SCIENCE OF PAIN

SCIENCE OF PAIN

Editors

Dr Allan I. Basbaum
University of California, San Francisco, CA, USA

Dr M. Catherine Bushnell
McGill University, Montreal, Quebec, Canada

ELSEVIER

AMSTERDAM   BOSTON   HEIDELBERG   LONDON   NEW YORK   OXFORD
PARIS   SAN DIEGO   SAN FRANCISCO   SINGAPORE   SYDNEY   TOKYO

Academic Press is an imprint of Elsevier
The Boulevard, Langford Lane, Kidlington, Oxford OX5 1GB, UK
525 B Street, Suite 1900, San Diego, CA 92101-4495, USA

First edition 2009

Notice
No responsibility is assumed by the publisher for any injury and/or damage to persons
or property as a matter of products liability, negligence or otherwise, or from any use
or operation of any methods, products, instructions or ideas contained in the material
herein, Because of rapid advances in the medical sciences, in particular, independent
verfication of diagnoses and drug dosages should be made

British Library Cataloguing in Publication Data
A catalogue record for this book is available from the British Library

Library of Congress Catalog Number: 2008930958

ISBN: 978-012-374625-2

For information on all Elsevier publications
visit our website at books.elsevier.com

Printed and bound in Slovenia

07 08 09 10 11 10 9 8 7 6 5 4 3 2 1

Contents

Introduction

"There is no coming to consciousness without pain." Carl Gustav Jung

It has been argued that pain, unlike audition, vision, somatosensation, and olfaction, is not a primary sense, but instead is more of an emotional experience. Most researchers of pain, however, consider pain to be a complex perception evoked by noxious stimuli. Pain is probably far more complicated than the other perceptual modalities described in this series. For example, in the setting of tissue or nerve injury, where pain is persistent, the stimulus that evokes pain can change. In fact, under these conditions innocuous stimuli can readily evoke the perception of pain.

But even these unusual characteristics do not capture the features that make pain among the most complex of perceptions. The International Association for the Study of Pain defines pain as "An unpleasant sensory and emotional experience associated with actual or potential tissue damage, or described in terms of such damage." In other words, although there is a very discrete anatomical and physiological basis for the detection and transmission of messages that are interpreted as painful, what makes the experience of pain so special is that there is always a profound emotional quality to the experience.

Pain in general, and pain research in particular, is especially exciting as it brings together elements of so many disciplines. This volume is comprehensive. It includes a wealth of information on the molecular biology, anatomy, physiology, and biochemical bases of 'pain', both in the normal and injury setting. But this volume also addresses the critical cognitive component of the pain experience, including some of the most provocative cerebral imaging studies that for the first time are providing insights into the gestalt of brain activity that occurs when pain is experienced. There are chapters on the pharmacological basis of the placebo, on the utility of hypnosis for the treatment of pain, and even an essay on consciousness and pain.

This is not a 'how to treat' clinical textbook. Nevertheless, the editors are advocates of the new mantra in the field, namely that *chronic* pain is not a symptom of disease, but rather is a disease entity itself, a disease of nervous system function. Therefore, in addition to covering the fundamentals of acute 'pain' processing, from the nociceptor to cortical activation, we also cover, in depth, the changes that occur in the setting of injury, including molecular, structural, and biochemical alterations in the properties of nociceptors and central nervous system pathways. Some of the particularly intractable clinical pain conditions are discussed. These chapters not only provide insights into pathophysiology but also clues to pain management.

Of course, a variety of compendia have recently appeared, and many also provide comprehensive reviews of the field. With this in mind, the editors have made a concerted effort to produce a final product that is different. Too often the excitement that epitomizes the field of pain research is buried within, or indeed omitted from, the typical edited book. Some textbooks include the proverbial 'box' that highlights an interesting topic, but these are generally very limited. We wanted to bring these topics to the forefront. Our approach is to include, in association with each major chapter, at least one or two cameos that illustrate fascinating and provocative areas of basic and clinical neuroscience that intersect the study of pain.

A few years ago we knew almost nothing about the cortical mechanisms that underlie the pain experience. Today some scientists, albeit the minority, believe that cortical imaging can provide an objective measure of the pain experience. A few years ago, the tetrodotoxin-resistant subtype of voltage-gated sodium channel, NaV1.8, was considered the Holy Grail for the next breakthrough in pain management. How fast things change. The discovery that a loss of function mutation of NaV1.7 underlies a condition of congenital insensitivity to pain and that a gain of function mutation underlies the excruciatingly painful condition of erythromelalgia has dramatically altered the focus, not only of the science community but also of the pharmaceutical industry. The pace of discovery in pain research is indeed remarkable. We hope that this volume conveys the excitement inherent in this discovery process and, most importantly, that it stimulates the next generation of basic and clinical scientists to unravel the mystery of the pain experience.

Allan I. Basbaum and M. Catherine Bushnell

1 The Adequate Stimulus

R D Treede, Johannes Gutenberg-University, Mainz, Germany

Glossary

nociception The processes of encoding and processing of noxious stimuli by the nervous system.

nociceptive stimulus An actually or potentially tissue damaging event that is encoded by primary nociceptive afferents. Although actual or potential tissue damage is the common denominator of those stimuli that may cause pain, there are some types of tissue damage that are not detected by any afferents, and thus do not cause pain. Therefore, not all noxious stimuli are adequate stimuli of nociceptive afferents. The adequate stimuli of nociceptors are termed nociceptive stimuli, which is a subset of noxious stimuli.

nociceptor A primary afferent nerve fiber that is capable of encoding noxious stimuli. All non-nociceptive afferents (e.g., tactile receptors, warm receptors) do respond to noxious stimuli (mechanical or thermal, respectively), because these stimuli are way above their respective thresholds. But only nociceptors are capable of encoding the relevant properties of those stimuli (e.g., sharpness, heat intensity in the painful range).

noxious stimulus An actual or potential tissue damaging event. This was found to be the common denominator of those stimuli that may cause pain. But there are some types of tissue damage that are not detected by any afferents, and thus do not cause pain. See nociceptive stimulus.

The term adequate stimulus (Kandel, E. R. *et al.*, 2000) is used in sensory physiology to describe the class of environmental phenomena that requires the least amount of energy in order to elicit a percept mediated by a particular sensory system, for example, a visual percept can be elicited by a punch to the eye, but light from a television screen needs much less energy for this purpose. The implication from such observations in subjective sensory physiology is that the receptive organs of each sensory system are specialized to detect a corresponding class of environmental phenomena (i.e., photoreceptors in the eye are specialized to detect photons).

It was difficult to transfer this concept to the perception of pain and to the nociceptive system. Many different stimuli may cause pain (e.g., pin prick, burn injury, freeze injury, and inflammation), none of which needs particularly low amounts of energy. Sherrington is credited with identifying the common denominator of those stimuli as being tissue damage (in Greek: νοξη Noxe) or environmental phenomena that threaten to cause such damage. Hence, the adequate stimulus to elicit pain is traditionally called a noxious stimulus. It may be defined as follows.

1.1 Noxious Stimulus

A noxious stimulus is an actual or potential tissue damaging event. Noxious stimuli may belong to thermal, mechanical, or chemical modalities of energy supply. Therefore, the sensory organs of the nociceptive system, free nerve endings of thinly myelinated $A\delta$ fibers, and unmyelinated C fibers in the skin and most other tissues, have a characteristic property: most of them are polymodal, that is, they respond to more than one modality of stimulus energy. Some nociceptive nerve endings, particularly those of $A\delta$ fibers, are specialized to be most sensitive to one particular stimulus modality (e.g., either heat or pin prick). The differential sensitivity spectra of nociceptive afferents suggest that pain quality may be encoded by a similar population code in the nociceptive system as taste quality in the gustatory system (Treede, R. D. *et al.*, 1998).

Polymodality is not the result of a primitive non-specific responsivity to tissue damage. Instead, it is mediated by specialized signal transduction pathways, some of which have been elucidated at the molecular level (Julius, D. and Basbaum, A. I., 2001). One member of the gene family of transient receptor potential channels (TRPV1), for example, is specifically activated by moderate heat stimuli, low tissue pH, and a class of irritant substances called vanilloids. For most inflammatory mediators, specific receptor molecules are present in the membranes of nociceptive nerve endings.

The topic of this volume of the *Handbook of the Senses* is pain. The taxonomy of the International Association for the Study of Pain (IASP) defines pain as, "An unpleasant sensory and emotional experience associated with actual or potential tissue damage, or described in terms of such damage" (Merskey, H. *et al.*, 1979). This definition implicitly refers to the adequate stimulus as identified by Sherrington, in its extended form including potential tissue damage rather than calling for outright tissue damage. There are two good reasons for this extension. First, a system that only responds after tissue damage has occurred, cannot subserve a warning function for the organism. The sensory system that mediates pain sensation (the nociceptive system), however, provides even primitive organisms with an array of protective reflexes that can flexibly respond to threatening environmental challenges. Such a system needs to be sensitive enough to signal impending tissue damage before it occurs. Second, although primary nociceptive afferents, the sensory organs of the nociceptive system, encode different intensities of manifest tissue damage as graded action potential discharge rates (Raja, S. N. *et al.*, 1999), the nociceptor activation thresholds were found to be clearly lower than the intensity needed to damage the skin. In humans as well as many animal species, heating the skin to above 40 °C, cooling to below 30 °C, or punctate pressure around 1–5 bars activates nociceptors, but none of these stimuli damages the skin. In fact, for most people these stimuli are not even painful.

The relatively low peripheral nociceptor activation thresholds allow the nociceptive system to subserve its warning function in a highly flexible and plastic way. The peripheral input can be shut down by descending inhibition, if there is an *a priori* reason to ignore the warning signal. The system can also enhance its response to the warning signal by enhancing the efficacy of its central synapses. This process is called central sensitization and may be considered to be one of the phylogenetically oldest mechanisms of learning and memory (Woolf, C. J. and Walters, E. T., 1991). Primary nociceptive afferents may also enhance their responsiveness to their adequate stimuli, a process called peripheral sensitization (Raja, S. N. *et al.*, 1999). Both peripheral and central sensitization contribute to the warning function of the nociceptive system.

As a warning system, the nociceptive system has gaps in its sensitivity: there are some types of tissue damage that are not detected by any afferents, and thus do not cause pain or any protective behavior. This is a well-known phenomenon in internal organs such as the liver or the brain, where a malignant tumor may cause extensive damage without being noticed by the patient. There is an even more common phenomenon of tissue damage that goes unnoticed by the nociceptive system: damage by ultraviolet radiation (Figure 1). The pain of sunburn always comes too late, after the skin has

Figure 1 The concept of the adequate stimulus to activate the nociceptive system and to elicit pain can be illustrated by this scene. Ultraviolet radiation causes tissue damage, but this noxious stimulus is not detected by the nociceptive nerve endings. Only the inflammatory response of the skin leads to adequate activation of nociceptive nerve endings via chemical mediators, which serve as nociceptive stimuli.

already been damaged. This pain does not signal the initial damage but the body's response by an inflammatory reaction. The ensuing peripheral and central sensitization of the nociceptive system lead to pronounced hyperalgesia of the injured skin to heat and mechanical stimuli, which then serves to protect the skin from further damage. These observations have led to the introduction of a new term, nociceptive stimulus (Cervero, F. and Merskey, H., 1996).

1.2 Nociceptive Stimulus

A nociceptive stimulus is an actual or potential tissue damaging event that is encoded by primary nociceptive afferents. In summary, the adequate stimulus to activate the receptive organs of the nociceptive system consists of either actual or potential tissue damage (noxious stimulus). However, not all noxious stimuli are detected by the nociceptive system. Therefore, the adequate stimulus for this system in the strict sense is that subset of noxious stimuli that can be encoded by the nociceptive system (nociceptive stimuli). It is not unusual for a sensory system to encode only a part of the range of environmental phenomena that its receptive organs are specialized for: visual stimuli consist of a restricted range of wavelengths of electromagnetic waves, and auditory stimuli consist of a restricted frequency range of pressure waves in the air. Likewise, nociceptive stimuli consist of a restricted range of actually or potentially tissue-damaging events. Linguistically, the term nociceptive stimulus may appear to be a little odd, as it also implies reception rather than just stimulation, but for exactly that reason it fits well into the broader system of terms in sensory physiology.

According to the concept of the adequate stimulus, nociceptive stimuli are encoded by the nociceptive system, just as visual stimuli are encoded by the visual system and auditory stimuli are encoded by the auditory system. However, a full appreciation of our senses goes beyond these aspects of sensory physiology (Bieri, P., 1995; Metzinger, T., 2000): seeing comprises more than vision, hearing comprises more than audition, and pain comprises more than nociception. This is reflected in the definition of pain as given above that does not relate to any objective observable measure, but refers to the subjective experience of the person in pain.

References

Bieri, P. 1995. Pain a Case Study for the Mind–Body Problem. In: Pain and the Brain: From Nociception to Cognition (eds. B. Bromm and J. E. Desmedt), pp. 99–110. Raven Press.

Cervero, F. and Merskey, H. 1996. What is a noxious stimulus? Pain Forum 5, 157–161.

Julius, D. and Basbaum, A. I. 2001. Molecular mechanisms of nociception. Nature 413, 203–210.

Kandel, E. R, Schwartz, J. H, and Jessell, T. M 2000. Principles of Neural Science. 4th edn., p. 414. McGraw-Hill.

Merskey, H., Albe-Fessard, D., Bonica, J. J., Carmon, A., Dubner, R., Kerr, F. W. L., Lindblom, U., Mumford, J. M., Nathan, P. W., Noordenbos, W., Pagni, C. A., Renaer, M. J., Sternbach, R. A., and Sunderland, S. 1979. Pain terms: a list with definitions and notes on usage. Recommended by the IASP subcommittee on taxonomy. Pain 6, 249–252.

Metzinger, T. 2000. The Subjectivity of Subjective Experience: A Representationalist Analysis of the First-Person Perspective. In: Neural Correlates of Consciousness: Empirical and Conceptual Questions (ed. T. Metzinger), pp. 285–306. MIT Press.

Raja, S. N., Meyer, R. A., Ringkamp, M., and Campbell, J. N. 1999. Peripheral Neural Mechanisms of Nociception. In: Textbook of Pain (eds. P. D. Wall and R. Melzack), 4th edn, pp. 11–57. Churchill Livingstone.

Treede, R. D., Meyer, R. A., and Campbell, J. N. 1998. Myelinated mechanically insensitive afferents from monkey hairy skin: heat response properties. J. Neurophysiol. 80, 1082–1093.

Woolf, C. J. and Walters, E. T. 1991. Common patterns of plasticity contributing to nociceptive sensitization in mammals and aplysia. Trends Neurosci. 14, 74–78.

2 Pain Theories

F Cervero, McGill University, Montreal, QC, Canada

References	9

From the beginning of scientific enquiry there have been two opposing views on the biological meaning of pain. One view proposes that pain is a sense similar to vision or hearing, a component of the sensory repertoire of most animals that warns us of impending damage, gives accurate information to the brain about injuries, and helps us to heal. The inclusion of pain in *The Senses: A Comprehensive Reference,* alongside vision, hearing, or olfaction shows that this view is persuasive. But there has always been an alternative interpretation of pain that denies it being a sense like vision or hearing and attaches to both pain and its opposite pleasure fundamental roles in shaping the emotions and behaviors of the individual. Pain is seen as a trigger of emotional states, a behavioral drive, and a highly effective learning tool. Aristotle, who was the originator of this view, made it very clear: there are only five senses – vision, hearing, touch, taste, and smell. Pain and pleasure are not senses but passions of the soul.

Whether or not pain is a sense like the others is not just an academic exercise. The experimental paradigms used to study the nervous system can be fundamentally different depending on the theoretical approach to the object of the study. If pain is regarded as a sense, like vision or hearing, then we will look for sensors that are activated selectively by painful stimuli and for sensory pathways in the brain and spinal cord that carry pain information much in the same way as we identify photoreceptors in the retina and a visual pathway to the cortex. This approach has generated an interpretation of pain mechanisms known as the specificity theory, which maintains that there are elements of the peripheral and central nervous system (CNS) specifically and exclusively dedicated to the processing of pain-related information.

However, if pain is not a sense like vision or hearing, then we do not need to look for a specific neural machinery for its processing but for patterns of activation, spatial or temporal, in neurons that do not necessarily have time-locked responses to painful stimuli. If pain is a passion of the soul then we need distributed networks and interactive parallel processing and not a pain pathway. This view has been articulated throughout the last 100 years as the pattern theory of pain, denying the existence of dedicated sensory elements for pain processing and attributing the perception of pain to interactions between patterns of impulses in nonspecific neuronal networks. The influential gate theory of pain (Melzack, R. and Wall, P. D., 1965) is the best contemporary example of such an interpretation.

The specificity theory of pain was a natural development of the Doctrine of Specific Nerve Energies put forward by the German physiologist Johannes Müller in the nineteenth century. The basic proposal of Müller's doctrine is that each sensory modality is the result of the activation of a specific neural system in the brain. If we are able to perceive touch it is because we have a subset of cells in our peripheral and CNS that respond to touch; if an animal can perceive infrared light or an electromagnetic field then it must have a subset of neurons capable of sensing and processing these stimuli. As he put it: "Sensation is not the conduction of a quality or state of external bodies to consciousness, but the conduction of a quality or state of our nerves to consciousness, excited by an external cause" (Müller, J., 1835–1840). At the end of the nineteenth century, von Frey M. (1895) extended Müller's doctrine to pain sensation, thus reinforcing a strict sensory interpretation of pain. He proposed that the fine nerve endings of unmyelinated afferents were the pain receptors in the periphery and that there was a specific pain pathway taking their signals to the brain. This influential proposal is responsible for the well-known model of pain mechanisms often found in textbooks whereby a pain receptor in the periphery is activated by a noxious stimulus and sends impulses to the spinal cord and from there to the thalamus and cortex via a crossed spinothalamic pathway. In other words, a relatively simple and straight forward pain pathway (Figure 1(a)).

In contrast, the pattern theory of pain represents a development of the Aristotelian concept of pain as a passion of the soul and was articulated in modern form by Goldscheider A. (1898) at the turn of the nineteenth century. The basic proposal is that pain is the result of intense stimulation of peripheral receptors, regardless of modality and tissue origin. For the pattern theory the neural substrate of pain perception consists of sequences of impulses in peripheral and central neurons that lead to pain sensations when certain spatial and temporal patterns of activity are produced. The emphasis of this interpretation is on patterns of neural activity evoked in the brain by a painful stimulus. In the 1960s, the gate theory of pain proposed a specific neural mechanism located at the first afferent relay in the spinal cord to illustrate how patterns of impulses in sensory receptors could lead to the modulation of pain sensation (Figure 1(b)). The gate theory emphasized the dynamic and plastic components of pain sensations and drew attention to pain modulation as opposed to an interpretation of pain exclusively as an alarm system. It also focused the attention of many researches on the clinical aspects of pain, away from physiological pain, which contributed a surge of studies on the effects of neuropathic lesions and on the development of

animal models of inflammatory and pathological pain (see Cervero, F., 2005 for a recent discussion on the gate theory).

The considerable amount of new information about pain mechanisms gathered in the last 40 years has relegated the specificity versus pattern argument to a secondary role. It is now well established that there are specialized sensors in the skin, muscles, and viscera of most animals that are activated exclusively by stimuli that cause injury and whose excitation leads to the sensation of pain (Belmonte, C. and Cervero, F., 1996). It is also known that pain cannot be evoked, under normal circumstances, by changing the patterns of activation of tactile sensory receptors (Ochoa, J. and Torebjork, E., 1983; 1989). There are also substantial data showing that there are neurons in the spinal cord and brain driven mainly or exclusively by nociceptive stimuli (Hunt, S. P. and Mantyh, P. W., 2001). However, there is also significant experimental evidence in favor of plasticity in the nociceptive sensory channel and of the existence of dynamic processes that can alter profoundly the functional properties of peripheral nociceptors and of nociceptive central neurons (Treede, R. D. *et al.*, 1992; Hunt, S. P. and Mantyh, P. W., 2001; Julius, D. and Basbaum, A. I., 2001). It has also been demonstrated that following a peripheral injury or inflammation, the activation of tactile afferents from uninjured skin can evoke pain sensations (see Cervero,

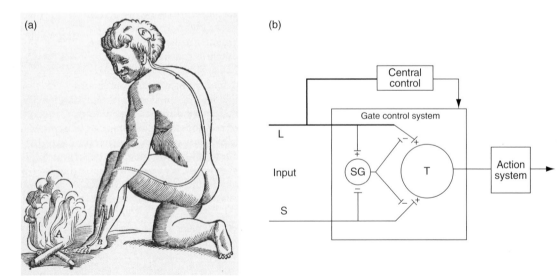

Figure 1   (a) Descartes' drawing of a pathway-oriented pain mechanism. The fire activates pain nerves in the foot of the child and these signals are transmitted to the brain where they are reflected into motor nerves that draw the foot away from the fire. (b) Schematic diagram of the gate theory of pain mechanisms. The diagram shows the presynaptic interaction model between large (L) and small (S) afferent fibers and the key role of substantia gelatinosa (SG) neurons controlling the activity of transmission (T) cells. (a) From Descartes, R. 1664. L'homme. Chez Jacques Le Gras. (b) Reprinted with permission from AAAS from Melzack, R. and Wall, P.D. 1965. Pain mechanisms: a new theory. Science 150, 971–979.

F. and Laird, J. M. A., 1996). Clearly, a point has been reached where neither a strict specificity nor a pattern interpretation can account for all that it is known about pain mechanisms.

We now realize that there are many different forms of pain (acute, traumatic, inflammatory, and neuropathic) and that pain is a dynamic process that cannot be explained with a single theory or a unique mechanism. One of the most striking expressions of the dynamic nature of pain sensation is its lack of adaptation. A continuous and uniform visual or auditory stimulus leads to sensory adaptation; we simply stop feeling this stimulus after a few seconds or minutes. However, the sensation of pain not only does not adapt to a continuous noxious stimulus but gets progressively worse so that, after a few minutes of persistent stimulation with a relatively mild painful stimulus, the sensation becomes unbearable. This change in pain sensitivity generates a state of pain amplification or hyperalgesia that is normally triggered and maintained by a persistent noxious stimulus but that can, under pathological circumstances, appear without an obvious cause so that the normal relationship between injury and pain is lost. Hyperalgesia and allodynia are the main symptoms of many chronic pain states and is the property of pain sensation that makes it particularly unpleasant and often unbearable.

In psychophysical terms a hyperalgesic state is represented by a leftward shift that occurs, following a peripheral injury, in the curve that relates stimulus intensity to pain sensation (Cervero, F. and Laird, J. M. A., 1996) (Figure 2). This shift causes the lower portion of the pain curve to fall in the innocuous stimulus intensity range (allodynia or pain produced by an innocuous stimulus) whereas the top portion shows an increased pain sensation to noxious stimuli (hyperalgesia proper or increased pain sensitivity to a noxious stimulus). Allodynia and hyperalgesia provide protective mechanisms to the organism, preventing the individual from stimulating an injured area and in so doing helping the healing process.

There are two forms of hyperalgesia: primary and secondary. Primary hyperalgesia is an increased pain sensitivity that occurs at the site of injury and it is the consequence of nociceptor sensitization, that is, the increased firing of peripheral nociceptors at the site of injury whose excitability has been increased by a number of locally released sensitizing agents. These sensitized nociceptors send enhanced afferent discharges to the CNS thus evoking increased pain from the primary hyperalgesic area and contributing

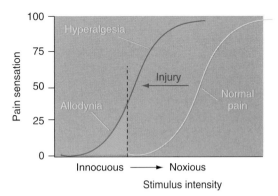

Figure 2 Diagram illustrating the changes in pain sensation induced by injury. The normal relationship between stimulus intensity and the magnitude of pain sensation is represented by the curve at the right-hand side of the figure. Pain sensation is only evoked by stimulus intensities in the noxious range (the vertical dotted line indicates the pain threshold). Injury provokes a leftward shift in the curve relating stimulus intensity to pain sensation. Under these conditions, innocuous stimuli evoke pain (allodynia). Reproduced From Cervero, F. and Laird, J. M. A. 1996. Mechanisms of touch-evoked pain (allodynia): a new model. Pain 68, 13–23, used with permission.

to the alterations in central processing that are, in turn, responsible for secondary hyperalgesia (Treede, R. D. et al., 1992).

Secondary hyperalgesia is defined as an increased sensitivity to pain occurring in areas adjacent or even remote to the site of injury. For instance, following an injury to the hand an area of hyperalgesia may develop covering the entire arm or an inflammation of the gastrointestinal tract or the bladder may produce an area of hyperalgesia in the abdominal or pelvic regions. Secondary hyperalgesia is the result of an alteration in the processing by the CNS of impulses from low-threshold mechanoreceptors, such that, these impulses are able to activate nociceptive neurons and evoke pain. This central alteration is initially triggered and later maintained by the enhanced afferent discharges from the primary hyperalgesic area (Treede, R. D. et al., 1992).

Different forms of pain are mediated by different mechanisms which participate in various ways in the generation of pain and hyperalgesia (Cervero, F. and Laird, J. M. A., 1991; Klein, T. et al., 2005). Generally speaking we can identify three different forms of pain taking into account the relationship between noxious stimulus and pain sensation: nociceptive, inflammatory, and neuropathic (also called phase 1, 2, and 3 pains by Cervero, F. and Laird, J. M. A., 1991) (Figure 3). Nociceptive pain refers to the processing

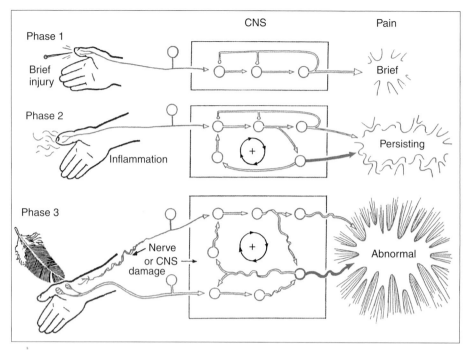

Figure 3 Diagram showing the three models of pain processing for the three main types of pain (nociceptive, inflammatory, and neuropathic or phases 1, 2, and 3). See text for further explanation. Reproduced from Cervero, F. and Laird, J. M. A. 1991 One pain or many pains? A new look at pain mechanisms. News Physiol. Sci. 6, 268–273, used with permission.

of brief noxious stimuli; inflammatory pain is the consequence of prolonged noxious stimulation leading to tissue damage; and neuropathic pain is the consequence of neurological damage, including peripheral neuropathies and central pain states.

Nociceptive pain is a protective sensation needed for the survival and well-being of the individual. The mechanisms subserving the processing of brief noxious stimuli can be viewed as a fairly simple pathway that carries impulses in peripheral nociceptors centrally toward the thalamus and cortex and leads to brief pain perception. In contrast, injury and tissue damage evoke an inflammatory reaction as part of the healing process and generate a pain state different from nociceptive pain as the response properties of the various components of the nociceptive system change. These changes include nociceptor sensitization and recruitment of populations of previously unresponsive receptors. In turn, CNS neurons show an amplification of their excitability expressed as increases in receptive field size and greater spontaneous and evoked firing. All of these changes indicate that the CNS has moved to a new, more excitable state as a result of the noxious input generated by tissue injury and inflammation. Under these conditions, an immediate correlation between

discharges in peripheral nociceptors and pain perception is lost.

Neuropathic pain syndromes are the consequence of damage to peripheral nerves or to the CNS itself and produce pain sensations well outside the range of the sensations produced by the normal nociceptive system, even after serious peripheral injury or inflammation. These include spontaneous pain, greatly reduced pain thresholds, and mechanical allodynia. Neuropathic pain states are characterized by an almost complete lack of correlation between peripheral noxious stimuli and pain sensation and are produced by neurological lesions that cause abnormal impulse activity generated in nerve sprouts, neuromas, or in dorsal root ganglion cells, ephaptic coupling between adjacent nerve fibers and abnormal responses of peripheral nociceptors and CNS neurons. Nociceptive and inflammatory pains are symptoms of peripheral injury, whereas neuropathic pain is a symptom of neurological disease.

Whereas the specificity theory can explain fairly well the simpler forms of pain, such as a pin prick or the acute pain of a minor burn, complex pain experiences, including hyperalgesic states require substantial peripheral and central plasticity such that low-intensity stimuli evoke pain (i.e., secondary

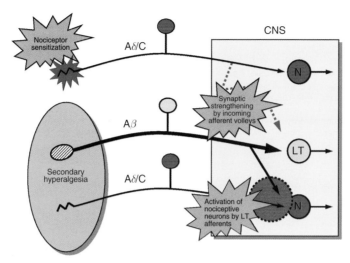

Figure 4 Diagram representing the basic mechanisms of primary and secondary hyperalgesia and the three key processes implicated in their generation. Primary hyperalgesia is produced by the stimulation of nociceptors connected to Aδ- and C-afferent fibers which activate nociceptive CNS pathways (N). Secondary hyperalgesia is produced by stimulation of tactile receptors connected to Aβ-afferents which normally activate low-threshold (LT) pathways but that as a consequence of the amplification of the nociceptive input from the injured area can now access nociceptive neurons (N). See text for further explanation.

hyperalgesia or touch-evoked allodynia). It is very likely that both specific and nonspecific nociceptive systems participate in the generation and maintenance of different pain states.

We can currently identify three key processes in the neurobiological approach to the mechanisms of pain and hyperalgesia: (1) the process of nociceptor activation and sensitization, responsible for the initial signaling of injury and the peripheral changes in the nociceptive system induced by a noxious stimulus; (2) the process of central amplification of nociceptive signals, known as central sensitization, generated by synaptic strengthening of connections between CNS neurons and responsible for the enhanced excitability that accompanies persistent pain states; and (3) the process whereby activity in low-threshold sensory receptors from undamaged peripheral areas can access the nociceptive system and evoke pain sensations and hyperalgesic states (e.g., touch-evoked pain, tactile allodynia) (Figure 4).

The analysis of these three processes supersedes interpretations of pain based on a specificity or pattern approach as all three contain elements that require the existence of groups of neurons dedicated to the processing of nociceptive signals as well as the expression of functional changes induced by afferent activity. Several of these changes can lead to major alterations of pain sensitivity through quite separate mechanisms. For instance, touch-evoked pain can be the result of sensitization of peripheral nociceptors leading to decreases in their activation threshold or be mediated by a purely central mechanism whereby low-threshold mechanoreceptors access nociceptive neurons. The emphasis has moved from theoretical interpretations based on a specific pain channel or on patterns of impulses to the in depth analysis of the functional processes that may cause the different pain and hyperalgesic states. Neither the specificity nor the pattern theories have been able to provide workable models for all forms of pain sensation. They still offer a useful theoretical framework for some of our observations but the weight of work has shifted to studying those processes that will help the development of effective pain relief therapies.

References

Belmonte, C. and Cervero, F. 1996. Neurobiology of Nociceptors. Oxford University Press.

Cervero, F. 2005. The Gate Theory Then and Now. In: The Paths of Pain (eds. H. Mershey, J. De Loeser, and R. Dubner), pp. 33–48. IASP Press.

Cervero, F. and Laird, J. M. A. 1991. One pain or many pains? A new look at pain mechanisms. News Physiol. Sci. 6, 268–273.

Cervero, F. and Laird, J. M. A. 1996. Mechanisms of touch-evoked pain (allodynia): a new model. Pain 68, 13–23.

Descartes, R. 1664. L'homme. Chez Jacques Le Gras.

Goldschaider, A. 1898. Über den Schmerz: Gesammelte Abhandlungen. Ambros.

Hunt, S. P. and Mantyh, P. W. 2001. The molecular dynamics of pain control. Nature Rev. Neurosci. 2, 83–91.

Julius, D. and Basbaum, A. I. 2001. Molecular mechanisms of nociception. Nature 413, 203–210.

Klein, T., Magerl, W., Rolke, R., and Treede, R. D. 2005. Human surrogate models of neuropathic pain. Pain 115, 227–233.

Melzack, R. and Wall, P. D. 1965. Pain mechanisms: a new theory. Science 150, 971–979.

Müller, J. 1835–1840. Handbuch der Physiologie des Menschen für Vorlesungen. J. Hölscher.

Ochoa, J. and Torebjork, E. 1983. Sensations evoked by intraneural microstimulation of single mechanoreceptor units innervating the human hand. J. Physiol. 342, 633–654.

Ochoa, J. and Torebjork, E. 1989. Sensations evoked by intraneural microstimulation of C nociceptor fibres in human skin nerves. J. Physiol. 415, 583–599.

Treede, R. D., Meyer, R. A., Raja, S. N., and Campbell, J. N. 1992. Peripheral and central mechanisms of cutaneous hyperalgesia. Prog. Neurobiol. 38, 397–421.

von Frey, M. 1895. Beitrage zur sinnesphysiologie der haut. Ber. Sachs. Ges. Wiss. 47, 166–184.

3 Anatomy of Nociceptors

S Mense, Institut für Anatomie und Zellbiologie, Universität Heidelberg, Heidelberg, Germany

Glossary

anterograde transport Intra-axonal transport in the distal direction (away from the soma) of substances synthesized in (or injected into) spinal or cranial ganglion cells. The transport is brought about by molecules (kinesin for anterograde, and dynein for retrograde, transport) that bind substances and vesicles and move along the microtubules in the axoplasm.

axon reflex Release of neuropeptides from a branch of a nociceptor when the branch is invaded antidromically (against the normal direction of propagation) by action potentials. The released neuropeptides influence the microcirculation in the vicinity of the ending.

deep somatic tissues All subcutaneous tissues that are not viscera, namely tendon, fascia, muscles, ligaments, and joints.

epitendineum (peritendineum externum) Dense connective tissue with blood vessels, lymphatic vessels, and nerve fibers surrounding a tendon.

group I–IV fibers Nomenclature for afferent fibers that originate in deep somatic tissues. Group IV fibers correspond to C-fibers from the skin, and group III to Aδ-fibers. Group I fibers comprise the primary endings from muscle spindles (Ia-fibers) and Ib-fibers from Golgi tendon organs. Group II fibers and cutaneous Aβ-fibers are largely identical. The denominations using roman numerals are based on the diameter of the nerve fibers, those using arabic letters on the conduction velocity.

noceffector This term was coined by Kruger L. (1988) for nociceptors in the tooth; it emphasizes the efferent function of the receptor. Efferent function means that the receptor is involved in the maintenance of the tissue under normal and pathological circumstances by releasing substances stored in the ending. One aspect of such a noceffector role in the skin would be regulation of Langerhans cell type expression and proliferation of keratinocytes. Under pathophysiological circumstances, the effector role of a nociceptor is reflected in the axon reflex (see above) and neurogenic inflammation.

paciniform corpuscle A rapidly adapting mechanoreceptor with a morphology similar to that of a Pacinian corpuscle. It consists of a layered capsule that surrounds a central receptive core. The capsule is derived from the perineurium; the central core contains the receptive portion of the axon with Schwann cells around it. Compared to the Pacinian corpuscle, the paciniform corpuscle is smaller and has fewer laminae in its capsule. In

skeletal muscle, it is often supplied by a group III fiber.

primary afferent unit The first sensory neuron in the body periphery. It includes the receptive ending, the afferent fiber, the soma in the spinal or cranial ganglion, and the central process in the dorsal root and the central nervous system (CNS).

receptor matrix An arrangement of cell organelles (mitochondria, vesicles, and axonal reticulum, i.e., a network of fluid-filled vacuoles) embedded in a granulated axonal cytoplasm.

Ruffini corpuscle A flat, encapsulated mechanoreceptor supplied by several myelinated nerve fibers. The fibers form a dense arborization within a capsule of connective tissue.

sensory terminal tree The preterminal axon and its branches. The term is part of the concept that both group III and IV endings do not have just one sensory site, but possess several branches with many receptive loci that together enhance the sensitivity of the free nerve ending as a sense organ.

varicosity An expansion of the preterminal axon of a slowly conducting sensory fiber (Aδ- or C-fiber). The varicosity contains mitochondria, vesicles, and other cell organelles. It is discussed as a receptive site. The varicosities of a free nerve ending are connected by thin stretches of axon; this arrangement causes the beaded appearance of a free nerve ending and the preterminal axon under the light microscope.

3.1 Introduction

Originally, pain was assumed to be an emotion like pleasure and fear. In the nineteenth century, this concept was replaced by the view that pain is due to activation of a set of specialized nerve endings. Von Frey M. (1894) was the first to link pain to fine nerve terminals in the skin. The functional term nociceptor is derived from the Latin word *noxius* for damaging or harmful. It denotes a sensory ending that detects actual or potential tissue damage. When activated, it may cause pain in humans and pain-related reflexes in animals. Another definition of a

nociceptor is an ending that by its discharge behavior is capable of distinguishing between an innocuous and a noxious stimulus. An important point for the understanding of the function of nociceptors is that their stimulation threshold is just below tissue-damaging intensity. The function of nociceptors is not to signal existing tissue damage, but to inform the central nervous system (CNS) when stimuli approach tissue-threatening intensities.

Nociceptive endings are present in almost all tissues and organs of the organism. However, the fact that lesions of the brain and parenchyma of the lung, liver, and cartilage are not painful suggests that in these tissues nociceptors are not present. Conversely, by stimulation of cornea, dura mater, and tooth pulp, pain is the predominant or only sensation that can be elicited. These observations led to the assumption that these tissues are equipped exclusively with nociceptors.

The available clinical and experimental evidence indicates that small-diameter afferent fibers have to be activated in order to elicit pain. These fibers conduct action potentials at a relatively slow velocity (below 30 m s^{-1} in the cat); histologically they comprise either thin myelinated (Aδ- or group III) fibers or nonmyelinated (C- or group IV) fibers. However, there are also nociceptors supplied by faster-conducting thick myelinated Aα/β-fibers: in the rat, approximately 20% of all Aα/β-fibers are nociceptive (for review, see Djouhri, L. and Lawson, S. N., 2004). The nomenclature with roman numerals (group I–IV fibers) was developed by Lloyd D. P. C. (1943) for muscle afferent fibers. It is now being used for afferent fibers from muscle, joint, tendons, and fascia (the so-called deep somatic tissues).

Generally, there is a negative relationship between mechanical threshold and conduction velocity (CV) of sensory afferents: the higher the CV, the lower the mechanical threshold (Burgess, R. P. and Perl, E. R., 1967). In the dorsal root ganglion (DRG), the correlation between soma diameter and axonal CV is weaker than generally thought. In Figure 1 (Hoheisel, U. and Mense, S., 1987), this correlation is shown for intracellularly stained DRG cells of the cat. For group III units (those having CVs between 2.5 and 30 ms^{-1}), there was no significant correlation between soma size and CV, and for group IV units (conducting at <2.5 ms^{-1}), the correlation was negative. Thus, from the soma size in the DRG, the diameter or CV of group III and IV afferent fibers cannot be inferred. Generally, the majority of nociceptive free nerve endings will originate from small-diameter DRG cells, but there are also relatively large somata that supply nociceptive terminals. Conversely, not all the small DRG cells supply nociceptive endings.

In morphological studies on the endings of unmyelinated fibers, the first hurdle to be taken is the distinction between efferent and afferent units, because postganglionic sympathetic fibers may look identical to unmyelinated afferent ones, except that the former have a larger mean diameter (Heppelmann, B. *et al.*, 1988). The presence of

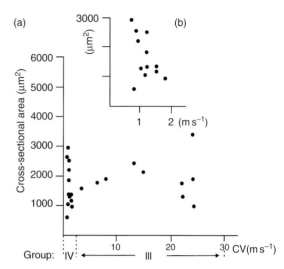

Figure 1 Relationship between cross-sectional area of dorsal root ganglion (DRG) cells and conduction velocity (CV) of the peripheral axon. (a) Data from 12 unmyelinated (C- or group IV) fibers and 10 thin myelinated (Aδ- or group III) fibers. (b) Data from the unmyelinated units at an expanded scale of the abscissa. There was no significant correlation between soma size and CV within group III fibers, and for group IV units the correlation was negative. (Modified from Hoheisel, U. and Mense, S., 1987, with permission).

neuropeptides may be helpful, since for instance CGRP is absent from postganglionic sympathetic fibers (Ju, G. *et al.*, 1987). However, postganglionic parasympathetic fibers have also been shown to contain neuropeptides. Therefore, if neuropeptide-immunoreactive (ir) free nerve endings are being studied, one has to assume that a certain (probably small) proportion of the endings originate from autonomic efferent fibers. Of course, the best way of eliminating this problem is to cut the autonomic supply to the studied body region.

Not all small-diameter, slowly conducting afferent fibers are nociceptive. In cutaneous nerves, there are thermoreceptors and low-threshold mechanoreceptors that have unmyelinated afferent fibers, and also in skeletal muscle low-threshold mechanoreceptors with group IV afferents can be found (Light, A. R. and Perl, E. R., 2003; Hoheisel, U. *et al.*, 2005). The presence of certain neuropeptides such as CGRP does not distinguish between high- and low-threshold mechanosensitive (HTM and LTM) group IV units, because the neuropeptide was found in both functional types (Hoheisel, U. *et al.*, 1994).

The development of myelinated nociceptors has been reported to depend upon the presence of trkA (tyrosin kinase A) and trkC receptors in the DRG, but not upon trkB receptors (Ichikawa, H. *et al.*, 2004). The trk receptors bind neurotrophins (NTs) selectively; the ligand for trkA is nerve growth factor (NGF), that for trkB, brain-derived neurotrophic factor (BDNF), and that for trkC, NT-3.

Even though several groups have studied and described the structure of putative nociceptors, clear morphological (light or electron microscopic) differences between the various functional types of nociceptor are not apparent. The functional differences are probably due to the equipment of the endings with different sets of receptor molecules (Cesare, P. and McNaughton, P., 1997).

In neurophysiological experiments on cats and rats, an unexpected finding was that many unmyelinated afferent units had two receptive fields (RFs), i.e., the same fiber could be activated by applying stimuli to more than one location. In the deep tissues of the cat tail, afferent units were found that had one RF in deep tissues (muscle, joint, and periosteum) and another one in the skin distal to the deep RF (Mense, S. *et al.*, 1981). Sometimes, the distance between the RFs could be several centimeters. The anatomical basis of this feature may be branching of the afferent fiber close to its area of termination. Anatomical and neurophysiological studies (Devor, M. *et al.*, 1984; Pierau, E.-K. *et al.*, 1984) have shown that afferent units with long branched axons are rare (a few percent of all afferent fibers), but they may be functionally relevant, since they are likely to reduce the spatial resolution of the nociceptive system, and thus could contribute to the diffuse nature of deep pain.

The data that are most important to this chapter were obtained in studies employing a combination of electrophysiological and neuroanatomical techniques, because otherwise one does not know if a given free nerve ending in a histological section has a nociceptive function. However, many purely morphological investigations on free nerve endings yielded results that are relevant to this chapter, even though the evidence for a nociceptive function is indirect. Therefore, many of these studies are included below.

The main emphasis of this chapter is on nociceptors or free nerve endings, respectively, in skin, muscle/joint, tooth pulp, and testis, because large sets of data are available from these tissues.

3.2 General Morphological Features of Nociceptors

3.2.1 Light Microscopic Structure

Morphologically, a nociceptor is a free nerve ending that is usually connected to the CNS through thin myelinated or unmyelinated nerve fibers. (As stated above, there are also nociceptors with thick myelinated fibers.) C- or group IV fibers are assumed to terminate exclusively in free nerve endings while group III fibers supply both free nerve endings and other types of receptors (for instance in muscle paciniform corpuscles; Barker, D., 1967; Stacey, M. J., 1969). A frequent location of free nerve endings is the wall of arterioles and the surrounding connective tissue; the capillaries proper do not have a rich supply with these endings (Stacey, M. J., 1969; Reinert, A. *et al.*, 1998). The marked sensitivity of free nerve endings to chemical stimuli, particularly to those associated with disturbances of the microcirculation, may be related to their location in or close to the wall of blood vessels.

The term free nerve ending or unencapsulated ending is derived from the fact that in the light microscope this type of ending lacks a visible (corpuscular) receptive structure (Stacey, M. J., 1969). Typically, such an ending consists of several branches or terminals that altogether form the receptor in the morphological sense. An afferent fiber together with its receptive ending is an afferent unit. In the fluorescence microscope, the receptive ending and the preterminal axon look like a string of beads with relatively wide diameter (so-called varicosities) connected by very thin stretches of axon. The diameter of a branch of a free nerve ending is 0.5–1.0 μm, i.e., at the limits of the resolution of the light microscope. Figure 2 shows a free nerve ending in a rat

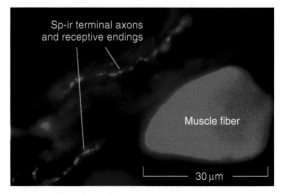

Figure 2 Histological section from rat gastrocnemius muscle showing two nerve fibers with varicosities visualized with fluorescent antibodies to substance P (SP; U. Hoheisel and S. Mense, unpublished data).

skeletal muscle after labeling with fluorescent antibodies to SP.

Most of the unmyelinated afferents enter the spinal cord via the dorsal root. However, there are also unmyelinated afferent fibers in the ventral root: approximately 30% of all ventral root fibers are unmyelinated and 50% of these fibers survive in the distal stump after cutting the ventral root. Therefore, they were assumed to be afferent with their soma located in the DRG (Applebaum, M. L. *et al.*, 1976).

In immunohistochemical studies, sensory free nerve endings have been described as either peptidergic or lectin-positive, the latter being largely peptide free. The lectin-positive endings possess membrane-associated glycoconjugates that bind the plant lectin *Griffonia simplicifolia* isolectin B4 (IB4). They are assumed to be identical to the fluoride-resistant acid phosphatase (FRAP) endings. Many of the IB4-positive endings express the purinergic P2X3 receptor. In contrast, the peptidergic endings are equipped with the vanilloid receptor TRPV1 (transient receptor potential subtype V1). These cytochemical differences led to the hypothesis that the IB4-positive and -negative endings represent two functionally distinct classes of nociceptors. (Snider, W. E. and McMahon, S. B., 1998; Stucky, C. L. and Lewin, G. R., 1999). Recent evidence indicates that the peptidergic endings express trkA receptors, are dependent on NGF in their development, and express predominantly tetrodotoxin (TTX)- sensitive Na$^+$ channels, whereas the nonpeptidergic IB4-positive endings depend on glial cell–derived neurotrophic factor (GDNF) in their development and express more TTX-resistant Na$^+$

channels than the IB4-negative fibers (Wu, Z.-Z. and Pan, H.-L., 2004). Another difference between these two types of putative nociceptors is that the IB4-positive endings terminate freely in the tissue without close spatial relationship to blood vessels, whereas the peptidergic fibers are often associated with blood vessels (Silverman, J. D. and Kruger, L., 1988; Figure 3). At present, however, the assumption prevails that the separation between these two postulated types is probably not that sharp and that IB4-positive endings may also express TRPV1 receptors.

In experiments employing a combination of electrophysiological and immunohistochemical techniques to study functionally identified DRG cells, Lawson S. N. *et al.* (1997) reported that cells terminating in cutaneous nociceptive endings showed a strong tendency to express SP, particularly if they had a slow CV or small somata in the DRG.

3.2.2 Ultrastructure

In electron microscopic studies, often uncertainties remain as to the identity of the ending. Even when in combined electrophysiological–morphological investigations an ending is first functionally characterized and then morphologically reconstructed, it is sometimes difficult to make sure that a given ending in a histological section is the one that was studied functionally.

The published findings demonstrated that free nerve endings are not free in the strict sense, because the majority are ensheathed by Schwann cells (therefore, the term free nerve ending is a misnomer). Exceptions to this rule are endings in the epidermis

(a)

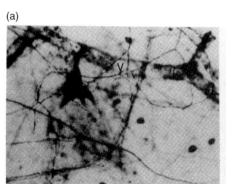

(b)

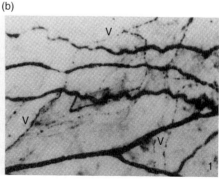

Figure 3 Testicular whole-mount preparation of the rat epididymal tunica vasculosa stained with Griffonia simplicifolia A-B4 lectin (a) and antibodies to calcitonin gene-related peptide (CGRP; (b)). Slender lectin-stained axonal bundles and terminals distribute freely throughout the tissue (a). The lectin-labeled axons bear little relation to blood vessels (V) whose endothelial basal lamina is also lectin-stained. In contrast, CGRP immunoreactivity is evident as heavily labeled coarse nerve bundles from which one can trace fine granular axons running closely adherent to blood vessels (V). (From Silverman, J. D. and Kruger, L., 1988, with permission.)

that lose their Schwann cell covering, when they pass the basal lamina of the epidermal layer (see below). Usually, also the receptive terminals are ensheathed by Schwann cells, and only small areas of the axonal membrane remain uncovered by Schwann cell processes. These areas are separated from the interstitial fluid only by the basal lamina that spans the exposed axonal membrane areas (Andres, K. H. *et al.*, 1985; Heppelmann, B. *et al.*, 1990a; 1990b; Messlinger, K., 1997). The varicosities in the exposed membrane areas are supplied with mitochondria and vesicles and show other structural specializations characteristic of receptive structures, and together with the end bulb they are assumed to be the site where external stimuli act. A reconstruction of such an ending is shown in Figure 4.

The arrangement of cell organelles (mitochondria, vesicles, and axonal reticulum, i.e., a network of fluid-filled vacuoles) embedded in a granulated axonal cytoplasm was called receptor matrix by Andres K. H. and von Düring M. (1973). Often, nociceptive endings exhibit granular or dense-core vesicles containing neuropeptides. However, vesicles containing catecholamines can have the same appearance, so that dense-core vesicles are not an unambiguous sign of a nociceptive ending. The function of the round clear vesicles in the peripheral ending is still obscure. They may contain the same transmitters as the central synaptic terminal (e.g., glutamate). Recent reports have shown that glutamate is an effective stimulant for nociceptive endings in the body periphery (Svensson, P. *et al.*, 2003). Therefore, it is conceivable that nociceptors enhance their own excitation by releasing glutamate when they are activated.

In a recent survey, Kruger L. *et al.* (2003a) list some features that were considered characteristic for nociceptive terminals in the body periphery:

(1) An axonal reticulum that may be derived from the smooth endoplasmic reticulum.
(2) Vesicle aggregates embedded in a granular axonal matrix that may be identical to the receptor matrix described by Andres K. H. and von Düring M. (1973).
(3) Exposed membrane areas that are not ensheathed by Schwann cells and directly contact the basal lamina.

The receptive terminal as a whole consists of several branches equipped with many receptive sites that have the appearance of varicosities. Together the

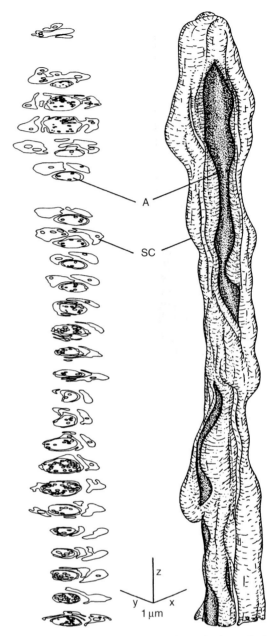

Figure 4 Reconstruction of an end branch of an articular group III ending. *Left*: Contours of the cross-sectioned nerve fiber at a distance of 1 μm each (computer-aided reconstruction). *Right*: Hand drawing based on the stack of sections showing the surface of the end branch. The bare axon areas (A) are densely dotted; SC, Schwann cell. (From Messlinger, K. *et al.*, 1995, with permission.)

branches constitute a sensory tree as shown in Figure 5 (Messlinger, K., 1996).

It is important to note that the above features cannot be found at the synaptic terminal of the same neuron, i.e., the central and distal terminals of

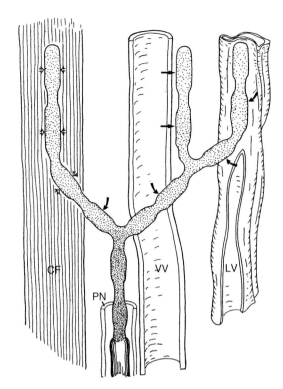

Figure 5 So-called sensory terminal tree of a group III fiber (schematic representation) as a possible morphological basis for sensitization of polymodal nociceptors. The summation of local receptive potentials created in perivascular (chemoreceptive) branches at venous vessels (VV) or lymphatic vessels (LV) in combination with the local potentials of mechanoreceptive branches in collagenous tissue (CF) may lead to a suprathreshold excitation at the basis of the sensory terminal tree within the perineurium (PN). (From Messlinger, K., 1996, with permission.)

a sensory neuron exhibit morphological differences. Small clear vesicles are present at both terminals, but at the peripheral terminal they show diversities in size, shape, and electron lucency, which are not present at the central terminal. The differences in vesicle content between the peripheral and the central terminal are difficult to explain, because if the vesicles are synthesized in the soma, as is generally assumed, a complicated sorting process is required. This is not impossible, because at least for channel proteins an exclusive transport to the peripheral terminal has been reported (Garcia-Anoveros, J. et al., 2001). Another explanation would be that the small clear vesicles in the distal terminal are synthesized in the ending itself, but at present this appears unlikely. As the origin and the contents of the small clear vesicles in the peripheral terminal are unknown, the function of these vesicles is likewise obscure.

One of the characteristic functional properties of a nociceptor is its relatively high mechanical stimulation threshold. The high mechanical threshold is surprising, considering the fact that a free nerve ending is a fragile structure with a semifluid membrane. One factor that may be responsible for the high mechanical threshold is a special mechanosensitive ion channel, the transient receptor potential subunit V4 (TRPV4; Liedtke, W., 2005).

3.3 Nociceptors in Various Tissues

A complete overview of all tissues is not attempted. Only those organs/tissues for which comprehensive sets of data are available are addressed below.

3.3.1 Skin

The cutaneous nociceptors are generally subdivided into several types according to the fibers supplying them, namely $A\delta/\beta$- or C-fibers. Among the receptors supplied by myelinated fibers are (data from monkey, after Djouhri, L. and Lawson, S. N., 2004):

- Type I A mechano-heat receptors (AMH, supplied by $A\delta$- and $A\beta$-fibers). They have a relatively low mechanical, but a high thermal, threshold. They are called moderate pressure receptors by some.
- Type II A mechano-heat nociceptors (AMHs, supplied by $A\delta$-fibers). They have high mechanical and low thermal thresholds.

Among the receptors supplied by C-fibers, which are by far more frequent, are

- C-fiber mechano-heat nociceptors (CMHs), which have little sensitivity to algesic agents
- Cold nociceptors
- Polymodal nociceptors that respond to all noxious stimuli including algesic chemical stimuli

In a cutaneous nerve of the cat, approximately 50% of all unmyelinated afferent fibers were found to have a high mechanical threshold in the noxious range (Bessou, P. and Perl, E. R., 1969). These units also responded to noxious heat and irritant chemicals and therefore were considered to be polymodal nociceptors.

The measured size of the RF of a single unmyelinated fiber supplying a polymodal nociceptor depends on the thickness of the probe used for mechanical stimulation: Thick probes yielded areas of 6–32 mm^2 (mean 18 mm^2) in the rabbit, whereas with small probes mean areas of 7.5 mm^2 were measured, which consisted of several mechanosensitive spots that could be activated independently from each other (Kenins, P., 1988).

3.3.1.1 Light microscopy

The nociceptive endings of the skin can extend in the epidermis to just beneath the surface. This explains why superficial abrasions of the skin can cause pain but no bleeding, because there are no blood vessels in the epidermis.

A study of the distribution of CGRP-ir fibers in the rat glabrous skin (Kruger, L. *et al.*, 1989) revealed a dense network in the subpapillary layer largely confined to dermal blood vessels and sweat glands. The intraepidermal fibers branched extensively in the stratum spinosum, but some extended nearly to the stratum corneum (Figures 6 and 7). These fibers were considered to fulfill a noceffector role rather than a nociceptive one. Noceffector means that these fibers have an efferent function and are involved in the maintenance of the tissue under normal and pathological circumstances (Kruger, L., 1988). One aspect of such a noceffector role in the skin would be regulation of Langerhans cell type expression and proliferation of keratinocytes (Kruger, L. and Halata, Z., 1996). In hairy skin of the rat, a similar arrangement of CGRP-ir fibers was found with the additional feature that the lower portion of each hair follicle exhibited an extensive innervation. Sebaceous glands and arrector pili muscles were not supplied with CGRP-ir fibers.

To find out if the effect of capsaicin treatment of skin pain (e.g., erythralgia) is due to a desensitization or destruction of nociceptive nerve fibers, Simone and coworkers (Simone, D. A. *et al.*, 1998) performed a light microscopic study of capsaicin effects in skin biopsies of human skin. In the epidermal layer of intact skin, mainly fibers ir for PGP 9.5 (protein gene product 9.5, a general marker for nerve fibers) were found, whereas in the dermis a dense innervation of CGRP- and SP-ir fibers was present. After topical capsaicin treatment, the CGRP- and PGP-ir fibers were almost completely gone, but the SP-ir fibers were less affected. Simultaneously, heat pain and pain produced by sharp objects were markedly reduced. This finding is surprising, because most SP-ir fibers are assumed to be nociceptive (Lawson, S. N. *et al.*, 1997).

Recently, unmyelinated fibers stained for PGP 9.5 in the skin of human biopsy material have been studied using confocal microscopy (Kennedy, W. R., 2004). Of interest in the context of the present chapter are fibers that penetrated the basal lamina and branched in the epidermis, because they can be assumed to be free nerve endings. The PGP 9.5-positive fibers in the epidermis had an appearance similar to that described in other reports: From a subepidermal plexus single fibers coursed in the epidermis, where they could be easily counted (Figure 8(a)). In skin biopsies of patients with diabetic neuropathy, the density of the epidermal

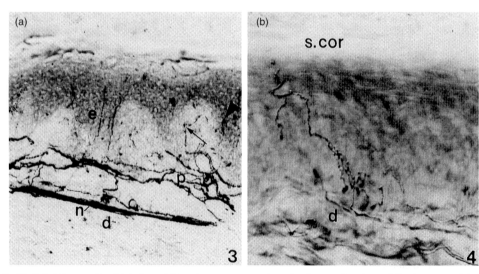

Figure 6 Digital glabrous skin. (a) A dermal (d) cutaneous nerve gives rise to an extensive calcitonin gene-related peptide-immunoreactive (CGRP-ir) subpapillary nerve plexus (p) from which individual axons can be traced into epidermal pegs (e) and dermal papillae (arrow). Some branches appear to end at varying depths within the stratum spinosum revealing terminal expansions (arrowhead) (×63). (b) Branched CGRP-ir axons in palmar glabrous skin extending from the dermis (d) nearly to the stratum corneum (s.cor) (×320). (From Kruger, L. *et al.*, 1989, with permission.)

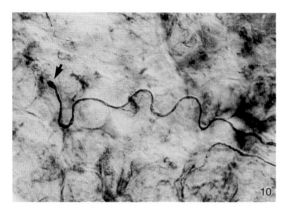

Figure 7 A dermal calcitonin gene-related peptide-immunoreactive (CGRP-ir) axon terminating as a free nerve ending with a bulbous enlargement (arrow) (×320). (From Kruger, L. *et al.*, 1989, with permission.)

branches was much less, and some fibers showed swellings that were interpreted as an indication of de- or regeneration (Figure 8(b)).

3.3.1.2 Electron microscopy

In a combined electrophysiological–electron microscopic study on mechanical nociceptors in the cat skin (Kruger, L. *et al.*, 1981), single receptive endings were first identified functionally in electrophysiological recordings and then examined morphologically. Mechano-nociceptors had RFs consisting of several sensitive spots with relatively insensitive areas in between. After functional identification, the location of the mechanosensitive spots was marked with fine steel pins and the tissue processed for light and electron microscopy. The study focused on afferent units conducting at 20–30 m s^{-1}, i.e., on myelinated mechano-nociceptors. Some of the thin myelinated fibers approached the papillary layer of the corium (dermis), lost their myelin sheath in this area but retained their Schwann cell layer, and penetrated the epidermal basal layer. The penetration site was considered to be the receptive portion of the ending; here the axons exhibited mitochondria, clear round and dense-core vesicles. Most of the axons that penetrated into the epidermal layer still had a Schwann cell covering (Figure 9), but in some cases the Schwann cell sheath was lost and the basal lamina of the axon fused with that of keratinocytes (Cauna, N., 1966).

Nociceptive endings originating from Aδ-fibers have been reported to terminate in the basal layer of the epidermis, i.e., they do not penetrate into the

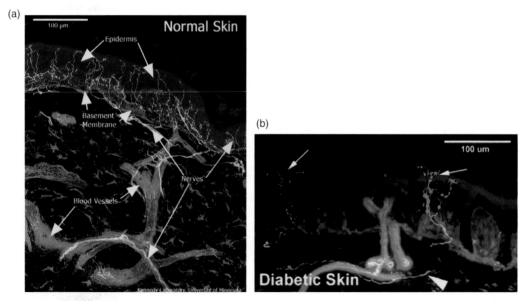

Figure 8 (a) Normal human epidermal and papillary dermis innervation. Nerve fibers (green and yellow) course in bundles through the dermis and branch into the papillary dermis to form the subepidermal neural plexus. Fibers from this plexus penetrate the epidermal–dermal basement membrane (red) to enter the epidermis (blue). Note that the basement membrane surrounding capillaries also appears red and some nonneuronal fibroblasts appear green. Epidermal nerve fibers are abundant and uniformly distributed. (b) Diabetic skin. Two tufts (arrows) arise from the subepidermal neural plexus (arrowhead). Nerve fibers comprising the right tuft have many swellings. Irregular distribution characterized by tufting and clustering of nerve fibers and the presence of swelling is distinctive of neuropathy. (From Kennedy, W. R., 2004, with permission.)

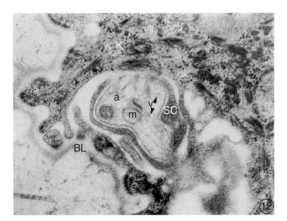

Figure 9 Cat hairy skin. An intradermal axon (a) containing mitochondria (m) and clear vesicles (v) (arrowheads) enclosed by a Schwann cell (SC). The basal lamina (BL) of the keratinocyte (K) is not shared by the Schwann cell, which lacks a basal lamina only within the epidermis (×40 000). (From Kruger, L. *et al.*, 1981, with permission.)

upper epidermal layer (Kruger, L. and Halata, Z., 1996). At present, these endings cannot be morphologically distinguished from cutaneous cold receptors that have the same location (Hensel, H. *et al.*, 1974).

Du J. *et al.* (2003) reported that nociceptive unmyelinated endings in the rat digital nerves exhibit *N*-methyl-D-aspartate (NMDA) receptor molecules type 1 (NMDAR1) and can be excited by glutamate. These data support those of Svensson P. *et al.* (2003). The NMDAR1 were assumed to contribute to the sensitization of nociceptors in inflamed tissue, because they showed an increased expression 7 days after induction of an experimental inflammation.

3.3.2   Cornea

Hundred years ago, von Frey assumed that from the cornea only pain can be evoked, but later several reports were published indicating that also sensation of touch, warmth, and cold can be elicited. The corneal epithelium is one of the tissues with the highest innervation density of the body and in this regard is comparable with the tooth pulp.

In an electrophysiological study on fine nerve fibers of the cornea *in vitro*, Aδ-fibers were described that were sensitive to mechanical and heat stimuli. These units were situated in the basal epithelial layer, whereas polymodal C-fiber endings were found in the superficial layers (MacIver, M. G. and Tanelian, D. L., 1993).

Recordings from single corneal sensory fibers demonstrated that the RFs of LTM endings were relatively large with a size of 50–200 mm^2 in the cat (Tower, S., 1940). Most of the slowly conducting (Aδ- or C-) fibers of the cornea were found to be polymodal and responded to mechanical, thermal, and chemical stimuli. They had a smaller RF size of ∼25 mm^2. Acidic solutions were used to test these units, and the endings responded readily to a pH of 5 to 4.5 with a discharge that was nearly proportional to the proton concentration. Capsaicin – a specific ligand for the TRPV1 receptor – was also excitatory for slowly conducting corneal afferents (Belmonte, C. *et al.*, 1991).

In contrast to most other body regions, the cornea is not supplied by blood vessels. Therefore, after a lesion has occurred, blood cells have no rapid access to the damaged area. In this case, lesion-induced sensitization of corneal nociceptors can occur through activation of tissue cyclooxygenase and lipoxygenase-generating prostaglandins (PGs) and other sensitizing agents.

An interesting feature of corneal sensory endings is that the continuous shedding of epithelial cells on the surface of the cornea changes the morphology of the nerve terminals. This results in a rearrangement of the arborization pattern of superficial terminals within 1 week. Also after a relatively mild injury, a remodeling of the corneal sensory ending takes place, and this remodeling is associated with an increased expression of the growth-associated protein (GAP)-43 in the corneal afferents (Belmonte, C. and Gallar, J., 1996). This protein is known to be involved in the development and regeneration of neurons; it is also present in growth cones of regenerating nerve fibers.

3.3.2.1   *Light microscopy*

Light microscopic investigations showed that corneal afferents were either thin myelinated or unmyelinated. As in the skin, they formed a dense plexus in the stroma from which branches ascended through the basal epithelial layer and terminated in bulb-like structures just underneath the corneal surface. In the electrophysiological study mentioned above (MacIver, M. G. and Tanelian, D. L., 1993) the fibers differed in their morphology: Aδ-fibers looked like leashes and polymodal C-fiber terminals like stranded endings under epifluorescence.

As in free nerve endings of other tissues, neuropeptides such as SP and CGRP are present in corneal terminals. Interestingly, in the cornea SP has a trophic effect on the corneal epithelium. The healing of a corneal wound is promoted by external administration of SP. Conversely, the healing process is delayed if SP is missing in the corneal terminals, e.g., after injection of capsaicin into the retrobulbar space to deplete SP-containing fibers. Generally, SP appears to promote mitosis in epithelial cells of the cornea, while CGRP inhibits it. After corneal irritation or damage, signs of neurogenic inflammation occur such as edema, conjunctivial vasodilatation, and photophobia. These symptoms are attributed to the release of neuropeptides from the sensory corneal terminals (e.g., SP and CGRP).

3.3.2.2 Electron microscopy

Electron microscopic studies showed that thin fibers entering the epithelium lost their Schwann cell covering, often branched and exhibited varicosities that contained mitochondria, neurofilaments and round clear vesicles (Figures 10 and 11; Belmonte, C. and Gallar, J., 1996; Müller, L. J. *et al.*, 1996). Terminals containing granular vesicles were numerous only at the limbus of the cornea. These endings disappeared after sympathectomy and, therefore, were considered to be adrenergic terminals. Sensory endings (i.e., those having their somata in the trigeminal ganglion) were often ir to SP and CGRP. Hoyes A. D. and Barber P. (1976) distinguished between two types of sensory free nerve endings that originated from unmyelinated corneal fibers:

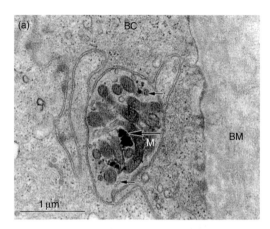

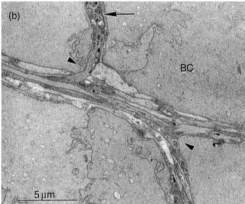

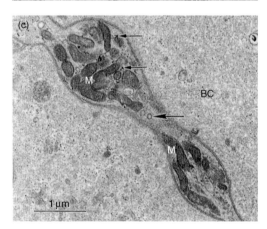

Figure 11 Nerve fibers in the epithelium of the human cornea. (a) Cross section of a single nerve fiber at the level of an axonal expansion (a bead), showing numerous mitochondria (M), glycogen particles (large arrow), and vesicles (small arrow) (×16576). (b) Frontal section of a nerve filament that bifurcates at the points indicated by the arrowheads. The single fiber running in the upper part of the picture has a bead filled with mitochondria (large arrow) (×3000). (c) Frontal section of a single fiber, showing two beads filled with numerous dark mitochondria (M), glycogen particles (small arrows), and vesicles (large arrow) (×12544). BC, basal cell; BM, Bowman's membrane. (From Müller, L. J. *et al.*, 1996, with permission.)

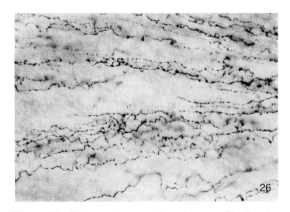

Figure 10 Flat mount preparation of cornea. The focal plane is through the base of the corneal epithelium, which is heavily innervated by varicose calcitonin gene-related peptide-immunoreactive (CGRP-ir) axons coursing radially inward from the limbus. Branches can be traced in more superficial focal planes ascending perpendicularly between epithelial cells to approach the epithelial surface (×200). (From Silverman, J. D. and Kruger, L., 1988, with permission.)

(1) Units that contained many mitochondria and were considered to be mechano- or thermo-receptors.

(2) Units exhibiting clear and dense-core vesicles. These units were assumed to be nociceptive afferents; they were concentrated in the central parts of the cornea where no blood vessels are present. The majority of this fiber type had an intraepithelial location.

Attempts to correlate morphology with function in corneal receptive endings did not yield unequivocal results. Therefore, in their review, Belmonte C. and Gallar J. (1996) stated that "no morphological specialization has been found in corneal nerve terminals associated with the various functional classes of corneal afferents."

3.3.3   Deep Somatic Tissues

3.3.3.1   *Muscle pain compared to cutaneous pain*

Muscle pain is less well localized than cutaneous pain. One reason for this difference may be the fact that muscle pain – but not cutaneous pain – is often referred to sites distant from the lesion. Another possible reason is the lower innervation density of muscle tissue. Direct quantitative comparisons between the innervation density of muscle and skin have not been published so far and are difficult to make, because muscle tissue is 3D and skin largely 2D.

Free nerve endings in muscle are partly nociceptive and partly nonnociceptive. The latter ones have a low mechanical threshold and are assumed to mediate pressure and tension sensations from muscle tissue. Another important function for these nonnociceptive endings in muscle is to control the adjustment of circulation and respiration to the requirements of muscle work. They are activated by physiological degrees of exercise and send their information to the circulatory and respiratory centers in the medulla (McCloskey, D. I. and Mitchell, J. H., 1972).

3.3.3.2   *Nociceptors in muscle and tendon*

Several studies on the response properties of unmyelinated afferent fibers from cat and dog muscle demonstrated that a subpopulation of these afferents are excited by stimuli (strong mechanical and algesic chemical) that are known to elicit pain (Kumazawa, T. and Mizumura, K., 1977; Mense, S., 1977; Mense, S. and Meyer, H., 1985).

Besides capsaicin, ATP (binding to the purinergic P2X3 receptor), NGF, and protons (acting through TRPV1 and acid-sensing ion channels (ASICs)) are effective stimulants for muscle nociceptors (Hoheisel, U. *et al.*, 2004; 2005). Similar to nociceptive afferents from the skin, afferent fibers from nociceptors in rat muscle are equipped with TTX-resistant (TTX-r) Na^+ channels (Steffens, H. *et al.*, 2003).

3.3.3.2.(i)   Light microscopy   In a quantitative evaluation of neuropeptide-ir free nerve endings and preterminal axons (both characterized by varicosities) in the gastrocnemius–soleus muscle of the rat, most endings were found around small blood vessels (arterioles or venules), whereas capillaries and the muscle cells were not supplied by these endings. Most numerous were the CGRP-ir endings followed by endings with SP-immunoreactivity (IR), VIP-IR, NGF-IR, and GAP-43-IR (Reinert, A. *et al.*, 1998). Many endings exhibited IR for more than one peptide, e.g., for SP and CGRP or SP and VIP (Figure 12).

After 12 days of an experimental myositis, the innervation density of the muscle with neuropeptide-ir free nerve endings was significantly increased. The effect was particularly marked for endings with SP-IR, GAP-43-IR, and NGF-IR (Reinert, A. *et al.*, 1998; Figure 13 (b and c)). The density of the SP-ir fibers doubled in inflamed muscle. Of course the question arose if this increase was due to inflammation-induced sprouting of the nerve fibers or due to an increase in the neuropeptide content of the individual fiber so that a higher proportion of fibers were above the detection threshold in the fluorescence microscope. The finding that the density of NGF-ir and GAP-43-ir fibers increased together with that of SP-ir endings was interpreted that sprouting had occurred, because NGF and GAP-43 are strongly expressed in growth cones.

Free nerve endings in the calcaneal tendon of the rat showed IR to the same neuropeptides, but the distribution of the fibers was different from muscle. In the epitendineum, the fibers formed a much denser network than in muscle, but the collagen fiber bundles in the center of the tendon were largely free from free nerve endings (Reinert, A. and Mense, S. unpublished data). This finding may explain why chronic mechanical stress to the epitendineum can be very painful, whereas ruptures of collagen bundles within the tendon may occur without pain.

3.3.3.2.(ii)   Electron microscopy
3.3.3.2.(ii).(a)   Muscle   The first comprehensive report on the morphology of free nerve endings

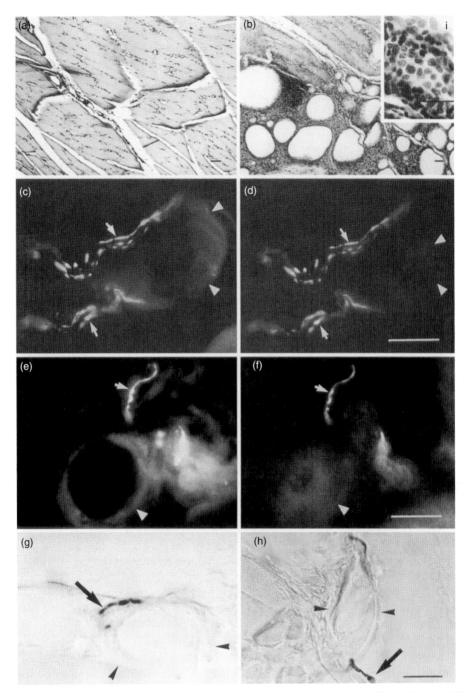

Figure 12   (a and b) Sections of intact (a) and persistently inflamed (b) muscle. Scale bar = 100 μm. The insert (i) in (b) shows lymphoid cells between vacuoles at a higher magnification. Scale bar = 50 μm. (c–f) Neuropeptide-ir nerve fibers in muscle. (c) Varicose SP-ir fibers (arrows) in the adventitia of an artery (arrowheads). (d) The fibers shown in (c) double-labeled for calcitonin gene-related peptide (CGRP). (e) Perivascular SP-ir fiber that is also ir for vasoactive intestinal polypeptide (VIP; (f)). (g) Perivascular fiber in inflamed muscle exhibiting immunoreactivity for NGF (arrow, same magnification as in (h)). (h) Growth-associated protein 43 (GAP 43)-ir fiber (arrow) in the wall of an artery (arrowheads). Scale bar = 25 μm (c–h). (From Reinert, A. *et al.*, 1998, with permission.)

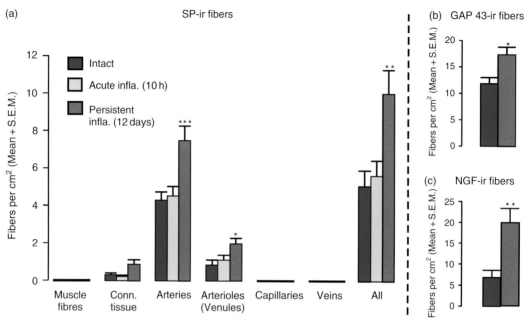

Figure 13 Innervation density of substance P-immunoreactive (SP-ir), growth-associated protein 43 (GAP 43)-ir, and nerve growth factor (NGF)-ir fibers in intact and inflamed muscle. Acute inflammation (yellow bars), duration 10 h; persistent inflammation (red bars), duration 12 days. (a) Distribution of SP-ir fibers in the components of muscle tissue. Note that no fibers were found close to muscle fibers and capillaries. Asterisks mark the level of significance compared to intact muscle. $^{*}p < 0.05$; $^{**}p < 0.01$; $^{***}p < 0.001$, U test. (Modified after Reinert, A. et al., 1998.)

in skeletal muscle of the cat was published by Stacey M. J. (1969). He used the silver impregnation technique in sympathectomized animals and focused on endings supplied by group III and IV fibers. The majority of the latter had a diameter of 0.35 μm; the unmyelinated afferents outnumbered the myelinated ones by a factor of 2. A prominent location of free nerve endings supplied by group IV fibers was the adventitia of arterioles and venules. Myelinated (group II and III) afferents generated not only free nerve endings but also paciniform corpuscles, whereas unmyelinated fibers terminated exclusively in free nerve endings.

In another study on sympathectomized cats, von Düring M. and Andres K. H. (1990) reported that the predominant location of group III and IV endings was in the perimysium surrounding larger or smaller bundles of muscle fibers. Other locations were the adventitia of arterioles, venules, and lymphatic vessels (LV), and finally the endoneurium of nerve fiber bundles. The latter terminals were assumed to belong to nervi nervorum. In muscle, the terminals of group III fibers were generally larger than those of group IV fibers, and they contained more mitochondria and a more distinct receptor matrix. Figure 14 shows a group IV terminal in the endomysium around the

precapillary segment of a blood vessel in the soleus muscle. The ending contained clear and dense-core vesicles and had exposed membrane areas. The authors suggested that those terminals (mainly originating from group III fibers) that had a close association with connective tissue may have a mechanoreceptive function mediating stretch or pressure, whereas endings that lacked this feature but had a spatial relation to mast cells were nociceptors.

3.3.3.2.(ii).(b) Tendon The electron microscopic reconstruction of free endings in the calcaneal tendon of the cat yielded various morphological types of free nerve endings connected to group III and IV afferent fibers (Andres, K. H. et al., 1985). Based on morphological criteria or the location in the tissue, the authors distinguished five types of free nerve endings supplied by group III fibers: Type 1 terminated in venous vessels (VV) (called the lanceolate type); it was special in that it had a flattened profile in cross sections and possessed exposed receptive areas on its edges. Type 2 ended in the wall of LV. Types 3 and 4 supplied the connective tissue around blood vessels, and one of these types had contacts to collagen fiber bundles. The collagen bundles were assumed to transfer mechanical forces to the receptive ending

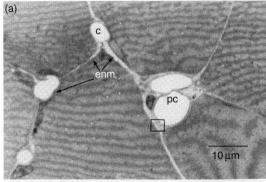

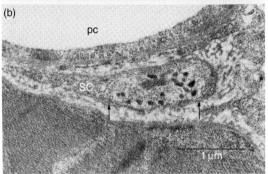

Figure 14 Nerve terminal of a group IV fiber in the soleus muscle of the cat. Rectangle in (a) represents the area of the receptor shown in (b). The terminal axon contains numerous dense-core and clear vesicles. pc, precapillary segment; c, capillary; SC, Schwann cell. Arrows indicate exposed axonal membrane areas. (From von Düring, M. and Andres, K. H., 1990, with permission.)

and, therefore, the endings were viewed as mechanoreceptors. The fifth type of group III ending innervated the endoneural connective tissues of small nerve fiber bundles (Figure 15).

In the same study, two types of free nerve endings of group IV fibers were described; both were located in the connective tissue of blood vessels, and some of the fibers contained granulated vesicles. The terminals showed penicillate formations (Figure 16).

3.3.3.3 Nociceptors in joints and ligaments

3.3.3.3.(i)   Pain from joints In humans, experimental joint pain has been evoked by administration of noxious mechanical and chemical stimuli to the articular capsule and ligaments. No pain is elicited by stimulation of cartilage, because cartilage is not innervated. Joint inflammation is characterized by hyperalgesia and persistent pain at rest. The discharge properties of joint nociceptors largely reflect these subjective phenomena of joint pain (Guilbaud, G. *et al.*, 1985;

Schaible, H.-G. and Schmidt, R. F., 1985; Schaible, H.-G. and Grubb, B. D., 1993; Schaible, H.-G., 2006). In inflamed joint, initially mechanoinsensitive afferents (silent nociceptors) are sensitized and may become mechanosensitive (Schaible, H.-G. and Schmidt, R. F., 1988).

3.3.3.3.(ii)   Ultrastructure of articular free nerve endings In a thorough investigation of articular endings, Heppelmann B. *et al.* (1990a) studied free nerve endings from the knee joint of the cat in sympathectomized animals. Figure 17 shows the reconstruction of group III and IV nerve endings from the joint capsule of the cat's knee joint. As in other tissues, free nerve endings in joint were ensheathed by Schwann cells, and only some sites were not covered. Usually, these exposed areas were located in the varicosities of the ending (the beads) and the end bulb, here the axon, often directly abutted the basal lamina. The exposed membrane areas were assumed to be the receptive sites of the fibers; they exhibited structures that are often associated with receptive sites: aggregates of mitochondria, glycogen particles, vesicles in the axoplasm, and an electron-dense filamentous structure of the axoplasm (the receptor matrix). The end bulb, i.e., the most distal beaded end of the fiber (Figure 18 shows the end bulb of one branch of the group III unit reconstructed in Figure 17), had generally the same structure as the axonal beads, except that the mitochondria showed a special arrangement: the organelles were not oriented in parallel as in the axonal beads but pointed with their axis in all possible directions.

In the cat's knee joint capsule, free nerve endings generated by group III fibers were smaller in axon diameter and had shorter branches (up to 200 μm) than those supplied by group IV fibers (more than 300 μm). The main difference between group III and IV endings was that the latter never had a neurofilament core in their axoplasm, which was present in all axons and axonal beads of group III fibers (but not in the end bulbs). The functional significance of the neurofilament core is obscure; one possible function is mechanical stabilization of the group III endings that usually have a more unfavorable quantitative relation between the fluid axoplasm and the more rigid axolemma.

The authors of these and similar data (Heppelmann, B. *et al.*, 1990a; 1990b; Messlinger, K., 1996) put forward the concept that both group III and IV fibers form sensory terminal trees that consist of the preterminal axon and its branches (Figure 5). The

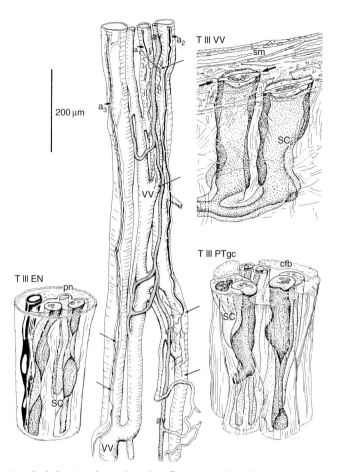

Figure 15 Group III fiber terminals in cat calcaneal tendon. Reconstruction of a series of sections over a length of 1 mm showing three different terminals of group III afferents. A fascicle of arterial vessels (av) and venous vessels (VV) with some arterio-venous anastomoses is accompanied by small nerve fiber bundles of which the myelinated axons (a_1, a_2, a_3) are reconstructed. The schematic drawings show segments of three different terminals at a higher magnification (T III VV, T III PTgc, and T III EN) with their typical relations to the surrounding tissue structures. Axon membrane segments without any Schwann cell (SC) covering are marked by higher contrast. The terminal of the a_1 axon is the T III VV type with two lanceolate terminals in the adventitia of the vessel. The arrows indicate the free edges of the terminal. Smooth muscle cell (sm). The terminal of the a_2 axon is type T III PTgc in the connective tissue of the vessel–nerve–fiber fascicle. The terminals are oriented to collagen fiber bundles (cfb). The terminal of the a_3 axon is the T III EN type. The terminals ramify within the endoneural space. Another small myelinated axon passes the terminals. Perineural sheath (pn). The length of the receptive areas of each myelinated axon is indicated in the reconstruction by arrows. (From Andres, K. H. *et al.*, 1985, with permission.)

terminal tree is not covered by myelin (in group III endings) and lacks a perineurial sheath (in both group III and IV endings); it is assumed to form the sensory ending proper.

Considerations on the space constant of free nerve endings prompted Heppelmann B. *et al.* (1990a; 1990b) to hypothesize that the distance from the end bulb as the main receptive region to the regenerative region – where the action potential originates – may be dangerously great for a safe transformation of the receptor potential into an action potential. Therefore, the

axonal beads may not only serve as receptive sites but also as amplifiers of the receptor potential.

The group IV endings terminating in the medial collateral ligament (Figure 19) exhibited the same features as the terminals in the joint capsule: They had exposed membrane areas and contained mitochondria, glycogen particles, and vesicles (Heppelmann, B. *et al.*, 1990b; Messlinger, K., 1996).

Group III and IV fibers of the joint terminate as free nerve endings in the fibrous capsule, adipose tissue, ligaments, menisci, and periosteum. Staining

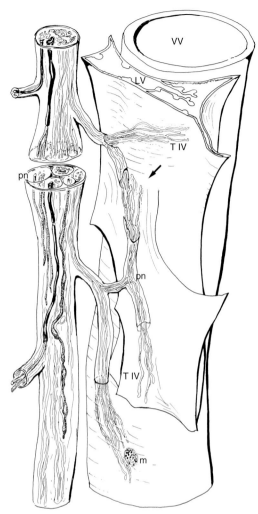

Figure 16 Group IV fiber terminals (T IV) in cat calcaneal tendon. Schematic drawing of group IV fibers and their sensory regions in the surrounding connective tissue of venous (VV) and lymphatic (LV) vessels. Perineural sheath (pn). The various parts of the scheme are not to scale. The terminal bundles end in penicillate formations. (From Andres, K. H. *et al.*, 1985, with permission.)

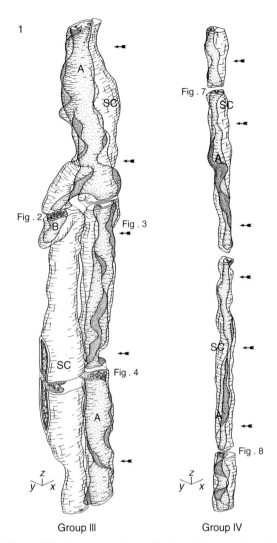

Group III   Group IV

Figure 17 Sensory endings in the knee joint capsule of the cat. Reconstructions based on ultrathin sections at 1-μm intervals. Areas of bare axolemma without Schwann cell covering are marked by dense stippling. A, sensory axon; SC, Schwann cell; arrows, thin segments between axonal beads (×6000). *x*, *y*, *z* bars, 1 μm. *Left*: group III fiber, length of reconstructed fiber 70 μm. *Right*: group IV fiber. The axonal beads are shorter and the thin segments longer than in group III fibers. Areas of bare axolemma are interrupted by axon segments with complete Schwann cell wrapping. (From Heppelmann, B. *et al.*, 1990b, with permission.)

for nerve fibers and neuropeptides also demonstrated endings in the synovial layer. Interestingly, a few group IV endings terminated in the perineurium (PN) of the articular nerve. Assuming that these endings are nociceptors, they might be responsible for neuropathic pain originating in the joint.

The major neuropeptides in small-diameter joint afferents were SP, CGRP, and somatostatin (SOM). Neurokinin (NK)-A, galanin, enkephalins, and neuropeptide Y have also been localized to joint afferents (Schaible, H.-G. and Grubb, B. D., 1993; Schaible, H.-G., 2006).

3.3.3.4 Tooth pulp

3.3.3.4.(i) Toothache In psychophysiological studies, electrical stimulation of the tooth pulp has been reported to elicit pain exclusively (Andersson, S. A. *et al.*, 1973). Therefore, the free nerve endings in the pulp were assumed to be either nociceptors or sympathetic postganglionic fibers. Rats show aversive behavior when capsaicin is applied to the exposed

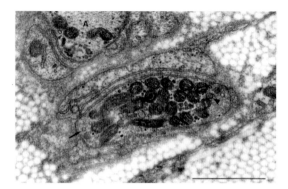

Figure 18 Axonal end bulb of an articular group III fiber in the cat. In addition to mitochondria and glycogen particles, the end bulb contains several clear vesicles (arrow) and larger dense-core vesicles (arrowhead). In contrast to more proximal axonal beads, microtubules and a neurofilament core are missing in the end bulb. (From Heppelmann, B. et al., 1990a, with permission.)

dentine of the incisors (Chidiac, J. J. et al., 2002; Rifai, K. et al., 2004). The latter observation indicates that nociceptive afferents of the tooth pulp are equipped with the TRPV1 receptor for which capsaicin is a specific ligand.

Functional studies of the receptive properties of single unmyelinated tooth pulp afferents are scarce. The published data show that slowly conducting afferent units from cat are sensitive to stimuli that are known to elicit pain in humans (hot and cold, local administration of bradykinin (BK), and mechanical stimulation) (Jyväsjärvi, E. and Kniffki, K. D., 1992). Many of the recorded units responded to more than one stimulus and thus behaved like polymodal nociceptors. In another study, the sensitivity to capsaicin of Aβ-, Aδ-, and C-fibers was compared in cat tooth pulp afferents. Not only C-fiber endings were excited but also Aδ-units, albeit at a smaller proportion. None of the Aβ-endings responded to the stimulant (Ikeda, H. et al., 1997).

No studies combining electrophysiological and morphological techniques were conducted on teeth. Therefore, in morphological investigations a possible nociceptive function of the free nerve endings can be discussed, but is not sure.

3.3.3.4.(ii) Light and electron microscopy

Data are available from many species including man. The morphology of the free nerve endings is similar to that described above for other tissues in more detail. In the following paragraph, possible differences between nerve fibers in mature and continuously growing teeth (e.g., mice incisors) are not considered.

(1) The great majority (56–79.6% in the cat mandibular canine tooth; Holland, G. R. and Robinson, P. P., 1983) of the fibers supplying the tooth pulp are unmyelinated with a mean diameter of 0.4 μm in man (Koling, A., 1985); they often branch and form a dense subodontoblastic plexus (the peripheral plexus of Raschkow, see Byers, M. R., 1984, for review) in the apex of the tooth. Between the plexus and the odontoblast layer, there is the cell-free zone of Weil. Figure 20 shows an overview of the afferent fibers in the root and coronal pulp of a rat molar. The data were obtained in experiments in which the trigeminal ganglion was injected with an anterogradely transported marker (WGA-HRP) to visualize afferent fibers only (Ibuki, T. et al., 1996). In the subodontoblastic region, axo-axonal appositions and appositions between axons and odontoblasts have been described by some authors, but others could not support this finding. The other pulpal afferents conduct in the Aδ-range; Aβ-fibers are sparse.

(2) Many axons are incompletely ensheathed by Schwann cells (Figure 21); they sometimes leave the Schwann cell sheath completely to terminate as naked axons in the extracellular space. In those fibers that do have a sheath, the Schwann cells often lack a basal lamina (Figure 22; Byers, M. R. et al., 1982). The majority of the afferent fibers form beaded free nerve endings in odontoblast layer, predentin, and in dentinal tubules, many of which are innervated. The intradentinal endings extend only 100–200 μm into the dentin, and sometimes more than one axon accompanies an odontoblast process (Figure 23). Gunji T. (1982) put forward the concept that the odontoblast processes and the axons in the dentinal tubules form a mechanoreceptive complex. In contrast to the intradentinal axons, the odontoblast processes reach the dentin–enamel border and could be deformed by mechanical stimuli. The mechanical movements or deformations of the processes may then excite the sensory axons that do not penetrate that far into the dentinal tubules.

(3) The varicosities of the free nerve endings exhibited specializations known from nociceptors in other tissue such as mitochondria, small clear and large dense-core vesicles, smooth endoplasmic reticulum, and few microtubules and neurofilaments (Figure 24; Byers, M. R., 1984; Ibuki, T. et al., 1996). Ibuki T. et al. (1996) distinguished between

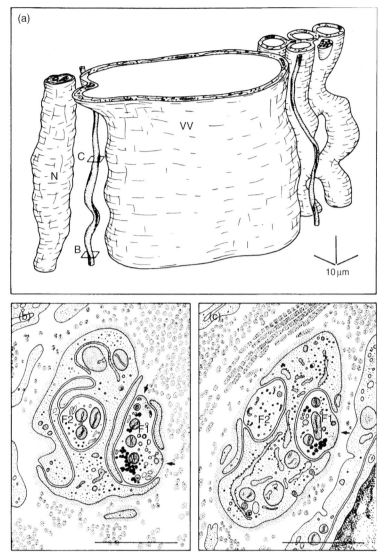

Figure 19 Group IV fiber endings at a venous vessel (VV) in the medial collateral ligament of the cat. (a) Reconstruction based on serial semi- and ultrathin sections. Besides a small peripheral nerve (N), two terminal nerve fiber bundles without perineurial sheath that enclose two group IV fibers (*left*) and one single fiber (*right*) run along the VV. (b and c) Cross sections of the nerve fiber bundle with two group IV fibers as marked in Figure 3(a) at high magnification. Drawings from original electron micrographs. Scale bar = 1 μm. (b) Cross section showing a bulge of one of the nerve fibers (FI) with bare areas of the axolemma (arrows), mitochondria, glycogen particles, and vesicles. The other nerve fiber (F2) is completely ensheathed by the Schwann cell. C, cross section 40 μm distal from that of Figure 3(b) showing the final thickening of the nerve fiber FI with bare area of axolemma (arrow). The second nerve fiber (F2) continues. (From Heppelmann, B. *et al.*, 1990b, with permission.)

two types of varicosity: one contained many mitochondria, while the other was characterized by the presence of large dense-core and small clear vesicles. Generally, a rich supply with mitochondria and glycogen particles is interpreted as an indication of high energy demands and, therefore is likely to occur in endings with a low mechanical threshold and high firing probability, while nociceptive

endings have a lower probability of being excited, and therefore need less mitochondria.

(4) Besides a nociceptive function many of the small-caliber tooth afferents may have a trophic influence on blood vessels in the pulp or on odontoblasts. They may be necessary for repair or normal growth, as in continuously growing teeth. The finding that the anterograde marker

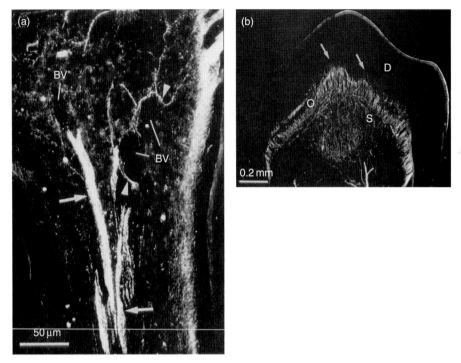

Figure 20   Innervation of root pulp (a) and coronal pulp (b) in a rat tooth. The fibers were labeled anterogradely from the trigeminal ganglion. (a) Bundles of labeled fibers (arrows) run along the blood vessels (BV) of the root pulp. Some fibers surround the blood vessels (arrowheads). (b) The fibers form a subodontoblast plexus (S) and penetrate the dentin (arrows). D, dentin; O, odontoblast. (From Ibuki, T. *et al.*, 1996, with permission.)

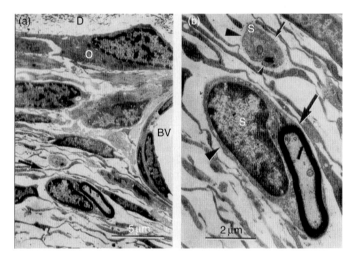

Figure 21   (a) Fibers labeled anterogradely with horseradish peroxidase in the subodontoblastic region of the tooth. BV, blood vessel; D, dentin; O, odontoblast. (b) High magnification of the area marked by an arrow in (a). Horseradish peroxidase reaction product is present in the myelinated Aδ-axon (large arrow) and unmyelinated axons (small arrow). The unmyelinated axon has partially lost its Schwann cell sheath (arrowheads). S, Schwann cell. (From Ibuki, T. *et al.*, 1996, with permission.)

(WGA-HRP) injected into the trigeminal ganglion by Ibuki T. *et al.* (1996) was found not only within the axons but also outside between the axons and the odontoblast processes (Figure 24) was assumed to reflect a release process that normally supplies the odontoblasts with trophic substances. This hypothesis resembles the nocef-fector concept of Kruger.

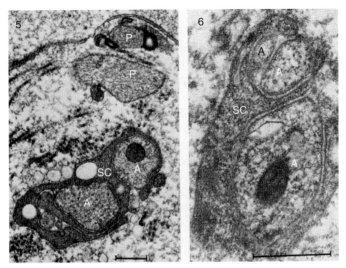

Figure 22 Nerve fibers in the cell-free zone of the tooth pulp. Unmyelinated axons (A) were identified in this zone when surrounded by a Schwann cell (SC), but no basal lamina accompanied the Schwann cell. Other cell processes (P) could be axons or fibroblasts. Scale bars, 0.5 μm (×42300). (From Byers, M. R. et al., 1982, with permission.)

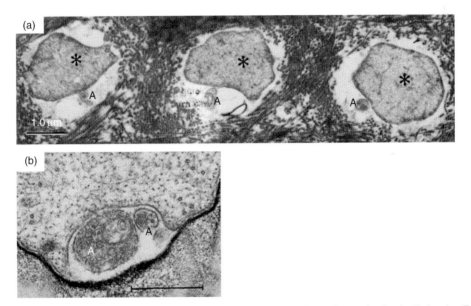

Figure 23 Axons in dentinal tubules. (a) A cluster of adjacent dentinal tubules at the predentin–dentin border. Each tubule contains an axon (A) and an odontoblast process (*) (×10350). (b) Two adjacent axons in the same dentinal tubule. (From Byers, M. R. et al., 1982, with permission.)

3.3.3.4.(iii) *Neuropeptide content and equipment with receptor molecules* Sensory dental axons contain many neuropeptides (SP, CGRP, VIP, enkephalin), the most interesting one being SP because of its relation to a nociceptive function. A particularly rich innervation with CGRP-ir fibers was found in the subodontoblast layer and the dentinal tubules (Kruger, L. and Halata, Z., 1996). These endings were thought to have a noceffector function (Kruger, L., 1988).

SP-ir fibers predominated in the center of the tooth pulp. The expression of SP in carious human teeth has been found to be increased, and the increase was larger in symptomatic than in asymptomatic teeth. This finding parallels the results obtained from inflamed muscle (see above) and suggests that

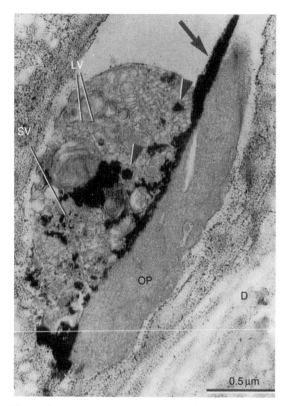

Figure 24 Anterogradely labeled sensory varicosity adjacent to an odontoblast process (OP) in dentin. The varicosity is full of large dense-core vesicles (LV) and small clear vesicles (SV). Some vesicles (arrowheads) contain the horseradish peroxidase reaction product used for anterograde labeling. The reaction product is also present in the extracellular space (arrow). D, dentin. (From Ibuki, T. *et al.*, 1996, with permission.)

SP plays an important role in the mediation of toothache (Rodd, H. D. and Boissonade, F. M., 2000). In one patient with hereditary sensory and autonomic neuropathy, CGRP- and SP-ir fibers were completely missing, which could explain the sensory deficit associated with this condition (Rodd, H. D. *et al.*, 1998). The lack of CGRP- and SP-ir fibers was accompanied by an upregulation of neuropeptide Y.

The nerve endings in the tooth pulp have been shown to be equipped with receptor molecules that are thought to be associated with a nociceptive function, e.g., P2X3, TRPV1 (Renton, T. *et al.*, 2003), and the vanilloid receptor-like protein, VRL1. In rat molar tooth pulp, fibers ir to trkB (the receptor for BDNF and NT-4) were found. These fibers included both unmyelinated and myelinated units. This and other findings were interpreted as indicating that BDNF and/or NT-4 are involved in the

development, differentiation, or regeneration of the tooth pulp innervation (Foster, E. *et al.*, 1995).

However, not all free nerve endings can be considered to be nociceptive; in the tooth pulp many endings appear to supply blood vessels (Zhang, J. Q. *et al.*, 1998). The vascular innervation with neuropeptide-ir nerve endings is largely restricted to arterioles; capillaries and lymphatics were mostly devoid of an innervation. This distribution pattern resembles that of skeletal muscle. Substances involved in the vascular and cellular regulation of the tooth pulp in the rat are SP, NK-A, and CGRP, which are thought to be released from nerve endings. This function was inferred from the finding that pulpal hard tissue cells, blood vessels, and fibroblasts were equipped with NK-1 and NK-2 receptors (for SP and NK-A, respectively) as well as CGRP1 receptors (Fristad, I. *et al.*, 2003).

3.3.3.5 Dura mater encephali

3.3.3.5.(i) Headache Generally, those types of headache that are associated with vascular changes in the meninges (e.g., migraine) are being attributed to the activation of para- or perivascular nociceptive endings by agents released from the nerve terminals. Among these substances are SP, CGRP, and NK-1. Observations supporting this assumption are that – when tested in awake subjects – meninges are most painful close to blood vessels, and that migraine pain is most intense during the phase of vasodilatation and increase of vascular permeability.

3.3.3.5.(ii) Light and electron microscopy In their light and electron microscopic study on receptive endings in the dura mater of the rat, Andres K. H. *et al.* (1987) described terminals of myelinated and unmyelinated fibers. The tissue around postcapillary blood vessels and the superior sagittal sinus exhibited a particularly high innervation density of endings. Contrary to former belief, the dura also contained LV (Figure 25). These vessels were innervated mainly by endings from unmyelinated fibers.

Some of the myelinated fibers terminated in Ruffini-like nerve endings, while most of the myelinated and all of the unmyelinated fibers terminated in free nerve endings. The Ruffini-like receptors were located close to collagen fiber bundles of the wall of the sagittal sinus and of other blood and LV. They were viewed as stretch receptors.

Among the free nerve endings formed by unmyelinated fibers, one type was located very close (at a distance of <1 μm) to the endothelial cells of the

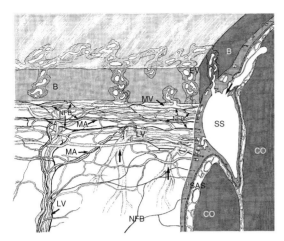

Figure 25 Vascular bed of the dura underneath the parietal bone (B) and dorsal to the sagittal sinus (SS). Arrows, entrance of the superior cerebral veins into the sagittal sinus; MA, branches of meningeal artery; MV, branches of meningeal vein; EV, emissary vein; LV, lymphatic vessels; NFB, nerve fiber bundles exhibiting a terminal plexus; CO, cortex; SAS, subarachnoidal space. (From Andres, K. H. *et al.*, 1987, with permission.)

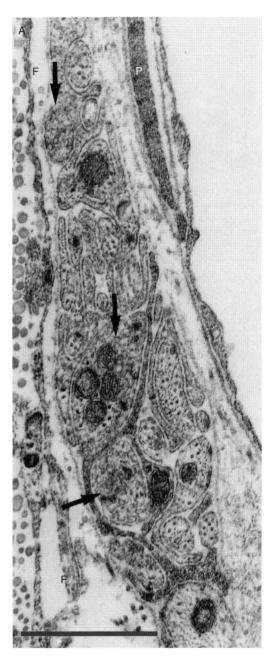

Figure 26 Nerve terminations in the dura at the wall of a venous vessel. Axon profiles with vesicles and receptor matrix are indicated (arrows); the other axons exhibit cytoskeletal elements (×45000). Scale bar = 1 μm. The gap between the terminal and the postcapillary venule with endothelial fenestration is <1 μm. F, process of fibrocyte; P, pericyte. (From Andres, K. H. *et al.*, 1987, with permission.)

postcapillary venules. It had the typical appearance of free nerve endings in other tissues with exposed membrane areas, mitochondria, vesicles, and a receptor matrix. The endings that were located close to the wall of blood vessels would be in a good position to monitor the composition of the interstitial fluid in the vicinity of the vessels (Figure 26). They could well perform nociceptive functions, for instance in cases of headache when sensitizing and pain-producing substances are released from the blood plasma and afferent nerve fibers.

Many of the nerve fibers accompanying blood vessels contained CGRP or SP (Figure 27; Hanesch, U., 1996). Like in other tissues, SP-ir fibers were less numerous than CGRP-ir ones. While SP-IR was located inside the axoplasm, the SP receptor NK-1 was found on the outer membrane of the Schwann cells surrounding the axons.

3.3.4 Visceral Organs

3.3.4.1 *Visceral pain*

The mechanisms of visceral pain are still poorly understood. Apparently, some inner organs are equipped with specific nociceptors (HTM and polymodal ones; Jänig W. (1996), for review), but their existence has not been proven for all viscera. It is conceivable that the pain from some viscera is mediated not by specific nociceptors but is encoded

in the pattern of discharge in populations of relatively unspecific visceral afferents. The clinical observation that a small cut into the gut is not painful, but spasm or ischemia of a larger area causes strong

(a)   (b)

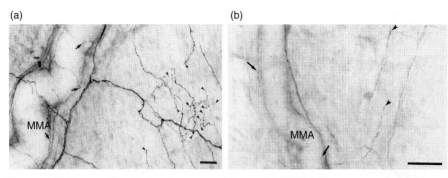

Figure 27 Neuropeptide-ir nerve fibers in the dura mater of the rat. (a) CGRP-ir fibers accompany and cross the middle meningeal artery (MMA). Small bundles of fibers branch off the main bundle and terminate in the connective tissue thereby forming tree-like endings (arrowheads). Some fibers end at the artery (arrows). (b) SP-ir fibers accompanying the MMA. Single beaded nerve fibers terminate at the vessel wall (arrows) or in the connective tissue between blood vessels (arrowheads). (From Hanesch, U., 1996, with permission.)

pain, is generally attributed to the low innervation density of the inner organs.

3.3.4.2 Testis

Electrophysiological studies by Kumazawa and coworkers (Kumazawa, T. and Mizumura, K., 1980; Kumazawa, T. *et al.*, 1987) have shown that in the tissue layers covering the seminiferous tubules in the rat, endings with unmyelinated afferent fibers are present that behave like polymodal nociceptors in that they can be activated by mechanical stimuli, pain-producing substances, and heat. The innervation density of the tissue is low, and the receptive endings are located directly underneath the tunica albuginea. These properties made it possible to relate the mechanosensitive spots found in electrophysiological experiments to structures in histological specimen.

Kruger L. and coworkers (2003b) used this approach to elucidate the ultrastructure of nociceptive endings of the testis. In the electrophysiological experiments, the location of the receptive ending was marked with insect pins. The endings were reconstructed from serial electron microscopic sections (Figure 28), and in general they showed the features described above for putative nociceptors in other tissues. The preterminal axon had expansions (varicosities) with axonal constrictions in between. Approaching the terminal proper, the axons often lost their microtubules and neurofilaments. Instead, a network of axonal reticulum (probably endoplasmic reticulum) appeared that was embedded in a granular matrix (similar to the receptor matrix of Andres, K. H. and von Düring, M., 1973). This arrangement of an axonal reticulum in a granular matrix (Figure 29) has not been described by other authors, but it is unclear if

this arrangement is special for testicular nociceptors, or if it had been overlooked in other tissues.

In the distal tip of the axon, the axolemma exhibited a greater number of exposed areas that directly contacted the basal lamina. The authors stress the presence of clusters of small spherical vesicles, aggregates of mitochondria, and axonal reticulum in a granular matrix in this part of the axon that apparently forms the receptive terminal (Figure 30). Interspersed between the spherical vesicles were so-called granular vesicles that differed somewhat from the dense-core vesicles of endings containing catecholamines and, therefore, could be distinguished from sympathetic nerve endings.

The functional interpretation of the ultrastructure of the terminal poses some problems: Even though it is characterized by special features (vesicle clusters, mitochondria, and exposed axolemma), the axonal zones close to the terminal likewise exhibited some of these specializations. Therefore, it is difficult to decide if only the ultimate terminal is the receptive site where the electrical receptor potential originates or if also other parts of the preterminal branches can be viewed as receptive, as in the sensory tree concept. An important question in this context is where the generator portion of the nociceptor axon is located and if it exhibits structural specializations different from those of the receptive terminal.

3.4 Neuropeptide Content of Nociceptors

3.4.1 General Remarks

In many cells, neurotransmitters coexist with neuropeptides, or one neuron contains more than

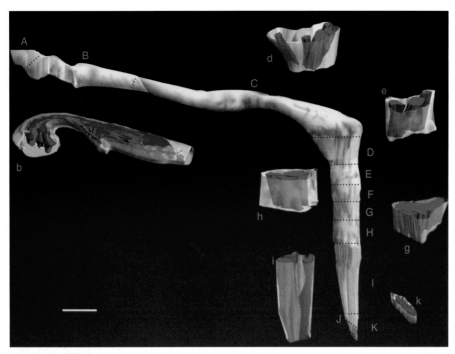

Figure 28 Three-dimensional reconstruction of a pin-marked receptive field (RF) identified from a physiologically characterized nociceptor polymodal fiber discharge in an *in vitro* canine testicular preparation based on a series of semithin (~1 μm) sections from which ultrathin sections were obtained in each of the sectors, labeled A–K. In the segments indicated in lower case, axons are shown in red and Schwann cell basal lamina boundaries in white. Scale bar = 1 μm. (From Kruger, L. *et al.*, 2003b, with permission.)

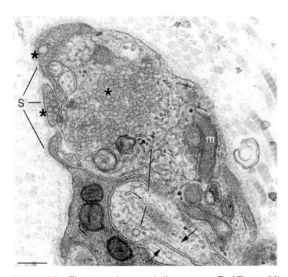

Figure 29 Electron micrograph (from sector E of Figure 28) displaying a putative receptive terminal containing a characteristic preterminal array of axonal reticulum (ar), microtubules (arrows), glycogen granules, and spherical vesicles (asterisks) embedded in a granular matrix and clustered below a surface ensheathed by dark Schwann cell processes (S) and containing vesicles. The bare upper axolemma is in direct contact with the basal lamina. Scale bar = 0.5 μm. (From Kruger, L. *et al.*, 2003, with permission.)

one neuropeptide. It is well known that a large proportion of SP-expressing DRG neurons also contain glutamate. Another example is that almost all DRG cells that are ir for CGRP also exhibit IR for SP and other neuropeptides (Ju, G. *et al.*, 1987). It is generally assumed that glutamate acts as the main neurotransmitter in nociceptive afferents, whereas the neuropeptides function as neuromodulators that – with few exceptions – enhance the central nervous effects of peripheral noxious stimuli (Hökfelt, T. *et al.*, 1980; Kow, L.-M. and Pfaff, D. W., 1988). The peptides are synthesized in the somata of the DRG or in ganglion cells of cranial nerves. They are transported to both the central and the peripheral terminal of the primary afferent unit.

3.4.2 Neuropeptides in Afferent Units of Different Tissues

No neuropeptide has been found that can be considered specific for afferent fibers from a particular tissue. DRG cells projecting in a cutaneous nerve

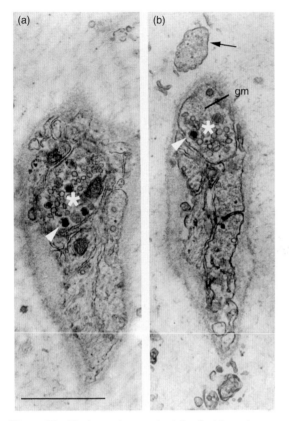

Figure 30 Electron micrograph of the final traced portion of the marked receptive field (RF) indicated in sector K of Figure 28 containing the final vesicle-rich axonal profiles, including several dense-core vesicles (arrowheads), all within a granular matrix (gm) exposed in b to the basal lamina. Arrow, additional unidentified profile. Scale bar = 1 μm. (From Kruger, L. et al., 2003b, with permission.)

have been shown to contain SP, CGRP, and SOM and thus present a peptide pattern similar to that of muscle nerves, whereas visceral afferent units lack IR to SOM (Molander, C. et al., 1987; O'Brien, C. et al., 1989). In comparison to skin nerves, muscle nerves have been reported to contain less SP. This finding has been teleologically explained by assuming that the vasodilatation and plasma extravasation caused by the release of SP and CGRP from afferent fibers would be deleterious for skeletal muscle, since it is surrounded by a tight fascia. Because of the fascia, an SP-induced muscle edema would result in a high increase in interstitial pressure and could cause muscle necrosis. On the other hand, there can be no doubt that a considerable proportion of free nerve endings in muscle exhibit IR to SP and CGRP (Reinert, A. et al., 1998). Therefore, SP and CGRP

are likely to be involved in painful alterations of muscle tissue.

3.4.3 Neuropeptides in Nociceptors and Other Types of Free Nerve Ending

Whether or not a particular neuropeptide or combination of neuropeptides is associated with a particular type of receptor is obscure. There is evidence from studies on DRG cells (Lawson, S. N. et al., 1997; Djouhri, L. and Lawson, S. N., 2004) indicating that SP-like IR (SP-LI) – and to a lesser extent, also CGRP-LI – was present predominantly in nociceptive units. Arguments supporting a nociceptive function of SP are that noxious stimulation is followed by a release of SP in the dorsal horn, and that the mRNA for SP and other tachykinins is upregulated under these circumstances. On the other hand, there is also evidence speaking against such a relation between the nociceptive function and the presence of SP. In a study by Leah J. D. et al. (1985), the great majority (10 out of 12) of individually identified nociceptive DRG cells of the cat did not exhibit SP-LI. Virtually all primary afferent units with SP-LI also exhibited CGRP-LI (Garry, M. G. et al., 1989); both peptides are presumably released together when the fiber is active. In the spinal cord, CGRP has been reported to prolong the action of SP by inhibiting its degradation and to facilitate synaptic transmission in general by enhancing the calcium influx into afferent fibers.

Data from experiments in which single DRG cells with receptive endings in muscle were first functionally identified and then injected with a dye, showed that at least some cell bodies whose peripheral processes terminated in muscle HTM (presumably nociceptive) receptors exhibited CGRP-LI (Hoheisel, U. et al., 1994). However, IR to SP, CGRP, and other neuropeptides was not only present in nociceptive units but also in other types of muscle receptors (e.g., in some muscle spindles and other LTM units). For instance, SP-LI has been found in LTM receptors, e.g., in nerve fibers from Krause corpuscles in the dog's tongue. Most of the cutaneous and deep receptor types studied so far showed CGRP-LI in some of their cell bodies in the DRG (Hoheisel, U. and Mense, S., 1987). The only condition for the presence of CGRP-LI seemed to be that the afferent unit had a small soma size and/ or a slow CV.

3.5   Efferent Function of Nociceptors

3.5.1   Release of Neuropeptides from the Nociceptive Ending

Neuropeptides (e.g., SP, CGRP, VIP, and SOM) are stored in vesicles in the varicosities of the peripheral terminal. Whenever a nociceptor is excited, it releases neuropeptides into the interstitial tissue. SP then releases histamine from mast cells, and together with CGRP these agents cause vasodilatation and an increase in vascular permeability of the blood vessels around the active ending. The result is a shift of blood plasma from the intravascular to the interstitial space. Outside the blood vessel, BK is cleaved from the plasma protein kallidin, and serotonin (5-HT) is set free from platelets and PGs (particularly PG E2) from endothelial and other tissue cells. All these substances sensitize nociceptors. Thus, the main tissue alteration induced by a nondestructive noxious mechanical stimulus is a localized region of vasodilatation, edema, and sensitized nociceptors.

3.5.2   The Axon Reflex

The release of neuropeptides from an activated nociceptor is an essential aspect of its function. A nociceptor is not a passive sensor for tissue-threatening stimuli; it actively influences the microcirculation and chemical composition of the interstitial space around it. The morphological basis of the axon reflex is that the branches of a single nociceptive ending can extend over a relatively large area (several hundred micrometers). If a noxious stimulus activates only parts of the ending, the action potentials originating in the regenerative region of the ending can invade antidromically (against the normal direction of propagation) those branches of the ending that were not excited by the stimulus. These antidromic action potentials release neuropeptides from the unstimulated branches. The whole process is called axon reflex. It is assumed to be the reason for the wheal and flare around a cutaneous lesion. The vascular permeability is increased mainly by SP (and by the NK-A and -B; Gamse, R. and Saria, A., 1985), whereas CGRP is assumed to act primarily as a vasodilatator. There is evidence showing that CGRP enhances the plasma extravasation induced by SP and NK-A and -B, but reduces the vasodilatory action of SP by desensitizing muscle arterioles to the peptide (Öhlén, A. *et al.*, 1988).

The area of wheal and flare after a localized damage to the skin – for instance around a needle prick – could be an indicator of the extent of the excited nociceptive ending. Of course, this is true only if a single ending is stimulated and if the distance covered by diffusion of the neuropeptides is constant.

The size of the RFs of cutaneous polymodal nociceptors was found to be $<2 \, mm^2$ in cat (Bessou, P. and Perl, E. R., 1969) and $6–32 \, mm^2$ in rabbit (Kenins, P., 1988). These figures are greater than the reported length of the branches of a nociceptor ending (a few hundred micrometers). The difference may be due to the fact that nociceptors can be excited from a certain distance, particularly when coarse probes are used. For a muscle nociceptor, the size of the RF can be determined only with limitations, particularly if it is located deep within the muscle. The reported sizes of superficially located fields or the projections of deep RFs on the muscle surface range from spotlike to several square centimeters in the gastrocnemius muscle of the cat and dog (Kumazawa, T. and Mizumura, K., 1977; Mense, S. and Meyer, H., 1985).

3.5.3   Neurogenic Inflammation

The release of SP, CGRP, NK-A, and other agents from nociceptors is the central factor in the cascade of events that lead to neurogenic inflammation in the periphery (Lembeck, F. and Holzer, P., 1979). Neurogenic inflammation is characterized by tissue edema and infiltration by immune cells after antidromic activation of nociceptive nerve endings, i.e., when action potentials are generated somewhere along the course of primary afferent units (spinal nerve or dorsal root). The action potentials propagate to both the CNS (causing pain) and the peripheral ending (causing release of neuropeptides and neurogenic inflammation). The published data indicate that vasodilatation can be elicited by antidromic stimulation of both $A\delta$- and C-fibers, but increase in vascular permeability and plasma extravasation by stimulation of C-fibers only. Neuropathies and radiculopathies and other pathological conditions that are associated with antidromic activity in sensory nerve fibers are examples of such events. Neurogenic inflammation is likely to increase the dysesthesias and pain of patients suffering from neuropathies.

Neuropeptides also influence immune cells (for a review, see Morley, J. E. *et al.*, 1987) and synoviocytes. These actions may be of particular importance for the development and maintenance of chronic arthritis and other inflammatory disorders of deep somatic tissues.

3.6 Receptor Molecules in the Membrane of Nociceptors

Classic stimulants for nociceptors are endogenous pain-producing (algesic) substances such as BK, 5-HT, and high concentrations of potassium ions. Injections of hypertonic NaCl solutions (4.5–6.0%) have been used for many decades to elicit pain from deep somatic tissues. Similar to potassium, the stimulating effects of hypertonic NaCl are considered to be unspecific (i.e., not mediated by a receptor–ligand interaction).

The effects of BK on nerve endings are mediated by G protein-coupled B1 and B2 receptors. Binding of the ligands to this and other receptors of this type results in changes of the state of activation of intracellular second messenger systems such as phospholipase C (PLC), cyclic adenosine monophosphate (cAMP), cyclic guanosin monophosphate (cGMP), and protein kinases (PK). One possible sequel of the activation of these messengers is the phosphorylation of ion channels. This increases the opening time or opening probability of the channels. Collectively, the processes lead either to direct excitation of the ending or to a changed reaction of the nerve ending to external stimuli (sensitization or desensitization).

In intact tissue, the discharges evoked in nerve endings by BK are mainly due to the activation of the BK receptor B2, whereas under pathological conditions the receptor B1 is the predominant one (for a review of receptor molecules mediating the effects of pain-producing substances, see Kumazawa, T., 1996). It is often overlooked that the above substances, particularly BK, excite not only nociceptors but also nonnociceptive endings with thin myelinated and unmyelinated afferent fibers. Therefore, BK cannot be considered a specific excitant of nociceptors. However, muscle receptors with group I and II afferents (e.g., muscle spindles and tendon organs) are not excited by BK.

The stimulating effects of serotonin on nociceptive terminals in the body periphery are largely mediated by the 5-HT3 receptor. Similar to PGE2, which binds to a G protein-coupled prostanoid receptor (EP2), serotonin sensitizes nociceptors rather than exciting them under (patho)physiological circumstances. The reason is that in damaged tissue concentrations of 5-HT and PGE2 are released, which are sufficient for sensitizing, but not for activating, nociceptors.

Novel stimulants for nociceptors are ATP, capsaicin, low pH, and NGF (Caterina, M. J. and David, J., 1999). ATP binds to purinergic receptors (in nociceptive endings, the P2X3 receptor; Ding, Y. et al., 2000) and opens a cation channel. Animal experiments have shown that nociceptors respond to ATP in concentrations that are present in tissue cells. This means that every time a cell is damaged (by trauma or a pathological process), it releases ATP in amounts that can cause pain. For this reason, ATP has been considered a general signal substance for tissue trauma (Cook, S. P. and McCleskey, E. W., 2002).

Capsaicin, the active ingredient of chilli peppers, is a natural stimulant for the vanilloid receptor VR1, now called TRPV1 (Caterina, M. J. and Julius, D., 2001). The receptor is also sensitive to an increase in H^+ concentration and to heat with a threshold of approximately $39\,^\circ C$. The sensitivity of this receptor to protons is important under conditions in which the pH of the tissue is lowered (exhaustive muscle work, ischemia, and inflammation). TRPV2 is present mainly in myelinated heat nociceptors that have a thermal threshold above $52\,^\circ C$, and TRPV4 is being discussed as the main receptor of mechano-nociceptors (Liedtke, W., 2005).

A special feature of nociceptive afferent fibers are TTX-r sodium channels. TTX is a neurotoxin that blocks TTX-sensitive sodium channel and thus inhibits conduction in nerve fibers that possess this type of Na^+ channel (mostly large-diameter fibers). Nociceptive fibers are equipped with TTX-r channels; therefore, TTX has no blocking action. Two TTX-r Na^+ channels that are important for nociception are the voltage-gated sodium channels (Na_v) 1.9 and 1.8. Na_v 1.9 has been found exclusively in nociceptive primary afferent neurons, whereas Na_v 1.8 is present in both nociceptive and nonnociceptive ones (for review, Djouhri, L. and Lawson, S. N., 2004).

References

Andersson, S. A., Ericson, T., Holmgren, E., and Lindqvist, G. 1973. Electro-acupuncture. Effect on pain threshold measured with electrical stimulation of teeth. Brain Res. 63, 393–396.

Andres, K. H. and von Düring, M. 1973. Morphology of cutaneous receptors. In: Handbook of Sensory Physiology, Vol II, Somatosensory System (ed. A. Iggo), pp. 3–28, Springer.

Andres, K. H., von Düring, M., Muszynski, K., and Schmidt, R. F. 1987. Nerve fibres and their terminals of the dura mater encephali in the rat. Anat. Embryol. 175, 289–301.

Andres, K. H., von Düring, M., and Schmidt, R. F. 1985. Sensory innervation of the Achilles tendon by group III and IV afferent fibers. Anat. Embryol. 172, 145–156.

Applebaum, M. L., Clifton, G. L., Coggeshall, R. E., Coulter, J. D., Vance, W. H., and Willis, W. D. 1976. Unmyelinated fibres in the sacral 3 and caudal 1 ventral root of the cat. J. Physiol. 256, 557–572.

Barker, D. 1967. The innervation of mammalian skeletal muscle. In: CIBA Foundation Symposium on Myotatic, Kinesthetic and Vestibular Mechanisms (eds. A. V. S. DeReuck and J. Knight), pp. 3–19, Little Brown.

Belmonte, C. and Gallar, J. 1996. Corneal nociceptors. In: Neurobiology of Nociceptors. (eds. C. Belmonte and F. Cervero), pp. 146–183, Oxford University Press.

Belmonte, C., Gallar, J., Pozo, M. A., and Rebollo, I. 1991. Excitation by irritant substances of sensory afferent units in the cat's cornea. J. Physiol. 437, 709–725.

Bessou, P. and Perl, E. R. 1969. Response of cutaneous sensory units with unmyelinated fibers to noxious stimuli. J. Neurophysiol. 32, 1025–1043.

Burgess, P. R. and Perl, E. R. 1967. Myelinated afferent fibers responding specifically to noxious stimulation of the skin. J. Physiol. 190, 541–562.

Byers, M. R. 1984. Dental sensory receptors. Intern. Rev. Neurobiol. 25, 39–94.

Byers, M. R., Neuhaus, S. J., and Gehrig, J. D. 1982. Dental sensory receptor structure in human teeth. Pain 13, 221–235.

Caterina, M. J. and David, J. 1999. Sense and specificity: a molecular identity for nociceptors. Curr. Opin. Neurobiol. 9, 525–530.

Caterina, M. J. and Julius, D. 2001. The vanilloid receptor: a molecular gateway to the pain pathway. Ann. Rev. Neurosci. 24, 487–517.

Cauna, N. 1966. Fine Structure of the Receptor Organs and Its Probable Functional Significance. In: Touch, Heat and Pain (eds. A. V. S. DeReuck and J. Knight), pp. 117–127, Churchill.

Cesare, P. and McNaughton, P. 1997. Peripheral pain mechanisms. Curr. Opin. Neurobiol. 7, 493–499.

Chidiac, J. J., Rifai, K., Hawwa, N. N., Massaad, C. A., Jurjus, A. R., Jabbur, S. J., and Saade, N. E. 2002. Nociceptive behaviour induced by dental application of irritants to rat incisors: a new model for tooth inflammatory pain. Eur. J. Pain 6, 55–67.

Cook, S. P. and McCleskey, E. W. 2002. Cell damage excites nociceptors through release of cytosolic ATP. Pain 95, 41–47.

Devor, M., Wall, P. D., and McMahon, S. B. 1984. Dichotomizing somatic nerve fibers exist in rats but they are rare. Neurosci. Lett. 49, 187–192.

Ding, Y., Cesare, P., Drew, L., Nikitaki, D., and Wood, J. N. 2000. ATP, P2X receptors, and pain pathways. J. Auton. Nerv. Syst. 81, 289–294.

Djouhri, L. and Lawson, S. N. 2004. Aβ-fiber nociceptive primary afferent neurons: a review of incidence and properties in relation to other afferent A-fiber neurons in mammals. Brain Res. Rev. 46, 131–145.

Du, J., Zhou, S., Coggeshall, R. E., and Carlton, S. M. 2003. N-methyl-D-aspartate-induced excitation and sensitization of normal and inflamed nociceptors. Neuroscience 118, 547–562.

von Düring, M. and Andres, K. H. 1990. Topography and Ultrastructure of Group III and IV Nerve Terminals of the Cat's Gastrocnemius-Soleus Muscle. In: The Primary Afferent Neuron (eds. W. Zenker and W. L. Neuhuber), pp. 35–41, Plenum Press.

Foster, E., Risling, M., and Fried, K. 1995. TrkB-like immunoreactivity in trigeminal sensory nerve fibers of the rat molar tooth pulp: development and response to nerve injury. Brain Res. 686, 123–133.

von Frey, M. 1894. Beiträge zur Physiologie des Schmerzsinns. Königl. Sächs. Ges. Wiss., Math. Phys. Klasse 46, 185–196.

Fristad, I., Vandevska-Radunovic, V., Fjeld, K., Wimalawansa, S. F., and Hals Kvinnsland, I. 2003. NK1, NK2, NK3 and CGRP1 receptors identified in rat oral soft tissues and in bone and dental hard tissue cells. Cell Tiss. Res. 311, 383–391.

Gamse, R. and Saria, A. 1985. Potentiation of tachykinin-induced plasma protein extravasation by calcitonin gene-related peptide. Eur. J. Pharmacol. 114, 61–66.

Garcia-Anoveros, J., Samad, T. A., Zuvela-Jelaska, L., Woolf, C. J., and Corey, D. P. 2001. Transport and localization of the DEG/ENaC ion channel BNaC1alpha to peripheral mechanosensory terminals of dorsal root ganglia neurons. J. Neurosci. 21, 2678–2686.

Garry, M. G., Miller, K. E., and Seybold, V. S. 1989. Lumbar dorsal root ganglia of the cat: a quantitative study of peptide immunoreactivity and cell size. J. Comp. Neurol. 284, 36–47.

Guilbaud, G., Iggo, A., and Tegner, R. 1985. Sensory receptors in ankle joint capsules of normal and arthritic rats. Exp. Brain Res. 58, 29–40.

Gunji, T. 1982. Morphological research on the sensitivity of dentin. Arch. Histol. Jpn. 45, 45–67.

Hanesch, U. 1996. Neuropeptides in Dural Fine Sensory Nerve Endings-Involvement in Neurogenic Inflammation? In: Progress in Brain Research, Vol. 113 (eds. T. Kumazawa, L. Kruger, and K. Mizumura), pp. 299–317. Elsevier Science.

Hensel, H., Andres, K. H., and von Düring, M. 1974. Structure and function of cold receptors. Pflügers Arch. 352, 1–10.

Heppelmann, B., Heuss, C., and Schmidt, R. F. 1988. Fiber size distribution of myelinated and unmyelinated axons in the medial and posterior articular nerves of the cat's knee joint. Somatosens. Res. 5, 273–281.

Heppelmann, B., Messlinger, K., Neiss, W., and Schmidt, R. F. 1990a. Ultrastructural three-dimensional reconstruction of group III and group IV sensory nerve endings (free nerve endings) in the knee joint capsule of the rat: evidence for multiple receptive sites. J. Comp. Neurol. 292, 103–116.

Heppelmann, B, Messlinger, K., Neiss, W. F., and Schmidt, R. F. 1990b. The Sensory Terminal Tree of 'Free Nerve Endings' in the Articular Capsule of the Knee. In: The Primary Afferent Neuron (eds. W. Zenker and W. L. Neuhuber), pp. 73–85, Plenum Press.

Hoheisel, U. and Mense, S. 1987. Observations on the morphology of axons and somata of slowly conducting dorsal root ganglion cells in the cat. Brain Res. 423, 269–278.

Hoheisel, U., Mense, S., and Scherotzke, R. 1994. Calcitonin gene-related peptide-immunoreactivity in functionally identified primary afferent neurones in the rat. Anat. Embryol. 189, 41–49.

Hoheisel, U., Reinöhl, J., Unger, T., and Mense, S. 2004. Acidic pH and capsaicin activate mechanosensitive group IV muscle receptors in the rat. Pain 110, 149–157.

Hoheisel, U., Unger, T., and Mense, S. 2005. Excitatory and modulatory effects of inflammatory cytokines and neurotrophins on mechanosensitive group IV muscle afferents in the rat. Pain 114, 168–176.

Hökfelt, T., Johansson, O., Ljungdahl, Å, Lundberg, J. M., and Schultzberg, M. 1980. Peptidergic neurones. Nature 284, 515–521.

Holland, G. R. and Robinson, P. P. 1983. The number and size of axons at the apex of the cat's canine tooth. Anat. Rec. 205, 215–222.

Hoyes, A. D. and Barber, P. 1976. Ultrastructure of the corneal nerves in the rat. Cell Tiss. Res. 172, 133–144.

Ibuki, T., Kido, M. A., Kiyoshima, T., Terada, Y., and Tanaka, T. 1996. An ultrastructural study of the relationship between sensory trigeminal nerves and odontoblasts in rat dentin/pulp as demonstrated by the anterograde transport of wheat germ agglutinin-horseradish peroxidase (WGA-HRP). J. Dent. Res. 75, 1963–1970.

Ichikawa, H., Matsuo, S., Silos-Santiago, I., Jacquin, M. F., and Sugimoto, T. 2004. The development of myelinated nociceptors is dependent upon trks in the trigeminal ganglion. Acta Histochem. 106, 337–343.

Ikeda, H., Tokita, Y., and Suda, H. 1997. Capsaicin-sensitive A delta fibers in cat tooth pulp. J. Dent. Res. 76, 1341–1349.

Jänig, W. 1996. Neurobiology of visceral afferent neurons: neuroanatomy, functions, organ regulations and sensations. Biol. Psychol. 42, 29–51.

Ju, G., Hökfelt, T., Brodin, E., Fahrenkrug, F. J., Fischer, J. A., Frey, P., Elde, R. P., and Brown, J. C. 1987. Primary sensory neurons of the rat showing calcitonin gene-related peptide immunoreactivity and their relation to substance P-, somatostatin-, galanin-, vasoactive intestinal polypeptide-, and cholecystokinin-immunoreactive ganglion cells. Cell Tiss. Res. 247, 417–431.

Jyväsjärvi, E. and Kniffki, K. D. 1992. Studies on the presence and functional properties of afferent C-fibers in the cat's dental pulp. Proc. Finn. Dent. Soc. 88(Suppl. 1), 533–542.

Kenins, P. 1988. The functional anatomy of the receptive fields of rabbit C polymodal nociceptors. J. Neurophysiol. 59, 1098–1115.

Kennedy, W. R. 2004. Opportunities afforded by the study of unmyelinated nerves in skin and other organs. Muscle Nerve 29, 756–767.

Koling, A. 1985. Freeze-fracture electron microscopy of non-myelinated nerve fibres in the human dental pulp. Arch. Oral Biol. 30, 685–690.

Kow, L.-M. and Pfaff, D. W. 1988. Neuromodulatory actions of peptides. Ann. Rev. Pharmacol. Toxicol. 28, 163–188.

Kruger, L. 1988. Morphological features of thin sensory afferent fibers: a new interpretation of 'nociceptor' function. Progr. Brain Res. 74, 253–257.

Kruger, L. and Halata, Z. 1996. Structure of nociceptor 'endings'. In: Neurobiology of Nociceptors (eds. C. Belmonte and F. Cervero), pp. 37–71, Oxford University Press.

Kruger, L., Kavookjian, A. M., Kumazawa, T., Light, A. R., and Mizumura, K. 2003b. Nociceptor structural specialization in canine and rodent testicular "free" nerve endings. J. Comp. Neurol. 463, 197–211.

Kruger, L., Light, A. R., and Schweizer, F. E. 2003a. Axonal terminals of sensory neurons and their morphological diversity. J. Neurocytol. 32, 205–216.

Kruger, L., Perl, E. R., and Sedivec, M. J. 1981. Fine structure of myelinated mechanical nociceptor endings in cat hairy skin. J. Comp. Neurol. 198, 137–154.

Kruger, L., Silverman, J. D., Mantyh, P. W., Sternini, C., and Brecha, N. C. 1989. Peripheral patterns of calcitonin-gene-related peptide general somatic sensory innervation: cutaneous and deep terminations. J. Comp. Neurol. 280, 291–302.

Kumazawa, T. 1996. The Polymodal Receptor Bio-Warning and Defense System. In: Progress in Brain Research, Vol. 113 (eds. T. Kumazawa, L. Kruger, and K. Mizumura), pp. 3–18, Elsevier Science.

Kumazawa, T. and Mizumura, K. 1977. Thin-fibre receptors responding to mechanical, chemical and thermal stimulation in the skeletal muscle of the dog. J. Physiol. 273, 179–194.

Kumazawa, T. and Mizumura, K. 1980. Chemical responses of polymodal receptors of the scrotal contents in dogs. J. Physiol. 299, 219–231.

Kumazawa, T., Mizumura, K., and Sato, J. 1987. Response properties of polymodal receptors studied using in vitro testis superior spermatic nerve preparations of dogs. J. Neurophysiol. 57, 702–711.

Lawson, S. N., Crepps, B. A., and Perl, E. R. 1997. Relationship of substance P to afferent characteristics of dorsal root ganglion neurons in guinea-pig. J. Physiol. 505, 177–191.

Leah, J. D., Cameron, A. A., and Snow, P. J. 1985. Neuropeptides in physiologically identified mammalian sensory neurones. Neurosci. Lett. 56, 257–263.

Lembeck, F. and Holzer, P. 1979. Substance P as neurogenic mediator of antidromic vasodilation and neurogenic plasma extravasation, Naunyn-Schmiedeb. Arch. Pharmacol. 310, 175–183.

Liedtke, W. 2005. TRPV4 plays an evolutionary conserved role in the transduction of osmotic and mechanical stimuli in live animals. J. Physiol. 567, 53–58.

Light, A. R. and Perl, E. R. 2003. Unmyelinated afferent fibers are not only for pain anymore. J. Comp. Neurol. 461, 137–139.

Lloyd, D. P. C. 1943. Neuron patterns controlling transmission of ipsilateral hind limb reflexes in cat. J. Neurophysiol. 6, 293–315.

MacIver, M. G. and Tanelian, D. L. 1993. Structural and functional specialization of Aδ and C fiber free nerve endings innervating rabbit corneal epithelium. J. Neurosci. 13, 4511–4524.

McCloskey, D. I. and Mitchell, J. H. 1972. Reflex cardiovascular and respiratory responses originating in exercising muscle. J. Physiol. 224, 173–186.

Mense, S. 1977. Nervous outflow from skeletal muscle following chemical noxious stimulation. J. Physiol. 267, 75–88.

Mense, S. and Meyer, H. 1985. Different types of slowly conducting afferent units in cat skeletal muscle and tendon. J. Physiol. 363, 403–417.

Mense, S., Light, A. R., and Perl, E. R. 1981. spinal Terminations of Subcutaneous High-Threshold Mechanoreceptors. In: Spinal Cord Sensation (eds. A. G. Brown and M. Réthelyi), pp. 79–86, Scottish Academic Press.

Messlinger, K. 1996. Functional Morphology of Nociceptive and Other Fine Sensory Endings (Free Nerve Endings) in Different Tissues. In: Progress in Brain Research, Vol. 113 (eds. T. Kumazawa, L. Kruger, and K. Mizumura), pp. 273–298. Elsevier Science.

Messlinger, K. 1997. Was ist ein Nozizeptor? Schmerz 5, 353–367.

Molander, C., Ygge, I., and Dalsgaard, C.-J. 1987. Substance P-, somatostatin-, and calcitonin gene-related peptide-like immunoreactivity and fluoride resistant acid phosphatase-activity in relation to retrogradely labeled cutaneous, muscular and visceral primary sensory neurons in the rat. Neurosci. Lett. 74, 37–42.

Morley, J. E., Kay, N. E., Solomon, G. F., and Plotnikoff, N. P. 1987. Neuropeptides: conductors of the immune orchestra. Life Sci. 41, 527–544.

Müller, L. J., Pels, L., and Vrensen, G. F. 1996. Ultrastructural organization of human corneal nerves. Invest. Ophthal. Vis. Sci. 37, 476–488.

O'Brien, C., Woolf, C. J., Fitzgerald, M., Lindsay, R. M., and Molander, C. 1989. Differences in the chemical expression of rat primary afferent neurons which innervate skin, muscle or joint. Neuroscience 32, 493–502.

Öhlén, A., Wiklund, N. P., Persson, M. G., and Hedqvist, P. 1988. Calcitonin gene-related peptide desensitizes skeletal muscle arterioles to substance P in vivo. Br. J. Pharmacol. 95, 673–674.

Pierau, F.-K., Fellmer, G., and Taylor, D. C. M. 1984. Somato-visceral convergence in cat dorsal root ganglion neurones demonstrated by double-labelling with fluorescent tracers. Brain Res. 321, 63–70.

Reinert, A., Kaske, A., and Mense, S. 1998. Inflammation-induced increase in the density of neuropeptide-immunoreactive nerve endings in rat skeletal muscle. Exp. Brain Res. 121, 174–180.

Renton, T., Yiangou, Y., Baecker, P. A., Ford, A. P., and Anand, P. 2003. Capsaicin receptor VR1 and ATP purinoceptor P2X3 in painful and nonpainful human tooth pulp. J. Orofac. Pain 17, 245–250.

Rifai, K., Chidiac, J. J., Hawwa, N., Baliki, M., Jabbur, S. J., and Saade, N. E. 2004. Occlusion of dentinal tubules and selective block of pulp innervation prevent the nociceptive behaviour induced in rats by intradental application of irritants. Arch. Oral Biol. 49, 457–468.

Rodd, H. D. and Boissonade, F. M. 2000. Substance P expression in human tooth pulp in relation to caries and pain experience. Eur. J. Oral Sci. 108, 467–474.

Rodd, H. D., Loescher, A. R., and Boissonade, F. M. 1998. Immunocytochemical and electron-microscopic features of tooth pulp innervation in hereditary sensory and autonomic neuropathy. Arch. Oral Biol. 43, 445–454.

Schaible, H.-G. 2005. Basic Mechanisms of Deep Somatic Pain. In: Textbook of Pain (eds. S. B. McMahon and M. Koltzenburg), pp. 621–633. Churchill Livingstone.

Schaible, H.-G. and Grubb, B. D. 1993. Afferent and spinal mechanisms of joint pain. Pain 55, 45–54.

Schaible, H.-G. and Schmidt, R. F. 1985. Effects of an experimental arthritis on the sensory properties of fine articular afferent units. J. Neurophysiol. 54, 1109–1122.

Schaible, H.-G. and Schmidt, R. F. 1988. Time course of mechanosensitivity changes in articular afferents during a developing experimental arthritis. J. Neurophysiol. 60, 2180–2195.

Silverman, J. D. and Kruger, L. 1988. Lectin and neuropeptide labeling of separate populations of dorsal root ganglion neurons and associated 'nociceptor' thin axons in rat testis and cornea whole-mount preparations. Somatosens. Res. 5, 259–267.

Simone, D. A., Nolano, M., Johnson, T., Wendelschafer-Crabb, G., and Kennedy, W. R. 1998. Intradermal injection of capsaicin in humans produces degeneration and subsequent reinnervation of epidermal nerve fibers: correlation with sensory function. J. Neurosci. 18, 8947–8959.

Snider, W. E. and McMahon, S. B. 1998. Tackling pain at the source: new ideas about nociceptors. Neuron 20, 629–632.

Stacey, M. J. 1969. Free nerve endings in skeletal muscle of the cat. J. Anat. 105, 231–254.

Steffens, H., Eek, B., Trudrung, P., and Mense, S. 2003. Tetrodotoxin block of A-fibre afferents from skin and muscle – A tool to study pure C-fibre effects in the spinal cord. Pflügers Arch. 445, 607–613.

Stucky, C. L. and Lewin, G. R. 1999. Isolectin B(4)-positive and –negative nociceptors are functionally distinct. J. Neurosci. 19, 6497–6505.

Svensson, P., Cairns, B. E., Wang, K., Hu, J. W., Graven-Nielsen, T., Arendt-Nielsen, L., and Sessle, B. J. 2003. Glutamate-evoked pain and mechanical allodynia in the human masseter muscle. Pain 101, 221–227.

Tower, S. 1940. Unit for sensory reception in cornea. With notes on nerve impulses from sclera, iris and lens. J. Neurophysiol. 3, 486–500.

Wu, Z.-Z. and Pan, H.-L. 2004. Tetrodotoxin-sensitive and – resistant Na+ channel currents in subsets of small sensory neurons of rats. Brain Res. 1029, 251–258.

Zhang, J. Q., Nagata, K., and Iijima, T. 1998. Scanning electron microscopy and immunohistochemical observations of the vascular nerve plexuses in the dental pulp of rat incisor. Anat. Rec. 251, 214–220.

Further Reading

Heppelmann, B, Messlinger, K., Neiss, W. F., and Schmidt, R. F. 1994. Mitochondria in fine afferent nerve fibres of the knee joint in the cat: a quantitative electron-microscopical examination. Cell Tiss. Res. 275, 493–501.

Heppelmann, B., Pfeffer, H.-G., Schaible, H.-G., and Schmidt, R. F. 1986. Effects of acetylsalicylic acid and indomethacin on single groups III and IV sensory units from acutely inflamed joints. Pain 26, 337–351.

Messlinger, K., Pawlak, M., Steinbach, H., Trost, B., and Schmidt, R. F. 1995. A new combination of methods for localization, identification, and three-dimensional reconstruction of the sensory endings of articular afferents characterized by electrophysiology. Cell Tiss. Res. 281, 283–294.

Yeh, Y. and Byers, M. R. 1983. Fine structure and axonal transport labeling of intraepithelial sensory nerve endings in anterior hard palate of the rat. Somatosens. Res. 1, 1–19.

4 Molecular Biology of the Nociceptor/Transduction

M S Gold, University of Pittsburgh, Pittsburgh PA, USA

M J Caterina, Johns Hopkins School of Medicine, Baltimore, MD, USA

Glossary

action potential A characteristic pattern of neuronal depolarization and hyperpolarization that mediates transmission of information along the length of the neuron and triggers communication with synaptically connected neurons.

afterpotential The change in membrane potential that follows the action potential.

cable properties Passive electrophysiological properties of a neuron that influence the movement of charge, and therefore the spread of a change in membrane potential, along a neuron. These properties include internal resistance (influenced by the diameter of the neural process), membrane resistance (influenced by the number of open ion channels in the membrane), and membrane capacitance (determined by the total membrane surface area).

generator potential Membrane depolarization resulting from the transduction of environmental stimuli (i.e., thermal, mechanical, or chemical). This transduction results in a change in membrane conductance that underlies the change in membrane potential.

G-protein-coupled receptors A diverse family of membrane proteins that trigger changes in cell function by signaling through intracellular guanine nucleotide-binding proteins.

hyperalgesia A state in which the sensitivity of nociceptive signaling pathways to environmental stimuli is enhanced.

inflammatory pain Spontaneous pain or hypersensitivity to environmental stimuli arising from inflammation in areas innervated by nociceptive neurons.

length constant A measure of the distance over which a current injected at a point source will spread within a neuron.

neurotrophins A diverse family of soluble signaling molecules that promote the differentiation, survival, and/or function of neurons.

nociceptor Specialized sensory neuron whose activation signals the presence of noxious, or potentially damaging stimuli. These neurons are selectively activated by noxious stimuli, under normal circumstances, or by molecules released once tissue damage has already occurred.

neuropathic pain Spontaneous pain or hypersensitivity to environmental stimuli arising from damage to neural tissue.

receptive field The area in or on the body within which adequate stimuli are capable of generating activity in a given neuron.

polymodality The ability of a molecule or neuron to be activated by more than one type (modality) of stimulus (e.g., mechanical and chemical).

tranduction A process in which environmental stimuli evoke conformational changes in the structure of specific proteins located on nociceptor terminals that directly or indirectly (i.e., via cellular signaling cascades) trigger the opening or closing of ion channels.

transient receptor potential A historically based name for a family of related iron channels that are prevalent in nociceptive neurons and other sensory neuron subtypes.

4.1   Introduction

The perception of pain is usually triggered when mechanical, chemical, or thermal stimuli of high intensity or a particular quality are detected by a specialized subpopulation of sensory neurons referred to as nociceptors. Although the term nociceptor has probably outlived its utility, we will use it here as a means of distinguishing afferents capable of responding to tissue damage from those that normally only encode innocuous stimuli. Nociceptor activation typically begins with transduction, a process in which environmental stimuli evoke conformational changes in the structure of specific proteins located on nociceptor terminals that directly or indirectly (i.e., via cellular signaling cascades) trigger the opening or closing of ion channels. The resulting changes in ionic flow across the plasma membrane produce a change in membrane potential, referred to as a generator potential. If the

generator potential reaches a critical threshold, action potential firing ensues. In order for these action potentials to trigger a perception of pain, however, they must be propagated to the central terminal of the sensory neuron and subsequently evoke neurotransmitter release in sufficient quantities to activate postsynaptic spinal cord neurons. Thus, there are multiple critical steps in the afferent transmission of nociceptive information, each of which involves a unique compilation of specialized proteins. We will begin this chapter by describing several important but sometimes overlooked anatomical and physiological features of nociceptive neurons. We will then introduce some of the major protein classes that contribute to transduction and, to a more limited extent, the ensuing neurophysiological processes that result in the transmission of nociceptive information to the spinal cord. Subsequently, we will provide a glimpse at how such variables as molecular heterogeneity, intermolecular interactions, posttranslational modifications, and ion channel distribution can vastly expand the sensory versatility of the nociceptive neurons. Finally, we will describe in greater detail a few of the changes that underlie nociceptor signaling plasticity in the context of inflammation and nerve injury. Although neurotransmission within the spinal cord dorsal horn represents another important function of nociceptors, we will defer a description of this process to other chapters of this volume.

4.1.1 Structure Dictates Function: The Anatomy of the Nociceptor

Sensory neurons are referred to as pseudounipolar neurons because they start out during development as bipolar neurons with a central process extending into the spinal or trigeminal dorsal horn and a peripheral process extending out to peripheral targets. During development, the initial segment of these two processes merges to form a T junction, leaving the cell body attached to peripheral and central processes via a single short length of axon (Fitzgerald, M., 2005) (Figure 1). Both central and peripheral processes terminate in a branching pattern referred to as a terminal arbor (Figure 1). The extent of the peripheral arbor depends on the afferent type and site of innervation. For example, afferents innervating the fingertips may innervate a square millimeter or less of skin, and therefore appear to have relatively few terminal branches (Darian-Smith, I. *et al.*, 1980; Andrew, D. and Greenspan, J. D., 1999). In contrast, afferents innervating a visceral structure such as the colon may have an extensive terminal arborization that extends over many centimeters of tissue (Berthoud, H. R. *et al.*, 2004). Afferents responsible for the transduction of low-threshold mechanical stimuli such as brush or vibration terminate in specialized structures such as Rufinni endings or Merkel discs that are critical for mediating the appropriate response to these stimuli (Mearow, K. M. and

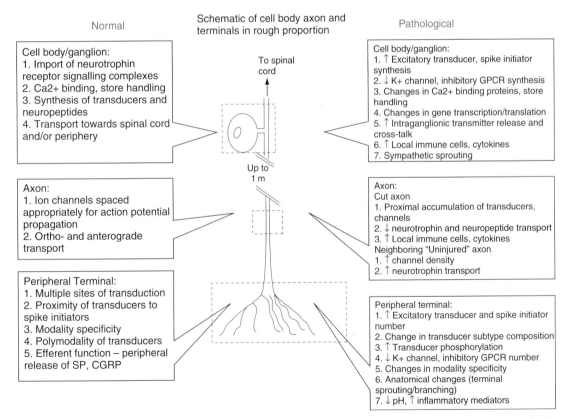

Figure 1 Schematic diagram of three major peripheral compactments of a nociceptive neuron: the cell body (top), axon (middle), and peripheral terminal arbor (bottom). Blowups at left: functional and anatomical properties of these compartments under normal conditions. Blowups at right: characteristic changes associated with nerve injury or inflammation.

Diamond, J., 1988; Johnson, K. O., 2001). In contrast, nociceptors are said to have free nerve endings, because peripheral terminals of these afferents do not appear to be associated with any specific cell type (Kruger, L. *et al.*, 2003a). Further description of this anatomical arrangment can be found in the chapter by Mense in this volume (Chapter Anatomy of Nociceptors).

There are at least three important implications of the observation that peripheral terminals of nociceptors are free. First, while recent evidence suggests that transduction may occur in cells surrounding nociceptive terminals (i.e., urothelial cells (Birder, L. A., 2005) or keratinocytes (Guler, A. D. *et al.*, 2002; Peier, A. M. *et al.*, 2002b; Khodorova, A. *et al.*, 2003; Wu, G. *et al.*, 2004a) (see below)), the fact that nociceptive terminals are generally not associated with specialized cells suggests that these neurons themselves possess the machinery necessary for transduction. Consistent with this suggestion, the isolated cell bodies of putative nociceptors *in vitro* have been show to respond to thermal (Cesare, P. and McNaughton, P., 1996; Reichling, D. B. and Levine, J. D., 1997; Reid, G. and Flonta, M., 2001; Thut, P. D. *et al.*, 2003), mechanical (McCarter, G. C. *et al.*, 1999), and various chemical (Baccaglini, P. I. and Hogan, P. G., 1983) stimuli. Second, if transduction machinery is present in nociceptive terminals, transduction may occur whenever and wherever a transducer is properly positioned in the cell membrane. The result is that transduction can and probably does occur at multiple sites in a peripheral arbor (Gover, T. D. *et al.*, 2003; Zhao, J. *et al.*, 2006). This has important implications for spike initiation and integration of stimuli applied throughout a receptive field (Darian-Smith, I. *et al.*, 1980). It also has important implications for the generation of ectopic activity in the presence of injury as there is evidence that the cut ends of axons may rapidly develop mechanical, thermal, and chemical sensitivity (Michaelis, M. *et al.*, 1999) and that spontaneous activity may arise from within sensory ganglia following nerve injury (Kajander, K. C. *et al.*, 1992). Third, because nerve endings are not only free of an association with a specific cell type, but free of even perineurial encapsulation (Devor, M., 1999), these endings are not only accessible to the chemical milieu of the tissue, but may be accessible to therapeutic interventions.

Both light and electron micrographic analyses of nociceptive terminals reveal complex anatomical structures. As suggested above, the structure of the terminal arbor varies with target of innovation. There is also evidence that subpopulations of nociceptive afferents have distinct terminal arbor patterns within the same structure. For example, the terminal arbor morphology of cutaneous C fibers varies according to whether the C fiber is peptidergic (i.e., expresses substance P or calcitonin gene-related peptide (CGRP)) or expresses the mas-related gene (mrg), MrgD (Zylka, M. J. *et al.*, 2005). Furthermore, in the cornea, the terminal arbor of C-fiber afferents is distinct from that of Aδ fibers (MacIver, M. B. and Tanelian, D. L., 1993). At an ultrastructural level, nociceptive terminals appear to have beaded endings with clusters of organelles such as mitochondria at sites of swelling (Kruger, L. *et al.*, 2003b). While the functional implications of this beaded structure remain to be determined, it has been suggested that swellings may be sites for transduction and/or the peripheral release of transmitter.

Finally, it is clear that active membrane properties, and therefore the nature of the afferent signal, are influenced by an interaction between the distribution, density, and biophysical properties of ion channels and transducers present in afferent terminals and the morphological features of the terminal arbor. Cable theory describes the impact of leak channels, terminal branch points, and irregularities in membrane shape such as those associated with the beaded pattern described above. As a first pass, all of these features, in addition to the relatively large surface-to-volume ratio associated with fine terminals suggest that the length constant, a measure of the distance over which a current injected at a point source will spread within the terminal, is quite small in nociceptive terminals. This implies that there are likely to be multiple sites of spike initiation throughout the terminal arbor. Importantly, it is likely that the cable properties, and therefore length constant, of nociceptive terminals are dynamically regulated. This notion is based on the observation that potassium leak currents, which determine membrane resistance, may be regulated by a number of factors such as transmitters and second messengers (Maingret, F. *et al.*, 2000). Furthermore, evidence from dorsal horn neurons (Mantyh, P. W. *et al.*, 1995) suggests that terminal morphology may be rapidly altered. Importantly, a change in a length constant may influence active membrane properties such as action potential threshold and spike accommodation.

4.1.2 Nociceptor Inputs and Outputs: More than Meets the Eye

The major focus of this chapter is on molecules involved in the initiation of fast orthograde electrical signaling by nociceptors to the central nervous system. However, several physiological features of nociceptors fall outside this conventional framework but are important within the larger context of sensory transduction. Therefore, before we dissect the mechanisms of nociceptor stimulus transduction, action potential generation, and action potential propagation in any detail, it is worth considering a few of these features.

4.1.2.1 The onset of nociceptor excitation: hyperpolarization versus depolarization

Sensory transduction in the mammalian nervous system can arise from either membrane hyper-polarization or membrane depolarization. The visual system utilizes the former while the somatosensory system utilizes the latter. Consequently, in the absence of stimuli, the resting membrane potential of nociceptive afferents is relatively hyperpolarized, with little, if any spontaneous activity (Rose, R. D. et al., 1986; Ritter, A. M. and Mendell, L. M., 1992). This resting membrane potential appears to reflect a relatively large K^+ conductance through both voltage-insensitive (Baumann, T. K. et al., 2004) and voltage-gated K^+ channels (Kirchhoff, C. et al., 1992). However, this K^+ conductance appears to be working against a relatively high membrane permeability to depolarizing cations (i.e., Na^+ and Ca^{2+}), at least in cutaneous afferents (Kirchhoff, C. et al., 1992). Data from studies of the isolated sensory neuron cell body in vitro suggest that a significant depolarizing conductance may be present in a relatively large percentage of putative nociceptive afferents (Viana, F. et al., 2002).

There are at least two implications of a significant depolarizing conductance in nociceptive afferents. First, it suggests that a viable mode of transduction may involve the closing of a K^+ channel. In the absence of a depolarizing conductance, closing a K^+ channel in a membrane in which K^+ conductance is the only major conductance responsible for the establishment of the resting membrane potential, will result in little, if any change in membrane potential. Consistent with the suggestion that there is a significant depolarizing conductance in putative nociceptive afferents, there is evidence that some forms of cold transduction reflect the closing of a K^+ channel (Reid, G. and Flonta, M., 2001; Viana, F.

et al., 2002). Second, the presence of a significant depolarizing conductance in nociceptive afferents suggests that the resting membrane potential will be relatively depolarized. This has been observed repeatedly in studies on dissociated sensory neurons, where the resting membrane potential of putative nociceptive afferents is more depolarized than that of putative non-nociceptive afferents (see Gold, M. S. et al., 1996a). There is evidence that intact nociceptive terminals in situ may be depolarized as well (Brock, J. A. et al., 2001). As mentioned above, resting membrane potential may influence not only transduction, given the voltage dependence of several transducers (see below), but also spike initiation.

4.1.2.2 Efferent function

It is clear that nociceptor afferent neurotransmission contributes to the perception of pain associated with tissue injury (Treede, R. D. et al., 1992). However, it is also clear there are tissues distributed throughout the body that are heavily innervated by nociceptive afferents, but that will never give rise to the sensation of pain in many of us. For example, even though pericardial tissue receives nociceptive innervation thought to be responsible for pain associated with cardiac ischemia (Immke, D. C. and McCleskey, E. W., 2001), the majority of us will never experience discomfort arising from this structure. What might be the function(s) of nociceptors in these locations? One potential answer to this question stems from the ability of many nociceptors to release bioactive substances with the potential to exert trophic and/or inflammatory effects on targets of innervation. For example, CGRP, which can be released by many nociceptive neurons, has been shown to regulate cellular activity underlying bone formation and resorption, two processes critical to the maintenance of healthy bone (Lerner, U. H., 2002). Such seemingly paradoxical efferent nociceptor activity is particularly important in tissues subjected to injury. Under these circumstances, the release of transmitters from the peripheral nociceptor terminals has been shown to contribute to local hyperemia and plasma extravasation (hallmarks of the so-called neurogenic inflammation), to the orchestration of tissue repair, and to the initiation and maintenance of pain and hyperalgesia (Willis, W. D., 1999). Much of the focus in this area has been on nociceptor release of CGRP and another peptide transmitter, substance P (Figure 1). However, it is now clear that glutamate is also released into the periphery, where it appears able to act in an autocrine fashion to further activate and/or sensitize nociceptive afferents (deGroot, J. *et al.*, 2000). Appreciation of these efferent capabilities has led to the suggestion that a better name for nociceptive sensory neurons might be noceffector afferents (Kruger, L., 1988).

Three mechanisms have been described that may contribute to the injury-induced efferent release of neurotransmitter. One is the axon reflex, thought to be mediated by action potential spread from one branch in a terminal arbor to other branches (Lewis, T., 1935). A second, relatively rare, mechanism is axonal coupling, in which axons traveling in the same bundle may, as a result of injury, come into such close proximity that activity in one axon is able to initiate activity in a second axon (Meyer, R. A. *et al.*, 1985). A third mechanism actually involves the initiation of action potentials in the central terminals of nociceptive afferents within the spinal cord dorsal horn that are propagated antidromically out into the periphery (Sluka, K. A. *et al.*, 1995). While the exact mechanism of this activation is still debated, there is evidence that in the presence of inflammation there is a change in presynaptic gamma-aminobutyric acid (GABA) ergic input, such that primary afferent-driven GABA release from interneurons is capable of action potential generation at central terminals (Sluka, K. A. *et al.*, 1995). A fourth mechanism, often overlooked in this discussion, involves the possible direct coupling between calcium-permeable transducers and transmitter release. Such a mechanism appears to contribute to the inflammation associated with capsaicin-induced activation of its receptor transient receptor potential vanilloid 1 (TRPV1) (see below).

4.1.2.3 Spike integration within the cell body

There is evidence indicating that once the T junction is formed within the sensory ganglia, action potentials need not invade the sensory neuron cell body (Amir, R. and Devor, M., 2003). This observation begs the question of why action potentials invade the cell body at all. One answer to this question appears to be that the invading action potential provides a mechanism for the recruitment of additional activity. The term cross-excitation has been used to describe the observation that activity in one axon can induce activity in neighboring axons (Devor, M. and Wall, P. D., 1990). Interestingly, the prevalence of cross-excitation increases in the presence of tissue injury, providing a mechanism for peripherally mediated secondary hyperalgesia (Devor, M. and

Wall, P. D., 1990). While the exact mechanisms underlying such cross talk have yet to be completely worked out (Amir, R. and Devor, M., 1996), there is evidence that the sensory neuron cell body is capable of releasing transmitter (Huang, L. Y. and Neher, E., 1996) and that peripheral nerve stimulation results in the release of transmitter within the ganglia (Matsuka, Y. *et al.*, 2001).

More relevant to the issue of transduction, however, is the observation that afferent activity is capable of driving changes in gene transcription (Figure 1). For example, there is evidence that electrical activity in sensory neurons is capable of changing Na^+ channel expression (Klein, J. P. *et al.*, 2003). Interestingly, the pattern of activity appears to influence the pattern of gene transcription (Fields, R. D. *et al.*, 2005). The result may be as profound as to determine whether or not sprouting axons are able to coalesce into a nerve. One mediator that appears to be critical for the integration of activity within the cell body is Ca^{2+}, acting through a number of Ca^{2+}-dependent molecules (Eshete, F. and Fields, R. D., 2001). Thus, the transduction of activity within the sensory neuron cell body involves the influx of Ca^{2+}, which then initiates a number of cellular processes including gene transcription.

4.1.2.4 Slow axonal transport

Another general mechanism critical to nociceptive signaling involves axonal transport to the cell body, rather than action potential propagation. This is a remarkable process, given the tremendous length of sensory nerve axons (up to 1 m), relative to their soma size ($\sim$18–30 μm). Among the primary functions of this form of signaling, which may take hours to days, are the maintenance of nociceptor survival during development, the maintenance of nociceptive phenotype and function in the adult, and orchestration of the afferent response to tissue injury (Koltzenburg, M., 1999). The best-studied example of signaling via slow axonal transport is that of neurotrophic factors bound to their receptors (Figure 1) (Zweifel, L. S. *et al.*, 2005). However, slow axonal transport can also lead to signaling in the distant sensory neuron cell body following ligand-dependent internalization of G-protein-coupled receptors (GPCRs) (e.g., those for neurokinins, opioids, and extracellular proteases) (Tseng, L. F. and Narita, M., 1995; Vergnolle, N. *et al.*, 2001). In addition, transcription factors, second messenger molecules, cell surface glycoproteins, and foreign particles such as plant lectins, toxins, and viruses can all be transported retrogradely from peripheral sensory neuron terminals toward the cell body (Wang, H. F. *et al.*, 1998; Vulchanova, L. *et al.*, 2001; Tarpley, J. W. *et al.*, 2004; Yeomans, D. C. *et al.*, 2005; Mata, M. *et al.*, 2006). In what has been a boon to researchers, there are some molecules that are selectively taken up by subpopulations of sensory neurons, and have therefore been used to characterize innervation patterns (see Wang, H. F. *et al.*, 1998). Sensory neurons also appear to be able to take up inflammatory mediators such as TNFα (Schafers, M. *et al.*, 2002). Exactly how sensory neurons use the information associated with the presence or absence of various molecules transported back to the cell body remains to be determined. However, the internalization and transport of molecules present at sensory neuron terminals appears to be a mechanism for sensory neurons to monitor tissue integrity and adjust to changes in the local environment.

4.2 The Molecular Toolbox of Sensory Neurons

Sensory neurons express a multitude of cell surface proteins that serve anatomically as markers of neuronal subtypes and functionally as mediators of the fundamental signaling and transmission functions outlined in the previous section. Three classes of sensory neuron cell surface proteins have been studied in the greatest detail: ion channels, metabotropic GPCRs, and receptors for neurotrophins or cytokines. Together, these proteins endow sensory afferents with responsiveness to a qualitatively and quantitatively diverse collection of environmental variables and dictate the passive and active properties essential for converting a generator potential into a signal that reaches the spinal cord. As detailed below, certain membrane proteins found in nociceptors (e.g., P2X3, TRPM8, mrgA, and NaV1.8) are expressed within subsets of sensory neurons at levels much higher than anywhere else in the body. Such enriched expression has provided reagents that can be used for sensory neuron- or nociceptor-specific gene disruption or transgene expression in the laboratory setting. More importantly, it provides at least the theoretical opportunity to develop antinociceptive drugs that have minimal effects on other cell types. A few of the more prominent classes of cell surface proteins expressed in sensory neurons are described below.

4.2.1 Ion Channels Involved in the Transduction of Extracellular Signals

4.2.1.1 P2X adenosine triphosphate-gated ion channels

Extracellular adenosine triphosphate (ATP), released upon cell damage, has been recognized as a nociceptive stimulus ever since the demonstration that application of this molecule to a blister base produces pain in human psychophysical studies (Bleehan, T. and Keele, C. A., 1977). At least one mechanism by which ATP activates nociceptors is via activation of ATP-gated ion channels of the P2X family. These channels are comprised of a heterotrimer of P2X subunits, each with two transmembrane domains and a large extracellular loop that contains the ATP binding site. Seven P2X subtypes have been cloned in mammals. They differ with respect to their sensitivity to ATP or ATP analogs, as well as the kinetics of their ATP-evoked currents. For example, homomeric P2X3 is responsive to α,β-methylene ATP, and exhibits a rapidly desensitizing current response, whereas homomeric P2X2 exhibits slowly desensitizing currents that are α,β-methylene ATP insensitive. When coassembled, these channels form an α,β-methylene sensitive, slowly desensitizing heteromultimeric channel. P2X channels are typically nonselective cation channels that exhibit considerable permeability to both Na^+ and Ca^{2+} ions. In several cases, however, (P2X2, P2X4, and P2X7) protracted stimulation of these channels can lead to an apparent dilation of the channel pore, such that much larger organic cations such as *N*-methyl D-glucamine are able to permeate the channel. The significance of this dynamic ion selectivity remains unclear (North, R. A., 2002).

Although mRNAs encoding P2X1 through P2X6 have all been detected in sensory neurons, P2X2 and P2X3 appear to be expressed most prominently in these cells. P2X2 is expressed in neurons with a wide range of cell body sizes. In contrast, at least in skin, P2X3 expression appears to be confined to a subpopulation of small-diameter sensory neurons that bind isolectin B4 (IB4) and that project to inner lamina II of the spinal cord dorsal horn (North, R. A., 2002). It is important to note, however, that the precise relationship between P2X3 expression and IB4 binding may vary between cutaneous afferents with their cell bodies in dorsal root ganglia and other nociceptor populations such as those in the trigeminal ganglion (Ambalavanar, R. *et al.*, 2005). Genetic disruption of both P2X3 and P2X2 results in elimination of all transient ATP-evoked currents and all but a small persistent ATP-evoked current in cultured sensory neurons (Cockayne, D. A. *et al.*, 2005).

P2X channels also contribute to pain sensation *in vivo*. For example, although application of P2X3 antagonists (Jarvis, M. F. *et al.*, 2002; McGaraughty, S. *et al.*, 2003) or antisense oligodeoxynucleotide-mediated knockdown of P2X3 (Barclay, J. *et al.*, 2002; Honore, P. *et al.*, 2002) have no apparent effect on acute nociceptive responses to mechanical or thermal stimuli, both interventions have been shown to reduce thermal and/or mechanical hyperalgesia associated with complete Freund's adjuvant (CFA) administration into the paw and several models of neuropathic pain, as well as reducing the extent of the first and second phases of paw licking following intraplantar formalin injection. Gene knockout of P2X3 or P2X2/3 receptors in mice has also been reported to diminish formalin-evoked nocifensive behavior. However, these knockout animals exhibit a perplexing mixture of pain-related phenotypes, including paradoxically enhanced behavioral avoidance of heat and cold stimuli (Cockayne, D. A. *et al.*, 2000; Souslova, V. *et al.*, 2000; Cockayne, D. A. *et al.*, 2005; Shimizu, I. *et al.*, 2005). In visceral structures, epithelial cells appear capable of the storage and Ca^{2+}-dependent release of ATP in response to mechanical and chemical stimuli. P2X3 and P2X2/3 receptors on primary afferent terminals enable visceral afferents to respond to this released ATP, perhaps contributing to visceral pain sensation (Cockayne, D. A. *et al.*, 2000; Vlaskovska, M. *et al.*, 2001; Andersson, K. E., 2002; Birder, L. A., 2005). A similar mechanism may underlie ATP-mediated stimulus transduction in cutaneous tissue in the presence of inflammation (Wu, G. *et al.*, 2004a). P2X receptors contribute to nociceptive neurotransmission in a number of other ways, as well. In the spinal cord, for example, activation of presynaptic P2X3-containing channels and postsynaptic P2X channels of other subtypes, modulates neurotransmitter release and excitability, respectively (Gu, J. G. and MacDermott, A. B., 1997; Nakatsuka, T. and Gu, J. G., 2001; Nakatsuka, T. *et al.*, 2003; Tsuzuki, K. *et al.*, 2003). In addition, although it appears not to be expressed within neurons themselves, P2X7 is necessary for normal inflammation, nerve injury-induced hyperalgesia (Chessell, I. P. *et al.*, 2005), and CFA-induced hyperalgesia in rats (Dell'Antonio, G. *et al.*, 2002a; 2002b), probably owing to its role in cytokine release from inflammatory cells (Labasi, J. M. *et al.*, 2002).

4.2.1.2 Proton-gated ion channels

Reductions in tissue pH, as might occur during ischemia or inflammation, can potently activate nociceptors and sensitize them to heat and other stimuli. *In vitro*, low pH activates cationic channels in sensory neurons that can exhibit transient, sustained, or biphasic kinetics (Bevan, S. and Yeats, J., 1991). At least some of these responses can be attributed to the acid-sensing ion channel (ASIC) family. These multimeric ion channels are comprised of subunits that topologically resemble the P2X channels, with two transmembrane domains each and an extracellular ligand (H^+) binding domain. Five ASIC subtypes have been cloned, of which four (ASIC1–4) are expressed in peripheral sensory neurons (Waldmann, R. *et al.*, 1999; Waldmann, R., 2001). Splice variants of several subtypes add to the overall complexity, and one such variant, ASIC1β, appears to be selectively enriched in sensory neurons (Chen, C. C. *et al.*, 1998). Upon activation by low pH (from <7 to <4, depending on the subtype) ASICs typically mediate a prominent and rapidly desensitizing Na^+ current. However, some subtypes (e.g., ASIC3), exhibit a biphasic current consisting of an initial desensitizing sodium current, followed by a more sustained current that includes Ca^{2+} ions (Waldmann, R. *et al.*, 1999; Waldmann, R., 2001).

Several lines of evidence support the participation of ASIC channels in acid responsiveness of sensory neurons. First, the kinetics of sensory neuron responses to a pH 6 stimulus are altered in neurons isolated from ASIC1, ASIC2, or ASIC3 knockout mice or in wild-type neurons exposed to APETx2, an ASIC3-selective antagonist from sea anemone (Price, M. P. *et al.*, 2001; Benson, C. J. *et al.*, 2002; Xie, J. *et al.*, 2002; Diochot, S. *et al.*, 2004). Second, pH 5-evoked C-fiber responses are reduced in *ex vivo* skin–nerve explants derived from ASIC3 knockout mice (Price, M. P. *et al.*, 2001). Third, mice lacking ASIC3 exhibit defects in the ipsilateral mechanical hyperalgesia that typically develops 4 h after injection of acid to the gastrocnemius muscle (Price, M. P. *et al.*, 2001) as well as defects in the bilateral hyperalgesia that develops after two intramuscular acid injections, 5 days apart (Sluka, K. A. *et al.*, 2003). Interestingly, no changes in behavioral responses to acute subcutaneous acid injection have been reported in these knockout studies. However, ASIC3 may be particularly important for sensing reductions in muscle pH during ischemia (Immke and McCleskey 2001).

In addition to acid detection, ASICs have also been implicated in mechanosensation. This hypothesis was initially based upon the recognition that ASICs bear structural and pharmacological similarities (i.e., amiloride sensitivity) to degenerins, cationic channels identified in genetic screens for mechanosensory defects in the nematode, *Cenorhabditis elegans* (O'Hagan, R. and Chalfie, M., 2005). The immunolocalization of ASIC2 and 3 to sensory afferent terminals (Price, M. P. *et al.*, 2000; Garcia-Anoveros, J. *et al.*, 2001; Price, M. P. *et al.*, 2001) and the systematic examination of mechanosensation in ASIC knockout mice have lent support to this notion. Mice lacking ASIC1 were recently reported to exhibit no detectable changes in mechanoreceptor properties in the skin–nerve preparation. However, there was an enhancement of mechanically evoked firing rate in gastrointestinal afferents in these mice, accompanied by a reduction in gastric emptying rate (Page, A. J. *et al.*, 2004). Studies of ASIC2 KO mice have yielded somewhat conflicting results. In one study, these mice were reported to exhibit reduced mechanically evoked firing rate of rapidly adapting Aβ mechanoreceptors (RAM), without a change in RAM threshold (Price, M. P. *et al.*, 2000). In another study, however, no change in RAM response characteristics was observed (Roza, C. *et al.*, 2004). ASIC3 KO mice were reported to exhibit decreased sensitivity of Aδ mechanoreceptors but increased sensitivity of RAM, together with increased behavioral withdrawal from mechanical stimulation after carageenan (Price, M. P. *et al.*, 2001).

In contrast to the situation for acid detection, it has been relatively difficult to correlate mechanically evoked currents with the presence of particular ASIC subunits. Indeed a recent *in vitro* study of mechanically evoked currents revealed no differences between wild type and ASIC2 knockout, ASIC3 knockout, or ASIC2/3 double knockout mice. Instead, the authors of this study observed that the cationic dye, ruthenium red, could inhibit mechanosensitive currents, suggesting the possible participation of transient receptor potential (TRP) channels (see below) (Drew, L. J. *et al.*, 2004).

4.2.1.3 Transient receptor potential ion channels

One early clue that nociceptive neurons might express signaling proteins distinct from those of other neuronal subtypes was the observation that exposure of neonatal rats to capsaicin, the main pungent ingredient in hot peppers, produces a lifelong depletion of small-diameter sensory neurons, with no such effects on larger-diameter neurons. As adults,

these animals exhibit diminished responsiveness not only to subsequent capsaicin challenge, but also to other painful stimuli (Jancso, G. *et al.*, 1977). The highly enriched expression of a vanilloid receptor in nociceptive neurons appears to underlie these incredibly selective effects of capsaicin and related vanilloid compounds (Szallasi, A. and Blumberg, P. M., 1999). This conclusion was corroborated by the molecular cloning of a capsaicin-gated ion channel named TRPV1 (Caterina, M. J. *et al.*, 1997).

TRPV1 belongs to the TRP family, which in mammals contains approximately 30 members. TRP channel subunits have six transmembrane domains and intracellular N- and C-termini, and apparently form tetrameric channels on the cell surface. A remarkable feature of TRPV1 is that, in addition to vanilloid compounds, it can be activated by a number of other stimuli, including protons (pH < 5) (Tominaga, M. *et al.*, 1998), certain endogenous arachidonic acid derivatives (anandamide, *N*-arachidonyl dopamine, and 12-HPETE) (Zygmunt, P. M. *et al.*, 1999; Hwang, S. W. *et al.*, 2000; Huang, S. M. *et al.*, 2002) and even noxious heat (>42 °C) (Caterina, M. J. *et al.*, 1997). Nonvanilloid exogenous chemicals capable of activating TRPV1 include ethanol (Trevisani, M. *et al.*, 2002), camphor (Xu, H. *et al.*, 2005), and 2-aminoethoxydiphenyl borate (2-APB) (Hu, H. Z. *et al.*, 2004).

Sensory neurons derived from TRPV1 knockout mice not only are devoid of vanilloid responsiveness but also exhibit deficits in current responses evoked by pH 5 or moderate heat (45–50 °C) stimuli. All *in vivo* vanilloid-evoked responses are correspondingly absent in TRPV1 knockout mice. These animals also exhibit modestly reduced (but not absent) behavioral responses to acute noxious heat. A far more striking nociceptive phenotype is apparent in these mice following cutaneous inflammation by mustard oil, carageenan, or complete Freund's adjuvant or following surgical incision of the skin. Under these conditions, almost none of the thermal hyperalgesia typically exhibited by wild-type mice is observed in mice lacking TRPV1, although mechanical hyperalgesia is unaffected (Caterina, M. J. *et al.*, 2000; Davis, J. B. *et al.*, 2000; Pogatzki-Zahn, E. M. *et al.*, 2005). More recent pharmacological studies have further supported a role for TRPV1 in inflammatory thermal hyperalgesia (Pomonis, J. D. *et al.*, 2003; Gavva, N. R. *et al.*, 2004; Honore, P. *et al.*, 2005). However, the degree to which this channel may or may not participate in neuropathic pain states is much less clear. Although some pharmacological studies using TRPV1 antagonists with significant potential for off-target effects have suggested a possible role (Pomonis, J. D. *et al.*, 2003; Walker, K. M. *et al.*, 2003), mice lacking TRPV1 exhibit apparently normal thermal hyperalgesia following partial ligation of the sciatic nerve (Caterina, M. J. *et al.*, 2000). The recent availability of more selective TRPV1 antagonists promises to resolve this issue (Rami, H. K. and Gunthorpe, M. J., 2004). One particularly promising recent finding was that a TRPV1-selective antagonist was shown to partially attenuate the mechanical hyperalgesia associated with experimental bone cancer pain (Ghilardi, J. R. *et al.*, 2005).

Following the identification of TRPV1, a host of other TRP channels were found to be expressed in sensory ganglia. Like TRPV1, many of these channels are polymodal, exhibiting responsiveness to a multiple physical and chemical stimuli (Jordt, S. E. *et al.*, 2003; Patapoutian, A. *et al.*, 2003). Whereas TRPV1 appears to be the only member of this family gated by capsaicin, TRPV2, TRPV3, and TRPV4, like TRPV1, can be activated by elevated ambient temperatures. TRPV2 is expressed most prominently in a subpopulation of neurons with medium- to large-diameter cell bodies. Based on their size, expression of CGRP and neurofilament proteins, and projection pattern within the spinal cord, many of these neurons are likely to correspond to Aδ neurons. However, some presumably Aβ neurons that project within the dorsal columns also appear to express TRPV2. Heterologously expressed TRPV2 can be activated by temperatures >52 °C, a thermal response pattern reminiscent of those exhibited by Type II Aδ fibers in peripheral nerve recording studies and by a subpopulation of medium-diameter, capsaicin-insensitive dorsal root ganglion (DRG) neurons in isolated cultures (Caterina, M. J. *et al.*, 1999). TRPV2 can alternatively be activated by hypo-osmolarity (Iwata, Y. *et al.*, 2003), cell stretch (Muraki, K. *et al.*, 2003), or 2-APB (Hu, H. Z. *et al.*, 2004).

TRPV3 is activated by warm temperatures (>34–37 °C), although its thermal responsiveness extends well into the noxious range (Peier, A. M. *et al.*, 2002b; Smith, G. D. *et al.*, 2002; Xu, H. *et al.*, 2002). This channel can, alternatively, be activated by certain exogenous chemical substances (camphor, 2-APB) (Chung, M. K. *et al.*, 2004b; Hu, H. Z. *et al.*, 2004; Moqrich, A. *et al.*, 2005). The expression pattern of TRPV3 remains somewhat controversial. Although there are reports of TRPV3 mRNA or protein expression in sensory neurons (Smith, G. D. *et al.*, 2002; Xu, H. *et al.*, 2002), its expression in these cells is far less clear than in skin keratinocytes (Peier, A. M.

et al., 2002b), where TRPV3-mediated current responses can be readily generated (Chung, M. K. *et al.*, 2004a; Moqrich, A. *et al.*, 2005). These findings have led to the notion that TRPV3 may signal the presence of cutaneous warmth via keratinocyte–sensory neuron communication. Support for a role of TRPV3 in thermosensation has come from the observation that in mice lacking this protein, heat-evoked nociception is blunted in the hot plate and tail immersion assays. In addition, on a thermal gradient, selection of preferred floor temperatures is slowed, but not absent, in these mice (Moqrich, A. *et al.*, 2005).

TRPV4 can be gated by warm temperatures (>32 °C), hypo-osmolarity, cytochrome P450 metabolites of arachidonic acid, or the synthetic phorbol ester, 4α-phorboldibutyryl didecanoate (Nilius, B. *et al.*, 2004). Like TRPV3, TRPV4 is expressed more prominently in keratinocytes than in sensory afferents (Guler, A. D. *et al.*, 2002), although there is evidence for low-level TRPV4 protein expression in a population of medium- to large-diameter sensory afferents (Liedtke, W. and Friedman, J. M., 2003). Mice lacking TRPV4 exhibit reduced sensitivity to noxious mechanical or hypertonic stimuli, prolonged tail immersion latencies at modestly hot temperatures, and a shift in thermal selection behavior toward floor temperatures warmer than those selected by wild-type mice (Liedtke, W. and Friedman, J. M., 2003; Suzuki, M. *et al.*, 2003; Alessandri-Haber, N. *et al.*, 2005; Lee, H. *et al.*, 2005b).

Just as heat can activate TRPV1–TRPV4, cold temperatures have been reported to activate at least two other TRP channels. TRPM8 was identified on the basis of its responsiveness to the cold-mimetic agent, menthol. Examination of its thermosensory responsiveness revealed, however, that it could alternatively be activated by modest reductions in ambient temperature (<22 °C) (McKemy, D. D. *et al.*, 2002; Peier, A. M. *et al.*, 2002a). Another cold-mimetic agent, icilin, can also activate TRPM8, but only when intracellular Ca^{2+} levels are elevated (McKemy, D. D. *et al.*, 2002; Chuang, H. H. *et al.*, 2004). TRPM8 is expressed in a subset of small-diameter neurons that lack expression of TRPV1, substance P or Isolectin B4 (Peier, A. M. *et al.*, 2002a).

Another icilin-sensitive TRP channel that has been implicated in cold sensation is TRPA1. This channel is expressed in a subpopulation of small-diameter sensory neurons and is commonly coexpressed with TRPV1. In some studies, recombinant TRPA1 has been reported to be responsive to intense cooling (<18 °C) (Story, G. M. *et al.*, 2003). Consistent with this finding, mice lacking TRPA1 were recently reported to exhibit reduced withdrawal behavior on a cold plate or in response to acetone-mediated cooling of the hind paw (Kwan, K. Y. *et al.*, 2006). However, the cold sensitivity of TRPA1 is not universally accepted, since other investigators have failed to detect such responses and have reported normal cold-evoked behavioral responses in a distinct line of TRPA1 null mice (Jordt, S. E. *et al.*, 2004). Another point of contention revolves around the potential contributions of TRPA1 to mechanosensation, which is supported by immunolocalization studies (Nagata, K. *et al.*, 2005) and by analysis of one knockout mouse line (Kwan, K. Y. *et al.*, 2006) but not another (Bautista, D. M. *et al.*, 2006). The basis for these apparent discrepancies remains unclear. However, there is uniform consensus that TRPA1 is a mediator of nociceptive responses to certain irritant chemicals, including mustard oil (allyl isothiocyanate), cinnamaldehyde, and acrolein and that it serves as a downstream target following the activation of bradykinin receptors (Bandell, M. *et al.*, 2004; Jordt, S. E. *et al.*, 2004; Bautista, D. M. *et al.*, 2006; Kwan, K. Y. *et al.*, 2006). Indeed, both lines of TRPA1 null mice were found to be defective in nocifensive behavioral responses to mustard oil injection, mustard oil-evoked allodynia, and bradykinin-evoked pain-like behavior (Bautista, D. M. *et al.*, 2006; Kwan, K. Y. *et al.*, 2006).

Thus far, the examination of TRP channels within sensory neurons has been confined largely to those that either exhibit preferentially high expression in these cells or have been identified as being responsive to chemical stimuli with recognizable sensory effects. However, other TRP channels are known to be expressed quite broadly in neuronal and non-neuronal cells, where they participate in numerous functions that are important to sensory neuron biology (e.g., store-operated Ca^{2+} entry, Mg^{2+} uptake, mechanosensation, axon guidance) (Ramsey, I. S. *et al.*, 2006). Therefore, studies aimed at the analysis of those channels in sensory ganglia may yield further important insights.

4.2.2 Ion Channels Influencing Passive Membrane Properties: Membrane Resistance

While a number of channels contribute to the establishment of input resistance, two general classes have been described at a molecular level. One class includes the two-pore K^+ channels. These channels have no intrinsic voltage sensitivity and are active at resting membrane potentials thereby contributing to resting input resistance. TWIK related potassium channel (TREK-1) was the first of these K^+ channels to be identified in sensory neurons. TWIK stands for tandem of P domains in a weak inward rectifier potassium channel (Patel, A. J. and Honore, E., 2001). TREK-1 was cloned in 1996 based on homology to TWIK. While TREK-1 was originally shown to be activated by arachidonic acid and inhibited by cAMP, it was subsequently demonstrated that TREK-1 was activated by membrane stretch, osmotic welling as well as molecules that caused membrane crenation. Two more mechanosensitive two-pore potassium channels, TREK-2 and TRAAK were subsequently described (Maingret, F. *et al.*, 1999; Bang, H. *et al.*, 2000), both of which are also activated by membrane stretch and crenators and both of which are present in sensory ganglia (Kang, D. and Kim, D., 2006). Nevertheless, it remains to be determined whether any of these channels underlies the mechanosensitivity of primary afferent neurons (Patel, A. J. and Honore, E., 2001).

A second class of ion channel that may also contribute to resting input resistance is a member of the voltage-gated Na^+ channel family. One member of this family, NaV1.9 appears to encode a voltage-gated channel with unique gating properties (Cummins, T. R. *et al.*, 1999). The channel has a very low threshold for activation, activates very slowly upon membrane depolarization, and undergoes very little inactivation with small membrane depolarizations. These gating properties have two important consequences. First, it is likely that this channel contributes little, if anything to the rapid upstroke of the action potential. Second, the channel should, in theory, contribute a depolarizing drive at resting membrane potentials (Herzog, R. I. *et al.*, 2001). The problem with this theory is that biophysical analysis of persistent Na^+ currents in sensory neurons (putative NaV1.9-mediated currents) (Priest, B. T. *et al.*, 2005) indicate that this current is subject to a profound slow inactivation (Cummins, T. R. *et al.*, 1999). This voltage and time-dependent process is associated with a channel transition from a closed state, from which it could be activated with membrane depolarization, to an inactivated state, from which the channel is no longer available for activation. The extent of slow inactivation of NaV1.9 is such that there is virtually no detectable persistent current evoked from a typical resting membrane potential (i.e., $-60\,mV$). There is evidence suggesting that the voltage dependence of slow inactivation is influenced by recording conditions, such that the use of fluoride-based electrode solutions results is a hyperpolarizing shift in the voltage dependence of this process (Coste, B. *et al.*, 2004). Consequently, the extent of inactivation may be overestimated when the whole-cell patch configuration is used, if stability of the internal milieu is as important for the stability of NaV1.9 as it is for other ion channels (Kepplinger, K. J. *et al.*, 2000). Either way, what is clear is that even a few NaV1.9 channels active at rest could have a profound influence on neuronal excitability (Herzog, R. I. *et al.*, 2001).

4.2.3 Ion Channels Underlying Active Membrane Processes

4.2.3.1 Action potential upstroke

Voltage-gated Na^+ channels are critical for the upstroke, or rapid depolarization, phase of the neuronal action potential. Typical voltage-gated Na^+ channels consist of an α-subunit and two β-subunits (Catterall, W. A., 1992). The α-subunit contains everything necessary for a functional voltage-gated channel, including the ion pore, voltage sensor, and inactivation gate. β-subunits appear to both influence channel gating as well as play a major role in channel trafficking (Isom, L. L., 2001). Recent evidence suggests that one of the β-subunits may also influence channel inactivation (Grieco, T. M. *et al.*, 2005). Nine α-subunits have been cloned and clearly demonstrated to underlie voltage-gated Na^+ currents (Catterall, W. A. *et al.*, 2005a). A tenth α-subunit has been identified that is distantly related to the other nine and has been suggested to function as a Na^+ transporter rather than a voltage-gated channel (Goldin, A. L. *et al.*, 2000). Four β-subunits have been cloned (Goldin, A. L. *et al.*, 2000; Yu, F. H. *et al.*, 2003). There is evidence that eight of the nine α-subunits and all four β-subunits are present in sensory neurons (Black, J. A. *et al.*, 1996; Shah, B. S. *et al.*, 2000; Renganathan, M. *et al.*, 2002; Takahashi, N. *et al.*, 2003). α-subunits are differentially distributed both between and within sensory neurons. For example, NaV1.7, 1.8, and 1.9 are preferentially expressed in putative nociceptive afferents (Fang, X. *et al.*, 2002; Djouhri, L. *et al.*, 2003a; 2003b), while NaV1.1 is preferentially expressed in non-nociceptive afferents (Black, J. A. *et al.*, 1996). Within nociceptive afferents, NaV1.8 appears to be normally localized to the cell body and terminal arbor (Ritter, A. M. and Mendell, L. M., 1992; Jeftinija, S., 1994; Gu, J. G. and

MacDermott, A. B., 1997; Brock, J. A. *et al.*, 1998; Khasar, S. G. *et al.*, 1998), while NaV1.6 is present along the axons (Rios, J. C. *et al.*, 2003). Furthermore, there is evidence that channel expression is developmentally regulated. For example, in the rat, expression levels of NaV1.3 and 1.5 are high during development, and then reduced to almost undetectable levels after birth (Waxman, S. G. *et al.*, 1994; Renganathan, M. *et al.*, 2002). Importantly, the pattern of both α- and β-subunit expression is altered both within and between neurons following injury, changes that are thought to play a significant role in pain and hyperalgesia associated with tissue injury.

Voltage-gated Na^+ currents in sensory neurons are generally described in terms of their sensitivity to tetrodotoxin (TTX) where currents are TTX-sensitive (NaV1.1, 1.2, 1.3, 1.6, and 1.7), TTX-insensitive (NaV1.5), or TTX-resistant (NaV1.8 and 1.9). The TTX-resistant currents are further subdivided into inactivating (NaV1.8) and persistent current (NaV1.9). Based on data from biophysical characterization of these subunits and toxin sensitivity, NaV1.8 appears to be largely responsible for spike initiation in the terminals and cell body of nociceptive afferents (Gold, M. S., 2000a), while the TTX-sensitive channels, NaV1.6 and NaV1.7, are likely to underlying action potential conduction. Recent evidence from the NaV1.7 null mutant mouse suggests that NaV1.7 may also play a significant role in action potential generation in nociceptive afferent terminals (Nassar, M. A. *et al.*, 2004). A small cohort of individuals was also recently identified who do not experience pain in response to noxious stimuli, even though they appear to be aware of stimuli that should produce pain, exhibiting few, if any detectable somatosensory abnormalities (Cox, J. J. *et al.*, 2006). All of these individuals possess loss of function mutations in NaV1.7, indicating that this channel is necessary for the stimulus evoked perception of pain.

4.2.3.2 Action potential downstroke

The falling phase of the action potential is the result of two processes. One is the inactivation of voltage-gated Na^+ channels and the second is the activation of K^+ channels. One class of channels that plays a major role in controlling the rate of membrane repolarization are the sustained (i.e., delayed rectifier type) voltage-gated K^+ channels (Werz, M. A. and MacDonald, R. L., 1983). There appear to be at least three types of sustained voltage-gated K^+ currents present in sensory neurons based on biophysical and pharmacological properties

(Gold, M. S. *et al.*, 1996c). While the K^+ channel subunits underlying these currents have yet to be conclusively identified, immunohistochemical data have implicated several possible candidates (Rasband, M. N. *et al.*, 1998).

Another class of channel that contributes to membrane repolarization is the large conductance Ca^{2+}-modulated K^+ channel, (i.e., BK or Maxi-K channel). These channels are encoded by a single α-subunit (*slo*, which has several splice variants), whose gating properties, Ca^{2+} sensitivity, and pharmacological sensitivity is influenced by the presence of one of four distinct β-subunits (Wallner, M. *et al.*, 1998). While these channels have intrinsic voltage sensitivity, the voltage dependence of activation is such that in the absence of Ca^{2+}, it is unlikely that these channels would be sufficiently activated to contribute to the downstroke of the action potential (Wallner, M. *et al.*, 1998). This suggests that there is sufficient Ca^{2+} entry via voltage-gated Ca^{2+} channels during the action potential (Blair, N. T. and Bean, B. P., 2002), to enable activation of BK channels. The nature of the association between BK channels and voltage-gated Ca^{2+} channels in afferent terminals has yet to be described. However, the timing of BK current activation in the cell body suggests that there is a relatively tight association between the two.

4.2.3.3 After potential

The after potential is the change in membrane potential that follows an action potential. This is generally a membrane hyperpolarization (more negative than resting membrane potential), but may also be a depolarizing after potential (DAP). Two different mechanisms have been described that contribute to the DAP. One mechanism involves a high density of low threshold, or T-type voltage-gated Ca^{2+} channels (White, G. *et al.*, 1989). The biophysical properties of these channels are such that if present at a high enough density, they are able to depolarize the membrane following an action potential, providing enough depolarizing drive to evoke a burst of action potentials. The second current suggested to contribute to a DAP is a Ca^{2+}-dependent Cl^- current (Mayer, M., 1985). As the name implies, activation of this current requires sufficient Ca^{2+} entry during the action potential. It also requires the intracellular Cl^- concentration is such that the equilibrium concentration for Cl^- is depolarized.

In the majority of nociceptive sensory neurons, the somatic action potential is followed by an after hyperpolarization (AHP). In fact, a distinguishing feature of nociceptive afferents is that the somatic action potential is associated with a relatively long AHP (McLachlan, E. M. *et al.*, 1993; Villiere, V. and McLachlan, E. M., 1996; Djouhri, L. *et al.*, 2003a). A number of distinct currents have been suggested to influence the magnitude and duration of the AHP. Furthermore, because of differences between currents with respect to biophysical properties of channel gating, changes in intracellular Ca^{2+} concentrations, etc., the relative contribution of different ion channels to the AHP changes with time. Because of this feature, investigators have often divided the AHP into fast (early), intermediate, and slow (late) AHP. Channels mediating the fast AHP are generally those that contributed to the downstroke of the action potential (i.e., sustained voltage-gated K^+ and BK currents). These channels generally deactivate (i.e., close) rapidly and therefore do not generally contribute to the AHP for more than several of milliseconds (Fowler, J. C. *et al.*, 1985a).

The hyperpolarization associated with the fast AHP may result in the activation of three additional types of currents that have opposing influences on membrane potential. First, there may be the activation of inward rectifying K^+ currents (Scroggs, R. S. *et al.*, 1994). These currents appear to activate and inactivate rapidly in sensory neurons and therefore appear to primarily influence the fast AHP. Second, there may be the activation of low-threshold inactivating or A-type voltage-gated K^+ currents (Cardenas, C. G. *et al.*, 1995). These channels are largely inactivated at rest. However, there may be enough membrane hyperpolarization associated with the fast AHP to enable recovery from inactivation of a number of A-type K^+ channels. Because these channels have a low threshold for activation, once recovered from inactivation, the channels will open and therefore prolong the decay of the AHP. Because A-type currents in sensory neurons may inactivate slowly (i.e., with time constants of decay will over 100 ms (Gold, M. S. *et al.*, 1996c)), these channels can and do contribute significantly to the intermediate AHP (Harriott, A. *et al.*, 2006). As such, these channels play a significant role in regulating interspike interval (Connor, J. A. and Stevens, C. F., 1971; Yoshimura, N. and de Groat, W. C., 1999). Importantly, persistent inflammation and nerve injury have been shown to result in a decrease in these currents, which is associated with a concomitant increase in membrane excitability (Yoshimura, N. and de Groat, W. C., 1999; Moore, B. A. *et al.*, 2002; Stewart, T. *et al.*, 2003; Dang, K. *et al.*, 2004; Harriott, A.

et al., 2006). Third, there may be the activation of hyperpolarization-activated cationic current (Ih). These currents, carried by family channels called hyperpolarization-activated cyclic nucleotide-gated (HCN) channels, are nonselective cationic currents (Robinson, R. B. and Siegelbaum, S. A., 2003). Consequently, activation of these channels results in a depolarizing current, which, if of sufficient magnitude, may contribute to spike initiation. A high density of these channels can result in membrane oscillations and spontaneous activity. Indeed, there is evidence that these currents are upregulated in the presence of injury, a change that has been suggested to contribute to spontaneous afferent activity observed following nerve injury (Chaplan, S. R. *et al.*, 2003). Furthermore, there is evidence that blocking these channels attenuates mechanical hypersensitivity observed following nerve injury (Lee, D. H. *et al.*, 2005a). These channels are also modulated by cyclic nucleotides (Ingram, S. L. and Williams, J. T., 1994; Robinson, R. B. and Siegelbaum, S. A., 2003), which suggests that they may contribute to the increased excitability observed in the presence of inflammation as well, given that many inflammatory mediators induce an increase in cAMP concentrations (Levine, J. D. *et al.*, 1993).

In a subpopulation of sensory neurons, a slow AHP has also been described (Fowler, J. C. *et al.*, 1985a; 1985b; Weinreich, D. and Wonderlin, W. F., 1987). This AHP apparently reflects the activation of a Ca^{2+}-dependent K^+ channel that takes several hundred milliseconds to activate, but once activated takes several seconds to close (Cordoba-Rodriguez, R. *et al.*, 1999). Consequently, activation of these channels has a profound influence of spike accommodation. The slow AHP has been well described in nodose ganglion neurons, where it appears to be a target for inflammatory mediators that are able to block it in a cAMP dependent manner. The result is a dramatic increase in excitability (Weinreich, D. and Wonderlin, W. F., 1987). While there is evidence that a similar channel may be present in DRG neurons, it does not appear to play a major role in inflammatory mediator-induced sensitization of nociceptive afferents (Gold, M. S. *et al.*, 1996d). Four Ca^{2+}-dependent but voltage-independent channels have been cloned and are referred to as SK1–4, in reference to their small single-channel conductance (Bond, C. T. *et al.*, 2005). There is immunohistochemical evidence for the presence of these SK1 in sensory neurons (Boettger, M. K. *et al.*, 2002) and we have data from a single-cell PCR

experiment suggesting that others are expressed as well (Gold, M. S., unpublished observation). It remains to be determined which if any of the SK channels mediates the slow AHP given several striking differences between the SK channels and the slow AHP in terms of pharmacological sensitivity, Ca^{2+} dependence and gating properties (Cordoba-Rodriguez, R. *et al.*, 1999).

4.2.3.4　Action potential threshold

The action potential threshold is where the various influences of all the ion channels in the membrane come into play. The magnitude and nature of leak conductance determines membrane resistance and the impact a generator current will have on the membrane potential. The biophysical properties of voltage-gated K^+ currents will determine the magnitude of the K^+ conductance associated with membrane depolarization as well as the impact of resting membrane potential on that value. Similarly, the biophysical properties of voltage-gated Na^+ currents will determine the voltage at which these channels begin to open, the rate at which they open and the rate at which they inactivate in the face of slow membrane depolarizations. Given the fact that there is considerable variation in the biophysical properties of Na^+ channels depending on which α- and β-subunits are present, the relative density of these subunits will have a significant impact on action potential threshold (Matzner, O. and Devor, M., 1992).

Another class of ion channels that may contribute to the action potential threshold are the low threshold (i.e., T-type) voltage-gated Ca^+ channels. As with voltage-gated Na^+ channels, T-type channels consist of an α-subunit that contains the channel-pore, voltage-sensing, and inactivation gates, and ancillary subunits that influence channel gating and trafficking (Catterall, W. A. *et al.*, 2005b). Three α-subunit genes have been cloned that encode for T-type currents. T-type currents appear to be present in a low density in putative nociceptive afferents (Scroggs, R. S. and Fox, A. P., 1992). While the activation rate of these channels does not appear to be sufficient to enable these channels to contribute to the action potential upstroke, the low threshold for activation enables these currents to amplify the effects of small membrane depolarizations. Consequently, an increase in the density of these channels can have a significant influence on membrane excitability. In support of this suggestion, these channels have recently been implicated in epileptogenisis (Czapinski, P. *et al.*, 2005). More recent evidence suggests that they may contribute to the sensitization of nociceptive afferents in the presence of tissue injury (Barton, M. E. *et al.*, 2005).

4.2.3.5 High-threshold voltage-gated Ca²⁺ current

The structure of high-threshold voltage-gated Ca^{2+} channels is analogous to that of voltage-gated Na^+ channels and T-type Ca^{2+} channels, with a single α-subunit that contains the ion pore, voltage-sensor, etc., and several different ancillary subunits that influence channel gating (i.e., the β-subunit) and/ or channel trafficking (i.e., the $\alpha2\delta$-subunit complex) (Catterall, W. A. *et al.*, 2005b). Seven α-subunits encoding high-threshold channels have been cloned, resulting in four currents that are distinguishable on the basis of unique pharmacological sensitivity. Molecular biological and pharmacological analyses indicate that all four current types are likely present in nociceptive afferents (Kostyuk, P. G. *et al.*, 1981; Scroggs, R. S. and Fox, A. P., 1992; Kim, D. S. *et al.*, 2001; Yusaf, S. P. *et al.*, 2001; Wu, Z. Z. *et al.*, 2004b). While these channels are critical for transmitter release, their role in direct regulation of excitability is less clear. On the one hand, they enable inward current that contributes to membrane depolarization and prolongation of the action potential. On the other hand, they enable activation of K^+ and Cl^- currents that may be inhibitory. Thus, the site of channel activation is critical. Of note, in studies involving the isolated sensory neuron cell body, there is evidence to suggest that inflammatory mediators decrease the density of high-threshold voltage-gated currents (Borgland, S. L. *et al.*, 2002), and in the presence of nerve injury, there is a decrease in the density of high-threshold voltage-gated currents as well (Baccei, M. L. and Kocsis, J. D., 2000; Hogan, Q. H. *et al.*, 2000; Kim, D. S. *et al.*, 2001). Authors of both the studies suggested that these changes may contribute to pain associated with tissue injury and all are certainly correct on this point.

4.2.4 Membrane Receptors that Indirectly Alter Nociceptor Function

4.2.4.1 G-protein-coupled receptors

Many of the cell surface proteins that define the functional and histological identity of sensory neurons are the so-called GPCRs. These proteins, which possess seven transmembrane domains, an extracellular amino terminus, and an intracellular carboxyl terminus, signal by catalyzing the dissociation of guanosine diphosphate (GDP) from the α-subunit of heterotrimeric G proteins associated with the intracellular face of the plasma membrane. As a consequence of its dissociation, GDP is replaced with guanosine triphosphate (GTP), causing the heterotrimer to dissociate into a GTP-bound α-subunit and a free $\beta\gamma$ dimer. Both the GTP-bound α-subunit and the free $\beta\gamma$ dimer have specific cellular targets through which they are able to activate or inhibit signaling cascades and/or directly modify the activity of specific ion channels (Gilman, A. G., 1987; Pierce, K. L. *et al.*, 2002). This versatile signal transduction strategy is used by at least 1000 different GPCRs expressed throughout the body, and dozens of these are expressed by nociceptors and other sensory neurons. A few notable examples include B1 and B2 receptors for bradykinin (Steranka, L. R. *et al.*, 1988), P2Y2 receptors for ATP (Tominaga, M. *et al.*, 2001), PAR2 protease-activated receptors (Vergnolle, N. *et al.*, 2001), EP1 EP3C and EP4 receptors for prostaglandin E2 (Ferreira, S. H. *et al.*, 1978; Sugimoto, Y. *et al.*, 1994; Oida, H. *et al.*, 1995), mu, kappa, and delta opioid receptors (Dado, R. J. *et al.*, 1993; Arvidsson, U. *et al.*, 1995), several subtypes of metabotropic glutamate receptor (Bhave, G. *et al.*, 2001; Yang, D. and Gereau, R. W. T., 2002), and at least seven subtypes of 5HT receptor (Nicholson, R. *et al.*, 2003). One recently discovered family of apparently pronociceptive GPCRs are those of the so-called mrg or sensory neuron specific (SNS) family (Dong, X. *et al.*, 2001; Lembo, P. M. *et al.*, 2002; Zylka, M. J. *et al.*, 2003). The expression pattern of this GPCR family exhibits remarkable species-specific heterogeneity, as there appear to be at least 50 members in mice, but only four in rat and only a few distantly related homolog in humans (Zylka, M. J. *et al.*, 2003).

GPCRs expressed in sensory ganglia can be broadly classified into those that enhance nociceptor excitability and those that diminish it. Sensitizing receptors tend to couple to either Gq (e.g., B2, P2Y2, PAR2, and EP1) or Gs (e.g., EP3C, EP4). The activated α-subunit of Gq activates phospholipase C (PLC), an enzyme that catalyzes cleavage of the membrane phospholipid phosphatidylinositol bisphosphate to inositol 1,3,5-trisphosphate (IP3) and diacylglycerol (DAG). IP3 evokes the release of Ca^{2+} ions from intracellular stores that, together with DAG, go on to trigger the plasma membrane localization and activation of protein kinase C (PKC) (Gilman, A. G., 1987). Multiple PKC isoforms (α, δ, ε, and μ) have been shown to facilitate nociceptor sensitization through their phosphorylation of ion channels and other target proteins (Cesare, P. *et al.*, 1999; Khasar, S. G. *et al.*, 1999; Olah, Z. *et al.*, 2002; Wang, Y. *et al.*, 2004). Ca^{2+} ions can also act in many other ways, such as by binding to proteins such as calmodulin that can allosterically regulate ion channels and intracellular enzymes in a Ca^{2+}-dependent manner (Numazaki, M. *et al.*, 2003). The GTP-bound $G\alpha s$ subunit activates the enzyme adenylyl cyclase, which catalyzes the generation of cyclic adenosine monophosphate (cAMP) from ATP (Gilman, A. G., 1987). cAMP, in turn, activates cAMP-dependent protein kinase (protein kinase A, PKA), which can also phosphorylate and augment the responsiveness of certain ligand-, heat-, or voltage-gated channels (Gold, M. S. *et al.*, 1998; Bhave, G. and Gereau, R. W. T., 2004). Another intracellular enzyme recently implicated in nociception is the cAMP-activated guanine nucleotide exchange protein, Epac, which couples intracellular cAMP generation to the PLC-dependent sensitization of TTX-resistant sodium channels (Hucho, T. B. *et al.*, 2005). Interestingly, although all of the actions of pronociceptive GPCRs on excitability are indirect, through their actions on downstream signaling mechanisms, activation of some (e.g., bradykinin receptors) is sufficient to trigger action potential firing, whereas activation of others (e.g., prostaglandin receptors) primarily enhances neuronal responsiveness to other exogenous stimuli (Birrell, G. J. *et al.*, 1991; 1993; Kumazawa, T. *et al.*, 1996). This distinction may reflect differences in signal amplification or differences in the proximity of a given GPCR, its target enzymes, and a given depolarizing ion channel.

In contrast to the pronociceptive receptors, GPCRs that couple to Gi or Go proteins (e.g., μ-opioid receptor, type II metabotropic glutamate receptors) tend to suppress neuronal activity. This outcome results from inhibition of adenylyl cyclase (to diminish PKA phosphorylation of target proteins), inhibition of voltage-gated Ca^{2+} channels, and/or activation of hyperpolarizing G-protein-coupled inwardly rectifying K^+ channels (GIRKs of the Kv4 family) (McFadzean, I., 1988; Bhave, G. and Gereau, R. W. T., 2004).

4.2.4.2   *Neurotrophin receptors*

As described in detail elsewhere in this chapter, neurotrophins are also important regulators of sensory neuron development, survival, and excitability. One major class of neurotrophic factors that carry out these functions is the nerve growth factor (NGF) family (NGF, BDNF, neurotrophin-3, and neurotrophin-4). The receptors for this family of nociceptors fall into two categories: a low-affinity receptor p75, which appears to bind all members of the family, and homo-dimeric high-affinity receptors that are specific for each of the family members (TrkA, TrkB, or TrkC) (Huang, E. J. and Reichardt, L. F., 2001). Interestingly, one p75 monomer and one Trk dimer can apparently bind the same neurotrophin dimer simultaneously (He, X. L. and Garcia, K. C., 2004). During development, all sensory neurons destined to become nociceptors are dependent on NGF for survival. However, with maturation, only a subset of nociceptive afferents continues to express the high-affinity NGF receptor, TrkA. In these neurons, continued access to NGF is no longer necessary for survival, but remains essential for the maintenance of the afferent phenotype, which includes peptide expression and the specific composition of ion channels and transducers. Afferents have continued access to NGF produced by target tissue, but the amount of NGF available is critical: a decrease in access to NGF appears to result in the loss of nociceptive phenotype, while increased access to NGF results in an increase in the expression of nociceptive markers (Koltzenburg, M., 1999). BDNF and neurotrophin-3 can also alter nociceptor function (Shu, X. Q. and Mendell, M. L., 1999).

Transduction of the signal arising from the binding of NGF family members to their receptors consists of both slow processes requiring internalization of the neurotrophin–receptor complex and rapid signaling processes that occur at the nociceptor terminal. In both cases, NGF binding drives the dimerization of the TrkA receptor, resulting in TrkA autophosphorylation and the recruitment of a complex of molecules that signal through distinct enzymatic pathways. These include, but are not limited to, pathways involving phospho-inositol 3 kinase (PI3K), ras-mitogen-associated protein kinase (MAPK), and PLC γ. The activated complex with NGF-bound TrkA at its core is internalized into a specialized endosomal vesicle and transported along microtubules to the cell body, where signaling from this complex can trigger changes in transcription, translation, and protein trafficking (Zweifel, L. S. *et al.*, 2005). For many years, this slow retrograde process was viewed as the predominant mode of neurotrophin signaling in sensory neurons. More recently, however, it has become evident that NGF–TrkA activation of PI3K, MAPK, and PLCγ

produces rapid effects at the nociceptor terminal by altering ion channel sensitivity and plasma membrane localization within seconds, chiefly via receptor phosphorylation (Shu, X. and Mendell, M. L., 2001; Bonnington, J. K. and McNaughton, P. A., 2003; Zhang, X. *et al.*, 2005). These events will be described in greater detail in the final section of this chapter. The p75 receptor is associated with the activation of a distinct set of second messenger pathways, including one involving the activation of sphingomyelinase and the liberation of ceramide (Zhang, Y. H. *et al.*, 2002; Zhang, Y. H. and Nicol, G. D., 2004).

Nociceptive afferents expressing TrkA constitute a subpopulation of nociceptive afferents. Other subpopulations of nociceptive afferents express receptors for other neurotrophic factors. For example, one subpopulation of nociceptive afferents that do not express the neuropeptides substance P or CGRP, lose their NGF dependence during early postnatal life and begin to express a receptor for glial cell-derived neurotrophic factor (GDNF) (Molliver, D. C. *et al.*, 1997). GDNF is a member of a second family of neurotrophins that includes neuturin, artemin, and persephin. As with NGF in TrkA-expressing neurons, the amount of GDNF to which these nonpeptidergic nociceptive afferents have access influences their phenotypic properties (see Fjell, J. *et al.*, 1999a). GDNF signaling involves a specific GDNF family receptor (GFR), which acts in conjunction with another receptor tyrosine kinase, ret, that is shared by all members of this family. Interestingly, GDNF signaling involves second messenger pathways similar to those underling the actions of NGF, including PLC-γ, MAPK, and PI3K (Airaksinen, M. S. and Saarma, M., 2002). Thus, the phenotype of subpopulations of nociceptive afferents is regulated by different receptor systems that appear to ultimately use similar transduction cascades.

4.3 Additive and Emergent Properties of Nociceptor Membrane Proteins

4.3.1 Introduction

It should be evident from the preceding section that sensory afferents have, at their disposal, an extensive repertoire of cell surface proteins with diverse stimulus specificities and signaling capabilities.

However, a full mechanistic understanding of nociceptive signal transduction, action potential generation, and spike propagation will require that pain biologists move beyond a mere itemization of the contents of this toolbox to an appreciation of the additive and emergent properties that can arise from the polymodal nature of some of these molecules, their posttranslational modulation, their interactions, and their spatial distribution. In this section, we attempt to illustrate these principles using a few specific examples, some well established, and others more speculative in nature.

4.3.2 Shared Modalities and Polymodality

One source of signaling complexity and heterogeneity among sensory neurons is the existence of multiple transducers for a given stimulus modality. For example, as described above, at least six TRP channels expressed in sensory neurons can be gated by heat or cold. Although each channel exhibits a characteristic temperature range over which it is thermally responsive, in many cases these ranges overlap, theoretically allowing continuous recognition of the thermal environment. At the same time, the largely nonoverlapping expression patterns of TRPM8, TRPV1, and TRPV2 lend some credence to the labeled-line hypothesis, which states that distinct sensory modalities (innocuous cold, pain, itch, etc.) are conveyed by anatomically distinct subsets of sensory afferents with specific central projection patterns (Craig, A. D., 2002). However, it should be noted that this segregation of thermotransduction mechanisms is not absolute. For example, some polymodal nociceptors exhibit paradoxical dual responsiveness to both intense heat and intense cold (Dodt, E. and Zotterman, Y., 1952). Even at the molecular level, TRPV1 and TRPA1 are coexpressed in a significant fraction of small-diameter sensory neurons. Though the implications for thermosensation remain controversial, such coexpression appears to be critical for full responsiveness to bradykinin, perhaps owing to the potentiation of TRPA1 by Ca^{2+} ions flowing through TRPV1 (Bautista, D. M. *et al.*, 2006). Thus, while the labeled line hypothesis may adequately address some aspects of thermosensation, the overall situation may be more complex.

Mechanotransduction represents an even more extreme example of likely complementarity and redundancy among transducing molecules. Members of both the ASIC and the TRP families have been implicated in this process, but only modest deficits in mechanotransduction have been reported in animals lacking any single putative mechanosensory protein (Price, M. P. et al., 2000; Liedtke, W. and Friedman, J. M., 2003; Suzuki, M. et al., 2003; Page, A. J. et al., 2004; Kwan, K. Y. et al., 2006). One interpretation of these findings is that mechanotransduction involves complexes of intracellular, transmembrane, and extracellular molecules, and that multiple interchangeable ion channel subunits can contribute to the ion permeation pathways of these transduction complexes.

Another source of complexity in sensory nerve signal transduction is responsiveness to multiple modalities within a single transduction molecule. One of the best-studied examples of this phenomenon is the capsaicin receptor, TRPV1. This channel protein can be activated by a diverse list of chemical or thermal stimuli (Caterina, M. J. and Julius, D., 2001). In addition, there is evidence that this channel or its splice variants may participate in osmotic or mechanical sensitivity in osmosensory regions of the brain and in non-neuronal cells of the urinary bladder (Birder, L. A., 2005; Naeini, R. S. et al., 2006). Such polymodal responsiveness provides ample opportunities for convergent stimuli to combine additively or even supraadditively to produce greater nociceptor activation than any single stimulus. For example, even a modest reduction in pH (to 6.4) can drop the apparent thermal threshold for TRPV1 activation to near body temperature, possibly leading to spontaneous pain in the context of inflammation or ischemia (Tominaga, M. et al., 1998).

4.3.3 Functional Interactions between Nociceptor Proteins

Interactions between different proteins on the surface of the nociceptor terminal can also have profound effects on nociceptors transduction and therefore function. Sometimes, these interactions are direct, as in the case of ion channel or GPCR heteromultimerization. In other cases, second messenger molecules and intracellular enzymes facilitate functional interactions among membrane proteins. Recently, for example, it was demonstrated that TRPM8 can activate Ca^{2+}-dependent phospholipases that, in turn, cleave the phosphatidyl inositol bisphosphate that inhibits or activates a host of channels, including inwardly rectifying potassium channels, as well as

TRPM8, itself (Rohacs, T. et al., 2005). Similarly, TRPV1 can activate MAPK signaling pathways (Zhuang, Z. Y. et al., 2004). An additional interaction mechanism that has been reported in other types of sensory neurons, but could also conceivably occur in nociceptors, is the stimulation of Ca^{2+}-activated K^+ channels by the Ca^{2+} ions that flow through ligand-gated ion channels (Yuhas, W. A. and Fuchs, P.A., 1999; Isaacson, J. S. and Murphy, G. J., 2001). This arrangement has the potential to shape the amplitude, duration, and even the direction of generator potential.

Active membrane properties are also strongly dependent upon the interaction, at multiple levels, between the ion channels and the other membrane proteins underlying transduction and those regulating action potential initiation and propagation. For example, even small changes in resting membrane potential or generator potential will dramatically influence the availability of different voltage-gated ion channels. Similarly, the second messenger pathways activated by GPCRs and neurotrophin receptors can impact many facets of action potential generation and propagation, as described in the final section of this chapter. The spatial distribution of ion channels underlying each of the steps in afferent signaling is another critical determinant of whether or not there will be successful transmission of sensory information. The ion channels underlying spike initiation must be close enough to the ion channels underlying stimulus transduction that the generator potential pushes the membrane potential above the action potential threshold (Figure 1). Similarly, the ion channels underlying action potential propagation must be close enough to the site of spike initiation, that the action potential is faithfully propagated toward the central nervous system. In addition, action potential threshold may be relatively high in nociceptors that utilize the high-threshold voltage-gated Na^+ channel NaV1.8 for spike initiation, whereas action potential burst duration may be relatively short in nociceptors in which Ca^{2+}-dependent K^+ channels are present in high densities.

One implication of the precise nature of these spatial relationships between of ion channels within sensory neurons is that cellular processes underlying the trafficking and anchoring of ion channels within sensory neurons must be tightly regulated. While there is much to be learned about the distribution and anchoring of ion channels in sensory neurons, the observation that it is possible to induce modality specific sensitization in polymodal nociceptors, raises the possibility that ion channels and/or transducers are differentially distributed throughout the terminal arbor. Adaptor/scaffold proteins can also shape nociceptor responsiveness by bridging ion channels or GPCRs with intracellular enzymes or with proteins that regulate intracellular transport. For example, PICK1, a PDZ domain-containing protein, appears to regulate the subcellular trafficking of ASIC channel subunits (Duggan, A. *et al.*, 2002), while annexin has been demonstrated to regulate surface expression and function of NaV1.8 (Okuse, K. *et al.*, 2002).

In summary, signaling heterogeneity, convergence of function, protein–protein interactions, and cytosolic signaling, as well as spatiotemporal organization of ion channels allow an already complex set of membrane proteins to vastly expand the sensory and coding capacity of nociceptive neurons.

4.4   Injury-Induced Plasticity in Transduction

4.4.1   Introduction

Research into the underlying mechanisms of pain associated with tissue injury has revealed that while there are a number of key players that are involved with a wide range of injury types the nature of the contribution of these players varies as a function of the type of injury. For example, acute, phosphorylation-dependent modulation of the voltage-gated Na^+ channel NaV1.8 (Fitzgerald, E. M. *et al.*, 1999), results in an increase in current that contributes to an inflammation-induced increase in nociceptors excitability (England, S. *et al.*, 1996; Gold, M. S. *et al.*, 1996b). In contrast, following traumatic nerve injury, redistribution of NaV1.8 to the axons of uninjured afferents appears to be necessary for the expression of mechanical hypersensitivity associated with nerve injury (Gold, M. S. *et al.*, 2003). There are many such examples, a number of which have been alluded to in the preceding discussion. However, while it is beyond the scope of the present chapter to detail the array of injury-induced changes in mechanisms underlying stimulus transduction, it is informative to consider several specific examples of modulatory events that occur in the context of inflammation and nerve injury. A few of these mechanisms are illustrated schematically in Figure 1.

4.4.2   Inflammation-Induced Changes in Nociceptor Sensory function – Inflammatory Pain

One of the best examples of injury-induced changes in the transduction properties of nociceptive neurons is the enhanced responsiveness to heat that occurs following tissue inflammation. As indicated in Section 4.2 of this chapter, the thermal hyperalgesia associated with several distinct inflammatory insults (CFA, carageenan, and mustard oil) or specific inflammatory mediators (bradykinin, NGF, ATP, proteases, and prostaglandin E2) depends strongly on the presence of TRPV1. Consistent with this observation, TRPV1 ion channel activity can be sensitized by these various agents in dissociated DRG neurons or heterologous expression systems (Lopshire, J. C. and Nicol, G. D., 1998; Shu,X. and Mendell, M. L., 2001; Vellani, V. *et al.*, 2001; Bonnington, J. K. and McNaughton, P. A., 2003). One lesson emerging from these studies, however, is that TRPV1 sensitization by inflammatory mediators can proceed via a number of alternative mechanisms.

The most direct mechanism by which inflammatory mediators can sensitize TRPV1 is by evoking the generation or liberation of TRPV1 coagonists, such as protons (Caterina, M. J. *et al.*, 1997) or arachidonic acid metabolites (Hwang, S. W. *et al.*, 2000; Huang, S. M. *et al.*, 2002). Convergent interactions of TRPV1 with multiple stimulators result in channel activity levels greater than those evoked by one agonist alone. Another extensively studied TRPV1 sensitization mechanism involves direct phosphorylation of this channel by serine/threonine kinases (PKC, PKA, Ca^{2+}/calmodulin-dependent protein kinase II) (Bhave, G. and Gereau, R. W. T., 2004) or the src tyrosine kinase (Zhang, X. *et al.*, 2005). Although there is some overlap in their recognition sequences, each kinase apparently targets a different collection of phosphorylation sites, resulting in the generation of distinct phospho-TRPV1 species. In the case of ser/thr kinases, phosphorylation appears to enhance TRPV1 sensitivity to its chemical and thermal agonists and/or reverse TRPV1 desensitization (Bhave, G. and Gereau, R. W. T., 2004). In contrast, TRPV1 phosphorylation by src kinase, which is activated by NGF–trkA signaling, leads within 10 min to a translocation of TRPV1 from an intracellular compartment to the plasma membrane, resulting in an increase in the overall number of TRPV1 channels on the cell surface (Zhang, X.

et al., 2005). The MAPK, p38, has also been shown to participate in NGF-stimulated facilitation of TRPV1 signaling during inflammation. Although the relevant target of p38 phosphorylation is not known, the end result is an apparent enhancement of TRPV1 translation and/or trafficking to peripheral nociceptor terminals that requires 1–2 days (Ji, R. R. *et al.*, 2002).

A third general mechanism by which TRPV1 can be modulated in the context of inflammation is by alterations in membrane phospholipid content. For example, it has been proposed that phosphatidylinositol bisphosphate exerts a tonic inhibition on TRPV1 activity and that its hydrolysis by PLC relieves that inhibition (Chuang, H. H. *et al.*, 2001; Prescott, E. D. and Julius, D., 2003). The effects of PIP2 on TRPV1 may be complex, however, given that this phospholipid is also required for resensitization of TRPV1 following agonist-dependent desensitization (Liu, B. *et al.*, 2005). More recently, work from several laboratories has provided evidence that phosphorylation of PIP2 to PIP3 by PI3 kinase to form PIP3 represents yet another mechanism for TRPV1 sensitization (Bonnington, J. K. and McNaughton, P. A., 2003; Zhuang, Z. Y. *et al.*, 2004; Zhang, X. *et al.*, 2005). Although the precise basis for this effect remains unclear, one intriguing suggestion is that it is due to the effects of PIP3 on src activity, thus potentially linking lipid-based and protein kinase-based sensitization events (Zhuang, Z. Y. *et al.*, 2004). Much remains to be learned about the relative importance of and interactions between these different processes. However, it is clear that nociceptors have developed multiple means of responding to tissue inflammation by becoming hypersensitive.

4.4.3 Changes in Nociceptor Function Following Traumatic Nerve Injury – Neuropathic Pain

Pain and hyperalgesia arising from direct damage to sensory neurons come under the heading of neuropathic pain. One of the most dramatic and common features of neuropathic pain is the presence of ongoing pain (Backonja, M. M. and Stacey, B., 2004). This feature is distinct from inflammatory pain, where it is often possible to relieve pain if it is possible to eliminate stimuli that impact the inflamed tissue. There is compelling evidence to suggest that ongoing pain associated with peripheral nerve injury is the result of aberrant and/or ongoing activity in

both injured afferents and their uninjured neighbors (Gold, M. S., 2000b). Most compelling are the observations that ongoing neuropathic pain may be blocked with the administration of local anesthetics (Gracely, R. H. *et al.*, 1992). However, there is also a very tight correlation between the development of nociceptive behaviors and the emergence of ongoing afferent activity in animal models of neuropathic pain (Liu. C. N. *et al.*, 2000). Because of this association between ongoing afferent activity and ongoing pain, considerable effort has been dedicated to the identification of mechanisms that may contribute to this activity. Central to this search is the question of whether the activity is driven by stimulus transduction, or whether there are changes in the relative contribution of ion channels resulting in membrane potential instability, the emergence or amplification of membrane potential oscillations that are capable of driving action potential generation. There is evidence for both sources of activity.

4.4.3.1 *Nerve injury-induced changes in transduction*

Several nerve injury-induced changes in transduction have been identified. These can be loosely grouped into those involving aberrant expression of transducers normally present in peripheral nerve, aberrant coupling between the transducers and cellular pathways and the emergence of inappropriate sources of stimulation. The list of tranducers aberrantly expressed following nerve injury has grown as fast as transducers have been identified. Aberrant expression comes in at least three forms. One is the expression of transducers in the wrong place within a neuron. The most dramatic example of this is the emergence of mechanical and thermal sensitivity at the cut end of a peripheral axon (Michaelis, M. *et al.*, 1999). More recently, it have been demonstrated that compression of a nerve is sufficient to alter the distribution of tranducers in the membrane (Ma, C. *et al.*, 2006). Importantly, aberrant expression patterns are not restricted to the point of compression or injury as a cut axon will result in increased mechanical sensitivity of the sensory neuron soma (Howe, J. F. *et al.*, 1977). Two consequences of these changes are the emergence of mechanical sensitivity at sites that are normally mechanically insensitive and mechanical allodynia, a situation in which a mechanical stimulus, such as that associated with movement, that is normally not painful, becomes a source of pain. Worse still, mechanical stimuli associated with physiological functions, such as movement of tissue associated with blood flow, may also become a source of stimuli for these transducers. A second form of aberrant expression is the emergence of transducers in the wrong neurons. For example, as indicated above, TRPV1 is primarily expressed in nociceptive afferents. However, following nerve injury, this transducer is expressed in what were previously low-threshold non-nociceptive afferents (Rashid, M. H. *et al.*, 2003). A third form of aberrant expression is a change in the levels of expression of transducers normally present in a neuron. There is evidence for both decreases in inhibitory transducers and increases in excitatory transducers. For example, nerve injury is associated with a decrease in the expression of opioid receptors (Kohno, T. *et al.*, 2005), which are the transducers for the inhibitory signal of both endogenous and exogenous opioids. The result is a decrease in the efficacy of mechanisms normally involved in attenuating pronociceptive processes. An example of an increase in an excitatory transducer is an increase in the expression of the ATP receptor, P2X3 (Fukuoka, T. *et al.*, 2002). The combination of these changes in expression levels is a net increase in afferent excitability.

Evidence in support of aberrant coupling between transducers and signaling pathways following nerve injury is less compelling but should be mentioned. The pathway that has received the greatest attention is the adrenergic system, largely as a result of the observation that pain associated with nerve injury may be sympathetically dependent. What most pain scientists agree on, is that in normal tissue, activation of the sympathetic nervous system is not painful. In fact, primary afferent neurons express adrenergic receptors that are critical for the antinociceptive actions of spinally administered adrenergic agonists. However, following nerve injury, adrenergic agonists may become excitatory (Sato, J. and Perl, E. R., 1991; Devor, M. *et al.*, 1994; Chen, Y. *et al.*, 1996; Liu, X. *et al.*, 1999; Moon, D. E. *et al.*, 1999). Mechanisms of this excitation are still debated, but there is evidence to support at least two mechanisms. One is that there is a change in adrenergic receptor expression and/or coupling such that activation of adrenergic receptors on primary afferent becomes excitatory (Davis, K. D. *et al.*, 1991; Birder, L. A. and Perl, E. R., 1999). A second is that there is a change in the signaling pathway underlying the actions of adrenergic agonists at autoreceptors on the sympathetic postganglionic neuron terminals: in the presence of elevated levels of intracellular Ca^{2+}, norepinephrine becomes capable of driving the release of arachidonic acid metabolites such as prostaglandin E2 (Gonzales, R. *et al.*, 1989; 1991), which then act directly on the primary afferent.

The emergence of inappropriate sources of stimulation may involve sources that are both expected and those that are much less so. Following nerve injury, there are a number of changes that occur in both the injured axon and the surrounding tissue that facilitate restoration of the normal innervation pattern (see Murinson, B. B. *et al.*, 2005). Injured nerves extend processes, which if scar tissue forms at the same time, may not be able to access former nerve tracks. The result may be the formation of a neuroma. Thus, an expected source of stimulation would be mechanical stimulation of the neuroma associated with traction of the surrounding scar tissue. Interestingly, nerve injury may also be associated with the sprouting of sympathic postganglionic terminals into the sensory ganglia (McLachlan, E. M. *et al.*, 1993). This striking observation provided anatomical evidence of sympathic/afferent coupling within the sensory ganglia. While many details about the mechanisms underlying this sprouting have yet to be determined, access to NGF appears to be a regulating factor (Ramer, M. S. and Bisby, M. A., 1999).

4.4.3.2 Changes in membrane stability

As emphasized above, the distribution, expression levels and biophysical properties of ion channels in a neural membrane are critical to the maintenance of a given level of membrane excitability. Change in any of these parameters may have profound effects. Not surprisingly, nerve injury is associated with changes in all three, with clear examples of changes in distribution (Gold, M. S. *et al.*, 2003), expression levels (Waxman, S. G. *et al.*, 1994; Dib-Hajj, S. *et al.*, 1996; Decosterd, I. *et al.*, 2002), and biophysical properties (Cummins, T. R. and Waxman, S. G., 1997) of ion channels in both injured neurons and their uninjured neighbors. One of the more striking ramifications of changes in all three of these properties is the emergence and/or amplification of membrane potential oscillations which, it has been argued, are sufficient to drive ongoing activity following nerve injury (Amir, R. *et al.*, 1999). Two critical players have been identified. One is a voltage-gated Na^+ channel with appropriate biophysical properties to sustain the upstroke of the oscillation. The second is a K^+ leak current that drives the downstroke of the oscillation (Amir, R. *et al.*, 1999). While the K^+ channels have yet to be identified, several different Na^+ channels have been implicated. One possibility is a TTX-resistant channel such as NaV1.8 or NaV1.9. This would be consistent with the observation that the upstroke of the oscillation in trigeminal neurons is TTX-resistant (Puil, E. *et al.*, 1989) as well as the biophysical properties of the current implicated in studies of isolated DRG neurons (Liu, C. N. *et al.*, 2002). However, spontaneous activity in injured afferents is most robust in sensory neurons with a large cell body diameter giving rise to myelinated axons (Liu, C. N. *et al.*, 2000); neither TTX-resistant channel is present in this subpopulation of sensory neurons and both are dramatically downregulated in injured neurons (Dib-Hajj, S. *et al.*, 1996; Decosterd, I. *et al.*, 2002). Another channel that has received a lot of attention is NaV1.3. This channel is upregulated in the presence of injury (Waxman, S. G. *et al.*, 1994; Black, J. A. *et al.*, 1999) and appears to have biophysical properties consistent with ongoing high-frequency activity (Cummins, T. R. and Waxman, S. G., 1997; Cummins, T. R. *et al.*, 2001). There is evidence for the accumulation of TTX-sensitive Na$^+$ channels at sites of nerve injury (Devor, M. *et al.*, 1993). Furthermore, there is evidence that at least some of the oscillatory activity in sensory neurons is sensitive to TTX (Amir, R. *et al.*, 1999). However, antisense knockdown of this channel has no impact on nociceptive behavior associated with nerve injury (Lindia, J. A. *et al.*, 2005).

Recent evidence from a cohort of individuals who suffer from a paroxysmal pain disorder raise the possibility that NaV1.7 may contribute to nerve injury-induced increases in afferent excitability. Sufferers of this disorder appear to share gain of function mutations in NaV1.7, decreasing the rate and/or extent of channel inactivation (Fertleman, C. R. *et al.*, 2006). While the link between neuropathic pain and paroxysmal pain disorder is tenuous, manifestations of this disorder share similarities with signs and symptoms neuropathic pain, most notably a sensitivity to the membrane stabilizer carbamazepine. A different set of NaV1.7 mutations appears to underlie primary erythermalgia (Yang, Y. *et al.*, 2004), a rare pain disorder associated with pain and redness in the feet and hands. Mutations in NaV1.7 associated with erythermalgia contribute to an increase in the excitability of primary afferent neurons (Rush, A. M. *et al.*, 2006), but they also appear to result in a decrease in the excitability of sympathetic postganglionic neurons, and it is likely the combination of these changes in excitability that appear to mediate the unique symptoms of the disorder. At least two lines of evidence argue against a role for NaV1.7 in pain associated with traumatic nerve injury. First, in patients with peripheral nerve injury NaV1.7 protein appears to decrease in the DRG neurons and remains undetectable in injured peripheral nerve (Coward, K. *et al.*, 2001). Second, there appears to be no deficit in the nerve injury-induced hypersensitivity in an NaV1.7 null mutant mouse (Nassar, M. A. *et al.*, 2005). Thus, while the specific channels underlying nerve injury-induced increases in afferent excitability have yet to be conclusively identified, it is clear that ion channels alone may become a generator of ongoing activity in injured neurons.

4.4.3.3 Mechanisms underlying nerve injury-induced changes in transduction

While many of the mechanistic details underlying nerve injury-induced changes in transducer and/or ion channels have yet to be identified, it is clear that changes in access to trophic factors plays a major role in the process (Waxman, S. G. *et al.*, 1999; Sah, D. W. *et al.*, 2003). Following nerve injury, different groups of afferents may either be deprived of access to neurotrophic factors or bathed in excess levels with those that are injured being deprived and those that remain being bathed in excess. The suggestion that access to neurotrophic factors plays a critical role in driving nerve injury-induced changes is supported by the observation that exogenous application of factors such as NGF or GDNF to injured neurons restores ion channel expression levels and may eliminate ongoing activity. Exogenous NGF restores expression levels of NaV1.8 and NaV1.9 and reduces expression levels of NaV1.3 (Black, J. A. *et al.*, 1997; Fjell, J. *et al.*, 1999b). GDNF has also been shown to differentially regulate expression levels of NaV1.3 (Boucher, T. J. *et al.*, 2000; Leffler, A. *et al.*, 2002). Similarly, the peripheral administration of NGF results in the sensitization of nociceptive afferents and increases in the expression of many proteins that are increased in uninjured afferents following nerve injury (e.g., TRPV1) (Ji, R. R. *et al.*, 2002; Rashid, M. H. *et al.*, 2003). Attenuation of neurotrophin signaling through antibodies or antagonist has also been shown to reverse inflammation-induced changes in afferent excitability (McMahon, S. B., 1996; Koltzenburg, M. *et al.*, 1999).

4.5 Conclusion

Adequate treatment of chronic pain in the absence of serious side effects remains an elusive goal for patients, clinicians, and basic researchers. What should be clear from the preceding discussion is that the identification of effective therapeutic interventions has been hampered by the realization that the system is far more complex than ever imagined. The redundancy of signaling molecules, the promiscuity of receptors, and the delicate balance of both inhibitory and facilitatory processes have all kept the silver bullet frustratingly out of reach. However, identification of proteins highly enriched in nociceptive afferents that are critical for the transmission of nociceptive information to the central nervous system, an understanding of how these molecules function within the larger context of molecules in and around the afferent, all coupled with the early success of selective blockers of these compounds in clinical trials, suggests that truly novel and effective therapeutic interventions may not be far off.

References

Airaksinen, M. S. and Saarma, M. 2002. The GDNF family: signalling, biological functions and therapeutic value. Nat. Rev. Neurosci. 3, 383–394.

Alessandri-Haber, N., Joseph, E., Dina, O. A., Liedtke, W., and Levine, J. D. 2005. TRPV4 mediates pain-related behavior induced by mild hypertonic stimuli in the presence of inflammatory mediator. Pain 118, 70–79.

Ambalavanar, R., Moritani, M., and Dessem, D. 2005. Trigeminal P2X3 receptor expression differs from dorsal root ganglion and is modulated by deep tissue inflammation. Pain 117, 280–291.

Amir, R. and Devor, M. 1996. Chemically mediated cross-excitation in rat dorsal root ganglia. J. Neurosci. 16, 4733–4741.

Amir, R. and Devor, M. 2003. Electrical excitability of the soma of sensory neurons is required for spike invasion of the soma, but not for through-conduction. Biophys. J. 84, 2181–2191.

Amir, R., Michaelis, M., and Devor, M. 1999. Membrane potential oscillations in dorsal root ganglion neurons: role in normal electrogenesis and neuropathic pain. J. Neurosci. 19, 8589–8596.

Andersson, K. E. 2002. Bladder activation: afferent mechanisms. Urology 59, 43–50.

Andrew, D. and Greenspan, J. D. 1999. Mechanical and heat sensitization of cutaneous nociceptors after peripheral inflammation in the rat. J. Neurophysiol. 82, 2649–2656.

Arvidsson, U., Dado, R. J., Riedl, M., Lee, J. H., Law, P. Y., Loh, H. H., Elde, R., and Wessendorf, M. W. 1995. Delta-opioid receptor immunoreactivity: distribution in brainstem and spinal cord, and relationship to biogenic amines and enkephalin. J. Neurosci. 15, 1215–1235.

Baccaglini, P. I. and Hogan, P. G. 1983. Some rat sensory neurons in culture express characteristics of differentiated pain sensory cells. Proc. Natl. Acad. Sci. U. S. A. 80, 594–598.

Baccei, M. L. and Kocsis, J. D. 2000. Voltage-gated calcium currents in axotomized adult rat cutaneous afferent neurons. J. Neurophysiol. 83, 2227–2238.

Backonja, M. M. and Stacey, B. 2004. Neuropathic pain symptoms relative to overall pain rating. J. Pain. 5, 491–497.

Bandell, M., Story, G. M., Hwang, S. W., Viswanath, V., Eid, S. R., Petrus, M. J., Earley, T. J., and Patapoutian, A. 2004. Noxious cold ion channel TRPA1 is activated by pungent compounds and bradykinin. Neuron 41, 849–857.

Bang, H., Kim, Y., and Kim, D. 2000. TREK-2, a new member of the mechanosensitive tandem-pore K^+ channel family. J. Biol. Chem. 275, 17412–17419.

Barclay, J., Patel, S., Dorn, G., Wotherspoon, G., Moffatt, S., Eunson, L., Abdel'al, S., Natt, F., Hall, J., Winter, J., et al. 2002. Functional downregulation of P2X3 receptor subunit in rat sensory neurons reveals a significant role in chronic neuropathic and inflammatory pain. J. Neurosci. 22, 8139–8147.

Barton, M. E., Eberle, E. L., and Shannon, H. E. 2005. The antihyperalgesic effects of the T-type calcium channel blockers ethosuximide, trimethadione, and mibefradil. Eur. J. Pharmacol. 521, 79–85.

Baumann, T. K., Chaudhary, P., and Martenson, M. E. 2004. Background potassium channel block and TRPV1 activation contribute to proton depolarization of sensory neurons from humans with neuropathic pain. Eur. J. Neurosci. 19, 1343–1351.

Bautista, D. M., Jordt, S. E., Nikai, T., Tsuruda, P. R., Read, A. J., Poblete, J., Yamoah, E. N., Basbaum, A. I., and Julius, D. 2006. TRPA1 mediates the inflammatory actions of environmental irritants and proalgesic agents. Cell 124, 1269–1282.

Benson, C. J., Xie, J., Wemmie, J. A., Price, M. P., Henss, J. M., Welsh, M. J., and Snyder, P. M. 2002. Heteromultimers of DEG/ENaC subunits form H^+-gated channels in mouse sensory neurons. Proc. Natl. Acad. Sci. U. S. A. 99, 2338–2343.

Berthoud, H. R., Blackshaw, L. A., Brookes, S. J., and Grundy, D. 2004. Neuroanatomy of extrinsic afferents supplying the gastrointestinal tract. Neurogastroenterol. Motil. 16(Suppl. 1), 28–33.

Bevan, S. and Yeats, J. 1991. Protons activate a cation conductance in a subpopulation of rat dorsal root ganglion neurones. J. Physiol. 433, 145–161.

Bhave, G. and Gereau, R. W. T. 2004. Posttranslational mechanisms of peripheral sensitization. J. Neurobiol. 61, 88–106.

Bhave, G., Karim, F., Carlton, S. M., and Gereau, R. W. T. 2001. Peripheral group I metabotropic glutamate receptors modulate nociception in mice. Nat. Neurosci. 4, 417–423.

Birder, L. A. 2005. More than just a barrier: urothelium as a drug target for urinary bladder pain. Am. J. Physiol. Renal Physiol. 289, F489–F495.

Birder, L. A. and Perl, E. R. 1999. Expression of alpha2-adrenergic receptors in rat primary afferent neurones after peripheral nerve injury or inflammation. J. Physiol. 515, 533–542.

Birrell, G. J., McQueen, D. S., Iggo, A., Coleman, R. A., and Grubb, B. D. 1991. PGI2-induced activation and sensitization of articular mechanonociceptors. Neurosci. Lett. 124, 5–8.

Birrell, G. J., McQueen, D. S., Iggo, A., and Grubb, B. D. 1993. Prostanoid-induced potentiation of the excitatory and sensitizing effects of bradykinin on articular mechanonociceptors in the rat ankle joint. Neuroscience 54, 537–544.

Black, J. A., Cummins, T. R., Plumpton, C., Chen, Y. H., Hormuzdiar, W., Clare, J. J., and Waxman, S. G. 1999. Upregulation of a silent sodium channel after peripheral, but not central, nerve injury in DRG neurons. J. Neurophysiol. 82, 2776–2785.

Black, J. A., Dib-Hajj, S., McNabola, K., Jeste, S., Rizzo, M. A., Kocsis, J. D., and Waxman, S. G. 1996. Spinal sensory neurons express multiple sodium channel alpha-subunit mRNAs. Brain. Res. Mol. Brain Res. 43, 117–131.

Black, J. A., Langworthy, K., Hinson, A. W., Dib-Hajj, S. D., and Waxman, S. G. 1997. NGF has opposing effects on Na^+ channel III and SNS gene expression in spinal sensory neurons. Neuroreport 8, 2331–2335.

Blair, N. T. and Bean, B. P. 2002. Roles of tetrodotoxin (TTX)-sensitive Na$^+$ current, TTX-resistant Na$^+$ current, and Ca^{2+} current in the action potentials of nociceptive sensory neurons. J. Neurosci. 22, 10277–10290.

Bleehan, T. and Keele, C. A. 1977. Observations on the algogenic actions of adenosine compounds on the human blister base preparation. Pain 3, 367–377.

Boettger, M. K., Till, S., Chen, M. X., Anand, U., Otto, W. R., Plumpton, C., Trezise, D. J., Tate, S. N., Bountra, C., Coward, K., *et al.* 2002. Calcium-activated potassium channel SK1- and IK1-like immunoreactivity in injured human sensory neurones and its regulation by neurotrophic factors. Brain 125, 252–263.

Bond, C. T., Maylie, J., and Adelman, J. P. 2005. SK channels in excitability, pacemaking and synaptic integration. Curr. Opin. Neurobiol. 15, 305–311.

Bonnington, J. K. and McNaughton, P. A. 2003. Signalling pathways involved in the sensitisation of mouse nociceptive neurones by nerve growth factor. J. Physiol. 551, 433–446.

Borgland, S. L., Connor, M., Ryan, R. M., Ball, H. J., and Christie, M. J. 2002. Prostaglandin E(2) inhibits calcium current in two sub-populations of acutely isolated mouse trigeminal sensory neurons. J. Physiol. 539, 433–444.

Boucher, T. J., Okuse, K., Bennett, D. L., Munson, J. B., Wood, J. N., and McMahon, S. B. 2000. Potent analgesic effects of GDNF in neuropathic pain states. Science 290, 124–127.

Brock, J. A., McLachlan, E. M., and Belmonte, C. 1998. Tetrodotoxin-resistant impulses in single nociceptor nerve terminals in guinea-pig cornea. J. Physiol. (Lond.) 512, 211–217.

Brock, J. A., Pianova, S., and Belmonte, C. 2001. Differences between nerve terminal impulses of polymodal nociceptors and cold sensory receptors of the guinea-pig cornea. J. Physiol. 533, 493–501.

Cardenas, C. G., Del Mar, L. P., and Scroggs, R. S. 1995. Variation in serotonergic inhibition of calcium channel currents in four types of rat sensory neurons differentiated by membrane properties. J. Neurophysiol. 74, 1870–1879.

Caterina, M. J. and Julius, D. 2001. The vanilloid receptor: a molecular gateway to the pain pathway. Ann. Rev. Neurosci. 24, 487–517.

Caterina, M. J., Leffler, A., Malmberg, A. B., Martin, W. J., Trafton, J., Petersen-Zeitz, K. R., Koltzenburg, M., Basbaum, A. I., and Julius, D. 2000. Impaired nociception and pain sensation in mice lacking the capsaicin receptor (see comments). Science 288, 306–313.

Caterina, M. J., Rosen, T. A., Tominaga, M., Brake, A. J., and Julius, D. 1999. A capsaicin receptor homologue with a high threshold for noxious heat. Nature 398, 436–441.

Caterina, M. J., Schumacher, M. A., Tominaga, M., Rosen, T. A., Levine, J. D., and Julius, D. 1997. The capsaicin receptor: a heat-activated ion channel in the pain pathway. Nature 389, 816–824.

Catterall, W. A. 1992. Cellular and molecular biology of voltage-gated sodium channels. Physiol. Rev. S15–S48.

Catterall, W. A., Goldin, A. L., and Waxman, S. G. 2005a. International Union of Pharmacology. XLVII. Nomenclature and structure–function relationships of voltage-gated sodium channels. Pharmacol. Rev. 57, 397–409.

Catterall, W. A., Perez-Reyes, E., Snutch, T. P., and Striessnig, J. 2005b. International Union of Pharmacology. XLVIII. Nomenclature and structure–function relationships of voltage-gated calcium channels. Pharmacol. Rev. 57, 411–425.

Cesare, P. and McNaughton, P. 1996. A novel heat-activated current in nociceptive neurons and its sensitization by bradykinin. Proc. Natl. Acad. Sci. U. S. A. 93, 15435–15439.

Cesare, P., Dekker, L. V., Sardini, A., Parker, P. J., and McNaughton, P. A. 1999. Specific involvement of PKC-epsilon in sensitization of the neuronal response to painful heat. Neuron 23, 617–624.

Chaplan, S. R., Guo, H. Q., Lee, D. H., Luo, L., Liu, C., Kuei, C., Velumian, A. A., Butler, M. P., Brown, S. M., and Dubin, A. E. 2003. Neuronal hyperpolarization-activated pacemaker channels drive neuropathic pain. J. Neurosci. 23, 1169–1178.

Chen, C. C., England, S., Akopian, A. N., and Wood, J. N. 1998. A sensory neuron-specific, proton gated ion channel. Proc. Natl. Acad. Sci. U. S. A. 95, 10240–10245.

Chen, Y., Michaelis, M., Janig, W., and Devor, M. 1996. Adrenoreceptor subtype mediating sympathetic-sensory coupling in injured sensory neurons. J. Neurophysiol. 76, 3721–3730.

Chessell, I. P., Hatcher, J. P., Bountra, C., Michel, A. D., Hughes, J. P., Green, P., Egerton, J., Murfin, M., Richardson, J., Peck, W. L., *et al.* 2005. Disruption of the P2X7 purinoceptor gene abolishes chronic inflammatory and neuropathic pain. Pain 114, 386–396.

Chuang, H. H., Neuhausser, W. M., and Julius, D. 2004. The super-cooling agent icilin reveals a mechanism of coincidence detection by a temperature-sensitive TRP channel. Neuron 43, 859–869.

Chuang, H. H., Prescott, E. D., Kong, H., Shields, S., Jordt, S. E., Basbaum, A. I., Chao, M. V., and Julius, D. 2001. Bradykinin and nerve growth factor release the capsaicin receptor from PtdIns(4,5)P2-mediated inhibition. Nature 411, 957–962.

Chung, M. K., Lee, H. A. M., Suzuki, M., and Caterina, M. J. 2004a. TRPV3 and TRPV4 mediate warmth-evoked currents in primary mouse keratinocytes. J. Biol. Chem. 279, 21569–21575.

Chung, M. K., Lee, H. A. M., Suzuki, M., and Caterina, M. J. 2004b. 2-aminoethoxydiphenyl borate activates and sensitizes the heat-gated ion channel, TRPV3. J. Neurosci. 24, 5177–5182.

Cockayne, D. A., Dunn, P. M., Zhong, Y., Rong, W., Hamilton, S. G., Knight, G. E., Ruan, H. Z., Ma, B., Yip, P., Nunn, P., *et al.* 2005. P2X2 knockout mice and P2X2/P2X3 double knockout mice reveal a role for the P2X2 receptor subunit in mediating multiple sensory effects of ATP. J. Physiol. 567, 621–639.

Cockayne, D. A., Hamilton, S. G., Zhu, Q. M., Dunn, P. M., Zhong, Y., Novakovic, S., Malmberg, A. B., Cain, G., Berson, A., Kassotakis, L., *et al.* 2000. Urinary bladder hyporeflexia and reduced pain-related behaviour in P2X3-deficient mice. Nature 407, 1011–1015.

Connor, J. A. and Stevens, C. F. 1971. Voltage clamp studies of a transient outward membrane current in gastropod neural somata. J. Physiol. 213, 21–30.

Cordoba-Rodriguez, R., Moore, K. A., Kao, J. P., and Weinreich, D. 1999. Calcium regulation of a slow post-spike hyperpolarization in vagal afferent neurons. Proc. Natl. Acad. Sci. U. S. A. 96, 7650–7657.

Coste, B., Osorio, N., Padilla, F., Crest, M., and Delmas, P. 2004. Gating and modulation of presumptive NaV1.9 channels in enteric and spinal sensory neurons. Mol. Cell Neurosci. 26, 123–134.

Coward, K., Aitken, A., Powell, A., Plumpton, C., Birch, R., Tate, S., Bountra, C., and Anand, P. 2001. Plasticity of TTX-sensitive sodium channels PN1 and brain III in injured human nerves. Neuroreport 12, 495–500.

Cox, J. J., Reimann, F., Nicholas, A. K., Thornton, G., Roberts, E., Springell, K., Karbani, G., Jafri, H., Mannan, J., Raashid, Y., *et al.* 2006. An SCN9A channelopathy causes congenital inability to experience pain. Nature 444, 894–898.

Craig, A. D. 2002. How do you feel? Interoception: the sense of the physiological condition of the body. Nat. Rev. Neurosci. 3, 655–666.

Cummins, T. R. and Waxman, S. G. 1997. Downregulation of tetrodotoxin-resistant sodium currents and upregulation of a rapidly repriming tetrodotoxin-sensitive sodium current in small spinal sensory neurons after nerve injury. J. Neurosci. 17, 3503–3514.

Cummins, T. R., Aglieco, F., Renganathan, M., Herzog, R. I., Dib-Hajj, S. D., and Waxman, S. G. 2001. Nav1.3 sodium channels: rapid repriming and slow closed-state inactivation display quantitative differences after expression in a mammalian cell line and in spinal sensory neurons. J. Neurosci. 21, 5952–5961.

Cummins, T. R., Dib-Hajj, S. D., Black, J. A., Akopian, A. N., Wood, J. N., and Waxman, S. G. 1999. A novel persistent tetrodotoxin-resistant sodium current in SNS-null and wild-type small primary sensory neurons. J. Neurosci. 19, RC43.

Czapinski, P., Blaszczyk, B., and Czuczwar, S. J. 2005. Mechanisms of action of antiepileptic drugs. Curr. Top. Med. Chem. 5, 3–14.

Dado, R. J., Law, P. Y., Loh, H. H., and Elde, R. 1993. Immunofluorescent identification of a delta (delta)-opioid receptor on primary afferent nerve terminals. Neuroreport 5, 341–344.

Dang, K., Bielefeldt, K., and Gebhart, G. F. 2004. Gastric ulcers reduce A-type potassium currents in rat gastric sensory ganglion neurons. Am. J. Physiol. Gastrointest. Liver. Physiol. 286, G573–G579.

Darian-Smith, I., Davidson, I., and Johnson, K. O. 1980. Peripheral neural representation of spatial dimensions of a textured surface moving across the monkey's finger pad. J. Physiol. 309, 135–146.

Davis, J. B., Gray, J., Gunthorpe, M. J., Hatcher, J. P., Davey, P. T., Overend, P., Harries, M. H., Latcham, J., Clapham, C., Atkinson, K., et al. 2000. Vanilloid receptor-1 is essential for inflammatory thermal hyperalgesia. Nature 405, 183–187.

Davis, K. D., Treede, R. D., Raja, S. N., Meyer, R. A., and Campbell, J. N. 1991. Topical application of clonidine relieves hyperalgesia in patients with sympathetically maintained pain. Pain 47, 309–317.

Decosterd, I., Ji, R. R., Abdi, S., Tate, S., and Woolf, C. J. 2002. The pattern of expression of the voltage-gated sodium channels Na(v)1.8 and Na(v)1.9 does not change in uninjured primary sensory neurons in experimental neuropathic pain models. Pain 96, 269–277.

De Groot, J., Zhou, S., and Carlton, S. M. 2000. Peripheral glutamate release in the hindpaw following low and high intensity sciatic stimulation. Neuroreport 11, 497–502.

Dell'Antonio, G., Quattrini, A., Cin, E. D., Fulgenzi, A., and Ferrero, M. E. 2002a. Relief of inflammatory pain in rats by local use of the selective P2X7 ATP receptor inhibitor, oxidized ATP. Arthritis Rheum. 46, 3378–3385.

Dell'Antonio, G., Quattrini, A., Dal Cin, E., Fulgenzi, A., and Ferrero, M. E. 2002b. Antinociceptive effect of a new P(2Z)/P2X7 antagonist, oxidized ATP, in arthritic rats. Neurosci. Lett. 327, 87–90.

Devor, M. 1999. Unexplained peculiarities of the dorsal root ganglion. Pain Suppl. 6, S27–S35.

Devor, M. and Wall, P. D. 1990. Cross-excitation in dorsal root ganglia of nerve-injured and intact rats. J. Neurophysiol. 64, 1733–1746.

Devor, M., Govrin, L. R., and Angelides, K. 1993. Na+ channel immunolocalization in peripheral mammalian axons and changes following nerve injury and neuroma formation. J. Neurosci. 13, 1976–1992.

Devor, M., Janig, W., and Michaelis, M. 1994. Modulation of activity in dorsal root ganglion neurons by sympathetic activation in nerve-injured rats. J. Neurophysiol. 71, 38–47.

Dib-Hajj, S., Black, J. A., Felts, P., and Waxman, S. G. 1996. Down-regulation of transcripts for Na channel alpha-SNS in

spinal sensory neurons following axotomy. Proc. Natl. Acad. Sci. U. S. A. 93, 14950–14954.

Diochot, S., Baron, A., Rash, L. D., Deval, E., Escoubas, P., Scarzello, S., Salinas, M., and Lazdunski, M. 2004. A new sea anemone peptide, APETx2, inhibits ASIC3, a major acid-sensitive channel in sensory neurons. EMBO J. 23, 1516–1525.

Djouhri, L., Fang, X., Okuse, K., Wood, J. N., Berry, C. M., and Lawson, S. N. 2003a. The TTX-resistant sodium channel Nav1.8 (SNS/PN3): expression and correlation with membrane properties in rat nociceptive primary afferent neurons. J. Physiol. 550, 739–752.

Djouhri, L., Newton, R., Levinson, S. R., Berry, C. M., Carruthers, B., and Lawson, S. N. 2003b. Sensory and electrophysiological properties of guinea-pig sensory neurones expressing Nav 1.7 (PN1) Na+ channel alpha subunit protein. J. Physiol. 546, 565–576.

Dodt, E. and Zotterman, Y. 1952. The discharge of specific cold fibers at high temperatures (paradoxical cold.). Acta Physiol. Scand. 26, 358–365.

Dong, X., Han, S., Zylka, M. J., Simon, M. I., and Anderson, D. J. 2001. A diverse family of GPCRs expressed in specific subsets of nociceptive sensory neurons. Cell 106, 619–632.

Drew, L. J., Rohrer, D. K., Price, M. P., Blaver, K. E., Cockayne, D. A., Cesare, P., and Wood, J. N. 2004. Acid-sensing ion channels ASIC2 and ASIC3 do not contribute to mechanically activated currents in mammalian sensory neurones. J. Physiol. 556, 691–710.

Duggan, A., Garcia-Anoveros, J., and Corey, D. P. 2002. The PDZ domain protein PICK1 and the sodium channel BNaC1 interact and localize at mechanosensory terminals of dorsal root ganglion neurons and dendrites of central neurons. J. Biol. Chem. 277, 5203–5208.

England, S., Bevan, S., and Docherty, R. J. 1996. PGE2 modulates the tetrodotoxin-resistant sodium current in neonatal rat dorsal root ganglion neurones via the cyclic AMP-protein kinase A cascade. J. Physiol. 495(Pt 2), 429–440.

Eshete, F. and Fields, R. D. 2001. Spike frequency decoding and autonomous activation of Ca^{2+}-calmodulin-dependent protein kinase II in dorsal root ganglion neurons. J. Neurosci. 21, 6694–6705.

Fang, X., Djouhri, L., Black, J. A., Dib-Hajj, S. D., Waxman, S. G., and Lawson, S. N. 2002. The presence and role of the tetrodotoxin-resistant sodium channel Na(v)1.9 (NaN) in nociceptive primary afferent neurons. J. Neurosci. 22, 7425–7433.

Ferreira, S. H., Nakamura, M., and de Abreu Castro, M. S. 1978. The hyperalgesic effects of prostacyclin and prostaglandin E2. Prostaglandins 16, 31–37.

Fertleman, C. R., Baker, M. D., Parker, K. A., Moffatt, S., Elmslie, F. V., Abrahamsen, B., Ostman, J., Klugbauer, N., Wood, J. N., Gardiner, R. M., and Rees, M. 2006. SCN9A mutations in paroxysmal extreme pain disorder: allelic variants underlie distinct channel defects and phenotypes. Neuron 52, 767–774.

Fields, R. D., Lee, P. R., and Cohen, J. E. 2005. Temporal integration of intracellular Ca^{2+} signaling networks in regulating gene expression by action potentials. Cell Calcium 37, 433–442.

Fitzgerald, M. 2005. The development of nociceptive circuits. Nat. Rev. Neurosci. 6, 507–520.

Fitzgerald, E. M., Okuse, K., Wood, J. N., Dolphin, A. C., and Moss, S. J. 1999. cAMP-dependent phosphorylation of the tetrodotoxin-resistant voltage-dependent sodium channel SNS. J. Physiol. 516(Pt 2), 433–446.

Fjell, J., Cummins, T. R., Dib-Hajj, S. D., Fried, K., Black, J. A., and Waxman, S. G. 1999a. Differential role of GDNF and NGF in the maintenance of two TTX-resistant sodium channels in adult DRG neurons. Brain. Res. Mol. Brain Res. 67, 267–282.

Fjell, J., Cummins, T. R., Fried, K., Black, J. A., and Waxman, S. G. 1999b. *In vivo* NGF deprivation reduces SNS expression and TTX-R sodium currents in IB4-negative DRG neurons. J. Neurophysiol. 81, 803–810.

Fowler, J. C., Greene, R., and Weinreich, D. 1985a. Two calcium-sensitive spike after-hyperpolarizations in visceral sensory neurones of the rabbit. J. Physiol. (Lond.) 365, 59–75.

Fowler, J. C., Wonderlin, W. F., and Weinreich, D. 1985b. Prostaglandins block Ca^{2+}-dependent slow afterhyperpolarization independent of effects on Ca^{2+} influx in visceral afferent neurons. Brain Res. 345, 345–349.

Fukuoka, T., Tokunaga, A., Tachibana, T., Dai, Y., Yamanaka, H., and Noguchi, K. 2002. VR1, but not P2X(3), increases in the spared L4 DRG in rats with L5 spinal nerve ligation. Pain 99, 111–120.

Garcia-Anoveros, J., Samad, T. A., Zuvela-Jelaska, L., Woolf, C. J., and Corey, D. P. 2001. Transport and localization of the DEG/ENaC ion channel BNaC1alpha to peripheral mechanosensory terminals of dorsal root ganglia neurons. J. Neurosci. 21, 2678–2686.

Gavva, N. R., Klionsky, L., Qu, Y., Shi, L., Tamir, R., Edenson, S., Zhang, T. J., Viswanadhan, V. N., Toth, A., Pearce, L. V., *et al.* 2004. Molecular determinants of vanilloid sensitivity in TRPV1. J. Biol. Chem. 279, 20283–20295.

Ghilardi, J. R., Rohrich, H., Lindsay, T. H., Sevcik, M. A., Schwei, M. J., Kubota, K., Halvorson, K. G., Poblete, J., Chaplan, S. R., Dubin, A. E., *et al.* 2005. Selective blockade of the capsaicin receptor TRPV1 attenuates bone cancer pain. J. Neurosci. 25, 3126–3131.

Gilman, A. G. 1987. G proteins: transducers of receptor-generated signals. Annu. Rev. Biochem. 56, 615–649.

Gold, M. S. 2000a. Sodium channels and pain therapy. Cur. Opin. Anaesthesiol. 13, 565–572.

Gold, M. S. 2000b. Spinal nerve ligation: what to blame for the pain and why. Pain 84, 117–120.

Gold, M. S., Dastmalchi, S., and Levine, J. D. 1996a. Co-expression of nociceptor properties in dorsal root ganglion neurons from the adult rat *in vitro*. Neuroscience 71, 265–275.

Gold, M. S., Levine, J. D., and Correa, A. M. 1998. Modulation of TTX-R INa by PKC and PKA and their role in PGE2-induced sensitization of rat sensory neurons *in vitro*. J. Neurosci. 18, 10345–10355.

Gold, M. S., Reichling, D. B., Shuster, M. J., and Levine, J. D. 1996b. Hyperalgesic agents increase a tetrodotoxin-resistant Na^+ current in nociceptors. Proc. Natl. Acad. Sci. U. S. A. 93, 1108–1112.

Gold, M. S., Shuster, M. J., and Levine, J. D. 1996c. Characterization of six voltage-gated K^+ currents in adult rat sensory neurons. J. Neurophysiol. 75, 2629–2646.

Gold, M. S., Shuster, M. J., and Levine, J. D. 1996d. Role of a slow Ca^{2+}-dependent slow afterhyperpolarization in prostaglandin E2-induced sensitization of cultured rat sensory neurons. Neurosci. Lett. 205, 161–164.

Gold, M. S., Weinreich, D., Kim, C. S., Wang, R., Treanor, J., Porreca, F., and Lai, J. 2003. Redistribution of Na(V)1.8 in uninjured axons enables neuropathic pain. J. Neurosci. 23, 158–166.

Goldin, A. L., Barchi, R. L., Caldwell, J. H., Hofmann, F., Howe, J. R., Hunter, J. C., Kallen, R. G., Mandel, G., Meisler, M. H., Netter, Y. B., *et al.* 2000. Nomenclature of voltage-gated sodium channels. Neuron 28, 365–368.

Gonzales, R., Goldyne, M. E., Taiwo, Y. O., and Levine, J. D. 1989. Production of hyperalgesic prostaglandins by sympathetic postganglionic neurons. J. Neurochem. 53, 1595–1598.

Gonzales, R., Sherbourne, C. D., Goldyne, M. E., and Levine, J. D. 1991. Noradrenaline-induced prostaglandin production by sympathetic postganglionic neurons is mediated by alpha$_2$-adrenergic receptors. J. Neurochem. 57, 1145–1150.

Gover, T. D., Kao, J. P., and Weinreich, D. 2003. Calcium signaling in single peripheral sensory nerve terminals. J. Neurosci. 23, 4793–4797.

Gracely, R. H., Lynch, S. A., and Bennett, G. J. 1992. Painful neuropathy: altered central processing maintained dynamically by peripheral input. Pain 51, 175–194.

Grieco, T. M., Malhotra, J. D., Chen, C., Isom, L. L., and Raman, I. M. 2005. Open-channel block by the cytoplasmic tail of sodium channel beta4 as a mechanism for resurgent sodium current. Neuron 45, 233–244.

Gu, J. G. and MacDermott, A. B. 1997. Activation of ATP P2X receptors elicits glutamate release from sensory neuron synapses. Nature 389, 749–753.

Guler, A. D., Lee, H., Iida, T., Shimizu, I., Tominaga, M., and Caterina, M. 2002. Heat-evoked activation of the ion channel, TRPV4. J. Neurosci. 22, 6408–6414.

Harriott, A., Dessem, D., and Gold, M. S. 2006. Inflammation increases the excitability of masseter muscle afferents. Neuroscience 141(1), 433–442.

He, X. L. and Garcia, K. C. 2004. Structure of nerve growth factor complexed with the shared neurotrophin receptor p75. Science 304, 870–875.

Herzog, R. I., Cummins, T. R., and Waxman, S. G. 2001. Persistent TTX-resistant Na+ current affects resting potential and response to depolarization in simulated spinal sensory neurons. J. Neurophysiol. 86, 1351–1364.

Hogan, Q. H., McCallum, J. B., Sarantopoulos, C., Aason, M., Mynlieff, M., Kwok, W., and Bosnjak, Z. J. 2000. Painful neuropathy decreases membrane calcium current in mammalian primary afferent neurons. Pain 86, 43–53.

Honore, P., Kage, K., Mikusa, J., Watt, A. T., Johnston, J. F., Wyatt, J. R., Faltynek, C. R., Jarvis, M. F., and Lynch, K. 2002. Analgesic profile of intrathecal P2X(3) antisense oligonucleotide treatment in chronic inflammatory and neuropathic pain states in rats. Pain 99, 11–19.

Honore, P., Wismer, C. T., Mikusa, J. P., Zhu, C. Z., Zhong, C., Gauvin, D. M., Gomtsyan, A., El Kouhen, R., Lee, C. H., Marsh, K., Sullivan, J. P., Faltynek, C. R., and Jarvis, M. F. 2005. A-425619, [1-isoquindin-5-yl-3-(4-trifluromethyl-benzyl)-urea], a novel transient receptor potential type V1 receptor antagonist, relieves pathophysiological pain associated with inflammation and tissue injury in rats. J. Pharmacol. Exp. Ther. 314(1), 410–421.

Howe, J. F., Loeser, J. D., and Calvin, W. H. 1977. Mechanosensitivity of dorsal root ganglia and chronically injured axons: a physiological basis for the radicular pain of nerve root compression. Pain 3, 25–41.

Hu, H. Z., Gu, Q., Wang, C., Colton, C. K., Tang, J., Kinoshita-Kawada, M., Lee, L. Y., Wood, J. D., and Zhu, M. X. 2004. 2-aminoethoxydiphenyl borate is a common activator of TRPV1, TRPV2, and TRPV3. J. Biol. Chem. 279, 35741–35748.

Huang, L. Y. and Neher, E. 1996. Ca(2+)-dependent exocytosis in the somata of dorsal root ganglion neurons. Neuron 17, 135–145.

Huang, E. J. and Reichardt, L. F. 2001. Neurotrophins: roles in neuronal development and function. Annu. Rev. Neurosci. 24, 677–736.

Huang, S. M., Bisogno, T., Trevisani, M., Al-Hayani, A., De Petrocellis, L., Fezza, F., Tognetto, M., Petros, T. J., Krey, J. F., Chu, C. J., *et al.* 2002. An endogenous capsaicin-like substance with high potency at recombinant and native vanilloid VR1 receptors. Proc. Natl. Acad. Sci. U. S. A. 99, 8400–8405.

Hucho, T. B., Dina, O. A., and Levine, J. D. 2005. Epac mediates a cAMP-to-PKC signaling in inflammatory pain: an isolectin

B4(+) neuron-specific mechanism. J. Neurosci. 25, 6119–6126.

Hwang, S. W., Cho, H., Kwak, J., Lee, S. Y., Kang, C. J., Jung, J., Cho, S., Min, K. H., Suh, Y. G., Kim, D., and Oh, U. 2000. Direct activation of capsaicin receptors by products of lipoxygenases: endogenous capsaicin-like substances. Proc. Natl. Acad. Sci. U. S. A. 97, 6155–6160.

Immke, D. C. and McCleskey, E. W. 2001. ASIC3: a lactic acid sensor for cardiac pain. ScientificWorldJournal 1, 510–512.

Ingram, S. L. and Williams, J. T. 1994. Opioid inhibition of Ih via adenylyl cyclase. Neuron 13, 179–186.

Isaacson, J. S. and Murphy, G. J. 2001. Glutamate-mediated extrasynaptic inhibition: direct coupling of NMDA receptors to Ca(2+)-activated K$^+$ channels. Neuron 31, 1027–1034.

Isom, L. L. 2001. Sodium channel beta subunits: anything but auxiliary. Neuroscientist 7, 42–54.

Iwata, Y., Katanosaka, Y., Arai, Y., Komamura, K., Miyatake, K., and Shigekawa, M. 2003. A novel mechanism of myocyte degeneration involving the Ca^{2+}-permeable growth factor-regulated channel. J. Cell. Biol. 161, 957–967.

Jancso, G., Kiraly, E., and Jancso-Gabor, A. 1977. Pharmacologically induced selective degeneration of chemosensitive primary sensory neurons. Nature 270, 741–743.

Jarvis, M. F., Burgard, E. C., McGaraughty, S., Honore, P., Lynch, K., Brennan, T. J., Subieta, A., Van Biesen, T., Cartmell, J., Bianchi, B., et al. 2002. A-317491, a novel potent and selective non-nucleotide antagonist of P2X3 and P2X2/3 receptors, reduces chronic inflammatory and neuropathic pain in the rat. Proc. Natl. Acad. Sci. U. S. A. 99, 17179–17184.

Jeftinija, S. 1994. The role of tetrodotoxin-resistant sodium channels of small primary afferent fibers. Brain Res. 639, 125–134.

Ji, R. R., Samad, T. A., Jin, S. X., Schmoll, R., and Woolf, C. J. 2002. p38 MAPK activation by NGF in primary sensory neurons after inflammation increases TRPV1 levels and maintains heat hyperalgesia. Neuron 36, 57–68.

Johnson, K. O. 2001. The roles and functions of cutaneous mechanoreceptors. Curr. Opin. Neurobiol. 11, 455–461.

Jordt, S. E., Bautista, D. M., Chuang, H. H., McKemy, D. D., Zygmunt, P. M., Hogestatt, E. D., Meng, I. D., and Julius, D. 2004. Mustard oils and cannabinoids excite sensory nerve fibres through the TRP channel ANKTM1. Nature 427, 260–265.

Jordt, S. E., McKemy, D. D., and Julius, D. 2003. Lessons from peppers and peppermint: the molecular logic of thermosensation. Curr. Opin. Neurobiol. 13, 487–492.

Kajander, K. C., Wakisaka, S., and Bennett, G. J. 1992. Spontaneous discharge originates in the dorsal root ganglion at the onset of a painful peripheral neuropathy in the rat. Neurosci. Lett. 138, 225–228.

Kang, D. and Kim, D. 2006. TREK-2(K2P10.1) and TRESK (K2P18.1) are major background K$^+$ channels in dorsal root ganglion neurons. Am. J. Physiol. Cell Physiol. 291(1), C138–C146.

Kepplinger, K. J., Forstner, G., Kahr, H., Leitner, K., Pammer, P., Groschner, K., Soldatov, N. M., and Romanin, C. 2000. Molecular determinant for run-down of L-type Ca^{2+} channels localized in the carboxyl terminus of the 1C subunit. J. Physiol. 529(Pt 1), 119–130.

Khasar, S. G., Gold, M. S., and Levine, J. D. 1998. A tetrodotoxin-resistant sodium current mediates inflammatory pain in the rat. Neurosci. Lett. 256, 17–20.

Khasar, S. G., Lin, Y. H., Martin, A., Dadgar, J., McMahon, T., Wang, D., Hundle, B., Aley, K. O., Isenberg, W., McCarter, G., et al. 1999. A novel nociceptor signaling pathway revealed in protein kinase C epsilon mutant mice. Neuron 24, 253–260.

Khodorova, A., Navarro, B., Jouaville, L. S., Murphy, J. E., Rice, F. L., Mazurkiewicz, J. E., Long-Woodward, D., Stoffel, M., Strichartz, G. R., Yukhananov, R., and Davar, G. 2003. Endothelin-B receptor activation triggers an endogenous analgesic cascade at sites of peripheral injury. Nat. Med. 9, 1055–1061.

Kim, D. S., Yoon, C. H., Lee, S. J., Park, S. Y., Yoo, H. J., and Cho, H. J. 2001. Changes in voltage-gated calcium channel alpha(1) gene expression in rat dorsal root ganglia following peripheral nerve injury. Brain Res. Mol. Brain Res. 96, 151–156.

Kirchhoff, C., Leah, J. D., Jung, S., and Reeh, P. W. 1992. Excitation of cutaneous senory nerve endings in the rat by 4-aminopyridine and tetraethylammonium. J. Neurophysiol. 67, 125–131.

Klein, J. P., Tendi, E. A., Dib-Hajj, S. D., Fields, R. D., and Waxman, S. G. 2003. Patterned electrical activity modulates sodium channel expression in sensory neurons. J. Neurosci. Res. 74, 192–198.

Kohno, T., Ji, R. R., Ito, N., Allchorne, A. J., Befort, K., Karchewski, L. A., and Woolf, C. J. 2005. Peripheral axonal injury results in reduced mu opioid receptor pre- and post-synaptic action in the spinal cord. Pain 117, 77–87.

Koltzenburg, M. 1999. The changing sensitivity in the life of the nociceptor. Pain Suppl. 6, S93–S102.

Koltzenburg, M., Bennett, D. L., Shelton, D. L., and McMahon, S. B. 1999. Neutralization of endogenous NGF prevents the sensitization of nociceptors supplying inflamed skin. Eur. J. Neurosci. 11, 1698–1704.

Kostyuk, P. G., Veselovsky, N. S., and Fedulova, S. A. 1981. Ionic currents in the somatic membrane of rat dorsal root ganglion neurons-II. Calcium currents. Neuroscience 6, 2431–2437.

Kruger, L. 1988. Morphological features of thin sensory afferent fibers: a new interpretation of 'nociceptor' function. Prog. Brain Res. 74, 253–257.

Kruger, L., Kavookjian, A. M., Kumazawa, T., Light, A. R., and Mizumura, K. 2003a. Nociceptor structural specialization in canine and rodent testicular "free" nerve endings. J. Comp. Neurol. 463, 197–211.

Kruger, L., Light, A. R., and Schweizer, F. E. 2003b. Axonal terminals of sensory neurons and their morphological diversity. J. Neurocytol. 32, 205–216.

Kumazawa, T., Mizumura, K., Koda, H., and Fukusako, H. 1996. EP receptor subtypes implicated in the PGE2-induced sensitization of polymodal receptors in response to bradykinin and heat. J. Neurophysiol. 75, 2361–2368.

Kwan, K. Y., Allchorne, A. J., Vollrath, M. A., Christensen, A. P., Zhang, D. S., and Corey, D. P. 2006. TRPA1 contributes to cold, mechanical, and chemical nociception, but is not essential for hair cell transduction. Neuron 50, 277–289.

Labasi, J. M., Petrushova, N., Donovan, C., McCurdy, S., Lira, P., Payette, M. M., Brissette, W., Wicks, J. R., Audoly, L., and Gabel, C. A. 2002. Absence of the P2X7 receptor alters leukocyte function and attenuates an inflammatory response. J. Immunol. 168, 6436–6445.

Lee, D. H., Chang, L., Sorkin, L. S., and Chaplan, S. R. 2005a. Hyperpolarization-activated, cation-nonselective, cyclic nucleotide-modulated channel blockade alleviates mechanical allodynia and suppresses ectopic discharge in spinal nerve ligated rats. J. Pain 6, 417–424.

Lee, H., Iida, T., Mizuno, A., Suzuki, M., and Caterina, M. J. 2005b. Altered thermal selection behavior in mice lacking transient receptor potential vanilloid 4. J. Neurosci. 25, 1304–1310.

Leffler, A., Cummins, T. R., Dib-Hajj, S. D., Hormuzdiar, W. N., Black, J. A., and Waxman, S. G. 2002. GDNF and NGF reverse changes in repriming of TTX-sensitive Na(+) currents following axotomy of dorsal root ganglion neurons. J. Neurophysiol. 88, 650–658.

Lembo, P. M., Grazzini, E., Groblewski, T., O'Donnell, D., Roy, M. O., Zhang, J., Hoffert, C., Cao, J., Schmidt, R., Pelletier, M., et al. 2002. Proenkephalin A gene products activate a new family of sensory neuron-specific GPCRs. Nat. Neurosci. 5, 201–209.

Lerner, U. H. 2002. Neuropeptidergic regulation of bone resorption and bone formation. J. Musculoskelet. Neuronal Interact 2, 440–447.

Levine, J. D., Basbaum, A. I., and Fields, H. L. 1993. Peptides and primary afferent nociceptors. J. Neurosci. 13, 2273–2286.

Lewis, T. 1935. Experiments relating to cunateous hyperalgesia and its spread through somatic fibers. Clinical Sci. 2, 373–423.

Liedtke, W. and Friedman, J. M. 2003. Abnormal osmotic regulation in trpv4-/- mice. Proc. Natl. Acad. Sci. U. S. A. 100, 13698–13703.

Lindia, J. A., Kohler, M. G., Martin, W. J., and Abbadie, C. 2005. Relationship between sodium channel Na(V)1.3 expression and neuropathic pain behavior in rats. Pain 117, 145–153.

Liu, X., Chung, K., and Chung, J. M. 1999. Ectopic discharges and adrenergic sensitivity of sensory neurons after spinal nerve injury. Brain Res. 849, 244–247.

Liu, C. N., Devor, M., Waxman, S. G., and Kocsis, J. D. 2002. Subthreshold oscillations induced by spinal nerve injury in dissociated muscle and cutaneous afferents of mouse DRG. J. Neurophysiol. 87, 2009–2017.

Liu, C. N., Wall, P. D., Ben-Dor, E., Michaelis, M., Amir, R., and Devor, M. 2000. Tactile allodynia in the absence of C-fiber activation: altered firing properties of DRG neurons following spinal nerve injury. Pain 85, 503–521.

Liu, B., Zhang, C., and Qin, F. 2005. Functional recovery from desensitization of vanilloid receptor TRPV1 requires resynthesis of phosphatidylinositol 4,5-bisphosphate. J. Neurosci. 25, 4835–4843.

Lopshire, J. C. and Nicol, G. D. 1998. The cAMP transduction cascade mediates the prostaglandin E_2 enhancement of the capsaicin-elicited current in rat sensory neurons: whole cell and single-channel studies. J. Neurosci. 18, 6081–6092.

Ma, C., Greenquist, K. W., and Lamotte, R. H. 2006. Inflammatory mediators enhance the excitability of chronically compressed dorsal root ganglion neurons. J. Neurophysiol. 95, 2098–2107.

MacIver, M. B. and Tanelian, D. L. 1993. Free nerve ending terminal morphology is fiber type specific for A delta and C fibers innervating rabbit corneal epithelium. J. Neurophysiol. 69, 1779–1783.

Maingret, F., Fosset, M., Lesage, F., Lazdunski, M., and Honore, E. 1999. TRAAK is a mammalian neuronal mechano-gated K^+ channel. J. Biol. Chem. 274, 1381–1387.

Maingret, F., Patel, A. J., Lesage, F., Lazdunski, M., and Honore, E. 2000. Lysophospholipids open the two-pore domain mechano-gated K(+) channels TREK-1 and TRAAK. J. Biol. Chem. 275, 10128–10133.

Mantyh, P. W., DeMaster, E., Malhotra, A., Ghilardi, J. R., Rogers, S. D., Mantyh, C. R., Liu, H., Basbaum, A. I., Vigna, S. R., Maggio, J. E., et al. 1995. Receptor endocytosis and dendrite reshaping in spinal neurons after somatosensory stimulation. Science 268, 1629–1632.

Mata, M., Chattopadhyay, M., and Fink, D. J. 2006. Gene therapy for the treatment of sensory neuropathy. Expert. Opin. Biol. Ther. 6, 499–507.

Matsuka, Y., Neubert, J. K., Maidment, N. T., and Spigelman, I. 2001. Concurrent release of ATP and substance P within guinea pig trigeminal ganglia in vivo. Brain Res. 915, 248–255.

Matzner, O. and Devor, M. 1992. Na+ conductance and the threshold for repetitive neuronal firing. Brain Res. 597, 92–98.

Mayer, M. 1985. A calcium-activated chloride current generates the after-depolarization of rat sensory neurons in culture. J. Physiol. 364, 217–239.

McCarter, G. C., Reichling, D. B., and Levine, J. D. 1999. Mechanical transduction by rat dorsal root ganglion neurons in vitro. Neurosci. Lett. 273, 179–182.

McFadzean, I. 1988. The ionic mechanisms underlying opioid actions. Neuropeptides 11, 173–180.

McGaraughty, S., Wismer, C. T., Zhu, C. Z., Mikusa, J., Honore, P., Chu, K. L., Lee, C. H., Faltynek, C. R., and Jarvis, M. F. 2003. Effects of A-317491, a novel and selective P2X3/P2X2/3 receptor antagonist, on neuropathic, inflammatory and chemogenic nociception following intrathecal and intraplantar administration. Br. J. Pharmacol. 140, 1381–1388.

McKemy, D. D., Neuhausser, V. M., and Julius, D. 2002. Identification of a cold receptor reveals a general role for TRP channels in thermosensation. Nature 416, 52–58.

McLachlan, E. M., Jang, W., Devor, M., and Michaelis, M. 1993. Peripheral nerve injury triggers noradrenergic sprouting within dorsal root ganglia. Nature 363, 543–546.

McMahon, S. B. 1996. NGF as a mediator of inflammatory pain. Philos. Trans. R. Soc. Lond. B Biol. Sci. 351, 431–440.

Mearow, K. M. and Diamond, J. 1988. Merkel cells and the mechanosensitivity of normal and regenerating nerves in Xenopus skin. Neuroscience 26, 695–708.

Meyer, R. A., Raja, S. N., and Campbell, J. N. 1985. Coupling of action potential activity between unmyelinated fibers in the peripheral nerve of monkey. Science 227, 184–187.

Michaelis, M., Blenk, K. H., Vogel, C., and Janig, W. 1999. Distribution of sensory properties among axotomized cutaneous C-fibres in adult rats. Neuroscience 94, 7–10.

Molliver, D. C., Wright, D. E., Leitner, M. L., Parsadanian, A. S., Doster, K., Wen, D., Yan, Q., and W.D. S. 1997. IB4-binding neurons switch from NGF to GDNF dependence in early postnatal life. Neuron 19, 849–861.

Moon, D. E., Lee, D. H., Han, H. C., Xie, J., Coggeshall, R. E., and Chung, J. M. 1999. Adrenergic sensitivity of the sensory receptors modulating mechanical allodynia in a rat neuropathic pain model. Pain 80, 589–595.

Moore, B. A., Stewart, T. M., Hill, C., and Vanner, S. J. 2002. TNBS ileitis evokes hyperexcitability and changes in ionic membrane properties of nociceptive DRG neurons. Am. J. Physiol. Gastrointest. Liver Physiol. 282, G1045–G1051.

Moqrich, A., Hwang, S. W., Earley, T. J., Petrus, M. J., Murray, A. N., Spencer, K. S., Andahazy, M., Story, G. M., and Patapoutian, A. 2005. Impaired thermosensation in mice lacking TRPV3, a heat and camphor sensor in the skin. Science 307, 1468–1472.

Muraki, K., Iwata, Y., Katanosaka, Y., Ito, T., Ohya, S., Shigekawa, M., and Imaizumi, Y. 2003. TRPV2 is a component of osmotically sensitive cation channels in murine aortic myocytes. Circ. Res. 93(9), 829–838.

Murinson, B. B., Archer, D. R., Li, Y., and Griffin, J. W. 2005. Degeneration of myelinated efferent fibers prompts mitosis in remak Schwann cells of uninjured C-fiber afferents. J. Neurosci. 25, 1179–1187.

Naeini, R. S., Witty, M. F., Seguela, P., and Bourque, C. W. 2006. An N-terminal variant of Trpv1 channel is required for osmosensory transduction. Nat. Neurosci. 9, 93–98.

Nagata, K., Duggan, A., Kumar, G., and Garcia-Anoveros, J. 2005. Nociceptor and hair cell transducer properties of TRPA1, a channel for pain and hearing. J. Neurosci. 25, 4052–4061.

Nakatsuka, T. and Gu, J. G. 2001. ATP P2X receptor-mediated enhancement of glutamate release and evoked EPSCs in dorsal horn neurons of the rat spinal cord. J. Neurosci. 21, 6522–6531.

Nakatsuka, T., Tsuzuki, K., Ling, J. X., Sonobe, H., and Gu, J. G. 2003. Distinct roles of P2X receptors in modulating glutamate release at different primary sensory synapses in rat spinal cord. J. Neurophysiol. 89, 3243–3252.

Nassar, M. A., Levato, A., Stirling, L. C., and Wood, J. N. 2005. Neuropathic pain develops normally in mice lacking both Nav1.7 and Nav1.8. Mol. Pain 1, 24.

Nassar, M. A., Stirling, L. C., Forlani, G., Baker, M. D., Matthews, E. A., Dickenson, A. H., and Wood, J. N. 2004. Nociceptor-specific gene deletion reveals a major role for Nav1.7 (PN1) in acute and inflammatory pain. Proc. Natl. Acad. Sci. U. S. A. 101, 12706–12711.

Nicholson, R., Small, J., Dixon, A. K., Spanswick, D., and Lee, K. 2003. Serotonin receptor mRNA expression in rat dorsal root ganglion neurons. Neurosci. Lett. 337, 119–122.

Nilius, B., Vriens, J., Prenen, J., Droogmans, G., and Voets, T. 2004. TRPV4 calcium entry channel: a paradigm for gating diversity. Am. J. Physiol. Cell Physiol. 286, C195–C205.

North, R. A. 2002. Molecular physiology of P2X receptors. Physiol. Rev. 82, 1013–1067.

Numazaki, M., Tominaga, T., Takeuchi, K., Murayama, N., Toyooka, H., and Tominaga, M. 2003. Structural determinant of TRPV1 desensitization interacts with calmodulin. Proc. Natl. Acad. Sci. U. S. A. 100, 8002–8006.

O'Hagan, R. and Chalfie, M. 2005. Mechanosensation in *Caenorhabditis elegans*. Int. Rev. Neurobiol. 69, 169–203.

Oida, H., Namba, T., Sugimoto, Y., Ushikubi, F., Ohishi, H., Ichikawa, A., and Narumiya, S. 1995. *In situ* hybridization studies of prostacyclin receptor mRNA expression in various mouse organs. Br. J. Pharmacol. 116, 2828–2837.

Okuse, K., Malik-Hall, M., Baker, M. D., Poon, W. Y., Kong, H., Chao, M. V., and Wood, J. N. 2002. Annexin II light chain regulates sensory neuron-specific sodium channel expression. Nature 417, 653–656.

Olah, Z., Karai, L., and Iadarola, M. J. 2002. Protein kinase C(alpha) is required for vanilloid receptor 1 activation. Evidence for multiple signaling pathways. J. Biol. Chem. 277, 35752–35759.

Page, A. J., Brierley, S. M., Martin, C. M., Martinez-Salgado, C., Wemmie, J. A., Brennan, T. J., Symonds, E., Omari, T., Lewin, G. R., Welsh, M. J., and Blackshaw, L. A. 2004. The ion channel ASIC1 contributes to visceral but not cutaneous mechanoreceptor function. Gastroenterology 127, 1739–1747.

Patapoutian, A., Peier, A. M., Story, G. M., and Viswanath, V. 2003. ThermoTRP channels and beyond: mechanisms of temperature sensation. Nat. Rev. Neurosci. 4, 529–539.

Patel, A. J. and Honore, E. 2001. Properties and modulation of mammalian 2P domain K$^+$ channels. Trends Neurosci. 24, 339–346.

Peier, A. M., Moqrich, A., Hergarden, A. C., Reeve, A. J., Andersson, D. A., Story, G. M., Earley, T. J., Dragoni, I., McIntyre, P., Bevan, S., and Patapoutian, A. 2002a. A TRP channel that senses cold stimuli and menthol. Cell 108, 705–715.

Peier, A. M., Reeve, A. J., Andersson, D. A., Moqrich, A., Earley, T. J., Hergarden, A. C., Story, G. M., Colley, S., Hogenesch, J. B., McIntyre, P., et al. 2002b. A heat-sensitive TRP channel expressed in keratinocytes. Science 296, 2046–2049.

Pierce, K. L., Premont, R. T., and Lefkowitz, R. J. 2002. Seven-transmembrane receptors. Nat. Rev. Mol. Cell Biol. 3, 639–650.

Pogatzki-Zahn, E. M., Shimizu, I., Caterina, M., and Raja, S. N. 2005. Heat hyperalgesia after incision requires TRPV1 and is distinct from pure inflammatory pain. Pain 115, 296–307.

Pomonis, J. D., Harrison, J. E., Mark, L., Bristol, D. R., Valenzano, K. J., and Walker, K. 2003. *N*-(4-Tertiarybutylphenyl)-4-(3-cholorphyridin-2-yl)tetrahydropyrazine-1(2H)-carbox-amide (BCTC), a novel, orally effective vanilloid receptor 1 antagonist with analgesic

properties: II. *in vivo* characterization in rat models of inflammatory and neuropathic pain. J. Pharmacol. Exp. Ther. 306, 387–393.

Prescott, E. D. and Julius, D. 2003. A modular PIP2 binding site as a determinant of capsaicin receptor sensitivity. Science 300, 1284–1288.

Price, M. P., Lewin, G. R., McIlwrath, S. L., Cheng, C., Xie, J., Heppenstall, P. A., Stucky, C. L., Mannsfeldt, A. G., Brennan, T. J., Drummond, H. A., et al. 2000. The mammalian sodium channel BNC1 is required for normal touch sensation. Nature 407, 1007–1011.

Price, M. P., McIlwrath, S. L., Xie, J., Cheng, C., Qiao, J., Tarr, D. E., Sluka, K. A., Brennan, T. J., Lewin, G. R., and Welsh, M. J. 2001. The DRASIC cation channel contributes to the detection of cutaneous touch and acid stimuli in mice. Neuron 32, 1071–1083.

Priest, B. T., Murphy, B. A., Lindia, J. A., Diaz, C., Abbadie, C., Ritter, A. M., Liberator, P., Iyer, L. M., Kash, S. F., Kohler, M. G., et al. 2005. Contribution of the tetrodotoxin-resistant voltage-gated sodium channel NaV1.9 to sensory transmission and nociceptive behavior. Proc. Natl. Acad. Sci. U. S. A. 102, 9382–9387.

Puil, E., Miura, R. M., and Spigelman, I. 1989. Consequences of 4-aminopyridine applications to trigeminal root ganglion neurons. J. Neurophysiol. 62, 810–820.

Ramer, M. S. and Bisby, M. A. 1999. Adrenergic innervation of rat sensory ganglia following proximal or distal painful sciatic neuropathy: distinct mechanisms revealed by anti-NGF treatment. Eur. J. Neurosci. 11, 837–846.

Rami, H. K. and Gunthorpe, M. J. 2004. The therapeutic potential of TRPV1 (VR1) antagonists: clinical answers await. Drug Discov. Today: Therapeutic Strategies. DOI: 10.1016/j.ddstr.2004.08.020.

Ramsey, I. S., Delling, M., and Clapham, D. E. 2006. An introduction to trp channels. Annu. Rev. Physiol. 68, 619–647.

Rasband, M. N., Trimmer, J. S., Schwarz, T. L., Levinson, S. R., Ellisman, M. H., Schachner, M., and Shrager, P. 1998. Potassium channel distribution, clustering, and function in remyelinating rat axons. J. Neurosci. 18, 36–47.

Rashid, M. H., Inoue, M., Kondo, S., Kawashima, T., Bakoshi, S., and Ueda, H. 2003. Novel expression of vanilloid receptor 1 on capsaicin-insensitive fibers accounts for the analgesic effect of capsaicin cream in neuropathic pain. J. Pharmacol. Exp. Ther. 304, 940–948.

Reichling, D. B. and Levine, J. D. 1997. Heat transduction in rat sensory neurons by calcium-dependent activation of a cation channel. Proc. Natl. Acad. Sci. U. S. A. 94, 7006–7011.

Reid, G. and Flonta, M. 2001. Cold transduction by inhibition of a background potassium conductance in rat primary sensory neurones. Neurosci. Lett. 297, 171–174.

Renganathan, M., Dib-Hajj, S., and Waxman, S. G. 2002. Na(v)1.5 underlies the 'third TTX-R sodium current' in rat small DRG neurons. Mol. Brain Res. 106, 70–82.

Rios, J. C., Rubin, M., St Martin, M., Downey, R. T., Einheber, S., Rosenbluth, J., Levinson, S. R., Bhat, M., and Salzer, J. L. 2003. Paranodal interactions regulate expression of sodium channel subtypes and provide a diffusion barrier for the node of Ranvier. J. Neurosci. 23, 7001–7011.

Ritter, A. M. and Mendell, L. M. 1992. Somal membrane properties of physiologically identified sensory neurons in the rat: effects of nerve growth factor. J. Neurophysiol. 68, 2033–2041.

Robinson, R. B. and Siegelbaum, S. A. 2003. Hyperpolarization-activated cation currents: from molecules to physiological function. Annu. Rev. Physiol. 65, 453–480.

Rohacs, T., Lopes, C. M., Michailidis, I., and Logothetis, D. E. 2005. PI(4, 5)P2 regulates the activation and desensitization of TRPM8 channels through the TRP domain. Nat. Neurosci. 8, 626–634.

Rose, R. D., Koerber, H. R., Sedivec, M. J., and Mendell, L. M. 1986. Somal action potential duration differs in identified primary afferents. Neurosci. Lett. 63, 259–264.

Roza, C., Puel, J. L., Kress, M., Baron, A., Diochot, S., Lazdunski, M., and Waldmann, R. 2004. Knockout of the ASIC2 channel in mice does not impair cutaneous mechanosensation, visceral mechanonociception and hearing. J. Physiol. 558, 659–669.

Rush, A. M., Dib-Hajj, S. D., Liu, S., Cummins, T. R., Black, J. A., and Waxman, S. G. 2006. A single sodium channel mutation produces hyper- or hypoexcitability in different types of neurons. Proc. Natl. Acad. Sci. U. S. A. 103, 8245–8250.

Sah, D. W., Ossipo, M. H., and Porreca, F. 2003. Neurotrophic factors as novel therapeutics for neuropathic pain. Nat. Rev. Drug Discov. 2, 460–472.

Sato, J. and Perl, E. R. 1991. Adrenergic excitation of cutaneous pain receptors induced by peripheral nerve injury. Science 251, 1608–1610.

Schafers, M., Geis, C., Brors, D., Yaksh, T. L., and Sommer, C. 2002. Anterograde transport of tumor necrosis factor-alpha in the intact and injured rat sciatic nerve. J. Neurosci. 22, 536–545.

Scroggs, R. S. and Fox, A. P. 1992. Calcium current variation between acutely isolated adult rat dorsal root ganglion neurons of different size. J. Physiol. 445, 639–658.

Scroggs, R. S., Todorovic, S. M., Anderson, E. G., and Fox, A. P. 1994. Variation in IH, IIR, and ILEAK between acutely isolated adult rat dorsal root ganglion neurons of different size. J. Neurophysiol. 71, 271–279.

Shah, B. S., Stevens, E. B., Gonzalez, M. I., Bramwell, S., Pinnock, R. D., Lee, K., and Dixon, A. K. 2000. beta3, a novel auxiliary subunit for the voltage-gated sodium channel, is expressed preferentially in sensory neurons and is upregulated in the chronic constriction injury model of neuropathic pain. Eur. J. Neurosci. 12, 3985–3990.

Shimizu, I., Iida, T., Guan, Y., Zhao, C., Raja, S. N., Jarvis, M. F., Cockayne, D. A., and Caterina, M. J. 2005. Enhanced thermal avoidance in mice lacking the ATP receptor P2X3. Pain 116, 96–108.

Shu, X. and Mendell, L. M. 2001. Acute sensitization by NGF of the response of small-diameter sensory neurons to capsaicin. J. Neurophysiol. 86, 2931–2938.

Shu, X. Q. and Mendell, L. M. 1999. Neurotrophins and hyperalgesia. Proc. Natl. Acad. Sci. U. S. A. 96, 7693–7696.

Sluka, K. A., Price, M. P., Breese, N. M., Stucky, C. L., Wemmie, J. A., and Welsh, M. J. 2003. Chronic hyperalgesia induced by repeated acid injections in muscle is abolished by the loss of ASIC3, but not ASIC1. Pain 106, 229–239.

Sluka, K. A., Willis, W. D., and Westlund, K. N. 1995. The role of dorsal root reflexes in neurogenic inflammation. Pain Forum 4, 141–149.

Smith, G. D., Gunthorpe, M. J., Kelsell, R. E., Hayes, P. D., Reilly, P., Facer, P., Wright, J. E., Jerman, J. C., Walhin, J. P., Ooi, L., et al. 2002. TRPV3 is a temperature-sensitive vanilloid receptor-like protein. Nature 418, 186–190.

Souslova, V., Cesare, P., Ding, Y., Akopian, A. N., Stanfa, L., Suzuki, R., Carpenter, K., Dickenson, A., Boyce, S., Hill, R., et al. 2000. Warm-coding deficits and aberrant inflammatory pain in mice lacking P2X3 receptors. Nature 407, 1015–1017.

Steranka, L. R., Manning, D. C., DeHaas, C. J., Ferkany, J. W., Borosky, S. A., Connor, J. R., Vavrek, R. J., Stewart, J. M., and Snyder, S. H. 1988. Bradykinin as a pain mediator: receptors are localized to sensory neurons, and antagonists

have analgesic actions. Proc. Natl. Acad. Sci. U. S. A. 85, 3245–3249.

Stewart, T., Beyak, M. J., and Vanner, S. 2003. Ileitis modulates potassium and sodium currents in guinea pig dorsal root ganglia sensory neurons. J. Physiol. 552, 797–807.

Story, G. M., Peier, A. M., Reeve, A. J., Eid, S. R., Mosbacher, J., Hricik, T. R., Earley, T. J., Hergarden, A. C., Andersson, D. A., Hwang, S. W., et al. 2003. ANKTM1, a TRP-like channel expressed in nociceptive neurons, is activated by cold temperatures. Cell 112, 819–829.

Sugimoto, Y., Shigemoto, R., Namba, T., Negishi, M., Mizuno, N., Narumiya, S., and Ichikawa, A. 1994. Distribution of the messenger RNA for the prostaglandin E receptor subtype EP3 in the mouse nervous system. Neuroscience 62, 919–928.

Suzuki, M., Mizuno, A., Kodaira, K., and Imai, M. 2003. Impaired pressure sensation in mice lacking TRPV4. J. Biol. Chem. 278, 22664–22668.

Szallasi, A. and Blumberg, P. M. 1999. Vanilloid (Capsaicin) receptors and mechanisms. Pharmacol. Rev. 51, 159–212.

Takahashi, N., Kikuchi, S., Dai, Y., Kobayashi, K., Fukuoka, T., and Noguchi, K. 2003. Expression of auxiliary beta subunits of sodium channels in primary afferent neurons and the effect of nerve injury. Neuroscience 121, 441–450.

Tarpley, J. W., Kohler, M. G., and Martin, W. J. 2004. The behavioral and neuroanatomical effects of IB4-saporin treatment in rat models of nociceptive and neuropathic pain. Brain Res. 1029, 65–76.

Thut, P. D., Wrigley, D., and Gold, M. S. 2003. Cold transduction in rat trigeminal ganglia neurons in vitro. Neuroscience 119, 1071–1083.

Tominaga, M., Caterina, M. J., Malmberg, A. B., Rosen, T. A., Gilbert, H., Skinner, K., Raumann, B. E., Basbaum, A. I., and Julius, D. 1998. The cloned capsaicin receptor integrates multiple pain-producing stimuli. Neuron 21, 1–20.

Tominaga, M., Wada, M., and Masu, M. 2001. Potentiation of capsaicin receptor activity by metabotropic ATP receptors as a possible mechanism for ATP-evoked pain and hyperalgesia. Proc. Natl. Acad. Sci. U. S. A. 98, 6951–6956.

Treede, R. D., Davis, K. D., Campbell, J. N., and Raja, S. N. 1992. The plasticity of cutaneous hyperalgesia during sympathetic ganglion blockade in patients with neuropathic pain. Brain 115(Pt 2), 607–621.

Trevisani, M., Smart, D., Gunthorpe, M. J., Tognetto, M., Barbieri, M., Campi, B., Amadesi, S., Gray, J., Jerman, J. C., Brough, S. J., et al. 2002. Ethanol elicits and potentiates nociceptor responses via the vanilloid receptor-1. Nat. Neurosci. 5, 546–551.

Tseng, L. F. and Narita, M. 1995. Use of d-antisense oligo in the study of turnover of d-opioid receptors in the spinal cord of the mice. Soc. Neurosci. Abstr. 21, 1363.

Tsuzuki, K., Ase, A., Seguela, P., Nakatsuka, T., Wang, C. Y., She, J. X., and Gu, J. G. 2003. TNP-ATP-resistant P2X ionic current on the central terminals and somata of rat primary sensory neurons. J. Neurophysiol. 89, 3235–3242.

Vellani, V., Mapplebeck, S., Moriondo, A., Davis, J. B., and McNaughton, P. A. 2001. Protein kinase C activation potentiates gating of the vanilloid receptor VR1 by capsaicin, protons, heat and anandamide. J. Physiol. (Lond.) 543, 813–825.

Vergnolle, N., Wallace, J. L., Bunnett, N. W., and Hollenberg, M. D. 2001. Protease-activated receptors in inflammation, neuronal signaling and pain. Trends Pharmacol. Sci. 22, 146–152.

Viana, F., de la Pena, E., and Belmonte, C. 2002. Specificity of cold thermotransduction is determined by differential ionic channel expression. Nat. Neurosci. 5, 254–260.

Villiere, V. and McLachlan, E. M. 1996. Electrophysiological properties of neurons in intact rat dorsal root ganglia classified by conduction velocity and action potential duration. J. Neurophysiol. 76, 1924–1941.

Vlaskovska, M., Kasakov, L., Rong, W., Bodin, P., Bardini, M., Cockayne, D. A., Ford, A. P., and Burnstock, G. 2001. P2X3 knock-out mice reveal a major sensory role for urothelially released ATP. J. Neurosci. 21, 5670–5677.

Vulchanova, L., Olson, T. H., Stone, L. S., Riedl, M. S., Elde, R., and Honda, C. N. 2001. Cytotoxic targeting of isolectin IB4-binding sensory neurons. Neuroscience 108, 143–155.

Waldmann, R. 2001. Proton-gated cation channels – neuronal acid sensors in the central and peripheral nervous system. Adv. Exp. Med. Biol. 502, 293–304.

Waldmann, R., Champigny, G., Lingueglia, E., De Weille, J. R., Heurteaux, C., and Lazdunski, M. 1999. H(+)-gated cation channels. Ann. N. Y. Acad. Sci. 868, 67–76.

Walker, K. M., Urban, L., Medhurst, S. J., Patel, S., Panesar, M., Fox, A. J., and McIntyre, P. 2003. The VR1 antagonist capsazepine reverses mechanical hyperalgesia in models of inflammatory and neuropathic pain. J. Pharmacol. Exp. Ther. 304, 56–62.

Wallner, M., Pratap, M., and Toro, L. 1998. Ca^{2+}-Dependent K^+ Channels in Muscle and Brain: Molecular Structure, Function and Diseases. In: Current Topics in Membranes, Vol. 46, Potassium Ion Channels: Molecular Structure, Function and Diseases (eds. Y. Kurachi, L. Y. Jan, and M. Lazdunski), pp. 117–140. Academic Press.

Wang, Y., Kedei, N., Wang, M., Wang, Q. J., Huppler, A. R., Toth, A., Tran, R., and Blumberg, P. M. 2004. Interaction between protein kinase Cmu and the vanilloid receptor type 1. J. Biol. Chem. 279(51), 53674–53682.

Wang, H. F., Shortland, P., Park, M. J., and Grant, G. 1998. Retrograde and transganglionic transport of horseradish peroxidase-conjugated cholera toxin B subunit, wheatgerm agglutinin and isolectin B4 from Griffonia simplicifolia I in primary afferent neurons innervating the rat urinary bladder. Neuroscience 87, 275–288.

Waxman, S. G., Dib-Hajj, S., Cummins, T. R., and Black, J. A. 1999. Sodium channels and pain. Proc. Natl. Acad. Sci. U. S. A. 96, 7635–7639.

Waxman, S. G., Kocsis, J. D., and Black, J. A. 1994. Type III sodium channel mRNA is expressed in embryonic but not adult spinal sensory neurons, and is reexpressed following axotomy. J. Neurophysiol. 72, 466–470.

Weinreich, D. and Wonderlin, W. F. 1987. Inhibition of calcium-dependent spike after-hyperpolarization increases excitability of rabbit visceral sensory neurones. J. Physiol. 394, 415–427.

Werz, M. A. and MacDonald, R. L. 1983. Opioid peptides selective for Mu- and Delta- opiate receptors reduce calcium-dependent action potential duration by increasing potassium conductance. Neurosci. Lett. 42, 173–178.

White, G., Lovinger, D. M., and Weight, F. F. 1989. Transient low-threshold Ca^{2+} current triggers burst firing through an afterdepolarizing potential in an adult mammalian neuron. Proc. Natl. Acad. Sci. 86, 6802–6806.

Willis, W. D., Jr. 1999. Dorsal root potentials and dorsal root reflexes: a double-edged sword. Exp. Brain Res. 124, 395–421.

Wu, Z. Z., Chen, S. R., and Pan, H. L. 2004b. Differential sensitivity of N- and P/Q-type Ca^{2+} channel currents to a mu opioid in isolectin B4-positive and -negative dorsal root ganglion neurons. J. Pharmacol. Exp. Ther. 311, 939–947.

Wu, G., Whiteside, G. T., Lee, G., Nolan, S., Niosi, M., Pearson, M. S., and Ilyin, V. I. 2004a. A-317491, a selective P2X3/P2X2/3 receptor antagonist, reverses inflammatory mechanical hyperalgesia through action at peripheral receptors in rats. Eur. J. Pharmacol. 504, 45–53.

Xie, J., Price, M. P., Berger, A. L., and Welsh, M. J. 2002. DRASIC contributes to pH-gated currents in large dorsal root ganglion sensory neurons by forming heteromultimeric channels. J. Neurophysiol. 87, 2835–2843.

Xu, H., Blair, N. T., and Clapham, D. E. 2005. Camphor activates and strongly desensitizes the transient receptor potential

vanilloid subtype 1 channel in a vanilloid-independent mechanism. J. Neurosci. 25, 8924–8937.

Xu, H., Ramsey, I. S., Kotecha, S. A., Moran, M. M., Chong, J. A., Lawson, D., Ge, P., Lilly, J., Silos-Santiago, I., Xie, Y., et al. 2002. TRPV3 is a calcium-permeable temperature-sensitive cation channel. Nature 418, 181–186.

Yang, D. and Gereau, R. W. T. 2002. Peripheral group II metabotropic glutamate receptors (mGluR2/3) regulate prostaglandin E2-mediated sensitization of capsaicin responses and thermal nociception. J. Neurosci. 22, 6388–6393.

Yang, Y., Wang, Y., Li, S., Xu, Z., Li, H., Ma, L., Fan, J., Bu, D., Liu, B., Fan, Z., et al. 2004. Mutations in SCN9A, encoding a sodium channel alpha subunit, in patients with primary erythermalgia. J. Med. Genet. 41, 171–174.

Yeomans, D. C., Levinson, S. R., Peters, M. C., Koszowski, A. G., Tzabazis, A. Z., Gilly, W. F., and Wilson, S. P. 2005. Decrease in inflammatory hyperalgesia by herpes vector-mediated knockdown of Nav1.7 sodium channels in primary afferents. Hum. Gene. Ther. 16, 271–277.

Yoshimura, N. and de Groat, W. C. 1999. Increased excitability of afferent neurons innervating rat urinary bladder after chronic bladder inflammation. J. Neurosci. 19, 4644–4653.

Yu, F. H., Westenbroek, R. E., Silos-Santiago, I., McCormick, K. A., Lawson, D., Ge, P., Ferriera, H., Lilly, J., DiStefano, P. S., Catterall, W. A., et al. 2003. Sodium channel beta4, a new disulfide-linked auxiliary subunit with similarity to beta2. J. Neurosci. 23, 7577–7585.

Yuhas, W. A. and Fuchs, P. A. 1999. Apamin-sensitive, small-conductance, calcium-activated potassium channels mediate cholinergic inhibition of chick auditory hair cells. J. Comp. Physiol. [A] 185, 455–462.

Yusaf, S. P., Goodman, J., Pinnock, R. D., Dixon, A. K., and Lee, K. 2001. Expression of voltage-gated calcium channel subunits in rat dorsal root ganglion neurons. Neurosci. Lett. 311, 137–141.

Zhang, Y. H. and Nicol, G. D. 2004. NGF-mediated sensitization of the excitability of rat sensory neurons is prevented by a blocking antibody to the p75 neurotrophin receptor. Neurosci. Lett. 366, 187–192.

Zhang, X., Huang, J., and McNaughton, P. A. 2005. NGF rapidly increases membrane expression of TRPV1 heat-gated ion channels. EMBO J. 24, 4211–4223.

Zhang, Y. H., Vasko, M. R., and Nicol, G. D. 2002. Ceramide, a putative second messenger for nerve growth factor, modulates the TTX-resistant Na(+) current and delayed rectifier K(+) current in rat sensory neurons. J. Physiol. 544, 385–402.

Zhao, J., Gover, T. D., Muralidharan, S., Auston, D. A., Weinreich, D., and Kao, J. P. 2006. Caged vanilloid ligands for activation of TRPV1 receptors by 1- and 2-photon excitation. Biochemistry 45, 4915–4926.

Zhuang, Z. Y., Xu, H., Clapham, D. E., and Ji, R. R. 2004. Phosphatidylinositol 3-kinase activates ERK in primary sensory neurons and mediates inflammatory heat hyperalgesia through TRPV1 sensitization. J. Neurosci. 24, 8300–8309.

Zweifel, L. S., Kuruvilla, R., and Ginty, D. D. 2005. Functions and mechanisms of retrograde neurotrophin signalling. Nat. Rev. Neurosci. 6, 615–625.

Zygmunt, P. M., Petersson, J., Andersson, D. A., Chuang, H., Sorgard, M., Di Marzo, V., Julius, D., and Hogestatt, E. D. 1999. Vanilloid receptors on sensory nerves mediate the vasodilator action of anandamide. Nature 400, 452–457.

Zylka, M. J., Dong, X., Southwell, A. L., and Anderson, D. J. 2003. Atypical expansion in mice of the sensory neuron-specific Mrg G protein-coupled receptor family. Proc. Natl. Acad. Sci. U. S. A. 100, 10043–10048.

Zylka, M. J., Rice, F. L., and Anderson, D. J. 2005. Topographically distinct epidermal nociceptive circuits revealed by axonal tracers targeted to Mrgprd. Neuron 45, 17–25.

5 Zoster-Associated Pain and Nociceptors

H Maija, Helsinki University Hospital, Helsinki, Finland

Glossary

allodynia Painful response to normally pain-free stimulus on the skin.

herpes zoster (HZ) Usually painful reactivation of varicella-zoster virus.

postherpetic neuralgia (PHN) Prolonged pain after herpes zoster.

varicella-zoster virus (VZV) The causative agent of herpes zoster.

5.1 Epidemiology and Course of Herpes Zoster and Postherpetic Neuralgia

Postherpetic neuralgia (PHN) is a painful aftermath of herpes zoster (HZ), an acute inflammation of spinal nerves due to reactivation of varicella-zoster virus (VZV) residing dormant in sensory ganglia. HZ is the most common neurological disorder with a cumulative lifetime incidence of 10–20% in general population, and as high as 50% of a cohort surviving to 85 years of age (Dworkin, R. H. and Schmader, K. E., 2001). Surveys from two United Kingdom general-practice populations reported annual incidence rates of 3.4–4.8 per 1000 persons (McGregor, R. M., 1957; Hope-Simpson, R. E., 1965). Two population-based studies of epidemiology of HZ have been published from Rochester, Minnesota (Ragozzino, M. W. *et al.*, 1982; Donahue, J. G. *et al.*, 1995) and reported the annual age-adjusted incidences of 1.3 and 2.1 per 1000. About 50% of the HZ cases are thoracic, whereas trigeminal, cervical, and lumbar locations represent each 10–15% of the cases and sacral dermatomes about 5% of the cases. The distribution of HZ reflects the initial virus burden to sensory ganglia (Hope-Simpson, R. E., 1965).

In HZ, viral replication first causes ganglionitis followed by infection of the corresponding nerve and skin dermatome. Pain and unilateral dermatomal rash are the typical symptoms, often accompanied by paresthesias and itching. Patients with HZ affecting the extremities can have complex regional painlike symptoms (color changes, skin temperature changes, edema, weakness, and extradermatomal involvement), and the risk of PHN is high in these cases (Berry, J. D. *et al.*, 2004). Most patients experience pain during HZ, with only about 10% remaining pain-free. Pain precedes rash in 75% of cases (Haanpää, M. *et al.*, 1999a; Chidiac, C. *et al.*, 2001), lasting usually for some days, but in some cases weeks and even months before the eruption of rash (Gilden, D. H. *et al.*, 1991). Zoster pain tends to resolve spontaneously with time, but a minority of patients continue to have pain after the healing of rash. Retrospective epidemiological studies based on help-seeking behavior reported that about 10% of the patients have pain after the healing of rash, 5% of the patients have pain at 3 months, and 2% have pain at 1 year (Burgoon, C. F. *et al.*, 1957; Hope-Simpson, R. E., 1965; Ragozzino, M. W. *et al.*, 1982). A prospective long-term follow-up study from Iceland reported that 19% of the patients have pain at 1 month, 7% at 3 months, and 3% at 1 year (Helgason, S. *et al.*, 2000). A prospective observational community study reported pain in 30% of the patients at 6 weeks, 27% at 12 weeks, 16% at 6 months, and 9% at 1 year (Scott, F. T. *et al.*, 2003). The most important risk factors for PHN are old age and severe acute pain. Other predictors of PHN are severe rash, presence of prodromal pain, viremia at presentation (Scott, F. T. *et al.*, 2003), and more pronounced immune response (Sato-Takeda, M. *et al.*, 2004), which reflect more severe acute disease. Mechanical allodynia, that is, painful response to normally pain-free mechanical stimulus on the skin, and pinprick hypoesthesia in acute HZ are associated with PHN at 3 months, but due to their insufficient sensitivity and specificity, they cannot be used as predictors of PHN for an individual patient (Haanpää, M. *et al.*, 2000). Abnormal findings in quantitative sensory testing (QST) are common in HZ: warm and cold thresholds are elevated in 20%, heat pain thresholds are lowered in 10%, and tactile thresholds are elevated in 25% of the patients with HZ. All somatosensory abnormalities tend to normalize with time. The abnormal QST findings are associated neither with the severity of acute pain nor with the development of PHN (Haanpää, M. *et al.*, 1999b).

HZ has been regarded as a simple model of peripheral unilateral neuritis caused by VZV, but evidence from neuropathological reports suggest that viral invasion and inflammation may extend to the central nervous system (CNS) as well (Denny-Brown, D. *et al.*, 1944; Watson, C. P. N. *et al.*, 1991). In a study with 56 immunocompetent HZ patients without clinical symptoms of CNS infection, cerebrospinal fluid (CSF) samples were obtained from 46 patients on days 1–18 from the eruption of rash, and 16 consecutive patients with cranial or cervical HZ underwent magnetic resonance imaging (MRI) 1–5 weeks following rash (Haanpää, M. *et al.*, 1998). In 35% of the patients there was evidence of VZV in the CSF either in the form of a positive polymerase chain reaction (PCR) or anti-VZV-immunoglobulin G (IgG). Leucocytosis (range 5–1440 μL^{-1}) was found in 46% of the patients. These changes were more common in patients with complications (e.g., peripheral motor paresis or ophthalmic complications) but they did not predict PHN. Zoster-related MRI changes were found in the brainstem in 56% of the patients and in the cervical cord in two. Three patients had enhancement of the trigeminal nerve in addition to the brainstem lesions. The presence of brainstem lesions in MRI was associated with the development of PHN at 3 months but not with the severity of acute pain.

Those who continue to suffer from prolonged pain after HZ describe three types of pain: a constant deep aching or burning pain, an intermittent pain with a lancinating quality, and allodynia of the skin. In an individual patient, any component can be the most distressing feature of the pain. Itching and dysesthetic sensations may also be present, being in some patients even more annoying than the pain itself (Oaklander, A. L. *et al.*, 2003). Normal sensory function is often altered in PHN. Mechanical allodynia is reported in 65–87% of patients with PHN (Watson, C. P. N. *et al.*, 1988; Nurmikko, T. and Bowsher, D., 1990; Pappagallo, M. *et al.*, 2000), and some patients have cold allodynia (Bowsher, D., 1996). The patients may experience paradoxical sensations in the affected skin area: hot may be felt cold or vice versa, and although light touch may be unpleasant or painful, firm pressure may attenuate pain (Nurmikko, T., 1994). Noordenbos W. (1959) investigated the skin of the patients with PHN and noted that there were areas of major sensory loss as well as minimal sensory loss in the affected region. Others have suggested that extreme allodynia to light touch is often restricted to areas surrounding scarred skin or lying at the border of affected and unaffected dermatomes. Allodynic areas may be often warm termographically, while scarred areas with maximal sensory loss are cool (Rowbotham, M. C. and Fields, H. L., 1989). When tested with conventional clinical methods, hypoesthesia to touch was present in 90% and hypoesthesia to pinprick in 92% of the patients in a pain clinic series (Watson, C. P. N. *et al.*, 1988). In another case series, 94% of the patients had impaired temperature and/or pinprick sensation and/or allodynia, and about 33% had also tactile deficit (Bowsher, D., 1996).

In QST, marked threshold elevations within the affected dermatome compared with the mirror-image site have been found for touch, pinprick, vibration, warmth, and cold (Bjerring, P. *et al.*, 1990; Nurmikko, T. and Bowsher, D., 1990; Eide, P. K. *et al.*, 1994; Rowbotham, M. C. and Fields, H. L., 1996; Rowbotham, M. C. *et al.*, 1996a; Choi, B. and Rowbotham, M. C., 1997; Attal N. *et al.*, 1999; Pappagallo, M. *et al.*, 2000). This indicates damage to unmyelinated and myelinated sensory fibers. Heat pain hypoesthesia is common in patients with PHN, but a minority of patients has heat pain hyperalgesia (Nurmikko, T. and Bowsher, D., 1990; Rowbotham, M. C. and Fields, H. L., 1996; Rowbotham, M. C. *et al.*, 1996a; Choi, B. and Rowbotham, M. C., 1997). In some patients with PHN, no asymmetry was found in warmth or heat pain thresholds, while the greatest threshold difference observed between the affected and contralateral dermatomes was as high as 12.2 °C for warmth and 9.6 °C for heat pain (Choi, B. and Rowbotham, M. C., 1997). However, noticeable asymmetry in thresholds between the affected site and its mirror-image site was found in different modalities from 5% to 13% of HZ patients not developing PHN (Nurmikko, T. and Bowsher, D., 1990). Hence, it seems possible to have sensory threshold abnormality without PHN, and PHN without sensory threshold abnormality in the affected dermatome after HZ.

5.2 Neuropathology of Herpes Zoster and Postherpetic Neuralgia

The classic report of Head H. and Campbell A. W. (1900) is the most comprehensive study of the neuropathology of HZ, but it does not report the possible presence of PHN. Despite it and later studies, our understanding of the pathophysiology of PHN is far from complete. In acute HZ the dorsal root ganglion shows hemorrhagic inflammation and subsequent scarring (Head, H. and Campbell, A. W., 1900; Denny-Brown, D. *et al.*, 1944; Reske-Nielsen, E. *et al.*, 1986; Schmidbauer, M. *et al.*, 1992). Typically, only one ganglion is severely affected, while mild patchy inflammatory changes can be found in the neighboring ganglia (Denny-Brown, D. *et al.*, 1944; Schmidbauer, M. *et al.*, 1992). Ganglion cell necrosis is followed by secondary degeneration and fibrosis in the sensory root and the peripheral nerve. Direct inflammation can also damage the root and nerve (Head, H. and Campbell, A. W., 1900; Denny-Brown, D. *et al.*, 1944; Zacks, S. L. *et al.*, 1964; Watson, C. P. N. *et al.*, 1991; Schmidbauer, M. *et al.*, 1992). VZV has been shown in ganglial and satellite cells; in damaged nerve roots; in the vasa nervorum; in the distal peripheral nerve endoneurally, perineurally, and epineurally; and in Schwann cells (Schmidbauer, M. *et al.*, 1992). Inflammation in peripheral nerves may persist from weeks to months, which can eventually lead to noticeable fibrosis (Head, H. and Campbell, A. W., 1900; Zacks, S. L. *et al.*, 1964; Watson, C. P. N. *et al.*, 1991).

Only a few studies compare the neuropathological findings of those who recover with the finding of those who have PHN. In four patients who had HZ and went on to develop PHN, atrophy of the dorsal horn was found at autopsy whereas those without PHN did not show similar changes (Watson, C. P. N. *et al.*, 1991). However, one patient without PHN showed subacute myelopathy. The investigators found no difference in substance P, calcitonin gene-related peptide, a marker for norepinephrine, or in opioid receptor binding studies (Watson, C. P. N.

et al., 1991). After HZ, a preponderance of small-diameter fibers in the surviving fiber population in peripheral nerves is suggested (Noordenbos, W., 1959; Zacks, S. L. *et al.*, 1964; Watson, C. P. N. *et al.*, 1991), but the significance of this finding is ambiguous, as the fibers in question have not been fully identified. Loss of myelin combined with the preservation of axons has also been described (Fabian, V. A. *et al.*, 1997). Loss of axons and myelin in the peripheral nerve is similar in cases with and without PHN (Zacks, S. L. *et al.*, 1964).

Three studies using skin biopsies report that HZ lowers the innervation density of skin (Rowbotham, M. C. *et al.*, 1996a; Oaklander, A. L. *et al.*, 1998; Oaklander, A. L., 2001). These changes are more profound in patients with PHN, possibly because the absence of pain requires the preservation of a minimum density of primary nociceptive neurons (Oaklander, A. L., 2001). The severity of allodynia is nevertheless associated with well-preserved fiber density (Rowbotham, M. C. *et al.*, 1996a). The results of Oaklander A. L. *et al.* (1998) suggest that unilateral HZ can cause bilateral segmental damage to primary sensory neurons. It is of interest that changes in the epidermal innervation in the contralateral unaffected skin in patients with PHN correlate with the presence and severity of pain, but the pathophysiological explanation for this is unknown.

5.3 Mechanisms of Acute Zoster Pain

Mechanisms of acute zoster pain are not fully established. It is obvious that acute zoster pain represents a combination of nociceptive and neuropathic pain. The nociceptive component is caused by inflammation of the skin, which leads to excitation and sensitization of nociceptors. Increased afferent activity transmitted to the spinal cord creates a state of central hyperexcitability in the dorsal horn neurons, which resolves gradually in concert with the healing of tissue damage in a majority of patients. The neuropathic component is caused by direct viral damage and inflammation in the dorsal ganglion, peripheral nerve, and in some cases also in the dorsal root and spinal cord. According to the study of acute HZ, preherpetic pain is not associated with the severity of rash or QST changes, which suggests that virus-induced irritation in the afferent pathways rather than dermal inflammation or axonal damage is more salient in preherpetic pain (Haanpää, M. *et al.*, 1999a).

Because the abnormal QST findings are not associated with the severity of acute zoster pain, dysfunction of the thin fibers per se seems not to be the dominating factor in acute zoster pain. The presence of severe acute pain in acute HZ is associated with the presence of mechanical allodynia but not with severe rash, which implies a neuropathic character of allodynia in acute HZ rather than hyperesthesia or hyperalgesia due to cutaneous inflammation (Haanpää, M. *et al.*, 2000). It is not known to what extent the mechanisms of allodynia in acute HZ are the same as the mechanisms of allodynia in PHN. The extension of allodynia and somatosensory changes outside the affected dermatome is not uncommon in acute HZ, which indicates that the CNS is quite frequently involved even in mild cases of HZ (Haanpää, M. *et al.*, 1999a; 1999b; 2000). Although QST changes are common in HZ, they are not associated with the development of PHN. Hence, the prolonged pain in HZ cannot be explained by as a simple consequence of either axonal damage (represented as threshold elevation) or sensitization of the nociceptive system (represented as heat pain hyperalgesia). Transition from acute zoster-associated pain to PHN is likely to be a multifactorial process, perhaps involving both peripheral and central changes. The genetic factors regulating the inflammatory responses may play an important role in this process (Sato-Takeda, M. *et al.*, 2004).

5.4 Pathophysiology of Postherpetic Neuralgia

Patients with PHN have been studied using quantitative testing of the primary afferent function, skin biopsies, and controlled treatment trials. In studies of patients with allodynic PHN, the results concerning the role of C-nociceptors in PHN are conflicting. Using indirect assessment of C-fibers by quantifying the axon reflex reactions induced by histamine iontophoresis, Baron R. and Saguer M. (1993) found that histamine responses were reduced or abolished within allodynic areas, indicating degeneration of C-fibers. Patients, who had recovered from HZ without PHN, had bilaterally identical histamine responses, indicating complete recovery of C-fibers. Additionally, impairment of the C-fiber function correlated positively with the intensity of postherpetic pain. In contrast, Rowbotham M. C. and Fields H. L. (1996) found, using QST measurements, that in the allodynic area of PHN patients, pain severity correlated with the preservation of thermal sensory function. The same conclusion was drawn from the study using skin punch biopsies; the loss of cutaneous innervation correlated inversely with allodynia, and pain intensity was strongly related to the severity of allodynia (Rowbotham, M. C. *et al.*, 1996a). In another study, significant correlation was found between the intensity of ongoing pain and mechanical allodynia in patients with short-lasting (<1 year) but not in patients with long-lasting (≥1 year) PHN (Pappagallo, M. *et al.*, 2000). According to these results, the underlying pathophysiology may vary from patient to patient and with the duration of PHN.

In PHN, three subtypes have been suggested (Table 1): the irritable nociceptor subtype, the deafferented allodynic subtype, and the deafferented nonallodynic subtype (Fields, H. L. *et al.*, 1998; Petersen, K. *et al.*, 2000; Rowbotham, M. and Petersen, K., 2001). In patients with anatomically intact primary afferent nociceptors, pain and allodynia may be due to abnormal hyperactivity of the nociceptors and subsequent central sensitization, respectively (irritable nociceptor type). The topical capsaicin application test can be used to identify PHN patients in whom the mechanism of pain involves sensitization of peripheral afferents (Petersen, K. *et al.*, 2000;

Table 1 Proposed subtypes of PHN

Subtype	Thermal sensory deficit	Allodynia	Anesthetic infiltration	Capsaicin sensitivity
Irritable nociceptors	Minimal deficit or hyperalgesic to heat pain	Marked	Prolonged relief	Severe burning pain
Deafferented nonallodynic subtype	Marked to all modalities	None	No change	No sensation
Deafferented allodynic subtype	Marked to all modalities	Variable	Relief of allodynia short duration	Variably reduced

Fields, H. L., Rowbotham, M., and Baron, R. 1998. Postherpetic neuralgia: irritable nociceptors and deafferentation. Neurobiol. Dis. 5, 209–227 and Rowbotham, M. C., Petersen, K. L., and Feilds, H. L. Is postherpetic neuralgia more than one disorder. 1999. IASP New. Let. 3–7.

Petersen, K. *et al.*, 2002). In patients with loss of cutaneous C-nociceptor function, evaluated either by using QST of thermal thresholds (Nurmikko, T. and Bowsher, D., 1990; Nurmikko, T., 1994; Choi, B. and Rowbotham, M. C., 1997), C-fiber axon reflex (Baron, R. and Saquer, M., 1993), or capsaicin-induced flare (Morris, G. C. *et al.*, 1995), deafferentation and central reorganization may explain the PHN pain (Fields, H. L. *et al.*, 1998). If tactile ($A\beta$) fibers are preserved, allodynia may result from sprouting of the spinal terminals of $A\beta$ mechanoreceptors into contact with the pain transmission neurons. These cases represent the deafferented allodynic subtype. In patients of deafferented nonallodynic subtype, pain may result from the activation of the CNS pain transmission neurons (Nurmikko, T. and Bowsher, D., 1990; Fields, H. L. *et al.*, 1998). In a study of 63 PHN patients, 25% of the cases were of irritable nociceptor type, 50% were of deafferentation with allodynia type, and 25% were of deafferentation without allodynia type (Pappagallo, M. *et al.*, 2000).

Pain relief of peripherally acting treatments, that is, topical agents, supports the role of peripheral nociceptors in PHN. Topical lidocaine, when used as a 5% gel under an occlusive dressing (Rowbotham, M. C. *et al.*, 1995) or as 5% patch (Rowbotham, M. C. *et al.*, 1996b; Galer, B. *et al.*, 1999) reduces pain and allodynia in PHN. Because it produces clinically insignificant serum levels, it is thought to relieve pain by reducing ectopic discharges in the affected peripheral afferent fibers by blocking upregulated sodium channels. Topical capsaicin has been studied in two randomized controlled trials (Bernstein, J. E. *et al.*, 1989; Watson, C. P. N. *et al.*, 1993). Capsaicin binds to nociceptors in the skin, causing excitation of neurons and a period of enhanced sensitivity perceived as itching, pricking, or burning, with cutaneous vasodilatation. This is followed by a refractory period with reduced sensitivity to, and after repeated applications, persistent desensitization (Nolano, M. *et al.*, 1999). The efficacy of topical capsaicin appears limited: the decrease of pain intensity was a maximum of 23% in one trial (Watson, C. P. *et al.*, 1993) and 30% in the other (Bernstein, J. E. *et al.*, 1989).

The question of the role of the sympathetic nervous system in PHN is ambiguous. In a double-blind study, the cutaneous infiltration of an adrenergic agonist into PHN skin caused only a modest increase of pain and allodynia severity (Choi, B. and Rowbotham, M. C., 1997). The role of sympathetic blocks in HZ and PHN remains controversial, and direct evidence of a major sympathetic contribution to PHN is nonexistent (Wu, C. L. *et al.*, 2000).

The treatment of HZ aims at minimizing the neural damage with antiviral drugs, maximal pain relief with analgesic drugs, and prevention of PHN. For those who continue to have prolonged pain, the current armamentarium of medical treatments, that is, tricyclic antidepressants, gabapentinoids, opioids, and topical lidocaine should be tested (Dubinsky, R. M. *et al.*, 2004). The possibility of multiple mechanisms in the same patient may explain why response to any single treatment is so often partial and forms a logic base for polytherapy. The detailed information of the efficacy of the treatments on the various components of PHN is lacking.

References

Attal, N., Brasseur, L., Chauvin, M., and Bouhassira, D. 1999. Effects of a single and repeated applications of a eutectic mixture of local anaesthetics (EMLA^R) cream on spontaneous and evoked pain in post-herpetic neuralgia. Pain 81, 203–209.

Baron, R. and Saguer, M. 1993. Postherpetic neuralgia: are C-nociceptors involved in signalling and maintenance of tactile allodynia? Brain 116, 1477–1496.

Bernstein, J. E., Korman, N. J., Bickers, D. R., Dahl, M. V., and Millikan, L. E. 1989. Topical capsaicin treatment of chronic postherpetic neuralgia. J. Am. Acad. Dermatol. 21, 265–270.

Berry, J. D., Rowbotham, M. C., and Petersen, K. L. 2004. Complex regional pain syndrome-like symptoms during herpes zoster. Pain 110, 1–12.

Bjerring, P., Arendt-Nielsen, L., and Soderberg, U. 1990. Argon laser induced cutaneous sensory and pain thresholds in post-herpetic neuralgia. Quantitative modulation by topical capsaicin. Acta Derm. Venereol. (Stockholm) 70, 121–125.

Bowsher, D. 1996. Postherpetic neuralgia and its treatment: a retrospective survey of 191 patients. J. Pain Symptom Manage. 12, 290–299.

Burgoon, C. F., Burgoon, J. S., and Baldridge, C. D. 1957. The natural history of herpes zoster. JAMA 164, 265–269.

Chidiac, C., Bruxelle, J., Daures, J. P., Hoang-Xuan, T., Morel, P., Leplege, A., El Hasnaoui, A., and de Labareyre, C. 2001. Characteristics of patients with herpes zoster on presentation to practitioners in France. Clin. Infect. Dis. 33, 62–69.

Choi, B. and Rowbotham, M. C. 1997. Effect of adrenergic receptor activation on post-herpetic neuralgia pain and sensory disturbances. Pain 69, 55–63.

Denny-Brown, D., Adams, R. D., and Fitzgerald, P. J. 1944. Pathologic features of herpes zoster. A note on geniculate herpes. Arch. Neurol. Psychiatry 51, 216–231.

Donahue, J. G., Choo, P. W., Manson, J. E., and Platt, R. 1995. The incidence of herpes zoster. Arch. Intern. Med. 155, 1605–1609.

Dubinsky, R. M., Kabbani, H., El-Chami, Z., Boutwell, C., and Ali, H. 2004. Practice parameter: treatment of postherpetic neuralgia. An evidence based report of the Quality Standards Subcommittee of the American Academy of Neurology. Neurology 63, 959–965.

Dworkin, R. H. and Schmader, K. E. 2001. The Epidemiology and Natural History of Herpes Zoster and Postherpetic Neuralgia. In: Herpes Zoster and Postherpetic Neuralgia, 2nd edn. (eds. C. P. N. Watson and A. A. Gershorn), pp. 39–64. Elsevier.

Eide, P. K., Jorum, E., Stubhaug, A., Bremnes, J., and Breivik, H. 1994. Relief of post-herpetic neuralgia with the N-methyl-D-aspartic acid receptor antagonist ketamine: a double-blind, cross-over comparison with morphine and placebo. Pain 58, 347–354.

Fabian, V. A., Wood, B., Crowley, B., and Kakulas, B. A. 1997. Herpes zoster brachial neuritis. Clin. Neuropathol. 16, 61–64.

Fields, H. L., Rowbotham, M., and Baron, R. 1998. Postherpetic neuralgia: irritable nociceptors and deafferentation. Neurobiol. Dis. 5, 209–227.

Galer, B. S., Rowbotham, M. C., Perander, J., and Friedman, E. 1999. Topical lidocaine patch relieves postherpetic neuralgia more effectively than a vehicle topical patch: results of an enriched enrollment study. Pain 80, 533–538.

Gilden, D. H., Dueland, A. N., Cohrs, R., Martin, J. R., Kleinschmidt-DeMasters, B. K., and Mahalingham, R. 1991. Preherpetic neuralgia. Neurology 41, 1215–1218.

Haanpää, M., Dastidar, P., Weinberg, A., Levin, M., Miettinen, A., Lapinlampi, A., Laippala, P., and Nurmikko, T. 1998. CSF and MRI findings in patients with acute herpes zoster. Neurology 51, 1405–1411.

Haanpää, M., Laippala, P., and Nurmikko, T. 1999a. Pain and somatosensory dysfunction in acute herpes zoster. Clin. J. Pain 15, 78–84.

Haanpää, M., Laippala, P., and Nurmikko, T. 1999b. Thermal and tactile perception thresholds in patients with herpes zoster. Eur. J. Pain 3, 375–386.

Haanpää, M., Laippala, P., and Nurmikko, T. 2000. Allodynia and pinprick hypoaesthesia in acute herpes zoster, and the development of postherpetic neuralgia. J. Pain Symptom Manage. 20, 50–58.

Head, H. and Campbell, A. W. 1900. The pathology of herpes zoster and its bearing on sensory localisation. Brain 23, 353–523.

Hope-Simpson, R. E. 1965. The nature of herpes zoster: a long-term study and a new hypothesis. Proc. R. Soc. Med. 58, 9–20.

McGregor, R. M. 1957. Herpes zoster, chicken pox and cancer in general practice. BMJ 1, 84–87.

Morris, G. C., Gibson, S. J., and Helme, R. D. 1995. Capsaicin-induced flare and vasodilatation in patients with post-herpetic neuralgia. Pain 63, 93–101.

Nolano, M., Simone, D. A., Wenderscharfer-Crabb, G., Johnson, T., Hazen, E., and Kennedy, W. R. 1999. Topical capsaicin in humans: parallel loss of epidermal nerve fibers and pain sensation. Pain 81, 135–145.

Noordenbos, W. 1959. Pain: Problems Pertaining to the Transmission of Nerve Impulses Which Give Rise to Pain. Elsevier.

Nurmikko, T. 1994. Sensory Dysfunction in Postherpetic Neuralgia. In: Touch, Temperature and Pain in Health and Disease, Progress in Pain Research and Management, Vol. 3 (eds. J. Boivie, P. Hansson, and U. Lindblom), pp. 133–141. IASP Press.

Nurmikko, T. and Bowsher, D. 1990. Somatosensory findings in postherpetic neuralgia. J. Neurol. Neurosurg. Psychiatry 53, 135–141.

Oaklander, A. L., Romans, K., Horasek, S., Stocks, A., Hauer, P., and Meyer, R. A. 1998. Unilateral postherpetic neuralgia is associated with bilateral sensory neuron damage. Ann. Neurol. 44, 789–795.

Oaklander, A. L. 2001. The density of remaining nerve endings in human skin with and without postherpetic neuralgia after shingles. Pain 92, 139–145.

Oaklander, A. L., Bowsher, D., Galer, B., Haanpää, M., and Jensen, M. P. 2003. Herpes zoster itch: preliminary epidemiological data. J. Pain 6, 338–343.

Pappagallo, M., Oaklander, A., Quatrano-Piacentini, A., Clark, M., and Raja, S. 2000. Heterogenous patterns of sensory dysfunction in postherpetic neuralgia suggest multiple pathophysiologic mechanisms. Anesthesiology 92, 691–698.

Petersen, K., Fields, H., Brennum, J., Sandroni, P., and Rowbotham, M. 2000. Capsaicin evoked pain and allodynia in post-herpetic neuralgia. Pain 88, 125–133.

Petersen, K. L., Rice, F. L., Suess, F., Berro, M., and Rowbotham, M. C. 2002. Relief of post-herpetic neuralgia by surgical removal of painful skin. Pain 98, 119–126.

Ragozzino, M. W., Melton, L. J., Kurland, L. T., Chu, C. P., and Perry, H. O. 1982. Population-based study on herpes zoster and its sequelae. Medicine 61, 310–316.

Reske-Nielsen, E., Oster, S., and Pedersen, B. 1986. Herpes zoster ophthalmicus and the mesencephalic nucleus. Acta Pathol. Microbiol. Immunol. Scand. 94, 263–269.

Rowbotham, M. and Petersen, K. 2001. Zoster-associated pain and neural dysfunction. Pain 93, 1–5.

Rowbotham, M. C. and Fields, H. L. 1989. Post-herpetic neuralgia: the relation of pain complaint, sensory disturbance, and skin temperature. Pain 39, 129–144.

Rowbotham, M. C. and Fields, H. L. 1996a. The relationship of pain, allodynia and thermal sensation in post-herpetic neuralgia. Brain 119, 347–354.

Rowbotham, M. C., Davies, P. S., and Fields, H. L. 1995. Topical lidocaine gel relieves postherpetic neuralgia. Ann. Neurol. 37, 246–253.

Rowbotham, M. C., Davies, P. S., Verkempinck, C., and Galer, B. S. 1996b. Lidocaine patch: double-blind controlled study of a new treatment method for post-herpetic neuralgia. Pain 65, 39–44.

Rowbotham, M. C., Yosipovitch, G., Connolly, M. K., Finlay, D., Forde, G., and Fields, H. L. 1996. Cutaneous innervation density in the allodynic form of postherpetic neuralgia. Neurobiol. Dis. 3, 205–214.

Rowbotham, M. C., Petersen, K. L., and Feilds, H. L. 1999. Is postherpetic neuralgia more than one disorder. IASP New. Let. 3–7.

Sato-Takeda, M., Ihn, H., Ohashi, J., Tsuchiya, N., Satake, M., Arita, H., Tamaki, K., Hanaoka, K., Tokunaga, K., and Yabe, T. 2004. The human histocompatibility leukocyte antigen (HLA) haplotype is associated with the onset of postherpetic neuralgia after herpes zoster. Pain 110, 329–336.

Schmidbauer, M., Budka, H., Pilz, P., Kurata, T., and Hondo, R. 1992. Presence, distribution and spread of productive varicella zoster virus infection in nervous tissue. Brain 115, 383–398.

Scott, F. T., Leedham-Green, M. E., Barrett-Muir, W. Y., Hawrami, K., Gallagher, W. J., Johnson, R., and Breuer, J. 2003. A study of shingles and the development of postherpetic neuralgia in East London. J. Med. Virol. 70 Suppl 1, S24–S30.

Watson, C. P. N., Evans, R., Watt, V. R., and Birkett, N. 1988. Post-herpetic neuralgia: 208 cases. Pain 35, 289–297.

Watson, C. P. N., Deck, H. J., Morshead, C., Van der Kooy, D., and Evans, R. J. 1991. Post-herpetic neuralgia: further post-mortem studies of cases with and without pain. Pain 44, 105–117.

Watson, C. P., Tyler, K. L., Bickers, D. R., Millikan, L. E., Smith, S., and Coleman, E. 1993. A randomized vehicle-controlled trial of topical capsaicin in the treatment of postherpetic neuralgia. Clin. Ther. 15, 510–526.

Wu, C. L., Marsh, A., and Dworkin, R. H. 2000. The role of sympathetic blocks in herpes zoster and postherpetic neuralgia. Pain 87, 121–129.

Zacks, S. L., Langfitt, T. W., and Elliott, F. A. 1964. Herpetic neuritis: a light and electron microscopic study. Neurology 14, 744–750.

Further Reading

Helgason, S., Petursson, G., Gudmundsson, S., and
Sigurdsson, J. A. 2000. Prevalence of postherpetic neuralgia
after a first episode of herpes zoster: prospective study with
long term follow up. BMJ 321, 794–796.

Relevant Websites

http://www.iasp-pain.org – International Association for
the Study of Pain.
http://www.vzvfoundation.org – VZN Research
Foundation.

6 Ectopic Generators

M Devor, Hebrew University of Jerusalem, Jerusalem, Israel

Glossary

ectopic generator Pathophysiological site of electrical impulse generation in the peripheral nervous system (PNS) or central nervous system (CNS). Ectopic excludes locations of impulse generation in normal, healthy tissue. Examples are nerve-end neuromas and dorsal root ganglia (DRGs) in the event of nerve injury.

electrogenesis The process of electrical impulse generation in nervous system tissue. Electrogenesis usually includes two phases: formation of a generator depolarization (sensory transduction in the case of afferent endings) and encoding of the generator depolarization into an impulse train.

microneurographic recording An electrophysiological technique in which a fine wire microelectrode is inserted transcutaneously into the peripheral nerve, permitting recording of impulse activity in individual nerve fibers. This minimally invasive procedure can be applied to awake humans.

6.1 Introduction

Normally, sensation originates in impulse discharge generated at sensory endings in skin, viscera, and other peripheral tissues. When there has been injury to peripheral nerves, neuropathic pain may arise as a result of abnormal impulse discharge generated at ectopic sites. The most common locations of ectopic impulse generation are the site of injury and associated dorsal root ganglia (DRGs). This chapter will touch on the characteristics of ectopic impulse generation (ectopia), the ways in which ectopia contributes to neuropathic pain sensation, and the cellular mechanisms that underlie it.

6.2 Nerve Injury and Disease Trigger Ectopia

6.2.1 Precipitating Factors

Ectopic impulse generation develops secondary to frank injury of the axon or soma of sensory neurons, or disruption of the myelin sheath that surrounds many axons (dysmyelination and demyelination). Disruption of the Schwann cell sheath that embraces bundles of nonmyelinated axons (Remak bundles) may also be a cause. It is fairly obvious why damage or disease that disrupts the ability of axons to conduct nerve impulses should cause negative sensory abnormalities such as hypoesthesia and numbness. It is much less obvious why neural pathology causes positive sensory symptoms and signs such as ongoing and evoked paresthesias, dysesthesias, and pain. The link is the emergence of ectopia. If the axotomy happens suddenly, there may be a brief period of injury discharge. In general, however, ectopia results from secondary changes that develop over time (Devor, M., 2006a). The pathophysiological changes that induce ectopia can be precipitated by a wide variety of events including trauma (frequently iatrogenic), nerve entrapment, infection (bacterial or viral), sterile inflammation, metabolic abnormalities, malnutrition, vascular abnormalities, neurotoxins (including many chemotherapeutic agents), radiation, inherited mutations, and autoimmune attack.

6.2.2 Spontaneous Ectopic Discharge

A considerable fraction of injured sensory neurons, as many as one-third in some experimental preparations, begin to fire spontaneously in the hours and days following injury. Microneurographic recordings in patients with nerve injury have revealed similar results (Nordin, M. *et al.*, 1984; Devor, M., 2006a). The origin of the activity has been identified as the swollen end-bulbs that form in the neuroma just proximal to sites of axonal transection (Fried, K. *et al.*, 1991), outgrowing sprouts, plaques of demyelination, and the cell soma in the DRG. In the event of sudden traumatic neuropathy activity usually begins earliest in myelinated axons (A-fibers), sometimes as soon as 16 h after axotomy, with a preference for muscle over cutaneous afferents. Activity in unmyelinated (C-) fibers tends to appear in earnest later, after a few weeks. Specific parameters

vary with the type of injury, source (axon or soma), and strain and species of animal.

Firing pattern is usually tonic-rhythmic (i.e., continuous), bursty, or irregular, with the occasional neuron showing complex, often cyclic variations in firing pattern. The instantaneous firing rate of tonic-rhythmic and bursty activity is typically 15–40 impulses per second (ips), with bursts typically occurring at intervals of between a few per second and one every few seconds (Figure 1). Irregular firing of single spikes occurs at the same relatively slow rate as bursting. The reason for these patterns is now known (Amir, R. *et al.*, 2002b). The basic clock driving ectopia triggers activity at the slow rate. Some cells render only singlet spikes. In others, an iterative process based on postspike depolarizing afterpotentials (DAPs) yields a short rhythmic burst upon each trigger event, resulting in bursty discharge. Tonic autorhythmicity reflects an unending burst, which can occur in cells that generate DAPs but in which the mechanism for ending bursts is relatively ineffective. Interestingly, observations in nerves in which only a part of the axons have been severed show that spontaneous ectopic activity may also develop in neighboring uninjured axons, particularly in C-fibers (Wu, G. *et al.*, 2001; Shim, B. *et al.*, 2005).

6.2.3 Ectopic Mechanosensitivity

Gentle percussion over sites of nerve injury, areas of entrapment or neuromas for example, typically evokes an intense stabbing or electric shocklike sensation. This is the Tinel sign. Similar sensations can be evoked by other maneuvers that apply mechanical force to injured

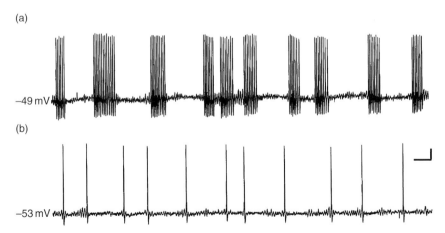

Figure 1 Typical patterns of spike activity generated at ectopic pacemaker sites. (a) Bursty discharge. (b) Irregular singlet firing. Subthreshold oscillations are visible in the baseline traces. Calibrations: 5 mV/100 ms.

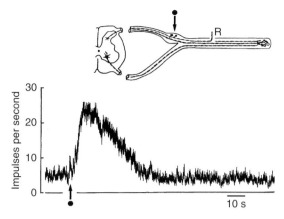

Figure 2 Mechanical afterdischarge. A weak, momentary stimulus to the surface of the DRG triggered a prolonged, self-sustained ectopic discharge burst.

nerves or spinal roots such as straight leg lifting in sciatica (Lasègue's sign), and the signs of Spurling and L'hermitte. Momentary mechanical stimulation of this sort frequently evokes ectopic discharge that outlasts the stimulus itself (afterdischarge), with corresponding aftersensation (Figure 2). This indicates that the underlying process is not just activation of sensitized nociceptors, or signal amplification, but the triggering of autonomous repetitive firing at ectopic pacemaker sites. The nature of this local electrical hyperexcitability is discussed below.

In the event of injuries that leave the nerve in-continuity, such as crush or freeze lesions, there is usually a second Tinel sign that advances with time, marking the farthest position reached by regenerating sprouts. In some conditions (e.g., diabetic neuropathy) paresthesias and pain may occur on tapping anywhere along the course of a nerve. This occurs when there are scattered outgrowing sprouts, demyelination plaques, or mechanosensitive end-bulbs resulting from axonal dying back. Although direct evidence has not yet been sought systematically in humans, it follows logically from studies in animals that pain evoked by deep palpation at tender spots, such as in fibromyalgia, might reflect ectopic mechanosensitivity at locations where small nerve branches cross through fascial planes, under tendons, or are otherwise at risk of being locally pinched.

6.2.4 Abnormal Response to Other Stimuli

Ectopic pacemaker sites also develop abnormal sensitivity to other depolarizing stimuli. Notable among these are response to circulating catecholamines and noradrenalin released from nearby (injured) postganglionic sympathetic axons (Devor, M. and Janig, W., 1981). Ectopic adrenosensitivity of sensory neurons yields sympathetic–sensory coupling, a likely substrate of sympathetically maintained chronic pain states. Afferent response to local and circulating inflammatory mediators is a second example of ectopic chemosensitivity. Abnormal discharge may also arise from temperature changes, ischemia, hypoxia, hypoglycemia, and other conditions capable of locally depolarizing afferent neurons at the sites at which they have developed local ectopic pacemaker capability (Devor, M., 2006a). Whereas intact sensory endings may be sensitive to these chemical and physical stimuli, axons in healthy nerve trunks and cell bodies in sensory ganglia are not. The key change in neuropathy is not the presence of excitatory stimuli but rather the ectopic emergence of sensitivity to them.

6.3 Ectopic Firing Contributes to Pain Sensation in Several Ways

6.3.1 Ongoing Pain, Tender-Points, and Other Sensitivities

The presence of ongoing ectopic discharge under circumstances in which ongoing dysesthesias and pain are present, and their simultaneous exacerbation by the same factors (e.g., mechanical stimulation, sympathetic efferent activity), constitutes *a priori* evidence that the ectopia is the cause of the pain. This conclusion is further supported by the observation that (diagnostic) nerve blocks using local anesthetics reliably obtund pain sensation in cases where the block is placed central to the site of nerve injury (Devor, M., 2006a). An interesting test case often raised as a counterexample is phantom limb pain in amputees. Although this pain is frequently eliminated (transiently) by blocking stump neuromas, there are reports of phantom limb pain persisting despite peripheral nerve block, and this observation has formed the basis for the widely held belief that phantom limb pain reflects abnormal neural activity originating in the cerebral cortex (Melzack, R., 1989; Flor, H. *et al.*, 1995). This conclusion is potentially flawed, however, because it ignores ectopic electrical impulse generation (electrogenesis) in sensory cell somata in the DRG. Amputation stump revision surgery is a common procedure, and is usually done under foramenal or spinal block. Although little published documentation is available (Tessler, M. J. and

Kleiman, S. J., 1994), experienced clinicians insist that these blocks reliably eliminate phantom limb pain for the duration of the pharmacological action of the local anesthetic agent. This would not occur if the pain signal originated supraspinally.

6.3.2 Allodynia – Pain Evoked by Stimulation of Residual Uninjured Afferents

Pain in response to light touch of the skin, tactile allodynia, is a common symptom in neuropathy. The simplest explanation is reduced response threshold in nociceptive C-fiber afferents, the classic excitable nociceptor hypothesis. This explanation is probably valid for heat allodynia. However, while neuropathy (and pronociceptive inflammatory mediators) can significantly reduce mechanical thresholds of sensory endings, few fibers that were originally nociceptors come to respond to the very weak tactile stimuli that are typically sufficient to evoke allodynia (Banik, R. K. and Brennan, T. J., 2004; Schlegel, T. *et al.*, 2004; Shim, B. *et al.*, 2005). Moreover, the response to painful touch in allodynic skin is rapid. There is no indication of the long delay between stimulus and response that would be expected from conduction in sensitized C-fibers, although sensitized Aδ nociceptors, if they exist, could contribute.

There is considerable evidence that cutaneous tactile allodynia in neuropathy (and inflammation) is primarily a sensory response to impulse activity in low-threshold mechanosensitive Aβ touch afferents abnormally amplified in the spinal cord (Campbell, J. N. *et al.*, 1988; Torebjork, H. *et al.*, 1992). The altered signal processing responsible for this Aβ pain is believed to result from one or a combination of numerous central nervous system (CNS) changes collectively called central sensitization (Woolf, C. J., 1983; Devor, M., 2006b). Central sensitization, which also amplifies ectopic afferent signals (spontaneous and evoked), is a dynamic process triggered and maintained by ongoing nociceptive input from the periphery. When the maintaining peripheral drive is eliminated, central sensitization fades, along with tactile allodynia. In neuropathy, sustained ectopic discharge can maintain central sensitization indefinitely, and with it Aβ pain and tactile allodynia (Gracely, R. *et al.*, 1992; Koltzenburg, M. *et al.*, 1994). This constitutes a second role of ectopia, in addition to providing a primary pain signal.

In contrast to the skin, it is likely that peripherally sensitized nociceptors play an important role in pain evoked by mechanical stimuli in inflamed viscera and deep somatic tissues such as muscles and joints. Reduced mechanical thresholds, including the recruitment of previously silent C-nociceptors, render the sensory endings of these afferents responsive to the (substantial) forces present during weight bearing, for example, or joint rotation (Schmidt, R., 1996). There is also good evidence for a contribution of central sensitization in deep tissue pain (Banic, B. *et al.*, 2004). Relatively little is known about changes in the response of injured afferent neurons, or the sensory endings of residual noninjured afferents, in the presence of neuropathy affecting nerves that supply deep tissues.

6.4 Cellular Mechanisms of Ectopic Impulse Generation

6.4.1 Resonance

The emergence of ectopic hyperexcitability in neuropathy is not a simple matter of lowered spike threshold. Rather, it reflects a more fundamental, qualitative change in the injured neuron's electrical characteristics. Most intact sensory neurons are incapable of generating sustained impulse discharge in midnerve or from within sensory ganglia, even in the presence of a strong sustained depolarizing stimulus. They are designed to fire exclusively in response to stimuli applied to the sensory ending in skin, muscle, etc., where the neuronal membrane is specialized for electrogenesis. Alterations in three cellular processes, gene expression, protein trafficking, and channel kinetics (below), actually create repetitive firing capability *de novo* at ectopic locations.

Ectopic repetitive firing capability (pacemaker capability) derives from electrical resonance that emerges in the cell membrane of the injured afferents, and resultant subthreshold oscillations in their normally stable resting potential (Amir, R. *et al.*, 1999; Wu, N. *et al.*, 2001). Subthreshold oscillations drive sustained ectopic discharge. This takes the form of irregular singlet spikes in some cells, and as repetitive bursting in others, depending on the ability of the cell to generate DAPs (Figure 1). Due to subthreshold oscillations, ectopic generator sites become responsive to slow onset and tonic depolarizing stimuli, which previously did not evoke sustained discharge. Some begin to fire spontaneously at rest, in the absence of stimulation. The biophysical processes that lead to membrane resonance, repetitive firing capability, ectopic discharge, and neuropathic pain are slowly coming into focus. Briefly, resonance in neuropathy appears to reflect reciprocation between inward Na$^+$

current carried by fast-activating, inactivating, and repriming tetrodotoxin (TTX)-sensitive Na$^+$ channels, and K$^+$ current passing through one or more voltage-insensitive K$^+$ leak channels, perhaps of the KCNK K2p family (Abdulla, F. A. and Smith, P. A., 2002; Amir, R. *et al.*, 2002a; Waxman, S. G., 2002). Voltage-sensitive K$^+$ channels tend to hold resonance in check, and their downregulation after axotomy is probably an important adjunct to ectopia (Kocsis, J. D. and Devor, M., 2000). Other channel types/conductances that might contribute to resonance characteristics of primary sensory neurons include certain Ca^{2+} channels, HCN (the Ih pacemaker channel), KCNQ-type K$^+$ channels, the β4 subunit-associated resurgent Na$^+$ current, and a persistent Na$^+$ conductance (gNaP), but their role, at this stage, is speculative. Many of the first-line analgesic agents used in the treatment of neuropathic pain (McQuay, M. and Moore, A., 1998), including (certain) anticonvulsants, tricyclic antidepressants, systemic local anesthetics, and (certain) antiarrhythmics are thought to act by stabilizing membrane resonance (Catterall, W. A., 1987; Deffois, A. *et al.*, 1996).

6.4.2 The Development of Ectopic Pacemaker Capability

The cascade of cellular events that lead to pacemaker capability and ectopic neuronal discharge begins with the injured primary sensory neuron. Our current understanding of the process is as follows. Axonal transection blocks the normal retrograde flow of neurotrophic signaling molecules (glial cell-derived neurotrophic factor (GDNF), nerve growth factor (NGF), and perhaps others) between the periphery and the sensory cell body. This triggers a change in the quantity of various proteins expressed by the cell body and exported by anterograde axoplasmic transport to both peripheral and central axon endings; some are upregulated and others downregulated (Boucher, T. J. and McMahon, S. B., 2001; Costigan, M. *et al.*, 2002; Xiao, H. S *et al.*, 2002). These changes alter the phenotype of the injured neuron, increasing its intrinsic resonance and supporting the emergence of ectopic electrogenesis.

In addition to changes in gene expression in injured sensory neurons, the delivery (trafficking) of transported molecules is disrupted. The most important change of this sort is the accumulation or depletion of molecules of excitability at ectopic generator sites (Figure 3), notably the accumulation of Na$^+$ channels (Devor, M., *et al.*, 1989; Devor, M.,

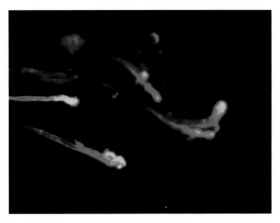

Figure 3 Immunolabeling shows the accumulation of Na$^+$ channels at the chronic cut end of hyperexcitable injured axons in an experimental nerve-end neuroma (for details see Devor, M. *et al.*, 1989).

2006a). Na$^+$ channels, for example, accumulate in neuroma endbulbs and sprouts, and in areas of demyelination. A third process that contributes to ectopic electrogenesis is altered channel kinetics. It is not ion channel proteins themselves that determine cell excitability but the ionic currents they carry. Increasing mean channel open time, for example, can have much the same effect on excitability as increasing the number of channels present. Proinflammatory cytokines and other mediators associated with neuropathy can affect channel kinetics in this way. For example, cAMP-dependent phosphorylation of Na$^+$ channel molecules reduces Na$^+$ current, while dephosphorylation returns it to normal (e.g., Li, M. *et al.*, 1992; Gold, M. S. *et al.*, 1996). Because certain hormones, trophic factors, cytokines, and other inflammatory mediators (notably prostaglandins) can activate protein kinases (PKA, PKC), they are positioned to affect the basic excitability of afferents, and not just to depolarize and excite sensory endings.

References

Abdulla, F. A. and Smith, P. A. 2002. Changes in Na$^+$ channel currents of rat dorsal root ganglion neurons following axotomy and axotomy-induced autotomy. J. Neurophysiol. 88, 2518–2529.

Amir, R., Liu, C. N., Kocsis, J. D., and Devor, M. 2002a. Oscillatory mechanism in primary sensory neurones. Brain 125, 421–435.

Amir, R., Michaelis, M., and Devor, M. 1999. Membrane potential oscillations in dorsal root ganglion neurons: role in normal electrogenesis and in neuropathic pain. J. Neurosci. 19, 8589–8596.

Amir, R., Michaelis, M., and Devor, M. 2002b. Burst discharge in primary sensory neurons: triggered by subthreshold oscillations, maintained by depolarizing afterpotentials. J. Neurosci. 22, 1187–1198.

Banic, B., Petersen-Felix, S., Andersen, O. K., Radanov, B. P., Villiger, P. M., Arendt-Nielsen, L., and Curatolo, M. 2004. Evidence for spinal cord hypersensitivity in chronic pain after whiplash injury and in fibromyalgia. Pain 107, 15–17.

Banik, R. K. and Brennan, T. J. 2004. Spontaneous discharge and increased heat sensitivity of rat C-fiber nociceptors are present *in vitro* after plantar incision. Pain 112, 204–213.

Boucher, T. J. and McMahon, S. B. 2001. Neurotrophic factors and neuropathic pain. Curr. Opin. Pharmacol. 1, 66–72.

Campbell, J. N., Raja, S. N., Meyer, R. A., and MacKinnon, S. E. 1988. Myelinated afferents signal the hyperalgesia associated with nerve injury. Pain 32, 89–94.

Catterall, W. A. 1987. Common modes of drug action on Na$^+$ channels: local anaesthetics, antiarrhythmics and anticonvulsants. Trends Pharmacol. Sci. 8, 57–65.

Costigan, M., Befort, K., Karchewski, L., Griffin, R. S., D'Urso, D., Allchorne, A., Sitarski, J., Mannion, J. W., Pratt, R. E., and Woolf, C. J. 2002. Replicate high-density rat genome oligonucleotide microarrays reveal hundreds of regulated genes in the dorsal root ganglion after peripheral nerve injury. BMC Neurosci. 3, 16–28.

Deffois, A., Fage, D., and Carter, C. 1996. Inhibition of synaptosomal veratridine-induced sodium influx by antidepressants and neuroleptics used in chronic pain. Neurosci. Lett. 220, 117–120.

Devor, M. 2006a. Response of Nerves to Injury in Relation to Neuropathic Pain. In: Wall and Melzack's Textbook of Pain, 5th edn (eds. S. L. McMahon and M. Koltzenburg), Chapter 58, pp. 905–927. Churchill Livingstone.

Devor, M. 2006b. Central Changes after Peripheral Nerve Injury. In: Encyclopedia of Pain (eds. R. F. Schmidt and W. M. Willis), pp. 306–311. Springer.

Devor, M. and Janig, W. 1981. Activation of myelinated afferents ending in a neuroma by stimulation of the sympathetic supply in the rat. Neurosci. Lett. 24, 43–47.

Devor, M., Keller, C. H., Deerinck, T., Levinson, S. R., and Ellisman, M. H. 1989. Na$^+$ channel accumulation on axolemma of afferents in nerve end neuromas in Apteronotus. Neurosci. Lett. 102, 149–154.

Flor, H., Elbert, T., Knecht, S., Wienbruch, C., Pantev, C., Birbaumer, N., Larbig, W., and Taub, E. 1995. Phantom-limb pain as a perceptual correlate of cortical reorganization following arm amputation. Nature 375, 482–484.

Fried, K., Govrin-Lippmann, R., Rosenthal, F., Ellisman, M., and Devor, M. 1991. Ultra-structure of afferent axon endings in a neuroma. J. Neurocytol. 20, 682–701.

Gold, M. S., Reichling, D. B., Shuster, M. J., and Levine, J. D. 1996. Hyperalgesic agents increase a tetrodotoxin-resistant Na+ current in nociceptors. Proc. Nat. Acad. Sci. U. S. A. 93, 1108–1112.

Gracely, R., Lynch, S., and Bennett, G. 1992. Painful neuropathy: altered central processing, maintained dynamically by peripheral input. Pain 51, 175–194.

Kocsis, J. D. and Devor, M. 2000. Altered Excitability of Large Diameter Cutaneous Afferents Following Nerve Injury: Consequences For Chronic Pain. In: Proceedings of the 9th World Congress on Pain. Progress in Pain Research and Management (eds. M. Devor, M. C. Rowbotham, and Z. Wiesenfeld-Hallin). Vol. 16, pp. 119–135. IASP Press.

Koltzenburg, M., Torebjork, H., and Wahren, L. 1994. Nociceptor modulated central sensitization causes mechanical hyperalgesia in acute chemogenic and chronic neuropathic pain. Brain 117, 579–591.

Li, M., West, J. W., Lai, Y., Scheuer, Y., and Catterall, W. A. 1992. Functional modulation of brain sodium channels by cAMP-dependent phosphorylation. Neuron 8, 1151–1159.

McQuay, M. and Moore, A. 1998. An Evidence-Based Resource for Pain Relief, p. 264. Oxford University Press.

Melzack, R. 1989. Phantom limbs, the self and the brain. Can. Psychol. 30, 1–16.

Nordin, M., Nystrom, B., Wallin, U., and Hagbarth, K.-E. 1984. Ectopic sensory discharges and paresthesiae in patients with disorders of peripheral nerves, dorsal roots and dorsal columns. Pain 20, 231–245.

Schlegel, T., Sauer, S. K., Handwerker, H. O., and Reeh, P. W. 2004. Responsiveness of C-fiber nociceptors to punctate force-controlled stimuli in isolated rat skin: lack of modulation by inflammatory mediators and flurbiprofen. Neurosci. Lett. 361, 163–167.

Schmidt, R. 1996. The articular polymodal nociceptor in health and disease. Prog. Brain Res. 113, 53–81.

Shim, B., Kim, D. W., Kim, B. H., Nam, T. S., Leem, J. W., and Chung, J. M. 2005. Mechanical and heat sensitization of cutaneous nociceptors in rats with experimental peripheral neuropathy. Neuroscience 132, 193–201.

Tessler, M. J. and Kleiman, S. J. 1994. Spinal anaesthesia for patients with previous lower limb amputations. Anaesthesia 49, 439–441.

Torebjork, H., Lundberg, L., and LaMotte, R. 1992. Central changes in processing of mechanoreceptive input in capsaicin-induced secondary hyperalgesia in humans. J. Physiol. (Lond.) 448, 765–780.

Waxman, S. G. (ed.) 2002. Sodium channels and neuronal hyperexcitability. Novartis Found. Symp. 241, 232.

Woolf, C. J. 1983. Evidence for a central component of postinjury pain hypersensitivity. Nature 306, 686–688.

Wu, G., Ringkamp, M., Hartke, T. V., Murinson, B. B., Campbell, J. N., Griffin, J. W., and Meyer, R. A. 2001. Early onset of spontaneous activity in uninjured C-fiber nociceptors after injury to neighboring nerve fibers. J. Neurosci. 21, RC140.

Wu, N., Hsiao, C.-F., and Chandler, S. 2001. Membrane resonance and subthreshold membrane oscillations in mesencephalic V neurons: participants in burst generation. J. Neurosci. 21, 3729–3739.

Xiao, H. S., Huang, Q. H., Zhang, F. X., Bao, L., Lu, Y. J., Guo, C., Yang, L., Huang, W. J., Fu, G., Xu, S. H., Cheng, X. P., Yan, Q., Zhu, Z. D., Zhang, X., Chen, Z., Han, Z. G., and Zhang, X. 2002. Identification of gene expression profile of dorsal root ganglion in the rat peripheral axotomy model of neuropathic pain. Proc. Natl. Acad. Sci. U. S. A. 99, 8360–8365.

7 Sodium Channels

John N Wood, University College London, London, UK

7.1 How Do Sodium Channels Work?

All sodium channels are single proteins made up of α-subunits that contain four homologous domains and form a voltage-gated sodium-selective aqueous pore (Figure 1). They comprise a family of nine structurally related α-subunits (Table 1). They are more than 75% identical over the amino acid sequences comprising the transmembrane and extracellular domains (Yu, F. H. and Catterall, W. A., 2003). The α-subunits show distinct patterns of expression, and are associated with accessory β-subunits that modify channel properties and interact with cytoskeletal and extracellular matrix proteins. The mechanism of channel gating involves the movement of voltage sensors (shown with positive charges in Figure 1) in response to changes in membrane potential. Lysine or arginine residues are found at intervals of three amino acids to produce a linear array of positive charges in the α-helical S4 transmembrane segments, and the movement of these charged areas of the channel opens the sodium-selective pore, defined by residues within the channel atrium. An intracellular loop lying between transmembrane domains 3 and 4 containing the tripeptide isoleucine phenylalanine methionine (IFM) is subsequently responsible for channel inactivation on a millisecond timescale. Sodium channel subtypes can be pharmacologically distinguished by their susceptibility to block by tetrodotoxin (TTX), a toxin isolated from puffer fish, which binds to the ion selectivity pore of some channels (Hille, B., 2002). Most sodium channels are blocked by nanomolar concentrations of TTX and are defined as TTX-sensitive (TTXs) whilst others

are resistant to micromolar TTX and are defined as TTX-resistant (TTXr).

A short sequence found upstream of neuronal sodium channel genes (as well as other neuronal genes) was identified and named neuron-restricted silencing element (NRSE) or repressor element 1(RE-1). Transcription factors that bound to the motif were found to act as inhibitors of gene expression in nonneuronal cells. These proteins were named RE-1 silencing transcription factor (REST) or neuron-restrictive silencer factor (NRSF) and mutational analysis identified a single zinc finger motif in the carboxyl-terminal domain of the factor as essential for repressing sodium channel-derived reporter genes (Kraner, S. D. *et al.*, 1992). The inhibitory activity of the complex can be further modulated by double-stranded RNA molecules that have the same sequence as NRSE/RE-1, and are found in developing neuronal precursors. These regulatory RNA molecules are able to switch the repressor function of the complex to an ac activator role (Kuwabara, T. *et al.*, 2004). In this way, the assumption of a neuronal phenotype seems to depend in part upon regulatory RNAs driving gene expression downstream of NRSE/RE1 motifs. Sodium channels are known to be expressed at the very earliest stages of the appearance of a neuronal phenotype in the mouse. These studies highlight the significance of sodium channel expression in neuronal function throughout development.

The α-subunits of the voltage-gated sodium channel that form the sodium pore are associated with accessory β-subunits that modulate channel properties and interact with cytoskeletal and extracellular

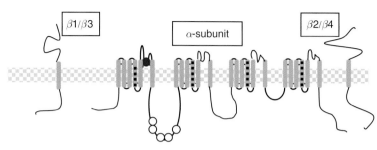

Figure 1 Structure of sodium channels. Voltage-gated sodium channels comprise a single functional α-subunit, which is tethered by associated accessory subunits to subcellular locations determined by both extracellular and intracellular proteins. Phosphorylation sites are shown on intracellular loop 2, and the serine residues in domain 1 that confers tetrodotoxin sensitivity is highlighted.

Table 1 Names and properties of voltage-gated sodium channel α-subunits

Type	Gene symbol	Name	Tissue	TTX-sensitive
$Na_V1.1$	SCN1A	Type I	CNS/PNS	+
$Na_V1.2$	SCN2A	Type II	CNS	+
$Na_V1.3$	SCN3A	Type III	CNS	+
$Na_V1.4$	SCN4A	SKM1	Skeletal muscle	+
$Na_V1.5$	SCN5A	SKMII	Heart/CNS	−
$Na_V1.6$	SCN8A	Nach6	CNS/glia	+
$Na_V1.7$	SCN9A	PN1	DRG	+
$Na_V1.8$	SCN10	SNS	DRG	−
$Na_V1.9$	SCN11	NAN	DRG	−

The nomenclature for sodium channel α-subunits, combined with their main site of expression (CNS – central nervous system, PNS – peripheral nervous system) and their sensitivity to block by tetrodotoxin (TTX), is shown.

matrix proteins. β-subunits are single transmembrane domain proteins that contain an extracellular domain that is homologous to the V-set of the immunoglobulin superfamily that includes cell adhesion molecules (Isom, L. L., 2002). The biochemical purification of two proteins, β1 and β2, that associate with α-subunits allowed the properties of these accessory factors to be investigated in heterologous expression systems. Molecular cloning has identified a β1-like member of the family named β3 and β2-like subunit named β4, as well as a splice variant of β1, β1A.

β1 and β1A are structurally homologous in the extracellular immunoglobulin-like loop region. β1A mRNA is expressed during embryonic development, and becomes undetectable after birth, concomitant with the onset of β1 expression. Immunocytochemical analysis of β1A expression revealed selective expression in brain and spinal cord neurons, with high expression in heart and all dorsal root ganglia neurons. Coexpression of $Na_V1.2A$ and β1A-subunits results in a 2.5-fold increase in sodium current density as a result of an increase in the level of functional sodium channels in expressing cells. β1A-expressing cell lines also revealed subtle differences in sodium

channel activation and inactivation. β3 most closely related to β1 is the product of a separate gene localized to human chromosome 11q23.3. The expression pattern of β3 differs significantly from β1.

Coexpression of the β-subunits with sodium channel α-subunits in *Xenopus* oocytes or mammalian cell lines accelerates channel activation, and shifts the voltage dependence of inactivation to more negative potentials, indicating that β-subunits are important in the assembly, expression, and functional modulation of the rat brain sodium channel heterotrimeric complex (Isom, L. L., 2002). However, β-subunits do not appear to modulate the biophysical properties of all sodium channel α-subunits. $Na_V1.8$ α-subunits show aberrant properties when expressed in heterologous cells, but coexpression with β1, β1A, or β3 does not recapitulate normal channel behavior. The existence of factors other than accessory β-subunits that can alter inactivation kinetics has also been suggested by studies in which sodium channel α-subunits are expressed in cells lacking endogenous sodium channel β-subunits. Expression of rat cardiac $Na_V1.5$ and rat brain $Na_V1.2$ sodium channels in mammalian cell lines results in sodium channels with rapid activation and inactivation characteristic of native neuronal sodium channel.

An additional function of β-subunits as cell adhesion molecules has been demonstrated by the ability of β-subunits to mediate homophilic interactions via their extracellular immunoglobulin-like repeats (Isom, L. L., 2002). Interacting β-subunits also recruit the cytoskeletal protein ankyrin-G to the cell surface via an interaction with their cytoplasmic domains, suggesting that these subunits link sodium channel α-subunits indirectly both to the cytoskeleton and to the extracellular matrix proteins, such as tenascin-R. Tenascin-R is an extracellular matrix molecule that is secreted by oligodendrocytes during myelination and that binds F3-contactin. Transfected cells stably expressing $\beta1$- or $\beta2$-subunits initially recognize and then are repelled from substrates containing tenascin-R. The cysteine-rich amino-terminal domain of tenascin-R appears to be responsible for the repellent effect on β-subunit-expressing cells. Application of the cysteine-rich N-terminal domain of tenascin-R to channels expressed in *Xenopus* oocytes potentiates expressed sodium currents. The binding of neuronal sodium channels to extracellular matrix molecules such as tenascin-R may thus play a role both in functional regulation and in localizing sodium channels in high density at nodes of Ranvier.

β-subunits also interact with the ankyrin-G membrane skeletal protein that is associated with spectrin–actin networks and also binds to integral membrane proteins including the L1 CAM family of cell adhesion molecules. Ankyrin-G is highly concentrated along with the L1CAM family members neurofascin and NrCAM at nodes of Ranvier and axon initial segments. Voltage-gated sodium channels are thus likely to associate in a complex containing neurofascin/NrCAM and ankyrin-G in myelinated neurons.

As well as the accessory subunits that link sodium channels to cytoskeletal and extracellular elements, other proteins are associated with sodium channels. Voltage-gated sodium channels in brain neurons were found to associate with receptor protein tyrosine phosphatase-β (RPTP-β) (Catterall 200. Both the extracellular domain and the intracellular catalytic domain of RPTP-β interact with sodium channels. Sodium channels were found to be tyrosine phosphorylated and dephosphorylation slowed $Na_v1.1$ inactivation, positively shifting voltage dependence and increasing whole-cell sodium current.

Regulation of sodium channels by protein kinases (PK) has already been demonstrated in a number of systems. Protein kinase A (PKA) can decrease or increase sodium currents by phosphorylation at consensus sites in the cytoplasmic domain I–II linker. Effects on membrane trafficking may also be mediated through phosphorylation. In the case of the cardiac channel, activation of PKA causes a slow increase in sodium current in *Xenopus* oocytes. Chloroquine and monensin, both compounds that disrupt plasma membrane recycling, reduce sodium currents. Preincubation with these agents also prevented the PKA-mediated rise in sodium current, indicating that this effect likely resulted from an increased number of channels in the plasma membrane.

There is strong evidence for specialized roles of the various sodium channel isoforms based on human genetic studies as well as the production of knockout mice. A perinatal lethal phenotype occurs in the $Na_v1.2$ sodium channel mouse knockout, which shows severe hypoxia as a result of brain stem apoptosis. $Na_v1.8$ channel knockouts have selective deficiencies in pain pathways (Akopian, A. N. *et al.*, 1999). Nax, a non-voltage-gated sodium channel homologue has been shown to play an important role in salt homeostasis, whilst loss of $Na_v1.6$ in the natural *med* mutant has been shown to lead to ataxia, dystonia, and paralysis. The deletion of accessory β-subunits, already known to play an important role in regulating functional channel expression and tethering α-subunits to specific subcellular locations, leads to a complex phenotype resulting in seizures and epileptogenic activity.

Naturally occurring mutations in both sodium channel α- and β-subunits have been implicated in various inherited disorders, many of which have been characterized in muscle sodium channel isoforms. Brugada syndrome, characterized by ventricular fibrillation, heart failure, and sudden death, is associated with mutations in the cardiac sodium channel $Na_v1.5$. Hyperkalemic periodic paralysis is associated with mutations in the skeletal muscle channel $Na_v1.4$. However, there is also evidence of a role for neuronal sodium channels and β-subunits in a variety of pathologies, including epilepsy, autism, and pain perception. Generalized epilepsy with febrile seizures (GEFS) is associated with mutations in either the $\beta1$-subunit (type I) or the α-subunit $Na_v1.1$ (type II). Some forms of autism associate with mutations in SCN1-3A sodium channel genes. The peripheral neuron-associated sodium channel $Na_v1.7$ encoded by gene SCN9A has been shown to be the site of mutations that lead to the chronic familial pain disorder primary erythermalgia (Yang, Y. *et al.*, 2004). This dominant condition is characterized by burning in the extremities, pain evoked by standing and a generalized inflammatory phenotype. Interestingly, systemic treatment with the sodium channel blockers lidocaine or mexiletine leads to dramatic improvements in the symptoms associated with this disease.

Neuronal excitability depends to a considerable extent on sodium channel trafficking, distribution and density, as well as the intrinsic properties of the channels themselves in terms of thresholds of activation and repriming characteristics. The repertoire of sodium channels expressed by sensory neurons and the alterations that occur in pain states are now well documented (Waxman, *et al.*, 2002, Wood, J. N. *et al.*, 2004). Unmyelinated C-fiber-associated dorsal root ganglion (DRG) neurons are usually nociceptive whilst thinly myelinated Aδ-fibers also mediate fast pain, and a subset of myelinated Aβ-fibers are also nociceptive. TTXr sodium channels are present on C-fibers and seem to play a significant role in nociception (Akopian, A. N. *et al.*, 1999).

7.2 Sodium Channels and Pain Pathways

Acute, inflammatory, and neuropathic pain can all be attenuated or abolished by local treatment with sodium channel blockers such as lidocaine. The peripheral input that drives pain perception thus depends on the presence of functional voltage-gated sodium channels. Remarkably, two voltage-gated sodium channel genes (Na$_v$1.8 and Na$_v$1.9) are expressed selectively in damage-sensing peripheral neurons, whilst a third channel (Na$_v$1.7) is found predominantly in sensory and sympathetic neurons. An embryonic channel (Na$_v$1.3) is also upregulated in damaged peripheral nerves and associated with increased electrical excitability in neuropathic pain states. A combination of antisense and knockout studies support a specialized role for these sodium channels in pain pathways.

Small-diameter sensory neurons express a variety of sodium channel transcripts as well as a repertoire of electrophysiological and pharmacologically distinct type of sodium currents. Persistent sodium channels that are resistant to TTX are probably encoded for by Na$_v$1.9 channels, whilst the major transient TTXr channel isotype present predominantly in nociceptors is Na$_v$1.8. TTXs currents are encoded by a number of genes including Na$_v$1.1, Na$_v$1.7, and Na$_v$1.6. The current–voltage relationship of these different sorts of sodium channels found in DRG sensory neurons is shown in Figure 2. The persistent current activates at voltages close to the resting potential and is likely to play a role in setting thresholds of activation.

Evidence that altered sodium channel activity in peripheral neurons is associated with the development of inflammatory and neuropathic pain is strong. Altered patterns of sodium channel transcripts as well as changes in posttranslational modifications have been observed. Neuropathic pain that results from direct damage to peripheral nerves is the most problematic condition in terms of analgesic therapy. The pain evoked by these conditions seems to result initially from enhanced neuronal excitability that can be blocked by low-dose TTX.

7.3 Na$_v$1.3 and Neuropathic Pain

Na$_v$1.3 is widely expressed in the adult human central nervous system but is normally present at low levels in adult peripheral nervous system. Axotomy or other forms of nerve damage lead to the reexpression of Na$_v$1.3 and the associated β3-subunit in sensory neurons, but not in primary motor neurons (Waxman, S. G. *et al.*, 1994). This event can be reversed *in vitro* and *in vivo* by treatment with high levels of exogenous glial cell-derived neurotrophic factor (GDNF). Na$_v$1.3 is known to recover (reprime) rapidly from inactivation (Cummins *et al.*, 2001). Axotomy has been shown to induce the expression of rapidly repriming TTXs sodium channels in damaged neurons, and this event can also be reversed by the combined actions of GDNF and nerve growth factor (NGF). Concomitant with the reversal of Na$_v$1.3 expression by GDNF, ectopic action potential generation is diminished and thermal and mechanical pain-related behavior in a rat chronic constriction injury model is reversed. Moreover, Na$_v$1.3 is upregulated in dorsal horn neurons following experimental spinal cord injury and this upregulation is associated with hyperexcitability of these nociceptive neurons and pain; antisense knockdown of Na$_v$1.3 (and probably other sodium channels) attenuates the dorsal horn neuron hyperexcitability and the pain behaviors in spinal cord-injured animals (Hains, B. C. *et al.*, 2003). It therefore seems likely that Na$_v$1.3 reexpression may play a role in increasing neuronal excitability that contributes to neuropathic pain after nerve and spinal cord injury.

TTXs and TTXr Na⁺ currents in small-diameter DRG neurons

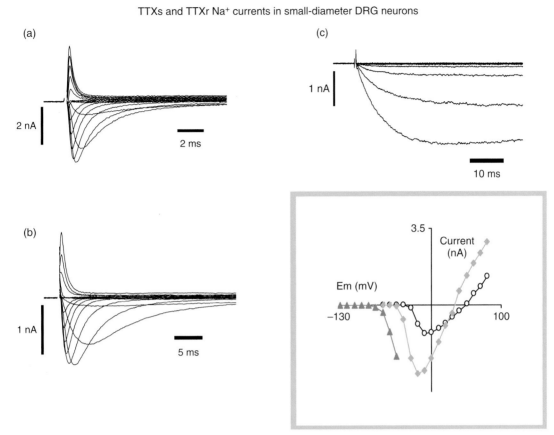

Figure 2 (a) TTX-sensitive (b) -resistant, and (c) persistent currents are shown. Note the different timescales for inactivation of the currents. The voltage at which the different types of channels activate is shown in panel 4 (red channel (c), blue channel (a), and black channel (b)). Although those channels that activate at the most negative potentials might seem to render other channels redundant, the ability to recover from inactivation is also an important determinant of neuronal excitability.

7.4 Na$_v$1.7 and Inflammation

Na$_v$1.7 encodes a TTXs channel found predominantly in peripheral sensory and sympathetic neurons. Interestingly, this channel is located at the terminal of sensory neurons, and is regulated in its expression by inflammatory mediators such as NGF, probably through mechanisms that involve the control of trafficking. The fact that a chronic inflammatory dominant human disease, primary erythermalgia, maps to this gene SCN10A suggests that this channel may have an important role in inflammatory pain (Yang, Y. *et al.*, 2004). Tissue-specific deletion in mouse nociceptors shows that Na$_v$1.7 plays a major role in inflammatory pain (Nassar, M. A. *et al.*, 2004). Low-dose lidocaine both inhibits the pain associated with primary erythermalgia and unmasks the phenotype of the Na$_v$1.8 null, further supporting a role for Na$_v$1.7 in inflammatory pain. Na$_v$1.7 mRNA contributes to an increase in TTXs sodium current in caraggeenan-inflamed rats, further implying a role in altering inflammatory pain thresholds. Recently humans with a loss of functional Na$_v$1.7 have been found to be completely pain free (Cox, J. J. *et al.*, 2006). Not surprisingly, this has stimulated major interest in developing an Na$_v$1.7-specific channel blocker.

7.5 Na$_v$1.8 and Nociception

Na$_v$1.8 is a TTXr sensory neuron-specific channel mainly expressed in nociceptive neurons. This channel contributes much of the sodium current underlying the depolarizing phase of the action potential in cells in which it is present and knockout mice lose all the inactivating TTXr sodium channel activity in DRG neurons (Akopian, A. N. *et al.*, 1999).

Na$_V$1.8 thus underlies the inactivating TTXr sodium currents that have been found to play a critical role in many aspects of nociceptor function. Functional expression of TTXr currents encoded by this channel is regulated by inflammatory mediators, including NGF. Both antisense and knockout studies support a role for the channel in contributing to inflammatory pain, as PGE$_2$-induced hyperalgesia is inhibited by antisense oligonucleotides. Antisense studies have also suggested a role for this protein in the development of neuropathic pain (Lai, J. *et al.*, 2002). A late-onset deficit in ectopic action propagation has been described in the Na$_V$1.8-null mutant mouse together with deficits in mechanohypersensitivity. A redistribution of immunoreactive Na$_V$1.8-like material has also been described in animal models of neuropathic pain, further supporting the lobby for Na$_V$1.8 as a key molecule in the pathogenesis of neuropathic pain. However, neuropathic pain behavior at early time points seems to be normal in the Na$_V$1.8-null mutant mouse.

Inflammatory mediators are known to alter the level of TTXr channel expression, and shift their threshold of activation to more negative potentials through the activation of PKA. Prostaglandin E$_2$, adenosine, and serotonin all increased the magnitude of Na$_V$1.8-mediated currents, shifted its conductance–voltage relation in a hyperpolarizing direction, and increased the rates of activation and inactivation of sodium channels in small-diameter sensory neurons in culture. These phenomena can be replicated *in vitro* with heterologously expressed Na$_V$1.8 channels, implicating this TTXr channel in setting peripheral pain thresholds. In support of a role for the Na$_V$1.8 channel in pain pathways, an Na$_V$1.8-null mutant shows deficits in inflammatory pain processing, in contrast to its responses to neuropathic damage (Akopian, A. N. *et al.*, 1999).

This channel also appears to have an important role in visceral pain. TTXr sodium currents were present in every colonic DRG neuron studied (Gold, *et al.*, 2002). Modulation of TTXr I(Na) in colonic afferents may be an underlying mechanism of hyperalgesia and pain associated with inflammation of the colon. Consistent with this, the Na$_V$1.8-null mutant has deficits in visceral pain and referred hyperalgesia.

Several strands of evidence also support a role for Na$_V$1.9 as an important contributor to the control of excitability in nociceptive afferents. The gene is expressed predominantly in neurons functionally defined as nociceptors. However, the phenotype of a mouse Na$_V$1.9 knockout is essentially normal.

7.6 The Diversity of Sodium Channel Subtypes

Given the fairly similar properties of voltage-gated sodium channels why are there so many different types of α-subunits present in nociceptive sensory neurons? A number of explanations are possible:

Firstly, the trafficking in sensory neurons of different sodium channels to distinct cellular locations (terminals, nodes of Ranvier, etc.) and the regulation of this process may provide a number of options to control neuronal excitability in different physiological contexts. Thus trafficking of Na$_V$1.8 into the cell membrane through its interaction with p11, which is massively upregulated by NGF, could play an important role in inflammatory pain – other channels do not show this behavior. It is important to remember that there are no external cues such as nodes of Ranvier to define the topological organization of sodium channels in C-fibers – the relatively even distribution of sodium channels to promote electrical propagation along the length of a very long axon is a process that is not understood, and altered channel density particularly at C-fiber terminals can have important effects on excitability. The two sodium channels that are uniquely associated with C-fibers may be involved in these processes.

Secondly, different structural features of the channels means that their response to posttranslational modification by enzymes such as PK is quite distinct. Phosphorylation on intracellular serines in Na$_V$1.8 increases peak current density, whilst in other channels present in DRG neurons, for example Na$_V$1.7, peak current diminishes. Primary sequence also determines the repriming characteristics of the different channels as well as their threshold of activation, which are crucial determinants of peripheral pain thresholds.

Thirdly, the topological relationship between different sodium channels and channels involved in sensing noxious stimuli may also play a role in setting thresholds of activation and thus pain thresholds. As noxious mechanosensation but not heat sensing is deficient in Na$_V$1.8-null mutant mice, we may argue that this channel is closely apposed to potential mechanosensors through a sequence-specific interaction with a macromolecular complex, and local depolarizations cause this channel to activate first to generate action potentials. Heat sensors may be linked with other sodium channel subtypes. If true, this would suggest that molecules that disrupt specialized nociceptive complexes at the terminals of sensory neurons may have a useful analgesic effect. There is already evidence that disruption of the cytoskeleton has antihyperalgesic effects.

7.6.1 Selective Sodium Channel Blockers as Drugs

Tetrodotoxin has proved invaluable in discriminating between two classes of sodium channels. The $Na_v1.5$ channel, as well as the peripheral neuron-specific channel $Na_v1.8$, is a subfamily of TTXr sodium channels, including the evolutionarily earliest sodium channel $Na_v1.9$. However, little progress has been made in developing specific subtype selective channel blockers. Conotoxins provide a possible route to this goal. The venom from marine predatory cone shell snails (*Conus*) contains a complex cocktail of neurotoxic disulfide-rich peptides, termed conotoxins. Conotoxins produce their toxic effects via the targeted modulation of specific ion channels, and there is some evidence that they target specific sodium channels.

Sodium channel subtypes are thus differentially regulated in different pain states. New subtype selective blockers should therefore prove useful as both research tools and potential analgesic drugs. The presence of DRG-specific sodium channel isoforms presents an exciting opportunity to develop new classes of analgesic drugs that may have quite novel properties from present-day analgesics with utility in some conditions of intractable or chronic pain.

Acknowledgments

We thank the Wellcome Trust, the BBSRC, and the MRC for support.

References

Akopian, A. N., Souslova, V., England, S., Okuse, K., Ogata, N., Ure, J., Smith, A., Kerr, B. J., McMahon, S. B., Boyce, S., Hill, R., Stanfa, L. C., Dickenson, A. H., and Wood, J. N. 1999. The TTX-R sodium channel SNS has a specialized function in pain pathways. Nat. Neurosci. 2, 541–5489.

Cox, J. J., Reimann, F., Nicholas, A. K., Thornton, G., Roberts, E., Springell, K., Karbani, G., Jafri, H., Mannan, J., Raashid, Y., Al-Gazali, L., Hamamy, H., Valente, E. M., Gorman, S., Williams, R., McHale, D. P., Wood, J. N.,

Gribble, F. M., and Woods, C. G. 2006. An SCN9A channelopathy causes congenital inability to experience pain. Nature 444(7121), 894–898.

Cummins, T. R., Aglieco, F., Renganathan, M., Herzog, R. I., Dib-Hajj, S. D., and Waxman, S. G. 2001. Nav1.3 sodium channels: rapid repriming and slow closed-state inactivation display quantitative differences after expression in a mammalian cell line and in spinal sensory neurons. J. Neurosci. 21(16), 5952–5961.

Gold, M. S., Zhang, L., Wrigley, D. L., and Traub, R. J. 2002. Prostaglandin E(2) modulates TTX-RI(Na) in rat colonic sensory neurons. J. Neurophysiol. 88(3), 1512–1522.

Hains, B. C., Klein, J. P., Saab, C. Y., Craner, M. J., Black, J. A., and Waxman, S. G. 2003. Upregulation of sodium channel Nav1.3 and functional involvement in neuronal hyperexcitability associated with central neuropathic pain after spinal cord injury. J. Neurosci. 23(26), 8881–8892.

Hille, B. 2002. Ion Channels of Excitable Membranes. Sinauer.

Isom, L. L. 2002. Beta subunits: players in neuronal hyperexcitability? Novartis Found. Symp. 241, 124–138; discussion 138–143, 226–232.

Kraner, S. D., Chong, J. A., Tsay, H. J., and Mandel, G. 1992. Silencing the type II sodium channel gene: a model for neural-specific gene regulation. Neuron 9(1), 37–44.

Kuwabara, T., Hsieh, J., Nakashima, K., Taira, K., and Gage, F. H. 2004. A small modulatory dsRNA specifies the fate of adult neural stem cells. Cell 116(6), 779–793.

Lai, J., Lai, J., Gold, M. S., Kim, C. S., Bian, D., Ossipov, M. H., Hunter, J. C., and Porreca, F. 2002. Inhibition of neuropathic pain by decreased expression of the tetrodotoxin-resistant sodium channel, Nav1.8. Pain 95(1–2), 143–152.

Nassar, M. A., Stirling, L. C., Forlani, G., Baker, M. D., Matthews, E. A., Dickenson, A. H., and Wood, J. N. 2004. Nociceptor-specific gene deletion reveals a major role for Nav1.7 (PN1) in acute and inflammatory pain. Proc. Natl. Acad. Sci. U. S. A. 101(34), 12706–12711.

Ratcliffe, C. F., Qu, Y., McCormick, K. A., Tibbs, V. C., Dixon, J. E., Scheuer, T., and Catterall, W. A. 2000. A sodium channel signaling complex: Modulation by associated receptor protein tyrosine phosphatase beta. Nat. Neurosci. 3(5), 437–444.

Waxman, S. G., Cummins, T. R., Dib-Hajj, S. D., and Black, J. A. 2000. Voltage-gated sodium channels and the molecular pathogenesis of pain: A review. J. Rehabil. Res. Dev. 37(5), 517–528.

Waxman, S. G., Kocsis, J. D., and Black, J. A. 1994. Type III sodium channel mRNA is expressed in embryonic but not adult spinal sensory neurons, and is re-expressed following axotomy. J. Neurophysiol. 72(1), 466–470.

Wood, J. N., Boorman, J. P., Okuse, K., and Baker, M. D. 2004. Voltage-gated sodium channels and pain pathways. J. Neurobiol. 61(1), 55–71.

Yang, Y., Wang, Y., Li, S., Xu, Z., Li, H., Ma, L., Fan, J., Bu, D., Liu, B., Fan, Z., Wu, G., Jin, J., Ding, B., Zhu, X., and Shen, Y. 2004. Mutations in SCN9A, encoding a sodium channel alpha subunit, in patients with primary erythermalgia. J. Med. Genet. 41(3), 171–174.

Yu, F. H. and Catterall, W. A. 2003. Overview of the voltage-gated sodium channel family. Genome Biol. 4(3), 207.

Further Reading

Fauchier, L., Babuty, D., and Cosnay, P. 2000. Epilepsy, Brugada syndrome and the risk of sudden unexpected death. J. Neurol. 247(8), 643–644.

8 Physiology of Nociceptors

M Ringkamp and R A Meyer, Johns Hopkins University, Baltimore, MD, USA

Glossary

AMH A-fiber mechanoheat-sensitive nociceptor responsive to mechanical and heat stimuli.

central sensitization An enhanced neural response that occurs in the central nervous system (CNS).

CMH C-fiber mechanoheat-sensitive nociceptor responsive to mechanical and heat stimuli.

fatigue Decrement in response to repeated stimuli.

hyperalgesia Enhanced stimulus-evoked pain.

MIA Mechanically insensitive afferent.

MSA Mechanically sensitive afferent.

nociceptor Neuron responsive to noxious or injurious stimuli.

primary hyperalgesia Hyperalgesia that develops at the site of injury. Mechanical and thermal hyperalgesia are present at the zone of primary hyperalgesia following a cutaneous injury.

receptive field Area over which neuron is responsive to mechanical, thermal, chemical, or electrical stimulation.

secondary hyperalgesia Hyperalgesia that develops in an uninjured tissue surrounding a site of injury. Only mechanical hyperalgesia is present at the zone of secondary hyperalgesia.

sensitization Enhanced neural response.

8.1 Introduction

A vital function of the sensory nervous system is to signal the threat or occurrence of tissue injury. A specialized class of afferents performs this function. These afferents respond preferentially to noxious or injurious stimuli and are called nociceptors. Nociceptors are found in virtually every organ (except the brain), and activity in nociceptive afferents is thought to lead to the percept of pain. In this chapter, we will focus on nociceptors innervating skin. Nociceptors from other somatic tissues have similar properties.

The skin contains a number of distinct sensory receptors that respond selectively to different types of innocuous stimuli and encode different aspects of the sensory experience. For example, several types of low-threshold mechanoreceptors signal different aspects of touch sensation. Receptors that are exquisitely sensitive to gentle warming or gentle cooling are responsible for the percepts of warmth and cooling, respectively.

Nociceptors are distinguished from these innocuous receptors by their relatively high threshold for activation. Unlike innocuous receptors, nociceptors often respond to more than one energy form – that is, they are polymodal. Nociceptors are often subclassified based on the conduction velocity of their parent axon. C-fiber nociceptors are associated with unmyelinated fibers and have conduction velocities less than $2\,\mathrm{m\,s^{-1}}$. A-fiber nociceptors are associated with myelinated fibers and have conduction velocities greater than $2\,\mathrm{m\,s^{-1}}$.

We will first consider the properties of nociceptors from uninjured skin and how their response is related to sensations of acute pain. We will then consider the effects of skin injury on the response of nociceptors and how this is related to enhanced pain sensitivity called hyperalgesia. Where possible, electrophysiological responses to noxious stimuli will be compared to psychophysical measures of pain.

8.2 Nociceptors in Uninjured Skin

In experimental studies of nociceptors, the first task is to identify where on the skin the afferent is responsive. This area is called the receptive field. The receptive field is often located based on its response to mechanical stimuli (e.g., gentle pinching), and then the border is mapped with flexible small-diameter cylindrical probes that exert a relatively constant force when they bend (called von Frey probes). Quantitative

stimuli of different energy modalities are then applied to this mechanical receptive field. In early studies, the response of nociceptors to heat and mechanical stimuli was tested leading to the nomenclature of C-fiber (CMH) or A-fiber mechanoheat nociceptors (AMH) corresponding to C-fiber or A-fiber nociceptors responsive to heat and mechanical stimuli. Most of these nociceptors also respond to chemical and/or cold stimuli and therefore are polymodal.

Nociceptors that respond to mechanical stimuli are also called mechanically sensitive nociceptive afferents (MSAs). Recently, it has become apparent that afferents exist that either are insensitive to mechanical stimuli or have extremely high mechanical thresholds; these afferents are called mechanically insensitive afferents (MIAs) or silent nociceptors. These afferents were missed in earlier studies of nociceptors that relied on mechanical techniques to search for the receptive field. MIAs have been identified in many different tissues including skin, cornea, knee joint, and viscera. In the skin, electrocutaneous stimulation can be used to search for the receptive field of MIAs (see Box 1). As will become apparent below, cutaneous MIAs differ from MSAs not only in their response to mechanical stimuli but also in their response to other stimulus modalities.

8.2.1 C-Fiber Nociceptors

The response of a typical CMH to a stepped increase in temperature in the receptive field is shown in Figure 1. For stimuli applied to the skin, there are five processes that occur to achieve this heat response. First, the heat stimulus needs to be absorbed by the skin. Second, the energy must be transmitted to the receptor terminal that sits below the skin surface. Third, the energy must be converted (or transduced) into a change in membrane potential at the terminal of the nociceptor. Molecular mechanisms of heat transduction are described in Chapter Molecular Biology of the Nociceptor/Transduction. Fourth, the increase in membrane potential must be sufficient to generate an action potential. Finally, the action potential must propagate orthodromically to the central nervous system (CNS) (and to the recording electrode).

Thermal modeling studies combined with electrophysiological recordings indicate that the heat threshold of CMHs in monkeys depends on the temperature at the depth of the receptor terminal and not on the rate of temperature change (Tillman, D. B. et al., 1995). Based on this analysis, the majority of CMHs have heat thresholds between 39 and 41 °C. However, the depth of the heat-responsive terminal

---

Box 1   Mapping mechanically insensitive afferents

Mechanically insensitive afferents (MIAs) either are completely unresponsive to mechanical stimuli or have such a high threshold for activation that they cannot be activated by light pinch stimuli applied to the skin. Mechanical stimuli are inadequate to locate the cutaneous receptive field of these afferents. Furthermore, repetitive noxious mechanical and heat stimuli could sensitize the nociceptive endings over the course of an electrophysiological experiment. Also, the use of mechanical and thermal stimuli bias the study toward those afferents that are responsive to these stimuli. Because of these reasons, a different search strategy has been developed that allows the receptive fields of afferents, including MIAs, to be localized in an unbiased manner. Instead of using natural stimuli, an electrical stimulus is applied to the skin (see adjacent figure).

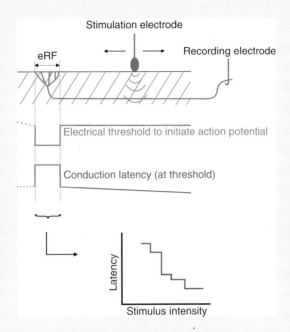

A saline-soaked cotton swap is used as a monopolar cathotic search electrode to locate the receptive field. Electrical stimuli are applied through the cotton swap at the skin over the presumed path of the cutaneous nerve from which the recordings are obtained. In response to the electrical stimulus, the afferent nerve fiber is excited and an action potential is generated and conducted to the recording electrode. The latency depends on the conduction velocity of the nerve fiber and the conduction distance. When the stimulation electrode is moved further distally along the peripheral nerve, the conduction latency (green line) of the unit increases due to the increase in conduction length. By moving the electrode further distally, the path of the unit can be traced peripherally toward its receptive field. The receptive field has been located when three phenomena occur. First, the threshold electrical stimulus intensity (pink line) that is needed to excite the afferent transcutaneously drops to its minimum. This drop is most likely due to the fact that the afferent nerve fiber ends in peripheral terminals that are located in the superficial layers of the skin. Therefore, the distance between the stimulation electrode and the nerve endings is minimized. Second, at stimulation threshold, the conduction latency of the action potential increases to its maximum, since the action potential is generated at the most distal part of the axon, that is, its peripheral ending. This latency increase is not continuous but a step increase. Third, within the electrical receptive field (eRF), the latency of the action potential decreases in discrete steps as the stimulus intensity increases. These sudden latency jumps are thought to be due to a switch of the action potential initiation site, for example, from a distal site of the axon (like the most distal ending) to a more proximal site (like a branch point) when the stimulus intensity is increased. Skin sites at which the conduction latency decreases in discrete steps with increasing stimulus intensity belong to the eRF of the unit under study. The described technique can therefore be used to locate the peripheral receptive field and to delineate its borders.

---

varies widely (from 20 to 570 μm below the skin surface). Since the tissue at the depth of the receptor heats more slowly than the skin surface due to thermal inertia, the heat threshold as measured by the surface temperature of the skin can vary greatly.

The heat response of CMHs increases monotonically with skin temperature above threshold. This suprathreshold response varies directly with the rate of temperature change at the receptor – slow temperature changes give lower instantaneous discharge frequencies than fast temperature changes.

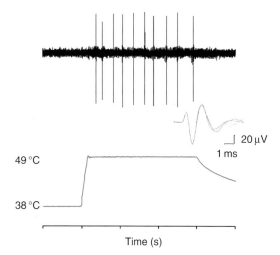

Figure 1 Response of a typical C-fiber mechanoheat-sensitive afferent (CMH) to a heat stimulus. A laser thermal stimulator was used to deliver a constant temperature stimulus (from a 38 °C baseline to 49 °C for 3 s) to the receptive field of a CMH on the hairy skin of the hand of an anesthetized monkey. The large action potential waveform corresponds to a CMH that responded with 10 action potentials to this stimulus. Inset: High-resolution recording of action potential waveforms.

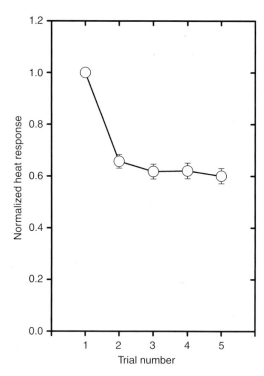

Figure 2 C-fiber mechanoheat-sensitive afferents (CMHs) exhibit fatigue to repeated heat stimuli. Identical heat stimuli were presented every 60 s to the receptive field of CMHs recorded from the anesthetized monkey. The response decreased significantly to the second stimulus and reached a plateau after several stimuli. Data were normalized by dividing by the response to the first stimulus. Adapted from Peng, Y. B., Ringkamp, M., Meyer, R. A., and Campbell, J. 2003. Fatigue and paradoxical enhancement of heat response in C-fiber nociceptors from cross-modal excitation. J. Neurosci. 23, 4766–4774, copyright 2003 by the society for Neuroscience, with permission.

The heat response of CMHs is strongly influenced by stimulus history. For example, when identical heat stimuli are repeated every 60 s, the response to the second stimulus is about 66% of the response to the first (Figure 2). This decrement in response is called fatigue. The amount of fatigue is inversely related to the interstimulus interval (i.e., shorter intervals lead to more fatigue) and directly related to the magnitude of the preceding stimulus (i.e., the larger the preceding stimulus, the greater the fatigue). As a result of fatigue, the shape of the stimulus–response function can also vary with stimulus history. Fatigue can also greatly influence threshold measurements. Complete recovery from fatigue takes more than 10 min.

A decrease in the heat response can also be produced by mechanical stimulation of the receptive field or electrical stimulation of the parent axon (Peng, Y. B. *et al.*, 2003). Thus, fatigue to one stimulus modality can be produced by a stimulus of a different modality. Interestingly, recovery from cross-modal fatigue is much faster than recovery from fatigue produced by the same stimulus modality. This suggests that cross-modal fatigue does not effect the stimulus transduction apparatus in the same way as unimodal fatigue.

In contrast to fatigue, which leads to a decreased response, the response of nociceptors can also be enhanced, for example, by tissue injury. This sensitization may account for aspects of hyperalgesia and is discussed in more detail later in this chapter.

CMHs do not respond to mechanical stimulation with blunt objects but respond well to application of sharp objects. Their mean mechanical threshold to von Frey probes is 2.2 bars. The receptive field size is around 20 mm² in monkeys but can be much larger in humans. The receptive field typically consists of multiple punctate regions of high mechanical sensitivity surrounded by areas of lesser sensitivity. Presumably these hot spots correspond to the locations of the receptor terminals.

For cylindrical probes, the response at a given force increases as the size of the probe decreases (Garell, P. C. *et al.*, 1996). This relationship scales inversely with a linear dimension of the probe (i.e., probe circumference) not probe area. Thus, the adequate stimulus for activating nociceptors appears to be related to force per unit length, and activation of the receptor terminals appears to occur at the edge of the probe where the strain is the greatest.

C-fiber MIAs are by definition insensitive to mechanical stimuli but are more responsive to chemical stimuli than CMHs and may be considered as chemosensors. For example, microneurographic recordings in human subjects (see Box 2) have shown that capsaicin (the active ingredient in hot peppers) evokes a vigorous response in C-fiber MIAs. Histamine also vigorously activates a subpopulation of C-MIAs. In contrast, histamine and capsaicin evoke weak responses in CMHs. Interestingly, MIAs can develop a response to mechanical stimuli after skin injury (discussed in more detail below).

C-MIAs also differ from C-MSAs with regard to their conductive properties. Most unmyelinated fibers

exhibit a slowing of conduction with repetitive electrical stimulation that is called activity-dependent slowing. C-MIAs show significantly more slowing than C-MSAs (Weidner, C. *et al.*, 1999). One possible explanation for this finding is that the relative distribution of ion channels involved with conduction is different in these two classes of afferents.

C-fiber afferents with extremely low thresholds to mechanical stimuli have been identified in the hairy skin of rodents, monkeys, and humans (Nordin, M., 1990). These afferents are responsive to light stroking of the skin but not to heat. They are thought to be involved in the sensation of pleasant touch and therefore may play a role in affiliative behavior.

C-fiber afferents with extremely low thresholds to heat are prevalent in all species. These afferents do not respond to mechanical stimuli but have ongoing activity at room temperature and respond to gentle warming of the skin (Darian-Smith, I. *et al.*, 1979).

Box 2   Marking technique

The marking technique is used in recordings that are hampered by a small signal-to-noise ratio that makes the identification of an action potential based on its waveform impractical. The marking technique is commonly used in human microneurography experiments, in which the signal-to-noise ratio is usually small. To use the marking technique, the primary afferent is electrically activated at a constant frequency (e.g., one stimulus every 4 s) through a pair of needle electrodes that are inserted intradermally close to the receptive field of the unit. In response to the ongoing low-frequency electrical stimulation, an action potential can be recorded at the recording electrode at a constant, fixed conduction latency (see the following figure).

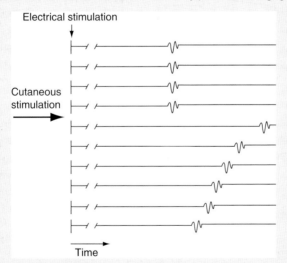

If the unit is also activated by a cutaneous (mechanical, thermal, and chemical) stimulus (see arrow), the conduction latency of the following electrically induced action potential will increase. This increase in conduction latency marks the unit as being activated by the stimulus applied to the receptive field. With ongoing electrical stimulation, the conduction latency slowly recovers to the original latency. Marking obviously indicates slowing of conduction along the course of the axon and is thought to be due to the additional action potentials that were initiated by the mechanical (or heat or chemical) stimulus and were conducted along the axon. Due to these additional action potentials, more sodium channels along the axonal membrane will be in an inactivated state when the next action potential induced by the ongoing electrical stimulation is generated and conducted. As a consequence of this inactivation, the conduction velocity for this action potential is decreased, leading to an increase in conduction latency.

Table 1 Properties of type I and type II A-fiber nociceptors

Property	Type I	Type II
Heat threshold (short-duration stimuli)	High	Low
Heat threshold (long-duration stimuli)	Low	Low
Response to heat	Slow onset, slowly increasing	Fast onset, rapidly adapting
Mechanical threshold	Low (mostly MSAs)	High (mostly MIAs)
Conduction velocity	Aδ and Aβ fibers	Aδ fibers
Sensitization after heat injury	Yes	No
Location	Hairy and glabrous skin	Hairy skin

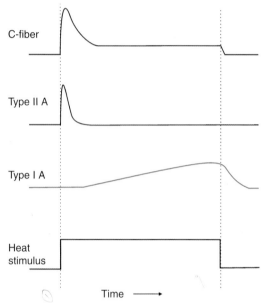

Figure 3 Response of nociceptors to a long-duration intense heat stimulus. The response of C-fiber nociceptors is quick and slowly adapts. The response of type II A-fibers is also quick but quickly adapts. The response of type I A-fibers is slow and builds during the stimulus.

The ongoing activity is suppressed by cooling. These afferents are thought to serve the percept of warmth and are often referred to as warm fibers.

8.2.2 A-Fiber Nociceptors

Two distinct types of heat-sensitive A-fiber nociceptors have been described (Table 1) (Treede, R.-D. et al., 1998). Type I fibers are responsive to mechanical and chemical stimuli but have relatively high heat thresholds when short-duration stimuli are used. For this reason, they have also been called high-threshold mechanoreceptors (HTMs). However, most type I A-fibers respond to long-duration heat stimuli (Figure 3), and their thresholds to long-duration heat stimuli are low. In addition, they develop a response to short-duration heat stimuli after injury. Heat transduction in type I afferents is thought to be subserved by the transient receptor potential channel V2 (TRPV2) ion channel. The mechanical threshold and receptive field size for type I AMHs are comparable to those of CMHs. The mean conduction velocity of type I fibers is around $25\,\mathrm{m\,s^{-1}}$ with some fibers having conduction velocities as high as $55\,\mathrm{m\,s^{-1}}$. Thus, their conduction velocity overlaps the traditional Aβ and Aδ myelinated fiber range. Type I AMHs are found in both hairy and glabrous skin.

Type II A-fiber nociceptors are characterized by a fast response to heat stimuli that quickly adapts (Figure 3). Their heat thresholds to short-duration stimuli are much lower than those of type I AMHs. Most type II fibers have very high mechanical thresholds or do not respond to mechanical stimuli and therefore are MIAs. Their mean conduction velocity ($15\,\mathrm{m\,s^{-1}}$) places them within the Aδ myelinated fiber range. In monkeys, they are only found in hairy skin.

Similar to CMHs, A-fiber nociceptors do not respond to mechanical stimulation with blunt object but respond well to sharp objects. The response to mechanical stimuli varies with position in the receptive field. When a $100\,\mu m$ wide blade is systematically moved to different positions within the receptive field, highly reproducible receptive field response maps can be obtained that contain multiple peaks and valleys (Slugg, R. M. et al., 2004). As the force increases, the response and width of the peaks increase, the response in the valleys increases, and new peaks emerge (Figure 4(a)). Although the response increases with force in most positions in the receptive field, there are positions where the response reaches a plateau at higher forces (Figure 4(b)). The averaged response across the receptive field provides an estimate for the response in the population of nociceptors that innervates a particular location (Figure 4(c)). The population response function for A- and C-fiber nociceptors

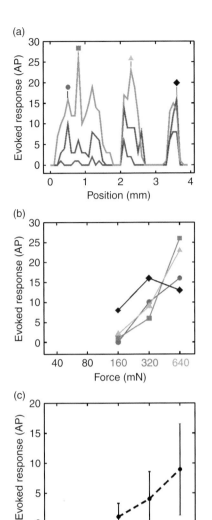

Figure 4 Response of a type I A-fiber nociceptor to a blade stimulus. (a) Maps of response as a function of position and force. A 100 μm wide blade was systematically applied to every position across the receptive field. The response shows peaks and valleys that grow in height and width as the stimulus strength increased (forces: 160, 320, and 640 mN). (b) Stimulus–response function varied with position in the receptive field. In most positions, the response increased monotonically with force, but in one position shown the response reached a plateau at the higher forces. (Symbols in (a) indicate the positions plotted.) (c) The average response across the receptive field is plotted. This provides an estimate of how the population of nociceptors that innervate a single spot on the skin would respond to the stimulus. Adapted from Slugg, R. M., Campbell, J. N., and Meyer, R. A. 2004. The population response of A- and C-fiber nociceptors in monkey ncodes high-intensity mechanical stimuli. J. Neurosci. 24, 4649–4656, copyright 2004 by the Society for Neuroscience, with permission.

increases monotonically with force. The main differences between A- and C-fiber nociceptors are (1) A-fibers in general exhibit a larger response than C-fibers, (2) the population response of A-fibers quadruples with every doubling of force, whereas the response of C-fibers doubles with every doubling of force, and (3) the A-fibers exhibit a better differential response with respect to different probe sizes and forces (Garell, P. C. *et al.*, 1996; Slugg, R. M. *et al.*, 2000).

Aδ-fiber afferents with ongoing activity are present in monkeys and humans. The neural activity in these afferents increases with gentle cooling and decreases with gentle warming (Darian-Smith, I. *et al.*, 1973). These afferents do not respond to mechanical stimulation with thermally neutral probes. These afferents are thought to subserve the percept of cool.

8.2.3 Classification of Nociceptors Based on Molecular Markers

In addition to a classification that is based on receptive properties, nociceptive afferents can be categorized using molecular markers that are expressed in their cell bodies or on their cell membrane. Markers that have been used for this purpose include neuropeptides, enzymes, and proteins involved in signal transduction, receptors for growth factors, and other molecules on the cell surface.

The somata of slowly conducting nociceptive afferents are small. In contrast to large cells that can be labeled with RT97, an antibody that is directed against neurofilament protein (NF200), only some of the small cells can be labeled with RT97. Small, RT97-positive neurons most likely represent the somata of Aδ nociceptive fibers. RT97-negative cells likely represent unmyelinated neurons.

Nociceptive afferents are commonly classified as peptidergic or nonpeptidergic (Figure 5). Peptidergic cells express substance P (SP) and/or calcitonin gene-related peptide (CGRP) or somatostatin (SOM). In rats, about 50% of neurons are peptidergic (Lawson, S. N., 1992). Peptidergic cells also express the tyrosine kinase receptor A (TrkA), the receptor for nerve growth factor (NGF), suggesting that they are NGF dependent. In addition, peptidergic cells express the TRPV1 channel that is activated by noxious heat stimuli and protons and also mediates the responsiveness to vanilloids (capsaicin and resiniferatoxin). Centrally, peptidergic neurons project to lamina I and the outer lamina II of the dorsal horn.

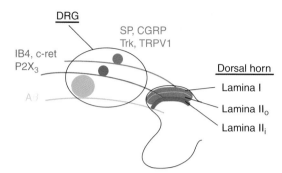

Figure 5 Small-diameter dorsal root ganglion neurons can be divided into two classes based on molecular markers in their cell bodies. The neurons in one class (shown in pink) are tyrosine kinase receptor A (TrkA), transient receptor potential V1 (TRPV1) channel, calcitonin gene-related peptide (CGRP), and substance P (SP)-positive and project to lamina I and lamina II$_o$ of the dorsal horn. The neurons in the other class (shown in blue) are IB4, c-ret, and P2X$_3$ positive and project to lamina II$_i$.

Nonpeptidergic cells contain fluoride-resistant acid phosphatase (FRAP) and bind the plant lectin IB4 from *Griffonia simplicifolia*. About 50% of dorsal root ganglion cells are IB4 positive (IB4$^+$). In contrast to peptidergic cells, IB4$^+$ neurons depend on glial cell-derived neurotrophic factor (GDNF) and they express the necessary receptors, tyrosine kinase Ret (c-ret), and glial cell-derived neurotrophic factor receptors (GDNFRα1–4). IB4$^+$ cells express the P2X$_3$ receptor, a purinergic ligand-gated ion channel that is activated by adenosine triphospate (ATP). The central terminals of IB4$^+$ neurons terminate in the inner sheet of lamina II. Some IB4$^+$ neurons also express *mas-related genes* A and D (mrgA and mrgD) or *mas-related G protein-coupled receptors* A and D (mrgprA and mrgprD) (Dong, X. *et al.*, 2001). IB4$^+$ and mrgA$^+$ neurons are mostly P2X$_3^-$ and express c-ret only at a low level. In contrast, IB4$^+$ and mrgD$^+$ neurons are also P2X$_3^+$ and express high levels of c-ret (Dong, X. *et al.*, 2001). IB4$^+$ fibers innervate many tissues of the body, including muscle and visceral organs. However, IB4$^+$/mrgpd$^+$ fibers are exclusively found in skin, where they terminate in the stratum granulosum of the epidermis, whereas CGRP-containing nerve fibers end less superficial in the stratum spinosum (Zylka, M. J. *et al.*, 2005). In addition, IB4$^+$/mrgpd$^+$ neurons appear to project to a distinct lamina of the dorsal horn, that is, between the outer and the inner sheet of lamina II. Thus peptidergic and nonpeptidergic nociceptive neurons appear to differ in growth factor dependency, in ion channel expression involved in signal transduction,

in their peripheral terminations, and in their central, dorsal horn projections.

While the classification into peptidergic and nonpeptidergic cells is commonly used, it should be emphasized that there is an overlap of marker expression between peptidergic and nonpeptidergic neurons. For example, some IB4$^+$ neurons express TRPV1or CGRP. Furthermore, there are considerable species differences in the expression and coexpression of these markers. For example, in mice only a minority of IB4$^+$ cells express TRPV1 (Zwick, M. *et al.*, 2002), whereas in rats the majority of IB4$^+$ and IB4$^-$ cells express TRPV1 (Michael, G. JC. and Priestley, J. V., 1999). Finally, the expression of these markers is not rigid but changes during development and after injury. Thus, peripheral inflammation increases the proportion of cells expressing neuropeptides, whereas following nerve injury, the proportion of cells expressing neuropeptides decreases.

8.3 Efferent and Trophic Functions of Nociceptors

In addition to serving the afferent function of signaling pain sensation, nociceptors also mediate efferent effects such as the flare that surrounds an injury site. This flare is thought to be due to an axonreflex mechanism, since the flare is absent in denervated tissue but present when a cutaneous nerve is blocked at a proximal location. When an action potential is initiated in the cutaneous terminals, it propagates not only orthodromically up the axons toward the CNS but also antidromically into peripheral branches of the nerve fiber where it causes the release of neuropeptides that are present in the terminals (Figure 6). Since afferent and efferent paths of this reflex arc are within the same axon, this reflex has been coined an axon reflex. The peripherally released neuropeptides include SP, neurokinin A (NKA), CGRP, vasoactive intestinal polypeptide (VIP), and SOM.

In animals, SP and CGRP are released from capsaicin-sensitive nerve endings and may induce vasodilation and plasma extravasation. These neuropeptides are also thought to stimulate epidermal cells (e.g., Langerhans' cells and kerotinocytes) and to play a role in immunological processes such as the migration of leukocytes at sites of tissue injury. In human skin, the antidromic release of neuropeptides is thought to lead to vasodilation, but not plasma extravasation (Schmelz, M. and Petersen, L. J., 2001). The area of the axon reflexive flare in humans is much

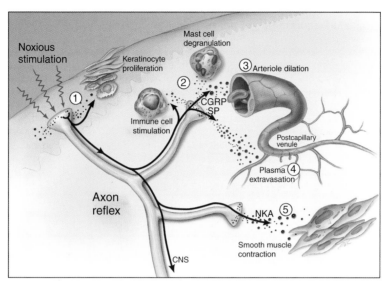

Figure 6 Nociceptors have an efferent function. Noxious stimulation of the nociceptor terminals leads to action potential propagation not only to the CNS but also into other peripheral branches of the nociceptor where they cause the release of neuropeptides such as calcitonin gene-related peptide (CGRP), SP, or NKA. The release of neuropeptides from nociceptors can lead to (1) proliferation of keratinocytes, (2) stimulation of immune cells or degranulation of mast cells, (3) vasodilation, (4) plasma extravasation, and (5) smooth muscle contraction. Artwork by Ian Suk, Johns Hopkins University. Reproduced from Meyer, R. A., Ringkamp, R., Campbell, J. N., and Raja, S. N. 2005. Peripheral Mechanisms of Cutaneous Nociception. In: Wall and Melzack's Textbook of Pain, 5 edn. (*ed.* M. K. S. McMahon), pp. 3–34. Elsevier, with permission from Elsevier.

Table 2   Roles of different afferents in acute pain sensation

Percept	Afferents responsible
Heat pain on glabrous skin	CMHs
First pain to heat on hairy skin	Type II A-fibers
Second pain to heat on hairy skin	CMHs
Pain to sustained heat stimulus	Type I AMHs
Pain to sharp mechanical stimulus	A-fiber nociceptors
Pain to sustained pressure	C-MIAs
Pain to capsaicin	C-MIAs, A-MIAs

larger than the receptive field size of CMHs. C-MIAs have large receptive field sizes and could therefore account for the large area of flare. The high threshold needed to induce flare by electrocutaneous stimulation is also consistent with mediation by MIAs (Schmelz, M. *et al.*, 2000).

8.4 Acute Pain Sensations and Nociceptor Activity

We now address the role of different classes of nociceptors in acute pain sensation (Table 2). To do this, psychophysical measurements of pain intensity in

human subjects are compared to the neuronal response in nociceptors when both are exposed to identical stimuli. A number of different scaling techniques can be used by humans including a visual analog scale (VAS) and the magnitude estimation technique (see Box 3)

8.4.1 Pain from Heat Stimuli

8.4.1.1 Heat pain from glabrous skin is signaled by C-fiber nociceptors

Near pain threshold, only two classes of fibers respond to heat stimuli applied to the glabrous skin, warm fibers, and CMHs. Type II A-fiber nociceptors are not found on glabrous skin, and type I AMHs have high heat thresholds. Warm fibers respond to gentle warming, but their response reaches a peak at temperatures at or below the threshold for pain (Figure 7) and then decreases for higher temperatures. In contrast, the response of CMHs starts at or below the heat pain threshold, and the response increases with temperature into the painful range. The stimulus–response functions for CMHs and for human pain ratings match well (Figure 7, Box 3).

There are a number of additional lines of corroborating evidence that CMHs encode heat pain sensation from glabrous skin: First, the long reaction time in human subjects to step increases in temperature is

Box 3   Psychophysical measurements of pain

A number of different techniques exist to measure the intensity of pain in human subjects. One of the most commonly used techniques is the visual analog scale (VAS) in which descriptors of pain are placed along the edge of the scale. This technique is very useful for measuring moderate levels of pain but suffers from ceiling effects for high levels of pain (i.e., once the subjects rates a given stimulus near the top of the scale, there is little room left for stimuli that may be much more intense).

Another method that is extremely useful for correlations with neural responses is the technique of magnitude estimation. In this procedure, the subjects are presented a moderately painful stimulus and asked to assign a number (e.g., 100) to their perceived intensity of that stimulus. Subsequent stimuli are then rated relative to this modulus. For example, if the next stimulus is twice as intense, the rating would be 200; half as intense, the rating would be 50. This technique does not suffer from ceiling effects, since the subject can use whatever numbers are needed to indicate the relative intensity of the stimulus. This is basically a ratio scale, since all ratings are made relative to the modulus. Ideally, the modulus is repeated during the course of the experiment to refresh the subject's memory.

An example of the use of the magnitude estimation technique to measure the intensity of heat pain in human subjects is shown in the figure.

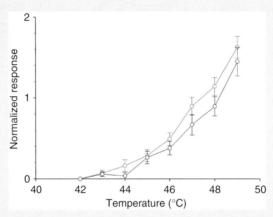

In this experiment, the glabrous skin of the hand was exposed to a random presentation of temperatures ranging from 41 to 49 °C in 1 °C increments. The random sequence was preceded by the presentation of a 45 °C stimulus that was used by the subjects to establish the reference rating (i.e., their modulus). To combine data across subjects who used different numbers for their modulus, data were normalized by dividing the ratings to a given stimulus by the ratings to the first stimulus. An identical sequence of temperatures was presented to the receptive field of CMHs recorded from the anesthetized monkey. The data across the population of CMHs were normalized by dividing the response of the nociceptors to a given temperature by the response to the initial 45 °C stimulus. The close correlation between the psychophysical data and the electrophysiological data provides strong evidence that CMHs encode heat pain from glabrous skin. Figure reproduced from Meyer, R. A. and Campbell, J. N. 1981b. Peripheral neural coding of pain sensation. Johns Hopkins APL Tech. Dig. 2, 164–171, with permission from APLTD.

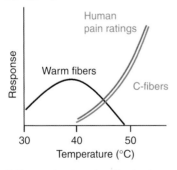

Figure 7   C-fiber nociceptors signal heat pain sensation from the glabrous skin. The response of warm fibers reaches a peak at relatively low temperatures and decreases as the temperature approaches the painful range. The response of C-fiber nociceptors starts near the pain threshold and increases with temperature. The pain ratings of human subjects closely matches the response profile of C-fiber nociceptors.

consistent with peripheral action potential propagation in the slowly conducting unmyelinated fibers. Second, selective blockade of peripheral conduction in myelinated fibers does not abolish heat pain sensation. Third, the fatigue to repeated heat stimuli described above for CMHs is also observed in psychophysical studies of pain to repeated heat stimuli. Fourth, in human microneurography experiments, intraneural electrical stimulation of presu-mably single CMHs elicits pain sensations.

8.4.1.2   *First pain to heat on hairy skin is signaled by type II A-fiber nociceptors*

In the hairy skin of the arm, stepped increases in temperature lead to a double pain sensation; an initial pricking pain is followed after a short lull by a

burning sensation (Figure 8). The reaction time to the pricking pain sensation is too short for conduction by slowly conducting unmyelinated fibers in the periphery. In addition, first pain sensation is blocked during a selective blockade of myelinated fibers.

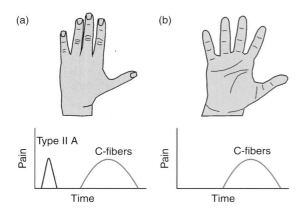

Figure 8 Heat pain on hairy and glabrous skin. (a) A heat stimulus on hairy skin leads to a dual pain sensation, an initial pricking pain followed by a burning pain. This is consistent with the finding that type II A-fiber nociceptors are found on hairy skin. (b) A heat stimulus on glabrous skin leads to burning pain only. Type II A-fibers are not found on glabrous skin.

These findings suggest that the first pain sensation to heat is mediated by myelinated afferents.

The type II A-fiber nociceptors are ideally suited to signal first pain sensation. The quick burst in action potential activity at the onset of a heat stimulus seen in type II A-fiber nociceptors is consistent with the percept of a short-lasting pricking sensation. The heat threshold of type II A-fiber nociceptors is also near the heat threshold for first pain sensation. The lack of type II fibers in glabrous skin is consistent with the absence of first pain to heat on the glabrous skin (Figure 8).

Figure 9 illustrates the neural responses to stepped increases in temperature in hairy skin. The heat stimulus activates a population of CMHs and type II A-fiber nociceptors with receptive fields that overlap the area of the heat stimulus. At the cutaneous terminals, both classes of fibers respond relatively quickly to the heat stimulus, but the response of the type II A-fibers rapidly adapts. As the neural signal propagates up to the CNS, these responses separate in time due to the conduction velocity differences of the myelinated and unmyelinated fibers. The signal from the type II fibers arrives at the CNS much sooner than the response of the

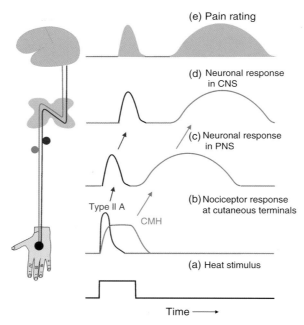

Figure 9 First pain to heat on hairy skin is signaled by A-fiber nociceptors. A stepped heat stimulus on hairy skin (a) leads to a dual pain sensation, a sharp pain followed after a lull by a burning pain (e). Heat stimuli near pain threshold activate two classes of nociceptors on hairy skin, type II A-fibers and C-fiber mechanoheat-sensitive afferents (CMHs) (b). The signal that reaches the central nervous system (CNS) is both delayed and broader than the signal at the receptor terminals. The A-fibers have a faster conduction velocity than the C-fibers and therefore their signals arrive at the CNS with less delay. The broadening of the signal is due to the variation in conduction velocity across the population of fibers that respond.

CMHs. In addition, the signals from each class of afferents become broader because of the variation in conduction velocity in each population of fibers that respond to the stimulus. The double pain response reflects the different arrival times at the CNS of neural signals from these two classes of afferents. This difference in arrival time is more apparent for stimuli applied to the distal extremities where the conduction lengths are long. A double pain sensation is not as apparent at more rostral locations (e.g., the face) where the conduction distances are short.

8.4.1.3 Pain to prolonged heat stimuli is signaled by type I A-fiber nociceptors

A long-duration heat stimulus applied to the glabrous skin produces pain that lasts through out the stimulus. However, the response of CMHs adapts to a long heat stimulus. In contrast, the response of type I A-fiber nociceptors builds during the course of a prolonged heat stimulus (Figure 3). Therefore, type I A-fiber nociceptors are thought to account for the pain to sustained high-intensity heat stimuli.

8.4.2 Pain from Controlled Mechanical Stimuli

8.4.2.1 Sharp pain is signaled by A-fiber nociceptors

The force thresholds for sharpness and pain from cylindrical objects increase in direct proportion with the circumference of the probe (Greenspan, J. D. and McGillis, S. L. B., 1991). As noted above, a similar relationship exists between probe circumference and the response of nociceptors. Sharp pain sensation is dramatically reduced in the presence of a selective blockade of A-fiber conduction. In addition, the reaction time for perception of sharp pain is short. Thus, A-fiber nociceptors appear to be primarily responsible for encoding the percept of sharp pain. Prolonged topical application of capsaicin leads to a substantial decrease in heat pain sensitivity but does not greatly affect mechanical pain sensibility (Magerl, W. et al., 2001). This suggests that the A-fibers involved in sharp pain are insensitive to capsaicin (likely, type I AMHs).

8.4.2.2 Pain from tonic pressure is signaled by C-fiber MIAs

Firm pressure does not hurt if applied for short periods of time but becomes painful when sustained for long periods of time. The response of mechanically sensitive nociceptors adapts to a constant mechanical stimulus.

In contrast, C-fiber MIAs develop a response during a maintained stimulus (Schmidt, R. et al., 2000). In addition, the pain to tonic pressure does not disappear during a selective A-fiber block (Andrew, D. and Greenspan, J. D., 1999). Thus, it appears that C-fiber MIAs encode the pain to tonic pressure.

8.4.3 Cold Pain Sensation

The threshold for cold pain sensation is around 14 °C on hairy skin. This is much farther from resting skin temperature (33 °C) than the threshold for heat pain (around 45 °C). In addition, the slope of the stimulus–response function for cold pain is shallower than that for heat pain (Morin, C. and Bushnell, M. C., 1998). Cold pain appears to be signaled, at least in part, from deep receptors since topical anesthetics do not abolish it and the latency for the perception of cold pain is long.

The majority of nociceptors respond to ice applied to the receptive field. A nonselective cation channel (TRPA1) has been identified that has an activation threshold (18 °C) near the cold pain threshold and is found in a subset of nociceptors (Story, G. M. et al., 2003). However, the role of this channel in signaling the pain to a noxious cold stimulus is controversial (Bautista, D. M. et al., 2006; Kwan, K. Y. et al., 2006). In the presence of an A-fiber block, gentle cooling stimuli become painful. It is thought that cold-evoked activity in A-delta cold fibers normally inhibits pain produced by activation of nociceptors.

8.4.4 Capsaicin-Evoked Pain

Capsaicin, the active ingredient in hot peppers, produces a burning percept upon contact with mucous membranes and produces intense pain that lasts for several minutes when injected into the skin. Topical or intradermal capsaicin can also lead to a marked desensitization to heat stimuli. The TRPV1 channel expressed by many nociceptors is thought to be responsible for the pungency of capsaicin. Injection of capsaicin into the receptive field of CMHs produces a weak response and can lead to a desensitization of the afferent to heat. Injection of capsaicin into the receptive field of A-fiber and C-fiber MIAs produces a long-lasting, strong response, suggesting that these afferents are responsible for the pain from capsaicin.

8.5 Hyperalgesia

A tissue injury can lead to an enhancement of pain sensibility called hyperalgesia. Hyperalgesia is characterized by a leftward shift in the stimulus–response function relating pain intensity to stimulus intensity (Figure 10(a)). There is a decrease in the threshold for pain, an increase in the pain to suprathreshold stimuli, and ongoing pain. The electrophysiological correlate for hyperalgesia is sensitization of nociceptive neurons (Figure 10(b)), either of peripheral nociceptors (peripheral sensitization) or of central neurons involved in processing nociceptive input (central sensitization) or both. Sensitization is characterized by a leftward shift in the neural response at different stimulus intensities. There is a decrease in the threshold for eliciting a response, an increase in the response to suprathreshold stimuli, and spontaneous activity.

Two zones of hyperalgesia are readily identified (Figure 11). Primary hyperalgesia occurs at the site of injury and is characterized by hyperalgesia to both heat and mechanical stimuli. Secondary hyperalgesia occurs in the uninjured tissue surrounding the site of injury and is characterized by hyperalgesia to mechanical but not heat stimuli. This dichotomy provides evidence that the mechanisms for primary and secondary hyperalgesia differ.

8.5.1 Primary Hyperalgesia

8.5.1.1 Hyperalgesia to heat is mediated by primary afferent sensitization

Hyperalgesia to heat is prevalent after a cutaneous injury and after inflammation. Substantial evidence indicates that sensitization of primary afferent nociceptors to heat accounts for this hyperalgesia. For example, a burn injury to the glabrous skin of the hand leads to marked hyperalgesia to heat and sensitization of type I AMHs to heat (Figure 12). Although a burn to the glabrous skin does not lead to sensitization of CMHs, a burn injury to the hairy skin does. This suggests that the propensity to sensitize is a function of skin type. Whether this is due to differences in the milieu or due to differences in the properties of the neurons is not known.

Hyperalgesia to heat can also develop after inflammation. When a cocktail of inflammatory mediators is injected into the receptive field of nociceptive afferents, marked sensitization to heat stimuli is observed.

8.5.1.2 Hyperalgesia to mechanical stimuli

Hyperalgesia to mechanical stimuli also occurs after a cutaneous injury and after inflammation. A number of possible mechanisms may account for this mechanical hyperalgesia.

Nociceptor sensitization. Although a burn injury leads to sensitization of nociceptors to heat stimuli, the mechanical threshold of nociceptors is not lowered, suggesting that this injury does not produce sensitization to mechanical stimuli. However, nociceptors are sensitized to mechanical stimuli in situations of inflammation. For example, injection into the receptive field of inflammatory substances leads to a sensitization of MIAs to mechanical stimuli. Inflammation of the knee joint also leads to a sensitization of MIAs to mechanical stimuli.

Spatial summation. When a burn injury is applied at the edge of the receptive field of a nociceptor, the receptive field expands into the area of injury. As a result, a mechanical stimulus in the area of injury evokes a response in a greater number of nociceptive

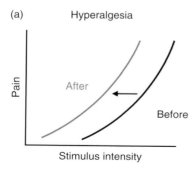

(a) Hyperalgesia

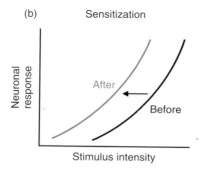

(b) Sensitization

Figure 10 Hyperalgesia and sensitization. (a) Hyperalgesia is characterized by a leftward shift in the stimulus function relating pain intensity to stimulus intensity. (b) Sensitization is characterized by a leftward shift in the stimulus function relating neural response to stimulus intensity.

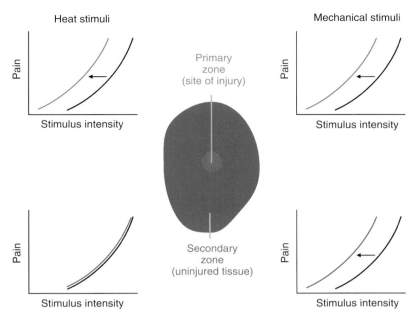

Figure 11   Primary and secondary hyperalgesia. There are two forms of hyperalgesia. Primary hyperalgesia occurs at the site of injury and is characterized by hyperalgesia to both mechanical and heat stimuli. Secondary hyperalgesia occurs in the uninjured tissue surrounding the injury site and is characterized by hyperalgesia to mechanical but not heat stimuli.

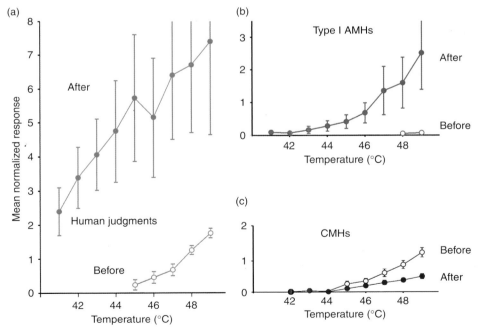

Figure 12   Primary hyperalgesia to heat is signaled by sensitization of primary afferent nociceptors. (a) A burn injury to the glabrous skin of human subjects leads to a dramatic increase in pain ratings to heat stimuli. (b) The same burn injury leads to a marked sensitization of type I AMHs in the anesthetized monkey. (c) This burn injury led to a desensitization of CMHs on glabrous skin. However, burn injuries to CMHs on hairy skin do lead to sensitization. Adapted from Meyer, R. A. and Campbell, J. N. 1981a. Myelinated nociceptive afferents account for the hyperalgesia that follows a burn to the hand. Science 213, 1527–1529, with permission from AAAS.

afferents. This will lead to spatial summation in the CNS that would be expected to produce more pain.

Central sensitization. Peripheral sensitization appears to account, at least in part, for hyperalgesia to mechanical stimuli. In addition, the central sensitization phenomena to be described below for secondary hyperalgesia may also play a role in primary hyperalgesia to mechanical stimuli.

8.5.1.3 Inflammatory mediators and nociceptor sensitization

Injury leads to the local release of various endogenous substances that excite or modulate the terminals of primary-afferent nociceptors (Figure 13). For example, bradykinin is released on tissue injury (e.g., from the plasma) and can excite and sensitize nociceptor terminals. Intradermal injection of bradykinin produces pain and hyperalgesia. Serotonin, histamine, and prostagladin can also be released from mast cells and alter

nociceptor terminal sensitivity. Tissue injury can also lead to the release or leakage of adenosine and its derivatives (e.g., ATP) into the extracellular space and could therefore activate sensory terminals. Low pH levels found in inflamed tissue activate and modulate nociceptor terminals. During inflammation, cytokines (e.g., interleukins and tumor necrosis factor alpha) are released by a variety of cells (e.g., macrophages) and are not only involved in regulating the inflammatory process but may also act as algesic substances either by directly activating nociceptive endings or by increasing their sensitivity to natural stimuli.

As illustrated in Figure 13, these agents can act directly at specialized receptors on the nociceptor terminal to excite the ending or to alter the sensitivity of the ending. The molecular mechanisms of transduction and sensitization are discussed in more detail in Chapter Molecular Biology of the Nociceptor/Transduction.

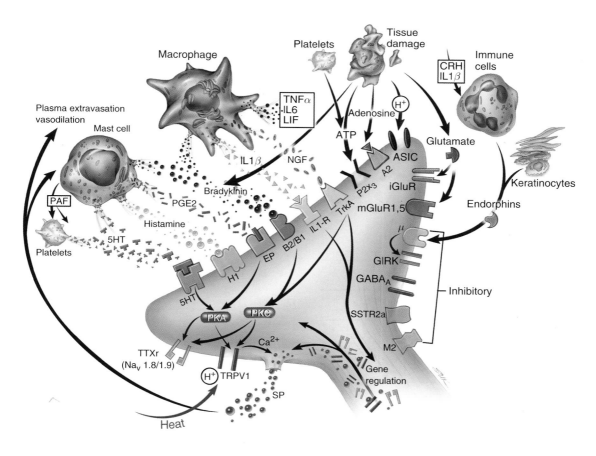

Figure 13 Inflammatory mediators act on specific receptors located at the nociceptor terminals. Artwork by Ian suk, Johns Hopkins University. Reproduced from Meyer, R. A., Ringkamp, R., Campbell, J. N., and Raja, S. N. 2005. Peripheral Mechanisms of Cutaneous Nociception. In: Wall and Melzack's Textbook of Pain, 5 edn. (*ed.* M. K. S. McMahon), pp. 3–34. Elsevier.

8.5.2   Secondary Hyperalgesia

Secondary hyperalgesia is characterized by hyperalgesia to mechanical but not heat stimuli. Two distinct forms of mechanical hyperalgesia have been reported (Figure 14). Punctate hyperalgesia (or static hyperalgesia) is characterized by hyperalgesia to sharp stimuli that normally cause pain (e.g., stiff von Frey probes and pin prick). Stroking hyperalgesia (or dynamic hyperalgesia or allodynia) is characterized by pain to a light moving touch (e.g., stroking of the skin with a soft brush) but not blunt pressure.

8.5.2.1   Sensitization of primary afferent nociceptors does not occur

One possible explanation for secondary hyperalgesia is that the sensitization process that occurs at the site of injury spreads to adjacent uninjured tissue. In a manner similar to that described above for the axon-reflexive flare response, the nerve endings could release not only vasoactive substances but also substances that produce sensitization of nociceptive terminals. However, a number of lines of evidence suggest that spreading sensitization does not occur. For example, burn or mechanical injuries or mustard oil application to one half of the receptive field of nociceptors lead to sensitization of that part of the receptive field but not of the uninjured part of the receptive field. In addition, antidromic electrical stimulation of nociceptors, which should lead to the release of these putative sensitizing substances from the terminals, does not lead to sensitization. Also, the area of flare is generally smaller than the area of hyperalgesia. In addition, flare can be induced by substances (e.g., histamine injection) that do not lead to the development of hyperalgesia. Finally, flare can spread across the midline of the body, but hyperalgesia does not.

8.5.2.2   Central mechanisms of secondary hyperalgesia

The preponderance of experimental data indicate that secondary hyperalgesia is the result of plasticity in the CNS often called central sensitization (Figure 15). This central sensitization is initiated by

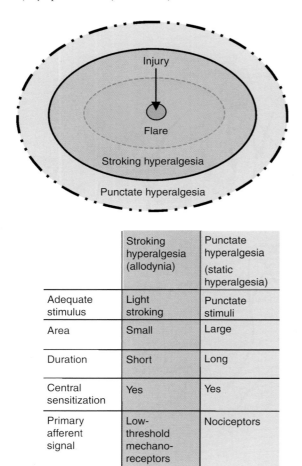

	Stroking hyperalgesia (allodynia)	Punctate hyperalgesia (static hyperalgesia)
Adequate stimulus	Light stroking	Punctate stimuli
Area	Small	Large
Duration	Short	Long
Central sensitization	Yes	Yes
Primary afferent signal	Low-threshold mechano-receptors	Nociceptors

Figure 14   Two forms of secondary hyperalgesia to mechanical stimuli. The properties of punctate and stroking hyperalgesia are different, suggesting that the neural mechanisms may differ.

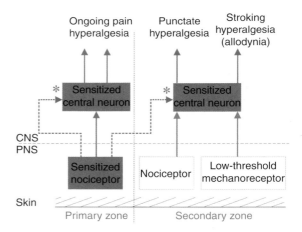

Figure 15   Secondary hyperalgesia is due to central sensitization. An injury or inflammation leads to activation and sensitization of nociceptors in the zone of primary hyperalgesia. Activity in the peripheral nociceptors produces a state of central plasticity in the CNS called central sensitization (*). This sensitization spreads to central neurons with input from adjacent, uninjured afferents. Sensitization to input from nociceptors accounts for punctate hyperalgesia. Sensitization to input from low-threshold mechanoreceptors accounts for stroking hyperalgesia (or allodynia).

neural activity in primary-afferent nociceptors and is already apparent at the level of the dorsal horn.

Compelling evidence for central sensitization comes from an experiment in which capsaicin was injected into both the forearms of human volunteers (LaMotte, R. H. *et al.*, 1991). In the control arm, the capsaicin injection produced intense pain and led to the development of a flare and a large zone of secondary hyperalgesia. In the test arm, the nerve supplying the forearm was anesthetized at a proximal site prior to the injection. The injection did not hurt but did lead to the development of an axon-reflexive flare, indicating that the nociceptors had been activated and were able to release vasoactive substances (and also putative sensitizing substances). Three hours after the capsaicin injection, there was no hyperalgesia on the test arm, but the control arm still had a large zone of secondary hyperalgesia.

The characteristics of stroking and punctate hyperalgesia are different, suggesting that the neural mechanisms are different (Figure 14). For example, the area of punctate hyperalgesia is typically larger than the area of stroking hyperalgesia. In addition, the duration of punctate hyperalgesia is longer than the duration of stroking hyperalgesia. Stroking hyperalgesia appears to be more dependent on ongoing input from nociceptors, since cooling or anesthetizing the site of injury substantially reduces the zone of stroking hyperalgesia without greatly affecting the zone of punctate hyperalgesia. As described below, both forms of secondary hyperalgesia are thought to be due to central sensitization mechanisms.

Stroking hyperalgesia appears to be due to central sensitization to input from low-threshold myelinated mechanoreceptors (Figure 15). The light stroking stimulus that produces pain in the secondary zone readily activates low-threshold mechanoreceptors but not nociceptors. Compelling evidence comes from an experiment in which intraneural microstimulation was performed in human volunteers (Torebjörk, H. E. *et al.*, 1992). A low-intensity, fixed-frequency electrical stimulus produced a tactile percept referred to a small area on the foot. Capsaicin was then injected adjacent to this area and produced secondary hyperalgesia that spread into this area. When the electrical stimulation was repeated at the same intensity and frequency, it now produced pain. This electrical stimulation of myelinated fibers bypassed the natural transduction apparatus and indicates that there was central sensitization to input from myelinated fibers. Additional evidence comes from studies which show that selective peripheral blockade of myelinated fibers (with conduction maintained in unmyelinated fibers) leads to the disappearance of stroking hyperalgesia.

Punctate hyperalgesia appears to be due to central sensitization to input from nociceptors (Figure 15). Punctate hyperalgesia, and not stroking hyperalgesia, developed when capsaicin was injected into the arm of a patient with a selective, large-fiber neuropathy, indicating that small myelinated or unmyelinated fibers are involved. In addition, enhanced pain to woolen fabrics is observed in the zone of secondary hyperalgesia. Since the prickliness of woolen fabrics is encoded by activity in nociceptors, not mechano-receptors, activity in nociceptors likely accounts for this hyperalgesia. Selective nerve blocks revealed that punctate hyperalgesia disappeared when $A\delta$ nerve fibers were blocked, suggesting that $A\delta$ nociceptors are important. The nociceptors appear to be capsaicin insensitive since topical application of capsaicin produced desensitization to heat but did not eliminate punctate hyperalgesia produced by a subsequent injection of capsaicin adjacent to the treated area (Magerl, W. *et al.*, 2001).

Acknowledgments

This research was supported in part by the National Institutes of Health (NS-14447 and NS-026363) and by a Stuart S. Janney Fellowship from the Applied Physics Laboratory (RAM).

References

Andrew, D. and Greenspan, J. D. 1999. Peripheral coding of tonic mechanical cutaneous pain: comparison of nociceptor activity in rat and human psychophysics. J. Neurophysiol. 82, 2641–2648.

Bautista, D. M., Jordt, S. E., Nikai, T., Tsuruda, P. R., Read, A. J., Poblete, J., Yamoah, E. N., Basbaum, A. I., and Julius, D. 2006. TRPA1 mediates the inflammatory actions of environmental irritants and proalgesic agents. Cell 124, 1269–1282.

Darian-Smith, I., Johnson, K. O., and Dykes, R. 1973. "Cold" fiber population innervating palmar and digital skin of the monkey: responses to cooling pulses. J. Neurophysiol. 36, 325–346.

Darian-Smith, I., Johnson, K. O., LaMotte, C., Shigenaga, Y., Kenins, P., and Champness, P. 1979. Warm fibers innervating palmar and digital skin of the monkey: responses to thermal stimuli. J. Neurophysiol. 42, 1297–1315.

Dong, X., Han, S., Zylka, M. J., Simon, M. I., and Anderson, D. J. 2001. A diverse family of GPCRs expressed in specific subsets of nociceptive sensory neurons. Cell 106, 619–632.

Garell, P. C., McGillis, S. L. B., and Greenspan, J. D. 1996. Mechanical response properties of nociceptors innervating feline hairy skin. J. Neurophysiol. 75, 1177–1189.

Greenspan, J. D. and McGillis, S. L. B. 1991. Stimulus features relevant to the perception of sharpness and mechanically evoked cutaneous pain. Somatosens. Motor Res. 8, 137–147.

Kwan, K. Y., Allchorne, A. J., Vollrath, M. A., Christensen, A. P., Zhang, D. S., Woolf, C. J., and Corey, D. P. 2006. TRPA1 contributes to cold, mechanical, and chemical nociception but is not essential for hair-cell transduction. Neuron 50, 277–289.

LaMotte, R. H., Shain, C. N., Simone, D. A., and Tsai, E.-F. P. 1991. Neurogenic hyperalgesia: psychophysical studies of underlying mechanisms. J. Neurophysiol. 66, 190–211.

Lawson, S. N. 1992. Morphological and Biochemical Cell Types of Sensory Neurons. In: Sensory Neurons (ed. S. A. Scott), pp. 27–59. Oxford University Press.

Magerl, W., Fuchs, P. N., Meyer, R. A., and Treede, R.-D. 2001. Roles of capsaicin-insensitive nociceptors in cutaneous pain and secondary hyperalgesia. Brain 124, 1754–1764.

Meyer, R. A. and Campbell, J. N. 1981a. Myelinated nociceptive afferents account for the hyperalgesia that follows a burn to the hand. Science 213, 1527–1529.

Meyer, R. A. and Campbell, J. N. 1981b. Peripheral neural coding of pain sensation. Johns Hopkins APL Tech. Dig. 2, 164–171.

Meyer, R. A., Ringkamp, R., Campbell, J. N., and Raja, S. N. 2005. Peripheral Mechanisms of Cutaneous Nociception. In: Wall and Melzack's Textbook of Pain, 5 edn. (ed. M. K. S. McMahon), pp. 3–34. Elsevier.

Michael, G. J. and Priestley, J. V. 1999. Differential expression of the mRNA for the vanilloid receptor subtype 1 in cells of the adult rat dorsal root and nodose ganglia and its downregulation by axotomy. J. Neurosci. 19, 1844–1854.

Morin, C. and Bushnell, M. C. 1998. Temporal and qualitative properties of cold pain and heat pain: a psychophysical study. Pain 74, 67–73.

Nordin, M. 1990. Low-threshold mechanoreceptive and nociceptive units with unmyelinated (C) fibres in the human supraorbital nerve. J. Physiol. (Lond.) 426, 229–240.

Peng, Y. B., Ringkamp, M., Meyer, R. A., and Campbell, J. 2003. Fatigue and paradoxical enhancement of heat response in C-fiber nociceptors from cross-modal excitation. J. Neurosci. 23, 4766–4774.

Schmelz, M. and Petersen, L. J. 2001. Neurogenic inflammation in human and rodent skin. News Physiol. Sci. 16, 33–37.

Schmelz, M., Michael, K., Weidner, C., Schmidt, R., Torebörk, H. E., and Handwerker, H. O. 2000. Which nerve fibers mediate the axon reflex flare in human skin? Neuroreport 11, 645–648.

Schmidt, R., Schmelz, M., Torebjörk, H. E., and Handwerker, H. O. 2000. Mechano-insensitive nociceptors encode pain evoked by tonic pressure to human skin. Neuroscience 98, 793–800.

Slugg, R. M., Campbell, J. N., and Meyer, R. A. 2004. The population response of A- and C-fiber nociceptors in monkey encodes high-intensity mechanical stimuli. J. Neurosci. 24, 4649–4656.

Slugg, R. M., Meyer, R. A., and Campbell, J. N. 2000. Response of cutaneous A- and C-fiber nociceptors in the monkey to controlled-force stimuli. J. Neurophysiol. 83, 2179–2191.

Story, G. M., Peier, A. M., Reeve, A. J., Eid, S. R., Mosbacher, J., Hricik, T. R., Earley, T. J., Hergarden, A. C., Andersson, D. A., Hwang, S. W., McIntyre, P., Jegla, T., Bevan, S., and Patapoutian, A. 2003. ANKTM1, a TRP-like channel expressed in nociceptive neurons, is activated by cold temperatures. Cell 112, 819–829.

Tillman, D. B., Treede, R.-D., Meyer, R. A., and Campbell, J. N. 1995. Response of C fibre nociceptors in the anaesthetized monkey to heat stimuli: estimates of receptor depth and threshold. J. Physiol. 485(Pt. 3), 753–765.

Torebjörk, H. E., Lundberg, L. E. R., and LaMotte, R. H. 1992. Central changes in processing of mechanoreceptive input in capsaicin-induced secondary hyperalgesia in humans. J. Physiol. (Lond.) 448, 765–780.

Treede, R.-D., Meyer, R. A., and Campbell, J. N. 1998. Myelinated mechanically insensitive afferents from monkey hairy skin: heat-response properties. J. Neurophysiol. 80, 1082–1093.

Weidner, C., Schmelz, M., Schmidt, R., Hansson, B., Handwerker, H. O., and Torebjork, H. E. 1999. Functional attributes discriminating mechano-insensitive and mechano-responsive C nociceptors in human skin. J. Neurosci. 19, 10184–10190.

Zwick, M., Davis, B. M., Woodbury, C. J., Burkett, J. N., Koerber, H. R., Simpson, J. F., and Albers, K. M. 2002. Glial cell line-derived neurotrophic factor is a survival factor for isolectin B4-positive, but not vanilloid receptor 1-positive, neurons in the mouse. J. Neurosci. 22, 4057–4065.

Zylka, M. J., Rice, F. L., and Anderson, D. J. 2005. Topographically distinct epidermal nociceptive circuits revealed by axonal tracers targeted to Mrgprd. Neuron 45, 17–25.

9 Itch

E Carstens, University of California, Davis, CA, USA

Glossary

allodynia Pain due to a stimulus that does not normally provoke pain.

alloknesis Itch elicited by light touching or stroking.

hyperalgesia An increased response to a stimulus which is normally painful.

hyperknesis Itch elicited by strong mechanical stimulation.

itch Unpleasant sensation associated with the desire to scratch.

nociceptor A receptor preferentially sensitive to a noxious stimulus or to a stimulus, which would become noxious if prolonged.

pain An unpleasant sensory and emotional experience associated with actual or potential tissue damage, or described in terms of such damage.

pruriceptor A sensory receptor preferentially sensitive to a pruritic stimulus.

9.1 Introduction and Definition of Itch

Itch is widely considered to be an unpleasant sensation associated with the desire to scratch. The definition of itch has recently been refined to include the following subcategories: pruriceptive (originating from skin damage), neuropathic (originating from nerve damage), neurogenic (originating centrally), and psychogenic (Yosipovitch, G. *et al.*, 2003). Itch is frequently associated with common insect bites or exposure to plants, providing a warning signal to remove the offending stimulus by scratching or rubbing the affected skin area. Itch can be experimentally elicited by cutaneous histamine and other chemicals (Schmelz, M. *et al.*, 2003b) and by electrical surface stimulation of the skin (Tuckett, R. P., 1982; Ikoma, A. *et al.*, 2005). However, itch also commonly occurs under a variety of skin conditions such as dry skin and contact or atopic dermatoses (eczema). Moreover, a variety of systemic disorders also have itching as a symptom (Krajnik, M. and Zbigniew, Z., 2001), in particular, liver dysfunctions such as cholestasis or biliary cirrhosis (Bergasa, N. V. *et al.*, 2000), and renal failure (Murphy, M. and Carmichael, A. J., 2000). Neuropathic itch is less commonly associated with lesions of the spinal cord (Dey, D. D. *et al.*, 2005) or other central structures. While itch associated with, for example, urticaria (hives) can be relieved by antihistamines acting at the H1 histamine receptor, itch associated with other skin conditions and systemic diseases is often refractory to antihistamines. This suggests that there may be other itch mediators and mechanisms in addition to histamine, the prototypical itch mediator in normal human skin.

Itch has historically been considered by some to be a low level of pain, and perhaps for this reason the experimental study of itch has until now received much less attention compared to pain. Nevertheless, recent studies have provided exciting new evidence for the existence of itch-selective peripheral receptors and central pathways separate from those mediating pain (for recent reviews, see Stander, S. *et al.*, 2003; Twycross, R. *et al.*, 2003; Yosipovitch, G. *et al.*, 2003; 2004; Biro, T. *et al.*, 2005). This vignette will describe some of the recent advances in our understanding of itch mechanisms. An overview of the neural basis for itch versus pain is provided first, followed by consideration of animal itch models and, finally, possible itch mediators and mechanisms based on human and animal research.

9.2 Is Itch Distinct from Pain?

The concept that itch is on a continuum with pain prompted the frequency theory (von Frey, M., 1922), depicted schematically in Figure 1(a). A low firing rate in cutaneous nociceptors connected to a common central pathway elicits a sensation of itch, while high-frequency firing elicits pain. This idea is supported by observations that numerous stimuli, including histamine (Broadbent, J. L., 1955) elicit itch at low intensities and pain at higher intensities, and that both pain and itch are conveyed via the spinothalamic tract (White, J. C. and Sweet, W. H., 1969). The notion that itch and pain are conveyed by a common set of neurons, however, is

belied by the following observations. Most importantly, itch can be elicited by focal electrical stimulation of the skin surface (Tuckett, R. P., 1982; Ikoma, A. *et al.*, 2005) or by intraneural microstimulation (Schmidt, R. *et al.*, 1993) and the itch becomes more intense but never takes on a painful quality as the stimulus frequency is increased. Conversely, pain elicited by intraneural stimulation never transforms to itch at lower frequencies (Ochoa, J. and Torebjörk, E., 1989; Handwerker, H. O. *et al.*, 1991). Furthermore, itch and pain rarely coexist and concurrent noxious stimulation, such as scratching, is well known to inhibit itch. Morphine and other opioids depress pain but frequently elicit or exacerbate itch, as discussed further below. Also, the behavioral manifestation of itch, scratching, is very different from manifestations of pain that often include withdrawal of the stimulated limb and/or escape reactions. Withdrawal from a noxious stimulus serves a protective function to prevent tissue damage, whereas scratching may represent an attempt to remove a pruritic stimulus that has already impinged upon tissue. In this regard, humans can localize the site of itch elicited by focal histamine application with errors of less than 10 mm, comparable to localization of punctate painful stimuli and only slightly worse than localization of tactile stimuli (Koltzenburg, M. *et al.*, 1993). This would aid the accuracy of scratches directed to an itchy site. However, two-point discrimination for histamine-induced itch is fairly poor, approximately 12–15 cm on forearm skin (Wahlgren, C. F and Ekblom, A., 1996) consistent with the very large

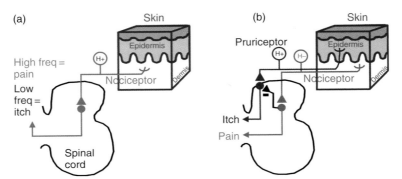

Figure 1 Theories of itch transmission. (a) Frequency theory. Both pruritic and noxious stimuli excite polymodal nociceptors in skin, which project to a common population of neurons in the spinal cord. These neurons signal itch at a low firing rate, and pain at a higher firing rate. H+: responsive to histamine. (b) Specificity (or selectivity) theory. Pruritic stimuli excite a selective population of pruriceptors in superficial epidermis that are histamine-sensitive (H+), and that project to a select population of spinal neurons (blue) in the superficial dorsal horn. Activity in this pathway signals itch. Dermal nociceptors project to a separate population of pain-signaling spinal neurons (red). Incorporated in this schematic is an inhibitory connection (black) from pain- to itch-transmitting neurons, reflecting the ability of noxious counterstimuli to inhibit itch.

innervation territories of putative pruriceptors (Schmelz, M. *et al.*, 1997) (see below).

For the reasons noted above, it has long been postulated that itch and pain are conveyed via separate central pathways, with pain capable of inhibiting itch transmission. This concept is depicted in Figure 1(b). The separate pathways for itch and pain constitute the classic specificity theory. Given recent evidence showing a relative, rather than absolute, sensitivity of potential itch-signaling fibers to prurities, the term selectivity is more appropriate. The ability of pain to inhibit itch is incorporated by the black inhibitory neuron shown in Figure 1(b), although interactions between nociceptive and pruriceptive systems may conceivably occur at supraspinal as well as spinal levels.

9.3 Pain Inhibition of Itch

Numerous experimental studies have demonstrated suppression of itch in humans by a variety of counter-stimuli including cooling, mustard oil, and electrical cutaneous field stimulation (Bickford, R. G., 1937; Gammon, G. D. and Starr, I., 1941; Murray, F. S. and Weaver, M. M., 1975; Frustorfer, H. *et al.*, 1986; Ward, L. *et al.*, 1996; Nilsson, H. J. *et al.*, 1997). This suppression undoubtedly underlies the antipruritic effect of scratching. Opioids acting at the μ-opioid receptor depress pain, yet frequently induce or exacerbate itch in human patients (Bromage, P., 1981; Bromage, P. R. *et al.*, 1982; Morgan, M., 1987; Ballantyne, J. C. *et al.*, 1988), suggesting that inhibition of tonic nociceptive activity facilitates itch transmission. Consistent with this, μ-opioid antagonists reduce histamine-induced itch (Heyer, G. *et al.*, 1997a). Intradermal injection of histamine at a site previously anesthetized by local anesthetic in human subjects resulted in markedly enhanced itch (Atanassoff, P. G. *et al.*, 1999). The authors postulated that histamine excites two types of peripheral receptor, one conveying itch and the other being antipruritic (i.e., exciting central itch-inhibitory neurons). Preferential local anesthetic blockade of the antipruritic fibers would unmask pruriceptive input leading to increased itch.

Scratching behavior in mice and monkeys induced by μ-opioids is reduced by κ-opioid agonists in mice (Togashi, Y. *et al.*, 2002; Umeuchi, H. *et al.*, 2003) and monkeys (Ko, M. C. *et al.*, 2003). The κ agonists are also analgesic, suggesting a complex interaction between pro- and antipruritic opioid receptor-

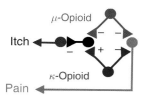

Figure 2 Schematic of opioid interactions with itch- and pain-signaling spinal neurons. μ-Opioid agonists (purple) inhibit pain and can induce or exacerbate itch, possibly by inhibiting interneurons (black) that tonically inhibit itch-transmitting neurons (blue), i.e., disinhibition. κ-Opioid agonists (dark green) also inhibit pain and may excite the inhibitory interneuron (black), thus inhibiting itch.

mediated central mechanisms. One possible scenario consistent with this interplay between μ- and κ-opioids is shown in Figure 2.

9.4 Peripheral and Spinal Mechanisms of Itch

Recent studies have provided evidence favoring the selectivity theory of itch depicted in Figure 1(b). Human microneurographic studies have uncovered a class of mechanically insensitive cutaneous receptors with slowly conducting unmyelinated afferent fibers that respond to cutaneous histamine over a time course that closely matches that of concomitant itch sensation (Figure 3; Schmelz, M. *et al.*, 1997). These potential itch receptors are called pruriceptors. Such histamine-responsive fibers also responded more weakly to prostaglandin E$_2$, serotonin, and acetylcholine, which were capable of inducing weak itch when delivered intradermally via microdialysis (Schmelz, M. *et al.*, 2003b). However, the fibers were not exclusively pruriceptive, as they also responded to capsaicin and bradykinin, which induce burning pain (Schmelz, M. *et al.*, 2003b), indicating that these fibers are itch-selective rather than itch-specific. These fibers could also be excited by electrical skin stimulation; however, the current intensities were much higher compared to electrical currents eliciting itch from the wrist (Ikoma, A. *et al*, 2005) suggesting that there may be additional pruriceptor subtypes.

Following the discovery of histamine-selective pruriceptors, a subpopulation of mechanically insensitive spinothalamic tract neurons in the cat superficial dorsal horn (lamina I) with temporally similar responses to histamine was identified (Andrew, D. and Craig, A. D., 2001) (Figure 3). Chemical selectivity was tested in four cases, with two additionally responding to mustard oil (which elicits burning pain) while the

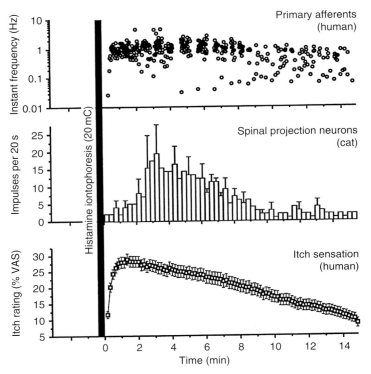

Figure 3 Time course of histamine-evoked response of an unmyelinated primary afferent fiber (upper panel), and mean responses of lamina I spinothalamic tract neurons (middle panel) in relation to itch sensation (lower panel). Vertical bar indicates time of delivery of histamine by iontophoresis to the skin. Adapted from Schmelz, M., Schmidt, R., Bickel, A., Forster, C., Handwerker, H. O., and Torebjork, H. E. 1997. Specific C-receptors for itch in human skin. J. Neurosci. 17, 8003–8008 and Andrew, D. and Craig, A. D. 2001. Spinothalamic lamina I neurons selectively sensitive to histamine: a central neural pathway for itch. Nat. Neurosci. 4, 72–76, with permission from Nature Publishing Group.

other two did not. These latter two units may be histamine-selective, while the units sensitive to both histamine and mustard oil resemble lamina I units in rats that gave prolonged response to serotonin (5-HT; also a pruritic) but usually also exhibited phasic responses to mustard oil, capsaicin, and other algogens (Jinks, S. L. and Carstens, E., 2002). An older study reported a fraction of wide dynamic range (WDR)-type neurons to respond to cutaneous application of cowhage (Wei, J. Y. and Tuckett, R. P., 1991). A recent study of identified spinothalamic tract neurons in primate found that about half responded to intra-dermal histamine, with all additionally responding to capsaicin (Simone, D. A. *et al.*, 2004). Moreover, neuronal responses to light mechanical stroking were unchanged after histamine, arguing against a role in alloknesis for these neurons. The authors conclude that such neurons are more likely to signal pain than itch. Similar findings were reported in the rat for neurons not identified by ascending projections (Carstens, E., 1997). Intradermal histamine produced an *N*-methyl-D-aspartic acid (NMDA) receptor-

dependent expansion of lumbar neuronal receptive fields (Jinks, S. L. and Carstens, E., 1998) possibly reflecting central sensitization of pain rather than itch, as histamine may be algesic rather than pruritic in rats (see below; Jinks, S. L. and Carstens, E., 2002). Thus, the bulk of current evidence favors the concept of an itch-selective pathway, originating from mechanically insensitive, pruritogen-selective C-fiber afferents which project to lamina I spinothalamic tract neurons that, in turn, convey itch signals to the lateral thalamus and cortex (Figure 1(b)).

An alternative view is that spinal WDR neurons may also contribute to itch, much as they are thought to contribute to signaling pain together with nociceptive-specific neurons. Both pruritic (serotonin, histamine) and algesic chemicals (capsaicin, nicotine, formalin) excited neurons in overlapping regions of the superficial dorsal horn (laminae I–II) as assessed by c-fos (Jinks, S. L. *et al.*, 2002; Nojima, H. *et al.*, 2003b). The vast majority of single WDR neurons recorded in rat superficial dorsal horn responded to both pruritic and algesic chemical stimuli as well as noxious heat (Jinks, S. L. and Carstens, E., 2000; 2002). Importantly, superficial dorsal horn neurons gave prolonged responses to serotonin that corresponded with the time course of serotonin-evoked scratching behavior in rats (Jinks, S. L. and Carstens, E., 2002) (Figure 4). The neuronal discharge rate during the response to serotonin was much lower compared to the phasic responses evoked by capsaicin and other algesic stimuli. Serotonin elicits dose-related scratching in rodents that was significantly reduced by opioid antagonists (Yamaguchi, T. *et al.*, 1999; Nojima, H. and Carstens, E., 2003a; Nojima, H. *et al.*, 2003b). Furthermore, c-fos expression in the superficial dorsal horn elicited by intradermal capsaicin, but not serotonin, was attenuated by morphine (Nojima, H. *et al.*, 2003b), consistent with serotonin being pruritic rather than algesic. If this is the case, then one must consider the possibility that pruriceptive information is being signaled by superficial WDR neurons, consistent with the frequency theory of itch.

In conclusion, exciting new information strongly favors the existence of an itch-selective pathway, but it would probably be wise not to completely discount the possible added contribution of nonselective neurons until more complete details about itch signaling have emerged. Future efforts to characterize itch-signaling neurons should test the following properties consistent with itch:

- Responses inhibited by scratching and other antipruritic counterstimuli.
- Responses facilitated by morphine (Jinks, S. L. and Carstens, E., 2000).

- Morphine facilitation blocked by κ-opioid antagonists.
- Responses inhibited by μ-opioid antagonists.
- Increased mechanical sensitivity consistent with alloknesis.

9.5 Alloknesis (Itchy Skin) and Sensitization

Intradermal injection of histamine elicits itch sensation, a wheal at the injection site, and a surrounding flare reaction due to histamine-evoked release of vasodilatory peptides (substance P (SP), calcitonin gene-related peptide). Light stroking in a broad area of normal skin surrounding the histamine injection can elicit itch, a phenomenon called itchy skin or alloknesis (Simone, D. A. *et al.*, 1991). The spread of alloknesis was prevented by local anesthesic block adjacent to the site of histamine injection, indicating that alloknesis is of neurogenic origin (Simone, D. A. *et al.*, 1991). Noxious punctuate stimulation in and around the region of alloknesis also elicits itch (hyperknesis) (Atanassoff, P. G. *et al.*, 1999).

The chronic itch of skin conditions such as atopic dermatitis suggests the possibility of peripheral or central sensitization along the itch-signaling pathway (Ikoma, A. *et al.*, 2004). Moreover, noxious stimuli such as acetylcholine, acid, heat, and pinprick, which normally evoke pain, instead elicit itch in atopic dermatitis patients (Vogelsang, M. *et al.*, 1995; Ikoma, A. *et al.*, 2004). A peripheral sensitization of

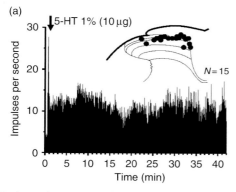
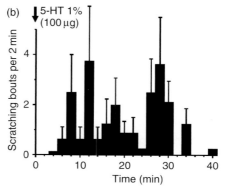

Figure 4 Prolonged responses of superficial dorsal horn neurons to 5-HT compared with scratching. (a) Averaged peristimulus–time histogram (PSTH, bin width: 1 s) of responses of superficial dorsal horn neurons to intradermal microinjection of 5-HT (60 mM; 1 μl) at arrow in pentobarbital-anesthetized rats. Inset shows recordings sites (dots) compiled on section through L5 dorsal horn. (b) Graph plots mean number of hindlimb scratching bouts (error bars: SEM) directed toward the site of injection of 5-HT in the nape of the neck in Sprague-Dawley rats. Scratching bouts were counted in 2 min intervals. Adapted from Jinks, S. L. and Carstens, E. 2002. Prolonged responses of superficial dorsal horn neurons to intracutaneous serotonin and other irritant chemicals: comparison with scratching behavior. J. Neurophysiol. 87, 1280–1289.

pruriceptors is suggested by microneurographic recordings from an atopic dermatitis patient, revealing a subset of mechanically insensitive C-fiber afferents that exhibited irregular patterns of spontaneous activity; two of these were activated by pruritic stimulation (Schmelz, M. *et al.*, 2003a). Since mechanically insensitive, histamine-responsive C-fibers normally do not exhibit any spontaneous activity (Schmelz, M. *et al.*, 1997), their increased firing in atopic dermatitis may represent sensitization that contributes to spontaneous itch. Central sensitization is also likely to contribute to pathological itch (Ikoma, A. *et al.*, 2004). Central sensitization of pain is mediated partly via excessive excitation of spinal neurons by nociceptors releasing glutamate and SP, which act at postsynaptic NMDA and neurokinin-1 receptors to initiate Ca^{2+}-dependent intracellular events resulting in enhanced neuronal excitability (Ji, R. R. *et al.*, 2003). Central sensitization is associated with hyperalgesia (increased pain), allodynia (pain from a normally nonpainful stimulus), and expansion of mechanosensitive receptive fields. It is conceivable that conditions associated with chronic itch may induce central sensitization of itch-signaling neurons via mechanisms similar to those for central sensitization of pain (Ikoma, A. *et al.*, 2004).

9.6 Central Representation of Itch

Using positron emission tomography (PET), histamine injection in the right upper arm differentially activated the contralateral anterior cingulate cortex, ipsilateral inferior parietal lobule, as well as bilateral supplementary and premotor cortical areas (Hsieh, J. C. *et al.*, 1994). Activation of the ipsilateral premotor cortical areas was suggested to represent an urge to scratch. A more recent PET study using histamine pin prick on the right arm reported dose-related activation in premotor and supplementary motor areas and anterior cingulate gyrus, as well as in contralateral primary (but not secondary) somatosensory and primary motor cortex, prefrontal cortex and contralateral insula, and prefrontal and superior temporal cortex (Drzezga, A. *et al.*, 2001). While there was some overlap in cortical areas activated by itch versus pain, there were notable differences, particularly in the activation of secondary somatosensory cortex and thalamus by pain but not itch (Drzezga, A. *et al.*, 2001). A third recent PET study also reported dose-related activation in many of the same cortical areas by iontophoretically applied histamine, but notably also in thalamus. Simultaneous

application of a noxious cold ($5\,^{\circ}$C) stimulus reduced the degree of histamine-evoked cortical activation. Interestingly, coapplication of noxious cold and histamine revealed significant activation in midbrain periaqueductal gray (PAG). This suggests that in addition to the well-known role of the PAG in pain modulation, it may also play a role in modulation of itch (Carstens, E., 1997; Mochizuki, H. *et al.*, 2003). These studies indicate that noxious and pruritic stimuli activate overlapping as well as separate cortical and subcortical regions, lending partial support for separate central pathways mediating itch versus pain.

9.7 Animal Models of Itch

Scratching directed toward a site of pruritic stimulation in rodents is thought to reflect itch sensation (Berendsen, H. H. G. and Broekkamp, C. L. E., 1991; Kuraishi, Y. *et al.*, 1995). However, scratching is a normal grooming behavior and in some cases may reflect pain rather than itch (DeCastro-Costa, M. *et al.*, 1987). Moreover, hindlimb scratching is elicited by intrathecal injection of numerous agents including neurokinins, capsaicin, and morphine (Frenk, H. *et al.*, 1988; Wilcox, G. L., 1988; Ko, M. C. and Naughton, N. N., 2000), even in spinalized animals (Bossut, D. *et al.*, 1988) suggesting direct activation of motor scratch reflex circuits. Facial scratching is also elicited by intracerebroventricular and intramedullary injection of opiates (Königstein, H., 1948; Thomas, D. A. *et al.*, 1993; Thomas, D. A. and Hammond, D. L., 1995; Tohda, C. *et al.*, 1997b; Johnson, M. D. *et al.*, 1999; Kuraishi, Y. *et al.*, 2000) and other substances including adrenocorticotropic hormone (ACTH) and many neuropeptides (Share, N. N. and Rackham, A., 1981; Van Wimersma Greidanus, T. B., 1986; Van Wimersma Greidanus, T. B. *et al.*, 1987; Van Wimersma Greidanus, T. B. and Maigret, C., 1988; Van Wimersma Greidanus, T. B. *et al.*, 1988; Van Wimersma Greidanus, T. B. and Maigret, C., 1991; Johnson, M. D. *et al.*, 1999).

Despite the caveats noted above, there is good evidence that hindlimb scratching may distinguish between itch and pain. Substances that induce itch in humans, including SP (Hagermark, O. *et al.*, 1978), the mast cell degranulator compound 48/80 (Fjellner, B. and Hägermark, O., 1982), serotonin (5-HT; Weisshaar, E. *et al.*, 1997), and platelet activating factor (PAF, Fjellner, B. and Hägermark, O., 1985), elicit dose-related scratching in rodents (Berendsen, H. H. G and Broekkamp, C. L. E, 1991; Kuraishi, Y. *et al.*, 1995;

Woodward, D. F. *et al.*, 1995; Yamaguchi, T. *et al.*, 1999; Thomsen J. S. *et al.*, 2001; Jinks, S. L. and Carstens, E., 2002). In contrast, capsaicin and formalin, which induce burning pain, elicit little or no scratching (Kuraishi, Y. *et al.*, 1995; Yamaguchi, T. *et al.*, 1999; Jinks, S. L. and Carstens, E., 2002). One notable difference is histamine, which is the definitive pruritogen in humans (Simone, D. A. *et al.*, 1987; 1991; Hagermark, O., 1995) yet does not elicit scratching in several strains of mice (Kuraishi, Y. *et al.*, 1995; Yamaguchi, T. *et al.*, 1999; Inagaki, N. *et al.*, 2001) or Sprague-Dawley rats (Thomsen, J. S. *et al.*, 2001; Jinks, S. L. and Carstens, E., 2002). This might be explained by the much lower concentration of histamine in cutaneous mast cells of rodents (Gustafsson, B., 1980; Graziano, F. M., 1988; Purcell, W. M. *et al.*, 1989) in favor of 5-HT which induces scratching via the 5-HT$_2$ receptor (Yamaguchi, T. *et al.*, 1999; Nojima, H. and Carstens, E., 2003b). However, other rodent strains including the ICR mouse (Maekawa, T. *et al.*, 2000) and hairless guinea-pig (Woodward, D. F. *et al.*, 1995) exhibit dose-related scratching to histamine, indicating the existence of strain differences in the scratch-inducing potency of itch mediators.

Importantly, opiate antagonists significantly attenuated hindlimb scratching elicited by 5-HT (Yamaguchi, T. *et al.*, 1999; Nojima, H. and Carstens, E., 2003a; Nojima, H. *et al.*, 2003b) or SP (Andoh, T. *et al.*, 1998). This is consistent with the report that opiate antagonists reduce histamine-induced itch in humans (Heyer, G. *et al.*, 1997a) and further supports scratching as an itch-related response.

Additional animal models of acute itch include hindlimb scratching directed toward the eye in albino guinea-pigs (Woodward, D. F. *et al.*, 1995). In this model, histamine, 5-HT, PAF, and prostaglandin E$_2$ elicited significant, dose-related ocular scratching while even high concentrations of the algesic agents bradykinin, acetic acid, or saline did not. Another model is biting directed toward the site of 5-HT injected into the hind paw of mice (Hagiwara, K. *et al.*, 1999). 5-HT (60–100 nmol) elicited approximately equal numbers of licks and bites, but only biting was significantly attenuated by both naloxone and the 5-HT antagonist, methysergide. In contrast, formalin elicited hind paw licking but no biting. These results suggest that biting represents a means of noxious counterstimulation of the paw to relieve itch.

Animal models of acute allergic (e.g., Inagaki, N. *et al.*, 2000; Ohtsuka, E. *et al.*, 2001; Nojima, H. and Carstens, E., 2003b) and dry-skin itch (Miyamoto, T. *et al.*, 2002b; Nojima, H. *et al.*, 2003a; Nojima, H. *et al.*,

2004) have also been developed. One model that most people can relate to is mosquito-bite-induced itch. Initial exposure of ICR mice to mosquito bites did not result in scratching. However, twice-weekly exposure of the mice to mosquito bites resulted in a progressive increase in scratching. Naïve mice sensitized by injection of female mosquito salivary gland extract (the presumed antigen) exhibited marked scratching upon the first mosquito bite, indicating that it triggered an acute allergic reaction and associated itch (Ohtskua, E. *et al.*, 2001).

A potential model of chronic itch is the NC/jic mouse (Tohda, C. *et al.*, 1997a; Yamaguchi, T. *et al.*, 2001). Within 2–6 months after birth, a majority of these animals spontaneously develop skin lesions (eczema, bleeding, alopecia) associated with excessive scratching that is naloxone-sensitive. These animals did not develop skin lesions, scratching, or increased plasma immonoglobulin G levels when reared in a specific pathogen-free environment, but did so within 4 weeks after being transferred to a conventional environment. The NC/jic mouse may represent an animal model of itch associated with chronic dermatitis, affording the possibility to develop novel antipruritic treatment strategies for clinical cases of itch that are poorly treated by antihistamines.

9.8 Itch Mediators

Histamine is the most important mediator of normal human itch. It is localized in cutaneous mast cells along with a variety of other chemicals including serotonin, chymase, and tryptase. Intradermal histamine primarily elicits a dose-dependent sensation of itch (Melton, F. M. and Shelly, W. P., 1950; Keele, C. A. and Armstrong, D., 1964; Handwerker, H. O. *et al.* 1987; Simone, D. A. *et al.* 1987; 1991). Microdialysis of histamine in the skin elicited almost exclusively, itch sensation and strongly excited mechanically insensitive C-fibers that are considered to be pruriceptors as discussed earlier (Schmelz, M. *et al.*, 2003b). The itch sensation evoked by histamine and other pruritic chemicals (e.g., serotonin, SP, PAF, vasoactive intestinal polypeptide) is in nearly all cases prevented by pretreatment with H$_1$ receptor antagonists (Hagermark, O., 1992; Heyer, G. and Magerl, W., 1995), suggesting a common action via histamine released by mast cell degranulation. However, other mediators might act independently of histamine release to induce itch, as discussed further below.

More than 50 years ago, Arthur R. P. and Shelley W. B. (1955) reported the itch-inducing effects of proteases including mucunain (Shelley W. B. and Arthur, R. P., 1955), a peptidase extracted from spicules of the velvet bean plant, cowhage, which elicit itch upon contact with skin. The serine proteases trypsin and chymase and the cystein protease papain induce itch that may be independent of histamine (Rajka, G., 1969; Hagermark, O. *et al.*, 1972). Mast cell chymase elicited scratching in mice, and a chymase inhibitor reduced allergic scratching (Imada, T. *et al.*, 2002). An exciting new development is the discovery of the proteinase-activated receptor (PAR). One subtype, PAR-2, is targeted by trypsin and mast cell tryptase, is expressed in sensory neurons and other tissues (Steinhoff, M. and Roosterman, D., 2005) and may be involved in itch (Steinhoff, M. *et al.*, 2003). In particular, PAR-2 expression is increased fourfold in patients with atopic dermatitis, and these patients experience itch upon intralesional application of PAR-2 agonists that is histamine-independent (Steinhoff, M. *et al.*, 2003). Thus, PAR-2 represents a novel target for antipruritic drugs, particularly for conditions like atopic dermatitis in which the itch is refractory to antihistamines.

Serotonin (5-HT) delivered to the skin induces weak itch in humans (Hagermark, O., 1992; Weisshaar, E. *et al.*, 1997) and may be involved in the pruritus of polycythemia vera (Fjellner, B. and Hagermark, O., 1979; Fitzsimons, E. J. *et al.*, 1981) and cholestasis (Schworer, H. *et al.*, 1995; Weisshaar, E. *et al.*, 1999; Jones, E. A. and Bergasa, N. V., 2000). Intradermal microdialysis of 5-HT elicited weak-to-moderate itch in some subjects and pain in others, but excited mechanically insensitive, histamine-sensitive C-fiber pruriceptors (Schmelz, M. *et al.*, 2000b; 2003a) in support of a role as itch mediator. As noted earlier, 5-HT potently induces scratching in rodents via the 5-HT_2 receptor (Yamaguchi, T. *et al.*, 1999; Nojima, H. and Carstens, E., 2003b), whereas the 5-HT_3 receptor may be involved under certain chronic itch conditions in humans (Schworer, H. *et al.*, 1995; Weisshaar, E. *et al.*, 1999).

Intradermal injection (Hagermark, O. and Strandberg, K., 1977) or microdialysis (Neisius, U. *et al.*, 2002; Schmelz, M. *et al.*, 2003) of prostaglandin E_2 also elicited weak itch in some subjects and strongly excited histamine-sensitive C-fiber pruriceptors (Schmelz, M. *et al.*, 2003).

Acetylcholine normally elicits pain, but instead induced itch in patients with atopic dermatitis (Heyer, G. *et al.*, 1997b). Intradermal microdialysis of acetylcholine induced itch in some subjects but more frequently pain, and excited most histamine-sensitive C-fiber pruriceptors tested (Schmelz, M. *et al.*, 2003). Cholinergic agonists elicited dose-related scratching in mice that was reduced by opiate and muscarinic M3 receptor antagonists; nicotine did not elicit scratching (Miyamoto, T. *et al.*, 2002).

Intradermal injection of the neuropeptide SP elicits itch (Hagermark, O. *et al.*, 1978; Thomsen, D. A. *et al.*, 2002) via mast cell degranulation (Hagermark, O. *et al.*, 1978). However, a recent study reported no itch or pain sensations with up to 10^{-5} M SP delivered by intradermal microdialysis, and concluded that physiological levels of SP are insufficient to induce mast cell degranulation (Weidner, C. *et al.*, 2000). In rodents, SP elicited dose-dependent scratching that was reduced by opiate and neurokinin-1 antagonists, and was at least partly independent of mast cell degranulation (Andoh, T. *et al.*, 1998).

A variety of other mediators may also have some role in itch. PAF elicits histamine-dependent itch in humans (Fjellner, B. and Hagermark, O., 1985), and scratching behavior in guinea-pigs, which, however, was reduced by PAF antagonists but not by histamine H1 antagonists (Woodward, D. F. *et al.*, 1995). Leukotriene B_4 (LTB_4) elicited scratching in mice that was attenuated by an LTB_4 antagonist (Andoh, T. and Kuraishi, Y., 1998). LTB_4 also plays a role in scratching elicited by intradermal injection of SP (Andoh, T. *et al.*, 2001) and nociceptin (Andoh, T. *et al.*, 2004). Endothelin delivered by intradermal microdialysis or injection induced short-lasting burning itch accompanied by release of histamine and nitric oxide (Wenzel, R. R *et al.*, 1998; Katugampola, R. *et al.*, 2000). Interleukin-2 (IL-2), but not tumor necrosis factor-alpha, given intradermally induced weak itch (Wahlgren, C. F. *et al.*, 1995; Darsow, U. *et al.*, 1997) and there is indirect evidence that IL-4 and IL-6 may also have some role in pathological itch (Nordlind, K. *et al.*, 1996; Chan, L. S. *et al.*, 2001).

The preceding list is by no means comprehensive, but certainly we have now at least scratched the surface in terms of identifying peripheral mediators, receptors, and neural pathways involved in signaling the enigmatic sensory quality of itch.

Acknowledgments

The author's work has been supported by research grants from the NIH (#DE13685) and the State of California Tobacco-Related Disease Research Program (#11RT-0053).

References

Andoh, T. and Kuraishi, Y. 1998. Intradermal leukotriene B_4, but not prostaglandin E_2, induces itch-associated responses in mice. Eur. J. Pharmacol. 353, 93–96.

Andoh, T., Katsube, N., Maruyama, M., and Kuraishi, Y. 2001. Involvement of leukotriene B4 in substance P-induced itch-associated response in mice. J. Invest. Dermatol. 117, 1621–1626.

Andoh, T., Nagasawa, T., Satoh, M., and Kuraishi, Y. 1998. Substance P induction of itch-associated response mediated by cutaneous NK1 tachykinin receptors in mice. J. Pharmacol. Exp. Ther. 286, 1140–1145.

Andoh, T., Yageta, Y., Takeshima, H., and Kuraishi, Y. 2004. Intradermal nociceptin elicits itch-associated responses through leukotriene B(4) in mice. J. Invest. Dermatol. 123, 196–201.

Andrew, D. and Craig, A. D. 2001. Spinothalamic lamina I neurons selectively sensitive to histamine: a central neural pathway for itch. Nat. Neurosci. 4, 72–76.

Arthur, R. P. and Shelley, W. B. 1955. The role of proteolytic enzymes in the production of pruritus in man. J. Invest. Dermatol. 25, 341–346.

Atanassoff, P. G., Brull, S. J., Zhang, J., Greenquist, K., Silverman, D. G., and LaMotte, R. H. 1999. Enhancement of experimental pruritus and mechanically evoked dysesthesiae with local anesthesia. Somatosens. Motor Res. 16, 291–298.

Ballantyne, J. C., Loach, A. B., and Carr, D. B. 1988. Itching after epidual and spinal opiates. Pain 33, 149–160.

Berendsen, H. H. G. and Broekkamp, C. L. E. 1991. A peripheral 5-HT$_{1D}$-like receptor involved in serotonergic induced hindlimb scratching in rats. Eur. J. Pharmacol. 194, 201–208.

Bergasa, N. V., Mehlman, J. K., and Jones, E. A. 2000. Pruritus and fatigue in primary biliary cirrhosis. Bailliere's Best Pract. Res. Clin. Gastroenterol. 14, 643–655.

Bickford, R. G. 1937. Experiments relating to the itch sensation, its peripheral mechanism and central pathway. Clin. Sci. 3, 377–386.

Biro, T., Ko, M. C., Bromm, B., Wei, E. T., Bigliardi, P., Siebenhaar, F., Hashizume, H., Misery, L., Bergasa, N. V., Kamei, C., Schouenborg, J., Roostermann, D., Szabo, T., Maurer, M., Bigliardi-Qi, M., Meingassner, J. G., Hossen, M. A., Schmelz, M., and Steinhoff, M. 2005. How best to fight that nasty itch – from new insights into the neuroimmunological, neuroendocrine, and neurophysiological bases of pruritus to novel therapeutic approaches. Exp. Dermatol. 14, 225–240.

Bossut, D., Frenk, H., and Mayer, D. J. 1988. Is substance P a primary afferent neurotransmitter for nociceptive input? II. Spinalization does not reduce and intrathecal morphine potentiates behavioral responses to substance P. Brain Res. 445, 232–239.

Broadbent, J. L. 1955. Observations on histamine-induced pruritis and pain. Br. J. Pharmacol. 10, 183–185.

Bromage, P. 1981. The price of intraspinal narcotic analgesia: basic constraints. Anesth. Analg. 60, 461–463.

Bromage, P. R., Camporesi, E. M., Durant, P. A., and Nielsen, C. H. 1982. Nonrespiratory side effects of epidural morphine. Anesth. Analg. 61, 490–495.

Carstens, E. 1997. Responses of rat spinal dorsal horn neurons to intracutaneous microinjection of histamine, capsaicin, and other irritants. J. Neurophysiol. 77, 2499–2514.

Chan, L. S., Robinson, N., and Xu, L. 2001. Expression of interleukin-4 in the epidermis of transgenic mice results in a pruritic inflammatory skin disease: an experimental animal model to study atopic dermatitis. J. Invest. Dermatol. 117, 977–983.

Darsow, U., Scharein, E., Bromm, B., and Ring, J. 1997. Skin testing of the pruritogenic activity of histamine and cytokines (interleukin-2 and tumour necrosis factor-alpha) at the dermal–epidermal junction. Br. J. Dermatol. 137, 415–417.

DeCastro-Costa, M., Gybels, J., Kupers, R., and Van Hees, J. 1987. Scratching behaviour in arthritic rats: a sign of chronic pain or itch? Pain 29, 123–131.

Dey, D. D., Landrum, O., and Oaklander, A. L. 2005. Central neuropathic itch from spinal-cord cavernous hemangioma: a human case, a possible animal model, and hypotheses about pathogenesis. Pain 113, 233–237.

Drzezga, A., Darsow, U., Treede, R. D., Siebner, H., Frisch, M., Munz, F., Weilke, F., Ring, J., Schwaiger, M., and Bartenstein, P. 2001. Central activation by histamine-induced itch: analogies to pain processing: a correlational analysis of O-15 H_2O positron emission tomography studies. Pain 92, 295–305.

Fitzsimons, E. J., Dagg, J. H., and McAllister, E. J. 1981. Pruritus of polycythaemia vera: a place for pizotifen? Br. Med. J. 283, 277.

Fjellner, B. and Hagermark, O. 1979. Pruritus in polycythemia vera: treatment with aspirin and possibility of platelet involvement. Acta Derm. Venereol. 59, 505–512.

Fjellner, B. and Hägermark, O. 1982. Influence of ultraviolet light on itch and flare reactions in human skin induced by histamine and the histamine liberator compound 48/80. Acta Derm. Venereol. 62, 137–140.

Fjellner, B. and Hägermark, O. 1985. Experimental pruritus evoked by platelet activating factor (PAF-acether) in human skin. Acta Derm. Venereol. 65, 409–412.

Frenk, H., Bossut, D., Urca, G., and Mayer, D. J. 1988. Is substance P a primary afferent neurotransmitter for nociceptive input? I. Analysis of pain-related behaviors resulting from intrathecal administration of substance P and 6 excitatory compounds. Brain Res. 455, 223–231.

Frustorfer, H, Hermanns, M., and Latzke, L. 1986. The effects of thermal stimulation on clinical and experimental itch. Pain 24, 259–269.

Gammon, G. D. and Starr, I. 1941. Studies on the relief of pain by counterirritation. J. Clin. Invest. 20, 13–20.

Graziano, F. M. 1988. Mast cells and mast cell products. Methods Enzymol. 162, 501–522.

Gustafsson, B. 1980. Cytofluorometric analysis of anaphylactic secretion of 5-hydroxytryptamine and heparin from rat mast cells. Int. Arch Allergy Appl. Immunol. 63, 121–128.

Hagermark, O. 1992. Peripheral and central mediators of itch. Skin Pharmacol. 5, 1–8.

Hägermark, O. 1995. Itch mediators. Sem. Dermatol. 14, 271–276.

Hagermark, O. and Strandberg, K. 1977. Pruritogenic activity of prostaglandin E2. Acta Derm. Venereol. 57, 37–43.

Hagermark, O., Hokfelt, T., and Pernow, B. 1978. Flare and itch induced by substance P in human skin. J. Invest. Dermatol. 71, 233–235.

Hagermark, O., Rajka, G., and Bergvist, U. 1972. Experimental itch in human skin elicited by rat mast cell chymase. Acta Derm. Venereol. 152, 25–128.

Hagiwara, K, Nojima, H., and Kuraishi, Y. 1999. Serotonin-induced biting of the hind paw is itch-related response in mice. Pain Res. 14, 53–59.

Handwerker, H. O., Forster, C., and Kirchoff, C. 1991. Discharge patterns of human C-fibers induced by itching and burning stimuli. J. Neurophysiol. 66, 307–315.

Handwerker, H. O., Magerl, W., Klemm, F., Land, E., and Westerman, R. A. 1987. Quantitative Evaluation of Itch Sensation. In: Fine Afferent Nerve Fibers and Pain (eds. R. F. Schmidt, H.-G. Schaible, and C. Vahle-Hinz), pp. 462–473. VCH.

Heyer, G. and Magerl, W. 1995. Skin Reactions and Sensations Induced by Intradermal Injection of Substance P into

Compound 48/80 Pretreated Skin. In: Wound Healing and Skin Physiology (*ed*. M. Altmeyer), pp. 335–344. Springer.

Heyer, G., Dotzer, M., Diepgen, T. L., and Handwerker, H. O. 1997a. Opiate and H1 antagonist effects on histamine induced pruritus and alloknesis. Pain 73, 239–243.

Heyer, G., Vogelgsang, M., and Hornstein, O. P. 1997b. Acetylcholine is an inducer of itching in patients with atopic eczema. J. Dermatol. 24, 621–625.

Hsieh, J. C., Hagermark, O., Stahle-Backdahl, M., Ericson, K., Eriksson, L., Stone-Elander, S., and Ingvar, M. 1994. Urge to scratch represented in the human cerebral cortex during itch. J. Neurophysiol. 72, 3004–3008.

Ikoma, A., Fartasch, M., Heyer, G., Miyachi, Y., Handwerker, H., and Schmelz, M. 2004. Painful stimuli evoke itch in patients with chronic pruritus: central sensitization for itch. Neurology 62, 212–217.

Ikoma, A., Handwerker, H., Miyachi, Y., and Schmelz, M. 2005. Electrically evoked itch in humans. Pain 113, 148–154.

Imada, T., Komorita, N., Kobayashi, F., Naito, K., Yoshikawa, T., Miyazaki, M., Nakamura, N., and Kondo, T. 2002. Therapeutic potential of a specific chymase inhibitor in atopic dermatitis. Jpn. J. Pharmacol. 90, 214–217.

Inagaki, N., Nagao, M., Igeta, K., Kawasaki, H., Kim, J. F., and Nagai, H. 2001. Scratching behavior in various strains of mice. Skin Pharmacol. Appl. Skin Physiol. 14, 87–96.

Inagaki, N., Nakamura, N., Nagao, M., Kawasaki, H., and Nagai, H. 2000. Inhibition of passive cutaneous anaphylaxis-associated scratching behavior by mu-opioid receptor antagonists in ICR mice. Int. Arch. Allergy Immunol. 123(4), 365–368.

Ji, R. R., Kohno, T., Moore, K. A., and Woolf, C. J. 2003. Central sensitization and LTP: do pain and memory share similar mechanisms? Trends Neurosci. 26, 696–705.

Jinks, S. L. and Carstens, E. 1998. Spinal NMDA receptor involvement in expansion of dorsal horn neuronal receptive field area produced by intracutaneous histamine. J. Neurophysiol. 79, 1613–1618.

Jinks, S. L. and Carstens, E. 2000. Superficial dorsal horn neurons identified by intracutaneous histamine: chemonociceptive responses and modulation by morphine. J. Neurophysiol. 84, 616–627.

Jinks, S. L. and Carstens, E. 2002. Prolonged responses of superficial dorsal horn neurons to intracutaneous serotonin and other irritant chemicals: comparison with scratching behavior. J. Neurophysiol. 87, 1280–1289.

Jinks, S. L., Simons, C. T., Dessirier, J. M., Carstens, M. I., Antognini, J. F., and Carstens, E. 2002. C-fos induction in rat superficial dorsal horn following cutaneous application of noxious chemical or mechanical stimuli. Exp. Brain Res. 145, 261–269.

Johnson, M. D., Ko, M., Choo, K. S., Traynor, J. R., Mosberg, H. I., Naughton, N. N., and Woods, J. H. 1999. The effects of the phyllolitorin analogue [desTrp(3), Leu(8)]phyllolitorin on scratching induced by bombesin and related peptides in rats. Brain Res. 839, 194–198.

Jones, E. A. and Bergasa, N. V. 2000. Evolving concepts of the pathogenesis and treatment of the pruritus of cholestasis. Can. J. Gastroenterol. 14, 33–40.

Katugampola, R., Church, M. K., and Clough, G. F. 2000. The neurogenic vasodilator response to endothelin-1: a study in human skin *in vivo*. Exp. Physiol. 85, 839–844.

Keele, C. A. and Armstrong, D. 1964. Substances Producing Pain and Itch. Williams and Wilkins.

Ko, M. C. and Naughton, N. N. 2000. An experimental itch model in monkeys: characterization of intrathecal morphine-induced scratching and antinociception. Anesthesiology 92, 795–805.

Ko, M. C., Lee, H., Song, M. S., Sobczyk-Kojiro, K., Mosberg, H. I., Kishioka, S., Woods, J. H., and

Naughton, N. N. 2003. Activation of kappa-opioid receptors inhibits pruritus evoked by subcutaneous or intrathecal administration of morphine in monkeys. J. Pharmacol. Exp. Ther. 305, 173–179.

Koltzenburg, M., Handwerker, H. O., and Torebjörk, H. E. 1993. The ability of humans to localise noxious stimuli. Neurosci. Lett. 150, 219–222.

Königstein, H. 1948. Experimental study of itch stimuli in animals. Arch. Derm. Syphilol. 57, 829–849.

Krajnik, M. and Zbigniew, Z. 2001. Understanding pruritis in systemic disease. J. Pain Symptom Manage. 21, 151–168.

Kuraishi, Y., Nagasawa, T., Hayashi, K., and Satoh, M. 1995. Scratching behavior induced by pruritogenic but not algesiogenic agents in mice. Eur. J. Pharmacol. 275, 229–233.

Kuraishi, Y., Yamaguchi, T., and Miyamoto, T. 2000. Itch-scratch responses induced by opioids through central mu opioid receptors in mice. J. Biomed. Sci. 7, 248–252.

Maekawa, T., Nojima, H., and Kurahishi, Y. 2000. Itch-associated responses of afferent nerve innervating the murine skin: different effects of histamine and serotonin in ICR and ddY mice. Jpn. J. Pharmacol. 84, 462–466.

Melton, F. M. and Shelly, W. P. 1950. The effect of topical antipruritic therapy on experimentally induced pruritis in man. J. Invest. Dermatol. 15, 325–332.

Miyamoto, T., Nojima, H., and Kuraishi, Y. 2002a. Cholinergic agonists induce itch-associated response via M3 muscarinic acetylcholine receptors in mice. Jpn. J. Pharmacol. 88, 351–354.

Miyamoto, T., Nojima, H., Shinkado, T., Nakahashi, T., and Kuraishi, Y. 2002b. Itch-associated response induced by experimental dry skin in mice. Jpn. J. Pharmacol. 88, 285–292.

Mochizuki, H., Tashiro, M., Kano, M., Sakurada, Y., Itoh, M., and Yanai, K. 2003. Imaging of central itch modulation in the human brain using positron emission tomography. Pain 105, 339–346.

Morgan, M. 1987. Epidural and intrathecal opioids. Anaesth. Intensive Care 15, 60–67.

Murphy, M. and Carmichael, A. J. 2000. Renal itch. Clin. Exp. Dermatol. 25, 103–106.

Murray, F. S. and Weaver, M. M. 1975. Effects of ipsilateral and contralateral counterirritation on experimentally induced itch in human beings. J. Comp. Physiol. Psychol. 89, 819–826.

Neisius, U., Olsson, R., Rukwied, R., Lischetzki, G., and Schmelz, M. 2002. Prostaglandin E2 induces vasodilation and pruritus, but no protein extravasation in atopic dermatitis and controls. J. Am. Acad. Dermatol. 47, 28–32.

Nilsson, H. J., Levinsson, A., and Schouenborg, J. 1997. Cutaneous field stimulation (CFS): a new powerful method to combat itch. Pain 71, 49–55.

Nojima, H. and Carstens, E. 2003a. Quantitative assessment of directed hind limb scratching behavior as a rodent itch model. J. Neurosci. Methods 126, 137–143.

Nojima, H. and Carstens, E. 2003b. Serotonin 5-HT$_2$ receptor involvement in acute 5-HT-evoked scratching but not in allergic pruritus induced by dinitrofluorobenzene in rats. J. Pharmacol. Exp. Ther. 306, 245–252.

Nojima, H., Cuellar, J. M., Simons, C. T., Iodi Carstens, M., and Carstens, E. 2004. Spinal *c-fos* expression associated with spontaneous biting in a mouse model of dry skin pruritus. Neurosci. Lett. 361, 79–82.

Nojima, H., Iodi Carstens, M., and Carstens, E. 2003a. *c-fos* expression in superficial dorsal horn of cervical spinal cord associated with spontaneous scratching in rats with dry skin. Neurosci. Lett. 347, 62–64.

Nojima, H., Simons, C. T., Cuellar, J. M., Carstens, M. I., Moore, J. A., and Carstens, E. 2003b. Opioid modulation of scratching and spinal c-fos expression evoked by intradermal serotonin. J. Neurosci. 23, 10784–10790.

Nordlind, K., Chin, L. B., Ahmed, A. A., Brakenhoff, J., Theodorsson, E., and Liden, S. 1996. Immunohistochemical localization of interleukin-6-like immunoreactivity to peripheral nerve-like structures in normal and inflamed human skin. Arch. Dermatol. Res. 288, 431–435.

Ochoa, J. and Torebjork, E. 1989. Sensations evoked by intraneural microstimulation of C nociceptor fibres in human skin nerves. J. Physiol. Lond. 415, 583–599.

Ohtsuka, E., Kawai, S., Ichikawa, T., Nojima, H., Kitagawa, K., Shirai, Y., Kamimura, K., and Kuraishi, Y. 2001. Roles of mast cells and histamine in mosquito bite-induced allergic itch-associated responses in mice. Jpn. J. Pharmacol. 86, 97–105.

Purcell, W. M., Cohen, D. L., and Hanahoe, T. H. 1989. Comparison of histamine and 5-hydroxytryptamine content and secretion in rat mast cells isolated from different anatomical locations. Int. Arch. Allergy Applied Immunol. 90, 382–386.

Rajka, G. 1969. Latency and duration of pruritus elicited by trypsin in aged patients with itching eczema and psoriasis. Acta. Derm. Venereol. 49, 401–403.

Schmelz, M., Hilliges, M., Schmidt, R., Orstavik, K., Vahlquist, C., Weidner, C., Handwerker, H. O., and Torebjork, H. E. 2003a. Active "itch fibers" in chronic pruritus. Neurology 61, 564–566.

Schmelz, M., Schmidt, R., Bickel, A., Forster, C., Handwerker, H. O., and Torebjork, H. E. 1997. Specific C-receptors for itch in human skin. J. Neurosci. 17, 8003–8008.

Schmelz, M., Schmidt, R., Weidner, C., Hilliges, M., Torebjork, H. E., and Handwerker, H. O. 2003b. Chemical response pattern of different classes of C-nociceptors to pruritogens and algogens. J. Neurophysiol. 89, 2441–2448.

Schmidt, R., Torebjork, E., and Jorum, E. 1993. Pain and itch from intraneural microstimulation. Abstracts of the 7th World Congress on Pain, Paris, 143.

Schworer, H., Hartmann, H., and Ranadori, G. 1995. Relief of cholestatic pruritus by a novel class of drugs: 5-hydroxytryptamine type 3 (5-HT₃) receptor antagonists: effectiveness of ondansetron. Pain 61, 33–37.

Share, N. N. and Rackham, A. 1981. Intracerebral substance P in mice: behavioral effects and narcotic agents. Brain Res. 211, 379–386.

Shelley, W. B. and Arthur, R. P. 1955. Mucunain, the active pruritogenic proteinase of cowhage. Science 122, 469–470.

Simone, D. A., Alreja, M., and LaMotte, R. H. 1991. Psychophysical studies of the itch sensation and itchy skin ("alloknesis") produced by intracutaneous injection of histamine. Somatosens. Motor Res. 8, 271–279.

Simone, D. A., Nigeow, J. Y. F., Whitehouse, J., Becerra-Cabal, L., Putterman, G. J., and LaMotte, R. H. 1987. The magnitude and duration of itch produced by intracutaneous injections of histamine. Somatosens. Res. 5, 81–92.

Simone, D. A., Zhang, X., Li, J., Zhang, J. M., Honda, C. N., LaMotte, R. H., and Giesler, G. J., Jr. 2004. Comparison of responses of primate spinothalamic tract neurons to pruritic and algogenic stimuli. J. Neurophysiol. 91(1), 213–222.

Stander, S., Steinhoff, M., Schmelz, M., Weisshaar, E., Metze, D., and Luger, T. 2003. Neurophysiology of pruritus: cutaneous elicitation of itch. Arch. Dermatol. 139, 1463–1470.

Steinhoff, M. and Roosterman, D. 2005. Viewpoint 5. Exp. Dermatol. 14, 231–233.

Steinhoff, M., Neisius, U., Ikoma, A., Fartasch, M., Heyer, G., Skov, P. S., Luger, T. A., and Schmelz, M. 2003. Proteinase-activated receptor-2 mediates itch: a novel pathway for pruritus in human skin. J. Neurosci. 23, 6176–6180.

Thomas, D. A. and Hammond, D. L. 1995. Microinjection of morphine into the rat medullary dorsal horn produces a dose-dependent increase in facial scratching. Brain Res. 695, 267–270.

Thomas, D. A., Williams, G. M., Iwata, K., Kenshalo, D. R., Jr., and Dubner, R. 1993. The medullary dorsal horn. A site of action of morphine in producing facial scratching in monkeys. Anesthesiology 9, 548–554.

Thomsen, J. S., Petersen, M. B., Benfeldt, E., Jensen, S. B., and Serup, J. 2001. Scratch induction in the rat by intradermal serotonin: a model for pruritus. Acta Derm. Venereol. 81, 250–254.

Thomsen, J. S., Sonne, M., Benfeldt, E., Jensen, S. B., Serup, J., and Menne, T. 2002. Experimental itch in sodium lauryl sulphate-inflamed and normal skin in humans: a randomized, double-blind, placebo-controlled study of histamine and other inducers of itch. Br. J. Dermatol. 146, 792–800.

Togashi, Y., Umeuchi, H., Okano, K., Ando, N., Yoshizawa, Y., Honda, T., Kawamura, K., Endoh, T., Utsumi, J., Kamei, J., Tanaka, T., and Nagase, H. 2002. Antipruritic activity of the kappa-opioid receptor agonist, TRK-820. Eur. J. Pharmacol. 435, 259–264.

Tohda, C., Yamaguchi, T., and Kuraishi, Y. 1997a. Increased expression of mRNA for myocyte-specific enhancer binding factor (MEF) 2C in the cerebral cortex of the itching mouse. Neurosci. Res. 29, 209–215.

Tohda, C., Yamaguchi, T., and Kuraishi, Y. 1997b. Intracisternal injection of opioids induces itch-associated response through mu-opioid receptors in mice. Jpn. J. Pharmacol. 74, 77–82.

Tuckett, R. P. 1982. Itch evoked by electrical stimulation of the skin. J. Invest. Dermatol. 79, 368–373.

Twycross, R., Greaves, M. W., Handwerker, H., Jones, E. A., Libretto, S. E., Szepietowski, J. C., and Zylicz, Z. 2003. Itch: scratching more than the surface. QJM 96, 7–26.

Umeuchi, H., Togashi, Y., Honda, T., Nakao, K., Okano, K., Tanaka, T., and Nagase, H. 2003. Involvement of central mu-opioid system in the scratching behavior in mice, and the suppression of it by the activation of kappa-opioid system. Eur. J. Pharmacol. 477, 29–35.

Van Wimersma Greidanus, T. B. 1986. Effects of naloxone and neurotensin on excessive grooming behavior of rats induced by bombesin, beta-endorphin and ACTH. NIDA Res Monogr. 75, 477–480.

Van Wimersma Greidanus, T. B. and Maigret, C. 1988. Grooming behavior induced by substance P. Eur. J. Pharmacol. 154, 217–220.

Van Wimersma Greidanus, T. B. and Maigret, C. 1991. Neuromedin-induced excessive grooming/scratching behavior is suppressed by naloxone, neurotensin and a dopamine D1 receptor antagonist. Eur. J. Pharmacol. 209, 57–61.

Van Wimersma Greidanus, T. B., Maigret, C., and Krechting, B. 1987. Excessive grooming induced by somatostatin or its analog SMS 201-995. Eur. J. Pharmacol. 144, 277–285.

Van Wimersma Greidanus, T. B., Maigret, C., Rinkel, G. J., Metzger, P., Panis, M., Van Zinnicq Bergmann, F. E., Poelman, J., and Colbern, D. L. 1988. Some characteristics of TRH-induced grooming behavior in rats. Peptides 9, 283–288.

Vogelsang, M., Heyer, G., and Hornstein, O. P. 1995. Acetylcholine induces different cutaneous sensations in atopic and non-atopic subjects. Acta Derm. Venereol. 75, 434–436.

Von Frey, M. 1922. Zur Physiologie der Juckempfindung. Arch. Neerland. Physiol. 7, 142–145.

Ward, L, Wright, E., and McMahon, S. B. 1996. A comparison of the effects of noxious and non-noxious counterstimuli on experimentally induced itch and pain. Pain 64, 129–138.

Wahlgren, C. F. and Ekblom, A. 1996. Two-point discrimination of itch in patients with atopic dermatitis and healthy subjects. Acta Derm. Venereol. 76, 48–51.

Wahlgren, C. F, Tengvall Linder, M., Hagermark, O., and Scheynius, A. 1995. Itch and inflammation induced by intradermally injected interleukin-2 in atopic dermatitis patients and healthy subjects. Arch. Dermatol. Res. 287, 572–580.

Wei, J. Y. and Tuckett, R. P. 1991. Response of cat ventrolateral spinal axons to an itch-producing stimulus (cowhage). Somatosens Mot. Res. 8, 227–239.

Weidner, C., Klede, M., Rukwied, R., Lischetzki, G., Neisius, U., Skov, P. S., Petersen, L. J., and Schmelz, M. 2000. Acute effects of substance P and calcitonin gene-related peptide in human skin – a microdialysis study. J. Invest. Dermatol. 115, 1015–1020.

Weisshaar, E., Ziethen, B., and Gollnick, H. 1997. Can a serotonin type 3 (5-HT3) receptor antagonist reduce experimentally-induced itch? Inflamm. Res. 46, 412–416.

Weisshaar, E., Ziethen, B., Rohl, F. W., and Gollnick, H. 1999. The antipruritic effect of a 5-HT$_3$ receptor antagonist (tropisetron) is dependent on mast cell depletion: an experimental study. Exp. Dermatol. 8, 254–260.

Wenzel, R. R., Zbinden, S., Noll, G., Meier, B., and Luscher, T. F. 1998. Endothelin-1 induces vasodilation in human skin by nociceptor fibres and release of nitric oxide. Br. J. Clin. Pharmacol. 45, 441–446.

White, J. C. and Sweet, W. H. 1969. Pain and the Neurosurgeon: A Forty Year Experience. Thomas.

Wilcox, G. L. 1988. Pharmacological studies of grooming and scratching behavior elicited by spinal substance P and excitatory amino acids. Ann. N. Y. Acad. Sci. 525, 228–236.

Woodward, D. F., Nieves, A. L., Spada, C. S., Williams, L. S., and Tuckett, R. P. 1995. Characterization of a behavioral model for peripherally evoked itch suggests platelet-activating factor as a potent pruritogen. J. Pharmacol. Exp. Ther. 272, 758–765.

Yamaguchi, T., Maekawa, T., Nishikawa, Y., Nojima, H., Kaneko, M., Kawakita, T., Miyamoto, T., and Kuraishi, Y. 2001. Characterization of itch-associated responses of NC mice with mite-induced chronic dermatitis. J. Dermatol. Sci. 25, 20–28.

Yamaguchi, T., Nagasawa, T., Satoh, M., and Kuraishi, Y. 1999. Itch-associated response induced by intradermal serotonin through 5-HT$_2$ receptors in mice. Neurosci. Res. 35, 77–83.

Yosipovitch, G., Greaves, M. W., Fleischer, A. B., and McGlone, F. 2004. Itch: Basic Mechanisms and Therapy. Marcel Dekker.

Yosipovitch, G., Greaves, M. W., and Schmelz, M. 2003. Itch. Lancet 361, 690–694.

10 Thermal Sensation (Cold and Heat) through Thermosensitive TRP Channel Activation

Makoto Tominaga, National Institutes of Natural Sciences, Okazaki, Japan

10.1 Introduction

We feel a wide range of temperatures spanning from cold to heat. Within this range, temperatures over about 43 °C and below about 15 °C evoke not only a thermal sensation, but also a feeling of pain. The neurons that allow us to sense temperatures are located in the dorsal root ganglia (DRG) and within cranial nerve ganglia such as trigeminal ganglion (TG). It has been hypothesized that cutaneous nociceptor endings detect temperature and other physical stimuli by means of ion channels responsive to these stimuli. Insight into the molecular nature of the temperature-activated channels came with the cloning of the capsaicin receptor, TRPV1 (Caterina, M. J. *et al.*, 1997). TRPV1 is a member of the TRP (transient receptor potential) superfamily of ion channels, whose prototypical member, TRP, was found to be deficient in a *Drosophila* mutant exhibiting abnormal responsiveness to continuous light (Montell, C., 2005). Three other TRPV channels, TRPV2, TRPV3, and TRPV4, have been cloned and characterized as heat thermosensors (Jordt, S. E. *et al.*, 2003; Patapoutian, A. *et al.*, 2003). The threshold temperatures for activation of these channels range from relatively warm (TRPV3 and TRPV4) to extremely hot (TRPV2). In contrast to the four thermosensitive TRPV channels, two other TRP channels, TRPM8 and TRPA1, have been found to be activated by cold stimuli (Jordt, S. E. *et al.*, 2003; Patapoutian, A. *et al.*, 2003). Between the two channels, TRPA1 is perhaps more likely to be involved in nociception because the temperature threshold for its activation is about 17 °C, close to the reported threshold for cold nociceptors.

10.2 Heat Receptors

10.2.1 TRPV1 (VR1)

10.2.1.1 Cloning of TRPV1 and its characterization

One characteristic shared by many nociceptive neurons is sensitivity to capsaicin, the main pungent ingredient of hot chili peppers (Szallasi, A. and Blumberg, P. M., 1999). A capsaicin receptor was isolated in 1997 and designated vanilloid receptor subtype 1 (VR1) (Caterina, M. J. *et al.*, 1997), and now it is called TRPV1 in TRPV subfamily of a large TRP ion channel superfamily (Figure 1). Patch-clamp recordings (Caterina, M. J. *et al.*, 1997; Tominaga, M. *et al.*, 1998) revealed that TRPV1 is a nonselective cation channel with high Ca^{2+} permeability and that TRPV1 can be activated not only by capsaicin but also by protons and heat over 43 °C, all of which are known to cause pain *in vivo*. Single channel openings of TRPV1 by any of the three

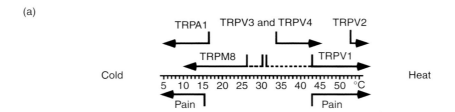

Figure 1 (a) Temperatures causing pain and activating six thermosensitive transient receptor potential (TRP) channels. Dotted lines indicate that threshold temperatures for activation of TRPV1 and TRPM8 are not fixed but changeable in the presence of other stimuli (see text). (b) Comparison of the mammalian six thermosensitive TRP channels.

stimuli in a membrane-delimited fashion indicate that the three stimuli gate TRPV1 directly and that TRPV1 is, itself, a heat sensor. These stimuli are likely to work in concert to regulate the activity of TRPV1 *in vivo*, especially under pathological conditions where tissue acidosis and elevated temperature may come into play. TRPV1 transcript and protein were found to be most highly expressed in sensory neurons (Caterina, M. J. *et al.*, 1997; Tominaga, M. *et al.*, 1998). Moreover, *in situ* hybridization and immunostaining revealed that, within DRG and TG, TRPV1 expression predominated in small-diameter cell bodies, most of which give rise to unmyelinated C fibers.

To determine whether TRPV1 really contributes to the detection of these noxious stimuli *in vivo*, mice lacking this protein were generated and analyzed for nociceptive function. Sensory neurons from mice lacking TRPV1 were deficient in their responses to each of the reported noxious stimuli – capsaicin, proton, and heat (Caterina, M. J. *et al.*, 2000). Consistent with this observation, behavioral responses to capsaicin were absent, and responses to acute thermal stimuli were diminished in these mice. The most prominent feature

of the knockout mouse thermosensory phenotype was a virtual absence of thermal hypersensitivity in the setting of inflammation. These findings indicate that TRPV1 is essential for selective modalities of pain sensation and for tissue injury-induced thermal hyperalgesia.

10.2.1.2 Reduction of temperature threshold for TRPV1 activation: A mechanism for inflammatory pain

Inflammatory pain is characterized by hypersensitivity both at the site of damage and in the adjacent tissue, and one underlying mechanism is the modulation (sensitization) of ion channels such as TRPV1. Sensitization is triggered by extracellular inflammatory mediators that are released *in vivo* from surrounding damaged or inflamed tissues and from nociceptive neurons themselves (Julius, D. and Basbaum, A. I., 2001). Tissue acidification is also induced in pathological conditions such as ischemia or inflammation, and such acidification exacerbates or causes pain. Among the inflammatory mediators, extracellular ATP, bradykinin, prostaglandins (PGE_2 and PGI_2), or tryptase or trypsin have been reported to potentiate TRPV1 responses through their

Gq-coupled $P2Y_2$, B_2, EP_1, IP, or PAR2 (proteinase activated receptor 2) receptors, respectively, mainly in a protein kinase C (PKC)-dependent manner in both a heterologous expression system and native DRG neurons (Tominaga, M. and Caterina, M. J., 2004; Moriyama, T. *et al.*, 2005). In addition to potentiating capsaicin- or proton-evoked currents, in the presence of the inflammatory mediators, the threshold temperature for heat activation of TRPV1 was reduced to as low as 30 °C, such that normally nonpainful thermal stimuli (i.e., normal body temperature) are capable of activating TRPV1. Under these circumstances, the inflammatory mediators seem to act as direct activators of TRPV1. This represents a novel mechanism through which the large amounts of the mediators released from different cells in inflammation might trigger a sensation of pain. Interactions between the Gq-coupled receptors and TRPV1 in relation to inflammatory pain have been proven at behavioral level using the Gq-coupled receptor- or TRPV1-deficient mice (Tominaga, M. and Caterina, M. J., 2004; Moriyama, T. *et al.*, 2005). A PKA-dependent and PIP_2-mediated pathways also seem to be involved in TRPV1 sensitization (Tominaga, M. and Caterina, M. J., 2004).

Tissue acidification is induced in pathological conditions such as ischemia or inflammation, and such acidification exacerbates or causes pain. In addition to the direct activation of TRPV1, acidification also shifts temperature–response curve of TRPV1 to the left so that the channel can be activated at lower temperatures (lower than body temperature) and responses to heat are bigger at a given suprathreshold temperature (Tominaga, M. *et al.*, 1998). This phenomenon might also contribute to inflammatory pain.

10.2.2 TRPV2 (VRL-1)

A protein, which is 49% identical to TRPV1, was isolated and designated VRL-1 and later renamed TRPV2 (Caterina, M. J. *et al.*, 1999; Jordt, S. E. *et al.*, 2003; Patapoutian, A. *et al.*, 2003; Tominaga, M. and Caterina, M. J., 2004) (Figure 1). TRPV2 can be activated by high temperatures with a threshold of ~52 °C. TRPV2 currents showed similar properties to those of TRPV1 such as high Ca^{2+} permeability.

Intense TRPV2 immunoreactivity was observed in medium- to large-diameter cells in rat DRG neurons (Caterina, M. J. *et al.*, 1999). Together with the fact that many of the TRPV2 immunoreactive cells in rat DRG costained with the marker for myelinated neurons, it was concluded that TRPV2 is preferentially expressed in Aδ fibers. TRPV2 transcript and protein were found not only in sensory neurons but also in motoneurons and in many non-neuronal tissues that are unlikely to be exposed to temperatures above 50 °C (Caterina, M. J. *et al.*, 1999). These results indicate that TRPV2 undoubtedly contributes to numerous functions in addition to nociceptive processing.

10.2.3 TRPV3 and TRPV4

TRPV3 and TRPV4 have been found to be activated by warm temperatures, ~34–38 °C for TRPV3 and ~27–35 °C for TRPV4, in heterologous expression systems and mouse keratinocytes, and to be expressed in multiple tissues, including, among others, sensory and hypothalamic neurons and keratinocytes (Jordt, S. E. *et al.*, 2003; Patapoutian, A. *et al.*, 2003; Tominaga, M. and Caterina, M. J., 2004) (Figure 1). Among the tissues, both TRPV3 and TRPV4 were most predominantly expressed in the skin keratinocytes and have been reported to be involved in thermosenation by keratinocytes, based on the studies using wildtype and TRPV3- or TRPV4-deficient keratinocytes (Chung, M. K. *et al.*, 2004; Lee, H. *et al.*, 2005; Moqrich, A. *et al.*, 2005). Thermosensing ability of TRPV3 or TRPV4 were proven at animal level because both TRPV3- and TRPV4-deficient mice exhibited some different thermal preference from that in wildtype mice. TRPV4 has also been shown to be involved in mechanical stimulus- and hypotonicity-induced nociception in rodents at baseline or following hypersensitivity induced by prostaglandin injection or taxol hypersensitivity. Furthermore, TRPV4-deficient mice reportedly exhibit deficits in inflammation-induced thermal hyperalgesia (Todaka, H. *et al.*, 2004). TRPV3 has been reported to function as a camphor receptor (Moqrich, A. *et al.*, 2005).

10.3 Cold Receptors

10.3.1 TRPM8

The cooling sensation of menthol, a chemical agent found in mint, is well established, and both cooling and menthol have been suggested to be transduced through a nonselective cation channel in DRG neurons (Figure 1). Two groups independently cloned and characterized a cold receptor, TRPM8, which can be also activated by menthol (McKemy, D. D.

et al., 2002; Peier, A. M. *et al.*, 2002). In heterologous expression systems, TRPM8 could be activated by menthol or by cooling, with an activation temperature of ~25–28 °C. TRPM8 could alternatively be activated by other cooling compounds, such as menthone, eucalyptol, and icilin. There also appears to be interaction between effective stimuli for TRPM8, in that subthreshold concentrations of menthol increased the temperature threshold for TRPM8 activation from 25 °C to 30 °C. This is reminiscent of TRPV1, whose activation temperature is reduced under mildly acidic conditions that do not open TRPV1 alone (Tominaga, M. *et al.*, 1998). Whole-cell recording in HEK293 cells expressing TRPM8 revealed that TRPM8 is a nonselective cation channel with relatively high Ca^{2+} permeability.

TRPM8 is expressed in a subset of DRG and TG neurons that can be classified as small-diameter C fiber-containing neurons (McKemy, D. D. *et al.*, 2002; Peier, A. M. *et al.*, 2002). Interestingly, however, TRPM8 is not coexpressed with TRPV1.

10.3.2 TRPA1

TRPA1 was reported as a distantly related TRP channel which is activated by cold with a lower activation threshold as compared with TRPM8 (Story, G. M. *et al.*, 2003) (Figure 1). In heterologous expression systems, TRPA1 was activated by cold stimuli with an activation temperature of about 17 °C, which is close to the reported noxious cold threshold. This finding led to the suggestion that TRPA1 is involved in cold nociception. Whole-cell recording in Chinese hamster ovary (CHO) cells expressing TRPA1 revealed cationic permeability with similar preferences for monovalent and divalent cations. Whether TRPA1 is gated directly by cold remains to be elucidated. A study from another group failed to reproduce cold responsiveness in TRPA1 (Jordt, S. E. *et al.*, 2004). The reason for this apparent discrepancy is unclear. However, both groups have demonstrated that TRPA1 can be activated by pungent isothiocyanate compounds such as those found in wasabi, horseradish, cinnamon, and mustard oil (Tominaga, M. and Caterina, M. J., 2004). Thus, several of the thermosensitive TRP channels likely to be involved in nociception can be activated by stimuli other than temperature.

Unlike TRPM8, TRPA1 is specifically expressed in a subset of sensory neurons that express the nociceptive markers CGRP and substance P (Story, G. M.

et al., 2003). Furthermore, TRPA1 is frequently coexpressed with TRPV1, raising the possibility that TRPA1 and TRPV1 mediate the function of a class of polymodal nociceptors. Such coexpression might also explain the paradoxical hot sensation experienced when one is exposed to a very cold stimulus.

10.4 Thermonociception by Non-TRP Proteins

The discovery of thermosensitive TRP channels indicates that thermosensation is mediated in part by a common molecular mechanism involving ion channels such as TRP channels as primary transducers of thermal stimuli. However, it is still possible that other proteins are involved in thermosensation and thermonociception. Indeed, TREK-1, a member of a family of mammalian two-pore domain K^+ channels, is expressed in mouse DRG neurons and activated by heat. Because opening of TREK-1 causes hyperpolarization of the membrane potential by efflux of K^+ ions, TREK-1 might be involved in pain relief through the reduction of firing upon noxious heat stimulus. Alternatively, inhibition of TREK-1 by cold stimuli might lead to nociception through depolarization. Further studies will be necessary to determine whether TREK-1 is involved in thermonociception. In addition to K^+ channels, other membrane proteins such as Na^+/K^+ ATPase, members of degenerin/epithelial sodium channels (DEG/ENaC), and $P2X_3$ receptors have been reported to be thermosensative. However, the involvement of those proteins in thermonociception remains to be established.

10.5 Conclusion

Significant advances in thermonociception research have been made in the last several years with the cloning and characterization of thermosensitive TRP channels. It is not known how multiple thermosensitive TRP channels are used within the same or different sensory neurons to provide us with the ability to evaluate temperature precisely over a broad range. Central integration of the thermal information acquired by peripheral nerve endings is almost certainly a critical component of this process. Furthermore, regulatory mechanisms similar to those emerging for TRPV1 are likely to exist for most of the thermosensitive TRP channels. Therefore, an understanding of the molecules surrounding thermo-sensitive TRP channels from the periphery to the central nervous system will be indispensable to a complete understanding of thermonociception.

References

Caterina, M. J., Leffler, A., Malmberg, A. B., Martin, W. J., Trafton, J., Petersen-Zeitz, K. R., Koltzenburg, M., Basbaum, A. I., and Julius, D. 2000. Impaired nociception and pain sensation in mice lacking the capsaicin receptor. Science 288, 306.

Caterina, M. J., Rosen, T. A., Tominaga, M., Brake, A. J., and Julius, D. 1999. A capsaicin-receptor homologue with a high threshold for noxious heat. Nature 398, 436.

Caterina, M. J., Schumacher, M. A., Tominaga, M., Rosen, T. A., Levine, J. D., and Julius, D. 1997. The capsaicin receptor: a heat-activated ion channel in the pain pathway. Nature 389, 816.

Chung, M. K., Lee, H., Mizuno, A., Suzuki, M., and Caterina, M. J. 2004. TRPV3 and TRPV4 mediate warmth-evoked currents in primary mouse keratinocytes. J. Biol. Chem. 279, 21569.

Jordt, S. E., Bautista, D. M., Chuang, H. H., McKemy, D. D., Zygmunt, P. M., Hogestatt, E. D., Meng, I. D., and Julius, D. 2004. Mustard oils and cannabinoids excite sensory nerve fibres through the TRP channel ANKTM1. Nature 427, 260.

Jordt, S. E., McKemy, D. D., and Julius, D. 2003. Lessons from peppers and peppermint: the molecular logic of thermosensation. Curr. Opin. Neurobiol. 13, 487.

Julius, D. and Basbaum, A. I. 2001. Molecular mechanisms of nociception. Nature 413, 203.

Lee, H., Iida, T., Mizuno, A., Suzuki, M., and Caterina, M. J. 2005. Altered thermal selection behavior in mice lacking transient receptor potential vanilloid 4. J. Neurosci. 25, 1304.

McKemy, D. D., Neuhausser, W. M., and Julius, D. 2002. Identification of a cold receptor reveals a general role for TRP channels in thermosensation. Nature 416, 52.

Montell, C. 2005. The TRP superfamily of cation channels. Sci. STKE 2005, re3.

Moqrich, A., Hwang, S. W., Earley, T. J., Petrus, M. J., Murray, A. N., Spencer, K. S., Andahazy, M., Story, G. M., and Patapoutian, A. 2005. Impaired thermosensation in mice lacking TRPV3, a heat and camphor sensor in the skin. Science 307, 1468.

Moriyama, T., Higashi, T., Togashi, K., Iida, T., Segi, E., Sugimoto, Y., Tominaga, T., Narumiya, S., and Tominaga, M. 2005. Sensitization of TRPV1 by EP1 and IP reveals peripheral nociceptive mechanism of prostaglandins. Mol. Pain 1, 3.

Patapoutian, A., Peier, A. M., Story, G. M., and Viswanath, V. 2003. ThermoTRP channels and beyond: mechanisms of temperature sensation. Nat. Rev. Neurosci. 4, 529.

Peier, A. M., Moqrich, A., Hergarden, A. C., Reeve, A. J., Andersson, D. A., Story, G. M., Earley, T. J., Dragoni, I., McIntyre, P., Bevan, S., and Patapoutian, A. 2002. A TRP channel that senses cold stimuli and menthol. Cell 108, 705.

Story, G. M., Peier, A. M., Reeve, A. J., Eid, S. R., Mosbacher, J., Hricik, T. R., Earley, T. J., Hergarden, A. C., Andersson, D. A., Hwang, S. W., McIntyre, P., Jegla, T., Bevan, S., and Patapoutian, A. 2003. ANKTM1, a TRP-like channel expressed in nociceptive neurons, is activated by cold temperatures. Cell 112, 819.

Szallasi, A. and Blumberg, P. M. 1999. Vanilloid (Capsaicin) receptors and mechanisms. Pharmacol. Rev. 51, 159.

Todaka, H., Taniguchi, J., Satoh, J., Mizuno, A., and Suzuki, M. 2004. Warm temperature-sensitive transient receptor potential vanilloid 4 (TRPV4) plays an essential role in thermal hyperalgesia. J. Biol. Chem. 279, 35133.

Tominaga, M. and Caterina, M. J. 2004. Thermosensation and pain. J. Neurobiol. 61, 3.

Tominaga, M., Caterina, M. J., Malmberg, A. B., Rosen, T. A., Gilbert, H., Skinner, K., Raumann, B. E., Basbaum, A. I., and Julius, D. 1998. The cloned capsaicin receptor integrates multiple pain-producing stimuli. Neuron 21, 531.

11 The Development of Nociceptive Systems

G J Hathway and M F Fitzgerald, University College London, London, UK

Glossary

A fibers Large-diameter myelinated primary afferent sensory fibers with large cell bodies in the dorsal root ganglion. The largest diameter Aβ fibers are mainly low-threshold mechanoceptors and the smaller Aδ fibers are both mechanoreceptors and nociceptors.

C fibers Small-diameter unmyelinated primary afferent sensory fibers with small cell bodies in the dorsal root ganglion. The majority are nociceptors. They divide into a neuropeptide containing, Trk-receptor expressing group and a lectin IB4 binding group although the functional implications of this are still unclear.

embryonic (E) age These are dated from the time of fertilization. Rat gestation is 21.5 days, mouse a little shorter. Rats are born relatively early in terms of CNS development and the early postnatal period is often paralleled with the final gestation of development in man.

receptive field The area on the body surface that when stimulated evokes action potentials in a given neuron.

periaqueductal gray An area of the brainstem that surrounds the aqueduct connecting the third and fourth ventricles. This area projects to the medullary raphe region, which in turn sends projections down the dorsolateral funiculus of the spinal cord to the dorsal horn. This pathway is known to strongly modulate spinal pain processing.

receptor subunits Ion channels are generally made up of several glycoprotein subunits around a central pore. These subunits can confer special characteristics upon a channel, such as increased calcium permeability or longer opening times. The subunits of many channels change with development, thereby altering the channel properties.

11.1 The Development of Nociceptors and Their Peripheral and Spinal Connections

Pain is usually, although not always, triggered by tissue damage and consequently a discussion of developing nociceptive pathways begins with the maturation of primary sensory neurons in the dorsal root ganglia (DRG) and their peripheral and central connections.

The two main subtypes of primary sensory neurons, A cells and C cells (future nociceptors) are specified at an early stage of development under the control of specific transcription factors. In the rat, lumbar DRG cells are born in two waves from embryonic day (E)12–E15 (gestation 21.5 days); C cells are born in the second wave, later than the larger-diameter A cells. There is evidence also that the future peptidergic nociceptors are born before the nonpeptidergic group (Hall, A. K.

et al., 1997; Jackman, A. and Fitzgerald, M., 2000). Young C cells immediately send out centrally and peripherally directed processes, again lagging some days behind A cells. However, from the time that they innervate peripheral skin, nociceptors are capable of transmitting information about noxious cutaneous stimulation (Box 1).

The study of sensory neurogenesis, axon outgrowth, and target finding is a fast-moving and exciting area of developmental biology. Not only are there a number of key regulatory genes controlling these events but the many aspects of nociceptor development are dependent upon access to neurotrophic factors (Fitzgerald, M., 2005). A key neurotrophic factor affecting nociceptor development is nerve growth factor, NGF, which not only determines nociceptor fate but is also critically required for nociceptor survival (Davies, A. M., 2000; Fitzgerald, M., 2005) during embryonic life and continues to influence nociceptor development postnatally by

1. NGF regulates axon outgrowth, independently of cell survival (Markus, A. *et al.*, 2002).
2. NGF determines nociceptor physiological properties. NGF signaling in the early postnatal period is required for C nociceptors to respond to noxious heat and for the normal development of myelinated high-threshold mechanoreceptors (Aδ fibers) (Ritter, A. M. *et al.*, 1991; Lewin, G. R. and Mendell, L. M., 1994).
3. NGF influences nociceptor expression of the neuropeptides substance P (SP) and calcitonin-gene-related peptide (CGRP) (Davies, A. M., 2000; Hall, A. K. *et al.*, 2001).
4. NGF regulates skin innervation density. Excess epidermal NGF leads to hyperinnervation (Davis, B. M. *et al.*, 1997).

The growth of C-fiber terminals into the spinal dorsal horn occurs just before birth in the rat and from the outset, these fibers terminate in a somatotopically precise manner in laminae I–II of the dorsal horn. Neuropeptide containing nociceptive terminals in laminae I and II can be detected prenatally, steadily increasing in density over the first 10 postnatal days, while IB4+ve terminals appear a few days later, suggesting that these fibers may form central connections later than peptidergic fibers and indeed be involved in a separate central pain pathway (Braz, J. M. *et al.*, 2005).

The immature C-fiber terminals are functional at birth but neurotransmitter release is of low frequency and the synapses are too immature to evoke substantial spike activity *in vivo* until the second postnatal week. This in contrast to the A fiber which, when stimulated, can evoke robust action potentials in dorsal horn cells from before birth (Fitzgerald, M. and Jennings, E., 1999; Baccei, M. L. *et al.*, 2003). As in the adult, immature C fibers terminate exclusively in laminae I and II. However, in the first postnatal weeks, they share this termination area with large-diameter Aβ afferents whose terminals extend dorsally into laminae II and I and form transient synaptic connections. This is followed by a gradual withdrawal from the superficial laminae over the first 3 postnatal weeks (Fitzgerald, M. *et al.*, 1994). The postnatal withdrawal of Aβ fibers from the superficial dorsal horn is an activity-dependent process, which that can be prevented by intrathecal application of the NMDA receptor antagonist MK801 (Beggs, S. *et al.*, 2002). The functional implications of the presence of A fibers in newborn lamina II are not known but they may contribute to the larger receptive fields and increased excitability to tactile stimulation described below.

Box 1   Postnatal development of C nociceptor properties

Although all of the broad functional classes of primary afferents are present in the rat hindlimb at birth, the maturation process varies according to the individual stimulus modality (Koltzenburg, M., 1999). From birth, polymodal nociceptors respond to noxious chemical, mechanical, and thermal stimuli with thresholds and firing frequencies that are characteristic of mature C fibers (Fitzgerald, M., 1987; Keller, A. F. *et al.*, 2004). This is consistent with early expression of TRPV1, critical for the detection of painful thermal and chemical stimuli, and the ATP-gated P2X$_3$ receptor (Guo, A. *et al.*, 2001) that plays an important role in thermal and mechanical hyperalgesia following inflammation or nerve injury and the tetrodotoxin-resistant sodium channel Nav1.8, necessary for the normal detection of noxious mechanical and thermal stimuli and the sensitization of nociceptive sensory neurons (Benn, S. C. *et al.*, 2001).

Nociceptive cells require activation of trkA receptors by nerve growth factor (NGF) for survival during the embryonic period, but during the postnatal period, a subset of these neurons downregulate trkA and become dependent on glial-cell-line-derived neurotrophic factor (GDNF) via the expression of the GDNF receptor Ret (Bennett, D. L. *et al.*, 1996; Molliver, D. C. *et al.*, 1997). Thus the clear distinction between (1) the trkA+ve, neuropeptide containing C fibers and (2) the IB4+ve, nonpeptidergic nociceptors appears postnatally, although the functional implications of this in terms of nociception is not clear.

11.2 The Development of Spinal Nociceptive Circuits

At birth, strong spinal nociceptive reflexes can be evoked by mechanical stimulation, which are exaggerated compared to the adult for the first 2–3 postnatal weeks (Ekholm, J., 1967; Fitzgerald, M. et al., 1988) with lower thresholds and larger, more synchronized and prolonged muscle contractions. A prick on the hind foot in the newborn rat can evoke limb withdrawal, wriggling, rolling, and simultaneous responses from all four limbs. Receptive fields of hindlimb flexor muscles are large and disorganized such that the limb withdrawal can be evoked that is not always appropriate to the stimulus (Holmberg, H. and Schouenborg, J., 1996). As the pup matures, the response decreases in magnitude and becomes restricted to an isolated leg or foot movement. Thresholds for withdrawal from heat stimuli are also lower in younger animals. In contrast with mechanical reflexes, but consistent with the slow synaptic maturation of C-fiber-evoked responses, cutaneous application of the selective C-fiber irritant, mustard oil, evokes little or no spinal reflex response until postnatal day 10 (Baccei, M. L. and Fitzgerald, M., 2005).

The excitability of mechanical neonatal cutaneous reflexes is attributable to underlying changes in dorsal horn sensory processing and is likely to be due to absence of fine-tuning of sensory and nociceptive neuronal circuitry in the newborn. The earlier maturation and widespread presence of functional A-fiber terminals in both the substantia gelatinosa and deeper laminae in the first weeks of life appear to adequately compensate for the weak C-fiber input and dominate the physiological responses of dorsal horn neurons, fewer of which are nociceptive in the newborn cord, especially of the wide dynamic range (WDR) variety. Receptive field areas of dorsal horn cells, mapped by natural mechanical stimulation of skin, are large at birth and decrease in size with age. Pinching or brushing the receptive field of many neonatal dorsal horn cells results in long-lasting after-discharges (30–90 s) which on days 0–3 may be more pronounced than the initial evoked response but which decrease in duration and amplitude of these responses with age (Fitzgerald, M. and Jennings, E., 1999; Torsney, C. and Fitzgerald, M., 2002). This predominantly A-fiber-induced spike activity appear to have a critical role in the formation of nociceptive circuits. Newborn rat pups have a high error rate in the direction of a tail flick on noxious stimulation than adults, which gradually improves over 3 weeks (Waldenstrom, A. et al., 2003). The maturation is unaffected by daily noxious stimulation but blocking low-intensity tactile inputs from the tail with local anesthetic, during the critical 10-day period, prevents this postnatal tuning. Low-intensity tactile input from, for example, spontaneous twitching during sleep could play a role in shaping nociceptive circuits in early life (Petersson, P. et al., 2003).

Postnatal maturation and tuning of nociceptive circuits critically depends upon the postnatal development of excitatory and inhibitory neurotransmitter/receptor function in the neonatal dorsal horn. The developmental pharmacology of spinal pain pathways has been an area of intense recent interest (Pattinson, D. and Fitzgerald, M., 2004; Fitzgerald, M., 2005) and some important features are shown in Box 2. The properties of immature

---

Box 2   The development of excitatory and inhibitory transmission in developing spinal pain circuits

Glutamate AMPA and NMDA receptors are highly expressed in the newborn dorsal horn and their expression decreases and becomes spatially organized with age. This is accompanied by changes in subunit composition leading to greater AMPA-dependent Ca^{2+} influx, altered receptor kinetics, greater NMDA receptor affinity, and increased Mg^{2+} sensitivity in neonatal spinal neurons (Watanabe, M. et al., 1994; Jakowec, M. W. et al., 1995; Albuquerque, C. et al., 1999; Bardoni, R., 2001; Green, G. M. and Gibb, A. J., 2001; Stegenga, S. L. and Kalb, R. G., 2001). Presynaptic Ca^{2+}-permeable KA receptors found on IB4+ve C cells switch to a Ca^{2+}-impermeable form early in the first postnatal week (Lee, C. J. et al., 2001).

Despite early expression of GABA and glycine and their receptors, newborn lamina II neurons exhibit only GABA mIPSCs (Baccei, M. L. and Fitzgerald, M., 2004) probably due to low levels of gephyrin, the postsynaptic scaffolding protein (Bremner et al., 2005). There is also a fourfold acceleration in the decay rate of GABA mIPSCs in laminae I–II neurons between P8 and P23 triggered by local neurosteroid release (Keller, A. F. et al., 2004). $GABA_A$R activation in the developing dorsal horn can be excitatory rather than inhibitory, due to the high intracellular Cl^- concentration in immature neurons producing a positive E_{Cl}. $GABA_A$R activation depolarizes some lamina II cells but the response becomes exclusively hyperpolarizing by P6–P7 (Baccei, M. L. and Fitzgerald, M., 2004). This is probably due to the onset of expression of the K^+–Cl^- cotransporter KCC2 which decreases $[Cl]_i$ over the postnatal period. However, E_{Cl} is always more negative than action potential threshold at P0–P1 suggesting that GABA will still inhibit the firing of newborn superficial dorsal horn neurons (albeit less effectively than in the adult).

Both presynaptic and postsynaptic $GABA_B$Rs are functional from birth in the neonatal dorsal horn, but seem to be regulated differently over the early postnatal period (Baccei, M. L. and Fitzgerald, M., 2004).

synaptic transmission tends to enhance calcium influx into developing dorsal horn neurons compared to adult neurons, which may be important for a variety of developmental processes including synapse formation and remodeling.

11.3   The Development of Ascending Pain Pathways

Following integration in the dorsal horn, sensory and nociceptive information is conducted to supraspinal centers via ascending tracts to targets in the brainstem, midbrain, hypothalamus, thalamus, and amygdala. A key ascending pathway involved in pain and hyperalgesia arises from lamina I of the dorsal horn of the spinal cord, particularly neurokinin 1 (NK1) expressing neurons receiving Aδ- and C-fiber nociceptive input, which project to the parabrachial area of the brainstem. These neurons then project to the amygdala (affective components of pain) and the hypothalamus (autonomic functions such as changes in heart rate and blood pressure). Outputs from these structures are passed via the periaqueductal gray (PAG), which in turn sends descending projections back to the dorsal horn in a spinobulbar spinal loop. Descending projections to the dorsal horn arise from the brainstem nuclei, nucleus raphe magnus (NRM), raphe dorsalis, reticularis, paragigantocellularis (NRPG), as well as locus coeruleus, PAG, and mesencephalic reticular formation (Todd, A. J. *et al.*, 2000; Mantyh, P. W. and Hunt, S. P., 2004; Suzuki, R. *et al.*, 2004; Ding, Y. Q. *et al.*, 1995; Marshall, G. E. *et al.*, 1996).

Little is known about the development of ascending spinothalamic and spinoparabrachial tracts. In the rat, neurogenesis of dorsal horn projection neurons is complete by E16 having followed a systematic pattern of migration and settlement in the dorsal horn related to their axon trajectories (Nissen, U. V. *et al.*, 2005). The projecting axons grow up the spinal cord prenatally (Bice, T. N. and Beal, J. A., 1997); spinothalamic tract fibers reach the thalamus just before birth at E19. In the rat, afferents appear to reach the thalamus by E19 (Higashi, S. *et al.*, 2002) while in the sheep, projection fibers from dorsal horn cells are in the thalamus approximately two-thirds through gestation (Rees, S. *et al.*, 1994; Ding, Y. Q. *et al.*, 1995). The thalamocortical axon (TCA) projection originates in dorsal thalamus, conveys sensory input to the neocortex, and has a critical role in cortical development. Information about the development of these pathways is entirely from visual and

somatosensory (whisker) pathways in rodents and their relevance to nociceptive processing can only be inferred. Recently, however, a study by Man S. H. *et al.* (2005) has shown the presence at birth of the NK1+ve lamina 1 ascending projection from the DH to the parabrachial area of the brainstem using retrograde tracing and c-fos immunostaining. The outgrowth of axons from the thalamus to the cortex begins at E13–E14 in rodents just before the arrival of the first cells migrating into the preplate zone. The migration of cortical neurons from the proliferative neuroepithelium lining the telencephalic ventricle to the cortical plate is highly complex and tightly regulated (Meyer, G., 2001). Thalamic axons and preplate fibers meet in the lateral part of the internal capsule at E15 and grow in association with early corticofugal axons, right up to the cortical subplate. The topography of thalamic axons is maintained throughout the pathway and they reach the cortex by associating with the projections of a number of preexisting cells, including the preplate scaffold. These cells invade the cortical plate 24–48 h later and form functional synaptic connections in the cortical plate around E18–E19. At birth, thalamic axons have extended into cortical layers IV, V, and VI in a topographically precise manner and exhibit a periphery-related pattern in layer IV (Schlaggar, B. L. and O'Leary, D. D., 1994; Molnar, Z. *et al.*, 2003). The spatial distribution of thalamic synapses onto cortical neurons (predominantly found at dendritic spines) is established at least as early as P11 in the somatosensory cortex of the mouse (Lev, D. L. *et al.*, 2002).

11.4   Pain and the Developing Cortex

The development of specific cortical areas is dependent upon both extrinsic influences from thalamocortical input and intrinsic genetic controls such as the transcription factors Emx2 and Pax6 (O'Leary, D. D. and Nakagawa, Y., 2002). This process extends over several months in man compared to the mouse cortex that requires a week (Angevine, J. B. and Sidman, R. L., 1961). In man, the generation and migration of cortical neurons begins in the fifth gestational week (O'Rahilly, R. and Muller, F., 1994) and continues until week 22 (Sidman, R. L. and Rakic, P., 1973). Following the arrival of neurons at their correct destination, they begin to differentiate and to extend dendritic fields. It is in the last 2 months of gestation that the characteristic folding of the cortex into sulci and gyri takes place (Fees-Higgins, A. and Larroche, J. C., 1987).

One area of interest is the anterior cingulate cortex (ACC) that appears to play an important role in the affective and cognitive aspects of pain, but little is known of the development of this region. Ultrasonic scans taken from preterm neonates have confirmed that at week 28 the ACC is fully developed (Huang, C. C., 1991).

Subplate neurons (SPn) play an important role in the formation of thalamocortical connections during early development and show glutamatergic and GABAergic spontaneous synaptic activity. Whole-cell recordings from SPn in somatosensory cortical slices of postnatal day 0–3 rats, showed that SPn receive distinct functional synaptic inputs arising from the thalamus, cortical plate, and other SPn mediated by AMPA and NMDA receptors (Hanganu, I. L. et al., 2002; Luhmann, H. J. et al., 2003). GABA activity is depolarizing at this age (see Box 2). Electrophysiological analysis of cortical cells at P7 shows them to be organized in columns as in the adult but to have larger receptive fields, similar to that discussed above for the spinal cord (Armstrong-James, M. A., 1970; Armstrong-James, M., 1975). Human somatosensory potentials evoked by electrical stimulation are observed at 29 weeks (Klimach, V. J. and Cooke, R. W., 1988) while preliminary studies using infrared spectroscopy during routine clinical heel lances suggest that a cortical blood flow response is present as early as 25 weeks (Slater, R. et al., 2006).

While axonal growth and early topographic arrangement of thalamocortical fiber pathways do not rely on activity-dependent mechanisms requiring evoked neurotransmitter release (Molnar, Z. et al., 2002), the formation of cortical circuits is highly dependent upon normal sensory experience. The postnatal plasticity of cortical circuits has been studied in great detail for the visual (Yao, H. and Dan, Y., 2005) and somatosensory whisker barrel cortex (Fox, K., 2002), but little is known about the effect of nociceptive activity upon cortical development (Anand, K. J., 2000).

11.5 Pain and the Developing Limbic System

The amygdala is a part of the limbic system and plays a key role in the emotional evaluation of sensory stimuli (Davis, M., 1998a; Aggleton, J. P., 1999; Gallagher, M. and Schoenbaum, G., 1999; Maren, S., 1999; Ledoux, J. E., 2000b; Cardinal, R. N. et al., 2002; Davidson, R. J., 2002; Zald, D. H., 2003). In the rat, the amygdala is a complex of 13 nuclei and cortical regions and their subdivisions. There is some debate as to whether the amygdala should be considered a single entity or as a collection of functionally diverse nuclei that share some anatomical and physiological similarities (Aggleton, J. P. and Saunders, R. C., 2000). There are both direct and indirect nociceptive inputs to the laterocapsular part of the central nucleus of the amygdala (the nociceptive amygdala) and the whole central area supplies the major output forming widespread connections with forebrain areas, the hypothalamus and the brainstem (Pitkanen, A. et al., 1998; Davis, M., 1998b; Aggleton, J. P. and Saunders, R. C., 2000; Ledoux, J. E., 2000a; Bourgeais, S. et al., 2001; Neugebauer, V. and Li, W., 2002; 2003; Li, W. and Neugebauer, V., 2004).

Neurons in the amygdala of the rhesus monkey are generated during a 3-week period very early in gestation (E30–E50 of a 165-day gestation; Kordower, J. H. et al., 1992). This neurogenesis appears to occur simultaneously within all the amygdalal nuclei. This pattern has also been reported in humans (Humphrey, T., 1968; Nikolic, I. and Kostovic, I., 1986). In the rodent, LOM-homeodomain genes have been shown to play a role in the segregation of embryonic amygdalal cells into appropriate groupings and this family of genes plays a fundamental role in the development of every single amygdalal nucleus (Remedios, R. et al., 2004). The distribution of opiate receptors in the amygdala of a 1-week-old monkey appears the same as in the adult indicating early neurochemical maturation (Bachevalier, J. et al., 1986). Widespread connections to and from the monkey amygdala at 1 week of age appear to be pruned over the postnatal period (Webster, M. J. et al., 1991). The same connections are likely to occur over the first 2 postnatal weeks in rats.

11.6 The Development of Descending Pain Pathways

Brainstem descending pathways form a major mechanism in the control of pain transmission (Pitkanen, A. et al., 1998; Ren, K. and Dubner, R., 2002; Gebhart, G. F., 2004) and there is more known about the development of these descending controls compared to the ascending systems.

Brainstem nuclei differentiate between E11 and E16 in the rat and all other major brain systems are morphologically identifiable by E18 (Altman, J. and Bayer, S. A., 1984; Momose-Sato, Y. *et al.*, 2001). In the rat embryo, the parabrachial area can be visualized histochemically from at least E17 (Cholley, B. *et al.*, 1989) (Figure 1).

As with ascending tracts, axons from brainstem nuclei appear to grow down the spinal cord well before birth (Cabana, T. and Martin, G. F., 1984). Injections of horseradish peroxidase into lumbosacral spinal cord of the neonatal rat label brainstem nuclei with similar density to that seen in the adult (Leong, S. K., 1983; Fitzgerald, M. *et al.*, 1987) but it is unclear at what stage these axons begin producing collaterals that innervate the dorsal horn and make synapses with their target cells. Mapping degenerating axons and synaptic endings following thoracic hemisections

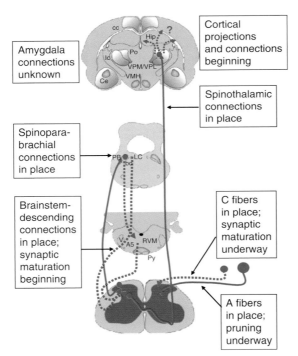

Figure 1 Diagram summarizing the state of development of ascending and descending pathways involved in the transmission of pain in the newborn rat. It is important to remember that changes in the anatomy of the pathways depicted here are also accompanied by changes in the physiology and neurochemistry of these pathways also. Additionally, the maturation of human nociceptive pathways occurs over different and as yet unknown time courses from those discussed here. Adapted from Hunt, S. and Mantyh, P. 2001. The molecular dynamics of pain control. Nat. Rev. Neurosci. 2, 83–91.

of the spinal cord at various postnatal stages suggests that descending axon collaterals innervate the intermediate and central gray from birth but that they are not found in the dorsal horn until P15 (Gilbert, M. and Stelzner, D. J., 1979).

Descending inhibitory pathways traveling from the brainstem via the dorsolateral funiculus of the spinal cord to the dorsal horn grow down the spinal cord early in fetal life, but they do not extend collateral branches into the dorsal horn for some time. Stimulation of the dorsolateral funiculi (DLF) fails to produce descending inhibition of C-fiber-evoked activity in neonatal rats before P9, but high-intensity stimulation produced descending inhibition at P18. By P22–P24 more typical adult responses were observed (Fitzgerald, M. and Koltzenburg, M., 1986). Similarly, the inhibition of C and Aδ activity in the dorsal horn by PAG stimulation via the DLF that is clearly observed in adults (Basbaum, A. I. and Fields, H. L., 1978) cannot be produced in rat pups before P21 (van Praag, H. and Frenk, H., 1991). Diffuse noxious inhibitory control (DNIC), measured as a reduction in noxious-evoked dorsal horn c-fos expression in the presence of an inflammatory stimulus elsewhere on the body, also develops between P12 and P21 (Boucher, T. *et al.*, 1998).

The delayed postnatal onset of functional descending inhibition, despite the presence of DLF terminals in the dorsal horn may be due to immaturity of neurotransmitter/receptor interactions (see below) or delayed maturation of critical interneurons. It has been suggested that the maturation of descending inhibition is dependent upon afferent C-fiber activity, because rats treated with capsaicin at birth have reduced inhibitory controls as adults (Cervero, F. and Plenderleith, M. B., 1985). The lack of descending inhibition in the neonatal dorsal horn means that there is no endogenous analgesic system to dampen noxious inputs as they enter the CNS and their effects may therefore be more profound than in the adult. It also explains why stimulus-produced analgesia from the PAG is not effective until P21 in rats (van Praag, H. and Frenk, H., 1991).

11.7 Development of Neurotransmitter/Receptor Function in Descending Pathways

It is widely accepted that 5-HT is the most relevant neurotransmitter with respect to the descending control of spinal cord excitability. Descending serotonergic pathways predominantly mediate descending inhibitory control over spinal cord excitability, but 5-HT is involved in both antinociceptive and pronociceptive effects depending upon the subtype of receptor that it is acting upon (Millan, M. J., 2002; Gebhart, G. F., 2004). An excitatory, pronociceptive role has been attributed to the 5-HT$_3$ receptor (Suzuki, R. et al., 2002). In the adult, descending inhibition and excitation are conveyed rostrocaudally via different spinal pathways, the DLF and ventral/ventrolateral funiculi (VLF), respectively (Millan, M. J., 2002), and since both of these tracts appear in the embryonic period, both types of control may also be possible in the newborn. Neurons that produce 5-HT (serotonin) are amongst the earliest to be detected in the developing CNS (Rubenstein, J. L., 1998) with the mesencephalon containing detectable levels by E12 (E12; Aitken, A. R. and Tork, I., 1988). 5-HT containing descending neurons is present in the rat between E15 and E19 (Lidov, H. G. and Molliver, M. E., 1982; Lakke, E. A., 1997), those arising from the raphe nuclei are present in the cervical cord by E14 and in the lumbar cord by E16–E17. 5-HT fibers were not found to innervate the dorsal horn until E19 (Rajaofetra, N. et al., 1989). Examination of the pre- and postnatal development of the descending serotonergic pathways to the spinal cord show that the maturation of 5-HT terminals occurs in a ventral to dorsal direction such that innervation of the lumbar dorsal horn does not occur until P14 and the adult pattern of high-density terminals in lamina II at P21 (Bregman, B. S., 1987). Direct application of 5-HT to the neonatal spinal cord in vitro is effective in very young superficial dorsal horn cells, suggesting that at least some 5-HT receptors are functional from birth (Li, P. and Zhuo, M., 1998).

Although 5-HT is the most important neurotransmitter, other monoaminergic and peptidergic compounds must also be considered with regard to the descending control of spinal cord excitability. The major noradrenergic input to the dorsal horn is from the locus coeruleus (LC) and adjacent noradrenergic nuclei in the brainstem. Activation of this pathway leads to the release of noradrenaline that acts postsynaptically on α_2-adrenergic receptors to produce a decrease in excitability and a concomitant analgesia (Sagen, J. and Proudfit, H. K., 1984; Jones, B. E. 1991; Jasmin, L. et al., 2002). Noradrenaline containing terminals appear earlier in the dorsal horn than serotonin (Loizou, L. A., 1972; Commissiong, J. W., 1983) and levels are generally mature by 2 weeks

postnatally. As with 5-HT, the development of the noradrenergic system in the ventral horn precedes that in the dorsal horn. Expression of the α2A receptor mRNA is observed in high levels in the dorsal horn from E19 and is downregulated postnatally (Winzer-Serhan, U. H. et al., 1997), which has led to the proposal that noradrenaline has a trophic role in the developing cord (Huang, Y. et al., 2002). Recent functional studies have shown that the local spinal cord application dexmedetomidine, the potent α2A agonist, has antinociceptive and antihyperalgesic actions from at least P3 (Walker, S. 2005).

Dopamine is also released by descending tracts; the main source is from the A11 (posterior) region of the hypothalamus (Commissiong, J. W. et al., 1978; Commissiong, J. W. and Neff, N. H., 1979; Swanson, L. W. and Kuypers, H. G., 1980), which provide extensive networks of dopamine-containing fibers in the dorsal horn and lamina X of the spinal cord (Skagerberg, G. et al., 1982; 1988). In the rat, dopamine containing neurons appear in the mesencephalon at E12–E13 (Olson, L. and Seiger, A., 1972) and levels of dopmaine appear to increase throughout development with particularly high levels being found in the solitary tract in the brainstem (Moore, R. Y. and Bloom, F. E., 1978).

Numerous other nonmonoaminergic transmitters are found in the terminals of descending pathways in the spinal cord including histamine, vasopressin, and oxytocin. Other transmitters released by descending projections are also present in intrinsic DH neurons. Of these, acetylcholine is the most prominent, having both a pro- and antinociceptive effect depending on the subtype of the receptor activated. Other neurotransmitters in the category are GABA and glycine. Cholinergic neurons in the brainstem generally develop later than catecholaminergic systems with levels of cholinergic markers such as choline acetyltransferase being only a few percent of the levels seen in the adult rat at birth but start to increase rapidly after P10 (Diebler, M. F. et al., 1979).

A major peptidergic descending input from the brainstem to the dorsal horn of the spinal cord is via enkephalin and other opioid peptides (pre-proenkephalin, pre-prodynorphin, pre-proendomorphin, pre-pro-opiomelanocortin (POMC), and orphaninFQ). Opioids play an important role in pain transmission both by initiating the descending controls from the brainstem and by directly reducing the evoked activity of C and Aδ fibers in the dorsal horn via presynaptic opiate receptors. There is considerable postnatal development of this receptor expression, with neonatal rats not only having a higher density but also a less selective distribution of μ-opioid receptors on DRG cells (Nandi, R. *et al.*, 2004). Morphine has a markedly higher analgesic efficacy in young rats in mechanical but not thermal noxious tests but it is not known whether this results from central or peripheral regulation of opioid pathways (Nandi, R. and Fitzgerald, M., 2005).

11.8   The Development of Persistent Pain

While it is essential to understand the underlying development of nociception, tissue injury, inflammation, and nerve damage can lead to more persistent pain that lasts for days, weeks, or months. Such pains arise from central sensitization and changes in gene expression as triggered by intense and sensitized peripheral nociceptor inputs (Woolf, C. J. and Salter, M. W., 2000).

Cutaneous mechanical thresholds fall in neonatal rat pups following local carrageenan injection (Marsh, D. *et al.*, 1999) and mustard oil application (Jiang, M. C. and Gebhart, G. F., 1998) but the effect is less at P3 than at P21. This may partly reflect peripheral nociceptor terminal immaturity. C fibers are capable of peripheral sensitization from before birth (Koltzenburg M., and Lewin, G. R., 1997) but peripheral C-fiber terminals do not release sufficient SP to produce neurogenic extravasation until P10 (Fitzgerald, M., and Gibson, S., 1984). Electrophysiological studies *in vivo* show that sensitization of dorsal horn cells can be observed 2–5 h after carageenan inflammation of the hindpaw at all ages although the exact pattern of effects is age dependent (Fitzgerald, M. and Jennings, E., 1999; Torsney, C. and Fitzgerald, M., 2003). In slice preparations, NMDA-dependent C-fiber-evoked depolarization of spinal cord cells and windup of cells to repeated C-fiber stimulation has been demonstrated in the young (8–14 days) spinal cord (Sivilotti, L. G, *et al.*, 1993). Recently, the role of descending brainstem excitatory pathways in maintaining persistent pain have been emphasized (Gebhart, G. F., 2004), but the developmental regulation of this system has not been investigated.

In adults, partial nerve injury can trigger neuropathic pain, charcterized by a lowered mechanical sensitivity or allodynia. In postnatal animal models of neuropathic pain, this allodynia is not observed until the animals are over 3 weeks old (Howard, R. F. *et al.*,

2001). This is interesting in view of the substantial cell death that occurs in the DRG of axons damaged in early life compared to adults (Yip, H. K. *et al.*, 1984; Benn, S. C. *et al.*, 2002) followed by transganglionic degeneration and subsequent withdrawal of central terminals (Bondok, A. A. and Sansone, F. M., 1984; Aldskogius, H. and Risling, M., 1989), sprouting of adjacent, intact axon collaterals into the denervated region (Fitzgerald, M., 1985; Fitzgerald, M. *et al.*, 1990; Shortland, P. and Fitzgerald, M., 1994) and disruption of the somatotopic organization of central terminals within the dorsal horn and cortex (Kaas, J. H. *et al.*, 1983). Clearly these changes do not trigger pain. One possibility is that the central microglial and immune responses that follow nerve injury in the adult and may contribute to neuropathic pain (Marchand, F. *et al.*, 2005) are not activated in the immature spinal cord.

There is also evidence that injury at a critical period of development may lead to long-term alterations in sensory processing resulting in hypo- or hyperalgesia (Lidow, M. S., 2002). The mechanisms underlying this are poorly understood. It is possible that early tissue damage, which is known to release excess neurotrophins into the wounded area (Constantinou, J. *et al.*, 1994) not only sensitizes peripheral nociceptors, but also permanently influences nociceptor development as described above. Alternatively central changes in activity-dependent connectivity are produced because of the increased barrage of C-fiber nociceptor input at a time when synapses are still forming (Fitzgerald, M. and Walker, S. M., 2003). Central changes in transmitter–receptor expression may also occur. Early inflammatory pain leads to an increase in expression in the PAG of serotonin receptors (5HT1A, 5HT1D, 5HT2A, 5HT2C, 5HT4), NR2D and NR3A subunits of the NMDA receptor, $GABA_A\gamma2$ and $GABA_A\theta$ subunits, the NK1, and the CCK-2 receptor. As well as being hypoalgesic, these animals exhibited decreased anxiety and greater stress-coping ability; this was indexed by better performance on the elevated plus maze and decreased basal and stress-induced neuroendocrine markers of anxiety and stress (Anseloni, V. C. *et al.*, 2005). For further review of this area, see Baccei M. L. and Fitzgerald M. (2005).

11.9   Concluding Remarks

In this chapter we have attempted to briefly review what is currently known about the development and maturation of pain pathways. While our knowledge of

Box 3   Development of pain pathways in humans

Because human infants cannot report pain, we have to use indirect physiological and behavioral methods to assess its existence and severity and there is considerable debate on how best to measure pain in infants (Stevens, B. J. and Franck, L. S., 2001).

From 23 weeks of gestational age, human fetuses mount a hormonal response when needles are inserted into the innervated intrahepatic vein (Giannakoulopoulos, X. et al., 1994). Noxious activation of nerve endings stimulates central pathways (most likely through the fetal hypothalamic–pituitary–adrenal axis), producing a biochemical stress response, although the link between perceived pain and the hormonal stress response is unpredictable, even in adults. Hormonal responses are also used to measure pain in the youngest preterm infants (~24–26 weeks), along with many other physiological measures such as heart rate variability, crying and palmar sweating and motor activity (Stevens, B. J. and Franck, L. S., 2001), and characteristic facial expressions (Craig, K. D. et al., 2002). Many of these measures can be applied to pain related to a particular noxious event, such as a needle prick, but are harder to use for persistent pain. Studies using cutaneous reflexes evoked by stimulation in and around an area of injury, show that infants display cutaneous sensitization or hyperalgesia lasting days and weeks (Andrews, K. and Fitzgerald, M., 2002; Andrews, K. et al., 2002), consistent with laboratory studies. Nothing is known of the maturation of descending brainstem controls in human infants.

True pain experience requires functional maturation of higher brain centers and recent studies using infrared spectroscopy during routine clinical heel lances suggest that a cortical blood flow response is present as early as 25 weeks (Slater, R. et al., 2006).

the development of spinal nociceptive pathways has increased considerably in recent years (Fitzgerald, M., 2005), the understanding of the structural and functional maturation of supraspinal centers involved in the affective and nociceptive components of pain is severely lacking. With rapid improvements in medical knowledge and technologies, it is now possible to treat children born at increasing premature stages of development. This requires long periods of hospitalization and large numbers of often invasive procedures that in an adult or older child would be considered painful (Box 3), leading to the administration of analgesic and sedative compounds. We know that neonates respond to pain but the pathways differ in important ways from those in adults. Until we understand more of the functional development and maturation of the central neuronal elements involved in the experience of pain, the best course of analgesic action in infancy remains unclear.

Acknowledgments

The authors would like to thank The Wellcome Trust, The Wellcome London Pain Consortium, and Professor S. P. Hunt (UCL).

References

Aggleton, J. P. 1999. Mapping recognition memory in the primate brain: why it's sometimes right to be wrong. Brain Res. Bull. 50, 447–448.

Aggleton, J. P. and Saunders, R. C. 2000. The Amygdala, 2nd edn. Oxford University Press.

Aitken, A. R. and Tork, I. 1988. Early development of serotonin-containing neurons and pathways as seen in wholemount preparations of the fetal rat brain. J. Comp. Neurol. 274, 32–47.

Albuquerque, C., Lee, C. J., Jackson, A. C., and MacDermott, A. B. 1999. Subpopulations of GABAergic and non-GABAergic rat dorsal horn neurons express Ca^{2+}-permeable AMPA receptors. Eur. J. Neurosci. 11, 2758–2766.

Aldskogius, H. and Risling, M. 1989. Number of dorsal root ganglion neurons and axons in cats of different ages. Exp. Neurol. 106, 70–73.

Altman, J. and Bayer, S. A. 1984. The development of the rat spinal cord. Adv. Anat. Embryol. Cell Biol. 85, 1–164.

Anand, K. J. 2000. Pain, plasticity, and premature birth: a prescription for permanent suffering? Nat. Med. 6, 971–973.

Andrews, K. and Fitzgerald, M. 2002. Wound sensitivity as a measure of analgesic effects following surgery in human neonates and infants. Pain 99, 185–195.

Andrews, K. A., Desai, D., Dhillon, H. K., Wilcox, D. T., and Fitzgerald, M. 2002. Abdominal sensitivity in the first year of life: comparison of infants with and without prenatally diagnosed unilateral hydronephrosis. Pain 100, 35–46.

Angevine, J. B. and Sidman, R. L. 1961. Autoradiogrpahic study of cell migration suring histogenesis of cerebral cortex in the mouse. Nature 192, 766–768.

Anseloni, V. C., He, F., Novikova, S. I., Turnbach, R. M., Lidow, I. A., Ennis, M., and Lidow, M. S. 2005. Alterations in stress-associated behaviors and neurochemical markers in adult rats after neonatal short-lasting local inflammatory insult. Neuroscience 131, 635–645.

Armstrong-James, M. A. 1970. Spontaneous and evoked single unit activity in 7-day rat cerebral cortex. J. Physiol. 208, 10P–11P.

Armstrong-James, M. 1975. The functional status and columnar organization of single cells responding to cutaneous stimulation in neonatal rat somatosensory cortex S1. J. Physiol. 246, 501–538.

Baccei, M. L. and Fitzgerald, M. 2004. Development of GABAergic and glycinergic transmission in the neonatal rat dorsal horn. J. Neurosci. 24, 4749–4757.

Baccei, M. L. and Fitzgerald, M. 2005. In: The Textbook of Pain, 5th edn. (eds McMahon and Koltzenburg), Chapter 8, pp.143–158. Elsevier.

Baccei, M. L., Bardoni, R., and Fitzgerald, M. 2003. Development of nociceptive synaptic inputs to the neonatal rat dorsal horn: glutamate release by capsaicin and menthol. J. Physiol. 549, 231–242.

Bachevalier, J., Ungerleider, L. G., O'Neill, J. B., and Friedman, D. P. 1986. Regional distribution of [3H]naloxone binding in the brain of a newborn rhesus monkey. Brain Res. 390, 302–308.

Bardoni, R. 2001. Excitatory synaptic transmission in neonatal dorsal horn: NMDA and ATP receptors. News Physiol. Sci. 16, 95–100.

Basbaum, A. I. and Fields, H. L. 1978. Endogenous pain control mechanisms: review and hypothesis. Ann. Neurol. 4, 451–462.

Beggs, S., Torsney, C., Drew, L. J., and Fitzgerald, M. 2002. The postnatal reorganization of primary afferent input and dorsal horn cell receptive fields in the rat spinal cord is an activity-dependent process. Eur. J. Neurosci. 16, 1249–1258.

Benn, S. C., Costigan, M., Tate, S., Fitzgerald, M., and Woolf, C. J. 2001. Developmental expression of the TTX-resistant voltage-gated sodium channels Nav1.8 (SNS) and Nav1.9 (SNS2) in primary sensory neurons. J. Neurosci. 21, 6077–6085.

Benn, S. C., Perrelet, D., Kato, A. C., Scholz, J., Decosterd, I., Mannion, R. J., Bakowska, J. C., and Woolf, C. J. 2002. Hsp27 upregulation and phosphorylation is required for injured sensory and motor neuron survival. Neuron 36, 45–56.

Bennett, D. L., Averill, S., Clary, D. O., Priestley, J. V., and McMahon, S. B. 1996. Postnatal changes in the expression of the trkA high-affinity NGF receptor in primary sensory neurons. Eur. J. Neurosci. 8, 2204–2208.

Bice, T. N. and Beal, J. A. 1997. Quantitative and neurogenic analysis of neurons with supraspinal projections in the superficial dorsal horn of the rat lumbar spinal cord. J. Comp. Neurol. 388, 565–574.

Bondok, A. A. and Sansone, F. M. 1984. Retrograde and transganglionic degeneration of sensory neurons after a peripheral nerve lesion at birth. Exp. Neurol. 86, 322–330.

Boucher, T., Jennings, E., and Fitzgerald, M. 1998. The onset of diffuse noxious inhibitory controls in postnatal rat pups: a C-Fos study. Neurosci. Lett. 257, 9–12.

Bourgeais, L., Gauriau, C., and Bernard, J. F. 2001. Projections from the nociceptive area of the central nucleus of the amygdala to the forebrain: a PHA-L study in the rat. Eur. J. Neurosci. 14, 229–255.

Braz, J. M., Nassar, M. A., Wood, J. N., and Basbaum, A. I. 2005. Parallel "pain" pathways arise from subpopulations of primary afferent nociceptor. Neuron 47, 787–793.

Bregman, B. S. 1987. Development of serotonin immunoreactivity in the rat spinal cord and its plasticity after neonatal spinal cord lesions. Brain Res. 431, 245–263.

Bremner, L., Fitzgerald, M., and Baccei, M. 2006. The functional GABA(A)-receptor mediated inhibition in the neonatal dorsal horn. J. Neurophysiol. 95(6), 3893–3897.

Cabana, T. and Martin, G. F. 1984. Developmental sequence in the origin of descending spinal pathways. Studies using retrograde transport techniques in the North American opossum (Didelphis virginiana). Brain Res. 317, 247–263.

Cardinal, R. N., Parkinson, J. A., Lachenal, G., Halkerston, K. M., Rudarakanchana, N., Hall, J., Morrison, C. H., Howes, S. R., Robbins, T. W., and Everitt, B. J. 2002. Effects of selective excitotoxic lesions of the nucleus accumbens core, anterior cingulate cortex, and central nucleus of the amygdala on autoshaping performance in rats. Behav. Neurosci. 116, 553–567.

Cervero, F. and Plenderleith, M. B. 1985. C-fibre excitation and tonic descending inhibition of dorsal horn neurones in adult rats treated at birth with capsaicin. J. Physiol. 365, 223–237.

Cholley, B., Wassef, M., Arsénio-Nunes, L., Brehier, A., and Sotelo, C. 1989. Proximal trajectory of the brachium conjunctivum in rat fetuses and its early association with the parabrachial nucleus. A study combining in vitro HRP anterograde axonal tracing and immunocytochemistry. Brain Res. Dev. Brain Res. 45, 185–202.

Commissiong, J. W. 1983. The development of catecholaminergic nerves in the spinal cord of rat. II. Regional development. Brain Res. 313, 75–92.

Commissiong, J. W. and Neff, N. H. 1979. Current status of dopamine in the mammalian spinal cord. Biochem. Pharmacol. 28, 1569–1573.

Commissiong, J. W., Galli, C. L., and Neff, N. H. 1978. Differentiation of dopaminergic and noradrenergic neurons in rat spinal cord. J. Neurochem. 30, 1095–1099.

Constantinou, J., Reynolds, M. L., Woolf, C. J., Safieh-Garabedian, B., and Fitzgerald, M. 1994. Nerve growth factor levels in developing rat skin: upregulation following skin wounding. Neuroreport 5, 2281–2284.

Craig, K. D., Korol, C. T., and Pillai, R. R. 2002. Challenges of judging pain in vulnerable infants. Clin. Perinatol. 29, 445–457.

Davidson, R. J. 2002. Anxiety and affective style: role of prefrontal cortex and amygdala. Biol. Psychiatry 51, 68–80.

Davis, M. 1998a. Are different parts of the extended amygdala involved in fear versus anxiety? Biol. Psychiatry 44, 1239–1247.

Davis, M. 1998b. Are different parts of the extended amygdala involved in fear versus anxiety? Biol. Psychiatry 44, 1239–1247.

Davies, A. M. 2000. Neurotrophins: more to NGF than just survival. Curr. Biol. 10, R374–R376.

Davis, B. M., Fundin, B. T., Albers, K. M., Goodness, T. P., Cronk, K. M., and Rice, F. L. 1997. Overexpression of nerve growth factor in skin causes preferential increases among innervation to specific sensory targets. J. Comp. Neurol. 387, 489–506.

Diebler, M. F., Farkas-Bargeton, E., and Wehrle, R. 1979. Developmental changes of enzymes associated with energy metabolism and the synthesis of some neurotransmitters in discrete areas of human neocortex. J. Neurochem. 32, 429–435.

Ding, Y. Q., Takada, M., Shigemoto, R., and Mizumo, N. 1995. Spinoparabrachial tract neurons showing substance P receptor-like immunoreactivity in the lumbar spinal cord of the rat. Brain Res. 674, 336–340.

Ekholm, J. 1967. Postnatal changes in cutaneous reflexes and in the discharge pattern of cutaneous and articular sense organs. A morphological and physiological study in the cat. Acta. Physiol. Scand. Suppl. 297, 1–130.

Fees-Higgins, A. and Larroche, J. C. 1987. Development of The Human Foetal Brain: An Anatomical Atlas. INSERM.

Fitzgerald, M. 1985. The post-natal development of cutaneous afferent fibre input and receptive field organization in the rat dorsal horn. J. Physiol. 364, 1–18.

Fitzgerald, M. 1987. Cutaneous primary afferent properties in the hind limb of the neonatal rat. J. Physiol. 383, 79–92.

Fitzgerald, M. 2005. The development of nociceptive circuits. Nat. Rev. Neurosci. 6, 507–520.

Fitzgerald, M. and Gibson, S. 1984. The postnatal physiological and neurochemical development of peripheral sensory C fibres. Neuroscience 13, 933–944.

Fitzgerald, M. and Jennings, E. 1999. The postnatal development of spinal sensory processing. Proc. Natl. Acad. Sci. U. S. A. 96, 7719–7722.

Fitzgerald, M. and Koltzenburg, M. 1986. The functional development of descending inhibitory pathways in the dorsolateral funiculus of the newborn rat spinal cord. Brain Res. 389, 261–270.

Fitzgerald, M. and Walker, S. M. 2003. The Role of Activity in Developing Pain Pathways. In: Proceedings of the 10th

World Congress on Pain (eds. J. Dostrovsky, D. Carr, and M. Koltzenburg), pp. 185–196. IASP Press.

Fitzgerald, M., Butcher, T., and Shortland, P. 1994. Developmental changes in the laminar termination of A fibre cutaneous sensory afferents in the rat spinal cord dorsal horn. J. Comp. Neurol. 348, 225–233.

Fitzgerald, M., King, A. E., Thompson, S. W., and Woolf, C. J. 1987. The postnatal development of the ventral root reflex in the rat; a comparative in vivo and in vitro study. Neurosci. Lett. 78, 41–45.

Fitzgerald, M., Shaw, A., and MacIntosh, N. 1988. Postnatal development of the cutaneous flexor reflex: comparative study of preterm infants and newborn rat pups. Dev. Med. Child Neurol. 30, 520–526.

Fitzgerald, M., Woolf, C. J., and Shortland, P. 1990. Collateral sprouting of the central terminals of cutaneous primary afferent neurons in the rat spinal cord: pattern, morphology, and influence of targets. J. Comp. Neurol. 300, 370–385.

Fox, K. 2002. Anatomical pathways and molecular mechanisms for plasticity in the barrel cortex. Neuroscience 111, 799–814.

Gallagher, M. and Schoenbaum, G. 1999. Functions of the amygdala and related forebrain areas in attention and cognition. Ann. N. Y. Acad. Sci. 877, 397–411.

Gebhart, G. F. 2004. Descending modulation of pain. Neurosci. Biobehav. Rev. 27, 729–737.

Giannakoulopoulos, X., Sepulveda, W., Kourtis, P., Glover, V., and Fisk, N. M. 1994. Fetal plasma cortisol and beta-endorphin response to intrauterine needling. Lancet 344, 77–81.

Gilbert, M. and Stelzner, D. J. 1979. The development of descending and dorsal root connections in the lumbosacral spinal cord of the postnatal rat. J. Comp. Neurol. 184, 821–838.

Green, G. M. and Gibb, A. J. 2001. Characterization of the single-channel properties of NMDA receptors in laminae I and II of the dorsal horn of neonatal rat spinal cord. Eur. J. Neurosci. 14, 1590–1602.

Guo, A., Simone, D. A., Stone, L. S., Fairbanks, C. A., Wang, J., and Elde, R. 2001. Developmental shift of vanilloid receptor 1 (VR1) terminals into deeper regions of the superficial dorsal horn: correlation with a shift from TrkA to Ret expression by dorsal root ganglion neurons. Eur. J. Neurosci. 14, 293–304.

Hall, A. K., Ai, X., Hickman, G. E., MacPhedran, S. E., Nduaguba, C. O., and Robertson, C. P. 1997. The generation of neuronal heterogeneity in a rat sensory ganglion. J. Neurosci. 17, 2775–2784.

Hall, A. K., Dinsio, K. J., and Cappuzzello, J. 2001. Skin cell induction of calcitonin gene-related peptide in embryonic sensory neurons in vitro involves activin. Dev. Biol. 229, 263–270.

Hanganu, I. L., Kilb, W., and Luhmann, H. J. 2002. Functional synaptic projections onto subplate neurons in neonatal rat somatosensory cortex. J. Neurosci. 22, 7165–7176.

Higashi, S., Molnar, Z., Kurotani, T., and Toyama, K. 2002. Prenatal development of neural excitation in rat thalamocortical projections studied by optical recording. Neuroscience 115, 1231–1246.

Holmberg, H. and Schouenborg, J. 1996. Postnatal development of the nociceptive withdrawal reflexes in the rat: a behavioural and electromyographic study. J. Physiol. 493(Pt 1), 239–252.

Howard, R. F., Hatch, D. J., Cole, T. J., and Fitzgerald, M. 2001. Inflammatory pain and hypersensitivity are selectively reversed by epidural bupivacaine and are developmentally regulated. Anesthesiology 95, 421–427.

Huang, C. C. 1991. Sonographic cerebral sulcal development in premature newborns. Brain Dev. 13, 27–31.

Huang, Y., Stamer, W. D., Anthony, T. L., Kumar, D. V., St, J. P., and Regan, J. W. 2002. Expression of alpha(2)-adrenergic receptor subtypes in prenatal rat spinal cord. Brain Res. Dev. Brain Res. 133, 93–104.

Humphrey, T. 1968. The development of the human amygdala during early embryonic life. J. Comp. Neurol. 132, 135–165.

Hunt, S. and Mantyh, P. 2001. The molecular dynamics of pain control. Nat. Rev. Neurosci. 2, 83–91.

Jackman, A. and Fitzgerald, M. 2000. Development of peripheral hindlimb and central spinal cord innervation by subpopulations of dorsal root ganglion cells in the embryonic rat. J. Comp. Neurol. 418, 281–298.

Jakowec, M. W., Fox, A. J., Martin, L. J., and Kalb, R. G. 1995. Quantitative and qualitative changes in AMPA receptor expression during spinal cord development. Neuroscience 67, 893–907.

Jasmin, L., Tien, D., Weinshenker, D., Palmiter, R. D., Green, P. G., Janni, G., and Ohara, P. T. 2002. The NK1 receptor mediates both the hyperalgesia and the resistance to morphine in mice lacking noradrenaline. Proc. Natl. Acad. Sci. U. S. A. 99, 1029–1034.

Jiang, M. C. and Gebhart, G. F. 1998. Development of mustard oil-induced hyperalgesia in rats. Pain 77, 305–313.

Jones, B. E. 1991. Noradrenergic locus coeruleus neurons: their distant connections and their relationship to neighboring (including cholinergic and GABAergic) neurons of the central gray and reticular formation. Prog. Brain Res. 88, 15–30.

Kaas, J. H., Merzenich, M. M., and Killackey, H. P. 1983. The reorganization of somatosensory cortex following peripheral nerve damage in adult and developing mammals. Annu. Rev. Neurosci. 6, 325–356.

Keller, A. F., Breton, J. D., Schlichter, R., and Poisbeau, P. 2004. Production of 5alpha-reduced neurosteroids is developmentally regulated and shapes GABA(A) miniature IPSCs in lamina II of the spinal cord. J. Neurosci. 24, 907–915.

Klimach, V. J. and Cooke, R. W. 1988. Maturation of the neonatal somatosensory evoked response in preterm infants. Dev. Med. Child Neurol. 30, 208–214.

Koltzenburg, M. 1999. The changing sensitivity in the life of the nociceptor. Pain Suppl. 6, S93–S102.

Koltzenburg, M. and Lewin, G. R. 1997. Receptive properties of embryonic chick sensory neurons innervating skin. J. Neurophysiol. 78, 2560–2568.

Kordower, J. H., Piecinski, P., and Rakic, P. 1992. Neurogenesis of the amygdaloid nuclear complex in the rhesus monkey. Brain Res. Dev. Brain Res. 68, 9–15.

Lakke, E. A. 1997. The projections to the spinal cord of the rat during development: a timetable of descent. Adv. Anat. Embryol. Cell Biol. 135, 1–143.

Ledoux, J. E. 2000a. Emotion circuits in the brain. Annu. Rev. Neurosci. 23, 155–184.

Ledoux, J. E. 2000b. Emotion circuits in the brain. Annu. Rev. Neurosci. 23, 155–184.

Lee, C. J., Kong, H., Manzini, M. C., Albuquerque, C., Chao, M. V., and MacDermott, A. B. 2001. Kainate receptors expressed by a subpopulation of developing nociceptors rapidly switch from high to low Ca^{2+} permeability. J. Neurosci. 21, 4572–4581.

Leong, S. K. 1983. Localizing the corticospinal neurons in neonatal, developing and mature albino rat. Brain Res. 265, 1–9.

Lev, D. L., Weinfeld, E., and White, E. L. 2002. Synaptic patterns of thalamocortical afferents in mouse barrels at postnatal day. J. Comp. Neurol. 442, 63–77.

Lewin, G. R. and Mendell, L. M. 1994. Regulation of cutaneous C-fibre heat nociceptors by nerve growth factor in the developing rat. J. Neurophysiol. 71, 941–949.

Li, W. and Neugebauer, V. 2004. Differential roles of mGluR1 and mGluR5 in brief and prolonged nociceptive processing in central amygdala neurons. J. Neurophysiol. 91, 13–24.

Li, P. and Zhuo, M. 1998. Silent glutamatergic synapses and nociception in mammalian spinal cord. Nature 393, 695–698.

Lidov, H. G. and Molliver, M. E. 1982. Immunohistochemical study of the development of serotonergic neurons in the rat CNS. Brain Res. Bull. 9, 559–604.

Lidow, M. S. 2002. Long-term effects of neonatal pain on nociceptive systems. Pain 99, 377–383.

Loizou, L. A. 1972. The postnatal ontogeny of monoamine-containing neurones in the central nervous system of the albino rat. Brain Res. 40, 395–418.

Luhmann, H. J., Hanganu, I., and Kilb, W. 2003. Cellular physiology of the neonatal rat cerebral cortex. Brain Res. Bull. 60, 345–353.

Man, S. H., Wong, Y. M., Fitzgerald, M. F., and Hunt, S. P. 2005. Postnatal development of excitatory link between lamina I and the brainstem. Proceedings of the International Society for the Study of Pain. Abstract.

Mantyh, P. W. and Hunt, S. P. 2004. Setting the tone: superficial dorsal horn projection neurons regulate pain sensitivity. Trends Neurosci. 27, 582–584.

Marchand, F., Perretti, M., and McMahon, S. B. 2005. Role of the immune system in chronic pain. Nat. Rev. Neurosci. 6, 521–532.

Maren, S. 1999. Long-term potentiation in the amygdala: a mechanism for emotional learning and memory. Trends Neurosci. 22, 561–567.

Markus, A., Zhong, J., and Snider, W. D. 2002. Raf and akt mediate distinct aspects of sensory axon growth. Neuron 35, 65–76.

Marsh, D., Dickenson, A., Hatch, D., and Fitzgerald, M. 1999. Epidural opioid analgesia in infant rats II: responses to carrageenan and capsaicin. Pain 82, 33–38.

Marshall, G. E., Shehab, S. A., Spike, R. C., and Todd, A. J. 1996. Neurokinin-1 receptors on lumbar spinothalamic neurons in the rat. Neuroscience 72, 255–263.

Meyer, G. 2001. Human neocortical development: the importance of embryonic and early fetal events. Neuroscientist 7, 303–314.

Millan, M. J. 2002. Descending control of pain. Prog. Neurobiol. 66, 355–474.

Molliver, D. C., Wright, D. E., Leitner, M. L., Parsadanian, A. S., Doster, K., Wen, D., Yan, Q., and Snider, W. D. 1997. IB4-binding DRG neurons switch from NGF to GDNF dependence in early postnatal life. Neuron 19, 849–861.

Molnar, Z., Kurotani, T., Higashi, S., Yamamoto, N., and Toyama, K. 2003. Development of functional thalamocortical synapses studied with current source-density analysis in whole forebrain slices in the rat. Brain Res. Bull. 60, 355–371.

Molnar, Z., Lopez-Bendito, G., Small, J., Partridge, L. D., Blakemore, C., and Wilson, M. C. 2002. Normal development of embryonic thalamocortical connectivity in the absence of evoked synaptic activity. J. Neurosci. 22, 10313–10323.

Momose-Sato, Y., Sato, K., and Kamino, K. 2001. Optical approaches to embryonic development of neural functions in the brainstem. Prog. Neurobiol. 63, 151–197.

Moore, R. Y. and Bloom, F. E. 1978. Central catecholamine neuron systems: anatomy and physiology of the dopamine systems. Annu. Rev. Neurosci. 1, 129–169.

Nandi, R. and Fitzgerald, M. 2005. Opioid analgesia in the newborn. Eur. J. Pain 9, 105–108.

Nandi, R., Beacham, D., Middleton, J., Koltzenburg, M., Howard, R. F., and Fitzgerald, M. 2004. The functional expression of mu opioid receptors on sensory neurons is developmentally regulated; morphine analgesia is less selective in the neonate. Pain 111, 38–50.

Neugebauer, V. and Li, W. 2002. Processing of nociceptive mechanical and thermal information in central amygdala neurons with knee-joint input. J. Neurophysiol. 87, 103–112.

Neugebauer, V. and Li, W. 2003. Differential sensitization of amygdala neurons to afferent inputs in a model of arthritic pain. J. Neurophysiol. 89, 716–727.

Nikolic, I. and Kostovic, I. 1986. Development of the lateral amygdaloid nucleus in the human fetus: transient presence of discrete cytoarchitectonic units. Anat. Embryol. (Berl.) 174, 355–360.

Nissen, U. V., Mochida, H., and Glover, J. C. 2005. Development of projection-specific interneurons and projection neurons in the embryonic mouse and rat spinal cord. J. Comp. Neurol. 483, 30–47.

O'Leary, D. D. and Nakagawa, Y. 2002. Patterning centers, regulatory genes and extrinsic mechanisms controlling arealization of the neocortex. Curr. Opin. Neurobiol. 12, 14–25.

Olson, L. and Seiger, A. 1972. Early prenatal ontogeny of central monoamine neurons in the rat: fluorescence histochemical observations. Z. Anat. Entwicklungsgesch 137, 301–316.

O'Rahilly, R. and Muller, F. 1994. Neurulation in the normal human embryo. Ciba Found. Symp. 181, 70–82.

Pattinson, D. and Fitzgerald, M. 2004. The neurobiology of infant pain: development of excitatory and inhibitory neurotransmission in the spinal dorsal horn. Reg. Anesth. Pain Med. 29, 36–44.

van Praag, H. and Frenk, H. 1991. The development of stimulation-produced analgesia (SPA) in the rat. Brain Res. Dev. Brain Res. 64, 71–76.

Petersson, P., Waldenstrom, A., Fahraeus, C., and Schouenborg, J. 2003. Spontaneous muscle twitches during sleep guide spinal self-organization. Nature 424, 72–75.

Pitkanen, A., Savander, V., and Ledoux, J. E. 1998. Organization of intra-amygdaloid circuitries in the rat: an emerging framework for understanding functions of the amygdala. Trends Neurosci. 20(11), 517–523 Abstract.

Rajaofetra, N., Sandillon, F., Geffard, M., and Privat, A. 1989. Pre- and post-natal ontogeny of serotonergic projections to the rat spinal cord. J. Neurosci. Res. 22, 305–321.

Rees, S., Nitsos, I., and Rawson, J. 1994. The development of cutaneous afferent pathways in fetal sheep: a structural and functional study. Brain Res. 661, 207–222.

Remedios, R., Subramanian, L., and Tole, S. 2004. LIM genes parcellate the embryonic amygdala and regulate its development. J. Neurosci. 24, 6986–6990.

Ren, K. and Dubner, R. 2002. Descending modulation in persistent pain: an update. Pain 100, 1–6.

Ritter, A. M., Lewin, G. R., Kremer, N. E., and Mendell, L. M. 1991. Requirement for nerve growth factor in the development of myelinated nociceptors in vivo. Nature 350, 500–502.

Rubenstein, J. L. 1998. Development of serotonergic neurons and their projections. Biol. Psychiatry 44, 145–150.

Sagen, J. and Proudfit, H. K. 1984. Effect of intrathecally administered noradrenergic antagonists on nociception in the rat. Brain Res. 310, 295–301.

Schlaggar, B. L. and O'Leary, D. D. 1994. Early development of the somatotopic map and barrel patterning in rat somatosensory cortex. J. Comp. Neurol. 346, 80–96.

Shortland, P. and Fitzgerald, M. 1994. Neonatal sciatic nerve section results in a rearrangement of the central terminals of saphenous and axotomized sciatic nerve afferents in the dorsal horn of the spinal cord of the adult rat. Eur. J. Neurosci. 6, 75–86.

Sidman, R. L. and Rakic, P. 1973. Neuronal migration, with special reference to developing human brain: a review. Brain Res. 62, 1–35.

Sivilotti, L. G., Thompson, S. W., and Woolf, C. J. 1993. Rate of rise of the cumulative depolarization evoked by repetitive stimulation of small-caliber afferents is a predictor of action potential windup in rat spinal neurons in vitro. J. Neurophysiol. 69, 1621–1631.

Skagerberg, G., Bjorklund, A., Lindvall, O., and Schmidt, R. H. 1982. Origin and termination of the diencephalo-spinal dopamine system in the rat. Brain Res. Bull. 9, 237–244.

Skagerberg, G., Meister, B., Hokfelt, T., Lindvall, O., Goldstein, M., Joh, T., and Cuello, A. C. 1988. Studies on dopamine-, tyrosine hydroxylase- and aromatic L-amino acid decarboxylase-containing cells in the rat diencephalon: comparison between formaldehyde-induced histofluorescence and immunofluorescence. Neuroscience 24, 605–620.

Slater, R., Cantarella, A., Gallella, S., Worley, A., Boyd, S., Meek, J., and Fitzgerald, M. 2006. Cortical pain responses in human infants. J. Neurosci. 26(14), 3662–3666.

Stegenga, S. L. and Kalb, R. G. 2001. Developmental regulation of N-methyl-D-aspartate- and kainate-type glutamate receptor expression in the rat spinal cord. Neuroscience 105, 499–507.

Stevens, B. J. and Franck, L. S. 2001. Assessment and management of pain in neonates. Paediatr. Drugs 3, 539–558.

Suzuki, R., Morcuende, S., Webber, M., Hunt, S. P., and Dickenson, A. H. 2002. Superficial NK1-expressing neurons control spinal excitability through activation of descending pathways. Nat. Neurosci. 5, 1319–1326.

Suzuki, R., Rygh, L. J., and Dickenson, A. H. 2004. Bad news from the brain: descending 5-HT pathways that control spinal pain processing. Trends Pharmacol. Sci. 25, 613–617.

Swanson, L. W. and Kuypers, H. G. 1980. The paraventricular nucleus of the hypothalamus: cytoarchitectonic subdivisions and organization of projections to the pituitary, dorsal vagal complex, and spinal cord as demonstrated by retrograde fluorescence double-labeling methods. J. Comp. Neurol. 194, 555–570.

Todd, A. J., McGill, M. M., and Shehab, S. A. 2000. Neurokinin 1 receptor expression by neurons in laminae I, III and IV of the rat spinal dorsal horn that project to the brainstem. Eur. J. Neurosci. 12, 689–700.

Torsney, C. and Fitzgerald, M. 2002. Age-dependent effects of peripheral inflammation on the electrophysiological properties of neonatal rat dorsal horn neurons. J. Neurophysiol. 87, 1311–1317.

Torsney, C. and Fitzgerald, M. 2003. Spinal dorsal horn cell receptive field size is increased in adult rats following neonatal hindpaw skin injury. J. Physiol. 550, 255–261.

Waldenstrom, A., Thelin, J., Thimansson, E., Levinsson, A., and Schouenborg, J. 2003. Developmental learning in a pain-related system: evidence for a cross-modality mechanism. J. Neurosci. 23, 7719–7725.

Walker, S. 2005. Modulation of pain by alpha 2 agonists – effect of developmental age. Ph.D. thesis, University College London.

Watanabe, M., Mishina, M., and Inoue, Y. 1994. Distinct spatiotemporal distributions of the N-methyl-D-aspartate receptor channel subunit mRNAs in the mouse cervical cord. J. Comp. Neurol. 345, 314–319.

Webster, M. J., Ungerleider, L. G., and Bachevalier, J. 1991. Lesions of inferior temporal area TE in infant monkeys alter cortico-amygdalar projections. Neuroreport 2, 769–772.

Winzer-Serhan, U. H., Raymon, H. K., Broide, R. S., Chen, Y., and Leslie, F. M. 1997. Expression of alpha 2 adrenoceptors during rat brain development – I. Alpha 2A messenger RNA expression. Neuroscience 76, 241–260.

Woolf, C. J. and Salter, M. W. 2000. Neuronal plasticity: increasing the gain in pain. Science 288, 1765–1769.

Yao, H. and Dan, Y. 2005. Synaptic learning rules, cortical circuits, and visual function. Neuroscientist 11, 206–216.

Yip, H. K., Rich, K. M., Lampe, P. A., and Johnson, E. M., Jr. 1984. The effects of nerve growth factor and its antiserum on the postnatal development and survival after injury of sensory neurons in rat dorsal root ganglia. J. Neurosci. 4, 2986–2992.

Zald, D. H. 2003. The human amygdala and the emotional evaluation of sensory stimuli. Brain Res. Brain Res. Rev. 41, 88–123.

12 Appropriate/Inappropriate Developed Pain Paths

J Schouenborg, Lund University, Lund, Sweden

12.1 General Background

The topographical organization of the somatosensory input to the spinal cord has traditionally been regarded as a somatotopically organized map of the body (Molander, C. and Grant, G., 1985; Mirnics, K. and Koerber, H. R., 1995; Silos-Santiago, I. *et al.*, 1995; Wilson, P. and Kitchener, P. D., 1996; Brown, P. B. *et al.*, 1997; 2004; 2005), with different body parts represented in different central nervous system (CNS) areas and the size of the representation being a function of receptor density and, to some extent, practise (see Blake, D. T., *et al.*, 2002). Most studies on the body representation have been focused on the sensory discriminative aspects (Brown, P. B. *et al.*, 2004; 2005). However, a major function of the spinal cord is to use the sensory information to control skeletal muscles. It has therefore been suggested that the somatosensory input may be primarily organized in a motor frame of reference (Levinsson, A. *et al.*, 2002). Such a view might explain some of the oddities with the known topography. For example, in the dorsal horn, there is an extensive overlap of afferent terminals from different body parts, and the sequence of the representation from distal to proximal body parts is neither continuous nor sequential (Levinsson, A. *et al.*, 2002). Moreover, the phenomenon of somatotopically inappropriate terminations, that is, primary afferents terminating outside the main projection area, might be explained if different motor systems requiring sensory information from same body region have different locations in the spinal cord.

To be useful in motor control, somatosensory information must be encoded with respect to body anatomy and movement patterns produced by the sensorimotor circuits. This is a computationally demanding task since the multisensory information (nociception, pressure, temperature, joint angles, muscle force, and length) arises from a complex body constitution. Understanding how sensory information is processed by spinal sensorimotor circuits and how the adult organization emerges during development is therefore a major task. This review will deal with the current literature on the modular organization of the spinal cord and the role of somatosensory imprinting in establishing an action-based sensory encoding in these modules.

12.2 Modular Organization of Sensorimotor Circuits in the Spinal Cord

During the last 10 years, the concept of a modular organization of the spinal cord has grown stronger. The idea of a modular organization of motor circuits in CNS is not new and, for the spinal cord, was proposed already in 1981 by Grillner, S. for locomotor circuits (unit bursters causing rhythmic activity around a joint) and later for circuits controlling scratch reflexes in the turtle (Mortin, L. I. *et al.*, 1985; Stein, P. S. *et al.*, 1998; Berkowitz, A., 2001). However, although a modular organization is conceivable, it is not yet clear exactly what constitutes a module in these rhythm-generating systems, the extent to which the different modules overlap in space, and how sensory information is related to the function of these modules (Tresch, M. C. *et al.*, 2002).

A modular type of reflex organization in the mammalian spinal cord was first demonstrated for the nociceptive withdrawal reflex (NWR) system in the rat (Schouenborg, J. and Kalliomäki, J., 1990; Schouenborg, J. *et al.*, 1992), but subsequently also in cats (Levinsson, A. *et al.*, 1999), mice (Thelin, J. M. and Schouenborg, J., 2003), and humans

(Sonnenborg, F. A. *et al.*, 2000). Here, the word modular is used synonymous to the term functional unit of a system, that is, not alluding to the existence of different motor systems, such as stepping, standing, scratching, or withdrawal reflex systems. For the NWR system, each excitatory module preferentially acts on a single muscle and performs a detailed sensorimotor transformation resulting in a graded withdrawal of the limb (or part of the limb) from its receptive field. For each excitatory NWR module, the input strength has a characteristic pattern on the skin that mimics the pattern of withdrawal efficacy in a standing position when the principal output muscle of the module contracts (Figure 1; Schouenborg, J. and Weng, H.-R., 1994). In a sense, the pattern of withdrawal (or unloading) efficacy is imprinted on the receptive field of the module. Neurones that encode sensory input in this motor frame of reference, termed reflex encoders (Schouenborg, J. *et al.*, 1995), are mainly located in deep dorsal horn laminae IV–VI. These neurones receive and weight the multisensory input from both tactile Aβ and nociceptive Aδ and C afferent fibers in proportion to the withdrawal/unloading action of the modules in a standing position (Figure 2). It is worth noting that, although the nociceptive input is stronger than that from tactile input, the strength of the input mediated by different types of afferent fibers is weighted in the same way. In fact, a large proportion of the wide dynamic range neurones, that is, neurones receiving a convergent input from tactile and nociceptive receptors, often referred to as wide dynamic range neurons, in the deep layers of the fifth lumbar segment, appear to be of the reflex encoder type. Within the L4–5 segments, reflex encoders for the interossei, flexor digitorum longus, gastrocnemius, peronei, and extensor digitorum longus muscles are located in a

mediolateral sequence reminiscent of the corresponding topographical organization of the motoneuron columns in the ventral horn. Hence, the reflex encoders appear to be located in discrete pools that have a musculotopic organization.

A corresponding set of inhibitory reflex modules also exists. In this case, the receptive fields correspond to the graded movement of the skin area toward external stimulation (i.e., increase in load) on contraction of the muscle in the module (Weng, H.-R. and Schouenborg, J., 1996). As a result of this organization, the excitatory and inhibitory modules are engaged to a degree that is proportional to their respective withdrawal/loading efficacy on skin stimulation.

The same principles of sensorimotor transformation as found for rat withdrawal reflexes also apply to the cat, mice, and humans (Levinsson, A. *et al.*, 1999 in cats; Sonnenborg, F. A. *et al.*, 2000; Thelin, J. M. and Schouenborg, J., 2003). Interestingly, muscles with different function in two different species exhibit corresponding differences in the receptive fields of the withdrawal reflexes. Thus, withdrawal reflexes elicited in dorsiflexors are much weaker in the cat than in the rat and vice versa for reflexes to plantar flexors. Cats stand on their digits, whereas rats stand on the whole plantar surface. In the standing position, the dorsiflexors of cat digits are much less effective in withdrawing the plantar skin than the corresponding muscles in rats and vice versa for plantar flexors. These findings are consistent with the view that these reflexes consist of adaptive modules (see Section 12.4).

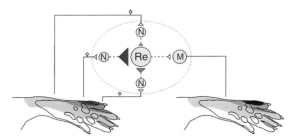

Figure 2 Schematic circuitry organization of a withdrawal reflex module. Nociceptive afferents from the receptive field (right) synapse with nociceptive neurones (N) in the superficial laminae of the dorsal horn. Reflex encoder neurones (Re) receive a weighted nociceptive input from superficially located neurones. Re project either directly or indirectly to motoneurones essentially controlling the action of a single muscle. Withdrawal efficacy of the module is shown to the right. Sensitivity in strength of primary afferent input and sensitivity in receptive fields is coded by color intensity. Tactile input is omitted for clarity.

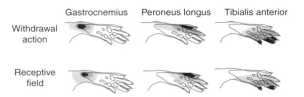

Figure 1 Action-based sensory encoding. Typical distributions of sensitivity and corresponding withdrawal action for three muscles are shown. Low, medium, and high dot density indicates areas of the skin from which the evoked responses were 0–30%, 30–70%, and 70–100% of maximal response, respectively. Adapted from Schouenborg, J. and Weng, H.-R. 1994. Sensorimotor transformation in a spinal motor system. Exp. Brain. Res. 100, 170–174.

Based on microstimulation in the dorsal horn of the spinal cord, a somewhat different modular organization of sensorimotor circuits acting on synergistic muscle groups was later proposed in frogs and rats (Tresch, M. C. and Bizzi, E., 1999; Bizzi, E. *et al.*, 2000, Saltiel, P. *et al.*, 2001). According to these authors, electrical and glutamatergic stimulation of the deep dorsal horn often results in a movement toward an equilibrium point, independent of the starting position of the limb. Here, the output of a module would include activity in two or more different muscles. Whether these findings reflect the existence of a fundamentally different type of reflex system than that of the withdrawal reflex system or the outcome of a combined stimulation of many withdrawal reflex modules is not clear at present. In a recent study, Avella and colleagues suggested, by analyzing electromyography (EMG) in frog hindlimbs during kicking, that a combination of a limited number of synergistic units was used for this behaviour (Avella, A. *et al.*, 2003).

12.3 Action-Based Sensory Encoding

By mapping tactile input to the dorsal horn, longitudinal cigar-shaped zones that weight the monosynaptic cutaneous input as a function of the unloading pattern caused by individual modules were found (Figure 3; Levinsson, A. *et al.*, 2002). This was a surprising finding in that it indicates that sensory input to the spinal cord gets organized and weighted in a motor frame of reference already at the level of the first-order synapses and that there may, consequently, not be a true 'map' of the body in the spinal cord as previously thought (Molander, C. and Grant, G., 1985; Mirnics, K. and Koerber, H. R., 1995; Silos-Santiago, I. *et al.*, 1995; Brown, P. B. *et al.*, 1997). Given that a main function of the spinal cord is to translate sensory information into movement corrections, an action-based sensory encoding do, however, appear to be an economical and fast way of processing the sensory information.

There are many reasons to think that action-based sensory encoding is fundamental also for higher-order motor systems. For example, the C1, C3, and X zones in the anterior cerebellar lobe is divided into microzones, where each microzone is defined by its climbing fibre input from a specific receptive field on the skin (Garwicz, M., 2002; Garwicz, M. *et al.*, 2002; Apps, R. and Garwicz, M., 2005). These climbing fibers encode the sensory input in the same way as individual spinal modules, that is, as a function of the unloading pattern caused by single muscle contraction.

12.4 Functional Adaptation of Sensorimotor Circuits During Development

Given that the adult sensorimotor transformations performed by the spinal withdrawal reflex circuits reflect precisely weighted connections in modules, how can this weighting be achieved during development? While the gross topographical organization of interneurons of the spinal cord is likely to be guided by gradients of trophic substances during the development (Albright, T. D. *et al.*, 2000, Chen, A. I. *et al.*, 2006; Chen, C. L. *et al.*, 2006; Kramer, I. *et al.*, 2006), it is difficult to see how such mechanisms could encode the detailed strength of every connection in the networks. Recent studies on the NWR system provide some clues to this problem. The sensorimotor transformations performed by its modules are functionally adapted during the first postnatal weeks in the rat (Holmberg, H. and Schouenborg, J., 1996a). During this time, the strength of the erroneous (or inappropriate) connections becomes weaker, whereas the strength of the adequate connections becomes proportional to the unloading effect on the skin of muscle contraction. Moreover, during this time, NWR can adapt to both altered innervation of the skin caused by nerve sections (Holmberg, H. and Schouenborg, J., 1996b) and to altered movement patterns caused by tendon transfer in the neonatal rat (Holmberg, H. *et al.*, 1997). Blocking the sensory input during the time of functional adaptation abolishes the learning (Waldenström, A. *et al.*, 2003), indicating a role for sensory input. Notably, a selective block of the nociceptive input by EMLA salve did not have an effect as compared with vehicle treatment. The functional adaptation is not blocked unless tactile input is also blocked, indicating a key role for tactile input in tuning nociceptive connections (Figure 4(a)). This conclusion is strongly supported by the finding that tactile feedback ensuing on spontaneous motility in spinal sensorimotor circuits alters the nociceptive connection strengths in NWR modules during postnatal development (Figures 4(b)–4(c); Petersson, P. *et al.*, 2003). Thus, the pattern of tactile inputs arriving in conjunction with the spontaneous movements has an instructive

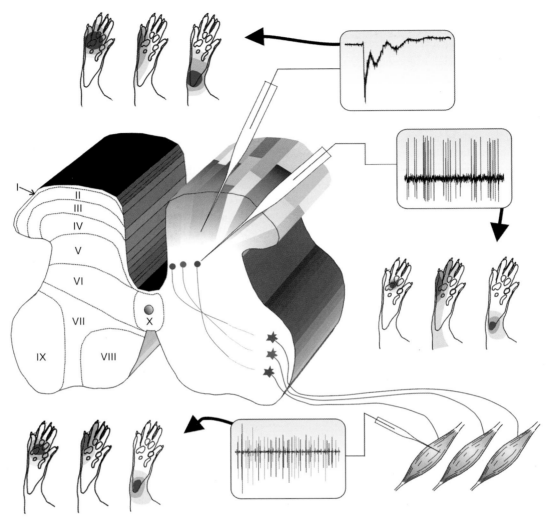

Figure 3 Schematic figure of proposed modular organization of the spinal cord. Columns of the dorsal horn receive a cutaneous input that has a specific weight distribution. This weight distribution is the same as that of nociceptive input to reflex encoders (interneurones that can encode the withdrawal reflex strength of individual modules) in deep dorsal horn. The reflex encoders are assumed to project to single muscles and weight the input according to the withdrawal efficacy of the output muscle. Sensitivity in strength of primary afferent input and sensitivity in receptive fields is coded by color intensity. Upper recordings, receptive fields of monosynaptic tactile field potentials; middle recordings, receptive fields of reflex encoders, bottom recordings, receptive fields of single muscles. Schematic indication of Rexed's laminae to the left.

role in the functional adaptation of the reflex modules. Uncorrelated input (given at random time points) does not cause a learning effect. Since this learning process results in an imprint of the withdrawal efficacy on the reflex modules, it was termed somatosensory imprinting (Holmberg, H. *et al.*, 1997; Petersson, P. *et al.*, 2003).

The finding that the strength of the nociceptive connections to the NWR network can be functionally adapted by tactile input discloses a hitherto unknown form of cross-modality learning in the spinal cord that

solves the puzzle of how a pain-related system can be learned during postnatal development, despite the rare occurrence of noxious events. Notably, this novel form of unsupervised learning occurs during active sleep, characterized by atonia in the musculature. This state may be particularly advantageous for learning since the sensory feedback on muscle contraction stands out from a more or less silent background.

A hitherto neglected problem is how the myriad of different types of somatosensory primary afferents, conveying information from different sensory

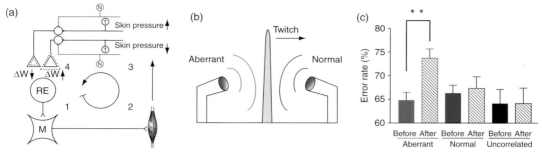

Figure 4 Somatosensory imprinting. (a) Proposed learning cycle underlying somatosensory imprinting. A learning cycle consists of the following chain of events. (1) Spontaneous bursts in reflex encoders (REs). (2) Motoneuron (M) activation leading to a muscle twitch that causes skin movement. (3) Increased or decreased skin pressure resulting in altered sensory input to pre-reflex encoder interneurons. Black lines (T) represent afferents from skin areas from where an increase and decrease, respectively, in low-threshold mechanoreceptor input would occur. Red lines (N) symbolize nociceptive afferents. (4) The strengths of erroneous connections (receiving increased tactile input) between pre-RE interneurons and REs are weakened and that of appropriate ones (receiving reduced tactile input) is strengthened (Dw = change in connection strength) (modified from Waldenström, A. et al., 2003). (b) Behavioural experiments. Schematic training set-up depicting aberrant and normal training protocol. Distal tail of a sleeping rat pup is shown. Air puffs directed to the skin surface moving toward or away from the air stream is termed aberrant or normal conditioning, respectively. One session of conditioning stimulation lasted for 2 h. Arrow indicates direction of spontaneous tail flick (modified from Petersson, P. et al., 2003). (c) The nociceptive withdrawal reflex (NWR) error rate (%) before and after aberrant or normal tactile stimulation (mean SE). The NWR error rate was significantly increased (Kruskal–Wallis and Dunnett's test $P < 0.01$) in rats given aberrant tactile feedback for 2 h (47 training sessions). Normal air-puff stimulation (53 training sessions) had no effect. Control experiments, using random air-puff stimulation (23 training sessions) or no stimulation (23 training sessions) did not affect NWR adaptation (Wilcoxon signed rank test $P = 0.65$ and $P = 0.37$, respectively), ruling out direct effects of the set-up, test stimulations, or of uncorrelated stimulation. (Adapted from Petersson, P., Waldenström, A., Fåhreaus, C., and Schouenborg, J. 2003. Spontaneous muscle twitches during sleep guide spinal self-organization. Nature 424(6244), 72–75.

modalities and locations, gets organized in the dorsal horn so that spinal multisensory modules receive topographically ordered information. Considering the apparent lack of somatotopic organization in the dorsal root ganglia and that different nerve fibre types enter the spinal cord by different routes and at different times during development, this is not a trivial problem. Indeed, recent data indicate that the adult body representation in the dorsal horn arises from an initially floating and plastic organization with many inappropriate connections through profound activity-dependent rewiring, involving both sprouting and elimination of afferent connections. In this process, tactile input seems to guide the alignment of tactile and nociceptive termination since disturbed tactile input patterns results in disturbed termination patterns of both tactile and nociceptive afferent fibers (Granmo, G. et al., 2002; Petersson, P. et al., 2004; Granmo, G. et al., 2005; 2007). The activity-dependent changes in the organization of the primary afferent termination in the spinal cord including changes in laminar termination patterns of tactile afferents are paralleled in time by reflex sensorimotor transformations, suggesting that the emerging action-based body representation is at least partly imprinted on the first-order synapses.

Importantly, these findings suggest that the adult action-based sensory encoding emerges by sensorimotor circuits probing the patterns of tactile input that ensue on movements caused by the effector muscles.

Spontaneous movements are a ubiquitous phenomenon during embryonic development in all vertebrates and mammals (De Vries, J. I. et al., 1982; Blumberg, M. S. and Lucas, D. E., 2000). However, while a role in maturation of motoneurones, motoneuronal axonal path finding, and contacts with skeletal muscles appears likely (Hanson, M. G. and Landmesser, L. T., 2004; 2006), their role in the CNS and for sensorimotor learning in particular has, however, not been known. The activity appears to be caused by spontaneous endogenous activity in neuronal circuits in the spinal cord and brain stem (Blumberg, M. S. and Lucas, D. E., 2000). Although present classifications tend to lump the spontaneous motility broadly into a few categories, detailed studies in humans distinguished 16 different types (De Vries, J. I. et al., 1982). The prevalence and complexity of these movements lead us to suggest that all major spinal motor systems contribute to the spontaneous movements during development. Furthermore, since somatosensory imprinting is highly effective, it may well be that all major groups of spinal motor systems learn relevant aspects of the body anatomy and biomechanics during development by probing the sensory feedback after spontaneous endogenous activation. It has also been suggested that the sensory input arising after spontaneous movements guide the tuning of the somatotopic organization of the somatosensory cortex during development (Khazipov, R. *et al.*, 2004). It may thus be proposed that sensory input arising as a consequence of spontaneous movements during development plays a key role also in the organization of higher-order sensorimotor systems.

As yet, little is known about the molecular mechanisms underlying somatosensory imprinting. While calcium transients (Perrier, J. F. and Hounsgaard, J., 2000; Russier, M. *et al.*, 2003) and/or spontaneous fluctuations of glutamate release (O'Donovan, M. J. *et al.*, 1998; O'Donovan, M. J., 1999) have been suggested to play a role for the initiation of spontaneous movements, *N*-methyl-D-aspartate (NMDA) receptors may also be involved in the learning mechanisms since the rewiring of the tactile afferent fiber termination in the dorsal horn during development is blocked by topically applied MK-801 (Mendelson, B., 1994; Beggs, S. *et al.*, 2002; Granmo, G. *et al.*, 2002; 2005). Moreover, mice with a disturbed long-term potentiation (LTP) in the hippocampus exhibit disturbed receptive fields of NWR, suggesting that LTP-like mechanisms in the spinal cord may be involved in somatosensory imprinting (Thelin, J. M., 2005). To further unravel the mechanisms that underlie the somatosensory imprinting is clearly an important task in future.

12.5 Concluding Remarks

The embryonic development of the spinal cord is characterized by a ventral to dorsal temporal sequence. Appearance of motoneurons precedes that of interneurons which in turn precedes primary afferent ingrowth (Altman, J. and Bayer, S. A., 1984). It is thus of interest that there is a specific spatial relation between the topography of motoneuron pools and lamina V reflex encoders (Schouenborg, J. *et al.*, 1995) and between the weight distribution of cutaneous input to the dorsal horn and the receptive fields of the reflex encoders. Taken together, these findings indicate an instrumental role of motoneurons and the patterns of sensory feedback caused by spontaneous motility during development in the

functional assembly of the dorsal horn. The elimination of inappropriate connections and strengthening of appropriate connections that normally occurs during development is likely to be a reflection of such a learning-dependent process.

Besides providing insights into how the wiring of the spinal cord is accomplished, the reviewed findings may well have clinical relevance. While the repair of the spinal cord after injury has so far been relatively unsuccessful, some advances indicate that regeneration across spinal lesions is possible and that some functional recovery may occur after such regeneration (Li, Y. *et al.*, 1997; Zgraggen, W. J. *et al.*, 1998; Levinsson, A. *et al.*, 2000). A major problem, once the obstacle of limited regeneration has been overcome, will be to eliminate the erroneous connections arising from regeneration (Bareyre, F. M. *et al.*, 2004). Such connections may be devastating for the patient. For example, patients with partially transected spinal cords often report chronic pain that is very difficult to alleviate (Sjölund, B. H., 2002). If adaptive mechanisms, such as somatosensory imprinting, which decrease the gain in erroneous connections and increase the gain in adequate connections can be utilized, true functional recovery may be attainable also from chaotically regenerated connections. A combination therapy based on training using feedback stimulation to tune the regenerated connections and pharmacological treatment that disinhibits regeneration might thus prove to be extremely useful.

Acknowledgments

This work was supported by Swedish MRC project no 01013 and a Linné grant, Knut and Alice Wallenbergs Foundation, Kocks Foundation.

References

Albright, T. D., Jessell, T. M., Kandel, E. R., and Posner, M. I. 2000. Neural science: a century of progress and the mysteries that remain. Neuron 25, (Suppl.), S1–S55. Review.

Altman, J. and Bayer, S. A. 1984. The development of the rat spinal cord. Adv. Anat. Embryol. Cell Biol. 85, 1–166.

Apps, R. and Garwicz, M. 2005. Anatomical and physiological foundations of cerebellar information processing. Nat. Rev. Neurosci. Apr: 6(4), 297–311. Review.

Avella, A., Saltiel, P., and Bizzi, E. 2003. Combinations of muscle synergies in the construction of a natural motor behavior. Nat. Neurosci. 6, 300–308.

Bareyre, F. M., Kerschensteiner, M., Raineteau, O., Mettenleiter, T. C., Weinmann, O., and Schwab, M. E. 2004.

The injured spinal cord spontaneously forms a new intraspinal circuit in adult rats. Nat. Neurosci. 7, 269–277.

Beggs, S., Torsney, C., Drew, L. J., and Fitzgerald, M. 2002. The postnatal reorganization of primary afferent input and dorsal horn cell receptive fields in the rat spinal cord is an activity-dependent process. Eur. J. Neurosci. 16, 1249–1258.

Berkowitz, A. 2001. Broadly tuned spinal neurons for each form of fictive scratching in spinal turtles. J. Neurophysiol. 86, 1017–1025.

Bizzi, E., Tresch, M. C., Saltiel, P., and d'Avella, A. 2000. New perspectives on spinal motor systems. Nat. Rev. Neurosci. 1, 101–108.

Blake, D. T., Byl, N. N., and Merzenich, M. M. 2002. Representation of the hand in the cerebral cortex. Behav. Brain Res. 135(1–2), 179–184.

Blumberg, M. S. and Lucas, D. E. 2000. Dual mechanisms of twitching during sleep in neonatal rats. Behav. Neurosci. 108, 1196–1202.

Brown, P. B., Harton, P., Millecchia, R., Lawson, J. J., Brown, A. G., Koerber, H. R., Culberson, J., and Stephens, S. 2005. From innervation density to tactile acuity 2: Embryonic and adult pre- and postsynaptic somatotopy in the dorsal horn. Brain Res. 1055, 36–59.

Brown, P. B., Koerber, H. R., and Millecchia, R. 1997. Assembly of the dorsal horn somatotopic map. Somatosens. Mot. Res. 14, 93–106.

Brown, P. B., Koerber, H. R., and Millecchia, R. 2004. From innervation density to tactile acuity. 1. Spatial representation. Brain Res. 1011, 14–32.

Chen, A. I., de Nooij, J. C., and Jessell, T. M. 2006. Graded activity of transcription factor Runx3 specifies the laminar termination pattern of sensory axons in the developing spinal cord. Neuron 49, 395–408.

Chen, C. L., et al. 2006. Runx1 determines nociceptive sensory neuron phenotype and is required for thermal and neuropathic pain. Neuron 49, 365–377.

De Vries, J. I., Visser, G. H., and Prechtl, H. F. 1982. The emergence of fetal behaviour. I. Qualitative aspects. Early Hum. Dev., 7, 301–322.

Garwicz, M. 2002. Spinal reflexes provide motor error signals to cerebellar modules – relevance for motor coordination. Brain Res. Rev. 40, 152–165.

Garwicz, M., Levinsson, A., and Schouenborg, J. 2002. Common principles of sensory encoding spinal reflex modules and cerebellar climbing fibres. J. Physiol. 540, 1061–1069.

Granmo, G., Jensen, T., Lind, G., and Schouenborg, J. 2005. NMDA dependent changes in dorsal horn somatotopy in the developing rat. IASP 11th World Congress on Pain, Sydney. Abstract.

Granmo, M., Levinsson, A., and Schouenborg, J. 2002. Postnatal changes in dorsal horn somatotopy in the rat. IASP 10th World Congress on Pain, 77–P73. Abstract.

Granmo, G., Petersson, P., and Schouenborg, J. Establishment of an action based body representation in the spinal cord. Submitted for publication.

Grillner, S. 1981. Control of locomotion in bipeds, tetrapods, and fish. In: Handbook of Physiology (ed. V. B. Brooks), Vol. 2, American Physiological Society, Bethesda, pp. 1179–1236, Section 1.

Hanson, M. G. and Landmesser, L. T. 2004. Normal patterns of spontaneous activity are required for correct motor axon guidance and the expression of specific guidance molecules. Neuron 43(5), 687–701.

Hanson, M. G. and Landmesser, L. T. 2006. Increasing the frequency of spontaneous rhythmic activity disrupts pool-specific axon fasciculation and pathfinding of embryonic spinal motoneurons. J. Neurosci. 26(49), 12769–12780.

Holmberg, H. and Schouenborg, J. 1996a. Postnatal development of the nociceptive withdrawal reflexes in the

rat: a behavioural and electromyographic study. J. Physiol. 493, 239–252.

Holmberg, H. and Schouenborg, J. 1996b. Developmental adaptation of withdrawal reflexes to early alteration of peripheral innervation in the rat. J. Physiol. 495, 399–409.

Holmberg, H., Schouenborg, J., Yu, Y., and Weng, H.-R. 1997. Developmental adaptation of rat nociceptive withdrawal reflexes after neonatal tendon transfer. J. Neurosci. 17, 2071–2078.

Khazipov, R., Sirota, A., Leinekugel, X., Holmes, G. L., Ben-Ari, Y., and Buzsaki, G. 2004. Early motor activity drives spindle bursts in the developing somatosensory cortex. Nature 432(7018), 758–761.

Kramer, I., et al. 2006. A role for Runx transcription factor signaling in dorsal root ganglion sensory neuron diversification. Neuron 49, 379–393.

Levinsson, A., Garwicz, M., and Schouenborg, J. 1999. Sensorimotor transformation in cat nociceptive withdrawal reflex system. Eur. J. Neurosci. 11, 4327–4332.

Levinsson, A., Holmberg, H., Broman, J., Zhang, M., and Schouenborg, J. 2002. Spinal sensorimotor transformation: Relation between cutaneous somatotopy and a reflex network. J. Neurosci. 22, 8170–8182.

Levinsson, A., Holmberg, H., Schouenborg, J., Seiger, A., Aldskogius, H., and Kozlova, E. N. 2000. Functional connections are established in the deafferented rat spinal cord by peripherally transplanted human embryonic sensory neurons. Eur. J. Neurosci. 12, 3589–3595.

Li, Y., Field, P. M., and Raisman, G. 1997. Repair of adult rat corticospinal tract by transplants of olfactory ensheating cells. Science 277, 2000–2002.

Mendelson, B. 1994. Chronic embryonic MK-801 exposure disrupts the somatotopic organization of cutaneous nerve projections in the chick spinal cord. Brain Res. Dev. Brain Res. 82, 152–166.

Mirnics, K. and Koerber, H. R. 1995. Prenatal development of rat primary afferent fibers: II. Central projections. J. Comp. Neurol. 355, 601–614.

Molander, C. and Grant, G. 1985. Cutaneous projections from the rat hindlimb foot to the substantia gelatinosa of the spinal cord studied by transganglionic transport of WGA-HRP conjugate. J. Comp. Neurol. 237, 476–484.

Mortin, L. I., Keifer, J., and Stein, P. S. G. 1985. Three forms of the scratch reflex in the spinal turtle: movement analyses. J. Neurophysiol., 53, 1501–1516.

O'Donovan, M. J. 1999. The origin of spontaneous activity in developing networks of the vertebrate nervous system. Curr. Opin. Neurobiol., 9, 94–104.

O'Donovan, M. J., Wenner, P., Chub, N., Tabak, J., and Rinzel, J. 1998. Mechanisms of spontaneous activity in the developing spinal cord and their relevance to locomotion. Ann. NY Acad. Sci. 860, 130–141.

Perrier, J. F. and Hounsgaard, J. 2000. Development and regulation of response properties in spinal cord motoneurones. Brain Res. Bull. 53, 529–535.

Petersson, P., Mundt-Petersen, K. and Schouenborg, J. 2004. Changes in primary afferent termination patterns accompany functional adaptation of nociceptive withdrawal reflexes. Society for Neuroscience Congress, Abstract 177.3.

Petersson, P., Waldenström, A., Fåhreaus, C., and Schouenborg, J. 2003. Spontaneous muscle twitches during sleep guide spinal self-organization. Nature 424(6944), 72–75.

Russier, M., Carlier, E., Ankri, N., Fronzaroli, L., and Debanne, D. 2003. A-, T-, and H-type currents shape intrinsic firing of developing rat abducens motoneurons. J. Physiol. (Lond.) 549, 21–36.

Saltiel, P., Wyler-Duda, K., D'Avella, A., Tresch, M. C., and Bizzi, E. 2001. Muscle synergies encoded within the spinal cord: evidence from focal intraspinal NMDA iontophoresis in the frog. J. Neurophysiol. 85, 605–619.

Schouenborg, J. and Kalliomäki, J. 1990. Functional organization of the nociceptive withdrawal reflexes. I. Activation of hindlimb muscles in the rat. Exp. Brain Res. 83, 67–78.

Schouenborg, J. and Weng, H.-R. 1994. Sensorimotor transformation in a spinal motor system. Exp. Brain. Res. 100, 170–174.

Schouenborg, J., Holmberg, H., and Weng, H. R. 1992. Functional organization of the nociceptive withdrawal reflexes. II. Changes of excitability and receptive fields after spinalization in the rat. Exp. Brain Res. 90, 469–478.

Schouenborg, J., Weng, H. R., Kalliomäki, J., and Holmberg, H. 1995. A survey of spinal dorsal horn neurones encoding the spatial organization of withdrawal reflexes in the rat. Exp. Brain Res. 106, 19–27.

Silos-Santiago, I., Jeng, B., and Snider, W. D. 1995. Sensory afferents show appropriate somatotopy at the earliest stage of projection to dorsal horn. Neuroreport 6, 861–865.

Sjölund, B. H. 2002. Pain and rehabilitation after spinal cord injury – the case of sensory spasticity? Brain Res. Rev. 40, 250–256.

Sonnenborg, F. A., Andersen, O. K., and Arendt-Nielsen, L. 2000. Modular organization of excitatory and inhibitory reflex receptive fields elicited by electrical stimulation of the foot sole in man. Clin. Neurophysiol. 111(12), 2160–2169.

Stein, P. S., McCullough, M. L., and Currie, S. N. 1998. Spinal motor patterns in the turtle. Ann. NY Acad. Sci. 16, 142–154.

Thelin, J. M. 2005. Plasticity in mice nociceptive spinal circuits – role of cell adhesion molecules Lund University, Faculty of Medicine Doctoral Dissertation Series, 94 ISSN 1652-8220, ISBN: 91-85439-95-9.

Thelin, J. M. and Schouenborg, J. 2003. Postnatal learning in nociceptive withdrawal reflexes in different mouse strains. Program No. 260.18. 2003. Abstract Viewer/Itinerary Planner. Washington, DC: Society for Neuroscience Congress.

Tresch, M. C. and Bizzi, E. 1999. Responses to spinal microstimulation in the chronically spinalized rat and their relationship to spinal systems activated by low threshold cutaneous stimulation. Exp. Brain Res. 129, 401–416.

Tresch, M. C., Saltiel, P., D'Avella, A., and Bizzi, E. 2002. Coordination and localization in spinal motor systems. Brain Res. Rev. 40, 66–79.

Waldenström, A., Thelin, J., Thimansson, E., Levinsson, A., and Schouenborg, J. 2003. Developmental learning in a pain-related system: evidence for a cross-modality mechanism. J. Neurosci., 23, 7719–7725.

Weng, H.-R. and Schouenborg, J. 1996. Cutaneous inhibitory receptive fields of withdrawal reflexes in the decerebrate spinal rat. J. Physiol. (Lond.) 493, 253–265.

Wilson, P. and Kitchener, P. D. 1996. Plasticity of cutaneous primary afferent projections to the spinal dorsal horn. Prog. Neurobiol. 48, 105–129.

Zgraggen, W. J., Metz, G. A. S., Kartje, G. L., Thallmair, M., and Schwab, M. E. 1998. Functional recovery and enhanced corticofugal plasticity after unilateral pyramidal tract lesion and blockade of myelin-associated neurite growth inhibitors in adult rats. J. Neurosci., 18, 4744–4757.

Further Reading

Brown, P. B., Harton, P., Millecchia, R., Lawson, J. J., Kunjara-na-ayudhya, T., Stephens, S., Miller, M. A., Hicks, L., and Culberson, J. 2000. Spatial convergence and divergence between cutaneous afferent axons and dorsal horn cells are not constant. J. Comp. Neurol. 420, 277–290.

Schouenborg, J. 2003. Somatosensory imprinting in spinal reflex modules. J. Rehab. Med. 41(Suppl.), 73–80 Review.

13 Pain Control: A Child-Centered Approach

Patricia A McGrath, The University of Toronto, Toronto, ON, Canada

Glossary

behavioral factors These factors refer to the specific behaviors of children, parents, and staff during pain episodes and also to parents' and children's broader actions in response to a chronic pain problem.

cognitive factors These factors include parents' and children's understanding of the pain problem, knowledge of effective therapies, and expectations for continuing pain or pain relief.

emotional factors These factors include parent's and children's feelings about the pain itself or responsible health condition and its adverse impact on the family, as well as any associated emotional disorders (such as anxiety or depression).

situational factors These factors represent a unique interaction between the child and the context in which the pain is experienced; they can vary dynamically, depending on the specific circumstances and setting.

13.1 Introduction

During the last two decades, unprecedented scientific and clinical attention has focused on the unique pain problems of infants, children, and adolescents (Schechter, N. L. *et al.*, 2003; McGrath, P. A., 2005; Schechter, N. L., 2007). Our knowledge of how children perceive pain and how we can alleviate their suffering has dramatically improved. Like adults, children's pain is modified both by ascending systems activated by sensory stimuli and by descending pain-suppressing systems activated by psychological factors. However, children's pain seems more plastic or modifiable in comparison to adults', so that environmental and psychological factors may exert a more powerful influence on children's pain perceptions (McGrath, P. A. and Dade, L. A., 2004).

Plasticity, from both a biological and a psychological perspective, is an essential concept in treating a child's pain. Our treatment approach has shifted from an almost exclusive disease-centered focus – detecting and treating the putative source of tissue damage – to a broader child-centered perspective, assessing the child with pain, identifying contributing psychological and contextual factors, and then targeting interventions accordingly. This chapter describes such a treatment approach, based on the unique factors that modify a child's pain.

13.2 Modifying a Child's Pain

Certain modifying factors are relevant for all children's pain experiences, irrespective of etiology. The model shown in Figure 1 lists the key factors that can increase pain, intensify distress, and exacerbate pain-related disability. A child's characteristics, such as cognitive level, sex, gender, temperament, previous pain experience, family, and cultural background (listed in the shaded box), generally shape how children interpret and cope with pain (Blount, R. L. *et al.*, 1991; Schechter, N. L. *et al.*, 1991; Bennett-Branson, S. and Craig, K., 1993; Schanberg, L. E. *et al.*, 1998; Chambers, C. T., 2003, Uman, L. S. and Chambers, C. T., 2007). However, other factors (listed in the open boxes) represent a

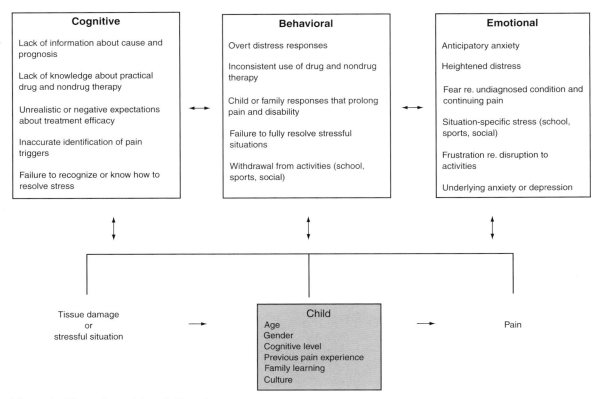

Figure 1 Biopsychosocial model for pain.

unique interaction between the child and the context in which the pain is experienced. These situational factors can vary dynamically, depending on the specific circumstances and setting. For example, a child receiving treatment for cancer may have repeated injections, central venous port access, or lumbar punctures, any of which can cause pain. Yet, even when tissue damage from these procedures is consistent, the situational factors present for a particular treatment are unique for each child.

The beliefs, behaviors, and emotional states of the children, parents, and clinicians all play important roles in modifying children's pain. Certain factors can intensify pain and distress, whereas others can eventually trigger pain episodes, prolong pain-related disability, or maintain the cycle of repeated pain episodes for children with recurrent pain syndromes (McGrath, P. A. and Hillier, L. M., 2003). While they may be unable to change the more stable child characteristics, health care providers can dramatically improve children's pain by modifying situational factors. Cognitive factors include parents' and children's understanding of the pain problem, knowledge of effective therapies, and expectations for continuing pain or pain relief.

Behavioral factors refer to the specific behaviors of children, parents, and staff during pain episodes and also to parents' and children's broader actions in response to a chronic pain problem. Emotional factors include parent's and children's feelings about the pain itself or responsible health condition and its adverse impact on the family, as well as any associated emotional disorders (such as anxiety or depression). Parents' understanding of the cause of pain, possible treatments, and long-term prognosis guides their behaviors toward children and shapes children's emotional responses to the pain problem.

Situational factors may affect children even more than adults. Adults typically have experienced a wide variety of pains differing in etiology, intensity, and quality; their diverse experiences create a broad base of knowledge, realistic expectations, and versatile coping behaviors. In contrast, children, especially very young children, have considerably limited pain experience. They rely mainly on information conveyed within the immediate situation to interpret pain and to respond appropriately. Thus, children are particularly sensitive to environmental cues and to the behaviors of the adults who are present. A child's pain is not predetermined by the extent of injury, disease progression, or invasiveness of a medical procedure. When health care providers ignore the powerful mediating impact of situational factors, they can easily, albeit inadvertently, increase a child's distress and pain. Generally, accurate information, realistic expectations, choices, control, and the use of some independent pain-reducing strategies will result in less pain.

13.3   Guidelines for Treating Acute Pain

Although distraction and parental support can minimize the pain caused by minor bumps and scrapes of childhood, pharmacological therapy (analgesics, adjunct analgesics, anesthetics, anxiolytics, and sedatives, as reviewed in other chapters) is the foundation for controlling children's acute pain (Finley, G. A. and McGrath, P. J., 2001; Dalens, B., 2003; Krane, E. J. *et al.*, 2003; Maunuksela, E.-L. and Olkkola, K. T., 2003; Yaster, M., 2003; Wilder, R. T., 2003; Brown, S. C. *et al.*, 2005; Bean-Lijewski, J. D., 2007; Schneeweiss, S., 2007). Children require appropriate analgesics, i.e., drugs selected according to type and severity of pain, doses determined by a child's weight based on the drug's duration of action to provide consistent pain relief and prevent breakthrough pain throughout the recovery period, and doses administered at regular dosing intervals. (For dosing guidelines, please see Finley, G. A. and McGrath, P. J., 2001; Krane, E. J.

et al., 2003; Brown, S. C. 2006.) Premature and term newborns show reduced clearance of most opioids (Wong, C. M. *et al.*, 2003). The differences in pharmacokinetics and pharmacodynamics among neonates, preterm infants, and full-term infants warrant special dosing considerations for infants and close monitoring when they receive opioids.

Recommended starting doses for analgesic medications for children are listed in Tables 1 and 2. Starting doses for adjuvant analgesic medications to control pain, drug-related side effects, and other symptoms are listed in Table 3.

Analgesics should be complemented by a practical cognitive-behavioral approach to ensure optimal pain relief, especially when treating acute pain from invasive medical procedures. First, health care providers should appreciate that children learn about pain because of their own injury-related acute pains. They perceive acute pains more as the occasionally inevitable result of daily activities than as something to fear. They quickly learn that pain is caused by injury, often easily visible, is relatively brief, and is relieved by many drug and nondrug interventions. Thus, they usually have accurate developmentally appropriate information about why something hurts and positive expectations for pain relief. In contrast, when children experience pain during invasive medical and dental procedures or after surgery, they often feel that they have no control, they may not know what to expect, and they may not know any simple pain control methods that they can use in these

Table 1   Nonopioid drugs to control pain in children

Drug	Dosage	Comments
Acetaminophen	10–15 mg kg^{-1} PO, every 4–6 h	Lacks gastrointestinal and hematological side effects; lacks anti-inflammatory effects (may mask infection-associated fever)
		Dose limit of 65 mg kg^{-1} day^{-1} or 4 g day^{-1}, whichever is less
Ibuprofen	5–10 mg kg^{-1} PO, every 6–8 h	Anti-inflammatory activity
		Use with caution in patients with hepatic or renal impairment, compromised cardiac function or hypertension (may cause fluid retention, edema), history of GI bleeding or ulcers, may inhibit platelet aggregation
		Dose limit of 40 mg kg^{-1} day^{-1}; maximum dose of 2400 mg day^{-1}
Naproxen	10–20 mg kg^{-1} day^{-1} PO, divided every 12 h	Anti-inflammatory activity. Use with caution and monitor closely in patients with impaired renal function. Avoid in patients with severe renal impairment
		Dose limit of 1 g day^{-1}
Diclofenac	1 mg kg^{-1} PO, every 8–12 h	Anti-inflammatory activity. Similar GI, renal, and hepatic precautions as noted above for ibuprofen and naproxen
		Dose limit of 50 mg dose^{-1}

Note: Increasing the dose of nonopioids beyond the recommended therapeutic level produces a ceiling effect, in that there is no additional analgesia but there are major increases in toxicity and side effects.
PO, by mouth; GI, gastrointestinal.
Brown, S. C. and McGrath, P. A. In Press. Pain Control in Children. In: Oxford Textbook of Palliative Medicine 4/e (*eds.* G. Hanks, N. Cherny, N. Christakis, M. Fallon, S. Kaasa, R. Portenoy). Oxford University Press.

Table 2   Opioid analgesics – usual starting doses for children

Drug	Equianalgesic dose (parenteral)	Starting Dose IV	IV:PO ratio	Starting dose PO/ transdermal	Duration of action
Morphine	10 mg	Bolus dose = 0.05– 0.1 mg kg^{-1} every 2–4 h Continuous infusion = 0.01– 0.04 mg kg^{-1} h^{-1}	1:3	0.15–0.3 mg kg^{-1} dose^{-1} every 4 h	3–4 h
Hydromorphone	1.5 mg	0.015–0.02 mg kg^{-1} every 4 h	1:5	0.06 mg kg^{-1} every 3–4 h	2–4 h
Codeine	120 mg	Not recommended		1.0 mg kg^{-1} every 4 h (dose limit 1.5 mg kg^{-1} dose^{-1})	3–4 h
Oxycodone	5–10 mg	Not recommended		0.1–0.2 mg kg^{-1} every 3–4 h	3–4 h
Meperidine[a]	75 mg	0.5–1.0 mg kg^{-1} every 3–4 h	1:4	1.0–2.0 mg kg^{-1} every 3–4 h (dose limit 150 mg)	1–3 h
Fentanyl[b]	100 μg	1–2 μg kg^{-1} h^{-1} as continuous infusion		25 μg patch	72 h (patch)

[a]Avoid use in renal impairment. Metabolite may cause seizures.
[b]Potentially highly toxic. Not for use in acute pain control.
Principles of opioid administration:
1. If inadequate pain relief and no toxicity at peak onset of opioid action, increase dose in 50% increments.
2. Avoid IM administration.
3. Whenever using continuous infusion, plan for hourly rescue doses with short onset opioids if needed. Rescue dose is usually 50–200% of continuous hourly dose. If greater than 6 rescues are necessary in 24-h period, increase daily infusion total by the total amount of rescues for previous 24 h/24. An alternative is to increase infusion by 50%.
4. To change opioids because of incomplete cross-tolerance: if changing between opioids with short duration of action, start new opioid at 50% of equianalgesic dose. Titrate to effect.
5. To taper opioids – anyone on opioids over 1 week must be tapered to avoid withdrawal: taper by 50% for 2 days, and then decrease by 25% every 2 days. When dose is equianalgesic to an oral morphine dose of 0.6 mg kg^{-1} day^{-1}, it may be stopped. Some patients on opioids for prolonged periods may require much slower weaning.
PO, by mouth; IV, intravenous.
Brown, S. C. and McGrath, P. A. In Press. Pain Control in Children. In: Oxford Textbook of Palliative Medicine 4/e (eds. G. Hanks, N. Cherny, N. Christakis, M. Fallon, S. Kaasa, and R. Portenoy). Oxford University Press.
NOTE: Doses are for opioid-naive patients. For infants under 6 months, start at one-quarter to one-third the suggested dose and titrate to effect.

new situations. In fact, health care providers may instruct children how to cope (i.e., behave) during a painful procedure in a manner that makes it easier for the person administering the procedure, but does not necessarily make it better for the child.

Health care providers often focus on explaining to parents and inadvertently neglect the child. Time constraints necessitate juggling who has priority and, understandably, concerned parents are a priority. Yet, children should remain the number 1 priority because it is their pain that is directly modified. Every word and action conveys powerful information to children. As an example, a young boy became very distressed when he learned that a local anesthetic should freeze his arm so he wouldn't feel anything. Finally, he explained that he was afraid his arm would become like a Popsicle. Health care providers have tremendous power in influencing what children know, how they can respond, and the relevance of the situation (i.e., injury, disease, treatment) and consequent pain. The manner in which health care providers introduce themselves, the words they choose to explain what will happen, and whether they confidently show children what they could do to lessen the pain profoundly impact what children will experience.

These situational factors can intensify children's pain and distress, even when children receive anxiolytics and analgesics. As an example, when children with cancer at our center first started receiving topical anesthetic creams before intravenous insertions and portacather injections, considerably more children were referred to our pain clinic. We learned that busy clinical staff no longer focused on an individual child in the manner that they had previously – encouraging the child to understand from their specific perspective (i.e., age, gender, culture) and use their preferred 'close your pain gate' methods.

Table 3 Adjuvant analgesics – doses for children

Drug category	Drug, dosage	Indications	Comments
Antidepressants	Amitriptyline, 0.2–0.5 mg kg^{-1} PO. Titrate upward by 0.25 mg kg^{-1} every 2–3 days Maintenance: 0.2–3.0 mg kg^{-1} Alternatives: nortriptyline, doxepin, imipramine	Neuropathic pain (i.e., vincristine-induced, radiation plexopathy, tumor invasion, CRPS-1) Insomnia	Usually improved sleep and pain relief within 3–5 days Anticholinergic side effects are dose-limiting. Use with caution for children with increased risk for cardiac dysfunction
Anticonvulsants	Gabapentin, 5 mg kg^{-1} day^{-1} PO. Titrate upward over 3–7 days. Maintenance: up to 15–50 mg kg^{-1} day^{-1} PO divided TID. Carbamazepine, initial dosing: 10 mg kg^{-1} day^{-1} PO divided OD or BID. Maintenance: up to 20–30 mg kg^{-1} day^{-1} PO divided every 8 h. Increase dose gradually over 2–4 weeks. Alternatives: phenytoin, clonazepam	Neuropathic pain, especially shooting, stabbing pain.	Side effects: gastrointestinal upset, ataxia, dizziness, disorientation, somnolence Monitor for hematological, hepatic, and allergic reactions with carbamazepine

(Continued)

Table 3 (Continued)

Drug category	Drug, dosage	Indications	Comments
Sedatives, hypnotics, anxiolytics	Diazepam, 0.025–0.2 mg kg^{-1} PO every 6 h.	Acute anxiety, muscle spasm	Sedative effect may limit opioid use. Other side effects include depression and dependence with prolonged use
	Lorazepam, 0.05 mg kg^{-1} dose^{-1} SL	Premedication for painful procedures	
	Midazolam, 0.5 mg kg^{-1} dose^{-1} PO administered 15–30 min prior to procedure; 0.05 mg kg^{-1} dose^{-1} IV for sedation		
Antihistamines	Hydroxyzine, 0.5 mg kg^{-1} PO every 6 h.	Opioid-induced pruritus, anxiety, nausea	Sedative side effects may be helpful
	Diphenhydramine, 0.5–1.0 mg kg^{-1} PO/IV every 6 h		
Psychostimulants	Dextroamphetamine, methylphenidate, 0.1–0.2 mg kg^{-1} BID.	Opioid-induced somnolence	Side effects include agitation, sleep disturbance, and anorexia.
	Escalate to 0.3–0.5 mg kg^{-1} as needed	Potentiation of opioid analgesia	Administer second dose in afternoon to avoid sleep disturbances
Corticosteroids	Prednisone, prednisolone, and dexamethasone dosage depends on clinical situation (i.e. dexamethasone initial dosing: 0.2 mg kg^{-1} IV. Dose limit 10 mg. Subsequent dose 0.3 mg kg^{-1}day^{-1} IV divided every 6 h)	Headache from increased intracranial pressure, spinal or nerve compression; widespread metastases	Side effects include edema, dyspeptic symptoms, and occasional gastrointestinal bleeding

CRPS-1, complex regional pain syndrome, type 1; PO, by mouth; IV, intravenous; SL, sublingual.
Brown, S. C. and McGrath, P. A. In Press. Pain Control in Children. In: Oxford Textbook of Palliative Medicine 4/e (eds. G. Hanks, N. Cherny, N. Christakis, M. Fallon, S. Kaasa, and R. Portenoy). Oxford University Press.

Instead, they applied the magic cream and sent children to a play area until it was time to conduct the invasive procedure. Despite the proven anesthetic properties of the cream, many children reported increased pain and exhibited increased distress in comparison to their pretopical procedures (McGrath, P. A., 1990).

Although psychological therapies are reviewed in other chapters, even the simple provision of information about what is happening, why the procedure is conducted in a certain manner, what equipment will be used, who will be present, the nature of the child's sensations during different phases of the procedure, and some straightforward instructions about how a child might lessen pain will dramatically reduce the pain caused by any invasive procedure to minimize anxiety and pain. As listed in Table 4, health care providers should explain what will happen and emphasize what sensations a child may feel, e.g., warmth, coolness, pressure, rather than the frightening label of pain. They should teach children a few basic attention and distraction methods to reduce pain and guide families to recognize the particular circumstances that exacerbate pain and distress.

Children who receive multiple invasive procedures over a prolonged period are at particular risk for developing anxiety and fear. Inaccurate understanding of the disease or injury, parents' anxiety about the underlying condition and painful treatments, inconsistency in treatment administration, lack of control and coping strategies, and fears about the effects of the disease or side effects of treatment can intensify a child's treatment-related pain. Regrettably, some inconsistencies in how procedures are conducted are common in clinical practice. The more procedures a child has, the more likely that they will experience a difficult procedure – more pain, more nausea, etc. A child then becomes fearful and expects that another difficult procedure is inevitable and a cycle of heightened anxiety, increased pain, and emotional distress ensues.

Many cognitive-behavioral programs have been designed to complement drug therapies for children with cancer (Hilgard, J. R. and LeBaron, S., 1984; LeBaron, S. and Zeltzer, L. K., 1996; Olness, K. and Kohen, D. P., 1996; Liossi, C., 2002; Collins, J. and Weisman, S. J., 2003; Kuttner, L. and Solomon, R., 2003). These child-centered programs usually combine basic education with psychological therapies (i.e., attention and distraction, hypnosis or hypnotic-like suggestions for analgesia, desensitization procedures, behavioral management, relaxation exercises, and biofeedback) to minimize children's anxiety and pain. The specific program components vary for different children because therapies are selected to target the unique child, family, and situational factors that were identified from the child's pain assessment as the factors that intensified children's pain, anxiety, and distress. Generally, as children receive developmentally appropriate accurate information, gain realistic expectations, have more choices and control, and use independent pain control strategies, they have less pain and distress.

13.4 Guidelines for Treating Chronic Pain

Effective treatment of chronic pain usually requires an integrated treatment approach comprising drug therapy, physical therapy, and psychological counseling, rather than a single therapy. Chronic pain generally has multiple sources affecting both peripheral and central nervous systems, with both nociceptive and neuropathic components. Certain cognitive, behavioral, and emotional factors may intensify pain, increase distress, and exacerbate a

Table 4   Preparing children for invasive procedures

Remember that anxiety and distress can increase pain.

Explain what will happen clearly, in simple developmentally appropriate language.

Emphasize the qualities of the sensations they may experience such as cold, sharp, tingling, and pressure so that children focus on what they are feeling, rather than only on the hurting aspect.

If the procedure will hurt, describe what children might feel using more familiar examples of pains that they have experienced during play and sports.

Use versatile pain control methods that involve children such as attention and distraction, deep breathing, counting exercises, or guided imagery.

Give children as many choices and as much control as possible such as choosing which arm for injections, whether to watch or look away, and which pain control method to use.

child's disability. The pain adversely affects all aspects of children's lives. Parents are distressed by the pain itself, its implications for their children's future, its life-threatening potential (if any), and the prospect of continuing pain and progressive disability. Parents and adolescents tend to emphasize and fear the future consequences of children's physical condition, whereas young children are more preoccupied by the immediate consequences and disruption to their daily activities. The dynamics within the family (both for siblings and for extended family members) inevitably change as chronic pain prevents children from pursing their normal activities and as family schedules adjust accordingly to the health care needs of the child with pain (for review, please see McGrath, P. J. and Finley, G. A., 1999; Berde, C. B. *et al.*, 2003; McGrath, P. A. and Dade, L. A., 2004).

In order to treat chronic pain effectively, health care providers should carefully determine the physical etiology (if possible) and also critically appraise the unique situational factors that may influence pain. Interdisciplinary teams, comprising physicians, nurses, psychologists, psychiatrists, and physical therapists, are better equipped to conduct a more comprehensive biopsychosocial assessment than a single pain specialist. After consultation among members, the team should provide a diagnosis that includes both the etiology and the relevant factors influencing children's pain. The biopsychosocial model shown in Figure 1 provides a concrete framework for explaining these factors and subsequently presenting the rationale for an integrated treatment approach comprising pharmacology, physiotherapy, and psychology. The specific drugs and therapies within this child-centered approach derive from the assessment results and are selected to target the responsible central and peripheral mechanisms and to mitigate the pain-exacerbating impact of situational factors.

NSAIDs and opioids in effective doses and regular dosing schedules can relieve pain with a primary nociceptive etiology. Drugs should be selected according to their analgesic potency based on the child's pain level – acetaminophen to control mild pain, codeine to control moderate pain, and morphine for strong pain (World Health Organization, 1998). Even when children require opioid analgesics, they should continue to receive acetaminophen (and NSAIDs, if appropriate) as supplemental analgesics. Health care providers should monitor a child's pain regularly and adjust analgesic doses as necessary to control the pain. The effective opioid dose to relieve pain varies widely among different children or in the same child at different times. To relieve ongoing pain, opioid doses should be increased steadily until comfort is achieved, unless the child experiences unacceptable side effects such as somnolence or respiratory depression. Adjuvant analgesics (especially gabapentin and tricyclic antidepressants) have become the main drugs for treating neuropathic pain (Berde, C. B. *et al.*, 2003).

Some children, irrespective of what may have caused their pain initially, develop significant functional disability that far exceeds their reported pain level. These children do not attend school, participate in sports, or socialize with peers. Families are frustrated because they may have tried to follow treatment recommendations and reintegrate children back into their normal activities. Parents may believe that health care providers do not believe the child and do not understand the severity of the pain problem. Families believe that children will return to normal (even when children have been disabled for years) as soon as health care providers cure the pain problem. A futile cycle ensues, with health care providers attempting to restore normal functioning and families refusing to cooperate until the pain is relieved. Children with severe disability constitute a unique clinical challenge. Gradually, I have learned to identify the problem to families as twofold – a chronic pain problem and a disability problem. Usually disability is the more serious problem. Thus, treatment should immediately focus on the both aspects – we must deal with the pain, but, concurrently, we must deal with the increasing disability. Almost all children who have a major disability problem will require additional psychological treatment and assistance to reenter school and resume their previous activities.

For any child with chronic pain, health care provides should recognize the common situational factors that influence parents' responses to children's continuing pain complaints, especially when pain is not caused by active disease or injury. Parents usually try to understand their child's pain from an acute disease perspective, where pain is due to a single cause and can be relieved by a single treatment. They do not understand that chronic pain, unlike most pains that they have already experienced, may have several interrelated causes. Thus, parents request further medical investigations as they search for the clear-cut physical etiology and concentrate on finding one treatment that will immediately stop the pain. Parents may reject potentially effective treatments after only one attempt, even though the treatment would address some of the causes and might lessen children's pain over time. Some parents expect that their child will remain very disabled until the one right treatment is found and believe that they have no control in changing the future course of continuing pain and disability.

To mitigate the factors that can intensify a child's chronic pain or exacerbate disability, health care providers should "Stop, Look, & Listen": Stop an exclusive disease focus; Look at the big picture, i.e., assess the child with pain; and Listen to the child's pain history, appreciating the potential impact of situational factors (McGrath, P. A. and Dade, L. A., 2004). The diagnosis is a critical component of pain management. Even when some aspects of the pain condition are puzzling, health care providers should honestly describe what they know and explain what they need to explore further in a straightforward and reassuring manner. Accurate information about the pain source(s) can mitigate the increased distress caused by a family's misinterpretations and anxieties. Moreover, since expectancies can profoundly affect treatment effectiveness, health care providers should build on this knowledge and directly address their patients' expectancies within the diagnostic appointment.

An accurate diagnosis should include information about both the primary causes and the secondary contributing factors so that treatments can be targeted at each relevant factor. Since these factors may vary over time, pain assessment remains a critical component of any treatment regimen. The focus of continuing assessment is the child with pain – not only the pain features but also the situational factors that modulate pain. Even when pain is due to actively progressing disease such as for children with advanced cancer, many other factors can affect their pain and suffering. Health care providers should assess these factors and target treatments accordingly. The treatment rationale is on the multifactorial etiology of pain and recommendations for a multimodal treatment approach. This is in direct contrast to the single cause and single treatment approach normally adequate for relieving acute pain.

References

Bean-Lijewski, J. D. 2007. Acute Pain in Children – Post-operative Pain. In: Encyclopedic Reference of Pain (eds. R. F. Schmidt and W. D. Willis), pp. 16–28. Springer-Verlag.

Bennett-Branson, S. and Craig, K. 1993. Post-operative pain in children: developmental and family influences on spontaneous coping strategies. Can. J. Behav. Sci. 25, 355–383.

Berde, C. B., Lebel, A. A., and Olsson, G. 2003. Neuropathic Pain in Children. In: Pain in Infants, Children, and Adolescents (eds. N. L. Schechter, C. B. Berde, and M. Yaster), 2nd edn., pp. 620–638. Lippincott Williams and Wilkins.

Blount, R. L., Landolf-Fritsche, B., Powers, S. W., and Sturges, J. W. 1991. Differences between high and low coping children and between parent and staff behaviors during painful medical procedures. J. Pediatr. Psychol. 16, 795–809.

Brown, S. C. 2006. Cancer pain and palliative care in children. In: Encyclopedic Reference of Pain (eds. R. F. Schmidt and W. D. Willis), pp. 220–224. Springer-Verlag.

Brown, S. C., McGrath, P. A., and Krmpotic, K. 2005. Pain in Children. In: Neurologic Basis of Pain (ed. M. Pappagallo), pp. 225–242. McGraw-Hill.

Chambers, C. T. 2003. The Role of Family Factors in Pediatric Pain. In: Pediatric Pain: Biological and Social Context, Progress in Pain Research and Management, Vol. 26. (eds. P. J. McGrath and G. A. Finley), pp. 99–130. IASP Press.

Collins, J. and Weisman, S. J. 2003. Management of Pain in Childhood Cancer. In: Pain in Infants, Children, and Adolescents (eds. N. L. Schechter, C. B. Berde, and M. Yaster), 2nd edn., pp. 517–538. Lippincott Williams and Wilkins.

Dalens, B. 2003. Peripheral Nerve Blockade in the Management of Postoperative Pain in Infants, Children, and Adolescents (eds. N. L. Schechter, C. B. Berde, and M. Yaster), 2nd edn., pp. 363–395. Lippincott Williams and Wilkins.

Finley, G. A. and McGrath, P. J. (eds.) 2001. Acute and Procedure Pain in Infants and Children. Progress in Pain Research and Management. IASP Press.

Hilgard, J. R. and LeBaron, S. 1984. Hypnotherapy of Pain in Children with Cancer. William Kaufman.

Krane, E. J. et al. 2003. Treatment of Pediatric Pain with Nonconventional Analgesics. In: Pain in Infants, Children, and Adolescents (eds. N. L. Schechter, C. B. Berde, and M. Yaster), 2nd edn., pp. 225–240. Lippincott Williams and Wilkins.

Kuttner, L. and Solomon, R. 2003. Hypnotherapy and Imagery for Managing Children's Pain. In: Pain in Infants, Children, and Adolescents (eds. N. L. Schechter, C. B. Berde, and M. Yaster), 2nd edn., pp. 317–328. Lippincott Williams and Wilkins.

LeBaron, S. and Zeltzer, L. K. 1996. Children in Pain. In: Hypnosis and Suggestion in the Treatment of Pain: A Clinical Guide (eds. J. Barber and C. J. Bejenke), pp. 305–340. W. W. Norton.

Liossi, C. 2002. Procedure-Related Cancer Pain in Children. Radcliffe Medical Press Ltd.

Maunuksela, E.-L. and Olkkola, K. T. 2003. Nonsteroidal Anti-inflammatory Drugs in Pediatric Pain Management. In: Pain in Infants, Children, and Adolescents (eds. C. B. Berde, N. L. Schechter, and, M. Yaster), pp. 171–180. Lippincott Williams and Wilkins.

McGrath, P. A. 1990. Pain in Children: Nature, Assessment and Treatment. Guilford Press.

McGrath, P. A. 2005. Children – Not Simply "Little Adults". In: Paths of Pain 1975–2005 (eds. H. Merskey, J. D. Loeser, and R. Dubner), pp. 433–446. IASP Press.

McGrath, P. A. and Dade, L. A. 2004. Effective Strategies to Decrease Pain and Minimize Disability. In: Psychological Modulation of Pain: Integrating Basic Science and Clinical Perspectives (eds. D. D. Price and M. C. Bushnell), pp. 73–96. IASP Press.

McGrath, P. J. and Finley, G. A. 1999. Chronic and Recurrent Pain in Children and Adolescents. IASP Press.

McGrath, P. A. and Hillier, L. M. 2003. Modifying the Psychological Factors that Intensify Children's Pain and Prolong Disability. In: Pain in Infants, Children, and Adolescents (eds. N. L. Schechter, C. B. Berde, and M. Yaster), 2nd edn., pp. 85–104. Lippincott Williams and Wilkins.

Olness, K. and Kohen, D. P. 1996. Hypnosis and Hypnotherapy with Children, 3rd edn. Guilford Press.

Schanberg, L. E., Keefe, F. J., Lefebvre, J. C., Kredich, D. W., and Gil, K. M. 1998. Social context of pain in children with juvenile primary fibromyalgia syndrome: parental pain history and family environment. Clin. J. Pain 14, 107–115.

Schechter, N. L. 2007. Evolution of Pediatric Pain Treatment. In: Encyclopedic Reference of Pain (eds. R. F. Schmidt and W. D. Willis), pp. 749–752. Springer-Verlag.

Schechter, N. L., Berde C. B., and Yaster, M. (eds.) 2003. Pain in Infants, Children, and Adolescents, 2nd edn. Lippincott Williams and Wilkins.

Schechter, N. L., Bernstein, B. A., Beck, A., Hart, L., and Scherzer, L. 1991. Individual differences in children's response to pain: role of temperament and parental characteristics. Pediatrics 87, 171–177.

Schneeweiss, S. 2007. Pain and Sedation of Children in the Emergency Setting. In: Encyclopedic Reference of Pain (eds. R. F. Schmidt and W. D. Willis), pp. 1637–1641. New York. Springer-Verlag.

Uman, L. S. and Chambers, C. T. 2007. Impact of Familial Factors on Children's Chronic Pain. In: Encyclopedic Reference of Pain (eds. R. F. Schmidt and W. D. Willis), pp. 961–964. Springer-Verlag.

Wilder, R. T. 2003. Regional Anesthetic Techniques for Chronic Pain Management. In: Pain in Infants, Children, and Adolescents (eds. N. L. Schechter, C. B. Berde, and M. Yaster), 2nd edn., pp. 396–416. Lippincott Williams and Wilkins.

Wong, C. M., et al. 2003. The Pain (and Stress) in Infants in a Neonatal Intensive Care Unit. In: Pain in Infants, Children, and Adolescents (eds. N. L. Schechter, C. B. Berde, and M. Yaster), 2nd edn., pp. 669–692. Lippincott Williams and Wilkins.

World Health Organization 1998. Cancer Pain Relief and Palliative Care in Children. World Health Organization, Geneva.

Yaster, M. 2003. Clinical Pharmacology. In: Pain in Infants, Children and Adolescents (eds. N. L. Schechter, C. B. Berde, and M. Yaster), pp. 71–84. Lippincott Williams and Wilkins.

Further Reading

Brown, S. C. and McGrath, P. A. (In Press). Pain Control in Children. In: Oxford Textbook of Palliative Medicine 4/e (eds. G. Hanks, N. Cherny, N. Christakis, M. Fallon, S. Kaasa, and R. Portenoy). Oxford University Press.

Schanberg, L. E., Gil, K. M., Anthony, K. K., Yow, E., and Rochon, J. 2005. Pain, stiffness, and fatigue in juvenile polyarticular arthritis: contemporaneous stressful events and mood as predictors. Arthritis Rheum. 52(4), 1196–1204.

14 Assaying Pain-Related Genes: Preclinical and Clinical Correlates

V E Scott, R Davis-Taber, and P Honore, Global Pharmaceutical Research and Development, Abbott Park, IL, USA

14.1 Introduction

Microarray gene chip technology is a powerful tool to evaluate changes in multiple transcript levels simultaneously in different tissue preparations. In the pain field, efforts have focused on characterizing gene transcript changes that occur in chronic pain states such as neuropathic pain (Wang, H. *et al.*, 2002; Xiao, H. S. *et al.*, 2002; Davis-Taber, R. *et al.*, 2003; Thimmapaya, R. *et al.*, 2003; Valder, C. R. *et al.*, 2003; Davis-Taber, R. *et al.*, 2004) and this will be the main focus of this chapter. Neuropathic pain is initiated or caused by a primary lesion or dysfunction of the nervous system. As this technology emerged, it was anticipated that it could be used to determine the key molecular players in both the development and maintenance of neuropathic pain, which would enhance our understanding of this disease and provide insights into the identification of potential therapeutic targets for the treatment of pain. Prior to microarray approaches, studies were limited to evaluation of a set of specific preselected genes/proteins in samples from animal pain models or human tissues using techniques such as quantitative reverse transcriptase polymerase chain reaction (RT-PCR), Western blot analysis, and immunocytochemistry. Dependent upon the gene chips used, this analysis permits evaluation of both known and unknown gene transcripts (expressed sequence tags (ESTs)) using relatively small amounts of RNA. In each of the studies highlighted in this chapter a wealth of transcriptional information was obtained. However, determination of which gene product would offer the best therapeutic potential will require extensive additional follow-up studies. Microarray studies are clearly just the starting point. This chapter will be divided into sections detailing the studies that have been reported evaluating gene expression changes from neuropathic pain models using microarray gene chip analysis, including discussion in each section on study design, sample selection, isolation of RNA, choice of microarray gene chip platforms, and data analysis. The follow-up to microarray studies outlining the expectations and limitations of these studies will also be described. Lastly, the use of microarray gene chips for pinpointing genetic variation in disease states that will direct suitable therapeutics based upon genetic predispositions for specific patient populations will be discussed.

14.2 Microarray Analysis of Whole Dorsal Root Ganglion from the Spinal Nerve Ligation Model

Several different studies have evaluated the expression changes that occur in the rat whole dorsal root ganglion (DRG; Wang, H. *et al.*, 2002; Thimmapaya, R. *et al.*, 2003; Valder, C. R. *et al.*, 2003) and dorsal horn of the spinal cord (Sun, H. *et al.*, 2002; Wang, H. *et al.*, 2002) following spinal nerve ligation (SNL) using microarray analysis. The overall goal of each of these studies was to obtain a better understanding of the molecular pathophysiology of neuropathic pain and through this analysis identify potential new targets for the treatment of this chronic pain state. In some studies following tight ligation of the lumbar 5 (L5) and lumbar 6 (L6) DRG, the development of allodynia was shown to develop 3 days following surgery and remained established for weeks ipsilateral to the insult (Wang, H. *et al.*, 2002; Valder, C. R. *et al.*, 2003). In these studies each animal was assessed for the development of tactile allodynia by behavioral testing prior to being included/excluded from the study. This was performed using von Frey filaments and to be included in the study animals needed to have a paw withdrawal threshold of less than 3 g in the neuropathic rats and greater than 10 g in the control rats. While some studies focused on evaluating the changes that occur both during the development and in fully established neuropathic pain (Wang, H. *et al.*, 2002), others assessed only established pain (Sun, H. *et al.*, 2002; Thimmapaya, R. *et al.*, 2003; Valder, C. R. *et al.*, 2003; Davis-Taber *et al.*, 2003). Several experimental design differences exist in each of these reports. In some studies since the contralateral paw did not display significant allodynia similarly to the sham animals, the L4/L5/L6 (Wang, H. *et al.*, 2002) or L5/L6 (Valder, C. R. *et al.*, 2003) DRG contralateral to the injury were selected for comparison with the equivalent injured DRGs from the same animals. In a parallel study to avoid any potential issues that could arise during surgery to alter gene expression contralateral to the injury and not revealed by allodynia assessment, the L5/L6 ipsilateral to the surgery were compared with the L5/L6 from sham-operated animals (Thimmapaya, R. *et al.*, 2003).

Another interesting comparison examined the expression differences between the injured L5/L6 DRG from Harlan and Holtzman Sprague-Dawley rats (Valder, C. R. *et al.*, 2003). The Holtzman rats were selected for this analysis as they do not develop the tactile allodynia observed in Harlan rats following SNL (Luo, Z. D. *et al.*, 1999). The goal of this study was to compare the expression profiles between these two strains and then extend these analyses to assess changes that were regulated following injury in one or both strains. This study did identify differentially expressed genes in the two strains, providing new information on neuroplasticity in DRG of these rats in response to nerve injury (Valder, C. R. *et al.*, 2003) in addition to those associated with neuropathic pain. However, cautious analysis of these data is important since similar regulation of gene changes in both strains is not an eliminating factor in defining whether the altered gene transcript is related to neuropathic pain development. Several transcripts that are known to be centrally involved in pain processing did change in

both strains, that is, Nav1.8 and protein kinase C epsilon highlighting the limitations of this study.

A critical consideration when designing these experiments is to ensure that sufficient samples are collected to reflect the true changes that arise due to neuropathic pain and animal variation accurately. To this end assessment of DRG from L5/L6 region was performed using at least ten animals that were pooled prior to isolation of RNA for each of these studies. The comparison was made between the injured L5/L6 and either the contralateral L5/L6 (Wang, H. *et al.*, 2002) or sham L5/L6 (Thimmapaya, R. *et al.*, 2003). In one of the spinal cord studies nine animals were divided into three groups and in each group the dorsal and ventral spinal cords were pooled separately to form each pair of samples (three pairs). The comparison that was made in this tissue was between the dorsal and ventral spinal cord of neuropathic rats (Wang, H. *et al.*, 2002). In the other study, the global gene expression in the ipsilateral lumbar spinal cord was compared with the contralateral lumbar spinal cord from the neuropathic rats and to the sham control (Sun, H. *et al.*, 2002).

Typically protocols for isolating total RNA have been comparable throughout these studies using the TRizol protocol (Life Technologies/Invitrogen; Carlsbad, CA, USA) followed by RNeasy (Qiagen; Valencia, CA, USA) RNA clean-up step. The RNA quality was assessed either by denatured gel electrophoresis or capillary electrophoresis on the Agilent 2100 Bioanalyzer (Applied Biosystems; Foster City, CA, USA) to ensure that the 28S:18S ribosomal RNA (rRNA) ratio was greater than 1.0 for each sample, indicating that the sample was pure.

The microarray gene chip platform that was selected for each of these studies was the Affymetrix rat RG-U34a GeneChip®. On these microarray chips, each gene transcript is represented by probe sets containing 16 oligonucleotides, eight of which are matched and eight mismatched to the sequence of the transcript of interest. The oligonucleotides are 20 base pairs long, with a 3′ bias and may be to the coding and noncoding regions of the gene. Hybridization probes from each of these samples were generated and hybridized to the chips as recommended by Affymetrix. The microarray data was analyzed using the SAFER algorithm (Holder, D. *et al.*, 2002), Microarray Suite Expression Algorithm (Affymetrix; Santa Clara, CA, USA) or Rosetta Resolver (Rosetta Biosoftware; Seattle, WA, USA). Several microarray analysis software packages are available. Some utilize only a statistical algorithm (Affymetrix analysis software) while other analysis packages incorporate error models for normalization of intensity between arrays and adjustments for probe-specific biases (SAFER and Resolver). Following a paired *t*-test and analysis of variance (ANOVA) analysis to determine a ratio for the expression changes between both samples a *P* value is generated and represents the probability that an observed gene expression change was due to a measurement error. To be considered significant, the change in gene expression had to occur in each microarray experiment and had to exhibit a value of $P \leq 0.05$ within each experiment. Another consideration that has been shown in each of these studies is how much does the transcript need to change to be considered significant enough to play a role in pain? In all the studies discussed in this chapter a cut-off of twofold was selected. This arbitrary cut-off was chosen since a large number of significant changes were observed but it is known that the expression of P2X$_3$, a validated pain target, only changed by 1.8-fold and would be excluded from this analysis. Therefore it is important not to exclude all small changes due to magnitude but if their changes are significant with $P \leq 0.05$, further analysis is warranted.

14.3 Microarray Data Analysis and Follow-up Studies

Following microarray analysis, the changes that were observed must be confirmed using either quantitative RT-PCR (Q-RTPCR), *in situ* hybridization, or, if reagents are available, immunocytochemistry. The latter two are preferable as the microarray studies described so far have all examined heterogeneous populations of cells (neurons and glial cells) and evaluation of specific tissue sections would provide a more precise profile of where the expression changes have occurred. Following confirmation of a subset of transcripts identified by microarray, data analysis has focused on classifying the gene transcripts into groups based upon the function of the gene product if known. An example of such a classification comes from the studies examining the changes in the L5/L6 DRG following SNL compared to sham L5/L6 DRG whereby gene products were grouped into cellular homeostasis and remodeling, ion channels and receptors, enzymes, immune response, neuronal markers, synaptic plasticity, transcription factors, and a miscellaneous category (Thimmapaya, R. *et al.*, 2003). Following this classification, it became evident that several examples of genes involved in synaptic plasticity (e.g., Narp and synaptobrevin) were downregulated, whereas genes involved in the immune response (e.g., MHCII) and cell regeneration/proliferation (e.g., Reg-2 and GADD45) were upregulated. A similar pattern of changes has been observed in the different neuropathic pain studies reported (Wang, H. *et al.*, 2002; Xiao, H. S. *et al.*, 2002; Valder, C. R. *et al.*, 2003).

14.4 Microarray Analysis of Subpopulations of Neurons Isolated from Dorsal Root Ganglion

All the studies described above examine the gene changes that occur in the entire DRG, which is composed of heterogenous subpopulations of cells (neurons and glial cells). Examining specific subpopulations of neurons would provide more precise information on the transcript expression changes. To accomplish this, LCM was used. This is a novel technology that permits the isolation and capture of specific cells from a tissue such as DRG in a native state. Determining the gene expression profile of different cell types has been hampered until recently by the inability to recover enough high-quality RNA from these cells for use in microarray experiments. Application of a new technique, antisense RNA (aRNA) amplification, has increased the yield of RNA to accommodate the requirements necessary for complementary DNA (cDNA) spotted and high-density oligonucleotide microarrays. DRG are composed of multiple neuronal and glial cell types and in particular, small-diameter neurons mediate nociceptive responses whereas large neurons facilitate mechanosensation. Two studies have profiled the differential gene expression of these nociceptive and mechanosensitive naive neurons by microarray (Luo, L. *et al.*, 1999; Davis-Taber, R. *et al.*, 2003). DRGs were flash frozen in cryomolds, sectioned to ~8 μm thickness and visualized with a Nissl (cresyl violet) stain to identify the neurons. Naive small (<25 μm and an identifiable nucleus) and large (>40 μm) diameter neurons were captured (PixCell II LCM system; Arcturus, Mountain View, CA, USA) and total RNA extracted from cellular material adherent to the thermoplastic film on the LCM caps. Multiple RNA isolation and amplification reagents and kits are commercially available from a variety of vendors (e.g., Arcturus, Mountain View, CA, USA; Stratagene, San Diego, CA, USA; Epicentre Technologies, Madison, WI, USA). aRNA was amplified two to three times prior to hybridization onto a custom cDNA spotted array or an Affymetrix high-density oligonucleotide GeneChip® array (Luo, L. *et al.*, 1999; Davis-Taber, R. *et al.*, 2003).

There are several aspects of experimental design that should be considered prior to initiation of an LCM microarray study. First, a sufficient number of cells should be captured. For each of the LCM studies discussed, approximately 1000 small and 1000 large DRG neurons were isolated for microarray profiling. An alternative to capturing a greater amount of cellular material is increasing the number of rounds of RNA amplification. One trade-off to higher yields is a shortening of the RNA transcripts with each successive round of amplification. One other consideration is the 3′ bias of the aRNA transcripts generated. This is important when determining the microarray platform to use. Many high-density oligonucleotide arrays already incorporate a 3′ bias in their probe design (e.g., Affymetrix) but this issue may be more significant when using custom spotted arrays.

The initial studies profiling naive small and large DRG neurons observed a differential gene expression pattern for these two neuronal subpopulations. In the first study, naive cervical DRG aRNA from adult rats were hybridized to a custom cDNA spotted microarray (Luo, L. *et al.*, 1999). The microarray chip had 477 cDNAs that were obtained from two differential display experiments: (1) identified genes uniquely expressed in DRG compared to brain, liver, and kidney; and (2) genes that were differentially expressed in L5/L6 between ipsilateral (SNL) and contralateral (control). Approximately 40 genes were found to be enriched in either neuronal subpopulation with a greater majority detected in the small DRG neurons. The second study collected similar sized neuronal cells from rat lumbar L5/L6 DRG and hybridized the aRNA to the Affymetrix Neurobiology GeneChip® (Davis-Taber, R. *et al.*, 2003). In the small and large DRG neurons, ~100 genes were detected and classified into functional groups (channels and receptors, enzymes, cell signaling, and cytoskeletal). In the categories of cell signaling and channels and receptors, there was an equal number of genes differentially enriched in each neuronal subpopulation, however, large DRG neurons predominantly expressed cytoskeletal proteins and enzymes. Examining the results reported in both experiments has shown that similar expression patterns and fold enrichment were observed for both neuronal subpopulations. A lack of glial-specific gene expression suggests that there was minimal glial contamination in the capturing process.

To understand the gene expression changes that occur in neuropathic pain states further, neuronal subpopulations were isolated from DRG following SNL or sham surgery and profiled by microarray. Similar to the naive DRG studies, small (~1000 cells) and large (~1000 cells) neurons were captured from SNL and sham L5/L6 ipsilateral DRGs (~4000 total) by LCM and evaluated by Affymetrix rat RG-U34a GeneChip® arrays (Davis-Taber, R. *et al.*, 2004). Microarray analysis showed altered expression patterns between SNL and sham animals especially in large DRG neurons where the majority of gene changes occurred. Although some genes showed a common expression signature across both small and large neurons, a number of genes were dysregulated in only one neuronal subtype. A comparison of the whole DRG SNL microarray data (Thimmapaya, R. *et al.*, 2003) with the SNL small and large DRG neuron study revealed only a small number of genes that were commonly altered in all three samples. Also, only a small percentage of gene expression changes observed in the small (35%) or large (22%) neurons were dysregulated in the whole DRG microarray study. This suggests that within the individual neuronal subtypes, the distinct expression patterns that were observed may have been masked or minimized in the whole DRG microarray study. Microarray analysis of the small and large DRG neuronal subpopulations isolated from the injured DRG following SNL has provided a deeper understanding of the differential gene expression associated with neuropathic pain (Davis-Taber, R. *et al.*, 2004).

A potential limitation of the LCM microarray studies is the ability to collect a pure subpopulation of neurons. The time constraints of standard immunohistochemical staining procedures would allow tissue sections to rehydrate which could hinder precise capture of individual cells so a modified cresyl violet staining protocol developed by Arcturus Engineering, Mountain View, CA, USA, was used to visualize both small and large diameter DRG neurons. Without a way to distinguish the different small DRG neuron subtypes, a mixed population of peptidergic and non-peptidergic neurons were captured and pooled together. TRPV1 expression is an example of this limitation whereby it was shown to be downregulated in the L5/L6 small DRG neurons in the LCM SNL study but was not detected in the naive, L5/L6 small DRG neurons. Localization studies have shown that TRPV1 is found on peptidergic small-diameter neurons and the receptor's inconsistent detection may result from the level of sensitivity of RNA amplification and microarray techniques from a mixed population of small DRG neurons. However, evaluation of the expression levels of RNA from subpopulations of neurons has provided a depth of information on transcripts that was previously not possible when whole DRG samples were assessed.

14.5 Evaluation of Additional Neuropathic Pain Models by Microarray

In addition to studies on the expression changes in DRG from SNL model of neuropathic pain, microarray analysis has also been performed on DRG following peripheral axotomy. In one study a custom microarray chip was generated that contained genes that were overexpressed in DRG 14 days following axotomy (Xiao, H. S. *et al.*, 2002). The L4/L5 DRG were then probed against this chip at various timepoints to assess the changes in expression that occur both during the development and following established pain. Some genes changed expression immediately by day 2 (neuropeptide Y, heat shock protein 27, LIM domain protein CLP36) while others showed a more significant change at 14 days (neuronal nitric oxide synthase (nNOS), calcitonin gene-related peptide (CGRP)) following axotomy. The changes were confirmed for a subset of genes by *in situ* hybridization. Interestingly many of the genes that changed expression were the same as those observed in the SNL model (Table 1) indicating that these neuropathic pain models share some underlying molecular components. A second study used the Affymetrix rat RG-U34a GeneChip® gene chip and also examined the L4/L5 DRG following sciatic nerve axotomy over a time course of 1–14 days postinjury (Costigan, M. *et al.*, 2002). This

Table 1   Representative gene expression changes identified by microarray analysis from rat L5/L6 DRG following peripheral nerve injury

Accession number	Description	SNL (L5/L6) (Thimmapaya, R. et al., 2003)	Published changes: preclinical neuropathic models
X59864	ASM15	4.84	↑ Valder C. R. *et al.* (2003)
AI009268	Ca^{2+}/calmodulin-dependent protein kinase 2 delta	2.29	↑ Xiao H. S. *et al.* (2002)
X81193	Cysteine-rich protein 3 (muscle LIM protein)	42.57	↑ Wang H. *et al.* (2002); ↑ Xiao H. S. *et al.* (2002); ↑ Valder C. R. *et al.* (2003)
L21192	GAP43	2.76	↑ Wang H. *et al.* (2002); ↑ Xiao H. S. *et al.* (2002); ↑ Valder C. R. *et al.* (2003)
AF028784	Glial fibrillary acidic protein (GFAP)	5.54	↑ Valder C. R. *et al.* (2003)
M98049	Pancreatitis-associated protein (Reg-2)	20.36	↑ Wang H. *et al.* (2002); ↑ Valder C. R. *et al.* (2003)
AA799645	Phospholemman chloride channel	2.45	↑ Costigan M. *et al.* (2002)
X80290	Pituitary adenylate cyclase activating peptide (PACAP)	3.04	↑ Costigan M. *et al.* (2002); ↑ Wang H. *et al.* (2002); ↑ Xiao H. S. *et al.* (2002)
M18331	Protein kinase C epsilon	−5.55	↑ Xiao H. S. *et al.* (2002); ↓ Valder C. R. *et al.* (2003)
U48246	Protein kinase C-binding protein NELL1	−12.15	↓ Wang H. *et al.* (2002); ↑ Xiao H. S. *et al.* (2002); ↓ Valder C. R. *et al.* (2003)
AB003991	SNAP-25A	−2.50	↓ Costigan M. *et al.* (2002); ↓ Wang H. *et al.* (2002); ↓ Xiao H. S. *et al.* (2002); ↓ Valder C. R. *et al.* (2003)
AF059030	Sodium voltage-gated channel, Na$_v$1.9	−23.85	↓ Wang H. *et al.* (2002); ↓ Valder C. R. *et al.* (2003)
X52772	Synaptotagmin I	−3.03	↔ Xiao H. S. *et al.* (2002)
X59737	Ubiquitous mitochondrial creatine kinase	−6.33	↓ Wang H. *et al.* (2002)

Genes that show enhanced or diminished expression from injured dorsal root ganglia (DRG) versus sham-operated DRG with a twofold change and $P \leq 0.05$. The fold increases and decreases in gene expression observed with DRG from spinal nerve ligated (SNL) rats are given, together with published microarray analysis of gene changes from preclinical neuropathic pain models (Xiao, H. S. *et al.*, 2002, axotomy; Wang, H. *et al.*, 2002, SNL; Costigan, M. *et al.*, 2002, axotomy; Valder, C. R. *et al.*, 2003, SNL). Genes were upregulated (↑), downregulated (↓), or unchanged (↔).

analysis compared naive DRG to the injured DRG and focused on the evaluation of data following this comparison. The data presented supported the use of a $P < 0.05$ significance threshold for detecting regulated genes. When a cut-off of a twofold change in expression was applied to select gene transcripts to assess, this led to an estimated error rate of 16%. Combining a more than 1.5-fold expression change and a $P < 0.05$ reduced the estimated error to 5%. Using the latter analysis, 240 genes changed expression significantly and many had not been previously described as changing following axotomy. Changes in expression of a subset of the genes were confirmed by *in situ* and Northern blots analysis. This study also identified increases in gene expression of genes mostly involved in immune responses and inflammation and decreases in those involved in neurotransmission similar to that reported for the SNL model of neuropathic pain. Taken together the microarray studies identified genes that are dysregulated in neuropathic pain states from different pain models providing an opportunity to unveil potential neuropathic molecular signatures that are common between these models.

14.6 The Strengths and Limitations of Microarray Analysis for Pain Research

Our knowledge of the complexity of neuropathic pain has heightened following the use of microarray gene chip technology. This approach provided opportunities to examine multiple gene transcripts simultaneously in pain states. Interestingly in these different studies similar gene changes have been observed in the DRG from diverse peripheral nerve injury models. A subset of these changes are given in Table 1, which shows that in most cases the gene products were dysregulated in a similar direction. However, in a couple of examples the transcripts were altered differently between the different models (Nell1; protein kinase C epsilon). The studies that have been performed have focused on gene changes that occur in the spinal cord and DRG but it is also known that areas of the brain are important for pain processing. Changes in gene expression can also occur in brain areas such as the rostral ventral medulla and periaqueductal gray, which participate in ascending and descending nociceptive signaling (Miki, K. *et al.*, 2002) in chronic pain states.

In addition to the strengths of microarray analysis detailed above there are also limitations. These assays are not suitable for detection of post-translational modifications of functional proteins including phosphorylation/dephosphorylation of ion channels, membrane proteins, and transcription factors. Functional modifications of pain-relevant proteins that do not alter their transcriptional levels include activation of *N*-methyl-D-aspartate receptors by injury-induced protein kinase C-mediated phosphorylation (Chen, L. and Huang, L. Y., 1992) and the downstream regulatory element antagonistic modulator (DREAM)-mediated derepression of prodynorphin expression that is regulated by elevated intracellular calcium and protein kinase A activation (Carrion, A. M. *et al.*, 1998). DREAM-deficient mice have been shown to have elevated spinal cord prodynorphin levels and reduced inflammatory pain behaviors (Cheng, H. Y. *et al.*, 2002).

Additionally, translation of the data arising from these studies to identify potential novel therapeutic targets is also a challenge. The approach undertaken in the studies highlighted in this chapter, has been to examine the literature for any data that would indicate that the protein encoded by a dysregulated transcript plays a role in pain. A complicating factor studying nociception and associated gene regulation in pain states is that unlike metabolic diseases where pathways are well understood, little is known about the specific pathways underlying pain. Application of programs that assess gene ontology (i.e., GenMAPP, GoSurfer, GOTM, WebGestalt, GoMiner, OntoExpress, and ErmineJ) and known pathways (i.e., Pathway Assist) can provide some insights into nociceptive signaling.

As mentioned earlier, microarray studies have generated a wealth of information but to understand the role of each gene product in pain will require extensive follow-up studies. Another approach that could be taken would be to evaluate the effects of administering target specific agonists or antagonists in preclinical pain models to determine if these proteins offer potential opportunities for novel therapeutic intervention if such agents already exist. Gene silencing using antisense or small interfering RNA (siRNA) approaches would also be other ways to establish if a protein that was dysregulated in a pain state would be a useful pain target (Hemmings-Mieszczak, M. *et al.*, 2003; Tan, P. H. *et al.*, 2005). Pain microarray studies as predicted have deepened our understanding of key players in neuropathic pain. However, deciphering the precise role of specific gene products in pain and identification of novel therapeutic targets awaits further investigation and is eagerly anticipated.

14.7 The Genetics of Pain

The pain microarray gene chip analysis detailed earlier in this chapter provides a snapshot of the changes in gene expression that occur in chronic pain states with the ultimate goal of understanding pain mechanisms and identifying potential novel therapeutic targets for pain treatment. Additional efforts to identify human genes associated with pain have examined congenital insensitivity to pain (CIP) that has been linked to the gene encoding serine palmitoyl-transferase (Nagasako, E. M. *et al.*, 2003). A subtype of this disorder CIP IV has been linked to single changes in nucleotide sequence (single nucleotide polymorphisms (SNPs)) of the TrkA receptor rendering the protein nonfunctional (Indo, Y. *et al.*, 1996). While nerve growth factor binding to the TrkA receptor is known to contribute to the survival and regulation of small diameter neurons, loss-of-function phenotypes observed in these individuals with SNPs are similar to the knockout animals. Another example is the gene responsible for familial hemiplegic migraine that has been linked to human chromosome 19p13 and more recently identified as the gene encoding a calcium channel α-subunit (CACNA1A; Joutel, A. *et al.*, 1993). Specific patients suffering from migraines have been shown to have missense mutations in this gene that lead to gain-of-function phenotypes. This calcium channel has been shown to be associated with serotonin release, magnesium levels, and the phenomenon of cortical spreading depression, each of which contribute to the pathophysiology of migraine (Ophoff, R. A. *et al.*, 1996).

Differences in human pain perception have also been associated to a common SNP (V158M) in the gene of the enzyme catechol-*o*-methyltransferase (COMT). Individuals homozygous for the Met[158] allele of COMT showed diminished regional μ-opioid responses to pain compared with heterozygotes, which was opposite to the effects observed with the Val[158] homozygote (Zubieta, J. K. *et al.*, 2003). Individuals with the more active form of the enzyme (Val[158]) are less susceptible to pain than people with Met[158] at the same position. COMT activity is also associated with a pathological pain condition known as temporomandibular joint disorder (TMD) that impacts 5–15% of the adult population. The relationship between COMT polymorphism, pain sensitivity, and risk of TMD development has recently been reported (Diatchenko, L. *et al.*, 2005). Three genetic variants (haplotypes) of the COMT gene, designated as low-pain sensitivity, average-pain sensitivity, and high-pain sensitivity were identified, that encompass 96% of the examined genotypes and lead to altered enzyme activity. The more active the COMT enzyme inversely correlated with pain susceptibility and the lower risk of developing TMD (Diatchenko, L. *et al.*, 2005).

Another interesting aspect of human genetics examined the sex differences in pain and analgesic sensitivity has been an area of research focus and increasing amounts of evidence suggest that sex differences may reflect the activation of gender-specific neural mechanisms. Since clinically κ-opioid receptor agonists show more efficacy in women than men, a recent report investigated the possible sex specificity of the genetic mediation of κ-opioid analgesia in mice using quantitative trait locus (QTL) mapping (Mogil, J. S. *et al.*, 2003). QTL mapping is a technique in which the co-inheritance of continuous traits and polymorphic DNA markers (e.g., microsatellites or SNPs) can be used to broadly identify the chromosomal location of genes responsible for trait variability (Lander, E. S. and Schork, N. J., 1994). Three convergent lines of evidence point to the mouse melanocortin-1 (*MCR1*) gene as the female-specific QTL. Furthermore analysis of women with red hair and fair skin containing two variant *MCR1* alleles displayed significantly greater analgesia from the κ-opioid pentazocine than all other groups (Mogil, J. S. *et al.*, 2003).

With the completion of the human genome project and ongoing efforts to sequence the genomes of a variety of other species, a plethora of sequence information is now available. New microarray applications and chip platforms are being developed to pinpoint genetic variation in disease states, isolate biomarkers that may indicate a predisposition for a given disease and identify treatment strategies and/or drug candidates that would have maximal benefit for a patient population with a specific genetic make-up.

The first of these new array platforms are comparative genomic hybridization (CGH) arrays. These microarrays are designed across large areas of sequence (i.e., a single chromosome) and even an entire genome to provide broad coverage for genotyping. CGH arrays have been generated from bacterial artificial chromosome (BAC) clones containing large genomic inserts and require a reference genome for comparison to the sample genome (Herr, A. *et al.*, 2005). These arrays have been used to determine differences in copy number of particular genes and identify chromosomal imbalances found in cancer, genetic disorders, and deletion mutants with unique phenotypes used for scientific research (Hughes, T. R. *et al.*, 2000; Bruder, C. E. G. *et al.*, 2001; Herr, A. *et al.*, 2005). In contrast, to CGH arrays, resequencing and genome tiling arrays are designed to smaller, specific regions of a chromosome or genomic sequence and have been used to detect unknown mutations or to narrow down the area where a mutation may occur (Chee, M. *et al.*, 1996; Galvin, P. *et al.*, 2004; Lin, S. M. *et al.*, 2006). Resequencing arrays have probes designed to a particular sequence which when combined together will encompass all of the sequence within a specified region. Tiling arrays are designed to walk along a section of sequence with each probe representing one of the four nucleotides at a single nucleotide position. The probes are designed in sets of four. At a certain nucleotide position in the middle of the probe, each of the four probes will represent a different nucleotide (A, C, G, or T). The next four overlapping probes will have a mismatch (A, C, G, or T) at the next nucleotide position in succession from the mismatch in the previous probe set. The overall result from both these array platforms is a lack of hybridization and a subsequent decrease in signal intensity for the probes surrounding a mutation or polymorphism.

Linkage analysis studies using restriction-fragment length polymorphisms (RFLP) and polymorphic microsatellites have been useful in understanding mendelian-linked diseases. However, these tools have proven less consistent for mapping more complex disorders (i.e., schizophrenia, manic depression, and multiple sclerosis) that may have greater genetic heterogeneity or nonmendelian inheritance (Risch, N. J., 2000). More recently, SNPs, single nucleotide changes in DNA sequence, have been identified and catalogued in a number of genomic databases. Several SNP microarray chips have been developed (e.g., Affymetrix GeneChip Human Mapping 10K array) and used in concert with high-density microsatellite linkage sets to identify susceptibility or candidate genes that may be associated with the phenotype of a specific disease (Riley, J. H. *et al.*, 2000; Risch, N. J., 2000; Roses, A. D., 2000; Sawcer, S. J. *et al.*, 2004; Schaid, D. J. *et al.*, 2004).

Microarray can be a powerful tool, but also a hindrance. This technology can quickly and efficiently generate large amounts of data. However, excavating through this mass of information can be a time-consuming and possibly fruitless process

without the proper analysis tools. With the implementation of these new array platforms, software has also been designed to aid in genotyping and linkage mapping studies. In the future, microarray may deepen our understanding of the genomic landscape and perhaps shed light on the origins of disease states. This genetic-driven trend could lead to individualized medical diagnoses and therapeutic treatment strategies.

References

Bruder, C. E. G., Hirvela, C., Tapia-Paez, I., Fransson, I., Segraves, R., Hamilton, G., Zhang, X. X., Evans, D. G., Wallace, A. J., Baser, M. E., Zucman-Rossi, J., Hergersberg, M., Boltshauser, E., Papi, L., Rouleau, G. A., Poptodorov, G., Jordanova, A., Rask-Andersen, H., Kluwe, L., Mautner, V., Sainio, M., Hung, G., Mathiesen, T., Moller, C., Pulst, S. M., Harder, H., Heiberg, A., Honda, M., Niimura, M., Sahlen, S., Blennow, E., Albertson, D. G., Pinkel, D., and Dumanski, J. P. 2001. High resolution deletion analysis of constitutional DNA from neurofibromatosis type 2 (NF2) patients using microarray-CGH. Hum. Mol. Genet. 10, 271–282.

Carrion, A. M., Mellstrom, B., and Naranjo, J. R. 1998. Protein kinase A-dependent derepression of the human prodynorphin gene via differential binding to an intragenic silencer element. Mol. Cell. Biol. 18, 6921–6929.

Chee, M., Yang, R., Hubbell, E., Berno, A., Huang, X. C., Stern, D., Winkler, J., Lockhart, D. J., Morris, M. S., and Fodor, S. P. A. 1996. Accessing genetic information with high-density DNA arrays. Science 274, 610–614.

Chen, L. and Huang, L. Y. 1992. Protein kinase C reduces Mg^{2+} block of NMDA-receptor channels as a mechanism of modulation. Nature 356, 521–523.

Cheng, H. Y., Pitcher, G. M., Laviolette, S. R., Whishaw, I. Q., Tong, K. I., Kockeritz, L. K., Wada, T., Joza, N. A., Crackower, M., Goncalves, J., Sarosi, I., Woodgett, J. R., Oliveira-dos-Santos, A. J., Ikura, M., van der Kooy, D., Salter, M. W., and Penninger, J. M. 2002. DREAM is a critical transcriptional repressor for pain modulation. Cell 108, 31–43.

Costigan, M., Befort, K., Karchewski, L., Griffin, R. S., D'Urso, D., Allchorne, A., Sitarski, J., Mannion, J. W., Pratt, R. E., and Woolf, C. J. 2002. Replicate high-density rat genome oligonucleotide microarrays reveal hundreds of regulated genes in the dorsal root ganglion after peripheral nerve injury. BMC Neurosci. 3, 16.

Davis-Taber, R., Choi, W., Kroeger, P., Seiler, F., Gagne, G., Kage, K., Faltynek, C., Gopalakrishnan, M., and Scott, V. 2003. Differential gene expression in small and large diameter DRG neurons isolated by laser capture microdissection. Am. Pain Soc. 2003; poster .

Davis-Taber, R., Choi, W., Vos, M., Zhu, C., Seiler, F., Gagne, G., Kage, K., Kroeger, K., Honore, P., Fagerland, J., Faltynek, C., Gopalakrishnan, M., Surowy, C., and Scott, V. E. 2004. Transcriptional profiling in small and large DRG neurons from a neuropathic pain model using LCM. Am. Pain Soc. 2004; poster .

Diatchenko, L., Slade, G. D., Nackley, A. G., Bhalang, K., Sigurdsson, A., Belfer, I., Goldman, D., Xu, K., Shabalina, S. A., Shagin, D., Max, M. B., Makarov, S. S., and Maixner, W. 2005. Genetic basis for individual variations in pain perception and the development of a chronic pain condition. Hum. Mol. Genet. 14(1), 135–143.

Galvin, P., Clarke, L. A., Harvey, S., and Amaral, M. D. 2004. Microarray analysis in cystic fibrosis. J. Cyst Fibros. 3 (Suppl 2), 29–33.

Hemmings-Mieszczak, M., Dorn, G., Natt, F. J., Hall, J., and Wishart, W. L. 2003. Independent combinatorial effect of antisense oligonucleotides and RNAi-mediated specific inhibition of the recombinant rat P2X3 receptor. Nucl. Acids Res. 31, 2117–2126.

Herr, A., Grutzmann, R., Matthaei, A., Artelt, J., Schrock, E., Rump, A., and Pilarsky, C. 2005. High-resolution analysis of chromosomal imbalances using the Affymetrix 10K SNP genotyping chip. Genomics 85, 392–400.

Holder, D., Pikounis, V., Raubertas, R., Svetnik, V., and Soper, K. 2002. Statistical Analysis of High Density Oligonucleotide Arrays: A SAFER Approach. In: Proceedings of the American Statistical Association, Atlanta, Georgia.

Hughes, T. R., Roberts, C. J., Dai, H., Jones, A. R., Meyer, M. R., Slade, D., Burchard, J., Dow, S., Ward, T. R., Kidd, M. J., Friend, S. H., and Marton, M. J. 2000. Widespread aneuploidy revealed by DNA microarray expression profiling. Nature Genet. 25, 333–337.

Indo, Y., Tsuruta, M., Hayashida, Y., Karim, M. A., Ohta, K., Kawano, T., Mitsubuchi, H., Tonoki, H., Awaya, Y., and Matsuda, I. 1996. Mutations in the TRKA/NGF receptor gene in patients with congenital insensitivity to pain with anhidrosis. Nat. Genet. 13(4), 485–488.

Joutel, A., Bousser, M. G., Biousse, V., Labauge, P., Chabriat, H., Nibbio, A., Maciazek, J., Meyer, B., Bach, M. A., Weissenbach, J., et al. 1993. A gene for familial hemiplegic migraine maps to chromosome 19. Nat. Genet. 5(1), 40–45.

Lander, E. S. and Schork, N. J. 1994. Genetic dissection of complex traits. Science 265(5181), 2037–2048.

Lin, S. M., Devakumar, J., and Kibbe, W. A. 2006. Improved prediction of treatment response using microarrays and existing biological knowledge. Pharmacogenomics 7, 495–501.

Luo, L., Salunga, C., Guo, H., Bittner, A., Joy, K., Galindo, J., Xiao, H., Rogers, K., Wan, J., Jackson, M., and Erlander, M. 1999. Gene expression profiles of laser-captured adjacent neuronal subtypes. Nature Med. 5, 117–122.

Luo, Z. D., Chaplan, S. R., Scott, B. P., Cizkova, D., Calcutt, N. A., and Yaksh, T. L. 1999. Neuronal nitric oxide synthase mRNA upregulation in rat sensory neurons after spinal nerve ligation: lack of a role in allodynia development. J. Neurosci. 19, 9201–9208.

Miki, K., Zhou, Q. Q., Guo, W., Guan, Y., Terayama, R., Dubner, R., and Ren, K. 2002. Changes in gene expression and neuronal phenotype in brain stem pain modulatory circuitry after inflammation. J. Neurophysiol. 87, 750–760.

Mogil, J. S., Wilson, S. G., Chesler, E. J., Rankin, A. L., Nemmani, K. V., Lariviere, W. R., Groce, M. K., Wallace, M. R., Kaplan, L., Staud, R., Ness, T. J., Glover, T. L., Stankova, M., Mayorov, A., Hruby, V. J., Grisel, J. E., and Fillingim, R. B. 2003. The melanocortin-1 receptor gene mediates female-specific mechanisms of analgesia in mice and humans. Proc. Natl. Acad. Sci. U. S. A. 100(8), 4867–4872.

Nagasako, E. M., Oaklander, A. L., and Dworkin, R. H. 2003. Congenital insensitivity to pain: an update. Pain 101(3), 213–219.

Ophoff, R. A., Terwindt, G. M., Vergouwe, M. N., van Eijk, R., Oefner, P. J., Hoffman, S. M., Lamerdin, J. E.,

Mohrenweiser, H. W., Bulman, D. E., Ferrari, M., Haan, J., Lindhout, D., van Ommen, G. J., Hofker, M. H., Ferrari, M. D., and Frants, R. R. 1996. Familial hemiplegic migraine and episodic ataxia type-2 are caused by mutations in the Ca^{2+} channel gene CACNL1A4. Cell 87(3), 543–552.

Riley, J. H., Allan, C. J., Lai, E., and Roses, A. 2000. The use of single nucleotide polymorphisms in the isolation of common disease genes. Pharmacogenomics 1, 39–47.

Risch, N. J. 2000. Searching for genetic determinants in the new millennium. Nature 405, 847–856.

Roses, A. D. 2000. Pharmacogenetics and the practice of medicine. Nature 405, 857–865.

Sawcer, S. J., Maranian, M., Singlehurst, S., Yeo, T., Compston, A., Daly, M. J., De Jager, P. L., Gabriel, S., Hafler, D. A., Ivinson, A. J., Lander, E. S., Rioux, J. D., Walsh, E., Gregory, S. G., Schmidt, S., Pericak-Vance, M. A., Barcellos, L., Hauser, S. L., Oksenberg, J. R., Kenealy, S. J., and Haines, J. L. 2004. Enhancing linkage analysis of complex disorders: an evaluation of high density genotyping. Hum. Mol. Genet. 13, 1943–1949.

Schaid, D. J., Guenther, J. C., Christensen, G. B., Hebbring, S., Rosenow, C., Hilker, C. A., McDonnell, S. K., Cunningham, J. M., Slager, S. L., Blute, M. L., and Thibodeau, S. N. 2004. Comparison of microsatellites versus single-nucleotide polymorphisms in a genome linkage screen for prostate cancer-susceptibility loci. Am. J. Hum. Genet. 75, 948–965.

Sun, H., Xu, J., Della Penna, K. B., Benz, R. J., Kinose, F., Holder, D. J., Koblan, K. S., Gerhold, D. L., and Wang, H. 2002. Dorsal horn-enriched genes identified by DNA microarray, in situ hybridization and immunohistochemistry BMC Neurosci. 3, 11–23.

Tan, P. H., Yang, L. C., Shih, H. C., Lan, K. C., and Cheng, J. T. 2005. Gene knockdown with intrathecal siRNA of NMDA receptor NR2B subunit reduces formalin-induced nociception in the rat. Gene Ther. 12, 59–66.

Thimmapaya, R., Davis-Taber, R., Choi, W., Zhu, C., Gubbins, E., Vos, M., Helfrich, R., Kage, K., Daza, A., Donnelly-Roberts, D., Harris, R., Surowy, C., Honore, P., Kroeger, P., Faltynek, C., Gopalakrishnan, M., and Scott, V. E. 2003. Gene expression profiling in the spinal nerve ligation model of neuropathic pain using microarray technology. Soc. Neurosci. 2003; poster .

Valder, C. R., Liu, J. J., Song, Y. H., and Luo, Z. D. 2003. Coupling gene chip analyses and rat genetic variances in identifying potential target genes that may contribute to neuropathic allodynia development. J. Neurochem. 87, 560–573.

Wang, H., Sun, H., Della Penna, K., Benz, R. J., Xu, J., Gerhold, D. L., Holder, D. J., and Koblan, K. S. 2002. Chronic neuropathic pain is accompanied by global changes in gene expression and shares pathobiology with neurodegenerative diseases. Neuroscience 114, 529–546.

Xiao, H. S., Huang, Q. H., Zhang, F. X., Bao, L., Lu, Y. J., Guo, C., Yang, L., Huang, W. J., Fu, G., Xu, S. H., Cheng, X. P., Yan, Q., Zhu, Z. D., Zhang, X., Chen, Z., Han, Z. G., and Zhang, X. 2002. Identification of gene expression profile of dorsal root ganglion in the rat peripheral axotomy model of neuropathic pain. Proc. Natl. Acad. Sci. U. S. A. 99, 8360–8365.

Zubieta, J. K., Heitzeg, M. M., Smith, Y. R., Bueller, J. A., Xu, K., Xu, Y., Koeppe, R. A., Stohler, C. S., and Goldman, D. 2003. COMT val158met genotype affects mu-opioid neurotransmitter responses to a pain stressor. Science 299(5610), 1240–3.

15 Evolutionary Aspects of Pain

E T Walters, University of Texas at Houston, Medical School, Houston, TX, USA

Glossary

CREB Ca^{2+} and cAMP response element-binding protein, a transcription factor that binds to specific enhancer elements in DNA and enhances the expression of corresponding genes when certain Ca^{2+}- or cAMP-activated protein kinases, such as CaMK or PKA, phosphorylate the CREB protein.

degenerin/epithelial Na^+ (DEG/ENaC) channels A family of ion channels that conduct Na^+ ions and can be constitutively active or gated by mechanical distention or protons. Channels that are gated by protons are called acid-sensing ion channels (ASICs).

MAPK (ERK) Mitogen-activated protein kinase, a family of protein kinases that includes the extracellular signal-regulated kinase (ERK). ERK has many functions related to neural plasticity, both in the cytoplasm and nucleus.

MRG Mas-related genes, a family of genes encoding G-protein-coupled receptors, some of which are expressed selectively in mammalian nociceptors. Also called sensory neuron-specific G-protein-coupled receptors (SNSRs).

NMDA-receptor-dependent LTP N-Methyl-D-asparate-receptor-dependent long-term synaptic potentiation, a form of use-dependent synaptic plasticity induced by Ca^{2+} influx through the class of glutamate receptors that is selectively activated (pharmacologically) by the drug, NMDA.

RFamide peptides A family of peptide neurotransmitters sharing the same C terminal RFamide (arginine-phenylalanine-NH_2) sequence.

transient receptor potential (TRP) A family of ion channels, many of which conduct cations nonselectively. Different channels in this family are opened by various inputs, including heat, cooling, protons, and intracellular signals.

TTX-resistant, voltage-gated Na^+ channels Members of the voltage-gated Na^+ channel family that are relatively insensitive to blockade by the specific antagonist, tetrodotoxin (TTX). These channels (NaV1.8 and NaV1.9) are expressed selectively in nociceptive dorsal root ganglion neurons.

15.1 Introduction

Pain, according to a widely accepted definition (Merskey, H. and Bogduk, N., 1994), is an unpleasant sensory and emotional experience associated with actual or potential tissue damage. This definition is intended to capture essential features of human experiences linked to tissue damage and makes no claims about the evolutionary roots or possible biological functions of pain. Nonetheless, it raises two distinct sets of evolutionary questions. First, how did sensory responses to actual or potential tissue damage evolve? What biological adaptations do these responses represent, and to what extent are the underlying mechanisms conserved across the animal kingdom? Second, how did the unpleasant emotional components of pain evolve? This essay considers comparative functional and molecular findings from diverse species in the animal kingdom that begin to address these questions. Because this task requires reference to far more reports than can be cited in a brief account, citations are made primarily to review articles or to recent reports that provide additional references to this diverse literature. Although no recent, comprehensive review of nociceptive biology in invertebrates exists, highly comprehensive reviews of nociceptive mechanisms in mammals can be found in other chapters in this volume, and in a recent book (Willis, W. D. and Coggeshall, R. E., 2004).

15.2 Evolution of Nociceptive Mechanisms

Because an animal's reproductive success is clearly enhanced if it can detect and avoid pervasive threats to its structural integrity that otherwise would impair or end its ability to reproduce, selection pressures for mechanisms that recognize noxious stimuli and trigger adaptive physiological and behavioral responses must be very strong (Walters, E. T., 1994). Indeed, animals in several phyla have now been shown to possess nociceptors connected to neural networks controlling defensive responses, and these varied nociceptive systems share numerous features including distinctive forms of plasticity. Common properties in diverse species are of interest for evolutionary questions because they point to (1) common selection pressures that have shaped similar but independently derived (analogous) adaptations; and/or

(2) the conservation of homologous mechanisms descending from common ancestors. Figure 1 shows a simplified view of evolutionary relationships among selected species mentioned in this essay that have been used to study nociceptive biology. Those that have had their genome completely sequenced are indicated by asterisks (human, mouse, fruit fly, and roundworm). The current explosion of molecular genetic information from a growing number of species promises to illuminate evolutionary relationships among nociceptive mechanisms with a level of detail and precision that could only be dreamed of a few years ago.

15.2.1 Nociceptors

Nociceptors are defined as sensory neurons that are activated preferentially by tissue-damaging stimuli, either immediately or when a noxious stimulus is sufficiently prolonged. Because of the medical, social, and economic importance of pain, most studies of nociceptors have been in selected mammalian species rather than animals more distantly related to humans. Thus, the comparative physiological and molecular information needed to discern evolutionary patterns in nociceptive function is still quite limited. However, the availability of extensive genomic and proteomic data from a growing number of species should soon shed light on the evolution of specific molecular components of nociceptive systems.

15.2.1.1 Identification of nociceptors in diverse phyla

Physiological identification and characterization of nociceptors is challenging because the somata are usually dispersed among those of many other types of neurons, and the fine peripheral and central fibers are essentially inaccessible for direct intracellular examination in living preparations. Nevertheless, nociceptors have now been demonstrated convincingly in representatives of all the major animal phyla (Figure 1), including nematode worms (Tobin, D. M. and Bargmann, C. I., 2004), leeches (Nicholls, J. G. and Baylor, D. A., 1968), insects (Tracey, W. D. Jr., *et al.*, 2003), mollusks (Illich, P. A. and Walters, E. T., 1997), and various chordates, including fish (Sneddon, L. U. *et al.*, 2003), amphibians (Lynn, B. and O'Shea, N. R., 1998), reptiles (Liang, Y. F. *et al.*, 1995), birds (Holloway, J. A. *et al.*, 1980), and mammals (Light, A. R., 1992). Additional species, including snails (Malyshev, A. Y. and Balaban, P. M.,

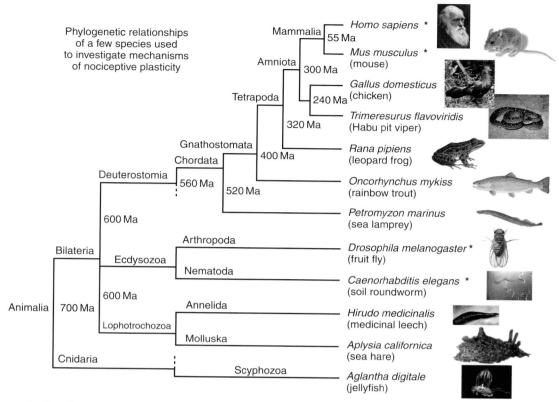

Figure 1 Greatly simplified phylogeny of a few selected species (right) that have been used to investigate nociceptive biology. Numbers indicate approximate time of divergence (millions of years ago, Ma) of the indicated groups (Dawkins, R., 2004). Groupings reflect recent molecular phylogenetics rather than traditional views based on morphological evidence. The traditional phyla indicated on this chart are chordata, arthropoda, nematoda, annelida, molluska, and cnidaria. Because of the paucity of fossil evidence prior to the Cambrian period (more than 540 Ma) and uncertainties in molecular clocks over longer periods, the older times of divergence on the chart are much rougher estimates than the more recent ones. Asterisks indicate species for which complete genome sequence information is currently available.

2002) and lampreys (Christenson, J. et al., 1988) have mechanosensory neurons that are likely to be nociceptive, despite their relatively low mechanical thresholds, because the tissue they innervate is delicate enough to be damaged by modest pressures (see Illich, P. A. and Walters, E. T., 1997).

15.2.1.2 Conserved mechanisms of nociceptive transduction

Tissue damage can be produced by various mechanical forces and chemical agents as well as by excessive temperature changes, so one would expect that nociceptors utilize diverse stimulus transduction mechanisms. Combined molecular and physiological analyses of nociceptive transduction have revealed important mechanisms in a few species. Many forms of mechanotransduction, perhaps including the transduction of noxious forces in some cells, depend upon depolarizing current entry through channels in the degenerin/epithelial Na^+ channel (DEG/ENaC) family. Several of these channels are required for activation of low-threshold mechanosensory neurons in the nematode, *C. elegans* (Tobin, D. M. and Bargmann, C. I., 2004), while in mice some channel subunits within this same family are acid sensitive (DRASIC/ASIC3) and contribute to the responses of some nociceptive neurons both to acid and to noxious pinch (Price, M. P. *et al.*, 2001). Another family of channels, the transient receptor potential (TRP) superfamily, has demonstrated its importance for nociceptive transduction in both vertebrates and invertebrates. TRPV channels in *C. elegans* are required for nociceptive, osmosensory, mechanosensory, and some chemosensory responses (Tobin, D. M. and Bargmann, C. I., 2004). In mammals, TRPV1 channels in sensory neurons are activated by heat, acid, capsaicin (the pungent vanilloid ingredient of chili peppers), and other chemicals (Krause, J. E. *et al.*, 2005). Mice lacking TRPV1 channels show impaired sensory and behavioral responses to heat, vanilloids, and inflammation (Caterina, M. J. *et al.*, 2000). TRP channels also transduce noxious stimuli in the fly, *Drosophila*; mutations of the painless gene (which encodes a TRPA family member) nearly eliminate nociceptive behavioral responses of larvae to noxious heat or mechanical stimuli (Tracey, W. D., Jr., *et al.*, 2003). TRPA family channels are found in vertebrate nociceptors as well, where they are activated by noxious cold, various pungent compounds, and bradykinin (Bandell, M. *et al.*, 2004). These recent results show that nociceptive transduction in at least three different phyla makes use of shared families of ion channels, notably TRPV and TRPA channels, and perhaps DEG/ENaC channels. However, these channel families are not specific to nociception; each family also has members that are involved in innocuous mechanosensation (including hearing in *Drosophila*) and various forms of chemosensation (Tobin, D. M. and Bargmann, C. I., 2004). Thus, nociception has exploited a number of ancient and versatile ion channel families that have also been used during evolution for additional forms of sensory transduction.

15.2.1.3 *Nociceptor-specific gene expression*

Although evolution is opportunistic and tends to find multiple uses for individual gene products, some remain functionally specialized. Several genes have been found to be expressed relatively selectively in mammalian nociceptors, suggesting specialized functions related to nociceptive processing. These include the TTX-resistant, voltage-gated Na^+ channels, NaV1.8 and NaV1.9, which show alterations after nerve injury that are likely to contribute to neuropathic pain (Lai, J. *et al.*, 2004). They also include a family of receptors called Mas-related genes (MRG) (Dong, X. *et al.*, 2001) or sensory neuron-specific G-protein-coupled receptors (SNSR) (Lembo, P. M. et al., 2002). These sensory neuron receptors have displayed exceptionally rapid recent evolution in some mammals (Choi, S. S. and Lahn, B. T., 2003), and might have first appeared in the chordate lineage. Rapid changes within the MRG family are illustrated dramatically by the variation in nociceptor-specific MRG genes within a single mammalian order; mice have 22 MRGA and 14 MRGC genes, whereas gerbils, and rats only have one of each (Zylka, M. J. *et al.*, 2003). Only six of the 13 MRG genes in humans are clearly orthologous (displaying sequence homology across species) with mice MRG genes. Little is known about the functions of these receptors, although the diversity of ligands implicated for them include a variety of RFamide peptides thought to be antinociceptive (Dong, X. *et al.*, 2001 and see below). The recency and rapidity of change in the MRG family suggests that strong selection pressures may be acting on antinociceptive modulation of nociceptors in vertebrates, including primates (Choi, S. S. and Lahn, B. T., 2003; Zylka, M. J. *et al.*, 2003; Zhang, L. *et al.*, 2005). Further investigation of the distribution – across diverse taxa – of genes found to be preferentially expressed in mammalian nociceptors should provide interesting insights into the evolution of nociceptive systems.

15.2.2 Nociceptive Networks and Antinociceptive Modulation

The capability for rapid detection of noxious stimuli could only have been selected during evolution if systems to translate this information into adaptive behavior had also been selected. Almost every animal that has been examined, from jellyfish to humans, displays withdrawal and/or escape responses to noxious stimulation (Walters, E. T., 1994), although these responses are inhibited under some circumstances (see further ahead).

15.2.2.1 *Common nociceptive functions mediated by diverse neural circuits*

Nociceptive responses are mediated by neural circuits that range enormously in complexity across different animals. In cnidarians (hydra, jellyfish, sea anemones), some simple withdrawal reflexes may be mediated in part by just a two-cell pathway involving a multifunction sensory-motor neuron synaptically connected to epithelial-muscular cells. However, even in cnidarians, which lack a brain or central ganglia, relatively complex withdrawal and escape responses can be mediated by a neural net containing sensory, motor, and interneurons (e.g., Mackie, G. O., 2004). At the other end of the behavioral spectrum, nociceptive circuits in mammals involve not only diverse types of sensory, motor, and interneurons at the spinal level, but vast numbers of neurons in higher brain structures (e.g., Craig, A. D., 2003). Different populations of neurons process nociceptive sensory information, generate defensive motor patterns, and modulate nociceptive responses depending upon the state of the organism and its previous experience. These basic nociceptive/defensive tasks are also mediated by functionally similar, but much simpler, neural circuits in several invertebrate species that have been investigated, including *C. elegans* (Wittenburg, N. and Baumeister, R., 1999), the medicinal leech (Shaw, B. K. and Kristan, W. B., Jr., 1995), and several gastropod mollusks – *Aplysia* (Cleary, L. J. et al., 1995), *Tritonia* (Getting, P. A., 1977), and *Helix* (Balaban, P. M., 2002). Although the organization of these circuits differs greatly among the mammalian and invertebrate species examined, common mechanisms of nociceptive plasticity appear to be shared by these diverse animals (see further ahead).

15.2.2.2 Widespread occurrence of antinociceptive mechanisms

One of the most important functions of nociceptive circuits is to modulate nociceptive responses, so that the threshold, selection, and intensity of defensive responses are well matched to the nature and context of a noxious threat. The form of modulation that has received the most experimental attention in nociceptive systems is antinociception, which produces analgesia or hypoalgesia in humans. Many authors have argued that it is adaptive for an animal to inhibit responses, including pain and recuperative responses evoked by injury, during active defense in life-threatening encounters with predators or aggressive conspecifics (see Walters, E. T., 1994). Indeed, most animals examined thus far, including various invertebrates, transiently suppress appetitive, recuperative, and other nondefensive behaviors that might compete with urgent fight-or-flight responses evoked immediately by noxious stimuli. In mammals (Light, A. R., 1992) and at least one invertebrate, *Aplysia* (Cleary, L. J. et al., 1995), inhibitory effects of noxious stimulation are exerted at multiple stages within sensory-motor pathways. Strong pre- and/or postsynaptic inhibition during noxious stimulation occurs at the first synapse in these pathways (from somatic primary afferent neurons) in various species, including mammals (Light, A. R., 1992) and *Aplysia* (Mackey, S. L. et al., 1987).

15.2.2.3 Recent evolution of opiate systems

There has been much interest in the phylogenetic distribution of antinociceptive neuropeptides, and especially the opioid family because of its well-known analgesic functions in humans. Pharmacological, immunocytochemical, and biochemical studies have suggested that opiate-like substances and opiate receptors occur in nervous and immune systems of diverse species, including invertebrates and even unicellular organisms (Stefano, G. B. et al., 2003). However, genes within the opioid/orphanin family have not been detected in the genomes of *C. elegans* or *Drosophila* (nor have genes within the corresponding melanocortin receptor family), which supports the hypothesis that this gene family first appeared in a deuterostome ancestor around the time that the chordate lineage diverged (Dores, R. M. et al., 2002) (Figure 1). The only invertebrate opioids that have been sequenced are proteins related to pro-opiomelanocortin (POMC), which have been found in immunocytes in a leech (Salzet, M. et al., 1997) and in the mussel, *Mytilus edulis* (Stefano, G. B. et al., 1999). Paradoxical features of these POMC sequences led to the suggestion that the vertebrate-like POMC in these invertebrates may have come somehow by cross-phylum gene transfer (Dores, R. M. et al., 2002), and these unusual features reinforce indications that antinociceptive opioids may be a chordate innovation. In contrast, another neuropeptide family having antinociceptive functions is widespread in the animal kingdom: the RFamide family which is expressed in many phyla, including cnidarians, nematodes, arthropods, mollusks, and chordates (Li, C. et al., 1999; Nichols, R., 2003). In mammals some RFamides activate MRG receptors (Dong, X. et al., 2001) (and see above), while in *Aplysia* FMRFamide can produce both short-term and long-term inhibition of synaptic transmission from nociceptive sensory neurons (Montarolo, P. G. et al., 1988). However, RFamides also have pronociceptive actions (Yudin, Y. K. et al., 2004) as well as many actions unrelated to nociceptive modulation.

15.2.3 Nociceptive Plasticity and Sensitization

As just discussed, many animals show a transient reduction in responsiveness to somatic stimuli during active defensive behavior. However, if an animal is injured and escapes, it often displays enhanced responsiveness to somatosensory stimuli while it recuperates. This nociceptive sensitization (and hyperalgesia in some species) is targeted to the injured region and can be very long lasting. It facilitates defensive responses, helping to protect an injured animal during a period when it may be weakened and emitting wound-related signals that can attract the attention of predators, parasites, and aggressive competitors (Walters, E. T., 1994). The persistence of nociceptive sensitization during recuperation periods that can last weeks or longer raises the possibility that nociceptive sensitization may utilize mechanisms that are shared with other long-lasting neural alterations, such as learning and memory. Demonstrations that the same plasticity mechanisms contribute both to nociceptive sensitization and to learning and memory, and evidence that these mechanisms have common functions in very distantly related animal species, indicate that evolutionary links between injury and memory may be quite primitive (Walters, E. T., 1994; Weragoda, R. M. *et al.*, 2004). Of several apparent commonalities between injury-induced and learning-related mechanisms in diverse phyla, three will be considered here.

15.2.3.1 N-Methyl-D-asparate-receptor-dependent long-term synaptic potentiation

As in mammals (Ji, R. R. *et al.*, 2003), somatosensory afferents in several intensively studied invertebrate species appear to release glutamate, which acts upon both NMDA and non-NMDA postsynaptic receptors. These invertebrates include *C. elegans* (Brockie, P. J. *et al.*, 2001), *Aplysia* (Dale, N. and Kandel, E. R., 1993), *Helix* (another snail) (Bravarenko, N. I. *et al.*, 2003), and the leech (Burrell, B. D. and Sahley, C. L., 2004). In mammals (Ji, R. R. *et al.*, 2003), *Aplysia* (Lin, X. Y. and Glanzman, D. L., 1994), *Helix* (Nikitin, V. P. *et al.*, 2002), and the leech (Burrell, B. D. and Sahley, C. L., 2004), intense activation of a sensory pathway (which under normal conditions would occur during noxious stimulation of the periphery) strongly depolarizes second-order neurons and activates NMDA receptors, leading to LTP in the nociceptive pathway, which contributes to nociceptive sensitization. As in mammals (Martin, S. J. *et al.*, 2000), in *Aplysia* (Murphy, G. G. and Glanzman, D. L., 1997), *Helix* (Nikitin, V. P. *et al.*, 2002), and the leech (Burrell, B. D. and Sahley, C. L., 2004), NMDA-receptor-dependent LTP has been implicated as a mechanism of learning and memory.

15.2.3.2 Activation of the transcription factor, Ca^{2+} and cAMP Response Element-Binding Protein

A second commonality is a role for a transcription factor, the Ca^{2+} and cAMP response element-binding protein (CREB), in inducing long-term neuronal alterations that require changes in gene transcription and protein synthesis. A key role for this transcription factor was first implicated in long-term memory formation in *Aplysia*, *Drosophila*, mice, and rats

(Silva, A. J. *et al.*, 1998; Kandel, E. R., 2001). Later, hyperalgesia was found to be accompanied by CREB phosphorylation in the rat spinal cord (Ji, R. R. and Rupp, F., 1997), and interference with CREB binding to DNA was shown to attenuate sensitized behavioral responses in a rat neuropathic pain model (Ma, W. et al., 2003). Interestingly, the sensory neurons from *Aplysia* originally used to demonstrate possible roles of CREB in long-term memory formation (Kandel, E. R., 2001) are nociceptive and exhibit long-term hyperexcitability following noxious cutaneous stimulation, and this hyperexcitability is blocked by injection of CRE oligonucleotides to prevent CREB binding (Lewin, M. R. and Walters, E. T., 1999). Cellular signals involved in activating CREB are also common to both memory and nociceptive sensitization/hyperalgesia in various phyla. Particular attention has been paid to the roles of PKA and MAPK (ERK) in CREB-related plasticity in mammalian hippocampus and forebrain (Waltereit, R. and Weller, M., 2003), mammalian spinal cord (Kawasaki, Y. *et al.*, 2004), and *Aplysia* sensory neurons (Kandel, E. R., 2001).

15.2.3.3 Local protein synthesis and primitive memory mechanisms

The observations mentioned above suggest that evolution made use of common basic mechanisms, including NMDA-receptor-dependent LTP and CREB-dependent transcriptional regulation, to induce long-lasting neuronal alterations underlying quite different functions: nociceptive sensitization and learning. Because strong selection pressures would have favored adaptive responses to peripheral injury at the earliest stages of animal evolution (prior to the evolution of brains complex enough for associative learning), it has been suggested that adaptive responses to peripheral injury, including nociceptive sensitization, emerged first as an adaptation to tissue damage and later evolved into mechanisms of learning and memory (Walters, E. T., 1994). In ancient, soft-bodied animals that were ancestral to most contemporary animals (Figure 1), some of these mechanisms may have functioned initially at sites of peripheral injury in axons and terminals. A vestige of early memory-like mechanisms linked to peripheral injury might be represented today by localized, long-term hyperexcitability of peripheral axon segments of nociceptive sensory neurons found in *Aplysia* (Weragoda, R. M. *et al.*, 2004). This sensitizing axonal alteration is induced by nerve crush or local depolarization and, like the late phase of LTP in mammalian hippocampus (Tang, S. J. *et al.*, 2002), it requires local, rapamycin-sensitive protein synthesis (Weragoda, R. M. *et al.*, 2004). It has not yet been determined whether this apparently primitive mechanism contributes to neuropathic pain in mammals, but evidence for intra-axonal protein synthesis following peripheral nerve injury has been found in axons of mammalian DRG neurons (e.g. Zheng, J. Q. *et al.*, 2001).

15.3 Evolution of Emotional Components of Pain

Perhaps the most commonly asked question related to the evolution of pain is one that has yet to be answered by scientists: Which animals feel pain? This question is critical for animal welfare issues, and is also of considerable philosophical interest (Allen, C., 2004). Many philosophers of science have argued that a human observer cannot know for sure whether another being or species has subjective, conscious experiences such as pain. The existence of emotional pain in other species may defy proof, but evolutionary and neurobiological considerations indicate strongly that these components of pain did not emerge *de novo* in humans, while also suggesting that the capacity to experience subjective pain is not a universal animal trait.

15.3.1 Emotional Awareness is not Required for Nociceptive Responses

As discussed above, most animals have nociceptive systems that detect noxious stimuli and activate defensive responses. Nociceptive reflexes and even nociceptive plasticity, however, can almost certainly operate without conscious, emotional experience because these responses are expressed not only in the simplest animals but also in greatly simplified physiological preparations, such as a spinalized rat or an *Aplysia* ganglion connected to part of the animal's tail. Moreover, in human patients nociceptive reflexes can be expressed without conscious awareness below the level of a spinal transection.

15.3.2 Even Snails Display Complex Motivational Effects after Noxious Stimulation

In addition to the abilities to detect noxious stimuli and activate defensive responses, adaptations to actual or incipient tissue damage include motivational

responses, and these too can be observed throughout the animal kingdom. Pain motivates humans and other mammals to avoid contexts or actions associated with the onset of the pain, and similar motivational effects of noxious stimulation (often electric shock that activates nociceptors) have been used experimentally to produce aversive learning in a wide variety of invertebrates, including flies, worms, and snails (reviewed by Krasne, F. B. and Glanzman, D. L., 1995; Waddell, S. and Quinn, W. G., 2001; Rankin, C. H., 2004). Some of these motivational effects appear remarkably similar to those produced by noxious stimuli in mammals. For example, pairing an odor with strong cutaneous shock causes the previously neutral odor to evoke a conditioned fear-like state in *Aplysia*, so that the odor comes to evoke a freezing response, inhibit appetitive responses, and facilitate defensive responses to tactile stimuli (Walters, E. T. *et al.*, 1981).

15.3.3 Cognitive and Emotional Responses to Noxious Stimulation Differ across Species

Although relatively simple invertebrates show functional similarities to vertebrates (sometimes with conserved molecular mechanisms) in detecting noxious stimuli, activating defensive behavior, and motivating learned and unlearned responses to a noxious event, vast differences in the size and structure of their brains indicate that major differences exist between most invertebrates and many vertebrates in the cognitive and possible emotional processing of information about tissue damage. For example, in *Aplysia*, the body surface is represented by a somatotopic map formed by nociceptor somata in central ganglia (Walters, E. T. *et al.*, 2004), and peripheral injury causes nociceptor soma hyperexcitability selectively in nociceptors representing the injured region. These and other sensory changes are sufficient to alter reflexive responses to stimuli received by the injured region, but they do not cause these normally silent sensory neurons to become spontaneously active. If one assumes that ongoing pain requires ongoing neural activity somewhere in a nervous system, then an absence of ongoing activity may be taken as evidence that tissue damage fails to produce long-lasting, pain-like states. In relatively simple invertebrates, one can systematically sample the activity of individual neurons and use optical imaging methods to sample the activity of populations of neurons to test this possibility following strong noxious stimulation. Although many more neurons need to be sampled before firm conclusions can be made, preliminary observations in *Aplysia* suggest that a persistent (lasting days or longer) increase in ongoing neural activity within the CNS does not accompany long-term, nociceptive sensitization of defensive reflexes. Instead, sensitization lasting days or weeks appears to be maintained largely by synaptic facilitation and hyperexcitability of sensory neurons, which remain silent in the absence of peripheral stimulation. An apparent lack of ongoing activity in the *Aplysia* CNS would contrast with increases in neural activity that probably occur in some regions of the human brain during persistent pain states (Casey, K. L., 2000).

Investigating the evolution of emotional aspects of pain is an enormous challenge because the biological functions of conscious pain (as opposed to unconscious nociceptive responses) are not obvious and the neural substrates are very difficult to define. Nevertheless, there are many intriguing observations in these areas. For example, much research has been stimulated by the suggestion that conscious awareness of an animal's own bodily states might help it predict the behavioral states (read the mind) of other animals (Keenan, J. et al., 2003). Conscious awareness of bodily states, including the emotional awareness of pain, involves dynamic neural representations of the body (e.g. Craig, A. D., 2003). Even the most complex brains of invertebrates, such as the octopus or honeybee, as well as the brains of non-mammalian vertebrates, have much smaller numbers of neurons and less obviously organized brain structures than found in behaviorally sophisticated mammals such as primates and cetaceans, and thus presumably less capacity for such representations. Indeed, extremely complex neuroanatomical substrates support the multiple representations and meta-representations of the body that are associated with emotional awareness and suffering in humans, which has suggested to some that these aspects of pain may not occur in animals lacking, for example, an equivalent to the anterior insular cortex of primates (Craig, A. D., 2003). Although direct measurement of subjective pain in animals is not possible, clues about the evolution of this capacity may still come from comparing across a broad range of species possible functions of emotional awareness as well as the neural substrates critical for expressing signs of such awareness.

15.4 Conclusions

Nociceptive mechanisms that permit an animal to detect, respond to, and remember peripheral injury are found throughout the animal kingdom. Some of these are likely to represent convergent (analogous) evolutionary responses to similar injury-related selection pressures in diverse organisms, which may depend upon different molecular substrates in different animal groups. For example, while most animals show transient, antinociceptive responses to noxious stimulation, the use of opiates and melanocortin receptors for antinociception may largely be restricted to vertebrates. On the other hand, emerging molecular genetic information indicates that other nociceptive mechanisms have been highly conserved from mechanisms that must have already been present around 600 million years ago in worm-like animals that were the last common ancestors of modern roundworms, flies, leeches, snails, mice, and humans (Figure 1). Conserved (homologous) nociceptive mechanisms that have been revealed thus far are related to nociceptive transduction (e.g., TRP channels), plasticity of nociceptor synapses (e.g., NMDA-receptor-dependent LTP), and regulation of gene transcription in nociceptive pathways (e.g., CREB activation). Because of dramatic increases in molecular genetic information from all the major phyla and the coupling of this information with continuing physiological studies of various species, it seems likely that knowledge gained from investigating highly conserved mechanisms in disparate creatures will provide insights that might help control intransigent pain states in humans. Comparative studies may also shed some light on the selection pressures that shaped the evolution of emotional aspects of pain, even though the subjective experience of pain may not be amenable to direct experimental investigation in other species.

References

Allen, C. 2004. Animal pain. Nous 38, 617–643.
Balaban, P. M. 2002. Cellular mechanisms of behavioral plasticity in terrestrial snail. Neurosci. Biobehav. Rev. 26, 597–630.
Bandell, M., Story, G. M., Hwang, S. W., Viswanath, V., Eid, S. R., Petrus, M. J., Earley, T. J., and Patapoutian, A. 2004. Noxious cold ion channel TRPA1 is activated by pungent compounds and Bradykinin. Neuron 41, 849–57.
Bravarenko, N. I., Korshunova, T. A., Malyshev, A. Y., and Balaban, P. M. 2003. Synaptic contact between mechanosensory neuron and withdrawal interneuron in

terrestrial snail is mediated by L-Glutamate-like transmitter. Neurosci. Lett. 341, 237–240.

Brockie, P. J., Madsen, D. M., Zheng, Y., Mellem, J., and Maricq, A. V. 2001. Differential expression of glutamate receptor subunits in the nervous system of Caenorhabditis elegans and their regulation by the homeodomain protein unc-42. J. Neurosci. 21, 1510–1522.

Burrell, B. D. and Sahley, C. L. 2004. Multiple forms of long-term potentiation and long-term depression converge on a single interneuron in the leech cns. J. Neurosci. 24, 4011–4019.

Casey, K. L. 2000. Concepts of pain mechanisms: the contribution of functional imaging of the human brain. Prog. Brain. Res. 129, 277–287.

Caterina, M. J., Leffler, A., Malmberg, A. B., Martin, W. J., Trafton, J., Petersen-Zeitz, K. R., Koltzenburg, M., Basbaum, A. I., and Julius, D. 2000. Impaired nociception and pain sensation in mice lacking the capsaicin receptor. Science 288, 306–313.

Choi, S. S. and Lahn, B. T. 2003. Adaptive evolution of MRG, a neuron-specific gene family implicated in nociception. Genome Res. 13, 2252–2259.

Christenson, J., Boman, A., Lagerback, P., and Grillner, S. 1988. The dorsal cell, one class of primary sensory neuron in the lamprey spinal cord. i. touch, pressure but no nociception - a physiological study. Brain Res. 440, 1–8.

Cleary, L. J., Byrne, J. H., and Frost, W. N. 1995. Role of interneurons in defensive withdrawal reflexes of *Aplysia*. Learn. Mem. 2, 133–151.

Craig, A. D. 2003. Pain mechanisms: labeled lines versus convergence in central processing. Annu. Rev. Neurosci. 26, 1–30.

Dale, N. and Kandel, E. R. 1993. L-Glutamate may be the fast excitatory transmitter of aplysia sensory neurons. Proc. Natl. Acad. Sci. USA 90, 7163–7167.

Dawkins, R. 2004. The Ancestor's Tale: a Pilgrimage to the Dawn of Evolution, Houghton Mifflin.

Dong, X., Han, S., Zylka, M. J., Simon, M. I., and Anderson, D. J. 2001. A diverse family of gpcrs expressed in specific subsets of nociceptive sensory neurons. Cell 106, 619–632.

Dores, R. M., Lecaude, S., Bauer, D., and Danielson, P. B. 2002. Analyzing the evolution of the opioid/orphanin gene family. Mass Spectrom Rev. 21, 220–243.

Getting, P. A. 1977. Neuronal organization of escape swimming in *Tritonia*. J. Comp. Physiol. 121, 325–342.

Holloway, J. A., Trouth, C. O., Wright, L. E., and Keyser, G. F. 1980. Cutaneous receptive field characteristics of primary afferents and dorsal horn cells in the avian (*Gallus domesticus*). Exp. Neurol. 68, 477–488.

Illich, P. A. and Walters, E. T. 1997. Mechanosensory neurons innervating *Aplysia* siphon encode noxious stimuli and display nociceptive sensitization. J. Neurosci. 17, 459–469.

Ji, R. R. and Rupp, F. 1997. Phosphorylation of transcription factor CREB in rat spinal cord after formalin-induced hyperalgesia: relationship to c-fos induction. J. Neurosci. 17, 1776–1785.

Ji, R. R., Kohno, T., Moore, K. A., and Woolf, C. J. 2003. Central sensitization and ltp: do pain and memory share similar mechanisms? Trends Neurosci. 26, 696–705.

Kandel, E. R. 2001. The molecular biology of memory storage: a dialogue between genes and synapses. science 294, 1030–1038.

Kawasaki, Y., Kohno, T., Zhuang, Z. Y., Brenner, G. J., Wang, H., Van Der Meer, C., Befort, K., Woolf, C. J., and Ji, R. R. 2004. ionotropic and metabotropic receptors, protein kinase a, protein kinase c, and src contribute to c-fiber-induced ERK activation and cAMP response element-binding protein phosphorylation in dorsal horn neurons, leading to central sensitization. J. Neurosci. 24, 8310–8321.

Keenan, J., Gallup, G. G., and Falk, D. 2003. The Face in the Mirror: The Search for the Origins of Consciousness. HarperCollins Publishers.

Krasne, F. B. and Glanzman, D. L. 1995. What we can learn from invertebrate learning. Annu. Rev. Psychol. 46, 585–624.

Krause, J. E., Chenard, B. L., and Cortright, D. N. 2005. Transient receptor potential ion channels as targets for the discovery of pain therapeutics. Curr. Opin. Investig. Drugs 6, 48–57.

Lai, J., Porreca, F., Hunter, J. C., and Gold, M. S. 2004. Voltage-gated sodium channels and hyperalgesia. Annu. Rev. Pharmacol. Toxicol. 44, 371–97.

Lembo, P. M., Grazzini, E., Groblewski, T., O'Donnell, D., Roy, M. O., Zhang, J., Hoffert, C., Cao, J., Schmidt, R., Pelletier, M., Labarre, M., Gosselin, M., Fortin, Y., Banville, D., Shen, S. H., Strom, P., Payza, K., Dray, A., Walker, P., and Ahmad, S. 2002. Proenkephalin a gene products activate a new family of sensory neuron–specific Gpcrs. Nat. Neurosci. 5, 201–209.

Lewin, M. R. and Walters, E. T. 1999. Cyclic GMP pathway is critical for inducing long-term sensitization of nociceptive sensory neurons. Nature Neurosci. 2, 18–23.

Li, C., Kim, K., and Nelson, L. S. 1999. FMRFamide-related neuropeptide gene family in caenorhabditis elegans. Brain Res. 848, 26–34.

Liang, Y. F., Terashima, S., and Zhu, A. Q. 1995. Distinct morphological characteristics of touch, temperature, and mechanical nociceptive neurons in the crotaline trigeminal ganglia. J. Comp. Neurol. 360, 621–633.

Light, A. R. 1992. The Initial Processing of Pain and Its Descending Control: Spinal and Trigeminal Systems Karger.

Lin, X. Y. and Glanzman, D. L. 1994. Hebbian induction of long-term potentiation of *Aplysia* synapses: partial requirement for activation of an NMDA-related receptor. Proc. R. Soc. Lond. B 255, 215–221.

Lynn, B. and O'Shea, N. R. 1998. Inhibition of forskolin-induced sensitisation of frog skin nociceptors by the cyclic AMP-Dependent Protein Kinase A Antagonist H-89. Brain Res. 780, 360–362.

Ma, W., Hatzis, C., and Eisenach, J. C. 2003. Intrathecal injection of cAMP response element binding protein (CREB) antisense oligonucleotide attenuates tactile allodynia caused by partial sciatic nerve ligation. Brain Res. 988, 97–104.

Mackey, S. L., Glanzman, D. L., Small, S. A., Dyke, A. M., Kandel, E. R., and Hawkins, R. D. 1987. Tail shock produces inhibition as well as sensitization of the siphon-withdrawal reflex of Aplysia: possible behavioral role for presynaptic inhibition mediated by the peptide Phe-Met-Arg-Phe-Nh2. Proc. Natl. Acad. Sci. USA 84, 8730–8734.

Mackie, G. O. 2004. Central neural circuitry in the jellyfish aglantha: a model 'simple nervous system'. Neurosignals 13, 5–19.

Malyshev, A. Y. and Balaban, P. M. 2002. Identification of mechanoafferent neurons in terrestrial snail: response properties and synaptic connections. J. Neurophysiol. 87, 2364–2371.

Martin, S. J., Grimwood, P. D., and Morris, R. G. 2000. Synaptic plasticity and memory: an evaluation of the hypothesis. Annu. Rev. Neurosci. 23, 649–711.

Merskey, H. and Bogduk, N. 1994. Classification of Chronic Pain, Second Edition, IASP Task Force on Taxonomy. IASP Press.

Montarolo, P. G., Kandel, E. R., and Schacher, S. 1988. Long-Term Heterosynaptic Inhibition in Aplysia. Nature 333, 171–174.

Murphy, G. G. and Glanzman, D. L. 1997. Mediation of classical conditioning in aplysia californica by long-term potentiation of sensorimotor synapses. Science, 278, 467–471.

Nicholls, J. G. and Baylor, D. A. 1968. Specific modalities and receptive fields of sensory neurons in cns of the leech. J. Neurophysiol. 31, 740–756.

Nichols, R. 2003. Signaling pathways and physiological functions of Drosophila Melanogaster FMRFamide-related Peptides. Annu. Rev. Entomol. 48, 485–503.

Nikitin, V. P., Koryzev, S. A., and Shevelkin, A. V. 2002. Selective effects of an NMDA glutamate receptor antagonist on the sensory input from chemoreceptors in the snail's head during acquisition of nociceptive sensitization. Neurosci. Behav. Physiol. 32, 129–134.

Price, M. P., McIlwrath, S. L., Xie, J., Cheng, C., Qiao, J., Tarr, D. E., Sluka, K. A., Brennan, T. J., Lewin, G. R., and Welsh, M. J. 2001. The DRASIC cation channel contributes to the detection of cutaneous touch and acid stimuli in mice. Neuron 32, 1071–1083.

Rankin, C. H. 2004. Invertebrate learning: what can't a worm learn? Curr. Biol. 14, R617–R618.

Salzet, M., Salzet-Raveillon, B., Cocquerelle, C., Verger-Bocquet, M., Pryor, S. C., Rialas, C. M., Laurent, V., and Stefano, G. B. 1997. Leech immunocytes contain proopiomelanocortin: nitric oxide mediates hemolymph proopiomelanocortin processing. J. Immunol. 159, 5400–5411.

Shaw, B. K. and Kristan, W. B., Jr. 1995. The whole-body shortening reflex of the medicinal leech: motor pattern, sensory basis, and interneuronal pathways. J. Comp. Physiol. A 177, 667–681.

Silva, A. J., Kogan, J. H., Frankland, P. W., and Kida, S. 1998. CREB and memory. Annu. Rev. Neurosci. 21, 127–148.

Sneddon, L. U., Braithwaite, V. A., and Gentle, M. J. 2003. Do fishes have nociceptors? evidence for the evolution of a vertebrate sensory system. Proc. R. Soc. Lond. B. Biol. Sci. 270, 1115–1121.

Stefano, G. B., Salzet-Raveillon, B., and Salzet, M. 1999. Mytilus edulis hemolymph contains pro-opiomelanocortin: LPS and morphine stimulate differential processing. Brain Res. Mol. Brain Res. 63, 340–350.

Stefano, G. B., Cadet, P., Rialas, C. M., Mantione, K., Casares, F., Goumon, Y., and Zhu, W. 2003. Invertebrate opiate immune and neural signaling. Adv. Exp. Med. Biol. 521, 126–147.

Tang, S. J., Reis, G., Kang, H., Gingras, A. C., Sonenberg, N., and Schuman, E. M. 2002. A rapamycin-sensitive signaling pathway contributes to long-term synaptic plasticity in the hippocampus. Proc. Natl. Acad. Sci. USA 99, 467–472.

Tobin, D. M. and Bargmann, C. I. 2004. Invertebrate nociception: behaviors, neurons and molecules. J. Neurobiol. 61, 161–174.

Tracey, W. D., Jr., Wilson, R. I., Laurent, G., and Benzer, S. 2003. Painless, a Drosophila gene essential for nociception. Cell 113, 261–273.

Waddell, S. and Quinn, W. G. 2001. Flies, genes, and learning. Annu. Rev. Neurosci. 24, 1283–1309.

Waltereit, R. and Weller, M. 2003. Signaling from cAMP/PKA to MAPK and synaptic plasticity. Mol. Neurobiol. 27, 99–106.

Walters, E. T. 1994. Injury-related behavior and neuronal plasticity: an evolutionary perspective on sensitization, hyperalgesia and analgesia. Int. Rev. Neurobiol. 36, 325–427.

Walters, E. T., Bodnarova, M., Billy, A. J., Dulin, M. F., Diaz-Rios, M., Miller, M. W., and Moroz, L. L. 2004. Somatotopic organization and functional properties of mechanosensory neurons expressing sensorin-A mRNA in Aplysia californica. J. Comp. Neurol. 471, 219–240.

Walters, E. T., Carew, T. J., and Kandel, E. R. 1981. Associative learning in Aplysia: evidence for conditioned fear in an invertebrate. Science 211, 504–506.

Weragoda, R. M., Ferrer, E., and Walters, E. T. 2004. Memory-like alterations in Aplysia axons after nerve injury or localized depolarization. J. Neurosci. 24, 10393–10401.

Willis, W. D. and Coggeshall, R. E. 2004. Sensory Mechanisms of the Spinal Cord: Primary Afferent Neurons and the Spinal Dorsal Horn. Kluwer Academic/Plenum Publishers.

Wittenburg, N. and Baumeister, R. 1999. Thermal avoidance in Caenorhabditis elegans: an approach to the study of nociception. Proc. Natl. Acad. Sci. U. S. A. 96, 10477–10482.

Yudin, Y. K., Tamarova, Z. A., Ostrovskaya, O. I., Moroz, L. L., and Krishtal, O. A. 2004. RFa-related peptides are algogenic: evidence in vitro and in vivo. Eur. J. Neurosci. 20, 1419–1423.

Zhang, L., Taylor, N., Xie, Y., Ford, R., Johnson, J., Paulsen, J. E., and Bates, D. 2005. Cloning and expression of MRG receptors in macaque, mouse, and human. Brain Res. Mol. Brain Res. 133, 187–197.

Zheng, J. Q., Kelly, T. K., Chang, B., Ryazantsev, S., Rajasekaran, A. K., Martin, K. C., and Twiss, J. L. 2001. A functional role for intra-axonal protein synthesis during axonal regeneration from adult sensory neurons. J. Neurosci. 21, 9291–9303.

Zylka, M. J., Dong, X., Southwell, A. L., and Anderson, D. J. 2003. Atypical expansion in mice of the sensory neuron-specific MRG G protein-coupled receptor family. Proc. Natl. Acad. Sci. USA 100, 10043–10048.

16 Redheads and Pain

J S Mogil, McGill University, Montreal, QC, Canada

Glossary

allele One possible form (or variant) of a gene, defined by its particular nucleotide sequence.

genotype The inherited alleles (one from each parent) at a genetic locus (e.g., a gene).

phenotype An observable or measurable biological trait.

pleiotropy The involvement of a single gene in more than one phenotype

polymorphism see Allele.

16.1 Finding Genes Associated with Phenotypic Variability

With recent developments in classical and molecular genetic techniques, identifying genes, and the DNA sequence variants in or near them, responsible for disease susceptibility and normal variation in biological phenotypes is becoming increasingly tractable. Two main methodologies are employed to this end: genetic linkage mapping and genetic association studies (see Lander, E. S. and Schork, N. J., 1994). These approaches have begun to be applied to the study of pain (see Mogil, J. S., 2004). For example, variable sensitivity of humans to experimental and clinical pain has been provisionally associated with the genes *COMT* (catechol-*O*-methyltransferase) (Zubieta, J. K. *et al.*, 2003; Diatchenko, L. *et al.*, 2005), *OPRD* (δ-opioid receptor; Kim, H. *et al.*, 2004), *TRPV1* (transient receptor potential vanilloid type 1; Kim, H. *et al.*, 2004), *OPRM* (μ-opioid receptor gene) (Klepstad, P. *et al.*, 2004; Fillingim, R. B. *et al.*, 2005), *IL6* (interleukin-6; Noponen-Hietala, N. *et al.*, 2005), *IL1* (interleukin-1; Solovieva, S. *et al.*, 2004), and others. These association studies were inspired in some cases by a mouse linkage mapping finding, in others by the phenotype of a null mutant (transgenic knockout) mouse, and in others simply by the known involvement of the protein in pain processing.

There are a finite number of human genes, current estimates are less than 30 000, but a virtually unlimited number of biological phenotypes. It is thus necessarily true that every gene will eventually be shown to be genetically associated with multiple (and possibly unrelated) traits. This principle is known as pleiotropy. Currently, one of the best-known examples of a gene with pleiotropic effects is *MC1R*, the gene coding for the melanocortin-1 receptor (MC1R).

16.2 MC1R and the Melanocortin-1 Receptor

The melanocortin receptor family consists of five subtypes, MC1R–MC5R. These receptors bind the melanocortin peptides, including α-melanocyte-stimulating hormone (α-MSH), β-MSH, γ-MSH, and adrenocorticotrophic hormone (ACTH), all cleaved from the proopiomelanocortin (*POMC*) gene along with β-endorphin (see Tatro, J. B., 1996; Wikberg, J. E. S., 1999; Abdel-Malek, Z. A., 2001; Starowicz, K. and Przewlocka, B., 2003). The receptors are transcribed from five separate genes, and each receptor has differential tissue distributions and affinities for the melanocortins. All melanocortin receptor subtypes couple positively to adenylate cyclase when activated (Chhajlani, V. and Wikberg, J. E. S., 1992). The MC1R, previously known as the α-MSH receptor, actually has equal affinity for ACTH, and is found primarily in melanocytes of the skin. It is also been demonstrated in a variety of glial cells, the testes, placenta, pituitary (see Tatro, J. B., 1996; Wikberg, J. E. S., 1999; Abdel-Malek, Z. A., 2001; Starowicz, K. and Przewlocka, B., 2003), and in neurons of the midbrain periaqueductal gray (Xia, Y. et al., 1995). The role of α-MSH acting on melanocyte MC1Rs to regulate pigmentation of skin and hair by stimulation of eumelanin synthesis has been understood in great detail for some time (see Eberle, A. N., 1988). Other α-MSH effects are also well known, and have been attributed to actions at the MC1R. For example, α-MSH has potent anti-inflammatory and antipyretic activity (see Tatro, J. B., 1996), and may play a role in nerve regeneration (see Starowicz, K. and Przewlocka, B., 2003) and female sexual behavior (Pfaus, J. G. et al., 2004). Injection of α-MSH (or ACTH) into the brain has been reported to affect nociceptive sensitivity (see Starowicz, K. and Przewlocka, B., 2003), usually in a manner opposed to the actions of endogenous opioids, although the literature is contradictory. It is possible that some or all of these effects are mediated at spinal MC4Rs, since the MC4R-selective antagonist, SHU9119, reduces neuropathic hypersensitivity states when injected into the spinal cord (Vrinten, D. H. et al., 2001).

Given the known role of MC1Rs in pigmentation and the highly polymorphic nature of *MC1R*, the hypothesis that the *MC1R* gene might be associated with variability in human hair color and skin tone was palpable. Indeed, Valverde P. et al. (1995) demonstrated that the inheritance of loss-of-function *MC1R* alleles was genetically associated with red hair and fair skin in humans. Subsequent work has determined that approximately 60–85% of natural redheads are so because of their inheritance of variant alleles of *MC1R*, especially those producing the very common Arg151Cys, Arg160Trp, and Asp294His amino acid substitutions (see Schaffer, J. V. and Bolognia, J. L., 2001). Red hair is much more likely if multiple alleles are inherited, and the number of variant alleles is important in determining the precise shade of red. In addition, it has been subsequently demonstrated that *MC1R* is the major gene responsible for solar lentigines (i.e., freckles; Bastiaens, M. et al., 2001a), reduced tanning ability (Healy, E. et al., 2000), and skin cancer (malignant melanoma, squamous cell, and basal cell carcinoma) susceptibility even in individuals without red hair (Palmer, J. S. et al., 2000; Bastiaens, M. et al., 2001b). It should be noted that *MC1R* genetics is rather complex, with a large number of known rare (<1% allele frequency) mutations and at least three common variants (Val60Leu, Val92Met, and Arg163Gln) that are only partially penetrant or completely silent, producing no impairment of MC1R functioning (see Schaffer, J. V. and Bolognia, J. L., 2001).

16.3 MC1R and Analgesia

Using genetic linkage (formally, quantitative trait locus or QTL) mapping, a gene affecting the magnitude of swim stress-induced analgesia (SIA) in female but not male mice was localized to distal mouse chromosome 8 (Mogil, J. S. et al., 1997). This SIA was previously demonstrated to be sex-specific in its neurochemical mediation (e.g., Mogil, J. S. et al., 1993; Kavaliers, M. and Galea, L. A. M., 1995). Specifically, pharmacological blockade of the N-methyl-D-aspartate (NMDA) excitatory amino acid receptor abolishes SIA in male rodents, but does not do so (at any dose or in any estrous phase) in females. Kavaliers M. and Choleris E. (1997) found that analgesia from a κ-opioid-selective agonist, NPC12626, was similarly sexually dimorphic in its neurochemical mediation, being blocked by NMDA antagonists only in males. The possible sex-specificity of κ-opioid analgesia was then a hot topic, as data from Gear R. W. et al. (1996) suggested that κ-opioid-acting drugs may only produce efficacious analgesia, in the third molar extraction model, in women. Mogil J. S. et al. (2003) hypothesized that if SIA and κ-opioid analgesia were both sexually dimorphic, and a female-specific linkage on chromosome 8 had been already demonstrated for SIA, one would likely be seen for κ-opioid analgesia as well. This was quickly confirmed using the κ-opioid-selective compound, U50,488 (Mogil, J. S. et al., 2003), but the confidence interval containing the gene(s) responsible for the linkage was quite large. Of the $\sim$200 murine genes in this region, however, one emerged as a high-probability candidate: *Mc1r*. Evidence for its candidacy included the possible role of α-MSH and/or ACTH as antiopioid peptides (see above) and the demonstration of neuronal MC1Rs in the periaqueductal gray (Xia, Y. et al., 1995), a supraspinal region of well-known importance in pain modulation (see Basbaum, A. I. and Fields, H. L., 1984). It should also be noted that dynorphin can bind with reasonable affinity in *in vitro* assays to MC1R (Quillan, J. M. and Sadee, W., 1997).

Using a three-pronged strategy, Mogil J. S. and colleagues have provided considerable supporting evidence that the mouse *Mc1r* gene is indeed the female-specific QTL for U50,488 analgesia. First, null mutant recessive yellow (C57BL/6-*Mc1r*$^{e/e}$, or *e/e*) mice, which arose spontaneously rather than being engineered using transgenic technology (see Cone, R. D. *et al.*, 1996), were tested for U50,488 analgesia along with their C57BL/6 wild-type controls. We found that female but not male *e/e* mutants displayed significantly greater U50,488 analgesia (see Figure 1(a)),

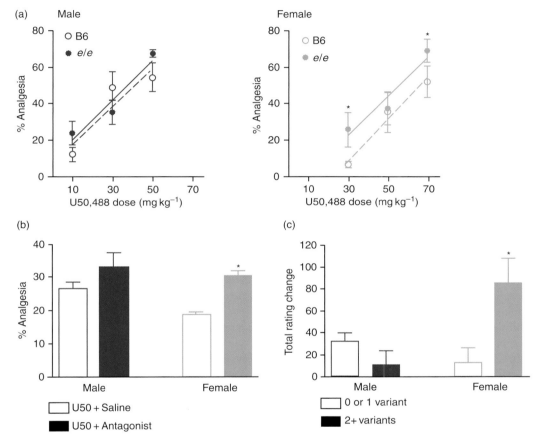

Figure 1 Female-specific effect of MC1R functional status on κ-opioid analgesia in mice and humans. (a) Response on the 49 °C tail-withdrawal test to systemic injection of various doses (10–70 mg kg^{-1}, subcutaneous (s.c.)) of the selective κ-opioid agonist, U50,488, in male and female spontaneous mutant recessive yellow (*e/e*) mice and their wild-type counterparts, C57BL/6 (B6). The *e/e* mutants of both sexes have nonfunctional MC1Rs due to a frameshift mutation at position 183. Symbols represent mean ± standard error of the mean (SEM) percent of total analgesia (based on area under the time–effect curve) over a 60 min testing period postdrug administration. (b) Female-specific potentiation of U50,488 (U50) analgesia (70 mg kg^{-1}, s.c.), on the 49 °C tail-withdrawal test, by injection of a peptidergic MC1R antagonist, Ac-Nle-Asp-Trp-DPhe-Nle-Trp-Lys-NH$_2$ (20 µg/mouse in 2.5 µl), into the lateral ventricles of outbred mice. Bars represent mean ± SEM percent of total analgesia (based on area under the time–effect curve) over a 60-min testing period postdrug administration. (c) Female-specific influence of *MC1R* genotype on pentazocine (0.5 mg kg^{-1}, intravenous) analgesia in humans. Subjects of both sexes were studied in two experimental sessions as described in detail in Mogil J. S. *et al.* (2003). Bars represent mean ± SEM difference scores for summed pain intensity ratings (0–100; 5 ×) during the ischemic pain procedure, computed by subtracting postdrug from predrug ratings, in subjects with consensus sequence or one *MC1R* variant (0 or 1 variant) and subjects with two or more *MC1R* variants (2+ variants). The 2+ variant subjects possessed nonfunctional MC1Rs, and were unanimously redheaded. $^{*}P < 0.05$ compared to other condition within-sex. Adapted from Mogil, J. S. *et al.* 2003. The melanocortin-1 receptor gene mediates female-specific mechanisms of analgesia in mice and humans. Proc. Natl. Acad. Sci. U. S. A. 100, 4867–4872.

and perhaps more importantly, that this analgesia was sensitive to NMDA blockade in the female mutants, suggesting that they had switched over to the malelike system (Mogil, J. S. *et al.*, 2003). This switching has been observed before and since with other hormonal disruptions of the female system (Mogil, J. S. *et al.*, 1993; Kavaliers, M. and Galea, L. A. M., 1995; Sternberg, W. F. *et al.*, 2004). Second, injection of a selective peptidergic MC1R antagonist, Ac-Nle-Asp-Trp-DPhe-Nle-Trp-Lys-NH$_2$, into the cerebral ventricles of (outbred) male and female mice produced a potentiation of U50,488 analgesia only in females (see Figure 1(b)), and furthermore rendered that analgesia newly sensitive to inhibition by an NMDA antagonist (Mogil, J. S. *et al.*, 2003). Finally, Austin and Mogil have demonstrated sex and strain differences in periaqueductal gray expression of the *Mc1r* gene using quantitative polymerase chain reaction (J. S. Austin and J. S. Mogil, unpublished data) in directions entirely consistent with the hypothesis that lower MC1R functionality is associated with higher κ-opioid analgesia.

These data, supportive of the sex-specific involvement of *Mc1r* in κ-opioid analgesia in the mouse, inspired a small association study in humans designed to test the same hypothesis in our species. College-age subjects were tested for thermal and ischemic pain before and after administration of pentazocine (Talwin®; 0.5 mg kg^{-1}) or saline, and then genotyped at the human *MC1R* gene by direct sequencing of exons containing common variants. We found that pentazocine analgesia was modest in all male subjects, and also in women with functional MC1Rs (i.e., those with the consensus DNA sequence or only one variant). Women with nonfunctional MC1Rs (i.e., those with two or more variants; all phenotypic redheads) displayed robust pentazocine analgesia (Mogil, J. S. *et al.*, 2003; see Figure 1(c)).

These data represent the first time a genetic finding in the mouse has ever been directly translated to humans within a single published study. More importantly, though, the data are the first to suggest that, like the mouse, humans possess qualitatively (i.e., neurochemically) distinct mechanisms of pain modulation, since the functional status of the MC1R affected analgesic output in one sex but not the other. A follow-up study was performed, in order to determine whether *Mc1r* and *MC1R* play a role in μ-opioid analgesia in mice and humans, respectively. μ-Opioid analgesia is of much greater clinical relevance than κ-opioid analgesia due to the high μ-selectivity of most clinically used opiate drugs. In addition to differences in pain sensitivity *per se* (see

Section 16.5 below), we found that *e/e* mice were significantly more sensitive to analgesia from both morphine and its bioactive metabolite, morphine-6β-glucuronide (Mogil, J. S. *et al.*, 2005a). Curiously, the effect of MC1R functional status on μ-opioid analgesia did not appear to be sex-specific, with both male and female mutants showing increased sensitivity (see Figure 2(b)). Again, data collected in humans were entirely analogous, with redheads of both sexes showing robustly higher morphine-6β-glucuronide inhibition of noxious electrical stimuli (Mogil, J. S. *et al.*, 2005a; see Figure 2(d)).

16.4 MC1R and Anesthesia

It has long been clinical lore among anesthesiologists that redheads are harder to anesthetize with inhalant anesthetics than others, but these anecdotal impressions were never tested experimentally until recently. Liem E. B. *et al.* (2004) recruited 20 young adult women (10 with red hair, 10 with dark hair), and studied the minimum alveolar concentration of desflurane required to prevent movement in response to intradermal electrical stimulation of the anterior thigh. They found that red-haired women required almost 20% more anesthetic to abolish responding. This finding regarding anesthetic sensitivity cannot easily be reconciled with the aforementioned findings regarding analgesia. If the greater response of *MC1R* mutants (i.e., redheads) to κ- and μ-opioid analgesics is reflective of higher endogenous activity of a tonically active analgesic system, one might have expected red-haired women to require less anesthetic, not more. Unfortunately, only women were tested in the study by Liem E. B. *et al.* (2004), so sex differences could not be evaluated. In apparent support of this finding, this same group tested *e/e* mutants for sensitivity to four inhaled anesthetic compounds (isoflurane, sevoflurane, desflurane, and halothane), and found that although mutant mice were not significantly less sensitive to any one anesthetic compared to controls, when data from all anesthetics were combined the genotypic difference achieved statistical significance ($P = 0.023$; Xing, Y. *et al.*, 2004).

16.5 MC1R and Pain

One limitation of the study by Liem E. B. *et al.* (2004) is that no attempt was made to study nociceptive sensitivity of the subjects in the absence of anesthetic, so it was impossible to tell whether the differences observed involved differential sensitivity among genotypes to the anesthetic, or differential sensitivity to the noxious stimulus itself. It has recently been demonstrated in mice that the latter can easily be confused for the former (Mogil, J. S. *et al.*, 2005b). In a recently published follow-up study by their group (Liem, E. B. *et al.*, 2005), women (30 with red hair, 30 with dark hair) were tested for sensory thresholds, pain thresholds, and pain tolerance to electrical and thermal stimulation. No differences were seen in sensitivity to electrical stimulation, but heat and cold pain perception was higher in redheads. Redheads were also found to be resistant to subcutaneous (but not liposomal) lidocaine anesthesia against electrical stimulation (Liem, E. B. *et al.*, 2005).

These data are in contrast with an earlier study, showing no *MC1R* genotypic differences in thermal pain sensitivity (Mogil, J. S. *et al.*, 2003), although that study was probably too lightly powered to assess the issue. However, the data of Liem E. B. *et al.* (2005) are also entirely contradictory with a larger follow-up study (Mogil, J. S. *et al.*, 2005a), which observed in *e/e* mutant mice of both sexes (across a broad range of noxious stimulus modalities) and humans of both sexes a decreased sensitivity to noxious electrical stimulation (see Figures 2(a) and 2(b)). There are a number of possible reasons for the discrepancy. Methodological differences can be found between the three experiments, of course. It is difficult to blind experimenters to hair color, and it is possible that this lack of blinding was a greater issue in one

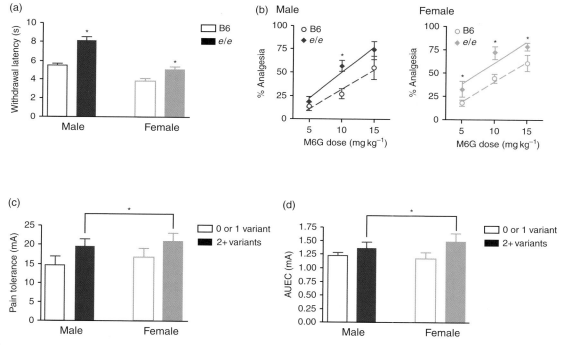

Figure 2 Effect of MC1R functional status on pain sensitivity and μ-opioid analgesia in mice and humans of both sexes. (a) Sensitivity of male and female spontaneous mutant recessive yellow (*e/e*) mice and their wildtype counterparts, C57BL/6 (B6), on the thermal paw-withdrawal test. Bars represent mean $\pm$ standard error of the mean (SEM) latency (in s) to withdraw from a radiant heat stimulus positioned under the ventral hind paws (right and left hind paw data combined). (b) Response of B6 and *e/e* mutants on the 49 °C tail-withdrawal test to systemic injection of various doses (5–15 mg kg^{-1}, subcutaneous) of the selective μ-opioid agonist (and morphine metabolite), morphine-6β-glucuronide (M6G). Symbols represent mean $\pm$ SEM percent of total analgesia (based on area under the time–effect curve) over a 300-min testing period postdrug administration. (c) Baseline tolerance to electrical current (10 Hz; 0.1 ms; increased by 0.5-mA steps) applied to the skin in men and women with functional (0 or 1 variant) or nonfunctional (2+ variants) MC1Rs. Bars represent mean $\pm$ SEM tolerated stimulus intensity. (d) Influence of *MC1R* genotype on M6G analgesia (0.3 mg kg^{-1}, intravenous) in men and women. Subjects were studied as described in detail in Mogil J. S. *et al.* (2005a). Bars represent mean $\pm$ SEM analgesia as expressed as area under the time–effect curve (AUEC relative to baseline current shown in (c)). *P < 0.05 compared to other genotype. Adapted from Mogil, J. S. *et al.* 2005a. Melanocortin-1 receptor gene variants affect pain and μ-opioid analgesia in mice and humans. *J. Med. Genet.* 42, 583–587.

study than another. Liem E. B. and colleagues studied women only, whereas Mogil J. S. and colleagues studied men and women. Most importantly, perhaps, Liem E. B. and colleagues compared phenotypes (women with dark or red hair) whereas Mogil J. S. and colleagues compared *MC1R* genotypes (subjects with functional or nonfunctional MC1Rs). It is not the case that all redheads are so because of inheritance of *MC1R* variants (see Section 16.2 above); that needs to be determined empirically by sequencing. Thus it is possible that some subjects in Liem E. B. *et al.* (2005) were misclassified in their inferred genotype.

Finally, a human genetic association study has revealed a possible association with *MC1R* variants and vulvar vestibulitis, with sufferers 3.5-fold more likely to carry loss-of-function alleles at the *MC1R* (Foster, D. C. *et al.*, 2004). It is unclear whether this risk is related to altered pain processing or, as suggested by the authors, to the known anti-inflammatory role of MC1Rs.

16.6   Future Directions

It is obvious that future work is needed to resolve the true directions of *MC1R* genotypic effect on pain and pain inhibition phenotypes, and to clarify which phenotypes are sex-specific and which are not. Nonetheless, the fact that a gene like *MC1R* would be involved in pain, analgesia, or anesthesia should be considered quite a surprise. It does not appear on any lists of highly probable candidate genes for mediating pain traits (Belfer, I. *et al.*, 2004), and was considered by many to be a gene whose function was already annotated, or solved. The unearthing of *MC1R*'s role in this particular biological domain highlights the advantages for gene discovery of both mouse genetic linkage mapping and bedside-to-bench strategies.

References

Abdel-Malek, Z. A. 2001. Melanocortin receptors: their functions and regulation by physiological agonists and antagonists. Cell. Mol. Life Sci. 58, 434–441.

Basbaum, A. I. and Fields, H. L. 1984. Endogenous pain control systems: brainstem spinal pathways and endorphin circuitry. Annu. Rev. Neurosci. 7, 309–338.

Bastiaens, M., ter Huurne, J., Gruis, N., Bergman, W., Westendorp, R., Vermeer, B. J., and Bavinck, J. N. B. 2001a. The melanocortin-1-receptor gene is the major freckle gene. Hum. Mol. Genet. 10, 1701–1708.

Bastiaens, M. T., ter Huurne, J. A. C., Kielich, C., Gruis, N. A., Westendorp, R. G. J., Vermeer, B. J., and Bavinck, J. N. B. 2001b. Melanocortin-1 receptor gene variants determine the risk of nonmelanoma skin cancer independently of fair skin and red hair. Am. J. Hum. Genet. 68, 884–894.

Belfer, I., Wu, T., Kingman, A., Krishnaraju, R. K., Goldman, D., and Max, M. B. 2004. Candidate gene studies of human pain mechanisms: a method for optimizing choice of polymorphisms and sample size. Anesthesiology 100, 1562–1572.

Chhajlani, V. and Wikberg, J. E. S. 1992. Molecular cloning and expression of the human melanocyte stimulating hormone receptor. FEBS Lett. 309, 417–420.

Cone, R. D., Lu, D., Koppula, S., Vage, D. I., Klungland, H., Boston, B., Chen, W., Orth, D. N., Pouton, C., and Kesterson, R. A. 1996. The melanocortin receptors: agonists, antagonists, and the hormonal control of pigmentation. Rec. Prog. Horm. Res. 51, 287–317.

Diatchenko, L., Slade, G. D., Nackley, A. G., Bhalang, K., Sigurdsson, A., Belfer, I., Goldman, D., Xu, K., Shabalina, S. A., Shagin, D., Max, M. B., Makarov, S. S., and Maixner, W. 2005. Genetic basis for individual variations in pain perception and the development of a chronic pain condition. Hum. Mol. Genet. 14, 135–143.

Eberle, A. N. 1988. The Melanotropins. Chemistry, Physiology and Mechanisms of Action. S. Karger.

Fillingim, R. B., Kaplan, L., Staud, R., Ness, T. J., Glover, T. L., Campbell, C. M., Mogil, J. S., and Wallace, M. R. 2005. The A118G single nucleotide polymorphism of the mu-opioid receptor gene (OPRM1) is associated with pressure pain sensitivity in humans. J. Pain 6, 159–167.

Foster, D. C., Sazenski, T. M., and Stodgell, C. J. 2004. Impact of genetic variation in interleukin-1 receptor antagonist and melanocortin-1 receptor genes on vulvar vestibulitis syndrome. J. Reprod. Med. 49, 503–509.

Gear, R. W., Miaskowski, C., Gordon, N. C., Paul, S. M., Heller, P. H., and Levine, J. D. 1996. Kappa-opioids produce significantly greater analgesia in women than in men. Nature Med. 2, 1248–1250.

Healy, E., Flanagan, N., Ray, A., Todd, C., Jackson, I. J., Matthews, J. N. S., Birch-Machin, M. A., and Rees, J. L. 2000. Melanocortin-1-receptor gene and sun sensitivity in individuals without red hair. Lancet 355, 1072–1073.

Kavaliers, M. and Choleris, E. 1997. Sex differences in N-methyl-D-aspartate involvement in κ opioid and non-opioid predator-induced analgesia in mice. Brain Res. 768, 30–36.

Kavaliers, M. and Galea, L. A. M. 1995. Sex differences in the expression and antagonism of swim stress-induced analgesia in deer mice vary with the breeding season. Pain 63, 327–334.

Kim, H., Neubert, J. K., San Miguel, A., Xu, K., Krishnaraju, R. K., Iadarola, M. J., Goldman, D., and Dionne, R. A. 2004. Genetic influence on variability in human acute experimental pain sensitivity associated with gender, ethnicity and psychological temperament. Pain 109, 488–496.

Klepstad, P., Rakvag, T. T., Kaasa, S., Holthe, M., Dale, O., Borchgrevink, P. C., Baar, C., Vikan, T., Krokan, H. E., and Skorpen, F. 2004. The 118 A > G polymorphism in the human μ-opioid receptor gene may increase morphine requirements in patients with pain caused by malignant disease. Acta Anaesthesiol. Scand. 48, 1232–1239.

Lander, E. S. and Schork, N. J. 1994. Genetic dissection of complex traits. Science 265, 2037–2048.

Liem, E. B., Joiner, T. V., Tsueda, K., and Sessler, D. I. 2005. Increased sensitivity to thermal pain and reduced subcutaneous lidocaine efficacy in redheads. Anesthesiology 102, 509–514.

Liem, E. B., Lin, C. M., Suleman, M. I., Doufas, A. G., Gregg, R. G., Veauthier, J. M., Loyd, G., and Sessler, D. I.

2004. Anesthetic requirement is increased in redheads. Anesthesiology 101, 279–283.

Mogil, J. S. (*ed*.) 2004. The Genetics of Pain, p. 349. IASP Press.

Mogil, J. S., Richards, S. P., O'Toole, L. A., Helms, M. L., Mitchell, S. R., Kest, B., and Belknap, J. K. 1997. Identification of a sex-specific quantitative trait locus mediating nonopioid stress-induced analgesia in female mice. J. Neurosci. 17, 7995–8002.

Mogil, J. S., Ritchie, J., Smith, S. B., Strasburg, K., Kaplan, L., Wallace, M. R., Romberg, R. R., Bijl, H., Sarton, E. Y., Fillingim, R. B., and Dahan, A. 2005a. Melanocortin-1 receptor gene variants affect pain and μ-opioid analgesia in mice and humans. J. Med. Genet. 42, 583–587.

Mogil, J. S., Smith, S. B., O'Reilly, M. K., and Plourde, G. 2005b. Influence of nociception and stress-induced antinociception on genetic variation in isoflurane anesthetic potency among mouse strains. Anesthesiology 103, 751–758.

Mogil, J. S., Sternberg, W. F., Kest, B., Marek, P., and Liebeskind, J. C. 1993. Sex differences in the antagonism of swim stress-induced analgesia: effects of gonadectomy and estrogen replacement. Pain 53, 17–25.

Mogil, J. S., Wilson, S. G., Chesler, E. J., Rankin, A. L., Nemmani, K. V. S., Lariviere, W. R., Groce, M. K., Wallace, M. R., Kaplan, L., Staud, R., Ness, T. J., Glover, T. L., Stankova, M., Mayorov, A., Hruby, V. J., Grisel, J. E., and Fillingim, R. B. 2003. The melanocortin-1 receptor gene mediates female-specific mechanisms of analgesia in mice and humans. Proc. Natl. Acad. Sci. U. S. A. 100, 4867–4872.

Noponen-Hietala, N., Virtanen, I., Karttunen, R., Schwenke, S., Jakkula, E., Li, H., Merikivi, R., Barral, S., Ott, J., Karppinen, J., and Ala-Kokko, L. 2005. Genetic variations in *IL6* associate with intervertebral disc disease characterized by sciatica. Pain 114, 186–194.

Palmer, J. S., Duffy, D. L., Box, N. F., Aitken, J. F., O'Gorman, L. E., Green, A. C., Hayward, N. K., Martin, N. G., and Sturm, R. A. 2000. Melanocortin-1 receptor polymorphisms and risk of melanoma: is the association explained solely by pigmentation phenotype. Am. J. Hum. Genet. 66, 176–186.

Pfaus, J. G., Shadiack, A., Van Soest, T., Tse, M., and Molinoff, P. B. 2004. Selective facilitation of sexual solicitation in the female rat by a melanocortin receptor agonist. Proc. Natl. Acad. Sci. U. S. A. 101, 10201–10204.

Quillan, J. M. and Sadee, W. 1997. Dynorphin peptides: antagonists of melanocortin receptors. Pharm. Res. 14, 713–719.

Schaffer, J. V. and Bolognia, J. L. 2001. The melanocortin-1 receptor: red hair and beyond. Arch. Dermatol. 137, 1477–1485.

Solovieva, S., Leino-Arjas, P., Saarela, J., Luoma, K., Raininko, R., and Riihimaki, H. 2004. Possible association of interleukin 1 gene locus polymorphisms with low back pain. Pain 109, 8–19.

Starowicz, K. and Przewlocka, B. 2003. The role of melanocortins and their receptors in inflammatory processes, nerve regeneration and nociception. Life Sci. 73, 823–847.

Sternberg, W. F., Chesler, E. J., Wilson, S. G., and Mogil, J. S. 2004. Acute progesterone can recruit sex-specific neurochemical mechanisms mediating swim stress induced and kappa-opioid analgesia in mice. Horm. Behav. 46, 467–473.

Tatro, J. B. 1996. Receptor biology of the melanocortins, a family of neuroimmunomodulatory peptides. Neuroimmunomodulation 3, 259–284.

Valverde, P., Healy, E., Jackson, I., Rees, J. L., and Thody, A. J. 1995. Variants of the melanocyte-stimulating hormone receptor gene are associated with red hair and fair skin in humans. Nat. Genet. 11, 328–330.

Vrinten, D. H., Adan, R. A. H., Groen, G. J., and Gispen, W. H. 2001. Chronic blockade of melanocortin receptors alleviates allodynia in rats with neuropathic pain. Anesth. Analg. 93, 1572–1577.

Wikberg, J. E. S. 1999. Melanocortin receptors: perspectives for novel drugs. Eur. J. Pharmacol. 375, 295–310.

Xia, Y., Wikberg, J. E. S., and Chhajlani, V. 1995. Expression of melanocortin 1 receptor in periaqueductal gray matter. Neuroreport 6, 2193–2196.

Xing, Y., Sonner, J. M., Eger, E. I., II, Cascio, M., and Sessler, D. I. 2004. Mice with a melanocortin 1 receptor mutation have a slightly greater minimum alveolar concentration than control mice. Anesthesiology 101, 544–546.

Zubieta, J. K., Heitzeg, M. M., Smith, Y. R., Bueller, J. A., Xu, K., Xu, Y., Koeppe, R. A., Stohler, C. S., and Goldman, D. 2003. COMT *val158met* genotype affects μ-opioid neurotransmitter responses to a pain stressor. Science 299, 1240–1243.

17 Autonomic Nervous System and Pain

Wilfrid Jänig, Physiologisches Institut, Christian-Albrechts-Universität zu Kiel, Germany

Glossary

allodynia, hyperalgesia Hyperalgesia denotes increased pain generated by a stimulus that is normally painful and excites nociceptors. It has a peripheral (sensitization of nociceptors) and/or a central component (sensitization of central neurons, e.g., in the dorsal horn of the spinal cord). Allodynia is pain generated by stimuli that activate low-threshold mechanoreceptors (mechanical allodynia) or cold receptors (cold allodynia). The mechanism of allodynia is central (central sensitization generated by persistent excitation of nociceptors). Secondary allodynia is pain elicited by stimulation of low-threshold mechanoreceptors in an area of skin that surrounds a territory with sensitized nociceptors (e.g., generated by inflammation) (Merskey, H. and Bogduk, H., 1994; Meyer, R. A. *et al.*, 2005).

causalgia See Chapter Complex Regional Pain Syndromes.

complex regional pain syndrome See Chapter Neuropathic Pain: Clinical.

enteric nervous system The enteric nervous system (Jänig, W., 2006).

homeostasis The maintenance of physiological parameters such as concentration of ions, blood glucose, arterial blood gases, and body core temperature in a narrow range and around predetermined set-points is called homeostasis. Homeostatic regulation involves autonomic systems, endocrine systems, and the respiratory system (Jänig, W., 2006).

neuropathic pain See Chapters Neuropathic Pain: Basic Mechanisms (Animal) and Complex Regional Pain Syndromes.

referred pain Pain from a particular visceral organ or a particular deep somatic structure is perceived to arise from the body surface (skin), from deep somatic structures, or from other viscera. The area or deep body structures where the pain is referred to is innervated by nerves of the same spinal segments as the affected deep body structures (Giamberardino, M. A., 1999; Jänig, W., 1993; Jänig, W. and Häbler, H. J., 2002).

sympathetic block, sympathectomy See Chapter Sprouting in Dorsal Root Ganglia.

sympathetically maintained pain (SMP) See Chapter Complex Regional Pain Syndromes.

sympatho-adrenal system The sympatho-adrenal system is the adrenal medulla and its innervation by sympathetic preganglionic neurons. The adrenal medulla consists of cells that release either adrenaline or noradrenaline upon impulses in the preganglionic neurons. The sympatho-adrenal system is functionally distinct from the various types of sympatho-neural systems (Jänig, W., 2006).

viscero-somatic convergent neurons (dorsal horn of the spinal cord) See Chapter Spinal Cord Mechanisms of Hyperalgesia and Allodynia.

17.1 Introduction

Research on the relationship between the autonomic nervous system and pain is a controversial field with many interesting facets that are important from the biological, pathobiological, and clinical point of view. (I prefer to use the terms biology and pathobiology, instead of physiology and pathophysiology, because they include physiological, morphological, biochemical, and molecular aspects.) I will discuss this field in a broad context and focus mainly on the sympathetic nervous system. I will argue that the peripheral sympathetic noradrenergic neuron may have, in addition to its conventional function to transmit signals generated in the brain to peripheral target cells (e.g., smooth muscle cells, secretory epithelia, heart cells, and neurons of the enteric nervous system), quite different functions that are directly or indirectly related to protection of body tissues and pain. These functions have not been studied as extensively as those related to the regulation of autonomic target cells (Jänig, W., 2006). The parasympathetic nervous system is not involved in the generation of pain, yet it may also be important in protection of body tissues (e.g., the gastrointestinal tract (GIT)). Vagal afferents are involved in integrative aspects of pain, hyperalgesia, and inflammation. This is discussed elsewhere (Jänig, W., 2005; Jänig, W. and Levine, J. D., 2006) and summarized in Jänig W. (2007).

The topic *Autonomic Nervous System and Pain* cannot be reduced to peripheral local processes in which peripheral nociceptive afferent fibers (sensitization of nociceptors and generation of ectopic impulses after nerve lesion), sympathetic fibers, and nonneural cells (e.g., cells of the immune system and the vasculature) are involved. The question to be asked is: How are the cellular and subcellular processes that lead to ongoing pain, allodynia, and hyperalgesia in neuropathic pain and inflammatory pain orchestrated by the brain in the continuous protection of the body against agents from outside as well as from within the body? The brain is an important player to organize and control these protective body reactions via neuroendocrine (hypothalamo-pituitary-adrenocortical and sympatho-adrenal) and autonomic (preferentially the sympathetic) systems. Feedback to the brain occurs via various signaling molecules (e.g., interleukines), hard-wired nociceptive afferents, and other afferent neurons (Figure 1). Table 1 lists different functions of the sympathetic nervous system related to pain and protection of body tissues in biological and pathobiological

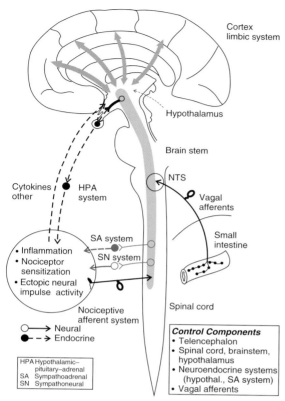

Figure 1 Spinal cord, brainstem, and hypothalamus contain neuronal circuits (pink shaded area) that modulate nociceptor sensitivity, inflammation, and ectopic activity in lesioned afferent neurons in the periphery of the body via the sympatho-adrenal (SA) system, the sympatho-neural system, and the hypothalamic–pituitary–adrenal (HPA) system. Feedback information from the peripheral inflammatory process occurs via nociceptive primary afferent neurons and cytokines. Abdominal vagal afferents signal events from the inner defense line of the body (GALT, gut-associated lymphoid tissue) to the lower brainstem (NTS, nucleus tractus solitarii). The telencephalon controls inflammation and sensitivity of nociceptors via the circuits in the neuraxis (see shaded double arrows).

conditions. It shows that peripheral as well as central mechanisms have to be considered and that the topic includes not only pain and hyperalgesia but also neural and neuroendocrine regulation of inflammation and the immune system.

The integration of the control of protective neural, endocrine, and immune mechanisms occurs in the brain (brainstem, hypothalamus, limbic system, and neocortex). Perception of sensations, feeling of emotions, autonomic, endocrine, and somatomotor responses are coordinated and are, therefore, parallel readouts of the central representations, although the evolvement of

feelings of emotions may require explicitly the afferent feedback from body responses generated by the efferent neural autonomic and somatomotor signals (Damasio, A., 2003). These central representations act back on the peripheral tissues, the immune and nociceptive primary afferent neurons, via the endocrine and autonomic nervous systems (Figure 1). The central circuits are also the origin of illness responses that include aversive feelings, pain, and hyperalgesia. This central integration is related or identical to the integrative processes, involving neuroendocrine, immune, and neural systems, occurring during environmental challenges such as viral and bacterial infection (Arkins, S. *et al.*, 2001; Miller, A. H. *et al.*, 2001). Thus, the autonomic nervous system is one component in the generation of pain and protective body reactions that are organized and regulated by the brain and disease occurs when this neural regulation fails.

I will distinguish three aspects of the topic Sympathetic nervous system and pain (Table 1):

1. Reactions of the sympathetic nervous system during pain. This includes the autonomic integrated and adaptive reactions during pain, reflexes to noxious stimulation, and reactions mediated by sympathetic systems in referred zones.
2. Coupling (cross-talk) from sympathetic (noradrenergic) neurons to afferent neurons in the generation of pain. Four broad categories of coupling will be discussed: Direct coupling related to excitation of the sympathetic postganglionic neurons and release of noradrenaline; indirect coupling related to influence of inflammatory processes by the sympathetic nervous system; coupling independent of excitation and noradrenaline release by sympathetic neurons. A fourth category is the effect of adrenaline released by the adrenal medulla on the sensitivity of nociceptors.
3. Sympathetic nervous system and central mechanisms. Based on observations made on patients and animal experimentation, it is likely that activity in sympathetic noradrenergic neurons or in the sympatho-adrenal system, both generated in the brain, contributes directly or indirectly to pain. This central aspect includes the various forms of sympathetic afferent coupling discussed before.

The role of the sympathetic nervous system and its transmitters in chronic inflammation (such as rheumatoid diseases) or in systemic chronic pain diseases (e.g., fibromyalgia) and its interaction with the peptidergic afferent innervation as well as the

Table 1 Sympathetic nervous system and pain

Reactions of the sympathetic nervous system in pain[a]
  Protective spinal reflexes
  Fight, flight, and quiescence organized at the level of the periaqueductal gray
  Hyperalgesia and sympathetically mediated changes in referred zones during visceral pain
Role of the sympathetic nervous system in the generation of pain[b]
  Sympathetic–afferent coupling in the periphery
    Coupling after nerve lesion (noradrenaline, α-adrenoceptors)
    Coupling via the micromilieu of the nociceptor and the vascular bed
    Sensitization of nociceptors mediated by sympathetic terminals independent of excitation and release of noradrenaline
    Sensitization of nociceptors initiated by cytokines or nerve growth factor and mediated by sympathetic terminals
    Sympatho-adrenal system and nociceptor sensitization
Sympathetic nervous system and central mechanisms[c]
  Control of inflammation and hyperalgesia by sympathetic and neuroendocrine mechanisms
    Complex regional pain syndrome and sympathetic nervous system
    Immune system and sympathetic nervous system
    Rheumatic diseases and sympathetic nervous system
    Systemic chronic pain (e.g., fibromyalgia) and sympathetic nervous system

[a]Bandler R. and Shipley M. T. (1994); Bandler R. *et al.* (2000a; 2000b); Jänig W. (1993; 2006).
[b]Jänig W. and Baron R. (2001); Jänig W. (2005); Jänig W. and Häbler H. J. (2000b); Jänig W. and Koltzenburg M. (1991); Jänig W. and Levine J. D. (2005).
[c]Jänig W. and Baron R. (2002; 2003; 2004); Jänig W. and Häbler H. J. (2000a); Jänig W. *et al.* (2000); Jänig W. and Levine J. D. (2005); Straub R. H. and Härle P. (2005); Straub R. H. *et al.* (2005); Vierck C. J. Jr. (2006).

hypothalamo-pituitary-adrenal (HPA) system will not be discussed. We are at the beginning of the experimental investigation of these control mechanisms (Straub, R. H. and Härle, P., 2005; Straub, R. H. *et al.*, 2005; Vierck, C. J. Jr., 2006).

17.2 Neurobiology of the Autonomic Nervous System and Pain

The peripheral autonomic nervous system is by definition an efferent system consisting of many functionally distinct pathways that transmit the impulse activity from the spinal cord or brainstem to the effector cells. These pathways are separate from each other and functionally distinct with respect to the effector cells, as is the impulse transmission in the autonomic ganglia and at the neuroeffector junctions to the effector cells. Each peripheral sympathetic or parasympathetic pathway is connected to distinct neuronal networks in the spinal cord, brainstem, and hypothalamus that generate the typical discharge patterns in the autonomic neurons and are responsible for the precise regulation of the target organs (cardiovascular regulation, thermoregulation, regulation of pelvic organs, regulation of GIT, and so forth). The autonomic regulations are represented in these central networks. The anatomical, histochemical, and functional organization of the autonomic nervous system,

in particular of the sympathetic nervous system, in the periphery and in the brain, mostly receives insufficient attention in studies of the functioning of this system under pathobiological conditions. I emphasize three points (Jänig, W. and McLachlan, E. M., 2002; Jänig, W., 2006): First, the sympathetic nervous system is functionally as well differentiated as the parasympathetic nervous system. It does not react as a unitary system under normal physiological conditions. Second, the terms sympathetic and parasympathetic are defined anatomically and not functionally. Thus, to use these terms functionally is misleading. Third, the long-held view that sympathetic and parasympathetic nervous system function universally antagonistically is also misleading. Both systems complement each other in their functions.

17.2.1 Reactions of the Sympathetic Nervous System in Pain

Any acute, but possibly also chronic tissue-damaging stimulus affects the sympathetic nervous system. Neurons of sympathetic systems exhibit generalized and specific reactions to these stimuli. The generalized reactions probably occur only in certain types of sympathetic system (e.g., muscle vasoconstrictor, visceral vasoconstrictor, sudomotor neurons, or sympathetic cardiomotor neurons) but are weak or absent in other systems (e.g., sympathetic systems to pelvic organs). They are organized in spinal cord, brainstem (medulla oblongata and mesencephalon), and hypothalamus and can be conceptualized as component parts of the different patterns of defense behavior, such as confrontational defense, flight, or quiescence (Figure 2). Confrontational defense and flight are typical of an active defense strategy when animals encounter threatening stimuli that are potentially injurious for the body, the former potentially leading to fight and the latter to forward avoidance. Both patterns are represented in the lateral and dorsolateral periaqueductal gray of the mesencephalon, activated from the body surface or cortex, and associated with endogenous nonopioid analgesia, hypertension, and tachycardia. Quiescence is similar to the natural reactions of mammals to serious injury

and chronic pain occurring particularly in the deep and visceral body domains. It is represented in the ventrolateral periaqueductal gray, activated from the deep (somatic and visceral) body domains, and consists of hyporeactivity, hypotension, bradycardia, and an endogenous opioid analgesia. These stereotyped preprogrammed elementary behaviors and their association with the endogenous control of analgesia enable the organism to cope with dangerous situations that are always accompanied by pain or impending pain. The dorsolateral, lateral, and ventrolateral columns of the periaqueductal gray have distinct reciprocal connections with the autonomic centers in the lower brainstem and hypothalamus that differentially regulate the activity in neurons of the autonomic pathways. They are under differential

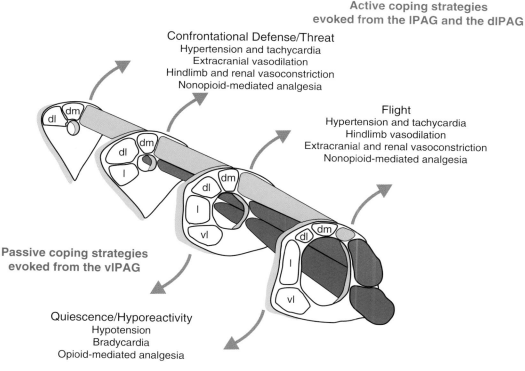

Figure 2 Representation of defense behaviors in the dorsolateral, lateral, and ventrolateral periaqueductal gray. Schematic illustration of the dorsolateral (dl), lateral (l), and ventrolateral (vl) columns within the rostral, intermediate, and caudal periaqueductal gray (PAG). The dorsomedial neural PAG columns are indicated too. Stimulation of neuron populations in the dlPAG, lPAG, and vlPAG by microinjections of the excitatory amino acid glutamate that excites only cell bodies of neurons but not axons evokes distinct defense behaviors: Confrontational defense is elicited from the rostral portion of the dlPAG and lPAG; flight is elicited from the caudal part of the dlPAG and lPAG; quiescence (cessation of spontaneous motor activity) is elicited from the vlPAG in the caudal portion of the PAG. These defense behaviors include typical autonomic cardiovascular reactions (changes of blood pressure, heart rate, and blood flows) and sensory changes (nonopioid- or opioid-mediated analgesia). The representations of confrontational defense and flight are the basis for active coping strategies produced by the cortex. The representation of quiescence is the basis for passive coping strategies produced by the cortex. Modified from Bandler R. and Shipley M. T. (1994) and Bandler R. *et al.* (2000b).

control of the medial and orbital prefrontal cortex and other telencephalic centers (Bandler, R. and Shipley, M. T., 1994; Bandler, R. *et al.*, 2000a; 2000b; Keay, K. A. and Bandler, R., 2004).

There also exist more localized reactions of the sympathetic nervous system to noxious stimuli that are organized within the spinal cord and trigeminal nucleus and in the periphery. These localized reactions are reflected in various somato-sympathetic and viscero-sympathetic reflexes, the afferent neurons of these reflexes are nociceptive (skin, deep somatic tissues, and viscera) and the efferent neurons sympathetic innervating blood vessels (in skin, deep somatic tissues, and viscera), sweat glands, erector pili muscles, enteric nervous system, or other effectors in the viscera. Both afferent neurons and efferent pathways are synaptically connected by various groups of interneurons, establishing circuits that probably are the basis of the changes occurring in referred zones during chronic stimulation of deep nociceptive primary afferents (see below). For example, cutaneous vasoconstrictor neurons exhibit distinct inhibitory reflexes to noxious stimuli of the territories innervated by these neurons.

The hypothalamo-mesencephalic and the spinal level of integration are presumably protective under normal biological conditions and are associated with the activation of the hypothalamo-pituitary adrenocortical system (Jänig, W., 2006).

17.2.2 Visceral Afferents, Autonomic Nervous System, and Pain

Visceral organs in the thoracic, abdominal, and pelvic cavities are innervated by two sets of extrinsic primary afferent neurons: spinal visceral afferent neurons and vagal visceral afferent neurons (the latter also including afferent neurons from arterial baro- and chemoreceptors that project through the glossopharyngeal nerve). Spinal visceral afferent neurons have their cell bodies in the dorsal root ganglia (DRG) and vagal afferent neurons in the inferior (nodose) ganglion of the vagus nerve (a few in the superior (jugular) ganglion of the vagus nerve and some arterial baro- and chemoreceptor afferents in the petrosal ganglion). Visceral afferent neurons are involved in specific organ regulations, multiple organ reflexes, neuroendocrine regulations, and visceral sensations (including visceral pain), shaping and eliciting emotional feelings and other functions. They monitor the inner state of the body and serve to maintain homeostasis and to adapt the internal

milieu and the regulation of the organs to the behavior of the organism. They are anatomically and functionally closely associated with the autonomic nervous system. However, they should not be labeled sympathetic, parasympathetic, or autonomic afferent neurons. These functional labels are misleading since they imply that the visceral afferent neurons have functions that uniquely pertain to the sympathetic or parasympathetic autonomic nervous system, the exception being that afferent neurons of the enteric nervous system are per definition enteric afferent neurons. No functional, morphological, histochemical, or other criteria do exist to associate any type of visceral afferent neuron with only one autonomic system (Jänig, W., 2006).

Usually, vagal afferent neurons are described and characterized by adequate physiological stimuli exciting them and by the main reflexes associated with the cardiovascular system, respiratory system, GIT, or other regulatory systems (Undem, B. J. and Weinreich, D., 2005). However, several observations show that groups of vagal afferents innervating visceral organs in the thoracic and abdominal cavity are involved in the control of nociception and pain, in neuroendocrine control of nociceptors, in the control of inflammation, and in the control of general protective body reactions including illness responses. Subgroups of vagal afferent neurons may signal events from visceral organs, in particular those occurring at the internal defense line of the body in the GIT represented in the liver and in the small intestine (which contains the largest immune system of the body, the gut associated lymphoid tissue (GALT)), to the central nervous system that triggers protective body reactions not only in the viscera but also in the superficial and deep somatic body tissues (see Chapters Visceral Pain; Vagal Afferent Neurons and Pain; Jänig, W., 2005). These vagal afferent neurons may be activated or sensitized by cellular processes of these inner body defense lines and their signaling molecules. Stimulation of vagal afferents may generate sickness behavior that is characterized by several protective illness responses (immobility, decreased social interaction, decrease in food intake and of digestion, formation of taste aversion against novel food, loss of weight, fever, increase of sleep, change in endocrine functions (e.g., of the activation of the hypothalamic–pituitary system), malaise; Watkins, L. R. *et al.*, 1995; Maier, S. F. and Watkins, L. R., 1998; Watkins, L. R. and Maier, S. F., 1999; 2000; Goehler, L. E. *et al.*, 2000), activates the inhibitory control of nociceptive impulse transmission in the spinal and trigeminal dorsal horn (Foreman, R. D., 1989; Randich, A. and Gebhart, G. F., 1992; Jänig, W., 2005), inhibits nociceptive–neuroendocrine reflex circuits that control inflammation (via the sympatho-adrenal system and the hypothalamic–pituitary system (Green, P. G. *et al.*, 1995; 1997; Jänig, W. *et al.*, 2000; Miao, F. J.-P. *et al.*, 2000; 2001)), and inhibits the central circuits that control via the adrenal medulla and release of adrenaline the sensitivity of nociceptors (Khasar, S. G. *et al.*, 1998a; 1998b; Jänig, W. *et al.*, 2000; Khasar, S. G. *et al.*, 2003; Jänig, W. *et al.*, 2006; see below). For these hypothetic functions of vagal afferents involving neuroendocrine and autonomic systems, in the protection of the body, an umbrella concept in the frame of body protection, pain, and inflammation has to be formulated. This has to include the central circuits (see Jänig, W., 2005; Jänig, W. and Levine, J. D., 2006).

17.2.3 Hyperalgesia and Sympathetically Mediated Changes in Referred Zones During Visceral Pain

Chronic stimulation of spinal afferents from visceral organs may elicit pain, hyperalgesia, and allodynia in the referred zones of the body surface and of the deep somatic domain (skin, subcutaneous tissue, musculature, tendons, fascia, and bones). Hyperalgesia and allodynia that are elicited in the superficial and deep somatic domains by visceral diseases are equivalent to secondary hyperalgesia and allodynia in the skin caused by ongoing activity in polymodal C-nociceptors (Cervero, F. and Jänig, W., 1992; Vecchiet, L. *et al.*, 1993). The degree and spatial spread of referred pain, hyperalgesia, and allodynia depend on the intensity and duration of the visceral stimulation and the type of visceral organ. They are abolished after blockade of the impulse traffic in spinal visceral afferents to the spinal cord. They may be generated by sensitization of neurons in the spinal cord that are involved in the generation of sensations (e.g., spino-thalamic tract neurons), in the descending control of spinal circuits (e.g., spinal neurons projecting to the brainstem or interneurons mediating signals of the endogenous control system), and in the regulation of sympathetic target organs and skeletal musculature constituting spinal and supraspinal viscero-sympathetic and viscero-somatic reflex pathways (Figure 3). The most likely candidate of these neurons is the broad class of viscero-somatic convergent neurons in the spinal cord (Cervero, F. and Jänig, W., 1992; Cervero, F., 1995; Bielefeldt, T. K. and Gebhart, G. F., 2006).

Little attention has been given to the generation of autonomic changes (reflected in changes of blood flow, sweating, and piloerection) and of trophic changes in the skin and its appendages, subcutis, joint capsules, and fascia of the referred hyperalgesic (Head's) zones during visceral diseases (Jänig, W. and Morrison, S. F., 1986). These changes are very variable between individuals but reported to be relatively stable for each individual. The clinical diagnosis of a visceral disease (e.g., in the GIT, of the evacuative organs, or of the heart) may be predicted from these changes with a probability of about 70% (Beal, M. C., 1983; Cox, J. M. *et al.*, 1983; Beal, M. C., 1985; Nicholas, A. S. *et al.*, 1985).

It is important to note that the trophic changes occur in regions that are outside of the site of injury or trauma, that the first sign of the trophic changes is probably the edema, and that the severe signs of these changes develop relatively slowly and late. These changes are believed to be generated by the segmental sympathetic outflow to the visceral reference zones at the trunk. However, the trophic changes in the referred zones may also or additionally be related to changes in the retrograde axonal transport in afferent neurons that innervate the referred zones (see discussion in Jänig, W., 1993). We need more reliable and quantitative data about the changes occurring in the referred somatic zones in patients with various visceral diseases to develop testable hypotheses.

The mechanisms explaining referred visceral pain and the changes observed in the referred zones that probably are associated with the sympathetic neurons are barely understood. Classical central theories trying to explain referred visceral pain are the convergence-projection theory of Ruch T. C. (1965) and the convergence-facilitation theory of MacKenzie, J. (1920). (The convergence-projection theory of Ruch T. C. (1965) is based on the concept that spinal neurons in the dorsal horn projecting to the thalamus obtain convergent synaptic input from primary afferent neurons innervating viscera or somatic body tissues. The telencephalon localizes the pain generated by activity in visceral afferents erroneously into somatic tissues of the corresponding dermatomes, myotomes, and sklerotomes. The convergence-facilitation theory of MacKenzie J. (1920) proposes that afferent inputs from the viscera to the spinal cord are not directly related to pain sensation but that they produce a generalized increase of excitability in the area of the spinal cord, which is related to the respective visceral organ generating

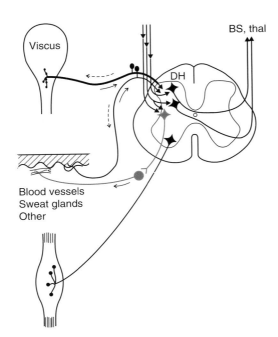

Observed phenomena:

1) Referred pain,
   hyperalgesia, allodynia
2) Autonomic changes:
   blood vessels, sweat glands
3) Muscle tension
4) Trophic changes

Possible mechanisms:

1) Sensitization of
   spinal neurons
2) Changes of balance
   between spinal circuits
   and descending
   control systems
3) Viscerosympathetic
   reflexes
4) Viscerosomatic
   reflexes
5) Changes in regulation of
   peripheral microcirculation
6) Changes of retrograde
   axonal transport in
   afferent neurons

Figure 3   Observed phenomena generated in referred zones during visceral trauma (e.g., inflammation) and the putative mechanisms underlying these changes. Sympathetic pathways (indicated in red) mediate changes of blood flow, sweating, or piloerection and possibly changes of tissue structure (edema, trophic changes). The mechanisms underlying the latter are not understood. Continuous arrows, excitation; dashed arrows, retrograde axonal transport. BS, brainstem; DH, dorsal horn; thal, thalamus. Modified from Jänig W. (1993).

an irritable focus. This was to promote activity in neurons projecting to the thalamus and obtaining convergent synaptic input from somatic tissue (leading to referred pain and referred hyperalgesia), in motoneurons (leading to muscle spasms), and in sympathetic neurons (leading to autonomic changes in the referred zones) (Ness, T. J., 1995).) However, both theories explain the mechanisms underlying referred pain rather incompletely and the mechanisms of the other changes in the referred zones in which the sympathetic nervous system is involved not at all (see Ness, T. J. and Gebhart, G. F., 1990; Jänig, W. and Koltzenburg, M., 1993; Vecchiet, L. *et al.*, 1993; Ness, T. J., 1995; Giamberardino, M. A., 1999, for discussion). It is suggested that persistent sensitization and activation of visceral afferent neurons (e.g., during chronic inflammation) sensitize the second-order viscero-somatic convergent neurons in the dorsal horn of the spinal cord that can be synaptically activated from viscera as well as from somatic tissues (skin and deep somatic tissues). However, it has not been convincingly demonstrated that these viscero-somatic convergent neurons can be sensitized by

activity in visceral afferents (Herrero, J. F. *et al.*, 2000; Jänig, W. and Häbler, H. J., 2002).

This invites speculation that not only spinal mechanisms but also, or in particular, supraspinal mechanisms are responsible to explain the phenomenology of referred pain. Supraspinal mechanisms controlling synaptic impulse transmission in the dorsal horn are inhibitory or excitatory (Urban, M. O. and Gebhart, G. F., 1999; Porreca, F. *et al.*, 2002). These mechanisms are represented in the periaqueductal gray, the caudal raphe nuclei, the ventromedial medulla, and the dorsolateral pontine tegmentum (Fields, H. L. *et al.*, 2006). They do control not only nociceptive impulse transmission in the dorsal horn but also various autonomic regulations and possibly neuroendocrine regulations (Mason, P., 2001) and are in turn under the control of the cerebral hemispheres. Thus, the reference of visceral affections (e.g., chronic inflammation in the viscera) into somatic zones may be dependent on these endogenous protective control systems. These referred phenomena (that include sensory, autonomic, and motor changes) may therefore be the biological expression of protective mechanisms operating in spinal cord and brainstem in the somatic body domains.

17.3 Role of the Sympathetic Nervous System in the Generation of Pain

17.3.1 Clinical Background: Sympathetically Maintained Pain

Pain being dependent on activity in sympathetic neurons is called sympathetically maintained pain (SMP). SMP is a symptom and includes generically spontaneous pain and pain evoked by mechanical or thermal stimuli. It may be present in the complex regional pain syndrome (CRPS) type I and type II and in other pain syndromes. The idea about the involvement of the (efferent) sympathetic nervous system in the generation of pain is based on various clinical observations that have been documented in the literature since tens of years (Bonica, J. J., 1990). Representative for these multiple observations on patients with pain in which the sympathetic nervous system is causally involved are experimental investigation on patients with CRPS who had SMP (1–4) and other less well-controlled observations on patients with pain.

1. Spontaneous pain, mechanical allodynia, and cold allodynia in the upper extremity of patients with CRPS II, which are alleviated by stellate ganglion block, can be rekindled, under the condition of sympathetic block, by injection of noradrenaline into the skin area that was painful before sympathetic block (Figure 4; Torebjörk, H. E. et al., 1995). By the same token, it has been shown in patients with causalgia, who had been successfully treated either by sympathectomy or by transcutaneous nerve stimulation, that ionophoretic application of noradrenaline to the previously causalgic skin area rekindles spontaneous pain and mechanical allodynia (Wallin, G. et al., 1976). Rekindling of spontaneous and evoked pain or enhancement of both by noradrenaline injected into the skin area, which was painful before sympathetic block or before successful treatment, took 10 min or longer to develop (Figure 4b). This long latency indicates that noradrenaline does not act possibly directly on the afferent fiber terminals, but by some unknown mechanism via changes in the micromilieu of the nociceptors.

2. Intradermal injection of noradrenaline in physiologically relevant doses in patients with SMP evokes greater pain in the affected limb regions than in the contralateral unaffected limb or in control subjects. Most patients whose pain increased after injection of noradrenaline in the affected extremity report a decrease in pain following systemic injection of the α-adrenoceptor blocker phentolamine (Ali, Z. et al., 2000; see also Arnér, S., 1991).

3. In CRPS type I patients with SMP, pain is relieved after blockade of sympathetic activity to the affected extremity by a local anesthetic applied to the appropriate sympathetic paravertebral ganglia. This pain relief lasts significantly longer than the short-lasting pain-relieving effect of saline injected close to the same paravertebral ganglia (control injection) in the same group of CRPS patients showing that the real pain-relieving effect generated by blockade of sympathetic activity can be discriminated from the placebo effect and the duration of pain relief outlasts the duration of conduction block generated by the local anesthetic by an order of magnitude (Price, D. D. et al., 1998; see Figure 15).

4. Pain (spontaneous, mechanical allodynia) in the hand of chronic CRPS I patients with SMP, which is clamped to a temperature of about $36\,^\circ C$, is enhanced when activity in cutaneous vasoconstrictor neurons is increased by central cooling (Figure 5), yet not in chronic CRPS I patients without SMP (Baron, R. et al., 2002). This experiment demonstrates that an increase of activity in sympathetic neurons by a physiological intervention (central cooling) can increase pain, possibly independent of a vascular component. An important side effect of this experimental investigation is that the peripheral mechanisms underlying SMP in CRPS I patients occur mainly in deep somatic tissues.

5. Injection of adrenaline around a stump neuroma that has developed long after limb amputation may be intensely painful (Chabal, C. et al., 1992).

6. In patients with CRPS II, but not in those with hyperhidrosis, intraoperative electrical stimulation of the sympathetic chain may increase spontaneous pain (Walker, A. E. and Nulsen, F., 1948; White, J. C. and Sweet, W. H., 1969).

7. In patients with SMP phentolamine (α-adrenoceptor blocker) given intravenously may relieve pain but not propranolol (β-adrenoceptor blocker) (Arnér, S., 1991; Raja, S. N. et al., 1991; Dellemijn, P. L. et al., 1994).

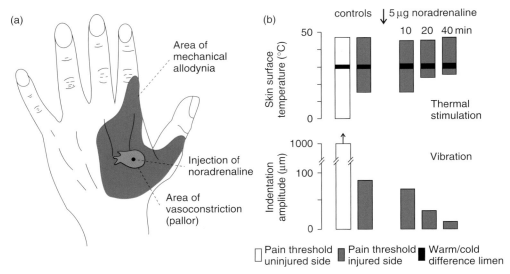

Figure 4 Mechanical and cold allodynia in CRPS type II patients with sympathetically maintained pain (SPM) elicited by intradermal injection of noradrenaline. Noradrenaline was injected intradermally at the site of mechanical and cold allodynia when the patients were temporarily relieved from their pain by a stellate ganglion block with lidocaine (1%) or bupivacaine (0.25%). (a) Area of pallor and mechanical allodynia 20 min after intradermal injection of 2 μg noradrenaline. (b) Sequential quantitative sensory testing of patients before and after intradermal injection of 5 μg noradrenaline during a time without blockade of the stellate ganglion. This patient had mechanical allodynia (indentation amplitude of a mechanical vibration stimulus to evoke pain about 90 μm, *lower panel*) and cold hyperalgesia/allodynia (cold pain threshold about 15 °C, *upper panel*). The mechanical allodynia (further decrease of indentation amplitude to elicit pain) and cold allodynia (further decrease of cold pain threshold) progressively worsened starting approximately 20 min after intradermal injection of 5 μg noradrenaline. Modified from Torebjörk H. E. *et al.* (1995) with permission.

8. In some patients with postherpetic neuralgia, spontaneous pain and mechanical allodynia are enhanced after intracutaneous injection of adrenaline or the α1-adrenoceptor agonist phenylephrine (Choi, B. and Robotham, M. C., 1997).

I recommend to read Mitchell S. W. (1872), Livingston W. K. (1943/1976; here in particular chapters 5, 6, 14, and 15), and Bonica J. J. (1953; Chapter 28). These texts are not only of historical interest (here the literature before 1953 is discussed) but also the reader will find (1) a wealth of carefully reported clinical observations and (2) ideas developed on the basis of these clinical observations that are still quite modern, showing that the progress in this field is not as impressive as it appears to be. These key experiments conducted on human patients with SMP demonstrate that (1) activation of sympathetic postganglionic neurons can produce pain, (2) blockade of sympathetic activity can relieve the pain, (3) noradrenaline injected intracutaneously rekindles the pain, and (4) α-adrenoceptor blockers or guanethidine (which depletes noradrenaline from its stores) can relieve the pain (Jänig, W. and Baron, R., 2001; 2002; Jänig, W., 2002; Jänig, W. and Baron, R., 2003).

The interpretation of these data is that the nociceptors are excited and possibly sensitized by noradrenaline released by the sympathetic fibers. Either the nociceptors have expressed adrenoceptors or the excitatory effect is generated indirectly, e.g., via changes in blood flow and subsequent ischemia or by other components modulated by sympathetic fibers. The ways of sympathetic afferent coupling to be discussed are illustrated in Figure 6. Sympathetically maintained activity in nociceptive neurons may generate and maintain a state of central sensitization/ hyperexcitability, leading to spontaneous pain and secondary evoked pain (mechanical and possibly cold allodynia). This is schematically illustrated in Figure 7.

17.3.2 Sympathetically Maintained Pain Following Nerve Lesion Simulated in Behavioral Animal Models

In animals (rats), signs of pain are measured by quantifiable behavioral components, e.g., paw-withdrawal latency to thermal noxious stimulation (heat and cold), frequency and threshold of paw withdrawal to mechanical stimulation, degree of self-mutilation

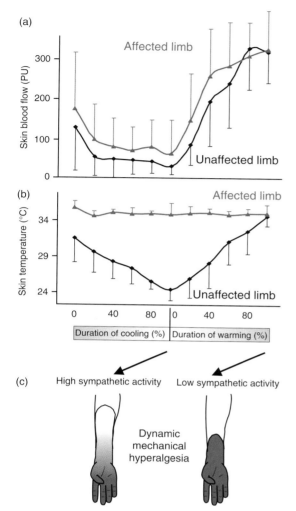

Figure 5 Experimental changes of activity in cutaneous sympathetic vasoconstrictor neurons by physiological thermoregulatory reflex stimuli in 13 complex regional pain syndrome (CRPS) patients. With the help of a thermal suit, whole-body cooling and warming was performed to alter sympathetic skin nerve activity. The subjects were lying in a suit supplied by tubes, in which running water at 12 °C and 50 °C, respectively (inflow temperature), was used to cool or warm the whole body. By these means, sympathetic activity can be switched on and off. (a) High sympathetic vasoconstrictor activity during cooling induces considerable drop in skin blood flow on the affected and unaffected extremity (laser Doppler flowmetry). Measurements were taken at 5-min intervals (mean + SD). (b) On the unaffected side, a secondary decrease of skin temperature was documented. On the affected side, the forearm temperature was clamped at 35 °C by a feedback-controlled heat lamp to exclude temperature effects on the sensory receptor level. Measurements were taken at 5-min intervals (mean + SD). (c) Effect of cutaneous sympathetic vasoconstrictor activity on dynamic mechanical hyperalgesia in one CRPS patient with sympathetically maintained pain (SMP). Activation of sympathetic neurons (during cooling) leads to an increase of the area of dynamic mechanical hyperalgesia. Reprinted with permission from Baron R. *et al.* (2002).

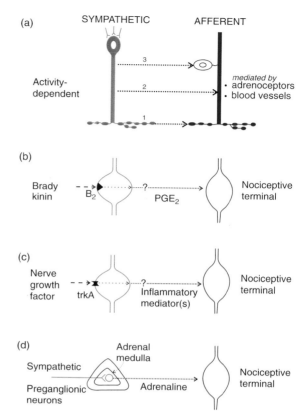

Figure 6 Ways hypothesized to couple sympathetic and primary afferent neurons following peripheral nerve lesion (a) or during inflammation (b–d). (a) These types of coupling depend on the activity in the sympathetic neurons and on the expression of functional adrenoceptors by the afferent neurons or mediation indirectly via the blood vessels (blood flow). It can occur in the periphery (1), in the dorsal root ganglion (3), or possibly also in the lesioned nerve (2). (b) The inflammatory mediator bradykinin acts at B_2 receptors in the membrane of the sympathetic varicosities or in cells upstream of these varicosities, inducing release of prostaglandin E_2 (PGE_2) and sensitization of nociceptors. This way of coupling is probably not dependent on activity in the sympathetic neurons. (c) Nerve growth factor released during an experimental inflammation reacts with the high-affinity receptor trkA, in the membrane of the sympathetic varicosities, inducing release of an inflammatory mediator or inflammatory mediators and sensitization of nociceptors. This effect is probably not dependent on activity in the sympathetic neurons. (d) Activation of the adrenal medulla by sympathetic preganglionic neurons leads to release of adrenaline that generates sensitization of nociceptors. The ? in (b) and (c) indicates that PGE_2 or other inflammatory mediators may be released by cells other than the sympathetic varicosities. Modified from Jänig W. and Häbler H. J. (2000b).

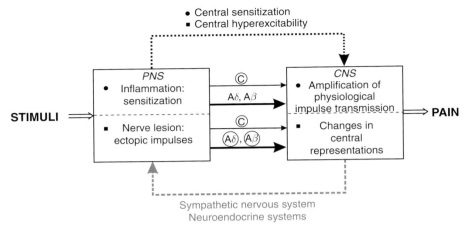

Figure 7 Concept of generation of peripheral and central hyperexcitability during inflammatory pain and neuropathic pain. The upper interrupted arrow indicates that the central changes are generated (and possibly maintained) (a) by persistent activation of nociceptors with C-fibers (e.g., during chronic inflammation) called here central sensitization or (b) after trauma with nerve lesion by ectopic activity and other changes in lesioned afferent neurons called here central hyperexcitability. The lower interrupted arrow indicates the efferent feedback via the sympathetic nervous system and neuroendocrine systems (e.g., the sympatho-adrenal system). Primary afferent nociceptive neurons (in particular those with C-fibers) are sensitized during inflammation. After nerve lesion, all lesioned primary afferent neurons (unmyelinated as well as myelinated ones) undergo biochemical, physiological, and morphological changes that become irreversible with time. These peripheral changes entail changes of the central representation (of the somato-sensory system) that become irreversible if no regeneration of primary afferent neurons to their target tissue occurs. The central changes, induced by persistent activity in afferent nociceptive neurons during inflammation or induced after nerve lesion, are also reflected in the efferent feedback systems that may establish positive feedback loops to the primary afferent neurons.

(autotomy), or spontaneous lifting and licking of paw. These reactions are interpreted as equivalent to mechanical or thermal hyperalgesia or allodynia or spontaneous pain in humans. Behavioral models allow to test whether the sympathetic nervous system contributes to these hyperalgesic/allodynic behaviors or to spontaneous pain behavior. Animals are given clinically relevant nerve lesions, and behavioral experiments are performed before and after interventions aimed at the sympathetic supply of the affected extremity. The strength of these models includes the range of possible experimental manipulations performed at the sympathetic nervous system and the use of different strains of animals or transgenic animals. Quantitative data obtained in these behavioral experiments lead to the formulation of hypotheses that can be tested in reduced experiments *in vivo* and *in vitro* using various methodical approaches.

Several rat models of neuropathic pain involving controlled nerve lesions have been used to determine the contribution of the sympathetic nervous system to neuropathic pain. The results obtained in these experiments are summarized in Table 2. The interventions include surgical or chemical sympathectomy, systemic or local application of adrenoceptor blockers

(e.g., phentolamine (α_1,α_2), prazosin (α_1), or yohimbin (α_2)) or of adrenoceptor agonists (e.g., phenylephrine (α_1) and clonidine (α_2)), and intraperitoneal application of guanethidine (that is taken up by the noradrenergic terminals and depletes noradrenaline).

Overall, the results obtained on these rat behavioral models of nerve injury are somewhat disappointing since they did not lead to straightforward answers. Thus, the controversial results on these models provide conflicting evidence on the role of the sympathetic nervous system in generating pain behavior, indicating that behavioral animal models must be designed with utmost care. Small changes in the experimental procedures may create major behavioral changes. Behavioral testing should be performed in a blinded fashion if possible, and the models must be designed in close association with the clinical situation to be certain to test what is intended to be tested. The results of systemic pharmacological interventions (e.g., application of adrenoceptor agonists or antagonists, guanethidine, or chemical sympathectomy with 5-hydroxy-dopamine) do not necessarily allow functional interpretations, partly because of the widespread systemic effects of these agents. Finally, the controversial results obtained from the animal behavior

Table 2   Nerve lesion animal (rat) models to study sympathetically maintained pain (SMP)

Model	Nerve lesion	Behavioral measurements	Interventions at the sympathetic supply	Results
1. Autotomy model[a]	Ligation and transection of sciatic and saphenous nerve	Autotomy of affected paw (self-mutilation)	Intraperitoneal guanethidine	Autotomy ↓
			Subcutaneous noradrenaline	Autonomy ↑
2. Partial lesion of the sciatic nerve[b]	Ligation of 33–50% of sciatic nerve	Heat HA/ mechanical HA	Intraperitoneal guanethidine, Noradrenaline local	Heat HA ↓, Mechanical HA ø, Heat HA ↑, Mechanical HA ↑
3. Chronic constriction injury of the sciatic nerve[c]	Loose ligation of sciatic nerve	Heat HA/ mechanical HA	Surgical sympathectomy, intraperitoneal guanethidine	Heat HA ↓, Mechanical HA (↓), ø
4. Spinal nerve ligation model[d]	Ligation and transection of spinal nerve L5, L5/L6	Mechanical HA/ allodynia heat HA	Surgical sympathectomy	Mechanical HA ↓ or ø[f] heat HA ↓
5. Cryolysis of the sciatic nerve[e]	Cryolysis of sciatic nerve	Mechanical allodynia	Surgical sympathectomy	Mechanical allodynia ø

[a]Wall P. D. *et al.* (1979a; 1979b); Coderre T. J. *et al.* (1984; 1986); Coderre T. J. and Melzack R. (1986).
[b]Seltzer Z. *et al.* (1990); Shir Y. and Seltzer Z. (1991); Tracey D. J. *et al.* (1995).
[c]Bennett G. J. and Xie Y. K. (1988); Neil A. *et al.* (1991); Desmeules J. A. *et al.* (1995); Kim K. J. *et al.* (1997).
[d]Kim S. H. and Chung J. M. (1991); Kim S. H. *et al.* (1993); Kim K. J. *et al.* (1997); Kinnman E. and Levine J. D. (1995); Ringkamp M. *et al.* (1999).
[e]Willenbring S. *et al.* (1995).
[f]The negative effects of surgical sympathectomy reported by Ringkamp M. *et al.* (1999: Pain 79, 142) were obtained independently by two different groups.
HA, hyperalgesia; ↓, decrease; ↑, increase; ø, no effect.

models may argue that the models do not reflect SMP as defined clinically.

This controversial situation does not mean that behavioral animal models are useless and cannot simulate the clinical situation. On the contrary, we are dependent on the design of these animal models if we want to unravel the mechanisms behind the different types of pathological pain in which the sympathetic nervous system is involved. However, the same model cannot be expected to represent SMP in different groups of patients. Furthermore, the animal models so far tested are related to nerve lesions. Rat models simulating SMP in patients with pain states developing after trauma without nerve lesion (Table 1) practically do not exist (e.g., for patients with CRPS I).

17.3.3   Direct Involvement of the Sympathetic Nervous System in the Generation of Pain Following Nerve Trauma

Under physiological conditions, there exists almost no acute influence of sympathetic postganglionic neurons on sensory neurons projecting to skin or deep somatic tissues in mammals. The effects that have been measured under experimental conditions on sensory receptors with myelinated or unmyelinated axons were weak and can in part be explained by changes of the effector organs (erector pili muscles and blood vessels) induced by the activation of sympathetic neurons. These rather negative results do not rule out that noradrenaline or colocalized substances released by the postganglionic terminals have secondary long-term effects on the excitability of sensory receptors although we have no experimental evidence for this (Jänig, W. and Koltzenburg, M., 1991).

17.3.3.1   Coupling between lesioned postganglionic and afferent nerve terminals

Coupling may occur between sympathetic fibers and afferent terminals in a neuroma, following nerve section or ligation. Some myelinated as well as unmyelinated nerve fibers in the neuroma can be excited following electrical stimulation of the sympathetic supply or by noradrenaline or adrenaline injected systemically. The coupling has been observed in young neuromas but less so in old ones, weeks and months after nerve lesion. Furthermore, the frequency of electrical stimulation applied to the sympathetic trunk in order to excite the lesioned afferent nerve fibers was high, generating activity in the sympathetic neurons that does not occurs under physiological conditions (Jänig, W., 2006). This corresponds to the clinical experience showing that neuroma pain is usually not dependent on sympathetic activity. It also corresponds to histological observations showing that catecholamine-containing axon profiles are rare within the neuroma and for several centimeters proximal to it but may be present in the scar tissue around the neuroma), many weeks after cutting and ligating the nerve (McLachlan, E. M., unpublished data). Thus, this coupling is chemical and occurs via noradrenaline acting on α-adrenoceptors, although other mediator substances may also be involved (Jänig, W., 1990). Ephaptic coupling between sympathetic fibers and afferent fibers has so far not been observed in a neuroma (Devor, M. and Jänig, W., 1981; Blumberg, H. and Jänig, W., 1982; 1984; Jänig, W. and Koltzenburg, M., 1991).

The situation is different when afferent and sympathetic fibers are allowed to regenerate to the target tissue. This has been shown experimentally for the chronic situation more than a year after cross-union of nerves (e.g., adaptation of the proximal stump of the sural or superficial peroneal nerve to the distal stump of the tibial nerve, in the cat) and after reinnervation of appropriate and inappropriate target tissues. Now unmyelinated afferent fibers may be vigorously excited by electrical low-frequency stimulation of the sympathetic supply (Häbler, H. J. *et al.*, 1987). Also this excitation is adrenoceptor-mediated (Figure 8). It can be mimicked by adrenaline or noradrenaline and blocked by α-adrenoceptor blockers (e.g., phentolamine).

17.3.3.2   Coupling between unlesioned postganglionic and afferent nerve terminals following partial nerve lesion

Intact C-fiber polymodal nociceptors in the skin may develop sensitivity to catecholamines following partial nerve injury. Sympathetic nerve stimulation and noradrenaline may excite the polymodal nociceptors or sensitize them for heat stimuli. The activation and sensitization is already seen 4–10 days after partial nerve lesion and is maintained for at least 150 days. This sympathetic afferent coupling apparently involves the nonlesioned polymodal nociceptive

afferent axons that project through the lesion site and nonlesioned postganglionic axons. It is suggested that the unlesioned unmyelinated afferents develop hyper-reactivity to catecholamines following degeneration of the sympathetic postganglionic axons. Expression of adrenoceptors in afferent fibers may be triggered by collateral sprouting of both afferent fibers and postganglionic fibers in the target tissue (Sato, J. and Perl, E. R., 1991). However, sympathetic afferent coupling leading to activation or sensitization of nociceptors requires sympathetic stimulation at high nonphysiological frequencies.

17.3.3.3   Coupling in the dorsal root ganglion and collateral sprouting following peripheral nerve lesion

Chemical sympathetic afferent coupling following peripheral nerve lesion (e.g., a spinal nerve, the sciatic nerve, or another hindlimb nerve) may occur proximally to the injury site in the nerve or in the DRG (Figure 6a3). Sympathetic postganglionic fibers reach the spinal nerves via gray rami. Most of these fibers project distally to peripheral target cells, and others project proximally (e.g., to the DRG) and are normally found along blood vessels. This situation changes after an experimental nerve lesion (e.g., transection and ligation of the sciatic nerve in rats). Now many perivascular catecholamine-containing axons of the unlesioned proximally projecting neurons start to sprout in the DRGs that contain somata with lesioned axons. The extent of this collateral sprouting increases with time after nerve lesion. Some somata are partially or almost completely surrounded by varicose catecholaminergic terminals several weeks after nerve lesion; the frequency of these catecholamine-fluorescent structures increases for more than 70 days after nerve lesion. The noradrenergic axons sprout preferentially to axotomized large-diameter afferent cell bodies (McLachlan, E. M. *et al.*, 1993; Chapter Sprouting in Dorsal Root Ganglia).

Neurophysiological experiments on rats with nerve lesion (ligation and transection of a spinal nerve, the sciatic nerve, or other hindlimb nerves) show that afferent neurons projecting in the lesioned nerve can be excited or depressed in their activity affected via the DRG by electrical stimulation of sympathetic neurons and by catecholamines: (1) In the first 2 to 3 weeks after nerve lesion, most A-fiber neurons with spontaneous activity and only very few silent A-fiber neurons are excited. At later time periods, when the catecholaminergic sprouting in the

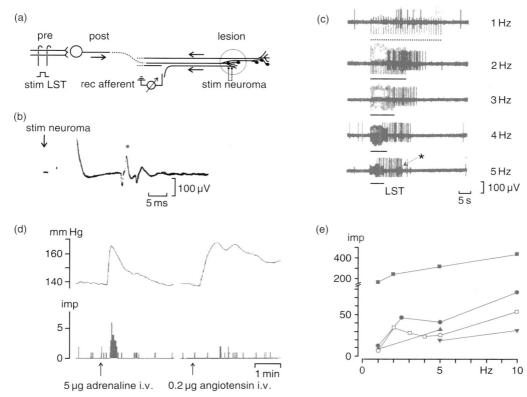

Figure 8 Sympathetic afferent coupling after nerve lesion with subsequent fiber regeneration. Excitation of unmyelinated afferent fibers by electrical stimulation of sympathetic fibers following nerve injury. Unmyelinated primary afferents were recorded in cats 11–20 months following a nerve lesion. The central cut stump of a cutaneous nerve innervating hairy skin (sural or superficial peroneal nerve) had been imperfectly adapted to the distal stump of a transected mixed nerve (tibial nerve). This preparation was designed to mimic the consequences of lesion of a mixed nerve. A neuroma-in-continuity at the lesion site was present and cutaneous nerve fibers had regenerated into skin and deep somatic tissue supplied by the mixed nerve. (a) Experimental setup. post, postganglionic; pre, preganglionic; LST, lumbar sympathetic trunk. (b) The afferent fibers were identified as unmyelinated by electrical stimulation of the neuroma with single impulses. The signal indicated by the dot was recorded from the same afferent fiber as in (c) and (d); the afferent fiber conducted at 1.3 m s^{-1}. (c) Record from a single unmyelinated afferent unit. Supramaximal stimulation of the LST with trains of 30 pulses at 1–5 Hz (trains and stimulation artifacts indicated by bars). Note that the afferent unit had some low rate of ongoing activity (impulses before the trains at 1 and 4 Hz) and a second fiber was activated at 5 Hz stimulation (marked by an asterisk). (d) Adrenaline (5 μg injected i.v.) activated the fiber. Angiotensin (0.2 μg injected i.v.) generated a large increase of blood pressure (MAP, mean arterial blood pressure) but did not activate the afferent fiber. (e) Stimulus response curves for the single unit (□) and four filaments containing two to three (●, ▲/▼) or more than five (■) afferent units. Ordinate scale is the total number of impulses exceeding ongoing activity in response to variable stimulation frequency of the LST. Modified from Häbler H. J. et al. (1987).

DRG is more prominent, most spontaneously active A-fiber neurons are inhibited during stimulation of the sympathetic supply but rarely excited. (2) Only very few afferent neurons with unmyelinated axons respond to electrical stimulation of the sympathetic trunk; most of them are inhibited in their activity (McLachlan, E. M. et al., 1993; Devor, M. et al., 1994; Michaelis, M. et al., 1996). (3) The coupling from sympathetic postganglionic fibers to the afferent cell bodies in the DRG probably occurs only to spontaneously active muscle afferent neurons with Aδ-fibers (Michaelis, M. et al., 2000). (4) Activation of the afferent neurons, via the DRG, requires high-frequency stimulation of the sympathetic neurons at 5 to 20 Hz. Sympathetic pre- and postganglionic neurons rarely discharge at these frequencies under physiological conditions. (5) Mechanical allodynic and hyperalgesic behavior shown by rats with L5 spinal nerve injury (transection and ligation) is not dependent on the innervation of the DRG by sympathetic neurons (Ringkamp, M. et al., 1999; see point 4 in Table 2). The rate of ectopic activity originating in the DRG, 15 to 45 days after spinal nerve injury, is not maintained by activity in sympathetic neurons (Liu, X. et al., 2000). (6) Afferent neurons in the DRG may be activated indirectly by ischemia generated by decrease of blood flow during vasoconstriction in the DRG following electrical stimulation of the sympathetic trunk and not directly via adrenoceptors expressed in the DRG cells (Häbler, H.-J. et al., 2000). (7) Axotomized medium- to large-diameter DRG neurons express α_{2A}-adrenoceptors after sciatic nerve injury (Birder, L. A. and Perl, E. R., 1999). However, these upregulated adreceptors disappear after several weeks as perineural catecholaminergic sprouting develops and only small depolarizations can be generated by high concentrations of noradrenaline in axotomized DRG somata (Lopez de Armentia, M. et al., 2003).

The mechanism leading to the collateral sprouting of postganglionic noradrenergic fibers in the DRG is possibly related to neurotrophic signals and their receptors generated by afferent cells with lesioned axons in the DRG and the surrounding satellite glia cells (see Chapter Sprouting in Dorsal Root Ganglia). Whether these aberrant pathological connections can account for spontaneous pain and allodynia in some patients after peripheral nerve lesions is doubtful and awaits further investigations (Jänig, W. et al., 1996).

17.3.3.4 Adrenoceptors involved in chemical sympathetic afferent coupling following nerve lesion

Excitation or depression of lesioned primary afferent neurons (in the DRG, in their lesioned terminals in the neuroma or in their regenerating sprouts), or of unlesioned collaterally sprouting primary afferents generated by activation of the sympathetic innervation, is mimicked by systemic injection of noradrenaline or adrenaline and prevented by blockade of α-adrenoceptors (e.g., phentolamine). Thus, both excitation and depression are suggested to be mediated by α-adrenoceptors. The cellular mechanisms underlying the increased sensitivity are unknown. Novel expression or upregulation of adrenoceptors occurs; alternatively, normally present adrenoreceptors that are not functional become uncovered and effective during the response to damage. The subtype of α-adrenoreceptor being involved in the sympathetic–afferent coupling in the different rat models is predominantly α_2 (Chen, Y. et al., 1996). Knowledge about the subtypes of adrenoceptor following nerve trauma may turn out to be useful in the design of more specific treatment modalities for neuropathic pain conditions involving sympathetic efferent activity (Jänig, W. et al., 1996; Jänig, W. and Häbler, H. J., 2000b).

17.3.3.5 Synopsis

Peripheral trauma with nerve injury may lead to sensitization and activation of nociceptive and activation of other primary afferent neurons. These processes may depend on and are maintained by the sympathetic nervous system. Several ways of coupling between sympathetic postganglionic neurons and primary afferent neurons are possible and have been worked out experimentally on animal models with controlled nerve lesions, showing that there may develop intimate relationships between sympathetic and afferent neurons under pathophysiological conditions (Figure 9):

1. The sympathetic postganglionic neuron may develop chemically mediated cross talk to primary afferent neurons. Whether this occurs probably depends on the time after nerve lesion as well as on the type of nerve lesion (partial or complete).
2. Following peripheral nerve lesion, remote collateral sprouting of unlesioned postganglionic fibers occurs in the DRG, preferentially toward the large-diameter sensory cells. Collateral sprouting of unlesioned postganglionic fibers may also occur in the peripheral target tissue (in particular after partial nerve lesion).
3. The transmission from sympathetic neurons to afferent neurons is mediated by noradrenaline, but additional mediator substances cannot be excluded.
4. Primary afferent neurons express functional adrenoceptors. The type of adrenoceptor involved is preferentially α_2 in animal models.
5. The signals that initiate the sprouting and the functional expression of adrenoceptors in the afferent neurons may be related to neurotrophic substances that derive from the Schwann cells or other cells. Their upregulation is related to the plastic changes of the primary afferent and sympathetic postganglionic neurons and possibly to decrease of the density of the noradrenergic innervation and/or to activity in the afferent and sympathetic neurons (see Chapter Sprouting in Dorsal Root Ganglia).

These plastic changes of primary afferent and sympathetic postganglionic neurons following peripheral trauma with nerve lesion may explain some of the

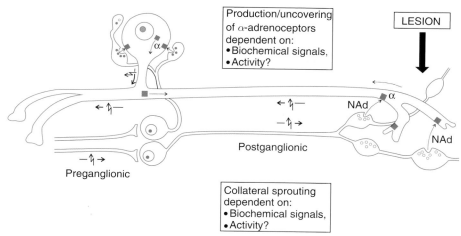

Figure 9 Relation between afferent and sympathetic neurons following peripheral nerve lesion. Collateral sprouting of unlesioned sympathetic neurons in the dorsal root ganglion (DRG) and in the peripheral target tissue. Upregulation or uncovering of functional adrenoceptors (α) by afferent neurons after nerve lesion. It is unclear in which way these processes are related to the biochemical signals (e.g., neurotrophins) synthesized by neurons, Schwann cells, and other cells in the DRG and the expression of their receptors. In which way are these processes dependent on activity in the afferent neurons, on the presence/absence of postganglionic noradrenergic neurons, or on the activity in the postganglionic neurons? NAd, noradrenaline. Modified from Jänig W. et al. (1996).

clinical sensory phenomena in patients with SMP, e.g., in CRPS type II. However, they do not explain (1) the slow development of spontaneous pain and allodynia after noradrenaline is injected into the skin that was painful before sympathetic block or before sympathectomy (Torebjörk, H. E. et al., 1995) (Figure 4); (2) the long-lasting pain relief in patients with SMP after sympathetic blocks Figure 15 and (3) the mechanism of SMP in patients without nerve lesion, e.g., in patients with CRPS type I.

17.3.4 Indirect Involvement of the Sympathetic Nervous System in the Generation of Pain

Nociceptive afferents are embedded in a complex micromilieu (Figure 10). The state of this micromilieu surrounding the receptive terminals depends on mediator substances that are released during inflammatory processes following trauma from non-neural cells such as mast cells, polymorphonuclear leucocytes, macrophages, fibroblasts, endothelial cells, or other cells. The microcirculation is under neural control of sympathetic vasoconstrictor neurons. Moreover, activation of subgroups of nociceptive primary afferents causes not only orthodromic impulse traffic but also arteriolar (precapillary) vasodilation and (in most tissues, but not in human skin) venular plasma extravasation by release of

neuropeptides from the receptive terminals (e.g., substance P (SP) and calcitonin-gene-related peptide (CGRP)). Both afferent-induced changes are called neurogenic inflammation (McDonald, D. M., 1997; Holzer, P., 1998). Thus, there are possibilities for indirect coupling between sympathetic and afferent nerve terminals (Figure 10): (1) Vascular perfusion of the micromilieu surrounding the nociceptors after nerve trauma may change as consequences of denervation and reinnervation by postganglionic vasoconstrictor neurons and afferent nociceptive neurons and hyper-reactivity of blood vessels to nerve impulses and circulating catecholamines may develop (Jobling, P. et al., 1992; Koltzenburg, M. et al., 1995). (2) Nociceptors may be sensitized by inflammatory mediators such as prostaglandins and interleukins, the release of which from nonneuronal cells of the micromilieu may be mediated by noradrenergic nerve terminals.

17.3.4.1 Changes of neurovascular transmission and development of hyperreactivity of blood vessels

Neural control of blood vessels can change dramatically after trauma with nerve lesion, but possibly also after trauma without (Koltzenburg, M. et al., 1995):

1. Cutaneous blood vessels that are reinnervated after a nerve lesion exhibit stronger than normal

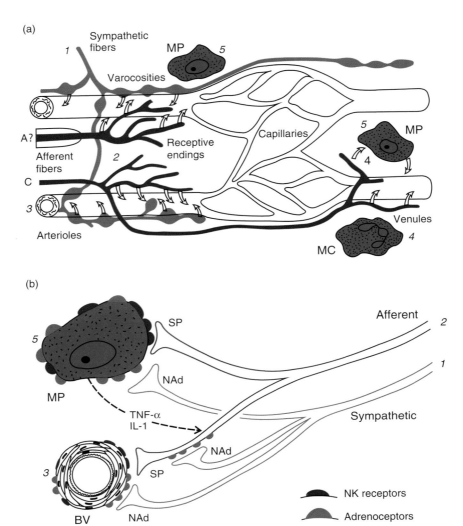

Figure 10 (a) The micromilieu of nociceptors. The microenvironment of primary afferents is thought to affect the properties of the receptive endings of myelinated (Aδ) and unmyelinated (C) afferent fibers. This has been particularly documented for inflammatory processes, but one may speculate that pathological changes in the direct surroundings of primary afferents may contribute to other pain states as well. The vascular bed consists of arterioles (directly innervated by sympathetic and afferent (Aδ, C) fibers), capillaries (not innervated and not influenced by nerve fibers), and venules (not directly innervated but influenced by nerve fibers). The micromilieu depends on several interacting components: Neural activity in postganglionic noradrenergic fibers (1) supplying blood vessels (3, BV) causes release of noradrenaline (NAd) and possibly other substances causing vasoconstriction. Excitation of primary afferents (Aδ- and C-fibers) (2) causes vasodilation in precapillary arterioles and plasma extravasation in postcapillary venules (C-fibers only) by the release of SP and other vasoactive compounds (e.g., calcitonin gene-related peptide, CGRP). Some of these effects may be mediated by nonneuronal cells such as mast cells (MC, 4) and macrophages (MP, 5). Other factors that affect the control of the microcirculation are the myogenic properties of arterioles (3) and more global environmental influences such as a change of the temperature and the metabolic state of the tissue (modified from Jänig, W. and Koltzenburg, M., 1991). (b) Hypothetical relation between sympathetic noradrenergic nerve fibers (1), peptidergic afferent nerve fibers (2), blood vessels (3, BV), and macrophages (4, MP). The activated and sensitized afferent nerve fibers activate macrophages (via SP release). The immune cells start to release cytokines, such as tumor necrosis factor α (TNF-α) and interleukin 1 (IL-1), which further activate afferent fibers by enhancing sodium influx into the cells. SP (and CGRP) released from the afferent nerve fibers reacts with neurokinin 1 (NK1) receptors (CGRP receptors) in the blood vessels (arteriolar vasodilation, venular plasma extravasation; neurogenic inflammation). The sympathetic nerve fibers interact with this system potentially on three levels: (1) via adrenoceptors (mainly α) on the blood vessels (vasoconstriction), (2) via adrenoceptors (mainly β) on macrophages (further release of cytokines), and (3) via adrenoceptors (mainly α) on afferents (further sensitization of these fibers). Modified from Jänig W. and Baron R. (2003).

vasoconstriction to impulses in sympathetic neurons.

2. The sympathetically reinnervated cutaneous blood vessels show stronger than normal vasoconstrictions to systemic catecholamines and appear to be hyper-reactive.

3. Blood vessels exhibit changes in vasodilation to antidromic activation of reinnervated peptidergic unmyelinated afferents.

The reinnervated blood vessels may therefore be under stronger than normal vasoconstrictor influence that can no longer be counteracted by an afferent-mediated vasodilation (see Häbler, H. J., *et al.*, 1997). *In vitro* investigation of the rat tail artery has shown that the functional recovery of neurovascular transmission is permanently disturbed after a nerve lesion (Jobling, P. *et al.*, 1992).

The altered neural and nonneural control of blood vessels, following trauma with nerve lesion, can contribute to the abnormal regulation of blood flow through skin and deep somatic tissues and possibly to the trophic changes (including the edema) that are seen in patients with CRPS. They may furthermore be a permissive factor in the generation of afferent nociceptive impulse activity and therefore in the sensitization of nociceptors and in the generation of pain.

Finally, it is commonly believed that skin temperature changes observed in painful disorders following trauma with or without nerve lesion (e.g., in CRPS I or II) reflect changes of activity in cutaneous vasoconstrictor neurons, in the sense that cold skin is associated with a high level and warm skin with a low level of activity in these neurons. This is probably a misconception, and there is no proof for this relation between skin temperature and activity in sympathetic neurons under these pathophysiological conditions.

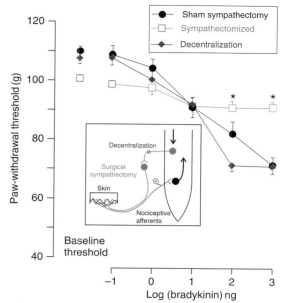

Figure 11 Mechanical hyperalgesic behavior and sympathetic innervation. Cutaneous nociceptors on the dorsum of the paw are stimulated by a linearly increasing mechanical force. Paw-withdrawal threshold (PWT) is defined as the mean ($\pm$SEM) minimum force (in grams, ordinate scale) at which the rat withdraws its paw. Abscissa scale: log dose of bradykinin (BK, in nanograms) injected in a volume of 2.5 μl of saline into the dermis of the skin. Decrease of PWT to mechanical stimulation induced by intradermal injection of BK in sham-sympathectomized (control) rats ($\bullet$, 6 hindpaws), in surgically sympathectomized rats (lumbar paravertebral ganglia L2 – L4 removed 8 days before; $\square$, 12 hindpaws), and in rats with decentralized lumbar sympathetic chains (white rami with preganglionic axons to paravertebral ganglia L2 and L3 and chain rostral to ganglion L2 transected 8 days before; $\blacklozenge$, 10 paws). Abscissa scale: log dose of BK (in nanograms). Sham sympathectomy and sympathetic decentralization were both significantly different from sympathectomy groups at doses of BK of 10^2 and 10^3 ng ($*$, $P < 0.01$). Modified from Khasar S. G. *et al.* (1998a).

17.3.4.2 Sensitization of nociceptors mediated by sympathetic terminals independent of excitation and release of noradrenaline

Withdrawal threshold to stimulation of the rat hindpaw with a linearly increasing mechanical stimulus applied to the dorsum of the paw decreases dose-dependently after intradermal injection of the inflammatory mediator bradykinin (an octapeptide cleaved from plasma α_2-globulins, by kallikreins circulating in the plasma) (Figure 11). Following a

single injection of bradykinin, this decrease lasts for more than 1 h for mechanical stimulation. This type of mechanical hyperalgesic behavior is mediated by the B_2 bradykinin-receptor and is not present when bradykinin is injected subcutaneously (Khasar, S. G. *et al.*, 1993; 1995; Jänig, W. and Häbler, H. J., 2000b). In normal rat, bradykinin-induced hyperalgesic behavior is blocked by the cyclooxygenase inhibitor indomethacin and therefore mediated by a prostaglandin (probably PGE_2) that sensitizes nociceptors for mechanical stimulation. However, in vagotomized rats in which the bradykinin-induced mechanical hyperalgesia is significantly enhanced (by activation of the adrenal medulla, see below and Figure 13a), indomethacin has almost no effect on bradykinin-induced hyperalgesia. This failure is not related to a switch from B2- to B1-receptor subtype because the selective B1-receptor agonist des-Arg9-bradykinin fails to produce hyperalgesia in vagotomized rats (Khasar, S. G. *et al.*, 1998a). The same applies to rats that were sound-stressed for 4 days (Khasar, S. G. *et al.*, 2005). Thus, bradykinin-induced hyperalgesia may not be mediated by prostanoids in vagotomized or sound-stressed rats. Furthermore, convincing sensitization to mechanical stimulation by bradykinin has only been demonstrated for afferent fibers innervating the cat knee joint (Neugebauer, V. *et al.*, 1989) or the cat skeletal muscle (Mense, S. and Meyer, H., 1988). Manning D. C. *et al.* (1991) demonstrated in humans that intradermal injection of bradykinin (100 ng to 10 μg) generates hyperalgesia to heat stimulation but not to mechanical stimulation.

The decrease in paw-withdrawal threshold to mechanical stimulation generated by intradermal injection of bradykinin is significantly reduced after surgical sympathectomy. Decentralization of the lumbar paravertebral sympathetic ganglia (denervating the postganglionic neurons by cutting the preganglionic sympathetic axons) does not abolish the bradykinin-induced mechanical hyperalgesic behavior (Figure 11). This indicates that the sensitizing effect of bradykinin is not dependent on activity in the sympathetic neurons innervating skin and therefore not on release of noradrenaline (Khasar, S. G. *et al.*, 1998a) and that bradykinin stimulates the release of prostaglandin from the sympathetic terminals or from other cells in association with the sympathetic terminals in the skin (Figure 6b).

Mechanical hyperalgesic behavior generated by intracutaneous injection of bradykinin and its dependence on the sympathetic innervation of the skin is an interesting phenomenon. However, that mechanical hyperalgesia is due to prostanoids sensitizing nociceptors since indomethacin prevents it appears to be too simple. The reasons are (1) the finding that sympathetic fibers mediate this hyperalgesia independent of neural activity and release of noradrenaline; (2) indomethacin does not block this behavior under certain conditions (e.g., when the adrenal medullae are activated after vagotomy (see below) or after sound stress); (3) sensitization of nociceptive afferents innervating deep somatic tissues by bradykinin to mechanical stimulation has been demonstrated in neurophysiological

experiments in the cat, not yet in humans for cutaneous pain (Manning, D. C. *et al.*, 1991); (4) in skin there is a spatial dissociation between afferent nociceptive fibers and sympathetic fibers, many of the former ending in the epidermis and subepidermal plexus, whereas most of the latter terminate in deeper dermal layer, raising the question as to the nature of the interaction between both groups of fibers.

These points argue that the dependence of bradykinin-induced cutaneous mechanical behavior on the sympathetic innervation has to be reinvestigated using a rigorous experimental approach. Neurophysiological experiments are required to demonstrate that sensitization of nociceptors by bradykinin is dependent on the presence of sympathetic terminals.

17.3.4.3 Sensitization of nociceptors initiated by nerve growth factor or cytokines possibly mediated by sympathetic terminals

17.3.4.3.(i) Nerve growth factor Systemic injection of nerve growth factor (NGF) is followed by a transient thermal and mechanical hyperalgesia in rats (Lewin, G. R. *et al.*, 1993; 1994) and humans (Petty, B. G. *et al.*, 1994). During experimental inflammation (evoked by Freund's adjuvant in the rat hindpaw), NGF increases in the inflamed tissue paralleled by the development of thermal and mechanical hyperalgesia. Both are prevented by anti-NGF antibodies (Lewin, G. R. *et al.*, 1994). The mechanisms responsible are sensitization of nociceptors via high-affinity NGF-receptors (trkA receptors) and an increased synthesis of CGRP and SP in the afferent cell bodies induced by NGF taken up by the afferent terminals and transported to the cell bodies. The NGF-induced sensitization of nociceptors or part of it seems to be mediated indirectly by the sympathetic postganglionic terminals. In rats, heat and mechanical hyperalgesic behavior generated by local injection of NGF into the skin is prevented or significantly reduced after chemical or surgical sympathectomy (Woolf, C. J. *et al.*, 1996). These experiments suggest that the NGF released during inflammation by inflammatory cells acts on the sympathetic terminals via high-affinity trkA receptors, inducing the release of inflammatory mediators (the source of which is unknown) and subsequently sensitization of nociceptors for mechanical and heat stimuli (Figure 6c; McMahon, S. B., 1996; Woolf, C. J., 1996; Jänig, W. and Häbler, H. J., 2000b). It is unclear whether this sensitization of nociceptors mediated by terminal sympathetic nerve fibers (1) is dependent on activity in the sympathetic neurons and release of noradrenaline and on adrenoceptors expressed in the nociceptive afferent neurons or (2) is independent of neural activity and release of noradrenaline.

17.3.4.3.(ii) *Proinflammatory cytokines*

Based on behavioral experiments conducted on rats (studying mechanical and heat hyperalgesia), it has been shown that tissue injury, injection of the bacterial cell wall endotoxin lipopolysaccharide, or injection of carrageenan (a plant polysaccharide) stimulates tissue inflammation and leads to sensitization of nociceptors. Systematic pharmacological interventions using blockers or inhibitors of the various mediators demonstrate that the proinflammatory cytokines, tumor necrosis factor α (TNF-α), interleukin-1 (IL-1), IL-6, and IL-8 may be involved in this process of nociceptor sensitization and therefore in the generation of hyperalgesia (Woolf, C. J. et al., 1996; 1997; Poole, S. et al., 1999). Pathogenic stimuli lead to activation of resident cells and release of the inflammatory mediator bradykinin and other mediators. The inflammatory mediators and the pathogenic stimuli themselves activate macrophages, monocytes, and other immune-related cells that release TNF-alpha. Nociceptors are believed to be sensitized by two possible pathways (Figure 12): (1) TNF-α induces production of IL-6 and IL-1β by immune cells, whereby IL-6 enhances the production of IL-1β. These interleukins stimulate cyclooxygenase 2 (COX 2) and the production of prostaglandins (PGE$_2$, PGI$_2$) that in turn react with the nociceptive terminal via E-type prostaglandin receptors. (2) TNF-α induces the release of IL-8 from endothelial cells and macrophages. IL-8 reacts with the sympathetic terminals that are supposed to mediate sensitization of nociceptive afferent terminals by release of noradrenaline to act via β_2-adrenoceptors. These two peripheral pathways by which nociceptive afferents are hypothesized to be sensitized involving cytokines are under inhibitory control of circulating glucocorticoids (indicated by asterisks in Figure 12) and of other, anti-inflammatory, interleukins (e.g., IL-4 and IL-10 indicated by # in Figure 12).

It is important to emphasize that the mechanisms of sensitization of nociceptive afferents, involving NGF, proinflammatory cytokines, and noradrenergic sympathetic postganglionic fibers, have been deduced on the basis of behavioral and pharmacological experiments. Proof of such interaction by directly assessing the activity of nociceptors with electrophysiological techniques and of the effect of noradrenergic nerve

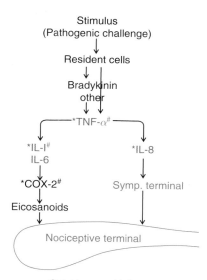

Cytokines and inflammatory hyperalgesia
(Sensitization of nociceptors)

Figure 12 Hypothetical role of cytokines in sensitization of nociceptors during inflammation and the underlying putative peripheral mechanisms leading to hyperalgesia. Pathogenic stimuli activate resident cells and lead to release of inflammatory mediators (such as bradykinin). Proinflammatory cytokines are synthesized and released by macrophages and other immune or immune-related cells. Nociceptors are postulated to be sensitized by two pathways involving the cytokines: (1) Tumor necrosis factor α (TNF-α) induces synthesis and release of interleukin 1 (IL-1) and IL-6 which, in turn, induce the release of eicosanoids (prostaglandin E$_2$ and I$_2$ (PGE$_2$, PGI$_2$)) by activating cyclooxygenase-2 (COX-2). (2) TNF-α induces synthesis and release of IL-8. IL-8 activates sympathetic terminals that sensitize nociceptors via β_2-adrenoceptors. Glucocorticoids inhibit the synthesis of the cytokines and the activation of COX-2 (indicated by asterisks). Anti-inflammatory cytokines (such as IL-4 and IL-10) that are also synthesized and released by immune cells inhibit the synthesis and release of proinflammatory cytokines (indicated by #). This scheme is fully dependent on behavioral experiments and pharmacological interventions. The different steps will need to be verified experimentally using neurophysiological experiments. Modified from Poole S. et al. (1999).

fibers on the afferent nerve fibers are lacking. These experiments have to be done.

17.3.4.4 *Sympatho-adrenal system and nociceptor sensitization*

Activity of the sympatho-adrenal system (adrenal medulla) is regulated by central circuits that are different from those regulating other sympathetic systems (Morrison, S. F. and Cao, W. H., 2000;

Morrison, S. F., 2001; Jänig, W., 2006). Activation of this system (e.g., by release of the central sympathetic circuits, connected to preganglionic neurons that innervate the adrenal medulla, from vagal inhibition by subdiaphragmatic vagotomy (Figure 13c; Khasar, S. G. *et al.*, 1998a; 1998b), but not of the sympatho-neural system, generates mechanical hyperalgesia and enhances bradykinin-induced mechanical hyperalgesia (decrease of paw-withdrawal threshold to mechanical stimulation of the skin, compare closed circles with open triangles in Figure 13a). Both develop slowly over 7 to 14 days during continuous activation of the adrenal medullae (e.g., induced by subdiaphragmatic vagotomy) and are maintained over 5 weeks tested (open triangles in Figure 13b). In rats with denervated adrenal medullae (adrenal nerves with preganglionic axons innervating the cells of the adrenal medulla sectioned (Figure 13c)), subdiaphragmatic

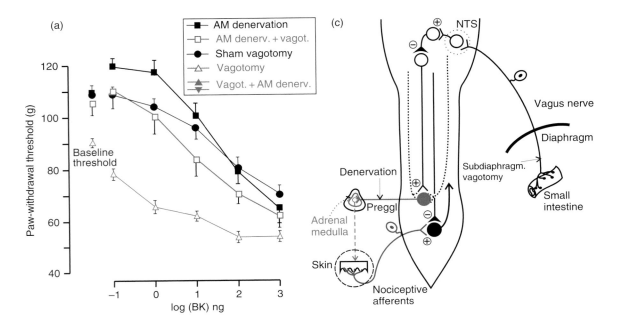

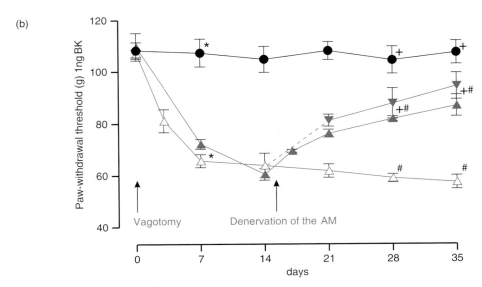

vagotomy has only a small effect (compare open squares with closed squares in Figure 13a). This decrease of paw-withdrawal threshold can be explained by removal of inhibition of nociceptive impulse transmission in the spinal dorsal horn (Figure 13c; see Randich, A. and Gebhart, G. F., 1992; see Chapter Descending Control Mechanisms). Denervation of the adrenal medullae, 14 days after vagotomy when the decrease of paw-withdrawal threshold is maximal, is followed by a slow increase of paw-withdrawal threshold taking more than 3 weeks (compare closed triangles with open triangles in Figure 13b). Thus, the decrease of paw-withdrawal threshold following vagotomy is reversed after denervation of the adrenal medullae. Application of adrenaline through a minipump for 3 to 14 days slowly enhances the development of mechanical hyperalgesia induced by bradykinin injected intradermally. Furthermore, continuous application of a β_2-adrenoceptor blocker over 7–14 days by way of a minipump prevents the enhancement of this hyperalgesic behavior generated by activation of the adrenal medullae (Khasar, S. G. et al., 2003).

The results of this experiment are interpreted in the following way: Adrenaline released by persistent activation of the adrenal medullae sensitizes cutaneous nociceptors for mechanical stimuli. This sensitization of nociceptors and its reversal are slow and take days to develop. The slow time course implies that the nociceptor sensitization cannot be acutely blocked by an adrenoceptor antagonist given intracutaneously or systemically. Adrenaline probably does not act directly on the cutaneous nociceptors but on cells in the microenvironment of the nociceptors inducing slow changes, which result in nociceptor sensitization. Candidate cells may be mast cells, macrophages, or keratinocytes, which then release substances that generate sensitization (Khasar, S. G. et al., 1998b; Jänig, W. et al., 2000; Jänig, W. and Häbler, H. J., 2000b). The change of sensitivity of a population of nociceptors generated by adrenaline released by the adrenal medullae that is regulated by the brain would be a novel mechanism of sensitization. This novel mechanism would be different from mechanisms that lead to activation and/or sensitization of nociceptors by sympathetic afferent coupling as discussed above (Figure 6d).

This peripheral mechanism of pain and hyperalgesia involving adrenaline released by the sympathoadrenal system and acting on nociceptors may operate in ill-defined pain syndromes such as irritable bowel syndrome, functional dyspepsia, fibromyalgia, and chronic fatigue syndrome (Wolfe, F. et al., 1990; Mayer, E. A. et al., 1995; Clauw, D. J. and Chrousos, G. P., 1997). The conclusions about long-term sensitization of nociceptors for mechanical stimulation by adrenaline fully rest on behavioral experiments. Neurophysiological experiments in vivo on primary afferent nociceptive neurons are required to test directly whether all or only a subpopulation of nociceptors are sensitized.

Figure 13 Long-term enhancement of bradykinin (BK)-induced mechanical hyperalgesia after vagotomy and its disappearance after denervation of the adrenal medullae (AM). (a) Decrease of paw withdrawal threshold (PWT) to mechanical stimulation of the dorsum of the rat hindpaw induced by BK injected intradermally (BK-induced behavioral mechanical hyperalgesia), in sham-vagotomized rats (●, $n = 18$), in vagotomized rats (△, $n = 16$; 7 days after subdiaphragmatic vagotomy), in rats with denervated adrenal medullae (■, $n = 6$), and in vagotomized rats with denervated AM (□, $n = 6$). For details see legend of Figure 11. (b) Total change of PWT (baseline plus BK-induced) in response to intradermal injection of 1 ng BK in rats before and 7 to 35 days after vagotomy (△, $n = 6$), before and 7 to 35 days after sham vagotomy (●, $n = 8$) and in rats whose AM are denervated 14 days after vagotomy and measurements taken up to 35 days after initial surgery (▲, rats tested after vagotomy and denervation of the AM ($n = 6$); ▼, rats only tested after additional denervation of the AM ($n = 4$)). Note that 1ng BK injected intradermally does not significantly change the PWT in rats with intact vagus nerves (see (a)). $^*P < 0.01$, ▲/△ versus ● on day 7; $^+P > 0.05$ ▲/▼ versus ● on days 28 and 35; $^\#P < 0.01$, ▲/▼ versus △ on days 28 and 35. (c) Schematic diagram showing the proposed neural circuits in spinal cord and brainstem which modulate nociceptor sensitivity via the sympatho-adrenal system. Sensitivity of cutaneous nociceptors for mechanical stimulation is modulated by adrenaline released by the AM. Activation of the AM increases the sensitivity of the nociceptors. Activity in preganglionic neurons innervating the AM depends on activity in vagal afferents from the small intestine that has an inhibitory influence on the central pathways to these preganglionic (preggl) neurons. Thus, interruption of the vagal afferents leads to activation of the AM. It is hypothesized that these neuronal (reflex) circuits in the brainstem are under the control of upper brainstem, hypothalamus, and telencephalon. Dotted thin lines: Axons of sympathetic premotor neurons in the brainstem that project through the dorsolateral funiculi of the spinal cord to the preganglionic neurons of the AM. ⊕, excitation; ⊖, inhibition. Modified from Khasar S. G. et al. (1998b).

17.4   Sympathetic Nervous System and Central Integrative Mechanisms in the Control of Hyperalgesia and Inflammation

The brain is suggested to modulate inflammation via the sympatho-neural and/or the sympatho-adrenal system although the mechanisms underlying these influences are little understood. Any influence of this type will affect indirectly the sensitivity of nociceptors. Here I will briefly discuss the hypothetical role of both sympathetic systems: (1) in CRPS that is characterized by a peripheral inflammatory process (in addition to changes in the central nervous system) suggested to be dependent on the sympathetic nervous system; (2) in experimental inflammation (bradykinin-induced synovial plasma extravasation); and (3) in the potential modulation of the immune tissue.

17.4.1   The Complex Regional Pain Syndrome Type I as Model

CRPS are painful disorders that may develop as a consequence of trauma typically affecting the limbs. Clinically, they are characterized by pain (spontaneous, hyperalgesia, and allodynia), active and passive movement disorders including an increased physiological tremor, abnormal regulation of blood flow and sweating, edema of skin and subcutaneous tissues, and trophic changes of skin, appendages of skin, and subcutaneous tissues (Stanton-Hicks, M. et al., 1995; Jänig, W. and Stanton-Hicks, M., 1996; Harden, R. N. et al., 2001; Jänig, W. and Baron, R., 2002; 2003). CRPS type I (previously called reflex sympathetic dystrophy) usually develops after minor trauma with a small or no obvious nerve lesion at an extremity (e.g., bone fracture, sprains, bruises or skin lesions, and surgeries) and rarely also after remote trauma in the visceral domain or after a lesion of the central nervous system (e.g., stroke). An important feature of CRPS I is that the severity and combination of clinical symptoms are disproportionate to the severity and type of trauma with a tendency to spread in the affected distal limb. The symptoms are not confined to the innervation zone of an individual nerve. CRPS type II (previously called causalgia) develops after trauma with a mostly large nerve lesion (Stanton-Hicks, M. et al., 1995; Jänig, W. and Stanton-Hicks, M., 1996; Harden, R. N. et al., 2001).

On the basis of clinical observations and research on human patients and animals, the hypothesis has been put forward that CRPS (in particular type I) is a systemic disease involving the central and peripheral nervous system (Figure 14). Various traumas can trigger variable combinations of clinical phenomena in which the somatosensory system, the sympathetic nervous system, the somatomotor system, and peripheral (vascular and inflammatory) systems are involved (Harden, R. N. et al., 2001; Jänig, W. and Baron, R., 2002; 2003). The central changes are reflected in changes of somatic sensations (increase of detection thresholds for mechanical, cold, warm, and heat stimuli), of motor performances, and of neural regulation of sympathetic effector systems (vasculature, sweat glands, inflammatory cells, etc.). Thus, it is hypothesized that the central representations of the sensory, motor, and sympathetic systems are changed. The peripheral changes consist of inflammation involving blood vessels, inflammatory cells, peptidergic afferent nerve fibers, and sympathetic afferent coupling (see above and Figures 6 and 10). The peripheral changes cannot be seen independently of the central ones; both interact with each other via afferent and efferent signals. Furthermore, the mechanisms that underlie CRPS cannot be reduced to one system or to one mechanism only (e.g., to sympathetic afferent coupling, to an adrenoceptor disease, to a peripheral inflammatory disease, and to a psychogenic disease) (Harden, R. N. et al., 2001; Jänig, W. and Baron, R., 2002; 2003).

The centrally generated activity in sympathetic neurons may be involved in various ways in the pathogenesis of CRPS type I, the main argument being that the peripheral changes are reversed or aggravated after intervention at the peripheral sympathetic nervous system (sympathetic blocks, stimulation of the sympathetic innervation; Figure 14). Three points that are relevant in the present context of sympathetic nervous system, pain, and body protection need to be emphasized for CRPS type I with SMP:

● In a subgroup of CRPS I patients, pain or a component of pain is obviously dependent on activity in sympathetic neurons and related to activation or sensitization of nociceptors by noradrenaline released by the sympathetic fibers (SMP). The nociceptors have expressed adrenoceptors and/or the excitatory effect is generated indirectly, e.g., by way of changes in blood flow in deep somatic tissues of an extremity (Baron, R. et al., 2002). Sympathetically

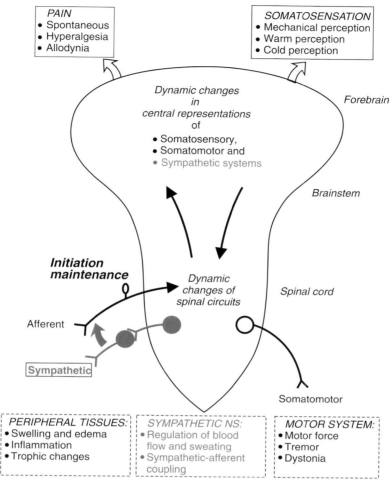

Figure 14 Schematic diagram summarizing the sensory, autonomic, and somatomotor changes in complex regional pain syndrome I (CRPS I) patients. The figure symbolizes the CNS (forebrain, brainstem, and spinal cord). Changes occur in the central representations of the somatosensory, the motor, and the sympathetic nervous system (which include the spinal circuits) and are reflected in the changes of the sensory painful and nonpainful perceptions, of cutaneous blood flow and sweating, and of motor performances. They are triggered and possibly maintained by the nociceptive afferent input from the somatic and visceral body domains. It is unclear whether these central changes are reversible in chronic CRPS I patients. These central changes affect the endogenous control system of nociceptive impulse transmission possibly too. Coupling between the sympathetic neurons and the afferent neurons in the periphery (see red arrow) is one component of the pain in CRPS I patients with sympathetically maintained pain (SMP). However, it seems to be unimportant in CRPS I patients without SMP. Modified from Jänig W. and Baron R. (2002; 2003).

maintained activity in nociceptive neurons may generate a state of central sensitization/hyperexcitability that is responsible for spontaneous pain and secondary evoked pain (mechanical and cold allodynia; see above and Figures 5 and 7).

• Swelling (edema) and inflammation may be generated by sympathetic and peptidergic afferent fibers interacting with each other at the arteriolar site (influencing blood flow) and venular site (influencing plasma extravasation) of the vascular bed. Sympathetic fibers are suggested to influence

inflammatory cells (macrophages and mast cells) by way of noradrenaline and adrenoceptors on the inflammatory cells (Figure 10) (Jänig, W. and Baron, R., 2002; 2003).

• Trophic changes in skin, appendages of skin, and subcutaneous tissues (including joints and so forth) are believed to be dependent, at least in part, on the sympathetic innervation. The nature of this neural influence on the tissue structure is unknown (Jänig, W. and Stanton-Hicks, M., 1996; Harden, R. N. *et al.*, 2001; Jänig, W. and Baron, R., 2003).

These three groups of observation made on patients with CRPS I (who have no nerve lesion and therefore by definition not neuropathic pain) argue that the change of activity in sympathetic neurons entails peripheral changes which in turn may result in secondary activation and/or sensitization of primary afferent nociceptive neurons. The underlying mechanism(s) of this activation/sensitization of nociceptive afferent neurons may involve short- and long-term processes as discussed above

As already mentioned above, an important observation, in CRPS I patients with SMP (but possibly also in CRPS II patients with SMP), is that pain relief following conduction block of sympathetic neurons by a local anesthetic applied to the sympathetic chain mostly outlasts the conduction block by at least an order of magnitude (days) (Figure 15; Price, D. D. et al., 1998). The long-lasting pain-relieving effect of sympathetic blocks suggests that activity in sympathetic neurons, which is of central origin, maintains a positive feedback circuit via the primary afferent neurons that are probably nociceptive in function. This positive feedback circuit maintains a central state of hyperexcitability (e.g., of neurons in the spinal dorsal horn; Figure 7), via excitation of afferent neurons started by an intense noxious event. The persistent afferent activity needed to maintain such a state of central hyperexcitability is switched off during temporary block of conduction in the sympathetic chain lasting only a few hours and cannot be immediately switched on again when the conduction block wears off and the activity in the sympathetic postganglionic neurons (and therefore probably also the sympathetically induced activity in afferent neurons) returns. It is hypothesized that the afferent activity has to act over a long time period to initiate and maintain the central state of hyperexcitability (via the positive feedback). The mechanism underlying this important and interesting phenomenon needs to be studied experimentally in patients with SMP as well as in animal models that have to be designed.

17.4.2 Sympathetic Nervous System and Acute Experimental Inflammation

Inflammation of body tissues is accompanied by sensitization of nociceptors leading to ongoing pain and hyperalgesia (Figure 7). The sensitization of nociceptors is generated by various signaling molecules released by the inflammatory cells including cells of the immune system (Figures 10 and 12; McMahon, S. B.

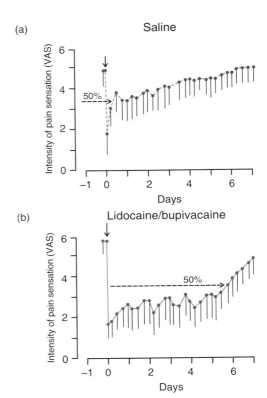

Figure 15 Effect of sympathetic blocks with a local anesthetic (lidocaine/bupivacaine) or of injection of saline close to the corresponding paravertebral sympathetic ganglia on pain in seven patients with CRPS I. Double-blind crossover study. Effect on pain following both interventions at the sympathetic supply was measured in the same group of CRPS I patients. Pain was systematically measured repeatedly using the visual analogue scale (VAS) on the day of the injection and on 7 days after the injection. Both interventions produced pain relief (see 50% value of pain relief). However, the mean relief of pain to injection of the local anesthetic lasted for 6 days and was significantly longer than the mean pain relief following local injection of saline that lasted for 6 h (placebo block). The initial maximal peaks of analgesia were statistically not different. Means + SEM. Modified from Price D. D. et al. (1998).

et al., 2006; Meyer, R. A. et al., 2005). Inflammation is potentially controlled by the brain via the HPA system, the sympatho-neural system, and the sympatho-adrenal system (Figure 16c), resulting in an enhancement or inhibition of the inflammatory (protective) responses depending on the behavioral state of the organism. Any change in neural and neuroendocrine control of inflammation should indirectly affect the sensitization of nociceptors innervating the inflamed tissue and therefore ongoing pain and hyperalgesia. Although we know relatively little about the role of both sympathetic systems in the

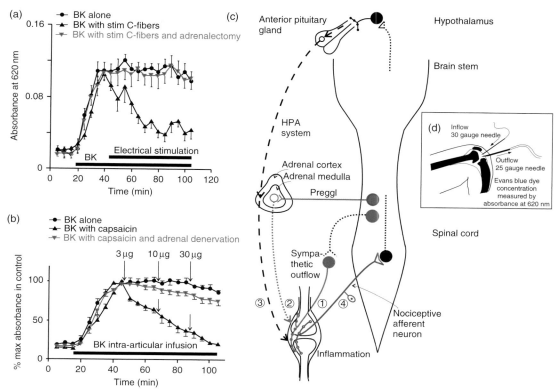

Figure 16 Experimental inflammation is controlled by the hypothalamo-pituitary system and the sympatho-adrenal system. Bradykinin (BK)-induced plasma extravasation (PE) was determined in the knee joint of anesthetized rats (d). Skin overlying the knee was excised to expose the joint capsule, and rats were then given an intravenous injection of Evans blue dye (50 mg kg^{-1}). The 30-gauge inflow and 25-gauge outflow needles were then inserted into the knee joint cavity for the inflow of perfusion fluid (0.9% saline at 250 μl min^{-1}). Perfusion fluid was collected over 5-min periods for up to 120 min and dye concentration determined spectrophotometrically at 620 nm. The absorbance at this wavelength is linearly related to the dye concentration and therefore to the degree of PE of the synovia (see ordinate scales in (a) and (b)). Following collection of the first three samples to establish baseline PE levels when the joint cavity is perfused with saline, BK (160 nM; and other substances) is (are) added to the perfusing fluid. This produces a rapid and sustained increase in the magnitude of PE (● in (a) ($n = 8$) and (b) ($n = 16$)). (a) Noxious transcutaneous electrical stimulation (25 mA, 3 Hz, 0.25 ms pulses) exciting C-fibers applied to the contralateral hindpaw after sample 8 produces a rapid decrease in the magnitude of PE (▲, $n = 7$). This inhibition produced by noxious electrical stimulation is prevented in rats in which the adrenal gland had been removed 1 week prior to the perfusion study (▼, $n = 7$). (b) Stimulation of cutaneous nociceptors by capsaicin injected into the palmar skin of the rat forepaw (3, 10, 30 μg in 2.5 μl) decreased the magnitude of PE (▲, $n = 8$). This inhibition produced by capsaicin was prevented in rats with denervated adrenal medullae (section of the adrenal nerves; ▼, $n = 8$). The rats were vagotomized. (c) Schematic diagram showing the neural and neuroendocrine components being involved in the control of BK-induced synovial PE: (*1*) Terminals of sympathetic postganglionic neurons mediate 60–70% of BK-induced synovial PE and the inhibitory effect of corticosterone on BK-induced synovial PE. (*2, 3*) Control of BK-induced PE by the HPA system and the sympatho-adrenal system. (*4*) Change of sensitivity of nociceptive afferents during inflammation as well as modulation of experimental inflammation by peptides released these afferents. The central nociceptive–neuroendocrine reflex circuits are not shown (see Green, P. G. *et al.*, 1995; Miao, F. J.-P. *et al.*, 2000; 2001). Mean + SEM in (a) and (b). (a) after Green P. G. *et al.* (1995); (b) after Miao F. J.-P. *et al.* (2000); (d) after Jänig W. *et al.* (2006).

control of inflammation (see Straub, R. H. *et al.*, 2005; Straub, R. H. and Härle, P., 2005), I will summarize the role of both systems in the control of acute experimental inflammation generated in the rat knee joint perfused with the inflammatory mediator bradykinin (bradykinin-induced plasma extravasation).

Continuous perfusion of bradykinin through the rat knee joint (Figure 16d) produces a large, sustained increase in plasma extravasation with a decay of about 10% over 1–2 h (Figure 16a and b, filled circles). This bradykinin-induced plasma extravasation response is largely dependent on the postganglionic sympathetic neuron terminal: surgical sympathectomy reduces the magnitude of response by 60–70%, but acute sympathetic decentralization (section of the preganglionic axons) has no significant effect. Similarly, acute interruption of the lumbar sympathetic chains during ongoing bradykinin-induced plasma extravasation does not reduce this extravasation. Finally, the sodium channel blocker tetrodotoxin coperfused with bradykinin into the knee joint cavity does not reduce the plasma extravasation (Miao, F. J.-P. et al., 1996a; 1996b). Thus, bradykinin-induced synovial plasma extravasation is dependent on the presence of sympathetic terminals in the synovia, but not on activity in these terminals and not on release of noradrenaline (function 1 in Figure 16c). This dependence on the sympathetic terminals is particularly large at bradykinin concentrations that have been measured in inflamed tissues (10^{-8} to 10^{-7} M; Hargreaves, K. M. et al., 1993; Swift, J. Q. et al., 1993). This is a novel peripheral function of postganglionic sympathetic fibers, at least in the context of the model of synovial inflammation, produced by the potent endogenous inflammatory mediator bradykinin. This function occurs at the postcapillary venules by release of a chemical substance (possibly prostaglandin E2 and/or related substances), which is independent of the electrical activity in the sympathetic neurons. Whether this substance is released by the sympathetic terminals or by other cells in association with these terminals is unknown. Future investigations will have to show whether this unusual function of sympathetic postganglionic terminals also applies to other inflammatory models.

Bradykinin-induced synovial plasma extravasation is under control of the brain mediated by the sympatho-adrenal system and the HPA system (functions 2 and 3 in Figure 16c). Both the neuroendocrine control systems can be reflex activated by noxious stimuli. The HPA system is most efficiently activated by continuous transcutaneous electrical stimulation of afferent C-fibers and the sympatho-adrenal system during stimulation of nociceptors by capsaicin (e.g., injected into the plantar or palmar skin or into the peritoneal cavity).

The differential activation of both neuroendocrine systems, the hypothalamic–pituitary system and the sympatho-adrenal system, generated by the modes of afferent stimulation is interesting but puzzling. During transcutaneous electrical stimulation, C-fibers are stimulated continuously and synchronously at 1–4 Hz together with A-fibers. This activation of C-fibers is certainly unphysiological.

Capsaicin injected into the skin or into the peritoneal cavity excites only C-fibers (and a few Aδ-fibers; but not Aβ-fibers) asynchronously. This activation is similar to physiological activation, e.g., by heat. These modes of afferent stimulation can be used to differentially activate both neuroendocrine systems experimentally whatever the central underlying mechanisms are for this differential activation.) The nociceptive–neuroendocrine reflex circuits in spinal cord and brainstem are enhanced after subdiaphragmatic vagotomy; thus, they are under inhibitory control of central circuits activated by abdominal vagal afferents. Both the central nociceptive–neuroendocrine reflex circuits and their inhibitory control exerted by activity in abdominal vagal afferents are described elsewhere (Miao, F. J.-P. et al., 1997a; 1997b; 2000; 2001; 2003).

Transcutaneous electrical stimulation of C-fibers inhibits bradykinin-induced plasma extravasation (normal triangles in Figure 16a). This inhibition is mediated by the HPA system since it can be prevented by removal of the adrenal glands (inverted triangles in Figure 16a), by hypophysectomy, by blockade of the corticosterone metabolism, and by blockade of corticosterone receptors. It can be mimicked by intravenous infusion of corticosterone (Green, P. G. et al., 1997). The inhibition of bradykinin-induced plasma extravasation generated by activation of the HPA system or by intravenous infusion of corticosterone is dependent on the presence of sympathetic postganglionic nerve fibers in the synovia: The bradykinin-induced plasma extravasation remaining after sympathectomy (removal of paravertebral ganglia) cannot any longer be inhibited by activation of the hypothalamic–pituitary system or by corticosterone injected intravenously. This shows that the sympathetic fibers are in a strategic, yet unexpected, position in the control of synovial plasma extravasation by the brain (Green, P. G. et al., 1997).

Reflex inhibition of bradykinin-induced plasma extravasation generated by stimulation of nociceptors by capsaicin (normal triangles in Figure 16b) is prevented by denervation of the adrenal medullae (inverted triangles in Figure 16b). Thus, it is mediated by adrenaline released by the adrenal medullae.

The mechanisms by which corticosterone released by the adrenal cortex and adrenaline released by the adrenal medulla depress bradykinin-induced synovial plasma extravasation and the role of the sympathetic nerve fibers innervating the synovia in this depression are unknown. Corticosterone seems to act via the peptide annexin possibly released by inflammatory cells in the synovia (Green, P. G. *et al.*, 1998). Here it is important to emphasize that the control of experimental synovial inflammation by both neuroendocrine systems affects indirectly the sensitivity of nociceptors innervating the knee joint and, therefore, also pain and hyperalgesia and that this change of sensitivity is therefore under the control of the brain.

17.4.3 Sympathetic Nervous System and Immune System

The hypothalamus can influence the immune system by way of the sympathetic nervous system and can therefore control protective mechanisms of the body at the cellular level (Besedovsky, H. O. and del Rey, A., 1995; Hori, T. *et al.*, 1995; Madden, K. S. and Felten, D. L., 1995; Madden, K. S. *et al.*, 1995). The parameters of the immune tissues potentially controlled are proliferation, trafficking and circulation of lymphocytes, functional activity of lymphoid cells (e.g., activity of natural killer cells) and cytokine production, hematopoesis of bone marrow, mucosal immunity, thymocyte development, etc. (for details of potential mechanisms, see Elenkov, I. J. *et al.*, 2000). The mechanisms of this influence remain largely unsolved (Ader, A. and Cohen, N., 1993; Saphier, D., 1993; Besedovsky, H. O. and del Rey, A., 1995). It is unknown (1) whether there exists a functionally distinct sympathetic system from the brain to the immune system or (2) whether the modulation of the immune system is a general function of the sympathetic nervous system. The more likely hypothesis favored by me is that the immune system is modulated by the brain by way of a functionally and anatomically distinct sympathetic pathway (Figure 17; for discussion, see Jänig, W. and Häbler, H. J., 2000a; Jänig, W., 2006). The reflex discharge characteristics of the neurons of this hypothetical sympathetic pathway innervating the immune tissue and the neural circuits in spinal cord, brainstem, and hypothalamus being involved in regulation of the activity of this pathway should be different from those of other sympathetic pathways (Jänig, W., 2006). In fact a hypothalamo-sympathetic neural system that controls the immune system has been postulated (Hori, T. *et al.*, 1995). In both cases it is hypothesized that activity in the sympathetic neurons supplying the immune system entails indirect modulation of the sensitivity of nociceptors involving various types of immune-related cells and their signaling molecules (e.g., cytokines, see above and Jänig, W. and Levine, J. D., 2006).

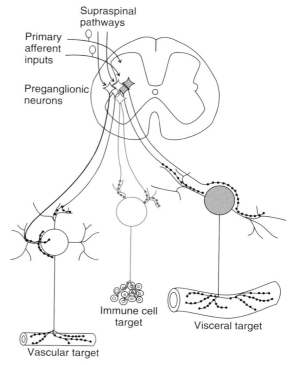

Figure 17 A separate sympathetic pathway to the immune tissue: a hypothesis. Demonstrated are three functional types of sympathetic pathway, one each to the vasculature (e.g., to blood vessels in skeletal muscle or skin or viscera), to the viscera (e.g., to the gastrointestinal tract being involved in the regulation of motility or secretion), or to the immune tissue. This neural link to the immune tissue may indirectly lead to sensitization of nociceptors and pain (modified from Jänig, W., 2006).

Acknowledgment

Supported by the Deutsche Forschungsgemeinschaft and the German Ministry of Research and Education (German Research Network on Neuropathic Pain).

References

Ader, A. and Cohen, N. 1993. Psychoneuroendocrinology: conditioning and stress. Annu. Rev. Physiol. 44, 53–85.

Ali, Z., Raja, S. N., Wesselmann, U., Fuchs, P. N., Meyer, R. A., and Campbell, J. N. 2000. Intradermal injection of norepinephrine evokes pain in patients with sympathetically maintained pain. Pain 88, 161–168.

Arkins, S., Johnson, R. W., Minshall, C., Dantzer, R., and Kelley, K. W. 2001. Immunophysiology: The Interaction of Hormones, Lymphohemopoietic Cytokines, and the Neuroimmune Axis. In: Handbook of Physiology. Section 7: The Endocrine System. Vol. IV: Coping with the Environment: Neural and Neuroendocrine Mechanisms (*ed.* B. S. McEwen), pp. 469–495. Oxford University Press.

Arnér, S. 1991. Intravenous phentolamine test: diagnostic and prognostic use in reflex sympathetic dystrophy. Pain 46, 17–22.

Bandler, R. and Shipley, M. T. 1994. Columnar organization in the midbrain periaqueductal gray: modules for emotional expression? Trends Neurosci. 17, 379–389.

Bandler, R., Keay, K. A., Floyd, N., and Price, J. 2000a. Central circuits mediating patterned autonomic activity during active vs. passive emotional coping. Brain Res. Bull. 53, 95–104.

Bandler, R., Price, J. L., and Keay, K. A. 2000b. Brain mediation of active and passive emotional coping. Prog. Brain Res. 122, 333–349.

Gebhart G. Mechanisms Underlying Complex Regional Pain Syndrome. Role of the Sympathetic Nervous System. In: Autonomic Failure (eds. C.J. Mathias and R. Bannister). Oxford University Press.

Baron, R., Schattschneider, J., Binder, A., Siebrecht, D., and Wasner, G. 2002. Relation between sympathetic vasoconstrictor activity and pain and hyperalgesia in complex regional pain syndromes: a case-control study. Lancet 359, 1655–1660.

Beal, M. C. 1983. Palpatory testing for somatic dysfunction in patients with cardiovascular disease. J. Am. Osteopath. Assoc. 82, 822–831.

Beal, M. C. 1985. Viscerosomatic reflexes: a review. J. Am. Osteopath. Assoc. 85, 786–801.

Bennett, G. J. and Xie, Y. K. 1988. A peripheral mononeuropathy in rat that produces disorders of pain sensation like those seen in man. Pain 33, 87–107.

Besedovsky, H. O. and del Rey, A. 1995. Immune-neuroendocrine interactions: facts and hypotheses. Endocr. Rev. 17, 64–102.

Bielefeldt, T. K. and Gebhart, G. F. 2006. Visceral Pain: Basic Mechanisms. In: Wall and Melzack's Textbook of Pain, 5th edn (eds. S. B. McMahon and M. Koltzenburg), pp. 721–736. Elsevier Churchill Livingstone.

Birder, L. A. and Perl, E. R. 1999. Expression of alpha2-adrenergic receptors in rat primary afferent neurones after peripheral nerve injury or inflammation. J. Physiol. 515, 533–542.

Blumberg, H. and Jänig, W. 1982. Activation of fibers via experimentally produced stump neuromas of skin nerves: ephaptic transmission or retrograde sprouting? Exp. Neurol. 76, 468–482.

Blumberg, H. and Jänig, W. 1984. Discharge pattern of afferent fibers from a neuroma. Pain 20, 335–353.

Bonica, J. J. 1953. Causalgia and Other Reflex Sympathetic Dystrophies. In: The Management of Pain, Chapter 28 (ed. J. J. Bonica), pp. 913–978. Lea and Febiger.

Bonica, J. J. 1990. Causalgia and Other Reflex Sympathetic Dystrophies. In: The Management of Pain, 2nd edn (ed. J. J. Bonica), pp. 220–243. Lea and Febiger.

Cervero, F. 1995. Mechanisms of Visceral Pain: Past and Present. In: Visceral Pain (ed. G. F. Gebhart), pp. 25–41. IASP Press.

Cervero, F. and Jänig, W. 1992. Visceral nociceptors: a new world order? Trends Neurosci. 15, 374–378.

Chabal, C., Jacobson, L., Russell, L. C., and Burchiel, K. J. 1992. Pain response to perineuromal injection of normal saline, epinephrine, and lidocaine in humans. Pain 49, 9–12.

Chen, Y., Michaelis, M., Jänig, W., and Devor, M. 1996. Adrenoreceptor subtype mediating sympathetic-sensory coupling in injured sensory neurons. J. Neurophysiol. 76, 3721–3730.

Choi, B. and Rowbotham, M. C. 1997. Effect of adrenergic receptor activation on post-herpetic neuralgia pain and sensory disturbances. Pain 69, 55–63.

Clauw, D. J. and Chrousos, G. P. 1997. Chronic pain and fatigue syndromes: overlapping clinical and neuroendocrine

features and potential pathogenic mechanisms. Neuroimmunomodulation 4, 134–153.

Coderre, T. J. and Melzack, R. 1986. Procedures which increase acute pain sensitivity also increase autotomy. Exp. Neurol. 92, 713–722.

Coderre, T. J., Abbott, F. V., and Melzack, R. 1984. Effects of peripheral antisympathetic treatments in the tail-flick, formalin and autotomy tests. Pain 18, 13–23.

Coderre, T. J., Grimes, R. W., and Melzack, R. 1986. Deafferentation and chronic pain in animals: an evaluation of evidence suggesting autotomy is related to pain. Pain 26, 61–84.

Cox, J. M., Gorbis, S., Dick, L. M., Rogers, J. C., and Rogers, F. J. 1983. Palpable musculoskeletal findings in coronary artery disease: results of a double-blind study. J. Am. Osteopath. Assoc. 82, 832–836.

Damasio, A. 2003. Looking for Spinoza. Joy, Sorrow, and the Feeling Brain. Harvest Books, Harcourt, Inc.

Dellemijn, P. L., Fields, H. L., Allen, R. R., McKay, W. R., and Rowbotham, M. C. 1994. The interpretation of pain relief and sensory changes following sympathetic blockade. Brain 117, 1475–1487.

Desmeules, J. A., Kayser, V., Weil-Fuggaza, J., Bertrand, A., and Guilbaud, G. 1995. Influence of the sympathetic nervous system in the development of abnormal pain-related behaviours in a rat model of neuropathic pain. Neuroscience 67, 941–951.

Devor, M. and Jänig, W. 1981. Activation of myelinated afferents ending in a neuroma by stimulation of the sympathetic supply in the rat. Neurosci. Lett. 24, 43–47.

Devor, M., Jänig, W., and Michaelis, M. 1994. Modulation of activity in dorsal root ganglion neurons by sympathetic activation in nerve-injured rats. J. Neurophysiol. 71, 38–47.

Elenkov, I. J., Wilder, R. L., Chrousos, G. P., and Vizi, E. S. 2000. The sympathetic nerve – an integrative interface between two supersystems: the brain and the immune system. Pharmacol. Rev. 52, 595–638.

Fields, H. L., Basbaum, A. I., and Heinricher, M. M. 2006. In: Central nervous system mechanisms of pain modulation (eds. S. B. McMahon and M. Koltzenburg), pp. 125–142. Elsevier Churchill Livingstone.

Foreman, R. D. 1989. Organization of the spinothalamic tract as a relay for cardiopulmonary sympathetic afferent fiber activity. Prog. Sensory Physiol. 9, 1–51.

Gebhart, G. F. 2007. In: The Senses: A Comprehensive References. In: Pain, Vol. 3 (eds. C. Bushnell and A. I. Basbaum). Elsevier.

Giamberardino, M. A. 1999. Recent and forgotten aspects of visceral pain. Eur. J. Pain 3, 77–92.

Goehler, L. E., Gaykema, R. P., Hansen, M. K., Anderson, K., Maier, S. F., and Watkins, L. R. 2000. Vagal immune-to-brain communication: a visceral chemosensory pathway. Auton. Neurosci. 85, 49–59.

Green, P. G., Jänig, W., and Levine, J. D. 1997. Sympathetic terminal: target for negative feedback neuroendocrine control of inflammatory response in the rat. J. Neurosci. 17, 3234–3238.

Green, P. G., Miao, F. J.-P., Jänig, W., and Levine, J. D. 1995. Negative feedback neuroendocrine control of the inflammatory response in rats. J. Neurosci. 15, 4678–4686.

Green, P. G., Strausbaugh, H. J., and Levine, J. D. 1998. Annexin I is a local mediator in neural-endocrine feedback control of inflammation. J. Neurophysiol. 80, 3120–3126.

Häbler, H.-J., Eschenfelder, S., Liu, X.-G., and Jänig, W. 2000. Sympathetic-sensory coupling after L5 spinal nerve lesion in the rat and its relation to changes in dorsal root ganglion blood flow. Pain 87, 335–345.

Häbler, H. J., Jänig, W., and Koltzenburg, M. 1987. Activation of unmyelinated afferents in chronically lesioned nerves by adrenaline and excitation of sympathetic efferents in the cat. Neurosci. Lett. 82, 35–40.

Häbler, H. J., Wasner, G., and Jänig, W. 1997. Interaction of sympathetic vasoconstriction and antidromic vasodilatation in the control of skin blood flow. Exp. Brain Res. 113, 402–410.

Harden, R.N., Baron, R., and Jänig, W. (eds.). 2001 Complex Regional Pain Syndrome. IASP Press.

Hargreaves, K. M., Roszkowski, M. T., and Swift, J. Q. 1993. Bradykinin and inflammatory pain. Agents Actions Suppl. 41, 65–73.

Herrero, J. F., Laird, J. M., and Lopez-Garcia, J. A. 2000. Wind-up of spinal cord neurones and pain sensation: much ado about something? Prog. Neurobiol. 61, 169–203.

Holzer, P. 1998. Neurogenic vasodilatation and plasma leakage in the skin. Gen. Pharmacol. 30, 5–11.

Hori, T., Katafuchi, T., Take, S., Shimizu, N., and Niijima, A. 1995. The autonomic nervous system as a communication channel between the brain and the immune system. Neuroimmunomodulation 2, 203–215.

Jänig, W. 1990. Activation of afferent fibers ending in an old neuroma by sympathetic stimulation in the rat. Neurosci. Lett. 111, 309–314.

Jänig, W. 1993. Spinal Visceral Afferents, Sympathetic Nervous System and Referred Pain. In: New Trends in Referred Pain and Hyperalgesia. Pain Research and Clinical Management, Vol. 7 (eds. L. Vecchiet, D. Albe-Fessard, U. Lindblom, and M. A. Giamberardino), pp. 83–89. Elsevier Science Publishers.

Jänig, W. 2002. Pain in the Sympathetic Nervous System: Pathophysiological Mechanisms. In: Autonomic Failure, 4th edn (eds. C. J. Mathias and R. Bannister), pp. 99–108. Oxford University Press.

Jänig, W. 2005. Vagal Afferents and Visceral Pain. Advances in Vagal Afferent Neurobiology (eds. B. Undem and D. Weinreich), pp. 465–493. CRC Press.

Jänig, W. 2006. The Integrative Action of the Autonomic Nervous System. Neurobiology of Homeostasis. Cambridge University Press.

Jänig, W. and Baron, R. 2001. The Role of the Sympathetic Nervous System in Neuropathic Pain: Clinical Observations and Animal Models. In: Neuropathic Pain: Pathophysiology and Treatment (eds. P. T. Hansson, H. L. Fields, R. G. Hill, and P. Marchettini), pp. 125–149. IASP Press.

Jänig, W. and Baron, R. 2002. Complex regional pain syndrome is a disease of the central nervous system. Clin. Auton. Res. 12, 150–164.

Jänig, W. and Baron, R. 2003. Complex regional pain syndrome: mystery explained? Lancet Neurol. 2, 687–697.

Janig, W. and Baron, R. 2004. Experimental approach to CRPS. Pain 108, 3–7.

Jänig, W. and Häbler, H. J. 2000a. Specificity in the organization of the autonomic nervous system: a basis for precise neural regulation of homeostatic and protective body functions. Prog. Brain Res. 122, 351–367.

Jänig, W. and Häbler, H. J. 2000b. Sympathetic nervous system: contribution to chronic pain. Prog. Brain Res. 129, 451–468.

Jänig, W. and Häbler, H. J. 2002. Physiologie und Pathophysiologie viszeraler Schmerzen [Physiology and pathophysiology of visceral pain]. Schmerz 16, 429–446.

Jänig, W. and Koltzenburg, M. 1991. What Is the Interaction Between the Sympathetic Terminal and the Primary Afferent Fiber? In: Towards a New Pharmacotherapy of Pain (eds. A. I. Basbaum and J. -M. Besson), pp. 331–352. Dahlem Workshop Reports. John Wiley and Sons.

Jänig, W. and Koltzenburg, M. 1993. Pain Arising from the Urogenital Tract. In: Nervous Control of the Urogenital Tract.

Vol. 3. The Autonomic Nervous System (eds. G. Burnstock and C. A. Maggi), pp. 523–576. Harwood Academic Publisher.

Jänig, W. and Levine, J. D. 2006. Autonomic–endocrine–immure interactions in acute nd chronic pain. In: Wall and Melzack's Textbook of Pain, 5th edn (eds. S. B. McMahon and M. Koltzenburg), pp. 205–218. Elsevier. Churchill Livingstone.

Jänig, W. and McLachlan, E. M. 2002. Neurobiology of the Autonomic Nervous System. In: Autonomic Failure, 5th edn (eds. C. J. Mathias and R. Bannister), pp. 3–15. Oxford University Press.

Jänig, W. and Morrison, J. F. B. 1986. Functional properties of spinal visceral afferents supplying abdominal and pelvic organs, with special emphasis on visceral nociception. Prog. Brain Res. 67, 87–114.

Jänig, W. and Stanton-Hicks, M. (eds.). 1996 Reflex Sympathetic Dystrophy – A Reappraisal. IASP Press.

Jänig, W., Chapman, C. R., and Green, P. G. 2006. Pain and Body Protection: Sensory, Autonomic, Neuroendocrine and Behavioral Mechanisms. In: The Control of Inflammation and Hyperalgesia. In Proceedings of the 11th World Congress on Pain (eds. H. Flor, E. Kalso, and J. O. Dostrovsky), pp. 331–347. IASP Press.

Jänig, W., Khasar, S. G., Levine, J. D., and Miao, F. J.-P. 2000. The role of vagal visceral afferents in the control of nociception. Prog. Brain Res. 122, 273–287.

Jänig, W., Levine, J. D., and Michaelis, M. 1996. Interactions of sympathetic and primary afferent neurons following nerve injury and tissue trauma. Prog. Brain Res. 112, 161–184.

Jobling, P., McLachlan, E. M., Jänig, W., and Anderson, C. R. 1992. Electrophysiological responses in the rat tail artery during reinnervation following lesions of the sympathetic supply. J. Physiol. 454, 107–128.

Keay, K. A. and Bandler, R. 2004. Periaqueductal gray. In: The Rat Nervous System, 3rd edn (ed. G. Paxinos), pp. 243–257. Academic Press.

Khasar, S. G., Green, P. G., and Levine, J. D. 1993. Comparison of intradermal and subcutaneous hyperalgesic effects of inflammatory mediators in the rat. Neurosci. Lett. 153, 215–218.

Khasar, S. G., Green, P. G., and Levine, J. D. 2005. Repeated sound stress enhances inflammatory pain in the rat. Pain 116, 79–86.

Khasar, S. G., Green, P. G., Miao, F. J.-P., and Levine, J. D. 2003. Vagal modulation of nociception is mediated by adrenomedullary epinephrine in the rat. Eur. J. Neurosci. 17, 909–915.

Khasar, S. G., Miao, F. J.-P., Jänig, W., and Levine, J. D. 1998a. Modulation of bradykinin-induced mechanical hyperalgesia in the rat skin by activity in the abdominal vagal afferents. Eur. J. Neurosci. 10, 435–444.

Khasar, S. G., Miao, F. J.-P., Jänig, W., and Levine, J. D. 1998b. Vagotomy-induced enhancement of mechanical hyperalgesia in the rat is sympathoadrenal-mediated. J. Neurosci. 18, 3043–3049.

Khasar, S. G., Miao, F. J.-P., and Levine, J. D. 1995. Inflammation modulates the contribution of receptor-subtypes to bradykinin-induced hyperalgesia in the rat. Neuroscience 69, 685–690.

Kim, S. H. and Chung, J. M. 1991. Sympathectomy alleviates mechanical allodynia in an experimental animal model for neuropathy in the rat. Neurosci. Lett. 134, 131–134.

Kim, S. H., Na, H. S., Sheen, K., and Chung, J. M. 1993. Effects of sympathectomy on a rat model of peripheral neuropathy. Pain 55, 85–92.

Kim, K. J., Yoon, Y. W., and Chung, J. M. 1997. Comparison of three rodent neuropathic pain models. Exp. Brain Res. 113, 200–206.

Kinnman, E. and Levine, J. D. 1995. Sensory and sympathetic contributions to nerve injury-induced sensory abnormalities in the rat. Neurosci. 64, 751–767.

Koltzenburg, M., Häbler, H.-J., and Jänig, W. 1995. Functional reinnervation of the vasculature of the adult cat paw pad by axons originally innervating vessels in hairy skin. Neuroscience 67, 245–252.

Lewin, G. R., Ritter, A. M., and Mendell, L. M. 1993. Nerve growth factor-induced hyperalgesia in the neonatal and adult rat. J. Neurosci. 13, 2136–2148.

Lewin, G. R., Rueff, A., and Mendell, L. M. 1994. Peripheral and central mechanisms of NGF-induced hyperalgesia. Eur. J. Neurosci. 6, 1903–1912.

Lopez de Armentia, M., Leeson, A. H., Stebbing, M. J., Urban, L., and McLachlan, E. M. 2003. Responses to sympathomimetics in rat sensory neurones after nerve transection. Neuroreport 14, 9–13.

Livingston, W. K. 1943. Pain Mechanisms. A Physiological Interpretation of Causalgia and Its Related States. Macmillan (reprinted by Plenum Press [1976]).

Liu, X., Eschenfelder, S., Blenk, K. H., Jänig, W., and Häbler, H. 2000. Spontaneous activity of axotomized afferent neurons after L5 spinal nerve injury in rats. Pain 84, 309–318.

MacKenzie, J. 1920. Symptoms and Their Interpretation, 4th edn. Shaws and Sons.

Madden, K. S. and Felten, D. L. 1995. Experimental basis for neural-immune interactions. Physiol. Rev. 75, 77–106.

Madden, K. S., Sanders, K., and Felten, D. L. 1995. Catecholamine influences and sympathetic modulation of immune responsiveness. Rev. Pharmacol. Toxicol. 35, 417–448.

Maier, S. F. and Watkins, L. R. 1998. Cytokines for psychologists: implications of bidirectional immune-to-brain communication for understanding behavior, mood, and cognition. Psychol. Rev. 105, 83–107.

Manning, D. C., Raja, S. N., Meyer, R. A., and Campbell, J. N. 1991. Pain and hyperalgesia after intradermal injection of bradykinin in humans. Clin. Pharmacol. Ther. 50, 721–729.

Mason, P. 2001. Contributions of the medullary raphe and ventromedial reticular region to pain modulation and other homeostatic functions. Annu. Rev. Neurosci. 24, 737–777.

Mayer, E. A., Munakata, J., Mertz, H., Lembo, T., and Bernstein, C. N. 1995. Visceral Hyperalgesia and Irritable Bowel Syndrome. In: Visceral Pain (ed. G. F. Gebhart), pp. 429–468. IASP Press.

McDonald, D. M. 1997. Neurogenic Inflammation in the Airways. In: Autonomic Control of the Respiratory System (ed. P. J. Barnes), pp. 249–289. Harwood Academic Publishers GmbH.

McLachlan, E. M., Jänig, W., Devor, M., and Michaelis, M. 1993. Peripheral nerve injury triggers noradrenergic sprouting within dorsal root ganglia. Nature 363, 543–546.

McMahon, S. B. 1996. NGF as a mediator of inflammatory pain. Philos. Trans. R. Soc. Lond. B Biol. Sci. 351, 431–440.

McMahon, S. B., Bennett, D. L. H., and Bevan, S. 2006. Inflammatory Mediators and Modulators of Pain. In: Wall and Melzack's Textbook of Pain, 5th edn (eds. S. B. McMahon and M. Koltzenburg), pp. 49–72. Elsevier Churchill Livingstone.

Mense, S. and Meyer, H. 1988. Bradykinin-induced modulation of the response behaviour of different types of feline group III and IV muscle receptors. J. Physiol. 398, 49–63.

Merskey, H. and Bogduk, H. 1994. Classification of Chronic Pain: Description of Chronic Pain Syndromes and Definition of Pain Terms, 2nd edn. IASP Press.

Meyer, R. A., Ringkamp, M., Campbell, J. N., and Raja, S. N. 2006. Peripheral Mechanisms of Cutaneous Nociception. In: Wall and Melzack's Textbook of Pain, 5th edn

(eds. S. B. McMahon and M. Koltzenburg), pp. 3–34. Elsevier Churchill Livingstone.

Miao, F. J.-P, Green, P. G., Coderre, T. J., Jänig, W., and Levine, J. D. 1996a. Sympathetic-dependence in bradykinin-induced synovial plasma extravasation is dose-related. Neurosci. Lett. 205, 165–168.

Miao, F. J.-P., Jänig, W., Green, P. G., and Levine, J. D. 1997a. Inhibition of bradykinin-induced synovial plasma extravasation produced by noxious cutaneous and visceral stimuli and its modulation by activity in the vagal nerve. J. Neurophysiol. 78, 1285–1292.

Miao, F. J.-P., Jänig, W., Jasmin, L., and Levine, J. D. 2001. Spino-bulbo-spinal pathway mediating vagal modulation of nociceptive-neuroendocrine control of inflammation in the rat. J. Physiol. 532, 811–822.

Miao, F. J.-P., Jänig, W., Jasmin, L., and Levine, J. D. 2003. Blockade of nociceptive inhibition of plasma extravasation by opioid stimulation of the periaqueductal gray and its interaction with vagus-induced inhibition in the rat. Neuroscience 119, 875–885.

Miao, F. J.-P., Jänig, W., and Levine, J. D. 1996b. Role of sympathetic postganglionic neurons in synovial plasma extravasation induced by bradykinin. J. Neurophysiol. 75, 715–724.

Miao, F. J.-P., Jänig, W., and Levine, J. D. 1997b. Vagal branches involved in inhibition of bradykinin-induced synovial plasma extravasation by intrathecal nicotine and noxious stimulation in the rat. J. Physiol. 498, 473–481.

Miao, F. J.-P., Jänig, W., and Levine, J. D. 2000. Nociceptive-neuroendocrine negative feedback control of neurogenic inflammation activated by capsaicin in the skin: role of the adrenal medulla. J. Physiol. 527, 601–610.

Michaelis, M., Devor, M., and Jänig, W. 1996. Sympathetic modulation of activity in rat dorsal root ganglion neurons changes over time following peripheral nerve injury. J. Neurophysiol. 76, 753–763.

Michaelis, M., Liu, X., and Jänig, W. 2000. Axotomized and intact muscle afferents but no skin afferents develop ongoing discharges of dorsal root ganglion origin after peripheral nerve lesion. J. Neurosci. 20, 2742–2748.

Miller, A. H., Pearce, B. D., Ruzek, M. C., and Biron, C. A. 2001. Interactions Between the Hypothalamic-Pituitary-Adrenal Axis and Immune System During Viral Infection: Pathways for Environmental Effects on Disease Expression. In: Handbook of Physiology. Section 7: The Endocrine System. Vol. IV: Coping with the Environment: Neural and Neuroendocrine Mechanisms (ed. B. S. McEwen), pp. 425–450. Oxford University Press.

Mitchell, S. W. 1872. Injuries of Nerves and Their Consequences. J. P. Lippincott.

Morrison, S. F. 2001. Differential control of sympathetic outflow. Am. J. Physiol. Regul. Integr. Comp. Physiol. 281, R683–R698.

Morrison, S. F. and Cao, W. H. 2000. Different adrenal sympathetic preganglionic neurons regulate epinephrine and norepinephrine secretion. Am. J. Physiol. Regul. Integr. Comp. Physiol. 279, R1763–R1775.

Neil, A., Attal, N., and Guilbaud, G. 1991. Effects of guanethidine on sensitization to natural stimuli and self-mutilating behaviour in rats with a peripheral neuropathy. Brain Res. 565, 237–246.

Ness, T. J. 1995. Historical and Clinical Perspectives of Visceral Pain. In: Visceral Pain. Progress in Pain Research and Management, Vol. 5 (ed. G. F. Gebhart), pp. 3–23. IASP Press.

Ness, T. J. and Gebhart, G. F. 1990. Visceral pain: a review of experimental studies. Pain 41, 167–234.

Neugebauer, V., Schaible, H. G., and Schmidt, R. F. 1989. Sensitization of articular afferents to mechanical stimuli by bradykinin. Pflügers Arch. 415, 330–335.

Nicholas, A. S., DeBias, D. A., Ehrenfeuchter, W., England, K. M., England, R. W., Greene, C. H., *et al.* 1985. A somatic component to myocardial infarction. Br. Med. J. 291, 13–17.

Petty, B. G., Cornblath, D. R., Adornato, B. T., Chaudhry, V., Flexner, C., Wachsman, M., *et al.* 1994. The effect of systemically administered recombinant human nerve growth factor in healthy human subjects. Ann. Neurol. 36, 244–246.

Poole, S., Cunha, F. Q., and Ferreira, S. H. 1999. Hyperalgesia from Subcutaneous Cytokines. In: Cytokines and Pain (*eds*. L. R. Watkins and S. F. Maier), pp. 59–87. Birkhäuser Verlag.

Porreca, F., Ossipov, M. H., and Gebhart, G. F. 2002. Chronic pain and medullary descending facilitation. Trends Neurosci. 25, 319–325.

Price, D. D., Long, S., Wilsey, B., and Rafii, A. 1998. Analysis of peak magnitude and duration of analgesia produced by local anesthetics injected into sympathetic ganglia of complex regional pain syndrome patients. Clin. J. Pain 14, 216–226.

Raja, S. N., Treede, R. D., Davis, K. D., and Campbell, J. N. 1991. Systemic alpha-adrenergic blockade with phentolamine: a diagnostic test for sympathetically maintained pain. Anesthesiology 74, 691–698.

Randich, A. and Gebhart, G. F. 1992. Vagal afferent modulation of nociception. Brain Res. Rev. 17, 77–99.

Ringkamp, M., Eschenfelder, S., Grethel, E. J., Häbler, H.-J., Meyer, R. A., Jänig, W., and Raja, S. N. 1999. Lumbar sympathectomy failed to reverse mechanical allodynia- and hyperalgesia-like behavior in rats with L5 spinal nerve injury. Pain 79, 143–153.

Ruch, T. C. 1965. Pathophysiology of Pain. In: Physiology and Biophysics (*eds*. T. C. Ruch and H. D. Patton), pp. 345–363. Saunders.

Saphier, D. 1993. Psychoimmunology: The Missing Link. In: Hormonally Induced Changes in Mind and Brain (*ed*. J. Schulkin), pp. 191–224. Academic Press.

Sato, J. and Perl, E. R. 1991. Adrenergic excitation of cutaneous pain receptors induced by peripheral nerve injury. Science 251, 1608–1610.

Seltzer, Z., Dubner, R., and Shir, Y. 1990. A novel behavioral model of neuropathic pain disorders produced in rats by partial sciatic nerve injury. Pain 43, 205–218.

Shir, Y. and Seltzer, Z. 1991. Effects of sympathectomy in a model of causalgiform pain produced by partial sciatic nerve injury in rats. Pain 45, 309–320.

Stanton-Hicks, M., Jänig, W., Hassenbusch, S., Haddox, J. D., Boas, R., and Wilson, P. 1995. Reflex sympathetic dystrophy, changing concepts and taxonomy. Pain 63, 127–133.

Straub, R. H. and Härle, P. 2005. Sympathetic neurotransmitters in joint inflammation. Rheum. Dis. Clin. North Am. 31, 43–59.

Straub, R. H., Baerwald, C. G., Wahle, M., and Jänig, W. 2005. Autonomic dysfunction in rheumatic diseases. Rheum. Dis. Clin. North Am. 31, 61–75.

Swift, J. Q., Garry, M. G., Roszkowski, M. T., and Hargreaves, K. M. 1993. Effect of flurbiprofen on tissue levels on immunoreactive bradykinin and acute postoperative pain. J. Oral. Max. Surg. 51, 112–117.

Torebjörk, H. E., Wahren, L. K., Wallin, B. G., Hallin, R., and Koltzenburg, M. 1995. Noradrenaline-evoked pain in neuralgia. Pain 63, 11–20.

Tracey, D. J., Cunningham, J. E., and Romm, M. A. 1995. Peripheral hyperalgesia in experimental neuropathy: mediation by alpha 2-adrenoreceptors on post-ganglionic sympathetic terminals. Pain 60, 317–327.

Undem, B. J. and Weinreich, D. (*eds*.). 2005. Advance in Vagal Afferent Neurobiology. CRC Press.

Urban, M. O. and Gebhart, G. F. 1999. Supraspinal contributions to hyperalgesia. Proc. Natl. Acad. Sci. U. S. A. 96, 7687–7692.

Vecchiet, L., Albe-Fessard, D., Lindblom, U., and Giamberardino, M.A. (*eds*.). 1993. New Trends in Referred Pain and Hyperalgesia. Pain Research and Clinical Management, Vol. 7. Elsevier Science Publishers.

Vierck, C. J., Jr. 2006. Mechanisms underlying development of spatially distributed chronic pain (fibromyalgia). Pain 124, 242–263.

Walker, A. E. and Nulsen, F. 1948. Electrical stimulation of the upper thoracic portion of the sympathetic chain in man. Arch. Neurol. Psychiatry 59, 559–560.

Wall, P. D., Devor, M., Inbal, R., Scadding, J. W., Schonfeld, D., Seltzer, Z., and Tomkiewicz, M. M. 1979a. Autotomy following peripheral nerve lesions: experimental anaesthesia dolorosa. Pain 7, 103–111.

Wall, P. D., Scadding, J. W., and Tomkiewicz, M. M. 1979b. The production and prevention of experimental anesthesia dolorosa. Pain 6, 175–182.

Wallin, G., Torebjörk, E., and Hallin, R. 1976. Preliminary Observations on the Pathophysiology of Hyperalgesia in the Causalgic Pain Syndrome. In: Sensory Functions of the Skin in Primates (*ed*. Y. Zottermann), pp. 489–499. Pergamon Press.

Watkins, L.R. and Maier, S.F. (*eds*.). 1999. Cytokines and Pain, Birkhäuser Verlag.

Watkins, L. R. and Maier, S. F. 2000. The pain of being sick: implications of immune-to-brain communication for understanding pain. Annu. Rev. Psychol. 51, 29–57.

Watkins, L. R., Maier, S. F., and Goehler, L. E. 1995. Immune activation: the role of pro-inflammatory cytokines in inflammation, illness responses and pathological pain states. Pain 63, 289–302.

White, J. C. and Sweet, W. H. 1969. Pain and the Neurosurgeon: A Forty Year Experience. Springfield.

Willenbring, S., Beauprie, I. G., and DeLeo, J. A. 1995. Sciatic cryoneurolysis in rats: a model of sympathetically independent pain. Part 1: Effects of sympathectomy. Anesth. Analg. 81, 544–548.

Wolfe, F., Smythe, H. A., Yunus, M. B., Bennett, R. M., Bombardier, C., Goldenberg, D. L., *et al.* 1990. The American College of Rheumatology (1990) Criteria for the Classification of Fibromyalgia. Report of the Multicenter Criteria Committee. Arthritis Rheum. 33, 160–172.

Woolf, C. J. 1996. Phenotypic modification of primary sensory neurons: the role of nerve growth factor in the production of persistent pain. Phil. Trans. R. Soc. Lond. B 351, 441–448.

Woolf, C. J., Allchorne, A., Safieh-Garabedian, B., and Poole, S. 1997. Cytokines, nerve growth factor and inflammatory hyperalgesia: the contribution of tumour necrosis factor alpha. Br. J. Pharmacol. 121, 417–424.

Woolf, C. J., Ma, Q.-P., Allchorne, A., and Poole, S. 1996. Peripheral cell types contributing to the hyperalgesic action of nerve growth factor in inflammation. J. Neurosci. 16, 2716–2723.

Further Reading

Miao, F. J.-P., Green, P. G., and Levine, J. D. 2004. Mechanosensitive duodenal afferents contribute to vagal modulation of inflammation in the rat. J. Physiol. 554, 227–235.

18 Sympathetic Blocks for Pain

A Sharma, Columbia University, New York, NY, USA

J N Campbell and S N Raja, Johns Hopkins University, Baltimore, MD, USA

18.1 Introduction

Traditionally, the sympathetic nervous system (SNS) is considered as an efferent system that controls peripheral blood flow, sweating, and piloerection. This autonomic part of the nervous system, however, may play a pivotal role in certain pain conditions. Over the last decade, much knowledge has been gained on the mechanisms of interaction between the sensory afferent and sympathetic efferent systems. Pain dependent on the discharge of sympathetic nerves is referred to as sympathetically maintained pain (SMP). Insights into the pathophysiology of SMP help its treatment. The concept of SMP stemmed from clinical observations that certain patients with persistent pain after traumatic nerve lesions received dramatic pain relief from blockade of sympathetic ganglion with local anesthetic agents or perivascular sympathectomy.

A pain syndrome often associated with SMP is now termed complex regional pain syndrome (CRPS). Only subsets of patients with CRPS have SMP. Formerly termed as reflex sympathetic dystrophy (RSD), CRPS is an incompletely understood pain disorder that usually results from a traumatic injury and clinically presents as nondermatomal distribution of pain and sensory abnormalities in an extremity. Symptoms and signs include hyperalgesia and allodynia (enhanced pain to natural stimuli such as a pin prick or light stroking of the skin, respectively), autonomic features, and motor dysfunction. Some patients in addition have a motor impairment not readily explained by a lesion in the peripheral nervous system. Edema at the time of presentation or at an earlier time point is a common feature. CRPS may be present in cases with (CRPS type II) or without a nerve lesion (CRPS type I). After the initial description of this disease by Weir Mitchell in 1864 (who called it causalgia) and decades later by Paul Sudeck (Sudeck's syndrome), Leriche advocated its treatment by sympathetic interruption in 1916. Leriche (vicious cycle hypothesis, 1939) and Mandl (1947) were pioneers in proposing the connection between pain and SNS. The concept that the SNS was involved in the mechanisms of pain in CRPS was further propagated by Evans, who coined the term reflex sympathetic dystrophy. In view of the explosion of research regarding mechanisms of pain in RSD and lack of uniform diagnostic criteria, major taxonomical changes were proposed by the International Association for the Study of Pain (IASP) Consensus Team, and the term complex regional pain syndrome was introduced (Stanton-Hicks, M. et al., 1995). According to the IASP classification, two distinct subtypes are now identified. Patients in whom the above-described conglomeration of symptoms and signs develop after injury to a major peripheral nerve are considered to have CRPS type II (previously known as causalgia). The remaining patients in whom injuries to a limb or lesions in remote body areas precede the onset of symptoms without any identifiable nerve damage are classified as CRPS type I (previously termed reflex sympathetic dystrophy) (Merskey, H. and Bogduk, N., 1995). Since the initial proposal of diagnostic criteria for CRPS, further modifications have been made (Harden, R. N. and Bruehl, S. P., 2006). Presently, the clinical signs and symptoms of CRPS are grouped into the following categories:

1. Positive sensory abnormalities (spontaneous pain, mechanical, somatic or deep somatic hyperalgesia).
2. Vascular abnormalities (vasodilation or vasoconstriction, skin temperature asymmetries, or skin color changes).
3. Sudomotor abnormalities (swelling, hyper-, or hyperhidrosis).
4. Motor changes (weakness, tremor, dystonia, or coordination deficits) and trophic changes (nail or hair changes, skin atrophy, joint stiffness, or soft-tissue changes).

The criteria are being further validated as no gold standard diagnostic tool has been established for CRPS (Baron, R. and Jänig, W., 2004; Harden, R. N. and Bruehl, S. P., 2006). The 1993 IASP consensus team also acknowledged the term sympathetically maintained pain (SMP) and defined it as "Pain that is maintained by sympathetic efferent innervation or by circulating catecholamines." A role for circulating catecholamines has not been clearly established though. SMP is traditionally diagnosed by relief of allodynia, hyperalgesia, ongoing pain, and a selective blockade of sympathetic function. Notably, patients with CRPS, as previously indicated, may have both SMP and sympathetically independent pain (SIP) in varying proportions (Stanton-Hicks, M. et al., 1995).

Even after decades of animal and human research, the pathophysiology of CRPS remains uncertain. Though it is common to refer to CRPS as a neuropathic pain disorder, it is not clear that pathology of the nervous system is at the root of this disorder as patients with or without nervous system lesions may manifest with symptoms of CRPS (Jänig, W. and Baron, R., 2002; 2003; 2004). Moreover, it is clear that CRPS is a diagnosis of exclusion, as other diseases (e.g., nerve entrapment and diabetic neuropathy) may present with similar signs and symptoms.

18.1.1 Sympathetically Maintained Pain

The clinical picture of CRPS with vascular (vasodilation or vasoconstriction, skin temperature asymmetries, or skin color changes) and sudomotor abnormalities (swelling, hyper-, or hypohidrosis) perhaps promoted the initial idea of SNS dysfunction as the primary etiology of the disease. Several lines of evidence suggest a dysregulation of autonomic function in the initial phase of the disease. Furthermore, enhancement of spontaneous pain and hyperalgesia after physiological activation of SNS (Baron, R. et al., 2002; Drummond, P. D. and Finch, P. M., 2004) and resolution of pain, physiological tremor, and swelling in some patients after sympathetic blocks strengthen this view. The detection of autoantibodies against the autonomic nervous system in certain CRPS patients provides further evidence to this concept.

The relief of pain by blockade of the SNS initially invited speculation of an increased SNS discharge as a potential mechanism for SMP. Several observations were against this hypothesis. First, patients with SMP, particularly in the early period of the disease, may present with warm rather than cold skin. Under conditions of increased sympathetic discharge, vasoconstriction should lead to a cold skin. Also others noted that catecholamine levels in the venous return of the affected extremity were below those in normal subjects or their unaffected side.

In 1948, Walker et al. made an important clinical observation that provided convincing evidence for SMP. At that time, sympathectomy was performed to treat limb ischemia and also to relieve pain in patients with causalgia. Walker and colleagues performed a preganglionic sympathectomy and placed electrodes on the ganglion. Postoperatively, the ganglion was stimulated. Since the ganglion was dissociated from the spinal cord, electrical stimulation would only induce an efferent discharge in the peripheral sympathetic fibers. Stimulation induced pain in patients with causalgia, but not in patients with limb ischemia. This indicated that SMP was due to efferent function of the SNS. In a further experiment performed decades later, norepinephrine was injected in patients with SMP. In these patients, the sympathetic ganglia were blocked with local anesthetic. If SNS dysfunction does exist, does it involve under- or overactivity? Earlier description of the disease promoted the concept of distinct phases with an initial phase involving warm, erythematous skin (stage I or acute stage) presumably due to SNS underactivity, an intermittent phase of both warm and cold sensation (stage II or dystrophic stage) followed by chronic cold sensation (stage III or atrophic stage) due to SNS hyperactivity. Although the existence of distinct phases of CRPS is debated at present (Bruehl, S. *et al.*, 2002), many studies have shown that vasomotor tone is reduced in very early stages of CRPS (Kurvers, H. A. *et al.*, 1995; Wasner, G. *et al.*, 2001), but not in later stages. The finding of reduced plasma norepinephrine and neuropeptide Y concentration on the affected extremity in early phase of the disease confirms this proposition (Raja, S. N. *et al.*, 1995). It seems that the apparent increase in vasomotor tone in later stages of CRPS (as evident by vasoconstriction or cold skin) is due to increased sensitivity (hyper- or supersensitivity) of skin microvessels to catecholamines (Kurvers, H. A. *et al.*, 1995) or secondary changes in neurovascular transmission while actual sympathetic tone is still depressed. Possible mechanisms for this supersensitivity might be decreased neuronal uptake of norepinephrine in the sympathetic neuroeffector junction (Raja, S. N. *et al.*, 1995) or increased number of peripheral α-receptors (Davis, K. D. *et al.*, 1991; Ali, Z. *et al.*, 2000). Studies in experimental animals report an increased expression of α-adrenoceptors in the neuromas that develop after nerve damage. Norepinephrine causes increased excitation of the neuroma afferents. Peripheral nerve injury may also result in sprouting of adrenergic fibers into the dorsal root ganglion and forms contact with sensory neurons. Thus, sympathetic afferent interactions might exist in the vicinity of the sensory receptors in partially denervated peripheral tissues, the peripheral nerve at the site of injury, and the dorsal root ganglion where cell bodies of injured sensory nerves are located.

Although SNS dysfunction is a fascinating concept, it does not explain all features of CRPS. Inflammatory and central origin of pain in CRPS patients has been suggested, and significant animal and human research support their contributions. The readers are advised to review the articles by Jänig W. and Baron R. (2003; 2004) for further understanding in this matter. Readers are also encouraged to read the conflicting views of Ochoa J. and Verdugo R. J. (2001) on the existence of SMP as well as the value of diagnostic phentolamine test or sympathetic blocks. The authors have negated the existence of SMP based on their own studies (Verdugo, R. J. and Ochoa, J. L., 1994), which in our opinion, had methodological deficiencies. A detailed discussion of these mechanisms and opposing views is beyond the scope of this chapter.

18.2 Techniques of Sympathetic Blockade

The techniques to achieve sympathetic blockade include the following:

1. Local anesthetic regional block (LASB) of sympathetic ganglion or trunk, for example, stellate ganglion block or lumbar sympathetic block.
2. Neuraxial techniques, for example, epidural or spinal block.
3. Intravenous regional sympathetic blockade (IRSB) with phentolamine, guanethidine, bretylium, clonidine, or lidocaine using tourniquet (Bier block technique).
4. Intravenous infusion of systemic alpha-adrenergic antagonists.
5. Radiofrequency denervation of sympathetic ganglion or trunk.
6. Surgical sympathectomy.

While LASB, IRSB with sympatholytic or local anesthetic medications, neuraxial techniques, and intravenous infusion of adrenergic antagonist can be used for both diagnostic and therapeutic benefits, neurolytic procedures (e.g., 5 and 6) are reserved for treatment of selected patients who show consistent but short-lasting response to initial diagnostic modalities.

18.3 Sympathetic Block for Diagnosis

The effectiveness of any diagnostic test or procedure is measured by its sensitivity and specificity. A good diagnostic test, when performed appropriately, should give consistent results in a given patient and should help guide therapy or predict outcome. Before any meaningful conclusions can be drawn from a sympathetic block by any technique, optimum inhibition of sympathetic outflow to the involved extremity needs to be achieved. This requires knowledge of the anatomical distribution of sympathetic innervation and the common pitfalls associated with the conduct and interpretation of these procedures. For instance, in performing stellate ganglion block to achieve an upper extremity sympathetic blockade, it is imperative to block contributions from upper thoracic ganglions that are anatomically separate from the stellate ganglion. This could be achieved by modifications in the previous technique. Similarly, injection of local anesthetic solution into nearby vascular structures could cause failure to achieve sympatholysis. This can be avoided with the use of intravenous contrast agent and live fluoroscopy to rule out intravascular placement of needle tip.

Following sympathetic blockade, determination of adequacy of sympatholysis is imperative. Monitoring skin temperature change is commonly used in clinical practice. Adequate sympathetic blockade is inferred by approximation of core body temperature and skin temperature of the ipsilateral limb. Accepting a rise in skin temperature by an arbitrary number in comparison with the contralateral limb leads to higher false-positive blocks (Schurmann, M. *et al.*, 2001), probably due to bilateral increase in blood flow following sympathetic block techniques. Many authors recommend further assessment of sympathetic function by sweat test (Stevens, R. A. *et al.*, 1998) or skin conductance response.

18.3.1 Local Anesthetic Sympathetic Blocks

LASBs are technically challenging procedures that require training. They have the advantage of selectively blocking stellate or lumbar sympathetic ganglion and/or chain with a small dose of local anesthetic solution without concomitant blockade of sensory or motor fibers (Table 1). Using fluoroscopic or CT guidance, the needle tip is positioned in close proximity to the sympathetic ganglia and a local anesthetic solution is injected. Although numerous complications of LASB have been reported in the literature, these techniques are safe in experienced hands. Efficacy of the block is assessed as previously described by monitoring objective evidence of sympathetic blockade, for example, cutaneous temperature monitoring. The patient is asked to report the intensity of pain using a numeric

Table 1   Comparison of diagnostic modalities for sympathetically maintained pain

Technique	*Advantages*	*Disadvantages*

Interventional regional sympathetic block with local anesthetic agents (LASBs)	1. Well-localized sympathetic ganglion or chain can be anesthetized (blocked). Objective signs of sympathectomy are seen with small doses of LA
	2. Systemic side effects from intravenous medications are avoided
		1. Requires technical expertise and has higher cost
		2. Requires fluoroscopy or CT guidance
		3. Complication from contrast or medication allergy
		4. Complication from technical reasons; e.g., infection, bleeding, and nerve damage
		5. Poorly tolerated in pediatric population and patients with needle phobia
		6. Sedatives used during the interventional technique might hinder with postprocedure pain responses
		7. False negative
		  • LA fails to anesthetize the sympathetic ganglion adequately
		  • Intravascular injection
		  • Central sympathetic–nociceptive interactions may be missed
		8. False positive
		  • LA anesthetizing somatic afferent fibers in the vicinity and causing pain relief from concurrent somatic blockade
		  • Regional ischemic pain will improve with sympathetic block
		  • Placebo effect

Intravenous regional sympathetic blockade (IRSB)	1. Easy to perform 2. Better patient acceptance 3. Low cost 4. Intravenous medications like phentolamine can cross blood–brain barrier and act on spinal cord α-2 adrenergic receptors, which are involved in pain modulation	1. Side effects of medications, e.g., phentolamine can cause positive chrontropic and inotrophic effects 2. Achieving complete sympathectomy at clinically tolerable doses might be difficult 3. Placebo effect cannot be ruled out

pain score before and after sympathetic blockade. The response is usually considered to be positive when a 50% or greater improvement in pain scores is reported in the light of adequate sympathectomy. LASB is traditionally considered as the gold standard against which other techniques have been tested in past (Dellemijn, P. L. *et al.*, 1994). Beyond CRPS, LASBs are also used to block efferent sympathetic fibers at thoracic sympathetic trunk, celiac plexus, superior hypogastric plexus, and ganglion impar to confirm improvement in pain prior to more invasive therapeutic blocks. These invasive procedures are used in the treatment of cancer pain, which may be somatic or visceral, and neuropathic or mixed in nature. A positive response to a temporary diagnostic local anesthetic block is a useful predictor of the success of subsequent neurodestructive procedures. Differential spinal or epidural blocks have also been described to distinguish somatic or central mechanisms from sympathetic etiologies. However, it is difficult to achieve selective sympathetic blockade with these techniques.

18.3.2 Intravenous Regional Sympathetic Blockade

Intravenous regional anesthesia, a technique described by August Bier in 1908, involves injection of a local anesthetic solution in a limb, isolated by tourniquet. Based on this concept, Hannington-Kiff proposed IRSB using guanethidine. Initially, guanethidine (with or without a local anesthetic agent) was used to perform this block and later other medications were tried. These include sympatholytics like reserpine, bretylium, or clonidine; local anesthetics like prilocaine or lidocaine; and others like kitanserin (5-HT2 antagonist). Although IRSBs have been used therapeutically in SMP, their use in diagnostic blocks has been limited. The ischemic tourniquet used during the procedure can itself relieve hyperalgesia in neuropathic pain patient (Campbell, J. N. *et al.*, 1988) and cause false-positive results. Certain patients tolerate tourniquet poorly. With IRSBs, guanethidine is often coinjected with local anesthetics since the release of norepinephrine by guanethidine is usually associated with the exacerbation of pain. As a result, diagnosis is often confounded by the effects of local anesthetics. Sudden release of tourniquet might release high doses of medications in systemic circulation and cause significant side effects. This technique is, hence, not useful as a diagnostic tool.

18.3.3 Systemic Alpha-Adrenergic Blockade

Intravenous alpha-adrenergic blockade with phentolamine (Phentolamine block, PhB) has been proposed as a safe and effective method of diagnosing SMP (Raja, S. N. *et al.*, 1991; Dellemijn, P. L. *et al.*, 1994). This technique is different from IRSB as no tourniquet is used to isolate the affected limb. In the initial study by Raja S. N. *et al.* (1991), 20 patients were randomly assigned to 25–35 mg of intravenous phentolamine (being administered in a double-blinded fashion) or LASB with 0.25% bupivacaine, and similar pain relief patterns were observed. Although 5–7 °C temperature difference was seen following LASB, the changes in cutaneous temperature after PhB were variable (1.8 ± 0.7 °C). Similar results were obtained by Dellemijn P. L. *et al.* (1994) who concluded that phentolamine infusion is less sensitive but more specific test of SMP than LASB. Both groups (Raja, S. N. *et al.*, 1991; Dellemijn, P. L. *et al.*, 1994) mentioned that the mechanism of pain relief during sympathetic blockade was independent of cutaneous vasodilatation or skin temperature changes. In an attempt to find optimum test dose, Raja S. N. *et al.* (1996) then compared the effect of two different doses of phentolamine (0.5 and 1 mg kg^{-1}) on skin blood flow (using a laser Doppler blood flow monitor) and sympathetically mediated vasoconstrictor response (a reflex decrease in peripheral blood flow in response to deep inhalation) (Raja, S. N. *et al.*, 1996). They concluded that the higher dose of phentolamine (1 mg kg^{-1} over 10 min) resulted in more complete adrenoceptor blockade and recommended this higher dose with cutaneous temperature monitoring to evaluate the sympathetically mediated component in neuropathic pain states.

There is a paucity of research in recognition of placebo responsiveness to sympathetic blocks. Randomization of patients for interventional techniques not only is difficult but also has ethical implications. Clinical observation frequently shows prolonged benefit in some patients beyond the duration of action of local anesthetic agents. Although reversal of central sensitization in the spinal cord and reduced coupling of sympathetic and sensory neurons have been put forward as a possible explanation, other possibilities need further consideration. In order to reduce placebo responsiveness, patients should have different diagnostic procedures like LASB and PhB on separate days. These two procedures provide complementary information and both may be needed to confirm the diagnosis of SMP (Dellemijn, P. L. *et al.*, 1994).

18.4 Sympathetic Block for Therapy of Chronic Pain States

In the past, once diagnosis of SMP was established, patients were often offered a series of regional or intravenous blocks in attempt to reduce pain and vasomotor symptoms. At present, they are considered as an element of multidisciplinary treatment plans. Patients who get significant benefit from initial blocks are often subjected to more aggressive interventional therapies. These treatment modalities are presumed to have enduring sympathectomy or modulation of SNS. They include spinal cord (dorsal column) stimulation, chemical neurolytic procedures, radiofrequency denervation of sympathetic trunk, or even surgical sympathectomy.

18.4.1 Local Anesthetic Sympathetic Blocks and Intravenous Regional Sympathetic Blockade

LASBs provide short-term benefits in relieving spontaneous and evoked pain and reducing vasomotor symptoms. Their long-term efficacy is unproven. In contrast, IRSB does not require any special expertise, is easier to perform and, comparatively, has better patient acceptance (Table 1). These procedures have the risk of tourniquet-related pain (if used) and significant side effects from the medications. Both techniques can get false-positive results from placebo effect.

Randomized controlled trials (RCTs) are considered as gold standard to evaluate the efficacy of any therapeutic modality, but not many of these studies have been conducted to assess the usefulness of LASB in SMP. Most randomized trials have been done in CRPS patients (includes both SMP and SIP patients) and not specifically in SMP states. SIP patients are not expected to get any long-term benefits from sympathetic blocks and their inclusion in the study biases the results. Cepeda M. S. *et al.* (2002) attempted a systematic review of RCTs of the effect of LASB in CRPS (not SMP) patients but were unable to pool data due to significant differences in their design. The authors then reviewed nonrandomized controlled studies, case series, and RCTs with acceptable designs, published in English-language peer review journals from 1916 through 1999. Only 29 studies met the inclusion criteria of a sample size of at least 10 patients undergoing LASB with convincible evidence of CRPS. Interestingly, of these 29 studies, only 10 studies evaluated the technical success of block and only two of those assessed pain and outcome. Based on the data from 14 studies (454 patients) that quantified the magnitude of patient's responses and also reported the number of patients with different degrees of responses, authors found that 29% patients undergoing LASB obtain full pain relief (>75% improvement) while 41% obtain partial relief (25–75% improvement). Authors concluded that less than one-third positive response (full pain relief) is consistent with placebo effect and that the efficacy of sympathetic blocks for treatment of CRPS is inconclusive. Many of these 14 studies were published prior to 1960 and only two actually identified technical success of block and pain scores.

Long-term usefulness of IRSB is even less certain. Jadad A. R. *et al.* (1995) attempted a small RCT (nine patients) of the effects of repeated intravenous guanethidine blocks on pain intensity and relief, adverse effects, mood, duration of analgesia, and global scores. The trial was stopped prematurely because of the severity of the adverse effects (hypotension). No significant difference was found between guanethidine and placebo on any of the outcome measures. The authors also presented a systematic review of seven studies on the effects of IRSB in CRPS. Two of these studies, one using ketanserin and one bretylium, with 17 patients in total, showed some advantage over control but few other RCTs showed lack of efficacy of guanethidine. Authors concluded that the use of guanethidine in IRSBs for patients with CRPS was not supported by any available literature. A recent study by Livingstone *et al.* again showed no significant analgesic advantage of guanethidine over a normal saline placebo block in the treatment of early CRPS type I.

Difficulty in conducting RCTs for LASBs or IRSBs for treatment of CRPS or SMP is explicable. Randomization to placebo is almost impossible. Patients or physicians cannot be blinded as outcomes of adequate sympathectomy are obvious. Also, ethical issues might arise in performing technically challenging procedures like stellate ganglion block with saline. But randomization can be done in a different fashion. The new concept of expertise-based RCTs might be useful in this condition when patient outcomes can be compared between physicians who perform repeat LASBs or IVRBs to those who solely impose conservative management. But again, such studies should compare patients with documented diagnosis of SMP and not CRPS. Based on available data in literature, efficacy and duration of LASB is variable and, indeed, unpredictable. Short duration of pain relief and reduction in vasomotor symptoms is often seen in clinical practice and should be utilized to improve mobility, range of motion, and motor strength by physiotherapy. Repeated blocks are beneficial in selected patients with willingness to actively participate in physiotherapy and are showing signs of continuing improvement.

18.4.2 Spinal Cord Stimulation

Spinal cord stimulation (SCS) has recently gained acceptance for a wide variety of neuropathic pain syndromes. A recent RCT (Kemler, M. A. *et al.*, 2004) and meta-analysis of literature (Grabow, T. S. *et al.*, 2003) showed that SCS results in a long-term pain reduction and health-related quality-of-life improvement, but no clinically important improvement of functional status, in chronic RSD. The technique is more effective and less expensive when compared with the standard treatment protocol for chronic CRPS (Mekhail, N. A. *et al.*, 2004). The mechanism of action by which the device provides pain relief is unclear. In general, patients with good response to sympathetic blocks are likely to have long-term pain relief (Hord, E. D. *et al.*, 2003) with SCS, although it does not affect skin microcirculation. The readers are encouraged to read a recent review on this subject for a detailed discussion (Meyerson, B. and Linderoth, B., 2001).

18.4.3 Chemical Neurolysis and Radiofrequency Denervation Techniques

Patients who show good response to initial sympathetic blocks are often subjected to radiofrequency denervation or chemical neurolytic destruction of sympathetic innervation. The technique of radiofrequency denervation has gone through various modifications by Noe *et al.*, Rocco, and Sluitjer. At present, destruction of T2 and T3 ganglions is recommended for upper extremity involvement while guidelines for lower extremity vary. Most authors suggest destruction of at least the first three ganglions (L1, L2, and L3), whereas others believe that the L4 and even L5 must be included for good long-term results. Complications of neurodestructive procedures include postsympathectomy sympathalgia, compensatory hyperhidrosis, Horner's

syndrome, wound infection, and spinal cord injury. There is a theoretical advantage of lesser incidence of postsympathetic neuralgia and more precise sympatholysis with radiofrequency technique in comparison with chemical neurolysis. Outcome data for these techniques vary. While Wilkinson H. A. (1996) reported almost 90% partial or complete evidence of sympathectomy after 2 years, most authors report sustained pain relief in less than two-thirds of patients at 2 years, and about one-third at 5 years. In a systematic review of the effects of percutaneous neurodestructive procedures for neuropathic pain, Mailis A. and Furlan A. (2003) concluded that the practice of surgical and chemical sympathectomy is based on poor quality evidence, uncontrolled studies, and personal experience. Therefore, more clinical trials of sympathectomy are required to establish the overall effectiveness and potential risks of this procedure. Because of the limited long-term outcomes, consideration should be given to neuromodulatory methods for SMP involving extremities. For other SMP syndromes, like visceral neuropathic pain in pancreatic cancer, chemical neurolyis continues to be the procedure of choice.

18.4.4 Surgical Sympathectomy

Surgical sympathectomy has been tried in patients who achieve good pain relief with a series of sympathetic block. Better outcomes are predicted when sympathectomy is performed early in the disease. Concerns about recurrence or development of new CRPS or disabling compensatory sweating syndrome have been raised after surgical sympathectomy, but current data suggests low incidence (7%) of these complications (Bandyk, D. F. *et al.*, 2002). Other disturbing complication includes transient (<3 months) postprocedural sympathalgia (in one-third of the patients following cervicodorsal sympathectomy and 20% of the patients after lumbar sympathectomy). At 1 year postprocedure, one-quarter of the patients are expected to continue experience significant improvement (pain severity score <3) and an additional 50% of the patients would have continued but reduced pain severity and an increase in daily/work activities (Bandyk, D. F. *et al.*, 2002). Many surgeons prefer video-assisted minimally invasive techniques instead of open sympathectomy to reduce other complication rates and promote early postoperative recovery. With the increasing popularity of radiofrequency procedure, rates of surgical sympathectomy have decreased.

18.5   Conclusions

SNS plays an intriguing role in maintaining pain in certain conditions. Complex regional pain syndromes are prime example of such states. Selective sympathetic blocks are commonly used to diagnose a subset of patients with a predominant sympathetically maintained pain state. In order to avoid false-positive diagnosis, at least two different diagnostic tests should be implicated on separate days to confirm the diagnosis. Treatment options that have shown some promising results include local anesthetic sympathetic blocks, SCS, radiofrequency techniques, and surgical sympathectomy in early stages. Visceral pain of pancreatic origin responds well to chemical neurolytic blocks. More randomized trials are warranted to prove long-term efficacy of neuromodulation as well as neurodestructive techniques.

References

Ali, Z., Raja, S. N., Wesselmann, U., Fuchs, P. N., Meyer, R. A., and Campbell, J. N. 2000. Intradermal injection of norepinephrine evokes pain in patients with sympathetically maintained pain. Pain 88(2), 161–168.

Bandyk, D. F., Johnson, B. L., Kirkpatrick, A. F., Novotney, M. L., and Back, M. R. 2002. Surgical sympathectomy for reflex sympathetic dystrophy syndromes. J. Vasc. Surg. 35(2), 269–277.

Baron, R. and Jänig, W. 2004. Complex regional pain syndromes – how do we escape the diagnostic trap? Lancet 364(9447), 1739–1741.

Baron, R., Schattschneider, J., Binder, A., Siebrecht, D., and Wasner, G. 2002. Relation between sympathetic vasoconstrictor activity and pain and hyperalgesia in complex regional pain syndromes: a case-control study. Lancet 359(9318), 1655–1660.

Bruehl, S., Harden, R. N., Galer, B. S., Saltz, S., Backonja, M., and Stanton-Hicks, M. 2002. Complex regional pain syndrome: are there distinct subtypes and sequential stages of the syndrome? Pain 95(1–2), 119–124.

Campbell, J. N., Raja, S. N., Meyer, R. A., and Mackinnon, S. E. 1988. Myelinated afferents signal the hyperalgesia associated with nerve injury. Pain 32(1), 89–94.

Cepeda, M. S., Lau, J., and Carr, D. B. 2002. Defining the therapeutic role of local anesthetic sympathetic blockade in complex regional pain syndrome: a narrative and systematic review. Clin. J. Pain 18(4), 216–233.

Davis, K. D., Treede, R. D., Raja, S. N., Meyer, R. A., and Campbell, J. N. 1991. Topical application of clonidine relieves hyperalgesia in patients with sympathetically maintained pain. Pain 47(3), 309–317.

Dellemijn, P. L., Fields, H. L., Allen, R. R., McKay, W. R., and Rowbotham, M. C. 1994. The interpretation of pain relief and sensory changes following sympathetic blockade. Brain 117(Pt 6), 1475–1487.

Drummond, P. D. and Finch, P. M. 2004. Persistence of pain induced by startle and forehead cooling after sympathetic blockade in patients with complex regional pain syndrome. J. Neurol. Neurosurg. Psychiatry 75(1), 98–102.

Grabow, T. S., Tella, P. K., and Raja, S. N. 2003. Spinal cord stimulation for complex regional pain syndrome: an evidence-based medicine review of the literature. Clin. J. Pain 19(6), 371–383.

Harden, R. N. and Bruehl, S. 2006. Diagnosis of complex regional pain syndrome: signs, symptons, and new empirically derived diagnostic criteria. Clin. J. Pain 22(5), 415–419.

Hord, E. D., Cohen, S. P., Cosgrone, G. R., Ahmed, S. U., Vallejo, R., Chang, Y., and Stojanovic, M. P. 2003. The predictive value of sympathetic block for the success of spinal cord stimulation. Neurosurgery 53(3), 626–632.

Jadad, A. R., Carroll, D., Glynn, C. J., and McQuay, H. J. 1995. Intravenous regional sympathetic blockade for pain relief in reflex sympathetic dystrophy: a systematic review and a randomized, double-blind crossover study. J. Pain Symp. Manage. 10(1), 13–20.

Jänig, W. and Baron, R. 2002. Complex regional pain syndrome is a disease of the central nervous system. Clin. Auton. Res. 12(3), 150–164.

Jänig, W. and Baron, R. 2003. Complex regional pain syndrome: mystery explained? Lancet Neurol. 2(11), 687–697.

Jänig, W. and Baron, R. 2004. Experimental approach to CRPS. Pain Mar. 108(1–2), 3–7.

Kemler, M. A., De Vet, H. C., Barendse, G. A., Van Den Wildenberg, F. A., and Van Kleef, M. 2004. The effect of spinal cord stimulation in patients with chronic reflex sympathetic dystrophy: two years' follow-up of the randomized controlled trial. Ann. Neurol. 55(1), 13–18.

Kurvers, H. A., Jacobs, M. J., Beuk, R. J., Van den Wildenberg, F. A., Kitslaar, P. J., Slaaf, D. W., and Reneman, R. S. 1995. Reflex sympathetic dystrophy: evolution of microcirculatory disturbances in time. Pain 60(3), 333–340.

Mailis, A. and Furlan, A. 2003. Sympathectomy for neuropathic pain. Cochrane Database Syst. Rev. (2), CD002918.

Mekhail, N. A., Aeschbach, A., and Stanton-Hicks, M. 2004. Cost benefit analysis of neurostimulation for chronic pain. Clin. J. Pain 20(6), 462–468.

Merskey, H. and Bogduk, N. 1995. Classification of Chronic Pain: Descriptions of Chronic Pain Syndromes and Definition of Terms. IASP Press.

Meyerson, B. and Linderoth, B. 2001. Spinal Cord Stimulation. In: Bonica's Management of Pain, 3rd edn. (eds. J. D. Loeser, S. D. Butler, C. R. Chapman, et al.), pp. 1857–1876. Lippincott Williams and Wilkins.

Ochoa, J. and Verdugo, R. J. 2001. Mechanisms of neuropathic pain: nerve, brain, and psyche: perhaps the dorsal horn but not the sympathetic system. Clin. Auton. Res. 11(6), 335–339.

Raja, S. N., Choi, Y., Asano, Y., Holmes, C., and Goldstein, D. S. 1995. Arteriovenous differences in plasma concentrations of catechols in rats with neuropathic pain. Anesthesiology 83(5), 1000–1008.

Raja, S. N., Treede, R. D., Davis, K. D., and Campbell, J. N. 1991. Systemic alpha-adrenergic blockade with phentolamine: a diagnostic test for sympathetically maintained pain. Anesthesiology 74(4), 691–698.

Raja, S. N., Turnquist, J. L., mEleka, S., and Campbell, J. N. 1996. Monitoring adequacy of alpha-adrenoceptor blockade following systemic phentolamine administration. Pain 64(1), 197–204.

Schurmann, M., Gradl, G., Wizgal, I., Tutic, M., Moser, C., Azad, S., and Beyer, A. 2001. Clinical and physiologic evaluation of stellate ganglion blockade for complex regional pain syndrome type I. Clin. J. Pain 17(1), 94–100.

Stanton-Hicks, M., Jänig, W., Hassenbusch, S., Haddox, J. D., Boas, R., and Wilson, P. 1995. Reflex sympathetic dystrophy: changing concepts and taxonomy. Pain 63(1), 127–133.

Stevens, R. A., Stotz, A., Kao, T. C., Powar, M., Burgen, S., and Kleinman, B. 1998. The relative increase in skin temperature after stellate ganglion block is predictive of a complete sympathectomy of the hand. Reg. Anesth. Pain Med. 23(3), 266–270.

Verdugo, R. J and Ochoa, J. L. 1994. 'Sympathetically maintained pain'. I. Phentolamine block questions the concept. Neurology 44(6), 1003–1010.

Verdugo, R. J., Campero, M., and Ochoa, J. L. 1994. Phentolamine sympathetic block in painful polyneuropathies. II. Further questioning of the concept of 'sympathetically maintained pain'. Neurology 44(6), 1010–1014.

Wasner, G., Schattschneider, J., Heckmann, K., Maier, C., and Baron, R. 2001. Vascular abnormalities in reflex sympathetic dystrophy (CRPS I): mechanisms and diagnostic value. Brain. 124(Pt 3), 587–599.

Wilkinson, H. A. 1996. Percutaneous radiofrequency upper thoracic sympathectomy. Neurosurgery 38(4), 715–725.

Further Reading

Jänig, W., Levine, J. D., and Michaelis, M. 1996. Interactions of sympathetic and primary afferent neurons following nerve injury and tissue trauma. Prog. Brain Res. 113, 161–184.

Nathan, P. W. 1983. Pain and the sympathetic system. J. Auton. Nerv. Syst. 7(3–4), 363–370.

19 Sprouting in Dorsal Root Ganglia

E M McLachlan, Prince of Wales Medical Research Institute, Randwick, NSW, Australia

Glossary

Aβ nociceptors They have often been ignored but there is accumulating evidence that a subpopulation of large-diameter fast-conducting (A) sensory neurons are nociceptive (Djouhri, L. and Lawson, S. N., 2004).

calcitonin gene-related peptide (CGRP) It is a neuropeptide present in the majority of peptidergic primary afferent nociceptor neurons. CGRP is also expressed in some medium to large diameter afferent neurons and appears *de novo* in others. including gracile-projecting mechanosensitive neurons, after nerve injuries (Ma, W. *et al.*, 1999).

galanin It is a neuropeptide present in a subpopulation of CGRP-containing primary afferent nociceptor neurons. It is upregulated both in these neurons and in sympathetic postganglionic neurons after a peripheral nerve injury.

neuropeptide Y (NPY) It is a neuropeptide that is present in many sympathetic postganglionic vasoconstrictor neurons in most species. It is upregulated in gracile-projecting low-threshold mechanosensitive primary afferent neurons after peripheral nerve lesions.

nonpeptidergic sensory neurons These are small-diameter dorsal root ganglion neurons that lack any known neuropeptide but express GFRα1-2 and c-RET, components of the receptor complex for glial-derived neurotrophic factor, and bind isolectin B4 (derived from *Griffonia simplicifolia*) (Bennett, D. L. *et al.*, 1998), neither of which are features of peptidergic sensory neurons. Nonpeptidergic sensory neurons are usually nociceptive (Lawson, S. N., 2002).

p75 It is the low-affinity nerve growth factor (NGF) receptor recognized by NGF, brain-derived neurotrophic factor (BDNF), neurotrophin-3 (NT-3), and neurotrophin-4/5 (NT-4/5). p75 interacts with each neurotrophin but with slightly different binding properties. It has striking structural homology with a family of receptors, including tumor necrosis factor (TNF) receptors, CD40, and Fas, consistent with it having many functions.

peptidergic sensory neurons These are small-diameter dorsal root ganglion neurons that contain a known neuropeptide (such as substance P, calcitonin gene-related peptide, galanin, somatostatin, and vasoactive intestinal polypeptide) and are usually nociceptors (Lawson, S. N., 2002).

perineuronal rings or baskets These are arrangements of varicose nerve terminals formed by the endings of sympathetic neurons and peptidergic and nonpeptidergic sensory neurons within dorsal root ganglia that project into a lesioned nerve trunk.

retrograde reaction Refers to the response in the soma of a neuron when its axon is severed. Following axotomy, either a positive signal of injury or the arrest of the normal retrograde traffic from the axon terminals reaches the soma within a few hours via the retrograde axoplasmic transport mechanism. The neuron soma swells and the nucleus may assume an eccentric position. These structural changes are referred to chromatolysis as the chromophilic Nissl bodies are disrupted. Protein synthesis is modified so that synthesis of neurotransmitter substances is reduced and other

activities such as regeneration are supported. The satellite glia around the soma also respond and begin to release cytokines and chemokines. They proliferate and start to synthesize neurotrophins.
satellite glia These are the support cells that surround the cell bodies of primary afferent neurons within dorsal root ganglia. They normally express low levels of glial fibrillary acidic protein (GFAP) which

is upregulated after nerve injury prior to the cells proliferating around damaged neurons.
substance P (SP) It is a neuropeptide present in a subpopulation of peptidergic primary afferent nociceptor neurons. It is upregulated in gracile-projecting low-threshold mechanosensitive primary afferent neurons after peripheral nerve lesions.

19.1   The Discovery

Clinicians observe that, in a subgroup of patients with chronic neuropathic pain after peripheral nerve injury of varying degrees of severity, sympathetic activity exacerbates or initiates pain, usually described as burning or stabbing (Bonica, J. J., 1990). The pain can be referred to superficial and/or deep sites and is often associated with allodynia and hyperalgesia. It has been postulated that, after the injury, nociceptors are activated by norepinephrine released by sympathetic activity, for example, during emotional stress or exposure to low environmental temperatures. This idea was based on evidence that pain was relieved by blockade of (1) the activity of sympathetic postganglionic neurons by injecting local anesthetics into paravertebral ganglia supplying the affected limb (Bonica, J. J., 1990), (2) norepinephrine release by injecting guanethidine into the limb (Wahren, L. K. *et al.*, 1991), or (3) the effects of norepinephrine by injecting α-adrenoceptor antagonists such phentolamine intravenously (Raja, S. N. *et al.*, 1991). Sympathetically maintained pain usually develops days to weeks after the injury.

These observations are mysterious because sympathetic nerve terminals are not co-located with nociceptor neurons or their terminals. In the skin, sympathetic axons terminate on arteriolar vessels deep within the dermis whereas nociceptor terminals lie below and within the epidermis. Sensory and sympathetic terminals do not usually co-locate in deeper tissues, other than rare axons containing substance P (SP) and calcitonin gene-related peptide (CGRP) that run among the sympathetic perivascular terminals on blood vessels (Holzer, P., 1992). There is no evidence for α-adrenoceptors on nociceptor terminals in normal skin and subcutaneous introduction of norepinephrine produces blanching but no pain.

Following early studies triggered by military injuries, an explosion of research in the last 25 years has utilized mainly rodents with injuries to one sciatic nerve that lead to diverse alterations in the damaged neurons, their supporting glia, the tissues they previously supplied, and the related vasculature. In addition inflammatory cells invade, not only the lesion site, but also retrogradely along the nerve trunk and within L4/5 dorsal root ganglia (DRGs), L3/4 sympathetic paravertebral ganglia, and the ventral horn of L4/5 spinal cord (Aldskogius, H. and Kozlova, E. N., 1998; Hu, P. and McLachlan, E. M., 2002; 2004). These changes persist for many months unless the damaged axons can regenerate to their target tissues.

Of relevance to injury-induced spontaneous pain is the generation of spontaneous activity in 20–30% of primary afferents within the damaged nerve. This activity is thought to contribute to central sensitization in the dorsal horn leading to abnormal discharge of nociceptive pathways (Mannion, R. J. and Woolf, C. J., 2000). The ectopic activity arises at two sites: in unmyelinated (C) neurons, it is initiated at the site of the injury (Michaelis, M. *et al.*, 1995); in myelinated (A) neurons, it arises from oscillations in the membrane potential of the somata in the DRG (Kajander, K. C. *et al.*, 1992; Liu, C. N. *et al.*, 2000). When the DRG and its connections are intact *in vivo*, activity in A axons arises in muscle but not cutaneous afferents (Michaelis, M. *et al.*, 2000).

The ectopic discharge in some DRG neurons can be generated or modulated by sympathetic activity. Wilfrid Jänig recorded from single myelinated afferent axons projecting down a ligated sciatic nerve (McLachlan, E. M. *et al.*, 1993). Stimulation of the sympathetic supply at high frequency (50 Hz for 10 s) excited a subgroup of afferents. Intravenous application of phentolamine completely inhibited the response, confirming that the sympathetic-sensory link was located proximally. Histochemical demonstration of norepinephrine revealed perivascular terminals outside the ganglion sprouting through the DRG (McLachlan, E. M. *et al.*, 1993). After a few weeks, some sprouting axons had formed interwoven baskets or rings of varicose terminals around a few large-diameter neurons. These novel structures appeared to be the anatomical basis for a functional link between sympathetic and sensory systems after injury.

19.2 Sprouting of Sympathetic Axons within Dorsal Root Ganglia

Sprouting of norepinephrine-containing terminals into DRGs after injury was soon confirmed in many laboratories (e.g., Chung, K. *et al.*, 1996; Kim, H. J. *et al.*, 1996; Ramer, M. S. and Bisby, M. A., 1997), mainly using immunohistochemistry to demonstrate tyrosine hydroxylase (TH), the rate-limiting enzyme for the synthesis of norepinephrine. The time course of development of sprouts varies with the site and nature of the lesion (Ramer, M. S. *et al.*, 1999). In rats, the sprouts develop over a few weeks after sciatic transection and form increasing numbers of perineuronal rings after ~4 weeks. The number of rings reaches a maximum (~2.5% of neurons or ~7.5% of A neurons) between 8 and 12 weeks and then gradually declines over 1 year or more (Hu, P. and McLachlan, E. M., 2001). While sympathetic sprouts appear in normal aged animals (Ramer, M. S. and Bisby, M. A., 1998b), this cannot account for the raised proportion of rings on the side ipsilateral to the lesion. Some perineuronal rings may persist indefinitely after injury, probably accounting for their presence in humans as described by Ramon y Cajal (Garcia-Poblete, E. *et al.*, 2003). The extent of sprouting is generally greater if the lesion is closer to the DRG (Kim, H. J. *et al.*, 1996; Ramer, M. S. and Bisby, M. A., 1998a) and if more axons are cut (Kim, H. J. *et al.*, 2001), although the relationships are not linear and may depend on the proportions of cutaneous and muscle axons involved (Hu, P. and McLachlan, E. M., 2003). Sprouting of sympathetic terminals into DRGs after sciatic transection does not occur in guinea-pigs that lack sympathetic terminals on nearby vessels (Hu, P. and McLachlan, E. M., 2000).

Sympathetic sprouts are already quite extensive after 7 days if the spinal nerve is transected (Chung, K. *et al.*, 1996; Ramer, M. S. *et al.*, 1998). Cutting L5 spinal nerve leads to hyperalgesia within a few days in the territory of the intact root (L4). Sympathetic

sprouts appear sooner and are more prolific when the sciatic nerve is only partially transected (Shir, Y. and Seltzer, Z., 1991) or subjected to chronic constriction (CCI) (Bennett, G. J. and Xie, Y. K., 1988) rather than transection. These injuries, like spinal nerve transection, leave intact axons lying in regions of Wallerian degeneration and the ensuing inflammatory reaction damages more axons. Such injuries produce mechanical allodynia and thermal hyperalgesia within a few days.

19.3 Sprouting of Primary Afferent Neurons within Dorsal Root Ganglia

One possible trigger for sprouting of perivascular terminals is an increased local concentration of neurotrophins, such as nerve growth factor (NGF) (Herzberg, U. *et al.*, 1997; see below). Consistent with this, sensory neurons containing SP and CGRP which bear the appropriate receptors (trkA) sprout within the lesioned DRG, forming similar numbers of perineuronal rings (McLachlan, E. M. and Hu, P., 1998). Galanin appears in some sprouts of both sympathetic and peptidergic origin (Hu, P. and McLachlan, E. M., 2001).

There is one report (Li, L. and Zhou, X. F., 2001) that the nonpeptidergic primary afferent axons that bind *Griffonia simplicifolia* I isolectin B4 (IB4) also sprout within L5 DRG after spinal nerve transection. Increasingly from 5 weeks after the lesion, IB4+ axons were found to encircle mainly large-diameter neuron somata where they were intermingled with proliferated satellite cells that also bound IB4. In some cases, TH+ axons were present around the same neurons. What is amazing is that tenfold more neurons received IB4+ perineuronal rings than TH+, CGRP+, or galanin+ rings at the same postoperative time (McLachlan, E. M. and Hu, P., 1998; Hu, P. and McLachlan, E. M., 2001). However, only a few of the IB4+ axons also stained for protein gene product (PGP) 9.5, a pan-neuronal marker (ubiquitin C-terminal hydroxylase).

19.4 Formation of Perineuronal Rings or Baskets

Perineuronal rings form more often at either pole of the DRG where large-diameter neurons are clustered (McLachlan, E. M. *et al.*, 1993). The neurons that receive them can be axotomized or intact (Jones, M. G. *et al.*, 1999; Hu, P. and

(a)   (b)

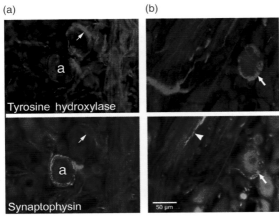

Figure 1   Perineuronal rings in rat L5 dorsal root ganglia 10 weeks after sciatic nerve transection and ligation. Two pairs of micrographs (a) and (b) show immunofluorescence for tyrosine hydroxylase (TH, green) and synaptophysin (Syn, red). (a) TH+ terminals that lack Syn (thin arrows) wrap around a neuron soma adjacent to another (A) that bears a dense basket of Syn+ terminals. (b) A neuron receives a TH+/Syn+ perineuronal ring (thick arrows). TH+ axons in the small nerve bundle to the left in (b) do not express Syn but a sprouting axon in the main axon bundle (arrowhead) is Syn+.

McLachlan, E. M., 2003). The varicose terminals intertwine around the somata but normally remain within the layers of proliferated satellite glia, very rarely forming contact with neuron somata (Chung, K. *et al.*, 1997; Shinder, V. *et al.*, 1999).

Sometimes isolated neurons are targeted, or more than one type of axon can encircle the same neuron (McLachlan, E. M. and Hu, P., 1998) or, in other cases, adjacent neurons receive a ring from one or other axon type (Figure 1). Synaptophysin, a non-specific marker of sprouting axons, appears in rings at a similar density to those with TH, CGRP, or galanin (Hu, P. and McLachlan, E. M., 2003). This may result because synaptophysin is only partly co-located with these substances or because of nonpeptidergic axon sprouts. As synaptophysin is also a synaptic vesicle protein, its presence in some but not all varicosities within rings supports them being release sites. Unfortunately, it is likely to be impossible to measure release directly from these terminals.

19.5   Underlying Cellular Mechanisms

Systemic lidocaine is successful in treating some forms of neuropathic pain (Wallace, M. S. *et al.*, 2000). Delivery of lidocaine by osmotic pump to

rats with nerve transection reduced the amount of sympathetic sprouting and the number of TH+ rings (Zhang, J. M. *et al.*, 2004), as did its application locally at the time of the lesion. These data suggest that activity derived from the neuroma is important for initiating the sprouting response.

Sympathetic axons sprout from two sources: (1) from intact perivascular terminals, which occur after a distant lesion, and are dependent on NGF and (2) by retrograde growth along the spinal nerve, which occurs when anterograde growth is obstructed at the lesion, and is NGF-independent. The former predominates after a distant injury, and the latter after spinal nerve ligation (Ramer, M. S. and Bisby, M. A., 1999). However, a common response to axotomy is the sprouting of collaterals from a transected axon within a short distance from the soma (Kelly, M. E. M. *et al.*, 1989) and this may occur after lesions to either sympathetic or sensory axons.

Upregulation of neurotrophin messenger ribonucleic acid (mRNA) by satellite glia leads to their local production within the DRG, stimulating growth of axons bearing appropriate receptors. This has been shown for NGF and neurotrophin-3 (NT-3) which are synthesized by the proliferated satellite glia that surround large-diameter neurons. Treatment with antibodies to NGF or NT-3 reduces the length of TH+ sprouts after a spinal nerve lesion (Ramer, M. S. and Bisby, M. A., 1999; Zhou, X. F. *et al.*, 1999) as both trkA and trkC are present on sympathetic and sensory axons. Brain-derived neurotrophic factor (BDNF) is upregulated in lesioned DRG neurons and antibodies to BDNF also inhibit sympathetic sprouting (Deng, Y. S. *et al.*, 2000). These approaches do not seem to have been applied to examine sprouting of sensory neurons. Glial cell-derived neurotrophic factor (GDNF) mRNA is present in satellite cells in DRGs and is upregulated after injury (Hammarberg, H. *et al.*, 1996), providing a basis for the sprouting of IB4+ axons (Li, L. and Zhou, X. F., 2001). Intrathecal GDNF does not elicit sympathetic sprouts (Jones, M. G. *et al.*, 1999) but leukemia inhibitory factor does (Thompson, S. W. and Majithia, A. A., 1998), perhaps by an effect that modifies NGF production.

The formation of perineuronal rings has been studied in trigeminal ganglia of transgenic mice. When these animals overexpress NGF (Davis, B. M. *et al.*, 1994), TH+ terminals preferentially target trkA+ CGRP+ sensory neurons that contain NGF (Walsh, G. S. and Kawaja, M. D., 1998; Walsh, G. S. *et al.*, 1999). The data indicate that CGRP+ axons do not sprout in these animals. If NGF or trkA is deleted, the TH+ axons do not enter the ganglion. If p75 is deleted but NGF is overexpressed, abundant TH+ sprouts wander through the DRG without associating with specific neuron somata (Walsh, G. S. *et al.*, 1999).

These elegant experiments do not help to explain sprouting or the formation of rings after nerve injury in normal adult animals. As mentioned above, NGF is synthesized by satellite glia that express p75 whereas trkA is downregulated on neuron somata (Verge, V. M. *et al.*, 1989). Further, no sprouts appear in the rat trigeminal ganglion after transection of the infraorbital or inferior alveolar nerve quite close to the ganglion (Bongenhielm, U. *et al.*, 1999) or chronic constriction injury (CCI) of the mental nerve (Grelik, C. *et al.*, 2005). This seems surprising as sympathetic axons join the nerve immediately distal to the trigeminal ganglion as for the DRG. Perhaps the absence of large-diameter proprioceptive neurons (located in the mesencephalic nucleus V) prevents the changes that trigger sympathetic sprouting in lesioned DRGs. Whether peptidergic neurons in trigeminal ganglia sprout after nerve lesions does not seem to have been examined.

Nevertheless it seems likely that the expression of p75 on proliferated glia in rat DRGs is important for ring formation. In guinea-pigs, in which sympathetic sprouting does not occur (see above), peptidergic axons sprout into the DRG but do not form perineuronal rings after sciatic transection. This might be explained by both the absence of glial reaction to axotomy (i.e., no upregulation of glial fibrillary acidic protein) and the failure of glia to express p75 in this species (Hu, P. and McLachlan, E. M., 2000).

What initiates neurotrophin production? It might follow the retrograde reaction in satellite glia and the subsequent invasion of the DRG by macrophages and lymphocytes (Hu, P. and McLachlan, E. M., 2002). Following sciatic transection, macrophage and T-cell density peaks after 1 week and remains raised for several months. Lymphocytes and macrophages secrete neurotrophins and proinflammatory cytokines. Neurotrophins trigger sprouting and cytokines sensitize DRG neurons. Further, T cells contribute to injury-induced pain (Moalem, G. *et al.*, 2004).

19.6 Other Functional Considerations

The somata around which perineuronal rings form have been thought to be low-threshold mechanoreceptors. The absence of sympathetic rings around peptidergic neurons (Ramer, M. S. and Bisby, M. A., 1998c) excludes the gracile-projecting A neurons which express SP, CGRP, NPY, and galanin several weeks after a peripheral injury (Zhang, X. *et al.*, 1993; Noguchi, K. *et al.*, 1995; Ma, W. and Bisby, M. A., 1997; Ma, W. *et al.*, 1999). Varicose peptidergic rings do not form around peptidergic neurons (McLachlan, E. M. and Hu, P., 1998). However, ~20% of A neurons in rat L5 DRG are Aβ-nociceptors (Djouhri, L. and Lawson, S. N., 2004) that is, 6% of all neurons, and at least some have large diameters. Allowing that TH+ terminals make up only ~50% of rings, rings may occur on 6% of neurons, making it feasible that Aβ-nociceptor neurons are the targets. While trkA is currently the only specific marker of these Aβ-neurons (Fang, X. *et al.*, 2005), this is downregulated after axotomy (Verge, V. M. *et al.*, 1989; Li, L. *et al.*, 2000) so that it may not be possible to test this idea.

Despite evidence that α-adrenoceptors are expressed *de novo* by a small proportion of largely medium-diameter neurons in lesioned DRGs (Birder, L. A. and Perl, E. R., 1999), there is limited evidence that norepinephrine can excite lesioned A neurons. Although norepinephrine can depolarize a small proportion of both control and damaged neurons in intact DRGs *in vitro* (Jones, M. G. *et al.*, 1999), this does not seem to be α-adrenoceptor mediated (Lopez de Armentia, M. *et al.*, 2003; cf. Xing, J. L. *et al.*, 2003). For neuropeptides, neuron somata with myelinated axons lack CGRP binding sites (Segond von Banchet, G. *et al.*, 2002), although they can develop sensitivity to galanin and SP after axotomy (Xu, Z. Q. *et al.*, 1997; Abdulla, F. A. *et al.*, 2001). This has not been demonstrated *in situ*.

19.7 Other Forms of Sympathetic Plasticity after Nerve Injury

Is there another explanation for the enhanced responses to sympathetic activation after injury? The normal consequence of sympathetic activity is vasoconstriction which is abolished by phentolamine. Stimulation paradigms that excite DRG neurons (>10 Hz) elicit marked vasoconstriction in most vascular beds. Stimulation of the lumbar sympathetic chain augmented discharge in a proportion of myelinated dorsal root axons from spinal nerve-injured rats only when vascular resistance in the

DRG was near maximal (Häbler, H. J. *et al.*, 2000). Discharge was also increased by vasoconstrictor agents and blockade of nitric oxide synthase. These data imply that enhanced neurovascular responses after a lesion jeopardize perfusion of the DRG, and raise the possibility that ischemia may be responsible for increasing excitability of sensory neurons.

Finally, the phenomenon of sympathetic sprouting in lesioned DRGs has had attention because of its possible relation to sympathetically maintained neuropathic pain. It is, however, important to note the failure of sympathectomy to modify ectopic activity in anesthetized rats (Liu, X. G. *et al.*, 2000) or mechanical allodynia (Ringkamp, M. *et al.*, 1999), and the lack of relation between the extent of sympathetic sprouting and pain behavior (Kim, H. J. *et al.*, 1999). Thus, although sympathetic perineuronal rings have been postulated to underlie ectopic activity, evidence for this in experimental animals is not convincing. Nevertheless the experiments on sprouting have provided insights into plasticity in the damaged nervous system.

Acknowledgments

Work in the author's laboratory was supported by the National Health & Medical Research Council and the Motor Accidents Authority of New South Wales.

References

Abdulla, F. A., Stebbing, M. J., and Smith, P. A. 2001. Effects of substance P on excitability and ionic currents of normal and axotomized rat dorsal root ganglion neurons. Eur. J. Neurosci. 13, 545–562.

Aldskogius, H. and Kozlova, E. N. 1998. Central neuron–glial and glial–glial interactions following axon injury. Prog. Neurobiol. 55, 1–26.

Bennett, D. L., Michael, G. J., Ramachandran, N., Munson, J. B., Averill, S., Yan, Q., McMahon, S. B., and Priestley, J. V. 1998. A distinct subgroup of small DRG cells express GDNF receptor components and GDNF is protective for these neurons after nerve injury. J. Neurosci. 18, 3059–3072.

Bennett, G. J. and Xie, Y. K. 1988. A peripheral mononeuropathy in rat that produces disorders of pain sensation like those seen in man. Pain 33, 87–107.

Birder, L. A. and Perl, E. R. 1999. Expression of alpha2-adrenergic receptors in rat primary afferent neurons after peripheral nerve injury or inflammation. J. Physiol. 515, 533–542.

Bongenhielm, U., Biossonade, F. M., Westermark, A., Robinson, P. P., and Fried, K. 1999. Sympathetic nerve sprouting fails to occur in the trigeminal ganglion after peripheral nerve injury in the rat. Pain 82, 283–288.

Bonica, J. J. 1990. The Management of Pain. Lea and Febiger.

Chung, K., Lee, B. H., Yoon, Y. W., and Chung, J. M. 1996. Sympathetic sprouting in the dorsal root ganglia of the injured peripheral nerve in a rat neuropathic pain model. J. Comp. Neurol. 376, 241–252.

Chung, K., Yoon, Y. W., and Chung, J. M. 1997. Sprouting sympathetic fibers form synaptic varicosities in the dorsal root ganglion of the rat with neuropathic injury. Brain Res. 751, 275–280.

Davis, B. M., Albers, K. M., Seroogy, K. B., and Katz, D. M. 1994. Overexpression of nerve growth factor in transgenic mice induces novel sympathetic projections to primary sensory neurons. J. Comp. Neurol. 349, 464–474.

Deng, Y. S., Zhong, J. H., and Zhou, X. F. 2000. Effects of endogenous neurotrophins on sympathetic sprouting in the dorsal root ganglia and allodynia following spinal nerve injury. Exp. Neurol. 164, 344–350.

Djouhri, L. and Lawson, S. N. 2004. Aβ-fiber nociceptive primary afferent neurons: a review of incidence and properties in relation to other afferent A-fiber neurons in mammals. Brain Res. Rev. 46, 131–145.

Fang, X., Djouhri, L., McMullan, S., Berry, C., Okuse, K., Waxman, S. G., and Lawson, S. N. 2005. trkA is expressed in nociceptive neurons and influences electrophysiological properties via Nav1.8 expression in rapidly conducting nociceptors. J. Neurosci. 25, 4868–4878.

Garcia-Poblete, E., Fernandez-Garcia, H., Moro-Rodriguez, E., Catala-Rodriguez, M., Rico-Morales, M. L., Garcia-Gomez-de-las-Heras, S., and Palomar-Gallego, M. A. 2003. Sympathetic sprouting in dorsal root ganglia (DRG): a recent histological finding? Histol. Histopathol. 18, 575–586.

Grelik, C., Bennett, G. J., and Ribeiro-da-Silva, A. 2005. Autonomic fibre sprouting and changes in nociceptive sensory innervation in the rat lower lip skin following chronic constriction injury. Eur. J. Neurosci. 21, 2475–2487.

Häbler, H. J., Eschenfelder, S., Liu, X. G., and Jänig, W. 2000. Sympathetic-sensory coupling after L5 spinal nerve lesion in the rat and its relation to changes in dorsal root ganglion blood flow. Pain 87, 335–345.

Hammarberg, H., Piehl, F., Cullheim, S., Fjell, J., Hokfelt, T., and Fried, K. 1996. GDNF mRNA in Schwann cells and DRG satellite cells after chronic sciatic nerve injury. Neuroreport 7, 857–860.

Herzberg, U., Eliav, E., Dorsey, J. M., Gracely, R. H., and Kopin, I. J. 1997. NGF involvement in pain induced by chronic constriction injury of the rat sciatic nerve. Neuroreport 8, 1613–1618.

Holzer, P. 1992. Peptidergic sensory neurons in the control of vascular functions: mechanisms and significance in the cutaneous and splanchnic vascular beds. Rev. Physiol. Biochem. Pharmacol. 121, 49–146.

Hu, P. and McLachlan, E. M. 2000. Distinct sprouting responses of sympathetic and peptidergic sensory axons proximal to a sciatic nerve transection in guinea pigs and rats. Neurosci. Lett. 295, 59–63.

Hu, P. and McLachlan, E. M. 2001. Long-term changes in the distribution of galanin in dorsal root ganglia after sciatic or spinal nerve transection in rats. Neuroscience 103, 1059–1071.

Hu, P. and McLachlan, E. M. 2002. Macrophage and lymphocyte invasion of dorsal root ganglia after peripheral nerve lesions in the rat. Neuroscience 112, 23–38.

Hu, P. and McLachlan, E. M. 2003. Selective reactions of cutaneous and muscle afferent neurons to peripheral nerve transection in rats. J. Neurosci. 23, 10559–10567.

Hu, P. and McLachlan, E. M. 2004. Inflammation in sympathetic ganglia proximal to sciatic nerve transection in rats. Neurosci. Lett. 365, 39–42.

Jones, M. G., Munson, J. B., and Thompson, S. W. 1999. A role for nerve growth factor in sympathetic sprouting in rat dorsal root ganglia. Pain 79, 21–29.

Kajander, K. C., Wakisaka, S., and Bennett, G. J. 1992. Spontaneous discharge originates in the dorsal root ganglion at the onset of a painful peripheral neuropathy in the rat. Neurosci. Lett. 138, 225–228.

Kelly, M. E. M., Bulloch, A. G. M., Lukowiak, K., and Bisby, M. A. 1989. Regeneration of frog sympathetic neurons is accompanied by sprouting and retraction of intraganglionic neurites. Brain Res. 477, 363–368.

Kim, H. J., Na, H. S., Back, S. K., and Hong, S. K. 2001. Sympathetic sprouting in sensory ganglia depends on the number of injured neurons. Neuroreport 12, 3529–3532.

Kim, H. J., Na, H. S., Nam, H. J., Park, K. A., Hong, S. K., and Kang, B. S. 1996. Sprouting of sympathetic nerve fibers into the dorsal root ganglion following peripheral nerve injury depends on the injury site. Neurosci. Lett. 212, 191–194.

Kim, H. J., Na, H. S., Sung, B., Nam, H. J., Chung, Y. J., and Hong, S. K. 1999. Is sympathetic sprouting in the dorsal root ganglia responsible for the production of neuropathic pain in a rat model? Neurosci. Lett. 269, 103–106.

Lawson, S. N. 2002. Phenotype and function of somatic primary afferent nociceptive neurons with C-, Adelta- or Aalpha/beta-fibres. Exp. Physiol. 87, 239–244.

Li, L. and Zhou, X. F. 2001. Pericellular Griffonia simplicifolia I isolectin B4-binding ring structures in the dorsal root ganglia following peripheral nerve injury in rats. J. Comp. Neurol. 439, 259–274.

Li, L., Deng, Y. S., and Zhou, X. F. 2000. Downregulation of TrkA expression in primary sensory neurons after unilateral lumbar spinal nerve transection and some rescuing effects of nerve growth factor infusion. Neurosci. Res. 38, 183–191.

Liu, C. N., Michaelis, M., Amir, R., and Devor, M. 2000. Spinal nerve injury enhances subthreshold membrane potential oscillations in DRG neurons: relation to neuropathic pain. J. Neurophysiol. 84, 205–215.

Liu, X. G., Eschenfelder, S., Blenk, K. H., Jänig, W., and Häbler, H. J. 2000. Spontaneous activity of axotomized afferent neurons after L5 spinal nerve injury in the rats. Pain 84, 309–318.

Lopez de Armentia, M., Leeson, A. H., Stebbing, M. J., Urban, L., and McLachlan, E. M. 2003. Responses to sympathomimetics in rat sensory neurons after nerve transection. Neuroreport 14, 9–13.

Ma, W. and Bisby, M. A. 1997. Differential expression of galanin immunoreactivities in the primary sensory neurons following partial and complete sciatic nerve injuries. Neuroscience 79, 1183–1195.

Ma, W., Ramer, M. S., and Bisby, M. A. 1999. Increased calcitonin gene-related peptide immunoreactivity in gracile nucleus after partial sciatic nerve injury: age-dependent and originating from spared sensory neurons. Exp. Neurol. 159, 459–473.

Mannion, R. J. and Woolf, C. J. 2000. Pain mechanisms and management: a central perspective. Clin. J. Pain 16, S144–S156.

McLachlan, E. M. and Hu, P. 1998. Axonal sprouts containing calcitonin gene-related peptide and substance P form pericellular baskets around large diameter neurons after sciatic nerve transection in the rat. Neuroscience 84, 961–965.

McLachlan, E. M., Jänig, W., Devor, M., and Michaelis, M. 1993. Peripheral nerve injury triggers noradrenergic sprouting within dorsal root ganglia. Nature 363, 543–546.

Michaelis, M., Blenk, K. H., Janig, W., and Vogel, C. 1995. Development of spontaneous activity and mechanosensitivity in axotomized afferent nerve fibers during the first hours after nerve transection in rats. J. Neurophysiol. 74, 1020–1027.

Michaelis, M., Liu, X., and Jänig, W. 2000. Axotomized and intact muscle afferents but no skin afferents develop ongoing discharges of dorsal root origin after peripheral nerve lesion. J. Neurosci. 20, 2742–2748.

Moalem, G., Xu, K., and Yu, L. 2004. T lymphocytes play a role in neuropathic pain following peripheral nerve injury in rats. Neuroscience 129, 767–777.

Noguchi, K., Kawai, Y., Fukuoka, T., Senba, E., and Miki, K. 1995. Substance P induced by peripheral nerve injury in primary afferent sensory neurons and its effect on dorsal column nucleus neurons. J. Neurosci. 15, 7633–7643.

Raja, S. N., Treede, R. D., Davis, K. D., and Campbell, J. N. 1991. Systemic alpha-adrenergic blockade with phentolamine: a diagnostic test for sympathetically maintained pain. Anesthesiology 74, 691–698.

Ramer, M. S. and Bisby, M. A. 1997. Rapid sprouting of sympathetic axons in dorsal root ganglia of rats with a chronic constriction injury. Pain 70, 237–244.

Ramer, M. S. and Bisby, M. A. 1998a. Differences in sympathetic innervation of mouse DRG following proximal or distal nerve lesions. Exp. Neurol. 152, 197–207.

Ramer, M. S. and Bisby, M. A. 1998b. Normal and injury-induced sympathetic innervation of rat dorsal root ganglia increases with age. J. Comp. Neurol. 394, 38–47.

Ramer, M. S. and Bisby, M. A. 1998c. Sympathetic axons surround neuropeptide-negative axotomized sensory neurons. Neuroreport 9, 3109–3113.

Ramer, M. S. and Bisby, M. A. 1999. Adrenergic innervation of rat sensory ganglia following proximal or distal painful sciatic neuropathy: distinct mechanisms revealed by anti-NGF treatment. Eur. J. Neurosci. 11, 837–846.

Ramer, M. S., Murphy, P. G., Richardson, P. M., and Bisby, M. A. 1998. Spinal nerve lesion-induced mechanoallodynia and adrenergic sprouting in sensory ganglia are attenuated in interleukin-6 knockout mice. Pain 78, 115–121.

Ramer, M. S., Thompson, S. W. N., and McMahon, S. B. 1999. Causes and consequences of sympathetic basket formation in dorsal root ganglia. Pain 6, S111–S120.

Ringkamp, M., Eschenfelder, S., Grethel, E. J., Häbler, H. J., Meyer, R. A., Jänig, W., and Raja, S. N. 1999. Lumbar sympathectomy failed to reverse mechanical allodynia- and hyperalgesia-like behavior in rats with L5 spinal nerve injury. Pain 79, 143–153.

Segond von Banchet, G., Pastor, A., Biskup, C., Schlegel, K., Benndorf, K., and Schaible, H. G. 2002. Localization of functional calcitonin gene-related peptide binding sites in a subpopulation of cultured dorsal root ganglion neurons. Neuroscience 110, 131–145.

Shinder, V., Govrin-Lippmann, R., Cohen, S., Belenky, M., Ilin, P., Fried, K., Wilkinson, H. A., and Devor, M. 1999. Structural basis of sympathetic-sensory coupling in rat and human dorsal root ganglia following peripheral nerve injury. J. Neurocytol. 28, 743–761.

Shir, Y. and Seltzer, Z. 1991. Effects of sympathectomy in a model of causalgiform pain produced by partial sciatic nerve injury in rats. Pain 45, 309–320.

Thompson, S. W. and Majithia, A. A. 1998. Leukemia inhibitory factor induces sympathetic sprouting in intact dorsal root ganglia in the adult rat in vivo. J. Physiol. 506, 809–816.

Verge, V. M., Riopelle, R. J., and Richardson, P. M. 1989. Nerve growth factor receptors on normal and injured sensory neurons. J. Neurosci. 9, 914–922.

Wahren, L. K., Torebjork, E., and Nystrom, B. 1991. Quantitative sensory testing before and after regional guanethidine block in patients with neuralgia in the hand. Pain 46, 23–30.

Wallace, M. S., Ridgeway, B. M., Leung, A. Y., Gerayli, A., and Yaksh, T. L. 2000. Concentration-effect relationship of intravenous lidocaine on the allodynia of complex regional pain syndrome types I and II. Anesthesiology 92, 75–83.

Walsh, G. S. and Kawaja, M. D. 1998. Sympathetic axons surround nerve growth factor-immunoreactive trigeminal neurons: observations in mice overexpressing nerve growth factor. J. Neurobiol. 34, 347–360.

Walsh, G. S., Krol, K. M., and Kawaja, M. D. 1999. Absence of the p75 neurotrophin receptor alters the pattern of sympathosensory sprouting in the trigeminal ganglia of mice overexpressing nerve growth factor. J. Neurosci. 19, 258–273.

Xing, J. L., Hu, S. J., Jian, Z., and Duan, J. H. 2003. Subthreshold membrane potential oscillation mediates the excitatory effect of norepinephrine in chronically compressed dorsal root ganglion neurons in the rat. Pain 105, 177–183.

Xu, Z. Q., Zhang, X., Grillner, S., and Hökfelt, T. 1997. Electrophysiological studies on rat dorsal root ganglion neurons after peripheral axotomy: changes in responses to neuropeptides. Proc. Nat. Acad. Sci. U. S. A. 94, 13262–13266.

Zhang, J. M., Li, H., and Munir, M. A. 2004. Decreasing sympathetic sprouting in pathologic sensory ganglia: a new mechanism for treating neuropathic pain using lidocaine. Pain 109, 143–149.

Zhang, X., Meister, B., Elde, R., Verge, V. M., and Hokfelt, T. 1993. Large calibre primary afferent neurons projecting to the gracile nucleus express neuropeptide Y after sciatic nerve lesions: an immunohistochemical and in situ hybridization study in rats. Eur. J. Neurosci. 5, 1510–1519.

Zhou, X. F., Deng, Y. S., Chie, E., Xue, Q., Zhong, J. H., McLachlan, E. M., Rush, R. A., and Xian, C. 1999. Satellite cell-derived nerve growth factor and neurotrophin-3 are involved in noradrenergic sprouting in the dorsal root ganglia following peripheral nerve injury in the rat. Eur. J. Neurosci. 11, 1711–1722.

20   Vagal Afferent Neurons and Pain

W Jänig, Christian-Albrechts-Universität zu Kiel, Kiel, Germany

Glossary

allodynia, hyperalgesia See Chapter Autonomic Nervous System and Pain.

angina (pectoris) A paroxysmal thoracic pain, with a feeling of suffocation and impending death, due, most often but not always, to anoxia of the myocardium.

GALT (gut-associated lymphoid tisue) Immune system of the gastrointestinal tract, in particular in the small intestine. It includes the immune cells of Peyer's patches, M-cells of the epithelial lining modified entestinal epithelial cells (enterocytes), lyphocytes in the almina propria, macrophages, and mast cells. Its function is to defend the body against invading antigens from food, bacteria, parasites, and toxins.

neurogenic inflammation Peptidergic primary afferent neurons with unmyelinated fibers (some with small-diameter myelinated (A delta) fibers) generate upon excitation precapillary (arteriolar) vasodilation (by release of calcitonin gene-related peptide (CGRP) supported by release of substance P and neurokinin A) and postcapillary (venular) plasma extravasation (by release of substance P and neurokinin A). Originally venular plasma extravasation was called neurogenic inflammation, but both precapillary vasodilation and postganglionic plasma extravasation are collectively now called neurogenic inflammation.

referred pain See Chapter Autonomic Nervous System and Pain.

vagal afferents Afferents projecting through the vagal nerves. Most of their cell bodies are located in the inferior vagal (nodose) ganglion and a few in the superior vagal (jugular) ganglion.

1. Vagal afferents innervating the heart may be involved in pain during angina referred to the upper cervical dermatomes and myotomes.

2. Vagal afferents innervating the stomach and being excited by acid are involved in nociceptive protective reactions, but probably not in conscious perception of pain.

3. Transmission of nociceptive impulses in the spinal or trigeminal dorsal horn is under inhibitory control generated by activity in cardiopulmonary and/or abdominal vagal afferents.

4. Vagal afferents innervating the liver (thus passing through the hepatic branch of the abdominal vagus nerves) are involved in illness responses generated by lipopolysaccharides (LPS) (bacterial antigens) or proinflammatory cytokines (e.g., interleukin-1β (IL-1β) or tumor necrosis factor α (TNF-α)) into the peritoneal cavity. One component of the illness responses is hyperalgesic behavior.

5. Activity in abdominal vagal afferents innervating the small intestine and probably sensing toxic and other events in the gastrointestinal tract (GIT) may be involved in reflex modulation of inflammation and sensitivity of nociceptors in somatic tissues via the sympatho-adrenal system (releasing adrenaline) and possibly the hypothalamo-pituitary–adrenal system.

The response properties of vagal afferents to physiological stimuli being involved in the reflex modulation of nociception, pain, hyperalgesia, inflammation, and illness responses are unknown. It is likely that not one functionally specific type of vagal afferent neuron is involved but several different types.

20.1   Introduction

The vagus nerves contain 80–85% visceral afferent fibers, most of them innervating the GIT, and 15–20% preganglionic parasympathetic fibers, at least in rats (Precht, J. C. and Powley, T. L., 1990; Berthoud, H. R. *et al.*, 1991). These afferent neurons are involved not only in organ regulation by the autonomic nervous system (Jänig, W., 2006) but also in the control of protective body reactions that include nociception and pain. These functions are inferred from various types of experimental investigations and from some clinical observations. It is unclear whether functionally specific groups of vagal afferents are involved in protective body reactions or whether vagal afferents being involved in organ regulation are also responsible for protective body reactions. In this chapter I will summarize:

1. data related to the role of vagal afferents in generating pain and in modulating central processing of nociceptive impulse activity and
2. data about the role of vagal afferents in protective reactions of the body including inflammation and hyperalgesia (Jänig, W., 2005; Jänig, W. and Levine, J. D., 2006).

20.2   Vagal Afferents, Pain, and Nociception

20.2.1   Thoracic Visceral Organs

Pain elicited from the proximal esophagus and proximal trachea is probably mediated by vagal afferents innervating the mucosa of these organs. These afferents are peptidergic (i.e., contain calcitonin gene related peptide (CGRP) and/or substance P); their cell bodies are probably located in the jugular ganglion (superior vagal ganglion; see Berthoud, H. R. and Neuhuber, W. L., 2000) cells in the nodose ganglion (inferior vagal ganglion) are almost exclusively peptide-negative, whereas neurons in the jugular ganglion may contain peptides). Activation of these afferents generates neurogenic inflammation in the mucosa (venular plasma extravasation and vasodilation (McDonald, D. M. *et al.*, 1988; McDonald, D. M., 1990)). Pain elicited from the more distal sections of

esophagus and trachea as well as from the bronchi may be generated by stimulation of spinal visceral afferent neurons yet not of vagal afferents. However, this situation is unclear and needs further experimentation.

Cardiac pain (e.g., during ischemic heart disease) is considered to be mediated by activation of spinal visceral afferents having their cell bodies in the dorsal root ganglia C8–T9 (mainly T2–6). However, attempts to relieve cardiac angina pain by surgical interventions (cervico-thoracic sympathectomy, dorsal rhizotomy, injection of alcohol into the sympathetic chain) consistently showed that only 50–60% of patients report complete relief from angina while the remaining patients report either partial relief or no relief at all. Some failures to relief pain may have been attributed to incomplete spinal denervation of the heart; but vagal afferents innervating particularly the inferior–posterior part of the heart may mediate cardiac pain too (see Meller, S. T. and Gebhart, G. F., 1992).

Electrical stimulation of cervical vagal afferent fibers in rats modulates the nociceptive tail reflex as well as the responses of dorsal horn neurons in the lumbar spinal cord. Low-intensity electrical vagal stimulation, probably exciting mainly myelinated afferents, facilitates both whereas high-intensity stimulation, probably exciting mainly C-fibers, attenuates both (Randich, A. and Gebhart, G. F., 1992; Ren, K. *et al.*, 1993). Nociceptive reflexes in neurons of the trigeminal nucleus caudalis or oralis of rats were mainly suppressed during electrical stimulation of cervical vagal afferents (Bossut, D. F. and Maixner, W., 1996; Takeda, M. *et al.*, 1998). Neurophysiological investigations in monkeys and rats show that some spino-thalamic tract (STT) neurons in the superficial dorsal horn and deeper laminae of the cervical segments C1–C2 (C3) can be synaptically activated by electrical stimulation of cardiopulmonary spinal and vagal afferents or by injection of algogenic chemicals in the pericardial sac via both afferent pathways. The activation of the STT neurons by vagal afferents is relayed through the nucleus tractus solitarii. These STT neurons are also synaptically activated by mechanical stimulation of the somatic receptive fields in the corresponding segments from the head, jaw, neck, and shoulder (dermatomes, myotomes); this is fully in line with clinical observations showing that cardiac pain may be referred to neck, shoulder, and jaw (Lindgren, I. and Olivecrona, H., 1947; White, J. C. and Bland, E. F., 1948; Meller, S. T. and Gebhart, G. F., 1992). The dorsal horn of the same high cervical spinal segments contains neurons with similar convergent synaptic inputs from vagal, spinal visceral, and somatic afferents that project to more caudal spinal thoracic, lumbar, and sacral segments. These dorsal horn neurons are involved in the inhibitory control of nociceptive impulse transmission (see Foreman, R. D., 1999; Chandler, M. J. *et al.*, 2002; see also Section 20.3, point 2).

Experimental studies on animals showing that nociceptive behavior and inhibition of nociceptive impulse transmission in the spinal and trigeminal dorsal horn is inhibited by electrical stimulation of the cervical vagus nerve have led to the idea that continuous electrical stimulation of vagal afferents could be used to treat chronic pain syndromes in patients. Electrical stimulation of the vagus nerve via a pair of electrodes implanted at the left cervical vagus nerve has been used recently in patients with some success to treat drug-resistant refractory epilepsy. The mechanisms underlying this antiepileptic effect are unknown (Schachter, S. C., 2005). A few experimental studies have been conducted on these patients to test whether experimental pain can be attenuated by vagal nerve stimulation. Furthermore, in patients with chronic headache (migraine, tension-type headache, cluster headache) electrodes have been implanted at the left cervical vagus nerve. In these experiments, in pilot studies, as well as in some case reports it has been shown that electrical vagal nerve stimulation may reduce experimental pain or relieve chronic headache (for literature and discussion see Multon, S. and Schoenen, J., 2005).

20.2.2 Pelvic and Abdominal Organs

Visceral pain elicited from pelvic and most abdominal organs is triggered by stimulation of spinal visceral afferent neurons and not by stimulation of vagal afferent neurons (Cervero, F., 1994; see Chapter Spinal Cord Mechanisms of Hyperalgesia and Allodynia). However, the situation is not entirely clear for the gastroduodenal section of the GIT. Whether patients with complete interruption of spinal ascending impulse transmission at the thoracic level T1 or at a more rostral segmental level can experience pain from the gastroduodenal section (e.g., during gastritis, a peptic ulcer, or distension of the stomach) has never been systematically investigated. Patients with complete lesion of the cervical spinal cord experience abdominal hunger, dread, and nausea (Crawford, J. P. and Frankel, H. L., 1971). Furthermore, these patients may experience vague sensations of fullness after a hot

meal. But acid reflux or an obstructed or distended viscus usually does not generate discomfort and pain (Juler, G. L. and Eltorai, I. M., 1985; Strauther, G. R. *et al.*, 1999). An acute perforation of a duodenal ulcer may be accompanied by violent pain in the right or left shoulder (page 284 in Guttmann, L., 1976), indicating that vagal afferents may be involved.

It is generally accepted that stimulation of vagal afferents innervating the stomach elicits emesis, bloating, and nausea, all three being protective reactions. Experiments on rats show that influx of acid into and other chemical insults of the gastroduodenal mucosa lead to a host of locally and centrally organized protective reactions that are mediated by spinal visceral and vagal afferent neurons. Activation of vagal afferents by these chemical stimuli excites neurons in the nucleus tractus solitarii, area postrema, lateral parabrachial nucleus, subceruleus nucleus, thalamic and hypothalamic paraventricular nuclei, supraoptic nucleus, and central amygdala, but not in the insular cortex (the major central representation area of the stomach) as shown by the expression of the marker protein Fos (Michl, T. *et al.*, 2001). Furthermore, neurons in the spinal dorsal horn do not seem to be activated. Thus, vagal afferents innervating the gastroduodenal mucosa (Holzer, P. and Maggi, C. A., 1998; Holzer, P.; 2002; 2003; 2006) (1) seem to be involved in chemonociception, local protective reactions, and generation of autonomic, endocrine, and behavioral protective reactions but (2) not in pain perception, yet in the emotional aspect of pain and therefore indirectly in upper abdominal hyperalgesia.

These fascinating ideas, formulated by Holzer, need verification by further experimentation. For example, it is unclear in which way (1) activity in spinal visceral afferents and vagal afferents is centrally integrated *in vivo* so as to elicit the protective reflexes, protective behavior, and pain sensations including visceral hyperalgesia and (2) activity in vagal afferents is responsible for the emotional aspects of visceral pain, but spinal visceral afferents for the conscious perception of visceral pain.

20.3 Abdominal Vagal Afferents, Protection of the Body, and Illness Responses

The small intestine and liver are very vulnerable portals of entries into the body. Both have potent local defense systems and serve as internal defense lines of the body. The small intestine contains a powerful immune system (the gut associated lymphoid tissue (GALT) (Shanahan, F., 1994)) that is innervated by vagal afferents projecting through the celiac branches of the abdominal vagus nerve. Specific modulation of activity in these afferents in conjunction with the reaction of the GALT may act as an early warning system to the rest of the body by transmitting important information to the brain about toxic events and agents in the small intestine that are dangerous for the organism (Berthoud, H. R. and Neuhuber, W. L., 2000). Thus, particularly vagal afferents in the celiac branches of the abdominal vagus nerves that innervate the small intestine (in addition to the distal duodenum and the proximal colon) and vagal afferents in the hepatic branch that innervate the liver (in addition to the proximal duodenum, pancreas, and pylorus) may be important for protective functions of the GIT and the body:

1. Vagal afferents in the celiac branches of the abdominal vagus nerve monitor chemical and mechanical events that occur in the intestine under physiological and pathophysiological conditions (for review see Grundy, D. and Scratcherd, T., 1989), that is, are related to meals, ingestion of toxic substances, inflammation, obstruction. These afferents respond to distension or contraction of the small intestine or to intraluminal chemical stimulation (e.g., maltose, glucose, protein products of long-chain lipids, or intraluminal osmotic stimuli). The responses to chemical stimuli are mediated by enterochromaffin cells releasing 5-HT and by 5-HT_3 receptors on the terminals of the vagal afferents (Zhu, J. X. et al., 2001) or enteroendocrine cells releasing cholecystokinin (CCK) and the CCK_A receptor on the terminals of the vagal afferent neurons (Richards, W. et al., 1996; Lal, S. et al., 2001). Vagal afferent neurons may be associated with the GALT and may be excited by inflammatory processes and toxic processes, also probably mediated by enterochromaffin cells releasing serotonin (5-HT), enteroendocrine cells releasing CCK, or mast cells releasing histamine and other compounds (Kirkup, A. J. et al., 2001; Kreis, M. E. et al., 2002). The functional specificity of these afferents with respect to the different types of intraluminal stimuli is unknown. The vagal afferents are important for preabsorptive detection of energy-yielding molecules and probably for other properties of nutrient solutions that are toxic and

deleterious for the GIT and for the body (Walls, E. K. et al., 1995a; 1995b). Additionally, some vagal afferents that innervate the small intestine and the liver respond to cytokines (e.g., IL-1β); these afferents may encode events that are related to the immune system of the GIT and liver (Niijima, A., 1996).

2. Electrical stimulation of abdominal vagal afferents exerts inhibition or facilitation of central nociceptive impulse transmission in the spinal dorsal horn and depresses nociceptive behavior depending on as to whether unmyelinated or myelinated vagal afferents are excited (Randich, A. and Gebhart, G. F., 1992). Electrical stimulation of cervical vagal afferents in monkeys suppresses transmission of impulse activity in STT neurons with nociceptive function at all levels of the spinal cord. Electrical stimulation of subdiaphragmatic vagal afferents has no effect on STT neurons in this species, arguing that particularly cardiopulmonary vagal afferents are involved in this inhibitory control. The central pathways mediating the inhibitory effect are neurons in the subceruleus–parabrachial complex (noradrenergic) and neurons in the nucleus raphe magnus of the rostroventromedial medulla (serotonergic) that project to the spinal cord (Foreman, R. D., 1989). The central pathways mediating the facilitatory effect are mediated by suprapontine pathways (Gebhart, G. F. and Randich, A., 1992; Randich, A. and Gebhart, G. F., 1992). The functional types of vagal afferents being involved in inhibitory and facilitatory control of nociceptive impulse transmission are unknown.

3. Activity in vagal afferents innervating the small intestine (i.e., projecting through the celiac branches of the abdominal vagus nerves) is important in reflex modulation of inflammatory processes (e.g., in the knee joint) and mechanical hyperalgesic behavior (by sensitization of nociceptors) in remote body tissues involving the sympatho-adrenal system and possibly the hypothalamo-pituitary–adrenal system (Green, P. G. et al., 1995; 1997; Miao, F. J.-P. et al., 1997a;1997b; Khasar, S. G. et al., 1998a; 1998b; Jänig, W. et al., 2000; Miao, F. J. -P. et al., 2000, Miao; F. J. -P.et al., 2001; Khasar, S. G. et al., 2003; Miao, F. J. -P. et al., 2003a; 2003b; see Section 5.17.3.4.4 and 5.17.4 in Chapter Autonomic Nervous System and Pain.

4. Intraperitoneal injection of illness-inducing agents, such as the bacterial cell wall endotoxin LPS in rats, produces behavioral hyperalgesia.

This is mediated by activity in subdiaphragmatic vagal afferents, specifically afferents running in the hepatic branch. It is suggested that LPS activates hepatic macrophages (Kupffer cells), dendritic cells, and leukocytes that release IL-1β and TNF-α. This in turn activates vagal afferents from the liver. IL-1β and TNF-α injected intraperitoneally themselves generate behavioral hyperalgesia that is abolished by vagotomy (Watkins, L. R. *et al.*, 1994a; 1994b; 1995a; 1995b; 1995c; 1995d). The proinflammatory cytokines either activate the vagal afferents directly or bind to glomus cells in the abdominal paraganglia that are innervated by vagal afferents; the activated vagal afferents signal the peripheral events to the brain, leading to hyperalgesia (as shown by an enhanced thermal nociceptive tail-flick reflex).

5. Pain behavior mediated by vagal afferents, activated by intraperitoneal injection of LPS or of proinflammatory cytokines, is part of a general sickness behavior characterized by various protective illness responses (e.g., immobility, decreased social interaction, decrease in food intake, formation of taste aversion to novel foods, decrease of digestion, loss of weight (anorexia), fever, increase of sleep, change in endocrine functions (activation of the hypothalamo–pituitary–adrenal axis), malaise) and is correlated with marked alterations of brain functions. For example, food aversion and anorexia generated in rats by intraperitoneal TNF-α are abolished or attenuated by subdiaphragmatic vagotomy (Bret-Dibat, J. L. *et al.*, 1995; Bernstein, I. L., 1996). Endotoxin, IL-1β, or TNF-α injected intraperitoneally activate nodose ganglion cells (Gaykema, R. P. *et al.*, 1998) and various brain areas in rodents (nucleus tractus solitarii, parabrachial nuclei, and hypothalamic supraoptic and paraventricular nuclei) as well as induce IL-1β mRNA in the pituitary gland, hypothalamus, and hippocampus. These changes do not occur or are significantly attenuated in subdiaphragmatically vagotomized animals (Ericsson, A. *et al.*, 1994; Wan, W. *et al.*, 1994; Layé, S. *et al.*, 1995; see Sawchenko, P. E. *et al.*, 1996; Dantzer, R. *et al.*, 1998; 2000).

6. Watkins, Maier, and coworkers have developed the idea that vagal abdominal afferents projecting through the hepatic branch of the abdominal vagus nerves form an important neural interface between the immune system and the brain. Activation of these afferents by signals from the immune system (proinflammatory cytokines IL-1β, TNF-α, IL-6) trigger via different centers in brain stem and hypothalamus illness responses, one component being pain and hyperalgesia. The underlying mechanisms of these illness responses have been discussed (Maier, S. F. and Watkins, L. R., 1998; Watkins, L. R. and Maier, S. F., 1999, Goehler, L. E. *et al.*, 2000; Watkins, L. R. and Maier, S. F., 2000). The physiology of the vagal afferents involved in the communication between the immune system, which operates as a diffuse sensory system to detect chemical constituents associated with dangerous microorganisms and their toxins, and the brain has to be worked out.

Acknowledgments

Supported by the Deutsche Forschungsgemeinschaft and the German Ministry of Research and Education (German Research Network on Neuropathic Pain).

References

Bernstein, I. L. 1996. Neural mediation of food aversions and anorexia induced by tumor necrosis factor and tumors. Neurosci. Biobehav. Rev. 20, 177–181.

Berthoud, H. R. and Neuhuber, W. L. 2000. Functional and chemical anatomy of the afferent vagal system. Auton. Neurosci. 85, 1–17.

Berthoud, H. R., Carlson, N. R., and Powley, T. L. 1991. Topography of efferent vagal innervation of the rat gastrointestinal tract. Am. J. Physiol. 260, R200–R207.

Bossut, D. F. and Maixner, W. 1996. Effects of cardiac vagal afferent electrostimulation on the responses of trigeminal and trigeminothalamic neurons to noxious orofacial stimulation. Pain 65, 101–109.

Bret-Dibat, J. L., Bluthe, R. M., Kent, S., Kelley, K. W., and Dantzer, R. 1995. Lipopolysaccharide and interleukin-1 depress food-motivated behavior in mice by a vagal-mediated mechanism. Brain Behav. Immun. 9, 242–246.

Cervero, F. 1994. Sensory innervation of the viscera: peripheral basis of visceral pain. Physiol. Rev. 74, 95–138.

Chandler, M. J., Zhang, J., Qin, C., and Foreman, R. D. 2002. Spinal inhibitory effects of cardiopulmonary afferent inputs in monkeys: neuronal processing in high cervical segments. J. Neurophysiol. 87, 1290–1302.

Crawford, J. P. and Frankel, H. L. 1971. Abdominal 'visceral' sensation in human tetraplegia. Paraplegia 9, 153–158.

Dantzer, R., Bluthe, R. M., Gheusi, G., Cremona, S., Laye, S., Parnet, P., and Kelley, K. W. 1998. Molecular basis of sickness behavior. Ann. N. Y. Acad. Sci. 856, 132–138.

Dantzer, R., Konsman, J. P., Bluthe, R. M., and Kelley, K. W. 2000. Neural and humoral pathways of communication from the immune system to the brain: parallel or convergent? Auton. Neurosci. 85, 60–65.

Ericsson, A., Kovacs, J. C., and Sawchenko, P. E. 1994. A functional anatomical analysis of central pathways subserving the effects of interleukin-1 on stress-related neuroendocrine neurons. J. Neurosci. 14, 897–913.

Foreman, R. D. 1989. Organization of the spinothalamic tract as a relay for cardiopulmonary sympathetic afferent fiber activity. Prog. Sens. Physiol. 9, 1–51.

Foreman, R. D. 1999. Mechanisms of cardiac pain. Annu. Rev. Physiol. 61, 143–167.

Gaykema, R. P., Goehler, L. E., Tilders, F. J., Bol, J. G., McGorry, M., Fleshner, M., Maier, S. F., and Watlans, L. R. 1998. Bacterial endotoxin induces fos immunoreactivity in primary afferent neurons of the vagus nerve. Neuroimmunomodulation 5, 234–240.

Gebhart, G. F. and Randich, A. 1992. Vagal modulation of nociception. APS J. 1, 26–32.

Goehler, L. E., Gaykema, R. P., Hansen, M. K., Anderson, K., Maier, S. F., and Watkins, L. R. 2000. Vagal immune-to-brain communication: a visceral chemosensory pathway. Auton. Neurosci. 85, 49–59.

Green, P. G., Jänig, W., and Levine, J. D. 1997. Sympathetic terminal: target for negative feedback neuroendocrine control of inflammatory response in the rat. J. Neurosci. 17, 3234–3238.

Green, P. G., Miao, F. J. P., Jänig, W., and Levine, J. D. 1995. Negative feedback neuroendocrine control of the inflammatory response in rats. J. Neurosci. 15, 4678–4686.

Grundy, D. and Scratcherd, T. 1989. Sensory Afferents from the Gastrointestinal Tract. In: Handbook of Physiology, The Gastrointestinal System (ed. J. D. Wood), pp. 593–620. Am. Physiol. Soc.

Guttmann, L. 1976. Spinal Cord Injuries. Comprehensive Management and Research. Blackwell Scientific Publications.

Holzer, P. 2002. Sensory neurone responses to mucosal noxae in the upper gut: relevance to mucosal integrity and gastrointestinal pain. Neurogastroenterol. Motil. 14, 459–475.

Holzer, P. 2003. Afferent signalling of gastric acid challenge. J. Physiol. Pharmacol. 54(Suppl. 4), 43–53.

Holzer, P. 2006. Efferent-like roles of afferent neurons in the gut: Blood flow regulation and tissue protection. Auton. Neurosci. 125, 70–75.

Holzer, P. and Maggi, C. A. 1998. Dissociation of dorsal root ganglion neurons into afferent and efferent-like neurons. J. Neurosci. 86, 389–398.

Jänig, W. 2005. Vagal Afferents and Visceral Pain. In: Advances in Vagal Afferent Neurobiology (eds. B. Undem and D. Weinreich), pp. 465–493. CRC Press.

Jänig, W. 2006. The Integrative Action of the Autonomic Nervous System. Neurobiology of Homeostasis. Cambridge University Press.

Jänig, W. and Levine, J. D. 2006. Autonomic-Neuroendocrine-Immune Responses in Acute and Chronic Pain. In: Wall & Melzack's Textbook of Pain, 5th edn. (eds. S. B. McMahon and M. Koltzenburg), pp. 205–218. Elsevier Churchill Livingstone.

Jänig, W., Khasar, S. G., Levine, J. D., and Miao, F. J.-P. 2000. The role of vagal visceral afferents in the control of nociception. Prog. Brain Res. 122, 273–287.

Juler, G. L. and Eltorai, I. M. 1985. The acute abdomen in spinal cord injury patients. Paraplegia 23, 118–123.

Khasar, S. G., Green, P. G., Miao, F. J.-P., and Levine, J. D. 2003. Vagal modulation of nociception is mediated by adrenomedullary epinephrine in the rat. Eur. J. Neurosci. 17, 909–915.

Khasar, S. G., Miao, F. J.-P., Jänig, W., and Levine, J. D. 1998a. Modulation of bradykinin-induced mechanical hyperalgesia in the rat skin by activity in the abdominal vagal afferents. Eur. J. Neurosci. 10, 435–444.

Khasar, S. G., Miao, F. J.-P., Jänig, W., and Levine, J. D. 1998b. Vagotomy-induced enhancement of mechanical hyperalgesia in the rat is sympathoadrenal-mediated. J. Neurosci. 18, 3043–3049.

Kirkup, A. J., Brunsden, A. M., and Grundy, D. 2001. Receptors and transmission in the brain-gut axis: potential for novel therapies. I. Receptors on visceral afferents. Am. J. Physiol. Gastrointest. Liver Physiol. 280, G787–G794.

Kreis, M. E., Jiang, W., Kirkup, A. J., and Grundy, D. 2002. Cosensitivity of vagal mucosal afferents to histamine and 5-HT in the rat jejunum. Am. J. Physiol. Gastrointest. Liver Physiol. 283, G612–G617.

Lal, S., Kirkup, A. J., Brunsden, A. M., Thompson, D. G., and Grundy, D. 2001. Vagal afferent responses to fatty acids of different chain length in the rat. Am. J. Physiol. Gastrointest. Liver Physiol. 281, G907–G915.

Layé, S., Bluthe, R. M., Kent, S., Combe, C., Medina, C., Parnet, P., Kelley, K. and Dantzer, R. 1995. Subdiaphragmatic vagotomy blocks induction of IL-1 beta mRNA in mice brain in response to peripheral LPS. Am. J. Physiol. 268, R1327–R1331.

Lindgren, I. and Olivecrona, H. 1947. Surgical treatment of angina pectoris. J. Neurosurg. 4, 19–39.

Maier, S. F. and Watkins, L. R. 1998. Cytokines for psychologists: implications of bidirectional immune-to-brain communication for understanding behavior, mood, and cognition. Psychol. Rev. 105, 83–107.

McDonald, D. M. 1990. The ultrastructure and permeability of tracheobronchial blood vessels in health and disease. Eur. Respir. J. (Suppl.)12, 572s–585s.

McDonald, D. M., Mitchell, R. A., Gabella, G., and Haskell, A. 1988. Neurogenic inflammation in the rat trachea. II. Identity and distribution of nerves mediating the increase in vascular permeability. J. Neurocytol. 17, 605–628.

Meller, S. T. and Gebhart, G. F. 1992. A critical review of the afferent pathways and the potential chemical mediators involved in cardiac pain. J. Neurosci. 48, 501–524.

Miao, F. J.-P., Green, P. G., and Levine, J. D. 2003b. Mechano-sensitive duodenal afferents contribute to vagal modulation of inflammation in the rat. J. Physiol. (Lond.) 554, 227–235.

Miao, F. J.-P., Jänig, W., and Levine, J. D. 1997a. Vagal branches involved in inhibition of bradykinin-induced synovial plasma extravasation by intrathecal nicotine and noxious stimulation in the rat. J. Physiol. (Lond.) 498, 473–481.

Miao, F. J.-P., Jänig, W., and Levine, J. D. 2000. Nociceptive-neuroendocrine negative feedback control of neurogenic inflammation activated by capsaicin in the skin: role of the adrenal medulla. J. Physiol. (Lond.) 527, 601–610.

Miao, F. J.-P., Jänig, W., Green, P. G., and Levine, J. D. 1997b. Inhibition of bradykinin-induced synovial plasma extravasation produced by noxious cutaneous and visceral stimuli and its modulation by activity in the vagal nerve. J. Neurophysiol. 78, 1285–1292.

Miao, F. J.-P., Jänig, W., Jasmin, L., and Levine, J. D. 2001. Spino-bulbo-spinal pathway mediating vagal modulation of nociceptive-neuroendocrine control of inflammation in the rat. J. Physiol. (Lond.) 532, 811–822.

Miao, F. J.-P., Jänig, W., Jasmin, L., and Levine, J. D. 2003a. Blockade of nociceptive inhibition of plasma extravasation by opioid stimulation of the periaqueductal gray and its interaction with vagus-induced inhibition in the rat. Neurosci. 119, 875–885.

Michl, T., Jocic, M., Heinemann, A., Schuligoi, R., and Holzer, P. 2001. Vagal afferent signaling of a gastric mucosal acid insult to medullary, pontine, thalamic, hypothalamic and limbic, but not cortical, nuclei of the rat brain. Pain 92, 19–27.

Multon, S. and Schoenen, J. 2005. Pain control by vagus nerve stimulation: from animal to man . . . and back. Acta Neurol. Belg. 105, 62–67.

Niijima, A. 1996. The afferent discharge from sensors for interleukin-1 beta in the hepatoportal system in the anesthetized rat. J. Auton. Nerv. Syst. 61, 287–291.

Precht, J. C. and Powley, T. L. 1990. The fibre composition of the abdominal vagus in the rat. Anat. and Embryol. 181, 101–115.

Randich, A. and Gebhart, G. F. 1992. Vagal afferent modulation of nociception. Brain Res. Rev. 17, 77–99.

Ren, K., Zhuo, M., Randich, A., and Gebhart, G. F. 1993. Vagal afferent stimulation-produced effects on nociception in capsaicin-treated rats. J. Neurophysiol. 69, 1530–1540.

Richards, W., Hillsley, K., Eastwood, C., and Grundy, D. 1996. Sensitivity of vagal mucosal afferents to cholecystokinin and its role in afferent signal transduction in the rat. J. Physiol. (Lond.) 497, 473–481.

Sawchenko, P. E., Brown, E. R., Chan, R. K., Ericsson, A., Li, H. Y., Roland, B. L., and Kovacs, K. J. 1996. The paraventricular nucleus of the hypothalamus and the functional neuroanatomy of visceromotor responses to stress. Prog. Brain Res. 107, 201–222.

Schachter, S. C. 2005. Electrical Stimulation of the Vagus Nerve for the Treatment of Epilepsy. In: Advances in Vagal Afferent Neurobiology (eds. B. J. Undem and D. Weinreich), pp. 495–510. CRC Press.

Shanahan, F. 1994. The Intestinal Immune System. Physiology of the Gastrointestinal Tract (ed. L. R. Johnson), pp. 643–684. Raven Press.

Strauther, G. R., Longo, W. E., Virgo, K. S., and Johnson, F. E. 1999. Appendicitis in patients with previous spinal cord injury. Am. J. Surg. 178, 403–405.

Takeda, M., Tanimoto, T., Ojima, K., and Matsumoto, S. 1998. Suppressive effect of vagal afferents on the activity of the trigeminal spinal neurons related to the jaw-opening reflex in rats: involvement of the endogenous opioid system. Brain Res. Bull. 47, 49–56.

Walls, E. K., Phillips, R. J., Wang, F. B., Holst, M. C., and Powley, T. L. 1995b. Suppression of meal size by intestinal nutrients is eliminated by celia vagal deafferentation. Am. J. Physiol. 269, R1410–1419.

Walls, E. K., Wang, F. B., Holst, M. C., Phillips, R. J., Voreis, J. S., Perkins, A. R., Pollard, L. E. and Powley, T. L.

1995a. Selective vagal rhizotomies: a new dorsal surgical approach used for intestinal deafferentions. Am. J. Physiol. 269, R1279–1288.

Wan, W., Wetmore, L., Sorensen, C. M., Greenberg, A. H., and Nance, D. M. 1994. Neural and biochemical mediators of endotoxin and stress-induced c-fos expression in the rat brain. Brain Res. Bull. 34, 7–14.

Watkins, L.R. and Maier, S. F. (eds.) 1999. Cytokines and Pain. Birkhäuser Verlag.

Watkins, L. R. and Maier, S. F. 2000. The pain of being sick: implications of immune-to-brain communication for understanding pain. Annu. Rev. Psychol. 51, 29–57.

Watkins, L. R., Goehler, L. E., Relton, J., Brewer, M. T., and Maier, S. F. 1995a. Mechanisms of tumor necrosis factor-alpha (TNF-alpha) hyperalgesia. Brain Res. 692, 244–250.

Watkins, L. R., Goehler, L. E., Relton, J. K., Tartaglia, N., Silbert, L., Martin, D., and Maier, S. F. 1995b. Blockade of interleukin-1 induced hyperthermia by sub-diaphragmatic vagotom: evidence for vagal mediation of immune-brain communication. Neurosci. Lett. 183, 27–31.

Watkins, L. R., Maier, S. F., and Goehler, L. E. 1995c. Cytokine-to-brain communication: a review & analysis of alternative mechanisms. Life Sci. 57, 1011–1026.

Watkins, L. R., Maier, S. F., and Goehler, L. E. 1995d. Immune activation: the role of pro-inflammatory cytokines in inflammation, illness responses and pathological pain states. Pain 63, 289–302.

Watkins, L. R., Wiertelak, E. P., Goehler, L. E., Mooney-Heiberger, K., Martinez, J., Furness, L. Smith, K. P. and Maier, S. F. 1994a. Neurocircuitry of illness-induced hyperalgesia. Brain Res. 639, 283–299.

Watkins, L. R., Wiertelak, E. P., Goehler, L. E., Smith, K. P., Martin, D., and Maier, S. F. 1994b. Characterization of cytokine-induced hyperalgesia. Brain Res. 654, 15–26.

White, J. C. and Bland, E. F. 1948. The surgical relief of severe angina pectoris. Medicine 27, 1–42.

Zhu, J. X., Zhu, X. Y., Owyang, C., and Li, Y. 2001. Intestinal serotonin acts as a paracrine substance to mediate vagal signal transmission evoked by luminal factors in the rat. J. Physiol. (Lond.) 530, 431–442.

21 Sex, Gender, and Pain

R B Fillingim, University of Florida College of Dentistry, Community Dentistry and Behavioral Science
Gainesville, FL, USA

Glossary

biopsychosocial model A model of pain that posits that the experience of and response to pain results from complex, dynamic interactions among biological, psychological, and sociocultural variables.

gender A person's self-representation as male or female, or how that person is responded to by social institutions on the basis of that individual's gender presentation.

sex The classification of living things, generally as male or female, according to their reproductive organs and functions assigned by the chromosomal complement.

21.1 Sex Differences in Pain Responses

Considerable evidence suggests that females and males differ in their experience of both clinical and experimentally induced pain (for reviews, see Fillingim, R. B. and Maixner, W., 1995; Unruh, A. M., 1996; Berkley, K. J., 1997; LeResche, L., 1999; Fillingim, R. B., 2003). For example, several chronic pain disorders are more prevalent among women, and in population surveys women report more pain than men (Fillingim, R. B. and Maixner, W., 1995; Unruh, A. M., 1996; Berkley, K. J., 1997). Moreover, some evidence indicates that clinical pain may be more severe for women than men. For example, women have reported greater postsurgical pain than men in some (Savedra, M. C. *et al.*, 1993; Averbuch, M. and Katzper, M., 2000; Coulthard, P. *et al.*, 2000; Taenzer, A. H. *et al.*, 2000; Morin, C. *et al.*, 2000; Averbuch, M. and Katzper, M., 2001; Kalkman, C. J. *et al.*, 2003) but not all (e.g., Gordon, N. C. *et al.*, 1995; Gear, R. W. *et al.*, 1996a; Chia, Y. Y. *et al.*, 2002) studies. Women have also reported higher pain ratings in acute cancer-related pain (Cepeda, M. S. *et al.*, 2003), procedural pain such as colonoscopy (Froehlich, F. *et al.*, 1997), and conditions presented in emergency rooms (Boccardi, L. and Verde, M., 2003). In addition, women display greater perceptual responses to laboratory-induced pain relative to men, although the magnitude of these differences varies across studies, perhaps due to methodological differences (e.g., pain assay, stimulus characteristics, sample selection; Fillingim, R. B. and Maixner, W., 1995; Riley, J. L. *et al.*, 1998). Interestingly, consistent with the human literature, greater behavioral responses to noxious stimuli have been reported among female compared to male rodents (e.g., Barrett, A. C. *et al.*, 2002b; 2003; Terner, J. M. *et al.*, 2003; also for reviews, see Bodnar, R. J. *et al.*, 1988; Berkley, K. J., 1997), with some exceptions (Mogil, J. S. *et al.*, 1993; Kayser, V. *et al.*, 1996). In general, these findings suggest that females exhibit greater perceptual responses to pain than males; however, the mechanisms underlying these sex differences have not been determined.

21.2 Sex Differences in Responses to Pain Treatment

In addition to basal pain sensitivity, sex-related influences on responses to pharmacologic and non-pharmacologic pain treatments have received increasing attention in recent years. The rodent literature fairly consistently demonstrates greater analgesic responses to opioid agonists among males versus female animals (Bodnar, R. J. *et al.*, 1988; Kepler, K. L. *et al.*, 1989; 1991; Kiefel, J. M. and Bodnar. R. J., 1992; Islam, A. K. *et al.*, 1993; Cicero, T. J. *et al.*, 1996a; 1997; Kest, B. *et al.*, 2000; Craft, R. M., 2003a; 2003b). However, the human literature regarding sex differences in analgesic responses has shown mixed results. Some clinical studies demonstrate greater analgesic responses among women, others show no sex differences, and others report greater analgesia among men (Fillingim, R. B. and Gear, R. W., 2004). Among the most consistent findings is a series of studies indicating more robust analgesic responses κ-opioid-acting drugs after oral surgery (Gear, R. W. *et al.*, 1996a; 1996b; 1999; 2003). Laboratory studies in humans have shown mixed results, as well (Sarton, E. *et al.*, 2000; Fillingim, R. B. *et al.*, 2004; Romberg, R. *et al.*, 2004; Olofsen, E. *et al.*, 2005; Fillingim, R. B. *et al.*, 2005c). Sex differences in response to nonpharmacologic interventions for pain has received less empirical attention, though a few laboratory (Keogh, E. *et al.*, 2000; Sternberg, W. F. *et al.*, 2001) and clinical (Hansen, F. R. *et al.*, 1993; Krogstad, B. S. *et al.*, 1996; Jensen, I. B. *et al.*, 2001) studies suggest that sex differences in treatment responses may emerge. It is possible that the consistent sex differences in analgesic responses are specific to the oral surgery pain model, perhaps due to its inflammatory component or based on interactions with other medications administered perioperatively. The clinical relevance of these results beyond this acute pain model remains unknown, and this is an area ripe for future investigation.

21.3 Mechanisms Underlying Sex Differences in Pain

It is important to recognize that sex differences in pain are inevitably determined by interactions among multiple biopsychosocial processes. While these mechanisms are typically described as either psychosocial or biological, this dualistic conceptualization is artificial and potentially counterproductive. For instance, sex differences in expression of pain are often attributed to the effects of stereotypic sex roles, typically considered a psychosocial issue. However, neurobiological and hormonal correlates of masculine versus feminine sex roles may well contribute to differences in nociceptive processing. Thus, various psychosocial and biological mechanisms underlying sex differences in pain responses could refer to the same fundamental processes described at different levels of analysis.

Several biological variables have been implicated in sex differences in both clinical and experimental pain responses. Abundant preclinical and clinical evidence suggests that gonadal hormones are important, because hormonal influences on basal pain sensitivity as well as responses to analgesic compounds have been well documented (Riley, J. L. III. *et al.*, 1999; Fillingim, R. B. and Ness, T. J., 2000; Bodnar. R. J. *et al.*, 2002). Also, endogenous pain inhibitory systems may function differently in females and males, including the endogenous opioid system. Zubieta and colleagues have reported sex differences in both basal and pain-evoked μ-opioid receptor binding in certain brain regions (Zubieta, J. K. *et al.*, 1999; 2002). Genetic factors may also contribute to sex differences in pain. In rodents, sex differences in nociceptive and antinociceptive responses are often strain-dependent (Mogil, J. S. *et al.*, 2000; Barrett, A. C. *et al.*, 2002a; Terner, J. M. *et al.*, 2003), and pain-related quantitative trait loci (QTLs) are often sex-specific (Mogil, J. S. *et al.*, 1997; Mogil, J. S., 2000). Recent findings provide corroborating results in humans (Kim, H. *et al.*, 2004). Also, Mogil J. S. *et al.* (2003) demonstrated in both mice and humans that the melanocortin-1-receptor gene (*MC1R*) was associated with analgesic responses to κ-opioids among females but not males. Also, the A118G single nucleotide polymorphism of the μ-opioid receptor gene (*OPRM1*) was significantly associated with pressure pain thresholds, but this association was only significant among males (Fillingim, R. B. *et al.*, 2005b). Thus, several biological factors may contribute to sex differences in pain.

Numerous psychosocial variables also contribute to sex differences in pain responses. Anxiety and other negative emotions have been more strongly related with both clinical and experimental pain responses among men than women (Fillingim, R. B. *et al.*, 1996; Edwards, R. R. *et al.*, 2000; Riley, J. L., III *et al.*, 2001; Jones, A. and Zachariae, R., 2002; Robinson, M. E. *et al.*, 2005) Moreover, positive affect predicted lower experimental pain sensitivity and fewer side effects in responses to pentazocine (a mixed action opioid) among men, but not women (Fillingim, R. B. *et al.*, 2005a). Additional psychosocial factors that may contribute to sex differences in pain include cognitive and behavioral pain coping strategies (Affleck, G. *et al.*, 1999; Unruh, A. M. *et al.*, 1999; Keefe, F. J. *et al.*, 2000; Mercado, A. C. *et al.*, 2000; Osman, A. *et al.*, 2000), gender roles (Otto, M. W. and

Dougher, M. J., 1985; Myers, C. D. *et al.*, 2001; Robinson, M. E. *et al.*, 2003), and family history of pain (Edwards, P. W. *et al.*, 1985; Fillingim, R. B. *et al.*, 2000).

21.4 Conclusions and Future Directions

The evidence reviewed indicates consistent sex differences in pain-related responses, to which multiple biopsychosocial factors contribute. The literature focuses more heavily on quantitative sex differences (i.e., does the magnitude of pain or analgesia differ in women vs. men?); however, of equal if not greater importance are qualitative sex differences (i.e., are the mediators of pain and analgesia different in women vs. men?). Some examples are provided above, including findings demonstrating sex-specific genetic and psychosocial associations with pain. The clearest challenge for the future is to determine the practical significance of these sex differences and to integrate this information into clinical practice. It seems plausible that expanding research into qualitative sex differences may be more informative in this regard, because such differences imply that some components of the pain processing system may be fundamentally different in women versus men, which calls for sex-specific treatment approaches. It is hoped that continued research into sex, gender and pain will ultimately enhance pain management for both women and men.

Acknowledgment

This work was supported by NIH grant NS41670.

References

Affleck, G., Tennen, H., Keefe, F. J., Lefebvre, J. C., Kashikar-Zuck, S., Wright, K., Starr, K., and Caldwell, D. S. 1999. Everyday life with osteoarthritis or rheumatoid arthritis: independent effects of disease and gender on daily pain, mood, and coping. Pain 83, 601–609.

Averbuch, M. and Katzper, M. 2000. A search for sex differences in response to analgesia. Arch. Intern. Med. 160, 3424–3428.

Averbuch, M. and Katzper, M. 2001. Gender and the placebo analgesic effect in acute pain. Clin. Pharmacol. Ther. 70, 287–291.

Barrett, A. C., Cook, C. D., Terner, J. M., Roach, E. L., Syvanthong, C., and Picker, M. J. 2002a. Sex and rat strain determine sensitivity to kappa opioid-induced antinociception. Psychopharmacology (Berl.) 160, 170–181.

Barrett, A., Smith, E., and Picker, M. 2002b. Sex-related differences in mechanical nociception and antinociception produced by mu- and kappa-opioid receptor agonists in rats. Eur. J. Pharmacol. 452, 163.

Barrett, A. C., Smith, E. S., and Picker, M. J. 2003. Capsaicin-induced hyperalgesia and μ-opioid-induced antihyperalgesia in male and female Fischer 344 rats. J. Pharmacol. Exp. Ther. 307, 237–245.

Berkley, K. J. 1997. Sex Differences in Pain. Behav. Brain Sci. 20, 371–380.

Boccardi, L. and Verde, M. 2003. Gender differences in the clinical presentation to the emergency department for chest pain. Ital. Heart J. 4, 371–373.

Bodnar, R. J., Commons, K., and Pfaff, D. W. 2002. Central Neural States Relating Sex and Pain. Johns Hopkins University Press.

Bodnar, R. J., Romero, M. T., and Kramer, E. 1988. Organismic variables and pain inhibition: roles of gender and aging. Brain Res. Bull. 21, 947–953.

Cepeda, M. S., Africano, J. M., Polo, R., Alcala, R., and Carr, D. B. 2003. Agreement between percentage pain reductions calculated from numeric rating scores of pain intensity and those reported by patients with acute or cancer pain. Pain 106, 439–442.

Chia, Y. Y., Chow, L. H., Hung, C. C., Liu, K., Ger, L. P., and Wang, P. N. 2002. Gender and pain upon movement are associated with the requirements for postoperative patient-controlled iv analgesia: a prospective survey of 2,298 Chinese patients. Can. J. Anaesth. 49, 249–255.

Cicero, T. J., Nock, B., and Meyer, E. R. 1996. Gender-related differences in the antinociceptive properties of morphine. J. Pharmacol. Exp. Ther. 279, 767–773.

Cicero, T. J., Nock, B., and Meyer, E. R. 1997. Sex-related differences in morphine's antinociceptive activity: relationship to serum and brain morphine concentrations. J. Pharmacol. Exp. Ther. 282, 939–944.

Coulthard, P., Pleuvry, B. J., Dobson, M., and Price, M. 2000. Behavioural measurement of postoperative pain after oral surgery. Br. J. Oral Maxillofac. Surg. 38, 127–131.

Craft, R. M. 2003a. Sex differences in drug- and non-drug-induced analgesia. Life Sci. 72, 2675–2688.

Craft, R. M. 2003b. Sex differences in opioid analgesia: from mouse to man. Clin. J. Pain 19, 175–186.

Edwards, R. R., Augustson, E., and Fillingim, R. B. 2000. Sex-specific effects of pain-related anxiety on adjustment to chronic pain. Clin. J. Pain 16, 46–53.

Edwards, P. W., Zeichner, A., Kuczmierczyk, A. R., and Boczkowski, J. 1985. Familial pain models: the relationship between family history of pain and current pain experience. Pain 21, 379–384.

Fillingim, R. B. 2003. Sex-related influences on pain: a review of mechanisms and clinical implications. Rehabil. Psychol. 48, 165–174.

Fillingim, R. B. and Gear, R. W. 2004. Sex differences in opioid analgesia: clinical and experimental findings. Eur. J. Pain 8, 413–425.

Fillingim, R. B. and Maixner, W. 1995. Gender differences in the responses to noxious stimuli. Pain Forum 4, 209–221.

Fillingim, R. B. and Ness, T. J. 2000. Sex-related hormonal influences on pain and analgesic responses. Neurosci. Biobehav. Rev. 24, 485–501.

Fillingim, R. B., Edwards, R. R., and Powell, T. 2000. Sex-dependent effects of reported familial pain history on clinical and experimental pain responses. Pain 86, 87–94.

Fillingim, R. B., Hastie, B. A., Ness, T. J., Glover, T. L., Campbell, C. M., and Staud, R. 2005a. Sex-related psychological predictors of baseline pain perception and analgesic responses to pentazocine. Biol. Psychol. 69, 97–112.

Fillingim, R. B., Kaplan, L., Staud, R., Ness, T. J., Glover, T. L., Campbell, C. M., Mogil, J. S., and Wallace, M. R. 2005b. The A118G single nucleotide polymorphism of the mu-opioid receptor gene (OPRM1) is associated with pressure pain sensitivity in humans. J. Pain 6, 159–167.

Fillingim, R. B., Keefe, F. J., Light, K. C., Booker, D. K., and Maixner, W. 1996. The influence of gender and psychological factors on pain perception. J. Gender Cult. Health 1, 21–36.

Fillingim, R. B., Ness, T. J., Glover, T. L., Campbell, C. M., Hastie, B. A., Price, D. D., and Staud, R. 2005c. Morphine responses and experimental pain: sex differences in side effects and cardiovascular responses but not analgesia. J. Pain 6, 116–124.

Fillingim, R. B., Ness, T. J., Glover, T. L., Campbell, C. M., Price, D. D., and Staud, R. 2004. Experimental pain models reveal no sex differences in pentazocine analgesia in humans. Anesthesiology 100, 1263–1270.

Froehlich, F., Thorens, J., Schwizer, W., Preisig, M., Kohler, M., Hays, R. D, Fried, M., and Gonvers, J. J. 1997. Sedation and analgesia for colonoscopy: patient tolerance, pain, and cardiorespiratory parameters. Gastrointest. Endosc. 45, 1–9.

Gear, R. W., Gordon, N. C., Heller, P. H., Paul, S., Miaskowski, C., and Levine, J. D. 1996a. Gender difference in analgesic response to the kappa-opioid pentazocine. Neurosci. Lett. 205, 207–209.

Gear, R. W., Gordon, N. C., Miaskowski, C., Paul, S. M., Heller, P. H., and Levine, J. D. 2003. Sexual dimorphism in very low dose nalbuphine postoperative analgesia. Neurosci. Lett. 339, 1–4.

Gear, R. W., Miaskowski, C., Gordon, N. C., Paul, S. M., Heller, P. H., and Levine, J. D. 1999. The kappa opioid nalbuphine produces gender- and dose-dependent analgesia and antianalgesia in patients with postoperative pain. Pain 83, 339–345.

Gear, R. W., Miaskowski, C., Gordon, N. C., Paul, S. M., Heller, P. H., and Levine, J. D. 1996b. Kappa-opioids produce significantly greater analgesia in women than in men. Nat. Med. 2, 1248–1250.

Gordon, N. C., Gear, R. W., Heller, P. H., Paul, S., Miaskowski, C., and Levine, J. D. 1995. Enhancement of morphine analgesia by the GABAB agonist baclofen. Neuroscience 69, 345–349.

Hansen, F. R., Bendix, T., Skov, P., Jensen, C. V., Kristensen, J. H., Krohn, L., and Schioeler, H. 1993. Intensive, dynamic back-muscle exercises, conventional physiotherapy, or placebo-control treatment of low-back pain. A randomized, observer-blind trial. Spine 18, 98–108.

Islam, A. K., Beczkowska, I. W., and Bodnar, R. J. 1993. Interactions among aging, gender, and gonadectomy effects upon naloxone hypophagia in rats. Physiol. Behav. 54, 981–992.

Jensen, I. B., Bergstrom, G., Ljungquist, T., Bodin, L., and Nygren, A. L. 2001. A randomized controlled component analysis of a behavioral medicine rehabilitation program for chronic spinal pain: are the effects dependent on gender? Pain 91, 65–78.

Jones, A. and Zachariae, R. 2002. Gender, anxiety, and experimental pain sensitivity: an overview. J. Am. Med. Womens Assoc. 57, 91–94.

Kalkman, C. J., Visser, K., Moen, J., Bonsel, G. J., Grobbee, D. E., and Moons, K. G. 2003. Preoperative prediction of severe postoperative pain. Pain 105, 415–423.

Kayser, V., Berkley, K. J., Keita, H., Gautron, M., and Guilbaud, G. 1996. Estrous and sex variations in vocalization thresholds to hindpaw and tail pressure stimulation in the rat. Brain Res. 742, 352–354.

Keefe, F. J., Lefebvre, J. C., Egert, J. R., Affleck, G., Sullivan, M. J., and Caldwell, D. S. 2000. The relationship of gender to pain, pain behavior, and disability in osteoarthritis patients: the role of catastrophizing. Pain 87, 325–334.

Keogh, E., Hatton, K., and Ellery, D. 2000. Avoidance versus focused attention and the perception of pain: differential effects for men and women. Pain 85, 225–230.

Kepler, K. L., Kest, B., Kiefel, J. M., Cooper, M. L., and Bodnar, R. J. 1989. Roles of gender, gonadectomy and estrous phase in the analgesic effects of intracerebroventricular morphine in rats. Pharmacol. Biochem. Behav. 34, 119–127.

Kepler, K. L., Standifer, K. M., Paul, D., Kest, B., Pasternak, G. W., and Bodnar, R. J. 1991. Gender effects and central opioid analgesia. Pain 45, 87–94.

Kest, B., Sarton, E., and Dahan, A. 2000. Gender differences in opioid-mediated analgesia: animal and human studies. Anesthesiology 93, 539–547.

Kiefel, J. M. and Bodnar, R. J. 1992. Roles of gender and gonadectomy in pilocarpine and clonidine analgesia in rats. Pharmacol. Biochem. Behav. 41, 153–158.

Kim, H., Neubert, J. K., San, M. A., Xu, K., Krishnaraju, R. K., Iadarola, M. J., Goldman, D., and Dionne, R. A. 2004. Genetic influence on variability in human acute experimental pain sensitivity associated with gender, ethnicity and psychological temperament. Pain 109, 488–496.

Krogstad, B. S., Jokstad, A., Dahl, B. L., and Vassend, O. 1996. The reporting of pain, somatic complaints, and anxiety in a group of patients with TMD before and 2 years after treatment: sex differences. J. Orofac. Pain 10, 263–269.

LeResche, L. 1999. Gender Considerations in the Epidemiology of Chronic Pain. In: Epidemiology of Pain (ed. I. K. Crombie), pp. 43–52. IASP Press.

Mercado, A. C., Carroll, L. J., Cassidy, J. D., and Cote, P. 2000. Coping with neck and low back pain in the general population. Health Psychol. 19, 333–338.

Mogil, J. S. 2000. Interactions Between Sex and Genotype in the Mediation and Modulation of Nociception in Rodents. In: Sex, Gender, and Pain (ed. R. B. Fillingim), pp. 25–40. IASP Press.

Mogil, J. S., Chesler, E. J., Wilson, S. G., Juraska, J. M., and Sternberg, W. F. 2000. Sex differences in thermal nociception and morphine antinociception in rodents depend on genotype. Neurosci. Biobehav. Rev. 24, 375–389.

Mogil, J. S., Richards, S. P., O'Toole, L. A., Helms, M. L., Mitchell, S. R., and Belknap, J. K. 1997. Genetic sensitivity to hot-plate nociception in DBA/2J and C57BL/6J inbred mouse strains: possible sex-specific mediation by delta2-opioid receptors. Pain 70, 267–277.

Mogil, J. S., Sternberg, W. F., Kest, B., Marek, P., and Liebeskind, J. C. 1993. Sex differences in the antagonism of stress-induced analgesia: effects of gonadectomy and estrogen replacement. Pain 53, 17–25.

Mogil, J. S., Wilson, S. G., Chesler, E. J., Rankin, A. L., Nemmani, K. V., Lariviere, W. R., Groce, M. K., Wallace, M. R., Kaplan, L., Staud, R., Ness, T. J., Glover, T. L., Stankova, M., Mayorov, A., Hruby, V. J., Grisel, J. E., and Fillingim, R. B. 2003. The melanocortin-1 receptor gene mediates female-specific mechanisms of analgesia in mice and humans. Proc. Natl. Acad. Sci. U. S. A. 100, 4867–4762.

Morin, C., Lund, J. P., Villarroel, T., Clokie, C. M., and Feine, J. S. 2000. Differences between the sexes in post-surgical pain. Pain 85, 79–85.

Myers, C. D., Robinson, M. E., Riley, J. L., III, and Sheffield, D. 2001. Sex, gender, and blood pressure: contributions to experimental pain report. Psychosom. Med. 63, 545–550.

Olofsen, E., Romberg, R., Bijl, H., Mooren, R., Engbers, F., Kest, B., and Dahan, A. 2005. Alfentanil and placebo analgesia: no sex differences detected in models of experimental pain. Anesthesiology 103, 130–139.

Osman, A., Barrios, F. X., Gutierrez, P. M., Kopper, B. A., Merrifield, T., and Grittmann, L. 2000. The Pain Catastrophizing Scale: further psychometric evaluation with adult samples. J. Behav. Med. 23, 351–365.

Otto, M. W. and Dougher, M. J. 1985. Sex differences and personality factors in responsivity to pain. Percept. Mot. Skills 61, 383–390.

Riley, J. L., III, Robinson, M. E., Wade, J. B., Myers, C. D., and Price, D. D. 2001. Sex differences in negative emotional responses to chronic pain. J. Pain 2, 354–359.

Riley, J. L., III, Robinson, M. E., Wise, E. A., Myers, C. D., and Fillingim, R. B. 1998. Sex differences in the perception of noxious experimental stimuli: a meta-analysis. Pain 74, 181–187.

Riley, J. L. I., Robinson, M. E., Wise, E. A., and Price, D. D. 1999. A meta-analytic review of pain perception across the menstrual cycle. Pain 81, 225–235.

Robinson, M. E., Dannecker, E. A., George, S. Z., Otis, J., Atchison, J. W., and Fillingim, R. B. 2005. Sex differences in the associations among psychological factors and pain report: a novel psychophysical study of patients with chronic low back pain. J. Pain 6, 463–470.

Robinson, M. E., Gagnon, C. M., Riley, J. L., III, and Price, D. D. 2003. Altering gender role expectations: effects on pain tolerance, pain threshold, and pain ratings. J. Pain 4, 284–288.

Romberg, R., Olofsen, E., Sarton, E., den, Hartigh J., Taschner, P. E., and Dahan, A. 2004. Pharmacokinetic–pharmacodynamic modeling of morphine-6-glucuronide-induced analgesia in healthy volunteers: absence of sex differences. Anesthesiology 100, 120–133.

Sarton, E., Olofsen, E., Romberg, R., den Hartigh, J., Kest, B., Nieuwenhuijs, D., Burm, A., Teppema, L., and Dahan, A. 2000. Sex differences in morphine analgesia: an experimental study in healthy volunteers. Anesthesiology 93, 1245–1254.

Savedra, M. C., Holzemer, W. L., Tesler, M. D., and Wilkie, D. J. 1993. Assessment of postoperation pain in children and adolescents using the adolescent pediatric pain tool. Nurs. Res. 42, 5–9.

Sternberg, W. F., Bokat, C., Kass, L., Alboyadjian, A., and Gracely, R. H. 2001. Sex-dependent components of the analgesia produced by athletic competition. J. Pain 2, 65–74.

Taenzer, A. H., Clark, C., and Curry, C. S. 2000. Gender affects report of pain and function after arthroscopic anterior cruciate ligament reconstruction. Anesthesiology 93, 670–675.

Terner, J. M., Lomas, L. M., Smith, E. S., Barrett, A. C., and Picker, M. J. 2003. Pharmacogenetic analysis of sex differences in opioid antinociception in rats. Pain 106, 381–391.

Unruh, A. M. 1996. Gender variations in clinical pain experience. Pain 65, 123–167.

Unruh, A. M., Ritchie, J., and Merskey, H. 1999. Does gender affect appraisal of pain and pain coping strategies? Clin J. Pain 15, 31–40.

Zubieta, J. K., Dannals, R. F., and Frost, J. J. 1999. Gender and age influences on human brain mu-opioid receptor binding measured by PET. Am. J. Psychiatry 156, 842–848.

Zubieta, J. K., Smith, Y. R., Bueller, J. A., Xu, Y., Kilbourn, M. R., Jewett, D. M., Meyer, C. R., Koeppe, R. A., and Stohler, C. S. 2002. mu-Opioid receptor-mediated antinociceptive responses differ in men and women. J. Neurosci. 22, 5100–5107.

22 Neurotrophins and Pain

Lorne M Mendell, State University of New York, Stony Brook, NY, USA

Glossary

immunoadhesin A chimeric molecule consisting of a IgG backbone (Fc region) linked to the binding site of a receptor that binds to a molecule to be sequestered, e.g., trkA-IgG binds nerve growth factor (NGF).

keratinocyte Immune-competent cell in the epidermis.

mast cell Immune-competent cell that stores inflammatory mediators.

neutrophil A type of white blood cell that phagocytoses microorganisms (i.e., due to infection) and breaks them down.

skin nerve preparation *In vitro* preparation in which the skin is removed with its nerve supply (e.g., ankle skin and saphenous nerve) and placed in a chamber. The nerve is led into a second chamber separated from the skin chamber so that the skin can be maintained in oxygenated Ringer, whereas the nerve can be placed in oil for extracellular recording.

tachyphylaxis A process whereby the response to an agent applied repeatedly becomes smaller and smaller.

22.1 The Neurotrophin Family and Its Receptors

Neurotrophin molecules are 13.5-kDa proteins that come in four major forms that show a substantial overlap in their primary structure. These are known as nerve growth factor (NGF), brain-derived neurotrophic factor (BDNF), neurotrophin-3 (NT-3), and neurotrophin-4 (also known neurotrophin-5 or neurotrophin-4/5, NT-4/5). The molecular biology underlying the synthesis of these molecules has been reviewed elsewhere (Teng, K. K. and Hempstead, B. L., 2004). Neurotrophins bind to two classes of receptors: the tropomyosin-related kinase (trk) receptors, trkA, trkB, and trkC, as well as the p75 receptor that is a member of the tumor necrosis factor (TNF) superfamily. Although these two receptor types are often considered to function together as a complex (Chao, M. V. and Hempstead, B. L., 1995; He, X. L. and Garcia, K. C., 2004), there is also evidence that they can function independently (see Section 22.7).

The trk receptors are known as the high-affinity neurotrophin receptors and selectively bind the different neurotrophin molecules. NGF binds selectively to trkA, BDNF and NT-4/5 to trkB, and NT-3 to trkC (Huang, E. J. and Reichardt, L. F., 2001). Some cross-binding with lower affinity also occurs (Barbacid, M., 1994). The p75 receptor is generally known as the low-affinity receptor, although it actually binds neurotrophins with high affinity but with rapid kinetics. It binds all known neurotrophins with similar affinity. Thus, the selective expression of trk receptors in different cell types (MCMahon, S. B. *et al.*, 1994) is a major factor underlying the selective effects of neurotrophins in the nervous system.

Ligands binding to the extracellular component of trk receptors induce autophosphorylation of the trk molecule, which leads to downstream effects via multiple signaling pathways (Kaplan, D. R. and Miller, F. D., 2000). For example, phospholipase Cγ (PLCγ) is phosphorylated subsequent to its recruitment to the cell membrane as a consequence of phosphorylation of tyrosine 785 on trkA (Obermeier, A. *et al.*, 1993). This is very likely an important mechanism in mediating neurotrophin-induced effects (see below). However, other intracellular signaling pathways are also activated by neurotrophin binding to trk receptors including the Ras–Raf–Mek–Erk pathway and the PI-3K pathway (Kaplan, D. R. and Cooper, E., 2001). Some of these have also been implicated in mediating the functional effects of neurotrophins (see Section 22.6).

22.2 Neurotrophins and Development of Nociceptors

The early history of the discovery of NGF has been extensively reviewed (Cowan, W. M., 2001). It began with the observation that extracts from mouse salivary glands induce proliferation of processes of explanted dorsal root ganglion (DRG) and sympathetic ganglion cells. Later, biochemical studies revealed that the active factor is a peptide that became known as nerve growth factor. NGF was subsequently identified as an important factor assuring survival of developing sensory neurons since its absence induced by treating the animal with an NGF antibody resulted in the survival of fewer cells in the DRG as well as in sympathetic ganglia. This led to the neurotrophic hypothesis (Davies, A. M., 1996), which states that a soluble trophic factor is present in limiting amounts in the target tissue of a population of neurons, which require it in order to survive. Nerve terminals that are successful in internalizing this factor and transporting it to the cell body for its metabolic use survive; the unsuccessful ones die. More detailed studies of the DRG revealed that selective neutralization of NGF via its antibody results in depletion of a limited population of DRG cells, specifically those that express peptides such as substance P and calcitonin gene-related peptide (CGRP) and project into superficial laminae of the spinal cord. i.e., those that are identified with nociception (reviewed in Mendell, L. M., 1995). Animals treated in this way exhibit a higher threshold for painful stimuli. Genetically modified animals have proven very useful in demonstrating these relationships with those under- or overexpressing NGF in the skin being hypo- or hyperalgesic, respectively (Davis, B. M. *et al.*, 1993), and those underexpressing the NGF receptor trkA being hypoalgesic (Smeyne, R. J. *et al.*, 1994).

In all these manipulations, NGF was considered to affect the number of nociceptors, permitting more or fewer to survive. However, it has become apparent that NGF influences the properties of nociceptors in the immediate postnatal period, beyond the time it can affect DRG cell number. Ritter A. M. *et al.* (1991; see also Lewin, G. R. *et al.*, 1992a) demonstrated that a class of nociceptors known as Aδ high-threshold mechanoreceptors (HTMRs) is severely depleted in rats treated postnatally with an antibody to NGF (anti-NGF). The proportion of Aδ low-threshold mechanoreceptors known as down hair receptors (D hairs) is elevated by a similar proportion.

Significantly, the proportion of a third population of afferent fibers also conducting in the Aδ range and innervating subcutaneous tissues is unchanged. Thus, HTMRs are not selectively depleted since if this occurs, the proportion of all other fiber types would be expected to increase. Furthermore, when cell counts were undertaken in the DRG, no change in cell number was noted. Consistent with these findings, the magnitude of the vasodilator response elicited by antidromic activation of cutaneous nerves is reduced in proportion to the depletion of Aδ HTMRs (Lewin, G. R. et al., 1992b).

These findings suggest that the postnatal development of nociceptors depends on the availability of NGF and that in its absence sensory neurons normally destined to be HTMRs become D hairs. This has been referred to as a phenotypic shift (Lewin, G. R. et al., 1992a). Cell number in the DRG declines if the antibody is delivered between P0 and P2, whereas the phenotypic shift from HTMR to D hair occurs after anti-NGF treatment between P4 and P11. Attention was drawn to evidence that sensory axons normally first innervate the epidermis, with many terminals later withdrawing to the dermal layer as hair follicles and other structures develop (Reynolds, M. L. et al., 1991). NGF is expressed by a limited population of cells in the epidermis, for example in basal keratinocytes (Di Marco, E. et al., 1991), and it was hypothesized that the availability of NGF there would cause trkA-expressing neurons to differentiate into nociceptors with epidermal terminals. In the absence of NGF due to anti-NGF treatment, these terminals would recede into the dermal layers of the skin, as occurred with most non-trkA-expressing neurons (reviewed in Lewin, G. R. and Mendell L. M., 1993; Mendell, L. M. et al., 1999).

Exogenous NGF delivered postnatally does not elevate the relative proportion of HTMRs (Lewin, G. R. et al., 1993). However, enhancing NGF levels in the skin beginning during development (E11) in transgenic mice via the keratin promoter increases the number of sensory neurons well above normal levels (Albers, K. M. et al., 1994). Electrophysiological studies have confirmed that nociceptors are selectively rescued by the availability of the additional NGF (Stucky, C. L. et al., 1999).

The DRG cell population undergoes substantial changes in the expression of trkA postnatally. A substantial number lose trkA expression over the first 3 postnatal weeks, and there is a parallel increase in the number of cells exhibiting receptors for glial-derived neurotrophic factor (GDNF) (Molliver, D. C. et al.,

1997). Cells undergoing this conversion do not express p75 even in neonates, i.e., they are p75$^-$/trkA$^+$ as distinct from cells that are p75$^+$/trkA$^+$ (Figure 1). Receptors for GDNF consist of an intracellular signaling receptor domain RET, which is a tyrosine kinase receptor. There are four different ligand-binding domains, GFR1–GFR4, that bind different members of the GDNF family. GDNF itself binds to GFRα1. Other agonists are neurturin (NTN) binding to GFRα2, artemin binding to GFRα3 and persepherin that binds to GFRα4 (Saarma, M.

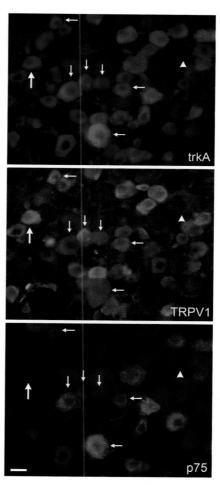

Figure 1 Dorsal root ganglion (DRG) tissue section from 10-day-old rat immunostained for trkA (red), TRPV1 (green), and p75 (blue). Most small cells are at least weakly positive for trkA, TRPV1, and p75 (small down and horizontal arrows). In one case (large up arrow), the cell was trkA$^+$/TRPV1$^+$/p75$^-$; this cell was likely to lose its trkA expression and gain receptors for GDNF. The cell denoted by an arrowhead is negative for all three markers used. Scale bar (lower left) is 20 μm. Data from Petruska J. C. and Mendell L. M. (unpublished).

and Sariola, H., 1999). In rats, the population expressing cRet also express the lectin IB4 (in adults trkA$^+$ cells are IB4$^-$). In mice, this separation is not as distinct. In rats, IB4$^+$/cret$^+$/trkA$^-$ cells have been shown to project more deeply into lamina II (lamina IIi) and are considered to be sensitive to nonnoxious stimulation. These neurons, unlike trkA$^+$/IB4$^-$ neurons, do not contain CGRP. GDNF has actions on GFR/Ret-expressing nociceptors that only partially overlap those of NGF on trkA-expressing nociceptors (see Section 22.8).

22.3 Nerve Growth Factor Affects Electrophysiological Properties of Nociceptors that Express TrkA

The finding that treatment with anti-NGF affects properties of nociceptors suggested that trkA is expressed selectively in nociceptors. This has been confirmed in a number of electrophysiological studies (Ritter, A. M. and Mendell, L. M., 1992; Djouhri, L. *et al.*, 2001) where application of NGF or anti-NGF was found to affect properties of the action potential in nociceptors only, regardless of axon size (Aβ, Aδ, or C). Interestingly, both these studies found effects on action potential fall time although the effects differed, probably because of the very different conditions of the experiments. Both studies also revealed that NGF does not affect the duration of after hyperpolarization. The selective expression of trkA in electrophysiologically characterized nociceptors has recently been confirmed directly by Fang X. *et al.* (2005).

22.4 Exogenous Nerve Growth Factor Elicits Hyperalgesia

22.4.1 Chronic Changes

Neonatal rats are more sensitive to being handled after an initial injection of NGF (Lewin, G. R. *et al.*, 1993). It was surmised that changes had occurred in the nociceptive system. Behavioral studies in 5-week-old rats treated with NGF every other day from birth revealed both mechanical and thermal hyperalgesia (Lewin, G. R. *et al.*, 1993). Additional studies with serial observations on rats treated either as neonates (0–2 weeks old) or as juveniles (2–5 weeks old) demonstrated that the sensitizing effect of NGF lasts several weeks after the end of treatment.

Individual Aδ HTMRs in rats treated systemically over the first 2 postnatal weeks with NGF exhibit a significantly diminished mechanical threshold (i.e., are more sensitive) up to 9 weeks of age after which it returns to control values (Lewin, G. R. *et al.*, 1993). Neonatal treatment with anti-NGF elevates mechanical threshold (Lewin, G. R. *et al.*, 1992). Treatment of older rats with NGF has no measurable effect on mechanical threshold. A similar study on C-fibers (Lewin, G. R. and Mendell, L. M., 1994) has revealed that treatment with NGF increases the fraction sensitive to noxious heat, whereas anti-NGF reduces it. Unlike the changes in mechanical sensitivity in Aδ-fibers, these changes are apparently permanent. The mechanical sensitivity of these C-fiber units is unchanged, although a population of mechanically sensitive units with an unusually low threshold is observed in anti-NGF-treated preparations. Reducing NGF levels in the skin in adult rats using the immunoadhesin trkA-IgG 10–12 days before recording (Bennett, D. L. *et al.*, 1998a) also causes a substantial decline in the response of individual units to noxious heat with no change in the response to mechanical stimulation. The number of units sensitive to bradykinin also decreases after trkA-IgG treatment, suggesting an important contribution of NGF to bradykinin sensitivity of nociceptors (see Section 22.8.4).

22.4.2 Acute Changes

Examination of the acute effects of a single administration of NGF (i.p.) in adult rats revealed that thermal hyperalgesia becomes significant virtually immediately, unlike mechanical hyperalgesia that develops gradually and reaches significance only after a few hours. Hyperalgesia elicited by a single i.p. injection of NGF lasts several days. Localized small injections of NGF into the skin elicit thermal and mechanical hyperalgesia of much shorter duration (Woolf, C. J. *et al.*, 1994; Shu, X. *et al.*, 1999). These local injections indicate the likelihood that NGF can act directly on cells located in the skin. NGF itself does not evoke any behavioral response; it merely sensitizes the response to subsequent nociceptive stimuli.

Although administration of NGF to the skin elicits hyperalgesia, it does not cause inflammation. On the other hand, treatments or injuries that elicit inflammation, such as complete Freund's adjuvant (CFA), enhance NGF levels in the skin (Weskamp, G. and Otten, U., 1987; Aloe, L. *et al.*, 1992).

Inflammatory cytokines, specifically TNFα and IL-1β, known to be expressed during inflammation, enhance NGF levels when delivered separately into the skin (Woolf, C. J. *et al.*, 1997). These studies have suggested a temporal sequence: TNFα → IL-1β → NGF (Figure 2). Cells responsible for producing NGF include keratinocytes (Di Marco, E. *et al.*, 1991) and mast cells (Leon, A. *et al.*, 1994).

The requirement for NGF as an essential intermediate in causing inflammatory hyperalgesia has been demonstrated in experiments where its levels are experimentally diminished using an antibody to NGF (Lewin, G. R. *et al.*, 1994; Woolf, C. J. *et al.*, 1994) or an immunoadhesin such as trkA-IgG (McMahon, S. B. *et al.*, 1995; Koltzenburg, M. *et al.*, 1999), both of which deplete endogenous NGF. These agents prevent the development of hyperalgesia to mechanical or thermal stimuli in response to inflammatory agents such as CFA. In the case of thermal hyperalgesia, these agents can also substantially reduce ongoing inflammatory pain within an hour. However, mechanical hyperalgesia is reduced only after 24–48 h (Woolf, C. J. *et al.*, 1994; 1996). This difference between thermal and mechanical hyperalgesia is consistent with the conclusion, based on the latency of onset after exposure to NGF, that thermal hyperalgesia is due to a peripheral action, whereas the mechanical hyperalgesia is indirect, probably involving transcriptional changes at the level of the DRG (but see Section 22.7).

Rueff A. and Mendell L. M. (1996) recorded the activity of single nociceptor axons in an excised skin nerve preparation with the corium side up where NGF could be applied directly to the electrophysiologically identified receptive field of the fiber. They tested the response to noxious heat and mechanical stimulation under control conditions and 10 min later with NGF applied to the corium surface of the unit or under control conditions (no NGF). They observed a significant decline in threshold to noxious heat, but no acute change in threshold to mechanical stimulation. The sensitivity to NGF was abolished if mast cells were degranulated over a period of several days prior to treatment (see Section 22.4.3).

22.4.3 Mechanisms of Nerve Growth Factor Action in the Skin

A number of pharmacological approaches have been undertaken to explore the identity of cells and mechanisms underlying sensitization of nociceptive responses by NGF. A contribution by the sympathetic nervous system to the sensitizing effect of NGF has been identified since guanethidine treatment substantially reduces NGF's acute hyperalgesic effect (Andreev, N. Yu. *et al.*, 1995) by delaying its onset (Woolf, C. J. *et al.*, 1996).

Mast cells also participate in NGF-induced sensitization, particularly of noxious thermal stimuli. Mast cells express trkA receptors (Horigome, K. *et al.*, 1993), and NGF causes them to degranulate their contents consisting of serotonin, histamine, and NGF itself (reviewed in Mendell, L. M. *et al.*, 1999). Degranulation of mast cells in advance of NGF administration delays the development of thermal hyperalgesia for about 3 h but does not prevent the later phase of thermal hyperalgesia, nor does it abolish mechanical hyperalgesia (Lewin, G. R. *et al.*, 1994; see also Woolf, C. J. *et al.*, 1996). Behavioral experiments using pharmacological antagonists indicate that serotonin is an important contributor to mast cell-mediated sensitization initiated by NGF (Lewin, G. R. *et al.*, 1994). Under some special conditions, NGF can elicit sensitization in the absence of mast cell activity (Amann, R. *et al.*, 1996a), suggesting the existence of other pathways (Figure 2), or perhaps that NGF acts directly on trkA receptors expressed by other cells, for example, the nociceptors themselves (Averill, S. *et al.*, 1995; Galoyan, S. M. *et al.*, 2003; see Section 22.5).

Another action of NGF is enhancement of the level of leukotrienes, specifically leukotriene B$_4$ (LTB$_4$), via upregulation of the enzyme 5-lipoxygenase (Amann, R. *et al.*, 1996b; Bennett, G. *et al.*, 1998b). Leukotrienes are known to sensitize nociceptors (Martin, H. A. *et al.*, 1988). This occurs as a result of neutrophil accumulation and their subsequent

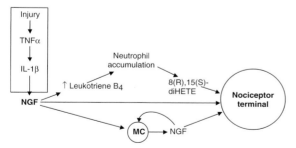

Figure 2 Major mechanisms underlying acute action of nerve growth factor (NGF) on sensory neurons. Exogenous NGF sensitizes sensory neurons via neutrophils, mast cells (MCs), and perhaps directly. Other possible mechanisms are mentioned in the text. NGF is released endogenously after injury via the action of cytokines TNFα and IL-1β (rectangle).

release of 8(R), 15(S)-diHETE. If neutrophils are depleted pharmacologically, NGF-induced hyperalgesia is abolished (Bennett, G. *et al.*, 1998b).

Keratinocytes may also play an important role in NGF-induced hyperalgesia. These cells are known to express trkA receptors and to contain NGF (Terenghi, G. *et al.*, 1997) and thus may act in a manner similar to that described for mast cells (see above). They are also known to release ATP (Koizumi, S. *et al.*, 2004) that activates nociceptor terminals (Cook, S. P. and McCleskey, E. W., 2002). Enhancement of NGF expression in these cells in genetically modified mice enhances nociceptive behavior (Davis, B. M. *et al.*, 1993).

The multiplicity of cell types that express trkA receptors and the number of substances capable of sensitizing nociceptors released by these cells in response to NGF indicate the complexity of peripheral sensitization by neurotrophins. They also indicate the difficulty of using behavioral methods and preparations such as the skin nerve preparation, where all these cells remain in close contiguity, to arrive at cellular mechanisms of sensitization. Thus, many studies have made use of dissociated DRG cells to examine the effects of NGF directly on trkA-expressing neurons. Such studies have the obvious advantage of providing a relatively unobstructed view of cellular mechanisms but at the expense of the intercellular interactions that take place *in vivo*.

22.5 Acute Sensitization by Nerve Growth Factor of the Response of Dissociated Nociceptors

Shu X. Q. and Mendell L. M. (1999a; 1999b) have investigated the effect of NGF on the response of small (<30 μm) dissociated DRG cells to capsaicin, known to elicit the sensation of noxious heat (Lamotte, R. H. *et al.*, 1992) via the VR1 (now referred to as TRPV1) receptor (Caterina, M. J. *et al.*, 1997). NGF sensitizes the response to capsaicin (Shu, X. and Mendell, L. M., 1999b), measured as an increased inward current in the presence of NGF, in a dose-dependent manner over the range of 2–100 ng ml^{-1}. Its effect is blocked by the protein kinase inhibitor K252a known to antagonize the response of trkA, although not in a highly selective manner (Knusel, B. and Hefti, F., 1992). Not all cells tested are sensitized by NGF, indicating that not all these small cells express the trkA receptor, as predicted from immunohistochemical studies (Averill, S. *et al.*, 1995; Bennett, D. L. *et al.*, 1996a; Figures 1 and 3).

The duration of NGF action is difficult to determine because the test input, capsaicin, is associated with long-lasting tachyphylaxis. To circumvent this problem, Shu X. and Mendell L. M. (2001) have examined the duration of NGF action under conditions of reduced Ca^{2+} that eliminates tachyphylaxis

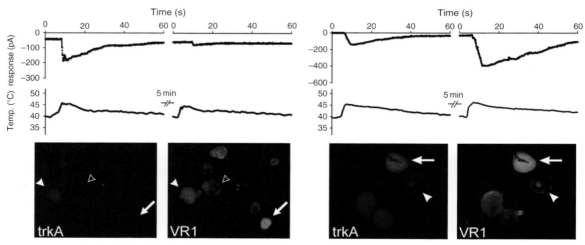

Figure 3 The response of two dissociated dorsal root ganglion (DRG) cells to administration of nerve growth factor (NGF) and demonstration that sensitization of the response requires expression of trkA. Cell on left was tested with two heat pulses (*bottom traces*). The second response 5 min later exhibited tachyphylaxis despite the presence of NGF in the bathing solution for 5 min between heat pulses. The cell (diameter 26 μm; thick arrow) was recovered and immunostained for TRPV1 and trkA. It was classified as TRPV1$^+$/trkA$^-$. Another cell in the same field denoted by white arrowhead is trkA$^+$/TRPV1$^+$ and a third cell denoted by white-bordered arrowhead is classified trkA$^-$/TRPV1$^-$. Cell on right showed sensitization of the response to heat after 5 min exposure to NGF. The cell (diameter 25 μm; arrow) was recovered, immunostained, and classified trkA$^+$/TRPV1$^+$. Cell denoted by arrowhead was trkA$^-$/TRPV1$^-$. Adapted from Galoyan *et al.* (2003) with permission.

(Koplas, P. A. *et al.*, 1997). They found that a 10-min exposure to NGF sensitizes a neuron to repeated puffs of capsaicin for as long as 70 min, suggesting that the effect of NGF involves more than a direct action on the conductance of a membrane channel or the sensitivity of a receptor (see Section 22.6).

Galoyan S. M. *et al.* (2003) have carried out similar experiments in small dissociated DRG cells from adult rats using noxious heat instead of capsaicin as the test stimulus (Figure 3). One advantage of the noxious heat stimulus over capsaicin is the ease with which threshold can be established. These cells respond an inward current whose threshold is 43–44 °C which is characteristic of the TRPV1 receptor expressed in hetrologous cells (caterina, M. J., *et al.*, 1997). NGF-induced sensitization of the response to noxious heat has properties that are similar to sensitization of the response to capsaicin. NGF does not alter temperature threshold; its major effect is to enhance the magnitude of the heat evoked current. Recent studies suggest that the acute increase of TRPV1 current elicited by exposure to NGF is due to increased trafficking and subsequent insertion of the additional TRPV1 receptors into the plasma membrane rather than an increase in channel unitary conductance (Zhang, X., *et al.*, 2005; Stein, A. T., *et al.*, 2006). These cells respond with a threshold of about 43–44 °C, which is characteristic of the TRPV1 receptor (Caterina, M. J. *et al.*, 1997). NGF-induced sensitization of the response to noxious heat has properties that are similar to sensitization of the response to capsaicin. One of the advantages of noxious heat over capsaicin as the stimulus is the ability to establish a threshold value. This has permitted the conclusion that NGF does not alter temperature threshold; its major effect is to enhance the magnitude of the heat-evoked current.

As observed with capsaicin, NGF does not sensitize the heat-evoked response in some neurons. Immunohistochemical characterization of the recorded cells has revealed that the NGF-induced sensitization of the heat-evoked inward current is closely associated with the expression of trkA (Figure 3). This is not unexpected since studies in heterologous cells have demonstrated that the ability for NGF to sensitize the TRPV1 response requires that the cell express trkA (Chuang, H. H. *et al.*, 2001).

Zhu W. *et al.* (2004) have recently demonstrated that the capsaicin- or noxious heat-induced TRPV1 response in DRG cells isolated from neonatal rats cannot be sensitized by NGF unlike the response in cells from adults. Thus, in neonates, there is no close correspondence between the expression of trkA and the ability of NGF to sensitize the TRPV1 response, i.e., trkA is expressed in cells that fail to be sensitized by NGF. However, cells in neonates failing to be sensitized by NGF can be sensitized by bradykinin, i.e., the failure for NGF to sensitize the cells is not a generalized failure for sensitization to occur. It is not presently known what factor develops during the neonatal period that permits NGF to sensitize the heat-evoked response.

NGF-induced sensitization develops gradually from P4 to P10 (Zhu, W. *et al.*, 2004). This is the time at which trkA-expressing cells lose their requirement for NGF in order to survive (Lewin, G. R. *et al.*, 1992). It has been hypothesized that the inability for NGF to sensitize cells in neonates may be a protective mechanism to prevent undue nociceptive input due to sensitization at a time that NGF is required to assure survival of nociceptors (Zhu, W. *et al.*, 2004).

GDNF, like NGF, acutely elevates cytosolic Ca^{2+} in DRG cells (Lamb, K. and Bielefeldt, K., 2003). Recent experiments suggest that exposure of dissociated sensory neurons to GDNF, neurturin or artemin for several minutes rapidly enhances the sensitivity of TRPV1 receptors to capsaicin in isolated DRG cells (Malin, S. A., *et al.*, 2006). Artemin mRNA, but not neurturin or GDNF, is upregulated in the skin in response to inflammatory conditions; its expression increases faster, and to a greater extent, than that of NGF. Administration of these agents elicits thermal hyperalgesia lasting several hours which is substantially prolonged if NGF is co-injected. These findings broaden our understanding of the role of trophic factors in sensitization of nociceptors.

22.6 TrkA Signaling Pathways Responsible for Acute Effects of Nerve Growth Factor

Signaling initiated by trk receptor activation is diverse, and for a general treatment of this subject from a more molecular perspective, the reader is referred to a recent review (e.g., Huang, E. J. and Reichardt, L. F., 2001; Nicol, G. D. and Vasko, M. R., 2007). Many studies carried out on NGF-induced signaling have been directed toward determining the signaling pathways involved in cell survival and axon outgrowth. Here, we provide a more focused treatment related to NGF's role as a sensitizing agent in sensory neurons.

Chuang H. H. *et al.* (2001) reported that phosphatidyl-4.5-inositol biphosphate (PtdIns(4,5)P$_2$ – PIP$_2$) exerts a tonic inhibitory effect on the inward current produced by TRPV1 via its interaction with the C-terminal domain of TRPV1. They suggested that NGF sensitizes TRPV1 by activating PLC which inhibits PIP$_2$, thereby disinhibiting TRPV1, a mechanism corroborated by the demonstration that the PLC inhibitor U-73122 antagonizes the sensitizing effect of NGF in adult rat DRG neurons (Galoyan, S. M., *et al.*, 2003). This mechanism has not been confirmed in subsequent studies (see below). Furthermore, recent findings suggest that PIP$_2$ facilitates rather than inhibits TRPV1 receptor activity (Stein, A. T. *et al.*, 2006). A related finding indicates that PIP$_2$ is required for recovery from desensitization (Lin, B., *et al.*, 2005) known to occur after TRPV1 receptor activation.

Chuang H. H. and coworkers pointed out that the activation of PLC would also be expected to enhance levels of protein kinase C (PKC) via diacylglycerol (DAG). However, they found that inhibitors of PKC fail to abolish NGF-induced sensitization of TRPV1. This has been confirmed by Shu X. and Mendell L. M. (2001) in adult rat DRG cells using a different pair of PKC blockers. Interestingly, other investigators working in dissociated DRG neurons have obtained very different results on the signaling pathways involved in NGF-induced sensitization. Bonnington J. K. and McNaughton P. A. (2003), working in neonatal mouse DRG cells, used a different inhibitor of PLC, neomycin, and found no inhibition of NGF-induced sensitization of the capsaicin response and also that the most robust inhibition of NGF-induced potentiation occurred by blocking the activity of PI-3K and PKC, the latter in contrast to the earlier findings (see above). Zhuang Z.-Y. *et al.* (2004), working in adult rat (see also Zhu, W. and Oxford, G. S., 2007) have confirmed the role for PI-3K in NGF-induced sensitization, agreeing with Bonnington J. K. and McNaughton P. A., but also reported that the MEK inhibitor PD98059 blocked the effect of NGF unlike the findings of both Bonnington J. K. and McNaughton P. A. (2003) and Shu X. and Mendell L. M. (2001).

It is important to note that there are several differences among these experiments including the use of rats and mice of different strains, cell culture conditions, protocols (e.g., number of capsaicin pulses), different assays such as Ca^{2+} imaging, and membrane current. These studies point out the complexity of studies on signaling pathways, specifically the need to define the conditions under which they are investigated. It may be necessary to carry out such studies *in vivo* where conditions are more tightly regulated in order to understand which of the many possible pathways are operating under physiological conditions.

22.7 p75 Receptor Influence on Sensory Neuron Function

The p75 receptor is expressed in all cells that also express trkA and trkB receptors, but in only about half the cells expressing trkC. Cells not expressing trk do not express p75 (Wright, D. E. and Snider, W. D., 1995). p75 binds all neurotrophins (see reviews by Patapoutian, A. and Reichardt, L. F., 2001; Huang, E. J. and Reichardt, L. F., 2001). The functional role of trk and p75 in mediating the effects of neurotrophins has received greatest attention in developmental studies as determinants of neuronal survival and apoptosis (Hempstead, B. L. *et al.*, 1991; Miller, F. D. and Kaplan, D. R., 2001). For example, p75 acts as a mediator of sympathetic neuron apoptosis, whereas trkA mediates cell survival (Majdan, M. *et al.*, 2001).

Relatively little attention was paid to the p75 receptor initially in evaluating the physiological effects of NGF because of evidence that sensitization to noxious heat by NGF is normal in p75 knockouts (Bergmann, I. *et al.*, 1998, but see von Schack, D. *et al.*, 2001, for evidence that this knockout is not complete). However, the finding that activation of p75 enhances ceramide via activation of the sphingomyelin pathway (Dobrowsky, R. T. *et al.*, 1994) prompted a search for physiological effects elicited by NGF binding to the p75 receptor (reviewed by Nicol, G. D. and Vasko, M. R., 2007). Zhang Y. H. *et al.* (2002) reported that NGF-induced activation of the ceramide pathway via the p75 receptor enhances membrane excitability (membrane gain-number of APs per nanoampere of current applied across the cell membrane) in capsaicin-sensitive DRG neurons (presumptive nociceptors) by enhancing a TTX-insensitive Na$^+$ conductance and decreasing the magnitude of the K$^+$-rectifier. These experiments were carried out first by short circuiting p75 by administering ceramide itself, and second by demonstrating that the effects of NGF could be blocked by administering glutathione, an inhibitor of neutral sphingomyelinase, the enzyme responsible for producing ceramide (Liu, B. and Hannun, Y. A., 1997). A p75 blocking antibody also prevents ceramide facilitation of membrane gain (Zhang, Y. H. and Nicol, G. D., 2004). Thus, NGF might sensitize the response of the cell by enhancing its depolarization (trkA) and independently by increasing the number of action potentials produced per unit of depolarization (p75) (Mendell, L. M., 2002). In this context, it is interesting that one of the major deficits in p75-knockout mice is a substantial reduction of discharge frequency of certain mechanoreceptors (Stucky, C. L. and Koltzenburg, M., 1997).

Cellular mechanisms for interaction between p75 and trkA have also been described. Working in human neuroblastoma cells, Plo I. *et al.* (2004) reported that activation of trkA reduces sphingomyelinase levels and thus reduces ceramide production (i.e., reduced p75 signaling) via PKC. Similarly, Bilderback T. R. *et al.* (2001) demonstrated in PC12 cells that PI-3K, activated by NGF via trkA, acutely inhibits acidic sphingomyelinase activated by NGF via the p75 receptor. Since PI-3K under certain conditions has also been shown to play a role in NGF-induced activation of TRPV1 via trkA (see Section 22.6), NGF appears to have the potential to affect cell discharge in a complex manner involving interaction of the signaling pathways activated by p75 and trkA. Also in PC12 cells, MacPhee I. and Barker P. A. (1997; 1999) demonstrated a reciprocal action of NGF, i.e., brief activation of p75 acutely reduces trkA activity via a ceramide-dependent mechanism, whereas extended p75 activation facilitates trkA. Another recent report (He, X. L. and Garcia, K. C., 2004) suggests that NGF can bind to p75 in a 2:1 molecular ratio and that this complex might bind to trkA. Such a scheme would predict cooperativity between p75 and trkA at the level of the cell surface in addition to intracellular signaling. Furthermore, there is now evidence that p75 can modulate the action of the trk receptors by increasing trk receptor autophosphorylation (Berg, M. M. *et al.*, 1991; Verdi, J. M. *et al.*, 1994), by increasing the concentration of the neurotrophin in the vicinity of its trk receptor (Barker, P. A. and Shooter, E. M., 1994), and/or by increasing the selectivity of the trk receptor for its ligand (Rodriguez-Tebar, A. *et al.*, 1992; Benedetti, M. *et al.*, 1993). Thus, even though there is evidence for independent effects of neurotrophins via trkA and p75, the possibility that these pathways could interact in nociceptive neurons must be considered.

22.8 Effects of Nerve Growth Factor on Expression of Channels and Receptors in Nociceptors

The acute changes in sensory neuron function considered heretofore are very rapid and hence considered to be posttranscriptional changes. However, there is now substantial evidence that NGF can also regulate the expression of numerous receptors and channels via transcriptional mechanisms and in this way exert much more profound and long-lasting effects on the properties of nociceptors (Figure 4). Furthermore, in the case of several of the channels (TRPV1, BK, P2X3, and TTX-resistant Na^+ channels), there is evidence for differential regulation by NGF and GDNF.

22.8.1 TRPV1 Receptors

Chronic exposure of DRG cells in culture to NGF increases their response to capsaicin (Bevan, S. and Winter, J., 1995; Ganju, P. *et al.*, 1998); this is attributed to an increase in expression of TRPV1 associated with enhanced levels of TRPV1 mRNA (Winston, J. *et al.*, 2001). *In vivo* studies have documented that peripheral administration of NGF enhances expression of TRPV1 via activation of p^{38} MAP kinase in trkA+ sensory neurons in the DRG and in sciatic nerve and especially in the skin by selective centrifugal transport (Ji, R. R. *et al.*, 2002). A recent report (Amaya, F. *et al.*, 2004) has shown that the elevated expression of TRPV1 begins 1 day after CFA administration into the paw and lasts for up to 1 week, returning to normal at 2 weeks. Both NGF and GDNF levels increase in DRG neurons after CFA treatment, with NGF

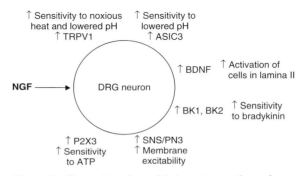

Figure 4 The variety of possible long-term actions of nerve growth factor (NGF) on expression of various channels, receptors, and neurotrophins in nociceptors. In each case, the major functional effect is listed. Further details in text.

exhibiting a relatively rapid rise and fall (peak at 1 day) and GDNF reaching its peak level at 1 week and falling to control levels by the second week. Selectively interfering with NGF and GDNF upregulation using antibodies has clarified that NGF and GDNF elevate TRPV1 expression in different populations of nociceptive sensory neurons, trkA$^+$ and IB4$^+$ neurons, respectively. Furthermore, restricting elevation of TRPV1 expression to either population of DRG cells reduces the thermal nociception elicited during CFA-induced inflammation.

22.8.2　P2X3 Receptors

NGF also enhances expression of the P2X3 receptor (Bradbury, E. J. et al., 1998; Ramer, M. S. et al., 2001) considered important in pain produced by ATP (Cook, S. P. and McCleskey, E. W., 2002). One week after intrathecal administration, GDNF elevates P2X3 expression in spinal terminals of Ret/IB4$^+$ neurons in which it is normally expressed. NGF has a more profound effect since it causes novel expression of P2X3 in neurons terminating in more superficial laminae of the spinal cord where P2X3 is normally not expressed. Although these latter changes are relatively small, it was concluded that they might be a significant component in NGF-mediated inflammatory pain.

22.8.3　ASIC Channels

ASIC channels are Na channels gated by H$^+$. The ASIC3 isoform is expressed in DRG cells identified as nociceptors by coexpression of markers characteristic of such neurons. Along with TRPV1 receptors, ASIC3 receptors are considered to be important mediators of the pain associated with tissue acidosis (Voilley, N. et al., 2001). Mamet J. et al. (2003) have shown in culture that NGF enhances the expression of the ASIC3 isoform in DRG cells. They describe a physiological control mechanism, suggesting that low levels of NGF regulate ASIC3 expression via trkA and the PLC/PKC signaling pathway. During inflammation, NGF levels rise and the increased activation of p75 removes inhibition of the downstream effects of JNK and p38 that are also activated by NGF via trkA (Ji, R. R. et al., 2002). Since these downstream effects include expression of ASIC3, this mechanism can act as a switch whereby the enhanced level of NGF switches on production of ASIC3 by removing tonic inhibition of JNK and p38.

22.8.4　BK Receptors

Another receptor upregulated by NGF is a member of the bradykinin (BK) receptor family. Nociceptors express the B2 form of the BK receptor constitutively; the B1 form is expressed only after inflammation (Dray, A. and Perkins, M., 1993). McNaughton P. A. and coworkers have reported that freshly dissociated mouse DRG cells undergo upregulation of B2 mRNA (confirmed at the protein level immunocytochemically) when exposed to NGF for 3 days; GDNF is much less effective (Lee, Y. J. et al., 2002). These studies are consistent with previous observations that cultured cells are more responsive to bradykinin when cultured in the presence of NGF (Petersen, M. et al., 1998). The identity of the receptor responsible for this action of NGF (trkA or p75) is controversial (see discussion in Lee, Y. J. et al., 2002). Interestingly, expression of the B1 receptor can be enhanced in cultured DRG neurons in the presence of GDNF but not NGF (Vellani, V. et al., 2004). However, in vivo the effects of NGF may be much more complex as suggested by Rueff A. et al. (1996) who hypothesized that NGF might indirectly cause upregulation of B1 receptors on DRG neurons via release of cytokines such as TNFα and IL-1β from inflammatory cells.

22.8.5　Na$^+$ Channels

Waxman, S. G., Kocsis, J. D., and coworkers have shown that NGF can affect Na$^+$ and K$^+$ channel expression of nociceptive afferent fibers (Fjell, J. et al., 1999; Everill, B. and Kocsis, J. D., 2000). Fjell J. and coworkers noted changes in the properties of small-diameter DRG cells after peripheral axotomy that could be reversed by administration of NGF. Specifically, they observed that axotomy diminishes the TTX-resistant Na current in these neurons and that provision of NGF partially reverses this decline due to an effect on SNS/PN3 TTX-resistant Na$^+$ channel (Na$_v$ 1.8). NGF is unable to reverse the axotomy-induced attenuation of activity in NaN, another TTX-resistant Na$^+$ channel (Na$_v$ 1.9). However, GDNF is able to reverse the decline in both NaN and SNS/PN3. Interestingly, GDNF but not NGF enhances TTX-resistant currents in small DRG neurons studied in culture, suggesting that the ability to restore function after axotomy does not predict effects under all conditions. However, given the likely importance of TTX-resistant Na$^+$ channels in the function of nociceptors (Ritter, A. M. and

Mendell, L. M., 1992; Fang, X. *et al.*, 2005), the finding that neurotrophins can influence their function makes this system a potentially important target in achieving pain control.

22.8.6 PGE2 Receptors

The multiplicity of channels affected by NGF suggests that this growth factor acts as a master switch in regulating the sensitivity of sensory neurons to nociceptive stimuli. However, its control is not universal since NGF does not control expression of the PGE2 receptor (Southall, M. D. and Vasko, M. R., 2000) that mediates the hyperalgesia caused by prostaglandins.

22.9 Neurotrophins and Sensitization of Tissues Other than Skin

The skin has attracted most of the attention in studies of the peripheral effects of neurotrophins. This is in keeping with the focus on this organ in investigations of somatosensory mechanisms. However, there is considerable evidence that NGF contributes to the sensitization of other tissues, specifically viscera, muscle, and joints, all of which are sites where inflammatory pain can occur.

22.9.1 Visceral Structures

Several investigators have observed enhanced levels of NGF in the bladder in models of interstitial cystitis, a normally painful condition (Oddiah, D. *et al.*, 1998; Toma, H. *et al.*, 2000; Bielefeldt, K. *et al.*, 2003). In human studies, the level of NGF determined in bladder biopsies was found to be higher in individuals whose condition is painful (Lowe, E. M. *et al.*, 1997). Using viral transfer methods, Lamb K. *et al.* (2004) demonstrated that enhancing NGF levels in rat bladder leads to overactivity measured cystometrographically. If NGF is neutralized using immunological methods, the bladder overactivity produced by external irritants is reduced (Lamb, K. *et al.*, 2003). NGF has also been demonstrated to be upregulated in other visceral organs such as the bowel in association with inflammatory conditions (di Mola, F. F. *et al.*, 2000) or in an animal model of pancreatitis (Toma, H. *et al.*, 2000).

22.9.2 Muscle Afferents

Recent studies have indicated that neurotrophins can influence the discharge of small-diameter muscle afferents. Hoheisel U. *et al.* (2005) evaluated the effect of intramuscular NGF on the discharge of small-diameter (Group IV) muscle afferents and found that only units with high mechanical threshold are facilitated. They reported that threshold is unchanged, but discharge frequency in response to adequate stimulation increases virtually immediately after NGF (0.8 μM) administration. Interestingly, they also observed that administration of BDNF (1 μM) causes a decline in discharge frequency, but only several minutes after administration.

The sensitization of high-threshold muscle afferents is consistent with the sensitization observed in response to intramuscular administration of NGF to the masseter muscle in humans (Svensson, P. *et al.*, 2003). Changes were observed 1 and 7 days after administration but not after 1 h and were described as consisting of both mechanical allodynia and hyperalgesia. Under pathological conditions, it seems likely that the cytokine TNFα is upregulated in muscle and that this leads to production of NGF (Schafers, M. *et al.*, 2003).

22.9.3 Joint Receptors

Synovial fluid from arthritic joints in humans expresses elevated levels of NGF (Falcini, F. *et al.*, 1996; Halliday, D. A. *et al.*, 1998), which has been attributed to elevated levels of cytokines such as TNFα and IL1-β (Manni, L. *et al.*, 2003), much as has been reported in skin (see Section 22.4.3). There have been no reports of enhanced sensitivity of joint afferents to mechanical stimulation in response to NGF although sensitization is known to occur in models of experimental arthritis (Schaible, H. G. and Schmidt, R. F., 1985). Joints treated with CFA have a more extensive innervation (Shinoda, M. *et al.*, 2003), possibly as a result of the enhanced levels of NGF known to promote terminal axonal sprouting in the skin (Diamond, J. *et al.*, 1992). An enhancement in peripheral axon terminals of trkA$^+$ nociceptive afferents might be expected to increase sensitivity to peripheral stimuli as a result of increased numbers of receptor molecules. Recently, Shelton D. L. *et al.* (2005) have demonstrated that administration of a monoclonal antibody against NGF (mAb 911, Rinat Neuroscience) can eliminate the pain associated with an experimental autoimmune arthritic condition in rats.

22.10   Nerve Growth Factor, Enhances Peptide Expression in the Dorsal Root Ganglion, and Central Sensitization

The ability of NGF to elicit rapid sensitization via posttranscriptional changes and to increase the expression of receptors and channels involved in setting the sensitivity and activity of nociceptors is indicative of its ability to enhance peripheral sensitization. However, NGF also elevates expression of peptides in the DRG ganglion and these are believed to enhance the central effects of glutamate released by nociceptors, i.e., to contribute to central sensitization (McMahon, S. B. *et al.*, 1993). Lindsay R. M. and Harmar A. J. (1989) were among the first to demonstrate that DRG cells maintained in culture exhibit enhanced levels of substance P and CGRP after exposure to NGF in the medium. These peptides are considered to be important mediators in central sensitization because of the very long-lasting depolarization that they produce in postsynaptic cells, leading to activation of the NMDA receptor (McMahon, S. B. *et al.*, 1993). Both NGF and GDNF enhance capsaicin-stimulated release of CGRP from cultured trigeminal ganglion neurons; however, the effect of GDNF is due entirely to the greater survival of these neurons in its presence (i.e., no change in response of individual neurons to capsaicin), whereas the effect of NGF is attributed in part to increased CGRP release per neuron, i.e., NGF sensitizes the response to capsaicin (Price, T. J. *et al.*, 2005).

The possibility for direct participation of neurotrophins in central sensitization was initially advanced by the finding *in vivo* that BDNF is also upregulated in the cell body of trkA receptor-expressing DRG cells several hours to days after exposure to NGF (Apfel, S. C. *et al.*, 1996; Michael, G. J. *et al.*, 1997). This suggested that BDNF would be released into the spinal cord, a prediction that has been borne out experimentally (Lever, I. J. *et al.*, 2001). The high-affinity receptor for BDNF, trkB, is found on neurons throughout the superficial dorsal horn (Bradbury, E. J. *et al.*, 1998; Garraway, S. M. *et al.*, 2003). A physiological role for BDNF has been established in models of inflammatory pain by demonstrating that sequestration of spinal BDNF using intrathecally delivered trkB-IgG diminishes the response to noxious heat (Kerr, B. J. *et al.*, 1999; Mannion, R. J. *et al.*, 1999; Pezet, S. *et al.*, 2002). In these experiments, it was also demonstrated that NGF has no central action, i.e.,

different neurotrophins act to promote peripheral and central sensitization (but see Section 22.11, indicating that BDNF and NT-4 can also elicit peripheral sensitization).

Direct physiological evidence for the action of BDNF in the superficial dorsal horn has come from experiments in spinal slices where it has been reported that BDNF facilitates the synaptic response of cells in lamina II to electrical stimulation of C-fibers both in neonates (Garraway, S. M. *et al.*, 2003) and in adults (Garraway, S. M. *et al.*, 2005). This facilitation of AMPA/kainate responses (measured in voltage clamp) is virtually immediate after superfusion of BDNF $(100\,\mathrm{ng\,ml^{-1}})$ and lasts for at least a few hours after washout of the BDNF. The effect of BDNF requires participation of NMDA receptors since it is blocked by APV. Additional pharmacological experiments have demonstrated a role for PKC in this process and also that administration of BDNF results in phosphorylation of the NMDA receptor in neurons in the superficial dorsal horn (Slack, S. E. *et al.*, 2004). It is important to note that BDNF itself does not elicit any detectable effect on lamina II neurons; it only acts to sensitize the response to stimulation of small-diameter sensory afferents. The neurotrophin effect is specific; superfusion of NT-3 does not sensitize this response (Garraway, S. M. *et al.*, 2003).

The requirement for functional NMDA receptors in order for BDNF to be effective in sensitizing the response of lamina II neurons to C-fiber input is consistent with the behavioral finding that pretreating rats for several days with systemic MK-801, an antagonist of NMDA receptors, selectively blocks the late phase of thermal and mechanical hyperalgesia elicited by NGF (Lewin, G. R. *et al.*, 1994). This late phase was attributed to central sensitization, whereas the acute phase, not affected by MK-801 treatment, was identified with peripheral sensitization. Thus, NGF exerts its effects by peripheral and central sensitization which together have been advanced as mechanisms to explain different components of hyperalgesia (McMahon, S. B. *et al.*, 1993).

22.11   Role of Brain-Derived Neurotrophic Factor, Neurotrophin-4, and Neurotrophin-3 in Peripheral Sensitization

There is increasing evidence that acute sensitization of nociceptors can be affected by neurotrophins other than NGF. Specifically, BDNF and NT-4 sensitize nociceptive function acutely at the behavioral level (Shu, X. *et al.*, 1999), at the level of nociceptor discharge measured in isolated skin nerve preparations (Rueff, A. and Mendell, L. M., 1996; Shu, X. *et al.*, 1999), and at the level of capsaicin-activated TRPV1 currents measured in dissociated cells (Shu, X. and Mendell, L. M., 1999b). The proportion of small cells ($d \leq 30 \, \mu m$) sensitized by NT-4 is similar to that sensitized by NGF. Therefore, it seems unlikely that trkA expression is required for cells to be sensitized by NT-4 since as a population, relatively few (10%) small trkA$^+$ sensory neurons also express trkB (Karchewski, L. A. *et al.*, 1999). Thus, activation of trkB alone can sensitize the response to capsaicin.

It is not known where trkB ligands would come from naturally in the periphery since there is no evidence at present that BDNF or NT-4 is upregulated in the skin immediately after inflammation. However, BDNF is upregulated in nerves undergoing Wallerian degeneration, opening the possibility that trkB ligands sensitize neurons after damage to neighboring nerve trunks. Another possibility is that BDNF, which is known to be released from the central terminals of nociceptors, would also be released from their peripheral terminals in the skin and that this would be the normal source of trkB ligands. Such a line of reasoning suggests that BDNF might act peripherally to reinforce the central sensitization it also elicits.

NT-3 does not elicit any acute sensitization of the response to noxious heat measured either electrophysiologically or behaviorally (Shu, X. *et al.*, 1999; Shu, X. and Mendell, L. M., 1999b) despite evidence that it can bind to trkA (Lindsay, R. M., 1996). However, 7-day exposure of trkA$^+$/trkC$^-$ sensory neurons to NT-3 results in a substantial reduction in trkA receptor expression (Gratto, K. A. and Verge, V. M., 2003). Accompanying this loss of trkA is a decline in substance P levels. These studies suggest that NT-3 might have therapeutic value in diminishing painful states initiated by the activation of trkA.

22.12 Neurotrophins and Neuropathic Pain

Most considerations of neurotrophins and pain have centered on their role in inflammatory pain since they are naturally upregulated in the both periphery and DRG and play a prominent role in peripheral and central sensitization. However, there is also evidence that neurotrophins may contribute to the pain

associated with nerve manipulations that are accepted as models for neuropathic pain. Thus, Zhou X. F. *et al.* (2000) have demonstrated that antibodies to NGF, BDNF, and NT-3 delivered to the DRG can reduce the pain associated with spinal nerve axotomy. Interestingly, antibodies to NGF elicit their effect in the early stages after injury, whereas BDNF and NT-3 antibodies exert their effects only later. In addition, Zhou X. F. *et al.* demonstrated that BDNF and NGF, but not NT-3, delivered to the DRG in intact preparations elicit mechanical allodynia. It is worth contrasting these findings with those of Kerr B. J. *et al.* (1999; see above) who showed that NGF delivered intrathecally elicits no nociceptive behavior. These differences presumably reflect the differences in the expression of trkA that is known to be present in the DRG, but not in the spinal cord. Theodosiou M. *et al.* (1999) carried out similar experiments with neurotrophin antibodies in rats subject to chronic partial axotomy of the sciatic nerve. They observed both mechanical and thermal hyperalgesia in these rats that could be relieved by antibodies to NGF or BDNF but not NT-3.

Fukuoka T. *et al.* (2001) have studied uninjured neurons in the L4 DRG in rats subjected to L5 spinal nerve ligation (Chung model). They observed an increase in BDNF mRNA levels in L4 associated with increased levels of expression as well as recruitment of additional neurons into the population of DRG neurons expressing BDNF mRNA. They also found that the enhanced sensitivity to noxious heat associated with this injury could be diminished temporarily by intrathecal application of an antibody to BDNF. Significantly, NGF appears to play a role in this enhancement of BDNF levels since NGF is increased in the DRG whose spinal nerve is ligated, and an antibody to NGF delivered to the injured DRG can prevent the upregulation of BDNF in the neighboring ganglion as well as the associated thermal hyperalgesia.

A recent study (Wilson-Gerwing, T. D. *et al.*, 2005) indicates that NT-3 can prevent the long-lasting increased expression of TRPV1 observed in the chronic constriction injury (CCI) model of neuropathic pain. They further demonstrated that NT-3 reduces expression of p38 MAPK shown to be a primary factor in the delayed elevation of TRPV1 expression in the skin (Ji, R. R. *et al.*, 2002). Treatment with NT-3-reduced thermal hyperalgesia, but not the mechanical hypersensitivity associated with CCI.

A different role for neurotrophins in neuropathic pain was reported by Bennett D. L. *et al.* (1996b) who demonstrated that sprouting of peripherally axotomized Aβ-fibers into the superficial laminae of the spinal cord, considered to be a possible mechanism for neuropathic pain after such damage (Woolf, C. J. et al., 1995), is reduced by application of NGF to the damaged nerve. They argued that this was not a direct effect of NGF on the sprouting of A-fibers since it would imply that sprouting is paradoxically reduced by administration of NGF. Also, large non-nociceptive Aβ-fibers do not express trkA, and so it is not clear how NGF would affect their sprouting. They proposed that NGF elicits sprouting of axotomized C-fibers that would reverse the A-fiber sprouting caused initially by the peripheral axotomy.

The pain associated with peripheral nerve damage has been shown to also have a microglial component. ATP released from damaged nociceptive afferents activates microglial P2 × 4 receptors; the activated microglia release BDNF which inhibits the KCC2 chloride transporter in lamin I cells (Coull, J. A., et al., 2005). This shifts the chloride equilibrium potential in the depolarizing direction which changes the sign of inhibitory input toward excitation. This increases activity in nociceptive pathways contributing to pain. Thus, microglial BDNF as well as neuronal (DRG) BDNF can contribute to central sensitization. Interaction between these two sources of BDNF remains to be discovered.

Despite the lack of consistency, these experiments indicate several potential roles for neurotrophins in neuropathic pain. These are based on their ability to directly sensitize neuronal responses as well as their ability to elicit structural changes in the sensory nervous system.

22.13   Neurotrophins and the Clinic

The very strong sensitizing effect of neurotrophins on nociceptive pathways both peripherally and in the spinal cord has given them a very high profile in pain research. One of the earliest indications of a clinical effect emerged from Phase I clinical trials to determine the safety of NGF delivered systemically (Petty, B. G. et al., 1994). NGF was administered subcutaneously or intramuscularly in human volunteers in amounts ranging from 0.03 to 1.0 μg kg^{-1}, dosages that are much lower than typically used in rats (1 μg g^{-1}). These subjects experienced myalgia in truncal and bulbar muscles as well as hyperalgesia at the site of injection. The latter could persist for up to 7 weeks. Similar human trials using single

localized injections of NGF (1–3 μg per patient) in the arm resulted in localized allodynia and thermal hyperalgesia lasting several weeks (Dyck, P. J. et al., 1997). These experiments in humans with verbal reports provide significant confirmation of the results from animal experiments. They also suggest that the use of NGF to increase cell survival or for other purposes will require measures to overcome the nociceptive side effects.

As discussed above, peripheral NGF levels are upregulated as a consequence of inflammation resulting in hyperalgesia. Furthermore, the hyperalgesia associated with inflammation can be largely eliminated by reducing NGF levels using either a sequestering antibody or a trkA immunoadhesin. This has prompted the development of NGF antibodies for eventual use in humans. Recently, Sevcik M. A. et al. (2005) reported that an antibody mAb 911 (Rinat Neuroscience Corporation) is effective in attenuating the pain in a mouse model of bone cancer whose sarcoma tumor cells are a source of high levels of NGF. As shown above (Section 22.9), bladder pain associated with models of cystitis is also associated with elevated expression of NGF, as is arthritic pain. Thus, we can expect that clinical trials using will be undertaken as soon as antibodies have been rendered safe for human use.

Several clinical conditions reported to be associated with alterations in NGF or trkA expression have been reviewed by Anand P. (2004). The trkA receptor is dysfunctional in individuals with congenital insensitivity to noxious stimulation, probably as a result of a mutation. Leprosy, a condition associated with hypoalgesia, is accompanied by reduced expression of NGF. Similar findings have been made in diabetic neuropathy. Conversely, the pain associated with injury to nerves has been correlated with enhanced expression of NGF in the associated nerve trunks. Together, these findings suggest that manipulation of neurotrophin levels may turn out to be a useful therapeutic approach to various conditions associated with disorders of nociception.

22.14   Discussion, Conclusions, and Open Questions

The past 15 years have seen an important expansion in our understanding of the role of neurotrophins as it has become generally accepted that they function throughout the life of the animal, not only during development. With regard to their effect on nociception, it is clear that the survival of nociceptors is dependent on the integrity of the NGF/trkA system during the perinatal period. Beyond this period, the influence of NGF on nociceptors is very wide ranging (Petruska, J. C. and Mendell, L. M., 2004). NGF sensitizes nociceptors both acutely via post-transcriptional mechanisms (Figure 2) and chronically by enhancing the expression of numerous channels associated with nociceptors (Figure 4). These effects are exerted on nociceptors innervating many different tissues, including skin, muscle, joints, and viscera although at this time it is not yet known whether these neurons are regulated identically by NGF.

The discovery of the capsaicin receptor VR1/TRPV1 (Caterina, M. J. et al., 1997), the revelation that this receptor mediated the response to noxious heat (Tominaga, M. et al., 1998), and the finding that NGF could sensitize the response to capsaicin (Shu, X. and Mendell, L. M., 1999b) or to noxious heat (Galoyan, S. M. et al., 2003) seemed to be entirely consistent with the behavioral evidence that NGF sensitizes the response to noxious heat. However, several experiments have raised difficulties with this simple picture. A substantial fraction of cell membrane patches from small-diameter sensory neurons were found to be sensitive to capsaicin (indicating TRPV1 expression) or noxious heat, but not both (Nagy, I. and Rang, H. P., 1999), suggesting that the receptor for capsaicin and for noxious heat are not the same. Furthermore, although two studies in dissociated cells from TRPV1$^{-/-}$ mice revealed the expected deficit in noxious heat sensitivity (Caterina, M. J. et al., 2000; Davis, J. B. et al., 2000), the same reports demonstrated that less reduced preparations (skin nerve or *in vivo* preparations) displayed considerable residual sensitivity to noxious heat. This raises questions concerning our model of how NGF acts to sensitize the response to noxious heat – specifically whether other noxious heat receptors exist (Woodbury, C. J. et al., 2004) and whether these are affected by neurotrophins.

The finding that activation of the p75 receptor enhances membrane gain via by increasing TTX-insensitive Na current and decreasing the strength of K-rectification (see Section 22.7) expands possible mechanisms for sensitization by NGF. It also opens the possibility for sensitizing effects of other neurotrophins since p75, unlike trkA, trkB, and trkC, binds all the neurotrophins. The role of p75 in mediating the functional effects of NGF is still largely unknown.

Finally, the fact that NGF has so many different possible actions on nociceptive function in terms of the variety of cells that express trkA (Section 22.4.3) as well as the multiple effects that NGF can elicit in individual nociceptors (Sections 22.8 and 22.10) raises important questions about how all these mechanisms interact to cause sensitization. If the scheme outlined in Figure 2 is a complete representation of the peripheral mechanism of NGF action, one would expect that elimination of mast cells would reduce but not abolish NGF-induced sensitization. However, acute sensitization is abolished (Lewin, G. R. et al., 1994), suggesting that NGF activation of the neutrophil pathway (Bennett, G. et al., 1998b) or direct activation of nociceptor terminals is not sufficient to mediate the acute phase of NGF-induced hyperalgesia. Similarly, inhibiting neutrophil accumulation completely eliminates the hyperalgesic effect of NGF (Bennett, G. et al., 1998b), indicating that the mast cell or direct effects are not sufficient either. Together, these findings suggest an interaction between these different mechanisms, and it will be important to dissect these interactions in order to more fully understand how NGF (and neurotrophins more generally) acts as a sensitizing agent for nociceptors. In view of increasing interest in manipulating NGF levels in a clinical setting, it will be necessary to understand its physiology at a more integrative level.

Acknowledgment

I thank Dr. Jeff Petruska for useful comments on an early draft of the manuscript and for help with figures. The author's research was supported by NIH (NS39420; NS16996) and the Christopher Reeve Paralysis Foundation Consortium on Spinal Injury.

References

Albers, K. M., Wright, D. E., and Davis, B. M. 1994. Overexpression of nerve growth factor in epidermis of transgenic mice causes hypertrophy of the peripheral nervous system. J. Neurosci. 14, 1422–1432.

Aloe, L., Tuveri, M. A., and Levi-Montalcini, R. 1992. Studies on carrageenan-induced arthritis in adult rats: presence of nerve growth factor and role of sympathetic innervation. Rheumatol. Internat. 12, 213–216.

Amann, R., Schuligoi, R., Herzeg, G., and Donnerer, J. 1996a. Intraplantar injection of nerve growth factor into the rat hind paw: local edema and effects on thermal nociceptive threshold. Pain 64, 323–329.

Amann, R., Schuligoi, R., Lanz, I., and Peskar, B. A. 1996b. Effect of a 5-lipoxygenase inhibitor on nerve growth factor-induced thermal hyperalgesia in the rat. Eur. J. Pharmacol. 306, 89–91.

Amaya, F., Shimosato, G., Nagano, M., Ueda, M., Hashimoto, S., Tanaka, Y., Suzuki, H., and Tanaka, M. 2004. NGF and GDNF differentially regulate TRPV1 expression that contributes to development of inflammatory thermal hyperalgesia. Eur. J. Neurosci. 20, 2303–2310.

Anand, P. 2004. Neurotrophic factors and their receptors in human sensory neuropathies. Prog. Brain Res. 146, 477–492.

Andreev, N. Yu., Dimitrieva, N., Koltzenburg, M., and McMahon, S. B. 1995. Peripheral administration of nerve growth factor in the adult rat produces a thermal hyperalgesia that requires the presence of sympathetic post-ganglionic neurones. Pain 63, 109–115.

Apfel, S. C., Wright, D. E., Wiideman, A. M., Dormia, C., Snider, W. D., and Kessler, J. A. 1996. Nerve growth factor regulates the expression of brain-derived neurotrophic factor mRNA in the peripheral nervous system. Mol. Cell. Neurosci. 7, 134–142.

Averill, S., McMahon, S. B., Clary, D. O., Reichardt, L. F., and Priestley, J. V. 1995. Immunocytochemical localization of trkA receptors in chemically identified subgroups of adult rat sensory neurons. Eur. J. Neurosci. 7, 1484–1494.

Barbacid, M. 1994. The Trk family of neurotrophin receptors. J. Neurobiol. 25, 1386–1403.

Barker, P. A. and Shooter, E. M. 1994. Disruption of NGF binding to the low affinity neurotrophin receptor p75LNTR reduces NGF binding to TrkA on PC12 cells. Neuron 13, 203–215.

Benedetti, M., Levi, A., and Chao, M. V. 1993. Differential expression of nerve growth factor receptors leads to altered binding affinity and neurotrophin responsiveness. Proc. Natl. Acad. Sci. U. S. A. 90, 7859–7863.

Bennett, D. L., Averill, S., Clary, D. O., Priestley, J. V., and McMahon, S. B. 1996a. Postnatal changes in the expression of the trkA high-affinity NGF receptor in primary sensory neurons. Eur. J. Neurosci. 8, 2204–2208.

Bennett, D. L., French, J., Priestley, J. V., and McMahon, S. B. 1996b. NGF but not NT-3 or BDNF prevents the A fiber sprouting into lamina II of the spinal cord that occurs following axotomy. Mol. Cell Neurosci. 8, 211–220.

Bennett, D. L., Koltzenburg, M., Priestley, J. V., Shelton, D. L., and McMahon, S. B. 1998a. Endogenous nerve growth factor regulates the sensitivity of nociceptors in the adult rat. Eur. J. Neurosci. 10, 1282–1291.

Bennett, G., al-Rashed, S., Hoult, J. R., and Brain, S. D. 1998b. Nerve growth factor induced hyperalgesia in the rat hind paw is dependent on circulating neutrophils. Pain 77, 315–322.

Berg, M. M., Sternberg, D. W., Hempstead, B. L., and Chao, M. V. 1991. The low-affinity p75 nerve growth factor (NGF) receptor mediates NGF-induced tyrosine phosphorylation. Proc. Natl. Acad. Sci. U. S. A. 88, 7106–7110.

Bergmann, I., Reiter, R., Toyka, K. V., and Koltzenburg, M. 1998. Nerve growth factor evokes hyperalgesia in mice lacking the low-affinity neurotrophin receptor p75. Neurosci. Lett. 255, 87–90.

Bevan, S. and Winter, J. 1995. Nerve growth factor (NGF) differentially regulates the chemosensitivity of adult rat cultured sensory neurons. J. Neurosci. 15, 4918–4926.

Bielefeldt, K., Ozaki, N., and Gebhart, G. F. 2003. Role of nerve growth factor in modulation of gastric afferent neurons in the rat. Am. J. Physiol. Gastrointest. Liver Physiol. 284, G499–G507.

Bilderback, T. R., Gazula, V. R., and Dobrowsky, R. T. 2001. Phosphoinositide 3-kinase regulates crosstalk between Trk A tyrosine kinase and p75(NTR)-dependent sphingolipid signaling pathways. J. Neurochem. 76, 1540–1551.

Bonnington, J. K. and McNaughton, P. A. 2003. Signalling pathways involved in the sensitisation of mouse nociceptive neurones by nerve growth factor. J. Physiol. 551(Pt 2), 433–446.

Bradbury, E. J., King, V., Simmons, L. J., Priestley, J. V., and McMahon, S. B. 1998. NT-3, but not BDNF, prevents atrophy and death of axotomized spinal cord projection neurons. Eur. J. Neurosci. 10, 3058–3068.

Caterina, M. J., Lefler, A., Malmberg, A. B., Martin, W. J., Trafton, J., Petersen-Zeitz, K. R., Koltzenburg, M., Basbaum, A. I., and Julius, D. 2000. Impaired nociception and pain sensation in mice lacking the capsaicin receptor. Science 288, 306–313.

Caterina, M. J., Schumacher, M. A., Tominaga, M., Rosen, T. A., Levine, J. D., and Julius, D. 1997. The capsaicin receptor: a heat activated ion channel in the pain pathway. Nature 389, 816–824.

Chao, M. V. and Hempstead, B. L. 1995. p75 and Trk: a two-receptor system. Trends Neurosci. 18, 321–326.

Chuang, H. H., Prescott, E. D., Kong, H., Shields, S., Jordt, S. E., Basbaum, A. I., Chao, M. V., and Julius, D. 2001. Bradykinin and nerve growth factor release the capsaicin receptor from PtdIns(4,5)P2-mediated inhibition. Nature 411, 957–962.

Cook, S. P. and McCleskey, E. W. 2002. Cell damage excites nociceptors through release of cytosolic ATP. Pain 95, 41–47.

Coull, J. A., Beggs, S., Boudreau, D., Boivin, D., Tsuda, M., Inoue, K., Gravel, C., Salter, M. W., and De Koninck, Y. 2005. BDNF from microglia causes the shift in neuronal anion gradient underlying neuropathic pain. Nature 438, 1017–1021.

Cowan, W. M. 2001. Viktor Hamburger and Rita Levi-Montalcini: the path to the discovery of nerve growth factor. Annu. Rev. Neurosci. 24, 551–600.

Davies, A. M. 1996. The neurotrophic hypothesis: where does it stand? Philos. Trans. R. Soc. Lond. B Biol. Sci. 351, 389–394.

Davis, J. B., Gray, J., Gunthorpe, M. J., Hatcher, J. P., Davey, P. T., Overend, P., Harries, M. H., Latcham, J., Clapham, C., Atkinson, K., Hughes, S. A., Rance, K., Grau, E., Harper, A. J., Pugh, P. L., Rogers, D. C., Bingham, S., Randall, A., and Sheardown, S. A. 2000. Vanilloid receptor-1 is essential for inflammatory thermal hyperalgesia. Nature 405, 183–187.

Davis, B. M., Lewin, G. R., Mendell, L. M., Jones, M. E., and Albers, K. M. 1993. Altered expression of nerve growth factor in the skin of transgenic mice leads to changes in response to mechanical stimuli. Neuroscience 56, 789–792.

Di Marco, E., Marchisio, P. C., Bondanza, S., Franzi, A. T., Cancedda, R., and De Luca, M. 1991. Growth-regulated synthesis and secretion of biologically active nerve growth factor by human keratinocytes. J. Biol. Chem. 266, 21718–21722.

Diamond, J., Holmes, M., and Coughlin, M. 1992. Endogenous NGF and nerve impulses regulate the collateral sprouting of sensory axons in the skin of the adult rat. J. Neurosci. 12, 1454–1466.

Djouhri, L., Dawbarn, D., Robertson, A., Newton, R., and Lawson, S. N. 2001. Time course and nerve growth factor dependence of inflammation-induced alterations in electrophysiological membrane properties in nociceptive primary afferent neurons. J. Neurosci. 21, 8722–8733.

Dobrowsky, R. T., Werner, M. H., Castellino, A. M., Chao, M. V., and Hannun, Y. A. 1994. Activation of the sphingomyelin cycle through the low-affinity neurotrophin receptor. Science 265, 1596–1599.

Dray, A. and Perkins, M. 1993. Bradykinin and inflammatory pain. Trends Neurosci. 16, 99–104.

Dyck, P. J., Peroutka, S., Rask, C., Burton, E., Baker, M. K., Lehman, K. A., Gillen, D. A., Hokanson, J. L., and O'Brien, P. C. 1997. Intradermal recombinant human nerve growth factor induces pressure allodynia and lowered heat-pain threshold in humans. Neurology 48, 501–505.

Everill, B. and Kocsis, J. D. 2000. Nerve growth factor maintains potassium conductance after nerve injury in adult cutaneous afferent dorsal root ganglion neurons. Neuroscience 100, 417–422.

Falcini, F., Matucci Cerinic, M., Lombardi, A., Generini, S., Pignone, A., Tirassa, P., Ermini, M., Lepore, L., Partsch, G., and Aloe, L. 1996. Increased circulating nerve growth factor is directly correlated with disease activity in juvenile chronic arthritis. Ann. Rheum. Dis. 55, 745–748.

Fang, X., Djouhri, L., McMullan, S., Berry, C., Okuse, K., Waxman, S. G., and Lawson, S. N. 2005. trkA is expressed in nociceptive neurons and influences electrophysiological properties via Nav1.8 expression in rapidly conducting nociceptors. J. Neurosci. 25, 4868–4878.

Fjell, J., Cummins, T. R., Dib-Hajj, S. D., Fried, K., Black, J. A., and Waxman, S. G. 1999. Differential role of GDNF and NGF in the maintenance of two TTX-resistant sodium channels in adult DRG neurons. Brain Res. Mol. Brain Res. 67, 267–282.

Fukuoka, T., Kondo, E., Dai, Y., Hashimoto, N., and Noguchi, K. 2001. Brain-derived neurotrophic factor increases in the uninjured dorsal root ganglion neurons in selective spinal nerve ligation model. J. Neurosci. 21, 4891–4900.

Galoyan, S. M., Petruska, J. C., and Mendell, L. M. 2003. Mechanisms of sensitization of the response of single dorsal root ganglion cells from adult rat to noxious heat. Eur. J. Neurosci. 18, 535–541.

Ganju, P., O'Bryan, J. P., Der, C., Winter, J., and James, I. F. 1998. Differential regulation of SHC proteins by nerve growth factor in sensory neurons and PC12 cells. Eur. J. Neurosci. 10, 1995–2008.

Garraway, S. M., Anderson, A. J., and Mendell, L. M. 2005. BDNF-induced facilitation of afferent evoked responses in lamina II neurons is reduced following neonatal spinal cord contusion injury. J. Neurophysiol. [Epub ahead of print].

Garraway, S. M., Petruska, J. C., and Mendell, L. M. 2003. BDNF sensitizes the response of lamina II neurons to high threshold primary afferent inputs. Eur. J. Neurosci. 18, 2467–2476.

Gratto, K. A. and Verge, V. M. 2003. Neurotrophin-3 down-regulates trkA mRNA, NGF high-affinity binding sites, and associated phenotype in adult DRG neurons. Eur. J. Neurosci. 18, 1535–1548.

Halliday, D. A., Zettler, C., Rush, R. A., Scicchitano, R., and McNeil, J. D. 1998. Elevated nerve growth factor levels in the synovial fluid of patients with inflammatory joint disease. Neurochem. Res. 23, 919–922.

He, X. L. and Garcia, K. C. 2004. Structure of nerve growth factor complexed with the shared neurotrophin receptor p75. Science 304, 870–875.

Hempstead, B. L., Martin-Zanca, D., Kaplan, D. R., Parada, L. F., and Chao, M. V. 1991. High-affinity NGF binding requires coexpression of the trk proto-oncogene and the low-affinity NGF receptor. Nature 350, 678–683.

Hoheisel, U., Unger, T., and Mense, S. 2005. Excitatory and modulatory effects of inflammatory cytokines and neurotrophins on mechanosensitive group IV muscle afferents in the rat. Pain 114, 168–176.

Horigome, K., Pryor, J. C., Bullock, E. D., and Johnson, E. M., Jr. 1993. Mediator release from mast cells by nerve growth factor. J. Biol. Chem. 268, 14881–14887.

Huang, E. J. and Reichardt, L. F. 2001. Neurotrophins: roles in neuronal development and function. Annu. Rev. Neurosci. 24, 677–736.

Ji, R. R., Samad, T. A., Jin, S. X., Schmoll, R., and Woolf, C. J. 2002. p38 MAPK activation by NGF in primary sensory neurons after inflammation increases TRPV1 levels and maintains heat hyperalgesia. Neuron 36, 57–68.

Kaplan, D. R. and Cooper, E. 2001. PI-3 kinase and IP3: partners in NT3-induced synaptic transmission. Nat. Neurosci. 4, 5–7.

Kaplan, D. R. and Miller, F. D. 2000. Neurotrophin signal transduction in the nervous system. Curr. Opin. Neurobiol. 10, 381–391.

Karchewski, L. A., Kim, F. A., Johnston, J., McKnight, R. M., and Verge, V. M. 1999. Anatomical evidence supporting the potential for modulation by multiple neurotrophins in the majority of adult lumbar sensory neurons. J. Comp. Neurol. 413, 327–341.

Kerr, B. J., Bradbury, E. J., Bennett, D. L., Trivedi, P. M., Dassan, P., French, J., Shelton, D. B., McMahon, S. B., and Thompson, S. W. 1999. Brain-derived neurotrophic factor modulates nociceptive sensory inputs and NMDA-evoked responses in the rat spinal cord. J. Neurosci. 19, 5138–5148.

Knusel, B. and Hefti, F. 1992. K-252 compounds: modulators of neurotrophin signal transduction. J. Neurochem. 59, 1987–1996.

Koizumi, S., Fujishita, K., Inoue, K., Shigemoto-Mogami, Y., Tsuda, M., and Inoue, K. 2004. Ca^{2+} waves in keratinocytes are transmitted to sensory neurons: the involvement of extracellular ATP and P2Y2 receptor activation. Biochem. J. 380, 329–338.

Koltzenburg, M., Bennett, D. L., Shelton, D. L., and McMahon, S. B. 1999. Neutralization of endogenous NGF prevents the sensitization of nociceptors supplying inflamed skin. Eur. J. Neurosci. 11, 1698–1704.

Koplas, P. A., Rosenberg, R. L., and Oxford, G. S. 1997. The role of calcium in the desensitization of capsaicin responses in rat dorsal root ganglion neurons. J. Neurosci. 17, 3525–3537.

Lamb, K. and Bielefeldt, K. 2003. Rapid effects of neurotrophic factors on calcium homeostasis in rat visceral afferent neurons. Neurosci. Lett. 336, 9–12.

Lamb, K., Gebhart, G. F., and Bielefeldt, K. 2004. Increased nerve growth factor expression triggers bladder overactivity. J. Pain 5, 150–156.

Lamb, K., Kang, Y. M., Gebhart, G. F., and Bielefeldt, K. 2003. Nerve growth factor and gastric hyperalgesia in the rat. Neurogastroenterol. Motil. 15, 355–361.

LaMotte, R. H., Lundberg, L. E., and Torebjork, H. E. 1992. Pain, hyperalgesia and activity in nociceptive C units in humans after intradermal injection of capsaicin. J. Physiol. 448, 749–764.

Lee, Y. J., Zachrisson, O., Tonge, D. A., and McNaughton, P. A. 2002. Upregulation of bradykinin B2 receptor expression by neurotrophic factors and nerve injury in mouse sensory neurons. Mol. Cell Neurosci. 19, 186–200.

Leon, A., Buriani, A., Dal Toso, R., Fabris, M., Romanello, S., Aloe, L., and Levi-Montalcini, R. 1994. Mast cells synthesize, store, and release nerve growth factor. Proc. Natl. Acad. Sci. U. S. A. 91, 3739–3743.

Lever, I. J., Bradbury, E. J., Cunningham, J. R., Adelson, D. W., Jones, M. G., McMahon, S. B., Marvizon, J. C., and Malcangio, M. 2001. Brain-derived neurotrophic factor is released in the dorsal horn by distinctive patterns of afferent fiber stimulation. J. Neurosci. 21, 4469–4477.

Lewin, G. R. and Mendell, L. M. 1993. Nerve growth factor and nociception. Trends Neurosci. 16, 353–359.

Lewin, G. R. and Mendell, L. M. 1994. Regulation of cutaneous C-fibre nociceptors by nerve growth factor in the developing rat. J. Neurophysiol. 71, 941–949.

Lewin, G. R., Lisney, S. J., and Mendell, L. M. 1992b. Neonatal anti-NGF treatment reduces the A delta- and C-fibre evoked vasodilator responses in rat skin: evidence that nociceptor afferents mediate antidromic vasodilatation. Eur. J. Neurosci. 4, 1213–1218.

Lewin, G. R., Ritter, A. M., and Mendell, L. M. 1992a. On the role of nerve growth factor in the development of myelinated nociceptors. J. Neurosci. 12, 1896–1905.

Lewin, G. R., Ritter, A. M., and Mendell, L. M. 1993. Nerve growth factor-induced hyperalgesia in the neonatal and adult rat. J. Neurosci. 13, 2136–2148.

Lewin, G. R., Rueff, A., and Mendell, L. M. 1994. Peripheral and central mechanisms of NGF-induced hyperalgesia. Eur. J. Neurosci. 6, 1903–1912.

Lindsay, R. M. 1996. Role of neurotrophins and trk receptors in the development and maintenance of sensory neurons: an overview. Philos. Trans. R. Soc. Lond. B Biol. Sci. 351, 365–373.

Lindsay, R. M. and Harmar, A. J. 1989. Nerve growth factor regulates expression of neuropeptide genes in adult sensory neurons. Nature 337, 362–364.

Liu, B. and Hannun, Y. A. 1997. Inhibition of the neutral magnesium-dependent sphingomyelinase by glutathione. J. Biol. Chem. 272, 16281–16287.

Liu, B., Zhang, C., and Qin, F. 2005. Functional recovery from desensitization of vanilloid receptor TRPV1 requires resynthesis of phosphatidylinositol 4,5-bisphosphate. J. Neurosci. 25, 4835–4843.

Lowe, E. M., Anand, P., Terenghi, G., Williams-Chestnut, R. E., Sinicropi, D. V., and Osborne, J. L. 1997. Increased nerve growth factor levels in the urinary bladder of women with idiopathic sensory urgency and interstitial cystitis. Br. J. Urol. 79, 572–577.

MacPhee, I. J. and Barker, P. A. 1997. Brain-derived neurotrophic factor binding to the p75 neurotrophin receptor reduces TrkA signaling while increasing serine phosphorylation in the TrkA intracellular domain. J. Biol. Chem. 272, 23547–23551.

MacPhee, I. and Barker, P. A. 1999. Extended ceramide exposure activates the trkA receptor by increasing receptor homodimer formation. J. Neurochem. 72, 1423–1430.

Majdan, M., Walsh, G. S., Aloyz, R., and Miller, F. D. 2001. TrkA mediates developmental sympathetic neuron survival in vivo by silencing an ongoing p75NTR-mediated death signal. J. Cell Biol. 155, 1275–1285.

Malin, S. A., Molliver, D. C., Koerber, H. R., Cornuet, P., Frye, R., Albers, K. M., and Davis, B. M. 2006. Glial cell line-derived neurotrophic factor family members sensitize nociceptors in vitro and produce thermal hyperalgesia *in vivo*. J. Neurosci. 26, 8588–8599.

Mamet, J., Lazdunski, M., and Voilley, N. 2003. How nerve growth factor drives physiological and inflammatory expressions of acid-sensing ion channel in sensory neurons. J. Biol. Chem. 278, 48907–48913.

Manni, L., Lundeberg, T., Fiorito, S., Bonini, S., Vigneti, E., and Aloe, L. 2003. Nerve growth factor release by human synovial fibroblasts prior to and following exposure to tumor necrosis factor-alpha, interleukin-1 beta and cholecystokinin-8: the possible role of NGF in the inflammatory response. Clin. Exp. Rheumatol. 21, 617–624.

Mannion, R. J., Costigan, M., Decosterd, I., Amaya, F., Ma, Q. P., Holstege, J. C., Ji, R. R., Acheson, A., Lindsay, R. M., Wilkinson, G. A., and Woolf, C. J. 1999. Neurotrophins: peripherally and centrally acting modulators of tactile stimulus-induced inflammatory pain hypersensitivity. Proc. Natl. Acad. Sci. U. S. A. 96, 9385–9390.

Martin, H. A., Basbaum, A. I., Goetzl, E. J., and Levine, J. D. 1988. Leukotriene B4 decreases the mechanical and thermal thresholds of C-fiber nociceptors in the hairy skin of the rat. J. Neurophysiol. 60, 438–445.

di Mola, F. F., Friess, H., Zhu, Z. W., Koliopanos, A., Bley, T., Di Sebastiano, P., Innocenti, P., Zimmermann, A., and Buchler, M. W. 2000. Nerve growth factor and Trk high affinity receptor (TrkA) gene expression in inflammatory bowel disease. Gut 46, 670–679.

McMahon, S. B., Armanini, M. P., Ling, L. H., and Phillips, H. S. 1994. Expression and coexpression of trk receptors in subpopulations of adult primary sensory neurons projecting to identified peripheral targets. Neuron 12, 1161–1171.

McMahon, S. B., Bennett, D. L., Priestley, J. V., and Shelton, D. L. 1995. The biological effects of endogenous nerve growth factor on adult sensory neurons revealed by a trkA-IgG fusion molecule. Nat. Med. 1, 774–780.

McMahon, S. B., Lewin, G. R., and Wall, P. D. 1993. Central hyperexcitability triggered by noxious inputs. Curr. Opin. Neurobiol. 3, 602–610.

Mendell, L. M. 1995. Neurotrophic factors and the specification of neural function. Neuroscientist 1, 26–34.

Mendell, L. M. 2002. Does NGF binding to p75 and trkA receptors activate independent signaling pathways to sensitize nociceptors? J. Physiol. (Perspective) 544, 333.

Mendell, L. M., Albers, K. M., and Davis, B. M. 1999. Neurotrophins, nociceptors and pain. Microsc. Res. Tech. 45, 252–261.

Michael, G. J., Averill, S., Nitkunan, A., Rattray, M., Bennett, D. L., Yan, Q., and Priestley, J. V. 1997. Nerve growth factor treatment increases brain-derived neurotrophic factor selectively in TrkA-expressing dorsal root ganglion cells and in their central terminations within the spinal cord. J. Neurosci. 17, 8476–8490.

Miller, F. D. and Kaplan, D. R. 2001. Neurotrophin signalling pathways regulating neuronal apoptosis. Cell Mol. Life Sci. 58, 1045–1053.

Molliver, D. C., Wright, D. E., Leitner, M. L., Parsadanian, A. S., Doster, K., Wen, D., Yan, Q., and Snider, W. D. 1997. IB4-binding DRG neurons switch from NGF to GDNF dependence in early postnatal life. Neuron 19, 849–861.

Nagy, I. and Rang, H. P. 1999. Similarities and differences between the responses of rat sensory neurons to noxious heat and capsaicin. J. Neurosci. 19, 647–655.

Nicol, G. D. and Vasko, M. R. 2007. Unraveling the story of NGF-mediated sensitization of nociceptive sensory neurons: ON or OFF the Trks? Mol Interv. 7, 26–41.

Obermeier, A., Halfter, H., Wiesmuller, K. H., Jung, G., Schlessinger, J., and Ullrich, A. 1993. Tyrosine 785 is a major determinant of Trk–substrate interaction. EMBO J. 12, 933–941.

Oddiah, D., Anand, P., McMahon, S. B., and Rattray, M. 1998. Rapid increase of NGF, BDNF and NT-3 mRNAs in inflamed bladder. Neuroreport 9, 1455–1458.

Patapoutian, A. and Reichardt, L. F. 2001. Trk receptors: mediators of neurotrophin action. Curr. Opin. Neurobiol. 11, 272–280.

Petersen, M., Klusch, A., and Eckert, A. 1998. The proportion of isolated rat dorsal root ganglion neurones responding to bradykinin increases with time in culture. Neurosci. Lett. 252, 143–146.

Petruska, J. C. and Mendell, L. M. 2004. The many functions of nerve growth factor: multiple actions on nociceptors. Neurosci. Lett. 361, 168–171.

Petty, B. G., Cornblath, D. R., Aldornato, B. T., Chaudhry, V., Flexner, C., Wachsman, M., Sinicropi, D., Burton, L. E., and Peroutka, S. J. 1994. The effect of systemically administered recombinant human nerve growth factor in healthy human subjects. Ann. Neurol. 36, 244–246.

Pezet, S., Cunningham, J., Patel, J., Grist, J., Gavazzi, I., Lever, I., and Malcangio, M. 2002. BDNF modulates sensory neuron synaptic activity by a facilitation of GABA transmission in the dorsal horn. Mol. Cell. Neurosci. 21, 51–62.

Plo, I., Bono, F., Bezombes, C., Alam, A., Bruno, A., and Laurent, G. 2004. Nerve growth factor-induced protein kinase C stimulation contributes to TrkA-dependent inhibition of p75 neurotrophin receptor sphingolipid signaling. J. Neurosci. Res. 77, 465–474.

Price, T. J., Louria, M. D., Candelario-Soto, D., Dussor, G. O., Jeske, N. A., Patwardhan, A. M., Diogenes, A., Trott, A. A., Hargreaves, K. M., and Flores, C. M. 2005. Treatment of trigeminal ganglion neurons in vitro with NGF, GDNF or BDNF: effects on neuronal survival, neurochemical properties and TRPV1-mediated neuropeptide secretion. BMC Neurosci. 6, 4.

Ramer, M. S., Bradbury, E. J., and McMahon, S. B. 2001. Nerve growth factor induces P2X(3) expression in sensory neurons. J. Neurochem. 77, 864–875.

Ramer, M. S., Bradbury, E. J., Michael, G. J., Lever, I. J., and McMahon, S. B. 2003. Glial cell line-derived neurotrophic factor increases calcitonin gene-related peptide immunoreactivity in sensory and motoneurons in vivo. Eur. J. Neurosci. 18, 2713–2721.

Reynolds, M. L., Fitzgerald, M., and Benowitz, L. I. 1991. GAP-43 expression in developing cutaneous and muscle nerves in the rat hindlimb. Neuroscience 41, 201–211.

Ritter, A. M. and Mendell, L. M. 1992. Somal membrane properties of physiologically identified sensory neurons in the rat: effects of nerve growth factor. J. Neurophysiol. 68, 2033–2041.

Ritter, A. M., Lewin, G. R., Kremer, N. E., and Mendell, L. M. 1991. Requirement for nerve growth factor in the development of myelinated nociceptors in vivo. Nature 350, 500–502.

Rodriguez-Tebar, A., Dechant, G., Gotz, R., and Barde, Y. A. 1992. Binding of neurotrophin-3 to its neuronal receptors and interactions with nerve growth factor and brain-derived neurotrophic factor. EMBO J. 11, 917–922.

Rueff, A. and Mendell, L. M. 1996. Nerve growth factor and NT-5 induce increased thermal sensitivity of cutaneous nociceptors in vitro. J. Neurophysiol. 76, 3593–3596.

Rueff, A., Dawson, A. J., and Mendell, L. M. 1996. Characteristics of nerve growth factor induced hyperalgesia in adult rats: dependence on enhanced bradykinin-1 receptor activity but not neurokinin-1 receptor activation. Pain 66, 359–372.

Saarma, M. and Sariola, H. 1999. Other neurotrophic factors: glial cell line-derived neurotrophic factor (GDNF). Microsc. Res. Tech. 45, 292–302.

Stein, A. T., Ufret-Vincenty, C. A., Hua, L., Santana, L. F., and Gordon, S. E. 2006. Phosphoinositide 3-kinase binds to TRPV1 and mediates NGF-stimulated TRPV1 trafficking to the plasma membrane. J. Gen. Physiol. 128, 509–522.

von Schack, D., Casademunt, E., Schweigreiter, R., Meyer, M., Bibel, M., and Dechant, G. 2001. Complete ablation of the neurotrophin receptor p75NTR causes defects both in the nervous and the vascular system. Nat. Neurosci. 4, 977–978.

Schafers, M., Sorkin, L. S., and Sommer, C. 2003. Intramuscular injection of tumor necrosis factor-alpha induces muscle hyperalgesia in rats. Pain 104, 579–588.

Schaible, H. G. and Schmidt, R. F. 1985. Effects of an experimental arthritis on the sensory properties of fine articular afferent units. J. Neurophysiol. 54, 1109–1122.

Sevcik, M. A., Ghilardi, J. R., Peters, C. M., Lindsay, T. H., Halvorson, K. G., Jonas, B. M., Kubota, K., Kuskowski, M. A., Boustany, L., Shelton, D. L., and Mantyh, P. W. 2005. Anti-NGF therapy profoundly reduces bone cancer pain and the accompanying increase in markers of peripheral and central sensitization. Pain 115, 128–141.

Shelton, D. L., Zeller, J., Ho, W. H., Pons, J., and Rosenthal, A. 2005. Nerve growth factor mediates hyperalgesia and cachexia in auto-immune arthritis. Pain. [Epub ahead of print].

Shinoda, M., Honda, T., Ozaki, N., Hattori, H., Mizutani, H., Ueda, M., and Sugiura, Y. 2003. Nerve terminals extend into the temporomandibular joint of adjuvant arthritic rats. Eur. J. Pain. 7, 493–505.

Shu, X. Q. and Mendell, L. M. 1999a. Neurotrophins and hyperalgesia. Proc. Natl. Acad. Sci. U. S. A. 96, 7693–7696.

Shu, X. and Mendell, L. M. 1999b. Nerve growth factor acutely sensitizes the response of adult rat sensory neurons to capsaicin. Neurosci. Lett. 274, 158–162.

Shu, X. and Mendell, L. M. 2001. Acute sensitization by NGF of the response of small-diameter sensory neurons to capsaicin. J. Neurophysiol. 86, 2931–2938.

Shu, X., Llinas, A., and Mendell, L. M. 1999. Effects of trkB and trkC neurotrophin receptor agonists on thermal nociception: a behavioural and electrophysiological study. Pain 80, 463–470.

Slack, S. E., Pezet, S., McMahon, S. B, Thompson, S. W., and Malcangio, M. 2004. Brain-derived neurotrophic factor induces NMDA receptor subunit one phosphorylation via ERK and PKC in the rat spinal cord. Eur. J. Neurosci. 20, 1769–1778.

Smeyne, R. J., Klein, R., Schnapp, A., Long, L. K., Bryant, S., Lewin, A., Lira, S. A., and Barbacid, M. 1994. Severe sensory and sympathetic neuropathies in mice carrying a disrupted Trk/NGF receptor gene. Nature 368, 246–249.

Southall, M. D. and Vasko, M. R. 2000. Prostaglandin E(2)-mediated sensitization of rat sensory neurons is not altered by nerve growth factor. Neurosci. Lett. 287, 33–36.

Stucky, C. L. and Koltzenburg, M. 1997. The low-affinity neurotrophin receptor p75 regulates the function but not the selective survival of specific subpopulations of sensory neurons. J. Neurosci. 11, 4398–4405.

Stucky, C. L., Koltzenburg, M., Schneider, M., Engle, M. G., Albers, K. M., and Davis, B. M. 1999. Overexpression of nerve growth factor in skin selectively affects the survival and functional properties of nociceptors. J. Neurosci. 19, 8509–8516.

Svensson, P., Cairns, B. E., Wang, K., and Arendt-Nielsen, L. 2003. Injection of nerve growth factor into human masseter muscle evokes long-lasting mechanical allodynia and hyperalgesia. Pain 104, 241–247.

Teng, K. K. and Hempstead, B. L. 2004. Neurotrophins and their receptors: signaling trios in complex biological systems. Cell Mol. Life Sci. 61, 35–48.

Terenghi, G., Mann, D., Kopelman, P. G., and Anand, P. 1997. trkA and trkC expression is increased in human diabetic skin. Neurosci. Lett. 228, 33–36.

Theodosiou, M., Rush, R. A., Zhou, X. F., Hu, D., Walker, J. S., and Tracey, D. J. 1999. Hyperalgesia due to nerve damage: role of nerve growth factor. Pain 181, 245–255.

Toma, H., Winston, J., Micci, M. A., Shenoy, M., and Pasricha, P. J. 2000. Nerve growth factor expression is up-regulated in the rat model of L-arginine-induced acute pancreatitis. Gastroenterology 119, 1373–1381.

Tominaga, M., Caterina, M. J., Malmberg, A. B., Rosen, T. A., Gilbert, H., Skinner, K., Raumann, B. E., Basbaum, A. I., and Julius, D. 1998. The cloned capsaicin receptor integrates multiple pain-producing stimuli. Neuron 21, 531–543.

Vellani, V., Zachrisson, O., and McNaughton, P. A. 2004. Functional bradykinin B1 receptors are expressed in nociceptive neurones and are upregulated by the neurotrophin GDNF. J. Physiol. 560, 391–401.

Verdi, J. M., Birren, S. J., Ibanez, C. F., Persson, H., Kaplan, D. R., Benedetti, M., Chao, M. V., and Anderson, D. J. 1994. p75LNGFR regulates Trk signal transduction and NGF-induced neuronal differentiation in MAH cells. Neuron 12, 733–745.

Voilley, N., de Weille, J., Mamet, J., and Lazdunski, M. 2001. Nonsteroid anti-inflammatory drugs inhibit both the activity and the inflammation-induced expression of acid-sensing ion channels in nociceptors. J. Neurosci. 21, 8026–8033.

Weskamp, G. and Otten, U. 1987. An enzyme-linked immunoassay for nerve growth factor (NGF): a tool for studying regulatory mechanisms involved in NGF production in brain and in peripheral tissues. J. Neurochem. 48, 1779–1786.

Wilson-Gerwing, T. D., Dmyterko, M. V., Zochodne, D. W., Johnston, J. M., and Verge, V. M. 2005. Neurotrophin-3 suppresses thermal hyperalgesia associated with neuropathic pain and attenuates transient receptor potential vanilloid receptor-1 expression in adult sensory neurons. J. Neurosci. 25, 758–767.

Winston, J., Toma, H., Shenoy, M., and Pasricha, P. J. 2001. Nerve growth factor regulates VR-1mRNA levels in cultures of adult dorsal root ganglion neurons. Pain 89, 181–186.

Woodbury, C. J., Zwick, M., Wang, S., Lawson, J. J., Caterina, M. J., Koltzenburg, M., Albers, K. M., Koerber, H. R., and Davis, B. M. 2004. Nociceptors lacking TRPV1 and TRPV2 have normal heat responses. J. Neurosci. 24, 6410–6415.

Woolf, C. J., Allchorne, A., Safieh-Garabedian, B., and Poole, S. 1997. Cytokines, nerve growth factor and inflammatory hyperalgesia: the contribution of tumour necrosis factor alpha. Br. J. Pharmacol. 121, 417–424.

Woolf, C. J., Ma, Q. P., Allchorne, A., and Poole, S. 1996. Peripheral cell types contributing to the hyperalgesic action of nerve growth factor in inflammation. J. Neurosci. 16, 2716–2723.

Woolf, C. J., Safieh-Garabedian, B., Ma, Q.-P., Crilly, P., and Winter, J. 1994. Nerve growth factor contributes to the generation of inflammatory sensory hypersensibility. Neuroscience 62, 327–331.

Woolf, C. J., Shortland, P., Reynolds, M., Ridings, J., Doubell, T., and Coggeshall, R. E. 1995. Reorganization of central terminals of myelinated primary afferents in the rat dorsal horn following peripheral axotomy. J. Comp. Neurol. 360, 121–134.

Wright, D. E. and Snider, W. D. 1995. Neurotrophin receptor mRNA expression defines distinct populations of neurons in rat dorsal root ganglia. J. Comp. Neurol. 351, 329–338.

Zhang, Y. H. and Nicol, G. D. 2004. NGF-mediated sensitization of the excitability of rat sensory neurons is prevented by a blocking antibody to the p75 neurotrophin receptor. Neurosci. Lett. 366, 187–192.

Zhang, X., Huang, J., and McNaughton, P. A. 2005. NGF rapidly increases membrane expression of TRPV1 heat-gated ion channels. EMBO J. 24, 4211–4223.

Zhang, Y. H., Vasko, M. R., and Nicol, G. D. 2002. Ceramide, a putative second messenger for NGF, modulates the TTX-resistant Na current and delayed rectifier K current in rat sensory neurons. J. Physiol. (Lond.) 544, 385–402.

Zhou, X. F., Deng, Y. S., Xian, C. J., and Zhong, J. H. 2000. Neurotrophins from dorsal root ganglia trigger allodynia after spinal nerve injury in rats. Eur. J. Neurosci. 12, 100–105.

Zhu, W. and Oxford, G. S. 2007. Phosphoinositide-3-kinase and mitogen activated protein kinase signaling pathways mediate acute NGF sensitization of TRPV1. Mol. Cell. Neurosci. 34, 689–700.

Zhu, W., Galoyan, S. M., Petruska, J. C., Oxford, G. S., and Mendell, L. M. 2004. A developmental switch in acute sensitization of small dorsal root ganglion (DRG) neurons to capsaicin or noxious heating by NGF. J. Neurophysiol. 92, 3148–3152.

Zhuang, Z.-Y., Xu, H., Clapham, D. E., and Ji, R.-R. 2004. Phosphatidylinositol 3-kinase activates ERK in primary sensory neurons and mediates inflammatory heat hyperalgesia through TRPV1 sensitization. J. Neurosci. 24, 8300–8309.

23 Morphological and Neurochemical Organization of the Spinal Dorsal Horn

A Ribeiro-da-Silva, McGill University, Montreal, QC, Canada

Y De Koninck, Centre de recherche Université Laval Robert-Giffard, Québec, QC, Canada

Glossary

bouton Dilated portion of an axon possessing synaptic vesicles and establishing synapses with adjacent dendrites or other neuronal components. Synonyms – axonal varicosity or axon terminal.

presynaptic dendrite Dendrite possessing synaptic vesicles, and which is presynaptic to other neuronal processes, mainly dendrites. In the dorsal horn, presynaptic dendrites occur in islet cells.

substantia gelatinosa (of Rolando) Synonym of lamina II of Rexed. So named because of the translucent and gelatinous appearance when examined in fresh tissue.

superficial dorsal horn Region of the dorsal horn corresponding to laminae I and II of Rexed.

synaptic glomerulus (of the spinal cord) Complex synaptic arrangement in which a central axonal bouton, of primary sensory afferent origin, establishes synaptic contacts with several processes from dorsal horn neurons.

23.1   Introduction

The superficial laminae of the dorsal horn of the spinal cord, particularly the marginal layer (or lamina I of Rexed) and the substantia gelatinosa (or lamina II of Rexed), represent the area of the central nervous system (CNS) where the first modulation of pain-related information occurs. Although some progress has been made in recent years concerning our knowledge of the anatomical and neurochemical characteristics of the relevant cells and systems, our understanding of this area is far from complete. In particular, despite significant recent progress, our understanding of the synaptic circuitry and how neurotransmitters/neuromodulators in this region interact with their receptors is still incomplete, and even less is known about the changes that occur in acute and chronic pain (for recent review see Todd, A. J. and Ribeiro-da-Silva, A., 2005). This chapter presents an overview of what is known concerning the morphology and neurochemistry of the spinal dorsal horn, attempting to extract an emerging integrated view of the morphological and neurochemical organization in this area of the CNS from the available literature.

Because most of the work on the structure of the spinal cord has been carried out on the rat, we have based our description on this species, but when possible have made comparison to other species. We have focused on the superficial dorsal horn because of its relevance in the transmission and modulation of pain-related information.

23.2   Overall Organization of Dorsal Horn: Rexed Lamination

The first subdivision of the dorsal horn into horizontal laminae, based on the morphological properties of the cells in a Nissl-type staining, was performed in the cat dorsal horn (Rexed, B., 1952; 1954). It was later verified that this lamination could be adapted with minor modifications to almost any mammalian species. In particular, it has been adapted to the monkey (Ralston, H. J., III, 1979), human (Harmann, P. A. *et al.*, 1988), rat (Steiner, T. J. and Turner, L. M., 1972; Molander, C. *et al.*, 1984; 1989), and mouse (Ma, W. Y. *et al.*, 1995), besides nonmammalian species such as the pigeon (Leonard, R. B. and Cohen, D. H., 1975). Through a correlation of light and electron microscopy, it has been possible since

the 1970s to establish criteria to identify the laminae at the ultrastructural level (for reviews see Light, A. R., 1992; Ribeiro-da-Silva, A., 2004).

Despite some recent criticism (Woodbury, C. J. *et al.*, 2000), the Rexed lamination remains the reference. Unfortunately, it is not always followed accurately, and many published micrographs show laminar limits that were based more on speculation rather than on objective evaluation using the proper approaches. This is true particularly for ultrastructural studies and also for those using radioactive ligand binding. This is unfortunate because, as we will discuss below, it is easy to set up the limits of the laminae with acceptable accuracy (see Figures 1 and 2 and below).

Besides neuroglia and vessels, the spinal dorsal horn contains several elements of neuronal origin: (a) the final arborization and termination of primary afferent fibers; (b) local circuit neurons, some excitatory and others inhibitory; (c) projection neurons to the brain; (d) propriospinal neurons, which interconnect different levels of the spinal cord; and (e) descending axons from several supraspinal sources.

Lamina I of the dorsal horn is also known as the marginal layer and lamina II as the substantia gelatinosa (of Rolando). These two laminae together represent what is frequently called the superficial dorsal horn and are of particular importance for the spinal processing of pain-related information and its forwarding to higher levels. This is true in particular for lamina I and the outer 2/3 of lamina II, as the inner 1/3 is mostly nonnociceptive. In contrast, laminae III–IV represent what is called the nucleus proprius, which was thought to represent a nonnociceptive region of the spinal cord. While this view is mostly valid, this region does contain neurons that are known to respond to noxious stimuli and project to the brain, which were initially thought to occur only in lamina V (De Koninck, Y. *et al.*, 1992; Ma, W. *et al.*, 1996).

Figure 1 shows the laminar organization of the rat spinal cord at the C4 level, as detected using a Nissl method on thick sections, and the correlation with laminae detected on semithin plastic sections. Although the criteria are quite different, there is a very good match between the laminae when identified by the Nissl approach and the semithin sections approach. When using a Nissl staining, lamina I can be recognized by a rather low cell density (Figure 1(a) and (1b)), with an occasional large neuron among mostly small cells. Lamina II is a layer with a higher density of cells than lamina I; these cells

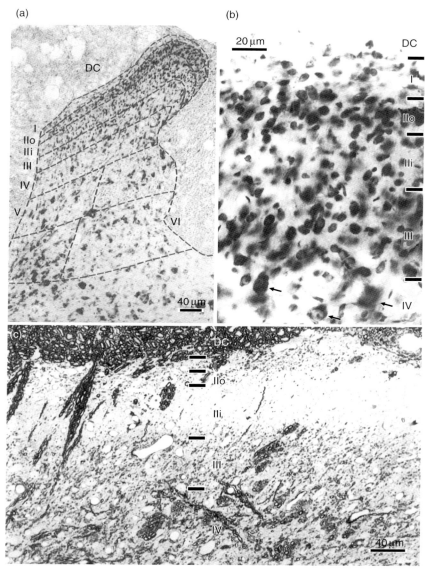

Figure 1 Rexed's laminae at the C4 level of the adult rat spinal cord. (a) and (b) represent micrographs from frozen cross sections, 50 μm in thickness, stained with toluidine blue as described by Rexed B. (1952). In (a), note the overall laminar organization of the dorsal horn. In (b), note that lamina I (I) has a low density of cells, whereas outer lamina II (LIIo) stands out because of the clustering of small neurons, which differentiate it from the inner part of lamina II (IIi). Both LIIo and inner lamina II (LIIi) have numerous small neurons, of rather uniform size. The presence of some slightly larger neurons separates lamina III from LIIi, whereas lamina IV can be easily distinguished from lamina III by the lower cell density and the presence of some large neurons. (c) represents a micrograph from an Epon-embedded semithin (2 μm thick) transverse section of the rat dorsal horn at the C4 level, stained with toluidine blue and Azur II, to illustrate that Rexed's laminae can be identified in these sections, which are used as reference for ultrastructural observations; notably, lamina IIi can be clearly delineated from lamina III by the absence of myelin (see text for details). DC, dorsal columns.

are rather small in size and occur in a considerably higher density in outer lamina II (LIIo) than in inner lamina II (LIIi). It should be noted that in many articles the subdivision of lamina II into outer and inner parts does not follow a cytoarchitectonic scheme as it should and some authors simply consider LIIo as the dorsal half of lamina II. Although in rat we have used lamina IIA to refer to LIIo and lamina IIB to refer to LIIi (see Ribeiro-da-Silva, A., 2004, for rationale), in the current chapter we will use the nomenclature followed in most of the literature. The border between LIIi and lamina III is difficult to distinguish in Nissl-type preparations, as the main features are similar except for the presence of some larger cells in lamina III. Lamina IV, in contrast, is easy to distinguish from lamina III because of the lower cell density and the presence of some rather large neurons (Figure 1(b)).

The superficial laminae can also be recognized in Epon-embedded 1–2-μm-thick semithin sections stained with toluidine blue, as this approach has been validated by integrated light and electron microscopic studies (Ribeiro-da-Silva, A. and Coimbra, A., 1982). In these preparations, lamina I can be distinguished from LIIo by the presence of a higher density of small myelinated fibers, whereas LIIi is virtually devoid of myelin (Figure 1(c)). In contrast, lamina III is easily distinguished from IIi by the appearance of numerous thinly myelinated fibers (Figure 1(c)).

At mid-lumbar levels, which represent the region innervating the hind limbs and which has been studied extensively, the thickness of the superficial laminae is greater in the medial 2/3 than in the lateral third. Overall the thickness of LI and LIIo is twice the value of mid-cervical levels. Figure 2 illustrates the laminar subdivisions at the level of the L5 segment of the young rat spinal cord.

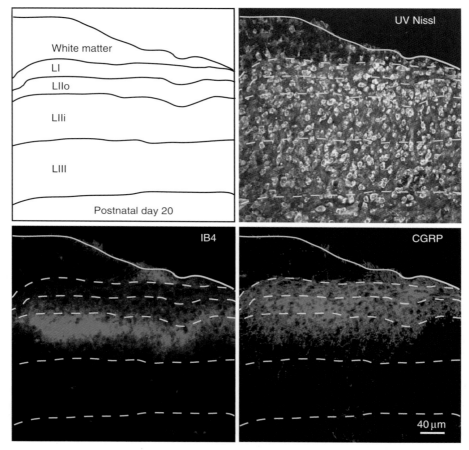

Figure 2 Rexed's laminae at the L5 level of the spinal cord of a young rat (20-day old). Micrographs represent confocal images that result from the fusion of several optical sections obtained using a multitrack approach on a 50-μm-thick transverse section. Section was immunostained for calcitonin gene-related peptide (CGRP) and incubated for IB4 binding and a Nissl-type labeling using an ultraviolet fluorescent marker. The Nissl-type labeling was used to identify the Rexed's laminae and to create an overlay, which was applied to the other images, and is shown isolated in the upper left picture. This figure clearly illustrates that CGRP immunostaining occurs in lamina I (LI), outer lamina II (LIIo), and part of inner lamina II (LIIi), whereas IB4 labeling is most intense in the dorsal part of LIIi. Because the thickness of each lamina is rather constant at the same level of the spinal cord, overlays created like this are useful to identify the laminae even in material not stained with a Nissl method.

23.3 Primary Afferent Fibers

23.3.1 Types of Afferents, Neurochemistry, and Termination in Spinal Cord

The great majority of afferents that transmit pain-related information (nociception) are of small diameter and have unmyelinated (C) or thinly myelinated (Aδ) axons. Most such small-diameter fibers convey pain-related information and, therefore, are often named nociceptors. These small-diameter fibers represent the central processes of pseudounipolar neurons of small or medium size located in the dorsal root ganglia (DRG) and terminate in the superficial laminae of the dorsal horn of the spinal cord (for reviews see Grant, G. and Robertson, B., 2004; Ribeiro-da-Silva, A., 2004). The smaller-diameter (unmyelinated) nociceptive afferents can be divided into two major subpopulations, the peptidergic and the nonpeptidergic. The peptidergic expresses substance P (SP) and calcitonin gene-related peptide (CGRP), and the nonpeptidergic possesses fluoride-resistant acid phosphatase (FRAP) activity, binds the lectin GSA-IB4, and expresses purinergic P2X$_3$ receptors (Hunt, S. P. and Rossi, J., 1985; Alvarez, F. J. and Fyffe, R. E., 2000; Hunt, S. P. and Mantyh, P. W., 2001). These two populations differ in neurotrophic support in the adult. In fact, during development, both populations require nerve growth factor (NGF) for survival, but shortly after birth only the peptidergic continues to respond to NGF, whereas the nonpeptidergic population starts to respond instead to glial cell line-derived neurotrophic factor (GDNF) (Bennett, D. L. H. et al., 1998). Accordingly, the peptidergic population expresses the NGF high-affinity receptor, trkA, whereas the nonpeptidergic expresses GDNF receptors. It has also been shown that the latter population expresses the purinergic receptor P2X$_3$ (Bradbury, E. J. et al., 1998; Snider, W. D. and McMahon, S. B., 1998). Although the distinction between two populations of primary sensory fibers, peptidergic and nonpeptidergic, seems attractive, it is not fully accurate as a small proportion of peptidergic sensory fibers (those that colocalize CGRP and somatostatin (SOM)) do not respond to NGF in the adult and bind the lectin GSA-IB4 (Alvarez, F. J. and Fyffe, R. E., 2000; Priestley, J. V. et al., 2002). It should be noted that, in all of the above putative nociceptive fibers, the classical synaptic transmitter is very likely glutamate (Battaglia, G. and Rustioni, A., 1988; De Biasi, S. and Rustioni, A., 1988;

Merighi, A. et al., 1991). In the spinal cord, sensory fibers that express SP and CGRP terminate mostly in laminae I, outer II, and lamina V; those that colocalize CGRP and SOM terminate in laminae I and II, and those that contain FRAP, bind the lectin GSA-IB4 or express the purinergic P2X$_3$ receptor, terminate mostly in the middle third of lamina II (see Figure 2; for reviews see Bradbury, E. J. et al., 1998; Ribeiro-da-Silva, A., 2004). Not all the nociceptive afferents that express SP are C fibers; some are thinly myelinated (Aδ) afferents. There is a significant population of nociceptive fibers that are Aδ and were shown to represent high-threshold mechanoreceptors (HTM) after being characterized electrophysiologically and filled intracellularly with horseradish peroxidase in cat and monkey; they were shown to terminate in laminae I and V (Light, A. R. and Perl, E. R., 1979). The morphological and neurochemical identification of these Aδ HTM fibers is difficult because they do not contain peptides, do not bind IB4, and do not express P2X$_3$ receptors. However, there is evidence from an *in situ* hybridization study that the 5-HT3 may be highly expressed by Aδ HTM fibers (Zeitz, K. P. et al., 2002). Although the prevailing view is that Aβ fibers are nonnociceptive, it has been suggested that some of the myelinated nociceptors conduct in the Aβ range. In rat, these Aβ nociceptors might represent up to 20% of all Aβ somatic afferents according to some reports (for review see Djouhri, L and Lawson, S. N., 2004).

Although a small number of C fibers are nonnociceptive and convey either innocuous mechanical or thermal information, most nonnociceptive mechanosensitive afferents are either thinly myelinated (Aδ) or thick afferents (Aβ) (Alvarez, F. J. and Fyffe, R. E., 2000). Most innocuous Aδ afferents innervate down hairs (D hair afferents) and terminate in the deeper part of lamina II and in lamina III, a distribution that contrasts with the preferential termination of nociceptive Aδ afferents in laminae I and V (Light, A. R. and Perl, E. R., 1979). Like the nociceptive fibers, the nonnociceptive afferents are glutamatergic (Battaglia, G. and Rustioni, A., 1988; De Biasi, S. and Rustioni, A., 1988; Merighi, A. et al., 1991). Curiously, the T-type calcium channel Cav3.2 seems to be expressed exclusively by D hair afferents, as detected by in situ hybridization at the level of the DRG (Shin, J. B. et al., 2003).

It should be noted that other neuropeptides have been localized in the fibers that contain SP and CGRP: neurokinin A (Dalsgaard, C. J. et al., 1985), galanin (Ju, G. et al., 1987; Zhang, X. et al., 1993), and

Table 1 Overview of the classical transmitter and peptidergic innervation of the spinal dorsal horn of the rat, as described in text

	Origin of neurotransmitter/neuropeptide		
>*Primary afferents*	*Local circuit neurons*	*Neurons projecting to brain*	*Descending fibers*
Glu + SP[a] + CGRP + End-2 + (Gal[b]) + (CCK[c]) + (VIP[c])	Glu + SP + ENK	Glu + SP + ENK	5-HT + SP + TRH
Glu + SOM + CGRP	Glu + NKB	Glu + dynorphins (?)	5-HT + ENK
Glu + dynorphin B (at sacral levels) + ?	Glu + SOM	Glu + Gal + CCK	GABA
Glu + ?	Endomorphin-1 + ?		GABA + Gly
Glu + (NPY[c])	GABA + ENK		Noradrenaline
	GABA + Ach		
	GABA + NPY		
	Glu + Gal + CCK (?)		
	Glu + dynorphins (?)		
	Glu + neurotensin		

[a]All neurons that express SP colocalize NKA in rat. Neurochemical in parenthesis are not expressed or expressed in a reduced number of afferents in the absence of lesion.
[b]Galanin is strongly upregulated after nerve lesion.
[c]These peptides are expressed only in a significant number of primary afferents following peripheral nerve lesion.
This table does not intend to be a complete list. It is restricted to the peptides mentioned in the text.
Abbreviations: Ach, acetylcholine; CCK, cholecystokinin; CGRP, calcitonin gene-related peptide; End-2, endomorphin 2; ENK, enkephalin; Gal, galanin; Glu, glutamate; NKA, neurokinin A; NKB, neurokinin B; NPY, neuropeptide Y; SP, substance P; SOM, somatostatin; TRH, thyrotropin-releasing hormone; VIP, vasoactive intestinal peptide.

the opioid peptide endomorphin-2 (Martin-Schild, S. *et al.*, 1997; Martin-Schild, S. *et al.*, 1998). Dynorphin B has been detected in some primary sensory fibers from visceral afferents from pelvic viscera at sacral levels in the cat (Basbaum, A. I. *et al.*, 1986). Some neuropeptides are not expressed in any significant amount in primary sensory fibers in naïve rats but are upregulated following peripheral axotomy: vasoactive intestinal peptide (VIP) and cholecystokinin (CCK) appear on small, peptidergic afferents, whereas neuropeptide Y (NPY) appears in large-diameter afferents (for review see Hökfelt, T. *et al.*, 1994). Table 1 gives an overview of classical transmitters and peptides in the dorsal horn, including those in the terminals of primary sensory fibers.

23.3.2 Synaptic Arrangements of Primary Afferents

In the dorsal horn, primary afferent boutons establish mostly simple axodendritic, and to a lesser extent, axosomatic synapses. However, a significant proportion of primary afferent endings participate in complex arrangements named synaptic glomeruli, where they form the central terminal (Coimbra, A. *et al.*, 1974; Ribeiro-da-Silva, A. and Coimbra, A., 1982). Glomeruli represent multiplier systems, i.e., devices that transmit sensory information to several

dorsal horn neurons. They also constitute important modulatory devices as the primary afferent terminals are often postsynaptic to other neuronal profiles. Since the central terminals of glomeruli are all of primary afferent origin, as has been shown by their complete disappearance following multiple dorsal rhizotomies (Coimbra, A. *et al.*, 1984), they provide a very useful means of identifying these afferents with electron microscopy, without the need for transported or immunocytochemical markers. The detailed features of synaptic glomeruli in the rat have been reviewed in a recent publication (Ribeiro-da-Silva, A., 2004). To be classified as a synaptic glomerulus, a synaptic arrangement must meet simultaneously all of following criteria when viewed in a single ultrathin section: (a) it should possess a central terminal (C) with round synaptic vesicles; (b) this C terminal has to be surrounded by at least four neuronal profiles, representing either dendritic profiles or axonal bouton profiles; (c) there have to be at least two visible synaptic contacts between the central bouton profiles and the peripheral profiles. Two main types of synaptic glomeruli have been identified in rat by Ribeiro-da-Silva A. and Coimbra A. (1982), and are named type I and type II. Glomeruli of type I have a central bouton of scalloped contour, with an electrondense matrix, with rather densely packed synaptic vesicles and a low density of mitochondria, whereas type II glomeruli are larger and possess a central terminal, which has a rounder contour, a lighter cytoplasmatic matrix, synaptic vesicles, which are more uniform in size and less densely packed, and a higher density of mitochondria (Ribeiro-da-Silva, A. and Coimbra, A., 1982). This classification is applicable to the mouse and, with some modifications, to the cat and the monkey (unpublished observations). A comparable classification was established for the monkey dorsal horn (Knyihár-Csillik, E. *et al.*, 1982b), in which three glomerular types were described, based on the properties of their central terminals. Of the three types described in the monkey by Knyihár-Csillik E. *et al.* (1982b), the dense sinusoid axon (DSA) type corresponds to the type I of Ribeiro-da-Silva A. and Coimbra A. (1982) and the regular synaptic vesicles (RSV) type corresponds to type II. Although the large dense-core vesicle (LDCV) type of Knyihár-Csillik E. *et al.* (1982b) has apparently no equivalent in the classification of Ribeiro-da-Silva A. and Coimbra A. (1982), this type has been considered in rat as a subtype of type I (subtype Ib), as in this species most of their central terminals share the dark axoplasm, the sinuous contour, and small size with type I central boutons, although they display numerous dense-core vesicles (Ribeiro-da-Silva, A. *et al.*, 1989; Ribeiro-da-Silva, A., 2004). However, the incorporation of these glomeruli with dense-core vesicles in type I is not legitimate in cat and monkey, as in these species they have larger central boutons, with more mitochondria and a lighter matrix than in rat. Therefore, for the sake of simplicity, we propose that in cat and primates, glomeruli can be classified into type I, type II, and peptide-type (because all the central boutons of this later type express neuropeptides – see below). In rat both type Ia and type II glomeruli have complex synaptic arrangements in which the central bouton is postsynaptic to presynaptic dendrites and axonal boutons and presynaptic to both normal dendrites and presynaptic dendrites (see Figure 3 for details). One main feature of peptide-type synaptic glomeruli (type Ib in rat) is that their central terminal is seldom postsynaptic to presynaptic dendrites and peripheral axons.

Of the nociceptive afferents, the nonpeptidergic, IB4-binding afferents terminate mostly as central boutons of synaptic glomeruli of type I (or more precisely in rat of subtype Ia), whereas the peptidergic endings are glomerular in only a small proportion, particularly in rat (Ribeiro-da-Silva, A. *et al.*, 1989); when they are glomerular, the central boutons (subtype Ib) contain dense-core vesicles and peptide immunoreactivity (CGRP, together with either SP or SOM; Ribeiro-da-

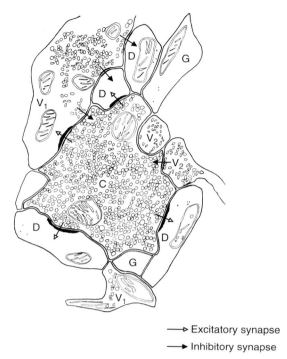

→ Excitatory synapse

→ Inhibitory synapse

Figure 3 Diagram of a synaptic glomerulus of type Ia, with representation of the complex synaptic arrangements of the triadic type. The drawing was inspired by an electron micrograph of a type Ia glomerulus and shows the morphological features and the synaptic circuits involving the central bouton, of primary sensory origin, and the peripheral profiles. C, Central bouton; D, regular dendrite; V₁, presynaptic dendrite; V₂, peripheral axonal bouton; G, glial profile.

Silva, A., 1995). Concerning nonnociceptive boutons, many terminate as central boutons of type II synaptic glomeruli (e.g., the D-hair afferents), but many are nonglomerular, including the ones corresponding to the termination of G-hair afferents, based on the data obtained in the cat (for review see Maxwell, D. J. and Réthelyi, M., 1987). Figure 4 shows a diagrammatic representation of the different types and subtypes of synaptic glomeruli in rat. There are considerable species differences in the type of synaptic arrangements of primary sensory fibers. For instance, lamina I and LIIo are virtually devoid of synaptic glomeruli in rat (Ribeiro-da-Silva, A. and Coimbra, A., 1982), but not in cat or monkey; in this latter species, there are, in laminae I and IIo, numerous type II glomeruli corresponding to the termination of the HTM myelinated nociceptors (Réthelyi, M. *et al.*, 1982), which are nonglomerular in rat, as well as a considerable number of peptide-type glomeruli (Knyihár-Csillik, E. *et al.*, 1982b), which are sparse in rat.

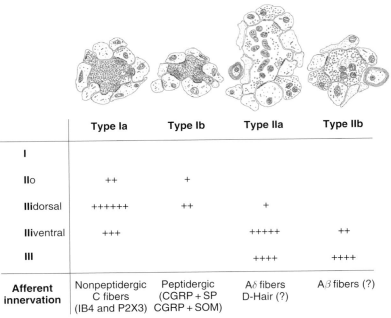

	Type Ia	Type Ib	Type IIa	Type IIb
I				
IIo	++	+		
IIdorsal	++++++	++	+	
IIventral	+++		+++++	++
III			++++	++++
Afferent innervation	Nonpeptidergic C fibers (IB4 and P2X3)	Peptidergic (CGRP + SP CGRP + SOM)	Aδ fibers D-Hair (?)	Aβ fibers (?)

Figure 4 Types of synaptic glomeruli and their laminar incidence in laminae I–III of the rat dorsal horn. The drawings on top were inspired by actual electron micrographs of representative samples of each glomerular type. Note that in rat, lamina I (I) is virtually devoid of glomeruli and that outer lamina II (LIIo) has only a low number of them. This is a considerable difference when comparing to the cat and primate, which have glomeruli in significant numbers in lamina I and outer II. In contrast, the other laminae have abundant synaptic glomeruli. Most type Ia glomeruli are condensed in a rather narrow band (~20 μm thick at cervical levels), which represents the dorsal part of inner lamina II (IIi$_{dorsal}$). Most of the inner lamina II (IIi$_{ventral}$) has a predominance of type IIa glomeruli, whereas lamina III is normally devoid of type I glomeruli and has similar numbers of both subtypes of type II glomeruli. The data concerning laminar incidence of glomeruli are semiquantitative and are based on the data collected from other publications (Ribeiro-da-Silva, A. and Coimbra, A., 1982; Ribeiro-da-Silva, A. et al., 1989). The information on the types of afferent fibers, which terminate as central boutons of each type of glomerulus, is either based on evidence obtained in rat (for type I glomeruli) or extrapolated from the data obtained in cat and primate. See text for details on glomerular classification.

It is important to stress that of the populations of nociceptive primary sensory fibers, the peptidergic is not postsynaptic to GABAergic neurons in the species studied (Alvarez, F. J. et al., 1993; Ribeiro-da-Silva, A., 2004). In contrast, in rat, the IB4-binding population (which terminates as central boutons of glomeruli of type Ia) is postsynaptic to GABAergic neurons (Todd, A. J. and Lochhead, V., 1990; Ribeiro-da-Silva, A., 2004). In cat and primates, the HTM myelinated nociceptors that terminate in synaptic glomeruli are also postsynaptic to dorsal horn neurons at synapses, which have the morphological properties of inhibitory junctions or in which the profiles have been shown to express GABA (Réthelyi, M. et al., 1982; Alvarez, F. J. et al., 1992). There seem to be species differences though, as a primate study has shown that the terminals that have the morphology of the IB4 nonpeptidergic population of the rat did not show them to be involved in axo- or dendroaxonic synapses (Knyihár-Csillik, E.

et al., 1982a). Similarly, in rat the HTM myelinated nociceptors are not glomerular and not postsynaptic to other structures. The above indicates different presynaptic control of the different type of nociceptive inputs, as well as likely species differences in such presynaptic control.

23.3.3 Receptors on Primary Sensory Fibers

Receptors expressed on primary afferent terminals in the spinal dorsal horn include neurotransmitter/modulator receptors and neurotrophic factor receptors. It is important to point out that most of the past studies of the neurotransmitter/modulator receptors on primary afferents have been performed in the dorsal root ganglia. Information on localization of receptors on the terminals of primary sensory fibers in the spinal cord is recent and still rather limited as it required the development of specific antibodies against the neurotransmitter/modulator receptors.

Concerning glutamate receptors, α-amino-3-hydroxy-5-methylisoxazole-4-propionic acid (AMPA) receptors have a widespread distribution in DRG neurons (Tachibana, M. *et al.*, 1994; Lu, C. R. *et al.*, 2002). It has been shown at the light and ultrastructural levels that N-methyl-D-aspartic acid (NMDA) receptors occur in sensory fibers in the dorsal horn (Liu, H. *et al.*, 1994b). A presynaptic localization of sensory fibers in the dorsal horn has also been shown for kainate (Hwang, S. J. *et al.*, 2001) and metabotropic glutamate receptors (Ohishi, H. *et al.*, 1995). It is important to note that while ionotropic glutamate receptors are typically viewed as excitatory, their activation on primary afferent terminals leads to inhibition of transmitter release, likely via primary afferent terminal depolarization (PAD) akin to the action of presynaptic GABA$_A$ receptors (Kerchner, G. A. *et al.*, 2001; Lee, C. J. *et al.*, 2002; Bardoni, R. *et al.*, 2004). This contrasts with the apparent excitatory role that glutamate receptors appear to play at the peripheral endings of primary afferents (Cairns, B. E. *et al.*, 1998; Svensson, P. *et al.*, 2003). To further complicate the matter, PAD appears to have an opposite effect on spontaneous transmitter release (facilitation) versus evoked release (inhibition) (for review see Engelman, H. S. and MacDermott, A. B., 2004). Subunits of the GABA$_A$ receptor have also been identified by *in situ* hybridization in DRG neurons (Persohn, E. *et al.*, 1991), consistent with the functional evidence of a role for these receptors in presynaptic inhibition of primary afferents (for review see Rudomin, P. and Schmidt R. F., 1999). Intriguingly, however, to date it has not yet been possible to immunocytochemically label GABA$_A$ receptor subunits at axoaxonic synapses in the dorsal horn, although they have been labeled in dorsal horn neurons (Alvarez, F. J. *et al.*, 1996; Todd, A. J. *et al.*, 1996). There also appears to be a difference in the strength of GABA$_A$ receptor-mediated responses between subtypes of small-diameter afferents (Labrakakis, C. *et al.*, 2003). Subunits of the metabotropic GABA$_B$ receptor have been identified in DRG neurons and, in the dorsal horn, in dendrites and central boutons of glomeruli (Poorkhalkali, N. *et al.*, 2000; Ribeiro-da-Silva and Shigemoto, unpublished observations).

Concerning opioid receptors, it is well established that they occur in primary sensory fibers. The evidence originates from studies using radioactive ligand binding, which show that both μ- and δ-opioid receptors in the dorsal horn are reduced following dorsal rhizotomy (Fields, H. L. *et al.*, 1980), and from

in situ hybridization studies, which have shown that mRNA for each of the main opioid receptors (μ, δ, and κ) occurs in DRG neurons (Minami, M. *et al.*, 1995).

There is both immunocytochemical and *in situ* hybridization evidence that the SP receptor, the neurokinin 1 receptor (NK-1r), occurs in DRG cells and on sensory fibers in the periphery (Ruocco, I. *et al.*, 1997; Li, H. S. and Zhao, Z. Q., 1998). Surprisingly, the NK-1r cannot be detected by immunocytochemistry on terminals of primary sensory fibers in the spinal dorsal horn, an observation that suggests that the NK-1r synthesized in DRG neurons is targeted preferentially toward the periphery. In contrast, the NK-1r occurs in well-defined populations of dorsal horn neurons, as described below.

The capsaicin receptor, TRPV1, previously named VR1 (Caterina, M. J., 2003), occurs in small diameter primary sensory afferents. Unfortunately, there is conflicting evidence in the literature regarding which afferent populations express the receptor. The *in situ* hybridization data of Michael G. J. and Priestley J. V. (1999) concluded that it is expressed by the majority of small- and medium-sized dorsal root ganglion neurons, which label with the lectin IB4 or are CGRP-immunoreactive (IR). In contrast, using immunocytochemistry Guo A. *et al.* (1999) concluded that TRPV1 occurred mainly in the nonpeptidergic, IB4-binding afferents, and to a minor extent in the peptidergic as well. An important observation is that there was a phenotypic switch in the postnatal period leading to the progressive expression of TRPV1 in DRG neurons that bind the lectin IB4, whereas the population that colocalized peptides did not change (Guo, A. *et al.*, 2001). Furthermore, a very significant species difference does occur because in mouse TRPV1 is not expressed by the IB4-binding population of afferents but rather by the peptidergic population (Zwick, M. *et al.*, 2002), suggesting that the phenotypic switch, which occurs postnatally in the rat, does not take place in the mouse. This is a particularly important point as it suggests that the results obtained with TRPV1 knockout mice are not transposable to the rat.

The purinergic receptor P2X$_3$ has been detected in the nonpeptidergic population of nociceptive afferents (Bradbury, E. J. *et al.*, 1998) and, in agreement with this, in the central boutons of glomeruli of type I (Llewellyn-Smith, I. J. and Burnstock, G., 1998). Because of the variability with the conditions of incubation in a number of afferents that are labeled by IB4 binding (see Alvarez, F. J. and Fyffe, R. E.,

2000, for a discussion), P2X$_3$ immunostaining may represent a better way of identifying the nonpeptidergic subpopulation of C fibers.

Concerning cholinergic receptors on primary sensory fibers, both nicotinic and muscarinic receptors have been identified (Flores, C. M. *et al.*, 1996; Haberberger, R. *et al.*, 1999). Concerning adrenergic receptors, both $\alpha 1$ and $\alpha 2$ receptors have also been localized in the DRG neurons (Kinnman, E. *et al.*, 1997; Birder, L. A. and Perl, E. R., 1999; Xie, J. *et al.*, 2001), and there is also some evidence that they occur in nociceptive primary afferents in the spinal cord (Stone, L. S. *et al.*, 1998). Because α-adrenergic receptors in the DRG are upregulated after peripheral nerve lesion, these receptors may be implicated in the sensitization of primary afferents in sympathetically maintained pain.

There is abundant literature on the expression of trophic factor receptors by primary sensory fibers (for review see Priestley, J. V. *et al.*, 2002). All three high-affinity neurotrophin receptors, trkA, trkB. and trkC, have been detected in DRG neurons. Of these, trkB and trkC occur in large-diameter neurons, corresponding to low-threshold mechanoreceptors, whereas trkA is found in nociceptive neurons that express the neuropeptides CGRP and SP. The low-affinity neurotrophin receptor, p75, has been identified in the populations of DRG neurons that express one of the high-affinity neurotrophin receptors. The nonpeptidergic, IB4-binding population, including its subpopulation that colocalizes CGRP and SOM, expresses in the adult components of the GDNF receptor family, but not in neurotrophin receptors (Alvarez, F. J. and Fyffe, R. E., 2000).

23.3.4 Expression of Voltage-Gated Sodium Channels by Sensory Fibers

In recent years, there has been a developing interest on voltage-gated sodium channels expressed by primary sensory fibers, as changes in some have been linked to pain states (for review see Wood, J. N. *et al.*, 2004). Channels with α subunits coded by the Na$_v$1.8 and Na$_v$1.9 genes are sensory neuron-specific and commonly referred to as tetrodotoxin (TTX)-resistant because higher doses of TTX are required to block them (Amaya, F. *et al.*, 2000). Although the two channels are mostly colocalized in the same DRG neurons in the rat (Amaya, F. *et al.*, 2000), the Na$_v$1.8 channel also occurs in some larger-diameter afferents, as was demonstrated in a study in which the physiological properties of the neurons expressing it was studied in

the DRG (Djouhri, L. *et al.*, 2003) and by immunocytochemistry, which shows that Na$_v$1.9 channel immunoreactivity is restricted to laminae I–II whereas some Na$_v$1.8 signal was detected in the deeper laminae as well (Amaya, F. *et al.*, 2000). Regarding small-diameter afferents that express these channels, both channels seem to be expressed equally by peptidergic and nonpeptidergic (IB4-positive) afferents, although such expression was present only in about 50% of trkA-positive and 50% IB4-positive neurons (Amaya, F. *et al.*, 2000). However, a recent study in the mouse indicated that, in this species, the Na$_v$1.8 channel is expressed by 80% of the IB4-binding population of sensory fibers and only by 20% of the CGRP-IR (Braz, J. M. *et al.*, 2005), which would suggest that there are significant species differences in the relative expression of these channels. Of the TTX-sensitive channels, the Na$_v$1.7 channel is also expressed selectively in nociceptive, small-diameter afferents, although it also occurs in sympathetic neurons, and seems to play a major role in acute and inflammatory pain (Nassar, M. A. *et al.*, 2004). Interestingly, mutations that cause loss of function of Na$_v$1.7 were recently found in three consanguineous families from northern Pakistan characterized by complete inability to sense pain (Cox, J. J. *et al.*, 2006).

23.4 Dorsal Horn Neurons

The great majority of neurons in each lamina of the dorsal horn are local circuit neurons (interneurons), with axons that remain in the spinal cord. It should be noted that laminae I and V also possess predominantly interneurons, despite the fact that they are normally regarded as the origin of major ascending nociceptive pathways to the brain. These interneurons participate in synaptic circuits, which unfortunately are still very incompletely known. To complicate matters, there is evidence of considerable interspecies differences. These differences affect the distribution and synaptic connections of primary sensory fibers as well as the neurotransmitter/neuropeptide colocalization and the ascending pathways. It certainly does not help that most immunocytochemical studies on physiologically characterized, intracellularly filled neurons were performed in the cat, whereas most of those based on a combination of tract tracing and immunocytochemistry are from the rat and the monkey, and studies on transgenic animals were performed in the mouse.

The primary sensory fibers terminate in the dorsal horn in contact with both interneurons and projection neurons. Unfortunately, our knowledge concerning the details of the termination of specific fiber populations on identified dorsal horn neurons is extremely limited, because a systematic study combining intracellular recordings and labeling of sensory fibers with labeling of dorsal horn neurons has never been done for the nociceptive afferents. This lack of information from intracellular techniques is not a major issue for the SP afferents that innervate neurons that express the NK-1r because the available antibodies label the NK-1r-IR dorsal horn neurons extremely well, in a Golgi-like manner, which allows their morphological characterization. Because of this, we have rather detailed information on NK-1r-positive neurons innervated directly by SP, including their supraspinal projection sites (for recent review see Todd, A. J. and Ribeiro-da-Silva, A., 2005). Unfortunately, the same is not true for the termination of the nonpeptidergic population of nociceptive fibers in the dorsal horn, despite the fact that their endings can be identified directly on electron micrographs (EM) without the need of any label, because of their unique morphology as type I glomeruli central boutons (Ribeiro-da-Silva, A., 2004). In fact, it is still unknown if the central boutons of type I glomeruli synapse directly on the projection neurons or if there is an interposed interneuron. Recent evidence demonstrates a pathway for nonpeptidergic afferents connecting to lamina I neurons via lamina IIi interneurons (Lu, Y. and Perl, E. R., 2005) (given that the great majority of the C fibers that innervate lamina IIi are from the nonpeptidergic subpopulation; see also below). A challenging alternative has been proposed recently in a study using a transgenic mouse endogenously expressing a tracer transported transsynaptically in the nonpeptidergic afferents that express the $Na_V1.8$ channel. This study has shown that these afferents are connected to lamina II interneurons that link primarily to lamina V projection neurons that, in turn, target fourth-order neurons in the amygdala, hypothalamus, bed nucleus of the stria terminalis, and globus pallidus (Braz, J. M. et al., 2005). The $Na_V1.8$ subclass of nonpeptidergic, IB4-binding afferents demonstrated in mice would thus likely be distinct from the putative subclass that would connect to lamina I via lamina IIi based on the studies by Lu Y. and Perl E. R. (2003; 2005) in the rat. It remains to be determined whether this is a species difference and how these pathways transpose to other species.

23.4.1 Lamina I Neurons

23.4.1.1 Morphological classification of lamina I neurons

Four morphological types of neurons have been described in the rat by Lima D. and Coimbra A. (1983; 1986): fusiform, flattened, multipolar, and pyramidal (Figure 5). Most of these neurons have their perikarya and dendritic arborization extending predominantly in the rostrocaudal axis, but with some processes extending along the mediolateral axis. For most cells, the dendrites remain within the limits of lamina I. However, some specific cell subtypes have been described with dendrites extending into laminae II and III (Figure 6); these represent a small proportion of the total number of cells that may represent local circuit neurons (Yu, X. H. et al., 2005), although the possibility remains that they project to another supraspinal site different from the thalamus and the parabrachial nucleus (Lima, D. and Coimbra, A., 1990; Almarestani, L. et al., 2007). Antidromic recording and retrograde labeling with horseradish peroxidase or cholera toxin subunit b have demonstrated that lamina I neurons project to several supraspinal areas including the thalamus, the periaqueductal gray, and the parabrachial nucleus (for reviews see Lima, D. and Coimbra, A., 1988; Spike, R. C. et al., 2003; Yu, X. H. et al., 2005). A comparable classification of lamina I projection neurons has been carried out in the cat (Zhang, E. T. et al., 1996) and the monkey (Zhang, E. T. and Craig, A. D., 1997; Yu, X. H. et al., 1999) based on the analysis of horizontal sections (Figure 5). In this latter classification, the multipolar and flattened types were merged into one category (multipolar neurons). Because of the correlation with classifications done in other species and correlation with function (see below), we have found it more appropriate to classify lamina I neurons in rat into the general types of fusiform, multipolar, and pyramidal (Yu, X. H. et al., 1999; 2005; Almarestani, L. et al., 2007) (Figure 6). While to a large extent these cell types can be recognized in horizontal sections (Figure 5), complete classification of these cells require a 3D reconstruction of their dendritic tree (Figures 5 and 6) (Yu, X. H. et al., 1999; 2005).

23.4.1.2 Correspondence with function

Recent evidence suggests a correspondence between the morphological characteristics of lamina I neurons and their functional properties (Han, Z. S. et al., 1998). Three main functional types of lamina I neurons

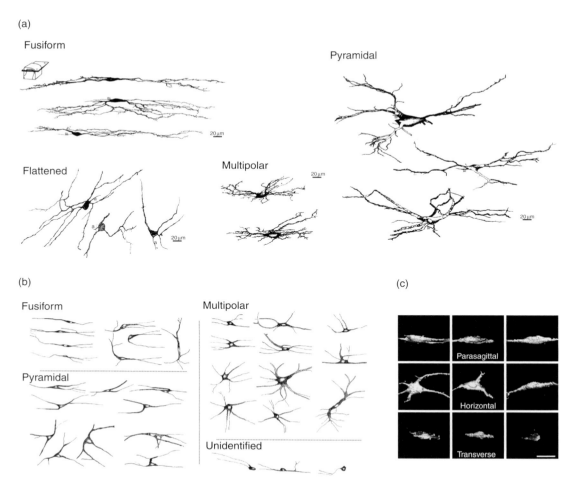

Figure 5 Representative drawings of lamina I neurons in the rat and the monkey spinal dorsal horn. The drawings in (a) are based on camera lucida reconstructions from Golgi-impregnated cells viewed in the horizontal plane (modified from Lima, D. and Coimbra, A., 1986). (b) The basic scheme was extended to higher species, such as the monkey, with minor nomenclature adjustments (modified from Zhang, E. T. and Craig, A. D., 1997; see text). (c) Three-dimensional reconstructions from serial confocal images of retrogradely labeled lamina I spinothalamic neurons in rats (from Yu, X. H. *et al.*, 2005). Note that lamina I projection neurons in rat (Yu, X. H. *et al.*, 2005), cat (Zhang, E. T. *et al.*, 1996; Han, Z. S. *et al.*, 1998), and monkey (Zhang, E. T. and Craig, A. D., 1997) do not appear to have ventrally directed dendrites (see also text) and are thus properly identified only in the horizontal plane as shown in (c) (modified from Yu, X. H. *et al.*, 2005).

have been identified on the basis of their response to cutaneous sensory inputs: nociceptive-specific (NS) neurons, responsive only to noxious heat and pinch; polymodal nociceptive neurons, responsive to noxious heat and pinch as well as to innocuous and noxious cold (HPC); and innocuous thermoreceptive neurons responsive only to innocuous cooling (COLD) (Craig, A. D. and Kniffki, A. I., 1985; Craig, A. D. and Bushnell, M. C., 1994; Dostrovsky, J. O. and Craig, A. D., 1996; Han, Z. S. *et al.*, 1998). A correspondence between these functional classes of lamina I and the morphological types (fusiform, pyramidal, and multipolar) has been found in the cat whereby

fusiform cells were all NS, pyramidal cells were all COLD, and multipolar cells were divided between HPC and NS (Han, Z. S. *et al.*, 1998). These results are also consistent with previous reports that many pyramidal and multipolar cells possess large axons in the cat (Gobel, S., 1978a) and the rat (Lima, D. and Coimbra, A., 1986). In contrast, fusiform cells are characterized by thin, likely unmyelinated axons (Lima, D. and Coimbra, A., 1986), consistent with the observation that HPC and COLD cells have faster antidromic conduction velocities (Craig, A. D. and Kniffki, K. D., 1985) than NS neurons (Craig, A. D. and Kniffki, K. D., 1985). Recent evidence using

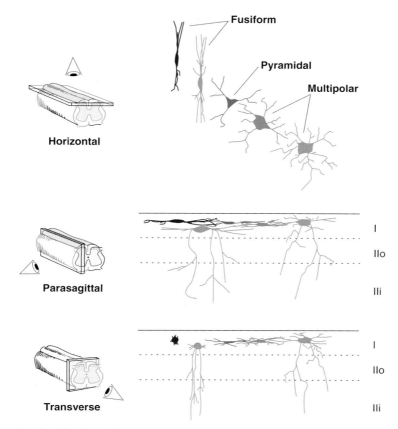

Figure 6 Morphological classification of the neuronal populations in lamina I. The classification is mainly based on cell body shape and primary dendrite orientation. Adaptation of the original classification by Lima D. and Coimbra A. (1986) to take into account recent findings on projection neurons (see Figure 5 and text) and interneurons (see text). Most cells types (especially the projection neurons) are best identified in the horizontal plane. Sagittal and transverse views may be useful to test for ventrally directed dendrites.

whole cell recording in slices indicates the presence of distinct classes of lamina I neurons on the basis of their intrinsic firing properties (Prescott, S. A. and De Koninck, Y., 2002; Ruscheweyh, R. *et al.*, 2004). In addition, a correspondence between intrinsic electrophysiological properties and morphology in lamina I has been shown in rat (Prescott, S. A. and De Koninck, Y., 2002) whereby tonic cells were typically fusiform, phasic cells were pyramidal, and delayed onset and single spike cells were multipolar. In the latter study, the multipolar cell subtype probably correspond to the small multipolar subtype with ventrally directed dendrites and therefore are likely not projection neurons.

23.4.1.3 Expression of NK-1r in subpopulations of lamina I neurons

The finding that pyramidal neurons corresponded to a class of lamina I neurons that only respond to innocuous thermal stimuli was of particular interest because of the evidence that nonnociceptive neurons fail to respond to SP (Henry, J. L., 1976) and do not receive significant amount of SP input (De Koninck, Y. *et al.*, 1992; Ma, W. *et al.*, 1996). We therefore decided to test whether pyramidal neurons express NK-1r (Yu, X. H. *et al.*, 1999). The results indicated that, while normally one-third of lamina I neurons are pyramidal in the cervical and lumbar enlargements of the spinal cord (Zhang, E. T. *et al.*, 1996; Zhang, E. T. and Craig, A. D., 1997; Yu, X. H. *et al.*, 1999), only ∼5% of the NK-1r-expressing neurons in these areas were pyramidal, most being either fusiform or multipolar. This type of differential distribution gives further support to the concept that there is a correlation between the functional properties of neurons and their morphological features. In a recent study (Yu, X. H. *et al.*, 2005), we obtained data in the rat consistent with that in the monkey whereby retrogradely labeled lamina I spinothalamic tract (STT) cells were evenly distributed among the three main cell types (Zhang, E. T. and Craig, A. D., 1997; Yu, X. H. *et al.*, 1999), and NK-1r immunoreactivity was also preferentially expressed in fusiform and multipolar lamina I STT neurons.

A high proportion (60–80%), but not all, of lamina I projection neurons are IR for NK-1r (Marshall, G. E. *et al.*, 1996; Yu, X. H. *et al.*, 1999; 2005). In contrast, however, only a subset of NK-1r-IR neurons are projection neurons (49% after the injection of tracer in the thalamus; Yu, X. H. *et al.*, 2005). A large proportion of lamina I neurons project to the parabrachial nucleus. Importantly, although it had been reported that very few lamina I neurons in rat project to the thalamus at lumbar levels (Spike, R. C. *et al.*, 2003), we found that a significant number, about 9%, of lamina I neurons project to the thalamus (Yu, X. H. *et al.*, 2005), a figure comparable to that found in primate by us (Yu, X. H. *et al.*, 1999) and others (Andrew, D. *et al.*, 2003). Our recent data on spinoparabrachial neurons also indicate that most pyramidal cells in this subpopulation do not express NK-1r (Almarestani, L. *et al.*, 2007).

23.4.2   Lamina II Neurons

Lamina II neurons have been classified originally by Ramón y Cajal (1909) into two main types, the central cell, which occurs throughout the lamina, and the limiting cell, which is restricted to a region close to the laminae I–II border. These same morphological types were identified in the cat by Gobel S. (1975; 1978b), and were renamed islet cells and stalked cells, respectively. A Golgi method study by Todd A. J. and Lewis S. G. (1986) identified cells with the characteristics of stalked and islet cells, as described by Gobel, in lamina II of the rat (Figure 7). Concerning incidence of these cells, the stalked cells were shown to represent about half of the stained cells in LIIo; in contrast, islet cells were found throughout lamina II and corresponded to about one-third of the entire stained neuronal population (Todd, A. J. and Lewis, S. G., 1986). It should be noted that these authors reported other cell types in lamina II; in fact, about half of the cells in the ventralmost part of lamina II did not fit in either the stalked or islet cell categories, although they could be subdivided into groups based on their dendritic arborization. The axons of these cells either crossed to lamina III or remained within the limits of lamina II. Some of these neurons may correspond to the stellate and laminae II–III border

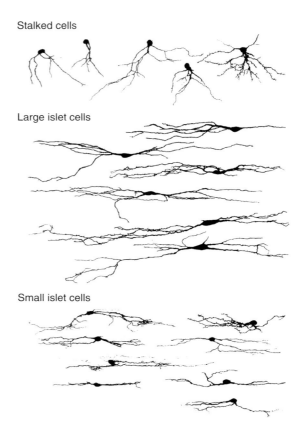

Stalked cells

Large islet cells

Small islet cells

Figure 7   Representative drawings of lamina II neurons in the rat spinal dorsal horn. Three main cell types are represented in these camera lucida drawings of Golgi-impregnated cells: the stalked, small islet, and large islet cells (see text for further details). Modified from Todd A. J. and Spike R. C. (1993).

cells described in other species (Todd, A. J. and Lewis, S. G., 1986).

Recently, Grudt T. J. and Perl E. R. (2002) performed a detailed morphological analysis of the different cells types in lamina II following intracellular labeling in spinal cord slices of the hamster. They identified five morphological types of lamina II neurons: islet, central, medial–lateral, radial, and vertical (Figure 8). Most of lamina II neurons had dendritic arbors that were primarily orientated in the rostrocaudal dimension and relatively flattened in the mediolateral and dorsoventral directions (consistent with Scheibel, M. E. and Scheibel, A. B., 1968). The islet cells of Grudt T. J. and Perl E. R. (2002) appear to correspond well with the islet cells of Gobel S. (1975; 1978b) and the large islet cells of Todd A. J. and Lewis S. G. (1986) (Figures 7 and 8); the radial cells resembled the star-shape neurons of Bicknell H. R. and Beal J. A. (1984); or the stellate

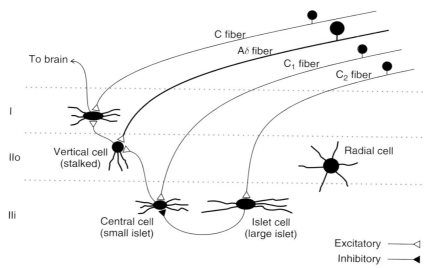

Figure 8 Diagrammatic representation of morphological types of neuron in lamina II and their primary afferent input and interconnections. A correspondence between the classification by Todd A. J. and Spike R. C. (1993) (words in parenthesis) and that of Grudt T. J. and Perl E. R. (2002) and Lu Y. and Perl E. R. (2003; 2005) is attempted as described in the text. Scheme for interconnection is based on Lu Y. and Perl E. R. (2003; 2005) (see text for a complete description). Note that distinct classes of C fibers appear to contact the two classes of islet cells based on the conduction velocity data. Note also the feed-forward inhibition between the islet cell and the central cell, which is unidirectional (Lu, Y. and Perl, E. R., 2003).

neurons described by Schoenen J. (1982). The central cells were similar to islet cells, but with a much smaller dendritic expansion in the rostrocaudal direction, thus appearing to correspond to Ramon y Cajal's central cells and Todd's small islet cells (Figures 7 and 8). The medial–lateral cells were the sole type with dendrites extending substantially in the mediolateral axis. The vertical cells appear to correspond to the limitroph neurons of Ramon y Cajal and stalked cells of Gobel.

Using paired recordings in slices and stimulation of primary afferents in rat spinal cord slices, Lu Y. and Perl E. R. (2003; 2005) were able to describe a specific connection path between subclasses of C-fiber afferents and the network of large islet cells, small islet cells (central), and stalked cells (vertical), as summarized in Figure 8. Notably, their results show that distinct subclasses of C fibers connect to different laminae II neurons subtypes and that stalked cells are indeed feed-forward excitatory cells to lamina I neurons, confirming what had been postulated in previous reports (Bennett, G. J. et al., 1980; Gobel, S. et al., 1980).

Concerning islet cells, it has been found that their electrophysiological properties differed depending on their localization in lamina II, as those situated in deep lamina II were nonnociceptive, while those in LIIo were nociceptive (Bennett, G. J. et al., 1980;

Gobel, S. et al., 1980). These findings concur with previous studies in the cat by Light A. R. et al. (1979). It interesting to note that in the cat, according to Light A. R. and collaborators (1979), the type of response elicited from lamina II cells, either nociceptive or nonnociceptive, had little correlation with their morphology but was related to where the dendritic tree arborized. This observation was interpreted as a consequence of the type of primary sensory fibers, which contacted the dendrites of the cells, as the afferents that terminated in deep lamina II did not seem to be nociceptive in the cat (Light, A. R. and Perl, E. R., 1979).

23.4.3 Laminae III–VI

Laminae III–VI possess both interneurons and projection neurons and have a rather heterogeneous population of cells with dendrites that cut across these laminae (Brown, A. G., 1982). Cells of some of the major ascending pathways (spinothalamic, spinocervical, and dorsal column postsynaptic pathway) have cell bodies located in these laminae (Brown, A. G., 1982). Lamina III can be considered as a continuation of the ventralmost part of lamina II, with some small neurons with characteristics similar to those of that part of lamina II, as described in one study (Wall, P. D. et al., 1979). In cat, cells in lamina

III have been described as nonnociceptive and heterogeneous (Maxwell, D. J. *et al.*, 1983) based on intracellular injections of physiologically characterized neurons. However, this concept that all lamina III cells are nonnociceptive is not correct based on the detection of wide dynamic range cells in this laminae III–IV of the cat, which possess dorsally oriented dendrites branching in the superficial laminae (De Koninck, Y. *et al.*, 1992; Ribeiro-da-Silva, A. *et al.*, 1992; Ma, W. *et al.*, 1996; 1997), as well as on the detection of neurons in the same laminae of the rat, which express the NK-1r and possess dorsally oriented dendrites that branch into laminae I and II (Liu, H. *et al.*, 1994a; Brown, J. L. *et al.*, 1995; Littlewood, N. K. *et al.*, 1995; Naim, M. *et al.*, 1997).

23.4.4   Neurochemistry of Dorsal Horn Neurons

In spite of a large number of studies that identified neurochemicals in dorsal horn neurons, the available information is less than satisfactory because, with a few exceptions, the information on the morphological properties of the neurons is either incomplete or simply lacking. This is a consequence of the fact that, in most cases, little more is known than the size and the localization of the neurochemically characterized cell bodies. There are some exceptions, however. The most striking exception is likely the excellent characterization of the population that expresses the NK-1r because a Golgi-like staining of the cells is obtained. With most neurochemicals, either the staining is inadequate to allow a proper characterization of the cell (e.g., cell body faintly labeled) or the number of immunolabeled cells is so high that it is impossible to characterize individual cells. In the case of *in situ* hybridization studies, only cell bodies are labeled, and therefore the information obtained with the use of this approach is significant but limited. The only approaches that can deliver an integrative view of morphological and neurochemical properties of the cells are those that involve the combination of intracellular filling of neurons with immunocytochemistry, but they have been sparsely used to date. Therefore, in the text below little correlation to cell populations described above will be made because such data are simply lacking. Table 1 gives an overview of classical transmitter/neuropeptide colocalizations in the dorsal horn, as described in the text.

23.4.4.1   *Neurokinins*

All three main mammalian neurokinins (SP, neurokinin A, and neurokinin B (NKB)) occur in dorsal horn neurons. SP-containing cell bodies have been detected in laminae I and II, both with immunocytochemistry (Ljungdahl, A. *et al.*, 1978; Ribeiro-da-Silva, A. *et al.*, 1991) and *in situ* hybridization (Warden, M. K. and Young, W. S., 1988). It is interesting to note that most, if not all, SP-IR cell bodies in the rat dorsal horn colocalize enkephalin (ENK) immunoreactivity (Ribeiro-da-Silva, A. *et al.*, 1991), and they seem to be glutamatergic as well because they express the vesicular glutamate transporter 2 (VGLUT2) (Todd, A. J. *et al.*, 2003). It has been speculated in the past that some of these neurons colocalizing SP and enkephalin immunoreactivities represent stalked cells, based on their laminar localization in LIIo, absence of presynaptic dendrites, and synapses on lamina I neurons (Ribeiro-da-Silva, A. *et al.*, 1991). However, neurons colocalizing SP and enkephalin likely belong to other cell types as well and some of these project to higher levels. In particular, SP-IR lamina I and lamina V cells have been shown to project to the thalamus in rat and cat (Battaglia, G. and Rustioni, A., 1992), and it is possible that they also project to other targets such as the midbrain parabrachial nucleus. Based on the finding that spinal cord SP-IR neurons normally colocalize enkephalin and glutamate, we can speculate that such projection neurons colocalize all three neurochemicals.

Virtually all SP-containing neurons in the rat express neurokinin A (NKA) as well (Carter, M. S. and Krause, J. E., 1990), which means that their distribution is essentially the same. This applies to the primary sensory fibers that also express SP, which represent the main source of SP in the dorsal horn (see Section Types of Afferents, Neurochemistry, and Termination in Spinal Cord). Another source of SP and NKA in the dorsal horn are axons originating from cell bodies located in the brain stem (Hökfelt, T. *et al.*, 1978; Gilbert, R. F. *et al.*, 1982); although most such fibers terminate in the ventral horn, some terminate in the dorsal horn.

NKB derives from a different precursor, and in contrast to SP and NKA, does not occur in primary sensory neurons (Ogawa, T. *et al.*, 1985). A light and electron microscopic immunocytochemical study using a marker of NKB neurons revealed that the signal occurs in axon terminals in laminae I–II, with a peak in lamina IIi, and in cell bodies and dendrites mostly in lamina IIi (McLeod, A. L. *et al.*, 2000).

Lamina III showed much less immunolabeling. Signal for NKB was detected in dendrites of type I glomeruli, suggesting a participation in the modulation of nociception (McLeod, A. L. *et al.*, 2000). Neurons expressing NKB seem to be excitatory and glutamatergic (Polgár, E. *et al.*, 2006).

23.4.4.2   Neurokinin receptors

Originally, NK-1r immunoreactivity was described in neurons with cell bodies residing in lamina I and in laminae III–IV, but not in lamina II (Liu, H. *et al.*, 1994a; Nakaya, Y. *et al.*, 1994). More recent studies, however, have identified NK-1r immunoreactivity in some neurons in LIIo (McLeod, A.L. *et al.*, 1998; Ribeiro-da-Silva, A. *et al.*, 2000). Most of the NK-1r-IR neurons project to higher levels, having targets such as the thalamus (Marshall, G. E. *et al.*, 1996), the parabrachial nucleus (Ding, Y.-Q. *et al.*, 1995; Todd, A. J. *et al.*, 2000), the lateral reticular nucleus, the dorsal part of caudal medulla, and, to a minor extent, the periaqueductal gray (Todd, A. J. *et al.*, 2000).

The receptor for neurokinin A (the neurokinin-2 receptor) hardly occurs in the CNS (for review see Ribeiro-da-Silva, A. *et al.*, 2000). Immunoreactivity for the neurokinin-2 receptor in the superficial spinal dorsal horn was rather weak and restricted to a narrow band in the lateral part of lamina I, and seemed to be localized in glial cells (Zerari, F. *et al.*, 1998). This is an unexpected finding that indicates that either neurokinin A acts through another receptor in the CNS or that it acts mainly in the periphery. There is evidence that NKA coreleased with SP from primary sensory fibers acts through the NK-1r and may be an important component of signaling through peptidergic afferent fibers (Trafton, J. A. *et al.*, 2001).

The preferential receptor for NKB, the neurokinin-3 receptor, was identified in cell bodies located in lamina I and, mostly, in lamina II (Ding, Y. Q. *et al.*, 1996; Zerari, F. *et al.*, 1997; Mileusnic, D. *et al.*, 1999; Ding, Y. Q. *et al.*, 2002). Interestingly, the cell populations that express the neurokinin-3 receptor coexpress either the μ-opioid receptor or GABA and nitric oxide synthase (NOS) (Ding, Y. Q. *et al.*, 2002).

23.4.4.3   CGRP and CGRP receptors

Immunoreactivity for CGRP in the spinal dorsal horn is restricted to primary sensory fibers in rodents, and likely in others species as well. Concerning CGRP receptors, their distribution has been studied with ligand binding and have been shown to occur in high density in lamina I and in deeper laminae but

were sparse in lamina II (Yashpal, K. *et al.*, 1992). Curiously, following peripheral nerve lesion, CGRP receptors were upregulated in lamina II, indicating that lamina II cells have the capacity to synthesize the receptor (Kar, S. *et al.*, 1994).

23.4.4.4   Somatostatin

The neuropeptide SOM occurs both in primary sensory fibers (see above) and in dorsal horn neurons (Hökfelt, T. *et al.*, 1976; Alvarez, F. J. and Priestley, J. V., 1990; Ribeiro-da-Silva, A. and Cuello, A. C., 1990b). Cells bodies from SOM-IR neurons occur mainly in lamina II (Alvarez, F. J. and Priestley, J. V., 1990; Ribeiro-da-Silva, A. and Cuello, A. C., 1990b) and seem to be glutamatergic as they also express VGLUT2 (Todd, A. J. *et al.*, 2003). The receptors for SOM are G protein-coupled and form a family of five receptors (sst1 to sst5) (Dournaud, P. *et al.*, 2000). The distribution of these receptor subtypes has been studied with immunocytochemistry, and they have been found in cell bodies and processes in the superficial dorsal horn (Schulz, S. *et al.*, 1998; Von Banchet, G. S. *et al.*, 1999).

23.4.4.5   Opioid peptides

23.4.4.5.(i)   Enkephalins   Although there are two main enkephalins, met- and leu-enkephalin, no differences in distribution have been detected with antibodies generated against each of the two, and as they share a common precursor, *in situ* hybridization studies do not reveal differences in distribution either. Therefore, as in most studies, we refer to enkephalin immunoreactivity, not discriminating whether studies were done with antibodies raised against leu- or met-enkephalin. Enkephalin immunoreactivity is intense in axons and cell processes in laminae I and II (Hökfelt, T. *et al.*, 1977; Hunt, S. P. *et al.*, 1981) and has also been detected in neuronal cell bodies in laminae I–III (Del Fiacco, M. and Cuello, A. C., 1980; Hunt, S. P. *et al.*, 1981; Miller and Seybold, 1989). *In situ* hybridization studies have not detected any significant evidence for the presence of the enkephalin precursor signal in DRG neurons, beyond a few isolated cells (Ruda, M. A. *et al.*, 1989), and sensory deafferentation does not result in any decrease of enkephalin levels in the dorsal horn (Ruda, M. A. *et al.*, 1986). Therefore, almost all the enkephalin in the dorsal horn seems to be from local circuit neurons, with a minor component colocalized in axons from serotonergic neurons of the raphe nuclei that project to the spinal cord (Hökfelt, T. *et al.*, 1979); most such axons, however, terminate in the ventral horn (Tashiro, T. *et al.*, 1988; Menétrey, D. and Basbaum, A. I., 1987). A small percentage of enkephalinergic neurons project to the brain (Standaert, D. G. *et al.*, 1986). Concerning enkephalin in neurons with cell bodies located in the dorsal horn, it occurs in two different populations of neurons: (a) cells that are SP-IR (Tashiro, T. *et al.*, 1987; Senba, E. *et al.*, 1988; Ribeiro-da-Silva, A. *et al.*, 1991) and that likely are glutamatergic (Todd, A. J. *et al.*, 2003); and (b) cells that are GABAergic (Todd, A. J. *et al.*, 1992b). An interesting conclusion is that as all SP-IR neurons in the dorsal horn seem to possess enkephalin immunoreactivity, the colocalization of both peptides can be used as a marker to identify SP-IR fibers that are intrinsic to the dorsal horn (Ribeiro-da-Silva, A., 2004). The ultrastructural studies of SP and enkephalin double labeling have confirmed the colocalization of the two peptides in both rat and cat and provided evidence that enkephalin-IR boutons or presynaptic dendrites are never presynaptic to SP-IR boutons (Ribeiro-da-Silva, A. *et al.*, 1991; Ma, W. *et al.*, 1997). In the cat, enkephalin-IR boutons have been shown to contact spinothalamic neurons (Ruda, M. A. *et al.*, 1984) and neurons of the dorsal column postsynaptic pathway (Nishikawa, N. *et al.*, 1983). Our own data on physiologically characterized, intracellularly filled dorsal horn neurons, some of which likely projected to the brain, has shown that these cells had appositions from boutons colocalizing SP and enkephalin immunoreactivities and from boutons IR for only one of the peptides, which likely represented the population colocalizing enkephalin and GABA and the population colocalizing SP and CGRP (primary sensory in origin) (Ma, W. *et al.*, 1997).

23.4.4.5.(ii) Dynorphins Most of the dynorphin in the dorsal horn is from spinal cord neurons, with a minor component at sacral levels from visceral afferents (see above). The distribution of the dynorphins in the dorsal horn has been studied both by immunocytochemistry with antibodies raised against dynorphin A and dynorphin B as well as by *in situ* hybridization using probes for the dynorphin precursor. The available data indicate that the dynorphins occur in axons and cells bodies in laminae I, IIo, and V, in both rat and cat (Cruz, L. and Basbaum, A. I., 1985; Miller, K. E. and Seybold, V. S., 1987; Leah, J. *et al.*, 1988; Ruda, M. A. *et al.*, 1988; Cho, H. J and Basbaum, A. I., 1989). These neurons have been reported as forming an heterogeneous population, although some neurons in LIIo resembled stalked

cells (Cruz, L. and Basbaum, A. I., 1985; Cho, H. J. and Basbaum, A. I., 1989). The dynorphins are not restricted to local circuits neurons, as some of the neurons that express them project to the parabrachial nucleus (Standaert, D. G. *et al.*, 1986).

23.4.4.5.(iii) Endormorphins Of the two endomorphins, endomorphin-2 is restricted to primary sensory fibers, where it is colocalized with SP (Martin-Schild, S. *et al.*, 1998; Sanderson Nydahl K., *et al.*, 2004). In contrast, endomorphin-1 is intrinsic to the CNS and occurs in fibers in laminae I and II (Martin-Schild, S. *et al.*, 1999).

23.4.4.6 Opioid receptors

Besides occurring on primary sensory fibers (see above), opioid receptors are located on dorsal horn neurons as well. *In situ* hybridization cytochemistry has confirmed the occurrence of opioid receptors in dorsal horn neurons (Minami, M. and Satoh, M., 1995). The δ-opioid receptor occurs both on cell bodies and dendrites of dorsal horn neurons and in axon terminals (Cheng, P. Y. *et al.*, 1995; 1997); some of the axon terminals expressing the receptor colocalize enkephalin (Cheng, P. Y. *et al.*, 1995). In contrast, the κ receptor has been localized mostly in dendrites and cell bodies (Arvidsson, U. *et al.*, 1995). Immunoreactivity for the μ receptor occurs mostly in lamina II (Honda, C. N. and Arvidsson, U., 1995; Kemp, T. *et al.*, 1996), axon terminals, dendritic profiles, and cell bodies (Cheng, P. Y. *et al.*, 1996; 1997). Most of the cell bodies displaying μ-opioid receptor immunoreactivity are located in lamina II and do not colocalize GABA or glycine immunoreactivities, which suggests that the neurons that express the μ receptor may be mostly excitatory interneurons (Kemp, T. *et al.*, 1996).

23.4.4.7 Glutamate

It has been known for a long time that a high percentage of neurons in the dorsal horn are glutamatergic. However, the evidence was mostly indirect, as glutamate immunostaining is not a reliable marker of glutamatergic cells bodies (Ottersen, O. P. and Storm-Mathisen, J., 1984). Currently, the most reliable way of identifying glutamatergic dorsal horn neurons is the immunocytochemical detection of one of the vesicular glutamate transporters. However, because the immunoreactivity is located mostly in axonal terminals and is low in dendrites and cell bodies, we have to rely on colocalization studies to identify the glutamatergic neurons.

Colocalization studies have revealed that most non-primary sensory axonal varicosities in the dorsal horn that expressed SP, SOM, or neurotensin were IR for the vesicular glutamate transporter (VGLUT2), strongly indicating that these cell populations are glutamatergic (Todd, A. J. *et al.*, 2003). It is also assumed that most of the projection neurons, including the cells of lamina I, are glutamatergic, although direct immunocytochemical evidence is lacking. Systematic studies combining *in situ* hybridization for vesicular glutamate transporters are required to confirm this point.

23.4.4.8 Glutamate receptors

Besides occurring in primary sensory fibers (see section Receptors on Primary Sensory Fibers above), glutamate receptors have also been identified in dorsal horn neurons. Concerning the ionotropic glutamate receptors, their distribution has been studied with both *in situ* hybridization and immunocytochemistry. The predominant subunits of AMPA receptors in the dorsal horn are GluR1 and GluR2 (Furuyama, T. *et al.*, 1993; Jakowec, M. W. *et al.*, 1995a; 1995b), and the labeling of many cell bodies has been detected by immunocytochemistry. An interesting observation is that the GluR1 subunit is particularly associated with GABAergic neurons, while the GluR2 is mainly found in cells that lack GABA (Spike, R. C. *et al.*, 1998). Concerning NMDA receptors, the NR1 subunit is highly expressed in dorsal horn neurons (Tölle, T. R. *et al.*, 1993; Watanabe, M. *et al.*, 1994) and may be expressed by all cells. Less is known about the expression of the NR2 subunit of the NMDA receptor and of kainate receptors (Furuyama, T. *et al.*, 1993; Tölle, T. R. *et al.*, 1993; Watanabe, M. *et al.*, 1994). Metabotropic glutamate receptors are also expressed by dorsal horn neurons and of these the mGluR5 is the most prevalent (Berthele, A. *et al.*, 1999; Jia, H. *et al.*, 1999).

23.4.4.9 Inhibitory amino acids

23.4.4.9.(i) GABA The distribution of GABAergic neurons in the dorsal horn was originally studied by autoradiography using ³H-GABA high-affinity uptake and by immunocytochemistry using anti-glutamic acid decarboxylase (GAD) antibodies (Ribeiro-da-Silva, A. and Coimbra, A., 1980; Hunt, S. P. *et al.*, 1981; Barber, R. P. *et al.*, 1982). Later studies used immunoreactivity for GABA itself (Todd, A. J. and McKenzie, J., 1989). A study in rat showed that GABAergic neurons were evenly distributed in laminae I–III, where they represented 24–33% or 28–

46% of the overall neuronal population (Todd, A. J. and McKenzie, J., 1989; Todd, A. J. and Sullivan, A. C., 1990). Through a combination of GABA immunostaining and the Golgi method, some of the GABAergic neurons of lamina II were identified as islet cells, but never as stalked cells (Todd, A. J. and McKenzie, J., 1989). This observation is consistent with the observation of islet cells with presynaptic dendrites (Gobel, S. *et al.*, 1980; Spike, R. C. and Todd, A. J., 1992), which indicate an inhibitory function. Results from Lu Y. and Perl E. R. (2003; 2005) suggest that only the large islet cells of Todd A. J. (1986) (Figures 7 and 8) are GABAergic while small islet cells (central cell) are glutamatergic. Large islet cells appear to act as feed-forward inhibitory neurons to small islet cells (Lu, Y. and Perl, E. R., 2003) (Figure 8).

In the spinal cord, GABA participates in both presynaptic inhibition on terminals of primary afferents and postsynaptic inhibition of dorsal horn neurons. Morphological support for the involvement of GABAergic interneurons in presynaptic inhibition was obtained by demonstrating that these are presynaptic to primary sensory fibers; this evidence was obtained by means of a combination of GAD immunocytochemistry with nerve lesions in the rat (Barber, R. P. *et al.*, 1978) or with the intracellular filling of identified afferent fibers in the cat (Maxwell, D. J. and Noble, R., 1987). The identification of both presynaptic dendrites and peripheral axons in glomeruli presynaptic to their central bouton, of known primary sensory origin, provided further morphological support to the participation of GABAergic dorsal horn neurons in the presynaptic inhibition of primary sensory fibers (Todd, A. J., 1996; Ribeiro-da-Silva, A., 2004). In these presynaptic structures, GABA is colocalized with other neurochemicals, such as enkephalin, acetylcholine (Ribeiro-da-Silva, A., 2004), glycine (Todd, A. J., 1991; 1996), NPY, or galanin (Polgár, E. *et al.*, 1999b). A minor component of the GABAergic terminals in the dorsal horn originates from the brain stem (Antal, M. *et al.*, 1996), and a recent *in vivo* electrophysiological study provides evidence for descending monosynaptic GABAergic and/or glycinergic inputs from the rostral ventromedial medulla to the substantia gelatinosa (Kato, G. *et al.*, 2006).

23.4.4.9.(ii) Glycine Glycinergic neurons are found in most laminae of the dorsal horn; in laminae I–III, their proportion in the overall is considerably higher in lamina III (30%) than in lamina I (9%) or in lamina II (14%) (Todd, A. J. and Sullivan, A. C., 1990). Virtually all glycinergic neurons in laminae I–III are also GABAergic (Todd, A. J. and Sullivan, A. C., 1990; Todd, A. J., 1991); however, only about half of the GABAergic cells colocalize glycine immunoreactivity (Todd, A. J. and Sullivan, A. C., 1990; Todd, A. J., 1991). Terminals colocalizing GABA and glycine immunoreactivities are presynaptic mostly to dendrites and cell bodies, except in LIIi and deeper dorsal horn where they are also presynaptic to primary sensory fibers in type II but not in type I glomeruli (Todd, A. J., 1996). There is indirect evidence that glycine is likely colocalized with GABA in some descending fibers from the brain stem (Kato, G. et al., 2006).

23.4.4.10 GABA and glycine receptors

$GABA_A$ receptors seem to be ubiquitous in dorsal horn neurons, where their subunit precursors have been identified by in situ hybridization (Persohn, E. et al., 1991) and their proteins by immunocytochemistry (Alvarez, F. J. et al., 1996; Bohlhalter, S. et al., 1996). The combination of $\alpha 3$, $\beta 3$, and $\gamma 2$ subunits appears to occur in cells throughout the dorsal horn, while $\alpha 1$ subunits are expressed by neurons in deeper laminae (Bohlhalter, S. et al., 1996). An electron microscopic study using an antibody-recognizing subunits $\beta 2$–$\beta 3$ of the $GABA_A$ receptor revealed predominantly dendrite and cell body staining, with no direct evidence of labeling of sensory fibers (Alvarez, F. J. et al., 1996).

A study of the distribution of immunoreactivity for the two subunits of the metabotropic $GABA_B$ receptor, $GABA_B R1$ and $GABA_B R2$, revealed that they occurred in highest concentrations in laminae I and II, where they were detected both in cell bodies and in the neuropile (Margeta-Mitrovic, M. et al., 1999). The localization of the cell bodies that express the subunits of the $GABA_B$ receptor was confirmed by in situ hybridization (Towers, S. et al., 2000).

Interestingly, in the dorsal horn, glycine receptors seem restricted to intrinsic neurons. Glycine receptors are ionotropic receptors made up of a combination of α and β subunits. The initial studies were performed with antibodies against gephyrin (a glycine- and GABA receptor-associated protein). These studies were performed at the time when it was thought that this protein was associated only with the glycine receptor. Surprisingly, these studies were rather accurate, as in the spinal cord, gephyrin-IR sites match the distribution of glycine receptors more closely than that of $GABA_A$ receptors (see e.g.

Mitchell, K. et al., 1993). One of the glycine receptor subunits (GlyR $\alpha 3$) has been found to be specifically expressed in the superficial laminae of the spinal dorsal horn, particularly in lamina II (Harvey, R. J. et al., 2004), although the most abundant subunit in the adult spinal cord is the $\alpha 1$ (Becker, C. M. et al., 1988). One interesting feature relative to the distribution of GABA and glycine receptors in the spinal cord is that many synapses colocalize GABA and glycine, and the specificity of the mixed GABA/glycine synapses seems to be determined by the expression, properties, and the subsynaptic localization of the $GABA_A$, $GABA_B$, and glycine receptors in the target cells (Chéry, N. and De Koninck, Y., 1999; 2000) and their changes during development (Keller, A. F. et al., 2001).

It should be kept in mind that the great majority of GABA- and glycine-containing terminals are presynaptic to spinal neurons, not to primary afferent terminals (Todd, A. J. and Spike, R. C., 1993), and GABAergic and glycinergic interneurons control an important network of polysynaptic connections between primary afferents and dorsal horn neurons. For example, blockade of GABAergic and glycinergic inhibition unmasks low-threshold input to dorsal horn neurons that do not receive monosynaptic input from low-threshold afferents (Baba, H. et al., 2003; Torsney, C. and MacDermott, A. B., 2006). Impaired inhibition is thus a likely substrate for cross talk between sensory modalities and appear to underlie allodynia characteristic of neuropathic pain syndromes (Moore, K. A. et al., 2002; Coull, J. A. et al., 2003; Coull, J. A. et al., 2005).

23.4.4.11 Other classical transmitters and other neuropeptides

One of the better-defined transmitter-specific systems in the dorsal horn is the cholinergic one. In fact, neuronal perikarya IR for choline acetyltransferase (ChAT) have been identified in the spinal dorsal horn (Kimura, H. et al., 1981; Barber, R. P. et al., 1984; Todd, A. J., 1991). These neurons are not numerous, occur mainly in laminae III–IV, and are presynaptic to primary sensory fibers in synaptic glomeruli and to neurons (Ribeiro-da-Silva, A. and Cuello, A. C., 1990a). Interestingly, as mentioned above, most, if not all, of these cholinergic cell bodies and boutons colocalize GABA.

Regarding other transmitters, spinal serotonin originates from neurons with cell bodies in the brain stem (for review see Ruda, M. A. et al., 1986). Interestingly, in the cat, these serotonin-IR profiles have been shown to synapse on projection neurons (Ruda, M. A., 1986). These fibers colocalizes other transmitters such as enkephalin (Hökfelt, T. *et al.*, 1979) (see above) or SP and thyrotropin-releasing hormone (TRH) (Johansson, O. *et al.*, 1981). Another transmitter-specific system originating from the brain stem is the noradrenergic one. Although their light microscopic distribution and their origin in the brain stem are well known (Westlund, K. N. *et al.*, 1983; Fritschy, J. M. and Grzanna, R., 1990), there is still not much information about these noradrenergic fibers except that they are presynaptic to spinal dorsal horn neurons (Hagihira, S. *et al.*, 1990; Doyle, C. A. and Maxwell, D. J., 1991). Interestingly, noradrenergic analgesia appears to involve presynaptic inhibition of glutamate release from α2c-adrenergic terminals of local excitatory interneurons (Olave, M. J. and Maxwell, D. J., 2002; 2003a) contacting NK-1r-IR projection neurons (Olave, M. J. and Maxwell, D. J., 2003b). Neurotensin immunoreactivity occurs in neurons in laminae I and II (Hunt, S. P. *et al.*, 1981; Seybold, V. S. and Elde, R. P., 1982), which do not colocalize GABA (Todd, A. J. *et al.*, 1992a) but do express VGLUT2, and are likely glutamatergic (Todd, A. J. *et al.*, 2003). Furthermore, most of the neurotensin-IR neurons in LIIi express protein kinase C gamma (Polgár, E. *et al.*, 1999a).

23.5 Identified Neuronal Circuits

While our knowledge of synaptic circuits in the dorsal horn remains limited, some patterns are emerging based on recent findings. Figure 9 attempts to summarize some observations described above. It should be noted here that there are significant interspecies differences in the neurochemical and anatomical organization of the region. Thus some of the proposed circuits may not hold across all species.

One should stress here that, while we propose well-defined circuits, not all chemical communication involves hard-wired synapses (Zoli, M. *et al.*, 1999). Nonsynaptic or paracrine transmission is thought to occur particularly with peptides and monoamines; however, it probably occurs for amino acid transmitters also acting both on their ionotropic and metabotropic receptors (Rossi, D. and Hamann, M., 1998; Chéry, N. and De Koninck, Y., 1999; 2000). Whereas under physiological conditions, most interneuronal communication in the dorsal horn probably takes place either at

synapses or close to the site of release, diffusion of neurochemicals over longer distances cannot be excluded, particularly in the case of neuropeptides. Nevertheless, it should be kept in mind that the neuropile represents a significant diffusion barrier, and it is unlikely that even peptides diffuse large distances when released at physiological concentrations. Therefore, a detailed description of the chemical neuroanatomy of the dorsal horn circuits is important to understand sensory processing.

As most synapses in the dorsal horn are axodendritic, it is not surprising that the majority of interneurons appear to act on their targets through a postsynaptic mechanism; however, some cells clearly participate in presynaptic interactions. GABAergic axons, which are thought to originate from local interneurons, establish axoaxonic synapses on the central boutons of synaptic glomeruli, except for those in peptidergic (type Ib) glomeruli that are seldom postsynaptic to other profiles. There is a fundamental difference between the types of GABAergic neurons that innervate type Ia and type II glomeruli, because boutons containing GABA and glycine are presynaptic to the central boutons of type II, but not to type Ia glomeruli (Todd, A. J., 1996). In contrast, neurons colocalizing GABA and enkephalin immunoreactivities seem to innervate central axons of type I and II glomeruli equally (through both axoaxonic and dendroaxonic synapses) (Ribeiro-da-Silva, A., 2004). The population of neurons with GABA and ChAT is relatively small (Todd, A. J., 1991); however, their axons contribute a significant contingent of axoaxonic synapses in type II, and to a lesser extent in type Ia, glomeruli (Ribeiro-da-Silva, A. and Cuello, A. C., 1990a; Ribeiro-da-Silva, A., 2004). The neurons with GABA and acetylcholine (revealed by ChAT immunoreactivity) have cell bodies in laminae III and IV and dendrites that extend superficially, where they receive synaptic input from primary afferents in both type Ia and type II glomeruli. Cells that colocalize GABA and NPY appear to selectively target NK-1r-IR laminae III/IV projection neurons, with which they make numerous axodendritic and axosomatic synapses (Polgár, E. *et al.*, 1999b).

As discussed above, most (if not all) of the SP-containing neurons in the superficial laminae are also enkephalin-IR (Ribeiro-da-Silva, A. *et al.*, 1991). They are located in laminae I and II and receive synaptic input from nonpeptidergic C fibers in type Ia glomeruli and from SP-containing afferents outside glomeruli. Interestingly, based on evidence

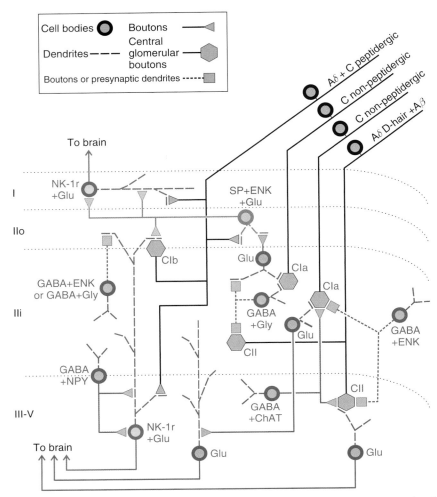

Figure 9 Summary diagram of neuronal circuits in the dorsal horn. Different synaptic arrangements involve the different classes of primary afferents endings in the dorsal horn (glomerular and nonglomerular). In particular, the glomerular structures are exquisite sites of presynaptic modulation of afferent inputs. Yet, it should be kept in mind that, while this diagram highlights the different presynaptic arrangements in which the different populations of GABAergic interneurons are involved, most GABAergic and glycinergic synapses remain axodendritic and axosomatic in the dorsal horn. While physiological studies indicate that Aδ high-threshold mechanoreceptors (HTM) provide input to laminae I and V, further information on the synaptic arrangements they are involved in is unavailable due to the lack of specific markers for these types of afferents. Peptidergic Aδ and C nociceptors are known to contact spinal projection neurons that express the NK-1 receptor both in laminae I and V. The substance P (SP)–enkephalin (ENK)–Glu interneuron positioned in lamina IIo represents a prototype stalked cell, which, for the sake of simplicity, accounts for both the circuitry illustrated in Figure 8 based on the physiological data (see Lu, Y. and Perl, E. R., 2003; 2005) and that based on what is known of the peptidergic circuitry (Ribeiro-da-Silva, A. et al., 1991). It remains to be confirmed, however, whether a single type of stalked cell, expressing SP, ENK, and glutamate, constitutes a common pathway to lamina I projection neurons or whether parallel circuits are involved. Two nonpeptidergic afferent pathways are illustrated to account for the possibility that parallel circuits with no cross talk at the level of the dorsal horn may connect to ascending pathways (see text for further details; Braz, J. M. et al., 2005).

from studies in the cat (Ma, W. et al., 1997), these neurons seem to preferentially innervate nociceptive neurons in lamina I and in the deep dorsal horn, which probably express the NK-1r (Figure 9).

Finally, these findings, together with those of Lu Y. and Perl E. R. (2003; 2005) and the evidence of a parallel pathway via laminae II–V (Braz, J. M. et al., 2005) (see above) that exclude NK-1r-expressing cells, suggest that distinct subclasses of nonpeptidergic afferents connect to laminae I and V ascending pathways with virtually no cross talk at the level of the dorsal horn (Figure 9).

23.6   Conclusion and Future Directions

An important conclusion of this review is that, despite all the progress, we are still unable to answer a few very basic questions regarding the synaptic organization of the dorsal horn. This is in part a consequence of the heterogeneity of neuronal populations in each layer and therefore the difficulty to properly identify subtypes of neurons in functional studies (in contrast with highly organized structures like the cerebellum, hippocampus, and even neocortex).

For example, while a significant amount of data on the organization of inhibitory circuits is available from purely morphological and immunocytochemical studies, little is known at the functional level, in part because of the difficulty to identify GABAergic and glycinergic interneurons for targeted electrophysiological recording. This is likely to change, thanks to the growing availability of transgenic mice that express enhanced green fluorescent protein (GFP) in specific subpopulations of inhibitory interneurons (e.g., under the control of the promoter for glutamic acid decarboxylase or the glycine transporter 2) (Hantman, A. W. *et al.*, 2004; Heinke, B. *et al.*, 2004; Dougherty, K. J. *et al.*, 2005; Zeilhofer, H. U. *et al.*, 2005). The same is true for subpopulations of interneurons that are associated with other specific markers (e.g., neuropeptides and calbindin).

Emerging classification schemes, based on combinations of morphological features, transmitter and receptor phenotypes, and electrophysiological properties (Thomson, A. M. *et al.*, 1989; Yoshimura, M. and Jessell, T. M. 1989; Lopez-Garcia, J. A. and King, A. E., 1994; Han, Z. S. *et al.*, 1998; Jo, Y. H. *et al.*, 1998; Yu, X. H. *et al.*, 1999; Grudt, T. J. and Perl, E. R., 2002; Prescott, S. A. and De Koninck, Y., 2002; Ruscheweyh, R. and Sandkuhler, J., 2002; Ruscheweyh, R. *et al.*, 2004; Hantman, A. W. and Perl, E. R., 2005; Yu, X. H. *et al.*, 2005), will also be critical to better understand the organization of neuronal circuits in the dorsal horn. Indeed, studies combining electrophysiological recording and intracellular labeling with immunocytochemistry remain critical to obtain a finer identification of neuronal subpopulations and their function as well as to obtain a more complete picture of the afferent and efferent connections of transmitter-specific interneurons; however, such approaches are technically demanding and remain insufficiently used (De Koninck, Y. *et al.*,

1992; Ribeiro-da-Silva, A. *et al.*, 1992; Ma, W. *et al.*, 1996; 1997; Lawson, S. N. *et al.*, 1997; 2002).

An important challenge for the future will also be to better unravel the functional significance of the multiple transmitter signals originating from the same cells and the same synaptic terminals. For example, the functional meaning of the corelease of GABA and glycine (Chéry, N. and De Koninck, Y., 1999) – two very similar neurotransmitters – from the same terminals remains elusive in a large part. Several synapses also appear to use opposing transmitters, such as SP and enkephalin (Ribeiro-da-Silva, A. *et al.*, 1991) or GABA and ATP (Jo, Y. H. and Schlichter, R., 1999); whether these involve local feedback systems, differential release conditions, or nonlinear modes of action (Kupfermann, I., 1991; Nusbaum, M. P. *et al.*, 2001) remains to be determined. Several of these questions continue to be elusive simply because the modes of action of several neuropeptides remain incompletely understood. Target-selective expression of receptors may determine the functional meaning of the mixed transmitter signals released from the same neuron. For example, when GABA and glycine are colocalized in axoaxonic contacts with primary afferent terminals, only $GABA_A$ and $GABA_B$ are expressed on those terminals; the glycinergic signal must only be meaningful at other synapses from the same presynaptic neuron or acting on receptors located on dendrites neighboring the release site (Rossi, D. and Hamann, M., 1998). In the same line, fine subcellular segregation of receptors in the target cells may have very significant impact on how mixed transmitter signals are decoded by the postsynaptic cell (see e.g., Chéry, N. and De Koninck, Y., 1999; 2000; Keller, A. F. *et al.*, 2001).

While important information has arisen from the use of markers of specific primary afferent subclasses, it is necessary to develop more refined classification schemes. For example, both on anatomical and physiological grounds, the current classification schemes of peptidergic and nonpeptidergic C-fiber afferents (e.g., binding IB4 or expressing CGRP) remains too general to account for the existing diversity of subclasses. The search for newer cell surface markers or the development of transgenic models expressing GFP in specific subpopulations of cells will be important to unravel the synaptic circuitry and functional role of each subclass of afferents and their spinal dorsal horn targets.

An important issue to keep in mind that may be an important limitation in our current understanding is the occurrence of interspecies differences. For example, many of the original studies on neurons characterized physiologically and labeled intracellularly with injections of markers have been performed in the cat, whereas tract-tracing and immunocytochemical studies have generally been carried out in the rat and the monkey, while studies relying on transgenic tools have to be conducted in the mouse. Many of these species differences have been described earlier in this chapter. One difference that may be worth stressing again is the fact that the TRPV1 receptor is restricted to the peptidergic afferents in the mouse, but not in the rat (see above), raising concerns about the use of mouse transgenic models to study the function of this and other receptors.

While currently much emphasis is naturally placed on studying direct interactions between pairs of neurons to unravel the synaptic circuitry in the dorsal horn, a better understanding of the organization of the network will eventually require the integrative study of polysynaptic pathways (e.g., from primary afferents to spinal projection neurons). For example, cross talk between ascending pathways by means of polysynaptic connections regulated by local inhibitory circuits in the dorsal horn is likely a critical element underlying plasticity and pathology of the pain system (Yaksh, T. L., 1989; Sherman, S. E. and Loomis, C.W., 1994; Baba, H. *et al.*, 2003; Torsney, C. and MacDermott, A. B., 2006).

In summary, novel tools are emerging to allow us to improve our understanding of the structural and functional organization of the dorsal horn. These include novel markers, especially markers compatible with live cell studies such as transgenic animals expressing GFP in specific cell types (Hantman, A. W. *et al.*, 2004; Heinke, B. *et al.*, 2004; Dougherty, K. J. *et al.*, 2005; Zeilhofer, H. U. *et al.*, 2005) and the use of specific fluorescent ligands compatible with live cell studies (see e.g., Labrakakis, C. *et al.*, 2003; Labrakakis, C. and MacDermott, A. B., 2003). Novel methods are also emerging to identify neuronal pathways such as transsynaptic markers (Jasmin, L. *et al.*, 1997; Kissa, K. *et al.*, 2002; Braz, J. M. *et al.*, 2005). It also remains critical to combine these tools with other approaches to establish correlations between different identifiers of cell structure and function (e.g., morphology, transmitter/receptor phenotype, and electrophysiological properties). These steps are important to obtain a better grasp of the organization of the dorsal horn, which remains an essential condition for the understanding of nociceptive

processing as well as of the neural basis of altered nociception in chronic pain conditions.

References

Almarestani, L., Waters, S. M., Krause, J. E., Bennett, G. J., and Ribeiro-da-Silva, A. 2007. Morphological characterization of spinal cord dorsal horn neurons projecting to the parabrachial nucleus in the rat. J. Comp. Neurol., in press.

Alvarez, F. J. and Fyffe, R. E. 2000. Nociceptors for the 21st century. Curr. Rev. Pain 4, 451–458.

Alvarez, F. J. and Priestley, J. V. 1990. Anatomy of somatostatin-immunoreactive fibres and cell bodies in the rat trigeminal subnucleus caudalis. Neuroscience 38, 343–357.

Alvarez, F. J., Kavookjian, A. M., and Light, A. R. 1992. Synaptic interactions between GABA-immunoreactive profiles and the terminals of functionally defined myelinated nociceptors in the monkey and cat spinal cord. J. Neurosci. 12, 2901–2917.

Alvarez, F. J., Kavookjian, A. M., and Light, A. R. 1993. Ultrastructural morphology, synaptic relationships, and CGRP immunoreactivity of physiologically identified C-fiber terminals in the monkey spinal cord. J. Comp. Neurol. 329, 472–490.

Alvarez, F. J., Taylor-Blake, B., Fyffe, R. E. W., de Blas, A. L., and Light, A. R. 1996. Distribution of immunoreactivity for the β_2 and β_3 subunits of the GABA$_A$ receptor the mammalian spinal cord. J. Comp. Neurol. 365, 392–412.

Amaya, F., Decosterd, I., Samad, T. A., Plumpton, C., Tate, S., Mannion, R. J., Costigan, M., and Woolf, C. J. 2000. Diversity of expression of the sensory neuron-specific TTX-resistant voltage-gated sodium ion channels SNS and SNS2. Mol. Cell Neurosci. 15, 331–342.

Andrew, D., Krout, K. E., and Craig, A. D. 2003. Differentiation of lamina I spinomedullary and spinothalamic neurons in the cat. J. Comp. Neurol. 458, 257–271.

Antal, M., Petko, M., Polgár, E., Heizmann, C. W., and Storm-Mathisen, J. 1996. Direct evidence of an extensive GABAergic innervation of the spinal dorsal horn by fibres descending from the rostral ventromedial medulla. Neuroscience 73, 509–518.

Arvidsson, U., Riedl, M., Chakrabarti, S., Vulchanova, L., Lee, J.-H., Nakano, A. H., Lin, X., Loh, H. H., Law, P.-Y., Wessendorf, M. W., and Elde, R. 1995. The kappa-opioid receptor is primarily postsynaptic: combined immunohistochemical localization of the receptor and endogenous opioids. Proc. Natl. Acad. Sci. U. S. A. 92, 5062–5066.

Baba, H., Ji, R. R., Kohno, T., Moore, K. A., Ataka, T., Wakai, A., Okamoto, M., and Woolf, C. J. 2003. Removal of GABAergic inhibition facilitates polysynaptic A fiber-mediated excitatory transmission to the superficial spinal dorsal horn. Mol. Cell Neurosci. 24, 818–830.

Barber, R. P., Phelps, P. E., Houser, C. R., Crawford, G. D., Salvaterra, P. M., and Vaughn, J. E. 1984. The morphology and distribution of neurons containing choline acetyltransferase in the adult rat spinal cord: an immunohistochemical study. J. Comp. Neurol. 229, 329–346.

Barber, R. P., Vaughn, J. E., and Roberts, E. 1982. The cytoarchitecture of GABAergic neurons in rat spinal cord. Brain Res. 238, 305–328.

Barber, R. P., Vaughn, J. E., Saito, K., McLaughlin, B. J., and Roberts, E. 1978. GABAergic terminals are presynaptic to primary afferent terminals in the substantia gelatinosa of the rat spinal cord. Brain Res. 141, 35–55.

Bardoni, R., Torsney, C., Tong, C. K., Prandini, M., and MacDermott, A. B. 2004. Presynaptic NMDA receptors modulate glutamate release from primary sensory neurons in rat spinal cord dorsal horn. J. Neurosci. 24, 2774–2781.

Basbaum, A. I., Cruz, L., and Weber, E. 1986. Immunoreactive dynorphin B in sacral primary afferent fibers of the cat. J. Neurosci. 6, 127–133.

Battaglia, G. and Rustioni, A. 1988. Coexistence of glutamate and substance P in dorsal root ganglion neurons of the rat and monkey. J. Comp. Neurol. 277, 302–312.

Battaglia, G. and Rustioni, A. 1992. Substance P innervation of the rat and cat thalamus. II. Cells of origin in the spinal cord. J. Comp. Neurol. 315, 473–486.

Becker, C. M., Hoch, W., and Betz, H. 1988. Glycine receptor heterogeneity in rat spinal cord during postnatal development. EMBO J. 7, 3717–3726.

Bennett, G. J., Abdelmoumene, M., Hayashi, H., and Dubner, R. 1980. Physiology and morphology of substantia gelatinosa neurons intracellularly stained with horseradish peroxidase. J. Comp. Neurol. 194, 809–827.

Bennett, D. L. H., Michael, G. J., Ramachandran, N., Munson, J. B., Averill, S., Yan, Q., McMahon, S. B., and Priestley, J. V. 1998. A distinct subgroup of small DRG cells express GDNF receptor components and GDNF is protective for these neurons after nerve injury. J. Neurosci. 18, 3059–3072.

Berthele, A., Boxall, S. J., Urban, A., Anneser, J. M., Zieglgänsberger, W., Urban, L., and Tölle, T. R. 1999. Distribution and developmental changes in metabotropic glutamate receptor messenger RNA expression in the rat lumbar spinal cord. Dev. Brain Res. 112, 39–53.

Bicknell, H. R. and Beal, J. A. 1984. Axonal and dendritic development of substantia gelatinosa neurons in the lumbosacral spinal cord of the rat. J. Comp. Neurol. 226, 508–522.

Birder, L. A. and Perl, E. R. 1999. Expression of α_2-adrenergic receptors in rat primary afferent neurones after peripheral nerve injury or inflammation. J. Physiol. (Lond.) 515, 533–542.

Bohlhalter, S., Weinmann, O., Mohler, H., and Fritschy, J. M. 1996. Laminar compartmentalization of GABA$_A$-receptor subtypes in the spinal cord: an immunohistochemical study. J. Neurosci. 16, 283–297.

Bradbury, E. J., Burnstock, G., and McMahon, S. B. 1998. The expression of P2X$_3$ purinoreceptors in sensory neurons: Effects of axotomy and glial-derived neurotrophic factor. Mol. Cell. Neurosci. 12, 256–268.

Braz, J. M., Nassar, M. A., Wood, J. N., and Basbaum, A. I. 2005. Parallel "pain" pathways arise from subpopulations of primary afferent nociceptor. Neuron 47, 787–793.

Brown, A. G. 1982. The dorsal horn of the spinal cord. Q. J. Exp. Physiol. 67, 193–212.

Brown, J. L., Liu, H., Maggio, J. E., Vigna, S. R., Mantyh, P. W., and Basbaum, A. I. 1995. Morphological characterization of substance P receptor-immunoreactive neurons in the rat spinal cord and trigeminal nucleus caudalis. J. Comp. Neurol. 356, 327–344.

Cairns, B. E., Sessle, B. J., and Hu, J. W. 1998. Evidence that excitatory amino acid receptors within the temporomandibular joint region are involved in the reflex activation of the jaw muscles. J. Neurosci. 18, 8056–8064.

Carter, M. S. and Krause, J. E. 1990. Structure, expression, and some regulatory mechanisms of the rat preprotachykinin gene encoding substance P, neurokinin A, neuropeptide K, and neuropeptide gamma. J. Neurosci. 10, 2203–2214.

Caterina, M. J. 2003. Vanilloid receptors take a TRP beyond the sensory afferent. Pain 105, 5–9.

Cheng, P. Y., Liu-Chen, L. Y., and Pickel, V. M. 1997. Dual ultrastructural immunocytochemical labeling of μ and δ opioid receptors in the superficial layers of the rat cervical spinal cord. Brain Res. 778, 367–380.

Cheng, P. Y., Moriwaki, A., Wang, J. B., Uhl, G. R., and Pickel, V. M. 1996. Ultrastructural localization of μ-opioid receptors in the superficial layers of the rat cervical spinal cord: Extrasynaptic localization and proximity to Leu^5-enkephalin. Brain Res. 731, 141–154.

Cheng, P. Y., Svingos, A. L., Wang, H., Clarke, C. L., Jenab, S., Beczkowska, I. W., Inturrisi, C. E., and Pickel, V. M. 1995. Ultrastructural immunolabeling shows prominent presynaptic vesicular localization of δ-opioid receptor within both enkephalin- and nonenkephalin-containing axon terminals in the superficial layers of the rat cervical spinal cord. J. Neurosci. 15, 5976–5988.

Chéry, N. and De Koninck, Y. 1999. Junctional versus extrajunctional glycine and GABA$_A$ receptor-mediated IPSCs in identified lamina I neurons of the adult rat spinal cord. J. Neurosci. 19, 7342–7355.

Chéry, N. and De Koninck, Y. 2000. GABA$_B$ receptors are the first target of released GABA at lamina I inhibitory synapses in the adult rat spinal cord. J. Neurophysiol. 84, 1006–1011.

Cho, H. J. and Basbaum, A. I. 1989. Ultrastructural analysis of dynorphin B-immunoreactive cells and terminals in the superficial dorsal horn of the deafferented spinal cord of the rat. J. Comp. Neurol. 281, 193–205.

Coimbra, A., Ribeiro-da-Silva, A., and Pignatelli, D. 1984. Effects of dorsal rhizotomy on the several types of primary afferent terminals in laminae I–III of the rat spinal cord. An electron microscope study. Anat. Embryol. (Berl.) 170, 279–287.

Coimbra, A., Sodré-Borges, B. P., and Magalhães, M. M. 1974. The substantia gelatinosa Rolandi of the rat. Fine structure, cytochemistry (acid phosphatase) and changes after dorsal root section. J. Neurocytol. 3, 199–217.

Coull, J. A., Beggs, S., Boudreau, D., Boivin, D., Tsuda, M., Inoue, K., Gravel, C., Salter, M. W., and De Koninck, Y. 2005. BDNF from microglia causes the shift in neuronal anion gradient underlying neuropathic pain. Nature 438, 1017–1021.

Coull, J. A., Boudreau, D., Bachand, K., Prescott, S. A., Nault, F., Sik, A., De Koninck, P., and De Koninck, Y. 2003. Trans-synaptic shift in anion gradient in spinal lamina I neurons as a mechanism of neuropathic pain. Nature 424, 938–942.

Cox, J. J., Reimann, F., Nicholas, A. K., Thornton, G., Roberts, E., Springell, K., Karbani, G., Jafri, H., Mannan, J., Raashid, Y., Al-Gazali, L., Hamamy, H., Valente, E. M., Gorman, S., Williams, R., McHale, D. P., Wood, J. N., Gribble, F. M., and Woods, C. G. 2006. An SCN9A channelopathy causes congenital inability to experience pain. Nature 444, 894–898.

Craig, A. D. and Bushnell, M. C. 1994. The thermal grill illusion: unmasking the burn of cold pain. Science 265, 252–255.

Craig, A. D. and Kniffki, K. D. 1985. Spinothalamic lumbosacral lamina I cells responsive to skin and muscle stimulation in the cat. J. Physiol. (Lond.) 365, 197–221.

Cruz, L. and Basbaum, A. I. 1985. Multiple opioid peptides and the modulation of pain: immunohistochemical analysis of dynorphin and enkephalin in the trigeminal nucleus caudalis and spinal cord of the cat. J. Comp. Neurol. 240, 331–348.

Dalsgaard, C. J., Haegerstrand, A., Theodorsson-Norheim, E., Brodin, E., and Hökfelt, T. 1985. Neurokinin A-like immunoreactivity in rat primary sensory neurons; coexistence with substance P. Histochemistry 83, 37–39.

De Biasi, S. and Rustioni, A. 1988. Glutamate and substance P coexist in primary afferent terminals in the superficial laminae of spinal cord. Proc. Natl. Acad. Sci. U. S. A. 85, 7820–7824.

De Koninck, Y., Ribeiro-da-Silva, A., Henry, J. L., and Cuello, A. C. 1992. Spinal neurons exhibiting a specific nociceptive response receive abundant substance P-containing synaptic contacts. Proc. Natl. Acad. Sci. U. S. A. 89, 5073–5077.

Del Fiacco, M. and Cuello, A. C. 1980. Substance P-and enkephalin-containing neurones in the rat trigeminal system. Neuroscience 5, 803–815.

Ding, Y. Q., Lu, C. R., Wang, H., Su, C. J., Chen, L. W., Zhang, Y. Q., and Ju, G. 2002. Two major distinct subpopulations of neurokinin-3 receptor-expressing neurons in the superficial dorsal horn of the rat spinal cord. Eur. J. Neurosci. 16, 551–556.

Ding, Y. Q., Shigemoto, R., Takada, M., Ohishi, H., Nakanishi, S., and Mizuno, N. 1996. Localization of the neuromedin K receptor (NK3) in the central nervous system of the rat. J. Comp. Neurol. 364, 290–310.

Ding, Y.-Q., Takada, M., Shigemoto, R., and Mizuno, N. 1995. Spinoparabrachial tract neurons showing substance P receptor- like immunoreactivity in the lumbar spinal cord of the rat. Brain Res. 674, 336–340.

Djouhri, L. and Lawson, S. N. 2004. A beta-fiber nociceptive primary afferent neurons: a review of incidence and properties in relation to other afferent A-fiber neurons in mammals. Brain Res. Rev. 46, 131–145.

Djouhri, L., Fang, X., Okuse, K., Wood, J. N., Berry, C. M., and Lawson, S. N. 2003. The TTX-resistant sodium channel Nav1.8 (SNS/PN3): expression and correlation with membrane properties in rat nociceptive primary afferent neurons. J. Physiol 550, 739–752.

Dostrovsky, J. O. and Craig, A. D. 1996. Cooling-specific spinothalamic neurons in the monkey. J. Neurophysiol. 76, 3656–3665.

Dougherty, K. J., Sawchuk, M. A., and Hochman, S. 2005. Properties of mouse spinal lamina I GABAergic interneurons. J. Neurophysiol. 94, 3221–3227.

Dournaud, P., Slama, A., Beaudet, A., and Epelbaum, J. 2000. Somatostatin Receptors. In: Peptide Receptors, Part 1 (eds. R. Quirion, A. Björklund, and T. Hökfelt), pp. 1–43. Elsevier Science.

Doyle, C. A. and Maxwell, D. J. 1991. Ultrastructural analysis of noradrenergic nerve terminals in the cat lumbosacral spinal dorsal horn: A dopamine-β- hydroxylase immunocytochemical study. Brain Res. 563, 329–333.

Engelman, H. S. and MacDermott, A. B. 2004. Presynaptic ionotropic receptors and control of transmitter release. Nat. Rev. Neurosci. 5, 135–145.

Fields, H. L., Emson, P. C., Leigh, B. K., Gilbert, R. F., and Iversen, L. L. 1980. Multiple opiate receptor sites on primary afferent fibres. Nature 284, 351–353.

Flores, C. M., DeCamp, R. M., Kilo, S., Rogers, S. W., and Hargreaves, K. M. 1996. Neuronal nicotinic receptor expression in sensory neurons of the rat trigeminal ganglion: demonstration of $\alpha3\beta4$, a novel subtype in the mammalian nervous system. J. Neurosci. 16, 7892–7901.

Fritschy, J. M. and Grzanna, R. 1990. Demonstration of two separate descending noradrenergic pathways to the rat spinal cord: evidence for an intragriseal trajectory of locus coeruleus axons in the superficial layers of the dorsal horn. J. Comp. Neurol. 291, 553–582.

Furuyama, T., Kiyama, H., Sato, K., Park, H. T., Maeno, H., Takagi, H., and Tohyama, M. 1993. Region-specific expression of subunits of ionotropic glutamate receptors (AMPA-type, KA-type and NMDA receptors) in the rat spinal cord with special reference to nociception. Mol. Brain Res. 18, 141–151.

Gilbert, R. F., Emson, P. C., Hunt, S. P., Bennett, G. W., Marsden, C. A., Sandberg, B. E., Steinbusch, H. W., and Verhofstad, A. A. 1982. The effects of monoamine

neurotoxins on peptides in the rat spinal cord. Neuroscience 7, 69–87.

Gobel, S. 1975. Golgi studies of the substantia gelatinosa neurons in the spinal trigeminal nucleus. J. Comp. Neurol. 162, 397–416.

Gobel, S. 1978a. Golgi studies of the neurons in layer I of the dorsal horn of the medulla (trigeminal nucleus caudalis). J. Comp. Neurol. 180, 375–393.

Gobel, S. 1978b. Golgi studies of the neurons in layer II of the dorsal horn of the medulla (trigeminal nucleus caudalis). J. Comp. Neurol. 180, 395–414.

Gobel, S., Falls, W. M., Bennett, G. J., Abdelmoumene, M., Hayashi, H., and Humphrey, E. 1980. An EM analysis of the synaptic connections of horseradish peroxidase-filled stalked cells and islet cells in the substantia gelatinosa of the adult cat spinal cord. J. Comp. Neurol. 194, 781–807.

Grant, G. and Robertson, B. 2004. Primary Afferent Projections to the Spinal Cord. In: The Rat Nervous System (ed. G. Paxinos), pp. 111–119. Elsevier Academic Press.

Grudt, T. J. and Perl, E. R. 2002. Correlations between neuronal morphology and electrophysiological features in the rodent superficial dorsal horn. J. Physiol. (Lond.) 540, 189–207.

Guo, A., Simone, D. A., Stone, L. S., Fairbanks, C. A., Wang, J., and Elde, R. 2001. Developmental shift of vanilloid receptor 1 (VR1) terminals into deeper regions of the superficial dorsal horn: correlation with a shift from TrkA to Ret expression by dorsal root ganglion neurons. Eur. J. Neurosci. 14, 293–304.

Guo, A., Vulchanova, L., Wang, J., Li, X., and Elde, R. 1999. Immunocytochemical localization of the vanilloid receptor 1 (VR1): relationship to neuropeptides, the $P2X_3$ purinoceptor and IB4 binding sites. Eur. J. Neurosci. 11, 946–958.

Haberberger, R., Henrich, M., Couraud, J. Y., and Kummer, W. 1999. Muscarinic M2-receptors in rat thoracic dorsal root ganglia. Neurosci. Lett. 266, 177–180.

Hagihira, S., Senba, E., Yoshida, S., Tohyama, M., and Yoshiya, I. 1990. Fine structure of noradrenergic terminals and their synapses in the rat spinal dorsal horn: an immunohistochemical study. Brain Res. 526, 73–80.

Han, Z. S., Zhang, E. T., and Craig, A. D. 1998. Nociceptive and thermoreceptive lamina I neurons are anatomically distinct. Nat. Neurosci. 1, 218–225.

Hantman, A. W. and Perl, E. R. 2005. Molecular and genetic features of a labeled class of spinal substantia gelatinosa neurons in a transgenic mouse. J. Comp. Neurol. 492, 90–100.

Hantman, A. W., van den Pol, A. N., and Perl, E. R. 2004. Morphological and physiological features of a set of spinal substantia gelatinosa neurons defined by green fluorescent protein expression. J. Neurosci. 24, 836–842.

Harmann, P. A., Chung, K., Briner, R. P., Westlund, K. N., and Carlton, S. M. 1988. Calcitonin gene-related peptide (CGRP) in the human spinal cord: a light and electron microscopic analysis. J. Comp. Neurol. 269, 371–380.

Harvey, R. J., Depner, U. B., Wassle, H., Ahmadi, S., Heindl, C., Reinold, H., Smart, T. G., Harvey, K., Schutz, B., bo-Salem, O. M., Zimmer, A., Poisbeau, P., Welzl, H., Wolfer, D. P., Betz, H., Zeilhofer, H. U., and Muller, U. 2004. GlyR alpha3: an essential target for spinal PGE2-mediated inflammatory pain sensitization. Science 304, 884–887.

Heinke, B., Ruscheweyh, R., Forsthuber, L., Wunderbaldinger, G., and Sandkuhler, J. 2004. Physiological, neurochemical and morphological properties of a subgroup of GABAergic spinal lamina II neurones identified by expression of green fluorescent protein in mice. J. Physiol. 560, 249–266.

Henry, J. L. 1976. Effects of substance P on functionally identified units in cat spinal cord. Brain Res. 114, 439–451.

Hökfelt, T., Elde, R., Johansson, O., Luft, R., Nilsson, G., and Arimura, A. 1976. Immunohistochemical evidence for separate populations of somatostatin-containing and substance P-containing primary afferent neurons in the rat. Neuroscience 1, 131–136.

Hökfelt, T., Ljungdahl, A., Steinbusch, H. W., Verhofstad, A. N., Nilsson, G., Brodin, E., Pernow, B., and Goldstein, M. 1978. Immunohistochemical evidence of substance P-like immunoreactivity in some 5-hydroxytryptamine-containing neurons in the rat central nervous system. Neuroscience 3, 517–538.

Hökfelt, T., Ljungdahl, A., Terenius, L., Elde, R., and Nilsson, G. 1977. Immunohistochemical analysis of peptide pathways possibly related to pain and analgesia: enkephalin and substance P. Proc. Natl. Acad. Sci. USA 74, 3081–3085.

Hökfelt, T., Terenius, L., Kuypers, H. G. J. M., and Dann, O. 1979. Evidence for enkephalin immunoreactive neurons in the medulla oblongata projecting to the spinal cord. Neurosci. Lett. 14, 55–60.

Hökfelt, T., Zhang, X., and Wiesenfeld-Hallin, Z. 1994. Messenger plasticity in primary sensory neurons following axotomy and its functional implications. Trends Neurosci. 17, 22–30.

Honda, C. N. and Arvidsson, U. 1995. Immunohistochemical localization of delta- and mu-opioid receptors in primate spinal cord. NeuroReport 6, 1025–1028.

Hunt, S. P. and Mantyh, P. W. 2001. The molecular dynamics of pain control. Nat. Rev. Neurosci. 2, 83–91.

Hunt, S. P. and Rossi, J. 1985. Peptide- and non-peptide-containing unmyelinated primary sensory afferents: the parallel processing of nociceptive information. Philos. Trans. R. Soc. Lond. [Biol.] 308, 283–289.

Hunt, S. P., Kelly, J. S., Emson, P. C., Kimmel, J. R., Miller, R. J., and Wu, J.-Y. 1981. An immunohistochemical study of neuronal populations containing neuropeptides or gamma-aminobutyrate within the superficial layers of the rat dorsal horn. Neuroscience 6, 1883–1898.

Hwang, S. J., Pagliardini, S., Rustioni, A., and Valtschanoff, J. G. 2001. Presynaptic kainate receptors in primary afferents to the superficial laminae of the rat spinal cord. J. Comp. Neurol. 436, 275–289.

Jakowec, M. W., Fox, A. J., Martin, L. J., and Kalb, R. G. 1995a. Quantitative and qualitative changes in AMPA receptor expression during spinal cord development. Neuroscience 67, 893–907.

Jakowec, M. W., Yen, L., and Kalb, R. G. 1995b. In situ hybridization analysis of AMPA receptor subunit gene expression in the developing rat spinal cord. Neuroscience 67, 909–920.

Jasmin, L., Burkey, A. R., Card, J. P., and Basbaum, A. I. 1997. Transneuronal labeling of a nociceptive pathway, the spino-(trigemino-)parabrachio-amygdaloid, in the rat. J. Neurosci. 17, 3751–3765.

Jia, H., Rustioni, A., and Valtschanoff, J. G. 1999. Metabotropic glutamate receptors in superficial laminae of the rat dorsal horn. J. Comp. Neurol. 410, 627–642.

Jo, Y. H. and Schlichter, R. 1999. Synaptic corelease of ATP and GABA in cultured spinal neurons. Nat. Neurosci. 2, 241–245.

Jo, Y. H., Stoeckel, M. E., and Schlichter, R. 1998. Electrophysiological properties of cultured neonatal rat dorsal horn neurons containing GABA and met-enkephalin-like immunoreactivity. J. Neurophysiol. 79, 1583–1586.

Johansson, O., Hökfelt, T., Pernow, B., Jeffcoate, S. L., White, N., Steinbusch, H. W., Verhofstad, A. A., Emson, P. C., and Spindel, E. 1981. Immunohistochemical support for three putative transmitters in one neuron: coexistence of 5-hydroxytryptamine, substance P- and thyrotropin releasing hormone-like immunoreactivity in medullary neurons projecting to the spinal cord. Neuroscience 6, 1857–1881.

Ju, G., Hökfelt, T., Brodin, E., Fahrenkrug, J., Fischer, J. A., Frey, P., Elde, R. P., and Brown, J. C. 1987. Primary sensory neurons of the rat showing calcitonin gene-related peptide immunoreactivity and their relation to substance P-, somatostatin-, galanin-, vasoactive intestinal polypeptide-and cholecystokinin-immunoreactive ganglion cells. Cell Tissue Res. 247, 417–431.

Kar, S., Rees, R. G., and Quirion, R. 1994. Altered calcitonin gene-related peptide, substance P and enkephalin immunoreactivities and receptor binding sites in the dorsal spinal cord of the polyarthritic rat. Eur. J. Neurosci. 6, 345–354.

Kato, G., Yasaka, T., Katafuchi, T., Furue, H., Mizuno, M., Iwamoto, Y., and Yoshimura, M. 2006. Direct GABAergic and glycinergic inhibition of the substantia gelatinosa from the rostral ventromedial medulla revealed by in vivo patch-clamp analysis in rats. J. Neurosci. 26, 1787–1794.

Keller, A. F., Coull, J. A., Chéry, N., Poisbeau, P., and De Koninck, Y. 2001. Region-specific developmental specialization of GABA-glycine cosynapses in laminas I–II of the rat spinal dorsal horn. J. Neurosci. 21, 7871–7880.

Kemp, T., Spike, R. C., Watt, C., and Todd, A. J. 1996. The μ-opioid receptor (MOR1) is mainly restricted to neurons that do not contain GABA or glycine in the superficial dorsal horn of the rat spinal cord. Neuroscience 75, 1231–1238.

Kerchner, G. A., Wilding, T. J., Li, P., Zhuo, M., and Huettner, J. E. 2001. Presynaptic kainate receptors regulate spinal sensory transmission. J. Neurosci. 21, 59–66.

Kimura, H., McGeer, P. L., Peng, J. H., and McGeer, E. G. 1981. The central cholinergic system studied by choline acetyltransferase immunohistochemistry in the cat. J. Comp. Neurol. 200, 151–201.

Kinnman, E., Nygårds, E. B., and Hansson, P. 1997. Peripheral α-adrenoreceptors are involved in the development of capsaicin induced ongoing and stimulus evoked pain in humans. Pain 69, 79–85.

Kissa, K., Mordelet, E., Soudais, C., Kremer, E. J., Demeneix, B. A., Brulet, P., and Coen, L. 2002. In vivo neuronal tracing with GFP-TTC gene delivery. Mol. Cell Neurosci. 20, 627–637.

Knyihár-Csillik, E., Csillik, B., and Rakic, P. 1982a. Periterminal synaptology of dorsal root glomerular terminals in the substantia gelatinosa of the spinal cord in the rhesus monkey. J. Comp. Neurol. 210, 376–399.

Knyihár-Csillik, E., Csillik, B., and Rakic, P. 1982b. Ultrastructure of normal and degenerating glomerular terminals of dorsal root axons in the substantia gelatinosa of the rhesus monkey. J. Comp. Neurol. 210, 357–375.

Kupfermann, I. 1991. Functional studies of cotransmission. Physiol. Rev. 71, 683–732.

Labrakakis, C. and MacDermott, A. B. 2003. Neurokinin receptor 1-expressing spinal cord neurons in lamina I and III/IV of postnatal rats receive inputs from capsaicin sensitive fibers. Neurosci. Lett. 352, 121–124.

Labrakakis, C., Tong, C. K., Weissman, T., Torsney, C., and MacDermott, A. B. 2003. Localization and function of ATP and GABA$_A$ receptors expressed by nociceptors and other postnatal sensory neurons in rat. J. Physiol. 549, 131–142.

Lawson, S. N., Crepps, B. A., and Perl, E. R. 1997. Relationship of substance P to afferent characteristics of dorsal root ganglion neurones in guinea-pig. J. Physiol. (Lond.) 505, 177–191.

Lawson, S. N., Crepps, B., and Perl, E. R. 2002. Calcitonin gene-related peptide immunoreactivity and afferent receptive properties of dorsal root ganglion neurones in guinea-pigs. J. Physiol. 540, 989–1002.

Leah, J., Menétrey, D., and De Pommery, J. 1988. Neuropeptides in long ascending spinal tract cells in the rat: evidence for parallel processing of ascending information. Neuroscience 24, 195–207.

Lee, C. J., Bardoni, R., Tong, C. K., Engelman, H. S., Joseph, D. J., Magherini, P. C., and MacDermott, A. B. 2002. Functional expression of AMPA receptors on central terminals of rat dorsal root ganglion neurons and presynaptic inhibition of glutamate release. Neuron 35, 135–146.

Leonard, R. B. and Cohen, D. H. 1975. A cytoarchitectonic analysis of the spinal cord of the pigeon (Columba livia). J. Comp. Neurol. 163, 159–180.

Li, H. S. and Zhao, Z. Q. 1998. Small sensory neurons in the rat dorsal root ganglia express functional NK-1 tachykinin receptor. Eur. J. Neurosci. 10, 1292–1299.

Light, A. R. 1992. Organization of the Spinal Cord. In: The Initial Processing of Pain and its Descending Control: Spinal and Trigeminal Systems. Pain and Headache Series, Vol.12 (Series ed. P. L. Gildenberg) (ed. A. R. Light), pp. 75–86. Basel.

Light, A. R. and Perl, E. R. 1979. Spinal termination of functionally identified primary afferent neurons with slowly conducting myelinated fibers. J. Comp. Neurol. 186, 133–150.

Light, A. R., Trevino, D. L., and Perl, E. R. 1979. Morphological features of functionally defined neurons in the marginal zone and substantia gelatinosa of the spinal dorsal horn. J. Comp. Neurol. 186, 151–172.

Lima, D. and Coimbra, A. 1983. The neuronal population of the marginal zone (lamina I) of the rat spinal cord. A study based on reconstructions of serially sectioned cells. Anat. Embryol. (Berl.) 167, 273–288.

Lima, D. and Coimbra, A. 1986. A Golgi study of the neuronal population of the marginal zone (lamina I) of the rat spinal cord. J. Comp. Neurol. 244, 53–71.

Lima, D. and Coimbra, A. 1988. The spinothalamic system of the rat: structural types of retrogradely labelled neurons in the marginal zone (lamina I). Neuroscience 27, 215–230.

Lima, D. and Coimbra, A. 1990. Structural types of marginal (lamina I) neurons projecting to the dorsal reticular nucleus of the medulla oblongata. Neuroscience 27, 591–606.

Littlewood, N. K., Todd, A. J., Spike, R. C., Watt, C., and Shehab, S. A. S. 1995. The types of neuron in spinal dorsal horn which possess neurokinin-1 receptors. Neuroscience 66, 597–608.

Liu, H., Brown, J. L., Jasmin, L., Maggio, J. E., Vigna, S. R., Mantyh, P. W., and Basbaum, A. I. 1994a. Synaptic relationship between substance P and the substance P receptor: Light and electron microscopic characterization of the mismatch between neuropeptides and their receptors. Proc. Natl. Acad. Sci. U. S. A. 91, 1009–1013.

Liu, H., Wang, H., Sheng, M., Jan, L. Y., Jan, Y. N., and Basbaum, A. I. 1994b. Evidence for presynaptic N-methyl-D-aspartate autoreceptors in the spinal cord dorsal horn. Proc. Natl. Acad. Sci. U. S. A. 91, 8383–8387.

Ljungdahl, A., Hökfelt, T., and Nilsson, G. 1978. Distribution of substance P-like immunoreactivity in the central nervous system of the rat - I. Cell bodies and nerve terminals. Neuroscience 3, 861–943.

Llewellyn-Smith, I. J. and Burnstock, G. 1998. Ultrastructural localization of P2X$_3$ receptors in rat sensory neurons. NeuroReport 9, 2545–2550.

Lopez-Garcia, J. A. and King, A. E. 1994. Membrane properties of physiologically classified rat dorsal horn neurons in vitro: correlation with cutaneous sensory afferent input. Eur. J. Neurosci. 6, 998–1007.

Lu, Y. and Perl, E. R. 2003. A specific inhibitory pathway between substantia gelatinosa neurons receiving direct C-fiber input. J. Neurosci. 23, 8752–8758.

Lu, Y. and Perl, E. R. 2005. Modular organization of excitatory circuits between neurons of the spinal superficial dorsal horn (laminae I and II). J. Neurosci. 25, 3900–3907.

Lu, C. R., Hwang, S. J., Phend, K. D., Rustioni, A., and Valtschanoff, J. G. 2002. Primary afferent terminals in spinal cord express presynaptic AMPA receptors. J. Neurosci. 22, 9522–9529.

Ma, W., Ribeiro-da-Silva, A., De Koninck, Y., Radhakrishnan, V., Cuello, A. C., and Henry, J. L. 1997. Substance P and enkephalin immunoreactivities in axonal boutons presynaptic to physiologically identified dorsal horn neurons. An ultrastructural multiple-labelling study in the cat. Neuroscience 77, 793–811.

Ma, W., Ribeiro-da-Silva, A., De Koninck, Y., Radhakrishnan, V., Henry, J. L., and Cuello, A. C. 1996. Quantitative analysis of substance P immunoreactive boutons on physiologically characterized dorsal horn neurons in the cat lumbar spinal cord. J. Comp. Neurol. 376, 45–64.

Ma, W. Y., Ribeiro-da-Silva, A., Noel, G., Julien, J.-P., and Cuello, A. C. 1995. Ectopic substance P and calcitonin gene-related peptide immunoreactive fibres in the spinal cord of transgenic mice over-expressing nerve growth factor. Eur. J. Neurosci. 7, 2021–2035.

Margeta-Mitrovic, M., Mitrovic, I., Riley, R. C., Jan, L. Y., and Basbaum, A. I. 1999. Immunohistochemical localization of GABA$_B$ receptors in the rat central nervous system. J. Comp. Neurol. 405, 299–321.

Marshall, G. E., Shehab, S. A. S., Spike, R. C., and Todd, A. J. 1996. Neurokinin-1 receptors on lumbar spinothalamic neurons in the rat. Neuroscience 72, 255–263.

Martin-Schild, S., Gerall, A. A., Kastin, A. J., and Zadina, J. E. 1998. Endomorphin-2 is an endogenous opioid in primary sensory afferent fibers. Peptides 19, 1783–1789.

Martin-Schild, S., Gerall, A. A., Kastin, A. J., and Zadina, J. E. 1999. Differential distribution of endomorphin 1- and endomorphin 2-like immunoreactivities in the CNS of the rodent. J. Comp. Neurol. 405, 450–471.

Martin-Schild, S., Zadina, J. E., Gerall, A. A., Vigh, S., and Kastin, A. J. 1997. Localization of endomorphin-2-like immunoreactivity in the rat medulla and spinal cord. Peptides 18, 1641–1649.

Maxwell, D. J. and Noble, R. 1987. Relationship between hair-follicle afferent terminations and glutamic acid decarboxylase-containing boutons in cat's spinal cord. Brain Res. 408, 308–312.

Maxwell, D. J. and Réthelyi, M. 1987. Ultrastructure and synaptic connections of cutaneous afferent fibres in the spinal cord. Trends Neurosci. 10, 117–123.

Maxwell, D. J., Fyffe, R. E. W., and Réthelyi, M. 1983. Morphological properties of physiologically characterized lamina III neurones in the cat spinal cord. Neuroscience 10, 1–22.

McLeod, A. L., Krause, J. E., Cuello, A. C., and Ribeiro-da-Silva, A. 1998. Preferential synaptic relationships between substance P-immunoreactive boutons and neurokinin 1 receptor sites in the rat spinal cord. Proc. Natl. Acad. Sci. U. S. A. 95, 15775–15780.

McLeod, A. L., Krause, J. E., and Ribeiro-da-Silva, A. 2000. Immunocytochemical localization of neurokinin B in the rat spinal dorsal horn and its association with substance P and GABA: an electron microscopic study. J. Comp. Neurol. 420, 349–362.

Ménétrey, D. and Basbaum, A. I. 1987. The distribution of substance P-, enkephalin- and dynorphin-immunoreactive neurons in the medulla of the rat and their contribution to bulbospinal pathways. Neuroscience 23, 173–187.

Merighi, A., Polak, J. M., and Theodosis, D. T. 1991. Ultrastructural visualization of glutamate and aspartate immunoreactivities in the rat dorsal horn, with special reference to the co-localization of glutamate, substance P and calcitonin- gene related peptide. Neuroscience 40, 67–80.

Michael, G. J. and Priestley, J. V. 1999. Differential expression of the mRNA for the vanilloid receptor subtype 1 in cells of the adult rat dorsal root and nodose ganglia and its downregulation by axotomy. J. Neurosci. 19, 18444–18854.

Mileusnic, D., Lee, J. M., Magnuson, D. J., Hejna, M. J., Krause, J. E., Lorens, J. B., and Lorens, S. A. 1999. Neurokinin-3 receptor distribution in rat and human brain: an immunohistochemical study. Neuroscience 89, 12269–12290.

Miller, K. E. and Seybold, V. S. 1987. Comparison of met-enkephalin-, dynorphin A-, and neurotensin immunoreactive neurons in the cat and rat spinal cords: II. Lumbar cord. J. Comp. Neurol. 255, 293–304.

Miller, K. E. and Seybold, V. S. 1989. Comparison of met-enkephalin, dynorphin A, and neurotensin immunoreactive neurons in the cat and rat spinal cords: III. Segmental differences in the marginal zone. J. Comp. Neurol. 279, 619–628.

Minami, M. and Satoh, M. 1995. Molecular biology of the opioid receptors: Structures, functions and distributions. Neurosci. Res. 23, 121–145.

Minami, M., Maekawa, K., Yabuuchi, K., and Satoh, M. 1995. Double in situ hybridization study on coexistence of μ-, δ- and kappa-opioid receptor mRNAs with preprotachykinin A mRNA in the rat dorsal root ganglia. Mol. Brain Res. 30, 203–210.

Mitchell, K., Spike, R. C., and Todd, A. J. 1993. An immunocytochemical study of glycine receptor and GABA in laminae I–III of the rat spinal dorsal horn. J. Neurosci. 13, 2371–2381.

Molander, C., Xu, Q., and Grant, G. 1984. The cytoarchitectonic organization of the spinal cord in the rat: II. The lower thoracic and lumbosacral cord. J. Comp. Neurol. 230, 133–141.

Molander, C., Xu, Q., Rivero-Melian, C., and Grant, G. 1989. Cytoarchitectonic organization of the spinal cord in the rat: III. The cervical and upper thoracic spinal cord. J. Comp. Neurol. 289, 375–385.

Moore, K. A., Kohno, T., Karchewski, L. A., Scholz, J., Baba, H., and Woolf, C. J. 2002. Partial peripheral nerve injury promotes a selective loss of GABAergic inhibition in the superficial dorsal horn of the spinal cord. J. Neurosci. 22, 6724–6731.

Naim, M., Spike, R. C., Watt, C., Shehab, S. A., and Todd, A. J. 1997. Cells in laminae III and IV of the rat spinal cord that possess the neurokinin-1 receptor and have dorsally directed dendrites receive a major synaptic input from tachykinin-containing primary afferents. J. Neurosci. 17, 5536–5548.

Nakaya, Y., Kaneko, T., Shigemoto, R., Nakanishi, S., and Mizuno, N. 1994. Immunohistochemical localization of substance P receptor in the central nervous system of the adult rat. J. Comp. Neurol. 347, 2249–2774.

Nassar, M. A., Stirling, L. C., Forlani, G., Baker, M. D., Matthews, E. A., Dickenson, A. H., and Wood, J. N. 2004. Nociceptor-specific gene deletion reveals a major role for Nav1.7 (PN1) in acute and inflammatory pain. Proc. Natl. Acad. Sci. U. S. A. 101, 127096–127101.

Nishikawa, N., Bennett, G. J., Ruda, M. A., Lu, G. W., and Dubner, R. 1983. Immunocytochemical evidence for a serotoninergic innervation of dorsal column postsynaptic neurons in cat and monkey: light- and electron-microscopic observations. Neuroscience 10, 1333–1340.

Nusbaum, M. P., Blitz, D. M., Swensen, A. M., Wood, D., and Marder, E. 2001. The roles of co-transmission in neural network modulation. Trends Neurosci. 24, 146–154.

Ogawa, T., Kanazawa, I., and Kimura, S. 1985. Regional distribution of substance P, neurokinin α and neurokinin β in rat spinal cord, nerve roots and dorsal root ganglia, and the effects of dorsal root section or spinal transection. Brain Res. 359, 152–157.

Ohishi, H., Nomura, S., Ding, Y. Q., Shigemoto, R., Wada, E., Kinoshita, A., Li, J. L., Neki, A., Nakanishi, S., and Mizuno, N.

1995. Presynaptic localization of a metabotropic glutamate receptor, mGluR7, in the primary afferent neurons: An immunohistochemical study in the rat. Neurosci. Lett. 202, 85–88.

Olave, M. J. and Maxwell, D. J. 2002. An investigation of neurones that possess the alpha 2C adrenergic receptor in the rat dorsal horn. Neuroscience 115, 31–40.

Olave, M. J. and Maxwell, D. J. 2003a. Axon terminals possessing the alpha 2c adrenergic receptor in the rat dorsal horn are predominantly excitatory. Brain Res. 965, 269–273.

Olave, M. J. and Maxwell, D. J. 2003b. Neurokinin-1 projection cells in the rat dorsal horn receive synaptic contacts from axons that possess alpha2C-adrenergic receptors. J. Neurosci. 23, 6837–6846.

Ottersen, O. P. and Storm-Mathisen, J. 1984. Glutamate- and GABA-containing neurons in the mouse and rat brain, as demonstrated with a new immunocytochemical technique. J. Comp. Neurol. 229, 374–392.

Persohn, E., Malherbe, P., and Richards, J. G. 1991. In situ hybridization histochemistry reveals a diversity of GABA$_A$ receptor subunit mRNAs in neurons of the rat spinal cord and dorsal root ganglia. Neuroscience 42, 497–507.

Polgár, E., Fowler, J. H., McGill, M. M., and Todd, A. J. 1999a. The types of neuron which contain protein kinase C gamma in rat spinal cord. Brain Res. 833, 71–80.

Polgár, E., Furuta, T., Kaneko, T., and Todd, A. J. 2006. Characterization of neurons that express preprotachykinin B in the dorsal horn of the rat spinal cord. Neuroscience 1–27

Polgár, E., Shehab, S. A. S., Watt, C., and Todd, A. J. 1999b. GABAergic neurons that contain neuropeptide Y selectively target cells with the neurokinin 1 receptor in laminae III and IV of the rat spinal cord. J. Neurosci. 19, 2637–2646.

Poorkhalkali, N., Juneblad, K., Jönsson, A. C., Lindberg, M., Karlsson, O., Wallbrandt, P., Ekstrand, J., and Lehmann, A. 2000. Immunocytochemical distribution of the GABA$_B$ receptor splice variants GABA$_B$ R1a and R1b in the rat CNS and dorsal root ganglia. Anat. Embryol. (Berl.) 201, 1–13.

Prescott, S. A. and De Koninck, Y. 2002. Four cell types with distinctive membrane properties and morphologies in lamina II of the spinal dorsal horn of the adult rat. J. Physiol. 539, 817–836.

Priestley, J. V., Michael, G. J., Averill, S., Liu, M., and Willmott, N. 2002. Regulation of nociceptive neurons by nerve growth factor and glial cell derived neurotrophic factor. Can. J. Physiol. Pharmacol. 80, 495–505.

Ralston, H. J., III 1979. The fine structure of laminae I, II and III of the macaque spinal cord. J. Comp. Neurol. 184, 619–642.

Ramón y Cajal, S. 1909. Histologie du Système Nerveux de l'Homme et des Vertébrés. Vol. 1. Transl. L. Azoulay. A. Maloine.

Réthelyi, M., Light, A. R., and Perl, E. R. 1982. Synaptic complexes formed by functionally defined primary sensory afferent units with fine myelinated fibers. J. Comp. Neurol. 207, 381–393.

Rexed, B. 1952. The cytoarchitectonic organization of the spinal cord in the cat. J. Comp. Neurol. 96, 415–495.

Rexed, B. 1954. A cytoarchitectonic atlas of the spinal cord in the cat. J. Comp. Neurol. 100, 297–397.

Ribeiro-da-Silva, A. 1995. Ultrastructural features of the colocalization of calcitonin gene related peptide with substance P or somatostatin in the dorsal horn of the spinal cord. Can. J. Physiol. Pharmacol. 73, 940–944.

Ribeiro-da-Silva, A. 2004. Substantia Gelatinosa of the Spinal Cord. In: The Rat Nervous System (ed. G. Paxinos), pp. 129–148. Elsevier Academic Press.

Ribeiro-da-Silva, A. and Coimbra, A. 1980. Neuronal uptake of ^{3}H-GABA and ^{3}H-glycine in laminae I–III substantia gelatinosa Rolandi) of the rat spinal cord. An autoradiographic study. Brain Res. 188, 449–464.

Ribeiro-da-Silva, A. and Coimbra, A. 1982. Two types of synaptic glomeruli and their distribution in laminae I–III of the rat spinal cord. J. Comp. Neurol. 209, 176–186.

Ribeiro-da-Silva, A. and Cuello, A. C. 1990a. Choline acetyltransferase-immunoreactive profiles are presynaptic to primary sensory fibers in the rat superficial dorsal horn. J. Comp. Neurol. 295, 370–384.

Ribeiro-da-Silva, A. and Cuello, A. C. 1990b. Ultrastructural evidence for the occurrence of two distinct somatostatin-containing systems in the substantia gelatinosa of rat spinal cord. J. Chem. Neuroanat. 3, 141–153.

Ribeiro-da-Silva, A., De Koninck, Y., Cuello, A. C., and Henry, J. L. 1992. Enkephalin-immunoreactive nociceptive neurons in the cat spinal cord. NeuroReport 3, 25–28.

Ribeiro-da-Silva, A., McLeod, A. L., and Krause, J. E. 2000. Neurokinin Receptors in the CNS. In: Handbook of Chemical Neuroanatomy, Vol. 16: Peptide Receptors, Part I (eds. R. Quirion, A. Björklund, and T. Hökfelt), pp. 195–240. Elsevier Science.

Ribeiro-da-Silva, A., Pioro, E. P., and Cuello, A. C. 1991. Substance P- and enkephalin-like immunoreactivities are colocalized in certain neurons of the substantia gelatinosa of the rat spinal cord. An ultrastructural double-labeling study. J. Neurosci. 11, 1068–1080.

Ribeiro-da-Silva, A., Tagari, P., and Cuello, A. C. 1989. Morphological characterization of substance P-like immunoreactive glomeruli in the superficial dorsal horn of the rat spinal cord and trigeminal subnucleus caudalis: a quantitative study. J. Comp. Neurol. 281, 497–415.

Rossi, D. J. and Hamann, M. 1998. Spillover-mediated transmission at inhibitory synapses promoted by high affinity α6 subunit GABA$_A$ receptors and glomerular geometry. Neuron 20, 783–795.

Ruda, M. A. 1986. The Pattern and Place of Nociceptive Modulation in the Dorsal Horn. A Discussion of the Anatomically Characterized Neural Circuitry of Enkephalin, Serotonin, and Substance P. In: Spinal Afferent Processing (ed. T. L. Yaksh), pp. 141–164. Plenum Press.

Ruda, M. A., Bennett, G. J., and Dubner, R. 1986. Neurochemistry and neural circuitry in the dorsal horn. Prog. Brain Res. 66, 219–268.

Ruda, M. A., Coffield, J., and Dubner, R. 1984. Demonstration of postsynaptic opioid modulation of thalamic projection neurons by the combined techniques of retrograde horseradish peroxidase and enkephalin immunocytochemistry. J. Neurosci. 4, 2117–2132.

Ruda, M. A., Cohen, L. V., Shiosaka, S., Takahashi, T., Allen, B., Humphrey, E., and Iadarola, M. J. 1989. In situ Hybridization Histochemical and Immunocytochemical Analysis of Opioid Gene Products in a Rat Model of Peripheral Inflammation. In: Processing of Sensory Information in the Superficial Dorsal Horn of the Spinal Cord (eds. F. Cervero, G. J. Bennett, and P. M. Headley), pp. 383–394. Plenum Press.

Ruda, M. A., Iadarola, M. J., Cohen, L. V., and Young, W. S. 1988. In situ hybridization histochemistry and immunocytochemistry reveal an increase in spinal dynorphin biosynthesis in a rat model of peripheral inflammation and hyperalgesia. Proc. Natl. Acad. Sci. U. S. A. 85, 622–626.

Rudomin, P. and Schmidt, R. F. 1999. Presynaptic inhibition in the vertebrate spinal cord revisited. Exp. Brain Res. 129, 1–37.

Ruocco, I., Krause, J. E., and Ribeiro-da-Silva, A. 1997. Anatomical localization of substance P and the neurokinin-1 receptor in the skin of the rat lower lip: an electron microscopy study. Soc. Neurosci. Abst. 23, 1490. Ref Type: Abstract

Ruscheweyh, R. and Sandkuhler, J. 2002. Lamina-specific membrane and discharge properties of rat spinal dorsal horn neurones in vitro. J. Physiol. 541, 231–244.

Ruscheweyh, R., Ikeda, H., Heinke, B., and Sandkuhler, J. 2004. Distinctive membrane and discharge properties of rat spinal lamina I projection neurones in vitro. J Physiol 555, 527–543.

Sanderson Nydahl K., Skinner, K., Julius, D., and Basbaum, A. I. 2004. Co-localization of endomorphin-2 and substance P in primary afferent nociceptors and effects of injury: a light and electron microscopic study in the rat. Eur. J. Neurosci. 19, 1789–1799.

Scheibel, M. E. and Scheibel, A. B. 1968. Terminal axonal patterns in the cat spinal cord. II. The dorsal horn. Brain Res. 9, 32–58.

Schoenen, J. 1982. The dendritic organization of the human spinal cord: the dorsal horn. Neuroscience 7, 2057–2087.

Schulz, S., Schreff, M., Schmidt, H., Händel, M., Przewlocki, R., and Höllt, V. 1998. Immunocytochemical localization of somatostatin receptor sst$_{2A}$ in the rat spinal cord and dorsal root ganglia. Eur. J. Neurosci. 10, 3700–3708.

Senba, E., Yanaihara, C., Yanaihara, N., and Tohyama, M. 1988. Co-localization of substance P and Met-enkephalin-Arg^6-Gly^7-Leu^8 in the intraspinal neurons of the rat, with special reference to the neurons in the substantia gelatinosa. Brain Res. 453, 110–116.

Seybold, V. S. and Elde, R. P. 1982. Neurotensin immunoreactivity in the superficial laminae of the dorsal horn of the rat: I. Light microscopic studies of cell bodies and proximal dendrites. J. Comp. Neurol. 205, 89–100.

Sherman, S. E. and Loomis, C. W. 1994. Morphine insensitive allodynia is produced by intrathecal strychnine in the lightly anesthetized rat. Pain 56, 17–29.

Shin, J. B., Martinez-Salgado, C., Heppenstall, P. A., and Lewin, G. R. 2003. A T-type calcium channel required for normal function of a mammalian mechanoreceptor. Nat. Neurosci. 6, 724–730.

Snider, W. D. and McMahon, S. B. 1998. Tackling pain at the source: new ideas about nociceptors. Neuron 20, 629–632.

Spike, R. C. and Todd, A. J. 1992. Ultrastructural and immunocytochemical study of lamina II islet cells in rat spinal dorsal horn. J. Comp. Neurol. 323, 359–369.

Spike, R. C., Kerr, R., Maxwell, D. J., and Todd, A. J. 1998. GluR1 and GluR2/3 subunits of the AMPA-type glutamate receptor are associated with particular types of neurone in laminae I–III of the spinal dorsal horn of the rat. Eur. J. Neurosci. 10, 324–333.

Spike, R. C., Puskár, Z., Andrew, D., and Todd, A. J. 2003. A quantitative and morphological study of projection neurons in lamina I of the rat lumbar spinal cord. Eur. J. Neurosci. 18, 2433–2448.

Standaert, D. G., Watson, S. J., Houghten, R. A., and Saper, C. B. 1986. Opioid peptide immunoreactivity in spinal and trigeminal dorsal horn neurons projecting to the parabrachial nucleus in the rat. J. Neurosci. 6, 1220–1226.

Steiner, T. J. and Turner, L. M. 1972. Cytoarchitecture of the rat spinal cord. J. Physiol. (Lond.) 222, 123P–125P.

Stone, L. S., Broberger, C., Vulchanova, L., Wilcox, G. L., Hökfelt, T., Riedl, M. S., and Elde, R. 1998. Differential distribution of α_{2A} and α_{2C} adrenergic receptor immunoreactivity in the rat spinal cord. J. Neurosci. 18, 5928–5937.

Svensson, P., Cairns, B. E., Wang, K., Hu, J. W., Graven-Nielsen, T., rendt-Nielsen, L., and Sessle, B. J. 2003. Glutamate-evoked pain and mechanical allodynia in the human masseter muscle. Pain 101, 221–227.

Tachibana, M., Wenthold, R. J., Morioka, H., and Petralia, R. S. 1994. Light and electron microscopic immunocytochemical localization of AMPA-selective glutamate receptors in the rat spinal cord. J. Comp. Neurol. 344, 431–454.

Tashiro, T., Satoda, T., Takahashi, O., Matsushima, R., and Mizuno, N. 1988. Distribution of axons exhibiting both enkephalin- and serotonin- like immunoreactivities in the lumbar cord segments: an immunohistochemical study in the cat. Brain Res. 440, 357–362.

Tashiro, T., Takahashi, O., Satoda, T., Matsushima, R., and Mizuno, N. 1987. Immunohistochemical demonstration of coexistence of enkephalin- and substance P-like immunoreactivities in axonal components in the lumbar segments of cat spinal cord. Brain Res. 424, 391–395.

Thomson, A. M., West, D. C., and Headley, P. M. 1989. Membrane characteristics and synaptic responsiveness of superficial dorsal horn neurons in a slice preparation of adult rat spinal cord. Eur. J. Neurosci. 1, 479–488.

Todd, A. J. 1991. Immunohistochemical evidence that acetylcholine and glycine exist in different populations of GABAergic neurons in lamina III of rat spinal dorsal horn. Neuroscience 44, 741–746.

Todd, A. J. 1996. GABA and glycine in synaptic glomeruli of the rat spinal dorsal horn. Eur. J. Neurosci. 8, 2492–2498.

Todd, A. J. and Lewis, S. G. 1986. The morphology of Golgi-stained neurons in lamina II of the rat spinal cord. J. Anat. 149, 113–119.

Todd, A. J. and Lochhead, V. 1990. GABA-like immunoreactivity in type I glomeruli of rat substantia gelatinosa. Brain Res. 514, 171–174.

Todd, A. J. and McKenzie, J. 1989. GABA-immunoreactive neurons in the dorsal horn of the rat spinal cord. Neuroscience 31, 799–806.

Todd, A. J. and Ribeiro-da-Silva, A. 2005. Molecular Architecture of the Dorsal Horn. In: Neurobiology of Pain (eds. S. P. Hunt and M. Koltzenburg), pp. 65–94. Oxford University Press.

Todd, A. J. and Spike, R. C. 1993. The localization of classical transmitters and neuropeptides within neurons in laminae I–III of the mammalian spinal dorsal horn. Prog. Neurobiol. 41, 609–638.

Todd, A. J. and Sullivan, A. C. 1990. Light microscope study of the coexistence of GABA-like and glycine-like immunoreactivities in the spinal cord of the rat. J. Comp. Neurol. 296, 496–505.

Todd, A. J., Hughes, D. I., Polgár, E., Nagy, G. G., Mackie, M., Ottersen, O. P., and Maxwell, D. J. 2003. The expression of vesicular glutamate transporters VGLUT1 and VGLUT2 in neurochemically defined axonal populations in the rat spinal cord with emphasis on the dorsal horn. Eur. J. Neurosci. 17, 13–27.

Todd, A. J., McGill, M. M., and Shehab, S. A. 2000. Neurokinin 1 receptor expression by neurons in laminae I, III and IV of the rat spinal dorsal horn that project to the brainstem. Eur. J. Neurosci. 12, 689–700.

Todd, A. J., Russell, G., and Spike, R. C. 1992a. Immunocytochemical evidence that GABA and neurotensin exist in different neurons in laminae II and III of rat spinal dorsal horn. Neuroscience 47, 685–691.

Todd, A. J., Spike, R. C., Russell, G., and Johnston, H. M. 1992b. Immunohistochemical evidence that Met-enkephalin and GABA coexist in some neurones in rat dorsal horn. Brain Res. 584, 149–156.

Todd, A. J., Watt, C., Spike, R. C., and Sieghart, W. 1996. Colocalization of GABA, glycine, and their receptors at synapses in the rat spinal cord. J. Neurosci. 16, 974–982.

Tölle, T. R., Berthele, A., Zieglgänsberger, W., Seeburg, P. H., and Wisden, W. 1993. The differential expression of 16 NMDA and non-NMDA receptor subunits in the rat spinal cord and in periaqueductal gray. J. Neurosci. 13, 5009–5028.

Torsney, C. and MacDermott, A. B. 2006. Disinhibition opens the gate to pathological pain signaling in superficial neurokinin 1 receptor-expressing neurons in rat spinal cord. J. Neurosci. 26, 1833–1843.

Towers, S., Princivalle, A., Billinton, A., Edmunds, M., Bettler, B., Urban, L., Castro-Lopes, J., and Bowery, N. G. 2000. GABAB receptor protein and mRNA distribution in rat spinal cord and dorsal root ganglia. Eur. J. Neurosci. 12, 3201–3210.

Trafton, J. A., Abbadie, C., and Basbaum, A. I. 2001. Differential contribution of substance P and neurokinin A to spinal cord neurokinin-1 receptor signaling in the rat. J. Neurosci. 21, 3656–3664.

Von Banchet, G. S., Schindler, M., Hervieu, G. J., Beckmann, B., Emson, P. C., and Heppelmann, B. 1999. Distribution of somatostatin receptor subtypes in rat lumbar spinal cord examined with gold-labelled somatostatin and anti-receptor antibodies. Brain Res. 816, 254–257.

Wall, P. D., Merrill, E. G., and Yaksh, T. L. 1979. Responses of single units in laminae 2 and 3 of cat spinal cord. Brain Res. 160, 245–260.

Warden, M. K. and Young, W. S. 1988. Distribution of cells containing mRNAs encoding substance P and neurokinin B in the rat central nervous system. J. Comp. Neurol. 272, 90–113.

Watanabe, M., Mishina, M., and Inoue, Y. 1994. Distinct spatiotemporal distributions of the N-methyl-D-aspartate receptor channel subunit mRNAs in the mouse cervical cord. J. Comp. Neurol. 345, 314–319.

Westlund, K. N., Bowker, R. M., Ziegler, M. G., and Coulter, J. D. 1983. Noradrenergic projections to the spinal cord of the rat. Brain Res. 263, 15–31.

Wood, J. N., Boorman, J. P., Okuse, K., and Baker, M. D. 2004. Voltage-gated sodium channels and pain pathways. J. Neurobiol. 61, 55–71.

Woodbury, C. J., Ritter, A. M., and Koerber, H. R. 2000. On the problem of lamination in the superficial dorsal horn of mammals: a reappraisal of the substantia gelatinosa in postnatal rat. J. Comp. Neurol. 417, 88–102.

Xie, J., Ho, L. Y., Wang, C., Mo, C. J., and Chung, K. 2001. Differential expression of alpha1-adrenoceptor subtype mRNAs in the dorsal root ganglion after spinal nerve ligation. Mol. Brain Res. 93, 164–172.

Yaksh, T. L. 1989. Behavioral and autonomic correlates of the tactile evoked allodynia produced by spinal glycine inhibition: effects of modulatory receptor systems and excitatory amino acid antagonists. Pain 37, 111–123.

Yashpal, K., Kar, S., Dennis, T., and Quirion, R. 1992. Quantitative autoradiographic distribution of calcitonin gene-related peptide (hCGRPα) binding sites in the rat and monkey spinal cord. J. Comp. Neurol. 322, 224–232.

Yoshimura, M. and Jessell, T. M. 1989. Primary afferent evoked synaptic responses and slow potential generation in rat substantia gelatinosa neurons in vitro. J. Neurophysiol. 62, 96–108.

Yu, X. H., Ribeiro-da-Silva, A., and De Koninck, Y. 2005. Morphology and NK-1 receptor expression of spinothalamic lamina I neurons in the rat spinal cord. J. Comp. Neurol. 491, 56–68.

Yu, X. H., Zhang, E. T., Craig, A. D., Shigemoto, R., Ribeiro-da-Silva, A., and De Koninck, Y. 1999. NK-1 receptor immunoreactivity in distinct morphological types of lamina I neurons of the primate spinal cord. J. Neurosci. 19, 3545–3555.

Zeilhofer, H. U., Studler, B., Arabadzisz, D., Schweizer, C., Ahmadi, S., Layh, B., Bosl, M. R., and Fritschy, J. M. 2005. Glycinergic neurons expressing enhanced green fluorescent protein in bacterial artificial chromosome transgenic mice. J. Comp. Neurol. 482, 123–141.

Zeitz, K. P., Guy, N., Malmberg, A. B., Dirajlal, S., Martin, W. J., Sun, L., Bonhaus, D. W., Stucky, C. L., Julius, D., and Basbaum, A. I. 2002. The 5-HT3 subtype of serotonin receptor contributes to nociceptive processing via a novel subset of myelinated and unmyelinated nociceptors. J. Neurosci. 22, 1010–1019.

Zerari, F., Karpitskiy, V., Krause, J., Descarries, L., and Couture, R. 1997. Immunoelectron microscopic localization of NK-3 receptor in the rat spinal cord. NeuroReport 8, 2661–2664.

Zerari, F., Karpitskiy, V., Krause, J., Descarries, L., and Couture, R. 1998. Astroglial distribution of neurokinin-2 receptor immunoreactivity in the rat spinal cord. Neuroscience 84, 1233–1246.

Zhang, E. T. and Craig, A. D. 1997. Morphology and distribution of spinothalamic lamina I neurons in the monkey. J. Neurosci. 17, 3274–3284.

Zhang, E. T., Han, Z. S., and Craig, A. D. 1996. Morphological classes of spinothalamic lamina I neurons in the cat. J. Comp. Neurol. 367, 537–549.

Zhang, X., Nicholas, A. P., and Hökfelt, T. 1993. Ultrastructural studies on peptides in the dorsal horn of the spinal cord-II. Co-existence of galanin with other peptides in primary afferents in normal rats. Neuroscience 57, 365–384.

Zoli, M., Jansson, A., Sykova, E., Agnati, L. F., and Fuxe, K. 1999. Volume transmission in the CNS and its relevance for neuropsychopharmacology. Trends Pharmacol. Sci. 20, 142–150.

Zwick, M., Davis, B. M., Woodbury, C. J., Burkett, J. N., Koerber, H. R., Simpson, J. F., and Albers, K. M. 2002. Glial cell line-derived neurotrophic factor is a survival factor for isolectin B4-positive, but not vanilloid receptor 1-positive, neurons in the mouse. J. Neurosci. 22, 4057–4065.

24 Spinal Cord Physiology of Nociception

A R Light, University of Utah, Salt Lake City, UT, USA

S Lee, Korea Institute of Science and Technology, Seoul, Korea

24.1 Organization of the Spinal Cord

The spinal cord consists of a wrapping of axons inside of which are neurons comprising the first relay for sensory information and motoneurons with axons that connect directly with skeletal muscles. In addition to sensory and motor integration, considerable sympathetic nervous system integration occurs in this structure. The neurons of the spinal cord are clustered together in a butterflylike arrangement, with the each of the two wings containing neurons representing or controlling muscles and skin in the right or left side of the body, respectively. The more posterior (or dorsal) portion of each wing is specialized for sensory function and is called the dorsal horn, while the more anterior (or ventral) portion is specialized for motor function and is called the ventral horn (see Figures 1 and 2).

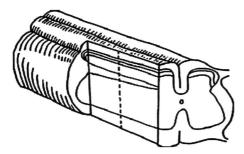

Figure 1   Cut-away view of lumbar spinal cord. From Light, A.R. 1992. Pain and Headache (Vol. 12): the Initial Processing of Pain and its Descending Control: Spinal and Trigeminal Systems. Karger, with permission.

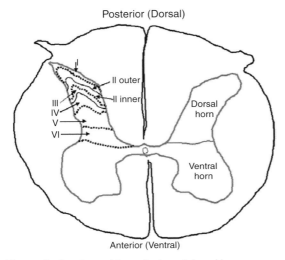

Figure 2   Laminae of the spinal cord dorsal horn.

24.2   Ventral Horn: Motor

The ventral horn contains all of the motoneurons whose axons leave the ventral horn in the ventral roots and directly innervate skeletal muscles of the body. However, most of the neurons in the ventral horn have axons that remain in the spinal cord and are, therefore, interneurons and are involved in motor integration. For the purposes of this chapter, the ventral horn is most important for the integration of reflexes. The most important of which are nociceptive reflexes. The most notable of the nociceptive reflexes is the flexion reflex which is often used as a surrogate for a response to pain, although the flexion reflex can often be regulated in different ways than the perception of pain

(Mauderli, A. P. *et al.*, 2000). This chapter will not consider the ventral horn in detail. The axons surrounding the ventral gray matter are of interest in the physiology of pain systems, as the axons of many nociceptive (and non-nociceptive) neurons in the dorsal horn travel in the white matter of the ventral horn. Some ascend all the way to the thalamus, or even hypothalamus.

24.3   Dorsal Horn: Sensory

The dorsal horn consists of the terminations of primary afferent neurons whose cell bodies lie in the dorsal root ganglia, and many interneurons with terminations within the spinal cord. The dorsal horn also contains many transmission neurons with axons that leave the spinal cord and terminate in the brainstem (medulla and pons), thalamus, and hypothalamus.

The dorsal horn also receives many inputs from higher brain centers that can modify, or in some cases, completely inhibit transmission from primary afferent neurons to other brain centers, including the ventral horn.

24.4   Laminar Organization

Many scientists have described the organization of the dorsal horn as laminar. Rexed B. (1952) divided the dorsal horn into six layers or laminae defined by the sizes and shapes of the cells within them (see Figure 2). Other defining characteristics of these laminae are the orientation of axons and dendrites of the neurons within them, and the amount of myelin that coats the axons in each layer (Clarke, J. L., 1859). The most dorsal, or superficial laminae (laminae I and II) as well as one of the deepest layers (lamina V) are most often associated with nociception (Light, A. R., 1992). While these laminae were originally defined using anatomical criteria they prove to also segregate some of the physiology of sensory systems of the spinal cord.

24.5   Laminae I and II (the Substantia Gelatinosa)

Because of its peculiar appearance in fresh tissue, the gelatinous translucent region that caps the dorsal horn has been called the substantia gelatinosa. This region was defined as lamina II by Rexed B. (1952). Capping the substantia gelatinosa is the even more superficial lamina I, also called the substantia spongiosa or marginal zone (Waldeyer, W., 1889; Ramon Y Cajal, S., 1909). These regions have been associated with nociception since at least the time of Ranson (Ranson, S. W., 1913; 1914; 1915; Ranson, S. W. and von Hess, C. L., 1915; Ranson, S. W. and Billingsley, P. R., 1916; Ranson, S. W. and Davenport, H. K., 1931). Ranson made the association between the translucent appearance of these regions and the lack of myelinated axon terminations in these regions and the likely termination of unmyelinated axons, which even then he associated with pain.

Fifty years later, lamina I and II were definitively shown to contain neurons that specifically responded to noxious (noxious meaning tissue damaging) stimuli and others responding to innocuous thermal stimuli (Christensen, B. N. and Perl, E. R., 1969; 1970; Mosso, J. A. and Kruger, L., 1973). Some of these neurons were shown to project through the anterolateral tract to the thalamus (Kumazawa, T. et al., 1971; 1975; Craig, A. D. and Kniffki, K. D., 1985). However, these investigators also showed that laminae I and II integrated more than purely nociceptive inputs with innocuous thermal neurons found in lamina I and innocuous mechanical neurons found in lamina II.

24.6 The Adequate Stimulus

Physiological descriptions of nociceptive neurons in the spinal cord dorsal horn are often based on the adequate stimulus. We define this as the best (most effective) natural (most likely to occur in an environment not manipulated by humans) stimulus that excites (evokes action potentials) in a neuron. For example, while punctate pressure that does not cause immediate harm to the skin may occasionally evoke an action potential in a spinal neuron, if that neuron fires a train of action potentials at 100 Hz to punctuate pressure that damages the skin, then the adequate stimulus is presumed to be noxious mechanical stimulation. If this same neuron responds to electrical shock, its adequate stimulus is not presumed to be electrical stimulation. Electrical shock is unlikely to occur in a world without human intervention.

24.7 Physiology of Neurons in the Dorsal Horn by Laminae

Being largely sensory in nature, the types of peripheral stimuli that activate the intrinsic neurons of the dorsal horn are strongly related to the physiological types of primary afferents that synapse on the spinal neurons. A discussion of the properties of nociceptive primary afferents and their responses to peripheral stimuli of various types is found in chapter 4. Integration of these primary afferent inputs occurs to various degrees in the spinal neurons. Some neurons retain the modality selectivity of the primary afferents, while others integrate the input of more than one type of primary afferent.

24.8 Lamina I

24.8.1 Modality of Input to Lamina I

Most investigators agree that a substantial portion of lamina I (see Figure 2 for location in spinal cord) neurons respond specifically or preferentially to various forms of noxious stimuli (Light, A. R., 1992). In studies using cats, Light A. R. found that 63% of neurons identified as lamina I neurons by intracellular labeling were nociceptive while 12% responded to innocuous cooling (4%) or warming (cooling and warming neurons are referred to collectively as thermoreceptive neurons); 14% responded to innocuous mechanical stimuli, and 5% responded to both innocuous mechanical stimuli and noxious stimuli. Of the neurons with nociceptive inputs, 28% responded only to noxious mechanical stimuli while 35% responded to noxious mechanical stimuli, and also to noxious heat and chemical stimuli. Only some of these neurons were tested with noxious cold. Figure 3 shows some characteristic responses of lamina I cells.

Andrew D. and Craig A. D. (2002) determined that the types of inputs observed for lamina I neurons projecting to the thalamus were similar to those described for unidentified lamina I neurons. Some differences for identified spinothalamic neurons were a disproportionate percentage of neurons with thermal inputs in the spinothalamic population. Of 125 lamina I spinothalamic neurons, 61% were nociceptive and 39% were thermoreceptive (most responded to cooling, 35%, with the rest responding to innocuous warming, 4%). Andrew D. and Craig A. D. also found that the nociceptive neurons were classifiable into

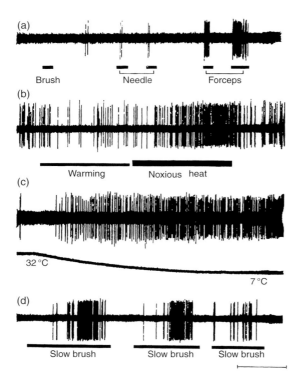

Figure 3 Responses of lamina I neurons (a) Neuron activated only by noxious mechanical stimulation of skin. (b) Neuron activated by both noxious mechanical stimuli (not shown) and noxious heating of the skin. (c) Neuron activated only by innocuous cooling of the skin. (d) Neurons activated only by innocuous mechanical stimulation of the skin. This neuron responded only to brushing at less than 10 cm s^{-1}. Scale bar = 10 s for (a), (b), and (d); 20 s for (c). From Light, A. R. 1992. Pain and Headache (Vol. 12): the Initial Processing of Pain and its Descending Control: Spinal and Trigeminal Systems. Karger. Reprinted with permission from Light, A. R., Trevino, D. L., and Perl, E. R. 1979. Morphological features of functionally defined neurons in the marginal zone and substantia gelatinosa of the spinal dorsal horn. J. Comp. Neurol. 186, 151–171. Copyright 1979, Wiley-Liss of John Wiley & Sons, Inc.

somewhat different groups than those summarized by Light A. R. (1992). Of the 61% nociceptive neurons, a minority (43%) were described as nociceptive specific. Of the nociceptive-specific neurons, most (80%) were responsive to both noxious mechanical and noxious heat stimuli, with only 20% responsive to only mechanical stimuli. The majority (57%) of the nociceptive neurons was classified as responding to heat, pressure, and cold (HPC). These neurons responded not only to noxious mechanical pressure, but also responded to innocuous cooling temperatures. Craig (Andrew, D. and Craig, A. D., 2002) has also determined the quantitative response properties of lamina I

nociceptive neurons (see Figure 4). The data presented in these publications clearly indicate that lamina I neurons are capable of reliably detecting and transmitting precise quantitative information about noxious pressure to higher centers (Craig, A. D. and Andrew, D., 2002). These authors showed the quantitative responses of lamina I neurons to activation by noxious heat. The responses correlate very well with human pain ratings of the same stimuli (see Figure 5).

In addition to nociceptive and thermal inputs, lamina I neurons also encode stimuli interpreted as itch and metabolic signals produced by contracting muscles.

Craig's laboratory reported recording neurons in lamina I that specifically respond to stimuli provoking itch (Andrew, D. and Craig, A. D., 2001; see Figure 6), and others that responded only to actively contracting muscles (Wilson, L. B. et al., 2002; see Figure 7). Some of the neurons responding only to actively contracting muscles may represent metaboreceptors that encode the metabolites produced by actively contracting muscles, and may be the peripheral receptors that are capable of encoding the signals interpreted as muscle fatigue, and/or muscle pain. These neurons may also be the sensory arm of sympathetic reflexes known to be activated by actively contracting muscles (Rotto, D. M. and Kaufman, M. P., 1988).

While the detailed neuronal properties presented here were obtained from recordings in cats, similar percentages of these various types of neurons in lamina I have been found in monkeys (Kumazawa, T. et al., 1975; Kumazawa, T. and Perl, E. R., 1978; Light, A. R. et al., 1979; Price, D. D. et al., 1979; Chung, J. M. et al., 1986; Craig, A. D., 2004) and rats (Giesler, G. J., Jr. et al., 1976; Menetrey, D. et al., 1977; Light, A. R. and Willcockson, H. H., 1999) although some of those referenced here report many more neurons responding to both noxious and non-noxious stimuli in the wide-dynamic-range (WDR) fashion. Very recently, similar types of neurons have also been reported in lamina I of the mouse, in an *ex vivo* preparation in mouse (Koerber, H. R., personal communica- tion, 2007).

24.8.2 Receptive Field Characteristics

The receptive fields of lamina I neurons range from quite small (1 mm^2) to quite large (1000 mm^2) depending on the body part encoded. The smallest receptive fields are found on the digits, while the largest are found on the back and hip. The receptive

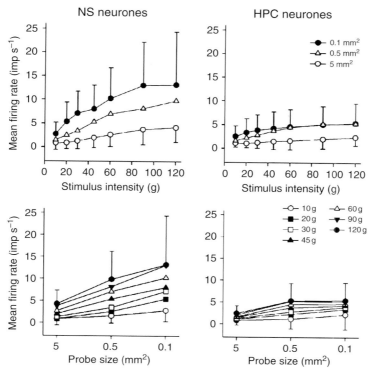

Figure 4 Quantitative responses of lamina I neurons Mean population stimulus–response curves to graded mechanical stimulation (10–120 g) for NS (n = 20) and HPC (n = 19) neurones (top row). The same data replotted with respect to toprobe size are shown in the bottom row. NS, nociceptive-specific neurons; HPC, heat, pressure, and cold. From Andrew, D. and Craig, A. D. 2002. Quantitative responses of spinothalamic lamina I neurones to graded mechanical stimulation in the cat. J. Physiol. 545, 913–931, reprinted with permission.

field sizes and pattern indicates considerable convergence from primary afferent neurons. Thus, the smallest excitatory receptive fields for lamina I neurons can approximate the size of C-primary afferent polymodal nociceptors (<1 mm). The largest excitatory receptive fields, however, are many times the size of the largest primary afferent receptive field. Lamina I neurons with A-fiber nociceptive input do not have the spotlike receptive fields of the primary afferents, but have a more uniform receptive field, indicating convergence of several A-nociceptors onto the lamina I neurons.

Excitatory receptive fields often have an inhibitory surround region, stimulation of which with noxious stimuli causes inhibitory post synaptic potentials (IPSPs)in the recorded neuron and can inhibit activation induced by stimulation of the more central excitatory receptive field (see Figure 8 for explanation of how these receptive fields are formed).

24.9 Projections and Function of Lamina I Neurons

While many of the neurons in lamina I most likely function as integrative interneurons within lamina I, a large proportion of these neurons project to other spinal cord and brain regions.

Many of the neurons in lamina I have been shown to project to higher brain centers including the thalamus, the parabrachial region, the medullary reticular formation, and even the hypothalamus. A much larger proportion of lamina I cells have been shown to project to higher brain centers than neurons located in laminae II-IV. Craig A. D. (1996; 2003) has also found lamina I projection neurons responding specifically to itch inputs, and others responding specifically to ergoreception (inputs from actively contracting muscles; see above). He has attempted to generalize these inputs as homeostasis inputs, that is, inputs necessary to maintain body integrity.

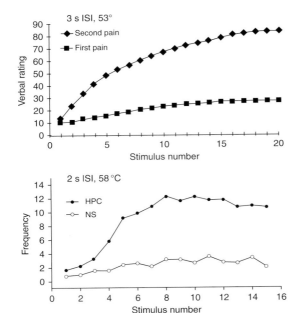

Figure 5 Comparison of the mean response curves of heat, pressure, and cold (HPC) and nociceptive-specific (NS) lamina I spinothalamic tract cells (bottom) to the maximal repeated brief contact heat stimulus trial with the mean report of human subjects to the maximal trial in the study of Vierck *et al.* (1997) (top). In the top panel, 53 DEG = 53°C; ISI = Interstimulus interval. From Craig, A. D. and Andrew, D. 2002. Responses of spinothalamic lamina I neurons to repeated brief contact heat stimulation in the cat. J. Neurophysiol. 87, 1902–1914, reprinted with permission.

Thus, one of the major functions of lamina I is to convey information about nociceptive inputs, itch, and thermal inputs, to higher brain centers . Lamina I also integrates and relays inputs necessary for sympathetic functions. Craig has grouped these functions as homeostatic and suggests the view that lamina I is critical for relatively objective information about body integrity necessary to maintain homeostasis, and thus, life (Craig, A. D., 1996; 2003).

24.10 Lamina II

Almost 30 years after the first definitive recordings were made from the substantia gelatinosa, the function of this region is still not understood. Recordings combined with intracellular labeling suggested that the anatomically observable division of lamina II into an inner and outer region had an apparent functional correlation.

24.11 Outer Lamina II

24.11.1 Modality of Input to Outer Lamina II

The outer region contains mostly nociceptive neurons while the inner region contains mostly innocuous mechanically responsive neurons (see Figure 9 for recordings of common responses). This relationship has been shown in cats, monkeys, and rats (Kumazawa, T. and Perl, E. R., 1978; Bennett, G. J. *et al.*, 1979; Light, A. R. *et al.*, 1979; Bennett, G. J. *et al.*, 1980; Light, A. R. *et al.*, 1981; Light, A. R. and Kavookjian, A. M., 1984; Steedman, W. M. *et al.*, 1985; Light, A. R. and Willcockson, H. H., 1999).

In a summary of data from experiments in the Light and Perl laboratory (Light, A. R., 1992), 70% of the neurons identified as having cell bodies in lamina IIo were nociceptive. Forty-eight per cent responded only to mechanical nociceptive inputs, while 32% responded to at least noxious heat and noxious mechanical inputs, and many that were tested, also responded to noxious chemicals, making them polymodal nociceptive inputs. Of the remainder, 20% responded to both innocuous and noxious mechanical inputs and 5% each were thermoreceptive (most to cooling) and 5% responded only to innocuous mechanoreceptive stimuli.

The neurons in the outer region of lamina II have small receptive fields, and neurons can receive only noxious mechanical inputs, only noxious heat or cold inputs, and mixtures of mechanical and heat, and noxious chemical inputs. Summarizing data from several publications (Light, A. R. *et al.*, 1979; Light, A. R. and Kavookjian, A. M., 1984; Light, A. R. and Kavookjian, A. M., 1988; Light, A. R. and Willcockson, H. H., 1999), some 70% of lamina IIo neurons are nociceptive, 5% are thermoreceptive, 5% are responsive to innocuous mechanical stimuli, and 20% respond to both innocuous mechanical stimuli and nociceptive stimuli. Similar to lamina I neurons, some of the nociceptive neurons respond selectively to noxious mechanical inputs, with perhaps more of these than other types. Other nociceptive neurons respond to noxious heat as well as to noxious mechanical stimuli.

24.11.2 Receptive Field Characteristics

The receptive fields of lamina IIo neurons tend to be smaller than those of the overlying lamina I

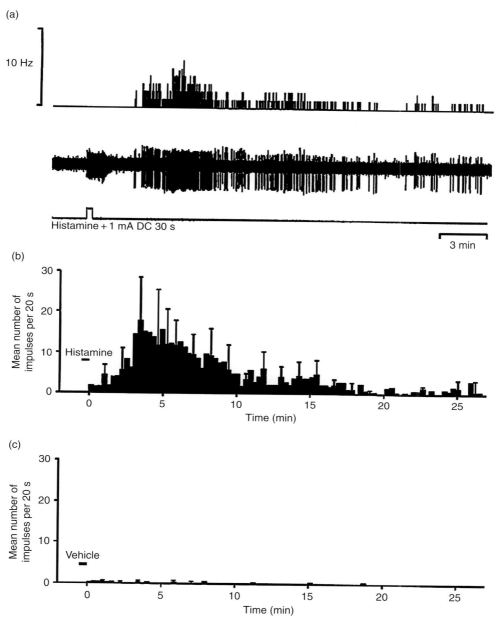

Figure 6 Histamine- and vehicle-evoked responses from histamine-sensitive lamina I spinothalamic tract neurons. (a) The response of a single neuron to histamine. Top histogram, binned firing rate of the neuron (1-s bins). The middle trace shows the analog record of neuronal activity; the thickening of the baseline in this record during the iontophoresis (indicated by the lower trace) is due to the current-evoked activation of several neighboring neurons. (b) Mean response of all 10 histamine-sensitive lamina I spinothalamic tract neurons to histamine application. (c) Mean responses of the same 10 neurons to vehicle application. Error bars, 1 standard deviation; bin size, 20 s; DC, direct current. With permission from Andrew, D. and Craig, A. D. 2001. Spinothalamic lamina I neurons selectively sensitive to histamine: a central neural pathway for itch. Nat. Neurosci. 4, 72–77.

neurons (Price, D. D. *et al.*, 1979). This may be due to the less extensive mediolateral spread of dendrites of lamina II neurons. Lamina II neurons may also have fewer complex receptive fields, sometimes lacking an inhibitory surround region. These two observations, along with some

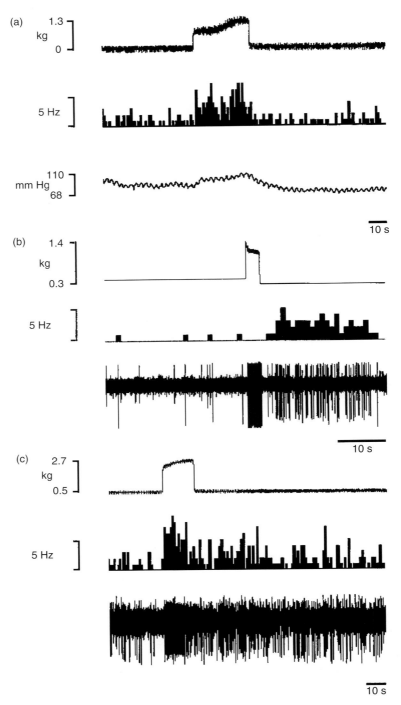

Figure 7 Examples of the variety of responses of lamina I neurons that were excited by muscle contraction. The traces show from the top downward: the tension developed in the triceps surae muscle, the discharge of the single neuron histogrammed in 1 s bins and either mean blood pressure (a) or the neural recording ((b) and (c)). (a) Response from a heat, pressure, and cold (HPC) lamina I neuron with unidentified projections that was excited by contraction that paralleled the tension developed in the muscle and the centrally evoked cardiovascular reflex. (b) Response of another HPC lamina I neuron with unidentified projections that was only excited after the contraction. (c) Response of a spinobulbar lamina I neuron (central conduction velocity 2.2 m s^{-1}) that was excited during and after the contraction; this neuron had no identifiable cutaneous receptive field but responded to stretching the triceps muscle and also to rhythmic contractions at 5 Hz. From Wilson, L. B., Andrew, D., and Craig, A. D. 2002. Activation of spinobulbar lamina I neurons by static muscle contraction. J. Neurophysiol. 87, 1641–1645, reprinted with permission.

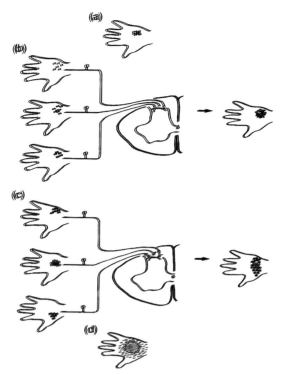

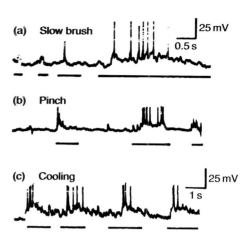

Figure 9 Low-gain, direct-current patch-clamp recordings *in vivo* from the most common types of responsive neurons encountered in the substantia gelatinosa of rats. Examples of slow brush (a), pinch (b), and innocuous cooling (c) (ethyl chloride spray applied to brush and brushed on skin) responsive neurons. From Light, A. R. and Willcockson, H. H. 1999. Spinal laminae I–II neurons in rat recorded in vivo in whole cell, tight seal configuration: properties and opioid responses. J. Neurophysiol. 82, 3316–3326, reprinted with permission.

Figure 8 Diagram of mechanisms of integration of nociceptive inputs in lamina I neurons. (a) Indicates the receptive field of a single primary afferent nociceptor with an Aδ axon. Each spot represents a point at which noxious mechanical stimulation evokes a discharge. (b) Left panel indicates the receptive fields of three mechanical nociceptors with overlapping receptive fields. The three primary afferents terminate on the same lamina I neuron in the middle panel. The right panel indicates the resulting receptive field. The receptive field is somewhat larger, and more responsive points are found within the responsive area. Thus, unlike primary afferent receptors, spinal nociceptive neurons have nearly continuous receptive fields with less unresponsive regions between responsive spots. (c) Left panel indicates the receptive fields of three nonoverlapping primary afferent mechanical nociceptors. The middle panel indicates that the lamina I neuron receives excitatory input from the axons whose receptive field is indicated with open circles. The other two axons excite inhibitory neurons (this is a simplification), and in all cases, both excitatory and inhibitory neurons are excited by all primary afferent neurons). The result is that the receptive field of the lamina I neuron, shown at right, consists of inhibitory inputs (shown with filled symbols) and excitatory inputs (shown with open symbols). (d) This receptive field is typical for spinal cord nociceptive neurons and indicates both of the processes described in (a)–(c). The central crosshatched region is the excitatory receptive field in which noxious mechanical stimulation causes excitation of the spinal neuron. Noxious mechanical stimulation of the region outside the excitatory receptive field causes inhibition and is indicated by minus signs. This inhibitory field fades out gradually as the distance from the excitatory field increases. Adapted from Light, A. R. 1992. Pain and Headache (Vol. 12): the Initial Processing of Pain and its Descending Control: Spinal and Trigeminal Systems. Karger, reprinted with permission.

anatomical findings suggest that at least some lamina II neurons project inputs to overlying lamina I neurons.

24.12 Projections and Function of Outer Lamina II Neurons

In addition to projections to lamina I, lamina IIo neurons have been shown to project to deeper laminae, including laminae III, IV, and V (Light, A. R. and Kavookjian, A. M., 1988; Eckert, W. A., III, *et al.*, 2003). Substantial intralaminar projects have also been noted for many years (Szentagothai, J., 1964; Matsushita, M., 1969; Rethelyi, M. and Szentagothai, J., 1969; Rethelyi, M. *et al.*, 1981; 1983; 1989). Both inhibitory and excitatory intralaminar, as well as extralaminar connections, are possible (Lu, Y. and Perl, E. R., 2005). A few neurons in lamina II have also been shown to project long connections to the thalamus (Giesler, G. J., Jr. *et al.*, 1978; Willis, W. D. *et al.*, 1978). Whether these projections represent ectopic lamina I neurons, or more projections of this variety have yet to be detected is unknown.

Thus, lamina IIo neurons may function both as inhibitory neurons helping to create the inhibition surround receptive fields described above, and as excitatory relay neurons for convergence of receptive fields and possible amplification of the excitatory signal (see Figure 8).

24.12.1   Inner Lamina II (II$_i$)

The inner division of lamina II has many neurons that respond to innocuous mechanical stimuli (Light, A. R. et al., 1979; Price, D. D. et al., 1979; Woolf, C. J. and Fitzgerald, M., 1983; Light, A. R. and Kavookjian, A. M., 1988; Light, A. R. and Willcockson, H. H., 1999). A detailed analysis of the types of inputs to this lamina has not been made. One survey (Light, A. R., 1992) found that between 50% and 80% of the neurons responded only to innocuous mechanical stimuli. A very common response was a preferential response to a very slowly moving stimulus (e.g., see Figure 9(a)). The response to mechanical stimulation often adapts with very few repetitions of the stimulus, sometimes after only a single stimulation. Responses of this type do not relate directly to any known type of primary afferent input. Intracellular analysis of these response properties identified a prominent IPSP evoked by innocuous mechanical stimulation that was capable of inhibiting nearly all action potential generation (Light, A. R. et al., 1979). Neurons with these types of inputs seem to dominate this sublamina.

However, in addition to innocuous mechanical inputs, considerable circumstantial evidence suggests that nociceptive inputs should also reach inner lamina II. First, a few neurons in this layer have been found to be activated preferentially by nociceptive stimuli. Second, primary afferents labeling with isolectin B4 (IB4) terminate heavily in inner lamina II. At least some of the IB4-positive afferents have been identified as nociceptive, and eliminating IB4-containing primary afferents reduced inflammatory pain in rats. vanilloid receptor like 1 (VRL-1 also called TRPV2)also labels inner lamina II suggesting that a population of myelinated nociceptor may selectively terminate in this region (Lewinter, R. D. et al., 2004). Neurons in lamina II$_i$ show increases in protein kinase Cγ (PKCγ) following inflammation which results in allodynia (Martin, W. J. et al., 1999; 2001). At least a few nociceptive primary afferent fibers have shown extensive terminations in lamina II$_i$ (Sugiura, Y. et al., 1986; Koerber H. R. et al., personal communication, 2004). Finally, either muscle nociceptive input to laminae II$_i$ cells, and/or muscle metaboreceptor input to this layer is

predicted by the terminations of single group IV muscle afferents in this layer (Ling, L. J. et al., 2003).

Potentially then, nociceptive neurons, and possibly, metaboreceptive neurons should be found in inner lamina II. Whether these neurons represent special types of nociceptive inputs (such as inputs sensitive to inflammation) or have nociceptive inputs (such as muscle nociceptive inputs) that have not been found in recording studies thus far remains to be examined.

24.12.1.1   Projections and function of inner lamina II neurons

Like other neurons in lamina II, lamina II$_i$ has substantial axonal projections to other regions of lamina II (Szentagothai, J., 1964; Matsushita, M., 1969; Rethelyi, M. and Szentagothai, J., 1969; Rethelyi, M. et al., 1981; 1983; 1989). In addition, at least some of these neurons project axons to deeper laminae, where they synapse with other spinal cord neurons (Light, A. R. and Kavookjian, A. M., 1988; Eckert, W. A., III, et al., 2003). Basbaum's laboratory has also recently proposed that neurons in this region project to lamina V neurons which subsequently project to higher brain regions, not usually considered in nociceptive processing (Braz, J. M. et al., 2005).

The innocuous responses of lamina II$_i$ neurons seems most related to sensuous touch (Olausson, H. et al., 2002) or tickle. As these neurons have not been shown to project in large numbers to the thalamus or other brain regions rostral to the spinal cord, it is unknown how these inputs are relayed to higher brain centers. However, it is interesting that some have noted that tickle is abolished following lesions of the anterolateral tract (Lahuerta, J. et al., 1990).

24.12.2   Lamina III

Lamina III mostly contains neurons that receive only innocuous mechanical inputs (Pomeranz, B. et al., 1968; Price, D. D. and Mayer, D. J., 1974; Wall, P. D. et al., 1979; Light, A. R. and Durkovic, R. G., 1984). However, at least some nociceptive neurons can be found here. Nociceptive neurons in lamina III may respond only to noxious mechanical stimuli, or to noxious thermal and chemical stimuli as well as to noxious mechanical stimuli (Brown, A. G. and Fyffe, R. E., 1981; Maxwell, D. J. et al., 1983; Maxwell, D. J. and Koerber, H. R., 1987). Most investigators have reported that neurons that do respond to noxious stimuli, also respond to innocuous stimuli, albeit with much lower action potential frequency. The importance of lamina III in nociception is unknown. In cats, many lamina III neurons project to the lateral cervical nucleus. In rats and primates, at least some of these neurons project to the dorsal column nuclei (Rustioni, A. *et al.*, 1979; Giesler, G. J., Jr. *et al.*, 1984). As with lamina IIi, PKCγ is upregulated in lamina III neurons following nerve injury and inflammation (Polgar, E. *et al.*, 1999; Miletic, V. *et al.*, 2000), suggesting at least inflammatory inputs, if not nociceptive inputs to this lamina.

24.12.3   Lamina IV

Lamina IV neurons are also dominated by responses to innocuous mechanical stimuli (Armett, C. J. *et al.*, 1962; Pomeranz, B. *et al.*, 1968; Price, D. D. and Browe, A. C., 1973). Many other neurons, however, respond more vigorously to noxious stimuli (both mechanical and thermal) than to innocuous stimuli (the WDR response). In cats, for example, Light A. R. and Durkovic R. G. (1984) found that 56% of the recorded neurons responded only to innocuous mechanical stimuli, while 37% responded to innocuous stimuli, but increased their response rate to noxious stimuli. Only 6% of the neurons responded only to noxious stimuli. In cats, a substantial number of lamina IV neurons project to the lateral cervical nucleus and are, therefore, spinocervical neurons (Brown, A. G., 1968; Brown, A. G. and Franz, D. N., 1969; Bryan, R. N. *et al.*, 1973; 1974). At least some of these neurons encode some aspects of noxious stimuli. The proportion of spinocervical neurons in primates appears to be less than in cats (Bryan, R. N. *et al.*, 1974). Again some of these neurons encode noxious stimuli (Brown, A. G., 1968; Brown, A. G. and Franz, D. N., 1969; Bryan, R. N. *et al.*, 1973; 1974; Kajander, K. C. and Giesler, G. J., Jr., 1987). Specific functions for noxious inputs relayed via the spinocervical tract are unknown, but may represent an alternative pathway for pain transmission, separate from the spinothalamic tract (Willis, W. D. and Westlund, K. N., 1997). Finally, at least some lamina IV neurons project to the thalamus (Trevino, D. L. *et al.*, 1973; Albe-Fessard, D. *et al.*, 1974; Trevino, D. L. *et al.*, 1974; Willis, W. D. *et al.*, 1974; Price, D. D. and Mayer, D. J., 1975).

24.12.4   Lamina V

A very large literature exists on the nociceptive properties of neurons in lamina V (Wall, P. D., 1967; Mendell, L. M. and Wall, P. D., 1965; Hillman, P. and Wall, P. D., 1969; Wall, P. D., 1973; Price, D. D. and Dubner, R., 1977; Dubner, R., 1978; Dubner, R. and Bennett, G. J., 1983; Willis, W. D., 1985; 1988).

This is partly due to the relative ease of recording the larger neurons located in this layer, and partly due to its known large projection to the thalamus, making it a large contributor to the spinothalamic tract and this tract has long been implicated as a pain transmission pathway.

Lamina V contains many neurons that respond to both noxious and innocuous stimuli, and also neurons that respond selectively to noxious stimuli. Innocuous mechanoreceptive neurons are also found in this lamina (Mendell, L. M. and Wall, P. D., 1965; Hillman, P. and Wall, P. D., 1969; Light, A. R. and Durkovic, R. G., 1984).

The receptive fields of lamina V neurons tend to be larger than neurons with similar modalities in lamina I–IV. A common feature of the receptive fields, like those in lamina I, is a zone of inhibitory influence surrounding the excitatory receptive field (inhibition surround). Typically, the most effective stimulus for eliciting inhibition is the same as that of the excitatory region (Hillman, P. and Wall, P. D., 1969).

The response characteristics of at least some lamina V neurons correlate well with human estimates of pain intensity and with reflex response of animals to noxious stimuli (Dubner, R. *et al.*, 1977; 1980; Casey, K. L., 1982; Dubner, R. *et al.*, 1989; e.g., see figure 10). This, along with the fact that many lamina V neurons project in the anterolateral tract to the thalamus, both of which are implicated in pain transmission, indicate that lamina V neurons are important in at least some aspects of pain perception. The likely contribution of lamina V neurons to the flexion reflex pathway probably contributes to the observations that lamina V neuron responses are similar to nociceptive reflexes (Light, A. R. and Durkovic, R. G., 1984; Morgan, M. M., 1998).

The presence of innocuous activation of many lamina V neurons in addition to nociceptive activation requires additional mechanisms to allow discrimination between innocuous and noxious stimuli.

24.13   The Ventral Horn

The ventral horn contains many neurons that project to the thalamus. However, the properties of these spinal neurons suggest a role in proprioception, rather than in nociceptive processing (Menetrey, D. *et al.*, 1984; Ammons, W. S., 1987; Craig, A. D., Jr. *et al.*, 1989).

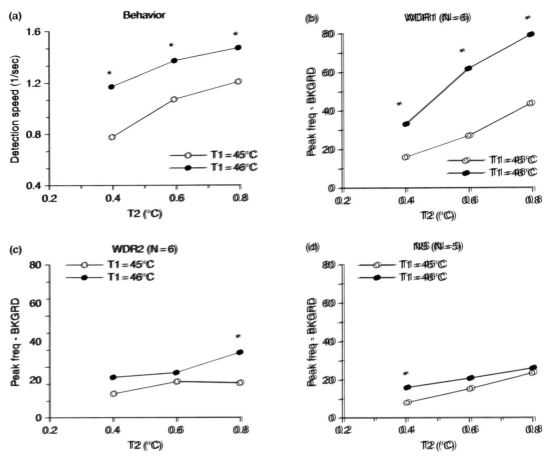

Figure 10 (a) Detection speeds in response to 0.4, 0.6, and 0.8 °C T2 stimuli from T1s of 45 and 46 °C while recording from wide-dynamic range (WDR)1, WDR2, and nociceptive-specific (NS) neurons. There was a significant increase in detection speed for each T2 temperature when T1 was 46 °C (P < 0.01). Each point represents the average of four trials from each of 17 neurons. (b)–(d) Responses of WDR1 (b), WDR2 (c), and NS (d) neurons to the same stimuli as in (a). Asterisks indicate significant increases (P < 0.01) in neuronal discharge when T1 was 46 °C. Each point represents the average of four trials per neuron. From Dubner, R., Kenshalo, D.R., Jr., Maixner, W., Bushnell, M. C., and Oliveras, J. L. 1989. The correlation of monkey medullary dorsal horn neuronal activity and the perceived intensity of noxious heat stimuli. J. Neurophysiol. 62, 450–457, reprinted with permission.

24.14 Somatotopic Organization

Submodality is encoded by neurons in different laminae. However, spatial encoding of the body surface is encoded in the medial–lateral and rostral–caudal planes (Brown, P. B. and Fuchs, J. L., 1975; Koerber, H. R. and Brown, P. B., 1977; Light, A. R. and Durkovic, R. G., 1984; Ritz, L. A. et al., 1985; Brown, P. B. et al., 1997; 2005). The overall map of the body surface is similar in all laminae. Figures 11 and 12 show the overall somatotopic map and more detail of the lumbar region. Overall, the map follows the segmental body plan. However, the map is distorted in the regions that encode the highly innervated digits and feet with a much greater representation of these body regions than the neck and trunk regions. The distortions in this map also represent the density of primary afferent innervation of the skin, with digits being much more densely innervated than proximal limb regions and trunk regions. These distortions are correlated to nociceptive discrimination of these same body regions with greater discrimination possible at the tips of the digits, and less on the trunk.

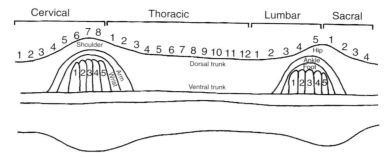

Figure 11 Somatotopic map of representation of body surface in primal spinal cord gray matter. Cervical, thoracic, lumber, and sacral segments are indicated. Maps of this sort are somewhat misleading in that receptive fields often encompass more than one of the body regions indicated on the map. For example, in the lumbar cord, receptive fields notes as hip may be restricted to the hip, but others may include a strip that includes the entire leg including the toes as well as the hip. From Light, A. R. 1992. Pain and Headache (Vol. 12): the Initial Processing of Pain and its Descending Control: Spinal and Trigeminal Systems. Karger, reprinted with permission.

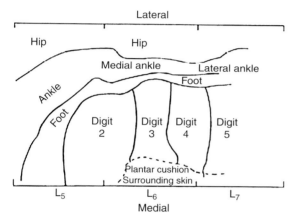

Figure 12 Somatopic organization of the foot representation in the spinal cord of cats. Most medial toe designated as toe 2; most lateral toe designated as toe 5. Map depicts density centers and does not indicate the overlapping of cell groups that have different receptive field locations. From Light, A. R. and Kavookjian, A. M. 1984. Substantia gelatinosa neurons with axons projecting to other dorsal horn laminae. Soc. Neurosci. Abstr. 10, 489, reprinted with permission.

24.15 Physiological Properties of Spinal Neurons

Considerable detail of the physiological properties of the intrinsic neurons of the spinal cord has been obtained by using recordings from spinal cord slices. These preparations have the advantage of a known a stable milieu surrounding the neurons. In addition, access to neurons in either the transverse or horizontal slice is less problematic because less white matter tissue must be traversed with a microelectrode before

attaining recordings from the cell bodies. Because access is easier, whole-cell patch clamp of the spinal cord neurons can be readily accomplished in slice preparations. Recordings in this mode are more stable, details of synaptic inputs more easily observed, and membrane currents rather than voltage changes can be measured.

The major disadvantage, however, is that except for very recent recordings (Koerber, personal communication, 2006) the inputs to the spinal cord are removed. Thus, the adequate stimulus for the neurons is unknown in slice recordings. Since neurons are heterologous with respect to their inputs in all areas of the spinal cord, the physiological properties obtained from slice preparations can only be related to the types of inputs they receive in a broad fashion, drawing on what is know about the most common types of neurons found in the specific laminae. With these limitations in mind, still, much useful knowledge has been obtained from slice preparations. Where possible, we relate the properties observed in slice preparations to those obtained from intact preparations.

24.16 Intrinsic Properties of Spinal Dorsal Horn Neurons

In the spinal dorsal horn, using sharp electrode and whole-cell patch clamp recording several types of neurons have been distinguished on the basis of their intrinsic firing properties. However, these firing types and the relative numbers of each type vary depending on the experimental conditions and species. Seven types of firing have been identified. Tonic, burst, delayed onset, gap, phasic, and single spike. These are shown in Figure 13.

Tonically firing neurons have repetitive action potentials with little frequency adaptation and little amplitude attenuation during sustained depolarization. (Figures 13 (Aa) and 13 (Ca)). The majority of tonic neurons are either WDR or nociceptive-specific neurons (Lopez-Garcia, J. A. and King, A. E., 1994). The ion channels expressed by tonic neurons that

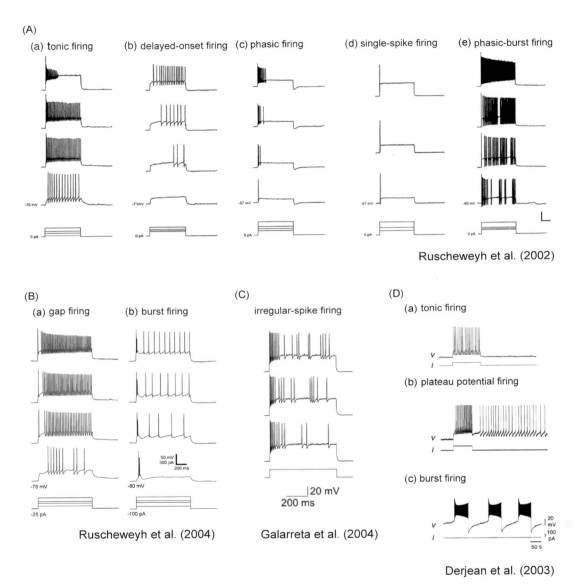

Figure 13 Firing patterns of neurons. (A) Five patterns of rat spinal dorsal horn neurons. Firing patterns were obtained in response to 1-s injections of depolarizing current (25–350 pA, 25-pA steps) at resting membrane potential. Bottom traces in (a)–(e), injected currents. (B) Typical firing patterns of lamina I projection neurons. Firing patterns were obtained in response to depolarizing current injection from hyperpolarized holding potentials. C: Deep dorsal horn neurons show three modes of firing. A: from Ruscheweyh, R. and Sandkuhler, J. 2002. Lamina-specific membrane and discharge properties of rat spinal dorsal horn neurones in vitro. J. Physiol. 541, 231–244, reprinted with permission; B: from Dahlhaus, A., Ruscheweyh, R., and Sandkuhler, J. 2005. Synaptic input of rat spinal lamina I projection and unidentified neurones in vitro. J. Physiol. 566(Pt 2), 355–368, reprinted with permission; C: from Derjean, D., Bertrand, S., Le Masson, G., Landry, M., Morisset, V., and Nagy, F. 2003. Dynamic balance of metabotropic inputs causes dorsal horn neurons to switch functional states. Nat. Neurosci. 6(3), 274–281, reprinted with permission.

shape this response pattern are only partly understood. It has been suggested that a voltage gated Na^+ current in combination with a pronounced delayed rectifier K^+ (K_{DR}) current and absent or weak A-type K^+ (K_A) current help to generate the basic tonic firing pattern (Ruscheweyh, R. and Sandkuhler, J., 2002; Melnick, I. V. et al., 2004b). A Ca^{2+}-dependent K^+ (K_{Ca}) conductance that is sensitive to apamin regulates the discharge frequency, prolonging interspike intervals, and stabilizing firing pattern evoked by sustained membrane depolarization (Melnick, I. V. et al., 2004a).

Tonic firing neurons in lamina I display prolonged excitatory postsynaptic potentials (EPSPs). The mechanisms for these prolonged EPSPs appears not to be a property of the presynaptic release, but are mediated by a persistent sodium current, $I_{Na,p}$ that amplifies and prolongs the depolarization caused by brief stimulation, combined with a persistent calcium current, $I_{Ca,p}$, that contributes to the prolongation, but not the amplification (Prescott, S. A. and De Koninck, Y., 2005). The net result is that these currents can increase integration time and encourage temporal summation of inputs, at the expense of reduced ability to encode the timing of synaptic inputs. Thus, the tonic firing neurons operate as integrators of synaptic inputs.

24.16.1   Burst Firing Neurons

Burst firing neurons exhibit a high-frequency barrage of action potentials that ride on the peak of a low-threshold Ca^{2+} spike (Figure 13 (Bb)). In the superficial dorsal horn (laminae I and II), the burst firing pattern was characterized by a short burst of two to four action potentials riding on a slow depolarizing wave at the onset of a depolarizing current pulse (Ruscheweyh, R. et al., 2004; see Figure 13 (Bb)). With stronger current injections, this pattern could be converted to the tonic pattern (Ruscheweyh, R. and Sandkuhler, J., 2002). The burst pattern was evident only if the neuron was held more negative than −60 mV in whole-cell recordings. The current causing the bursting pattern was not affected by tetrodotoxin, but was partially blocked by Ni^{2+} and Cd^{2+}. Interestingly, the burst firing occurred almost exclusively in projection neurons and was nearly absent in nonprojection neurons, and found more often in neurons projecting to the periaqueductal gray than neurons projecting to the parabrachial region (Ruscheweyh, R. et al., 2004).

The potential function of burst firing for dorsal horn nociceptive neurons is not currently understood. However, some have suggested that this pattern could contribute to the distortion of nociceptive processing in the dorsal horn that is associated with persistent pathological pain (Herrero, J. F. et al., 2000). An example of an animal model of this is that increased burst firing of neurons in the spinal cord was correlated with the induction of recurrent autotomy in nerve injured rats (Asada, H. et al., 1996).

24.16.2   Delayed-Onset Firing Neurons

Delayed-onset firing neurons demonstrate a slow ramp depolarization in response to subthreshold depolarizing current pulses (Yoshimura, M. and Jessell, T. M., 1989; Grudt, T. J. and Perl, E. R., 2002; Prescott, S. A. and De Koninck, Y., 2002; Ruscheweyh, R. and Sandkuhler, J., 2002). With higher current injections, a delay between the onset of the depolarizing current and the first action potentials gradually shortens. Repetitive firing is less regular than in tonic firing neurons, and often, the firing rate increases during the current pulse (Prescott, S. A. and De Koninck, Y., 2002; Ruscheweyh, R. and Sandkuhler, J., 2002). Yoshimura M. and Jessell T. M. (1989a) and Ruscheweyh R. and Sandkuhler J. (2002) suggested that the transient outward rectifying current (I_A) was involved in the response of neurons to sudden depolarization and served to delay the onset of the first action potentials.

Delayed-onset neurons were found almost entirely in lamina I and II, and not in deeper laminae. Whether these neurons are nociceptive, and the functional relevance of this firing pattern is not known.

24.16.3   Gap Firing Neurons

Gap firing neurons, also called late spiking neurons, are characterized by a considerably delayed-onset first spike flowed by tonic firing after week current pulses. With stronger current pulses, a long interval occurs between the first and second spike apparently due to a transient hyperpolarization occurring after the onset of the first depolarizing impulse (see Figure 13 (Ba)). Ruscheweyh R. et al. (2004) suggested that this firing pattern could be evoked only if the neuron was held at a negative potential more negative than −75 mV. They also suggested that gap firing and burst firing are the most common patterns of projection neurons with less excitable membranes having a broad action potential duel to a hump in the falling phase of the action potential. They determined that gap firing neurons preferentially project to the parabrachial region of the brainstem, with fewer projecting to the periaqueductal gray region. Apparently, a slow I_A is required for the appearance of gap firing (Ruscheweyh, R. *et al.*, 2004).

24.16.4   Phasic Firing Neurons

Phasic firing neurons (Figure 13Ac) generate a short burst of action potentials during sustained depolarization. They have been reported by several groups (Thomson, A. M. *et al.*, 1989; Lopez-Garcia, J. A. and King, A. E., 1994; Prescott, S. A. and De Koninck, Y., 2002; Ruscheweyh, R. and Sandkuhler, J., 2002; Melnick, I. V. *et al.*, 2004b). The phasic pattern may be related to the density of Na^+ channels, and their voltage dependency, which may be shifted towards more positive potentials. Melnick I. V. *et al.*, (2004a) found that phasic neurons were found in the lateral part of lamina II. Lopez-Garcia J. A. and King A. E. (1994) classified most phasic neurons as nociceptive specific.

24.16.5   Single-Spike Firing Neurons

Single-spike firing neurons (Figure 13(Ad)) typically generate only one action potential at the onset of a current pulse, regardless of the intensity of the current pulse. A transient, voltage-dependent outward current may be involved in generating the single spike pattern. 4-aminopyridine (a K_v blocker) converts single-spike firing to tonic firing (Ruscheweyh, R. and Sandkuhler, J., 2002).

24.17   Relationship between Intrinsic Firing Patterns and Physiological Types of Dorsal Horn Neurons

Very few investigators have attempted to relate the firing patterns described above to the types of adequate stimuli that normally activate these neurons. Figure 14 demonstrates the data from one such attempt in rats. As this figure indicates Lopez-Garcia J. A. and King A. E. (1994) determined that the majority (81%) of WDR neurons responded with tonic firing, while nociceptive-specific neurons were mostly phasic firing neurons. Low-threshold mechanoreceptor (LTM) neurons were of both the tonic and phasic firing types. Neurons that were not fired by adequate stimuli, but demonstrated subthreshold inputs were either tonic firing or single spike type of neurons. In contrast to their conclusions, Graham B. A. *et al.* (2004) using whole-cell recordings in adult mice *in vivo* did not

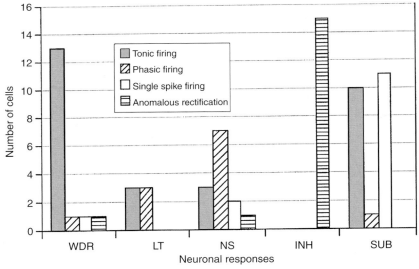

Figure 14 Distribution of electrophysiological groups from a hemisected slice preparation from rats with hindlimb attached. The number of physiologically classified wide-dynamic range neurons (WDR), low-threshold neurons (LT), nociceptive-specific neurons (NS), and neurons displaying inhibitory events (INH) are shown along with the firing type. SUB, subthreshold for firing action potentials. Adapted from Lopez-Garcia, J. A. and King, A. E. 1994. Membrane properties of physiologically classified rat dorsal horn neurons in vitro: correlation with cutaneous sensory afferent input. Eur. J. Neurosci. 6, 998–1007, with permission.

find a consistent relationship between any of the firing pattern types of neurons and physiological inputs.

Lu Y. and Perl E. R. (2003) also found that all lamina II neurons receiving inhibitory inputs from other lamina II neurons were of the phasic firing type, indicating at least some functional organization is related to the firing pattern of lamina II neurons.

Clearly, the firing patterns described here reflect specific currents mediated by specific ion channels. These specific ion channels can be altered by a number of pharmacologic agents, endogenous neurotransmitters and neuromodulators. The specificity of neurons for these channels could convey specificity of drug action on particular functional types of neurons. Some of this is the topic of Chapter 20, and much of this will be the subject of future investigations into the rich modulatory capacity of sensory neurons in the dorsal horn of the spinal cord.

24.18　Summary

The spinal cord dorsal horn contains the first relay for afferent inputs from the skin, muscles, and viscera of the body. The superficial dorsal horn contains at least some neurons that maintain the selectivity for modalities encoded by the primary afferent endings. These include a number of different modalities including innocuous warming, innocuous cooling, noxious mechanical, noxious thermal, noxious chemical, prurigenic (itch producing), metabolic (non-noxious signals from actively contracting muscles), and innocuous mechanical stimuli of various forms as well as other inputs from autonomic sensory neurons.

In the majority of neurons in the dorsal horn, various combinations of these inputs are integrated to allow detection of features (such as edges, speed of movement, location on the skin, noxious hot objects, etc.). Such integration allows for useful motor outputs to be generated to compensate for the inputs received, e.g., moving a limb away from a hot object, scratching a biting insect, or wiping away an insect crawling on the skin.

Many neurons in the dorsal horn also transmit selective afferent inputs, or integrated inputs to other regions of the central nervous system, including other parts of the spinal cord, the medulla, midbrain, and thalamus as well as the hypothalamus. Thus, the dorsal horn serves as the first integration and relay center for nociceptive as well as a variety of non-nociceptive inputs.

Dorsal horn neurons are modulated by inputs from higher brain centers. Inputs from other centers can greatly modify the amplitude of signals relayed from primary afferent neurons.

Dorsal horn neurons are quite heterogeneous with respect to the types of inputs that they encode, and with respect to their intrinsic membrane properties. In at least some cases, it appears that the intrinsic membrane properties are associated with the types of inputs they receive, and these intrinsic properties may allow the neurons to better integrate certain types of inputs.

References

Albe-Fessard, D., Levante, A., and Lamour, Y. 1974. Origin of spino-thalamic tract in monkeys. Brain Res. 65, 503–509.

Ammons, W. S. 1987. Characteristics of spinoreticular and spinothalamic neurons with renal input. J. Neurophysiol. 58, 480–495.

Andrew, D. and Craig, A. D. 2001. Spinothalamic lamina I neurons selectively sensitive to histamine: a central neural pathway for itch. Nat. Neurosci. 4, 72–77.

Andrew, D. and Craig, A. D. 2002. Quantitative responses of spinothalamic lamina I neurones to graded mechanical stimulation in the cat. J. Physiol. 545, 913–931.

Armett, C. J., Gray, J. A. B., Hunsperger, R. W., and Lal, K. B. 1962. The transmission of information in primary receptor neurones and second-order neurones of a phasic system. J. Physiol. 164, 395–421.

Asada, H., Yamaguchi, Y., Tsunoda, S., and Fukuda, Y. 1996. The role of spinal cord activation before neurectomy in the development of autotomy. Pain 64, 161–167.

Bennett, G. J., Abdelmoumene, M., Hayashi, H., and Dubner, R. 1980. Physiology and morphology of substantia gelatinosa neurons intracellularly stained with horseradish peroxidase. J. Comp. Neurol. 194, 809–827.

Bennett, G. J., Hayashi, H., Abdelmoumene, M., and Dubner, R. 1979. Physiological properties of stalked cells of the substantia gelatinosa intracellularly stained with horseradish peroxidase. Brain Res. 164, 285–289.

Braz, J. M., Nassar, M. A., Wood, J. N., and Basbaum, A. I. 2005. Parallel "pain" pathways arise from subpopulations of primary afferent nociceptor. Neuron 47, 787–793.

Brown, A. G. 1968. Single unit activity in the spinocervical tract of the cat. J. Physiol. (Lond.) 198, 62passim–62pas63p.

Brown, A. G. and Franz, D. N. 1969. Responses of spinocervical tract neurones to natural stimulation of identified cutaneous receptors. Exp. Brain Res. 7, 231–249.

Brown, A. G. and Fyffe, R. E. 1981. Form and function of dorsal horn neurones with axons ascending the dorsal columns in cat. J. Physiol. (Lond.) 321, 31–47.

Brown, P. B. and Fuchs, J. L. 1975. Somatotopic representation of hindlimb skin in cat dorsal horn. J. Neurophysiol. 38, 1–9.

Brown, P. B., Koerber, H. R., and Millecchia, R. 1997. Assembly of the dorsal horn somatotopic map. Somatosens. Mot. Res. 14, 93–106.

Brown, P. B., Millecchia, R., Lawson, J. J., Brown, A. G., Koerber, H. R., Culberson, J., and Stephens, S. 2005. From innervation density to tactile acuity 2: embryonic and adult pre- and postsynaptic somatotopy in the dorsal horn. Brain Res. 1055, 36–59.

Bryan, R. N., Coulter, J. D., and Willis, W. D. 1974. Cells of origin of the spinocervical tract in the monkey. Exp. Neurol. 42, 574–586.

Bryan, R. N., Trevino, D. L., Coulter, J. D., and Willis, W. D. 1973. Location and somatotopic organization of the cells of origin of the spino-cervical tract. Exp. Brain Res. 17, 177–189.

Casey, K. L. 1982. Neural mechanisms of pain: an overview. Acta Anaesthesiol. Scand. Suppl. 74, 13–20.

Christensen, B. N. and Perl, E. R. 1969. Unique excitation of cells at the margin of the spinal dorsal horn by cutaneous nociceptors. Fed. Proc. 28, 276.

Christensen, B. N. and Perl, E. R. 1970. Spinal neurons specifically excited by noxious or thermal stimuli: marginal zone of the dorsal horn. J. Neurophysiol. 33, 293–307.

Chung, J. M., Surmeier, D. J., Lee, K. H., Sorkin, L. S., Honda, C. N., Tsong, Y., and Willis, W. D. 1986. Classification of primate spinothalamic and somatosensory thalamic neurons based on cluster analysis. J. Neurophysiol. 56, 308–327.

Clarke, J. L. 1859. Further researches on the grey substance of the spinal cord. Phil. Trans. 149, 437–467.

Craig, A. D. 1996. An ascending general homeostatic afferent pathway originating in lamina I. Prog. Brain Res. 107, 225–242 [Review; 78 references]

Craig, A. D. 2003. A new view of pain as a homeostatic emotion. Trends Neurosci. 26, 303–307.

Craig, A. D. 2004. Lamina I, but not lamina V, spinothalamic neurons exhibit responses that correspond with burning pain. J. Neurophysiol. 92, 2604–2609.

Craig, A. D. and Andrew, D. 2002. Responses of spinothalamic lamina I neurons to repeated brief contact heat stimulation in the cat. J. Neurophysiol. 87, 1902–1914.

Craig, A. D. and Kniffki, K. D. 1985. Spinothalamic lumbosacral lamina I cells responsive to skin and muscle stimulation in the cat. J. Physiol. (Lond.) 365, 197–221.

Craig, A. D., Jr., Linington, A. J., and Kniffki, K. D. 1989. Cells of origin of spinothalamic tract projections to the medial and lateral thalamus in the cat. J. Comp. Neurol. 289, 568–585.

Dahlhaus, A., Ruscheweyh, R., and Sandkuhler, J. 2005. Synaptic input of rat spinal lamina I projection and unidentified neurones in vitro. J. Physiol. 566(Pt 2), 355–368.

Derjean, D., Bertrand, S., Le Masson, G., Landry, M., Morisset, V., and Nagy, F. 2003. Dynamic balance of metabotropic inputs causes dorsal horn neurons to switch functional states. Nat. Neurosci. 6(3), 274–281.

Dubner, R. 1978. Neurophysiology of pain. Dent. Clin. North. Am. 22, 11–30.

Dubner, R. and Bennett, G. J. 1983. Spinal and trigeminal mechanisms of nociception. Ann. Rev. Neurosci. 381–418.

Dubner, R., Hayes, R. L., and Hoffman, D. S. 1980. Neural and behavioral correlates of pain in the trigeminal system. Res. Publ. Assoc. Res. Nerv. Ment. Dis. 58, 63–72.

Dubner, R., Kenshalo, D. R., Jr., Maixner, W., Bushnell, M. C., and Oliveras, J. L. 1989. The correlation of monkey medullary dorsal horn neuronal activity and the perceived intensity of noxious heat stimuli. J. Neurophysiol. 62, 450–457.

Dubner, R., Price, D. D., Beitel, R. E., and Hu, J. W. 1977. Peripheral Neural Correlates of Behavior in Monkey and Human Related to Sensory-Discriminative Aspects of Pain. In: Pain in the Trigeminal Region (eds. D. J. Anderson and B. Matthews), pp. 57–66. Elsevier.

Eckert, W. A., III, McNaughton, K. K., and Light, A. R. 2003. Morphology and axonal arborization of rat spinal inner lamina II neurons hyperpolarized by mu-opioid-selective agonists. J. Comp. Neurol. 458, 240–256.

Giesler, G. J., Jr., Cannon, J. T., Urca, G., and Liebeskind, J. C. 1978. Long ascending projections from substantia

gelatinosa Rolandi and the subjacent dorsal horn in the rat. Science 202, 984–986.

Giesler, G. J., Menetrey, D., Guilbaud, G., and Besson, J. M. 1976. Lumbar cord neurons at the origin of the spinothalamic tract in the rat. Brain Res. 118, 320–324.

Giesler, G. J., Jr., Nahin, R. L., and Madsen, A. M. 1984. Postsynaptic dorsal column pathway of the rat. I. Anatomical studies. J. Neurophysiol. 51, 260–275.

Graham, B. A., Brichta, A. M., and Callister, R. J. 2004. An in vivo mouse spinal cord preparation for patch-clamp analysis of nociceptive processing. J. Neurosci. Methods 136, 221–228.

Grudt, T. J. and Perl, E. R. 2002. Correlations between neuronal morphology and electrophysiological features in the rodent superficial dorsal horn. J. Physiol. 540, 189–207.

Herrero, J. F., Laird, J. M., and Lopez-Garcia, J. A. 2000. Wind-up of spinal cord neurones and pain sensation: much ado about something? Prog. Neurobiol. 61, 169–203.

Hillman, P. and Wall, P. D. 1969. Inhibitory and excitatory factors influencing the receptive fields of lamina 5 spinal cord cells. Exp. Brain Res. 9, 284–306.

Kajander, K. C. and Giesler, G. J., Jr. 1987. Responses of neurons in the lateral cervical nucleus of the cat to noxious cutaneous stimulation. J. Neurophysiol. 57, 1686–1704.

Koerber, H. R. and Brown, P. B. 1977. Somatotopic organization of the brachial cord of cat. Soc. Neurosci. Abstr. 3, 485.

Kumazawa, T. and Perl, E. R. 1978. Excitation of marginal and substantia gelatinosa neurons in the primate spinal cord: indications of their place in dorsal horn functional organization. J. Comp. Neurol. 177, 417–434.

Kumazawa, T., Perl, E. R., Burgess, P. R., and Whitehorn, D. 1971. Excitation of posteromarginal cells (lamina I) in monkey and their projection in lateral spinal tracts. Proc. Intl. Union Physiol. Sci. 9, 328.

Kumazawa, T., Perl, E. R., Burgess, P. R., and Whitehorn, D. 1975. Ascending projections from marginal zone (lamina I) neurons of the spinal dorsal horn. J. Comp. Neurol. 162, 1–11.

Lahuerta, J., Bowsher, D., Campbell, J., and Lipton, S. 1990. Clinical and instrumental evaluation of sensory function before and after percutaneous anterolateral cordotomy at cervical level in man. Pain 42, 23–30.

Lewinter, R. D., Skinner, K., Julius, D., and Basbaum, A. I. 2004. Immunoreactive TRPV-2 (VRL-1), a capsaicin receptor homolog, in the spinal cord of the rat. J. Comp. Neurol. 470, 400–408.

Light, A. R. 1992. Pain and Headache (Vol. 12): the Initial Processing of Pain and its Descending Control: Spinal and Trigeminal Systems. Karger.

Light, A. R. and Durkovic, R. G. 1984. Features of laminar and somatotopic organization of lumbar spinal cord units receiving cutaneous inputs from hindlimb receptive fields. J. Neurophysiol. 52, 449–458.

Light, A. R. and Kavookjian, A. M. 1984. Substantia gelatinosa neurons with axons projecting to other dorsal horn laminae. Soc. Neurosci. Abstr. 10, 489.

Light, A. R. and Kavookjian, A. M. 1988. Morphology and ultrastructure of physiologically identified substantia gelatinosa (lamina II) neurons with axons that terminate in deeper dorsal horn laminae (III–V). J. Comp. Neurol. 267, 172–189.

Light, A. R., Rethelyi, M., and Perl, E. R. 1981. Ultrastructure of Functionally Identified Neurones in the Marginal Zone and the Substantia Gelatinosa. In: Spinal Cord Sensation (eds. A. G. Brown and M. Rethelyi), pp. 97–102. Scottish Academic Press.

Light, A. R., Trevino, D. L., and Perl, E. R. 1979. Morphological features of functionally defined neurons in the marginal zone and substantia gelatinosa of the spinal dorsal horn. J. Comp. Neurol. 186, 151–171.

Light, A. R. and Willcockson, H. H. 1999. Spinal laminae I–II neurons in rat recorded in vivo in whole cell, tight seal configuration: properties and opioid responses. J. Neurophysiol. 82, 3316–3326.

Ling, L. J., Honda, T., Shimada, Y., Ozaki, N., Shiraishi, Y., and Sugiura, Y. 2003. Central projection of unmyelinated (C) primary afferent fibers from gastrocnemius muscle in the guinea pig. J. Comp. Neurol. 461, 140–150.

Lopez-Garcia, J. A. and King, A. E. 1994. Membrane properties of physiologically classified rat dorsal horn neurons in vitro: correlation with cutaneous sensory afferent input. Eur. J. Neurosci. 6, 998–1007.

Lu, Y. and Perl, E. R. 2003. A specific inhibitory pathway between substantia gelatinosa neurons receiving direct C-fiber input. J. Neurosci. 23, 8752–8758.

Lu, Y. and Perl, E. R. 2005. Modular organization of excitatory circuits between neurons of the spinal superficial dorsal horn (laminae I and II). J. Neurosci. 25, 3900–3907.

Martin, W. J., Liu, H., Wang, H., Malmberg, A. B., and Basbaum, A. I. 1999. Inflammation-induced up-regulation of protein kinase Cγ immunoreactivity in rat spinal cord correlates with enhanced nociceptive processing. Neuroscience 88, 1267–1274.

Martin, W. J., Malmberg, A. B., and Basbaum, A. I. 2001. PKCgamma contributes to a subset of the NMDA-dependent spinal circuits that underlie injury-induced persistent pain. J. Neurosci. 21, 5321–5327.

Matsushita, M. 1969. Some aspects of interneuronal connections cat's spinal gray matter. J. Comp. Neurol. 136, 57–80.

Mauderli, A. P., Acosta-Rua, A., and Vierck, C. J. 2000. An operant assay of thermal pain in conscious, unrestrained rats. J. Neurosci. Methods 97, 19–29.

Maxwell, D. J. and Koerber, H. R. 1987. Morphologically unusual feline spinocervical tract neurons. Exp. Neurol. 95, 521–524.

Maxwell, D. J., Fyffe, R. E., and Rethelyi, M. 1983. Morphological properties of physiologically characterized lamina III neurones in the cat spinal cord. Neuroscience 10, 1–22.

Melnick, I. V., Santos, S. F., and Safronov, B. V. 2004a. Mechanism of spike frequency adaptation in substantia gelatinosa neurones of rat. J. Physiol. 559, 383–395.

Melnick, I. V., Santos, S. F., Szokol, K., Szucs, P., and Safronov, B. V. 2004b. Ionic basis of tonic firing in spinal substantia gelatinosa neurons of rat. J. Neurophysiol. 91, 646–655.

Mendell, L. M. and Wall, P. D. 1965. Responses of single dorsal cord cells to peripheral cutaneous unmyelinated fibres. Nature 206, 97–99.

Menetrey, D., de Pommery, J., and Roudier, F. 1984. Properties of deep spinothalamic tract cells in the rat, with special reference to ventromedial zone of lumbar dorsal horn. J. Neurophysiol. 52, 612–624.

Menetrey, D., Giesler, G. J., Jr., and Besson, J. M. 1977. An analysis of response properties of spinal cord dorsal horn neurones to nonnoxious and noxious stimuli in the spinal rat. Exp. Brain Res. 27, 15–33.

Miletic, V., Bowen, K. K., and Miletic, G. 2000. Loose ligation of the rat sciatic nerve is accompanied by changes in the subcellular content of protein kinase C beta II and gamma in the spinal dorsal horn. Neurosci. Lett. 288, 199–202.

Morgan, M. M. 1998. Direct comparison of heat-evoked activity of nociceptive neurons in the dorsal horn with the hindpaw withdrawal reflex in the rat. J. Neurophysiol. 79, 174–180.

Mosso, J. A. and Kruger, L. 1973. Receptor categories represented in spinal trigeminal nucleus caudalis. J. Neurophysiol. 36, 472–488.

Olausson, H., Lamarre, Y., Backlund, H., Morin, C., Wallin, B. G., Starck, G., Ekholm, S., Strigo, I., Worsley, K.,

Vallbo, A. B., and Bushnell, M. C. 2002. Unmyelinated tactile afferents signal touch and project to insular cortex. Nat. Neurosci. 5, 900–904.

Polgar, E., Fowler, J. H., McGill, M. M., and Todd, A. J. 1999. The types of neuron which contain protein kinase C gamma in rat spinal cord. Brain Res. 833, 71–80.

Pomeranz, B., Wall, P. D., and Weber, W. V. 1968. Cord cells responding to fine myelinated afferents from viscera, muscle, and skin. J. Physiol. (Lond.) 199, 511–532.

Prescott, S. A. and De Koninck, Y. 2002. Four cell types with distinctive membrane properties and morphologies in lamina I of the spinal dorsal horn of the adult rat. J. Physiol. 539, 817–836.

Prescott, S. A. and De Koninck, Y. 2005. Integration time in a subset of spinal lamina I neurons is lengthened by sodium and calcium currents acting synergistically to prolong subthreshold depolarization. J. Neurosci. 25, 4743–4754.

Price, D. D. and Browe, A. C. 1973. Responses of spinal cord neurons to graded noxious and non-noxious stimuli. Brain Res. 64, 425–429.

Price, D. D. and Dubner, R. 1977. Neurons that subserve the sensory-discriminative aspects of pain. Pain 3, 307–338.

Price, D. D. and Mayer, D. J. 1974. Physiological laminar organization of the dorsal horn of M. mulatta. Brain Res. 79, 321–325.

Price, D. D. and Mayer, D. J. 1975. Neurophysiological characterization of the anterolateral quadrant neurons subserving pain in M. mulatta. Pain 1, 59–72.

Price, D. D., Hayashi, H., Dubner, R., and Ruda, M. A. 1979. Functional relationships between neurons of marginal and substantia gelatinosa layers of primate dorsal horn. J. Neurophysiol. 42, 1590–1608.

Ramon Y, and Cajal, S. 1909. Histologie du Systeme Nerveux de l'Homme et des Vertebres. Maloine.

Ranson, S. W. 1913. The course within the spinal cord of the non-medullated fibers of the dorsal roots: a study of Lissauer's tract in the cat. J. Comp. Neurol. 23, 259–281.

Ranson, S. W. 1914. The tract of Lissauer and the substantia gelatinosa Rolandi. Am. J. Anat. 16, 97–126.

Ranson, S. W. 1915. Unmyelinated nerve-fibres as conductors of protopathic sensation. Brain 38, 381–389.

Ranson, S. W. and Billingsley, P. R. 1916. The conduction of painful afferent impulses in the spinal nerves. Am. J. Physiol. 40, 571–584.

Ranson, S. W. and Davenport, H. K. 1931. Sensory unmyelinated fibers in the spinal nerves. Am. J. Anat. 48, 331–353.

Ranson, S. W. and von Hess, C. L. 1915. The conduction within the spinal cord of the afferent impulses producing pain and the vasomotor reflexes. Am. J. Physiol. 38, 128–152.

Rethelyi, M. and Szentagothai, J. 1969. The large synaptic complexes of the substantia gelatinosa. Exp. Brain Res. 7, 258–274.

Rethelyi, M., Light, A. R., and Perl, E. R. 1981. Synapses made by nociceptive laminae I and II neurones. Pain suppl. S127.

Rethelyi, M., Light, A. R., and Perl, E. R. 1983. Synapses made by nociceptive laminae I and II neurons. In: (eds. J. J. Bonica, U. Lindblom, and A. Iggo),Advances in Pain Research and Therapy, Vol. 5, pp. 111–118. Raven Press.

Rethelyi, M., Light, A. R., and Perl, E. R. 1989. Synaptic ultrastructure of functionally and morphologically characterized neurons of the superficial spinal dorsal horn of cat. J. Neurosci. 9, 1846–1863.

Rexed, B. 1952. The cytoarchitectonic organization of the spinal cord in the cat. J. Comp. Neurol. 96, 415–466.

Ritz, L. A., Culberson, J. L., and Brown, P. B. 1985. Somatotopic organization in cat spinal cord segments with fused dorsal horns: caudal and thoracic levels. J. Neurophysiol. 54, 1167–1177.

Rotto, D. M. and Kaufman, M. P. 1988. Effect of metabolic products of muscular contraction on discharge of group III and IV afferents. J. Appl. Physiol. 64, 2306–2313.

Ruscheweyh, R. and Sandkuhler, J. 2002. Lamina-specific membrane and discharge properties of rat spinal dorsal horn neurones in vitro. J. Physiol. 541, 231–244.

Ruscheweyh, R., Ikeda, H., Heinke, B., and Sandkuhler, J. 2004. Distinctive membrane and discharge properties of rat spinal lamina I projection neurones *in vitro*. J. Physiol. 555, 527–543.

Rustioni, A., Hayes, N. L., and O'Neill, S. 1979. Dorsal column nuclei and ascending spinal afferents in macaques. Brain 102, 95–125.

Steedman, W. M., Molony, V., and Iggo, A. 1985. Nociceptive neurones in the superficial dorsal horn of cat lumbar spinal cord and their primary afferent inputs. Exp. Brain Res. 58, 171–182.

Sugiura, Y., Lee, C. L., and Perl, E. R. 1986. Central projections of identified, unmyelinated (C) afferent fibers innervating mammalian skin. Science 234, 358–361.

Szentagothai, J. 1964. Neuronal and synaptic arrangement in the substantia gelatinosa Rolandi. J. Comp. Neurol. 122, 219–239.

Thomson, A. M., West, D. C., and Headley, P. M. 1989. Membrane characteristics and synaptic responsiveness of superficial dorsal horn neurons in a slice preparation of adult rat spinal cord. Eur. J. Neurosci. 1, 479–488.

Trevino, D. L., Coulter, J. D., and Willis, W. D. 1973. Location of cells of origin of spinothalamic tract in lumbar enlargement of the monkey. J. Neurophysiol. 36, 750–761.

Trevino, D. L., Coulter, J. D., Maunz, R. A., and Willis, W. D. 1974. Location and functional properties of spinothalamic cells in the monkey. Adv. Neurol. 4, 167–170.

Vierck, C. J., Jr., Cannon, R. L., Fry, G., Maixner, W., and Whitsel, B. L. 1997. Characteristics of temporal summation of second pain sensations elicited by brief contact of glabrous skin by a preheated thermode. J. Neurophysiol. 78, 992–1002.

Waldeyer, W. 1889. Das Gorilla-Ruckenmark. Acedemie der Wissenschaften.

Wall, P. D. 1967. The laminar organization of dorsal horn and effects of descending impulses. J. Physiol. (Lond.) 188, 403–423.

Wall, P. D. 1973. Dorsal Horn Electrophysiology. In: Handbook of Sensory Physiology. Somatosensory System (*ed*. A. Iggo), pp. 253–270. Springer-Verlag.

Wall, P. D., Merrill, E. G., and Yaksh, T. L. 1979. Responses of single units in laminae 2 and 3 of cat spinal cord. Brain Res. 160, 245–260.

Willis, W. D. 1985. The Pain System: the Neural Basis of Nociceptive Transmission in the Mammalian Nervous System. Karger.

Willis, W. D., Jr. 1988. Dorsal horn neurophysiology of pain. Ann. N. Y. Acad. Sci. 531, 76–89.

Willis, W. D. and Westlund, K. N. 1997. Neuroanatomy of the pain system and of the pathways that modulate pain. J. Clin. Neurophysiol. 14, 2–31. [Review; 393 references]

Willis, W. D., Leonard, R. B., and Kenshalo, D. R., Jr. 1978. Spinothalamic tract neurons in the substantia gelatinosa. Science 202, 986–988.

Willis, W. D., Trevino, D. L., Coulter, J. D., and Maunz, R. A. 1974. Responses of primate spinothalamic tract neurons to natural stimulation of hindlimb. J. Neurophysiol. 37, 358–372.

Wilson, L. B., Andrew, D., and Craig, A. D. 2002. Activation of spinobulbar lamina I neurons by static muscle contraction. J. Neurophysiol. 87, 1641–1645.

Woolf, C. J. and Fitzgerald, M. 1983. The properties of neurones recorded in the superficial dorsal horn of the rat spinal cord. J. Comp. Neurol. 221, 313–328.

Yoshimura, M. and Jessell, T. M. 1989. Primary afferent-evoked synaptic responses and slow potential generation in rat substantia gelatinosa neurons *in vitro*. J. Neurophysiol. 62, 96–108.

25 What is a Wide-Dynamic-Range Cell?

D Le Bars, INSERM U-713, Paris, France

S W Cadden, University of Dundee, Dundee, UK

Glossary

allodynia Pain caused by stimuli which would not normally cause pain (i.e., non-noxious stimuli).

angina pectoris Chest pain from cardiac ischemia.

basic somesthetic activity Description of the on-going activity in somatosensory pathways in the absence of any deliberate stimuli.

bistable A mechanism which can essentially be in one of two states (e.g., on or off)

body image Conscious mental representation of the body.

body schema Unconscious model of the body in the brain.

bulbo-spinal control A neural control mechanism originating in the brainstem and projecting to the spinal cord.

dermatome Area of skin innervated by nerves from a given segment of the spinal cord.

diffuse noxious inhibitory controls (DNIC) Neural controls triggered by noxious stimulation of widespread areas of the body which exert an inhibitory influence on wide-dynamic-range neurons.

dorsal column stimulation Technique for inducing analgesia/hypoalgesia by electrically stimulating the dorsal columns of the spinal cord.

excitatory receptive field Area of the body (surface or interior) which when stimulated will produce excitation of a given neuron.

hyperalgesia Increased level of pain felt in response to a noxious stimulus (cf. allodynia).

hypoalgesic Pertaining to a reduction of pain.

inhibitory receptive field Area of the body (surface or interior) which when stimulated will produce inhibitory effects on the activity of a given neuron.

microelectrophoretic application Application of a substance in its ionic form from a microelectrode by the application of electrical current.

monoarthritis Inflammation of a single joint.

nociceptive stimulus Stimulus which activates nociceptive afferents (which could be a noxious stimulus (see below) or an electrical stimulus).

noxious stimulus Stimulus which causes or threatens to cause tissue damage.

polyarthritis Inflammation of several joints.

propriospinal mechanism Neural mechanism mediated entirely within the spinal cord.

radiating pain Pain which spreads from the area of stimulation to adjacent areas.

referred pain Pain which is felt in areas remote to the area of stimulation (as well as or instead of from the area of stimulation).

> **supraspinal mechanism** Neural mechanism which at least in part, involves structures above the spinal cord.
> **transcutaneous electrical nerve stimulation** Technique for inducing analgesia/hypoalgesia by electrically stimulating peripheral nerves through the skin (i.e., with electrodes placed on the skin).
> **trigeminal nucleus caudalis** Most caudal part of the trigeminal sensory nuclear complex (or of the trigeminal spinal nucleus); also sometimes called the medullary dorsal horn.

25.1   Introduction

Wide-dynamic-range (WDR) neurons (various other names have been used by different authors for these neurones, e.g., trigger, lamina V-type, class 2, multi-receptive, and convergent neurones) are found in the spinal dorsal horn (and its medullary homolog, trigeminal nucleus caudalis), both in the superficial layers and more so in deeper areas centered on lamina V. They are activated by both nociceptive and non-nociceptive stimuli and include interneurons involved in polysynaptic reflexes and neurons projecting in pathways such as the spinothalamic and spinoreticular tracts (Willis, W. D. and Coggeshall, R. E., 1991). For various reasons, these neurons are believed to play a key role in pain. For example, many procedures used to alleviate pain in man (e.g., systemic or intrathecal morphine, transcutaneous electrical nerve stimulation, and dorsal column stimulation), also result in a reduction in the responses of WDR cells to nociceptive stimuli. Indeed, some WDR neurons show patterns of activity which are more reminiscent of certain types of pain than activity in peripheral nociceptive nerves (Adriaensen, H. *et al.*, 1984; Cervero, F. *et al.*, 1988). For example, a WDR neuron may remain active after removal of a stimulus from its peripheral excitatory field (e.g., Wagman, I. H. and Price, D. D., 1969) while conversely it may remain inactive following the brief activation of inhibitory mechanisms even if there is still a stimulus on the excitatory field (Cadden, S. W., 1993). Such bistable patterns of activity might be explained by inherent properties of WDR neurons (e.g., Morisset, V. and Nagy, F., 1996) and/or their involvement in positive feedback circuits (see Cadden, S. W., 1993).

WDR neurons also constitute a site where excitatory and inhibitory influences converge. This chapter focuses on the properties of WDR neurons under physiological conditions, and most particularly, on the concept that they exhibit whole body receptive fields. One consequence of this unusual property is that it necessitates consideration of the output of the whole population of these neurons (Le Bars, D., 2002). The extent of a nociceptive focus can differ greatly depending on its cause, and consequently the global volume of information sent to the brain will also vary. Although *a priori* this volume of information is determined by the intensity, duration, and area being stimulated, it can also be altered profoundly by past and present experiences.

25.2   Segmental Excitatory Receptive Fields

25.2.1   General

WDR neurons receive information from all three groups of cutaneous afferents, that is, $A\beta$-, $A\delta$-, and C-fibers (e.g., Wagman, I. H. and Price, D. D., 1969; Menétrey, D. *et al.*, 1977). The excitatory receptive fields from which this information arises are fairly restricted in size (Figure 1). Studies in conscious monkeys illustrate that these receptive fields can expand or contract depending on the attentional state of the animal (e.g., Dubner, R. *et al.*, 1981). Although the receptive fields are usually larger than those of nociceptive-specific neurons, they can still provide useful information regarding stimulus location. Usually, these receptive fields exhibit a gradient of sensitivity: in the center even light mechanical stimuli can activate the neuron, while at the periphery only intense stimuli elicit responses (e.g., Menétrey, D. *et al.*, 1977).

When increasing intensities of mechanical or thermal stimuli are applied to the center of the receptive field, the rate and duration of neuronal firing increases (see insert in Figure 1). In this respect, these neurons seem better than nociceptive-specific neurons at encoding intensities of

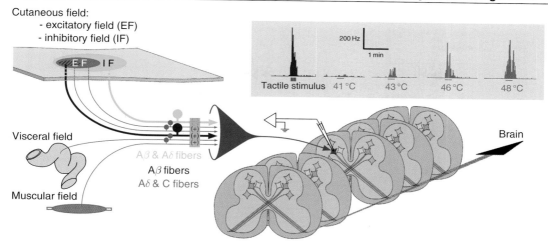

Figure 1 Schematic organization of the segmental receptive field of a wide-dynamic-range (WDR) neuron. The principal part on the skin is made up of an excitatory (EF) and an inhibitory (IF) field. The center of the EF can be activated by both innocuous and nociceptive stimuli (blue and red stripes, respectively). The periphery of the EF is activated only by nociceptive stimuli (red). The inhibitory field (yellow) is activated mainly by non-nociceptive stimuli. As shown, the receptive field may also include visceral and/or muscular components thus completing a convergence of excitatory (+) and inhibitory (−) information onto a single neuron. A ratemeter recording from a WDR neuron in the rat lumbar cord (insert, top right) shows responses to stimuli applied to the center of the excitatory field: repetitive innocuous (light pressure) stimuli produced a high level of firing while applications of radiant heat (indicated by bars with temperatures achieved) evoked responses which increased with temperature within the noxious range. Adapted from Le Bars, D. and Chitour, D. 1983. Do convergent neurons in the spinal cord discriminate nociceptive from non-nociceptive information? Pain 17, 1–19.

noxious stimuli (Hoffman, D. S. *et al.*, 1981). It is often believed that WDR neurons respond more strongly to noxious stimulation than to any form of innocuous stimulation. However, this is not quite true. In fact, the responses evoked in WDR neurons by noxious heat can be exceeded by repetitively applied innocuous mechanical stimuli (insert of Figure 1) (see Le Bars, D. and Chitour, D., 1983). Thus, firing frequency alone cannot distinguish between non-noxious and noxious stimuli. How can this fact be reconciled with WDR neurons having a central role in pain processing? It seems that the meaningful signal is the pattern of activity from the whole population of WDR cells and not simply the signals from each individual neuron. As their receptive fields overlap one another, the organization of spatial convergence from WDR neurons must be taken into account. This is illustrated by the fact that when a tactile stimulus is applied to a given territory, it activates only the few WDR neurons whose excitatory fields have centers in the territory. However, a noxious stimulus applied in the same place, will activate not only the same WDR neurons, but also those whose excitatory fields have peripheries in the territory (Figure 2).

Moreover, some WDR neurons are activated from more than one type of tissue (e.g., cutaneous and visceral or cutaneous and muscular). This convergence helps explain clinical curiosities such as radiating or referred pain, for example, to the left arm in angina pectoris. Indeed, WDR neurons can capture all the information coming from interfaces with both the external environment (the skin) and the internal milieu (viscera, muscles, etc.). This information constitutes a basic somesthetic activity. Thus, it is possible that these neurons are not involved exclusively in pain and, by transmitting this basic somesthetic activity, inform the brain that the organism is suffering no special perturbation from either external or internal sources.

25.2.2 Plasticity

The convergence onto a single WDR neuron is actually greater than is observed under normal conditions. This discrepancy is such that on fringes of the excitatory field, there is a zone which when stimulated will trigger a depolarization of the WDR neuron (Figure 2(c)) which normally will be insufficient to elicit action potentials. This is consistent with the observation that cutaneous receptive fields of WDR neurons are apparently smaller when mapped using natural as opposed to electrical stimulation (e.g., Mendell, L. M., 1984), given that the electrical stimuli produce more synchronized afferent activity which results in more effective

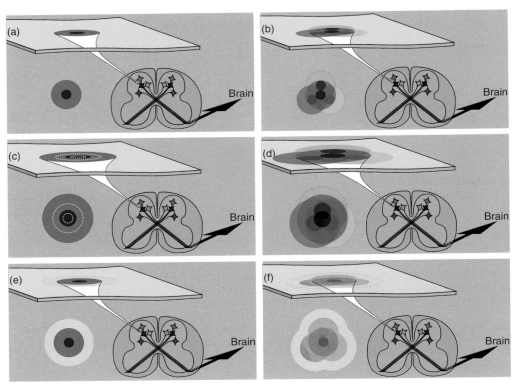

Figure 2 (a) When recordings are made from an individual wide-dynamic-range (WDR) neuron (the blue neuron in the diagram), activation by noxious stimuli applied to any part of its excitatory receptive field can be observed (blue surface on the skin). Its center (blue dark) is also sensitive to low-intensity stimuli. (b) The excitatory fields of adjacent WDR neurons overlap one another, so that a noxious stimulus will activate not only the center of some receptive fields (blue area) but also the edges of others (pale surfaces colored red, purple, and yellow). Thus, when applied to a given surface area (here the center of the blue field), a noxious stimulus will activate many neurons (four in this case) whereas a non-noxious stimulus will activate far fewer (here only one). (c) In fact the surface potentially covered by the excitatory receptive field of a WDR neuron is larger than that observed under physiological conditions. If the neuron (blue) from which recordings are being made is partially depolarized (e.g., by microelectrophoretically applying an excitatory amino acid to its membrane), an enlargement of the excitatory receptive field can be observed. This means that the span of the excitatory receptive fields is potentially much greater than shown in (a) (as indicated by the broken white lines). This is probably what occurs under inflammatory conditions. (d) Thus, considering four neurons, it can be seen that a non-noxious stimulus applied to the center of the blue field will now be able to activate the red, purple, and yellow neurons as well (cf. (b)). (e) At the periphery of their excitatory fields, WDR neurons have an inhibitory field (white zone). When mechanical stimuli are applied to this area, the neuron is inhibited. (f) The inhibitory fields of a population of adjacent WDR neurons substantially overlap one another. Thus in the center of the red field, mechanical stimuli activate the red neuron but inhibit the purple and yellow neurons. When tactile stimuli are applied to a large surface, they will not only activate the centers of excitatory fields and potentially generate a false nociceptive signal, they will also activate inhibitory fields with a resulting attenuation of the global response. This is the basis for some physical methods of pain-relief (see text).

summation of excitatory postsynaptic potentials. The concept of quiescent synapses (e.g., Wall, P. D., 1977) developed with the idea that in some conditions, this fringe of the receptive field can produce significant information transfer, that is, trigger action potentials. Such synapses can be made more efficacious experimentally by partially depolarizing a neuron using microelectrophoretic application an excitatory amino acid (Zieglgänsberger, W. and Herz, A., 1971) or pharmacologically blocking inhibitory transmitter mechanisms (e.g., Yokota, T. and

Nishikawa, Y., 1979). Under these circumstances, the receptive fields enlarge and this is also seen after transient electrical stimulation of C-fibers, injury, noxious pinch, intradermal injection of capsaicin, or noxious stimulation of viscera or muscle.

Such plasticity may be triggered naturally by peripheral or central pathologies. In general, sensitization of excitatory mechanisms or a deficiency of inhibitory ones will produce both increased activity in the neuronal population associated with a painful focus and a growth of this population (Figure 2(d)). This is illustrated by the increased size of excitatory receptive fields during acute or chronic inflammation (e.g., Grubb, B. D. *et al.*, 1993). The resulting increased information is transmitted to, and decoded in, the brain where it constitutes a form of hyperalgesia. This increase in nociceptive information varies with the volume of the body affected by the lesion (i.e., in a three-dimensional way) and also has a temporal dimension.

25.3 Segmental Inhibitory Receptive Field

25.3.1 General

WDR neurons frequently exhibit not only excitatory cutaneous receptive fields, but also inhibitory fields that are often situated proximally (e.g., Hillman, P. and Wall, P. D., 1969). When applied to this area, most mechanical stimuli, notably those of low intensity, inhibit the activity of WDR neurons. It is often implied that these phenomena are triggered only by the activation of $A\beta$-fibers. However, numerous studies have demonstrated that activation of $A\delta$-fibers results in more powerful inhibitions (e.g., Chung, J. M. *et al.*, 1984). Thus the activation of a pool of $A\delta$-fiber terminals will both activate those WDR neurons with excitatory receptive fields in this particular area and inhibit those with overlapping inhibitory receptive fields. As described below, the final output from these neurons constitutes a contrasted picture reminiscent of the lateral inhibition described in other systems (e.g., visual and touch).

25.3.2 Plasticity

It is important to consider the overlap of excitatory and inhibitory fields of neighboring WDR neurons (Figures 2(e) and 2(f)). The organization of the inhibitory fields offers an explanation for how, at least in nonpathological conditions, the high levels of activity produced in individual neurons by application of multiple non-nociceptive stimuli to large surface areas (incorporating numerous centers of excitatory fields) do not constitute a nociceptive signal. Such an eventuality is prevented by the concomitant stimulation of numerous inhibitory fields which attenuates the overall response of the neuronal population. in contrast, a deficit of such inhibitory effects will result in a greater amount of activity being produced by harmless stimuli. Once transmitted to and decoded by the brain,

such information could constitute a form of allodynia such as that triggered by brushing the skin with cotton wool, which is characteristic of some neuropathies.

The convergence of excitatory and inhibitory influences may also explain the hypoalgesic effects of stimulating parts of the body close to a painful focus, for example, by rubbing or transcutaneous electrical nerve stimulation (TENS). Indeed, the fact that large and fine cutaneous afferent fibers interact with this potential effect has been known for a long time (e.g., Head, H. *et al.*, 1905; Noordenbos, W., 1959; Melzack, R. and Wall, P. D., 1965).

25.4 Perturbance of the Basic Somesthetic Activity by a Painful Focus

In addition to the segmental inhibitory influences discussed above, WDR neurons are inhibited by nociceptive stimulation of areas of the body remote from their segmental receptive fields. These effects involve propriospinal and supraspinal mechanisms. The former can be observed in the dorsal horn of spinal animals (e.g., Gerhart, K. D. *et al.*, 1981) but the underlying mechanisms have not been investigated in detail. The supraspinal mechanisms are observed in intact but not in spinalized animals and have been studied more extensively (see Chapter Diffuse Noxious Inhibitory Controls (DNIC)).

In fact, any noxious stimulus will activate descending bulbo-spinal controls and these will strongly depress the activity of neurons in the spinal dorsal horn and its trigeminal homolog. In several species, the vast majority of WDR neurons (and some nociceptive-specific neurons) are profoundly inhibited by noxious stimulation of any part of the body provided it is not the neuron's own excitatory receptive field. In effect, the receptive fields of these neurons are made up by the whole body – mainly in the form of an inhibitory receptive field except in its own neural segment where both inhibitory and excitatory fields exist. These whole body phenomena which have been termed diffuse noxious inhibitory controls (DNIC) work by hyperpolarizing WDR neurons (see Chapter Diffuse Noxious Inhibitory Controls (DNIC)).

In summary, a noxious focus activates a segmental subset of WDR neurons and inhibits all other WDR neurons (Figure 3(b)). This mechanism effectively improves the signal-to-noise ratio by increasing the contrast between the activities of the excited neurons and the depressed activity of the remainder.

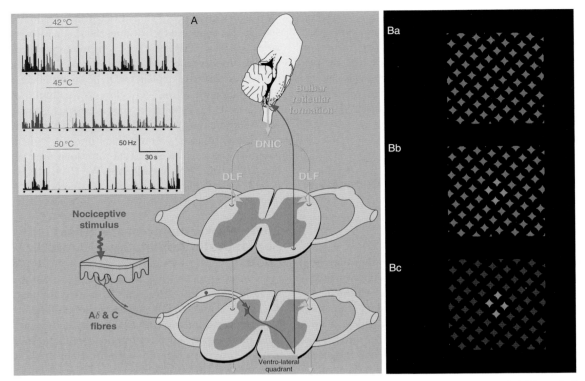

Figure 3 (A) Triggering of descending inhibitory controls by nociceptive stimuli. One consequence of a painful focus is the modification of activity of a whole population of WDR neurons. Nociceptive signals pass up the ventrolateral quadrant to activate structures in the bulbar reticular formation. These structures in turn send signals down the dorsolateral cord to trigger powerful inhibitory mechanisms in all segments not concerned with the initial stimulus. The insert (top left) shows a recording from a WDR neuron in the rat lumbar cord whose receptive field on the extremity of a hindpaw, was mechanically stimulated by a paint brush once every 10 s (indicated by black spots). The resulting activity was inhibited by the application of remote noxious stimuli (immersion of the tail in 42, 45, or 50 °C water). This inhibition increased with the water temperature. (B) Schematization of the activity of all WDR neurons: (Ba) in the absence of a painful focus, these neurons produce the basic somaesthetic information, consisting of signals from the body about its immediate environment; (Bb) as a consequence of a painful focus, some of these neurons are activated (four of them in the center of this diagram), but it is difficult to extract this information because of the background noise; (Bc) however the inhibition of the other neurons by the mechanisms described in (a), makes it possible for the brain to extract the information about the painful focus due to the contrast between the activated and the inhibited neurons. DLF, dorsolateral funiculus; DNFC, diffuse noxious inhibitory control.

Interestingly, DNIC are exacerbated in experimental models of clinical pain (e.g., monoarthritis, polyarthritis, and peripheral mononeuropathies; e.g., Danziger, N. *et al.*, 2000).

25.5   Multiple or Vast Painful Foci

If it is accepted that WDR neurons are important in pain, then interactions between nociceptive signals from separate parts of the body must result in interactions between the corresponding pains. In fact, a painful stimulus can decrease or even mask pain arising from another part of the body (see Chapter Diffuse Noxious Inhibitory Controls (DNIC)).

The size of a painful focus can vary greatly depending on the underlying pathology and consequently the volume of nociceptive information sent to the brain will also vary. How do WDR neurons behave when noxious stimuli are applied to increasingly large surface areas? In fact, two opposite effects are observed: up to a particular area of stimulation – approximately twice that of the neuron's excitatory field – there is an accelerating transfer function; however for larger surface areas, the responses gradually diminish (Bouhassira, D. *et al.*, 1995). This is consistent with findings in humans that noxious thermal stimulation of increasingly large areas of skin causes an increasingly painful sensation only over a restricted range of surface areas (e.g., Hardy, J. D.

et al., 1952) and that painful thermal stimuli evoke pleasant warmth when applied to larger areas (Melzack, R. *et al.*, 1962). The classical poor correlation between the extent of an injury and the intensity of pain may also involve such mechanisms.

Thus it seems that a painful focus activates a negative feedback loop affecting WDR neurons only when it generates a sufficient volume of nociceptive information. This volume is undoubtedly determined by the intensity of the stimulus, the area to which it is applied, and its duration. In this respect, the role of temporal summation in triggering inhibitory bulbo-spinal mechanisms must be addressed (Gozariu, M. *et al.*, 1997).

25.6 Summary and Conclusions

The global population of WDR neurons is activated continuously but unpredictably by all the non-noxious stimuli provided by the environment. The resulting basic somaesthetic activity can constitute either noise from which the brain may have difficulty extracting a clear signal, or a significant message. According to modern neuropsychological concepts, this on-going information from the whole body, along with that from other sensory systems, notably the vestibular and visual systems, is integrated to generate a permanent, unconscious, representation of the physiological reality of self. At the beginning of the twentieth century, the great neurologist Henry Head proposed that our brains contain two schemas – "a postural model of ourselves and another schema allowing the recognition of the locality of the stimulated spot" (Head, H. and Holmes, G., 1911). However, following Schilder P. (1950), the simple term body schema is usually used to refer to both models. Even though the terms are sometimes used interchangeably, it is advisable to distinguish body schema which is an unconscious cerebral model of the body, from body image which is a conscious mental representation. WDR neurons are good candidates for constructing and endlessly rebuilding the memory of one's physical self (Le Bars, D., 2002). Previous experiences, be they neutral, pleasant, or unpleasant (e.g., pain), contribute to the building of this memory; all will have activated WDR neurons. Intense or long-lasting pains are probably among the physical causes which most commonly disturb the body schema: they generate a painful body by focusing attention on one part of the body to the detriment of the others. The body schema then emerges from

relative unconsciousness and the painful focus becomes pre-eminent and relatively oversized. Spino-bulbo-spinal mechanisms could underlie this over-representation by providing a filter which imposes the nociceptive signal to the detriment of information from the rest of the body.

After all, this is what happens within any large gathering: silence is necessary if a speaker is to be heard. DNIC establish such silence among spinal neurons. Conversely, a hubbub does not reduce the speaker to silence, but drowns his speech to the extent that he cannot be heard. This is one way in which morphine works – and DNIC are extremely sensitive to low doses of morphine (see Le Bars, D. *et al.*, 1995).

In summary, WDR neurons do not constitute a single-channel transmission network. Nociceptive information is processed by complex mechanisms which can modify the gain of the system. The modulation of this gain is determined by the three-dimensional characteristics of the painful focus – not only its intensity and duration but also its surface area. Beyond a critical level, an increase in the number of activated noci-responsive neurons is associated with a decrease in their individual responses.

A complete analysis of nociceptive processes must consider not only a large population of neurons but also their interactions. Pain in clinical practice is not punctuate and involves a large number of excitatory receptive fields of peripheral fibers and central neurons. When several foci coexist, the global signal from WDR neurons may be polymorphic with the resulting pain showing fluctuations, instability, or even versatility due to the resulting interactions.

References

Adriaensen, H., Gybels, J., Handwerker, H. O., and Van Hees, J. 1984. Nociceptor discharges and sensations due to prolonged noxious mechanical stimulation: a paradox. Human Neurobiol. 3, 53–58.

Bouhassira, D., Gall, O., Chitour, D., and Le Bars, D. 1995. Dorsal horn convergent neurones: negative feed-back triggered by spatial summation of nociceptive afferents. Pain 62, 195–200.

Cadden, S. W. 1993. The ability of inhibitory controls to 'switch-off' activity in dorsal horn convergent neurones in the rat. Brain Res. 628, 65–71.

Cervero, F., Handwerker, H. O., and Laird, J. M. 1988. Prolonged noxious mechanical stimulation of the rat's tail: responses and encoding properties of dorsal horn neurones. J. Physiol. 404, 419–436.

Chung, J. M., Fang, Z. R., Hori, Y., Lee, K. H., and Willis, W. D. 1984. Prolonged inhibition of primate spinothalamic tract cells by peripheral nerve stimulation. Pain 19, 259–275.

Danziger, N., Le Bars, D., and Bouhassira, D. 2000. Diffuse Noxious Inhibitory Controls and Arthritis in the Rat. In: Pain and Neuroimmune Interactions (*eds*. N. E. Saadé, A. V. Apkarian, and S. J. Jabbur), pp. 69–78. Kluwer Academic Press.

Dubner, R., Hoffman, D. S., and Hayes, R. L. 1981. Neuronal activity in medullary dorsal horn of awake monkeys trained in a thermal discrimination task. III. Task-related responses and their functional role. J. Neurophysiol. 46, 444–464.

Gerhart, K. D., Yezierski, R. P., Giesler, G. J., Jr., and Willis, W. D. 1981. Inhibitory receptive fields of primate spinothalamic tract cells. J. Neurophysiol. 46, 1309–1325.

Gozariu, M., Bragard, D., Willer, J. C., and Le Bars, D. 1997. Temporal summation of C-fiber afferent inputs: competition between facilitatory and inhibitory effects on C-fiber reflex in the rat. J. Neurophysiol. 78, 3165–3179.

Grubb, B. D., Stiller, R. U., and Schaible, H. G. 1993. Dynamic changes in the receptive field properties of spinal cord neurons with ankle input in rats with chronic unilateral inflammation in the ankle region. Exp. Brain Res. 92, 441–452.

Hardy, J. D., Woolf, H. G., and Goodel, H. 1952. Pain Sensations and Reactions. William and Wilkins.

Head, H. and Holmes, G. 1911. Sensory disturbances from cerebral lesions. Brain 34, 102–245.

Head, H., Rivers, W. H. R., and Sherren, J. 1905. The afferent nervous system from a new aspect. Brain 28, 99–115.

Hillman, P. and Wall, P. D. 1969. Inhibitory and excitatory factors influencing the receptive fields of lamina 5 spinal cord cells. Exp. Brain Res. 9, 284–306.

Hoffman, D. S., Dubner, R., Hayes, R. L., and Medlin, T. P. 1981. Neuronal activity in medullary dorsal horn of awake monkeys trained in a thermal discrimination task. I. Responses to innocuous and noxious thermal stimuli. J. Neurophysiol. 46, 409–427.

Le Bars, D. 2002. The whole body receptive field of dorsal horn multireceptive neurones. Brain Res. Rev. 40, 29–44.

Le Bars, D. and Chitour, D. 1983. Do convergent neurons in the spinal cord discriminate nociceptive from non-nociceptive information? Pain 17, 1–19.

Le Bars, D., Bouhassira, D., and Villanueva, L. 1995. Opioids and Diffuse Noxious Inhibitory Controls in the Rat. In: Pain and the Brain: From Nociceptor to Cortical Activity (*eds*. B. Bromm and J. E. Desmedt), Advances in Pain Research and Therapy, Vol. 22, pp. 517–539. Raven Press.

Melzack, R. and Wall, P. D. 1965. Pain mechanisms: a new theory. Science 150, 971–979.

Melzack, R., Rose, G., and McGinty, D. 1962. Skin sensitivity to thermal stimuli. Exp. Neurol. 6, 300–314.

Mendell, L. M. 1984. Modifiability of sipnal synapses. Physiol. Rev. 64, 260–324.

Menétrey, D., Giesler, G. J., Jr., and Besson, J. M. 1977. An analysis of response properties of spinal cord dorsal horn neurones to nonnoxious and noxious stimuli in the spinal rat. Exp. Brain Res. 27, 15–33.

Morisset, V. and Nagy, F. 1996. Modulation of regenerative membrane properties by stimulation of metabotropic glutamate receptors in rat deep dorsal horn neurons. J. Neurophysiol. 76, 2794–2798.

Noordenbos, W. 1959. Pain. Elsevier.

Schilder, P. 1950. The Image and Appearance of the Human Body. International Universities Press.

Wagman, I. H. and Price, D. D. 1969. Responses of dorsal horn cells of *M. Mulatta* to cutaneous and sural A and C fiber stimuli. J. Neurophysiol. 32, 803–817.

Wall, P. D. 1977. The presence of ineffective synapses and the circumstances which unmask them. Philos. Trans. R Soc. London Ser. B 278, 361–372.

Willis, W. D. and Coggeshall, R. E. 1991. Sensory Mechanisms of the Spinal Cord. Plenum.

Yokota, T. and Nishikawa, Y. 1979. Actions of picrotoxin upon trigeminal subnucleus caudalis neurons in the monkey. Brain Res. 171, 369–373.

Zieglgänsberger, W. and Herz, A. 1971. Changes of cutaneous receptive fields of spino-cervical tract neurones and other dorsal horn neurones by microelectrophoretically administered amino acids. Exp. Brain Res. 13, 111–126.

26 Spinal Cord Mechanisms of Hyperalgesia and Allodynia

T J Coderre, McGill University, Montreal, QC, Canada

Glossary

adaptor protein A protein which is accessory to main proteins in a signal transduction pathway. These proteins often lack any intrinsic enzymatic activity themselves, but instead mediate specific protein-protein interactions that drive the formation of protein complexes.

allodynia Pain due to a stimulus that does not normally provoke pain.

antisense oligonucleotides Single strands of DNA or RNA that are complementary to a chosen sequence. Antisense RNAs prevent translation of complementary RNA strands by binding to them.

apoptosis An orchestrated series of biochemical events leading to cell death.

carrageenan A linear sulphated polysaccharide extracted from red seaweeds that is used as an inflammatory agent.

central sensitization Activity- or injury-dependent changes in the excitability of central nervous system (CNS) neurons, typically in pain pathways, and is reflected by increased spontaneous activity, reduced thresholds or increased responsiveness to afferent inputs, prolonged after- discharges to repeated stimulation, and the expansion of the peripheral receptive fields of central neurons.

complete Freund's adjuvant An antigen solution composed of inactivated and dried *Mycobacterium tuberculosis* emulsified in mineral oil, used as an immunopotentiator (booster of the immune system) and inflammatory agent.

2-deoxy-D-glucose A marker for tissue glucose use.

excitatory amino acids Amino acid neurotransmitters that include aspartate and glutamate. Glutamate is the most abundant fast excitatory neurotransmitter in the mammalian nervous system.

excitatory postsynaptic potential (EPSP) Is a temporary increase in postsynaptic membrane potential caused by the flow of positively charged ions into the postsynaptic cell.

glial cells Commonly called neuroglia or simply glia, are non-neuronal cells that provide support and nutrition, maintain homeostasis, form myelin, and participate in signal transmission in the nervous system.

homosynaptic sensitization Facilitation of activity occuring at the synapses that connect two neurons.

hyperalgesia An increased response to a stimulus which is normally painful. Primary hyperalgesia occurs at the site of injury, and secondary hyperalgesia in adjacent, or sometimes remote, uninjured tissue.

inhibitory postsynaptic potential (IPSP) Is the change in membrane voltage of a postsynaptic neuron which results from synaptic activation of inhibitory neurotransmitter receptors.

ion channels Pore-forming proteins that help to establish and control the small voltage gradient across the plasma membrane of all living cells by allowing the flow of ions down their electrochemical gradient.

long-term potentiation (LTP) An increase in the strength of a chemical synapse that lasts from minutes to several days.

neuropeptides A variety of peptides found in neural tissue.

neurotrophins Also called "neurotrophic factors", are a family of protein which induce the survival of neurons. They belong to a family of growth factors, secreted proteins which are capable of signaling particular cells to survive, or differentiate, or grow.

phenotype switch A process by which neurons change their phenotype and begin to produce and release substances they normally do not produce.

protein kinase Is a kinase enzyme that modifies other proteins by chemically adding phosphate groups to them (phosphorylation).

protein tyrosine phophatase (PTP) An enzyme that remove phosphate groups from phosphorylated tyrosine residues on proteins.

scaffold proteins Proteins that bring together various other proteins in a signaling pathway and allows for their interaction. They recruit downstream effectors in a pathway and enhance specificity of the signal.

secondary messengers Secondary messengers are a component of signal transduction cascades that utilize receptors on the surface of the plasma membrane which are generally coupled to a kinase on the interior surface of the membrane. The kinase then phosphorylates another molecule which carries out another action.

transcription factor A protein that works in concert with other proteins to either increase or decrease the transcription of genes.

transporters Membrane-bound pumps that move biochemicals and other atomic or molecular substances across membranes.

wide dynamic range A property of spinal cord dorsal horn neurons (typically in lamina V) that exhibit graded responses to non-noxious and noxious stimuli since they receive convergent input from non-nociceptive and nociceptive primary afferent fibers.

wind-up Cumulative depolarization, action potential bursting and afterdischarges in spinal cord dorsal horn neurons following high frequency, C-fiber strength electrical stimulation or intensely noxious natural stimuli.

26.1 Characteristics of Hyperalgesia

26.1.1 Definition and Historical Background

The term hyperalgesia was first coined by Gowers in the 1800s, and is often still used as a term to define what Hardy J. D. (1956) described as "a state of increased pain sensation induced by either noxious or ordinarily nonnoxious stimulation of peripheral tissue." The International Association for the Study of Pain (IASP) currently uses separate terms for the two reactions Hardy described. Thus, hyperalgesia is defined as "an increased response to a stimulus which is normally painful," while "pain due to a stimulus that does not normally provoke pain" is known as allodynia (IASP website). Over the years, hyperalgesia (not distinguished from allodynia) has been further separated into primary and secondary hyperalgesia; primary hyperalgesia occurs at the site of injury, and secondary hyperalgesia in adjacent, or sometimes remote, uninjured tissue. Primary hyperalgesia is reflected by a hypersensitivity to thermal and mechanical stimuli, and is driven largely by sensitization of peripheral nociceptors. Secondary hyperalgesia is predominantly characterized by hypersensitivity to mechanical stimuli, and likely involves central sensitization. Central sensitization refers to activity- or injury-dependent changes in the excitability of central nervous system (CNS) neurons, typically in pain pathways, and is reflected by increased spontaneous activity, reduced thresholds or increased responsiveness to afferent inputs, prolonged after-discharges to repeated stimulation, and the expansion of the peripheral receptive fields of central neurons (Woolf, C. J. *et al.*, 1988; Coderre, T. J. *et al.*, 1993). Although central sensitization has been demonstrated in neurons throughout the somatosensory nervous system, the mechanisms underlying central sensitization have been largely generated from studies concentrating on spinal cord dorsal horn neurons. The characteristics and underlying mechanisms of central sensitization in spinal cord dorsal horn neurons will be the main focus of the remaining sections of this chapter, and will be a recurring theme in various chapters of this volume that address alterations in pain processing associated with injury or disease of peripheral tissue, primary afferent nerves, or CNS structures.

26.1.2 Precipitating Conditions for Hyperalgesia and Evidence of Central Sensitization

26.1.2.1 Cutaneous injury

26.1.2.1.(i) *Heat injury* Considerable evidence demonstrates that heat injury of the skin produces a sensitization of peripheral nociceptors. Peripheral sensitization occurs both in Aδ and C-fiber nociceptors (Beitel, R. E. and Dubner, R., 1976; Fitzgerald, M. and Lynn, B., 1977), and largely accounts for primary hyperalgesia (Meyer, R. A. and Campbell, J. N., 1981; Torebjörk, H. E. *et al.*, 1984). In addition, dorsal horn neurons fire with increasing frequency in response to repeated applications of a noxious heat stimulus (Perl, E. R. *et al.*, 1976, Kenshalo, D. R. *et al.*, 1979) or following thermal injury (Price, D. D. *et al.*, 1978; Kenshalo, D. R. *et al.*, 1982), indicative of central sensitization. In addition to the sensitization and prolonged excitation of dorsal horn cells, noxious stimulation associated with tissue injury also produces an expansion of the receptive fields of dorsal horn neurons. Thus, neurons in the dorsal horn of the spinal cord with receptive fields adjacent to a cutaneous heat injury expand their receptive fields to incorporate the site of injury (McMahon, S. B. and Wall, P. D., 1984).

Evidence that central sensitization contributes to secondary hyperalgesia after heat injury was supported initially by reports that heat injury-induced cutaneous hyperalgesia spreads well beyond the area of nociceptor sensitization. Thalhammer J. G. and Lamotte R. H. (1982) found that a heat injury in one half of a cutaneous nociceptor's receptive field did not produce heat sensitization in the other half, despite the fact that hyperalgesia spread into this area. Typically, nociceptor sensitization associated with injury is restricted to about 5–10 mm of the site of injury (Fitzgerald, M., 1979) while, in contrast, cutaneous hyperalgesia spreads as far as 10–20 cm beyond the site of injury (Hardy, J. D. *et al.*, 1950).

Although it is commonly reported that there is mechanical, but not heat hyperalgesia in the secondary zone (Raja, S. N. *et al.*, 1984), secondary heat hyperalgesia has been reported in some cases (Pedersen, J. L. and Kehlet, H. 1998). Importantly, secondary heat hyperalgesia is significantly reduced when lidocaine is administered i.v. prior to heat injury (Holthusen, H. *et al.*, 2000), suggesting that central sensitization plays a role in its initiation. Animal behavioral studies support this conclusion, as the spread of hyperalgesia to the rat hind paw contralateral to the paw that received a thermal injury is unaffected by either deafferentation or anesthetic blocks of the injured hind paw following the injury, but is prevented if deafferentation or anesthetic block precedes the injury (Coderre, T. J. and Melzack, R., 1985; 1987).

26.1.2.1.(ii) *Mechanical injury* Mechanical injury of the skin also produces both primary and secondary hyperalgesia, as has been demonstrated in the rat plantar incision model (Brennan, T. J. *et al.*, 1996), as well as following incision of human skin (Kawamata, M. *et al.*, 2002a). Although both pre- and postincisional administration of intrathecal (i.t.) morphine or bupivacaine reduce secondary hyperalgesia in the rat model (Zahn, P. K. and Brennan, T. J., 1999), only preincisional administration of s.c. or i.v. lidocaine reduce secondary hyperalgesia in humans (Kawamata, M. *et al.*, 2002a). These results again support a role for central sensitization in the initiation of secondary hyperalgesia after mechanical skin injury. This conclusion is also consistent with the finding that rat plantar incisions or other mechanical injuries produce an increase in the responsiveness of spinal wide dynamic range (WDR) neurons to innocuous and noxious stimuli, as well as increases in their receptive fields (Cervero, F. *et al.*, 1988; Kawamata, M. *et al.*, 2002b).

26.1.2.1.(iii) *Capsaicin* Capsaicin, the active ingredient of hot chilli peppers, produces hyperalgesia in both animals (Gilchrist, H. D. *et al.*, 1996) and humans (Carpenter, S. E. and Lynn, B., 1981) when injected or applied to the skin. Evidence that capsaicin-induced hyperalgesia depends on central sensitization initially came from animal studies that showed that intradermal injections of capsaicin produced enhanced excitability of spinal cord dorsal horn neurons in cats (Simone, D. A. *et al.*, 1989) and monkeys (Dougherty, P. M. and Willis, W. D., 1992). In both monkeys (Baumann, T. K. *et al.*, 1991) and humans (Lamotte, R. H. *et al.*, 1992), capsaicin injections produce extensive, spreading hyperalgesia without producing the same degree of spreading sensitization of primary afferent nociceptors. Indeed, capsaicin caused the spread of both brush allodynia and pinprick hyperalgesia beyond the borders of the territory of the nerve which innervates the injected skin area (Sang, C. N. *et al.*, 1996), suggesting that there is extraterritorial pain that is dependent on central sensitization. Furthermore, in humans, hyperalgesia to punctate mechanical stimuli, which

develops after intradermal injection of capsaicin, is maintained even after anesthetizing the region where capsaicin was injected (Lamotte, R. H. *et al.*, 1991). However, if the skin region is anesthetized prior to capsaicin injection, cutaneous hyperalgesia does not develop. Furthermore, hyperalgesic responses to capsaicin can be prevented if the area of skin where the injection is made is rendered anesthetic by a proximal anesthetic block of the peripheral nerve which innervates it. Thus, for hyperalgesia to develop it is critical that initial inputs from the injury reach the CNS. However, once hyperalgesia is established, it does not need to be maintained by inputs from the injured peripheral tissue. In support of this, Torebjörk H. E. *et al.* (1992) have shown that pain thresholds to intraneural electrical stimulation of afferent fibers are dramatically reduced following intradermal capsaicin injection in the skin from which the stimulated nerve emanates: neural stimulation which was felt as tactile before administration of capsaicin, was painful after capsaicin. Importantly, this reduced pain threshold is evident even when the sensory projected field of the afferent nerve is anesthetized after the capsaicin injection. Again, a state of central sensitization is indicated since once they have established their effects, inputs from the injured region are not required to maintain the lowered threshold.

26.1.2.2 Peripheral nerve stimulation

Much of the early evidence that spinal mechanisms contribute to hyperalgesia and allodynia came from studies that used peripheral nerve stimulation as a conditioning stimulus, and assessed the influence of the conditioning on spinal function or the function of spinal output neurons. Early studies showed that repeated high-frequency C-fiber afferent stimulation sequentially increases dorsal horn activity resulting in a prolonged discharge of the cell, lasting from seconds to minutes poststimulation (Mendell, L. M. and Wall, P. D., 1965; Mendell, L. M., 1966). This phenomenon, which has been labeled as windup, results from a homosynaptic sensitization (Woolf, C. J. *et al.*, 1988) of dorsal horn neurons, and likely represents a short-term sensitization of dorsal horn neurons that underlies the intense reactions and lingering after sensations to stimuli at the pain tolerance level. Although windup is not synonymous with central sensitization (Woolf, C. J., 1996), stimulation that causes windup can produce some of the classical characteristics of central sensitization including expansion of dorsal horn neuronal receptive fields and enhanced spinal neuronal responses to C-fiber, but not A-fiber electrical stimulation (Li, J. et al., 1999).

In early studies of central mechanisms of hyperalgesia, it was shown that noxious electrical stimulation of cutaneous and muscle afferent nerves (Woolf, C. J., 1983; Wall, P. D. and Woolf, C. J., 1984), as well as cutaneous (Woolf, C. J., 1983) and deep (Woolf, C. J. and McMahon, S. B., 1985) tissue injury, produce an increase in the excitability of the ipsilateral and contralateral flexor efferent nerves in response to noxious mechanical stimulation of the hind paw. Because the increased excitability in the contralateral flexor efferent nerve is maintained even after inputs from the injured paw are blocked by local anesthesia, the results suggest that central, not peripheral, changes underlie this effect.

Although typically associated with hippocampal neurons, evidence suggests that the phenomenon long-term potentiation (LTP) can also be elicited in spinal cord dorsal horn neurons after stimulation of primary afferents in transverse spinal cord slices (Randic, M. et al., 1993; Lozier, A. P. and Kendig, J. J., 1995), as well as in vivo after electrical nerve stimulation (Svendsen, F. et al., 1997; 1999) or intense natural noxious stimulation (Rygh, L. J. et al., 1999). Importantly, LTP induced by electrical nerve stimulation or noxious stimulation is maintained after lidocaine block distal to the stimulation site in the sciatic nerve (Svendsen, F. et al., 1999) or proximal to the site of noxious stimulation (Rygh, L. J. et al., 1999).

26.1.2.3 Inflammation

26.1.2.3.(i) Subcutaneous inflammation Subcutaneous injections of inflammatory substances such as carrageenan and complete Freund's adjuvant (CFA) in the rat hind paw produce thermal hyperalgesia and mechanical allodynia (Buritova, J. et al., 1996; Arevalo, M. I. et al., 2003), as well as a sensitization of peripheral nociceptors (Kocher, L. et al., 1987; Andrew, D. and Greenspan, J. D., 1999). There is also evidence of central sensitization associated with peripheral inflammation. Subcutaneous carrageenan injections produce increased excitability of rat spinal cord neurons as shown by increased responses to mechanical and electrical stimulation, enhanced windup, expanded receptive fields (Hedo, G. et al., 1999; Torsney, C. and Fitzgerald, M., 2002), and increased flexor efferent neuron responsiveness (Xu, X.-J. et al., 1995). Subcutaneous injections of CFA in rats also produce central sensitization, as evidenced by the enhancement of dorsal horn neurons spontaneous activity and increased responsiveness to thermal, mechanical and

electrical nerve stimulation (Hylden, J. L. K. et al., 1989; Ren, K. et al., 1992). After subcutaneous CFA, there is also an increased ability to induce LTP (Vikman, K. S. et al., 2003), expanded receptive fields in rat dorsal horn neurons (Hylden, J. L. K. et al., 1989; Ren, K. et al., 1992), and an increase in the excitability of flexor efferent neurons (Seybold, V. S. et al., 2003).

26.1.2.3.(ii) Joint inflammation Injection of kaolin and carrageenan into the rat knee joint produces limping and guarding of the limb (Sluka, K. A. and Westlund, K. N., 1993), as well as a sensitization of peripheral nociceptors (Neugebauer, V. and Schaible, H. G., 1988). The same stimulus produces enhanced responsiveness of spinal cord dorsal horn neurons to normal movement or tactile stimulation of the knee joint (Neugebauer, V. and Schaible, H. G., 1988; 1990), as well as an expansion of the receptive field of dorsal horn neurons that respond to joint input (Neugebauer, V. and Schaible, H. G., 1990). There is also an expansion of dorsal horn neuron receptive fields following the induction of polyarthritis in multiple joints three weeks or more following CFA injections to the rat tail (Ménétrey, D. and Besson, J. M. 1982; Calvino, B. et al., 1987).

26.1.2.4 Peripheral nerve injury

Traumatic (Bennett, G. J. and Xie, Y. K., 1988; Seltzer, Z. et al., 1990; Kim, S. H. and Chung, J. M., 1992; Decosterd, I. and Woolf, C. J., 2000), inflammatory (Eliav, E. et al., 1999; Milligan, E. D. et al., 2003), or chemical (Tanner, K. D. et al., 1998; Polomano, R. C. et al., 2000) injury of peripheral nerves induces heat and/or mechanical hyperalgesia, or mechanical and/or cold allodynia in animal models. While there is an established contribution of peripheral pathology to hyperalgesia and allodynia in these models (Chen, Y. and Devor, M., 1998; Liu, X. et al., 1999), it is also clear that central sensitization plays a critical role. Thus, 9–11 days after chronic constriction injury (CCI) of the rat sciatic nerve, lumbar spinal cord dorsal horn neurons exhibit abnormal characteristics including spontaneous activity, exaggerated, and prolonged responses to noxious and innocuous mechanical stimulation (Laird, J. M. and Bennett, G. J., 1993), with prolonged after-discharges more prevalent in WDR neurons, as opposed to nociceptive-specific neurons (Sotgiu, M. L. et al., 1995; Pitcher, G. M. and Henry, J. L., 2000). Two weeks after L5/L6 spinal nerve ligation (SNL), rat dorsal horn neurons also show increased spontaneous firing with irregular firing patterns (Suzuki, R. and Dickenson, A. H., 2005).

Thresholds for excitatory postsynaptic potentials (EPSPs) of lamina II neurons are dramatically lower after CCI, and spared nerve injury (Kohno, T. *et al.*, 2003), and there is a long-lasting posttetanic potentiation of sciatic-evoked A-fiber superficial dorsal horn field potentials that is present over the same time course as thermal hyperalgesia in CCI rats (Draganic, P. *et al.*, 2001). Tight ligation of the L5/L6 spinal nerve also results in a lower threshold for activation of C-fiber-driven dorsal horn field potentials, and for the induction of LTP in dorsal horn neurons (Xing, G. *et al.*, 2003).

Increases in spontaneous and evoked activity in dorsal horn neurons of CCI rats are paralleled by increases in $[^{14}C]$-2-deoxyglucose metabolic activity throughout the lumbar spinal dorsal horn, with greater increases in the medial portion of the ipsilateral dorsal horn (Mao, J. *et al.*, 1992a). Fos protein expression, often used as a measure of neuronal activation, is also increased in the spinal cord dorsal horn of CCI rats (Ro, L. S. *et al.*, 2004). Importantly, Fos protein expression is elevated throughout the dorsal horn (lamina I–VI) of the entire lumbar spinal cord, and like $[^{14}C]$-2-deoxyglucose activity, significant but lower increases were found in the contralateral dorsal horn. Thus, for both the $[^{14}C]$-2-deoxyglucose and Fos protein expression measures, there is a spread of neuronal activity to spinal regions outside of the normal central terminations of the sciatic nerve. This may explain why abnormal pain sensations are sometimes exhibited contralaterally, or in body areas that do not correspond with the terminations of an injured peripheral nerve (Koltzenburg, M. *et al.*, 1999; Ro, L. S. *et al.*, 2004).

In addition to enhanced spontaneous and evoked activity, spinal cord dorsal horn neurons of CCI rats are also found to have expanded receptive fields (Pitcher, G. M. and Henry, J. L., 2000). Rats with diabetic neuropathy following treatment with streptozotocin are also found to have dorsal horn neurons with high level of spontaneous activity and with expanded receptive fields (Chen, S. R. and Pan, H. L., 2002). A recent study measured dorsal horn neuronal activity in neuropathic pain patients undergoing surgery for ablation of the dorsal root entry zone (Guenot, M. *et al.*, 2003). These authors found similar alterations in dorsal horn neuron activity including bursting activity (in patients with deafferentation pain after brachial plexus avulsion) and nonrandom patterns of spontaneous discharge (in patients with peripheral nerve injury).

The enhanced spontaneous or evoked activity of dorsal horn neurons likely contribute to hyperalgesia and allodynia since systemic low-dose lidocaine reduces both spontaneous activity in WDR neurons (Sotgiu, M. L. *et al.*, 1992) and the thermal hyperalgesia (Abram, S. E. and Yaksh, T. L., 1994) in CCI rats. Mechanical allodynia induced by partial sciatic nerve lesion or SNL is also attenuated by i.t. treatment with very low doses of lidocaine that are not effective when given systemically (Ma, W. *et al.*, 2003a). The hyperalgesia after CCI depends partly on an injury discharge at the time of surgery, since the onset of thermal hyperalgesia in CCI rats was delayed by treating the nerve with bupivacaine 15 min before, but not 15 min after nerve constriction (Yamamoto, T. *et al.*, 1993).

26.1.2.5 Spinal cord injury

Animal models of central pain have been generated following traumatic, inflammatory, or excitotoxic injury of the spinal cord (see Vierck, C. J. and Light, A. R., 2000 or Yezierski, R. P., 2000 for reviews). Traumatic (Christensen, M. D. *et al.*, 1996; Christensen, M. D. and Hulsebosch, C. E., 1997; Bruce, J. C. *et al.*, 2002), inflammatory (Cahill, C. M. *et al.*, 1998), and excitotoxic (Yezierski, R. P. *et al.*, 1998) injury of spinal cord produces mechanical and thermal allodynia and/or hyperalgesia in rat hind paws. Spinal cord injury also results in a sensitization of spinal cord dorsal horn neurons as indicated by increased spontaneous activity, increased evoked responses to mechanical stimuli, and increased duration of after-discharges and enhanced windup (Yezierski, R. P. and Park, S. H., 1993; Christensen, M. D. and Hulsebosch, C. E., 1997; Wang, J. *et al.*, 2005; Zhang, H. *et al.*, 2005). It was recently demonstrated that spinal sensitization (enhanced spontaneous and evoked activity and enhanced windup in dorsal horn neurons), as well as hind paw thermal and mechanical allodynia, are reduced in spinal cord injured rats that are treated spinally with bupivacaine at the time of the injury (Zhang, H. *et al.*, 2005), suggesting that spinal sensitization underlies the development of allodynia/hyperalgesia after spinal cord injury. Spinal treatment with morphine also reduces the enhanced mechanically evoked responses of WDR neurons in the spinal cord dorsal horn of spinal cord injured rats (Wang, J. *et al.*, 2005). Consistent with these findings, it has recently been shown that central pain (spontaneous pain, mechanical allodynia, and/or mechanical hyperalgesia) associated with spinal cord injury in rats is reduced by intravenous lidocaine administration (Attal, N. *et al.*, 2000; Finnerup, N. B. *et al.*, 2005).

26.1.2.6 Illness-induced hyperalgesia

Thermal hyperalgesia is also induced by intraperitoneal (i.p.) injection of agents that induce illness including lithium chloride and the bacterial cell wall endotoxin lipopolysaccharide (LPS) (Wiertelak, E. P. *et al.*, 1994; Watkins, L. R. *et al.*, 1994a). Illness-induced hyperalgesia has been shown to be dependent on spinal mechanisms that are activated by a descending pain facilitatory pathway (see below) that projects from the nucleus raphe magnus (NRM) through the dorsolateral funiculus to the spinal cord dorsal horn. The NRM has also been shown to be activated by brain regions that receive inputs from hepatic vagal afferent nerves (Watkins, L. R. *et al.*, 1994b).

26.1.2.7 Chronic opiate exposure

Thermal and/or mechanical hyperalgesia are observed in rats after repeated high-dose morphine treatment (Kayan, S. *et al.*, 1971; Mao, J. and Mayer, D. J., 2001) or in some cases after administration of low doses of morphine (Holtman, J. R., Jr. and Wala, E. P., 2005). This phenomenon mirrors a trend that has been observed in clinics in which chronic pain patients are treated with morphine (Jacobsen *et al.*, 1995). Mao J. *et al.* (1995a) and Mayer D. J. *et al.* (1999) have noted a parallel between hyperalgesia that occurs after repeated morphine exposure and that produced by neuropathic injury, each causing the activation of similar intracellular signaling pathways and the appearance of dark neurons in spinal cord dorsal horn. It is also significant that expression of the transcription factor Fos, which is commonly used as an indicator of neuronal activation, is upregulated in the spinal dorsal horn of rats that are morphine dependent and have withdrawal precipitated by injection of naltrexone (Rohde, D. S. *et al.*, 1996).

26.2 Alterations in Spinal Cord Dorsal Horn

26.2.1 Alterations in Transmitter or Growth Factor Expression or Release

26.2.1.1 Neuropeptides

Several lines of evidence suggest that C-fiber neuropeptides are involved in triggering CNS plasticity following injury or noxious stimulation. A role for C-fiber neuropeptides in nociception is suggested as noxious stimulation or peripheral inflammation causes the release of substance P (SP) (Go, V. L. and Yaksh, T. L., 1987; Oku, R. *et al.*, 1987; Duggan, A. W. *et al.*, 1988), neurokinin (NK) A (Hua, X. Y. *et al.*, 1986; Duggan, A. W. *et al.*, 1990), somatostatin (Kuraishi, Y. *et al.*, 1985; Morton, C. R. *et al.*, 1988), calcitonin gene-related peptide (CGRP) (Saria, A. *et al.*, 1986), dynorphin 1–17 (Parra, M. C. *et al.*, 2002), and galanin (Morton, C. R. and Hutchison, W. D., 1990; Colvin, L. A. and Duggan, A. W. *et al.*, 1998) in spinal cord dorsal horn. Spontaneous and noxious stimulus-evoked release of neuropeptide Y (NPY) is enhanced in the dorsal horn of rats with CCI of the sciatic nerve (Mark, M. A. *et al.*, 1998). Furthermore, the iontophoretic application of SP and other NKs produces an excitation of dorsal horn neurons (Henry, J. L., 1976; Willcockson, W. S. *et al.*, 1984a), while i.t. treatment produces behavioral hyperalgesia (Moochhala, S. M. and Sawynok, J., 1984; Cridland, R. A. and Henry, J. L., 1986) or nociceptive behaviors (Hylden, J. L., and Wilcox, G. L., 1981; Seybold, V. S. *et al.*, 1982; Gamse, R. and Saria, A., 1986) in rodents.

There is a decrease in immunoreactive staining for the neuropeptides SP and CGRP in the superficial dorsal horn of rats with experimental arthritis (Sluka, K. A. *et al.*, 1992). While the decrease in these neuropeptides occurs early (4–8 h) after knee inflammation, substance P and CRGP are increased after 5–7 days in the dorsal root ganglia and spinal cord of rats with inflammation of the knee (Sluka, K. A. and Westlund, K. N., 1993) or hind paw (Donnerer, J. *et al.*, 1993). Conversely, and spinal cord dorsal horn substance P and CGRP are decreased in the weeks after CCI of the sciatic nerve (Cameron, A. A. *et al.*, 1991). There is also a decrease in both mRNA and peptide levels of SP, somatostatin and CGRP in the DRG after peripheral nerve section (Jessell, T. M. *et al.*, 1979; Nielsch, V. *et al.*, 1987; Noguchi, K. *et al.*, 1989, 1990, 1993; Henken, D. B. *et al.*, 1990; Baranowsli, A. P. *et al.*, 1993), whereas the levels of cholecystokinin (CCK), galanin, pituitary adenylate cyclase activating peptide (PACAP), vasoactive intestinal polypeptide (VIP) and NPY are increased (Shehab, S. A. and Atkinson, M. E., 1986; Hokfelt, T. *et al.*, 1987; Villar, M. J. *et al.*, 1989; Noguchi, K. *et al.*, 1989, 1993; Wakisaka, S. *et al.*, 1991, 1992; Zhang, X. *et al.*, 1993b, Zhang, Y. Z., 1996). Dynorphin mRNA is also increased in the spinal dorsal horn following traumatic spinal cord injury (Tachibana, T. *et al.*, 1998).

For SP, the decreased immunoreactivity may reflect increased neuropeptide release since *in vivo* microdialysis studies show that spinal SP is increased following partial sciatic nerve ligation (Wallin, J. and Schott, E., 2002). The increased levels of galanin may play an antinociceptive role as galanin overexpressing mice have attenuated mechanical and heat hyperalgesia after partial ischemic injury (Hygge-Blakeman, K. *et al.*, 2004) or CCI (Eaton, M. J. *et al.*, 1999) of the sciatic nerve. Spinal administration of galanin also produces an inhibition of dorsal horn neuronal responses to natural and electrical stimuli in rats with SNL (Flatters, S. J. *et al.*, 2002). In contrast, galanin knockout mice have attenuated responses to hind paw formalin injections (Kerr, B. J. *et al.*, 2001a), and have reduced thermal hyperalgesia after intraplantar injection of carrageenan (Kerr, B. J. *et al.*, 2001a) or partial sciatic nerve transection (Kerr, B. J. *et al.*, 2000). In addition, both windup and the facilitation of spinal reflexes following conditioning stimulation are significantly reduced in galanin knockout mice with sciatic nerve transections (Kerr, B. J. *et al.*, 2001b). Upregulated CCK may contribute to mechanical hypersensitivity as allodynia induced by spinal ischemic injury is reduced following administration of a CCK-B receptor antagonist (Wiesenfeld-Hallin, Z. *et al.*, 1997). Prodynorphin knockout mice or mice receiving dynorphin A antisera do not develop mechanical and thermal hyperalgesia after SNL (Gardell, L. R. *et al.*, 2004), suggesting that upregulated dynorphin contributes to nerve injury-induced hyperalgesia. Indeed, elevated spinal dynorphin content associated with SNL has been demonstrated to be coincident with the onset of mechanical allodynia and thermal hyperalgesia (Malan, T. P. *et al.*, 2000). Although the release of both VIP and NPY are elevated in the deep dorsal horn of rats with spinal nerve transection, the increases do not appear to be related to neuropathic pain as increases were observed even in rats that did not exhibit behavioral signs of hyperalgesia (Kim, H. J. *et al.*, 2003).

A role of C-fiber neuropeptides in noxious stimulus-induced plasticity is further suggested by several findings. Repetitive stimulation of dorsal roots elicits a slow depolarization in dorsal horn neurons which is mimicked by SP (Murase, K. and Randic, M., 1984), NK A (Murase, K. *et al.*, 1989), CGRP (Ryu, P. D. *et al.*, 1988), VIP (Urban, L. and Randic, M., 1984), or CCK (Murase, K. *et al.*, 1987), and is blocked by SP antagonists or capsaicin applied to the tissue bath (Urban, L. and Randic, M., 1984). Iontophoretic application of neuropeptides, such as SP, produces enhanced dorsal horn neuron responses to noxious thermal and mechanical stimulation (Henry, J. L.,

1976; Randic, M. and Miletic, V., 1977). Dorn horn neuronal LTP induced *in vivo* by high-intensity, high-frequency tetanic afferent nerve stimulation is blocked by i.v. administration of NK-1) or NK-2 antagonists (Liu, X. and Sandkuhler, J., 1997). Knockout of the preprotackykinin-A gene, from which SP and NKA are derived, results in a reduction in the time course over which dorsal horn WDR neurons are sensitized following mustard oil application to the mice hind paw, as well as a reduction in the magnitude of poststimulus discharges of WDR neurons (Martin, W. J. *et al.*, 2004).

The increased excitability in flexor efferents, induced either by C-fiber electrical stimulation or by the application of chemical irritants, is blocked by pretreatment of the sciatic and saphenous nerves with the C-fiber neurotoxin capsaicin (Woolf, C. J. and Wall, P. D., 1986) or by the SP antagonist spantide II (Wiesenfeld-Hallin, Z. *et al.*, 1990). I.t. application of the C-fiber neuropeptides SP (Woolf, C. J. and Wiesenfeld-Hallin, Z., 1986), NK A (Xu, X.-J. *et al.*, 1991), CGRP (Woolf, C. J. and Wiesenfeld-Hallin, Z., 1986), VIP (Wiesenfeld-Hallin, Z., 1987), somatostatin (Wiesenfeld-Hallin, Z., 1985), and galanin (Wiesenfeld-Hallin, Z. *et al.*, 1989) produces prolonged enhancements in the excitability of the flexion reflex.

In addition, the hyperalgesia that develops in the hind paw contralateral to a thermal injury is mimicked by i.t. treatment with SP and NKA, and reversed by pretreatment with a SP antagonist (Coderre, T. J. and Melzack, R., 1991). Subcutaneous injection of formalin, which elicits a persistent nociceptive response associated with central changes (Coderre, T. J. *et al.*, 1990), evokes an immediate, intense barrage of C-fiber afferent activity (McCall, W. D. *et al.*, 1996), and produces an increase in SP in the cerebrospinal fluid (Kuraishi, Y. *et al.*, 1989). Furthermore, nociceptive responses to formalin are significantly suppressed in rats pretreated with peptide (Murray, C. W. *et al.*, 1991) and nonpeptide (Yamamoto, T. and Yaksh, T. L., 1991; Yashpal, K. *et al.*, 1993) NK-1 receptor antagonists, as well as the C-fiber neurotoxin capsaicin (Dray A. and Dickenson, A. H., 1991). Finally, various nonpeptide NK-1 receptor antagonists attenuate allodynia and hyperalgesia that is evoked by CCI of the sciatic nerve (Cumberbatch, M. J. *et al.*, 1998; Cahill, C. M. and Coderre, T. J., 2002).

26.2.1.2 Excitatory amino acids

Additional evidence implicates a contribution of excitatory amino acids (EAAs) to injury-induced neuroplasticity. EAAs have widespread activity in the CNS including the spinal cord (Watkins, J. C. and Evans, R. H., 1981; Davies, J. and Watkins, J. C., 1983) and thalamus (Eaton, S. A. and Salt, T. E., 1990). The role of EAAs in nociception is suggested since noxious stimulation or peripheral inflammation causes the release of glutamate and aspartate in spinal cord dorsal horn (Skilling, S. R. *et al.*, 1988; Sorkin, L. S. *et al.*, 1992). The dorsal horn synaptosomal level of EAAs is increased in rats with CCI of the sciatic nerve (Somers, D. L. and Clemente, F. R., 2002), and dorsal horn levels of EAAs are increased in rats with a plantar incision (Zahn, P. K. *et al.*, 2002) or CCI of the sciatic nerve (al-Ghoul, W. M. *et al.*, 1993; Kawamata, M. and Omote, K., 1996). Furthermore, iontophoretic application of EAAs produces an excitation of dorsal horn neurons (Curtis, D. R. and Watkins, J. C., 1960; Willcockson, W. S. *et al.*, 1984b; Schneider, S. P. and Perl, E. R., 1988), while i.t. treatment produces both behavioral hyperalgesia and spontaneous nociceptive behaviors (Aanonsen, L. M. and Wilcox, G. L., 1986; 1987).

A role of EAAs in noxious stimulus-induced plasticity is further suggested by several findings. The windup of dorsal horn neuron activity after frequency C-fiber stimulation is mimicked by the application of L-glutamate or N-methyl-D-aspartate (NMDA) (Gerber, G. and Randic, M., 1989), and blocked by application of either competitive (Dickenson, A. H. and Sullivan, A. F., 1987; Thompson, S. W. *et al.*, 1990) or noncompetitive (Davies, S. N. and Lodge, D., 1987; Thompson, S. W. *et al.*, 1990) NMDA antagonists. Iontophoretic application of EAAs produces receptive field changes in dorsal horn neurons (Zieglgansberger, W. and Herz, A., 1971), as well as enhanced dorsal horn neuron responses to nonnoxious and noxious mechanical stimulation (Aanonsen, L. M. *et al.*, 1990; Dougherty, P. M. and Willis, W. D., 1991). Dorsal horn neurons that are sensitized following peripheral tissue injury/inflammation show increased responsiveness to the iontophoretic application of EAAs (Dougherty, P. M. and Willis, W. D., 1992; Dougherty, P. M. *et al.*, 1992b), and exhibit a reduction in responsiveness or sensitization following intravenous administration of ketamine or iontophoretic application of NMDA antagonists (Schaible, H.-G. *et al.*, 1991), or the administration of α-amino-3-hydroxy-5-methylisoxazole-4-propionic acid (AMPA) or NMDA antagonists to dorsal horn neurons by reverse microdialysis (Dougherty, P. M. *et al.*, 1992a).

I.t. administration of the EAAs L-glutamate or L-aspartate produces an increase in the excitability of flexor efferents (Woolf, C. J. and Wiesenfeld-Hallin, Z., 1986), while competitive or noncompetitve NMDA antagonists reduce the facilitation of flexion reflexes induced by electrical (C-fiber) stimulation or cutaneous application of the chemical irritant mustard oil (Woolf, C. J. and Thompson, S. W., 1991). Hyperalgesia that develops in the hind paw contralateral to a thermal injury is both mimicked following i.t. treatment with NMDA, and reversed by an NMDA receptor antagonist (Coderre, T. J. and Melzack, R. 1991). Studies in humans (Kristensen, J. D. *et al.*, 1992), show that i.t. treatment with the competitive NMDA receptor antagonist 3-(2-carboxypiperazin-4-yl)-propyl-1-phosphonic acid (CPP) abolished after-discharges and spreading pain and hyperalgesia (symptoms proposed to be associated with windup) in a patient with neuropathic pain.

A contribution of ionotropic glutamate receptors (iGluRs) to nociceptive processing was originally suggested by the ability of AMPA/kainate, and particularly NMDA, receptor antagonists to produce analgesic effects in both phasic and tonic nociceptive tests in rats (Cahusac, P. M. *et al.*, 1984; Nasstrom, J. *et al.*, 1992). In addition, AMPA antagonists have been found to produce an inhibition of miniature endplate potentials in dorsal horn neurons (Hori, Y. and Endo, K., 1992) and dorsal root potentials evoked by single shock C-fiber stimulation (Thompson, S. W. *et al.*, 1992). Furthermore, tetanic electrical stimulation of the primary afferent fibers *in vitro* (Randic, M. *et al.*, 1993) or *in vivo* (Liu, X. G. and Sandkuhler, J., 1995) induces a LTP of the C-fiber-evoked potentials in dorsal horn neurons that is prevented by superfusion with NMDA antagonists.

NMDA antagonists have been particularly effective at reducing persistent pain associated with central sensitization. The noncompetitive NMDA antagonist MK-801 reduces the hyperalgesia which develops in rats with peripheral neuropathy (Davar, G. *et al.*, 1991; Mao, J. *et al.*, 1992c) or adjuvant-induced inflammation (Ren, K. *et al.*, 1992) and reduces autotomy behavior in rats with peripheral nerve sections (Seltzer, Z. *et al.*, 1991). Various competitive and noncompetitive NMDA receptor antagonists reduce the enhanced after-discharges, windup, and thermal- and mechanical-evoked responses of spinal dorsal horn neurons in rats with SNL (Leem, J. W. *et al.*, 1996; Suzuki, R. *et al.*, 2001). MK-801 has been found to reduce the increases in intracellular calcium ion (Ca^{2+}) concentration in spinal dorsal horn slices of rats which had CCI of the sciatic nerve (Kawamata, M. and Omote, K., 1996), as well as reducing increased Fos protein expression in response to nonnoxious mechanical stimulation of the hind paw of CCI rats (Kosai, K. *et al.*, 2001). MK-801 also reduces the adjuvant inflammation-induced expansion of the receptive fields of nociceptive neurons in spinal cord dorsal horn (Dubner, R. and Ruda, M. A., 1992). In humans, ischemic and postoperative pain is suppressed by subanesthetic doses of ketamine (Maurset, A. *et al.*, 1989), while in the rat the increased activity in dorsal horn in response to ischemia associated with femoral artery occlusion is inhibited by i.t. application of an NMDA antagonist (Sher, G. and Mitchell, D., 1990). Subcutaneous injection of formalin evokes an increased release of glutamate and aspartate in spinal cord dorsal horn (Skilling, S. R. *et al.*, 1988), while the sustained responses of spinal nociceptive cells to noxious peripheral stimulation produced by subcutaneous formalin injection are reduced by i.t. administration of selective NMDA antagonists (Haley, J. E. *et al.*, 1990). Nociceptive responses to formalin are also both enhanced by pretreatment with the EAA agonists, L-glutamate and L-aspartate, as well as combinations of NMDA + AMPA or NMDA + trans-1-aminocyclopentane-1,3-dicarboxylate (ACPD) (Coderre, T. J. and Melzack, R., 1992a), and suppressed by pretreatment with NMDA antagonists (Murray, C. W. *et al.*, 1991; Coderre, T. J. and Melzack, R., 1992a).

Evidence suggests that in addition to iGluRs, metabotropic glumate receptors (mGluRs) also play a significant role in nociception. Unlike iGluRs, which gate cation channels, mGluRs are coupled by G proteins to various intracellular messengers (Pin, J. P. and Duvoisin, R., 1995). Molecular cloning techniques have identified eight subtypes of mGluRs to date, some of which include splice variants (Pin, J. P. *et al.*, 1992; Hollmann, M. and Heinemann, S. F., 1994). The eight subtypes of mGluRs are divided into three groups based on sequence homology, signal transduction mechanisms, and ligand selectivites (Nakanishi, S., 1992). Group I includes mGluR1 and 5 which stimulate phospholipase C (PLC), leading to protein kinase C (PKC) activation, phosphoinositide (PI) hydrolysis, and intracellular Ca^{2+} mobilization (Houamed, K. M. *et al.*, 1991; Masu, M. *et al.*, 1991; Abe, T. *et al.*, 1992).

The nonselective mGluR antagonist L-AP3 attenuates rat dorsal horn neuronal activity associated with repeated mustard oil application (Young, M. R. *et al.*, 1994; 1995) and knee joint inflammation (Neugebauer, V. *et al.*, 1994). Furthermore, the nonselective mGluR agonists trans-ACPD or 1S,3R-ACPD produce a depolarization (Young, M.

R. *et al.*, 1994; 1995) or increased evoked responses (Palecek, J. *et al.*, 1994) in dorsal horn neurons. 1S,3R-ACPD also produces a ventral root (VR) depolarization in neonatal rat spinal cord *in vitro* (Boxall, S. J. *et al.*, 1996; 1998). The VR depolarization produced by 1S,3R-ACPD and windup after stimulation of dorsal roots, or capsaicin application, are blocked by the mGluR antagonist (+/−)-alpha-methyl-4-carboxphenylglycine (MCPG) (Boxall, S. J. *et al.*, 1996).

Selective group I mGluR antagonists also reduce rat dorsal horn neuronal activity evoked by hind paw mustard oil (Boxall, S. J. *et al.*, 1996) or capsaicin (Neugebauer, V. *et al.*, 1999). Furthermore, dorsal horn neuronal activity is increased by spinal administration of the group I mGluR agonist dihydroxyphenylglycine (DHPG) (Young, M. R. *et al.*, 1997; Neugebauer, V. *et al.*, 1999); although see Chen Y. *et al.* (2000). DHPG may act primarily at mGluR1, as agonists of mGluR5 (*trans*-ADA, CHPG) have either less (Young, M. R. *et al.*, 1997) or no (Neugebauer, V. *et al.*, 1999) excitatory effects. However, 1S,3R-ACPD, DHPG, and CHPG all produce a long-lasting potentiation of polysynaptic EPSPs in dorsal horn cells receiving C-fiber afferent input (Zhong, J. *et al.*, 2000). The neuronal activation evoked by spinal DHPG, or hind paw application of mustard oil or capsaicin, is also reduced by mGluR1-selective antagonists (Young, M. R. *et al.*, 1997; Neugebauer, V. *et al.*, 1999). However, a selective mGluR5 antagonist 2-methyl-6-(phenylethynyl)-pyridine (MPEP) attenuates shock-induced VR potentials and windup in isolated spinal cords (Bordi, F. and Ugolini, A., 2000), as well as reducing the enhanced spontaneous and noxious-stimulus-evoked activity in spinal cord neurons of rats with CCI of the sciatic nerve (Sotgiu, M. L. *et al.*, 2003). Alternatively, group II and III mGluR agonists depress spinal reflexes or activity in isolated spinal cords (Saitoh, T. *et al.*, 1998; Gerber, G. *et al.* 2000; Krieger, P. and El Manira, A., 2002), and reverse the hypersensitivity of dorsal horn neurons after skin capsaicin treatment (Neugebauer, V. *et al.*, 2000) or SNL (Chen, S. R. and Pan, H. L., 2005).

Behavioral studies also suggest mGluRs influence nociception. I.t. quisqualate, an mGluR and AMPA receptor agonist, produces hyperalgesia (Aanonsen, L. M. *et al.*, 1990; Kolhekar, R. and Gebhart, G. F., 1994), as well as nociceptive behaviors (Sun, X. and Larson, A. A. 1991) in rodents. The mGluR agonist 1S,3R-ACPD (Fisher, K. and Coderre, T. J., 1996a) produces similar behaviors, implicating mGluRs in these responses. Furthermore, nociceptive behaviors and hyperalgesia are also produced by i.t. DHPG (Fisher, K. and Coderre, T. J., 1996a; 1998; Dolan, S. and Nolan, A. M., 2000; Hama, A. T., 2003), but importantly, not by the selective mGluR5 agonist trans-ADA (Fisher, K. and Coderre, T. J., 1996a). This suggests a pronociceptive role of spinal mGluR1. Conversely, i.t. administration of trans-ACPD, and a group II mGluR agonist (L-CCG-I), produce increased mechanical withdrawal thresholds, that are blocked by a group II mGluR antagonist (Chen, S. R. and Pan, H. L., 2005). This suggests an antinociceptive role for group II mGluRs.

Group I mGluRs are also implicated in the development of hyperalgesia and persistent pain associated with inflammatory or chemical injury. Thus, nonselective and group I selective mGluR antagonists (Walker, K. *et al.*, 2001; Zhang, L. *et al.*, 2002; Zhu, C. Z. *et al.*, 2004), and group I mGuR antisense oligonucleotides (Fundytus, M. E. *et al.*, 2002) reduce hyperalgesia in rats with hind paw inflammation. Furthermore, formalin nociception is enhanced by spinal i.t. 1*S*,3*R*-ACPD (Coderre, T. J. and Melzack, R., 1992a) and DHPG (Fisher, K. and Coderre, T. J., 1996b), and nociception associated with hind paw formalin (Fisher, K. and Coderre, T. J., 1996b; Varty, G. B. *et al.*, 2005) and i.p. acetic acid injections are reduced by group I mGluR antagonists (Chen, Y. *et al.*, 2000; Zhu, C. Z. *et al.*, 2004). Formalin nociception is also reduced by group II and III mGluR agonists (Fisher, K. and Coderre, T. J., 1996b; Simmons, R. M. A. *et al.*, 2002), and by enhancing an endogenous group II mGluR agonist (Yamamoto, T. *et al.*, 2004). Thus, while Group I activation is nociceptive, activation of Group II or III mGluRs is antinociceptive. Indeed, group I mGluR-linked intracellular messengers (i.e., PKC, IP$_3$ (inositol trisphosphate), and $_i$Ca^{2+}) contribute to nociception (Igwe, O. J. and Ning, L., 1994; Mao, J. *et al.*, 1995b), while events linked to group II/III mGluR (i.e., decreased cyclic adenosine monophosphate (cAMP) production) produce antinociception (Gereau, R. W. and Conn, P. J., 1994; Shen, J. *et al.*, 2000).

mGluRs also influence hyperalgesia induced by nerve injury. Group I mGluR antagonists reduce hyperalgesia in CCI-injured rats (Fisher, K. *et al.*, 1998). CCI-induced hyperalgesia is also reduced by selective mGluR1 and mGluR5 antagonists (Fisher, K. *et al.*, 2002; Zhu, C. Z. *et al.*, 2004), as well as group II (2R,4R-APDC) and III (L-AP4)

mGluR agonists (Fisher, K. *et al.*, 2002), and by antisense oligonucleotides to mGluR1 and mGluR5 (Fundytus, M. E. *et al.*, 2001). Hyperalgesia after SNL is reduced by group I mGluR antagonists (Dogrul, A. *et al.*, 2000; Zhu, C. Z. *et al.*, 2004; Varty, G. B. *et al.*, 2005) and group II and III agonists (Simmons, R. M. A. *et al.*, 2002; Chen, S. R. and Pan, H. L., 2005). Tactile allodynia after spinal cord injury is also reduced by the mGluR1 antagonist (RS)-1-aminoindan-1,5-dicarboxylic acid (AIDA) (Mills, C. D. *et al.*, 2000), as well as group II and III mGluR agonists (Mills, C. D. *et al.*, 2002). Group II (mGluR3,4) and group III (mGluR4, 6, 7, 8) result in reduced production of cyclic adenosine monophosphate (cAMP) (Schoepp, D. D. *et al.*, 1995; Tanabe, Y. et al., 1993).

26.2.1.3 Other transmitters

The expression of a variety of other transmitters are affected following inflammatory or nerve injuries that produce hyperalgesia and allodynia. Thus, there is a reduction in the number of gamma-aminobutyric acid (GABA)- and glutamate decarboxylase (GAD)-, the synthetic enzyme for GABA, immunoreactive cells in the spinal cord of rats with CCI of the sciatic nerve (Eaton, M. J. *et al.*, 1998). There is also a reduction in the level of GABA in spinal cord microdialysates taken form rats with a partial constriction of the sciatic nerve that exhibit allodynia (Stiller, C. O., *et al.*, 1996). Recent studies also indicate that there is a reduction in endomorphin-2 immunoreactivity in the superficial dorsal horn of mice with a partial ligation of the sciatic nerve, which paralleled the development of thermal hyperalgesia (Smith, R. R. *et al.*, 2001).

26.2.1.4 Neurotrophins

Neurotrophins (NTs) are growth factors which stimulate neuronal growth in embryonic development, and maintain neuronal viability in adult tissues. NTs include nerve growth factor (NGF) that acts selectively at tyrosine kinase receptor A (TrkA), brain-derived neurotrophic factor (BDNF), NT-4, and NT-5, which act at TrkB receptors, and NT-3 that acts at TrkC receptors in spinal cord. An additional NT, glial-derived neurotrophic factor (GDNF), binds to the receptor GFR-alpha1. Small diameter unmyelinated primary afferent fibers generally fall into two groups: one which contains CGRP, SP, and TrkA receptors, and depends on NGF for its development; a second contains the lectin IB-4 and TrkB receptors, and is dependent on GDNF for its development (McMahon, S. B. *et al.*, 1994). Importantly, these neurotrophic factors are not only required for neuronal development; they continue to be produced by numerous types of cells, produce excitatory effects on primary afferent fibers, and are required to maintain normal neuronal function of primary afferent fibers. BDNF has been found to be expressed in primary afferent fibers (particularly C fibers containing CGRP and SP) (Barakat-Walter, I., 1996), and to act in the spinal cord to enhance neuronal excitability (Kerr, B. J. *et al.*, 1999; Thompson, S. W. *et al.*, 1999).

It has been hypothesized that hyperalgesia after inflammatory injury is influenced by an upregulation of BDNF in primary afferent fibers that occurs in response to peripheral stimulation of primary afferent fibers with NGF (Thompson, S. W. *et al.*, 1999). Thus, inflammation leads to increased production of NGF in peripheral tissue, which stimulates TrkA receptors on the first group of primary afferent fibers (peptidergic fibers) described above (Apfel, S. C., 2000). The stimulation of TrkA receptors results in increased BDNF production in this group of C fibers (Apfel, S. C. *et al.*, 1996; Michael, G. J. *et al.*, 1997), and precipitates enhanced BDNF release, and a subsequent sensitization of dorsal horn neurons (Thompson, S. W. *et al.*, 1999). Noxious thermal, mechanical, and chemical stimuli, but not innocuous stimuli, also causes activation of spinal TrkB receptors as shown by increased Trk phosphorylation and increased extracellular signal-regulated kinase (ERK) activation (see below) in dorsal horn, which is blocked by sequestering BDNF with a TrkB–IgG fusion molecule (Pezet, S. *et al.*, 2002).

Importantly, BDNF protein and mRNA expression is increased in rat DRG and spinal dorsal horn following hind paw injection of CFA (Cho, H. J. *et al.*, 1997a; 1997b). BDNF protein levels are increased in DRG and spinal cord dorsal after CCI of the sciatic nerve (Ha, S. O. *et al.*, 2001; Miletic, G. and Miletic, V., 2002), while BDNF mRNA and or protein is also increased in uninjured L4 DRG and dorsal horn following L5 SNL (Fukuoka, T. *et al.*, 2001; Ha, S. O. *et al.*, 2001). I.t. administration of antibodies to BDNF attenuate thermal hyperalgesia in rats with L5 SNL (Fukuoka, T. *et al.*, 2001), while BDNF antisense olgionucleotides and the BDNF-scavenging protein TrkB–IgG attenuates carrageenan-induced thermal hyperalgesia (Kerr, B. J. *et al.*, 1999; Groth, R. and Aanonsen, L. M., 2002). Conversely, i.t. administration of BDNF and NT-4/5, which act at TrkB, but not the TrkC agonist NT-3,

produce thermal hyperalgesia, while antisense to TrkB attenuates carrageenan-induced thermal hyperalgesia in rats (Groth, R. and Aanonsen, L. M., 2002). In contrast, antisense oliginucleotides to NT-3 have been found to attenuate mechanical allodynia in rats with partial sciatic nerve ligation (White, D. M., 2000). Alternatively, rats intrathecally treated with adenovirus to produce a spinal cord overexpression of BDNF (Eaton, M. J. *et al.*, 2002), or transgenic mice that overexpress NT-3 in muscle (Gandhi, R. *et al.*, 2004), each exhibit reduced mechanical allodynia and thermal hyperalgesia, after CCI of the sciatic nerve or acid injection into the gastrocnemius muscle, respectively. Protein and mRNA levels of GDNF and GFRα-1 are significantly increased in DRGs of rats with CCI of the sciatic nerve, and thermal hyperalgesia in CCI rats is significantly enhanced by i.t. administration of GFRα-1 antisense oligonucleotides, suggesting that GDNF may act at GFRα-1 to produce antihyperalgesic effects (Dong, Z. Q. *et al.*, 2005).

26.2.1.5 Phenotype switch

Evidence suggests that specific primary afferent fibers may be able to switch their phenotype following intense noxious stimulation. Thus, it has been shown that after inflammatory injury Aβ fibers begin to produce and release SP (which they normally do not contain) (Neumann, S. *et al.*, 1996); a phenomenon that has important implications for the development of mechanical allodynia (which is believed to mediated largely by Aβ fibers). Evidence suggests that growth factors or NTs may play a critical role in the development of these injury-induced phenotype changes (Woolf, C. J., 1996). The NT BDNF itself has been shown to undergo a phenotype switch after nerve injury as there is a shift in the size distribution of rat DRG neurons that contain BDNF after sciatic nerve lesions, so that large neurons, which do not initially express BDNF, start to express more than small neurons (Zhou, X. F. *et al.*, 1999a). Large DRG neurons also exhibit a phenotype switch when they start to express both BDNF (Ohtori, S. *et al.*, 2002) and CGRP (Ohtori, S. *et al.*, 2001) after CFA-induced inflammation of the lumbar facet joint.

26.2.2 Expression of Second Messengers and Transcription Factors

26.2.2.1 Second messengers

A growing body of literature suggests that intracellular messengers linked with various ionotropic and metabotropic receptors contribute to synaptic plasticity in spinal cord dorsal horn. Ca^{2+} is a key intracellular messenger that is elevated in the cytoplasm of neurons following influx through receptor-operated and voltage-gated Ca^{2+} channels, or following release from intracellular stores. Glutamate and aspartate stimulate the influx of Ca^{2+} through NMDA receptor-operated channels (MacDermott, A. B. *et al.*, 1986). SP produces an elevation in intracellular Ca^{2+} by mobilizing its release from intracellular stores (Womack, M. D. *et al.*, 1988), while both SP (Womack, M. D. *et al.*, 1989) and CGRP (Oku, R. *et al.*, 1988) increase Ca^{2+} influx through voltage-gated Ca^{2+} channels. Activity at either NK-1 receptors by SP (Mantyh, P. W. *et al.*, 1984), or at metabotropic EAA receptors by glutamate and aspartate (Sugiyama, H. *et al.*, 1987), stimulates the hydrolysis of inositol phospholipids by activating a polyphosphoinositide-specific PLC. PLC is an enzyme which catalyzes the hydrolysis of polyphosphatidylinositol into the intracellular messengers IP_3 and diacylglycerol (DAG). Following its production, IP_3 stimulates the release of Ca^{2+} from internal stores; on the other hand, DAG stimulates the translocation and activation of PKC.

Gerber G. *et al.* (1989) showed that activators of PKC enhance the basal and evoked release of glutamate and aspartate in the spinal cord slice, as well as the depolarizing responses of dorsal horn neurons to exogenous glutamate and NMDA. Furthermore, Chen L. and Huang L.-Y. M. (1992) demonstrated that PKC increases NMDA-activated currents in isolated trigeminal cells by increasing the probability of channel openings and by reducing the voltage-dependent Mg^{2+} block of NMDA receptor channels. PKC is involved in synaptic plasticity causing a phosphorylation of both group I mGluR (Alaluf, S. *et al.*, 1995) and NMDA receptors (Chen, L. and Huang, L.-Y. M., 1992). Enhanced spinal PKC activity is observed following inflammatory (Yashpal, K. *et al.*, 1995; 2001; Sweitzer, S. M. *et al.*, 2004b) and nerve injury (Mao, J. *et al.*, 1992b; Miletic, V. *et al.*, 2000; Fundytus, M. E. *et al.*, 2001; Yashpal, K. *et al.*, 2001) and PKC inhibitors reduce pain induced by hind paw injection of noxious chemicals (Yashpal, K. *et al.*, 1995; 2001; Li, K. C. and Chen, J., 2003; Sweitzer, S. M. *et al.*, 2004b) and nerve injury (Hua, X. Y. *et al.*, 1999b; Miletic, V. *et al.*, 2000; Yajima, Y. *et al.*, 2003). PKC also enhances NMDA receptor activity (Liao, G. Y. *et al.*, 2001; Skeberdis, V. A. *et al.*, 2001) and NMDA receptor trafficking to neuronal plasma membranes (Lan, J. Y. *et al.*, 2001). PKC has several different isoforms, and evidence suggests

that $PKC\gamma$ is increased in the rat spinal cord dorsal horn after CCI of the sciatic nerve (Mao, J. *et al.*, 1995b), and the synaptosomal membrane fraction of both $PKC\gamma$ and $PKC\beta$-II is increased in the spinal cord dorsal of CCI rats (Miletic, V. *et al.*, 2000). Importantly, $PKC\gamma$ knockout mice exhibit reduced mechanical allodynia and thermal hyperalgesia after partial sciatic nerve ligation (Malmberg, A. B. *et al.*, 1997a). $PKC\varepsilon$ is also upregulated in rat DRG following hind paw injections of carrageenan or CFA, in a manner that is correlated with the development of inflammatory hyperalgesia (Zhou, Y. *et al.*, 2003).

Increased neuronal Ca^{2+} infux also activates phospholipase A2 that triggers the production of arachidonic acid (Dumuis, A. *et al.*, 1988; 1990), which is metabolized into various eicosanoids, including prostaglandins (PGs). Further increases in arachidonic acid occur following the breakdown of DAG (Gammon, C. M. *et al.*, 1989), which is produced following stimulation of PLC with activity at various metabotropic receptors. Spinal administration of PGE_2 enhances the responsiveness and receptive field sizes of rat spinal dorsal horn neurons (Vasquez, E. *et al.*, 2001), as well as producing mechanical allodynia and thermal hyperalgesia in mice (Minami, T. *et al.*, 1994a; 1994b). There is an increase in the levels of PGE_2 in CSF of rat after knee joint injection of carrageenan/kaolin (Yang, L. C. *et al.*, 1996) or i.t. injection of SP (Hua, X. Y. *et al.*, 1999a). The breakdown of arachidonic acid into PGs in inhibited by nonsteroidal anti-inflammatory drugs (NSAIDs), and i.t. administration of NSAIDs produces analgesia in animals subjected to i.p. acetic acid or hind paw formalin injections (Yaksh, T. L., 1982; Malmberg, A. B. and Yaksh, T. L., 1992a), as well as reducing enhanced flexor reflexes in rats with hind paw injections of CFA (Seybold, V. S. *et al.*, 2003). I.t. NSAIDs block thermal hyperalgesia induced by i.t. treatment with NMDA or SP (Malmberg, A. B. and Yaksh, T. L., 1992b). There is also an elevation of the enzyme cyclooxygenase-2 (COX-2) in the spinal CSF of rats with hind paw carrageenan injections (Ibuki, T. *et al.*, 2003) or in the spared nerve injury model (Broom, D. C., *et al.*, 2004). Furthermore, thermal hyperalgesia associated with carrageenan injection is attenuated by i.t. treatment with a selective inhibitor of COX-2 (Dirig, D. M. *et al.*, 1998; Ibuki, T. *et al.*, 2003), but not COX-1 (Yaksh, T. L. *et al.*, 2001). I.t. COX-2 inhibitors are found to relieve hyperalgesia in a model of hind paw inflammation, but not in the spared nerve injury model (Broom *et al.*, 2004).

Elevation in intracellular Ca^{2+} in neurons also leads to increases in additional protein kinases such as Ca^{2+}/calmodulin-dependent protein kinase (CaMK) (Kennedy, M. B. et al., 1983). There is an increased expression and phosphorylation of CaMK-II in rat spinal cord dorsal horn after hind paw injection of capsaicin, while i.t. administration of a CaMK-II inhibitor reduces enhanced dorsal horn neuronal responsiveness and behavioral responses after hind paw capsaicin (Fang, L. et al., 2002), as well as reducing mechanical allodynia and thermal hyperalgesia in CCI mice (Garry, E. M. et al., 2003). Mice with a mutation of the alpha-CaMK-II gene, which prevents autophosphorylation of CaMK-II, also have diminished nociceptive responses to hind paw formalin injections (Zeitz, K. P. et al., 2004).

The binding of Ca^{2+} to calmodulin activates nitric oxide (NO) synthase, which generates NO from free L-arginine (Garthwaite, J. et al., 1988), which in turn activates soluable guanylate cyclase and increases cGMP (Southam, E. et al., 1991). Intracellularly the Ca^{2+} binds to calmodulin which activates NO synthase. NO synthase converts L-arginine to NO, and the released NO stimulates the production of cGMP from soluable guanylate cyclase (Synder, S. H., 1992). The levels of neuronal NO synthase, as well as NADPHd and [^{3}H]citrulline staining (markers of NO activity), are increased in the rat DRG and spinal cord dorsal horn after CCI or transection of the sciatic nerve (Cizkova, D. et al., 2002a; Lukacova, N. et al., 2003), SNL (Steel, J. H., et al., 1994), or hind paw formalin injection (Lam, H. H. et al., 1996). There is also increased cGMP in the spinal dorsal horn of rats with CCI of the sciatic nerve that parallels the development of mechanical allodynia and thermal hyperalgesia (Siegan, J. B. et al., 1996). In addition, NO synthase inhibitors or an inhibitor of soluable guanylate cyclase reduce the thermal hyperalgesia that develops in rats with chronic nerve constriction injury (Meller, S. T. et al., 1992) or hind paw carrageenan (Meller, S. T. et al., 1994; Osborne, M. G. and Coderre, T. J., 1999).

Mediators such as NO, which increase neuronal cGMP, stimulate the production of cAMP-dependent protein kinase (PKG) (Scott, J. D., 1991). The spinal dorsal horn levels of PKG1α are increased in response to hind paw injection of formalin, and i.t. administration of PKG inhibitors reduces nociceptive responses to formalin (Tao, Y. X. et al., 2000; Schmidtko, A. et al., 2003). I.t. treatment with a PKG inhibitor also reverses mechanical allodynia observed in rats with a hind paw injection of capsaicin (Sluka,

K. A. and Willis, W. D., 1997), and thermal hyperalgesia induced by i.t. injections of NMDA or an NO donor (Tao, Y. X. and Johns, R. A., 2000). I.t. administration of a PKG activator produces mechanical allodynia which is absent in PKG-I knockout mice, further implicating a role for spinal PKG in allodynia (Tegeder, I. et al., 2004a).

Activity at various neuropeptide receptors leads to increased levels of adenylate cyclase in spinal cord neurons (see Millan, M. J., 1999 for a review). The enzyme adenylate cyclase increases cAMP and its associated kinase, protein kinase A (PKA), which causes membrane phosphorylation that enhances Na^+ and Ca^{2+} currents, and decreases K^+ currents in primary afferent fibers (Vanegas, H. and Schaible, H. G., 2001). Enhancing the spinal cord levels of PKA leads to a sensitization of dorsal horn neurons to noxious mechanical stimulation (Lin, Q. et al., 2002), and i.t. administration of PKA inhibitors attenuate the mechanical allodynia induced by hind paw injection of capsaicin (Sluka, K. A. and Willis, W. D., 1997). There is also a reduction in thermal hyperalgesia associated with hind paw inflammation in mice with a target mutation of the type I regulatory subunit of PKA (Malmberg, A. B. et al., 1997b).

Activity of growth factors and various hormones leads to the activation of a mitogen-activated protein kinase (MAPK) pathway through stimulation of a GTP-binding protein Ras and protein kinases, such as PKC (Davis, R. J., 1993). A growing body of literature suggests that MAPK, including ERK, p38 MAPK, and c-Jun N-terminal kinase (JNK) contribute to spinal nociceptive processing. Recent studies have shown ERK activity is observed in lamina I and IIo of the ipsilateral dorsal horn after noxious thermal stimulation of the hind paw or C-fiber electrical nerve stimulation, but not after nonnoxious thermal or A fiber electrical stimulation (Ji, R. R. et al., 1999). ERK activity is also observed in rodent DRG or spinal cord dorsal horn in response to hind paw injections of carrageenan (Galan, A. et al., 2002), CFA (Ji, R. R. et al., 2002a, Obata, K. et al., 2003), formalin (Ji, R. R. et al., 1999, Karim, F. et al., 2001), or capsaicin (Ji, R. R. et al., 1999), or to sciatic nerve transection (Obata, K. et al., 2003), CCI (Obata, K. et al., 2004a) partial sciatic nerve ligation (Ma, W. and Quirion, R., 2002) or L5 SNL (Obata, K. et al., 2004b). Inhibitors of ERK have been found to reduce nociception induced by formalin (Ji, R. R. et al., 1999, Karim, F. et al., 2001), and hyperalgesia or allodynia after hind paw injection of carrageenan (Sammons, M. J. et al., 2000), CFA (Ji, R. R. et al., 2002a) or capsaicin (Kawasaki, Y. *et al.*, 2004), and in rats with CCI of the sciatic nerve (Ciruela, A. *et al.*, 2003) or L5 SNL (Obata, K. *et al.*, 2004b). Recently, it has been shown that stimulation of group I mGluRs leads to activation of ERK in spinal cord dorsal horn, and that antagonists of group I mGluRs produce a reduction in formalin-stimulated ERK activation (Karim, F. *et al.*, 2001). Adwanikar, H. *et al.* (2004) has also shown that ERK inhibitors significantly reduce DHPG-induced spontaneous nociceptive behaviors (SNBs) in mice that have received a prior hind paw injection of CFA.

Recent evidence indicates there is an increase in phosphorylated p38 MAPK in DRG neurons and microglia of the DRG and spinal cord following hind paw injection of formalin (Kim, S. Y. *et al.*, 2002) or capsaicin (Sweitzer, S. M. *et al.*, 2004a; Mizushima, T. *et al.*, 2005) or CCI of the sciatic nerve (Kim, S. Y. *et al.*, 2002; Obata, K. *et al.*, 2004a). I.t. treatment with a p38 MAPK inhibitor reverses thermal hyperalgesia produced by hind paw capsaicin injection (Sweitzer, S. M. *et al.*, 2004a; Mizushima, T. *et al.*, 2005) or in CCI rats (Obata, K. *et al.*, 2004a). There is also increased p38 MAPK activity in neurons of both the injured L5 and the uninjured L4 DRG after L5 SNL, and i.t. treatment with a p38 MAPK inhibitor reverses mechanical allodynia and thermal hyperalgesia in L5 SNL rats (Obata, K. *et al.*, 2004b). In contrast, JNK activity is increased only in L5 DRG neurons after L5 SNL, and i.t. administration of a JNK inhibitor reverses only mechanical allodynia in rats with L5 SNL (Obata, K. *et al.*, 2004b). JNK activity is also increased in the spinal cord dorsal horn of rats with partial sciatic nerve ligation (Ma, W. and Quirion, R., 2002).

It is now evident that activity of the NMDA receptor is regulated by the protein tyrosine kinases (PTKs), Src and Fyn, which phosphorylate the NMDA channel and enhance its activity (Yu, X. M. *et al.*, 1997; Tezuka, T. *et al.*, 1999; Yu, X. M. and Salter, M. W., 1999). Evidence that this process might contribute to hyperalgesia after nerve injury is suggested as phosphotyrosine staining is increased in the spinal cord dorsal horn following sciatic nerve transection (Eckert, W. A. *et al.*, 1994). Furthermore, i.t. administration of nonselective PTK inhibitors, reduce hind paw mechanical and thermal hyperalgesia produced by injection of carrageenan and kaolin to the base of the rat tail, or following i.t. injection of NMDA (Sato, E. *et al.*, 2003), as well as reducing thermal hyperalgesia in rats with partial sciatic nerve ligation (Yajima, Y. *et al.*, 2002). I.t.

treatment with a selective Src inhibitor delays the onset of mechanical allodynia and hyperalgesia induced by hind paw injection of CFA (Guo, W. *et al.*, 2002), as well as reducing capsaicin-induced ERK activation in spinal cord slices (Kawasaki, Y. *et al.*, 2004). In contrast to the effects of PTKs, NMDA receptors currents have been shown to be inhibited by protein tyrosine phosphatases (PTPs) (Wang, Y. T. *et al.*, 1996). Recent studies have shown that i.t. administration of PTP inhibitors produces an enhancement of mechanical allodynia and hyperalgesia induced by hind paw injection of capsaicin (Zhang, X. *et al.*, 2003). Studies using selective inhibitors or antisense oligonucleotides have also implicated various other kinases including: IkappaB kinase (Tegeder, I. *et al.*, 2004b), cyclin-dependent kinase 5 (Cdk5) (Wang, C. H. *et al.*, 2005), and Rho kinase (Tatsumi, S. *et al.*, 2005) in nociception or hyperalgesia associated with hind paw formalin or zymosan injection or CCI of the sciatic nerve.

26.2.2.2 Transcription factors

Noxious stimulation leads to the expression of proto-oncogenes and their protein products which act as transcription factors. Hunt S. P. *et al.* (1987) first demonstrated that the c-fos protein product Fos is expressed in postsynaptic dorsal horn neurons following noxious thermal or chemical stimulation of the skin. The expression of Fos has also been demonstrated in rat spinal dorsal horn in response to noxious pinch of the hind paws (Bullitt, E., 1989), the injection of formalin (Presley, R. W. *et al.*, 1990) or carrageenan (Draisci, G. and Iadorola, M. J., 1989) into a hind paw or sodium urate crystals into joints (Menétrey, D. *et al.*, 1989), the injection of acetic acid into viscera (Menétrey, D. *et al.*, 1989), the induction of polyarthritis with CFA (Menétrey, D. *et al.*, 1989). Fos expression in the rat spinal dorsal horn is also increased after transection (Chi, S. I. *et al.*, 1993), CCI (Kajander, K. C. *et al.*, 1996; Catheline, G. *et al.*, 1999) or partial ligation (Delander, G. E. *et al.*, 1997) of the sciatic nerve. Importantly, there is a strong correlation between pain behavior and the number of cells expressing Fos (Presley, R. W. *et al.*, 1990). The time course of Fos expression after peripheral inflammation (Draisci, G. and Iadorola, M. J., 1989) or nerve injury (Delander, G. E. *et al.*, 1997) has been found to be correlated with the development of hyperalgesia. Moreover, morphine pretreatment produces a dose-dependent suppression of Fos expression which corresponds with its analgesic effects (Presley, R. W. *et al.*, 1990; Tolle, T. R. *et al.*, 1990).

An increase in phosphorylated cyclic AMP response element binding (pCREB) protein is observed in the spinal cord dorsal horn of rats with partial ligation or CCI of the sciatic nerve (Ma, W. and Quirion, R., 2001; Miletic, G. *et al.*, 2002, 2004b) or hind paw injections of capsaicin (Kawasaki, Y. *et al.*, 2004; Fang, L. *et al.*, 2005). Elevation in spinal pCREB associated with nerve injury, hind paw capsaicin, or electrical nerve stimulation has been found to be dependent on the activation of PKA, PKC, ERK, or CaMK (Kawasaki, Y. *et al.*, 2004; Miletic, G. *et al.*, 2004b; Fang, L. *et al.*, 2005; Miyabe, T. and Miletic, V., 2005). I.t. treatment with antisense oligonucleotides to CREB reduces the mechanical allodynia in rats with partial sciatic nerve ligation (Ma, W. *et al.*, 2003b).

Noxious stimulation, inflammatory, or nerve injury also leads to the spinal cord expression of other transcription factors, including Fos B, Jun, Jun B, Jun C, Jun D (Herdegen, T. *et al.*, 1991a; 1991b; Delander, G. E. *et al.*, 1997), NGFI-A/Krox 24, NGFI-B, serum response factor (Herdegen, T. *et al.*, 1990; Wisden, W. *et al.*, 1990), FRA-1, FRA-2 (Munglani, R. and Hunt, S. P., 1995), AP-1, nuclear factor kappa B, octamer factors (Chan, C. F. *et al.*, 2000; Pollock, G. *et al.*, 2005), activating transcription factor 3 (Tsujino, H. *et al.*, 2000), cyclic AMP responsive element modulator (Naranjo, J. R. *et al.*, 1997), and peroxisome proliferators-activated receptor-alpha (Benani, A. *et al.*, 2004).

26.2.3 Alterations in Receptor and Ion Channel Expression, Activity, or Cellular Location

26.2.3.1 Neuropeptide receptors

Recent studies show that there is an increase in the internalization of NK-1 receptors in the spinal cord dorsal horn following electrical nerve or noxious peripheral stimulation, and that this internalization is enhanced in rats with inflammatory or nerve injuries (Allen, B. J. *et al.*, 1999). There is also an upregulation of NK-1 receptors in the spinal cord dorsal horn in response to hind paw inflammation induced by CFA and following sciatic nerve transection (Aanonsen, L. M. *et al.*, 1992). Importantly, upregulation of dorsal horn NK-1 mRNA after partial sciatic nerve ligation is significantly correlated with the development of thermal hyperalgesia (Taylor, B. K. and McCarson, K. E., 2004). The importance of NK-1 receptors to the development of hyperalgesia after tissue injury is also highlighted

by the finding that destruction of neurons with NK-1 receptors with the toxin saporin bound to SP abolishes the development of mechanical and thermal hyperalgesia following intraplantar capsaicin injection (Khasabov, S. G. *et al.*, 2002). NK-1 knockout mice also have attenuated dorsal horn neuronal responses to suprathreshold stimuli and a reduction of windup of deep dorsal horn neurons after C-fiber electrical nerve stimulation (Suzuki, R. *et al.*, 2003).

As for other neuropeptide receptors, it has been shown that sciatic nerve transection produces a marked increase if CCK-B receptor mRNA in the dorsal root ganglia (Zhang, X. *et al.*, 1993a). Although there is no change in the expression of the VIP/PACAP receptor PAC1, there is a decrease in VPAC1 and an increase in VPAC2 in the dorsal horn following CCI of the sciatic nerve.

26.2.3.2 Excitatory amino acid receptors

The NMDA receptor has two subunits NMDAR1 (NR1) and NMDAR2 (NR2), with four splice variants of the NR2 subunit (NR2A–D). Two of the NR2 subunits are heavily expressed in the spinal cord dorsal horn and have been studied in respect to plasticity associated with nociception. Recent evidence indicates that there is enhanced phosphorylation of the NR1 subunit of rat dorsal horn neurons after intradermal injection of capsaicin (Zou, X. *et al.*, 2000) or after noxious heat stimulation of the hind paw (Brenner, G. J., *et al.*, 2004), effects which depends on the activation of kinases, including PKA and/or PKC. Elevated levels of phosphorylated NR1 is also found in the spinal dorsal horn of rats with an excitotoxic injury of the spinal cord that causes central neuropathic pain (Caudle, R. M. *et al.*, 2003), or following hind paw inflammation with carrageenan (Caudle, R. M. *et al.*, 2005).

There is also a prolonged increase in the phosphorylation of NR2B subunits in rat spinal dorsal horn after inflammation induced by hind paw injection of CFA, that is dependent on activation of PKC and Src kinase (Guo, W. *et al.*, 2002), although decreases in rat spinal NR2B phosphorylation have also been observed after hind paw inflammation (Caudle, R. M. *et al.*, 2005). Increased phosphorylation of NR2B in spinal dorsal horn has also been observed in mice with transected sciatic nerves. The increased phosphorylated NR2B was attenuated by administration of a NR2B selective antagonist and in mice lacking Fyn kinase, a Src-family kinase (Abe, T. *et al.*, 2005). Recently, the C-fiber-stimulated windup of spinal dorsal horn neurons was shown to be attenuated by i.v. administration of a selective NR2B antagonist (Kovacs, G. *et al.*, 2004). The importance of phosphorylation of spinal NR subunits to central sensitization is based on now long established evidence that such phosphorylation produces an enhancement of Ca^{2+} currents at NMDA receptor channels (Chen, L. and Huang, L.-Y. M., 1992). Thus, PKC activation, initiated by glutamate- or aspartate-induced PLC activity and Ca^{2+} influx through NMDA channels, leads to further influx of Ca^{2+} through NMDA channels, creating a positive feedback loop for glutamate and aspartate neurotransmission. As described above, a similar mechanism has been demonstrated for pathways that involve Src or Fyn kinase (Yu, X. M. *et al.*, 1997; Tezuka, T. *et al.*, 1999).

A recent question concerning synaptic activity at glutamate synapses is the issue of possible alterations in the trafficking of glutamate receptor to the plasma membrane and synapses of postsynaptic neurons. In recent years it has been established that glutamate receptors move between the cell cytoplasm and the plasma membrane, altering the number of receptors that are available to respond to glutamate and aspartate. Generally referred to as receptor trafficking, glutamate receptors are inserted into and removed from the plasma membrane by exocytosis and endocytosis (or internalization), respectively, and diffused laterally within the plasma membrane (Collingridge, G. L. *et al.*, 2004). Recent studies have implicated trafficking of AMPA (Malinow, R. R., and Malenka, R. C., 2002), kainite (Jaskolski, F. *et al.*, 2005), and NMDA (Carroll, R. C. and Zukin, R. S., 2002) receptors as key processes in regulating synaptic plasticity.

Alterations in the activity of AMPA receptors are implicated in plasticity associated with nociception. AMPA receptors in lamina III–IV of the dorsal horn are upregulated following dorsal rhizotomy (Carlton, S. M. *et al.*, 1998). There are four subunits of AMPA receptors, known as GluR1–4, some of which have splice variants, and GluR1 and GluR2 are predominant in the spinal cord (Nagy, G. G. *et al.*, 2004). It has recently been shown that there is a PKA- and PKC-dependent upregulation of phosphorylated GluR1 in the spinal cord dorsal horn of rats that have received a hind paw injection of capsaicin or noxious stimulation (Fang, L. *et al.*, 2003; Nagy, G. G. *et al.*, 2004). Recent evidence also indicates that GluR1 receptors are trafficked to the plasma membrane of dorsal horn neurons in response to visceral inflammation (Galan, A. *et al.*, 2004), via a CaMK-II-dependent mechanism. This raises the possibility that noxious stimulation may lead to a recruitment of functional AMPA receptors to the postsynaptic membrane and synapses of spinal neurons that normally do not have functional AMPA receptors. The phenomenom of silent synapses has been demonstrated in cultured hippocampal neurons exposed to stimuli that produce LTP (Liao, D. *et al.*, 1995), and may contribute to hyperexcitability of dorsal horn neurons in persistent pain conditions (Kerchner, G. A. *et al.*, 1999).

Recent studies have demonstrated alterations in the expression of various mGluRs in the DRG and dorsal horn of rats following nerve injury or persistent pain. There is a dramatic downregulation of mGluR1 mRNA in rat L5 DRG cells following tibial nerve axotomy (Hofmann, H. A. *et al.*, 2001). In contrast, mGluR5 immunoreactivity is increased in L4 and L5 DRGs of rats with L5 SNL (Hudson, L. J. *et al.*, 2002), and in the rat spinal cord dorsal horn after hind paw carrageenan injection (Dolan, S. *et al.*, 2003). There is also increased mGluR3 mRNA in the dorsal horn of rats with their hind paws exposed to ultraviolet irradiation (Boxall, S. J. *et al.*, 1998), and an upregulation of both mGluR3 and mGluR5 mRNA in the dorsal horn of sheep with chronic hindlimb inflammation (Dolan, S. *et al.*, 2003). Upregulations of mGluR2, 3, and 5 mRNA in the dorsal horn have also been reported a sheep model of postsurgical pain (Dolan, S. *et al.*, 2004). Spinal mGluR1, but not mGluR5, mRNA is also upregulated in response to spinal cord injury, while group II mGluRs are downregulated (Mills, C. D. *et al.*, 2001).

26.2.3.3 Other receptors

In addition to effects on neuropeptides and EAA receptors, inflammatory and nerve injuries that produce hyperalgesia are known to produce alterations in various other transmitter receptors and ion channels in neurons in spinal cord dorsal horn or DRG neurons. Sciatic nerve transection produces an upregulation of GABA-A and a downregulation of GABA-B receptors in the rat spinal cord dorsal horn (Castro-Lopes, J. M. *et al.*, 1995). A similar downregulation of GABA-B receptors occurs after joint inflammation induced by CFA injection in the rat knee, but no change in GABA-A receptors was found (Castro-Lopes, J. M. *et al.*, 1995). There is also a downregulation of μ-opioid receptor immunoreactivity in the rat and monkey spinal cord dorsal horn following peripheral nerve axotomy (Zhang, X. *et al.*, 1998). In addition, there is an increase in phosphorylated μ-opioid receptors following sciatic nerve ligation (Narita, M. *et al.*, 2004). The reduction in number of μ-opioid receptors or the phosphorylation of remaining receptors may explain why morphine-induced antinociception (Ossipov, M. H. *et al.*, 1995; Narita, M. *et al.*, 2004) and the presynaptic inhibitory effects of μ-opioids (Kohno, T. *et al.*, 2005) are reduced in mice and rats with peripheral nerve injury, although see also Suzuki R. *et al.* (1999). There is, however, also an increase in the levels of phosphorylated κ-opioid receptors in the spinal cord dorsal horn of mice with a partial sciatic nerve lesion, an effect that is not shown in prodynorphin knockout mice that have enhanced tactile allodynia and thermal hyperalgesia (Xu, M. *et al.*, 2004). These findings suggest that the loss of endogenous dynorphin activity at κ-opioid receptors may also contribute to allodynia and hyperalgesia after nerve injury.

In contrast to the effects of opioid receptors, CCI of the sciatic nerve leads to a MAPK-dependent upregulation of spinal cannabinoid-1 (CB-1) receptors (Lim, G. *et al.*, 2003). The upregulation of spinal CB-1 was determined to be responsible for the anti-allodynic and antihyperalgesic effects of a CB-1 agonist in CCI rats, as the reduction in mechanical allodynia and thermal allodynia produced by Win 55,212-2 was lost in rats pretreated with an ERK inhibitor (PD98059) that blocked the spinal CB-1 upregulation (Lim, G. *et al.*, 2003). CCI of the rat sciatic nerve also leads to an upregulation of P2X3 receptors in spinal cord dorsal horn (Novakovic, S. D. *et al.*, 1999). This upregulation may contribute to hyperalgesia and allodynia, as an adenosine kinase inhibitor produces a greater inhibition of post-discharge, windup, and C-fiber-evoked responses in spinal cord neurons of rat with SNLs than in sham controls. Recent studies also indicate that there is an upregulation of α3 and α5 spinal nicotinic acetylcholine (nAcH) receptors following rat SNL (Vincler, M. and Eisenach, J. C., 2004), and an increase in spinal melanocortin (MC) receptors after CCI of the rat sciatic nerve (Vrinten, D. H. *et al.*, 2000). It is expected that upregulated α5 nAcH and MC receptors may contribute to allodynia after nerve injuries, as an α5 nAcH antisense olgionucleotides and an MC receptor antagonist significantly attenuate mechanical allodynia in SNL and CCI rats, respectively (Vrinten, D. H. *et al.*, 2000; Vincler, M. A. and Eisenach, J. C., 2005).

26.2.3.4 Scaffolding and adaptor proteins

NMDA receptors bind directly to membrane-associated guanylate kinases (MAGUKs) that regulate surface and synaptic NMDAR trafficking in the CNS (Ponting, C. P. *et al.*, 1997). MAGUKs include the adaptor proteins known as postsynaptic density proteins 93 and 95 (PSD-93, PSD-95), which are abundantly expressed in the spinal cord dorsal horn (Garry, E. M. *et al.* 2003; Tao, Y. X. *et al.*, 2003). Knockout of PSD-93 reduces surface NR2A and NR2B, lowers NMDA receptor-mediated postsynaptic currents and potentials, and reduces hyperalgesia and/or allodynia in mice with SNL or hind paw injection of CFA (Tao, Y. X. *et al.*, 2003). Antisense knockdown of spinal cord PSD-93 also reduces mechanical allodynia and thermal hyperalgesia in rats with hind paw CFA injection or following peripheral nerve injury (Zhang, B. *et al.*, 2003). Knockout or knockdown of PSD-95 also reduces hyperalgesia and allodynia in CCI mice or in rats with L5 SNL, respectively (Tao, F. *et al.*, 2001; Garry, E. M. *et al.*, 2003).

Homer is an intracellular scaffolding protein that is implicated in the intracellular retention and trafficking of group I mGluRs (Roche, K. W. *et al.*, 1999; Ango, F. *et al.*, 2000). Homer proteins also interact with the postsynaptic density protein Shank, a PDZ domain-containing protein binding to the PSD-95/NMDA receptor complex (Naisbitt, S. *et al.*, 1999; Tu, J. C. *et al.*, 1999). There are increases in the levels of both Homer and Shank in spinal cord dorsal horn in rats with CCI of the sciatic nerve at the time points when they exhibit mechanical allodynia and thermal hyperalgesia (Miletic, G. *et al.*, 2005). Another scaffolding protein, neurofilament light chain (NFL) protein, is downregulated in the DRG and spinal dorsal horn of rats with inflammation after a hind paw injection of zymosan (Kunz, S. *et al.*, 2004), the breakdown of NFL protein and inflammatory hyperalgesia were blocked with an inhibitor of the protease calpain. The cell adhesion molecule E-cadherin is dramatically downregulated in spinal cord dorsal horn in response to sciatic nerve transection (Brock, J. H. *et al.*, 2004), while the microtubule associated protein 1B is upregulated (Soares, S. *et al.*, 2002). The presynaptic vesicle protein synaptophysin is increased in spinal dorsal horn in parallel with thermal hyperalgesia in CCI rats (Chou, A. K. *et al.*, 2002).

26.2.3.5 Ion channels

Inflammatory and nerve injuries lead to alterations in sensory neuron ion channel expression, distribution and function that contribute to hyperalgesia and allodynia. In particular, evidence indicates that various sodium (Na^+) channels subtypes are altered in association with inflammation or neuropathy. There is an upregulation of the Na^+ channel subunit

$Na_V 1.3$ in DRG neurons or spinal cord dorsal horn neurons following peripheral nerve axotomy (Waxman, S. G. et al., 1994), traumatic spinal cord injury (Hains, B. C. et al., 2003) and CCI of the sciatic nerve (Hains, B. C. et al., 2004), as well as hind paw injection of carrageenan (Black, J. A. et al., 2004). In the case of CCI, a spinal upregulation likely contributes to hyperalgesia/allodynia, as i.t. administration of antisense oligonucleotides to $Na_V 1.3$ decreased expression to normal levels and decreased the enhanced spontaneous activity and evoked responses of dorsal horn neurons, as well as reducing thermal hyperalgesia and mechanical allodynia in CCI rats (Hains, B. C. et al., 2004).

There is also an upregulation of $Na_V 1.7$ and $Na_V 1.8$, but not $Na_V 1.9$, in DRG in response to hind paw injection of carrageenan (Tanaka, M. et al., 1998; Black, J. A. et al., 2004). Hyperalgesia in response to hind paw inflammation is reduced or delayed in rats treated with $Na_V 1.8$ antisense oligonucleotides (Khasar, S. G. et al., 1998) or in $Na_V 1.8$ knockout mice (Akopian, A. N. et al., 1999). Although $Na_V 1.8$ channels are not upregulated in DRG after nerve injury (Okuse, K. et al., 1997; Novakovic, S. D. et al., 1998), antisense oligonucleotide knockdown of $Na_V 1.8$ attenuates neuropathic pain in rats with SNL (Lai, J. et al., 2002; Gold, M. S. et al., 2003). The effectiveness of the knockdown may be explained by findings that there is a redistribution of $Na_V 1.8$ channels in injured nerves with an accumulation of these channels at the site of the nerve injury (Novakovic, S. D. et al., 1998; Gold, M. S. et al, 2003).

As for $Na_V 1.9$, it has been shown that knockdown of its protein has no influence on allodynia or hyperalgesia in SNL rats (Porreca, F. et al., 1999), while disruption of the SCN11A gene, which encodes $Na_V 1.9$, attenuates thermal hyperalgesia and spontaneous pain after peripheral inflammation (Priest, B. T. et al., 2005).

Ca^{2+} channels also show changes in expression following nerve injury. Thus, after axotomy or partial nerve ligation of the sciatic nerve there is an upregulation of $Ca_V 1.2$ (L-type $\alpha 1C$) and $Ca_V 2.3$ (R-type $\alpha 1E$) (Yang, L. et al., 2004; Dobremez, E. et al., 2005). Evidence suggests that L-type Ca^{2+} channels contribute to tonic nociception (Del Pozo, E. et al., 1987; Coderre, T. J. and Melzack, R., 1992b), and recent evidence indicates that $Ca_V 2.3$ knockout mice have reduced responses to inflammatory pain (Saegusa, H. et al., 2000), suggesting that R-type Ca^{2+} channels may also play a role in hyperalgesia/allodynia. There is also an upregulation of $Ca_V 2.2$ (N-type Ca^{2+} channels $\alpha 1B$) in both DRG and spinal dorsal horn after hind paw injection of carrageenan (Yokoyama, K. et al., 2003) or CCI of the sciatic nerve (Cizkova, D. et al., 2002b). Many preclinical studies have shown that antagonists of N-type Ca^{2+} channels reduce allodynia or hyperalgesia associated with inflammatory or nerve injuries, and suggest that $Ca_V 2.2$ is a key target (see Winquist, R. J. et al., 2005 for a review). There is also a reduction of tactile allodynia and thermal hyperalgesia associated with SNL in $Ca_V 2.2$ knockout mice (Saegusa, H. et al., 2001). Although the P-type Ca^{2+} channel have also been implicated in the development of hyperalgesia, the effects of their antagonists have been less effective (Sluka, K. A., 1998; Matthews, E. A. and Dickenson, A. H., 2001), and a selective Q-type antagonist was ineffective (Nebe, J. et al., 1997). In addition, others have found in response to axotomy or CCI of the sciatic nerve there is a downregulation of $Ca_V 1.2$ (L-type $\alpha 1C$), $Ca_V 1.3$ (L-type $\alpha 1D$), $Ca_V 3.2$ (T-type $\alpha 1H$) and $Ca_V 3.3$ (T-type $\alpha 1I$) mRNA expression in rat DRG (Kim, D. S. et al., 2001). A role for T-type channels in hyperalgesia is suggested since antisense knockdown of $Ca_V 3.2$ reduces both hyperalgesia and allodynia in rats with CCI of the sciatic nerve (Bourinet, E. et al., 2004).

Recent studies also show there is an upregulation of $\alpha_2 \delta$ Ca^{2+} channel auxiliary subunits found in the DRG and spinal dorsal horn of rats after SNL and to a lesser extent CCI of the sciatic nerve (Luo, Z. D. et al., 2001; 2002). The upregulation of $\alpha_2 \delta$ subunits is temporally related to the development of tactile allodynia after SNL, and i.t. treatment of antisense oligonucleotide to $\alpha_2 \delta$ diminishes the tactile allodynia in SNL rats (Li, C. Y. et al., 2004). The importance of this subunit to hyperalgesia and allodynia is also highlighted by the fact that the $\alpha_2 \delta$ subunit is the only known binding site for gabapentin and pregabalin which are effective antihyperalgesic agents for many neuropathic pain patients (Frampton, J. E. and Foster, R. H., 2005; Wiffen, P. J. et al., 2005).

There is considerable evidence that sensory neuron vanilloid receptors, including TRPV1 (VR-1) and TRPV2 (VRL-1), play a critical role in the transduction of noxious heat (Caterina, M. J. and Julius, D., 2001). An up-regulation of TRPV1 has also been found in the spared L4 DRG following L5 SNL (Hudson, L. J. et al., 2001) and CCI (Wilson-Gerwing, T. D. et al., 2005). Importantly, the upregulation of L4 DRG TRPV1 correlates temporally with the development and maintenance of thermal hyperalgesia after L5 SNL (Fukuoka, T.

et al., 2002). I.t. administration of NT-3 reduces thermal hyperalgesia as well as reducing the elevated DRG TRPV1 expression in CCI rats (Wilson-Gerwing, T. D. *et al.*, 2005). Systemic treatment with a TRPV1 antagonist reduces thermal hyperalgesia induced by hindpaw capsaicin, carrageenan or CFA, as well as mechanical hyperalgesia after partial sciatic nerve lesion in the guinea pig (Walker, K. M. *et al.*, 2003), and carrageenan-induced thermal hyperalgesia is absent in TRPV1 knock-out mice (Davis, J. B. *et al.*, 2000). In contrast, i.t. administration of a TRPV1 antagonist is less effective at reducing noxious stimulus-evoked dorsal horn neuronal responses in SNL rats as compared to sham controls (Kelly, S. and Chapman, V., 2002). Evidence suggests that there is also an up-regulation of TRPV1 in DRG associated with peripheral inflammation, and that the upregulation is associated with thermal hyperalgesia and depends on p38 MAPK activation (Ji, R. R. *et al.*, 2002b). Recent evidence indicates that while the levels of TRPV2 receptors in DRG are not affected by peripheral nerve axotomy, their levels in sympathetic post-ganglionic neurons are upregulated after nerve sections (Gaudet, A. D. *et al.*, 2004).

26.2.4 Alterations in Transporter Activity

26.2.4.1 *Glutamate transporters*

Glutamate transporters are critical for the reuptake of transmitter in glutamate synapses, and the major ones include vesicular glutamate transporters VGluT1-3 and EAAC1 and glial glutamate transporters (GLAST and GLT-1) (Shigeri, Y. *et al.*, 2004). There is a downregulation of EAAC1, GLAST, and GLT-1 is the spinal cord dorsal horn of CCI rats at time points when they exhibit mechanical allodynia and thermal hyperalgesia (Sung, B. *et al.*, 2003). The EAAC1, GLAST, and GLT-1 downregulation in rats with taxol-induced neuropathy is paralleled by increases in spontaneous activity, prolonged after-discharges, and windup in dorsal horn neurons (Cata, J. P. *et al.*, 2005; Weng, H. R. *et al.*, 2005). Dorsal horn EAAC1, GLAST, and GLT-1 are also all downregulated in chronic morphine-treated rats in a manner that is temporally correlated with the development of thermal hyperalgesia (Mao, J. *et al.*, 2002a; Lim, G. *et al.*, 2005). I.t. administration of inhibitors of glutamate transporters has also been shown to produce mechanical allodynia and thermal hyperalgesia (Mao, J. *et al.*, 2002a; Liaw, W. J. *et al.*, 2005). Finally, using *in vivo* voltametry and pressure-ejected glutamate, it has been shown that

the uptake of glutamate is significantly reduced in the spinal cord dorsal horn of rats with SNL (Binns, B. C. *et al.*, 2005), illustrating the impact of glutamate transporter suppression.

26.2.4.2 *Cation chloride transporters*

Recent studies suggest that alterations in the function of cation chloride cotransporters may play a role in hyperalgesia states by modulating the chloride concentration inside nociceptive cells. Galan A. and Cervero F. (2005) report that there is a 50% increase in the $Na^+-Cl^--K^+$ isoform 1 (NKCC1) in the dorsal horn of mice after intracolonic capsaicin injection, as well as an increase in the phosphorylation of NKCC1. In addition, NKCC1 knockout mice have a reduction in mechanical allodynia to stroking induced by intradermal capsaicin (Laird, J. M. *et al.*, 2004). They propose that increased NKCC-1 shifts the anion gradient and alters normal presynaptic inhibition in dorsal horn by producing an increase in the intracellular concentration of chloride and an enhanced GABA-mediated primary afferent depolarization that leads to mechanical allodynia (Galan, A. and Cervero, F., 2005). Thus, GABAergic depolarization in primary afferents will lead to a cross excitation between low- and high-threshold afferents that produce allodynia by reducing GABA-mediated presynaptic inhibition (Price, T. J. *et al.*, 2005).

Furthermore, Coull J. A. *et al.* (2003) found a reduction in the potassium chloride exporter (KCC2) in the spinal dorsal horn of CCI rats. This decrease was paralleled by a shift in the transmembrane anion gradient that caused normally inhibitory anion synaptic currents to be excitatory, and an increase in the net excitability of lamina I dorsal horn neurons in CCI rats (Coull, J. A. *et al.*, 2003). In addition, blockade or knockdown of spinal KCC2 in intact rats produced mechanical allodynia, suggesting that disruption of anion homeostatis could be responsible for allodynia in neuropathic rats (Coull, J. A. *et al.*, 2003). The decrease in spinal KCC2 is proposed then to result in a loss of GABA-/glycinergic inhibitory tone and, in some cases, inverting its action into net excitation that causes allodynia after nerve injury (Price, T. J. *et al.*, 2005).

26.2.5 Glial Cell Alterations

26.2.5.1 *Microglia*

Activation of microglial and the release of proinflammatory cytokines from these cells was originally found to play a significant role in illness-induced hyperalgesia, including that produced by bacterial endotoxins and viruses such as HIV-1 (see Watkins, L. R. and Maier, S. F., 1999 for a review). There is also an increase in activated microglia, as indicated by increased complement receptor C3bi (OX-42) immunoreactivity in rat spinal cord dorsal horn following sciatic nerve transection (Eriksson, N. P. *et al.*, 1993). There is an increase in immunoreactivity for microglial markers OX-42 and histocompatibility complex II (OX-6), as well, in the dorsal horn of rats with partial sciatic nerve ligation (Coyle, D. E., 1998), with a linear correlation between increased mechanical allodynia and increased OX-42 staining. OX-42 is also increased in dorsal horn following injury of spinal nerve roots, and is accompanied by increases in expression of the interleukin 1β (IL-1β) (Hashizume, H. *et al.*, 2000) and the fractalkine receptor X3CR1 in dorsal horn microglia (Verge, G. M. *et al.*, 2004; Lindia, J. A. *et al.*, 2005). Peripheral inflammation induced by hind paw CFA injections also leads to increases in microglial markers, including OX-42, Mac-1, TLR4, and CD14 in spinal dorsal horn (Raghavendra, V. *et al.*, 2004). I.t. administration of an inhibitor of microglial cell activation attenuates mechanical allodynia in rats with sciatic nerve inflammation, as well as reducing the expression of the proinflammatory cytokines IL-1β and tumor necrosis factor alpha (TNFα) and their converting enzymes in spinal dorsal horn after sciatic nerve inflammation (Milligan, E. D. *et al.*, 2003; Ledeboer, A. *et al.*, 2005). A glial metabolic inhibitor also reduces C-fiber afferent-stimulated LTP in spinal cord dorsal horn neurons (Ikeda, H. and Murase, K., 2004).

Mechanical allodynia in ipsilateral and contralateral hind paws of rats with a unilateral inflammation of the sciatic nerve is also reduced following i.t. treatment with antagonists of the proinflammatory cytokines IL-1β, IL-6, and TNFα (Milligan, E. D. *et al.*, 2003). I.t. administration of the cytokine fractalkine produces mechanical allodynia and thermal hyperalgesia that is blocked by an inhibitor of microglial activation, as well as an antagonist of IL-1 and antibodies to IL-6 (Milligan, E. D. *et al.*, 2004; 2005). In contrast, a neutralizing antibody of the fractalkine receptor (CX3CR1) delays the development of mechanical allodynia and thermal hyperalgesia in rats with CCI or inflammatory injury of the sciatic nerve (Milligan, E. D. *et al.*, 2004). Increased cytokine activity in spinal nerve injured rats leads to a phosphorylation of p38 MAPK in dorsal horn microglia, and a p38 MAPK inhibitor reduces mechanical allodynia in neuropathic rats (Inoue, K. *et al.*, 2003; Tsuda, M. *et al.*, 2004). The expression of the P2X4 receptor, a

subtype of ATP receptors, is also enhanced in rat spinal microglia after SNL, and ATP acting at P2X4 receptor on microglia leads to the release of cytokines which activate p38 MAPK (Inoue, K., 2006). Pharmacological blockade or knockdown of P2X4 receptors reverse mechanical allodynia after SNL in rats, while i.t. administration of microglia activated by P2X4 receptor stimulation produces mechanical allodynia in naïve rats (Tsuda, M. *et al.*, 2003).

26.2.5.2 Astrocytes

There is also an increase in activated astrocytes as indicated by increased staining of glial fibrillary acidic protein (GFAP) in rat spinal cord dorsal horn following sciatic nerve transection (Eriksson, N. P. *et al.*, 1993) or partial sciatic nerve ligation (Ma, W. and Quirion, R., 2002). The degree of thermal hyperalgesia or mechanical allodynia in rats with CCI of the sciatic nerve (Garrison, C. J. *et al.*, 1991) and injury of spinal nerve roots (Hashizume, H. *et al.*, 2000) or partial sciatic nerve injury (Coyle, D. E., 1998), respectively, is correlated with the magnitude of GFAP staining in the spinal cord dorsal horn. GFAP is also increased in dorsal horn following SNL, and is accompanied by increases expression the cytokine fibroblast growth factor-2 (Madiai, F. *et al.*, 2003) and increases in fractalkine (CX3CL1) in astrocytes (Lindia, J. A. *et al.*, 2005). Hind paw injection of CFA also produces spinal astrocyte activation as indicated by increased GFAP and S100B expression in the dorsal horn (Raghavendra, V. *et al.*, 2004). There is also an increase in MAPK activity in astrocytes as indicated by increases in phosphorylated ERK and JNK in cells that colocalize GFAP (Ma, W. and Quirion, R., 2002).

26.2.5.3 Neuron–glial interactions

Many of the factors released by glial cells discussed in the two above sections (particularly cytokines) act to sensitize neurons and represent neuron–glial interactions that contribute to hyperalgesia and/or allodynia. A recent study by Coull J. A. *et al.* (2005) provides an excellent example of a neuronal–glial interactive pathway that has been implicated in the development of allodynia and hyperalgesia in nerve-injured rats. These authors have shown that ATP-stimulated microglia cause a shift in the anion reversal potential in spinal lamina I neurons similar to that observe in rats with sciatic nerve injury. The shift inverts the polarity of GABA currents so that GABA produces excitation rather than the normal inhibition of lamina I neurons. In addition, they showed that applying BDNF mimics the alteration in anion gradient, that ATP stimulation evokes the release of BDNF from microglia, and that blocking BDNF effects on TrkB receptors reverses allodynia and anion gradient shifts that follow nerve injury or administration of ATP-stimulated microglia (Coull, J. A. et al., 2005). This study illustrates the importance of neuron–glial interactions to allodynia in neuropathic rats, and indicates that BDNF is a crucial signalling molecule between microglia and neurons.

26.2.6 Descending Facilitation

Sensitization of spinal cord dorsal horn neurons is not only enabled by increased input from primary afferent nociceptors, but also by a decrease in descending inhibitory controls and a phenomenon known as descending facilitation. The role of descending facilitation was reported by Watkins L. R. et al. (1994b) who found lesions of the NRM reduced illness-induced hyperalgesia. Pertovaara, A. (1998) further reported that mustard oil application to the rat hind paw produced a facilitation of evoked nociceptive responses of dorsal horn neurons that was attenuated by spinal transection or lidocaine block of the rostroventromedulla (RVM). Descending facilitation of spinal neurons has also been observed through a pathway that involves the nucleus reticularis gigantocellularis (Zhuo, M. and Gebhart, G. F. 1992). Spinal cord transection or RVM blocks, as well as inhibition of RVM NMDA receptors or NO synthase, also prevent the development of secondary hyperalgesia in rats after topical mustard oil, carrageenan inflammation and SNL (see Urban, M. O. and Gebhart, G. F., 1999). Mechanical allodynia after SNL in rats is reversed by ipsilateral dorsal column lesions of by administration of lidocaine into the ipsilateral nucleus gracilis, suggesting that descending facilitation from supraspinal sites depends on afferent signals transmitted by large myelinated fibers in the dorsal column pathway (Sun, H. et al., 2001). Importantly, large myelinated fibers projecting to the nucleus gracilis have been found to change their phenotype and express SP following sciatic nerve transection, and NK-1 receptor antagonists suppress stimulus-induced Fos expression in the gracile nucleus in sciatic nerve lesioned rats (Noguchi, K. et al., 1995). It has also been suggested that a descending facilitatory serotonergic pathway is activated by superficial dorsal horn neurons that express NK-1 (Suzuki, R. et al., 2002).

26.2.7 Anatomical Changes

26.2.7.1 Cell death and apoptotic genes

26.2.7.1.(i) Cell death
Transection of the sciatic nerve results in transganglionic degeneration atrophy and signs of Wallerian degeneration in the monkey spinal cord (Knyihar-Csillik, E. et al., 1987). In rats, there is a dramatic increase of terminal deoxynucleotidyl transferase-mediated biotinylated UTP nick end labeling (TUNEL)-stained interneurons in superficial (Azkue, J. J. et al., 1998) and deep (Oliveira, A. L. et al., 1997) dorsal horn neurons after sciatic nerve transection, indicating that apoptotic mechanisms are involved in the cell death process. CCI of the rat sciatic nerve also leads to degenerative changes in spinal cord with the appearance of hyperchromatic or dark neurons (Sugimoto, T. et al., 1990; Nachemson, A. K. and Bennett, G. J., 1993) and TUNEL-labeled neurons (Whiteside, G. T. and Munglani, R., 2001; Maione, S. et al., 2002) in the dorsal horn. The apoptotic degeneration after nerve injury appears to depend on EAA-induced excitotoxicity since premptive treatment with NMDA or mGluR5 antagonists reduces TUNEL staining after transection (Azkue, J. J. et al., 1998) or CCI (Whiteside, G. T. and Munglani, R., 2001; de Novellis, V. et al., 2004) of the sciatic nerve. Reducing neuronal apopotosis in spinal dorsal horn by administration of erythroprotein also prevents mechanical allodynia in rats with spinal nerve root crush injury (Sekiguchi, Y. et al., 2003).

Recent evidence suggest that there is a decrease in GABA-A-mediated inhibitory postsynaptic potentials (IPSPs) in rats with spared nerve injury or CCI (Moore, K. A. et al., 2002), suggesting a reduction in the presynaptic release of GABA after nerve injury. Partial nerve injured was also found to produce a reduction in the GABA synthesizing enzyme glutamic acid decarboxylase (GAD) as well as increased TUNEL staining indicative of apoptosis in dorsal horn (Moore, K. A. et al., 2002), although it has been recently suggested that most TUNEL staining is in microglia and not neurons (Polgar, E. et al., 2005). Blocking apoptosis with a caspase inhibitor prevents the loss of spinal GABAergic interneurons, the reduction in spinal IPSCs, and mechanical and cold allodynia after CCI or partial sciatic nerve injury (Scholz, J. et al., 2005). Furthermore, rats with SNL have alterations in the normal inhibition of C-fiber-evoked activity by GABA-A agonists (Kontinen, V. K. and Dickenson, A. H., 2000; Sokal, D. M. and Chapman, V., 2003), and spinal cord injured rats with allodynia do not exhibit the normal increases in evoked responses of dorsal horn neurons after ionotophorectic application of GABA-A antagonists (Drew, G. M. *et al.*, 2004). In addition, a reduction in the glycinergic inhibitory control of spinal neurons occurs after CFA-induced inflammation of the hind paw as reflected by a reduced mean frequency of glycine-mediated miniature IPSCs in lamina I neurons (Muller, F. *et al.*, 2003). All these findings point to a potentially important contribution to hyperalgesia and/or allodynia of cell death of inhibitory neurons.

26.2.7.1.(ii) *Apoptotic genes*

Recent research suggests that chronic pain produces alterations in apoptotic and antiapoptotic genes either following gene transcription or by posttranslational mechanisms. CCI of the sciatic nerve leads to alterations of dorsal horn levels of various members of Bcl-2 gene family, including an upregulation of the proapoptotic gene Bax and the antiapoptotic genes Bcl-2 and Bcl-xL, and a downregulation of Bcl-xS (Maione, S. *et al.*, 2002; de Novellis, V. *et al.*, 2004). Importantly, spinal neuron apoptosis indicated by TUNEL staining, thermal, and mechanical hyperalgesia and the alterations in apoptotic genes were all reversed by treatment with an mGluR5 antagonist (de Novellis, V. *et al.*, 2004). There is also an upregulation of the proapoptotic protein Bax and caspase 3, and a downregulation of the antiapopotic protein Bcl-2 in chronic morphine-treated rats that exhibit thermal hyperalgesia (Mao, J. *et al.*, 2002b). Neuronal apopotosis in dorsal horn and thermal hyperalgesia in chronic morphine-treated rats are attenuated by i.t. administration of a nonselective caspase and a caspase-3-selective inhibitor (Mao, J. *et al.*, 2002b). I.t. quisqualic acid-induced excitotoxic spinal cord injury produces pain-related behaviors and increased expression of CD-95 or Fas ligand and TNF-related apoptotis-inducing ligand (TRAIL), and i.t. treatment with IL-10, an anti-inflammatory cytokine, reduced pain-related behaviors in spinal injured rats (Plunkett, J. A. *et al.*, 2001). Furthermore, inhibition of activator (1,2,8, and 9) and effector (3) caspases attenuates neuropathic pain induced in rats by cancer and HIV/AIDS chemotherapeutic agents, as well as pain-related behavior induced by TNFα (Joseph, E. K. and Levine, J. D., 2004). A caspase inhibitor also reduces mechanical and cold allodynia in rats with partial nerve injury (Scholz, J. *et al.*, 2005), while an inhibitor of poly(ADP-ribose) synthetase, an enzyme involved in glutamate-induced

neurotoxicity, reduces the expression of dark neurons and allodynia and hyperalgesia in CCI rats (Mao, J. *et al.*, 1997). Galectin-1, a member of the family of β-galactoside-binding animal lectins, has been shown to be a proapoptotic factor, and is increased in spinal cord dorsal horn after peripheral nerve transection (Imbe, H. *et al.*, 2003; McGraw, J. *et al.*, 2005). I.t. treatment with antibodies to galectin-1 attenuates the mechanical allodynia in the spared nerve injury model (Imbe, H. *et al.*, 2003).

26.2.7.2 *Sprouting*

26.2.7.2.(i) *Sympathetic fibers*

Although controversial, there is long-standing evidence that alterations in sympathetic nervous system function plays a role in hyperalgesia and allodynia following nerve injury (Janig, W. *et al.*, 1996). Sympathetic activity may influence inputs to spinal cord dorsal horn following the development of abnormal connectivity between sympathetic efferent fibers and primary afferent fibers. McLachlan E. M. *et al.* (1993) showed that sympathetic axons sprout into the DRGs of rats with sciatic nerve transection and form basket-like structures around large neurons. Similar sympathetic fiber sprouting is observed after SNL (Chung, K. *et al.*, 1993) and partial ligation (Lee, B. H. *et al.*, 1998) or CCI of the sciatic nerve (Ramer, M. S. and Bisby, M. A., 1997). Mechanical and allodynia in SNL rats is abolished after sympathectomy, which eliminates most sprouting fibers (Chung, K. *et al.*, 1996). Sympathetic sprouting, mechanical allodynia, and thermal hyperalgesia after CCI depends on the occurrence of Wallerian degeneration (Ramer, M. S. *et al.*, 1997) and an upregulation of NGF (Ramer, M. S. and Bisby, M. A., 1997) in DRG neurons. There is also and upregulation of NGF and NT-3 in glial cells of the DRG in SNL rats, and antibodies to NGF, BDNF, and NT-3 reduce sympathetic sprouting in rat DRG and thermal hyperalgesia after SNL (Zhou, X. F. *et al.*, 1999b; Deng, Y. S. *et al.*, 2000). Sympathetic sprouting and mechanical allodynia after SNL are attenuated in IL-6 knockout mice, implicating cytokines as key factors contributing to these changes (Ramer, M. S. *et al.*, 1998).

26.2.7.2.(ii) *Aβ fibers*

It it difficult to explain the development of allodynia in response to activity-drive plasticity as Aβ fibers, which transmit low-threshold touch inputs generally do not terminate on spinal nociceptive neurons. Evidence generated over the past 15 years suggests, however, that allodynia may be partly explained the nerve injury-induced sprouting of the terminals of Aβ fibers into dorsal horn lamina I and IIo where they are able to contact spinal nociceptive neurons (Woolf, C. J. *et al.*, 1992; Shortland, P. and Woolf, C. J., 1993). It has been suggested that regenerating Aβ fibers exhibit collateral sprouting into superficial dorsal horn lamina after the terminal arbors of small Aδ and C fibers degenerate following peripheral nerve injury (Wilson, P. and Kitchener, P. D., 1996), a theory that is supported by evidence of similar Aβ fiber sprouting after local application of the C-fiber neurotoxin capsaicin to the sciatic nerve (Mannion, R. J. *et al.*, 1996). Aβ fiber sprouting into rat spinal lamina II has also been observed in rats with SNL or CCI of the sciatic nerve (Shortland, P. *et al.*, 1997; Nakamura, S. and Myers, R. R., 1999) or even chronic inflammation of the rat hind paw (Ma, Q. P. and Tian, L., 2002). Lekan H. A. *et al.* (1997) also found increased Aβ fibers in dorsal roots despite overall decreases in rat DRG cells after SNL. Both the sprouting of Aβ fibers into lamina II and mechanical allodynia after sciatic nerve injury are absent in mice lacking membrane-type 5 matrix metalloproteinase (Komori, K. *et al.*, 2004), and the sprouting is accompanied by *de novo* expression of SP, somatostatin, and PACAP in deeper lamina (Jongsma, H. *et al.*, 2000; Swamydas, M. *et al.*, 2004). The conclusions of these studies have been questioned, however, by evidence that after nerve injury Aδ fibers (Ma, Q. P. and Tian, L., 2001) or even C fibers (Tong, Y. G. *et al.*, 1999; Shehab, S. A. *et al.*, 2003) may begin to uptake the cholera toxin B subunit that has been used in the previous studies as a selective marker of Aβ fibers.

26.3 Conclusion

Hyperalgesia and allodynia are consequences of cutaneous or deep tissue and nerve injury and depend on both peripheral and central sensitization. Central sensitization relies on various alterations in both presynaptic DRG cells, interneurons and postsynaptic spinal cord dorsal horn neurons, as well as satellite cells in the dorsal horn and DRG. There are changes in the expression of receptors and ion channels, as well as alterations in their cellular location within neurons. There is evidence that changes in scaffolding proteins, kinases and phosphatases, and transporters influence the synaptic activity and anion gradients of pre- and postsynaptic neurons in the spinal cord dorsal horn. Activation of microglial

and astrocytes and the release of mediators such as cytokines and neurotrophins result in neuron–glial interactions that also contribute to increased excitability in the spinal cord. Finally there is evidence of cell death, trans-synaptic degeneration, and the sprouting of sympathetic efferent fibers in DRG and A-fiber primary afferent fibers in the dorsal horn which contribute to central sensitization and the development and/or maintenance of hyperalgesia and allodynia.

Acknowledgments

This work is supported by grants form CIHR, NSERC, and FRSQ to T. J. C.

References

Aanonsen, L. M. and Wilcox, G. L. 1986. Phencyclidine selectively blocks a spinal action of N-methyl-D-aspartate in mice. Neurosci. Lett. 67, 191–197.

Aanonsen, L. M. and Wilcox, G. L. 1987. Nociceptive action of excitatory amino acids in the mouse: effects of spinally administered opioids, phencyclidine and sigma agonists. J. Pharmacol. Exp. Ther. 243, 9–19.

Aanonsen, L. M., Lei, S., and Wilcox, G. L. 1990. Excitatory amino acid receptors and nociceptive neuro-transmission in rat spinal cord. Pain 41, 309–321.

Aanonsen, L. M., Kajander, K. C., Bennett, G. J., and Seybold, V. S. 1992. Autoradiographic analysis of 125I-substance P binding in rat spinal cord following chronic constriction injury of the sciatic nerve. Brain Res. 596, 259–268.

Abe, T., Matsumura, S., Katano, T., Mabuchi, T., Takagi, K., Xu, L., Yamamoto, A., Hattori, K., Yagi, T., Watanabe, M., Nakazawa, T., Yamamoto, T., Mishina, M., Nakai, Y., and Ito, S. 2005. Fyn kinase-mediated phosphorylation of NMDA receptor NR2B subunit at Tyr1472 is essential for maintenance of neuropathic pain. Eur. J. Neurosci. 22, 1445–1454.

Abe, T., Sugihara, H., Nawa, H., Shigemoto, R., Mizuno, N., and Nakanishi, S. 1992. Molecular characterization of a novel metabotropic glutamate receptor mGluR5 coupled to inositol phosphate/Ca²⁺ signal transduction. J. Biol. Chem. 267, 13361–13368.

Abram, S. E. and Yaksh, T. L. 1994. Systemic lidocaine blocks nerve injury-induced hyperalgesia and nociceptor-driven spinal sensitization in the rat. Anesthesiology 80, 383–391.

Adwanikar, H., Karim, F., and Gereau, R. W., 4th 2004. Inflammation persistently enhances nocifensive behaviors mediated by spinal group I mglurs through sustained ERK activation. pain 111, 125–135.

Akopian, A. N., Souslova, V., England, S., Okuse, K., Ogata, N., Ure, J., Smith, A., Kerr, B. J., McMahon, S. B., Boyce, S., Hill, R., Stanfa, L. C., Dickenson, A. H., and Wood, J. N. 1999. The tetrodotoxin-resistant sodium channel SNS has a specialized function in pain pathways. Nat. Neurosci. 2, 541–548.

Alaluf, S., Mulvihill, E. R., and McIlhinney, A. J. 1995. Rapid agonist mediated phosphorylation of the metabotropic glutamate receptor 1a by protein kinase C in permanently transfected BHK cells. FEBS Lett. 367, 301–305.

Allen, B. J., Li, J., Menning, P. M., Rogers, S. D., Ghilardi, J., Mantyh, P. W., and Simone, D. A. 1999. Primary afferent fibers that contribute to increased substance P receptor internalization in the spinal cord after injury. J. Neurophysiol. 81, 1379–1390.

Andrew, D. and Greenspan, J. D. 1999. Mechanical and heat sensitization of cutaneous nociceptors after peripheral inflammation in the rat. J. Neurophysiol. 82, 2649–2656.

Ango, F., Pin, J. P., Tu, J. C., Xiao, B., Worley, P. F., Bockaert, J., and Fagni, L. 2000. Dendritic and axonal targeting of type 5 metabotropic glutamate receptor is regulated by homer1 proteins and neuronal excitation. J. Neurosci. 20, 8710–8716.

Apfel, S. C. 2000. Neurotrophic factors and pain. Clin. J. Pain 16, S7–S11.

Apfel, S. C., Wright, D. E., Wiideman, A. M., Dormia, C., Snider, W. D., and Kessler, J. A. 1996. Nerve growth factor regulates the expression of brain-derived neurotrophic factor mRNA in the peripheral nervous system. Mol. Cell. Neurosci. 7, 134–142.

Arevalo, M. I., Escribano, E., Caleno, A., Domenech, J., and Queralt, J. 2003. Thermal hyperalgesia and light touch allodynia after intradermal *Mycobacterium butyricum* administration in rat. Inflammation 27, 293–299.

Attal, N., Gaude, V., Brasseur, L., Dupuy, M., Guirimand, F., Parker, F., and Bouhassira, D. 2000. Intravenous lidocaine in central pain: a double-blind, placebo-controlled, psychophysical study. Neurology 54, 564–574.

Azkue, J. J., Zimmermann, M., Hsieh, T. F., and Herdegen, T. 1998. Peripheral nerve insult induces NMDA receptor-mediated, delayed degeneration in spinal neurons. Eur. J. Neurosci. 10, 2204–2206.

Barakat-Walter, I. 1996. Brain-derived neurotrophic factor-like immunoreactivity is localized mainly in small sensory neurons of rat dorsal root ganglia. J. Neurosci. Methods 68, 281–288.

Baranowsli, A. P., Priestley, J. V., and McMahon, S. 1993. Substance P in cutaneous primary sensory neurons-a comparison of models of nerve injury that allow varying degrees of regeneration. Neuroscience 55, 1025–1036.

Baumann, T. K., Simone, D. A., Shain, C. N., and LaMotte, R. H. 1991. Neurogenic hyperalgesia: the search for the primary cutaneous afferent fibers that contribute to capsaicin-induced pain and hyperalgesia. J. Neurophysiol. 66, 212–227.

Beitel, R. E. and Dubner, R. 1976. Response of unmyelinated (C) polymodal nociceptors to thermal stimuli applied to monkey's face. J. Neurophysiol. 39, 1160–1175.

Benani, A., Heurtaux, T., Netter, P., and Minn, A. 2004. Activation of peroxisome proliferator-activated receptor alpha in rat spinal cord after peripheral noxious stimulation. Neurosci. Lett. 369, 59–63.

Bennett, G. J. and Xie, Y. K. 1988. A peripheral mononeuropathy in rat produces disorders of pain sensation like those seen in man. Pain 33, 87–107.

Binns, B. C., Huang, Y., Goettl, V. M., Hackshaw, K. V., and Stephens, R. L. Jr. 2005. Glutamate uptake is attenuated in spinal deep dorsal and ventral horn in the rat spinal nerve ligation model. Brain Res. 1041, 38–47.

Black, J. A., Liu, S., Tanaka, M., Cummins, T. R., and Waxman, S. G. 2004. Changes in the expression of tetrodotoxin-sensitive sodium channels within dorsal root ganglia neurons in inflammatory pain. Pain 108, 237–247.

Bordi, F. and Ugolini, A. 2000. Involvement of mGluR(5) on acute nociceptive transmission. Brain Res. 871, 223–233.

Bourinet, E., Alloui, A., Monteil, A., Barrere, C., Couette, B., Poirot, O., Pages, A., McRory, J., Snutch, T. P., Eschalier, A., and Nargeot, J. 2004. Silencing of the Cav3.2

T-type calcium channel gene in sensory neurons demonstrates its major role in nociception. EMBO J. 24, 315–324.

Boxall, S. J., Berthele, A., Laurie, D. J., Sommer, B., Zieglgansberger, W., Urban, L., and Tolle, T. R. 1998. Enhanced expression of metabotropic glutamate receptor 3 messenger RNA in the rat spinal cord during ultraviolet irradiation induced peripheral inflammation. Neuroscience 82, 591–602.

Boxall, S. J., Thompson, S. W., Dray, A., Dickenson, A. H., and Urban, L. 1996. Metabotropic glutamate receptor activation contributes to nociceptive reflex activity in the rat spinal cord in vitro. Neuroscience 74, 13–20.

Brennan, T. J., Vandermeulen, E. P., and Gebhart, G. F. 1996. Characterization of a rat model of incisional pain. Pain 64, 493–501.

Brenner, G. J., Ji, R. R., Shaffer, S., and Woolf, C. J. 2004. Peripheral noxious stimulation induces phosphorylation of the NMDA receptor NR1 subunit at the PISC-dependent site, serine-896, in spinal cord dorsal horn neurons. Eur. J. Neurosci. 20, 375–384.

Brock, J. H., Elste, A., and Huntley, G. W. 2004. Distribution and injury-induced plasticity of cadherins in relationship to identified synaptic circuitry in adult rat spinal cord. J. Neurosci. 24, 8806–8817.

Broom, D. C., Samad, T. A., Kohno, T., Tegeder, I., Geisslinger, G., and Woolf, C. J. 2004. Cyclooxygenase 2 expression in the spared nerve injury model of neuropathic pain. Neuroscience 124, 891–900.

Bruce, J. C., Oatway, M. A., and Weaver, L. C. 2002. Chronic pain after clip-compression injury of the rat spinal cord. Exp. Neurol. 178, 33–48.

Bullitt, E. 1989. Induction of c-fos-like protein within the lumbar spinal cord and thalamus of the rat following peripheral stimulation. Brain Res. 493, 391–397.

Buritova, J., Fletcher, D., Honore, P., and Besson, J. M. 1996. Effects of local anaesthetics on carrageenan-evoked inflammatory nociceptive processing in the rat. Br. J. Anaesth. 77, 645–652.

Cahill, C. M. and Coderre, T. J. 2002. Attenuation of hyperalgesia in a rat model of neuropathic pain after intrathecal pre- or post-treatment with a neurokinin-1 antagonist. Pain 95, 277–285.

Cahill, C. M., Dray, A., and Coderre, T. J. 1998. Priming enhances endotoxin-induced thermal hyperalgesia and mechanical allodynia in rats. Brain Res. 808, 13–22.

Cahusac, P. M., Evans, R. H., Hill, R. G., Rodriquez, R. E., and Smith, D. A. 1984. The behavioural effects of an *N*-methylaspartate receptor antagonist following application to the lumbar spinal cord of conscious rats. Neuropharmacology 23, 719–724.

Calvino, B., Villanueva, L., and LeBars, D. 1987. Dorsal horn (convergent) neurones in the intact anaesthetized arthritic rat. II. Heterotopic inhibitory influences. Pain 28, 81–98.

Cameron, A. A., Cliffer, K. D., Dougherty, P. M., Willis, W. D., and Carlton, S. M. 1991. Changes in lectin, GAP-43 and neuropeptide staining in the rat dorsal horn following experimental peripheral neuropathy. Neurosci. Lett. 131, 249–252.

Carlton, S. M., Hargett, G. L., and Coggeshall, R. E. 1998. Plasticity in alpha-amino-3-hydroxy-5-methyl-4-isoxazolepropionic acid receptor subunits in the rat dorsal horn following deafferentation. Neurosci. Lett. 242, 21–24.

Carpenter, S. E. and Lynn, B. 1981. Vascular and sensory responses of human skin to mild injury after topical treatment with Capsaicin. Br. J. Pharmacol. 73, 755–758.

Carroll, R. C. and Zukin, R. S. 2002. NMDA-receptor trafficking and targeting: implications for synaptic transmission and plasticity. Trends Neurosci. 25, 571–577.

Castro-Lopes, J. M., Malcangio, M., Pan, B. H., and Bowery, N. G. 1995. Complex changes of GABAA and GABAB receptor binding in the spinal cord dorsal horn following peripheral inflammation or neurectomy. Brain Res. 679, 289–297.

Cata, J. P., Weng, H. R., Chen, J. H., and Dougherty, P. M. 2006. Altered discharges of spinal wide dynamic range neurons and down-regulation of glutamate transporter expression in rats with paclitaxel-induced hyperalgesia. Neuroscience 138, 329–338.

Caterina, M. J. and Julius, D. 2001. The vanilloid receptor: a molecular gateway to the pain pathway. Annu. Rev. Neurosci. 24, 487–517.

Catheline, G., Le Guen, S., Honore, P., and Besson, J. M. 1999. Are there long-term changes in the basal or evoked Fos expression in the dorsal horn of the spinal cord of the mononeuropathic rat? Pain 80, 347–357.

Caudle, R. M., Perez, F. M., Del Valle-Pinero, A. Y., and Iadarola, M. J. 2005. Spinal cord NR1 serine phosphorylation and NR2B subunit suppression following peripheral inflammation. Mol. Pain 1, 25.

Caudle, R. M., Perez, F. M., King, C., Yu, C. G., and Yezierski, R. P. 2003. N-methyl-ᴅ-aspartate receptor subunit expression and phosphorylation following excitotoxic spinal cord injury in rats. Neurosci. Lett. 349, 37–40.

Cervero, F., Handwerker, H. O., and Laird, J. M. A. 1998. Prolonged noxious stimulation of the rat's tail: responses and encoding properties of dorsal horn neurons. J. Physiol. (Lond.) 404, 419–436.

Chan, C. F., Sun, W. Z., Lin, J. K., and Lin-Shiau, S. Y. 2000. Activation of transcription factors of nuclear factor kappa B, activator protein-1 and octamer factors in hyperalgesia. Eur. J. Pharmacol. 402, 61–68.

Chen, Y. and Devor, M. 1998. Ectopic mechanosensitivity in injured sensory axons arises from the site of spontaneous electrogenesis. Eur. J. Pain 2, 165–178.

Chen, L. and Huang, L.-Y. M. 1992. Protein kinase C reduces Mg^{2+} block of NMDA-receptor channels as a mechanism of modulation. Nature 356, 521–523.

Chen, S. R. and Pan, H. L. 2002. Hypersensitivity of spinothalamic tract neurons with diabetic neuropathic pain in rats. J. Neurophysiol. 87, 2726–2733.

Chen, S. R. and Pan, H. L. 2005. Distinct roles of group III metabotropic receptors in control of nociception and dorsal horn neurons in normal and nerve-injured rats. J. Pharmacol. Exp. Ther. 312, 120–126.

Chen, Y., Bacon, G., Sher, E., Clark, B. P., Kallman, M. J., Wright, R. A., Johnson, B. G., Schoepp, D. D., and Kingston, A. E. 2000. Evaluation of the activity of a novel metabotropic glutamate receptor antagonist (+/−)-2-amino-2-(3-cis and trans-carboxycyclobutyl-3-(9-thioxanthyl)propionic acid) in the in vitro neonatal spinal cord and in an in vivo pain model. Neuroscience 95, 787–793.

Chi, S. I., Levine, J. D., and Basbaum, A. I. 1993. Peripheral and central contributions to the persistent expression of spinal cord fos-like immunoreactivity produced by sciatic nerve transection in the rat. Brain Res. 617, 225–237.

Cho, H. J., Kim, S. Y., Park, M. J., Kim, D. S., Kim, J. K., and Chu, M. Y. 1997a. Expression of mRNA for brain-derived neurotrophic factor in the dorsal root ganglion following peripheral inflammation. Brain Res. 749, 358–362.

Cho, H. J., Kim, J. K., Zhou, X. F., and Rush, R. A. 1997b. Increased brain-derived neurotrophic factor immunoreactivity

in rat dorsal root ganglia and spinal cord following peripheral inflammation. Brain Res. 764, 269–272.

Chou, A. K., Muhammad, R., Huang, S. M., Chen, J. T., Wu, C. L., Lin, C. R., Lee, T. H., Lin, S. H., Lu, C. Y., and Yang, L. C. 2002. Altered synaptophysin expression in the rat spinal cord after chronic constriction injury of sciatic nerve. Neurosci. Lett. 333, 155–158.

Christensen, M. D. and Hulsebosch, C. E. 1997. Chronic central pain after spinal injury. J. Neurotrauma 14, 517–537.

Christensen, M. D., Everhart, A. W., Pickelman, J. T., and Hulsebosch, C. E. 1996. Mechanical and thermal allodynia in chronic central pain following spinal cord injury. Pain 68, 97–107.

Chung, K., Kim, H. J., Na, H. S., Park, M. J., and Chung, J. M. 1993. Abnormalities of sympathetic innervation in the area of an injured peripheral nerve in a rat model of neuropathic pain. Neurosci. Lett. 162, 85–88.

Chung, K., Lee, B. H., Yoon, Y. W., and Chung, J. M. 1996. Sympathetic sprouting in the dorsal root ganglia of the injured peripheral nerve in a rat neuropathic pain model. J. Comp. Neurol. 376, 241–252.

Ciruela, A., Dixon, A. K., Bramwell, S., Gonzalez, M. I., Pinnock, R. D., and Lee, K. 2003. Identification of MEK1 as a novel target for the treatment of neuropathic pain. Br. J. Pharmacol. 138, 751–756.

Cizkova, D., Lukacova, N., Marsala, M., and Marsala, J. 2002a. Neuropathic pain is associated with alterations of nitric oxide synthase immunoreactivity and catalytic activity in dorsal root ganglia and spinal dorsal horn. Brain Res. Bull. 58, 161–171.

Cizkova, D., Marsala, J., Lukacova, N., Marsala, M., Jergova, S., Orendacova, J., and Yaksh, T. L. 2002b. Localization of N-type Ca2+ channels in the rat spinal cord following chronic constrictive nerve injury. Exp. Brain Res. 147, 456–463.

Coderre, T. J. and Melzack, R. 1985. Increased pain sensitivity following heat injury involves a central mechanism. Behav. Brain Res. 15, 259–262.

Coderre, T. J. and Melzack, R. 1987. Cutaneous hyperalgesia: contributions of the peripheral and central nervous systems to the increase in pain sensitivity after injury. Brain Res. 404, 95–106.

Coderre, T. J. and Melzack, R. 1991. Central neural mediators of secondary hyperalgesia following heat injury in rats: neuropeptides and excitatory amino acids. Neurosci. Lett. 131, 71–74.

Coderre, T. J. and Melzack, R. 1992a. The contribution of excitatory amino acids to central sensitization and persistent nociception after formalin-induced tissue injury. J. Neurosci. 12, 3665–3670.

Coderre, T. J. and Melzack, R. 1992b. The role of NMDA receptor-operated calcium channels in persistent nociception after formalin-induced tissue injury. J. Neurosci. 12, 3671–3675.

Coderre, T. J., Katz, J., Vaccarino, A. L., and Melzack, R. 1993. Contribution of central neuroplasticity to pathological pain: review of clinical and experimental evidence. Pain 52, 259–285.

Coderre, T. J., Vaccarino, A. L., and Melzack, R. 1990. Central nervous system plasticity in the tonic pain response to subcutaneous formalin injection. Brain Res. 535, 155–158.

Collingridge, G. L., Isaac, J. T. R., and Wang, Y. T. 2004. Receptor trafficking and synaptic plasticity. Nat. Rev. Neurosci. 5, 952–961.

Colvin, L. A. and Duggan, A. W. 1998. Primary afferent-evoked release of immunoreactive galanin in the spinal cord of the neuropathic rat. Br. J. Anaesth. 81, 436–443.

Coull, J. A., Beggs, S., Boudreau, D., Boivin, D., Tsuda, M., Inoue, K., Gravel, C., Salter, M. W., and De Koninck, Y. 2005.

BDNF from microglia causes the shift in neuronal anion gradient underlying neuropathic pain. Nature 438, 1017–1021.

Coull, J. A., Boudreau, D., Bachand, K., Prescott, S. A., Nault, F., Sik, A., De Koninck, P., and De Koninck, Y. 2003. Trans-synaptic shift in anion gradient in spinal lamina I neurons as a mechanism of neuropathic pain. Nature 424, 938–942.

Coyle, D. E. 1998. Partial peripheral nerve injury leads to activation of astroglia and microglia which parallels the development of allodynic behavior. Glia 23, 75–83.

Cridland, R. A. and Henry, J. L. 1986. Comparison of the effects of substance P, neurokinin A, physalaemin and eledoisin in facilitating a nociceptive reflex in the rat. Brain Res. 381, 93–99.

Cumberbatch, M. J., Carlson, E., Wyatt, A., Boyce, S., Hill, R. G., and Rupniak, N. M. 1998. Reversal of behavioural and electrophysiological correlates of experimental peripheral neuropathy by the NK1 receptor antagonist GR205171 in rats. Neuropharmacology 37, 1535–1543.

Curtis, D. R. and Watkins, J. C. 1960. The excitation and depression of spinal neurones by structurally related amino acids. J. Neurochem. 6, 117–141.

Davar, G., Hama, A., Deykin, A., Vos, B., and Maciewicz, R. 1991. MK-801 blocks the development of thermal hyperalgesia in a rat model of experimental painful neuropathy. Brain Res. 553, 327–330.

Davies, S. N. and Lodge, D. 1987. Evidence for involvement of N-methylaspartate receptors in 'wind-up' of class 2 neurones in the dorsal horn of the rat. Brain Res. 424, 402–406.

Davies, J. and Watkins, J. C. 1983. Role of excitatory amino acid receptors in mono- and polysynaptic excitation in the cat spinal cord. Exp. Brain Res. 49, 280–290.

Davis, R. J. 1993. The mitogen-activated protein kinase signal transduction pathway. J. Biol. Chem. 268, 14553–14556.

Davis, J. B., Gray, J., Gunthorpe, M. J., Hatcher, J. P., Davey, P. T., Overend, P., Harries, M. H., Latcham, J., Clapham, C., Atkinson, K., Hughes, S. A., Rance, K., Grau, E., Harper, A. J., Pugh, P. L., Rogers, D. C., Bingham, S., Randall, A., and Sheardown, S. A. 2000. Vanilloid receptor-1 is essential for inflammatory thermal hyperalgesia. Nature 405, 183–187.

Decosterd, I. and Woolf, C. J. 2000. Spared nerve injury: an animal model of persistent peripheral neuropathic pain. Pain 87, 149–158.

Del Pozo, E., Caro, G., and Baeyens, J. M. 1987. Analgesic effects of several calcium channel blockers in mice. Eur. J. Pharmacol. 137, 155–160.

Delander, G. E., Schott, E., Brodin, E., and Fredholm, B. B. 1997. Spinal expression of mRNA for immediate early genes in a model of chronic pain. Acta Physiol. Scand. 161, 517–525.

Deng, Y. S., Zhong, J. H., and Zhou, X. F. 2000. Effects of endogenous neurotrophins on sympathetic sprouting in the dorsal root ganglia and allodynia following spinal nerve injury. Exp. Neurol. 164, 344–350.

Dickenson, A. H. and Sullivan, A. F. 1987. Evidence for a role of the NMDA receptor in the frequency dependent potentiation of deep rat dorsal horn nociceptive neurones following C fibre stimulation. Neuropharmacology 26, 1235–1238.

Dirig, D. M., Isakson, P. C., and Yaksh, T. L. 1998. Effect of COX-1 and COX-2 inhibition on induction and maintenance of carrageenan-evoked thermal hyperalgesia in rats. J. Pharmacol. Exp. Ther. 285, 1031–1038.

Dobremez, E., Bouali-Benazzouz, R., Fossat, P., Monteils, L., Dulluc, J., Nagy, F., and Landry, M. 2005. Distribution and regulation of L-type calcium channels in deep dorsal horn neurons after sciatic nerve injury in rats. Eur. J. Neurosci. 21, 3321–3333.

Dogrul, A., Ossipov, M. H., Lai, J., Malan, T. P., Jr., and Porreca, F. 2000. Peripheral and spinal antihyperalgesic activity of SIB-1757, a metabotropic glutamate receptor (mGLUR(5)) antagonist, in experimental neuropathic pain in rats. Neurosci. Lett. 292, 115–118.

Dolan, S. and Nolan, A. M. 2000. Behavioural evidence supporting a differential role for group I and II metabotropic glutamate receptors in spinal nociceptive transmission. Neuropharmacology 39, 1132–1138.

Dolan, S., Kelly, J. G., Monteiro, A. M., and Nolan, A. M. 2003. Up-regulation of metabotropic glutamate receptor subtypes 3 and 5 in spinal cord in a clinical model of persistent inflammation and hyperalgesia. Pain 106, 501–512.

Dolan, S., Kelly, J. G., Monteiro, A. M., and Nolan, A. M. 2004. Differential expression of central metabotropic glutamate receptor (mGluR) subtypes in a clinical model of post-surgical pain. Pain 110, 369–377.

Dong, Z. Q., Ma, F., Xie, H., Wang, Y. Q., and Wu, G. C. 2006. Down-regulation of GFRalpha-1 expression by antisense oligodeoxynucleotide attenuates electroacupuncture analgesia on heat hyperalgesia in a rat model of neuropathic pain. Brain Res. Bull. 69, 30–36.

Donnerer, J., Schuligoi, R., Stein, C., and Amann, R. 1993. Upregulation, release and axonal transport of substance P and calcitonin gene-related peptide in adjuvant inflammation and regulatory function of nerve growth factor. Regul Pept 46, 50–154.

Dougherty, P. M. and Willis, W. D. 1991. Enhancement of spinothalamic neuron responses to chemical and mechanical stimuli following combined microiontophoretic application of N-methyl-D-aspartic acid and substance P. Pain 47, 85–93.

Dougherty, P. M. and Willis, W. D. 1992. Enhanced responses of spinothalamic tract neurons to excitatory amino acids accompany capsaicin-induced sensitization in the monkey. J. Neurosci. 12, 883–894.

Dougherty, P. M., Palecek, J., Paleckova, V., Sorkin, L. S., and Willis, W. D. 1992a. The role of NMDA and non-NMDA excitatory amino acid receptors in the excitation of primate spinothalamic tract neurons by mechanical, thermal and electrical stimuli. J. Neurosci. 12, 3025–3041.

Dougherty, P. M., Sluka, K. A., Sorkin, K. N., Westlund, K. N., and Willis, W. D. 1992b. Neural changes in acute arthritis in monkeys. I. Parallel enhancement of responses of spinothalamic tract neurons to mechanical stimulation and excitatory amino acids. Brain Res. Rev. 17, 1–13.

Draganic, P., Miletic, G., and Miletic, V. 2001. Changes in post-tetanic potentiation of A-fiber dorsal horn field potentials parallel the development and disappearance of neuropathic pain after siatic nerve ligation in rats. Neurosci. Lett. 301, 127–130.

Draisci, G. and Iadarola, M. J. 1989. Temporal analysis of increases in c-fos, preprodynorphin and preproenkephalin mRNAs in rat spinal cord. Brain Res. Mol. Brain Res. 6, 31–37.

Dray, A. and Dickenson, A. 1991. Systemic capsaicin and olvanil reduce the acute algogenic and the late inflammatory phase following formalin injection into rodent paw. Pain 47, 79–83.

Drew, G. M., Siddall, P. J., and Duggan, A. W. 2004. Mechanical allodynia following contusion injury of the rat spinal cord is associated with loss of GABAergic inhibition in the dorsal horn. Pain 109, 379–388.

Dubner, R. and Ruda, M. A. 1992. Activity-dependent neuronal plasticity following tissue injury and inflammation. Trends Neurosci. 15, 96–103.

Duggan, A. W., Hendry, I. A., Mortom, C. R., Hutchison, W. D., and Zhao, Z. Q. 1988. Cutaneous stimuli releasing immunoreactive substance P in the dorsal horn of the cat. Brain Res. 451, 261–273.

Duggan, A. W., Hope, P. J., Jarrott, B., Schaible, H. -G., and Fleetwood-Walker, S. M. 1990. Release, spread and persistence of immunoreactive neurokinin A in the dorsal horn of the cat following noxious cutaneous stimulation. Studies with antibody microprobes. Neuroscience 35, 195–202.

Dumuis, A., Pin, J. -P., Oomagari, K., Sebben, M., and Bockaert, J. 1990. Arachidonic acid released from striatal neurons by joint stimulation of ionotropic and metabotropic quisqualate receptors. Nature 347, 182–184.

Dumuis, A., Sebben, M., Haynes, L., Pin, J. -P., and Bockaert, J. 1988. NMDA receptors activate the arachidonic acid cascade system in striatal neurons. Nature 336, 68–70.

Eaton, S. A. and Salt, T. E. 1990. Thalamic NMDA receptors and nociceptive sensory synaptic transmission. Neurosci. Lett. 110, 297–302.

Eaton, M. J., Blits, B., Ruitenberg, M. J., Verhaagen, J., and Oudega, M. 2002. Amelioration of chronic neuropathic pain after partial nerve injury by adeno-associated viral (AAV) vector-mediated over-expression of BDNF in the rat spinal cord. Gene Ther. 9, 1387–1395.

Eaton, M. J., Karmally, S., Martinez, M. A., Plunkett, J. A., Lopez, T., and Cejas, P. J. 1999. Lumbar transplant of neurons genetically modified to secrete galanin reverse pain-like behaviors after partial sciatic nerve injury. J. Peripher. Nerv. Syst. 4, 245–257.

Eaton, M. J., Plunkett, J. A., Karmally, S., Martinez, M. A., and Montanez, K. 1998. Changes in GAD- and GABA-immunoreactivity in the spinal dorsal horn after peripheral nerve injury and promotion of recovery by lumbar transplant of immortalized serotonergic precursors. J. Chem. Neuroanat. 16, 57–72.

Eckert, W. A., Valtschanoff, J. G., Otey, C. A., Rustioni, A., and Weinberg, R. J. 1994. Tyrosine phosphorylation in rat spinal cord after sciatic nerve transection. Neuroreport 5, 1289–1292.

Eliav, E., Herzberg, U., Ruda, M. A., and Bennet, G. J. 1999. Neuropathic pain from an experimental neuritis of the rat sciatic nerve. Pain 83, 169–182.

Eriksson, N. P., Persson, J. K., Svensson, M., Arvidsson, J., Molander, C., and Aldskogius, H. 1993. A quantitative analysis of the microglial cell reaction in central primary sensory projection territories following peripheral nerve injury in the adult rat. Exp. Brain Res. 96, 19–27.

Fang, L., Wu, J., Lin, Q., and Willis, W. D. 2002. Calcium-calmodulin-dependent protein kinase II contributes to spinal cord central sensitization. J. Neurosci. 22, 4196–4204.

Fang, L., Wu, J., Lin, Q., and Willis, W. D. 2003. Protein kinases regulate the phosphorylation of the GluR1 subunit of AMPA receptors of spinal cord in rats following noxious stimulation. Brain Res. Mol. Brain Res. 118, 160–165.

Fang, L., Wu, J., Zhang, X., Lin, Q., and Willis, W. D. 2005. Calcium/calmodulin dependent protein kinase II regulates the phosphorylation of cyclic AMP-responsive element-binding protein of spinal cord in rats following noxious stimulation. Neurosci. Lett. 374, 1–4.

Finnerup, N. B., Biering-Sorensen, F., Johannesen, I. L., Terkelsen, A. J., Juhl, G. I., Kristensen, A. D., Sindrup, S. H., Bach, F. W., and Jensen, T. S. 2005. Intravenous lidocaine relieves spinal cord injuriy pain: a randomized controlled trial. Anesthesiology 105, 1023–1030.

Fisher, K. and Coderre, T. J. 1996a. Comparison of nociceptive effects produced by intrathecal administration of mGluR agonists. Neuroreport 7, 2743–2747.

Fisher, K. and Coderre, T. J. 1996b. The contribution of metabotropic glutamate receptors (mGluRs) to formalin-induced nociception. Pain 68, 255–263.

Fisher, K. and Coderre, T. J. 1998. Hyperalgesia and allodynia induced by intrathecal (RS)-dihydroxyphenylglycine in rats. Neuroreport 9, 1169–1172.

Fisher, K., Fundytus, M. E., Cahill, C. M., and Coderre, T. J. 1998. Intrathecal administration of the mGluR compound, (S)-4CPG, attenuates hyperalgesia and allodynia associated with sciatic nerve constriction injury in rats. Pain 77, 59–66.

Fisher, K., Lefebvre, C. D., and Coderre, T. J. 2002. The effects of intrathecal administration of selective metabotropic glutamate receptor compounds in a rat model of neuropathic pain. Pharmacol. Biochem. Behav. 73, 411–418.

Fitzgerald, M. 1979. The spread of sensitization of polymodal nociceptors in the rabbit from nearby injury and by antidromic nerve stimulation. J. Physiol. 297, 207–216.

Fitzgerald, M. and Lynn, B. 1977. The sensitization of high threshold mechanoreceptors with myelinated axons by repeated heating. J. Physiol. 265, 549–563.

Flatters, S. J., Fox, A. J., and Dickenson, A. H. 2002. Nerve injury induces plasticity that results in spinal inhibitory effects of galanin. Pain 98, 249–258.

Frampton, J. E. and Foster, R. H. 2005. Pregabalin: in the treatment of postherpetic neuralgia. Drugs 65, 111–118; discussion 119–120.

Fukuoka, T., Kondo, E., Dai, Y., Hashimoto, N., and Noguchi, K. 2001. Brain-derived neurotrophic factor increases in the uninjured dorsal root ganglion neurons in selective spinal nerve ligation model. J. Neurosci. 21, 4891–4900.

Fundytus, M. E., Osborne, M. G., Henry, J. L., Coderre, T. J., and Dray, A. 2002. Antisense oligonucleotide knockdown of mGluR1 alleviates hyperalgesia and allodynia associated with chronic inflammation. Pharmacol. Biochem. Behav. 73, 401–410.

Fundytus, M. E., Yashpal, K., Chabot, J. G., Osborne, M. G., Lefebvre, C. D., Dray, A., Henry, J. L., and Coderre, T. J. 2001. Knockdown of spinal metabotropic glutamate receptor 1 (mGluR(1)) alleviates pain and restores opioid efficacy after nerve injury in rats. Br. J. Pharmacol. 132, 354–367.

Galan, A. and Cervero, F. 2005. Painful stimuli induce in vivo phosphorylation and membrane mobilization of mouse spinal cord NKCC1 co-transporter. Neuroscience 133, 245–252.

Galan, A., Laird, J. M., and Cervero, F. 2004. In vivo recruitment by painful stimuli of AMPA receptor subunits to the plasma membrane of spinal cord neurons. Pain 112, 315–323.

Galan, A., Lopez-Garcia, J. A., Cervero, F., and Laird, J. M. 2002. Activation of spinal extracellular signaling-regulated kinase-1 and -2 by intraplantar carrageenan in rodents. Neurosci. Lett. 322, 37–40.

Gammon, C. M., Allen, A. C. C., and Morell, P. 1989. Bradykinin stimulates phosphoinositide hydrolysis and mobilization of arachidonic acid in dorsal root ganglion neurons. J. Neurochem. 53, 95–101.

Gamse, R. and Saria, A. 1986. Nociceptive behavior after intrathecal injections of substance P, neurokinin A and calcitonin gene-related peptide in mice. Neurosci. Lett. 70, 143–147.

Gandhi, R., Ryals, J. M., and Wright, D. E. 2004. Neurotrophin-3 reverses chronic mechanical hyperalgesia induced by intramuscular acid injection. J. Neurosci. 24, 9405–9413.

Gardell, L. R., Ibrahim, M., Wang, R., Wang, Z., Ossipov, M. H., Malan, T. P., Jr., Porreca, F., and Lai, J. 2004. Mouse strains that lack spinal dynorphin upregulation after peripheral nerve injury do not develop neuropathic pain. Neuroscience 123, 43–52.

Garrison, C. J., Dougherty, P. M., Kajander, K. C., and Carlton, S. M. 1991. Staining of glial fibrillary acidic protein (GFAP) in lumbar spinal cord increases following a sciatic nerve constriction injury. Brain Res. 565, 1–7.

Garry, E. M., Moss, A., Delaney, A., O'Neill, F., Blakemore, J., Bowen, J., Husi, H., Mitchell, R., Grant, S. G., and Fleetwood-Walker, S. M. 2003. Neuropathic sensitization of behavioral reflexes and spinal NMDA receptor/CaM kinase II interactions are disrupted in PSD-95 mutant mice. Curr. Biol. 13, 321–328.

Garthwaite, J., Charles, S. L., and Chess-Williams, R. 1988. Endothelium-derived relaxing factors released on activation of NMDA receptors suggests role as intracellular messenger in the brain. Nature 336, 385–388.

Gaudet, A. D., Williams, S. J., Hwi, L. P., and Ramer, M. S. 2004. Regulation of TRPV2 by axotomy in sympathetic, but not sensory neurons. Brain Res. 1017, 155–162.

Gerber, G. and Randic, M. 1989. Participation of excitatory amino acid receptors in the slow excitatory synaptic transmission in the rat spinal cord in vitro. Neurosci. Lett. 106, 220–228.

Gerber, G., kangrga, I., Ryu, P-D., Larew, J. S., and Randic, M. 1989. Multiple effect of phorbol esters in the rat spinal dorsal horn. J. Neurosci. 9, 3606–3617.

Gerber, G., Zhong, J., Youn, D., and Randic, M. 2000. Group II and group III metabotropic glutamate receptor agonists depress synoptic transmission in the rat spinal cord dorsal horn. Neuroscience 100, 393–406.

Gereau, R. W. and Conn, P. J. 1994. Potentiation of cAMP responses by metabotropic glutamate receptors depresses excitatory synaptic transmission by a kinase-independent mechanism. Neuron 12, 1121–1129.

al-Ghoul, W. M., Volsi, G. L., Weinberg, R. J., and Rustioni, A. 1993. Glutamate immunocytochemistry in the dorsal horn after injury or stimulation of the sciatic nerve of rats. Brain Res. Bull. 30, 453–459.

Gilchrist, H. D., Allard, B. L., and Simone, D. A. 1996. Enhanced withdrawal responses to heat and mechanical stimuli following interplantar injection of capsaicin. Pain 67, 179–188.

Go, V. L. and Yaksh, T. L. 1987. Release of substance P from the cat spinal cord. J. Physiol. 391, 141–167.

Gold, M. S., Weinreich, D., Kim, C. S., Wang, R., Treanor, J., Porreca, F., and Lai, J. 2003. Redistribution of Na(V)1.8 in uninjured axons enables neuropathic pain. J. Neurosci. 23, 158–166.

Groth, R. and Aanonsen, L. 2002. Spinal brain-derived neurotrophic factor (BDNF) produces hyperalgesia in normal mice while antisense directed against either BDNF or trkB, prevent inflammation-induced hyperalgesia. Pain 100, 171–181.

Guenot, M., Bullier, J., Rospars, J. P., Lansky, P., Mertens, P., and Sindou, M. 2003. Single-unit analysis of the spinal dorsal horn in patients with neuropathic pain. J. Clin. Neurophysiol. 20, 143–150.

Guo, W., Zou, S., Guan, Y., Ikeda, T., Tal, M., Dubner, R., and Ren, K. 2002. Tyrosine phosphorylation of the NR2B subunit of the NMDA receptor in the spinal cord during the development and maintenance of inflammatory hyperalgesia. J. Neurosci. 22, 6208–6217.

Ha, S. O., Kim, J. K., Hong, H. S., Kim, D. S., and Cho, H. J. 2001. Expression of brain-derived neurotrophic factor in rat dorsal root ganglia, spinal cord and gracile nuclei in experimental models of neuropathic pain. Neuroscience 107, 301–309.

Hains, B. C., Klein, J. P., Saab, C. Y., Craner, M. J., Black, J. A., and Waxman, S. G. 2003. Upregulation of sodium channel Nav1.3 and functional involvement in neuronal hyperexcitability associated with central neuropathic pain after spinal cord injury. J. Neurosci. 23, 8881–8892.

Hains, B. C., Saab, C. Y., Klein, J. P., Craner, M. J., and Waxman, S. G. 2004. Altered sodium channel expression in second-order spinal sensory neurons contributes to pain after peripheral nerve injury. J. Neurosci. 24, 4832–4839.

Haley, J. E., Sullivan, A. F., and Dickenson, A. H. 1990. Evidence for spinal N-methyl-D-aspartate receptor involvement in prolonged chemical nociception in the rat. Brain Res. 518, 218–226.

Hama, A. T. 2003. Acute activation of the spinal cord metabotropic glutamate subtype-5 receptor leads to cold hypersensitivity in the rat. Neuropharmacology 44, 423–430.

Hardy, J. D. 1956. The nature of pain. J. Chronic. Dis. 4, 22–51.

Hardy, J. D., Wolff, H. G., and Goodell, H. 1950. Experimental evidence on the nature of cutaneous hyperalgesia. J. Clin. Invest. 29, 115–140.

Hashizume, H., DeLeo, J. A., Colburn, R. W., and Weinstein, J. N. 2000. Spinal glial activation and cytokine expression after lumbar root injury in the rat. Spine 25, 1206–1217.

Hedo, G., Laird, J. M., and Lopez-Garcia, J. A. 1999. Time-course of spinal sensitization following carrageenan-induced inflammation in the young rat: a comparative electrophysiological and behavioural study in vitro and in vivo. Neuroscience 92, 309–318.

Henken, D. B., Battisti, W. P., Chesselet, M. E., Murray, M., and Tessler, A. 1990. Expression of B-preprotachykinin mRNA and tachykinins in rat dorsal root ganglion cells following peripheral or central axotomy. Neuroscience 39, 733–742.

Henry, J. L. 1976. Effects of substance P on functionally identified units in cat spinal cord. Brain Res. 114, 439–451.

Herdegen, T., Leah, J. D., Manisali, A., Bravo, R., and Zimmermann, M. 1991a. c-JUN-like immunoreactivity in the CNS of the adult rat: basal and transynaptically induced expression of an immediate-early gene. Neuroscience 41, 643–654.

Herdegen, T., Tolle, T. R., Bravo, R., Zieglgansberger, W., and Zimmermann, M. 1991b. Sequential expression of JUN B, JUN D and FOS B proteins in rat spinal neurons: cascade of transcriptional operations during nociception. Neurosci. Lett. 129, 1–4.

Herdegen, T., Walker, T., Leah, J. D., Bravo, R., and Zimmermann, M. 1990. The KROX-24 protein, a new transcription regulating factor: expression in the rat central nervous system following afferent somatosensory stimulation. Neurosci. Lett. 120, 21–24.

Hofmann, H. A., Siegling, A., Denzer, D., Spreyer, P., and De Vry, J. 2001. Metabotropic glutamate mGluR1 receptor mRNA expression in dorsal root ganglia of rats after peripheral nerve injury. Eur. J. Pharmacol. 429, 135–138.

Hokfelt, T., Wiesenfeld-Hallin, X., Villar, M. J., and Melander, T. 1987. Increase of galanin-like immunoreactivity in rat dorsal root ganglion cells after peripheral axotomy. Neurosci. Lett. 83: 217–220.

Hollmann, M. and Heinemann, S. F. 1994. Cloned glutamate receptors. Annu. Rev. Neurosci. 17, 31–108.

Holthusen, H., Irsfeld, S., and Lipfert, P. 2000. Effect of pre- or post-traumatically applied i.v. lidocaine on primary and secondary hyperalgesia after experimental heat trauma in humans. Pain 88, 295–302.

Holtman, J. R., Jr. and Wala, E. P. 2005. Characterization of morphine-induced hyperalgesia in male and female rats. Pain 114, 62–70.

Hori, Y. and Endo, K. 1992. Miniature postsynaptic currents recorded from identified rat spinal dorsal horn projection neurons in thin-slice preparations. Neurosci. Lett. 142, 191–195.

Houamed, K. M., Kuijper, J. L., Gilbert, T. L., Haldeman, B. A., O'Hara, P. J., Mulvihill, E. R., Almers, W., and Hagen, F. S. 1991. Cloning, expression, and gene structure of a G protein-coupled glutamate receptor from rat brain. Science 252, 1318–1321.

Hua, X. Y., Chen, P., Marsala, M., and Yaksh, T. L. 1999a. Intrathecal substance P-induced thermal hyperalgesia and spinal release of prostaglandin E2 and amino acids. Neuroscience 89, 525–534.

Hua, X. Y., Chen, P., and Yaksh, T. L. 1999b. Inhibition of spinal protein kinase C reduces nerve injury-induced tactile allodynia in neuropathic rats. Neurosci. Lett. 276, 99–102.

Hua, X. Y., Saria, A., Gamse, R., Theodorsson-Norheim, E., Brodin, E., and Lundberg, J. M. 1986. Capsaicin-induced released of multiple tachykinins (substance P, neurokinin A and eledoisin-like material) from guinea-pig spinal cord and ureter. Neuroscience 19, 313–319.

Hudson, L. J., Bevan, S., McNair, K., Gentry, C., Fox, A., Kuhn, R., and Winter, J. 2002. Metabotropic receptor 5 upregulation in A-fibers after spinal nerve injury: 2-methyl-6-(phenylethynyl)-pyridine (MPEP) reverses the induced thermal hyperalgesia. J. Neurosci. 22, 2660–2668.

Hudson, L. J., Bevan, S., Wotherspoon, G., Gentry, C., Fox, A., and Winter, J. 2001. VR1 protein expression increases in undamaged DRG neurons after partial nerve injury. Eur. J. Neurosci. 13, 2105–2114.

Hunt, S. P., Pini, A., and Evan, G. 1987. Induction of c-fos-like protein in spinal cord neurones following sensory stimulation. Nature 328, 632–634.

Hygge-Blakeman, K., Brumovsky, P., Hao, J. X., Xu, X. J., Hokfelt, T., Crawley, J. N., and Wiesenfeld-Hallin, Z. 2004. Galanin over-expression decreases the development of neuropathic pain-like behaviors in mice after partial sciatic nerve injury. Brain Res. 1025, 152–158.

Hylden, J. L. and Wilcox, G. L. 1981. Intrathecal substance P elicits a caudally-directed biting and scratching behavior in mice. Brain Res. 217, 212–215.

Hylden, J. L. K., Nahin, R. L., Traub, R. J., and Dubner, R. 1989. Expansion of receptive fields of spinal lamina 1 projection neurones in rats with unilateral adjuvant-induced inflammation: the contribution of dorsal horn mechanisms. Pain 37, 229–243.

Ibuki, T., Matsumura, K., Yamazaki, Y., Nozaki, T., Tanaka, Y., and Kobayashi, S. 2003. Cyclooxygenase-2 is induced in the endothelial cells throughout the central nervous system during carrageenan-induced hind paw inflammation; its possible role in hyperalgesia. J. Neurochem. 86, 318–328.

Igwe, O. J. and Ning, L. 1994. Regulation of the second-messenger systems in the rat spinal cord during prolonged peripheral inflammation. Pain 58, 63–75.

Ikeda, H. and Murase, K. 2004. Glial nitric oxide-mediated long-term presynaptic facilitation revealed by optical imaging in rat spinal dorsal horn. J. Neurosci. 24, 9888–9896.

Imbe, H., Okamoto, K., Kadoya, T., Horie, H., and Senba, E. 2003. Galectin-1 is involved in the potentiation of neuropathic pain in the dorsal horn. Brain Res. 993, 72–83.

Inoue, K. 2006. The function of microglia through purinergic receptors: neuropathic pain and cytokine release. Pharmacol. Ther. 109, 210–226.

Inoue, K., Koizumi, S., Tsuda, M., and Shigemoto-Mogami, Y. 2003. Signaling of ATP receptors in glia-neuron interaction and pain. Life Sci. 74, 189–197.

Jacobsen, L. S., Olsen, A. K., Sjogren, P., and Jensen, N. H. 1995. Morphine-induced hyperalgesia, allodynia and myoclonus – new side-effects of morphine? Vgeskr Laeger 157, 3307–3310.

Janig, W., Levine, J. D., and Michaelis, M. 1996. Interactions of sympathetic and primary afferent neurons following nerve injury and tissue trauma. Prog. Brain Res. 113, 161–184.

Jaskolski, F., Coussen, F., and Mulle, C. 2005. Subcellular localization and trafficking of kainite receptors. Trends Pharmacol. Sci. 26, 20–26.

Jessell, T., Tsunoo, A., Kanazawa, I., and Otsuka, M. 1979. Substance P: depletion in the dorsal horn of the rat spinal cord after section of the peripheral processes of primary sensory neurons. Brain Res. 168, 247–259.

Ji, R. R., Baba, H., Brenner, G. J., and Woolf, C. J. 1999. Nociceptive-specific activation of ERK in spinal neurons contributes to pain hypersensitivity. Nat. Neurosci. 2, 1114–1119.

Ji, R. R., Befort, K., Brenner, G. J., and Woolf, C. J. 2002a. ERK/MAP kinase activation in superficial spinal cord neurons induces prodynorphin and NK-1 upregulation and contributes to persistent inflammatory pain hypersensitivity. J. Neurosci. 22, 478–485.

Ji, R. R., Samad, T. A., Jin, S. X., Schmoll, R., and Woolf, C. J. 2002b. p38 MAPK activation by NGF in primary sensory neurons after inflammation increases TRPV1 levels and maintains heat hyperalgesia. Neuron 36, 57–68.

Jongsma, H., Danielsen, N., Sundler, F., and Kanje, M. 2000. Alteration of PACAP distribution and PACAP receptor binding in the rat sensory nervous system following sciatic nerve transection. Brain Res. 853, 186–196.

Joseph, E. K. and Levine, J. D. 2004. Caspase signalling in neuropathic and inflammatory pain in the rat. Eur. J. Neurosci. 20, 2896–2902.

Kajander, K. C., Madsen, A. M., Iadarola, M. J., Draisci, G., and Wakisaka, S. 1996. Fos-like immunoreactivity increases in the lumbar spinal cord following a chronic constriction injury to the sciatic nerve of rat. Neurosci. Lett. 206, 9–12.

Karim, F., Wang, C. C., and Gereau, R. W. 2001. Metabotropic glutamate receptor subtypes 1 and 5 are activators of extracellular signal-regulated kinase signaling required for inflammatory pain in mice. J. Neurosci. 21, 3771–3779.

Kawamata, M. and Omote, K. 1996. Involvement of increased excitatory amino acids and intracellular Ca^{2+} concentration in the spinal dorsal horn in an animal model of neuropathic pain. Pain 68, 85–96.

Kawamata, M., Watanabe, H., Nishikawa, K., Takahashi, T., Kozuka, Y., Kawamata, T., Omote, K., and Namiki, A. 2002a. Different mechanisms of development and maintenance of experimental incision-induced hyperalgesia in human skin. Anesthesiology 97, 550–559.

Kawamata, M., Takahashi, T., Kozuka, Y., Nawa, Y., Nishikawa, K., Narimatsa, E., Watanabe, H., and Namiki, A. 2002b. Experimental incision-induced pain in human skin: effects of systemic lidocaine on flare formation and hyperalgesia. Pain 100, 77–89.

Kawasaki, Y., Kohno, T., Zhuang, Z. Y., Brenner, G. J., Wang, H., Van Der Meer, C., Befort, K., Woolf, C. J., and Ji, R. R. 2004. Ionotropic and metabotropic receptors, protein kinase A, protein kinase C, and Src contribute to C-fiber-induced ERK activation and cAMP response element-binding protein phosphorylation in dorsal horn neurons, leading to central sensitization. J. Neurosci. 24, 8310–8321.

Kayan, S., Woods, L. A., and Mitchell, C. L. 1971. Morphine-induced hyperalgesia in rats tested on the hot plate. J. Pharmacol. Exp. Ther. 177, 509–513.

Kelly, S. and Chapman, V. 2002. Effects of peripheral nerve injury on functional spinal VR1 receptors. Neuroreport 13, 1147–1150.

Kennedy, M. B., McGuinness, T., and Greengard, P. 1983. A calcium/calmodulin-dependent protein kinase from mammalian brain that phosphorylates Synapsin I: partial purification and characterization. J. Neurosci. 3, 818–831.

Kenshalo, D. R. Jr., Leonard, R. B., Chung, J. M., and Willis, W. D. 1979. Responses of primate spinothalamic neurons to graded and to repeated noxious heat stimuli. J. Neurophysiol. 42, 1370–1389.

Kenshalo, D. R., Jr., Leonard, R. B., Chung, J. M., and Willis, W. D. 1982. Facilitation of the responses of primate spinothalamic cells to cold and mechanical stimuli by noxious heating of the skin. Pain 12, 141–152.

Kerchner, G. A., Li, P., and Zhuo, M. 1999. Speaking out of turn: a role for silent synapses in pain. IUBMB Life 48, 251–256.

Kerr, B. J., Bradbury, E. J., Bennett, D. L., Trivedi, P. M., Dassan, P., French, J., Shelton, D. B., McMahon, S. B., and Thompson, S. W. 1999. Brain-derived neurotrophic factor modulates nociceptive sensory inputs and NMDA-evoked responses in the rat spinal cord. J. Neurosci. 19, 5138–5148.

Kerr, B. J., Cafferty, W. B., Gupta, Y. K., Bacon, A., Wynick, D., McMahon, S. B., and Thompson, S. W. 2000. Galanin knockout mice reveal nociceptive deficits following peripheral nerve injury. Eur. J. Neurosci. 12, 793–802.

Kerr, B. J., Gupta, Y., Pope, R., Thompson, S. W., Wynick, D., and McMahon, S. B. 2001a. Endogenous galanin potentiates spinal nociceptive processing following inflammation. Pain 93, 267–277.

Kerr, B. J., Thompson, S. W., Wynick, D., and McMahon, S. B. 2001b. Endogenous galanin is required for the full expression of central sensitization following peripheral nerve injury. Neuroreport 12, 3331–3334.

Khasabov, S. G., Rogers, S. D., Ghilardi, J. R., Peters, C. M., Mantyh, P. W., and Simone, D. A. 2002. Spinal neurons that possess the substance P receptor are required for the development of central sensitization. J. Neurosci. 22, 9086–9098.

Khasar, S. G., Gold, M. S., and Levine, J. D. 1998. A tetrodotoxin-resistant sodium current mediates inflammatory pain in the rat. Neurosci. Lett. 256, 17–20.

Kim, S. H. and Chung, J. M. 1992. An experimental model for peripheral neuropathy produced by segmental spinal nerve ligation in the rat. Pain 50, 355–363.

Kim, H. J., Back, S. K., Kim, J., Sung, B., Hong, S. K., and Na, H. S. 2003. Increases in spinal vasoactive intestinal polypeptide and neuropeptide Y are not sufficient for the genesis of neuropathic pain in rats. Neurosci. Lett. 342, 109–113.

Kim, S. Y., Bae, J. C., Kim, J. Y., Lee, H. L., Lee, K. M., Kim, D. S., and Cho, H. J. 2002. Activation of p38 MAP kinase in the rat dorsal root ganglia and spinal cord following peripheral inflammation and nerve injury. Neuroreport 13, 2483–2486.

Kim, D. S., Yoon, C. H., Lee, S. J., Park, S. Y., Yoo, H. J., and Cho, H. J. 2001. Changes in voltage-gated calcium channel alpha(1) gene expression in rat dorsal root ganglia following peripheral nerve injury. Brain. Res. Mol. Brain Res. 96, 151–156.

Knyihar-Csillik, E., Rakic, P., and Csillik, B. 1987. Transganglionic degenerative atrophy in the substantia gelatinosa of the spinal cord after peripheral nerve transection in rhesus monkeys. Cell Tissue Res. 247, 599–604.

Kocher, L., Anton, F., Reeh, P. W., and Handwerker, H. O. 1987. The effect of carrageenan-induced inflammation on the sensitivity of unmyelinated skin nociceptors in the rat. Pain 29, 363–373.

Kohno, T., Ji, R. R., Ito, N., Allchorne, A. J., Befort, K., Karchewski, L. A., and Woolf, C. J. 2005. Peripheral axonal injury results in reduced mu opioid receptor pre- and post-synaptic action in the spinal cord. Pain 117, 77–87.

Kohno, T., Moore, K. A., Baba, H., and Woolf, C. J. 2003. Peripheral nerve injury alters excitatory synaptic transmission in lamina II of the rat dorsal horn. J. Physiol. 548, 131–138.

Kolhekar, R. and Gebhart, G. F. 1994. NMDA and quisqualate modulation of visceral nociception in the rat. Brain Res. 651, 215–226.

Koltzenburg, M., Wall, P. D., and McMahon, S. B. 1999. Does the right side know what the left is doing? Trends Neurosci. 22, 122–127.

Komori, K., Nonaka, T., Okada, A., Kinoh, H., Hayashita-Kinoh, H., Yoshida, N., Yana, I., and Seiki, M. 2004. Absence of mechanical allodynia and Abeta-fiber sprouting after sciatic nerve injury in mice lacking membrane-type 5 matrix metalloproteinase. FEBS Lett. 557, 125–128.

Kontinen, V. K. and Dickenson, A. H. 2000. Effects of midazolam in the spinal nerve ligation model of neuropathic pain in rats. Pain 85, 425–431.

Kosai, K., Tateyama, S., Ikeda, T., Uno, T., Nishimori, T., and Takasaki, M. 2001. MK-801 reduces non-noxious stimulus-evoked Fos-like immunoreactivity in the spinal cord of rats with chronic constriction nerve injury. Brain Res. 910, 12–18.

Kovacs, G., Kocsis, P., Tarnow, J., Horvath, C., Szombathelyi, Z., and Farisas, S 2004. NR2B containing NMDA receptor dependent windup of single spinal neurons. Neuropharmacology 46, 23–30.

Krieger, P. and El Manira, A. 2002. Group II mGluR-mediated depression of sensory synaptic transmission. Brain Res. 937, 41–44.

Kristensen, J. D., Svensson, B., and Gordh, T., Jr. 1992. The NMDA-receptor antagonist CPP abolishes neurogenic 'wind-up pain' after intrathecal administration in humans. Pain 51, 249–253.

Kunz, S., Niederberger, E., Ehnert, C., Coste, O., Pfenninger, A., Kruip, J., Wendrich, T. M., Schmidtko, A., Tegeder, I., and Geisslinger, G. 2004. The calpain inhibitor MDL 28170 prevents inflammation-induced neurofilament light chain breakdown in the spinal cord and reduces thermal hyperalgesia. Pain 110, 409–418.

Kuraishi, Y., Hirota, N., Sato, Y., Hanashima, N., Takagi, H., and Satoh, M. 1989. Stimulus specificity of peripherally evoked substance P release from the rabbit dorsal horn in situ. Neuroscience 30, 241–250.

Kuraishi, Y., Hirota, N., Sato, Y., Hino, Y., Satoh, M., and Takagi, H. 1985. Evidence that substance P and somatostatin transmit separate information related to pain in the spinal dorsal horn. Brain Res. 325, 294–298.

Lai, J., Gold, M. S., Kim, C. S., Bian, D., Ossipov, M. H., Hunter, J. C., and Porreca, F. 2002. Inhibition of neuropathic pain by decreased expression of the tetrodotoxin-resistant sodium channel, NaV1.8. Pain 95, 143–152.

Laird, J. M. and Bennett, G. J. 1993. An electrophysiological study of dorsal horn neurons in the spinal cord of rats with an experimental peripheral neuropathy. J. Neurophysiol. 69, 2072–2085.

Laird, J. M., Garcia-Nicas, E., Delpire, E. J., and Cervero, F. 2004. Presynaptic inhibition and spinal pain processing in mice: a possible role of the NKCC1 cation-chloride co-transporter in hyperalgesia. Neurosci. Lett. 361, 200–203.

Lam, H. H., Hanley, D. F., Trapp, B. D., Saito, S., Raja, S., Dawson, T. M., and Yamaguchi, H. 1996. Induction of spinal cord neuronal nitric oxide synthase (NOS) after formalin injection in the rat hind paw. Neurosci. Lett. 210, 201–204.

LaMotte, R. H., Lundberg, L. E. R., and Torebjörk, H. E. 1992. Pain, hyperalgesia and activity in nociceptive C units in humans after intradermal injection of capsaicin. J. Physiol. 448, 749–764.

LaMotte, R. H., Shain, C. N., Simone, D. A., and Tsai, E. F. 1991. Neurogenic hyperalgesia: psychophysical studies of underlying mechanisms. J. Neurophysiol. 66, 190–211.

Lan, J. Y., Skeberdis, V. A., Jover, T., Grooms, S. Y., Lin, Y., Araneda, R. C., Zheng, X., Bennett, M. V., and Zukin, R. S. 2001. Protein kinase C modulates NMDA receptor trafficking and gating. Nat. Neurosci. 4, 382–390.

Ledeboer, A., Sloane, E. M., Milligan, E. D., Frank, M. G., Mahony, J. H., Maier, S. F., and Watkins, L. R. 2005. Minocycline attenuates mechanical allodynia and proinflammatory cytokine expression in rat models of pain facilitation. Pain 115, 71–83.

Lee, B. H., Yoon, Y. W., Chung, K., and Chung, J. M. 1998. Comparison of sympathetic sprouting in sensory ganglia in three animal models of neuropathic pain. Exp. Brain Res. 120, 432–438.

Leem, J. W., Choi, E. J., Park, E. S., and Paik, K. S. 1996. N-methyl-D-aspartate (NMDA) and non-NMDA glutamate receptor antagonists differentially suppress dorsal horn neuron responses to mechanical stimuli in rats with peripheral nerve injury. Neurosci. Lett. 211, 37–40.

Lekan, H. A., Chung, K., Yoon, Y. W., Chung, J. M., and Coggeshall, R. E. 1997. Loss of dorsal root ganglion cells concomitant with dorsal root axon sprouting following segmental nerve lesions. Neuroscience 81, 527–534.

Li, C. Y., Song, Y. H., Higuera, E. S., and Luo, Z. D. 2004. Spinal dorsal horn calcium channel alpha2delta-1 subunit upregulation contributes to peripheral nerve injury-induced tactile allodynia. J. Neurosci. 24, 8494–8499.

Li, J., Simone, D. A., and Larson, A. A. 1999. Windup leads to characteristics of central sensitization. Pain 79, 75–82.

Li, K. C. and Chen, J. 2003. Differential roles of spinal protein kinases C and A in development of primary heat and mechanical hypersensitivity induced by subcutaneous bee venom chemical injury in the rat. Neurosignals 12, 292–301.

Liao, D., Hessler, N. A., and Malinow, R. 1995. Activation of postsynaptically silent synapses during pairing-induced LTP in CA1 region of hippocampal slice. Nature 375, 400–404.

Liao, G. Y., Wagner, D. A., Hsu, M. H., and Leonard, J. P. 2001. Evidence for direct protein kinase-C mediated modulation of N-methyl-D-aspartate receptor current. Mol. Pharmacol. 59, 960–964.

Liaw, W. J., Stephens, R. L. Jr., Binns, B. C., Chu, Y., Sepkuty, J. P., Johns, R. A., Rothstein, J. D., and Tao, Y. X. 2005. Spinal glutamate uptake is critical for maintaining normal sensory transmission in rat spinal cord. Pain 115, 60–70.

Lim, G., Sung, B., Ji, R. R., and Mao, J. 2003. Upregulation of spinal cannabinoid-1-receptors following nerve injury enhances the effects of Win 55,212-2 on neuropathic pain behaviors in rats. Pain 105, 275–283.

Lim, G., Wang, S., and Mao, J. 2005. cAMP and protein kinase A contribute to the downregulation of spinal glutamate transporters after chronic morphine. Neurosci. Lett. 376, 9–13.

Lin, Q., Wu, J., and Willis, W. D. 2002. Effects of protein kinase a activation on the responses of primate spinothalamic tract neurons to mechanical stimuli. J. Neurophysiol. 88, 214–221.

Lindia, J. A., McGowan, E., Jochnowitz, N., and Abbadie, C. 2005. Induction of CX3CL1 expression in astrocytes and CX3CR1 in microglia in the spinal cord of a rat model of neuropathic pain. J. Pain 6, 434–438.

Liu, X. G. and Sandkuhler, J. 1995. Long-term potentiation of C-fiber-evoked potentials in the rat spinal dorsal horn is prevented by spinal N-methyl-D-aspartic acid receptor blockage. Neurosci. Lett. 191, 43–46.

Liu, X. and Sandkuhler, J. 1997. Characterization of long-term potentiation of C-fiber-evoked potentials in spinal dorsal horn of adult rat: essential role of NK1 and NK2 receptors. J. Neurophysiol. 78, 1973–1982.

Liu, X., Chung, K., and Chung, J. M. 1999. Ectopic discharges and adrenergic sensitivity of sensory neurons after spinal nerve injury. Brain Res. 849, 244–247.

Lozier, A. P. and Kendig, J. J. 1995. Long-term potentiation in an isolated peripheral nerve–spinal cord preparation. J. Neurophysiol. 74, 1001–1009.

Lukacova, N., Cizkova, D., Krizanova, O., Pavel, J., Marsala, M., and Marsala, J. 2003. Peripheral axotomy affects nicotinamide adenine dinucleotide phosphate diaphorase and nitric oxide synthases in the spinal cord of the rabbit. J. Neurosci. Res. 71, 300–313.

Luo, Z. D., Calcutt, N. A., Higuera, E. S., Valder, C. R., Song, Y. H., Svensson, C. I., and Myers, R. R. 2002. Injury type-specific calcium channel alpha 2 delta-1 subunit up-regulation in rat neuropathic pain models correlates with antiallodynic effects of gabapentin. J. Pharmacol. Exp. Ther. 303, 1199–1205.

Luo, Z. D., Chaplan, S. R., Higuera, E. S., Sorkin, L. S., Stauderman, K. A., Williams, M. E., and Yaksh, T. L. 2001. Upregulation of dorsal root ganglion (alpha)2(delta) calcium channel subunit and its correlation with allodynia in spinal nerve-injured rats. J. Neurosci. 21, 1868–1875.

Ma, Q. P. and Tian, L. 2001. A-fibres sprouting from lamina I into lamina II of spinal dorsal horn after peripheral nerve injury in rats. Brain Res. 904, 137–140.

Ma, Q. P. and Tian, L. 2002. Cholera toxin B subunit labeling in lamina II of spinal cord dorsal horn following chronic inflammation in rats. Neurosci. Lett. 327, 161–164.

Ma, W. and Quirion, R. 2001. Increased phosphorylation of cyclic AMP response element-binding protein (CREB) in the superficial dorsal horn neurons following partial sciatic nerve ligation. Pain 93, 295–301.

Ma, W. and Quirion, R. 2002. Partial sciatic nerve ligation induces increase in the phosphorylation of extracellular signal-regulated kinase (ERK) and c-Jun N-terminal kinase (JNK) in astrocytes in the lumbar spinal dorsal horn and the gracile nucleus. Pain 99, 175–184.

Ma, W., Du, W., and Eisenach, J. C. 2003a. Intrathecal lidocaine reverses tactile allodynia caused by nerve injuries and potentiates the antiallodynic effect of the COX inhibitor ketorolac. Anesthesiology 98, 203–208.

Ma, W., Hatzis, C., and Eisenach, J. C. 2003b. Intrathecal injection of cAMP response element binding protein (CREB) antisense oligonucleotide attenuates tactile allodynia caused by partial sciatic nerve ligation. Brain Res. 988, 97–104.

MacDermott, A. B., Mayer, M. L., Westbrook, G. L., Smith, S. J., and Barker, J. L. 1986. NMDA-receptor activation increases cytoplasmic calcium concentration in cultured spinal cord neurons. Nature 321, 519–522.

Madiai, F., Hussain, S. R., Goettl, V. M., Burry, R. W., Stephens, R. L., Jr., and Hackshaw, K. V. 2003. Upregulation of FGF-2 in reactive spinal cord astrocytes following unilateral lumbar spinal nerve ligation. Exp. Brain Res. 148, 366–376.

Maione, S., Siniscalco, D., Galderisi, U., de Novellis, V., Uliano, R., Di Bernardo, G., Berrino, L., Cascino, A., and Rossi, F. 2002. Apoptotic genes expression in the lumbar dorsal horn in a model neuropathic pain in rat. Neuroreport 13, 101–106.

Malan, T. P., Ossipov, M. H., Gardell, L. R., Ibrahim, M., Bian, D., Lai, J., and Porreca, F. 2000. Extraterritorial neuropathic pain correlates with multisegmental elevation of spinal dynorphin in nerve-injured rats. Pain 86, 185–194.

Malinow, R. and Malenka, R. C. 2002. AMPA receptor trafficking and synaptic plasticity. Annu. Rev. Neurosci. 25, 103–126.

Malmberg, A. B. and Yaksh, T. L. 1992a. Antinociceptive actions of spinal nonsteroidal anti-inflammatory agents on the formalin test in the rat. J. Pharmacol. Exp. Ther. 263, 136–146.

Malmberg, A. B. and Yaksh, T. L. 1992b. Hyperalgesia mediated by spinal glutamate or substance P receptor blocked by spinal cyclooxygenase inhibition. Science 257, 1276–1279.

Malmberg, A. B., Brandon, E. P., Idzerda, R. L., Liu, H., McKnight, G. S., and Basbaum, A. I. 1997a. Diminished inflammation and nociceptive pain with preservation of neuropathic pain in mice with a targeted mutation of the type I regulatory subunit of cAMP-dependent protein kinase. J. Neurosci. 17, 7462–7470.

Malmberg, A. B., Chen, C., Tonegawa, S., and Basbaum, A. I. 1997b. Preserved acute pain and reduced neuropathic pain in mice lacking PKCgamma. Science 278, 279–283.

Mannion, R. J., Doubell, T. P., Coggeshall, R. E., and Woolf, C. J. 1996. Collateral sprouting of uninjured primary afferent A-fibers into the superficial dorsal horn of the adult rat spinal cord after topical capsaicin treatment to the sciatic nerve. J. Neurosci. 16, 5189–5195.

Mantyh, P. W., Pinnock, R. D., Downes, C. P., Goedert, M., and Hunt, S. P. 1984. Correlation between inositol phospholipid hydrolysis and substance P receptors in rat CNS. Nature 309, 795–797.

Mao, J. and Mayer, D. J. 2001. Spinal cord neuroplasticity following repeated opioid exposure and its relation to pathological pain. Ann. N. Y. Acad. Sci. 933, 175–184.

Mao, J., Price, D. D., Coghill, R. C., Mayer, D. J., and Hayes, R. L. 1992a. Spatial patterns of spinal cord [^{14}C]-2-deoxyglucose metabolic activity in a rat model of painful peripheral mononeuropathy. Pain 50, 89–100. Erratum in: Pain 1992 51, 389.

Mao, J., Price, D. D., and Mayer, D. J. 1995a. Mechanisms of hyperalgesia and morphine tolerance: a current view of their possible interactions. Pain 62, 259–274.

Mao, J., Price, D. D., Mayer, D. J., and Hayes, R. L. 1992b. Pain-related increases in spinal cord membrane-bound protein kinase C following peripheral nerve injury. Brain Res. 588, 144–149.

Mao, J., Price, D. D., Mayer, D. J., Lu, J., and Hayes, R. L. 1992c. Intrathecal MK-801 and local nerve anesthesia synergistically reduce nociceptive behaviours in rats with experimental peripheral mononeuropathy. Brain Res. 576, 254–262.

Mao, J., Price, D. D., Phillips, L. L., Lu, J., and Mayer, D. J. 1995b. Increases in protein kinase C gamma immunoreactivity in the spinal dorsal horn of rats with painful mononeuropathy. Neurosci. Lett. 198, 75–78.

Mao, J., Price, D. D., Zhu, J., Lu, J., and Mayer, D. J. 1997. The inhibition of nitric oxide-activated poly(ADP-ribose) synthetase attenuates transsynaptic alteration of spinal cord dorsal horn neurons and neuropathic pain in the rat. Pain 72, 355–366.

Mao, J., Sung, B., Ji, R. R., and Lim, G. 2002a. Chronic morphine induces downregulation of spinal glutamate transporters: implications in morphine tolerance and abnormal pain sensitivity. J. Neurosci. 22, 8312–8323.

Mao, J., Sung, B., Ji, R. R., and Lim, G. 2002b. Neuronal apoptosis associated with morphine tolerance: evidence for an opioid-induced neurotoxic mechanism. J. Neurosci. 22, 7650–7661.

Mark, M. A., Colvin, L. A., and Duggan, A. W. 1998. Spontaneous release of immunoreactive neuropeptide Y from the central terminals of large diameter primary afferents of rats with peripheral nerve injury. Neuroscience 83, 581–589.

Martin, W. J., Cao, Y., and Basbaum, A. I. 2004. Characterization of wide dynamic range neurons in the deep dorsal horn of the spinal cord in preprotachykinin-a null mice in vivo. J. Neurophysiol. 91, 1945–1954.

Masu, M., Tanabe, Y., Tsuchida, K., Shigemoto, R., and Nakanishi, S. 1991. Sequence and expression of a metabotropic glutamate receptor. Nature 349, 760–765.

Matthews, E. A. and Dickenson, A. H. 2001. Effects of spinally delivered N- and P-type voltage-dependent calcium channel antagonists on dorsal horn neuronal responses in a rat model of neuropathy. Pain 92, 235–246.

Maurset, A., Skoglund, L. A., Hustveit, O., and Oye, I. 1989. Comparison of ketamine and pethidine in experimental and postoperative pain. Pain 36, 37–41.

Mayer, D. J., Mao, J., Holt, J., and Price, D. D. 1999. Cellular mechanisms of neuropathic pain, morphine tolerance, and their interactions. Proc. Natl. Acad. Sci. U. S. A. 96, 7731–7736.

McCall, W. D., Tanner, K. D., and Levine, J. D. 1996. Formalin induces biphasic activity in C-fibers in the rat. Neurosci. Lett. 208, 45–48.

McGraw, J., Gaudet, A. D., Oschipok, L. W., Kadoya, T., Horie, H., Steeves, J. D., Tetzlaff, W., and Ramer, M. S. 2005. Regulation of neuronal and glial galectin-1 expression by peripheral and central axotomy of rat primary afferent neurons. Exp. Neurol. 195, 103–114.

McLachlan, E. M., Janig, W., Devor, M., and Michaelis, M. 1993. Peripheral nerve injury triggers noradrenergic sprouting within dorsal root ganglia. Nature 363, 543–546.

McMahon, S. B. and Wall, P. D. 1984. Receptive fields of rat lamina 1 projection cells move to incorporate a nearby region of injury. Pain 19, 235–247.

McMahon, S. B., Armanini, M. P., Ling, L. H., and Phillips, H. S. 1994. Expression and coexpression of Trk receptors in subpopulations of adult primary sensory neurons projecting to identified peripheral targets. Neuron 12, 1161–1171.

Meller, S. T., Cummings, C. P., Traub, R. J., and Gebhart, G. F. 1994. The role of nitric oxide in the development and maintenance of the hyperalgesia produced by intraplanter injection of carrageenan in the rat. Neuroscience 60, 367–374.

Meller, S. T., Pechman, P. S., Gebhart, G. F., and Maues, T. J. 1992. Nitric oxide mediates the thermal hyperalgesia produced in a model of neuropathic pain in the rat. Neuroscience 50, 7–10.

Mendell, L. M. 1966. Physiological properties of unmyelinated fiber projections to the spinal cord. Exp. Neurol. 16, 316–332.

Mendell, L. M. and Wall, P. D. 1965. Responses of single dorsal cord cells to peripheral cutaneous unmyelinated fibres. Nature 206, 97–99.

Ménétrey, D. and Besson, J. M. 1982. Electrophysiological characteristics of dorsal horn cells in rats with cutaneous inflammation resulting from chronic arthritis. Pain 13, 343–364.

Ménétrey, D., Gannon, A., Levine, J. D., and Basbaum, A. I. 1989. Expression of c-fos protein in interneurons and projection neurons of the rat spinal cord in response to noxious somatic, articular, and visceral stimulation. J. Comp. Neurol. 285, 177–195.

Meyer, R. A. and Campbell, J. N. 1981. Myelinated nociceptive afferents account for the hyperalgesia that follows a burn to the hand. Science 213, 1527–1529.

Michael, G. J., Averill, S., Nitkunan, A., Rattray, M., Bennett, D. L., Yan, Q., and Priestley, J. V. 1997. Nerve growth factor treatment increases brain-derived neurotrophic factor selectively in TrkA-expressing dorsal root ganglion cells and in their central terminations within the spinal cord. J. Neurosci. 17, 8476–8490.

Miletic, G. and Miletic, V. 2002. Increases in the concentration of brain derived neurotrophic factor in the lumbar spinal dorsal horn are associated with pain behavior following chronic constriction injury in rats. Neurosci. Lett. 319, 137–140.

Miletic, G., Hanson, E. N., Savagian, C. A., and Miletic, V. 2004b. Protein kinase A contributes to sciatic ligation-associated early activation of cyclic AMP response element binding protein in the rat spinal dorsal horn. Neurosci. Lett. 360, 149–152.

Miletic, G., Miyabe, T., Gebhardt, K. J., and Miletic, V. 2005. Increased levels of Homer1b/c and Shank1a in the post-synaptic density of spinal dorsal horn neurons are associated with neuropathic pain in rats. Neurosci. Lett. 386, 189–193.

Miletic, G., Pankratz, M. T., and Miletic, V. 2002. Increases in the phosphorylation of cyclic AMP response element binding protein (CREB) and decreases in the content of calcineurin accompany thermal hyperalgesia following chronic constriction injury in rats. Pain 99, 493–500.

Miletic, V., Bowen, K. K., and Miletic, G. 2000. Loose ligation of the rat sciatic nerve is accompanied by changes in the subcellular content of protein kinase C beta II and gamma in the spinal dorsal horn. Neurosci. Lett. 288, 199–202.

Mills, C. D., Johnson, K. M., and Hulsebosch, C. E. 2002. Role of group II and group III metabotropic glutamate receptors in spinal cord injury. Exp. Neurol. 173, 153–167.

Millan, M. J. 1999. The induction of pain: an integrative review. Prog. Neurobiol. 57, 1–164.

Milligan, E. D., Twining, C., Chacur, M., Biedenkapp, J., O'Connor, K., Poole, S., Tracey, K., Martin, D., Maier, S. F., and Watkins, L. R. 2003. Spinal glia and proinflammatory cytokines mediate mirror-image neuropathic pain in rats. J. Neurosci. 23, 1026–1040.

Milligan, E. D., Zapata, V., Chacur, M., Schoeniger, D., Biedenkapp, J., O'Connor, K. A., Verge, G. M., Chapman, G., Green, P., Foster, A. C., Naeve, G. S., Maier, S. F., and Watkins, L. R. 2004. Evidence that exogenous and endogenous fractalkine can induce spinal nociceptive facilitation in rats. Eur. J. Neurosci. 20, 2294–2302.

Milligan, E. D., Zapata, V., Schoeniger, D., Chacur, M., Green, P., Poole, S., Martin, D., Maier, S. F., and Watkins, L. R. 2005. An initial investigation of spinal mechanisms underlying pain enhancement induced by fractalkine, a neuronally released chemokine. Eur. J. Neurosci. 22, 2775–2782.

Mills, C. D., Fullwood, S. D., and Hulsebosch, C. E. 2001. Changes in metabotropic glutamate receptor expression following spinal cord injury. Exp. Neurol. 170, 244–257.

Mills, C. D., Xu, G. Y., Johnson, K. M., McAdoo, D. J., and Hulsebosch, C. E. 2000. AIDA reduces glutamate release and attenuates mechanical allodynia after spinal cord injury. Neuroreport 11, 3067–3070.

Minami, T., Nishihara, I., Uda, R., Ito, S., Hyodo, M., and Hayaishi, O. 1994a. Characterization of EP-receptor subtypes involved in allodynia and hyperalgesia induced by intrathecal administration of prostaglandin E2 to mice. Br. J. Pharmacol. 112, 735–740.

Minami, T., Uda, R., Horiguchi, S., Ito, S., Hyodo, M., and Hayaishi, O. 1994b. Allodynia evoked by intrathecal administration of prostaglandin E2 to conscious mice. Pain 57, 217–223.

Miyabe, T. and Miletic, V. 2005. Multiple kinase pathways mediate the early sciatic ligation-associated activation of CREB in the rat spinal dorsal horn. Neurosci. Lett. 381, 80–85.

Mizushima, T., Obata, K., Yamanaka, H., Dai, Y., Fukuoka, T., Tokunaga, A., Mashimo, T., and Noguchi, K. 2005. Activation of p38 MAPK in primary afferent neurons by noxious stimulation and its involvement in the development of thermal hyperalgesia. Pain 113, 51–60.

Moochhala, S. M. and Sawynok, J. 1984. Hyperalgesia produced by intrathecal substance P and related peptides: desensitization and cross desensitization. Br. J. Pharmacol. 82, 381–388.

Moore, K. A., Kohno, T., Karchewski, L. A., Scholz, J., Baba, H., and Woolf, C. J. 2002. Partial peripheral nerve injury promotes a selective loss of GABAergic inhibition in the superficial dorsal horn of the spinal cord. J. Neurosci. 22, 6724–6731.

Morton, C. R. and Hutchison, W. D. 1990. Release of sensory neuropeptides in the spinal cord: studies with calcitonin gene-related peptide and galanin. Neuroscience 31, 807–815.

Morton, C. R., Hutchison, W. D., and Hendry, I. A. 1988. Release of immunoreactive somatostatin in the spinal dorsal horn of the cat. Neuropeptides 12, 189–197.

Muller, F., Heinke, B., and Sandkuhler, J. 2003. Reduction of glycine receptor-mediated miniature inhibitory postsynaptic currents in rat spinal lamina I neurons after peripheral inflammation. Neuroscience 122, 799–805.

Munglani, R. and Hunt, S. P. 1995. Molecular biology of pain. Br. J. Anaesth. 75, 186–192.

Murase, K. and Randic, M. 1984. Actions of substance P on rat spinal dorsal horn neurones. J. Physiol. 346, 203–217.

Murase, K., Randic, M., Ryu, P. D., and Usui, S. 1987. Substance P and Cholecystokinin Octapeptide Modify a Slow Inward Calcium-Sensitive Current Relaxation in Rat Spinal Cord Dorsal Horn Neurones. In: Fine Afferent Nerve Fibers and Pain (eds. R. F. Schmidt, H. G. Schaible, and C. Vahle-Hinz), pp. 265–271. VCH Verlagsgesellschaft.

Murase, K., Ryu, P. D., and Randic, M. 1989. Tachykinins modulate multiple ionic conductances in voltage-clamped rat spinal dorsal horn neurons. J. Neurophysiol. 61, 854–865.

Murray, C. W., Cowan, A., and Larson, A. A. 1991. Neurokinin and NMDA antagonists (but not a kainic acid antagonist) are antinociceptive in the mouse formalin model. Pain 44, 179–185.

Nachemson, A. K. and Bennett, G. J. 1993. Does pain damage spinal cord neurons? Transsynaptic degeneration in rat following a surgical incision. Neurosci. Lett. 162, 78–80.

Nagy, G. G., Al-Ayyan, M., Andrew, D., Fukaya, M., Watanabe, M., and Todd, A. J. 2004. Widespread expression of the AMPA receptor GluR2 subunit at glutamatergic synapses in the rat spinal cord and phosphorylation of GluR1 in response to noxious stimulation revealed with an antigen-unmasking method. J. Neurosci. 24, 5766–5777.

Naisbitt, S., Kim, E., Tu, J. C., Xiao, B., Sala, C., Valtschanoff, J., Weinberg, R. J., Worley, P. F., and Sheng, M. 1999. Shank, a novel family of postsynaptic density proteins that binds to the NMDA receptor/PSD-95/GKAP complex and cortactin. Neuron 23, 569–582.

Nakamura, S. and Myers, R. R. 1999. Myelinated afferents sprout into lamina II of L3-5 dorsal horn following chronic constriction nerve injury in rats. Brain Res. 818, 285–290.

Nakanishi, S. 1992. Molecular diversity of glutamate receptors and implications for brain function. Science 258, 579–603.

Naranjo, J. R., Mellstrom, B., Carrion, A. M., Lucas, J. J., Foulkes, N. S., and Sassone-Corsi, P. 1997. Peripheral noxious stimulation induces CREM expression in dorsal horn: involvement of glutamate. Eur. J. Neurosci. 9, 2778–2783.

Narita, M., Kuzumaki, N., Suzuki, M., Narita, M., Oe, K., Yamazaki, M., Yajima, Y., and Suzuki, T. 2004. Increased phosphorylated-mu-opioid receptor immunoreactivity in the mouse spinal cord following sciatic nerve ligation. Neurosci. Lett. 354, 148–152.

Näsström, J., Karlsson, U., and Post, C. 1992. Antinociceptive actions of different classes of excitatory amino acid receptor antagonists in mice. Eur. J. Pharmacol. 212, 21–29.

Nebe, J., Vanegas, H., Neugebauer, V., and Schaible, H. G. 1997. Omega-agatoxin IVA, a P-type calcium channel antagonist, reduces nociceptive processing in spinal cord neurons with input from the inflamed but not from the normal knee joint – an electrophysiological study in the rat in vivo. Eur. J. Neurosci. 9, 2193–2201.

Neugebauer, V. and Schaible, H. G. 1988. Peripheral and spinal components of the sensitization of spinal neurons during an acute experimental arthritis. Agents Actions 25, 234–236.

Neugebauer, V. and Schaible, H. G. 1990. Evidence for a central component in the sensitization of spinal neurons with joint input during development of acute arthritis in cat's knee. J. Neurophysiol. 64, 299–311.

Neugebauer, V., Chen, P. S., and Willis, W. D. 1999. Role of metabotropic glutamate receptor subtype mGluR1 in brief nociception and central sensitization of primate STT cells. J. Neurophysiol. 82, 272–282.

Neugebauer, V., Chen, P. S., and Willis, W. D. 2000. Groups II and III metabotropic glutamate receptors differentially modulate brief and prolonged nociception in primate STT cells. J. Neurophysiol. 84, 2998–3009.

Neugebauer, V., Lucke, T., and Schaible, H. G. 1994. Requirement of metabotropic glutamate receptors for the generation of inflammation-evoked hyperexcitability in rat spinal cord neurons. Eur. J. Neurosci. 6, 1179–1186.

Neumann, S., Doubell, T. P., Leslie, T., and Woolf, C. J. 1996. Inflammatory pain hypersensitivity mediated by phenotypic switch in myelinated primary sensory neurons. Nature 384, 360–364.

Nielsch, V., Bisby, M. A., and Keen, P. 1987. Effect of cutting or crushing the rat sciatic nerve on synthesis of substance P by isolated L5 dorsal root ganglia. Neuropeptides 10, 137–145.

Noguchi, K., De Leon, M., Nahin, R. L., Senba, E., and Ruda, M. A. 1993. Quantification of axotomy-induced alteration of neuropeptide mRNAs in dorsal root ganglion neurons with special reference to neuropeptide Y mRNA and the effects of neonatal capsaicin treatment. J. Neurosci. Res. 35, 54–66.

Noguchi, K., Kawai, Y., Fukuoka, T., Senba, E., and Miki, K. 1995. Substance P induced by peripheral nerve injury in primary afferent sensory neurons and its effect on dorsal column nucleus neurons. J. Neurosci. 15, 7633–7643.

Noguchi, K., Senba, E., Morita, Y., Sato, M., and Tohyama, M. 1989. Prepro-VIP and preprotachykinin mRNAs in the rat dorsal root ganglion cells following peripheral axotomy. Mol. Brain Res. 6, 327–330.

Noguchi, K., Senba, E., Morita, Y., Sato, M., and Tohyama, M. 1990. Alpha-CGRP and beta-CGRP mRNAs are differentially regulated in the rat spinal cord and dorsal root ganglion. Mol. Brain Res. 7, 299–304.

Novakovic, S. D., Kassotakis, L. C., Oglesby, I. B., Smith, J. A., Eglen, R. M., Ford, A. P., and Hunter, J. C. 1999. Immunocytochemical localization of P2X3 purinoceptors in sensory neurons in naive rats and following neuropathic injury. Pain 80, 273–282.

Novakovic, S. D., Tzoumaka, E., McGivern, J. G., Haraguchi, M., Sangameswaran, L., Gogas, K. R., Eglen, R. M., and Hunter, J. C. 1998. Distribution of the tetrodotoxin-resistant sodium channel PN3 in rat sensory neurons in normal and neuropathic conditions. J. Neurosci. 18, 2174–2187.

de Novellis, V., Siniscalco, D., Galderisi, U., Fuccio, C., Nolano, M., Santoro, L., Cascino, A., Roth, K. A., Rossi, F., and Maione, S. 2004. Blockade of glutamate mGlu5 receptors in a rat model of neuropathic pain prevents early over-expression of pro-apoptotic genes and morphological changes in dorsal horn lamina II. Neuropharmacology 46, 468–479.

Obata, K., Yamanaka, H., Dai, Y., Mizushima, T., Fukuoka, T., Tokunaga, A., and Noguchi, K. 2004a. Differential activation of MAPK in injured and uninjured DRG neurons following chronic constriction injury of the sciatic nerve in rats. Eur. J. Neurosci. 20, 2881–2895.

Obata, K., Yamanaka, H., Dai, Y., Tachibana, T., Fukuoka, T., Tokunaga, A., Yoshikawa, H., and Noguchi, K. 2003. Differential activation of extracellular signal-regulated protein kinase in primary afferent neurons regulates brain-derived neurotrophic factor expression after peripheral inflammation and nerve injury. J. Neurosci. 23, 4117–4126.

Obata, K., Yamanaka, H., Kobayashi, K., Dai, Y., Mizushima, T., Katsura, H., Fukuoka, T., Tokunaga, A., and Noguchi, K. 2004b. Role of mitogen-activated protein kinase activation in

injured and intact primary afferent neurons for mechanical and heat hypersensitivity after spinal nerve ligation. J. Neurosci. 24, 10211–10222.

Ohtori, S., Takahashi, K., Chiba, T., Yamagata, M., Sameda, H., and Moriya, H. 2001. Phenotypic inflammation switch in rats shown by calcitonin gene-related peptide immunoreactive dorsal root ganglion neurons innervating the lumbar facet joints. Spine 26, 1009–1013.

Ohtori, S., Takahashi, K., and Moriya, H. 2002. Inflammatory pain mediated by a phenotypic switch in brain-derived neurotrophic factor-immunoreactive dorsal root ganglion neurons innervating the lumbar facet joints in rats. Neurosci. Lett. 323, 129–132.

Oku, R., Nanayama, T., and Sato, M. 1988. Calcitonin gene-related peptide modulates calcium mobilization in synaptosomes of rat spinal dorsal horn. Brain Res. 475, 356–360.

Oku, R., Satoh, M., and Takagi, H. 1987. Release of substance P from the spinal dorsal horn is enhanced in polyarthritic rats. Neurosci. Lett. 74, 315–319.

Okuse, K., Chaplan, S. R., McMahon, S. B., Luo, Z. D., Calcutt, N. A., Scott, B. P., Akopian, A. N., and Wood, J. N. 1997. Regulation of expression of the sensory neuron-specific sodium channel SNS in inflammatory and neuropathic pain. Mol. Cell Neurosci. 10, 196–207.

Oliveira, A. L., Risling, M., Deckner, M., Lindholm, T., Langone, F., and Cullheim, S. 1997. Neonatal sciatic nerve transection induces TUNEL labeling of neurons in the rat spinal cord and DRG. Neuroreport 8, 2837–2840.

Osborne, M. G. and Coderre, T. J. 1999. Effects of intrathecal administration of nitric oxide synthase inhibitors on carrageenan-induced thermal hyperalgesia. Br. J. Pharmacol. 126, 1840–1846.

Ossipov, M. H., Lopez, Y., Nichols, M. L., Bian, D., and Porreca, F. 1995. Inhibition by spinal morphine of the tail-flick response is attenuated in rats with nerve ligation injury. Neurosci. Lett. 199, 83–86.

Palecek, J., Paleckova, V., Dougherty, P. M., and Willis, W. D. 1994. The effect of trans-ACPD, a metabotropic excitatory amino acid receptor agonist, on the responses of primate spinothalamic tract neurons. Pain 56, 261–269.

Parra, M. C., Nguyen, T. N., Hurley, R. W., and Hammond, D. L. 2002. Persistent inflammatory nociception increases levels of dynorphin 1-17 in the spinal cord, but not in supraspinal nuclei involved in pain modulation. J. Pain 3, 330–336.

Pedersen, J. L. and Kehlet, H. 1998. Secondary hyperalgesia to heat stimuli after burn injury in man. Pain 76, 377–384.

Perl, E. R., Kumazawa, T., Lynn, B., and Kenins, P. 1976. Sensitization of high threshold receptors with unmyelinated (C) afferent fibers. Prog. Brain Res. 43, 263–277.

Pertovaara, A. 1998. A neuronal correlate of secondary hyperalgesia in the rat spinal dorsal horn is submodality selective and facilitated by supraspinal influence. Exp. Neurol. 149, 193–202.

Pezet, S., Malcangio, M., Lever, I. J., Perkinton, M. S., Thompson, S. W., Williams, R. J., and McMahon, S. B. 2002. Noxious stimulation induces Trk receptor and downstream ERK phosphorylation in spinal dorsal horn. Mol. Cell Neurosci. 21, 684–695.

Pin, J. P. and Duvoisin, R. 1995. The metabotropic glutamate receptors: structure and functions. Neuropharmacology 34, 1–26.

Pin, J. P., Waeber, C., Prezeau, L., Bockaert, J., and Heinemann, S. F. 1992. Alternative splicing generates metabotropic glutamate receptors inducing different patterns of calcium release in Xenopus oocytes. Proc. Natl. Acad. Sci. U. S. A. 89, 10331–10335.

Pitcher, G. M. and Henry, J. L. 2000. Cellular mechanisms of hyperalgesia and spontaneous pain in a spinalized rat model of peripheral neuropathy: changes in myelinated afferent inputs implicated. Eur. J. Neurosci. 12, 2006–2020.

Plunkett, J. A., Yu, C. G., Easton, J. M., Bethea, J. R., and Yezierski, R. P. 2001. Effects of interleukin-10 (IL-10) on pain behavior and gene expression following excitotoxic spinal cord injury in the rat. Exp Neurol 168, 144–154.

Polgar, E., Hughes, D. I., Arham, A. Z., and Todd, A. J. 2005. Loss of neurons from laminas I-III of the spinal dorsal horn is not required for development of tactile allodynia in the spared nerve injury model of neuropathic pain. J. Neurosci. 25, 6658–6666.

Pollock, G., Pennypacker, K. R., Memet, S., Israel, A., and Saporta, S. 2005. Activation of NF-kappaB in the mouse spinal cord following sciatic nerve transection. Exp. Brain Res. 165, 470–477.

Polomano, R. C., Mannes, A. J., Clark, U. S., and Bennett, G. J. 2000. A painful peripheral neuropathy in the rat produced by the chemotherapeutic drug, paclitaxel. Pain 94, 293–304.

Ponting, C. P., Phillips, C., Davies, K. E., and Blake, D. J. 1997. PDZ domains: targeting signalling molecules to sub-membranous sites. Bioessays 19, 469–479.

Porreca, F., Lai, J., Bian, D., Wegert, S., Ossipov, M. H., Eglen, R. M., Kassotakis, L., Novakovic, S., Rabert, D. K., Sangameswaran, L., and Hunter, J. C. 1999. A comparison of the potential role of the tetrodotoxin-insensitive sodium channels, PN3/SNS and NaN/SNS2, in rat models of chronic pain. Proc. Natl. Acad. Sci. U. S. A. 96, 7640–7644. Erratum in: Proc. Natl. Acad. Sci. U. S. A. 96, 10548.

Presley, R. W., Menétrey, D., Levine, J. D., and Basbaum, A. I. 1990. Systemic morphine supresses noxious stimulus-evoked Fos protein-like immunoreactivity in the rat spinal cord. J. Neurosci. 10, 323–335.

Price, D. D., Hayes, R. L., Ruda, M., and Dubner, R. 1978. Spatial and temporal transformations of input to spinothalamic tract neurons and their relation to somatic sensations. J. Neurophysiol. 41, 933–946.

Price, T. J., Cervero, F., and de Koninck, Y. 2005. Role of cation-chloride-cotransporters (CCC) in pain and hyperalgesia. Curr. Top. Med. Chem. 5, 547–555.

Priest, B. T., Murphy, B. A., Lindia, J. A., Diaz, C., Abbadie, C., Ritter, A. M., Liberator, P., Iyer, L. M., Kash, S. F., Kohler, M. G., Kaczorowski, G. J., MacIntyre, D. E., and Martin, W. J. 2005. Contribution of the tetrodotoxin-resistant voltage-gated sodium channel NaV1.9 to sensory transmission and nociceptive behavior. Proc. Natl. Acad. Sci. U. S. A. 102, 9382–9387.

Raghavendra, V., Tanga, F. Y., and DeLeo, J. A. 2004. Complete Freund's adjuvant-induced peripheral inflammation evokes glial activation and proinflammatory cytokine expression in the CNS. Eur. J. Neurosci. 20, 467–473.

Raja, S. N., Campbell, J. N., and Meyer, R. A. 1984. Evidence for different mechanisms of primary and secondary hyperalgesia following heat injury to the glabrous skin. Brain 107, 1179–1188.

Ramer, M. S. and Bisby, M. A. 1997. Rapid sprouting of sympathetic axons in dorsal root ganglia of rats with a chronic constriction injury. Pain 70, 237–244.

Ramer, M. S., French, G. D., and Bisby, M. A. 1997. Wallerian degeneration is required for both neuropathic pain and sympathetic sprouting into the DRG. Pain 72, 71–78.

Ramer, M. S., Murphy, P. G., Richardson, P. M., and Bisby, M. A. 1998. Spinal nerve lesion-induced mechanoallodynia and adrenergic sprouting in sensory ganglia are attenuated in interleukin-6 knockout mice. Pain 78, 115–121.

Randic, M. and Miletic, V. 1977. Effect of substance P in cat dorsal horn neurones activated by noxious stimuli. Brain Res. 128, 164–169.

Randic, M., Jiang, M. C., and Cerne, R. 1993. Long-term potentiation and long-term depression of primary afferent neurotransmission in the rat spinal cord. J. Neurosci. 13, 5228–5241.

Ren, K., Hylden, J. L. K., Williams, G. M., Ruda, M. A., and Dubner, R. 1992. The effects of a non-competitive NMDA receptor antagonist, MK-801, on behavioral hyperalgesia and dorsal horn neuronal activity in rats with unilateral inflammation. Pain 50, 331–344.

Ro, L. S., Li, H. Y., Huang, K. F., and Chen, S. T. 2004. Territorial and extra-territorial distribution of Fos protein in the lumbar spinal dorsal horn neurons in rats with chronic constriction nerve injuries. Brain Res. 1004, 177–187.

Roche, K. W., Tu, J. C., Petralia, R. S., Xiao, B., Wenthold, R. J., and Worley, P. F. 1999. Homer 1b regulates the trafficking of group I metabotropic glutamate receptors. J. Biol. Chem. 274, 25953–25957.

Rohde, D. S., Detweiler, D. J., and Basbaum, A. I. 1996. Spinal cord mechanisms of opioid tolerance and dependence: fos-like immunoreactivity increases in subpopulations of spinal cord neurons during withdrawal. Neuroscience 72, 233–242.

Rygh, L. J., Svendsen, F., Hole, K., and Tjolsen, A. 1999. Natural noxious stimulation can induce long-term increase of spinal nociceptive responses. Pain 82, 305–310.

Ryu, P. D., Gerber, G., Murase, K., and Randic, M. 1988. Calcitonin gene related peptide enhances calcium current of rat dorsal root ganglion neurons and spinal excitatory synaptic transmission. Neurosci. Lett. 89, 305–312.

Saegusa, H., Kurihara, T., Zong, S., Kazuno, A., Matsuda, Y., Nonaka, T., Han, W., Toriyama, H., and Tanabe, T. 2001. Suppression of inflammatory and neuropathic pain symptoms in mice lacking the N-type Ca^{2+} channel. EMBO J. 20, 2349–2356.

Saegusa, H., Kurihara, T., Zong, S., Minowa, O., Kazuno, A., Han, W., Matsuda, Y., Yamanaka, H., Osanai, M., Noda, T., and Tanabe, T. 2000. Altered pain responses in mice lacking alpha 1E subunit of the voltage-dependent Ca^{2+} channel. Proc. Natl. Acad. Sci. U. S. A. 97, 6132–6137.

Saitoh, T., Ishida, M., and Shinozaki, H. 1998. Potentiation by DL-alpha-aminopimelate of the inhibitory action of a novel mGluR agonist (L-F2CCG-I) on monosynaptic excitation in the rat spinal cord. Br. J. Pharmacol. 123, 771–779.

Sammons, M. J., Raval, P., Davey, P. T., Rogers, D., Parsons, A. A., and Bingham, S. 2000. Carrageenan-induced thermal hyperalgesia in the mouse: role of nerve growth factor and the mitogen-activated protein kinase pathway. Brain Res. 876, 48–54.

Sang, C. N., Gracely, R. H., Max, M. B., and Bennett, G. J. 1996. Capsaicin-evoked mechanical allodynia and hyperalgesia cross nerve territories. Evidence for a central mechanism. Anesthesiology 85, 491–496.

Saria, A., Gamse, R., Petermann, J., Fischer, J. A., Theorodorsson-Norheim, E., and Lundeberg, J. M. 1986. Stimultaneous release of several tachykinins and calcitonin gene-related peptide from rat spinal cord. Neurosci. Lett. 63, 310–314.

Sato, E., Takano, Y., Kuno, Y., Takano, M., and Sato, I. 2003. Involvement of spinal tyrosine kinase in inflammatory and N-methyl-D-aspartate-induced hyperalgesia in rats. Eur. J. Pharmacol. 468, 191–198.

Schaible, H.-G., Grubb, B. D., Neugebauer, V., and Oppmann, M. 1991. The effects of NMDA antagonists on neuronal activity in cat spinal cord evoked by acute inflammation in the knee joint. Eur. J. Neurosci. 3, 981–991.

Schmidtko, A., Ruth, P., Geisslinger, G., and Tegeder, I. 2003. Inhibition of cyclic guanosine 5′-monophosphate-dependent protein kinase I (PKG-I) in lumbar spinal cord reduces formalin-induced hyperalgesia and PKG upregulation. Nitric Oxide 8, 89–94.

Schneider, S. P. and Perl, E. R. 1988. Comparison of primary afferent and glutamate excitation of neurons in the mammalian spinal dorsal horn. J. Neurosci. 8, 2062–2073.

Schoepp, D. D., Johnson, B. G., Salhoff, C. R., Valli, M. J., Desai, M. A., Burnett, J. P., Mayne, N. G., and Monn, J. A. 1995. Selective inhibition of forskolin-stimulated cyclic AMP formation in rat hippocampus by a novel mGluR agonist, 2R, 4R-4-amino-pyrrolidine-2,4- dicarboxylate. Neuropharmacology 34, 843–850.

Scholz, J., Broom, D. C., Youn, D. H., Mills, C. D., Kohno, T., Suter, M. R., Moore, K. A., Decosterd, I., Coggeshall, R. E., and Woolf, C. J. 2005. Blocking caspase activity prevents transsynaptic neuronal apoptosis and the loss of inhibition in lamina II of the dorsal horn after peripheral nerve injury. J. Neurosci. 25, 7317–7323.

Scott, J. D. 1991. Cyclic nucleotide-dependent protein kinases. Pharmacol. Ther. 50, 123–145.

Sekiguchi, Y., Kikuchi, S., Myers, R. R., and Campana, W. M. 2003. ISSLS prize winner: erythropoietin inhibits spinal neuronal apoptosis and pain following nerve root crush. Spine 28, 2577–2584.

Seltzer, Z., Cohn, S., Ginzburg, R., and Beilin, B. Z. 1991. Modulation of neuropathic pain behavior in rats by spinal disinhibition and NMDA receptor blockade of injury discharge. Pain 45, 69–75.

Seltzer, Z., Dubner, R., and Shir, Y. 1990. A novel behavioral model of neuropathic pain disorders produced in rats by partial sciatic nerve injury. Pain 43, 205–218.

Seybold, V. S., Hylden, J. K. L., and Wilcox, G. L. 1982. Intrathecal substance P and somatostatin in rats: behaviors indicative of sensation. Peptides 3, 49–54.

Seybold, V. S., Jia, Y. P., and Abrahams, L. G. 2003. Cyclo-oxygenase-2 contributes to central sensitization in rats with peripheral inflammation. Pain 105, 47–55.

Shehab, S. A. and Atkinson, M. E. 1986. Vasoactive intestinal polypeptide (VIP) increases in the spinal cord after peripheral axotomy of the sciatic nerve originate from primary afferent neurons. Brain Res. 372, 37–44.

Shehab, S. A., Spike, R. C., and Todd, A. J. 2003. Evidence against cholera toxin B subunit as a reliable tracer for sprouting of primary afferents following peripheral nerve injury. Brain Res. 964, 218–227.

Shen, J., Benedict Gomes, A., Gallagher, A., Stafford, K., and Yoburn, B. C. 2000. Role of cAMP-dependent protein kinase (PKA) in opioid agonist-induced mu-opioid receptor downregulation and tolerance in mice. Synapse 38, 322–327.

Sher, G. and Mitchell, D. 1990. N-methyl-D-aspartate receptors mediate responses of rat dorsal horn neurons to hindlimb ischemia. Brain Res. 522, 55–62.

Shigeri, Y., Seal, R. P., and Shimamoto, K. 2004. Molecular pharmacology of glutamate transporters, EAATs and VGLUTs. Brain Res. Brain Res. Rev. 45, 250–265.

Shortland, P. and Woolf, C. J. 1993. Chronic peripheral nerve section results in a rearrangement of the central axonal arborizations of axotomized A beta primary afferent neurons in the rat spinal cord. J. Comp. Neurol. 330, 65–82.

Shortland, P., Kinman, E., and Molander, C. 1997. Sprouting of A-fibre primary afferents into lamina II in two rat models of neuropathic pain. Eur. J. Pain 1, 215–227.

Siegan, J. B., Hama, A. T., and Sagen, J. 1996. Alterations in rat spinal cord cGMP by peripheral nerve injury and adrenal medullary transplantation. Neurosci. Lett. 215, 49–52.

Simmons, R. M. A., Webster, A. A., Kalra, A. B., and Iyenger, S. 2002. Group II mGluR receptor agonists are effective I persistent and neuropathic pain models in rats. Pharmacol. Biochem. Behav. 73, 419–427.

Simone, D. A., Baumann, T. K., Collins, J. G., and LaMotte, R. H. 1989. Sensitization of cat dorsal horn neurons to innocuous mechanical stimulation after intradermal injection of capsaicin. Brain Res. 486, 185–189.

Skeberdis, V. A., Lan, J., Opitz, T., Zheng, X., Bennett, M. V., and Zukin, R. S. 2001. mGluR1-mediated potentiation of NMDA receptors involves a rise in intracellular calcium and activation of protein kinase C. Neuropharmacology 40, 856–865.

Skilling, S. R., Smullin, D. H., and Larson, A. A. 1988. Extracellular amino acid concentrations in the dorsal spinal cord of freely moving rats following veratridine and nociceptive stimulation. J. Neurochem. 51, 127–132.

Sluka, K. A. 1998. Blockade of N- and P/Q-type calcium channels reduces the secondary heat hyperalgesia induced by acute inflammation. J. Pharmacol. Exp. Ther. 287, 232–237.

Sluka, K. A. and Westlund, K. N. 1993. Behavioral and immunohistochemical changes in an experimental arthritis model in rats. Pain 55, 367–377.

Sluka, K. A. and Willis, W. D. 1997. The effects of G-protein and protein kinase inhibitors on the behavioral responses of rats to intradermal injection of capsaicin. Pain 71, 165–178.

Sluka, K. A., Dougherty, P. M., Sorkin, L. S., Willis, W. D., and Westlund, K. N. 1992. Neural changes in acute arthritis in monkeys III. Changes in substance P, calcitonin gene-related peptide and glutamate in the dorsal horn of the spinal cord. Brain Res. Rev. 17, 29–38.

Smith, R. R., Martin-Schild, S., Kastin, A. J., and Zadina, J. E. 2001. Decreases in endomorphin-2-like immunoreactivity concomitant with chronic pain after nerve injury. Neuroscience 105, 773–778.

Soares, S., von Boxberg, Y., Lombard, M. C., Ravaille-Veron, M., Fischer, I., Eyer, J., and Nothias, F. 2002. Phosphorylated MAP1B is induced in central sprouting of primary afferents in response to peripheral injury but not in response to rhizotomy. Eur. J. Neurosci. 16, 593–606.

Sokal, D. M. and Chapman, V. 2003. Effects of spinal administration of muscimol on C- and A-fibre evoked neuronal responses of spinal dorsal horn neurones in control and nerve injured rats. Brain Res. 962, 213–220.

Somers, D. L. and Clemente, F. R. 2002. Dorsal horn synaptosomal content of aspartate, glutamate, glycine and GABA are differentially altered following chronic constriction injury to the rat sciatic nerve. Neurosci. Lett. 323, 171–174.

Sorkin, L. S., Westlund, K. N., Sluka, K. A., Dougherty, P. M., and Willis, W. D. 1992. Neural changes in acute arthritis in monkeys. IV. Time course of amino acid release into the lumbar dorsal horn. Brain Res. Rev. 17, 39–50.

Sotgiu, M. L., Bellomi, P., and Biella, G. E. 2003. The mGluR5 selective antagonist 6-methyl-2-(phenylethynyl)-pyridine reduces the spinal neuron pain-related activity in mononeuropathic rats. Neurosci. Lett. 342, 85–88.

Sotgiu, M. L., Biella, G., and Riva, L. 1995. Poststimulus afterdischarges of spinal WDR and NS units in rats with chronic nerve constriction. Neuroreport 6, 1021–1024.

Sotgiu, M. L., Lacerenza, M., and Marchettini, P. 1992. Effect of systemic lidocaine on dorsal horn neuron hyperactivity following chronic peripheral nerve injury in rats. Somatosens. Mot. Res. 9, 227–233.

Southam, E., East, S. J., and Garthwaite, J. 1991. Excitatory amino acid receptors coupled to the nitric oxide: cyclic GMP pathway in rat cerebellum during development. J. Neurochem. 56, 2072–2081.

Stiller, C. O., Cui, J. G., O'Connor, W. T., Brodin, E., Meyerson, B. A., and Linderoth, B. 1996. Release of gamma-aminobutyric acid in the dorsal horn and suppression of tactile allodynia by spinal cord stimulation in mononeuropathic rats. Neurosurgery 39, 367–374.

Steel, J. H., Terenghi, G., Chung, J. M., Na, H. S., Carlton, S. M., and Polak, J. M. 1994. Increased nitric oxide synthase immunoreactivity in rat dorsal root ganglia in a neuropathic pain model. Neurosci. Lett. 169, 81–84.

Sugimoto, T., Bennett, G. J., and Kajander, K. C. 1990. Transsynaptic degeneration in the superficial dorsal horn after sciatic nerve injury: effects of a chronic constriction injury, transection, and strychnine. Pain 42, 205–213.

Sugiyama, H., Ito, I., and Hirono, C. 1987. A new type of glutamate receptor linked to inositol phospholipid metabolism. Nature 325, 531–533.

Sun, X. and Larson, A. A. 1991. Behavioral sensitization to kainic acid and quisqualic acid in mice: comparison to NMDA and substance P responses. J. Neurosci. 11, 3111–3123.

Sun, H., Ren, K., Zhong, C. M., Ossipov, M. H., Malan, T. P., Lai, J., and Porreca, F. 2001. Nerve injury-induced tactile allodynia is mediated via ascending spinal dorsal column projections. Pain 90, 105–111.

Sung, B., Lim, G., and Mao, J. 2003. Altered expression and uptake activity of spinal glutamate transporters after nerve injury contribute to the pathogenesis of neuropathic pain in rats. J. Neurosci. 23, 2899–2910.

Suzuki, R. and Dickenson, A. H. 2006. Differential pharmacological modulation of the spontaneous stimulus-independent activity in the rat spinal cord following peripheral nerve injury. Exp. Neurol. 198, 72–80.

Suzuki, R., Chapman, V., and Dickenson, A. H. 1999. The effectiveness of spinal and systemic morphine on rat dorsal horn neuronal responses in the spinal nerve ligation model of neuropathic pain. Pain 80, 215–228.

Suzuki, R., Matthews, E. A., and Dickenson, A. H. 2001. Comparison of the effects of MK-801, ketamine and memantine on responses of spinal dorsal horn neurones in a rat model of mononeuropathy. Pain 91, 101–109.

Suzuki, R., Morcuende, S., Webber, M., Hunt, S. P., and Dickenson, A. H. 2002. Superficial NK1-expressing neurons control spinal excitability through activation of descending pathways. Nat. Neurosci. 5, 1319–1326.

Suzuki, R., Hunt, S. P., and Diusenson, A. H. 2003. The coding of noxious mechanical and thermal stimuli of deep dorsal horn neurones is attenuated in NK1 knockout mice. Neuropharmacology 45, 1093–1100.

Svendsen, F., Tjolsen, A., Gjerstad, J., and Hole, K. 1999. Long term potentiation of single WDR neurons in spinalized rats. Brain Res. 816, 487–492.

Svendsen, F., Tjolsen, A., and Hole, K. 1997. LTP of spinal A beta and C-fibre evoked responses after electrical sciatic nerve stimulation. Neuroreport 8, 3427–3430.

Swamydas, M., Skoff, A. M., and Adler, J. E. 2004. Partial sciatic nerve transection causes redistribution of pain-related peptides and lowers withdrawal threshold. Exp. Neurol. 188, 444–451.

Sweitzer, S. M., Peters, M. C., Ma, J. Y., Kerr, I., Mangadu, R., Chakravarty, S., Dugar, S., Medicherla, S., Protter, A. A., and Yeomans, D. C. 2004a. Peripheral and central p38 MAPK mediates capsaicin-induced hyperalgesia. Pain 111, 278–285.

Sweitzer, S. M., Wong, S. M., Peters, M. C., Mochly-Rosen, D., Yeomans, D. C., and Kendig, J. J. 2004b. Protein kinase C epsilon and gamma: involvement in formalin-induced nociception in neonatal rats. J. Pharmacol. Exp. Ther. 309, 616–625.

Synder, S. H. 1992. Nitric oxide: first in a new class of neurotransmitters. Science 257, 494–496.

Tachibana, T., Miki, K., Fukuoka, T., Arakawa, A., Taniguchi, M., Maruo, S., and Noguchi, K. 1998. Dynorphin mRNA expression in dorsal horn neurons after traumatic spinal cord injury: temporal and spatial analysis using in situ hybridization. J. Neurotrauma 15, 485–494.

Tanabe, Y., Nomura, A., Masu, M., Shigemoto, R., Mizuno, N., and Nakanishi, S. 1993. Signal transduction, pharmacological properties, and expression patterns of two rat metabotropic glutamate receptors, mGluR3 and mGluR4. J. Neurosci. 13, 1372–1378.

Tanaka, M., Cummins, T. R., Ishikawa, K., Dib-Hajj, S. D., Black, J. A., and Waxman, S. G. 1998. SNS Na$^+$ channel expression increases in dorsal root ganglion neurons in the carrageenan inflammatory pain model. Neuroreport 9, 967–972.

Tanner, K. D., Reichling, D. B., and Levine, J. D. 1998. Nociceptor hyper-responsiveness during vincristine-induced painful peripheral neuropathy in the rat. J. Neurosci. 18, 6480–6491.

Tao, Y. X. and Johns, R. A. 2000. Activation of cGMP-dependent protein kinase Ialpha is required for N-methyl-D-aspartate- or nitric oxide-produced spinal thermal hyperalgesia. Eur. J. Pharmacol. 392, 141–145.

Tao, Y. X., Hassan, A., Haddad, E., and Johns, R. A. 2000. Expression and action of cyclic GMP-dependent protein kinase Ialpha in inflammatory hyperalgesia in rat spinal cord. Neuroscience 95, 525–533.

Tao, Y. X., Rumbaugh, G., Wang, G. D., Petralia, R. S., Zhao, C., Kauer, F. W., Tao, F., Zhuo, M., Wenthold, R. J., Raja, S. N., Huganir, R. L., Bredt, D. S., and Johns, R. A. 2003. Impaired NMDA receptor-mediated postsynaptic function and blunted NMDA receptor-dependent persistent pain in mice lacking postsynaptic density-93 protein. J. Neurosci. 23, 6703–6712.

Tao, F., Tao, Y. X., Gonzalez, J. A., Fang, M., Mao, P., and Johns, R. A. 2001. Knockdown of PSD-95/SAP90 delays the development of neuropathic pain in rats. Neuroreport 12, 3251–3255.

Tatsumi, S., Mabuchi, T., Katano, T., Matsumura, S., Abe, T., Hidaka, H., Suzuki, M., Sasaki, Y., Minami, T., and Ito, S. 2005. Involvement of Rho-kinase in inflammatory and neuropathic pain through phosphorylation of myristoylated alanine-rich C-kinase substrate (MARCKS). Neuroscience 131, 491–498.

Taylor, B. K. and McCarson, K. E. 2004. Neurokinin-1 receptor gene expression in the mouse dorsal horn increases with neuropathic pain. J. Pain 5, 71–76.

Tegeder, I., Del Turco, D., Schmidtko, A., Sausbier, M., Feil, R., Hofmann, F., Deller, T., Ruth, P., and Geisslinger, G. 2004a. Reduced inflammatory hyperalgesia with preservation of acute thermal nociception in mice lacking cGMP-dependent protein kinase I. Proc. Natl. Acad. Sci. U. S. A. 101, 3253–3257.

Tegeder, I., Niederberger, E., Schmidt, R., Kunz, S., Guhring, H., Ritzeler, O., Michaelis, M., and Geisslinger, G. 2004b. Specific Inhibition of IkappaB kinase reduces hyperalgesia in inflammatory and neuropathic pain models in rats. J. Neurosci. 24, 1637–1645.

Tezuka, T., Umemori, H., Akiyama, T., Nakanishi, S., and Yamamoto, T. 1999. PSD-95 promotes Fyn-mediated tyrosine phosphorylation of the N-methyl-D-aspartate receptor subunit NR2A. Proc. Natl. Acad. Sci. U. S. A. 96, 435–440.

Thalhammer, J. G. and LaMotte, R. H. 1982. Spatial properties of nociceptor sensitization following heat injury of the skin. Brain Res. 231, 257–265.

Thompson, S. W., Bennett, D. L., Kerr, B. J., Bradbury, E. J., and McMahon, S. B. 1999. Brain-derived neurotrophic factor is an endogenous modulator of nociceptive responses in the spinal cord. Proc. Natl. Acad. Sci. U. S. A. 96, 7714–7718.

Thompson, S. W., Gerber, G., Sivilotti, L. G., and Woolf, C. J. 1992. Long duration ventral root potentials in the neonatal rat spinal cord in vitro; the effects of ionotropic and metabotropic excitatory amino acid receptor antagonists. Brain Res. 595, 87–97.

Thompson, S. W. N., King, A. E., and Woolf, C. J. 1990. Activity-dependent changes in rat ventral horn neurones in vitro; summation of prolonged afferent evoked postsynaptic depolarizations produce a d-APV sensitive windup. Eur. J. Neurosci. 2, 638–649.

Tolle, T. R., Castro-Lopes, J. M., Coimbra, A., and Zieglgansberger, W. 1990. Opiates modify induction of c-fos proto-oncogene in the spinal cord of the rat following noxious stimulation. Neurosci. Lett. 111, 46–51. Erratum in: Neurosci. Lett. 1990, 114, 239.

Tong, Y. G., Wang, H. F., Ju, G., Grant, G., Hokfelt, T., and Zhang, X. 1999. Increased uptake and transport of cholera toxin B-subunit in dorsal root ganglion neurons after peripheral axotomy: possible implications for sensory sprouting. J. Comp. Neurol. 404, 143–158.

Torebjörk, H. E., LaMotte, R. H., and Robinson, C. J. 1984. Peripheral neural correlation of magnitude of cutaneous pain and hyperalgesia: simultaneous recordings in humans of sensory judgements of pain and evoked response in nociceptors with C-fibres. J. Neurophysiol. 51, 325.

Torebjörk, H. E., Lundeberg, L., and LaMotte, R. H. 1992. Central changes in processing of mechanoreceptive input in capsaicin-induced secondary hyperalgesia in humans. J. Physiol. 448, 765–780.

Torsney, C. and Fitzgerald, M. 2002. Age-dependent effects of peripheral inflammation on the electrophysiological properties of neonatal rat dorsal horn neurons. J. Neurophysiol. 87, 1311–1137.

Tsuda, M., Mizokoshi, A., Shigemoto-Mogami, Y., Koizumi, S., and Inoue, K. 2004. Activation of p38 mitogen-activated protein kinase in spinal hyperactive microglia contributes to pain hypersensitivity following peripheral nerve injury. Glia 45, 89–95.

Tsuda, M., Shigemoto-Mogami, Y., Koizumi, S., Mizokoshi, A., Kohsaka, S., Salter, M. W., and Inoue, K. 2003. P2X4 receptors induced in spinal microglia gate tactile allodynia after nerve injury. Nature 424, 778–783.

Tsujino, H., Kondo, E., Fukuoka, T., Dai, Y., Tokunaga, A., Miki, K., Yonenobu, K., Ochi, T., and Noguchi, K. 2000. Activating transcription factor 3 (ATF3) induction by axotomy in sensory and motoneurons: a novel neuronal marker of nerve injury. Mol. Cell Neurosci. 15, 170–182.

Tu, J. C., Xiao, B., Naisbitt, S., Yuan, J. P., Petralia, R. S., Brakeman, P., Doan, A., Aakalu, V. K., Lanahan, A. A., Sheng, M., and Worley, P. F. 1999. Coupling of mGluR/Homer and PSD-95 complexes by the Shank family of postsynaptic density proteins. Neuron 23, 583–592.

Urban, M. O. and Gebhart, G. F. 1999. Supraspinal contributions to hyperalgesia. Proc. Natl. Acad. Sci. U. S. A. 96, 7687–7692.

Urban, L. and Randic, M. 1984. Slow excitatory transmission in rat dorsal horn: possible mediation by peptides. Brain Res. 290, 336–341.

Vanegas, H. and Schaible, H. G. 2001. Prostaglandins and cyclooxygenases [correction of cycloxygenases] in the spinal cord. Prog. Neurobiol. 64, 327–363. Erratum in: Prog. Neurobiol. 2001 65, 609.

Varty, G. B., Grilli, M., Forlani, A., Fredduzzi, S., Grzelak, M. E., Guthrie, D. H., Hodgson, R. A., Lu, S. X., Nicolussi, E., Pond, A. J., Parker, E. M., Hunter, J. C., Higgins, G. A., Reggiani, A., and Bertorelli, R. 2005. The antinociceptive and anxiolytic-like effects of the metabotropic glutamate receptor 5 (mGluR5) antagonists, MPEP and MTEP, and the mGluR1 antagonist, LY456236, in rodents: a comparison of efficacy and side-effect profiles. Psychopharmacology (Berl.) 179, 207–217.

Vasquez, E., Bar, K. J., Ebersberger, A., Klein, B., Vanegas, H., and Schaible, H. G. 2001. Spinal prostaglandins are involved in the development but not the maintenance of inflammation-induced spinal hyperexcitability. J. Neurosci. 21, 9001–9008.

Verge, G. M., Milligan, E. D., Maier, S. F., Watkins, L. R., Naeve, G. S., and Foster, A. C. 2004. Fractalkine (CX3CL1) and fractalkine receptor (CX3CR1) distribution in spinal cord and dorsal root ganglia under basal and neuropathic pain conditions. Eur. J. Neurosci. 20, 1150–1160.

Vierck, C. J., Jr. and Light, A. R. 2000. Allodynia and hyperalgesia within dermatomes caudal to a spinal cord injury in primates and rodents. Prog. Brain Res. 129, 411–428.

Vikman, K. S., Duggan, A. W., and Siddall, P. J. 2003. Increased ability to induce long-term potentiation of spinal dorsal horn neurones in monoarthritic rats. Brain Res. 990, 51–57.

Villar, M. J., Cortes, R., Theodorsson, E., Wiesenfeld-Hallin, Z., Schalling, M., Fahrenkrug, J., Emson, P. C., and Hokfelt, T. 1989. Neuropeptide expression in rat dorsal root ganglion cells and spinal cord after peripheral nerve injury with special reference to galanin. Neuroscience 33, 587–604.

Vincler, M. and Eisenach, J. C. 2004. Plasticity of spinal nicotinic acetylcholine receptors following spinal nerve ligation. Neurosci. Res. 48, 139–145.

Vincler, M. A. and Eisenach, J. C. 2005. Knock down of the alpha 5 nicotinic acetylcholine receptor in spinal nerve-ligated rats alleviates mechanical allodynia. Pharmacol. Biochem. Behav. 80, 135–143.

Vrinten, D. H., Gispen, W. H., Groen, G. J., and Adan, R. A. 2000. Antagonism of the melanocortin system reduces cold and mechanical allodynia in mononeuropathic rats. J. Neurosci. 20, 8131–8137.

Wakisaka, S., Kajander, K. C., and Bennett, G. J. 1991. Increased neuropeptide Y (NPY)-like immunoreactivity in rat sensory neurons following peripheral axotomy. Neurosci. Lett. 124, 200–203.

Wakisaka, S., Kajander, K. C., and Bennett, G. J. 1992. Effects of peripheral nerve injuries and tissue inflammation on the levels of neuropeptide Y-like immunoreactivity in rat primary afferent neurons. Brain Res. 598, 349–352.

Walker, K., Bowes, M., Panesar, M., Davis, A., Gentry, C., Kesingland, A., Gasparini, F., Spooren, W., Stoehr, N., Pagano, A., Flor, P. J., Vranesic, I., Lingenhoehl, K., Johnson, E. C., Varney, M., Urban, L., and Kuhn, R. 2001. mGluR5 receptors and nociceptive function I. Selective blockade of mGluR5 receptors in models of acute, persistent and chronic pain. Neuropharmacology 40, 1–9.

Walker, K. M., Urban, L., Medhurst, S. J., Patel, S., Panesar, M., Fox, A. J., and McIntyre, P. 2003. The VR1 antagonist capsazepine reverses mechanical hyperalgesia in models of inflammatory and neuropathic pain. J. Pharmacol. Exp. Ther. 304, 56–62.

Wall, P. D. and Woolf, C. J. 1984. Muscle but not cutaneous C-afferent input produces prolonged increases in the excitability of the flexion reflex in the rat. J. Physiol. (Lond.) 356, 443–458.

Wallin, J. and Schott, E. 2002. Substance P release in the spinal dorsal horn following peripheral nerve injury. Neuropeptides. 36, 252–256.

Wang, C. H., Chou, W. Y., Hung, K. S., Jawan, B., Lu, C. N., Liu, J. K., Hung, Y. P., and Lee, T. H. 2005. Intrathecal administration of roscovitine inhibits Cdk5 activity and attenuates formalin-induced nociceptive response in rats. Acta Pharmacol. Sin. 26, 46–50.

Wang, J., Kawamata, M., and Namiki, A. 2005. Changes in properties of spinal dorsal horn neurons and their sensitivity to morphine after spinal cord injury in the rat. Anesthesiology 102, 152–164.

Wang, Y. T., Yu, X. M., and Salter, M. W. 1996. Ca(2+)-independent reduction of N-methyl-D-aspartate channel activity by protein tyrosine phosphatase. Proc. Natl. Acad. Sci. U. S. A. 93, 1721–1725.

Watkins, J. C. and Evans, R. H. 1981. Excitatory amino acid transmitters. Annu. Rev. Pharmacol. Toxicol. 21, 165–204.

Watkins, L. R. and Maier, S. F. 1999. Implications of immune-to-brain communication for sickness and pain. Proc. Natl. Acad. Sci. U. S. A. 96, 7710–7713.

Watkins, L. R., Wiertelak, E. P., Furness, L. E., and Maier, S. F. 1994a. Illness-induced hyperalgesia is mediated by spinal neuropeptides and excitatory amino acids. Brain. Res. 664, 17–24.

Watkins, L. R., Wiertelak, E. P., Goehler, L. E., Mooney-Heiberger, K., Martinez, J., Furness, L., Smith, K. P., and Maier, S. F. 1994b. Neurocircuitry of illness-induced hyperalgesia. Brain Res. 639, 283–299.

Waxman, S. G., Kocsis, J. D., and Black, J. A. 1994. Type III sodium channel mRNA is expressed in embryonic but not adult spinal sensory neurons, and is reexpressed following axotomy. J. Neurophysiol. 72, 466–470.

Weng, H. R., Aravindan, N., Cata, J. P., Chen, J. H., Shaw, A. D., and Dougherty, P. M. 2005. Spinal glial glutamate transporters downregulate in rats with taxol-induced hyperalgesia. Neurosci. Lett. 386, 18–22.

White, D. M. 2000. Neurotrophin-3 antisense oligonucleotide attenuates nerve injury-induced Abeta-fibre sprouting. Brain Res. 885, 79–86.

Whiteside, G. T. and Munglani, R. 2001. Cell death in the superficial dorsal horn in a model of neuropathic pain. J. Neurosci. Res. 64, 168–173.

Wiertelak, E. P., Furness, L. E., Watkins, L. R., and Maier, S. F. 1994. Illness-induced hyperalgesia is mediated by a spinal NMDA–nitric oxide cascade. Brain Res. 664, 9–16.

Wiesenfeld-Hallin, Z. 1985. Intrathecal somatostatin modulates spinal sensory and reflex mechanisms: behavioral and electrophysiological studies in the rat. Neurosci. Lett. 62, 69–74.

Wiesenfeld-Hallin, Z. 1987. Intrathecal vasoactive intestinal polypeptide modulates spinal reflex excitability primarily to cutaneous thermal stimuli in rats. Neurosci. Lett. 80, 293–297.

Wiesenfeld-Hallin, Z., Aldskogius, H., Grant, G., Hao, J. X., Hokfelt, T., and Xu, X. J. 1997. Central inhibitory dysfunctions: mechanisms and clinical implications. Behav. Brain. Sci. 20, 420–425.

Wiesenfeld-Hallin, Z., Villar, M. J., and Hokfelt, T. 1989. The effects of intrathecal galanin and C-fibre stimulation on the flexor reflex in the rat. Brain Res. 486, 205–213.

Wiesenfeld-Hallin, Z., Xu, X. J., Hakanson, R., Feng, D. M., and Folkers, K. 1990. Plasticity of the peptidergic mediation of spinal reflex facilitation after peripheral nerve section in the rat. Neurosci. Lett. 116, 293–298.

Wiffen, P. J., McQuay, H. J., Edwards, J. E., and Moore, R. A. 2005. Gabapentin for acute and chronic pain. Cochrane Database Syst. Rev. 20, CD005452. Review.

Willcockson, W. S., Chung, J. M., Hori, Y., Lee, K. H., and Willis, W. D. 1984a. Effects of iontophoretically released peptides on primate spinothalamic tract cells. J. Neurosci. 4, 741–750.

Willcockson, W. S., Chung, J. M., Hori, Y., Lee, K. H., and Willis, W. D. 1984b. Effects of iontophoretically released amino acids and amines on primate spinothalamic tract cells. J. Neurosci. 4, 732–740.

Wilson, P. and Kitchener, P. D. 1996. Plasticity of cutaneous primary afferent projections to the spinal dorsal horn. Prog. Neurobiol. 48, 105–129.

Wilson-Gerwing, T. D., Dmyterko, M. V., Zochodne, D. W., Johnston, J. M., and Verge, V. M. 2005. Neurotrophin-3 suppresses thermal hyperalgesia associated with neuropathic pain and attenuates transient receptor potential vanilloid receptor-1 expression in adult sensory neurons. J. Neurosci. 25, 758–767.

Winquist, R. J., Pan, J. Q., and Gribkoff, V. K. 2005. Use-dependent blockade of Cav2.2 voltage-gated calcium channels for neuropathic pain. Biochem. Pharmacol. 70, 489–499.

Wisden, W., Errington, M. L., Williams, S., Dunnett, S. B., Waters, C., Hitchcock, D., Evan, G., Bliss, T. V., and Hunt, S. P. 1990. Differential expression of immediate early genes in the hippocampus and spinal cord. Neuron 4, 603–614.

Womack, M. D., MacDermott, A. B., and Jessell, T. M. 1988. Sensory transmitters regulate intracellular calcium in dorsal horn neurons. Nature 334, 351–353.

Womack, M. D., MacDermott, A. B., and Jessell, T. M. 1989. Substance P increases [Ca2+]i in dorsal horn neurons via two distinct mechanisms. Soc. Neurosci. Abstr. 15, 184.

Woolf, C. J. 1983. Evidence for a central component of post-injury pain hypersensitivity. Nature 306, 686–688.

Woolf, C. J. 1996. Windup and central sensitization are not equivalent. Pain 66, 105–108.

Woolf, C. J. and McMahon, S. B. 1985. Injury-induced plasticity of the flexor reflex in chronic decerebrate rats. Neuroscience 16, 395–404.

Woolf, C. J. and Thompson, S. W. 1991. The induction and maintenance of central sensitization is dependent on *N*-methyl-D-aspartic acid receptor activation; implications for the treatment of post-injury pain hypersensitivity states. Pain 44, 293–299.

Woolf, C. J. and Wall, P. D. 1986. Relative effectiveness of C primary afferent fibers of different origins in evoking a prolonged facilitation of the flexor reflex in the rat. J. Neurosci. 6, 1433–1442.

Woolf, C. J. and Wiesenfeld-Hallin, Z. 1986. Substance P and calcitonin gene-related peptide synergistically modulate the gain of the nociceptive flexor withdrawal reflex in the rat. Neurosci. Lett. 66, 226–230.

Woolf, C. J., Shortland, P., and Coggeshall, R. E. 1992. Peripheral nerve injury triggers central sprouting of myelinated afferents. Nature 355, 75–78.

Woolf, C. J., Thompson, S. W., and King, A. E. 1988–1989. Prolonged primary afferent induced alterations in dorsal horn neurones, an intracellular analysis *in vivo* and *in vitro*. J. Physiol. (Paris) 83, 255–266.

Xing, G., Liu, F., Yao, L., Wan, Y., and Han, J. 2003. Changes in long-term synaptic plasticity in the spinal dorsal horn of neuropathic pain rats. Beijing Da Xue Xue Bao 35, 226–230.

Xu, X. J., Elfvin, A., and Wiesenfeld-Hallin, Z. 1995. Subcutaneous carrageenan, but not formalin, increases the excitability of the nociceptive flexor reflex in the rat. Neurosci. Lett. 196, 116–118.

Xu, X.-J., Maggi, C. A., and Wiesenfeld-Hallin, Z. 1991. On the role of NK-2 tachykinin receptors in the mediation of spinal reflex excitability in the rat. Neuroscience 44, 483–490.

Xu, M., Petraschka, M., McLaughlin, J. P., Westenbroek, R. E., Caron, M. G., Lefkowitz, R. J., Czyzyk, T. A., Pintar, J. E., Terman, G. W., and Chavkin, C. 2004. Neuropathic pain activates the endogenous kappa opioid system in mouse spinal cord and induces opioid receptor tolerance. J. Neurosci. 24, 4576–4584.

Yajima, Y., Narita, M., Narita, M., Matsumoto, N., and Suzuki, T. 2002. Involvement of a spinal brain-derived neurotrophic factor/full-length TrkB pathway in the development of nerve injury-induced thermal hyperalgesia in mice. Brain Res. 958, 338–346.

Yajima, Y., Narita, M., Shimamura, M., Narita, M., Kubota, C., and Suzuki, T. 2003. Differential involvement of spinal protein kinase C and protein kinase A in neuropathic and inflammatory pain in mice. Brain Res. 992, 288–293.

Yaksh, T. L. 1982. Central and Peripheral Mechanisms for the Antialgesic Action of Acetylsalicylic Acid. In: Acetylsalicylic

Acid: New Uses for an Old Drug (eds. H. J. M. Barett, J. Hirsh, and J. F. Mustard), pp. 137–151. Raven Press.

Yaksh, T. L., Dirig, D. M., Conway, C. M., Svensson, C., Luo, Z. D., and Isakson, P. C. 2001. The acute antihyperalgesic action of nonsteroidal, anti-inflammatory drugs and release of spinal prostaglandin E2 is mediated by the inhibition of constitutive spinal cyclooxygenase-2 (COX-2) but not COX-1. J. Neurosci. 21, 5847–5853.

Yamamoto, T. and Yaksh, T. L. 1991. Stereospecific effects of a nonpeptide NK-1 selective antagonist, CP-96,345: antinociception in the absence of motor dysfunction. Life Sci. 49, 1955–1963.

Yamamoto, T., Hirasawa, S., Wroblewska, B., Grajkowska, E., Zhou, J., Kozikowski, A., Wroblewski, J., and Neale, J. H. 2004. Antinociceptive effects of N-acetylaspartylglutamate (NAAG) peptidase inhibitors ZJ-11, ZJ-17 and ZJ-43 in the rat formalin test and in the rat neuropathic pain model. Eur. J. Neurosci. 20, 484–494.

Yamamoto, T., Shimoyama, N., and Mizuguchi, T. 1993. Role of the injury discharge in the development of thermal hyperesthesia after sciatic nerve constriction injury in the rat. Anesthesiology 79, 993–1002.

Yang, L., Zhang, F. X., Huang, F., Lu, Y. J., Li, G. D., Bao, L., Xiao, H. S., and Zhang, X. 2004. Peripheral nerve injury induces trans-synaptic modification of channels, receptors and signal pathways in rat dorsal spinal cord. Eur. J. Neurosci. 19, 871–883.

Yang, L. C., Marsala, M., and Yaksh, T. L. 1996. Characterization of time course of spinal amino acids, citrulline and PGE2 release after carrageenan/kaolin-induced knee joint inflammation: a chronic microdialysis study. Pain 67, 345–354.

Yashpal, K., Fisher, K., Chabot, J.-G., and Coderre, T. J. 2001. Differential effects of NMDA and group I mGluR antagonists on both nociception and spinal cord protein kinase C translocation in the formalin test and a model of neuropathic pain in rats. Pain 94, 17–29.

Yashpal, K., Pitcher, G. M., Parent, A., Quirion, R., and Coderre, T. J. 1995. Noxious thermal and chemical stimulation induce increases in 3H-Phorbol 12, 13-dibutyrate binding in spinal cord dorsal horn as well as persistent pain and hyperalgesia, which is reduced by inhibition of protein kinase C. J. Neurosci. 15, 3263–3272.

Yashpal, K., Radhakrishnan, V., Coderre, T. J., and Henry, J. L. 1993. CP-96,345, but not its stereoisomer, CP-96,344, blocks the nociceptive responses to intrathecally administered substance P and to noxious thermal and chemical stimuli in the rat. Neuroscience 52, 1039–1047.

Yezierski, R. P. 2000. Pain following spinal cord injury: pathophysiology and central mechanisms. Prog. Brain Res. 129, 429–449.

Yezierski, R. P. and Park, S. H. 1993. The mechanosensitivity of spinal sensory neurons following intraspinal injections of quisqualic acid in the rat. Neurosci. Lett. 157, 115–119.

Yezierski, R. P., Liu, S., Ruenes, G. L., Kajander, K. J., and Brewer, K. L. 1998. Excitotoxic spinal cord injury: behavioral and morphological characteristics of a central pain model. Pain 75, 141–155.

Yokoyama, K., Kurihara, T., Makita, K., and Tanabe, T. 2003. Plastic change of N-type Ca channel expression after preconditioning is responsible for prostaglandin E2-induced long-lasting allodynia. Anesthesiology 99, 1364–1370.

Young, M. R., Fleetwood-Walker, S. M., Dickinson, T., Blackburn-Munro, G., Sparrow, H., Birch, P. J., and Bountra, C. 1997. Behavioural and electro-physiological evidence supporting a role for group I metabotropic glutamate receptors in the mediation of nociceptive inputs to the rat spinal cord. Brain Res. 777, 161–169.

Young, M. R., Fleetwood-Walker, S. M., Mitchell, R., and Dickinson, T. 1995. The involvement of metabotropic glutamate receptors and their intracellular signalling pathways in sustained nociceptive transmission in rat dorsal horn neurons. Neuropharmacology 34, 1033–1041.

Young, M. R., Fleetwood-Walker, S. M., Mitchell, R., and Munro, F. E. 1994. Evidence for a role of metabotropic glutamate receptors in sustained nociceptive inputs to rat dorsal horn neurons. Neuropharmacology 33, 141–144.

Yu, X. M. and Salter, M. W. 1999. Src, a molecular switch governing gain control of synaptic transmission mediated by N-methyl-D-aspartate receptors. Proc. Natl. Acad. Sci. U. S. A. 96, 7697–7704.

Yu, X. M., Askalan, R., Keil, G. J., 2nd, and Salter, M. W. 1997. NMDA channel regulation by channel-associated protein tyrosine kinase Src. Science 275, 674–678.

Zahn, P. K. and Brennan, T. J. 1999. Primary and secondary hyperalgesia in a rat model for human postoperative pain. Anesthesiology 90, 863–872.

Zahn, P. K., Sluka, K. A., and Brennan, T. J. 2002. Excitatory amino acid release in the spinal cord caused by plantar incision in the rat. Pain 100, 65–76.

Zeitz, K. P., Giese, K. P., Silva, A. J., and Basbaum, A. I. 2004. The contribution of autophosphorylated alpha-calcium-calmodulin kinase II to injury-induced persistent pain. Neuroscience 128, 889–898.

Zhang, X., Dagerlind, A., Elde, R. P., Castel, M. N., Broberger, C., Wiesenfeld-Hallin, Z., and Hokfelt, T. 1993a. Marked increase in cholecystokinin B receptor messenger RNA levels in rat dorsal root ganglia after peripheral axotomy. Neuroscience 57, 227–233.

Zhang, X., Bao, L., Shi, T. J., Ju, G., Elde, R., and Hokfelt, T. 1998. Down-regulation of mu-opioid receptors in rat and monkey dorsal root ganglion neurons and spinal cord after peripheral axotomy. Neuroscience 82, 223–240.

Zhang, Y. Z., Hannibal, J., Zhao, Q., Moller, K., Danielsen, N., Fahrenkrug, J., and Sundler, F. 1996. Pituitary adenylate cyclase activating peptide expression in the rat dorsal root ganglia: up-regulation after peripheral nerve injury. Neuroscience 74, 1099–1110.

Zhang, L., Lu, Y., Chen, Y., and Westlund, K. N. 2002. Group I metabotropic glutamate receptor antagonists block secondary thermal hyperalgesia in rats with knee joint inflammation. J. Pharmacol. Exp. Ther. 300, 149–156.

Zhang, X., Meister, B., Elde, R., Verge, V. M., and Hokfelt, T. 1993b. Large calibre primary afferent neurons projecting to the gracile nucleus express neuropeptide Y after sciatic nerve lesions: an immunohistochemical and in situ hybridization study in rats. Eur. J. Neurosci. 5, 1510–1519.

Zhang, B., Tao, F., Liaw, W. J., Bredt, D. S., Johns, R. A., and Tao, Y. X. 2003. Effect of knock down of spinal cord PSD-93/chapsin-110 on persistent pain induced by complete Freund's adjuvant and peripheral nerve injury. Pain 106, 187–196.

Zhang, X., Wu, J., Fang, L., and Willis, W. D. 2003. The effects of protein phosphatase inhibitors on nociceptive behavioral responses of rats following intradermal injection of capsaicin. Pain 106, 443–451.

Zhang, H., Xie, W., and Xie, Y. 2005. Spinal cord injury triggers sensitization of wide dynamic range dorsal horn neurons in segments rostral to the injury. Brain Res. 1055, 103–110.

Zhong, J., Gerber, G., Kojic, L., and Randic, M. 2000. Dual modulation of excitatory synaptic transmission by agonists at group I metabotropic glutamate receptors in the rat spinal dorsal horn. Brain Res. 887, 359–377.

Zhou, X. F., Chie, E. T., Deng, Y. S., Zhong, J. H., Xue, Q., Rush, R. A., and Xian, C. J. 1999a. Injured primary sensory neurons switch phenotype for brain-derived neurotrophic factor in the rat. Neuroscience 92, 841–853.

Zhou, X. F., Deng, Y. S., Chie, E., Xue, Q., Zhong, J. H., McLachlan, E. M., Rush, R. A., and Xian, C. J., 1999b. Satellite-cell-derived nerve growth factor and neurotrophin-3 are involved in noradrenergic sprouting in the dorsal root ganglia following peripheral nerve injury in the rat. Eur. J. Neurosci. 11, 1711–1722.

Zhou, Y., Li, G. D., and Zhao, Z. Q. 2003. State-dependent phosphorylation of epsilon-isozyme of protein kinase C in adult rat dorsal root ganglia after inflammation and nerve injury. J. Neurochem. 85, 571–580.

Zhu, C. Z., Wilson, S. G., Mikusa, J. P., Wismer, C. T., Gauvin, D. M., Lynch, J. J., III, Wade, C. L, Decker, M. W., and Honore, P. 2004. Assessing the role of metabotropic glutamate receptor 5 in multiple nociceptive modalities. Eur. J. Pharmacol. 506, 107–118.

Zhuo, M. and Gebhart, G. F. 1992. Characterization of descending facilitation and inhibition of spinal nociceptive transmission from the nuclei reticularis gigantocellularis and gigantocellularis pars alpha in the rat. J. Neurophysiol. 67, 1599–1614.

Zieglgansberger, W. and Herz, A. 1971. Changes of cutaneous receptive fields of spino-cervical-tract neurones and other dorsal horn neurones by microelectrophoretically administered amino acids. Exp. Brain Res. 13, 111–126.

Zou, X., Lin, Q., and Willis, W. D. 2000. Enhanced phosphorylation of NMDA receptor 1 subunits in spinal cord dorsal horn and spinothalamic tract neurons after intradermal injection of capsaicin in rats. J. Neurosci. 20, 6989–6997.

Further Reading

Abbadie, C., Brown, J. L., Mantyh, P. W., and Basbaum, A. I. 1996. Spinal cord substance P receptor immunoreactivity increases in both inflammatory and nerve injury models of persistent pain. Neuroscience 70, 201–209.

Barbour, B., Szatkowski, M., Ingledew, N., and Attwell, D. 1989. Arachidonic acid induces a prolonged inhibition of glutamate uptake into glial cells. Nature 342, 918–920.

Miletic, G., Hanson, E. N., and Miletic, V. 2004a. Brain-derived neurotrophic factor-elicited or sciatic ligation-associated phosphorylation of cyclic AMP response element binding protein in the rat spinal dorsal horn is reduced by block of tyrosine kinase receptors. Neurosci. Lett. 361, 269–271.

Watkins, L. R., Martin, D., Ulrich, P., Tracey, K. J., and Maier, S. F. 1997. Evidence for the involvement of spinal cord glia in subcutaneous formalin induced hyperalgesia in the rat. Pain 71, 225–235.

Relevant Website

http://www.iasp-pain.org/terms-p.html

27 Glycine Receptors

H U Zeilhofer, University of Zurich, Zurich, Switzerland

Glossary

GABA Gamma aminobutyric acid; fast inhibitory neurotransmitter in the mammalian CNS.

GABA$_A$ receptor Chloride channel-coupled GABA receptor blocked by bicuculline.

glycine Fast inhibitory neurotransmitter in the spinal cord and brain stem.

gephyrin Postsynaptic protein mediating the clustering of GABA$_A$ and glycine receptors in the postsynaptic membrane.

cyclooxygenase-2 Inducible enzyme mediating the production of prostaglandin precursors.

EGFP Enhanced green fluorescent protein used, for example to label cells *in vitro* and *in vivo*.

KCC2 Potassium/chloride exporter protein, needed for central neurons to reduce their intracellular chloride concentration to physiological levels.

27.1 Molecular Architecture of Glycine Receptors

Glycine is the major inhibitory neurotransmitter in the spinal cord and brain stem. In addition, glycine is an obligatory co-agonist at excitatory glutamate receptors of the *N*-methyl-D-aspartate (NMDA) subtype. This chapter focuses on inhibitory (strychnine-sensitive) glycine receptors. These glycine receptors are heteropentameric chloride-permeable ion channels composed of α and β subunits (Legendre, P., 2001). Four different genes (*glra1–glra4*) encode for the glycine-binding α subunits and one gene (*glrb*) encodes for the so-called structural β subunit. α Subunits carry the glycine binding site and are capable of forming functional homomeric receptor channels. The β subunit permits postsynaptic clustering of glycine receptors through an interaction with the postsynaptic density protein gephyrin. Most glycine receptors in the adult are heteromeric receptors composed of α1 and β subunits. GlyRα2 forms homomeric receptor channels during embryonic development and in early postnatal life. In most organs GlyRα2 homomers are replaced by GlyRα1/β heteromers during postnatal maturation. GlyRα3 is another adult glycine receptor subunit but much less abundant than GlyRα1. GlyRα4 is probably not expressed in humans because of a premature stop codon in the *GLRA4* gene.

27.2 Glycinergic Innervation in the Spinal Cord Dorsal Horn

Immunofluorescence studies have shown that markers of glycinergic innervation are abundant in the spinal cord dorsal horn (Todd, A. J. and Sullivan, A. C., 1990). Glycine receptors are found throughout the dorsal horn. As in other regions of the central nervous system (CNS), most glycine receptors in the adult spinal cord are α1/β heteromeric receptor channels. However, in the superficial layers of the dorsal horn, where most nociceptive afferents terminate, glycine receptors contain GlyRα3 in addition to GlyRα1 and GlyRβ (Figure 1) (Harvey, R. J. *et al.*, 2004). The prominent

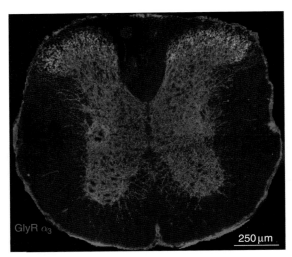

GlyR α3

250 μm

Figure 1 Expression of GlyRα3 in the spinal cord dorsal horn. GlyRα3 (green) is distinctly expressed in the superficial layers where nociceptive afferents terminate, whereas gephyrin (red), a marker for glycinergic and GABAergic synapses, is found throughout the spinal cord. Adapted from Harvey, R. J., Depner, U. B., Wässle, H., Ahmadi, S., Heindl, C., Reinold, H., Smart, T. G., Harvey, K., Schütz, B., Abo-Salem, O. M., Zimmer, A., Poisbeau, P., Welzl, H., Wolfer, D. P., Betz, H., Zeilhofer, H. U., and Müller, U. 2004. GlyR α3: an essential target for spinal inflammatory pain sensitization. Science 304, 884–887; with permission.

expression of glycine receptors in the spinal cord is accompanied by dense glycinergic innervation. Glycinergic axon terminals, characterized by the expression of the neuronal glycine transporter isoform GlyT2, and glycinergic cell bodies, identified by their immunoreactivity against glycine, are widely distributed in the dorsal horn. Recently, mice expressing enhanced green fluorescent protein (EGFP) under the transcriptional control of the GlyT2 gene have been generated (Zeilhofer, H. U. *et al.*, 2005). In the dorsal horn of these mice, numerous glycinergic (EGFP/GlyT2-positive) neurons are found in the deeper dorsal horn (laminae III–V) and in lamina I, whereas relatively few glycinergic neurons are seen in lamina II. In addition to these local inhibitory interneurons, descending antinociceptive glycinergic fiber tracts (Antal, M. *et al.*, 1996; Kato, G. *et al.*, 2006) probably contribute to the glycinergic innervation of the dorsal horn.

27.3 Inhibitory Glycine Receptors and Spinal Pain Control

Plenty of information meanwhile indicates that relief from synaptic inhibition by spinal blockade of glycine (and also of GABA$_A$) receptors can elicit and exaggerate nociceptive responses (for a recent review see Zeilhofer, H. U., 2005). Several groups have meanwhile demonstrated that relief from inhibition also occurs endogenously during inflammation and possibly also after peripheral nerve injury.

27.3.1 Glycine Receptors in the Spinal Control of Inflammatory Pain

Inflammatory pain originates to a large extent from prostaglandins, which are produced in response to inflammation and tissue damage mainly by inducible cyclooxygenase-2 (COX-2). The pivotal role of prostaglandins for pain sensitization is obvious from our everyday experience that profound analgesia can be achieved through inhibition of prostanoid formation either by the classical nonspecific cyclooxygenase (COX) inhibitors (aspirin and related drugs) or by the more recently developed COX-2-specific inhibitors (Brune, K. and Zeilhofer, H. U., 2005).

For a long time it was widely believed that prostaglandins sensitize the nociceptive system only at the level of the peripheral nociceptor. This long-held view has changed significantly since the mid-1990s when several groups demonstrated that inflammation induces COX-2 expression not only in the peripheral inflamed tissue, but also in the CNS, in particular in the spinal cord dorsal horn. A central action of prostaglandins and accordingly also a central mode of action for the analgesic action of COX inhibitors are meanwhile generally accepted (Malmberg, A. B. and Yaksh, T. L., 1992; Samad, T. A. *et al.*, 2002).

Electrophysiological experiments have shown that PGE$_2$ reduces inhibitory glycinergic neurotransmission in the spinal cord dorsal horn, but does not interfere with GABAergic or glutamatergic neurotransmission (Figure 2) (Ahmadi, S. *et al.*, 2002). This inhibition is (at least in adult animals) specific for neurons in the superficial layers of the dorsal horn and for PGE$_2$ (PGD$_2$, PGI$_2$, and PGF$_{2\alpha}$ are without effect). It involves activation of postsynaptic prostaglandin E receptors of the EP2 subtype and a specific protein kinase A (PKA)-dependent phosphorylation of GlyRα3-containing receptors, which are in the spinal cord distinctly expressed in the superficial layers (Harvey, R. J. *et al.*, 2004). Subsequent experiments using site-directed mutagenesis have shown that GlyRα3 is phosphorylated at serine 346 in the long intracellular loop between transmembrane regions 3 and 4. Figure 3 shows a schematic

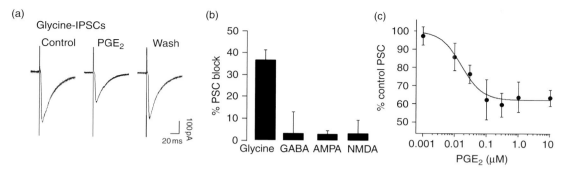

Figure 2 PGE$_2$ selectively inhibits glycinergic neurotransmission onto spinal cord dorsal horn neurons. (a) Averaged inhibitory postsynaptic current (IPSC) traces recorded from a neuron in the superficial dorsal horn under control conditions, in the presence of PGE$_2$, and after its removal. (b) Statistical analysis (mean ± SEM). Significant inhibition by PGE$_2$ occurred only for glycinergic IPSCs, GABAergic IPSCs, and excitatory postsynaptic currents mediated by glutamate receptors of the alpha-amino-3-hydroxy-5-methyl-4-isoxazolepropionic acid (AMPA) or N-methyl-D-aspartate (NMDA) type remained unchanged. (c) Concentration–response curve. Significant inhibition of glycinergic IPSC occurred at low nanomolar concentrations. Adapted from Ahmadi, S., Lipross, S., Neuhuber, W. L., and Zeilhofer, H. U. 2002. PGE$_2$ selectively blocks inhibitory glycinergic neurotransmission onto rat superficial dorsal horn neurons. Nat. Neurosci. 5, 34–40, with permission from Nature Publishing Group.

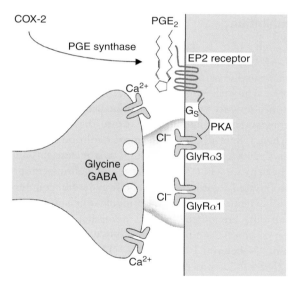

Figure 3 Schematic representation of the intracellular pathway leading to inhibition of glycinergic neurotransmission by PGE$_2$. PGE$_2$ activates E-type prostaglandin receptors of the EP2 subtype, which subsequently lead to an increase in intracellular camp and activate protein kinase A (PKA). PKA finally phosphorylates and inhibits GlyRα3. Adapted from Progress in Pain research and Management, vol. 30.

representation of the intracellular pathway that underlies inhibition of glycine receptors by PGE$_2$. Unlike GlyRα3-containing receptors, homomeric glycine receptors formed by GlyRα1 and heteromeric channels composed of GlyRα1 and GlyRβ are not inhibited by PKA activation and mice

carrying a loss-of-function mutation in the GlyRα1 subunit lack a nociceptive phenotype.

Mice lacking GlyRα3 (or EP2 receptors) not only exhibit a nearly complete loss of pain sensitization by spinal PGE$_2$ but also display nearly identical nociceptive phenotypes in models of inflammatory pain (Figure 4) (Harvey, R. J. *et al.*, 2004, Reinold, H. *et al.*, 2005; see also Zeilhofer, H. U., 2005). Both types of knockout mice develop normal initial thermal and mechanical sensitization, but recover much faster from hyperalgesia than wild-type mice. Inhibition of glycinergic neurotransmission by PGE$_2$ thus is a key event in the generation of inflammatory pain.

27.3.2 Glycine Receptors in the Spinal Control of Neuropathic Pain

A role for cyclooxygenase products has also been proposed for neuropathic pain originating from peripheral nerve injury. However, a significant contribution of prostaglandins to this pain form is discussed controversially and other mechanisms leading to a similar relief from inhibition may be more important (Hösl, K. *et al.*, 2006). Possible mechanisms include an inhibition of synaptic glycine release, reduced responsiveness of postsynaptic glycine receptors, a reduction in the transmembrane chloride gradient rendering the inhibitory tone of GABAergic and glycinergic synaptic input less efficient, and a loss of inhibitory innervation, for

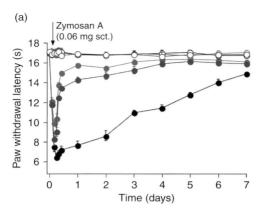 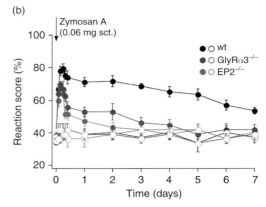

Figure 4 Inflammatory pain sensitization in mice lacking PGE$_2$-mediated inhibition of glycinergic neurotransmission. An inflammatory response was induced by subcutaneous injection by a yeast extract (zymosan A) into one hind paw. Inflammatory hyperalgesia was assessed by stimulating the inflamed paw either with noxious heat or mechanically with calibrated von Frey filaments. Both mice lacking the EP2 receptor (EP2$^{-/-}$ mice) and mice deficient in GlyRα3 (GlyRα3$^{-/-}$ mice) recovered quickly from inflammatory hyperalgesia. Adapted from Zeilhofer, H. U. 2005. The glycinergic control of spinal pain processing. Cell Mol. Life Sci. 62, 2027–2035.

example due to a selective death of GABAergic or glycinergic interneurons.

There is indeed experimental evidence for all three possibilities, but their contribution to neuropathic pain is not clear. A reduction in the transmembrane chloride gradient in dorsal horn neurons following peripheral nerve injury has been reported by Coull J. A. *et al.* (2003). Peripheral nerve trauma reduced the expression of the potassium chloride exporter KCC2 in dorsal horn neurons. This induced a shift of the chloride equilibrium potential to more depolarized values and rendered inhibitory GABAergic and glycinergic input less efficient. Another extensively discussed report suggests that peripheral nerve injury induces a specific loss of spinal inhibitory GABAergic neurotransmission in the dorsal horn of rats in the chronic constriction injury model and the spared nerve injury model of neuropathic pain (Moore, K. A. *et al.*, 2002). This original report has proposed that the loss of GABAergic input was due to the selective apoptotic death of GABAergic interneurons. Subsequent studies have however suggested that such a loss may not be necessary for the development of hyperalgesia in the chronic nerve injury model of neuropathic pain (Polgar, E. *et al.*, 2003; 2004). Finally, the neuropeptide nocistatin is a specific inhibitor of GABA and glycine release in the spinal cord dorsal horn (Zeilhofer, H. U. *et al.*, 2000) and it increases nociceptive responses in the rat formalin test and in the chronic constriction injury of neuropathic pain after intrathecal injection.

27.3.3 Transmitters Facilitating Glycinergic Inhibition

Given that a relief from glycinergic and/or GABAergic inhibition underlies inflammatory and neuropathic pain, one might speculate that transmitters, which facilitate inhibitory transmission might contribute to endogenous antinociception. Several transmitters including ATP (Jang, I. S. *et al.*, 2001; Rhee, J. S. *et al.*, 2000) and norepinephrine (Baba, H. *et al.*, 2000a; 2000b) facilitate glycine release through the activation of P2X receptors and α2 adrenoceptors in the spinal cord, respectively. In rat sacral commissural neurons (lamina X) glycinergic membrane currents are potentiated by norepinephrine acting on α2 adrenoceptors through a postsynaptic mechanism (Nabekura, J. *et al.*, 1999). These and possibly other similar processes may contribute to endogenous pain-controlling processes. Unfortunately, only very few drugs target glycine receptors and most of them lack the isoform specificity (Laube, B. *et al.*, 2002).

References

Ahmadi, S., Lippross, S., Neuhuber, W. L., and Zeilhofer, H. U. 2002. PGE$_2$ selectively blocks inhibitory glycinergic neurotransmission onto rat superficial dorsal horn neurons. Nat. Neurosci. 5, 34–40.

Antal, M., Petko, M., Polgar, E., Heizmann, C. W., and Storm-Mathisen, J. 1996. Direct evidence of an extensive GABAergic innervation of the spinal dorsal horn by fibres descending from the rostral ventromedial medulla. Neuroscience 73, 509–518.

Baba, H., Goldstein, P. A., Okamoto, M., Kohno, T., Ataka, T., Yoshimura, M., and Shimoji K. 2000a. Norepinephrine facilitates inhibitory transmission in substantia gelatinosa of adult rat spinal cord (part 2): effects on somatodendritic sites of GABAergic neurons. Anesthesiology 92, 485–492.

Baba, H., Shimoji, K., and Yoshimura, M. 2000b. Norepinephrine facilitates inhibitory transmission in substantia gelatinosa of adult rat spinal cord (part 1): effects on axon terminals of GABAergic and glycinergic neurons. Anesthesiology 92, 473–484.

Brune, K. and Zeilhofer, H. U. 2005. Antipyretic Analgesics – Basic Aspects. In: Melzack and Wall's Textbook of Pain, 5th edn. (eds. S. McMahon and M. Koltzenburg), pp. 459–469. Churchill Livingstone.

Coull, J. A., Boudreau, D., Bachand, K., Prescott, S. A., Nault, F., Sik, A., De Koninck, P., and De Koninck, Y. 2003. Trans-synaptic shift in anion gradient in spinal lamina I neurons as a mechanism of neuropathic pain. Nature 424, 938–942.

Harvey, R. J., Depner, U. B., Wässle, H., Ahmadi, S., Heindl, C., Reinold, H., Smart, T. G., Harvey, K., Schütz, B., Abo-Salem, O. M., Zimmer, A., Poisbeau, P., Welzl, H., Wolfer, D. P., Betz, H., Zeilhofer, H. U., and Müller, U. 2004. GlyR α3: an essential target for spinal inflammatory pain sensitization. Science 304, 884–887.

Hösl, K., Reinold, H., Harvey, R. J., Müller, U., Narumiya, S., and Zeilhofer, H. U. 2006. Spinal prostaglandin E receptors of the EP2 subtype and the glycine receptor α3 subunit, which mediate central inflammatory hyperalgesia, do not contribute to pain after peripheral nerve injury or formalin injection. Pain 126, 46–53.

Jang, I. S., Rhee, J. S., Kubota, H., Akaike, N., and Akaike, N. 2001. Developmental changes in P2X purinoceptors on glycinergic presynaptic nerve terminals projecting to rat substantia gelatinosa neurones. J. Physiol. 536, 505–519.

Kato, G., Yasaka, T., Katafuchi, T., Furue, H., Mizuno, M., Iwamoto, Y., and Yoshimura, M. 2006. Direct GABAergic and glycinergic inhibition of the substantia gelatinosa from the rostral ventromedial medulla revealed by in vivo patch-clamp analysis in rats. J. Neurosci. 26, 1787–1794.

Laube, B., Maksay, G., Schemm, R., and Betz, H. 2002. Modulation of glycine receptor function: a novel approach for therapeutic intervention at inhibitory synapses? Trends Pharmacol. Sci. 23, 519–527.

Legendre, P. 2001. The glycinergic inhibitory synapse. Cell. Mol. Life Sci. 58, 760–793.

Malmberg, A. B. and Yaksh, T. L. 1992. Hyperalgesia mediated by spinal glutamate or substance P receptor blocked by spinal cyclooxygenase inhibition. Science 257, 1276–1279.

Moore, K. A., Kohno, T., Karchewski, L. A., Scholz, J., Baba, H., and Woolf, C. J. 2002. Partial peripheral nerve injury promotes a selective loss of GABAergic inhibition in the superficial dorsal horn of the spinal cord. J. Neurosci. 22, 6724–6731.

Nabekura, J., Xu, T. L., Rhee, J. S., Li, J. S., and Akaike, N. 1999. Alpha2-adrenoceptor-mediated enhancement of glycine response in rat sacral dorsal commissural neurons. Neuroscience 9, 29–41.

Polgar, E., Gray, S., Riddell, J. S., and Todd, A. J. 2004. Lack of evidence for significant neuronal loss in laminae I–III of the spinal dorsal horn of the rat in the chronic constriction injury model. Pain 111, 144–150.

Polgar, E., Hughes, D. I., Riddell, J. S., Maxwell, D. J., Puskar, Z., and Todd, A. J. 2003. Selective loss of spinal GABAergic or glycinergic neurons is not necessary for development of thermal hyperalgesia in the chronic constriction injury model of neuropathic pain. Pain 104, 229–239.

Reinold, H., Ahmadi, S., Depner, U. B., Layh, B., Heindl, C., Hamza, M., Pahl, A., Brune, K., Narumiya, S., Müller, U., and Zeilhofer, H. U. 2005. Spinal inflammatory hyperalgesia is mediated by prostaglandin E receptors of the EP2 subtype. J. Clin. Invest. 115, 673–679.

Rhee, J. S., Wang, Z. M., Nabekura, J., Inoue, K., and Akaike, N. 2000. ATP facilitates spontaneous glycinergic IPSC frequency at dissociated rat dorsal horn interneuron synapses. J. Physiol. 524, 471–483.

Samad, T. A., Sapirstein, A., and Woolf, C. J. 2002. Prostanoids and pain: unraveling mechanisms and revealing therapeutic targets. Trends Mol. Med. 8, 390–396.

Todd, A. J. and Sullivan, A. C. 1990. Light microscope study of the coexistence of GABA-like and glycine-like immunoreactivities in the spinal cord of the rat. J. Comp. Neurol. 296, 496–505.

Zeilhofer, H. U. 2005. The glycinergic control of spinal pain processing. Cell Mol. Life Sci. 62, 2027–2035.

Zeilhofer, H. U., Muth-Selbach, U., Gühring, H., Erb, K., and Ahmadi, S. 2000. Selective suppression of inhibitory synaptic transmission by nocistatin in the rat spinal cord dorsal horn. J. Neurosci. 20, 4922–4999.

Zeilhofer, H. U., Studler, B., Arabadzisz, D., Schweizer, C., Ahmadi, S., Layh, B., Bösl, M. R., and Fritschy, J. M. 2005. Glycinergic neurons expressing enhanced green fluorescent protein in bacterial artificial chromosome transgenic mice. J. Comp. Neurol. 482, 123–141.

Further Reading

Baba, H., Kohno, T., Moore, K. A., and Woolf, C. J. 2001. Direct activation of rat spinal dorsal horn neurons by prostaglandin E2. J. Neurosci. 21, 1750–1756.

Sherman, S. E., Luo, L., and Dostrovsky, J. O. 1997. Spinal strychnine alters response properties of nociceptive-specific neurons in rat medial thalamus. J. Neurophysiol. 78, 628–637.

Sivilotti, L. and Woolf, C. J. 1994. The contribution of GABAA and glycine receptors to central sensitization: disinhibition and touch-evoked allodynia in the spinal cord. J. Neurophysiol. 72, 169–179.

28 Pain Following Spinal Cord Injury

R P Yezierski, Comprehensive Center for Pain Research and The McKnight Brain Institute, University of Florida, Gainesville, FL, USA

Glossary

at-level pain Pain condition associated with spinal cord injury. Located in dermatomes associated with spinal segments immediately adjacent to the site of injury. Characteristics typically include elevated sensitivity to mechanical and thermal stimuli. Spontaneous sensations (i.e., tingling, numbness, burning) can also be associated with this condition of abnormal sensation. Believed to have a predominately spinal mechanism.

below-level pain Pain condition associated with spinal cord injury. Located in dermatomes associated with spinal segments below the level of injury. Pain is typically spontaneous with characteristics of burning, stabbing, and shooting sensations. Mechanism of below-level pain is believed to involve spinal as well as supraspinal mechanisms.

central injury cascade Pathological events triggered by injury to the central nervous system. Initial events include the extracellular release of excitatory amino acids and the initiation of excitotoxic and inflammatory cascades. Included in this cascade are neurochemical and molecular changes that ultimately lead to changes in the functional properties of spinal and supraspinal sensory neurons.

central sensitization Change in the functional state of central neurons secondary to persistent input or injury-induced elevations of excitatory amino acids leading to the activation of cellular signaling pathways. The end result of this change in functional state is an increase in excitability that can lead to clinical symptoms of allodynia, hyperalgesia, or pain.

pattern generators of pain Dysfunctional neurons with increased excitability are capable of producing abnormal discharges. Collectively the discharges generated from a large population of neurons can form the substrate responsible for the perception of pain.

secondary injury Following an initial mechanical or vascular insult to the brain or spinal cord there is a series of events that are components of the central injury cascade. These events lead to the spread of pathological damage of central neurons beyond the primary insult.

28.1 The Condition of Pain Following Spinal Cord Injury

Pain following spinal cord injury (SCI) is one of many challenges associated with the life-threatening consequences of SCI. Over the past 15 years clinical and preclinical studies related to at- and below-level pain have provided an understanding of the spinal and supraspinal mechanism(s) responsible for these conditions. In spite of the fact that loss of sensory and motor function below the level of injury are regarded as the most significant consequences of SCI, the condition of pain has a direct relationship with the ability of patients to regain an optimal level of activity, return to work, and achieve an acceptable quality of life. The impact of spinal injury pain on the healthcare community is evident from studies reporting the incidence of painful sensations at a rate of 60–80% for all SCI patients with nearly 40% reporting severe pain to the extent they would trade any chance of recovering motor function for relief of pain (Nepomuceuno, C. *et al.*, 1979; Beric, A., 1990; Britell, C. W. and Mariano, A. J., 1991; Tasker, R. R. *et al.*, 1991; Beric, A., 1992; Widerstrom-Noga, E. G. *et al.*, 1999a). The functional impact of pain following SCI is further demonstrated by a report describing the challenge of dealing with pain is only surpassed by the decreased ability to walk, loss of sexual function, and bladder and bowel dysfunction (Widerstrom-Noga, E. G. *et al.*, 1999b).

28.2 Clinical Characteristics of Spinal Injury Pain

Pain associated with SCI is typically bilateral and perceived in anesthetic regions at and below the level of injury. Although changes in sensation above the injury have been reported, there is little evidence supporting the clinical significance of this type of altered sensation. The more common at- and below-level conditions are often referred to as deafferentation pain, dysesthetic pain, or central dysesthesia syndrome (Beric, A. *et al.*, 1988; Bonica, J. J., 1991; Yezierski, R. P., 2002). A characteristic of SCI pain that has been consistently documented is the detrimental impact across multiple quality of life domains, including life satisfaction, physical health, and overall handicap. Commonly used descriptors of SCI pain include burning, tingling, numbness, aching, and throbbing. Although it is often difficult to classify different categories of pain, below-level neuropathic pain is the most common and most difficult to treat (Finnerup, N. M. *et al.*, 2002). When discussing different classifications of pain it is helpful to have a standard taxonomy. A Task Force of the International Association for the Study of Pain proposed a taxonomy for SCI pain based on the type of pain, location, and putative mechanism of pain onset (Siddall, P. J. *et al.*, 2002; Table 1). Having a taxonomy that is accepted and used by clinical and basic research specialists in the fields of pain and spinal injury is essential to achieving a better understanding of different pain conditions.

Table 1 International Association for the Study of Pain proposed classification of spinal cord injury pain

Broad type (tier one)	Broad system (tier two)	Specific structures/pathology (tier three)
Nociceptive	Musculoskeletal	Bone, joint, muscle trauma, or inflammation
		Mechanical instability
		Muscle spasm
		Secondary overuse syndromes
	Visceral	For example, Renal calculus, bowel dysfunction
		Dysreflexic headache
Neuropathic	(above level)	Compressive mononeuropathies
		Complex regional pain syndromes
	(at level)	Nerve root compression, cauda equina
		Syringomyelia
		Spinal cord trauma/ischaemia
		Dual level cord and root trauma
	(below level)	Spinal cord trauma/ischaemia

Adapted from Siddall, P. J., Yezierski, R. P., and Loeser, J. D. 2002. Taxonomy and Epidemiology of Spinal Cord Injury. In: Spinal Cord Injury Pain: Assessment, Mechanisms, Management (*eds*. R. P. Yezierski and K. Burchiel), pp. 9–24. IASP Press, with permission from International Association for the Study of pain.

The condition of pain following spinal injury has a variable onset depending on the type of pain. Siddall P. J. *et al.* (2003) described the prevalence and onset of different types of pain in a cohort of 100 patients. Musculoskeletal pain (59%) was the most common type of pain followed by at-level (41%) and below-level (34%) neuropathic pain, with visceral pain (5%) the least common type of pain experienced. The mean onset time for any type of pain was 1.6 years with at-level and musculoskeletal pain having the shortest onset times, 1.2 and 1.3 years, respectively. Below-level pain generally had a longer onset time (1.8 years) and visceral pain had the longest (4.2 years). While there has been considerable effort to identify physical factors such as completeness and level of injury that correlate with pain onset, few consistent predictors have been identified (Siddall, P. J. *et al.*, 1999a; Werhagen, L. *et al.*, 2004). A positive relationship between the higher incidence of pain in patients with thoroacolumbar and incomplete lesions has been described (Demirel, G. *et al.*, 1998).

28.3 The Research Challenge of Spinal Injury Pain

A major challenge in the study of altered sensation following SCI is the development of appropriate injury models that parallel the human condition. The study of SCI pain has seen the use of several different models, including mechanical trauma, isolated lesions, complete transection, chemical lesions, and ischemic injury, each with pathological components found in the human condition (Wiesenfeld-Hallin, Z. *et al.*, 1994; Siddall, P. J. *et al.*, 1995; Christensen, M. D. and Hulsebosch, C. E., 1997; Yezierski, R. P. *et al.*, 1998; Bruce, J. C. *et al.*, 2002; Scheifer, C. *et al.*, 2002). Behavioral outcome measures used to evaluate injury-induced changes in sensation represents another critical factor in the study of spinal and supraspinal mechanisms of different pain conditions. The majority of behavioral measures used in the study of spinal injury pain rely on responses to mechanical and thermal stimuli to evaluate changes in nociceptive sensitivity. These studies have used a variety of nociceptive reflexes, such as hindpaw withdrawal, that are regulated by segmentally organized spinal mechanisms and are present in spinalized animals. Lick and guard responses have also been used and these responses to nociceptive input depend on spino-bulbo-spinal circuits and are present in decerebrate animals (Woolf, C. J., 1984). Reflex responses are important measures of

nociception that have the potential to reveal changes in excitability of spinal circuits, an important consequence of spinal injury. Enhancement of flexion/withdrawal reflexes as reported in the spastic syndrome, however, can be dissociated from chronic below-level pain in individual cases of subtotal SCI (Vierck, C. J. *et al.*, 2002a; 2002). These responses do not represent the component of pain perception that requires cerebral processing of sensation (Mauderli, A. P. *et al.*, 2000) and are more likely to apply to the study of at-level pain that is thought to depend on a spinal mechanism. A misconception in the preclinical study of SCI pain is that stimulation of peripheral dermatomes below the lesion to elicit a reflex response is thought to qualify as the study of below-level pain. The fact is below-level pain is likely to involve supraspinal mechanisms and therefore cannot be studied with reflex-based outcome measures.

Another strategy of behavioral assessment is the use of operant tests that require discrimination between nociceptive and non-nociceptive stimulation and the organization of complex behavioral adaptations to specific environmental contingencies. This approach is believed to provide a more accurate assessment of the pain experience (Vierck, C. J. *et al.*, 2002). The challenge of selecting appropriate behavioral measures is especially evident in the study of below-level pain. If it is assumed that this type of pain requires cortical involvement then to study this type of pain effectively a behavioral measure that relies on activation of cortical structures must be used. The only measures to fit these criteria are operant behaviors that rely on cortical processing for the perception of nociceptive stimuli as well as the planning and execution of learned responses. Acceptance of the differences between responses obtained with operant versus reflex measures represents a major challenge in the study of spinal injury pain, especially studies related to at- and below-level pain.

By far the most significant sensory consequence of spinal injury is below-level pain. This condition has been difficult to study in preclinical models as the pain is typically spontaneous and occurs in parts of the body that are anesthetic. Below-level pain has been attributed to white matter damage, although evidence supports the presence of abnormal activity within the gray matter contributing to below-level hyperalgesia and spontaneous pain behavior. This was demonstrated in experiments with rats and monkeys in which ischemic involvement of spinal gray matter was a critical factor in the development of hyperalgesia after interruption of the spinothalamic tract (Vierck, C. J. *et al.*, 2000). Similarly, quisqualic acid injections in the spinal gray matter produce below-level hyperalgesia and spontaneous pain behavior without damage to surrounding gray matter (Yezierski, R. P. *et al.*, 1998; King, C. D. *et al.*, 2006). Clinically, results of a magnetic resonance imaging study in patients with and without below-level pain showed the most common pathological feature of patients with below-level pain was the presence of lesions that included the central core of spinal gray matter (Finnerup, N. M. *et al.*, 2003b).

28.4   Pathophysiology of Spinal Cord Injury

In order to understand the mechanisms responsible for SCI pain it is important to acknowledge that associated with spinal injury is a cascade of biochemical, molecular, and anatomical changes that collectively

impact the survivability and functional integrity of spinal neurons (Figure 1). The clinical consequences associated with SCI, including at- and below-level pain, are believed to result from the influence of this cascade on the functional state of spinal and supraspinal neurons. Therefore to understand the end result (i.e., pain), it is important to appreciate the basic biology of secondary events that precede this condition.

The most obvious pathological changes associated with traumatic or ischemic injury to the spinal cord include, but are by no means limited to, the loss of neurons, damage to surrounding white matter, astrocytic scarring, syrinx formation, and breakdown of the spinal blood–brain barrier (Kakulas, B. A. *et al.*, 1990; Bunge, R. P. *et al.*, 1993). Also contributing to the progression of tissue damage are secondary injury cascades that include excitotoxic and inflammatory processes (Young, W., 1987; Tator, C. H. and

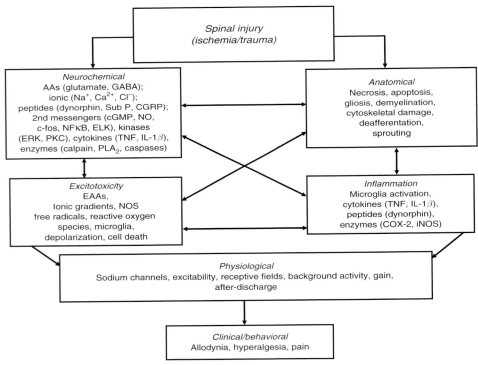

Figure 1   Components of the spinal injury cascade that contribute to the mechanism responsible for the development of pain. Evidence supporting the involvement of this cascade comes from results of clinical and experimental studies involving mechanical, ischemic, and/or chemical models of spinal injury (see text for details). The four major components of the cascade (neurochemical, excitotoxicity, anatomical, and inflammation) are represented as interactive and collectively lead to conditions producing functional changes in spinal and supraspinal neurons. The endpoint of the cascade is the onset of clinical and behavioral symptoms, e.g., allodynia, hyperalgesia, and pain. AAs, amino acids; cGMP, cyclic guanidine monophosphate; CGRP, calcium gene-related peptide; COX-2, cyclo-oxygenase-2; EAA, excitatory amino acid; ELK, ets-like gene; ERK, extracellular signal-regulated kinase; GABA, gamma-aminobutyric acid; IL-1β, interleukin-1β; iNOS, inducible nitric oxide synthase; NFkB, nuclear factor kappa B; NO, nitric oxide; PKC, protein kinase C; PLA$_2$, phospholipase A2; Sub P, substance P; TNF, tumor necrosis factor. Reproduced from Yezierski, R. P. 2006. Pain Following Spinal Cord Injury: Central Mechanisms. In: Handbook of Clinical Neurology (*eds.* F. Cervero and T. S. Jensen), Vol. 81, pp. 293–307, with permission from Elsevier.

Fehlings, M. G., 1991; Hsu, C. Y. *et al.*, 1994; Regan, R. and Choi, D. W., 1994; Bethea, J. R. *et al.*, 1999). Excitotoxic events following spinal injury are initiated by the well-documented effects of excitatory amino acids (EAAs) that are involved in the secondary pathology, including neuronal degeneration (Faden, A. I. and Simon, R. P., 1988; Nag, S. and Riopelle, R. J., 1990; Hao, J. X. *et al.*, 1991; Yezierski, R. P. *et al.*, 1993; Regan, R. and Choi, D. W., 1994; Wrathall, J. R. *et al.*, 1994). For this reason glutamate has been viewed as one of several putative chemical mediators contributing to the central cascade of secondary pathological changes following injury (Tator, C. H. and Fehlings, M. G., 1991). Inflammation is another major contributor to secondary injury (Bethea, J. R. and Dietrich, W. D., 2002). With the aid of specific histological and immunohistological stains the temporal profile of glia activation (astrocytes, microglial) and the infiltration of macrophages and other inflammatory cell types can be followed. Upregulation of messenger RNA (mRNA) for c-fos, tumor necrosis factor (TNF)-α and dynorphin have also been described following SCI (Yakovlev, A. G. and Faden, A. I., 1994). From this discussion it should be clear that the pathological sequela of SCI is by no means simple neither is it restricted to the site of insult as pathological consequences of SCI have been observed throughout the full extent of the neuraxis (Brewer, K. L. *et al.*, 1997; Jain, N. *et al.*, 1998; Morrow, T. J. *et al.*, 2000).

One of the earliest responses to spinal injury is the breakdown of cellular membranes and subsequent activation of phospholipases. Within minutes of injury, lipid peroxidation and membrane hydrolysis release polyunsaturated fatty acids capable of causing damage, through arachidonic acid and eicosanoid pathways (Anderson, D. and Hall, E., 1993; Tator, C., 1995). Reactive oxygen species (ROS) likewise trigger excitotoxic reactions (glutamate and quinolinate); catecholamine oxidation; mitochondrial breakdown; oxidation of extravasated hemoglobin; neutrophil infiltration; microglial activation; and release of nitric oxide, cytokines, and growth factors. Thus, ROS contribute to the development of a permissive environment for pain following SCI (Kim, H. K. *et al.*, 2004; Hains, B. C. and Waxman, S. G., 2006; Peng, X. M. *et al.*, 2006). Other changes associated with different models of SCI include: afferent sprouting in distant segments (Ondarza, A. B. *et al.*, 2003), upregulation of vanilloid receptor expression (Zhou, Y. *et al.*, 2002), changes in expression of metabotropic glutamate receptors (Mills, C. D. *et al.*, 2001), activation of protein kinases and transcription factors

associated with the MAPK signaling pathway (Crown, E. D. *et al.*, 2005; Yu, C. G. and Yezierski, R. P., 2005a), increased NR1 serine phosphorylation of the NMDA receptor (Caudle, R. M. *et al.*, 2003), changes in galanin immunoreactivity (Zvarova, K. *et al.*, 2004), and increased expression of c-fos mRNA (Yakovlev, A. G. and Faden, A. I., 1994; Siddall, P. J. *et al.*, 1999b; Abraham, K. E. and Brewer, K. L., 2001). Furthermore, analysis of molecular events using DNA microarray analysis has shown that 1 h after spinal trauma there are changes in mRNA levels for 165 genes (Nesic, O. *et al.*, 2002; 2005), including those regulating transcription factors, inflammatory processes, cell survival, and membrane excitability. In spite of the wide range of changes associated with spinal injury, the challenge remains to determine which ones have a causal relationship with the mechanism of pain onset, as opposed to being secondary events associated with the process of injury.

An important consideration in the progression of pathological changes associated with spinal injury is the longitudinal extent over which spinal tissue is influenced by the central injury cascade. In a study using agents with neuroprotective properties it was shown that limiting the spread of injury had a significant impact on the expression of pain-related behaviors (Yu, C. G. *et al.*, 2003). This study led to the proposal of a spatial threshold in which a critical distance of spinal tissue must be affected by the secondary injury cascade in order for pain behaviors to develop. Therefore, a critical determinant underlying the expression of pain following SCI may be a combination of specific injury related events together with the longitudinal spread of these events within the cord (Yezierski, R. P., 2000; Gorman, A. L. *et al.*, 2001). Identifying the mechanism responsible for the spread of injury may provide valuable insight into the development of different pain conditions. Given the permissive microenvironment for the development of pain associated with the wide spectrum of molecular, anatomical, neurochemical, and functional changes that occur following spinal injury it is not surprising that pain is a common clinical finding following injury.

28.5 Mechanism(s) of Pain Following Spinal Injury

The precise mechanism(s) underlying different types of neuropathic pain (at- and below-level) associated with spinal injury remain to be determined. While below-level pain is considered to have a central mechanism involving cortical and subcortical structures, at-level pain may have peripheral and central components. Unfortunately, little is known about the potential peripheral contribution to spinal injury pain. Little is also known about the role of the autonomic nervous system (ANS) in the development of different pain conditions following injury. Given the likely damage to sympathetic and parasympathetic neurons following injury and the contribution of the ANS to chronic pain states, there is a need for studies focusing on the role of autonomic dysfunction in the development of spinal injury pain.

Results of experimental studies have provided evidence to construct a working model for at- and below-level behavioral effects of SCI. This model includes a combination of six conditions: (1) loss of spinal inhibitory neurons, (2) emergence of spinal generators and supraspinal amplifiers of abnormal activity, (3) dependence on a population of NK-1R expressing neurons in the superficial dorsal horn, (4) longitudinal

progression of a secondary injury cascade, (5) reorganization of supraspinal structures, and (6) activation of the MAPK cell signaling pathway (Figure 2).

28.5.1 Loss of Inhibitory Tone

A critical factor in the onset of painlike behaviors following SCI is the loss of inhibitory influences within the injured spinal cord (Wiesenfeld-Hallin, Z., et al. 1994). The loss of inhibitory control allows for the recruitment of surrounding neurons and intensification and spread of abnormal sensations, including pain. Support for this conclusion comes from four lines of evidence: (1) following ischemic injury there is an increased excitability of WDR neurons that is reversed by the GABA$_B$ receptor agonist baclofen (Hao, J. X. et al., 1992), (2) hypersensitivity to peripheral stimuli can be produced in normal animals by the intrathecal administration of gamma-aminobutyric acid (GABA) receptor antagonists (Yaksh, T. L., 1989; Hao, J.-X. et al., 1994),

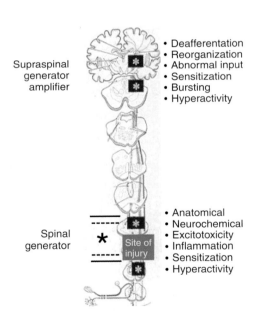

Supraspinal generator amplifier
- Deafferentation
- Reorganization
- Abnormal input
- Sensitization
- Bursting
- Hyperactivity

Spinal generator
* Site of injury
- Anatomical
- Neurochemical
- Excitotoxicity
- Inflammation
- Sensitization
- Hyperactivity

1. Loss of spinal inhibitory mechanisms.

2. Spinal and supraspinal generators and/or amplifiers.

3. NK-1 receptor expressing neurons in lamina I.

4. Longitudinal progression of injury cascade.

5. Functional reorganization of supraspinal structures.

6. Injury-induced activation of cell signaling pathways.

Figure 2 Summary of six different injury-induced changes contributing to the development of at- and below-level pain associated with spinal cord injury. A spinal pain generator evolves due to the collective impact of anatomical, neurochemical, inflammatory, and excitotoxic changes that result in the increased excitability of spinal sensory neurons. Specific changes contributing to the development of the spinal generator include loss of inhibitory mechanisms, the longitudinal progression of the spinal injury cascade, and the injury-induced activation of cell-signaling pathways. At supraspinal levels the presence of a generator/amplifier mechanism emerges as a result of the loss of afferent input from spinal segments below the level of injury (deafferentation). The functional impact of this and other injury-induced changes at supraspinal levels (e.g., sprouting, unmasking of connections) results in the activation of supraspinal regions by input from somatotopically inappropriate regions of the cord and the referral of pain sensations to dermatomes below the level of injury. Reproduced from Yezierski, R. P. 2006. Pain Following Spinal Cord Injury: Central Mechanisms. In: Handbook of Clinical Neurology (eds. F. Cervero and T. S. Jensen), Vol. 81, pp. 293–307, with permission from Elsevier.

(3) decreased numbers of GABA-positive neurons following ischemic spinal injury (Zhang, A.-L. *et al.*, 1999), and (4) administration of drugs that prolong the action of inhibitory neurotransmitters (i.e., tricyclic antidepressants) are effective in the short-term treatment of spinal injury pain (Tunks, E., 1986; Lejon, G. and Boivie, J., 1991; Boivie, J., 1994). Although existing evidence supports the decreased inhibitory influence of GABAergic neurotransmission in altering the functional properties of neurons in the injured cord, not to be overlooked in this process is a decreased influence of supraspinal inhibitory pathways that may be disrupted following injury. Alterations in descending serotonergic projections above and below the site of injury have led to the suggestion that loss of inhibitory influences below the injury may contribute to the increased excitability of neurons while an increase in serotonergic projections above the injury may contribute, through descending facilitation, to a change in functional state of neurons above the level of injury (Oatway, M. A. *et al.*, 2004).

28.5.2 Pattern Generators of Pain

Another component of the pathophysiological sequela of SCI that is thought to contribute to the onset of altered sensations is the emergence of a pattern generating mechanism (spinal and supraspinal). Several observations are consistent with the pattern generating hypothesis: (1) existence of focal regions of hyperactivity in the spinal cord and thalamus of spinal injured patients (Loeser, J. D. *et al.* 1968; Edgar, R. E. *et al.*, 1993; Lenz, F. A. *et al.*, 2000); (2) effectiveness of spinal anesthesia in alleviating pain when delivered proximal to the site of injury (Pollock, L. J. *et al.*, 1951; Botterell, E. H. *et al.*, 1953; Loubser, P. G. and Donovan, W. H., 1991); and (3) sensitization and prolonged after-discharges of spinal and supraspinal sensory neurons following SCI (Hao, J. X. *et al.*, 1992; Yezierski, R. P. and Park, S. H., 1993; Lenz, F. A. *et al.*, 2000; Drew, G. M. *et al.*, 2001; Hains, B. C. *et al.*, 2003b; Hoheisel, U. *et al.*, 2003; Hao, J. X. *et al.*, 2004; Hains, B. C. *et al.*, 2005). Consistent with an involvement of neuronal hyperexcitability as a component of the spinal and possibly supraspinal mechanism of SCI pain are results of pharmacological studies. For example, lidocaine and ketamine, two drugs that reduce membrane excitability and glutamate receptor activation are effective in treating SCI pain (Loubser, P. G. and Donovan, W. H., 1991; Eide, P. K. *et al.*, 1995; Attal, N. *et al.*, 2000). Efforts to increase inhibition with either baclofen or

propofol are also effective (Herman, R. M. *et al.*, 1992; Canavero, S. *et al.*, 1995). The anticonvulsant lamotrigine that affects sodium channels involved in hyperexcitability is also suggested to be effective in SCI patients with spontaneous and evoked pain (Finnerup, N. M. *et al.*, 2002).

Consistent with the excitability hypothesis and involvement of neurons in the superficial dorsal horn in spontaneous and evoked pain behaviors, are clinical reports of abnormal focal hyperactivity within the superficial laminae of the injured cord (Edgar, R. E. *et al.*, 1993). Microcoagulation of these hyperactive areas in the dorsal root entry zone (DREZ) at depths of 1–2 mm, representing Rexed's laminae I–II, results in a significant diminution of pain. A variant of the DREZ technique utilizing intramedullary recordings of C-fiber evoked responses to guide DREZ lesioning allows for the somatotopic mapping of the DREZ with regard to the generation of central deafferentation pain. In 25 patients where this technique was used, 84% received complete pain relief (Falci, S. *et al.*, 2002). An involvement of this region in the generation of pain is further supported by a report where elimination of NK-1R-expressing neurons in the superficial dorsal horn resulted in not only the prevention of spontaneous pain behavior, but also the reversal of the behavior once it had been initiated (Yezierski, R. P. *et al.*, 2004).

Injury-induced changes contributing to increased excitability of spinal circuitry has also been attributed to an increased expression of the $Na_v1.3$ sodium channel in spinal segments L3–L5 following contusion injury at T9 (Hains, B. C. *et al.*, 2003a). Upregulation of $Na_v1.3$ expression at sites remote from the epicenter of injury was shown to have a significant role in producing increased excitability of spinal multireceptive neurons recorded in L3–L5. Efforts to block these changes after injury showed that responses of neurons in animals treated intrathecally with antisense oligodeoxynucleotides targeting $Na_v1.3$ decreased expression of $Na_v1.3$ mRNA and protein, reduced hyperexcitability of multireceptive neurons, and attenuated mechanical allodynia and thermal hyperalgesia after SCI. Similarly, upregulation of $Na_v1.3$ sodium channels in thalamic neurons has also been shown to contribute to changes in the electrophysiological properties of these neurons (Hains, B. C. *et al.*, 2005). Thus, changes in the functional state of sensory neurons secondary to upregulation of sodium channels provides the mechanism for a pain generator in the injured cord as well as an amplifier of abnormal activity in thalamus.

The contribution of abnormal spinal input together with the effects of deafferentation (secondary to loss of spinal projection neurons) and upregulation of sodium channels could result in the development of abnormal generators or amplifiers of spinal input at supraspinal sites (Vierck, C. J., Jr., 1991). The transient relief of below-level pain in humans following application of local anesthetic to the proximal stump of a spinal lesion (Pollock, L. J. et al., 1951) suggests that the presence of a spinal generator plays a significant role in maintaining chronic below-level pain. Thus, a combination of supraspinal changes secondary to deafferentation, and the emergence of a spinal generator of abnormal activity are two significant events that may contribute to the development and maintenance of below-level pain. Recently, the report of elevated blood flow in thalamic nuclei (possibly reflecting changes in the functional state of neurons) lend additional support to the thalamic involvement in the supraspinal response to SCI and to the mechanism of injury-induced pain (Morrow, T. J. et al., 2000).

28.5.3 Cell Signaling Pathways: Synaptic Plasticity and Functional State

Another consideration for the mechanism of chronic pain following spinal injury is synaptic plasticity and changes in the functional state of spinal neurons. A significant response to injury of central or peripheral origin is the cellular response believed to contribute to the sensitization of spinal neurons (Woolf, C. J. and Csotigan, M., 1999; Woolf, C. J., 1992; Willis, W. D., 2002). Synaptic plasticity in the central nervous system is thought to be part of central sensitization and together with the long-term changes in functional state of spinal sensory neurons represents a potential mechanism for persistent pain (Ji, R. R. et al., 2003). Unfortunately, we are only at the beginning of understanding the biochemical and molecular events underlying long-term functional plasticity and their role in chronic pain. Included in the events thought to be involved in producing long-term changes are: (1) phosphorylation of regulatory proteins (as well as dephosphorylation of others), (2) positive and negative regulation of gene transcription, (3) injury-induced synthesis of proteins (as well as reduced synthesis of others), (4) strengthening of some and weakening of other connections (synaptic long-term potentiation, sprouting, and pruning), and (5) the death or rescuing of neurons. These neuronal changes share initiating events such as glutamate-induced elevations in calcium, maintaining events (nitric oxide synthesis, activation of protein kinases), perpetuating forces (activation of cyclic AMP response element-binding (CREB), nuclear factor kappa B (NF-κB), and terminal results (altered synaptic efficacy). In contrast to cellular events underlying persistent pain of peripheral origin, the mechanism underlying long-term plasticity associated with pain of central origin remains largely unexplored.

Important events contributing to the development of injury-induced pain are found in a construct of central sensitization and plasticity (Woolf, C. J. and Csotigan, M., 1999; Ji, R. R. and Woolf, C. J., 2001; Ji, R. R. et al., 2003). Hypersensitivity is believed to be a consequence of early posttranslational changes as well as later transcription-dependent changes in effector genes. Neuroplasticity resulting from these events includes activity-dependent changes in neurons and specific signaling molecules in signal transduction pathways. These changes likely contribute to alterations in the basal state of excitability that ultimately contributes to the development of spinal pain generators and possibly supraspinal amplifiers.

Examples of cell signaling events following spinal injury were described in a study of changes in the mitogen-activated protein kinase (MAPK) signaling pathway following excitotoxic spinal injury including: (1) increased phosphorylation of ERK1/2, (2) increased activation of NF-κB and phosphorylation of ELK-1, and (3) increased gene expression for the NK-1 receptor and NR1 and NR-2A subunits of the NMDA receptor (Yu, C. G. and Yezierski, R. P., 2005a). Blockade of the MAPK cascade with the MEK inhibitor PD98059 inhibited phosphorylation of ELK-1, activation of NF-κB, and gene expression of NR1, NR-2A, and NK-1R, and prevented the development of spontaneous pain behavior. Injury-induced elevations in spinal levels of EAAs thus lead to the activation of the ERK → ELK-1 and NF-κB signaling cascades and the transcriptional regulation of receptors important in the development of chronic pain. Blockade of these intracellular kinase cascades prevents the onset of injury-induced pain behavior. Similar changes in different components of the MAPK signaling pathway have been described in the diencephalon following injury (Yu, C.-G. and Yezierski, R. P., 2005b). The persistent upregulation of CREB in spinothalamic tract cells is another example of how cell signaling events are affected by SCI (Crown, E. D. et al., 2005).

The above results suggest that many of the same molecular changes described as activity-dependent following peripheral nerve and tissue injury can also be induced following central injury, that is, injury dependent. This suggests that the mechanism responsible for the increased excitability of spinal neurons following spinal injury may be similar to the well-documented activity-dependent mechanism induced by damage to peripheral tissue and/or nerves, a mechanism resulting in the activation of intracellular kinase cascades and ultimately long-term changes in synaptic efficacy and neuronal excitability. The end result of these events is the emergence of a hyperexcitable pain-generating mechanism.

28.5.4 Supraspinal Changes Following Spinal Cord Injury

One cannot discuss SCI without acknowledging the most fundamental of all anatomical facts and that is the connection between the brain and spinal cord. Following injury to the cord it has been well documented that there are multiple changes (e.g., anatomical, molecular, functional) that occur within cortical and subcortical structures. Changes in supraspinal structures could potentially contribute to the central mechanism responsible for the onset and progression of injury-induced pain behaviors. Studies carried out include an evaluation of changes in forebrain blood flow (Morrow, T. J. et al., 2000), and evaluation of opiate transmitters in selected supraspinal sites (Abraham, K. E. et al., 2001). In the study by Morrow T. J. et al. (2000) significant increases in regional cerebral blood flow were found in seven of 22 supraspinal structures examined, including the arcuate nucleus, hindlimb region of S1 cortex, parietal cortex and the thalamic posterior, ventral posterior medial and lateral nuclei. All of these structures have well-documented roles in sensory processing. These studies complement clinical observations showing similar changes in thalamic blood flow following SCI (Ness, T. J. et al., 1998), alterations in the chemical profile of ventroposterolateral thalamus in patients with SCI pain (Pattany, P. M. et al., 2002), and reports of hyperactive foci of thalamic activity in SCI patients with spontaneous burning pain (Lenz, F. A. et al., 2000; Ohara, S. et al., 2002).

Additional support for supraspinal changes contributing to the mechanism of spinal injury pain is the work from Brewer and colleagues focusing on changes in peptidergic transmitter systems at supraspinal levels following excitotoxic SCI (Abraham, K. E. et al., 2001). Preproenkephalin (PPE) and preprodynorphin (PPD) expression was shown to increase in cortical regions associated with nociceptive function; PPE in the cingulate cortex and PPD in the parietal cortex, both ipsilateral and contralaterally at various timepoints following injury. Increases in PPD were significant in animals that developed spontaneous pain behaviors versus those that did not. PPE expression in the anterior cingulate cortex and PPD expression in the contralateral parietal cortex were significantly higher in animals with spontaneous pain behaviors versus those without pain behaviors. The important conclusion from these reports is that following spinal injury there are significant changes in selected supraspinal sites including somatosensory structures putatively involved in pain processing. These results are consistent with a recent report in humans using proton magnetic resonance spectroscopy showing changes in the thalamic concentrations of selected metabolites in SCI patients with pain (Pattany, P. M. et al., 2002). Thus there is likely to be a significant supraspinal component to the mechanism responsible for the development of below-level SCI pain. Understanding the role of these supraspinal changes in the development of pain, especially below-level, remains a challenge for future studies.

28.6 Future Directions

An important realization of clinical and experimental studies directed toward understanding the mechanism(s) of SCI pain is the importance of identifying the critical events responsible for the onset of this condition. Continued studies focusing on different components of secondary excitotoxic and inflammatory cascades ultimately responsible for the dysfunction of spinal and supraspinal neurons are essential to the design of more effective interventions. The condition of pain associated with spinal injury is unlikely to have a simple mechanism that relies exclusively on any one consequence of injury. It is more probable that several components of a central injury cascade contribute to the development and maintenance of different pain states. The precise mechanism is likely to depend on the nature of the injury and progression of pathological and biochemical changes along the rostrocaudal axis of the cord, and include anatomical, molecular, and functional changes occurring in supraspinal structures. The mechanism is likely to include the combination of deafferentation resulting from spinal lesions combined with tonic excitatory input from a generator of abnormal activity in or around the epicenter of gray matter damage, that is, penumbral region. This input reaches deafferented supraspinal processing centers (via undamaged projections or multisynaptic pathways) and contributes to the abnormal discharge and/or amplification of activity ultimately leading to the perception of pain (at- or below-level).

In conclusion, well-established animal models indicate that at-level SCI pain results from excitotoxic/ischemic damage to the spinal gray and white matter. Experimental models have identified putative mechanisms for generation of at-level pain and have suggested a number of potential therapeutic approaches, including pharmacological strategies targeting inhibitory systems or spinal mediators of cell survival. Of special importance is control over secondary signaling pathways along with injury-induced chemical and molecular changes that modulate neuronal excitability. Additionally, current evidence suggests that restricting the extent of excitotoxic injury after traumatic SCI might prevent the development of below-level neuropathic pain, both by reducing excitatory influences and by limiting gray matter damage. In contrast, below-level neuropathic pain does not appear to result only from interruption of spinothalamic projections to rostral targets, but is potentiated by interruption and/or activation of other pathways, including propriospinal systems. Important questions related to different pain conditions associated with SCI include the role of glia activation and descending modulation (i.e., facilitation) in the initiation and maintenance of different pain states. It will also be important to determine if there are autonomic and peripheral components to the mechanism of SCI pain. Central and/or peripheral contributions to injury-induced visceral pain is another area requiring additional study. Continued use of available animal models of at- and below-level pain should provide an increasingly precise definition of mechanism(s) and potential therapeutic targets for these conditions.

Acknowledgments

This work was supported by NS40096 (RPY).

References

Abraham, K. E. and Brewer, K. L. 2001. Expression of c-fos mRNA is increased and related to dynorphin mRNA expression following excitotoxic spinal cord injury in the rat. Neurosci. Lett. 307, 187–191.

Anderson, D. and Hall, E. 1993. Pathophysiology of spinal cord trauma. Ann. Emerg. Med. 22, 987–992.

Attal, N., Gaude, V., Brasseur, L., Dupuy, M., Guirimand, F., Parker, F., and Bouhassira, D. 2000. Intravenous lidocaine in central pain: a double-blind, placebo controlled, psychophysical study. Neurology 54, 564–574.

Beric, A. 1990. Altered Sensation and Pain in Spinal Cord Injury. In: Recent Achievements in Restorative Neurology (eds. M. R. Dimitrijevic, P. D. Wall, and U. Lindblom), pp. 27–36. Karger.

Beric, A. 1992. Pain in Spinal Cord Injury. In: Spinal Cord Dysfunction (ed. L. S. Illis), pp. 156–165. Oxford University Press.

Beric, A., Dimitrijevic, M., and Lindblom, U. 1988. Central dysesthesia syndrome in SCI patients. Pain 34, 109–116.

Bethea, J. R. and Dietrich, W. D. 2002. Targeting the host inflammatory response in traumatic spinal cord injury. Curr. Opin. Neurol. 15, 355–360.

Bethea, J. R., Yu, C.-G., Plunkett, J. A., and Yezierski, R. P. 1999. Effects of interleukin-10 (IL-10) on pain behaviors following excitotoxic spinal cord injury. Soc. Neurosci. Abstr. 25, 1443.

Boivie, J. 1994. Central Pain. In: Textbook of Pain, 4th edn. (eds. P. D. Wall and R. Melzack), Churchill Livingston. 871–902.

Bonica, J. J. 1991. Semantic, Epidemiologic and Educational Issues of Central Pain. In: Pain and Central Nervous System Disease: the Central Pain Syndromes (ed. K. L. Casey), pp. 13–29. Raven Press.

Botterell, E. H., Callaghan, H. C., and Jousse, A. T. 1953. Pain in paraplegia: clinical management and surgical treatment. Proc. R. Soc. Med. 47, 281–288.

Brewer, K., Yezierski, R. P., and Bethea, J. R. 1997. Excitotoxic spinal cord injury induces diencephalic changes in gene expression. Soc. Neurosci. Abstr. 23, 438.

Britell, C. W. and Mariano, A. J. 1991. Chronic pain in spinal cord injury. Phys. Med. Rehab. 5, 71–82.

Bruce, J. C., Oatway, M. A., and Weaver, L. C. 2002. Chronic pain after clip-compression injury of the rat spinal cord. Exp. Neurol. 178, 33–48.

Bunge, R. P., Puckett, W. R., Becerra, J. L., Marcillo, A., and Quencer, R. M. 1993. Observations on the pathology of human spinal cord injury. A review and classification of 22 new cases with details from a case of chronic cord compression with extensive focal demyelination. Adv. Neurol. 59, 75–89.

Canavero, S., Bonicalzi, V., Pagni, C. A., et al. 1995. Propofol analgesia in central pain: preliminary clinical observations. J. Neurol. 242, 561–567.

Caudle, R. M., Perez, F. M., King, C., Yu, C. G., and Yezierski, R. P. 2003. N-methyl-D-aspartate receptor subunit expression and phosphorylation following excitotoxic spinal cord injury in rats. Neurosci. Lett. 349, 37–40.

Christensen, M. D. and Hulsebosch, C. E. 1997. Chronic central pain after spinal cord injury. J. Neurotrauma 14, 517–537.

Crown, E. D., Ye, Z., Johnson, K. M., Xu, G. Y., McAdoo, D. J., Westlund, K. N., and Hulsebosch, C. E. 2005. Upregulation of the phosphorylated form of CREB in spinothalamic tract cells following spinal cord injury: relation to central neuropathic pain. Neurosci. Lett. 384, 139–144.

Demirel, G., Yllmas, H., Gencosmanoglu, B., and Kesiktas, N. 1998. Pain following spinal cord injury. Spinal Cord. 36, 25–28.

Drew, G. M., Siddall, P. J., and Duggan, A. W. 2001. Responses of spinal neurons to cutaneous and dorsal root stimuli in rats with mechanical allodynia after contusive spinal cord injury. Br. Res. 893, 59–69.

Edgar, R. E., Best, L. G., Quail, P. A., and Obert, A. D. 1993. Computer-assisted DREZ microcoagulation: posttraumatic spinal deafferentation pain. J. Spinal Dis. 6, 48–56.

Eide, P. K., Stubhaug, A., and Stenehjem, A. E. 1995. Central dysesthesia pain after traumatic spinal cord injury is dependent on N-methyl-D-aspartate receptor activation, Neurosurgery 37, 1080–1087.

Faden, A. I. and Simon, R. P. 1988. A potential role for excitotoxins in the pathophysiology of spinal cord injury. Ann. Neurol, 24, 623–626.

Falci, S., Best, L., Bayles, R., Lammertse, D., and Starnes, C. 2002. Dorsal root entry zone microcoagulation for spinal cord injury-related central pain: operative intramedullary electrophysiological guidance and clinical outcome. J. Neurosurg. 97, 193–200.

Finnerup, N. M., Gyldensted, C., Nielsen, E., Kristensen, A. D., Back, F. W., and Jensen, T. S. 2003. MRI in chronic spinal cord injury patients with and without central pain. Neurology 61, 1569–1575.

Finnerup, N. B., Johannesen, I. L., Sindrup, S. H., Bach, F. W., and Jensen, T. S. 2002. Pharmacological Treatment of Spinal Cord Injury Pain. In: Spinal Cord Injury Pain: Assessment, Mechanisms, Management (eds. R. P. Yezierski and K. Burchiel), pp. 341–351. IASP Press.

Gorman, A. L., Yu, C. G., Sanchez, D., Ruenes, G. R., Daniels, L., and Yezierski, R. P. 2001. Conditions affecting the onset, severity, and progression of a spontaneous pain-like behavior after excitotoxic spinal cord injury. J. Pain 2, 229–240.

Hains, B. C. and Waxman, S. G. 2006. Activated microglia contribute to the maintenance of chronic pain after spinal cord injury. J. Neurosci. 26, 4308–4317.

Hains, B. C., Klein, J. P., Saab, C. Y., Craner, M. J., Black, J. A., and Waxman, S. G. 2003a. Upregulation of sodium channel Na$_v$1.3 and functional involvement in neuronal hyperexcitability associated with central neuropathic pain after spinal cord injury. J. Neurosci. 23, 8881–8892.

Hains, B. C., Saab, C. Y., and Waxman, S. G. 2005. Changes in electrophysiological properties and sodium channel Nav1.3 expression in thalamic neurons after spinal cord injury. Brain 128, 2359–2371.

Hains, B. C., Willis, W. D., and Hulsebosch, C. E. 2003b. Temporal plasticity of dorsal horn somatosensory neurons after acute and chronic spinal cord hemisection in rat. Brain Res. 970, 238–241.

Hao, J. X., Xu, X. J., Aldskogius, H., Seiger, Å., and Wiesenfeld-Hallin, Z. 1991. The excitatory amino acid receptor antagonist MK-801 prevents the hypersensitivity induced by spinal cord ischemia in the rat. Exp. Neurol. 114, 182–191.

Hao, J.-X., Xu, X.-J., and Wiesenfeld-Hallin, Z. 1994. Intrathecal-aminobutyric acidB (GABA$_B$) receptor antagonist CGP 45448 induces hypersensitivity to mechanical stimuli in the rat. Neurosci. Lett. 182, 299–302.

Hao, J. X., Xu, X. J., Yu, Y. X, Seiger, Å., and Wiesenfeld-Hallin, Z. 1992. Baclofen reverses the hypersensitivity of dorsal horn wide dynamic range neurons to mechanical stimulation after transient spinal cord ischemia: implications for a tonic GABAergic inhibitory control of myelinated fiber input. Neurophysiology 68, 392–396.

Hao, J. X., Kupers, R. C., and Xu, X. J. 2004. Response characteristics of spinal cord dorsal horn neurons in chronic allodynic rats after spinal cord injury. J. Neurophysiol. 92, 1391–1399.

Herman, R. M., D'Luzansky, S. C., and Ippoliti, R. 1992. Intrathecal baclofen suppresses central pain in patients with spinal lesions: a pilot study. Clin. J. Pain 8, 338–345.

Hoheisel, U., Scheifer, C., Trudrung, P., Unger, T., and Mense, S. 2003. Pathophysiological activity in rat dorsal horn neurones in segments rostral to a chronic spinal cord injury. Br. Res. 974, 134–145.

Hsu, C. Y., Lin, T. N., Xu, J., Chao, J., and Hogan, E. L. 1994. Kinins and Related Inflammatory Mediators in Central Nervous System Injury. In: The Neurobiology of Central Nervous System Trauma (eds. S. K. Salzman and A. L. Faden), pp. 145–154. Oxford Press.

Jain, N., Florence, S., and Kaas, J. H. 1998. Reorganization of somatosensory cortex after nerve and spinal cord injury. News Physiol. Sci. 14, 144–149.

Ji, R. R. and Woolf, C. J. 2001. Neuronal plasticity and signal transduction in nociceptive neurons: implications for the initiation and maintenance of pathological pain. Neurobiol. Dis. 8, 1–10.

Ji, R. R., Kohno, T., Moore, K. A., and Woolf, C. J. 2003. Central sensitization and LTP: do pain and memory share similar mechanisms? Trends Neurosci. 26, 696–705.

Kakulas, B. A., Smith, E., Gaekwad, U. F., Kaelan, C., and Jacobsen, P. 1990. The Neuropathology of Pain and Abnormal Sensations in Human Spinal Cord Injury Derived from the Clinicopathological Data Base at the Royal Perth Hospital. In: Recent Achievements in Restorative Neurology, (eds. M. R. Dimitrijevic, P. D. Wall, and U. Lindblom) Karger, pp. 37–41.

Kim, H. K., Kwon, P. S., Zhou, J. L., Taglialatela, G., Kyungsoon, C., Coggeshall, R. E., and Chung, J. M. 2004. Reactive oxygen species (ROS) play an important role in a rat model of neuropathic pain. Pain 111, 116–124.

King, C. D., Vierck, C. J., Berens, S., and Yezierski, R. P. 2006. Acute stress enhances below-level pain after spinal cord injury. Soc. Neurosci. Meeting, Atlanta, GA.

Leijon, G. and Boivie, J. 1991. Pharmacological Treatment of Central Pain. In: Pain and the Central Nervous System Disease: The Central Pain Syndromes (ed. K. L. Casey), pp. 257–266. Raven Press.

Lenz, F. A., Lee, J. I., and Dougherty, P. M. 2000. Human Thalamus Reorganization Related to Nervous System Injury and Dystonia. In: Nervous System Plasticity and Chronic Pain (eds. J. Sandkühler, B. Bromm, and G. F. Gebhart),Prog. Brain Res. Vol. 129, pp. 259–273. Elsevier.

Loeser, J. D., Ward, A. A., and White, L. E. 1968. Chronic deafferentation of human spinal cord neurons. J. Neurosurg. 29, 48–50.

Loubser, P. G. and Donovan, W. H. 1991. Diagnostic spinal anesthesia in chronic spinal cord injury pain. Paraplegia 29, 25–36.

Mauderli, A. P., Acosta-Rua, A., and Vierck, C. J. 2000. A conscious behavioral assay of thermal pain in rodents. J. Neurosci. Methods 97, 19–29.

Mills, C. D., Fullwood, S. D., and Hulsebosch, C. E. 2001. Changes in metabotropic receptor expression following spinal cord injury. Exp. Neurol. 170, 244–257.

Morrow, T. J., Paulson, P. E., Brewer, K. L., Yezierski, R. P., and Casey, K. L. 2000. Chronic, selective forebrain responses to excitotoxic dorsal horn injury. Exp. Neurol. 161, 220–226.

Nag, S. and Riopelle, R. J. 1990. Spinal neuronal pathology associated with continuous intrathecal infusion of N-methyl-D-aspartate in the rat. Acta Neuropathol. 81, 7–13.

Nepomuceno, C., Fine, P. R., Richards, J. S., Gowens, H., Stover, S. L., Rantanuabol, U., and Houston, R. 1979. Pain in patients with spinal cord injury. Arch. Phys. Med. Rehabil. 60, 605–609.

Nesic, O., Svrakic, N. M., Xu, G. Y., McAdoo, D., Westlund, K. N., Hulsebosch, C. E., Galante, A., Soteropoulos, P., Tolias, P., Young, W., Hart, R. P., and Perez-Polo, J. R. 2002. DNA microarray analysis of the contused spinal cord: effect of NMDA receptor inhibition. J. Neurosci. Res. 68, 406–423.

Nesic, O., Lee, J., Johnson, K. M., Ye, Z., Xu, G. Y., Unaia, G. C., Wood, T. G., McAdoo, D. J., Westlund, K. N., Hulsebosch, C. E., and Perez-Polo, J. R. 2005. Transcriptional profiling of spinal cord injury-induced central neuropathic pain. J. Neurochem. 95, 998–1014.

Ness, T. J., Pedro, E. C. S., Richards, J. S., Kezar, L., Liu, H.-G., and Mountz, J. M. 1998. A case of spinal cord injury-related pain with baseline rCBF brain SPECT imaging and beneficial response to gabapentin. Pain 78, 139–143.

Ohara, S., Garonzik, I., Hua, S., and Lenz, F. A. 2002. Microelectrode Studies of the Thalamus in Patients with Central Pain and in Control Patients with Movement Disorders. In: Spinal Cord Injury Pain: Assessment, Mechanisms, Management, Prog. Pain Res. Manage., Vol. 23, pp. 219–236 IASP Press.

Oatway, M. A., Chen, Y., and Weaver, L. C. 2004. The 5-HT$_3$ receptor facilitates at-level mechanical allodynia following spinal cord injury. Pain 110, 259–268.

Ondarza, A. B., Ye, Z., and Hulsebosch, C. E. 2003. Direct evidence of primary afferent sprouting in distant segments following spinal cord injury in rat: colocalization of GAP-43 and CGRP. Exp. Neurol. 184, 373–380.

Pattany, P. M., Yezierski, R. P., Widerstrom-Noga, E. G., Bowen, B. C., Martinez-Arizala, A., Garcia, B. R., and Quencer, R. M. 2002. Proton magnetic resonance spectroscopy of the thalamus: evaluation of patients with chronic neuropathic pain following spinal cord injury. Am. J. Neuroradiology 23, 901–905.

Peng, X. M., Zhou, Z. G., Glorioso, J. C., Fink, D. J., and Mata, M. 2006. Tumor necrosis factor-alpha contributes to below-level neuropathic pain after spinal injury. Ann. Neurol. 59, 843–851.

Pollock, L. J., Brown, M., Boshes, B., Finkelman, I., Chor, H., Arieff, A. J., and Finkel, J. R. 1951. Pain below the level of injury of the spinal cord. AMA Arch. Neurol. Psychiat. 65, 319–322.

Regan, R. and Choi, D. W. 1994. Excitoxicity and Central Nervous System Trauma. In: The Neurobiology of Central Nervous System Trauma (eds. S. K. Salzman and A. L. Faden), pp. 173–181. Oxford Press.

Scheifer, C., Hoheisel, U., Trudrung, P., Unger, T., and Mense, S. 2002. Rats with chronic spinal cord transection as a possible model for the at-level pain of paraplegic patients. Neurosci. Lett. 323, 117–120.

Siddall, P., Xu, C. L., and Cousins, M. 1995. Allodynia following traumatic spinal cord injury in the rat. Neuroreport 6, 1241–1244.

Siddall, P. J., McClelland, J. M., Rutkowski, S. B., and Cousins, M. J. 2003. A longitudinal study of the prevalence and characteristics of pain in the first five years following spinal cord injury. Pain 103, 249–257.

Siddall, P. J., Taylor, D., McClelland, J., Rutkowski, S., and Cousins, M. 1999a. Pain report and the relationship of pain to physical factors in the first 6 months following spinal cord injury. Pain 81, 187–197.

Siddall, P. J., Xu, C. L., Floyd, N., and Keay, K. A. 1999b. C-fos expression in the spinal cord in rats exhibiting allodynia following contusive spinal cord injury. Brain Res. 851, 281–286.

Siddall, P. J., Yezierski, R. P., and Loeser, J. D. 2002. Taxonomy and Epidemiology of Spinal Cord Injury. In: Spinal Cord Injury Pain: Assessment, Mechanisms, Management (eds. R. P. Yezierski and K. Burchiel), pp. 9–24. IASP Press.

Tator, C. H. and Fehlings, M. G. 1991. Review of the secondary injury theory of acute spinal cord trauma with emphasis on vascular mechanisms. J. Neurosurg. 75, 15–26.

Tator, C. 1995. Update on the pathophysiology and pathology of acute spinal cord injury. Brain Pathol. 5, 407–413.

Tasker, R. R., de Carvalho, G., and Dostrovsky, J. O. 1991. The History of Central Pain Syndromes, with Observations Concerning Pathophysiology and Treatment. In: Pain and Central Nervous System Disease: The Central Pain Syndromes (ed. K. L. Casey), pp. 31–58. Raven Press.

Tunks, E. 1986. Pain in Spinal Cord Injured Patients. In: Management of Spinal Cord Injuries (eds. R. F. Bloch and M. Basbaum), pp. 180–211. William and Wilkins.

Vierck, C. J., Jr. 1991. Can Mechanisms of Central Pain Syndromes be Investigated in Animal Models? In: Pain and Central Nervous System Disease: The Central Pain Syndromes (ed. K. L. Casey), pp. 129–141. Raven Press.

Vierck, C. J., Siddall, P. J., and Yezierski, R. P. 2000. Pain following spinal cord injury: animal models and mechanistic studies. Pain 89, 1–5.

Vierck, C. J., Cannon, R. L., Stevens, K. A., Acosta-Ru, A. J., and Wirth, E. D. 2002. Mechanisms of Increased Pain Sensitivity Within Dermatomes Remote from an Injured Segment of the Spinal Cord. In: Spinal Cord Injury Pain: Assessment, Mechanisms, Management (eds. R. P. Yezierski and K. Burchiel), pp. 155–173. IASP Press.

Werhagen, L., Budh, C. N., Hutling, C., and Molander, C. 2004. Neuropathic pain after traumatic spinal cord injury – relations to gender, spinal, level, completeness, and age at the time of injury. Spinal Cord 42, 665–673.

Widerstrom-Noga, E. G., Cuervo, E., Broton, J. G., Duncan, R. C., and Yezierski, R. P. 1999a. Self-reported consequences of spinal cord injury: results of a postal survey. Arch. Phys. Med. Rehab. 80, 580–586.

Widerstrom-Noga, E. G., Cuevo, E. F., Broton, J. G., Duncan, R. C., and Yezierski, R. P. 1999b. Perceived difficulty in dealing with consequences of SCI. Arch. Phys. Med. Rehab. 80, 580–586.

Wiesenfeld-Hallin, Z., Hao, J. X., Aldskogius, H., Seiger, Å., and Xu, X. J. 1994. Allodynia-Like Symptoms in Rats After Spinal Cord Ischemia: An Animal Model of Central Pain. In: Touch, Temperature and Pain in Health and Disease: Mechanisms and Assessments (eds. J. Boivie, P. Hansson, and U. Lindblom), Vol. 4, pp. 455–472. IASP Press.

Willis, W. D. 2002. Long-term potentiation in spinothalamic neurons. Brain Res. Rev. 40, 202–214.

Woolf, C. J. 1984. Long term alterations in the excitability of the flexion reflex produced by peripheral tissue injury in the chronic decerebrate rat. Pain 18, 325–343.

Woolf, C. J. and Csotigan, M. 1999. Transcriptional and posttranslational plasticity and the generation of inflammatory pain. Proc. Natl. Acad. Sci. U. S. A. 96, 7723–7730.

Woolf, C. J., Shortland, P., and Loggeshall, R. E. 1992. Peripheral nerve injury triggen central sprouting of myelinated afferents. Nature 355, 75–78.

Wrathall, J. R., Choiniere, D., and Teng, Y. D. 1994. Dose dependent reduction of tissue loss and functional impairment after spinal cord trauma with the AMPA/kainate antagonist NBQX. J. Neurosci. 14, 6598–6607.

Yakovlev, A. G. and Faden, A. I. 1994. Sequential expression of c-fos protooncogene, TNF-alpha, and dynorphin genes in spinal cord following experimental traumatic injury. Mol. Chem. Neuropath. 24, 179–190.

Yaksh, T. L. 1989. Behavioral and autonomic correlates of the tactile evoked allodynia produced by spinal glycine inhibition: effects of modulatory receptor systems and excitatory amino acid antagonists. Pain 47, 111–123.

Yezierski, R. P. 2000. Pain Following Spinal Cord Injury: Pathophysiology and Central Mechanisms. In: Nervous System Plasticity and Chronic Pain (eds. J. Sandkühler, B. Bromm, and G. F. Gebhart), Vol. 129, pp. 429–449. Elsevier.

Yezierski, R. P. 2002. Central Neuropathic Pain. In: Surgical Management of Pain (ed. K. Burchiel), pp. 42–64. Thieme Medical Publishers.

Yezierski, R. P. 2006. Pain Following Spinal Cord Injury: Central Mechanisms. In: Handbook of Clinical Neurology (eds. F. Cervero and T. S. Jensen), Vol. 81, pp. 293–307. Elsevier.

Yezierski, R. P. and Park, S. H. 1993. The mechanosensitivity of spinal sensory neurons following intraspinal injections of quisqualic acid in the rat. Neurosci. Lett. 157, 115–119.

Yezierski, R. P., Liu, S., Ruenes, G. L., Kajander, K. J., and Brewer, K. L. 1998. Excitotoxic spinal cord injury: behavioral and morphological characteristics of a central pain model. Pain 75, 141–155.

Yezierski, R. P., Santana, M., Park, D. H., and Madsen, P. W. 1993. Neuronal degeneration and spinal cavitation following intraspinal injections of quisqualic acid in the rat. J. Neurotrauma 10, 445–456.

Yezierski, R. P., Yu, C. G., Mantyh, P. W., Vierck, C. J., and Lappi, D. A. 2004. Spinal neurons involved in the generation of at-level pain following spinal injury in the rat. Neurosci. Lett. 361, 232–236.

Young, W. 1987. The post-injury responses in trauma and ischemia: secondary injury or protective mechanism. CNS Trauma 4, 27–52.

Yu, C. G. and Yezierski, R. P. 2005a. Activation of the ERK 1/2 signaling cascade by excitotoxic spinal cord injury. Brain Res. Mol. Brain Res. 138, 244–255.

Yu, C.-G. and Yezierski, R. P. 2005b. Effects of Excitotoxic Spinal Cord Injury on Thalamic MAPK and GABA$_A$ Receptors, IASP Meeting, Sydney, Australia.

Yu, C. G., Fairbanks, C. A., Wilcox, G. L., and Yezierski, R. P. 2003. Effects of agmatine, interleukin-10 and cyclosporin on spontaneous pain behavior following excitotoxic spinal cord injury in rats. J. Pain 4, 129–140.

Zhang, A.-L., Hao, J.-X., Seiger, Å, Xu, X.-J., Wiesenfeld-Hallin, Z., Grant, G., and Aldskogius, H. 1999. Decreased GABA immunoreactivity in spinal cord dorsal horn neurons after transient spinal cord ischemia in the rat. Brain Res. 656, 187–190.

Zhou, Y., Wang, Y., Abdelhady, M., Mourad, M. S., and Hassouna, M. M. 2002. Change of vanilloid receptor 1 following neuromodulation in rats with spinal cord injury. J. Surg. Res. 107, 140–144.

Zvarova, K., Murray, E., and Vizzard, M. A. 2004. Changes in galanin immunoreactivity in rat lumbosacral spinal cord and dorsal root ganglia after spinal cord injury. J. Comp. Neurol. 475, 590–603.

Further Reading

Aimone, J. B., Leasure, J. L., Perreau, V. M., and Thallmair, M. 2004. Spatial and temporal gene expression profiling of the contused rat spinal cord. Exp. Neurol. 189, 204–221.

Bethea, J., Castro, M., Lee, T. T., Dietrich, W. D., and Yezierski, R. 1998. Traumatic spinal cord injury induces nuclear factor kappa B activation. J. Neurosci. 18, 3251–3260.

Brewer, K. L., Bethea, J. R., and Yezierski, R. P. 1999. Neuroprotective effects of interleukin-10 following excitotoxic spinal cord injury. Exp. Neurol. 159, 484–493.

Brewer, K. L., McMillan, D., Nolan, T., and Shum, K. 2003. Cortical changes in cholecystokinin mRNA are related to spontaneous pain behaviors following spinal cord injury in the rat. Molec. Brain Res. 118, 171–174.

Christensen, M. D., Everhart, A. W., Pickeman, J., and Hulsebosch, C. E. 1996. Mechanical and thermal allodynia in chronic central pain following spinal cord injury. Pain 68, 97–107.

Davidoff, G. and Roth, E. 1991. Clinical Characteristics of Central (Dysesthetic) Pain in Spinal Cord Injury Patients. In: Pain and Central Nervous System Disease: The Central Pain Syndromes (ed. K. L. Casey), pp. 77–83. Raven Press.

Davidoff, G., Roth, E., Guarracini, M., Sliwa, J., and Yarkony, G. 1987. Function limiting dysesthetic pain syndrome among traumatic SCI patients: a cross sectional study. Pain 29, 39–48.

Eide, P. K. 1998. Pathophysiological mechanisms of central neuropathic pain after spinal cord injury. Spinal Cord 36, 601–612.

Finnerup, N. M. and Jensen, T. S. 2004. Spinal cord injury pain – mechanisms and treatment. Eur. J. Neurol. 11, 73–82.

Finnerup, N. B., Johannesen, I. L., Fuglsang-Frederiksen, A., Bach, F. W., and Jensen, T. S. 2003a. Sensory function in spinal cord injury patients with and without central pain. Brain 126, 57–70.

Gerke, M. B., Duggan, A. W., and Siddall, P. J. 2003. Thalamic neuronal activity in rats with mechanical allodynia following contusive spinal cord injury. Neuroscience 117, 715–722.

Gris, D., Marsh, M. A., Oatway, M. A., Chen, Y., Hamilton, E. F., Dekaban, G. A., and Weaver, L. C. 2004. Transient blockade of the CD11d/CD18 integrin reduces secondary damage after spinal cord injury, improving sensory, autonomic and motor function. J. Neurosci. 24, 4043–4051.

Hains, B. C., Everhart, A. W., Fullwood, S. T., and Hulsebosch, C. E. 2002. Changes in serotonin, serotonin transporter expression and serotonin denervation supersensitivity: involvement in chronic central pain after spinal hemisection in the rat. Exp. Neurol. 175, 347–362.

Hains, B. C., Saab, C. Y., and Waxman, S. G. 2006. Alterations in burst firing of thalamic VPL neurons and reversal by Na$_v$1.3 antisense following spinal cord injury. J. Neurophysiol. 95, 3343–3352.

Hayashi, M., Ueyama, T., Nemoto, K., Tamaki, T., and Senba, E. 2000. Sequential mRNA expression for immediate early genes, cytokines and neurotrophins in spinal cord injury. J. Neurotrauma 17, 203–218.

Koetzner, L., Hua, X. Y., Lai, J., Porreca, F., and Yaksh, T. 2004. Nonopioid actions of intrathecal dynorphin evoke spinal excitatory amino acid and prostaglandin E2 release mediated by cyclooxygnease-1 and -2. J. Neurosci. 24, 1451–1458.

Melzack, R. and Loeser, J. D. 1978. Phantom body pain in paraplegics: evidence for a central "pattern generating mechanism" for pain. Pain 4, 195–210.

Moore, C. I., Stern, C. E., Dunbar, C., Kostyk, S. K., Gehi, A., and Corkin, S. 2000. Referred phantom sensations and cortical reorganization after spinal cord injury in humans. Proc. Natl. Acad. Sci. U. S. A. 97, 14703–14708.

Plunkett, J. A., Yu, C. G., Bethea, J. R., and Yezierski, R. P. 2001. Effects of interleukin-10 (IL-10) on pain behavior and gene expression following excitotoxic spinal cord injury in the rat. Exp. Neurol. 169, 144–154.

Porreca, F., Ossipov, M. H., and Gebhart, G. F. 2002. Chronic pain and medullary descending facilitation. Trends Neurosci. 25, 319–325.

Rintala, D. H., Loubser, P. G., Castro, J., Hart, K. A., and Fuhrer, M. J. 1998. A comprehensive assessment of chronic pain in a community-based sample of men with spinal cord injury. J. Phys. Med. Rehab. 79, 604–614.

Siddall, P. J. 2002. Spinal Drug Administration in the Treatment of Spinal Cord Injury Pain. In: Spinal Cord Injury Pain: Assessment, Mechanisms, Management (eds. R. P. Yezierski and K. Burchiel), pp. 353–364. IASP Press.

Summers, J. D., Rapoff, M. A., Varghese, G., Porter, K., and Palmer, R. E. 1991. Psychosocial factors in chronic spinal cord injury pain. Pain 47, 183–189.

Vierck, C. J. and Light, A. R. 2002. Assessment of Pain Sensitivity in Dermatomes Caudal to Spinal Cord Injury in Rats. In: Spinal Cord Injury Pain: Assessment, Mechanisms, Management (eds. R. P. Yezierski and K. Burchiel), pp. 137–154. IASP Press.

Vierck, C. J., Jr. and Light, A. R. 2000. Allodynia and Hyperalgesia within Dermatomes Caudal to a Spinal Cord Injury in Primates and Rodents. In: Nervous System Plasticity and Chronic Pain (eds. J. Sandkuhler, B. Bromm, and G. Gebhart), Vol. 129, pp. 411–428. Elsevier.

Xu, X. J., Hao, J. X., and Wiesenfeld-Hallin, Z. 2002. Physiological and Pharmacological Characterization of a Rat Model of Spinal Cord Injury Pain After Spinal Ischemia. In: Spinal Cord Injury Pain: Assessment, Mechanisms, Management (*eds*. R. P. Yezierski and K. Burchiel), pp. 175–187. IASP Press.

Yezierski, R. P. 2002. Pathophysiology and Animals Models of Spinal Cord Injury Pain. In: Spinal Cord Injury Pain: Assessment, Mechanisms, Management (*eds*. R. P. Yezierski and K. Burchiel), Prog. Pain Res. Manage. Vol. 23, pp. 117–136. IASP Press.

R. P. Yezierski and K. J. Burchiel (*eds*). 2002. Spinal Cord Injury Pain: Assessment, Mechanisms, Management IASP Press.

Zhang, A. L., Hao, J. X., Seiger, Å., Xu, X. J., Wiesenfeld-Hallin, Z., Grant, G., and Aldskogius, H. 1999. Decreased GABA immunoreactivity in spinal cord dorsal horn neurons after transient spinal cord ischemia in the rat. Brain Res. 656, 187–190.

29 Long-Term Potentiation in Pain Pathways

J Sandkühler, Medical University of Vienna, Vienna, Austria

Glossary

synaptic efficacy, syn. synaptic strength Magnitude of postsynaptic current or potential in response to a presynaptic action potential. Synaptic strength may be modified by pre- and/or postsynaptic modulation.

long-term potentiation (LTP) Enduring increase in synaptic efficacy.

long-term depression (LTD) Enduring decrease in synaptic efficacy of a putatively nonpotentiated synapse.

depotentiation Enduring decrease of synaptic efficacy of a previously potentiated synapse, that is, reversal of LTP.

LTP induction Cellular signal transduction pathways which lead to LTP.

LTP maintenance Cellular signal transduction pathways which are involved in stabilizing established LTP.

spinal field potential Extracelluar electrical field which is mainly generated by simultaneous postsynaptic currents in spinal cord neurons in response to a brief stimulus, similar to evoked potentials in cerebral cortex.

29.1 Introduction

Long-term potentiation (LTP) is defined as the persistent increase in synaptic strength (Bliss, T. V. P. and Collingridge, G. L., 1993; Malenka, R. C. and Nicoll, R. A., 1999). Repetitive use of a synapse often leads to LTP which may last for a few hours only or for days to months. Synaptic strength may also be reduced for prolonged periods of time which is called long-term depression (LTD) when induced in a normal synapse or depotentiation when elicited in a previously potentiated synapse, that is, in the maintenance phase of LTP (Huang, C. C. and Hsu, K. S., 2001).

LTP has been most intensively studied in hippocampus where it is considered a fundamental cellular model of learning and memory formation (Bliss, T. V. P. and Collingridge, G. L., 1993; Malenka, R. C. and Nicoll, R. A., 1999). LTP has also been shown in pain pathways at synapses between primary afferent Aδ- or C-fibers and spinal dorsal horn neurons and is one of the proposed cellular mechanisms of pain amplification (Moore, K. A. *et al.*, 2000; Sandkühler, J., 2000). The potential consequences of LTP in pain pathways are manifold: (1) after LTP is established, suprathreshold excitatory input will now evoke stronger excitation of nociceptive neurons. Thus, LTP at nociceptive synapses may underlie hyperalgesia; (2) previously subthreshold excitatory input, from the subliminal fringe of a neurons' receptive field may now elicit action potential firing. This may widen painful areas and contribute to hyperalgesia as well; and (3) some spinal dorsal horn neurons receive subthreshold input from somatotopically inappropriate body areas, for example, the contralateral body half. LTP anywhere along this subliminal input path may cause mirror image pain or radiating pain. LTP at nociceptive synapses may, however, not significantly lower pain thresholds and is therefore not considered a likely mechanism of allodynia.

The most direct way to ascertain LTP in pain pathways is to measure monosynaptically evoked postsynaptic currents or potentials. So far this has only been done *in vitro* by recording Aδ- or C-fiber-evoked excitatory postsynaptic currents or potentials from dorsal horn neurons in response to electrical stimulation of dorsal roots. When postsynaptic currents are elicited synchronously in a sufficient number of neighboring neurons, electrical field potential are generated which are strong enough to be detected with extracellular microelectrodes (Schouenborg, J., 1984). Thus, to evaluate changes in synaptic strength, C-fiber-evoked field potentials can be recorded in superficial spinal dorsal horn *in vivo* and *in vitro* in response to supramaximal electrical stimulation of a dorsal root or a sensory nerve, for example, the sciatic nerve. If, however, events downstream to postsynaptic currents are recorded, such as action potential discharges, polysynaptically evoked responses, withdrawal reflexes or pain ratings, LTP can no longer be assessed directly. These measures are, however, indispensable to evaluate the functional consequences of LTP in pain pathways. This is discussed in greater detail elsewhere (Sandkühler, J., 2007).

29.2 Induction of Long-Term Potentiation in Pain Pathways

In superficial spinal lamina I, LTP is selectively induced in a group of nociceptive specific neurons which express the NK1 receptor for substance P and which project to the brain (see Figure 1; Ikeda, H. *et al.*, 2003; Ikeda, H. *et al.*, 2006). This group of neurons plays a key role in hyperalgesia following inflammation and nerve injury (Mantyh, P. W. *et al.*, 1997). In spinal cord LTP can be induced in different ways, for example, by electrical nerve stimulation , by nerve injury, or peripheral noxious events. Stimulation within the innervation territory of a sensory nerve, for example, with chemical irritants, or by inflammation may also trigger LTP at C-fiber synapses. All conditioning stimuli tested so far which induce LTP at Aδ-synapses (Randic, M. *et al.*, 1993) and/ or C-fiber synapses (Liu, X. G. and Sandkühler, J., 1995) may also cause hyperalgesia in experimental animals.

1. Electrical nerve stimulation: LTP has been induced by conditioning high intensity, high frequency burst-like stimulation (HFS, typically given as 50–100 Hz bursts for 1 s at C-fiber strength) *in vitro* (Randic, M. *et al.*, 1993; Hamba, M. *et al.*, 2000; Ikeda, H. *et al.*, 2000; Ikeda, H. *et al.*, 2003) and *in vivo* (Liu, X. G. and Sandkühler, J., 1997; Miletic, G. and Miletic, V., 2000; Ma, J. Y. and Zhao, Z. Q., 2001; Zhang, H. M. *et al.*, 2001). Stimulation intensity must be sufficient to recruit C-fibers; stimulation at A-fiber intensity is ineffective to induce LTP at C-fiber synapses (Liu, X. G. and Sandkühler, J., 1997). C-fibers do, however, discharge at such high frequencies only rarely. Thus, the pathological relevance of HFS-induced LTP can be challenged.

2. Recent studies show that not only HFS but also conditioning low frequency stimuli at 2–10 Hz which is well within the frequency band of C-fiber discharges during inflammation or tissue injury causes robust LTP, both, *in vivo* and *in vitro* (Terman, G. W. *et al.*, 2001; Ikeda, H. *et al.*, 2006)/electrical stimulation of sciatic nerve at C-fiber intensity (2 Hz, 2 min) induces robust LTP to 300% of control values which is not reversible within the recording periods of up to 18 h (Figure 1; Ikeda, H. *et al.*, 2006).

3. Electrical stimulation of nerves evokes regular and highly synchronous discharges which is not typical for naturally occurring, irregular nonsynchronous discharge patterns. Thus, it was important to show that LTP can also be induced by excitation of sensory nerve endings, for example, by subcutaneous injection of either capsaicin or formalin. In animals with spinal cord and descending pathways intact, intraplantar, s.c. injections of capsaicin or formalin induce slowly rising LTP (Figure 1; Ikeda, H. *et al.*, 2006).

4. Finally, with descending spinal pathways interrupted noxious skin heating and acute nerve injury also causes LTP at C-fiber synapses (Sandkühler, J. and Liu, X., 1998). This demonstrates that naturally occurring, low frequency, irregular, and nonsynchronous afferent barrage can cause synaptic plasticity in pain pathways.

LTP of C-fiber-evoked spinal field potentials can also be induced pharmacologically. In spinalised rats, topical application of *N*-methyl-D-aspartate (NMDA), substance P or neurokinin A induces LTP (Liu, X. G. and Sandkühler, J., 1998). In intact animals, spinal

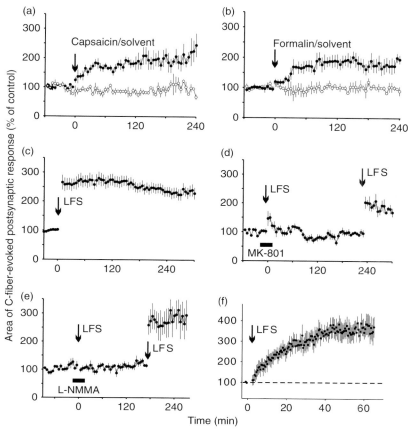

Figure 1 LTP can be induced by natural, low-frequency afferent barrage evoked by inflammation of peripheral tissue *in vivo* and by low-frequency stimulation of dorsal root afferents *in vitro*. Mean time courses of C-fiber-evoked field potentials recorded extracellulary in superficial spinal dorsal horn in response to electrical stimulation of left sciatic nerve of deeply anaesthetized adult rats with spinal cords and afferent nerves intact (a)–(e). Subcutaneous injections of transient receptor potential vanilloid 1 channel agonist capsaicin (1%, 100 μl, $n = 5$, (a) or formalin (5%, 100 μl, $n = 6$), (b) into the glabrous skin at the ipsilateral hind paw, within the innervation territory of the sciatic nerve at time zero (arrows) induced LTP (closed circles), while injections of the respective solvents (open circles) had no effects ($n = 3$ in each group). Conditioning electrical LFS (2 Hz, 2 min at C-fiber intensity) of sciatic nerve at time zero (arrow) also induced LTP ($n = 28$), (c) which was prevented by NMDA receptor antagonist MK-801 (3 mg kg^{-1}, i.v.-infusion over 30 min: horizontal bar, $n = 5$); (d) a second conditioning LFS four hours later (arrow) was partially effective in inducing LTP. NOS inhibitor L-NMMA (100 mg kg^{-1} h^{-1}, i.v., horizontal bar, $n = 5$); (e) also blocked LTP induction. This block was fully reversible as shown by a second LFS 3 h later (arrow). Similarly, LFS of dorsal roots at C-fibre strength induces robust LTP at synapses between C-fibre afferents and lamina I neurons with a projection to the midbrain periaqueductal gray (PAG) in a rat spinal cord–dorsal root preparation ($n = 18$). Adapted from Ikeda, H., Stark, J., Fischer, H., Wagner, M., Drdla, R., Jäger, T., and Sandkühler, J. 2006. Synaptic amplifier of inflammatory pain in the spinal dorsal horn. Science 312, 1659–1662.

application of dopamine D1/D5 or protein kinase A activators but not D2 receptor agonists selectively induces protein synthesis-dependent late phase of LTP (Yang, H. W. *et al.*, 2005).

29.3 Signal Transduction Pathways

The signal transduction pathways which lead to LTP in spinal dorsal horn largely overlap with those involved in

hyperalgesia in behaving animals. At C-fiber synapses, LTP induction requires co-activation of ionotropic glutamate receptors of the NMDA type (Randic, M. *et al.*, 1993; Liu, X. G. and Sandkühler, J., 1995; Ikeda, H. *et al.*, 2003; 2006; see, however, Hamba, M. *et al.* 2000) and group I but not group II or group III metabotropic glutamate receptors (mGluRs) (Hamba, M. *et al.*, 2000; Zhong, J. *et al.*, 2000; Azkue, J. J. *et al.*, 2003). LTP-inducing conditioning stimuli release substance P in spinal cord (Afrah, A. W. *et al.*, 2002) and activation of

NK1 receptors for substance P and NK2 receptors for neurokinin A are required for LTP induction (Liu, X. G. and Sandkühler, J., 1997). Opening of T-type voltage gated calcium channels is also necessary (Ikeda, H. *et al.*, 2003; 2006). All this leads to an activity-dependent rise in intracellular calcium ion concentration and activation of calcium-dependent signal transduction. The magnitude of HFS-induced LTP of C-fiber-evoked responses is linearly correlated with the rise in postsynaptic $[Ca^{2+}]_i$ (Ikeda, H. *et al.*, 2003). Blockade of calcium rise in the postsynaptic neuron or calcium-calmodulin-dependent protein kinase II prevents LTP induction (Ikeda, H. *et al.*, 2003; see, however, Hamba, M. *et al.*, 2000; Yang, H. W. *et al.*, 2004). Activation of group I mGluRs may activate phospholipase C (PLC) pathway leading to either protein kinase C activation or formation of inositol-1,4,5-trisphosphate (IP_3). Blockade of either PLC, PKC, or IP_3 receptors all abolish LTP induction (Ikeda, H. *et al.*, 2003).

Activation of NMDA, NK1, or NK2 receptors is not necessary for maintenance of LTP (Liu, X. G. and Sandkühler, J., 1997). Maintenance of the late phase of LTP which starts about 3 h after induction requires, however, *de novo* protein synthesis (Hu, N. W. *et al.*, 2003).

29.3.1 Long-Term Potentiation Induction is Context Sensitive

Induction of LTP not only depends upon the activity patterns at the presynaptic terminals, but also on the functional state of the postsynaptic neurons, namely the level of postsynaptic inhibition. Under conditions of impaired inhibition, previously ineffective presynaptic activity may now induce LTP and protocols which normally induce LTD may then cause LTP. *In vitro*, both, Aδ-fiber-evoked (Randic, M. *et al.*, 1993) and C-fiber-evoked (Ikeda, H. *et al.*, 2003) responses are potentiated by conditioning 100 Hz stimulation of dorsal roots when postsynaptic neurons are mildly depolarized to -70 to -60 mV. The same conditioning stimulation induces, however, LTD of Aδ-fiber-evoked responses if cells are hyperpolarized to -85 mV, suggesting that the direction of synaptic plasticity is voltage dependent (Randic, M. *et al.*, 1993).

Spinal Aβ-/Aδ-fiber-evoked field potentials are depressed by conditioning 50 Hz stimulation of sciatic nerve. After blockade of $GABA_A$ receptors with bicuculline the same conditioning stimulus now produces LTP rather than LTD (Miletic, G. and Miletic, V., 2001). In animals with a nerve injury,

inhibition in spinal cord may be impaired. Conditioning stimulation at 50 Hz produces a short lasting potentiation followed by LTD in control animals but LTP in animals with a chronic constriction injury of sciatic nerve (Miletic, G. and Miletic, V., 2000). Topical application of muscimol (10 μg) to spinal cord prevents tetanus-induced LTP of A-fiber-evoked field potentials in animals with a nerve injury. In the presence of $GABA_A$ receptor agonist muscimol the same conditioning stimulation evokes an LTD rather than an LTP (Miletic, G. *et al.*, 2003).

Pharmacological activation of spinal NMDA receptors and NK1 or NK2 receptors is insufficient to induce LTP in intact animals. When descending pathways, which are mainly inhibitory in acute preparations, are cut, then the same receptor agonists now trigger LTP (Liu, X. G. and Sandkühler, J., 1998). This confirms that level of postsynaptic inhibition regulates induction and polarity of synaptic plasticity. Some pharmacological induction protocols do not require any presynaptic stimulation which suggests that LTP can be induced in the absence of presynaptic activity. This may be relevant when neuromodulators such as substance P diffuse to distant, inactive synapses (volume transmission), which is a potential mechanism of spreading synaptic plasticity in the spinal cord.

Functional blockade of glial cells by i.t. administration of fluorocitrate changes the direction of HFS induced plasticity of C-fiber-evoked potentials. When HFS is given 1 h, but not 3 h after fluorocitrate LTD rather than LTP is induced (Ma, J. Y. and Zhao, Z. Q., 2002). The mechanism of this switch in polarity of synaptic plasticity is presently unknown but recent evidence suggests that glial cell-derived nitric oxide is liberated via activation of type 1 mGluRs and triggers LTP (Ikeda, H. and Murase, K., 2004).

29.4 Prevention of Long-Term Potentiation

LTP of C-fiber-evoked field potentials *in vivo* is not affected by deep (surgical) level of anesthesia with urethane, isoflurane, or sevoflurane in mature rats (Benrath, J. *et al.*, 2004). LTP is, however, prevented by low-dose intravenous infusion of μ-opioid receptor agonist fentanyl (Benrath, J. *et al.*, 2004). Similarly, spinal field potentials elicited by stimulation in the tract of Lissauer in spinal cord slices is blocked by DAMGO, a more specific agonist at these receptors

(Terman, G. W. *et al.*, 2001). Spinal application of Zn^{2+} also blocked LTP induction *in vivo* (Ma, J. Y. and Zhao, Z. Q., 2001).

29.5 Reversal of Long-Term Potentiation

Brief, high-frequency conditioning of sciatic nerve at Aδ-fiber intensity partially and temporarily reverses LTP of C-fiber-evoked field potentials when given 15 or 60 min but not 3 h after LTP induction (Liu, X. G. *et al.*, 1998;Zhang, H. M., *et al.*, 2001). Similarly, blockade of either protein kinase C, calcium-calmodulin-dependent protein kinase II or protein kinase A before or up to 30–60 min after conditioning stimulation abolishes LTP induction. Kinase blockers are, however, ineffective, when applied 3 h after induction. Inhibition of *de novo* protein synthesis in spinal dorsal horn during high frequency stimulation of sciatic nerve selectively blocked late phase of LTP but not its early phase lasting for about 3 h (Hu, N. W. *et al.*, 2003).

29.6 Functional Consequences of Long-Term Potentiation in Pain Pathways

A number of studies provide convergent evidence that LTP at spinal synapses of nociceptive Aδ- or C-fibers facilitates downstream processing of pain-related information. This includes action potential firing in superficial and deep dorsal horn neurons (Svendsen, F. *et al.*, 1999; Willis, W. D., 2002; Wallin, J. *et al.*, 2003; Twining, C. M. *et al.*, 2004; Pedersen, L. M. *et al.*, 2005) and pain perception in human subjects (Klein, T. *et al.*, 2004). Induction protocols signal transduction pathways and pharmacology of LTP and facilitation of C-fiber-evoked discharges largely overlap, which is in line with the hypothesis that LTP underlies these facilitations. Under some pathological conditions facilitation of C-fiber-evoked action potential discharges is enhanced, for example, when descending pathways are interrupted (Gjerstad, J. *et al.*, 2001) or if a joint is inflamed (Vikman, K. S. *et al.*, 2003). Induction of LTP of A-fiber-evoked field potentials by high frequency conditioning stimulation is facilitated in animals with a loose ligation of sciatic nerve (Miletic, G. and Miletic, V., 2000). One study reported reduced facilitation in animals with spinal nerve ligations (Rygh, L. J. *et al.*, 2000).

29.7 Conclusions

Induction protocols signal transduction pathways and pharmacology of LTP and some forms of hyperalgesia are virtually identical, rendering LTP in spinal dorsal horn an attractive cellular model of hyperalgesia.

References

Afrah, A. W., Fiska, A., Gjerstad, J., Gustafsson, H., Tjølsen, A., Olgart, L., Stiller, C. O., Hole, K., and Brodin, E. 2002. Spinal substance P release *in vivo* during the induction of long-term potentiation in dorsal horn neurons. Pain 96, 49–55.

Azkue, J. J., Liu, X. G., Zimmermann, M., and Sandkühler, J. 2003. Induction of long-term potentiation of C fibre-evoked spinal field potentials requires recruitment of group I, but not group II/III metabotropic glutamate receptors. Pain 106, 373–379.

Benrath, J., Brechtel, C., Martin, E., and Sandkühler, J. 2004. Low doses of fentanyl block central sensitization in the rat spinal cord *in vivo*. Anesthesiology 100, 1545–1551.

Bliss, T. V. P. and Collingridge, G. L. 1993. A synaptic model of memory: long-term potentiation in the hippocampus. Nature 361, 31–39.

Gjerstad, J., Tjølsen, A., and Hole, K. 2001. Induction of long-term potentiation of single wide dynamic range neurones in the dorsal horn is inhibited by descending pathways. Pain 91, 263–268.

Hamba, M., Onodera, K., and Takahashi, T. 2000. Long-term potentiation of primary afferent neurotransmission at trigeminal synapses of juvenile rats. Eur. J. Neurosci. 12, 1128–1134.

Hu, N. W., Zhang, H. M., Hu, X. D., Li, M. T., Zhang, T, Zhou, L. J., and Liu, X. G. 2003. Protein synthesis inhibition blocks the late-phase LTP of C-fiber evoked field potentials in rat spinal dorsal horn. J. Neurophysiol. 89, 2354–2359.

Huang, C. C. and Hsu, K. S. 2001. Progress in understanding the factors regulating reversibility of long-term potentiation. Rev. Neurosci. 12, 51–68.

Ikeda, H. and Murase, K. 2004. Glial nitric oxide-mediated long-term presynaptic facilitation revealed by optical imaging in rat spinal dorsal horn. J. Neurosci. 24, 9888–9896.

Ikeda, H., Asai, T., and Murase, K. 2000. Robust changes of afferent-induced excitation in the rat spinal dorsal horn after conditioning high-frequency stimulation. J. Neurophysiol. 83, 2412–2420.

Ikeda, H., Heinke, B., Ruscheweyh, R., and Sandkühler, J. 2003. Synaptic plasticity in spinal lamina I projection neurons that mediate hyperalgesia. Science 299, 1237–1240.

Ikeda, H., Stark, J., Fischer, H., Wagner, M., Drdla, R., Jäger, T., and Sandkühler, J. 2006. Synaptic amplifier of inflammatory pain in the spinal dorsal horn. Science 312, 1659–1662.

Klein, T., Magerl, W., Hopf, H. C., Sandkühler, J., and Treede, R. D. 2004. Perceptual correlates of nociceptive long-term potentiation and long-term depression in humans. J. Neurosci. 24, 964–971.

Liu, X. G. and Sandkühler, J. 1995. Long-term potentiation of C-fiber-evoked potentials in the rat spinal dorsal horn is prevented by spinal *N*-methyl-D-aspartic acid receptor blockage. Neurosci. Lett. 191, 43–46.

Liu, X. G. and Sandkühler, J. 1997. Characterization of long-term potentiation of C-fiber-evoked potentials in spinal dorsal horn of adult rat: essential role of NK1 and NK2 receptors. J. Neurophysiol. 78, 1973–1982.

Liu, X. G. and Sandkühler, J. 1998. Activation of spinal *N*-methyl-ᴅ-aspartate or neurokinin receptors induces long-term potentiation of spinal C-fibre-evoked potentials. Neuroscience 86, 1209–1216.

Liu, X. G., Morton, C. R., Azkue, J. J., Zimmermann, M., and Sandkühler, J. 1998. of C-fibre-evoked spinal field potentials by stimulation of primary afferent Aδ-fibres in the adult rat. Eur. J. Neurosci. 10, 3069–3075.

Ma, J. Y. and Zhao, Z. Q. 2001. The effects of Zn^{2+} on long-term potentiation of C fiber-evoked potentials in the rat spinal dorsal horn. Brain Res. Bull. 56, 575–579.

Ma, J. Y. and Zhao, Z. Q. 2002. The involvement of glia in long-term plasticity in the spinal dorsal horn of the rat. Neuroreport 13, 1781–1784.

Malenka, R. C. and Nicoll, R. A. 1999. Long-term potentiation– a decade of progress? Science 285, 1870–1874.

Mantyh, P. W., Rogers, S. D., Honoré, P., Allen, B. J., Ghilardi, J. R., Li, J., Daughters, R. S., Lappi, D. A., Wiley, R. G., and Simone, D. A. 1997. Inhibition of hyperalgesia by ablation of lamina I spinal neurons expressing the substance P receptor. Science 278, 275–279.

Miletic, G. and Miletic, V. 2000. Long-term changes in sciatic-evoked A-fiber dorsal horn field potentials accompany loose ligation of the sciatic nerve in rats. Pain 84, 353–359.

Miletic, G. and Miletic, V. 2001. Contribution of GABA-A receptors to metaplasticity in the spinal dorsal horn. Pain 90, 157–162.

Miletic, G., Draganic, P., Pankratz, M. T., and Miletic, V. 2003. Muscimol prevents long-lasting potentiation of dorsal horn field potentials in rats with chronic constriction injury exhibiting decreased levels of the GABA transporter GAT-1. Pain 105, 347–353.

Moore, K. A., Baba, H., and Woolf, C. J. 2000. Synaptic Transmission and Plasticity in the Superficial Dorsal Horn. In: Nervous System Plasticity and Chronic Pain, (*eds*. J. Sandkühler, B. Bromm, and G. F. Gebhart), Vol. 129, pp. 63–81. Elsevier.

Pedersen, L. M., Lien, G. F., Bollerud, I., and Gjerstad, J. 2005. Induction of long-term potentiation in single nociceptive dorsal horn neurons is blocked by the CaMKII inhibitor AIP. Brain Res. 1041, 66–71.

Randic, M., Jiang, M. C., and Cerne, R. 1993. Long-term potentiation and long-term depression of primary afferent neurotransmission in the rat spinal cord. J. Neurosci. 13, 5228–5241.

Rygh, L. J., Kontinen, V. K., Suzuki, R., and Dickenson, A. H. 2000. Different increase in C-fibre evoked responses after nociceptive conditioning stimulation in sham-operated and neuropathic rats. Neurosci. Lett. 288, 99–102.

Sandkühler, J. 2000. Learning and memory in pain pathways. Pain 88, 113–118.

Sandkühler, J. 2007. Understanding LTP in pain pathways. Mol. Pain 3, 9.

Sandkühler, J. and Liu, X. 1998. Induction of long-term potentiation at spinal synapses by noxious stimulation or nerve injury. Eur. J. Neurosci. 10, 2476–2480.

Schouenborg, J. 1984. Functional and topographical properties of field potentials evoked in rat dorsal horn by cutaneous C-fibre stimulation. J. Physiol. 356, 169–192.

Svendsen, F., Tjølsen, A., Rygh, L. J., and Hole, K. 1999. Expression of long-term potentiation in single wide dynamic range neurons in the rat is sensitive to blockade of glutamate receptors. Neurosci. Lett. 259, 25–28.

Terman, G. W., Eastman, C. L., and Chavkin, C. 2001. Mu opiates inhibit long-term potentiation induction in the spinal cord slice. J. Neurophysiol. 85, 485–494.

Twining, C. M., Sloane, E. M., Milligan, E. D., Chacur, M., Martin, D., Poole, S., Marsh, H., Maier, S. F., and Watkins, L. R. 2004. Peri-sciatic proinflammatory cytokines, reactive oxygen species, and complement induce mirror-image neuropathic pain in rats. Pain 110, 299–309.

Vikman, K. S., Duggan, A. W., and Siddall, P. J. 2003. Increased ability to induce long-term potentiation of spinal dorsal horn neurones in monoarthritic rats. Brain Res. 990, 51–57.

Wallin, J., Fiska, A., Tjølsen, A., Linderoth, B., and Hole, K. 2003. Spinal cord stimulation inhibits long-term potentiation of spinal wide dynamic range neurons. Brain Res. 973, 39–43.

Willis, W. D. 2002. Long-term potentiation in spinothalamic neurons. Brain Res. Rev. 40, 202–214.

Yang, H. W., Zhou, L. J., Hu, N. W., Xin, W. J., and Liu, X. G. 2005. Activation of spinal D1/D5 receptors induces latep hase LTP of c-fiber evoked field potentials in rat spinal dorsal horn. J. Neurophysiol. 94, 961–967.

Yang, H. W., Hu, X. D., Zhang, H. M., Xin, W. J., Li, M. T., Zhang, T., Zhou, L. J., and Liu, X. G. 2004. Roles of CaMKII, PKA and PKC in the induction and maintenance of LTP of C-Fiber evoked field potentials in rat spinal dorsal horn. J. Neurophysiol. 91, 1122–1133.

Zhang, H. M., Qi, Y. J., Xiang, X. Y., Zhang, T., and Liu, X. G. 2001. Time-dependent plasticity of synaptic transmission produced by long-term potentiation of C-fiber evoked field potentials in rat spinal dorsal horn. Neurosci. Lett. 315, 81–84.

Zhong, J., Gerber, G., Kojic, L., and Randic, M. 2000. Dual modulation of excitatory synaptic transmission by agonists at group I metabotropic glutamate receptors in the rat spinal dorsal horn. Brain Res. 887, 359–377.

30 Immune System, Pain and Analgesia

H L Rittner, H Machelska, and C Stein, Charité – Universitätsmedizin Berlin, Campus Benjamin Franklin, Berlin, Germany

Glossary

CFA Complete Freund's adjuvant.
CCR/CXCR Chemokine receptor.
CRF Corticotropin releasing factor.
COX Cyclooxygenase.
CXCL1 KC keratinocyte-derived chemokine.
CXCL2/3 MIP-2 macrophage inflammatory protein.
CXCL8 IL-8.
DRG Dorsal root ganglion.
GDNF Glial cell-line-derived neurotrophic factor.
ICAM-1 Intercellular adhesion molecule-1.
IL Interleukin.
IL-1ra IL-1 receptor antagonist.
NGF Nerve growth factor.

NMDA N-methyl-D-aspartate.
PC Prohormone convertase.
PECAM-1 Platelet-endothelial cell adhesion molecule-1.
PENK Proenkephalin.
POMC Proopiomelanocortin.
SRIF Somatotropin release inhibiting factor (somatostatin).
TNF-α Tumor necrosis factor-α.
TrkA Tyrosine receptor kinase A.
TRPV1 Vanilloid receptor-1 (transient receptor potential vanilloid 1).
TTX Tetrodotoxin.

30.1   Introduction

Tissue destruction or inflammation are accompanied by an invasion of immune cells and liberation of numerous mediators. While these mediators contribute to the body's ability to counteract the destruction of tissue integrity, they also elicit pain by activation of specialized primary afferent neurons called nociceptors (reviewed in Julius, D. and Basbaum, A. I., 2001). They are defined as "receptors preferentially sensitive to a noxious stimulus or to a stimulus which would become noxious if prolonged" (definition of the International Association for the Study of Pain, IASP) (Merskey, H. and Bogduk, N., 1994). Trigeminal and dorsal root ganglia (DRG) contain nociceptor cell bodies which give rise to myelinated Aδ and unmyelinated C fibers. Aδ and C fibers transduce and propagate noxious stimuli to the dorsal horn of the spinal cord from where these stimuli are transmitted to the brain. At the level of the spinal cord and at supraspinal sites, various neurotransmitters come into plays which, together with environmental and cognitive factors, contribute to the eventual sensation of pain (Scholz, J. and Woolf, C. J., 2002). Strictly speaking, analgesia is defined as the inhibition of pain in man, while antinociception is defined as the inhibition of behavioral responses to noxious stimuli in animals. Although central mechanisms also play a prominent role, the following chapter will focus on the peripheral injured tissue itself.

30.2   Proalgesic Mechanisms

Inflammatory pain is characterized by an increased response to mechanical and heat stimuli, which are normally painful (Merskey, H. and Bogduk, N., 1994). This is called mechanical or thermal hyperalgesia, respectively. After tissue injury, local tissue macrophages and dendritic cells are activated. The inflammatory response is amplified by migration of leukocytes into the inflamed tissue, by production of inflammatory mediators as well as tissue acidification. Several agents are used experimentally to induce nonspecific inflammation in experimental animals. The most widely used are carrageenan and complete Freund's adjuvant (CFA) injected into the hindpaw of rodents. Carrageenan is used for short-term inflammation, while CFA is used to model longer time courses of tissue injury. Less frequently used agents include lipopolysaccharide, zymosan injection, or peritoneal inflammation using acetic acid or glycogen.

Neuropathic pain can arise following injury of peripheral nerves, when damaged or neighboring undamaged nerve fibers are sensitized or fire ectopically. It is also characterized by mechanical and thermal hyperalgesia. In addition, animals with neuropathy are sensitive to stimuli that do not evoke pain behavior under normal conditions, e.g., to touch, cooling, or warming. This is called allodynia (Merskey, H. and Bogduk, N., 1994). The most common models of neuropathic pain are tight ligation of spinal nerves and tight or loose ligation of the sciatic nerve. These models differ in the degree of neuronal damage but all resemble human neuropathy resulting from trauma-induced injury of peripheral nerves (e.g. entrapment, chronic nerve compression) (Bennet, G. J., 1994).

In the following section the effects of different mediators including cytokines, chemokines, nerve growth factor (NGF), and bradykinin are discussed in three aspects: injection of the mediator into non-inflamed tissue, blocking the mediator in models of inflammatory pain, and the role of the mediator in neuropathic pain.

30.2.1   Proinflammatory Cytokines

Cytokines are small proteins produced in the inflamed tissue as well as in lymphoid organs with stimulatory or inhibitory properties on immune function. Cytokines act through high-affinity receptors. Among the many different cytokines, the type I cytokines have a similar four α helical bundle structure. Their receptors correspondingly share characteristic features that have led to their description as the cytokine receptor superfamily, or type I cytokine receptors. Cytokine receptors in general are characterized by low abundance and a multisubunit structure yielding low- and high-affinity receptor forms. They are integral membrane glycoproteins with the N-terminal outside and a single membrane-spanning domain. Unlike growth factors that contain an instrinsic tyrosine kinase in their cytoplasmic domain, cytokine receptors induce tyrosine phosphorylation via additional molecules. Multiple signaling pathways/molecules have been observed for various cytokines. Collectively, these include the Jak-STAT pathway, the Ras–mitogen-activated protein (MAP) kinase pathway, Src and ZAP70 and related proteins, phosphatidylinositol 3-kinase

(PI3K), insulin receptor substrates 1 and 2 (IRS-1 and IRS-2), and phosphatases.

Cytokines are produced on demand and travel only over short distances. *In vivo* concentrations are in the range of a few picograms to nanograms per milliliter. Cytokines are pleiotropic and redundant, they influence mostly immune cells and the inflammatory response. Recently, however, cytokines have been shown to link the immune system and the nervous system and they seem to modulate pain and hyperalgesia.

30.2.1.1 Tumor necrosis factor-α

The tumor necrosis factor (TNF) recptor superfamily is comprised of at least 19 genes encoding 20 type II (i.e., intracellular N-terminus, extracellular C-terminus) transmembrane proteins. Included are several well-known ligands such as TNF-α, TNF-β (lymphotoxin-α, LT-α), Fas ligand, and CD40 ligand, as well as an increasing number of newly described mediators. TNF-α is the prototypic proinflammatory cytokine due to its role in initiating a cascade of cytokines and growth factors in the inflammatory response. Intraplantar injection of TNF-α into non-inflamed paws produced a short-lived dose-dependent mechanical hyperalgesia in rats (Cunha, F. Q. *et al.*, 1992). TNF-α seems to have indirect as well as direct hyperalgesic properties. Indirect action is supported by the following evidence: inhibitors of prostaglandin synthesis via cyclooxygenase (COX) inhibitors, and antagonists of adrenergic receptors can attenuate TNF-α-induced hyperalgesia in rats. In mice, TNF-α action is only dependent on the prostaglandin pathway (Cunha, T. M. *et al.*, 2005). Interestingly, the actions of TNF-α seem to be longer lasting if TNF-α is administered repetitively. Such treatment resulted in hyperalgesia lasting up to 30 days after cessation of the daily TNF-α injections in rats (Sachs, D. *et al.*, 2002). TNF-α-induced hyperalgesia in this study was again dependent on release of prostaglandins, as well as sympathetic amines, because blockade of either reduced hyperalgesia. Other groups have demonstrated a role of NGF in TNF-α-induced hyperalgesia. Pretreatment of rats with anti-NGF antiserum significantly attenuated TNF-α-induced mechanical and thermal hyperalgesia (Woolf, C. J. *et al.*, 1997). A direct action on nociceptors was postulated in electrophysiological studies (Sorkin, L. S. *et al.*, 1997; Junger, H. and Sorkin, L. S., 2000). TNF-α applied onto the sciatic nerve or injected subcutaneously into normal paws produced aberrant electrophysiological activity and dose dependently sensitized nociceptors to mechanical stimuli. This effect was restricted to Aδ and C fibers and no effect was seen in Aβ mechanoreceptive fibers. Interestingly, however, higher doses of TNF-α produced lower responses (U-shaped dose–response). The authors postulated that this was related to TNF-α-induced release of anti-inflammatory cytokines (Sorkin, L. S. *et al.*, 1997; Junger, H. and Sorkin, L. S., 2000).

TNF-α exerts its effects through two known receptors, TNFR1 and TNFR2. Expression of both was evaluated in DRG from rats (Li, Y. *et al.*, 2004): TNFR1 mRNA was expressed in virtually all neurons and in nonneuronal cells with increased levels in models of peripheral inflammation. TNFR2 was exclusively expressed and regulated in nonneuronal cells. *In situ* hybridization revealed varying levels of TNFR1 mRNA in virtually all DRG neurons including putative nociceptive neurons coding for calcitonin gene-related peptide, substance P, or vanilloid receptor 1 (transient receptor potential vanilloid 1; TRPV1). Similarly, expression of TNFR1 and TNFR2 protein was seen in rat DRG neurons and their axons and increased in models of neuropathic pain (Shubayev, V. I. and Myers, R. R., 2001; Schäfers, M. *et al.*, 2003c). In contrast, TNFR1 is not expressed in DRG neurons from normal mice but is seen in mice with nerve injury (Ohtori, S. *et al.*, 2004). TNFR localization on peripheral endings has not been examined but seems likely because of expression in DRG and its presence in axons.

Some studies are available regarding the role of endogenous TNF-α in inflammatory hyperalgesia. In one study, antisera against TNF-α could completely block hyperalgesia induced by intraplantar carrageenan injection (Cunha, F. Q. *et al.*, 1992). Similar results are seen in lipopolysaccharide-induced inflammation (Ferreira, S. H. *et al.*, 1993). A third study examined TNF-α in CFA-induced hindpaw inflammation (Woolf, C. J. *et al.*, 1997). Here, a single injection of anti-TNF-α serum before CFA application significantly delayed the onset of the resultant inflammatory hyperalgesia and reduced interleukin (IL)-1β but not NGF levels measured 24 h later. A role for TNF-α in inflammatory pain has also been confirmed in mice using TNFR1 knockout mice (Cunha, T. M. *et al.*, 2005). The intensity of hyperalgesia after intraplantar carrageenan injection was significantly lower in these animals. Therefore, TNF-α seems to be an endogenous mediator in inflammatory pain in mice and in rats. Monocytes and tissue macrophages are the primary sources for TNF-α synthesis. TNF-α synthesis is stimulated by a wide variety of agents. In macrophages, it is induced by biological, chemical, and physical stimuli such as viruses, bacterial and parasitic products, tumor cells, complement, cytokines, ischemia, trauma, and irradiation (Figure 1) (Cunha, F. Q. and Ferreira, S. H., 2003; Cunha, T. M. *et al.*, 2005). In peripheral tissues of rats, TNF-α activity is followed by increases in IL-6 and IL-8, and IL-6 is followed by IL-1β and then by NGF. All of these finally mediate their effects through prostaglandins or sympathetic amines. This cascade is also seen with few modifications in mice (Cunha, T. M. *et al.*, 2005).

The role of TNF-α has also been studied in neuropathic pain. TNF-α mRNA was upregulated in the sciatic nerve at 2 weeks after its loose ligation (a chronic constriction injury model) (Okamoto, K. *et al.*, 2001). Also, TNF-α protein was detected in inflammatory cells accumulating at the ligated sciatic nerve (Cui, J. G. *et al.*, 2000; Schäfers, M. *et al.*, 2003a). Treatment of animals with neuropathic injury using anti-TNF-α antibodies and especially with the combination of anti-TNF-α and anti-IL-1β antibodies significantly reduced signs of neuropathic pain (Schäfers, M. *et al.*, 2001). In another model, treatment starting 2 days before spinal nerve ligation with the TNF-α receptor antagonist etanercept attenuated

mechanical allodynia (Schäfers, M. *et al.*, 2003d). *In vivo* studies have recently been substantiated by electrophysiological studies. *In vitro* perfusion of DRG cells with TNF-α elicits neuronal discharges in Aδ and C fibers, which are markedly higher and longer lasting after nerve injury (Schäfers, M. *et al.*, 2003b). This demonstrates an increased sensitivity of injured neurons to TNF-α. TNF-α-induced hyperalgesia in this model is mediated via TNFR1 (Schäfers, M. *et al.*, 2003c). Downstream effectors of TNF-α include the p38 MAP kinase pathway. Intrathecal infusion of a p38 inhibitor reduced neuropathic pain if it was started before but not 7 days after spinal nerve ligation. Cessation of therapy resulted in increased allodynia.

In summary, TNF-α has hyperalgesic properties if injected exogenously. It also has endogenous hyperalgesic actions in inflammatory and neuropathic pain. The effect seems to be mediated in part indirectly via prostaglandins and sympathetic amines. Recent data also suggest direct sensitizing effects on sensory neurons.

30.2.1.2 *Interleukin-1β*

IL-1 and its family members are primarily proinflammatory cytokines. They can stimulate the expression of genes associated with inflammation and autoimmune diseases. The most salient and relevant properties of IL-1 in inflammation are the initiation of COX, type 2 phospholipase A, and inducible nitric oxide synthase. Hyperalgesic effects of IL-1 have first been demonstrated by injection into hindpaws of rats without inflammation or nerve injury (Ferreira, S. H. *et al.*, 1988): in contrast to IL-1α, IL-1β is able to sensitize nociceptors. Several mechanisms have been proposed to mediate this indirect action, including prostaglandins, nitric oxide, NGF, and bradykinin. The hyperalgesic effect of IL-1β is blocked by treatment with indomethacin, a nonselective COX inhibitor reducing the formation of prostaglandins (Figure 1) (Ferreira, S. H. *et al.*, 1988). In mice, IL-1β also induces prostaglandin formation mediating hyperalgesia (Cunha, T. M. *et al.*, 2005). Injection of IL-1β into normal tissue induces thermal hyperalgesia and the production of NGF. The thermal hyperalgesia but not the NGF production can be prevented by pretreatment with recombinant IL-1 receptor antagonist (IL-1ra). There is also evidence of a direct action of IL-1β on nociceptors: intraplantar injection of IL-1β potentiates action potentials in rat DRG neurons in response to thermal or mechanical stimuli, and a decrease of the mechanical threshold for nerve firing was observed

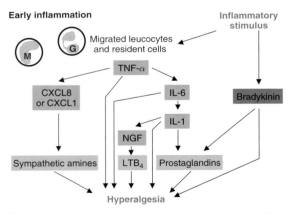

Figure 1 Hyperalgesic mechanisms in early inflammation. An inflammatory agent stimulates the migration of leukocytes, e.g., granulocytes (G) and monocytes (M), into the inflamed tissue. In the tissue, these leukocytes as well as resident cells initiate a cascade of cytokines including tumor necrosis factor-α (TNF-α) and interleukins (ILs), chemokines (CXCL8, CXCL1), nerve growth factor (NGF), bradykinin, and secondary mediators like sympathetic amines, leukotriene B$_4$ (LTB$_4$), and prostaglandins culminating in hyperalgesia. TNF-α, IL-6, IL-1, and bradykinin can also have direct hyperalgesic effects on nociceptors.

after local injection of IL-1β (Fukuoka, H. *et al.*, 1994). In the *in vitro* skin–nerve preparation brief exposure to IL-1β facilitates the heat-induced release of calcitonin gene-related peptide (a pronociceptive mediator) from peptidergic sensory neurons (Opree, A. and Kress, M., 2000). This was confirmed by a follow-up study (Obreja, O. *et al.* 2002a). IL-1 potentiated heat-activated inward currents and shifted the activation threshold towards lower temperature without altering intracellular calcium currents.

Intracellular mediators of these effects included protein tyrosine kinases and protein kinase C. One of the IL-1 receptors subtypes, the IL-1RI, was expressed in DRG cells as shown by polymerase chain reaction and *in situ* hybridization. The authors propose that recruitment of tyrosine kinases and downstream activation of protein kinase C phosphorylate TRPV1 to cause heat hyperalgesia. Both mRNA and IL-1RI protein are found in DRG neurons and glial cells (Li, M. *et al.*, 2005) of rats with and without paw inflammation. However IL-1RI localization has not been shown on peripheral nerve terminals. Using autoradiography, we did not detect IL-1 binding on nerves in either inflamed or noninflamed paws, but on immune cells in inflamed paws (Mousa, S. A. *et al.*, 1996). Therefore, IL-1 receptor expression in the periphery requires further studies.

IL-1β also has an endogenous hyperalgesic role in inflammatory as well as in neuropathic pain. In hindpaw inflammation induced by carrageenan or lipopolysaccharide IL-1ra can reduce, by not completely block, hyperalgesia in rats and in mice (Ferreira, S. H. *et al.*, 1993; Cunha, J. M. *et al.*, 2000; Cunha, T. M. *et al.*, 2005). The same effect can be seen in CFA-induced hindpaw inflammation or in mice after intraperitoneal injection of acetic acid (Safieh-Garabedian, B. *et al.*, 1995; Cunha, J. M. *et al.*, 2000). In CFA-induced inflammation, the authors propose that the effect is mediated by NGF, because treatment with IL-1ra also reduced the local production of NGF (Safieh-Garabedian, B. *et al.*, 1995). Interestingly, our group has shown analgesic effects of IL-1β via release of opioid peptides from immune cells in CFA-induced inflammation (see Analgesic Effects of Corticotropin Releasing Factor, Interleukin-1β, and Noradreline for discussion).

In neuropathic pain models, the role of IL-1 has not been thoroughly examined. Expression of IL-1β is seen in inflammatory cells accumulating at the sciatic nerve in ligation and inflammatory models of neuropathy as assesed by immunohistochemistry and by polymerase chain reaction (Gazda, L. S. *et al.*,

2001; Okamoto, K. *et al.*, 2001). Blocking of IL-1R in mice with chronic constriction injury reduces thermal hyperalgesia, mechanical allodynia, and also immunoreactivity for TNF-α (Sommer, C. *et al.*, 1999). However, the mechanisms remain unclear. So far, no study has demonstrated IL-1R expression in neuropathy. In summary, IL-1β seems to have hyperalgesic actions in different models of pain. The mechanism – direct or indirect – still needs to be clarified.

30.2.1.3 Interleukin-6

IL-6 is a one of the major proinflammatory cytokines. Its level is below detection under physiological conditions but is rapidly and strongly upregulated in early inflammation and can be detected early in the serum. Intraplantar injection of IL-6 induces hyperalgesia in normal rats (Poole, S. *et al.*, 1995). An indirect mechanism of this effect is supported by data showing inhibition of IL-6-induced hyperalgesia when animals are treated with COX blockers (Cunha, F. Q. *et al.*, 1992) (Figure 1). However, so far it is not known which cells and receptors are involved.

Direct effects have been proposed by other groups: In the rat *in vitro* skin–nerve preparation basal and heat-evoked release of calcitonin gene-related peptide from nociceptors was measured. IL-6 did not induce heat sensitization, when applied alone, but was effective in the presence of soluble IL-6 receptor (sIL-6R) (Opree, A. and Kress, M., 2000). sIL-6R is needed for cells that express only the signaling receptor unit gp130. The complex of sIL-6R and IL-6 can then induce a signal to activate cells. This so-called *trans*-signaling expands the spectrum of cells responsive to IL-6 to include neurons. IL-6, in combination with its soluble receptor, can sensitize nociceptors to heat, providing evidence for the constitutive expression of gp130, but not of the IL-6-membrane-bound (specific) receptor, in nociceptors (Obreja, O. *et al.*, 2002b). Hyper-IL-6 (HIL-6), a fusion protein of IL-6 and sIL-6R, was designed to mimic the effects of the IL-6–sIL-6R complex. *In vitro* exposure of DRG neurons to HIL-6 potentiated heat-activated inward currents and shifted the activation thresholds towards lower temperatures without affecting intracellular calcium levels (Obreja, O. *et al.*, 2005). This effect was reduced by a Janus tyrosine kinase inhibitor, a selective protein kinase C inhibitor as well as a selective blocker of the protein kinase C delta isoform, but not by the COX inhibitor indomethacin. *In situ* hybridization and reverse transcription-polymerase chain reaction revealed expression of the signal-transducing beta subunit of gp130 in neuronal somata but not in satellite cells in DRG from normal rats (Gadient, R. A. and Otten, U., 1996; Obreja, O. *et al.*, 2005). These data suggest a direct activation of sensory neurons by IL-6–sIL-6R.

As seen for IL-1β and TNF-α, endogenous IL-6 is also important for the development of hyperalgesia: Antisera neutralizing IL-6 inhibited lipopolysaccharide-induced hyperalgesia in rats (Ferreira, S. H. *et al.*, 1993). IL-6 knockout mice were shown to have reduced thermal and mechanical hyperalgesia after injection of carrageenan into the hindpaw (Xu, X. J. *et al.*, 1997). In this study, however, IL-6 knockout mice without inflammation displayed baseline hyperalgesia, which could not be explained by the authors.

In neuropathic mice, IL-6 mRNA levels were upregulated in DRG and correlated well with the development of nerve injury-induced thermal hyperalgesia and mechanical allodynia (Murphy, P. G. *et al.*, 1999). IL-6 knockout mice had reduced thermal and mechanical hyperalgesia in the chronic constriction injury model. In summary, IL-6 belongs to the group of proinflammatory hyperalgesic cytokines with direct as well as indirect effects on neurons, similar to IL-1β and TNF-α.

30.2.1.4   *Interleukin-18*

IL-18 is a member of the IL-1 ligand superfamily and is expressed in various cell types including macrophages and dendritic cells. Its biological functions include stimulation of activated T cells and enhancement of natural killer cell lytic activity. Furthermore, IL-18 activates and attracts neutrophils by inducing the production of TNF-α and subsequently leukotriene B$_4$. Injection of IL-18 into normal paws caused time and dose-dependent mechanical hyperalgesia (Verri, W. A., Jr. *et al.*, 2004) Interestingly, only morphine and dexamethasone blocked IL-18-induced hyperalgesia, whereas blockade of the prostaglandin or sympathetic pathway had no effect. In fact, the hyperalgesia elicited by IL-18 seems to be mediated via endothelin receptors, because it was abolished by blockage of endothelin receptor B (Verri, W. A., Jr. *et al.*, 2004).

30.2.2   Chemokines

Chemokines (CC and CXC) are chemotatic mediators produced in inflamed tissue. They control trafficking of leukocytes under physiological and inflammatory conditions, angiogenesis, and wound healing. They can be divided into subfamilies based on their structural motifs (the position of cysteine residues located near the N-terminus of the protein: CC, CXC, and CX$_3$C). Chemokines bind to a group of chemokine receptors (CCR, CXCR, and CX$_3$CR; seven transmembrane-spanning G$_i$ protein-coupled receptors) expressed on leukocytes. Although each receptor subtype typically binds multiple chemokines, the specificity is defined by a chemokine subfamily. Recently, chemokines and their receptors have been detected on cells in the central and peripheral nervous system.

Several observations indicate a role of chemokines in hyperalgesia. Chemokine receptors such as CXCR4 and CCR4 are expressed on subpopulations of DRG neurons and their corresponding ligands can induce calcium influx (Oh, S. B. *et al.*, 2001). However, in neonates the percentage of neurons responding to chemokine stimulation was significantly lower than the percentage responding to bradykinin or capsaicin. Direct intraplantar injection of some chemokines (CCL5, CXCL12, and CCL22) induces hyperalgesia in normal animals (Oh, S. B. *et al.*, 2001). In addition, some chemokines (human CXCL8 (IL-8) or rat CXCL1 (KC keratinocyte-derived chemokine)) can indirectly cause hyperalgesia through release of sympathetic amines when applied subcutaneously into the hindpaw (Lorenzetti, B. B. *et al.*, 2002; Cunha, F. Q. and Ferreira, S. H., 2003). Hyperalgesic effects of CXCL1 were further examined in mice (Cunha, T. M. *et al.*, 2005). Intraplantar injection of a small dose of CXCL1 induced hyperalgesia with a peak between 3 and 5 h returning to normal after 24 h. Pretreatment of mice with indomethacin, a COX inhibitor, or guanethidine, a blocker of sympathetic amines, reduced the hyperalgesia. An almost complete blockade was achieved if both agents were combined. Interestingly, CXCL1-induced hyperalgesia could also be inhibited by IL-1ra but no change was seen in TNFR1 knockout mice. Therefore, IL-1β seems to be downstream of CXCL1, whereas TNF-α is upstream. In contrast, our own unpublished data suggest that intraplantar injection of CXCL1 as well as CXCL2/3 (MIP-2 macrophage inflammatory protein) in normal rats does not evoke thermal or mechanical hyperalgesia for up to 12 h postinjection despite significant recruitment of granulocytes. Differences could be explained by the doses or the behavioral test used (see below). Doses producing maximal chemokine-induced granulocyte recruitment do not induce hyperalgesia in our model Rittner, H. L., *et al.* 2006.

In a recent study, a connection between chemokines and TRPV1 was shown (Zhang, N. *et al.*, 2005). CCL3 (macrophage inflammatory protein 1α) that binds to CCR1 was tested *in vitro* and *in vivo*. Activation of CCR1 resulted in increases in TRPV1-mediated Ca^{2+} influx and increased sensitivity of TRPV1 to its ligand capsaicin. *In vivo* the latency in the mouse hot plate test was reduced for up to 45 min after a local injection of CCL3. Thus, CCL3 seems to be capable of enhancing the sensitivity of TRPV1, possibly through a G protein-dependent signaling pathway.

The endogenous role of chemokines was studied in rats and mice in carrageenan-induced hindpaw inflammation. In studies by Ferreira *et al.*, human CXCL8 and rat CXCL1 were shown to contribute to carrageenan-induced hyperalgesia in rats (Cunha, F. Q. *et al.*, 1991; Lorenzetti, B. B. *et al.*, 2002; Sachs, D. *et al.*, 2002). Similarly, injection of carrageenan elicited hyperalgesia, which could be reduced by pretreatment with anti-CXCL1 antibodies in mice (Cunha, T. M. *et al.*, 2005). Therefore, part of the hyperalgesia seen in carrageenan inflammation is dependent directly or indirectly on certain chemokines. However, our own studies in CFA-induced inflammation suggest an additional role of chemokines in counteracting inflammatory pain. In this model, concomitant injection of CXCL2/3 and CFA induced a threefold increase in the number of neutrophils in the inflamed paw after 2 h, while mechanical and thermal hyperalgesia was not changed (Brack, A. *et al.*, 2004d). Furthermore, inhibition of CXCL1 and CXCL2/3 in inflamed paws had no effect on mechanical hyperalgesia despite significant reduction of the number of infiltrating neutrophils after 2 h (Brack, A. *et al.*, 2004c) (see also Analgesic Effects of Corticotropin Releasing Factor, Interleukin-1β, and Noradreline and Endogenous Opoid Analgesia). The differences between our studies and those by Cunha *et al.*, might be related to the model of inflammation (CFA versus carrageenan) or to the type of behavioral test used. While we determined the amount of mechanical pressure required for the rat to withdraw its paw (modified Randall–Selitto method), Cunha *et al.*, used a substantially more painful stimulus that induced sympathetic activation with a freezing reaction and apnea.

In neuropathic pain the role of chemokines has not been thoroughly examined. Disruption of the gene for CCR2 prevented mechanical allodynia in the chronic constriction injury model (Abbadie, C. *et al.*, 2003). CCR2, activated by CCL2 (monocyte chemotrattractant protein-1), is responsible for attracting monocytes into inflamed tissue. Therefore the authors conclude that the decreased number of monocytes is responsible for the reduced pain in this model. In summary, the contribution of chemokines to hyperalgesia may depend on their type, the model of pain, the time frame and the dose of chemokine used. Some of the indirect effects are mediated by the ability of chemokines to attract monocytes.

30.2.3 Nerve Growth Factor

NGF belongs to the family of neurotrophic proteins, together with brain-derived neurotrophic factor (BDNF), neurotrophin-3 (NT-3), NT-4/5, and NT-6. NGF governs the innervation of target tissues during development and plays an important role in neuronal survival and maintenance of connectivity. Besides its action on neurons, NGF also has immunomodulatory effects in different inflammatory diseases including asthma, arthritis, and psoriasis. NGF is constitutively expressed in keratinocytes in normal skin. However, many cells including all subtypes of leukocytes as well as sensory neurons can express NGF in inflammation (Vega, J. A. *et al.*, 2003). After binding to its high-affinity receptor tyrosine receptor kinase A (trkA) on peripheral terminals of sensory neurons, NGF is internalized and retrogradely transported to the somata in the DRG. NGF is a potent regulator of gene expression of neuropeptides like calcitonin gene-related peptide and substance P, of receptors like TRPV1 and of ion channels like tetrodotoxin (TTX)-resistant sodium channels. The role of endogenous NGF has mostly been examined using proteins that block the bioactivity of NGF because deletion of NGF or trkA in mice is lethal within the first week of life.

Injection of NGF into the normal rat paw induces a long-lasting thermal as well as mechanical hyperalgesia (Lewin, G. R. *et al.*, 1994). These effects seem to be in part dependent on mast cells and on a central component involving *N*-methyl-D-aspartate (NMDA) receptors. Nociceptor sensitization was also observed *in vitro* in a skin–nerve preparation (Rueff, A. and Mendell, L. M., 1996). Infusion of trkA–IgG into the noninflamed rat paw to block local NGF diminished thermal hyperalgesia and decreased calcitonin gene-related peptide content in DRG neurons (McMahon, S. B. *et al.*, 1995). The detection of sensitizing properties of NGF has been corroborated by the analysis of transgenic mice overexpressing NGF in the epidermis (Stucky, C. L. *et al.*, 1999). These animals developed hyperalgesia without any signs of inflammation of the skin. In electrophysiological studies unmyelinated noci-ceptors responded with a fourfold intensity to heat stimuli in comparison to wild-type mice. Moreover, almost every C fiber responded to heat in NGF over-expressing animals in contrast to only half of the C fibers in wild-type animals. No enhanced response to mechanical stimuli was seen. Together, this provides supporting evidence for the heat-sensitizing property of NGF. Importantly, endogenous NGF seems to be responsible for maintenance of the sensitivity of noci-ceptors as application of trkA–IgG fusion molecule to the innervation territory of the cutaneous saphenous nerve resulted in a reduction of innervation density of the epidermis and of the percentage of neurons responding to heat or bradykinin (Bennett, D. L. *et al.*, 1998).

Blocking of NGF by trkA–IgG in carrageenan-induced inflammation prevents the development of thermal hyperalgesia and the sensitization of nocicep-tors despite normal development of inflammation as measured by tissue edema (McMahon, S. B. *et al.*, 1995; Koltzenburg, M. *et al.*, 1999). Similarly, injection of anti-NGF antibodies blocks the development of thermal hyperalgesia in CFA-induced inflammation (Lewin, G. R. *et al.*, 1994). Thus, NGF seems to be part of the inflammatory cascade involving initially inflammatory cytokines like IL-1β and TNF-α and subsequently NGF (Figure 1) (Cunha, T. M. *et al.*, 2005).

Neuropathic pain is also influenced by NGF; however, the data seem to be controversial. Treatment with anti-NGF in the chronic constric-tion injury model reduced hyperalgesia but the effect showed a delayed onset, short duration, and depen-dency on the dosage (Ro, L. S. *et al.*, 1999). In a recent paper the role of NGF and glial cell-line-derived neurotrophic factor (GDNF) on nociceptor function was examined (Zwick, M. *et al.*, 2003). Two lines of transgenic mice that contained an increased number of either NGF- or GDNF-dependent neurons were used. These mice were tested in a model of inflam-matory pain (CFA) and neuropathic pain (spinal nerve ligation model). Contrary to expectations, neither line of transgenic mice became more hype-ralgesic following induction of persistent pain. In fact, NGF-overexpressing mice recovered more rapidly from initial inflammatory hyperalgesia and became hypoalgesic despite extensive paw swelling in the inflammatory model. In the neuropathic model, only wild-type mice became hyperalgesic. mRNA expression of opioid receptors, NMDA receptors, metabolic glutamate receptor, sodium channels and vanilloid receptors showed marked differences in

DRG neurons but not in the spinal cord between wild-type and NGF- or GDNF-overexpressing animals. This suggested that the somatosensory sys-tem is capable of producing normal pain behavior despite significant modification of nociceptors.

NGF blockade seems to produce analgesia also in other pain states. In a model of bone cancer pain, osteosarcoma tumor cells are injected into the intramedullary space of the mouse femur. In a recent study, the authors focused on a novel NGF-sequestering antibody and demonstrated that two administrations produce a profound reduction in both ongoing and movement-evoked bone cancer pain-related behaviors that was greater than that achieved with acute administration of 10 or 30 mg kg^{-1} of morphine (Sevcik, M. A. *et al.*, 2005). In summary, neutralization of NGF produces analge-sia in most pain models studied.

30.2.4   Other Mediators: Bradykinin

Bradykinin is a nonapeptide generated from high-molecular-weight kininogen through cleavage by kal-likrein under inflammatory conditions in plasma. It is also released from mast cells during asthma attacks and within the gut as a gastrointestinal vasodilator. Bradykinin is rapidly generated after tissue injury and seems to modulate most of the events observed during the inflammatory process including vasodilatation, increase of vascular permeability, plasma extravasation, and cell migration (reviewed in Calixto, J. B. *et al.*, 2000). It is also one of the most potent endogenous proalgesic mediators (Ferreira, S. H. *et al.*, 1993). It directly activates bradykinin receptors on primary afferent neurons or sensitizes nociceptors indirectly through release of prostaglandins, nitric oxide, neuro-kinins, calcitonin gene-related peptide, cytokines, and histamine (Figure 1), or via alterations of vanilloid receptor channel gating (Chuang, H. H. *et al.*, 2001). In fact, lipopolysaccharide- or carrageenan-induced inflammation triggers the TNF-α-driven cytokine cas-cade via the release of bradykinin (references in Cunha, F. Q. and Ferreira, S. H., 2003). Bradykinin is released at the beginning of the inflammation because delayed treatment with bradykinin receptor antago-nists 2 h after injection of lipopolysaccharide or carrageenan has no effect on hyperalgesia or inflammation.

Bradykinin also plays a role in neuropathic pain. Ablation of the gene for the bradykinin receptor 1 resulted in a significant reduction of mechanical allodynia and thermal hyperalgesia in early stages of nerve injury (Ferreira, J. *et al.*, 2005). Furthermore, systemic treatment with the bradykinin receptor 1-selective antagonist des-Arg9-[Leu8]-bradykinin reduced the established mechanical allodynia observed 7–28 days after nerve lesion in wild-type mice. Nonpeptidic orally active bradykinin receptor 1 and 2 antagonists are currently being developed as peripheral analgesics (Sawynok, J., 2003; Marceau, F. and Regoli, D., 2004).

30.2.5   Clinical Implications and Perspectives

A myriad of peripheral mediators is involved in inflammatory as well as neuropathic pain. The most frequently used pain medications are selective or nonselective inhibitors of COX. These nonsteroidal anti-inflammatory drugs inhibit the common final pathway of most proalgesic compounds, the production of prostaglandins. However, these drugs have serious side effects such as gastrointestinal ulcer formation, bleeding and/or, thromboembolic events. Because of the multitude of proinflammatory and proalgesic mediators the development of new analgesics is difficult. Inhibiting one pathway may still leave other mediators available. On the other hand, cytokine antagonists such as the TNF-α receptor antagonist etanercept have been developed for treatment of autoimmune disease like rheumatoid arthritis and have been shown to produce analgesia in patients (Sommer, C. *et al.*, 2001). Despite the almost innumerable number of mediators new treatments are under current investigation including, e.g., the blockade of the bradykinin system (Sawynok, J., 2003; Marceau, F. and Regoli, D., 2004). The development of peripherally acting drugs would have advantages such as the avoidance of central side effects. For example, the localized (e.g., topical) administration of drugs can potentially optimize drug concentrations at the site of injury without systemically active drug levels and fewer adverse systemic effects, fewer drug interactions, and no need to titrate doses into a therapeutic range compared with systemic administration.

30.3   Analgesic Mechanisms

Leukocytes are the source not only of hyperalgesic mediators but also of antinociceptive mediators. The best-characterized and clinically tested compounds are the endogenous opioid peptides. Other analgesic

mediators include anti-inflammatory cytokines, somatostatin (somatotropin release inhibiting factor; SRIF), and the endocannabinoids.

30.3.1   Anti-Inflammatory Cytokines

In later stages of inflammation cytokines are produced, which limit inflammation and counteract hyperalgesia (references in Cunha, F. Q. and Ferreira, S. H., 2003). For example, IL-4, IL-10, IL-13, and IL-1ra can produce analgesia (Figure 2) (Cunha, F. Q. *et al.*, 1999; Cunha, J. M. *et al.*, 2000; Cunha, F. Q. and Ferreira, S. H., 2003). IL-4 and IL-13 are produced by Th$_2$ lymphocytes and mast cells and suppress the production of proinflammatory mediators such as IL-1, TNF-α, interferon-γ, and CXCL8. In addition, inhibition of COX-2 is observed. IL-10 is a product of T-lymphocytes and monocytes and inhibits the production of Th$_1$ proinflammatory cytokines including IL-1, IL-6, IL-8, and TNF-α. Besides, it upregulates IL-1ra. These effects are also seen in models of pain. Pretreatment with IL-4, IL-10, and IL-13 dose dependently blocked hyperalgesia induced by carrageenan, bradykinin, and TNF-α but did not affect hyperalgesia induced by CXCL8 and prostaglandins. Furthermore, the endogenous role of anti-inflammatory cytokines to limit hyperalgesia was demonstrated by application of antisera against IL-4, IL-10, and IL-13. They potentiated hyperalgesia induced by carrageenan, bradykinin,

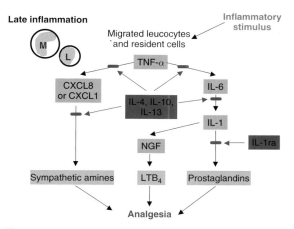

Figure 2   Analgesic cytokines in late inflammation. During ongoing inflammation leukocytes, e.g., lymphocytes (L) and monocytes/macrophages (M), start to produce anti-inflammatory cytokines like interleukin-4 (IL-4), IL-10, IL-13, and IL-1 receptor antagonist (IL-1ra). These cytokines inhibit the proinflammatory cytokines such as TNF-α, IL-1, and IL-6, and block the cascade.

and TNF-α. This effect was not seen in athymic and mast-cell-depleted rats. Thus, it seems that endogenous sources of IL-4 and IL-13 are mast cells and lymphocytes, respectively. Analgesic actions of IL-4, IL-10, and IL-13 are not only seen in paw inflammation but also in models of peritonitis and knee joint incapacitation induced by zymosan. Analgesic properties of anti-inflammatory cytokines are independent of endogenous production of opioid peptides (Vale, M. L. *et al.*, 2003). In summary, during ongoing inflammation analgesic cytokines are produced which counteract the effects of the proinflammatory hyperalgesic cytokines generated in the early stages.

30.3.2   Opioid Peptides

30.3.2.1   *Peripheral opioid receptors*

30.3.2.1.(i)   Localization   Opioid peptides are the natural ligands at opioid receptors. Three cDNAs and their genes have been identified, encoding the μ-, δ-, and κ-opioid receptor (MOP, DOP, and KOP), respectively (Kieffer, B. L. and Gaveriaux-Ruff, C., 2002). All three receptors can mediate pain inhibition and they are found throughout the nervous system, including somatic and visceral sensory neurons, spinal cord projection and interneurons, midbrain, and cortex. Recent interest has focused on the identification of opioid receptors on peripheral processes of sensory neurons. The cell bodies of these neurons in DRG express all three mRNAs and receptor proteins (Stein, C. *et al.*, 2001). Opioid receptors are intraaxonally transported into the neuronal processes and are detectable on peripheral sensory nerve terminals in animals and in humans. Colocalization studies confirmed the presence of opioid receptors on neurons expressing substance P and/or calcitonin gene-related peptide in DRG small- and medium-sized neurons and in their central and peripheral terminals (Dado, R. J. *et al.*, 1993; Zhang, X. *et al.*, 1998; Wenk, H. N. and Honda, C. N., 1999; Guan, J. S. *et al.*, 2005). Further, anatomical and electrophysiological studies showed expression of opioid receptors on unmyelinated C and on myelinated $A\delta$ fibers (i.e., nociceptors) (Beland, B. and Fitzgerald, M., 2001; Pare, M. *et al.*, 2001; Suzuki, R. and Dickenson, A. H., 2002) but not on large myelinated $A\beta$ fibers (Arvidsson, U. *et al.*, 1995; Beland, B. and Fitzgerald, M., 2001). However, after nerve injury morphine depressed responses of $A\beta$ fibers (Suzuki, R. and Dickenson, A. H., 2002), and expression δ-receptors on the $A\alpha\beta$ fibers in Meissner corpuscles in glaborus skin has been recently reported (Pare, M. *et al.*, 2001). The binding characteristics (affinity) of peripheral and central opioid receptors are similar but the molecular masses of peripheral and central μ-receptors appear to be different (references in Stein, C. *et al.*, 2003). If these findings are confirmed, a search for selective ligands at such distinct receptors may be warranted.

It has been suggested that opioid receptors are also located on sympathetic postganglionic neuron terminals. However, there are reports arguing against this notion and studies attempting the direct demonstration of opioid receptor mRNA in sympathetic ganglia have produced negative results. In addition, thorough morphological investigations have clearly demonstrated the presence of δ-receptors on unmyelinated primary afferent neurons and the absence of such receptors on postganglionic sympathetic neurons in skin, lip, and cornea. Moreover, chemical sympathectomy with 6-hydroxydopamine does not change the expression of opioid receptors in the DRG or the peripheral analgesic effects of μ-, δ-, and κ-receptor agonists in a model of inflammatory pain. Together, these findings have corroborated the notion that peripheral opioid receptors mediating analgesia are exclusively localized on primary sensory neurons (references in Stein, C. *et al.*, 2001).

Opioid binding sites and the expression of opioid receptor transcripts have also been demonstrated in immune cells. Opioid-mediated modulation of the proliferation of these cells and of their functions (e.g., chemotaxis, superoxide and cytokine production, mast cell degranulation) has been reported. These immunomodulatory actions can be stimulatory as well as inhibitory and have been ascribed to the activation of opioid receptors (Sacerdote, P. *et al.*, 2003; Sharp, B. M., 2003). However, the significance of such effects with regard to pain transmission has not been investigated as yet.

30.3.2.1.(ii)   Opioid receptor signaling in sensory neurons   All three types of opioid receptors mediate the inhibition of high-voltage-activated calcium currents in cultured primary afferent neurons. These effects are transduced by G proteins (G_i and/or G_o) (references in Stein, C. *et al.*, 2001; 2003). Although it is well-known that opioids induce membrane hyperpolarization due to increased potassium currents in central neurons, this could not be detected in DRG neurons so far (Akins, P. T. *et al.*, 1993). Thus, the modulation of calcium channels appears to be the primary mechanism for the inhibitory effects of opioids on peripheral sensory neurons. In addition, opioids – via inhibition of adenylyl cyclase – suppress TTX-resistant sodium- and nonselective cation currents stimulated by the inflammatory agent prostaglandin E_2 (Ingram, S. L. and Williams, J. T., 1994; Gold, M. S. and Levine, J. D., 1996). Interestingly, TTX-resistant sodium channels are selectively expressed in nociceptors, they are important for impulse initiation and action potential conductance, they mediate spontaneous activity in sensitized nociceptors and they accumulate at the site of injury in damaged nerves leading to ectopic impulse generation (Porreca, F. et al., 1999; Laird, J. M. et al., 2002). The latter observations may explain the notable efficacy of peripheral opioids in inflammatory and neuropathic pain (Stein, C. et al., 2001; 2003). Consistent with their effects on ion channels, opioids attenuate the excitability of peripheral nociceptor terminals, the propagation of action potentials, the release of excitatory proinflammatory neuropeptides (substance P, calcitonin gene-related peptide) from peripheral sensory nerve endings, and vasodilatation evoked by stimulation of C fibers (references in Stein, C. et al., 2001). All of these mechanisms result in analgesia and/or anti-inflammatory actions.

30.3.2.1.(iii) Peripheral opioid receptors and inflammation

Peripheral opioid analgesic effects are augmented under conditions of tissue injury such as inflammation, neuropathy, or bone damage (Stein, C. et al., 2001; 2003). One underlying mechanism is an increased number (upregulation) of peripheral opioid receptors. In neuronal cell cultures, μ-receptor transcription is upregulated by the cytokine IL-4 through binding of STAT6 transcription factors to the μ-receptor gene promoter (Kraus, J. et al., 2001). In DRG, the synthesis and expression of opioid receptors can be increased by peripheral tissue inflammation (Ji, R. R. et al., 1995; Zöllner, C. et al., 2003). Subsequently, the axonal transport of opioid receptors is greatly enhanced leading to their upregulation and to enhanced agonist efficacy at peripheral nerve terminals (Jeanjean, A. P. et al., 1995; Mousa, S. A. et al.,

2001). In addition, the specific milieu (low pH, prostanoid release) of inflamed tissue may increase opioid agonist efficacy by enhanced G protein coupling and by increased neuronal cyclic adenosine monophosphate levels (Selley, D. E. et al., 1993; Ingram, S. L. and Williams, J. T., 1994; Zöllner, C. et al., 2003). Inflammation also leads to an increase in the number of sensory nerve terminals (sprouting) and disrupts the perineurial barrier, thus facilitating the access of opioid agonists to their receptors (Stein, C. et al., 2001; 2003).

30.3.2.2 Opioid peptides produced by immune cells

Three families of opioid peptides are well characterized in the central nervous and neuroendocrine systems (Akil, H. et al., 1998). Each family derives from a distinct gene and from one of the three precursor proteins proopiomelanocortin (POMC), proenkephalin (PENK), or prodynorphin, respectively. Appropriate processing yields their respective representative opioid peptides, the endorphins, enkephalins, and dynorphins (Table 1). These peptides exhibit different affinity and selectivity for the three opioid receptors μ (endorphins, enkephalins), δ (enkephalins, endorphins), and κ (dynorphin). Two additional endogenous opioid peptides have been isolated from bovine brain: endomorphin-1 and endomorphin-2 (Table 1). Both peptides are considered highly selective μ-receptor ligands. Their precursors are not known yet. All of these opioid peptides have been detected in immune cells but the POMC and PENK families have been studied most extensively.

30.3.2.2.(i) Proopiomelanocortin-derived opioid peptides

POMC-related opioid peptides have been found in immune cells of many vertebrates and nonvertebrates (references in Machelska, H. et al., 2002). To determine whether these immunecompetent cells actually synthesize POMC rather than simply absorb related peptides from plasma,

Table 1 Structure of endogenous opioid peptides

β-Endorphin	Tyr-Gly-Gly-Phe-Met-Thr-Ser-Glu-Lys-Ser-Gln-Thr-Pro-Leu-Val-Thr-Leu-Phe-Lys-Asn-Ala-Ile-Val-Lys-Asn-Ala-His-Lys-Lys-Gly-Gln-OH
Methionine-enkephalin	Tyr-Gly-Gly-Phe-Met-OH
Leucine-enkephalin	Tyr-Gly-Gly-Phe-Leu-OH
Dynorphin A	Tyr-Gly-Gly-Phe-Leu-Arg-Arg-Ile-Arg-Pro-Lys-Leu-Lys-Tyr-Asp-Asn-Gln-OH
Endomorphin-1	Tyr-Pro-Trp-Phe-OH
Endomorphin-2	Tyr-Pro-Phe-Phe-OH

mRNA encoding POMC was sought for and demonstrated in many of these studies. The pituitary POMC gene is organized into three exons separated by intervening sequences which are removed during processing following transcription to produce the full-length, 1200-nt transcript. Initially, truncated POMC transcripts were found in leukocytes (references in Machelska, H. *et al.*, 2002). Lyons P. D. and Blalock J. E. (1997) reexamined the question of POMC mRNA expression using novel polymerase chain reaction procedures. With this exacting and sensitive methodology, they found expression of full-length transcripts encoding all three POMC exons in rat mononuclear leukocytes. This POMC transcript is spliced in the same way as the pituitary transcript and contains the sequence for the signal peptide, which is necessary for the correct routing into the regulated secretory pathway. The POMC protein is also proteolytically processed in a way consistent with the pituitary gland (Lyons, P. D. and Blalock, J. E., 1997; Mousa, S. A. *et al.*, 2004). These results unequivocally demonstrate that immune cells can produce full-length POMC transcripts. Apparently, this production is stimulated by various immune and inflammatory mediators (Smith, E. M., 2003).

30.3.2.2.(ii) *Proenkephalin-derived opioid peptides* PENK-derived opioid peptides have also been detected in human and rodent immune cells. Both the mRNA and methionine (Met)-enkephalin protein were detected. Preproenkephalin mRNA was found in T- and B-cells, macrophages, and mast cells. In subpopulations of immune cells this mRNA was shown to be highly homologous to brain PENK mRNA, abundant and apparently translated, as immunoreactive enkephalin is present and/or released. The appropriate enzymes necessary for posttranslational processing of PENK have also been identified in immune cells (references in Machelska, H. *et al.*, 2002).

30.3.2.2.(iii) *Immune-derived opioid peptides in inflammation* Immune-derived opioid peptides apparently play a substantial role in the modulation of inflammatory pain (Machelska, H. *et al.*, 2002; Stein, C. *et al.*, 2003). POMC mRNA and β-endorphin, as well as Met-enkephalin and dynorphin were found in cells derived from lymph nodes, and in leukocytes in the blood and within inflamed tissue in CFA model of inflammatory pain (Cabot, P. J. *et al.*, 1997; 2001; Machelska, H. *et al.*, 2003; Mousa, S. A. *et al.*, 2004).

In the pituitary, POMC processing begins as the nascent polypeptide enters the endoplasmic reticulum directed by the signal peptide, and POMC cleavage begins in the *trans*-Golgi network (references in Mousa, S. A. *et al.*, 2004). The POMC prohormone is directed to the regulated secretory pathway at the *trans*-Golgi network by binding to a sorting receptor, membrane-bound carboxypeptidase E. The prohormone convertases PC1 (also called PC1/3) and PC2 cleave POMC within the *trans*-Golgi network. PC1 mediates the initial cleavage into adrenocorticotropic and β-lipotropic hormones. The inactive pro-PC2 is bound to 7B2 (a chaperone-like binding protein) and is transported from the endoplasmic reticulum to later compartments of the secretory pathway, where it matures to active PC2; thereafter, PC2 converts β-lipotropic hormone into β-melanocyte stimulating hormone and β-endorphin (references in Mousa, S. A. *et al.*, 2004). Recently, we detected β-endorphin and POMC alone and colocalized with PC1, PC2, carboxypeptidase E, and 7B2 in leukocytes in the blood and within inflamed paw tissue (Mousa, S. A. *et al.*, 2004). This demonstrates that immune cells express the entire machinery required for POMC processing into functionally active β-endorphin.

In this CFA model, mRNAs encoding POMC and PENK and the corresponding opioid peptides β-endorphin and Met-enkephalin are abundant in cells of inflamed but not in noninflamed tissue (Przewlocki, R. *et al.*, 1992). Histomorphological procedures and flow cytometry have identified the opioid-containing cells as T- and B-lymphocytes, granulocytes, and monocytes/macrophages (Przewlocki, R. *et al.*, 1992; Rittner, H. L. *et al.*, 2001). Also, it was shown that β-endorphin is present in activated/memory T-cells within inflamed tissue (Cabot, P. J. *et al.*, 1997; Mousa, S. A. *et al.*, 2001). Thus, opioid peptides are processed and present both in circulating and inflammatory cells infiltrating injured tissue.

30.3.2.3 *Migration of opioid-containing immune cells to inflamed tissue*

The recruitment of leukocytes from the circulation into areas of inflammation begins with the attachment of these cells to vascular endothelium, followed by their transmigration into the tissue. This is mediated by specific cell adhesion and chemoattractant/activator molecules. Leukocytes are recruited from the circulation by a well-orchestrated set of events. They undergo multiple attachments to and detachments from the vessel's endothelial cells prior to transendothelial migration. This includes slowing and rolling along the endothelial cell wall that is mediated predominantly by the interaction of selectins expressed on leukocytes (L-selectin) and on endothelial cells (P- and E-selectin) with their ligands on endothelium or immune cells, respectively. The rolling leukocytes can then be activated by chemoattractants such as chemokines released from inflammatory cells and presented on endothelium. This leads to the upregulation and increased avidity of integrins. These mediate the firm adhesion of leukocytes to endothelial cells via ligands of the immunoglobulin superfamily, e.g., intercellular adhesion molecule-1 (ICAM-1). Finally, the immune cells transmigrate through the endothelial wall mediated by immunoglobulin superfamily members (e.g., platelet-endothelial adhesion molecule-1; PECAM-1) and are directed to the sites of inflammation to initiate a host defense (reviewed by Petruzzelli, L. *et al.*, 1999; von Andrian, U. H. and Mackay, C. R., 2000).

Recent findings indicate that these events can also be involved in the endogenous control of inflammatory pain. In the model of unilateral hind paw inflammation with CFA we have shown that integrin α_4 and the chemokines CXCL2 and CXCL1 are expressed by leukocytes, while adhesion molecules such as P- and E-selectins, ICAM-1, and PECAM-1 are upregulated on endothelium in inflamed paw tissue (Mousa, S. A. *et al.*, 2000; Machelska, H. *et al.*, 2002; Brack, A. *et al.*, 2004; Machelska, H. *et al.*, 2004) (Figure 3). Expression of CXCL1 and CXCL2 mRNAs and protein contents significantly increased in inflamed tissue during the course of inflammation (Brack, A. *et al.*, 2004). Importantly, L-selectin, integrins β_2, and the CXC chemokine receptor 2 (CXCR2) are coexpressed by opioid-containing leukocytes, which have migrated to inflamed subcutaneous paw tissue (Mousa, S. A. *et al.*, 2000; Brack, A. *et al.*, 2004; Machelska, H. *et al.*, 2004). Furthermore, pretreatment of rats with a selectin blocker (fucoidin), selective antibodies against

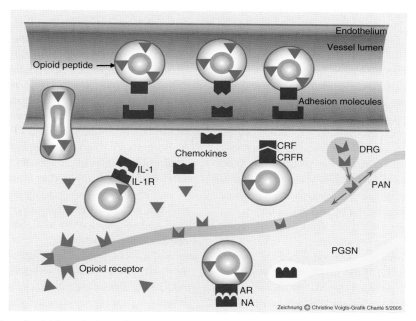

Figure 3 Migration of opioid-producing cells and opioid secretion within inflamed tissue. Adhesion molecules are upregulated on vascular endothelium and are coexpressed by circulating immune cells producing opioid peptides. These cells coexpress receptors for chemokines, which are presented on endothelium. Certain adhesion molecules mediate rolling of opioid-containing leukocytes along the endothelium. The rolling leukocytes can then be activated by chemokines, which upregulate another class of adhesion molecules mediating adhesion. Finally cells transmigrate through the endothelium. Once extravasated, these cells can be stimulated by stress or releasing agents such as corticotropin releasing factor (CRF), interleukin-1β (IL-1β), chemokines, and noradrenaline (NA) to secrete opioid peptides. CRF, IL-1, and NA (the latter derived from postganglionic sympathetic neurons; PGSN) elicit opioid release by activating CRF receptors (CRFR), IL-1 receptors (IL-1R), and adrenergic receptors (AR) on immune cells, respectively. Opioids bind to peripheral opioid receptors (produced in dorsal root ganglia (DRG) and transported to peripheral endings of primary afferent neurons; PAN) and lead to analgesia.

ICAM-1, integrins α_4 and β_2, or chemokines CXCL1 and CXCL2 substantially decreases the number of opioid-containing leukocytes infiltrating the inflamed tissue (Machelska, H. *et al.*, 1998; Machelska, H. *et al.*, 2002; Brack, A. *et al.*, 2004; Machelska, H. *et al.*, 2004) and in consequence abolishes endogenous peripheral opioid analgesia (see Endogenous Opoid Analgesia). This suggests that circulating opioid-producing immune cells home to inflamed tissue where they secrete the opioids to inhibit pain. Afterward, they travel to the regional lymph nodes, depleted of the opioid peptides (Cabot, P. J. *et al.*, 1997; 2001). Thus, local signals apparently not only stimulate the synthesis of opioid peptides in resident inflammatory cells but also attract opioid-containing cells from the circulation to the site of injury to reduce pain. This is controlled by specific chemotactic and adhesive mechanisms.

30.3.2.4   *Release of opioid peptides from immune cells*

Regulated secretion of peptides requires secretory granules deriving from the Golgi network for transport to the cell membrane. As discussed in Immune-Derived Opoid Peptides in Inflammation immune cells in the blood and in inflamed tissue coexpress the entire machinery required for POMC processing into functionally active β-endorphin (Mousa, S. A. *et al.*, 2004). Furthermore, our ultra-structural observations show that β-endorphin-expressing macrophages, monocytes, granulocytes, and lymphocytes, contain rough endoplasmic reticulum and Golgi apparatus, similar to pituitary cells. Immunostaining of β-endorphin was detectable in secretory granules, which were grouped in small or large membranous vesicular structures. The smaller β-endorphin-immunoreactive secretory granules were localized within cytoplasm, and the larger ones were arranged at the cell periphery ready for the exocytosis, similar to the pituitary (Mousa, S. A. *et al.*, 2004). In the pituitary, β-endorphin and other POMC-derived peptides are released by corticotropin releasing factor (CRF) and IL-1β (references in Schäfer, M., 2003). Similar mechanisms can trigger opioid release within peripheral inflamed tissue (Figure 3). CRF is present in immune cells, fibroblasts and vascular endothelium and peripheral CRF expression is enhanced in inflamed synovial and subcutaneous tissue of animals and humans (Schäfer, M. *et al.*, 1996). CRF and IL-1β receptors and their upregulation were demonstrated in inflamed lymph nodes and paw tissue in CFA model. Their pharmacological characteristics were similar to the high-affinity CRF- and IL-1β-binding sites in the pituitary (Mousa, S. A. *et al.*, 1996). Furthermore, the coexpression of CRF-1 and CRF-2 receptors with β-endorphin in monocytes/macrophages, granulocytes and lymphocytes were recently shown in the blood and in inflamed paw tissue (Mousa, S. A. *et al.*, 2003). In contrast, CRF receptors were not detected on peripheral sensory nerves. (Mousa, S. A. *et al.*, 1996; Mousa, S. A. *et al.*, 2003). The activation of CRF and IL-1β receptors on cells from lymph nodes results in the secretion of opioid peptides *in vitro*. In those studies β-endorphin, Met-enkephalin, and dynorphin were dose dependently released by CRF, while IL-1β released β-endorphin and dynorphin but not Met-enkephalin. These effects were specific to CRF and IL-1β receptors (Schäfer, M. *et al.*, 1994; Cabot, P. J. *et al.*, 1997; 2001; Mamet, J. *et al.*, 2002). Moreover, this release of opioid peptides was calcium-dependent and mimicked by elevated extracellular concentrations of potassium. This is consistent with a regulated pathway of release from secretory vesicles, as in neurons and endocrine cells (Cabot, P. J. *et al.*, 1997; 2001).

Adrenergic receptor agonists have also been shown to secrete β-endorphin from human peripheral blood mononuclear cells (Kavelaars, A. *et al.*, 1990). Similar mechanisms are involved in β-endorphin release from leukocytes in the rat CFA model of inflammatory pain (Binder, W. *et al.*, 2004) (Figure 3). Double labeling demonstrated adrenergic α_1-, β_2-, and, to a lesser degree, α_2 receptors expressed on β-endorphin-containing leukocytes in inflamed paw tissue. β-endorphin-containing and adrenergic α_1- and β_2-receptor-expressing immune cells were localized in close proximity to sympathetic nerve fibers, and chemical ablation of these fibers abolished intrinsic opioid analgesia. Finally, noradrenaline induced adrenergic receptor-specific release of β-endorphin from immune cells *in vitro* (Binder, W. *et al.*, 2004). Taken together, CRF, IL-1β, and sympathetic neuron-derived noradrenaline can act on their respective receptors on immune cells resulting in release of opioid peptides (Figure 3).

30.3.2.5   *Analgesia produced by immune-derived opioid peptides*

30.3.2.5.(i) *Analgesic effects of corticotropin releasing factor, interleukin-1β, and noradrenaline*

CRF-, IL-1β-, and noradrenaline-induced release of opioids from immune cells also occurs *in vivo*

(Figure 3). CRF and IL-1β injected into inflamed paws in CFA model produce dose-dependent analgesia reversible by their respective antagonists (Schäfer, M. *et al.*, 1994). CRF can produce analgesia both in early and late stages of inflammation (2 h and 6 days) and, in accord with anatomical findings (see Release of Opoid Peptides from Immune Cells), CRF analgesia involves both CRF-1 and CRF-2 receptors (Machelska, H. *et al.*, 2003; Mousa, S. A. *et al.*, 2003; Brack, A. *et al.*, 2004c). Intravenous administration of these agents in locally effective doses does not change pain thresholds, demonstrating a peripheral site of action (Schäfer, M. *et al.*, 1994; Machelska, H. *et al.*, 2003; Mousa, S. A. *et al.*, 2003; Brack, A. *et al.*, 2004c). Leukocytes apparently are the target for CRF and IL-1β because immunosuppression with cyclosporine A, depletion of granulocytes, blockade of chemokines (CXCL1 and CXCL2/3), as well as anti-selectin and anti-ICAM-1 treatments result in a significant reduction of opioid-containing immune cells and of CRF- or IL-1β-induced analgesia (Schäfer, M. *et al.*, 1994; Machelska, H. *et al.*, 1998; 2002; Brack, A. *et al.*, 2004c). Also, cyclosporine A-induced attenuation of CRF analgesia could be restored by injection of activated lymphocytes (Hermanussen, S. *et al.*, 2004). CRF- and IL-1β-induced analgesia is blocked by an antibody against β-endorphin, suggesting that this opioid plays a major role. In addition, Met-enkephalin appears to be involved in CRF-, and dynorphin in IL-1β-induced analgesia (Schäfer, M. *et al.*, 1994).

These results are in line with other studies on local analgesic effects of CRF (Hargreaves, K. M. *et al.*, 1989; McLoon, L. K. *et al.*, 2002) and of the cytokines IL-6 and TNF-α (Czlonkowski, A. *et al.*, 1993) but are in contrast to hyperalgesia mediated by IL-1β, IL-6, and TNF-α discussed in Cunha F. Q. and Ferreira S. H. (2003). Importantly, however, the latter hyperalgesic effects of exogenous cytokines were observed after injections into noninflamed tissue (Cunha, F. Q. *et al.*, 1992; Poole, S. *et al.*, 1995; Safieh-Garabedian, B. *et al.*, 1995; Woolf, C. J. *et al.*, 1997; Cunha, T. M. *et al.*, 2005). Apparently, hyperalgesia in highly inflamed tissue (4 days after CFA) has already reached a ceiling effect (Machelska, H. *et al.*, 2003) and therefore cannot be further increased. Instead, in such tissue opioid-containing immune cells became the predominant target for cytokines to produce analgesia. Contribution of endogenous proinflammatory cytokines to the generation of pain was mostly observed either in early stages (3–6 h) when cytokine blockade preceded induction

of CFA inflammation (Woolf, C. J. *et al.*, 1997) or in short-lasting (30 min to 3 h) inflammation, e.g., induced with carrageenan, lipopolysaccharide, or acetic acid (Cunha, F. Q. *et al.*, 1992; Ferreira, S. H. *et al.*, 1993; Cunha, J. M. *et al.*, 2000; Cunha, T. M. *et al.*, 2005). Thus, the presence or absence of inflammation, the duration and/or model of inflammatory pain are factors to be taken into consideration.

Noradrenaline administered directly into inflamed tissue has been shown to produce analgesia, reversible by α_1-, α_2-, and β_2-adrenergic receptor antagonists in CFA model. Further, this effect was dose dependently blocked by μ and δ receptor antagonists and antibody against β-endorphin (Binder, W. *et al.*, 2004). These data suggest that this catecholamine produces analgesia via opioid peptides activating peripheral opioid receptors. In noninflamed tissue, noradrenaline did not influence pain behavior in this model, consistent with the lack of opioid-containing cells and with the scarcity of adrenergic receptors (Binder, W. *et al.*, 2004). However, the role of peripheral adrenergic receptors in nociception appears controversial. In noninflamed tissue, noradrenaline has been shown to produce hyperalgesia via α_2-adrenergic receptors. Others found that α_2-agonists could also produce peripheral analgesia. It has even been postulated that hyperalgesia is mediated via α_{2B} and analgesia is mediated via α_{2C} receptors (references in Binder, W. *et al.*, 2004). Thus, different receptor subtypes, receptor localization, microenvironment, and the presence or absence of inflammation are important parameters to be considered.

30.3.2.5.(ii) Endogenous opioid analgesia Stress is a natural stimulus triggering inhibition of pain (Willer, J. C. *et al.*, 1981; Terman, G. W. *et al.*, 1984). In rats, stress induced by cold water (4 °C) swim (for 1 min) elicits potent analgesia in inflamed but not in noninflamed paws in CFA model (Stein, C. *et al.*, 1990a; Machelska, H. *et al.*, 2003). In early inflammation (6 h), this swim stress-induced analgesia was only partially attenuated by peripherally selective doses of different opioid peptide antibodies and receptor antagonists but was fully reversed by centrally acting doses of an opioid receptor antagonist (Machelska, H. *et al.*, 2003). At later stages of inflammation (4–6 days), swim stress analgesia was completely abolished by peripherally selective doses of antibody against β-endorphin and by μ and δ-antagonists (Stein, C. *et al.*, 1990a). Together, these data indicate that at early stages of CFA-induced inflammation all three families of opioid peptides and receptors are involved, while at later stages β-endorphin acting at μ- and δ-receptors dominates. Whereas at early stages both peripheral and central opioid receptors are involved, at later stages endogenous analgesia is mediated exclusively by peripheral opioid receptors. Thus, peripheral opioid mechanisms of pain control become more prevalent with the duration and severity of inflammation. Endogenous triggers of swim stress-induced analgesia are locally produced CRF and sympathetic nerve-derived catecholamines because this effect is abolished by local neutralization of CRF (Schäfer, M. *et al.*, 1996; Machelska, H. *et al.*, 2003), and by sympathetic blockade (Binder, W. *et al.*, 2004).

Various types of immune cells are the source of opioids as demonstrated by the abolishment of stress-induced analgesia by immunosuppression with cyclosporine A or whole-body irradiation and by depletion of monocytes/macrophages (Stein, C. *et al.*, 1990b; Przewlocki, R. *et al.*, 1992; Brack, A. *et al.*, 2004a). Moreover, this effect is also extinguished by inhibiting the extravasation of β-endorphin-containing immune cells following blockade of L- and P-selectins, α_4 and β_2 integrins or of ICAM-1 (Machelska, H. *et al.*, 1998; 2002; 2004). These adhesion molecules apparently regulate the migration of opioid-containing immune cells and the subsequent generation of intrinsic pain control in inflamed tissue (Figure 3).

A future challenge is to identify factors that will increase homing of opioid-containing cells to injured tissue and will enhance analgesia. To this end, we have shown that hematopoetic growth factors strongly mobilized granulocytes in the blood but resulted only in a minor increase in the number of opioid-containing leukocytes in inflamed paws and in no change of CRF- and swim stress-induced analgesia (Brack, A. *et al.*, 2004b). Another approach was to increase the migration of opioid-containing cells to inflamed tissue with local injections of CXCL2. However, this did not result in stronger CRF- or swim stress-induced analgesia either, most probably as a result of the small number of peripheral opioid receptors at this very early (2 h) stage of inflammation (Brack, A. *et al.*, 2004d). Indeed, previous studies had shown that intrinsic analgesia increases with the duration of inflammation (2 h to 4 days), in parallel with the number of opioid-containing leukocytes, with the number of peripheral opioid receptors and with the efficacy of opioid receptor–G protein coupling in sensory neurons

(Mousa, S. A. *et al.*, 2001; Rittner, H. L. *et al.*, 2001; Zöllner, C. *et al.*, 2003).

30.3.2.6 Clinical implications

Peripheral endogenous opioid analgesia has found many clinical applications. Opioid receptors have been demonstrated on peripheral terminals of sensory nerves in human synovia (Stein, C. *et al.*, 1996) and these receptors mediate analgesia in patients with various types of pain (e.g., in chronic rheumatoid arthritis and osteoarthritis, bone pain, after dental, laparoscopic, urinary bladder, and knee surgery) (references in Stein, C. *et al.*, 2003). Opioid peptides are found in human synovial lining cells, mast cells, lymphocytes, and macrophages. The prevailing peptides are β-endorphin and Met-enkephalin, while only minor amounts of dynorphin are detectable (Stein, C. *et al.*, 1993; 1996). The interaction of synovial opioids with peripheral opioid receptors was examined in patients undergoing knee surgery. Blocking intraarticular opioid receptors by the local administration of naloxone resulted in significantly increased postoperative pain (Stein, C. *et al.*, 1993). Taken together, these findings suggest that in a stressful (e.g., postoperative) situation, opioids are tonically released from inflamed tissue and activate peripheral opioid receptors to attenuate clinical pain. Importantly, these endogenous opioids do not interfere with exogenous morphine, i.e., intraarticular morphine is an equally potent analgesic in patients with and without opioid-producing inflammatory synovial cells (Stein, C. *et al.*, 1996). This suggests that, in contrast to the rapid development of tolerance in the central nervous system, the immune cell-derived opioids do not readily produce cross-tolerance to morphine at peripheral opioid receptors.

30.3.2.7 Perspectives

Effective control of inflammatory pain can result from interactions between leukocyte-derived opioid peptides and their receptors on peripheral sensory neurons. These findings provide new insights into intrinsic mechanisms of pain control and open strategies to develop new drugs and alternative approaches to treatment of pain. Immunocompromised patients (e.g., in AIDS, cancer, diabetes) frequently suffer from painful neuropathies, which can be associated with intra- and perineural inflammation, with reduced intraepidermal nerve fiber density and with low $CD4^+$ lymphocyte counts (Polydefkis, M. *et al.*, 2002). Thus it may be interesting to investigate the opioid production/release and the migration of opioid-containing leukocytes in these patients. The important role of certain adhesion molecules and chemokines in the trafficking of opioid-containing cells to injured tissues indicates that antiadhesion or antichemokine strategies for the treatment of inflammatory diseases may in fact carry a significant risk to exacerbate pain. It would be highly desirable to identify stimulating factors and strategies that selectively attract opioid-producing cells and increase peripheral opioid receptor numbers in damaged tissue. Augmenting the synthesis and/or secretion of opioid peptides and opioid receptor numbers within injured tissue may be accomplished by gene therapy: delivery of PENK, POMC, and of μ receptor cDNAs have been shown to decrease chronic pain and inflammation (Braz, J. *et al.*, 2001; Lu, C. Y. *et al.*, 2002; Xu, Y. *et al.*, 2003).

Importantly, opioid analgesia resulting from neuroimmune interactions occurs in peripheral tissues and therefore is devoid of central opioid side effects (such as depression of breathing, nausea, clouding of consciousness, addiction, and high rate of tolerance) and of typical side effects produced by COX inhibitors (such as gastric erosions, ulcers, bleeding, diarrhea, thromboembolic events, and renal toxicity). Many efforts are currently undertaken to develop peripherally acting analgesics by aiming at individual excitatory receptors or channels on sensory neurons (Simonin, F. and Kieffer, B. L., 2002). The major advantage of targeting opioid receptors is their mechanism of action: the inhibition of calcium (and possibly sodium) channels simply renders the nociceptor less excitable to the plethora of stimulating molecules expressed in damaged tissue. Thus, peripherally acting opioids can prevent and reverse the action of multiple excitatory agents simultaneously, in contrast to blocking only one single noxious stimulus. Uncovering mechanisms that can enhance the availability of endogenous opioids within injured tissue and the signal transduction of peripheral opioid receptors will open exciting possibilities for pain research and therapy.

References

Abbadie, C., Lindia, J. A., Cumiskey, A. M., Peterson, L. B., Mudgett, J. S., Bayne, E. K., DeMartino, J. A., MacIntyre, D. E., and Forrest, M. J. 2003. Impaired neuropathic pain responses in mice lacking the chemokine receptor CCR2. Proc. Natl. Acad. Sci. U. S. A. 100, 7947–7952.

Akil, H., Owens, C., Gutstein, H., Taylor, L., Curran, E., and Watson, S. 1998. Endogenous opioids: overview and current issues. Drug Alcohol Depend. 51, 127–140.

Akins, P. T. and McCleskey, E. W. 1993. Characterization of potassium currents in adult rat sensory neurons and modulation by opioids and cyclic AMP. Neuroscience 56, 759–769.

von Andrian, U. H. and Mackay, C. R. 2000. T-cell function and migration. Two sides of the same coin. N. Engl. J. Med. 343, 1020–1034.

Arvidsson, U., Riedl, M., Chakrabarti, S., Lee, J. H., Nakano, A. H., Dado, R. J., Loh, H. H., Law, P. Y., Wessendorf, M. W., and Elde, R. 1995. Distribution and targeting of a mu-opioid receptor (MOR1) in brain and spinal cord. J. Neurosci. 15, 3328–3341.

Beland, B. and Fitzgerald, M. 2001. Mu- and delta-opioid receptors are downregulated in the largest diameter primary sensory neurons during postnatal development in rats. Pain 90, 143–150.

Bennet, G. J. 1994. Animal Models of Neuropathic Pain. In: Proceedings of the 7th World Congress on Pain, Progress in Pain Research and Management (eds. G. F. Gebhart, D. L. Hammond, and T. S. Jensen), pp. 495–510. IASP Press.

Bennett, D. L., Koltzenburg, M., Priestley, J. V., Shelton, D. L., and McMahon, S. B. 1998. Endogenous nerve growth factor regulates the sensitivity of nociceptors in the adult rat. Eur. J. Neurosci. 10, 1282–1291.

Binder, W., Mousa, S. A., Sitte, N., Kaiser, M., Stein, C., and Schäfer, M. 2004. Sympathetic activation triggers endogenous opioid release and analgesia within peripheral inflamed tissue. Eur. J. Neurosci. 20, 92–100.

Brack, A., Labuz, D., Schiltz, A., Rittner, H. L., Machelska, H., Schäfer, M., Reszka, R., and Stein, C. 2004a. Tissue monocytes/macrophages in inflammation: hyperalgesia versus opioid-mediated peripheral antinociception. Anesthesiology 101, 204–211.

Brack, A., Rittner, H. L., Machelska, H., Beschmann, K., Sitte, N., Schäfer, M., and Stein, C. 2004b. Mobilization of opioid-containing polymorphonuclear cells by hematopoetic growth factors and influence on inflammatory pain. Anesthesiology 100, 149–157.

Brack, A., Rittner, H. L., Machelska, H., Shaqura, M., Mousa, S. A., Labuz, D., Zöllner, C., Schäfer, M., and Stein, C. 2004c. Control of inflammatory pain by chemokine-mediated recruitment of opioid-containing polymorphonuclear cells. Pain 112, 229–238.

Brack, A., Rittner, H. L., Machelska, H., Shaqura, M., Mousa, S. A., Labuz, D., Zöllner, C., Schäfer, M., and Stein, C. 2004d. Endogenous peripheral antinociception in early inflammation is not limited by the number of opioid-containing leukocytes but by opioid receptor expression. Pain 108, 67–75.

Braz, J., Beaufour, C., Coutaux, A., Epstein, A. L., Cesselin, F., Hamon, M., and Pohl, M. 2001. Therapeutic efficacy in experimental polyarthritis of viral-driven enkephalin overproduction in sensory neurons. J. Neurosci. 21, 7881–7888.

Cabot, P. J., Carter, L., Gaiddon, C., Zhang, Q., Schäfer, M., Loeffler, J. P., and Stein, C. 1997. Immune cell-derived beta-endorphin. Production, release, and control of inflammatory pain in rats. J. Clin. Invest. 100, 142–148.

Cabot, P. J., Carter, L., Schäfer, M., and Stein, C. 2001. Methionine-enkephalin- and Dynorphin A-release from immune cells and control of inflammatory pain. Pain 93, 207–212.

Calixto, J. B., Cabrini, D. A., Ferreira, J., and Campos, M. M. 2000. Kinins in pain and inflammation. Pain 87, 1–5.

Chuang, H. H., Prescott, E. D., Kong, H., Shields, S., Jordt, S. E., Basbaum, A. I., Chao, M. V., and Julius, D. 2001.

Bradykinin and nerve growth factor release the capsaicin receptor from PtdIns(4,5)P2-mediated inhibition. Nature 411, 957–962.

Cui, J. G., Holmin, S., Mathiesen, T., Meyerson, B. A., and Linderoth, B. 2000. Possible role of inflammatory mediators in tactile hypersensitivity in rat models of mononeuropathy. Pain 88, 239–248.

Cunha, F. Q. and Ferreira, S. H. 2003. Peripheral hyperalgesic cytokines. Adv. Exp. Med. Biol. 521, 22–39.

Cunha, J. M., Cunha, F. Q., Poole, S., and Ferreira, S. H. 2000. Cytokine-mediated inflammatory hyperalgesia limited by interleukin-1 receptor antagonist. Br. J. Pharmacol. 130, 1418–1424.

Cunha, F. Q., Lorenzetti, B. B., Poole, S., and Ferreira, S. H. 1991. Interleukin-8 as a mediator of sympathetic pain. Br. J. Pharmacol. 104, 765–767.

Cunha, F. Q., Poole, S., Lorenzetti, B. B., and Ferreira, S. H. 1992. The pivotal role of tumour necrosis factor alpha in the development of inflammatory hyperalgesia. Br. J. Pharmacol. 107, 660–664.

Cunha, F. Q., Poole, S., Lorenzetti, B. B., Veiga, F. H., and Ferreira, S. H. 1999. Cytokine-mediated inflammatory hyperalgesia limited by interleukin-4. Br. J. Pharmacol. 126, 45–50.

Cunha, T. M., Verri, W. A., Jr., Silva, J. S., Poole, S., Cunha, F. Q., and Ferreira, S. H. 2005. A cascade of cytokines mediates mechanical inflammatory hypernociception in mice. Proc. Natl. Acad. Sci. U. S. A. 102, 1755–1760.

Czlonkowski, A., Stein, C., and Herz, A. 1993. Peripheral mechanisms of opioid antinociception in inflammation: involvement of cytokines. Eur. J. Pharmacol. 242, 229–235.

Dado, R. J., Law, P. Y., Loh, H. H., and Elde, R. 1993. Immunofluorescent identification of a delta (delta)-opioid receptor on primary afferent nerve terminals. Neuroreport 5, 341–344.

Ferreira, J., Beirith, A., Mori, M. A., Araujo, R. C., Bader, M., Pesquero, J. B., and Calixto, J. B. 2005. Reduced nerve injury-induced neuropathic pain in kinin B1 receptor knock-out mice. J. Neurosci. 25, 2405–2412.

Ferreira, S. H., Lorenzetti, B. B., Bristow, A. F., and Poole, S. 1988. Interleukin-1 beta as a potent hyperalgesic agent antagonized by a tripeptide analogue. Nature 334, 698–700.

Ferreira, S. H., Lorenzetti, B. B., and Poole, S. 1993. Bradykinin initiates cytokine-mediated inflammatory hyperalgesia. Br. J. Pharmacol. 110, 1227–1231.

Fukuoka, H., Kawatani, M., Hisamitsu, T., and Takeshige, C. 1994. Cutaneous hyperalgesia induced by peripheral injection of interleukin-1 beta in the rat. Brain Res. 657, 133–140.

Gadient, R. A. and Otten, U. 1996. Postnatal expression of interleukin-6 (IL-6) and IL-6 receptor (IL-6R) mRNAs in rat sympathetic and sensory ganglia. Brain Res. 724, 41–46.

Gazda, L. S., Milligan, E. D., Hansen, M. K., Twining, C. M., Poulos, N. M., Chacur, M., KA, O. C., Armstrong, C., Maier, S. F., Watkins, L. R., and Myers, R. R. 2001. Sciatic inflammatory neuritis (SIN): behavioral allodynia is paralleled by peri-sciatic proinflammatory cytokine and superoxide production. J. Peripher. Nerv. Syst. 6, 111–129.

Gold, M. S. and Levine, J. D. 1996. DAMGO inhibits prostaglandin E2-induced potentiation of a TTX-resistant Na$^+$ current in rat sensory neurons in vitro. Neurosci. Lett. 212, 83–86.

Guan, J. S., Xu, Z. Z., Gao, H., He, S. Q., Ma, G. Q., Sun, T., Wang, L. H., Zhang, Z. N., Lena, I., Kitchen, I., Elde, R., Zimmer, A., He, C., Pei, G., Bao, L., and Zhang, X. 2005. Interaction with vesicle luminal protachykinin regulates surface expression of delta-opioid receptors and opioid analgesia. Cell 122, 619–631.

Hargreaves, K. M., Dubner, R., and Costello, A. H. 1989. Corticotropin releasing factor (CRF) has a peripheral site of action for antinociception. Eur. J. Pharmacol. 170, 275–279.

Hermanussen, S., Do, M., and Cabot, P. J. 2004. Reduction of beta-endorphin-containing immune cells in inflamed paw tissue corresponds with a reduction in immune-derived antinociception: reversible by donor activated lymphocytes. Anesth. Analg. 98, 723–729.

Ingram, S. L. and Williams, J. T. 1994. Opioid inhibition of Ih via adenylyl cyclase. Neuron 13, 179–186.

Jeanjean, A. P., Moussaoui, S. M., Maloteaux, J. M., and Laduron, P. M. 1995. Interleukin-1 beta induces long-term increase of axonally transported opiate receptors and substance P. Neuroscience 68, 151–157.

Ji, R. R., Zhang, Q., Law, P. Y., Low, H. H., Elde, R., and Hokfelt, T. 1995. Expression of mu-, delta-, and kappa-opioid receptor-like immunoreactivities in rat dorsal root ganglia after carrageenan-induced inflammation. J. Neurosci. 15, 8156–8166.

Julius, D. and Basbaum, A. I. 2001. Molecular mechanisms of nociception. Nature 413, 203–210.

Junger, H. and Sorkin, L. S. 2000. Nociceptive and inflammatory effects of subcutaneous TNFalpha. Pain 85, 145–151.

Kavelaars, A., Ballieux, R. E., and Heijnen, C. J. 1990. In vitro beta-adrenergic stimulation of lymphocytes induces the release of immunoreactive beta-endorphin. Endocrinology 126, 3028–3032.

Kieffer, B. L. and Gaveriaux-Ruff, C. 2002. Exploring the opioid system by gene knockout. Prog. Neurobiol. 66, 285–306.

Koltzenburg, M., Bennett, D. L., Shelton, D. L., and McMahon, S. B. 1999. Neutralization of endogenous NGF prevents the sensitization of nociceptors supplying inflamed skin. Eur. J. Neurosci. 11, 1698–1704.

Kraus, J., Borner, C., Giannini, E., Hickfang, K., Braun, H., Mayer, P., Hoehe, M. R., Ambrosch, A., Konig, W., and Hollt, V. 2001. Regulation of mu-opioid receptor gene transcription by interleukin-4 and influence of an allelic variation within a STAT6 transcription factor binding site. J. Biol. Chem. 276, 43901–43908.

Laird, J. M., Souslova, V., Wood, J. N., and Cervero, F. 2002. Deficits in visceral pain and referred hyperalgesia in Nav1.8 (SNS/PN3)-null mice. J. Neurosci. 22, 8352–8356.

Lewin, G. R., Rueff, A., and Mendell, L. M. 1994. Peripheral and central mechanisms of NGF-induced hyperalgesia. Eur. J. Neurosci. 6, 1903–1912.

Li, Y., Ji, A., Weihe, E., and Schafer, M. K. 2004. Cell-specific expression and lipopolysaccharide-induced regulation of tumor necrosis factor alpha (TNFalpha) and TNF receptors in rat dorsal root ganglion. J. Neurosci. 24, 9623–9631.

Li, M., Shi, J., Tang, J. R., Chen, D., Ai, B., Chen, J., Wang, L. N., Cao, F. Y., Li, L. L., Lin, C. Y., and Guan, X. M. 2005. Effects of complete Freund's adjuvant on immunohistochemical distribution of IL-1beta and IL-1R I in neurons and glia cells of dorsal root ganglion. Acta Pharmacol. Sin. 26, 192–198.

Lorenzetti, B. B., Veiga, F. H., Canetti, C. A., Poole, S., Cunha, F. Q., and Ferreira, S. H. 2002. CINC-1 mediates the sympathetic component of inflammatory mechanical hypersensitivity in rats. Eur. Cytokine Netw. 13, 456–461.

Lu, C. Y., Chou, A. K., Wu, C. L., Yang, C. H., Chen, J. T., Wu, P. C., Lin, S. H., Muhammad, R., and Yang, L. C. 2002. Gene-gun particle with pro-opiomelanocortin cDNA produces analgesia against formalin-induced pain in rats. Gene Ther. 9, 1008–1014.

Lyons, P. D. and Blalock, J. E. 1997. Pro-opiomelanocortin gene expression and protein processing in rat mononuclear leukocytes. J. Neuroimmunol. 78, 47–56.

Machelska, H., Brack, A., Mousa, S. A., Schopohl, J. K., Rittner, H. L., Schäfer, M., and Stein, C. 2004. Selectins and integrins but not platelet-endothelial cell adhesion molecule-1 regulate opioid inhibition of inflammatory pain. Br. J. Pharmacol. 142, 772–780.

Machelska, H., Cabot, P. J., Mousa, S. A., Zhang, Q., and Stein, C. 1998. Pain control in inflammation governed by selectins. Nat. Med. 4, 1425–1428.

Machelska, H., Mousa, S. A., Brack, A., Schopohl, J. K., Rittner, H. L., Schäfer, M., and Stein, C. 2002. Opioid control of inflammatory pain regulated by intercellular adhesion molecule-1. J. Neurosci. 22, 5588–5596.

Machelska, H., Schopohl, J. K., Moussa, S. A., Schäfer, M., and Stein, C. 2003. Different mechanisms of intrinsic pain inhibition in early and late inflammation. J. Neuroimmuno 14, 30–39.

Mamet, J., Baron, A., Lazdunski, M., and Voilley, N. 2002. Proinflammatory mediators, stimulators of sensory neuron excitability via the expression of acid-sensing ion channels. J. Neurosci. 22, 10662–10670.

Marceau, F. and Regoli, D. 2004. Bradykinin receptor ligands: therapeutic perspectives. Nat. Rev. Drug. Discov. 3, 845–852.

McLoon, L. K., Sandnas, A. M., Nockleby, K. J., and Wirtschafter, J. D. 2002. Reduction in vesicant-induced cellular inflammation and hyperalgesia by local injection of corticotropin releasing factor in rabbit eyelid. Inflamm. Res. 51, 16–23.

McMahon, S. B., Bennett, D. L., Priestley, J. V., and Shelton, D. L. 1995. The biological effects of endogenous nerve growth factor on adult sensory neurons revealed by a trkA–IgG fusion molecule. Nat. Med. 1, 774–780.

Merskey, H. and Bogduk, N. 1994. Pain Terminology. In: Classification of Chronic Pain. IASP Task Force on Taxonomy (eds. H. Merskey and N. Bogduk), pp. 209–214. IASP Press.

Mousa, S. A., Bopaiah, C. P., Stein, C., and Schäfer, M. 2003. Involvement of corticotropin-releasing hormone receptor subtypes 1 and 2 in peripheral opioid-mediated inhibition of inflammatory pain. Pain 106, 297–307.

Mousa, S. A., Machelska, H., Schafer, M., and Stein, C. 2000. Co-expression of beta-endorphin with adhesion molecules in a model of inflammatory pain. J. Neuroimmunol. 108, 160–170.

Mousa, S. A., Schafer, M., Mitchell, W. M., Hassan, A. H., and Stein, C. 1996. Local upregulation of corticotropin-releasing hormone and interleukin-1 receptors in rats with painful hindlimb inflammation. Eur. J. Pharmacol. 311, 221–231.

Mousa, S. A., Shakibaei, M., Sitte, N., Schafer, M., and Stein, C. 2004. Subcellular pathways of beta-endorphin synthesis, processing, and release from immunocytes in inflammatory pain. Endocrinology 145, 1331–1341. Epub 2003 Nov 1320.

Mousa, S. A., Zhang, Q., Sitte, N., Ji, R., and Stein, C. 2001. beta-Endorphin-containing memory-cells and mu-opioid receptors undergo transport to peripheral inflamed tissue. J. Neuroimmunol. 115, 71–78.

Murphy, P. G., Ramer, M. S., Borthwick, L., Gauldie, J., Richardson, P. M., and Bisby, M. A. 1999. Endogenous interleukin-6 contributes to hypersensitivity to cutaneous stimuli and changes in neuropeptides associated with chronic nerve constriction in mice. Eur. J. Neurosci. 11, 2243–2253.

Obreja, O., Biasio, W., Andratsch, M., Lips, K. S., Rathee, P. K., Ludwig, A., Rose-John, S., and Kress, M. 2005. Fast modulation of heat-activated ionic current by proinflammatory interleukin 6 in rat sensory neurons. Brain 128, 1634–1641.

Obreja, O., Rathee, P. K., Lips, K. S., Distler, C., and Kress, M. 2002a. IL-1 beta potentiates heat-activated currents in rat sensory neurons: involvement of IL-1RI, tyrosine kinase, and protein kinase C. FASEB J. 16, 1497–1503.

Obreja, O., Schmelz, M., Poole, S., and Kress, M. 2002b. Interleukin-6 in combination with its soluble IL-6 receptor sensitises rat skin nociceptors to heat, in vivo. Pain 96, 57–62.

Oh, S. B., Tran, P. B., Gillard, S. E., Hurley, R. W., Hammond, D. L., and Miller, R. J. 2001. Chemokines and glycoprotein120 produce pain hypersensitivity by directly exciting primary nociceptive neurons. J. Neurosci. 21, 5027–5035.

Ohtori, S., Takahashi, K., Moriya, H., and Myers, R. R. 2004. TNF-alpha and TNF-alpha receptor type 1 upregulation in glia and neurons after peripheral nerve injury: studies in murine DRG and spinal cord. Spine 29, 1082–1088.

Okamoto, K., Martin, D. P., Schmelzer, J. D., Mitsui, Y., and Low, P. A. 2001. Pro- and anti-inflammatory cytokine gene expression in rat sciatic nerve chronic constriction injury model of neuropathic pain. Exp. Neurol. 169, 386–391.

Opree, A. and Kress, M. 2000. Involvement of the proinflammatory cytokines tumor necrosis factor-alpha, IL-1 beta, and IL-6 but not IL-8 in the development of heat hyperalgesia: effects on heat-evoked calcitonin gene-related peptide release from rat skin. J. Neurosci. 20, 6289–6293.

Pare, M., Elde, R., Mazurkiewicz, J. E., Smith, A. M., and Rice, F. L. 2001. The Meissner corpuscle revised: a multiafferented mechanoreceptor with nociceptor immunochemical properties. J. Neurosci. 21, 7236–7246.

Petruzzelli, L., Takami, M., and Humes, H. D. 1999. Structure and function of cell adhesion molecules. Am. J. Med. 106, 467–476.

Polydefkis, M., et al., 2002. Reduced intraepidermal nerve fiber density in HIV-associated sensory neuropathy. Neurology 58, 115–119.

Poole, S., Cunha, F. Q., Selkirk, S., Lorenzetti, B. B., and Ferreira, S. H. 1995. Cytokine-mediated inflammatory hyperalgesia limited by interleukin-10. Br. J. Pharmacol. 115, 684–688.

Porreca, F., Lai, J., Bian, D., Wegert, S., Ossipov, M. H., Eglen, R. M., Kassotakis, L., Novakovic, S., Rabert, D. K., Sangameswaran, L., and Hunter, J. C. 1999. A comparison of the potential role of the tetrodotoxin-insensitive sodium channels, PN3/SNS and NaN/SNS2, in rat models of chronic pain. Proc. Natl. Acad. Sci. U. S. A. 96, 7640–7644.

Przewlocki, R., Hassan, A. H., Lason, W., Epplen, C., Herz, A., and Stein, C. 1992. Gene expression and localization of opioid peptides in immune cells of inflamed tissue: functional role in antinociception. Neuroscience 48, 491–500.

Rittner, H. L., Brack, A., Machelska, H., Mousa, S. A., Bauer, M., Schäfer, M., and Stein, C. 2001. Opioid peptide-expressing leukocytes: identification, recruitment, and simultaneously increasing inhibition of inflammatory pain. Anesthesiology 95, 500–508.

Rittner, H. L., Mousa, S. A., Labuz, D., Beschmann, K., Schäfer, M., Stein, C., and Brack, A. 2006. Selective local PMN recruitment by CXCL1 or CXCL2/3 injection does not cause inflammatory pain. J. Leuk. Biol. 79, 1022–1032.

Ro, L. S., Chen, S. T., Tang, L. M., and Jacobs, J. M. 1999. Effect of NGF and anti-NGF on neuropathic pain in rats following chronic constriction injury of the sciatic nerve. Pain 79, 265–274.

Rueff, A. and Mendell, L. M. 1996. Nerve growth factor NT-5 induce increased thermal sensitivity of cutaneous nociceptors in vitro. J. Neurophysiol. 76, 3593–3596.

Sacerdote, P., Limiroli, E., and Gaspani, L. 2003. Experimental evidence for immunomodulatory effects of opioids. Adv. Exp. Med. Biol. 521, 106–116.

Sachs, D., Cunha, F. Q., Poole, S., and Ferreira, S. H. 2002. Tumour necrosis factor-alpha, interleukin-1beta and interleukin-8 induce persistent mechanical nociceptor hypersensitivity. Pain 96, 89–97.

Safieh-Garabedian, B., Poole, S., Allchorne, A., Winter, J., and Woolf, C. J. 1995. Contribution of interleukin-1 beta to the inflammation-induced increase in nerve growth factor levels and inflammatory hyperalgesia. Br. J. Pharmacol. 115, 1265–1275.

Sawynok, J. 2003. Topical and peripherally acting analgesics. Pharmacol. Rev. 55, 1–20.

Schäfer, M. 2003. Cytokines and peripheral analgesia. Adv. Exp. Med. Biol. 521, 40–50.

Schäfer, M., Carter, L., and Stein, C. 1994. Interleukin 1 beta and corticotropin-releasing factor inhibit pain by releasing opioids from immune cells in inflamed tissue. Proc. Natl. Acad. Sci. U. S. A. 91, 4219–4223.

Schäfer, M., Mousa, S. A., Zhang, Q., Carter, L., and Stein, C. 1996. Expression of corticotropin-releasing factor in inflamed tissue is required for intrinsic peripheral opioid analgesia. Proc. Natl. Acad. Sci. U. S. A. 93, 6096–6100.

Schäfers, M., Brinkhoff, J., Neukirchen, S., Marziniak, M., and Sommer, C. 2001. Combined epineurial therapy with neutralizing antibodies to tumor necrosis factor-alpha and interleukin-1 receptor has an additive effect in reducing neuropathic pain in mice. Neurosci. Lett. 310, 113–116.

Schäfers, M., Geis, C., Svensson, C. I., Luo, Z. D., and Sommer, C. 2003a. Selective increase of tumour necrosis factor-alpha in injured and spared myelinated primary afferents after chronic constrictive injury of rat sciatic nerve. Eur. J. Neurosci. 17, 791–804.

Schäfers, M., Lee, D. H., Brors, D., Yaksh, T. L., and Sorkin, L. S. 2003b. Increased sensitivity of injured and adjacent uninjured rat primary sensory neurons to exogenous tumor necrosis factor-alpha after spinal nerve ligation. J. Neurosci. 23, 3028–3038.

Schäfers, M., Sorkin, L. S., Geis, C., and Shubayev, V. I. 2003c. Spinal nerve ligation induces transient upregulation of tumor necrosis factor receptors 1 and 2 in injured and adjacent uninjured dorsal root ganglia in the rat. Neurosci. 347, 179–182.

Schäfers, M., Svensson, C. I., Sommer, C., and Sorkin, L. S. 2003d. Tumor necrosis factor-alpha induces mechanical allodynia after spinal nerve ligation by activation of p38 MAPK in primary sensory neurons. J. Neurosci. 23, 2517–2521.

Scholz, J. and Woolf, C. J. 2002. Can we conquer pain? Nat. Neurosci. 5(Suppl.), 1062–1067.

Selley, D. E., Breivogel, C. S., and Childers, S. R. 1993. Modification of G protein-coupled functions by low-pH pretreatment of membranes from NG108-15 cells: increase in opioid agonist efficacy by decreased inactivation of G proteins. Mol. Pharmacol. 44, 731–741.

Sevcik, M. A., Ghilardi, J. R., Peters, C. M., Lindsay, T. H., Halvorson, K. G., Jonas, B. M., Kubota, K., Kuskowski, M. A., Boustany, L., Shelton, D. L., and Mantyh, P. W. 2005. Anti-NGF therapy profoundly reduces bone cancer pain and the accompanying increase in markers of peripheral and central sensitization. Pain 115, 128–141.

Sharp, B. M. 2003. Opioid receptor expression and intracellular signaling by cells involved in host defense and immunity. Adv. Exp. Med. Biol. 521, 98–105.

Shubayev, V. I. and Myers, R. R. 2001. Axonal transport of TNF-alpha in painful neuropathy: distribution of ligand tracer and TNF receptors. J. Neuroimmunol. 114, 48–56.

Simonin, F. and Kieffer, B. L. 2002. Two faces for an opioid peptide – and more receptors for pain research. Nature 5, 185–186.

Smith, E. M. 2003. Opioid peptides in immune cells. Adv. Exp. Med. Biol. 521, 51–68.

Sommer, C., Petrausch, S., Lindenlaub, T., and Toyka, K. V. 1999. Neutralizing antibodies to interleukin 1-receptor reduce pain associated behavior in mice with experimental neuropathy. Neurosci. Lett. 270, 25–28.

Sommer, C., Schäfers, M., Marziniak, M., and Toyka, K. V. 2001. Etanercept reduces hyperalgesia in experimental painful neuropathy. J. Peripher. Nerv. Syst. 6, 67–72.

Sorkin, L. S., Xiao, W. H., Wagner, R., and Myers, R. R. 1997. Tumour necrosis factor-alpha induces ectopic activity in nociceptive primary afferent fibres. Neuroscience 81, 255–262.

Stein, C., Gramsch, C., and Herz, A. 1990a. Intrinsic mechanisms of antinociception in inflammation: local opioid receptors and beta-endorphin. J. Neurosci. 10, 1292–1298.

Stein, C., Hassan, A. H., Lehrberger, K., Giefing, J., and Yassouridis, A. 1993. Local analgesic effect of endogenous opioid peptides. Lancet 342, 321–324.

Stein, C., Hassan, A. H., Przewlocki, R., Gramsch, C., Peter, K., and Herz, A. 1990b. Opioids from immunocytes interact with receptors on sensory nerves to inhibit nociception in inflammation. Proc. Natl. Acad. Sci. U. S. A. 87, 5935–5939.

Stein, C., Machelska, H., and Schafer, M. 2001. Peripheral analgesic and antiinflammatory effects of opioids. Z. Rheumatol. 60, 416–424.

Stein, C., Pfluger, M., Yassouridis, A., Hoelzl, J., Lehrberger, K., Welte, C., and Hassan, A. H. 1996. No tolerance to peripheral morphine analgesia in presence of opioid expression in inflamed synovia. J. Clin. Invest. 98, 793–799.

Stein, C., Schäfer, M., and Machelska, H. 2003. Attacking pain at its source: new perspectives on opioids. Nat. Med. 9, 1003–1008.

Stucky, C. L., Koltzenburg, M., Schneider, M., Engle, M. G., Albers, K. M., and Davis, B. M. 1999. Overexpression of nerve growth factor in skin selectively affects the survival and functional properties of nociceptors. J. Neurosci. 19, 8509–8516.

Suzuki, R. and Dickenson, A. H. 2002. Nerve injury-induced changes in opioid modulation of wide dynamic range dorsal column nuclei neurones. Neuroscience 111, 215–228.

Terman, G. W., Shavit, Y., Lewis, J. W., Cannon, J. T., and Liebeskind, J. C. 1984. Intrinsic mechanisms of pain inhibition: activation by stress. Science 226, 1270–1277.

Vale, M. L., Marques, J. B., Moreira, C. A., Rocha, F. A., Ferreira, S. H., Poole, S., Cunha, F. Q., and Ribeiro, R. A. 2003. Antinociceptive effects of interleukin-4, -10, and -13 on the writhing response in mice and zymosan-induced knee joint incapacitation in rats. J. Pharmacol. Exp. Ther. 304, 102–108.

Vega, J. A., Garcia-Suarez, O., Hannestad, J., Perez-Perez, M., and Germana, A. 2003. Neurotrophins and the immune system. J. Anat. 203, 1–19.

Verri, W. A., Jr., Schivo, I. R., Cunha, T. M., Liew, F. Y., Ferreira, S. H., and Cunha, F. Q. 2004. Interleukin-18 induces mechanical hypernociception in rats via endothelin acting on ETB receptors in a morphine-sensitive manner. J. Pharmacol. Exp. Ther. 310, 710–717.

Wenk, H. N. and Honda, C. N. 1999. Immunohistochemical localization of delta opioid receptors in peripheral tissues. J. Comp. Neurol. 408, 567–579.

Willer, J. C., Dehen, H., and Cambier, J. 1981. Stress-induced analgesia in humans: endogenous opioids and naloxone-reversible depression of pain reflexes. Science 212, 689–691.

Woolf, C. J., Allchorne, A., Safieh-Garabedian, B., and Poole, S. 1997. Cytokines, nerve growth factor and inflammatory hyperalgesia: the contribution of tumour necrosis factor alpha. Br. J. Pharmacol. 121, 417–424.

Xu, Y., Gu, Y., Xu, G. Y., Wu, P., Li, G. W., and Huang, L. Y. 2003. Adeno-associated viral transfer of opioid receptor gene to primary sensory neurons: a strategy to increase opioid antinociception. Proc. Natl. Acad. Sci. U. S. A. 100, 6204–6209.

Xu, X. J., Hao, J. X., Andell-Jonsson, S., Poli, V., Bartfai, T., and Wiesenfeld-Hallin, Z. 1997. Nociceptive responses in interleukin-6-deficient mice to peripheral inflammation and peripheral nerve section. Cytokine 9, 1028–1033.

Zhang, X., Bao, L., Arvidsson, U., Elde, R., and Hokfelt, T. 1998. Localization and regulation of the delta-opioid receptor in dorsal root ganglia and spinal cord of the rat and monkey: evidence for association with the membrane of large dense-core vesicles. Neuroscience 82, 1225–1242.

Zhang, N., Inan, S., Cowan, A., Sun, R., Wang, J. M., Rogers, T. J., Caterina, M., and Oppenheim, J. J. 2005. A proinflammatory chemokine, CCL3, sensitizes the heat- and capsaicin-gated ion channel TRPV1. Proc. Natl. Acad. Sci. U. S. A. 102, 4536–4541.

Zöllner, C., Shaqura, M. A., Bopaiah, C. P., Mousa, S. A., Stein, C., and Schäfer, M. 2003. Painful inflammation induced increase in μ opioid receptor binding and G-protein coupling in primary afferent neurons. Mol. Pharm. 64, 202–210.

Zwick, M., Molliver, D. C., Lindsay, J., Fairbanks, C. A., Sengoku, T., Albers, K. M., and Davis, B. M. 2003. Transgenic mice possessing increased numbers of nociceptors do not exhibit increased behavioral sensitivity in models of inflammatory and neuropathic pain. Pain 106, 491–500.

Further Reading

Machelska, H. and Stein, C. 2002. Immune mechanisms in pain control. Anesth. Analg. 95, 1002–1008.

Relevant Website

http://www.iasp.pain.org – International Association for the Study of Pain.

31 Mechanisms of Glial Activation after Nerve Injury

L R Watkins, E D Milligan, and S F Maier, University of Colorado at Boulder, Boulder, CO, USA

31.1 Introduction

It has been known since the early 1970s that peripheral nerve injury leads to the activation of glia in anatomically linked regions of the brain and spinal cord (Aldskogius, H. and Kozlova, E. N., 1998). In these studies, glial activation was inferred by upregulation of the so-called glial activation markers. These signs that glial activation has occurred can be visualized under the microscope using immunohistochemistry to detect cell-type specific changes (e.g., increased expression of glial fibrillary acidic protein by astrocytes; increased expression of complement type 2 receptors by microglia). Such studies revealed that damage of sensory afferent nerves activates glia in the region of the afferent terminations in the central nervous system (CNS). Likewise, damage of motor efferents activates glia in the CNS in the region of the involved motor neuron cell bodies (Aldskogius, H. and Kozlova, E. N., 1998). While this early literature documented the phenomenon of glial activation in CNS following peripheral nerve injury, the mechanisms underlying such activation remained obscure.

Two decades later, glial activation was linked to neuropathic pain. This link was again initially based on the upregulation of glial activation markers (Garrison, C. J. et al., 1991; 1994). Pharmacology, anatomy, and molecular biology studies that followed documented that microglia and then astrocytes are sequentially activated in response to inflammation and damage to peripheral nerves (Raghavendra, V. et al., 2003; Ledeboer, A. et al., 2005). Indeed, studies of p38 mitogen-activated protein (MAP) kinase activation demonstrated that microglial activation occurs with remarkable rapidity following nerve injury (Svensson, C. I. et al., 2004). Upon activation, neuroexcitatory products that glia release are importantly involved in the initiation and maintenance of neuropathic pain, via their actions on neurons of the pain pathway (Watkins, L. R. et al., 2001; Watkins, L. R. and Maier, S. F., 2003). Thus, converging lines of evidence utilizing a variety of techniques all support the conclusion that glia are key players in neuropathic pain.

However, what is known to date regarding glial involvement in neuropathic pain begs a fundamentally important question: how do glia know to become activated? That is, what message(s) do glia receive that trigger their transition from basal to activated states? This will be discussed first in the context of glial activation in response to injury or inflammation of peripheral nerves, and then more broadly in the context of glial activation in response to analgesics used clinically to treat chronic pain, such as morphine and methadone.

31.2 Glial Activation in Response to Neurotransmitters

Given that (1) glia become activated in response to inflammation and damage of peripheral tissues and peripheral nerves and (2) this activation occurs within the somatotopically appropriate region of spinal cord, this suggests that glia may be activated by neurotransmitters released by nocisponsive primary afferents. Indeed, calcitonin gene-related peptide (CGRP) (Haas, C. A. et al., 1991; Reddington, M. et al., 1995), adenosine triphosphate (ATP) (Araque, A. et al., 1999; Tsuda, M. et al., 2003), glutamate (Araque, A. et al., 1999; Tsuda, M. et al., 2003), and substance P

(Svensson, C. I. *et al.*, 2003; Marriott, I., 2004) can each activate microglia and astrocytes. Such neurotransmitter-driven glial activation can be relevant for pain facilitation. For example, pain facilitation that is normally produced upon intrathecal administration of the excitatory amino acid, *N*-methyl-D-aspartate (NMDA), is abolished by blocking microglial activation (Hua, X.-Y. *et al.*, 2005).

Regarding substance P, studies of spinal cord glia have demonstrated that it can activate both astrocytes and microglia. Astrocytes isolated from spinal cord (but not from other brain regions) release prostaglandins upon exposure to substance P (Marriott, D. R. *et al.*, 1991). Substance P can also stimulate the production of proinflammatory cytokines in both astrocytes (Martin, F. C. *et al.*, 1992; Palma, C. *et al.*, 1997; Lieb, K. *et al.*, 1998) and microglia (Martin, F. C. *et al.*, 1993; Luber-Narod, J. *et al.*, 1994). When substance P is injected intrathecally in rats, this activates spinal cord microglia as reflected by phosphorylation (activation) of microglial p38 MAP kinase, a signaling pathway linked to pain facilitation (Svensson, C. I. *et al.*, 2004). Interestingly, microglia can also produce substance P, suggesting that these glial cells may potentially contribute to substance P-mediated pain signaling (Lai, J. P. *et al.*, 2002).

ATP has also recently attracted attention as a neurotransmitter that can enhance pain via the activation of glia. Microglia (but not neurons or astrocytes) express an ionotropic ATP receptor subtype called P2X4 whose expression can become dramatically upregulated in response to peripheral nerve injury (Tsuda, M. *et al.*, 2003). Pharmacological blockade of spinal P2X4 receptors and knockdown of P2X4 receptors using antisense oligodeoxynucleotides each reverse neuropathy-induced tactile allodynia (Tsuda, M. *et al.*, 2003). Intriguingly, activation of microglia by ATP is sufficient to cause pain facilitation. That is, microglia that are first stimulated *in vitro* by ATP, and then injected intrathecally in rats, induces mechanical allodynia (Tsuda, M. *et al.*, 2003).

While, at first glance, it would seem logical that glia must become activated in response to neurotransmitters released as a consequence of peripheral nerve injury, one must remember the literature of the 1970s. That is, if one damages a pure motor nerve, glia are activated surrounding the motor neuron cell bodies within the CNS. In this case, there is no damage to sensory afferent fibers. So, despite there being no neurotransmitter release, glia are still being intensely activated. Thus, while glia do express receptors for neurotransmitters and can respond to them, they are not required for glial activation to take place. This predicts that other signals, beyond neurotransmitters, must exist which can trigger the activation of glia.

31.3 Glial Activation in Response to Neuromodulators

Neuromodulators, such as prostaglandins and nitric oxide (NO) are well documented to enhance pain (Woolf, C. J. and Salter, M. W., 2000). In addition to direct excitatory effects on neurons, these substances can also activate glia. Using NO as the example, NO induces calcium oscillations in astrocytes, a marker of activation in this cell type. In turn, NO-induced calcium oscillations causes a nondecrementing, rapid spread of excitation to surrounding astrocytes through cell-to-cell low-resistance connections called gap junctions (Heidemann, A. C. *et al.*, 2004; Sul, J. *et al.*, 2004). Thus NO-excited astrocytes lead to the excitation of a much larger population of astrocytes. Once activated, these glia can then begin releasing neuroexcitatory substances (Araque, A. *et al.*, 1999). In addition, NO induces the release of glutamate, ATP, and prostaglandins from astrocytes (Molina-Holgado, F. *et al.*, 1995; Bal-Price, A. *et al.*, 2002) and stimulates the production of proinflammatory cytokines in both microglia and astrocytes (Sun, D. *et al.*, 1998). Regarding actions in spinal cord, NO has recently been demonstrated to drive the production and release of proinflammatory cytokines in response to intrathecally administered gp120, a protein expressed on the surface of the acquired immune deficiency syndrome (AIDS) virus that activates glia (Holguin, A. *et al.*, 2004). As gp120-induced proinflammatory cytokines induce mechanical allodynia, it would be predicted that NO synthesis inhibitors would block such pain changes concomitant with blocking NO-induced production of these cytokines. Indeed, that was the result found (Holguin, A. *et al.*, 2004).

31.4 Glial Activation by Unique Neuron-to-Glia Signals

Beyond neurotransmitters and neuromodulators classically known for their pain-enhancing effects, glia can also be activated by substances novel to the pain field. This class includes fractalkine and substances released by damaged cells. Regarding fractalkine, it is a protein expressed on the extracellular surface of both sensory afferent neurons and neurons intrinsic to the spinal cord (Verge, G. M. et al., 2004; Lindia, J. A. et al., 2005). Fractalkine is tethered to the neurons in such a way as to be enzymatically released under strong neuroexcitatory conditions. Upon release, fractalkine acts as a diffusible signaling molecule (Chapman, G. et al., 2000). Intriguingly, under basal conditions, the receptor for fractalkine is expressed predominantly by spinal microglia (Verge, G. M. et al., 2004; Lindia, J. A. et al., 2005). Thus, upon insult to the body, released fractalkine serves as a neuron-to-microglial signal. In response to peripheral nerve inflammation and injury, there is an increase in fractalkine receptor expression by microglia in the dorsal horn and microglia remain the sole apparent source of fractalkine receptor expression (Verge, G. M. et al., 2004; Lindia, J. A. et al., 2005). Under some (but not all) neuropathic conditions, astrocytes may begin producing fractalkine, suggestive that astrocytes also stimulate microglia via the release of fractalkine, so to maintain microglial activation under such conditions (Verge, G. M. et al., 2004; Lindia, J. A. et al., 2005).

Regarding the role of fractalkine in pain facilitation, what is known to date is that fractalkine upregulates P2X4 receptors in dorsal spinal cord microglia (Wieseler-Frank, J. et al., 2004). As reviewed above, this microglia-specific ATP receptor is implicated in the induction of mechanical allodynia (Tsuda, M. et al., 2003). When fractalkine is injected intrathecally, it induces both mechanical allodynia and thermal hyperalgesia, via the release of proinflammatory cytokines (Johnston, I. N. et al., 2004; Milligan, E. D. et al., 2004; 2005). Importantly, endogenous fractalkine enhances pain. That is, blocking fractalkine actions both delays and reverses neuropathic pain (Milligan, E. D. et al., 2004). This suggests that peripheral nerve injury causes perseverative release of fractalkine and this release contributes to the induction and maintenance of neuropathic pain.

Lastly, glia can be activated by substances that signal neuronal damage. Microglia, like other immune cells, express evolutionarily conserved receptors for detecting cell damage, as well as for detecting pathogens such as bacteria and viruses. The genes for these receptors are homologous to the Toll gene in Drosophila and have therefore come to be referred to as Toll-like receptors (TLRs). One TLR subtype, TLR4, has been clearly implicated in pain facilitation (Tanga, F. Y. et al., 2005). TLR4 knockout mice show reduced neuropathic pain in response to L5 nerve transection, as do rats treated with TLR4 antisense

oligodeoxynucleotide to reduce the expression of spinal TLR4. This treatment also decreased expression of spinal microglial markers and proinflammatory cytokines, compared to controls (Tanga, F. Y. et al., 2005). While TLR4 is best known as the receptor that recognizes bacterial endotoxin (lipopolysaccharide), it also binds endogenous substances associated with cell damage. These include members of the heat shock protein family and proteoglycans.

The potential link to heat shock proteins is especially intriguing. Heat shock proteins are classically thought of as intracellular proteins that aid in cell survival after stress or trauma (Costigan, M. et al., 1998). However, heat shock proteins are also released by damaged and dying cells and such extracellular heat shock proteins can activate glia (Kakimura, J. et al., 2002; Takata, K. et al., 2003). Indeed, heat shock proteins are known to be upregulated for months in the spinal processes of sensory neurons whose peripheral axons are damaged (Costigan, M. et al., 1998). Preliminary data suggest that heat shock proteins can synergize with other glial excitatory stimuli to enhance the production of proinflammatory cytokines by dorsal spinal cord microglia (Wieseler-Frank, J. et al., 2004).

31.5 Beyond Neuropathic Pain: Glial Activation in Response to Opioids

While the focus to this point has been on endogenous signals that activate glia, it is also noteworthy that glia are activated by opioids, including morphine (Song, P. and Zhao, Z. Q., 2001; Raghavendra, V. et al., 2002; Johnston, I. N. et al., 2004; Shavit, Y. et al., 2004) and methadone (Hutchinson, M. R. et al., 2005). The results for morphine that have been reported across labs to date are remarkably consistent in demonstrating that spinal cord glia become progressively more activated by morphine upon repeated administration and that morphine-induced glial activation leads to the production and release of proinflammatory cytokines. This opioid-induced proinflammatory cytokine release has been linked, using diverse methodologies, to an attenuation of acute morphine analgesia, to the development of morphine tolerance, and to the expression of withdrawal-induced thermal hyperalgesia and mechanical allodynia (for a review, see Watkins, L. R. et al., 2005). What is important about this line of research is that it documents that glia are not only activated in response to peripheral injury and inflammation, but also by first-line opioids used for clinical pain control. Thus, glial activation becomes a double-edged sword, occurring both in response to situations that induce pathological pain and in response to drugs intended to treat it.

31.6 Conclusions

It is clear from the studies reviewed above that spinal cord glia can become activated in response to a wide array of chemical signals. Some are neurotransmitters and neuromodulators naturally released in spinal cord dorsal horn in response to inflammation and damage in the body, others are unique neuron-to-glia signals released by activated and damaged neurons, and surprisingly some are drugs commonly used for clinical pain control. Regardless of what the activating stimulus is, the end result is strikingly similar. That is, activated glia begin producing and releasing an array of neuroexcitatory substances, including proinflammatory cytokines. Taken together, these studies clearly predict that targeting spinal cord glia and glial proinflammatory products should suppress pain facilitation induced by inflammation and damage to peripheral tissues and enhance the clinical efficacy of opioids.

References

Aldskogius, H. and Kozlova, E. N. 1998. Central neuron–glial and glial–glial interactions following axon injury. Prog. Neurobiol. 55, 1–26.

Araque, A., Parpura, V., Sanzgiri, R. P., and Haydon, P. G. 1999. Tripartite synapses: glia, the unacknowledged partner. Trends Neurosci. 22, 208–215.

Bal-Price, A., Moneer, Z., and Brown, G. C. 2002. Nitric oxide induces rapid, calcium-dependent release of vesicular glutamate and ATP from cultured rat astrocytes. Glia 40, 312–323.

Chapman, G., Moores, K., Harrison, D., Campbell, C. A., Stewart, B. R., and Strijbos, P. J. L. M. 2000. Fractalkine cleavage from neuronal membranes represents an acute event in the inflammatory response to excitotoxic brain damage. J. Neurosci. 20(RC87), 1–5.

Costigan, M., Mannion, R. J., Kendall, G., Lewis, S. E., Campagna, J. A., Coggeshall, R. E., Meridith-Middleton, J., Tate, S., and Woolf, C. J. 1998. Heat shock protein 27: developmental regulation and expression after peripheral nerve injury. J. Neurosci. 18, 5891–5900.

Garrison, C. J., Dougherty, P. M., and Carlton, S. M. 1994. GFAP expression in lumbar spinal cord of naive and neuropathic rats treated with MK-801. Exp. Neurol. 129, 237–243.

Garrison, C. J., Dougherty, P. M., Kajander, K. C., and Carlton, S. M. 1991. Staining of glial fibrillary acidic protein (GFAP) in lumbar spinal cord increases following a sciatic nerve constriction injury. Brain Res. 565, 1–7.

Haas, C. A., Reddington, M., and Kreutzberg, G. W. 1991. Calcitonin gene-related peptide stimulates the induction of

c-fos gene expression in rat astrocyte cultures. Eur. J. Neurosci. 3, 708–712.

Heidemann, A. C., Schipke, C. G., Peters, O., and Kettenmann, H. 2004. Nitric oxide triggers repetitive calcium transients in astrocytes in mouse cortical slices. Proc. Soc. Neurosci. 34, 405–407.

Holguin, A., Biedenkapp, J., Campisi, J., Wieseler-Frank, J., O'Connor, K. A., Milligan, E. D., Maksimova, E., Bravmann, C., Hansen, M. K., Martin, D., Fleshner, M., Maier, S. F., and Watkins, L. R. 2004. HIV-1 gp120 stimulates proinflammatory cytokine-mediated pathological pain via activation of nitric oxide synthase-I (nNOS). Pain 110, 517–530.

Hua, X.-Y., Svensson, C. I., Matsui, T., Fitzsimmons, B., Yaksh, T. L., and Webb, M. 2005. Intrathecal minocycline attenuates peripheral inflammation-induced hyperalesia by inhibiting p38 MAPK in spinal microglia. Eur. J. Neurosci. 22, 2431–2440.

Hutchinson, M. R., Milligan, E. D., Maier, S. F., and Watkins, L. R. 2005. Interleukin-1 receptor antagonist (IL1ra) "unmasks" analgesia following both R- & S-methadone: evidence for induction of spinal proinflammatory cytokines (PICs) via non-classical opioid receptors. Proc. Soc. Neurosci. abstract no. 49.11.

Johnston, I. N., Milligan, E. D., Wieseler-Frank, J., Frank, M. G., Zapata, V., Campisi, J., Langer, S., Martin, D., Green, P., Fleshner, M., Leinwand, L., Maier, S. F., and Watkins, L. R. 2004. A role for proinflammatory cytokines and fractalkine in analgesia, tolerance, and subsequent pain facilitation induced by chronic intrathecal morphine. J. Neurosci. 24, 7353–7365.

Kakimura, J., Kitamura, Y., Takata, K., Umeki, M., Suzuki, S., Shibagaki, K., Taniguchi, T., Nomura, Y., Gebicke-Haerter, P. J., Smith, M. A., Perry, G., and Shimohama, S. 2002. Microglial activation and amyloid-beta clearance induced by exogenous heat-shock proteins. FASEB J. 16, 601–603.

Lai, J. P., Douglas, S. D., Shaheen, F., Pleasure, D. E., and Ho, W. Z. 2002. Quantification of substance p mRNA in human immune cells by real-time reverse transcriptase PCR assay. Clin. Diagn. Lab. Immunol. 9, 138–143.

Ledeboer, A., Sloane, E. M., Milligan, E. D., Frank, M., Maier, S. F., and Watkins, L. R. 2005. Selective inhibition of spinal cord microglial activation attenuates mechanical allodynia and proinflammatory cytokine expression in rat models of pain facilitation. Pain 115, 71–83.

Lieb, K., Schaller, H., Bauer, J., Berger, M., Schulze-Osthoff, K., and Fiebich, B. L. 1998. Substance P and histamine induce interleukin-6 expression in human astrocytoma cells by a mechanism involving protein kinase C and nuclear factor-IL-6. J. Neurochem. 70, 1577–1583.

Lindia, J. A., McGowan, E., Jochnowitz, N., and Abbadie, C. 2005. Induction of CX3CL1 expression in astrocytes and CX3CR1 in microglia in the spinal cord of a rat model of neuropathic pain. J. Pain 6, 434–438.

Luber-Narod, J., Kage, R., and Leeman, S. E. 1994. Substance P enhances the secretion of tumor necrosis factor-alpha from neuroglial cells stimulated with lipopolysaccharide. J. Immunol. 152, 819–824.

Marriott, I. 2004. The role of tachykinins in central nervous system inflammatory responses. Front. Biosci. 9, 2153–2165.

Marriott, D. R., Wilkin, G. P., and Wood, J. N. 1991. Substance P-induced release of prostaglandins from astrocytes: regional specialisation and correlation with phosphoinositol metabolism. J. Neurochem. 56, 259–265.

Martin, F. C., Anton, P. A., Gornbein, J. A., Shanahan, F., and Merrill, J. E. 1993. Production of interleukin-1 by microglia in response to substance P: role for a non-classical NK-1 receptor. J. Neuroimmunol. 42, 53–60.

Martin, F. C., Charles, A. C., Sanderson, M. J., and Merrill, J. E. 1992. Substance P stimulates IL-1 production by astrocytes via intracellular calcium. Brain Res. 599, 13–18.

Milligan, E. D., Zapata, V., Chacur, M., Schoeniger, D., Biedenkapp, J., O'Connor, K., Verge, G. M., Chapman, G., Green, P., Foster, A. C., Naeve, G. S., Maier, S. F., and Watkins, L. R. 2004. Evidence that exogenous and endogenous fractalkine can induce spinal nociceptive facilitation. Eur. J. Neurosci. 20, 2294–2302.

Milligan, E. D., Zapata, V., Schoeniger, D., Chacur, M., Green, P., Poole, S., Martin, D., Maier, S. F., and Watkins, L. R. 2005. An initial investigation of spinal mechanisms underlying pain enhancement induced by fractalkine, a neuronally-released chemokine. Eur. J. Neurosci. 22, 2775–2782.

Molina-Holgado, F., Lledo, A., and Guaza, C. 1995. Evidence for cyclooxygenase activation by nitric oxide in astrocytes. Glia 15, 167–172.

Palma, C., Minghetti, L., Atolfi, M., Ambrosini, E., Silberstein, F. C., manzini, S., Levi, G., and Aloisi, F. 1997. Functional characterization of substance P receptors on cultured human spinal cord astrocytes: synergism of substance P with cytokines in inducing interleukin-6 and prostaglandin E2 production. Glia 21, 183–193.

Raghavendra, V., Rutkowski, M. D., and DeLeo, J. A. 2002. The role of spinal neuroimmune activation in morphine tolerance/hyperalgesia in neuropathic and sham-operated rats. J. Neurosci. 22, 9980–9989.

Raghavendra, V., Tanga, F., and DeLeo, J. A. 2003. Inhibition of microglial activation attenuates the development but not existing hypersensitivity in a rat model of neuropathy. J. Pharmacol. Exp. Ther. 306, 624–630.

Reddington, M., Priller, J., Treichel, J., Haas, C., and Kreutzberg, G. W. 1995. Astrocytes and microglia as potential targets for calcitonin gene related peptide in the central nervous system. Can. J. Physiol. Pharmacol. 73, 1047–1049.

Shavit, Y., Wolf, G., Goshen, I., Livshits, D., and Yirmiya, R. 2004. Interleukin-1 antagonizes morphine analgesia and underlies morphine tolerance. Pain 115, 50–59.

Song, P. and Zhao, Z. Q. 2001. The involvement of glial cells in the development of morphine tolerance. Neurosci. Res. 39, 281–286.

Sul, J., Legresley, A., Ellis-Davies, G., and Haydon, P. G. 2004. Photoreleased nitric oxide regulates glial calcium signaling. Proc. Soc. Neurosci., 34, 405–409.

Sun, D., Coleclough, C., Cao, L., Hu, X., Sun, S., and Whitaker, J. N. 1998. Reciprocal stimulation between TNF-alpha and nitric oxide may exacerbate CNS inflammation in experimental autoimmune encephalomyelitis. J. Neuroimmunol. 89, 122–130.

Svensson, C. I., Fitzsimmons, B., Azizi, S., Powell, H. C., Hua, X.-Y., and Yaksh, T. L. 2004. Spinal p38beta isoform mediates tissue injury-induced hyperalgesia and spinal sensitization. J. Neurochem. 92, 1508–1520.

Svensson, C. I., Marsala, M., Westerlund, A., Calcutt, N. A., Campana, W. M., Freshwater, J. D., Catalano, R., Feng, Y., Protter, A. A., Scott, B., and Yaksh, T. L. 2003. Activation of p38 mitogen-activated protein kinase in spinal microglia is a critical link in inflammation-induced spinal pain processing. J. Neurochem. 86, 1534–1544.

Takata, K., Kitamura, Y., Tsuchiya, D., Kawasaki, T., Taniguchi, T., and Shimohama, S. 2003. Heat shock protein-90-induced microglial clearance of exogenous amyloid-beta1-42 in rat hippocampus in vivo. Neurosci. Lett. 344, 87–90.

Tanga, F. Y., Nutile-McMenemy, N., and DeLeo, J. A. 2005. The CNS role of Toll-like receptor 4 in innate neuroimmunity and painful neuropathy. Proc. Natl. Acad. Sci. U. S. A. 102, 5856–5861.

Tsuda, M., Shigemoto-Mogami, Y., Koizumi, S., Mizokoshi, A., Kohsaka, S., Slater, M. W., and Inoue, K. 2003. P2X4 receptors induced in spinal microglia gate tactile allodynia after nerve injury. Nature 424, 778–783.

Verge, G. M., Milligan, E. D., Maier, S. F., Watkins, L. R., Naeve, G. S., and Foster, A. C. 2004. Fractalkine (CX3CL1) and fractalkine receptor (CX3CR1) distribution in spinal cord and dorsal root ganglia under basal versus neuropathic pain conditions. Eur. J. Neurosci. 20, 1150–1160.

Watkins, L. R. and Maier, S. F. 2003. Glia: a novel drug discovery target for clinical pain. Nat. Rev. Drug. Discov. 2, 973–985.

Watkins, L. R., Hutchinson, M. R., Johnston, I., and Maier, S. F. 2005. Glia: novel counter-regulators of opioid analgesia. Trends Neurosci. 28, 661–669.

Watkins, L. R., Milligan, E. D., and Maier, S. F. 2001. Glial activation: a driving force for pathological pain. Trends Neurosci. 24, 450–455.

Wieseler-Frank, J., Kwiecien, O., Jekich, B., Maier, S. F., and Watkins, L. R. 2004. Putative neuron-to-glia signals synergize to enhance interleukin-1 production by rat dorsal spinal cord glial cells in vitro. Proc. Soc. Neurosci. abstract no. 518.11.

Woolf, C. J., and Salter, M. W. 2000. Neuronal plasticity: increasing the gain in pain. Science 288, 1765–1769.

32 Trigeminal Mechanisms of Nociception: Peripheral and Brainstem Organization

D A Bereiter, University of Minnesota, Minneapolis, MN, USA

K M Hargreaves, University of Texas Health Science Center, San Antonio, TX, USA

J W Hu, University of Toronto, Toronto, ON, Canada

32.1 Introduction

At first glance the trigeminal nerve (Vn), the largest cranial nerve, serves many of the same functions as spinal nerves for lower parts of the body by relaying innocuous and noxious mechanical, thermal and chemical sensory information from all tissues of the head and oral cavity to the brain (Figure 1). Although there are many similarities, peripheral and central aspects of the Vn system are organized quite differently from the spinal system (see Table 1). A primary function of the Vn system is to detect damaging or potentially damaging environmental challenges to the head and oral cavity and one obvious distinction from spinal systems is the diversity of specialized craniofacial tissues innervated by the Vn that provide such detection. Tissues such as the dental pulp, cornea, and dura lack appreciable innervation by large diameter afferent fibers and generally evoke only pain sensation to natural stimuli (Feindel, W. *et al.*, 1960; Kenshalo, D. R., 1960; Beuerman, R. W. and Tanelian, D. L., 1979; Trowbridge, H. O. *et al.*, 1980). Sensory afferents that innervate mucosal tissues of the eye, nose, and oral cavity have a high sensitivity to airborne chemicals (Cometto-Muniz, J. E. and Cain, W. S., 1995) and are thought to underlie the common chemical sense in mammals (see Bryant, B. and Silver, W. L., 2000). The Vn also supplies all specialized receptor regions of the head including the fungiform taste papillas (Whitehead, M. C. *et al.*, 1999), cochlea (Vass, Z. *et al.*, 1997) and the eye (Belmonte, C. *et al.*, 1997). In some mammals and lower vertebrates the Vn relays sensory information unrelated to nociception from unique peripheral structures over distinct central pathways to meet environmental demands. For example, Vn afferents that supply the pit organ of crotaline snakes, a highly sensitive thermal receptor with broad spectral tuning that complements visual input, do not respond to capsaicin (Moiseenkova, V. *et al.*, 2003) and project to separate regions of the trigeminal brainstem complex (Molenaar, G. J., 1978). The skin of the bill of the platypus contains some 30 000–40 000 mucous sensory glands innervated by small myelinated trigeminal afferent fibers that detect electromagnetic radiation emitted by prey and used for underwater navigation (Proske, U. *et al.*, 1998). The whisking of mystacial vibrissas on the face of rodents provides essential textural and perceptual information for exploration and food-gathering in a frequency-dependent manner (Petersen, C. C., 2003; Moore, C. I., 2004), whereas the trigeminal innervation of the snout of the coati mundi (Barker, D. J. and Welker,

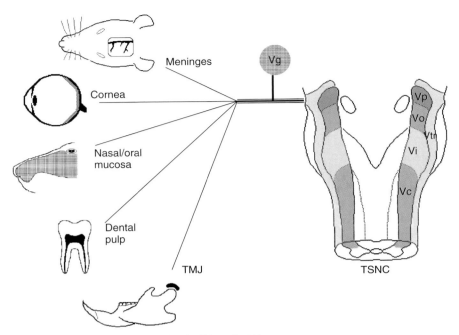

Figure 1 Unique craniofacial tissues supplied by the Trigeminal Nerve.

W. I., 1969) and Eimer's organ of the appendages of the star-nosed mole (Catania, K. C., 1999) relay highly refined tactile information and appear more analogous to somatosensory receptors of the hands than face of primates. Thus, a significant feature of the peripheral Vn is its association with highly specialized sensory functions of craniofacial and oral tissues.

The organization of the central Vn system also displays unusual features distinct from spinal systems that may be relevant to pain processing. At the level of the initial synapse, the trigeminal sensory brainstem nuclear complex (TSNC), craniofacial tissues are represented somatotopically at multiple, and in some cases discontinuous levels, while spinal nerves project to a few dominant contiguous segments (see Renehan, W. E. and Jacquin, M. F., 1993; Bereiter, D. A. *et al.*, 2000; Sessle, B. J., 2000). The importance of multiple representation remains uncertain; however, based on encoding properties and efferent projections, this feature may reflect the segregation of neuronal populations that serve different aspects of trigeminal function. Although the representation of craniofacial tissues accounts for a relatively large area of sensory thalamus and somatosensory cortex, a feature that has long been appreciated since the classic studies of Penfield W. and Rasmussen T. (1950), the cortical representation of some specialized craniofacial tissues

is organized in unusual patterns (e.g., teeth) and not always proportional to innervation density (e.g., ocular surface). In lower mammals, peripheral and central specializations can lead to remarkable magnification of the cortical representation of orofacial tissues such as the nasal appendages of the star-nosed mole which accounts for over 50% of the area of primary somatosensory cortex (SmI; Catania, K. C. and Kaas, J. H., 1997). Similarly, the representation of the enlarged incisors of the naked mole rat used for tunneling and underground navigation accounts for over 30% of SmI (Catania, K. C. and Remple, M. S., 2002). In the albino rat the face occupies more than 60% of SmI with a large area devoted to the cortical barrels and mystacial vibrissae (Welker, C., 1971), whereas a significant portion of the lateral somatosensory cortex is devoted to the teeth (Remple, M. S. *et al.*, 2003). Cortical representation of craniofacial tissues in primates is much reduced compared to lower mammals; however, in select regions such as area 3b, the representation of the face and oral cavity remains unexpectedly large, though curiously, the ocular surface with its high innervation density is represented only weakly (Manger, P. R. *et al.*, 1995; Jain, N. *et al.*, 2001). Beyond the biological importance of protecting the head and oral cavity from damaging stimuli, the Vn is critical for preserving normal eating and sleeping habits, while the emotional

Table 1   General distinctive features of the trigeminal sensory system

Factor	Details	References
Peripheral trigeminal system		
Sensory innervation of specialized tissues	Cavernous sinus, cornea/conjunctiva, dental pulp, lacrimal gland, nasal mucosa, pineal gland, salivary glands, temporomandibular joint, tongue, vibrissae	Rozsa A. J. and Beuerman R. W. (1982), Anggard A. et al. (1983), Byers M. R. (1984), Kido M. A. et al. (1995), Bleys R. L. et al. (1996), Reuss S. (1999), Cheng S. B. et al. (2000), Muller L. J. et al. (2003), Kawabata A. et al. (2004)
Somatotopy within TG	TG neurons from major Vn branches (ophthalmic, maxillary and mandibular) distributed from anteromedial to posterolateral, respectively	Kerr F. W. L. (1963), Marfurt C. F. (1981), Arvidsson J. et al. (1992), Aigner M. et al. (2000)
Mesencephalic nucleus of V (Vmes)	Vmes within the CNS; jaw muscle spindle and periodontal ligament afferents only	Cody F. W. et al. (1972), Linden R. W. (1978)
Muscle innervation	Sparse or misshaped muscle spindles in lip, extraocular and jaw opening muscles of humans	Lennartsson B. (1979), Bruenech J. R. and Ruskell G. L. (2001)
Embryonic origin	Ophthalmic division of TG from neurogenic placodes maxillary, mandibular divisions of TG from neural crest	Noden D. M. (1991), Artinger K. B. et al. (1998), Baker C. V. et al. (2002)
Sensory axon diameter	% A-δ fibers > % C-fibers	Young R. F. and King R. B. (1973), Holland G. R. and Robinson P. P. (1992)
Sensory fiber bifurcation	Descending branch of Vtr: % C-fibers > A-δ-fibers	Tashiro T. et al. (1984)
Sensory fiber termination	Direct sensory fiber projection to autonomic relay nuclei (e.g., NTS)	Jacquin M. F. et al. (1983), Marfurt C. F. and Rajchert D. M. (1991), Panneton W. M. et al. (1994)
Linkage to parasympathetic outflow	Cutaneous vasodilatation, hypotension and bradycardia to TG stimuli	Kumada M. et al. (1977), Drummond P. D. (1992), Ramien M. et al. (2004)
Sympathetic fibers	Vn << spinal nerves	Hoffmann K. D. and Matthews M. A. (1990)
Central trigeminal system		
Facial dermatomes	Representation of orofacial tissues at multiple levels of TSNC	Marfurt C. F. (1981), Panneton W. M. and Burton H. (1981), Shigenaga Y. et al. (1986b), Ma P. M. (1991)
Onion skin organization	Perioral and midline tissues represented more rostral than lateral tissues	Jacquin M. F. et al. (1986), Shigenaga Y. et al. (1986a)
Sensory convergence	Termination of cranial nerves V, VII, IX, X, and upper cervical rootlets in caudal Vc	Kerr F. W. L. (1961), Beckstead R. M. and Norgren R. (1979), Hu J. W. et al. (2005)
Intersubnuclear connections	Lissauer's tract equivalent plus deep bundles	Gobel S. and Purvis M. B. (1972), Kruger L. et al. (1977), Ikeda M. et al. (1984), Jacquin M. F. et al. (1990)

CNS, central nervous system; NTS, nucleus tractus solitarius; TG, trigeminal ganglion; TSNC, trigeminal sensory brainstem nuclear complex.

significance of protecting one's face and self-image suggests that the Vn system has evolved features that set it apart from spinal systems.

Chronic orofacial pain is a significant public health concern that affects more than 20% of the population in western societies (Macfarlane, T. V. et al., 2002b). Persistent orofacial pain has profound effects on eating and sleeping patterns, a strong association with depression, and a reduced sense of wellbeing (Korszun, A., 2002). Many common forms of chronic

orofacial pain such as headache, temporomandibular disorders (TMD), odontalgia, and dry eye are difficult to treat clinically and may be associated with lower pain thresholds elsewhere in the body (Burstein, R. et al., 2000; Sarlani, E. and Greenspan, J. D., 2003; Van Bijsterveld, O. P., et al., 2003). Moreover, epidemiological studies indicate that many chronic craniofacial pain conditions are more prevalent and occur with greater intensity in women than men (Lipton, J. A. et al., 1993; Sandstedt, P. and Sorensen, S., 1995;

LeResche, L., 1997; Rasmussen, B. K., 2001; Wolf, E. *et al.*, 2001). This review highlights several neurobiological aspects of the Vn system that differ from spinal systems and may contribute to the pattern and magnitude of craniofacial pain.

32.2   The Peripheral Trigeminal Nerve System

32.2.1   General Features

The sensory portion of the Vn consists of three main nerve branches: ophthalmic, maxillary, and mandibular. Proprioceptive afferents from select craniofacial muscles and periodontal ligaments travel within the Vn, however; the cell bodies are located in the mesencephalic nucleus of V (Vme) in the rostral pons (Cody, F. W. *et al.*, 1972; Linden, R. W., 1978). Unlike the spinal dorsal root ganglion (DRG) that has no apparent intraganglionic organization, the somata of neurons giving rise to the three branches of Vn are somatotopically arranged within the trigeminal ganglion (TG; Kerr, F. W. L., 1963; Marfurt, C. F., 1981; Arvidsson, J. *et al.*, 1992; Aigner, M. *et al.*, 2000). Table 1 lists several features of the peripheral Vn system that differ from the spinal system. Differences relevant to pain processing include the innervation of specialized tissues, composition, and central termination of afferents; relationship of sensory nerves to sympathetic efferent outflow; and neurochemical markers in ganglion cells. The Vn innervates a heterogeneous and unique group of craniofacial tissues well associated with pain sensation. The cornea is the most densely innervated tissue of the body, supplied exclusively by small diameter fibers (Rozsa, A. J. and Beuerman, R. W., 1982); is the only tissue in which nerve fibers penetrate the outer epithelial layers (Hoyes, A. D. and Barber, P., 1976; Muller, L. J. *et al.*, 1996); and responds to very small changes in ocular surface temperature, moisture status, or foreign bodies (Belmonte, C. *et al.*, 1997). The oral mucosal epithelium also is densely innervated by polymodal small diameter Vn afferents with relatively large receptive fields (RFs) compared to cutaneous receptors (Toda, K. *et al.*, 1997) that provide proprioceptive information on lip/tongue position necessary for eating and speech as well as detecting noxious stimuli within the mouth (Trulsson, M. and Johansson, R. S., 2002). By contrast, the dental pulp is well insulated from the exterior environment and, while pain is the predominant sensation relayed by pulpal afferents, dental afferents respond best to natural stimuli after injury (Byers, M. R., 1984; Hildebrand, C. *et al.*, 1995).

Specialized transduction processes also are associated with these unique tissues such as the proposed hydrodynamic movement of fluid as the basis for dentinal pain (Brannstrom, M., 1986).

Peripheral Vn afferents are subject to frequent injury due to trauma, infection, or iatrogenic damage following surgery that can lead to peripheral sensitization. Indeed, one of the first demonstrations of enhanced peripheral C-fiber activity due to injury was reported for facial afferents in the monkey (Beitel, R. E. and Dubner, R., 1976). Since peripheral sensitization of small myelinated A-δ versus unmyelinated C-fibers does not occur uniformly, but rather varies for different tissue types and depends on the nature of the injury (see Meyer, R. A. *et al.*, 2006), it may be significant that the sensory root of the Vn has a greater percentage of A-δ fibers than C-fibers compared to spinal nerves as determined by counts of axons (Young, R. F. and King, R. B., 1973; Holland, G. R. and Robinson, P. P., 1992) or lectin staining for C-fibers (Ambalavanar, R. and Morris, R., 1992; Wang, H. *et al.*, 1994). Correspondingly, corneal (Belmonte, C. *et al.*, 1997) and dural afferents (Levy, D. and Strassman, A. M., 2002) conducting in the A-δ range display greater sensitization than more slowly conducting C-fibers.

Complex regional pain syndrome (CRSP), formerly referred to as causalgia, is a neuropathic pain condition with sympathetic involvement, often resulting from trauma or bone fracture to the distal limbs (see Wasner, G. *et al.*, 1998). The incidence of CRSP-like symptoms following trauma to craniofacial tissues is relatively low compared to injury of other tissues (Matthews, B., 1989). The exact reason for this observation is not certain; however, fewer sympathetic efferent fibers course within sensory branches of Vn compared to spinal nerves, and instead these fibers travel along the arteries that supply the head and oral cavity (Hoffmann, K. D. and Matthews, M. A., 1990; Maklad, A. *et al.*, 2001). Spinal nerve injury induces sprouting of noradrenergic fibers into the DRG (McLachlan, E. M. *et al.*, 1993), whereas comparable injury to Vn produces no such sprouting into the TG (Bongenhielm, U. *et al.*, 1999; Benoliel, R. *et al.*, 2001). Furthermore, cervical sympathectomy does not alter the development of ectopic discharge or increased mechanical sensitivity after inferior alveolar nerve transection (Bongenhielm, U. *et al.*, 1998). Since the basis for sprouting into the DRG following nerve injury is thought to involve neurotrophins (Davis, B. M. *et al.*, 1994), it is interesting to note that the number of trkA-positive corneal and dental sensory TG neurons is significantly less (< 10%, Mosconi, T. et al., 2001) than for spinal DRG neurons (40%, Molliver, D. C. and Snider, W. D., 1997). Failure to observe sympathetic sprouting may be specific for nerve injury, since central administration of nerve growth factor (NGF) induces significant sprouting into the TG and spinal DRG (Nauta, H. J. et al., 1999).

32.2.2 Neurochemical and Molecular Properties of Trigeminal Ganglion Neurons

Although this review emphasizes comparisons of the TG and DRG systems in the adult, significant developmental differences also exist. While the functional significance of developmental differences (e.g., neurogenic placode vs. neural crest) may not be obvious in the healthy adult, many responses to injury recapitulate developmental differences and thus may have implications in the injured adult nervous system. Studies reporting similarities in development of the TG and DRG systems have included: P2X3 receptor expression (Ruan, H. Z. et al., 2004); phenotype expression in trkB knockout mice (Klein, R. et al., 1993; Gonzalez-Martinez, T. et al., 2004); vesicular monoamine transporter levels (Hansson, S. R. et al., 1998); the apoptotic enzyme, CPP32/apopain (Mukasa, T. et al., 1997); neuronal tyrosine hydroxylase (Son et al., 1996; Kim, S. J. et al., 1997); six 4/ARECC3 mRNA (Esteve, P. and Bovolenta, P., 1999); the transcription factor, retinoid-X receptor γ (Georgiades, P. et al., 1998); guanosine triphosphate (GTP)-binding protein G-α z (Kelleher, K. L. et al., 1998); the fetal plasma glycoprotein, fetuin (Kitchener, P. D. et al., 1997); FGF receptor mRNA (Wanaka, A. et al., 1991); and the large zinc finger protein, KRC (Hicar, M. D. et al., 2002). By contrast, studies reporting significant differences in the developing TG and DRG systems have included: expression of trkC at day 12.5 (DRG: yes, TG: no; Elkabes, S. et al., 1994); requirement for NGF at day E16.5 (DRG > TG; Goedert, M. et al., 1984); and sensitivity to cadmium-induced neurotoxicity (TG > DRG; Arvidson, B., 1983). Thus, developmental biology indicates some similarities with notable differences in the responsiveness to certain neurotrophic factors.

Table 2 summarizes results from studies comparing adult TG and DRG systems under basal (naive) conditions. These studies reveal many similarities among markers associated with nociception such as the percentage of TRPV1-positive neurons (Ambalavanar, R. et al., 2005) and expression of P2X

receptor subtypes (Collo, G. et al., 1996; Cook, S. P. et al., 1997; Xiang, Z. et al., 1998). However, substantial differences also are observed that include: the percentage of neurons stained for substance P and somatostatin (TG > DRG; Kai-Kai, M. A., 1989), trkA receptor (DRG > TG; Mosconi, T. et al., 2001), galectin-1 (DRG >> TG; Akazawa, C. et al., 2004), μ- and δ-opioid receptors (DRG > TG; Buzas, B. and Cox, B. M., 1997), CCK (DRG > TG; Ghilardi, J. R. et al., 1992), cytokeratin (TG >> DRG; Okabe, H. et al., 1997), and neuropeptide Y (NPY)-binding sites (TG > DRG; Mantyh, P. W. et al., 1994). In addition, TG and DRG systems display a differential sensitivity to ganglion cell labeling by selected anatomical tracers in which the TG system has a greater uptake of Fluoro-Gold than DRG but not of Fast Blue (Yoshimura, N. et al., 1994).

Differences in cellular properties between the TG and DRG systems under naive conditions may contribute to differential responses to tissue injury. For example, sprouting of sympathetic nerve terminals into the DRG (McLachlan, E. M. et al., 1993), but not the TG after nerve injury (Bongenhielm, U. et al., 1999; Benoliel, R. et al., 2001) is consistent with findings that trkA-positive neurons are more numerous in the DRG than TG (Mosconi, T. et al., 2001). Comparison of nerve injury-induced changes for a majority of neuropeptides associated nociception such as substance P, TRPV1, P2X3, and NPY appear similar for TG and DRG systems (Zhang, X. et al., 1996; Okuse, K. et al., 1997; Eriksson, J. et al., 1998; Elcock, C. et al., 2001; Tsuzuki, K. et al., 2001; Stenholm, E. et al., 2002; Tsuzuki, K. et al., 2003). However, species differences have been reported such as a decrease in galanin in TG of ferret (Elcock, C. et al., 2001) compared to an increase in rat (Zhang, X. et al., 1996) after Vn injury, similar to the increase in galanin in rat DRG after spinal nerve injury (Villar, M. J. et al., 1989). Nerve injury causes higher spontaneous discharge rates and greater rhythmic firing patterns in DRGs than TGs (Tal, M. and Devor, M., 1992), effects often associated with sodium channel activity (see Wood, J. N. et al., 2004). However, the basis for this difference is not certain since changes in the expression of the tetrodotoxin (TTX)-resistant sodium channel, NaV1.8, appear comparable for DRG and TG systems (Dib-Hajj, S., et al., 1996; Bongenhielm, U. et al., 2000). Susceptibility to infection also may differ between TG and DRG systems, since injection of herpes simplex virus (HSV) to the left ear pinna in

Table 2   Comparison of trigeminal (TG) and dorsal root ganglion (DRG) systems under naive conditions

Factor	Details	References
Substance P	TG $\geq$ DRG	Kai-Kai M. A. (1989)
Somatostatin	TG $\geq$ DRG	Kai-Kai M. A. (1989)
P2X$_1$, P2X$_2$, P2X$_3$, P2X$_4$, P2X$_5$, P2X$_6$ mRNA	TG $\sim$ DRG; no P2X2, P2X3 in Vme	Collo G. et al. (1996), Cook S. P. et al. (1997) Xiang Z. et al. (1998)
P2X$_3$-isolectin B4 co-expression	DRG $>>$ TG	Ambalavanar R. et al. (2005)
TRPV1	TG $\sim$ DRG	Guo A. et al. (1999)
trkA	DRG $>$ TG; $\sim$ 40% DRG versus 10–15% of TG neurons innervating pulp or cornea	Mosconi T. et al. (2001)
Arginine vasopressin	DRG $>$ TG About 40% AVP is in capsaicin-sensitive neurons in both ganglia	Kai-Kai M. A. and Che Y. M. (1995)
5-HT$_{1d}$	TG $\sim$ DRG	Potrebic S. et al. (2003)
Oxytocin	TG $>$ DRG	Kai-Kai M. A. (1989)
NADPH-diaphorase	DRG: T5–L1 $>>$ C1-T4 $=$ L2–S $=$ TG	Aimi Y. et al. (1991)
CCK mRNA in monkey	DRG $=$ 20%; TG $=$ 10% of neurons	Verge V. M. et al. (1993)
CCK(B) receptor	Rat, rabbit: TG $\sim$ DRG Monkey: DRG $>$ TG (nondetectable)	Ghilardi J. R. et al. (1992)
PYY binding sites (NPY receptor)	TG $\geq$ DRG	Mantyh P. W. et al. (1994)
Galectin-1 mRNA	DRG $>>$ TG (nondetectable)	Akazawa C. et al. (2004)
Oncostatin M (OSM-β)	DRG $>$ TG (OSMr-β coexpressed in TRPV1-positive neurons)	Tamura S. et al. (2003)
glycogen phosphorylase	DRG $>$ TG	Pfeiffer B. et al. (1995)
Cytokeratin (AE1 and CAM5.2)	TG $>>$ DRG	Okabe H. et al. (1997)
Glucocorticoid receptor (GR)	TG: GR expressed in substance P, CGRP, but not galanin-positive neurons DRG: GR expressed in substance P, CGRP, and galanin-positive neurons	DeLeon M. et al. (1994)
MOR mRNA	DRG: lumbar $>$ thoracic $\sim$ cervical $>$ TG	Buzas B. and Cox B. M. (1997)
DOR mRNA	DRG: lumbar $\sim$ thoracic $\sim$ cervical $>$ TG	Buzas B. and Cox B. M. (1997)
TGF-α mitogenic effect in vitro	DRG: Yes TG: No	Chalazonitis A. et al. (1992)
Parvalbumin	TG smaller size than DRG, but both populations have high expression of carbonic anhydrase and low expression of CGRP	Ichikawa H. et al. (1994)
Osteocalcin $\pm$ parvalbumin	TG: 25% of neurons (31% express parvalbumin) Vme: 63% of neurons ($>$90% express parvalbumin) DRG: 16% of neurons ($>$90% express parvalbumin)	Ichikawa H. et al. (1999)
Osteocalcin and TRPV1 coexpression	TG $=$ 14% DRG $=$ none	Ichikawa H. and Sugimoto T. (2002)
Calretinin	TG: neurons mostly $<$ 800 μm^2 and 34% positive for tachykinin DRG: neurons mostly $>$ 800 μm^2 and 7% positive for tachykinin	Ichikawa H. et al. (1993)
S100 calcium-binding protein	TG: 59% ($>$ 90% coexpress parvalbumin and calbindin D-28k) DRG: 44% ($>$ 90% coexpress parvalbumin, none with calbindin D-28k)	Ichikawa H. et al. (1997)

(*Continued*)

Table 2   (Continued)

Factor	Details	References
tr7kC at day E12.5 of development	DRG: Yes TG: No	Elkabes S. *et al.* (1994)
NT-3 overexpression	TG > DRG for responsiveness	Albers K. M. *et al.* (1996)
Neonatal anti-NGF antisera at E16.5	DRG > TG for loss of substance P- and somatostatin-positive neurons	Goedert M. *et al.* (1984)
Uptake of fluorescent dyes	TG ~ DRG uptake of Fast Blue TG > DRG uptake of Fluoro-Gold	Yoshimura N. *et al.* (1994)

5-HT$_{1d}$, serotonin receptor subtype; CCK, cholecystokinin; CGRP, calcitonin gene-related peptide; DOR, δ-opioid receptor; DRG, dorsal root ganglion; MOR, μ-opioid receptor, messenger RNA; NADPH, nicotinamide adenine dinucleotide phosphate hydrogen; ; NGF, nerve growth factor; NPY, neuropeptide Y; P2X, ATP receptor; PYY, polypeptide Y; TG, trigeminal ganglion; TGF, tumor growth factor; trkA, tyrosine kinase A receptor subtype; TRPV1, vanilloid receptor.

mice produces 100% HSV infection in both the ipsilateral TG and cervical DRG. Interestingly, 70% of the TGs contralateral to injection, while only 10% of contralateral DRGs were infected (Thackray, A. M. and Field, H. J., 1996). Interhemispheric neural communication likely contributes to the progression of joint-related pain in trigeminal (see Bereiter, D. A. *et al.*, 2005b) as well as spinal systems (Levine, J. D. *et al.*, 1985; Shenker, N. *et al.*, 2003). The responsiveness to viral vectors in the Vn system has prompted recent efforts to deliver targeted transgene-derived products for control of trigeminal neuropathic pain (Meunier, A. *et al.*, 2005). Collectively, these studies reveal significant differences in peripheral Vn and spinal systems under naïve and injured conditions, differences that could not be predicted on the basis of results from spinal sensory systems alone.

32.3 Central Aspects of Trigeminal Organization

Noxious sensory information is relayed from Vn afferents to second-order neurons in the TSNC and the upper cervical spinal cord. The TSNC is the initial site of synaptic integration for sensory input from the head and oral cavity and shares this feature with the spinal dorsal horn and dorsal column nuclei that receive sensory input from the rest of the body (Figure 1). However, unlike the spinal cord, the TSNC is comprised of several cell groups with distinct cytological and organizational features (see Darian-Smith, I., 1973; Kruger, L. and Young, R. F., 1981) yet each cell group receives direct primary afferent projections from specific craniofacial tissues. The TSNC consists of: the principal nucleus (Vp), supratrigeminal region (Vsup) lying

dorsal to Vp, an elongated spinal nucleus (Vsp) extending from the pons to the upper cervical spinal cord, and the interstitial islands or the paratrigeminal region (Pa5) embedded within the spinal trigeminal tract, dorsal and lateral to the caudal Vsp (Figure 2; see Kruger, L. and Young, R. F., 1981; Renehan, W. E. and Jacquin, M. F., 1993). The Vsp is further subdivided, from rostral to caudal, into subnucleus oralis (Vo), subnucleus interpolaris (Vi), and a laminated trigeminal subnucleus caudalis (Vc) as described originally by Olszewski J. (1950). Although nociceptive neurons in the caudal laminated portion of the TSNC, Vc, display properties similar to those at spinal levels (Price, D. D. *et al.*, 1976; Dubner, R. and Bennett, G. J., 1983) consistent with a prominent role in nociceptive processing (see also Bereiter, D. A. *et al.*, 2000; Sessle, B. J., 2000), the contribution of rostral portions of the TSNC to orofacial pain is less certain.

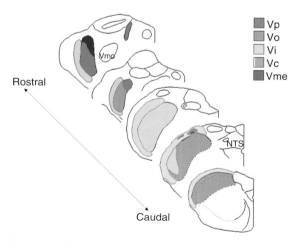

Figure 2   Trigeminal brainstem sensory complex.

32.3.1 Somatotopy

Somatotopy is a key feature of the trigeminal system and is seen within the TG as well as the TSNC distinct from the spinal cord. Craniofacial tissues are represented at multiple levels of the TSNC, while sensory afferents from other body loci terminate at several contiguous spinal segments of the spinal dorsal horn. Also, at caudal levels of the TSNC craniofacial tissues are represented in a series of semicircular bands that converge at the rostral midline of the face, often referred to as an onion-skin arrangement and by a medial-lateral representation in which the head is inverted (Jacquin, M. F. et al., 1986; Shigenaga, Y. et al., 1986a). Although somatotopy along the mediolateral axis is preserved at all levels of the TSNC, the onion-skin arrangement is most apparent in Vc. The implications of this organization for facial pain have been debated since an early report by Sjoqvist O. (1938) that trigeminal tractotomy at the level of rostral Vc reduced the pain of trigeminal neuralgia, while preserving the sense of temperature and touch on the face. It is understood that rostral portions of TSNC play a more prominent role in dental and intraoral pain than caudal portions (Young, R. F. and Perryman, K. M., 1984; see Sessle, B. J., 2000); however, the importance of multiple representation of extraoral tissues for other forms of craniofacial pain is not well understood. One reason for this uncertainty may be due to the fact that many craniofacial tissues are represented in a discontinuous manner along the rostrocaudal extent of the TSNC as summarized in Table 3. Although a small percentage of Vn afferents projects to both rostral and caudal portions of the TSNC (Silverman, J. D. and Kruger, L., 1985; Li, Y. Q. et al., 1992), most fibers either ascend in a short sensory root to terminate in Vp or descend to give off branches to Vo, Vi, Vc, and the upper cervical dorsal horn. Several aspects of the afferent input pattern to TSNC are notable. First, the superficial laminas in Vc receive substantial input from all specialized tissues well associated with craniofacial pain conditions (e.g., cornea, dura, teeth, temporomandibular joint (TMJ)), while afferents from structures with no known relationship to pain perception (vibrissas) do not project to this region. Second, dental pulp afferents are the

Table 3 Summary of the relative density of trigeminal primary afferent terminals within different portions of the sensory trigeminal sensory nuclear complex.

| | Vp | | Vo | | Vi | dPa5 | Vi/Vc | Vc | | | References |
	dm	vl	dm	vl				I–II	III–IV	V	
Cornea	−	+	−	+	+	−	+++	++	−	−	Panneton W. M. and Burton H. (1981), Marfurt C. F. and del Toro D. R. (1987), Marfurt C. F. and Echtenkamp S. F. (1988)
Nasal mucosa	−	+	+	+	++	−	+++	+++	+	+	Anton F. and Peppel P. (1991); Panneton W. M. (1991)
Dura	−	−	−	−	+	−	+++	+++	+	−	Arbab M. A. et al. (1988), Liu Y. et al. (2004)
Teeth	++	+	+++	+	+	+++	−	++	−	+	Marfurt C. F. and Turner D. F. (1984), Shigenaga Y. et al. (1986c); Takemura M. et al. (1993)
Masseter muscle	+	−	+	−	++	++	−	++	−	+	Nishimori T. et al. (1986), Shigenaga Y. et al. (1988), Arvidsson J. and Raappana P. (1989)
TMJ	+	−	+	−	++	++	−	+++	−	+	Jacquin M. F. et al. (1983); Shigenaga Y. et al. (1986a), (1986b), Capra N. F. (1987), Takemura M. et al. (1987)
Vibrissae	−	+++	−	+	+++	−	−	−	++	−	Jacquin M. F. et al. (1986), Nomura S. et al. (1986), Arvidsson J. (1982)

−, very few or none; +,++,+++, weak, moderate and dense terminal distribution; dPa5, dorsal paratrigeminal region; TMJ, temporomandibular joint; Vc, subnucleus caudalis; Vi, subnucleus interpolaris; Vo, subnucleus oralis; Vp, principal sensory nucleus.

only group of putative nociceptive fibers with a substantial projection to rostral portions of the TSNC, namely, to the dorsomedial portions of Vp and Vo, thus supporting the role for these regions in intraoral pain (Marfurt, C. F. and Turner, D. F., 1984; Shigenaga, Y. *et al.*, 1986c). Also, compared to other orofacial tissues associated with pain sensation, the termination pattern of dental afferents is particularly widespread along the rostrocaudal extent of the TSNC. Third, input from structures supplied by the ophthalmic branch of the Vn (e.g., cornea, nasal dura) project only sparsely to rostral portions of the TSNC suggesting that these regions play a lesser role than Vc in mediating pain due to uveitis, dry eye, sinusitis, or headache. Fourth, discontinuous representation is not unique to tissues supplied by the ophthalmic division (e.g., cornea, dura, nasal cavity) since the auriculotemporal nerve, a major source of innervation for the TMJ region also displays an uneven terminal distribution in the TSNC (Shigenaga, Y. *et al.*, 1986a; 1986b). In rodents vibrissae afferents project to most levels of the TSNC (Arvidsson, J., 1982; Nomura, S. *et al.*, 1986); however, the architectonic representation of the vibrissae fields, the so-called barrelettes, are well delineated in Vp, Vi, and Vc, but not Vo (Ma, P. M., 1991), supporting the notion that different levels of the TSNC mediate different aspects of sensory processing of innocuous as well as noxious inputs.

32.3.2 Intersubnuclear Connections

Rostral and caudal portions of the TSNC are connected by a rich longitudinal fiber network coursing within spinal trigeminal tract and through the deep bundles that extend from Vp to the upper cervical spinal cord (Gobel, S. and Purvis, M. B., 1972; Kruger, L. *et al.*, 1977; Ikeda, M. *et al.*, 1984; Jacquin, M. F. *et al.*, 1990). Although propriospinal-like connections of the TSNC share anatomical similarities with those at lower spinal levels, intersubnuclear connections in the TSNC link spatially distinct brainstem regions with common somatotopic representation of facial fields. The implications of this organization for facial pain remain to be determined; however, results from numerous animal studies support the clinical findings of Sjoqvist O. (1938) and indicate that ascending connections from caudal Vc generally facilitate the activity of neurons in more rostral portions within the TSNC. Lesion or chemical blockade of caudal Vc reduced the excitability of rostral trigeminal neurons responsive to noxious stimulation of tooth pulp

(Greenwood, L. F. and Sessle, B. J., 1976; Chiang, C. Y. *et al.*, 2002), dura (Davis, K. D. and Dostrovsky, J. O., 1988), and cornea (Hirata, H. *et al.*, 2003). By contrast, innocuous sensory input from facial skin (Greenwood, L. F. and Sessle, B. J., 1976) and vibrissae (Hallas, B. H. and Jacquin, M. F., 1990) often was enhanced. Fewer studies have assessed the influence of rostral TSNC regions on caudal Vc neural activity, though in a recent study, muscimol blockade of the Vi/Vc transition region facilitated cornea-responsive neurons in laminas I–II of caudal Vc (Hirata, H. *et al.*, 2003). These results suggest that ascending as well as descending connections within the TSNC contribute to the integration of sensory inputs relevant for craniofacial pain.

32.3.3 Relationship to the Autonomic Nervous System

The trigeminal system is closely linked to brain regions that control autonomic outflow, especially parasympathetic outflow and vagus nerve activity. This linkage likely contributes to craniofacial pain conditions such as primary headache (Edvinsson, L. and Uddman, R., 2005) and dry eye (Hocevar, A. *et al.*, 2003), sudden bradycardia and asystole during maxillofacial surgery (Schaller, B., 2004), and the so-called diving reflex in infant humans (Goksor, E. *et al.*, 2002) and aquatic mammals (Butler, P. J. and Jones, D. R., 1997). Even facial skin differs from other cutaneous regions in that it is well supplied by parasympathetic fibers (Ramien, M. *et al.*, 2004). Under experimental conditions noxious stimulation of craniofacial tissues in humans evokes long-lasting vasodilatation in orofacial regions (Drummond, P. D., 1992; Izumi, H., 1999) that differs from responses evoked by stimulation of other body regions consistent with the existence of specialized trigeminal vasodilator reflex mechanisms (Kemppainen, P. *et al.*, 2001). Two aspects of the relationship between autonomic nerves and the trigeminal system are distinct from spinal cord and deserve special mention. Unlike at spinal levels where nearly all sensory nerves relay initially in the dorsal horn or dorsal column nuclei, many Vn afferents, especially those from the mandibular branch, project directly to brainstem nuclei that control autonomic outflow such as the nucleus tractus solitarius (NTS), parabrachial complex, and ventrolateral medulla (Kerr, F. W. L., 1961; Jacquin, M. F. *et al.*, 1983; Marfurt, C. F. and Rajchert, D. M., 1991; Panneton, W. M., 1991; Panneton, W. M. *et al.*, 1994). Also, unlike lower portions of the spinal cord, there is an extensive convergence of Vn, facial, glossopharyngeal, and vagal afferents to common laminae of the lower TSNC and upper cervical dorsal horn (Denny-Brown, D. and Yanagisawa, N., 1973; Beckstead, R. M. and Norgren, R., 1979; Contreras, R. J. *et al.*, 1982; Altschuler, S. M. *et al.*, 1989; McNeill, D. L. *et al.*, 1991). The dorsal paratrigeminal region (dPa5) also receives inputs from multiple cranial nerves and upper cervical rootlets, has widespread connections to central autonomic pathways and other regions of the TSNC, and may be a key site of somatic–autonomic integration for cutaneous and visceral sensory input and control of homeostasis (Panneton, W. M. and Burton, H., 1985; Saxon, D. W. and Hopkins, D. A., 1998; Caous, C. A. *et al.*, 2001). Considerable evidence suggests that the Vc/upper cervical cord (Vc/C2) junction region differs from lower spinal cord. In addition to receiving convergent input from multiple cranial nerves and upper cervical rootlets (Pfaller, K. and Arvidsson, J., 1988; Neuhuber, W. L. and Zenker, W., 1989), second-order Vc/C2 neurons have widespread ascending connections to the hypothalamus (Burstein, R. *et al.*, 1990) and periaqueductal gray (PAG; Keay, K. A. *et al.*, 1997), brainstem regions well associated with control of autonomic outflow, and endogenous pain modulation circuits. The Vc/C2 junction also sends long-range descending projections to the lower spinal cord and is a critical region for visceral sensory, particularly vagus nerve, modulation of somatic input to lower spinal segments (Chandler, M. J. *et al.*, 2002). Although vagus nerve stimulation generally is associated with antinociception in humans (Kirchner, A. *et al.*, 2000) and animals (Randich, A. and Gebhart, G. F., 1992; Khasar, S. G. *et al.*, 1998), the relationship between vagus nerve activity and facial pain may be more complex. For example, increased vagal afferent activity has been suggested as one source of facial pain referred from the lung in cancer patients (see Sarlani, E. *et al.*, 2003), whereas in animal studies increased vagal activity inhibits painlike behavior and c-fos expression after formalin injection into facial skin (Bohotin, C. *et al.*, 2003) and tooth pulp-evoked activity of Vc/C2 dorsal horn neurons (Tanimoto, T. *et al.*, 2002). The effects of vagus nerve stimulation have been tested mainly on neurons in the caudal portions of the TSNC; however, vagal stimulation also inhibits tooth pulp-evoked digastric reflexes (Bossut, D. F. *et al.*, 1992) suggesting that modulation of neurons in rostral portions of the TSNC is possible. Given the extensive convergence and efferent projections of second-order neurons at the dPa5 and Vc/C2 regions, it is tempting to speculate that trigeminal–vagal interactions play a significant role in modulating pain above as well as below the neck.

32.3.4 Neurochemical Markers

The role of central neurons in mediating the various aspects of pain (e.g., sensation, autonomic control, motor reflexes) can be predicted, in part, on the basis of key factors such as the nature of the sensory input, encoding properties, response to analgesic agents, efferent projections (Price, D. D. and Dubner, R., 1977; Price, D. D. *et al.*, 2003), and, more recently, the distribution of a growing list of neurochemical markers associated with nociceptive processing (Woolf, C. J. and Salter, M. W., 2000; Julius, D. and Basbaum, A. I., 2001; Lewin, G. R. *et al.*, 2004). The data provided in Tables 4 (neurochemical markers) and 5 (efferent projections) are consistent with the notion that different portions of the TSNC contribute to different aspects of craniofacial pain. However, these data derive from results in naïve animals and do not specifically indicate which regions respond with phenotypic or long-term structural changes during chronic pain (see Basbaum, A. I., 1999; Hunt, S. P. and Mantyh, P. W., 2001; Scholz, J. and Woolf, C. J., 2002), changes that likely occur unequally in different portions of the TSNC.

Table 4 summarizes the pattern of distribution of selected neurochemical markers with known association to nociceptive processing and/or its modulation. The most striking aspect of these data is the dense and almost universal distribution of all markers within the superficial laminae of Vc, while only weak labeling is seen in laminae not associated with nociceptive processing (laminae III–IV). In rostral portions of the TSNC the density of different markers is more varied than in Vc. For example, the dorsomedial portions of Vp and Vo, regions that have a high density of afferent terminals from tooth pulp nerves, also display intense labeling for calcitonin gene-related peptide (CGRP) and trkA, whereas labeling for substance P and inositol 1,4,5 triphosphate (IP3) receptor are relatively weak and that for MOR1, the μ-opioid receptor, is absent. Interestingly, selective agonists for the $5-HT_{1B}$ receptor, a serotonergic receptor subtype that binds sumatriptan, an effective therapeutic agent for migraine, displays moderate density in the dorsomedial portions of Vp and Vo, yet neither region receives significant input from meningeal afferents. Although the laminar distribution of most neurochemical markers in Vc and the spinal dorsal horn are similar, significant differences have been reported for IB4 (Sugimoto, T. *et al.*, 1997a) and TRPV1 (Bae, Y. C. *et al.*, 2004), where spinal lamina IIi contains a greater expression of both markers than lamina IIo, while the reverse is seen in Vc. Also, the distribution of CGRP and substance P appears

Table 4   Summary of the distribution of neurochemical markers associated with nociceptive processing in different portions of the trigeminal sensory nuclear complex

	Vp		Vo					Vc			
	dm	vl	dm	vl	Vi	dPa5	Vi/Vc	I–II	III–IV	V	References
IB4	+	−	+	−	+	++	−	+++	−	−	Sugimoto T. *et al.* (1997a)
TRPV1	+	−	+	−	+	+++	++	+++	−	−	Bae Y. C. *et al.* (2004)
SP	+	−	+	−	+	++	++	+++	−	+	Boissonade F. M. *et al.* (1993), Sugimoto T. *et al.* (1997b)
NK1	+	−	+	−	+	+	+	+++	+	+	Nakaya Y. *et al.* (1994)
CGRP	++	−	+++	−	+	++	++	+++	−	+	Henry M. A. *et al.* (1996), Kruger L. *et al.* (1988), Sugimoto T. *et al.* (1997b)
P2X$_2$	+	+	+	++	++	+	++	++	+	+	Kanjhan R. *et al.* (1999)
trkA	+++	+	++	+	++	++	++	+++	+	−	Pioro E. P. and Cuello A. C. (1990), Sobreviela T. *et al.* (1994)
BDNF	+	−	+	−	+	++	++	+++	−	−	Connor J. M. *et al.* (1997)
trkB	++	++	++	++	++	++	++	+++	+	+	Yan Q. *et al.* (1997)
EP3	−	−	−	−	−	−	−	+++	−	−	Nakamura K. *et al.* (2000)
ChAT	+	−	+	−	+	−	−	++	−	−	Tatehata T. *et al.* (1987)
nNOS	+	−	+	−	+	++	+	+++	+	−	Dohrn C. S. *et al.* (1994), Rodrigo J. *et al.* (1994)
NR1	++	+	++	+	++	+++	++	+++	+	+	Petralia R. S. *et al.* (1994)
mGluR2	+	+	+	+	+	++	+	++	+	+	Ohishi H. *et al.* (1998)
GABA	+	+	++	+	+	+	++	+++	++	+	Ginestal E. and Matute C. (1993)
GABAaR	+	+	+	+	++	+	++	++	+	++	Fritschy J. M. and Mohler H. (1995), Pirker S. *et al.* (2000)
GABAbR	+	+	+	+	+	++	++	+++	+	+	Margeta-Mitrovic M. *et al.* (1999)
NE/DA	−	+	+	+	+	−	+	+	+	+	Kitahama K. *et al.* (2000), Levitt P. and Moore R. Y. (1979)
ARα/β	+	+	+	+	+	+	+	++	+	+	Talley E. M. *et al.* (1996)
5-HT	+	+	+	+	+	+	++	+++	+	+	Harding A. *et al.* (2004), Steinbusch H. W. M. (1981)
5-HT$_{1B/1D}$	++	−	++	−	+	++	+	+++	+	+	Potrebic S. *et al.* (2003, Thor K. B. *et al.* (1992)
IP$_3$R	−	+	−	+	+	−	+	++	−	−	Rodrigo J. *et al.* (1993)
Calcineurin	−	+	−	+	+	−	+	+++	+	−	Strack S. *et al.* (1996)
Osteocalcin	−	+	−	+	+	++	+	+++	+	−	Ichikawa H. and Sugimoto T. (2002)
Endo2	−	−	+	−	+	++	+	+++	−	−	Martin–Schild S. *et al.* (1999)
MOR1	−	−	−	−	−	++	+	+++	−	−	Bereiter D. A. and Bereiter D. F. (2000), Ding Y. Q. *et al.* (1996)
ER$_\alpha$	−	−	−	−	−			−	+++	+	Bereiter D. A. *et al.* (2005a)

−, very weak or no staining; +,++,+++, weak, moderate and dense staining; 5-HT, serotonin; 5HT$_{1/2}$, serotonin receptor subtypes; AR$_{\alpha/\beta}$, adrenergic receptor subtypes; BDNF, brain-derived neurotrophic factor; CGRP, calcitonin gene-related peptide; ChAT, choline acetyltransferase; dPa5, dorsal paratrigeminal region; Endo2, endomorphin 2; EP3, prostaglandin receptor; ER$_\alpha$, estrogen receptor alpha subtype; GABA, gamma-aminobutyric acid; GABAaR, GABA receptor α subtype, b2/3 subunit; GABAbR, GABA receptor β subtype, R1a/b subunit; IB4, isolectin B4; IP$_3$R, inositol triphosphate receptor; MOR1, μ-opioid receptor; NE/DA, norepinephrine/ dopamine; NK1, neurokinin 1 receptor; nNOS, neuronal nitric oxide synthase; NR1, N-methyl-D-aspartic acid (NMDA) receptor subunit; P2X$_2$, ATP receptor; SP, substance P; trkA, tyrosine kinase A receptor subtype; trkB, tyrosine kinase B receptor subtype; TRPV1, vanilloid receptor; Vc, subnucleus caudalis; Vi, subnucleus interpolaris; Vo, subnucleus oralis; Vp, principal sensory nucleus.

more widely distributed across laminae I–III in Vc than in spinal dorsal horn where both neuropeptides are restricted to laminae I and IIo. The functional significance of these differences are not certain; however, the high degree of convergence of afferents from multiple sensory ganglion sources (e.g., trigeminal, nodose, cervical dorsal root) to the caudal Vc may underlie the poor localization and spreading of pain for many craniofacial pain conditions.

32.3.5 Efferent Projections

Table 5 summarizes major efferent projection targets of trigeminal neurons in different portions of the

Table 5 Summary of efferent projections from different portions of the trigeminal sensory nuclear complex to thalamic, pontine and medullary targets associated with various aspects of nociception

	Vp		Vo		Vi	dPa5	Vi/Vc	Vc			References
	dm	vl	dm	vl				I–II	III–IV	III–IV	
Thalamus											
VPM	+++	+++	+	+	++	+	+	++	+	++	Shigenaga Y. et al. (1983), Bruce L. L. et al. (1987); Mantle-St. John L. A. and Tracey D. J. (1987)
PO	−	+	−	++	++	−	+	+++	+	+	Dado R. J. and Giesler G. J. (1990), Guy N. et al. (2005)
SM	−	−	−	−	+	−	+++	++	−	−	Craig A. D. and Burton H. (1981), Dado R. J. and Giesler G. J. (1990), Yoshida A. et al. (1991)
Pf	+	−	+	−	+	−	+	−	−	++	Krout K. E. et al. (2002)
Hypothalamus											
VMH	−	−	+	+	+	−	++	+++	+	++	Malick A. and Burstein R. (1998)
LH	+	−	+	−	+	++	++	++	+	++	Malick A. and Burstein R. (1998), Ikeda T. et al. (2003)
Tectum											
APT	−	+	−	+	++	−	++	−	−	−	Yoshida A. et al. (1992)
SC	−	++	−	++	++	−	++	++	+	−	Bruce L. L. et al. (1987), Ndiaye A. et al. (2002)
PAG	−	−	−	−	−	−	−	++	−	+	Beitz A. J. (1982), Mantyh P. W. (1982), Wiberg M et al. (1986), Keay K. A. et al. (1997)
PBA	+	−	++	+	+	+++	++	+++	−	++	Panneton W. M. et al. (1994), Feil K. and Herbert H. (1995), Allen G. V. et al. (1996)
NTS	+	−	+	+	+	+++	+	++	−	−	Menetrey D. and Basbaum A. I. (1987), Zerari-Mailly F. et al. (2005)
ION	−	−	−	+	++	−	+	+	+	−	Huerta M. F. et al. (1985), Van Ham J. J. and Yeo C. H. (1992), Yatim N. et al. (1996)

−, very weak or no staining; +,++,+++, weak, moderate, and dense staining; APT, anterior pretectal nucleus; dPa5, dorsal paratrigeminal region; ION, inferior olivary nucleus; LH, lateral hypothalamic area; NTS, nucleus tractus solitarius; PAG, periaqueductal gray region; PBA, parabrachial region; Pf, medial and lateral parafascicular nuclei; PO, posterior thalamic nucleus; SC, superior colliculus; SM, nucleus submedius of thalamus; Vc, subnucleus caudalis; VMH, ventromedial hypothalamic area; Vi, subnucleus interpolaris; Vo, subnucleus oralis; Vp, principal sensory nucleus; VPM, ventroposteromedial nucleus of thalamus.

TSNC independent of the functional class (i.e., nociceptive versus innocuous sensory encoding). These data derive largely from studies in rodents; however, the qualitative pattern of efferent projections from the TSNC is similar across species, although there are significant quantitative species differences for some targets. Efferent projections from second-order neurons to the sensory thalamus have commanded considerable attention. The main sources of projections from the TSNC to the ventral posterior medial nucleus (VPM) arise from Vp and Vi, which in the case of rodents and most carnivores, is due to a heavy projection from vibrissae-driven rather than nociceptive neurons (Veinante, P. et al., 2000). Nociceptive neurons in Vc that project to VPM originate mainly in laminae I and V (Ikeda, M. et al., 2003) and moreover, the majority of projection neurons are found in rostral Vc rather than near the Vc/C2 junction region (Guy, N. et al., 2005). Lamina I of Vc projects heavily to the posterior thalamus (PO); however, ventrolateral portions of Vo also provide a significant input to PO (Guy, N. et al., 2005). Both Vc and spinal dorsal horn lamina I cells project to similar though adjacent regions of PO (Gauriau, C. and Bernard, J. F., 2004). Projections to parafascicular thalamic nuclei from Vc are sparse compared to spinal dorsal horn (Craig, A. D., 2004; Gauriau, C. and Bernard, J. F., 2004). Trigeminal projections to thalamic nucleus submedius (SM) reveal a unique pattern that differs markedly from spinal cord and exhibits significant species differences. In the rat the majority of SM projections arise from the ventrolateral portion of the Vi/Vc transition region with only weak input from lamina I and V of caudal Vc, while spinal projections to SM originate mainly from laminae V–VII (Dado, R. J. and Giesler, G. J., 1990; Yoshida, A. et al., 1991). By contrast, in cat and monkey lamina I cells in Vc and spinal cord provide a significant direct input to SM with Vc displaying a somewhat more extensive projection (Craig A. D. and Burton, H., 1981). Trigeminal projections to autonomic control regions such as hypothalamus, parabrachial area, and NTS derive mainly from the dPa5, the Vi/Vc transition and lamina I of Vc, while more rostral regions of the TSNC provide relatively sparse input. This pattern is consistent with the dPa5 and lamina I of Vc receiving direct input from the vagus nerve. Spinal lamina I cells also project heavily to similar autonomic control regions of the brainstem (Cechetto, D. F. et al., 1985; Menetrey, D. and Basbaum, A. I., 1987; Westlund, K. N. and Craig, A. D., 1996). Although nociceptive neurons in Vc

(Sessle, B. J. et al., 1981; Chiang, C. Y. et al., 1994) and Vo (Chiang, C. Y. et al., 1989) are markedly inhibited by direct stimulation of PAG or rostral ventromedial medulla (RVM), the afferent pathways from second-order TSNC neurons to these endogenous pain control regions are not well defined. Compared to the significant input from upper cervical levels of spinal cord (Keay, K. A. et al., 1997) projections from TSNC to PAG are sparse (Beitz, A. J., 1982). These results add substantial support to the notion that laminae I–II of Vc are critical regions for processing nociceptive information relevant for multiple aspects of craniofacial pain. Behavioral evidence, though less extensively tested compared to spinal pain models, indicates that the Vc is necessary for opioid modulation of cutaneous facial pain (Oliveras, J. L. et al., 1986) and that levels of attention markedly influence Vc nociceptive neurons and behavioral responsiveness (Hayes, R. L. et al., 1981).

32.4 Functional Considerations

Recent developments in methods that assess neural activity and encoding properties provide the strongest evidence regarding the functional role of different regions of the TSNC in craniofacial pain. Advances in neuroimaging can distinguish somatotopic and simultaneous activation of brainstem and cortical responses to trigeminal stimuli in conscious humans (DaSilva, A. F. et al., 2002), though resolution is not yet sufficient to discern the relative activation of different portions of the TSNC. Nociceptive neurons in the TSNC have been identified and their properties determined mainly on the basis of electrophysiological recording and, more recently, immediate early gene expression such as c-fos. Immunostaining for Fos, the protein product of c-fos, is a reliable method to identify populations of nociceptive central neurons at the single cell level (see Bullitt, E., 1990), an advantage not readily achieved by electrophysiology. Although there are examples of mismatches, properly designed c-fos studies generally complement electrophysiological results and have shed new light on long-standing controversies in trigeminal physiology. For example, a role for Vo in dental pain is suggested by: a dense terminal pattern in dorsomedial Vo from tooth pulp afferents (Marfurt, C. F. and Turner, D. F., 1984), a moderate-to-high density of CGRP staining (Sugimoto, T. et al., 1997b), and behavioral studies revealing preservation of dental pain after trigeminal tractotomy at the level of caudal Vi (Young, R. F. and Perryman, K. M., 1984). By contrast, few Vo neurons can be driven by natural stimulation of the tooth pulp compared to Vc (Hu, J. W. and Sessle, B. J., 1984) and few Fos-positive neurons are found in Vo after acute thermal stimulation of teeth (Chattipakorn, S. C. *et al.*, 1999). Thus, despite the fact that many Vo cells can be classified as nociceptive on the basis of cutaneous RF properties (Dallel, R. *et al.*, 1990), display wind-up to repeated cutaneous stimulation (Dallel, R. *et al.*, 1999), and are inhibited by systemic morphine (Dallel, R. *et al.*, 1996), the role of Vo in acute dental pain remains uncertain. It is possible that the Vo acts as a silent pain relay in the TSNC and becomes active only after persistent tissue damage since Fos-positive cells first appear in Vo only several days after molar tooth pulp exposure (Byers, M. R. *et al.*, 2000) and Vo neurons display sensitization provided input from Vc remains intact (Chiang, C. Y. *et al.*, 2002; Hu, B. *et al.*, 2002). Alternatively, rostral TSNC contributions to intraoral sensation and homeostasis may involve nonpulpal tissues (e.g., periodontal, muscosal receptors) and mediate select aspects of craniofacial pain (e.g., somatomotor reflexes).

32.4.1   Ocular Pain Processing

Ascribing a role for different portions of the TSNC in ocular pain appears more straightforward than for dental pain.

Corneal afferents terminate mainly in ventrolateral portions of caudal Vi and Vc with few fibers projecting to more rostral regions (Panneton, W. M. and Burton, H., 1981; Marfurt, C. F. and del Toro, D. R., 1987; Marfurt, C. F. and Echtenkamp, S. F., 1988). Acute stimulation of the ocular surface in the rat evokes a high density of Fos-positive cells at the Vi/Vc transition and caudal Vc and none in rostral regions of TSNC (Lu, J. *et al.*, 1993; Strassman, A. M. and Vos, B. P., 1993; Bereiter, D. A. *et al.*, 1994; Meng, I. D. and Bereiter, D. A., 1996). Converging lines of evidence support the notion that the Vi/Vc transition and caudal Vc serve different aspects of ocular pain.

All ocular cells in laminae I–II at the Vc/C2 junction are classified as nociceptive (wide-dynamic range (WDR), nociceptive specific (NS)), while many cells at the Vi/Vc transition have no cutaneous RF (Meng, I. D. *et al.*, 1997; Hirata, H. *et al.*, 1999). Many cells at the Vi/Vc transition are sensitive to the moisture status of the ocular surface, while few such neurons are found in caudal Vc, suggesting that this

region is critical for reflex lacrimation (Hirata, H. *et al.*, 2004).

Repeated ocular surface stimulation evokes a wind-up like response among caudal Vc units, while Vi/Vc cells rapidly become desensitized (Meng, I. D. *et al.*, 1997). In a model for endotoxin-induced uveitis, at 7 days after ocular inflammation convergent cutaneous RF areas become enlarged and responsiveness to ocular surface stimulation is enhanced among caudal Vc neurons, while Vi/Vc transition cells display no evidence of hyperalgesia (Bereiter, D. A. *et al.*, 2005c).

Systemic morphine inhibits all ocular cells at the caudal Vc, while nearly 30% of Vi/Vc cells are enhanced; an effect that can be produced by microinjection of μ-opioid receptor agonists directly into the caudal Vc (Meng, I. D. *et al.*, 1998; Hirata, H. *et al.*, 2000).

The modality of ocular units at the Vi/Vc and caudal Vc predicts, in part, the efferent projections to PO or salivatory nucleus in the brainstem (Hirata, H. *et al.*, 2000). The Vi/Vc transition is unique among TSNC regions and is the main source of ascending projections to SM (Yoshida, A. *et al.*, 1991; Ikeda, M. *et al.*, 2003). These data suggest that the caudal Vc underlies the sensory-discriminative aspects of ocular pain and modulation of ocular cells in more rostral regions via intersubnuclear connections. By contrast, the Vi/Vc transition appears to play a significant role in mediating ocular-specific reflexes (e.g., lacrimation, eye blink). Projections to SM, coupled with the finding that many Vi/Vc neurons display enhanced responsiveness after morphine, suggests that this region may be part of the neural circuit that recruits endogenous pain controls in response to craniofacial tissue injury. Since a high percentage of ocular cells at each region also respond to meningeal stimulation (Strassman, A. M. *et al.*, 1994; Burstein, R. *et al.*, 1998; Schepelmann, K. *et al.*, 1999), it is proposed that rostral and caudal portions of Vc mediate different aspects of headache as well as ocular pain.

32.5   Chronic Craniofacial Pain

Chronic pain involving craniofacial tissues is a significant public health concern and a recognized research priority for the National Institutes of Health (NIH; e.g., PA 03-173: *Neurobiology of Persistent Pain Mediated by the Trigeminal Nerve*). The classification, diagnosis, and management of chronic craniofacial pain remain difficult since the mechanisms for many of these conditions are not well understood and animal models, though instructive, do not mimic the clinical state (Vos, B. P. *et al.*, 1994; Roveroni, R. C. *et al.*, 2001). Considerable progress has been in delineating the long-term changes in peripheral and central neural circuits that mediate pain sensation following tissue injury (Treede, R. D. *et al.*, 1992; Woolf, C. J. and Salter, M. W., 2000; Hunt, S. P. and Mantyh, P. W., 2001; Julius, D. and Basbaum, A. I., 2001; Lewin, G. R. *et al.*, 2004). Although similar cellular and molecular mechanisms may contribute to chronic pain due to Vn damage (Lavigne, G. *et al.*, 2005), many chronic craniofacial pain conditions such as TMD, migraine, or chronic daily headache and trigeminal neuralgia present with no overt signs of tissue injury. Indeed, it has long been appreciated that the correlation between tissue injury and magnitude of pain sensation may be weak (Wall, P. D., 1979). This has lead to proposals of emotional or neuropsychological (Tenenbaum, H. C. *et al.*, 2001; Korszun, A., 2002) and genetic factors as significant determinants of some forms of chronic craniofacial pain (TMD, Diatchenko, L. *et al.*, 2005; migraine, Wessman, M. *et al.*, 2004; trigeminal neuralgia, Duff, J. M., *et al.*, 1999; dry eye in Sjogren's syndrome, Takei, M. *et al.*, 2005). Chronic craniofacial pain can be broadly classified according to the pattern and origin of pain episodes: chronic/recurrent (TMD, migraine headache, trigeminal neuralgia), chronic/persistent (burning mouth, dry eye syndromes), and chronic deafferentation pain (postherpetic neuralgia, posttraumatic neuralgia, phantom tooth). The mechanisms that underlie these diverse conditions involve markedly different tissues and are likely quite heterogeneous. However, three features of many chronic craniofacial pain conditions are notable and provide evidence of commonality.

First, the prevalence of most chronic craniofacial pain conditions is higher in women than men (Dao, T. T. and LeResche, L., 2000; Macfarlane, T. V. *et al.*, 2002a). This is especially apparent for the chronic/recurrent conditions of TMD and migraine headache (LeResche, L., 1997; Rasmussen, B. K., 2001) and somewhat less so for trigeminal neuralgia (Kitt, C. A. *et al.*, 2000; Manzoni, G. C. and Torelli, P., 2005). Women also are more likely to develop burning mouth (Bergdahl, M. and Bergdahl, J., 1999; Grushka, M. *et al.*, 2003) and dry eye syndromes (Yazdani, C. *et al.*, 2001) and report greater sensory disturbances after Vn damage (Sandstedt, P. and

Sorensen, S., 1995). Among TMD (Isselee, H. *et al.*, 2002; LeResche, L. *et al.*, 2003) and migraine patients (Rasmussen, B. K., 1993; MacGregor, E. A. and Hackshaw, A., 2004) pain severity and symptoms vary over the menstrual cycle suggesting a significant interaction with factors related to sex hormone status. Correspondingly, recent anatomical (Amandusson, A. *et al.*, 1996; Pajot, J. *et al.*, 2003; Bereiter, D. A. *et al.*, 2005a; 2005b; Puri, V. *et al.*, 2005) and electrophysiological evidence (Okamoto, K. *et al.*, 2003; Flake, N. M. *et al.*, 2005) from animal models support the notion that estrogen preferentially enhances the excitability of trigeminal neurons that contribute to craniofacial pain. Furthermore, within the TSNC, estrogen receptor-positive neurons are found almost exclusively within the superficial laminae of Vc and not in more rostral portions of the complex (Bereiter, D. A. *et al.*, 2005a) suggesting that this region plays a key role in differential processing of orofacial sensory information under different sex hormone conditions.

Second, spreading and referral of pain and sensitization evoked from outside the affected dermatomal region are common features of many chronic craniofacial pain conditions. Sensory disturbances occurring outside the affected region have been well documented in clinical studies of TMD (Maixner, W. *et al.*, 1998; Sarlani, E. and Greenspan, J. D., 2003; 2005), migraine (Burstein, R. *et al.*, 2000; Katsarava, Z., *et al.*, 2002; Goadsby, P. J., 2005), trigeminal neuralgia (Dubner, R. *et al.*, 1987; Nurmikko, T. J. and Eldridge, P. R., 2001; Devor, M. *et al.*, 2002), and burning mouth syndrome (Svensson, P. *et al.*, 1993; Ito, M. *et al.*, 2002). Patients with postherpetic neuralgia involving trigeminal dermatomes had lower thermal warm and cool thresholds, while those with infection of spinal dermatomes had elevated thresholds (Pappagallo, M. *et al.*, 2000) suggesting different underlying pathologies for neuropathic pain in trigeminal and spinal systems. Mechanical allodynia and increased temporal summation are consistent with the notion that central neural mechanisms maintain chronic craniofacial pain, while peripheral mechanisms are required mainly for initiation of the pain state.

Third, many chronic craniofacial pain conditions are accompanied by significant disturbances of the autonomic and/or endocrine systems. Chronic TMD patients display altered secretion of stress hormones (Jones, D. A. *et al.*, 1997; Korszun, A. *et al.*, 2002) and elevated levels of neuropeptides and proinflammatory cytokines that could affect blood flow to joints (Kopp, S., 2001). The relationship between vascular reactivity and migraine has long been considered a critical variable (Janig, W., 2003; Goadsby, P. J., 2005), whereas stimulus-evoked oral mucosal blood flow is greater in patients with burning mouth syndrome (Heckmann, S. M. *et al.*, 2001). Evidence from animal models indicate that contributions from vagus nerve activity (Khasar, S. G. *et al.*, 2001) and adrenal medullary outflow (Green, P. G. *et al.*, 2001) have marked sex-related effects on cutaneous pain behavior. Although treatments to reduce sympathetically maintained pain generally have a poor outcome for posttraumatic trigeminal neuralgic patients (Gregg, J. M., 1990), the influence of the autonomic nervous system on other forms chronic craniofacial pain has not been adequately explored. It may be significant that brainstem regions critically involved in control of autonomic outflow also are densely stained for estrogen receptor-positive neurons (Shughrue, P. J. *et al.*, 1997; Simonian, S. X. *et al.*, 1998; Merchenthaler, I. *et al.*, 2004). Collectively, these features underscore the hypothesis that an interaction between sex hormone status and autonomic outflow occurs within the central nervous system to alter the expression of chronic craniofacial facial pain.

We recognize that chronic craniofacial pain is a complex problem with varying etiologies and possible contributions from genetic, neuropsychological, and neurobiological factors. However, regardless of the origin and relative contribution of these factors on pain circuits within the brain, a greater understanding of the unique organizational features of the trigeminal system may provide new perspectives and strategies to manage chronic craniofacial pain that would not otherwise be apparent from studies conducted only at the spinal level.

References

Aigner, M., Robert, Lukas J., Denk, M., Ziya-Ghazvini, F., Kaider, A., and Mayr, R. 2000. Somatotopic organization of primary afferent perikarya of the guinea-pig extraocular muscles in the trigeminal ganglion: a post-mortem DiI-tracing study. Exp. Eye Res. 70, 411–418.

Aimi, Y., Fujimura, M., Vincent, S. R., and Kimura, H. 1991. Localization of NADPH-diaphorase-containing neurons in sensory ganglia of the rat. J. Comp. Neurol. 306, 382–392.

Akazawa, C., Nakamura, Y., Sango, K., Horie, H., and Kohsaka, S. 2004. Distribution of the galectin-1 mRNA in the rat nervous system: its transient upregulation in rat facial motor neurons after facial nerve axotomy. Neuroscience 125, 171–178.

Albers, K. M., Perrone, T. N., Goodness, T. P., Jones, M. E., Green, M. A., and Davis, B. M. 1996. Cutaneous overexpression of NT-3 increases sensory and sympathetic neuron number and enhances touch dome and hair follicle innervation. J. Cell Biol. 134, 487–497.

Allen, G. V., Barbrick, B., and Esser, M. J. 1996. Trigeminal-parabrachial connections: possible pathway for nociception-induced cardiovascular reflex responses. Brain Res. 715, 125–135.

Altschuler, S. M., Bao, X., Bieger, D., Hopkins, D. A., and Miselis, R. R. 1989. Viscerotopic representation of the upper alimentary tract in the rat: sensory ganglia and nuclei of the solitary and spinal trigeminal tracts. J. Comp. Neurol. 283, 248–268.

Amandusson, A., Hermanson, O., and Blomqvist, A. 1996. Colocalization of oestrogen receptor immunoreactivity and preproenkephalin mRNA expression to neurons in the superficial laminae of the spinal and medullary dorsal horn of rats. Eur. J. Neurosci. 8, 2440–2445.

Ambalavanar, R. and Morris, R. 1992. The distribution of binding by isolectin I-B4 from *Griffonia simplicifolia* in the trigeminal ganglion and brainstem trigeminal nuclei in the rat. Neuroscience 47, 421–429.

Ambalavanar, R., Moritani, M., and Dessem, D. 2005. Trigeminal P2X(3) receptor expression differs from dorsal root ganglion and is modulated by deep tissue inflammation. Pain 117, 280–291.

Anggard, A., Lundberg, J. M., and Lundblad, L. 1983. Nasal autonomic innervation with special reference to peptidergic nerves. Eur. J. Respir. Dis. Suppl. 128, 143–149.

Anton, F. and Peppel, P. 1991. Central projections of trigeminal primary afferents innervating the nasal mucosa: a horseradish peroxidase study in the rat. Neuroscience 41, 617–628.

Arbab, M. A., Delgado, T., Wiklund, L., and Svendgaard, N. A. 1988. Brain stem terminations of the trigeminal and upper spinal ganglia innervation of the cerebrovascular system: WGA-HRP transganglionic study. J. Cereb. Blood Flow Metab. 8, 54–63.

Artinger, K. B., Fedtsova, N., Rhee, J. M., Bronner-Fraser, M., and Turner, E. 1998. Placodal origin of Brn-3-expressing cranial sensory neurons. J. Neurobiol. 36, 572–585.

Arvidson, J. 1982. Somatotopic organization of vibrissae afferents in the trigeminal sensory nuclei of the rat studied by transganglionic transport of HRP. J. Comp. Neurol. 211, 84–92.

Arvidson, B. 1983. Influence of age on the development of cadmium-induced vascular lesions in rat sensory ganglia. Environ. Res. 32, 240–246.

Arvidsson, J. and Raappana, P. 1989. An HRP study of the central projections from primary sensory neurons innervating the rat masseter muscle. Brain Res. 480, 111–118.

Arvidsson, J., Pfaller, K., and Gmeiner, S. 1992. The ganglionic origins and central projections of primary sensory neurons innervating the upper and lower lips in the rat. Somatosensory Motor Res. 9, 199–209.

Bae, Y. C., Oh, J. M., Hwang, S. J., Shigenaga, Y., and Valtschanoff, J. G. 2004. Expression of vanilloid receptor TRPV1 in the rat trigeminal sensory nuclei. J. Comp. Neurol. 478, 62–71.

Baker, C. V., Stark, M. R., and Bronner-Fraser, M. 2002. Pax3-expressing trigeminal placode cells can localize to trunk neural crest sites but are committed to a cutaneous sensory neuron fate. Dev. Biol. 249, 219–236.

Barker, D. J. and Welker, W. I. 1969. Receptive fields of first-order somatic sensory neurons innervating rhinarium in coati and raccoon. Brain Res. 14, 367–386.

Basbaum, A. I. 1999. Distinct neurochemical features of acute and persistent pain. Proc. Natl. Acad. Sci. U. S. A. 96, 7739–7743.

Beckstead, R. M. and Norgren, R. 1979. An autoradiographic examination of the central distribution of the trigeminal, facial, glossopharyngeal, and vagal nerves in the monkey. J. Comp. Neurol. 184, 455–472.

Beitel, R. E. and Dubner, R. 1976. Response of unmyelinated (C) polymodal nociceptors to thermal stimuli applied to monkey's face. J. Neurophysiol. 39, 1160–1175.

Beitz, A. J. 1982. The organization of afferent projections to the midbrain periaqueductal gray of the rat. Neuroscience 7, 133–159.

Belmonte, C., Garcia-Hirschfeld, J., and Gallar, J. 1997. Neurobiology of ocular pain. Prog. Retin. Eye Res. 16, 117–156.

Benoliel, R., Eliav, E., and Tal, M. 2001. No sympathetic nerve sprouting in rat trigeminal ganglion following painful and non-painful infraorbital nerve neuropathy. Neurosci. Lett. 297, 151–154.

Bereiter, D. A. and Bereiter, D. F. 2000. Morphine and NMDA receptor antagonism reduce c-fos expression in spinal trigeminal nucleus produced by acute injury to the TMJ region. Pain 85, 65–77.

Bereiter, D. A., Cioffi, J. L., and Bereiter, D. F. 2005a. Oestrogen receptor-immunoreactive neurons in the trigeminal sensory system of male and cycling female rats. Arch. Oral. Biol. 50, 971–979.

Bereiter, D. A., Hathaway, C. B., and Benetti, A. P. 1994. Caudal portions of the spinal trigeminal complex are necessary for autonomic responses and display Fos-like immunoreactivity after corneal stimulation in the cat. Brain Res. 657, 73–82.

Bereiter, D. A., Hirata, H., and Hu, J. W. 2000. Trigeminal subnucleus caudalis: beyond homologies with the spinal dorsal horn. Pain 88, 221–224.

Bereiter, D. A., Okamoto, K., and Bereiter, D. F. 2005b. Effect of persistent monoarthritis of the temporomandibular joint region on acute mustard oil-induced excitation of trigeminal subnucleus caudalis neurons in male and female rats. Pain 117, 58–67.

Bereiter, D. A., Okamoto, K., Tashiro, A., and Hirata, H. 2005c. Endotoxin-induced uveitis causes long-term changes in trigeminal subnucleus caudalis neurons. J. Neurophysiol. 94, 3815–3825.

Bergdahl, M. and Bergdahl, J. 1999. Burning mouth syndrome: prevalence and associated factors. J. Oral Pathol. Med. 28, 350–354.

Beuerman, R. W. and Tanelian, D. L. 1979. Corneal pain evoked by thermal stimulation. Pain 7, I–I4.

Bleys, R. L., Groen, G. J., and Hommersom, R. F. 1996. Neural connections in and around the cavernous sinus in rat, with special reference to cerebrovascular innervation. J. Comp. Neurol. 369, 277–291.

Bohotin, C., Scholsem, M., Multon, S., Martin, D., Bohotin, V., and Schoenen, J. 2003. Vagus nerve stimulation in awake rats reduces formalin-induced nociceptive behaviour and fos-immunoreactivity in trigeminal nucleus caudalis. Pain 101, 3–12.

Boissonade, F. M., Sharkey, K. A., and Lucier, G. E. 1993. Trigeminal nuclear complex of the ferret: anatomical and immunohistochemical studies. J. Comp. Neurol. 329: 291–312.

Bongenhielm, U., Boissonade, F. M., Westermark, A., Robinson, P. P., and Fried, K. 1999. Sympathetic nerve sprouting fails to occur in the trigeminal ganglion after peripheral nerve injury in the rat. Pain 82, 283–288.

Bongenhielm, U., Nosrat, C. A., Nosrat, I., Eriksson, J., Fjell, J., and Fried, K. 2000. Expression of sodium channel SNS/PN3 and ankyrin(G) mRNAs in the trigeminal ganglion after inferior alveolar nerve injury in the rat. Exp. Neurol. 164, 384–395.

Bongenhielm, U., Yates, J. M., Fried, K., and Robinson, P. P. 1998. Sympathectomy does not affect the early ectopic discharge from myelinated fibres in ferret inferior alveolar nerve neuromas. Neurosci. Lett. 245, 89–92.

Bossut, D. F., Whitsel, E. A., and Maixner, W. 1992. A parametric analysis of the effects of cardiopulmonary vagal electrostimulation on the digastric reflex in cats. Brain Res. 579, 253–260.

Brannstrom, M. 1986. The hydrodynamic theory of dentinal pain: sensation in preparations, caries, and the dentinal crack syndrome. J. Endod. 12, 453–457.

Bruce, L. L., McHaffie, J. G., and Stein, B. E. 1987. The organization of trigeminotectal and trigeminothalamic neurons in rodents: a double-labeling study with fluorescent dyes. J. Comp. Neurol. 262, 315–330.

Bruenech, J. R. and Ruskell, G. L. 2001. Muscle spindles in extraocular muscles of human infants. Cells Tissues Organs 169, 388–394.

Bryant, B. and Silver, W. L. 2000. Chemesthesis: the Common Chemical Sense. In: Neurobiology of Taste and Smell, 2nd edn (eds. T. E. Finger, W. L. Silver, and D. Restrepo), pp. 73–100. Wiley-Liss.

Bullitt, E. 1990. Expression of c-fos-like protein as a marker for neuronal activity following noxious stimulation in the rat. J. Comp. Neurol. 296, 517–530.

Burstein, R., Yamamura, H., Malick, A., and Strassman, A. M. 1998. Chemical stimulation of the intracranial dura induces enhanced responses to facial stimulation in brain stem trigeminal neurons. J. Neurophysiol. 79, 964–982.

Burstein, R., Yarnitsky, D., Goor-Aryeh, I., Ransil, B. J., and Bajwa, Z. H. 2000. An association between migraine and cutaneous allodynia. Ann. Neurol. 47, 614–624.

Butler, P. J. and Jones, D. R. 1997. Physiology of diving of birds and mammals. Physiol. Rev. 77, 837–899.

Buzas, B. and Cox, B. M. 1997. Quantitative analysis of mu and delta opioid receptor gene expression in rat brain and peripheral ganglia using competitive polymerase chain reaction. Neuroscience 76, 479–489.

Byers, M. R. 1984. Dental sensory receptors. Int. Rev. Neurobiol. 25, 39–94.

Byers, M. R., Chudler, E. H., and Iadorola, M. J. 2000. Chronic tooth pulp inflammation causes transient and persistent expression of Fos in dynorphin-rich regions of rat brainstem. Brain Res. 861, 191–207.

Caous, C. A., de Sousa Buck, H., and Lindsey, C. J. 2001. Neuronal connections of the paratrigeminal nucleus: a topographic analysis of neurons projecting to bulbar, pontine and thalamic nuclei related to cardiovascular, respiratory and sensory functions. Autonomic Neurosci. 94, 14–24.

Capra, N. F. 1987. Localization and central projections of primary afferent neurons that innervate the temporomandibular joint in cats. Somatosensory Motor Res. 4, 201–213.

Catania, K. C. 1999. A nose that looks like a hand and acts like an eye: the unusual mechanosensory system of the star-nosed mole. J. Comp. Physiol. A 185, 367–372.

Catania, K. C. and Kaas, J. H. 1997. Somatosensory fovea in the star-nosed mole: behavioral use of the star in relation to innervation patterns and cortical representation. J. Comp. Neurol. 387, 215–233.

Catania, K. C. and Remple, M. S. 2002. Somatosensory cortex dominated by the representation of teeth in the naked mole-rat brain. Proc. Natl. Acad. Sci. U. S. A. 99, 5692–5697.

Cechetto, D. F., Standaert, D. S., and Saper, C. B. 1985. Spinal and trigeminal dorsal horn projections to the parabrachial nucleus in the rat. J. Comp. Neurol. 240, 153–162.

Chalazonitis, A., Kessler, J. A., Twardzik, D. R., and Morrison, R. S. 1992. Transforming growth factor alpha, but not epidermal growth factor, promotes the survival of sensory neurons in vitro. J. Neurosci. 12, 583–594.

Chandler, M. J., Zhang, J., Qin, C., and Foreman, R. D. 2002. Spinal inhibitory effects of cardiovascular inputs in monkeys:

neuronal processing in high cervical segments. J. Neurophysiol. 87, 1290–1302.

Chattipakorn, S. C., Light, A. R., Willcockson, H. H., Nahri, M., and Maixner, W. 1999. The effect of fentanyl on c-fos expression in the trigeminal brainstem complex produced by pulpal heat stimulation in the ferret. Pain 82, 207–215.

Cheng, S. B., Kuchiiwa, S., Kuchiiwa, T., and Nakagawa, S. 2000. Three novel pathways to the lacrimal glands of the cat: an investigation with cholera toxin B subunit as a retrograde tracer. Brain Res. 873, 160–164.

Chiang, C. Y., Hu, B., Hu, J. W., Dostrovsky, J. O., and Sessle, B. J. 2002. Central sensitization of nociceptive neurons in trigeminal subnucleus orlais depends on the integrity of subnucleus caudalis. J. Neurophysiol. 88, 256–264.

Chiang, C. Y., Hu, J. W., Dostrovsky, J. O., and Sessle, B. J. 1989. Changes in mechanoreceptive field properties of somatosensory brainstem neurons induced by stimulation of nucleus raphe magnus in cats. Brain Res. 485, 371–381.

Chiang, C. Y., Hu, J. W., and Sessle, B. J. 1994. Parabrachial area and nucleus raphe magnus-induced modulation of nociceptive and nonnociceptive trigeminal subnucleus caudalis neurons activated by cutaneous or deep inputs. J. Neurophysiol. 71, 2430–2445.

Cody, F. W., Lee, R. W., and Taylor, A. 1972. A functional analysis of the components of the mesencephalic nucleus of the fifth nerve in the cat. J. Physiol. 226, 249–261.

Collo, G., North, R. A., Kawashima, E., Merlo-Pich, E., Neidhart, S., Surprenant, A., and Buell, G. 1996. Cloning of P2X5 and P2X6 receptors and the distribution and properties of an extended family of ATP-gated ion channels. J. Neurosci. 16, 2495–2507.

Cometto-Muniz, J. E. and Cain, W. S. 1995. Relative sensitivity of the ocular trigeminal, nasal trigeminal and olfactory systems to airborne chemicals. Chem. Senses 20, 191–198.

Connor, J. M., Lauterborn, J. C., Yan, Q., Gall, C. M., and Varon, S. 1997. Distribution of brain-derived neurotrophic factor (BDNF) protein and mRNA in the normal adult rat CNS: evidence for anterograde axonal transport. J. Neurosci. 17, 2295–2313.

Contreras, R. J., Beckstead, R. M., and Norgren, R. 1982. The central projections of the trigeminal, facial, glossopharyngeal and vagus nerves: an autoradiographic study in the rat. J. Autonomic Nerv. Syst. 6, 303–322.

Cook, S. P., Vulchanova, L., Hargreaves, K. M., Elde, R., and McCleskey, E. W. 1997. Distinct ATP receptors on pain-sensing and stretch-sensing neurons. Nature 387, 505–508.

Craig, A. D. 2004. Distribution of trigeminothalamic and spinothalamic lamina I terminations in the macaque monkey. J. Comp. Neurol. 477, 119–148.

Craig, A. D. and Burton, H. 1981. Spinal and medullary lamina I projection to nucleus submedius in medial thalamus: a possible pain center. J. Neurophysiol. 45, 443–466.

Dado, R. J. and Giesler, G. J. 1990. Afferent input to nucleus submedius in rats: retrograde labeling of neurons in the spinal cord and caudal medulla. J. Neurosci. 10, 2672–2686.

Dallel, R., Duale, C., Luccarini, P., and Molat, J. L. 1999. Stimulus-function, wind-up and modulation by diffuse noxious inhibitory controls of responses of convergent neurons of the spinal trigeminal nucleus oralis. Eur. J. Neurosci. 11, 31–40.

Dallel, R., Luccarini, P., Molat, J. L., and Woda, A. 1996. Effects of systemic morphine on the activity of convergent neurons of spinal trigeminal nucleus oralis in the rat. Eur. J. Pharmacol. 314, 19–25.

Dallel, R., Raboisson, P., Woda, A., and Sessle, B. J. 1990. Properties of nociceptive and non-nociceptive neurons in trigeminal subnucleus oralis of the rat. Brain Res. 521, 95–106.

Dao, T. T. and LeResche, L. 2000. Gender differences in pain. J. Orofacial Pain 14, 169–184.

DaSilva, A. F., Becerra, L., Makris, N., Strassman, A. M., Gonzalez, R. G., Geatrakis, N., and Borsook, D. 2002. Somatotopic activation in the human trigeminal pain pathway. J. Neurosci. 22, 8183–8192.

Davis, K. D. and Dostrovsky, J. O. 1988. Effect of trigeminal subnucleus caudalis cold block on the cerebrovascular-evoked responses of rostral trigeminal complex neurons. Neurosci. Lett. 94, 303–308.

Davis, B. M., Albers, K. M., Seroogy, K. B., and Katz, D. M. 1994. Overexpression of nerve growth factor in transgenic mice induces novel sympathetic projections to primary sensory neurons. J. Comp. Neurol. 349, 464–474.

DeLeon, M., Covenas, R., Chadi, G., Narvaez, J. A., Fuxe, K., and Cintra, A. 1994. Subpopulations of primary sensory neurons show coexistence of neuropeptides and glucocorticoid receptors in the rat spinal and trigeminal ganglia. Brain Res. 636, 338–342.

Denny-Brown, D. and Yanagisawa, N. 1973. The function of the descending root of the fifth nerve. Brain 96, 783–814.

Devor, M., Amir, R., and Rappaport, Z. H. 2002. Pathophysiology of trigeminal neuralgia: the ignition hypothesis. Clin. J. Pain 18, 4–13.

Diatchenko, L., Slade, G. D., Nackley, A. G., Bhalang, K., Sigurdsson, A., Belfer, I., Goldman, D., Shabalina, S. A., Shagin, D., Max, M. B., Makarov, S. S., and Maixner, W. 2005. Genetic basis for individual variations in pain perception and the development of a chronic pain condition. Human Mol. Genet. 14, 135–143.

Dib-Hajj, S., Black, J. A., Felts, P., and Waxman, S. G. 1996. Down-regulation of transcripts for Na channel alpha-SNS in spinal sensory neurons following axotomy. Proc. Natl. Acad. Sci. U. S. A. 93, 14950–14954.

Ding, Y. Q., Kaneko, T., Nomura, S., and Mizuno, N. 1996. Immunohistochemical localization of μ-opioid receptors in the central nervous system of the rat. J. Comp. Neurol. 367, 375–402.

Dohrn, C. S., Mullett, M. A., Price, R. H., and Beitz, A. J. 1994. Distribution of nitric oxide synthase-immunoreactive interneurons in the spinal trigeminal nucleus. J. Comp. Neurol. 346, 449–460.

Drummond, P. D. 1992. The mechanism of facial sweating and cutaneous vascular responses to painful stimulation of the eye. Brain 115, 1417–1428.

Dubner, R. and Bennett, G. J. 1983. Spinal and trigeminal mechanisms of nociception. Annu. Rev. Neurosci. 6, 38I–418.

Dubner, R., Sharav, Y., Gracely, R. H., and Price, D. D. 1987. Idiopathic trigeminal neuralgia: sensory features and pain mechanisms. Pain 31, 23–33.

Duff, J. M., Spinner, R. J., Lindor, N. M., Dodick, D. W., and Atkinson, J. L. 1999. Familial trigeminal neuralgia and contralateral hemifacial spasm. Neurology 53, 216–218.

Edvinsson, L. and Uddman, R. 2005. Neurobiology in primary headaches. Brain Res. Brain Res. Rev 48, 438–456.

Elcock, C., Boissonade, F. M., and Robinson, P. P. 2001. Changes in neuropeptide expression in the trigeminal ganglion following inferior alveolar nerve section in the ferret. Neuroscience 102, 655–667.

Elkabes, S., Dreyfus, C. F., Schaar, D. G., and Black, I. B. 1994. Embryonic sensory development: local expression of neurotrophin-3 and target expression of nerve growth factor. J. Comp. Neurol. 341, 204–213.

Eriksson, J., Bongenhielm, U., Kidd, E., Matthews, B., and Fried, K. 1998. Distribution of P2X3 receptors in the rat trigeminal ganglion after inferior alveolar nerve injury. Neurosci. Lett. 254, 37–40.

Esteve, P. and Bovolenta, P. 1999. cSix4, a member of the six gene family of transcription factors, is expressed during placode and somite development. Mech. Dev. 85, 161–165.

Feil, K. and Herbert, H. 1995. Topographic organization of spinal and trigeminal somatosensory pathways to the rat parabrachial and Kolliker-Fuse nuclei. J. Comp. Neurol. 353, 506–528.

Feindel, W., Penfield, W., and McNaughton, F. 1960. The tentorial nerves and localization of intracranial pain in man. Neurology 10, 555–563.

Flake, N. M., Bonebreak, D. B., and Gold, M. S. 2005. Estrogen and inflammation increase the excitability of rat temporomandibular joint afferent neurons. J. Neurophysiol. 93, 1585–1597.

Fritschy, J. M. and Mohler, H. 1995. GABAA-receptor heterogeneity in the adult rat brain: differential regional and cellular distribution of seven major subunits. J. Comp. Neurol. 359, 154–194.

Gauriau, C. and Bernard, J. F. 2004. A comparative reappraisal of projections from the superficial laminae of the dorsal horn in the rat: the forebrain. J. Comp. Neurol. 468, 24–56.

Georgiades, P., Wood, J., and Brickell, P. M. 1998. Retinoid X receptor-gamma gene expression is developmentally regulated in the embryonic rodent peripheral nervous system. Anat. Embryol. (Berl.) 197, 477–484.

Ghilardi, J. R., Allen, C. J., Vigna, S. R., McVey, D. C., and Mantyh, P. W. 1992. Trigeminal and dorsal root ganglion neurons express CCK receptor binding sites in the rat, rabbit, and monkey: possible site of opiate-CCK analgesic interactions. J. Neurosci. 12, 4854–4866.

Ginestal, E. and Matute, C. 1993. Gamma-aminobutyric acid-immunoreactive neurons in the rat trigeminal nuclei. Histochemistry 99, 49–55.

Goadsby, P. J. 2005. Trigeminal autonomic cephalgias: fancy term or constructive change to the IHS classification? J. Neurol. Neurosurg. Psychiatry 76, 301–305.

Gobel, S. and Purvis, M. B. 1972. Anatomical studies of the organization of the spinal V nucleus: the deep bundles and the spinal V tract. Brain Res. 48, 27–44.

Goedert, M., Otten, U., Hunt, S. P., Bond, A., Chapman, D., Schlumpf, M., and Lichtensteiger, W. 1984. Biochemical and anatomical effects of antibodies against nerve growth factor on developing rat sensory ganglia. Proc. Natl. Acad. Sci. U. S. A. 81, 1580–1584.

Goksor, E., Rosengren, L., and Wennergren, G. 2002. Bradycardic response during submersion in infant swimming. Acta Paediatr. 91, 307–312.

Gonzalez-Martinez, T., Germana, G. P., Monjil, D. F., Silos-Santiago, I., de Carlos, F., Germana, G., Cobo, J., and Vega, J. A. 2004. Absence of Meissner corpuscles in the digital pads of mice lacking functional TrkB. Brain Res. 1002, 120–128.

Green, P. G., Dahlqvist, S. R., Isenberg, W. M., Mao, F. J., and Levine, J. D. 2001. Role of adrenal medulla in development of sexual dimorphism in inflammation. Eur. J. Neurosci. 14, 1436–1444.

Greenwood, L. F. and Sessle, B. J. 1976. Inputs to trigeminal brain stem neurons from facial, oral, tooth pulp and pharyngolarygeal tissues: II. Role of trigeminal nucleus caudalis in modulating responses to innocuous and noxious stimuli. Brain Res. 117, 227–238.

Gregg, J. M. 1990. Studies of traumatic neuralgia in the maxillofacial region: symptom complexes and response to microsurgery. J. Oral Maxillofac. Surg. 48, 135–140; discussion 141.

Grushka, M., Epstein, J. B., and Gorsky, M. 2003. Burning mouth syndrome and other oral sensory disorders: a unifying hypthesis. Pain Res. Manag. 8, 133–135.

Guo, A., Vulchanova, L., Wang, J., Li, X, and Elde, R. 1999. Immunocytochemical localization of the vanilloid receptor 1 (VR1): relationship to neuropeptides, the P2X3 purinoceptor and IB4 binding sites. Eur. J. Neurosci. 11, 946–958.

Guy, N., Chalus, M., Dallel, R., and Voisin, D. L. 2005. Both oral and caudal parts of the spinal trigeminal nucleus project to the somatosensory thalamus in the rat. Eur. J. Neurosci. 21, 741–754.

Hallas, B. H. and Jacquin, M. F. 1990. Structure–function relationships in rat brain stem subnucleus interpolaris. IX. inputs from subnucleus caudalis. J. Neurophysiol. 64, 28–45.

Hansson, S. R., Mezey, E., and Hoffman, B. J. 1998. Ontogeny of vesicular monoamine transporter mRNAs VMAT1 and VMAT2. II. Expression in neural crest derivatives and their target sites in the rat. Brain Res. Dev. Brain Res. 110, 159–174.

Harding, A., Paxinos, G., and Halliday, G. 2004. The serotonin and tachykinin systems. In: Paxinos, G. (ed.), The Rat Nervous System, pp.1205–256, Elsevier.

Hayes, R. L., Price, D. D., Ruda, M. A., and Dubner, R. 1981. Neuronal activity in medullary dorsal horn of awake monkeys trained in a thermal discrimination task. II. Behavioral modulation of responses to thermal and mechanical stimuli. J. Neurophysiol. 46, 428–443.

Heckmann, S. M., Heckmann, J. G., Hilz, M. J., Popp, M., Marthol, H., Neundorfer, B., Henry, M. A., Johnson, L. R., Nousek-Goebl, N., and Westrum, L. E. 1996. Light microscopic localization of calcitonin gene-related peptide in the normal feline trigeminal system and following retrogasserian rhizotomy. J. Comp. Neurol. 365, 526–540.

Henry, M. A., Westrum, L. E., Bothwell, M., and Johnson, L. R. 1993. Nerve growth factor receptor (p75)-immunoreactivity in the normal adult feline trigeminal system and following retrogasserian rhizotomy. J. Comp. Neurol. 335, 425–436.

Hicar, M. D., Robinson, M. L., and Wu, L. C. 2002. Embryonic expression and regulation of the large zinc finger protein KRC. Genesis 33, 8–20.

Hildebrand, C., Fried, K., Tuisku, F., and Johansson, C. S. 1995. Teeth and tooth nerves. Prog. Neurobiol. 45, 165–222.

Hirata, H., Hu, J. W., and Bereiter, D. A. 1999. Responses of medullary dorsal horn neurons to corneal stimulation by CO_2 pulses in the rat. J. Neurophysiol. 82, 2092–2107.

Hirata, H., Okamoto, K., and Bereiter, D. A. 2003. GABAA receptor activation modulates corneal unit activity in rostral and caudal portions of trigeminal subnucleus caudalis. J. Neurophysiol. 90, 2837–2849.

Hirata, H., Okamoto, K., Tashiro, A., and Bereiter, D. A. 2004. A novel class of neurons at the trigeminal subnucleus interpolaris/caudalis transition region monitors ocular surface fluid status and modulates tear production. J. Neurosci. 24, 4224–4232.

Hirata, H., Takeshita, S., Hu, J. W., and Bereiter, D. A. 2000. Cornea–responsive medullary dorsal horn neurons: modulation by local opioid agonists and projections to thalamus and brainstem. J. Neurophysiol. 84, 1050–1061.

Hocevar, A., Tomsic, M., Praprotnik, S., Hojnik, M., Kveder, T., and Rozman, B. 2003. Parasympathetic nervous system dysfunction in primary Sjogren's syndrome. Ann. Rheum. Dis. 62, 702–704.

Hoffmann, K. D. and Matthews, M. A. 1990. Comparison of sympathetic neurons in orofacial and upper extremity nerves: implications for causalgia. J. Oral Maxillofac. Surg. 48, 720–726; discussion 727.

Holland, G. R. and Robinson, P. P. 1992. Axon populations in cat lingual and chorda tympani nerves. J. Dent. Res. 71, 1468–1472.

Hoyes, A. D. and Barber, P. 1976. Ultrastructure of the corneal nerves in the rat. Cell Tissue Res. 172, 133–144.

Hu, B., Chiang, C. Y., Hu, J. W., Dostrovsky, J. O., and Sessle, B. J. 2002. P2X receptors in trigeminal subnucleus caudalis modulate central sensitization in trigeminal subnucleus oralis. J. Neurophysiol. 88, 1614–1624.

Hu, J. W. and Sessle, B. J. 1984. Comparison of responses of cutaneous nociceptive and nonnociceptive brain stem neurons in trigeminal subnucleus caudalis (medullary dorsal horn) and subnucleus oralis to natural and electrical stimulation of tooth pulp. J. Neurophysiol. 52, 39–53.

Huerta, M. F., Hashikawa, T., Gayoso, M. J., and Harting, J. K. 1985. The trigemino-olivary projection in the cat: contributions of individual subnuclei. J. Comp. Neurol. 241, 180–190.

Hu, J. W., Sun, K. Q., Vernon, H., and Sessle, B. J. 2005. Craniofacial inputs to upper cervical dorsal horn: implications for somatosensory information processing. Brain Res. 1044, 93–106.

Hunt, S. P. and Mantyh, P. W. 2001. The molecular dynamics of pain control. Nat. Rev. Neurosci. 2, 83–91.

Ichikawa, H. and Sugimoto, T. 2002. The difference of osteocalcin-immunoreactive neurons in the rat dorsal root and trigeminal ganglia: co-expression with nociceptive transducers and central projection. Brain Res. 958, 459–462.

Ichikawa, H., Deguchi, T., Nakago, T., Jacobowitz, D. M., and Sugimoto, T. 1994. Parvalbumin, calretinin and carbonic anhydrase in the trigeminal and spinal primary neurons of the rat. Brain Res. 655, 241–245.

Ichikawa, H., Itota, T., Torii, Y., Inoue, K., and Sugimoto, T. 1999. Osteocalcin-immunoreactive primary sensory neurons in the rat spinal and trigeminal nervous systems. Brain Res. 838, 205–209.

Ichikawa, H., Jacobowitz, D. M., and Sugimoto, T. 1993. Calretinin-immunoreactive neurons in the trigeminal and dorsal root ganglia of the rat. Brain Res. 617, 96–102.

Ichikawa, H., Jacobowitz, D. M., and Sugimoto, T. 1997. S100 protein-immunoreactive primary sensory neurons in the trigeminal and dorsal root ganglia of the rat. Brain Res. 748, 253–257.

Ikeda, M., Tanami, T., and Matsushita, M. 1984. Ascending and descending internuclear connections of the trigeminal sensory nuclei in the cat. A study with the retrograde and anterograde horseradish peroxidase technique. Neuroscience 12, 1243–1260.

Ikeda, T., Terayama, R., Jue, S., Sugiyo, S., Dubner, R., and Ren, K. 2003. Differential rostral projections of caudal brainstem neurons receiving trigeminal input after masseter inflammation. J. Comp. Neurol. 465, 220–233.

Isselee, H., Laat, A. D., Mot, B. D., and Lysens, R. 2002. Pressure-pain threshold variation in temporomandibular disorder myalgia over the course of the menstrual cycle. J. Orofac. Pain 16, 105–117.

Ito, M., Kurita, K., Ito, T., and Arao, M. 2002. Pain threshold and pain recovery after experimental stimulation in patients with burning mouth syndrome. Psychiatry Clin. Neurosci. 56, 161–168.

Izumi, H. 1999. Nervous control of blood flow in the orofacial region. Pharmacol. Ther. 81, 141–161.

Jacquin, M. F., Chiaia, N. L., Haring, J. H., and Rhoades, R. W. 1990. Intersubnuclear connections within the rat trigeminal brainstem complex. Somatosensory Motor Res. 7, 399–420.

Jacquin, M. F., Renehan, W. E., Mooney, R. D., and Rhoades, R. W. 1986. Structure–function relationships in rat medullary and cervical dorsal horns. I. Trigeminal primary afferents. J. Neurophysiol. 55, 1153–1186.

Jacquin, M. F., Semba, K., Egger, M. D., and Rhoades, R. W. 1983. Organization of HRP-labeled trigeminal mandibular primary afferent neurons in the rat. J. Comp. Neurol. 215, 397–420.

Jain, N., Qi, H. X., Catania, K. C., and Kaas, J. H. 2001. Anatomic correlates of the face and oral cavity

representations in the somatosensory cortical area 3b of monkeys. J. Comp. Neurol. 429, 455–468.

Janig, W. 2003. Relationship between pain and autonomic phenomena in headache and other pain conditions. Cephalalgia 23, 43–48.

Jones, D. A., Rollman, G. B., and Brooke, R. I. 1997. The cortisol response to psychological stress in temporomandibular dysfunction. Pain 72, 171–182.

Julius, D. and Basbaum, A. I. 2001. Molecular mechanisms of nociception. Nature 413, 203–210.

Kai-Kai, M. A. 1989. Cytochemistry of the trigeminal and dorsal root ganglia and spinal cord of the rat. Comp. Biochem. Physiol. 93A, 183–193.

Kai-Kai, M. A. and Che, Y. M. 1995. Distribution of arginine-vasopressin in the trigeminal, dorsal root ganglia and spinal cord of the rat: depletion by capsaicin. Comp. Biochem. Physiol. A Physiol. 110, 71–78.

Kanjhan, R., Housley, G. D., Burton, L. D., Christie, D. L., Kippenberger, A., Thorne, P. R., Luo, L., and Ryan, A. F. 1999. Distribution of the P2X2 receptor subunit of the ATP-gated ion channels in the rat central nervous system. J. Comp. Neurol. 407, 11–32.

Katsarava, Z., Lehnerdt, G., Duda, B., Ellrich, J., Diener, H. C., and Kaube, H. 2002. Sensitization of trigeminal nociception specific for migraine but not pain of sinusitis. Neurology 59, 1450–1453.

Kawabata, A., Itoh, H., Kawao, N., Kuroda, R., Sekiguchi, F., Masuko, T., Iwata, K., and Ogawa, A. 2004. Activation of trigeminal nociceptive neurons by parotid PAR-2 activation in rats. Neuroreport 15, 1617–1621.

Keay, K. A., Feil, K., Gordan, B. D., Herbert, H., and Bandler, R. 1997. Spinal afferents to functionally distinct periaqueductal gray columns in the rat: an anterograde and retrograde tracing study. J. Comp. Neurol. 385, 207–229.

Kelleher, K. L., Matthaei, K. I., Leck, K. J., and Hendry, I. A. 1998. Developmental expression of messenger RNA levels of the alpha subunit of the GTP-binding protein, Gz, in the mouse nervous system. Brain Res. Dev Brain Res. 107, 247–253.

Kemppainen, P., Forster, C., and Handwerker, H. O. 2001. The importance of stimulus site and intensity in differences of pain-induced vascular reflexes in human orofacial regions. Pain 91, 331–338.

Kenshalo, D. R. 1960. Comparison of thermal sensitivity of the forehead, lip, conjunctiva and cornea. J. Appl. Physiol. 15, 987–991.

Kerr, F. W. L. 1961. Structural relation of trigeminal spinal tract to upper cervical roots and solitary nucleus in cat. Exp. Neurol. 4, 134–146.

Kerr, F. W. L. 1963. The divisional organization of afferent fibres of the trigeminal nerve. Brain 86, 721–732.

Khasar, S. G., Isenberg, W. M., Miao, F. J., Gear, R. W., Green, P. G., and Levine, J. D. 2001. Gender and gonadal hormone effects on vagal modulation of tonic nociception. J. Pain 2, 91–100.

Khasar, S. G., Miao, F. J., Janig, W., and Levine, J. D. 1998. Vagotomy-induced enhancement of mechanical hyperalgesia in the rat is sympathoadrenal-mediated. J. Neurosci. 15, 3043–3049.

Kido, M. A., Kiyoshima, T., Ibuki, T., Shimizu, S., Kondo, T., Terada, Y., and Tanaka, T. 1995. A topographical and ultrastructural study of sensory trigeminal nerve endings in the rat temporomandibular joint as demonstrated by anterograde transport of wheat germ agglutinin-horseradish peroxidase (WGA-HRP). J. Dental Res. 74, 1353–1359.

Kim, S. J., Lee, J. W., Chun, H. S., Joh, T. H., and Son, J. H. 1997. Monitoring catecholamine differentiation in the embryonic brain and peripheral neurons using *E. coli* lacZ as a reporter gene. Mol. Cells 7, 394–398.

Kirchner, A., Birklein, F., Stefan, H., and Handwerker, H. O. 2000. Left vagus nerve stimulation suppresses experimentally induced pain. Neurology 55, 1167–1171.

Kitahama, K., Nagatsu, I., Geffard, M., and Maeda, T. 2000. Distribution of dopamine-immunoreactive fibers in the rat brainstem. J. Chem. Neuroanat. 18, 1–9.

Kitchener, P. D., Dziegielewska, K. M., Knott, G. W., Miller, J. M., Nawratil, P., Potter, A. E., and Saunders, N. R. Fetuin expression in the dorsal root ganglia and trigeminal ganglia of perinatal rats. Int. J. Dev. Neurosci. 15, 717–727.

Kitt, C. A., Gruber, K., Davis, M., Woolf, C. J., and Levine, J. D. 2000. Trigeminal neuralgia: opportunities for research and treatment. Pain 85, 3–7.

Klein, R., Smeyne, R. J., Wurst, W., Long, L. K., Auerbach, B. A., Joyner, A. L., and Barbacid, M. 1993. Targeted disruption of the trkB neurotrophin receptor gene results in nervous system lesions and neonatal death. Cell 75, 113–122.

Kopp, S. 2001. Neuroendocrine, immune, and local responses related to temporomandibular disorders. J. Orofacial Pain 15, 9–28.

Korszun, A. 2002. Facial pain, depression and stress-connections and directions. J. Oral Pathol. Med. 31, 615–619.

Korszun, A., Young, E. A., Singer, K., Carlson, N. E., Brown, M. B., and Crofford, L. 2002. Basal circadian cortisol secretion in women with temporomandibular disorders. J. Dent. Res. 81, 279–283.

Krout, K. E., Belzer, R. E., and Loewy, A. D. 2002. Brainstem projections to midline and intralaminar thalamic nuclei of the rat. J. Comp. Neurol. 448, 53–101.

Kruger, L. and Young, R. F. 1981. Specialized features of the trigeminal nerve and its central connections. In: The Cranial Nerves (*eds*. M. Samii and P. J. Jannetta), pp. 274–301. Springer-Verlag.

Kruger, L., Saporta, S., and Feldman, S. G. 1977. Axonal Transport Studies of the Sensory Trigeminal Complex. In: Pain in the Trigeminal Region (*eds*. D. J. Anderson and B. Matthews), pp. 191–201. Elsevier.

Kruger, L., Sternini, C., Brecha, N. C., and Mantyh, P. W. 1988. Distribution of calcitonin-gene-related peptide immunoreactivity in relation to the rat central somatosensory projection. J. Comp. Neurol. 273, 149–162.

Kumada, M., Dampney, R. A. L., and Reis, D. J. 1977. The trigeminal depressor response: a novel vasodepressor response originating from the trigeminal system. Brain Res. 119, 305–326.

Lavigne, G., Woda, A., Truelove, E., Ship, J. A., Dao, T., and Goulet, J. P. 2005. Mechanisms associated with unusual orofacial pain. J. Orofac. Pain 19, 9–21.

Lennartsson, B. 1979. Muscle spindles in the human anterior digastric muscle. Acta Odontol. Scand. 37, 329–333.

LeResche, L. 1997. Epidemiology of temporomandibular disorders: implications for the investigation of etiological factors. Crit. Rev. Oral Biol. Med. 8, 291–305.

LeResche, L., Mancl, L., Sherman, J. J., Gandara, B., and Dworkin, S. F. 2003. Changes in temporomandibular pain and other symptoms across the menstrual cycle. Pain 106, 253–261.

Levine, J. D., Collier, D. H., Basbaum, A. I., Moskowitz, M. A., and Helms, C. A. 1985. Hypothesis: the nervous system may contribute to the pathophysiology of rheumatoid arthritis. J. Rheumatol. 12, 406–411.

Levitt, P. and Moore, R. Y. 1979. Origin and organization of brainstem catecholamine innervation in the rat. J. Comp. Neurol. 186, 505–528.

Levy, D. and Strassman, A. M. 2002. Distinct sensitizing effects of the cAMP-PKA second messenger cascade on rat dural mechanonociceptors. J. Physiol. 538, 483–493.

Lewin, G. R., Lu, Y., and Park, T. J. 2004. A plethora of painful molecules. Curr. Opin. Neurobiol. 14, 443–449.

Li, Y. Q., Takada, M., Ohishi, H., Shinonaga, Y., and Mizuno, N. 1992. Trigeminal ganglion neurons which project by way of axon collaterals to both the caudal spinal trigeminal and the principal sensory trigeminal nuclei. Brain Res. 594, 155–159.

Linden, R. W. 1978. Properties of intraoral mechanoreceptors represented in the mesencephalic nucleus of the fifth nerve in the cat. J. Physiol. 279, 395–408.

Lipton, J. A., Ship, J. A., and Larach-Robinson, D. 1993. Estimated prevalence and distribution of reported orofacial pain in the United States. J. Am. Dent. Assoc. 124, 115–121.

Liu, Y., Broman, J., and Edvinsson, L. 2004. Central projections of sensory innervation of the rat superior sagittal sinus. Neuroscience 129, 431–437.

Lu, J., Hathaway, C. B., and Bereiter, D. A. 1993. Adrenalectomy enhances Fos-like immunoreactivity within the spinal trigeminal nucleus induced by noxious thermal stimulation of the cornea. Neuroscience 54, 809–818.

Ma, P. M. 1991. The barrelettes – architectonic vibrissal representations in the brainstem trigeminal complex of the mouse. I. Normal structural organization. J. Comp. Neurol. 309, 161–199.

Macfarlane, T. V., Blinkhorn, A. S., Davies, R. M., Kincey, J., and Worthington, H. V. 2002a. Association between female hormonal factors and oro-facial pain: study in the community. Pain 97, 5–10.

Macfarlane, T. V., Blinkhorn, A. S., Davies, R. M., Kincey, J., and Worthington, H. V. 2002b. Oro-facial pain in the community: prevalence and associated impact. Comm. Dent. Oral Epidemiol. 30, 52–60.

MacGregor, E. A. and Hackshaw, A. 2004. Prevalence of migraine on each day of the natural menstrual cycle. Neurology 63, 351–353.

Maixner, W., Fillingim, R., Sigurdsson, A., Kincaid, S., and Silva, S. 1998. Sensitivity of patients with temporomandibular disorders to experimentally evoked pain: evidence for altered temporal summation of pain. Pain 76, 71–81.

Maklad, A., Quinn, T., and Fritzsch, B. 2001. Intracranial distribution of the sympathetic system in mice: DiI tracing and immunocytochemical labeling. Anat. Rec. 263, 99–111.

Malick, A. and Burstein, R. 1998. Cells of origin of the trigeminohypothalamic tract in the rat. J. Comp. Neurol. 400, 125–144.

Manger, P. R., Woods, T. M., and Jones, E. G. 1995. Representation of the face and intraoral structures in area 3b of the squirrel monkey (*Saimiri sciureus*) somatosensory cortex, with special reference to the ipsilateral representation. J. Comp. Neurol. 362, 597–607.

Mantle-St, John, L. A. and Tracey, D. J. 1987. Somatosensory nuclei in the brainstem of the rat: independent projections to the thalamus and cerebellum. J. Comp. Neurol. 255, 259–271.

Mantyh, P. W. 1982. The ascending input to the midbrain periaqueductal gray of the primate. J. Comp. Neurol. 211, 50–64.

Mantyh, P. W., Allen, C. J., Rogers, S., DeMaster, E., Ghilardi, J. R., Mosconi, T., Kruger, L., Mannon, P. J., Taylor, I. L., and Vigna, S. R. 1994. Some sensory neurons express neuropeptide Y receptors: potential paracrine inhibition of primary afferent nociceptors following peripheral nerve injury. J. Neurosci. 14, 3958–3968.

Manzoni, G. C. and Torelli, P. 2005. Epidemiology of typical and atypical craniofacial neuralgias. Neurol. Sci. 26, S65–S67.

Marfurt, C. F. 1981. The somatotopic organization of the cat trigeminal ganglion as determined by the horseradish peroxidase technique. Anat. Rec. 201, 105–118.

Marfurt, C. F. and del Toro, D. R. 1987. Corneal sensory pathway in the rat: a horseradish peroxidase tracing study. J. Comp. Neurol. 261, 450–459.

Marfurt, C. F. and Echtenkamp, S. F. 1988. Central projections and trigeminal ganglion location of corneal afferent neurons in the monkey, *Macaca fascicularis*. J. Comp. Neurol. 272, 370–382.

Marfurt, C. F. and Rajchert, D. M. 1991. Trigeminal primary afferent projections to "non-trigeminal" areas of the rat central nervous system. J. Comp. Neurol. 303, 489–511.

Marfurt, C. F. and Turner, D. F. 1984. The central projections of tooth pulp afferent neurons in the rat as determined by the transganglionic transport of horseradish peroxidase. J. Comp. Neurol. 223, 535–547.

Margeta-Mitrovic, M., Mitrovic, I., Riley, R. C., Jan, L. Y., and Basbaum, A. I. 1999. Immunohistochemical localization of GABAB receptors in the rat central nervous system. J. Comp. Neurol. 405, 299–321.

Martin-Schild, S., Gerall, A. A., Kastin, A. J., and Zadina, J. E. 1999. Differential distribution of endomorphin 1- and endomorphin2-like immunoreactivities in the CNS of the rodent. J. Comp. Neurol. 405, 450–471.

Matthews, B. 1989. Autonomic mechanisms in oral sensations. Proc. Finn. Dent. Soc. 85, 365–373.

McLachlan, E. M., Janig, W., Devor, M., and Michaelis, M. 1993. Peripheral nerve injury triggers noradrenergic sprouting within dorsal root ganglia. Nature 363, 543–546.

McNeill, D. L., Chandler, M. J., Fu, Q. G., and Foreman, R. D. 1991. Projection of nodose ganglion cells to the rat upper cervical spinal cord in the rat. Brain Res. Bull. 27, 151–155.

Menetrey, D. and Basbaum, A. I. 1987. Spinal and trigeminal projections to the nucleus of the solitary tract: a possible substrate for somatovisceral and viscerovisceral reflex activation. J. Comp. Neurol. 255, 439–450.

Meng, I. D. and Bereiter, D. A. 1996. Differential distribution of Fos-like immunoreactivity in the spinal trigeminal nucleus after noxious and innocuous thermal and chemical stimulation of rat cornea. Neuroscience 72, 243–254.

Meng, I. D., Hu, J. W., and Bereiter, D. A. 1998. Differential effects of morphine on corneal-responsive neurons in rostral versus caudal regions of spinal trigeminal nucleus in the rat. J. Neurophysiol. 79, 2593–2602.

Meng, I. D., Hu, J. W., Benetti, A. P., and Bereiter, D. A. 1997. Encoding of corneal input in two distinct regions of the spinal trigeminal nucleus in the rat: cutaneous receptive field properties, responses to thermal and chemical stimulation, modulation by diffuse noxious inhibitory controls, and projections to the parabrachial area. J. Neurophysiol. 77, 43–56.

Merchenthaler, I., Lane, M. V., Numan, S., and Dellovade, T. L. 2004. Distribution of estrogen receptor a and b in the mouse central nervous system: *in vivo* autoradiographic and immunocytochemical analyses. J. Comp. Neurol. 473, 270–291.

Meunier, A., Latremoliere, A., Mauborgne, A., Bourgoin, S., Kayser, V., Cesselin, F., Hamon, M., and Pohl, M. 2005. Attenuation of pain-related behavior in a rat model of trigeminal neuropathic pain by viral-driven enkephalin overproduction in trigeminal ganglion neurons. Mol. Ther. 11, 608–616.

Meyer, R. A., Ringkamp, M., Campbell, J. N., and Raja, S. N. 2006. Peripheral Mechanisms of Cutaneous Nociception. In: Wall and Melzack's Textbook of Pain (*eds*. S. B. McMahon and M. Koltzenburg), pp. 3–34. Elsevier.

Moiseenkova, V., Bell, B., Motamedi, M., Wozniak, E., and Christensen, B. 2003. Wide-band spectral tuning of heat receptors in the pit organ of the copperhead snake (*Crotalinae*). Am. J. Physiol. Regul. Integr. Comp. Physiol. 284, R598–R606.

Molenaar, G. J. 1978. The sensory trigeminal system of a snake in the possession of infrared receptors. I. The sensory trigeminal nuclei. J. Comp. Neurol. 179, 123–135.

Molliver, D. C. and Snider, W. D. 1997. Nerve growth factor receptor TrkA is down-regulated during postnatal development by a subset of dorsal root ganglion neurons. J. Comp. Neurol. 381, 428–438.

Moore, C. I. 2004. Frequency-dependent processing in the vibrissa sensory system. J. Neurophysiol. 91, 2390–2399.

Mosconi, T., Snider, W. D., and Jacquin, M. F. 2001. Neurotrophin receptor expression in retrogradely labeled trigeminal nociceptors – comparisons with spinal nociceptors. Somatosens. Mot. Res. 18, 312–321.

Mukasa, T., Urase, K., Momoi, M. Y., Kimura, I., and Momoi, T. 1997. Specific expression of CPP32 in sensory neurons of mouse embryos and activation of CPP32 in the apoptosis induced by a withdrawal of NGF. Biochem. Biophys. Res. Commun. 231, 770–774.

Muller, L. J., Marfurt, C. F., Kruse, F., and Tervo, T. M. 2003. Corneal nerves: structure, contents and function. Exp. Eye Res. 76, 521–542.

Nakamura, K., Kaneko, T., Yamashita, Y., Hasegawa, H., Katoh, H., and Negishi, M. 2000. Immunohistochemical localization of prostaglandin EP3 receptor in the rat nervous system. J. Comp. Neurol. 421, 543–569.

Nakaya, Y., Kaneko, T., Shigemoto, R., Nakanishi, S., and Mizuno, N. 1994. Immunohistochemical localization of substance P receptor in the central nervous system of the adult rat. J. Comp. Neurol. 347, 249–274.

Nauta, H. J., Wehman, J. C., Koliatsos, V. E., Terrell, M. A., and Chung, K. 1999. Intraventricular infusion of nerve growth factor as the cause of sympathetic fiber sprouting in sensory ganglia. J. Neurosurg. 91, 447–453.

Ndiaye, A., Pinganaud, G., Buisseret-Delmas, C., Buisseret, P., and Vanderwerf, F. 2002. Organization of trigeminocollicular connections and their relations to the sensory innervation of the eyelids in the rat. J. Comp. Neurol. 448, 373–387.

Neuhuber, W. L. and Zenker, W. 1989. Central distribution of cervical primary afferents in the rat, with emphasis on proprioceptive projections to vestibular, perihypoglossal, and upper thoracic spinal nuclei. J. Comp. Neurol. 280, 231–253.

Nishimori, T., Sera, M., Suemune, S., Yoshida, A., Tsuru, K., Tsuiki, Y., Akisaka, T., Okamoto, T., Dateoka, Y., and Shigenaga, Y. 1986. The distribution of muscle primary afferents from the masseter nerve to the trigeminal sensory nuclei. Brain Res. 372, 375–381.

Noden, D. M. 1991. Vertebrate craniofacial development: the relation between ontogenetic process and morphological outcome. Brain Behav. Evol. 38, 190–225.

Nomura, S., Itoh, K., Sugimoto, T., Yasui, Y., Kamiya, H., and Mizuno, N. 1986. Mystacial vibrissae representation within the trigeminal sensory nuclei of the cat. J. Comp. Neurol. 253, 121–133.

Nurmikko, T. J. and Eldridge, P. R. 2001. Trigeminal neuralgia – pathophysiology, diagnosis and current treatment. Br. J. Anaesth. 87, 117–132.

Ohishi, H., Neki, A., and Mizuno, N. 1998. Distribution of a metabotropic glutamate receptor, mGluR2, in the central nervous system of the rat and mouse: an immunohistochemical study with a monoclonal antibody. Neurosci. Res. 30, 65–82.

Okabe, H., Okubo, T., Adachi, H., Ishikawa, T., and Ochi, Y. 1997. Immunohistochemical demonstration of cytokeratin in human embryonic neurons arising from placodes. Brain Dev. 19, 347–352.

Okamoto, K., Hirata, H., Takeshita, S., and Bereiter, D. A. 2003. Response properties of TMJ neurons in superficial laminae at the spinomedullary junction of female rats vary over the estrous cycle. J. Neurophysiol. 89, 1467–1477.

Okuse, K., Chaplan, S. R., McMahon, S. B., Luo, Z. D., Calcutt, N. A., Scott, B. P., Akopian, A. N., and Wood, J. N. 1997. Regulation of expression of the sensory neuron-specific sodium channel SNS in inflammatory and neuropathic pain. Mol. Cell Neurosci. 10, 196–207.

Oliveras, J. L., Maixner, W., Dubner, R., Bushnell, M. C., Jr., Kenshalo, D. R., Duncan, G. H., Thomas, D. A., and Bates, R. 1986. The medullary dorsal horn: a target for the expression of opiate effects on the perceived intensity of noxious heat. J. Neurosci. 6, 3086–3093.

Olszewski, J. 1950. On the anatomical and functional organization of the trigeminal nucleus. J. Comp. Neurol. 92, 401–413.

Pajot, J., Ressot, C., Ngom, I., and Woda, A. 2003. Gonadectomy induces site-specific differences in nociception in rats. Pain 104, 367–373.

Panneton, W. M. 1991. Primary afferent projections from the upper respiratory tract in the muskrat. J. Comp. Neurol. 308, 51–65.

Panneton, W. M. and Burton, H. 1981. Corneal and periocular representation within the trigeminal sensory complex in the cat studied with transganglionic transport of horseradish peroxidase. J. Comp. Neurol. 199, 327–344.

Panneton, W. M. and Burton, H. 1985. Projections from the paratrigeminal nucleus and the medullary and spinal dorsal horns to the peribrachial area in the cat. Neuroscience 15, 779–797.

Panneton, W. M., Johnson, S. N., and Christensen, N. D. 1994. Trigeminal projections to the peribrachial region in the muskrat. Neuroscience 58, 605–625.

Pappagallo, M., Oaklander, A. L., Quatrano-Piacentini, A. L., Clark, M. R., and Raja, S. N. 2000. Heterogenous patterns of sensory dysfunction in postherpetic neuralgia suggest multiple pathophysiologic mechanisms. Anesthesiology 92, 691–698.

Penfield, W. and Rasmussen, T. 1950. The Cerebral Cortex of Man: a Clinical Study of Localization of Function. Macmillan.

Petersen, C. C. 2003. The barrel cortex – integrating molecular, cellular and systems physiology. Pflugers Arch. 447, 126–134.

Petralia, R. S., Yokotani, N., and Wenthold, R. J. 1994. Light and electron microscope distribution of the NMDA receptor subunit NMDAR1 in the rat nervous system using a selective anti-peptide antibody. J. Neurosci. 14, 667–696.

Pfaller, K. and Arvidsson, J. 1988. Central distribution of trigeminal and upper cervical primary afferents in the rat studied by anterograde transport of horseradish peroxidase conjugated to wheat germ agglutinin. J. Comp. Neurol. 268, 91–108.

Pfeiffer, B., Buse, E., Meyermann, R., and Hamprecht, B. 1995. Immunocytochemical localization of glycogen phosphorylase in primary sensory ganglia of the peripheral nervous system of the rat. Histochem. Cell Biol. 103, 69–74.

Pioro, E. P. and Cuello, A. C. 1990. Distribution of nerve growth factor receptor-like immunoreactivity in the adult rat central nervous system. Effect of colchicine and correlation with the cholinergic system – II. Brainstem, cerebellum and spinal cord. Neuroscience 34, 89–110.

Pirker, S., Schwarzer, C., Wieselthaler, A., Sieghart, W., and Sperk, G. 2000. GABA(A) receptors: immunocytochemical distribution of 13 subunits in the adult rat brain. Neuroscience 101, 815–850.

Potrebic, S., Ahn, A. H., Skinner, K., Fields, H. L., and Basbaum, A. I. 2003. Peptidergic nociceptors of both trigeminal and dorsal root ganglia express serotonin 1D receptors: implications for the selective antimigraine action of triptans. J. Neurosci. 23, 10988–10997.

Price, D. D. and Dubner, R. 1977. Neurones that subserve the sensori-discriminative aspects of pain. Pain 3, 307–338.

Price, D. D., Dubner, R., and Hu, J. W. 1976. Trigeminothalamic neurons in nucleus caudalis responsive to tactile, thermal, and nociceptive stimulation of monkey's face. J. Neurophysiol. 39, 936–953.

Price, D. D., Greenspan, J. D., and Dubner, R. 2003. Neurons involved in the exteroceptive function of pain. Pain 106, 215–219.

Proske, U., Gregory, J. E., and Iggo, A. 1998. Sensory receptors in monotremes. Phil. Trans. R. Soc. Lond. B Biol. Sci. 353, 1187–1198.

Puri, V., Cui, L., Liverman, C. S., Roby, K. F., Klein, R. M., Welch, K. M., and Berman, N. E. 2005. Ovarian steroids regulate neuropeptides in the trigeminal ganglion. Neuropeptides 39, 409–417.

Ramien, M., Roucco, I., Cuello, A. C., Louis, M. S., and Ribeiro-da-Silva, A. 2004. Parasympathetic nerve fibers invade the upper dermis following sensory denervation of the rat lower lip skin. J. Comp. Neurol. 469, 83–95.

Randich, A. and Gebhart, G. F. 1992. Vagal afferent modulation of nociception. Brain Res. Rev. 17, 77–99.

Remple, M. S., Henry, E. C., and Catania, K. C. 2003. Organization of somatosensory cortex in the laboratory rat (Rattus norvegicus): evidence for two lateral areas joined at the representation of the teeth. J. Comp. Neurol. 467, 105–118.

Renehan, W. E. and Jacquin, M. F. 1993. Anatomy of Central Nervous System Pathways Related to Head Pain. In: The Headaches (eds. J. Olesen, P. Tfelt-Hansen, and K. Welch), pp. 59–68. Raven.

Reuss, S. 1999. Trigeminal innervation of the mammalian pineal gland. Microsc. Res. Tech. 46, 305–309.

Rodrigo, J., Springall, D. R., Uttenthal, O., Bentura, M. L., Abadia-Molina, F., Riveros-Moreno, V., Martinez-Murillo, R., Polak, J. M., and Moncada, S. 1994. Localization of nitric oxide synthase in the adult rat brain. Phil. Trans. R. Soc. Lond. B Biol. Sci. 345, 175–221.

Rodrigo, J., Suburo, A. M., Bentura, M. L., Fernandez, T., Nakade, S., Mikoshiba, K., Martinez-Murillo, R., and Polak, J. M. 1993. Distribution of the inositol 1,4,5-trisphosphate receptor, P400, in adult rat brain. J. Comp. Neurol. 337, 493–517.

Roveroni, R. C., Prada, C. A., Cecilia, M., Veiga, F. A., and Tambeli, C. H. 2001. Development of a behavioral model of TMJ pain in rats: the TMJ formalin test. Pain 94, 185–191.

Rozsa, A. J. and Beuerman, R. W. 1982. Density and organization of free nerve endings in the corneal epithelium of the rabbit. Pain 14, 105–120.

Ruan, H. Z., Moules, E., and Burnstock, G. 2004. Changes in P2X3 purinoceptors in sensory ganglia of the mouse during embryonic and postnatal development. Histochem. Cell Biol. 122, 539–551.

Sandstedt, P. and Sorensen, S. 1995. Neurosensory disturbances of the trigeminal nerve: a long-term follow-up of traumatic injuries. J. Oral Maxillofac. Surg. 53, 498–505.

Sarlani, E. and Greenspan, J. D. 2003. Evidence for generalized hyperalgesia in temporomandibular disorders patients. Pain 102, 221–226.

Sarlani, E. and Greenspan, J. D. 2005. Why look in the brain for answers to temporomandibular disorder pain? Cells Tissues Organs 180, 69–75.

Sarlani, E., Schwartz, A. H., Greenspan, J. D., and Grace, E. G. 2003. Facial pain as first manifestation of lung cancer: a case of lung cancer-related cluster headache and a review of the literature. J. Orofac. Pain 17, 262–267.

Saxon, D. W. and Hopkins, D. A. 1998. Efferent and collateral organization of paratrigeminal nucleus projections: an anterograde and retrograde fluorescent tracer study in the rat. J. Comp. Neurol. 402, 93–110.

Schaller, B. 2004. Trigeminocardiac reflex. A clinical phenomenon or a new physiological entity? J. Neurol. 251, 658–665.

Schepelmann, K., Ebersberger, A., Pawlak, M., Oppmann, M., and Messlinger, K. 1999. Response properties of trigeminal brain stem neurons with input from dura mater encephali in the rat. Neuroscience 90, 543–554.

Scholz, J. and Woolf, C. J. 2002. Can we conquer pain? Nat. Neurosci. 5, 1062–1067.

Sessle, B. J. 2000. Acute and chronic craniofacial pain: brainstem mechanisms of nociceptive transmission and neuroplasticity, and their clinical correlates. Crit. Rev. Oral Biol. Med. 11, 57–91.

Sessle, B. J., Hu, J. W., Dubner, R., and Lucier, G. E. 1981. Functional properties of neurons in cat trigeminal subnucleus caudalis (medullary dorsal horn). II. Modulation of responses to noxious and nonnoxious stimuli by periaqueductal gray, nucleus raphe magnus, cerebral cortex, and afferent influences, and effect of naloxone. J Neurophysiol. 45, 193–207.

Shenker, N., Haigh, R., Roberts, E., Mapp, P., Harris, N., and Blake, D. 2003. A review of contralateral responses to a unilateral inflammatory lesion. Rheumatology 42, 1279–1286.

Shigenaga, Y., Chen, I. C., Suemune, S., Nishimori, T., Nasution, I. D., Yoshida, A., Sato, H., Okamoto, T., Sera, M., and Hosoi, M. 1986a. Oral and facial representation within the medullary and upper cervical dorsal horns in the cat. J. Comp. Neurol. 243, 388–408.

Shigenaga, Y., Nakatani, Z., Nishimori, T., Suemune, S., Kuroda, R., and Matano, S. 1983. The cells of origin of cat trigeminothalamic projections: especially in the caudal medulla. Brain Res. 277, 20l–222.

Shigenaga, Y., Okamoto, T., Nishimori, T., Suemune, S., Nasution, I. D., Chen, I. C., Tsuru, K., Yoshida, A., Tabuchi, K., Hosoi, M., and Tsuru, H. 1986b. Oral and facial representation in the trigeminal principal and rostral spinal nuclei of the cat. J. Comp. Neurol. 244, 1–18.

Shigenaga, Y., Sera, M., Nishimori, T., Suemune, S., Nishimura, M., Yoshida, A., and Tsuru, K. 1988. The central projection of masticatory afferent fibers to the trigeminal sensory nuclear complex and upper cervical spinal cord. J. Comp. Neurol. 268, 489–507.

Shigenaga, Y., Suemune, S., Nishimura, M., Nishimori, T., Sato, H., Ishidori, H., Yoshida, A., Tsuru, K., Tsuiki, Y., Dakeoka, Y., Nasution, I. D., and Hosoi, M. 1986c. Topographic representation of lower and upper teeth within the trigeminal sensory nuclei of adult cat as demonstrated by the transganglionic transport of horseradish peroxidase. J. Comp. Neurol. 251, 299–3l6.

Shughrue, P. J., Lane, M. V., and Merchenthaler, I. 1997. Comparative distribution of estrogen receptor-a and -b mRNA in the rat central nervous system. J. Comp. Neurol. 388, 507–525.

Silverman, J. D. and Kruger, L. 1985. Projections of the rat trigeminal sensory nuclear complex demonstrated by multiple fluorescent dye retrograde transport. Brain Res. 361, 383–388.

Simonian, S. X., Delaleu, B., Caraty, A., and Herbison, A. E. 1998. Estrogen receptor expression in brainstem noradrenergic neurons of the sheep. Neuroendocrinology 67, 392–402.

Sjoqvist, O. 1938. Studies on pain conduction in the trigeminal nerve. A contribution to the surgical treatment of facial pain. Acta Psychiat. Neurol. Scand. 1, 1–139.

Sobreviela, T., Clary, D. O., Reichardt, L. F., Brandabur, M. M., Kordower, J. H., and Mufson E. J. 1994. TrkA-immunoreactive profiles in the central nervous system: colocalization with neurons containing p75 nerve growth factor receptor, choline acetyltransferase, and serotonin. J. Comp. Neurol. 350, 587–611.

Son, J. H., Min, N., and Joh, T. H. 1996. Early ontogeny of catecholaminergic cell lineage in brain and peripheral neurons monitored by tyrosine hydroxylase-lacZ transgene. Brain Res. Mol. Brain Res. 36, 300–308.

Steinbusch, H. W. M. 1981. Distribution of serotonin-immunoreactivity in the central nervous system of the rat-cell bodies and terminals. Neuroscience 6, 557–618.

Stenholm, E., Bongenhielm, U., Ahlquist, M., and Fried, K. 2002. VRl- and VRL-l-like immunoreactivity in normal and injured trigeminal dental primary sensory neurons of the rat. Acta Odontol. Scand. 60, 72–79.

Strack, S., Wadzinski, B. E., and Ebner, F. F. 1996. Localization of the calcium/calmodulin-dependent protein phosphatase, calcineurin, in the hindbrain and spinal cord of the rat. J. Comp. Neurol. 375, 66–76.

Strassman, A. M. and Vos, B. P. 1993. Somatotopic and laminar organization of Fos-like immunoreactivity in the medullary and cervical dorsal horn induced by noxious facial stimulation in the rat. J. Comp. Neurol. 331, 495–516.

Strassman, A. M., Potrebic, S., and Maciewicz, R. J. 1994. Anatomical properties of brainstem trigeminal neurons that respond to electrical stimulation of dural blood vessels. J. Comp. Neurol. 346, 349–365.

Sugimoto, T., Fujiyoshi, Y., He, Y. F., Xiao, C., and Ichikawa, H. 1997a. Trigeminal primary projection to the rat brain stem sensory trigeminal nuclear complex and surrounding structures revealed by anterograde transport of cholera toxin B subunit-conjugated and *Bandeiraea simplicifolia* isolectin B4-conjugated horseradish peroxidase. Neurosci. Res. 28, 361–371.

Sugimoto, T., Fujiyoshi, Y., Xiao, C., He, Y. F., and Ichikawa, H. 1997b. Central projection of calcitonin gene-related peptide (CGRP)- and substance P (SP)-immunoreactive trigeminal primary neurons in the rat. J. Comp. Neurol. 378, 425–442.

Svensson, P., Bjerring, P., Arendt-Nielsen, L., and Kaaber, S. 1993. Sensory and pain thresholds to orofacial argon laser stimulation in patients with chronic burning mouth syndrome. Clin. J. Pain 9, 207–215.

Takemura, M., Nagase, Y., Yoshida, A., Yasuda, K., Kitamura, S., Shigenaga, Y., and Matano, S. 1993. The central projections of the monkey tooth pulp afferent neurons. Somatosens. Mot. Res. 10, 217–227.

Takemura, M., Sugimoto, T., and Sakai, A. 1987. Topographic organization of central terminal region of different sensory branches of the rat mandibular nerve. Exp. Neurol. 96, 540–557.

Takei, M., Shiraiwa, H., Azuma, T., Hayashi, Y., Seki, N., and Sawada, S. 2005. The possible etiopathogenic genes of Sjogren's syndrome. Autoimmun. Rev. 4, 479–484.

Tal, M. and Devor, M. 1992. Ectopic discharge in injured nerves: comparison of trigeminal and somatic afferents. Brain Res. 579, 148–151.

Talley, E. M., Rosen, D. I., Lee, A., Guyenet, P. G., and Lynch, K. R. 1996. Distribution of alpha 2A-adrenergic receptor-like immunoreactivity in the rat central nervous system. J. Comp. Neurol. 372, 111–134.

Tamura, S., Morikawa, Y., Miyajima, A., and Senba, E. 2003. Expression of oncostatin M receptor beta in a specific subset of nociceptive sensory neurons. Eur. J. Neurosci. 17, 2287–2298.

Tanimoto, T., Takeda, M., and Matsumoto, S. 2002. Suppressive effect of vagal afferents on cervical dorsal horn neurons responding to tooth pulp electrical stimulation in the rat. Exp. Brain Res. 145, 468–479.

Tashiro, T., Higo, S., and Matsuyama, T. 1984. Soma size comparison of the trigeminal ganglion cells giving rise to the ascending and descending tracts: a horseradish peroxidase study in the cat. Exp. Neurol. 84, 37–46.

Tatehata, T., Shiosaka, S., Wanaka, A., Rao, Z. R., and Tohyama, M. 1987. Immunocytochemical localization of the choline acetyltransferase containing neuron system in the rat lower brain stem. J. Hirnforsch. 28, 707–716.

Tenenbaum, H. C., Mock, D., Gordan, A. S., Goldberg, M. B., Grossi, M. L., Locker, D., and Davis, K. D. 2001. Sensory and affective components of orofacial pain: is it all in your brain? Crit. Rev. Oral Biol. Med. 12, 455–468.

Thackray, A. M. and Field, H. J. 1996. Differential effects of famciclovir and valaciclovir on the pathogenesis of herpes simplex virus in a murine infection model including reactivation from latency. J. Infect. Dis. 173, 291–299.

Thor, K. B., Blitz-Siebert, A., and Helke, C. J. 1992. Autoradiographic localization of 5HT1 binding sites in the medulla oblongata of the rat. Synapse 10, 185–205.

Toda, K., Ishii, N., and Nakamura, Y. 1997. Characteristics of mucosal nociceptors in the rat oral cavity: an in vitro study. Neurosci. Lett. 228, 95–98.

Treede, R. D., Meyer, R. A., Raja, S. N., and Campbell, J. N. 1992. Peripheral and central mechanisms of cutaneous hyperalgesia. Prog. Neurobiol. 38, 397–421.

Trowbridge, H. O., Franks, M., Korostoff, E., and Emling, R. 1980. Sensory response to thermal stimulation in human teeth. J. Endod. 6, 405–412.

Trulsson, M. and Johansson, R. S. 2002. Orofacial mechanoreceptors in humans: encoding characteristics and responses during natural orofacial behaviors. Behav. Brain Res. 135, 27–33.

Tsuzuki, K., Fukuoka, T., Sakagami, M., and Noguchi, K. 2003. Increase of preprotachykinin mRNA in the uninjured mandibular neurons after rat infraorbital nerve transection. Neurosci. Lett. 345, 57–60.

Tsuzuki, K., Kondo, E., Fukuoka, T., Yi, D., Tsujino, H., Sakagami, M., and Noguchi, K. 2001. Differential regulation of P2X(3) mRNA expression by peripheral nerve injury in intact and injured neurons in the rat sensory ganglia. Pain 91, 351–360.

Van Bijsterveld, O. P., Kruize, A. A., and Bleys, R. L. 2003. Central nervous system mechanisms in Sjogren's syndrome. Br. J. Ophthalmol. 87, 128–130.

Van Ham, J. J. and Yeo, C. H. 1992. Somatosensory trigeminal projections to the inferior olive, cerebellum and other precerebellar nuclei in rabbits. Eur. J. Neurosci. 4, 302–317.

Vass, Z., Shore, S. E., Nuttall, A. L., Jansco, G., Brechtelsbauer, P. B., and Miller, J. M. 1997. Trigeminal ganglion innervation of the cochlea – a retrograde transport study. Neuroscience 79, 605–615.

Veinante, P., Jacquin, M. F., and Deschenesm, M. 2000. Thalamic projections from the whisker-sensitive regions of the spinal trigeminal complex in the rat. J. Comp. Neurol. 420, 233–243.

Verge, V. M., Wiesenfeld-Hallin, Z., and Hokfelt, T. 1993. Cholecystokinin in mammalian primary sensory neurons and spinal cord: in situ hybridization studies in rat and monkey. Eur. J. Neurosci. 5, 240–250.

Villar, M. J., Cortes, R., Theodorsson, E., Wiesenfeld-Hallin, Z., Schalling, M., Fahrenkrug, J., Emson, P. C., and Hokfelt, T. 1989. Neuropeptide expression in rat dorsal root ganglion cells and spinal cord after peripheral nerve injury with special reference to galanin. Neuroscience 33, 587–604.

Vos, B. P., Strassman, A. M., and Maciewicz, R. J. 1994. Behavioral evidence of trigeminal neuropathic pain following chronic constriction injury to the rat's infraorbital nerve. J. Neurosci. 14, 2708–2723.

Wall, P. D. 1979. On the relation of injury to pain. Pain 6, 253–264.

Wanaka, A., Milbrandt, J., and Johnson, E. M., Jr. 1991. Expression of FGF receptor gene in rat development. Development 111, 455–468.

Wang, H., Rivero-Melian, C., Robertson, B., and Grant, G. 1994. Transganglionic transport and binding of the isolectin B4 from Griffonia simplicifolia I in rat primary sensory neurons. Neuroscience 62, 539–551.

Wasner, G., Backonja, M. M., and Baron, R. 1998. Traumatic neuralgias: complex regional pain syndromes (reflex sympathetic dystrophy and causalgia): clinical characteristics, pathophysiological mechanisms and therapy. Neurol. Clin. 16, 851–868.

Welker, C. 1971. Microelectrode delineation of fine grain somatotopic organization of (SmI) cerebral neocortex in albino rat. Brain Res. 26, 259–275.

Wessman, M., Kaunisto, M. A., Kallela, M., and Palotie, A. 2004. The molecular genetics of migraine. Ann. Med. 36, 462–473.

Westlund, K. N. and Craig, A. D. 1996. Association of spinal lamina I projections with brainstem catecholamine neurons in the monkey. Exp. Brain Res. 110, 151–162.

Whitehead, M. C., Ganchrow, J. R., Ganchrow, D., and Yao, B. 1999. Organization of geniculate and trigeminal ganglion cells innervating single fungiform taste papillae: a study with tetramethylrhodamine dextran amine labeling. Neuroscience 93, 931–941.

Wiberg, M., Westman, J., and Blomqvist, A. 1986. The projection to the mesencephalon from the sensory trigeminal nuclei. An anatomical study in the cat. Brain Res. 399, 51–68.

Wolf, E., Petersson, K., Petersson, A., and Nilner, M. 2001. Long-lasting orofacial pain – a study of 109 consecutive patients referred to a pain group. Swed. Dent. J. 25, 129–136.

Wood, J. N., Boorman, J. P., Okuse, K., and Baker, M. D. 2004. Voltage-gated sodium channels and pain pathways. J. Neurobiol. 61, 55–71.

Woolf, C. J. and Salter, M. W. 2000. Neuronal plasticity: increasing the gain in pain. Science 288, 1765–1768.

Xiang, Z., Bo, X., and Burnstock, G. 1998. Localization of ATP-gated P2X receptor immunoreactivity in rat sensory and sympathetic ganglia. Neurosci. Lett. 256, 105–108.

Yan, Q., Radeke, M. J., Matheson, C. R., Talvenheimo, J., Welcher, A. A., and Feinstein, S. C. 1997. Immunocytochemical localization of TrkB in the central nervous system of the adult rat. J. Comp. Neurol. 378, 135–157.

Yatim, N., Billig, I., Compoint, C., Buisseret, P., and Buisseret-Delmas, C. 1996. Trigeminocerebellar and trigemino-olivary projections in rats. Neurosci. Res. 25, 267–283.

Yazdani, C., McLaughlin, T., Smeeding, J. E., and Walt, J. 2001. Prevalence of treated dry eye disease in a managed care population. Clin. Ther. 23, 1672–1682.

Yoshida, A., Dostrovsky, J. O., Sessle, B. J., and Chiang, C. Y. 1991. Trigeminal projections to the nucleus submedius of the thalamus in the rat. J. Comp. Neurol. 307, 609–625.

Yoshimura, N., White, G., Weight, F. F., and de Groat, W. C. 1994. Patch-clamp recordings from subpopulations of autonomic and afferent neurons identified by axonal tracing techniques. J. Auton. Nerv. Syst. 49, 85–92.

Young, R. F. and King, R. B. 1973. Fiber spectrum of the trigeminal sensory root of the baboon determined by electron microscopy. J. Neurosurg. 38, 65–71.

Young, R. F. and Perryman, K. M. 1984. Pathways for orofacial pain sensation in the trigeminal brain-stem nuclear complex of the Macaque monkey. J. Neurosurg. 61, 563–568.

Zerari-Mailly, F., Buisseret-Delmas, P., and Nosjean, A. 2005. Trigemino-solitarii-facial pathway in rats. J. Comp. Neurol. 487, 176–189.

Zhang, X., Ji, R. R., Arvidsson, J., Lundberg, J. M., Bartfai, T., Bedecs, K., and Hokfelt, T. 1996. Expression of peptides, nitric oxide synthase and NPY receptor in trigeminal and nodose ganglia after nerve lesions. Exp. Brain Res. 111, 393–404.

Further Reading

Arvidsson, J. and Rice, F. L. 1991. Central projections of primary sensory neurons innervating different parts of the vibrissae follicles and intervibrissal skin on the mystacial pad of the rat. J. Comp. Neurol. 309, 1–16.

Crockett, D. P., Wang, L., Zhang, R. X., and Egger, M. D. 1999. Distribution of the low-affinity neurotrophin receptor (p75) in the developing trigeminal brainstem complex in the rat. Anat. Rec. 254, 549–565.

Darian-Smith, I. 1973. The Trigeminal System. In: Handbook of Sensory Physiology, vol. 2 Somatosensory System (ed. A. Iggo), pp. 271–314. Springer.

Hummel, T. 2001. Oral mucosal blood flow in patients with burning mouth syndrome. Pain 90, 281–286.

Kawabata, A., Feil, K., Gordan, B. D., Herbert, H., and Bandler, R. 1997. Spinal afferents to functionally distinct periaqueductal gray columns in the rat: an anterograde and retrograde tracing study. J. Comp. Neurol. 385, 207–229.

Lima, D., Mendes-Ribeiro, J. A., and Combra, A. 1991. The spino-latero-reticular system of the rat: Projections from the superficial dorsal horn and structural characterization of marginal neurons involved. Neuroscience 45, 137–152.

Luo, P. and Dessem, D. 1995. Inputs from identified jaw-muscle spindle afferents to trigeminothalamic neurons in the rat: a double-labeling study using retrograde HRP and intracellular biotinamide. J. Comp. Neurol. 353, 50–66.

Panneton, W. M., Klein, B. G., and Jacquin, M. F. 1991. Trigeminal projections to contralateral dorsal horn originate in midline hairy skin. Somatosensory Motor Res. 8, 165–173.

Pedersen, J., Reddy, H., Funch-Jensen, P., Arendt-Nielsen, L., Gregersen, H., and Drewes, A. M. 2004. Differences between male and female responses to painful thermal and mechanical stimulation of the human esophagus. Dig. Dis. Sci. 49, 1065–1074.

Rasmussen, B. K. 1993. Migraine and tension-type headache in a general population: precipitating factors, female hormones, sleep pattern and relation to lifestyle. Pain 53, 65–72.

Rasmussen, B. K. 2001. Epidemiology of headache. Cephalalgia 21, 774–777.

Sessle, B. J. and Hu, J. W. 1981. Raphe-induced suppression of the jaw-opening reflex and single neurons in trigeminal subnucleus oralis, and influence of naloxone and subnucleus caudalis. Pain 10, 19–36.

Shigenaga, Y., Nishimura, M., Suemune, S., Nishimori, T., Doe, K., and Tsuru, H. 1989. Somatotopic organization of tooth pulp primary afferent neurons in the cat. Brain Res. 477, 66–89.

Wiberg, M., Westman, J., and Blomqvist, A. 1987. Somatosensory projection to the mesencephalon: an anatomical study in the monkey. J. Comp. Neurol. 264, 92–117.

Willis, W. D., Jr., Zhang, X., Honda, C. N., and Giesler, G. J., Jr. 2001. Projections from the marginal zone and deep dorsal horn to the ventrobasal nuclei of the primate thalamus. Pain 92, 267–276.

Yoshida, A., Dostrovsky, J. O., and Chiang, C. Y. 1992. The afferent and efferent connections of the nucleus submedius in the rat. J. Comp. Neurol. 324, 115–133.

33 Migraine – A Disorder Involving Trigeminal Brainstem Mechanisms

P J Goadsby, University of California, San Francisco, CA, USA

33.1 Introduction

Headache in general, and in particular migraine (Goadsby, P. J. et al., 2002) and cluster headache (Goadsby, P. J., 2002), is better understood now than has been the case for the last four millennia (Lance, J. W. and Goadsby, P. J., 2005). Migraine is a common, disabling recurrent disorder of the central nervous system with a core manifestation involving activation, or the perception of activation, of trigeminal nociceptive afferents (Olesen, J. et al., 2005). Here how studies of the anatomy and physiology of the pain-producing innervation of the dura mater and large cranial vessels, the trigeminovascular system, has contributed to our current understanding of one of the most common maladies of humans will be explored.

33.2 Migraine – Explaining the Clinical Features

Migraine is in essence a familial episodic disorder whose key marker is headache with certain associated features (Table 1). It is these features that give clues to its pathophysiology, and ultimately will provide insights leading to new treatments.

The essential elements to be considered are:

Genetics of migraine;

Physiological basis for the aura;

Anatomy of head pain, particularly that of the trigeminovascular system;

Physiology and pharmacology of activation of the peripheral branches of ophthalmic branch of the trigeminal nerve;

Table 1 International Headache Society features of migraine (Headache Classification Committee of the International Headache Society, 2004)

Repeated episodic headache (4–72 h) with the following features:	
Any two of:	Any one of:
unilateral	nausea/vomiting
throbbing	photophobia and phonophobia
worsened by movement	
moderate or severe	

Physiology and pharmacology of the trigeminal nucleus, in particular its caudal most part, the trigeminocervical complex (TCC);

Brainstem and diencephalic modulatory systems that influence trigeminal pain transmission and other sensory modality processing.

33.3 Genetics of Migraine

One of the most important aspects of the pathophysiology of migraine is the inherited nature of the disorder. It is clear from clinical practice that many patients have first-degree relatives who also suffer from migraine (Lance, J. W. and Goadsby, P. J., 2005). Transmission of migraine from parents to children has been reported as early as the seventeenth century, and numerous published studies have reported a positive family history.

33.3.1 Genetic Epidemiology

Studies of twin pairs are the classical method to investigate the relative importance of genetic and environmental factors. A Danish study included 1013 monozygotic and 1667 dizygotic twin pairs of the same gender, obtained from a population-based twin register. The pairwise concordance rate was significantly higher among monozygotic than dizygotic twin pairs ($P < 0.05$). Several studies have attempted to analyze the possible mode of inheritance in migraine families and conflicting results have been obtained. Both twin studies and population-based epidemiological surveys strongly suggest that migraine without aura is a multifactorial disorder, caused by a combination of genetic and environmental factors.

33.3.2 Familial Hemiplegic Migraine

In approximately 50% of the reported families, familial hemiplegic migraine (FHM) has been assigned to chromosome 19p13. Few clinical differences have been found between chromosome 19-linked and unlinked FHM families. Indeed, the clinical phenotype does not associate particularly with the known mutations. The most striking exception is cerebellar ataxia, which occurs in approximately 50% of the chromosome 19-linked, but in none of the unlinked families. Another less striking difference includes the fact that patients from chromosome 19-linked families are more likely to have attacks that can be triggered by minor head trauma or are that associated by coma.

The biological basis for the linkage to chromosome 19 is mutations (Ophoff, R. A. *et al.*, 1996) involving the $Ca_v2.1$ (P/Q) type voltage-gated calcium channel *CACNA1A* gene. Now known as FHM-I, this mutation is responsible for about 50% of identified families. Mutations in the *ATP1A2* gene have been identified to be responsible for about 20% of FHM families. Interestingly, the phenotype of some FHM-II involves epilepsy, while it has also been suggested that alternating hemiplegia of childhood can be due to *ATP1A2* mutations. The latter cases are most unusual for migraine. Most recently mutations in the neuronal voltage-gated sodium channel *SCN1A* have been identified as the cause of FHM-III, thus continuing the ionopathic theme.

Taken together, the known mutations suggest that migraine, or at least the neurological manifestations currently called the aura, are caused by an ionopathy. Linking the channel disturbance for the first time to the aura process has demonstrated that human mutations expressed in a knockin mouse produce a reduced threshold for cortical spreading depression (CSD), which has some profound implications for understanding that process.

33.4 Migraine Aura

Migraine aura is defined as a focal neurological disturbance manifest as visual, sensory, or motor symptoms (Headache Classification Committee of the International Headache Society, 2004). It is seen in about 30% of patients, and it is clearly neurally driven. The case for the aura being the human equivalent of the CSD of Leao has been well made (Lauritzen, M., 1994). In humans, visual aura has been described as affecting the visual field, suggesting the visual cortex, and it starts at the center of the visual field and propagates to the periphery at a speed of $3\,\text{mm}\,\text{min}^{-1}$. This is very similar to spreading depression described in rabbits. Blood flow studies in patients have also shown that a focal hyperemia tends to precede the spreading oligemia, and again this is similar to what would be expected with spreading depression. After this passage of oligemia, the cerebrovascular response to hypercapnia in patients is blunted while autoregulation remains intact. Again this pattern is repeated with experimental spreading depression. Human observations have rendered the arguments reasonably sound that human aura has as its equivalent in animals in CSD. An area of controversy surrounds whether aura triggers the rest of the attack, and is indeed painful. Based on the available experimental and clinical data this author is not at all convinced that aura is painful *per se*, but this does not diminish its interest or the importance of understanding it. Indeed therapeutic developments may shed further light on these relationships, and studies are required to understand how in some patients aura is clearly not a sufficient trigger to pain.

33.5   Headache – Anatomy

33.5.1   The Trigeminal Innervation of Pain-Producing Intracranial Structures

Surrounding the large cerebral vessels, pial vessels, large venous sinuses, and dura mater is a plexus of largely unmyelinated fibers that arise from the ophthalmic division of the trigeminal ganglion and in the posterior fossa from the upper cervical dorsal roots. Trigeminal fibers innervating cerebral vessels arise from neurons in the trigeminal ganglion that contain substance P and calcitonin gene-related peptide (CGRP), both of which can be released when the trigeminal ganglion is stimulated either in humans or cats (Goadsby, P. J. *et al.*, 1988). Stimulation of the cranial vessels, such as the superior sagittal sinus (SSS), is certainly painful in humans (Wolff, H. G., 1948). Human dural nerves that innervate the cranial vessels largely consist of small diameter myelinated and unmyelinated fibers that almost certainly subserve a nociceptive function.

33.6   Headache Physiology – Peripheral Connections

33.6.1   Plasma Protein Extravasation

Moskowitz M. A. (1990) has provided a series of experiments to suggest that the pain of migraine may be a form of sterile neurogenic inflammation. Although this seems clinically implausible, the model system has been helpful in understanding some aspects of trigeminovascular physiology. Neurogenic plasma extravasation can be seen during electrical stimulation of the trigeminal ganglion in the rat. Plasma extravasation can be blocked by ergot alkaloids, indomethacin, acetylsalicylic acid, and the serotonin-5-$HT_{1B/1D}$ agonist, sumatriptan. The pharmacology of abortive antimigraine drugs has been reviewed in detail. In addition there are structural changes in the dura mater that are observed after trigeminal ganglion stimulation. These include mast cell degranulation and changes in postcapillary venules including platelet aggregation. While it is generally accepted that such changes, and particularly the initiation of a sterile inflammatory response, would cause pain, it is not clear whether this is sufficient of itself, or requires other stimulators, or promoters. Preclinical studies suggest that CSD may be a sufficient stimulus to activate trigeminal neurons, although this has been a controversial area.

Although plasma extravasation in the retina, which is blocked by sumatriptan, can be seen after trigeminal ganglion stimulation in experimental animals, no changes are seen with retinal angiography during acute attacks of migraine or cluster headache. A limitation of this study was the probable sampling of both retina and choroids elements in rats, given that choroidal vessels have fenestrated capillaries. Clearly, however, blockade of neurogenic plasma protein extravasation is not completely predictive of antimigraine efficacy in humans as evidenced by the failure in clinical trials of substance P, neurokinin-1 antagonists, specific plasma protein extravasation (PPE) blockers, CP122,288 and 4991w93, an endothelin antagonist, and a neurosteroid. The implications of these data have been recently reviewed (Peroutka, S. J., 2005).

33.6.2   Sensitization and Migraine

While it is highly doubtful that there is a significant sterile inflammatory response in the dura mater during migraine, it is clear that some form of sensitization takes place during migraine, since allodynia is common. About two-thirds of patients complain of pain from non-noxious stimuli, allodynia (Selby, G. and Lance, J. W., 1960). A particularly interesting aspect is the demonstration of allodynia in the upper limbs ipsilateral and contralateral to the pain. This finding is consistent with at least third-order neuronal sensitization, such as sensitization of thalamic neurons, and firmly places the pathophysiology within the central nervous system. Sensitization in migraine may be peripheral with local release of inflammatory markers, which would certainly activate trigeminal nociceptors. More likely in migraine is a form of central sensitization, which may be classical central sensitization, or a form of disinhibitory sensitization with dysfunction of descending modulatory pathways (Knight, Y. E. *et al.*, 2002). Just as dihydroergotamine (DHE) can block trigeminovascular nociceptive transmission, probably at least by a local effect in the TCC, DHE can block central sensitization associated with dural stimulation by an inflammatory soup, as can cyclo-oxygenase block both sensitization and trigeminocervical transmission.

33.6.3   Neuropeptide Studies

Electrical stimulation of the trigeminal ganglion in both humans and cats leads to increases in extracerebral blood flow and local release of both CGRP and SP (Goadsby, P. J. *et al.*, 1988). In the cat trigeminal ganglion stimulation also increases cerebral blood flow by a pathway traversing the greater superficial petrosal branch of the facial nerve (Goadsby, P. J. and Duckworth, J. W., 1987) again releasing a powerful vasodilator peptide, vasoactive intestinal polypeptide (VIP; May, A. and Goadsby, P. J., 1999). Interestingly, the VIPergic innervation of the cerebral vessels is predominantly anterior rather than posterior, and this may contribute to this region's vulnerability to spreading depression, explaining why the aura is so very often seen to commence posteriorly. Stimulation of the more specifically vascular pain-producing superior sagittal sinus increases cerebral blood flow and jugular vein CGRP levels. Human evidence that CGRP is elevated in the headache phase of migraine (Goadsby, P. J. *et al.*, 1990), supporting the view that the trigeminovascular system may be activated in a protective role in these conditions. Moreover, nitric oxide (NO)-donor-triggered migraine, which is in essence typical migraine, also results in increases in CGRP that are blocked by sumatriptan, just as in spontaneous migraine (Goadsby, P. J. and Edvinsson, L., 1993). It is of

interest in this regard that compounds that have not shown activity in migraine (Peroutka, S. J., 2005), notably the conformationally restricted analogue of sumatriptan, CP122,288, and the conformationally restricted analog of zolmitriptan, 4991w93, were both ineffective inhibitors of CGRP release after superior sagittal sinus in cats. The recent development of nonpeptide highly specific CGRP antagonists, and the announcement of proof-of-concept for a CGRP antagonist in acute migraine (Olesen, J. *et al.*, 2004), firmly establishes this as a novel and important new emerging principle for acute migraine. At the same time the lack of any effect of CGRP blockers on plasma protein extravasation, explains in some part why that model has proved inadequate at translation into human therapeutic approaches (Peroutka, S. J., 2005).

33.7   Headache Physiology – Central Connections

33.7.1   The Trigeminocervical Complex

Fos immunohistochemistry is a method for looking at activated cells by plotting the expression of Fos protein. After meningeal irritation with blood Fos expression is noted in the trigeminal nucleus caudalis, while after stimulation of the superior sagittal sinus Fos-like immunoreactivity is seen in the trigeminal nucleus caudalis and in the dorsal horn at the C_1 and C_2 levels in cats and monkey. These latter findings are in accord with similar data using 2-deoxyglucose measurements with superior sagittal sinus stimulation. Similarly, stimulation of a branch of C_2, the greater occipital nerve, increases metabolic activity in the same regions, i.e., trigeminal nucleus caudalis and $C_{1/2}$ dorsal horn, and fos expression can be elicited by injection of mustard oil into the occipital muscles. In experimental animals one can record directly from trigeminal neurons with both supratentorial trigeminal input and input from the greater occipital nerve, a branch of the C_2 dorsal root (Bartsch, T. and Goadsby, P. J., 2002). Stimulation of the greater occipital nerve for 5 min results in substantial increases in responses to supratentorial dural stimulation, which can last for over 1 h. Conversely, stimulation of the middle meningeal artery dura mater with the C-fiber irritant mustard oil sensitizes responses to occipital muscle stimulation. Taken together these data suggest convergence of cervical and ophthalmic inputs at the level of the second-order neuron. Moreover, stimulation of a lateralized structure, the middle meningeal artery, produces Fos expression bilaterally in both cat and monkey brains. This group of neurons from the superficial laminas of trigeminal nucleus caudalis and $C_{1/2}$ dorsal horns should be regarded functionally as the TCC.

These data demonstrate that trigeminovascular nociceptive information comes by way of the most caudal cells. This concept provides an anatomical explanation for the referral of pain to the back of the head in migraine. Moreover, experimental pharmacological evidence suggests that some abortive antimigraine drugs, such as, ergot derivatives, acetylsalicylic acid, sumatriptan, eletriptan, naratriptan, rizatriptan, and zolmitriptan can have actions at these

second-order neurons that reduce cell activity and suggest a further possible site for therapeutic intervention in migraine. This action can be dissected out to involve each of the 5-HT_{1B}, 5-HT_{1D}, and 5-HT_{1F} receptor subtypes, and are consistent with the localization of these receptors on peptidergic nociceptors. Interestingly, triptans also influence the CGRP promoter, and regulate CGRP secretion from neurons in culture, as well as perhaps require cell surface expression for their effect. Furthermore, the demonstration that some part of this action is postsynaptic with either 5-HT_{1B} or 5-HT_{1D} receptors located nonpresynatically offers a prospect of highly anatomically localized treatment options (Figure 1).

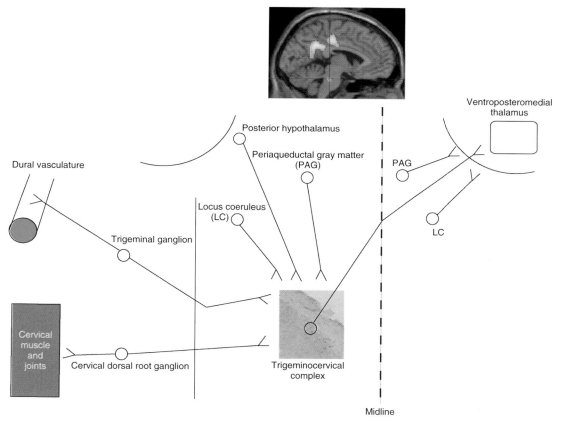

Figure 1 Illustration of the some elements of migraine biology. Patients inherit a dysfunction in brain control systems for pain and other afferent stimuli, which can be triggered and are in turn capable of activating the trigeminovascular system as the initiating event in a positive feedback of neurally driven vasodilatation. Nociceptive afferents from the cervical region terminate in the trigeminocervical complex (illustrated by Fos protein expression in the superficial laminas) and this accounts for the nontrigeminal distribution of pain in many patients. These afferents project to the thalamus, including ventroposteromedial thalamus, and are at least influenced by neurons in the posterior hypothalamic gray, the periaqueductal gray (PAG), and probably by neurons of the nucleus locus coeruleus in the pons. Functional brain imaging suggests that the brainstem, notably the pons as illustrated after Bahra, A., Matharu, M. S., Buchel, C., Frackowiak, R. S. J., and Goadsby, P. J. 2001. Brainstem activation specific to migraine headache. Lancet 357, 1016–1017, is a pivotal region in the migraine process.

Table 2   Neuroanatomical processing of vascular head pain

	Structure	Comments
Target innervation: Cranial vessels Dura mater	Ophthalmic branch of trigeminal nerve	
First	Trigeminal ganglion	Middle cranial fossa
Second	Trigeminal nucleus (quintothalamic tract)	Trigeminal nucleus caudalis and C_1/C_2 dorsal horns
Third	Thalamus	Ventrobasal complex Medial nucleus of posterior group Intralaminar complex
Modulatory	Midbrain Hypothalamus	Periaqueductal gray matter ?
Final	Cortex	Insulas Frontal cortex Anterior cingulate cortex Basal ganglia

33.7.2   Higher-Order Processing

Following transmission in the caudal brainstem and high cervical spinal cord information is relayed rostrally (Table 2).

33.7.3.1   *Thalamus*

Processing of vascular nociceptive signals in the thalamus occurs in the ventroposteromedial (VPM) thalamus, medial nucleus of the posterior complex, and in the intralaminar thalamus. It has been shown by application of capsaicin to the superior sagittal sinus that trigeminal projections with a high degree of nociceptive input are processed in neurons particularly in the ventroposteromedial thalamus and in its ventral periphery. These neurons in the VPM can be modulated by activation of gamma-aminobutyric acid (GABA)$_A$ inhibitory receptors, and perhaps of more direct clinical relevance by propranolol though a β_1-adrenoceptor mechanism (Shields, K. G. and Goadsby, P. J., 2005). Remarkably, triptans through $5\text{-HT}_{1B/1D}$ mechanisms can also inhibit VPM neurons locally, as demonstrated by microiontophoretic application, suggesting a hitherto unconsidered locus of action for triptans in acute migraine. Human imaging studies have confirmed activation of thalamus contralateral to pain in acute migraine (Bahra, A. *et al.*, 2001).

33.7.3.2   *Activation of modulatory regions*

Stimulation of nociceptive afferents by stimulation of the superior sagittal sinus in cats activates neurons in the ventrolateral periaqueductal gray matter (PAG). PAG activation in turn feeds back to the TCC with an inhibitory influence. PAG is clearly included in the area of activation seen in positron emission tomography (PET) studies in migraineurs. This typical negative feedback system will be further considered below as a possible mechanism for the symptomatic manifestations of migraine.

Another potentially modulatory region activated by stimulation of nociceptive trigeminovascular input is the posterior hypothalamic gray. This area is crucially involved in several primary headaches, notably cluster headache (Goadsby, P. J., 2002), Short-lasting unilateral neuralgiform headache attacks with conjunctival injection and tearing (SUNCT), paroxysmal hemicrania and hemicrania continua. Moreover, the clinical features of the premonitory phase, and other features of the disorder, suggest dopamine neuron involvement. Orexinergic neurons in the posterior hypothalamus can be both pro- and antinociceptive, offering a further possible region whose dysfunction might involve the perception of head pain.

33.8   Central Modulation of Trigeminal Pain

33.8.1   Brain Imaging in Humans

Functional brain imaging with PET has demonstrated activation of the dorsal midbrain, including the PAG, and in the dorsal pons, near the locus coeruleus, in studies during migraine without aura.

Dorsolateral pontine activation is seen with PET in spontaneous episodic and chronic migraine, and with nitrogylcerin-triggered attacks (Bahra, A. *et al.*, 2001; Afridi, S. *et al.*, 2005). These areas are active immediately after successful treatment of the headache but are not active interictally. The activation corresponds with the brain region that Raskin initially reported, and confirmed, to cause migrainelike headache when stimulated in patients with electrodes implanted for pain control. Similarly, excess iron in the PAG of patients with episodic and chronic migraine, and chronic migraine can develop after a bleed into a cavernoma in the region of the PAG, or with a lesion of the pons. What could dysfunction of these brain areas lead to?

33.8.2 Animal Experimental Studies of Sensory Modulation

It has been shown in experimental animals that stimulation of nucleus locus coeruleus, the main central noradrenergic nucleus, reduces cerebral blood flow in a frequency-dependent manner (Goadsby, P. J. *et al.*, 1982) through an α_2-adrenoceptor-linked mechanism. This reduction is maximal in the occipital cortex. While a 25% overall reduction in cerebral blood flow is seen, extracerebral vasodilatation occurs in parallel (Goadsby, P. J. *et al.*, 1982). In addition, the main serotonin-containing nucleus in the brainstem, the midbrain dorsal raphe nucleus, can increase cerebral blood flow when activated. Furthermore, stimulation of PAG will inhibit sagittal sinus-evoked trigeminal neuronal activity in cats, while blockade of P/Q-type voltage-gated Ca^{2+} channels in the PAG facilitates trigeminovascular nociceptive processing (Knight, Y. E. *et al.*, 2002) with the local GABAergic system in the PAG still intact.

33.8.3 Electrophysiology of Migraine in Humans

Studies of evoked potentials and event-related potentials provide some link between animal studies and human functional imaging. Authors have shown changes in neurophysiological measures of brain activation but there is much discussion as to how to interpret such changes (Schoenen, J. *et al.*, 2003). Perhaps the most reliable theme is that the migrainous brain does not habituate to signals in a normal way. Similarly, contingent negative variation (CNV), an event related potential, is abnormal in migraineurs compared to controls. Changes in CNV predict attacks and preventive therapies alter, normalize, such changes. Attempts to correlate clinical phenotypes with electrophysiological changes, may enhance further studies in this area.

33.9 What is Migraine?

Migraine is an inherited, episodic disorder involving sensory sensitivity. Patients complain of pain in the head that is throbbing, but there is no reliable relationship between vessel diameter and the pain, or its treatment. They complain of discomfort from normal lights and the unpleasantness of routine sounds. Some mention otherwise pleasant odors are unpleasant. The anatomical connections of, for example, the pain pathways are clear, the ophthalmic division of the trigeminal nerve subserves sensation within the cranium and explains why the top of the head is headache, and the maxillary division is facial pain. The convergence of cervical and trigeminal afferents explains why neck stiffness or pain is so common in primary headache. The genetics of channelopathies is opening up a plausible way to think about the episodic nature of migraine. However, where is the lesion, what is actually the pathology?

Migraine aura cannot be the trigger alone, there is no evidence at all after 4000 years that it occurs in more than 30% of migraine patients; aura can be experienced without pain at all, and is seen in the other primary headaches. Perhaps electrophysiological changes in the brain have been mislabeled as hyperexcitability whereas dyshabituation might be a simpler explanation. If migraine was basically an attentional problem with changes in cortical synchronization (Niebur, E. *et al.*, 2002), hypersynchronization, all its manifestations could be accounted for in a single over-arching pathophysiological hypothesis of a disturbance of subcortical sensory modulation systems. While it seems likely that the trigeminovascular system, and its cranial autonomic reflex connections, the trigeminal-autonomic reflex (May, A. and Goadsby, P. J., 1999), act as a feed-forward system to facilitate the acute attack, the fundamental problem in migraine is in the brain. Unraveling its basis will deliver great benefits to patients and considerable understanding of some very fundamental neurobiological processes.

Acknowledgment

The work of the author has been supported by the Migrane Trust.

References

Afridi, S., Matharu, M. S., Lee, L., Kaube, H., Friston, K. J., Frackowiak, R. S. J., and Goadsby, P. J. 2005. A PET study exploring the laterality of brainstem activation in migraine using glyceryl trinitrate. Brain 128, 932–939.

Bahra, A., Matharu, M. S., Buchel, C., Frackowiak, R. S. J., and Goadsby, P. J. 2001. Brainstem activation specific to migraine headache. Lancet 357, 1016–1017.

Bartsch, T. and Goadsby, P. J. 2002. Stimulation of the greater occipital nerve induces increased central excitability of dural afferent input. Brain 125, 1496–1509.

Goadsby, P. J. 2002. Pathophysiology of cluster headache: a trigeminal autonomic cephalgia. Lancet Neurol. 1, 37–43.

Goadsby, P. J. and Duckworth, J. W. 1987. Effect of stimulation of trigeminal ganglion on regional cerebral blood flow in cats. Am. J. Physiol. 253, R270–R274.

Goadsby, P. J. and Edvinsson, L. 1993. The trigeminovascular system and migraine: studies characterizing cerebrovascular and neuropeptide changes seen in humans and cats. Ann. Neurol. 33, 48–56.

Goadsby, P. J., Edvinsson, L., and Ekman, R. 1988. Release of vasoactive peptides in the extracerebral circulation of man and the cat during activation of the trigeminovascular system. Ann. Neurol. 23, 193–196.

Goadsby, P. J., Edvinsson, L., and Ekman, R. 1990. Vasoactive peptide release in the extracerebral circulation of humans during migraine headache. Ann. Neurol. 28, 183–187.

Goadsby, P. J., Lambert, G. A., and Lance, J. W. 1982. Differential effects on the internal and external carotid circulation of the monkey evoked by locus coeruleus stimulation. Brain Res. 249, 247–254.

Goadsby, P. J., Lipton, R. B., and Ferrari, M. D. 2002. Migraine – current understanding and treatment. New Engl. J. Med. 346, 257–270.

Headache Classification Committee of the International Headache Society. 2004. The International Classification of Headache Disorders (2nd edn.). Cephalalgia 24, 1–160.

Knight, Y. E., Bartsch, T., Kaube, H., and Goadsby, P. J. 2002. P/Q-type calcium channel blockade in the PAG facilitates trigeminal nociception: a functional genetic link for migraine? J. Neurosci. 22, 1–6.

Lance, J. W. and Goadsby, P. J. 2005. Mechanism and Management of Headache. Elsevier.

Lauritzen, M. 1994. Pathophysiology of the migraine aura. The spreading depression theory. Brain 117, 199–210.

May, A. and Goadsby, P. J. 1999. The trigeminovascular system in humans: pathophysiological implications for primary headache syndromes of the neural influences on the cerebral circulation. J. Cerebr. Blood Flow Metabol. 19, 115–127.

Moskowitz, M. A. 1990. Basic mechanisms in vascular headache. Neurolog. Clin. 8, 801–815.

Niebur, E., Hsiao, S. S., and Johnson, K. O. 2002. Synchrony: a neural mechanism for attentional selection? Curr. Opin. Neurobiol. 12, 190–194.

Olesen, J., Diener, H. C., Husstedt, I. W., Goadsby, P. J., Hall, D., Meier, U., Pollentier, S., and Lesko, L. M. 2004. Calcitonin gene-related peptide (CGRP) receptor antagonist BIBN4096BS is effective in the treatment of migraine attacks. New Engl. J. Med. 350, 1104–1110.

Olesen, J., Tfelt-Hansen, P., Ramadan, N., Goadsby, P. J., and Welch, K. M. A. 2005. The Headaches. Lippincott, Williams & Wilkins.

Ophoff, R. A., Terwindt, G. M., Vergouwe, M. N., van Eijk, R., Oefner, P. J., Hoffman, S. M. G., Lamerdin, J. E,, Mohrenweiser, H. W., Bulman, D. E., Ferrari, M., Haan, J., Lindhout, D., van Ommen, G. J., Hofker, M. H., Ferrari, M. D., and Frants, R. R. 1996. Familial hemiplegic migraine and episodic ataxia type-2 are caused by mutations in the Ca^{2+} channel gene CACNL1A4. Cell 87, 543–552.

Peroutka, S. J. 2005. Neurogenic inflammation and migraine: implications for therapeutics. Mol. Intervent. 5, 306–313.

Schoenen, J., Ambrosini, A., Sandor, P. S., and Maertens de Noordhout, A. 2003. Evoked potentials and transcranial magnetic stimulation in migraine: published data and viewpoint on their pathophysiologic significance. Clin. Neurophysiol. 114, 955–972.

Selby, G. and Lance, J. W. 1960. Observations on 500 cases of migraine and allied vascular headache. J. Neurol. Neurosurg. Psychiatry 23, 23–32.

Shields, K. G. and Goadsby, P. J. 2005. Propranolol modulates trigeminovascular responses in thalamic ventroposteromedial nucleus: a role in migraine? Brain 128, 86–97.

Wolff, H. G. 1948. Headache and Other Head Pain. Oxford University Press.

34 Tooth Pain

M R Byers, University of Washington, Seattle WA, USA

Glossary

ASIC receptors Acid-sensing ion channels that detect low pH, a typical condition of pulpitis.

atypical odontalgia A condition in which tooth pain derives from neuropathic or referred mechanisms. Tooth extractions or root canals do not relieve the pain.

BK receptors Activated by the inflammatory mediator, bradykinin.

convergence Many sensory afferents from multiple tissues project to individual central trigeminal neurons.

dentin Specialized, calcified, collagenous matrix that surrounds the pulp.

hot tooth An inflamed tooth that resists regional anesthesia and remains sensitive when neighboring teeth are numb.

hypersensitive dentin Sharp pain elicited from light touch to exposed dentin.

ionotropic receptors Activated by specific ligands causing ion flux through receptor pores.

mesencephalic trigeminal nucleus Location of cell bodies of primary sensory neurons that innervate stretch receptors in periodontal ligament, sutures, or mastication muscles.

metabotropic receptors Interaction with ligand activates G-protein intracellular signaling.

NK receptors Receptors for neurokinins such as substance P or neurokinin A.

nucleus caudalis Caudal region of spinal trigeminal subnuclei, specialized for processing, relaying and modulating orofacial pain.

odontalgia Tooth pain.

odontoblast Neural crest-derived cells that make dentin and regulate the pulp–dentin barrier.

P2 receptors Purinergic nucleotide receptors that respond to adenosine triphosphate (ATP).

periapex Region at base of root socket that surrounds the root apex, and includes periodontal ligament, neurovascular bundles and endings, and alveolar bone.

prepain The first sensation (tingling, vibration, or touch) elicited by electrical stimulation of teeth. It changes to sharp tooth pain at stronger levels of stimulation.

pulpitis Inflamed tooth pulp that can be healed locally (reversible), that can consume the pulp and spread into periapex (irreversible), or that can be undetected (silent) until it reaches the periapex.

receptive field Patch of tissue that activates a primary afferent or central neuron.

referred pain Pain that is felt at a different site from the neural activity that causes it.

Ruffini mechanoreceptors Complex stretch mechanoreceptors located in periodontal ligament.

Trk receptors Tyrosine kinase receptors that respond to neurotrophin factors.

TRP receptors Transient receptor potential channels (vanilloid receptor family) that are activated by capsaicin, heat, or low pH.

34.1 Introduction

Tooth pain is the most common type of orofacial pain (Lipton, J. et al., 1993; Hargreaves, K. M., 2002). Dental nerve fibers branch centrally to activate many neurons in the trigeminal brainstem complex or extratrigeminal relay sites, and those central neurons also receive extensive convergent input from other orofacial tissues, making location of tooth pain difficult. However, most of the time our teeth do not hurt, and most dental neural activity is unperceived. The sensory functions of tooth nerves are presented here, along with unusual features of central processing, mechanisms, perception, diagnosis, and treatment of tooth pain, which can be sharp or dull, focused or diffuse, episodic or relentless, referred or neuropathic (for reviews see: Närhi, M. V. O. et al., 1996; Olgart, L., 1996; Byers, M. R. and Närhi, M. V. O., 1999; Dionne, R. A. and Berthold, C. W., 2001; Byers, M. R. and Närhi, M. V. O., 2002; Hargreaves, K. M., 2002; Hu, J. W., 2004; Lavigne, G. et al., 2004; Truelove, E., 2004; Sessle, B. J., 2005; Henry, M. A. and Hargreaves, K. M., 2007).

34.2 Dental Sensory Mechanisms

34.2.1 Normal Teeth/Acute Pain

A-fibers respond especially well to acute stimuli that move fluid in dentin (Brännström, M and Åström, A., 1972) and C-fibers respond to acute heat stimuli or pulp damage (Närhi, M. V. O. et al., 1996; Närhi, M. V. O., 2005). Different factors affect axonal conduction in trigeminal nerves, ganglion, central tracts, and synaptic termination regions in the brainstem (Figure 1). Dental afferents project to low threshold mechanoreceptive, nociceptive-specific, and wide-dynamic-range central neurons, all of which receive a major input from other tissues, often from more than one trigeminal division (Sessle, B. J. et al., 1986).

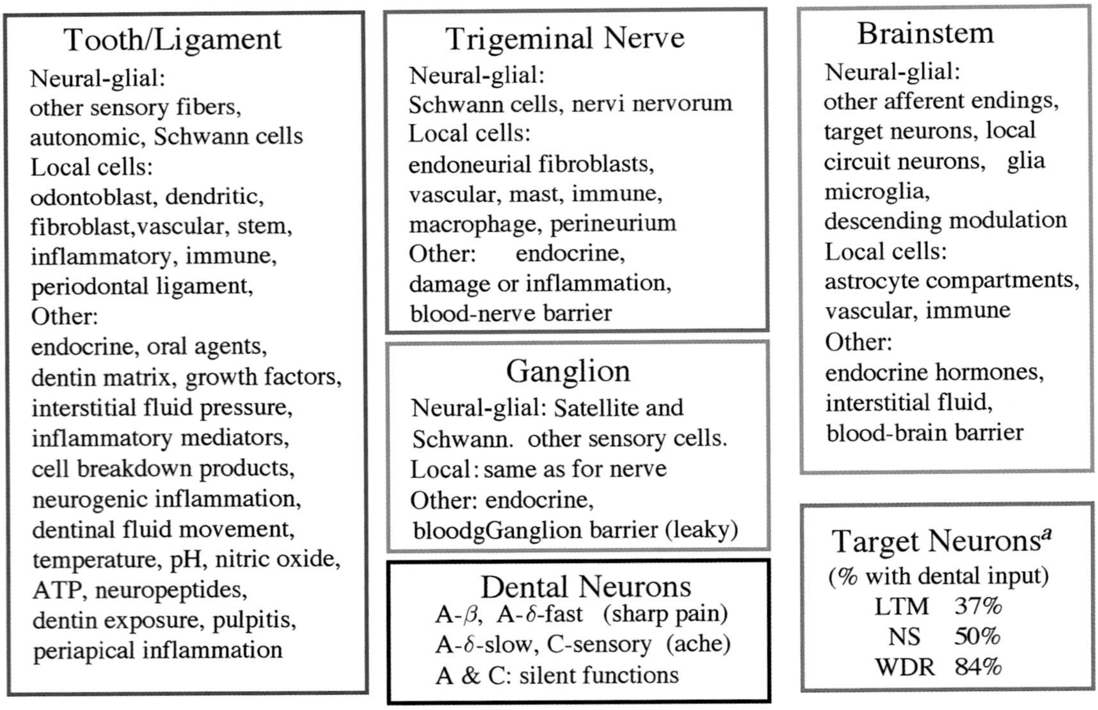

Tooth/Ligament	Trigeminal Nerve	Brainstem
Neural-glial: other sensory fibers, autonomic, Schwann cells	Neural-glial: Schwann cells, nervi nervorum	Neural-glial: other afferent endings, target neurons, local circuit neurons, glia microglia, descending modulation
Local cells: odontoblast, dendritic, fibroblast,vascular, stem, inflammatory, immune, periodontal ligament,	Local cells: endoneurial fibroblasts, vascular, mast, immune, macrophage, perineurium	Local cells: astrocyte compartments, vascular, immune
Other: endocrine, oral agents, dentin matrix, growth factors, interstitial fluid pressure, inflammatory mediators, cell breakdown products, neurogenic inflammation, dentinal fluid movement, temperature, pH, nitric oxide, ATP, neuropeptides, dentin exposure, pulpitis, periapical inflammation	Other: endocrine, damage or inflammation, blood-nerve barrier	Other: endocrine hormones, interstitial fluid, blood-brain barrier

Ganglion
Neural-glial: Satellite and Schwann. other sensory cells.
Local: same as for nerve
Other: endocrine, bloodgGanglion barrier (leaky)

Dental Neurons
A-β, A-δ-fast (sharp pain)
A-δ-slow, C-sensory (ache)
A & C: silent functions

Target Neurons[a]
(% with dental input)
LTM 37%
NS 50%
WDR 84%

[a] Inputs from teeth were most common for WDR neurons in nucleus caudalis of cats (Sessle, B. J. et al., 1986).

Figure 1 Interactions affecting dental neuronal function. ATP, adenosine triphosphate; LTM, low-threshold mechanoreceptor; NS, nociceptive specific; WDR, wide-dynamic neuron. Target Neurons: reprinted from Pain, 27, Sessele, B. J., Hu J. W., Amano, N., and Zhong, G., Convergence of cutaneous, tooth pulp, visceral, neck and muscle afferents onto nociceptive and non-nociceptive neurons in trigeminal subnucleus caudalis (medullary dorsal horn) and its implications for referred pain, 219–235, Copyright 1986, with permission from The International Association for the Study of Pain.

Most tooth pain perceptions are acute, and they derive from activation of neurons in nucleus caudalis, although rostral trigeminal nuclei are also involved, and there are other connection sites such as the paratrigeminal nucleus and reticular formation (Sessle, B. J., 2000; 2005).

Human studies show three kinds of evoked sensation from dental nerves: prepain, sharp pain, and dull ache. The first two depend on activation of fast A-β and A-δ-fibers and the latter involves polymodal capsaicin-sensitive slow A-δ and C-fibers (Närhi, M. V. O. *et al.*, 1996; Ikeda, H. *et al.*, 1997), some of which express neuropeptide receptors for autocrine modulation (Suzuki, H. *et al.*, 2002). Each of our teeth is innervated by many hundreds of highly branched trigeminal neurons, and the density of sensory nerve endings in coronal pulp and inner dentin is enormous (Figure 2). However, most intradental neural activity involves unperceived neural efferent functions or reflexes, such as vasodilatation by neuropeptides from sensory fibers that is counterbalanced by sympathetic-mediated vasoconstriction (Figure 3; Olgart, L. 1996; Fristad, I. *et al.*, 1997; Berggreen, E. and Heyeraas, K. J., 2000). Touch sensations during chewing come from Ruffini mechanoreceptors in the periodontal ligament outside the roots, while unconscious aspects of jaw reflexes involve intradental mechanoreceptive A-fibers and the periodontal endings of mesencephalic trigeminal neurons (Dong, W. K. *et al.*, 1993). Dental neurons can express a variety of ionotropic receptors (e.g., TRP-V1, TRP-V2, P2X$_3$, ASIC), metabotropic receptors (BK-1, BK-2, NK1-3, TrkA), ion channels, receptors for neuropeptides, neurotrophins, and opioid peptides that sensitize, inhibit, or modulate sensory neurons (Hu, J. W., 2004), and even immune regulators (Wadachi, R. and Hargreaves, K. M., 2006). The regulation of the pulpal milieu and the quality of tooth pain vary in relation to those activating and modulating systems.

Many dental nerve endings form close appositions with the odontoblasts (Figures 2(a) and 2(b)), while others end freely in pulp and dentin. Neuro-odontoblast interactions are not fully understood, and may involve odontoblast support for the free sensory endings, modulation of sensory activity, and/or specific sensory activity. The lack of synaptic contacts or gap

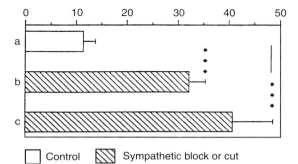

Figure 3 Laser Doppler demonstration of sensory nerve-mediated increased blood flow in rat incisor pulp after α-adrenergic block (b) or sympathectomy (c) compared to intact blood flow (a). All teeth received brief bipolar electrical stimulation of the intact tooth crown. Reproduced from Olgart, L. and Kerezoudis, N. P. 1994. Nerve–pulp interactions. Arch. Oral Biol. 39, 47S–54S, with permission.

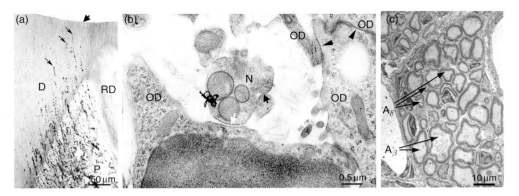

Figure 2 (a) Sensory endings in pulp (P) and dentin (D) (thin arrows) are shown by autoradiography of axonally transported ^{3}H-protein in adult rat molars. They avoid reparative dentin (RD) but come close to surface of tooth (wide arrow). Reprinted from Byers, M.R. 1984. Dental sensory receptors. Int. Rev. Neurobiol. 25, 39–94. (b) Electron microscopic autoradiography showed that transported ^{3}H-proteins (black coiled silver grain) are confined to sensory endings (N). Odontoblasts (OD) are connected by numerous gap junctions (arrowheads). A special apposition separates OD and N (white arrow). Reproduced from Byers, M. R. 1977. Fine structure of trigeminal receptors in rat molars. In: Pain in the Trigeminal Region (*eds*. D. J. Anderson and B. Matthews), pp. 13–24, with permission from Elsevier. (c) Mouse molar root nerves include A-β fibers that exceed 6 μm in diameter and several sizes of A-δ axons.

junctions between odontoblasts and nerves suggests a supportive role. However, demonstrations of neural-like ion channels (Guo, L. and Davidson, R. M., 1998) and TREK-1, a mechanosensitive potassium channel, in odontoblasts (Magloire, H. *et al.*, 2003) show that they are excitable and mechanosensitive. They also attract sensory nerves (Maurin, J. C. *et al.*, 2004) and express neurotrophin factors and receptors in developing, adult, and injured teeth (Fried, K. *et al.*, 2000; Woodnutt, D. A. *et al.*, 2000). It is still not clear whether odontoblast excitability directly affects tooth pain or just allows those cells to monitor dentinogenic requirements.

34.2.2 Inflammatory Tooth Pain

When pulp or periapex are inflamed, there are local cellular changes, nerve sprouting, peripheral and central sensitization, and neurochemical plasticity that alter the quality of tooth pain perceptions (Hargreaves, K. M., 2002; Hu, J. W., 2004). Many of the events in inflamed teeth are typical of any inflammation, but there are unusual nerve-sprouting reactions near the pulpitis, and those either subside if healing occurs, or they persist for months or years when lesions escape into the periapical region (Figures 4(a) and 4(b); Byers, M. R. and Närhi, M. V. O., 1999). Important changes occur in dental neuronal satellite cell glia (Figure 4(c)) when inflammation continues in teeth, such as expression of glial fibrillary acidic protein (Stephenson, J. L. and Byers, M. R., 1995). Specialized vascular reactions also occur (Olgart, L., 1996). Recording from individual fibers shows that there are expansions of receptive fields of the A-fiber afferents after 1–2 weeks of inflammation (Figure 5), that would further affect sensitization in the central neurons.

34.2.3 Dental Neuropathic Pain

Teeth are major players in referred orofacial pain, either as the source or the referral site, and that situation can lead to unnecessary multiple extractions. Teeth can also exhibit neuropathic symptoms such as hyperalgesia, allodynia, and spontaneous pain (Truelove, E., 2004; Lavigne, G. *et al.*, 2004). A principal mechanism for atypical odontalgia or dental neuropathic pain is the extensive convergence of inputs onto brainstem neurons from a wide area, often involving multiple trigeminal divisions and a variety of tissues (Sessle, B. J., 2000; 2005), and glial actions also affect pain quality (Xie, Y. F. *et al.*, 2006).

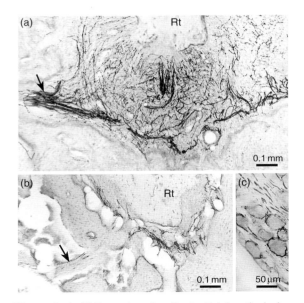

Figure 4 (a, b) Several weeks after tooth injury the lesion has destroyed the pulp and emerged from the root (Rt) into periapical tissue, where there is intense sprouting and immunoreactivity for calcitonin gene-related peptide (CGRP) nerve fibers compared to a normal molar (b). Arrows indicate sensory nerve in alveolar canal. (c) Satellite cells surrounding trigeminal cell bodies are shown by immunocytochemistry to express glial fibrillary acidic protein (black rings) after molar injury in adult rats. (a, b) Reproduced from Kimberly, C. L. and Byers, M. R. 1988. Inflammation of rat molar pulp and periodontium causes increased calcitonin gene-related peptide and axonal sprouting. Anat. Rec. 222, 289–300, with permission.

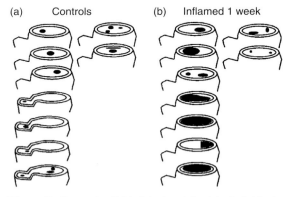

Figure 5 Receptive fields (black patches) for individual A-fibers were larger in dog teeth after 1 week of induced pulpitis compared to control teeth. Reproduced from Närhi, M. V. O., Yamamoto, H., and Ngassapa, D. 1996. Function of Intradental Nociceptors in Normal and Inflamed Teeth. In: Dentin/Pulp Complex (*eds*. M. Shimono, T. Maeda, H. Suda, and K. Takahashi), pp. 136–140. Quintessence Publishing Co, Inc., with permission.

34.3 Tooth Pain: Diagnosis and Management

Dentists routinely use evoked acute pain to diagnose tooth pathology, pulp vitality, and treatment (Bender, I. B., 2000). Dental procedures such as orthodontia or root canal treatment cause transient pain that usually decreases within a few days, as the local inflammation subsides, especially with nonsteroidal anti-inflammatory treatment (Dionne, R. A. and Berthold, C. W.,

2001). However, pain can persist weeks, months, or years, especially in individuals who have had long-term tooth pain or other chronic pain (Bender, I. B., 2000; Truelove, E., 2004). Tooth pain can expand to a wide area, making it difficult to locate the pathology. Competition from a second pain source can narrow the site to just one tooth, at least for a few minutes to aid diagnosis (Figure 6). This diffuse noxious inhibition implies substantial changes in central physiology for the widely dispersed tooth pains

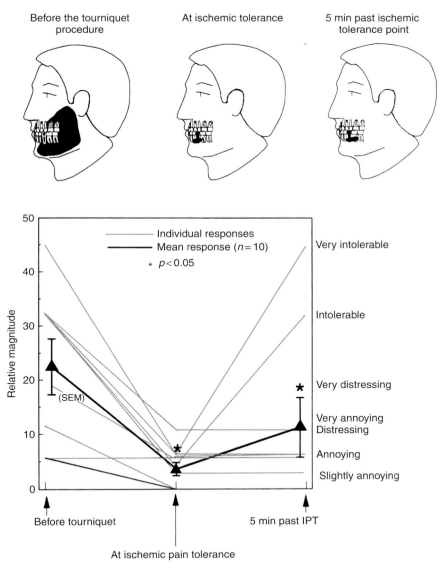

Figure 6 Prior to tourniquet-evoked arm pain, the area of dental pain was large. After maximal arm pain, the tooth pain area was reduced to an individual tooth. By 5 min later, the pain was still focused although the pain intensity (red lines) was returning to prestimulus levels. IPT, ischemic pain tolerance. Reprinted from Pain, 57, Sigurdsson, A. and Maxiner, W., Effects of experimental and clinical noxious counterirritants on pain perception, 265–275, Copyright 1994, with permission from The International Association for the Study of Pain.

(Sessle, B. J., 2005), though important changes in neural sodium channels occur in painful teeth (Renton, T. *et al.*, 2005; Henry, M. A. and Hargreaves, K. M., 2007) offering possible pharmacologic therapeutic targets.

34.4 Conclusions

Unusual features of tooth pain include: (1) difficulty locating the source of pain; (2) intense pain from stimulation of dentin; (3) hypersensitive dentin after loss of its protective enamel or periodontal covering, with concomitant pulpitis; (4) special neuro-pulpal interactions; (5) silent pulpal inflammation that only becomes painful when it invades periapical tissues; (6) the hot tooth problem, in which an inflamed painful tooth becomes difficult to anesthetize, even when the rest of the jaw is numb; and (7) referred pain. It is, perhaps, surprising that prolonged tooth pain is unusual, given the huge number of dental procedures every day that can cause tooth, nerve, or bone damage (i.e., dental implants that screw into the jaw, third molar extractions, delicate access for anesthetic injections to reach bone-encased nerves, root canals that remove pulp to eliminate infection). The keys to prevention of debilitating chronic tooth pain include avoiding nerve damage in the jaw, and removal of infected pulp (Bender, I. B., 2000; Hargreaves, K. M., 2002; Hu, J. W., 2004; Truelove, E., 2004; Sessle, B. J., 2005; Henry, M. A. and Hargreaves, K. M., 2007).

References

Bender, I. B. 2000. Pulp pain diagnosis – a review. J. Endodon. 26, 175–179.

Berggreen, E. and Heyeraas, K. J. 2000. Effect of the sensory neuropeptide antagonists h-CGRP (8-37) and SR 140.33 on pulpal and gingival blood flow in ferrets. Arch. Oral Biol. 45, 537–542.

Brännström, M and Åström, A. 1972. The hydrodynamics of dentine: its possible relationship to dentinal pain. Int. Dent. J. 22, 219–227.

Byers, M. R. 1978. Fine Structure of Trigeminal Receptors in Rat Molars. In: Pain in the Trigeminal Region (*eds*. D. J. Anderson and B. Matthews), pp. 13–24, Elsevier.

Byers, M. R. and Närhi, M. V. O. 1999. Dental injury models: experimental tools for understanding neuro-inflammatory interactions and polymodal nociceptor functions. Crit. Rev. Oral Biol. Med. 10, 4–39.

Byers, M. R. and Närhi, M. V. O. 2002. Nerve Supply of the Pulpodentin Complex and Responses to Injury. In: Seltzer and Bender's Dental Pulp (*eds*. K. M. Hargreaves and H. E. Goodis), pp. 151–179. Quintessence.

Dionne, R. A. and Berthold, C. W. 2001. Therapeutic uses of non-steroidal anti-inflammatory drugs in dentistry. Crit. Rev. Oral Biol. Med. 12, 315–330.

Dong, W. K., Shiwaku, T., Kawakami, Y., and Chudler, E. H. 1993. Static and dynamic responses of periodontal ligament mechanoreceptors and intradental mechanoreceptors. J. Neurophysiol. 69, 1567–1582.

Fried, K., Nosrat, C., Lillesaar, C., and Hildebrand, C. 2000. Molecular signaling and pulpal nerve development. Crit. Rev. Oral Biol. Med. 11, 318–332.

Fristad, I., Kvinnsland, I. H., Johnsson, R., and Heyeraas, K. J. 1997. Effect of intermittent long-lasting electrical tooth stimulation on pulpal blood flow and immunocompetent cells: a hemodynamic and immunohistochemical study in young rat molars. Exp. Neurol. 146, 230–239.

Guo, L. and Davidson, R. M. 1998. Potassium and chloride channels in freshly isolated odontoblasts. J. Dent. Res. 77, 341–350.

Hargreaves, K. M. 2002. Pain Mechanisms of the Pulpodentin Complex. In: Seltzer and Bender's Dental Pulp (*eds*. K. M. Hargreaves and H. E. Goodis), pp. 181–203. Quintessence.

Henry, M. A. and Hargreaves, K. M. 2007. Peripheral mechanisms of odontogenic pain. Dent. Clin. North Am. 51, 19–44.

Hu, J. W. 2004. Tooth Pulp. In: Clinical Oral Physiology (*eds*. T. S. Miles, B. Nauntofte, and P. Svensson), pp. 141–162. Quintessence.

Ikeda, H., Tokita, Y., and Suda, H. 1997. Capsaicin-sensitive A-delta fibers in cat tooth pulp. J. Dent. Res. 76, 1341–1349.

Kimberly, C. L. and Byers, M. R. 1988. Inflammation of rat molar pulp and periodontium causes increased calcitonin gene-related peptide and axonal sprouting. Anat. Rec. 222, 289–300.

Lavigne, G., Woda, A., Truelove, E., Ship, J. A., Dao, T., and Goulet, J. P. 2004. Mechanisms associated with unusual orofacial pain. J. Orofac. Pain 19, 9–21.

Lipton, J., Ship, J., and Larach-Robinson, D. 1993. Estimated prevalence and distribution of reported orofacial pain in the United States. J. Am. Dent. Assoc. 124, 115–121.

Magloire, H., Lesage, F., Couble, M. L., Lazdunski, M., and Bleicher, F. 2003. Expression and localization of TREK-1 K^+ channels in human odontoblasts. J. Dent. Res. 82, 542–545.

Maurin, J. C., Couble, M. L., Didier-Bazes, M., Brisson, C., Magloire, H., and Bleicher, F. 2004. Expression and localization of reelin in human odontoblasts. Matrix Biol. 23, 277–285.

Närhi, M. V. O., Yamamoto, H., and Ngassapa, D. 1996. Function of Intradental Nociceptors in Normal and Inflamed Teeth. In: Dentin/Pulp Complex (*eds*. M. Shimono, T. Maeda, H. Suda, and K. Takahashi), pp. 136–140. Quintessence.

Närhi, M. V. O. 2006. Nociceptors in the Dental Pulp. In: Pain Encyclopedia (*eds*. R. F. Schmidt and W. D. Willis), Springer.

Olgart, L. 1996. Neural control of pulpal blood flow. Crit. Rev. Oral Biol. Med. 7, 159–171.

Olgart, L. and Kerezoudis, N. P. 1994. Nerve–pulp interactions. Arch. Oral Biol. 39, 47S–54S.

Renton, T., Yiangou, Y., Plumpton, C., Tate, S., Bountra, C., and Anand, P. 2005. Sodium channel Nav1.8 immunoreactivity in painful human dental pulp. BMC Oral Health 5, 5.

Sessle, B. J. 2000. Acute and chronic craniofacial pain: brainstem mechanisms of nociceptive transmission and neuroplasticity, and their clinical correlates. Crit. Rev. Oral Biol. Med. 11, 57–91.

Sessle, B. J. 2005. Orofacial Pain. In: The Paths of Pain (*eds*. H. Merskey, J. D. Lowser, and R. Dubner), pp. 131–150. IASP Press.

Sessle, B. J., Hu, J. W., Amano, N., and Zhong, G. 1986. Convergence of cutaneous, tooth pulp, visceral, neck and muscle afferents onto nociceptive and non-nociceptive neurons in trigeminal subnucleus caudalis (medullary dorsal horn) and its implications for referred pain. Pain 27, 219–235.

Stephenson, J. L. and Byers, M. R. 1995. GFAP immunoreactivity in trigeminal ganglion satellite cells after tooth injury in rats. Exp. Neurol. 131, 11–22.

Sigurdsson, A. and Maixner, W. 1994. Effects of experimental and clinical noxious counterirritants on pain perception. Pain 57, 265–275.

Suzuki, H., Iwanaga, T., Yoshie, H., Li, J., Yamabe, K., Yanaihara, N., Langel, U., and Maeda, T. 2002. Expression of galanin receptor-1 (GALR1) in the rat trigeminal ganglia and molar teeth. Neurosci. Res. 42, 197–207.

Truelove, E. 2004. Management issues of neuropathic trigeminal pain from a dental perspective. J. Orofac. Pain 18, 374–380.

Wadachi, R. and Hargreaves, K. M. 2006. Trigeminal nociceptors express TLR-4 and CD-14; a mechanism for pain due to infection. J. Dent. Res. 85, 49–53.

Woodnutt, D. A., Wager-Miller, J., O'Neill, P. C., Bothwell, M., and Byers, M. R. 2000. Neurotrophin receptors and nerve growth factor are differentially expressed in adjacent nonneuronal cells of normal and injured tooth pulp. Cell Tissue Res. 299, 225–236.

Xie, Y. F., Zhang, S., Chiang, C. Y., Hu, J. W., Dostrovsky, J. O., and Sessle, B. J. 2006. Involvement of glia in central sensitization in trigeminal subnucleus cudalis (medullary dorsal horn). Brain Behav. Immun. (in press).

35 Ascending Pathways: Anatomy and Physiology

D Lima, Universidade do Porto, Porto, Portugal

Glossary

anterograde tracing Staining of axonal terminal arborizations with a substance (tracer) picked up by the neuronal soma and dendrites and transported through the axon up to its terminal structures.

antidromic activation Evoking neuronal spikes by electric activation of the axon terminal field.
antinociceptive action A neuronal effect that results in decrease of responses to pain

and inhibiton of neurons driven by nociceptive input.

ascending pathway Neuronal tract that conveys input from caudal to rostral areas along the spinal cord and brain.

contralateral pathway Pathway connecting gray matter regions (nuclei) located in opposite sides of the spinal cord or brain.

dendrites The branch units of a dendritic arbor.

dendritic arbor The receptive area of a neuron, organized from the perikarya as the ramifying branches of a tree.

dendritic spines Small protrusions of the dendritic surface that more often appear as a knob connected to the dendritic shaft by a short pedicle.

electrophysiological recording Recording of changes in the membrane potential or current in a neuron.

facilitatory loop Neuronal circuit playing a positive feedback action so that neuronal activity is enhanced.

fiber decussation Crossing of axons from one side to the other side of the spinal cord or the brain.

high-threshold neurons Neurons responsive solely to stimuli of high intensity (noxious).

ipsilateral pathway Pathway connecting gray matter regions (nuclei) located in the same side of the spinal cord or brain.

low-threshold neurons Neurons responsive solely to stimuli of low intensity (innocuous).

neuronal soma The central area of a neuron where the nucleus and most organelles are located.

neurotransmitters Molecules that functionally connect neurons at synapses by being delivered by the presynaptic element upon depolarization and acting upon ligand-gated receptors at the postsynaptic element.

nociresponsive Driven by the arrival of nociceptive input.

noxious stimulation Presentation of stimuli that represent a potential or effective aggression to a peripheral tissue.

pain inhibitory center Region in the central nervous system whose stimulation results in pain depression.

primary sensory neurons Neurons that convey input from the periphery to the central nervous system.

receptive field of a neuron The peripheral area whose stimulation elicits neuronal responses.

retrograde tracing Staining of the neuronal soma and dendritic arbors with a substance (tracer) picked up by the respective axonal boutons and transported back through the axon.

second-order neurons Neurons that transmit input from primary sensory neurons to higher order processing centers.

somatotopy Structural arrangement that correlates topographic organization at different sites, either at the periphery or in the central nervous system.

spinofugal transmission Input transmission away from the spinal cord.

target Area where the axon terminal arborization of a neuron is distributed.

transmitter receptors Plasma membrane molecules which, upon activation by a ligand (neurotransmitter), open ionic channels inducing alteration in the membrane potential.

wide dynamic-range neurons Neurons responsive to stimuli of graded intensities, from innocuous to noxious, but more intensively at the noxious range.

35.1 Introduction

35.1.1 Defining Nociceptive Ascending Pathways

Nociceptive information traveling from the periphery in primary sensory neurons is transmitted to second-order neurons located at the spinal cord and cranial sensory nuclei. From this first relay, various pathways distribute nociceptive input through higher processing centers so that pain is ultimately perceived in its multiple dimensions and adequate adaptive responses are generated. These ascending pathways are believed to terminate in the cortex with one or several relay stations in their way, although for most of the tracts, studies demonstrating that the intervening neurons are indeed serially connected from the spinal cord to the cortex are missing. The term 'nociceptive ascending pathways' is thus normally used in a restricted sense, to designate the neuronal tracts that connect the spinal cord with supraspinal regions, each pathway being named from the brain area at which it terminates.

The nociceptive nature of a pathway is classically demonstrated by recording responses to noxious stimuli from spinal neurons antidromically activated from the target (Perl, E. R. and Whitlock, D. G. 1961; Dilly, P. N. *et al.*, 1968). More recently, detection of molecular markers of nociceptive activation (Hunt, S. P. *et al.*, 1987) in conjunction with retrograde tracing has been largely used. By revealing large populations of putative nociceptive neurons, such a procedure allows an easy characterization of the location and morphology of neurons projecting in each pathway, and is particularly valuable to uncover varying population activation patterns as a function of stimulation conditions (Lima, D., 1998).

The structural features that best describe a pathway are the topographic and morphofunctional characteristics of the spinal neurons involved and the area of termination of their axons. Studies addressing the morphology of spinal projecting neurons are particularly few and mainly related to lamina I. One of the reasons for the restricted use of this kind of evaluation is the lack of systematic models of classification of spinal neurons apart from laminae I (Gobel, S. 1978a; Lima, D. and Coimbra, A., 1983; 1986) and II (Gobel, S., 1978b). Recent studies, however, call our attention to the importance of neuronal morphology, in particular, dendritic geometry, in defining the signal-processing properties of neurons (Prescott, S. A. and De Koninck, Y. 2002; Szucs, P. *et al.*, 2003; Mainen, Z. F. and Sejnowski, T. J. 2006), which justify taking these aspects into account when addressing the anatamophysiology of a pathway.

The multitude of ascending nociceptive pathways, together with the subtleness of the anatomical and physiological features that separate them as to their origin at the spinal cord, makes it difficult to attribute a particular functional meaning to each one. A tentative way of unraveling the role of each pathway in nociceptive processing has been the elucidation of the connections established by the target.

35.1.2   The Spinothalamic System

The classical view of the ascending nociceptive system puts particular emphasis on the dual distribution of nociceptive input through a lateral pathway responsible for sharp, well-localized short lasting

pain, and a medial pathway responsible for diffuse, poorly localized persisting pain (Figure 1). From the spinal cord, nociceptive input is conveyed both to the posterior lateral sensory nuclei of the thalamus, in the lateral or neospinothalamic pathway (Bowsher, D. 1957; Mehler, W. R. 1957; Willis, W. D. *et al.*, 1974; Giesler, G. J. *et al.*, 1976) and to medial thalamic nuclei, in the medial or paleospinothalamic pathway (Mehler, W. R. *et al.*, 1956; Bowsher, D. 1957; Mehler,

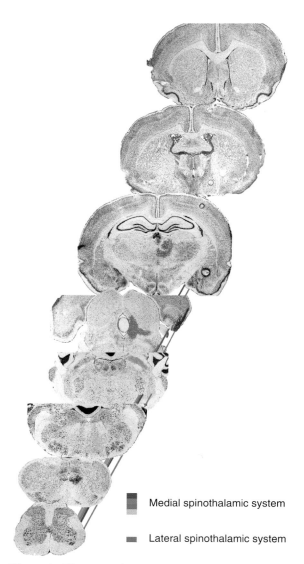

Medial spinothalamic system

Lateral spinothalamic system

Figure 1   Diagrammatic representation of the areas of termination of the lateral spinothalamic pathway and the medial spinothalamic pathway, and the spinoreticular pathways that serve as relays in the medial spinothalamic system. Brain and spinal cord photomicrographs were adapted from Paxinos, G. and Watson, C. 1998. The Rat Brain in Stereotaxic Coordinates, 4th edn. Academic Press.

W. R. 1957; Giesler, G. J. *et al.*, 1981b). While the first is monosynaptic, made up of spinal neurons projecting directly to the thalamus (Trevino, D. L. and Carstens, E. 1975; Willis, W. D. *et al.*, 1979; Giesler, G. J. *et al.*, 1979a), the second is either monosynaptic or polysynaptic with a variable number of relays along the brainstem (Johnson, F. H. 1954; Mehler, W. R. *et al.*, 1956; Rossi, G. F. and Brodal, A. 1957; Bowsher, D. 1957; Carstens, E. and Trevino, D. L. 1978b; Willis, W. D. *et al.*, 1979; Giesler, G. J. *et al.*, 1979a).

The lateral spinothalamic pathway was proposed to be involved in the discriminative aspects of pain (Bowsher, D. 1957; Melzack, R. and Casey, K. L. 1968) based on early studies on the perception deficits resulting from lesions of lateral thalamic nuclei (Dejerine, J. and Roussy, G. 1906; Melzack, R. and Casey, K. L. 1968). This was later supported by studies showing that both spinal and thalamic neurons of the lateral spinothalamic pathway present small receptive fields (Giesler, G. J. *et al.*, 1981b; Willis, W. D. 1988), are capable of encoding the extent of the stimulated area and the intensity of the stimulus(Kenshalo, D. R., Jr. *et al.*, 1979; Peschanski, M. *et al.*, 1980; Guilbaud, G. *et al.*, 1985; Willis, W. D. 1988; Guilbaud, G. and Kayser, V. 1988), and terminate following a somatotopic pattern in the thalamus and cortex respectively (Whitsel, B. L. *et al.*, 1978; Boivie, J. 1979). The medial spinothalamic pathway was claimed to deal with the affective and volitive aspects of pain (Bowsher, D. 1957; Melzack, R. and Casey, K. L. 1968) based on similar behavioral studies on human and experimental animals suffering from lesions of various brainstem or thalamic areas (Dejerine, J. and Roussy, G. 1906; Walker, A. E. 1942b; Hécaen, H. *et al.*, 1949). This view was again supported by studies revealing that spinal and thalamic neurons of the medial pathway present large receptive fields, responses unrelated to stimulus intensity (Giesler, G. J. *et al.*, 1981b; Guilbaud, G. *et al.*, 1985) and no topographical arrangement of their axonal terminal arborizations in both the thalamus (Boivie, J. 1979) and cortex (Morison, R. S. and Dempsey, E. W. 1942; Jones, E. G. and Leavitt, R. Y. 1974).

In the last decades, the use of very sensitive electrophysiological and tracing techniques revealed a wide variety of brainstem loci capable of contributing as relays to the medial spinothalamic system (reviewed by (Lima, D., 1997), and uncovered ascending pathways that bypass the thalamus to terminate directly on telencephalic areas, including the cortex (Cliffer, K. D. *et al.*, 1991). In this chapter, the various pathways involved in the spinofugal transmission of nociceptive input will be referred following a caudorostral sequence.

35.2 Spinocervical Pathway

The spinocervical pathway is one of the ascending putative nociceptive pathways uncovered the earliest, based on its capacity to activate cortical sensory areas (Catalano, J. V. and Lamarche, G., 1957; Mark, R. F. and Steiner, J., 1958) and on the finding that neurons of the lateral cervical nucleus (LCN) project to the ventrobasal complex of the thalamus (Ha, H. and Liu, C. N., 1966; Craig, A. D. and Burton, H., 1979; Baker, M. L. and Giesler, G. J., 1985; Giesler, G. J., *et al.*, 1987). It has been claimed, however, to be mostly dedicated to the processing of tactile, hair movement input (Lundberg, A. and Oscarsson, O., 1961; Taub, A. and Bishop, P. O., 1965) and be mainly represented in carnivores (Flink, R. and Westman, J., 1986). The identification of a neuronal column continuing the lateral cervical nucleus caudally along the entire length of the cord (the lateral spinal nucleus), together with the finding that their neurons also send projections to the thalamus as well as to many other brain areas (see ahead), raised the possibility that the lateral cervical nucleus is part of a more extensive spinal neuronal aggregation lying within the dorsolateral fasciculus, which serves as a relay in various ascending nociceptive pathways. Nonetheless, differences in the morphology and response properties of lateral cervical and lateral spinal neurons (Giesler, G. J. *et al.*, 1979b; Menétrey, D. *et al.*, 1980) point to a specific role of the former in conveying ascending input to the lateral thalamus.

35.2.1 Spinal Laminae of Origin and Sites of Termination

Neurons projecting to the lateral cervical nucleus are mainly located in the ipsilateral spinal cord at lamina IV (Figure 2) (Bryan, R. N. *et al.*, 1973; Craig, A. D., 1976; Brown, A. G. *et al.*, 1976; Cervero, F. *et al.*, 1977; Brown, A. G. *et al.*, 1977; Craig, A. D., 1978; Brown, A. G. *et al.*, 1980a; Baker, M. L. and Giesler, G. J., 1984), where they amount to 60% of the entire population, followed by lamina III (10%) (Brown, A. G. *et al.*, 1980a). Scattered neurons are present in other laminae such as laminae I, V, VI, and VII (Craig, A. D., 1976; Craig, A. D., 1978; Brown, A. G. *et al.*, 1980a;

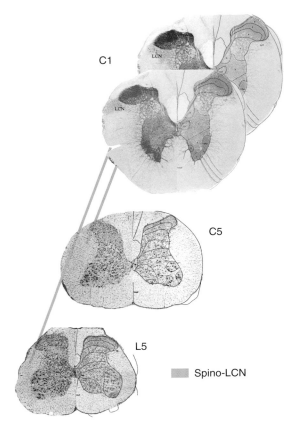

Figure 2 Diagram representing the spinal laminae of origin, ascendingb course in the spinal cord, and areas of termination of the spino-LCN pathway. Note the somatotopic arrangement of the axon terminal fields according to their rostrocaudal origin. (in this figure and figures 3, 7, 10, 12, 13, 15, 17, 19, 21 and 24, the left side is ipsilateral to the side of arrival of peripheral input) Brain and spinal cord photomicrographs were adapted from Paxinos, G. and Watson, C. 1998. The Rat Brain in Stereotaxic Coordinates, 4th edn. Academic Press.

Baker, M. L. and Giesler, G. J., 1984). Axonal terminals are somatotopically organized, fibers originated in caudal levels terminating in the dorsolateral aspect of the caudal portion of the lateral cervical nucleus, and fibers originated in rostral levels in the medial aspect of its rostral portion (Figure 2) (Svensson, B. A. et al., 1985).

35.2.2 Structural Types of Neurons Involved

Neurons targeting the lateral cervical nucleus are commonly described as small-sized (10–20 µm) (Craig, A. D., 1978; Baker, M. L. and Giesler, G. J.,

1984), but neurons with large round soma have also been observed in lamina IV (Craig, A. D., 1978). They present dorsally oriented dendrites (Brown, A. G. et al., 1976; Jankowska, E. et al., 1976; Brown, A. G. et al., 1977; Craig, A. D., 1978; Brown, A. G. et al., 1980b), which extend rostrocaudally for up to 2000 µm without penetrating lamina II (Brown, A. G. et al., 1977; Brown, A. G. et al., 1980b).

35.2.3 Spinal Location of Ascending Fibers

Spinal axons targeting the lateral cervical nucleus course ipsilaterally in the dorsal part of the lateral funiculus (Figure 2) (Brown, A. G. et al., 1977; Baker, M. L. and Giesler, G. J., 1984; Giesler, G. J. et al., 1988). According to Ha and Liu (Ha, H. and Liu, C. N., 1966), they are often collaterals of fibers coursing to more rostral levels.

35.2.4 Response properties

Nociceptive neurons at the origin of the spinocervical pathway present background activity in bursts (Brown, A. G. and Franz, D. N., 1969). They are activated by Aβ, Aδ, and C primary afferent fibers (Taub, A. and Bishop, P. O., 1965), A fibers exciting the entire neuronal population either alone (29%) or in convergence with C fibers (71%) (Brown, A. G. et al., 1975). They belong in the low-threshold (LT), wide-dynamic range (WDR) and high-threshold (NS) classes (Brown, A. G. and Franz, D. N., 1969; Bryan, R. N. et al., 1973; Bryan, R. N. et al., 1974; Cervero, F. et al., 1977; Downie, J. W. et al., 1988). Neurons receiving high-threshold input were reported to make up between 60% and 86% of the entire projecting population (Cervero, F. et al., 1977; Kajander, K. C. and Giesler, G. J., 1987). High-threshold input is generated either by pressure, pinch, heat, or cold (Brown, A. G. and Franz, D. N., 1969; Bryan, R. N. et al., 1973; Bryan, R. N. et al., 1974; Cervero, F. et al., 1977; Downie, J. W. et al., 1988), while low-threshold input is mainly originated in hair follicle afferent receptors (Brown, A. G. and Franz, D. N., 1969). Receptive fields are small and located in hairy as well as glabrous skin (Bryan, R. N. et al., 1974; Kunze, W. A. A. et al., 1987; Downie, J. W. et al., 1988). They are organized somatotopically so that cells located more laterally have receptive fields in the dorsal surface of the body and cells located more medially in the ventral surface (Bryan, R. N. et al., 1973; Bryan, R. N. et al., 1974; Brown, A. G. et al.,

1980a). Inhibitory receptive fields were described adjacent to the excitatory receptive field (Brown, A. G. *et al.*, 1987; Short, A. D. *et al.*, 1990) or in the contralateral limb (Brown, A. G. and Franz, D. N., 1969). Natural stimuli causing neuronal inhibition include hair movement, pressure, and squeezing (Brown, A. G. and Franz, D. N., 1969; Brown, A. G. *et al.*, 1987; Short, A. D. *et al.*, 1990). Convergence of cutaneous and deep tissues input has been reported (Kniffki, K. D. *et al.*, 1977; Hamann, W. C. *et al.*, 1978; Harrison, P. J. and Jankowska, E., 1984). Only cells responsive to both hair movement and skin pressure were shown to receive group III and IV muscle afferent input (Hamann, W. C. *et al.* 1978). Muscle and joint primary afferent activation can also elicit neuronal inhibition (Hamann, W. C. *et al.* 1978; Harrison, P. J. and Jankowska, E. 1984).

35.2.5 Pathways Driven at the Target

Axons from neurons in the lateral cervical nucleus join the medial lemniscus contralaterally and terminate in the midbrain, the ventral posterior lateral nucleus (VPL), and the posterior complex (PO) of the thalamus (Ha, H. and Liu, C. N., 1966). In the midbrain, the lateral part of the periaqueductal gray (PAG) receives afferents from the lateral two-thirds of the lateral cervical nucleus (Mouton, L. J. and Holstege, G., 2000; Mouton, L. J. *et al.*, 2004). In the VPL, fibers ascending from the lateral cervical nucleus are topographically arranged so that those originated in its dorsolateral portion terminate in the VPL, *pars lateralis*, and those originated in its ventromedial portion terminate in the VPL, *pars medialis* (Craig, A. D. and Burton, H., 1979).

35.3 Spinobulbar Pathways

Most spinobulbar pathways were uncovered relatively recently and are not as thoroughly studied as the more rostrally terminating pathways such as the spinomesencephalic and the spinothalamic. With the exception of the rostral ventromedial medulla (RVM), the majority of the studies dealing with spinobulbar pathways were performed in the rat. An anecdotic study addressing the spinal pathways terminating in the caudal medulla of the pigeon (Galhardo, V. *et al.*, 2000), however, reveals remarkable

similarities with the rat, pointing to a high degree of phylogenetic conservation.

35.3.1 Ventrolateral Reticular Formation

35.3.1.1 Spinal laminae of origin and sites of termination

Spinal cells projecting to the caudal ventrolateral reticular formation (VLM) are distributed through laminae I, II, IV–VII, VIII, X, and the lateral spinal nucleus (Figure 3) (Menétrey, D. *et al.*, 1983; Menétrey, D. *et al.*, 1984; Leah, J. *et al.*, 1988; Lima, D. *et al.*, 1991; Galhardo, V. *et al.*, 2000). Projections were first described to be mainly contralateral (Menétrey, D. *et al.*, 1983; Menétrey, D. *et al.*, 1984; Thies, R., 1985), but an important ipsilateral component was revealed by the use of more sensitive retrograde tracers (Menétrey, D. *et al.*, 1982; Lima, D. *et al.*, 1991; Menétrey, D. *et al.*, 1992b; Galhardo, V. *et al.*, 2000). There is, however, a large variability in the proportion of cells labeled in each spinal side, particularly with respect to cells located in the superficial dorsal horn. Both ipsilateral and contralateral predominance have been observed, in a few cases in the same animal at different rostrocaudal levels (Lima, D. *et al.*, 1991). The occurrence of subtle differences in the area of termination of each spinal side along the cord length was proposed as a tentative explanation, but experiments designed to clarify a putative somatotopic arrangement are missing.

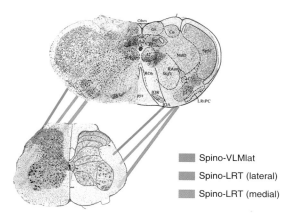

Spino-VLMlat
Spino-LRT (lateral)
Spino-LRT (medial)

Figure 3 Diagram representing the spinal laminae of origin, ascending course in the spinal cord, and areas of termination of the spino-VLM pathway. Note the contribution of lamina II neurons. Brain and spinal cord photomicrographs were adapted from Paxinos, G. and Watson, C. 1998. The Rat Brain in Stereotaxic Coordinates, 4th edn. Academic Press.

Spino-VLM axons appear to give axonal collaterals to various sites within the brainstem (Thies, R., 1985), such as the parabrachial nuclei (PBN) and the PAG (Spike, R. C. *et al.*, 2003).

The use of discrete injections, which became possible with the advent of particularly sensitive tracing methods, clearly demonstrated that the spino-VLM system is composed of three pathways originating in three distinct spinal regions (Figure 3).

The pathway originating in the superficial dorsal horn (laminae I–III) and lateral spinal nucleus was observed in the rat (Lima, D. and Coimbra, A., 1991; Tavares, I. *et al.*, 1993), cat, and monkey (Craig, A. D., 1995) to terminate in the lateral-most portion of the VLM (VLMlat), in between the spinal trigeminal nucleus and the ventral tip of the lateral reticular nucleus (LRt). The pathways originating in the deep dorsal horn (laminae IV–VI) and intermediate/ventral horn (laminae VII and X) terminate, respectively, in the lateral half and the medial half of the LRt (Lima, D. *et al.*, 1991).

A curious aspect of the spino-VLMlat projection is the participation of a large amount of cells of the substantia gelatinosa, or lamina II (Figure 4) (Lima, D. *et al.*, 1991; Lima, D. and Coimbra, A., 1991), an area that was not shown to participate significantly in any other ascending pathway and is therefore usually taken as involved in nociceptive modulatory local or propriospinal circuits (see Chapter Spinal Cord Physiology of Nociception). Lamina II neurons project, together with neurons in lamina I and the lateral spinal nucleus, to the VLMlat (Figure 3), contributing to almost one-third of the entire spinal-VLMlat projection in the cervical enlargement, and to one-fifth in the lumbar enlargement.

(Lima, D. and Coimbra, A., 1991). There are still no clues on the physiological meaning of this unique lamina II projection to the VLMlat. Studies aimed at clarifying whether these cells participate in inhibitory circuits similar to those described in the superficial dorsal horn failed to reveal GABA content in these neurons (Tavares, I. and Lima, D., 2002).

35.3.1.2 *Structural types of neurons involved*

Lamina I neurons projecting in the spino-VLM pathway were characterized in the rat, based on the comparison of the structural features of neurons retrogradely labeled with CTb at various spinal cord levels in the three anatomical planes (Lima, D. *et al.*, 1991). In the rat, as in the pigeon (Galhardo, V. *et al.*, 2000), spino-VLM neurons belong in three (fusiform,

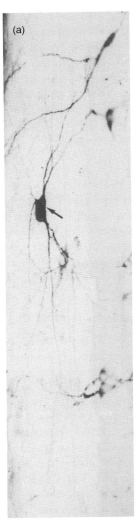

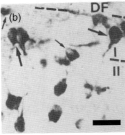

Figure 4 Superficial dorsal horn neurons labeled retrogradely with cholera toxin subunit B (CTb) from the VLMlat. (a) Lamina I neuron of the fusiform type, subtype B, in horizontal view. (b) Lamina I fusiforn neurons and lamina II neurons in transverse view. Note the ventrally oriented dendrites of fusiform B neurons (large arrows in (b)). DF – Dorsal funiculus. Scale bar = 30 μm. Adapted from figure 3 of Lima, D., Mendes-Ribeiro, J. A., and Coimbra, A. 1991. The spino-latero-reticular system of the rat: projections from the superficial dorsal horn and structural characterization of marginal neurons involved. Neuroscience 45, 137–152.

pyramidal, and flattened) of the four structural neuronal groups present in lamina I. The majority (around 80%) are of the fusiform type, a neuronal group whose main characteristic is the strict longitudinal spiny dendritic arbor. Some fusiform neurons present a few dendrites oriented ventrally and penetrating the entire width of lamina II (Figure 4). These neurons, classified as fusiform B by Lima and Coimbra (Lima, D. and Coimbra, A., 1986), represent a large fraction of VLMlat-projecting fusiform cells (20%) when compared to the small contribution of this neuronal subtype (6%) to the entire lamina I fusiform neuronal population (Lima, D. and Coimbra, A., 1986). Curiously, fusiform B neurons could not be observed in other spinofugal pathways. This finding is particularly interesting in the light of the contribution of lamina II neurons to this pathway (Figure 4). It may indicate that neurons of the fusiform B subtype cooperate with lamina II neurons in transmitting to the VLMlat primary input arriving at lamina II. The remaining 20% VLM-projecting lamina I neurons belong to the flattened and pyramidal types in similar amounts. These two cell groups have in common long practically aspiny dendrites coursing horizontally, parallel to the dorsal surface of lamina I. Pyramidal neurons have, in addition, dendrites that ramify inside the white matter overlying lamina I.

VLM-projecting lamina II neurons present ovoid, rostrocaudally elongated dendritic arbors that extend as parasagittal sheets through the entire width of lamina II (Lima, D. and Coimbra, A., 1991), resembling the central and the limiting cells of Ramón y Cajal (Ramón y Cajal, S., 1909).

35.3.1.3 Spinal location of ascending fibers

According to the intense fiber staining occurring in the dorsolateral fasciculus after retrograde tracer injections in theVLMlat (Lima, D. *et al.*, 1991), VLM projections from laminae I–III course in the dorsal portion of the lateral funiculus (Figure 3), as is the case of the spinomesencephalic (McMahon, S. B. and Wall, P. D., 1985; Hylden, J. L. K. *et al.*, 1986b) and spinothalamic (Apkarian, A. V. *et al.*, 1985; Apkarian, A. V. and Hodge, C. J., 1989c) pathways.

35.3.1.4 Surface receptors

Large numbers of lamina I neurons projecting to the VLM express neurokinin I (NK1) (Figure 5) (Todd, A. J. *et al.*, 2000; Spike, R. C. *et al.*, 2003; Castro, A. R. *et al.*, 2006) and GABA$_B$ receptors (Castro, A. R. *et al.*, 2006), the co-localization of both being relatively frequent (Castro, A. R. *et al.*, 2006).

In addition, 90% of the large deep dorsal horn neurons with dendrites entering superficial laminae and exhibiting the NK1 receptor project to the VLM (Todd, A. J. *et al.*, 2000). Appositions of serotonin (5HT) and noradrenalin-containing axonal boutons upon VLM-projecting lamina I neurons are common (Tavares, I. *et al.*, 1996a; Polgar, E. *et al.*, 2002). Serotoninergic boutons establish symmetrical synaptic contacts and are more abundant upon neurons expressing the NK1 receptor (Polgar, E. *et al.*, 2002).

35.3.1.5 Neurotransmitters

As in most ascending pathways conveying nociceptive input, studies addressing the neurochemical content of

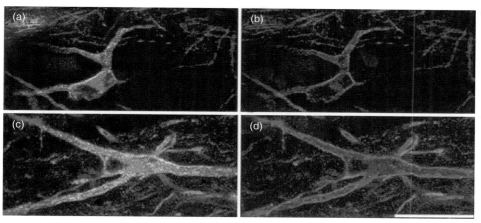

Figure 5 Flattened (a, b) and pyramidal (c, d) lamina I neurons, in horizontal view, retrogradely labeled with cholera toxin subunit B (CTb) from the VLM (red) and immunoreactive for the NK1 receptor (green). Scale bar = 50 μm. Adapted from figure 9 of Spike, R. C., Puskar, Z., Andrew, D., and Todd, A. J. 2003. A quantitative and morphological study of projection neurons in lamina I of the rat lumbar spinal cord. Eur. J. Neurosci. 18, 2433–2448.

the neurons of origin of the spino-VLM pathway have focused on neuropeptides. The exhaustive study of Leah and collaborators (Leah, J. *et al.*, 1988) revealed a relatively large amount of vasoactive intestinal peptide (VIP), bombesin, dynorphin, and substance P-immunoreactive VLM-projecting neurons in the lateral spinal nucleus. A few enkephalin immunoreactive neurons were observed in lamina X. Calbindin is present in particularly large numbers of lumbosacral cells projecting to the VLM-bilaterally, especially within lamina I and the lateral spinal nucleus, but also in lamina X (Menétrey, D. *et al.*, 1992b). Although fusiform neurons in lamina I (Lima, D. *et al.*, 1993) and neurons in lamina II (Todd, A. J. and Spike, R. C., 1993) are known to contain GABA, GABA-immunostaining of neurons retrogradely labeled from the VLM could not be observed (Tavares, I. and Lima, D., 2002).

35.3.1.6 Response properties

The response properties of neurons projecting to the ventrolateral reticular formation were studied in the rat (Menétrey, D. *et al.*, 1984) and cat (Thies, R., 1985). A high proportion of VLM-projecting neurons are spontaneously active (Menétrey, D. *et al.*, 1984; Thies, R., 1985). NS neurons make up about half of the entire population of VLM-projecting neurons (Menétrey, D. *et al.*, 1984; Thies, R., 1985). The remaining are either WDR or LT/proprioceptive neurons (Menétrey, D. *et al.*, 1984; Thies, R., 1985). They often present bilateral symmetrical receptive fields as well as cutaneous inhibitory receptive fields (Menétrey, D. *et al.*, 1984). Convergence of cutaneous, visceral, and muscle input is frequently observed (Thies, R., 1985). By monitoring noxious-evoked c-*fos* induction (Tavares, I. *et al.*, 1993), 10% to 20% of VLM-projecting neurons were shown to be activated by heat or mechanical stimulation in laminae I and IIo. In lamina IIi, neurons were activated in fewer numbers and only after thermal stimulation.

35.3.1.7 Pathways driven at the target

The VLM projects to several brain areas also receiving nociceptive input from the spinal cord and involved in pain control as well as cardiovascular, endocrine, or limbic functioning. Among these areas stand the RVM; the dorsal reticular nucleus (DRt); the A$_5$, A$_6$, and A$_7$ pontine noradrenergic groups (Tavares, I. *et al.*, 1996b); the hypothalamus (Calaresu, F. R. *et al.*, 1984; Malick, A. *et al.*, 2000); and the central nucleus of the amygdala (Zardettosmith, A. M. and Gray, T. S., 1995). Studies focused on the brainstem showed that the VLMlat is

the VLM region that establishes connections with other supraspinal pain inhibitory centers (Tavares, I. *et al.*, 1996b; Cobos, A. *et al.*, 2003), pointing to a role of the superficial dorsal horn-VLMlat pathway in driving the potent descending inhibition that can be elicited upon stimulation of VLM (Gebhart, G. F. and Ossipov, M. H., 1986). At the RVM and the pontine A$_5$ noradrenergic group, terminal arborizations of VLM axons appose spinal-projecting neurons (Figure 6), indicating that pain control actions from those areas are indeed under the control of the VLM (Tavares, I. *et al.*, 1996b).

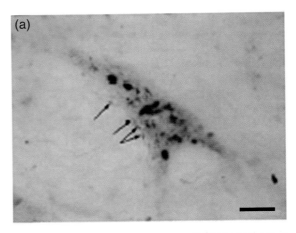

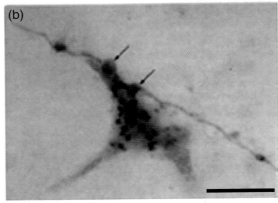

Figure 6 RVM (a) and A$_5$ (b) neurons retrogradely labeled with cholera toxin subunit B (CTb) from the spinal cord (brown granules) and receiving appositions (arrows) from axonal boutons anterogradely labeled with biotinilated dextran amine (BDA) from the VLM. Neuron in (b) is immunoreactive for dopamine-β-Hydroxylase (DBH; blue). Scale bar = 20 μm. (Adapted from figure 3 of Tavares, I., Lima, D., and Coimbra, A. 1996b. The ventrolateral medulla of the rat is connected with the spinal cord dorsal horn by an indirect descending pathway relayed in the A5 noradrenergic cell group. J. Comp. Neurol. 374, 84–95).

In the A_5, spinal-projecting neurons contacted by VLMlat fibers are noradrenergic (Figure 6) and post-synaptic in asymmetric, putative excitatory synaptic contacts (Tavares, I. *et al.*, 1996b). This was taken as suggestive that the spinal α_2 adrenoreceptor-mediated antinociceptive action triggered in the VLM is dependent on VLMlat activation and relayed in the A_5 group.

The VLM also sends direct descending projections to the spinal cord (Tavares, I. and Lima, D., 1994). Projections to both the superficial and deep dorsal horn originate in the VLMlat (Tavares, I. and Lima, D., 1994). Data from several studies indicate that noradrenaline is not used in the direct VLM-spinal pathway (Westlund, K. N. *et al.*, 1981; Westlund, K. N. *et al.*, 1983; Tavares, I. *et al.*, 1996b). VLMlat axons targeting lamina I make up a reciprocal closed VLM-spino-VLM loop which is entirely excitatory at the VLM level, and both excitatory and inhibitory at the spinal level (Tavares, I. *et al.*, 1998; Tavares, I. and Lima, D., 2002).

The cerebellar connections of the LRt (Cledenin, M. *et al.*, 1974; Parenti, R. *et al.*, 1996), together with the fact that, contrary to the VLMlat, the LRt does not contribute descending projections to the spinal cord dorsal horn (Tavares, I. and Lima, D., 1994), point to a role of the deep dorsal horn/ventral horn-LRt pathways in the control of motor activity in response to pain.

35.3.2 Dorsal Reticular Nucleus

35.3.2.1 Spinal laminae of origin and sites of termination

The medullary dorsal reticular nucleus (DRt) was first shown in the rat to be the site of termination of an important, mainly ipsilateral, pathway ascending from the spinal cord (Lima, D. and Coimbra, A., 1985; Lima, D., 1990; Villanueva, L. *et al.*, 1991). By the same time, neurons in the DRt were shown to be exclusively or preferentially activated by noxious stimuli from the skin and viscera (Villanueva, L. *et al.*, 1988). The spino-DRt pathway was later uncovered in other species such as the cat, monkey (Craig, A. D., 1995), and pigeon (Galhardo, V. *et al.*, 2000). It is constituted by a dorsal and ventral component differing in the area of termination within the DRt (Figure 7).

The dorsal pathway terminates at the dorsal-most portion of the DRt, immediately above the level of the central canal and surrounding the ventral border of the cuneate nucleus (Almeida, A. *et al.*, 1995; Almeida, A. *et al.*, 2000). It originates from the medial-most part of laminae I–III ipsilaterally, with a marked predominance of lamina I, and from lamina X, bilaterally (Figure 7) (Lima, D., 1990). The ventral pathway terminates ventrally to the area of termination of the dorsal pathway, within both sides of the DRt (Almeida, A. *et al.*, 1995; Almeida, A. *et al.*, 2000).

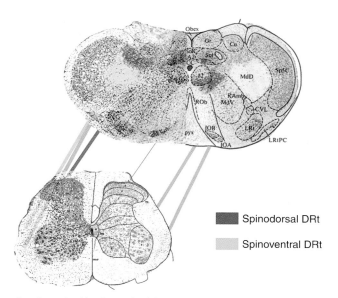

Figure 7 Diagram representing the spinal laminae of origin, ascending course in the spinal cord, and areas of termination of the spino-DRt pathway. Note the preferential medial location of dorsal horn neurons. Fibers of the spinodorsal DRt pathway course in the dorsal funiculus. Brain and spinal cord photomicrographs were adapted from Paxinos, G. and Watson, C. 1998. The Rat Brain in Stereotaxic Coordinates, 4th edn. Academic Press.

Its cells of origin prevail ipsilaterally in the medial portion of laminae IV–VI, with additional bilateral participation from laminae VII and X (Figure 7) (Lima, D., 1990). Interestingly, retrograde labeling from the superficial or deep dorsal horn demonstrated descending pathways arising, respectively, from the ipsilateral dorsal-most part of the DRt or from both sides of the ventral DRt (Tavares, I. and Lima, D., 1994). The DRt is thus part of two distinct reciprocal loops connecting its dorsal part with the superficial dorsal horn ipsilaterally, and ventral part with the deep dorsal horn bilaterally (Almeida, A. et al., 1993; Almeida, A. et al., 2000).

35.3.2.2 Structural types of neurons involved

Lamina I neurons projecting to the dorsal DRt were structurally characterized in the rat based on observations of neurons retrogradely filled with CTb in the three anatomical planes (Lima, D. and Coimbra, A., 1990). As to other spinal laminae, no data have been collected so far. In the rat, about 30% of lamina I neurons projecting to the DRt belong in the pyramidal and flattened types in similar amounts. The remaining 70% are of the multipolar type, a finding particularly curious taking into account that multipolar neurons were not seen to project in other ascending pathways. These cells are rather peculiar due to the typical lamina II crossing of the dendritic arbor, the bush pattern of the highly ramified proximal dendritic branches and the profusion and large variety of dendritic spines (Lima, D. and Coimbra, A., 1986). Another interesting feature of multipolar cells is their preferential location in the medial third of lamina I in the rat (Lima, D. and Coimbra, A., 1983; Lima, D. and Coimbra, A., 1986). The same structural types of lamina I neurons were seen to project to the DRt in the pigeon, although flattened and pyramidal neurons were relatively more abundant in this species (Galhardo, V. et al., 2000).

35.3.2.3 Spinal location of ascending fibers

An interesting aspect of the spino-DRt pathway, apparently only shared with the spino-NTS pathway (Gamboa-Esteves, F. O. et al., 2001c), is the course of their axons in the dorsal funiculus. This was first suggested by the dorsal orientation of axonal processes of lamina I neurons labeled retrogradely from the DRt and by comparing retrograde labeling rostrally and caudally to lesions of the dorsal funiculus (Lima, D., 1990). Retrograde labeling

was completely abolished in ipsilateral laminae I–III and markedly diminished in the deep dorsal horn at segments caudal to the lesion. Anterograde tracing later revealed that numerous fibers are labeled in the dorsal funiculus after injections in superficial laminae, and in the dorsolateral fasciculus after injections in deep dorsal horn laminae (Almeida, A. et al., 1995). Accordingly, injections in the dorsal funiculus resulted in DRt labeling restricted to its ipsilateral dorsal part, while injections in the dorsolateral fasciculus produced ipsilateral ventral DRt labeling (Almeida, A. et al., 1995). These findings not only confirm a dual ascending tract for the dorsal and ventral DRt pathways, but also indicate that fiber decussation in the ventral pathway takes place near the segment of origin, as in most other pathways.

35.3.2.4 Surface receptors

A significant amount of lamina I neurons projecting to the DRt express the NK1 receptor alone (Todd, A. J. et al., 2000; Castro, A. R. et al., 2006) or together with the GABA$_B$ receptor (Castro, A. R. et al., 2006). DRt-projecting neurons expressing only the GABA$_B$ receptor are, however, much more numerous (Castro, A. R. et al., 2006). In laminae III–IV, about 20% of NK1-expressing neurons project to the DRt (Todd, A. J. et al., 2000).

35.3.2.5 Neurotransmitters

So far, there are no studies addressing the possible neurotransmitters used in the spino-DRt pathway. Nevertheless, it is interesting to note that, although immunoreactions for GABA revealed immunostaining of lamina I neurons of the multipolar type, (Lima, D. et al., 1993), even in the early 2000s, GABA could not be detected in projecting spinal neurons (Gamboa-Esteves, F. O. et al., 2001b; Tavares, I. and Lima, D., 2002).

35.3.2.6 Response properties

The few data regarding the response characteristics of spino-DRt neurons rely on the induction of the c-fos proto-oncogene as a marker of activation of spinal neurons following noxious stimulation (Hunt, S. P. et al., 1987). This kind of approach only permitted to conclude on the activation by various kinds of cutaneous and visceral noxious stimulation of DRt-projecting lamina I neurons of all the three structural groups involved, namely multipolar, flattened, and pyramidal (Almeida, A. and Lima, D., 1997). The rate of activation of DRt-projecting lamina I neurons varied from 20% to 80% depending on the kind of stimulation applied and the neuronal cell group (Almeida, A. and Lima, D., 1997). In the deep dorsal horn, although high levels of c-*fos* expression were observed, the amount of activation of projecting neurons did not exceed 5%. Electrophysiological studies are absolutely needed to further characterize the response properties of spino-DRt neurons and to clarify the involvement of the deep dorsal horn in the transmission of nociceptive input to the DRt.

35.3.2.7 Pathways driven at the target

Projections from the DRt reach the parafascicular, ventromedial, and reunions thalamic nuclei (Figure 8) in the rat (Villanueva, L. *et al.*, 1998), which in turn are connected with telencephalic areas involved in emotional/affective and cognitive control. A putative spinoreticulothalamocortical projection relayed at the DRt has been proposed (Desbois, C. and Villanueva, L., 2001). Widespread projections to other brain targets of nociceptive ascending pathways have also been described. These include the VLM; the NTS; the rostral ventromedial medulla; the pontine noradrenergic cell groups A_5, A_6, and A_7 (PBN); the PAG; the posterior thalamus; the hypothalamus; the septal nuclei; the globus pallidus; and the amygdala (Figure 8) (Bernard, J. F. and Besson, J. M., 1990; Bernard, J. F. *et al.*, 1990; Villanueva, L. *et al.*, 1998; Leite-Almeida, H. *et al.*, 2006). Important projections to the orofacial motor nuclei (Bernard, J. F. *et al.*, 1990; Leite-Almeida, H. *et al.*, 2006) as well as to the deep cerebellar nuclei (Leite-Almeida, H. *et al.*, 2006) favor an important role in the organization of facial expressions and vocalization triggered by noxious stimulation.

The DRt also projects to the spinal cord superficial (Tavares, I. and Lima, D., 1994) and deep dorsal horn (Tavares, I. and Lima, D., 1994; Villanueva, L. *et al.*, 1995) as well as to the intermediate/ventral horn (Villanueva, L. *et al.*, 1995). DRt axons terminating in lamina I participate in a closed reciprocal spinodorsal DRt-spinal loop which, based on the asymmetric structure of synapses, is likely to be excitatory at both the spinal and medullary levels (Figure 9) (Almeida, A. *et al.*, 1993; Almeida, A. *et al.*, 2000; Lima, D. and Almeida, A., 2002). Nociceptive input arriving at the DRt is thus thought to drive a reverberating, lamina I centered pain facilitatory circuit, which is in accordance with the high proportion of c-*fos* activated spino-DRt neurons when compared to that of activated cells projecting to other targets

(Lima, D., 1998). A ventral DRt-deep dorsal horn pain facilitatory loop is also likely to be established as indicated by the increased responsiveness of WDR deep dorsal horn neurons upon DRt stimulation (Dugast, C. *et al.*, 2003).

35.3.3 Nucleus Tractus Solitarii

35.3.3.1 Spinal laminae of origin and sites of termination

The spino-NTS pathway was demonstrated in the rat (Menétrey, D. and Basbaum, A. I., 1987; Leah, J. *et al.*, 1988; Esteves, F. *et al.*, 1993) and pigeon (Galhardo, V. *et al.*, 2000). It originates in laminae I, IV–VI, VII, X, and the lateral spinal nucleus, mainly contralaterally, but with an important ipsilateral component (Figure 10) (Esteves, F. *et al.*, 1993).

In the rat, the thoracic and sacral autonomic cell columns were also shown to participate (Menétrey, D. and Basbaum, A. I., 1987; Menétrey, D. and DePommery, J., 1991). In addition, an important contribution from the superficial laminae of the spinal trigeminal nucleus, *pars caudalis*, was revealed (Menétrey, D. and Basbaum, A. I., 1987). The spino-NTS pathway terminates in the caudal part of the NTS, the general visceral zone (Figure 10) (Loewy, A. D., 1990). Based on restricted retrograde tracer injections, cells projecting to the lateral subnucleus are fewer than those targeting the medial NTS, and do not include lamina I cells (Figure 10) (Esteves, F. *et al.*, 1993). Anterograde tracing confirmed this finding by showing that fibers originating in the superficial dorsal horn terminate bilaterally in the medial part of the commissural subnucleus, while fibers originating in the deep dorsal horn terminate ipsilaterally in the lateral subnucleus, with a few fibers distributed to the dorsomedial subnucleus (Gamboa-Esteves, F. O. *et al.*, 2001c).

35.3.3.2 Structural types of neurons involved

Again, studies referring to the structural characteristics of spino-NTS neurons only addressed lamina I. Similar to what was observed for the VLM, lamina I neurons projecting to the NTS belong in the fusiform, pyramidal, and flattened groups. However, contrary to the VLM, fusiform neurons contribute a small fraction. In spite of constituting half of the lamina I neuronal population (Lima, D. and Coimbra, A., 1983; Lima, D. and Coimbra, A., 1986), fusiform neurons amount only to 25% of NTS-projecting lamina I neurons, whereas flattened and pyramidal neurons make up

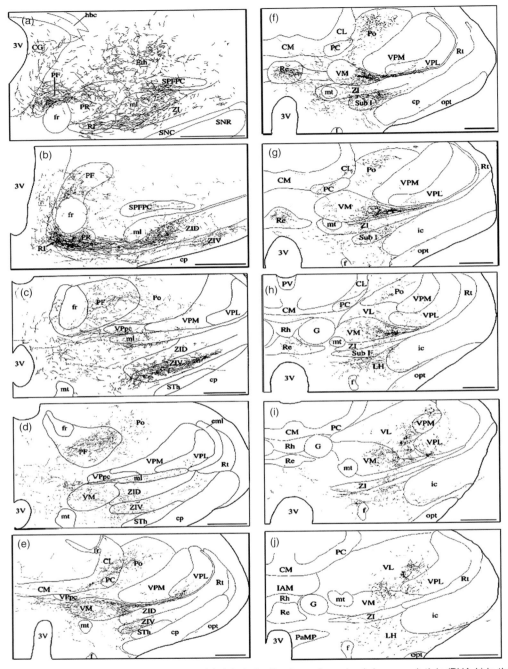

Figure 8 Distribution of DRt efferents anterogradely labeled with *phaseolus vulgaris* leucoagglutinin (PHA-L) in the diencephalon and telencephalon, in horizontal view. Adapted from figure 6 of Villanueva, L., Desbois, C., Le Bars, D., and Bernard, J. F. 1998. Organization of diencephalic projections from the medullary subnucleus reticularis dorsalis and the adjacent cuneate nucleus: a retrograde and anterograde tracer study in the rat. J. Comp. Neurol. 390, 133–160.

about 35–40% each. Similar relative amounts were observed in the pigeon, although pyramidal cells prevailed followed by flattened and fusiform cells (Galhardo, V. *et al.*, 2000). The abundant participation

of flattened cells in the spino-NTS pathway deserves particular attention since it may be related to the specific role of this pathway in pain processing. Flattened cells make up 10% of the entire lamina I

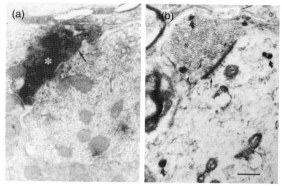

Figure 9 Anterogradely labeled axonal bouton (∗) in the DRt (a) and superficial dorsal horn (b) establishing asymmetric synaptic contacts (arrows) upon retrogradely labeled dendrites after injecting cholera toxin subunit B (CTb) in the superficial dorsal horn (a) or horseradish peroxidase (HRP) in the dorsal DRt (b). In (b), tracer deposits are pointed by cuved arrows. Scale bar −0.3 μm (a) Adapted from figure 3 of Almeida, A., Tavares, I., and Lima, D. 2000. Reciprocal connections between the medullary dorsal reticular nucleus and the spinal dorsal horn in the rat. Eur. J. Pain. 4, 373–387. (b) Adapted from figure 3 of Almeida, A., Tavares, I., Lima, D., and Coimbra, A. 1993. Descending projections from the medullary dorsal reticular nucleus make synaptic contacts with spinal cord lamina I cells projecting to that nucleus: an electron microscopic tracer study in the rat. Neuroscience 55, 1093–1106.

neuronal population and participate in a similar proportion in the spino-VLM (Lima, D. *et al.*, 1991) and the spino-DRt (Lima, D. and Coimbra, A., 1990) pathways. Their relative amount in the NTS system is, however, particularly high surpassing the one observed in the lateral spinothalamic system (25–30% at the spinal enlargements).

35.3.3.3 Spinal location of ascending fibers

By comparing anterograde tracing produced by injections centered in the dorsal funiculus or the dorsolateral fasciculus with that obtained from superficial or deep dorsal horn laminae, it was concluded that fibers from lamina I neurons course in the dorsal funiculus and fibers from deep dorsal horn neurons in the dorsolateral fasciculus (Figure 10) (Gamboa-Esteves, F. O. *et al.*, 2001c).

35.3.3.4 Neurotransmitters

A small fraction of the neurons projecting to the NTS was found, in immunocytochemical studies, to contain dynorphin in lamina I, VIP in the lateral spinal nucleus, and bombesin in lamina X (Leah, J. *et al.*, 1988). Calbindin was immunodetected in spinal NTS-projecting neurons located mainly in lamina I (Figure 11) and the lateral spinal nucleus (Menétrey, D. *et al.*, 1992b; Gamboa-Esteves, F. O. *et al.*, 2001a), as well as in lamina I of the spinal trigeminal nucleus, *pars caudalis* (Menétrey, D. *et al.*, 1992a). In lamina I, about 40% of fusiform NTS-projecting neurons contain calbindin. Pyramidal and flattened neurons also exhibit calbindin (Figure 11), but in smaller fractions (Gamboa-Esteves, F. O. *et al.*, 2001a). Glutamate also occurs mainly in fusiform neurons followed by pyramidal neurons, and co-localizes extensively with calbindin (Gamboa-Esteves, F. O. *et al.*, 2001a). Nitric oxide synthase is present in fewer cells, almost all of them the fusiform type (Gamboa-Esteves, F. O. *et al.*, 2001a). Calbindin-immunoreactive NTS-projecting neurons are c-*fos* activated by visceral and cutaneous stimulation (Gamboa-Esteves, F. O. *et al.*, 2001b). Glutamate-positive and nitric oxide synthase-positive neurons are c-*fos* activated only by visceral stimulation (Gamboa-Esteves, F. O. *et al.*, 2001b). About 5% neurons of NTS-projecting neurons of the pyramidal group are immunoreactive for substance P (Gamboa-Esteves, F. O. *et al.*, 2001a).

35.3.3.5 Response properties

By the use of the c-*fos* approach, spino-NTS neurons located in lamina I were shown to be activated by cutaneous and visceral noxious stimulation (Menétrey, D. and DePommery, J., 1991; Lima, D. *et al.*, 1994; Gamboa-Esteves, F. O. *et al.*, 2001b). They belong in the three neuronal groups that participate in the pathway (Esteves, F. *et al.*, 1993) irrespective of the kind of cutaneous or visceral noxious stimulation employed (Lima, D. *et al.*, 1994). However, cells activated by visceral input prevail over those activated by cutaneous input, and neurochemical differences between cutaneous- and visceral-activated cells were found (Gamboa-Esteves, F. O. *et al.*, 2001b).

35.3.3.6 Pathways driven at the target

The caudal NTS projects to several brain areas involved in pain processing such as the caudal ventrolateral reticular formation (Cobos, A. *et al.*, 2003), the dorsal reticular nucleus (Almeida, A. *et al.*, 2002), the rostral ventromedial medulla (Sim, L. J. and Joseph, S. A., 1994), the PBN (Cechetto, D. F. *et al.*, 1985), the PAG (Bandler, R. and Tork, I., 1987; Herbert, H. and Saper, C. B., 1992), and the hypothalamus (Reis, L. C. *et al.*, 2000). The NTS is also connected with the medullary vasopressor (Agarwal, S. K. and Calaresu,

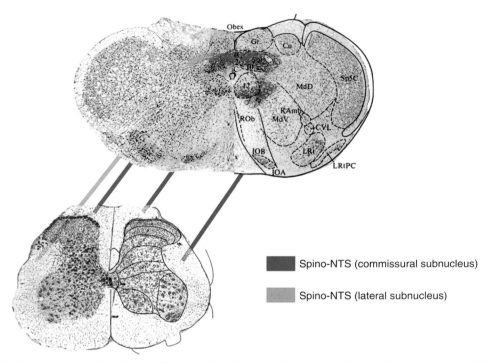

Figure 10   Diagram representing the spinal laminae of origin, ascending course in the spinal cord, and areas of termination of the spino-NTS pathway. Note the deep dorsal horn location of neurons terminating in the lateral subnucleus. Brain and spinal cord photomicrographs were adapted from Paxinos, G. and Watson, C. 1998. The Rat Brain in Stereotaxic Coordinates, 4th edn. Academic Press.

F. R., 1990) and vasodepressor (VLM) areas (Cobos, A. *et al.*, 2003). Although the NTS–VLM pathway is potentially involved in nociceptive/cardiovascular integration (Tavares, I. *et al.*, 1997), different VLM neurons are likely to mediate the antinociceptive and vasodepressive effects (Lima, D. *et al.*, 2001).

The caudal NTS also gives rise to projections descending directly to the spinal cord (Loewy, A. D. and Burton, H., 1978; Mtui, E. P. *et al.*, 1993; Tavares, I. and Lima, D., 1994). NTS fibers targeting the dorsal horn originate in the commissural subnucleus and terminate in superficial laminae (Tavares, I. and Lima, D., 1994).

35.3.4   Rostral Ventromedial Medulla

35.3.4.1   *Spinal laminae of origin and sites of termination*

Spinal cells projecting in the spino-RVM pathway predominate in laminae V, VII, VIII, and X (Figure 12) in the rat (Maunz, R. A. *et al.*, 1978; Kevetter, G. A. and Willis, W. D., 1982; Kevetter, G. A. *et al.*, 1982; Kevetter, G. A. and Willis, W. D., 1983; Chaouch, A. *et al.*, 1983; Peschanski, M. and Besson, J. M., 1984; Nahin, R. L. and Micevych, P. E., 1986;

Nahin, R. L. *et al.*, 1986), cat (Abols, I. A. and Basbaum, A. I., 1981; Ammons, W. S., 1987), and monkey (Haber, L. H. *et al.*, 1982; Kevetter, G. A. *et al.*, 1982).

The pathway is mainly contralateral, although a large number of ipsilaterally and bilaterally projecting cells have also been reported (Figure 12) (Haber, L. H. *et al.*, 1982; Thies, R. and Foreman, R. D., 1983; Foreman, R. D. *et al.*, 1984; Ammons, W. S., 1987). Cells are distributed throughout the entire length of the spinal cord, but are much more numerous at the upper cervical level due to an important contribution of the ipsilateral ventral horn (Kevetter, G. A. and Willis, W. D., 1983). Scattered neurons were observed in lamina I (Foreman, R. D. *et al.*, 1984; Leah, J. *et al.*, 1988) and the lateral spinal nucleus in the rat (Leah, J. *et al.*, 1988). The axonal termination domain distributes through the nucleus reticularis gigantocellularis and the nucleus paragigantocellularis (Figure 12) (Bowsher, D. and Westman, J., 1970; Kerr, F. W. L., 1975; Peschanski, M. and Besson, J. M., 1984). In the cat and monkey, fibers originating in lamina I were seen to extend medially through the reticular formation to terminate in nucleus raphe magnus (Craig, A. D., 1995).

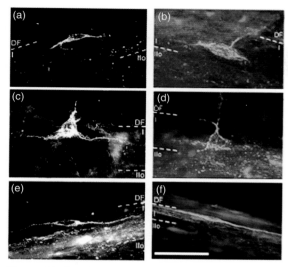

Figure 11 Lamina I neurons of the flattened (a, b), pyramidal (c, d), and fusiform (e, f) types retrogradely labeled with cholera toxin subunit B (CTb) from the NTS, in parasagittal view. In (b, d, and f), retrogradely labeled neurons present immunoreactivity for calbindin (green). Scale bar = 50 μm. (Adapted from figures 3 and 5 of Gamboa-Esteves, F. O., Kaye, J. C., McWilliam, P. N., Lima, D., and Batten, T. F. 2001a. Immunohistochemical profiles of spinal lamina I neurons retrogradely labelled from the nucleus tractus solitarii in rat suggest excitatory projections. Neuroscience 104, 523–538).

35.3.4.2   Structural types of neurons involved

Spino-RVM neurons located in lamina VII in the monkey were structurally relatively small as compared to spinothalamic neurons, with multipolar or, less frequently, fusiform or round perikarya and long dendrites across the lateromedial axis (Kevetter, G. A. et al., 1982).

35.3.4.3   Spinal location of ascending fibers

Spino-RVM axons course in the contralateral ventrolateral quadrant in the spinal cord (Figure 12) and follow ventrolaterally in the caudal medulla oblongata before they turn medially to reach their sites of termination (Rossi, G. F. and Brodal, A., 1957; Anderson, F. D. and Berry, C. M., 1959; Mehler, W. R. et al., 1960; Kerr, F. W. L., 1975; Nahin, R. L. et al., 1986).

35.3.4.4   Neurotransmitters

Immunocytochemical staining for enkephalin was observed in spino-RVM neurons located in laminae VII and X (Nahin, R. L. and Micevych, P. E., 1986).

Neurons in lamina X were also shown to contain cholecystokinin (CCK), somatostatin and, in fewer numbers, bombesin (Leah, J. et al., 1988). In the lateral spinal nucleus, a relatively large neuronal population is immunoreactive to VIP, while bombesin and dynorphin are expressed in just a few neurons (Leah, J. et al., 1988). Scarce somatostatin-immunoreactive neurons occur in lamina V (Leah, J. et al., 1988).

35.3.4.5   Response properties

Spino-RVM neurons are excited by stimulation of Aδ and C fibers from the skin, muscles, and viscera (Maunz, R. A. et al., 1978; Thies, R. and Foreman, R. D., 1983; Foreman, R. D. et al., 1984; Ammons, W. S., 1987). Both noxious and innocuous stimulation are effective, but the majority of RVM-projecting neurons are nociceptive (Maunz, R. A. et al., 1978; Haber, L. H. et al., 1982; Thies, R. and Foreman, R. D., 1983; Foreman, R. D. et al., 1984). Neurons responding only to cutaneous input belong in the LT, WDR, and NS classes (Fields, H. L. et al., 1977). Receptive fields vary from limited (Fields, H. L., et al., 1977) to large and complex, often including inhibitory regions (Fields, H. L. et al., 1977; Maunz, R. A. et al., 1978; Cervero, F. and Wolstencroft, J. H., 1984). Neurons responding to stimulation of deep structures are particularly numerous. They are mostly activated by deep noxious stimulation (Cervero, F. and Wolstencroft, J. H., 1984) and may receive convergent noxious or innocuous input from the skin (Fields, H. L. et al., 1977; Cervero, F. and Wolstencroft, J. H., 1984). Neurons responsive to cutaneous, deep, and visceral noxious stimulation have also been reported (Foreman, R. D. et al., 1984; Blair, R. W. et al., 1984a, 1984b; Ammons, W. S., 1987). The majority of these cells belong in the high-threshold class, the remaining being WDR neurons (Thies, R. and Foreman, R. D., 1983; Foreman, R. D. et al., 1984). Similar to pure somatic neurons, spinoreticular neurons with visceral input present either well-delimited or complex receptive fields, the latter including areas resulting in innocuous- or noxious-evoked inhibition (Thies, R. and Foreman, R. D., 1983; Foreman, R. D. et al., 1984). Ten percent to 20% of the neurons projecting to the RVM also target the lateral thalamus (Haber, L. H. et al., 1982; Foreman, R. D. et al., 1984).

35.3.4.6   Pathways driven at the target

The RVM was shown to project to the medial thalamus, including the intralaminar complex (Giesler, G.

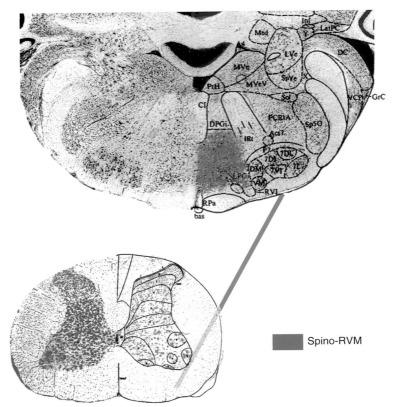

Figure 12 Diagram representing the spinal laminae of origin, ascending course in the spinal cord, and areas of termination of the spino-RVM pathway. Brain and spinal cord photomicrographs were adapted from Paxinos, G. and Watson, C. 1998. The Rat Brain in Stereotaxic Coordinates, 4th edn. Academic Press.

J. *et al.*, 1981b; Peschanski, M. and Besson, J. M., 1984), and hence was proposed to function as a relay station in a spinoreticulothalamic pathway involved in motor responses to noxious stimulation. It is also at the origin of an important pain control descending pathway that terminates in the superficial and deep dorsal horn and includes pain inhibitory and facilitatory neurons (Fields, H. L. *et al.*, 1995; Fields, H. L., 2000).

35.4 Spinopontine Pathways

The search for spinal fibers terminating in the brainstem (Craig, A. D., 1995), in particular in catecholaminergic nuclei (Westlund, K. N. and Craig, A. D., 1996), revealed various pontine spinal targets, which include the ventrolateral pons (A5), the locus coeruleus (A6), the subcoerulear region, the Kölliker–Fuse nucleus, and the PBN (Figure 13). These studies were, however, focused on lamina I ascending fibers, and, except for the PBN,

such putative nociceptive pathways were not thoroughly investigated. This section will, therefore, deal only with the spino-PBN system.

35.4.1 Parabrachial Nuclei

35.4.1.1 Spinal laminae of origin and sites of termination

Injections centered in the PBN in both the rat (Cechetto, D. F. *et al.*, 1985; Lima, D. and Coimbra, A., 1989; Hylden, J. L. *et al.*, 1989; Menétrey, D. and DePommery, J., 1991; Traub, R. J. and Murphy, A., 2002) and cat (Panneton, W. M. and Burton, H., 1985; Hylden, J. L. K. *et al.*, 1986a; 1986b) produce dense bilateral retrograde labeling in lamina I, mainly near the dorsal root entry zone, as well as in the lateral reticular portion of lamina V and laminae VIII and X (Figure 13). Additional labeling was observed in the intermediolateral column at thoracic levels, and in the parasympathetic column at sacral levels (Menétrey, D. and DePommery, J. 1991). Spino-PBN neurons with

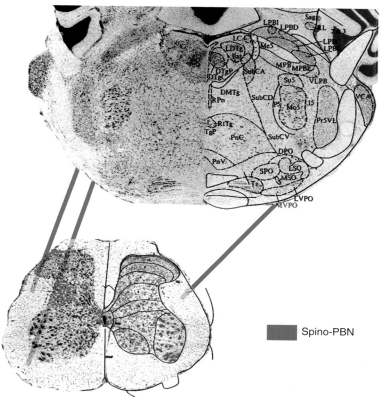

Spino-PBN

Figure 13 Diagram representing the spinal laminae of origin, ascending course in the spinal cord, and areas of termination of the spino-PBN pathway. Note the preferential lateral location of dorsal horn neurons. Spinal axonal termination areas in the locus coeruleus (LC), nucleus subcoeruleus (subCA) and the A_5 noradrenergic group are also represented. Brain and spinal cord photomicrographs were adapted from Paxinos, G. and Watson, C. 1998. The Rat Brain in Stereotaxic Coordinates, 4th edn. Academic Press.

axonal collaterals to the lateral thalamus were observed in all spinal areas with a marked prevalence in lamina I (Hylden, J. L. K. *et al.*, 1985; Hylden, J. L. *et al.*, 1989). A large percentage of spino-PBN neurons also project to the VLM (Spike, R. C. *et al.*, 2003).

The areas of termination of spinal fibers in the PBN are distributed bilaterally through the dorsal part of the lateral parabrachial nucleus, namely the dorsal, central, internal, and superior lateral subnuclei and the Kölliker–Fuse (Figure 13) (Cechetto, D. F. *et al.*, 1985; Blomqvist, A. *et al.*, 1989; Slugg, R. M. and Light, A. R., 1994). The medial and ventral lateral subnuclei are not targeted by spinal axons (Cechetto, D. F. *et al.*, 1985). No topographical arrangement has been disclosed (Blomqvist, A. *et al.*, 1989).

35.4.1.2 Structural types of neurons involved

Lamina I neurons labeled retrogradely in the rat following CTb injections centered in the PBN

(although extending to the cuneiform nucleus) belonged in the fusiform (65–70%) and pyramidal (30–35%) groups (Lima, D. and Coimbra, A., 1989). In the cat, lamina I neurons antidromically activated from mesencephalic sites with similar location and intracellularly stained resembled fusiform neurons (Figure 14) both from the description of their longitudinally extended spiny dendritic arbors and from their camera lucida drawings (Hylden, J. L. K. *et al.*, 1986a; see Figure 5).

35.4.1.3 Spinal location of ascending fibers

In the cat, spino-PBN fibers course bilaterally in the dorsolateral fasciculus and ipsilaterally in the ventrolateral and ventral funiculi (Figure 13) (Hylden, J. L. K. *et al.*, 1986b; Hylden, J. L. *et al..* 1989). Fibers originated in lamina I were shown to ascend through the dorsal aspect of the dorsolateral fasciculus (Hylden, J. L. K. *et al.*, 1986b). About one-fifth of

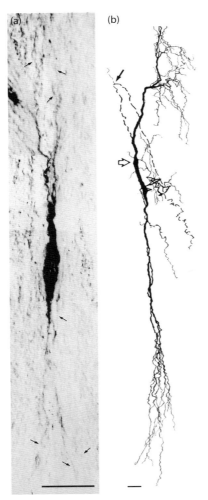

Figure 14 Lamina I PBN-projecting neurons retrogradely labeled with CTb (a) or intracellularly stained during antidromic activation (b), in horizontal (a) and parasagittal (b) views. In (a), arrows point to thin distal dendritic branches. In (b), the arrow points to the axon and the open arrow to the cell body. Scale bars = 30 μm. (a) Adapted from figure 4 of Lima, D. and Coimbra, A. 1989. Morphological types of spinomesencephalic neurons in the marginal zone (lamina I) of the rat spinal cord, as shown after retrograde labeling with cholera toxin subunit B. J. Comp. Neurol. 279, 327–339. (b) Adapted from figure 5 of Hylden, J. L. K., Hayashi, H., Dubner, R., and Bennett, G. J. 1986a. Physiology and morphology of the lamina I spinomesencephalic projections. J. Comp. Neurol. 247, 505–515).

lamina I neurons project bilaterally along the spinal cord (Hylden, J. L. K. *et al.*, 1986a).

35.4.1.4   Surface receptors

Most spino-PBN neurons located in lamina I express the NK1 receptor (Todd, A. J. *et al.*, 2000; Spike, R. C.

et al., 2003), and most lamina I neurons expressing the NK1 receptor and receiving synaptic contacts from TRPV1 immunoreactive primary afferent fibers project to the PBN (Hwang, S. J. *et al.*, 2003). Lamina I PBN-projecting Giant cells (lamina I cells three times larger than the remaining, amounting to about 5% in each lamina I structural group) (Lima, D. and Coimbra, A., 1983, 1986) of the pyramidal type were shown to lack the NK1 receptor and exhibit instead the glycine receptor-associated protein gephyrin (Puskar, Z. *et al.*, 2001). These cells are apposed by nitric oxide synthase and GABA-containing axonal boutons (Puskar, Z. *et al.*, 2001). In the deep dorsal horn, about 60% of NK1-immunoreacive neurons project to the PBN (Todd, A. J. *et al.*, 2000).

35.4.1.5   Neurotransmitters

Around half of the lamina I neuronal population projecting to the PBN immunostains for either dynorphin or enkephalin. Staining of sequential sections did not reveal co-localization of the two peptides (Standaert, D. *et al.*, 1986). According to Lima and colleagues (Lima, D. and Coimbra, A., 1989; Lima, D. *et al.*, 1993), lamina I enkephalinergic PBN-projecting cells should belong in the pyramidal group, whereas dynorphinergic cells could be either pyramidal or fusiform. Lumbosacral neurons projecting both ipsi- and contralaterally, mainly from lamina I and the lateral spinal nucleus, are immunoreactive for calbindin (Menétrey, D. *et al.*, 1992b).

35.4.1.6   Response properties

Neurons antidomically activated from the PBN in lamina I of the lumbar spinal cord of the rat (Bester, H. *et al.*, 2000) and cat (Hylden, J. L. K. *et al.*, 1985; Hylden, J. L. K. *et al.*, 1986a) belong mostly in the NS class (75–90%), the remaining being WDR neurons. They present extremely low spontaneous activity and small receptive fields, respond to stimulation of Aδ and C primary afferent fibers and conduct in the C–Aδ range (Hylden, J. L. K. *et al.*, 1986a; Bester, H. *et al.*, 2000). The large majority respond to both mechanical and heat-noxious stimulation and a few also to noxious cold stimulation. C-*fos* studies identified thoracolumbar spino-PBN neurons located preferentially in the superficial dorsal horn that were activated by visceral input (Menétrey, D. and DePommery, J., 1991; Traub, R. J. and Murphy, A., 2002). In whole-cell patch-clamp recordings, most lamina I neurons projecting to the PBN present a gap firing pattern, with a voltage-dependent delay in action potential firing, which was only shared by part of the neurons projecting to the PAG and not by neurons that were not labeled from these two sites (Ruscheweyh, R. *et al.*, 2004).

35.4.1.7 Pathways driven at the target

By combining antidromic activation from the thalamus with orthodromic activation from the periphery, Bourgeais and coworkers (Bourgeais, L. *et al.*, 2001b) demonstrated that neurons in the parabrachial internal lateral nucleus responding exclusively, with sustained firing, to noxious stimulation of large receptive fields, project to the paracentral thalamic nucleus. This spino-PBN-paracentral thalamic pathway was claimed to be responsible for triggering the aversive reactions to pain at the prefrontal cortex. By the use of a similar approach as well as by anatomical tracing, a spino-PBN-amygdaloid pathway with relay neurons in the external pontine parabrachial area and terminating in the lateral capsular division of the central nucleus of the amygdala was demonstrated (Ma, W. and Peschanski, M., 1988; Bernard, J. F. and Besson, J. M., 1990). This pathway was confirmed by retrograde transneuronal tracing from the amygdala with pseudorabies virus, and found to originate mainly in lamina I neurons (Jasmin, L. *et al.*, 1997).

35.5 Spinomesencephalic Pathways

Spinomesencephalic pathways target a multitude of regions located close to each other, which include the PAG, the intercollicular nucleus, the superior colliculus, the cuneiform nuclei, the posterior and anterior pretectal nuclei, and the nucleus of Darkschewitsch (Wiberg, M. and Blomqvist, A., 1984; Björkeland, M. and Boivie, J., 1984; Yezierski, R. P., 1988). The PAG itself, the major target of spinofugal mesencephalic pathways, has its spinal afferents distributed through several areas, each one playing particular integrative roles (Yezierski, R. P., 1988). Most retrograde tracing studies that refer to the spino-PAG pathway are based on injections that encompass different areas of the PAG as well as part of the above referred neighbor regions. Since the PAG is the principal site of termination of the spinomesencephalic tract, this chapter, will focus on the spino-PAG pathway without separating the various mesencephalic targets, as a large number of studies addressing this ascending system do. However, it should be kept in mind that it comprises several parallel systems that

are likely to deal with different aspects of pain processing.

35.5.1 Periaqueductal Gray

35.5.1.1 Spinal laminae of origin and sites of termination

Neurons of origin of the spino-PAG pathway are located in lamina I, the reticular region of laminae IV–V, laminae VI–VIII, lamina X, and the lateral spinal nucleus (Figure 15) in the rat (Menétrey, D. *et al.*, 1982; Beitz, A. J., 1982; Liu, R. P., 1983; Swett, J. E. *et al.*, 1985; Yezierski, R. P., 1988; Lima, D. and Coimbra, A., 1989; Yezierski, R. P. and Broton, J. G., 1991; Yezierski, R. P. and Mendez, C. M., 1991), cat (Wiberg, M. and Blomqvist, A., 1984; Yezierski, R. P., 1988), and monkey (Trevino, D. L., 1976; Mantyh, P. W., 1982; Wiberg, M. *et al.*, 1987; Yezierski, R. P., 1988; Zhang, D. *et al.*, 1990). In the cat, cells located in lamina I were found to account for the majority of spino-PAG neurons in the cervical and lumbar enlargements, but only to around 30% in the remaining spinal segments (Mouton, L. J. *et al.*, 2001). In the rat, an additional important projection, apparently exclusive of this pathway, originates from neurons located inside the white matter overlying lamina I, at the dorsal funiculus (Lima, D. and Coimbra, A., 1989).

Projections originated in the dorsal horn are mainly contralateral, especially from lamina I, the lateral spinal nucleus, and the dorsal funiculus (Trevino, D. L., 1976; Wiberg, M. *et al.*, 1987; Lima, D. and Coimbra, A., 1989), although a significant ipsilateral projection from lumbosacral spinal segments has been reported (Menétrey, D. *et al.*, 1992b). Projections originated in lamina X and the ventral horn are bilateral (Figure 15) (Trevino, D. L., 1976; Wiberg, M. *et al.*, 1987; Lima, D. and Coimbra, A., 1989). The upper cervical cord makes a major additional bilateral contribution both from the ventral horn and the lateral cervical nucleus (Yezierski, R. P. and Mendez, C. M., 1991; Mouton, L. J. and Holstege, G., 2000). Spino-PAG neurons were shown to leave axonal collaterals in the DRt, RVM, and locus coeruleus in the rat (McMahon, S. B. and Wall, P. D., 1985; Pechura, C. and Liu, R., 1986), and to collateralize a lot within the mesencephalon (Hylden, J. L. K. *et al.*, 1985; Yezierski, R. P. and Schwartz, R. H., 1986). Projections to both the mesencephalon and thalamus were reported in the rat (Harmann, P. A. *et al.*, 1988; Yezierski, R. P. and Mendez, C. M., 1991), cat (Hylden, J. L. K. *et al.*, 1986a; Yezierski, R. P. and Broton, J. G., 1991) and

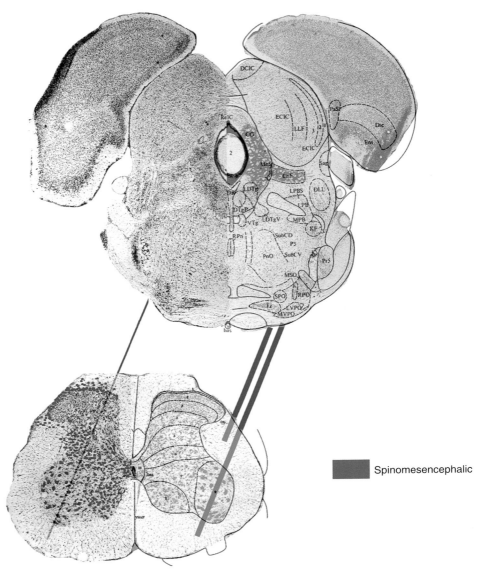

Figure 15 Diagram representing the spinal laminae of origin, ascending course in the spinal cord, and areas of termination of the spinomesencephalc pathway. (Only the caudal termination area is represented) Brain and spinal cord photomicrographs were adapted from Paxinos, G. and Watson, C. 1998. The Rat Brain in Stereotaxic Coordinates, 4th edn. Academic Press.

monkey (Price, D. D. *et al.*, 1978; Yezierski, R. P. *et al.*, 1987; Zhang, D. *et al.*, 1990). Thalamic sites of termination are mainly located in the ventrobasal complex, but collateralization to the posterior complex and medial thalamic nuclei has also been observed (Yezierski, R. P. and Mendez, C. M., 1991).

The mesencephalic sites of termination of the fibers ascending from the spinal cord and the spinal trigeminal nucleus were depicted in the rat (Yezierski, R. P., 1988), cat (Wiberg, M. and Blomqvist, A., 1984;

Björkeland, M. and Boivie, J., 1984; Wiberg, M. *et al.*, 1987; Yezierski, R. P., 1988), and monkey (Kerr, F. W. L., 1975; Wiberg, M. *et al.*, 1987; Yezierski, R. P., 1988). The termination pattern is very similar in the three species (Figure 15). With the exception of the nucleus of Darkschewitsch, terminal arborizations are sparse in the most rostral part of the mesencephalon (Yezierski, R. P., 1988; Lima, D. and Coimbra, A., 1989). Fibers are mainly distributed through the middle and caudal part of the PAG, nucleus cuneiformis, deep and intermediate gray layers of the superior colliculus, and intercollicular nucleus (Wiberg, M. *et al.*, 1987; Yezierski, R. P., 1988). In the caudalmost PAG of the monkey, but not in the rat and cat (Figure 16), spinal afferents contribute to two distinct dorsolateral and ventrolateral dense arborizations, while immediately rostrally, in the intercollicular region, they concentrate in a sole domain located laterally (Wiberg, M. *et al.*, 1987; Yezierski, R. P.,

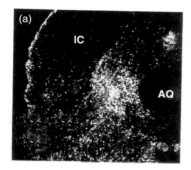

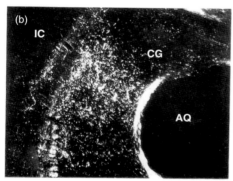

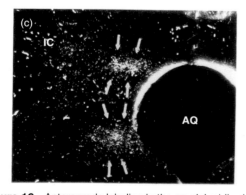

Figure 16 Anterograde labeling in the caudal midbrain following injection of wheat grem agglutinin–horseradish peroxidase (WGA-HRP) in the lumbosacral spinal cord of the rat (a), cat (b), and monkey (c). AQ, Cerebral aqueduct; CG , Periaqueductal gray; IC, Inferior colliculus. Adapted from figure 3 of Yezierski, R. P. 1988. Spinomesencephalic tract: projections from the lumbosacral spinal cord of the rat, cat, and monkey. J. Comp. Neurol. 267, 131–146.

1988), with additional labeling dorsally to the aqueduct (Yezierski, R. P., 1988).

This is an interesting finding in the light of data showing that the dorsolateral/lateral PAG is involved in aversive/defense behavior and vasopressor responses, and the vantrolateral PAG in immobility, positive reinforcing, and vasodepression (Lovick, T. A., 1993).

The possibility that the spino-PAG pathway is a composite of multiple pathways subserving the various functions in which the PAG is involved is supported by studies showing that different regions in the PAG receive afferents from distinct spinal neuronal populations (Keay, K. A. and Bandler, R., 1992; VanderHorst, V. G. J. M. *et al.*, 1996; Mouton, L. J. and Holstege, G.., 2000). Neurons at spinal segments C1–C3 that project to the lateral part of the PAG are mainly located in lamina I, whereas those projecting to the ventrolateral part of the PAG prevail in laminae VII–VIII (Keay, K. A. and Bandler, R., 1992). In the lumbosacral spinal cord, neurons in medial lamina VII and lamina VIII terminate in the lateral part of the lateral PAG and adjacent tegmentum, whereas neurons distributed thoughout laminae I and V terminate diffusely in the dorsal and lateral PAG (VanderHorst, V. G. J. M. *et al.*, 1996). In the cat, Mouton and Holstege (Mouton, L. J. and Holstege, G., 2000) described five distinct spinal neuronal groups based on their clustering pattern in the spinal cord and termination pattern in the PAG: (1) neurons located in laminae I and V along the entire length of the spinal cord and terminating in all parts of the intermediate and caudal PAG; (2) neurons located bilaterally in lateral laminae VI–VII and dorsolateral lamina VIII of segments C1–C3 and terminating in the ventrolateral and lateral part of the entire PAG and deep tectum; (3) neurons located in lamina X of the thoracic and upper lumbar cord and terminating in the ventrolateral and lateral PAG and deep tectum; (4) neurons located in medial laminae VI–VII of segments L5–S3 and terminating in the lateral and ventrolateral intermediate and caudal PAG; and (5) neurons located laterally in lamina I of segments L6–S2 and laminae V–VII and X of segments S1–S3 and terminating in the medial part of the ventrolateral intermediate and caudal PAG. According to electrophysiolgcal studies using antidromic activation (Yezierski, R. P. and Schwartz, R. H., 1986), spinal cells projecting to the rostral-most part of the PAG are located more ventrally, in laminae V–VII, than those projecting to the intercollicular and caudal levels, to where lamina I neurons project.

35.5.1.2 Structural types of neurons involved

Spinomesencephalic neurons were characterized as to the size and shape of the soma both in the rat and cat (Menétrey, D. *et al.*, 1980; VanderHorst, V. G. J. M. *et al.*, 1996). Neurons in lamina I are smaller than in other laminae and present oval to fusiform soma in transverse view (VanderHorst, V. G. J. M. *et al.*, 1996). Neurons in the lateral spinal nucleus and lamina X present oval to fusiform soma of variable sizes. Deep dorsal horn and ventral horn neurons are large and multipolar (Menétrey, D. *et al.*, 1980; VanderHorst, V. G. J. M. *et al.*, 1996).

A detailed structural characterization was obtained for lamina I neurons retrogradely labeled with CTb in the rat (Lima, D. and Coimbra, A., 1989). Large numbers of fusiform and pyramidal neurons were shown to participate. However, all injections were directed to the ventrolateral caudal PAG and encompassed part of the PBN. Injections hitting mainly the PBN failed to stain as many pyramidal neurons as those targeting the ventrolateral PAG, while still labeling a relatively large number of fusiform neurons. Pyramidal neurons were therefore taken as projecting mainly to the ventrolateral PAG and fusiform neurons to the PBN (Lima, D. and Coimbra, A., 1989). Some fusiform neurons projecting to the PAG were shown to have myelinated axons and give off collaterals inside lamina I (Hylden, J. L. K. *et al.*, 1986a).

35.5.1.3 Spinal location of ascending fibers

Fibers of the spinomesencephalic pathway are classically considered to travel in the ventrolateral quadrant of the spinal white matter (Mehler, W. R. *et al.*, 1960; Kerr, F. W. L., 1975). More recent data using antidromic activation revealed that fibers arising from lamina I decussate near their level of origin and course in the dorsal part of the dorsolateral fasciculus (McMahon, S. B. and Wall, P. D., 1985; Hylden, J. L. K. *et al.*, 1986b).

35.5.1.4 Surface receptors

The majority of spino-PAG neurons in lamina I, but not in the deep dorsal horn, express the NK1 receptor (Todd, A. J. *et al.*, 2000). Their amounts are, however, smaller than those of neurons projecting to the VLM or the PBN (Spike, R. C. *et al.*, 2003).

35.5.1.5 Neurotransmitters

In the lateral spinal nucleus and lamina X, neurons containing various neuropeptides and projecting to a mesencephalic area centered in the PAG (but extending to the parabrachial nuclei) were observed (Leah, J. *et al.*, 1988). Neurons in the lateral spinal nucleus were immunoreactive for VIP, bombesin, and substance P, while those located in lamina X were immunoreactive for bombesin and enkephalin. A few VIP-immunoreactive neurons were located in lamina I. At lumbosacral spinal levels, calbindin immunoreactive PAG-projecting neurons were observed bilaterally in all spinal areas of origin of the pathway, with a particularly high concentration in lamina I and the lateral spinal nucleus (Menétrey, D. *et al.*, 1992b).

35.5.1.6 Response properties

The response properties of spinal neurons projecting to the PAG were recorded in the rat (Menétrey, D. *et al.*, 1980), cat (Yezierski, R. P. and Schwartz, R. H., 1986; Yezierski, R. P. and Broton, J. G., 1991), and monkey (Yezierski, R. P. *et al.*, 1987). PAG-projecting neurons belong in the LT, WDR, and NS classes. Both WDR and NS neurons respond to mechanical and heat stimuli at the noxious range. WDR neurons largely prevail over the other neuronal classes, representing about half of the population recorded. Many WDR cells were found to respond to both cutaneous and visceral/deep tissue stimulation (Yezierski, R. P. and Schwartz, R. H., 1986; Yezierski, R. P. *et al.*, 1987). In the rat, NS neurons are predominant in lamina I, WDR neurons are distributed through both lamina I and the deep dorsal horn, and LT cells prevail in the deep dorsal horn (Menétrey, D. *et al.*, 1980). In the cat and monkey, neurons are distributed evenly through the dorsal horn and around the central canal irrespective of the physiological class they belong to (Yezierski, R. P. and Schwartz, R. H., 1986; Yezierski, R. P. *et al.*, 1987).

C-*fos* studies (Clement, C. I. *et al.*, 2000; Keay, K. A. *et al.*, 2002) revealed neurons at the thoracic spinal cord that project to the rostral ventrolateral PAG to be activated by noxious visceral stimulation. Neurons at both the lumbosacral and upper cervical spinal cord and projecting to the caudal ventrolateral PAG were activated by hind limb muscle noxious stimulation. Lamina I neurons projecting to the caudal ventrolateral PAG at the lumbar enlargement and expressing c-*fos* following either mechanical, thermal, or chemical noxious stimulation of the skin or noxious stimulation of the urinary bladder belong in both the fusiform and pyramidal groups (Lima, D. *et al.*, 1992; Lima, D., 1998).

Most spino-PAG neurons, including those few cells projecting to both the PAG and the ventrobasal complex of the thalamus, present small excitatory receptive fields confined to a single limb (Yezierski, R. P. *et al.*, 1987). However, neurons with extensive and complex receptive fields have also been observed, in particular in the upper cervical cord and in deep spinal laminae, including lamina X (Menétrey, D. *et al.*, 1980; Yezierski, R. P. and Schwartz, R. H., 1986; Yezierski, R. P., 1990; Yezierski, R. P. and Broton, J. G., 1991). Both groups present complex inhibitory receptive fields and include NS and WDR neurons (Yezierski, R. P. and Schwartz, R. H., 1986; Yezierski, R. P. *et al.*, 1987; Yezierski, R. P. and Broton, J. G., 1991).

Spino-PAG lamina I neurons have slow conducting velocities, at the $A\delta$ range, while those in the deep dorsal horn and ventral horn conduct at the low A_β range (Yezierski, R. P. *et al.*, 1987). Neurons in the lateral spinal nucleus present particularly slow axons, which belong in the unmyelinated and thin myelinated classes (Menétrey, D. *et al.*, 1980). A recent study using whole-cell patch-clamp in spinal slices showed that spino-PAG neurons present either gap-firing or burst-firing patterns, contrary to neurons that were not labeled from either the PAG or the PBN (Ruscheweyh, R. *et al.*, 2004).

35.5.1.7 *Pathways driven at the target*

The spinomesencephalic pathway was first uncovered as a relay station of the medial spinothalamic tract. Early anatomical studies revealed a projection from the mesencephalon to the intralaminar nuclei of the thalamus (Bowsher, D., 1957). Later anterograde tracing studies confirmed this connection as well as important projections to the hypothalamus, striatum, and amygdala (Eberhart, J. A. *et al.*, 1985; Meller, S. T. and Dennis, B. J., 1991). Discrete injections confined to different portions of the PAG in the rabbit showed that the ventral portion is the main source of afferent systems (Meller, S. T. and Dennis, B. J., 1991). Two distinct ascending systems were recognized: a periventricular system terminating in intralaminar and midline thalamic nuclei and along the hypothalamus, and a ventrolateral system terminating in the ventral tegmental area, ventral thalamus, zona incerta, amygdala, substantia innominata, lateral preopric nucleus, diagonal band of Broca, and the lateral septal nucleus. This multitude of pathways is likely to reflect the morphofunctional complexity of the PAG and is

taken as indicative of a PAG role in motor responses, escape/avoidance, aversive versus positive reinforcing, and neuroendocrine and autonomic responses to pain (Lovick, T. A., 1993).

A dense descending projection connects the ventrolateral portion of the caudal PAG with the ipsilateral nucleus raphe magnus (NRM) and adjacent reticular formation, locus coeruleus (LC), nucleus subcoeruleus, and the ventral reticular formation of the medulla (Meller, S. T. and Dennis, B. J., 1991). Sparse fibers originated in the dorsal PAG and superior colliculus also terminate in the locus coeruleus/subcoeruleus area (Cowie, R. J. and Holstege, G., 1992). Since PAG projections to the spinal cord are limited and restricted to laminae VII–VIII (Behbehani, M. M., 1995), the analgesic effects elicited from PAG stimulation are likely to be mediated by these PAG-pontine and PAG-medullary pathways (Lovick, T. A., 1993).

35.6 Spinodiencephalic Pathways

Of the pathways terminating in the diencephalon, the spinothalamic are by far those known for longer and therefore more thoroughly investigated. Although a large proportion of the studies dealing with the spinothalamic system address together the lateral and medial pathways, it turned clear from clinical (Dejerine, J. and Roussy, G., 1906; Walker, A. E., 1942a; Hécaen, H. *et al.*, 1949), electrophysiological (Kenshalo, D. R., Jr. *et al.*, 1979; Giesler, G. J. *et al.*, 1981b), and anatomical (Boivie, J., 1979) studies that each pathway is engaged in particular aspects of nociceptive processing. The medial pathway has been implicated in arousal, motivational, affective, and motor responses to pain, and the lateral pathway in stimulus discrimination. Accordingly, in this chapter, the two pathways will be dealt with separately in spite of the difficulties raised by being often assessed as a whole, particularly in retrograde tracing studies. There are, nonetheless, common aspects that will be more thoroughly described in the context of the lateral spinothalamic pathway. Also, a relatively large percentage of spinothalamic neurons (around 15%) of various species and different spinal laminae projects to both the lateral and medial thalamus (Giesler, G. J. *et al.*, 1981b; Kevetter, G. A. and Willis, W. D., 1983; Stevens, R. T. *et al.*, 1989; Craig, A. D. *et al.*, 1989). These neurons share, however, all the properties of lateral spinothalamic neurons

(Giesler, G. J. *et al.*, 1981b) and are therefore considered as part of the lateral spinothalamic pathway.

35.6.1 Lateral Thalamus

The lateral spinothalamic pathway was the first nociceptive spinofugal pathway described (Edinger, L., 1890). Its identification, at the turning of the nineteenth century, was based on the observation in necropsia tissue from humans, of degenerating profiles at the lateral sensory thalamus after ventrolateral cordotomy that disrupted pain sensation (Quensel, F., 1898; Kohnstamm, O., 1900; Thiele, F. H. and Horsley, V., 1901; Collier, J. and Buzzard, E. F., 1903; Foerster, O. and Gagel, O., 1932; Clark, W. E.

L., 1936). Only much later, with the advent of neurophysiology and tracing techniques, this pathway was revealed in detail. Nevertheless, early clinical studies (Dejerine, J. and Roussy, G., 1906; Melzack, R. and Casey, K. L., 1968) correlated the lateral spinothalamic pathway with the discriminative processing of nociceptive input.

35.6.1.1 *Spinal laminae of origin and sites of termination*

The lateral spinothalamic pathway takes origin on the contralateral spinal cord in laminae I, IV–VI, VII–VIII, X, and the lateral spinal nucleus (Figure 17) in the rat, cat, and monkey (Trevino, D. L. and Carstens, E., 1975; Carstens, E. and Trevino, D. L.,

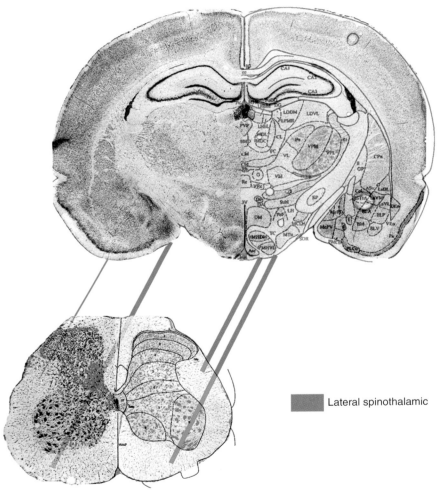

Lateral spinothalamic

Figure 17   Diagram representing the spinal laminae of origin, ascending course in the spinal cord and areas of termination of the lateral spinothalamic pathway. Note the cluster appearance of axon termination in the VPL. Brain and spinal cord photomicrographs were adapted from Paxinos, G. and Watson, C. 1998. The Rat Brain in Stereotaxic Coordinates, 4th edn. Academic Press.

1978b; Willis, W. D. *et al.*, 1979; Giesler, G. J. *et al.*, 1979a; Berkley, K. J., 1980; Leah, J. *et al.*, 1988; Lima.D. and Coimbra, A., 1988; Burstein, R. *et al.*, 1990b).

In laminae I and IV, a somatotopic arrangement has been described, neurons receiving input from extensor surfaces being located more laterally than neurons receiving input from flexor surfaces (Willis, W. D. *et al.*, 1974). The amount of spinothalamic neurons varies considerably along the rostrocaudal extent of the spinal cord due to additional intense labeling in particular areas at various spinal levels. This is the case of the deep dorsal horn contralaterally, and the intermediate/ventral horn, bilaterally, in spinal segments C1–C2 (Carstens, E. and Trevino, D. L., 1978a; Carstens, E. and Trevino, D. L., 1978b; Giesler, G. J. *et al.*, 1979a; Lima.D. and Coimbra, A., 1988; Burstein, R. *et al.*, 1990b), and the intermediate basilar nucleus of Cajal, contralaterally, in the rat (Giesler, G. J. *et al.*, 1979a; Lima.D. and Coimbra, A., 1988; Burstein, R. *et al.*, 1990b). Due to this regional variability, together with the fact that the multiple spinal groups projecting to the lateral thalamus also project to many other supraspinal targets, the erroneous assumption that thalamic projections from certain areas, such as lamina I, to the main lateral spinal target, the VPL, are not sufficiently relevant gained credit (Blomqvist, A. *et al.*, 2000; Craig, A. D. *et al.*, 2002; Klop, E. M. *et al.*, 2004). However, although the relative participation of lamina I neurons is below 10% in the cat (Klop, E. M. *et al.*, 2004) and rat (Burstein, R. *et al.*, 1990b) when the spinal cord is considered as a whole, small relative amounts are only found in segments where additional labeling occurs in particular spinal groups, as is the case of the upper cervical and lumbar cord (Burstein, R. *et al.*, 1990b). Notably, in the cervical enlargement of the rat, numbers of lamina I spinothalamic neurons equal those in the deep dorsal horn (Burstein, R. *et al.*, 1990b). Moreover, many lamina I neurons in the contralateral spinal and medullary dorsal horn of the monkey project to the ventrobasal complex of the thalamus, amounting to one-third of the entire dorsal horn labeled population (Willis, W. D. *et al.*, 2001).

Although the spinothalamic lateral pathway is classically considered to project contralaterally except for the ventral horn in segments C_1–C_2 (Trevino, D. L. and Carstens, E., 1975; Carstens, E. and Trevino, D. L., 1978a; Carstens, E. and Trevino, D. L., 1978b; Willis, W. D. *et al.*, 1979; Giesler, G. J. *et al.*, 1979a), the use of very sensitive tracers such as

CTb disclosed an important ipsilateral component (Lima.D. and Coimbra, A., 1988). Ipsilateral neurons amounted to about half the neurons labeled in the contralateral side in the lateral spinal nucleus and the deep dorsal horn in almost all the spinal segments examined. In the ventral horn, ipsilateral neurons equalized in number contralateral neurons, except in the cervical enlargement, where they were more abundant. Only in lamina I and the intermediate basilar nucleus of Cajal, contralateral neurons largely surpassed ipsilateral neurons. The need of using a very sensitive tracing technique to reveal the ipsilateral neuronal population suggests that these neurons may actually be contralaterally projecting neurons that send axonal collaterals to the ipsilateral thalamus. However, there is evidence from ventral spinal lesions in primates that nociceptive input is also conveyed supraspinally in the ipsilateral anterolateral quadrant (Vierck, C. J. and Luck, M. M., 1979).

The areas of spinal axon arborization in the lateral thalamus (Figure 17) were thoroughly studied in primates, including the monkey (Mehler, W. R. *et al.*, 1960; Bowsher, D., 1961; Mehler, W. R., 1966; Mehler, W. R., 1969; Kerr, F. W. L. and Lippman, H. H., 1974; Boivie, J., 1979; Berkley, K. J., 1980; Mantyh, P. W., 1983a; Apkarian, A. V. and Hodge, C. J., 1989a) and humans (Mehler, W. R., 1962; Mehler, W. R., 1974). In these species, as in the rat (Lund, R. D. and Webster, K. E., 1967; Mehler, W. R., 1969; Zemlan, F. P. *et al.*, 1978; Peschanski, M. *et al.*, 1983; Cliffer, K. D. *et al.*, 1991), the VPL is the major recipient of spinal fibers. In the VPL, spinal afferents are somatotopically arranged in rostrocaudally oriented clusters so that axons arriving from the lumbosacral spinal cord terminate in the lateral part of the nucleus and axons from the cervical enlargement terminate in the medial part (Boivie, J., 1979; Mantyh, P. W., 1983a). Such a somatotopic arrangement supports the ability of the lateral spinothalamic pathway to process spatial discrimination. Other important lateral thalamic areas of spinal termination are the posterior complex (PO), the ventral posteroinferior nucleus (VPI) and the zona incerta (ZI) ((Mehler, W. R., 1974; Boivie, J., 1979; Apkarian, A. V. and Hodge, C. J., 1989a; Cliffer, K. D. *et al.*, 1991). In the cat, fibers in the lateral spinothalamic pathway appear to be fewer and terminate in the ZI, the posterior complex, and in a shell area surrounding the VPL ventrolaterally (Boivie, J., 1971; Jones, E. G. and Burton, H., 1974; Berkley, K. J., 1980; Mantyh, P. W., 1983b; Craig, A. D. and Burton, H., 1985). Recently, Craig and colleagues claimed that a region located posteromedially to the VPL, which they called VMpo, is the site of termination of the lamina I spinothalamic fibers (Craig, A. D. *et al.*, 1994, 2002; Blomqvist, A. *et al.*, 2000). However, besides the fact that the so-called VMpo was most probably included in the area of termination of spinal and trigeminal thalamic afferents described by Mehler (Mehler, W. R., 1966) in humans, numerous retrograde and anterograde studies have proved that lamina I neurons project to many other areas in the thalamus (references given earlier), including a study by Craig (Craig, A. D., 2003) using anterograde tracing with *phaseolus vulgaris* leucoagglutinin. Moreover, this assumption was based on the dense calbindin-staining observed in the VMpo, but neither is the calbindin-immunoreactive region restricted to the VMpo lying within the medial aspect of the ventral posterior medial nucleus (Graziano, A. and Jones, E. G., 2004), nor are calbindin-immonoreactive projecting cells exclusively located in lamina I (Menétrey, D. *et al.*, 1992b).

35.6.1.2 *Structural types of neurons involved*

Spinothalamic cells were shown, in retrograde labeling studies, to have cell bodies that vary in shape from roundish or flattened to polygonal (Willis, W. D. *et al.*, 1979). They are mainly small in lamina I and include cells with fusiform, pyriform, and triangular shapes in transverse view, beyond the classic Waldeyer cells. In deep dorsal horn as well as the ventral horn, they are medium to large sized and polygonal in shape. Similar data were obtained by Apkarian and Hodge (Apkarian, A. V. and Hodge, C. J., 1989d), although in this case, tracer injections included the medial thalamus. Spinothalamic cells intracellularly stained in laminae IV–VIII of the cat (Meyers, D. E. R. and Snow, P. J., 1982) and monkey (Surmeier, D. J. *et al.*, 1988) presented long dendritic branches that could reach the lateral funiculus, lamina I and lamina X. In the intermediomedial gray matter, spinothalamic cells presented spheroidal cell bodies and narrow dendrites (Milne, R. J. *et al.*, 1982). Around the central canal, neuropeptide-containing spinothalamic cells had oval cell bodies and dendrites oriented transversely, reaching the central canal medially (Leah, J. *et al.*, 1988).

Injections of CTb confined to the VPL in the rat revealed that VPL-projecting lamina I neurons belong in the pyramidal and flattened groups (Lima.D. and Coimbra, A., 1988). Pyramidal cells prevailed over flattened cells in the cervical and lumbar enlargements (70% and 77%, respectively), the reverse occurring at

C1 and C2 (40%) (Lima.D. and Coimbra, A., 1988). In the cat (Zhang, E. T. *et al.*, 1996) and monkey (Zhang, E. T. and Craig, A. D., 1997), large tracer injections filling both the lateral and medial thalamus, resulted in labeling of fusiform neurons, beyond pyramidal and flattened neurons. Although the authors explained their labeling of fusiform neurons by putative species differences, fusiform neurons were most probably labeled from the medial thalamus (see further). It should be emphasized that, in the studies by Craig and co-workers (Zhang, E. T. *et al.*, 1996; Zhang, E. T. and Craig, A. D., 1997), flattened neurons were designated 'multipolar' due to their appearance in horizontal view, on which the authors based their observations. However, this designation is misleading and should be avoided since a neuronal group completely distinct in dendritic geometry and specializations was previously designated 'multipolar' in the rat (Lima, D. and Coimbra, A., 1986), and subsequently observed in the cat (Galhardo, V. and Lima, D., 1999), monkey (Lima, D. *et al.*, 2002), and pigeon (Galhardo, V. *et al.*, 2000) as well. When compared to flattened neurons, multipolar neurons have ventrally oriented rather than horizontal dendritic arbors and highly spiny rather than smooth dendritic branches. This kind of misuse of nomenclature already led some authors to disregard the occurrence of flattened neurons as an independent group in the cat and monkey, based on the results of retrograde labeling and immunostaining (Yu, X. H. *et al.*, 1999), despite the fact that flattened neurons were clearly identified in both species by the use of Golgi impregnation (Galhardo, V. and Lima, D., 1999; Lima, D. *et al.*, 2002).

35.6.1.3 *Spinal location of ascending fibers*

Axons of the lateral spinothalamic tract travel in the ventral, ventrolateral, and dorsolateral funiculi after decussating the spinal cord within a short distance from the cell body (Applebaum, A. E. *et al.*, 1975; Willis, W. D. *et al.*, 1979; Giesler, G. J. *et al.*, 1981a; Jones, M. W. *et al.*, 1985; Surmeier, D. J. *et al.*, 1988; Stevens, R. T. *et al.*, 1989; Apkarian, A. V. and Hodge, C. J., 1989a,b). Spinothalamic axons from lamina I cells were shown to project through the dorsolateral fasciculus in the cat (Apkarian, A. V. *et al.*, 1985; Stevens, R. T. *et al.*, 1989) and monkey (Apkarian, A. V. and Hodge, C. J., 1989a,b,c). Neurons located in the deep dorsal horn and ventral horn project through the ventrolateral fasciculus and ventral funiculus (Stevens, R. T. *et al.*, 1989; Apkarian, A. V. and Hodge, C. J., 1989b; Zhang, X. J. *et al.*, 2000). In the ventrolateral quadrant of the spinal cord white matter, axons are arranged somatotopically so that those originating in more caudal levels are located dorsolaterally to the more rostral ones (Horrax, G., 1929; Foerster, O. and Gagel, O., 1932; Hyndman, O. R. and Van Epps, C., 1939; Walker, A. E., 1940; Applebaum, A. E. *et al.*, 1975). The termination sites of the dorsolateral and ventrolateral fibers are equally distributed in the lateral thalamus, except for VPI and the ZI whose spinal afferents course mainly in the dorsolateral fasciculus and the ventral spinal quadrant, respectively (Apkarian, A. V. and Hodge, C. J., 1989a).

35.6.1.4 Surface receptors

Enkephalin immunoreactive varicosities were shown to establish asymmetric synaptic contacts upon medullary and spinal neurons retrogradely labeled from large HRP injections centered in the lateral thalamus of the cat and monkey (Ruda, M. A. *et al.*, 1984). These neurons make up 30% of lamina I and 50% of lamina V labeled neurons. In transverse sections, neurons present bipolar configuration in lamina I (equivalent to flattened neurons of Lima, D. and Coimbra, A., 1986) and multipolar configuration in lamina V (Ruda, M. A. *et al.*, 1984). Spinothalamic neurons exhibiting immunostaining for the NMDA receptor (Zou, X. Y. *et al.*, 2000) and metabotropic glutamate receptor subtype 1 (mGluR1) (Millis, C. D. and Hulsebosch, C. E., 2002) have been described in studies. A study focused on lamina I neurons retrogradely labeled from large injections comprising both the lateral and medial thalamus revealed NK1 receptors in flattened (called 'multipolar' by the authors) and pyramidal neurons, the latter being relatively few, however (Yu, X. H. *et al.*, 1999). This was taken as supporting the non-nociceptive nature of pyramidal neurons, although large amounts of pyramidal neurons expressing the NK1 receptor were observed by other authors (Todd, A. J. *et al.*, 2002), and pyramidal neurons are c-*fos*-activated following various kinds of noxious stimulation (Lima, D., 1998).

35.6.1.5 Neurotransmitters

Most studies addressing the neurochemical nature of spinothalamic neurons looked for the presence of neuropeptides in the rat. Neuropeptide-containing lateral spinothalamic neurons were preferentially observed in the lateral spinal nucleus and around the central canal (including lamina X). Both VIP (Nahin, R. L., 1988) and bombesin (Leah, J. *et al.*,

1988) were observed in the lateral spinal nucleus at the lumbar cord. In lamina X, neurons immunoreactive for bombesin (Leah, J. *et al.*, 1988), CCK (Ju, G. *et al.*, 1987; Leah, J. *et al.*, 1988), and galanin (Ju, G. *et al.*, 1987) were described. Galanin and CCK were seen to co-localize in lamina X neurons projecting to the VPL (Ju, G. *et al.*, 1987). Neurons containing glutamate or glutaminase were described in the lateral trigeminothalamic system in areas where WDR and low-threshold neurons predominate (Magnusson, K. R. *et al.*, 1987). Calbindin was claimed to be present in the majority of lamina I spinothalamic neurons of all structural groups projecting to the thalamus (Craig, A. D. *et al.*, 2002), although anterograde tracing combined with immunocytochemical staining failed to reveal calbindin-immunostaining in lamina I axons terminating in the thalamus (Graziano, A. and Jones, E. G., 2004).

35.6.1.6 Response properties

Spinothalamic cells were shown to present background activity at variable firing rates depending on the species and their laminar location in the spinal cord. Only few lamina I cells present background activity (Craig, A. D. and Kniffki, K. D., 1985) in the cat as compared to lamina I cells in the monkey (Ferrington, D. G. *et al.*, 1987) and to laminae IV–V cells in both species (Giesler, G. J. *et al.*, 1981b; Ferrington, D. G. *et al.*, 1986). Most spinothalamic cells respond to stimulation of C primary afferent fibers (Chung, J. M. *et al.*, 1979), but many of them respond to volleys in Aδ and Aβ fibers of somatic nerves as well (Foreman, R. D. *et al.*, 1975; Beall, J. E. *et al.*, 1977; Chung, J. M. *et al.*, 1979). Spinothalamic cells also respond to Aδ and C-fiber volleys in visceral nerves (Foreman, R. D. and Weber, R. N., 1980; Blair, R. W. *et al.*, 1981; Foreman, R. D. *et al.*, 1981, 1984; Rucker, H. K. and Holloway, J. A., 1982; Ammons, W. S., 1987) and to group II, III, and IV muscle afferents (Foreman, R. D. *et al.*, 1979). Lateral spinothalamic neurons were found to be activated by mechanical and/or thermal noxious stimulation of the skin in the rat (Giesler, G. J. *et al.*, 1976), cat (Fox, R. E. *et al.*, 1980; Ferrington, D. G. *et al.*, 1986), and monkey (Willis, W. D. *et al.*, 1974, 1979; Applebaum, A. E. *et al.*, 1975; Price, D. D. *et al.*, 1978; Kenshalo, D. R., Jr. *et al.*, 1979; Giesler, G. J. *et al.*, 1981b; Surmeier, D. J. *et al.*, 1986a,b; Ferrington, D. G. *et al.*, 1987) as well as by noxious chemical stimulation (Simone, D. A. *et al.*, 1991) and low-threshold mechanical (Willis, W. D. *et al.*, 1974; Applebaum, A. E. *et al.*, 1975; Giesler, G. J. *et al.*, 1976; Price, D. D. *et al.*, 1978) cooling

(Dostrovsky, J. O. and Craig, A. D., 1996; Craig, A. D. *et al.*, 2001) and warming (Andrew, D. and Craig, A. D., 2001) stimulation. Nociceptive neurons belong in both the NS and WDR classes (Willis, W. D. *et al.*, 1974; Giesler, G. J. *et al.*, 1981b; Ferrington, D. G. *et al.*, 1986). NS neurons are equally distributed throughout laminae I and IV–V, while WDR neurons predominate in laminae IV–V (Willis, W. D. *et al.*, 1974) and neurons responsive to cooling are located in lamina I (Craig, A. D. *et al.*, 2001). Thermal-responsive neurons belonging in either the NS or the WDR neuronal classes were shown to be capable of encoding noxious heat intensity (Figure 18) irrespective of their location in lamina I or the deep dorsal horn (Kenshalo, D. R., Jr. *et al.*, 1979; Ferrington, D. G. *et al.*, 1986; Surmeier, D. J. *et al.*, 1986a, b).

Neurons with convergent input from the skin and viscera also contribute to the lateral spinothalamic

pathway (Foreman, R. D. and Weber, R. N., 1980; Blair, R. W. *et al.*, 1981; Foreman, R. D. *et al.*, 1984; Ammons, W. S. *et al.*. 1984; Ammons, W. S., 1987; Al-Chaer, E. D. *et al.*, 1999; Chandler, M. J. *et al.*, 2000). Curiously, in lamina X, neurons of the postsynaptic dorsal column tract were shown to receive viscerosomatic input and reported to be more numerous than those of the spinothalamic tract (Al-Chaer, E. D. *et al.*, 1999; Dorsal Columns and Visceral Pain). These neurons belong in the WDR, NS, or high-threshold inhibitory classes, and present cutaneous receptive fields that occupy regions to which pain is frequently referred (Foreman, R. D. and Weber, R. N., 1980; Foreman, R. D. *et al.*, 1984; Ammons, W. S. *et al.*, 1984). Some of these neurons were shown to leave axonal collaterals in the medial medullary reticular formation (Foreman, R. D. *et al.*, 1984). Some spinothalamic neurons receive convergent input from the skin and deep tissues (Willis, W. D. *et al.*, 1974; Giesler, G. J. *et al.*, 1981b; Ferrington, D. G. *et al.*, 1986). These neurons are thought to have a proprioceptive function (Milne, R. J. *et al.*, 1982). They are mainly located in the lateral part of lamina V and intermediomedial gray matter (Stilling's nucleus) and respond to either weak or intense cutaneous stimulation (Willis, W. D. *et al.*, 1974).

Lateral spinothalamic neurons normally present ipsilateral receptive fields that vary from very small (less than one digit) to moderate (the entire limb) (Willis, W. D. *et al.*, 1974; Giesler, G. J. *et al.*, 1981b), as well as additional inhibitory receptive fields (Gerhart, K. D. *et al.*, 1981; Giesler, G. J. *et al.*, 1981b; Ammons, W. S., 1987). In a few cells, convergent inhibitory cutaneous or visceral receptive fields were reported (Willis, W. D. *et al.*, 1974; Blair, R. W. *et al.*, 1981; Milne, R. J. *et al.*, 1982). Receptive fields tend to be smaller in high-threshold neurons and in lamina I neurons (Applebaum, A. E. *et al.*, 1975; Giesler, G. J. *et al.*, 1981b; Ferrington, D. G. *et al.*, 1987). The relatively small size of the receptive fields of lateral spinothalamic neurons favors a role in discriminating the size of the stimulated area (Giesler, G. J. *et al.*, 1981b).

C-*fos* induction after noxious cutaneous or visceral stimulation was observed in spinothalamic neurons in laminae I, III–VII, and X (Palecek, J. *et al.*, 2003). Curiously, neurons projecting in the postsynaptic dorsal column pathway were activated in similar proportions by the noxious cutaneous stimuli and in even higher proportions by the visceral stimuli (Palecek, J. *et al.*, 2003). In lamina I, both flattened and pyramidal neurons projecting to the

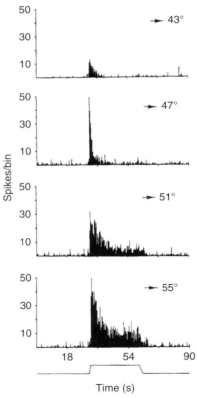

Figure 18 Responses of a lateral spinothalamic neuron to heat stimulation of increasing intensities. Adapted from figure 1 of Surmeier, D. J., Honda, C. N., and Willis, W. D. 1986a. Temporal features of the responses of primate spinothalamic neurons to noxious thermal stimulation of hairy and glabrous skin. J. Neurophysiol. 56, 351–369.

VPL were c-*fos*-activated following cutaneous mechanical, thermal, and chemical noxious stimulation, and, in smaller amounts, following visceral chemical stimulation (Lima, D. *et al.*, 1992; Lima, D., 1998).

35.6.1.7 Pathways driven at the target

The VPL is long-known for sending nociceptive input to the parietal somatosensory cortex (Burton, H. and Jones, E. G., 1976; Whitsel, B. L. *et al.*, 1978; Kenshalo, D. R., Jr. *et al.*, 1980). Combined anterograde tracing from the spinal cord and retrograde tracing from the somatosensory cortex in the monkey revealed that overlapping between spinal thalamic afferents and thalamic neurons projecting to cortical areas SI or SII occurs in the VPL, but also in the VPI and PO, the VPL being the area where the number of overlapping neurons is smaller (Stevens, R. T. *et al.*, 1993; Shi, T. and Apkarian, A. V., 1995). VPL afferents in cortical areas SI and SII are arranged in a somatotopic fashion (Burton, H. and Jones, E. G., 1976; Whitsel, B. L. *et al.*, 1978) so that the more posterior the thalamic source of afferents, the more posterior the cortical termination sites. Sparse cells in the lateral part of the VPL and the ventral part of the VPI target the cingulated cortex (Apkarian, A. V. and Shi, T., 1998).

The PO projects to the granular insular and retroinsular cortex in primates (Burton, H. and Jones, E. G., 1976) and was therefore proposed to be involved in nociceptive visceral processing (Cechetto, D. F. and Saper, C. B., 1987). PO neurons projecting to the anterior insula were seen to clearly overlap with spinothalamic axonal arborizations (Apkarian, A. V. and Shi, T., 1998). Neurons in the VPI also show insular projections, but with no overlapping. Moreover, neurons projecting to the insula do not overlap with neurons projecting to SI (Apkarian, A. V. and Shi, T., 1998).

35.6.2 Medial Thalamus

The medial spinothalamic pathway was described at the middle 1960s as an ascending nociceptive system especially devoted to the processing of the affective and motivational aspects of pain (Melzack, R. and Casey, K. L., 1968). While anatomical degeneration studies demonstrated the termination in medial thalamic nuclei of spinal axons ascending in the ventrolateral quadrant of the spinal cord (Bowsher, D., 1957, 1961; Mehler, W. R. *et al.*, 1960; Boivie, J., 1971), clinical studies revealed that, in patients with lesions centered in the medial thalamus, the painful poorly localized unpleasant feeling was abolished (Hécaen, H. *et al.*, 1949).

35.6.2.1 Spinal laminae of origin and sites of termination

Spinal neurons projecting to medial thalamic nuclei prevail in the contralateral intermediate and ventral gray matter (Figure 19) in the rat (Giesler, G. J. *et al.*, 1979a), cat (Carstens, E. and Trevino, D. L., 1978b), and monkey (Willis, W. D. *et al.*, 1979; Giesler, G. J. *et al.*, 1981b). In the cat, although a predominant location in laminae VII and VIII has been described by some authors (Carstens, E. and Trevino, D. L., 1978b; Comans, P. E. and Snow, P. J., 1981), others point to a laminar distribution similar to that observed for the lateral spinothalamic pathway, namely laminae I, IV–VI, and VII to X (Stevens, R. T. *et al.*, 1989; Craig, A. D. *et al.*, 1989). Lamina I was also shown to be the source of spinal afferents to the nucleus submedius in the cat and monkey (Craig, A. D. and Burton, H., 1981; Stevens, R. T. *et al.*, 1989) and to contribute to the spinal projections to the intralaminar complex in the monkey (Albe-Fessard, D. *et al.*, 1975; Ammons, W. S. *et al.*, 1985). In the rat, projections from laminae IV–VII were recently shown to target the central lateral nucleus, whereas the medial dorsal nucleus was found to be mainly innervated bilaterally by the lateral spinal nucleus and, to a lesser extent, by contralateral lamina I (Gauriau, C. and Bernard.J.F., 2004).

Spinal afferents ascending to the medial thalamus appear to be fewer than those reaching the lateral thalamus (Mehler, W. R. *et al.*, 1960; Apkarian, A. V. and Hodge, C. J., 1989a). They target similar medial thalamic regions in the rat (Lund, R. D. and Webster, K. E., 1967; Mehler, W. R., 1969; Zemlan, F. P. *et al.*, 1978; Cliffer, K. D. *et al.*, 1991), cat (Boivie, J., 1971; Berkley, K. J., 1980; Mantyh, P. W., 1983a; Craig, A. D. and Burton, H., 1985), monkey (Mehler, W. R. *et al.*, 1960; Boivie, J., 1979; Mantyh, P. W., 1983a; Apkarian, A. V. and Hodge, C. J., 1989a), and humans (Mehler, W. R., 1962; Mehler, W. R., 1974). These include the medial dorsal and paraventricular nuclei and the intralaminar complex, namely the central lateral, center median, paracentral, and parafascicular nuclei (Figure 19). In the monkey, the medial dorsal nucleus (except for its dorsomedial part) and the central lateral nucleus are the major sites of termination of spinal fibers (Apkarian, A. V. and Hodge, C. J., 1989a). The nucleus submedius is also a consistent site of spinal axon arborization in the cat (Boivie, J., 1971; Craig, A. D. and Burton, H., 1981; Mantyh, P.

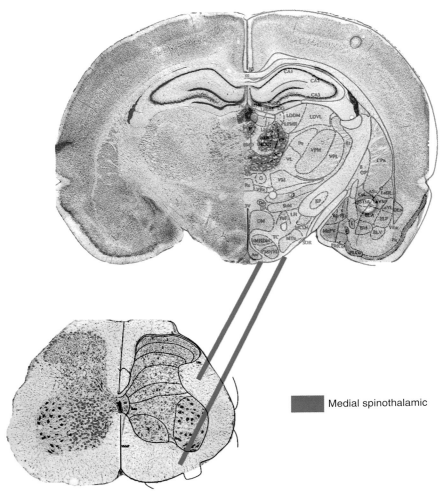

Medial spinothalamic

Figure 19 Diagram representing the spinal laminae of origin, ascending course in the spinal cord and areas of termination of the medial spinothalamic pathway. Brain and spinal cord photomicrographs were adapted from Paxinos, G. and Watson, C. 1998. The Rat Brain in Stereotaxic Coordinates, 4th edn. Academic Press.

W., 1983b; Craig, A. D. and Burton, H., 1985) and monkey (Boivie, J., 1979; Mantyh, P. W., 1983a; Apkarian, A. V. and Hodge, C. J., 1989a), but in relatively small amounts (Mantyh, P. W., 1983a; Apkarian, A. V. and Hodge, C. J., 1989a).

Contrary to the lateral thalamus, no somatotopic arrangement of spinal terminals could be detected in the intralaminar complex (Boivie, J., 1979), which agrees with electrophysiological studies (Giesler, G. J. *et al.*, 1981b; Guilbaud, G. *et al.*, 1985) as to the inadequacy of the medial spinothalamic system to process information related to stimulus discrimination. In the nucleus submedius, however, a somatotopic organization was described, with fibers from more rostral spinal levels terminating more rostrally in the nucleus (Craig, A. D. and Burton, H., 1985).

35.6.2.2 Structural types of neurons involved

Although there are no descriptions of the morphology of spinal neurons labeled from injections restricted to medial thalamic nuclei, data on laminae VI–X spinothalamic cells revealed medium to large-sized soma of polygonal or occasionally flattened configuration (Willis, W. D. *et al.*, 1979). Following injections filling both the lateral and medial thalamus in the cat and monkey (Zhang, E. T. *et al.*, 1996; Zhang, E. T. and Craig, A. D., 1997), lamina I neurons belonging in the fusiform, pyramidal, and flattened groups were labeled. As to pyramidal and flattened neurons, the possibility that labeled neurons picked up the tracer in the lateral thalamus can not be ruled out, since they were shown to project to the VPL in the rat (Lima, D. and Coimbra, A., 1988). Fusiform neurons, however, most probably project to medial thalamic nuclei as they could not be labeled from the VPL in the rat (Lima, D. and Coimbra, A., 1988).

35.6.2.3 Spinal location of ascending fibers

Spinal axons targeting the medial thalamus were early shown to course in the ventral quadrant of the spinal cord (Mehler, W. R. *et al.*, 1960; Bowsher, D., 1961; Mehler, W. R., 1969). In the rat this appears to be the way taken by all spinothalamic axons (Giesler, G. J. *et al.*, 1981a), whereas in the monkey axons were shown to distribute through the lateral funiculus also (Giesler, G. J. *et al.*, 1981b; Apkarian, A. V. and Hodge, C. J., 1989a). Studies using anterograde tracing after lesioning either the contralateral spinal ventral quadrant or the contralateral dorsolateral fasciculus (Apkarian, A. V. and Hodge, C. J., 1989a) revealed that the termination pattern of fibers coursing in both pathways is similar except for the fibers targeting the dorsolateral region of the medial dorsal nucleus, which course mostly in the ventral quadrant. The ventral quadrant also contributes with particularly large amounts of fibers to the central lateral nucleus afferents, mainly in its more anterior portion.

35.6.2.4 Neurotransmitters

Enkephalin (Coffield, J. A. and Miletiæ, V., 1987; Nahin, R. L., 1988) and dynorphin (Nahin, R. L., 1988) were immunodetected in deep dorsal horn and intermediate gray spinal neurons projecting to the medial thalamus. Enkephalin-immunoreactive neurons amount to 10% of the entire spinothalamic population (Coffield, J. A. and Miletiæ, V., 1987). Spinothalamic lamina X neurons immunoreactive for CCK (Ju, G. *et al.*, 1987; Leah, J. *et al.*, 1988) and galanin (Ju, G. *et al.*, 1987) are likely to terminate in the parafascicular nucleus, which was shown to contain fibers immunoreactive to both neuropeptides (Ju, G. *et al.*, 1987).

35.6.2.5 Response properties

Neurons projecting to the medial thalamus respond, although weakly, to activation of A (α to δ) and C primary afferent fibers with sustained after-discharges (Giesler, G. J. *et al.*, 1981b). Their background activity is practically nil. They conduct at relatively low velocities, which amount to about half those of lateral spinothalamic neurons (Giesler, G. J. *et al.*, 1981b). Most medial spinothalamic neurons belong in the NS class (around two-third vs. one-

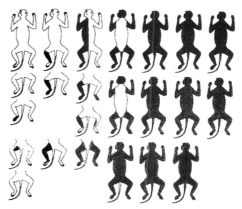

Figure 20 Receptive fields of spinal cells projecting to the medial thalamus in the monkey. Adapted from figure 3 of Giesler, G. J., Yezierski, R. P., Gerhart, K. D., and Willis, W. D. 1981b. Spinothalamic tract neurons that project to medial and/or lateral thalamic nuclei: evidence for a physiologically novel population of spinal cord neurons. J. Neurophysiol. 46, 1285–1308.

third in the lateral spinothalamic pathway). Some are WDR neurons and a few respond to deep tissue stimulation (Giesler, G. J. *et al.*, 1981b). They normally present very large, frequently complex, receptive fields that may encompass the entire surface of the body (Figure 20).

Inhibitory receptive fields are rare (Giesler, G. J. *et al.*, 1981b). Viscerosomatic convergence has been reported (Rucker, H. K. and Holloway, J. A., 1982).

35.6.2.6 Pathways driven at the target

Intralaminar nuclei, in particular the central lateral nucleus, project to widely distributed areas of the cerebral cortex, including sensorimotor areas, and to the basal ganglia (Jones, E. G. and Leavitt, R. Y., 1974). Ascending pathways terminating in the intralaminar nuclei have hence been proposed to be involved in motor and arousal nociceptive responses. Nociceptive pathways from the medial dorsal nucleus and the central lateral nucleus terminate, respectively, in the anterior cingulated cortex and frontal motor cortex (Wang, C. C. and Shyu, B. C., 2004). The posterior cingulated cortex also receives projections from restricted areas of both, the medial dorsal and the central lateral nuclei (Apkarian, A. V. and Shi, T., 1998). These connections are consonant with the role of the medial thalamic pathway in the emotional aspects of nociception (Wang, C. C. and Shyu, B. C., 2004). The medial dorsal nucleus and the nucleus submedius, which are similar in the response properties of their nociceptive neurons (Dostrovsky,

J. O. *et al.*, 1987), project to adjacent regions in the orbital cortex (Krettek, J. E. and Price, J. L., 1977; Craig, A. D. *et al.*, 1982; Yoshida, A. *et al.*, 1992).

35.6.3 Hypothalamus

35.6.3.1 *Spinal laminae of origin and sites of termination*

A direct spinohypothalamic pathway (Figure 21) was uncovered in the rat (Burstein, R. *et al.*, 1987; Burstein, R. *et al.*, 1990a; Burstein, R. *et al.*, 1990b; Menétrey, D. and DePommery, J., 1991) and cat (Katter, J. T. *et al.*, 1991) by the use of both electrophysiologic and tracing methods. It originates bilaterally, although with a slight contralateral prevalence, from the entire extent of the spinal cord (Figure 22) (Burstein, R. *et al.*, 1987, 1990a; 1990b) as well as from the spinal trigeminal nucleus, mainly the *pars caudalis* (Malick, A. and Burstein, R., 1998). Neurons in the deep dorsal horn make up the largest population of spinohypothalamic neurons, especially at upper cervical segments (around 50%), followed by the lateral spinal nucleus (around 30%), laminae I and X (around 10% each), and the intermediate ventral horn (Burstein, R. *et al.*, 1990a).

An additional contribution from the parasympathetic cell column was described in the rat (Burstein, R. *et al.*, 1990a; Menétrey, D. and DePommery, J., 1991). In the cat, the spinohypothalamic tract appears to be much smaller and to contain fewer lamina I

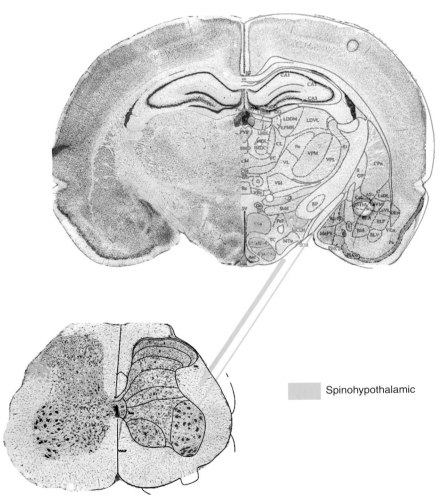

Spinohypothalamic

Figure 21   Diagram representing the spinal laminae of origin, ascending course in the spinal cord, and areas of termination of the spinohypothalamic pathway. Note that ipsilaterally terminating axons course contralaterally in the spinal cord. Brain and spinal cord photomicrographs were adapted from Paxinos, G. and Watson, C. 1998. The Rat Brain in Stereotaxic Coordinates, 4th edn. Academic Press.

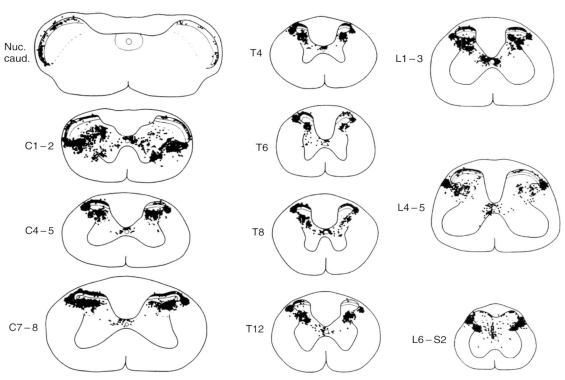

Figure 22 Distribution in the spinal cord and spinal trigeminal nucleus, *pars caudalis*, of neurons retrogradely labeled from injection of fluorogold encompassing the lateral and medial hypothalamus. (Adapted from figure 4 of Burstein, R., Cliffer, K. D., and Geisler, G. J. 1990b. Cells of origin of the spinohypothalamic tract in the rat. J. Comp. Neurol. 291, 329–344).

neurons than in the rat (Katter, J. T. *et al.*, 1991). No differences in spinal distribution were found between the medial and lateral hypothalamic pathways, except for the sacral parasympathetic nucleus, which appears to project mainly to medial nuclei (Burstein, R. *et al.*, 1990a). According to a recent study by Braz and coworkers (Braz, J. M. *et al.*, 2005) in which transgenic mice expressing a transneuronal tracer in a subset of nociceptors were used, nociceptive input conveyed by nonpeptidergic primary afferent neurons is relayed to lamina V neurons projecting to the ventromedial hypothalamus by spinal neurons located in the dorsal part of lamina II.

Consonant with retrograde and antidromic stimulation studies, terminal arborizations of spinal axons were shown to distribute massively through the lateral hypothalamus, although important labeling was also observed in the medial hypothalamus (Figure 21) (Burstein, R. *et al.*, 1987; Cliffer, K. D. *et al.*, 1991). In the lateral hypothalamus, fibers terminate bilaterally along its rostrocaudal extent, and throughout the course of the supraoptic decussation. In the medial hypothalamus, fibers terminate mainly contralaterally in the posterior and dorsal hypothalamic areas, the

dorsomedial, paraventricular and suprachiasmatic nuclei, and the preoptic area (Cliffer, K. D. *et al.*, 1991). Anterograde tracer injections restricted to the superficial or deep dorsal horn confirmed that the latter contributes much more fibers to the spinohypothalamic pathway, the main areas of termination being the posterior hypothalamic area, the posterior part of the lateral hypothalamic area, and the ventral part of the paraventricular hypothalamic nucleus (Gauriau, C. and Bernard. J. F., 2004).

35.6.3.2 Spinal location of ascending fibers

A few studies based on degeneration following spinal lesions revealed the termination in the lateral (Anderson, F. D. and Berry, C. M., 1959; Ring, G. and Ganchrow, D., 1983) and medial (Kerr, F. W. L., 1975) hypothalamus of spinal fibers coursing in the ventral funiculus. However, electrophysiological studies using antidromic activation from the supraoptic decussation showed that only 5% of the spinohypothalamic axons course in the ventral funiculus (Burstein, R. *et al.*, 1991). The remaining travel through the lateral funiculus, mainly in the dorsolateral fasciculus (57%), irrespective of their origin in the superficial or deep dorsal horn (Figure 21) (Burstein, R. *et al.*, 1991). Although spinal and trigeminal neurons project to the hypothalamus bilaterally, their axons ascend contralaterally in the spinal cord (Burstein, R. *et al.*, 1991) and brainstem (Kostarczyk, E. *et al.*, 1997). Those terminating ipsilaterally cross the midline within the supraoptic decussation (Burstein, R. *et al.*, 1991). Extensive collateralization along the entire brainstem has been reported (Kostarczyk, E. *et al.*, 1997).

35.6.3.3 Surface receptors

A particularly high number of spinal neurons projecting to the hypothalamus are apposed by profiles immunoreactive to nitric oxide synthase or to interferon-γ receptor in the lateral spinal nucleus (Kayalioglu, G. *et al.*, 1999).

35.6.3.4 Response properties

The majority of the neurons antidromically activated from the hypothalamus both in the spinal cord and the spinal trigeminal nucleus belong in the WDR and NS classes and also respond to noxious heat or cooling (Burstein, R. *et al.*, 1987, 1991; Malick, A. *et al.*, 2000; Zhang, X. J. *et al.*, 2002). Incremental responses to increasingly intense noxious heat stimulation were observed (Burstein, R. *et al.*, 1987). Low-threshold neurons make up 20% of the trigeminohypothalamic neurons (Malick, A. *et al.* 2000) and only 4% of the spinohypothalamic neurons (Burstein, R. *et al.*, 1991). The cutaneous receptive fields are particularly small indicating that this pathway may convey relatively precise information about the stimulated area (Burstein, R. *et al.*, 1991; Malick, A. *et al.*, 2000). About half of the spinohypothalamic neurons recorded in the thoracic spinal cord were activated by visceral distension with responses that increased with increasing stimulus intensities (Zhang, X. J. *et al.*, 2002). Eight percent of lumbar spinohypothalamic neurons respond to deep low-threshold input (Burstein, R. *et al.*, 1991).

35.6.3.5 Pathways driven at the target

Projections from the hypothalamus, namely the dorsomedial nucleus, descend through a dorsal pathway to the PAG, and through a ventral smaller pathway to the NTS (Thompson, R. H. *et al.*, 1996). According to c-*fos* studies (Snowball, R. K. *et al.*, 2000), neurons projecting to the ventrolateral PAG receive visceral input in the lateral hypothalamus, presumably in convergence with somatic input, and neurons

projecting to the NTS are concentrated in the posterolateral hypothalamus and the paraventricular nucleus. Projections connecting the paraventricular nucleus with the ventrolateral medulla were also reported (Hardy, S. G. P., 2001).

35.7 Spinothelencephalic Pathways

During the last decade, evidence accumulated about the existence of ascending nociceptive pathways that connect the spinal cord directly with various telencephalic regions in the rat. Although data are mainly based on anatomic retrograde and anterograde tracing studies, both the location of the spinal cells of origin, the demonstration of cells responding to noxious stimulation in areas such as the amygdala (Miyagama, T. *et al.*, 1986), and a recent study using genetic-controlled transneuronal tracing initiated in IB4-positive putative nociceptive primary afferent fibers (Braz, J. M. *et al.*, 2005) suggest that some of these areas are spinal targets of nociceptive input.

The data on spinotelencephalic pathways collected till the early years of 2000 are still few and each study deals with several systems. Therefore, notwithstanding their probable functional individuality, they will be described together although tentatively grouped according to their putative functions.

35.7.1 Thelencephalic Targets of Spinal Ascending Fibers

Studies using very sensitive anterograde tracers, such as *phaseolus vulgaris* leucoagglutinin or dextran (Burstein, R. *et al.*, 1987; Cliffer, K. D. *et al.*, 1991; Gauriau, C. and Bernard. J. F., 2004), revealed axonal terminal arborizations in various regions of the basal forebrain and cortex (Figure 23), which can be grouped as areas involved in motor control and areas of the limbic system.

The first group includes the globus pallidus, substantia nigra, and nucleus accumbens, in particular its medial part. A participation in the striatopallidal system as well as a role in innate motor patterns triggered by noxious stimuli has been proposed for the spinal–globus pallidus projection (Braz, J. M. *et al.*, 2005). Limbic spinal targets include nuclei of the septal complex, thought to be involved in motivation and emotion, but also in attention, arousal, learning, and memory (Burstein, R. *et al.*, 1987; Cliffer, K. D. *et al.*, 1991). Positive and negative reinforcement

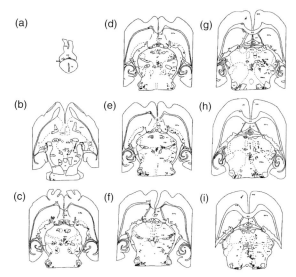

Figure 23 Distribution of spinal fibers labeled anterogradely with *phaseolus vulgaris* leucoagglutinin in the diencephalon and telencephalon, in horizontal view. Adapted from figure 3 of Cliffer, K. D., Burstein, R., and Giesler, G. J. 1991. Distributions of spinothalamic, spinohypothalamic, and spinotelencephalic fibers revealed by anterograde transport of PHA-L in rats. J. Neurosci. 11, 852–868.

associated with learning trials during nociceptive processing have been claimed to be accomplished by this pathway (Cliffer, K. D. *et al.*, 1991). The central nucleus of the amygdala together with what has been called the 'extended amygdala', namely the substantia innonimata and the Bed Nucleus of the stria terminalis, also receive direct projections from the spinal cord (Cliffer, K. D. *et al.*, 1991; Gauriau, C. and Bernard.J.F., 2004), as do the medial orbital and infralimbic cortices (Cliffer, K. D. *et al.*, 1991). Projections to the orbital cortex may affect autonomic, endocrine, and behavioral functions in relation to pain (Burstein, R. and Potrebic, S., 1993). The horizontal and vertical limbs of the diagonal band of Broca are also targeted by spinal axons (Burstein, R. *et al.*, 1987; Cliffer, K. D. *et al.*, 1991).

These direct spinotelencephalic projections appear to be paralleled by ascending polysynaptic pathways relaying not only in the brainstem, as is the case of the spinoparabrachialamygdaloid pathway (Ma, W. and Peschanski, M., 1988; Bernard, J. F. and Besson, J. M., 1990; Jasmin, L. *et al.*, 1997), but also within the telencephalon. Anterograde tracer injections in the central and basolateral anterior nuclei of the amygdala revealed projections from there to the substantia

innonimata and stria terminalis as well as the posterior hypothalamus (Bourgeais, L. *et al.*, 2001a).

35.7.2 Spinal Laminae of Origin

Based on anterograde labeling from restricted injections in various spinal laminae, Gauriau and Bernard (Gauriau, C. and Bernard. J. F., 2004) concluded that cells of origin of spinopallidal fibers are located in the deep dorsal horn, laminae VII and X and, in smaller amounts, in the superficial dorsal horn (Table 1). Marked labeling was obtained in the lateral aspect of the globus pallidus from neurons located in lamina V, as revealed by transneuronal tracing with wheat germ agglutinin synthesized by IB4-positive primary afferents in mice (Braz, J. M. *et al.*, 2005).

Spinal fibers projecting to the nucleus accumbens and the septal nuclei have similar bilateral origins at the reticulated portion of the deep dorsal horn, the lateral spinal nucleus, and lamina X throughout the entire length of the spinal cord (Burstein, R. and Giesler, G. J., 1989). Deep dorsal horn neurons account to half of the entire projection population, followed by the lateral spinal nucleus, and lamina X (about 15% each). Neurons in lamina I and the intermediate/ventral horn are very few, but the first are slightly more numerous in the spinoseptal pathway (Table 1).

Spinal neurons projecting to the amygdala present similar laminar distribution (Menétrey, D. and DePommery, J., 1991; Burstein, R. and Potrebic, S., 1993). Neurons in the reticulated region of the deep dorsal horn also make up about half of the entire spinoamygdala population, but they are located in its dorsal portion in cervical segments and ventral portion in thoracolumbar segments. The lateral spinal nucleus contributes to 25% of the projection, lamina X to 13% (mainly at upper lumbar segments), and the intermediate and ventral horn to 10%, at upper cervical and lumbar segments (Table 1). Neurons in lamina V receiving primary afferent input through lamina II neurons activated by IB4-positive primary afferents also send projections to the amygdala, as well as to the bed nucleus of stria terminalis (Braz, J. M. *et al.*, 2005). A projection to the dorsal part of substantia innonimata from the superficial and deep dorsal horn, contralaterally, and from the lateral spinal nucleus, ipsilaterally, has been described (Gauriau, C. and Bernard. J.F., 2004).

Table 1 Relative participation of spinal cord laminae in the various nociceptive ascending pathways

	LSN	I	II	III	IV	V	VI	VII	VIII	X
LCN		■		■ (10%)	■ (60%)	■	■	■		
VLMlat	■	■	■	■						■
LRt					■	■	■	■	■	■
DRt dorsal		■ (Medial)	■ (Medial)	■ (Medial)						■
DRt ventral					■ (Medial)	■ (Medial)	■ (Medial)	■		■
NTS	■	■			■	■	■			■
RVM						■	■		■	■
PBN		■ (Lateral)				■ (Lateral)			■	■
PAG	■	■			■ (Lateral)	■ (Lateral)	■	■	■	■
Lateral thalamus	■	■			■	■	■	■	■	■
Medial thalamus	■	■						■	■	
Hypothalamus	■ (30%)	■ (10%)			■	■ (----50%----)	■			■ (10%)
N. Pallidus		■			■	■	■	■		■
N. Accumbens	■ (25%)				■	■ (----50%----)	■			■ (10%)
Septal nucleus	■ (15%)	■			■	■ (----50%----)	■			■ (15%)
Amygdala	■ (25%)					■ (--50%--)	■	■ (-10%-)	■	■ (10%)
Subst. inonimata					■	■	■			
Orbital cortex	■ (15%)					■ (--62%--)	■	■ (-13%-)	■	■ (10%)

Relative amounts refer to each pathway and do not allow comparisons between pathways. Whenever quantified, the relative contribution to each pathway is referred between brackets. LSN, lateral spinal nucleus; LCN, lateral cervical nucleus; VLMlat, caudal ventrolateral reticular formation, lateral portion; LRt, lateral reticular nucleus; DRt dorsal, dorsal reticular nucleus, dorsal portion; DRt ventral, dorsal reticular nucleus – ventral portion; NTS, nucleus tractus solitarii; RVM, rostral ventromedial medulla; PBN, parabrachial nuclei; PAG, periaqueductal gray; I–X, spinal laminae.

The orbital cortex receives its major spinal projections from the contralateral reticulated area of the deep dorsal horn (62%) at the cervical level or its ventromedial aspect at the thoracic and lumbar levels. The lateral spinal nucleus contributes to 15% of the projection, but mainly at lumbar segments, while the intermediate/ventral horn and lamina X contribute to 13% and 10%, respectively, both at the upper cervical and lower thoracic/upper lumbar segments (Table 1) (Burstein, R. and Potrebic, S., 1993).

35.8 Discussion

35.8.1 Multiple Parallel Ascending Pathways

The main feature that stands out is the multiplicity of the ascending nociceptive system in terms of the variety of supraspinal regions that are targeted (Figure 24). This has been interpreted as the anatomical substrate for the triggering of a multitude of responses to the noxious event, from autonomic and motor reactions to affective and cognitive behaviors. Nevertheless, it is curious to note that, contrary to what was thought in the middle of the twentieth century, nociceptive input does not necessarily arrive to high processing motor, affective, and cognitive centers in the thelencephalon through multisynaptic chains capable of filtering information at various successive levels, but can reach those areas through direct spinofugal pathways.

Another aspect that emerges is that multiple polysynaptic pathways seem to parallel monosynaptic connections between the spinal cord and each supraspinal target. So far, this organization pattern was demonstrated only for a few systems, such as the medial paracentral spinothalamic and the spinoamygdaloid

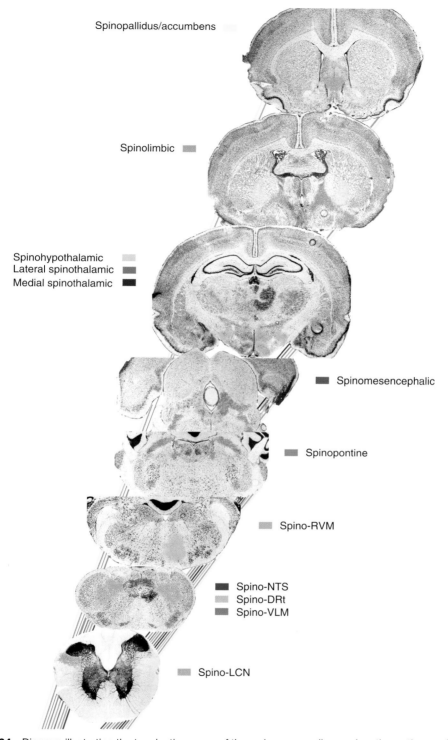

Spinopallidus/accumbens

Spinolimbic

Spinohypothalamic
Lateral spinothalamic
Medial spinothalamic

Spinomesencephalic

Spinopontine

Spino-RVM

Spino-NTS
Spino-DRt
Spino-VLM

Spino-LCN

Figure 24 Diagram illustrating the termination areas of the various ascending nociceptive pathways. The laterality of the ascending tracts and termination fields with respect to the side of arrival of primary afferent input is represented, the left side being ipsilateral and the right side contralateral. Brain and spinal cord photomicrographs were adapted from Paxinos, G. and Watson, C. 1998. The Rat Brain in Stereotaxic Coordinates, 4th edn. Academic Press.

systems, served by a direct pathway and a disynaptic pathway with a relay in the PBN (Ma, W. and Peschanski, M., 1988; Bernard, J. F. and Besson, J. M., 1990; Jasmin, L. *et al.*, 1997). However, the fact that most spinal targets send projections to other brain areas that are also targeted by spinal fibers strongly suggests a similar architecture for most systems, with the participation of parallel monosynaptic and multisynaptic chains of different lengths. Such an arrangement may imply that the responses to the noxious event generated at each site evolve along time according to postarriving of noxious-evoked input generated in other pain-processing centers. Notwithstanding the extensive anatomical and electrophysiological data that are still needed to corroborate this hypothesis, it is worth to take it into consideration in future investigation.

35.8.2 Spinal Neuronal Populations at the Origin of Nociceptive Ascending Pathways

When facing such a variety of nociceptive ascending pathways, it is tempting to assume that they differ by either channeling different sensory modalities to brain regions specifically dedicated to their processing, or by the responses they induce to whatever stimulus through activation of a particular supraspinal region. For the first assumption to be correct, the spinal source of input should differ between different pathways. The overview of the ascending nociceptive system here presented clearly shows that this is not the case. On the contrary, if one compares the contribution of the various spinal laminae to each pathway (Table 1), the first emerging picture is that of a strong similarity, which favors the second assumption that the functional properties of a pathway depend on the functional engagement of its target. However, in spite of a large overlap, there are subtle dissimilarities between pathways consisting of a preponderance of some laminae over others or differences in the location of the projecting neurons in each lamina (Table 1). Also the electrophysiological response properties of spinal neurons participating in the various pathways overlap considerably (Table 2): all pathways including nociceptive specific, wide-dynamic-range and low-threshold neurons, as well as neurons responding to cutaneous, visceral, and deep noxious stimuli. Again, subtle differences are likely to occur about the proportion of neurons of each kind that take part in each pathway (Table 2). Differences between pathways over the

Table 2 Relative participation of low threshold (LT), wide-dynamic-range (WDR) and nociceptive specific (NS) spinal cord neurons in the various nociceptive ascending pathways

	LT	*WDR*	*NS*
LCN	Mainly hair movement	60–86%	
VLM	25%	25%	50%
NTS	+	The majority	
RVM	+	The majority	
PBN	−	+	+ (75–90% in lamina I)
PAG	+ (DDH)	50% (Lamina I + DDH	+ (lamina I)
Lateral thalamus	+	+	35%
Medial thalamus	+	+	70%
Hypothalamus	4%	The majority	

DDH, deep dorsal horn. Other abbreviations as in Table 1.

amount of projecting neurons they involve have also been pointed out (Apkarian, A. V. and Hodge, C. J., 1989d; Burstein, R. *et al.*, 1990b; Mouton, L. J. and Holstege, G., 1998; Willis, W. D. *et al.*, 2001) but a complete picture of the quantitative variations is still hard to attain due to the use of tracers of variable sensitivity. As to the possibility that each pathway identity relies on a particular neurochemical architecture at the spinal relay, data are too scarce to allow any sort of considerations.

The difficulty in concealing this anatomofunctional organization with the well-known capacity of modality discrimination during acute physiological pain led some authors to ascribe discrimination capacity to a particular spinal region, leaving the remaining spinal cord with a secondary, largely unknown but eventually not important role in pain processing. The high concentration of nociceptive specific neurons in lamina I, together with the convergence of input of various nature and peripheral origin to this lamina and the easy separation of its structural neuronal groups (see Chapter Spinal Cord Physiology of Nociception) brought it into focus during the last decade. However, lamina I does not contribute to all spinofugal nociceptive pathways while participating similarly in many others (Table 1). Moreover, an appraisal of the participation of the various lamina I structural cell groups in a sample of nociceptive ascending pathways revealed

Table 3 Relative amounts of supraspinally projecting lamina I neurons

	Fusiform	*Flattened*	*Pyramidal*	*Multipolar*
VLM	80%	10%	10%	
DRt		10–30%	5–20%	60–85%
NTS	25%	40%	35%	
PBN	70%		30%	
PAG	30%		70%	
Thalamus VBC		25–30% (enlargements)	70–75% (enlargements)	

VBC, ventrobasal complex of the thalamus.

large superposition, although again slight differences as to their relative amount in each pathway and the specific involvement or noninvolvement of certain groups were detected (Table 3). What is noteworthy is that when taken together, the various electrophysiological studies on the response properties of lamina I neurons do not support a clear-cut structural–functional correlation based on stimulus-modality processing (see discussion in Galhardo, V. *et al.*, 2000). The possibility that stimulus characterization at the central nervous system depends on a combinatory activation process rather than on the activation of specific channels should be addressed in the future.

As a working hypothesis, it could be postulated at this point that each pathway has particular characteristics that depend on both, the kind of input it transmits (defined by the relative contribution of the various spinal neurons) and the functional properties of the target. Ultimately, for each noxious event, ascending transmission of nociceptive input would be the result of the relative activation of the various pathways.

35.8.3 Stimulus Discrimination

A clear separation between the lateral and the medial spinothalamic nociceptive pathways as to the ability of the former to discriminate between stimulus location and intensity has been established, based mainly on the electrophysiological properties of neurons projecting in each pathway. Amongst the supporting data stand out the relatively small size of the receptive fields, the stimulus intensity-encoding capacity, and the somatotopic organization of the spinal, thalamic, and cortical neurons in the lateral pathway, as opposed to the medial pathway. It should be noted, however, that early studies pointed out that the medial spinothalamic system conveys spinal input either directly or through a brainstem relay, but spinal

neurons projecting to brainstem regions connected to the medial thalamus do not necessarily share the same properties of the medial spinothalamic neurons. Receptive fields of spino-RVM and spino-PAG neurons vary from small, confined to a sole limb, to very large and complex. PBN-projecting spinal neurons were shown to present small receptive fields (Bester, H. *et al.*, 2000), contrary to PBN neurons projecting to the paracentral nucleus of the thalamus, which have large receptive fields (Bourgeais, L. *et al.*, 2001b). Also noteworthy in this respect is the fact that neurons in the RVM and PAG send axonal collaterals to the lateral thalamus, while neurons projecting through the lateral spinothalamic or the spinocervical pathways send collaterals to the medial thalamus. As a whole, the data suggest that, although the lateral spinothalamic pathway as well as the spinocervical and the spinohypothalamic pathways appear to be morphofuntionally organized to allow stimulus location and intensity discrimination, an extensive cross-talk between the various nociceptive ascending pathways is likely to take place.

35.8.4 Nociceptive Ascending Pathways as Part of a Complex Nociceptive Integration System

Although the ascending transmission system and the descending endogenous pain control system (see Chapter Descending Control Mechanisms) are normally dealt with separately, evidence has been accumulated to prove that they are both part of a sole nociceptive system buildup in such a way that information is treated at different brain levels in order to integrate nociception and various brain functions. The brain areas of termination of the ascending nociceptive pathways are those and the same from where pain-control actions are elicited upon local stimulation (Jones, S. L., 1992). These areas are intimately connected with each other, and each of them with the spinal cord dorsal horn, in most cases through direct descending projections that very often participate in spino-brain-spinal reciprocal loops (Lima, D. *et al.*, 1998). Descending pain-control actions appear to be triggered not only by arriving of nociceptive input, but also by the current state of processing of other functions carried out at those brain areas, as blood pressure variations at the caudal ventrolateral reticular formation (Tavares, I. and Lima, D., 2002).

The so-called ascending and descending systems are arranged as an intricate network of neuronal pathways that interchange information and dynamically control, through modulation of spinal dorsal horn activity, the level of activation of the perception and response centers. Pain perception and pain reactions are thus adapted to autonomic, affective, and cognitive states, which in turn are short- or long-term modified in response to the noxious event. The various nociceptive ascending pathways must not be viewed as static separate lines that allow the passage of nociceptive signals to distinct brain regions, but rather as an ensemble of spinal transmission neurons that tune pain perception and reactions according to peripheral and central conditions through a constant and dynamic interplay with the brain.

References

Abols, I. A. and Basbaum, A. I. 1981. Afferent connections of the rostral medulla of the cat: a neural-substrate for midbrain-medullary interactions in the modulation of pain. J. Comp. Neurol. 201, 285–297.

Agarwal, S. K. and Calaresu, F. R. 1990. Reciprocal connections between nucleus tractus solitarii and rostral ventrolateral medulla. Brain Res. 523, 305–308.

Al-Chaer, E. D., Feng, Y., and Willis, W. D. 1999. Comparative study of viscerosomatic input onto postsynaptic dorsal column and spinothalamic tract neurons in the primate. J. Neurophysiol. 82, 1876–1882.

Albe-Fessard, D., Boivie, J., Grant, G., and Levante, A. 1975. Labelling of cells in the medulla oblongata and the spinal cord of the monkey after injections of horseradish peroxidase in the thalamus. Neurosci. Lett. 1, 75–80.

Almeida, A. and Lima, D. 1997. Activation by cutaneous or visceral noxious stimulation of spinal neurons projecting to the medullary dorsal reticular nucleus in the rat: a *c-fos* study. Eur. J. Neurosci. 9, 686–695.

Almeida, A., Cobos, A., Tavares, A., and Lima, D. 2002. Brain afferents to the medullary dorsal reticular nucleus: a retrograde and anterograde tracing study in the rat. Eur. J. Neurosci. 16, 81–95.

Almeida, A., Tavares, I., and Lima, D. 1995. Projection sites of superficial or deep dorsal horn in the dorsal reticular nucleus. Neuroreport 6, 1245–1248.

Almeida, A., Tavares, I., and Lima, D. 2000. Reciprocal connections between the medullary dorsal reticular nucleus and the spinal dorsal horn in the rat. Eur. J. Pain. 4, 373–387.

Almeida, A., Tavares, I., Lima, D., and Coimbra, A. 1993. Descending projections from the medullary dorsal reticular nucleus make synaptic contacts with spinal cord lamina I cells projecting to that nucleus: an electron microscopic tracer study in the rat. Neuroscience 55, 1093–1106.

Ammons, W. S. 1987. Characteristics of spinoreticular and spinothalamic neurons with renal input. J. Neurophysiol. 50, 480–495.

Ammons, W. S., Blair, R. W., and Foreman, R. D. 1984. Responses of primate T1–T5 spinothalamic neurons to gallbladder distension. Am. J. Physiol. 247, R995–R1002.

Ammons, W. S., Girardot, M. N., and Foreman, R. D. 1985. T2–T5 spinothalamic neurons projecting to medial thalamus with viscerosomatic input. J. Neurophysiol. 54, 73–89.

Anderson, F. D. and Berry, C. M. 1959. Degeneration studies of long ascending fiber systems in the cat brain stem. J. Comp. Neurol. 111, 195–229.

Andrew, D. and Craig, A. D. 2001. Spinothalamic lamina I neurons selectively responsive to cutaneous warming in cats. J. Physiol. 537, 489–495.

Apkarian, A. V. and Hodge, C. J. 1989a. The primate spinothalamic pathways: III. Thalamic terminations of the dorsolateral and ventral spinothalamic pathways. J. Comp. Neurol. 288, 493–511.

Apkarian, A. V. and Hodge, C. J. 1989b. The primate spinothalamic pathways: II. The cells of origin of the dorsolateral and ventral spinothalamic pathways. J. Comp. Neurol. 288, 474–492.

Apkarian, A. V. and Hodge, C. J. 1989c. A dorsolateral spinothalamic tract in macaque monkey. Pain 37, 323–333.

Apkarian, A. V. and Hodge, C. J. 1989d. The primate spinothalamic pathways: I. A quantitative study of the cells of origin of the spinothalamic pathway. J. Comp. Neurol. 288, 447–473.

Apkarian, A. V. and Shi, T. 1998. Thalamocortical Connections of the Cingulate and Insula in Relation to Nociceptive Inputs to the Cortex. In: Pain Mechanisms and Management (*eds*. S. N. Ayrapetyan and A. V. Apkarian), pp. 212–220. IOS Press.

Apkarian, A. V., Stevens, R. T., and Hodge, C. J. 1985. Funicular location of ascending axons of lamina L cells in the cat spinal-cord. Brain Res. 334, 160–164.

Applebaum, A. E., Beall, J. E., Foreman, R. D., and Willis, W. D. 1975. Organization and receptive fields of primate spinothalamic tract neurons. J. Neurophysiol. 38, 572–586.

Baker, M. L. and Giesler, G. J. 1984. Anatomical studies of the spinocervical tract of the rat. Somatosens. Res. 2, 1–18.

Bandler, R. and Tork, I. 1987. Midbrain periaqueductal grey region in the cat has afferent and efferent connections with solitary tract nuclei. Neurosci. Lett. 74, 1–6.

Beall, J. E., Foreman, R. D., Willis, W. D., and Applebaum, A. E. 1977. Spinal cord potentials evoked by cutaneous afferents in the monkey. J. Neurophysiol. 40, 199–211.

Behbehani, M. M. 1995. Functional characteristics of the midbrain periaqueductal gray. Prog. Neurobiol. 46, 575–605.

Beitz, A. J. 1982. The organization of afferent-projections to the midbrain periaqueductal gray of the rat. Neuroscience 7, 133–159.

Berkley, K. J. 1980. Spatial relationships between the terminations of somatic sensory and motor pathways in the rostral brainstem of cats and monkeys. I. Ascending somatic sensory inputs to lateral diencephalon. J. Comp. Neurol. 193, 283–317.

Bernard, J. F. and Besson, J. M. 1990. The spino(trigemino)-pontoamygdaloid pathway: electrophysiological evidence for an involvement in pain processes. J. Neurophysiol. 63, 473–490.

Bernard, J. F., Villanueva, L., Carroué, J., and Le Bars, D. 1990. Efferent projections from the subnucleus reticularis dorsalis

(SRD): a *Phaseolus vulgaris* leucoagglutinin study in the rat. Neurosci. Lett. 116, 257–262.

Bester, H., Chapman, V., Besson, J. M., and Bernard, J. F. 2000. Physiological properties of the lamina I spinoparabrachial neurons in the rat. J. Neurophysiol. 83, 2239–2259.

Björkeland, M. and Boivie, J. 1984. The termination of spinomesencephalic fibers in cat. An experimental anatomical study. Anat. Embryol. 170, 265–277.

Blair, R. W., Ammons, W. S., and Foreman, R. D. 1984a. Responses of thoracic spinothalamic and spinoreticular cells to coronary artery occlusion. J. Neurophysiol. 51, 636–648.

Blair, R. W., Weber, R. N., and Foreman, R. D. 1981. Characteristics of primate spinothalamic tract neurons receiving viscerosomatic convergent inputs in the T3–T5 segments. J. Neurophysiol. 46, 797–811.

Blair, R. W., Weber, R. N., and Foreman, R. D. 1984b. Responses of thoracic spinoreticular and spinothalamic cells to intracardiac bradykinin. Am. J. Physiol. 246, H500–H507.

Blomqvist, A., Ma, W., and Berkley, K. J. 1989. Spinal input to the parabrachial nucleus in the cat. Brain Res. 480, 29–36.

Blomqvist, A., Zhang, E. T., and Craig, A. D. 2000. Cytoarchitectonic and immunohistochemical characterization of a specific pain and temperature relay, the posterior portion of the ventral medial nucleus, in the human thalamus. Brain 123, 601–619.

Boivie, J. 1971. The termination of the spinothalamic tract in the cat. An experimental study with silver degeneration methods. Exp. Brain Res. 12, 331–353.

Boivie, J. 1979. An anatomical reinvestigation of the termination of the spinothalamic tract in the monkey. J. Comp. Neurol. 186, 343–370.

Bourgeais, L., Gauriau, C., and Bernard, J. F. 2001a. Projections from the nociceptive area of the central nucleus of the amygdala to the forebrain: a PHA-L study in the rat. Eur. J. Neurosci. 14, 229–255.

Bourgeais, L., Monconduit, L., Villanueva, L., and Bernard, J. F. 2001b. Parabrachial internal lateral neurons convey nociceptive messages from the deep laminas of the dorsal horn to the intralaminar thalamus. J. Neurosci. 21, 2159–2165.

Bowsher, D. 1957. Termination of the central pain pathway in man: the conscious appreciation of pain. Brain 80, 606–622.

Bowsher, D. 1961. The termination of secondary somatosensory neurons within the thalamus of Macaca mulatta: an experimental degeneration study. J. Comp. Neurol. 117, 213–227.

Bowsher, D. and Westman, J. 1970. The gigantocellular reticular region and its spinal afferents: a light and electron microscopic study in the cat. J. Anat. 106, 23–36.

Braz, J. M., Nassar, M. A., Wood, J. N., and Basbaum, A. I. 2005. Parallel 'pain' pathways arise from subpopulations of primary afferent nociceptor. Neuron 47, 787–793.

Brown, A. G. and Franz, D. N. 1969. Responses of spinocervical tract neurones to natural stimulation of identified cutaneous receptors. Exp. Brain Res. 7, 231–249.

Brown, A. G., Fyffe, R. E., Noble, R., Rose, P. K., and Snow, P. J. 1980a. The density, distribution and topographical of spinocervical tract neurones in the cat. J. Physiol. 300, 409–428.

Brown, A. G., Hamann, W. C., and Martin, H. F. 1975. Effects of activity in non-myelinated afferent fibres on the spinocervical tract. Brain Res. 98, 243–259.

Brown, A. G., House, C. R., Rose, P. K., and Snow, P. J. 1976. The morphology of spinocervical tract neurones in the cat. J. Physiol. 260, 719–738.

Brown, A. G., Koerber, H. R., and Noble, R. 1987. An intracellular study of spinocervical tract cell responses to natural stimuli and single hair afferent fibres in cats. J. Physiol. 382, 331–354.

Brown, A. G., Rose, P. K., and Snow, P. J. 1977. The morphology of spinocervical tract neurones revealed by intracellular injection of horseradish peroxidase. J. Physiol. 270, 747–764.

Brown, A. G., Rose, P. K., and Snow, P. J. 1980b. Dendritic trees and cutaneous receptive fields of adjacent spinocervical tract neurones in the cat. J. Physiol. 300, 429–440.

Bryan, R. N., Coulter, J. D., and Willis, W. D. 1974. Cells of origin of the spinocervical tract in the monkey. Exp. Neurol. 42, 574–586.

Bryan, R. N., Trevino, D. L., Coulter, J. D., and Willis, W. D. 1973. Location and somatotopic organization of the cells of origin of the spino-cervical tract. Exp. Brain Res. 17, 177–189.

Burstein, R. and Giesler, G. J. 1989. Retrograde labeling of neurons in spinal cord that project directly to nucleus accumbens or the septal nuclei in the rat. Brain Res. 497, 149–154.

Burstein, R. and Potrebic, S. 1993. Retrograde labelling of neurons in the spinal cord that project directly to the amygdala or the orbital cortex in the rat. J. Comp. Neurol. 335, 469–485.

Burstein, R., Cliffer, K. D., and Giesler, G. J. Jr. 1987. Direct somatosensory projections from the spinal cord to the hypothalamus and telencephalon. J. Neurosci. 4159–4164.

Burstein, R., Cliffer, K. D., and Giesler, G. J. 1990a. Cells of origin of the spinohypothalamic tract in the rat. J. Comp. Neurol. 291, 329–344.

Burstein, R., Dado, R. J., Cliffer, K. D., and Giesler, G. J., Jr. 1991a. Physiological characterization of spinohypothalamic tract neurons in the lumbar enlargement of rats. J. Neurophysiol. 66, 261–284.

Burstein, R., Cliffer, K. D., and Geisler, G. J. 1990b. Cells of origin of the spinohypothalamic tract in the rat. J. Comp. Neurol. 291, 329–344.

Burton, H. and Jones, E. G. 1976. The posterior thalamic region and its cortical projection in new world and old world monkeys. J. Comp. Neurol. 168, 249–301.

Calaresu, F. R., Ciriello, J., Caverson, M. M., Chechetto, D. F., and Krukoff, T. L. 1984. Functional neuroanatomy of central pathways controlling the circulation. In: Hypertension and the brain (eds. G. P. Guthrie, Jr. and T. A. Kotchen), pp. 3–22. Futura Publishing.

Carstens, E. and Trevino, D. L. 1978a. Anatomical and physiological properties of ipsilaterally projecting spinothalamic neurons in the second cervical segment of the cat's spinal cord. J. Comp. Neurol. 182, 167–184.

Carstens, E. and Trevino, D. L. 1978b. Laminar origins of spinothalamic projections in the cat as determined by the retrograde transport of horseradish peroxidase. J. Comp. Neurol. 182, 151–166.

Castro, A. R., Morgado, C., Lima, D., and Tavares, I. 2006. Differential expression of NK1 and GABA_B receptors in spinal neurons projecting to antinociceptive or pronociceptive medullary centers. Brain Res Bull. 69, 266–275.

Catalano, J. V. and Lamarche, G. 1957. Central pathway for cutaneous impulses in the cat. Am. J. Physiol. 189, 141–144.

Cechetto, D. F. and Saper, C. B. 1987. Evidence for a viscerotopic sensory representation in the cortex and thalamus in the rat. J. Comp. Neurol. 262, 27–45.

Cechetto, D. F., Standaert, D. G., and Saper, C. B. 1985. Spinal and trigeminal dorsal horn projections to the parabrachial nucleus in the rat. J. Comp. Neurol. 240, 153–160.

Cervero, F. and Wolstencroft, J. H. 1984. A positive feeedback loop between spinal cord nociceptive pathways and antinociceptive areas of the cat's brain stem. Pain 20, 125–138.

Cervero, F., Iggo, A., and Molony, V. 1977. Responses of spinocervical tract neurones to noxious stimulation of the skin. J. Physiol. 267, 537–558.

Chandler, M. J., Zhang, J., Qin, C., Yuan, Y., and Foreman, R. D. 2000. Intrapericardiac injections of algogenic chemicals excite primate C1-C2 spinothalamic tract neurons. Am. J. Physiol. Regul. Integr. Comp. Physiol. 279, R560–R568.

Chaouch, A., Menétrey, D., Binder, D., and Besson, J. M. 1983. Neurons at the origin of the component of the bulbopontine spinoreticular tract in the rat: an anatomical study using horseradish peroxidase retrograde transport. J. Comp. Neurol. 214, 309–320.

Chung, J. M., Kenshalo, D. R., Gerhart, K. D., and Willis, W. D. 1979. Excitation of primate spinothalamic neurons by cutaneous C-fiber volleys. J. Neurophysiol. 42, 1354–1369.

Clark, W. E. L. 1936. The termination of ascending tracts in the thalamus of the thalamus of the macaque monkey. J. Anat. 71, 7–40.

Cledenin, M., Ekerot, C. F., Oscarsson, O., and Rosén, I. 1974. The lateral reticular nucleus in the cat. II: Organization of component activated from bilateral ventral flexor reflex tract (bVFRT). Exp. Brain Res. 21, 487–500.

Clement, C. I., Keay, K. A., Podzebenko, K., Gordon, B. D., and Bandler, R. 2000. Spinal sources of noxious visceral and noxious deep somatic afferent drive onto the ventrolateral periaqueductal gray of the rat. J. Comp. Neurol. 425, 323–344.

Cliffer, K. D., Burstein, R., and Giesler, G. J. 1991. Distributions of spinothalamic, spinohypothalamic, and spinotelencephalic fibers revealed by anterograde transport of PHA-L in rats. J. Neurosci. 11, 852–868.

Cobos, A., Lima, D., Almeida, A., and Tavares, I. 2003. Brain afferents to the lateral caudal ventrolateral medulla: a retrograde and anterograde tracing study in the rat. Neuroscience 120, 485–498.

Coffield, J. A. and Miletiæ, V. 1987. Immunoreactive enkephalin is contained within some trigeminal and spinal neurons projecting to the rat medial thalamus. Brain Res. 425, 380–383.

Collier, J. and Buzzard, E. F. 1903. The degenerations resulting from lesions of posterior nerve roots and away from transverse lesions of the spinal cord in man. Brain 26, 559–591.

Comans, P. E. and Snow, P. J. 1981. Ascending projections to nucleus parafascicularis of the cat. Brain Res. 230, 337–341.

Cowie, R. J. and Holstege, G. 1992. Dorsal mesencephalic projections to pons, medulla, and spinal-cord in the cat - limbic and nonlimbic components. J. Comp. Neurol. 319, 536–559.

Craig, A. D. 1976. Spinocervical tract cells in cat and dog, labeled by the retrogade transport of horseradish peroxidase. Neurosci. Lett. 3, 173–177.

Craig, A. D. 1978. Spinal and medullary input tothe lateral cervical nucleus. J. Comp. Neurol. 181, 729–744.

Craig, A. D. 1995. Distribution of brainstem projections from spinal lamina I neurons in the cat and the monkey. J. Comp. Neurol. 361, 225–248.

Craig, A. D. 2003. Distribution of trigeminothalamic and spinothalamic lamina I, terminations in the cat. Somatosens. Mot. Res. 20, 209–222.

Craig, A. D. and Burton, H. 1979. The lateral cervical nucleus in the cat: anatomic organization of cervico-thalamic neurons. J. Comp. Neurol. 185, 329–346.

Craig, A. D. and Burton, H. 1981. Spinal and medullary lamina I projection to nucleus submedius in medial thalamus: a possible pain center. J. Neurophysiol. 45, 443–466.

Craig, A. D. and Burton, H. 1985. The distribution and topographical organization in the thalamus of antegradely-transported horseradish peroxidase after spinal injections in cat and raccoon. Exp. Brain Res. 58, 227–254.

Craig, A. D. and Kniffki, K. D. 1985. Spinothalamic lumbosacral lamina I cells responsive to skin and muscle stimulation in the cat. J. Physiol. 365, 197–221.

Craig, A. D., Bushnell, M. C., Zhang, E. T., and Blomqvist, A. 1994. A thalamic nucleus specif for pain and temperature sensation. Nature 372, 770–773.

Craig, A. D., Krout, K., and Andrew, D. 2001. Quantitative response characteristics of thermoreceptive and nociceptive lamina I spinothalamic neurons in the cat. J. Neurophysiol. 86, 1459–1480.

Craig, A. D., Linington, A. J., and Kniffki, K. D. 1989. Cells of origin of spinothalamic tract projections to the medial and lateral thalamus in the cat. J. Comp. Neurol. 289, 568–585.

Craig, A. D., Wiegand, S. J., and Price, J. L. 1982. The thalamo-cortical projection of the nucleus submedius in the cat. J. Comp. Neurol. 206, 28–48.

Craig, A. D., Zhang, E. T., and Blomqvist, A. 2002. Association of spinothalamic lamina I neurons and their ascending axons with calbindin-immunoreactivity in monkey and human. Pain 97, 105–115.

Dejerine, J. and Roussy, G. 1906. Le syndrome thalamique. Rev. Neurol. 14, 521–532.

Desbois, C. and Villanueva, L. 2001. The organization of lateral ventromedial thalamic connecions in the rat: a link for the distribution of nociceptive signals to widespread cortical regions. Neuroscience 102, 885–898.

Dilly, P. N., Wall, P. D., and Webster, K. E. 1968. Cells of origin of the spinothalamic tract in the cat and rat. Exp. Neurol. 21, 550–562.

Dostrovsky, J. O. and Craig, A. D. 1996. Cooling-specific spinothalamic neurons in the monkey. J. Neurophysiol. 76, 3656–3665.

Dostrovsky, J. O., Guilbaud, G., and Gautron, M. 1987. Nociceptive neurons in nucleus submedius of the arthritic rat. Pain 4, S263.

Downie, J. W., Ferrington, D. G., Sorkin, L. S., and Willis, W. D. 1988. The primate spinocervicothalamic pathway: responses of cells of the lateral cervical nucleus and spinocervical tract to innocuous and noxious stimuli. J. Neurophysiol. 59, 861–885.

Dugast, C., Almeida, A., and Lima, D. 2003. The medullary dorsal reticular nucleus enhances the responsiveness of spinal nociceptive neurons to peripheral stimulation in the rat. Eur. J. Neurosci. 18, 580–588.

Eberhart, J. A., Morrell, J. I., Krieger, M. S., and Pfaff, D. W. 1985. An autoradiographic study of projections ascending from the midbrain central gray, and from the region lateral to it, in the rat. J. Comp. Neurol. 241, 285–310.

Edinger, L. 1890. Einiges vom Verlauf der Gefuehlsbahnen im centralen Nervensysteme. Deut. Med. Woch. 16, 421–426.

Esteves, F., Lima, D., and Coimbra, A. 1993. Structural types of spinal cord marginal (lamina I) neurons projecting to the nucleus of the tractus solitarius in the rat. Somatosens. Mot. Res. 10, 203–216.

Ferrington, D. G., Sorkin, L. S., and Willis, W. D. 1986. Responses of spinothalamic tract cells in the cat cervical spinal cord to innocuous and graded noxious stimuli. Somatosens. Res 3, 339–358.

Ferrington, D. G., Sorkin, L. S., and Willis, W. D. 1987. Responses of spinothalamic tract cells in the superficial dorsal horn of the primate lumbar spinal cord. J. Physiol. 388, 681–703.

Fields, H. L. 2000. Pain modulation: expectation, opioid analgesia and virtual pain. Prog. Brain Res. 122, 245–253.

Fields, H. L., Clanton, C. H., and Anderson, S. D. 1977. Somatosensory properties of spinoreticular neurons in the cat. Brain Res. 120, 49–66.

Fields, H. L., Malick, A., and Burstein, R. 1995. Dorsal horn projection targets of On and OFF cells in the rostral ventromedial medulla. J. Neurophysiol. 1742–1759.

Flink, R. and Westman, J. 1986. Different neuron populations in the feline lateral cervical nucleus: A light and electron microscopic study with the retrograde axonal transport technique. J. Comp. Neurol. 250, 265–281.

Foerster, O. and Gagel, O. 1932. Die Vorderseitenstrangdurchschneidung beim Menschens. Eine klinisch-patho-physiologisch-anatomische Studie. Z. Ges. Neurol. Psychiat. 138, 1–92.

Foreman, R. D. and Weber, R. N. 1980. Responses from neurons of the primate spinothalamic tract to electrical stimulation of afferents from the cardiopulmonary region and somatic structures. Brain Res. 186, 463–468.

Foreman, R. D., Applebaum, A. E., Beall, J. E., Trevino, D. L., and Willis, W. D. 1975. Responses of primate spinothalamic tract neurons to electrical stimulation of hindlimb peripheral nerve. J. Neurophysiol. 38, 132–145.

Foreman, R. D., Blair, R. W., and Weber, R. N. 1984. Viscerosomatic convergence onto T2-T4 spinoreticular, spinoreticular-spinothalamic, and spinothalamic tract neurons in the cat. Exp. Neurol. 85, 597–619.

Foreman, R. D., Hancock, M. B., and Willis, W. D. 1981. Responses of spinothalamic tract cells in the thoracic spinal cord of the monkey to cutaneous and visceral inputs. Pain 11, 149–162.

Foreman, R. D., Kenshalo, D. R., Jr., Schmidt, R. F., and Willis, W. D. 1979. Field potentials and excitation of primate spinothalamic neurons in response to volleys in muscle afferents. J. Physiol. 286, 197–213.

Fox, R. E., Holloway, J. A., Iggo, A., and Mokha, S. S. 1980. Spinothalamic neurones in the cat: some electrophysiological obhservations. Brain Res. 182, 186–190.

Galhardo, V. and Lima, D. 1999. Structural characterization of marginal (Lamina I) spinal cord neurons in the cat: A Golgi Study. J. Comp. Neurol. 414, 315–333.

Galhardo, V., Lima, D., and Necker, R. 2000. Spinomedullary pathways in the pigeon (Columba livia): Differential involvement of lamina I cell. J. Comp. Neurol. 423, 631–645.

Gamboa-Esteves, F. O., Kaye, J. C., McWilliam, P. N., Lima, D., and Batten, T. F. 2001a. Immunohistochemical profiles of spinal lamina I neurons retrogradely labelled from the nucleus tractus solitarii in rat suggest excitatory projections. Neuroscience 104, 523–538.

Gamboa-Esteves, F. O., Lima, D., and Batten, T. F. C. 2001b. Neurochemistry of superficial spinal neurones projecting to nucleus of the solitary tract that express c-fos on chemical somatic and visceral nociceptive input in the rat. Metab. Brain Dis. 16, 151–164.

Gamboa-Esteves, F. O., Tavares, I., Almeida, A., Batten, T. F., McWilliam, P. N., and Lima, D. 2001c. Projection sites of superficial and deep spinal dorsal horn cells in the nucleus tractus solitarii of the rat. Brain Res. 921, 195–205.

Gauriau, C. and Bernard, J. F. 2004. A comparative reappraisal of projections from the superficial laminae of the dorsal horn in the rat: the forebrain. J. Comp. Neurol. 468, 24–56.

Gebhart, G. F. and Ossipov, M. H. 1986. Characterization of inhibition of the spinal nociceptive tail-flick reflex in the rat from the medullary lateral reticular nucleus. J. Neurosci. 6, 701–713.

Gerhart, K. D., Yezierski, R. P., Giesler, G. J., and Willis, W. D. 1981. Inhibitory receptive fields of primate spinothalamic tract cells. J. Neurophysiol. 46, 1309–1325.

Giesler, G. J. and Elde, R. P. 1985. Immunocytochemical studies of the peptidergic content of fibers and terminals within the lateral spinal and lateral cervical nuclei. J. Neurosci. 5, 833–1841.

Giesler, G. J., Björkeland, M., Xu, Q., and Grant, G. 1988. Organization of the spinocervicothalamic pathways in the rat. J. Comp. Neurol. 268, 223–233.

Giesler, G. J., Ménétrey, D., and Basbaum, A. I. 1979a. Differential origins of spinothalamic tract projections to medial and lateral thalamus in the rat. J. Comp. Neurol. 184, 107–126.

Giesler, G. J., Ménétrey, D., Guilbaud, G., and Besson, J. M. 1976. Lumbar cord neurons at the origin of the spinothalamic tract in the rat. Brain Res. 118, 320–324.

Giesler, G. J., Miller, L. R., Madsen, A. M., and Katter, J. T. 1987. Evidence for the existence of a lateral cervical nucleus in mice, guinea pigs and rabbits. J. Comp. Neurol. 263, 106–112.

Giesler, G. J., Spiel, H. R., and Willis, W. D. 1981a. Organization of spinothalamic tract axons within the rat spinal cord. J. Comp. Neurol. 195, 243–252.

Giesler, G. J., Urca, G., Cannon, J. T., and Liebeskind, J. C. 1979b. Response properties of neurons of the lateral cervical nucleus in the rat. J. Comp. Neurol. 186, 65–78.

Giesler, G. J., Yezierski, R. P., Gerhart, K. D., and Willis, W. D. 1981b. Spinothalamic tract neurons that project to medial and/or lateral thalamic nuclei: evidence for a physiologically novel population of spinal cord neurons. J. Neurophysiol. 46, 1285–1308.

Gobel, S. 1978a. Golgi studies of the neurons in layer I of the dorsal horn of the medulla (trigeminal nucleus caudalis). J. Comp. Neurol. 180, 375–394.

Gobel, S. 1978b. Golgi studies of the neurons in layer II of the dorsal horn of the medulla (trigeminal nucleus caudalis). J. Comp. Neurol. 180, 395–414.

Graziano, A. and Jones, E. G. 2004. Widespread thalamic terminations of fibers arising in the superficial medullary dorsal horn of monkeys and their relation to calbindin immunoreactivity. J. Neurosci. 24, 248–256.

Guilbaud, G. and Kayser, V. 1988. New evidence for the involvement of the rat ventrobasal thalamic complex in nociception: electrophysiological and neuropharmacological studies. 93–107.

Guilbaud, G., Peschanski, M., and Besson, J. M. 1985. Experimental Data Related to Nociception and Pain at the Supraspinal Level. In: Textbook of Pain (eds. P. D. Wall and R. Melzack), pp. 110–118. Churchill Livingston.

Ha, H. and Liu, C. N. 1966. Organization of the spino-cervico-thalamic system. J. Comp. Neurol. 127, 445–470.

Haber, L. H., Moore, B. D., and Willis, W. D. 1982. Electrophysiological response properties of spinoreticular neurons in the monkey. J. Comp. Neurol. 207, 75–84.

Hamann, W. C., Hong, S. K., Kniffki, K. D., and Schmidt, R. F. 1978. Projections of primary afferent fibres from muscle to neurones of the spinocervical tract of the cat. J. Physiol. 283, 369–378.

Hardy, S. G. P. 2001. Hypothalamic projections to cardiovascular centers of the medulla. Brain Res. 894, 233–240.

Harmann, P. A., Carlton, S. M., and Willis, W. D. 1988. Collaterals of spinothalamic tract cells to the periaqueductal gray: a fluorescent double-labelling study in the rat. Brain Res. 441, 87–97.

Harrison, P. J. and Jankowska, E. 1984. An intracellular study of descending and non-cutaneous afferent input to spinocervical tract neurones in the cat. J. Physiol. 356, 245–261.

Hécaen, H., Talairach, J., David, M., and Dell, M. B. 1949. Coagulations limitees du thalamus dans les algies du syndrome thalamique – resultats therapeutiques et physiologiques. Rev. Neurol. 81, 917–931.

Herbert, H. and Saper, C. B. 1992. Organization of medullary adrenergic and noradrenergic projections to the periaqueductal gray matter in the rat. J. Comp. Neurol. 315, 34–52.

Horrax, G. 1929. Experiences with cordotomy. Arch. Surg. 1140–1164.

Hunt, S. P., Pini, A., and Evan, G. 1987. Induction of c-fos-like protein in spinal cord neurons following sensory stimulation. Nature 328, 632–634.

Hwang, S. J., Burette, A., and Valtschanoff, J. G. 2003. VR1-positive primary afferents contact NK1-positive spinoparabrachial neurons. J. Comp. Neurol. 460, 255–265.

Hylden, J. L. K., Anton, F., and Nahin, R. L. 1989. Spinal lamina I projection neurons in the rat: collateral innervation of parabrachial area and thalamus. Neuroscience 28, 27–37.

Hylden, J. L. K., Hayashi, H., and Bennett, G. J. 1986b. Lamina I spinomesencephalic neurons in the cat ascend via the dorsolateral funiculi. Somatosens. Res. 4, 31–41.

Hylden, J. L. K., Hayashi, H., Bennett, G. J., and Dubner, R. 1985. Spinal lamina I neurons projecting to the parabrachial area of the cat midbrain. Brain Res. 336, 195–198.

Hylden, J. L. K., Hayashi, H., Dubner, R., and Bennett, G. J. 1986a. Physiology and morphology of the lamina I spinomesencephalic projections. J. Comp. Neurol. 247, 505–515.

Hyndman, O. R. and Van Epps, C. 1939. Possibility of differential section of the spinothalamic tract. A clinical and histological study. Arch. Surg. 38, 1036–1053.

Jankowska, E., Rastad, J., and Westman, J. 1976. Intracellular application of horseradish peroxidase and its light and electron microscopical appearance in spinocervical tract cells. Brain Res. 105, 557–562.

Jasmin, L., Burkey, A. R., Card, J. P., and Basbaum, A. I. 1997. Transneuronal labeling of a nociceptive pathway, the spino-(trigemino-) parabrachio-amygdaloid, in the rat. J. Neurosci. 17, 3751–3765.

Johnson, F. H. 1954. Experimental study of spino-reticular connections in the cat. Anat. Rec. 118, 316.

Jones, E. G. and Burton, H. 1974. Cytoarchitecture and somatic sensory connectivity of thalamic nuclei other than the ventrobasal complex in the cat. J. Comp. Neurol. 154, 395–432.

Jones, E. G. and Leavitt, R. Y. 1974. Retrograde axonal-transport and demonstration of nonspecific projections to cerebral-cortex and striatum from thalamic intralaminar nuclei in rat, cat and monkey. J. Comp. Neurol. 154, 349–377.

Jones, M. W., Hodge, C. J., Apkarian, A. V., and Stevens, R. T. 1985. A dorsolateral spinothalamic pathway in cat. Brain Res. 335, 188–193.

Jones, S. L. 1992. Descending Control of Nociception. In: The Initial Processing of Pain and its Descending Control: Spinal and Trigeminal Systems (ed. A. R. Light), pp. 203–295. Karger.

Ju, G., Melander, T., Ceccatelli, S., Hökfelt, T., and Frey, P. 1987. Immunohistochemical evidence for a spinothalamic pathway co-containing cholecystokinin- and galanin-like immunoreactivities in the rat. Neuroscience 20, 439–456.

Kajander, K. C. and Giesler, G. J. 1987. Responses of neurons in the lateral cervical nucleus of the cat to noxious cutaneous stimulation. J. Neurophysiol. 57, 1686–1704.

Katter, J. T., Burstein, R., and Giesler, G. J. 1991. The cells of origin of the spinohypothalamic tract in cats. J. Comp. Neurol. 303, 101–112.

Kayalioglu, G., Robertson, B., Kristensson, K., and Grant, G. 1999. Nitric oxide synthase and interferon-gamma receptor immunoreactivities in relation to ascending spinal pathways to thalamus, hypothalamus, and the periaqueductal grey in the rat. Somatosens. Mot. Res. 16, 280–290.

Keay, K. A. and Bandler, R. 1992. Anatomical evidence for segregated input from the upper cervical spinal-cord to functionally distinct regions of the periaqueductal gray region of the cat. Neurosci. Lett. 139, 143–148.

Keay, K. A., Clement, C. I., Matar, W. M., Heslop, D. J., Henderson, L. A., and Bandler, R. 2002. Noxious activation of spinal or vagal afferents evokes distinct patterns of fos-like immunoreactivity in the ventrolateral periaqueductal gray of unanaesthetised rats. Brain Res. 948, 122–130.

Kenshalo, D. R., Jr., Giesler, G. J., Leonard, R. B., and Willis, W. D. 1980. Responses of neurons in the primate ventral posterior lateral nucleus to noxious stimuli. J. Neurophysiol. 43, 1594–1614.

Kenshalo, D. R., Jr., Leonard, R. B., Chung, J. M., and Willis, W. D. 1979. Responses of primate spinothalamic neurons to graded and to repeated noxious heat stimuli. J. Neurophysiol. 42, 1370–1389.

Kerr, F. W. L. 1975. The ventral spinothalamic tract and other ascending systems of the ventral funiculus of the spinal cord. J. Comp. Neurol. 159, 335–356.

Kerr, F. W. L. and Lippman, H. H. 1974. The primate spinothalamic tract as demonstrated by anterolateral cordotomy and commissural myelotomy. Adv. Neurol. 4, 147–156.

Kevetter, G. A. and Willis, W. D. 1982. Spinothalamic cells in the rat lumbar cord with collaterals to the medullary reticular formation. Brain Res. 238, 181–185.

Kevetter, G. A. and Willis, W. D. 1983. Collaterals of spinothalamic cells in the rat. J. Comp. Neurol. 215, 453–464.

Kevetter, G. A., Haber, L. H., Yezierski, R. P., Chung, J. M., Martin, R. F., and Willis, W. D. 1982. Cells of origin of the spinoreticular tract in the monkey. J. Comp. Neurol. 207, 61–74.

Klop, E. M., Mouton, L. J., and Holstege, G. 2004. Less than 15% of the spinothalamic fibers originate from neurons in lamina I in cat. Neurosci. Lett. 360, 125–128.

Kniffki, K. D., Mense, S., and Schmidt, R. F. 1977. The spinocervical tract as a possible pathway for muscular nociception. J. Physiol. (Paris) 73, 359–366.

Kohnstamm, O. 1900. Ueber die Coordinationskerne des Hirnstammes und die absteigenden Spinalbahnen. Nach den Ergebnissen der combinierten Degenerationsmethode. Mschr. Psychiat. Neurol. 8, 261–293.

Kostarczyk, E., Zhang, X. J., and Giesler, G. J. 1997. Spinohypothalamic tract neurons in the cervical enlargement of rats: Locations of antidromically identified ascending axons and their collateral branches in the contralateral brain. J. Neurophysiol. 77, 435–451.

Krettek, J. E. and Price, J. L. 1977. The cortical projections of the mediodorsal nucleus and adjacent thalamic nuclei in the rat. J. Comp. Neurol. 171, 157–191.

Kunze, W. A. A., Wilson, P., and Snow, P. J. 1987. Response of lumbar spinocervical tract cells to natural and electrical stimulation of the hindlimb footpads in cats. Neurosci. Lett. 75, 253–258.

Leah, J., Ménétrey, D., and De Pommery, J. 1988. Neuropeptides in long ascending spinal tract cells in the rat: evidence for parallel processing of ascending information. Neuroscience 24, 195–207.

Leite-Almeida, H., Valle-Fernandes, A., and Almeida, A. 2006. Brain projections from the medullary dorsal reticular nucleus: an anterograde and retrograde tracing study in the rat. Neuroscience 140, 577–595.

Lima, D. 1990. A spinomedullary projection terminating in the dorsal reticular nucleus of the rat. Neuroscience 34, 577–589.

Lima, D. 1997. Functional anatomy of spinofugal nociceptive pathways. Pain Rev. 4, 1–19.

Lima, D. 1998. Anatomical basis for the dynamic processing of nociceptive input. Eur. J. Pain. 2, 195–202.

Lima, D. and Almeida, A. 2002. The medullary dorsal reticular nucleus as a pronociceptive centre of the pain control system. Prog. Neurobiol. 66, 81–108.

Lima, D. and Coimbra, A. 1983. The neuronal population of the marginal zone (lamina I) of the rat spinal cord. A study based on reconstructions of serially sectioned cells. Anat. Embryol. 167, 273–288.

Lima, D. and Coimbra, A. 1985. Marginal neurons of the rat spinal cord at the origin of a spinobulboreticular projection. Neurosci. Lett. 22, S9.

Lima, D. and Coimbra, A. 1986. A Golgi study of the neuronal population of the marginal zone (lamina I) of the rat spinal cord. J. Comp. Neurol. 244, 53–71.

Lima, D. and Coimbra, A. 1988. The spinothalamic system of the rat: structural types of retrogradely labeled neurons in the marginal zone (lamina I). Neuroscience 27, 215–230.

Lima, D. and Coimbra, A. 1989. Morphological types of spinomesencephalic neurons in the marginal zone (lamina I) of the rat spinal cord, as shown after retrograde labeling with cholera toxin subunit B. J. Comp. Neurol. 279, 327–339.

Lima, D. and Coimbra, A. 1990. Structural types of marginal (Lamina I) neurones projecting to the dorsal reticular nucleus of the medulla oblongata. Neuroscience 34, 591–606.

Lima, D. and Coimbra, A. 1991. Neurons in the substantia gelatinosa rolandi (lamina II) project to the caudal ventrolateral reticular formation of the medulla oblongata in the rat. Neurosci. Lett. 132, 16–18.

Lima, D., Albino-Teixeira, A., and Tavares, I. 2001. The caudal medullary ventrolateral reticular formation in nociceptive-cardiovascular integration. An experimental study in the rat. Exp. Physiol. 87.2, 267–274.

Lima, D., Almeida, A., and Tavares, A. 1998. The endogenous pain control system: facilitating and inhibitory pathways. In: Pain Mechanisms and Management (eds. S. N. Ayrapetyan and A. V. Apkarian), pp. 192–210. IOS Press.

Lima, D., Avelino, A., and Coimbra, A. 1992. Differential participation of marginal cells in the transmission of nociceptive input to the thalamus and mesencephalon. Eur. J. Neurosci. Suppl. 5, 2057.

Lima, D., Avelino, A., and Coimbra, A. 1993. Morphological characterization of marginal (Lamina I) neurons immunoreactive for substance P, enkephalin, dynorphin and gamma-aminobutyric acid in the rat spinal cord. J. Chem. Neuroanat. 6, 43–52.

Lima, D., Castro, A. R., Castro-Lopes, J. M., and Galhardo, V. 2002. Differential expression of GABAB receptors in monkey lamina I neuronal types; a Golgi and immunocytochemical study. Society Neurosci Abst 258.4.

Lima, D., Esteves, F., and Coimbra, A. 1994. C-fos activation by Noxious Input of Spinal Neurons Projecting to the Nucleus of the Tractus Solitarius in the Rat. In: Proceedings of the 7th World Congress on Pain (ed. G. F. Gebhart), Vol. 2, pp. 423–434. IASP Press.

Lima, D., Mendes-Ribeiro, J. A., and Coimbra, A. 1991. The spino-latero-reticular system of the rat: projections from the superficial dorsal horn and structural characterization of marginal neurons involved. Neuroscience 45, 137–152.

Liu, R. P. 1983. Laminar origins of spinal projection to the periaqueductal gray of the rat. Brain Res. 264, 118–122.

Loewy, A. D. 1990. Central autonomic pathways. In: Central Regulation of Autonomic Functions (eds. A. D. Loewy and K. M. Spyer), pp. 88–103. Oxford University Press.

Loewy, A. D. and Burton, H. 1978. Nuclei of the solitary tract: Efferent projections to the lower brain stem and spinal cord of the cat. J. Comp. Neurol. 181, 421–450.

Lovick, T. A. 1993. Integrated activity of cardiovascular and pain integratory systems: Role in adaptive behaviour. Prog. Neurobiol. 40, 631–644.

Lund, R. D. and Webster, K. E. 1967. Thalamic afferents from the spinal cord and trigeminal nuclei. An experimental anatomical study in the rat. J. Comp. Neurol. 130, 313–328.

Lundberg, A. and Oscarsson, O. 1961. Three ascending spinal pathways in the dorsal part of the lateral funiculus. Acta. Physiol. Scand. 51, 1–16.

Ma, W. and Peschanski, M. 1988. Spinal and trigeminal projections to the parabrachial nucleus in the rat-electron-microscopic evidence of a spino-ponto-amygdalian somatosensory pathway. Somatosens. Res. 5, 247–257.

Magnusson, K. R., Magnusson, K. R., Clements, J. R., Larson, A. A., Madl, J. E., and Beitz, A. J. 1987. Localization of glutamate in trigeminothalamic projection neurons: a combined retrograde transport-immunohistochemical study. Somatosens. Res. 4, 177–190.

Mainen, Z. F. and Sejnowski, T. J. 2006. Influence of dendritic structure on firing pattern in model neocortical neurons. Nature 382, 363–366.

Malick, A. and Burstein, R. 1998. Cell of origin of the trigeminohypothalamic tract in the rat. J. Comp. Neurol. 400, 125–144.

Malick, A., Strassman, A. M., and Burstein, R. 2000. Trigeminohypothalamic and reticulohypothalamic tract in the upper cervical spinal cord and caudal medulla of the rat. J. Neurophysiol. 84, 2078–2112.

Mantyh, P. W. 1982. The ascending input to the midbrain periaqueductal gray of the primate. J. Comp. Neurol. 211, 50–64.

Mantyh, P. W. 1983a. The spinothalamic tract in the primate: a re-examination using wheatgerm agglutinin conjugated to horseradish peroxidase. Neuroscience 9, 847–862.

Mantyh, P. W. 1983b. The terminations of the spinothalamic tract in the cat. Neurosci. Lett. 38, 119–124.

Mark, R. F. and Steiner, J. 1958. Cortical projection of impulser in myelinated cutaneous afferent nerve fibres of the cat. J. Physiol. 142, 544–562.

Maunz, R. A., Pitts, N. G., and Peterson, B. W. 1978. Cat spinoreticular neurons: locations, responses and changes in responses during repetitive stimulation. Brain Res. 148, 365–379.

McMahon, S. B. and Wall, P. D. 1985. Electrophysiological mapping of brainstem projections of spinal cord lamina I cells in the rat. Brain Res. 19–26.

Mehler, W. R. 1957. The mammalian pain tract in phylogeny. Anat. Rec. 127, 332.

Mehler, W. R. 1962. The anatomy of the so-called 'pain tract' in man: an analysis of the course and distribution of the ascending fibers of the fasciculus anterolateralis. In: Basic Research in Paraplegia (eds. J. D. French and R. W. Porter), pp. 26–55. Thomas.

Mehler, W. R. 1966. Some observations on secondary ascending afferent systems in the central nervous system. In: Pain (eds. R. S. Knighton and P. R. Dumke), pp. 11–32. Little Brown.

Mehler, W. R. 1969. Some neurological species differences – a posteriori. Ann. NY Acad. Sci. 167, 424–468.

Mehler, W. R. 1974. Central pain and the spinothalamic tract. Adv. Neurol. 4, 127–146.

Mehler, W. R., Feferman, M. E., and Nauta, W. J. H. 1956. Ascending axon degeneration following anterolateral chordotomy in the monkey. Anat. Rec. 124, 332–333.

Mehler, W. R., Feferman, M. E., and Nauta, W. J. H. 1960. Ascending axon degeneration following anterolateral cordotomy. An experimental study in the monkey. Brain 83, 718–751.

Meller, S. T. and Dennis, B. J. 1991. Efferent projections of the periaqueductal gray in the rabbit. Neuroscience 40, 191–216.

Melzack, R. and Casey, K. L. 1968. Sensory, motivational and central control determinants of pain. pp. 423–443.

Menétrey, D. and Basbaum, A. I. 1987. Spinal and trigeminal projections to the nucleus of the solitary tract – a possible substrate for somatovisceral and viscerovisceral reflex activation. J. Comp. Neurol. 255, 439–450.

Menétrey, D. and DePommery, J. 1991. Origins of spinal ascending pathways that reach central areas involved in viscenoception and visceronociception in the rat. Eur. J. Neurosci. 3, 249–259.

Menétrey, D., Chaouch, A., and Besson, J. M. 1980. Location and properties of dorsal horn neurons at origin of spinoreticular tract in lumbar enlargement of the rat. J. Neurophysiol. 44, 862–877.

Menétrey, D., Chaouch, A., Binder, D., and Besson, J. M. 1982. The origin of the spinomesencephalic tract in the rat: an anatomical study using the retrograde transport of horseradish peroxidase. J. Comp. Neurol. 206, 193–207.

Menétrey, D., DePommery, J., Baimbridge, K. G., and Thomasset, M. 1992a. Calbindin-D28K (CaBP28k)-like immunoreactivity in ascending projections.1. Trigeminal nucleus caudalis and dorsal vagal complex projections. Eur. J. Neurosci. 4, 61–69.

Menétrey, D., DePommery, J., and Besson, J. M. 1984. Electrophysiological characteristics of lumbar spinal cord neurons backfired from lateral reticular nucleus in the rat. J. Neurophysiol. 52, 595–611.

Menétrey, D., DePommery, J., Thomasset, M., and Baimbridge, K. G. 1992b. Calbindin-D28K (CaBP28k)-like immunoreactivity in ascending projections. II. Spinal Projections to brain stem and mesencephalic areas. Eur. J. Neurosci. 4, 70–76.

Menétrey, D., Roudier, F., and Besson, J. M. 1983. Spinal neurons reaching the lateral reticular nucleus as studied in the rat by retrograde transport of horseradish peroxidase. J. Comp. Neurol. 220, 439–452.

Meyers, D. E. R. and Snow, P. J. 1982. The morphology of physiologically identified deep spinothalamic tract cells in the lumbar spinal cord of the cat. J. Physiol. 329, 373–388.

Millis, C. D. and Hulsebosch, C. E. 2002. Increased expression of metabotropic glutamate receptor subtype 1 on spinothalamic tract neurons following spinal cord injury in the rat. Neurosci. Lett. 319, 59–62.

Milne, R. J., Foreman, R. D., and Willis, W. D. 1982. Responses of primate spinothalamic neurons located in the sacral intermediomedial gray (Stilling's nucleus) to proprioceptive input from the tail. Brain Res. 234, 227–236.

Miyagawa, T., Ando, R., Sakurada, S., Sakurada, T., and Kisara, K. 1986. Effects of tooth pulp stimulation on single unit activity of the amygdala in cats. Nippon Yakurigaku Zasshi 88, 173–178.

Morison, R. S. and Dempsey, E. W. 1942. A study of thalamo-cortical relations. Am. J. Physiol. 135, 281–292.

Mouton, L. J. and Holstege, G. 1998. Three times as many lamina I neurons project to the periaqueductal gray than to the thalamus: a retrograde tracing study in the cat. Neurosci. Lett. 255, 107–110.

Mouton, L. J. and Holstege, G. 2000. Segmental and laminar organization of the spinal neurons projecting to the periaqueductal gray (PAG) in the cat suggests the existence of at least five separate clusters of Spino-PAG neurons. J. Comp. Neurol. 428, 389–410.

Mouton, L. J., Klop, E. M., and Holstege, G. 2001. Lamina I-periaqueductal gray (PAG) projections represent only a limited part of the total spinal and caudal medullary input to the PAG in the cat. Brain. Res. Bull. 54, 167–174.

Mouton, L. J., Klop, E. M., Broman, J., Zhang, M., and Holstege, G. 2004. Lateral cervical nucleus projections to periaqueductal gray matter in cat. J. Comp. Neurol. 471, 434–445.

Mtui, E. P., et al. 1993. Projections from the nucleus tractus solitarii to the spinal cord. J. Comp. Neurol. 337, 231–252.

Nahin, R. L. 1988. Immunocytochemical identification of long ascending, peptidergic lumbar spinal neurons terminating in either the medial or lateral thalamus in the rat. Brain Res. 443, 345–349.

Nahin, R. L. and Micevych, P. E. 1986. A long ascending pathway of enkephalin-like immunoreactive spinoreticular neurons in the rat. Neurosci. Lett. 65, 271–276.

Nahin, R. L., Madsen, A. M., and Giesler, G. J. 1986. Funicular location of the ascending axons of neurons adjacent to the spinal cord central canal in the rat. Brain Res. 384, 367–372.

Palecek, J., Paleckova, V., and Willis, W. D. 2003. Fos expression in spinothalamic and postsynaptic dorsal column neurons following noxious visceral and cutaneous stimuli. Pain 104, 249–257.

Panneton, W. M. and Burton, H. 1985. Projections from the paratrigeminal nucleus and the medullary and spinal dorsal horns to the peribrachial area in the cat. Neuroscience 15, 779–797.

Parenti, R., Cicirata, F., Panto, M. R., and Serapide, M. F. 1996. The projections of the lateral reticular nucleus to the deep cerebellar nuclei. An experimental analysis in the rat. Eur. J. Neurosci. 8, 2157–2167.

Paxinos, G. and Watson, C. 1998. The Rat Brain in Stereotaxic Coordinates, 4th ed. Academic Press.

Pechura, C. and Liu, R. 1986. Spinal neurons which project to the periaqueductal gray and the medullary reticular formation via axon collaterals: a double-label fluorescence study in the rat. Brain Res. 374, 357–361.

Perl, E. R. and Whitlock, D. G. 1961. Somatic stimuli exciting spinothalamic projections to thalamic neurons in cat and monkey. Exp. Neurol. 3, 256–296.

Peschanski, M. and Besson, J. M. 1984. A spino-reticulo-thalamic pathway in the rat: an anatomical study with reference to pain transmission. Neuroscience 12, 165–178.

Peschanski, M., Guilbaud, G., Gautron, M., and Besson, J. M. 1980. Encoding of noxious heat messages in neurons of the ventrobasal thalamic complex in the rat. Brain Res. 197, 401–413.

Peschanski, M., Mantyh, P. W., and Besson, J. M. 1983. Spinal afferents to the ventrobasal thalamic complex in the rat: an anatomical study using wheat-germ agglutinin conjugated to horseradish peroxidase. Brain Res. 278, 240–244.

Polgar, E., Puskár, Z., Watt, C., Matesz, C., and Todd, A. J. 2002. Selective innervation of lamina I projection neurones that possess the neurokinin 1 receptor by serotonin-containing axons in the rat spinal cord. Neuroscience 109, 799–809.

Prescott, S. A. and De Koninck, Y. 2002. Four cell types with distinctive membrane properties and morphologies in lamina I of the spinal dorsal horn of the adult rat. J. Physiol. 539, 817–836.

Price, D. D., Hayes, R. L., Rude, M., and Dubner, R. 1978. Spatial and temporal transformations of input to spinothalamic tract neurons and their relation to somatic sensations. J. Neurophysiol. 41, 933–947.

Puskar, Z., Polgar, E., and Todd, A. J. 2001. A population of large lamina I projection neurons with selective inhibitory input in rat spinal cord. Neuroscience 102, 167–176.

Quensel, F. 1898. Ein Fall von Sarcom der Dura spinalis. Beitrag zur Kenntnis der secundaeren Degenerationen nach Rueckenmarkscompression. Neurol. Zbl. 17, 482–493.

Ramón y Cajal, S. 1909. Histologie dusystème nerveux de l'homme et des vertébrés. Maloine.

Reis, L. C., Colombari, E., Canteras, N. S., and Cravo, S. L. 2000. Projections from nucleus tractus solitarii (NTS) and ventrolateral medulla (VLM) to paraventricular nucleus of the hypothalamus (PVH) and median preoptic nucleus (MnPO). Projections from nucleus tractus solitarii (NTS) and ventrolateral medulla (VLM) to paraventricular nucleus of the hypothalamus (PVH) and median preoptic nucleus (MnPO). FASEB J. 14, A66.

Ring, G. and Ganchrow, D. 1983. Projections of nucleus caudalis and spinal cord to brainstem and diencephalon in the hedgehog (*Erinaceus europaeus* and *Paraechinus aethiopicus*): A degeneration study. J. Comp. Neurol. 216, 132–151.

Rossi, G. F. and Brodal, A. 1957. Terminal distribution of spinoreticular fibers in the cat. Arch. Neurol. Psychiat. 78, 439–453.

Rucker, H. K. and Holloway, J. A. 1982. Viscerosomatic convergence onto spinothalamic tract neurons in the cat. Brain Res. 243, 155–157.

Ruda, M. A., Coffield, J., and Dubner, R. 1984. Demonstration of postsynaptic opioid modulation of thalamic projection neurons by the combined techniques of retrograde horseradish peroxidase and enkephalin immunocytochemistry. J. Neurosci. 4, 2117–2132.

Ruscheweyh, R., Ikeda, H., Heinke, B., and Sandkuhler, J. 2004. Distinctive membrane and discharge properties of rat spinal lamina I projection neurons in vitro. J. Physiol. 555, 527–543.

Shi, T. and Apkarian, A. V. 1995. Morphology of thalamocortical neurons projecting to the primary somatosensory cortex and their relationship to spinothalamic terminals in the squirrel monkey. J. Comp. Neurol. 361, 1–24.

Short, A. D., Brown, A. G., and Maxwell, D. J. 1990. Afferent inhibition and facilitation of transmission through the spinocervical tract in the anaesthetized cat. J. Physiol. 429, 511–528.

Sim, L. J. and Joseph, S. A. 1994. Efferent of the opiocortin-containing region of the commissural nucleus tractus solitarius. Peptides 15, 169–174.

Simone, D. A., Sorkin, L. S., Oh, U., Chung, J. M., Owens, C., LaMotte, R. H., and Willis, W. D. 1991. Neurogenic hyperalgesia: Central neural correlates in responses of spinothalamic tract neurons. J. Neurophysiol. 66, 228–246.

Slugg, R. M. and Light, A. R. 1994. Spinal cord and trigeminal projections to the pontine parabrachial region in the rat as demonstrated with phaseolus-vulgaris-leukoagglutinin. J. Comp. Neurol. 339, 49–61.

Snowball, R. K., Semenenko, F. M., and Lumb, B. M. 2000. Visceral inputs to neurons in the anterior hypothalamus including those that project to the periaqueductal gray: A functional anatomical and electrophysiological study. Neuroscience 99, 351–361.

Spike, R. C., Puskar, Z., Andrew, D., and Todd, A. J. 2003. A quantitative and morphological study of projection neurons in lamina I of the rat lumbar spinal cord. Eur. J. Neurosci. 18, 2433–2448.

Standaert, D., Watson, S., Houghten, R., and Saper, C. 1986. Opioid peptide immunoreactivity in spinal and trigeminal dorsal horn neurons projecting to the parabrachial nucleus in the rat. J. Neurosci. 6, 1220–1226.

Stevens, R. T., Hodge, C. J., and Apkarian, A. V. 1989. Medial, intralaminar and lateral terminations of lumbar spinothalamic tract neurons: a fluorescent double label study. Somatosens. Mot. Res. 6, 285–308.

Stevens, R. T., London, S. M., and Apkarian, A. V. 1993. Spinothalamocortical projections to the secondary somatosensory cortex (SII) in squirrel monkey. Brain Res. 631, 241–246.

Surmeier, D. J., Honda, C. N., and Willis, W. D. 1986a. Temporal features of the responses of primate spinothalamic neurons to noxious thermal stimulation of hairy and glabrous skin. J. Neurophysiol. 56, 351–369.

Surmeier, D. J., Honda, C. N., and Willis, W. D. 1986b. Responses of primate spinothalamic neurons to noxious thermal stimulation of glabrous and hairy skin. J. Neurophysiol. 56, 328–350.

Surmeier, D. J., Honda, C. N., and Willis, W. D. 1988. Natural grouping of primate spinothalamic neurons based on cutaneous stimulation. Physiological and anatomical features. J. Neurophysiol. 59, 833–860.

Svensson, B. A., Rastad, J., Westman, J., and Wiberg, M. 1985. Somatotopic termination of spinal afferents to the feline lateral cervical nucleus. Exp. Brain Res. 57, 576–584.

Swett, J. E., McMahon, S. B., and Wall, P. D. 1985. Long ascending projections to the midbrain from cells of lamina I and nucleus of the dorsolateral funiculus of the rat spinal cord. J. Comp. Neurol. 238, 401–416.

Szucs, P., Odeh, F., Szokol, K., and Antal, M. 2003. Neurons with distinctive firing patterns, morphology and distribution in lamina V-VII of the neonatal rat lumbar spinal cord. Eur. J. Neurosci. 17, 537–544.

Taub, A. and Bishop, P. O. 1965. The spinocervical tract: dorsal column linkage, conduction velocity, primary afferent spectrum. Exp. Neurol. 13, 1–21.

Tavares, I., Almeida, A., Albino-Teixeira, A., and Lima, D. 1997. Lesions of the caudal ventrolateral medulla block the hypertension-induced inhibition of noxious-evoked *c-fos* expression in the rat spinal cord. Eur. J. Pain 1, 149–160.

Tavares, I., Almeida, A., Esteves, F., Lima, D., and Coimbra, A. 1998. The caudal ventrolateral medullary reticular formation is reciprocally connected with the spinal cord. Society Neurosci Abst 24, 1132.

Tavares, I. and Lima, D. 1994. Descending projections from the caudal medulla oblongata to the superficial or deep dorsal horn of the rat spinal cord. Exp. Brain Res. 99, 455–463.

Tavares, I. and Lima, D. 2002. The caudal ventrolateral medulla as an important inhibitory modulator of pain transmission in the spinal cord. J. Pain. 3, 337–346.

Tavares, I., Lima, D., and Coimbra, A. 1993. Neurons in the superficial dorsal horn of the rat spinal cord projecting to the medullary ventrolateral reticular formation express *c-fos* after noxious stimulation of the skin. Brain Res. 623, 278–286.

Tavares, I., Lima, D., and Coimbra, A. 1996a. Catecholaminergic and serotonergic input to spinal dorsal horn neurons projecting to the caudal ventrolateral medulla of the rat. Abst 8th World Congress on Pain 448.

Tavares, I., Lima, D., and Coimbra, A. 1996b. The ventrolateral medulla of the rat is connected with the spinal cord dorsal horn by an indirect descending pathway relayed in the A5 noradrenergic cell group. J. Comp. Neurol. 374, 84–95.

Thiele, F. H. and Horsley, V. 1901. A study of the degenerations observed in the central nervous system in a case of fracture dislocation of the spine. Brain 24, 519–531.

Thies, R. 1985. Activation of lumbar spinoreticular neurons by stimulation of muscle, cutaneous and sympathetic afferents. Brain Res. 333, 151–155.

Thies, R. and Foreman, R. D. 1983. Inhibition and excitation of thoracic spinoreticular neurons by electrical stimulation of vagal afferent nerves. Exp. Neurol. 82, 1–16.

Thompson, R. H., Canteras, N. S., and Swanson, L. W. 1996. Organization of projections from the dorsomedial nucleus of the hypothalamus: A PHA-L study in the rat. J. Comp. Neurol. 376, 143–173.

Todd, A. J., McGill, M. M., and Shehad, S. A. S. 2000. Neurokinin 1 receptor expression by neurons in laminae I, III and IV of the rat spinal dorsal horn that project to the brainstem. Eur. J. Neurosci. 12, 689–700.

Todd, A. J., Puskar, Z., Spike, R. C., Hughes, C., Watt, C., and Forrest, L. 2002. Projection neurons in lamina I of rat spinal cord with the neurokinin 1 receptor are selectively innervated by substance p-containing afferents and respond to noxious stimulation. J. Neurosci. 22, 4103–4113.

Todd, A. J. and Spike, R. C. 1993. Localization of classical neurotransmitters and neuropeptides within neurons in lamina I-III of the mammalian spinal dorsal horn. Prog. Neurobiol. 41, 609–645.

Traub, R. J. and Murphy, A. 2002. Colonic inflammation induces fos expression in the thoracolumbar spinal cord increasing activity in the spinoparabrachial pathway. Pain 95, 93–102.

Trevino, D. L. 1976. The Origin and Pojections of a Spinal Nociceptive and Thermoreceptive Pathway. In: Sensory Functions of the Skin in Primates, with Special Reference to Man (ed. Y. Zotterman), pp. 367–376. Pergamon Press.

Trevino, D. L. and Carstens, E. 1975. Confirmation of the location of spinothalamic neurons in the cat and monkey by the retrograde transport of horseradish peroxidase. Brain Res. 98, 177–182.

VanderHorst, V. G. J. M., Mouton, L. J., Blok, B. F. M., and Holstege, G. 1996. Distinct cell groups in the lumbosacral cord of the cat project to different areas in the periaqueductal gray. J. Comp. Neurol. 376, 361–385.

Vierck, C. J. and Luck, M. M. 1979. Loss and recovery of reactivity to noxious stimulation in monkeys with primary spinothalamic cordotomies, followed by secundary and tertiary lesions of other cord sectors. Brain 102, 233–248.

Villanueva, L., Bernard, J. F., and Le Bars, D. 1995. Distribution of spinal cord projections from the medullary subnucleus reticularis dorsalis and the adjacent cuneate nucleus: a Phaseolus vulgaris-leucoagglutinin study in the rat. J. Comp. Neurol. 352, 11–32.

Villanueva, L., Bouhassira, D., Bing, Z., and Le Bars, D. 1988. Convergence of heterotopic nociceptive information onto subnucleus reticularis dorsalis (SRD) neurons in the rat medulla. J. Neurophysiol. 60, 980–1009.

Villanueva, L., Bouhassira, D., Bing, Z., and Le Bars, D. 1991. Spinal afferent projections to subnucleus reticularis dorsalis in the rat. J. Neurophysiol. 60, 980–1009.

Villanueva, L., Desbois, C., Le Bars, D., and Bernard, J. F. 1998. Organization of diencephalic projections from the medullary subnucleus reticularis dorsalis and the adjacent cuneate nucleus: a retrograde and anterograde tracer study in the rat. J. Comp. Neurol. 390, 133–160.

Walker, A. E. 1940. The spinothalamic tract in man. Arch. Neurol. Psychiat. 43, 284–298.

Walker, A. E. 1942a. Relief of pain by mesencephalic tractotomy. Arch. Neurol. Psychiat. 48, 865–880.

Walker, A. E. 1942b. Somatotopic localization of spino thalamic and sensory trigeminal tracts in mesencephalon. Arch. Neurol. Psychiat. 48, 884–889.

Wang, C. C. and Shyu, B. C. 2004. Differential projections from the mediodorsal and centrolateral thalamic nuclei to the frontal cortex in rats. Brain Res. 995, 226–235.

Westlund, K. N. and Craig, A. D. 1996. Association of spinal lamina I projections with brainstem catecholamine neurons in the monkey. Exp. Brain Res. 110, 151–162.

Westlund, K. N., Bowker, R. M., Ziegler, M. G., and Coulter, J. D. 1983. Noradrenergic projections to the spinal cord of the rat. Brain Res. 263, 15–31.

Westlund, K. N., Bowker, R. M., Ziegler, M. G., and Dan, J. 1981. Origins of spinal noradrenergic pathways demonstrated by retrograde transport of antibody to dopamine-â-hydroxylase. Neurosci. Lett. 25, 243–249.

Whitsel, B. L., Rustioni, A., Dreyer, D. A., Loe, P. R., Allen, E. E., and Metz, C. B. 1978. Thalamic projections to S-1 in macaque monkey. J. Comp. Neurol. 178, 385–409.

Wiberg, M. and Blomqvist, A. 1984. The spinomesencephalic tract in the cat: its cells of origin and termination pattern as demonstrated by the intraaxonal transport method. Brain Res. 291, 1–18.

Wiberg, M., Westman, J., and Blomqvist, A. 1987. Somatosensory projection to the mesencephalon: an anatomical study in the monkey. J. Comp. Neurol. 264, 92–117.

Willis, W. D. 1988. Nociceptive neurons in the primate ventral posterior lateral (VPL) nucleus. In: Cellular Thalamic Mechanisms (eds. M. Bentivoglio and R. Spreafico), pp. 77–92. Elsevier.

Willis, W. D., Kenshalo, D. R., Jr., and Leonard, R. B. 1979. The cells of origin of the primate spinothalamic tract. J. Comp. Neurol. 188, 543–574.

Willis, W. D., Trevino, D. L., Coulter, J. D., and Maunz, R. A. 1974. Responses of primate spinothalamic tract neurons to natural stimulation of hindlimb. J. Neurophysiol. 37, 358–372.

Willis, W. D., Zhang, X., Honda, C. N., and Giesler, G. J. 2001. Projections from the marginal zone and deep dorsal horn to the ventrobasal nuclei of the primate thalamus. Pain 92, 267–276.

Yezierski, R. P. 1988. Spinomesencephalic tract: projections from the lumbosacral spinal cord of the rat, cat, and monkey. J. Comp. Neurol. 267, 131–146.

Yezierski, R. P. 1990. The effects of midbrain and medullary stimulation on spinomesencephalic tract cells in the cat. J. Neurophysiol. 63, 240–255.

Yezierski, R. P. and Broton, J. G. 1991. Functional properties of spinomesencephalic tract (SMT) cells in the upper cervical spinal cord of the cat. Pain 45, 187–196.

Yezierski, R. P. and Mendez, C. M. 1991. Spinal distribution and collateral projections of rat spinomesencephalic tract cells. Neuroscience 44, 113–130.

Yezierski, R. P. and Schwartz, R. H. 1986. Response and receptive-field properties of spinomesencephalic tract cells in the cat. J. Neurophysiol. 55, 76–96.

Yezierski, R. P., Sorkin, L. S., and Willis, W. D. 1987. Response properties of spinal neurons projecting to midbrain or midbrain-thalamus in the monkey. Brain Res. 437, 165–170.

Yoshida, A., Dostrovsky, J. O., and Chiang, C. Y. 1992. The afferent and efferent connections of the nucleus submedius in the rat. J. Comp. Neurol. 324, 115–133.

Yu, X. H., et al. 1999. NK-1 receptor immunoreactivity in distinct morphological types of lamina I neurons of the primate spinal cord. J. Neurosci. 19, 3545–3555.

Zardettosmith, A. M. and Gray, T. S. 1995. Catecholamine and NPY efferents from the ventrolateral medulla to the amygdala in the rat. Brain Res. Bull. 38, 253–260.

Zemlan, F. P., Leonard, C. M., Kow, L. M., and Plaff, D. W. 1978. Ascending tracts of the lateral columns of the rat spinal cord: a study using the silver impregnation and horseradish peroxidase techniques. Exp. Neurol. 62, 298–334.

Zhang, D., Carlton, S. M., Sorkin, L. S., and Willis, W. D. 1990. Collaterals of primate spinothalamic tract neurons to the periaqueductal gray. J. Comp. Neurol. 296, 277–290.

Zhang, E. T. and Craig, A. D. 1997. Morphology and distribution of spinothalamic lamina I neurons in the monkey. J. Neurosci. 17, 3274–3284.

Zhang, E. T., Han, Z. S., and Craig, A. D. 1996. Morphological classes of spinothalamic lamina I neurons in the cat. J. Comp. Neurol. 367, 537–549.

Zhang, X. J., Gokin, A. P., and Giesler, G. J. 2002. Responses of spinohypothalamic tract neurons in the thoracic spinal cord of rats to somatic stimuli and to graded distention of the bile duct. Somatosens. Mot. Res. 19, 5–17.

Zhang, X. J., Wenk, H. N., Honda, C. N., and Giesler, G. J. 2000. Locations of spinothalamic tract axons in cervical and thoracic spinal cord white matter in monkeys. J. Neurophysiol. 83, 2869–2880.

Zou, X. Y., Lin, Q., and Willis, W. D. 2000. Enhanced phosphorylation of NMDA receptor 1 subunits in spinal cord dorsal horn and spinothalamic tract neurons after intradermal injection of capsaicin in rats. J. Neurosci. 20, 6989–6997.

36 Dorsal Columns and Visceral Pain

W D Willis Jr. and K N Westlund, University of Texas Medical Branch, Galveston, TX, USA

Glossary

activity box Apparatus used to determine the amount and time course of exploratory activity of an animal, such as a rat.

central neuropathic pain pain that develops following injury to the central nervous system.

CNQX 6-cyano-7-nitroquinoxaline-2,3-dione, a non-*N*-methyl-D-aspartic acid receptor antagonist.

kainic acid lesion damage produced by injection of kainic acid, a substance that produces excitotoxicity.

NMDA *N*-methyl-D-aspartic acid.

pain referral Projection of the source of pain to an area of the body distant to the actual area of injury.

paresthesias Unusual sensations, such as tingling or burning.

rhizotomy Transaction of one or more spinal roots.

viscerospecific responses Neuronal activity evoked by sensory input from a particular visceral organ.

36.1   Clinical Evidence Concerning Spinal Cord Pathways That Signal Visceral Pain

36.1.1   Spinothalamic Tract

36.1.1.1   *Anterolateral cordotomy*

It is well known that n is relieved (at least temporarily) and thermal sense is lost over the appropriate part of the contralateral body following an anterolateral cordotomy (Spiller, W. G., 1905; Spiller, W. G. and Martin, E., 1912; Foerster, O. and Gagel, G., 1932; reviewed in Gybels, J. M. and Sweet, W. H., 1989). The signals are transmitted from one side of the body to the contralateral thalamus (Kenshalo, D. R. *et al.*, 1980; Chung, J. M. *et al.*, 1986; Bushnell, M. C. *et al.*, 1993; Lee, J. I. *et al.*, 1999; 2005) and from there to the appropriate regions of the cerebral cortex (see Casey, K. L. and Bushnell, M. C., 2000). It has been presumed that the pain signals are conveyed chiefly by the spinothalamic tract, which decussates at the spinal cord level (Mehler, W. R., 1962), although other pathways that accompany the spinothalamic tract are also likely to contribute. These include the spinoreticular, spinoparabrachial, spinohypothalamic, and other tracts (Willis, W. D. and Westlund, K. N.,1997; Willis, W. D. and Coggeshall, R. E., 2004). Innocuous thermal signals are attributed just to the spinothalamic tract.

Anterolateral cordotomy can relieve superficial and deep somatic pain, as well as visceral pain (and can block thermal sense), provided that the lesion is extensive enough (Gybels, J. M. and Sweet, W. H., 1989; Nathan, P. W. *et al.*, 2001). For complete pain relief on the contralateral side of the body, a cordotomy at an upper cervical level needs to extend from just posterior to the denticulate ligament, across the remainder of the lateral funiculus and well into the anterior funiculus (Figure 1(a); Nathan, P. W. *et al.*, 2001).

However, even after an initially successful cordotomy, pain relief may not persist more than a few months to a year or so, although in some cases the pain relief is of very long duration (Gybels, J. M. and Sweet, W. H., 1989). It is unclear why the pain can recur. Some suggestions are that other pathways now convey the pain signals, the disease advances (e.g., metastatic cancer may activate spinothalamic neurons whose axons were not interrupted by the cordotomy), or central neuropathic pain develops. Because of the recurrence of pain after a large proportion of cordotomies, neurosurgeons have

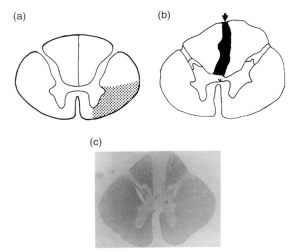

Figure 1   (a) Drawing of a cross section of the human spinal cord at the third cervical level. The dotted area shows the region that needs to be sectioned by a cordotomy in order to relieve contralateral pain completely. Note that the lesion must extend to a level that is posterior to the expected location of the denticulate ligament and anteromedially across the lateral funiculus and into the anterior funiculus. (b) Drawing of a cross section of the human spinal cord at a mid-thoracic level showing the location of a limited midline myelotomy that was used to relieve completely the pain of colon cancer for the duration of the 3-month survival period. No opioid analgesics were required after a tapering-off period. (c) Photomicrograph of a cross section of the human spinal cord at the level of a limited midline myelotomy that was placed at a mid-thoracic level in a patient with a painful presacral sarcoma. The lesion (shown by the area of demyelination near the midline of the dorsal columns) eliminated the pain and the need for narcotics for the duration of the patient's survival time, which was 3 years postsurgery. (a) From Nathan, P. W., Smith, M., and Deacon, P. 2001. The crossing of the spinothalamic tract. Brain 124, 793–803. (b) From Hirshberg, R. M., Al-Chaer, E. D., Lawand, N. B., Westlund, K. N., and Willis, W. D. 1996. Is there a pathway in the posterior funiculus that signals visceral pain? Pain 47, 291–305.

tended to limit the use of this procedure to patients with pain from a terminal disease, usually cancer, or to abandon the procedure altogether in favor of pharmacotherapy.

Visceral pain can be relieved by cordotomy, provided that the lesion extends into the anterior funiculus medially to the spinal cord gray matter (Gybels, J. M. and Sweet, W. H., 1989). A unilateral cordotomy can be effective if the visceral pain is restricted to one side. However, if the pain is bilateral, a bilateral cordotomy may be needed. Unfortunately, this can lead to undesired side effects, such as incontinence because of interruption of descending pathways that control bowel and bladder function. If the cordotomy is done at a high cervical level for pain originating from the upper part of the body, there is danger that the respiratory control pathways may be interrupted, leading to severe (potentially lethal) difficulty with breathing.

36.1.1.2 Commissural myelotomy

An alternative procedure called commissural myelotomy is directed at interrupting the axons of spinothalamic tract neurons of both sides as they cross the midline (Armour, D., 1927; Mansuy, L. et al., 1944; 1976; see Gybels, J. M. and Sweet, W. H., 1989). A lesion is made near the midline of the spinal cord and extending rostrocaudally for as many segments as needed to interrupt the appropriate projections. The lesion should be sufficiently deep to ensure that the crossing spinothalamic axons in the anterior white commissure are cut. Of course, the crossing axons of a number of tracts that accompany the spinothalamic tract, such as spinoreticular axons, are also interrupted, as are many axons in the posterior funiculi. According to the analysis of Dargent, M. et al. (1963), this procedure is apparently particularly effective for vaginal and visceral pain. Interestingly, commissural myelotomy often relieves clinical pain even if on testing there is no hypalgesia. Furthermore, pain and temperature sensation may be lost over a much greater extent of the body than could be predicted from the location and dimensions of the commissural myelotomy (Gybels, J. M. and Sweet, W. H., 1989). Based on evidence reviewed in the following, it now seems likely that the effect of a commissural myelotomy depends at least in part on the interruption of a visceral pain pathway that ascends in the posterior funiculi.

36.1.2 Posterior Column

36.1.2.1 Hitchcock procedure (stereotactic C1 central myelotomy)

Hitchcock developed a completely different approach to lesioning the spinal cord to relieve pain (Hitchcock, E., 1970; 1974; 1977; see also Papo, L. and Luongo, A., 1976; Eiras, J. et al., 1980; Schvarcz, J. R., 1977; 1978; Sourek, K., 1985). He placed a lesion in the central part of the upper cervical spinal cord using a stereotactic approach. The location of the lesion was sometimes determined by recordings of evoked potentials but more often by the sensory effects of electrical stimulation through the electrode

and by the changes in the sensory examination produced by a lesion. Stimulation in the central cord at C1 at 50 Hz typically resulted in a burning sensation in the chest or abdomen (Hitchcock, E., 1970; see Gybels, J. M. and Sweet, W. H., 1989). A lesion in the same location produced a change in pinprick sensation, either a loss of the ability to distinguish sharp from blunt or loss of a sensation of pain in response to pinprick. The analgesia was bilateral. Cancer pain was relieved in most cases, and the relief persisted until death or for over 5 years. Some patients (20%) failed to have a sensory loss and also did not experience pain relief.

In the surgical series done by Schvarcz J. R. (1977; 1978), lesions were made at a depth of 5 mm from the posterior surface of the spinal cord, reaching the base of the posterior column. Electrical stimulation produced paresthesias in the feet, legs, or trunk (and also in the trigeminal distribution). Relief of cancer pain persisted in 78% of cases for at least 0.5–2 years (most of the patients died by this time). Other types of pain, including peripheral and central neuropathic pain, were relieved for 0.5–4 years. No neurological deficits were seen. Several other clinical studies reported similar results (Papo, L. and Luongo, A., 1976; Eiras, J. et al., 1980; Sourek, K., 1985).

36.1.2.2 Limited midline myelotomy

Instead of a lesion at an upper cervical level, Gildenberg P. L. and Hirshberg R. M. (1984) made a midline lesion at T10 in patients with pelvic cancer pain. The location was chosen to be just above the level of entrance of primary afferent fibers from the pelvic viscera into the spinal cord. The procedure was termed a limited midline myelotomy. Eight cases done by Hirshberg, using a mechanical probe, were described further and a postmortem specimen showing the spinal cord lesion in one patient was illustrated (Figure 1(b)) in Hirshberg R. M. et al. (1996). The lesion interrupted the medial parts of the posterior funiculi but did not appear to intrude into the central gray matter or anterior white commissure. The results of the lesions in the eight cases were generally quite favorable, with pain relief that lasted as long as the patients survived and with little or no need for strong analgesic drugs.

The chief of neurosurgery at our institution, Dr. H. J. W. Nauta, made lesions using a technique similar to that described by Hirshberg in a series of patients who had pain from pelvic or other cancers. The results in each case were again generally favorable and the need for strong analgesics greatly reduced. In one case, a midline myelotomy was done in a patient with inflammatory bowel disease secondary to irradiation that had successfully eradicated a cervical cancer (Nauta, H. J. W. *et al.*, 1997; 2000). The patient had severe ongoing pain, and the pain intensified during bowel movements. As a consequence of the pain, the patient had lost a substantial fraction of her body weight, and it was judged that she would not survive much more weight loss. After midline myelotomy, this patient lived for almost 5 years with no further pelvic pain (although she did experience subdiaphragmatic pain because of peritonitis). She died from complications of diabetes.

In another of the cases reported by Nauta H. J. W. *et al.* (1997; 2000), pain was reduced from 10 to 2–3 on the visual analog scale, and narcotic medication was tapered from 30 mg of intravenously administered morphine per hour preoperatively to 5 mg per hour within 5 days postoperatively. Another patient had unbearable pain from a presacral sarcoma, despite several efforts to resect the tumor. A midline myelotomy was done in the area shown in Figure 1(c). The lesion resulted in dramatic pain relief for the remainder of the patient's life. He survived the surgery by 36 months, and for most of this time he did not require narcotic medication. In general, only in some cases were even minor and transient sensory side effects observed. Pelvic cancer pain was consistently reduced following midline myelotomy, and narcotic usage could be decreased on average by 83%.

Several other groups around the world have had similar experiences with midline myelotomies (Becker, R. *et al.*, 1999; Kim, Y. S. and Kwon, S. J., 2000; Filho, O. V. *et al.*, 2001; Hu, J. S. and Li, Y. J., 2002; Hwang, S. L. *et al.*, 2004), and the results have been reviewed by Becker, R. *et al.*, (2002).

36.2 Basic Science Evidence Concerning Spinal Cord Pathways That Signal Visceral Pain

36.2.1 Spinothalamic Tract

The spinothalamic tract conveys somatic but also some visceral nociceptive information to the thalamus (Figure 2(a)). Studies in which recordings were made from antidromically activated spinothalamic tract neurons have demonstrated that many of these cells can be activated (or in some instances inhibited) following stimulation of viscera. For example, Foreman and his group have shown that feline and

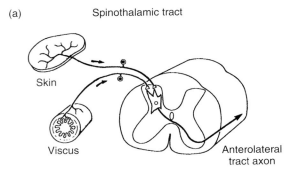

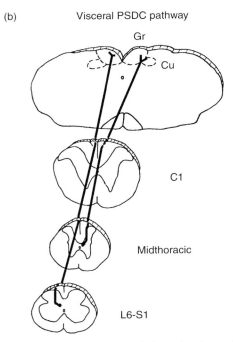

Figure 2 (a) Convergence of visceral and somatic afferent inputs on an anterolateral tract neuron. Such viscerosomatic convergence is thought to be responsible for referral of visceral pain to somatic structures. (b) Course followed by the axonal projections of postsynaptic dorsal column neurons found in the central region of the spinal cord gray matter at sacral and mid-thoracic levels. The anterograde tracer, phaseolus vulgaris leucoagglutinin (PHA-L), was injected and then several days later traced in axons of the dorsal column to the brainstem. The projection from the sacral cord ascended at the midline, whereas that from the mid-thoracic cord ascended more laterally, just ventral to the dorsal intermediate sulcus. Information about visceral nociception most likely travels with other sensory input to the thalamus in the medial lemniscus. Data from Wang, C. C., Willis, W. D., and Westlund, K. N. 1999. Ascending projections from the central, visceral processing region of the spinal cord: a PHA-L study in rats. J. Comp. Neurol. 415, 341–367; illustration from Willis, W. D., Al-Chaer, E. D., Quast M. J., and Westlund, K. N. 1999. A visceral pain pathway in the dorsal column of the spinal cord. Proc. Natl. Acad. Sci. U. S. A 96, 7675–7679.

primate spinothalamic tract cells can be excited by electrical stimulation of cardiopulmonary visceral afferent fibers, occlusion of a coronary artery, or injection of bradykinin into the coronary circulation (see review by Foreman, R. D., 1989). Similarly, spinothalamic tract cells have been shown to respond to electrical stimulation of the greater splanchnic nerve (Hancock, M. B. *et al.*, 1975; Foreman, R. D. *et al.*, 1981), distension of any of several hollow viscera, including the gall bladder (Ammons, W. S. *et al.*, 1984), kidney (Ammons, W. S., 1987), ureter (Ammons, W. S., 1989), urinary bladder (Milne, R. J. *et al.*, 1981), and colon (Al-Chaer, E. D. *et al.*, 1999) or noxious stimulation of the testicle (Milne, R. J. *et al.*, 1981). In general, spinothalamic tract and other neurons that are activated by visceral stimulation also respond to stimulation of somatic tissue. This convergence of visceral and somatic input onto spinal cord neurons, including nociceptive neurons that project to the brain, is thought to help account for the referral of visceral pain to the body wall in many clinical conditions (Figure 2(a); Head, H., 1893). Examples of the somatic receptive fields of viscerosensitive spinothalamic cells that correspond to the distribution of pain referral are noted in Foreman (Foreman, R. D., 1989) and Milne (Milne, R. J. *et al.*, 1981). Only rarely can a spinal neuron be found that seems to respond to just visceral stimuli and not to somatic ones. However, when the search stimulus is colon distension, rather than antidromic activation from the lateral thalamus, viscerospecific responses can be recorded from cells located rostrally in the central lateral nucleus of the intralaminar complex of the thalamus (Ren, Y. *et al.*, 2006). This region has been shown to receive direct spinal innervation from lamina X neurons (Wang, C. C. *et al.*, 1999) and is rich in opiates (Sar, M. *et al.*, 1978). The spinothalamic projection from lamina X is situated at the medialmost edge of the spinothalamic tract and shifts laterally as it ascends with the spinothalamic tract.

36.2.2 Spinoreticular, Spinoparabrachial, Spinoamygdalar, and Spinohypothalamic Tracts

Several studies have demonstrated the effectiveness of visceral stimulation in exciting (or inhibiting) spinoreticular tract neurons (Blair, R. W. *et al.* 1984; Hobbs, S. F. *et al.*, 1990), spinoparabrachial and spinoamygdalar neurons (Menetrey, D. and De Pommery, J., 1991; Bernard, J. F. *et al.*, 1994), and spinohypothalamic tract neurons (Katter, J. T. *et al.*,

1996; Zhang, X. *et al.*, 2002). A cordotomy or a commissural myelotomy is likely to interrupt these pathways, and thus the relief of visceral pain by such lesions is likely to be at least partially explained by this, as well as by interruption of the spinothalamic tract.

36.2.3 Postsynaptic Dorsal Column Path

36.2.3.1 Historical evidence for a visceral projection in the dorsal column

Recordings from the human posterior funiculus have revealed the presence of visceral afferents that respond to distention of the urinary bladder (Puletti, F. and Blomqvist, A. J., 1967). Responses at various levels of the dorsal column-medial lemniscus pathway have also been reported in animal experiments following electrical stimulation of visceral nerves (Amassian, V. E., 1951; Aidar, O. *et al.*, 1952; Rigamonti, D. D. and Hancock, M. B., 1974; 1978).

36.2.3.2 Effects of interruption of the dorsal column or a lesion of dorsal column nuclei on responses of brainstem and thalamic neurons to noxious visceral stimuli

To determine if there is a visceral nociceptive pathway in the dorsal columns of experimental animals, recordings were made from viscero-responsive neurons in the ventral posterior lateral nucleus of the thalamus in rats before and after surgical interruption of either the dorsal columns bilaterally or the ipsilateral ventrolateral column, which contains the spinothalamic and associated nociceptive pathways (Al-Chaer, E. D. *et al.*, 1996a). The dorsal column lesion reduced the responses of the thalamic neurons to colorectal distension on average by about 80% (Figure 3(Ae), cf. upper and middle rows of records). The remaining response was eliminated by the lesion of the ventrolateral quadrant of the spinal cord (Figure 3(Ae), lower row of records). Responses to weak mechanical stimuli applied to the skin were abolished by the dorsal column lesion, whereas the responses to the strongest mechanical stimuli were affected only by the ventrolateral quadrant lesion (Figure 3(Ad)). Similar effects were seen when the noxious visceral stimulus was inflammation of the colon by intraluminal injection of mustard oil (not illustrated). The mustard oil produced a progressive increase in the background activity of the viscerosensitive thalamic neurons. A dorsal column lesion was more effective in reducing the background activity than was a ventrolateral quadrant lesion.

Neurons in the rat ventral posterior lateral thalamic nucleus could also be found that responded to distention of the duodenum (Figure 3(Ba); Feng, Y. *et al.*, 1998). In these experiments, a lesion of the dorsal column placed at the midline failed to reduce the responses (Figure 3(Bb)). However, bilateral lesions just ventral to the dorsal intermediate sulcus at an upper cervical level were effective (Figure 3(Bc)). This shift in the location of the dorsal column lesion needed to interrupt nociceptive responses in the

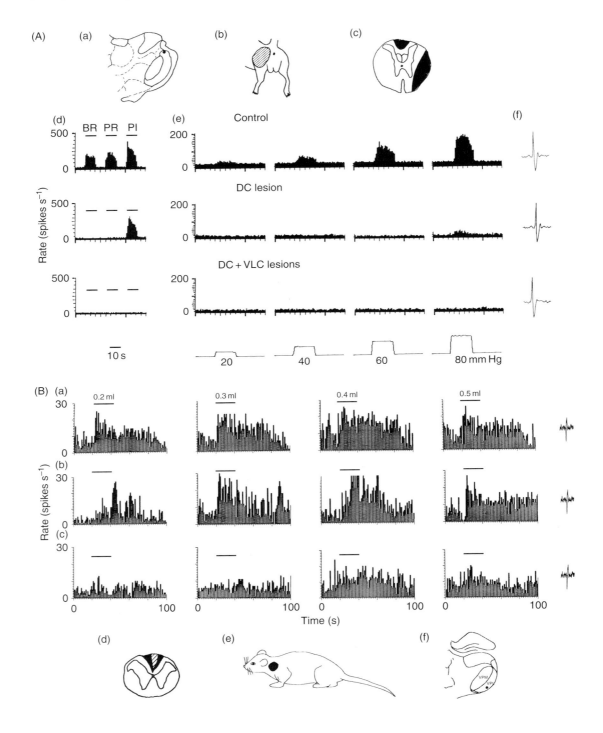

thalamus relates to the fact that the visceral input originated from an abdominal, rather than a pelvic organ (see Figure 2(b), discussion below, and Wang, C. C. *et al.*, 1999).

The off-midline part of the dorsal column pathway that carries visceral input from the thoracic level of the spinal cord was investigated in behavioral (see Figure 2(b) and discussion below) and electrophysiological studies after noxious chemical irritation of the pancreas (Houghton, A. K. *et al.*, 1997; 2001). Noxious stimulation of the pancreas by applications of bradykinin to its exposed surface in anesthetized rats resulted in excitation of neurons in the ventral posterior nucleus of the thalamus (Houghton, A. K. *et al.*, 2001). The mean firing rate of thalamic neurons was increased $412 \pm 120\%$ above baseline during the first 40 s after bradykinin application to the pancreas and was sustained for 3–6 min. Large dorsal column lesions effectively eliminated the increased responses of thalamic neurons to applications of bradykinin on the pancreas and baseline activity remained constant. Enhanced responses to skin pinch after pancreatic application of bradykinin dropped after a dorsal column lesion from 242% over baseline back to 100%. The effects of noxious stimulation of the pancreas with bradykinin involved a spinal synaptic relay rather than a vagal relay since thalamic activation was prevented by intrathecal administration of morphine. This action of morphine could be antagonized by naloxone.

Visceral responses can also be recorded from neurons in the ventral posterior lateral nucleus in monkeys (cf., Chandler, M. J. *et al.*, 1992; Brüggemann, J. *et al.*, 1994), as in rats. The effects of a dorsal column lesion on such responses in monkeys are consistent with those obtained in rats (Al-Chaer, E. D. *et al.*, 1998).

The thalamic responses to visceral input are relayed through the dorsal column nuclei. Visceral responses of neurons in the rat ventral posterior lateral thalamic nucleus could be reduced by small electrolytic or kainic acid lesions of the nucleus gracilis (Al-Chaer, E. D. *et al.*, 1997). Responses of cells in the nucleus gracilis have been recorded after noxious stimulation of either the colon or the pancreas (Al-Chaer, E. D. *et al.*, 1996b; 1997; Houghton, A. K. *et al.*, 2001).

36.2.3.3 Effects of a dorsal column lesion on behavioral responses

Changes observed in behavioral experiments were consistent with the electrophysiological findings. For example, when a rat is put into an unfamiliar plastic activity box, it moves freely around in the box until it adapts to this new environment. The movements can be tracked automatically by computer. Beams of ultraviolet light cross the plastic box, and whenever a beam is blocked by the animal, this is counted as a movement. Under normal conditions, the exploratory activity gradually decreases over about 45 min. However, if the animal experiences pain, for example, because of the presence of allodynia or hyperalgesia, the exploratory activity decreases more rapidly.

Figure 3 (A) shows the effects of sequential lesions of the dorsal columns (DC) and of the ventrolateral column (VLC) on one side on the responses of a neuron in the rat ventral posterior thalamic nucleus to a graded series of mechanical and visceral stimuli. The filled circle in (a) shows the recording site; the hatched area in (b) the somatic receptive field, and the black areas on a drawing of a transverse section of the spinal cord in (c) the extent of the lesions. Responses to brushing the skin (BR), application of pressure to a fold of skin with an arterial clip (PR) and pinching the skin with a stronger arterial clip (PI) are seen in (d), before and after the lesions. The responses to colorectal distentions of 20, 40, 60, and 80 mmHg are shown in (e), before and after the lesions. The action potential of the thalamic neuron at different times during the experiment is illustrated in (f) to indicate the stability of the recording. (B) Shows the effects of lesions of the dorsal columns placed in the upper cervical spinal cord at the midline or bilaterally just ventral to the dorsal intermediate sulci on the responses of a neuron in the rat ventral posterior lateral nucleus. In (a), upper row of records, are shown the responses of the neuron to graded distensions of the duodenum. The volumes of fluid injected into a balloon at the end of a catheter inserted into the duodenum are indicated in ml. The middle row of records (b) was taken after a midline lesion was made in the dorsal columns. The lesion had no clear effect. The lower row of records (c) shows that bilateral lesions of the dorsal columns ventral to the dorsal intermediate sulci substantially reduced the responses. The action potential of the thalamic neuron at different times during the experiment is shown in (b) to indicate the stability of the recording. The midline lesion is indicated by the hatched area and the more laterally placed lesions by the black area in (d). The receptive field of the neuron is shown in (e) and the recording site in (f) (A) From Al-Chaer, E. D., Lawand, N. B., Westlund, K. N., and Willis, W. D. 1996a. Visceral nociceptive input into the ventral posterolateral nucleus of the thalamus: a new function for the dorsal column pathway. J. Neurophysiol. 76, 2661–2674. (B) From Feng, Y., Cui, M., Al-Chaer, E. D., and Willis, W. D. 1998. Epigastric antinociception by cervical dorsal column lesions in rats. Anesthesiology 89, 411–420.

Exploratory activity is also reduced in rats with pancreatitis (Houghton, A. K. *et al.*, 1997), and this reduction is counteracted following a lesion of the dorsal columns. Similarly, infusion of bradykinin into the common bile and pancreatic duct in rats results in a reduction in rearing behavior, and this vertical exploratory behavior is increased after a dorsal column lesion (Houghton, A. K. *et al.*, 2001). Since the pancreas is an intra-abdominal organ, rather than a pelvic organ, the dorsal column had to be interrupted bilaterally and it included the areas ventral to the dorsal intermediate sulci (see below and Wang, C. C. *et al.*, 1999).

Feng Y. *et al.* (1998) inserted a balloon into the duodenum through a catheter. Distention of the balloon resulted in contractions of the abdominal muscles that were graded in force depending on the amount of balloon distention. Interruption of the dorsal columns resulted in a reduction in these abdominal muscle contractions.

The role of the dorsal column versus that of the spinothalamic tract and associated pathways was tested by Palecek J. *et al.* (2002). Exploratory activity of rats was shown to be reduced following either an intradermal injection of capsaicin (Figure 4(a)) or by inflammation of the colon with mustard oil, coupled with a mild degree of colorectal distention (Figure 4(c)). The effects of a capsaicin injection on exploratory activity could be reversed by dorsal rhizotomies on the side of the injection or by a lesion of the ventrolateral column of the spinal cord on the side opposite to the injection, but not by a bilateral lesion of the dorsal column (Figure 4(b)). By contrast, the effects of colon inflammation and distention on exploratory activity were eliminated by a bilateral lesion of the dorsal columns, and this change persisted for at least 90 days (Figure 4(d)).

36.2.3.4 Effects of a dorsal column lesion on the regional cerebral blood flow changes that result from colorectal distention in monkeys

In each of four monkeys anesthetized with isoflurane, it was possible to survey the changes in regional cerebral blood flow that were produced by colorectal distention, using a 4.7 T magnet for functional magnetic resonance imaging (Willis, W. D. *et al.*, 1999). The images were averaged over a standard period of time with or without visceral distention. The intensity of the colorectal distention was to 80 mm of Hg, well into the noxious range. Images through a coronal section of the brain at the level

of the posterior thalamus were compared, before and after surgery (Figure 5). In one animal, sham surgery was done over the dorsal column, but no lesion was made (Figure 5, left-most images). In the other three animals, the dorsal columns were interrupted at a mid-thoracic level. At various times after recovery from the surgery, the animals were re-anesthetized and fMRIs repeated at several intervals up to 4 months after the surgery. The sham surgery had no obvious effect on the regional cerebral blood flow produced by noxious colorectal distention. However, the lesion of the dorsal columns completely eliminated the blood flow changes in all three animals, and this effect of the lesion persisted for as long as the animals were followed (Figure 5).

36.2.3.5 Blockade of synaptic relay in sacral cord by morphine or 6-cyano-7-nitroquinoxaline-2,3-dione

The experiments described above indicate that the dorsal columns contain axons that signal visceral pain. However, a lesion of the dorsal columns interrupts not only the ascending branches of dorsal root ganglion cells that project to the dorsal column nuclei but also axons belonging to the postsynaptic dorsal column pathway (reviewed in Willis, W. D. and Coggeshall, R. E, 2004). An experiment was therefore designed to determine if noxious visceral signals are conveyed by the direct dorsal column projection or by postsynaptic dorsal column neurons (Al-Chaer, E. D. *et al.*, 1996b). The experimental arrangement is shown in Figure 6.

Recordings were made from the nucleus gracilis, rather than from the thalamus, to avoid the complication of visceral signals conveyed by the spinothalamic and accompanying pathways, as well as by the dorsal columns. The visceral stimulus was graded colorectal distention. The location of the part of the spinal cord that relayed the information to the medulla was restricted to the sacral segments by transecting the hypogastric nerves bilaterally. A microdialysis fiber was inserted across the sacral cord in order to introduce drugs into the dorsal horn that could block synaptic transmission but that would not interfere with direct nerve impulse transmission. The drugs included morphine and the non-N-methyl-D-aspartic acid (NMDA) glutamate receptor antagonist, 6-cyano-7-nitroquinoxaline-2,3-dione (CNQX). The prediction was that these drugs would prevent the responses of neurons of the gracile nucleus to colorectal distention if postsynaptic dorsal column neurons were responsible for the responses, but not

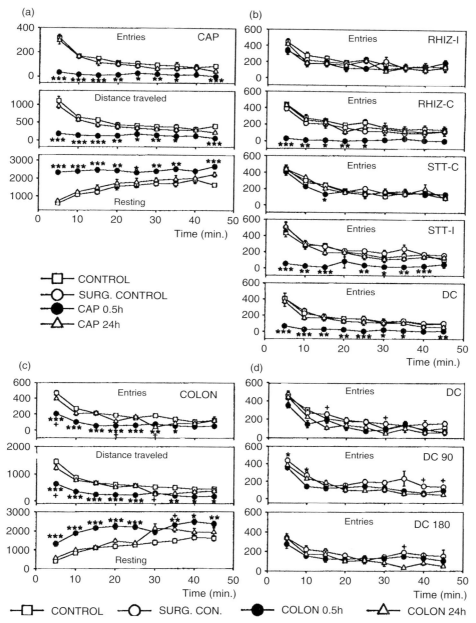

Figure 4 Time course of exploratory activity as monitored by a computer after introduction of rats into an activity box. The level of exploratory activity is determined by the number of interruptions of infrared light beams over a period of 45 min after the rat was placed in the activity box. The parameters included number of entries into a different zone, total distance traveled, and resting time. In (a) and (b), an intradermal injection of capsaicin was given, whereas in (c) and (d) mustard oil was injected into the colon and a latex balloon inserted and inflated to a pressure of 30 mmHg. At 30 min following either stimulus, exploratory activity was tested. Note that in (a), the capsaicin (CAP) injection resulted in a significant reduction in exploratory activity (and increased resting time). In (b), this was prevented by a prior extensive dorsal rhizotomy on the side ipsilateral to the injection (RHIZ-I), but was unaffected by a contralateral dorsal rhizotomy (RHIZ-C). A lesion that interrupted the spinothalamic tract contralateral to the injection (STT-C) also eliminated the reduction in exploratory activity following CAP, whereas an ipsilateral cordotomy (STT-I) had no effect; nor did a bilateral dorsal column (DC) lesion. In (c), colon inflammation and distention reduced the exploratory activity, and in (d) this change was eliminated by a dorsal column (DC) lesion. The effect lasted for 90 (DC 90) and 180 (DC 180) days. From Palecek, J., Paleckova, V., and Willis, W. D. 2002. The roles of pathways in the spinal cord lateral and dorsal funiculi in signaling nociceptive somatic and visceral stimuli in rats. Pain 96, 297–307.

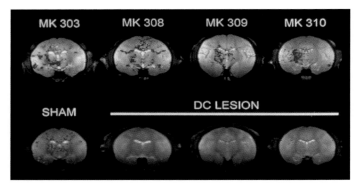

Figure 5   Functional magnetic resonance imaging showing the changes in regional cerebral blood flow in the brains of four different monkeys in coronal slices made at the level of the posterior thalamus. The blood flow changes were evoked by repeated colorectal distention to 80 mmHg under isoflurane anesthesia. The images in the upper row were made prior to surgery. Those in the lower row were from the same animals but after surgery. The animal whose images are shown at the left was subjected to sham surgery: the dorsal column was exposed at a mid-thoracic level, but no lesion was made. In the other three animals, the dorsal columns were interrupted at a mid-thoracic level. There was no obvious difference in regional blood flow changes in the animal with sham surgery, but the blood flow changes in the other animals were completely eliminated (previously unpublished data.).

if the direct dorsal column pathway were involved. This was important, since it was conceivable that noxious visceral signals could be transmitted to the dorsal column nuclei by unmyelinated, peptidergic branches of dorsal root ganglion neurons. Such fibers have been demonstrated, although the functions of these axons are still unknown (see Willis, W. D. and Coggeshall, R. E., 2004). The results of this experiment indicated that the postsynaptic dorsal column pathway is responsible for the visceral pain signals that ascend in the dorsal columns. Both morphine and CNQX blocked visceral nociceptive transmission from the colon to the gracile nucleus, and the effects of morphine were reversed by naloxone, indicating that the morphine acted on opiate receptors (Figure 7). The CNQX must have exerted its effect by blocking a critical synaptic relay that involves non-NMDA glutamate receptors.

36.2.3.6   Projections of the postsynaptic dorsal column pathway

The course of the ascending axons of postsynaptic dorsal column neurons in the sacral spinal cord was traced using the *Phaseolus vulgaris leukoagglutinin* (PHA-L) anterograde tracing technique and compared with the projections from postsynaptic dorsal column neurons in the mid-thoracic spinal cord (Wang, C. C. *et al.*, 1999). Injections of PHA-L were made into the central region of the spinal cord for several reasons. This area is known to receive visceral afferent input (Honda, C. N., 1985; Sugiura, Y. *et al.*, 1989), and recordings had been

made in this region from a number of postsynaptic dorsal column neurons that responded to noxious visceral stimuli (Al Chaer, E. D. *et al.*, 1996a; 1996b). Many cells around the central canal have ascending projections in the dorsal columns that are labeled following injections of retrograde tracer into the dorsal column nuclei or the dorsal column itself (Hirshberg, R. M. *et al.*, 1996). The nociceptive visceral information is likely to be relayed from the dorsal column nuclei to the lateral thalamus by way of the medial lemniscus.

However, recent observations indicate that many postsynaptic dorsal column neurons that are activated by noxious visceral stimuli are also located in the nucleus proprius of the spinal cord dorsal horn (Palecek, J. *et al.*, 2003b), presumably adjacent to postsynaptic dorsal column neurons that lack visceral input (Figure 8).

Postsynaptic dorsal column neurons in lamina X of the sacral cord were found to project their axons through the medial part of the fasciculi gracilis to the nuclei gracilis (Figure 2(b); Wang, C. C. *et al.*, 1999). Thus, the projections of these postsynaptic dorsal column cells are in exactly the position in the medial posterior columns that was targeted by the Hitchcock procedure (Hitchcock, E. 1970; 1974; 1977) and by limited midline myelotomy (see Hirshberg, R. M. *et al.*, 1996; Nauta, H. J. W. *et al.*, 1997; 2000). These axons would also be interrupted by a commissural myelotomy. However, axons that ascend from postsynaptic dorsal column neurons in the mid-thoracic cord to the lateral part of the gracilis

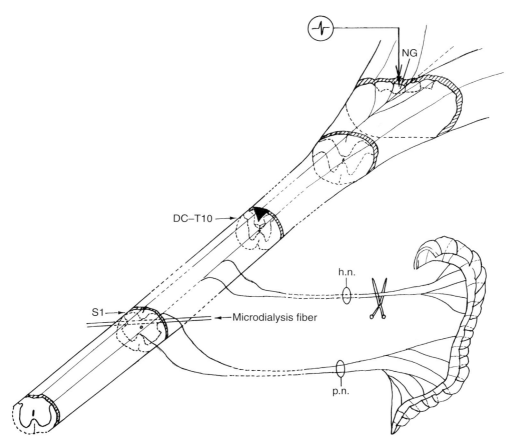

Figure 6 Drawing of the experimental arrangement. Recordings were made of the responses of neurons in the nucleus gracilis (NG, upper right) in response to colorectal distension. This avoided the problem that would have been encountered with thalamic recordings, since neurons of the ventral posterior lateral nucleus could be affected by either the dorsal column pathway or the spinothalamic tract. The input from the colon to the spinal cord was limited to just that traversing the pelvic nerve and ending in the sacral spinal cord. This was done by sectioning the hypogastric nerves prior to the experiment. A microdialysis fiber was placed across the spinal cord at S1 so that drugs that are likely to block synaptic transmission in the dorsal horn could be administered locally. Access to the dorsal column (DC) at T10 permitted a test of the effect of a dorsal column lesion. From Al-Chaer, E. D., Lawand, N. B., Westlund, K. N., and Willis, W. D. 1996a. Visceral nociceptive input into the ventral posterolateral nucleus of the thalamus: a new function for the dorsal column pathway. J. Neurophysiol. 76, 2661–2674.

nucleus and the medial part of the cuneate nucleus do so near the dorsal intermediate septum (Figure 2(b); Wang, C. C. *et al.*, 1999). Thus, there appears to be a viscerotopic organization of this visceral pathway similar to that of the somatosensory part of the postsynaptic dorsal column pathway (see Willis, W. D. and Coggeshall, R. E., 2004). The results of the studies of stimulation of internal organs of the abdomen described above by Houghton A. K. *et al.* (1997; 2001) and Feng Y. *et al.*(1998) are consistent with this observation. In these studies, distention or chemical irritation of abdominal viscera was used to activate neurons in the rat ventral posterior lateral nucleus. Lesions placed in the dorsal column at the midline

failed to influence the responses. However, lesions lateral to the midline in the area of the dorsal intermediate septum were effective. Clinical studies from a group in Korea (Kim, Y. S. and Kwon, S. J., 2000) report favorable results in patients experiencing stomach cancer pain when lesions were made near the incoming thoracic dorsal root fibers rather than at the midline. This indicates that the viscerotopic organization is similar in humans and animals and that off-midline myelotomies will offer better relief of thoracic or upper abdominal visceral pain than do midline myelotomies.

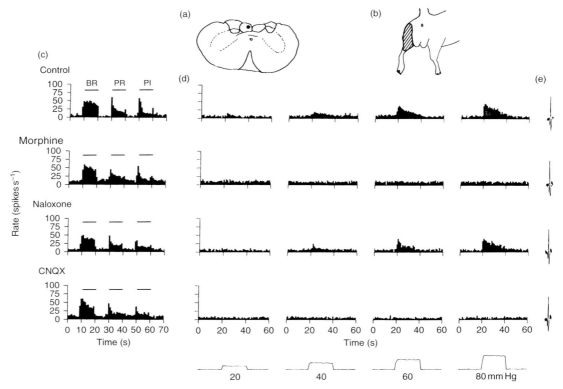

Figure 7 Effects of morphine and CNQX administered into the sacral cord by microdialysis on the responses of a neuron in the gracile nucleus to graded somatic and visceral stimuli. (a) Shows the recording site in the gracile nucleus. (b) Indicates the receptive field of the gracile neuron. In (c) are the responses of the gracile neuron to BR, PR, and PI stimuli applied to the cutaneous receptive field. (d) Shows the responses to graded colorectal distention to 20, 40, 60, and 80 mmHg. The upper row of recordings in (c) and (d) are the control responses taken before administration of drugs. The second row of recordings shows the effects of morphine given by microdialysis. There was little, if any, change in the responses to the somatic stimuli, but the responses to colorectal distention were eliminated. The third row of recordings shows that systemically administered naloxone counteracted the effects of morphine. The lowest row of recordings shows the effect of CNQX in blocking the responses to colorectal distention. In (e) are action potentials recorded at different times during the experiment to show the stability of the recordings. From Al-Chaer, E. D., Lawand, N. B., Westlund, K. N., and Willis, W. D. 1996b. Pelvic visceral input into the nucleus gracilis is largely mediated by the postsynaptic dorsal column pathway. J. Neurophysiol. 76, 2675–2690.

36.2.3.7 Upregulation of NK1 receptors in PSDC neurons after colon inflammation

The cells of origin of several nociceptive pathways that ascend from the spinal cord to the brain are known to contain neurokinin-1 (NK-1) receptors (Todd, A. J. et al., 2000). These pathways include the spinothalamic and spinoreticular tracts. However, postsynaptic dorsal horn neurons do not contain NK-1 receptors under normal conditions, although an up-regulation of these receptors does occur in some of these neurons following visceral inflammation (Palecek, J. et al., 2003a).

36.2.3.8 Fos expression in PSDC neurons after noxious visceral stimulation

A comparison was made of the numbers and locations of neurons belonging to the spinothalamic tract and to the postsynaptic dorsal column pathway that express Fos, the product of the immediate/early gene, c-fos, following either an intradermal injection of capsaicin or distention of the ureter (see Figure 8; Palecek, J. et al., 2003b). The spinothalamic neurons were retrogradely labeled from the contralateral thalamus and the postsynaptic dorsal column neurons from the ipsilateral nucleus gracilis. Ureter distention was accomplished by tightening a ligature that had previously been placed around the ureter. The ureter was distended proximal to the site where the ligature blocked the flow of urine. Fos-labeled neurons were distributed over a considerable length of the spinal cord, from T11 to L6. This study confirmed that spinothalamic tract cells can be activated by a noxious visceral stimulus and that postsynaptic dorsal column neurons are also activated by such stimuli.

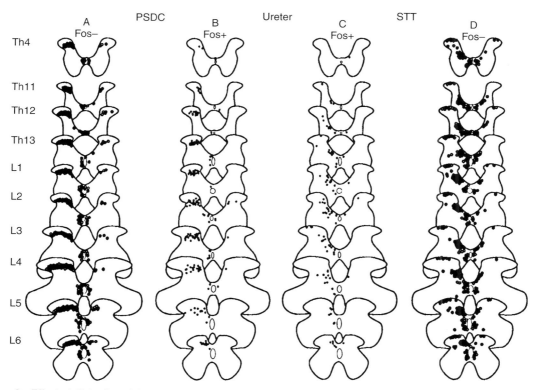

Figure 8 Effect of distention of the ureter in evoking expression of Fos protein in identified projection neurons in rats. Postsynaptic dorsal column neurons were labeled retrogradely following injection of a tracer into the nucleus gracilis (left-most column), and spinothalamic neurons were identified following injection of a different retrograde tracer into the ventral posterior lateral nucleus of the thalamus (right-most column). The middle columns show neurons identified by retrograde labeling that were also labeled for Fos protein following distention of the ureter. From Palecek, J., Paleckova, V., and Willis, W. D. 2003b. Fos expression in spinothalamic and postsynaptic dorsal column neurons following noxious visceral and cutaneous stimuli. Pain 104, 249–257.

36.3 Descending Facilitation

It has been shown that the responses of spinal cord neurons to visceral stimulation depend in large part on a spino-bulbo-spinal circuit (Cervero, F. and Wolstencroft, J. H., 1984; Tattersall, J. E. *et al.*, 1986; Zhuo, M. and Gebhart, G. F., 2002) that when activated facilitates visceral responses at the spinal cord level. Our anterograde tract tracing studies of the projections from lamina X demonstrate an ascending component that joins the medial edge of the spinothalamic tract (Wang, C. C. *et al.*, 1999). This ascending pathway has terminations in the nucleus raphe magnus, rostral ventromedial medulla, and medial reticular formation. This pathway is the only reported direct input to the nucleus raphe magnus from the spinal cord and is likely to be important for the activation of brainstem facilitatory neurons with axons that descend into the spinal cord to enhance the responses of spinal cord neurons to noxious visceral stimuli.

Palecek J. and Willis W. D. (2003) confirmed that a dorsal column lesion did not affect visceromotor reflex responses under normal conditions (cf., Ness, T. J. and Gebhart, G. F., 1988). However, the enhanced visceromotor responses to graded colorectal distention that followed colon inflammation with mustard oil were reduced or prevented by bilateral lesions of the dorsal columns. This suggests that the visceral reflex is enhanced because of activity induced by the colon inflammation and conveyed to the brainstem by way of the dorsal column. A lesion of the ventrolateral spinal cord eliminated the visceromotor reflex, presumably by interrupting a facilitatory pathway that descends from the brainstem and/or the ascending spinothalamic projection from lamina X.

It is our working hypothesis that interruption of the postsynaptic dorsal column pathway by a lesion placed in the dorsal columns reduces visceral pain not only because such a lesion prevents visceral pain signals from reaching the thalamus by way of the dorsal column nuclei and medial lemniscus, but also because of a reduction of descending facilitation affecting visceral nociceptive neurons in the spinal cord, including post-synaptic dorsal column and spinothalamic tract neurons.

References

Aidar, O., Geohegan, W. A., and Ungewitter, L. H. 1952. Splanchnic afferent pathways in the central nervous system. J. Neurophysiol. 15, 131–138.

Al-Chaer, E. D., Feng, Y., and Willis, W. D. 1998. A role for the dorsal column in nociceptive visceral input into the thalamus of primates. J. Neurophysiol. 79, 3143–3150.

Al-Chaer, E. D., Feng, Y., and Willis, W. D. 1999. Comparative study of viscerosomatic input onto postsynaptic dorsal column and spinothalamic tract neurons in the primate. J. Neurophysiol. 82, 1876–1882.

Al-Chaer, E. D., Lawand, N. B., Westlund, K. N., and Willis, W. D. 1996a. Visceral nociceptive input into the ventral posterolateral nucleus of the thalamus: a new function for the dorsal column pathway. J. Neurophysiol. 76, 2661–2674.

Al-Chaer, E. D., Lawand, N. B., Westlund, K. N., and Willis, W. D. 1996b. Pelvic visceral input into the nucleus gracilis is largely mediated by the postsynaptic dorsal column pathway. J. Neurophysiol. 76, 2675–2690.

Al-Chaer, E. D., Westlund, K. N., and Willis, W. D. 1997. Nucleus gracilis: an integrator for visceral and somatic information. J. Neurophysiol. 78, 521–527.

Amassian, V. E. 1951. Fiber groups and spinal pathways of cortically represented visceral afferents. J. Neurophysiol. 14, 445–460.

Ammons, W. S. 1987. Characteristics of spinoreticular and spinothalamic neurons with renal input. J. Neurophysiol. 50, 480–495.

Ammons, W. S. 1989. Primate spinothalamic cell responses to ureteral occlusion. Brain Res. 496, 124–130.

Ammons, W. S., Blair, R. W., and Foreman, R. D. 1984. Responses of primate T1-T5 spinothalamic neurons to gallbladder distension. Am. J. Physiol. 247, R995–R1002.

Armour, D. 1927. Surgery of the spinal cord and its membranes. Lancet i, 691–697.

Becker, R., Gatscher, S., Sure, U., and Bertalanffy, H. 2002. The punctuate midline myelotomy concept for visceral cancer pain control- case report and review of the literature. Acta Neurochir. Suppl. 79, 77–78.

Becker, R., Sure, U., and Bertalanffy, H. 1999. Punctate midline myelotomy. A new approach in the management of visceral pain. Acta Neurochir. (Wien) 141, 881–883.

Bernard, J. F., Huang, G. F., and Besson, J. M. 1994. The parabrachial area: electrophysiological evidence for an involvement in visceral nociceptive processes. J. Neurophysiol. 71, 1646–1660.

Blair, R. W., Ammons, W. S., and Foreman, R. D. 1984. Responses of thoracic spinothalamic and spinoreticular cells to coronary artery occlusion. J. Neurophysiol. 51, 636–648.

Brüggemann, J., Shi, T., and Apkarian, A. V. 1994. Squirrel monkey lateral thalamus. II. Viscero-somatic convergent representation of urinary bladder, colon and esophagus. J. Neurosci. 14, 6796–6814.

Bushnell, M. C., Duncan, G. H., and Tremblay, N. 1993. Thalamic VPM nucleus in the behaving monkey. I.

Multimodal and discriminative properties of thermosensitive neurons. J. Neurophysiol. 69, 739–752.

Casey, K. L. and Bushnell, M.C. (eds.) (2000) Pain Imaging. IASP Press.

Cervero, F. and Wolstencroft, J. H. 1984. A positive feedback loop between spinal cord nociceptive pathways and antinociceptive areas of the cat's brain stem. Pain 20, 125–138.

Chandler, M. J., Hobbs, S. F., Qing-Gong, Fu., Kenshalo, D. R., Blair, R. W., and Foreman, R. D. 1992. Responses of neurons in ventroposterolateral nucleus of primate thalamus to urinary bladder distension. Brain Res. 571, 26–34.

Chung, J. M., Lee, K. H., Surmeier, D. J., Sorkin, L. S., Kim, J., and Willis, W. D. 1986. Response characteristics of neurons in the ventral posterior lateral nucleus of the monkey thalamus. J. Neurophysiol. 56, 370–390.

Dargent, M., Mansuy, L., Cohen, J., and De Rougemont, J. 1963. Les problems poses par la douleur dans l'evolution des cancers gynécologiques. Lyon chirurg. 59, 62–83.

Eiras, J., Garcia, J., Gomez, J., Carvavilla, L. I., and Ucar, S. 1980. First results with extralemniscal myelotomy. Acta Neurochir. Suppl. 30, 377–381.

Feng, Y., Cui, M., Al-Chaer, E. D., and Willis, W. D. 1998. Epigastric antinociception by cervical dorsal column lesions in rats. Anesthesiology 89, 411–420.

Filho, O. V., Araujo, M. R., Florencio, R. S., Silva, M. A. C., and Silveira, M. T. 2001. CT-guided percutaneous punctuate midline myelotomy for the treatment of intractable visceral pain: a technical note. Stereotact. Funct. Neurosurg. 77, 177–182.

Foerster, O. and Gagel, G. 1932. Die Vorderseitenstrangdurchschneidung beim Menschen: Eine klinisch-pathophysioligisch-anatomische Studie. Z. Ges. Neurol. Psychiat. 138, 1–92.

Foreman, R. D. 1989. Organization of the spinothalamic tract as a relay for cardiopulmonary sympathetic afferent fiber activity. In: Progress in Sensory Physiology (ed. D. Ottoson), Vol. 9, pp. 1–51. Springer.

Foreman, R. D., Hancock, M. B., and Willis, W. D. 1981. Responses of spinothalamic tract cells in the thoracic spinal cord of the monkey to cutaneous and visceral inputs. Pain 11, 149–162.

Gildenberg, P. L. and Hirshberg, R. M. 1984. Limited myelotomy for the treatment of intractable cancer pain. J. Neurol. Neurosurg. Psychiat. 47, 94–96.

Gybels, J. M. and Sweet, W. H. 1989. Neurosurgical Treatment of Persistent Pain. Karger.

Hancock, M. B., Foreman, R. D., and Willis, W. D. 1975. Convergence of visceral and cutaneous input onto spinothalamic tract cells in the thoracic spinal cord of the cat. Exp. Neurol. 47, 240–248.

Head, H. 1893. On disturbances of sensation with especial reference to the pain of visceral disease. Brain 16, 1–132.

Hirshberg, R. M., Al-Chaer, E. D., Lawand, N. B., Westlund, K. N., and Willis, W. D. 1996. Is there a pathway in the posterior funiculus that signals visceral pain? Pain 47, 291–305.

Hitchcock, E. 1970. Stereotactic cervical myelotomy. J. Neurol. Neurosurg. Psychiat. 33, 224–230.

Hitchcock, E. 1974. Stereotactic myelotomy. Proc. R. Soc. Med. 67, 771–772.

Hitchcock, E. 1977. Stereotactic spinal surgery. Neurol. Surg. 433, 271–280.

Hobbs, S. F., Oh, U. T., Brennan, T. J., Chandler, M. J., Kim, K. S., and Foreman, R. D. 1990. Urinary bladder and hindlimb stimuli inhibit T1–T6 spinothalamic and spinoreticular cells. Am. J. Physiol. 258, R10–R20.

Honda, C. N. 1985. Visceral and somatic afferent convergence onto neurons near the central canal in the sacral spinal cord of the cat. J. Neurophysiol. 53, 1059–1078.

Houghton, A. K., Kadura, S., and Westlund, K. N. 1997. Dorsal column lesions reverse the reduction in homecage activity in rats with pancreatitis. NeuroReport 8, 3795–3800.

Houghton, A. K., Wang, C. C., and Westlund, K. N. 2001. Do nociceptive signals from the pancreas travel in the dorsal column? Pain 89, 207–220.

Hu, J. S. and Li, Y. J. 2002. Clinical application of midline myelotomy to treat visceral cancer pain. Natl. Med. J. China 82, 856–867 (in Chinese).

Hwang, S. L., Lin, C. L., Lieu, A. S., Kuo, T. H., Yu, K. L. K., Ou-Yang, F., Wang, S. N., Lee, K. T., and Howng, S. L. 2004. Punctate midline myelotomy for intractable visceral pain caused by hepatobiliary or pancreatic cancer. J. Pain Symptom Management 27, 79–84.

Katter, J. T., Dado, R. J., Kostarczyk, E., and Giesler, G. J. 1996. Spinothalamic and spinohypothalamic tract neurons in the sacral spinal cord of rats. II. Responses to cutaneous and visceral stimuli. J. Neurophysiol. 75, 2606–2628.

Kenshalo, D. R., Giesler, G. J., Leonard, R. B., and Willis, W. D. 1980. Responses of neurons in primate ventral posterior lateral nucleus to noxious stimuli. J. Neurophysiol. 43, 1594–1614.

Kim, Y. S. and Kwon, S. J. 2000. High thoracic midline dorsal column myelotomy for severe visceral pain due to advanced stomach cancer. Neurosurgery 46, 85–92.

Lee, J. I., Antezanna, D., Dougherty, P. M., and Lenz, F. A. 1999. Responses of neurons in the region of the thalamic somatosensory nucleus to mechanical and thermal stimuli graded into the painful range. J. Comp. Neurol. 410, 541–555.

Lee, J. I., Ohara, S., Dougherty, P. M., and Lenz, F. A. 2005. Pain and temperature encoding in the human thalamic somatic sensory nucleus (ventral caudal): inhibition-related bursting evoked by somatic stimuli. J. Neurophysiol. 94, 1676–1687.

Mansuy, L., Lecuire, J., and Acassat, L. 1944. Technique de la myélotomie commissurale postérieure. J. Chir. 60, 206–213.

Mansuy, L., Sindou, M., Fischer, G., and Brunon, J. 1976. La cortotomie spino-thalamique dans les douleurs cancéreuses. Résultats d'une série de 124 malades operas par abord direct postérieur. Neuro-chirurgie 22, 437–444.

Mehler, W. R. 1962. The anatomy of the so-called "pain tract" in man: analysis of the course and distribution of the ascending fibers of the fasciculus anterolateralis. In: Basic Research in Paraplegia (eds. J. D. French and R. W. Porter), pp. 26–55. Thomas.

Menetrey, D. and De Pommery, J. 1991. Origins of spinal ascending pathways that rech central areas involved in visceroception and visceronociception in the rat. Eur. J. Neurosci. 3, 249–259.

Milne, R. J., Foreman, R. D., Giesler, G. J., and Willis, W. D. 1981. Convergence of cutaneous and pelvic visceral nociceptive input onto primate spinothalamic neurons. Pain 11, 163–183.

Nathan, P. W., Smith, M., and Deacon, P. 2001. The crossing of the spinothalamic tract. Brain 124, 793–803.

Nauta, H. J. W., Hewitt, E., Westlund, K. N., and Willis, W. D. 1997. Surgical interruption of a midline dorsal column visceral pain pathway: case report and review of the literature. J. Neurosurg. 86, 538–542.

Nauta, H. J. W., Soukup, V. M., Fabian, R. H., Lin, J. H. T., Grady, J. J., Williams, C. G. A., Campbell, G. A., Westlund, K. N., and Willis, W. D. 2000. Punctate mid-line myelotomy for the relief of visceral cancer pain. J. Neurosurg. (Spine 1) 92, 125–130.

Ness, T. J. and Gebhart, G. F. 1988. Colorectal distention as a noxious visceral stimulus: physiologic and pharmacologic characterization of pseudoaffective reflexes in the rat. Brain Res. 450, 153–169.

Palecek, J. and Willis, W. D. 2003. The dorsal column pathway facilitates visceromotor responses to colorectal distention after colon inflammation in rats. Pain 104, 501–507.

Palecek, J., Paleckova, V., and Willis, W. D. 2002. The roles of pathways in the spinal cord lateral and dorsal funiculi in signaling nociceptive somatic and visceral stimuli in rats. Pain 96, 297–307.

Palecek, J., Paleckova, V., and Willis, W. D. 2003a. Postsynaptic dorsal column neurons express NK1 receptors following colon inflammation. Neuroscience 116, 565–572.

Palecek, J., Paleckova, V., and Willis, W. D. 2003b. Fos expression in spinothalamic and postsynaptic dorsal column neurons following noxious visceral and cutaneous stimuli. Pain 104, 249–257.

Papo, L. and Luongo, A. 1976. High cervical commissural myelotomy in the treatment of pain. J. Neurol. Neurosurg. Psychiat. 39, 705–710.

Puletti, F. and Blomqvist, A. J. 1967. Single neuron activity in posterior columns of the human spinal cord. J. Neurosurg. 27, 255–259.

Ren, Y., Lu, Y., Yang, H., and Westlund, K. N. Responses of neurons in central lateral thalamic nucleus to mechanical and chemical visceral stimulation in rats. APS Abstracts, American Pain Society, 2006.

Rigamonti, D. D. and Hancock, M. B. 1974. Analysis of field potentials elicited in the dorsal column nuclei by splanchnic nerve A-beta afferents. Brain Res. 77, 326–329.

Rigamonti, D. D. and Hancock, M. B. 1978. Viscerosomatic convergence in the dorsal column nuclei of the cat. Exp. Neurol. 61, 337–348.

Sar, M., Stumpf, W. E., Miller, R. J., Chang, K.-J., and Cuatrecasas, P. 1978. Immunohistochemical localization of enkephalin in rat brain and spinal cord. J Comp. Neurol. 182, 17–38.

Schvarcz, J. R. 1977. Functional exploration of the spinomedullary junction. Acta Neurochir. Suppl. 24, 179–185.

Schvarcz, J. R. 1978. Spinal cord stereotactic techniques re trigeminal nucleotomy and extralemniscal myelotomy. Appl. Neurophysiol. 41, 99–112.

Sourek, K. 1985. Central thermromyelo-coagulation for intractable chronic pain and opioid peptides. In: 8th International Congress of Neurological Surgery, Toronto, 1985, Abstract 221.

Spiller, W. G. 1905. The occasional clinical resemblance between caries of the vertebrae and lumbothoracic syringomyelia and the location within the spinal cord of the fibres for the sensations of pain and temperature. Univ. Pennsylvania Med. Bull. 18, 147–154.

Spiller, W. G. and Martin, E. 1912. The treatment of persistent pain of organic origin in the lower part of the body by division of the anterolateral column of the spinal cord. JAMA 58, 1489–1490.

Sugiura, Y., Terui, N., and Hossoya, Y. 1989. Difference in distribution of central terminals between visceral and somatic unmyelinated (C) primary afferent fibers. J. Neurophysiol. 62, 834–840.

Tattersall, J. E., Cervero, F., and Lumb, B. M. 1986. Effects of reversible spinalization on the visceral input to viscerosomatic neurons in the lower thoracic spinal cord of the cat. J. Neurophysiol. 56, 785–796.

Todd, A. J., McGill, M. M., and Shehab, S. A. 2000. Neurokinin 1 receptor expression by neurons in laminae I, III and IV of the rat spinal dorsal horn that project to the brainstem. Eur. J. Neurosci. 12, 689–700.

Wang, C. C., Willis, W. D., and Westlund, K. N. 1999. Ascending projections from the central, visceral processing region of the spinal cord: a PHHA-L study in rats. J. Comp. Neurol. 415, 341–367.

Willis, W. D. and Coggeshall, R. E. 2004. Sensory Mechanisms of the Spinal Cord, 3rd edn. Kluwer Academic/Plenum.

Willis, W. D. and Westlund, K. N. 1997. Neuroanatomy of the pain system and of the pathways that modulate pain. J. Clin. Neurophysiol. 14, 2–31.

Willis, W. D., Al-Chaer, E. D., Quast, M. J., and Westlund, K. N. 1999. A visceral pain pathway in the dorsal column of the spinal cord. Proc. Natl. Acad. Sci. U. S. A. 96, 7675–7679.

Zhang, X., Gokin, A. P., and Giesler, G. J. 2002. Responses of spinohypothalamic tract neurons in the thoracic spinal cord of rats to somatic stimuli and to graded distention of the bile duct. Somatosens. Mot. Res. 19, 5–17.

Zhuo, M. and Gebhart, G. F. 2002. Facilitation and attenuation of a visceral nociceptive reflex from the rostroventral medulla in the rat. Gastroenterology 122, 1007–1019.

37 Visceral Pain

G F Gebhart and K Bielefeldt, University of Pittsburgh, Pittsburgh, PA, USA

Glossary

referred visceral pain Visceral pain is not generally felt at the source, but rather is perceived as arising from other tissues (skin and/or muscle) innervated by non-visceral nerves having input onto second-order spinal neurons that converges with input from visceral organs onto the same spinal neurons. Accordingly, visceral pain is commonly referred (or transferred, an older terminology) to non-visceral tissues.

nociceptor sensitization Sensitization of nociceptors reflects a change in their excitability expressed as an increase in response magnitude to a noxious intensity of stimulation accompanied by a reduction in threshold stimulus intensity required to activate the nociceptor.

visceral nociceptors Sensory endings in viscera which, when activated by an adequate stimulus (stretch/tension, ischemia, inflammatory mediators), commonly give rise to sensations of discomfort and pain.

visceral hypersensitivity Increased sensitivity or response to a visceral stimulus, which can arise from sensitization of visceral sensory endings (e.g., visceral mechanoreceptors), sensitization of central nervous system neurons receiving increased input from sensitized visceral sensory endings, or a combination of both. Clinically, visceral hypersensitivity is associated with discomfort or pain produced by normally non-pain-producing stimuli (e.g., ingestion of food, normal bladder fillling) and commonly includes increased sensitivity to probing in the areas of referred visceral sensation, which moreover are increased in size (area).

37.1 Introduction

Of all possible sources of pain within the body, pain arising from internal organs commonly generates the greatest autonomic and emotional responses. This is linked to several characteristics of visceral pain, the most apparent being that the source of pain is neither visible nor typically felt at the source. Visceral pain also is commonly diffuse in character and poorly localized. In addition, it is also referred (or transferred) to nonvisceral, somatic structures (like skin and muscle), and there is considerable overlap in areas of referred pain from adjacent organs. For example, sites of referral from thoracic and abdominal organs (esophagus, gallbladder, and heart) overlap, making it difficult for both physicians and patients to determine the source of pain. That diffuse substernal chest pain has high emotional valence is readily appreciated, given the potential significance of such pain. Similarly, pelvic genitourinary organs have overlapping patterns of referral with the distal colon. It is also characteristic of visceral pain that the areas of referred sensation increase in size and sensitivity in organ inflammation and disease. For example, repetitive episodes of angina lead to increased tenderness (hyperalgesia) of chest and shoulder skin and abdominal tenderness spreads and increases in area in interstitial cystitis and irritable bowel syndrome (IBS).

Distinct from cutaneous, muscle, or joint pain, tissue-damaging stimuli do not commonly produce pain when applied to the viscera. Tissue crushing, cutting, or burning stimuli applied to visceral tissue generally produce little conscious sensation and rarely produce pain. Instead, stretch of the smooth muscle layers of hollow organs (e.g., by distension), traction on the mesentery, ischemia, and inflammation are more reliably uncomfortable or pain-producing in viscera. The bases of the above attributes of visceral pain and unique features that distinguish visceral from nonvisceral pain are the focus of this chapter.

37.2 Visceral Sensation

The viscera are invested with a wide range of sensory receptors, although when activated in healthy organs they do not generally give rise to conscious sensation. This raises the issue of whether these neurons are afferent or sensory, which is considered below in

Section 37.3. We are generally unaware of movement of air in and out of the lungs, blood flowing through vessels, beating of the heart, or food in the stomach despite the fact that sensory receptors in the bronchial tree, lungs, vessels, heart, esophagus, and stomach are activated by these events. Information from activation of visceral sensory neurons is faithfully transmitted to the central nervous system, but rarely is perceived. Rather, activation of most visceral sensory receptors plays important roles in the regulation and maintenance of many essential functions (e.g., respiration, blood flow, food digestion), which requires rapid feedback to adjust to changes in demand due to exertion, food intake or other activities. While autonomic functions associated with the visceral afferent innervation can influence affective and cognitive processes and are, in turn, influenced by affective and cognitive processes, most of the afferent input from the viscera never reaches consciousness.

Visceral events, of course, do lead to conscious sensations. Under physiological conditions, humans perceive a sense of fullness related to organs involved in food intake (i.e., stomach) or waste elimination (rectum and bladder). Aside from these sensations related to functions that trigger intentional changes in behavior, most consciously perceived input from the viscera carries negative connotations, such as nausea, palpitation, dyspnea, discomfort (e.g., overfilling, gas, bloating), and pain. Normally, visceral pain is infrequent, but in cases of functional gastrointestinal disorders (e.g., nonulcer dyspepsia, IBS (see Chapter Irritable Bowel Syndrome), noncardiac chest pain, functional abdominal pain syndrome), interstitial cystitis/painful bladder syndrome, inflammatory visceral disorders (e.g., pancreatitis, inflammatory bowel disease), neoplasias, etc., discomfort and pain are common, persistent, and typically difficult to treat.

The so-called functional visceral disorders are characterized by pain and discomfort and enhanced sensitivity to stimuli in the absence of a demonstrable organic cause (i.e., there are no apparent structural, biochemical, or inflammatory conditions to explain the symptoms). Such functional disorders constitute a significant socioeconomic burden worldwide; the estimated prevalence of abdominal pain syndromes can range widely for different specific syndromes, but prevalence rates typically range between 22% and 28% of the adult population (Halder, S. L. and Locke, G. R., 2007). Characteristically, these and other visceral disorders are characterized by hypersensitivity, meaning that normally non-pain-producing visceral events are perceived as uncomfortable or painful, which is likely contributed to by both peripheral and central nervous system mechanisms.

37.2.1 Organization of Visceral Sensory Innervation

The viscera are unique among all tissues in the body. Each internal organ is innervated by two nerves, which share some functions, but are not physiologically redundant and include important functional distinctions. Anatomically, the visceral sensory innervation is physically associated with the sympathetic and parasympathetic divisions of the autonomic nervous system. Correspondingly, Langley J. N. (1921) termed visceral afferent fibers associated with sympathetic nerves afferent sympathetic fibers. This sometimes confusing nomenclature has been replaced by use of the name of the nerve (e.g., inferior cardiac, lumbar splanchnic, pelvic), which in addition avoids assigning putative function(s) to the visceral innervation. Given the physical association of the visceral innervation with the autonomic nervous system, visceral afferent fibers are thus contained in nerves that terminate in the spinal cord (spinal nerves) or in the brainstem (vagus nerves).

The bilateral vagus nerves are perhaps the most widely distributed nerves in the body. They innervate most internal organs, including all of the thoracic viscera (proximal esophagus, heart, bronchopulmonary system), most of the abdominal viscera (distal esophagus, stomach, small and proximal large intestines, liver, pancreas, etc.), and some of the pelvic viscera. The vagus nerves are mixed and contain both efferent and afferent axons, but most axons in the vagus nerves ($\geq 80\%$) are sensory (afferent) fibers and the remainder parasympathetic efferents. The nodose (primarily) and smaller, rostrally located jugular ganglia contain the cell bodies of vagal afferent fibers, most of which terminate bilaterally with some viscerotopic organization in the medullary brainstem nuclei of the solitary tract (NTS). Some vagal afferents, at least in nonhuman primates and rats, project to the upper cervical spinal cord (C1–2), where they may contribute to referred sensations and modulate nociceptive processing within the spinal cord (Foreman, R., 1999).

The cell bodies of spinal afferents innervating the viscera are located bilaterally in dorsal root ganglia (DRG) from the cervical to sacral spinal cord. As mentioned above, their axons travel with sympathetic efferent fibers (except for the pelvic nerves)

and, thus, traverse pre- and paravertebral ganglia *en route* to the spinal cord (Figure 1). In prevertebral ganglia, some afferent axons give rise to collateral branches that synapse on intraganglionic neurons and influence organ function (e.g., motility, secretion).

37.2.2 Visceral Nociceptors and Sensory Endings

Because tissue-damaging crushing, cutting, or burning stimuli that generate pain from skin do not commonly produce pain when applied to the viscera, it was long argued that the viscera were not innervated, and when visceral innervation was established, it was argued that the viscera were insensate. It is now well established that stimuli adequate for activation of visceral nociceptors differ from those adequate for activation of cutaneous nociceptors. They include distension of hollow organs, traction on the mesentery, ischemia, and inflammation. For example, experimental hollow organ distension reproduces in patients the localization, intensity, and quality of sensations associated with their visceral disorder. Because distension of hollow organs, as well as traction on the mesentery, are mechanical events (stretch), it is assumed that many visceral nociceptors are mechanoreceptors with receptive endings in visceral smooth muscle layers or serosa where they transduce the mechanical energy of tension/stretch into electrical activity (nerve membrane depolarization, action potentials). There are chemo-nociceptors in epithelial and subepithelial layers of visceral organs, but they have not been studied as extensively as have mechanoreceptors and their receptive endings have not been identified as they have for some mechanoreceptors.

Virtually all visceral sensory neuron axons are slowly conducting, thinly myelinated Aδ- or unmyelinated C-fibers. Because mechanical events are common in most organs, visceral mechanosensation and endings in muscle layers of hollow organs that respond to tension/stretch have been the focus of most studies. Unfortunately, the morphology of spinal nerve peripheral terminals in organs is virtually unknown; they are assumed to be free without structural specialization. However, two morphologically distinct endings associated principally with vagal afferent endings have been described. Intraganglionic laminar endings (IGLEs), which have been shown to be mechanosensitive, and intramuscular arrays (IMAs) are located in the upper gastrointestinal tract (Zagorodnyuk, V. P. and

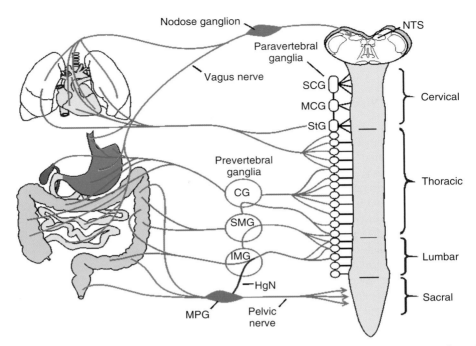

Figure 1 Cartoon illustrating innervation of the viscera. The vagus nerve innervates organs in the thoracic and abdominal cavities. Visceral afferent nerves that terminate in the spinal cord also innervate organs in the thoracic and abdominal cavities as well as those in the pelvic floor, including the genitalia (not shown). The cell bodies of vagal sensory neurons are contained in the nodose and jugular (not shown) ganglia with central terminations bilaterally in the nucleus tractus solitarii (NTS). The cell bodies of spinal visceral sensory neurons are contained in dorsal root ganglia interposed between the paravertebral ganglia (the sympathetic chain) and the spinal cord and are not illustrated in this cartoon. CG, coeliac ganglion; HgN, hypogastric nerve; IMG and SMG, inferior and superior mesenteric ganglia, respectively; MPG, major pelvic ganglion. SCG, superior cervical ganglia; MCG, middle cervical ganglia; StG, stellate cervical ganglia.

Brookes, S. J. H., 2000; Zagorodnyuk, V. P. *et al.*, 2003). As their name indicates, vagal IGLEs appear as branching endings surrounding neurons within a ganglion of the myenteric plexus (Phillips, R. J. and Powley, T. L., 2000; Powley, T. L. and Phillips, R. J., 2002). IGLEs associated with vagal afferent terminals are present throughout much of the gastrointestinal tract, but are most densely distributed in the esophagus, stomach, and duodenum (Wang, F. B. and Powley, T. L., 2000). Like IGLEs, IMAs are also thought to be mechanosensitive, but differ in morphology and distribution from IGLEs. IMAs are long arrays of terminals, typically running parallel with each other and parallel to the orientation of smooth muscle cells. They bridge between the parallel terminals within one of the muscle layers of an organ and form a lattice-like network. IMA distribution within the gastrointestinal tract is more restricted than that of IGLEs; IMAs associated with vagal afferent terminals are present in the stomach (cardia) and sphincters of the upper gastrointestinal tract and are uncommon in the intestines. Although terminal end organs have not

been described for visceral spinal nerves, an IGLE-like low threshold, slowly adapting mechanoreceptor has been reported in the colonic and rectal innervation of the guinea-pig (Lynn, P. A. *et al.*, 2003).

Whether activation of either IGLEs or IMAs give rise to visceral pain is uncertain because it would appear, based on their essentially proximal and distal distribution within the gastrointestinal tract and low response threshold, that neither are particularly well suited to that function. Powley T. L. and Phillips R. J. (2002) have speculated that IGLEs detect rhythmic motor activity, which requires response to muscle/organ tension; IMAs also respond to muscle/organ stretch. However, the association of IMAs with sphincters, and because IGLEs and IMAs have heretofore been best described and characterized in association with vagal afferent endings, suggests a role for these mechanosensors in normal physiological processes (e.g., food intake, defecation) rather than nociception. Recent studies suggest that varicose branch points of spinal afferents within the serosa or deeper layers of the gastrointestinal tract may correspond to fibers activated by high-intensity, noxious mechanical stimuli (Blackshaw, L. A. *et al.*, 2007).

37.2.3 Chemical Character of Visceral Sensory Neurons

During ontogeny, sensory neurons develop different phenotypes dependent on their differential sensitivity and exposure to growth factors. As mentioned above, essentially all visceral sensory neurons have unmyelinated or thinly myelinated axons. In DRG, this population of sensory neurons can be separated into two largely distinct groups based on the expression of growth factor receptors or other neurochemical markers (Molliver, D. C. *et al.*, 1997). Small-diameter peptide-containing DRG neurons express the high-affinity receptor for nerve growth factor trkA, the temperature-, proton-, and capsaicin-sensitive transient receptor potential vanilloid ion channel (TRPV)1 and the neuropeptides substance P and/or calcitonin gene-related peptide (CGRP). The second group of small-diameter DRG cells depends on glial cell line-derived neurotrophic factor (GDNF) during embryonic development, binds the plant lectin isolectin B4 (IB4) and releases neurotransmitters such as adenosine triphosphate (ATP) rather than peptides as transmitters. Immuno- histochemical studies have consistently shown that about 80% of visceral sensory neurons within the DRG are peptidergic and contain CGRP or substance P. Visceral sensory neurons in the placode-derived nodose ganglia, however, depend on different growth factors during development and show a different distribution of neurochemical markers. While the expression of neuropeptides CGRP and/or substance P is closely associated with the presence of TRPV1 channels in DRG neurons, this does not hold for nodose ganglion neurons; less than 10% express neuropeptides CGRP or substance P, whereas about 50% of nodose neurons exhibit TRPV1-like immunoreactivity.

In studies of cutaneous or muscle sensory neurons, expression of TRPV1, CGRP, or the presence of IB4 binding are often used as surrogate markers to classify neurons as nociceptive. As discussed below, 70–80% of mechanosensitive spinal afferents innervating abdominal or pelvic viscera are activated by innocuous intensities of stimulation (i.e., are low-threshold mechanosensors). Most studies of mechanosensitive vagal afferents, however, reveal an even more homogeneous response profile; high-threshold mechanosensitive fibers are rarely found. Thus, peptide expression is a poor predictor of physiological properties of visceral sensory neurons, arguing against the use of neurochemical content as a surrogate indicator of function.

Subsequent studies have established by immunohistochemistry, *in situ* hybridization, and/or current or voltage response to ligands applied directly to cell soma that some visceral sensory neurons also contain other peptides (e.g., vasoactive intestinal polypeptide, cholecystokinin) and express ligand-gated channels and receptors (e.g., different members of the family of transient receptor potential channels, including TRPV1, acid-sensing ion channels (ASICs), both ionotropic and metabotropic glutamate, purinergic, serotoninergic, γ-aminobutyric acid (GABA)-ergic, and opioid receptors). As information continues to accumulate, it is apparent that there is no marker at present that distinguishes visceral sensory neurons from nonvisceral sensory neurons or visceral nociceptors from nonnociceptors.

It could be argued that visceral sensory neuron terminals are generally exposed to a richer chemical milieu than other sensory neurons (and thus express a wider array of receptors and channels, although this has not been quantified). The environments in which visceral sensory endings reside include exposure to acids, digestive enzymes, hormones, nutrients, toxins (ingested or secreted by gut flora, for example), metabolites of xenobiotics, and bioactive chemicals released from nearby enteric nervous system neurons, sympathetic nerve terminals, epithelial cells like enterochromaffin cells in the gut, and urothelial cells in the urinary bladder and mast cells. Visceral sensory neurons have been shown to respond to products released in these environments, including protons, norepinephrine, serotonin, histamine, tryptase, ghrelin, cholecystokinin, and other gastric and duodenal secretions that initiate digestion and motility.

In the gastrointestinal tract, serotonin plays an especially important role as a signaling molecule. More than 90% of the body's serotonin is found within the intestine with most stored in secretory granules on the basolateral side of enteroendocrine cells. These enteroendocrine cells are specialized epithelial cells that can release into the lamina propria a variety of mediators, including serotonin, upon mechanical or chemical stimulation, which secondarily activate extrinsic and intrinsic sensory neurons. Interestingly, the number of enteroendocrine cells increases after intestinal inflammation and remains elevated in patients with continuing complaints despite resolution of the inciting infection (Spiller, R. C. et al., 2000), suggesting that serotonin release from enteroendocrine cells is important in the pathogenesis of functional bowel disorders.

An example illustrating the importance of luminal factors within the gastrointestinal tract is the interaction between the PAR_2 receptor and the pancreatic enzyme trypsin, which is especially relevant under pathologic conditions. The PAR_2 receptor is a member of the proteinase-activated receptor (PAR) family, a group of G-protein-coupled receptors characterized by a tethered peptide–ligand that is part of the molecule and can activate the receptor once cleaved from the molecule by specific proteases. The pancreatic protease trypsin is one of the endogenous proteases that cleaves the ligand and thus activates the receptor. Pancreatic acinar cells secrete precursor forms of proteolytic enzymes, such as trypsin, that require activation within the lumen of the small intestine. During pancreatitis, however, activated trypsin is also present in the pancreatic duct and parenchyma. Experimental infusion of trypsin into the pancreatic duct activates pancreatic sensory neurons, most of which express PAR_2 immunoreactivity, and triggers aversive behavior in animals. Selective experimental activation of PAR_2 receptors mimics this effect, suggesting an important role of PAR_2 in pain during acute pancreatitis (Hoogerwerf, W. A. et al., 2001; 2004). Similarly, mouse colon sensory neurons contain immunoreactivity for the PAR_2 receptor and PAR_2 agonists depolarize those neurons and cause a sustained hyperexcitability (Ahmed, K. et al., 2007). Relevant to IBS, biopsy samples from IBS patients release greater amounts of proteases than do samples from controls (Cenac, N. et al., 2007). Supernatants from colonic biopsies of IBS patients produce visceral hypersensitivity when given intracolonically in control, but not PAR_2 knockout mice. Mast cells, which are closely associated with nerves in the human gastrointestinal mucosa (Stead, R. H. et al., 1989), are also a source of tryptase as well as other potential mediators and modulators of visceral sensation (Bueno, L. et al., 1997; Barbara, G. et al., 2006). Mucosal biopsies from IBS patients contained mediators that increased activity in rat mesenteric afferents, an effect inhibited by histamine H_1 receptor blockade and serine protease inactivation (Barbara, G. et al., 2007).

In the airway system, similarly specialized cells have been identified within the epithelium of the intrapulmonary bronchial tree. Based on their ultrastructural and neurochemical characteristics with large, dense core vesicles and neuropeptide expression, they are referred to as neuroendocrine cells or neuroendocrine bodies when clustered in small groups (Cutz, E., 1982). Histological studies combining immunohistochemistry, retrograde labeling, and/or vagotomy reveal a complex innervation with vagal and spinal afferents projecting to these cells or cell clusters, where nerve terminals branch with formation of varicosities at branch points (Adriaensen, D. I. et al., 2003). Neuroepithelial bodies likely function as oxygen sensors within the airways as hypoxia inhibits potassium currents (Fu, X. W. et al., 2000). However, acute allergic inflammation triggers depletion of peptides from airway neuroepithelial cells, suggesting a role in broader signaling under physiologic and pathophysiologic conditions (Dakhama, A. et al., 2002).

While the expression of one or another receptor or ligand-gated channel of a particular subunit composition may be determined at some future time to identify all or a subset of visceral sensory neurons, one feature of visceral sensory neurons may distinguish them from nonvisceral counterparts, and that is cell size. Cutaneous nociceptors are on the whole associated with thinly myelinated and unmyelinated axons and it is generally the case that these nociceptors have small-diameter cell bodies (e.g., ≤ 15–$20\,\mu m$). Virtually all visceral sensory neurons have thinly myelinated or unmyelinated axons, yet their cell bodies tend to be medium in size (e.g., $\geq 25\,\mu m$ in diameter), at least those associated with the gastrointestinal tract and urinary bladder. This is of importance when nociceptors are studied and identified principally on the basis of cell diameter because visceral nociceptors are not generally contained in the population of small-diameter sensory neurons assumed to be nociceptors solely because of small diameter.

The foregoing addresses, at least indirectly, identification of and assignment of function of sensory neurons, including visceral sensory neurons, based on different properties of the cells: size (diameter), cell content (e.g., peptides), expression of one or another channel or receptor, dependence on exposure during development to different growth factors. Assignment of function based upon one or more of these properties is unfortunately relatively common in the literature, and is both confusing to nondiscriminating readers and misleading. It is not possible at present to assign function (e.g., nociceptor) to a sensory neuron based on any of the above properties. As discussed below, nociceptors, including visceral nociceptors, are defined by response to an adequate stimulus, and even then assignment of function can be difficult, if not contentious, if the quality of the stimulus is uncertain, unknown or of limited physiologic relevance, such as punctate stimulation of viscera.

37.2.4 Central Organization of Visceral Pain

37.2.4.1 *Vagus nerves*

Although they extensively innervate the viscera, the vagus nerves are generally considered not to play a role in nociception. However, they may contribute to the complex sensory experience of visceral pain, which includes strong autonomic reactions, including nausea or dyspnea. Vagal afferents project to the bilaterally located nuclei of the solitary tract in the dorsal medulla, an important relay station for visceral input. Vagal input reaches more rostral brain structures involved in autonomic regulation, such as the hypothalamus, supraoptic nucleus, and – via the parabrachial nucleus and ventro-medial thalamus – the insular cortex, anterior cingulate cortex, and amygdala, regions that play a role the regulation of affective responses to different stimuli including pain.

37.2.4.2 *Spinal nerves*

Because vagotomy was found to be generally ineffective in relieving visceral pain, whereas spinal nerve transaction or destruction of sympathetic prevertebral ganglia often provided pain relief (at least for a limited period of time), the spinal visceral nerves (sympathetic afferents) were inferred to be the conveyors of nociceptive information from the organs to the spinal cord. Their termination in the spinal cord is noteworthy on several counts. Firstly, spinal visceral afferent fibers terminate in a pattern that largely overlaps with terminations of cutaneous nociceptors: superficial layers of the spinal dorsal horn, deeper in lamina V and dorsal to the central canal, an area often referred to as lamina X. Visceral afferent fibers also terminate within the interomediolateral cell column/sacral parasympathetic nucleus where afferent input influences efferent output back to the same as well as to other organs (Figure 2). Secondly, spinal visceral afferents represent less than 10%, and probably closer to 5%, of the total afferent input into the spinal cord from all tissues. In compensation for this relatively sparse input, the central projection of a single visceral afferent fiber bifurcates at its spinal segment of entry in the dorsal root into caudally and rostrally directed main branches that can extend in either the dorsal funiculus or Lissauer's tract for two or three spinal segments before penetrating the spinal dorsal horn. In addition, during their longitudinal journey these main branches give off multiple collateral branches into the spinal dorsal horn (superficial laminae and laminae V and X, including contralateral laminae V and X) where, moreover, their number of terminal swellings are greater than found on cutaneous C-fiber

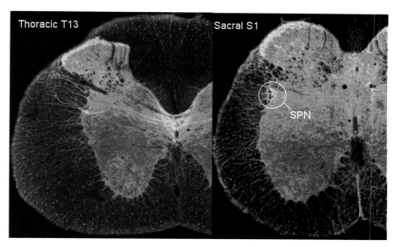

Figure 2 Distribution of visceral afferent terminals in the thoracic and sacral spinal cord. Photomicrographs illustrate internalization of the substance P receptor (yellow) in the superficial dorsal horn and area dorsal to the central canal in rat T13 and S1 spinal cord sections. Internalization of the substance P receptor was produced by distension of the colon in the rat (Honoré, P., Kamp, E. H., Rogers, S. D., Gebhart, G. F., and Mantyh, P. W. 2002. Activation of lamina I spinal cord neurons that express the substance p receptor in visceral nociception and hyperalgesia. J. Pain 3, 3–11). SPN, sacral parasympathetic nucleus.

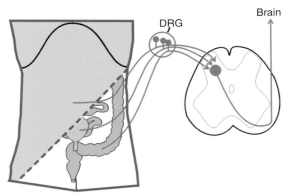

Figure 3 Illustration of viscerosomatic and viscerovisceral convergence of inputs onto a second-order spinal neuron. Illustrated is input from the abdominal skin, urinary bladder, and distal colon onto the same spinal neuron. Activation of the second-order neuron is illustrated here as being conveyed to the brain via the anterolateral ascending pathway; not illustrated is a postsynaptic dorsal column pathway, which also conveys visceral sensory information to the brain. DRG, dorsal root ganglion.

terminations within the spinal dorsal horn (which are typically limited to the spinal segment of entry). These anatomic characteristics of visceral afferent spinal terminations surely contribute to the diffuse nature of visceral pain and difficulty in localizing its source.

A final notable and defining characteristic of spinal visceral input that also contributes to poor localization is convergence. It is a feature of virtually all second-order spinal neurons upon which visceral afferent fibers terminate that convergent inputs from somatic and/or other visceral organs are also received. Typically, a single dorsal horn neuron that receives a visceral input (e.g., from colon) has a convergent cutaneous receptive field and also receives input from another viscus (e.g., urinary bladder, uterus) (Figure 3). Thus, viscerosomatic and viscerovisceral convergence upon second-order spinal neurons is the general rule (rather than the exception), and further compromises localization of visceral inputs.

37.2.4.3 Spinal pathways and supraspinal terminations

Spinal visceral afferent input is further transmitted rostrally and distributed widely in the brainstem (largely associated with reflex functions; e.g., micturition), hypothalamus, and thalamus and then to areas of the cerebral cortex where discriminative and affective components of visceral information assumes

consciousness and triggers responses/behavior. Interestingly, thalamic and cortical neurons that receive a visceral input also exhibit viscerosomatic and viscerovisceral convergence (e.g., Apkarian, A. et al., 1995) (Figure 4). The ascending pathways in spinal cord white matter that convey visceral information to the brain are contained principally in the anterolateral quadrants of the spinal cord and in the dorsal columns. Thus, there are spinomedullary, spinopontine, spinomesencephalic, spinohypothalamic, and spinothalamic tracts that ascend in the anterolateral quadrant of the spinal cord, predominantly contralateral to the side of input. The postsynaptic dorsal column paths are ipsilateral to the side of input and ascend to the nuclei gracilis and cuneatus in the medulla, where they cross and ascend to the contralateral ventrobasal thalamus (e.g., Palecek, J. et al., 2002). In addition to these principal pathways that convey information to supraspinal sites, there are intraspinal propriospinal pathways that are less well understood and presumably have modulatory functions and another pathway from lamina I neurons to pontine parabrachial nuclei and thence to the amygdala, principally the central nucleus. This latter pathway is believed to specifically convey visceral information of a noxious character to the amygdala, which importantly is associated with anxiety and affective dimensions of pain (Phelps, E. A. and LeDoux, J, E., 2005).

Functional imaging studies have complemented and expanded our appreciation of the central representation of visceral inputs. Although the current spatial resolution of imaging methods has limited ability to study the initial processing of visceral input in the brainstem, reports to date suggest that cutaneous and visceral stimuli activate similar brain regions. Pain-induced anxiety, which tends to be more significant during visceral stimulation, is associated with stronger activation of the midbrain periaqueductal gray (PAG) (Dunckley, P. et al., 2005). Similarly, activation of cortical structures does not show a pattern during visceral pain distinct from somatic pain, with the anterior cingulate cortex (ACC), the anterior insular cortex and – less consistently – the primary and secondary sensory cortices activated by both visceral and somatic pain (Hobson, A. R. and Aziz, Q., 2004). Despite the overall similarity between brain patterns of activation produced by cutaneous and visceral stimuli, subtle differences have emerged. Compared to noxious cutaneous stimulation, visceral pain tends to cause a stronger activation of the ACC and typically activates the

Amadesi, S., Creminon, C., Lanthorn, T., Geppetti, P., Bunnett, N. W., and Mayer, E. A. 2001. Role of peripheral N-methyl-D-aspartate (NMDA) receptors in visceral nociception in rats. Gastroenterology 120, 1737–1748.

Meller, S. T. and Gebhart, G. F. 1992. A critical review of the afferent pathways and the potential chemical mediators involved in cardiac pain. Neuroscience 48, 501–524.

Miranda, A., Peles, S., Rudolph, C., Shaker, R., and Sengupta, J. N. 2004. Altered visceral sensation in response to somatic pain in the rat. Gastroenterology 126, 1082–1089.

Molliver, D. C., Wright, D. E., Leitner, M. L., Parsadanian, A. S., Doster, K., Wen, D., Yan, O., and Snider, W. D. 1997. Ib4-binding drg neurons switch from ngf to gdnf dependence in early postnatal life. Neuron 19, 849–861.

Mukerji, G., Yiangou, Y., Agarwal, S. K., and Anand, P. 2006. Transient receptor potential vanilloid receptor subtype 1 in painful bladder syndrome and its correlation with pain. J. Urol. 176, 797–801.

Munakata, J., Naliboff, B., Harraf, F., Kodner, A., Lembo, T., Chang, L., Silverman, D. H., and Mayer, E. A. 1997. Repetitive sigmoid stimulation induces rectal hyperalgesia in patients with irritable bowel syndrome. Gastroenterology 112, 55–63.

Ness, T. J., Metcalf, A. M., and Gebhart, G. F. 1990. A psychophysiological study in humans using phasic colonic distension as a noxious visceral stimulus. Pain 43, 377–386.

Page, A. J., Brierley, S. M., Martin, C. M., Price, M. P., Symonds, E., Butler, R., Wemmie, J. A., and Blackshaw, L. A. 2005. Different contributions of asic channels 1a, 2, and 3 in gastrointestinal mechanosensory function. Gut 54, 1408–1415.

Page, A. J., Martin, C. M., and Blackshaw, L. A. 2002. Vagal mechanoreceptors and chemoreceptors in mouse stomach and esophagus. J. Neurophysiol. 87, 2095–2103.

Palecek, J., Paleckova, V., and Willis, W. D. 2002. The roles of pathways in the spinal cord lateral and dorsal funiculi in signaling nociceptive somatic and visceral stimuli in rats. Pain 96, 297–307.

Pan, H. and Chen, S. R. 2002. Myocardial ischemia recruits mechanically insensitive cardiac sympathetic afferents in cats. J. Neurophysiol. 87, 660–668.

Pan, H. L., Longhurst, J. C., Eisenach, J. C., and Chen, S. R. 1999. Role of protons in activation of cardiac sympathetic C-fibre afferents during ischaemia in cats. J. Physiol. 518, 857–866.

Perl, E. 1996. Pain and the Discovery of Nociceptors. In: Neurobiology of Nociceptors (eds. C. Belmonte and F. Cervero), pp. 5–36. Oxford University Press.

Pezzone, M. A., Liang, R., and Fraser, M. O. 2005. A model of neural cross-talk and irritation in the pelvis: implications for the overlap of chronic pelvic pain disorders. Gastroenterology 128, 1953–1964.

Phelps, E. A. and LeDoux, J. E. 2005. Contributions of the amygdala to emotion processing: from animal models to human behavior. Neuron 48, 175–187.

Phillips, R. J. and Powley, T. L. 2000. Tension and stretch receptors in gastrointestinal smooth muscle: re-evaluating vagal mechanoreceptor electrophysiology. Brain Res. Rev. 34, 1–26.

Porreca, F., Ossipov, M. H., and Gebhart, G. F. 2002. Tonic descending facilitation as a mechanism of chronic pain. Trends Neurosci. 25, 319–325.

Posserud, I., Agerforz, P., Ekman, R., Bjornsson, E. S., Abrahamsson, H., and Simren, M. 2004. Altered visceral perceptual and neuroendocrine response in patients with irritable bowel syndrome during mental stress. Gut 53, 1102–1108.

Powley, T. L. and Phillips, R. J. 2002. Musings on the wanderer: what's new in our understanding of vago-vagal reflexes?: I.

Morphology and topography of vagal afferents innervating the GI tract. Am. J. Physiol. 283, G1217–G1225.

Price, D. D., Craggs, J., Verne, N., Perlstein, W. M., and Robinson, M. E. 2007a. Placebo analgesia is accompanied by large reductions in pain-related brain activity in irritable bowel syndrome patients. Pain 127, 63–72.

Price, D. D., Robinson, M. E., and Verne, G. N. 2007b. Measuring Pain and Hyperalgesia in Persistent Pain Conditions with a Special Emphasis on Irritable Bowel Syndrome. In: Chronic Abdominal and Visceral Pain (eds. P. Pasricha, W. Willis, and G. F. Gebhart), pp. 127–140. Informa Healthcare.

Randich, A., Uzzell, T., DeBerry, J. J., and Ness, T. J. 2006. Neonatal urinary bladder inflammation produces adult bladder hypersensitivity. J. Pain 7, 469–479.

Ricco, M. M., Kummer, W., Biglari, B., Myers, A. C., and Undem, B. J. 1996. Interganglionic segregation of distinct vagal afferent fibre phenotypes in guinea-pig airways. J. Physiol 496, 521–530.

Robinson, D. R., McNaughton, P. A., Evans, M. L., and Hicks, G. A. 2004. Characterization of the primary spinal afferent innervation of the mouse colon using retrograde labelling. Neurogastroenterol. Motil. 16, 113–124.

Rong, W., Hillsley, K., Davis, J. B., Hicks, G., Winchester, W. J., and Grundy, D. 2004. Jejunal afferent nerve sensitivity in wild-type and TRPV1 knockout mice. J. Physiol. 560, 867–881.

Sarkar, S., Aziz, Q., Woolf, C. J., Hobson, A. R., and Thompson, D. G. 2000. Contribution of central sensitisation to the development of non-cardiac chest pain. Lancet 356, 1154–1159.

Sarkar, S., Hobson, A. R., Furlong, P. L., Woolf, C. J., Thompson, D. G., and Aziz, Q. 2001. Central neural mechanisms mediating human visceral hypersensitivity. Am. J. Physiol. 281, G1196–G1202.

Schmidt, R. F. 1996. The Articular Polymodal Nociceptor in Health and Disease. In: ThePolymodal Receptor: A Gateway to Pathological Pain, Progress in Brain Research (eds.T. Kumazawa, L. Kruger, and K. Mizumura), Vol. 113, pp. 53–81. Elsevier.

Schuligoi, R., Joci, M., Heinemann, A., Schoninkle, E., Pabst, M. A., and Holzer, P. 1998. Gastric acid-evoked c-fos messenger RNA expression in rat brainstem is signaled by capsaicin-resistant vagal afferents. Gastroenterology 115, 649–660.

Sharma, A. and Aziz, Q. 2007. Mechanisms of Visceral Sensitization in Humans. In: Chronic Abdominal and Visceral Pain (eds. P. Pasricha, W. Willis, and G. F. Gebhart), pp. 141–160. Informa Healthcare.

Sherrington, C. S. 1900. Cutaneous Sensations. In: Textbook of Physiology (ed. E. A. Schäffer), pp. 920–1001. Pentland.

Sherrington, C. S. 1906. The Integrative Action of the Nervous System. Scribner.

Sidhu, H., Kern, M., and Shaker, R. 2004. Absence of increasing cortical FMRI activity volume in response to increasing visceral stimulation in IBS patients. Am. J. Physiol. 287, G425–G435.

Song, G. H., Venkatraman, V., Ho, K. Y., Chee, M. W., Yeoh, K. G., and Wilder-Smith, C. H. 2006. Cortical effects of anticipation and endogenous modulation of visceral pain assessed by functional brain MRI in irritable bowel syndrome patients and healthy controls. Pain 126, 79–90.

Spencer, S. J., Hyland, N. P., Sharkey, K. A., and Pittman, Q. J. 2007. Neonatal immune challenge exacerbates experimental colitis in adult rats: potential role for TNF-α. Am. J. Physiol. 292, R308–R315.

Spiller, R. C., Jenkins, D., Thornley, J. P., Hebden, J. M., Wright, T., Skinner, M., and Neal, K. R. 2000. Increased rectal mucosal enteroendocrine cells, T lymphocytes and increased gut permeability following acute campylobacter enteritis and in post-dysenteric irritable bowel syndrome. Gut 47, 804–811.

Stead, R. H., Dixon, M. F., Bramwell, N. H., Riddell, R. H., and Bienenstock, J. 1989. Mast cells are closely apposed to nerves in the human gastrointestinal mucosa. Gastroenterology 97, 575–585.

Stewart, T. M., Beyak, M. J., and Vanner, S. J. 2003. Ileitis modulates potassium and sodium currents in guinea pig dorsal root ganglia neurons. J. Physiol. 552, 797–807.

Strigo, I. A., Duncan, G. H., Boivin, M., and Bushnell, M. C. 2003. Differentiation of visceral and cutaneous pain in the human brain. J. Neurophysiol. 89, 3294–3303.

Su, X., Sengupta, J. N., and Gebhart, G. F. 1997. Effects of opioids on mechanosensitive pelvic nerve afferent fibers innervating the urinary bladder of the rat. J. Neurophysiol. 77, 1566–1580.

Sugiura, T., Bielefeldt, K., and Gebhart, G. F. 2004. TRPV1 function in mouse colon sensory neurons is enhanced by metabotropic 5-hydroxytryptamine receptor activation. J. Neurosci. 24, 9521–9530.

Sugiura, T., Bielefeldt, K., and Gebhart, G. F. 2007. Mouse colon sensory neurons detect extracellular acidosis via TRPV1. Am. J. Physiol. 292, C1768–C1774.

Sugiura, T., Dang, K., Lamb, K., Bielefeldt, K., and Gebhart, G. F. 2005. Acid-sensing properties in rat gastric sensory neurons from normal and ulcerated stomach. J. Neurosci. 25, 2617–2627.

Taché, Y., Martinez, V., Wang, L., and Million, M. 2004. CRF1 receptor signaling pathways are involved in stress-related alterations of colonic function and viscerosensitivity: implications for irritable bowel syndrome. Br. J. Pharmacol. 141, 1321–1330.

Torebjörk, H. E., Schmelz, M., and Handwerker, H. O. 1996. Functional Properties of Human Cutaneous Nociceptors and their Role in Pain and Hyperalgesia. In: Neurobiology of Nociceptors (eds. C. Belmonte and F. Cervero), pp. 349–369. Oxford University Press.

Undem, B. J., Carr, M. J., and Kollarik, M. 2002. Physiology and plasticity of putative cough fibres in the Guinea pig. Pulm. Pharmacol. Ther. 15, 217–219.

Undem, B. J., Chuaychoo, B., Lee, M. G, Weinreich, D., Myers, A. C., and Kollarik, M. 2004. Sybtypes of vagal afferent c-fibres in guinea-pig lungs. J. Physiol. 556, 905–917.

Vandenberghe, J., Vos, R., Persoons, P., Demyttenaere, K., Janssens, J., and Tack, J. 2005. Dyspeptic patients with visceral hypersensitivity: sensitisation of pain specific or multimodal pathways? Gut 54, 914–919.

Vera-Portocarrero, L. P., Yie, J., Kowal, J., Ossipov, M., King, T., and Porreca, F. 2006. Descending facilitation from the rostral ventromedial medulla maintains visceral pain in rats with experimental pancreatitis. Gastroenterology 130, 2155–2164.

Wang, F. B. and Powley, T. L. 2000. Topographic inventories of vagal afferents in gastrointestinal muscle. J. Comp. Neurol. 421, 302–324.

Wemmie, J., Price, M., and Welsh, M. 2006. Acid-sensing ion channels: advances, questions and therapeutic opportunities. Trends Neurosci. 29, 578–586.

Whitehead, W., Palsson, O., and Jones, K. 2002. Systematic review of the comorbidity of irritable bowel syndrome with other disorders: What are the causes and implications? Gastroenterology 122, 1140–1156.

Willert, R. P., Woolf, C. J., Hobson, A. R., Delaney, C., Thompson, D. G., and Aziz, Q. 2004. The development and maintenance of human visceral pain hypersensitivity is dependent on the N-methyl-D-aspartate receptor. Gastroenterology 126, 683–692.

Winnard, K. P., Dmitrieva, N., and Berkley, K. 2006. Cross-organ interactions between reproductive, gastrointestinal, and urinary tracts: modulation by estrous stage and involvement of the hypogastric nerve. Am. J. Physiol. 291, R1592–R1601.

Woolf, C. J. and Thompson, S. W. 1991. The induction and maintenance of central sensitization is dependent on N-methyl-D-aspartic acid receptor activation: implications for the treatment of post-injury pain hypersensitivity states. Pain 44, 293–299.

Wynn, G., Ma, B., Ruan, H. Z., and Burnstock, G. 2004. Purinergic component of mechanosensory transduction is increased in a rat model of colitis. Am. J. Physiol. 287, G647–G657.

Wynn, G., Rong, W., Xiang, Z., and Burnstock, G. 2003. Purinergic mechanisms contribute to mechanosensory transduction in the rat colorectum. Gastroenterology 25, 1398–1409.

Xu, G., Winston, J., Shenoy, M., Yin, H., and Pasricha, P. J. 2006. Enhanced excitability and suppression of A-type K^+ current of pancreas-specific afferent neurons in a rat model of chronic pancreatitis. Am. J. Physiol. 291, G424–G431.

Yáguez, L., Coen, S., Gregory, L. J., Amaro, E., Jr., Altman, C., Brammer, M. J., Bullmore, E. T., Williams, S. C. R., and Aziz, Q. 2005. Brain response to visceral aversive conditioning: a functional magnetic resonance imaging study. Gastroenterology 128, 1819–1829.

Yoshimura, N. and de Groat, W. C. 1999. Increased excitability of afferent neurons innervating rat urinary bladder after chronic bladder inflammation. J. Neurosci. 19, 4644–4653.

Yoshimura, N., Seki, S., Erickson, K. A., Erickson, V. L., Chancellor, M. B., and de Groat, W. C. 2003. Histological and electrical properties of rat dorsal root ganglion neurons innervating the lower urinary tract. J. Neurosci. 23, 4355–4361.

Yoshimura, N., Seki, S., Novakovic, S. D., Tzoumaka, E., Erickson, V. L., Erickson, K. A., Chancellor, M. B., and de Groat, W. C. 2001. The involvement of the tetrodotoxin-resistant sodium channel nav1.8 (pn3/sns) in a rat model of visceral pain. J. Neurosci. 21, 8690–8696.

Yu, S., Undem, B. J., and Kollarik, M. 2005. Vagal afferent nerves with nociceptive properties in guinea-pig oesophagus. J. Physiol. 563, 831–842.

Zagorodnyuk, V. P. and Brookes, S. J. H. 2000. Transduction sites of vagal mechanoreceptors in the guinea pig esophagus. J. Neurosci. 20, 6249–6255.

Zagorodnyuk, V. P., Chen, B. N., Costa, M., and Brookes, J. H. 2003. Mechanotransduction by intraganglionic laminar endings of vagal tension receptors in the guinea-pig oesophagus. J. Physiol. 553, 575–587.

Zhuo, M. and Gebhart, G. F. 2002. Facilitation and attenuation of a visceral nociceptive reflex from the rostroventral medulla in the rat. Gastroenterology 122, 1007–1019.

Zhuo, M., Sengupta, J. N., and Gebhart, G. F. 2002. Biphasic modulation of spinal visceral nociceptive transmission from the rostroventral medial medulla in the rat. J. Neurophysiol. 87, 2225–2236.

Further Reading

Berkley, K. J., Robbins, A., and Sato, Y. 1988. Afferent fibers supplying the uterus in the rat. J. Neurophysiol. 59, 142–163.

Brune, K. and Handwerker, H. O. (eds.) 2004. Hyperalgesia: Molecular Mechanisms and Clinical Implications, Progress in Pain Research and Management,, Vol. 30. IASP Press.

Carr, M. J. and Undem, B. J. 2003. Bronchopulmonary afferent nerves. Respirology 8, 291–301.

Cervero, F. 1994. Sensory innervation of the viscera: peripheral basis of visceral pain. Physiol. Rev. 74, 95–138.

Francis, C. Y., Duffy, J. N., Whorwell, P. J., and Morris, J. 1997. High prevalence of irritable bowel syndrome in patients attending urological outpatient departments. Dig. Dis. Sci. 42, 404–407.

Gebhart, G. F. (ed.) 1995. Visceral Pain, Progress in Pain Research and Management, , Vol. 5. IASP Press.

Jänig, W. 2006. The Integrative Action of the Autonomic Nervous System. Cambridge University Press.

Kollarik, M. and Undem, B. J. 2002. Mechanisms of acid-induced activation of airway afferent nerve fibres in guinea-pig. J. Physiol. 543, 591–600.

Kollarik, M. and Undem, B. J. 2003. Activation of bronchopulmonary vagal afferent nerves with bradykinin, acid and vanilloid receptor agonists in wild-type and TRPV1$^{-/-}$ mice. J. Physiol. 555, 115–123.

Kreis, M. E., Jiang, W., Kirkup, A. J., and Grundy, D. 2002. Cosensitivity of vagal mucosal afferents to histamine and 5-HT in the rat jejunum. Am. J. Physiol. 283, G612–G617.

Linden, D. R., Chen, J.-X., Gershon, M. D., Sharkey, K. A., and Mawe, G. M. 2003. Serotonin availability is increased in mucosa of guinea pigs with TNBS-induced colitis. Am. J. Physiol. 285, G207–G216.

Ness, T. J. and Gebhart, G. F. 1990. Visceral pain: a review of experimental studies. Pain 41, 167–234.

Pan, H. L. and Longhurst, J. C. 1996. Ischaemia-sensitive sympathetic afferents innervating the gastrointestinal tract function as nociceptors in cats. J. Physiol. 492, 841–850.

Pasricha, P. J., Willis, W. D.,, and Gebhart, G. F. (eds.) 2007. Chronic Abdominal and Visceral Pain., Informa Healthcare.

Spiller, R. and Grundy, D. (eds.) 2004. Pathophysiology of the Enteric Nervous System., Blackwell Publishing.

38 Irritable Bowel Syndrome

S Bradesi and E A Mayer, University of California, Los Angeles, CA, USA

I Schwetz, Medical University, Graz, Austria

38.1 Introduction

Recurrent abdominal pain or discomfort in the absence of detectable structural or biochemical abnormalities, associated with alterations in bowel habits, are the principal symptoms of irritable bowel syndrome (IBS) (Drossman, D. A. *et al.*, 2002). Due to the likely heterogeneity of the syndrome (Mayer, E. A. and Collins, S. M., 2002) and the lack of reliable organic markers, the development of a unifying, and generally agreed upon, hypothesis of the pathophysiology of IBS has remained an elusive goal. Many investigators in the field agree that an enhanced perception of physiologically occurring, or experimentally generated visceral events (visceral hypersensitivity) (Mayer, E. A. and Gebhart, G. F., 1994; see Vagal Afferent Neurons and Pain and Visceral Pain) in conjunction with alterations in gastrointestinal motility and secretory function, are key pathophysiological mechanisms underlying the cardinal clinical features of IBS. In contrast, many other alterations reported in IBS patients over the past decades, including altered mucus production and altered gastrointestinal motility, have turned out to be epiphenomena which are unlikely to be essential for symptom generation. Considerable evidence supports the role of psychosocial (Bennett, E. J. *et al.*, 1998; Collins, S. M., 2002) and physical (Gwee, K. A. *et al.*, 1996) (i.e., acute gastroenteric infections) stressors as central and peripheral triggers of first symptom onset or symptom exacerbation of long-standing IBS (Mayer, E. A. *et al.*, 2001), and an IBS hypothesis of hyperresponsiveness of central stress circuits has been proposed (Mayer, E. A. and Collins, S. M., 2002).

38.2 Clinical Presentation and Epidemiology

IBS is one of the most common and most thoroughly studied functional disorders of the gastrointestinal tract (Drossman, D. A. *et al.*, 2002). In additional to chronic abdominal pain, the clinical presentation of IBS typically also includes nonpainful abdominal discomfort (sensations of urgency, bloating, fullness, gas and constipation) (Lembo, T. J. and Fink, R. N., 2002) and visible abdominal distension. These gastrointestinal symptoms are frequently associated with extraintestinal symptoms such as fatigue, decreased energy level, impaired sleep, depression, and anxiety (Zimmerman, J., 2003). In the absence of generally agreed upon reliable biological markers, the diagnosis of IBS remains one based on the presence of the so-called symptom criteria (Thompson, W. G. *et al.*, 1999). The most widely accepted diagnostic criteria are the Rome II Criteria that evolved from the Rome I and the Manning criteria initially defined about a decade ago (Longstreth, G. F., 2005). Different IBS patient subtypes have been identified based on their bowel habit predominance and have been classified as: constipation, diarrhea, or alternating periods of both. In a large US sample, approximately 50% of IBS patients present alternating bowel habit (IBS-A),

30% diarrhea (IBS-D), and 20% constipation (IBS-C) (Tillisch, K. *et al.*, 2005).

Large population based studies in the US have found a prevalence of IBS of about 14% with a greater proportion of women affected (female-to-male ratio approximately 3:1 to 3:2). The socioeconomic consequences of IBS are considerable with a large impact on work productivity and absenteeism (Dean, B. B. *et al.*, 2005). In addition, a substantial reduction in health-related quality of life (HRQoL) has been observed in IBS patients with moderate to severe symptoms who are seen in a referral setting compared with healthy controls (Spiegel, B. M. *et al.*, 2004). Interestingly, bowel habit alterations contributed little to the HRQoL impairment in this study, while extraintestinal manifestations such as loss of energy, fatigue, and excessive worry were important determinants. The annual cost of IBS in the US has been estimated to be between $1.7 billion and $10 billion in direct medical cost (repetitive use of multiple healthcare resources) and $20 billion for indirect costs (work absenteeism and impaired productivity) representing a high socioeconomic burden on society (Inadomi, J. M. *et al.*, 2003).

While research over the past few years has provided significant advances in the understanding of IBS pathophysiology, the precise mechanism(s) underlying symptom generation remains incompletely understood, generating considerable controversy among investigators. As a result, the development of effective IBS therapies has been slow and disappointing (Bradesi, S. *et al.*, 2006). IBS pathophysiology is often viewed within the so-called biopsychosocial model in which altered physiology (gastrointestinal motility and secretion, enhanced perception of visceral stimuli (visceral hypersensitivity) and psychosocial factors interact and determine the clinical expression of IBS (Schwetz, I. *et al.*, 2004). From a biological perspective, IBS symptomatology can be viewed as the manifestation of alterations in the brain–gut axis, specifically as a dysregulation in the complex interplay between events occurring within the gut, the enteric nervous system, and the central nervous system (Mulak, A. and Bonaz, B., 2004).

38.3   The Bidirectional Brain–Gut Axis Model

Brain-gut interactions play a prominent role in the modulation of gut function in health and disease (Mayer, E. A. *et al.*, 2001; Taché, Y. *et al.*, 2001).

Therefore, every conceptual model of IBS has to take into account that neither the central nervous system nor the gastrointestinal tract function in isolation, but that both systems interact with each other under normal conditions and particularly during perturbations of homeostasis. Afferent signals arising from the lumen of the gut are transmitted via various visceral afferent pathways (enteric, spinal, vagal) to the central nervous system. Homeostatic reflexes, which generate appropriate gut responses to physiological as well as pathological visceral stimuli, occur at the level of the enteric nervous system, the spinal cord, and pontomedullary nuclei and limbic regions. Vagal visceral afferent inputs may also play an important role in such diverse functions as modulation of emotion, pain sensitivity, satiety, and immune response (Jänig, W. and Habler, H.-J., 2000). Whereas the reflex circuits within the enteric nervous system in principle can regulate and synchronize all basic gastrointestinal functions (motility, secretion, blood flow), coordination of gut functions with the overall homeostatic state of the organism requires continuous communication between the central nervous system and the gastrointestinal tract (Mayer, E. A. and Collins, S. M., 2002). Descending cortico-limbic influences can set the gain and responsiveness of these reflexes, or impose distinct patterns of motor responses on lower circuits, should the overall condition of the body make it necessary. Such top-down override of local reflex function occurs during sleep, during the stress response, or during strong emotions such as fear and anger (Ito, M., 2002; Mayer, E. A., 2000a; Taché, Y. *et al.*, 2001; Welgan, P. *et al.*, 1988). While the great majority of homeostatic afferent input from the gut (as well as other viscera) to the central nervous system is not consciously perceived, there are both peripheral and central adaptive mechanisms which can result in enhanced perception and altered reflex responses to visceral stimuli (Mayer, E. A. and Gebhart, G. F., 1994).

38.4   Visceral Hypersensitivity

The initial clinical observations that led to the hypothesis that IBS patients exhibit visceral hypersensitivity include recurring abdominal pain, tenderness during palpation of the sigmoid colon during physical examination, and excessive pain during endoscopic evaluation of the sigmoid colon (Mayer, E. A. and Gebhart, G. F., 1994). Multiple human experimental studies using barostat-controlled balloon distension paradigms have reported lowered colorectal perceptual thresholds, increased sensory ratings and viscerosomatic referral areas in IBS patients compared to healthy individuals (Bouin, M. *et al.*, 2002; Chang, L. *et al.*, 2000a; Mertz, H. *et al.*, 1995). Despite the uncertainty about the underlying mechanism(s), this kind of experimentally induced visceral hypersensitivity has been considered a pathophysiological hallmark of the disease. Within the framework of homeostatic afferents, the finding of chronic, conscious awareness of unpleasant visceral sensations, together with the presence of altered reflex regulation in the gut, could be explained by several different mechanisms, including the following: (1) a peripheral up-regulation of the sensitivity of visceral afferent pathways which may be related to alterations in the activity of various effector cells within the gut (enteric nerves, enterochromaffin cells, immune cells); (2) a spinal or brainstem alteration in the sensitivity to incoming visceral afferents which could be a consequence of primary peripheral or central inputs; or (3) a primary central amplification of perceptual and reflex responses to incoming visceral afferent signals. In addition, the frequent presence of compromised vital functions such as sleep, mood, sexual drive, and affect, suggest a possible involvement of other homeostatic mechanisms, such as tonic serotonergic pontomedullary systems (Mason, P., 2005). It remains to be determined if these various abnormalities occur in distinct subsets of patients, at different stages of the disorder, or if various combinations of them generate a heterogeneous patient population.

38.4.1 Peripheral Up-Regulation of Visceral Afferent Sensitivity

The concept that gut mucosal alterations may play a role in the pathophysiology of IBS has been prompted by several lines of clinical and preclinical evidence including: (1) alterations in the number of immune cells, particularly mast cells, within intestinal biopsies of patients meeting diagnostic criteria for IBS (Barbara, G. *et al.*, 2004; O'Sullivan, M. *et al.*, 2000; Weston, A. P. *et al.*, 1993), and the demonstration in rodent studies that mucosal mast cells play a prominent role of transducers within the brain–gut axis (Santos, J. *et al.*, 2005; Soderholm, J. D. *et al.*, 2002; Wilson, L. M. and Baldwin, A. L., 1999); (2) demonstration of visceral hypersensitivity in different animal models of gut immune system activation (Barbara, G. *et al.*, 1997; Barreau, F. *et al.*, 2004; La, J. H. *et al.*, 2003;

Lamb, K. *et al.*, 2006); (3) preliminary evidence that supernatants from human mucosal biopsies (taken from IBS patients) have a unique effect on visceral afferent function in rodent bioassays (Barbara, G. *et al.*, 2005a; Barbara, G. *et al.*, 2005b); (4) the development of IBS-like symptoms in a small number of individuals following an episode of acute gastroenteritis (Gwee, K. A. *et al.*, 1998; Spiller, R. C., 2005) and in a subset of patients with inflammatory bowel disease (Bernstein, C. N. *et al.*, 1996); (5) evidence for alteration in the normal gut flora (including bacterial overgrowth) (Balsari, A. *et al.*, 1982; O'Leary, C. and Quigley, E. M., 2003; Swidsinski, A. *et al.*, 1999); and (6) therapeutic responses to treatment with antibiotics and probiotic treatments (Madden, J. A. and Hunter, J. O., 2002; Pimentel, M. *et al.*, 2000, 2003). Without discussing this large body of evidence supporting each of these points (Barbara, G. *et al.*, 2006; Crowell, M. D. *et al.*, 2005; Schwetz, I. *et al.*, 2003; Spiller, R. C., 2005), it is assumed, by many investigators focusing on peripheral etiologies of visceral hypersensitivity, that the mucosal immune activation plays some role in maintaining a chronic state of visceral afferent sensitization. Based on recent preclinical evidence, it is conceivable that transient peripheral sensitization may result in long-lasting up-regulation of spino-bulbo-spinal pain amplification mechanisms (Suzuki, R. *et al.*, 2005). While such peripheral mechanisms (in particular an increase in mucosal mast cells and intestinal enterochromaffin cells) may play an important role in a subset of IBS patients, or in the mediation of certain types of IBS symptoms, a series of observations argues against a simple relationship between peripheral mucosal events and IBS symptoms. For example, it has long been known that some patients in remission from inflammatory bowel disease, especially ulcerative colitis, report symptoms similar to those of IBS patients (Isgar, B. *et al.*, 1983; Simren, M. *et al.*, 2000). In contrast, inflammatory bowel disease patients (without associated IBS features) who have chronically recurring gut inflammation do not exhibit visceral hypersensitivity (Chang, L. *et al.*, 2000b) or enhanced central nervous system responses to visceral distension during periods of remission (Loennig-Baucke, V. *et al.*, 1989; Rao, S. S. C. *et al.*, 1987). In addition, clinical states characterized by chronic inflammation of the esophagus (gastroesophageal reflux disease) and stomach (*Helicobacter pylori* chronic gastritis) are also not associated with visceral mechanical hyperalgesia (Fass, R. *et al.*, 1998; Mertz, H. *et al.*, 1998). On the contrary, it has been reported that

Crohn's patients with isolated inflammation in the small bowel have elevated discomfort thresholds to controlled distension of the rectum (Bernstein, C. N. et al., 1996), and that patients with ulcerative colitis do not show sensitization following repetitive noxious distension of the sigmoid colon (Chang, L. et al., 2000b). Taken together, these data are consistent with the interpretation that chronic mucosal inflammatory changes in the esophagus, stomach, or colon by themselves do not necessarily result in mechanical visceral hypersensitivity. Chronic intestinal inflammation in inflammatory bowel disease seems to be associated with the activation of counterregulatory antinociceptive systems, inhibition of pain facilitatory pathways or both, resulting in a reduced perception of visceral afferent information (Chang, L. et al., 2000b). One may speculate that genetic polymorphisms recently identified as being associated with greater pain sensitivity may be related to such compromised endogenous pain modulation systems (Diatchenko, L. et al., 2005).

38.4.2 Spinal and Supraspinal Up-Regulation of Visceral Afferent Sensitivity

A growing body of literature supports the concept of an enhanced stress responsiveness playing a role in the development of IBS symptoms in a subset of patients (reviewed in Mayer, E. A., 2000b; Mayer, E. A. et al., 2001). An individual's response to stress (perturbation of homeostasis) is generated by a central network comprised of integrative brain structures, referred to as the emotional motor system (EMS) (for a detailed review see Mayer, E. A., 2000b); Mayer, E. A. et al., 2001). The main output systems of the EMS are ascending monoaminergic pathways, the autonomic nervous system, the hypothalamic–pituitary–adrenal axis and endogenous pain modulatory systems. The neuropeptide, corticotropin-releasing factor (CRF), plays a prominent role in integrating these various outputs in response to physical and psychological stressors (Bale, T. L., 2005).

The responsiveness of the EMS and its various output pathways is under partial genetic control (Pezawas, L. et al., 2005) and is programmed by prenatal (Matthews, S. G., 2002) and postnatal (Ladd, C. O. et al., 2000) aversive events and by certain types of pathological stress (Fuchs, E. and Fluegge, G., 1995). In humans, early (pre- and postnatal) life adverse experiences can lead to long-lasting stress hyperresponsiveness, which in turn has been associated with a wide range of health impairments in adult life (Whitehead, M. and Holland, P., 2003). Long-term consequences of early aversive events also include an increased vulnerability for stress-sensitive disorders such as IBS and posttraumatic stress disorder (Lowman, B. C. et al., 1987). Stress hyperresponsiveness is related in part to permanent, stress-induced hypersecretion of CRF (Heim, C. et al., 2000). A number of preclinical studies support the concept of centrally mediated alteration of visceral perception. Different stress paradigms were found to lead to enhanced visceral response to colonic distension (Bradesi, S. et al., 2002; Gue, M. et al., 1997; Schwetz, I. et al., 2005), which can be mimicked or abolished by central administration of pharmacological agents. For example, central CRF injection was found to mimic the effect of stress on visceral sensitivity, whereas stress-induced visceral hyperalgesia may be reduced by central injection of a CRF receptor subtype 1 antagonist (Taché, Y. et al., 2004). Similarly, a neurokinin-1 receptor antagonist injected into the spinal cord in stress-sensitized guinea pigs (Greenwood-Van Meerveld, B. et al., 2003) or rats (Bradesi, S. et al., 2005) abolished the stress-induced visceral hyperalgesia to colorectal distension. Together, the data available suggest the role of central stress circuits in the alteration of visceral nociceptive response in animal models for IBS. In contrast to the well-known analgesic response to severe acute stressors (stress-induced somatic analgesia), prolonged and milder stressors that are associated with anxiety-like states are commonly associated with hyperalgesic responses (Boccalon, S. et al., 2006; Vendruscolo, L. F. et al., 2004).

Suggestive evidence for an alteration in central pain modulation mechanisms in IBS patients comes from a series of functional brain-imaging studies. In response to rectal balloon distension, IBS patients have shown increased activation of subregions of the dorsal anterior cingulate cortex (ACC) (Naliboff, B. D. et al., 2001b; Porreca, F. et al., 2002; Verne, G. N. et al., 2003), a brain region involved in attentional and emotional modulation of stimulus perception (Petrovic, P. and Ingvar, M., 2002). Dorsal ACC subregions provide input to subcortical endogenous pain inhibitory circuits (Petrovic, P. and Ingvar, M., 2002), as well as to subcortical pain facilitatory circuits (Zhuo, M. and Gebhart, G. F., 2002). The relative balance between these simultaneously activated pain modulation systems determines the overall modulation of perception (Porreca, F. et al., 2002). Thus, the finding of greater dorsal ACC activation by a visceral stimulus in IBS may be related to greater attention and possibly associated activation of pain facilitation circuits to a visceral stimulus in IBS. On the other hand, the lesser activation of the periaqueductal gray (PAG) reported in some studies (Naliboff, B. D. *et al.*, 2001b) is consistent with possible deficiencies in the cortical activation of endogenous pain inhibition systems. Several recent studies (Chang, L., 2005) provide further support for the hypothesis of altered endogenous pain modulation in IBS patients. In one study (Mayer, E. A. *et al.*, 2005), IBS patients were compared to patients with ulcerative colitis and with healthy control subjects. IBS patients showed consistently greater activation of limbic/paralimbic brain regions (amygdala, hypothalamus, ventral/rostral ACC, dorsomedial prefrontal cortex) suggestive of increased activation of arousal circuits by a visceral stimulus. In addition, the results showed activation in the ulcerative colitis and control subjects, but not in IBS patients, in the lateral prefrontal regions and a midbrain region including the PAG. A connectivity analysis using structural equation modeling supported these regions acting as part of a pain inhibition network that involves lateral and medial prefrontal influences on the PAG. Another study provided evidence for the abnormal activation of diffuse noxious inhibitory control systems in response to a noxious stimulus in IBS patients (Wilder-Smith, C. H. *et al.*, 2004).

Several lines of evidence indicate that patients with IBS and other functional disorders have hypervigilance for symptom relevant sensations (Berman, S. M. *et al.*, 2002). In a recent longitudinal study of IBS patients exposed to six sessions of rectal inflations over a 1-year period, regional cerebral blood flow to the inflations and anticipation of inflations using $H_2^{15}O$ positron emission tomography (PET) at the first and last session were evaluated (Naliboff, B. D. *et al.*, 2001a). Subjective ratings of the rectal inflations normalized over the course of the study consistent with decreased vigilance towards the visceral stimuli. Stable activation of the central pain matrix (including thalamus and insula) by the rectal stimulus was observed over the 12-month period, while activity in limbic, paralimbic, and pontine regions decreased. An analysis examining the co-variation of these brain regions supported the hypothesis of changes in an arousal network including limbic, pontine, and cortical areas underlying the decreased perception seen over the multiple stimulation studies.

38.5 Treatment Options

Despite its high prevalence and considerable impact on patients' HRQoL, treatment options for IBS continue to be limited and there are few well designed studies to support the effectiveness of some of the most commonly used therapies (for review see Camilleri, M., 2004). Traditional treatment algorithms have primarily been aimed at peripheral targets and are largely symptom-based (e.g., laxatives, antidiarrheals, prokinetics). In addition, centrally targeted therapies include various forms of cognitive behavioral therapy (Hutton, J., 2005), low-dose tricyclic antidepressants (Drossman, D. A. *et al.*, 2003) and, in patients with co-morbid psychiatric conditions, full dose serotonin reuptake inhibitors (Creed, F. *et al.*, 2003). A series of compounds are currently in development, which is targeted at central circuits involved in stress responsiveness and pain modulation (Bradesi, S. *et al.*, 2006).

38.6 Summary and Conclusions

Despite its high prevalence, and considerable burden of illness, treatment of IBS remains unsatisfactory. However, considerable progress has been made to identify alterations at different levels of the brain–gut axis which may contribute to characteristic symptoms. In the absence of generally agreed upon animal models of the syndrome, functional brain-imaging techniques in well defined human patient populations has a high promise to identify central abnormalities related to altered stress responsiveness and associated pain modulation circuits. A number of novel treatment approaches aimed at these central abnormalities are currently under development.

References

Bale, T. L. 2005. Sensitivity to stress: Dysregulation of CRF pathways and disease development. Horm. Behav. 48, 1–10.

Balsari, A., Ceccarelli, A., Dubini, F., Fesce, E., and Poli, G. 1982. The fecal microbial population in the irritable bowel syndrome. Microbiologica 5, 185–194.

Barbara, G., Cremon, C., Riccardo, V., Elisabetta, C., Balestra, B., Di Nardo, G., De Giorgio, R., Stanghellini, V., Corinaldesi, R., and Tonini, M. 2005a. Colonic mucosal mast cell mediators from patients with irritable bowel syndrome excite enteric cholinergic motor neurons. Gastroenterology 128, A-58 (abstract).

Barbara, G., Stanghellini, V., De Giorgio, R., and Corinaldesi, R. 2006. Functional gastrointestinal disorders and mast cells: Implications for therapy. Neurogastroenterol. Motil. 18, 6–17.

Barbara, G., Stanghellini, V., De Giorgio, R., Cremon, C., Cottrell, G. S., Santini, D., Pasquinelli, G., Morselli-Labate, A. M., Grady, E. F., Bunnett, N. W., Collins, S. M., and Corinaldesi, R. 2004. Activated mast cells in proximity to colonic nerves correlate with abdominal pain in irritable bowel syndrome. Gastroenterology 126, 693–702.

Barbara, G., Vallance, B. A., and Collins, S. M. 1997. Persistent intestinal neuromuscular dysfunction after acute nematode infection in mice. Gastroenterology 113, 1224–1232.

Barbara, G., Wang, B., Grundy, D., Cremon, C., De Giorgio, R., Di Nardo, G., Trevisani, M., Campi, B., Geppetti, P., Tonini, M., Stanghellini, V., and Corinaldesi, R. 2005b. Mast cells are increased in the colonic mucosa of patients with irritable bowel syndrome and excite visceral sensory pathways. Gastroenterology 128, A-626 (abstract).

Barreau, F., Cartier, C., Ferrier, L., Fioramonti, J., and Bueno, L. 2004. Nerve growth factor mediates alterations of colonic sensitivity and mucosal barrier induced by neonatal stress in rats. Gastroenterology 127, 524–534.

Bennett, E. J., Tennant, C. C., Piesse, C., Badcock, C. A., and Kellow, J. E. 1998. Level of chronic life stress predicts clinical outcome in irritable bowel syndrome. Gut 43, 256–261.

Berman, S. M., Naliboff, B. D., Chang, L., FitzGerald, L., Antolin, T., Camplone, A., and Mayer, E. A. 2002. Enhanced preattentive central nervous system reactivity in irritable bowel syndrome. Am. J. Gastroenterol. 97, 2791–2797.

Bernstein, C. N., Niazi, N., Robert, M., Mertz, H., Kodner, A., Munakata, J., Naliboff, B., and Mayer, E. A. 1996. Rectal afferent function in patients with inflammatory and functional intestinal disorders. Pain 66, 151–161.

Boccalon, S., Scaggiante, B., and Perissin, L. 2006. Anxiety stress and nociceptive responses in mice. Life Sci. 78, 1225–1230.

Bouin, M., Plourde, V., Boivin, M., Riberdy, M., Lupien, F., Laganiere, M., Verrier, P., and Poitras, P. 2002. Rectal distension testing in patients with irritable bowel syndrome: Sensitivity, specificity, and predictive values of pain sensory thresholds. Gastroenterology 122, 1771–1777.

Bradesi, S., Eutamene, H., Garcia-Villar, R., Fioramonti, J., and Bueno, L. 2002. Acute and chronic stress differently affect visceral sensitivity to rectal distension in female rats. Neurogastroenterol. Motil. 14, 75–82.

Bradesi, S., Kokkotou, E., Song, B., Marvizon, J. C., Mittal, Y., McRoberts, J. A., Ennes, H. S., Pothoulakis, C., Ohning, G., and Mayer, E. A. 2005. Sustained visceral hyperalgesia following chronic psychological stress in rats involves up-regulation of spinal NK1 receptors. Gastroenterology 128, A-494 (abstract).

Bradesi, S., Tillisch, K., and Mayer, E. A. 2006. Emerging drugs for IBS. Expert Opin. Emerg. Drugs 11, 293–313.

Camilleri, M. 2004. Treating irritable bowel syndrome: overview, perspective and future therapies. Br. J. Pharmacol. 141, 1237–1248.

Chang, L. 2005. Brain responses to visceral and somatic stimuli in irritable bowel syndrome: A central nervous system disorder? Gastroenterol. Clin. North. Am. 34, 271–279.

Chang, L., Mayer, E. A., Johnson, T., FitzGerald, L., and Naliboff, B. 2000a. Differences in somatic perception in female patients with irritable bowel syndrome with and without fibromyalgia. Pain 84, 297–307.

Chang, L., Munakata, J., Mayer, E. A., Schmulson, M. J., Johnson, T. D., Bernstein, C. N., Saba, L., Naliboff, B., Anton, P. A., and Matin, K. 2000b. Perceptual responses in patients with inflammatory and functional bowel disease. Gut 47, 497–505.

Collins, S. M. 2002. A case for an immunological basis for irritable bowel syndrome. Gastroenterology 122, 2078–2080.

Creed, F., Fernandes, L., Guthrie, E., Palmer, S., Ratcliffe, J., Read, N., Rigby, C., Thompson, D., Tomenson, B. North of England IBS Research Group. 2003. The cost-effectiveness of psychotherapy and paroxetine for severe irritable bowel syndrome. Gastroenterology 124, 303–317.

Crowell, M. D., Harris, L., Jones, M. P., and Chang, L. 2005. New insights into the pathophysiology of irritable bowel syndrome: Implications for future treatments. Curr. Gastroenterol. Rep. 7, 272–279.

Dean, B. B., Aguilar, D., Barghout, V., Kahler, K. H., Frech, F., Groves, D., and Ofman, J. J. 2005. Impairment in work productivity and health-related quality of life in patients with IBS. Am. J. Manag. Care. 11, S17–S26.

Diatchenko, L., Slade, G. D., Nackley, A. G., Bhalang, K., Sigurdsson, A., Belfer, I., Goldman, D., Xu, K., Shabalina, S. A., Shagin, D., Max, M. B., Makarov, S. S., and Maixner, W. 2005. Genetic basis for individual variations in pain perception and the development of chronic pain condition. Hum. Mol. Gen. 14, 135–143.

Drossman, D. A., Camilleri, M., Mayer, E. A., and Whitehead, W. E. 2002. AGA technical review on irritable bowel syndrome. Gastroenterology 123, 2108–2131.

Drossman, D. A., Toner, B. B., Whitehead, W. E., Diamant, N. E., Dalton, C. B., Duncan, S., Emmott, S., Proffitt, V., Akman, D., Frusciante, K., Le, T., Meyer, K., Bradshaw, B., Mikula, K., Morris, C. B., Blackman, C. J., Hu, Y., Jia, H., Li, J. Z., Koch, G. G., and Bangdiwala, S. I. 2003. Cognitive-behavioral therapy versus education and desipramine versus placebo for moderate to severe functional bowel disorders. Gastroenterology 125, 19–31.

Fass, R., Naliboff, B., Higa, L., Johnson, C., Kodner, A., Munakata, J., Ngo, J., and Mayer, E. A. 1998. Differential effect of long-term esophageal acid exposure on mechanosensitivity and chemosensitivity in humans. Gastroenterology 115, 1363–1373.

Fuchs, E. and Fluegge, G. 1995. Modulation of binding sites for corticotropin-releasing hormone by chronic psychosocial stress. Psychoneuroendocrinology 20, 33–51.

Greenwood-Van Meerveld, B., Gibson, M. S., Johnson, A. C., Venkova, K., and Sutkowski-Markmann, D. 2003. NK1 receptor-mediated mechanisms regulate colonic hypersensitivity in the guinea pig. Pharmacol. Biochem. Behav. 74, 1005–1013.

Gue, M., Del Rio-Lacheze, C., Eutamene, H., Theodorou, V., Fioramonti, J., and Bueno, L. 1997. Stress-induced visceral hypersensitivity to rectal distension in rats: Role of CRF and mast cells. Neurogastroenterol. Motil. 9, 271–279.

Gwee, K. A., Collins, S. M., Marshall, J. S., Underwood, J. E., Moochala, S. M., and Read, N. W. 1998. Evidence of inflammatory pathogenesis in post-infectious irritable bowel syndrome. Gastroenterology 114, 758 (abstract).

Gwee, K. A., Graham, J. C., McKendrick, M. W., Collins, S. M., Marshall, J. S., Walters, S. J., and Read, N. W. 1996. Psychometric scores and persistence of irritable bowel after infectious diarrhoea. Lancet 347, 150–153.

Heim, C., Ehlert, U., and Hellhammer, D. H. 2000. The potential role of hypercortisolism in the pathophysiology of stress-related bodily disorders. Psychoneuroendocrinology 25, 1–35.

Hutton, J. 2005. Cognitive behaviour therapy for irritable bowel syndrome. Eur. J. Gastroenterol. Hepatol. 17, 11–14.

Inadomi, J. M., Fennerty, M. B., and Bjorkman, D. 2003. Systematic review: The economic impact of irritable bowel syndrome. Aliment. Pharmacol. Ther. 18, 671–682.

Isgar, B., Harman, M., Kaye, M. D., and Whorwell, P. J. 1983. Symptoms of irritable bowel syndrome in ulcerative colitis in remission. Gut 24, 190–192.

Ito, M. 2002. Controller-regulator model of the central nervous system. J. Integr. Neurosci. 1, 129–143.

Jänig, W. and Habler, H.-J. 2000. Specificity in the Organization of the Autonomic Nervous System: A Basis for Precise

Neural Regulation of Homeostatic and Protective Body Functions. In: The Biological Basis for Mind Body Interactions (ed. E. A. Mayer and C. B. Saper), pp. 351–367. Elsevier Science.

La, J. H., Kim, T. W., Sung, T. S., Kang, J. W., Kim, H. J., and Yang, I. S. 2003. Visceral hypersensitivity and altered colonic motility after subsidence of inflammation in a rat model of colitis. World J. Gastroenterol. 9, 2791–2795.

Ladd, C. O., Huot, R. L., Thrivikraman, K. V., Nemeroff, C. B., Meaney, M. J., and Plotsky, P. 2000. Long-term Behavioral and Neuroendocrine Adaptations to Adverse Early Experience. In: The Biological Basis for Mind Body Interactions (ed. E. A. Mayer and C. B. Saper), pp. 81–103. Elsevier.

Lamb, K., Zhong, F., Gebhart, G. F., and Bielefeldt, K. 2006. Experimental colitis in mice and sensitization of converging visceral and somatic afferent pathways. Am. J. Physiol. Gastrointest. Liver Physiol. 290, G451–G457.

Lembo, T. J. and Fink, R. N. 2002. Clinical assessment of irritable bowel syndrome. J. Clin. Gastroenterol. 35, S31–S36.

Loennig-Baucke, V., Metcalf, A. M., and Shirazi, S. 1989. Anorectal manometry in active and quiescent ulcerative colitis. Am. J. Gastroenterol. 84, 892–897.

Longstreth, G. F. 2005. Definition and classification of irritable bowel syndrome: Current consensus and controversies. Gastroenterol. Clin. North. Am. 34, 173–187.

Lowman, B. C., Drossman, D. A., Cramer, E. M., and McKee, D. C. 1987. Recollection of childhood events in adults with irritable bowel syndrome. J. Clin. Gastroenterol. 9, 324–330.

Madden, J. A. and Hunter, J. O. 2002. A review of the role of the gut microflora in irritable bowel syndrome and the effects of probiotics. Br. J. Nutr. 88, S67–S72.

Mason, P. 2005. Deconstructing endogenous pain modulations. J. Neurophysiol. 94, 1659–1663.

Matthews, S. G. 2002. Early programming of the hypothalamo-pituitary-adrenal axis. Trends Endocrinol. Metab. 13, 373–380.

Mayer, E. A. 2000a. Spinal and supraspinal modulation of visceral sensation. Gut 47, iv69–iv72.

Mayer, E. A. 2000b. The neurobiology of stress and gastrointestinal disease. Gut 47, 861–869.

Mayer, E. A., Berman, S., Suyenobu, B., Labus, J., Mandelkern, M. A., Naliboff, B. D., and Chang, L. 2005. Differences in brain responses to visceral pain between patients with irritable bowel syndrome and ulcerative colitis. Pain 115, 398–409.

Mayer, E. A. and Collins, S. M. 2002. Evolving pathophysiologic models of functional gastrointestinal disorders. Gastroenterology 122, 2032–2048.

Mayer, E. A. and Gebhart, G. F. 1994. Basic and clinical aspects of visceral hyperalgesia. Gastroenterology 107, 271–293.

Mayer, E. A., Naliboff, B. D., Chang, L., and Coutinho, S. V. 2001. Stress and the gastrointestinal tract: V. Stress and irritable bowel syndrome. Am. J. Physiol. Gastrointest. Liver. Physiol. 280, G519–G524.

Mertz, H., Fullerton, S., Naliboff, B., and Mayer, E. A. 1998. Symptoms and visceral perception in severe functional and organic dyspepsia. Gut 42, 814–822.

Mertz, H., Naliboff, B., Munakata, J., Niazi, N., and Mayer, E. A. 1995. Altered rectal perception is a biological marker of patients with irritable bowel syndrome. Gastroenterology 109, 40–52.

Mulak, A. and Bonaz, B. 2004. Irritable bowel syndrome: A model of the brain-gut interactions. Med. Sci. Monit. 10, RA55–RA62.

Naliboff, B. D., Derbyshire, S. W. G., Berman, S. M., Mandelkern, M. A., Chang, L., Stains, J., and

Mayer, E. A. 2001a. Assessing the stability of brain activation to visceral stimulation in IBS. Gastroenterology 120, A133 (abstract).

Naliboff, B. D., Derbyshire, S. W. G., Munakata, J., Berman, S., Mandelkern, M., Chang, L., and Mayer, E. A. 2001b. Cerebral activation in irritable bowel syndrome patients and control subjects during rectosigmoid stimulation. Psychosom. Med. 63, 365–375.

O'Leary, C. and Quigley, E. M. 2003. Small bowel bacterial overgrowth, celiac disease, and IBS: What are the real associations? Am. J. Gastroenterol. 98, 720–722.

O'Sullivan, M., Clayton, N., Breslin, N. P., Harman, I., Bountra, C., McLaren, A., and O'Morain, C. A. 2000. Increased mast cells in the irritable bowel syndrome. Neurogastroenterol. Motil. 12, 449–457.

Petrovic, P. and Ingvar, M. 2002. Imaging cognitive modulation of pain processing. Pain 95, 1–5.

Pezawas, L., Meyer-Lindenberg, A., Drabant, E. M., Verchinski, B. A., Munoz, K. E., Kolachana, B. S., Egan, M. F., Mattay, V. S., Hariri, A. R., and Weinberger, D. R. 2005. 5-HTTLPR polymorphism impacts human cingulate-amygdala interactions: A genetic susceptibility mechanism for depression. Nat. Neurosci. 8, 828–834.

Pimentel, M., Chow, E. J., and Lin, H. C. 2000. Eradication of small intestinal bacterial overgrowth reduces symptoms of irritable bowel syndrome. Am. J. Gastroenterol. 95, 3503–3506.

Pimentel, M., Chow, E. J., and Lin, H. C. 2003. Normalization of lactulose breath testing correlates with symptom improvement in irritable bowel syndrome. A double-blind, randomized, placebo-controlled study. Am. J. Gastroenterol. 98, 412–419.

Porreca, F., Ossipov, M. H., and Gebhart, G. F. 2002. Chronic pain and medullary descending facilitation. Trends Neurosci. 25, 319–325.

Rao, S. S. C., Read, N. W., Davison, P. A., Bannister, J. J., and Holdsworth, C. D. 1987. Anorectal sensitivity and responses to rectal distention in patients with ulcerative colitis. Gastroenterology 93, 1270–1275.

Santos, J., Guilarte, M., Alonso, C., and Malagelada, J. R. 2005. Pathogenesis of irritable bowel syndrome: The mast cell connection. Scand. J. Gastroenterol. 40, 129–140.

Schwetz, I., Bradesi, S., and Mayer, E. A. 2003. Current insights into the pathophysiology of irritable bowel syndrome. Curr. Gastroenterol. Rep. 5, 331–336.

Schwetz, I., Bradesi, S., and Mayer, E. A. 2004. The pathophysiology of irritable bowel syndrome. Minerva. Med. 95, 419–426.

Schwetz, I., McRoberts, J. A., Coutinho, S. V., Bradesi, S., Gale, G., Fanselow, M., Million, M., Ohning, G., Taché, Y., Plotsky, P. M., and Mayer, E. A. 2005. Corticotropin-releasing factor receptor 1 mediates acute and delayed stress-induced visceral hyperalgesia in maternally separated Long Evans rats. Am. J. Physiol. Gastrointest. Liver Physiol. 289, G704–G712.

Simren, M., Axelsson, J., Abrahamsson, H., Svedlund, J., and Bjsrnsson, E. S. 2000. Symptoms of irritable bowel syndrome in inflammatory bowel disease in remission and relationship to psychological factors. Gastroenterology 118, A702. (Abstract).

Soderholm, J. D., Yang, P. C., Ceponis, P., Vohra, A., Riddell, R., Sherman, P. M., and Perdue, M. H. 2002. Chronic stress induces mast cell-dependent bacterial adherence and initiates mucosal inflammation in rat intestine. Gastroenterology 123, 1099–1108.

Spiegel, B. M., Gralnek, I. M., Bolus, R., Chang, L., Dulai, G. S., Mayer, E. A., and Naliboff, B. 2004. Clinical determinants of health-related quality of life in patients with irritable bowel syndrome. Arch. Intern. Med. 164, 1773–1780.

Spiller, R. C. 2005. Irritable bowel syndrome. Br. Med. Bull. 72, 15–29.

Suzuki, R., Rahman, W., Rygh, L. J., Webber, M., Hunt, S. P., and Dickenson, A. H. 2005. Spinal-supraspinal serotonergic circuits regulating neuropathic pain and its treatment with gabapentin. Pain 117, 292–303.

Swidsinski, A., Khilkin, M., Ortner, M., Swidsinski, S., Wirth, J., Weber, J., Schlien, P., Lochs, H. 1999. Alteration of bacterial concentration in colonic biopsies from patients with irritable bowel syndrome (IBS). Gastroenterology 116, A-1. (Abstract).

Taché, Y., Martinez, V., Million, M., and Wang, L. 2001. Stress and the gastrointestinal tract III. Stress-related alterations of gut motor function: Role of brain corticotropin-releasing factor receptors. Am. J. Physiol. Gastrointest. Liver Physiol. 280, G173–G177.

Taché, Y., Martinez, V., Wang, L., and Million, M. 2004. CRF1 receptor signaling pathways are involved in stress-related alterations of colonic function and viscerosensitivity: implications for irritable bowel syndrome. Br. J. Pharmacol. 141, 1321–1330.

Thompson, W. G., Longstreth, G. F., Drossman, D. A., Heaton, K. W., Irvine, E. J., and Müller-Lissner, S. A. 1999. Functional bowel disorders and functional abdominal pain. Gut 45, II43–II47.

Tillisch, K., Labus, J. S., Naliboff, B. D., Bolus, R., Shetzline, M., Mayer, E. A., and Chang, L. 2005. Characterization of the alternating bowel habit subtype in patients with irritable bowel syndrome. Am. J. Gastroenterol. 100, 896–904.

Vendruscolo, L. F., Pamplona, F. A., and Takahashi, R. N. 2004. Strain and sex differences in the expression of nociceptive behavior and stress-induced analgesia in rats. Brain Res. 1030, 277–283.

Verne, G. N., Himes, N. C., Robinson, M. E., Gopinath, K. S., Briggs, R. W., Crosson, B., and Price, D. D. 2003. Central representation of visceral and cutaneous hypersensitivity in the irritable bowel syndrome. Pain 103, 99–110.

Welgan, P., Meshkinpour, H., and Beeler, M. 1988. Effect of anger on colon motor and myoelectric activity in irritable bowel syndrome. Gastroenterology 94, 1150–1156.

Weston, A. P., Biddle, W. L., Bhatia, P. S., and Miner, P. B. J. 1993. Terminal ileal mucosal mast cells in irritable bowel syndrome. Dig. Dis. Sci. 38, 1590–1595.

Whitehead, M. and Holland, P. 2003. What puts children of lone parents at a health disadvantage? Lancet 361, 271.

Wilder-Smith, C. H., Schindler, D., Lovblad, K., Redmond, S. M., and Nirkko, A. 2004. Brain functional magnetic resonance imaging of rectal pain and activation of endogenous inhibitory mechanisms in irritable bowel syndrome patient subgroups and healthy controls. Gut 53, 1595–1601.

Wilson, L. M. and Baldwin, A. L. 1999. Environmental stress causes mast cell degranulation, endothelial and epithelial changes, and edema in the rat intestinal mucosa. Microcirculation 6, 189–198.

Zhuo, M. and Gebhart, G. F. 2002. Facilitation and attenuation of a visceral nociceptive reflex from the rostroventral medulla in the rat. Gastroenterology 122, 1007–1019.

Zimmerman, J. 2003. Extraintestinal symptoms in irritable bowel syndrome and inflammatory bowel diseases: Nature, severity, and relationship to gastrointestinal symptoms. Dig. Dis. Sci. 48, 743–749.

39 Pain in Childbirth

U Wesselmann, The Johns Hopkins University School of Medicine, Baltimore, MD, USA

Glossary

dysmenorrhea Painful cramping of the lower abdomen occurring before or during menses primarily as a result of endogenous prostaglandins, often accompanied by other symptoms such as sweating, tachycardia, headaches, nausea, vomiting, diarrhea, and tremulousness.

dyspareunia Painful intercourse.

epidural analgesia Regional analgesia produced by injection of local anesthetic solution into the peridural space.

postpartum pain Pain after childbirth; the postpartum period starts immediately following childbirth but the length of the period is not well defined.

referred pain Pain perceived in a region different from the injured area during the time of injury.

stages of labor pain There are three stages of labor pain. Stage 1 begins with regular uterine contractions and ends with complete cervical dilatation at 10 cm. Stage 2 begins once the cervix has completely dilated and ends with delivery of the fetus. Stage 3 lasts from the delivery of the fetus until the delivery of the placenta.

39.1 Introduction

The experience of pain during childbirth is a complex, multidimensional response to sensory stimuli generated during labor and delivery. Pain in childbirth occurs in the context of an individual woman's physiology and psychology, as well as in the context of the sociology of the culture and the health care system and its providers surrounding her. For the majority of women in all societies and cultures, natural childbirth is likely to be one of the most painful events in their lifetime. Average pain scores for labor pain are exceeded only by those for causalgia in chronic-pain patients and amputation of a digit in acute-pain patients (Melzack, R., 1984). Different from other acute and chronic pain experiences, pain during childbirth is not associated with a pathological process. It is surprising that this physiological process associated with the most basic and fundamental life experience causes severe pain, and this has been the subject of many philosophic and religious debates. The International Association for the Study of Pain (IASP) defines pain as an unpleasant sensory and emotional experience associated with actual or potential tissue damage or described in terms of such damage (Merskey, H., 1979). Traditionally, different approaches to pain management of the pain in childbirth have addressed either the sensory or the affective dimension of pain, while more recently a multidimensional approach has been advocated, addressing both dimensions of labor pain (Lowe, N. K., 2002; Melzack, R., 1984).

39.2   Physiological Aspects of Labor Pain

Compared to many other pain conditions, research on the neurophysiological mechanisms of pain in childbirth has been sparse, due to the difficulty of developing animal models that could be studied adequately using electrophysiological and neuroanatomical techniques. Studies examining the influence of pregnancy on somatosensory responses in animals and humans have shown hypoalgesia in late pregnancy prior to the onset of labor and this chance in nociceptive threshold is at least in part opioid-mediated (Bajaj, P. et al., 2002; Jarvis, S. et al., 1997). There are three distinct stages of labor pain related to uterine contractions, cervical stretching, and distension of the vaginal canal during fetal descend (for reviews see: McDonald, J. S., 2001; Rowlands, S. and Permezel, M., 1998). The first stage begins with regular uterine contractions and ends with complete cervical dilatation at 10 cm. Stage 1 has been further subdivided into an earlier latent phase and an ensuing active phase, which begins at about 3–4 cm of cervical dilatation and heralds a period of more rapid cervical dilation. Once the cervix has completely dilated, stage 2 of labor has begun. It ends with delivery of the fetus. Stage 3 of labor lasts from the delivery of the fetus until the delivery of the placenta. Pain during the first stage of labor (dilatation phase) is thought to be due to nociceptive stimuli arising from mechanical distension of the lower uterine segment and cervical dilatation. In addition, high-threshold mechanoreceptors in the myometrium may be activated in response to uterine contractions. Several chemical nociceptive mediators have been suggested, including bradykinines, leukotrienes, prostaglandins, serotonin, lactic acid, and Substance P. Pain of the first stage of labor is predominantly mediated by neural pathways involving the T10 to L1 spinal cord segments. Similar to other types of visceral pain, labor pain may present with referred pain to somatic structures in corresponding myotomes and dermatomes, including the abdominal wall, lumbosacral region, iliac crests, gluteal areas and thighs. Pain in the second and third stages of labor involves spinal cord segments S2 to S4, and is considered to be due to distension and traction on pelvic structures surrounding the vaginal vault and from distension of the pelvic floor and perineum. The mean intensity of pain in childbirth has been reported to be positively correlated with the intensity, duration, and frequency of uterine contractions and with the degree of cervical distension.

Pain associated with childbirth provokes a generalized stress response, which has widespread physiological and potentially adverse effects on the progress of labor and the well-being of the mother and the fetus (Brownridge, P., 1995; Beilin, Y., 2002). Respiratory effects include hyperventilation, which might lead to maternal hypocarbia and respiratory alkalosis, and rise in cardiac output, peripheral resistance, and blood pressure. Pain associated with uterine contraction results in stimulation of the release of stress-related hormones from the adrenal sympathetic axis and the hypophyseal–pituitary axis. Labor pain promotes maternal and fetal acidosis, which is due to a catecholamine-induced shift toward lipolytic metabolism, hyperventilation, physical exertion, starvation, and diminished buffering capacity secondary to respiratory alkalosis. While these effects may be largely innocuous during the course of uncomplicated labor, they present a great risk in the presence of certain medical and obstetric complications and in situations where fetal compromise already exists.

39.2.1   Variables Associated with the Severity of Labor Pain

There is a very high level of individual variability in the severity of labor pain and this has been correlated with several factors (Melzack, R., 1984; Brownridge, P., 1995). Primiparas and younger women report more pain than multiparas and older women. Women of higher socioeconomic status report less pain than women of lower socioeconomic status. In addition physical factors might play a role in increased pain ratings during labor, including increased fetal size and increased maternal body weight. Labor pain is influenced by maternal positions – women experience more pain when delivering in the horizontal position as compared to the upright position. Childbirth at night has been reported to be less painful than childbirth during the day. Women with a history of dysmenorrhea report higher pain levels during labor as compared to women, who do not have a history of menstrual pain. In contrast, a previous history of nongynecological pain is correlated with decreased labor pain.

Severe fear of pain associated with childbirth occurs in 6–10% of parturients and is highly correlated to pain levels reported during the first stage of labor (for review see Saisto, T. and Halmesmäki, E., 2003). It is not an isolated variable, but associated with the woman's personal characteristics including general anxiety and fear of pain in general, low self-esteem, depression, dissatisfaction with her partnership, and lack of support. Fear of labor pain is strongly associated with fear of pain in general, independent of parity and is one of the most common reasons for requesting a cesarean section. Fear of childbirth has been reported to complicate about 20% of pregnancies in developed countries.

Although it is often assumed that culture and ethnicity have a significant influence on the intensity of labor pain, numerous studies have documented that there is no difference in self-report pain intensity ratings (see Lowe, N. K., 2002 for review). However, pain behavior is significantly influenced by culture and ethnicity, due to learned values and attitudes to the perception and expression of acute pain.

The environmental influences on pain perception have been explored in two prospective studies from Scandinavia comparing low-risk women delivering at birth centers and at standard obstetrical hospital-based units (Skibsted, L. and Lange, A. P., 1992; Waldenström, U. and Nilsson, C. A., 1994). These studies suggest that the environment provided may affect a woman's ability to cope with pain. While women delivering in birth centers reported significantly higher pain intensities than women delivering in hospital settings, there were no differences between the two groups with respect to satisfaction with the quality of the birth experience. These results emphasize the importance of differentiation between pain intensity and the attitude toward the pain experienced in labor and delivery.

It has been postulated that the memory for pain associated with disease, trauma, or surgical and medical procedures can be more damaging than its initial experience (Song, S.-O. and Carr, D. B., 1999). Although pain in childbirth is one of the most intense pains many women experience in their lifetime, childbirth is one of the most positive events of life for most women. Review of the literature on memory for labor pain shows that while memory of the events of childbirth is very accurate, the accuracy of recalled labor pain remains in question. Memories of labor pain can evoke intense negative reactions in a few women, but are more likely to give rise to positive consequences related to coping, self-efficacy and self-esteem (Niven, C. A. and Murphy-Black, T., 2000).

It is important to emphasize that labor pain, although being a very prominent aspect of childbirth, is just one aspect of the childbirth experience. A recent systematic review of the literature on pain and women's satisfaction with the experience of childbirth demonstrated that the influences of pain, pain relief, and intra-partum medical interventions on subsequent satisfaction are not as powerful as the personal expectations, the amount of support from caregivers, the quality of the caregiver-patient relationship, and involvement of decision making (Hodnett, E. D., 2002).

39.2.2 Treatment of Pain in Childbirth

39.2.2.1 Analgesia for Labor and Delivery

Analgesia for labor and delivery is now safer than ever and can be offered during all stages of labor, targeted to the individual needs and wishes of the pregnant woman, without compromising the safety of her or her unborn child (for reviews see Beilin, Y., 2002; Caton, D. et al., 2002; Nystedt, A. et al., 2004). Anesthesia related maternal mortality has decreased from 4.3 million live births during the years 1979–81 to 1.7 per million live births during the years 1988–90. The increased use of regional anesthesia techniques is partially responsible for this decrease in mortality. The most popular method for analgesia during labor and delivery is epidural analgesia using a combination of local anesthetics and opioids. Its popularity is related to its efficacy and safety. Drugs can be applied as continuous epidural infusions or as patient-controlled epidural analgesia. During the early stages of labor dilute solutions of local anesthetic can be used to achieve analgesia. As labor progresses, more concentrated solutions of local anesthetics can be used and opioids can be added. Typically the epidural catheter is inserted to maintain a low dermatomal level of analgesia for vaginal delivery (T10 to L1). If a cesarean section is required, the dermatomal level can be raised to T4. Combined spinal epidural techniques offer the advantage of a very rapid onset of analgesia with minimal motor block. The future of obstetric anesthesia lies in refining currently available drugs and techniques to make obstetric anesthesia even safer and more efficacious.

39.2.2.2 Nonpharmacological, complementary and alternative therapies for relief of pain in childbirth

Many nonpharmacological, complementary and alternative medicine methods exist to relieve labor pain (for reviews see Simkin, P. P. and O'Hara M., 2002; Smith, C. A. *et al.*, 2003; Cluett, E. R. *et al.*, 2004; Cyna, A. M. *et al.*, 2004; Huntley, A. L. *et al.*, 2004; Lee, H. and Ernst, E., 2004). These methods appeal to women and caregivers who are interested to reduce labor pain without creating potentially serious side-effects and high costs. In addition many women appreciate the simplicity of these approaches and the sense of control they gain from actively managing their pain. However, few of these therapies have been subjected to proper scientific study. Meta-analysis of randomized controlled trials has indicated that acupuncture and hypnosis may be beneficial for the management of pain in childbirth. There is evidence that water immersion during the first stage of labor reduces the use of analgesia and reported pain intensity, without adverse outcomes on labor duration, operative delivery, or neonatal outcomes. The effects of water immersion during the later stages of labor are not clear. No differences were observed for women receiving aromatherapy, music, or audio analgesia.

39.3 Pain during Pregnancy

While the focus of this review is on pain associated with childbirth, it is important to note that pain is also a significant issue during pregnancy (Stuge, B. *et al.*, 2003). Approximately 50% of women experience back pain and/or pelvic pain during their pregnancy, and in up to 15% of the cases the pain is rated as severe. Several hypotheses have been suggested to explain the etiology of this pain, including increased weight, decreased stability of the pelvic girdle due to hormonal changes and referred visceral pain mechanisms. The role of the hormone relaxin and pelvic pain in pregnancy is controversial. Physical therapy is often recommended for the prevention and treatment of these pregnancy-related pains, but the effects of these interventions have not been systematically studied and thus it is not clear if they are of any benefit to the pregnant women.

39.4 Postpartum Pain

While most of the discussion of adverse sequelae of labor and delivery has focused on urinary and fecal incontinence, there is now increasing awareness that pain associated with childbirth is not only related to the process of labor and delivery. Women report significant pains in many sites of the body after delivery, a phenomenon which has been defined as postpartum pain. Postpartum pains may include urogenital, pelvic, back and breast pains as well as headaches, persisting for months to years after labor and delivery (Audit Commission, 1997). It has been hypothesized that anesthetic techniques during labor and delivery and gonadal hormonal changes may play a role in the etiology of these pain complaints. Forty-nine percent of women report significant dyspareunia when resuming sexual intercourse after childbirth (Buhling, K. J. *et al.*, 2005). Women with a history of operative vaginal delivery have the highest prevalence of severe perineal pain when resuming sexual intercourse. The persistence of dyspareunia for longer than 6 months after delivery ranges from 3.4% to 14% based on mode of delivery. Treatment approaches for early intervention to prevent persistent perineal pain after childbirth have not been explored in detail. A meta-analysis assessing the effectiveness of topically applied anesthetics to the perineal region in the early postpartum period showed no compelling evidence of pain reduction (Hedayati, H. *et al.*, 2005). However, there has been no evaluation of the long-term effects of topically applied local anesthetics.

39.5 Future Aspects

Pain in childbirth is a complex, multidimensional experience. While the focus of this chapter is on the pain aspect of childbirth, it is important to keep in mind that pain is an important aspect, but not the only aspect of the childbirth experience. The acknowledgment of the existence of pain associated with labor and delivery and the recognition of the severity of this pain by the medical community over the last 50 years has resulted in a broad spectrum of pharmacological and nonpharmacological pain management options that can be offered to the parturient today. There have been tremendous advancements in the pharmacological treatment of pain in labor and delivery over the last 25 years, mainly due to improved regional anesthesia techniques. The future of obstetric anesthesia lies in refining currently available drugs and techniques and in developing new drugs targeted at specific pathophysiological mechanisms to make obstetric anesthesia even safer and more efficacious. This will require translational research efforts ranging from basic science to clinical research. As a first step an animal model has recently been developed in the rat to study the neurophysiological and neuropharmacological mechanisms of pain associated with uterine cervical distension (Liu, B. *et al.*, 2005). Neuroanatomical studies of the sensory pathways from the uterine cervix have shown that P2X3-receptor-expressing sensory neurons might play a role during birthing and signal nociceptive information such as labor pain (Papka, R. E. *et al.*, 2005). Numerous nonpharmacological and alternative treatments are available for the treatment of pain in childbirth. While there is experience with some of those approaches already for centuries, many of these treatments have not been subjected to proper scientific study. Clinical studies assessing the effects of these interventions on pain in childbirth and on a woman's satisfaction with the childbirth experience, as well as studies assessing the effects on the safety for the pregnant woman and her child are urgently needed.

Acknowledgments

Ursula Wesselmann is supported by NIH grants DK57315 (NIDDK), DK066641 (NIDDK), and HD39699 (NICHD, Office of Research for Women's Health).

References

Audit Commission 1997. First Class Delivery: improving Maternity Services in England and Wales. Audit Commission.

Bajaj, P., Madsen, H., Moller, M., and Arendt-Nielsen, L. 2002. Antenatal women with or without pelvic pain can be characterized by generalized or segmental hypoalgesia in late pregnancy. J. Pain 3, 451–460.

Beilin, Y. 2002. Advances in labor analgesia. Mt. Sinai J. Med. 69, 38–44.

Brownridge, P. 1995. The nature and consequences of childbirth pain. Eur. J. Obstet. Gynecol. Reprod. Biol. 59 Suppl, S9–15.

Buhling, K. J., Schmidt, S., Robinson, J. N., Klapp, C., Siebert, G., and Dudenhausen, J. W. 2005. Rate of dyspareunia after delivery in primiparae according to mode of delivery. Eur. J. Obstet. Gynecol. Reprod. Biol.124, 42–46.

Caton, D., Corry, M. P., Frigoletto, F. D., Hopkins, D. P., Lieberman, E., Mayberry, L., Rooks, J. P., Rosenfield, A.,

Sakala, C., Simkin, P., and Young, D. 2002. The nature and management of labor pain: executive summary. Am. J. Obstet. Gynecol. 186, S1–15.

Cluett, E. R., Nikodem, V. C., McCandlish, R. E., and Burns, E. E. 2004. Immersion in water in pregnancy, labor and birth. Cochrane Database Syst. Rev. CD000111.

Cyna, A. M., McAuliffe, G. L., and Andrew, M. I. 2004. Hypnosis for pain relief in labor and childbirth: a systematic review. Br. J. Anaesth. 93, 505–511.

Hedayati, H., Parsons, J., and Crowther, C. A. 2005. Topically applied anaesthetics for treating perineal pain after childbirth. Cochrane Database Syst. Rev. CD004223.

Hodnett, E. D. 2002. Pain and women's satisfaction with the experience of childbirth: a systematic review. Am. J. Obstet. Gynecol. 186, S160–172.

Huntley, A. L., Coon, J. T., and Ernst, E. 2004. Complementary and alternative medicine for labor pain: a systematic review. Am. J. Obstet. Gynecol. 191, 36–44.

Jarvis, S., McLean, K. A., Chirnside, J., Deans, L. A., Calvert, S. K., Molony, V., and Lawrence, A. B. 1997. Opioid-mediated changes in nociceptive threshold during pregnancy and parturition in the sow. Pain 72, 153–159.

Lee, H. and Ernst, E. 2004. Acupuncture for labor pain management: A systematic review. Am. J. Obstet. Gynecol. 191, 1573–1579.

Liu, B., Eisenach, J. C., and Tong, C. 2005. Chronic estrogen sensitizes a subset of mechanosensitive afferents innervating the uterine cervix. J. Neurophysiol. 93, 2167–2173.

Lowe, N. K. 2002. The nature of labor pain. Am. J. Obstet. Gynecol. 186, S16–24.

McDonald, J. S. 2001. Pain of Childbirth. In: Bonica's Management of Pain, 3rd edn. (ed. J. D. Loeser), pp. 1388–1414. Lippincott Williams & Wilkins.

Melzack, R. 1984. The myth of painless childbirth (the John J. Bonica lecture). Pain 19, 321–337.

Merskey, H. 1979. Pain terms: a list with definitions and a note on usage. Recommended by the IASP Subcommittee on Taxonomy. Pain 6, 249–252.

Niven, C. A. and Murphy-Black, T. 2000. Memory for labor pain: a review of the literature. Birth 27, 244–253.

Nystedt, A., Edvardsson, D., and Willman, A. 2004. Epidural analgesia for pain relief in labor and childbirth – a review with a systematic approach. J. Clin. Nurs. 13, 455–466.

Papka, R. E., Hafemeister, J., and Storey-Workley, M. 2005. P2X receptors in the rat uterine cervix, lumbosacral dorsal root ganglia, and spinal cord during pregnancy. Cell Tissue Res. 321, 35–44.

Rowlands, S. and Permezel, M. 1998. Physiology of pain in labor. Baillieres Clin. Obstet. Gynaecol. 12, 347–362.

Saisto, T. and Halmesmaki, E. 2003. Fear of childbirth: a neglected dilemma. Acta Obstet. Gynecol. Scand. 82, 201–208.

Simkin, P. P. and O'Hara, M. 2002. Nonpharmacologic relief of pain during labor: systematic reviews of five methods. Am. J. Obstet. Gynecol. 186, S131–159.

Skibsted, L. and Lange, A. P. 1992. The need for pain relief in uncomplicated deliveries in an alternative birth center compared to an obstetric delivery ward. Pain 48, 183–186.

Smith, C. A., Collins, C. T., Cyna, A. M., and Crowther, C. A. 2003. Complementary and alternative therapies for pain management in labor. Cochrane Database Syst. Rev. CD003521.

Song, S-O. and Carr, D. B. 1999. Pain and memory. IASP Clin. Updates 7, 1–4.

Stuge, B., Hilde, G., and Vollestad, N. 2003. Physical therapy for pregnancy-related low back and pelvic pain: a systematic review. Acta Obstet. Gynecol. Scand. 82, 983–990.

Waldenström, U. and Nilsson, C. A. 1994. Experience of childbirth in birth center care. A randomized controlled study. Acta Obstet. Gynecol. Scand. 73, 547–554.

40 Urothelium as a Pain Organ

L A Birder, University of Pittsburgh School of Medicine, Pittsburgh, PA, USA

40.1 Urothelial Cells: Detectors of Mechanical, Chemical, and Thermal Stimuli

The urothelium is a specialized lining of the urinary tract, extending from the renal pelvis to the urethra. The urothelium is composed of at least three layers: a basal cell layer attached to a basement membrane, an intermediate layer, and a superficial apical layer with large hexagonal cells (diameters of 25–250 μm) which are also termed umbrella cells (Lewis, S. A., 2000; Apodaca, G., 2004). The umbrella cells, and, perhaps, intermediate cells may have projections to the basement membrane (Martin, B.F., 1972; Hicks M., 1975; Apodaca, G., 2004). Basal cells, which are thought to be precursors for other cell types, normally exhibit a low (3–6 months) turnover rate; however, accelerated proliferation can occur in pathology. For example, using a model (protamine sulfate) that selectively damages the umbrella cell layer has shown that the urothelium rapidly undergoes both functional and structural changes in order to restore the barrier following injury (Lavelle, J. et al., 2002).

While the urothelium has been historically viewed primarily as a barrier, it is becoming increasingly appreciated as a responsive structure capable of detecting physiological and chemical stimuli, and releasing a number of signaling molecules. A number of investigators have described the release of diffusible mediators from the urothelium which could influence urinary bladder function (Ferguson, D.R. et al., 1997; Hawthorn, M.H. et al., 2000; Burnstock, G., 2001; Chess-Williams, R., 2002). There is now abundant evidence which indicates that urothelial cells display a number of properties similar to sensory neurons (nociceptors/mechanoreceptors) (see Table 1; Lewis, S. A. and Hanrahan, J. W., 1985; Birder, L. et al., 1998; 2004; Smith, P. R. et al., 1998; Birder, L. et al., 2001; 2002a; 2002b; Chess-Williams, R., 2002; Beckel, J. et al., 2004; Birder, L., et al., 2004; Chess-Williams, R., 2004; Burnstock, G. and Knight, G. E., 2004; Murray, E. et al., 2004; Tempest, H. V. et al., 2004; Chopra, B. et al., 2005; Birder, L. et al., 2007) and that both types of cells use diverse signal transduction mechanisms to detect physiological stimuli.

40.1.1 Sensor Molecules Expressed in Urothelium Which Could Contribute to Bladder Pain

One example of a urothelial sensor molecule is the TRP channel, TRPV1, known to play a prominent role in nociception and in urinary bladder function (Szallasi, A., 2001). It is well established that painful sensations induced by capsaicin, the pungent substance in hot peppers, are caused by stimulation of TRPV1, an ion channel protein (Caterina, M. J. et al., 1997; Caterina, M. J., 2001) which is activated by capsaicin as well as by moderate heat, protons, and lipid metabolites such as anandamide (endogenous ligand of both cannabinoid and vanilloid receptors). TRPV1 is expressed throughout the afferent limb of the micturition reflex pathway, Figure 1, including urinary bladder unmyelinated (C-fiber) nerves that detect bladder distension or the presence of irritant chemicals (Chancellor M. B. and de Groat, W. C., 1999). In the urinary bladder, TRPV1 is not only expressed by afferent nerves or myofibroblasts that form close contact with urothelial cells, but also by the urothelial cells themselves (Birder, L. et al., 2001). Urothelial TRPV1 receptor expression correlates with the sensitivity to vanilloid compounds, as exogenous application of capsaicin or resiniferatoxin to cultured cells increases intracellular calcium and

Table 1　Examples of sensor molecules (i.e., receptors/ion channels) associated with neurons that have been identified in urothelial cells

Sensor function/stimuli	Urothelial sensor molecules	Neuronal sensor molecules
ATP	P2X/P2Y	P2X/P2Y
Capsaicin resiniferatoxin	TRPV1	TRPV1
Heat	TRPV1; TRPV2; TRPV4	TRPV1; TRPV2; TRPV3; TRPV4
Cold	TRPM8; TRPA1	TRPM8; TRPA1
H^+	TRPV1; ?	TRPV1; ASIC; DRASIC
Osmolarity	In part TRPV4	In part TRPV4
Bradykinin	B1; B2	B1; B2
Acetylcholine	Nicotinic/muscarinic	Nicotinic/muscarinic
Norepinephrine	α/β subtypes	α/β subtypes
Nerve growth factor	p75/trkA	p75/trkA
Mechanosensitivity	Amiloride sensitive Na^+ channels	Amiloride sensitive Na^+ channels

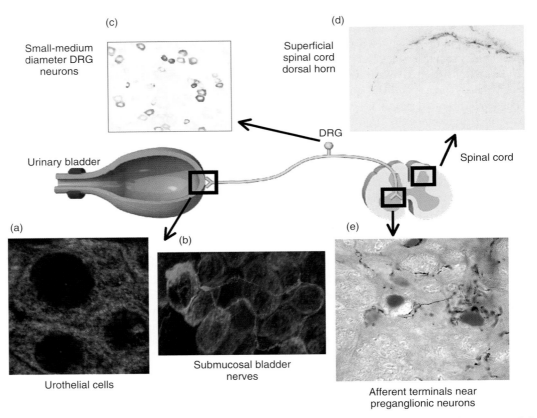

(c) Small-medium diameter DRG neurons

(d) Superficial spinal cord dorsal horn

DRG

Urinary bladder

Spinal cord

(a)

(b)

(e)

Urothelial cells

Submucosal bladder nerves

Afferent terminals near preganglionic neurons

Figure 1　TRPV1 is expressed throughout the afferent limb of the micturition reflex pathway. TRPV1-immunoreactivity (cy-3, red) in basal epithelial cells (cyt 17, FITC green) (a); in nerve fibers (cy-3, red) located in close proximity to basal cells (FITC, green) (b) (punctate TRPV1 staining in urothelial cells was electronically subtracted to facilitate imaging of the TRPV1-immunoreactive nerve fiber); in small to medium diameter dorsal root ganglion (DRG) neurons (c); and in superficial regions of the spinal cord dorsal horn (d) (staining indicate TRPV1 nerve fibers). (e) TRPV1-positive afferent terminals (in black) are localized in close proximity to pre-ganglionic neurons (PGN neurons labeled by injection of a tracer dye, fast blue, into the major pelvic ganglion). Reproduced from Birder, L. A. 2005. More than just a barrier: Urothelium as a drug target for urinary bladder pain. Am. J. Physiol. Renal. Physiol. 289(3): F489–F495, used with permission from the American Physiological Society.

evokes transmitter (NO, nitric oxide; ATP, adenosine triphosphate) release (Birder, L. *et al.*, 2001; 2002a). In neurons, TRPV1 is thought to integrate/amplify the response to various stimuli playing an essential role in the development of inflammation-induced hyperalgesia (Ghuang, H. H. *et al.*, 2001; Holzer, P., 2004). Thus, it seems likely that urothelial TRPV1 might participate in a similar manner, in the detection of irritant stimuli following bladder inflammation or infection.

Anatomically normal, TRPV1-null mice exhibited a number of alterations in bladder urothelial cell function including a reduction of *in vitro*, stretch-evoked ATP release and membrane capacitance as well as a decrease in hypotonic-evoked ATP release (Birder, L. *et al.*, 2002a). These findings demonstrate that the functional significance of TRPV1 in the bladder extends beyond pain sensation to include participation in normal bladder function, and is essential for normal mechanically evoked purinergic signaling by the urothelium.

In addition to TRPV1, urothelial cells also express additional TRP channels, including TRPV2, TRPV4, TRPM8, and TRPA1. In contrast to TRPV1, TRPV2, and TRPV4, which are detectors of warm temperatures (TRPV4 can also be gated by hypotonic stimuli) (Liedtke, W. *et al.*, 2000; Alessandri-Haber, N. *et al.*, 2003; Chung, M. K. *et al.*, 2003), TRPM8 and TRPA1 have been shown to be activated by cold (25–28 °C) temperatures as well as by cooling agents (menthol, icilin) and are expressed in a subset of sensory neurons as well as in non-neural cells. This expression suggests that these cells express a range of thermoreceptors underlying both cold and heat stimuli (Stein, R. J. *et al.*, 2004). While the functional role of these thermosensitive channels in urothelium remains to be clarified, it seems likely that a primary role for these proteins may be to recognize noxious stimuli in the bladder. However, the diversity of stimuli which can activate these proteins suggests a much broader sensory and/or cellular role. For example, TRPM8 expression has been shown to be increased in some epithelia in malignant disorders (prostate tumors), suggesting a role in proliferating cells (Tsavaler, L. *et al.*, 2001). Thus, further studies are needed to fully elucidate the role of TRP channels in urothelium and their influence on bladder function.

Purinergic receptors, which are activated by ATP and related nucleotides, are known to play an important role in bladder function and chronic pain (Burnstock, G., 2001; North, R. A., 2004). Two families of purinergic receptors have been identified, P2X and P2Y, both of which are expressed in urothelial cells (Lee, H.Y. *et al.*, 2000; Birder, L. *et al.*, 2004; Tempest, H. V. *et al.*, 2004). Although the function of purinergic receptors in nonexcitable cells is less clear than in afferent neurons, the presence of such receptors may be associated with cell proliferation, apoptosis, secretion, and sensory transduction (Coutinho-Silva R. *et al.*, 2005; Greig, A. V. *et al.*, 2003). In the urinary bladder, recent studies have shown that urothelial-derived ATP release can act as a trigger for exocytosis, in part, via autocrine activation of urothelial purinergic (P2X; P2Y) receptors (Wang, E. C. *et al.*, 2005). This type of signaling may be similar to that in airway epithelium, where nucleotides released to the epithelial surface may act in a paracrine/autocrine manner to regulate ion transport and/or other functions via interactions with luminal epithelial purinergic receptors (Poulsen, A. N. *et al.*, 2005).

40.1.2 Response to Stimuli: Transducer Function of Urothelial Cells

Release of chemical mediators (NO, ATP, acetylcholine (ACh); substance P; prostaglandins (PG)) (Ferguson, D. R. *et al.*, 1997; Birder, L. *et al.*, 1998; Burnstock, G., 2001; Birder, L. *et al.*, 2003; Chess-Williams, R., 2004) from urothelial cells suggests that these cells exhibit specialized sensory and signaling properties that could allow reciprocal communication with neighboring urothelial cells as well as nerves or other cells (i.e., immune, myofibroblasts, inflammatory) in the bladder wall. Recent studies have shown that both afferent as well as autonomic axons are located in close proximity to the urothelium (Wayabayashi, Y. *et al.*, 1995; Birder, L. *et al.*, 2001). For example, peptide and TRPV1 immunoreactive nerve fibers have been found localized throughout the urinary bladder musculature and in a plexus beneath and extending into the urothelium. Confocal microscopy revealed that TRPV1 immunoreactive nerve fibers are in close association with basal urothelial cells such that their fluorescent signals overlapped within 0.5 μm optical sections. This type of communication suggests that these cells may be targets for transmitters released from bladder nerves or other cells, or that chemicals released by urothelial cells may also alter the excitability of bladder nerves.

Studies have demonstrated that ATP released from urothelium during bladder stretch could contribute to activation of bladder afferents.

That ATP released from urothelial cells during stretch can activate a population of suburothelial bladder afferents expressing $P2X_3$ receptors, signaling changes in bladder fullness and pain supports this idea (Ferguson, D. R. *et al.*, 1997; Burnstock, G., 2001). Accordingly, $P2X_3$ null mice exhibit a urinary bladder hyporeflexia, suggesting that this receptor and neural–epithelial interactions are essential for normal bladder function (Cockayne, D. A. *et al.*, 2000). This type of regulation may be similar to epithelial-dependent secretion of mediators in airway epithelial cells which is thought to modulate submucosal nerves and bronchial smooth muscle tone and may play an important role in inflammation (Homolya, L. *et al.*, 2000; Jallat-Daloz, I. *et al.*, 2001). Thus, it is possible that activation of bladder nerves and urothelial cells can modulate bladder function directly or indirectly via the release of chemical factors in the urothelial layer.

40.2 How Might Urothelial Cells Influence Pain Processes?

Recent evidence has demonstrated that urothelial cells exhibit plasticity whereby inflammation or injury can alter the expression and/or sensitivity of a number of urothelial-sensor molecules. Examples include changes in urothelial expression of various receptors including tyrosine kinase (trk), low-affinity nerve growth factor (p75), bradykinin, TRPV1, protease-activated receptors (PPARs), and purinergic receptor (P2X and P2Y) subtypes in animal models as well as in patients diagnosed with a number of bladder disorders including neurogenic bladder and interstitial cystitis (IC) – a chronic clinical disease characterized by urgency, frequency, and bladder pain upon filling (an innocuous stimulus) (Murray, E. *et al.*, 2004; North, R. A., 2004; Tempest, H. V. *et al.*, 2004; Chopra, B. *et al.*, 2005; Dattilio, A. and Vizzard, M. A., 2005). Such changes could contribute to pain and hypersensitivity exhibited in these painful syndromes. Because urothelial cells appear to exhibit sensory function, it is possible that plasticity in urothelial receptors is then linked with pain in various bladder syndromes.

Sensitization can be triggered by various mediators (ATP, NO, nerve growth factor (NGF), PGE_2) which may be released by both neuronal and non-neuronal cells (urothelial cells, fibroblasts, mast cells) located near the bladder luminal surface. An important component of the inflammatory response is ATP release from various cell types including urothelium, which can initiate painful sensations by exciting purinergic (P2X) receptors on sensory fibers (Cockayne, D. A. *et al.*, 2000; Burnstock, G., 2001). Recently, it has been shown in sensory neurons that ATP can potentiate the response of vanilloids by lowering the threshold for protons, capsaicin, and heat (Tominaga, M. *et al.*, 2001). This represents a novel mechanism by which large amounts of ATP released from damaged or sensitized cells in response to injury or inflammation may trigger the sensation of pain. These findings have clinical significance and suggest that alterations in afferents or epithelial cells in pelvic viscera may contribute to the sensory abnormalities in a number of pelvic disorders, such as IC, which is consistent with augmented release of ATP in urothelial cells from some patients with IC (Sun, Y. *et al.*, 2001). A comparable disease in cats is termed feline interstitial cystitis (FIC) (Buffington, C. A. *et al.*, 1999; 2001), which is also accompanied by alterations in stretch-evoked release of urothelially derived ATP (Birder, L. *et al.*, 2003).

Though the urothelium maintains a tight barrier to ion and solute flux, a number of factors such as tissue pH, mechanical or chemical trauma, or bacterial infection can modulate this barrier function of the urothelium (Hicks, M., 1975; Anderson, G. *et al.*, 2003). When this function is compromised during injury or inflammation, it can result in the passage of toxic substances into the underlying tissue (neural/muscle layers) resulting in urgency, frequency, and pain during bladder distension. For example, inflammation, injury (spinal cord transection), or IC, all of which increase endogenously generated levels of NO, increase permeability to water/urea in addition to producing ultrastructural changes in the apical layer (Lavelle, J. *et al.*, 2000; Apodaca, G. *et al.*, 2003). Although the mechanism is unknown, these findings may be similar to that in other epithelia where excess production of NO has been linked to changes in epithelial integrity (Han, X. *et al.*, 2004).

Disruption of epithelial integrity may also be due to substances such as antiproliferative factor (APF), which has been shown to be secreted by bladder epithelial cells from IC patients and can inhibit epithelial proliferation thereby adversely affecting barrier function (Keay, S. *et al.*, 1999; 2004).

40.3 Potential Clinical Implications: Urothelial Receptors/Release Mechanisms as Targets for Drug Treatment

It is conceivable that the effectiveness of some agents currently used in the treatment of bladder disorders may involve urothelial receptors and/or release mechanisms. For example, intravesical instillation of vanilloids (capsaicin or resiniferatoxin) improves urodynamic parameters in patients with neurogenic detrusor overactivity and reduces bladder pain in patients with hypersensitivity disorders, presumably by desensitizing bladder nerves (Szallasi, A. and Fowler, C. J., 2002; Kim, J. H. *et al.*, 2003). This treatment could also target TRPV1 on urothelial cells, whereby a persistent activation might lead to receptor desensitization or depletion of urothelial transmitters. Recent studies have demonstrated that intradetrusor injection with botulinum neurotoxin type A (BoNTA) is an effective treatment for bladder hypersensitivity disorders including neurogenic detrusor overactivity (Harper, M. *et al.*, 2004; Reitz, A. and Schurch, B., 2004). Following injection, the toxin binds to bladder cholinergic nerve terminals and cleaves the protein, SNAP25, necessary for exocytosis and release of acetylcholine (Harper M. *et al.*, 2004). There is evidence the BoNTA can suppress the release of a number of mediators (acetylcholine, ATP, and neuropeptides) from both neural and non-neural cells (Morris, J. L. *et al.*, 2001). Suppression of neurotransmitter release from urothelium would serve to blunt afferent activity driven by urothelial-derived release of mediators in a number of lower urinary tract dysfunctions. These findings suggest that urothelial cells exhibit specialized sensory and signaling properties that could allow them to respond to their chemical and physical environments and to engage in reciprocal communication with neighboring urothelial cells as well as nerves within the bladder well. Taken together, pharmacologic interventions aimed at targeting urothelial receptor/ion channel expression or release mechanisms may provide a new strategy for the clinical management of bladder disorders.

Acknowledgments

This work was supported by grants to L.A.B. from the NIH (RO1 DK54824 and RO1 DK57284).

References

Alessandri-Haber, N., Yeh, J. J., Boyd, A. E., Parada, C. A., Chen, X., Reichling, D. B., and Levine, J. D. 2003. Hypotonicity induces TRPV4-mediated nociception in rat. Neuron 39, 497–511.

Anderson, G., Palermo, J., Schilling, J., Roth, R., Heuser, J., and Hultgren, S. 2003. Intracellular bacterial biofilm-like pods in urinary tract infections. Science 301, 105–107.

Apodaca, G. 2004. The uroepithelium: not just a passive barrier. Traffic 5, 1–12.

Apodaca, G., Kiss, S., Ruiz, W. G., Meyers, S., Zeidel, M. L., and Birder, L. 2003. Disruption of barrier epithelium function after spinal cord injury. Am. J. Physiol. 284, F966–F970.

Barrick, S. R., Lee, H., Meyers, S., Caterina, M. J., Kanai, A. J., Zeidel, M. L., Chopra, B., de Groat, W. C., and Birder, L. 2003 Expression and function of TRPV4 in urinary bladder urothelium. Soc. Neurosci. Abstr. 26, 608.6.

Beckel, J., Barrick, S. R., Keast, J.R., Meyers, S., Kanai, A. J., de Groat, W. C., Zeidel, M. L., and Birder, L. 2004. Expression and function of urothelial muscarinic receptors and interactions with bladder nerves. Soc. Neurosci. 26, 846.23.

Birder, L., Apodaca, G., de Groat, W. C., and Kanai, A. J. 1998. Adrenergic- and capsaicin-evoked nitric oxide release from urothelium and afferent nerves in urinary bladder. Am. J. Physiol. 275, F226–F229.

Birder, L., Kanai, A. J., de Groat, W. C., Kiss, S., Nealen, M. L., Burke, N. E., Dineley, K. E., Watkins, S., Reynolds, I. J., and Caterina, M. J. 2001. Vanilloid receptor expression suggests a sensory role for urinary bladder epithelial cells. Proc. Natl. Acad. Sci. U. S. A. 98, 13396–13401.

Birder, L., Kullmann, A., Lee, H., Barrick, S., de Groat, W., Kanai, A., and Caterina, M. 2007. Activation of Urothelial-TRPV4 by 4α-PDD contributes to altered bladder reflexes in the rat. JPET (in press).

Birder, L., Nakamura, Y., Kiss, S., Nealen, M. L., Barrick, S. R., Kanai, A. J., Wang, E., Ruiz, W. G., de Groat, W. C., Apodaca, G., Watkins, S., and Caterina, M. J. 2002a. Altered urinary bladder function in mice lacking the vanilloid receptor TRPV1. Nat. Neurosci. 5, 856–860.

Birder, L., Nealen, M. L., Kiss, S., de Groat, W. C., Caterina, M. J., Wang, E., Apodaca, G., and Kanai, A. J. 2002b. Beta-adrenoceptor agonists stimulate endothelial nitric oxide synthase in rat urinary bladder urothelial cells. J. Neurosci. 22, 8063–8070.

Birder, L., Barrick, S. R., Roppolo, J. R., Kanai, A. J., de Groat, W. C., Kiss, S., and Buffington, C. A. 2003. Feline interstitial cystitis results in mechanical hypersensitivity and altered ATP release from bladder urothelium. *American* J. Physiol. 285, F423–F429.

Birder, L., Ruan, H. Z., Chopra, B., Xiang, Z., Barrick, S. R., Buffington, C. A., Roppolo, J. R., Ford, A. P., de Groat, W. C., and Burnstock, G. 2004. Alterations in P2X and P2Y purinergic receptor expression in urinary bladder from normal cats and cats with interstitial cystitis. Am. J. Physiol. 287, F1084–F1091.

Buffington, C. A. 2001. Visceral pain in humans: lessons from animals. Curr. Pain Headache Rep. 5, 44–51.

Buffington, C. A., Chew, D. J., and Woodworth, B. E. 1999. Feline interstitial cystitis. J. Am. Vet. Med. Assoc. 215, 682–687.

Burnstock, G. 2001. Purine-mediated signalling in pain and visceral perception. Trends Pharmacol. Sci. 22, 182–188.

Burnstock, G. and Knight, G. E. 2004. Cellular distribution and functions of P2 receptor subtypes in different systems. Int. Rev. Cytol. 240, 301–304.

Caterina, M. J. 2001. The vanilloid receptor: a molecular gateway to the pain pathway. Ann. Rev. Neurosci. 24, 602–607.

Caterina, M. J., Schumacher, M. A., Tominaga, M., Rosen, T. A., Levine, J. D., and Julius, D. 1997. The capsaicin receptor: a heat activated ion channel in the pain pathway. Nature 389, 816–824.

Chancellor, M. B. and de Groat, W. C. 1999. Intravesical capsaicin and resiniferatoxin therapy: spicing up the ways to treat the overactive bladder. J. Urol. 162, 3–11.

Chess-Williams, R. 2002. Muscarinic receptors of the urinary bladder: detrusor, urothelial and prejunctional. Auton. Autacoid. Pharmacol. 22, 133–145.

Chess-Williams, R. 2004. Potential therapeutic targets for the treatment of detrusor overactivity. Expert Opin. Ther. Targets 8, 95–106.

Chopra, B., Barrick, S. R., Meyers, S., Beckel, J., Zeidel, M. L., Ford, A. P., de Groat, W. C., and Birder, L. A. 2005. Expression and function of bradykinin B1 and B2 receptors in normal and inflamed rat urinary bladder urothelium. J. Physiol. 562, 859–871.

Chung, M. K., Lee, H., and Caterina, M. J. 2003. Warm temperatures activate TRPV4 in mouse 308 keratinocytes. J. Biol. Chem. 278, 32037–32046.

Cockayne, D. A., Hamilton, S. G., Zhu, Q. M., Dunn, P. M., Zhong, Y., Novakovic, S., Malmberg, A. B., Cain, G., Berson, A., Kassotakis, L., Hedley, L., Lachnit, W. G., Burnstock, G., McMahon, S. B., and Ford, A. P. D. W. 2000. Urinary bladder hyporeflexia and reduced pain-related behaviour in P2X(3)-deficient mice. Nature 407, 1011–1015.

Coutinho-Silva, R., Stahl, L., Cheung, K. K., de Campos, N. E., de Oliveira Souza, C., Ojcius, D. M., and Burnstock, G. 2005. P2X and P2Y purinergic receptors on human intestinal epithelial carcinoma cells: effects of extracellular nucleotides on apoptosis and cell proliferation. Am. J. Physiol. 288, G1024–G1035.

Dattilio, A. and Vizzard, M. A. 2005. Up-regulation of protease activated receptors in bladder after cyclophosphamide induced cystitis and colocalization with capsaicin receptor (VR1) in bladder nerve fibers. J. Urol. 173, 635–639.

Ferguson, D. R., Kennedy, I., and Burton, T. J. 1997. ATP is released from rabbit urinary bladder epithelial cells by hydrostatic pressure changes – a possible sensory mechanism? J. Physiol. 505.2, 503–511.

Ghuang, H. H., Prescott, E. D., Kong, H., Shields, S., Jordt, S. E., Basbaum, A. I., Chao, M. V., and Julius, D. 2001. Bradykinin and nerve growth factor release the capsaicin receptor from PtdIns(4,5)P$_2$-mediated inhibition. Nature 411, 957–962.

Greig, A. V., Linge, C., Terenghi, G., McGrouther, D. A., and Burnstock, G. 2003. Purinergic receptors are part of a functional signaling system for proliferation and differentiation of human epidermal keratinocytes. J. Invest. Dermatol. 120, 1007–1015.

Han, X., Fink, M. P., Uchiyama, T., Yang, R., and Delude, R. L. 2004. Increased iNOS activity is essential for pulmonary epithelial tight junction dysfunction in endotoxemic mice. Am. J. Physiol. 286, L259–L267.

Harper, M., Fowler, C. J., and Dasgupta, P. 2004. Botulinum toxin and its applications in the lower urinary tract. BJU Int. 93, 702–706.

Hawthorn, M. H., Chapple, C. R., Cock, M., and Chess-Williams, R. 2000. Urothelium-derived inhibitory factor(s) influences on detrusor muscle contractility in vitro. Br. J. Pharmacol. 129, 416–419.

Hicks, M. 1975. The mammalian urinary bladder: an accommodating organ. Biol. Rev. 50, 215–246.

Holzer, P. 2004. TRPV1 and the gut: from a tasty receptor for a painful vanilloid to a key player in hyperalgesia. Eur. J. Pharmacol. 500, 231–241.

Homolya, L., Steinberg, T. H., and Boucher, R.. C. 2000. Cell to cell communication in response to mechanical stress via

bilateral release of ATP and UTP in polarized epithelia. J. Cell Biol. 150, 1349–1360.

Jallat-Daloz, I., Cognard, J. L., Badet, J. M., and Regnard, J. 2001. Neural-epithelial cell interplay: in vitro evidence that vagal mediators increase PGE$_2$ production by human nasal epithelial cells. Allerg. Asthma Proc. 22, 17–23.

Keay, S., Warren, J. W., Zhang, C. O., Tu, L. M., Gordon, D. A., and Whitmore, K. E. 1999. Antiproliferative activity is present in bladder but not renal pelvic urine from interstitial cystitis patients. J. Urol. 162, 1487–1489.

Keay, S. K., Szekely, Z., Conrads, T. P., Veenstra, T. D., Barchi, J. J., Zhang, C. O., Koch, K. R., and Michejda, C. J. 2004. An antiproliferative factor from interstitial cystitis patients is a frizzled 8 protein-related sialoglycopeptide. Proc. Natl. Acad. Sci. U. S. A. 101, 11803–11808.

Kim, J. H., Rivas, D. A., Shenot, P. J., Green, B., Kennelly, M., Erickson, J. R., O'Leary, M., Yoshimura, N., and Chancellor, M. B. 2003. Intravesical resiniferatoxin for refractory detrusor hyperreflexia: a multicenter, blinded, randomized, placebo-controlled trial. J. Spinal Cord Med. 26, 358–363.

Lavelle, J., Meyers, S., Ramage, R., Bastacky, S., Doty, D., Apodaca, G., and Zeidel, M. L. 2002. Bladder permeability barrier: recovery from selective injury of surface epithelial cells. Am. J. Physiol. 283, F242–F253.

Lavelle, J., Meyers, S., Ruiz, W. G., Buffington, C. A., Zeidel, M. L., and Apodaca, G. 2000. Urothelial pathophysiological changes in feline interstitial cystitis: a human model. Am. J. Physiol. 278, F540–F553.

Lee, H. Y., Bardini, M., and Burnstock, G. 2000. Distribution of P2X receptors in the urinary bladder and the ureter of the rat. J. Urol. 163, 2002–2007.

Lewis, S. A. 2000. Everything you wanted to know about the bladder epithelium but were afraid to ask. Am. J. Physiol. 278, F867–F874.

Lewis, S. A. and Hanrahan, J. W. 1985. Apical and basolateral membrane ionic channels in rabbit urinary bladder epithelium. Pflugers Arch 405, S83–S88.

Liedtke, W., Choe, Y., Marti-Renom, M. A., Bell, A. M., Denis, C. S., and Sali, A. 2000. Vanilloid receptor-related osmotically activated channel (VR-OAC), a candidate vertebrate osmoreceptor. Cell 103, 525–535.

Martin, B. F. 1972. Cell replacement and differentiation in transitional epithelium: a histological and autoradiographic study of the guinea-pig bladder and ureter. J. Anat. 112, 433–455.

Morris, J. L., Jobling, P., and Gibbins, I. L. 2001. Differential inhibition by botulinum neurotoxin A of cotransmitters released from autonomic vasodilator neurons. Am. J. Physiol. 281, H2124–H2127.

Murray, E., Malley, S. E., Qiao, L. Y., Hu, V. Y., and Vizzard, M. A. 2004. Cyclophosphamide induced cystitis alters neurotrophin and receptor kinase expression in pelvic ganglia and bladder. J. Urol. 172, 2434–2439.

North, R. A. 2004. P2X3 receptors and peripheral pain mechanisms. J. Physiol. 554, 301–308.

Poulsen, A. N., Klausen, T. L., Pedersen, P. S., Willumsen, N. J., and Frederidsen, O. 2005. Regulation of ion transport via apical purinergic receptors in intact rabbit airway epithelium. Pflugers Arch. 450(4), 227–235.

Reitz, A. and Schurch, B. 2004. Intravesical therapy options for neurogenic detrusor overactivity. Spinal Cord 42, 267–272.

Smith, P. R., Mackler, S. A., Weiser, P. C., Brooker, D. R., Ahn, Y. J., Harte, B. J., McNulty, K. A., and Kleyman, T. R. 1998. Expression and localization of epithelial sodium channel in mammalian urinary bladder. Am. J. Physiol. 274, F91–F96.

Stein, R. J., Santos, S., Nagatomi, J., Hayashi, Y., Minnery, B. S., Xavier, M., Patel, A. S., Nelson, J. B., Futrell, W. J.,

Yoshimura, N., Chancellor, M. B., and deMiguel, F. 2004. Cool (TRPM8) and hot (TRPV1) receptors in the bladder and male genital tract. J. Urol. 172, 1175–1178.

Sun, Y., Keay, S., DeDeyne, P. G., and Chai, T. C. 2001. Augmented stretch activated adenosine triphosphate release from bladder uroepithelial cells in patients with interstitial cystitis. J. Urol. 166, 1951–1956.

Szallasi, A. 2001. Vanilloid receptor ligands: hopes and realities for the future. Drugs Aging 18, 561–573.

Szallasi, A. and Fowler, C. J. 2002. After a decade of intravesical vanilloid therapy: still more questions than answers. Lancet Neurol. 1, 167–172.

Tempest, H. V., Dixon, A. K., Turner, W. H., Elneil, S., Sellers, L. A., and Ferguson, D. R. 2004. P2X and P2Y receptor expression in human bladder urothelium and changes in interstitial cystitis. BJU Int. 93, 1344–1388.

Tominaga, M., Wada, M., and Masu, M. 2001. Potentiation of capsaicin receptor activation by metabotropic ATP

receptors: a possible mechanism for ATP-evoked pain and hypersensitivity. Proc. Natl. Acad. Sci. USA 29, 820–825.

Tsavaler, L., Shapero, M. H., Morkowski, S., and Laus, R. 2001. TRP-p8, a novel prostate specific gene, is up regulated in prostate cancer and other malignancies and shares high homology with transient receptor potential calcium channel proteins. Cancer Res. 61, 3760–3769.

Wang, E. C., Lee, J. M., Ruiz, W. G., Balestreire, E. M., von Bodungen, M., Barrick, S., Cockayne, D. A., Birder, L. A., and Apodaca, G. 2005. ATP and purinergic-receptors dependent membrane traffic in bladder umbrella cells. J. Clin. Invert. 115, 2412–2422.

Wayabayashi, Y., Kojima, Y., Makiura, Y., Tomoyoshi, T., and Maeda, T. 1995. Acetylcholinesterase positive axons in the mucosa of urinary bladder of adult cats: retrograde tracing and degeneration studies. Histol. Histopathol. 10, 523–530.

41 The Brainstem and Nociceptive Modulation

M M Heinricher, Oregon Health & Science University, Portland, OR, USA

S L Ingram, Washington State University, Vancouver, WA, USA

Glossary

cyclooxygenase An enzyme–protein complex that catalyzes the production of prostaglandins from arachidonic acid. Cyclooxygenase activity is inhibited by nonsteroidal anti-inflammatory drugs, such as aspirin and ibuprofen.

GIRK G-protein-activated inwardly rectifying potassium channel. A potassium-selective channel that is gated by activated betagamma subunits of Gi/Go G proteins.

Kv channel Voltage-gated potassium channel.

12-lipoxygenase An enzyme that catalyzes the conversion of arachidonic acid to the hydroxyeicosenoic acid (12-HETE) structure.

NSAIDs Nonsteroidal anti-inflammatory drugs. Drugs that inhibit cyclooxygenases and the production of prostaglandins involved in the inflammatory response (e.g., aspirin, ibuprofen, naproxen).

phospholipase A$_2$ An enzyme that catalyes the formation of arachidonic acid from phospholipids.

41.1 The Periaqueductal Gray–Rostral Ventromedial Medulla Pain-Modulating System

The idea that there are well-defined central systems that selectively modulate nociception is usually traced to the demonstration that electrical stimulation in the periaqueductal gray (PAG) region of rats produced potent analgesia. This phenomenon came to be called stimulation-produced analgesia when subsequent work confirmed this finding using quantitative tests of nociception in animals, and when such stimulation was shown to produce clinical analgesia in humans. This was an important advance in documenting that the brain itself could regulate the processing of nociceptive information. Inspired by stimulation-produced analgesia, a significant research effort led to the definition of a brainstem pain modulatory network with critical links in the rostral ventromedial medulla (RVM) as well as the PAG (Figure 1). The antinociception resulting from stimulation in these structures is in large part due to regulation of nociceptive processing at the level of the spinal cord. The PAG–RVM system mediates the analgesic actions of opioids, and it is recruited by internal and environmental challenges. Accumulating evidence from neuroimaging studies supports a role for this system in top-down modulation of pain in humans, such as that produced by placebo or shifts in attention. An overview of descending control, including that mediated by the PAG–RVM system, is provided in Chapter Descending Control Mechanisms. The purpose of this chapter is to review the properties of pain-modulating neurons within the PAG and RVM, and to place this circuitry in a behavioral context.

41.1.1 Functional Characterization of the Periaqueductal Gray–Rostral Ventromedial Medulla Pain-Modulating System

The PAG is a cell-rich region surrounding the cerebral aqueduct in the midbrain. The RVM is defined functionally, rather than cytoarchitecturally, and includes the nucleus raphe magnus and adjacent reticular formation. Numerous behavioral studies have demonstrated that nonselective activation of neurons within the PAG or RVM produces a potent antinociception. Activation of PAG–RVM output neurons must mediate this antinociception, as direct excitation of either PAG or RVM neurons produces antinociception and inhibition of noxious-evoked activity of dorsal horn neurons, whereas inactivation of either the PAG or RVM does not. The PAG–RVM modulatory network is also an important substrate for opioid analgesia. Focal application of morphine or mu-opioid agonists at either site is sufficient to produce antinociception comparable to that resulting from systemic morphine administration (see Fields, H. L. *et al.*, 2005 or Heinricher, M. M. and Morgan, M. M., 1999 for reviews).

Along with the evidence that this system mediates the analgesic actions of exogenous and endogenous opioids, the fact that the net behavioral effect of nonselective experimental activation of the PAG or RVM is antinociception led to a general view of the PAG–RVM system as an analgesia system. This system is indeed activated by acute stress and opioid analgesics to inhibit spinal nociceptive processing. However, there is now increasing evidence that the system, especially the

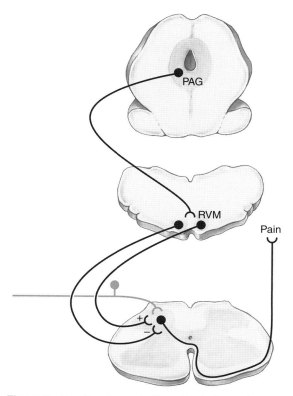

Figure 1 Functional organization of brainstem pain-modulating system with links in midbrain periaqueductal gray (PAG) and rostral ventromedial medulla (RVM). The PAG projects to the RVM. The RVM in turn regulates spinal nociceptive circuitry via a projection through the dorsolateral funiculus to the dorsal horn. This system exerts bidirectional control, and separate populations of RVM neurons mediate descending inhibition and descending facilitation. The PAG is reciprocally linked with forebrain structures including prefrontal cortex, amygdala, and hypothalamus. These substantial interconnections provide an anatomical substrate through which emotional and cognitive variables could influence nociception via the PAG–RVM system.

RVM, also facilitates nociception. The RVM has been implicated in hyperalgesia and allodynia associated with inflammation, nerve injury, acute opioid withdrawal, chronic opioid administration, and the sickness response (see Sections 41.2.1 and 41.3.2, also Porreca, F. *et al.*, 2002 and Heinricher, M. M. *et al.*, 2003 for recent reviews). The PAG–RVM circuit should therefore be viewed not specifically as an analgesia system, but more generally, as a pain-modulation system. From this perspective, the system has the potential for graded enhancement or inhibition of nociception under different conditions.

41.1.2 Pain Modulation as Part of Adaptive Responses to Behavioral and Physiological Challenges

One of the earliest hints that stimulation-produced analgesia was not an experimental curiosity was the demonstration of environmental or stress-induced analgesia. Stress-induced analgesia refers to the fact that any number of situations or experimental procedures that could be characterized as stressful induce behaviorally measurable, and in some cases quite potent, analgesia. For example, stress-induced analgesia can be produced by electric shock, forced swim, and centrifugal rotation as well as biologically relevant threat stimuli such as odors from stressed animals of the same species or exposure to a predator. Analgesia is also elicited as a conditioned response to cues that have been paired previously with noxious or aversive events. This analgesia can be opioid or nonopioid in nature and has been shown by a number of investigators to be mediated by the PAG–RVM system (Bodnar, R. J. *et al.*, 1980; Lewis, J. W. *et al.* 1980; Watkins, L. R. *et al.* 1982; Fanselow, M. S. 1986).

In addition to antinociception, electrical or chemical stimulation of the PAG also evokes autonomic changes commonly associated with cardiovascular aspects of defense, including hypertension (in anesthetized animals) and altered heart rate (Lovick, T. A., 1993; Bandler, R. *et al.*, 2000; Morgan, M. M. and Carrive, P., 2001). Patients in whom deep brain electrodes were implanted in the rostral midbrain/PAG region for relief of intractable pain often found the stimulation to be distinctly disquieting, with reports of a feeling of impending doom or a desire to flee when the stimulation was activated (Nashold, B. S. *et al.*, 1969). Consistent with these reports in humans, rats will work to terminate electrical stimulation in the dorsal PAG (Kiser, R. S. *et al.*, 1978). Moreover, stimulation-produced antinociception in rats is accompanied by species-specific defense behaviors, which include immobility and escape or attack (Bandler, R. and DePaulis, A., 1991; Fanselow, M. S. 1991; Morgan, M. M. and Carrive, P. 2001; Carrive, P. and Morgan, M. M. 2004).

Two factors contributed to the idea that antinociception is recruited as part of defense behaviors: the fact that antinociception is readily evoked by learned or innate danger signals and the observation that stress induces analgesia through activation of the PAG–RVM system. Pain behaviors must sometimes be inhibited in order to give higher precedence to more pressing needs such as escaping from an aggressor or avoiding detection by a predator. However, a more general view of pain modulation has now developed. Pain inhibition is currently viewed as one component of a number of organized responses that allow an organism to prioritize nociceptive behaviors relative to other internal and external demands. In addition to antinociception, such responses typically include autonomic, endocrine, and motor elements. One example of such an organized defense response would be antinociception as part of preparation for fight or flight when confronted by a predator. Although the circuitry through which the PAG–RVM system is brought into play in response to threat remains to be fully elucidated, inputs from the amygdala and hypothalamus are likely to be critical (Helmstetter, F. J., 1992; Lovick, T. A., 1993; Lumb, B. M. *et al.*, 2002).

There is also evidence that the PAG–RVM system is engaged as part of the response to general immune challenge, as occurs with systemic bacterial infection. Thus, systemic administration of lipopolysaccharide (LPS), a well-accepted model for bacterial infection and sickness, results in changes in nociception mediated at least in part by the RVM (Watkins, L. R. *et al.*, 1994; Romanovsky, A. A. *et al.*, 1996). Just as antinociception is viewed as an adaptive response to external threat, hyperalgesia associated with sickness may promote recuperative behaviors and facilitate healing. As with the defense response to external threat, the system is engaged during sickness to prioritize processing of nociceptive information in accord with other physiological and behavioral goals.

It is likely that the PAG–RVM system also mediates the more subtle shifts in pain processing that occur in the absence of extreme challenge. For example, reflexes or other more highly organized behaviors evoked by noxious stimuli would be expected to interfere with feeding or other homeostatic behaviors. Nociceptive threshold is increased in hungry cats given access to food (Casey, K. L. and Morrow, T. J. 1988; 1989). Moreover, there appears to be an equilibrium between responding to noxious inputs and the need to maintain energy balance. Feeding is suppressed in favor of pain behaviors during the first phase of the formalin response, generally thought to represent a relatively intense sensation. By contrast, pain behaviors are reduced in favor of feeding during the second, less intense, phase of the formalin response (LaGraize, S. C. *et al.*, 2004a). Similarly, paw withdrawal to noxious heat is attenuated during micturition, presumably allowing complete emptying of the bladder without interruption by noxious-evoked movements (Baez, M. A. *et al.* 2005).

Distraction or more global variables such as positive or negative mood also alter pain (see Villemure, C. and Bushnell, M. C., 2002 for a review). Similarly, expectation that a particular manipulation will relieve pain can lead to inhibition of pain (Price, D. D. *et al.*, 1999). Understanding the neural mechanisms through which these and other higher psychological factors modulate pain has been challenging, as these variables are at best difficult to study in nonhuman subjects. Moreover, some modulation of pain perception undoubtedly occurs within thalamocortical circuits. However, imaging studies in humans have now demonstrated that the PAG is activated in placebo and distraction paradigms. There are also changes in activation of this system and associated structures when subjects anticipate pain, and in functional pain disorders (see Tracey, I., in press for a recent review). Studies of opioid-dependent placebo have been most compelling. Petrovic P. *et al.* (2002) found that a placebo procedure (infusion of an inactive vehicle in subjects who had been given the powerful opiate remifentanil in a separate trial) resulted in reduced pain reports and activation of the PAG. This region had previously been activated by the exogenous opiate, suggesting that the placebo manipulation recruited endogenous opioid circuitry within the PAG. Wager T. D. *et al.* (2004) saw similar activation of the anterior cingulate and PAG in a conditioned placebo paradigm. These two studies suggest that recruitment of the PAG contributes to the pain inhibition produced by placebo. Similar studies investigating the effects of attention on pain responses also suggest that activation of the PAG–RVM system contributes to pain suppression when attention is redirected from the noxious stimulus to inputs that have greater behavioral significance (Tracey, I. *et al.*, 2002; Valet, M. *et al.*, 2004). Although correlative, these imaging studies suggest that the PAG–RVM system plays a role in pain modulation produced by placebo, shifts in attention and presumably other cognitive and affective variables.

41.2   The Periaqueductal Gray

The PAG comprises heterogeneous cell populations surrounding the cerebral aqueduct. It extends rostrally from the pericoerulear area of the pons to the opening of the third ventricle. As described above, electrical stimulation or focal application of opioids in the PAG produces behaviorally measurable antinociception. However, in addition to pain modulation, it integrates a variety of complex functions, including cardiovascular and other aspects of defense, general autonomic control, reproductive behaviors, and vocalization (Bandler, R. and Shipley, M. T., 1994).

41.2.1 The Periaqueductal Gray and Facilitation of Nociception

The idea that the PAG–RVM system exerts bidirectional control is now well accepted (Porreca, F. *et al.*, 2002; Heinricher, M. M. *et al.*, 2003). Although the focus of functional studies of descending facilitation has been primarily on the RVM (see below, Section 41.3.2), there is some evidence that the PAG also contributes to hyperalgesia under some conditions.

First, the PAG contributes to normal levels of responsiveness on the formalin test and to enhanced nociceptive behaviors after foot shock, suggesting a net facilitating output from this region in these paradigms (McLemore, S. *et al.*, 1999; Berrino, L. *et al.*, 2001). Dorsolateral PAG may be important in a generalized sensory sensitization induced by footshock (Crown, E. D. *et al.*, 2004), a facilitation that is recruited in parallel with the well-documented descending inhibition evoked by footshock. Second, focal application of capsaicin or bradykinin into the PAG has a pronociceptive effect (Burdin, T. A. *et al.*, 1992; McGaraughty, S. *et al.*, 2003). Third, prostaglandin release within the PAG apparently recruits descending facilitation, possibly under conditions of inflammation or systemic infection (Vanegas, H. and Tortorici, V. 2002). Thus, cyclooxygenase is constitutively expressed in PAG, and direct microinjection of cyclooxygenase inhibitors in this region reduces responding on cutaneous and visceral nociceptive tests. Moreover, direct application of prostaglandin E_2 in the PAG produces behavioral hyperalgesia and activates RVM neurons that facilitate nociception (Heinricher, M. M. *et al.*, 2004a). Taken together, these findings document the ability of the PAG to facilitate nociception. Consistent with this idea, a recent imaging study demonstrated that stimulation of an area of capsaicin-induced secondary hyperalgesia in human subjects was associated with activation of the PAG (Zambreanu, L. *et al.*, 2005).

41.2.2 Afferents to the Periaqueductal Gray

The PAG integrates information from all levels of the central nervous system. The spinal cord sends direct, somatotopically organized projections from the dorsal horn and the intermediate gray to the

ventrolateral and lateral PAG (Menetrey, D. *et al.*, 1982; Yezierski, R. P. and Mendez, C. M. 1991). These afferents convey innocuous and noxious information from cutaneous, musculoskeletal, and visceral structures (Wiberg, M. *et al.*, 1987; Yezierski, R. P., 1988; Yezierski, R. P. and Mendez, C. M., 1991; Clement, C. I. *et al.*, 2000). Inputs from the lumbosacral cord arise from a number of regions, including those receiving afferents from pelvic and pudendal nerves (Vanderhorst, V. G. *et al.*, 1996). These latter afferents are proposed to provide important sensory information contributing to PAG functions in sexual behavior and micturition. Approximately half of the spinal input to the PAG is from the upper cervical spinal cord, but the significance of this cervical predominance has not been determined (Vanderhorst, V. G. *et al.*, 2002).

The most substantial afferent inputs to the PAG are from forebrain, including prefrontal and agranular insular cortices as well as the amygdala and hypothalamus (Bandler, R. and Shipley, M. T., 1994; An, X. *et al.*, 1998; Floyd, N. S. *et al.*, 2000). Connections from medial prefrontal cortex are part of a medial prefrontal network associated with visceromotor control through links with the hypothalamus (Ongur, D. and Price, J. L., 2000). Stimulation in the anterior cingulate has been reported by different laboratories to facilitate or suppress nociceptive responses (Hardy, S. G., 1985; Fuchs, P. N. *et al.*, 1996; Calejesan, A. A. *et al.*, 2000). Lesions of the anterior cingulate are generally agreed to reduce nociception in animal models (Pastoriza, L. N. *et al.*, 1996; Donahue, R. R. *et al.*, 2001; LaGraize, S. C. *et al.*, 2004b; in press), consistent with reports in human patients (Foltz, E. L. and Lowell, E. W. 1962; Davis, K. D. *et al.*, 1994; Talbot, J. D. *et al.*, 1995). However, the antinociceptive effect of lesions is presumably due to interference with cortical processing, rather than to activation of descending control. Effects of experimental manipulation of the ventrolateral orbital cortex (VLO) on nociception are similarly complex. Stimulation of the VLO has been reported by one group of investigators to facilitate nociception (Hutchison, W. D. *et al.*, 1996) and another to suppress nociception (Zhang, Y. Q. *et al.*, 1997). The antinociceptive effect of VLO stimulation was blocked by lesion of the PAG, indicating that the connections from this cortex to the PAG are relevant to nociceptive modulation (Zhang, Y. Q. *et al.*, 1997).

Hypothalamic afferents to the PAG are predominantly from the lateral hypothalamus and the anterior hypothalamus/medial preoptic area (MPO)

(Shipley, M. T. *et al.*, 1991; Semenenko, F. M. and Lumb, B. M. 1992; Behbehani, M. M. and Da Costa Gomez, T. M., 1996). At least some of the anterior hypothalamic neurons projecting to the PAG are nociceptive (Lumb, B. M. *et al.*, 2002; Parry, D. M. *et al.*, 2002). The MPO has important roles in autonomic regulation including thermoregulation and fever, as well as in sleep, mating, and maternal behaviors. Electrical stimulation of the MPO produces *c-fos* labeling throughout the PAG (Rizvi, T. A. *et al.*, 1996) and activates PAG–RVM output neurons (Semenenko, F. M. and Lumb, B. M., 1999). Direct application of prostaglandin E_2 in the MPO at low doses produces hyperalgesia that is likely mediated by the PAG–RVM system (Hosoi, M. *et al.*, 1997; Abe, M. *et al.*, 2001; Heinricher, M. M. *et al.*, 2004b).

The central nucleus of the amygdala has massive reciprocal connections with the PAG (Rizvi, T. A. *et al.*, 1991; da Costa Gomez, T. M. and Behbehani, M. M., 1995). The central nucleus receives inputs from the basolateral nucleus, which is a major target of cortical afferents to the amygdala. The central nucleus of the amygdala also receives nociceptive input from the spinal cord, both directly and indirectly via the parabrachial nucleus (Burstein, R. and Potrebic, S., 1993; Gauriau, C. and Bernard, J. F., 2004). PAG neurons respond to stimulation in the amygdala (da Costa Gomez, T. M. *et al.*, 1996), and both stimulation and injection of morphine into the amygdala result in antinociception that is dependent on activation of the descending pathway from PAG to RVM (Pavlovic, Z. W. *et al.*, 1996; Helmstetter, F. J. *et al.*, 1998; McGaraughty, S. and Heinricher, M. M., 2002; 2004).

Afferents to the PAG from the brainstem arise from the medulla, including the RVM. Other brainstem sources of inputs to the PAG include the nucleus tractus solitarius, adjacent nucleus cuneiformis, pontine reticular formation, and the locus coeruleus and other catecholaminergic nuclei (Beitz, A. J., 1982; Herbert, H. and Saper, C. B., 1992). The strong input to the PAG from forebrain and hypothalamus provides an anatomical substrate through which emotional and cognitive variables could influence pain via the PAG. Information related to motivation and emotion thus converges with direct and indirect visceral and somatic afferent input in the PAG. This convergence is presumably important to pain modulation in allowing the organism to regulate nociceptive processing in accord with current needs and motivational state, particularly in response to environmental or internal challenge.

41.2.3 Efferents from the Periaqueductal Gray

The PAG has only a sparse projection to the spinal cord, but a dense projection to the RVM, which in turn projects to the dorsal horn via the dorsolateral funiculus. Inactivation of the RVM prevents the antinociceptive effects of PAG manipulations, indicating that the connection from the PAG to the RVM is the neuroanatomical basis for descending modulation of nociception by the PAG. PAG–RVM projection neurons express neuropeptides, excitatory amino acids, and serotonin. Functional studies implicate excitatory amino acids, serotonin, and endogenous opioids in recruitment of the RVM by the PAG to produce antinociception. A second relay through which the PAG is likely to influence spinal nociceptive processing is pontine catecholaminergic cell groups, as ultrastructural studies demonstrate that these neurons receive inputs from the ventrolateral PAG (see Fields, H. L. *et al.*, 2005 for a recent review).

41.2.4 Columnar Organization of the Periaqueductal Gray

Based on both connectivity and function, the PAG has been subdivided into rostrocaudally oriented columns, designated as dorsomedial (or dorsal), dorsolateral, lateral, and ventrolateral (Carrive, P. and Morgan, M. M., 2004). These columns roughly correspond with subdivisions based on cytoarchitecture that were proposed previously by Beitz A. J. (1985) and Beitz A. J. and Shepard R. D. (1985). Cortical, hypothalamic, and spinal afferents show a preferential distribution to different columns, although there is substantial spread across columnar boundaries (Rizvi, T. A. *et al.*, 1992; An, X. *et al.*, 1998; Floyd, N. S. *et al.*, 2000). Projections to the RVM arise in the dorsomedial, lateral, and ventrolateral columns, but not the dorsolateral column (Abols, I. A. and Basbaum, A. I., 1981; Van Bockstaele, E. J. *et al.*, 1991).

Pain modulation is not constrained by columnar boundaries, and electrical stimulation or microinjection of mu-opioid agonists or neuroexcitant agents throughout the dorsoventral extent of the PAG produces analgesia, particularly at more caudal levels (Waters, A. J. and Lumb, B. M., 1997 Carrive, P. and Morgan, M. M., 2004; Morgan, M. M. and Clayton, C. C. 2005). Antinociception resulting from stimulation in the ventrolateral PAG must be mediated by an endogenous opioid connection, as it is blocked by naloxone. By contrast, the antinociception resulting from electrical stimulation more dorsally is not mediated by endogenous opioids (Morgan, M. M. 1991).

Antinociception evoked by stimulation of the dorsal and lateral columns is often associated with escape-like behaviors or defensive posturing, hypertension, and tachycardia. By contrast, antinociception produced by stimulation of the ventrolateral PAG is often accompanied by immobility, bradycardia, and in anesthetized animals, hypotension (Depaulis, A. et al., 1994; Morgan, M. M. and Carrive, P., 2001). These findings gave rise to the proposal that the lateral/dorsolateral columns function specifically in active coping, particularly when confronted with an external threat. By contrast, the ventrolateral column was hypothesized to be important in passive coping and recuperation, for example, in response to visceral pain or some other challenge that cannot be controlled or escaped (Keay, K. A. and Bandler, R., 2001).

Two aspects of this framework for PAG organization have been controversial. The first issue is the interpretation of immobility evoked by stimulation in the ventrolateral column. It is not possible to determine from lack of movement alone whether an animal should be viewed as engaging in recuperative behaviors or whether it is instead freezing, that is, exhibiting an active cryptic defense that reduces the likelihood of detection by a predator (Blanchard, R. J. et al. 1986; Fanselow, M. S., 1991; Bittencourt, A. S. et al., 2004). The argument that immobility represents passive coping rather than a defensive response is based largely on observations that stimulation in the ventrolateral column in anesthetized animals produce hypotension (Keay, K. A. and Bandler, R., 2001). However, hypotension is not evoked by such stimulation in awake behaving animals (Morgan, M. M. and Carrive, P., 2001). Furthermore, hypotension may not imply a quiescence or recuperative state. For example, although the freezing in a conditioned fear paradigm is associated with hypertension (Carrive, P. 2000), defensive freezing to various threat stimuli can be associated with profound hypotension in various species of wild-trapped rodents (Hofer, M. A., 1970).

A second contentious issue is whether active offensive and escape behaviors should be associated specifically with the lateral/dorsolateral columns, whereas immobility is a function of the ventrolateral column. Immobility as well as escape can be evoked with stimulation of the lateral/dorsolateral columns. Notably, the current needed to induce immobility is lower than that required for active flight behaviors with stimulation in this column (Bittencourt, A. S. et al., 2004). Conversely, bursts of running are often interspersed with periods of inactivity during antinociception evoked from the ventrolateral column (Vianna, D. M. et al., 2001; Morgan, M. M. and Clayton, C. C., 2005). As in the lateral/dorsolateral columns, the current needed to elicit freezing with stimulation in the ventrolateral PAG is lower than that required to elicit jumping (Vianna, D. M. et al., 2001). Moreover, although lesions of the ventrolateral PAG reduce freezing and release running behavior in response to threat (Walker, P. and Carrive, P., 2003, Farook, J. M. et al., 2004), lesions of the dorsolateral PAG have no effect on either freezing or running evoked by exposure to a cat or to a context associated with footshock. In addition, changes in heart rate and activity in this paradigm are seen only when the animal is denied access to a safe hiding place (Dielenberg, R. A. et al., 2004; Farook, J. M. et al., 2004; Leman, S. et al., 2003). Thus, neither active fight or flight behaviors nor immobility appears to be a specific output of a particular column.

These observations suggest that although the idea that the lateral/dorsolateral and ventral columns of the PAG represent centers for active and passive coping has had great heuristic value, a full understanding of the functional significance of the PAG columns is likely to require sophisticated analyses of the neural circuitry of these regions in relevant behavioral paradigms.

41.2.5 Intrinsic Circuitry and Neurotransmitters in the Periaqueductal Gray

41.2.5.1 Endogenous opioids

All three opioid receptors, mu (MOR), delta (DOR), and kappa (KOR) opioid receptors, are moderately to densely expressed in the PAG (Mansour, A. et al., 1995; Gutstein H. B. et al., 1998; Kalyuzhny, A. E. and Wessendorf, M. W., 1998). The PAG is also rich in endogenous opioids (Mansour, A. et al., 1995). Enkephalin-like immunoreactivity is most dense in the ventrolateral PAG and enkephalin-containing terminals are found apposed to both gamma aminobutyric acid (GABA)ergic and non-GABAergic dendrites, including those of a small percentage of PAG–RVM neurons (Williams, F. G. and Beitz, A. J., 1990). Endomorphin-2 (Tyr-Pro-Phe-Phe-NH$_2$), an endogenous peptide with high selectivity for the MOR (Zadina, J. E. et al., 1997), is concentrated in the PAG as well as the dorsal horn of the spinal cord (Schreff, M. et al., 1998). Although the endomorphins have a high affinity for the MOR, it appears that these peptides are only partial agonists for the MOR in the PAG (Narita, M. *et al.*, 2000). Another endogenous opioid found in the PAG is β-endorphin. Discrete populations of β-endorphin-containing neurons in the ventromedial hypothalamus project to the PAG and have been implicated in analgesia produced by electrical stimulation and stress (Millan, M. J., 2002).

MOR agonists produce potent antinociception when applied directly in the PAG (Heinricher, M. M. and Fields, H. L., 2003). The behavioral antinociception produced by these agents is mediated by activation of output neurons projecting to the RVM (Sandkühler, J. and Gebhart, G. F., 1984; Tortorici, V. and Morgan, M. M., 2002). However, as the direct postsynaptic action of MOR agonists on PAG neurons is hyperpolarization (Chieng, B. and Christie, M. J., 1994a; Osborne, P. B. *et al.*, 1996), activation of the output neurons must be via disinhibition. MOR agonists are thought to act presynaptically to block GABAergic inhibition of PAG output neurons. Consistent with this hypothesis, blockade of GABA transmission within the PAG by microinjection of a GABA$_A$ receptor antagonist produces antinociception (Moreau, J. L. and Fields, H. L., 1986; Depaulis, A. *et al.*, 1987). Also consistent with this hypothesis are observations that MOR1 is frequently expressed in GABAergic neurons within the PAG. However, a substantial subset of PAG–RVM projection neurons do express MOR1 (Kalyuzhny, A. E. and Wessendorf, M. W., 1998; Commons, K. G. *et al.*, 2000). These MOR1-positive neurons may target nociceptive facilitating neurons in the RVM (Vanegas, H. *et al.*, 1984; Cheng, Z. F. *et al.*, 1986; Morgan, M. M. *et al.*, 1992;). If so, these MOR1-positive output neurons would be involved in descending facilitation rather than descending inhibition.

Postsynaptic effects of MOR agonists on PAG neurons include hyperpolarization (via activation of a postsynaptic G-protein-activated inwardly rectifying potassium conductance, GIRK) and inhibition of calcium channels (Connor, M. and Christie, M. J., 1998) (Figure 2). MOR agonists also have presynaptic effects, inhibiting GABA and glutamate release from terminals within the ventrolateral PAG (Chieng, B. and Christie, M. J., 1994b; Vaughan, C. W. and Christie, M. J. 1997; Vaughan, C. W. *et al.*, 1997). Presynaptic inhibition of GABAergic neurotransmission is through activation of the arachidonic acid–

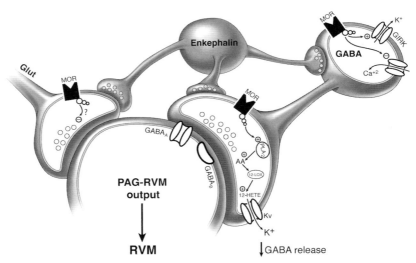

Figure 2 Cellular mechanisms of opioid action within the periaqueductal gray (PAG). Enkephalin-containing synapses are apposed to cell bodies as well as to GABA- and glutamate-containing terminals. The postsynaptic mu-uopoid receptor (MOR) activates G-protein-activated inwardly rectifying potassium channels (GIRKs) and inhibits voltage-gated Ca^{2+} channels to hyperpolarize cells and decrease cell activity. Presynaptic MORs inhibit both GABA and glutamate release on ventrolateral PAG neurons, and apparently use different signal transduction pathways in the terminals. MORs localized to GABA terminals are coupled to voltage-gated potassium channels via activation of the arachidonic acid/12-lipoxygenase (12-LOX) second messenger pathway. Hyperpolarization of the terminal decreases GABA release. The signal transduction pathway for MOR inhibition of glutamate release is currently unknown. MORs are found on GABA containing interneurons in the PAG, as well as on PAG output neurons projecting to the rostral ventromedial medulla (RVM). Activation of the descending antinociceptive pathway by opioids occurs via disinhibition. 12-HETE, 12-hydroxyeicosenoic acid; Kv, voltage-gated potassium; PLA$_2$, phospholipase A$_2$.

phospholipase A_2 second messenger pathway. Stimulation of this pathway results in activation of voltage-gated potassium channels (Kv channels) by metabolites of 12-lipoxygenase (Vaughan, C. W. and Christie, M. J., 1997; Vaughan, C. W. *et al.*, 1997). This pathway is independent of adenylyl cyclase, protein kinase A or protein kinase C activity (Vaughan, C. W. *et al.*, 1997). Further research is needed to determine the relevance of the various presynaptic versus post-synaptic opioid actions to the nociceptive modulatory function of the PAG. As already noted, there is good evidence that the behavioral antinociception produced by local application of MOR agonists involves activation of the PAG–RVM output neurons via dis-inhibition. Whether postsynaptic inhibition or the suppression of glutamatergic transmission plays a role in antinociception is unknown. However, at least in the PAG slice, the effect of presynaptic inhi-bition of GABAergic transmission apparently predominates, because bath application of the MOR agonist [D-Ala2, N-Met-Phe4, Gly-015] enkephalin (DAMGO) results in activation of neurons that are also directly hyperpolarized by the opioid (Chiou, L. C. and Huang, L. Y., 1999). Thus, under *in vitro* conditions, disinhibition has a relatively large impact, and the net effect of MOR agonist administration is neuronal activation. However, it is not known whether, or under what conditions, endogenous opioids are released to act simultaneously on presy-naptic and postsynaptic receptors.

In general, activation of DOR and KORs in the PAG does not result in significant antinociception, at least in rats. A number of groups have found no analgesic effect of microinjection of DOR agonists in the PAG in rats (Bodnar, R. J. *et al.*, 1988; Smith, D. J. *et al.*, 1988; Ossipov, M. H. *et al.*, 1995), although there is one report that microinjection of the delta$_2$ agonist deltorphin into the PAG produces a modest increase in latency of the tail flick response (Rossi, G. C. *et al.*, 1994). KOR agonists are also ineffective in producing analgesia in the PAG (Fang, F. G. *et al.*, 1989). At the cellular level, DOR and KOR agonists have no effect on presynaptic GABA release and in rat do not activate postsynaptic potassium channels (Vaughan, C. W. and Christie, M. J., 1997). However, in mice, KOR agonists inhibit presynaptic GABA release in the PAG, and agonists of all three opioid receptor subtypes activate postsynaptic potassium channels (Vaughan, C. W. *et al.*, 2003). The functions of the three receptors may therefore be different in the two species, and it would obviously be of interest

to test the behavioral effects of DOR- and KOR-selective agonists in the PAG in the mouse.

41.2.5.1.(i) Opioid/nonsteroidal anti-inflammatory drugs interactions Microinjections of nonsteroidal anti-inflammatory drugs (NSAIDs) into the PAG produce analgesia (Tortorici, V. and Vanegas, H., 1995). This antinociception is apparently mediated at least in part by an endogenous opioid peptide, as the analgesia produced by NSAIDs microinjected into the PAG and RVM is attenuated by the opioid antagonist naloxone (Vanegas, H. and Tortorici, V., 2002).

In addition to stimulating release of endogenous opioids, recent evidence at the cellular level suggests that NSAIDs may also augment the signaling pathway used by opiates, potentiating the actions of exogenous opioids (Figure 3). Coapplication of NSAIDs potenti-ates the inhibition of GABA release by the MOR partial agonist morphine, although NSAIDs have no effect on GABA release in the absence of morphine (Vaughan, C. W. *et al.*, 1997). NSAIDs primarily inhibit cyclooxygenases (COX-1 and COX-2), one of three types of enzymes (cyclooxygenases, 5-lipoxygenases, and 12-lipoxygenases) that metabolize arachidonic acid. One hypothesis proposed to explain the mechan-ism of increased analgesia with coapplication of opioids and NSAIDs is that blockade of COX-1 shunts arachidonic acid metabolism through the lipoxygenase pathways to increase the potency of opioid receptor agonists (Vaughan, C. W. *et al.*, 1997; Vaughan, C. W. 1998; Christie, M. J. *et al.*, 1999). The fact that inhibi-tors of 5-lipoxygenases also appear to potentiate the effects of opioid agonists in the PAG adds further weight to this proposal (Vaughan, C. W. *et al.*, 1997; Christie, M. J. *et al.*, 1999). These results are important in that the combined administration of NSAIDs and opiates may allow lower doses of morphine to be used to provide adequate analgesia while reducing the probability of the development of tolerance and side effects (such as respiratory depression) associated with high doses of opiates. Functional studies of NSAID/opioid interactions in the PAG would therefore be of great interest.

41.2.5.1.(ii) Opioid tolerance and dependence The PAG has been implicated in opioid tolerance as well as dependence. Opioid tolerance is the dimin-ished responsiveness to the antinociceptive actions of opioids with chronic administration. Dependence refers to the occurrence of withdrawal signs and/or rebound responses upon removal of the opioid or

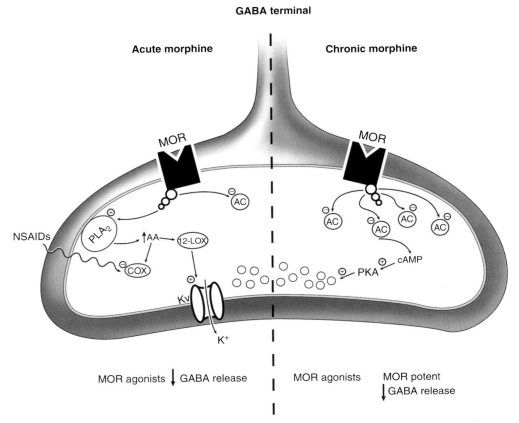

Figure 3 Mu-opioid receptor (MOR) coupling in presynaptic GABA terminals changes with chronic morphine administration. Acute administration of MOR agonists activates MORs coupled to phospholipase A_2 (PLA$_2$). Activation of PLA$_2$ increases production of arachidonic acid, which is further metabolized by 12-lipoxygenase (12-LOX). Lipoxygenase metabolites such as 12-HETE activate voltage-gated potassium channels (Kv channels) to hyperpolarize and decrease GABA release from the terminals. Nonsteroidal anti-inflammatory drugs (NSAIDs) potentiate this action of opioids by inhibiting cyclooxygenase (COX)-mediated arachidonic acid metabolism, thereby shunting arachidonic acid to the 12-LOX pathway. Activation of MORs presumably also acutely inhibits adenylyl cyclase (AC) activity in these terminals. Chronic morphine administration upregulates AC and protein kinase A (PKA) activity. After chronic, but not acute, opioid treatment, GABA release is enhanced by increased PKA activity. MOR agonists are more potent inhibitors of this PKA-dependent release, so that opioid removal or blockade of MORs by antagonists results in a rebound increase in GABA release. This increased GABA release may contribute to withdrawal behaviors mediated by the PAG.

administration of an opioid antagonist. Animals exhibit tolerance to the antinociceptive actions of MOR agonists when morphine or MOR agonists are applied directly in the PAG, especially in the caudal ventrolateral aspect (Morgan, M. M. *et al.*, 2005). This tolerance is not due to associative mechanisms, as continuous administration via an implanted pump leads to tolerance in the absence of cues associated with drug administration (Lane, D. A. *et al.*, 2004). The PAG is apparently critical for the development of antinociceptive tolerance when opioids are given systemically because animals chronically treated with morphine microinjected into the PAG display cross-tolerance with morphine given systemically

(Jacquet, Y. F. and Lajtha, A., 1976; Siuciak, J. A. and Advokat, C., 1987). Moreover, blockade of PAG opioid receptors using local microinjection of the opioid antagonist naltrexone prevents the development of tolerance to systemically administered morphine (Lane, D. A. *et al.*, 2005). Although RVM neurons lose their responsiveness to PAG morphine administration in tolerant animals (Tortorici, V. *et al.*, 2001), tolerance does not develop as readily when morphine is microinjected into the RVM itself (Morgan, M. M. *et al.*, 2005), and blockade of the RVM does not interfere with the development of tolerance when morphine is applied directly in the PAG (Lane, D. A. *et al.*, 2005). In addition, animals do not develop behavioral tolerance to direct activation of PAG output neurons that mediate antinociception, for example by microinjecting kainic acid (Morgan, M. M. *et al.*, 2003). These latter two findings demonstrate that tolerance is not a result of adaptations downstream from the opioid-sensitive PAG neuron, and indicate that the search for molecular and cellular mechanisms of opioid tolerance should focus on these opioid-sensitive neurons.

The idea that the PAG has a role in opioid dependence derives primarily from observations that blockade of PAG opioid receptors produces a number of withdrawal signs in tolerant animals, and that chronic administration of morphine in the PAG leads to the development of dependence (Bozarth, M. A. 1994). In addition, precipitated withdrawal is associated with increased expression of *c-fos* throughout the PAG, especially in the ventrolateral and lateral aspects (Chieng, B. *et al.*, 1995). A substantial number of neurons expressing *c-fos* after precipitated withdrawal are GABAergic (Chieng, B. *et al.*, 2005), and neurons positive for *c-fos* do not project to the RVM (Bellchambers, C. E. *et al.*, 1998). Thus, as was the case with tolerance, the locus for withdrawal signs in the PAG appears to be upstream from the output neuron projecting to the RVM.

Studies of the membrane properties of PAG neurons following chronic opioid treatment and withdrawal document a host of changes that could contribute to tolerance and/or dependence (Williams, J. T. *et al.*, 2001; Bailey, C. P. and Connor, M., 2005). These studies have emphasized altered effects of opioids on GABAergic inhibition and potassium channels. Opioid tolerance is widely assumed to involve a functional uncoupling between opioid receptors and their effectors. The decreased ability of opioids to inhibit the firing of PAG neurons after chronic morphine treatment is consistent with this idea (Bagley E. E. *et al.*, 2005; Chieng, B. and Christie, M. D., 1996). However, MOR activation of GIRK channels in PAG is not reduced in animals subjected to repeated intermittent morphine administration that is sufficient to induce antinociceptive tolerance (Ingram, S. L. *et al.*, 2007), and the ability of MOR agonists to inhibit GABA release is potentiated rather than reduced in slices from animals treated with chronic morphine. These enhanced actions of MOR agonists are mediated by upregulation of the protein kinase A pathway (Ingram S. L. *et al.*, 1998; Hack, S. P. *et al.*, 2003). Moreover, recent findings demonstrate that induction of a GABA transporter-mediated cation current in opioid-sensitive neurons in the PAG may

be the mechanism of the increased firing rate of these neurons during withdrawal (Bagley, E. E. *et al.*, 2005). These latter observations would argue that uncoupling of MOR from pre- or postsynaptic effectors does not underlie antinociceptive tolerance in the PAG. Rather, compensatory mechanisms may mask an underlying increase in the coupling of presynaptic MOR to its effectors that is revealed upon removal of the opioid, leading to withdrawal behaviors. Further work will be needed to bridge the gap between the cellular and behavioral analyses of tolerance and dependence.

41.2.5.2 Norepinephrine

The PAG has reciprocal connections with pontomedullary catecholaminergic cell groups (Herbert, H. and Saper, C. B. 1992; Bajic, D. *et al.*, 2001). Norepinephrine has multiple effects on PAG neurons that are mediated by α_1- and α_2-adrenergic receptors. Activation of α_1-receptors depolarizes all PAG neurons, whereas α_2-receptor activation hyperpolarizes all neurons, suggesting that both receptors are colocalized on PAG neurons (Vaughan, C. W. *et al.*, 1996). However, most neurons preferentially display either a hyperpolarizing or depolarizing response to norepinephrine itself. Neurons in ventrolateral PAG were more likely to be depolarized ($\sim$85%), whereas lateral PAG neurons were split equally in exhibiting depolarizing and hyperpolarizing responses. The α_2-mediated hyperpolarization is due to activation of a potassium conductance while the α_1-mediated depolarization depends on inhibition of a potassium conductance and an unidentified norepinephrine-sensitive conductance (Pan, Z. Z. *et al.*, 1994; Vaughan, C. W. *et al.*, 1996).

The α_2-adrenoceptor agonist clonidine also suppresses GABAergic synaptic transmission, and this suppression is enhanced after chronic morphine treatment (Ingram, S. L. *et al.*, 1998). This action of clonidine may be related to its ability to suppress many of the signs of the opioid withdrawal syndrome (Christie, M. J. *et al.*, 1997).

41.2.5.3 Substance P

NK1 receptors are found in the PAG (Commons, K. G. and Valentino, R. J., 2002), and substance P activates both opioid-sensitive and opioid-insensitive neurons in this region (Ogawa, S. *et al.*, 1992; Drew, G. M. *et al.*, 2005). In primary afferents and the dorsal horn, substance P has long been associated with nociceptive transmission, and noxious stimulation also increases the release of substance P in the

PAG. This observation would suggest that substance P plays a role in nociceptive transmission in the PAG as in the dorsal horn. However, local morphine administration also increases substance P release in the PAG, and microinjection of exogenous substance P in this region produces significant antinociception (Xin, L. *et al.*, 1997; Rosen, A. *et al.*, 2004). The role of substance P in nociceptive transmission or modulation thus remains to be determined. Substance P has also been implicated in defensive behaviors and aggression, and it produces a conditioned place aversion when microinjected into the dorsolateral PAG (Aguiar, M. S. and Brandao, M. L., 1994). These results point to a more general role for this peptide in defense. Thus substance P may apparently influence a number of functions in the PAG, and functional studies using antagonists are needed to determine the role of endogenous substance P in nociceptive processing and modulation.

41.2.5.4   Cannabinoids

Cannabinoids have been used for centuries to provide pain relief (Walker, J. M. and Huang S. M., 2002a) but are of limited utility today because of the significant psychoactive side effects and illegal status in the United States. CB1 receptors are dense throughout the PAG (Herkenham, M. *et al.*, 1991; Tsou, K. *et al.*, 1998). Cannabinoids, like opioids, can mediate antinociception by activating the PAG–RVM descending pathway (see Iversen, L. and Chapman, V., 2002 and Walker, J. M. and Huang, S. M., 2002b for reviews). Endocannabinoids, like endogenous opioids, contribute to the antinociceptive effect of activating the PAG–RVM system (Walker, J. M. *et al.*, 1999; Hough, L. B. *et al.*, 2002), and endocannabinoids in the PAG were recently shown to contribute to stress-induced analgesia (Hohmann, A. G. *et al.*, 2005). Activation of CB1 receptors inhibits both GABA and glutamate release via presynaptic mechanisms. However, cannabinoids have negligible postsynaptic actions on PAG neurons (Vaughan, C. W. *et al.*, 2000).

41.3   The Rostral Ventromedial Medulla

41.3.1   Connections of the Rostral Ventromedial Medulla

The RVM can be viewed as the brainstem output from the PAG–RVM system, receiving a dense innervation from the PAG and projecting to the dorsal horn through the dorsolateral funiculus.

Forebrain structures and the hypothalamus/MPO thus influence the RVM through the PAG, as well as by means of less dense connections directly to the RVM itself. In addition, RVM neurons are likely to receive spinothalamic inputs, either through direct connections to their widespread dendritic arbors or relayed via other brainstem regions such as nucleus reticularis gigantocellularis or the PAG (see Fields, H. L. *et al.*, 2005, for a review.)

In addition to a direct projection to the dorsal horn, the RVM can apparently influence spinal nociceptive processing via catecholamine cell groups in the pons, particularly the A7 cell group. Although there are no noradrenergic neurons within the RVM, antinociception produced by electrical or chemical stimulation in this region is frequently attenuated or blocked by intrathecal administration of noradrenergic antagonists (Hammond, D. L. and Yaksh, T. L., 1984; Proudfit, H. K., 1992). This indicates that part of the antinociceptive effect of RVM stimulation is relayed through one of the descending catecholaminergic pathways, most likely the A7 cell group in the mesopontine tegmentum. The RVM sends connections to the A7 region (Clark, F. M. and Proudfit, H. K., 1991), and these axons originate from a population of neurons distinct from that projecting to the dorsal horn (Buhler, A. V. *et al.*, 2004). This connection to A7 allows the RVM to engage spinally projecting noradrenergic neurons, which would parallel the direct projections from the RVM to the dorsal horn.

The primary outputs from the RVM are thus descending projections to the dorsal horn, both directly and via the mesopontine tegmentum. Ascending connections have also been demonstrated anatomically (Zagon, A. *et al.*, 1994; Hermann, D. M. *et al.*, 1996), but whether these rostral projections play a role in nociceptive modulation is unknown.

41.3.2   The Rostral Ventromedial Medulla and Facilitation of Nociception

Although the most robust effect of nonselective experimental activation of the RVM is antinociception, low-intensity electrical stimulation of the RVM and infusions of neuropeptides and *N*-methyl-D-aspartic acid (NMDA) in this region have been shown by various groups to facilitate nociceptive processing (Zhuo, M. and Gebhart, G. F., 1992; Urban, M. O. and Smith, D. J., 1993; Smith, D. J. *et al.*, 1997; Zhuo, M. and Gebhart, G. F., 1997; Kovelowski, C. J. *et al.*, 2000; Friedrich, A. E. and

Gebhart, G. F., 2003; Heinricher, M. M. and Neubert, M. J., 2004; Neubert, M. J. et al., 2004). Inactivation studies demonstrate that hyperalgesia produced by electrical stimulation is not merely an experimental curiosity: lesions implicate the RVM in a variety of models of hyperalgesia and persistent pain, including sickness, acute opiate withdrawal, chronic opioid-induced hyperalgesia, Freund's adjuvant, and mustard oil inflammatory hyperalgesia, following noxious stimulation of a remote body part, and in neuropathic pain models (Kaplan, H. and Fields, H. L., 1991; Morgan, M. M. et al., 1994; Watkins, L. R. et al., 1994; Pertovaara, A. et al., 1996; Ren, K. and Dubner, R., 1996; Urban, M. O. et al., 1996; Mansikka, H. and Pertovaara, A., 1997; Wiertelak, E. P. et al., 1997; Terayama, R. et al., 2000; Porreca, F. et al., 2001; Vanderah, T. W. et al., 2001a; 2001b).

These inactivation studies clearly demonstrate a role for descending facilitation in a variety of enhanced pain states, including inflammation. Nevertheless, there is evidence for an increase in descending inhibition from the RVM during inflammation (Ren, K. and Dubner, R., 2002), with potentiation of the analgesic effectiveness of opioids in the RVM (Hurley, R. W. and Hammond, D. L., 2000; 2001). One intriguing possibility is that inflammation increases the number of RVM neurons expressing MOR (Zhang, L. et al., 2004). Given both the clear demonstration that descending facilitation contributes to the hyperalgesia associated with inflammation and the apparently contradictory evidence for increased descending inhibition, an important question is whether different aspects of nociceptive transmission are differentially affected by inhibition and facilitation during inflammation (Vanegas, H., 2004; Vanegas, H. and Schaible, H.-G., 2004).

41.3.3 The Neural Basis for Bidirectional Control from the Rostral Ventromedial Medulla: On- and Off- Cells

41.3.3.1 Physiological classification of rostral ventromedial medulla neurons based on reflex-related activity

Initial electrophysiological approaches to the function of the RVM focused on the responses of RVM neurons to noxious stimulation (Anderson, S. D. et al., 1977; Behbehani, M. M. and Pomeroy, S. L., 1978; Guilbaud, G. et al., 1980; Heinricher, M. M. and Rosenfeld, J. P., 1985). In these experiments in deeply anesthetized animals, some neurons were excited,

some inhibited, and some unresponsive. The emphasis of most investigators at that time was on the subset of neurons excited by noxious inputs, at least in part because of the idea that pain inhibits pain. However, the notion of recruitment of the pain-inhibiting output from the RVM by noxious stimulation was difficult to reconcile with the observation that noxious-evoked firing of these neurons was suppressed by opioids given in doses thought sufficient to produce analgesia (Anderson, S. D. et al., 1977; Gebhart, G. F., 1982; Heinricher, M. M. and Rosenfeld, J. P., 1985). The on/off/neutral cell classification, introduced by Fields, H. L. et al. (1983a), was a significant advance in that it provided a defined framework for hypotheses relating RVM neuronal firing to the pain-modulating function of the region. This approach to RVM physiology was based on the recognition that pain-modulating neurons should discharge in relationship to variables implicated in pain modulation, rather than in response to noxious stimulation. Thus, pain-inhibiting neurons should be activated by analgesic drugs, and during periods of suppressed nociceptive responsiveness, such as that induced by intense stress. Conversely, firing of pain-facilitating neurons should be suppressed during analgesia, and enhanced during hyperalgesia (see Fields, H. M. and Heinricher, M. M., 1985 for an early discussion of the implications of the on/off/neutral cell classification).

In the on/off/neutral cell framework, RVM neurons are recorded in lightly anesthetized rats so that cell firing can be related to changes in nociceptive responding, as measured by spinal nocifensor reflexes. Off-cells are characterized by a cessation of firing during nociceptive reflexes (Figure 4). On-cells are defined by a burst of activity during nociceptive reflexes (Figure 4). Onset of the on-cell reflex-related burst and off-cell pause precede the reflex itself, as well as the thalamic response to the noxious stimulus (Hernandez, N. et al., 1989). Because nociceptive reflexes are generally suppressed when RVM neurons are experimentally activated using electrical stimulation or glutamate microinjection, the fact that off-cell activity ceases abruptly just prior to the execution of nociceptive reflexes suggested that off-cells are the antinociceptive output neurons of the RVM, and that the pause in firing permits responses to occur. Conversely, because on-cells are active during the animal's response to a noxious stimulus, it seemed unlikely that they exert a significant inhibitory effect on nociception. Rather, the reflex-related burst suggested that these neurons have

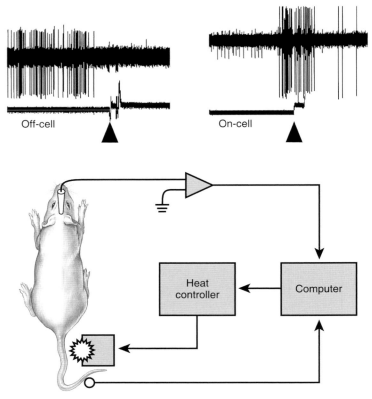

Figure 4 Two cell populations in the rostral ventromedial medulla (RVM), on- and off-cells, explain descending inhibition and facilitation from the RVM. Bottom: Experimental setup shows simultaneous brainstem recording and reflex testing. Top: Single oscilloscope sweeps show activity of a representative off- and on-cell during a tail heat trial. Upper trace is cell discharge, lower trace shows tail position, with arrowhead indicating tail movement. Heat is applied starting at the beginning of the trace. Each trace is 10 s.

a pronociceptive role. Both hypotheses have subsequently received ample confirmation in experiments using selective pharmacological manipulation of the different cell classes (see Section 41.3.3.2). At least some cells of both classes project to the dorsal horn (Fields, H. L. *et al.*, 1995).

The remaining cells in the RVM are classified as neutral cells because they show no change in activity associated with nociceptive reflexes. Whether neutral cells have a role in pain modulation remains to be determined (see Section 41.3.3.3).

Noxious pinch typically activates on-cells and suppresses off-cell firing. Cutaneous application of mustard oil, which produces an acute inflammation, evokes a strong activation of on-cells and suppresses off-cell firing, leaving neutral cells unaffected (Kincaid, W. *et al.*, in press). Occasional discrepancies between responses to heat and pinch are most likely explained by the fact that the heat-related classification is based on the reflex response evoked by a

stimulus at just threshold intensity (Fields, H. L. *et al.*, 1983a; Jinks, S. L. *et al.*, 2004a), whereas pinch stimuli in work published to date have not been limited to withdrawal threshold in intensity, and have not been linked to a withdrawal reflex (Leung, C. G. and Mason, P., 1998; Ellrich, J. *et al.*, 2000; 2001).

On- and off-cell firing is not constant over time. In animals lightly anesthetized with barbiturates, on- and off-cells frequently show alternations between periods of silence and spontaneous discharge. Simultaneous recordings from pairs of on- and off-cells show that excitability within each class varies across the population, but that firing within each class is in phase. Consequently, a subset of either the on- or off-cell class is active at any given time (Barbaro, N. M. *et al.*, 1989), but the two populations are not active simultaneously. By contrast, serotonergic neurons (which can be considered a specific division of neutral cells (Gao, K. *et al.*, 1997; Mason, P., 1997; Gao, K. *et al.*, 1998)) display regular, more or less constant activity. Nonserotonergic neutral cells vary in their firing, but are often continuously active for prolonged periods. If a withdrawal reflex is evoked when on-cells are in an active period, there is little or no discernible increase in firing (Barbaro N. M. *et al.*, 1986). Because on-cell firing apparently approaches this ceiling during active phases, definitive classification of a neuron as a neutral cell rather than an on-cell requires extensive characterization, with reflex trials delivered during periods of low spontaneous activity. Indeed, Barbaro N. M. *et al.* (1986) noted that earlier work from the Fields group had likely misclassified a number of on-cells as neutral cells.

On-, off-, and neutral cell classes can be identified in awake behaving animals (Oliveras, J. L. *et al.*, 1990; 1991a; 1991b; Leung, C. G. and Mason, P., 1999; Foo, H. and Mason, P., 2005). However, on-cells are more active during waking than sleep, firing in association with the animals' movements. Neurons of this class are also generally more responsive when animals are awake, and more easily activated by innocuous stimulation than when the animals are anesthetized or asleep. Off-cells are more active when the animal is sleeping, and are inhibited in association with movement in awake animals (Leung, C. G. and Mason, P., 1999; Foo, H. and Mason, P., 2005).

41.3.3.2 Role of on- and off-cells in pain modulation

Initial functional investigations revealed a number of correlations consistent with the idea that off-cells suppress, while on-cells facilitate, nociception. Both on- and off-cell classes respond to experimental manipulations of the PAG that produce behavioral analgesia (Vanegas, H. *et al.*, 1984; Hutchison, W. D. *et al.*, 1996; Tortorici, V. and Morgan, M. M., 2002). Withdrawal reflexes can generally be elicited at a lower threshold or with shorter latency if on-cells are active and off-cells inactive (Heinricher, M. M. *et al.*, 1989; Ramirez, F. and Vanegas, H., 1989; Bederson, J. B. *et al.*, 1990; Foo, H. and Mason, P., 2003). Antinociception produced by administration of morphine systemically or within the PAG is associated with a uniform activation of off-cells and inhibition of on-cells. Importantly, in animals in which opioid administration fails to produce behavioral analgesia, changes in cell firing are inconsistent. Thus, changes in on- and off-cell firing are related to antinociception rather than to opioid administration as such (Fields, H. L. *et al.*, 1983b; Cheng, Z. F. *et al.*, 1986.). Reduced nociceptive thresholds in acute opioid withdrawal and secondary hyperalgesia during noxious stimulation of a distant

body region are correlated with on-cell activation (Bederson, J. B. *et al.*, 1990; Morgan, M. M. and Fields, H. L., 1994). On-cells also facilitate escape responses to intense noxious stimulation, and probably contribute to secondary hyperalgesia during acute inflammation (Jinks, S. L. *et al.*, 2004b; Kincaid, W. *et al.*, in press). Finally, on-cells have been implicated in hyperalgesia associated with the sickness response. Hyperalgesia produced by systemic administration of bacterial endotoxin is blocked by lesion of the RVM (Watkins, L. R. *et al.*, 1994; Wiertelak, E. P. *et al.*, 1997). Although on-cells have not been studied in animals challenged with endotoxin, these neurons are activated by direct administration of prostaglandin E_2 in the MPO (Heinricher, M. M. *et al.*, 2004b), which can be considered a model for central components of the sickness response (Oka, T. *et al.*, 1994; Hosoi, M. *et al.*, 1997).

The obvious limitation of these observations is that they are correlative. The RVM plays an important role in a number of functions other than pain modulation, and many neurons in this region are known to exhibit correlations with various physiological parameters including EEG, blood pressure, and body temperature. Causal inferences relating on- or off-cell discharge to altered nociceptive processing thus require selective experimental manipulation of each cell class as a whole, for example, by microinjection of drugs that differentially alter the firing of the different classes. If altered firing of a particular class leads to modified nociceptive responses, one can conclude that the observed change in cell activity mediated the resulting change in behavior.

A series of studies using the above approach of manipulating RVM circuitry directly have demonstrated that activation of off-cells is sufficient to produce analgesia, and required for the antinociceptive actions of systemically administered morphine (Heinricher, M. M. and Tortorici, V., 1994; Heinricher, M. M. *et al.*, 1994; 1997; 1999; 2001a; 2001b; Meng, I. D. and Johansen, J. P., 2004; Neubert, M. J. *et al.*, 2004; Meng, I. D. *et al.*, 2005). Similar approaches have recently been extended to testing the effects of off-cell activation. Such studies show that on-cells do not contribute significantly to nociceptive threshold under basal conditions as suppression of the reflex-related firing of on-cells does not by itself reduce responding (Heinricher, M. M. and McGaraughty, S., 1998; Meng, I. D. *et al.*, 2005). However, direct activation of on-cells has a pronociceptive effect: microinjection of neurotensin or cholecystokinin (CCK) at doses that activate on-cells selectively produces thermal hyperalgesia (Heinricher, M. M. and Neubert, M. J., 2004; Neubert, M. J. *et al.*, 2004). Activation of on-cells by CCK likely contributes to nerve injury pain (Kovelowski, C. J. *et al.*, 2000; Porreca, F. *et al.*, 2001; Burgess, S. E. *et al.*, 2002). Microinjection of a CCK-2 receptor antagonist in the RVM also interferes with the paradoxical hyperalgesia associated with prolonged opioid administration (Xie, J. Y. *et al.* 2005), indicating that CCK-mediated activation of on-cells is also involved in this phenomenon. Viewed collectively, the above data provide strong evidence that off-cells suppress nociception, and that on-cells facilitate nociception. The balance of activity between the on- and off-cell populations therefore allows graded bidirectional control of nociceptive responding (Figure 5).

Functional studies demonstrate that the RVM exerts bidirectional control over visceral, as well as cutaneous, hyperalgesia (Coutinho, S. V. *et al.*, 1998). The behavioral response to colorectal distension is suppressed by microinjection of morphine in the RVM (Friedrich, A. E. and Gebhart, G. F., 2003), which is known to activate off-cells (Heinricher, M. M. *et al.*, 1994). Conversely, microinjection of CCK, at a dose that likely activates on-cells (Heinricher, M. M. and Neubert, M. J., 2004), enhances the behavioral response to colorectal distension (Friedrich, A. E. and Gebhart, G. F., 2003). Thus, on- and off-cells appear to modulate behavioral responses evoked by colorectal distension just as they modulate responses evoked by cutaneous heat. The lack of congruence between visceral inputs to RVM neurons and heat-evoked reflex-related responses of these neurons (Brink, T. S. and Mason, P., 2004) may be related to the inhibitory effect of colorectal distension on sensory processing of cutaneous stimuli (Bouhassira, D. *et al.*, 1994; 1998; Pertovaara, A. and Kalmari, J., 2002; Kalmari, J. and Pertovaara, A., 2004). More generally, a difference in

on- and off-cell firing associated with cutaneous versus visceral noxious-evoked reflexes emphasizes the basic principle that the functions of RVM neurons must be inferred from their outputs (i.e., from the behavioral or physiological effect of selective manipulation of the difference cell classes), rather from inputs.

41.3.3.3 *Role of neutral cells*

All cells that are not classified as on- or off-cells are grouped together under the heading of neutral cells. Whether these neurons have any role in pain modulation has long been an important question. By definition, the firing of neutral cells is unchanged during behavioral responses to noxious stimulation, and neutral cell firing is also unaffected by MOR or DOR agonists, norepinephrine, neurotensin, or CCK, all of which alter firing of on- and/or off-cells and nociceptive behavior when applied in the RVM (Heinricher, M. M. *et al.*, 1994; 1988; Harasawa, I. *et al.*, 2000; Heinricher, M. M. *et al.*, 2001a; Neubert, M. J., *et al.*, 2004). KOR agonists depress the firing of some neutral cells, but the functional implications of this depression have not been explored (Meng, I. D. *et al.*, 2005). The differential pharmacology of neutral cells supports their categorization as distinct from on- and off-cells. Neutral cells, or subpopulations of neutral cells, might be involved in some other functions of the RVM, for example thermoregulation (Nakamura, K. *et al.*, 2002; Madden, C. J. and Morrison, S. F., 2003).

Despite the lack of direct evidence that neutral cells have a role in pain modulation, interest in this cell class remains high. This is in part because a subset of neutral cells contains serotonin (Potrebic, S. B. *et al.*, 1994; Mason, P., 2001), and serotonin is implicated in descending facilitation as well as descending inhibition of nociception (Le Bars, D., 1988; Calejesan, A. A. *et al.*, 2000; Suzuki, R. *et al.*, 2004). However, there is significant controversy as to

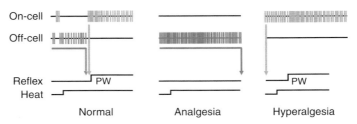

Figure 5 Balance of activity between on- and off-cell populations allows graded bidirectional control of nociceptive responding. *Left column, normal conditions*: Application of heat (arrow) to a rat's tail or paw evokes a paw withdrawal reflex. The off-cell pauses and on-cell becomes active just before the reflex occurs. *Middle column, analgesia*: Manipulations that cause off-cells to become continuously active also suppress on-cell firing and application of heat does not evoke a response. *Right column, hyperalgesia*: Reflexes occur with a shorter latency if heat is applied during a period when on-cells are activated (for example, as a consequence of a prior noxious stimulus).

the exact role of serotonergic neutral cells in pain modulation. Immunohistochemically identified serotonergic neurons express MOR, DOR, and KOR (Kalyuzhny, A. E. and Wessendorf, M. W., 1999; Wang, H. and Wessendorf, M.W., 1999; Marinelli, S. *et al.*, 2002). However, neutral cells, including putative serotonergic neutral cells, as a class do not respond to morphine or MOR agonists given systemically or microinjected into the RVM (Barbaro, N. M. *et al.*, 1986; Heinricher, M. M. *et al.*, 1994; Gao, K. *et al.*, 1998). Why neurons that express the MOR fail to respond to MOR agonists remains an important puzzle. A similar mismatch between *in vivo* functional data and immunohistochemical data arises with the neurotensin receptor. Some immunohistochemically identified serotonergic neurons express neurotensin receptor-like immunoreactivity (NT1-ir), and behavioral studies show that microinjection of high doses of neurotensin or a NT1 agonist produce antinociception (Smith, D. J. *et al.*, 1997; Neubert, M. J. *et al.*, 2004; Buhler, A. V. *et al.* 2005). However, physiologically identified neutral cells do not respond to neurotensin when microinjected into the RVM at a dose sufficient to produce behavioral antinociception (Neubert, M. J. *et al.*, 2004). Again, there is no obvious explanation for why neurons expressing the NT1 receptor show no change in firing following microinjection of neurotensin. Resolution of these discrepancies between immunohistochemical and electrophysiological findings will require immunohistochemical labeling of functionally identified neurons.

A second reason for continued interest in neutral cells is the report that the response properties of many neutral cells change to those of on- or off-cells over the course of hours during prolonged inflammation (Miki, K. *et al.*, 2002). A subset of neutral cells may therefore be involved in long-term facilitation or inhibition of nociception, or in other changes in behavior or physiology during inflammation. As discussed above however, the pharmacology of such cells would presumably differ from cells classified as on- or off-cells in the absence of inflammation.

41.3.4 Pharmacology of Nociceptive Modulation in the Rostral Ventromedial Medulla

Among the neuropeptides and neurotransmitters found in RVM neurons are serotonin, substance P, enkephalin, thyrotropin-releasing hormone, galanin, somatostatin, CCK, GABA, and acetylcholine (Bowker, R. M. *et al.*, 1983; Mantyh, P. W. and Hunt, S. P., 1984; Menetrey, D. and Basbaum, A. I., 1987; Millhorn, D. E. *et al.*, 1987a; 1987b; Bowker, R. M. and Abbott, L. C., 1988; Millhorn, D. E. *et al.*, 1989; Palkovits, M. and Horvath, S., 1994; Skinner, K. *et al.*, 1997). With the exception of serotonin (see above, Section 41.3.3.1), the relationship between physiological/functional class (on-, off-, and neutral cell classification) and expression of different neurotransmitters has yet to be examined.

41.3.4.1 Gamma-aminobutyric acid and glutamate: the off-cell pause and on-cell burst

One approach to identifying the neurotransmitter (s) mediating defined synaptic inputs is to attempt to block those inputs using iontophoretic application of neurotransmitter antagonists (Hicks, T. P., 1983). Iontophoresis of $GABA_A$ receptor antagonists selectively blocks the off-cell pause, demonstrating that this reflex-related inhibition of off-cell firing is mediated by GABA (Heinricher, M. M. *et al.*, 1991). Consistent with the idea that the off-cell pause removes descending inhibition and permits noxious information to be processed, microinjection of $GABA_A$ receptor antagonists in the RVM blocks the off-cell pause and produces behavioral antinociception (Drower, E. J. and Hammond, D. L., 1988; Heinricher, M. M. and Kaplan, H. J. 1991; Heinricher, M. M. and Tortorici, V., 1994; Gilbert, A. K. and Franklin, K. B., 2001). Interestingly, Nason M. W., Jr. and Mason P. (2004) recently suggested that the effect of RVM GABA receptor antagonism is selective for stimulation of the tail, with facilitation rather than inhibition of responses evoked by stimulation of the foot. However, this is difficult to reconcile with the antinociception observed by others using the hot plate and formalin tests, as both of these tests involve stimulation of the hindpaw (Gilbert, A. K. and Franklin, K. B., 2001). The source of GABAergic input to off-cells is unknown, but it is presumed to be the target of MOR agonists (see Section 41.3.4.2).

Although firing of both on-cells and neutral cells is inhibited by iontophoretically applied GABA, neither cell class displays significant disinhibition during iontophoresis of $GABA_A$ receptor antagonists. This observation implies that on-cells and neutral cells express GABA receptors, but that there is little ongoing GABAergic control of their firing, at least in the lightly anesthetized rat (Heinricher, M. M. *et al.*, 1991).

Iontophoretic application of excitatory amino acids excites all three classes of RVM neurons. However, antagonist studies specifically implicate excitatory amino acid transmitters in reflex-related activation of on-cells and opioid-induced activation of off-cells. Thus, iontophoretic application of the broad-spectrum excitatory amino acid antagonist kynurenate blocks the reflex-related on-cell burst, without affecting the activity of off-cells or neutral cells (Heinricher, M. M. and Roychowdhury, R., 1997). Kynurenate microinjected into the RVM at a dose that blocks on-cell activation does not result in increased withdrawal latency to noxious heat, indicating that the on-cell burst does not control nociceptive threshold, at least under normal conditions (Heinricher, M. M. and McGaraughty, S., 1998). However, it is possible that the firing of on-cells during and after the withdrawal modulates the magnitude or force of the reflex, or primes responses to subsequent stimuli (Ramirez, F. and Vanegas, H., 1989; Jinks, S. L. et al., 2004b).

A second important role for excitatory amino acid transmission in the RVM is the recruitment of off-cells by opioids (Heinricher, M. M. et al., 2001b). Disinhibition of off-cells leads to NMDA-mediated activation of these neurons, amplifying the effect of disinhibition (see the next Section 41.3.4.2).

41.3.4.2 Opioid actions in the rostral ventromedial medulla

All three opioid receptors, MOR, DOR, and KOR, are found in the RVM, although their expression is much less dense than in the PAG (Gutstein H. B. et al., 1998; Kalyuzhny, A. E. et al., 1996; Wang, H. and Wessendorf, M. W., 1999). The RVM, and specifically activation of off-cells, is required for the analgesic actions of systemically administered morphine (Dickenson, A. H. et al., 1979; Azami, J. et al., 1982; Mitchell, J. M. et al., 1998; Heinricher, M. M. et al., 2001a; 2001b; Gilbert, A. and Franklin, K., 2002). Like the PAG, the RVM supports MOR-mediated analgesia. Thus, direct local microinjection of morphine or MOR agonists in the RVM produces an antinociception equivalent to that produced by systemic morphine administration (Heinricher, M. M. and Morgan, M. M., 1999). Moreover, PAG and RVM interact in a synergistic fashion (Rossi, G. C. et al., 1994). This is at least in part because exogenous opioid administration at one site recruits endogenous opioid mechanisms at the other. Thus, microinjection of morphine in the PAG apparently evokes release of endogenous opioids in the RVM (Kiefel, J. M.

et al., 1993; Pan, Z. Z. and Fields, H. L., 1996; Roychowdhury, S. M. and Fields, H. L., 1996) to produce its antinociceptive effect.

The local actions of MOR agonists within the RVM are well understood at both the cellular and circuit levels. On-cells are directly inhibited by MOR agonists. Neither neutral cells nor off-cells are directly sensitive to MOR agonists, but off-cells are activated indirectly, most likely via disinhibition (Figure 6). Microinjection of MOR agonists in the RVM suppresses on-cell firing and activates off-cells, resulting in behavioral antinociception (Heinricher, M. M. et al., 1992; 1994). The activation of off-cells is required for the antinociceptive effect (Heinricher, M. M. et al., 1997). Suppression of on-cell firing is not necessary for RVM-mediated analgesia (Heinricher, M. M. et al., 1994; Neubert, M. J. et al., 2004), although it likely contributes. Selective inhibition of on-cell firing reduces the force, although not the latency, of withdrawal to noxious heat (Heinricher, M. M. and McGaraughty, S., 1998; Jinks, S. L. et al., 2004b).

Studies at the cellular level reveal that some neurons are inhibited directly by MOR agonists, whereas others are disinhibited via presynaptic inhibition of GABAergic transmission (Pan, Z. Z. et al., 1990). These findings are thus consistent with the direct inhibition of on-cells and indirect activation of off-cells observed *in vivo*. RVM neurons studied *in vitro* are classified as secondary cells, which are directly

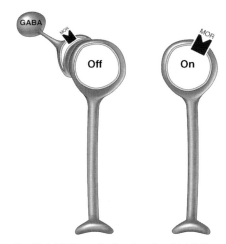

Figure 6 Disinhibition of off-cells, direct inhibition of on-cells mediates opioid analgesia. On-cells are inhibited, directly, by morphine and mu-opioid receptor (MOR) agonists. Off-cells are not inhibited by MOR agonists, but are activated, via disinhibition. MORs are thus presumed to be located on on-cells, and on gamma aminobutyric acid (GABA)ergic inputs to off-cells.

inhibited by MOR agonists, and primary cells, which are not. Primary cells do however receive opioid-sensitive GABAergic inputs, and are disinhibited by MOR agonists (Pan, Z. Z. *et al.*, 1990). Secondary cells presumably map to on-cells recorded *in vivo*, and primary cells to off-cells and neutral cells identified *in vivo*.

41.3.4.2.(i) *Delta opioid receptor*

Like MOR agonists, DOR agonists, particularly delta$_2$ agonists, produce behavioral hypoalgesia when microinjected into the RVM. The effects of delta$_1$ agonists are less robust than those of delta$_2$ agonists (see Heinricher, M. M. and Fields, H. L., 2003 for review of the role of the DOR in pain modulation in the RVM). *In vitro*, the DOR agonists deltorphin and [D-Pen2, D-Pen5] enkephalin (DPDPE) elicit hyperpolarizing currents in a substantial minority of RVM neurons (Marinelli S. *et al.*, 2005). Some, but not all, of these neurons were also hyperpolarized by MOR agonists. It thus appears that a subset of DOR-sensitive RVM neurons falls into the secondary cell category. In experiments *in vivo*, microinjection of deltorphin into the RVM increases the latency and reduces the amplitude of the reflex-related on-cell burst, reduces the duration of the reflex-related off-cell pause, and evokes a small but statistically significant increase in tail flick latency (Harasawa, I. *et al.*, 2000).

Molecular, cellular, and behavioral approaches point to interactions between MOR- and DOR-mediated processes, but experiments designed to uncover such interactions within the RVM reveal that the effects of MOR and DOR agonists in the RVM are independent under basal conditions (Hurley, R. W. and Hammond, D. L. 2001; Kalra, A. *et al.*, 2001). However, there is a time-dependent recruitment of supraspinal DOR function in inflammation, so that DOR agonists microinjected into the RVM are more effective in animals subjected to long-lasting inflammation compared to noninflamed controls. MOR agonists are also more potent in these animals (Hurley, R. W. and Hammond, D. L., 2000; 2001). This increased opioid action may reflect an increase in the release of endogenous opioids acting at the DOR (Williams F. G. *et al.*, 1995) and/or translocation of DOR from the cytoplasm to the plasma membrane (Commons, K. G. *et al.*, 2001; Commons, K. G., 2003). Coexpression of MOR and DOR on individual RVM neurons may also play a role (Marinelli, S. *et al.*, 2005).

41.3.4.2.(ii) *Kappa opioid receptor*

The role of the KOR in the pain-modulating functions of the RVM remains controversial. Electrophysiological studies in the RVM slice demonstrate pre- and post-synaptic effects of KOR agonists in the RVM. However, it is unclear whether neurons in which KOR agonists evoke an outward current are a separate population from those inhibited by MOR agonists (Pan, Z. Z. *et al.*, 1997; Ackley, M. A. *et al.*, 2001; Marinelli, S. *et al.*, 2002; Bie, B. and Pan, Z. Z., 2003). Microinjection of a KOR agonist into the RVM has been reported to either have no behavioral effect (Pan, Z. *et al.*, 2000; Meng, I. D. *et al.*, 2005) or to produce antinociception, depending on the nociceptive test as well as the sex of the animal (Tershner, S. A. *et al.*, 2000; Ackley, M. A. *et al.*, 2001). KOR agonists also block the antinociceptive effect of PAG microinjection of a MOR agonist or systemic morphine administration (Pan, Z. *et al.*, 2000; Bie, B. and Pan, Z. Z., 2003; Meng, I. D. *et al.*, 2005) as well as MOR-mediated conditioned hypoalgesia (Foo, H. and Helmstetter F. J., 2000). This antianalgesic effect of the KOR agonist is presumably due to a blockade of off-cell activation (Bie, B. and Pan, Z. Z., 2003; Meng, I. D *et al.*, 2005). KOR agonists can also block hyperalgesia when microinjected into the RVM, presumably via presynaptic blockade of excitatory inputs to on-cells (Ackley, M. A. *et al.*, 2001; Meng, I. D. *et al.*, 2005).

41.3.4.3 *Norepinephrine*

As with the KOR, norepinephrine likely influences both the nociceptive facilitating and inhibiting outputs from the RVM. Behavioral studies have reported that blockade of α_1-adrenergic receptors produces antinociception (Sagen, J. and Proudfit, H. K., 1981) and reduces hyperalgesia associated with acute opioid withdrawal (Bie, B. *et al.*, 2003). Agonists at the α_1 receptor produce hyperalgesia and conditioned place avoidance, and the increased nociception is dissociated from the negative affective state in these animals (Hirakawa, N. *et al.*, 2000).

Surprisingly, α_2 agonists have been reported to produce analgesia (Sagen, J. and Proudfit, H. K., 1985; Haws, C. W. *et al.*, 1990) as well as to block opioid analgesia (Bie, B. *et al.*, 2003) when microinjected into the RVM. The neural basis for these conflicting results is unclear, as electrophysiological studies are also inconsistent. *In vivo*, iontophoretically applied norepinephrine and α_2 agonists suppress the firing of on-cells, whilst α_1 agonists activate on-cells. Off-cells and neutral cells do not respond to iontophoretically applied norepinephrine or adrenergic agents (Heinricher, M. M. *et al.*, 1988). By contrast, in experiments in the RVM slice, α_2 agonists hyperpolarize primary cells (presumably off-cells and/or neutral cells), and α_1 agonists activate all cells (Bie, B. *et al.*, 2003). The findings from the *in vivo* study are thus more consistent with an antinociceptive effect of α_2 receptor activation, whereas the studies in the slice are more consistent with the antiopioid action of these agents.

41.3.4.4 Cannabinoids

As noted above (Section 41.2.5.4), the PAG–RVM system is now recognized to mediate at least part of the antinociceptive effect of cannabinoids. The RVM is required for the antinociceptive action of systemically administered cannabinoid agonists (Meng, I. D. *et al.*, 1998), and microinjection of CB1 agonists into the RVM produces moderate antinociception (Martin, W. J. *et al.*, 1998; Monhemius, R. *et al.*, 2001; Meng, I. D. and Johansen, J. P., 2004). Off-cells display a measurable increase in activity after local infusion of a CB1 agonist in the RVM, and the reflex-related activity of on-cells is significantly reduced (Meng, I. D. and Johansen, J. P., 2004). Both of these effects are presumed to reflect a presynaptic action, as cannabinoids block excitatory and inhibitory synaptic currents, but have negligible postsynaptic effects in RVM (Vaughan C. W. *et al.*, 1999).

41.3.4.5 Cholecystokinin

Endogenous CCK has long been recognized to counter opioid analgesia and to contribute to enhanced nociception, particularly in neuropathic pain states (Stanfa, L. *et al.*, 1994; Wiesenfeld-Hallin, Z. *et al.*, 1999). In humans, CCK antagonism increases the analgesic actions of morphine and placebo (Price, D. D. *et al.*, 1985; Benedetti, F. and Amanzio, M., 1997; McCleane, G. J., 1998). CCK is found in the RVM (Skinner, K. *et al.*, 1997), and has both pronociceptive and antiopioid actions in this region. CCK in the RVM is implicated in the expression of allodynia and thermal hyperalgesia in the spinal nerve ligation model of nerve injury pain as well as in hyperalgesia associated with chronic morphine administration (Kovelowski, C. J. *et al.*, 2000; Xie, J. Y. *et al.*, 2005). Furthermore, focal application of relatively high doses of exogenous CCK in the RVM produces cutaneous and visceral hyperalgesia (Kovelowski, C. J. *et al.*, 2000; Friedrich, A. E. and

Gebhart, G. F., 2003; Heinricher, M. M. and Neubert, M. J., 2004). Notably, CCK activates on-cells selectively when microinjected in a dose sufficient to produce behavioral hyperalgesia. This points to on-cells as critical mediators of CCK hyperalgesia in the RVM (Heinricher, M. M. and Neubert, M. J., 2004).

CCK also has antiopioid actions in the RVM, albeit through a different mechanism than that through which it produces hyperalgesia. Focal application of a low dose of CCK in the RVM attenuates both opioid activation of off-cells and opioid-induced antinociception without affecting baseline nociceptive responding or the activity of on-cells (Heinricher, M. M. *et al.*, 2001a). The signal transduction mechanisms that underlie this antiopioid action of CCK are unknown. In the hippocampus, CCK reduces opioid-induced disinhibition by increasing GABA release (Miller, K. K. *et al.*, 1997), raising the possibility that CCK acts presynaptically to block disinhibition of off-cells in the RVM.

41.3.4.6 Neurotensin

Like CCK, neurotensin has multiple effects in the RVM. Microinjection of this peptide within the RVM gives rise to a dose-related, bidirectional effect on nociceptive behaviors (tail flick, hot plate, and visceromotor responses to colorectal distension) and dorsal horn nociceptive neurons. Extremely low doses produce facilitation, whereas high doses produce antinociception. Intermediate doses are without effect (Urban, M. O. and Smith, D. J., 1993; Smith, D. J. *et al.*, 1997; Urban, M. O. and Gebhart, G. F., 1997; Urban, M. O. *et al.*, 1999; Neubert, M. J. *et al.*, 2004.). Behavioral hyperalgesia is mediated by selective activation of on-cells, whereas the hypoalgesia produced by higher doses is due to recruitment of off-cells. Neutral cells do not respond to neurotensin at any dose (Neubert, M. J. *et al.* 2004).

41.3.4.7 Nociceptin/orphanin FQ

The identification of an opioid-like receptor (ORL-1) that did not bind classical opioid peptides suggested that additional peptides might exist that could modulate nociception. However, despite its structural and functional similarities to the opioid receptors, activation of ORL-1 by its endogenous ligand, referred to as nociceptin or orphanin FQ (OFQ), does not produce antinociception when applied within the RVM. Rather nociceptin/OFQ potently inhibits all neurons in the RVM. As a consequence, focal application of nociception/OFQ in the RVM is effectively a functional lesion of the region, and interferes with both antinociceptive and pronociceptive outputs (see Heinricher, M. M., in press for a review).

41.3.5 Physiological Activation of Nociceptive Modulatory Neurons in the Rostral Ventromedial Medulla

Given a pronociceptive role for on-cells and an antinociceptive role for off-cells, the obvious question that arises is how and when these two classes of neurons with opposing functions are recruited to modulate nociception. This question remains a significant challenge, in part because a number of factors likely to influence nociception via the RVM involve higher-order neural structures and psychological processes that are difficult to study in rodents, particularly under anesthesia. Nevertheless, we do have some information as to how and when RVM nociceptive modulatory neurons are recruited to modulate nociceptive processing.

41.3.5.1 Response of RVM neurons to noxious stimulation: does pain inhibit pain?

Noxious-evoked withdrawals are associated with a period of on-cell activation and off-cell inhibition that can last from less than a second to many minutes. This period, during which the balance between the on- and off-cell populations is shifted toward on-cell activation, is associated with heightened nociceptive responsiveness that can be blocked by lesion of the RVM (Heinricher, M. M. *et al.* 1989; Ramirez, F. and Vanegas, H., 1989; Morgan, J. L. and Fields, H. L., 1994; Foo, H. and Mason, P., 2003). Thus noxious stimulation *per se* recruits the RVM to facilitate nociception. This conclusion is consistent with the early suggestion of Cervero F. and Wolstencroft J. H. (1984) that the RVM is part of a short-term positive feedback loop activated by noxious stimulation. Such a positive feedback process presumably prepares the organism to respond more briskly or at a lower threshold to subsequent damaging inputs. This view of on-cells as part of a positive feedback loop is further supported by evidence that on-cells are activated by acute cutaneous inflammation, and that the RVM is necessary for secondary hyperalgesia in this paradigm (Urban, M. O. *et al.*, 1996; Kincaid, W. *et al.*, in press).

The evidence for positive feedback mediated by the RVM is plainly at odds with the long-standing idea that pain inhibits pain, which arose from observations of an apparent lack of pain observed in some patients subjected to intense trauma (Wall, P. D. 1979; Melzack, R. *et al.*, 1982) and the from the ability of footshock or other noxious stimuli to suppress animals' responses to other painful events (Maier, S. F., 1986). Several factors must be considered in addressing this issue. First, other brainstem pathways, such as that through the medullary subnucleus reticularis dorsalis (Le Bars, D., 2002), act in parallel with the output from the RVM. The net effect of a given noxious stimulus on spinal nociceptive processing and pain sensation will reflect the actions of these pathways as well as that of the RVM. In addition, higher-order processes, such as fear or stress, are likely to be significant when an animal is subjected to an experimental noxious stimulus, particularly one over which it has no control. Such higher influences will be most prominent in awake behaving animals, and may shift the balance in the RVM towards activation of off-cells rather than on-cells. Such a shift is presumably the basis for stress-induced analgesia or antinociception associated with fear conditioning (Bodnar, R. J. *et al.*, 1980; Fanselow, M. S. 1986; Helmstetter, F. J. and Tershner, S. A., 1994). Conditioned fear, for example, produces antinociception mediated by the amygdala and the RVM, and direct opioid activation of the amygdala activates RVM off-cells via the PAG (Helmstetter, F. J. and Landeira-Fernandez, J., 1990; Helmstetter, F. J., 1992; Helmstetter, F. J. and Bellgowan, P. S., 1993; Helmstetter, F. J. and Tershner, S. A., 1994; Helmstetter, F. J. *et al.*, 1998; McGaraughty, S. and Heinricher, M. M., 2002; 2004).

Thus, because of higher-order inputs to the RVM and the actions of other descending pathways, the behavioral response to a given noxious stimulus may be enhanced or suppressed in the presence of pain. Which behavioral outcome predominates will depend on the stimulus history, the environment in which the stimulus is applied and the behavioral state of the animal.

41.3.5.2 Behavioral state control: anesthesia and sleep/waking cycle

The influence of behavioral state variables on nociceptive processing at the first central relay in the dorsal horn (Hayes, R. L. *et al.*, 1981; Bushnell, M. C. *et al.*, 1984; Soja, P. J. *et al.*, 1999) suggests that pathways descending from the brain to the dorsal horn contribute to the influence of behavioral state on nociception. One possibility that has been considered is that RVM neurons might mediate effects of general anesthesia or sleep/waking on nociceptive processing.

Correlative data suggest that on- and/or off-cell firing could contribute to isoflurane-induced motor depression at lighter planes of anesthesia (Jinks, S. L. *et al.*, 2004a). However, firing of both on- and off-cells is depressed when isoflurane concentration is increased to levels above that required to suppress nocifensor reflexes (Leung, C. G. and Mason, P., 1995). These neurons are thus unlikely to mediate the effects of a anesthesia at a surgical plane on nociceptive withdrawals.

Analysis of the role of on- and off-cells in mediating changes in nociception across the sleep–waking cycle is impeded by lack of a clear understanding of how nociception changes with sleep/waking. In cats, Kshatri A. M. *et al.* (1998) demonstrated that the latency of the tail flick reflex was slightly increased during slow-wave sleep compared to waking, but substantially longer during paradoxical sleep. Consistent with this, noxious-evoked activity of trigeminal neurons is depressed during paradoxical sleep, but not slow-wave sleep, compared to waking (Cairns, B. E. *et al.*, 1995; 1996). By contrast, Mason P. *et al.* (2001) reported that paw withdrawals evoked by heat were enhanced during slow-wave sleep compared to waking, suggesting a disinhibition of nociception during slow-wave sleep. Parallel recording studies from this group demonstrated that off-cells were more active during slow-wave sleep and on-cells and neutral cells were less active compared to waking (Leung, C. G. and Mason, P., 1999). It is therefore unlikely that on- or off-cells mediated the enhanced nociception seen in their behavioral analysis. However, these authors noted that rats return to sleep more quickly following noxious stimulation if the stimulus is delivered when the rats are sleeping rather than awake. They suggested that this could point to a role for off-cells in controlling arousal. However, it is not clear that the duration of waking triggered by the noxious stimulus could be differentiated from the ongoing waking pattern in their analysis. It remains to be determined whether off-cells control arousal, or mediate the influence of arousal on nociception.

A further question is whether on- and off-cells contribute to the atonia of paradoxical sleep. This has not been tested directly. However, although stimulation throughout the length of the medial pontomedullary reticular formation can produce atonia in decerebrate rats (Hajnik, T. *et al.*, 2000), the most effective sites are generally dorsal to the area defined as the RVM. Moreover, microinjection of kainate or morphine in the RVM in intact behaving rats does not produce loss of motor tone, even though locomotor activity in an open field is reduced (Morgan, M. M. and Whitney, P. K., 2000). Finally, the firing of on- and off-cells does not vary between slow wave sleep and paradoxical sleep (Foo, H. and Mason, P., 2003). Taken together, these data indicate that on- and off-cells are unlikely to mediate atonia during paradoxical sleep.

41.3.5.3　Micturition

Baez M. A. *et al.* (2005) recently reported that paw withdrawal to noxious heat is attenuated during micturition, presumably preventing potential interruption of bladder emptying by noxious stimulus-evoked movements. Consistent with the idea that off-cells inhibit and on-cells facilitate nociception during micturition, many presumptive off-cells were active during urine expulsion, whereas presumptive on-cells were inhibited. However, these authors propose a somewhat broader role for RVM neurons, as electrical stimulation in the RVM and morphine microinjection in the PAG are known to suppress bladder contractions (Sugaya, K. *et al.*, 1998; Matsumoto, S. *et al.*, 2004). Baez M. A. *et al.* (2005) therefore suggested that off-cell activation suppresses voiding, possibly by depressing transmission of bladder afferent information.

41.3.5.4　Environmental analgesia

Decreases in pain responsiveness can be induced by a wide range of experimental procedures, including electrical shock, forced swim, and centrifugal rotation (Watkins, L. R and Mayer, D. J., 1982; Bodnar, R. J., 1986; Maier, S. F., 1986). Biologically relevant threat stimuli such as odors from stressed animals of the same species or exposure to a predator also induce antinociception (Fanselow, M. S., 1985; Lester, L. S. and Fanselow, M. S., 1985; Kavaliers, M. 1988; Lichtman, A. H. and Fanselow, M. S., 1990). Activation of endogenous pain-modulating systems can also enhance the analgesic effects of exogenous opioids. Thus, presentation of a stressful shock potentiates morphine-induced antinociception, an effect that increases as the intensity of stress is increased (Grau, J. W. *et al.*, 1981; Rosellini, R. A. *et al.*, 1994). The PAG–RVM system is implicated in at least some of these environmentally induced changes in nociceptive responsiveness. Most important, some forms of environmental analgesia can be attenuated by lesioning the RVM (Prieto, G. J. *et al.*, 1983; Watkins, L. R. *et al.*, 1983b). Moreover, administration of a stressful foot shock blocks both the pause in firing that characterizes off-cells and the typical on-cell burst of activity (Friederich, M. W. and Walker, J. M., 1990). This indicates that foot-shock and possibly other stressors affect the activity of on- and off-cell classes in a manner consistent with their respective pronociceptive and antinociceptive roles.

Antinociception is also elicited as a conditioned response to previously neutral cues paired with noxious or aversive events (Fanselow, M. S., 1986). This latter phenomenon is termed conditioned antinociception, and has been widely adopted for experimental analysis of endogenous mechanisms involved in nociceptive modulation. In conditioned antinociception paradigms, antinociception is recruited in concert with other behaviors and autonomic adjustments as part of a defensive reaction (Williams, F. G. et al., 1990; Fanselow, M. S., 1991; Harris, J. A. and Westbrook, R. F., 1995). Benzodiazepine receptor inverse agonists have anxiogenic or fear-promoting effects, and yield antinociception after either intracerebral or systemic administration (Rodgers, R. J. and Randall, J. I., 1988; Helmstetter, H. J. et al., 1990; Fanselow, M. S. and Kim, J. J., 1992). Conversely, hypoalgesia is reduced by manipulations that would be expected to reduce fear, such as administration of anxiolytic agents (Fanselow, M. S. and Helmstetter, H. J., 1988).

The amygdala, a forebrain structure with a well-documented role in fear, stress and anxiety, is critical in organization of the fear-related processes described above (Davis, M., 1992a; 1992b). Notably, amygdala lesions attenuate freezing and analgesia in rats exposed to a cat, which is an innate fear stimulus for this species (Blanchard, D. C. and Blanchard, R. J., 1972; Fox, R. J. and Sorenson, C. A., 1994), and to intense, nonnoxious noise (Helmstetter, F. J. and Bellgowan, P. S., 1994; Bellgowan, P. S. and Helmstetter, F. J., 1996). Such lesions have no effect on baseline nociceptive responsiveness (Helmstetter, F. J. 1992; Fox, R. J. and Sorenson, C. A., 1994; Manning, B. H. and Mayer, D. J., 1995; Watkins, L. R. et al., 1993).

Antinociception produced by fear-related processes organized in the amygdala is mediated by the PAG–RVM system. Thus, conditioned antinociception involves suppression of nociceptive processing at the level of the spinal cord (Harris, J. A. et al., 1995), and is disrupted by lesions of the PAG or the RVM, or by infusion of an opiate antagonist into the PAG (Watkins, L. R. et al., 1983a; 1983b; Kinscheck, I. B. et al., 1984; Helmstetter, F. J. and Landeira-Fernandez, J., 1990; Helmstetter, F. J. and Tershner, S. A., 1994). The amygdala sends a sparse projection into the RVM region, raising the possibility that the amygdala could directly influence the RVM (Price, J. L. and Amaral, D. G., 1981; Hermann, D. M. et al., 1997). However, the projection from the amygdala to the PAG is much more robust, and the PAG is thought to be a necessary relay in the antinociception associated with conditioned fear. Amygdala projections to the PAG arise from the central nucleus, and to a lesser extent the medial nucleus (Krettek, J. E. and Price, J. L., 1978; Rizvi, T. A. et al., 1991; Canteras, N. S. et al., 1995). Terminations from the central nucleus are concentrated in PAG regions that in turn send projections to the RVM (Rizvi, T. A. et al., 1991). Activation of the central or basolateral nucleus of the amygdala alters the firing of PAG neurons, with approximately equal proportions of PAG neurons showing excitation and inhibition (Sandrew, B. B. and Poletti, C. E., 1984; da Costa Gomez, T. M. and Behbehani, M. M., 1995). Effects of basolateral stimulation are mediated primarily through the central nucleus (da Costa Gomez, T. M. et al., 1996). Finally, microinjection of morphine in the basolateral nucleus of the amygdala activates off-cells and suppresses on-cell firing, and these changes in on- and off-cell activity are associated with behavioral antinociception (McGaraughty, S. and Heinricher, M. M., 2002). Taken together, the above findings indicate that one way in which the modulatory circuitry of the PAG–RVM circuitry is engaged physiologically is via activation of the amygdala by stimuli that induce fear.

41.4 Conclusion

The effort to understand the neural basis of nociceptive modulation by the PAG–RVM system highlights the importance of studying functionally identified neurons. The RVM can both facilitate and inhibit nociception. Furthermore, this region is also implicated in a number of functions other than nociceptive modulation, including reproductive behaviors, cardiovascular and respiratory control, sleep–waking and arousal, thermoregulation, and behavioral suppression. A meaningful analysis of how the RVM contributes to enhanced pain states therefore requires functional identification of the neurons under study, so that mechanisms contributing to nociceptive facilitation can be distinguished from those involved in nociceptive inhibition or other functions. The on/off/neutral cell classification introduced over 20 years ago by Fields H. L. *et al.* (1983a) has been very productive. It has predicted the pharmacology of individual neurons (responses to opioids, norepinephrine, cannabinoids, neurotensin, and CCK) and allowed functional links between activation of on-cells and increased nociception, and between activation of off-cells and decreased nociception. Application of a similar analysis to the PAG, which is implicated in a host of functions from autonomic control to reproductive behavior and vocalization, would greatly advance our understanding of the neural circuitry of pain modulation in this region.

Numerous neurotransmitters and neuropeptides have been implicated in pain modulating functions of the PAG and RVM. Under what conditions are these substances released? Are they local or external to the PAG–RVM? Anatomical studies document substantial reciprocal connections between the PAG and limbic forebrain structures. Which of these connections are relevant to pain modulation, and under what conditions are they activated? These and similar questions must be addressed in order to meet the greater challenge of defining how and when this system is brought into play to enhance or inhibit pain.

Acknowledgments

MMH was supported by grants from NIDA (DA 05608) and NINDS (NS 40365), and S. L. I. by a NARSAD Young Investigator Award and a grant from NIDA (DA 015498).

References

Abe, M., Oka, T., Hori, T., and Takahashi, S. 2001. Prostanoids in the preoptic hypothalamus mediate systemic lipopolysaccharide-induced hyperalgesia in rats. Brain Res. 916, 41–49.

Abols, I. A. and Basbaum, A. I. 1981. Afferent connections of the rostral medulla of the cat: a neural substrate for midbrain-medullary interactions in the modulation of pain. J. Comp. Neurol. 201, 285–297.

Ackley, M. A., Hurley, R. W., Virnich, D. E., and Hammond, D. L. 2001. A cellular mechanism for the antinociceptive effect of a kappa opioid receptor agonist. Pain 91, 377–388.

Aguiar, M. S. and Brandao, M. L. 1994. Conditioned place aversion produced by microinjections of substance p into the periaqueductal gray of rats. Behav. Pharmacol. 5, 369–373.

An, X., Bandler, R., Ongur, D., and Price, J. L. 1998. Prefrontal cortical projections to longitudinal columns in the midbrain periaqueductal gray in macaque monkeys. J. Comp. Neurol. 401, 455–479.

Anderson, S. D., Basbaum, A. I., and Fields, H. L. 1977. Response of medullary raphe neurons to peripheral stimulation and to systemic opiates. Brain Res. 123, 363–368.

Azami, J., Llewelyn, M. B., and Roberts, M. H. 1982. The contribution of nucleus reticularis paragigantocellularis and nucleus raphe magnus to the analgesia produced by systemically administered morphine, investigated with the microinjection technique. Pain 12, 229–246.

Baez, M. A., Brink, T. S., and Mason, P. 2005. Roles for pain modulatory cells during micturition and continence. J. Neurosci. 25, 384–394.

Bagley, E. E., Chieng, B. C., Christie, M. J., and Connor, M. 2005. Opioid tolerance in periaqueductal gray neurons isolated from mice chronically treated with morphine. Br. J. Pharmacol. 146, 68–76

Bailey, C. P. and Connor, M. 2005. Opioids: cellular mechanisms of tolerance and physical dependence. Curr. Opin. Pharmacol. 5, 60–68.

Bajic, D., Van Bockstaele, E. J., and Proudfit, H. K. 2001. Ultrastructural analysis of ventrolateral periaqueductal gray projections to the a7 catecholamine cell group. Neuroscience 104, 181–197.

Bandler, R. and DePaulis, A. 1991. Midbrain Periaqueductal Gray Control of Defensive Behavior in Cat and Rat. In: Midbrain Periaqueductal Gray Matter (*eds*. A. DePaulis and R. Bandler), pp. 175–198. Plenum.

Bandler, R. and Shipley, M. T. 1994. Columnar organization in the midbrain periaqueductal gray: modules for emotional expression? Trends Neurosci. 17, 379–389.

Bandler, R., Keay, K. A., Floyd, N., and Price, J. 2000. Central circuits mediating patterned autonomic activity during active vs. passive emotional coping. Brain Res. Bull. 53, 95–104.

Barbaro, N. M., Heinricher, M. M., and Fields, H. L. 1986. Putative pain modulating neurons in the rostral ventral medulla: reflex-related activity predicts effects of morphine. Brain Res. 366, 203–210.

Barbaro, N. M., Heinricher, M. M., and Fields, H. L. 1989. Putative nociceptive modulatory neurons in the rostral ventromedial medulla of the rat display highly correlated firing patterns. Somatosens. Mot. Res. 6, 413–425.

Bederson, J. B., Fields, H. L., and Barbaro, N. M. 1990. Hyperalgesia during naloxone-precipitated withdrawal from morphine is associated with increased on-cell activity in the rostral ventromedial medulla. Somatosens. Mot. Res. 7, 185–203.

Behbehani, M. M. and Da Costa Gomez, T. M. 1996. Properties of a projection pathway from the medial preoptic nucleus to the midbrain periaqueductal gray of the rat and its role in the regulation of cardiovascular function. Brain Res. 740, 141–150.

Behbehani, M. M. and Pomeroy, S. L. 1978. Effect of morphine injected in periadueductal gray on the activity of single units in nucleus raphe magnus of the rat. Brain Res. 149, 266–269.

Beitz, A. J. 1982. The organization of afferent projections to the midbrain periaqueductal gray of the rat. Neuroscience 7, 133–159.

Beitz, A. J. 1985. The midbrain periaqueductal gray in the rat. I. Nuclear volume, cell number, density, orientation, and regional subdivisions. J. Comp. Neurol. 237, 445–459.

Beitz, A. J. and Shepard, R. D. 1985. The midbrain periaqueductal gray in the rat. Ii. A golgi analysis. J. Comp. Neurol. 237, 460–475.

Bellchambers, C. E., Chieng, B., Keay, K. A., and Christie, M. J. 1998. Swim-stress but not opioid withdrawal increases expression of c-fos immunoreactivity in rat periaqueductal gray neurons which project to the rostral ventromedial medulla. Neuroscience 83, 517–524.

Bellgowan, P. S. and Helmstetter, F. J. 1996. Neural systems for the expression of hypoalgesia during nonassociative fear. Behav. Neurosci. 110, 727–736.

Benedetti, F. and Amanzio, M. 1997. The neurobiology of placebo analgesia: from endogenous opioids to cholecystokinin. Prog. Neurobiol. 52, 109–125.

Berrino, L., Oliva, P., Rossi, F., Palazzo, E., Nobili, B., and Maione, S. 2001. Interaction between metabotropic and NMDA glutamate receptors in the periaqueductal grey pain modulatory system. Naunyn Schmiedebergs Arch. Pharmacol. 364, 437–443.

Bie, B., Fields, H. L., Williams, J. T., and Pan, Z. Z. 2003. Roles of α_1- and α_2-adrenoceptors in the nucleus raphe magnus in opioid analgesia and opioid abstinence-induced hyperalgesia. J. Neurosci. 23, 7950–7957.

Bie, B. and Pan, Z. Z. 2003. Presynaptic mechanism for anti-analgesic and anti-hyperalgesic actions of kappa-opioid receptors. J. Neurosci. 23, 7262–7268.

Bittencourt, A. S., Carobrez, A. P., Zamprogno, L. P., Tufik, S., and Schenberg, L. C. 2004. Organization of single components of defensive behaviors within distinct columns of periaqueductal gray matter of the rat: role of n-methyl-D-aspartic acid glutamate receptors. Neuroscience 125, 71–89.

Blanchard, D. C. and Blanchard, R. J. 1972. Innate and conditioned reactions to threat in rats with amygdaloid lesions. J. Comp. Physiol. Psychol. 81, 281–290.

Blanchard, R. J., Flannelly, K. J., and Blanchard, D. C. 1986. Defensive behaviors of laboratory and wild *Rattus norvegicus*. J. Comp. Psychol. 100, 101–107.

Bodnar, R. J. 1986. Neuropharmacological and neuroendocrine substrates of stress-induced analgesia. Ann. N. Y. Acad. Sci. 467, 345–360.

Bodnar, R. J., Kelly, D. D., Brutus, M., and Glusman, M. 1980. Stress-induced analgesia: neural and hormonal determinants. Neurosci. Biobehav. Rev. 4, 87–100.

Bodnar, R. J., Williams, C. L., Lee, S. J., and Pasternak, G. W. 1988. Role of mu 1-opiate receptors in supraspinal opiate analgesia: a microinjection study. Brain Res. 447, 25–34.

Bouhassira, D., Chollet, R., Coffin, B., Lemann, M., Le Bars, D., Willer, J. C., and Jian, R. 1994. Inhibition of a somatic nociceptive reflex by gastric distention in humans. Gastroenterology 107, 985–992.

Bouhassira, D., Sabate, J. M., Coffin, B., Le Bars, D., Willer, J. C., and Jian, R. 1998. Effects of rectal distensions on nociceptive flexion reflexes in humans. Am. J. Physiol. 275, G410–417.

Bowker, R. M. and Abbott, L. C. 1988. The origins and trajectories of somatostatin reticulospinal neurons: a potential neurotransmitter candidate of the dorsal reticulospinal pathway. Brain Res. 447, 398–403.

Bowker, R. M., Westlund, K. N., Sullivan, M. C., Wilber, J. F., and Coulter, J. D. 1983. Descending serotonergic, peptidergic and cholinergic pathways from the raphe nuclei: a multiple transmitter complex. Brain Res. 288, 33–48.

Bozarth, M. A. 1994. Physical dependence produced by central morphine infusions: an anatomical mapping study. Neurosci. Biobehav. Rev. 18, 373–383.

Brink, T. S. and Mason, P. 2004. Role for raphe magnus neuronal responses in the behavioral reactions to colorectal distension. J. Neurophysiol. 92, 2302–2311.

Buhler, A. V., Choi, J., Proudfit, H. K., and Gebhart, G. F. 2005. Neurotensin activation of the NTR1 on spinally-projecting serotonergic neurons in the rostral ventromedial medulla is antinociceptive. Pain 114, 285–294.

Buhler, A. V., Proudfit, H. K., and Gebhart, G. F. 2004. Separate populations of neurons in the rostral ventromedial medulla

project to the spinal cord and to the dorsolateral pons in the rat. Brain Res. 1016, 12–19.

Burdin, T. A., Graeff, F. G., and Pela, I. R. 1992. Opioid mediation of the antiaversive and hyperalgesic actions of bradykinin injected into the dorsal periaqueductal gray of the rat. Physiol. Behav. 52, 405–410.

Burgess, S. E., Gardell, L. R., Ossipov, M. H., Malan, T. P., Jr., Vanderah, T. W., Lai, J., and Porreca, F. 2002. Time-dependent descending facilitation from the rostral ventromedial medulla maintains, but does not initiate, neuropathic pain. J. Neurosci. 22, 5129–5136.

Burstein, R. and Potrebic, S. 1993. Retrograde labeling of neurons in the spinal cord that project directly to the amygdala or the orbital cortex in the rat. J. Comp. Neurol. 335, 469–485.

Bushnell, M. C., Duncan, G. H., Dubner, R., and He, L. F. 1984. Activity of trigeminothalamic neurons in medullary dorsal horn of awake monkeys trained in a thermal discrimination task. J. Neurophysiol. 52, 170–187.

Cairns, B. E., Fragoso, M. C., and Soja, P. J. 1995. Activity of rostral trigeminal sensory neurons in the cat during wakefulness and sleep. J. Neurophysiol. 73, 2486–2498.

Cairns, B. E., McErlane, S. A., Fragoso, M. C., Jia, W. G., and Soja, P. J. 1996. Spontaneous discharge and peripherally evoked orofacial responses of trigemino-thalamic tract neurons during wakefulness and sleep. J. Neurosci. 16, 8149–8159.

Calejesan, A. A., Kim, S. J., and Zhuo, M. 2000. Descending facilitatory modulation of a behavioral nociceptive response by stimulation in the adult rat anterior cingulate cortex. Eur. J. Pain 4, 83–96.

Canteras, N. S., Simerly, R. B., and Swanson, L. W. 1995. Organization of projections from the medial nucleus of the amygdala: a PHAL study in the rat. J. Comp. Neurol. 360, 213–245.

Carrive, P. 2000. Conditioned fear to environmental context: cardiovascular and behavioral components in the rat. Brain Res. 858, 440–445.

Carrive, P. and Morgan, M. M. 2004. Periaqueductal Gray. In: The Human Nervous System (eds. G. Paxinos and J. K. Mai), pp. 393–423. Elsevier.

Casey, K. L. and Morrow, T. J. 1988. Supraspinal nocifensive responses of cats: spinal cord pathways, monoamines, and modulation. J. Comp. Neurol. 270, 591–605.

Casey, K. L. and Morrow, T. J. 1989. Effect of medial bulboreticular and raphe nuclear lesions on the excitation and modulation of supraspinal nocifensive behaviors in the cat. Brain Res. 501, 150–161.

Cervero, F. and Wolstencroft, J. H. 1984. A positive feedback loop between spinal cord nociceptive pathways and antinociceptive areas of the cat's brain stem. Pain 20, 125–138.

Cheng, Z. F., Fields, H. L., and Heinricher, M. M. 1986. Morphine microinjected into the periaqueductal gray has differential effects on 3 classes of medullary neurons. Brain Res. 375, 57–65.

Chieng, B. and Christie, M. D. 1996. Local opioid withdrawal in rat single periaqueductal gray neurons in vitro. J. Neurosci. 16, 7128–7136.

Chieng, B. and Christie, M. J. 1994a. Hyperpolarization by opioids acting on mu-receptors of a sub-population of rat periaqueductal gray neurones in vitro. Br. J. Pharmacol. 113, 121–128.

Chieng, B. and Christie, M. J. 1994b. Inhibition by opioids acting on mu-receptors of gabaergic and glutamatergic postsynaptic potentials in single rat periaqueductal gray neurones in vitro. Br. J. Pharmacol. 113, 303–309.

Chieng, B. C., Hallberg, C., Nyberg, F. J., and Christie, M. J. 2005. Enhanced c-fos in periaqueductal grey gabaergic neurons during opioid withdrawal. Neuroreport 16, 1279–1283.

Chieng, B., Keay, K. A., and Christie, M. J. 1995. Increased fos-like immunoreactivity in the periaqueductal gray of anaesthetised rats during opiate withdrawal. Neurosci. Lett. 183, 79–82.

Chiou, L. C. and Huang, L. Y. 1999. Mechanism underlying increased neuronal activity in the rat ventrolateral periaqueductal grey by a mu-opioid. J. Physiol. 518 (Pt 2), 551–559.

Christie, M. J., Vaughan, C. W., and Ingram, S. L. 1999. Opioids, NSAIDs and 5-lipoxygenase inhibitors act synergistically in brain via arachidonic acid metabolism. Inflamm. Res. 48, 1–4.

Christie, M. J., Williams, J. T., Osborne, P. B., and Bellchambers, C. E. 1997. Where is the locus in opioid withdrawal? Trends Pharmacol. Sci. 18, 134–140.

Clark, F. M. and Proudfit, H. K. 1991. Projections of neurons in the ventromedial medulla to pontine catecholamine cell groups involved in the modulation of nociception. Brain Res. 540, 105–115.

Clement, C. I., Keay, K. A., Podzebenko, K., Gordon, B. D., and Bandler, R. 2000. Spinal sources of noxious visceral and noxious deep somatic afferent drive onto the ventrolateral periaqueductal gray of the rat. J. Comp. Neurol. 425, 323–344.

Commons, K. G. 2003. Translocation of presynaptic delta opioid receptors in the ventrolateral periaqueductal gray after swim stress. J. Comp. Neurol. 464, 197–207.

Commons, K. G. and Valentino, R. J. 2002. Cellular basis for the effects of substance P in the periaqueductal gray and dorsal raphe nucleus. J. Comp. Neurol. 447, 82–97.

Commons, K. G., Aicher, S. A., Kow, L. M., and Pfaff, D. W. 2000. Presynaptic and postsynaptic relations of mu-opioid receptors to gamma-aminobutyric acid-immunoreactive and medullary-projecting periaqueductal gray neurons. J. Comp. Neurol. 419, 532–542.

Commons, K. G., Beck, S. G., Rudoy, C., and Van Bockstaele, E. J. 2001. Anatomical evidence for presynaptic modulation by the delta opioid receptor in the ventrolateral periaqueductal gray of the rat. J. Comp. Neurol. 430, 200–208.

Connor, M. and Christie, M. J. 1998. Modulation of Ca^{2+} channel currents of acutely dissociated rat periaqueductal grey neurons. J. Physiol. 509, 47–58.

Coutinho, S. V., Urban, M. O., and Gebhart, G. F. 1998. Role of glutamate receptors and nitric oxide in the rostral ventromedial medulla in visceral hyperalgesia. Pain 78, 59–69.

Crown, E. D., Grau, J. W., and Meagher, M. W. 2004. Pain in a balance: noxious events engage opposing processes that concurrently modulate nociceptive reactivity. Behav. Neurosci. 118, 1418–1426.

da Costa Gomez, T. M. and Behbehani, M. M. 1995. An electrophysiological characterization of the projection from the central nucleus of the amygdala to the periaqueductal gray of the rat: the role of opioid receptors. Brain Res. 689, 21–31.

da Costa Gomez, T. M., Chandler, S. D., and Behbehani, M. M. 1996. The role of the basolateral nucleus of the amygdala in the pathway between the amygdala and the midbrain periaqueductal gray in the rat. Neurosci. Lett. 214, 5–8.

Davis, M. 1992a. The Role of the Amygdala in Conditioned Fear. In: The Amygdala: Neurobiological Aspects of Emotion, Memory and Mental Dysfunction (eds. J. P. Aggleton), pp. 255–305. Wiley–Liss.

Davis, M. 1992b. The role of the amygdala in fear and anxiety. Annu. Rev. Neurosci. 15, 353–375.

Davis, K. D., Hutchison, W. D., Lozano, A. M., and Dostrovsky, J. O. 1994. Altered pain and temperature perception following cingulotomy and capsulotomy in a patient with schizoaffective disorder. Pain 59, 189–199.

Depaulis, A., Keay, K. A., and Bandler, R. 1994. Quiescence and hyporeactivity evoked by activation of cell bodies in the ventrolateral midbrain periaqueductal gray of the rat. Exp. Brain Res. 99, 75–83.

Depaulis, A., Morgan, M. M., and Liebeskind, J. C. 1987. Gabaergic modulation of the analgesic effects of morphine microinjected in the ventral periaqueductal gray matter of the rat. Brain Res. 436, 223–228.

Dickenson, A. H., Oliveras, J. L., and Besson, J. M. 1979. Role of the nucleus raphe magnus in opiate analgesia as studied by the microinjection technique in the rat. Brain Res. 170, 95–111.

Dielenberg, R. A., Leman, S., and Carrive, P. 2004. Effect of dorsal periaqueductal gray lesions on cardiovascular and behavioral responses to cat odor exposure in rats. Behav. Brain Res. 153, 487–496.

Donahue, R. R., LaGraize, S. C., and Fuchs, P. N. 2001. Electrolytic lesion of the anterior cingulate cortex decreases inflammatory, but not neuropathic nociceptive behavior in rats. Brain Res. 897, 131–138.

Drew, G. M., Mitchell, V. A., and Vaughan, C. W. 2005. Postsynaptic actions of substance P on rat periaqueductal grey neurons in vitro. Neuropharmacology 49, 587–595.

Drower, E. J. and Hammond, D. L. 1988. Gabaergic modulation of nociceptive threshold: effects of thip and bicuculline microinjected in the ventral medulla of the rat. Brain Res. 450, 316–324.

Ellrich, J., Ulucan, C., and Schnell, C. 2000. Are 'neutral cells' in the rostral ventro-medial medulla subtypes of on- and off-cells? Neurosci. Res. 38, 419–423.

Ellrich, J., Ulucan, C., and Schnell, C. 2001. Is the response pattern of on- and off-cells in the rostral ventromedial medulla to noxious stimulation independent of stimulation site? Exp. Brain Res. 136, 394–399.

Fang, F. G., Haws, C. M., Drasner, K., Williamson, A., and Fields, H. L. 1989. Opioid peptides (dago-enkephalin, dynorphin a (1–13), bam 22p) microinjected into the rat brainstem: comparison of their antinociceptive effect and their effect on neuronal firing in the rostral ventromedial medulla. Brain Res. 501, 116–128.

Fanselow, M. S. 1985. Odors released by stressed rats produce opioid analgesia in unstressed rats. Behav. Neurosci. 99, 589–592.

Fanselow, M. S. 1986. Conditioned fear-induced opiate analgesia: a competing motivational state theory of stress analgesia. Ann. N. Y. Acad. Sci. 467, 40–54.

Fanselow, M. S. 1991. The Midbrain Periaqueductal Gray as a Coordinator of Action in Response to Fear and Anxiety. In: The Midbrain Periaqueductal Gray Matter (eds. A. DePaulis and R. Bandler), pp. 151–173. Plenum.

Fanselow, M. S. and Helmstetter, F. J. 1988. Conditional analgesia, defensive freezing, and benzodiazepines. Behav. Neurosci. 102, 233–243.

Fanselow, M. S. and Kim, J. J. 1992. The benzodiazepine inverse agonist dmcm as an unconditional stimulus for fear-induced analgesia: implications for the role of gaba_a receptors in fear-related behavior. Behav. Neurosci. 106, 336–344.

Farook, J. M., Wang, Q., Moochhala, S. M., Zhu, Z. Y., Lee, L., and Wong, P. T. 2004. Distinct regions of periaqueductal gray (PAG) are involved in freezing behavior in hooded pvg rats on the cat-freezing test apparatus. Neurosci. Lett. 354, 139–142.

Fields, H. L. and Heinricher, M. M. 1985. Anatomy and physiology of a nociceptive modulatory system. Philos. Trans. R. Soc. Lond. B. Biol. Sci. 308, 361–374.

Fields, H. L., Basbaum, A. I., and Heinricher, M. M. 2005. Central nervous system mechanisms of pain modulation. In: Wall and Melzack's textbook of pain, 5th ed. (ed. McMahon, S. and Koltzenburg, M. Elsevier.

Fields, H. L., Bry, J., Hentall, I., and Zorman, G. 1983a. The activity of neurons in the rostral medulla of the rat during withdrawal from noxious heat. J. Neurosci. 3, 2545–2552.

Fields, H. L., Malick, A., and Burstein, R. 1995. Dorsal horn projection targets of ON and OFF cells in the rostral ventromedial medulla. J. Neurophysiol. 74, 1742–1759.

Fields, H. L., Vanegas, H., Hentall, I. D., and Zorman, G. 1983b. Evidence that disinhibition of brain stem neurones contributes to morphine analgesia. Nature 306, 684–686.

Floyd, N. S., Price, J. L., Ferry, A. T., Keay, K. A., and Bandler, R. 2000. Orbitomedial prefrontal cortical projections to distinct longitudinal columns of the periaqueductal gray in the rat. J. Comp. Neurol. 422, 556–578.

Foltz, E. L. and Lowell, E. W. 1962. Pain "relief" by frontal cingulumotomy. J. Neurosurg. 19, 89–100.

Foo, H. and Helmstetter, F. J. 2000. Activation of kappa opioid receptors in the rostral ventromedial medulla blocks stress-induced antinociception. Neuroreport 11, 3349–3352.

Foo, H. and Mason, P. 2003. Discharge of raphe magnus ON and OFF cells is predictive of the motor facilitation evoked by repeated laser stimulation. J. Neurosci. 23, 1933–1940.

Foo, H. and Mason, P. 2005. Movement-related discharge of ventromedial medullary neurons. J. Neurophysiol. 93, 873–883.

Fox, R. J. and Sorenson, C. A. 1994. Bilateral lesions of the amygdala attenuate analgesia induced by diverse environmental challenges. Brain Res. 648, 215–221.

Friederich, M. W. and Walker, J. M. 1990. The effect of foot-shock on the noxious-evoked activity of neurons in the rostral ventral medulla. Brain Res. Bull. 24, 605–608.

Friedrich, A. E. and Gebhart, G. F. 2003. Modulation of visceral hyperalgesia by morphine and cholecystokinin from the rat rostroventral medial medulla. Pain 104, 93–101.

Fuchs, P. N., Balinsky, M., and Melzack, R. 1996. Electrical stimulation of the cingulum bundle and surrounding cortical tissue reduces formalin-test pain in the rat. Brain Res. 743, 116–123.

Gao, K., Chen, D. O., Genzen, J. R., and Mason, P. 1998. Activation of serotonergic neurons in the raphe magnus is not necessary for morphine analgesia. J. Neurosci. 18, 1860–1868.

Gao, K., Kim, Y. H., and Mason, P. 1997. Serotonergic pontomedullary neurons are not activated by antinociceptive stimulation in the periaqueductal gray. J. Neurosci. 17, 3285–3292.

Gauriau, C. and Bernard, J. F. 2004. A comparative reappraisal of projections from the superficial laminae of the dorsal horn in the rat: the forebrain. J. Comp. Neurol. 468, 24–56.

Gebhart, G. F. 1982. Opiate and opioid peptide effects on brain stem neurons: relevance to nociception and antinociceptive mechanisms. Pain 12, 93–140.

Gilbert, A. and Franklin, K. 2002. The role of descending fibers from the rostral ventromedial medulla in opioid analgesia in rats. Eur. J. Pharmacol. 449, 75.

Gilbert, A. K. and Franklin, K. B. 2001. GABAergic modulation of descending inhibitory systems from the rostral ventromedial medulla (RVM). Dose-response analysis of nociception and neurological deficits. Pain 90, 25–36.

Grau, J. W., Hyson, R. L., Maier, S. F., Madden, J. T., and Barchas, J. D. 1981. Long-term stress-induced analgesia and activation of the opiate system. Science 213, 1409–1411.

Guilbaud, G., Peschanski, M., Gautron, M., and Binder, D. 1980. Responses of neurons of the nucleus raphe magnus to noxious stimuli. Neurosci. Lett. 17, 149–154.

Gutstein, H. B., Mansour, A., Watson, S. J., Akil, H., and Fields, H. L. 1998. Mu and kappa opioid receptors in periaqueductal gray and rostral ventromedial medulla. Neuroreport 9, 1777–1781.

Hack, S. P., Vaughan, C. W., and Christie, M. J. 2003. Modulation of GABA release during morphine withdrawal in midbrain neurons in vitro. Neuropharmacology 45, 575–584.

Hajnik, T., Lai, Y. Y., and Siegel, J. M. 2000. Atonia-related regions in the rodent pons and medulla. J. Neurophysiol. 84, 1942–1948.

Hammond, D. L. and Yaksh, T. L. 1984. Antagonism of stimulation-produced antinociception by intrathecal administration of methysergide or phentolamine. Brain Res. 298, 329–337.

Harasawa, I., Fields, H. L., and Meng, I. D. 2000. Delta opioid receptor mediated actions in the rostral ventromedial medulla on tail flick latency and nociceptive modulatory neurons. Pain 85, 255–262.

Hardy, S. G. 1985. Analgesia elicited by prefrontal stimulation. Brain Res. 339, 281–284.

Harris, J. A. and Westbrook, R. F. 1995. Effects of benzodiazepine microinjection into the amygdala or periaqueductal gray on the expression of conditioned fear and hypoalgesia in rats. Behav. Neurosci. 109, 295–304.

Harris, J. A., Westbrook, R. F., Duffield, T. Q., and Bentivoglio, M. 1995. Fos expression in the spinal cord is suppressed in rats displaying conditioned hypoalgesia. Behav. Neurosci. 109, 320–328.

Haws, C. M., Heinricher, M. M., and Fields, H. L. 1990. Alpha-adrenergic receptor agonists, but not antagonists, alter the tail-flick latency when microinjected into the rostral ventromedial medulla of the lightly anesthetized rat. Brain Res. 533, 192–195.

Hayes, R. L., Dubner, R., and Hoffman, D. S. 1981. Neuronal activity in medullary dorsal horn of awake monkeys trained in a thermal discrimination task. Ii. Behavioral modulation of responses to thermal and mechanical stimuli. J. Neurophysiol. 46, 428–443.

Heinricher, M. M. 2005. Nociceptin/orphanin FQ: pain, stress and neural circuits. Life Sci. 77(25), 3127–3132.

Heinricher, M. M. and Fields, H. L. 2003. The Delta Opioid Receptor and Brain Pain-Modulating Circuits. In: The Delta Receptor: Molecular and Effect of Delta Opioid Compounds (eds. K. J. Chang, F. Porreca, and J. Woods), pp. 467–480. Marcel Dekker.

Heinricher, M. M. and Kaplan, H. J. 1991. GABA-mediated inhibition in rostral ventromedial medulla: role in nociceptive modulation in the lightly anesthetized rat. Pain 47, 105–113.

Heinricher, M. M. and McGaraughty, S. 1998. Analysis of excitatory amino acid transmission within the rostral ventromedial medulla: implications for circuitry. Pain 75, 247–255.

Heinricher, M. M. and Morgan, M. M. 1999. Supraspinal Mechanisms of Opioid Analgesia. In: Opioids and Pain Control (ed. C. Stein), pp. 46–69. Cambridge University Press.

Heinricher, M. M. and Neubert, M. J. 2004. Neural basis for the hyperalgesic action of cholecystokinin in the rostral ventromedial medulla. J. Neurophysiol. 92, 1982–1989.

Heinricher, M. M. and Rosenfeld, J. P. 1985. Microinjection of morphine into nucleus reticularis paragigantocellularis of the rat: suppression of noxious-evoked activity of nucleus raphe magnus neurons. Brain Res. 359, 388–391.

Heinricher, M. M. and Roychowdhury, S. 1997. Reflex-related activation of putative pain facilitating neurons in rostral ventromedial medulla (RVM) depends upon excitatory amino acid transmission. Neuroscience 78, 1159–1165.

Heinricher, M. M. and Tortorici, V. 1994. Interference with GABA transmission in the rostral ventromedial medulla: disinhibition of off-cells as a central mechanism in nociceptive modulation. Neuroscience 63, 533–546.

Heinricher, M. M., Barbaro, N. M., and Fields, H. L. 1989. Putative nociceptive modulating neurons in the rostral ventromedial medulla of the rat: firing of on- and off-cells is related to nociceptive responsiveness. Somatosens. Mot. Res. 6, 427–439.

Heinricher, M. M., Haws, C. M., and Fields, H. L. 1988. Opposing Actions of Norepinephrine and Clonidine on Single Pain-Modulating Neurons in Rostral Ventromedial Medulla. In: Pain Research and Clinical Management (eds. R. Dubner, G. F. Gebhart, and M. R. Bond), pp. 590–594. Elsevier.

Heinricher, M. M., Haws, C. M., and Fields, H. L. 1991. Evidence for GABA-mediated control of putative nociceptive modulating neurons in the rostral ventromedial medulla: iontophoresis of bicuculline eliminates the off-cell pause. Somatosens. Mot. Res. 8, 215–22.

Heinricher, M. M., Martenson, M. E., and Neubert, M. J. 2004a. Prostaglandin E_2 in the midbrain periaqueductal gray produces hyperalgesia and activates pain-modulating circuitry in the rostral ventromedial medulla. Pain 110, 419–426.

Heinricher, M. M., McGaraughty, S., and Farr, D. A. 1999. The role of excitatory amino acid transmission within the rostral ventromedial medulla in the antinociceptive actions of systemically administered morphine. Pain 81, 57–65.

Heinricher, M. M., McGaraughty, S., and Grandy, D. K. 1997. Circuitry underlying antiopioid actions of orphanin FQ in the rostral ventromedial medulla. J. Neurophysiol. 78, 3351–3358.

Heinricher, M. M., McGaraughty, S., and Tortorici, V. 2001a. Circuitry underlying antiopioid actions of cholecystokinin within the rostral ventromedial medulla. J. Neurophysiol. 85, 280–286.

Heinricher, M. M., Morgan, M. M., and Fields, H. L. 1992. Direct and indirect actions of morphine on medullary neurons that modulate nociception. Neuroscience 48, 533–543.

Heinricher, M. M., Morgan, M. M., Tortorici, V., and Fields, H. L. 1994. Disinhibition of off-cells and antinociception produced by an opioid action within the rostral ventromedial medulla. Neuroscience 63, 279–288.

Heinricher, M. M., Neubert, M. J., Martenson, M. E., and Gonçalves, L. 2004b. Prostaglandin E_2 in the medial preoptic area produces hyperalgesia and activates pain-modulating circuitry in the rostral ventromedial medulla. Neuroscience 128, 389–398.

Heinricher, M. M., Pertovaara, A., and Ossipov, M. H. 2003. Descending Modulation after Injury. In: Progress in Pain Research and Management, Vol. 24 (eds. J. O. Dostrovsky, D. B. Carr, and M. Koltzenburg), pp. 251–260. IASP Press.

Heinricher, M. M., Schouten, J. C., and Jobst, E. E. 2001b. Activation of brainstem n-methyl-D-aspartate receptors is required for the analgesic actions of morphine given systemically. Pain 92, 129–138.

Helmstetter, F. J. 1992. The amygdala is essential for the expression of conditional hypoalgesia. Behav. Neurosci. 106, 518–528.

Helmstetter, F. J. and Bellgowan, P. S. 1993. Lesions of the amygdala block conditional hypoalgesia on the tail flick test. Brain Res. 612, 253–257.

Helmstetter, F. J. and Bellgowan, P. S. 1994. Hypoalgesia in response to sensitization during acute noise stress. Behav. Neurosci. 108, 177–185.

Helmstetter, F. J. and Landeira-Fernandez, J. 1990. Conditional hypoalgesia is attenuated by naltrexone applied to the periaqueductal gray. Brain Res. 537, 88–92.

Helmstetter, F. J. and Tershner, S. A. 1994. Lesions of the periaqueductal gray and rostral ventromedial medulla disrupt antinociceptive but not cardiovascular aversive conditional responses. J. Neurosci. 14, 7099–7108.

Helmstetter, F. J., Calcagnetti, D. J., and Fanselow, M. S. 1990. The beta-carboline dmcm produces hypoalgesia after central administration. Psychobiology 18, 293–297.

Helmstetter, F. J., Tershner, S. A., Poore, L. H., and Bellgowan, P. S. 1998. Antinociception following opioid stimulation of the basolateral amygdala is expressed through the periaqueductal gray and rostral ventromedial medulla. Brain Res. 779, 104–118.

Herbert, H. and Saper, C. B. 1992. Organization of medullary adrenergic and noradrenergic projections to the periaqueductal gray matter in the rat. J. Comp. Neurol. 315, 34–52.

Herkenham, M., Lynn, A. B., Johnson, M. R., Melvin, L. S., de Costa, B. R., and Rice, K. C. 1991. Characterization and localization of cannabinoid receptors in rat brain: a quantitative in vitro autoradiographic study. J. Neurosci. 11, 563–583.

Hermann, D. M., Luppi, P. H., Peyron, C., Hinckel, P., and Jouvet, M. 1996. Forebrain projections of the rostral nucleus raphe magnus shown by iontophoretic application of choleratoxin b in rats. Neurosci. Lett. 216, 151–154.

Hermann, D. M., Luppi, P. H., Peyron, C., Hinckel, P., and Jouvet, M. 1997. Afferent projections to the rat nuclei raphe magnus, raphe pallidus and reticularis gigantocellularis pars alpha demonstrated by iontophoretic application of choleratoxin (subunit b). J. Chem. Neuroanat. 13, 1–21.

Hernandez, N., Lopez, Y., and Vanegas, H. 1989. Medullary on- and off-cell responses precede both segmental and thalamic responses to tail heating. Pain 39, 221–230.

Hicks, T. P. 1983. Antagonism of synaptic transmission in vivo: contributions of microiontophoresis. Brain Behav. Evol. 22, 1–12.

Hirakawa, N., Tershner, S. A., Fields, H. L., and Manning, B. H. 2000. Bi-directional changes in affective state elicited by manipulation of medullary pain-modulatory circuitry. Neuroscience 100, 861–871.

Hofer, M. A. 1970. Cardiac and respiratory function during sudden prolonged immobility in wild rodents. Psychosom. Med. 32, 633–647.

Hohmann, A. G., Suplita, R. L., Bolton, N. M., Neely, M. H., Fegley, D., Mangieri, R., Krey, J. F., Walker, J. M., Holmes, P. V., Crystal, J. D., Duranti, A., Tontini, A., Mor, M., Tarzia, G., and Piomelli, D. 2005. An endocannabinoid mechanism for stress-induced analgesia. Nature 435, 1108–1112.

Hosoi, M., Oka, T., and Hori, T. 1997. Prostaglandin e receptor ep3 subtype is involved in thermal hyperalgesia through its actions in the preoptic hypothalamus and the diagonal band of broca in rats. Pain 71, 303–311.

Hough, L. B., Nalwalk, J. W., Stadel, R., Timmerman, H., Leurs, R., Paria, B. C., Wang, X., and Dey, S. K. 2002. Inhibition of improgan antinociception by the cannabinoid (cb)₁ antagonist n- (piperidin-1-yl)-5- (4-chlorophenyl)-1- (2,4-dichlorophenyl)-4-methyl-1h-p yrazole-3-carboxamide (sr141716a): lack of obligatory role for endocannabinoids acting at cb (1) receptors. J. Pharmacol. Exp. Ther. 303, 314–322.

Hurley, R. W. and Hammond, D. L. 2000. The analgesic effects of supraspinal μ and δ opioid receptor agonists are potentiated during persistent inflammation. J. Neurosci. 20, 1249–1259.

Hurley, R. W. and Hammond, D. L. 2001. Contribution of endogenous enkephalins to the enhanced analgesic effects of supraspinal μ opioid receptor agonists after inflammatory injury. J. Neurosci 21, 2536–2545.

Hutchison, W. D., Harfa, L., and Dostrovsky, J. O. 1996. Ventrolateral orbital cortex and periaqueductal gray stimulation-induced effects on on- and off-cells in the rostral ventromedial medulla in the rat. Neuroscience 70, 391–407.

Ingram, S. L., Fossum, E., and Morgan, M. M. 2007. Behavioral and electrophysiological evidence for opioid tolerance in adolescent rats. Neuropsychopharmacology 32, 600–606.

Ingram, S. L., Vaughan, C. W., Bagley, E. E., Connor, M., and Christie, M. J. 1998. Enhanced opioid efficacy in opioid dependence is caused by an altered signal transduction pathway. J. Neurosci. 18, 10269–10276.

Iversen, L. and Chapman, V. 2002. Cannabinoids: a real prospect for pain relief? Curr. Opin. Pharmacol. 2, 50–55.

Jacquet, Y. F. and Lajtha, A. 1976. The periaqueductal gray: site of morphine analgesia and tolerance as shown by 2-way cross tolerance between systemic and intracerebral injections. Brain Res. 103, 501–513.

Jinks, S. L., Carstens, E., and Antognini, J. F. 2004a. Isoflurane differentially modulates medullary on and off neurons while suppressing hind-limb motor withdrawals. Anesthesiology 100, 1224–1234.

Jinks, S. L., Carstens, E., and Antognini, J. F. 2004b. Medullary on-cells facilitate multilimb movements elicited by intense noxious stimulation. Abstract Viewer/Itinerary Planner. Society for Neuroscience, Online, Program 296.297.

Kalmari, J. and Pertovaara, A. 2004. Colorectal distension-induced suppression of a nociceptive somatic reflex response in the rat: modulation by tissue injury or inflammation. Brain Res. 1018, 106–110.

Kalra, A., Urban, M. O., and Sluka, K. A. 2001. Blockade of opioid receptors in rostral ventral medulla prevents antihyperalgesia produced by transcutaneous electrical nerve stimulation (tens). J. Pharmacol. Exp. Ther. 298, 257–263.

Kalyuzhny, A. E. and Wessendorf, M. W. 1998. Relationship of μ- and δ-opioid receptors to GABAergic neurons in the central nervous system, including antinociceptive brainstem circuits. J. Comp. Neurol. 392, 528–547.

Kalyuzhny, A. E. and Wessendorf, M. W. 1999. Serotonergic and GABaergic neurons in the medial rostral ventral medulla express kappa-opioid receptor immunoreactivity. Neuroscience 90, 229–234.

Kalyuzhny, A. E., Arvidsson, U., Wu, W., and Wessendorf, M. W. 1996. M-opioid and δ-opioid receptors are expressed in brainstem antinociceptive circuits: studies using immunocytochemistry and retrograde tract-tracing. J. Neurosci. 16, 6490–6503.

Kaplan, H. and Fields, H. L. 1991. Hyperalgesia during acute opioid abstinence: evidence for a nociceptive facilitating function of the rostral ventromedial medulla. J. Neurosci. 11, 1433–1439.

Kavaliers, M. 1988. Brief exposure to a natural predator, the short-tailed weasel, induces benzodiazepine-sensitive analgesia in white-footed mice. Physiol. Behav. 43, 187–193.

Keay, K. A. and Bandler, R. 2001. Parallel circuits mediating distinct emotional coping reactions to different types of stress. Neurosci. Biobehav. Rev. 25, 669–678.

Kiefel, J. M., Rossi, G. C., and Bodnar, R. J. 1993. Medullary mu and delta opioid receptors modulate mesencephalic morphine analgesia in rats. Brain Res. 624, 151–161.

Kincaid, W., Neubert, M. J., Xu, M., Kim, C.-J., and Heinricher, M. M. 2006. Role for medullary pain facilitating neurons in secondary thermal hyperalgesia. J. Neurophysiol. 95, 33–41.

Kinscheck, I. B., Watkins, L. R., and Mayer, D. J. 1984. Fear is not critical to classically conditioned analgesia: the effects of periaqueductal gray lesions and administration of chlordiazepoxide. Brain Res. 298, 33–44.

Kiser, R. S., Lebovitz, R. M., and German, D. C. 1978. Anatomic and pharmacologic differences between two types of aversive midbrain stimulation. Brain Res. 155, 331–342.

Kovelowski, C. J., Ossipov, M. H., Sun, H., Lai, J., Malan, T. P., and Porreca, F. 2000. Supraspinal cholecystokinin may drive tonic descending facilitation mechanisms to maintain neuropathic pain in the rat. Pain 87, 265–273.

Krettek, J. E. and Price, J. L. 1978. Amygdaloid projections to subcortical structures within the basal forebrain and brainstem in the rat and cat. J. Comp. Neurol. 178, 225–254.

Kshatri, A. M., Baghdoyan, H. A., and Lydic, R. 1998. Cholinomimetics, but not morphine, increase antinociceptive behavior from pontine reticular regions regulating rapid-eye-movement sleep. Sleep 21, 677–685.

Lagraize, S. C., Borzan, J., Peng, Y. B., and Fuchs, P. N. 2006. Selective regulation of pain affect following activation of the opioid anterior cingulate cortex system. Exp. Neurol. 197(1), 22–30.

LaGraize, S. C., Borzan, J., Rinker, M. M., Kopp, J. L., and Fuchs, P. N. 2004a. Behavioral evidence for competing motivational drives of nociception and hunger. Neurosci. Lett. 372, 30–34.

LaGraize, S. C., Labuda, C. J., Rutledge, M. A., Jackson, R. L., and Fuchs, P. N. 2004b. Differential effect of anterior cingulate cortex lesion on mechanical hypersensitivity and escape/avoidance behavior in an animal model of neuropathic pain. Exp. Neurol. 188, 139–148.

Lane, D. A., Patel, P. A., and Morgan, M. M. 2005. Evidence for an intrinsic mechanism of antinociceptive tolerance within the ventrolateral periaqueductal gray of rats. Neuroscience 135, 227–234.

Lane, D. A., Tortorici, V., and Morgan, M. M. 2004. Behavioral and electrophysiological evidence for tolerance to continuous morphine administration into the ventrolateral periaqueductal gray. Neuroscience 125, 63–69.

Le Bars, D. 1988. Serotonin and Pain. In: Neuronal Serotonin (eds. N. N. Osborne and M. Hamon), pp. 171–226. Wiley.

Le Bars, D. 2002. The whole body receptive field of dorsal horn multireceptive neurones. Brain Res. Rev. 40, 29–44.

Leman, S., Dielenberg, R. A., and Carrive, P. 2003. Effect of dorsal periaqueductal gray lesion on cardiovascular and behavioural responses to contextual conditioned fear in rats. Behav. Brain Res. 143, 169–176.

Lester, L. S. and Fanselow, M. S. 1985. Exposure to a cat produces opioid analgesia in rats. Behav. Neurosci. 99, 756–759.

Leung, C. G. and Mason, P. 1995. Effects of isoflurane concentration on the activity of pontomedullary raphe and medial reticular neurons in the rat. Brain Res. 699, 71–82.

Leung, C. G. and Mason, P. 1998. Physiological survey of medullary raphe and magnocellular reticular neurons in the anesthetized rat. J. Neurophysiol. 80, 1630–1646.

Leung, C. G. and Mason, P. 1999. Physiological properties of raphe magnus neurons during sleep and waking. J. Neurophysiol. 81, 584–595.

Lewis, J. W., Cannon, J. T., and Liebeskind, J. C. 1980. Opioid and nonopioid mechanisms of stress analgesia. Science 208, 623–625.

Lichtman, A. H. and Fanselow, M. S. 1990. Cats produce analgesia in rats on the tail-flick test: naltrexone sensitivity is determined by the nociceptive test stimulus. Brain Res. 533, 91–94.

Lovick, T. A. 1993. Integrated activity of cardiovascular and pain regulatory systems: role in adaptive behavioural responses. Prog. Neurobiol. 40, 631–644.

Lumb, B. M., Parry, D. M., Semenenko, F. M., McMullan, S., and Simpson, D. A. 2002. C-nociceptor activation of hypothalamic neurones and the columnar organisation of their projections to the periaqueductal grey in the rat. Exp. Physiol. 87, 123–128.

Madden, C. J. and Morrison, S. F. 2003. Excitatory amino acid receptor activation in the raphe pallidus area mediates prostaglandin-evoked thermogenesis. Neuroscience 122, 5–15.

Maier, S. F. 1986. Stressor controllability and stress-induced analgesia. Ann. N. Y. Acad. Sci. 467, 55–72.

Manning, B. H. and Mayer, D. J. 1995. The central nucleus of the amygdala contributes to the production of morphine antinociception in the formalin test. Pain 63, 141–152.

Mansikka, H. and Pertovaara, A. 1997. Supraspinal influence on hindlimb withdrawal thresholds and mustard oil-induced secondary allodynia in rats. Brain Res. Bull. 42, 359–365.

Mansour, A., Fox, C. A., Akil, H., and Watson, S. J. 1995. Opioid-receptor mRNA expression in the rat CNS: anatomical and functional implications. Trends Neurosci. 18, 22–29.

Mantyh, P. W. and Hunt, S. P. 1984. Evidence for cholecystokinin-like immunoreactive neurons in the rat medulla oblongata which project to the spinal cord. Brain Res. 291, 49–54.

Marinelli, S., Connor, M., Schnell, S. A., Christie, M. J., Wessendorf, M. W., and Vaughan, C. W. 2005. Δ-opioid receptor-mediated actions on rostral ventromedial medulla neurons. Neuroscience 132, 239–244.

Marinelli, S., Vaughan, C. W., Schnell, S. A., Wessendorf, M. W., and Christie, M. J. 2002. Rostral ventromedial medulla neurons that project to the spinal cord express multiple opioid receptor phenotypes. J. Neurosci. 22, 10847–10855.

Martin, W. J., Tsou, K., and Walker, J. M. 1998. Cannabinoid receptor-mediated inhibition of the rat tail-flick reflex after microinjection into the rostral ventromedial medulla. Neurosci. Lett. 242, 33–36.

Mason, P. 1997. Physiological identification of pontomedullary serotonergic neurons in the rat. J. Neurophysiol. 77, 1087–1098.

Mason, P. 2001. Contributions of the medullary raphe and ventromedial reticular region to pain modulation and other homeostatic functions. Annu. Rev. Neurosci. 24, 737–777.

Mason, P., Escobedo, I., Burgin, C., Bergan, J., Lee, J. H., Last, E. J., and Holub, A. L. 2001. Nociceptive responsiveness during slow-wave sleep and waking in the rat. Sleep 24, 32–38.

Matsumoto, S., Levendusky, M. C., Longhurst, P. A., Levin, R. M., and Millington, W. R. 2004. Activation of mu opioid receptors in the ventrolateral periaqueductal gray inhibits reflex micturition in anesthetized rats. Neurosci. Lett. 363, 116–119.

McCleane, G. J. 1998. The cholecystokinin antagonist proglumide enhances the analgesic efficacy of morphine in humans with chronic benign pain. Anesth. Analg. 87, 1117–1120.

McGaraughty, S. and Heinricher, M. M. 2002. Microinjection of morphine into various amygdaloid nuclei differentially affects nociceptive responsiveness and RVM neuronal activity. Pain 96, 153–162.

McGaraughty, S. and Heinricher, M. M. 2004. Lesions of the periaqueductal gray disrupt input to the rostral ventromedial medulla following microinjections into the medial or basolateral nuclei of the amygdala. Brain Res. 1009, 223–227.

McGaraughty, S., Chu, K. L., Bitner, R. S., Martino, B., Kouhen, R. E., Han, P., Nikkel, A. L., Burgard, E. C., Faltynek, C. R., and Jarvis, M. F. 2003. Capsaicin infused into the PAG affects rat tail flick responses to noxious heat and alters neuronal firing in the RVM. J. Neurophysiol. 90, 2702–2710.

McLemore, S., Crown, E. D., Meagher, M. W., and Grau, J. W. 1999. Shock-induced hyperalgesia: Ii. Role of the dorsolateral periaqueductal gray. Behav. Neurosci. 113, 539–549.

Melzack, R., Wall, P. D., and Ty, T. C. 1982. Acute pain in an emergency clinic: latency of onset and descriptor patterns related to different injuries. Pain 14, 33–43.

Menetrey, D. and Basbaum, A. I. 1987. The distribution of substance p-, enkephalin- and dynorphin-immunoreactive neurons in the medulla of the rat and their contribution to bulbospinal pathways. Neuroscience 23, 173–187.

Menetrey, D., Chaouch, A., Binder, D., and Besson, J. M. 1982. The origin of the spinomesencephalic tract in the rat: an anatomical study using the retrograde transport of horseradish peroxidase. J. Comp. Neurol. 206, 193–207.

Meng, I. D. and Johansen, J. P. 2004. Antinociception and modulation of rostral ventromedial medulla neuronal activity by local microinfusion of a cannabinoid receptor agonist. Neuroscience 124, 685–693.

Meng, I. D., Johansen, J. P., Harasawa, I., and Fields, H. L. 2005. Kappa opioids inhibit physiologically identified medullary pain modulating neurons and reduce morphine antinociception. J. Neurophysiol. 93, 1138–1144.

Meng, I. D., Manning, B. H., Martin, W. J., and Fields, H. L. 1998. An analgesia circuit activated by cannabinoids. Nature 395, 381–383.

Miki, K., Zhou, Q. Q., Guo, W., Guan, Y., Terayama, R., Dubner, R., and Ren, K. 2002. Changes in gene expression and neuronal phenotype in brain stem pain modulatory circuitry after inflammation. J. Neurophysiol. 87, 750–760.

Millan, M. J. 2002. Descending control of pain. Prog. Neurobiol. 66, 355–474.

Miller, K. K., Hoffer, A., Svoboda, K. R., and Lupica, C. R. 1997. Cholecystokinin increases GABA release by inhibiting a resting K^+ conductance in hippocampal interneurons. J. Neurosci. 17, 4994–5003.

Millhorn, D. E., Hokfelt, T., Seroogy, K., Oertel, W., Verhofstad, A. A., and Wu, J. Y. 1987a. Immunohistochemical evidence for colocalization of gamma-aminobutyric acid and serotonin in neurons of the ventral medulla oblongata projecting to the spinal cord. Brain Res. 410, 179–185.

Millhorn, D. E., Hokfelt, T., Verhofstad, A. A., and Terenius, L. 1989. Individual cells in the raphe nuclei of the medulla oblongata in rat that contain immunoreactivities for both serotonin and enkephalin project to the spinal cord. Exp. Brain Res. 75, 536–542.

Millhorn, D. E., Seroogy, K., Hökfelt, T., Schmued, L. C., Terenius, L., Buchan, A., and Brown, J. C. 1987b. Neurons of the ventral medulla oblongata that contain both somatostatin and enkephalin immunoreactivities project to nucleus tractus solitarii and spinal cord. Brain Res. 424, 99–108.

Mitchell, J. M., Lowe, D., and Fields, H. L. 1998. The contribution of the rostral ventromedial medulla to the antinociceptive effects of systemic morphine in restrained and unrestrained rats. Neuroscience 87, 123–133.

Monhemius, R., Azami, J., Green, D. L., and Roberts, M. H. 2001. CB1 receptor mediated analgesia from the nucleus reticularis gigantocellularis pars alpha is activated in an animal model of neuropathic pain. Brain Res. 908, 67–74.

Moreau, J. L. and Fields, H. L. 1986. Evidence for GABA involvement in midbrain control of medullary neurons that modulate nociceptive transmission. Brain Res. 397, 37–46.

Morgan, M. M. 1991. Differences in Antinociception Evoked from Dorsal and Ventral Regions of the Caudal Periaqueductal Gray Matter. In: The Midbrain Periaqueductal Gray Matter (eds. A. DePaulis and R. Bandler), pp. 139–150. Plenum.

Morgan, M. M. and Carrive, P. 2001. Activation of the ventrolateral periaqueductal gray reduces locomotion but not mean arterial pressure in awake, freely moving rats. Neuroscience 102, 905–910.

Morgan, M. M. and Clayton, C. C. 2005. Defensive behaviors evoked from the ventrolateral periaqueductal gray of the rat: comparison of opioid and GABA disinhibition. Behav. Brain Res. 164, 61–66.

Morgan, M. M. and Fields, H. L. 1994. Pronounced changes in the activity of nociceptive modulatory neurons in the rostral ventromedial medulla in response to prolonged thermal noxious stimuli. J. Neurophysiol. 72, 1161–1170.

Morgan, M. M. and Whitney, P. K. 2000. Immobility accompanies the antinociception mediated by the rostral ventromedial medulla of the rat. Brain Res. 872, 276–281.

Morgan, M. M., Clayton, C. C., and Boyer-Quick, J. S. 2005. Differential susceptibility of the PAG and RVM to tolerance to the antinociceptive effect of morphine in the rat. Pain 113, 91–98.

Morgan, M. M., Clayton, C. C., and Lane, D. A. 2003. Behavioral evidence linking opioid-sensitive gabaergic neurons in the ventrolateral periaqueductal gray to morphine tolerance. Neuroscience 118, 227–232.

Morgan, M. M., Heinricher, M. M., and Fields, H. L. 1992. Circuitry linking opioid-sensitive nociceptive modulatory systems in periaqueductal gray and spinal cord with rostral ventromedial medulla. Neuroscience 47, 863–871.

Morgan, M. M., Heinricher, M. M., and Fields, H. L. 1994. Inhibition and facilitation of different nocifensor reflexes by spatially remote noxious stimuli. J. Neurophysiol. 72, 1152–1160.

Nakamura, K., Matsumura, K., Kaneko, T., Kobayashi, S., Katoh, H., and Negishi, M. 2002. The rostral raphe pallidus nucleus mediates pyrogenic transmission from the preoptic area. J. Neurosci. 22, 4600–4610.

Narita, M., Mizoguchi, H., Narita, M., Dun, N. J., Hwang, B. H., Endoh, T., Suzuki, T., Nagase, H., Suzuki, T., and Tseng, L. F. 2000. G protein activation by endomorphins in the mouse periaqueductal gray matter. J. Biomed. Sci. 7, 221–225.

Nashold, B. S., Wilson, W. P., and Slaughter, D. G. 1969. Sensations evoked by stimulation in the midbrain in man. J. Neurosurg. 30, 14–24.

Nason, M. W., Jr. and Mason, P. 2004. Modulation of sympathetic and somatomotor function by the ventromedial medulla. J. Neurophysiol. 92, 510–522.

Neubert, M. J., Kincaid, W., and Heinricher, M. M. 2004. Nociceptive facilitating neurons in the rostral ventromedial medulla. Pain 110, 158–165.

Ogawa, S., Kow, L. M., and Pfaff, D. W. 1992. Effects of lordosis-relevant neuropeptides on midbrain periaqueductal gray neuronal activity in vitro. Peptides 13, 965–975.

Oka, T., Aou, S., and Hori, T. 1994. Intracerebroventricular injection of prostaglandin E_2 induces thermal hyperalgesia in rats: the possible involvement of ep3 receptors. Brain Res. 663, 287–292.

Oliveras, J. L., Martin, G., and Montagne-Clavel, J. 1991a. Drastic changes of ventromedial medulla neuronal properties induced by barbiturate anesthesia. Ii. Modifications of the single-unit activity produced by brevital, a short-acting barbiturate in the awake, freely moving rat. Brain Res. 563, 251–260.

Oliveras, J. L., Martin, G., Montagne, J., and Vos, B. 1990. Single unit activity at ventromedial medulla level in the awake, freely moving rat: effects of noxious heat and light tactile stimuli onto convergent neurons. Brain Res. 506, 19–30.

Oliveras, J. L., Montagne-Clavel, J., and Martin, G. 1991b. Drastic changes of ventromedial medulla neuronal properties induced by barbiturate anesthesia. I. Comparison of the single-unit types in the same awake and pentobarbital-treated rats. Brain Res. 563, 241–250.

Ongur, D. and Price, J. L. 2000. The organization of networks within the orbital and medial prefrontal cortex of rats, monkeys and humans. Cereb Cortex 10, 206–219.

Osborne, P. B., Vaughan, C. W., Wilson, H. I., and Christie, M. J. 1996. Opioid inhibition of rat periaqueductal grey neurones with identified projections to rostral ventromedial medulla in vitro. J. Physiol. 490, 383–389.

Ossipov, M. H., Kovelowski, C. J., Nichols, M. L., Hruby, V. J., and Porreca, F. 1995. Characterization of supraspinal antinociceptive actions of opioid delta agonists in the rat. Pain 62, 287–293.

Palkovits, M. and Horvath, S. 1994. Galanin immunoreactive neurons in the medulla oblongata of rats. Acta Biol. Hung. 45, 399–417.

Pan, Z. Z. and Fields, H. L. 1996. Endogenous opioid-mediated inhibition of putative pain-modulating neurons in rat rostral ventromedial medulla. Neuroscience 74, 855–862.

Pan, Z. Z., Grudt, T. J., and Williams, J. T. 1994. Alpha 1-adrenoceptors in rat dorsal raphe neurons: regulation of two potassium conductances. J. Physiol. 478 Pt 3, 437–447.

Pan, Z., Hirakawa, N., and Fields, H. L. 2000. A cellular mechanism for the bidirectional pain-modulating actions of orphanin FQ/nociceptin. Neuron 26, 515–522.

Pan, Z. Z., Tershner, S. A., and Fields, H. L. 1997. Cellular mechanism for anti-analgesic action of agonists of the kappa- opioid receptor. Nature 389, 382–385.

Pan, Z. Z., Williams, J. T., and Osborne, P. B. 1990. Opioid actions on single nucleus raphe magnus neurons from rat and guinea-pig in vitro. J. Physiol. 427, 519–532.

Parry, D. M., Semenenko, F. M., Conley, R. K., and Lumb, B. M. 2002. Noxious somatic inputs to hypothalamic-midbrain projection neurones: a comparison of the columnar organisation of somatic and visceral inputs to the periaqueductal grey in the rat. Exp. Physiol. 87, 117–122.

Pastoriza, L. N., Morrow, T. J., and Casey, K. L. 1996. Medial frontal cortex lesions selectively attenuate the hot plate response: possible nocifensive apraxia in the rat. Pain 64, 11–17.

Pavlovic, Z. W., Cooper, M. L., and Bodnar, R. J. 1996. Opioid antagonists in the periaqueductal gray inhibit morphine and beta-endorphin analgesia elicited from the amygdala of rats. Brain Res. 741, 13–26.

Pertovaara, A. and Kalmari, J. 2002. Neuropathy reduces viscero-somatic inhibition via segmental mechanisms in rats. Neuroreport 13, 1047–1050.

Pertovaara, A., Wei, H., and Hamalainen, M. M. 1996. Lidocaine in the rostroventromedial medulla and the periaqueductal gray attenuates allodynia in neuropathic rats. Neurosci. Lett. 218, 127–130.

Petrovic, P., Kalso, E., Petersson, K. M., and Ingvar, M. 2002. Placebo and opioid analgesia – imaging a shared neuronal network. Science 295, 1737–1740.

Porreca, F., Burgess, S. E., Gardell, L. R., Vanderah, T. W., Malan, T. P., Jr., Ossipov, M. H., Lappi, D. A., and Lai, J. 2001. Inhibition of neuropathic pain by selective ablation of brainstem medullary cells expressing the μ-opioid receptor. J. Neurosci. 21, 5281–5288.

Porreca, F., Ossipov, M. H., and Gebhart, G. F. 2002. Chronic pain and medullary descending facilitation. Trends Neurosci. 25, 319–325.

Potrebic, S. B., Fields, H. L., and Mason, P. 1994. Serotonin immunoreactivity is contained in one physiological cell class in the rat rostral ventromedial medulla. J. Neurosci. 14, 1655–1665.

Price, J. L. and Amaral, D. G. 1981. An autoradiographic study of the projections of the central nucleus of the monkey amygdala. J. Neurosci. 1, 1242–1259.

Price, D. D., Milling, L. S., Kirsch, I., Duff, A., Montgomery, G. H., and Nicholls, S. S. 1999. An analysis of factors that contribute to the magnitude of placebo analgesia in an experimental paradigm. Pain 83, 147–156.

Price, D. D., von der Gruen, A., Miller, J., Rafii, A., and Price, C. 1985. Potentiation of systemic morphine analgesia in humans by proglumide, a cholecystokinin antagonist. Anesth. Analg. 64, 801–806.

Prieto, G. J., Cannon, J. T., and Liebeskind, J. C. 1983. N. Raphe magnus lesions disrupt stimulation-produced analgesia from ventral but not dorsal midbrain areas in the rat. Brain Res. 261, 53–57.

Proudfit, H. K. 1992. The Behavioural Pharmacology of the Noradrenergic System. In: Towards the Use of Noradrenergic Agonists for the Treatment of Pain (*eds*. J. M. Besson and G. Guilbaud), pp. 119–136. Elsevier.

Ramirez, F. and Vanegas, H. 1989. Tooth pulp stimulation advances both medullary off-cell pause and tail flick. Neurosci. Lett. 100, 153–156.

Ren, K. and Dubner, R. 1996. Enhanced descending modulation of nociception in rats with persistent hindpaw inflammation. J. Neurophysiol. 76, 3025–3037.

Ren, K. and Dubner, R. 2002. Descending modulation in persistent pain: an update. Pain 100, 1–6.

Rizvi, T. A., Ennis, M., Behbehani, M. M., and Shipley, M. T. 1991. Connections between the central nucleus of the amygdala and the midbrain periaqueductal gray: topography and reciprocity. J. Comp. Neurol. 303, 121–131.

Rizvi, T. A., Ennis, M., and Shipley, M. T. 1992. Reciprocal connections between the medial preoptic area and the midbrain periaqueductal gray in rat: a WGA-HRP and PHA-L study. J. Comp. Neurol. 315, 1–15.

Rizvi, T. A., Murphy, A. Z., Ennis, M., Behbehani, M. M., and Shipley, M. T. 1996. Medial preoptic area afferents to periaqueductal gray medullo-output neurons: a combined fos and tract tracing study. J. Neurosci. 16, 333–344.

Rodgers, R. J. and Randall, J. I. 1988. Environmentally Induced Analgesia: Situational Factors, Mechanisms and Significance. In: Endorphins, Opiates and Behavioral Processes. (*eds*. R. J. Rodgers and S. J. Cooper), pp. 107–144. Wiley.

Romanovsky, A. A., Kulchitsky, V. A., Akulich, N. V., Koulchitsky, S. V., Simons, C. T., Sessler, D. I., and Gourine, V. N. 1996. First and second phases of biphasic fever: two sequential stages of the sickness syndrome? Am. J. Physiol. 271, R244–253.

Rosellini, R. A., Abrahamsen, G. C., Stock, H. S., and Caldarone, B. J. 1994. Modulation of hypoalgesia by morphine and number of shock trials: covariation of a measure of context fear and hypoalgesia. Physiol. Behav. 56, 183–188.

Rosen, A., Zhang, Y. X., Lund, I., Lundeberg, T., and Yu, L. C. 2004. Substance p microinjected into the periaqueductal gray matter induces antinociception and is released following morphine administration. Brain Res. 1001, 87–94.

Rossi, G. C., Pasternak, G. W., and Bodnar, R. J. 1994. Mu and delta opioid synergy between the periaqueductal gray and the rostro-ventral medulla. Brain Res. 665, 85–93.

Roychowdhury, S. M. and Fields, H. L. 1996. Endogenous opioids acting at a medullary μ-opioid receptor contribute to the behavioral antinociception produced by GABA antagonism in the midbrain periaqueductal gray. Neuroscience 74, 863–872.

Sagen, J. and Proudfit, H. K. 1981. Hypoalgesia induced by blockade of noradrenergic projections to the raphe magnus: reversal by blockade of noradrenergic projections to the spinal cord. Brain Res. 223, 391–396.

Sagen, J. and Proudfit, H. K. 1985. Evidence for pain modulation by pre- and postsynaptic noradrenergic receptors in the medulla oblongata. Brain Res. 331, 285–293.

Sandkühler, J. and Gebhart, G. F. 1984. Relative contributions of the nucleus raphe magnus and adjacent medullary reticular formation to the inhibition by stimulation in the periaqueductal gray of a spinal nociceptive reflex in the pentobarbital-anesthetized rat. Brain Res. 305, 77–87.

Sandrew, B. B. and Poletti, C. E. 1984. Limbic influence on the periaqueductal gray: a single unit study in the awake squirrel monkey. Brain Res. 303, 77–86.

Schreff, M., Schulz, S., Wiborny, D., and Höllt, V. 1998. Immunofluorescent identification of endomorphin-2-containing nerve fibers and terminals in the rat brain and spinal cord. Neuroreport 9, 1031–1034.

Semenenko, F. M. and Lumb, B. M. 1992. Projections of anterior hypothalamic neurones to the dorsal and ventral periaqueductal grey in the rat. Brain Res. 582, 237–245.

Semenenko, F. M. and Lumb, B. M. 1999. Excitatory projections from the anterior hypothalamus to periaqueductal gray neurons that project to the medulla: a functional anatomical study. Neuroscience 94, 163–174.

Shipley, M. T., Ennis, M., Rizvi, T. A., and Behbehani, M. M. 1991. Topographical Specificity of Forebrain Inputs to the Midbrain Periaqueductal Gray: Evidence for Discrete Longitudinally Organized Input Columns. In: The Midbrain Periaqueductal Gray (*eds*. A. Depaulis and R. Bandler), pp. 417–448. Plenum.

Siuciak, J. A. and Advokat, C. 1987. Tolerance to morphine microinjections in the periaqueductal gray (PAG) induces tolerance to systemic, but not intrathecal morphine. Brain Res. 424, 311–319.

Skinner, K., Basbaum, A. I., and Fields, H. L. 1997. Cholecystokinin and enkephalin in brain stem pain modulating circuits. Neuroreport 8, 2995–2998.

Smith, D. J., Hawranko, A. A., Monroe, P. J., Gully, D., Urban, M. O., Craig, C. R., Smith, J. P., and Smith, D. L. 1997. Dose-dependent pain-facilitatory and -inhibitory actions of neurotensin are revealed by SR 48692, a nonpeptide neurotensin antagonist: influence on the antinociceptive effect of morphine. J. Pharmacol. Exp. Ther. 282, 899–908.

Smith, D. J., Perrotti, J. M., Crisp, T., Cabral, M. E., Long, J. T., and Scalzitti, J. M. 1988. The mu opiate receptor is responsible for descending pain inhibition originating in the periaqueductal gray region of the rat brain. Eur. J. Pharmacol. 156, 47–54.

Soja, P. J., Cairns, B. E., and Kristensen, M. P. 1999. Transmission through Ascending Trigeminal and Lumbar Sensory Pathways: Dependence on Behavioral State. In: Handbook of Behavioral State Control (*eds*. R. Lydic and H. A. Baghdoyan), pp. 521–544. CRC Press.

Stanfa, L., Dickenson, A., Xu, X. J., and Wiesenfeld-Hallin, Z. 1994. Cholecystokinin and morphine analgesia: variations on a theme. Trends Pharmacol. Sci. 15, 65–66.

Sugaya, K., Ogawa, Y., Hatano, T., Koyama, Y., Miyazato, T., and Oda, M. 1998. Evidence for involvement of the subcoeruleus nucleus and nucleus raphe magnus in urine storage and penile erection in decerebrate rats. J. Urol. 159, 2172–2176.

Suzuki, R., Rygh, L. J., and Dickenson, A. H. 2004. Bad news from the brain: descending 5-HT pathways that control spinal pain processing. Trends Pharmacol. Sci. 25, 613–617.

Talbot, J. D., Villemure, J. G., Bushnell, M. C., and Duncan, G. H. 1995. Evaluation of pain perception after anterior capsulotomy: a case report. Somatosens. Mot. Res. 12, 115–126.

Terayama, R., Guan, Y., Dubner, R., and Ren, K. 2000. Activity-induced plasticity in brain stem pain modulatory circuitry after inflammation. Neuroreport 11, 1915–1919.

Tershner, S. A., Mitchell, J. M., and Fields, H. L. 2000. Brainstem pain modulating circuitry is sexually dimorphic with respect to mu and kappa opioid receptor function. Pain 85, 153–159.

Tortorici, V. and Morgan, M. M. 2002. Comparison of morphine and kainic acid microinjections into identical PAG sites on the activity of RVM neurons. J. Neurophysiol. 88, 1707–1715.

Tortorici, V. and Vanegas, H. 1995. Anti-nociception induced by systemic or PAG-microinjected lysine–acetylsalicylate in rats. Effects on tail-flick related activity of medullary off- and on-cells. Eur. J. Neurosci. 7, 1857–1865.

Tortorici, V., Morgan, M. M., and Vanegas, H. 2001. Tolerance to repeated microinjection of morphine into the periaqueductal gray is associated with changes in the behavior of off- and on-cells in the rostral ventromedial medulla of rats. Pain 89, 237–244.

Tracey, I. 2005. Nociceptive processing in the human brain. Curr. Opin. Neurobiol. 15(4), 478–487.

Tracey, I., Ploghaus, A., Gati, J. S., Clare, S., Smith, S., Menon, R. S., and Matthews, P. M. 2002. Imaging attentional modulation of pain in the periaqueductal gray in humans. J. Neurosci. 22, 2748–2752.

Tsou, K., Brown, S., Sanudo-Pena, M. C., Mackie, K., and Walker, J. M. 1998. Immunohistochemical distribution of cannabinoid CB1 receptors in the rat central nervous system. Neuroscience 83, 393–411.

Urban, M. O. and Gebhart, G. F. 1997. Characterization of biphasic modulation of spinal nociceptive transmission by neurotensin in the rat rostral ventromedial medulla. J. Neurophysiol. 78, 1550–1562.

Urban, M. O. and Smith, D. J. 1993. Role of neurotensin in the nucleus raphe magnus in opioid-induced antinociception from the periaqueductal gray. J. Pharmacol. Exp. Ther. 265, 580–586.

Urban, M. O., Coutinho, S. V., and Gebhart, G. F. 1999. Biphasic modulation of visceral nociception by neurotensin in rat rostral ventromedial medulla. J. Pharmacol. Exp. Ther. 290, 207–213.

Urban, M. O., Jiang, M. C., and Gebhart, G. F. 1996. Participation of central descending nociceptive facilitatory systems in secondary hyperalgesia produced by mustard oil. Brain Res. 737, 83–91.

Valet, M., Sprenger, T., Boecker, H., Willoch, F., Rummeny, E., Conrad, B., Erhard, P., and Tolle, T. R. 2004. Distraction modulates connectivity of the cingulo-frontal cortex and the midbrain during pain – an FMRI analysis. Pain 109, 399–408.

Van Bockstaele, E. J., Aston-Jones, G., Pieribone, V. A., Ennis, M., and Shipley, M. T. 1991. Subregions of the periaqueductal gray topographically innervate the rostral ventral medulla in the rat. J. Comp. Neurol. 309, 305–327.

Vanderah, T. W., Ossipov, M. H., Lai, J., Malan, T. P., and Porreca, F. 2001a. Mechanisms of opioid-induced pain and antinociceptive tolerance: descending facilitation and spinal dynorphin. Pain 92, 5–9.

Vanderah, T. W., Suenaga, N. M., Ossipov, M. H., Malan, T. P., Jr., Lai, J., and Porreca, F. 2001b. Tonic descending facilitation from the rostral ventromedial medulla mediates opioid-induced abnormal pain and antinociceptive tolerance. J. Neurosci. 21, 279–286.

Vanderhorst, V. G., Mouton, L. J., Blok, B. F., and Holstege, G. 1996. Distinct cell groups in the lumbosacral cord of the cat project to different areas in the periaqueductal gray. J. Comp. Neurol. 376, 361–385.

Vanderhorst, V. G., Terasawa, E., and Ralston, H. J., 3rd 2002. Estrogen receptor-alpha immunoreactive neurons in the ventrolateral periaqueductal gray receive monosynaptic input from the lumbosacral cord in the rhesus monkey. J. Comp. Neurol. 443, 27–42.

Vanegas, H. 2004. To the descending pain-control system in rats, inflammation-induced primary and secondary hyperalgesia are two different things. Neurosci. Lett. 361, 225–228.

Vanegas, H. and Schaible, H.-G. 2004. Descending control of persistent pain: inhibitory or facilitatory? Brain Res. Rev. 46, 295–309.

Vanegas, H. and Tortorici, V. 2002. Opioidergic effects of nonopioid analgesics on the central nervous system. Cell. Mol. Neurobiol. 22, 655–661.

Vanegas, H., Barbaro, N. M., and Fields, H. L. 1984. Midbrain stimulation inhibits tail-flick only at currents sufficient to excite rostral medullary neurons. Brain Res. 321, 127–133.

Vaughan, C. W. 1998. Enhancement of opioid inhibition of GABAergic synaptic transmission by cyclo-oxygenase inhibitors in rat periaqueductal grey neurones. Br. J. Pharmacol. 123, 1479–1481.

Vaughan, C. W. and Christie, M. J. 1997. Presynaptic inhibitory action of opioids on synaptic transmission in the rat periaqueductal grey in vitro. J. Physiol. 498, 463–472.

Vaughan, C. W., Bagley, E. E., Drew, G. M., Schuller, A., Pintar, J. E., Hack, S. P., and Christie, M. J. 2003. Cellular actions of opioids on periaqueductal grey neurons from c57b16/j mice and mutant mice lacking mor-1. Br. J. Pharmacol. 139, 362–367.

Vaughan, C. W., Bandler, R., and Christie, M. J. 1996. Differential responses of lateral and ventrolateral rat periaqueductal grey neurones to noradrenaline in vitro. J. Physiol. 490, 373–381.

Vaughan, C. W., Connor, M., Bagley, E. E., and Christie, M. J. 2000. Actions of cannabinoids on membrane properties and synaptic transmission in rat periaqueductal gray neurons in vitro. Mol. Pharmacol. 57, 288–295.

Vaughan, C. W., Ingram, S. L., Connor, M. A., and Christie, M. J. 1997. How opioids inhibit GABA-mediated neurotransmission. Nature 390, 611–614.

Vaughan, C. W., McGregor, I. S., and Christie, M. J. 1999. Cannabinoid receptor activation inhibits GABAergic neurotransmission in rostral ventromedial medulla neurons in vitro. Br. J. Pharmacol. 127, 935–940.

Vianna, D. M., Graeff, F. G., Brandao, M. L., and Landeira-Fernandez, J. 2001. Defensive freezing evoked by electrical stimulation of the periaqueductal gray: comparison between dorsolateral and ventrolateral regions. Neuroreport 12, 4109–4112.

Villemure, C. and Bushnell, M. C. 2002. Cognitive modulation of pain: How do attention and emotion influence pain processing? Pain 95, 195–199.

Wager, T. D., Rilling, J. K., Smith, E. E., Sokolik, A., Casey, K. L., Davidson, R. J., Kosslyn, S. M., Rose, R. M., and Cohen, J. D. 2004. Placebo-induced changes in fMRI in the anticipation and experience of pain. Science 303, 1162–1167.

Walker, P. and Carrive, P. 2003. Role of ventrolateral periaqueductal gray neurons in the behavioral and cardiovascular responses to contextual conditioned fear and poststress recovery. Neuroscience 116, 897–912.

Walker, J. M. and Huang, S. M. 2002a. Cannabinoid analgesia. Pharmacol. Ther. 95, 127–135.

Walker, J. M. and Huang, S. M. 2002b. Endocannabinoids in pain modulation. Prostaglandins Leukot. Essent. Fatty Acids 66, 235–242.

Walker, J. M., Huang, S. M., Strangman, N. M., Tsou, K., and Sanudo-Pena, M. C. 1999. Pain modulation by release of the endogenous cannabinoid anandamide. Proc. Natl. Acad. Sci. U. S. A. 96, 12198–12203.

Wall, P. D. 1979. On the relation of injury to pain. Pain 6, 253–264.

Wang, H. and Wessendorf, M. W. 1999. Mu- and delta-opioid receptor mRNAs are expressed in spinally projecting serotonergic and nonserotonergic neurons of the rostral ventromedial medulla. J. Comp. Neurol. 404, 183–196.

Waters, A. J. and Lumb, B. M. 1997. Inhibitory effects evoked from both the lateral and ventrolateral periaqueductal grey are selective for the nociceptive responses of rat dorsal horn neurones. Brain Res. 752, 239–249.

Watkins, L. R. and Mayer, D. J. 1982. Organization of endogenous opiate and nonopiate pain control systems. Science 216, 1185–1192.

Watkins, L. R., Cobelli, D. A., and Mayer, D. J. 1982. Opiate vs non-opiate footshock induced analgesia (FSIA): descending and intraspinal components. Brain Res. 245, 97–106.

Watkins, L. R., Kinscheck, I. B., and Mayer, D. J. 1983a. The neural basis of footshock analgesia: the effect of periaqueductal gray lesions and decerebration. Brain Res. 276, 317–324.

Watkins, L. R., Wiertelak, E. P., Goehler, L. E., Mooney-Heiberger, K., Martinez, J., Furness, L., Smith, K. P., and Maier, S. F. 1994. Neurocircuitry of illness-induced hyperalgesia. Brain Res. 639, 283–299.

Watkins, L. R., Wiertelak, E. P., and Maier, S. F. 1993. The amygdala is necessary for the expression of conditioned but not unconditioned analgesia. Behav. Neurosci. 107, 402–405.

Watkins, L. R., Young, E. G., Kinscheck, I. B., and Mayer, D. J. 1983b. The neural basis of footshock analgesia: the role of specific ventral medullary nuclei. Brain Res. 276, 305–315.

Wiberg, M., Westman, J., and Blomqvist, A. 1987. Somatosensory projection to the mesencephalon: an anatomical study in the monkey. J. Comp. Neurol. 264, 92–117.

Wiertelak, E. P., Roemer, B., Maier, S. F., and Watkins, L. R. 1997. Comparison of the effects of nucleus tractus solitarius and ventral medial medulla lesions on illness-induced and subcutaneous formalin- induced hyperalgesias. Brain Res. 748, 143–150.

Wiesenfeld-Hallin, Z., de Arauja Lucas, G., Alster, P., Xu, X. J., and Hökfelt, T. 1999. Cholecystokinin/opioid interactions. Brain Res. 848, 78–89.

Williams, F. G. and Beitz, A. J. 1990. Ultrastructural morphometric analysis of GABA-immunoreactive terminals in the ventrocaudal periaqueductal grey: analysis of the relationship of GABA terminals and the GABA$_a$ receptor to periaqueductal grey-raphe magnus projection neurons. J. Neurocytol. 19, 686–696.

Williams, F. G., Mullet, M. A., and Beitz, A. J. 1995. Basal release of met-enkephalin and neurotensin in the ventrolateral periaqueductal gray matter of the rat: a microdialysis study of antinociceptive circuits. Brain Res. 690, 207–216.

Williams, J. L., Worland, P. D., and Smith, M. G. 1990. Defeat-induced hypoalgesia in the rat: effects of conditioned odors, naltrexone, and extinction. J. Exp. Psychol. Anim. Behav. Proc. 16, 345–357.

Williams, J. T., Christie, M. J., and Manzoni, O. 2001. Cellular and synaptic adaptations mediating opioid dependence. Physiol. Rev. 81, 299–343.

Xie, J. Y., Herman, D. S., Stiller, C. O., Gardell, L. R., Ossipov, M. H., Lai, J., Porreca, F., and Vanderah, T. W. 2005. Cholecystokinin in the rostral ventromedial medulla mediates opioid-induced hyperalgesia and antinociceptive tolerance. J. Neurosci. 25, 409–416.

Xin, L., Geller, E. B., Liu-Chen, L. Y., Chen, C., and Adler, M. W. 1997. Substance p release in the rat periaqueductal gray and preoptic anterior hypothalamus after noxious cold stimulation: effect of selective *mu* and *kappa* opioid agonists. J. Pharmacol. Exp. Ther. 282, 1055–1063.

Yezierski, R. P. 1988. Spinomesencephalic tract: projections from the lumbosacral spinal cord of the rat, cat, and monkey. J. Comp. Neurol. 267, 131–146.

Yezierski, R. P. and Mendez, C. M. 1991. Spinal distribution and collateral projections of rat spinomesencephalic tract cells. Neuroscience 44, 113–130.

Zadina, J. E., Hackler, L., Ge, L. J., and Kastin, A. J. 1997. A potent and selective endogenous agonist for the μ-opiate receptor. Nature 386, 499–502.

Zagon, A., Totterdell, S., and Jones, R. S. 1994. Direct projections from the ventrolateral medulla oblongata to the limbic forebrain: anterograde and retrograde tract-tracing studies in the rat. J. Comp. Neurol. 340, 445–468.

Zambreanu, L., Wise, R. G., Brooks, J. C., Iannetti, G. D., and Tracey, I. 2005. A role for the brainstem in central sensitisation in humans. Evidence from functional magnetic resonance imaging. Pain 114, 397–407.

Zhang, L., Sykes, K. T., and Hammond, D. L. 2004. Peripheral inflammation increases the proportion of spinally projecting neurons that express functional opioid receptors in the rostral ventromedial medulla (RVM). Abstract Viewer/Itinerary Planner. Society for Neuroscience. Online, Program 297.294.

Zhang, Y. Q., Tang, J. S., Yuan, B., and Jia, H. 1997. Inhibitory effects of electrically evoked activation of ventrolateral orbital cortex on the tail-flick reflex are mediated by periaqueductal gray in rats. Pain 72, 127–135.

Zhuo, M. and Gebhart, G. F. 1992. Characterization of descending facilitation and inhibition of spinal nociceptive transmission from the nuclei reticularis gigantocellularis and gigantocellularis pars alpha in the rat. J. Neurophysiol. 67, 1599–1614.

Zhuo, M. and Gebhart, G. F. 1997. Biphasic modulation of spinal nociceptive transmission from the medullary raphe nuclei in the rat. J. Neurophysiol. 78, 746–758.

42 Emotional and Behavioral Significance of the Pain Signal and the Role of the Midbrain Periaqueductal Gray (PAG)

K Keay and R Bandler, University of Sydney, Sydney, NSW, Australia

42.1 More than a Sensation?

In a provocative article entitled *On the Relation of Injury to Pain*, Wall P. D. (1979) suggested that "... pain is better classified as an awareness of a need state than as a sensation; that it has more in common with the phenomena of hunger and thirst than it has with seeing and hearing...." He also observed, somewhat anecdotally, that as pain changed from acute to chronic the emotional coping reaction also changed. Livingston W. K. (1998) in his posthumously published book *Pain and Suffering* stated a similar belief that "... pain is not strictly a physical sensation that can be defined simply by its anatomical and physiological substrates." Writing even earlier, Lewis T. (1942) observed that different emotional coping reactions were evoked by pain of different tissue origins, "... while painful sensations derived from the human skin are associated with brisk movements, with a rise of pulse rate and with a sense of invigoration ... painful sensations derived from deeper structures are associated often with quiescence, with slowing of the pulse, falling of the blood pressure, sweating and nausea." He went on to suggest that "... the difference in the quality of skin pain and of deep pain is so clear ... that it would perhaps seem unsafe to class both together under the one unqualified term pain and concluded that "... we should bear in mind the possibly serious fallacy of regarding both types (of pain) as represented in a common centre."

It is striking that these eminent researchers although starting from different perspectives arrive at a shared view, namely, that in addition to knowledge of the anatomy and physiology of its ascending sensory pathways, understanding how pain is localized in the brain also requires knowledge of the neurobiology of the emotional coping strategies triggered by different classes of pain.

42.2 Coping with Pain

42.2.1 Active versus Passive Emotional Coping

Animals (including humans) employ different strategies to cope with different classes of pain. Active emotional coping, which is characterized by engagement with the environment (i.e., fight or flight), is the usual response to escapable pain. Components of active coping include increased vigilance, hyperreactivity, increased somatomotor activity, and increased sympathetic activity. Passive emotional coping, is the antithesis of active coping. Instead of engagement with the environment, there is disengagement (e.g., conservation-withdrawal (Henry, J. P. and Stephens, P. M., 1977)). Components of passive coping include: decreased vigilance, hyporeactivity, quiescence, and falls in arterial pressure and heart rate. Passive emotional coping is the usual response to inescapable pain, i.e., pain of deep origin (muscle, joint, viscera) or any persistent/chronic pain (Lewis, T., 1942; Wall, P. D., 1979; Bandler, R. *et al.*, 2000; Keay, K. A. *et al.*, 2001).

42.2.2   Neural Representations of Active Emotional Coping

Research carried out in rats, cats, and primates has identified longitudinally oriented, neuronal columns within the midbrain periaqueductal gray (PAG) as critical substrates integrating active or passive emotional coping. As seen in Figure 1 active emotional coping responses (i.e., a coordinated reaction of freezing, fight or flight, hypertension, tachycardia, and a nonopioid-mediated analgesia) are evoked when excitatory amino acids are microinjected into either the dorsolateral (dlPAG) or lateral (lPAG) column. (Bandler, R. and Shipley, M. T., 1994). Anatomical and functional studies suggest, however, different neural circuits and triggers are involved.

42.2.2.1   *Dorsolateral periaqueductal gray*

The dlPAG is distinguished from the lPAG by the presence (e.g., nitric oxide synthase (NOS), cholecystokinin (CCK), acetylcholine (ACh), Met-enkephalin) or absence (e.g., cytochrome oxidase and Gly-2 transporter) of specific neurochemicals (for a review, see Keay, K. A. and Bandler, R., 2004). Anatomically, major descending inputs to dlPAG originate from medial prefrontal cortical (PFC) fields and select medial hypothalamic nuclei, but there is a striking absence of ascending afferents arising from either brainstem or spinal cord (Figure 2). The predominance of forebrain, primarily cortical inputs, led to the suggestion that when triggered by psychological (cortical) stimuli, active coping is mediated by the dlPAG (Bandler, R. *et al.*, 2000). Consistent with this suggestion, the potential threat signaled by the presence of a cat (sight and/or odor, but without physical contact) triggers active coping and a selective increase in immediate early gene (c-fos) expression in the dlPAG of the rat (Figure 2) (Canteras, N. S. and Goto, M., 1999).

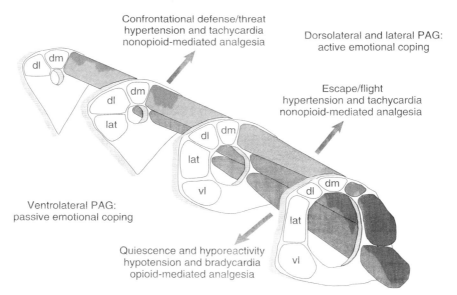

Figure 1   Schematic illustration of the dorsomedial (dm), dorsolateral (dl), lateral (lat), and ventrolateral (vl) neuronal columns within the periaqueductal gray (PAG). Injections of excitatory amino acids (EAA) within the dorsolateral and lateral PAG evoke active emotional coping strategies, whereas passive emotional coping strategies are evoked from the ventrolateral PAG. Specifically, EAA injections made within the dlPAG and lPAG columns evoke defensive reactions (confrontation, threat, escape, flight), hypertension, tachycardia, and a non-opioid-mediated analgesia. In contrast, EAA injections made within the vlPAG evoke cessation of spontaneous activity (quiescence), decreased responsiveness to the environment (hyporeactivity), hypotension, bradycardia, and an opioid-mediated analgesia. Adapted from Bandler, R. and Shipley, M. T. 1994. Columnar organization in the midbrain periaqueductal gray: modules for emotional expression? Trends Neurosci. 17, 379–389.

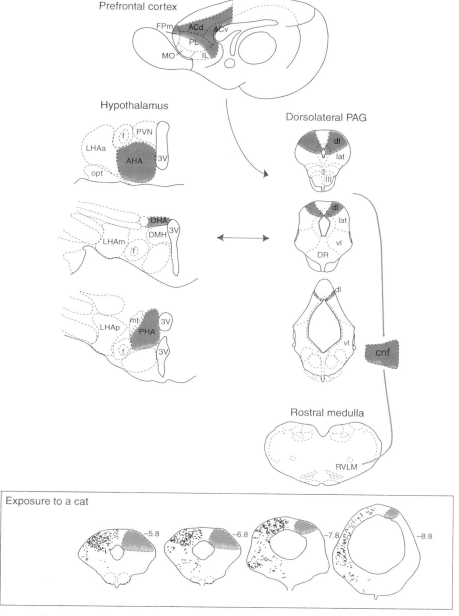

Figure 2 Schematic illustration of the major afferent and efferent projections of the dorsolateral periaqueductal gray (dlPAG). Major forebrain inputs to the dlPAG arise from dorsal, medial prefrontal cortical fields, and medial hypothalamic regions. The dlPAG does not project directly to the medulla, but can influence the rostral ventrolateral medulla via the cuneiform nucleus (cnf). The lower panel illustrates the pattern of neural activation (c-fos expression) evoked by a psychological stressor (exposure, without physical contact, of a rat to a cat), which evokes active emotional coping. The shaded area indicates the dorsolateral PAG column. RVLM, rostral ventrolateral medulla ACd, dorsal anterior cingulate cortex; ACv, ventral anterior cingulate cortex; AId, dorsal agranular insular cortex; AIp, posterior agranular insular cortex; DLO, dorsolateral orbital cortex; FPm, medial frontal pole; IL, infralimbic cortex; MO, medial orbital cortex; PL, prelimbic cortex; PRh, perirhinal cortex; VLOm, ventrolateral orbital cortex, medial part; VO, ventral orbital cortex AHA, anterior hypothalamic area; DHA, dorsal hypothalamic area; DMH, dorsomedial hypothalamic area; LHAa, lateral hypothalamic area, anterior; LHAm, lateral hypothalamic area, medial; LHAp, lateral hypothalamic area, posterior; PHA, posterior hypothalamic area; PVN, periventricular hypothalamic nucleus; f, fornix; mt, mamillothalmic tract; opt, optic tract; 3V, third ventricle; dl, dorsolateral PAG; lat, lateral PAG; vl, ventrolateral PAG, DR, dorsal raphe. Exposure to a cat: Adapted from Canteras, N. S. and Goto, M. 1999. Fos-like immunoreactivity in the periaqueductal gray of rats exposed to a natural predator. Neuroreport 10, 413–418.

42.2.2.2 *Lateral periaqueductal gray*

Although the boundary between the lPAG and dlPAG can be established on neurochemical grounds, the boundary between the lPAG and ventrolateral periaqueductal gray (vlPAG) rests on functional criteria (see Figure 1 and Section 42.2.3). Anatomically, the lPAG receives modest descending afferents of cortical and dorsal hypothalamic origins. However, in contrast to the dlPAG, the lPAG receives substantial and somatotopically organized ascending inputs from laminae I, II, IV, and V of the spinal cord, as well as the caudal, spinal trigeminal complex (Sp5) (Figure 3). The predominance of inputs arising from noci-responsive spinal and SpV regions suggests that active coping in response to physical stimuli (e.g., acute cutaneous pain) is mediated by the lPAG. In support of this hypothesis, brief applications of a noxious, cutaneous (thermal) stimulus evoked increased *Fos* expression within the lPAG, hypertension and active coping (Figure 3) (Keay, K. A. and Bandler, R. 1993; 2002).

42.2.3 Neural Representation of Passive Emotional Coping

Passive coping (i.e., a coordinated, conservation-withdrawal reaction of quiescence, hyporeactivity, hypotension, bradycardia, and an opioid-mediated analgesia) is evoked by microinjection of excitatory amino acids into the vlPAG column (Figure 1) (Bandler, R. and Shipley, M. T., 1994; Keay, K. A. and Bandler, R., 2001).

42.2.3.1 *Ventrolateral periaqueductal gray*

In common with the lPAG, substantial ascending inputs to the vlPAG originate from superficial and deep dorsal horn of spinal cord (and the transition zone between caudal and interpolar parts of Sp5, personal observations) (Keay, K. A. *et al.*, 1997). Double-label anatomical tracing studies indicate, however, that vlPAG and lPAG projections arise from distinct and separate populations of spinal neurons (Clement, C. I. *et al.*, 2000). Further, spino-vlPAG projections are not somatotopically organized (Keay, K. A. *et al.*, 1997). The vlPAG also receives substantial ascending inputs from nuclei of the solitary tract (NTS), as well as descending inputs from select orbital and insular cortical fields and the lateral hypothalamus (Figure 4). The convergence within the vlPAG of spinal, medullary, and forebrain afferents suggests that passive coping, whether evoked by physical or psychological stimuli, is mediated by the vlPAG. In support of this view, Fos expression is strongly evoked in the vlPAG when passive coping is triggered either (1) as a primary response to a deep or persistent noxious stimulus (see Figure 4) (Clement, C. I. *et al.*, 1996; Keay, K. A. *et al.*, 1997; 2001; Keay, K. A. and Bandler, R., 2002) or (2) in order to promote recovery and healing, as a delayed response to acute injury (for a discussion, see Wall, P. D., 1979; Keay, K. A. and Bandler, R., 2004).

42.2.4 Outputs Mediating Active and Passive Emotional Coping

42.2.4.1 *Lateral periaqueductal gray and ventrolateral periaqueductal gray*

Although they mediate distinct emotional coping strategies, the same ventromedial and ventrolateral medullary regions are projected upon by lPAG and vlPAG (see Figures 3 and 4) (Keay, K. A. and Bandler, R., 2001). Consistent with the opposing functional roles of lPAG and vlPAG, electrophysiological studies reveal opposite effects on medullary target neurons. For example, presympathetic rostral ventrolateral medulla (RVLM) vasopressor neurons are excited by lPAG stimulation, but inhibited by vlPAG stimulation (Verberne, A. J. and Boudier, H. R. S., 1991; Lovick, T. A., 1992a; 1992b; Verberne, A. and Guyenet, P., 1992). In contrast to shared descending targets, there exist distinct ascending projections to specific hypothalamic and midline and intralaminar thalamic fields (Floyd, N. S. *et al.*, 1996; 2000; Krout, K. E. and Loewy, A. D., 2000; Floyd, N. S. *et al.*, 2001).

42.2.4.2 *Dorsolateral periaqueductal gray*

Although it integrates an identical emotional coping strategy to the lPAG, the dlPAG has few direct medullary outputs. Its influence on the ventrolateral medulla is likely mediated indirectly, in part via a substantial projection to the cuneiform region (see Figure 2) (Redgrave, P. *et al.*, 1988; Mitchell, I. J. *et al.*, 1988a; 1988b). The dlPAG also projects to distinct thalamic and hypothalamic fields (Floyd, N. S. *et al.*, 1996; 2000; Krout, K. E. and Loewy, A. D., 2000; Floyd, N. S. *et al.*, 2001).

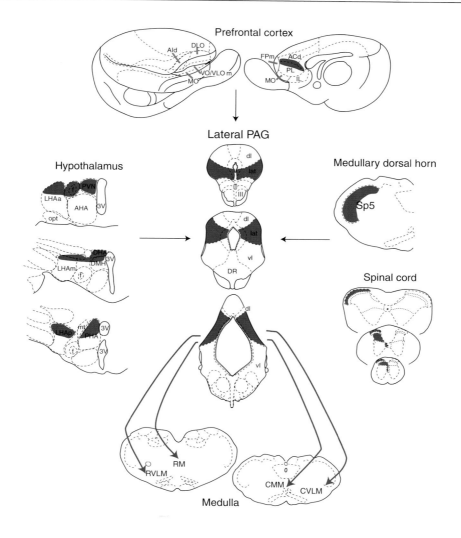

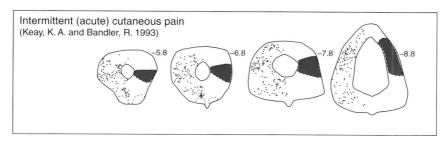

Figure 3 Schematic illustration of the major afferent and efferent projections of the lateral periaqueductal gray (IPAG). Major inputs to IPAG arise from spinal cord and the medullary dorsal horn (spinal trigeminal complex (SpV)). In addition, more modest inputs arise from a restricted part of medial prefrontal cortex as well as dorsal hypothalamic regions. The IPAG projects directly to both the rostral and caudal ventrolateral medulla (RVLM, CVLM), as well as the rostral and caudal ventromedial medulla (RM, CMM). The panel in the lower part of the figure shows the pattern of neural activation (c-fos expression) evoked by an acute cutaneous noxious stimulus (intermittent radiant heat), which evokes active emotional coping. The shaded area indicates the IPAG column. For abbreviations see caption to Figure 2.

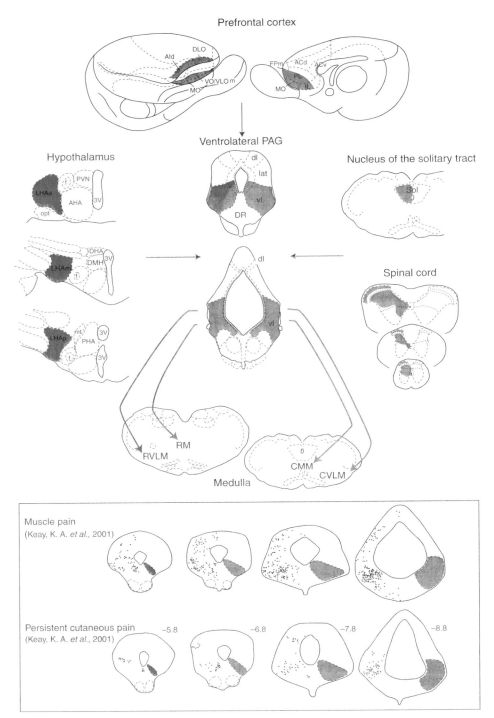

Figure 4 Schematic illustration of the major afferent and efferent projections of the ventrolateral periaqueductal gray (vlPAG). Major inputs to vlPAG arise from spinal cord and the nucleus of the solitary tract. In addition, significant inputs arise also from ventromedial prefrontal cortex and orbital/insular cortices. As well, substantial inputs arise from lateral hypothalamic fields. The vlPAG projects directly to both rostral and caudal ventrolateral medulla (RVLM, CVLM), as well as the rostral and caudal ventromedial medulla (RM, CMM). The panel in the lower part of the figure shows the patterns of neural activation (c-fos expression) evoked by an acute deep noxious stimulus (muscle pain: i.m. formalin) and a persistent cutaneous noxious stimulus (clip applied to dorsum of neck), each of which evoke passive emotional coping. The shaded area indicates the vlPAG column. For abbreviations see caption to Figure 2.

42.3 Conclusion

Animals (including humans) share the capacity to respond to escapable or inescapable pain with different emotional coping strategies. It has been established that the PAG is divisible into distinct longitudinal neuronal columns, which mediate distinct coping strategies (dlPAG/lPAG: active emotional coping; vlPAG: passive emotional coping). Further, each PAG column lies embedded within parallel, but distinct circuits, which extend rostrally to include specific PFC and hypothalamic regions (Bernard, J. F. and Bandler, J. F., 1998; Keay, K. A. and Bandler, R., 2001). On the output side, each PAG column projects either directly (lPAG/vlPAG) or indirectly (dlPAG) onto ventrolateral and ventromedial medullary regions containing (1) somatic and autonomic premotor neurons and (2) neurons mediating antinociception. A substantial body of evidence supports the view that the different PAG columns, and, their associated circuits, play critical roles in integrating the somatic, autonomic and antinociceptive components which characterize the distinct emotional coping reactions evoked by pain of different tissue origins and/or durations.

References

Bandler, R. and Shipley, M. T. 1994. Columnar organization in the midbrain periaqueductal gray: modules for emotional expression? Trends Neurosci. 17, 379–389.

Bandler, R., Keay, K. A., Floyd, N., and Price, J. 2000. Central circuits mediating patterned autonomic activity during active vs. passive emotional coping. Brain Res. Bull. 53, 95–104.

Bandler, R., Price, J. L., and Keay, K. A. 2000. Brain mediation of active and passive emotional coping. Progr. Brain Res. 122, 333–349.

Bernard, J. F. and Bandler, R. 1998. Parallel circuits for emotional coping behaviour: new pieces in the puzzle. J. Comp. Neurol. 401, 429–436.

Canteras, N. S. and Goto, M. 1999. Fos-like immunoreactivity in the periaqueductal gray of rats exposed to a natural predator. Neuroreport 10, 413–418.

Clement, C. I., Keay, K. A., Owler, B. K., and Bandler, R. 1996. Common patterns of increased and decreased fos expression in midbrain and pons evoked by noxious deep somatic and noxious visceral manipulations in the rat. J. Comp. Neurol. 366, 495–515.

Clement, C. I., Keay, K. A., Podzebenko, K., Gordon, B. D., and Bandler, R. 2000. Spinal sources of noxious visceral and noxious deep somatic afferent drive onto the ventrolateral periaqueductal gray of the rat. J. Comp. Neurol. 425, 323–344.

Floyd, N. S., Keay, K. A., and Bandler, R. 1996. A calbindin immunoreactive "deep pain" recipient thalamic nucleus in the rat. Neuroreport 7, 622–626.

Floyd, N. S., Price, J. L., Ferry, A. T., Keay, K. A., and Bandler, R. 2000. Orbitomedial prefrontal cortical projections to distinct longitudinal columns of the periaqueductal gray in the rat. J. Comp. Neurol. 422, 556–578.

Floyd, N. S., Price, J. L., Ferry, A. T., Keay, K. A., and Bandler, R. 2001. Orbitomedial prefrontal cortical projections to hypothalamus in the rat. J. Comp. Neurol. 432, 307–328.

Henry, J. P. and Stephens, P. M. 1977. Stress Health and the Social Environment: A Sociobiological Approach to Medicine. Springer.

Keay, K. A. and Bandler, R. 1993. Deep and superficial noxious stimulation increases fos-like immunoreactivity in different regions of the midbrain periaqueductal grey of the rat. Neurosci. Lett. 154, 23–26.

Keay, K. A. and Bandler, R. 2001. Parallel circuits mediating distinct emotional coping reactions to different types of stress. Neurosci. Biobehav. Rev. 25, 669–678.

Keay, K. A. and Bandler, R. 2002. Distinct central representations of inescapable and escapable pain: observations and speculation. Exp. Physiol. 87, 275–279.

Keay, K. A. and Bandler, R. 2004. Periaqueductal Gray. In: The Rat Nervous System, 3rd edn. (ed. G. Paxinos), pp. 243–257. Elsevier.

Keay, K. A., Clement, C. I., Depaulis, A., and Bandler, R. 2001. Different representations of inescapable noxious stimuli in the periaqueductal gray and upper cervical spinal cord of freely moving rats. Neurosci. Lett. 313, 17–20.

Keay, K. A., Clement, C. I., Owler, B., Depaulis, A., and Bandler, R. 1994. Convergence of deep somatic and visceral nociceptive information onto a discrete ventrolateral midbrain periaqueductal gray region. Neuroscience 61, 727–732.

Keay, K. A., Feil, K., Gordon, B. D., Herbert, H., and Bandler, R. 1997. Spinal afferents to functionally distinct periaqueductal gray columns in the rat: an anterograde and retrograde tracing study. J. Comp. Neurol. 385, 207–229.

Krout, K. E. and Loewy, A. D. 2000. Periaqueductal gray matter projections to midline and intralaminar thalamic nuclei of the rat. J. Comp. Neurol. 424, 111–141.

Lewis, T. 1942. Pain. MacMillan.

Livingston, W. K. 1998. Pain and Suffering. IASP Press.

Lovick, T. A. 1992a. Inhibitory modulation of the cardiovascular defence response by the ventrolateral periaqueductal grey matter in rats. Exp. Brain Res. 89, 133–139.

Lovick, T. A. 1992b. Midbrain influences on ventrolateral medullo-spinal neurones in the rat. Exp. Brain Res. 90, 147–152.

Mitchell, I. J., Dean, P., and Redgrave, P. 1988a. The projection from superior colliculus to cuneiform area in the rat. Ii. Defence-like responses to stimulation with glutamate in cuneiform nucleus and surrounding structures. Exp. Brain Res. 72, 626–639.

Mitchell, I. J., Redgrave, P., and Dean, P. 1988b. Plasticity of behavioural response to repeated injection of glutamate in cuneiform area of rat. Brain Res. 460, 394–397.

Redgrave, P., Dean, P., Mitchell, I. J., Odekunle, A., and Clark, A. 1988. The projection from superior colliculus to cuneiform area in the rat. I. Anatomical studies. Exp. Brain Res. 72, 611–625.

Verberne, A. and Guyenet, P. 1992. Midbrain central gray: influence on medullary sympathoexcitatory neurons and the baroreflex in rats. Am. J. Physiol. 263, R24–R33.

Verberne, A. J. and Boudier, H. R. S. 1991. Midbrain central grey: regional heamodynamic control and excitatory amino acidergic mechanisms. Brain Res. 550, 86–94.

Wall, P. D. 1979. On the relation of injury to pain. Pain 6, 253–264.

Further Reading

Bandler, R. and Keay, K. A. 1996. Columnar organization in the midbrain periaqueductal gray and the integration of emotional expression. Progr. Brain Res. 107, 285–300.

Keay, K. A., Clement, C. I., Matar, W. M., Heslop, D. J., Henderson, L. A., and Bandler, R. 1997. Noxious activation of spinal or vagal afferents evoked distinct patterns of fos-like immunoreactivity in the ventrolateral periaqueductal grey of unanaesthetised rats. Brain Res. 948, 122–130.

Lovick, T. and Bandler, R. 2005. The organization of the Midbrain Periaqueductal Grey and the Integration of Pain Behaviour. In: The Neurobiology of Pain (eds. S. Hunt and M. Koltzenburg), pp. 267–287. Oxford University Press.

43 The Thalamus and Nociceptive Processing

J O Dostrovsky, University of Toronto, Toronto, ON, Canada

A D Craig, Barrow Neurological Institute, Phoenix, AZ, USA

43.1 Introduction

The thalamus has been recognized to play a very important role in the higher-level processing of nociceptive inputs ever since the clinical observations by Dejerine J. and Roussy G. (1906) and Head H. and Holmes G. (1911) of pain resulting from strokes affecting the lateral thalamus. Modern-day anatomical and electrophysiological techniques have provided a wealth of information regarding the processing of nociceptive information at the thalamic level.

The concept that medial thalamus is involved in the affective/motivational aspect of pain and the lateral thalamus in the sensory/discriminative aspect of pain originated with the comprehensive analysis of 23 thalamic pain patients by Head and Holmes, and it was emphasized particularly by Melzack R. and Casey K. L. (1968). For many years, the available anatomical and physiological evidence suggested that these functions be ascribed to indirect spinoreticulothalamic (paleospinothalamic) input to medial thalamus and direct (neospinothalamic) spinothalamic tract (STT)-mediated input to somatosensory lateral thalamus, which was most prominent in humans. However, the recent evidence summarized below indicates that lamina I STT input to distinct portions of medial and lateral thalamus can be directly associated with these functions.

This chapter will review the major findings relating to the anatomy, physiology, and pharmacology of thalamus with respect to understanding its role in the mediation of acute and chronic pain. The chapter will start with an overview of thalamic anatomy and the termination sites of ascending somatosensory pathways focusing on the STT in the primate. Then the physiological findings in animals will be summarized, followed by discussions of findings in humans and a summary of thalamic pharmacology.

43.2 Comparative Anatomical Findings

In contrast to spinal cord, there are significant species differences in the anatomy of the pathways and the nuclei mediating nociception at the thalamic level. This chapter will focus on findings from primates including humans and will commence with a description of the thalamic termination sites of ascending pathways involved in the transmission of nociceptive information. The major pathway involved in the relay of nociceptive and thermoreceptive information is the STT and the functionally equivalent component of the trigeminothalamic tract (TTT) that originates in the medullary dorsal horn (subnucleus caudalis of the trigeminal spinal tract nucleus, also termed the medullary dorsal horn). Although the STT is frequently described as a single tract, this chapter will describe separately the terminations of its two components, the lateral and the ventral. The lateral STT originates largely from lamina I of the superficial dorsal horn and contains many neurons responding specifically to noxious stimuli and innocuous thermoreceptive neurons. In contrast, the ventral STT originates largely from neurons in deeper layers, most of which respond to innocuous tactile and proprioceptive inputs in addition to nociceptive inputs (Craig, A. D. and Dostrovsky, J. O., 1999; Craig, A. D., 2003).

43.2.1 Thalamic Nuclei Receiving Direct Spinothalamic Tract Inputs

There are six major regions of termination of the STT and TTT within the primate thalamus: the ventroposterior nucleus (VP), the posterior portion of the ventromedial nucleus (VMpo), the ventrolateral nucleus (VL), the central lateral nucleus (CL), the parafascicular nucleus (Pf), and the ventrocaudal portion of the medial dorsal nucleus (MDvc) (Mehler, W. R., 1969; Boivie, J., 1979; Berkley, K. J., 1980; Craig, A. D., 2003; 2004). Each of these is described in more detail below.

43.2.1.1 Ventroposterior nuclei

The VP that comprises the ventroposterior lateral (VPL) and ventroposterior medial (VPM) nuclei receives its major ascending afferent input from the medial lemniscus and is the main relay nucleus for tactile and kinesthetic information. This nucleus is also frequently termed the ventrobasal complex and in the human, the ventrocaudal nucleus (Vc) (Hassler, R., 1959; Mehler, W. R., 1966; Jones, E. G., 1990). However, it is well established that the STT and TTT also have terminations in this region. The STT terminations in VP form clusters (also referred to as archipelago), which are especially dense along its caudal border with the posterior group and pulvinar and along its rostral border with the VL and near the laminae that occur within VP (Stepniewska, I. et al., 2003). The terminations are roughly in register with the detailed somatotopic organization of the low-threshold mechanoreceptive (LTM) lemniscal terminations and thalamocortical neurons, with craniofacial inputs to VPM, arm/forelimb inputs to medial VPL, and leg/hindlimb inputs to lateral VPL. Some of the STT axons terminating in VP also send a collateral into the CL (Applebaum, A. E. et al., 1979; Giesler, G. J. Jr. et al., 1981).

The main source of STT and TTT terminations in VP is from the neurons in the deep dorsal horn (laminae IV and V) as shown in Figure 1(b). Interestingly, these terminations appear directed to VP neurons that are immunopositive for calbindin in contrast to the lemniscal afferents from the dorsal column nuclei (DCN) and principal trigeminal nucleus, which terminate in the vicinity of neurons that are immunopositive for parvalbumin (Rausell, E. and Jones, E. G., 1991). Furthermore, the calbindin-immunopositive neurons project to the superficial layers of cortex, whereas the parvalbumin neurons project to the middle layers (Gingold, S. I. et al., 1991; Rausell, E. and Jones, E. G., 1991). Another difference between spinothalamic and lemniscal terminations is that the latter but not the former form triadic synapses with GABAergic presynaptic dendrites (Ralston, H. J., III and Ralston, D. D., 1994). These differences suggest that the spinothalamic nociceptive inputs are processed differently than the innocuous mechanoreceptive inputs in VP.

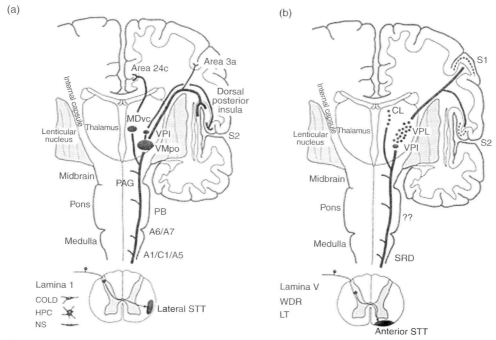

Figure 1 A schematic summary diagram of the ascending projections from the spinal cord to the thalamus and on to the cortex. (a) The lateral spinothalamic tract arising from nociceptive (HPC and NS) and innocuous thermoreceptive (COLD) neurons in lamina I. The major thalamic termination sites are in the posterior part of the ventromedial nucleus (VMpo), ventroposterior inferior nucleus (VPI), and the ventrocaudal part of the medial dorsal nucleus (MDvc). Also shown are the brainstem terminations in the ventrolateral medulla (A1/C1/A5), the dorsolateral pons (A6/A7), the parabrachial nucleus (PB), the periaqueductal gray (PAG), and the cortical projections of the thalamic sites receiving the lamina I inputs. (b) The ascending projections of lamina IV–V cells, primarily wide dynamic range (WDR) neurons and low-threshold mechanoreceptive (LTM) neurons, via the ventral spinothalamic tract (STT) to the VPI, ventroposterior lateral nucleus (VPL), ventrolateral nucleus (VL), and central lateral nucleus (CL). Terminations also occur in the brainstem in the subnucleus reticularis dorsalis (SRD) and other sites, probably including the reticular core. Adapted with permission from Craig, A. D. and Dostrovsky, J.O. 1999. Medulla to Thalamus. In: Textbook of Pain (eds. P. D. Wall and R. Melzack), pp. 183–214. Churchill-Livingstone.

Immediately ventral to VPL and VPM lies the ventroposterior inferior nucleus (VPI; roughly equivalent to the parvicellular part of the ventrocaudal nucleus (Vcpc) in humans). Curiously, this nucleus receives STT and TTT inputs from neurons located not only in laminae IV and V but also in lamina I, as well as from vestibular afferents (Ralston, H. J., III and Ralston, D. D., 1992). The VPI projects to secondary somatosensory cortex (SII) and retroinsular (vestibular) cortex (Friedman, D. P. and Murray, E. A., 1986; Stevens, R. T. *et al.*, 1993).

Species differences. In the rat, there are STT and TTT terminations throughout VP (Mehler, W. R., 1969; Peschanski, M., 1984), which originate from both superficial and deep layers of the dorsal horn. In contrast, in the cat, the STT and TTT terminate along the ventral aspect of VPL (and VPI and the basal part of the ventromedial nucleus (VMb)) and there are

almost no terminals within VP (Berkley, K. J., 1980; Craig, A. D. and Burton, H., 1985).

43.2.1.2 Posterior part of the ventromedial nucleus

Recent studies in the monkey have revealed that there is a very prominent and dense projection from STT and TTT neurons in lamina I to a region that has been termed the posterior ventromedial nucleus (VMpo), which lies immediately posterior and inferior to VP and is contiguous rostrally with the VMb (Craig, A. D., 2004). Of particular interest and importance are the findings that the STT and TTT terminations are topographically organized (Craig, A. D., 2003; 2004) and that the VMpo neurons project in a topographic manner to the dorsal posterior insula, a region that is consistently activated by thermal and nociceptive stimuli. The VMpo is the major projection target of the lamina I neurons in the primate, which comprise its almost exclusive ascending sensory input (Figure 1(a)). The trigeminal inputs terminate anteriorly and lumbar inputs most posteriorly. This anteroposterior topographic arrangement contrasts with the mediolateral topography of the VP.

The dense lamina I STT and TTT terminations in VMpo are clearly delineated with the use of immunohistochemical labeling for calbindin. A region of calbindin-positive terminal labeling has been observed in a comparable location in the human thalamus (Blomqvist, A. *et al.*, 2000) and corresponds with the area of dense STT terminations observed in thalamus following cordotomies performed to alleviate pain (Mehler, W. R., 1966). This region can also be delineated in monkeys and humans based on its different cytoarchitecture. Electron microscopy has revealed that the glutamatergic STT and TTT terminations form triadic synapses with the VMpo relay cell dendrites and GABAergic presynaptic dendrites (Beggs, J. *et al.*, 2003). This is similar to the termination of lemniscal afferents in VP but contrasts with the STT terminations in VP. The triadic contacts are believed to provide high synaptic security and temporal fidelity. VMpo was not specifically identified in earlier anatomical studies, but the region was included in the caudal VP (Mehler, W. R., 1966), the Vc portae (Hassler, R., 1970), and the posterior complex (Po) (Jones, E. G., 1990).

Species differences. In the cat, there is a sparse lamina I terminal field in the ventral VMb, and this may constitute a primordial homologue of the primate VMpo. In support of this notion is the fact that lesions to this part of VMb in the cat disrupt discriminative thermal sensation (Norrsell, U. and Craig, A. D., 1999) and that it projects to insular cortex (Clascá, F. *et al.*, 1997). A homologous region does not appear to exist in the rat.

43.2.1.3 Ventral lateral nucleus

Moderately dense STT terminations are observed extending rostrally from VP into caudal VL in cats and monkeys that overlap with inputs from the cerebellum (Berkley, K. J., 1980; Stepniewska, I. *et al.*, 2003). These inputs probably originate from neurons in the deep dorsal horn and ventral horn (laminae V and VII). VL projects to the motor cortex (Jones, E. G., 1985), and this STT component is most likely related with sensorimotor integration rather than nociception.

43.2.1.4 Parafascicular nucleus

The centré median (CM) and Pf regions are frequently cited as playing a major role in nociception. However, there is only a weak STT projection to Pf that originates from lamina I and V cells, and more recent anatomical studies fail to confirm earlier reports of STT terminations in CM. The Pf and CM nuclei appear to be involved in motor-related processing as their main connections are with the basal ganglia, substantia nigra, and motor cortex (Jones, E. G., 1985; Royce, G. J. *et al.*, 1989; Sadikot, A. F. *et al.*, 1992).

43.2.1.5 Medial dorsal nucleus

Recent studies in the monkey reveal moderately dense STT and TTT projections to the MDvc. The terminals are topographically organized along an anteroposterior axis, with trigeminal input located most posteriorly. The cells of origin of this projection are in lamina I of the spinal and medullary dorsal horn (Ganchrow, D., 1978; Craig, A. D., 2004). Cells in MDvc project to area 24c in the cortex at the fundus of the anterior cingulate sulcus (limbic motor cortex), rather than to the orbitofrontal and prefrontal cortex where the remainder of medial dorsal nucleus (MD) projects (Ray, J. P. and Price, J. L., 1993; Craig, A. D., 2003). It is likely that this region of MD plays an important role in mediating the affective/motivational aspect of pain. There appears to be no homologous region in the cat or rat (see Section 43.2.1.7).

43.2.1.6 Central lateral nucleus

Dense STT terminations are observed in the caudal part of CL and in some other portions of the nucleus. The cells of origin are located in laminae V and VII (Applebaum, A. E. *et al.*, 1979; Giesler, G. J., Jr. *et al.*, 1981; Craig, A. D., Jr. *et al.*, 1989). Projections from different regions of the spinal cord terminate in different cell clusters but do not appear to be topographically organized (Craig, A. D. and Burton, H., 1985). This nucleus also receives projections from the cerebellum, tectum, substantia nigra, and globus pallidus. Most of the neurons in this region project to the basal ganglia, but there are also projections to the superficial and deep layers of posterior parietal and motor cortices (Jones, E. G., 1985; Royce, G. J. *et al.*, 1989). The function of the STT input to this nucleus is unknown, but one can speculate that it may be involved in motor set, attention, and orientation. Similar STT terminations have been observed in cats, rats, and other vertebrates.

43.2.1.7 Nucleus submedius

In the cat, there are dense and topographically organized terminations of the TTT and STT arising from spinal and medullary dorsal horn lamina I neurons in the medial thalamic nucleus submedius (Sm). Although Sm originates developmentally from the pronucleus of MD, it projects to the ventrolateral orbital cortex rather than to the cingulate cortex and thus constitutes a major phylogenetic difference from the lamina I projection to MDvc in the primate (Craig, A. D., Jr. et al., 1982; Coffield, J. A. et al., 1992). In the cat, the spinal input to the anterior cingulate relays in the ventral VP (Musil, S. Y. and Olson, C. R., 1988; Yasui, Y. et al., 1988) and also indirectly by way of the parabrachial nucleus (PB) (Devinsky, O. et al., 1995). In the rat, input to Sm originates primarily from trigeminal cells at the junction of the caudalis and interpolaris subnuclei of the trigeminal spinal tract nucleus and from trigeminal and cervical lamina I cells, as well as other cells in the spinal cord (Dado, R. J. and Giesler, G. J., Jr., 1990; Yoshida, A. et al., 1992).

43.2.2 Indirect Nociceptive Pathways to the Thalamus

In addition to the STT, there are several polysynaptic pathways that may be involved in relaying nociceptive signals to the thalamus. First, the postsynaptic dorsal column pathway comprises dorsal horn neurons located primarily in laminae IV–VI and X whose axons ascend ipsilaterally in the superficial dorsolateral funiculus and deep dorsal columns and terminate in the ventral and rostral portions of the DCN (gracile and cuneate nuclei). These portions of the DCN have projections to motor-related regions of the brainstem rather than to VP (Brodal, A., 1982; Willis, W. D., 1985; Berkley, K. J. et al., 1986). Although most of the neurons in this pathway respond only to nonnoxious cutaneous mechanical stimuli, some respond also to noxious stimuli. Some of the nociceptive neurons in this pathway convey visceral activity in the rat and possibly also in the primate (Al-Chaer, E. D. et al., 1998; see also Villanueva, L. and Nathan, P., 2000). However, functional imaging studies of visceral pain in humans reveal a pattern of activation that does not appear consistent with the hypothesis that this pathway is important for visceral pain in humans (e.g., Strigo, I. A. et al., 2003). Second, the spinocervicothalamic tract comprises dorsal horn neurons whose axons ascend ipsilaterally in the

dorsolateral funiculus and terminate in the lateral cervical nucleus that is located at the level of C1-2 just lateral to the superficial dorsal horn (Boivie, J., 1983). The neurons of the lateral cervical nucleus project to the contralateral VP via the medial lemniscus. This pathway is prominent in the cat and raccoon but very small in the primate. Some of the neurons in this pathway respond to nociceptive stimuli although most respond only to innocuous stimuli (Kajander, K. C. and Giesler, G. J., Jr., 1987; Downie, J. W. et al., 1988). Third, evidence obtained in the rat indicates that spinal input to a dorsal reticular region in the caudal medulla is relayed by way of a portion of the ventromedial nucleus to layer 1 of widespread regions of the frontal cortex (Desbois, C. et al., 1999; Desbois, C. and Villanueva, L., 2001). This region also has a descending projection to the deep dorsal horn of the spinal cord. Modulation of activity, such as those associated with alerting responses, is suggested as a potential function, but whether this pathway exists in primates and humans is unknown. Fourth, a spinoreticulothalamic pathway (paleospinothalamic tract) was hypothesized long ago to accommodate clinical observations made in patients with thalamic pain syndrome (see below). Modern evidence indicates that spinal terminations in the brainstem reticular formation do not overlap with the locations of cells that project to thalamus, except within the PB (Blomqvist, A. and Berkley, K. J., 1992). The PB receives contralateral spinal input and ipsilateral trigeminal input that originates mainly from lamina I but also from lamina V cells. There is only a crude topography, and although there is considerable anatomical overlap with ascending homeostatic afferent input from the nucleus of the solitary tract, evidence in the rat indicates that nociceptive neurons are mainly modality selective (Bernard, J. F. and Besson, J. M., 1990). In primates, PB projects mainly to VMb in the thalamus, as well as to hypothalamus and amygdala, but in rats, it projects more broadly within medial thalamus and to various cortical regions associated with autonomic control.

43.3 Comparative Physiological Findings

The previous section described the pathways and thalamic termination sites of spinal and trigeminal neurons that include nociceptive neurons. On this basis, one would expect to find neurons in the thalamic termination sites that would also respond to noxious inputs, and this is particularly the case for the regions receiving inputs from lamina I. There have indeed been many reports of nociceptive neurons in lateral, posterior, and medial areas of the thalamus. However, it should be kept in mind that the existence of neurons responding to noxious stimuli in a given region of thalamus does not necessarily signify that the region is involved in mediating the sensations of pain that would be expected to be elicited by noxious stimuli giving rise to the noxious responses. For example, the responses could be related to arousal, attentional, and motor consequences of the noxious stimulus. The following section will summarize the physiological findings reported for each of the main thalamic regions receiving nociceptive inputs.

43.3.1   Ventroposterior Nuclei

As mentioned earlier, the VPL and VPM are the main thalamic targets for innocuous tactile information ascending via the medial lemniscus from the DCN and principal trigeminal nucleus. It is thus not surprising that most of the neurons in this region respond exclusively to innocuous low-threshold mechanical cutaneous inputs. These neurons are organized in a high-resolution somatotopic fashion and project primarily to areas 3b and 1 of primary somatosensory (SI) cortex. There is an anterior and dorsally located shell that contains neurons responding to deep, muscle and joint afferent inputs, and this proprioceptive information is relayed to areas 3a and 2 of SI cortex (Kaas, J. H. *et al.*, 1984; Jones, E. G., 1990).

It is generally considered that VP is the main relay nucleus for nociceptive inputs involved in mediating the sensory aspects of pain (localization and intensity discrimination) (Melzack, R. and Casey, K. L., 1968; Willis, W. D., 1985), since it receives STT inputs (primates and rodents, but not carnivores) and projects to the somatotopically organized SI cortex. Indeed, there have been many reports of nociceptive neurons within VP in the primate and rodent. In the monkey, about 10% of the neurons are nociceptive and most of these are of the wide dynamic range (WDR – responding to both LTM inputs and nociceptive inputs) type (Figure 2) (Willis, W. D., 1985; Apkarian, A. V. and Shi, T., 1994; Willis, W. D. and Westlund, K. N., 1997). They usually have moderately large receptive fields (e.g., more than half of the face or arm), and these are organized in a crude somatotopic manner roughly corresponding to the somatotopy of the intermingled

LTM neurons (Willis, W. D. and Westlund, K. N., 1997). In the monkey, these neurons are concentrated in the posterior part of VP and near the major fiber laminae, consistent with the predominant location of the terminations of the STT and TTT axons of lamina V neurons (Applebaum, A. E. *et al.*, 1979; Boivie, J., 1979; Kenshalo, D. R., Jr. *et al.*, 1980; Casey, K. L. and Morrow, T. J., 1983; Gingold, S. I. *et al.*, 1991; Bushnell, M. C. *et al.*, 1993). Anatomical studies have provided evidence suggesting that the nociceptive neurons project to areas 3b and 1 of SI cortex. Interestingly, these studies suggest that the cortical terminations of these nociceptive neurons are in the most superficial layers of cortex in contrast to the VP LTM neurons whose axons terminate in the middle layer of cortex (Rausell, E. *et al.*, 1992; Shi, T. *et al.*, 1993), indicating a possible modulatory role for these nociceptive inputs. Electrophysiological studies have confirmed that some of these WDR neurons within VP can be antidromically activated from areas 3b and 1 of SI (Kenshalo, D. R., Jr. *et al.*, 1980). Curiously, visceral noxious stimuli activate not only VP WDR neurons but also LTM neurons and do not appear to be somatotopically organized (Chandler, M. J. *et al.*, 1992; Al-Chaer, E. D. *et al.*, 1998).

Species differences. In the rat, nociceptive-specific (NS) and WDR nociceptive cells have been reported throughout VP intermixed in a roughly topographic manner with the LTM neurons (Guilbaud, G. *et al.*, 1980). The receptive fields of these nociceptive neurons are generally quite large and often bilateral. Some receive convergent visceral input. In rats with experimentally induced arthritis or neuropathies, increased numbers of WDR VP cells that project to the sensorimotor cortex have been reported (Guilbaud, G. *et al.*, 1990). In contrast to the primate and rodent, in the cat, there are virtually no nociceptive cells within VP proper. However, NS and WDR nociceptive neurons in this species can be found in the dorsal and ventral aspects of VP, the ventral aspect of VMb and VPI. These are the regions where STT and TTT terminations occur (see above). The nociceptive neurons within this shell-like region are organized in a crude mediolateral topographic pattern in parallel (but not in register) with the somatotopy of the adjacent VP. A portion of these neurons receive convergent input from skin, muscle, viscera, tooth pulp, or cranial vasculature (Davis, K. D. and Dostrovsky, J. O., 1988; Bruggemann, J. *et al.*, 1994). Some of these neurons have been shown to project to area 3a of SI, SII, and/or anterior cingulate cortex (Yasui, Y. *et al.*, 1988; Craig, A. D. and Dostrovsky, J. O., 2001).

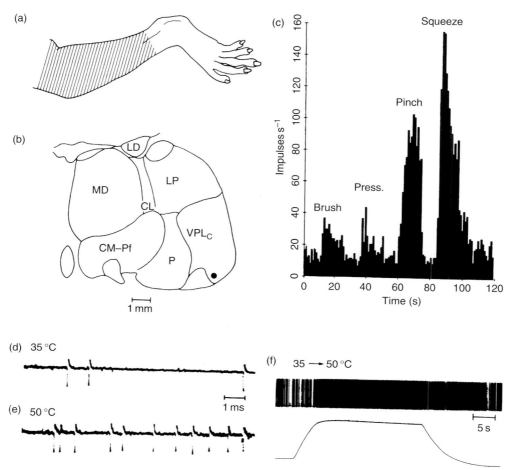

Figure 2 Example of responses of a wide-dynamic-range (WDR) neuron in ventroposterior nuclei (VP). The receptive field on the leg is shown in (a). The location of the recording site in VPLc (VP) is indicated in (b). The histogram (c) shows the responses to graded intensities of mechanical stimulation. This neuron also responded to noxious thermal stimulation (d, e, and f). Modified with kind permission from Kenshalo, D. R., Jr., Giesler, G. J., Jr., Leonard, R. B., and Willis, W. D. 1980. Responses of neurons in primate ventral posterior lateral nucleus to noxious stimuli. J. Neurophysiol. 43, 1594–1614.

43.3.2 Ventroposterior Inferior Nucleus

The neurons in the region located immediately inferior (ventral) to VPL and VPM (VPI) project to SII (Krubitzer, L. *et al.*, 1995) and to retroinsular (vestibular) cortex. Some of the neurons in this region are excited by nociceptive inputs (Figure 3). These WDR and NS neurons are arranged in a topographic representation within this region in the monkey (Apkarian, A. V. and Shi, T., 1994).

43.3.3 Posterior Thalamus and Posterior Part of the Ventromedial Nucleus

The posterior thalamus has frequently been implicated in pain processing and has been shown in

various species to contain nociceptive neurons (Albe-Fessard, D. *et al.*, 1985; Willis, W. D., 1985; Willis, W. D., Jr., 1997). Although it was formerly regarded as undifferentiated, more detailed and recent observations in monkeys and humans have described a distinct nucleus that has been termed VMpo. The VMpo has been shown to be the recipient of a major projection from lamina I STT and TTT neurons. In accordance with its lamina I inputs, it has been shown to contain NS and thermoreceptive neurons (Craig, A. D. *et al.*, 1994). Their responses are similar to those of lamina I STT and TTT neurons, and they have small receptive fields that are topographically organized in a rostrocaudal axis, in correspondence with the topographically organized terminations of the afferent inputs (Figure 4).

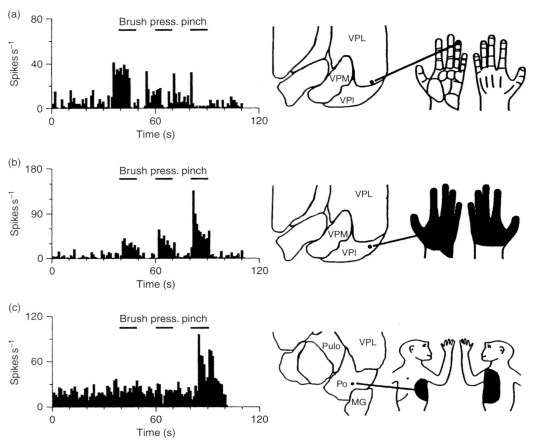

Figure 3 Example of responses of nociceptive neurons in ventroposterior inferior nucleus (VPI) and posterior complex (Po). Reproduced with kind permission from Apkarian, A. V. and Shi, T. 1994. Squirrel monkey lateral thalamus. I. Somatic nociresponsive neurons and their relation to spinothalamic terminals. J. Neurosci. 14, 6779–6795.

Neuronal recordings in awake monkeys in the presumed VMpo (originally assumed to be in VPM but subsequently re-interpreted from the original histology to be in VMpo (Craig, unpublished observations) revealed strong correlation between the neuronal activity and the behavioral detection of noxious and innocuous thermal stimuli applied to the face. Furthermore, lidocaine-induced block of activity in this region reduced the monkey's behavioral responses to these stimuli (Bushnell, M. C. *et al.*, 1993; Duncan, G. H. *et al.*, 1993).

Further support for the critical role of VMpo in relaying nociceptive and thermoreceptive information for perception of pain and temperature is the finding that it has a strong and topographic projection to the dorsal margin of the posterior insula (Craig, A. D., 2003), where preliminary fMRI observations in monkeys show strong activation with noxious stimuli (Craig, A. D., 2002). Furthermore, human imaging

studies have consistently found this region of insular cortex to be activated by noxious and thermal stimuli. This rostrocaudally organized topography is distinct from the mediolateral topography observed in SII and in the adjacent parietal operculum (Disbrow, E. *et al.*, 2000). There is also a minor projection from VMpo to area 3a of SI cortex.

Species differences. In the cat and rat where there is no clearly differentiated nociceptive and thermoreceptive relay site in the Po, there are nevertheless nociceptive neurons, but they tend to have very large receptive fields and are not arranged in a clear somatotopic fashion (Poggio, G. F. and Mountcastle, V. B., 1960; Carstens, E. and Yokota, T., 1980). This region does receive some lamina I inputs in the cat. In the rat, there is convergent STT input from laminae I and V to the posterior triangular nucleus, which projects to the region of SII and to the amygdala. It is possible that in the rat, nociceptive inputs also arise

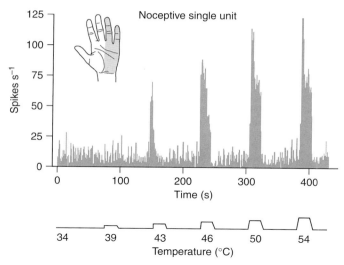

Figure 4 Example of the responses of a posterior part of the ventromedial (VMpo) nociceptive neuron. The histogram shows the graded responses of a single nociceptive-specific neuron to noxious heat pulses applied with a thermode applied to the receptive field on the contralateral ulnar hand. Adapted with kind permission from Craig, A. D., Bushnell, M. C., Zhang, E. -T., and Blomqvist, A. 1994. A thalamic nucleus specific for pain and temperature sensation. Nature 372, 770–773.

via a descending corticothalamic route (Diamond, M. E. *et al.*, 1992; Berkley, K. J. *et al.*, 1993). Furthermore, there is no direct projection to the portion of insular cortex homologous to that targeted by VMpo in the primate. The part of Po dorsal to VP in the cat, where some nociceptive neurons have been reported (Hutchison, W. D. *et al.*, 1992; Bruggemann, J. *et al.*, 1994), projects to area 5a (Jones, E. G., 1985). Nevertheless, in both the cat and the rat, there exists a narrow region along the ventral aspect of VMb that contains nociceptive and visceroceptive neurons and projects to the insular cortex adjacent to the gustatory cortex (Yasui, Y. *et al.*, 1991; Clasca, F. *et al.*, 1997; Norrsell, U. and Craig, A. D., 1999).

43.3.4 Intralaminar Nuclei

As described earlier, several regions in medial thalamus receive direct inputs from the spinal cord and the trigeminal nucleus, but there are also indirect inputs to these regions from the brainstem, and in particular from the PB.

Neurons responding to noxious electrical, mechanical, or heat stimuli have been recorded throughout the intralaminar thalamus, in particular in CL and CM–Pf. Most of these recordings have been obtained in anesthetized rats and cats although some recordings have been performed in monkeys (Albe-Fessard, D. *et al.*, 1985; Bushnell, M. C. *et al.*, 1993; Bruggemann, J. *et al.*, 1994). Most of the

nociceptive neurons in this region have large receptive fields that can be bilateral. Studies employing graded natural stimuli have revealed that some of the cells can code for stimulus intensity. Although natural stimulation was used in only some studies, cells with graded responses to noxious heat have been observed. In particular, a study in the awake monkey by Bushnell M. C. and Duncan G. H. (1989) reported the existence of nociceptive neurons in the CM–Pf and CL region that responded in a graded fashion to noxious thermal stimuli applied to the face (Figure 5). It is possible that the responses of at least some of these neurons may be related to attention and arousal rather than to perception of pain. It is interesting that many intralaminar neurons have been shown to respond to eye movements (Schlag, J. and Schlag-Rey, M., 1986), which is consistent with the termination of ascending inputs from the cerebellum and the superior colliculus. Indeed, it has been proposed that this region plays a role in gaze orientation (Jones, E. G., 1985).

43.3.5 Medial Dorsal Nucleus

As mentioned earlier, Craig and colleagues have recently described a direct lamina I input to the MDvc in the monkey. A particular interesting feature of this projection is that the MDvc contains neurons that project to the anterior cingulate cortex (Craig, A. D. and Zhang, E.-T., 1996), a region

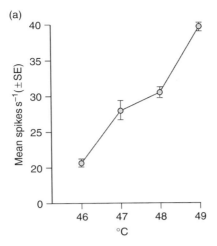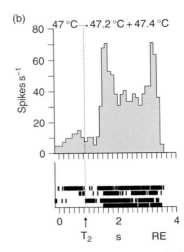

Figure 5 Responses of a neuron in posteromedial thalamus of an awake monkey to noxious thermal stimuli to the face. (a) The mean firing rate to increasing temperatures. (b) A peristimulus histogram with associated dot raster display showing increased firing when the temperature increased from 47 °C to 47.2 °C or 47.4 °C. In this particular example, the monkey did not respond until the end of the trial (RE button release). Adapted with kind permission from Bushnell, M. C. and Duncan, G. H. 1989. Sensory and affective aspects of pain perception: is medial thalamus restricted to emotional issues. Exp. Brain Res. 78, 415–418.

frequently implicated in pain. There is only very limited information regarding the properties of the neurons in MDvc; however, preliminary studies in anesthetized monkeys indicate that it contains a discrete group of NS neurons with large, sometimes bilateral receptive fields (Craig, A. D., 2003). Furthermore, this study reports that their spontaneous firing can be inhibited by innocuous thermal (cool, warm) stimuli. This phenomenon could provide an explanation for the well-known cold-induced inhibition of pain and the thermal grill illusion of pain (see below). It is possible that some of the recordings of nociceptive neurons in medial thalamus in the study of Bushnell M. C. and Duncan G. H. (1989) mentioned above may also have included neurons located in MDvc.

Species differences. Whereas in the monkey (and presumably the human), the STT and the TTT project to MDvc, in the cat and the rat, there is a projection that originates mainly in lamina I to the developmentally related Sm. Several studies have reported the existence of NS neurons in Sm in the rat (Dostrovsky, J. O. and Guilbaud, G., 1988; Coffield, J. A. and Miletic, V., 1993; Kawakita, K. *et al.*, 1993). Receptive fields are generally quite large and some have inputs from deep tissues. In the Freund's adjuvant-induced arthritic rat, many Sm cells were found to respond to joint movements, which are normally not effective in activating the Sm neurons

(Dostrovsky, J. O. and Guilbaud, G., 1988). Various studies in the rat that have examined the effects of lesioning or electrically and chemically induced excitation of Sm have provided evidence for a role of Sm and its cortical target VLO in the activation of descending antinociceptive pathways by way of the periaqueductal gray (PAG) (Roberts, V. J. and Dong, W. K., 1994; Zhang, S. *et al.*, 1998). Although it has been shown in the cat that there is also a projection from innocuous thermoreceptive cool lamina I STT neurons in addition to the NS neurons (Craig, A. D. and Dostrovsky, J. O., 2001), neurons excited by cooling stimuli have not been found. It is possible, however, that inputs from these neurons may provide a basis for the cold-induced inhibition of nociceptive processing, similar to MDvc of the primate (Ericson, A. C. *et al.*, 1996).

43.4 Direct Physiological Evidence in Humans

Functional stereotactic surgery for the treatment of chronic pain or tremor provides a unique opportunity to record and stimulate in the thalamus of awake patients. There have been several studies that have provided information of interest in terms of furthering our understanding of the role of thalamus in pain and these are described below.

43.4.1 Nociceptive Neurons

Lenz and colleagues have searched for neurons in VP (often termed Vc in humans) and the immediately adjacent posterior and inferior region of thalamus. Although most of the neurons in Vc respond only to nonnoxious mechanical stimulation of the skin (LTM neurons), they reported that some neurons also had an increased response to noxious mechanical stimuli and some responded weakly to noxious or innocuous thermal stimuli in addition to innocuous tactile stimuli (i.e., WDR neurons) (for review, see Lenz, F. A. and Dougherty, P. M., 1997; Lee J. *et al.*, 1999). These WDR neurons were primarily located in the posteroinferior portion of VP. In the adjoining posteroinferior area, which includes VMpo (Blomqvist, A. *et al.*, 2000), they found NS neurons that responded to noxious heat but no LTM neurons (Lenz, F. A. *et al.*, 1993a). Although nociceptive neurons have frequently been reported in medial thalamus of anesthetized animals, there have only been a few old reports of neurons responding to noxious stimuli in awake human medial thalamus. It is difficult to evaluate these old reports since few details were provided and more recent studies have not been able to confirm the findings (Lenz, F. A. *et al.*, 1997). However, there have been very few opportunities to record from neurons in medial thalamus, and the number of neurons that can be tested with noxious stimuli in the awake patient is limited, and this may account for the lack of definitive evidence.

43.4.2 Innocuous Cooling-Responsive Neurons

Neurons responding selectively to innocuous cooling of the skin have been found in the human thalamic region located medial and posteroinferior to VP, which likely corresponds to VMpo (Davis, K. D. *et al.*, 1999). These neurons had discrete receptive fields on the contralateral body and responses similar to those of lamina I spinal and trigeminal neurons (Figure 6(a)). This finding is consistent with the findings in animals, where cooling-specific neurons have only been found

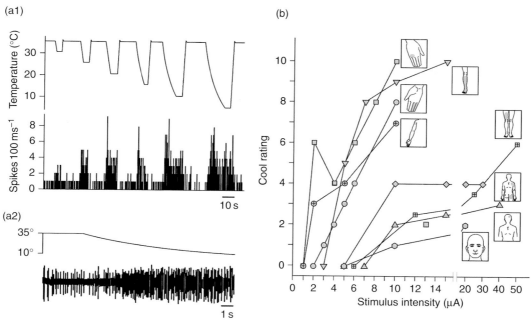

Figure 6 (a1) Recording from a single neuron in the presumed posterior ventromedial nucleus (VMpo) region in an awake patient showing responses to cooling stimuli applied to receptive field on the fifth digit. The top trace shows the temperature of the thermode with increasing cooling steps. The bottom trace is a histogram of the neuronal firing showing graded responses to the increasing cooling steps. (a2) Segment of raw trace of neuronal recording from (a1) showing response to first part of a cooling step. (b) Thalamic stimulation evoked cool sensations. Verbal ratings (0–10 scale) of the innocuous cool sensations evoked by threshold and suprathreshold intensities of thalamic microstimulation obtained in eight patients. Figurines adjacent to each line depict the location of the thalamic stimulation-evoked sensation at threshold. Reprinted with kind permission from Davis, K. D., Lozano, R. M., Manduch, M., Tasker, R. R., Kiss, Z. H., and Dostrovsky, J. O. 1999. Thalamic relay site for cold perception in humans. J. Neurophysiol. 81, 1970–1973.

in the cat medial VPM and in the monkey VMpo. Animal studies have also shown that cooling-specific neurons are found only in lamina I of the trigeminal medullary dorsal horn (nucleus caudalis) and spinal dorsal horn. Thus, the existence of cooling-specific neurons in the human VMpo is consistent with the evidence cited above that this region receives a major input from lamina I neurons. The close association of pain and temperature further supports an important role of VMpo in the relay of pain signals in the human.

43.4.3   Stimulation-Induced Pain and Temperature Sensations

The electrophysiological studies in human patients during stereotactic surgery provide a unique

opportunity to determine the sensations evoked by electrical stimulation within the brain, since these are performed with the patient awake. As might be expected, electrical stimulation (1 s trains of 100–300 Hz) within VP and adjacent regions of the thalamus usually evokes innocuous paresthesia. These sensations, which can be elicited by low stimulus intensities (e.g., 2 μA), are perceived as originating from a small region of the contralateral side of the body (Figure 7). The projected fields are usually in register with the receptive fields of the neurons recorded at the stimulation site. Increasing the intensity of stimulation results in an increase in the perceived intensity of sensation and usually also with an increase in the projected field size, but although the sensation can be very

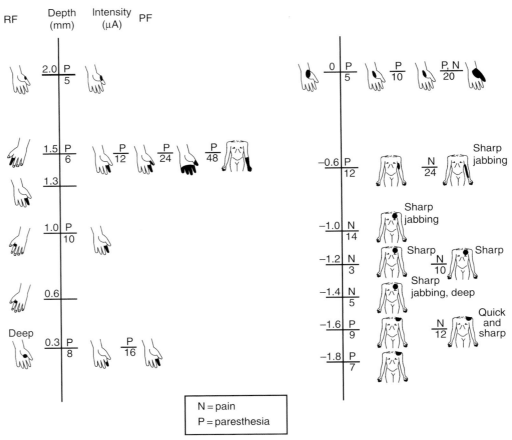

Figure 7   Stimulation-evoked pain in a chronic pain patient (causalgia left arm). The diagram shows a reconstruction of an electrode track through the ventrocaudal nucleus and on into the posteroinferior region, showing the locations of the receptive fields (RF) of the low-threshold mechanoreceptive neurons recorded at sites along the electrode track on the left of the vertical line and the projected fields (PF) induced by stimulation on the right side. The electrode track has been broken down into two contiguous segments. Depths of recordings are indicated in millimeters to the left of the line at each recording/stimulation site. While the electrode was in ventrocaudal nucleus (Vc), as evidenced by the low-threshold mechanoreceptive fields, stimulation induced nonpainful paresthesia (P). However, at the bottom of the Vc (0) and at deeper sites, stimulation induced painful sensations (N).

intense, it is not usually reported as painful. Nevertheless, stimulation at some sites in some patients is reported as eliciting a distinct painful and/or temperature sensation even at the threshold for sensation (Hassler, R. and Riechert, T., 1959; Halliday, A. M. and Logue, V., 1972; Tasker, R. R., 1984; Lenz, F. A. et al., 1993b; Davis, K. D. et al., 1996; Dostrovsky, J. O. et al., 2000). These sites are generally located near the posterior–inferior border of VP and extend several millimeters posterior, inferior, and medial to it (Figure 7). The incidence of sites where pain and/or thermal sensations can be evoked is much higher in the posteroinferior region than within VP (except in poststroke pain patients – see below). The sensations are always on the contralateral side of the body and can arise from a small region. The sensations are reported as being quite natural in contrast to the paresthesia (tingling or shock-like) sensations evoked within VP. There have been reports of cases where thalamic stimulation evoked pain arising from deep or visceral sites although these are rare (Lenz, F. A. et al., 1994; Davis, K. D. et al., 1995). Interestingly, Lenz and colleagues found that microstimulation within VP at sites where WDR neurons responding to noxious mechanical stimuli were found rarely elicited pain or temperature sensations, whereas at the sites in the region posteroinferior to VP where stimulation evoked pain, there was a high likelihood of finding nociceptive neurons (Lenz, F. A. and Dougherty, P. M., 1997). It seems likely that the sites in the region posteroinferior to VP where stimulation evokes pain and temperature are close to or in VMpo. It is also notable that at the sites where neurons responding selectively to cooling stimuli were found, microstimulation evoked cooling sensations arising from the same area of skin where the receptive fields of the neurons were located (Figure 6(b)). Increasing the stimulus intensity at such sites increased the intensity of the cold sensation (i.e., it felt colder).

There have been few studies of the effects of stimulation in medial thalamus. A few studies reported that stimulation in the posterior aspect of medial thalamus can evoke pain (Sano, K., 1979; Jeanmonod, D. et al., 1994). However, in most cases the stimuli were delivered from large-tipped electrodes and the sensations were only elicited at high current intensities, so current spread (e.g., to the STT) is an issue. Several more recent studies have failed to replicate these findings.

43.5 The Thalamus and Central Neuropathic Pain

43.5.1 Physiological Observations in Central Pain Patients

There have been many studies that have demonstrated the marked plasticity of the brain and in particular the alterations in neuronal properties and somatotopic organization at all levels of the somatosensory system subsequent to peripheral or central damage to the ascending pathways (Wall, J. T. et al., 2002). Although much less is known about the plasticity of the pain system, especially at the thalamic and cortical levels, it is reasonable to expect that marked changes will occur following damage to ascending nociceptive pathways. Indeed, there have been several animal studies that examined the effects of STT damage on thalamic physiology (Weng, H. R. et al., 2000; 2003). Studies in patients who have sustained deafferentation due, for example, to amputation or spinal cord injury provide evidence for plasticity (Lenz, F. A. et al., 1994; Davis, K. D. et al., 1998). For example, there is evidence suggesting expansion of the representation of intact regions into deafferented regions of VP thalamus. Micro-stimulation in such regions was found to evoke sensations arising from the phantom limb in amputees or to the deafferented body region in spinal injury patients (Lenz, F. A. et al., 1994; Davis, K. D. et al., 1998). Although these stimuli usually resulted in nonpainful paresthesia rather than pain, these types of changes may be involved in the development of chronic pain in these types of patients. Of particular interest have been the findings of two studies that found changes in the stimulation-evoked sensations in poststroke pain and some other types of chronic central pain patients. The incidence of stimulation sites in VP where stimulation evoked sensations in such patients was markedly elevated (Lenz, F. A. et al., 1998; Dostrovsky, J. O. et al., 2000). In addition, there was an increase in the number of sites in the posteroinferior region where pain sensations were evoked and a decrease in sites where innocuous thermal stimuli were evoked (Figure 8). These findings provide further evidence suggesting that there have been alterations in the thalamic and cortical processing of somatosensory inputs leading to increased perception of pain. It is interesting, however, that deep brain stimulation within thalamus (Siegfried, J., 1987; Kupers, R. C. and Gybels, J. M., 1993) can alleviate the pain in some of these chronic pain patients, possibly by disrupting the pathological patterns and balance of activity in these regions.

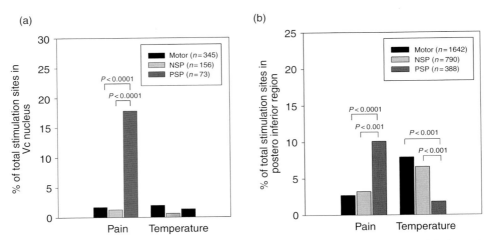

Figure 8 Bar graphs showing the incidence of sites in the ventrocaudal nucleus (a) and the posteroinferior region (b) where stimulation evoked sensations of pain or innocuous temperature in movement disorder patients (Motor), chronic nonpoststroke pain patients (NSP), and poststroke pain patients (PSP). Reprinted with kind permission from Dostrovsky, J. O., Manduch, M., Davis, K. D., Tasker, R. R., Lozano, A. M. 2000. Thalamic Stimulation-Evoked Pain and Temperature Sites in Pain and Non-pain Patients. In: Proceedings of the 9th World Congress on Pain (eds. M. Devor, M. C. Rowbotham, and Z. Wiesenfeld-Hallin), pp. 419–425. IASP Press.

43.5.2 Thalamic Bursting Activity

Thalamic neurons switch to a bursting mode when they are hyperpolarized. This bursting activity is generated by activation of low-threshold T-type calcium channels that give rise to a calcium spike and a burst of sodium channel-generated action potentials (so-called LTS bursting). This bursting activity is generally only observed during sleep. However, recordings in medial and lateral thalamus of chronic pain patients during the awake state revealed the existence of neurons firing in typical calcium spike-mediated bursts (Lenz, F. A. et al., 1989; Jeanmonod, D. et al., 1993; Lenz, F. A. et al., 1994). It was proposed that this activity may be related to the patients' chronic pain and in fact may mediate the spontaneous pain in these patients (Lenz, F. A. and Dougherty, P. M., 1997; Llinas, R. R. et al., 1999). However, similar LTS firing in similar regions of thalamus can also be observed in patients without pain and thus is not specific to pain (Jeanmonod, D. et al., 1996; Radhakrishnan, V. et al., 1999). More recently, Llinas R. R. et al. (1999) have proposed that LTS bursting in thalamus may be the cause of many different neurological diseases including chronic pain and have termed this condition thalamic dysrhythmia. Nevertheless, the role of LTS bursting in chronic central pain remains unclear.

It has been proposed that central neuropathic pain may be mediated by reduced GABAergic inhibition in the thalamus (Roberts, V. J. and Dong, W. K.,

1994). Studies in the primate have found that following chronic cervical rhizotomy there is a downregulation of thalamic GABA$_A$ receptors (Rausell, E. et al., 1992), providing some support for this suggestion. On the other hand, several positron emission tomography (PET) studies have obtained evidence suggesting that chronic neuropathic pain is associated with decreased blood flow, suggesting decreased activity (Apkarian, A. V. et al., 2005).

43.5.3 Effects of Thalamic Lesions on Pain

There have been many studies since the original pioneering studies of Dejerine and Roussy and Head and Holmes which have documented that strokes affecting the lateral thalamus can lead to central pain (poststroke pain and thalamic pain syndrome) (Pagni, C. A., 1998). In addition, infarcts and lesions that involve the posteroinferior region can also result in analgesia and thermanesthesia (Head, H. and Holmes, G., 1911; Hassler, R. and Riechert, T., 1959; Hassler, R., 1970; Tasker, R. R., 1984). Lesions have been purposefully made in medial thalamus of chronic pain patients for relief of their pain (Gybels, J. M. and Sweet, W. H., 1989). However, such procedures are rarely undertaken at the present time as most neurosurgeons do not believe that they are effective. It is possible, however, that in some of the reported successes the lesions may have included MDvc and/or VMpo (Jeanmonod, D. *et al.*, 1994; Lenz, F. A. and Dougherty, P. M., 1997). Interestingly, lesions in medial thalamus that spare the lateral thalamus and STT do not appear to cause central pain (Bogousslavsky, J. *et al.*, 1988).

43.5.4 The Thalamic Pain Syndrome

The poststroke central (thalamic) pain (CPSP) syndrome was first recognized by Dejerine and Roussy, and it has been comprehensively analyzed clinically by Head H. and Holmes G. (1911), Riddoch G. and Critchley Mc. D. (1937), Schott B. *et al.* (1986), Boivie J. (1994), Pagni C. A. (1998), and others. The first patients had lesions confined to the thalamus, and so the syndrome was first called thalamic pain. However, patients with central pain have lesions that interrupt the ascending lateral STT – lamina I – spinothalamocortical pathway by way of VMpo to the dorsal posterior insula at any level (Schmahmann, J. D. and Leifer, D., 1992; Pagni, C. A., 1998; Craig, A. D., 2003). Such lesions produce loss of pain and temperature sensation, but in about half of such cases, this disruption results, immediately or after a variable delay, in the paradoxical appearance of ongoing pain in the deafferented region. Head and Holmes inferred that this meant that pain sensation occurred in the thalamus, but we now know that the thalamus serves mainly as a relay for activity ascending to the cortex. This syndrome likely results from disruption of the interactions at cortical and subcortical levels between components of the pain processing network. One such interaction is the inhibition caused by cooling on pain, and it has been suggested that interruption of this interaction could cause central pain by disinhibition (Craig, A. D., 1998). The suggestion is based on similarities with the thermal grill illusion of pain, in which reduced activity in cooling-specific lamina I STT neurons can unmask the cold-activated burning pain elicited by polymodal nociceptive lamina I STT neurons (HPC) in the anterior cingulate cortex. A study of thermal sensation in a central pain patient directly supports that hypothesis (Morin, C. *et al.*, 2002). A recent imaging study of a central pain patient suggested that the posterolateral thalamic lesion did not involve VMpo but that conclusion was incompatible with the patient's loss of thermal sensation (cf. Craig, A. D. *et al.*, 2000; Montes, C. *et al.*, 2005). The indirect ascending nociceptive input to the medial thalamus by way of the PB may also play a role in central pain.

43.6 Pharmacology

The pharmacology of thalamic nociceptive neurons has not been extensively studied. However, it is reasonable to assume that they share many common features with those of other thalamic neurons about which much more is known. The two most important transmitters are glutamate and GABA (McCormick, D. A., 1992; Steriade, M. *et al.*, 1997; Sherman, S. M. and Guillery, R. W., 2001). Glutamate is released by the ascending axons of medial lemniscal and STT and TTT pathways to excite thalamocortical neurons. The released glutamate acts at both NMDA and non-NMDA (AMPA and/or kainate) ionotropic glutamate receptors. In the cat and primate, there are also many interneurons within the thalamus and these also receive glutamatergic inputs from these afferents that activate in addition to the ionotropic receptors and metabotropic glutamate receptors (Jones, E. G., 1985; Magnusson, K. R. *et al.*, 1987; Ericson, A. C. *et al.*, 1995; Blomqvist, A. *et al.*, 1996; Sherman, S. M. and Guillery, R. W., 2001). The thalamocortical neurons also receive a massive glutamatergic projection from layer 6 of the cortex, which activates both ionotropic and metabotropic receptors (Rustioni, A. *et al.*, 1983; Jones, E. G., 1987; Deschenes, M. and Hu, B., 1990; Eaton, S. A. and Salt, T. E., 1990; 1996; Salt, T. E. and Eaton, S. A., 1996). The thalamocortical neurons release glutamate at their cortical terminals (Kharazia, V. N. and Weinberg, R. J., 1994; Pirot, S. *et al.*, 1994).

GABA is the major inhibitory neurotransmitter in the thalamus, and the major source of these inputs is from the thalamic reticular nucleus (TRN). The TRN GABAergic neurons are topographically organized and receive excitatory inputs from collaterals of thalamocortical and corticothalamic neurons. In cats and primates, but not rats, there are also GABAergic interneurons within the thalamus (Houser, C. R. *et al.*, 1980; Rustioni, A. *et al.*, 1983; Jones, E. G., 1985; Steriade, M. *et al.*, 1997). The TRN produces both short- and long-latency inhibitory responses; the short-latency IPSPs are mediated by $GABA_A$ receptors, whereas the long-latency responses are mediated by the $GABA_B$ metabotropic receptors (Lee, S. M. *et al.*, 1994). Interestingly, activation of presynaptic glutamate metabotropic receptors reduces the release of GABA from the TRN receptors (Salt, T. E. and Eaton, S. A., 1995). Most of the studies of thalamic GABAergic mechanisms have involved nonnociceptive neurons; however, studies in Pf have shown that thalamic nociceptive neurons receive inhibitory inputs from the TRN and are inhibited by GABA (Reyes-Vazquez, C. and Dafny, N., 1983; Jia, H. *et al.*, 2004), and microinjection of the GABA$_A$ antagonist bicuculline or agonist muscimol into Sm enhanced or depressed, respectively, the antinociception induced by the prior microinjection of morphine into the rat Sm (Jia, H. *et al.*, 2004). There is also recent evidence for modulation of GABA$_B$ receptors in VP and Po by noxious inputs (inflammation) (Ferreira-Gomes, J. *et al.*, 2004).

The thalamic relay neurons as well as the TRN and inhibitory interneurons also receive projections from serotoninergic neurons in the dorsal raphe nucleus, and norardrenergic and cholinergic neurons in the peribrachial region (see reviews by McCormick, D. A., 1992; Steriade, M. *et al.*, 1997). Their role in nociception is, however, unclear. There have been recent studies in the rat that have demonstrated an involvement of serotoninergic mechanisms in modulation of transmission through the Sm, and these may play a role in a pain modulatory system proposed to relay in Sm (Xiao, D. Q. *et al.*, 2005). There is also evidence that 5HT modulates nociceptive processing in the rat Pf (Harte, S. E. *et al.*, 2005). Several neuropeptides have also been identified in thalamic afferents and neurons. For example, tachykinins, cholecystokinin, and enkephalins have been identified in spinothalamic fibers ending in the rat and human thalamus and are presumably coreleased with glutamate, but their function is not known (Coffield, J. A. and Miletic, V., 1987; Gall, C. *et al.*, 1987; Hirai, T. and Jones, E. G., 1989; Battaglia, G. *et al.*, 1992; Nishiyama, K. *et al.*, 1995).

It is usually assumed that the analgesia produced by opiates results from an action at the spinal and brainstem levels. However, there is increasing evidence for a role of opioids also at thalamic and cortical levels, and thus, a direct depressant effect of opioids on thalamic and cortical nociceptive neurons may contribute to the opioid-induced analgesia. Opioid receptors and enkephalinergic terminals have been identified in the thalamus (Mansour, A. *et al.*, 1987; Miletic, V. and Coffield, J. A., 1988). Systemic morphine inhibits thalamic nociceptive neurons (e.g., Benoist, J. M. *et al.*, 1983), but this does not prove that effect is due to an action at the thalamic level. However, there have been studies that have shown that local application of opioid agents can depress the responses of thalamic nociceptive neurons (He, L. F. *et al.*, 1991; Coffield, J. A. and Miletic, V., 1993) and produce antinociception when microinjected into Sm (Yang, Z. J. *et al.*, 2002). Furthermore, the use of radiolabeled μ-opioid agonists with PET has shown high opiate receptor levels in the human thalamus (see review in Apkarian, A. V. *et al.*, 2005), and evidence for their involvement in endogenous pain modulation.

References

Al-Chaer, E. D., Feng, Y., and Willis, W. D. 1998. A role for the dorsal column in nociceptive visceral input into the thalamus of primates. J. Neurophysiol. 79, 3143–3150.

Albe-Fessard, D., Berkley, K. J., Kruger, L., Ralston, H. J., III, and Willis, W. D., Jr. 1985. Diencephalic mechanisms of pain sensation. Brain Res. 356, 217–296.

Apkarian, A. V. and Shi, T. 1994. Squirrel monkey lateral thalamus. I. Somatic nociresponsive neurons and their relation to spinothalamic terminals. J. Neurosci. 14, 6779–6795.

Apkarian, A. V., Bushnell, M. C., Treede, R. D., and Zubieta, J. K. 2005. Human brain mechanisms of pain perception and regulation in health and disease. Eur. J. Pain 9, 463–484.

Applebaum, A. E., Leonard, R. B., Kenshalo, D. R., Jr., Martin, R. F., and Willis, W. D. 1979. Nuclei in which functionally identified spinothalamic tract neurons terminate. J. Comp. Neurol. 188, 575–585.

Battaglia, G., Spreafico, R., and Rustioni, A. 1992. Substance P innervation of the rat and cat thalamus. I. Distribution and relation to ascending spinal pathways. J. Comp. Neurol. 315, 457–472.

Beggs, J., Jordan, S., Ericson, A. C., Blomqvist, A., and Craig, A. D. 2003. Synaptology of trigemino- and spinothalamic lamina I terminations in the posterior ventral medial nucleus of the macaque. J. Comp. Neurol. 459, 334–354.

Benoist, J. M., Kayser, V., Gautron, M., and Guilbaud, G. 1983. Low dose of morphine strongly depresses responses of specific nociceptive neurones in the ventrobasal complex of the rat. Pain 15, 333–344.

Berkley, K. J. 1980. Spatial relationships between the terminations of somatic sensory and motor pathways in the rostral brainstem of cats and monkeys. I. Ascending somatic sensory inputs to lateral diencephalon. J. Comp. Neurol. 193, 283–317.

Berkley, K. J., Budell, R. J., Blomqvist, A., and Bull, M. 1986. Output systems of the dorsal column nuclei in the cat. Brain Res. Bull. 11, 199–225.

Berkley, K. J., Guilbaud, G., Benoist, J. M., and Gautron, M. 1993. Responses of neurons in and near the thalamic ventrobasal complex of the rat to stimulation of uterus, cervix, vagina, colon, and skin. J. Neurophysiol. 69, 557–568.

Bernard, J. F. and Besson, J. M. 1990. The spino(trigemino)pontoamygdaloid pathway: electrophysiological evidence for an involvement in pain processes. J. Neurophysiol. 63, 473–490.

Blomqvist, A. and Berkley, K. J. 1992. A re-examination of the spino-reticulo-diencephalic pathway in the cat. Brain Res. 579, 17–31.

Blomqvist, A., Ericson, A. C., Craig, A. D., and Broman, J. 1996. Evidence for glutamate as a neurotransmitter in spinothalamic tract terminals in the posterior region of owl monkeys. Exp. Brain Res. 108, 33–44.

Blomqvist, A., Zhang, E. T., and Craig, A. D. 2000. Cytoarchitectonic and immunohistochemical characterization of a specific pain and temperature relay, the posterior portion of the ventral medial nucleus, in the human thalamus. Brain 123 (Pt 3), 601–619.

Bogousslavsky, J., Regli, F., and Uske, A. 1988. Thalamic infarcts: clinical syndromes, etiology, and prognosis. Neurology 38, 837–848.

Boivie, J. 1979. An anatomical reinvestigation of the termination of the spinothalamic tract in the monkey. J. Comp. Neurol. 186, 343–370.

Boivie, J. 1983. Anatomic and Physiologic Features of the Spino-cervico-thalamic Pathway. In: Somatosensory Integration in the Thalamus (eds. G. Macchi, A. Rustioni, and R. Spreafico), pp. 63–106. Elsevier.

Boivie, J. 1994. Central Pain. In: Texbook of Pain (ed. P. D. Wall), pp. 871–902. Churchill Livingstone.

Brodal, A. 1982. Neurological Anatomy. Oxford.

Bruggemann, J., Vahle-Hinz, C., and Kniffki, K. D. 1994. Projections from the pelvic nerve to the periphery of the cat's thalamic ventral posterolateral nucleus and adjacent regions of the posterior complex. J. Neurophysiol. 72, 2237–2245.

Bushnell, M. C. and Duncan, G. H. 1989. Sensory and affective aspects of pain perception: is medial thalamus restricted to emotional issues. Exp. Brain Res. 78, 415–418.

Bushnell, M. C., Duncan, G. H., and Tremblay, N. 1993. Thalamic VPM nucleus in the behaving monkey. I. Multimodal and discriminative properties of thermosensitive neurons. J. Neurophysiol. 69, 739–752.

Carstens, E. and Yokota, T. 1980. Viscerosomatic convergence and responses to intestinal distension of neurons at the junction of midbrain and posterior thalamus in the cat. Exp. Neurol. 70, 392–402.

Casey, K. L. and Morrow, T. J. 1983. Ventral posterior thalamic neurons differentially responsive to noxious stimulation of the awake monkey. Science 221, 675–677.

Chandler, M. J., Hobbs, S. F., Fu, Q. G., Kenshalo, D. R., Jr., Blair, R. W., and Foreman, R. D. 1992. Responses of neurons in ventroposterolateral nucleus of primate thalamus to urinary bladder distension. Brain Res. 571, 26–34.

Clasca, F., Llamas, A., and Reinoso-Suarez, F. 1997. Insular cortex and neighboring fields in the cat: a redefinition based on cortical microarchitecture and connections with the thalamus. J. Comp. Neurol. 384, 456–482.

Coffield, J. A. and Miletic, V. 1987. Immunoreactive enkephalin is contained within some trigeminal and spinal neurons projecting to the rat medial thalamus. Brain Res. 425, 380–383.

Coffield, J. A. and Miletic, V. 1993. Responses of rat nucleus submedius neurons to enkephalins applied with micropressure. Brain Res. 630, 252–261.

Coffield, J. A., Bowen, K. K., and Miletic, V. 1992. Retrograde tracing of projections between the nucleus submedius, the ventrolateral orbital cortex, and the midbrain in the rat. J. Comp. Neurol. 321, 488–499.

Craig, A. D. 1998. A new version of the thalamic disinhibition hypothesis of central pain. Pain Forum 7, 1–14.

Craig, A. D. 2002. New and Old Thoughts on the Mechanisms of Pain Following Spinal Cord Injury. In: Spinal Cord Injury Pain: Assessment, Mechanisms, and Treatment (eds. R. P. Yezierski and K. J. Burchiel), pp. 237–261. IASP Press.

Craig, A. D. 2003. Pain mechanisms: labeled lines versus convergence in central processing. Annu. Rev. Neurosci. 26, 1–30.

Craig, A. D. 2004. Distribution of trigeminothalamic and spinothalamic lamina I terminations in the macaque monkey. J. Comp. Neurol. 477, 119–148.

Craig, A. D. and Burton, H. 1985. The distribution and topographical organization in the thalamus of anterogradely-transported horseradish peroxidase after spinal injections in cat and raccoon. Exp. Brain Res. 58, 227–254.

Craig, A. D., Chen, K., Bandy, D., and Reiman, E. M. 2000. Thermosensory activation of insular cortex. Nat. Neurosci. 3, 184–190.

Craig, A. D. and Dostrovsky, J. O. 1999. Medulla to Thalamus. In: Textbook of Pain (eds. P. D. Wall and R. Melzack), pp. 183–214. Churchill-Livingstone.

Craig, A. D. and Dostrovsky, J. O. 2001. Differential projections of thermoreceptive and nociceptive lamina I trigeminothalamic and spinothalamic neurons in the cat. J. Neurophysiol. 86, 856–870.

Craig, A. D. and Zhang, E.-T. 1996 Anterior cingulate projection from MDvc (a lamina I spinothalamic target in the medial thalamus of the monkey). Soc. Neurosci. Abstr. 22, 111.

Craig, A. D., Bushnell, M. C., Zhang, E.-T., and Blomqvist A. 1994. A thalamic nucleus specific for pain and temperature sensation. Nature 372, 770–773.

Craig, A. D., Jr., Linington, A. J., and Kniffki, K. D. 1989. Cells of origin of spinothalamic tract projections to the medial and lateral thalamus in the cat. J. Comp. Neurol. 289, 568–585.

Craig, A. D., Jr., Wiegand, S. J., and Price, J. L. 1982. The thalamo-cortical projection of the nucleus submedius in the cat. J. Comp. Neurol. 206, 28–48.

Dado, R. J. and Giesler, G. J., Jr. 1990. Afferent input to nucleus submedius in rats: retrograde labeling of neurons in the spinal cord and caudal medulla. J. Neurosci. 10, 2672–2686.

Davis, K. D. and Dostrovsky, J. O. 1988. Responses of feline trigeminal spinal tract nucleus neurons to stimulation of the middle meningeal artery and sagittal sinus. J. Neurophysiol. 59, 648–666.

Davis, K. D., Kiss, Z. H. T., Luo, L., Tasker, R. R., Lozano, A. M., and Dostrovsky, J. O. 1998. Phantom sensations generated by thalamic microstimulation. Nature 391, 385–387.

Davis, K. D., Kiss, Z. H. T., Tasker, R. R., and Dostrovsky, J. O. 1996. Thalamic stimulation-evoked sensations in chronic pain patients and in nonpain (movement disorder) patients. J. Neurophysiol. 75, 1026–1037.

Davis, K. D., Lozano, R. M., Manduch, M., Tasker, R. R., Kiss, Z. H., and Dostrovsky, J. O. 1999. Thalamic relay site for cold perception in humans. J. Neurophysiol. 81, 1970–1973.

Davis, K. D., Tasker, R. R., Kiss, Z. H., Hutchison, W. D., and Dostrovsky, J. O. 1995. Visceral pain evoked by thalamic microstimulation in humans. Neuroreport 6, 369–374.

Dejerine, J. and Roussy, G. 1906. La syndrome thalamique. Rev. Neurol. 14, 521–532.

Desbois, C. and Villanueva, L. 2001. The organization of lateral ventromedial thalamic connections in the rat: a link for the distribution of nociceptive signals to widespread cortical regions. Neuroscience 102, 885–898.

Desbois, C., Le, B. D., and Villanueva, L. 1999. Organization of cortical projections to the medullary subnucleus reticularis dorsalis: a retrograde and anterograde tracing study in the rat. J. Comp. Neurol. 410, 178–196.

Deschenes, M. and Hu, B. 1990. Electrophysiology and pharmacology of the corticothalamic input to lateral thalamic nuclei: an intracellular study in the cat. Eur. J. Neurosci. 2, 140–152.

Devinsky, O., Morrell, M. J., and Vogt, B. A. 1995. Contributions of anterior cingulate cortex to behaviour. Brain 118(Pt 1), 279–306.

Diamond, M. E., Armstrong-James, M., Budway, M. J., and Ebner, F. F. 1992. Somatic sensory responses in the rostral sector of the posterior group (POm) and in the ventral posterior medial nucleus (VPM) of the rat thalamus: dependence on the barrel field cortex. J. Comp. Neurol. 319, 66–84.

Disbrow, E., Roberts, T., and Krubitzer, L. 2000. Somatotopic organization of cortical fields in the lateral sulcus of Homo sapiens: evidence for SII and PV. J. Comp. Neurol. 418, 1–21.

Dostrovsky, J. O. and Guilbaud, G. 1988. Noxious stimuli excite neurons in nucleus submedius of the normal and arthritic rat. Brain Res. 460, 269–280.

Dostrovsky, J. O., Manduch, M., Davis, K. D., Tasker, R. R., and Lozano, A. M. 2000. Thalamic Stimulation-Evoked Pain and Temperature Sites in Pain and Non-Pain Patients. In: Proceedings of the 9th World Congress on Pain (eds. M. Devor, M. C. Rowbotham, and Z. Wiesenfeld-Hallin), pp. 419–425. IASP Press.

Downie, J. W., Ferrington, D. G., Sorkin, L. S., and Willis, W. D., Jr. 1988. The primate spinocervicothalamic pathway: responses of cells of the lateral cervical nucleus and spinocervical tract to innocuous and noxious stimuli. J. Neurophysiol. 59, 861–885.

Duncan, G. H., Bushnell, M. C., Oliveras, J. L., Bastrash, N., and Tremblay, N. 1993. Thalamic VPM nucleus in the behaving monkey. III. Effects of reversible inactivation by lidocaine on thermal and mechanical discrimination. J. Neurophysiol. 70, 2086–2096.

Eaton, S. A. and Salt, T. E. 1990. Thalamic NMDA receptors and nociceptive sensory synaptic transmission. Neurosci. Lett. 110, 297–302.

Eaton, S. A. and Salt, T. E. 1996. Role of N-methyl-D-aspartate and metabotropic glutamate receptors in corticothalamic excitatory postsynaptic potentials in vivo. Neuroscience 73, 1–5.

Ericson, A. C., Blomqvist, A., Craig, A. D., Ottersen, O. P., and Broman, J. 1995. Evidence for glutamate as neurotransmitter in trigemino-and spinothalamic tract terminals in the nucleus submedius of cats. Eur. J. Neurosci. 7, 305–317.

Ericson, A. C., Blomqvist, A., Krout, K., and Craig, A. D. 1996. Fine structural organization of spinothalamic and trigeminothalamic lamina I terminations in the nucleus submedius of the cat. J. Comp. Neurol. 371, 497–512.

Ferreira-Gomes, J., Neto, F. L., and Castro-Lopes, J. M. 2004. Differential expression of GABA(B(1b)) receptor mRNA in the thalamus of normal and monoarthritic animals. Biochem. Pharmacol. 68, 1603–1611.

Friedman, D. P. and Murray, E. A. 1986. Thalamic connectivity of the second somatosensory area and neighboring somatosensory fields of the lateral sulcus of the macaque. J. Comp. Neurol. 252, 348–373.

Gall, C., Lauterborn, J., Burks, D., and Seroogy, K. 1987. Co-localization of enkephalin and cholecystokinin in discrete areas of rat brain. Brain Res. 403, 403–408.

Ganchrow, D. 1978. Intratrigeminal and thalamic projections of nucleus caudalis in the squirrel monkey (Saimiri sciureus): a degeneration and autoradiographic study. J. Comp. Neurol. 178, 281–312.

Giesler, G. J., Jr., Yezierski, R. P., Gerhart, K. D., and Willis, W. D. 1981. Spinothalamic tract neurons that project to medial and/or lateral thalamic nuclei: evidence for a physiologically novel population of spinal cord neurons. J. Neurophysiol. 46, 1285–1308.

Gingold, S. I., Greenspan, J. D., and Apkarian, A. V. 1991. Anatomic evidence of nociceptive inputs to primary somatosensory cortex: relationship between spinothalamic terminals and thalamocortical cells in squirrel monkeys. J. Comp. Neurol. 308, 467–490.

Guilbaud, G., Benoist, J. M., Jazat, F., and Gautron, M. 1990. Neuronal responsiveness in the ventrobasal thalamic complex of rats with an experimental peripheral mononeuropathy. J. Neurophysiol. 64, 1537–1554.

Guilbaud, G., Peschanski, M., Gautron, M., and Binder, D. 1980. Neurones responding to noxious stimulation in VB complex and caudal adjacent regions in the thalamus of the rat. Pain 8, 303–318.

Gybels, J. M. and Sweet, W. H. 1989. Neurosurgical Treatment of Persistent Pain. Physiological and Pathological Mechanisms of Human Pain. Pain and Headache, Vol. 11. Karger.

Halliday, A. M. and Logue, V. 1972. Painful Sensations Evoked by Electrical Stimulation in the Thalamus. In: Neurophysiology Studied in Man (ed. G. G. Somjen), pp. 221–230. Excerpts Medica.

Harte, S. E., Kender, R. G., and Borszcz, G. S. 2005. Activation of 5-HT1A and 5-HT7 receptors in the parafascicular nucleus suppresses the affective reaction of rats to noxious stimulation. Pain 113, 405–415.

Hassler, R. 1959. Architechtronic Organization of the Thalamic Nuclei. In: An Introduction to Stereotaxis with an Atlas of the Human Brain (eds. G. Schaltenbrand and P. Bailey), pp. 142–180. Thieme.

Hassler, R. 1970. Dichotomy of Facial Pain Conduction in the Diencephalon. In: Trigeminal Neuralgia (eds. R. Hassler and A. E. Walker), pp. 123–138. Sanders.

Hassler, R. and Riechert, T. 1959. Klinische und anatomische Befunde bei stereotaktischen Schmerzoperationen im thalamus. Arch. Psychiatr. 200, 93–122.

He, L. F., Dong, W. Q., and Wang, M. Z. 1991. Effects of iontophoretic etorphine and naloxone, and electroacupuncture on nociceptive responses from thalamic neurones in rabbits. Pain 44, 89–95.

Head, H. and Holmes, G. 1911. Sensory disturbances from cerebral lesions. Brain 34, 102–254.

Hirai, T. and Jones, E. G. 1989. A new parcellation of the human thalamus on the basis of histochemical staining. Brain Res. Brain Res. Rev. 14, 1–34.

Houser, C. R., Vaughn, J. E., Barber, R. P., and Roberts, E. 1980. GABA neurons are the major cell type of the nucleus reticularis thalami. Brain Res. 200, 341–354.

Hutchison, W. D., Luhn, M. A., and Schmidt, R. F. 1992. Knee joint input into the peripheral region of the ventral posterior lateral nucleus of cat thalamus. J. Neurophysiol. 67, 1092–1104.

Jeanmonod, D., Magnin, M., and Morel, A. 1993. Thalamus and neurogenic pain: physiological, anatomical and clinical data. Neuroreport 4, 475–478.

Jeanmonod, D., Magnin, M., and Morel, A. 1994. A thalamic concept of neurogenic pain. Proc. 7th Cong. Pain 2, 767–787.

Jeanmonod, D., Magnin, M., and Morel, A. 1996. Low-threshold calcium spike bursts in the human thalamus. Common physiopathology for sensory, motor and limbic positive symptoms. Brain 119(Pt 2), 363–375.

Jia, H., Xie, Y. F., Xiao, D. Q., and Tang, J. S. 2004. Involvement of GABAergic modulation of the nucleus submedius (Sm) morphine-induced antinociception. Pain 108, 28–35.

Jones, E. G. 1985. The Thalamus. Plenum.

Jones, E. G. 1987. Immunocytochemical Studies on Thalamic Afferent Transmitters. In: Thalamus and Pain (eds. J. M. Besson, G. Guilbaud, and M. Peschanski), pp. 83–109. Excerpta Medica.

Jones, E. G. 1990. Correlation and revised nomenclature of ventral nuclei in the thalamus of human and monkey. Stereotact. Funct. Neurosurg. 54–55, 1–20.

Kaas, J. H., Nelson, R. J., Sur, M., Dykes, R. W., and Merzenich, M. M. 1984. The somatotopic organization of the ventroposterior thalamus of the squirrel monkey, Saimiri sciureus. J. Comp. Neurol. 226, 111–140.

Kajander, K. C. and Giesler, G. J., Jr. 1987. Responses of neurons in the lateral cervical nucleus of the cat to noxious cutaneous stimulation. J. Neurophysiol. 57, 1686–1704.

Kawakita, K., Dostrovsky, J. O., Tang, J. S., and Chiang, C. Y. 1993. Responses of neurons in the rat thalamic nucleus submedius to cutaneous, muscle and visceral nociceptive stimuli. Pain 55, 327–338.

Kenshalo, D. R., Jr., Giesler, G. J., Jr., Leonard, R. B., and Willis, W. D. 1980. Responses of neurons in primate ventral posterior lateral nucleus to noxious stimuli. J. Neurophysiol. 43, 1594–1614.

Kharazia, V. N. and Weinberg, R. J. 1994. Glutamate in thalamic fibers terminating in layer IV of primary sensory cortex. J. Neurosci. 14, 6021–6032.

Krubitzer, L., Clarey, J., Tweedale, R., Elston, G., and Calford, M. 1995. A redefinition of somatosensory areas in the lateral sulcus of macaque monkeys. J. Neurosci. 15, 3821–3839.

Kupers, R. C. and Gybels, J. M. 1993. Electrical stimulation of the ventroposterolateral thalamic nucleus (VPL) reduces mechanical allodynia in a rat model of neuropathic pain. Neurosci. Lett. 150, 95–98.

Lee, J., Dougherty, P. M., Antezana, D., and Lenz, F. A. 1999. Responses of neurons in the region of human thalamic principal somatic sensory nucleus to mechanical and thermal stimuli graded into the painful range. J. Comp. Neurol. 410, 541–555.

Lee, S. M., Friedberg, M. H., and Ebner, F. F. 1994. The role of GABA-mediated inhibition in the rat ventral posterior medial thalamus. II. Differential effects of GABAA and GABAB receptor antagonists on responses of VPM neurons. J. Neurophysiol. 71, 1716–1726.

Lenz, F. A. and Dougherty, P. M. 1997. Pain Processing in the Human Thalamus. In: Thalamus, Volume II, Experimental and Clincal Aspects (eds. M. Steriade, E. G. Jones, and D. McCormick), pp. 617–651. Elsevier.

Lenz, F. A., Gracely, R. H., Baker, F. H., Richardson, R. T., and Dougherty, P. M. 1998. Reorganization of sensory modalities evoked by microstimulation in region of the thalamic principal sensory nucleus in patients with pain due to nervous system injury. J. Comp. Neurol. 399, 125–138.

Lenz, F. A., Kwan, H. C., Dostrovsky, J. O., and Tasker, R. R. 1989. Characteristics of the bursting pattern of action potentials that occurs in the thalamus of patients with central pain. Brain Res. 496, 357–360.

Lenz, F. A., Kwan, H. C., Martin, R., Tasker, R., Richardson, R. T., and Dostrovsky, J. O. 1994. Characteristics of somatotopic organization and spontaneous neuronal activity in the region of the thalamic principal sensory nucleus in patients with spinal cord transection. J. Neurophysiol. 72, 1570–1587.

Lenz, F. A., Seike, M., Lin, Y. C., Baker, F. H., Rowland, L. H., Gracely, R. H., and Richardson, R. T. 1993a. Neurons in the area of human thalamic nucleus ventralis caudalis respond to painful heat stimuli. Brain Res. 623, 235–240.

Lenz, F. A., Seike, M., Richardson, R. T., Lin, Y. C., Baker, F. H., Khoja, I., Jaeger, C. J., and Gracely, R. H. 1993b. Thermal and pain sensations evoked by microstimulation in the area of human ventrocaudal nucleus. J. Neurophysiol. 70, 200–212.

Llinas, R. R., Ribary, U., Jeanmonod, D., Kronberg, E., and Mitra, P. P. 1999. Thalamocortical dysrhythmia: a neurological and neuropsychiatric syndrome characterized by magnetoencephalography. Proc. Natl. Acad. Sci. U. S. A. 96, 15222–15227.

Magnusson, K. R., Clements, J. R., Larson, A. A., Madl, J. E., and Beitz, A. J. 1987. Localization of glutamate in trigeminothalamic projection neurons: a combined retrograde transport-immunohistochemical study. Somatosens. Res. 4, 177–190.

Mansour, A., Khachaturian, H., Lewis, M. E., Akil, H., and Watson, S. J. 1987. Autoradiographic differentiation of mu, delta, and kappa opioid receptors in the rat forebrain and midbrain. J. Neurosci. 7, 2445–2464.

McCormick, D. A. 1992. Neurotransmitter actions in the thalamus and cerebral cortex and their role in neuromodulation of thalamocortical activity. Prog. Neurobiol. 39, 337–388.

Mehler, W. R. 1966. The posterior thalamic region in man. Confin. Neurol. 27, 18–29.

Mehler, W. R. 1969. Some neurological species differences – a posteriori. Ann. N. Y. Acad. Sci. 167, 424–468.

Melzack, R. and Casey, K. L. 1968. Sensory, Motivational, and Central Control Determinants of Pain. In: The Skin Senses (ed. D. R. Kenshalo), pp. 423–443. Thomas.

Miletic, V. and Coffield, J. A. 1988. Enkephalin-like immunoreactivity in the nucleus submedius of the cat and rat thalamus. Somatosens. Res. 5, 325–334.

Morin, C., Bushnell, M. C., Luskin, M. B., and Craig, A. D. 2002. Disruption of thermal perception in a multiple sclerosis patient with central pain. Clin. J. Pain 18, 191–195.

Montes, C., Magnin, M., Maarrow, J., Frot, M., Convers, P., Mauguiére, F., and Garcia-Larrea, L. 2005. Thalamic thermo-algesic transimission: ventral posterior (VP) complex versus VMpo in the light of a thalamic infarct with central pain. Pain 113, 223–232.

Musil, S. Y. and Olson, C. R. 1988. Organization of cortical and subcortical projections to anterior cingulate cortex in the cat. J. Comp. Neurol. 272, 203–218.

Nishiyama, K., Kwak, S., Murayama, S., and Kanazawa, I. 1995. Substance P is a possible neurotransmitter in the rat spinothalamic tract. Neurosci. Res. 21, 261–266.

Norrsell, U. and Craig, A. D. 1999. Behavioral thermosensitivity after lesions of thalamic target areas of a thermosensory spinothalamic pathway in the cat. J. Neurophysiol. 82, 611–625.

Pagni, C. A. 1998. Central Pain: A Neurosurgical Challenge. Edizioni Minerva Medica.

Peschanski, M. 1984. Trigeminal afferents to the diencephalon in the rat. Neuroscience 12, 465–487.

Pirot, S., Jay, T. M., Glowinski, J., and Thierry, A. M. 1994. Anatomical and electrophysiological evidence for an excitatory amino acid pathway from the thalamic mediodorsal nucleus to the prefrontal cortex in the rat. Eur. J. Neurosci. 6, 1225–1234.

Poggio, G. F. and Mountcastle, V. B. 1960. A study of the functional contributions of the lemniscal and spinothalamic systems to somatic sensibility. Bull. Johns Hopkins Hosp. 106, 266–316.

Radhakrishnan, V., Tsoukatos, J., Davis, K. D., Tasker, R. R., Lozano, A. M., and Dostrovsky, J. O. 1999. A comparison of the burst activity of lateral thalamic neurons in chronic pain and non-pain patients. Pain 80, 567–575.

Ralston, H. J., III and Ralston, D. D. 1992. The primate dorsal spinothalamic tract: evidence for a specific termination in the posterior nuclei (Po/SG) of the thalamus. Pain 48, 107–118.

Ralston, H. J., III, and Ralston, D. D. 1994. Medial lemniscal and spinal projections to the macaque thalamus: an electron microscopic study of differing GABAergic circuitry serving thalamic somatosensory mechanisms. J. Neurosci. 14, 2485–2502.

Rausell, E. and Jones, E. G. 1991. Histochemical and immunocytochemical compartments of the thalamic VPM nucleus in monkeys and their relationship to the representational map. J. Neurosci. 11, 210–225.

Rausell, E., Bae, C. S., Vinuela, A., Huntley, G. W., and Jones, E. G. 1992. Calbindin and parvalbumin cells in monkey VPL thalamic nucleus: distribution, laminar cortical projections, and relations to spinothalamic terminations. J. Neurosci. 12, 4088–4111.

Ray, J. P. and Price, J. L. 1993. The organization of projections from the mediodorsal nucleus of the thalamus to orbital and medial prefrontal cortex in macaque monkeys. J. Comp. Neurol. 337, 1–31.

Reyes-Vazquez, C. and Dafny, N. 1983. Microiontophoretically applied THIP effects upon nociceptive responses of neurons in medial thalamus. Appl. Neurophysiol. 46, 254–260.

Riddoch, G. and Critchley, Mc.D. 1937. La physiopatholgie de la douleru d'oriinge centrale. Rev. Neurol. 68, 77–104.

Roberts, V. J. and Dong, W. K. 1994. The effect of thalamic nucleus submedius lesions on nociceptive responding in rats. Pain 57, 341–349.

Royce, G. J., Bromley, S., Gracco, C., and Beckstead, R. M. 1989. Thalamocortical connections of the rostral intralaminar nuclei: an autoradiographic analysis in the cat. J. Comp. Neurol. 288, 555–582.

Rustioni, A., Schmechel, D. E., Spreafico, R., Cheema, S., and Cuenod, M. 1983. Excitatory and Inhibitory Amino Acid Putative Neurotransmitters in the Ventralis Posterior Complex: An Autoradiographic and Immunocytochemical Study in Rats and Cats. In: Somatosensory Integration in the Thalamus (eds. G. Machhi, A. Rustioni, and R. Spreafico), pp. 365–383. Elsevier.

Sadikot, A. F., Parent, A., and François, C. 1992. Efferent connections of the centromedian and parafascicular thalamic nuclei in the squirrel monkey: a PHA-L study of subcortical projections. J. Comp. Neurol. 315, 137–159.

Salt, T. E. and Eaton, S. A. 1995. Distinct presynaptic metabotropic receptors for L-AP4 and CCG1 on GABAergic terminals: pharmacological evidence using novel alpha-methyl derivative mGluR antagonists, MAP4 and MCCG, in the rat thalamus in vivo. Neuroscience 65, 5–13.

Salt, T. E. and Eaton, S. A. 1996. Functions of ionotropic and metabotropic glutamate receptors in sensory transmission in the mammalian thalamus. Prog. Neurobiol. 48, 55–72.

Sano, K. 1979. Stereotaxic Thalamolaminotomy and Posteromedial Hypothalamotomy for the Relief of Intractable Pain. In: Advances in Pain Research and Therapy (eds. J. J. Bonica and V. Ventafridda), pp. 475–485. Raven Press.

Schlag, J. and Schlag-Rey, M. 1986. Role of the central thalamus in gaze control. Prog. Brain Res. 64, 191–201.

Schmahmann, J. D. and Leifer, D. 1992. Parietal pseudothalamic pain syndrome. Clinical features and anatomic correlates. Arch. Neurol. 49, 1032–1037.

Schott, B., Laurent, B., and Mauguiere, F. 1986. Les douleurs thalamiques. Etude critique de 43 case. Rev. Neurol. 142, 308–315.

Sherman, S. M. and Guillery, R. W. 2001. Exploring the Thalamus. Academic Press.

Shi, T., Stevens, R. T., Tessier, J., and Apkarian, A. V. 1993. Spinothalamocortical inputs nonpreferentially innervate the superficial and deep cortical layers of SI. Neurosci. Lett. 160, 209–213.

Siegfried, J. 1987. Stimulation of Thalamic Nuclei in Human: Sensory and Therapeutical Aspects. In: Thalamus and Pain (eds. J.-M. Besson, G. Guilbaud, and M. Peschanski), pp. 271–278. Excerpta Medica.

Stepniewska, I., Sakai, S. T., Qi, H. X., and Kaas, J. H. 2003. Somatosensory input to the ventrolateral thalamic region in the macaque monkey: potential substrate for parkinsonian tremor. J. Comp. Neurol. 455, 378–395.

Steriade, M., Jones, E. G., and McCormick, D. A. 1997. Thalamus. Elsevier Science.

Stevens, R. T., London, S. M., and Apkarian, A. V. 1993. Spinothalamocortical projections to the secondary somatosensory cortex (SII) in squirrel monkey. Brain Res. 631, 241–246.

Strigo, I. A., Duncan, G. H., Boivin, M., and Bushnell, M. C. 2003. Differentiation of visceral and cutaneous pain in the human brain. J. Neurophysiol. 89, 3294–3303.

Tasker, R. R. 1984. Stereotaxic Surgery. In: Textbook of Pain (eds. P. D. Wall and R. Melzack), pp. 639–655. Churchill Livingstone.

Villanueva, L. and Nathan, P. 2000. Multiple Pain Pathways. In: Proceedings of the 9th World Congress on Pain (eds. M. Devor, M. C. Rowbotham, and Z. Wiesenfeld-Hallin), pp. 371–376. IASP Press.

Wall, J. T., Xu, J., and Wang, X. 2002. Human brain plasticity: an emerging view of the multiple substrates and mechanisms that cause cortical changes and related sensory dysfunctions after injuries of sensory inputs from the body. Brain Res. Brain Res. Rev. 39, 181–215.

Weng, H. R., Lee, J. I., Lenz, F. A., Schwartz, A., Vierck, C., Rowland, L., and Dougherty, P. M. 2000. Functional plasticity in primate somatosensory thalamus following chronic lesion of the ventral lateral spinal cord. Neuroscience 101, 393–401.

Weng, H. R., Lenz, F. A., Vierck, C., and Dougherty, P. M. 2003. Physiological changes in primate somatosensory thalamus induced by deafferentation are dependent on the spinal funiculi that are sectioned and time following injury. Neuroscience 116, 1149–1160.

Willis, W. D. 1985. The Pain System. Karger.

Willis, W. D., Jr. 1997. Nociceptive Functions of Thalamic Neurons. In: Thalamus, Volume II, Experimental and Clinical Aspects (eds. M. Steriade, E. G. Jones, and D. A. McCormick), pp. 373–424. Elsevier Science Ltd.

Willis, W. D. and Westlund, K. N. 1997. Neuroanatomy of the pain system and of the pathways that modulate pain. J. Clin. Neurophysiol. 14, 2–31.

Xiao, D. Q., Zhu, J. X., Tang, J. S., and Jia, H. 2005. 5-hydroxytryptamine 1A (5-HT1A) but not 5-HT3 receptor is involved in mediating the nucleus submedius 5-HT-evoked antinociception in the rat. Brain Res. 1046, 38–44.

Yang, Z. J., Tang, J. S., and Jia, H. 2002. Morphine microinjections into the rat nucleus submedius depress nociceptive behavior in the formalin test. Neurosci. Lett. 328, 141–144.

Yasui, Y., Breder, C. D., Saper, C. B., and Cechetto, D. F. 1991. Autonomic responses and efferent pathways from the insular cortex in the rat. J. Comp. Neurol. 303, 355–374.

Yasui, Y., Itoh, K., Kamiya, H., Ino, T., and Mizuno, N. 1988. Cingulate gyrus of the cat receives projection fibers from the thalamic region ventral to the ventral border of the ventrobasal complex. J. Comp. Neurol. 274, 91–100.

Yoshida, A., Dostrovsky, J. O., and Chiang, C. Y. 1992. The afferent and efferent connections of the nucleus submedius in the rat. J. Comp. Neurol. 324, 115–133.

Zhang, S., Tang, J. S., Yuan, B., and Jia, H. 1998. Inhibitory effects of glutamate-induced activation of thalamic nucleus submedius are mediated by ventrolateral orbital cortex and periaqueductal gray in rats. Eur. J. Pain 2, 153–163.

44 Psychophysics of Sensations Evoked by Stimulation of the Human Central Nervous System

S Ohara, C A Bagley, H C Lawson, and F A Lenz, Johns Hopkins Hospital, Baltimore, MD, USA

Glossary

dorsal columns A pathway carrying axons of fibers in the peripheral nerve through the posterior aspect of the spinal cord to the dorsal column nucleus at the posterior aspect of the medulla where they synapse and ascend through the medial lemnisus to the principal nucleus of thalamus. Classically, this pathway transmits low-threshold muscle and cutaneous afferents but recent studies demonstrate that afferents transmitting nociceptive, thermal, or visceral inputs are also transmitted.

nociceptive specific (NS) neurons Neurons responding only to stimuli which are noxious or painful.

wide dynamic range (WDR) neurons Neurons responding to stimuli across the intensive continuum into the painful or noxious range.

projected field in the somatosensory system A part of the body at which a sensation is evoked in response to stimulation of the central nervous system.

receptive field In the somatosensory system: skin region, stimulation of which can influence the discharge of the neuron under study

somatic sensory nuclear group of the thalamus Nuclei receiving input from the dorsal column nuclei, the spino- or the trigeminothalamic tract by either a direct or a transsynaptic route. Summary term for the specific somatosensory nuclei of the thalamus. Most commonly used taxonomy is for monkey thalamus: VPL (ventroposterolateral), VPM (ventroposteromedial), VPI (ventroposteroinferior) corresponding to humans: Vc (ventrocaudal, both medial and lateral). Adjacent, both posterior medial, to VPM ventral medial posterior (VMpo), a putative pain and temperature signaling nucleus may be located (Craig, A. D. et al., 1994, cf. Willis, W. D., Jr. et al., 2001, Graziano, A. and Jones, E. G., 2004; Lenz, F. A. et al., 2004). VMpo may be located medial to ventral caudal portae (Vcpor), while intralaminar nuclei mostly a thin layer of neurons located medial to Vim, Vc, and Vcpor (see Figure 1(a)).

spinothalamic tract (STT) The pain and temperature signaling pathway from neurons in the spinal dorsal horn to the lateral and anterolateral funiculi of the spinal cord and the brainstem to the lateral, posterior, medial, and intralaminar nuclei of thalamus.

spinothalamocortical pathways Cortical areas which receive input from thalamic nuclei that themselves receive STT input. These thalamocortical connections include those from Vc to primary somatosensory and secondary somatosensory cortex, from Vcpor, Vcpc, and VMpo to parietal opercular, insular, and retroinsular cortex, and finally from the medial dorsal nucleus to anterior cingulate cortex.

sylvian fissure Also called lateral sulcus, the sylvian fissure separates the temporal lobe from the frontal and parietal lobes (see Figure 6, left – middle and lower levels). The insula is a cortical surface, which is located deep to the lateral fissure and parallel to the cortical surface on coronal and axial sections (Figure 6, middle and right). The frontal,

parietal, and temporal opercula are folds of cortex that extend over the insula and meet each other over the insula. The sulcus between the frontal–parietal opercula and the temporal operculum is the sylvian fissure.

Talairach coordinates Initially developed for a specific stereotactic frame; based on one single brain; frequently used as common coordinate system; *x*: left–right, *y*: anterior–posterior, *z*: superior–inferior; the reference point (0, 0, 0) is the anterior commissure (Talairach, J. and Tournoux, P., 1988).

thalamus A subcortical gray matter structure characterized by reciprocal connections with the cortical mantle and by inputs from the periphery and nuclei of the brain, such as the pallidum.

44.1   Introduction

The dorsal column pathway and the spinothalamic tract (STT) are the two main somatosensory spinal tracts afferent to the thalamus. The dorsal column pathway is formed by the axons of low-threshold mechanoreceptors with cell bodies in the dorsal root ganglion. These axons terminate to the principal somatic sensory nucleus of the thalamus (human ventral caudal or Vc, monkey ventral posterior or VP) (Jones, E. G. *et al.*, 1982; Kaas, J. H. *et al.*, 1984; Lenz, F. A. *et al.*, 1988; Willis, W. D. and Coggeshall, R. E., 1991). The dorsal columns do contain a postsynaptic pain pathway signaling noxious visceral stimuli, which terminates in the Vc (Uddenberg, N., 1968; Rustioni, A. *et al.*, 1979; Willis, W. D., *et al.*, 1991; Al Chaer, E. D. *et al.*, 1996b). The STT is a pain pathway that originates from the spinal dorsal horn, and ascends in the anterolateral quadrant of the spinal cord before terminating in the thalamus (Willis, W. D., Jr. *et al.*, 2001). Stimulation of these thalamic nuclei in man reveals the psychophysical dimensions of these pathways which is the subject of this chapter.

44.2   The Spinothalamic Tract

Many neurons in the STT are characterized by their response to noxious or painful stimuli. Some of these respond to the somatic stimuli across the intensive

continuum into the noxious range (WDR – wide dynamic range). These cells arise from both superficial (lamina I) and deep laminae of the dorsal horn (laminae IV–V) (Kumazawa, T. and Perl, E. R., 1978; Willis, W. D., 1985; Ferrington, D. G. *et al.*, 1987). Some neurons in the superficial dorsal horn respond only to different stimuli including noxious stimuli (nociceptive specific (NS) cells), cold (Kumazawa, T. and Perl, E. R., 1978; Willis, W. D., 1985; Ferrington, D. G. *et al.*, 1987; Craig, A. D. *et al.*, 1994), injection of histamine (itch – Andrew, D. and Craig, A. D., 2001; Craig, A. D., 2003b), or visceral stimuli (Craig, A. D., 2003a). The deep and superficial laminae of the dorsal horn project to the brain, respectively, through the ventral lateral (ventral STT) and the dorsal lateral spinal funiculi (dorsal STT) (Apkarian, A. V. and Hodge, C. J., 1989; Cusick, C. G. *et al.*, 1989; Ralston, H. J. and Ralston, D. D., 1992; Craig, A. D., 1998, 2003b; Price, D. D. *et al.*, 2003).

The population of neurons in the superficial dorsal horn responding only to noxious stimuli has led to the description of the STT as a structure signaling only pain – a labeled line (Perl, E. R., 1984; Willis, W. D., 1985). A recent version of this hypothesis suggests that the STT is a series of labeled lines for cool and itch, as well as pain that jointly reflect the internal state of the body (interoception) (Craig, A. D., 2003a; 2003b). In this view, pain is the emotion produced by disequilibrium of the internal state. An alternate view is that pain is signaled by WDR neurons which transmit a graded signal, the strength of which might be decoded in the brain to identify the presence of a painful stimulus (Price, D. D. and Dubner, R., 1977; Willis, W. D., 1985; Price, D. D. *et al.*, 2003). This chapter examines recent evidence from stimulation of the human central nervous system as they impact these hypotheses.

One approach to identifying the pain pathway has been to stimulate the cord during cordotomy using paired-pulse stimulus parameters that selectively activate the axons arising in the superficial or deep laminae (Mayer, D. J. *et al.*, 1975). This interpulse threshold was more consistent with that of cells in the deep than the superficial dorsal horn (Price, D. D. and Mayer, D. J., 1975). The results suggest that the sensory aspect of pain is signaled through the axons in ventral STT which originate from the WDR neurons in the deep dorsal horn (Price, D. D. and Dubner, R., 1977; Dubner, R. *et al.*, 1989).

44.3 The Dorsal Column Pathway

The dorsal column pathway is formed by the axons of low-threshold mechanoreceptors that project through the dorsal column nuclei and medial lemniscus to the region of the principle somatic sensory nucleus (ventral caudal, Vc) (Jones, E. G. *et al.*, 1982; Kaas, J. H. *et al.*, 1984; Lenz, F. A. *et al.*, 1988; Willis, W. D. *et al.*, 1991). As shown in Figure 1 receptive fields are quite constant within a particular parasagittal plane in Vc. From medial to lateral planes, the sequence of neuronal cutaneous receptive fields progresses from intraoral through face, thumb, fingers (radial to ulnar), and arm to leg. Proximal parts of the limbs are represented dorsal to the corresponding digits (Lenz, F. A. *et al.*, 1988).

Substantial evidence demonstrates that the dorsal columns do contain a postsynaptic pain pathway signaling noxious visceral stimuli, rather than noxious somatic stimuli, which terminates in the VP (Uddenberg, N., 1968; Rustioni, A. *et al.*, 1979; Willis, W. D. *et al.*, 1991; Al Chaer, E. D. *et al.*, 1996b). This postsynaptic pathway has been demonstrated by infusion of neurotransmitter agonists/antagonists into the dorsal horn, which alters the response of neurons in the dorsal column nuclei to visceral stimuli. Furthermore, lesions of rat nucleus gracilis diminish the response of neurons in VP to visceral as well as cutaneous stimuli (Al-Chaer, E. D. *et al.*, 1997).

The cells of origin of this pathway are located just dorsal to the central canal, as demonstrated by antidromic invasion from nucleus gracilis (Al Chaer, E. D. *et al.*, 1996a), and by retrograde and anterograde tracer studies (Christensen, M. D. *et al.*, 1996). Retrograde tracer studies depositing marker in the nucleus gracilis (Christensen, M. D. *et al.*, 1996) label cell bodies of origin of the pathway in the medial base of the dorsal horn just above the central canal. Injection of tracer into this area resulted in fiber labeling of the dorsal column midline and in the medial aspect of the nucleus gracilis. Fibers originating in the thoracic cord terminate in the lateral part of nucleus gracilis and adjacent to the medial parts of nucleus cuneatus (Wang, C. C. *et al.*, 1999; Willis, W. D. *et al.*, 1999). Along with the STT, a major terminus of the postsynaptic dorsal horn pathway is the lateral thalamus.

44.4 Thalamic Nuclei

44.4.1 Lateral Thalamic Nuclei

The human principal sensory nucleus (Vc) (Hassler, R., 1959) is divided into a core area (equivalent to monkey VP, see Olszewski, J., 1952; Hirai, T. and Jones, E. G., 1989), posterior, and inferior regions. These two regions are defined relative to the most posterior and inferior cell with a response to non-painful, cutaneous stimuli (cell 57 in Figure 1(b)). In the core, the majority of cells respond to innocuous, mechanical, and cutaneous stimulation. This latter area corresponds to the posterior and inferior subnuclei of Vc, which are ventral caudal portae (Vcpor), ventral caudal parvocellular nucleus (Vcpc) (Mehler, W. R., 1966), the posterior nucleus, and the magnocellular medial geniculate (Mehler, W. R., 1962; Mehler, W. R., 1966; Lenz, F. A. *et al.*, 1993b) (see Vc and Vcpor in sagittal section in Figure 1(a)). Studies of patients at autopsy following lesions of the STT show terminations in all this nuclei (Bowsher, D., 1957; Mehler, W. R., 1966; Mehler, W. R., 1969). This area includes the ventral medial nucleus – posterior part (VMpo), which may receive STT inputs and may signal pain and temperature (Craig, A. D. *et al.*, 1994; Blomqvist, A. *et al.*, 2000; cf. Willis, W. D., Jr. *et al.*, 2001; Graziano, A. *et al.*, 2004). The physiology recapitulates this anatomy.

Cells in Vc responding to painful and thermal stimuli are of several types, including WDR and NS cells responding to painful thermal and mechanical stimuli (see figure 2 in Lee, J.-I. *et al.*, 1999). Figure 2(a) demonstrates the response of a single neuron with WDR properties to a painful 45 °C

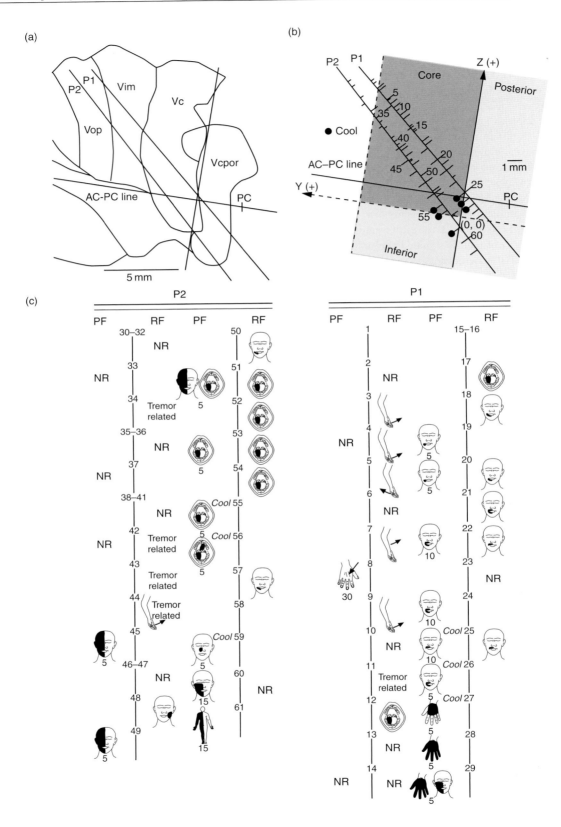

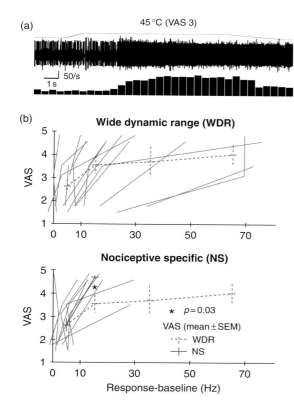

(a)   45 °C (VAS 3)

50/s

1 s

(b)

Wide dynamic range (WDR)

VAS

0   10   30   50   70

Nociceptive specific (NS)

VAS

★ $p = 0.03$

VAS (mean ± SEM)
- - - WDR
—+— NS

0   10   30   50   70

Response-baseline (Hz)

Figure 2   Activity of neurons in the region of Vc (ventrocaudal) responding to painful thermal stimuli. The response to nonpainful and painful heat/mechanical stimuli applied within the receptive field (RF) (Lenz, F. A. et al., 1994b; Lee, J.-I. et al., 1999) is compared with the visual analog scale of intensity (VAS) evoked by the same stimulus. (a) The response of a cell wide dynamic range (WDR) to painful heat. (b) VAS and firing rates for the response to painful stimuli are plotted for nociceptive (NS) cells which respond only to painful stimuli and WDR neurons, which respond in a graded fashion to nonpainful and painful stimuli. Average and one standard deviation scores by decade to 20 Hz and by 30 Hz steps from 20 to 80 Hz were compared by Mann–Whitney U test. Reprinted from Lenz F. A., Ohara, S., Gracely, R. H., Dougherty, P. M., and Patel, S. H. 2004. Pain encoding in the human forebrain: binary and analog exteroceptive channels. J. Neurosci. 24, 6540-6544 with permission ©2004 by the Society for Neuroscience).

stimulus. Responses to painful stimuli were characterized by the mean firing rate during painful stimulation, recorded through the microelectrode (Figure 2(b), x-axis), and by visual analog scale of intensity (VAS) ratings of microstimulation-evoked pain (y-axis). Some low-threshold cells respond to nonpainful mechanical and cold stimuli (Figure 2) (Lenz, F. A. et al., 1993a; Lenz, F. A. and Dougherty, P. M., 1998; Lee, J.-I. et al., 1999). Cells in the core and posterior region respond only to noxious heat stimuli (Lenz, F. A. et al., 1993a; 2004) and to noxious cold stimuli (Davis, K. D. et al., 1999).

Nociceptive cells in Vc appear to signal pain based on temporary lesioning and stimulation studies. Blockade of the activity in this region by injection of local anesthetic into monkey VP, corresponding to human Vc (Hirai, T. et al., 1989), significantly interferes with the monkey's ability to discriminate temperature in both the innocuous and noxious range (Duncan, G. H. et al., 1993). Stimulation within Vc and the regions posterior and inferior to it can evoke the sensations of pain (Hassler, R. and Reichert, T., 1959; Willis, W. D., 1985; Dostrovsky, J. O. et al., 1991; Lenz, F. A. et al., 1993b) and temperature (Lenz, F. A. et al., 1993b; Davis, et al., 1999).

The largest study of stimulation-evoked pain and temperature responses examined results of threshold microstimulation of the region of Vc in 124 thalami (116 patients), as summarized in Figure 3. The location of pain and temperature responses is defined relative to the posterior and inferior borders of the principal somatic sensory nucleus (Vc). Warm sensations were evoked more frequently in the posterior region (5.7%) than in the core (2.3%). Otherwise the proportions were not significantly different for cool or pain sensations between the core or the posterior region or both (cool 2.5%, 2.2%; pain 2.8%, 4.1%).

Figure 1   Map of receptive and projected fields for trajectories in the regions of the Vc (ventrocaudal) in a single patient (number 193.97). (a) Positions of the trajectories relative to nuclear boundaries as predicted radiologically from the position of the anterior commissure–posterior commissure (AC–PC) line. The AC–PC line is indicated by the horizontal line in the panel; the trajectories are shown by the two oblique lines. The positions of nuclei are inferred from the AC–PC line and therefore are only an approximate indicator of nuclear location. Scale as indicated. Abbreviations defined in the text. (b) Location of the cells, stimulation sites, and trajectories (P1 and P2) relative to the AC–PC line (a solid line) and the ventral border of the core of Vc (a dotted line). The locations of stimulation sites are indicated by ticks to the left of the trajectory; the locations of the cells are indicated by ticks to the right of the trajectory. Cells with receptive fields (RFs) are indicated by long ticks; those without are indicated by short ticks. The cold sensation evoked is indicated by filled circles at the end of the tick to the left of the trajectory. Scale is as indicated. Each site where a cell was recorded or stimulation was carried out or both is indicated by the same number in (b) and (c). The core of Vc and the regions posterior and inferior are as labelled. (c) P1 and P2 show the site number, PF, and RF for that site. The threshold (in microamperes) is indicated below the PF diagram. Reprinted from Ohara S. and Lenz F. A. (2003) Medial lateral extent of thermal and pain sensations evoked by microstimulation in somatic sensory nuclei of human thalamus. J. Neurophysiol. 90, 2367-2377, used with permission from the American physiological society.

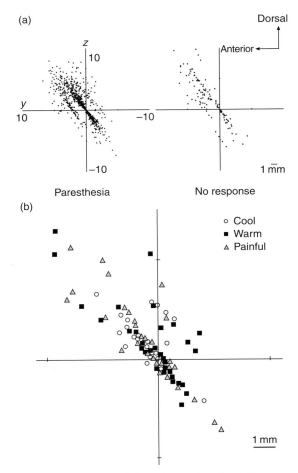

Figure 3 Locations of sites where microstimulation evoked paresthesias (a, left), no response (NR) (a, right), and thermal and pain sensations (b). Site location is shown relative to the posterior and inferior borders of the core of Vc. Note that thermal and pain sensations were evoked both in the core and posterior regions of Vc. Paresthesic sites are most dense where NR sites are least dense over the core and the posterior regions. Scale as indicated. Reprinted from Ohara S. and Lenz F. A. 2003. Medial lateral extent of thermal and pain sensations evoked by microstimulation in somatic sensory nuclei of human thalamus. J. Neurophysiol. 90, 2367-2377, used with permission from the American Physiological Society.

Warm sensations were evoked more frequently in the lateral plane (10.8%) than in the medial planes of the posterior region (3.9%) but no other significant medial lateral differences for any sensation were found in the core or posterior region or overall.

These results are in contrast to previous studies reporting a larger proportion of thermal/pain sites were evoked in the posterior and inferior regions (Lenz, F. A. *et al.*, 1993b; Davis, K. D. *et al.*, 1996).

These latter studies took the anterior commissure–posterior commissure line (ACPC) as the floor of Vc, contrary to atlas and physiologic maps (Schaltenbrand, G. and Bailey, P., 1959; Lenz, F. A. *et al.*, 1988). This recent study suggests that sites where thermal or pain sensations are evoked are located both within and posterior, inferior, and medial to Vc. If the proportion of such sites is larger in posterior and inferior regions, then those sites must be very close to the borders of the core.

Patterned stimulation at sites in the region of Vc, an STT terminal region, evokes sensations consistent with one of two pathways – one binary (pain+) and the other analog (pain−/+) (Figure 3). Specifically, current was applied at five frequencies (10, 20, 38, 100, and 200 Hz) in bursts of 4, 7, 20, 50, and 100 pulses in an ascending staircase protocol, the type of protocol commonly used in studies of pain (Gracely, R. H. *et al.*, 1988; Yarnitsky, D. and Sprecher, E., 1994), including our studies (Greenspan, J. D. *et al.*, 2004). Stimulation at pain+ sites evoked a constant high level of pain over large, often cutaneous, projected fields (PFs). These sites were characterized both by descriptors, which did not change along the staircase, or by more intense stimulation-evoked pain than that evoked at the pain−/+ sites (Figure 4). These results suggest that pain+ sites participate in a binary, exteroceptive, labeled line which signals the presence of a painful external stimulus.

The thalamic stimulation thresholds for nonpainful and painful sensations are not significantly different (Lenz, F. A. *et al.*, 1993b; Ohara, S. *et al.*, 2003) suggesting that pain−/+ sites did not result from activation of the system transmitting nonpainful sensations (largely medial lemniscal) before that transmitting painful sensations (largely STT) (Willis, W. D., 1985). In addition, the equivalence of current, pulse, and frequency thresholds for pain at both types of sites predicts that the neural elements, that is, possibly WDR and NS cells, should be activated together if they were found at the same site. At such sites, analog pain+ responses would be predicted to occur because the combination of binary plus analog neural elements at a site will have analog properties, given the assumption of linearity. However, analog pain+ sites were not observed. For all these reasons, it is plausible that our observations may be the result of selective activation of two functionally distinct pathways.

Figure 2(a) demonstrates the mean firing rate minus baseline (Figure 2(b), *x*-axis) for the response of multiple single WDR and NS neurons to painful stimuli versus VAS score (Figure 2(b), *y*-axis). There

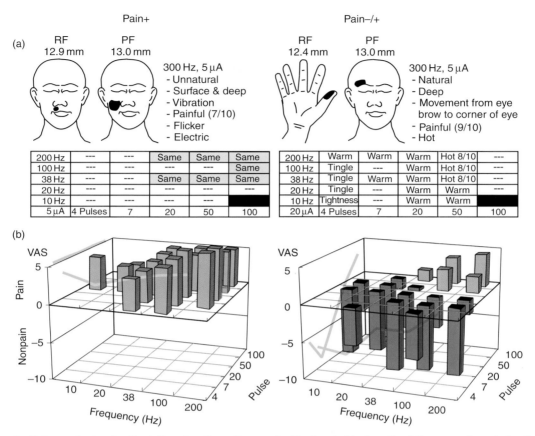

Figure 4 Pain+ and pain−/+ stimulation sites. Sensations evoked by threshold microstimulation were characterized by the PF and by descriptors from a validated questionnaire, and by a visual analog scale of intensity (VAS) (Lenz, F. A. *et al.*, 1993b; 1998a; Lenz, F. A. and Byl, N. N. 1999). (a, left) Site where stimulation at 300 Hz and 5 μA produced pain in the PF shown in the figurine and of the quality as described. Pain identical to that evoked by 300 Hz was evoked at most sites with trains ≥20 pulses and frequencies ≥200 Hz (yellow rectangle, see text). (a, right) Site where tightness was evoked in the first column at 10 Hz and then tingle at 20, 38, and 100 Hz. At 200 Hz and thereafter, warm was evoked at each step in the staircase, excepting 7 pulses – 10, 20, 100 Hz, until 50 pulses – 38 Hz. At this step and further up, the staircase painful heat was evoked. (b) Average VAS ratings across all pain+ and pain−/+ sites. Ratings were taken in response to pulse and frequency pairs ascending the staircase. The yellow lines along the outside surfaces of the 3D displays indicate the average VAS ratings across all sites by frequency and number of pulses. Reprinted from Lenz F. A., Ohara, S., Gracely, R. H., Dougherty, P. M., and Patel, S. H. 2004. Pain encoding in the human forebrain: binary and analog exteroceptive channels. J. Neurosci. 24, 6540-6544, with permission (©2004; by the Society for Neuroscience).

was a significantly steeper initial rise in VAS scores for the neurons that only responded to painful stimuli (NS neurons), than for WDR neurons. The steep initial rise of VAS with the firing rate of NS versus WDR neurons (Figure 2(b)) is consistent with the shorter dynamic range of thalamic NS cells (Apkarian, A. V. and Shi, T., 1994), and with the binary response to stimulation at pain+ sites (Figure 2(b)).

We suggest that the first pathway is characterized as a binary pain response signaling the presence/absence of painful stimuli, consistent with an alerting/alarm function (Becker, D. E. *et al.*, 1993;

Zaslansky, R. *et al.*, 1995). The second pathway may be an analog route in which activity is graded with intensity of the painful stimulus, consistent with STT neurons, which encode the properties of external stimuli (Willis, W. D., 1985; Price, D. D. *et al.*, 2003). Itch was rarely evoked and never in isolation. Emotion descriptors (e.g., nauseating, cruel, suffocating) were uncommonly endorsed at either pain+ and pain−/+ sites (cf. Lenz, F. A. *et al.*, 1995). Therefore, both painful responses to stimulation were described in terms usually applied to external stimuli (exteroception) rather than to internal or emotional phenomena (interoception). Exteroreceptive sensations can be associated with a strong affective dimension.

Pain with a strong unpleasant or affective component can be evoked by stimulation of the lateral thalamus in the region of Vc. These sensations have the character of memories of a previously experienced pain, unlike the pain sensations evoked by thalamic sensation which are not related to previous experience (see above) or a diffuse unpleasant sensation of the type evoked by stimulation of the medial thalamus (see below). In the first case, stimulation in Vcpc (Figure 5) evoked chest pain with an affective dimension in the case of a patient with coronary artery disease which had been effectively treated by balloon angioplasty (Lenz, F. A. et al., 1994a).

Microstimulation at site 49 (Figure 5) evoked an unnatural, painful (visual analog scale − 4.6/10), mechanical sensation in the flank and an unnatural nonpainful electrical sensation involving the left arm and leg. At sites 51 and 53 (Figure 5) microstimulation evoked a sensation described by the patient as "heart pain," which was "like what I took nitroglycerin for" except that "it starts and stops suddenly." It was not accompanied by dyspnea, diaphoresis, or after effects. The PF involved the precordium and left side of the chest from the sternum in the midline to the anterior axillary line. Microstimulation at site 51 also evoked a sensation of nonpainful surface, tingling in the left leg, which coincided with the stimulation-associated angina.

Characteristics of the patient's stimulation-associated angina and usual angina were measured by using a questionnaire. The same descriptors for stimulation-associated angina were chosen intraoperatively during stimulation at both sites 51 and 53 (Figure 5) including: natural, deep, painful (visual analog scale − 10/10), squeezing, frightful, fatiguing, and identical to her angina. The questionnaire was administered three times over several months postoperatively to describe the patient's usual angina. The following descriptors were chosen: natural (3/3 administrations), deep (3/3), painful (3/3), squeezing (3/3), frightful (2/3), suffocating (2/3), and fatiguing (2/3). Her usual angina involved the left side of the chest, arm, and neck and was associated with a surface (3/3), nonpainful (3/3), and tingling (3/3) in the left arm and hand. This coincidence of descriptors is unlikely to occur at random ($p < 10^{-6}$, combinatorial analysis).

Similar emotional responses, including crying in response to thalamic stimulation in the same region, have been reported in the case of atypical chest pain, dysparunia, and the pain of childbirth (Davis, K. D.

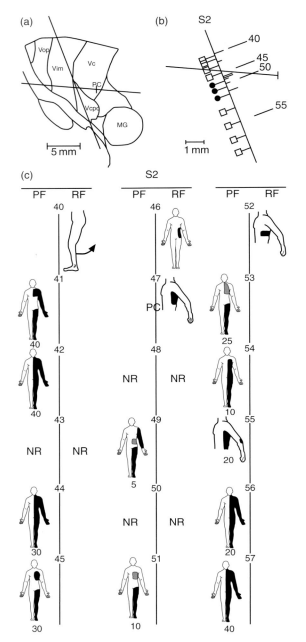

Figure 5 (a) Thalamic map of a patient with a history of angina pectoris successfully treated with coronary artery balloon angioplasty. (b) Dark circular balloons on ticks to the right of the line indicate sites (sites 49, 51, 53) where thalamic microstimulation evoked painful sensations in PFs indicated by stippling in the figurines (c). Open square balloons in (b) indicate sites where nonpainful, tingling sensations were evoked. All abbreviations and other conventions are as in the legend to Figure 1. Reproduced from Lenz F. A., Gracely, R. H., Hope, E. J., Baker, F. H., Rowland, L. H., Dougherty, P. M., and Richardson, R. T. 1994a. The sensation of angina can be evoked by stimulation of the human thalamus. Pain 59, 119-125, with permission from Elsevier.

et al., 1995; Lenz, F. A. *et al.*, 1995). Clinical criteria including a battery of cardiac tests (enzymes, EKGs, stress test) ruled out angina of cardiac origin in both these patients. Explorations in 50 patients without a history of angina found that stimulation-associated angina was not evoked at any of the 19 stimulation sites with PFs on the chest wall. PFs were located on the left chest wall at three sites and the right chest wall at 16. At one of these 19 sites an unnatural, sharp, mechanical, painful, and vibration was described in response to stimulation but emotional descriptors were not endorsed.

Preoperative pain was clearly of cardiac origin in the patient with angina (Lenz, F. A. *et al.*, 1994a), but clearly not of cardiac origin in the patient with panic disorder. The association of stimulation-associated angina and the affective dimension was not unexpected (Lenz, F. A. *et al.*, 1994a) as angina is often associated with a strong affective dimension, unlike other chest pains (Matthews, M. B., 1985; Braunwald, E., 1988; Procacci, P. and Zoppi, M., 1989; Pasternak, R. C. *et al.*, 1992). Stimulation-evoked sharp chest pain occurred without an affective dimension in a retrospective analysis of patients without prior experience of spontaneous chest pain with a strong affective dimension. Therefore, it is possible that in the case, stimulation-associated chest pain included an affective dimension as a result of conditioning by the prior experience of spontaneous chest pain with a strong affective dimension. In retrospect, the affective dimension of stimulation-associated angina might arise by similar conditioning.

44.4.2 Medial and Intralaminar Thalamic Nuclei

The medial and intralaminar nuclei also play a role in signaling pain sensations. Medial to Vc, the most dense STT terminal pattern is found in the intralaminar nucleus centralis lateralis (Mehler, W. R., 1962; 1969), while a much less dense termination is found in other interlaminar nuclei central medial, parafascicularis (Mehler, W. R., 1962), and the medial dorsal nucleus (Mehler, W. R., 1969). Nociceptive neurons have been identified in an area that apparently corresponds to the human central median nucleus (Ishijima, B. *et al.*, 1975; Tsubokawa, T. and Moriyasu, N., 1975; Rinaldi, P. C. *et al.*, 1991).

Milliampere-current-level stimulation at sites probably located in the parafascicular, limitans, and central medial (parvocellular part) nuclei was reported to evoke a diffuse, burning pain, or

sometimes to exacerbate the patient's ongoing pain (Sano, K., 1979). This pain was evoked in projected fields as large as the whole body or hemibody. Stimulation at sites, possibly the medial dorsal and periventricular nuclei evoked a generalized unpleasant sensation, not localized to a particular body part (Sano, K., 1979). Both responses may be consistent with the STT pathway or a multisynaptic pathway traversing the reticular formation (Willis, W. D., 1985).

44.5 Cortex

Functional imaging studies of the response to the application of painful stimuli (Jones, A. K. *et al.*, 1991; Talbot, J. D. *et al.*, 1991; Casey, K. L. *et al.*, 1994; Craig, A. D. *et al.*, 1996; Andrew, D. *et al.*, 2001) have identified three cortical areas with metabolic activation: primary somatosensory, parasylvian, and cingulate cortex. These cortical areas all receive input arising from nociceptors as demonstrated by cortical potentials evoked by cutaneous application of a laser (laser-evoked potentials – LEP) (Ohara, S. *et al.*, 2004b), which selectively activates nociceptors (Bromm, B. *et al.*, 1984). Pain-related responses to stimulation have been identified in relation to the parasylvian cortex.

44.5.1 Parasylvian Cortex and Pain Memory

Parasylvian cortex receives input from the nuclei around, and subnuclei within Vc (Van Buren, J. M. and Borke, R. C., 1972) such as Vcpc, Vcpor, and putative VMpo. Parasylvian receives nociceptive input as evidenced by the presence of neurons responding to noxious stimuli (Robinson, C. J. and Burton, H., 1980; Dong, W. K. *et al.*, 1989; 1994) and of LEP generators (Lenz, F. A. *et al.*, 1998c; Vogel, H. *et al.*, 2003; Ohara, S. *et al.*, 2004b). The LEP generator is anterior to primary auditory cortex, adjacent to the S2 (secondary somatosensory cortex) generator for vibratory SEPs (somatosensory-evoked potentials) (Hamalainen, H. *et al.*, 1990; Ohara, S. *et al.*, 2004b). Source modeling suggests that the LEP generator is in the dorsal insula–parietal operculum (Figure 6) (Lenz, F. A. *et al.*, 1998b; Vogel, H. *et al.*, 2003), different from the location of the local generator for the P3 event-related potentials in the temporal base (Lenz, F. A. *et al.*, 1998b). Differences between the locations of the late component of the LEP, the P2 wave, and the P3 suggest that the LEP P2 is not a

(a)　　　　　　(b)

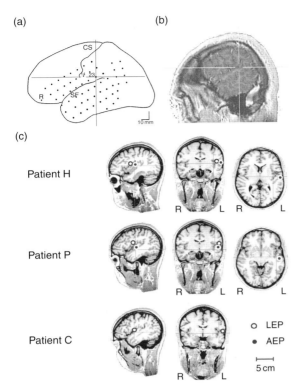

(c)

Patient H

Patient P

Patient C

○ LEP
● AEP

5 cm

R　　L

Figure 6 Verification of subdural grid electrode position in head model for dipole source analysis. Positions of all subdural grid electrodes were determined in a postoperative CT scan. These positions were entered into a spherical head model for dipole source analysis, from which they were projected back into a preoperative magnetic resonance imaging (MRI) of this patient's brain (patient H). (a) Intraoperative sketch shows the sylvian fissure with frontotemporal grid electrode 23 near the end of the fissure, and electrode 14 just posterior of the central sulcus. (b) Location of electrode 23 was projected from brain electromagnetic source analysis software (BESA) head coordinates into the preoperative MRI, showing excellent coincidence of the different coordinate systems used in this study. A similar correlation was found for electrode 14 and the central sulcus. (c) Projection of dipole source locations for LEP (○) and auditory – evoked potentials (AEP) (●). The projection was performed in Talairach space of a standard brain MRI. Left to right, panels show sagittal, coronal, and axial sections in patients H, P, and C, top to bottom. The LEP source was located above the sylvian fissure, and the AEP was located further posterior and below the fissure. The axial slice was aligned to pass through Heschl's gyrus (AEP generator) to illustrate the relative location of the source of the AEP (available only in patients H and P). Reprinted from Vogel H., Port, J. D., Lenz, F. A., Solaiyappan, M., Krauss, G., and Treede, R. D. 2003. Dipole source analysis of laser-evoked subdural potentials recorded from parasylvian cortex in humans. J. Neurophysiol. 89, 3051–3060, used with permission the American Physiological Society.

P3-like wave, signaling the alertness evoked by painful stimuli (Zaslansky, R. *et al.*, 1995; Lenz, F. A. *et al.*, 1998b). Rather, it is likely that the LEP P2 is primarily related to the sensation of pain (Ohara, S. *et al.*, 2004a).

The pain-related function of this area is consistent with decreased pain discrimination and tolerance with lesions of the parietal operculum and insula, respectively (Greenspan, J. D. *et al.*, 1999). Studies in which LEPs were recorded through depth electrodes implanted in S2 and insula did not show a phase reversal in the parietal operculum, that is, S2 (Frot, M. *et al.*, 1999). Stimulation through electrodes placed in the posterior, superior insula produced pains described as burning, stinging, electrical, and disabling sensations (Ostrowsky, K. *et al.*, 2002). In an earlier series, stimulation of the exposed insula during awake craniotomies ($n = 5$) produced pain uncommonly but did produce nausea, tastes, somatic sensations in the epigastric area, and rising sensations in the epigastric and umbilical areas (Penfield, W. and Jasper, H. 1954). Thus there is evidence from human stimulation studies that insula, possibly the posterior–superior portion, is involved in pain-related processes.

Pain with a strong, vivid, affective dimension evoked by stimulation of the region of Vc may be related activation of its parasylvian cortical projection zone (see Section 44.4.1) (Locke, S. *et al.*, 1961; Mehler, W. R., 1962; Van Buren, J. M. and Borke, R. C., 1972). These vivid memories are similar to those evoked by stimulation around the lateral sulcus in patients with epilepsy (Halgren, E. *et al.*, 1978; Gloor, P. *et al.*, 1982b; Gloor, P., 1990). These memories may be related to cortical activation rather than medial temporal structures as they are not fully formed memories like those described here (Halgren, E. *et al.*, 1978; Gloor, P. *et al.*, 1982a). Furthermore, fully formed memories can be evoked after removal of mesial temporal structures (Moriarity, J. L. *et al.*, 2001). Therefore, pain with a strong affective dimension in response to thalamic stimulation results might be related to the activation of limbic and associated cortical structures (Lenz, F. A. *et al.*, 1995). The role of the medial temporal lobe might be transiently involved in the formation of these memories, before being located in cortex, independent of medial temporal lobe structures (Mishkin, M., 1979; Friedman, D. P. *et al.*, 1986; Zola-Morgan, S. and Squire, L. R., 1990).

Painful stimuli sometimes lead to long-term changes in pain processing, as well as to signaling the presence of the stimulus. This appears to be the situation in the case of stimulation sites in and posterior to Vc, where stimulation can evoke complex, fully formed, pain in patients with angina or atypical chest pain. These memories seem to be due to long term changes in forebrain function, that is, conditioning, since memories of this type have not been reported in response to STT stimulation.

S2 and insular cortical areas involved in pain processing also satisfy criteria for areas involved in memory through corticolimbic connections (Mishkin, M., 1979). In monkeys, a nociceptive submodality selective area has been found within S2 (Dong, W. K. *et al.*, 1994; Willis, W. D., Jr. *et al.*, 2001). S2 cortex projects to insular areas that project to amygdala (Friedman, D. P. *et al.*, 1986). S2 and insular cortex have bilateral primary noxious sensory input (Chatrian, G. E. *et al.*, 1975), and cells in these areas responding to noxious stimuli have bilateral representation (Dong, W. K. *et al.*, 1994) and project to the medial temporal lobe (Chatrian, G. E. *et al.*, 1975; Dong, W. K. *et al.*, 1989). Parasylvian cortical areas receive input from nociceptive subnuclei within Vc, and from nuclei nearby (see Lateral Thalamic Nuclei) and so may be involved in memory for pain (Burton, H., 1986). This proposal is consistent with Mishkin's hypothesis of somatic sensory memory mediated through corticolimbic connections (Mishkin, M., 1979).

44.6 Conclusions

These stimulation results are consistent with those of functional imaging studies that have identified brain regions activated in a binary fashion by the application of a very specific, painful stimulus, while further increases in stimulus intensity do not produce increased activation. These brain regions may also mediate the long-term relationship of some intense pains to the strong affective dimension, which accompanies them. This relationship may also explain the affective dimension of chronic pain, which results from an intensely painful experience. These complex experiences of pain are not evoked by stimulation of the STT, again suggesting that these memories of pain involve conditioning of forebrain structures, rather than simple activation by a painful stimulus.

Acknowledgments

This work is supported by the National Institutes of Health – National Institute of Neurological Disorders and Stroke (NS38493 and NS40059 to F. A. L.).

References

Al Chaer, E. D., Lawand, N. B., Westlund, K. N., and Willis, W. D. 1996a. Pelvic visceral input into the nucleus gracilis is largely mediated by the postsynaptic dorsal column pathway. J. Neurophysiol. 76, 2675–2690.

Al Chaer, E. D., Lawand, N. B., Westlund, K. N., and Willis, W. D. 1996b. Visceral nociceptive input into the ventral posterolateral nucleus of the thalamus: a new function for the dorsal column pathway. J. Neurophysiol. 76, 2661–2674.

Al-Chaer, E. D., Westlund, K. N., and Willis, W. D. 1997. Nucleus gracilis: an integrator for visceral and somatic information. J. Neurophysiol. 78, 521–527.

Andrew, D. and Craig, A. D. 2001. Spinothalamic lamina I neurons selectively sensitive to histamine: a central neural pathway for itch. Nat. Neurosci. 4, 72–77.

Apkarian, A. V. and Hodge, C. J. 1989. Primate spinothalamic pathways. II. The cells of origin of the dorsolateral and ventral spinothalamic pathways. J. Comp. Neurol. 288, 474–492.

Apkarian, A. V. and Shi, T. 1994. Squirrel monkey lateral thalamus. I. Somatic nociresponsive neurons and their relation to spinothalamic terminals. J. Neurosci. 14, 6779–6795.

Becker, D. E., Yingling, C. D., and Fein, G. 1993. Identification of pain, intensity, and P300 components in the pain evoked potential. Electroencephalogr. Clin. Neurophysiol. 88, 290–301.

Blomqvist, A., Zhang, E. T., and Craig, A. D. 2000. Cytoarchitectonic and immunohistochemical characterization of a specific pain and temperature relay, the posterior portion of the ventral medial nucleus, in the human thalamus. Brain 123(Pt 3), 601–619.

Bowsher, D. 1957. Termination of the central pain pathway in man: the conscious appreciation of pain. Brain 80, 606–620.

Braunwald, E. 1988. The History. In: Heart Disease: A Textbook of Cardiovascular Medicine (ed. E. Braunwald), pp. 1–12. W. B. Saunders.

Bromm, B. and Treede, R. D. 1984. Nerve fibre discharges, cerebral potentials and sensations induced by CO_2 laser stimulation. Hum. Neurobiol. 3, 33–40.

Burton, H. 1986 Second Somatosensory Cortex and Related Areas. In: Cerebral Cortex, Sensory–Motor Areas and Aspects of Cortical Connectivity, (eds. E. G. Jones and A. Peters), Vol. 5. pp. 31–98. Plenum .

Casey, K. L., Minoshima, S., Berger, K. L., Koeppe, R. A., Morrow, T. J., and Frey, K. A. 1994. Positron emission tomographic analysis of cerebral structures activated specifically by repetitive noxious heat stimuli. J. Neurophysiol. 71, 802–807.

Chatrian, G. E., Canfield, R. C., Knauss, T. A., and Eegt, E. L. 1975. Cerebral responses to electrical tooth pulp stimulation in man. An objective correlate of acute experimental pain. Neurology 25, 745–757.

Christensen, M. D., Willis, W. D., and Westlund, K. N. 1996. Anatomical evidence for cells of origin of a postsynaptic dorsal column visceral pathway: sacral spinal cord cells innervating the medial nucleus gracilis. Society For Neuroscience Abstract 22, 109.

Craig, A. D. 1998. A new version of the thalamic disinhibition hypothesis of central pain. Pain Focus 7, 1–14.

Craig, A. D. 2003a. A new view of pain as a homeostatic emotion. Trends Neurosci. 26, 303–307.

Craig, A. D. 2003b. Pain mechanisms: labeled lines versus convergence in central processing. Annu. Rev. Neurosci. 26, 1–30.

Craig, A. D., Bushnell, M. C., Zhang, E. T., and Blomqvist, A. 1994. A thalamic nucleus specific for pain and temperature sensation. Nature 372, 770–773.

Craig, A. D., Reiman, E. M., Evans, A., and Bushnell, M. C. 1996. Functional imaging of an illusion of pain. Nature 384, 258–260.

Cusick, C. G., Wall, J. T., Felleman, D. J., and Kaas, J. H. 1989. Somatotopic organization of the lateral sulcus of owl monkeys: area 3b, S-II, and a ventral somatosensory area. J. Comp. Neurol. 282, 169–190

Davis, K. D., Kiss, Z. H. T., Tasker, R. R., and Dostrovsky, J. O. 1996. Thalamic stimulation-evoked sensations in chronic pain patients and nonpain (movement disorder) patients. J. Neurophysiol. 75, 1026–1037.

Davis, K. D., Lozano, A. M., Manduch, M., Tasker, R. R., Kiss, Z. H. T., and Dostovsky, J. O. 1999. Thalamic relay site for cold perception in humans. J. Neurophysiol. 81, 1970–1973.

Davis, K. D., Tasker, R. R., Kiss, Z. H. T., Hutchison, W. D., and Dostrovsky, J. O. 1995. Visceral pain evoked by thalamic microstimulation in humans. Neuroreport 6, 369–374.

Dong, W. K., Chudler, E. H., Sugiyama, K., Roberts, V. J., and Hayashi, T. 1994. Somatosensory, multisensory, and task-related neurons in cortical area 7b (PF) of unanesthetized monkeys. J. Neurophysiol. 72, 542–564.

Dong, W. K., Salonen, L. D., Kawakami, Y., Shiwaku, T., Kaukoranta, E. M., and Martin, R. F. 1989. Nociceptive responses of trigeminal neurons in SII-7b cortex of awake monkeys. Brain Res. 484, 314–324.

Dostrovsky, J. O., Wells, F. E. B., and Tasker, R. R. 1991. Pain Evoked by Stimulation in Human Thalamus. In: International Symposium on Processing Nociceptive Information (ed. Y. Sjigenaga), pp. 115–120. Elsevier.

Dubner, R., Kenshalo, D. R., Jr., Maixner, W., Bushnell, M. C., and Oliveras, J. L. 1989. The correlation of monkey medullary dorsal horn neuronal activity and the perceived intensity of noxious heat stimuli. J. Neurophysiol. 62, 450–457

Duncan, G. H., Bushnell, M. C., Oliveras, J. L., Bastrash, N., and Tremblay, N. 1993. Thalamic VPM nucleus in the behaving monkey. III. Effects of reversible inactivation by lidocaine on thermal and mechanical discrimination. J. Neurophysiol. 70, 2086–2096.

Ferrington, D. G., Sorkin, L. S., and Willis, W. D. 1987. Responses of spinothalamic tract cells in the superficial dorsal horn of the primate lumbar spinal cord. J. Physiol. (Lond.) 388, 681–703.

Friedman, D. P., Murray, E. A., O'Neill, J. B., and Mishkin, M. 1986. Cortical connections of the somatosensory fields of the lateral sulcus of macaques: evidence for a corticolimbic pathway for touch. J. Comp. Neurol. 252, 323–347.

Frot, M. and Mauguiere, F. 1999. Timing and spatial distribution of somatosensory responses recorded in the upper bank of the sylvian fissure (SII area) in humans. Cereb. Cortex 9, 854–863.

Gloor, P. 1990. Experiential phenomena of temporal lobe epilepsy: facts and hypotheses. Brain 113, 1673–1694.

Gloor, P., Olivier, A., Quesney, L. F., Andermann, F., and Horowitz, S. 1982a. The role of the limbic system in experiential phenomena of temporal lobe epilepsy. Ann. Neurol. 12, 129–144.

Gloor, P., Olivier, A., Quesney, L. F., Andermann, F., and Horowitz, S. 1982b. The role of the limbic system in experiential phenomena of temporal lobe epilepsy. Ann. Neurol. 12, 129–144.

Gracely, R. H., Lota, L., Walter, D. J., and Dubner, R. 1988. A multiple random staircase method of psychophysical pain assessment. Pain 32, 55–63.

Graziano, A. and Jones, E. G. 2004. Widespread thalamic terminations of fibers arising in the superficial medullary dorsal horn of monkeys and their relation to calbindin immunoreactivity. J. Neurosci. 24, 248–256.

Greenspan, J. D., Lee, R. R., and Lenz, F. A. 1999. Pain sensitivity alterations as a function of lesion location in the parasylvian cortex. Pain 81, 273–282.

Greenspan, J. D., Ohara, S., Sarlani, E., and Lenz, F. A. 2004. Allodynia in patients with post-stroke central pain (CPSP) studied by statistical quantitative sensory testing within individuals. Pain 109, 357–366.

Halgren, E., Walter, R. D., Cherlow, D. G., and Crandall, P. H. 1978. Mental phenomena evoked by electrical stimulation of the human hippocampal formation and amygdala. Brain 101, 83–117.

Hamalainen, H., Kekoni, J., Sams, M., Reinikainen, K., and Naatanen, R. 1990. Human somatosensory evoked potentials to mechanical pulses and vibration: contributions of SI and SII somatosensory cortices to P50 and P100 components. Electroencephalogr. Clin. Neurophysiol. 75, 13–21.

Hassler, R. 1959. Anatomy of the Thalamus. In: Introduction to Stereotaxis with an Atlas of the Human Brain (eds. G. Schaltenbrandand and P. Bailey), pp. 230–290. Thieme.

Hassler, R. and Reichert, T. 1959 Klinische und anatomische Befunde bei stereotaktischen Schmerzoperationen im Thalamus. Arch. Psychiat. Nerverkr. 200, 93–122.

Hirai, T. and Jones, E. G. 1989. A new parcellation of the human thalamus on the basis of histochemical staining. Brain Res. Rev. 14, 1–34.

Ishijima, B., Yoshimasu, N., Fukushima, T., Hori, T., Sekino, H., and Sano, K. 1975. Nociceptive neurons in the human thalamus. Confin. Neurol. 37, 99–106.

Jones, A. K., Brown, W. D., Friston, K. J., Qi, L. Y., and Frackowiak, R. S. 1991. Cortical and subcortical localization of response to pain in man using positron emission tomography. Proc. R. Soc. Lond. B Biol. Sci. 244, 39–44.

Jones, E. G., Friedman, D. P., and Hendry, S. H. 1982. Thalamic basis of place- and modality-specific columns in monkey somatosensory cortex: a correlative anatomical and physiological study. J. Neurophysiol. 48, 545–568.

Kaas, J. H., Nelson, R. J., Sur, M., Dykes, R. W., and Merzenich, M. M. 1984 The somatotopic organization of the ventroposterior thalamus of the squirrel monkey, Saimiri sciureus. J. Comp. Neurol. 226, 111–140.

Kumazawa, T. and Perl, E. R. 1978. Excitation of marginal and substantia gelatinosa neurons in the primate spinal cord: indications of their place in dorsal horn functional organization. J. Comp. Neurol. 177, 417–434

Lee, J.-I., Antezanna, D., Dougherty, P. M., and Lenz, F. A. 1999. Responses of neurons in the region of the thalamic somatosensory nucleus to mechanical and thermal stimuli graded into the painful range. J. Comp. Neurol. 410, 541–555

Lenz, F. A. and Byl, N. N. 1999. Reorganization in the cutaneous core of the human thalamic principal somatic sensory nucleus (ventral caudal) in patients with dystonia. J. Neurophysiol. 82, 3204–3212.

Lenz, F. A. and Dougherty, P. M. 1998. Cells in the human principal thalamic sensory nucleus (ventralis caudalis – Vc) respond to innocuous mechanical and cool stimuli. J. Neurophysiol. 79, 2227–2230.

Lenz, F. A., Dostrovsky, J. O., Tasker, R. R., Yamashiro, K., Kwan, H. C., and Murphy, J. T. 1988. Single-unit analysis of the human ventral thalamic nuclear group: somatosensory responses. J. Neurophysiol. 59, 299–316.

Lenz, F. A., Gracely, R. H., Baker, F. H., Richardson, R. T., and Dougherty, P. M. 1998a. Reorganization of sensory modalities evoked by stimulation in the region of the principal sensory nucleus (ventral caudal – Vc) in patients with pain secondary to neural injury. J. Comp. Neurol. 399, 125–138.

Lenz, F. A., Gracely, R. H., Hope, E. J., Baker, F. H., Rowland, L. H., Dougherty, P. M., and Richardson, R. T. 1994a. The sensation of angina can be evoked by stimulation of the human thalamus. Pain 59, 119–125.

Lenz, F. A., Gracely, R. H., Romanoski, A. J., Hope, E. J., Rowland, L. H., and Dougherty, P. M. 1995. Stimulation in the human somatosensory thalamus can reproduce both the affective and sensory dimensions of previously experienced pain. Nat. Med. 1, 910–913.

Lenz, F. A., Gracely, R. H., Rowland, L. H., and Dougherty, P. M. 1994b. A population of cells in the human thalamic principal sensory nucleus respond to painful mechanical stimuli. Neurosci. Lett. 180, 46–50.

Lenz, F. A., Ohara, S., Gracely, R. H., Dougherty, P. M., and Patel, S. H. 2004. Pain encoding in the human forebrain: binary and analog exteroceptive channels. J. Neurosci. 24, 6540–6544.

Lenz, F. A., Rios, M., Chau, D., Krauss, G. L., Zirh, T. A., and Lesser, R. P. 1998b. Painful stimuli evoke potentials recorded from the parasylvian cortex in humans. J. Neurophysiol. 80, 2077–2088.

Lenz, F. A., Rios, M. R., Zirh, T. A., Krauss, G., and Lesser, R. P. 1998c. Painful stimuli evoke potentials recorded over the human anterior cingulate gyrus. J. Neurophysiol. 79, 2231–2234.

Lenz, F. A., Seike, M., Lin, Y. C., Baker, F. H., Rowland, L. H., Gracely, R. H., and Richardson, R. T. 1993a. Neurons in the area of human thalamic nucleus ventralis caudalis respond to painful heat stimuli. Brain Res. 623, 235–240.

Lenz, F. A., Seike, M., Richardson, R. T., Lin, Y. C., Baker, F. H., Khoja, I., Jaeger, C. J., and Gracely, R. H. 1993b. Thermal and pain sensations evoked by microstimulation in the area of human ventrocaudal nucleus. J. Neurophysiol. 70, 200–212.

Locke, S., Angevine, J. B., and Marin, O. S. M. 1961. Projection of magnocellular medial geniculate nucleus in man. Anat. Rec. 139, 249–250.

Matthews, M. B. 1985. Clinical Diagnosis. In: Angina Pectoris (ed. D. G. Julian), pp. 62–83. Churchill Livingstone.

Mayer, D. J., Price, D. D., and Becker, D. P. 1975. Neurophysiological characterization of the anterolateral spinal cord neurons contributing to pain perception in man. Pain 1, 51–58.

Mehler, W. R. 1962. The Anatomy of the So-Called "Pain Tract" in Man: An Analysis of the Course and Distribution of the Ascending Fibers of the Fasciculus Anterolateralis. In: Basic Research in Paraplegia (eds. J. D. French and R. W. Porter), pp. 26–55. Thomas.

Mehler, W. R. 1966. The posterior thalamic region in man. Confin. Neurol. 27, 18–29.

Mehler, W. R. 1969. Some neurological species differences – a posteriori. Ann. N. Y. Acad. Sci. 167, 424–468.

Mishkin, M. 1979. Analogous neural models for tactual and visual learning. Neuropsychologia 17, 139–151.

Moriarity, J. L., Boatman, D., Krauss, G. L., Storm, P. B., and Lenz, F. A. 2001 Human "memories" can be evoked by stimulation of the lateral temporal cortex after ipsilateral medial temporal lobe resection. J. Neurol. Neurosurg. Psychiatry 71, 549–551.

Ohara, S. and Lenz, F. A. 2003. Medial lateral extent of thermal and pain sensations evoked by microstimulation in somatic sensory nuclei of human thalamus. J. Neurophysiol. 90, 2367–2377.

Ohara, S., Crone, N. E., Weiss, N., Treede, R. D., and Lenz, F. A. 2004a. Amplitudes of laser evoked potential recorded from primary somatosensory, parasylvian and medial frontal cortex are graded with stimulus intensity. Pain 110, 318–328.

Ohara, S., Crone, N. E., Weiss, N., Treede, R. D., and Lenz, F. A. 2004b. Cutaneous painful laser stimuli evoke responses recorded directly from primary somatosensory cortex in awake humans. J. Neurophysiol. 91, 2734–2746.

Olszewski, J. 1952. The Thalamus of Maccaca mulatta. Karger.

Ostrowsky, K., Magnin, M., Ryvlin, P., Isnard, J., Guenot, M., and Mauguiere, F. 2002. Representation of pain and somatic sensation in the human insula: a study of responses to direct electrical cortical stimulation. Cereb. Cortex 12, 376–385.

Pasternak, R. C., Braunwald, E., and Sobel, B. E. 1992 Acute Myocardial Infarction. In: Cardiac Disease (ed. E. Braunwald), pp. 1200–1291. W. B. Saunders.

Penfield, W. and Jasper, H. 1954. Epilepsy and the Functional Anatomy of the Human Brain. Little Brown.

Perl, E. R. 1984 Pain and Nociception. In: The Nervous System: Sensory Processes, Part 2 (eds. J. M. Brookhart, V. B. Mountcastle, I. Darian-Smith, and S. R. Geiger), pp. 915–975. American Physiological Society.

Price, D. D. and Dubner, R. 1977. Neurons that subserve the sensory-discriminative aspects of pain. Pain 3, 307–338.

Price, D. D., Greenspan, J. D., and Dubner, R. 2003. Neurons involved in the exteroceptive function of pain. Pain 106, 215–219.

Price, D. D. and Mayer, D. J. 1975. Neurophysiological characterization of the anterolateral quadrant neurons subserving pain in M. mulatta. Pain 1, 59–72.

Procacci, P. and Zoppi, M. 1989. Heart Pain. In: Textbook of Pain (eds. P. D. Wall and R. Melzack), pp. 410–419. Churchill Livingstone.

Ralston, H. J. and Ralston, D. D. 1992. The primate dorsal spinothalamic tract: evidence for a specific termination in the posterior nuclei [Po/SG] of the thalamus. Pain 48, 107–118.

Rinaldi, P. C., Young, R. F., Albe-Fessard, D. G., and Chodakiewitz, J. 1991. Spontaneous neuronal hyperactivity in the medial and intralaminar thalamic nuclei in patients with deafferentation pain. J. Neurosurg. 74, 415–421.

Robinson, C. J. and Burton, H. 1980. Somatic submodality distribution within the second somatosensory (SII), 7b, retroinsular, postauditory, and granular insular cortical areas of M. fascicularis. J. Comp. Neurol. 192, 93–108.

Rustioni, A., Hayes, N. L., and O'Neill, S. 1979. Dorsal column nuclei and ascending spinal afferents in macaques. Brain 102, 95–125.

Sano, K. 1979. Stereotaxic Thalamolaminotomy and Posteromedial Hypothalamotomy for the Relief of Intractable Pain. In: Advances in Pain Research and Therapy (eds. J. J. Bonica and V. Ventrafridda) Vol. 2, pp. 475–485. Raven Press.

Schaltenbrand, G. and Bailey, P. 1959. Introduction to Stereotaxis with an Atlas of the Human Brain. Thieme.

Talairach, J. and Tournoux, P. 1988. Co-Planar Stereotaxic Atlas of the Human Brain, 3D Proportional System: An Approach to Cerebral Imaging. Georg Thieme Verlag.

Talbot, J. D., Marrett, S., Evans, A. C., Meyer, E., Bushnell, M. C., and Duncan, G. H. 1991. Multiple representations of pain in human cerebral cortex. Science 251, 1355–1358.

Tsubokawa, T. and Moriyasu, N. 1975. Follow-up results of centre median thalamotomy for relief of intractable pain. A method of evaluating the effectiveness during operation. Confin. Neurol. 37, 280–284.

Uddenberg, N. 1968. Functional organization of long, second-order afferents in the dorsal funiculus. Exp. Brain Res. 4, 377–382.

Van Buren, J. M. and Borke, R. C. 1972. Variations and Connections of the Human Thalamus. Springer.

Vogel, H., Port, J. D., Lenz, F. A., Solaiyappan, M., Krauss, G., and Treede, R. D. 2003. Dipole source analysis of laser-evoked subdural potentials recorded from parasylvian cortex in humans. J. Neurophysiol. 89, 3051–3060.

Wang, C. C., Willis, W. D., and Westlund, K. N. 1999. Ascending projections from the area around the spinal cord central canal: a *Phaseolus vulgaris* leucoagglutinin study in rats. J. Comp. Neurol. 415, 341–367.

Willis, W. D. 1985 The Pain System. Karger.

Willis, W. D., Al Chaer, E. D., Quast, M. J., and Westlund, K. N. 1999. A visceral pain pathway in the dorsal column of the spinal cord. Proc. Natl. Acad. Sci. U. S. A. 96, 7675–7679.

Willis, W. D. and Coggeshall, R. E. 1991. Sensory Mechanisms of the Spinal Cord. Plenum.

Willis, W. D., Jr., Zhang, X., Honda, C. N., and Giesler, G. J., Jr. 2001. Projections from the marginal zone and deep dorsal horn to the ventrobasal nuclei of the primate thalamus. Pain 92, 267–276.

Yarnitsky, D. and Sprecher, E. 1994. Thermal testing: normative data and repeatability for various test algorithms. J. Neurol. Sci. 125, 39–45.

Zaslansky, R., Sprecher, E., Tenke, C. E., Hemli, J. A., and Yarnitsky, D. 1995. The P300 in pain evoked potentials. Pain 66, 39–49.

Zola-Morgan, S. and Squire, L. R. 1990. The primate hippocampal formation: evidence for a time-limited role in memory storage. Science 250, 288–290.

45 Nociceptive Processing in the Cerebral Cortex

R D Treede, Ruprecht-Karls-University Heidelberg, Heidelberg, Germany

A V Apkarian, Northwestern University, Chicago, IL, USA

Glossary

ACC anterior cingulate cortex (n.b.: some authors use this as a summary term for ACC and MCC)

CBP Chronic back pain

CRPS Complex regional pain syndrome

EEG Electro-encephalography

fMRI Functional magnetic resonance imaging

IBS Irritable bowel syndrome

IC Insular cortex

MCC mid-cingulate cortex (n.b.: some authors call this region the posterior part of ACC)

MEG Magneto-encephalography

MRS Magnetic resonance spectroscopy

OIC Operculo-insular cortex, consisting of insular cortex plus the frontal, parietal and temporal operculum.

PAG Periaqueductal grey

PET Positron emission tomography

PFC prefrontal cortex

PHN Postherpetic neuralgia

SI Primary somatosensory cortex

SII Secondary soamtosensory cortex

SCI Spinal cord injury

SPECT Single photon emission computed tomography

Th Thalamus

45.1　Introduction

Conscious perception of external stimuli requires encoding by sensory organs, processing within the respective sensory system, and activation of the appropriate sensory cortical areas. Based on a small case series of infra- and supratentorial brain lesions, Head H. and Holmes G. (1911) postulated that the sensation of pain is an exception to this rule and that its conscious perception occurs in the essential organ of the thalamus. In spite of evidence to the contrary from clinical reports (Marshall, J., 1951; Biemond, A., 1956), evoked potentials in humans (Spreng, M. and Ichioka, M., 1964; Duclaux, R. *et al.*, 1974; Carmon, A. *et al.*, 1976; Chen, A. C. N. *et al.*, 1979; Bromm, B. and Treede, R. D., 1984), single unit recordings in animals (Lamour, Y. *et al.*, 1982; Kenshalo, D. R. and Isensee, O., 1983), neuroanatomical tracing (Gingold, S. I. *et al.*, 1991), and some early PET studies (Buchsbaum, M. S. *et al.*, 1984), it was maintained for a long time that the cortical representation of pain is a *quantité négligable*.

This situation changed, when the modern neuroimaging techniques of positron emission tomography (PET) and later functional magnetic resonance imaging (fMRI) demonstrated systematic metabolic and perfusion changes in a large number of cortical areas following painful stimuli (Talbot, J. D. *et al.*, 1991; Jones, A. K. P. *et al.*, 1991a; Apkarian, A. V. *et al.*, 1992; Davis, K. D. *et al.*, 1995). These findings were supported by invasive and noninvasive electrophysiological studies in humans, using magnetoencephalography (MEG), electroencephalography (EEG), subdural recordings directly from the surface of the brain, and depth recordings during stereotactic procedures (for a systematic review see Apkarian, A. V. *et al.*, 2005).

Meanwhile it has been recognized that painful stimuli activate a vast network of cortical areas, including the primary and secondary somatosensory cortex (SI, SII), the insula, posterior parietal cortex, anterior and mid-cingulate cortex, and parts of the prefrontal cortex (PFC; Figure 1). These areas are involved in the generation of painful percepts as well as in the descending control of pain (for review see Kenshalo, D. R. and Willis, W. D., 1991; Treede, R. D. *et al.*, 1999; Price, D. D., 2000; Apkarian, A. V. *et al.*, 2005). Most of these areas are also involved in other sensory, emotional, cognitive, motor or autonomic functions. Hence, the nociceptive system converges with other systems for the generation of the conscious percept of pain. In that sense, the nociceptive system is not different from the visual system, for example. But it is still an open question, to what extent any cortical regions can be considered as nociceptive specific.

In this chapter we will briefly review the methods used to assess nociceptive processing in the human brain, present connectivity and functional properties of each of the principal cortical regions of the nociceptive system, and summarize the roles of the cerebral cortex in various aspects of pain perception and pain modulation.

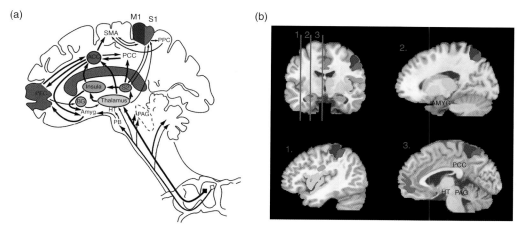

Figure 1 Cortical regions involved in pain perception, their interconnectivity and ascending pathways. Locations of brain regions involved in pain perception are color coded in a schematic drawing and in an example magnetic resonance image (MRI). (a) Schematic diagram shows the regions, their interconnectivity, and afferent pathways. (b) The areas corresponding to those shown in the schematic are shown in an anatomical MRI, on a coronal slice and three sagittal slices as indicated on the coronal slice: primary and secondary somatosensory cortices (S1, S2, red and orange), anterior and mid-cingulate cortex (ACC, green), insula (blue), thalamus (yellow), prefrontal cortex (PFC, purple), primary and supplementary motor cortex (M1 and SMA), posterior parietal cortex (PPC), posterior cingulate cortex (PCC), basal ganglia (BG, pink), hypothalamus (HT), amygdala (AMYG), parabrachial nuclei (PB), and periaqueductal gray (PAG). Reproduced from Apkarian, A. V., Bushnell, C., Treede, R. D., and Zubieta, J. K. 2005. Human brain mechanisms of pain perception and regulation in health and disease. Eur. J. Pain 9, 463–484.

45.2 Methods to Study Nociceptive Processing in the Human Cerebral Cortex

Table 1 summarizes properties of the different brain imaging techniques that have been used to define the nociceptive network in the human brain. The most direct approach to learn about the functions of cortical neurons is direct intracellular or extracellular recording of their electrical activity during sensory stimulation and in different contexts. This technique is mostly restricted to animal studies (Kenshalo, D. R. and Isensee, O., 1983; Dong, W. K. et al., 1994) and has rarely been possible in humans (Hutchison, W. D. et al., 1999). Field potentials within the brain invert their polarity, when an electrode track passes through or close to their generator source. This technique has been used in the course of presurgical epilepsy diagnostics (Frot, M. and Mauguière, F., 1999), but since the electrode tracks are related to the clinical indications, only few parts of the brain have been sampled that way. Presurgical epilepsy diagnostics using subdural electrode grids samples a much larger part of the brain surface, and dipole source analysis can be used to estimate the depths of the generators below the grids (Vogel, H. et al., 2003). All of these invasive recordings in the human brain need to be interpreted with caution, because the cortical pathology that provided the indication for the procedure (e.g., epilepsy, tumors) may have altered nociceptive signal processing.

EEG and MEG are noninvasive techniques for the direct assessment of electrical activity in the brain. Mathematical algorithms are available to estimate the location of the generators within the brain from the signals recorded at the surface of the head with an accuracy of about 10 mm (Scherg, M., 1992; Pascual-Marqui, R. D. et al., 1994; Hari, R. and Forss, N., 1999). EEG and MEG techniques provide accurate timing information. As a result, both methods have been used mainly to identify the arrival of information to various cortical regions (stimulus-evoked potentials). Spontaneous fluctuations in EEG and MEG would provide a view of the interactions between cortical areas. However, the application of the latter to painful states has remained minimal (Chen, A. C. N., 1993; Ohara, S. et al., 2006). MEG detects brain magnetic activity, a signal that is proportional and orthogonal to the local electrical activity. Depending on the orientation of a local generator source and the gyral geometry of the brain region, evoked potentials in different brain areas may be better detected by MEG or EEG. The main weakness of EEG and MEG methods is their limited spatial resolution (on the order of 1 cm for both methods).

PET, single photon emission-computed tomography (SPECT), and fMRI measure brain activity

Table 1 Brain mapping techniques, their properties, and application in pain studies

Method	Energy source	Spatial resolution (mm)	Temporal resolution (s)	Constraints	Output measured	Application in pain studies
EEG/MEG	Intrinsic electricity	10	0.001	Lack of unique localization	Electrophysiology of brain events	Increasing in use, mainly for detecting temporal sequences
fMRI	Radio waves, magnetic fields	4–5	4–10	Immobilization, loud, cooperation	Relative cerebral blood flow	Most used, mainly for localizing brain activity
MRS	Radio waves, magnetic fields	10	10–100	Immobilization, loud	Relative chemical concentrations	Recently used, for detecting long term changes in brain chemistry
Nuclear (PET/ SPECT)	Radiation	5–10	60–1000	Radiation limits, immobilization	Physiology, neurochemistry, absolute values	Decreasing in use, becoming limited to neurochemistry
Brain imaging techniques available but rarely or not yet used in pain studies						
Single or multiunit electrophysiology	Intrinsic electricity	0.01–1	0.001	Invasive, direct access to brain	Electrophysiology	
Near infrared spectroscopy and imaging	Infrared light	0.05	0.05	Immobilization, surface > depth, limited field of view	Relative cerebral blood flow	
Transcranial magnetic/ electric stimulation	Magnetic/ electric fields	10	0.01	Risk of seizures, immobilization, loud	Electrophysiology, conduction times	
Structural MRI	Radio waves, magnetic fields	1	N/A	Immobilization, loud	Structure, vasculature, white matter	
Postmortem	N/A	0.001	N/A	Postmortem	Microarchitecture, chemoarchitecture	

EEG, electroencephalography; MEG, magnetoencephalography; fMRI, functional magnetic resonance imaging; MRS, magnetic resonance spectroscopy; PET, positron emission tomography; SPECT, single photon emission-computed tomography; N/A, not applicable.

indirectly by imaging changes in blood flow, blood oxygenation, or local metabolic changes (Peyron, R. *et al.*, 2000; Davis, K. D., 2003). All three methods can provide similar spatial resolution, although PET and fMRI methodologies are now far more advanced than SPECT. The statistical models and experimental designs available for PET and fMRI are robust and very rich. Therefore, these two techniques are currently most extensively used for detecting brain circuitry underlying many cognitive states, including pain. The temporal resolution of PET and SPECT is in the order of tens of seconds, while for fMRI it is shorter. PET and SPECT provide the additional opportunity for examining *in vivo* biochemistry and pharmacology by imaging the distributions of specific neurotransmitters or receptors. Recent MRI methods, like magnetic resonance spectroscopy (MRS), have also provided the ability to examine brain biochemistry. This approach is

developing rapidly and has the potential to become a major method in the near future for studying brain chemistry. In addition, voxel based morphometry allows to image structural changes related to disease states (May, A. *et al.*, 1999).

45.3 Cortical Regions that are Part of the Nociceptive System

45.3.1 The Primary Somatosensory Cortex

The primary somatosensory cortex (SI) is located in the anterior part of the parietal lobe, where it constitutes the postcentral gyrus. It consists of Brodmann areas 1, 2, 3a, and 3b (Figure 2(a)). Areas 3b and 1 receive cutaneous tactile input, areas 3a and 2 proprioceptive input.

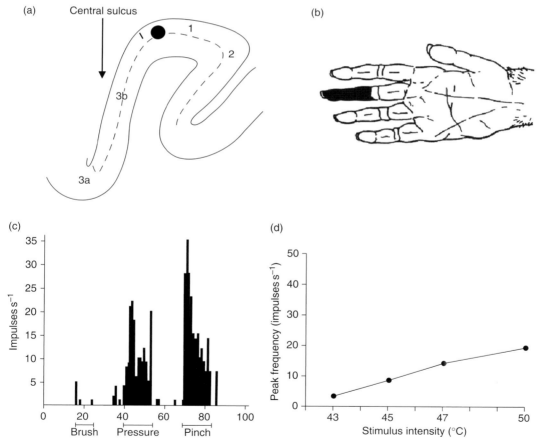

Figure 2 Nociceptive specific neuron in the primary somatosensory cortex (SI). (a) Left: SI consists of Brodmann areas 1, 2, 3a, and 3b in the postcentral gyrus. Black dot: location of the recorded neuron. (b) The small receptive field is consistent with a role in spatial discrimination. (c) Stimulus response function to painful mechanical stimuli. (d) Stimulus response function to painful heat stimuli. Modified from Kenshalo, D. R., Iwata, K., Sholas, M., and Thomas, D. A. 2000. Response properties and organization of nociceptive neurons in area 1 of monkey primary somatosensory cortex. J. Neurophysiol. 84, 719–729.

Nociceptive input to monkey SI was demonstrated anatomically. SI receives direct spino-thalamocortical input from the ventrobasal nuclei, in particular the ventro-posterolateral (VPL) nucleus (Gingold, S. I. *et al.*, 1991). Nociceptive neurons in SI are found in clusters, raising the possibility that SI may contain nociceptive specific columns (Lamour, Y. *et al.*, 1983). Since evidence for nociceptive neurons in the most superficial cortical layers is lacking, this hypothesis has not yet been confirmed. Nociceptive neurons are rare in monkey SI and have mainly been found in area 1 (Kenshalo, D. R. *et al.*, 2000), whereas optical imaging techniques have also suggested nociceptive input to area 3a (Tommerdahl, M. *et al.*, 1996). Thus, nociceptive signal processing within SI may be spatially distinct from tactile signal processing that is primarily directed to area 3b. There is also some EEG and MEG evidence in humans that nociceptive areas may be situated more medially within SI than tactile areas with the same receptive fields, suggesting that nociceptive and tactile signal processing may occur in different subareas of SI (Ploner, M. *et al.* 2000; Schlereth, T. *et al.*, 2003). Nociceptive input to human SI has been confirmed by subdural recordings (Kanda, M. *et al.*, 2000; Ohara, S. *et al.*, 2004). About 75% of the PET and fMRI studies reported activation of SI (Bushnell, M. C. *et al.*, 1999; Apkarian, A. V. *et al.*, 2005).

Nociceptive neurons in SI have small receptive fields (Figure 2(b)) that are somatotopically arranged, and hence are ideally suited to code for the location of nociceptive stimuli (Kenshalo, D. R. and Isensee, O., 1983). Somatotopy of nociceptive processing in the human SI has been confirmed by EEG and PET studies (Tarkka, I. M. and Treede, R. D., 1993; Andersson, J. L. R. *et al.*, 1997). Action potential discharges of nociceptive SI neurons in monkey are modulated by the intensity of both mechanical and heat stimuli (Figures 2(c) and 2(d)) and their discharges correlate with detection speed (Kenshalo, D. R. *et al.*, 1988). These findings suggest that nociceptive SI neurons are involved in the coding of pain intensity. This conclusion has been confirmed by a PET study of hypnotic modulation of perceived pain intensity that also modulated perfusion of SI (Hofbauer, R. K. *et al.*, 2001) and by correlation analysis (Timmermann, L. *et al.*, 2001).

45.3.2 Parasylvian Cortex: the Operculo-Insular Region

The parasylvian cortex has a complicated macroscopic structure and only some of its cytoarchitectonic areas have been charted in detail (Eickhoff, S. B. *et al.*, 2006). In lateral views of the brain, the Sylvian fissure runs above the temporal lobe and separates it from the parietal and frontal lobes above the fissure. Hidden deep inside the Sylvian fissure lies a further lobe of the brain: the insula (Figure 3(a)). The insula is covered by the

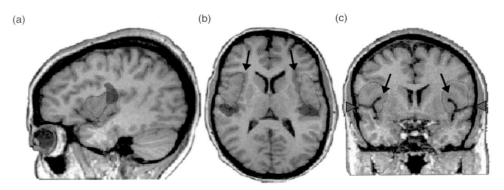

(a)   (b)   (c)

Figure 3 Nociceptive regions in parasylvian cortex. (a) Sagittal section shows the insula as a triangular region deep inside the Sylvian fissure. (b) Transaxial section illustrates the secondary somatosensory cortex (SII) as identified by tactile stimuli (blue) and regions responsive to nociceptive stimulation (orange) that extend further rostrally and medially. (c) Coronal section shows that the temporal operculum covers the insula below the Sylvian fissure, and frontal and parietal opercula cover the insula above the fissure. Frontal and parietal opercula consist of an outer part on the convexity of the brain, a horizontal part above the Sylvian Fissure, and an inner vertical part facing the insula across its circular sulcus. Arrowheads: Sylvian fissure (lateral sulcus). Arrows: circular sulcus of the insula. Modified from Treede, R. D., Baumgärtner, U., and Lenz, F. A. 2007. Nociceptive Processing in the Secondary Somatosensory Cortex. In: Encyclopedia of Pain (*eds*. R. F. Schmidt and W. D. Willis), pp. 1376–1379. Springer.

temporal, parietal and frontal opercula. Coronal sections reveal that the parasylvian cortex consists of the insula itself, the inner vertical surface of the opercula, the horizontal banks of the Sylvian fissure, and most laterally the outer surface of the convexity of the brain (Figure 3(c)).

The majority of human imaging studies showed consistent activation of the parasylvian cortex during painful stimulation, and this activation overlapped only partly with that by tactile stimuli (Treede, R. D. *et al.*, 2007). Lesions in parasylvian cortex cause deficits in pain perception (Greenspan, J. D. *et al.*, 1999), and intracortical electrical stimulation of this region is painful (Ostrowsky, K. *et al.*, 2002). Thus, this region is a good candidate to contain some nociceptive specific cortical areas, if they exist.

About 75% of the PET and fMRI studies reported activation of the SII region, and 94% found activation of the insula (Treede, R. D. *et al.*, 2000; Apkarian, A. V. *et al.*, 2005). But due to the curvature and oblique course of the Sylvian fissure (Özcan, M. *et al.*, 2005), activated areas are often misallocated, even across major sulci. Whereas the operculo-insular cortex in the parasylvian region has been recognized as one of the most important nociceptive cortical areas, its precise anatomical and functional organization has yet to be determined. In particular it is not yet known, whether insula and operculum subserve distinct functions or form one uniform area.

45.3.2.1 The secondary somatosensory cortex (SII)

The secondary somatosensory cortex is located in the superior bank of the Sylvian fissure, where it makes up a major part of the parietal operculum (Figures 1(b), 3(b), and 3(c)). Nociceptive input to monkey SII was demonstrated anatomically. SII receives direct spino-thalamo-cortical input from the ventrobasal nuclei, in particular the ventro-postero-inferior nucleus VPI (Stevens, R. T. *et al.*, 1993). Nociceptive input to human SII has been confirmed by subdural recordings (Lenz, F. A. *et al.*, 1998a).

Single neuron recordings in SII have largely focused on the tactile representation (Robinson, C. J. and Burton, H. 1980; Fitzgerald, P. J. *et al.*, 2006). The SII region contains multiple somatotopic representations of the body, suggesting the existence of several subregions (Disbrow, E. *et al.*, 2000, Fitzgerald, P. J. *et al.*, 2004). Although most neurons in SII have contralateral receptive fields, this is the first part of the somatosensory system with a sizable proportion of bilateral receptive fields. Hence, most imaging and electrophysiological studies in humans have shown a bilateral response to unilateral stimulation, with a contralateral preponderance. Functionally, SII is considered to play a role in tactile object recognition and memory (Seitz, R. J. *et al.*, 1991).

Evoked potential recordings in humans following brief laser heat stimuli showed that SII was activated simultaneously with or even earlier than SI (Ploner, M. *et al.*, 1999; Schlereth, T. *et al.*, 2003). Combined anterograde and retrograde tracer studies in monkey (Apkarian, A. V. and Shi, T., 1994) support the concept that nociceptive input reaches SII more directly than tactile input. Hence, SII has been supposed to be important for the recognition of painful stimuli as such.

In contrast to the abundance of evidence for nociceptive activation of the SII region from human studies, there are few single neuron recordings in this area showing specific nociceptive responses (Treede, R. D. *et al.*, 2000). In monkey, some convergence with visual input encoding the approach of a sharp object to the face has raised the possibility of a representation of threat. These neurons, however, were not in SII proper but in area 7b which is adjacent to SII in monkey but not in humans (Dong, W. K. *et al.*, 1994). These neurons are now considered to be part of the posterior parietal cortex (see below). Since neurons in all studies on SII were searched using mechanical skin stimulation, it is possible that these studies missed nociceptive specific neurons, because many primary nociceptive afferents are mechanically insensitive (Treede, R. D. *et al.*, 1998). Thus, an intriguing possibility is that tactile and nociceptive inputs are represented in different areas within the SII region.

45.3.2.2 The frontal operculum

Dipole source analysis of laser-evoked potentials (LEPs) in healthy volunteers, and subdural and depth recordings in patients undergoing epilepsy surgery have identified an area in the inner vertical surface of the frontal operculum (Figure 4(a)) that was activated by painful heat stimuli with a shorter latency (about 150 ms) than any other cortical area (Tarkka, I. M. and Treede, R. D., 1993; Valeriani, M. *et al.*, 1996; Ploner, M. *et al.*, 1999; Frot, M. and Mauguière, F., 2003; Schlereth, T. *et al.*, 2003; Vogel, H. *et al.*, 2003). This area anterior of the tactile SII area has a different somatotopic orientation (face: anterior, foot: posterior) than SII itself (face: lateral, foot: medial; Vogel, H. *et al.*, 2003). The thalamic source of nociceptive input to this region is not yet

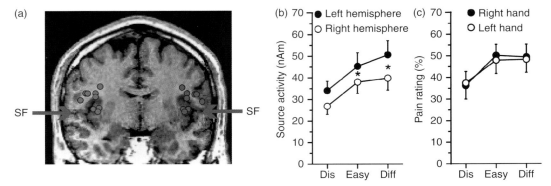

Figure 4   Nociceptive input to the frontal operculum. (a) Projection of dipole source locations for the first component of laser-evoked potentials (LEPs) onto a coronal magnetic resonance image slice at Talairach y = −6 mm. The distribution around the roof of the circular insular sulcus, ranging from the inner vertical face of the frontal operculum to the adjacent dorsal insula matches the projection area of the nociceptive thalamic nucleus VMpo. (b) Task effects and interhemispheric differences. (c) The hemispheric asymmetry of opercular activation (left hemisphere > right hemisphere) was not reflected or caused by different visual analog score pain ratings between both hands. Dis, distraction task; Easy, easy spatial and intensity discrimination tasks; diff, difficult discrimination tasks. ANOVA:* P < 0.05. Modified from Schlereth, T., Baumgärtner, U., Magerl, W., Stoeter, P., and Treede, R. D. 2003. Left-hemisphere dominance in early nociceptive processing in the human parasylvian cortex. Neuroimage 20, 441–454.

clear: it may be VPI like for the posteriorly adjacent SII, or it may be VMpo like for the medially adjacent dorsal insula. Somatotopy in the frontal operculum would be consistent with that of a VMpo projection target (Craig, A. D., 1995). Nociceptive input to the frontal operculum in humans has been confirmed by subdural and depth recordings (Lenz, F. A. et al., 1998a; Frot, M. et al., 1999). Responses in this area are modulated during spatial and intensity discrimination tasks and show a left-hemisphere dominance (Figures 4(b) and 4(c)).

45.3.2.3 The insula

The insula is located deep inside the Sylvian fissure, where it can be visualized as a triangular shape in sagittal sections (Figure 3(a)). It often contains two long sulci in its posterior part and three short sulci rostrally. Several functional subdivisions of the insula have been suggested (Dieterich, M. et al., 2003; Schweinhardt, P. et al., 2006), but there is no consensus yet. Parts of the insula subserve varied functions in the somatosensory, vestibular, gustatory and autonomic nervous system, which led to the suggestion that this region serves for a central representation of the internal state of the body (Craig, A. D., 2002). This concept is consistent with the interoceptive aspects of nociception.

Another source of nociceptive input into the parasylvian cortex is the posterior inferior part of the ventrobasal nucleus (Lenz, F. A. et al., 1993), a region designated as VMpo by some authors (Craig, A. D. et al., 1994). VMpo projects to the dorsal insula and the adjacent frontal operculum. Nociceptive input to the insula in humans has been confirmed by depth recordings (Frot, M. et al., 2003). The somatotopic representation of pain in the dorsal insula in monkey (face: anterior, foot: posterior) is orthogonal to that in SII (face: lateral, foot: medial; Baumgärtner, U. et al., 2006b). Direct electrical stimulation of the insula is painful with a strong affective component (Ostrowsky, K. et al., 2002).

45.3.3 The Posterior Parietal Cortex

The posterior parietal cortex is located adjacent and posterior to SI. It comprises Brodmann areas 5 and 7 (Figure 5(c)). Nociceptive input to this region is suggested by studies in monkey that reported short-latency responses to nociceptive stimuli in area 7b; the same neurons also responded to visual stimuli of sharp objects directed at their receptive field (Dong, W. K. et al., 1994). In the tactile system, this region is part of a dorsally directed stream involved in stimulus location, convergence with visual information and the generation of spatial information for motor control. Nociceptive input to this region in humans has not yet been explored with subdural recordings, but there is some evidence from EEG

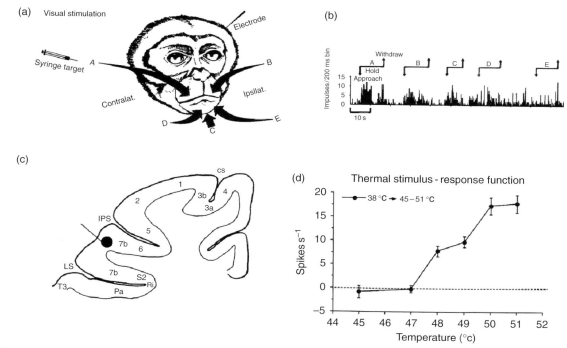

Figure 5 Nociceptive neuron in posterior parietal cortex (Area 7b threat neuron). (a) Bilateral receptive field in the orofacial region; seeing a syringe approach the receptive field was also an adequate stimulus. (b) Responses to the stimuli shown in (a). (c) Location of the recorded neuron on a coronal section that passes through both the central sulcus (CS) and the Sylvian fissure (LS: lateral sulcus). (d) Stimulus response function to painful heat stimuli. IPS, Intraparietal sulcus. Modified from Dong, W. K., Chudler, E. H., Sugiyama, K., Roberts, V. J., and Hayashi, T. 1994. Somatosensory, multisensory, and task-related neurons in cortical area 7b (PF) of unanesthetized monkeys. J. Neurophysiol. 72, 542–564.

and MEG studies in humans that nociceptive stimuli activate parietal lobe posterior of SI (Schlereth, T. *et al.*, 2003; Forss, N. *et al.*, 2005). Few fMRI studies have assessed this region (Kulkarni, B. *et al.*, 2005; Schmahl, C. *et al.*, 2006).

45.3.4 The Cingulate Cortex

The cingulate cortex is located above the corpus callosum and around its anterior knee (Figure 6). The anterior cingulate cortex (ACC) comprises Brodmann areas 24 and 32, whereas the posterior cingulate cortex (PCC) contains areas 23 and 31 (Vogt, B. A. *et al.*, 1995). ACC receives nociceptive thalamocortical input from the mediodorsal (MD) and parafascicular (Pf) nuclei (Vogt, B. A. *et al.*, 1979). ACC has been further subdivided into midcingulate cortex, which is associated with response selection and motor efferent functions, and ACC proper that is related to emotion and autonomic efferent functions (Vogt, B. A., 2005). Nociceptive

input to human cingulate cortex has been confirmed by subdural recordings and by intracortical recordings (Lenz, F. A. *et al.*, 1998b; Hutchison, W. D. *et al.*, 1999). About 87% of the PET and fMRI studies reported activation of the cingulate cortex (Apkarian, A. V. *et al.*, 2005), but no region of the cingulate cortex is considered to be nociceptive specific (Vogt, B. A., 2005).

Nociceptive neurons in ACC have large or even whole-body receptive fields (Figure 7, Sikes, R. W. and Vogt, B. A., 1992; Yamamura, H. *et al.*, 1996). For this reason it is unlikely that they contribute to the sensory dimension of pain. Monkey ACC neurons activate during pain avoidance behavior, reflecting anticipation, and response selection (Koyama, T. *et al.*, 1998; 2001). The cingulate cortex is supposed to participate in the affective-motivational dimension of pain. This conclusion has been confirmed by a PET study of hypnotic modulation of perceived pain affect that also modulated perfusion of ACC (Hofbauer, R. K. *et al.*, 2001) and by correlation analysis (Tölle, T. R. *et al.*, 1999).

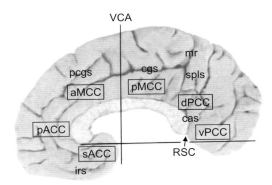

Figure 6 Distribution of cingulate cortex regions and subregions. Region borders are marked with arrows. Cross-hair shows vertical plane at the anterior commissure (VCA) and the anterior–posterior commissural line. A functional overview, derived from the analysis of a large volume of literature illustrates general regional function. aMCC, anterior mid-cingulate cortex; cas, callosal sulcus; cgs, cingulate sulcus; dPCC, dorsal posterior cingulate cortex; irs, inferior rostral sulcus; mr, marginal ramus of cgs; pACC, pregenual anterior cingulate cortex; pcgs, paracingulate sulcus; pMCC, posterior mid-cingulate cortex; RSC, retrosplenial cortex; sACC, subgenual anterior cingulate cortex; spls, splenial sulci; vPCC, ventral posterior cingulate cortex. Reproduced from Vogt, B. A. 2005. Pain and emotion interactions in subregions of the cingulate gyrus. Nat. Rev. Neurosci. 6, 533–544.

45.3.5 The Prefrontal Cortex

The PFC (including Brodmann areas 9, 10, 46) comprises the major part of the frontal lobe and is located anterior of the motor cortical areas. There is no evidence that it would receive a direct nociceptive thalamo-cortical input, but the PFC receives cortico-cortical input from the cingulate gyrus that may convey nociceptive information. About 55% of the PET and fMRI studies reported activation of the PRC in healthy subjects, and 81% of the studies in chronic pain patients (Apkarian, A. V. *et al.*, 2005). The PRC is assumed to participate in the cognitive-evaluative dimension of pain and in endogenous pain control (Lorenz, J. *et al.*, 2003; Schmahl, C. *et al.*, 2006).

45.4 Functional Roles of Cortical Nociceptive Signal Processing

Pain perception has been conceived to consist of sensory-discriminative, affective-motivational and cognitive-evaluative dimensions (Melzack, R. and Casey, K. L. 1968). The sensory-discriminative dimension includes intensity discrimination, pain qualities, stimulus localization and timing discrimination; this dimension is traditionally thought to involve lateral thalamic nuclei and the somatosensory cortices SI and SII. The affective-motivational dimension includes perception of the negative hedonic quality of pain, autonomic nervous system manifestations of emotions, and motivated behavioral responses; this dimension is traditionally thought to involve medial thalamic nuclei and the limbic cortices ACC and MCC. The insula has an intermediate position in that concept, receiving input from lateral thalamus but projecting into the limbic system. The cognitive-evaluative dimension includes interaction with previous experience, cognitive influence on perceived pain intensity and an overall evaluation of its salience; this dimension is traditionally thought to involve the PRC. Numerous neuroimaging studies have assessed various experimental paradigms derived from several psychological concepts that do not easily fit into the traditional three dimensions of pain. Therefore, we here report imaging evidence for involvement of cortical areas in specific functions instead of the dimensions of pain.

45.4.1 Location and Quality of Phasic Pain

Neuroimaging studies have examined brain regions activated by many types of painful stimulation, including noxious heat and cold, muscle stimulation using electric shock or hypertonic saline, topical and intradermal capsaicin, colonic distention, rectal distention, gastric distention, esophageal distention, ischemia, cutaneous electric shock, ascorbic acid, laser heat, as well as an illusion of pain evoked by combinations of innocuous temperatures (Apkarian, A. V. *et al.*, 2005; Bushnell, M. C. and Apkarian, A. V., 2005). Despite the differences in sensation, emotion and behavioral responses provoked by these different types of pain, individuals can easily identify each as being painful. Thus, there appears to be a common construct of pain with an underlying network of brain activity in the areas described above. Nevertheless, despite the similarities in pain experiences and similarities in neural activation patterns, each pain experience is unique. Subjects can usually differentiate noxious heat from noxious cold from noxious pressure. This ability to differentiate pains is particularly puzzling, since there is ubiquitous convergence of information from cutaneous, visceral and muscle tissue throughout the

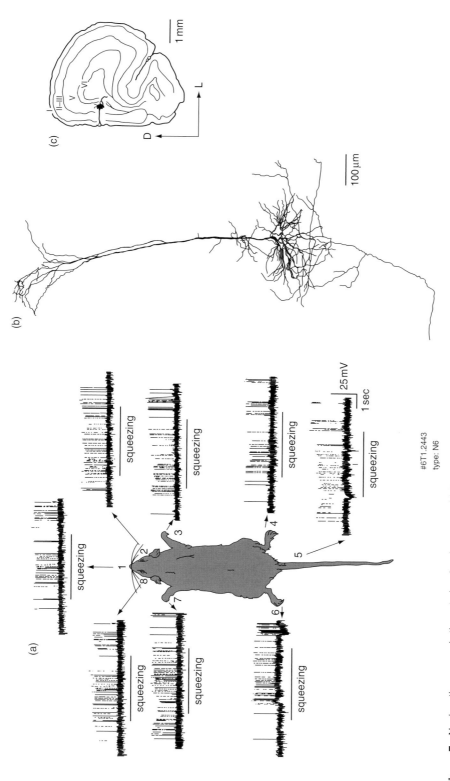

Figure 7 Nociceptive neuron in the anterior cingulate gyrus. (a) Responses to painful mechanical stimuli show a whole-body receptive field for this neuron. (b) Intracellular dye injection reveals a lamina V pyramidal neuron. (c) Location of the recorded neuron in the rat cingulate cortex. Reproduced from Yamamura, H., Iwata, K., Tsuboi, Y., Toda, K., Kitajima, K., Shimizu, N., Nomura, H., Hibiya, J., Fujita, S. and Sumino, R. 1996. Morphological and electrophysiological properties of ACCx nociceptive neurons in rats. Brain Res. 735, 83–92.

afferent nociceptive system (Willis, W. D. and Coggeshall, R. E., 2004). The convergence and the similarities in brain regions activated by different types of pain are consistent with phenomena such as referred pain, but cannot explain either the ability to identify the origin of pain or with contrasting behavioral reactions to cutaneous and visceral pain (withdrawal versus quiescence).

There is evidence from single neuron recordings, MEG, PET, and fMRI that neural activity in SI cortex could underlie the identification of the locus of cutaneous pain. Kenshalo and colleagues (Kenshalo, D. R. et al., 1988; Kenshalo, D. R. and Isensee, O., 1983) showed that SI nociceptive neurons have discrete receptive fields, so that different neurons respond to painful stimulation in different skin areas. Correspondingly, EEG, PET, and fMRI studies have shown a topographic organization of nociceptive responses in SI cortex similar to the organization of tactile responses, i.e., a medio-lateral organization of foot, hand, face, and intra-abdominal areas (Tarkka, I. M. and Treede, R. D., 1993; Andersson, J. L. R. et al., 1997; DaSilva, A. F. M. et al., 2002; Strigo, I. A. et al., 2003; Vogel, H. et al., 2003). Most imaging studies find little somatotopic organization of pain in other cortical areas (Tarkka, I. M. and Treede, R. D., 1993; Xu, X. P. et al., 1997), thus suggesting that responses in SI cortex may be most important for pain localization. More recently, a somatotopic organization has also been documented for operculo-insular cortex (Vogel, H. et al., 2003; Baumgärtner, U. et al., 2006a). A left hemisphere dominance has been reported for the sensory dimension of pain (Schlereth, T. et al., 2003), whereas right hemisphere dominance was observed for the affective dimension (Pauli, P. et al., 1999; Brooks, J. C. W. et al., 2002).

Strigo I. A. and colleagues (2003) directly compared brain activations produced by esophageal distension and cutaneous heat on the chest that were matched for pain intensity. They found that the two qualitatively different pains produced different primary loci of activation with insula, SI, motor and prefrontal cortices. Such local differences in responses within the nociceptive network might subserve our ability to distinguish visceral and cutaneous pain as well as the differential emotional, autonomic, and motor responses associated with these different sensations.

45.4.2   The Time Domain

Most information about the temporal sequence of pain-evoked brain activation comes from EEG or MEG studies. The dual pain sensation elicited by a single brief painful stimulus that is due to the different conduction times in nociceptive A- and C-fibers (about 1 s difference) is reflected in two sequential brain activations in EEG and MEG recordings from SI, SII, and MCC (Bromm, B. et al., 1983; Bragard, D. et al., 1996; Magerl, W. et al., 1999; Opsommer, E. et al., 2001; Iannetti, G. D. et al., 2003). EEG mapping studies (Kunde, V. and Treede, R. D., 1993; Miyazaki, M. et al., 1994), source analysis (Tarkka, I. M. and Treede, R. D., 1993; Valeriani, M. et al., 1996; Ploner, M. et al., 1999), and intracranial recordings (Lenz, F. A. et al., 1998a; Frot, M. et al., 1999) show that the earliest pain-induced brain activity originates in the vicinity of SII. In contrast, tactile stimuli activate this region only after processing in the primary somatosensory cortex (Ploner, M. et al., 2000). The adjacent dorsal insula is activated slightly but significantly later than the operculum (Frot, M. and Mauguière, F., 2003). These observations support the suggestion derived from anatomical studies that the SII region and adjacent insula are primary receiving areas for nociceptive input to the brain (Apkarian, A. V. and Shi, T., 1994; Craig, A. D., 2002).

45.4.3   Attention and Distraction Effects on Pain-Evoked Cortical Activity

Early human brain imaging studies examining the effects of attention and distraction show modulation of pain-evoked activity in a number of cortical regions, including sensory and limbic structures, as well as prefrontal areas (Bushnell, M. C. et al., 1999; Longe, S. E. et al., 2001; Bantick, S. J. et al., 2002; Schlereth, T. et al., 2003). These results generally show reduced activations in sensory regions of the cortex and some increased activity in more frontal regions, suggesting that attentional modulation is mediated through the latter structures resulting in reduced sensory processing, where the attentional distraction is usually reported resulting in reduced perceived magnitude of pain. A more recent study extends these notions by showing that during distraction there is a functional interaction between pregenual ACC and frontal cortex exerting a top-down modulation on periaqueductal gray (PAG) and thalamus to in turn reduce activity in cortical sensory regions and correspondingly decrease perception of pain (Petrovic, P. et al., 2000; Tracey, I. et al., 2002; Valet, M. et al., 2004). Given that ACC is implicated in attentional modulation as well as pain perception, a distraction study indicates that some portions of the pregenual ACC region are decreased with distraction while others are increased, consistent with these two different functions (Frankenstein, U. N. *et al.*, 2001).

45.4.4 Anticipation and Expectation

Anticipation or expectation of pain can activate many of the cortical areas related to perception of pain in the absence of a physical pain stimulus (Ploghaus, A. *et al.*, 1999; Hsieh, J. C. *et al.*, 1999b; Sawamoto, N. *et al.*, 2000; Porro, C. A. *et al.*, 2002). Two studies have attempted to identify the circuitry for modulation of pain by expectation. In one study MCC, caudate nucleus, cerebellum, and nucleus cuneiformis were modulated by systematic manipulation of pain intensity expectation by two different cues (Keltner, J. R. *et al.*, 2006), whereas pain intensity itself modulated somatosensory cortex, insula, and rostral ACC. In the second study expectancy was modulated by a placebo procedure, resulting mainly in modulation including MCC, PRC, cerebellum, pons, and parahippocampal gyrus (Kong, J. *et al.*, 2006). The latter study is complicated by the fact that the procedure is a combination of manipulation of expectancy and placebo acupuncture treatment. Generally, there remains a strong need for systematic studies to identify brain elements that modulate pain responses due to expectation.

45.4.5 Empathy

A provocative study opened the field regarding the interaction between pain and empathy, where the authors defined empathy as the ability to have an experience of another's pain. Using this definition and comparing brain activity for experiencing pain or knowing that their loved one, present in the same room, was experiencing the same pain, the authors showed many cortical regions similarly activated for both conditions, especially bilateral operculo-insular cortex and MCC (Singer, T. *et al.*, 2004). These results were interpreted as evidence for the affective component of pain being active in both empathy and pain, and thus concluded that empathy for pain involves the affective component, but not the sensory component, of pain. The study induced a flurry of activity in attempting to understand the relationship between empathy and pain. Multiple groups have replicated the main finding and proposed different underlying mechanisms (Morrison, I. *et al.*, 2004; Botvinick, M. *et al.*, 2005; Jackson, P. L. *et al.*, 2005), with multiple studies showing that at least MCC

activity reflects the pain experienced by others and that multiple cortical areas involved in sensory processing of pain are also activated.

The overall notion that empathy involves assessment of the pain experienced by others – pain mirroring – was tested directly in subjects with alexithymia, a cognitive and emotional deficit leading to difficulty in identifying one's own emotional state and also other people's emotional state. The study showed in fact reduced activity in PRC and MCC during a pain empathy condition in this patient population (Moriguchi, Y. *et al.*, 2006). Even though these results are internally consistent, their interpretation remains problematic. Simple introspection casts doubt on the notion that empathy means actually experiencing another person's pain. Instead, what is called empathy may be the assessment of the magnitude of negative emotion that the other person may be experiencing, i.e., a cognitive function of interpersonal communication. According to that concept, empathy may be defined as a complex form of psychological inference that enables us to understand the personal experience of another person through cognitive/evaluative and affective processes. A study in patients with congenital insensitivity to pain (Danziger, N. *et al.*, 2006) reported a deficit in rating pain-inducing events, but normal inference of pain from facial expressions (empathy), indicating that empathy for pain does not require an intact pain percept.

45.5 Pain Modulation

45.5.1 Psychological Modulation of Pain

The psychological modulation of pain has been observed very early on and studied in the clinical and laboratory settings (Beydoun, A. *et al.*, 1993; Villemure, C. and Bushnell, M. C., 2002). Modern brain imaging techniques now provide powerful tools with which mechanisms of these modulations can be documented and dissected. Given that these are cognitive/attentional modulations their effects should be observed at the cortical level.

45.5.1.1 Hypnosis and pain-evoked cortical activity

Hypnosis can alter pain perception. It has been used to differentially modulate sensory and affective dimensions of pain and thus distinguish the cortical regions involved in these dimensions. Such studies indicate that SI activity is preferentially modulated when the hypnotic instructions are directed to the intensity of pain, while MCC activity is preferentially modulated when hypnosis is directed to the unpleasantness of pain (Rainville, P. *et al.*, 1997, Hofbauer, R. K. *et al.*, 2001). Brain activity for hypnotically induced pain perception seem to be different from activity for imagined pain in sensory, limbic, and prefrontal activation patterns (Derbyshire, S. W. *et al.*, 2004). The sensory and limbic cortical activations for hypnotically induced and stimulation-induced pain seem relatively similar, the only region that may be differentiating them seems to be the medial PRC (Raij, T. T. *et al.*, 2005).

45.5.1.2 Mood and emotional states and pain-evoked cortical activity

Studies show that experimental procedures that improve mood generally reduce pain, while those that have a negative effect on mood increase pain (Zelman, D. C. *et al.*, 1991; Marchand, S. and Arsenault, P., 2002). One study showed that looking at fearful faces increased their level of anxiety and discomfort, which also resulted in enhanced esophageal stimulation-evoked activity in limbic regions like ACC and insula (Phillips, M. L. *et al.*, 2003).

45.5.1.3 Placebo and pain-evoked cortical activity

Placebo is a potent modulator of pain; it afflicts all clinical studies of pain pharmacology. Placebo effects have also been seen in depression and in Parkinson's disease and recent brain imaging studies show a robust brain and subcortical reward circuitry's involvement in these (Lidstone, S. C. and Stoessl, A. J., 2007). The first neurochemical evidence for opiate involvement of placebo was demonstrated about 30 years ago by showing that placebo analgesia can be blocked by naloxone (Levine, J. D. *et al.*, 1978). Consistent with this notion, changes in endogenous opiate release are shown to be involved in placebo-induced analgesia, where PRC (medial and lateral) as well as insula and ventral striatum seem to be involved, where high placebo responders increased opiate release in ventral striatum was positively correlated with pain ratings (Zubieta, J. K. *et al.*, 2005). Results generally consistent with this brain response pattern have been demonstrated by a number of other groups (Wager, T. D. *et al.*, 2004; Benedetti, F. *et al.*, 2005; Kong, J. *et al.*, 2006); the medial prefrontal/rostral ACC responses for placebo seem to recruit PAG and amygdala (Bingel, U. *et al.*, 2006); and involvement of PAG in placebo-induced analgesia

is observed in the above studies as well, which links opiate descending modulation with prefrontal cortical control of placebo analgesia. The correspondence between placebo analgesia and reward was directly studied and the results show a strong correspondence between brain regions involved in each (Petrovic, P. *et al.*, 2005).

45.5.2 Pharmacological Modulation of Pain

A league table of analgesic efficacy has been generated based on pain-related evoked potentials (Scharein, E. and Bromm, B., 1998). Since these studies used electrical stimuli that circumvent peripheral nociceptive transduction mechanisms, this table reflects central rather than peripheral analgesic actions, as evidenced, e.g., by the higher efficacy of the antidepressant imipramine than the nonsteroidal anti-inflammatory drug (NSAID) acetylsalicylic acid. Since dipole source analysis has not been applied in these EEG studies, possible cortical sites of actions were not differentiated. Combining fMRI and pharmacology promises to provide that type of information (Tracey, I., 2001; Borsook, D. *et al.*, 2006). In addition, PET techniques can be used for direct tracing of cortical distribution of a given drug, when it has been labeled with the positron emitting ^{11}C isotope.

45.5.2.1 Opiates

There is a vast literature regarding opiate-mediated descending modulation through the PAG and a similarly large literature on its effects on inhibitory interneurons in the spinal cord. At the cortical level, it has been noted that opiate receptors are present in many parts of the nociceptive system, with high specific binding in ACC, insula, and frontal operculum, and with moderate specific binding in MCC, SII, and SI (Jones, A. K. P. *et al.*, 1991b; Baumgärtner, U. *et al.*, 2006a).

Recent studies of opiate-mediated responses in the brain have used two approaches, examination of metabolic function in response to pharmacological agents and direct measurement of opiate receptor binding potential. Exogenous administration of μ-opioid receptor agonist drugs show dose-dependent increased metabolic activity in regions rich with μ-opioid receptors, which in the cortex are mainly localized to PRC and ACC (Firestone, L. L. *et al.*, 1996; Schlaepfer, T. E. *et al.*, 1998; Wagner, K. J. *et al.*, 2001). Also, μ-opioid agonist fentanyl on brain responses to painful stimuli have been explored, showing that most cortical responses to pain are reduced or eliminated, confirming analgesic effects of the opiate (Casey, K. L. *et al.*, 2000; Petrovic, P. *et al.*, 2002). Changes in endogenous opioid system is studied using a selective μ-opioid radiotracer, showing activation of opiate neurotransmission in rostral ACC, PRC, and insula during a tonic muscle pain (Zubieta, J. K. *et al.*, 2001).

45.5.2.2   Dopamine

Dopamine is best known for its role in motor, motivation, and pleasure control. There is accumulating evidence to suggest that dopamine acting at the level of the basal ganglia may also be involved in pain modulation. Human brain imaging studies document increased pain sensitivity to be associated with lower levels of endogenous dopamine (Pertovaara, A. *et al.*, 2004; Martikainen, I. K. *et al.*, 2005; Scott, D. J. *et al.*, 2006); and sustained experimental pain results in release of dopamine in the basal ganglia (Scott, D. J. *et al.*, 2006), and indicate an interaction between opiate activity and dopamine where alfentanil administration results in decreased mechanical pain and decreased release of dopamine in the basal ganglia (Hagelberg, N. *et al.*, 2002). Moreover, abnormal levels of dopamine in the basal ganglia have been associated with chronic pain in burning mouth syndrome and atypical facial pain (Jaaskelainen, S. K. *et al.*, 2001; Hagelberg, N. *et al.*, 2003a; 2003b), and perhaps in fibromyalgia (Wood, P. B. *et al.*, 2007). Patients with restless legs syndrome display a pronounced mechanical hyperalgesia to pinprick stimuli that is slowly reversed by dopaminergic agonists (Stiasny-Kolster, K. *et al.*, 2004), but this action is probably mediated by extrastriatal dopamine receptors.

45.5.2.3   Estrogen

Gender is one of the most important determinants of human health. Women far outnumber men in susceptibility to many autoimmune disorders, fibromyalgia, and chronic pain, differences in physiological responses to stress may potentially be an important risk factor for these disorders as physiologic responses to stress seem to differ according to gender, with phase of menstrual cycle, menopausal status and with pregnancy (Kajantie, E. and Phillips, D. I., 2006). Consistent with this idea recent fMRI study shows that brain activity in premenopausal women as studied for negative valence/high arousal in contrast to neutral visual stimuli show differences when the task is performed during early

follicular menstrual cycle phase compared to late follicular/mid-cycle; with greater activity found during early follicular phase in amygdala, hypothalamus, hippocampus, orbital frontal cortex, and ACC, suggesting that estrogen may attenuate arousal in women through cortical-subcortical control of hypothalamic–pituitary–adrenal circuitry (Goldstein, J. M. *et al.*, 2005). There is also growing evidence of gender differences in the anatomy of the brain, its connectivity, and in cognitive abilities (Hampson, E., 2002; Becker, J. B. *et al.*, 2005). Multiple studies have documented that threshold and tolerance for pain is lower for women (Wiesenfeld-Hallin, Z., 2005; Rolke, R. *et al.*, 2006; Wilson, J. F., 2006).

Gender differences in cortical activity for acute pain has been observed in early studies (Paulson, P. E. *et al.*, 1998). The association of sex hormones with pain perception and pain memory was studied by Zubieta J. K. and colleagues (Zubieta, J. K. *et al.*, 2002; Smith, Y. R. *et al.*, 2006). They scanned healthy women during their early follicular phase when estrogen levels are low and then repeated the scan during that same phase in another month after they had worn for a week an estrogen-releasing skin patch which increased their estrogen to levels normally seen in the menstrual cycle. These studies showed that more μ-opioid receptors were available in the presence of high estrogen levels, and women reported less pain in response to acute painful stimuli than when their estrogen levels were low. Moreover, estrogen-associated variations in the activity of μ-opioid neurotransmission correlated with individual ratings of the sensory and affective perceptions of pain and the subsequent recall of that experience. These data demonstrate a significant role of estrogen in modulating endogenous opioid neurotransmission and associated psychophysical responses to an acute pain stressor in humans. Approximately similar results have been reported by another group (De Leeuw, R. *et al.*, 2006).

45.6   Overview Regarding the Role of the Cortex in Acute Pain Perception

The above sections describe the contribution of modern imaging studies to our understanding of the involvement of the cortex in pain perception. Cortical activity is demonstrated to possess properties necessary for involvement in pain perception, like somatotopic representation of painful stimuli, correlation with stimulus intensity, modulation with attention, modulation with expectation and other psychological variables, and distinct brain regions showing differential activity for sensory and affective dimensions of pain, as well as attenuation of responses with analgesic drugs (Apkarian, A. V. *et al.*, 2005). Thus, human brain imaging studies have asserted the role of the cortex in acute pain.

However, because imaging studies identify brain responses in a correlative manner, they may all reflect secondary processes. Perception of pain automatically directs attention to the source of pain, results in autonomic responses, motor reflexes to escape from the pain, and other emotional and cognitive responses that undoubtedly are at least partially mediated through cortical processes. Therefore, the role of the cortex in pain perception in contrast to its activity as a consequence of these secondary responses remains unclear and needs to be properly addressed in future studies (Apkarian, A. V., 2004).

In fact, unpublished data from Apkarian's laboratory suggest that a large proportion of the brain network activated with acute pain may be responses that are commonly involved in general magnitude estimation for any sensory modality, and as a result are not specific for nociception (abstract Society for Neuroscience 2006), suggesting that the majority of cortical activity for acute pain are instead sensory, cognitive, emotional, and attentional responses to nociceptive inputs. Careful clinical and neuropsychological examination of patients with small brain lesions, combined with high-resolution structural and neuropharmacological neuroimaging in the same patients, will be needed to address the question what brain structures are necessary for acute pain perception. Anatomical tracing studies and single unit recordings should address the question, to what extent nociceptive specific neurons exist in these brain structures. For most parts of the nociceptive cortical network, as illustrated above, it is likely that they participate only partly in pain perception, by providing certain feature extraction functions, but they also participate in other functions in different contexts.

45.7   Clinical Applications

It should be emphasized that although the subjective phenomenon of being in pain can be considered an emergent phenomenon of cortical activity (Treede, R. D., 2001), there is currently no measure of brain activity that would objectively show whether or not a person is in pain. Therefore, neither EEG/MEG nor imaging with fMRI or PET can be used to verify the presence of ongoing spontaneous pain.

EEG and MEG recordings of evoked potentials, however, are sensitive enough to verify whether the ascending nociceptive pathways are intact in a given individual patient (Bromm, B. and Lorenz, J., 1998; Treede, R. D. *et al.*, 2003; Cruccu, G. *et al.*, 2004). A prerequisite for this use of EEG and MEG technology is a phasic adequate stimulus for nociceptor activation. Radiant heat pulses of a few milliseconds duration, as generated by infrared lasers, have been validated for this purpose (Plaghki, L. and Mouraux, A., 2003), and LEPs can thus be used to verify the presence of negative sensory signs of nociception (hypoalgesia).

Neither fMRI nor PET are sensitive enough to allow clinical assessment of nociceptive pathways in individual cases, since so far no activation paradigm has been developed that would reliably induce a particular cortical activation pattern in each and every healthy subject. Thus, negative findings with these techniques are inconclusive.

For the study of pathological nociceptive processing at the group level, however, fMRI and PET techniques are extremely powerful. These techniques have broadened our understanding of the pathophysiology of conditions with decreased pain perception such as afferent pathway lesions or borderline personality disorder, and conditions with increased pain perception such as neuropathic pain or fibromyalgia (Gracely, R. H. *et al.*, 2004; Maihöfner, C. *et al.*, 2005; Garcia-Larrea, L. *et al.*, 2006; Schmahl, C. *et al.*, 2006; Schweinhardt, P. *et al.*, 2006). In addition, PET allows direct estimation of pharmacological and biochemical processes in the brain, such as alterations in dopamine or opioid receptor availability (Hagelberg, N. *et al.*, 2003a; Willoch, F. *et al.*, 2004).

45.8   Chronic Pain

45.8.1   Studying Brain Activity in Chronic Pain with Nonspecific Painful Stimuli

Chronic pain might result from cortical processing of chronic nociceptive spinothalamic input according to the same mechanisms as in acute pain, or there might be specific changes in cortical processing of nociceptive input in patients with chronic pain. Such changes could then either be a causal factor for or a consequence of the chronicity of the pain condition.

A recent meta-analysis in fact shows that across some 100 studies one can establish statistically significant differences in incidence of different brain areas activated by experimental painful stimuli between acute and chronic pain conditions: PRC shows a stronger activation in chronic pain patients, whereas other nociceptive cortical areas and the thalamus show a weaker response (Table 2). A simple interpretation of these findings would be that nociceptive signal processing for experimental painful stimuli in chronic pain patients involves a reduced sensory discriminative component and an increased affective-motivational or cognitive-evaluational component. That interpretation would also be consistent with the stronger affective component of clinical pain as compared to experimental pain (Chen, A. C. N. and Treede, R. D., 1985). But there are further implications: Is the result a consequence of some trivial confounds or does it signify changes in the physiology of pain? One could construct a long list of confounds that may underlie the observation, from attentional shifts, to coping mechanisms, to effects of drug use, and heightened anxiety and depression.

The standard approach for studying brain activity for acute pain is to induce pain by a mechanical or thermal stimulus and determine brain regions modulated with the stimulus period and even with the various intensities used. Therefore, it is natural to carry the same technology to the clinical arena and apply it to chronic pain patients. As an example, we discuss one study which attempted to identify brain activity in complex regional pain syndrome (CRPS) patients using fMRI (Apkarian, A. V. et al., 2001a; 2001b).

The design of the study was to examine brain activity for thermal stimuli applied to the body part where CRPS pain was present, and compare brain responses to this stimulus between CRPS and healthy subjects. Moreover, as the pain in CRPS patients with sympathetically maintained pain (SMP) may be modulated by a sympathetic block, it was reasoned that one could decrease the patients' ongoing pain and then re-examine brain activity responses to the same stimulus. The study was done in a small group of patients and this by itself is an important weakness. The main observation was that thermal stimuli in CRPS evoked more prefrontal cortical activity than usually seen in healthy subjects, and this was reversed (became more similar in pattern to normal subjects' brain activity to thermal stimuli) following sympathetic blocks. The introduction of sympathetic blocks necessitated the use of the same procedure in healthy subjects as well, where its effects were minimal. The study also observed that when a placebo block resulted in decreased pain perception then the cortical response pattern changed similarly to that of effective blocks. These results show that brain activity may be distinct between CRPS and healthy subjects for thermal stimuli.

45.8.2 Clinical Pain Conditions Studied by Stimulation and the Role of the Cortex

A direct approach to studying clinical pain states is to provoke it and examine brain activity. This is doable by drugs in headaches and in cardiac pain. As a result there is growing literature in both fields. There is also now good evidence that migraine with aura is accompanied with decreased blood flow and decreased activity in the occipital cortex, and migraine with or

Table 2 Frequency of brain areas active during pain in normal subjects as compared to patients with clinical pain conditions

	ACC	*SI*	*SII*	*IC*	*Th*	*PFC*
Pain in normal subjects in 68 studies	47/54 (87%)	39/52 (75%)	38/51 (75%)	45/48 (94%)	28/35 (80%)	23/42 (55%)
Clinical pain conditions in 30 studies	13/29 (45%)	7/25 (28%)	5/25 (20%)	15/26 (58%)	16/27 (59%)	21/26 (81%)
Comparison between pain in normal subjects and in clinical conditions	$P < 0.001$	$P < 0.001$	$P < 0.001$	$P < 0.001$	$P = 0.095$	$P = 0.038$

Incidence values are based on positron emission tomography, single photon emission-computed tomography, and functional magnetic resonance imaging studies. For details see Apkarian et al., 2005. P values are based on Fisher's exact statistics contrasting incidence for each area. ACC, anterior cingulate cortex; IC, insular cortex; PFC, prefrontal cortex; SI, primary somatosensory cortex; SII, secondary somatosensory cortex; Th, thalamus.

without aura is associated with increased cortical thickness in visual cortical regions involved in motion detection (Granziera, C. *et al.*, 2006).

45.8.2.1 Migraine

Migraine attacks are characterized by unilateral severe headache often accompanied by nausea, phonophobia and photophobia. Activation of the trigeminovascular system is thought to be responsible for the pain itself, and cortical spreading depression (CSD) seems to underlie the aura symptoms. This view has been greatly advanced and substantiated by brain imaging studies. fMRI studies show CSD-typical cerebrovascular changes in the cortex of migraineurs while experiencing a visual aura (Hadjikhani, N. *et al.*, 2001). The subsequent decrease in fMRI signal is temporally correlated with the scotoma that follows the scintillations. These fMRI signal changes develop first in the extrastriate cortex, contralateral to the visual changes. It then slowly migrated towards more anterior regions of the visual cortex, representing peripheral visual fields, in agreement with the progressive movement of the scintillations and scotoma from the centre of vision towards the periphery. A recent study that analyzed visually triggered attacks showed hyperemia in the occipital cortex, independently of whether the headache was preceded by visual symptoms (Cao, Y. *et al.*, 1999). An alternative view considers migraine aura and headache as parallel rather than sequential processes, and proposes that the primary cause of migraine headache is an episodic dysfunction in brainstem nuclei that are involved in the central control of nociception (Goadsby, P. J. *et al.*, 2002).

45.8.2.2 Cluster headache

The pathophysiology of cluster headache is thought to involve multiple brain regions. Brain imaging studies imply that the associated excruciatingly severe unilateral pain is likely mediated by activation of the first (ophthalmic) division of the trigeminal nerve, while the autonomic symptoms are due to activation of the cranial parasympathetic out-flow from the VIIth cranial nerve. The circadian rhythmicity of cluster headache has led to the concept of a central origin for its initiation (Strittmatter, M. *et al.*, 1996).

Using PET in cluster headache patients, significant activations ascribable to the acute cluster headache were observed in the ipsilateral hypothalamic gray matter and in multiple cortical areas including cingulate and PRC. When compared to

the headache-free state only hypothalamic activity was distinct (May, A. *et al.*, 2000). This highly significant activation was not seen in cluster headache patients out of the bout when compared to the patients experiencing an acute cluster headache attack. In contrast to migraine, no brainstem activation was found during the acute attack compared to the resting state. Newer MRS results further substantiate this idea by showing reduced metabolites within the hypothalamus of cluster headache patients in contrast to healthy or migraine headache controls (Wang, S. J. *et al.*, 2006). These data suggest that while primary headaches such as migraine and cluster headache may share a common pain pathway, the trigeminovascular innervation, and activate similar cortical regions, the underlying pathogenesis may be quite different.

45.8.2.3 Cardiac pain

Cardiac pain and its variants have been studied by brain imaging using various drugs that bring about these symptoms (Rosen, S. D. *et al.*, 1996; 2002). In patients with myocardial ischemia the perception of angina is associated with activity in the hypothalamus, PAG, thalami, rostral ACC, and bilateral PRC. In patients with silent myocardial ischemia it seems that the silence is not due to impaired afferent signaling, but rather it is associated with a failure of transmission of signals from the thalamus to the frontal cortex. In contrast, in patients with syndrome X, a condition of chest pain with ischemiclike stress electrocardiography but entirely normal coronary angiogram, activity in the right anterior insula distinguished these patients from patients with known coronary disease. These patients appear to have a deficit in central pain habituation (Valeriani, M. *et al.*, 2005). Overall, these studies imply that difference between different cardiac pain conditions are due to central processing, e.g., syndrome X is interpreted as a cortical pain syndrome, a top-down process, in contrast with the bottom-up generation of a pain percept caused by myocardial ischemia in coronary artery disease.

45.8.2.4 Irritable bowel syndrome

Irritable bowel syndrome (IBS) is a disorder of abdominal pain or discomfort associated with bowel dysfunction. Hypersensitivity to visceral, but not somatic, stimuli has been demonstrated in IBS. A number of groups have examined brain activity in this condition mainly by monitoring responses to painful and nonpainful rectal distensions, as well as responses to the anticipation of painful distensions. Two studies are interesting since both show in normal subjects a significant positive correlation between cingulate cortex activity and subjective rating of rectal distension pain, and in both studies this relationship completely disappears in IBS patients (Silverman, D. H. *et al.*, 1997; Mertz, H. *et al.*, 2000). IBS is more prevalent in women than in men. Brain imaging studies have now shown gender differences in brain activity in IBS (Berman, S. M. *et al.*, 2000; Naliboff *et al.*, 2001; Berman, S. M. *et al.*, 2002b; Nakai, A. *et al.*, 2003). There are large differences between the studies, making their synthesis difficult (see Ringel, Y., 2006).

More recent studies show a hint for sensitization in IBS patients because subliminal and supraliminal distensions of rectal distension seem to indicate small differences between IBS and healthy controls in the total cortical volume activated or in regional activity as a function of distention volume (Andresen, V. *et al.*, 2005; Lawal, A. *et al.*, 2006). A study of IBS in contrast to healthy subjects examined thermal and visceral hyperalgesia and related brain activity (Verne, G. N. *et al.*, 2003). This seems the only study where besides pain intensity and unpleasantness measures, the authors also document fear and anxiety and show that all are rated higher by IBS for both heat and rectal distention, and not surprisingly these increased sensations and emotions give rise to larger cortical activations in IBS. The latter is most likely a reflection of a perceptual magnitude mismatch between the groups and says little as to the IBS cortical activity abnormalities. Such mismatches at least for fear and anxiety most likely are common in the majority of studies of IBS. One assumes that the simple introduction of a rectal balloon in IBS would result in increased anxiety, which undoubtedly effects cortical activity to visceral and somatic pain, yet its specific contribution has remained unexplored.

In a more elegant study the authors use perception-related ratings during rectal distention to evoke either urge to defecate or pain, and compared brain activity related to the ratings between IBS patients and healthy subjects. (Kwan, C. L. *et al.*, 2005). The approach is similar to the technique used in mapping brain activity for spontaneous pain in chronic back pain (CBP) and in postherpetic neuralgia (PHN; Baliki, M. N. *et al.*, 2006; Geha, P. Y. *et al.*, 2007). The results show large differences between the two groups contrasted, with far more extensive brain activations in the healthy subjects. The results are complicated by the fact that the authors do not take into consideration the influence of spontaneous pain. Still, this is perhaps one of the best-controlled IBS studies, and indicates distinct cortical areas involved in the urge and pain perceptions in each group.

There is now evidence that serotonin (5-HT) may be involved in IBS. One study (Nakai, A. *et al.*, 2003) examined serotonin synthesis in the brain and indicated greater brain regional serotonin synthesis in female IBS. There is also evidence that alosetron, a 5-HT_3 receptor antagonist, is clinically effective in treating some subtypes of IBS. Berman S. M. *et al.* (2002a) examined brain activity in a large population of IBS, before and after a randomized, placebo-controlled, 3-week use of alosetron. Treatment improved IBS symptoms and regional cerebral blood flow in brain regions rich with 5-HT_3 receptor and involved in emotional and aversive functions: amygdala, ventral striatum, and dorsal pons, implying that the therapeutic effects are due to central actions and not peripheral. Thus, generally the IBS studies show that brain responses are different to rectal stimuli in patients, and that these central events may be critical to IBS.

45.8.3 Spontaneous Pain as a Confound in Assessing Brain Activity

A person who has lived for years in the presence of pain, must have developed some coping mechanisms that aid in pursuing other everyday life interests in spite of the presence of the pain. How does this impact the brain? Can one consider the patient in chronic pain as composed of a brain-signaling pain together with a brain undertaking other tasks as in healthy subjects? Or, does the presence of ongoing pain interact and impact other processes as well? Certainly our cognitive and anatomic studies suggest that the latter is more likely. We have now direct evidence of the modulation that ongoing pain imposes on brain activity in general.

A recent study reported brain activity for spontaneous pain in PHN patients before and after topical lidocaine treatment (Geha, P. Y. *et al.*, 2007). The PHN patients were imaged before, after 6-hours and 2-weeks treatment with lidocaine. Behaviorally and based on questionnaires most participants showed a modest but significant decrease in their ongoing pain. The patients were scanned while they were either rating their ongoing pain or rating a visual bar that varied in time in a pattern that mimicked their ratings of pain (Figure 8). Thus, the

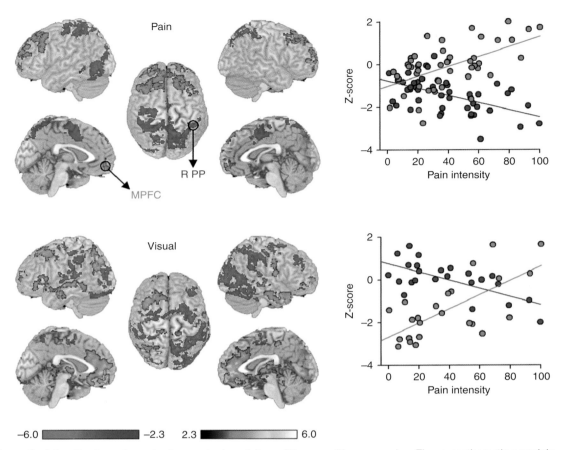

Figure 8   Intensity of ongoing pain changes brain activity and thus cognitive processing. Eleven postherpetic neuralgia patients were studied by functional magnetic resonance imaging (fMRI) once before and twice after lidocaine application on the painful skin. In all sessions, patients performed two different tasks: in the Pain task they continuously rated the fluctuations of their spontaneous pain, and brain activity related to this was identified using methods. In the Visual task they rated fluctuations of a bar varying in time, brain activity was determined with the same approach as for the pain task. The relationship between brain activity and intensity of ongoing pain was determined by using a covariate analysis, where for each fMRI scan its related pain intensity was used to determine the effect of this parameter on brain responses. Across-subject and across all scans average variation of brain activity is displayed for both tasks in the left. Red are brain regions that are positively correlated and blue regions that are negatively correlated with intensity of ongoing pain (normalized to z-values). The right scatter-grams show this effect for two brain regions (right posterior parietal cortex, R PP, x = 33 y = −45 z = 50; and medial prefrontal cortex, MPFC, x = 9 y = 50 z = −40, as respectively circled). Each dot represents a single patient's activity at a single time. Top scatter-gram is for Pain task; bottom for Visual task, red symbols and regression line are for MPFC; blue for R PP. MPFC exhibited significant positive correlations with pain intensity for pain ($r = 0.49$, $P < 0.05$) and visual ($r = 58$, $P < 0.01$) task, while R PP showed negative correlation for pain ($r = -0.48$, $P < 0.05$) and visual ($r = -0.64$, $P < 0.01$) task. Brain areas that show increased correlation with ongoing pain are interpreted as a functional compensation for the decreased attentional resources. Reproduced from Geha, P. Y., Baliki, M. N., Chialvo, D. R., Harden, R. N., Paice, J. A., and Apkarian, A. V. 2007. Brain activity for spontaneous pain of postherpetic neuralgia and its modulation by lidocaine patch therapy. Pain 128, 88–100.

latter is a control task that captures motor and cognitive parts of the task but, of course, it does not reflect the pain. Brain activity for both tasks was increasing from first to third session. This observation is similar to earlier reports that decrease in clinical pain in many cases results in increased brain activity. In this case, however, the internal control was also changing in a manner parallel to

the pain condition, hinting that the effects of decreased pain was modulating more than just pain-related circuitry.

To identify the role of spontaneous pain on brain activity in general, a correlation analysis was done for both tasks with mean spontaneous pain. Figure 8 shows the influence of pain intensity on across-sessions averaged brain activity for both tasks.

The resultant map is generally similar for both tasks: activity in medial and lateral prefrontal regions was positively correlated, while posterior parietal attentional areas were negatively correlated with mean pain intensity. This result shows that brain activity for both tasks is influenced by the level of spontaneous pain, implying that pain intensity influences task performance in general. This is in line with previous studies showing that ongoing pain may interfere with cognitive functions (Lorenz, J. et al., 1997).

This result reinforces the need for correcting brain activity by a control condition performed at the same pain level that is the necessity of subtracting the visual task from spontaneous pain rating task, at each treatment session. For both tasks, the fact that posterior parietal cortical activity was negatively correlated with mean ongoing pain suggests that the attentional abilities of patients are directly related to the intensity of their pain, which would in turn impact their abilities in performing anything that would demand concentration.

Moreover, multiple prefrontal regions were positively correlated to the mean pain, suggesting that the patients' brain regions underlying higher cognitive functions become more active as the pain intensity increases. The exact cognitive implications for these brain activity patterns remain unclear. In contrast, the finding indicates that the intensity of spontaneous pain impacts brain activity for any task that the subject attempts to perform, enhancing some aspects and inhibiting others. Therefore, the decreased brain activity reported for pain tasks in many clinical pain conditions (Peyron, R. et al., 2000; Derbyshire, S. W., 2003; Apkarian, A. V. et al., 2005; Kupers, R. and Kehlet, H., 2006) is most likely a reflection of the presence of the spontaneous pain, and is not specific to the task being investigated.

The fact that pain intensity seems to modulate brain activity in general has another powerful consequence. It suggests that simply studying brain activity, in tasks unrelated to pain, one should be able to identify the presence of pain and study its effects on sensory/cognitive/motor/attentional processing, an exciting prospect that remains to be pursued.

45.8.4 Functional Magnetic Resonance Imaging of Spontaneous Pain

Spontaneous pain is highly prevalent in clinical pain conditions, and is usually the primary drive for patients seeking medical care. Thus, understanding its related brain circuitry is both scientifically and therapeutically imperative. Cortical responses to standard mechanical or thermal stimulation are of limited value for understanding these clinical pain conditions. Spontaneous pain fluctuates unpredictably in the time scale of seconds to minutes, and these fluctuations have characteristic properties that differentiate between different chronic pain conditions such as PHN and CBP (Foss, J. M. et al., 2006). This variability (specific fractal dimension) can also be observed in fMRI signals when such patients rate their spontaneous pain. Therefore, this technique was applied to study brain activity in CBP (Baliki, M. N. et al., 2006) and PHN patients (Geha, P. Y. et al., 2007) in relation to their subjective report of fluctuations of spontaneous pain.

The combination of relating brain activity to spontaneous pain and correcting for confounds by subtracting brain activity for visual bar lengths, provides a robust approach with which clinical pain may be studied directly. Note that in this case the brain activity is related to exactly the event that the patient complains about. With this approach, in CBP patients (Baliki, M. N. et al., 2006) it was shown that the brain regions activated when the pain was increasing correspond to brain regions seen for acute pain in normal subjects. In contrast, for time periods when the pain was high and sustained, the brain activity was mainly limited to medial PRC, a region usually not activated for acute pain. The resultant brain activity was strongly correlated to the patients' reported pain intensity at the time of the scan, specifically with medial prefrontal activity. Also, the duration or chronicity of the pain was captured in the insular activity, a region activated only during increases in spontaneous pain. Thus, two fundamental properties of CBP its intensity and duration were directly reflected in the brain activity identified in these patients. By applying a thermal painful stimulus in the same patients (as well as in healthy subjects) the same study showed that brain regions reflecting the stimulus intensity were not related to that reflecting the intensity of spontaneous pain. In turn, the brain region that reflected spontaneous pain intensity was only activated for the latter and did not reflect thermal painful stimulus intensity. Therefore, at least in the patient group studied spontaneous pain involved a different brain activity pattern than acute pain.

45.8.5 Neuropathic Pain

Patients with neuropathic pain show decreased responses in the thalamus to experimental painful stimuli (Peyron, R. *et al.*, 2000). A MRS study showed a decrease in the level of *N*-acetyl-aspartate, a neuronal marker, in the thalami of patients with chronic neuropathic pain after spinal cord injury (SCI), when compared to patients with SCI but without pain (Pattany, P. M. *et al.*, 2002). Thus, neurochemical brain imaging provides evidence for the occurrence of long-term changes in the brain chemistry and morphology of chronic neuropathic pain patients. Thalamic activity in neuropathic patients was also reported to increase after pain relief (Hsieh, J. C. *et al.*, 1995), and to be significantly negatively correlated with the duration of the condition in CRPS patients (Fukumoto, M. *et al.*, 1999). Thus, the reduced activation of the thalamus may also be an altered functional state rather than an irreversible degeneration. Neuropathic pain patients in addition show a reduced availability of opioid receptor binding sites (Maarrawi, J. *et al.*, 2007). This reduction was symmetric in peripheral neuropathic pain, suggesting a possible release of endogenous opioids, but lateralized to the hemisphere contralateral to the pain in central pain patients, consistent with a loss of receptors (Jones, A. K. P. *et al.*, 2004, Willoch, F. *et al.*, 2004).

Brain activity differences between healthy subjects and patients in activation paradigms are difficult to interpret since they do not distinguish between brain activity specifically related to the clinical condition and abnormalities in sensory processing secondarily associated with the clinical state. Particularly in neuropathic pain, the accompanying sensory deficit may be reflected in the imaging results and not the pain. Reduced relevance of the acute stimulus to subjects who are already in pain may also account for much of the decreased regional brain activity in neuropathic pain. To overcome such nonspecific brain activity differences one needs to compare brain activity for stimuli where perceptual evaluation has been equated between patients and normal healthy subjects.

Three studies (Hsieh, J. C. *et al.*, 1995; 1999a; Apkarian, A. V. *et al.*, 2001b) have looked at the regions of the brain modulated by relief of chronic neuropathic pain: CRPS, peripheral neuropathy, and trigeminal neuropathy. Two of these studies show that the PRC activity is decreased, and all three studies report decreased rostral ACC activity, after successful pain relief. It is to be noted that in addition to those regions some areas were also less activated with pain relief such as the insula (Hsieh, J. C. *et al.*, 1995) and the anterior limbic thalamus (Hsieh, J. C. *et al.*, 1999a), whereas others were more activated

after pain relief like the medial PRC (Hsieh, J. C. *et al.*, 1999a). This heterogeneity is not surprising because pattern of brain activity may be specific to each neuropathic pain condition.

45.8.6   Low Back Pain and Fibromyalgia

As mentioned above, brain activity of healthy subjects and patients with increased pain sensitivity should be compared in such a way that perceived intensity has been matched across the two groups. A recent study used such a design and showed generally heightened brain activity for painful stimuli of equivalent perceptual intensity both in fibromyalgia and CBP patients as compared to healthy subjects (Gracely, R. H. *et al.*, 2002; Giesecke, T. *et al.*, 2004). Morphometric and neurochemical brain imaging studies provide evidence for the occurrence of long-term changes in the brain chemistry and morphology of chronic pain patients. The level of *N*-acetyl-aspartate, a neuronal marker, was decreased in the medial and lateral PRC of CBP patients compared to an age- and gender-matched control group (Grachev, I. D. *et al.*, 2000). A morphometric study in chronic pain showed also a decrease in gray matter density in the dorsolateral PRC and the thalamus of CBP patients when compared to matched controls (Apkarian, A. V. *et al.*, 2004). Furthermore, these long-term chemical and morphological changes are significantly correlated with different characteristics of pain such as pain duration (Apkarian, A. V. *et al.*, 2004), pain intensity (Pattany, P. M. *et al.*, 2002; Grachev, I. D. *et al.*, 2002; Apkarian, A. V. *et al.*, 2004), and sensory-affective components (Grachev, I. D. *et al.*, 2002). The morphometric and neurochemical studies imply an active role of the central nervous system in chronic pain, suggesting that supraspinal reorganization may be critical for chronic pain.

45.8.7   Overview Regarding the Role of the Cortex in Chronic Pain Perception

In spite of a plethora of data there remains a host of uncertainties about their significance. Overall, the clinical brain imaging studies indicate reduced information transmission through the thalamus to the cortex, and increased activity in PFC, mostly in medial PFC coupled with atrophy in dorsolateral PFC. The number of studies remain very small and hence our confidence as to the reproducibility of these changes remain minimal. Still, the observations regarding cortical and thalamic activity changes in chronic pain are in general consistent with the notion that chronic pain conditions preferentially engage brain areas involved in cognition/emotion and decreases activity in regions involved in sensory evaluation of nociceptive inputs.

Evidence has been presented that brain activity, chemistry, and morphology may be reorganized in chronic pain conditions. Does this evidence imply that there is supraspinal reorganization, above and beyond what is established in the periphery and spinal cord? That is, even if we establish a brain pattern of activity for some chronic pain condition, does this reflect some unique contribution of the brain to this state or is it simply a reflection of lower level reorganization? The answer is not straightforward. However, only by answering such questions will brain imaging be able to provide new information to the myriad mechanisms described for peripheral and spinal cord reorganization in chronic pain.

45.9 Conclusions and Outlook

The study of nociceptive processing in the cerebral cortex has come a long way. In contrast to earlier assumptions, the classical somatosensory cortex areas are not the only ones activated by painful stimuli. In addition, limbic areas such as the anterior and midcingulate cortex and the insula have also been recognized as part of the nociceptive network, and more recently also cognitive areas in the PRC. Limbic areas are usually considered to mediate emotional processes, but they are also involved in autonomic and motor functions. In this way, progress in understanding the cortical nociceptive network mirrors that in understanding the subcortical networks, which also include many connections to autonomic and motor nuclei as well as hypothalamus, cerebellum, and basal ganglia. Images of brain activation by painful stimuli leave the impression that at least half of the brain participates in processing nociceptive information. At other times, many of the same areas participate in visual, motor, emotional, cognitive, or other signal processing. In that sense, our current understanding of the nociceptive network in the brain is consistent with our current understanding of how the brain uses distributed processing for its many functions. It is not clear, however, to what extent any part of the cerebral cortex is specific for nociception. The best candidate region for such a

function lies in the parasylvian cortex, in the vicinity of SII and the dorsal insula. In chronic pain, nociceptive processing in the cerebral cortex is partly preserved and partly altered, in particular with respect to PRC functions. This reorganization may be a neuroplastic response to the chronicity of pain, it may reflect activation of antinociceptive processes, or it may even represent a predisposing factor for the development of chronic pain. The methods available for the study of nociceptive processing in the brain allow to address many of these open questions in the near future, and this part of pain research is bound to remain a very productive one.

References

Andersson, J. L. R., Lilja, A., Hartvig, P., Långström, B., Gordh, T., Handwerker, H., and Torebjörk, E. 1997. Somatotopic organization along the central sulcus, for pain localization in humans, as revealed by positron emission tomography. Exp. Brain Res. 117, 192–199.

Andresen, V., Bach, D. R., Poellinger, A., Tsrouya, C., Stroh, A., Foerschler, A., Georgiewa, P., Zimmer, C., and Monnikes, H. 2005. Brain activation responses to subliminal or supraliminal rectal stimuli and to auditory stimuli in irritable bowel syndrome. Neurogastroenterol. Motil. 17, 827–837.

Apkarian, A. V. 2004. Cortical pathophysiology of chronic pain. Novartis Found. Symp. 261, 239–245.

Apkarian, A. V. and Shi, T. 1994. Squirrel monkey lateral thalamus. I. Somatic nociresponsive neurons and their relation to spinothalamic terminals. J. Neurosci. 14, 6779–6795.

Apkarian, A. V., Bushnell, C., Treede, R. D., and Zubieta, J. K. 2005. Human brain mechanisms of pain perception and regulation in health and disease. Eur. J. Pain 9, 463–484.

Apkarian, A. V., Grachev, I. D., and Krauss, B. R. 2001a. Imaging brain pathophysiology of chronic CRPS pain. In: Complex Regional Pain Syndrome (eds. R. Harden, W. Janig, and J. C. Baron), pp. 209–227. IASP Press.

Apkarian, A. V., Sosa, Y., Sonty, S., Levy, R. E., Harden, R., Parrish, T., and Gitelman, D. 2004. Chronic back pain is associated with decreased prefrontal and thalamic gray matter density. J. Neurosci. 24, 10410–10415.

Apkarian, A. V., Stea, R. A., Manglos, S. H., Szeverenyi, N. M., King, R. B., and Thomas, F. D. 1992. Persistent pain inhibits contralateral somatosensory cortical activity in humans. Neurosci. Lett. 140, 141–147.

Apkarian, A. V., Thomas, P. S., Krauss, B. R., and Szeverenyi, N. M. 2001b. Prefrontal cortical hyperactivity in patients with sympathetically mediated chronic pain. Neurosci. Lett. 311, 193–197.

Baliki, M. N., Chialvo, D. R., Geha, P. Y., Levy, R. M., Harden, R. N., Parrish, T. B., and Apkarian, A. V. 2006. Chronic pain and the emotional brain: specific brain activity associated with spontaneous fluctuations of intensity of chronic back pain. J. Neurosci. 26, 12165–12173.

Bantick, S. J., Wise, R. G., Ploghaus, A., Clare, S., Smith, S. M., and Tracey, I. 2002. Imaging how attention modulates pain in humans using functional MRI. Brain 125, 310–319.

Baumgärtner, U., Buchholz, H. G., Bellosevich, A., Magerl, W., Siessmeier, T., Rolke, R., Höhnemann, S., Piel, M., Rösch, F., Wester, H. J., Henriksen, G., Stoeter, P., Bartenstein, P., Treede, R. D., and Schreckenberger, M. 2006a. High opiate receptor binding potential in the human lateral pain system. Neuroimage 30, 692–699.

Baumgärtner, U., Tiede, W., Treede, R. D., and Craig, A. D. 2006b. Laser-evoked potentials are graded and somatotopically organized anteroposteriorly in the operculoinsular cortex of anesthetized monkeys. J. Neurophysiol. 96, 2802–2808.

Becker, J. B., Arnold, A. P., Berkley, K. J., Blaustein, J. D., Eckel, L. A., Hampson, E., Herman, J. P., Marts, S., Sadee, W., Steiner, M., Taylor, J., and Young, E. 2005. Strategies and methods for research on sex differences in brain and behavior. Endocrinology 146, 1650–1673.

Benedetti, F., Mayberg, H. S., Wager, T. D., Stohler, C. S., and Zubieta, J. K. 2005. Neurobiological mechanisms of the placebo effect. J. Neurosci. 25, 10390–10402.

Berman, S. M., Chang, L., Suyenobu, B., Derbyshire, S. W., Stains, J., Fitzgerald, L., Mandelkern, M., Hamm, L., Vogt, B., Naliboff, B. D., and Mayer, E. A. 2002a. Condition-specific deactivation of brain regions by 5-HT3 receptor antagonist Alosetron. Gastroenterology 123, 969–977.

Berman, S., Munakata, J., Naliboff, B. D., Chang, L., Mandelkern, M., Silverman, D., Kovalik, E., and Mayer, E. A. 2000. Gender differences in regional brain response to visceral pressure in IBS patients. Eur. J. Pain 4, 157–172.

Berman, S. M., Naliboff, B. D., Chang, L., Fitzgerald, L., Antolin, T., Camplone, A., and Mayer, E. A. 2002b. Enhanced preattentive central nervous system reactivity in irritable bowel syndrome. Am. J. Gastroenterol. 97, 2791–2797.

Beydoun, A., Morrow, T. J., Shen, J. F., and Casey, K. L. 1993. Variability of laser-evoked potentials: attention, arousal and lateralized differences. Electroencephalogr. Clin. Neurophysiol. 88, 173–181.

Bingel, U., Lorenz, J., Schoell, E., Weiller, C., and Büchel, C. 2006. Mechanisms of placebo analgesia: rACC recruitment of a subcortical antinociceptive network. Pain 120, 8–15.

Borsook, D., Becerra, L., and Hargreaves, R. 2006. A role for fMRI in optimizing CNS drug development. Nat. Rev. Drug Discov. 5, 411–424.

Biemond, A. 1956. The conduction of pain above the level of the thalamus opticus. Arch. Neurol. Psychiat. 75, 231–244.

Botvinick, M., Jha, A. P., Bylsma, L. M., Fabian, S. A., Solomon, P. E., and Prkachin, K. M. 2005. Viewing facial expressions of pain engages cortical areas involved in the direct experience of pain. Neuroimage 25, 312–319.

Bragard, D., Chen, A. C. N., and Plaghki, L. 1996. Direct isolation of ultra-late (C-fibre) evoked brain potentials by CO2 laser stimulation of tiny cutaneous surface areas in man. Neurosci. Lett. 209, 81–84.

Bromm, B. and Lorenz, J. 1998. Neurophysiological evaluation of pain. Electroenceph. Clin. Neurophysiol. 107, 227–253.

Bromm, B. and Treede, R. D. 1984. Nerve fibre discharges, cerebral potentials and sensations induced by CO_2 laser stimulation. Human Neurobiol. 3, 33–40.

Bromm, B., Neitzel, H., Tecklenburg, A., and Treede, R. D. 1983. Evoked cerebral potential correlates of C-fibre activity in man. Neurosci. Lett. 43, 109–114.

Brooks, J. C. W., Nurmikko, T. J., Bimson, W. E., Singh, K. D., and Roberts, N. 2002. fMRI of thermal pain: effects of stimulus laterality and attention. Neuroimage 15, 293–301.

Buchsbaum, M. S., Kessler, R., King, A., Johnson, J., and Cappelletti, J. 1984. Simultaneous cerebral glucography with positron emission tomography and topographic electroencephalography. Prog. Brain Res. 62, 263–269.

Bushnell, M. C. and Apkarian, A. V. 2005. Representation of Pain in the Brains. In: Wall and Melzack's Textbook of Pain, Chapter 6, 5th Edition (eds. S. B. McMahon and M. Koltzenburg), pp. 107–124. Elsevier.

Bushnell, M. C., Duncan, G. H., Hofbauer, R. K., Ha, B., Chen, J. I., and Carrier, B. 1999. Pain perception: is there a role for primary somatosensory cortex? Proc. Natl. Acad. Sci. U. S. A. 96, 7705–7709.

Cao, Y., Welch, K. M. A., Aurora, S., and Vikingstad, E. M. 1999. Functional MRI-BOLD of visually triggered headache in patients with migraine. Arch. Neurol. 56, 548–554.

Carmon, A., Mor, J., and Goldberg, J. 1976. Evoked cerebral responses to noxious thermal stimuli in humans. Exp. Brain Res. 25, 103–107.

Casey, K. L., Svensson, P., Morrow, T. J., Raz, J., Jone, C., and Minoshima, S. 2000. Selective opiate modulation of nociceptive processing in the human brain. J. Neurophysiol. 84, 525–533.

Chen, A. C. N. 1993. Human brain measures of clinical pain: a review. I. Topographic mappings. Pain 54, 115–132.

Chen, A. C. N. and Treede, R. D. 1985. The McGill pain questionnaire in the assessment of phasic and tonic experimental pain: behavioral evaluation of the 'pain inhibiting pain' effect. Pain 22, 67–69.

Chen, A. C. N., Chapman, C. R., and Harkins, S. W. 1979. Brain evoked potentials are functional correlates of induced pain in man. Pain 6, 365–374.

Craig, A. D. 1995. Supraspinal Projections of Lamina I Neurons. In: Forebrain Areas Involved in Pain Processing (eds. J. M. Besson, G. Guilbaud, and H. Ollat), pp. 13–25. John Libbey Eurotext.

Craig, A. D. 2002. How do you feel? Interoception: the sense of the physiological condition of the body. Nat. Rev. 3, 655–666.

Craig, A. D., Bushnell, M. C., Zhang, E. T., and Blomqvist, A. 1994. A thalamic nucleus specific for pain and temperature sensation. Nature 372, 770–773.

Cruccu, G., Anand, P., Attal, N., Garcia-Larrea, L., Haanpää, M., Jørum, E., Serra, J., and Jensen, T. S. 2004. EFNS guidelines on neuropathic pain assessment. Eur. J. Neurol. 11, 153–162.

DaSilva, A. F. M., Becerra, L., Makris, N., Strassman, A. M., Gonzalez, R. G., Geatrakis, N., and Borsook, D. 2002. Somatotopic activation in the human trigeminal pain pathway. J. Neurosci. 22, 8183–8192.

Danziger, N., Prkachin, K. M., and Willer, J. C. 2006. Is pain the price of empathy? The perception of others' pain in patients with congenital insensitivity to pain. Brain 129, 2494–2507.

Davis, K. D. 2003. Neurophysiological and anatomical considerations in functional imaging of pain. Pain 105, 1–3.

Davis, K. D., Wood, M. L., Crawley, A. P., and Mikulis, D. J. 1995. fMRI of human somatosensory and cingulate cortex during painful electrical nerve stimulation. Neuroreport 7, 321–325.

De Leeuw, R., Albuquerque, R. J., Andersen, A. H., and Carlson, C. R. 2006. Influence of estrogen on brain activation during stimulation with painful heat. J. Oral Maxillofac. Surg. 64, 158–166.

Derbyshire, S. W. 2003. A systematic review of neuroimaging data during visceral stimulation. Am. J. Gastroenterol. 98, 12–20.

Derbyshire, S. W., Whalley, M. G., Stenger, V. A., and Oakley, D. A. 2004. Cerebral activation during hypnotically induced and imagined pain. Neuroimage 23, 392–401.

Dieterich, M., Bense, S., Lutz, S., Drzezga, A., Stephan, T., Bartenstein, P., and Brandt, T. 2003. Dominance for vestibular cortical function in the non-dominant hemisphere. Cereb. Cortex 13, 994–1007.

Disbrow, E., Roberts, T., and Krubitzer, L. 2000. Somatotopic organization of cortical fields in the lateral sulcus of homo sapiens: evidence for SII and PV. J. Comp. Neurol. 418, 1–21.

Dong, W. K., Chudler, E. H., Sugiyama, K., Roberts, V. J., and Hayashi, T. 1994. Somatosensory, multisensory, and task-related neurons in cortical area 7b (PF) of unanesthetized monkeys. J. Neurophysiol. 72, 542–564.

Duclaux, R., Franzen, O., Chatt, A. B., Kenshalo, D. R., and Stowell, H. 1974. Responses recorded from human scalp evoked by cutaneous thermal stimulation. Brain Res. 78, 279–290.

Eickhoff, S. B., Schleicher, A., Zilles, K., and Amunts, K. 2006. The human parietal operculum. I. Cytoarchitectonic mapping of subdivisions. Cereb. Cortex 16, 254–267.

Firestone, L. L., Gyulai, F., Mintun, M., Adler, L. J., Urso, K., and Winter, P. M. 1996. Human brain activity response to fentanyl imaged by positron emission tomography. Anesth. Analg. 82, 1247–1251.

Fitzgerald, P. J., Lane, J. W., Thakur, P. H., and Hsiao, S. S. 2004. Receptive field properties of the macaque second somatosensory cortex: evidence for multiple functional representations. J. Neurosci. 24, 11193–11204.

Fitzgerald, P. J., Lane, J. W., Thakur, P. H., and Hsiao, S. S. 2006. Receptive field (RF) properties of the macaque second somatosensory cortex: RF size, shape, and somatotopic organization. J. Neurosci. 26, 6485–6495.

Foss, J. M., Apkarian, A. V., and Chialvo, D. R. 2006. Dynamics of pain: fractal dimension of temporal variability of spontaneous pain differentiates between pain States. J. Neurophysiol. 95, 730–736.

Forss, N., Raij, T. T., Seppa, M., and Hari, R. 2005. Common cortical network for first and second pain. Neuroimage 24, 132–142.

Frankenstein, U. N., Richter, W., McIntyre, M. C., and Remy, F. 2001. Distraction modulates anterior cingulate gyrus activations during the cold pressor test. Neuroimage 14, 827–836.

Frot, M. and Mauguière, F. 1999. Operculo-insular responses to nociceptive skin stimulation in humans (A review). Neurophysiol. Clin. 29, 401–410.

Frot, M. and Mauguière, F. 2003. Dual representation of pain in the operculo-insular cortex in humans. Brain 126, 438–450.

Frot, M., Rambaud, L., Guénot, M., and Mauguière, F. 1999. Intracortical recordings of early pain-related CO2-laser evoked potentials in the human second somatosensory (SII) area. Clin. Neurophysiol. 110, 133–145.

Fukumoto, M., Ushida, T., Zinchuk, V. S., Yamamoto, H., and Yoshida, S. 1999. Contralateral thalamic perfusion in patients with reflex sympathetic dystrophy syndrome. Lancet 354, 1790–1791.

Garcia-Larrea, L., Maarrawi, J., Peyron, R., Costes, N., Mertens, P., Magnin, M., and Laurent, B. 2006. On the relation between sensory deafferentation, pain and thalamic activity in Wallenberg's syndrome: a PET-scan study before and after motor cortex stimulation. Eur. J. Pain 10, 677–688.

Geha, P. Y., Baliki, M. N., Chialvo, D. R., Harden, R. N., Paice, J. A., and Apkarian, A. V. 2007. Brain activity for spontaneous pain of postherpetic neuralgia and its modulation by lidocaine patch therapy. Pain 128, 88–100.

Giesecke, T., Gracely, R. H., Grant, M. A., Nachemson, A., Petzke, F., Williams, D. A., and Clauw, D. J. 2004. Evidence of augmented central pain processing in idiopathic chronic low back pain. Arthritis Rheum. 50, 613–623.

Gingold, S. I., Greenspan, J. D., and Apkarian, A. V. 1991. Anatomic evidence of nociceptive inputs to primary somatosensory cortex: relationship between spinothalamic terminals and thalamocortical cells in squirrel monkeys. J. Comp. Neurol. 308, 467–490.

Goadsby, P. J., Lipton, R. B., and Ferrari, M. D. 2002. Migraine – current understanding and treatment. N. Engl. J. Med. 346, 257–270.

Goldstein, J. M., Jerram, M., Poldrack, R., Ahern, T., Kennedy, D. N., Seidman, L. J., and Makris, N. 2005. Hormonal cycle modulates arousal circuitry in women using functional magnetic resonance imaging. J. Neurosci. 25, 9309–9316.

Gracely, R. H., Geisser, M. E., Giesecke, T., Grant, M. A., Petzke, F., Williams, D. A., and Clauw, D. J. 2004. Pain catastrophizing and neural responses to pain among persons with fibromyalgia. Brain 127, 835–843.

Gracely, R. H., Petzke, F., Wolf, J. M., and Clauw, D. J. 2002. Functional magnetic resonance imaging evidence of augmented pain processing in fibromyalgia. Arthritis Rheum. 46, 1333–1343.

Grachev, I. D., Fredrickson, B. E., and Apkarian, A. V. 2000. Abnormal brain chemistry in chronic back pain: an in vivo proton magnetic resonance spectroscopy study. Pain 89, 7–18.

Grachev, I. D., Fredrickson, B. E., and Apkarian, A. V. 2002. Brain chemistry reflects dual states of pain and anxiety in chronic low back pain. J. Neural Transm. 109, 1309–1334.

Granziera, C., DaSilva, A. F., Snyder, J., Tuch, D. S., and Hadjikhani, N. 2006. Anatomical alterations of the visual motion processing network in migraine with and without aura. PLoS Med. 3, e402.

Greenspan, J. D., Lee, R. R., and Lenz, F. A. 1999. Pain sensitivity alterations as a function of lesion location in the parasylvian cortex. Pain 81, 273–282.

Hadjikhani, N., Sanchez, d. R., Wu, O., Schwartz, D., Bakker, D., Fischl, B., Kwong, K. K., Cutrer, F. M., Rosen, B. R., Tootell, R. B., Sorensen, A. G., and Moskowitz, M. A. 2001. Mechanisms of migraine aura revealed by functional MRI in human visual cortex. Proc. Natl. Acad. Sci. U. S. A. 98, 4687–4692.

Hagelberg, N., Forssell, H., Aalto, S., Rinne, J. O., Scheinin, H., Taiminen, T., Nagren, K., Eskola, O., and Jaaskelainen, S. K. 2003a. Altered dopamine D2 receptor binding in atypical facial pain. Pain 106, 43–48.

Hagelberg, N., Forssell, H., Rinne, J. O., Scheinin, H., Taiminen, T., Aalto, S., Luutonen, S., Nagren, K., and Jaaskelainen, S. 2003b. Striatal dopamine D1 and D2 receptors in burning mouth syndrome. Pain 101, 149–154.

Hagelberg, N., Kajander, J. K., Nagren, K., Hinkka, S., Hietala, J., and Scheinin, H. 2002. Mu-receptor agonism with alfentanil increases striatal dopamine D2 receptor binding in man. Synapse 45, 25–30.

Hampson, E. 2002. Sex Differences in Human Brain and Cognition: the Influence of Sex Steroids in Early and Adult Life. In: Behavioral endocrinology, 2nd edn. (eds. J. B. Becker, S. M. Breedlove, D. Crews, and M. M. McCarthy) pp. 579–628. MIT Press.

Hari, R. and Forss, N. 1999. Magnetoencephalography in the study of human somatosensory cortical processing. Phil. Trans. R. Soc. Lond. B 354, 1145–1154.

Head, H. and Holmes, G. 1911. Sensory disturbances from cerebral lesions. Brain 34, 102–254.

Hofbauer, R. K., Rainville, P., Duncan, G. H., and Bushnell, M. C. 2001. Cortical representation of the sensory dimension of pain. J. Neurophysiol. 86, 402–411.

Hsieh, J. C., Belfrage, M., Stoneelander, S., Hansson, P., and Ingvar, M. 1995. Central representation of chronic ongoing neuropathic pain studied by positron emission tomography. Pain 63, 225–236.

Hsieh, J. C., Meyerson, B. A., and Ingvar, M. 1999a. PET study on central processing of pain in trigeminal neuropathy. Eur. J. Pain 3, 51–65.

Hsieh, J. C., Stone-Elander, S., and Ingvar, M. 1999b. Anticipatory coping of pain expressed in the human anterior cingulate cortex: a positron emission tomography study. Neurosci. Lett. 262, 61–64.

Hutchison, W. D., Davis, K. D., Lozano, A. M., Tasker, R. R., and Dostrovsky, J. O. 1999. Pain-related neurons in the human cingulate cortex. Nature Neurosci. 2, 403–405.

Iannetti, G. D., Truini, A., Romaniello, A., Galeotti, F., Rizzo, C., Manfredi, M., and Cruccu, G. 2003. Evidence of a specific spinal pathway for the sense of warmth in humans. J. Neurophysiol. 89, 562–570.

Jaaskelainen, S. K., Rinne, J. O., Forssell, H., Tenovuo, O., Kaasinen, V., Sonninen, P., and Bergman, J. 2001. Role of the dopaminergic system in chronic pain – a fluorodopa-PET study. Pain 90, 257–260.

Jackson, P. L., Meltzoff, A. N., and Decety, J. 2005. How do we perceive the pain of others? A window into the neural processes involved in empathy. Neuroimage 24, 771–779.

Jones, A. K. P., Brown, W. D., Friston, K. J., Qi, L. Y., and Frackowiak, R. S. J. 1991a. Cortical and subcortical localization of response to pain in man using positron emission tomography. Proc. R. Soc. Lond. B 244, 39–44.

Jones, A. K. P., Qi, L. Y., Fujiwara, T., Luthra, S. K., Ashburner, J., Bloomfield, P., Cunningham, V. J., Itoh, M., Fukuda, H., and Jones, T. 1991b. In vivo distribution of opioid receptors in man in relation to the cortical projections of the medial and lateral pain systems measured with positron emission tomography. Neurosci. Lett. 126, 25–28.

Jones, A. K. P., Watabe, H., Cunningham, V. J., and Jones, T. 2004. Cerebral decreases in opioid receptor binding in patients with central neuropathic pain measured by [^{11}C]diprenorphine binding and PET. Eur. J. Pain 8, 479–485.

Kajantie, E. and Phillips, D. I. 2006. The effects of sex and hormonal status on the physiological response to acute psychosocial stress. Psychoneuroendocrinology 31, 151–178.

Kanda, M., Nagamine, T., Ikeda, A., Ohara, S., Kunieda, T., Fujiwara, N., Yazawa, S., Sawamoto, N., Matsumoto, R., Taki, W., and Shibasaki, H. 2000. Primary somatosensory cortex is actively involved in pain processing in human. Brain Res. 853, 282–289.

Keltner, J. R., Furst, A., Fan, C., Redfern, R., Inglis, B., and Fields, H. L. 2006. Isolating the modulatory effect of expectation on pain transmission: a functional magnetic resonance imaging study. J. Neurosci. 26, 4437–4443.

Kenshalo, D. R. and Isensee, O. 1983. Responses of primate SI cortical neurons to noxious stimuli. J. Neurophysiol. 50, 1479–1496.

Kenshalo, D. R. and Willis, W. D. 1991. The Role of the Cerebral Cortex in Pain Sensation. In: Cerebral Cortex, Vol. 9 (ed. A. Peters), pp. 153–212. Plenum Press.

Kenshalo, D. R., Chudler, E. H., Anton, F., and Dubner, R. 1988. SI nociceptive neurons participate in the encoding process by which monkeys perceive the intensity of noxious thermal stimulation. Brain Res. 454, 378–382.

Kenshalo, D. R., Iwata, K., Sholas, M., and Thomas, D. A. 2000. Response properties and organization of nociceptive neurons in area 1 of monkey primary somatosensory cortex. J. Neurophysiol. 84, 719–729.

Kong, J., Gollub, R. L., Rosman, I. S., Webb, J. M., Vangel, M. G., Kirsch, I., and Kaptchuk, T. J. 2006. Brain activity associated with expectancy-enhanced placebo analgesia as measured by functional magnetic resonance imaging. J. Neurosci. 26, 381–388.

Koyama, T., Kato, K., Tanaka, Y. Z., and Mikami, A. 2001. Anterior cingulate activity during pain-avoidance and reward tasks in monkeys. Neurosci. Res. 39, 421–430.

Koyama, T., Tanaka, Y. Z., and Mikami, A. 1998. Nociceptive neurons in the macaque anterior cingulate activate during anticipation of pain. Neuroreport 9, 2663–2667.

Kunde, V. and Treede, R. D. 1993. Topography of middle-latency somatosensory evoked potentials following painful laser stimuli and non-painful electrical stimuli. Electroenceph. Clin. Neurophysiol. 88, 280–289.

Kulkarni, B., Bentley, D. E., Elliott, R., Youell, P., Watson, A., Derbyshire, S. W., Frackowiak, R. S., Friston, K. J., and Jones, A. K. 2005. Attention to pain localization and unpleasantness discriminates the functions of the medial and lateral pain systems. Eur. J. Neurosci. 21, 3133–3142.

Kupers, R. and Kehlet, H. 2006. Brain imaging of clinical pain states: a critical review and strategies for future studies. Lancet Neurol. 5, 1033–1044.

Kwan, C. L., Diamant, N. E., Pope, G., Mikula, K., Mikulis, D. J., and Davis, K. D. 2005. Abnormal forebrain activity in functional bowel disorder patients with chronic pain. Neurology 65, 1268–1277.

Lamour, Y., Guilbaud, G., and Willer, J. C. 1983. Rat somatosensory (SmI) cortex: II. Laminar and columnar organization of noxious and non-noxious inputs. Exp. Brain Res. 49, 46–54.

Lamour, Y., Willer, J. C., and Guilbaud, G. 1982. Neuronal responses to noxious stimulation in rat somatosensory cortex. Neurosci. Lett. 29, 35–40.

Lawal, A., Kern, M., Sidhu, H., Hofmann, C., and Shaker, R. 2006. Novel evidence for hypersensitivity of visceral sensory neural circuitry in irritable bowel syndrome patients. Gastroenterology 130, 26–33.

Lenz, F. A., Rios, M., Chau, D., Krauss, G. L., Zirh, T. A., and Lesser, R. P. 1998a. Painful stimuli evoke potentials recorded from the parasylvian cortex in humans. J. Neurophysiol. 80, 2077–2088.

Lenz, F. A., Seike, M., Richardson, R. T., Lin, Y. C., Baker, F. H., Khoja, I., Jaeger, C. J., and Gracely, R. H. 1993. Thermal and pain sensations evoked by microstimulation in the area of human ventrocaudal nucleus. J. Neurophysiol. 70, 200–212.

Lenz, F. A., Rios, M., Zirh, A., Chau, D., Krauss, G., and Lesser, R. P. 1998b. Painful stimuli evoke potentials recorded over the human anterior cingulate gyrus. J. Neurophysiol. 79, 2231–2234.

Levine, J. D., Gordon, N. C., and Fields, H. L. 1978. The mechanism of placebo analgesia. Lancet 2, 654–657.

Lidstone, S. C. and Stoessl, A. J. 2007. Understanding the placebo effect: contributions from neuroimaging. Mol. Imaging Biol. 9, 176–185.

Longe, S. E., Wise, R., Bantick, S., Lloyd, D., Johansen-Berg, H., McGlone, F., and Tracey, I. 2001. Counter-stimulatory effects on pain perception and processing are significantly altered by attention: an fMRI study. Neuroreport 12, 2021–2025.

Lorenz, J., Beck, H., and Bromm, B. 1997. Cognitive performance, mood and experimental pain before and during morphine-induced analgesia in patients with chronic non-malignant pain. Pain 73, 369–375.

Lorenz, J., Minoshima, S., and Casey, K. L. 2003. Keeping pain out of mind: the role of the dorsolateral prefrontal cortex in pain modulation. Brain 126, 1079–1091.

Maarrawi, J., Peyron, R., Mertens, P., Costes, N., Magnin, M., Sindou, M., Laurent, B., and Garcia-Larrea, L. 2007. Differential brain opioid receptor availability in central and peripheral neuropathic pain. Pain 127, 183–194.

Magerl, W., Ali, Z., Ellrich, J., Meyer, R. A., and Treede, R. D. 1999. C- and Ad-fiber components of heat-evoked cerebral potentials in healthy human subjects. Pain 82, 127–137.

Maihöfner, C., Forster, C., Birklein, F., Neundörfer, B., and Handwerker, H. O. 2005. Brain processing during mechanical hyperalgesia in complex regional pain syndrome: a functional MRI study. Pain 114, 93–103.

Marchand, S. and Arsenault, P. 2002. Odors modulate pain perception: a gender-specific effect. Physiol. Behav. 76, 251–256.

Martikainen, I. K., Hagelberg, N., Mansikka, H., Hietala, J., Nagren, K., Scheinin, H., and Pertovaara, A. 2005. Association of striatal dopamine D2/D3 receptor binding potential with pain but not tactile sensitivity or placebo analgesia. Neurosci. Lett. 376, 149–153.

Marshall, J. 1951. Sensory disturbances in cortical wounds with special reference to pain. J. Neurol. Neurosurg. Psychiatry 14, 187–204.

May, A., Ashburner, J., Büchel, C., McGonigle, D. J., Friston, K. J., Frackowiak, R. S., and Goadsby, P. J. 1999. Correlation between structural and functional changes in brain in an idiopathic headache syndrome. Nat. Med. 5, 836–838.

May, A., Bahra, A., Buchel, C., Frackowiak, R. S., and Goadsby, P. J. 2000. PET and MRA findings in cluster headache and MRA in experimental pain. Neurology 55, 1328–1335.

Melzack, R. and Casey, K. L. 1968. Sensory, Motivational, and Central Control Determinants of Pain. A New Conceptual Model. In: The Skin Senses. (eds. D. R Kenshalo and C. Charles), pp. 423–443. Thomas.

Mertz, H., Morgan, V., Tanner, G., Pickens, D., Price, R., Shyr, Y., and Kessler, R. 2000. Regional cerebral activation in irritable bowel syndrome and control subjects with painful and nonpainful rectal distention. Gastroenterology 118, 842–848.

Miyazaki, M., Shibasaki, H., Kanda, M., Xu, X., Shindo, K., Honda, M., Ikeda, A., Nagamine, T., Kaji, R., and Kimura, J. 1994. Generator mechanism of pain-related evoked potentials following CO2 laser stimulation of the hand: scalp topography and effect of predictive warning signal. J. Clin. Neurophysiol. 11, 242–254.

Moriguchi, Y., Decety, J., Ohnishi, T., Maeda, M., Mori, T., Nemoto, K., Matsuda, H., and Komaki, G. 2006. Empathy and judging other's pain: an fMRI study of alexithymia. Cereb. Cortex doi: 10.1093/cercor/bh1130.

Morrison, I., Lloyd, D., di Pellegrino, G., and Roberts, N. 2004. Vicarious responses to pain in anterior cingulate cortex: is empathy a multisensory issue? Cogn. Affect. Behav. Neurosci. 4, 270–278.

Nakai, A., Kumakura, Y., Boivin, M., Rosa, P., Diksic, M., D'Souza, D., and Kersey, K. 2003. Sex differences of brain serotonin synthesis in patients with irritable bowel syndrome using a [^{11}C]methyl-l-tryptophan, positron emission tomography and statistical parametric mapping. Can. J. Gastroenterol. 17, 191–196.

Özcan, M., Baumgärtner, U., Vucurevic, G., Stoeter, P., and Treede, R. D. 2005. Spatial resolution of fMRI in the human parasylvian cortex: comparison of somatosensory and auditory activation. Neuroimage 25, 877–887.

Ohara, S., Crone, N. E., Weiss, N., and Lenz, F. A. 2006. Analysis of synchrony demonstrates 'pain networks' defined by rapidly switching, task-specific, functional connectivity between pain-related cortical structures. Pain 123, 244–253.

Ohara, S., Crone, N. E., Weiss, N., Treede, R. D., and Lenz, F. A. 2004. Cutaneous painful laser stimuli evoke responses recorded directly from primary somatosensory cortex in awake humans. J. Neurophysiol. 91, 2734–2746.

Opsommer, E., Weiss, T., Plaghki, L., and Miltner, W. H. R. 2001. Dipole analysis of ultralate (C-fibres) evoked potentials after laser stimulation of tiny cutaneous surface areas in humans. Neurosci. Lett. 298, 41–44.

Ostrowsky, K., Magnin, M., Ryvlin, P., Isnard, J., Guenot, M., and Mauguière, F. 2002. Representation of pain and somatic sensation in the human insula: a study of responses to direct electrical cortical stimulation. Cereb. Cortex 12, 376–385.

Pascual-Marqui, R. D., Michel, C. M., and Lehmann, D. 1994. Low resolution electromagnetic tomography: a new method for localizing electrical activity in the brain. Int. J. Psychophysiol. 18, 49–65.

Pattany, P. M., Yezierski, R. P., Widerstrom-Noga, E. G., Bowen, B. C., Martinez-Arizala, A., Garcia, B. R., and Quencer, R. M. 2002. Proton magnetic resonance spectroscopy of the thalamus in patients with chronic neuropathic pain after spinal cord injury. AJNR Am. J. Neuroradiol. 23, 901–905.

Pauli, P., Wiedemann, G., and Nickola, M. 1999. Pain sensitivity, cerebral laterality, and negative affect. Pain 80, 359–364.

Paulson, P. E., Minoshima, S., Morrow, T. J., and Casey, K. L. 1998. Gender differences in pain perception and patterns of cerebral activation during noxious heat stimulation in humans. Pain 76, 223–229.

Pertovaara, A., Martikainen, I. K., Hagelberg, N., Mansikka, H., Nagren, K., Hietala, J., and Scheinin, H. 2004. Striatal dopamine D2/D3 receptor availability correlates with individual response characteristics to pain. Eur. J. Neurosci. 20, 1587–1592.

Petrovic, P., Dietrich, T., Fransson, P., Andersson, J., Carlsson, K., and Ingvar, M. 2005. Placebo in emotional processing-induced expectations of anxiety relief activate a generalized modulatory network. Neuron 46, 957–969.

Petrovic, P., Kalso, E., Petersson, K. M., and Ingvar, M. 2002. Placebo and opioid analgesia- imaging a shared neuronal network. Science 295, 1737–1740.

Petrovic, P., Petersson, K. M., Ghatan, P. H., Stone-Elander, S., and Ingvar, M. 2000. Pain-related cerebral activation is altered by a distracting cognitive task. Pain 85, 19–30.

Peyron, R., Laurent, B., and García-Larrea, L. 2000. Functional imaging of brain responses to pain. A review and meta-analysis 2000. Neurophysiol. Clin. 30, 263–288.

Phillips, M. L., Gregory, L. J., Cullen, S., Cohen, S., Ng, V., Andrew, C., Giampietro, V., Bullmore, E., Zelaya, F., Amaro, E., Thompson, D. G., Hobson, A. R., Williams, S. C., Brammer, M., and Aziz, Q. 2003. The effect of negative emotional context on neural and behavioural responses to oesophageal stimulation. Brain 126, 669–684.

Plaghki, L. and Mouraux, A. 2003. How do we selectively activate skin nociceptors with a high power infrared laser? Physiology and biophysics of laser stimulation. Neurophysiol. Clin. 33, 269–277.

Ploghaus, A., Tracey, I., Gati, J. S., Clare, S., Menon, R. S., Matthews, P. M., and Rawlins, J. N. 1999. Dissociating pain from its anticipation in the human brain. Science 284, 1979–1981.

Ploner, M., Schmitz, F., Freund, H. J., and Schnitzler, A. 1999. Parallel activation of primary and secondary somatosensory cortices in human pain processing. J. Neurophysiol. 81, 3100–3104.

Ploner, M., Schmitz, F., Freund, H. J., and Schnitzler, A. 2000. Differential organization of touch and pain in human primary somatosensory cortex. J. Neurophysiol. 83, 1770–1776.

Porro, C. A., Baraldi, P., Pagnoni, G., Serafini, M., Facchin, P., Maieron, M., and Nichelli, P. 2002. Does anticipation of pain affect cortical nociceptive systems? J. Neurosci. 22, 3206–3214.

Price, D. D. 2000. Psychological and neural mechanisms of the affective dimension of pain. Science 288, 1769–1772.

Raij, T. T., Numminen, J., Narvanen, S., Hiltunen, J., and Hari, R. 2005. Brain correlates of subjective reality of physically and psychologically induced pain. Proc. Natl. Acad. Sci. U. S. A. 102, 2147–2151.

Rainville, P., Duncan, G. H., Price, D. D., Carrier, B., and Bushnell, M. C. 1997. Pain affect encoded in human anterior cingulate but not somatosensory cortex. Science 277, 968–971.

Ringel, Y. 2006. New directions in brain imaging research in functional gastrointestinal disorders. Dig. Dis. 24, 278–285.

Robinson, C. J. and Burton, H. 1980. Somatic submodality distribution within the second somatosensory (SII), 7b, retroinsular, postauditory, and granular insular cortical areas of M. fascicularis. J. Comp. Neurol. 192, 93–108.

Rolke, R., Baron, R., Maier, C., Tölle, T. R., Treede, R. D., Beyer, A., Binder, A., Birbaumer, N., Birklein, F., Botefur, I. C., Braune, S., Flor, H., Huge, V., Klug, R., Landwehrmeyer, G. B., Magerl, W., Maihofner, C., Rolko, C., Schaub, C., Scherens, A., Sprenger, T., Valet, M., and Wasserka, B. 2006. Quantitative sensory testing in the German Research Network on Neuropathic Pain (DFNS): standardized protocol and reference values. Pain 123, 231–243.

Rosen, S. D., Paulesu, E., Nihoyannopoulos, P., Tousoulis, D., Frackowiak, R. S., Frith, C. D., Jones, T., and Camici, P. G. 1996. Silent ischemia as a central problem: regional brain activation compared in silent and painful myocardial ischemia. Ann. Intern. Med. 124, 939–949.

Rosen, S. D., Paulesu, E., Wise, R. J., and Camici, P. G. 2002. Central neural contribution to the perception of chest pain in cardiac syndrome X. Heart 87, 513–519.

Sawamoto, N., Honda, M., Okada, T., Hanakawa, T., Kanda, M., Fukuyama, H., Konishi, J., and Shibasaki, H. 2000. Expectation of pain enhances responses to nonpainful somatosensory stimulation in the anterior cingulate cortex and parietal operculum/posterior insula: an event-related functional magnetic resonance imaging study. J. Neurosci. 20, 7438–7445.

Scharein, E. and Bromm, B. 1998. The intracutaneous pain model in the assessment of analgesic efficacy. Pain Rev. 5, 216–246.

Scherg, M. 1992. Functional imaging and localization of electromagnetic brain activity. Brain Topogr. 5, 103–111.

Schlaepfer, T. E., Strain, E. C., Greenberg, B. D., Preston, K. L., Lancaster, E., Bigelow, G. E., Barta, P. E., and Pearlson, G. D. 1998. Site of opioid action in the human brain: mu and kappa agonists' subjective and cerebral blood flow effects. Am. J. Psychiatry 155, 470–473.

Schlereth, T., Baumgärtner, U., Magerl, W., Stoeter, P., and Treede, R. D. 2003. Left-hemisphere dominance in early nociceptive processing in the human parasylvian cortex. Neuroimage 20, 441–454.

Schmahl, C., Bohus, M., Esposito, F., Treede, R. D., Di Salle, F., Greffrath, W., Ludaescher, P., Jochims, A., Lieb, K., Scheffler, K., Hennig, J., and Seifritz, E. 2006. Neural correlates of antinociception in borderline personality disorder. Arch. Gen. Psychiatry 63, 659–667.

Schweinhardt, P., Glynn, C., Brooks, J., McQuay, H., Jack, T., Chessell, I., Bountra, C., and Tracey, I. 2006. An fMRI study of cerebral processing of brush-evoked allodynia in neuropathic pain patients. Neuroimage 32, 256–265.

Scott, D. J., Heitzeg, M. M., Koeppe, R. A., Stohler, C. S., and Zubieta, J. K. 2006. Variations in the human pain stress experience mediated by ventral and dorsal basal ganglia dopamine activity. J. Neurosci. 26, 10789–10795.

Seitz, R. J., Roland, P. E., Bohm, C., Greitz, T., and Stone-Elander, S. 1991. Somatosensory discrimination of shape: tactile exploration and cerebral activation. Eur. J. Neurosci. 3, 481–492.

Sikes, R. W. and Vogt, B. A. 1992. Nociceptive neurons in area 24 of rabbit cingulate cortex. J. Neurophysiol. 68, 1720–1732.

Silverman, D. H., Munakata, J. A., Ennes, H., Mandelkern, M. A., Hoh, C. K., and Mayer, E. A. 1997. Regional cerebral activity in normal and pathological perception of visceral pain. Gastroenterology 112, 64–72.

Singer, T., Seymour, B., O'Doherty, J., Kaube, H., Dolan, R. J., and Frith, C. D. 2004. Empathy for pain involves the affective but not sensory components of pain. Science 303, 1157–1162.

Smith, Y. R., Stohler, C. S., Nichols, T. E., Bueller, J. A., Koeppe, R. A., and Zubieta, J. K. 2006. Pronociceptive and antinociceptive effects of estradiol through endogenous opioid neurotransmission in women. J. Neurosci. 26, 5777–5785.

Spreng, M. and Ichioka, M. 1964. Langsame Rindenpotentiale bei Schmerzreizung am Menschen. Pflügers Arch. 279, 121–132.

Stevens, R. T., London, S. M., and Apkarian, A. V. 1993. Spinothalamocortical projections to the secondary somatosensory cortex (SII) in squirrel monkey. Brain Res. 631, 241–246.

Stiasny-Kolster, K., Magerl, W., Oertel, W. H., Möller, J. C., and Treede, R. D. 2004. Static mechanical hyperalgesia without dynamic tactile allodynia in patients with restless legs syndrome. Brain 127, 773–782.

Strigo, I. A., Duncan, G. H., Boivin, M., and Bushnell, M. C. 2003. Differentiation of visceral and cutaneous pain in the human brain. J. Neurophysiol. 89, 3294–3303.

Strittmatter, M., Hamann, G. F., Grauer, M., Fischer, C., Blaes, F., Hoffmann, K. H., and Schimrigk, K. 1996. Altered activity of the sympathetic nervous system and changes in the balance of hypophyseal, pituitary and adrenal hormones in patients with cluster headache. Neuroreport 7, 1229–1234.

Talbot, J. D., Marrett, S., Evans, A. C., Meyer, E., Bushnell, M. C., and Duncan, G. H. 1991. Multiple representations of pain in human cerebral cortex. Science 251, 1355–1358.

Tarkka, I. M. and Treede, R. D. 1993. Equivalent electrical source analysis of pain-related somatosensory evoked potentials elicited by a CO_2 laser. J. Clin. Neurophysiol. 10, 513–519.

Timmermann, L., Ploner, M., Haucke, K., Schmitz, F., Baltissen, R., and Schnitzler, A. 2001. Differential coding of pain intensity in the human primary and secondary somatosensory cortex. J. Neurophysiol. 86, 1499–1503.

Tölle, T. R., Kaufmann, T., Siessmeier, T., Lautenbacher, S., Berthele, A., Munz, F., Zieglgänsberger, W., Willoch, F., Schwaiger, M., Conrad, B., and Bartenstein, P. 1999. Region-specific encoding of sensory and affective components of pain in the human brain: a positron emission tomography correlation analysis. Ann. Neurol. 45, 40–47.

Tommerdahl, M., Delemos, K. A., Vierck, C. J., Favorov, O. V., and Whitsel, B. L. 1996. Anterior parietal cortical response to tactile and skin-heating stimuli applied to the same skin site. J. Neurophysiol. 75, 2662–2670.

Tracey, I. 2001. Prospects for human pharmacological functional magnetic resonance imaging (phMRI). J. Clin. Pharmacol. Suppl. 21S–28S.

Tracey, I., Ploghaus, A., Gati, J. S., Clare, S., Smith, S., Menon, R. S., and Matthews, P. M. 2002. Imaging attentional modulation of pain in the periaqueductal gray in humans. J. Neurosci. 22, 2748–2752.

Treede, R. D. 2001. Neural Basis of Pain. In: International Encyclopedia of the Social & Behavioral Sciences (eds. N. J. Smelser and P. B. Baltes), pp. 11000–11005. Elsevier.

Treede, R. D., Apkarian, A. V., Bromm, B., Greenspan, J. D., and Lenz, F. A. 2000. Cortical representation of pain: functional characterization of nociceptive areas near the lateral sulcus. Pain 87, 113–119.

Treede, R. D., Baumgärtner, U., and Lenz, F. A. 2007. Nociceptive Processing in the Secondary Somatosensory Cortex. In: Encyclopedia of Pain (eds. R. F. Schmidt and W. D. Willis), pp. 1376–1379. Springer.

Treede, R. D., Kenshalo, D. R., Gracely, R. H., and Jones, A. K. P. 1999. The cortical representation of pain. Pain 79, 105–111.

Treede, R. D., Meyer, R. A., and Campbell, J. N. 1998. Myelinated mechanically insensitive afferents from monkey hairy skin: heat response properties. J. Neurophysiol. 80, 1082–1093.

Treede, R. D., Lorenz, J., and Baumgärtner, U. 2003. Clinical usefulness of laser-evoked potentials. Neurophysiol. Clin. 33, 303–314.

Valeriani, M., Rambaud, L., and Mauguière, F. 1996. Scalp topography and dipolar source modelling of potentials evoked by CO_2 laser stimulation of the hand. Electroenceph. Clin. Neurophysiol. 100, 343–353.

Valeriani, M., Sestito, A., Le Pera, D., De Armas, L., Infusino, F., Maiese, T., Sgueglia, G. A., Tonali, P. A., Crea, F., Restuccia, D., and Lanza, G. A. 2005. Abnormal cortical pain processing in patients with cardiac syndrome X. Eur. Heart J. 26, 975–982.

Valet, M., Sprenger, T., Boecker, H., Willoch, F., Rummeny, E., Conrad, B., Erhard, P., and Tölle, T. R. 2004. Distraction modulates connectivity of the cingulo-frontal cortex and the midbrain during pain – an fMRI analysis. Pain 109, 399–408.

Verne, G. N., Himes, N. C., Robinson, M. E., Gopinath, K. S., Briggs, R. W., Crosson, B., and Price, D. D. 2003. Central representation of visceral and cutaneous hypersensitivity in the irritable bowel syndrome. Pain 103, 99–110.

Villemure, C. and Bushnell, M. C. 2002. Cognitive modulation of pain: how do attention and emotion influence pain processing? Pain 95, 195–199.

Vogel, H., Port, J. D., Lenz, F. A., Solaiyappan, M., Krauss, G., and Treede, R. D. 2003. Dipole source analysis of laser-evoked subdural potentials recorded from parasylvian cortex in humans. J. Neurophysiol. 89, 3051–3060.

Vogt, B. A. 2005. Pain and emotion interactions in subregions of the cingulate gyrus. Nat. Rev. Neurosci. 6, 533–544.

Vogt, B. A., Nimchinsky, E. A., Vogt, L. J., and Hof, P. R. 1995. Human cingulate cortex: surface features, flat maps, and cytoarchitecture. J. Comp. Neurol. 359, 490–506.

Vogt, B. A., Rosene, D. L., and Pandya, D. N. 1979. Thalamic and cortical afferents differentiate anterior from posterior cingulate cortex in the monkey. Science 204, 205–207.

Wager, T. D., Rilling, J. K., Smith, E. E., Sokolik, A., Casey, K. L., Davidson, R. J., Kosslyn, S. M., Rose, R. M., and Cohen, J. D. 2004. Placebo-induced changes in fMRI in the anticipation and experience of pain. Science 303, 1162–1167.

Wagner, K. J., Willoch, F., Kochs, E. F., Siessmeier, T., Tölle, T. R., Schwaiger, M., and Bartenstein, P. 2001. Dose-dependent regional cerebral blood flow changes during remifentanil infusion in humans: a positron emission tomography study. Anesthesiology 94, 732–739.

Wang, S. J., Lirng, J. F., Fuh, J. L., and Chen, J. J. 2006. Reduction in hypothalamic 1H-MRS metabolite ratios in patients with cluster headache. J. Neurol. Neurosurg. Psychiatry 77, 622–625.

Wiesenfeld-Hallin, Z. 2005. Sex differences in pain perception. Gend. Med. 2, 137–145.

Willis, W. D. and Coggeshall, R. E. 2004. Sensory Mechanisms of the Spinal Cord: Primary Afferent Neurons and the Spinal Dorsal Horn, Vol. 1. Kluwer Academic/Plenum Publishers.

Willoch, F., Schindler, F., Wester, H. E., Empl, M., Straube, A., Schwaiger, M., Conrad, B., and Tölle, T. R. 2004. Central poststroke pain and reduced opioid receptor binding within pain processing circuitries: a [C-11]diprenorphine PET study. Pain 108, 213–220.

Wilson, J. F. 2006. The pain divide between men and women. Ann. Intern. Med 144, 461–464.

Wood, P. B., Patterson, J. C., Sunderland, J. J., Tainter, K. H., Glabus, M. F., and Lilien, D. L. 2007. Reduced presynaptic dopamine activity in fibromyalgia syndrome demonstrated with positron emission tomography: a pilot study. J. Pain 8, 51–58.

Yamamura, H., Iwata, K., Tsuboi, Y., Toda, K., Kitajima, K., Shimizu, N., Nomura, H., Hibiya, J., Fujita, S., and Sumino, R. 1996. Morphological and electrophysiological properties of ACCx nociceptive neurons in rats. Brain Res. 735, 83–92.

Xu, X. P., Fukuyama, H., Yazawa, S., Mima, T., Hanakawa, T., Magata, Y., Kanda, M., Fujiwara, N., Shindo, K., Nagamine, T., and Shibasaki, H. 1997. Functional localization of pain perception in the human brain studied by PET. Neuroreport 8, 555–559.

Zelman, D. C., Howland, E. W., Nichols, S. N., and Cleeland, C. S. 1991. The effects of induced mood on laboratory pain. Pain 46, 105–111.

Zubieta, J. K., Smith, Y. R., Bueller, J. A., Xu, Y., Kilbourn, M. R., Jewett, D. M., Meyer, C. R., Koeppe, R. A., and Stohler, C. S. 2001. Regional mu opioid receptor regulation of sensory and affective dimensions of pain. Science 293, 311–315.

Zubieta, J. K., Bueller, J. A., Jackson, L. R., Scott, D. J., Xu, Y., Koeppe, R. A., Nichols, T. E., and Stohler, C. S. 2005. Placebo effects mediated by endogenous opioid activity on μ-opioid receptors. J. Neurosci. 25, 7754–7762.

Zubieta, J. K., Smith, Y. R., Bueller, J. A., Xu, Y., Kilbourn, M. R., Jewett, D. M., Meyer, C. R., Koeppe, R. A., and Stohler, C. S. 2002. μ-Opioid receptor-mediated antinociceptive responses differ in men and women. J. Neurosci. 22, 5100–5107.

46 Phantom Limb Pain

H Flor, Central Institute of Mental Health, Mannheim, Germany

Glossary

cortical reorganization Changes in the maps of the primary sensory or motor areas of the cortex related to injury or stimulation. Cortical representations can increase or decrease in size or they can shift to other locations on the sensory or motor map.

central sensitization Short- or long-term changes in the excitability of central neurons and their synaptic strength that lead to an enhanced or altered processing of peripheral sensory input resulting in hypersensibility.

neuroma When a limb is severed, a terminal swelling or endbulb is formed with axonal sprouting occurring. In the case of an amputation this sprouting and endbulb formation lead to neuroma, a tangled mass that is formed when the axons cannot reconnect or can only partially reconnect as is the case in partial lesions. These neuroma generate abnormal activity that is called ectopic because it does not originate from the nerve endings.

nonpainful phantom limbs and nonpainful phantom sensations These phenomena refer to the continued presence of the limb (corporeal awareness of the limb) and nonpainful sensations such as tingling or pressure sensations.

nonpainful residual limb sensations The residual limb can also have nonpainful sensations such as tingling or cramping sensations that can be distinguished from phantom sensation as well as from painful sensation.

phantom limb pain or phantom pain Pain in a body part that is no longer present. It may be related to a certain position or movement of the phantom and may be elicited or exacerbated by a range of physical (e.g., changes in weather or pressure on the residual limb) and psychological factors (e.g., emotional stress). It seems to be more intense in the distal portions of the phantom and may have a number of different qualities such as stabbing, throbbing, burning, or cramping. Phantom limb pain is often confused with pain in the area adjacent to the amputated body part.

postamputation pain Pain at the site of the wound that must be distinguished from pain in the residual limb and phantom limb pain that may all co-occur in the early phase after amputation.

preamputation pain Pain that occurred in the amputated body part before the amputation. It may be related to the incidence, type, and severity of phantom limb pain in the phase following the amputation.

residual limb pain or stump pain Pain in the body part adjacent to the amputation. It is usually positively correlated with phantom limb pain.

46.1 Definition

The amputation of a body part is often followed by the sensation that the deafferented body part is still present. These sensations may include not only the feeling of the continued presence of the limb but also nonpainful and painful phantom sensations such as the feeling of a specific position, shape, or movement of the missing limb, feelings of warmth or cold, itching, tingling or electric sensations, and other paresthesias.

Phantom limb pain (or phantom pain) is pain in the body part that is no longer present. This occurs to some degree in 50–80% of all amputees (Nikolajsen, L. and Jensen, T. S., 2006). Phantom limb pain may be related to a certain position or movement of the phantom and may be elicited or exacerbated by a range of physical (e.g., changes in weather or pressure on the residual limb) and psychological factors (e.g., emotional stress). It seems to be more intense in the distal portions of the phantom and may have a number of different qualities such as stabbing, throbbing, burning, or cramping (Hill, A., 1999). Phantom limb pain belongs to the neuropathic pain syndromes and is assumed to be related to damage to the axons of peripheral neurons with secondary changes induced in central neurons. Phantom body pain may also occur following spinal cord injury. Phantom limb pain is infrequent if the amputation occurred at a very young age. However, older children exhibit as high an incidence of phantom limb pain as adults. Although phantom sensations seem to occur occasionally in congenital amputees (i.e., those born without a limb), phantom limb pain seems to be very rare under these circumstances (Flor, H., 2002). The long-term course of phantom limb pain is unclear. While some authors report a slight decline in pain prevalence over the course of several years, others have described high prevalence rates also in long-term amputees. Both peripheral and central factors have been discussed as determinants of phantom limb pain. Psychological factors do not seem to be a primary cause, but they may well affect the course and the severity of the pain (Sherman, R. A., 1997). The general view today is that multiple changes along the neuraxis contribute to the experience of phantom limb pain.

46.2 Peripheral Mechanisms of Phantom Limb Pain

Peripheral changes that give rise to nociceptive input from the residual limb have been viewed as an important determinant of phantom limb pain. This is supported by the moderately high correlation between pain in the residual limb (stump) and phantom limb pain. Ectopic discharge from stump neuromas has been postulated as a potential source of such nociceptive input (Devor, M., 2006). When peripheral nerves are cut or injured, terminal swelling and regenerative sprouting of the injured axon end occur. In this process, neuromas form in the residual limb. The disorganized endings of C fibers and demyelinated A fibers in neuromas have

increased excitability and often show spontaneous impulse activity (ectopic discharge). Mechanical, chemical, and thermal stimulation may further exacerbate this ectopic discharge. The increased excitability of injured nerves that results in ectopic discharge seems to result from upregulation or novel expression, and altered trafficking, of molecules that are responsible for neuronal excitability, such as voltage-sensitive sodium channels (Devor, M., 2006). In addition, abnormal connections between injured axons, such as ephapses, may contribute to the spontaneous ectopic activity. Phantom limb pain is often present very soon after amputation before a palpable, swollen neuroma could have formed. However, ectopic discharge also appears rapidly, apparently originating at first in swollen endbulbs at the cut axon end rather than in outgrowing sprouts.

Local anesthesia of the stump does not eliminate phantom limb pain in all amputees (Birbaumer, N. et al., 1997). This fact motivated a search for other potential sources of ectopic input from the periphery. An additional site of ectopic discharge is the dorsal root ganglion (DRG). Ectopia originating in the DRG can summate with ectopia originating from neuromas in the stump. Indeed, processes such as cross-excitation can lead to the depolarization and activation of neighboring neurons, significantly amplifying the overall ectopic barrage (Devor, M., 2006). In experimental preparations and in humans it was found that anesthetic block of neuromas eliminates spontaneous and stimulation-induced nerve activity related to the stump, but not ectopic activity potentially originating in the DRG (Nystrom, B. and Hagbarth, K. E., 1981). Interestingly, there is evidence for genetic factors in the predisposition to develop ectopic neuroma, DRG discharge, and neuropathic pain. For example, Devor M. et al. (2005) presented evidence for the presence of several genes that predisposes to the pain behavior that follows peripheral neurectomy in rodents. This neuroma model of neuropathic pain has been considered to be a valid animal surrogate of phantom limb pain in humans.

Elevated sympathetic discharge, as well as increased levels of circulating epinephrine, can trigger and exacerbate ectopic neuronal activity from neuromas (Devor, M., 2006). In addition to such sympathetic–sensory coupling at the level of the neuroma, sympathetic–sensory coupling also occurs at the level of the DRG. This may account for the frequent exacerbation of phantom pain at times of emotional distress. Additional factors such as temperature, oxygenation level, and local inflammation in neuromas and associated DRGs may also play a role. The sympathetic maintenance of phantom limb pain in some patients is supported by evidence that systemic adrenergic blocking agents, and targeted chemical or surgical blockade of sympathetic nerves and ganglia, sometimes reduce phantom limb pain. Likewise, injections of epinephrine into stump neuromas have been shown to increase phantom limb pain and paresthesias in some amputees. Although sympathetically maintained pain does not necessarily covary with regional sympathetic abnormalities, in some patients, sympathetic dysregulation in the residual limb is apparent. Reduced near-surface blood flow to a limb has been implicated as a predictive physiological correlate of burning phantom limb pain. Correspondingly, onset and intensity of cramping and squeezing descriptions of phantom pain have been related to muscle tension in the residual limb. This relationship seems not to hold for any other descriptors of phantom pain. In addition, lower skin temperature in the residual limb in amputees with phantom limb pain and phantom sensation and a close relationship between phantom limb sensation and skin conductance responses in the residual limb can be viewed as evidence of a sympathetic-efferent somatic-afferent mechanism (for review see Sherman, R. A., 1997).

46.3 Central Factors

46.3.1 The Spinal Cord

Anecdotal evidence in human amputees first suggested that spinal mechanisms may play a role in phantom limb pain. For example, during spinal anesthesia, phantom pains have been reported to reoccur in patients who were pain-free at the time of the procedure. Experimental data in human amputees are lacking, but increasing amounts of evidence based on animal models of nerve injury are becoming available.

Increased activity in peripheral nociceptors leads to an enduring change in the synaptic responsiveness of neurons in the dorsal horn of the spinal cord, a process called central sensitization (Woolf, C. J. and Salter, M., 2006). Central sensitization is also triggered by nerve injury such as occurs during amputation. For example, spinal changes associated with nerve injury include increased firing of the dorsal horn neurons, structural changes at the central endings of the primary sensory neurons, and reduced inhibitory processes. Inhibitory GABAergic and glycinergic interneurons in the spinal cord may actually be destroyed by rapid ectopic discharge or other effects of axotomy, contributing to a hyperexcitable spinal cord (Scholz, J. and Woolf, C., 2006). Changes have also been noted in ascending projection neurons. The cascade of biological events that take place in the spinal cord after peripheral nerve damage may trigger abnormal firing of spinal origin. Part of this sensitization is due to facilitation of the response of N-methyl-D-aspartic acid (NMDA) receptors to the primary afferent neurotransmitter glutamate (Sandkühler, J., 2000). A remarkable effect of the spinal changes evoked by nerve injury is that low-threshold afferents may become functionally connected to ascending spinal projection neurons that carry nociceptive information. When this happens, normally innocuous A-fiber input from the periphery, ectopic input as well as input from residual intact low-threshold afferents, may contribute to phantom pain sensation.

A number of additional central nervous system (CNS) processes are thought to contribute to the hyperexcitability of spinal cord circuitry following major nerve damage. For example, there may be downregulation of opioid receptors, both on primary afferent endings and on intrinsic spinal neurons. This is expected to add to disinhibition due to the reduction of GABA and glycine activity. In addition, cholycystokinin, an endogenous inhibitor of the opiate receptor, is upregulated in injured tissue (for review see Woolf, C. J. and Salter, M., 2006). Another interesting example of changed gene expression after axotomy is the appearance of the neuropeptide substance P in low-threshold Aβ neurons. Substance P is normally expressed only by Aδ and C afferents, most of which are nociceptors. The injury-triggered expression of substance P by Aβ fibers may render them more like nociceptors. For example, it may permit ectopic or normal activity in Aβ fibers to trigger and maintain central sensitization. Changes such as these in gene expression in injured afferents (and in some postsynaptic spinal neurons), which result in a change in their functioning (i.e., their phenotype), are referred to as phenotypic switch. Recent work based on gene chip technology indicates that hundreds of genes are up- or downregulated in DRG and spinal neurons following peripheral nerve injury.

Following peripheral nerve injury, degeneration of central projection axon occurs. Massive deafferentation is observed when dorsal roots are injured, or avulsed from the spinal cord. Deafferentation may act hand-in-hand with the central effects of peripheral denervation to bring about the changes that contribute to spinal hyperexcitability. A mechanism that may be of special relevance to phantom phenomena is the invasion of regions of the spinal cord functionally vacated from injured afferents. For example, in the neuroma model in rats and cats, there is an expansion of receptive fields on skin adjacent to the denervated part of the limb, and a shift of activity from these adjacent areas into regions of the spinal cord that previously served the part of the limb that was functionally deafferented by the nerve lesion (Devor, M. and Wall, P. D., 1978). Such reorganization of the spinal map of the limb, which is probably due to unmasking of previously silent connections, is reflected in brainstem and cortical remapping also.

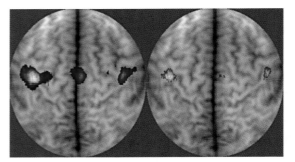

Figure 1 The representation of lip movements in amputees with phantom limb pain (left) and without phantom limb pain (right) in primary somatosensory cortex based on functional magnetic resonance imaging. Note that the amputees with phantom limb pain activate both the cortical hand and the mouth representation whereas the amputees without phantom limb pain activate only the cortical mouth representation.

46.3.2 Supraspinal Changes

Supraspinal changes related to phantom limb pain involve the brain stem, the thalamus, and the cerebral cortex (for review see Flor, H. *et al.*, 2006). New insights into phantom limb pain have come from studies demonstrating changes in the functional and structural architecture of primary somatosensory cortex subsequent to amputation and deafferentation in adult monkeys. In these studies, the amputation of digits in an adult owl monkey led to an invasion of adjacent areas into the representation zone of the deafferented fingers. Several imaging studies (for review see Flor, H., 2002) have reported that upper extremity amputees actually show a shift of the mouth into the hand representation in primary somatosensory and motor cortex (Figure 1). Flor H. *et al.* (1995) provided evidence that these cortical changes are less related to referred sensations but rather have a close association with phantom limb pain. The larger the shift of the mouth representation into the zone that formerly represented the amputated hand and arm the larger the phantom limb pain. These cortical changes could be reversed by the elimination of peripheral input from the amputation stump using brachial plexus anesthesia. Peripheral anesthesia completely eliminated cortical reorganization and phantom limb pain in half of the amputees that were studied. In the remaining half, both cortical reorganization and phantom limb pain remained unchanged (Birbaumer, N. *et al.*, 1997). This result suggests that in some amputees, cortical reorganization and phantom limb pain may be maintained by

peripheral input whereas in others central, potentially intracortical changes might be more important.

It is so far not known to what extent spinal changes contribute to these supraspinal alterations. It was shown that axonal sprouting in the cortex underlies the reorganizational changes observed in amputated monkeys (Florence, S. L. *et al.*, 1998). Thalamic stimulation and recordings in human amputees have revealed that reorganizational changes may also occur at the thalamic level and are closely related to the perception of phantom limbs and phantom limb pain (Davis, K. D. *et al.*, 1998). Studies in animals have shown that these changes may be relayed from the spinal and brain stem level; however, changes on the subcortical levels may also originate in the cortex, which has strong efferent connections to the thalamus and lower structures.

Sometimes pain in the phantom is similar to the pain that existed in the limb prior to amputation. The likelihood of this ranges from 10% to 79% in different reports and depends on the type and time of assessment (see Katz, J. and Melzack, R., 1990). The type and time of assessment and potential errors in retrospective reports are important determinants of the incidence of these pain memories. It has been proposed that pain memories established prior to the amputation are powerful elicitors of phantom limb pain (Katz, J. and Melzack, R., 1990; Flor, H., 2002). Pain memories may be implicit and not readily accessible to conscious recollection. The term implicit pain memory refers to central changes related to nociceptive input that lead to subsequent altered processing of the somatosensory system and do not require changes in conscious processing of the pain experience (Flor, H., 2002). In patients with chronic back pain it was shown that increasing chronicity is correlated with an increase in the representation zone of the back in primary somatosensory cortex, and it was also reported that the experience of acute pain alters the map in primary somatosensory cortex. These data suggest that long-lasting noxious input may lead to long-term changes at the central and especially at the cortical level. It has long been known that the primary somatosensory cortex is involved in the processing of pain and that it may be important for the sensory-discriminative aspects of the pain experience. There have also been reports that phantom limb pain was abolished after the surgical removal of portions of the primary somatosensory cortex and that stimulation of somatosensory cortex evoked phantom limb pain. If a somatosensory pain memory has been established with an important neural correlate in spinal and supraspinal structures, such as in primary somatosensory cortex, subsequent deafferentation and an invasion of the amputation zone by neighboring input may preferentially activate cortical neurons coding for pain. Since the cortical area coding input from the periphery seems to stay assigned to the original zone of input, the activation in the cortical zone representing the amputated limb is referred to this limb and the activation is interpreted as phantom sensation and phantom limb pain. It is likely that not only the areas involved in sensory-discriminative aspects of pain reorganize but also that those areas that mediate affective–motivational aspects of pain such as the insula and the anterior cingulate cortex undergo plastic changes that contribute to the experience of phantom pain. A prospective study (Larbig *et al.*, unpublished data) showed that the best predictor of phantom limb pain 12 months after an amputation is chronic pain before the amputation thus supporting the pain memory hypothesis. However, these authors tested a sample that did not include traumatic amputees but mainly amputees with long-standing prior pain problems. Further research is needed to better clarify these relationships.

46.4 Implications for the Treatment and Prevention of Phantom Limb Pain

Several studies, including large surveys of amputees, have shown that most treatments for phantom limb pain are ineffective and fail to consider the mechanisms underlying the production of the pain (Sindrup, S. H. and Jensen, T. S., 1999). Most studies are uncontrolled short-term assessments of small samples of phantom limb pain patients. The maximum benefit reported from a host of treatments such as local anesthesia, sympathectomy, dorsal root entry zone lesions, cordotomy and rhizotomy, neurostimulation methods, or pharmacological interventions such as anticonvulsants, barbiturates, antidepressants, neuroleptics, and muscle relaxants seems to be around 30%. This does not exceed the placebo effect reported in other studies. Controlled studies have been performed for opioids, calcitonin, ketamine, dextromethorphan, and gabapentin (see Nikolajsen, L. and Jensen, T. S., 2006), all of which were found to effectively reduce phantom limb pain. Mechanism-based treatments are rare but have been shown to be effective in a few small but mostly uncontrolled studies. Lidocaine was found to reduce phantom limb pain of patients with neuromas in two small-sample controlled studies. Biofeedback treatments resulting in vasodilatation of the residual limb or decreased muscle tension in the residual limb help to reduce phantom limb pain and seem promising in patients where peripheral factors contribute to the pain (Flor, H., 2002).

Based on the findings from neuroelectric and neuromagnetic source imaging, changes in cortical reorganization might influence phantom limb pain. Animal work on stimulation-induced plasticity would suggest that extensive behaviorally relevant stimulation of a body part leads to the expansion of its representation zone. It was shown that intensive use of a myoelectric prosthesis was positively correlated with both reduced phantom limb pain and reduced cortical reorganization. When cortical reorganization was partialled out, the relationship between prosthesis use and reduced phantom limb pain was no longer significant suggesting that cortical reorganization mediates this relationship. An alternative approach in patients where prosthesis use is not viable is the application of behaviorally relevant stimulation. A 2-week training that consisted of a discrimination training of electric stimuli to the stump for 2 h per day led to significant improvements in phantom limb pain and a significant reversal of cortical reorganization (Flor, H. *et al.*, 2001). A control group of patients who received standard medical treatment and general psychological counseling in this time period did not show similar changes in cortical reorganization and phantom limb pain. The basic idea of the treatment was to provide input into the amputation zone and thus undo the reorganizational changes that occurred subsequent to the amputation. Mirror treatment (where a mirror is used to trick the brain into perceiving movement of the phantom when the intact limb is moved) might be effective but has so far only been tested in an anecdotal manner.

Preemptive analgesia refers to the attempt to prevent chronic pain by early intervention before acute pain occurs, for example, before and during surgery. Based on the data on sensitization of spinal neurons by afferent barrage it has been suggested that general anesthesia should be complemented by peripheral anesthesia thus preventing peripheral nociceptive input from reaching the spinal cord and higher centers. However, preemptive analgesia that included both general and spinal anesthesia has not consistently been efficacious in preventing the onset of phantom limb pain (for review see Nikolajsen, L. and Jensen, T. S., 2006). A preexisting pain memory that has already led to central and especially cortical changes would not necessarily be affected by a short-term elimination of afferent barrage. Thus, it is possible that peripheral analgesia would eliminate new but not preexisting central changes in the perioperative phase. In addition, a short-term blockade is not sufficient to prevent discharges from severed peripheral nerves to reach the CNS. The perioperative use of NMDA receptor antagonists or GABA agonists might therefore be useful.

46.5 Future Developments

Both peripheral and central factors and their interaction need to be examined more closely in animal models of amputation-related pain and in human amputees. The role of spinal mechanisms has so far not been sufficiently elucidated. The detection of genes relevant for the development of phantom pain-like behaviors in animal is an important step and may aid in the identification of predisposing factors for phantom limb pain as well as in the development of new interventions. The development of more powerful treatments for phantom limb pain needs controlled treatment outcome, prospective and double-blind placebo-controlled outcome research. Only then effective evidence-based interventions will be available.

Acknowledgment

The preparation of this article was supported by a grant from the Deutsche Forschungsgemeinschaft (FL 156/29).

References

Birbaumer, N., Lutzenberger, W., Montoya, P., Larbig, W., Unertl, K., Töpfner, S., and Flor, H. 1997. Effects of regional anesthesia on phantom limb pain are mirrored in changes in cortical reorganization. J. Neurosci. 17, 5503–5508.

Davis, K. D., Kiss, Z. H. T., Luo, L., Tasker, R. R., Lozano, A. M., and Dostrovsky, J. O. 1998. Phantom sensations generated by thalamic microstimulation. Nature 391, 385–387.

Devor, M. 2006. Response of Nerves to Injury in Relation to Neuropathic Pain. In: Melzack and Wall's Textbook of Pain, 5th edn. (eds. S. McMahon and M. Koltzenburg), pp. 905–927. Churchill-Livingstone.

Devor, M. and Wall, P. D. 1978. Reorganization of the spinal cord sensory map after peripheral nerve injury. Nature 276, 76.

Devor, M., del Canho, S., and Raber, P. 2005. Heritability of symptoms in the neuroma model of neuropathic pain: replication and complementation analysis. Pain 116, 294–301.

Flor, H. 2002. Phantom limb pain: characteristics, aetiology and treatment. Lancet Neurology 3, 182–189.

Flor, H., Denke, C., Schaefer, M., and Grüsser, S. 2001. Sensory discrimination training alters both cortical reorganization and phantom limb pain. Lancet 357, 1763–1764.

Flor, H., Elbert, T., Knecht, S., Wienbruch, C., Pantev, C., Birbaumer, N., and Taub, E. 1995. Phantom limb pain as a perceptual correlate of cortical reorganization following arm amputation. Nature 357, 482–484.

Flor, H, Nikolajsen, L, and Jensen, T. 2006. Neuroplastic changes in phantom limbs and phantom limb pain. Nat. Neurosci. Rev. 1, 813–881.

Florence, S. L., Taub, H. B., and Kaas, J. H. 1998. Large-scale sprouting of cortical connections after peripheral injury in adult macaque monkeys. Science 282, 1117–1120.

Hill, A. 1999. Phantom limb pain: a review of the literature on attributes and potential mechanisms. J. Pain Sympt. Manage 17, 125–142.

Jensen, T. S. and Nikolajsen, L. 2000. Pre-emptive analgesia in postamputation pain: an update. Prog. Brain Res. 129, 493–503.

Katz, J. and Melzack, R. 1990. Pain 'memories' in phantom limbs: review and clinical observations. Pain 43, 319–336.

Nikolajsen, L. and Jensen, T. S. 2006. Phantom Limb and Other Postamputation Phenomena. In: Wall and Melzacks Textbook of Pain, 5th edn. (eds. M. Koltzenburg and S. McMahon), pp. 961–971. Churchill-Livingstone.

Nystrom, B. and Hagbarth, K. E. 1981. Microelectrode recordings from transsected nerves in amputees with phantom limb pain. Neurosci. Lett. 27, 211–216.

Sandkühler, J. 2000. Learning and memory in pain pathways. Pain 88, 113–118.

Sherman R. A. (ed.) 1997. Phantom Limb Pain. Plenum.

Sindrup, S. H. and Jensen, T. S. 1999. Efficacy of pharmacological treatments of neuropathic pain: an update and effect related to mechanism of drug action. Pain 83, 389–400.

Scholz, J. and Woolf, C. 2006. Neuropathic Pain: A Neurodegenerative Disease. In: Proceedings of the 11th

World Congress on Pain (eds. H. Flor, E. Kalso. and J. Dostrovsky), pp. 373–384. IASP Press.

Woolf, C. J. and Salter, M. 2006. Plasticity and Pain: Role of the Dorsal Horn. In: Melzack and Wall's Textbook of Pain, 5th edn. (eds. S. McMahon and M. Koltzenburg), pp. 91–105. Churchill-Livingstone.

Further Reading

Abraham, R. B., Marouani, N., and Weinbroum, A. A. 2003. Dextromethorphan mitigates phantom pain in cancer amputees. Ann. Surg. Oncol. 10, 268–274.

Appenzeller, O. and Bicknell, J. M. 1969. Effects of nervous system lesion on phantom experience in amputees. Neurology 19, 141–146.

Bone, M., Critchley, P., and Buggy, D. J. 2002. Gabapentin in postamputation phantom limb pain: a randomized, double-blind, placebo-controlled, cross-over study. Reg. Anesth. Pain Med. 27, 481–486.

Chabal, C., Jacobson, L., Russell, L. C., and Burchiel, K. J. 1989. Pain responses to perineuromal injection of normal saline, gallamine, and lidocaine in humans. Pain 36, 321–325.

Chabal, C., Jacobson, L., Russell, L. C., and Burchiel, K. J. 1992. Pain response to perineuromal injection of normal saline, epinephrine, and lidocaine in humans. Pain 49, 9–12.

Cronholm, B. 1951. Phantom limbs in amputees. A study of changes in the integration of centripetal impulses with special reference to referred sensations. Acta. Psychiatr. Neurol. Scand. Suppl. 72, 1–310.

Devor, M. and Govrin-Lippman, R. 1993. Angelides K. Na+ channels immunolocalization in peripheral mammalian axons and changes following nerve injury and neuroma formation. J. Neurosci. 13, 1976–1992.

Devor, M. and Raber, P. 1990. Heritability of symptoms in an experimental model of neuropathic pain. Pain 42, 51–67.

Devor, M., Jänig, W., and Michaelis, M. 1994. Modulation of activity in dorsal root ganglion neurones by sympathetic activation in nerve-injured rats. J. Neurophysiol. 71, 38–47.

Ehde, D. M., Czerniecki, J. M., Smith, D. G., Campbell, K. M., Edwards, W. T., et al. 2000. Chronic phantom sensations, phantom pain, residual limb pain, and other regional pain after lower limb amputation. Arch. Phys. Med. Rehab. 8, 1039–1044.

Ephraim, P. L., Wegener, S. T., Mackenzie, E. J., Dillingham, T. R., and Pezzin, L. E. 2005. Phantom pain, residual limb pain, and back pain in amputees: results of a national survey. Arch. Phys. Med. Rehabil. 86, 1910–1919.

Ergenzinger, E. R., Glasier, M. M., Hahm, J. O., and Pons, T. P. 1998. Cortically induced thalamic plasticity in the primate somatosensory system. Nat. Neurosci. 1, 226–229.

Flor, H., Braun, C., Elbert, T., and Birbaumer, N. 1997. Extensive reorganization of primary somatosensory cortex in chronic back pain patients. Neurosci. Lett. 224, 5–8.

Flor, H., Elbert, T., Mühlnickel, W., Pantev, C., Wienbruch, C., and Taub, E. 1998. Cortical reorganization and phantom phenomena in congenital and traumatic upper-extremity amputees. Exp. Brain Res. 119, 205–212.

Florence, S. L. and Kaas, J. H. 1995. Large-scale reorganization at multiple levels of the somatosensory pathway following therapeutic amputation of the hand in monkey. J. Neurosci. 15, 8083–8095.

Grüsser, S., Winter, C., Mühlnickel, W., Denke, C., Karl, A., Villringer, K., and Flor, H. 2001. The relationship of perceptual phenomena and cortical reorganization in upper extremity amputees. Neuroscience 102, 263–272.

Huse, E., Larbig, W., Flor, H., and Birbaumer, N. 2001. The effect of opioids on phantom limb pain and cortical reorganization. Pain 90, 47–55.

Jaeger, H. and Maier, C. 1992. Calcitonin in phantom limb pain: a double-blind study. Pain 48, 21–27.

Jensen, T. S., Krebs, B., Nielsen, J., and Rasmussen, P. 1985. Immediate and long-term phantom limb pain in amputees: incidence, clinical characteristics and relationship to preamputation pain. Pain 21, 267–278.

Jensen, T. S., Krebs, B., Nielsen, J., and Rasmussen, P. 2000. Immediate and long-term phantom limb pain in amputees: Incidence, clinical characteristics and relationship to pre-amputation of somatosensory and motor cortex after peripheral nerve or spinal cord injury in primates. Prog. Brain Res. 128, 173–179.

Katz, J. 1992. Pychophysical correlates of phantom limb experience. J. Neurol. Neurosurg. Psychiatry 55, 811–821.

Katz, J. and Melzack, R. A. 1991. Auricular transcutaneous electrical nerve stimulation (TENS) reduces phantom limb pain. J. Pain Sympt. Manage 6, 77–83.

Karl, A., Birbaumer, N., Lutzenberger, W., Cohen, L. G., and Flor, H. 2001. Reorganization of motor and somatosensory cortex in upper extremity amputees with phantom limb pain. J. Neurosci. 21, 3609–3618.

Kooijman, C. M., Dijkstra, P. U., Geertzen, J. H. B., Elzinga, A., and van der Schans, C. P. 2000. Phantom pain and phantom sensations in upper limb amputees: an epidemiological study. Pain 87, 33–41.

Krane, E. J. and Heller, L. B. 1995. The prevalence of phantom sensation and pain in pediatric amputees. J. Pain Sympt. Manage 10, 21–29.

Lotze, M., Grodd, W., Birbaumer, N., Erb, M., Huse, E., Larbig, W., and Flor, H. 1999. Does use of a myoelectric prosthesis prevent cortical reorganization and phantom limb pain? Nat. Neurosci. 2, 501–502.

Mackenzie, N. 1983. Phantom limb pain during spinal anaesthesia. Recurrence in amputees. Anaesthesia 38, 886–887.

MacLachlan, M., McDonald, D., and Waloch, J. 2004. Mirror treatment of lower limb phantom pain: a case study. Disabil. Rehabil. 26, 901–904.

Nathan, P. W. 1983. Pain and the sympathetic system. J. Auton. Nerv. Syst. 7, 363–370.

Nikolajsen, L. and Jensen, T. S. 2001. Phantom limb pain. Br. J. Anaesth. 87, 107–116.

Nikolajsen, L., Gottrup, H., Kristensen, A. G., and Jensen, T. S. 2000. Memantine (a N-methyl-D-aspartate receptor antagonist) in the treatment of neuropathic pain after amputation or surgery: a randomized, double-blinded, cross-over study. Anesth. Analg. 91, 960–966.

Nikolajsen, L., Hansen, C. L., Nielsen, J., Keller, J., Arendt-Nielsen, L., et al., 1996. The effect of ketamine on phantom limb pain: a central neuropathic disorder maintained by peripheral input. Pain 67, 69–77.

Nikolajsen, L., Ilkjaer, S., Christensen, J. H., Kroner, K., and Jensen, T. S. 1997a. Randomised trials of epidural bupivacaine and morphine in prevention of stump and phantom pain in lower-limb amputation. Lancet 350, 1353–1357.

Nikolajsen, L., Ilkjaer, S., and Jensen, T. S. 2000. Relationship between mechanical sensitivity and post-amputation pain: a perspective study. Eur. J. Pain 4, 327–334.

Nikolajsen, L., Ilkjaer, S., Kroner, K., Christensen, J. H., and Jensen, T. S. 1997b. The influence of preamputation pain on postamputation stump and phantom pain. Pain 72, 393–405.

Ramachandran, V. S., Rogers-Ramachandran, D. C., and Stewart, M. 1992. Perceptual correlates of massive cortical reorganization. Science 258, 1159–1160.

Sherman, R. A. 1989. Stump and phantom limb pain. Neurol. Clin. 7, 249–264.

Sherman, R. A. and Sherman, C. J. 1983. Prevalence and characteristics of chronic phantom limb pain. Results of a trial survey. Am. J. Phys. Med. 62, 227–238.

Soros, P., Knecht, S., Bantel, C., Imai, T., Wusten, R., *et al.* 2000. Functional reorganization of the human primary somatosensory cortex after acute pain demonstrated by magnetoencephalography. Neurosci. Lett. 298, 195–198.

Wei, F. and Zhuo, M. 2001. Potentiation of sensory responses in the anterior cingulate cortex following digit amputation in the anaesthetised rat. J. Physiol. 532, 823–833.

Wiech, K., Preissl, H., Kiefer, T., Töpfner, S., Pauli, P., *et al.* 2001. Prevention of phantom limb pain and cortical reorganization in the early phase after amputation in humans. Soc. Neurosci. Abstr. 28, 163.

Wilkins, K. L., McGrath, P. J., Finley, G. A., and Katz, J. 1998. Phantom limb sensations and phantom limb pain in child and adolescent amputees. Pain 78, 7–12.

47 Human Insular Recording and Stimulation

F Mauguière, Lyon I University and INSERM U879, Bron, France

M Frot, INSERM U879, Bron France

J Isnard, Lyon I University and INSERM U879, Bron, France

Glossary

cytoarchitectony (cytoarchitectonic)
Anatomical method based on microscopic observation of brain slices used to delineate cortical areas according to the organization of the different types of gray matter cells (neurons). This technique permits the drawing of cytoarchitectonic maps of the cortex, of which the most widely used is that of Korbinian Brodmann, first described in 1909, which counts 52 distinct areas in the human brain. The thickness of two different types of cortical cells layers (granule and pyramidal cells) and the number of layers are the two main parameters used for cytoarchitectonic mapping. Granule cells are over-represented in primary sensory (visual, auditory, and somatosensory) cortical areas (granular cortex), while large pyramidal cells are characteristic of the primary motor cortex (agranular cortex).

deep-brain stimulation Technique involving stimulating the brain with electric current pulses of low intensity using depth electrodes implanted through the skull. In humans, this technique can be used for therapeutic or diagnostic purposes. High-frequency continuous (or chronic) deep-brain stimulation is used as a treatment in various neurological diseases, e.g., Parkinson's disease. For diagnosis purposes, single pulses or trains of pulses are used mostly in presurgical assessment

of patients with drug-resistant focal epileptic seizures in order to map the seizure focus (or
epileptogenic zone) and the functional areas surrounding the focus. Chronically implanted
electrodes are painless; they also permit the
recording of ongoing brain activity and evoked
potentials.

***discharge (epileptic discharge, after-
discharge)*** Abnormal electroencephalographic
(EEG) activity with abrupt onset and termination
lasting at least several seconds seen during an
epileptic seizure; discharges can be recorded
either by scalp or by depth electrodes. They can
occur spontaneously or be triggered by deep-brain
stimulation and then named after-discharges.

electroencephalography (EEG) The record of
electrical activity of the brain taken by means of
electrodes placed on the surface of the scalp. First
applied to study the activity of the human brain by
Hans Berger in 1924, this noninvasive technique is
still used routinely for the diagnosis of epilepsy.

evoked potentials (EPs) Brain electric response
(wave or complex of waves) elicited by and time-
locked to a physiological or nonphysiological stimulus, the timing of which can be reliably
assessed, e.g., a short duration. Laser beam
applied to the skin surface. Because of their low
voltage most of EPs are not evident from the
ongoing electroencephalographic (EEG) activity
after a single stimulation. Repeated stimulations
and computer summation techniques are thus
needed for their detection. The number of single
responses to be summated is a function of the
signal-to-noise ratio (ratio between the EP and
ongoing EEG activity voltages); this number is
higher for scalp than for depth intracerebral
recordings.

somatotopy (somatotopic) Representation maps
of peripheral skin areas and joints in the somatosensory cortex that were first described in the
human brain by Wilder Penfield using electric stimulation of parietal cortex. The even existence of
such maps demonstrates that granule cells receiving inputs from the same peripheral areas are
grouped together in the somatosensory cortex.
Microelectrode recordings in mammalian brain
including monkeys have shown that separate
somatotopic maps of the peripheral skin and joint
receptors can be drawn in the somatosensory cortical areas.

somatosensory areas (S1, S2) Cortical areas
containing a full somatotopic map of the contralateral body half that have been identified in mammals
and human brain by means of electric cortical stimulation, evoked potentials, brain imaging
(functional magnetic resonance imaging), and unit
or multiple cell responses. The human primary
sensory area (S1) is located in the postcentral parietal gyrus and the second somatosensory area
(S2) is located in the upper bank of the sylvian
fissure (parietal operculum).

stereotaxy (stereotactic) The stereotactic space
first described by Jean Talairach is a three-dimensional representation of the human brain in which
each point in the brain can be located according to
its coordinates (x, y, and z) in a Cartesian space.
The space is defined by the intersection of the mid-
sagittal plane and the plane perpendicular to mid-
sagittal plane passing through the anterior and
posterior commissure of the brain (AC–PC). Using
depth electrodes implanted perpendicular to the
mid-sagittal plane, any given recording or stimulation site can be localized by its stereotactic
coordinates.

47.1   Anatomy, Connections, and Physiology of the Insula in Primates

47.1.1   The Insula as the Fifth Lobe of the Brain

In an article published in 1896 that was devoted to the
comparative anatomy of the insula Clark T. E. (1896)
quoted 39 synonyms used in anatomical literature to
name the fifth lobe of the brain buried in the lateral
sulcus and covered by the opercular parts of the frontal,
parietal, and temporal lobes, among which the term
insula, first proposed by Reil in 1804, has prevailed.

Due to its anatomical situation in depth of brain lateral
fissure (see Figure 1 and Tanriover, N. *et al.*, 2004), and
its cytoarchitectonic organization the insula has long be
considered by anatomists as a rather isolated lobe of the
brain mostly devoted to the processing of body and
visceral sensation including gustation, pain, and other
emotions, as well as to visceromotor and autonomic
control. In primates including humans, the insula
shows a caudorostral sequence of three distinct
cytoarchitectonic areas namely; (1) a large area of granular cortex, at its upper and posterior part, whose
structure is very similar to that of the second

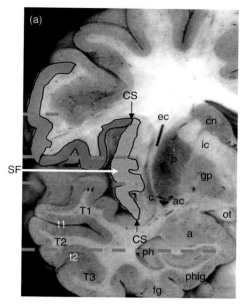 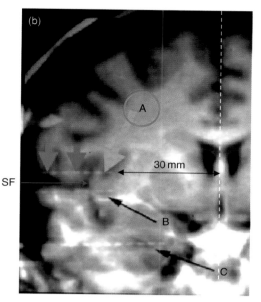

Figure 1 Anatomy of the human insular lobe (frontal slices). A) Anatomical fontal brain section showing the deep location of the insula (in yellow) in the sylvian fissure (SF, white arrow), which separates the frontal and parietal lobes (above) from the temporal lobe (below). The second somatosensory area (S2 in green), which occupies the superior bank of the SF, has common borders with the lower part of the primary somatosensory area (S1 in purple), and the insula. The circular sulcus (CS) separates the insula from the parietal operculum S2 and the first temporal gyrus (T1). The dashed red lines indicates the trajectory of depth electrodes used for insular, parietal, and temporal lobes recordings in patients with epilepsy of the temporal lobe and perisylvian cortex (see Isnard, J. et al., 2004). See also Figure 4 for sagittal view of the insula. B) Magnetic resonance imaging (MRI) slice showing the projection of the recording electrodes implanted horizontally perpendicular to the midline sagittal plane (dashed vertical white line). Recording contacts appears as small white segments along the electrodes trajectory. The deepest contacts of electrodes A and B are located in the insular cortex. The depths coordinates (x) of each contact is calculated from the midline sagittal plane in the human brain stereotactic space of Talairach. The x coordinates of insular recording sites vary from 27 mm to 36 mm from midline. Electrode A (red circle) contacts explore from depth to surface: the insula (yellow arrow), the S2 area (green arrow; $38 < x < 46$ mm) and the lower part of the S1 area (purple arrow; $48 < x < 63$ mm). Electrode B contacts explore the insula and the first temporal gyrus (T1 in Figure 1A). Electrode C contacts explore the second temporal gyrus (T2 in Figure 1B) and hippocampus. a, amygdala; ac, anterior commissure; c, claustrum; CS, circular sulcus; cn, caudate nucleus; ec, external capsule; fg, fusiform grus; gp, globus pallidus; ic, internal capsule; ot, optic tract; phig, parahippocamplal gyrus; SF, sylvian fissure; T1, first temporal gyrus; t1, first temporal sulcus; T2, second temporal gyrus; t2, second temporal sulcus; T3, third temporal gyrus.

somatosensory (S2) cortical area, which is involved mostly in the processing of somatosensory and pain sensation; (2) a transitional dysgranular field localized in its anterosuperior part involved in gustation and visceral sensation; and (3) an anteroventral agranular field, which is in continuity with the temporal pole and olfactory proisocortex, and related to olfactory and autonomic functions (see Tanriover, N. *et al.*, 2004 for a review).

47.1.2 The Insula as a Node in a Distributed Cortical Network

The insula is characterized by anatomical borders that are defined by a limiting sulcus (the circuminsular

fissure) but also by fuzzy cytoarchitectonic borders with neighboring cortical areas and by a dense network of corticocortical connections with adjacent or more distant cortical areas. Therefore its function cannot be sketched as an isolated functional center, as suggested by the term insula. A complete description of insular connections is given in the review by Augustine J. R. (1996) showing that the insula is connected to the limbic areas, amygdalar nucleus, basal ganglia, and all of the cortical lobes, except the occipital lobe. Several attempts have been made to identify functional networks in which the insula could play the role of a functional node. Among these networks the perisylvian–insular, the temporo–limbic–insular and the mesial–orbitofrontal–insular networks (Mesulam, M. M. and Mufson, E. J., 1982) deserve the attention of pain physiologists because they are involved in pain localization and intensity encoding, as well as in emotional and behavioral reaction to pain.

47.1.3 The Insula as a Polymodal Area

The question of insular physiology has been addressed by studies using neuroimaging, evoked potentials (EPs), and direct stimulation in humans as well microelectrode studies in monkeys (Augustine, J. R., 1996). These studies have confirmed that the insula is involved in somatosensory and pain sensation as assessed by numerous functional imaging studies in humans (see Peyron, R. *et al.*, 2002 for a review), as well as by the somatosensory and pain-evoked response recordings and direct electric stimulations of the insular cortex that are detailed below in this chapter. However, the insula also proved to be a highly organized lobe with multiple specific functions other than pain including in particular:

Visceral sensation, visceromotor control.

Cardiovascular function as demonstrated by insular stimulation in epileptic patients that produces changes in heart rate or blood pressure in 50% of cases (Oppenheimer, S. M. *et al.*, 1992), thus leading to suspect a role of insular discharges in cardiac arrhythmias causing sudden death during epileptic seizures.

Gustation as assessed first by the stimulation studies of Penfield W. and Faulk M. E. (1955) and further confirmed by neuroimaging studies and microelectrode recordings of insular neurons in monkeys. These physiological findings are consistent with the altered taste perception observed in patients with insular lesions.

Audition and language, in particular allocation of auditory attention, tuning in to novel auditory stimuli, temporal, and phonological processing of auditory stimuli. Furthermore several studies suggest that both right and left insulas are involved in the control of speech production.

47.1.4 The Insula as Part of the Mirror-Neuron System

Some recent studies suggest that the insular lobe could belong to the mirror-neuron system that characterizes regions of the brain that are able to respond when the subject performs an action and when the subject observes another individual doing a similar action

(see Rizzolatti, G. and Craighero, L., 2004 for a review). The concept also refers to regions of the brain that are able to encode for a sensation (or emotion) perceived by the subject and to respond to the observation of others experiencing that sensation (or emotion). In the human insula, the regions involved in visceral sensation or visceromotor responses are also responding to faces expressing disgust (Krolak-Salmon, P. *et al.*, 2003). Similarly the human insula respond to both pain perception and empathy for others' pain (Jackson, P. L. *et al.*, 2005).

47.2 Rational, Ethical Limitations and Procedure of Insular Recording and Stimulations in Humans

47.2.1 Recording and Stimulation of the Human Insula Are Justified only in the Context of Presurgical Evaluation of Epilepsy

In humans, electrical stimulation of the cortex, as well as invasive intracranial recordings of the cerebral activity, including surface recordings using electrodes placed on the cortex (electrocorticography) and depth recordings using electrodes implanted directly in the cortex (stereotactic electroencephalography or SEEG), aims at localizing the area that produces focal epileptic seizures. This area, referred to as the epileptogenic zone (EZ; Rosenow, F. and Luders, H. O., 2001), is that which is surgically removed to obtain seizure freedom in patients whose epilepsy cannot be controlled by conventional drug treatment. Ethically such recordings are justified only when the EZ cannot be localized noninvasively either by scalp recordings or neuroimaging techniques. In this context, invasive explorations aim mostly at recording spontaneous seizures. They are also used to perform a functional mapping of the cortex located in, or in the vicinity of, the EZ, in order to predict the functional consequences of the surgical treatment, which consists of removing the EZ cortex. This procedure, named cortectomy, is possible provided that the EZ is not located in an eloquent area involved in essential cortical functions such as language, memory, motor control, and vision. Both the recording of responses evoked by sensory stimulations and direct electrical stimulation of the cortex are commonly used to achieve this functional mapping in epilepsy surgery centers around the world. This context obviously entails some limitations in the use of depth cortical recordings and stimulation in studies of human brain physiology.

47.2.2 Depth Intracortical Electrodes Are Needed to Explore the Insula

Due to its deep location in the lateral fissure the insula can be explored either by electrocorticography during surgery in patients whose frontoparietal or temporal operculum has been removed by a previous cortectomy, as was the case in the few patients first explored in Montreal by Penfield W. and Faulk M. E. (1955), or by depth SEEG electrodes implanted perpendicular to the mid-sagittal plane through the opercular cortex (Figure 1). These electrodes usually have five to ten contacts, each of 2 mm in length separated by 1.5 mm and can be left in place chronically up to 15 days (Isnard, J. *et al.*, 2000; 2004). Only the deepest contacts of such electrodes reach the insular cortex while the superficial ones explore the frontoparietal or temporal opercular cortex. Therefore, concerning somatosensory responsive areas located close to the lateral fissure, it must be checked on individual brain magnetic resonance imaging (MRI) whether contacts are located in the suprasylvian external parietal cortex (area S1), in the parietal operculum cortex (area S2), or in the insula. Oblique electrodes trajectories can also been used to explore the insula, thus increasing the number of contacts located in the insular cortex.

47.2.3 The Number of Electrodes Is Limited by Ethical Issues

Only the minimal number of electrodes that are useful to diagnosis can be implanted. Consequently the spatial resolution of the mapping in each individual is low, and pooling interindividual data is necessary to draw topographic functional maps of the human insula based on depth electrode data.

47.2.4 The Site of Electrode Implantation Is Guided by Diagnostic Purposes

Implantation sites are guided by the hypothesis issued from noninvasive investigations concerning the most probable EZ location. Furthermore the trajectory of electrode tracts is restricted by the anatomy of blood vessels that are particularly dense in the lateral fissure. Therefore, some parts of the insula are rarely or never explored.

47.2.5 Only Data Recorded in Nonepileptogenic Cortex are Physiologically Relevant

Data from cortical recordings or stimulations can be considered as reflecting the normal physiology of the brain on the condition that they have been obtained in cortical areas that are not involved in the epileptogenic process and do not show increased excitability to peripheral inputs or local electric fields produced by stimulation. Two criteria are commonly used to check this, which are first the absence of insular epileptic discharge in the insula during spontaneous seizures and, second, the absence of sustained afterdischarge after electrical stimulation of the insular cortex.

47.3 Insular Recordings of Pain Evoked Responses in Humans

In what follows we will describe responses to pain and nonpainful cutaneous stimuli because the same contact in the insula most often detects both types of responses. Moreover pain intensity rating is subjective and corresponds to various levels of stimulus intensity between subjects. In complement to insular responses we will also describe responses recorded in the second somatosensory (S2) area (Figure 1), which is located in the superior bank of the lateral fissure, because it has a blurred anatomical border with the granular insular cortex and is also involved in building up pain sensation.

47.3.1 The Laser Stimulus

Most of the studies on pain-EPs in normal subjects and patients use a laser beam (mostly CO_2) applied on the skin surface as a stimulus. The laser beam is known to stimulate the endings of small diameters fibers and mostly those of A delta fibers. When the power output is fixed, the amount of thermal energy delivered depends on the duration of the pulse, which is in the order of a few milliseconds and, thus, permits an accurate timing for the analysis of the electrophysiologic response. The energy density of the laser beam is expressed in milli-Joules per mm^2 of skin surface ($mJ\,mm^{-2}$). Threshold values for pain show large interindividual variations between 30 and $40\,mJ\,mm^{-2}$ using a CO_2 laser beam. In most studies, the sensation perceived by the subject is that of sharp pinprick, without poststimulus pain, considered as characteristic of the sensations produced by the sti-mulation of A delta fibers. Although they are able to rate it on a visual analog scale of pain (usually at 4–7 on a 10 level scale), the subjects do not identify this sensation as a pain comparable to what they might have experienced in the past. Indeed no natural pain is provoked by selective activation of the A delta fibers, and this must be kept in mind when using laser stimulation in pain studies. In particular, the laser stimulus can be considered as adequate to assess the intensity coding of a pain stimulus, but is sub-optimal for studying of the emotional reaction to pain.

47.3.2 Cortical Surface Recordings (Electrocorticography)

Lenz F. A. *et al.* (1998) were the first to record cortical surface recordings by means of a subdural grid of electrodes placed on the surface of the parietal oper-culum with CO_2-laser-evoked potentials (LEPs) peaking between 160 and 340 ms after the stimulus. The spatial distribution of this response over the cortical surface of the perisylvian cortex was consid-ered as compatible with generators located in the parietal operculum and/or in the insular cortices. However subdural electrodes placed over the sylvian area do not allow direct recording neither in the deep aspect of the frontoparietal opercular cortex nor in the insula.

47.3.3 Intracortical S2 Recordings

In humans, LEPs at stimulus intensities below and above pain threshold are recorded in the suprasylvian opercular cortex corresponding to the human S2 area (Frot, M. *et al.*, 2001). These responses are not picked up in the other areas most often explored in presur-gical assessment of epilepsy including the amygdala, hippocampus, and orbitofrontal cortex. They show a biphasic negative–positive waveform peaking at 137 ± 13 ms (N140) and 172 ± 11 ms (P170), respec-tively, after stimulation of opposite hand (Figure 2). Similar responses are equally recorded in the homo-logous cortex, ipsilateral to the painful stimulus with a delay of 10–17 ms. The LEP voltage in S2 shows a significant increase as soon as the stimulus intensity reaches the sensory threshold as well as between sensory and pain thresholds while it rapidly reaches a plateau for intensities above pain threshold (Frot, M. *et al.*, 2007; Figure 3).

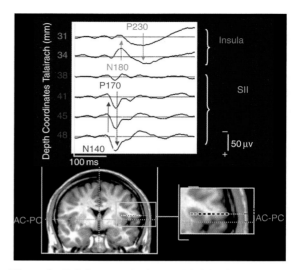

Figure 2　Pain laser-evoked potentials (LEPs) to skin laser stimulation of the dorsum of the hand. The lower part of the figure shows frontal magnetic resonance imaging slices of the right insula with projection of the contacts of a recording electrode exploring the insula (red) and S2 (blue). The vertical dashed blue line represent the midline sagittal plane, and the horizontal dashed blue line the anterior commissure–posterior commissure (AC–PC) horizontal plane used for stereotactic implantation of depth electrodes. Evoked potentials illustrated in the upper part of the figure are obtained by averaging responses to repeated stimulations of the left hand by a CO_2 laser beam. Figures along the vertical axis are the depth coordinates (*x*) in mm from midline of each recording contact. Note that distinct responses are obtained in both areas, which are separated by a delay of 50 ms.

47.3.4 Intracortical Insular Recordings

LEPs contralateral to stimulation recorded in the insular cortex itself (Frot, M. and Mauguière, F., 2003) consist in a N180 negative response (mean latency: 180 ± 16.5 ms) followed by a P230 positivity (mean latency: 226 ± 16 ms; Figure 2). As for S2, insular LEPs ipsilateral to stimulation peak 10–17 ms later than contralateral ones. LEP amplitudes increase between sensory and pain threshold intensities but, contrary to what is observed in S2, continue to increase significantly at intensities over the pain threshold (Figure 3).

The reason for the delay of ~50 ms (50 ± 16 ms between S2 N140 and insular N180 peaking laten-cies) observed between S2 and insular pain responses remains questionable. It is too long for a monosynap-tic transmission from S2 to the insula, the two areas being interconnected through direct projections. Alternatively, knowing that both S2 and the insula

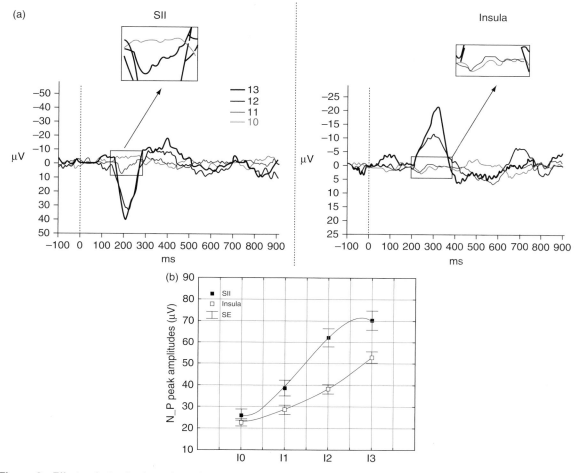

Figure 3 Effects of stimulus intensity on laser-evoked potential (LEP) amplitude in S2 and insula. A) LEP traces in S2 and insula to four different stimulus intensities are superimposed. CO_2 Laser pulses were applied at four different intensities in each subject. The power output being fixed, the amount of thermal energy delivered depends on the duration of the pulse. Pulse duration is set up according to subjects' subjective reports, rated on a visual analog scale (VAS) with an anchor point corresponding to the pain threshold. The printed scales consist of 10-cm horizontal lines where the left extreme is labeled no sensation and the right extreme maximal pain, and an anchored level 4 was at pain threshold (Lickert-type scale). The different stimuli and related subjective sensation are as follows: I0: below sensory threshold (pulse duration: 5–15 ms, mean energy density: 7 mJ mm^{-2}, no sensation); I1: above sensory threshold (pulse duration: 15–45 ms, mean energy density: 19 mJ mm^{-2}, producing a detectable nonpainful sensation reported for more than 90% of stimulations; For one-third of patients this sensation is a warmth sensation and for the others two-thirds a slight nonpainful pinprick sensation; VAS: 1.6 ± 1.09); I2: pain threshold (pulse duration: 25–80 ms, mean energy density: 33 mJ mm^{-2}, producing a pricking sensation, like a hair pulling or a drop of hot boiling water on the skin; VAS: 3.9 ± 1.46); I3: 20% above pain threshold (pulse duration: 35–110 ms, mean energy density: 46 mJ mm^{-2}, producing a pricking sensation described as clearly painful; VAS: 5.4 ± 1.6). In S2, a significant increase of the LEP is observed as soon as the stimulus intensity reaches the sensory threshold (between I0 and I1, $P < 0.001$), as well as between sensory and pain thresholds (I1 to I2, $P < 0.001$) while amplitudes rapidly reach a plateau for intensities above pain threshold (no significant amplitude difference between I2 and I3). In the insula, no significant amplitude increase is observed for low-stimulation intensities between I0 and I1, LEP amplitudes also increase between sensory and pain threshold intensities ($P < 0.05$ between I1 and I2) and continue to increase significantly at higher intensities over pain threshold ($P < 0.001$ between I2 and I3). B) Insula and S2 LEPs peaking amplitudes as a function of stimulus intensity. In these polynomial regression curves of the stimulus–response functions S2 LEPs amplitudes show a S-shaped profile with increasing stimulus intensities that reaches a plateau above pain threshold, while insular LEP amplitudes show an exponential profile. Laser stimuli above pain threshold trigger a maximal response in S2 and a pain level-related response in the insula (Modified from Frot, M., Magnin, M., Mauguière, F., and Garcia-Larrea, L. 2007. Human SII and posterior insula differently encode thermal laser stimuli. Cereb. Cortex 17, 610–620).

receive direct projections from the thalamus (see for a review Augustine, J. R., 1996) the explanation could be that the latter are triggered via thalamocortical fibers with a slower conduction than that of thalamic projections to the S2 area. To our knowledge, however, no electrophysiological demonstration of this hypothesis is hitherto available. A third hypothesis could be that the suprasylvian cortex and the insula are activated by inputs conveyed by peripheral fibers with different conduction velocities. Some studies have estimated the A delta conduction velocity in a large range of $7{-}20\,\mathrm{m\,s^{-1}}$ suggesting the existence of different A delta fiber subpopulations, with different conduction velocities. One can hypothesize that these different subpopulations of peripheral fibers could project in distinct cortical regions. However, to our knowledge, no electrophysiological study has been devoted to the identification of separate subpopulations of fibers with different conduction velocities in the spinothalamic tract or thalamocortical projections.

47.3.5 Topographic Specificity of Insular Laser-Evoked Potentials

Due to the anatomical proximity between the S2 area and the granular part of the insula the question arises whether insular responses might reflect the diffusion of S2 LEPs with a polarity reversal across the sylvian fissure, which is almost virtual in that region. The fact that when two contacts, or more, explore the insular cortex there is an amplitude increase from surface to depth of the N180–P230 response suggests that this is not so (see Frot, M. and Mauguière, F. 2003 for a complete discussion). However, the absolute proof that the N180–P230 is generated in the insular cortex would be its polarity reversal between the surface and the depth of the insular gray matter; unfortunately usual stereotactic recordings in patients do not have enough spatial resolution (2 mm interval between two neighboring contacts of 1.5 mm) to assess the distribution of potentials perpendicular to the cortical surface.

47.3.6 Interhemispheric Transmission of S2 and Insular Laser-Evoked Potentials

Insular and S2 responses to pain are bilateral; ipsilateral potentials peaking with a delay of 10–20 ms after contralateral ones. This delay is compatible with the interhemispheric transmission time through fibers of the corpus callosum as estimated by numerous studies (e.g., 15 ms between primary visual areas); it is in the same range as that measured between ipsi- and contralateral S2 magnetic fields evoked by electrical stimulation of the median nerve. The possibility remains, however, that responses ipsilateral to the stimulus could be triggered via ipsilateral thalamic fibers with slower conduction velocities. Only intracortical recordings of S2 or insular EPs to ipsilateral stimuli in patients with a lesion of the homologous areas in the opposite hemisphere could address this question directly.

47.4 Insular Stimulation

47.4.1 The Challenge of Insular Stimulation

In humans, it has long been a challenge to stimulate the insular cortex. Thus, only a few studies have reported nonnociceptive somesthetic symptoms, cardiovascular effects as well as visceromotor and viscerosensitive sensations consecutive to direct electrical stimulation of the insular cortex. Truly painful responses have not been reported during stimulation of any area of the cerebral cortex in humans by Penfield W. and Faulk M. E. (1955), who extensively stimulated the surface of all cortical areas of the human brain, including the insula, using surface electrodes. However, they did not explore the posterior part of the insula where electric stimulation evokes painful sensations.

Electric stimulations are produced by a current-regulated neurostimulator designed for a safe diagnostic stimulation of the human brain. Square pulses of current are applied between two adjacent contacts (bipolar stimulation) at a frequency of 50 Hz, pulse duration of 0.5 ms, train duration of 5 s, and intensity of 0.8–6.0 mA. These parameters are used to avoid any tissue injury (charge density per square pulse $<55\,\mu\mathrm{C\,cm^{-2}}$; Gordon, B. *et al.*, 1990). This stimulation paradigm ensures a good spatial specificity with respect to the desired structures to be stimulated. The study of current densities in the cortex for bipolar stimulation with a 10-mA stimulating current shows that the peak current density occurs in the region immediately beneath the bipolar electrodes ($0.05\,\mathrm{A\,cm^{-2}}$) and declines rapidly to $0.02\,\mathrm{A\,cm^{-2}}$ 0.5 cm away, and that the current density decreases in relation to the square of the distance into the cortex. In S1, S2, and the insula, the minimal stimulation intensity necessary to elicit a clinical response is 1.6–1.9 mA (Mazzola, L. *et al.*, 2006). With these stimuli there is virtually no current spread out of the stimulation current dipole as defined by the distance of 5.5 mm between the outer and inner limits of the superficial and deep-stimulating contacts, respectively (Ostrowsky, K. *et al.*, 2002). Moreover it has been demonstrated that no significant cerebral blood flow change occurs at the site of cortical stimulation, or in its close vicinity, in the absence of epileptic after discharge.

47.4.2 Pain Evoked by Insular Stimulation

Painful sensations in response to direct electric stimulation of the insula have been reported by Ostrowsky K. *et al.* (2000; 2002) and more recently by Mazzola L. *et al.* (2006). Qualities of the evoked pain are described as burning, stinging, disabling sensation, or electrical shock. Pain intensity varies from mild to intolerable but is not related to stimulation intensity. Pain disappears as soon as the stimulation is interrupted in most cases except a few where an intense pain is followed by a sore feeling that can last up to 1 min after the end of the stimulation. It is located contralateral to the stimulation site or bilaterally when midline parts of the body are involved, and affects large areas of the body (e.g., face, upper limb, whole half of the body), suggesting that receptive fields for pain in the insular cortex are more extended that somatosensory fields in the SI and SII cortex (Mazzola, L. *et al.*, 2006).

47.4.3 Anatomical Location of the Insular Pain Area

The area where pain can be elicited by direct electric stimulation occupies the posterosuperior part of the insular lobe (Figure 4). Surface electrodes cannot reach this area unless the lower precentral motor and postcentral somatosensory cortex be removed. This is the most probable reason why early electrocorticographic studies using that type of electrodes failed to produce pain by cortical stimulation in humans. Because only a few contacts can be placed in the insula of a given individual during stereotactic exploration of the brain with depth electrodes it is uncertain whether a pain somatotopic representation of the body, comparable to that demonstrated in the primary somatosensory cortex (S1), does exist in the pain insular area. Moreover due to the large skin extent of pains evoked by insular stimulation that the somatotopic pain representation is likely to be much more fuzzy, with overlapping receptive fields, in the insular pain area than in S1 or S2 somatosensory areas.

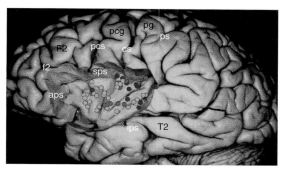

Figure 4 Pain insular area. The locations of stimulation sites in a population of patients are plotted on a sagittal view of the insula exposed by removal of the supra and infrasylvian cortex according to the type of somatosensory sensation reported by the subjects. Red dots: pain, Orange dots: nonpainful warm sensation, yellow dots: nonpainful paresthesias. Note that most of pain sensations are evoked by stimulation the posterior part of the insula situated behind the tip of the central sulcus (cs), while nonpainful responses are obtained over a larger and more anterior surface of the insular lobe. Aps, anterior peri-insular sulcus; cs, central sulcus; F2, second frontal gyrus; f2, second frontal sulcus; ips, inferior peri-insular sulcus; pcg, precentral gyrus; pg, postcentral gyrus; ps, parietal sulcus; sps, superior peri-insular sulcus; T2, second temporal gyrus.

References

Augustine, J. R. 1996. Circuitry and functional aspects of the insular lobe in primates including humans. Brain Res. Rev. 22, 229–244.

Clark, T. E. 1896. The comparative anatomy of the insula. J. Comp. Neurol. 6, 59–100.

Frot, M. and Mauguière, F. 2003. Dual representation of pain in the operculo-insular cortex in humans. Brain 126, 438–450.

Frot, M., Garcia-Larrea, L., Guénot, M., and Mauguière, F. 2001. Responses of the supra-sylvian (SII) cortex in humans to painful and innocuous stimuli. A study using intra-cerebral recordings. Pain 94, 65–73.

Frot, M., Magnin, M., Mauguière, F., and Garcia-Larrea, L. 2007. Human SII and posterior insula differently encode thermal laser stimuli. Cereb. Cortex 17, 610–620.

Gordon, B., Lesser, R. P., Rance, N. E., Hart, J., Jr., Webber, R., Uematsu, S., and Fisher, R. S. 1990. Parameters for direct cortical electrical stimulation in the human: histopathologic confirmation. Electroencephalogr. Clin. Neurophysiol. 75, 371–377.

Isnard, J., Guénot, M., Ostrowsky, K., Sindou, M., and Mauguière, F. 2000. The role of the insular cortex in temporal lobe epilepsy. Ann. Neurol. 48, 614–623.

Isnard, J., Guénot, M., Sindou, M., and Mauguière, F. 2004. Clinical manifestations of insular lobe epileptic discharges: a stereo-electroencephalographic study. Epilepsia 45, 1079–1790.

Jackson, P. L., Meltzoff, A. N., and Decety, J. 2005. How do we perceive the pain of the others? A window into the neural processes involved in empathy. Neuroimage 24, 771–779.

Krolak-Salmon, P., Henaff, M. A., Isnard, J., Tallon-Baudry, C., Guénot, M., Vighetto, A., Bertrand, O., and Mauguière, F. 2003. An attention modulated response to disgust in human ventral anterior insula. Ann. Neurol. 53, 446–453.

Lenz, F. A., Rios, M., Chau, D., Krauss, G. L., Zirh, T. A., and Lesser, R. P. 1998. Painful stimuli evoke potentials recorded from the parasylvian cortex in humans. J. Neurophysiol. 80, 2077–2088.

Mazzola, L., Isnard, J., and Mauguière, F. 2006. Somatosensory and pain responses to stimulation of the second somatosensory area (SII) in humans. A comparison with SI and insular responses. Cereb. Cortex 16, 960–968.

Mesulam, M. M. and Mufson, E. J. 1982. Insula of the old world monkey. I. Architectonics in the insulo orbito-temporal component of the paralimbic brain. J. Comp. Neurol. 212, 1–22.

Oppenheimer, S. M., Gelb, A., Girvin, J. P., and Hachinski, V. C. 1992. Cardio-vascular effects of human insular cortex stimulation. Neurology 42, 1727–1732.

Ostrowsky, K., Isnard, J., Ryvlin, P., Guenot, M., Fischer, C., and Mauguière, F. 2000. Functional mapping of the insular cortex: clinical implication in temporal lobe epilepsy. Epilepsia 41, 681–686.

Ostrowsky, K., Magnin, M., Ryvlin, P., Isnard, J., Guénot, M., and Mauguière, F. 2002. Representation of pain and somatic sensation in the human insula: a study of responses to direct electrical cortical stimulation. Cereb. Cortex 12, 376–385.

Penfield, W. and Faulk, M. E. 1955. The insula. Further observations on its function. Brain 78, 445–470.

Peyron, R., Frot, M., Schneider, F., Garcia-Larrea, L., Mertens, P., Barral, F. G., Sindou, M., Laurent, B., and Mauguière, F. 2002. Role of operculo-insular cortices in human pain processing. Converging evidence from PET, fMRI, dipole modeling and intra-cerebral recordings of evoked potentials. Neuroimage 17, 1336–1346.

Rizzolatti, G. and Craighero, L. 2004. The mirror-neuron system. Ann. Rev. Neurosci. 27, 169–192.

Rosenow, F. and Luders, H. O. 2001. Presurgical evaluation of epilepsy. Brain 124, 1683–1700.

Tanriover, N., Rhoton, A. L., Kawashima, M., Ulm, A. J., and Yasuda, A. 2004. Microsurgical anatomy of the insula and the sylvian fissure. J. Neurosurg. 100, 891–922.

Relevant Websites

http://imaging.mrc-cbu.cam.ac.uk – CBU Imaging

http://en.wikipedia.org

48 The Rostral Agranular Insular Cortex

L Jasmin, Neurosurgery and Gene Therapeutics Research Institute, Los Angeles, CA, USA

P T Ohara, University of California, San Francisco, CA, USA

Glossary

amygdala A group of neurons that is part of the limbic system. The amygdala is involved in many aspects of behavior such as memory, emotion, and motivation.

immunocytochemical staining A method that uses antibodies to identify and locate specific cellular components, such as neurotransmitters.

medial frontal, infralimbic, and cingulate cortices Different parts of the cerebral cortex that have been shown to be involved in different aspects of processing painful information.

nociception The perception of a stimulus that is potentially injurious to one's body.

nucleus accumbens A part of the limbic system involved in reward and addiction.

thalamus A part of the brain located at the center that processes sensory and motor information and sends the resulting information to the cerebral cortex.

48.1 What Is the Rostral Agranular Insular Cortex?

The rostral agranular insular cortex (RAIC) is a small area of the cerebral cortex located on the lateral aspect of the rat's cerebral hemisphere. The RAIC can be distinguished in terms of its location, cellular architecture, neurochemistry, and connectivity. Functionally, the RAIC is involved in processing pain information and, depending on which cells are activated, can act through connections with other brain areas to reduce or increase pain responses. The RAIC therefore is one of a limited number of cortical regions involved in the perception and response to pain.

48.2 Function of the Rostral Agranular Insular Cortex

Many parts of the cerebral cortex do not have precisely defined functions and usually do not have precisely defined borders. Until recently, the RAIC was one such region, but a picture indicating that the RAIC is involved in the behavioral responses to nociceptive (painful) stimuli is beginning to emerge from animal and human studies.

48.3 Anatomical Location

In the rat, the RAIC occupies a small area of the lateral aspect of cerebral cortex just above a groove,

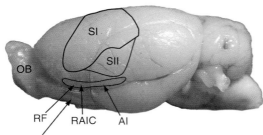

Primary olfactory cortex

Figure 1   Lateral view of the rat brain. The rhinal fissure (RF) is seen as a prominent groove starting near the ventral anterior cortex and extending toward the posterior cortex. The agranular insular cortex is a thin strip of cortex lying just above (dorsal to) the RF and the rostral part (RAIC) is indicated in red. The approximate location of the primary somatosensory cortex (SI) and the secondary somatosensory cortex (SII) are indicated for reference. OB, olfactory bulb.

which is one of the few surface features of the rat cortex, called the rhinal fissure (Figure 1) (Paxinos, G., 1995). Agranular cortex is defined by the absence of a layer 4 that gives the cortex a less granular appearance than the adjacent areas. The region was described by Krettek J. E. and Price J. L. (1977) as a strip of agranular cortex located immediately above the rhinal fissure and bounded dorsally by granular cortex. Rostrally, the agranular cortex was further subdivided into a ventral and a dorsal agranular region. This description was later modified (Cechetto, D. F. and Saper, C. B., 1987) to include a dysgranular field between the agranular and granular cortices. The dysgranular field had a layer 4 that was intermediate in prominence between the fully developed layer 4 of the granular cortex and the absent layer 4 of the agranular cortex. Later studies all distinguish agranular, dysgranular, and granular fields although the precise boundaries, particularly between the dorsal agranular and dysgranular fields, were delineated differently.

48.4   Neurochemical Signature

Although classically cortical regions are defined by cytoarchitecture and location, modern neuroanatomical techniques have refined our understanding of cortical organization based on connections and neurochemical profile. The RAIC has been shown to be distinguished from surrounding regions by a distinct pattern of immunocytochemical staining (Figure 2) (Ohara, P. T. *et al.*, 2003). The first feature is a

concentration of neurons in layer 5 of the RAIC that express the $GABA_B$ receptor (Jasmin, L. *et al.*, 2003), which is one of several types of receptors activated by the inhibitory neurotransmitter, gamma-aminobutyric acid (GABA). Similar concentrations of $GABA_B$ receptor-bearing neurons are also found in the insular cortex posterior to the RAIC and in a few other brain regions but are less concentrated in the remainder of the cortex. The neurons in layer 5 of the RAIC that express the $GABA_B$ receptor are large pyramidal projection neurons that have axons that leave the cortex and innervate other brain regions. This finding suggests that the neurons in the RAIC that have $GABA_B$ receptors play a specific role in how the RAIC communicates with other brain areas. A second notable feature of the RAIC is the concentration of nerve fibers containing the neurotransmitter dopamine. In the rat, there is only sparse dopamine innervation throughout most of the cortex dopamine fibers that are found in high concentration in then infralimbic cortex and in the RAIC (Figure 2). These dopamine fibers originate in a region of the brainstem called the ventral tegmental area (VTA) and probably are involved in determining the overall level of activity of the RAIC rather than conveying specific sensory information.

48.5   Connections of the Rostral Agranular Insular Cortex

The specific connections of a cortical region also help define the boundaries and give insight into the functions of the region. The RAIC shares the general pattern of connection common to most regions of cortex receiving its major subcortical afferents from the thalamus, having reciprocal connection with other cortical regions, and projecting to a number of subcortical sites (Jasmin, L. *et al.*, 2004). The picture that emerges from an analysis of the projections is that the RAIC plays a role in pain mechanisms. The inputs from the thalamus are from the nuclei that convey sensory information from the body. However, this input is not from thalamic nuclei that subserve the fine-detailed discriminative aspect of sensation (i.e., those that project to the primary somatosensory cortex), rather the input is from nuclei located in the midline of the thalamus that are generally associated with emotional or affective aspects of pain. The principal cortical connection of the RAIC is with areas such as the medial frontal, infralimbic, and cingulate cortices, which have long been known to

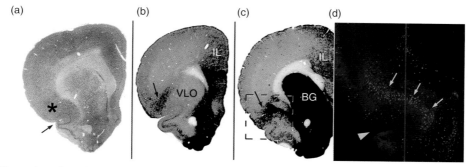

Figure 2 Coronal sections through the left anterior rat brain showing some of the characteristic features of the RAIC. (a) A section stained with a Nissl stain to show the general appearance and location of the RAIC. The rhinal fissure is indicated by the arrow and the asterisk indicates the location of the RAIC. (b, c) Sections stained immunocytochemically to show the location of fibers containing the neurotransmitter, dopamine. The contrast has been digitally enhanced to show the dopamine more clearly. The arrow indicates the region of dopamine staining that corresponds to the location of the RAIC. Intense staining is also present in the infralimbic cortex (IL), which is also a pain-related area, and in the basal ganglia (BG), which is a motor-related site. The ventrolateral orbital cortex (VLO), a site also associated with pain behavior, shows no dopamine staining. (d) A higher magnification of the boxed region indicated in (c). This section is stained to show the location of neurons that express the GABA$_B$ receptor (green-colored structures) in cortical layer 5. The arrowhead indicated the location of the rhinal fissure.

be associated with pain responses. The RAIC projects to a number of subcortical sites including the lateral hypothalamus, dorsal raphe, periaqueductal gray matter, pericerulear region, rostroventral medulla, and parabrachial nuclei – all of which play a role in pain inhibition mechanisms. Significant to the probable function of the RAIC are projections to the amygdala and the nucleus accumbens. These projections include a large component originating in the cells in layer 5 that express the GABA$_B$ receptor described above. The amygdala is known to be involved in several aspects of behavior including fear, anxiety, and attention (Williams, M. A. *et al.*, 2005).

48.6 Behavioral Studies

In addition to the connectional studies, there are a number of behavioral studies that give insight into the function of the RAIC. Direct injections of compounds into the RAIC have shown that morphine injection, increasing dopamine levels, and increasing GABA levels all result in behavioral antinociception. Because increasing GABA in the RAIC results in antinociception and the action of GABA is to inhibit neuronal activity, it suggests that activity in the RAIC is pronociceptive. Additional pharmacological experiments suggest that the neurons that express the GABA$_A$ and those that express the GABA$_B$ receptor change pain thresholds through different pathways. These antinociceptive effects mediated by cells that

possess GABA$_A$ receptors occur through projections to the brainstem and activation of descending pain inhibitory systems. However, as noted above, the RAIC is characterized by an abundance of cells that have GABA$_B$ receptors. The changes in pain behavior mediated by the GABA$_B$ receptor-bearing neurons point to projections to the amygdala playing a key role. The proposal that the activity in the RAIC leads to increased pain was tested by inhibiting GABA$_A$ receptor-bearing neurons and at the same time activating GABA$_B$ receptor-bearing neurons. This experiment resulted in an increase in pain behavior (Jasmin, L. *et al.*, 2003). These studies highlight two important issues. First, changes in activity of the RAIC can result in both increases and decreases in pain response and the overall effect of the RAIC results from a balance of pro- and antinociceptive activity. Second, the RAIC probably changes pain behavior both by activating descending inhibitory circuits and by changing activity in the amygdala and other cortical area to change the emotional/affective response to painful stimuli.

48.7 Relationship with Adjacent Cortical Regions

The location and small size of the RAIC makes it difficult to unequivocally isolate it from surrounding regions. The agranular insular cortex has been most associated with visceral and gustatory responses, and a number of studies on cells in the region have indicated that they respond to visceral input. These functions, however, seem more related to the more posterior regions of the agranular cortex or the dysgranular cortex immediately above the RAIC. Immediately anterior to the RAIC is the ventrolateral orbital cortex (VLO). This region of cortex in the rat is also associated with pain responses (Xie, Y. F. et al., 2004). As with all parts of the cerebral cortex, it is often difficult to define a clear boundary between any two regions. For this, and other technical reasons such as diffusion of drugs when injected into a specific site, it is possible that some studies previously carried out on the VLO may have encroached upon the RAIC and vice versa. There are both similarities and differences between the VLO and the RAIC. Morphine injected into the VLO and RAIC both produce antinociception, but the application of drugs that inhibit GABA has no effect in the RAIC but do have antinociceptive effects in the VLO. There is less dopamine innervation of VLO, and although the VLO contains GABA$_B$ receptor-bearing cells, they are not as numerous as in the RAIC (Ohara, P. T. et al., 2003). Whether the VLO and RAIC are really separate entities with specific, separate functions related to pain or whether both regions are parts of a larger forebrain nociceptive system is not clear.

48.8   Overview in the Rat

The major connections of the RAIC are with areas that have established roles in nociceptive information processing and behavior responses to nociceptive stimuli. RAIC projections to other cortical areas and subcortical sites such as the amygdala are likely to participate in the sensorimotor integration of nociceptive processing, while the hypothalamus and brainstem projections are most likely to contribute to descending pain inhibitory control.

48.9   From Rats to Humans

The agranular insular cortex is found in other mammals including cats, monkeys, and primates including humans (Augustine, J. R., 1996; Mesulam, M.-M., 2000). In primates, the divisions of the insular cortex are the same as in rats and the RAIC occupies an area immediately noticeable, that is, dorsal to the primary olfactory cortex.

The advent of high-resolution functional imaging has resulted in the availability of data relating to cortical function that has previously been difficult to obtain from animal studies (Figure 3). Specifically, it is possible to relate verbal descriptions and emotional responses to painful stimuli to specific cortical sites. In humans, imaging techniques and direct recording of field potentials show that painful stimuli activate the insular cortex (Coghill, R. C. et al., 1994; Brooks, J. C. et al., 2005). It has been possible to distinguish between activation of the anterior and of the posterior region of the agranular cortex located in the insular region (Strigo, I. A. et al., 2003). The anterior insula, which might be considered most equivalent to the RAIC, is activated when a strong emotional response is associated with the nociceptive stimulus. The posterior insula and the adjacent parietal cortex, in turn, are activated by all nociceptive stimuli, independently of their emotional content. These differences in function between the anterior and the posterior insula are explained by their neural connectivity. Notably, ascending nociceptive afferents from the spinal cord and brainstem directly activate the posterior insula

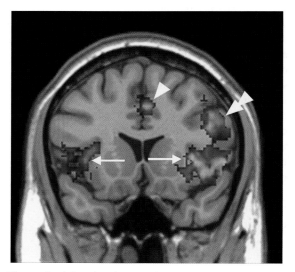

Figure 3 A functional magnetic resonance imaging blood oxygen level-dependent (fMRI BOLD) image averaged from nine normal healthy subjects who received a painful stimulus to the back. Note that the insular cortex is activated (arrows, red and yellow areas) on both sides of the cortex although more prominently on the left side. This bilateral activation of the insular region can be contrasted with the unilateral activation of other cortical regions such as the cingulate cortex (arrowhead) and dorsolateral prefrontal cortex (double arrowhead). Why the activation is lateralized in these latter areas in not known but is a common occurrence. Photo courtesy of Dr. Vania Apkarian, Northwestern University.

only. Accordingly, direct electrical stimulation of the posterior insula results in the generation of thermal sensations (Ostrowsky, K. *et al.*, 2002), and lesions of the posterior insula, and not the anterior insula, affect recognition of nociceptive heat (Greenspan, J. D. *et al.*, 1999). On the other hand, the anterior insula is implicated in the emotional awareness that occurs when a nociceptive stimuli endangers the self (Craig, A. D., 2002).

One aspect of function that has emerged from human studies and not fully explored in animal experiments is that in some cases activation of the anterior insula is lateralized. For example, air hunger activates mostly the right side (Banzett, R. B. *et al.*, 2000), and from this and other results, the right insular cortex has been assigned a greater role in the consciousness of the external world as well as interest in novelty (Manes, F. *et al.*, 1999). The left insula is more implicated in the regulation of cardiovascular function (Phillips, M. L. *et al.*, 2003).

48.10 Perspective

Our understating of cortical functioning, particularly with respect to pain, must be considered rudimentary. Although we refer to the RAIC as a distinct entity with a specific function in the rat, we need to remain aware of the close relationship it has with adjacent areas serving visceral functions and nearby regions such as the VLO involved in nociceptive processing. The most complete understanding of the RAIC will only come about when we understand how it is integrated into the other cortical and subcortical sites that appear to form a pain matrix. We also need to understand the relationship between the RAIC in the rat and the agranular insular cortex in the human. Although there are clear parallels between the two areas in terms of cytoarchitecture and function, it is not clear that they are strictly equivalent areas and our understanding of pain mechanisms in general will have to take into account such species-specific similarities and/or differences.

References

Augustine, J. R. 1996. Circuitry and functional aspects of the insular lobe in primates including humans. Brain Res. Brain Res. Rev. 22, 229–244.

Banzett, R. B., Mulnier, H. E., Murphy, K., Rosen, S. D., Wise, R. J., and Adams, L. 2000. Breathlessness in humans activates insular cortex. Neuroreport 11, 2117–2120.

Brooks, J. C., Zambreanu, L., Godinez, A., Craig, A. D., and Tracey, I. 2005. Somatotopic organisation of the human insula to painful heat studied with high resolution functional imaging. Neuroimage 27, 201–209.

Cechetto, D. F. and Saper, C. B. 1987. Evidence for a viscerotopic sensory representation in the cortex and thalamus in the rat. J. Comp. Neurol. 262, 27–45.

Coghill, R. C., Talbot, J. D., Evans, A. C., Meyer, E., Gjedde, A., Bushnell, M. C., and Duncan, G. H. 1994. Distributed processing of pain and vibration by the human brain. J. Neurosci. 14, 4095–4108.

Craig, A. D. 2002. How do you feel? Interoception: the sense of the physiological condition of the body. Nat. Rev. Neurosci. 3, 655–666.

Greenspan, J. D., Lee, R. R., and Lenz, F. A. 1999. Pain sensitivity alterations as a function of lesion location in the parasylvian cortex. Pain 81, 273–282.

Jasmin, L., Burkey, A. R., Granato, A., and Ohara, P. T. 2004. Rostral agranular insular cortex and pain areas of the central nervous system: a tract-tracing study in the rat. J. Comp. Neurol. 468, 425–440.

Jasmin, L., Rabkin, S. D., Granato, A., Boudah, A., and Ohara, P. T. 2003. Analgesia and hyperalgesia from GABA-mediated modulation of the cerebral cortex. Nature 424, 316–320.

Krettek, J. E. and Price, J. L. 1977. The cortical projections of the medio-dorsal nucleus and adjacent thalamic nuclei in the rat. J. Comp. Neurol. 171, 157–192.

Manes, F., Paradiso, S., Springer, J. A., Lamberty, G., and Robinson, R. G. 1999. Neglect after right insular cortex infarction. Stroke 30, 946–948.

Mesulam, M.-M. 2000. Principles of Behavioral and Cognitive Neurology. Oxford University Press.

Ohara, P. T., Granato, A., Moallem, T. M., Wang, B. R., Tillet, Y., and Jasmin, L. 2003. Dopaminergic input to GABAergic neurons in the rostral agranular insular cortex of the rat. J. Neurocytol. 32, 131–141.

Ostrowsky, K., Magnin, M., Ryvlin, P., Isnard, J., Guenot, M., and Mauguiere, F. 2002. Representation of pain and somatic sensation in the human insula: a study of responses to direct electrical cortical stimulation. Cereb. Cortex 12, 376–385.

Paxinos, G. (*ed.*) 1995. The Rat Nervous System. Academic Press.

Phillips, M. L., Gregory, L. J., Cullen, S., Coen, S., Ng, V., Andrew, C., Giampietro, V., Bullmore, E., Zelaya, F., Amaro, E., Thompson, D. G., Hobson, A. R., Williams, S. C., Brammer, M., and Aziz, Q. 2003. The effect of negative emotional context on neural and behavioural responses to oesophageal stimulation. Brain 126, 669–684.

Strigo, I. A., Duncan, G. H., Boivin, M., and Bushnell, M. C. 2003. Differentiation of visceral and cutaneous pain in the human brain. J. Neurophysiol. 89, 3294–3303.

Williams, M. A., McGlone, F., Abbott, D. F., and Mattingley, J. B. 2005. Differential amygdala responses to happy and fearful facial expressions depend on selective attention. Neuroimage 24, 417–425.

Xie, Y. F., Wang, J., Huo, F. Q., Jia, H., and Tang, J. S. 2004. Mu but not delta and kappa opioid receptor involvement in ventrolateral orbital cortex opioid-evoked antinociception in formalin test rats. Neuroscience 126, 717–726.

Further Reading

Allen, G. V., Saper, C. B., Hurley, K. M., and Cechetto, D. F. 1991. Organization of visceral and limbic connections in the insular cortex of the rat. J. Comp. Neurol. 311, 1–16.

Altier, N. and Stewart, J. 1998. Dopamine receptor antagonists in the nucleus accumbens attenuate analgesia induced by ventral tegmental area substance P or morphine and by nucleus accumbens amphetamine. J. Pharmacol. Exp. Ther. 285, 208–215.

Altier, N. and Stewart, J. 1999. The role of dopamine in the nucleus accumbens in analgesia. Life Sci. 65, 2269–2287.

Clasca, F., Llamas, A., and Reinoso-Suarez, F. 1997. Insular cortex and neighboring fields in the cat: a redefinition based on cortical microarchitecture and connections with the thalamus. J. Comp. Neurol. 384, 456–482.

Clasca, F., Llamas, A., and Reinoso-Suarez, F. 2000. Cortical connections of the insular and adjacent parieto-temporal fields in the cat. Cereb.. Cortex 10, 371–399.

Dupont, S., Bouilleret, V., Hasboun, D., Semah, F., and Baulac, M. 2003. Functional anatomy of the insula: new insights from imaging. Surg. Radiol. Anat. 25, 113–119.

Gear, R. W., Aley, K. O., and Levine, J. D. 1999. Pain-induced analgesia mediated by mesolimbic reward circuits. J. Neurosci. 19, 7175–7181.

van Groen, T., Kadish, I., and Wyss, J. M. 1999. Efferent connections of the anteromedial nucleus of the thalamus of the rat. Brain Res. Brain Res. Rev. 30, 1–26.

Hof, P. R., Mufson, E. J., and Morrison, J. H. 1995. Human orbitofrontal cortex: cytoarchitecture and quantitative immunohistochemical parcellation. J. Comp. Neurol. 359, 48–68.

Kawaguchi, Y. and Kondo, S. 2002. Parvalbumin, somatostatin and cholecystokinin as chemical markers for specific GABAergic interneuron types in the rat frontal cortex. J. Neurocytol. 31, 277–287.

Kawaguchi, Y. and Kubota, Y. 1998. Neurochemical features and synaptic connections of large physiologically-identified GABAergic cells in the rat frontal cortex. Neuroscience 85, 677–701.

Kawaguchi, Y. and Shindou, T. 1998. Noradrenergic excitation and inhibition of GABAergic cell types in rat frontal cortex. J. Neurosci. 18, 6963–6976.

Li, Y., Li, J. J., and Yu, L. C. 2002. Anti-nociceptive effect of neuropeptide Y in the nucleus accumbens of rats: an involvement of opioid receptors in the effect. Brain Res. 940, 69–78.

Li, N., Lundeberg, T., and Yu, L. C. 2001. Involvement of CGRP and CGRP1 receptor in nociception in the nucleus accumbens of rats. Brain Res. 901, 161–166.

Margeta-Mitrovic, M., Mitrovic, I., Riley, R. C., Jan, L. Y., and Basbaum, A. I. 1999. Immunohistochemical localization of GABA(B) receptors in the rat central nervous system. J. Comp. Neurol. 405, 299–321.

Meyer, S., Strittmatter, M., Fischer, C., Georg, T., and Schmitz, B. 2004. Lateralization in autonomic dysfunction in ischemic stroke involving the insular cortex. Neuroreport 15, 357–361.

Ray, J. P. and Price, J. L. 1992. The organization of the thalamocortical connections of the mediodorsal thalamic nucleus in the rat, related to the ventral forebrain-prefrontal cortex topography. J. Comp. Neurol. 323, 167–197.

Schmidt, B. L., Tambeli, C. H., Barletta, J., Luo, L., Green, P., Levine, J. D., and Gear, R. W. 2002. Altered nucleus accumbens circuitry mediates pain-induced antinociception in morphine-tolerant rats. J. Neurosci. 22, 6773–6780.

Schmidt, B. L., Tambeli, C. H., Gear, R. W., and Levine, J. D. 2001. Nicotine withdrawal hyperalgesia and opioid-mediated analgesia depend on nicotine receptors in nucleus accumbens. Neuroscience 106, 129–136.

Sengupta, J. N., Snider, A., Su, X., and Gebhart, G. F. 1999. Effects of kappa opioids in the inflamed rat colon. Pain 79, 175–185.

Suhara, T., Yasuno, F., Sudo, Y., Yamamoto, M., Inoue, M., Okubo, Y., and Suzuki, K. 2001. Dopamine D2 receptors in the insular cortex and the personality trait of novelty seeking. Neuroimage 13, 891–895.

Tambeli, C. H., Parada, C. A., Levine, J. D., and Gear, R. W. 2002. Inhibition of tonic spinal glutamatergic activity induces antinociception in the rat. Eur. J. Neurosci. 16, 1547–1553.

49 Descending Control Mechanisms

K Ren and R Dubner, University of Maryland, Baltimore, MD, USA

49.1 Introduction

Pain is a sensory process that reflects the level and intensity of noxious stimulus on one hand but is also closely controlled or modulated by the central nervous system (CNS). In fact, the intensity of the perceived pain is not necessarily proportional to the amount of stimulus under different conditions (Figure 1). Although psychological factors contribute to the variability of pain, the most important underlying mechanisms responsible for the variant pains are the existence of the endogenous pain modulatory systems, in which brainstem descending pathways play a fundamental role (Fields, H. L. and Basbaum, A. I., 1999; Millan, M. J., 2002). These pathways provide the neural mechanisms by which attentional, motivational, and cognitive variables filter ascending sensory information.

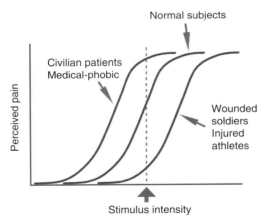

Figure 1 The variant pains. In this illustrative diagram, the intensity of the perceived pain is plotted against the stimulus intensity. The stimulus-response function curve of the normal subjects is located in the middle. For wounded soldiers and injured athletes, the curve may be shifted to the right and for civilian patients and medical-phobic, the curve may be shifted to the left. The shift of the curve is translated as changes in the sensation of pain. At certain stimulus intensity (arrow-dashed line), the civilian population may feel more pain and the wounded soldiers show less pain, when compared to the normal subjects. The endogenous pain modulatory system contributes to the variability of pain under different conditions. Adapted from Cervero, F. and Laird, J. M. 1996. Mechanisms of touch-evoked pain (allodynia): a new model. Pain 68, 13–23.

Our knowledge of the descending pain modulatory systems spans at least four decades. In 1965, Wall and Melzack in their gate control theory of pain questioned the concept of an invariant relationship between stimulus and sensation and a fixed, direct-line connection from the skin to the brain. They proposed that central signals arising from the brain could influence afferent input at the earliest synaptic levels of the somesthetic system, the spinal dorsal horn (Melzack, R. and Wall, P. D., 1965). Evidence supporting the concept was available even earlier when Hagbarth and colleagues (Hagbarth K. E. and Kerr, D. I., 1954; Hernandez-Peon, R. and Hagbarth, K. E., 1955) showed that sensory transmission through spinal cord and trigeminal sensory nuclei was influenced by descending fibers from the cerebral cortex. The gate control hypothesis placed this evidence within the framework of a theory of how pain modulation could account, in part, for the variant relationship between noxious stimuli and the sensations they produced. The first evidence to support endogenous pain control came from the study of Reynolds (Reynolds, D. V., 1969) who demonstrated that focal brain stimulation of the midbrain periaqueductal gray (PAG) produced sufficient analgesia to allow surgery in rats without the use of chemical anesthetics. Liebeskind and colleagues quickly confirmed this finding and concluded that stimulation of the PAG activated a normal function of the brain: pain inhibition (Mayer, D. J. *et al.*, 1971; Mayer, D. J. and Liebeskind, J. C., 1974). Later evidence indicates that pain modulation is not limited to inhibition. Descending pathways also facilitate pain transmission at the spinal level (Ren, K. *et al.*, 2000; Millan, M. J., 2002; Gebhart, G. F., 2004). Most recently, studies have been moved to address the role of pain modulatory circuitry in persistent, or chronic pain conditions. New lines of evidence suggest that the descending pathways exhibit dramatic plasticity and multiplicity and are actively involved in the development of persistent pain after tissue or nerve injury (Porreca, F. *et al.*, 2002; Ren, K. and Dubner, R., 2002; Gebhart, G. F., 2004; Vanegas, H. and Schaible, H.-G., 2004).

49.2 Organization of Descending Pathways

There are multiple brain sites and pathways that are involved in descending pain modulation, ranging from the cerebral cortex to caudal medulla (Fields, H. L. and Basbaum, A. I., 1978; 1999; Gebhart, G. F., 1988; Oliveras, J. L. and Besson, J. M., 1988; Sandkuhler, J., 1996; Willis, W. D. and Westlund, K. N., 1997; Millan, M. J., 2002) (Figure 2). To date, the most well characterized endogenous pain modulatory pathway involves a circuitry linking the midbrain PAG, rostral ventromedial medulla (RVM), and the spinal cord. The second important source of descending pain control is from the dorsolateral pontomesencephalic tegmentum (DLPT), which includes the cuneiform nucleus in the midbrain ventrolateral to PAG and locus coeruleus/subcoeruleus (LC/SC) together with other neighboring noradrenaline cell groups (Fields, H. L. and Basbaum, A. I., 1999). In the caudal medulla, ventrolateral medulla (CVLM) including lateral reticular nucleus (LRN), the subnucleus reticularis dorsalis (SRD), and the nucleus tractus solitarius (NTS) have been identified to play a role in descending pain control. Additionally, hypothalamus and other forebrain sites also participate in descending pain modulation.

49.2.1 Key Structures and Pathways

49.2.1.1 Midbrain PAG matter

PAG neurons are organized in longitudinal columns that function to integrate behavioral responses to noxious and stressful stimuli (Bandler, R. and Shipley, M. T., 1994; Cameron, A. A. *et al.*, 1995a; 1995b). The dorsal raphe nucleus ventromedial to the cerebral aqueduct and the cuneiform nucleus lateral to the ventrolateral PAG are also considered a functional entity with PAG for descending pain control (Jensen, T. S. and Gabhart, G. F., 1988; Fields, H. L. and Basbaum, A. I., 1999). The anatomical studies implicate the importance of the midbrain PAG in descending pain modulation. Ascending fibers from the spinal cord including axons of lamina I neurons terminate in the PAG (Menetrey, D. *et al.*, 1982; Hylden, J. L. K. *et al.*, 1986; Azkue, J. J. *et al.*, 1998). Afferent input from the cerebral cortex, hypothalamus, the cuneiform nucleus, caudal brainstem nuclei such as nucleus raphe magnus (NRM), also enters PAG (Beitz, A. J., 1982b). The PAG provides major input to most areas that are involved in endogenous

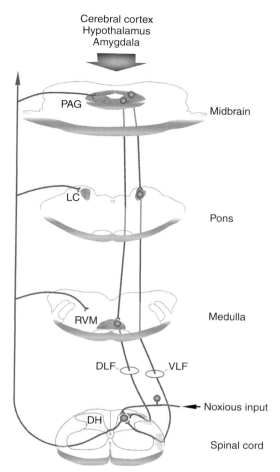

Figure 2 Two major brainstem descending pathways involved in pain modulation. The PAG in the midbrain has efferent projection to the rostral ventromedial medulla (RVM) and locus coeruleus (LC). The RVM and LC directly project to the spinal dorsal horn (DH) where nociceptive input is initially processed. The descending input from the RVM and LC travels in the dorsolateral (DLF) and ventrolateral funiculus (VLF) and modulates spinal pain transmission (only postsynaptic modulation of dorsal horn projection neuron is shown). Supraspinal projecting neurons also issue collaterals to reach pain modulatory structures and the PAG also receives input from forebrain structures including cerebral cortex, hypothalamus, and amygdala. See text for details of other key structures and pathways related to descending pain control. The drawings of brain tissue sections are adapted from Paxinos, G. and Watson, C. 2005. The Rat Brain in Stereotaxic Coordinates, 5th edn. Academic Press.

pain control, including RVM, LC/SC, A5, pontine parabrachial nuclei (PB), NTS, hypothalamus and amygdala (Jensen, T. S. and Gabhart, G. F., 1988; Fields, H. L. and Basbaum, A. I., 1999). The PAG also is a site for regulation of other homeostatic functions (Gauriau, C. and Bernard, J. F., 2002). For instance, PAG neurons are activated by cardiovascular input (Murphy, A. Z. *et al.*, 1995). Somatovisceral interaction may occur in distinct organized columns of neurons in the PAG.

Animal studies have demonstrated that electrical stimulation of PAG produces profound antinociception, or analgesia (Reynolds, D. V., 1969; Mayer, D. J. *et al.*, 1971, Mayer, D. J. and Liebeskind, J. C., 1974; Fardin, V. *et al.*, 1984a). The stimulation-produced analgesia was not due to a generalized sensory and motor deficit. The animals could move and were responsive to audiovisual stimuli during stimulation, while the responses to noxious stimuli were absent. Stimulation of PAG also selectively inhibits responses of dorsal horn and trigeminal sensory neurons to noxious stimulus (Sessle, B. J. *et al.*, 1981; Dostrovsky, J. O. *et al.*, 1983; Gebhart, G. F., 1988). The stimulus-produced analgesia has been successfully observed in humans with chronic intractable pain. Electrical stimulation of the midbrain periaqueductal and hypothalamus periventricular areas produces long-standing analgesia from very brief periods of stimulation. With implanted electrodes in the brain, some patients can self-stimulate for 10–15 min, two to three times per day for adequate pain control (Kumar, K. *et al.*, 1997). It has been observed in patients who have had failure of the implanted stimulation system, that the actual return of pain to its original pre-stimulation level takes 10 days to 2 weeks (Kumar, K. *et al.*, 1997). This would indicate that activation of the endogenous pain control system has a long-lasting effect.

49.2.1.2 PAG–RVM–spinal dorsal horn circuitry

Convergent lines of evidence indicate that PAG stimulation-produced analgesia is relayed through other brainstem nuclei. The RVM has been identified as the premier relay station between the PAG and spinal dorsal horn. The RVM is termed for collective structures that are functionally involved in descending pain modulation, mainly consisting of the midline nucleus raphe magnus (NRM) and the adjacent gigantocellular reticular nucleus, alpha part (GiA). Anatomically, the raphe pallidus nucleus (Rpa) and a portion of the medial lateral paragigantocellular nucleus (LPGi) also are located in this region (Zagon, A., 1995; Paxinos, G. and Watson, C., 2005). Extensive evidence supports the critical involvement of the PAG–RVM–spinal dorsal horn pathway in descending pain control.

In the rat, suppression of NRM neurotransmission by either lesions, local anesthesia, or microinjection of the *N*-methyl-D-aspartate (NMDA) receptor antagonist reverses the spinal inhibition of nocifensive behavior produced by activation of PAG neurons (Behbehani, M. M. and Fields, H. L., 1979; Aimone, L. D. and Gebhart, G. F., 1986). Lidocaine block of the NRM blocks PAG stimulation-produced inhibition of spinal nociceptive neurons in cat (Gebhart, G. F. *et al.*, 1983).

RVM does not receive direct input from the spinal cord, although indirect spinal input can reach RVM through the adjacent Gi (Basbaum, A. I. and Fields, H. L., 1984; Fields, H. L. and Basbaum, A. I., 1999). The major source of input to the RVM is from the PAG and cuneiformis nucleus (Fardin, V. *et al.*, 1984b; Fields, H. L. and Basbaum, A. I., 1999). The ventrolateral cell column of PAG projects to NRM and the dorsolateral PAG region projects to the GiA (Cameron, A. A., *et al.*, 1995b). However, Van Bockstaele E. J. *et al.* (1991) show that injections of tracers into the dorsal PAG resulted in anterograde labeling in the NRM.

Electrical stimulation and microinjection of an opioid receptor agonist into the PAG activate NRM neurons that project to the spinal cord (Behbehani, M. M. and Pomeroy, S. L., 1978; Fields, H. L. and Anderson, S. D., 1978). The excitation of NRM neurons by injection of glutamate into the PAG is associated with an increase in nociceptive response threshold (Behbehani, M. M. and Fields, H. L., 1979).

The RVM has major projections to the spinal cord. Neurons of all subregions of RVM, NRM, GiA, and LPGi send axons to the spinal cord via the dorsolateral funiculus (DLF) (Basbaum, A. I. and Fields, H. L., 1979; 1984). The densest terminal fields of RVM descending fibers are in superficial and deep dorsal horn where nociceptive information is processed.

Activation of RVM neurons produces antinociception. In the cat, electrical stimulation of NRM completely suppresses the behavioral responses to noxious pinch of the skin and modifies the jaw-opening reflex threshold to tooth pulp stimulation (Oliveras, J. L. *et al.*, 1975). Electrical stimulation of NRM or opioid microinjection inhibits activity of dorsal horn nociceptive neurons (Fields, H. L. and Anderson, S. D., 1978; Dickenson, A. H. *et al.*, 1979; Jones, S. L. and Light, A. R., 1990).

The DLF mediates descending modulation of dorsal horn substantia gelatinosa neurons (Cervero, F. et al., 1979). Spinal DLF lesions block antinociception produced by brainstem stimulation (Basbaum, A. I. et al., 1977).

Thus, although studies suggest that some PAG neurons project directly to the spinal cord and spinal trigeminal nucleus (Mantyh, P. W. and Peschanski, M., 1982; Mantyh, P. W., 1983; Skirboll, L. et al., 1983; Li, Y. Q. et al., 1993), the RVM relay is necessary for the analgesic effect of PAG. Whether PAG can modulate dorsal horn sensory transmission through a direct pathway to the spinal cord still needs to be determined.

49.2.1.3 Dorsolateral pontomesencephalic tegmentum

In addition to the PAG–RVM–dorsal horn circuitry, the components of the DLPT constitute separate parallel descending pathways responsible for pain modulation (Dubner, R. and Bennett, G. J., 1983; Gebhart, G. F., 1988; Fields, H. L. and Basbaum, A. I., 1999). One member of the DLPT, the cuneiform nucleus, is located in the midbrain lateral reticular formation neighboring ventrolateral PAG and shares anatomical and physiological similarities with PAG related to descending pain modulation. Electrical stimulation of the midbrain lateral reticular formation produces powerful inhibition of dorsal horn nociceptive neurons in cats (Carstens, E. et al., 1980). The other members of the DLPT consist of LC/SC, A7 cell group, PB and Kölliker-Fuse nuclei. The critical role of these pontine noradrenaline cell groups in descending control has been established.

49.2.1.3.(i) The PAG–LC/SC–spinal pathway
The LC in the dorsal pontine is the major noradrenergic nucleus originally designated the A6 cell group by Dahlström A. and Fuxe K. (1964). The SC is located immediately ventral to the LC and assigned to the A7 noradrenaline cell group by Westlund K. N. et al. (1983). Since the two structures are relatively small and the spinal projecting fibers from the LC are mainly from the ventral part, electrical stimulation will affect both nuclei (Jones, S. L., 1991). The following studies provide evidence that the LC/SC–spinal pathway represents an additional source of descending pain modulation.

The LC/SC neurons project to the spinal cord. Initially, bulbospinal noradrenergic fibers are thought to originate from the A1 cell group in the caudal medulla (Dahlström, A. and Fuxe, K., 1964). Using combined dopamine–beta-hydroxylase immunohistochemistry and retrograde tracing, Westlund K. N. et al. (1983) demonstrated that brainstem descending noradrenergic fibers are exclusively from the pontine noradrenergic cell groups. In the rat, the LC and SC provide major descending noradrenergic axons to the spinal cord. These findings form the anatomical basis for a role of the LC/SC in descending control. One concern on the spinal projection of LC/SC neurons is that they apparently terminate most heavily in laminae VII, VIII, and IX of the ventral horn. The termination of LC fibers is moderate in the deep dorsal horn and relatively sparse in the superficial dorsal horn (Clark, F. M. and Proudfit, H. K., 1991). However, there appears to be a substrain difference in LC/SC descending projection related to the commercial sources. More lamina I terminations are from the LC in rats raised in Harlan Sprague-Dawley while the SC of Sasco source provides more lamina I projection (Fritschy, J. M. and Grzanna, R., 1990; Sluka, K. A. and Westlund, K. N., 1992; Willis, W. D. et al., 1995; Proudfit, H. K., 2002).

LC/SC stimulation produces inhibition of dorsal horn nociceptive neurons in rats and cats (Mokha, S. S. et al., 1986; Jones, S. L. and Gebhart, G. F., 1986; Jones, S. L., 1991). Selective spinal lesions indicate that the effect of LC/SC is mediated by the ventrolateral funiculus (VLF) (Jones, S. L., 1991), whereas the NRM stimulation-produced inhibition is conveyed via the DLF (Basbaum, A. I. et al., 1977).

Unlike the PAG, the effect of LC on spinal nociceptive transmission is not mediated through RVM since local anesthesia of NRM does not block LC stimulation-produced inhibition of dorsal horn neurons (Jones, S. L., 1991).

Intrathecal yohimbine, an alpha 2 adrenoceptor antagonist, blocks the LC stimulation-produced inhibition of nociceptive reflex in rats (Gebhart, G. F., 1988).

The LC/SC may also relay the effect of PAG since they receive PAG input. In monkey, dense projection to the LC is identified from the PAG (Mantyh, P. W., 1983). In rat, the dorsolateral PAG cell column innervates LC/SC (Cameron, A. A. et al., 1995b), although another study suggests that PAG does not have major projections to the LC proper (Ennis, M. et al., 1991).

49.2.1.3.(ii) Other noradrenaline cell groups in the DLPT participating in descending modulation
49.2.1.3.(ii).(a) A7 cell column
The A7 cell group spans rostrocaudally the caudal midbrain and rostral

pons in the rat (Paxinos, G. and Watson, C., 2005). The rostral tip of the A7 group can be seen at the same transverse plane as the cuneiform nucleus and PAG. In the pons, the A7 is located medial to the dorsal part of the ventrolateral principal sensory trigeminal nucleus and ventral to the supratrigeminal and Kölliker-Fuse nuclei. The caudal A7 cells end before the appearance of the LC. The A7 cells also provide dense projection to the spinal cord (Westlund, K. N. et al., 1983). This cell group may play a unique role in the descending circuitry. The A7 group has reciprocal connections with RVM and may mediate some effect of RVM on spinal nociceptive transmission (Fields, H. L. and Basbaum, A. I., 1999; Buhler, A. V. et al., 2004).

49.2.1.3.(ii).(b) Parabrachial nucleus The PB consists of several distinct subnuclei located in the dorsolateral pons and belongs to the pontine cell groups that give rise to noradrenergic terminals in the spinal cord (Westlund, K. N. et al., 1983). The PB has been demonstrated as an important relay station in pain pathways. Spinal and subnucleus caudalis lamina I nociceptive neurons project to PB (Hylden, J. L. K. et al., 1989; Bester, H. et al., 2000). PAG neurons project to the PB in the rat (Krout, K. E. et al., 1998). The PB has further connections with the amygdala (Bernard, J. F. et al., 1993) and the hypothalamus (Bester, H. et al., 1997), and sends descending projections to the spinal and medullary dorsal horn (Yoshida, A. et al., 1997). The activation of the PB descending pathway modulates spinal and trigeminal nociceptive transmission (Girardot, M. N. et al., 1987; Chiang, C. Y. et al., 1994; Willis, W. D. and Westlund, K. N., 1997).

49.2.1.3.(ii).(c) The Kölliker-Fuse nucleus The Kölliker-Fuse nucleus is immediately dorsal to the A7 cell group (Paxinos, G. and Watson, C., 2005) and is the major source of spinopetal noradrenergic fibers in the cat (Stevens, R. T. et al., 1982). The Kölliker-Fuse nucleus is more developed in humans than in other animal species (Lavezzi, A. M. et al., 2004). Young R. F. et al. (1992) show that chronic stimulation of the Kölliker-Fuse nucleus region can relieve intractable pain in patients and the effect is comparable to that of PAG/periventricular stimulation.

49.2.1.3.(ii).(d) A5 cell column The A5 cell group spans the ventrolateral pontine tegmentum at the mid-pon level to the rostral ventrolateral medulla (RVLM) in the rat. This cell group shares many similarities with other noradrenergic cell groups in the DLPT such as spinal projection (Westlund, K. N. et al., 1983; Clark, F. M. and Proudfit, H. K., 1993) and extensive connections with other pain modulatory nuclei including PAG and CVLM (Cameron, A. A. et al., 1995b; Tavares, I. and Lima, D., 2002). Stimulation of the A5 produces inhibition of the nociceptive tail flick reflex in rats and the effect is mediated through spinal opioid and alpha adrenoceptors (Burnett, A. and Gebhart, G. F., 1991). Since A5 has been primarily considered as an important site for cardiovascular regulation, this cell group may play a role in somatoautonomic integration.

49.2.1.4   Caudal medulla

Several structures in the caudal medulla have emerged as important players in endogenous pain control. As these sites overlap with those critical in visceral and cardiovascular functions, the caudal medulla provides a point of convergence to facilitate interactions between somatosensory and autonomic regulation.

49.2.1.4.(i) Caudal ventrolateral medulla Anatomically, the CVLM is not a clearly defined structure, the center of which is about 2 mm ventral to the dorsal surface of the medulla and medial to the ventral pole of the spinal trigeminal nucleus in the rat. Rostrocaudally, the CVLM is within the most caudal segment of the medulla starting from approximately 0.5 mm rostral to the calamus scriptorius (Aicher, S. A. et al., 1996). This region is also referred as the lateral reticular formation and includes the A1 catecholamine cell group. The poorly defined ventromedial border of lamina V of the trigeminal subnucleus caudalis may fall into the territory of the CVLM, or vice versa. It has been consistently observed that a variety of noxious stimuli, including noxious heating of the hindpaw, peritoneal inflammation, urinary bladder inflammation, deep muscle irritation or inflammation, all induce Fos protein expression in the CVLM (see Ren, K., 2002).

49.2.1.4.(i).(a) Lateral reticular nucleus The LRN is a large nucleus located bilaterally in the ventrolateral medullary reticular formation. It receives input from a variety of structures including PAG and NRM and projects to the spinal cord (Janss, A. J. and Gebhart, G. F., 1987; Lee, H. S. and Mihailoff, G. A., 1999). Electrical or chemical stimulation of the LRN produces inhibition of the nociceptive reflex and spinal dorsal horn unit responses to noxious stimuli. The LRN-produced descending inhibition is partially mediated by bilateral DLF and spinal alpha 2 adrenergic and serotonin receptors (Janss, A. J. and Gebhart, G. F., 1987; 1988).

49.2.1.4.(i).(b) Lateral CVLM

The CVLMlat, a small area of the medullary reticular formation between the ventral pole of the spinal trigeminal nucleus and LRN, has recently been proposed to have a major role in descending pain inhibition (Tavares, I. and Lima, D., 2002). The CVLMlat has reciprocal connections with spinal lamina I neurons and output to the A5 cell group and RVM. These indirect spinopetal pathways from the CVLMlat via A5 and RVM may mediate noradrenergic and serotonergic antinociception produced by CVLM activation. Neurons in this region are reciprocally connected to the RVM and activated after orofacial tissue injury (Sugiyo, S. et al., 2005). A group of reticulohypothalamic tract neurons in the CVLM may overlap with the CVLMlat (Malick, A. et al., 2000). Most of these neurons respond to both innocuous and noxious stimulation of the head but only to noxious stimulation of extracephalic regions. Their response properties suggest that they participate in pain modulation.

49.2.1.4.(ii) Subnucleus reticularis dorsalis

The SRD includes a group of neurons distributed in the dorsal medullary reticular formation medial to the trigeminal subnucleus caudalis (also called medullary reticular nucleus, dorsal part, MdD) (Paxinos, G. and Watson, C., 2005). The SRD contributes to the spinoreticular–thalamic pathway by receiving nociceptive information from the spinal dorsal horn and relays it to the thalamus (Villaneuva, L., et al. 1998). The diffuse noxious inhibitory control (DNIC), proposed by Le Bars and colleagues, is mediated via the SRD and independent of the RVM-spinal pathway (Bouhassira, D. et al., 1992; 1993). The SRD is also proposed as a pronociceptive center (Lima, D. and Almeida, A., 2002).

49.2.1.4.(iii) Nucleus tractus solitarius

The NTS is a key structure in somatovisceral processing, since this nucleus receives somatic and visceral input and has extensive efferent connections with other nuclei (Randich, A. and Maixner, W., 1984; Ren, K. et al., 1990). Notably, PAG neurons project to NTS (Cameron, A. A. et al., 1995b). Activation of NTS neurons produces descending pain inhibition and the

vagal afferent modulation of nociception is mediated through NTS (Randich, A. and Gebhart, G. F., 1992). The function of NTS supports somatovisceral and somatoautonomic interactions.

49.2.1.5 Forebrain sites

A number of forebrain structures provide descending input to brainstem nuclei, particularly the PAG (Beitz, A. J., 1982b).

49.2.1.5.(i) Hypothalamus

The subregions of the hypothalamus provide the largest descending input to the PAG (Beitz, A. J., 1982b). The periventricular gray (PVG) of the hypothalamus is the rostral extension of PAG and rich in opioids. Axons of PVG neurons can reach the spinal cord (Jensen, T. S. and Gabhart, G. F., 1988; Van den Pol, A. N. and Collins, W .F., 1994). Stimulation of PVG can produce effective pain relief in patients (Kumar, K. et al., 1997). The lateral hyphthalamus (LH) has reciprocal connections with the PAG and NRM (Aimone, L. D. et al., 1988; Cameron, A. A. et al., 1995a; 1995b). Inhibition of spinal nociceptive transmission from the LH requires a bulbar relay in the NRM (Aimone, L. D. et al., 1988).

49.2.1.5.(ii) Amygdala

The ascending nociceptive input reaches the amygdala through a relay in the PB (Gauriau, C. and Bernard, J. F., 2002). The amygdala has reciprocal connections with PAG (Rizvi, T. A. et al., 1991). Activation of amygdala neurons produces analgesia that is mediated through PAG (see Fields, H. L. and Basbaum, A. I., 1999). The dorsal PAG interacts with the amygdala to affect fear and anxiety and the amygdala is an important site for interrelating sensory and affective dimensions of pain (Behbehani, M. M., 1995; Price, D. D., 2002).

49.2.1.5.(iii) Cerebral cortex

Recent brain imaging studies have generated a wealth of information on activation of the cortical areas during pain processing. It also shows that the cerebral cortex is not only the end point of the pain pathway, but also a source of descending input that contributes to the fine-tuning of pain threshold and modulation of the emotional aspect of pain (Jasmin, L. et al., 2003b; Petrovic, P. et al., 2004). One interesting pathway involves the anterior cingulate cortex (ACC) and amygdala in pain modulation. The ACC–amygdala pathway may modulate pain-related stress and aversive responses, as well as pain-related fear memory (Gao, Y. J. et al., 2004; Petrovic, P. et al.,

2004; Tang, J. *et al.*, 2005). Transcranial magnetic stimulation of the dorsolateral prefrontal cortex increased human pain tolerance level (Graff-Guerrero, A. *et al.*, 2005).

49.2.1.6 Summary

It is clear from the above overview that the pivotal anatomical substrates for descending pain control are located within the brainstem. In addition to the midbrain PAG, a number of important sites are in the medulla. The medulla oblongata is classically termed the vital center for life with its critical involvement in cardiovascular and respiratory controls. Since pain is arguably one of the most important body functions for survival, it is not surprising that the structures that are vital for pain modulation are also located in the medulla to facilitate interactions between somatosensory and autonomic regulation. The interplay of the descending system with visceral (NTS-PB), cardiovascular (CVLM, A5, PAG), respiratory (Kölliker-Fuse nucleus), micturation (RVM, pons), arousal (A6), and emotional (amygdala, ACC) regulation is extensive and serves to enhance survival (Randich, A. and Maixner, W., 1984; Randich, A. and Gebhart, G. F., 1992; Murphy, A. Z. *et al.*, 1995; Mason, P., 2001).

49.2.2 Neurotransmitters

A plethora of transmitters (also see Chapter Pharmacological Modulation of Pain) have been shown to be involved in descending pain control (Duggan, A. W., 1985; Fields, H. L. *et al.*, 1991; Millan M. J., 2002). The most studied and established candidates include endogenous opioid peptides, noradrenaline, and serotonin (5-HT). All these transmitters participate in inhibition and facilitation of pain transmission through their respective receptor subtypes. The major neurotransmitters that are implicated in descending pain control are briefly discussed below.

49.2.2.1 Endogenous opioids and related peptides

49.2.2.1.(i) Opioid peptides Tsou K. and Jang C. S. (1964) discovered that the most sensitive sites for the analgesic action of morphine were located in the PAG and the adjacent hypothalamic periventricular area. These early findings suggested that opiate drugs like morphine acted by binding to a receptor in the brain and that there likely were endogenous ligands or chemical mediators whose actions were mimicked by opiates. Opiate binding sites in the brain were subsequently demonstrated in early 1970s by radioligand binding assays (Pert, C. B. and Snyder, S. H., 1973; Simon, E. J. *et al.*, 1973; Terenius, L., 1973) and the first endogenous opioid peptides enkephalins were identified soon after (Hughes, J. *et al.*, 1975). To date, four classes of opioid peptides have been found in the brain: endorphins, enkephalins, dynorphins, and endomorphins (Hughes, J. *et al.*, 1975; Li, C. H. and Chung, D., 1976; Goldstein, A. *et al.*, 1979; Zadina, J. E. *et al.*, 1997). Martin W. R. *et al.* (1976) first proposed the existence of three subtypes of opioid receptors based on three different syndromes produced by congeners of morphine. All three opioid receptors, μ (MOP), δ (DOP), and κ (KOP), have now been cloned (Evans, C. J. *et al.*, 1992; Kieffer, B. L. *et al.*, 1992; Chen, Y. *et al.*, 1993; Meng, F. *et al.*, 1993). The endogenous opioid peptides and their respective precursors, amino acid sequence, and preferred receptor subtypes are summarized in Table 1. Recent studies also indicate the existence of alternative splicing (Pan, L. *et al.*, 2005), subtypes of the mu, delta, and kappa receptors (Fowler, C. J. and Fraser, G. L., 1994;

Table 1 Endogenous opioid peptides involved in pain modulation

Precursor	Name	Amino acid sequence	Receptor
POMC (pro-opiomelanocortin)	β-Endorphin	Tyr-Gly-Gly-Phe-Met-Thr-Ser-Glu-(human) Lys-Ser-Gln-Thr-Pro-Leu-Val-Thr-Leu-Phe-Lys-Asn-Ala-Ile-Ile-Lys-Asn-Ala-Tyr-Lys-Lys-Gly-Glu-OH	$\mu \geq \delta$
Proenkephalin	Leu-enkephalin	Tyr-Gly-Gly-Phe-Leu-OH	$\delta \geq \mu$
Proenkephalin	Met-enkephalin	Tyr-Gly-Gly-Phe-Met-OH	$\delta \geq \mu$
Prodynorphin	Dynorphin A	Tyr-Gly-Gly-Phe-Leu-Arg-Arg-Ile-Arg-Pro-Lys-Leu-Lys-Trp-Asp-Asn-Gln-OH	κ
Unidenitified	Endomorphin-1	Tyr-Pro-Trp-Phe-NH_2	μ
Unidenitified	Endomorphin-2	Tyr-Pro-Phe-Phe-NH_2	μ

Zaki, P. A. *et al.*, 1996; Pasternak, G. W., 2001), and opioid receptor gene polymorphism (Ikeda, K. *et al.*, 2005). The functional significance of the further molecular diversity of opioid receptor subtypes in descending pain control is poorly understood.

The endogenous opioids and their receptors are distributed in structures involved in descending pain control at all levels including the amygdala, hypothalamus, PAG, DLPT, RVM, and spinal cord (Basbaum, A. I. and Fields, H. L., 1984; Ruda, M. A., 1988; Mansour, A. *et al.*, 1995; Fields, H. L. and Basbaum, A. I., 1992; Wang, H. and Wessendorf, M. W., 2002). Notably, beta-endorphin terminals from hypothalamus descend along the borders of the third ventricle and reach PAG, DLPT, and NRM (Finley, J. C. *et al.*, 1981; Jensen, S. and Gabhart, G. F., 1988). Rostral ventromedial medulla neurons express multiple opioid receptor phenotypes (Marinelli, S. *et al.*, 2002). In the RVM, spinally projecting serotonergic and nonserotonergic neurons express mu- and delta-opioid receptor mRNAs (Wang, H. and Wessendorf, M. W., 1999). Enkephalinergic axons in the spinal cord synapse with thalamic projection neurons (Ruda, M. A., 1982), providing postsynaptic modulation of ascending nociceptive information.

Microinjection of morphine into the PAG produces behavioral analgesia (Camarata, P. J. and Yaksh, T. L., 1985). Brain stimulation-produced analgesia can be reversed by naloxone, an opioid receptor antagonist (Akil, H. *et al.*, 1976). Behavioral analgesia produced by an opioid receptor agonist is related to the inhibition of spinal nociceptive neurons (Yaksh, T. L., 1978). However, the effect of opioid agonists on dorsal horn nociceptive neurons is more variable and can be both inhibitory and facilitatory (Gebhart, G. F., 1982; Stamford, J. A., 1995). This paradox is likely related to the sampling limitation during single unit recording and cautions on predicting behavioral outcome from activity of single neurons. Using Fos immunoreactivity to study a population of activated neurons simultaneously, opioids produce behavioral analgesia that is correlated to an inhibition of dorsal horn neuronal activation as indicated by a reduction of noxious stimulus-induced Fos immunoreactivity (Gogas, K. R. *et al.*, 1991). Is the endogenous opioid system tonically active? The findings from opioid receptor knockout mice suggest a low endogenous opioid tone with each receptor subtype contributing differentially to the modulation of transient pain (Matthes, H. W. *et al.*, 1996; Sora, I. *et al.*, 1997; Gaveriaux-Ruff, C. and Kieffer, B. L., 2002).

Although there is no question that opioids produce potent pain inhibition through their respective receptors at all levels of the descending circuitry, the role of spinopetal opioidergic fibers in mediating descending modulation from supraspinal sites has yet to be convincingly demonstrated. Intrathecal administration of the opioid receptor antagonists in rats has been shown to reverse or attenuate analgesia produced by intracerebroventricular morphine (Gogas, K. R. *et al.*, 1996), intra-PAG injection of beta-endorphin (Tseng, L. F. and Tang, R., 1990), electrical stimulation of the RVM (Zorman, G. *et al.*, 1982; Lu, Y. *et al.*, 2004), or the hypothalamic arcuate nucleus (Wang, Q. *et al.*, 1990). However, the contribution of intrinsic opioidergic interneurons in the spinal dorsal horn to supraspinal-produced analgesia cannot be ruled out. In the cat, opioid peptides intrinsic to the spinal cord play a major role in inhibition of spinal nociceptive transmission (Duggan, A. W., 1985).

49.2.2.1.(ii) Orphanin FQ, nocistatin, and neuropeptideFF Orphanin FQ, or nociceptin, is a heptadecapeptide that was identified as the endogenous ligand for the orphan-opioid-receptor-like 1 (ORL-1, now nociceptin/orphanin FQ peptide receptor, NOP) receptor (Meunier, J. C. *et al.*, 1995; Reinscheid, R. K. *et al.*, 1995). Genetically, Orphanin FQ and NOP belong to the same family with the opioids and their receptors and they share homology with dynorphin A and the kappa opioid receptor, respectively. However, orphanin FQ does not interact with three classical opioid receptors, largely due to the replacement of the first tyrosine residue by phenylalanine. Most studies show that orphanin FQ produces analgesia at the spinal level and both pro- and anti-nociception supraspinally (Meunier, J. C., 1997; Henderson, G. and McKnight, A. T., 1997). However, intrathecal low dose of orphanin FQ also produces allodynia (Yamamoto, T. *et al.*, 1999). Orphanin FQ can suppress both pain facilitatory and inhibitory neurons in the RVM, which may explain its dual effects on pain processing under different conditions (Pan, Z. Z. *et al.*, 2000; Vaughan, C. W. *et al.*, 2001; Heinricher, M. M., 2005).

Nocistatin is liberated from the same orphanin FQ precursor protein, proorphanin, after posttranslational cleavage but binds to a receptor distinct from NOP (Okuda-Ashitaka, E. *et al.*, 1998). Nocistatin opposes action of orphanin FQ and can be considered a functional antagonist of orphanin FQ

(Yamamoto, T. *et al.*, 1999; Millan, M. J., 2002). Proorphanin also yields another heptadecapeptide, orphanin FQ2, which may participate in pain inhibition (Rossi, G. C. *et al.*, 1998).

The octapeptide neuropeptideFF (NPFF, Phe-Leu-Phe-Gln-Pro-Gln-Arg-Phe-NH₂) is originally isolated from the bovine brain and found to possess anti-morphine property (Yang, H. Y. *et al.*, 1985). NPFF fibers project to the PAG from hypothalamus and to the RVM from the NTS (Aarnisalo, A. A. and Panula, P., 1995). In the spinal cord, NPFF-immunoreactive cell bodies and terminals are localized to laminae I–IV and X and DLF, but a supraspinal origin of NPFF was not found (Kivipelto, L. and Panula, P., 1991). Available evidence indicates that NPFF produces antinociception at the spinal level and pronociception (anti-opioid) at the supraspinal level (Millan, M. J., 2002; Mollereau, C. *et al.*, 2005).

49.2.2.2 Monoamines

It is now well established that descending monoaminergic fibers mediate descending modulation from the supraspinal sites (Yaksh, T. L., 1979; Basbaum, A. I. and Fields, H. L., 1984; Gebhart, G. F., 1988; Tjolsen, A. *et al.*, 1991; Hung, K. C. *et al.*, 2003). The major pathways involve noradrenergic fibers from the DLPT and serotonergic fibers from the NRM, the major nucleus of RVM. The two pathways can be co-activated and may act synergistically to participate in descending pain modulation (Gebhart, G. F., 1988; Sawynok, J., 1989; Danzebrink, R. M. and Gebhart, G. F., 1991).

49.2.2.2.(i) Noradrenaline

Intrathecal administration of the alpha 2-adrenoceptor antagonist yohimbine, but not the nonselective opioid receptor antagonist naloxone and alpha 1 adrenoceptor antagonist prazosin, attenuates descending inhibition of nocifensive tail flick reflex produced by focal stimulation of the LC/SC region (Jones, S. L., 1991). The PAG-produced inhibition of dorsal horn neurons is reduced by microdialysis of alpha 2 adrenoceptor antagonists idazoxan and yohimbine into the dorsal horn (Peng, Y. B. *et al.*, 1996). Selective depletion of spinal noradrenaline by intrathecal DSP4 attenuates morphine analgesia (Zhong, F. X. *et al.*, 1985). These results clearly indicate a role of spinopetal noradrenergic fibers in descending modulation since there are no noradrenaline-producing cell bodies in the spinal cord (Swanson, L. W. and Hartman, B. K., 1975). Consistently, depletion of spinal NA by 6-hydroxydopamine decreases and

intrathecal NA elevates the nociceptive threshold (Reddy, S. V. *et al.*, 1980). Opioid analgesia is enhanced in norepinephrine transporter knockout mice and the effect is blocked by yohimbine (Bohn, L. M. *et al.*, 2000). Intrathecal administration of the adrenoceptor antagonists produces a decrease in nociceptive threshold in the rat, suggesting that the descending noradrenergic system is tonically active (Sagen, J. and Proudfit, H. K., 1984). The antinociceptive effect of NA in the spinal cord does not appear to require GABAergic, glycinergic, opioidergic, and serotonergic activity (Reddy, S. V. *et al.*, 1980), and may involve both pre- and postsynaptic mechanisms (Sonohata, M. *et al.*, 2004). *In vivo* whole-cell patch-clamp recordings from rat substantia gelatinosa neurons show that NA reduces EPSC and suppresses action potentials evoked by a noxious stimulus, an effect that is blocked by antagonism of alpha 2, but not alpha 1, adrenoceptors (Sonohata, M. *et al.*, 2004).

Studies using transgenic mice, such as D79N mice with a point mutation of the alpha 2A adrenoceptors, have examined the subtypes of spinal alpha 2 adrenoceptors involved in descending inhibition. Most studies support a role for the alpha 2A subtype in spinal analgesia (Hunter, J. C. *et al.*, 1997; Lakhlani, P. P. *et al.*, 1997; Stone, L. S. *et al.*, 1997; Kingery, W. S. *et al.*, 2000; Malmberg, A. B. *et al.*, 2001; Mansikka, H. *et al.*, 2004; Ozdogan, U. K. *et al.*, 2004). However, alpha 2C adrenoceptor-bearing terminals synapse with NK-1 projection cells in the rat dorsal horn (Olave, M. J. and Maxwell, D. J., 2003), suggesting a role in postsynaptic modulation. The alpha 2C adrenoceptors have been shown to be involved in spinal analgesia and adrenergic–opioid synergy (Fairbanks, C. A. *et al.*, 2002; Lahdesmaki, J. *et al.*, 2003). The alpha 2B adrenoceptors may play an important role in the pontine noradrenergic nuclei involved in analgesic action of nitrous oxide (Sawamura, S. *et al.*, 2000).

After peripheral nerve injury, there is a decrease in alpha 2A mRNA and protein immunoreactivity in the spinal cord of the rat (Stone, L. S. *et al.*, 1999; Leiphart, J. W. *et al.*, 2003); these findings suggesting a reduced noradrenergic transmission in these rats. However, there is also an increased descending noradrenergic innervation to the spinal dorsal horn after nerve injury, possibly serving to enhance inhibition (Ma, W. and Eisenach, J. C., 2003). Interestingly, pain hypersensitivity after nerve injury develops similarly in the wild-type mice and those with mutation of alpha 2A, 2B, or 2C adrenoceptors, suggesting that the net inhibitory tone is not affected by loss of alpha 2 adrenoceptors (Malmberg, A. B. *et al.*, 2001). It is worth noting that, different from previous observations in normal rats (Sagen, J. and Proudfit, H. K., 1984), the baseline nociceptive threshold is not affected in mice with mutation of either of the three alpha 2 subtypes (Malmberg, A. B. *et al.*, 2001), suggesting that compensatory mechanisms may have developed in the mutant mice to balance the loss of tonic noradrenergic inhibition.

In contrast to a predominant inhibitory role of the alpha 2 adrenoceptors, the alpha 1 adrenoceptor may be involved in descending pain facilitation (Holden, J. E. *et al.*, 1999; Nuseir, K. and Proudfit, H. K., 2000; Hedo, G. and Lopez-Garcia, J. A., 2001; Holden, J. E. and Naleway, E., 2001) and supraspinal pronociceptive action (Kingery, W. S. *et al.*, 2002). Intracellular recordings from adult rat spinal slice support the opposing actions of NA on substantia gelatinosa neurons. Noradrenaline induces hyperpolarization or increased spontaneous excitatory postsynaptic potentials in subpopulations of dorsal horn neurons, which is selectively blocked by alpha 2 (yohimbine) and alpha 1 (prazosin) adrenoceptor antagonists, respectively (North, R. A. and Yoshimura, M., 1984).

49.2.2.2.(ii) *Serotonin* There is dense serotonergic innervation in the spinal dorsal horn (Ruda, M. A. *et al.*, 1982; Ruda, M. A., 1988), which is almost exclusively derived from supraspinal sources (LaMotte, C. C., 1988; Fields, H. L. *et al.*, 1991; Mason, P., 2001; Millan, M. J., 2002). The NRM is the major structure that issues descending serotonergic fibers to the spinal cord (Bowker, R. M. *et al.*, 1988), although not all descending axons from the NRM are serotonergic (Jones, S. L. and Light, A. R., 1992; Wang, H. and Wessendorf, M. W., 1999). Descending inhibition produced by focal electrical stimulation in the PAG or NRM is mediated in part by descending serotonergic pathways acting on spinal serotonergic receptors. Microinjection of morphine into the PAG and stimulation of NRM evokes the release of serotonin from spinal cord (Yaksh, T. L. and Tyce, G. M., 1979; Rivot, J. P. *et al.*, 1982). Intrathecal methysergide, a nonselective 5-HT receptor antagonist, increases the intensity of stimulation in the PAG and the NRM for inhibition of the nocifensive tail flick reflex while naloxone has no effect (Gebhart, G. F., 1988). The role of the descending serotonergic pathway in descending pain inhibition is well established (Fields, H. L. and Basbaum, A. I., 1978; Basbaum, A. I. and Fields, H. L., 1984).

It has become clear that descending serotonergic fibers also participate in facilitation of nociception (Zhuo, M. and Gebhart, G. F., 1991; Millan, M. J., 2002; Suzuki, R. *et al.*, 2002b; 2004b) The nociceptive facilitation triggered by low intensity vagal afferent stimulation involves descending serotonergic pathways (Ren, K. *et al.*, 1991). Although spinal application of 5-HT generally produces an analgesic effect, excitation of dorsal horn neurons by serotonergic agonists is also reported (Todd, A. J. and Millar, J., 1984; Zhang, Y. *et al.*, 2001). In the spinal slice preparation, serotonin can recruit silent glutamatergic synapses in the superficial dorsal horn and enhance synaptic transmission (Li, P. and Zhuo, M., 1998). This result supports a facilitatory role of descending serotonergic fibers in spinal plasticity in response to injury. Indeed, intrathecal administration of methysergide has been shown to attenuate neuropathic pain in the rat (Pertovaara, A. *et al.*, 2001). In carrageenan-inflamed rats, the facilitation of dorsal horn neuronal responses to noxious stimuli by agonists of $5\text{-}HT_{1A}$ and $5\text{-}HT_{1B}$ receptors is further enhanced (Zhang, Y. *et al.*, 2001). Blockade of spinal $5\text{-}HT_3$ receptors reduces tissue injury-induced persistent pain (Zeitz, K. P. *et al.*, 2002).

At least 12 different subtypes of 5-HT receptors may be involved in pain modulation (Hoyer, D. *et al.*, 1994; Millan, M. J., 2002). However, their roles in descending pain control have only been evaluated in a few subtypes including $5\text{-}HT_{1A}$, $5\text{-}HT_{1B}$, $5\text{-}HT_{2A}$, $5\text{-}HT_3$, and $5\text{-}HT_4$ receptors (Millan, M. J., 2002). Recent studies have identified a pronociceptive role of the $5\text{-}HT_3$ subtype (Suzuki, R. *et al.*, 2002b; Zeitz, K. P. *et al.*, 2002). A number of studies have indicated a facilitatory role of the $5\text{-}HT_{1A}$ receptor while others pointed to an opposite effect (Millan, M. J., 2002). The equivocal literature on the role of 5-HT receptors in pain modulation suggests sophisticated differential involvement of the 5-HT receptor subtypes in pain modulation and transmission. Moreover, different experimental conditions including the chosen endpoints, stimulation modality, anesthetic state, and *in vivo* versus *in vitro* preparations are also confounding factors.

Despite a well-documented contribution of the descending serotonergic pathway to pain modulation, the specific role of RVM serotonergic neurons in the operation of descending circuitry is not understood. Stimulation of the PAG that produces analgesia does not excite RVM serotonergic neurons (Gao, K. *et al.*, 1997). Thus, these cells may not mediate the signal from PAG, at least not directly.

Stimulation of NRM that induces antinociception does not release 5-HT in the spinal cord, although further increasing the stimulation intensity results in 5-HT release (Sorkin, L. S. *et al.*, 1993). These results are consistent with earlier findings that both serotonergic and nonserotonergic pathways contribute to descending inhibition (Rivot, J. P. *et al.*, 1980). Careful analysis of the characteristics of RVM serotonergic neurons revealed that their levels of activity depend on the conscious state (Mason, P., 2001). Serotonergic cells exhibit the highest level of discharge during waking when nociceptive threshold is relatively high compared to the sleep state. One hypothesis is that the descending serotonergic modulation of pain transmission is state dependent and closely related to alterations in serotonergic tone.

49.2.2.2.(iii) Dopamine

Similar to noradrenaline and serotonin, spinal dopamine is virtually exclusively of supraspinal origin, mainly from the hypothalamus (see Millan, M. J., 2002). The dopamine receptor consists of D_1 and D_2 subtypes and they both have been localized in the spinal cord including the dorsal horn. The D_1 receptor may support opioid analgesia from PAG activation (Flores, J. A. *et al.*, 2004) and the activity of the D_2 subtype may oppose opioid analgesia (King, M. A. *et al.*, 2001). The contribution of the descending dopaminergic pathway to modulation of spinal nociceptive transmission remains to be determined.

49.2.2.3 Amino acids

Excitatory and inhibitory amino acid neurotransmitters are widely distributed in the pain modulatory circuitry including the PAG, DLPT, RVM, and spinal cord. They clearly participate in the process of descending pain control through actions on their respective receptors.

49.2.2.3.(i) Glutamate

Activation of supraspinal glutamate synapses is associated with antinociception. Glutamate-containing neurons are observed in the PAG of the cat (Barbaresi, P. *et al.*, 1997). Stimulation-produced analgesia from the PAG is mediated by excitatory amino acid receptors in the RVM (Aimone, L. D. and Gebhart, G. F., 1986; van Praag, H. and Frenk, H., 1990; Spinella, M. *et al.*, 1996). Stimulation of the cuneiform nucleus excites about 50% of NRM neurons, an effect blocked by an excitatory amino acid antagonist (Richter, R. C. and Behbehani, M. M., 1991). However, microinjection of glutamate into the discrete brainstem sites including some sites in the PAG can also induce pain-like behavior via activation of NMDA receptors (Jensen, T. S. and Yaksh, T. L., 1992). Thus, the glutamatergic transmission within the pain modulatory circuitry is also capable of bidirectional pain modulation.

Glutamate receptors consist of ionotropic and metabotropic (mGluR) subfamilies. The former includes NMDA, alpha-amino-3-hydroxy-5-methyl-4-isoxazole propionic acid (AMPA), and kainate (KA) receptor subtypes, and the eight subtypes of metabotropic glutamate receptors (mGluR's) are subdivided into three groups. Studies on the role of specific glutamate subtypes in descending pain control are still in the preliminary stage. In the PAG, group I mGluR may be involved in pain inhibition involving an interaction with NMDA receptors (Maione, S. *et al.*, 1998; Berrino, L. *et al.*, 2001). In the RVM, activation of group II mGluR produces descending inhibition of the nociceptive tail flick reflex (Kim, S. J. *et al.*, 2002). Activation of AMPA and NMDA receptors contributes to plasticity in the RVM in response to tissue injury (Urban, M. O. and Gebhart, G. F., 1999; Ren, K. and Dubner, R., 2002).

Unlike noradrenaline and serotonin, primary afferent fibers are rich in glutamate (Battaglia, G. and Rustioni, A., 1988). At the spinal level, glutamate receptors play a major role in the development of persistent pain. It is difficult to isolate the effect of descending glutamatergic input from that of primary afferent origin. Limited information is available on the anatomy of descending glutamatergic pathways. One study shows that descending pathways from the DLPT nuclei contain glutamatergic fibers (Liu, R. H. *et al.*, 1995). However, whether they contribute to descending pain modulation is unclear.

49.2.2.3.(ii) Gamma-aminobutyric acid and glycine

In general, gamma-aminobutyric acid (GABA) and glycine are used by inhibitory interneurons. In the PAG, GABAergic terminals are abundant (Reichling, D. B. and Basbaum, A. I., 1990). Blocking GABA receptors in the ventral PAG enhances morphine-produced analgesia in this region, consistent with the view that GABAergic neurons inhibit PAG output neurons involved in descending inhibition (Depaulis, A. *et al.*, 1987). GABAergic terminals from the PAG descend to the RVM (Reichling, D. B. and Basbaum, A. I., 1990).

RVM GABA-containing neurons and GABAergic terminals of PAG neurons both synapse with spinally projecting RVM neurons (Cho, H. J. and Basbaum, A. I., 1991). Inhibition of GABAergic transmission in the RVM leads to disinhibition of pain inhibitory activity. Apparently, the GABA system may set up the tone of descending inhibition and help to maintain an adequate level of pain sensitivity, which is essential to normal body function. Direct GABAergic innervation of the spinal dorsal horn by fibers descending from the RVM has been described (Antal, M. et al., 1996). However, confounded by the presence of abundant intrinsic GABAergic neurons in the spinal cord, the role of descending GABAergic fibers in descending control is unclear.

Alterations of GABAergic activity may result in either an increase or decrease in pain sensitivity. In the insular cortex of the rat, increasing and decreasing GABA transmission produces analgesia and hyperalgesia, respectively (Jasmin, L. et al., 2003b). Thus, GABAergic activity in this region appears to facilitate cortical output related to descending inhibition, which is in contrast to PAG and RVM. In the RVM, the GABA$_B$ agonist baclofen produces antinociception at low dose (0.1–1.0 ng) and hyperalgesia at high dose (30–150 ng) (Thomas, D. A. et al., 1995; Hammond, D. L. et al., 1998). Intrathecal methysergide antagonizes the baclofen-produced antinociception, suggesting that analgesia is related to disinhibition of bulbospinal serotonergic neurons (Hammond, D. L. et al., 1998). GABAergic neurons in the DLPT also exert bidirectional pain modulation by providing tonic inhibition of two populations of noradrenergic neurons in the A7 group (Nuseir, K. and Proudfit, H. K., 2000).

Stimulation of the PAG and NRM results in release of glycine in the spinal dorsal horn and inhibition of nociceptive neuronal activity (Sorkin, L. S. et al., 1993; Cui, M. et al., 1999). Iontophoresis of glycine and GABA receptor agonists suppresses responses of primate spinothalamic tract neurons to noxious stimuli (Lin, Q. et al., 1999). Clearly, disinhibition of glycinergic and GABAergic neurotransmission constitutes a mechanism of central sensitization and persistent pain (Cronin, J. N. et al., 2004; Zeilhofer, H. U., 2005), although to what extent the descending control intervenes in this process needs to be specifically addressed. Glycine may contribute to the development of hyperalgesia via an intriguing way. As a required co-agonist for NMDA receptor activation, glycine released from inhibitory interneurons in the superficial dorsal horn can escape the synaptic cleft and access and facilitate nearby NMDA receptors through spillover (Ahmadi, S. et al., 2003).

49.2.2.4 Others

A number of additional neurotransmitters also participate in descending pain control, although their roles are studied less extensively and a large portion of the study has focused on the different segmental levels. Some key observations are discussed below.

49.2.2.4.(i) Cannabinoids Endocannabinoids (endogenous marijuana-like compounds) include anandamide and 2-arachidonylglycerol, which act on two subclasses of cannabinoid receptors, CB$_1$ and CB$_2$ (Cravatt, B. F. and Lichtman, A. H., 2004; Hohmann, A. G. et al., 2005). The CB1 receptors are localized in the CNS including important structures of descending pain control (Tsou, K. et al., 1998). Endogenous cannabinoids produce analgesia via mechanisms apparently different from those of endogenous opioids, although they may act in similar neural circuitry involving the PAG and RVM (Walker, J. M. and Huang, S. M., 2002). Stress induces endocannabinoids in the PAG associated with analgesia that can be reversed by CB1 receptor blockade in the region (Hohmann, A. G. et al., 2005). In CB$_1$-mutant mice, the stress-induced analgesia is selectively affected, leaving the opioid analgesia unaffected (Valverde, O. et al., 2000). The analgesic effect of cannabinoids requires the RVM circuitry and cannabinoids modulate activity of RVM pain modulating neurons (Meng, I. D. et al., 1998). Although generally considered to be localized peripherally, increasing evidence indicates that the CB$_2$ receptor in the spinal cord plays a role in the persistent pain process (Zhang, J. et al., 2003).

49.2.2.4.(ii) Brain-derived neurotrophic factor In the adult mammalian brain, brain-derived neurotrophic factor (BDNF) facilitates excitatory synaptic transmission and long-term synaptic plasticity. Studies have shown that exogenously applied BDNF into the PAG produces analgesia in rats (Siuciak, J. A. et al., 1995; Frank, L. et al., 1997). Overexpression of BDNF in the spinal cord alleviates neuropathic pain (Cejas, P. J. et al., 2000; Eaton, M. J. et al., 2002). The analgesic effect of exogenous BDNF may be attributed to a downregulation of TrkB after a large dose of the BDNF treatment (Frank, L. et al., 1997; Chen, H. and Weber, A. J., 2004). If BDNF is microinjected into the RVM at a dose within the

physiological range (10–100 fmol), pain facilitation occurs (Guo, W. *et al.*, 2006). BDNF-containing neurons in the PAG project to and release BDNF in the RVM. Intra-RVM sequestration of BDNF by TrkB–IgG fusion protein or anti-BDNF antibody and knockdown of TrkB by RNAi attenuates pain after inflammation (Guo, W. *et al.*, 2006). The effect of BDNF in the RVM is dependent on the activation of NMDA receptors. BDNF induces tyrosine phosphorylation of the NMDA receptor NR2A subunit via a signal transduction cascade that involves IP3, PKC, and Src (Guo, W. *et al.*, 2006). Thus, supraspinal BDNF–TrkB signaling contributes to net descending pain facilitation through a mechanism involving NMDA receptor activity.

49.2.2.4.(iii) Substance P

Although substance P is a major neurotransmitter released from primary afferents at the spinal level, it is also widely distributed in supraspinal structures involved in descending pain control (Beitz, A. J., 1982a). A group of PAG neurons containing both substance P and cholecystokinin (CCK) projects to the spinal cord (Skirboll, L. *et al.*, 1983). However, the function of this descending pathway is unclear. A significant population of PAG neurons possesses neurokinin 1 (NK1) receptors, the primary binding site for substance P, and many NK1-positive PAG neurons also contain glutamate and to a lesser extent, enkephalin (Commons, K. G. and Valentino, R. J., 2002). This raises the possibility that activation of NK1 in the PAG may be upstream to glutamatergic activation and affect activity of the descending circuitry. Both pro- and antinociceptive properties have been attributed to the NK1 receptor. Holden J. E. *et al.* (2002) show that antinociception from lateral hypothalamic stimulation is mediated by NK1 receptors in the A7 cell group in rats. Suzuki R. *et al.* (2002b) demonstrate that excitation of spinal NK1-bearing projection neurons results in input to the RVM, which triggers descending pain facilitation.

49.2.2.4.(iv) Neurotensin

Microdialysis of morphine into the ventromedial PAG induces neurotensin release (Stiller, C. O. *et al.*, 1997). The neurotensinergic input from the PAG to the RVM has been described (Beitz, A. J., 1982c). Neurotensin is involved in bidirectional descending modulation from the RVM. Microinjection of low dose (30 fmol) of neurotensin into the RVM produces hyperalgesia and facilitation of dorsal horn neuronal responses to noxious stimuli and high doses of neurotensin (0.3–3 nmol) produces antinociception (Urban, M.

O. *et al.*, 1996; Urban, M. O. and Gebhart, G. F., 1997). Activation of the NT_1 receptor on spinally projecting serotonergic neurons in the RVM produces antinociception through release of serotonin in the spinal dorsal horn (Buhler, A. V. *et al.*, 2005).

49.2.2.4.(v) Cholecystokinin

Dense immunoreactive cholecystokinin (CCK) terminals are localized to PAG and RVM neurons (Skinner, K. *et al.*, 1997). However, CCK-containing neurons in the PAG do not appear to project to the NRM (Beitz, A. J. *et al.*, 1983). Most studies agree that CCK acts as an anti-opioid peptide through CCK_2 (CCK_B) receptors (Han, J. S., 1995; Fields, H. L. and Basbaum, A. I., 1999; Veraksits, A. *et al.*, 2003). Cholecystokinin in the RVM mediates opioid-induced hyperalgesia and antinociceptive tolerance (Xie, J. Y. *et al.*, 2005). Intra-RVM pretreatment with a selective CCK_2, but not CCK_1 (CCK_A), receptor antagonist significantly potentiates the antihyperalgesic effect of morphine in a model of visceral nociception (Friedrich, A. E. and Gebhart, G. F., 2003). Supraspinal CCK may drive tonic descending facilitation mechanisms to maintain neuropathic pain in the rat (Kovelowski, C. J. *et al.*, 2000). The cellular mechanisms of the pronociceptive and anti-opioid actions of CCK involve an action on pain modulatory neurons in the RVM (Heinricher, M. M. *et al.*, 2001; Heinricher, M. M. and Neubert, M. J., 2004).

49.2.2.4.(vi) Acetylcholine

The DLPT nuclei and RVM regions have been found to send descending cholinergic fibers to the spinal cord (Bowker, R. M. *et al.*, 1983; Jones, B. E. *et al.*, 1986). The cholinergic pedunculopontine tegmental nucleus (PPTg) in the DLPT, immediately ventral to the cuneiform nucleus, has attracted attention. The PPTg is one of the two most sensitive sites for intracranial nicotine-produced antinociception (Iwamoto, E. T., 1991). Neurons in the PPTg are responsive to noxious tail pinch (Carlson, J. D. *et al.*, 2004). Microinjection of nicotine into the NRM, the other sensitive site for the effect of nicotine, produces antinociception (Iwamoto, E. T., 1991). Intra-NRM ABT-594, a cholinergic channel modulator, produces analgesia, presumably through an action on alpha 4-containing nicotinic acetylcholine receptors coexpressed by serotonergic neurons (Bitner, R. S. *et al.*, 1998). The dorsal raphe nucleus is also a potential site for nicotinic analgesia (Cucchiaro, G. *et al.*, 2005). Spinal cholinergic receptors contribute to descending inhibition from the Gi/GiA region, although it is unclear whether the source of acetylcholine transmitters is supraspinal (Zhuo, M. and Gebhart, G. F., 1990). At the spinal level, acetylcholine tonically modulate serotonergic transmission (Cordero-Erausquin, M. and Changeux, J. P., 2001) and both inhibitory and excitatory neurons are responsive to nicotinic stimulation (Cordero-Erausquin, M. *et al.*, 2004).

49.2.2.4.(vii) *Histamine*

Central histamine is mainly derived from the tuberomammilllary nucleus of the hypothalamus and histaminergic terminals are found in the PAG, LC, RVM, and spinal cord (Millan, M. J., 2002). Among four classes of histamine receptors (H_{1-4}), H_{1-3} receptors have been localized to the CNS but their roles in descending pain control are unclear. Activation of central H_1 receptor induces behavioral hyperalgesia in mice involving the phospholipase C–protein kinase C pathway (Galeotti, N. *et al.*, 2004). Intracerebroventricular injection of H_3 agonist is hyperalgesic in rodents (Malmberg-Aiello, P. *et al.*, 1994). However, activation of spinal histamine H_3 receptors inhibits mechanical nociception (Cannon, K. E. *et al.*, 2003). It is clear that the effects of histamine-related agents may be mediated through nonhistamine receptors. A histamine-derived compound, improgan, produces analgesia exclusively through the supraspinal circuitry involving PAG and RVM through unknown receptor mechanisms (Hough, L. B. *et al.*, 1997; Nalwalk, J. W. *et al.*, 2004).

49.2.2.4.(viii) *Prostaglandin*

Recent observations suggest a role of centrally produced prostaglandins in descending pain modulation. Microinjection of prostaglandin E2 into the medial preoptic region and PAG regulates pain modulatory neurons in the RVM, the effects including activation of ON cells and suppression of OFF cells, and produces thermal hyperalgesia (Heinricher, M. M. *et al.*, 2004a; 2004b). These findings support the emerging view that the CNS glia-inflammatory cytokine network, in association with the cyclooxygenase 2-prostaglandin cascade, is involved in the development of hyperalgesia after tissue and nerve injury (Samad, T. A. *et al.*, 2001; Watkins, L. R. *et al.*, 2003). Such pathways may increase descending facilitation from the RVM circuitry and produce pain and hyperalgesia.

49.2.2.4.(ix) *Orexins (hypocretin)*

The novel neuropeptides orexin A and B are selectively synthesized in the lateral and posterior hypothalamus and originally identified as a regulator of feeding behavior (Sakurai, T. *et al.*, 1998). A number of studies have suggested a role of orexin in descending pain modulation. Hypothalamic descending axons containing orexin innervate all levels of the spinal cord including superficial dorsal horn (van den Pol, A. N., 1999). Orexin A-like immunoreactive fibers in the spinal cord are mostly confined to axon terminals containing dense-cored vesicles and mainly forming asymmetric synapses (Guan, J. L. *et al.*, 2003). Initial analysis indicates that orexin-B modulates superficial dorsal horn neurons in the rat spinal cord. Orexin-B increases spontaneous inhibitory postsynaptic currents in the majority of dorsal horn neurons and also produces excitatory effects on a smaller population of dorsal horn neurons (Grudt, T. J. *et al.*, 2002). Orexins produces analgesia through central but not peripheral mechanisms and do not appear to involve the endogenous opioids (Mobarakeh, J. I. *et al.*, 2005). Activation of spinal orexin-A receptor produces anti-allodynic effect in the rat with carrageenan-induced inflammation (Yamamoto, T. *et al.*, 2003). In the pre-pro-orexin (precursor of orexins) knockout mice, the baseline pain threshold is not altered but the knockout mice exhibit enhanced hyperalgesia after inflammation and reduced stress-induced analgesia (Watanabe, S. *et al.*, 2005). These results suggest that hypothalamic orexins and related descending pathways are activated by persistent pain and stress to produce pain inhibition.

49.2.2.5 *Summary*

The bulbospinal noradrenergic and serotonergic pathways are the most established systems with regard to descending pain modulation. The roles of other neurotransmitters and receptors in descending pain control are mainly documented at the segmental levels. The interactions between these transmitters and receptors, including receptor subtypes, are extremely extensive and sophisticated (see Chapter Pharmacological Modulation of Pain) (Millan, M. J., 2002), although their contributions are discussed individually here. A striking feature of the involvement of neurochemicals in descending pain control is that almost every substance has been shown to be involved in both inhibitory and facilitatory pain control, either through actions at the different receptor subtypes, different levels of the neuraxis, or different concentrations; and that the same transmitter may produce opposite modulatory effects at different, or even the same, segmental levels. This factor contributes to some conflicting results in this area and seriously complicates the interpretation of the findings derived from genetic approaches that involve manipulation of specific receptor(s) at the systemic level.

49.2.3 Neuronal Activity in Descending Pathways

49.2.3.1 Pain modulatory neurons

By comparing neuronal activity to the occurrence of nociceptive reflexes in lightly anesthetized rats, Fields H. L. *et al.* (1983a) identified three classes of neurons in the RVM. One type of cell typically shows a burst of activity immediately prior to the onset of the tail flick from a noxious thermal stimulus, thus named an ON cell. Another type of cell exhibits a pause in activity just before the tail flick and is called an OFF cell. The activity of the third class of cell, NEUTRAL cell, exhibits no clear relationship to nociceptive reflexes in response to transient stimuli. The firing profiles of ON and OFF cells suggest that they are pain modulatory neurons since their activity is correlated with facilitation and inhibition of nocifensive behaviors, respectively. In fact, both RVM ON and OFF cells project to the spinal cord, particularly laminae I, II, and V of the dorsal horn (Fields, H. L. *et al.*, 1995), and are modulated by PAG stimulation and opioids (Vanegas, H. *et al.*, 1984; Fields, H. L. *et al.*, 1983b; Fields, H. L. *et al.*, 1991). Although less studied, ON and OFF cells are also found elsewhere involved in pain modulation, including the PAG (Heinricher, M. M. *et al.*, 1987), cuneiformis nucleus, PB (Haws, C. M. *et al.*, 1989) and nucleus submedius of the thalamus (Fu, J. J. *et al.*, 2002). It should be clarified that OFF-cell activity may not be solely responsible for suppressing nocifensive withdrawals. In the unanesthetized rat during slow wave sleep, animals are able to execute withdrawal from noxious stimulation when OFF cell continues to fire (Leung, C. G. and Mason, P., 1999). Simultaneous recordings from spinal and RVM neurons show that ON- and OFF-cell firing reflects well input from the spinal dorsal horn, suggesting a role in gating ascending nociceptive information (Hernandez, N. *et al.*, 1989; Hernandez, N. and Vanegas, H., 2001). In addition to regulation of nocifensive responses, ON- and OFF-cell activity may also play a role in integrating cardiovascular, visceral, and somatosensory input (Thurston, C. L. and Randich, A., 1995) and regulating homeostatic functions (Mason, P., 2001).

In vitro recordings from medullary slices identified two distinct cell types in the NRM, primary and secondary cells (Pan, Z. Z. *et al.*, 1990). Primary cells exhibit large GABA-mediated synaptic potentials and are insensitive to opioids, while secondary cells have little or no GABAergic input and are hyperpolarized by opioids. Primary and secondary are likely corresponding to OFF and ON cells, respectively, based on their pharmacological characteristics, such that local iontophoretic of morphine directly inhibits ON cells but does not affect OFF cells (Fields, H. L. and Basbaum, A. I., 1999). Excitatory amino acid transmission is involved in modulating both ON and OFF cell activity (Heinricher, M. M. *et al.*, 1999), which is in line with the evidence that both primary and secondary cell types receive excitatory amino acid input. Models have been proposed to understand the functional significance of neuronal activity within the RVM circuitry. In the disinhibition model, opioids directly inhibit ON cells (corresponding to secondary cells) and release the GABAergic inhibition of OFF cells (corresponding to primary cells), resulting in pain inhibition (Fields, H. L. and Basbaum, A. I., 1999). RVM neurons are also subject to modulation by noradrenergic input (Hammond, D. L. *et al.*, 1980), which mainly originated from the DLPT nuclei (Tanaka, M. *et al.*, 1996; Meng, X. W. *et al.*, 1997). Opposite to the spinal cord, activation of alpha 2 adrenoceptors hyperpolarizes primary cells and attenuates opioid analgesia, and activation of alpha 1 adrenoceptors depolarizes primary cells and enhances analgesia (Bie, B. *et al.*, 2003).

The finding of ON and OFF cells provides a cellular mechanism for bidirectional descending pain control and favors our understanding of variability of pain (Fields, H. L., 1988). It should be emphasized that the descending facilitation parallels inhibition and is an active process. The enhanced nociceptive responsiveness in morphine-dependent rats after treatment with opioid receptor antagonists is accompanied by an increased RVM ON-cell firing (Bederson, J. B. *et al.*, 1990). In spite of a well-documented RVM–spinal serotonergic projection in descending pain control, Gao K. and Mason P. (2000) found that serotonergic RVM cells that respond to noxious tail heat are not ON or OFF cells. The transmitters released by axon terminals of RVM ON and OFF cells in the spinal cord remain to be verified.

Pan Z. Z. *et al.* (1997) show that activation of the kappa-receptor hyperpolarizes primary cells that are disinhibited through the inhibition of secondary cells by mu-opioid receptors. Thus, mu- and kappa-opioid receptors may produce opposing actions on RVM pain modulatory neurons. This kappa effect reverses the analgesic effect of mu-opioids (Bie, B. and Pan, Z. Z.,

2003). Activation of the kappa-opioid receptor in the RVM has also been shown to produce antinociception in a test modality-dependent manner. Microinjection of a kappa-receptor agonist U69593 into the RVM increased withdrawal latency of paw, but not the tail, to noxious thermal stimulus (Ackley, M. A. *et al.*, 2001). The opioid withdrawal-induced hyperalgesia is attenuated by intra-RVM kappa-agonist (Bie, B. and Pan, Z. Z., 2003). The cellular mechanisms of these differential actions of the kappa-opioid on RVM neurons and behavioral nociception are unclear and appear to involve a presynaptic inhibition of glutamatergic input to pain modulatory neurons (Ackley, M. A. *et al.*, 2001; Bie, B. and Pan, Z. Z., 2003) (also see Chapter The Brainstem and Nociceptive Modulation).

49.2.3.2 Descending influence on dorsal horn neurons

49.2.3.2.(i) Tonic effect Wall P. D. (1967) originally observed that blocking impulses descending from the brainstem induced an increased activity of laminae IV/V dorsal horn neurons in the cat. In contrast, the response of lamina VI neurons to movement was mainly reduced. This suggests a tonic brainstem descending inhibition on cutaneous sensory processing in the dorsal horn and a facilitatory control over proprioceptive activity. The tonic inhibition of dorsal horn neurons is confirmed by later studies (Besson, J. M. *et al.*, 1975; Cervero, F. *et al.*, 1976; Dickhaus, H. *et al.*, 1985; Ren, K. and Dubner, R., 1996).

It is worth noting that if a portion of the descending pathways is blocked or only NRM is anesthetized, dorsal horn nociceptive neurons do not show an increased activity (Cervero, F. *et al.*, 1979; Gebhart, G. F. *et al.*, 1983). Apparently, tonic suppression of dorsal horn activity is implemented by multiple descending pathways or supraspinal sites; and the loss of inhibition from one pathway can be compensated by others.

The findings of Wall also reveal the plastic nature of dorsal horn neurons. In lamina IV, many low threshold neurons became wide dynamic range neurons; and in lamina V, the receptive fields of nociceptive neurons were expanded during the spinal cord block. Collins J. G. and Ren K. (1987) showed that, after a low dose of pentobarbital in the intact awake cat, dorsal horn neurons (originally only responsive to noxious stimuli) became responsive to low-intensity stimuli, that is, the wide dynamic range response profile is unmasked. This effect is likely due to a release of inhibitory systems that regulate spinal

dorsal horn activity, consistent with Wall's hypothesis (Wall, P. D., 1967).

49.2.3.2.(ii) Inhibitory effect Descending input produces inhibition of dorsal horn neurons through multiple synaptic mechanisms (Light, A. R. *et al.*, 1986; Zhang, D. X. *et al.*, 1991; Fields, H. L. and Basbaum, A. I., 1999). As spinal serotonin and noradrenaline are almost exclusively of supraspinal origin, they provide postsynaptic control of spinal neurons. Westlund K. N. *et al.* (1990) demonstrated that spinothalamic neurons in primate receive direct catecholaminergic innervation. Stimulation of the NRM produces direct postsynaptic inhibition of spinothalamic tract neurons (Giesler, G. J. Jr. *et al.*, 1981). Descending input can also modulate secondary order neurons indirectly through action on interneurons. NRM and spinal stimulation increases glycine and GABA release into the spinal cord, suggesting activation of inhibitory interneurons (Sorkin, L. S. *et al.*, 1993; Cui, J. G. *et al.*, 1997). Since changes in activity of interneurons can lead to changes in presynaptic inhibition related to primary afferent depolarization, primary afferent fibers are also indirectly modulated by descending input (Martin, R. F. *et al.*, 1979).

49.2.3.2.(iii) Facilitatory effect Supraspinal structures also exert facilitatory influences on spinal nociceptive transmission. Tattersall J. E. *et al.* (1986) report that 44% of neurons recorded from the lower thoracic spinal cord of the cat show reduced response to visceral stimulation after reversible spinalization, suggesting that they are under descending excitatory control. Excitation and inhibition of dorsal horn neurons are produced by stimulation of the DLF of the spinal cord (Dubuisson, D. and Wall, P. D., 1979; McMahon, S. B. and Wall, P. D., 1988), NRM (Dubuisson, D. and Wall, P. D., 1980), and Gi (Haber, L. H. *et al.*, 1980). Even microinjection of morphine into the NRM produces facilitation of C-fiber responses of some dorsal horn nociceptive neurons (Le Bars, D. *et al.*, 1980). Although in most cases, it is unclear whether facilitation occurs on inhibitory interneurons, excitation of which may result in subsequent inhibition, Haber L. H. *et al.* (1980) demonstrate the Gi stimulation-produced excitation of primate spinothalamic tract neurons, indicating descending facilitation of nociceptive transmission. Several studies suggest a spinal–brainstem loop in the modulation of spinal pain processing. The activity of this loop involves descending inhibition or facilitation of dorsal horn neurons (Cervero, F.

and Wolstencroft, J. H., 1984; McMahon, S. B. and Wall, P. D., 1988; Suzuki, R. *et al.*, 2002b). Spinal 5-HT$_3$ receptors mediate descending facilitation of deep dorsal horn neurons, which involves a circuitry consisting of ascending NK1-expressing neurons in the superficial dorsal horn and descending fibers from RVM (Suzuki, R. *et al.*, 2002b).

49.2.4 Multiplicity of Descending Control

The bidirectional descending modulation of nociception has now been appreciated. It appears that some sites in the rostral medulla play an essential role in this balanced descending modulation. Depending on the intensities used and the specific site, stimulation of the Gi and RVM regions produces facilitation (5–25 µA) and inhibition (50–200 µA) of nocifensive reflexes and dorsal horn nociceptive neurons (Zhuo, M. and Gebhart, G. F., 1992; 1997). RVM neurons may exert bidirectional control of nociception through descending serotonergic pathways (Zhuo, M. and Gebhart, G. F., 1991) as well as via the A7 catecholamine cell group. Enkephalin-containing neurons in the RVM project to the A7 cell group and microinjection of morphine in the A7 cell group produces facilitation and inhibition of spinal nociception mediated by alpha 1- and alpha 2-adrenoceptors (Holden, J. E. *et al.*, 1999). The other site that is involved in descending facilitation is the subnucleus reticularid dorsalis (SRD) of the caudal medulla (Lima, D. and Almeida, A., 2002). Lesioning the SRD depresses nociceptive responses to acute and inflammatory pain, whereas stimulation produces pain facilitation. The SRD-mediated facilitation requires a reciprocal disynaptic putative excitatory circuit that links the SRD and the spinal dorsal horn. It is interesting that both facilitatory and inhibitory pathways can be triggered by activation of visceral afferents. Under certain circumstances, vagal afferent stimulation at relative low intensities produces facilitation of the nociceptive tail flick reflex as well as dorsal horn nociceptive neuronal activity (Randich, A. and Gebhart, G. F., 1992). The neural relays involved in vagal afferent-produced facilitation and inhibition also include sites within the RVM. Thus, spinal nociceptive processing is subject to bidirectional control from supraspinal sites so that the intensity of perceived pain can be fine-tuned by descending pathways.

In general, when biphasic pain modulation is observed at the same site, facilitation is produced by electrical stimulation given at lesser intensities or receptor agents administered at lower doses and inhibition is found at higher intensities of stimulation or larger doses of the receptor agents, with few exceptions (Thomas, D. A. *et al.*, 1995; Hammond, D. L. *et al.*, 1998). This may explain why descending facilitation was not recognized in many earlier reports because the facilitation might often be masked by more intense electrical stimulation or higher drug doses used to produce net descending inhibition. It remains to be explained how and why a manipulation at the same site by varying stimulation intensities would produce opposing effects on nociceptive transmission. Pain modulating neurons of different modalities intermingle with each other within the RVM and the membrane properties of ON and OFF cells are very similar (Zagon, A. *et al.*, 1997). One can only speculate that the net effect of excitation of brainstem nuclei is related to differential activation of subsets of neurons with different connections and different combinations of neurotransmitter receptors. In the case of BDNF, brain microinjection at a dose within the normal range, determined by measuring its concentration *in vivo*, produces nociceptive facilitation (Guo, W. *et al.*, 2006), but higher doses lead to antinociception (Siuciak, J. A. *et al.*, 1995; Frank, L. *et al.*, 1997; Guo, W. *et al.*, 2006). This suggests that activation of BDNFergic transmission at the physiological level results in facilitation and inhibition produced by larger doses may result from the recruitment of excessive synapses that are not normally active simultaneously. We should also be aware that supraspinal descending influences might exert differential effects on motor reflex and dorsal horn neuronal activity that are used by most animal studies as the endpoints (Willer, J. C. *et al.*, 1979). Future studies should validate pain facilitation and inhibition by including the affective component of pain and not merely measure reflex behaviors (Vierck, C. J. Jr. *et al.*, 2004; Neubert, J. K. *et al.*, 2005).

49.3 Plasticity of Descending Pathways

To be able to feel pain is a normal and essential function for living. Pain warns the body against injury and illness so that responses will be undertaken to protect the body. Normally, pain goes away when the damage is resolved and tissues are healed, and will not be long-lasting. However, chronic pain develops under a variety of conditions even without the presence of injury, and the suffering from excessive persistent pain has been a major health problem. Most earlier studies on the descending pain control mechanisms have focused on responses to acute or transient noxious stimuli, which has built up the fundamentals for our understanding of the system. Recent studies have turned attention to chronic pain conditions and examined the effects of pain that last for hours, days, or longer, following tissue damage or nerve injury. Persistent pain conditions are associated with prolonged functional changes in the nervous system, evidenced by the development of dorsal horn hyperexcitability, or activity-dependent plasticity, also commonly referred to as central sensitization. Evidence has accumulated to indicate that descending pathways also exhibit plasticity that contributes to altered pain processing after injury (Porreca, F. *et al.*, 2002; Ren, K. and Dubner, R., 2002; Vanegas, H. and Schaible, H.-G., 2004).

49.3.1 Descending Modulation in Persistent Pain

49.3.1.1 Responses of descending pathways to injury

In the rat inflammation model, intraplantar injection of carrageenan induces inflammation and hyperalgesia that is associated with an increased Fos protein expression, a marker of neuronal activation, in the PB and LC/SC (Bellavance, L. L. and Beitz, A. J., 1996; Buritova, J. *et al.*, 1998; Tsuruoka, M. *et al.*, 2003). In adjuvant-induced inflammation and the sciatic nerve ligation model, there is a selective increase in proneurotensin mRNA expression in the PAG (Williams, F. G. and Beitz, A. J., 1993). In contrast, veratridine-induced GABA release is decreased in the PAG after hind paw inflammation (Renno, W. M. and Beitz, A. J., 1999). Orofacial inflammation activates the coeruleotrigeminal antinociceptive pathway (Matsutani, K. *et al.*, 2003) and hind paw inflammation induces an increase in NA levels in the spinal dorsal horn, suggesting activation of descending noradrenergic pathways (Tsuruoka, M. *et al.*, 1999).

In rats receiving chronic constriction injury of the sciatic nerve (the CCI model), in addition to the sensory structures of the cortex and thalamus, changes in brain neural activity are observed in a variety of regions involved in descending control including the hypothalamic arcuate nucleus and central gray matter, LC, PB, and Gi (Mao, J. *et al.*, 1993). The activation of GiA-mediated descending

pain inhibition is shown in rats with partial sciatic nerve ligation (Azami, J. *et al.*, 2001). In the CCI model, there is an upregulation of spinal CB$_1$ receptors and an increased antihyperalgesic and anti-allodynic effect of Win 55,212-2, a CB receptor agonist (Lim, G. *et al.*, 2003), although it is unclear whether descending input contributes to this enhanced pain inhibition.

Intraplantar injection of formalin has been used extensively as a model of prolonged chemical nociception. In rats receiving formalin injection, there is an increase in noradrenaline and serotonin concentrations in the spinal dorsal horn. Intrathecal pretreatment with yohimbine, an alpha 2 adrenoceptor antagonist, and methysergide, a 5-HT receptor antagonist, significantly enhanced nocifensive behavior (Omote, K. *et al.*, 1998). These results suggest activation of descending monoaminergic pathways by formalin. In the PAG, formalin injection leads to a decrease in GABA release (Maione, S. *et al.*, 1999). A decreased GABAergic tone in the PAG presumably will disinhibit PAG neurons and result in activation of opioid interneurons in the RVM, leading to increased descending inhibition.

In humans with secondary hyperalgesia after heat/capsaicine sensitization, the PAG and the cuneiform nucleus activation is observed using whole-brain functional magnetic resonance imaging (Zambreanu, L. *et al.*, 2005). These multiple lines of evidence indicate that, not only can stimulation of pain modulatory circuitry attenuate hyperalgesia and dorsal horn neuronal hyperexcitability after inflammation and nerve injury (Morgan, M. M. *et al.*, 1991; Sotgui, M. L., 1993), descending pathways are also activated under persistent pain conditions.

49.3.1.2 Descending inhibition in persistent pain

There is now considerable evidence that inflammatory injury induces an enhanced net descending inhibition at sites of primary hyperalgesia (Ren, K. and Dubner, R., 2002). In cats with knee joint inflammation, descending inhibition is greater in neurons with input from the inflamed knee as revealed by reversible spinalization with a cold block and DNIC-induced inhibition (Schaible H. G. *et al.*, 1991; Danziger, N. *et al.*, 1999; 2001). Similar phenomenon is observed in rats with adjuvant-induced hind paw inflammation. Local anesthetic block of the thoracic spinal cord leads to an increased activity of dorsal horn nociceptive neurons that is greater in inflamed than that in noninflamed rats (Ren, K. and Dubner, R.,

1996). Studies using Fos protein expression as a marker of neuronal activation have reached a similar conclusion. There are more inflammation-induced Fos-expressing neurons in the spinal dorsal horn in spinally transected or DLF-lesioned rats, when compared to sham-operated inflamed rats (Ren, K. and Dubner, R., 2002). Kauppila P. *et al.* (1998) show that thermal but not mechanical nociceptive responses are further enhanced in hind paw-inflamed and spinal nerve-ligated rats after spinalization at the midthoracic level. Furthermore, behavioral hyperalgesia is intensified in rats with lesions of the dorsal lateral quadrant of the spinal cord after inflammation or formalin injection (Abbott, F. V. *et al.*, 1996; Ren, K. and Dubner, R., 1996). These studies reveal the net descending inhibitory effects after tissue injury. The enhanced descending net inhibition serves to dampen injury-induced dorsal horn hyperexcitability and hyperalgesia.

The source of the enhanced inhibition can be traced back to brainstem structures. Local anesthesia of the RVM results in a further increase in dorsal horn nociceptive neuronal activity in hind paw-inflamed rats (Ren, K. and Dubner, R., 1996). Focal lesions of the RVM and LC are followed by an increase in spinal Fos expression and hyperalgesia after inflammation (Tsuruoka, M. and Willis, W. D., 1996; Wei, F. *et al.*, 1999). When lidocaine was microinjected into the NRM in adjuvant-inflamed rats, increases in activity were observed in over 75% of the spinal dorsal horn neurons (Ren, K. and Dubner, R., 1996). Interestingly, there appears to be laminar selectivity in the effect of RVM serotoninergic and LC noradrenergic descending pathways. The descending serotoninergic and noradrenergic pathways differentially suppress the responses of spinal neurons, including spinoparabrachial neurons, in the deep and superficial dorsal horn, respectively (Ren, K. *et al.*, 2000). It appears that both RVM and LC descending pathways are major sources of enhanced net inhibition in inflamed animals.

49.3.1.3 Descending facilitation in persistent pain

Descending input also exerts a facilitatory effect following injury. Emerging evidence indicates that descending facilitation not only parallels inhibition but also can be a driving force for the development of persistent pain (Porreca, F. *et al.*, 2002; see Figure 3). The selective destruction of the Gi with a soma-selective neurotoxin, ibotenic acid, leads to an attenuation of hyperalgesia and a reduction of

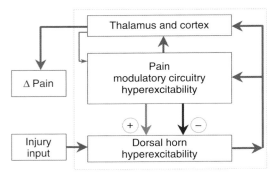

Figure 3 Diagram illustrating descending pain facilitation and inhibition in persistent pain. An enhanced primary afferent input after injury may lead to dorsal horn hyperexcitability and produce dynamic changes in excitability within brain stem pain modulatory circuitry. The net effect of descending modulation depends on a balance between inhibitory (−) and facilitatory (+) input to the spinal cord. An increase in descending net facilitatory modulation may lead to the development of persistent pain. Adapted from Ren, K., Zhuo, M., and Willis, W. D. 2000. Multiplicity and Plasticity of Descending Modulation of Nociception: Implications for Persistent Pain. In: Proceedings of the 9th World Congress on Pain (*eds*. M. Devor, M. C. Rowbotham, and Z. Wiesenfeld-Hallin), pp. 387–400. IASP Press.

inflammation-induced spinal Fos expression (Wei, F. *et al.*, 1999). The descending facilitatory effect may also originate from the medullary SRD (Lima, D. and Almeida, A., 2002) and other brain sites such as the anterior cingulate cortex (Calejesan, A. A. *et al.*, 2000).

Descending facilitation significantly contributes to the development of certain types of persistent pain, particularly those associated with secondary hyperalgesia or nerve injury. Spinalization blocks mustard oil-produced secondary allodynia and hyperexcitability of dorsal horn neurons (Mansikka, H. and Pertovaara, A., 1997). RVM lesions prevent formalin-induced nocifensive behavior (Wiertelak, E. P. *et al.*, 1997) and inhibit secondary hyperalgesia produced by topical application of mustard oil (Urban, M. O. and Gebhart, G. F., 1999). Masseter muscle inflammation produces mechanical hyperalgesia of the overlying skin that is blocked by RVM lesions, and likely includes secondary hyperalgesia of the skin and primary hyperalgesia of the muscle (Sugiyo, S. *et al.*, 2005). The same phenomenon occurs in models of neuropathic pain. Suzuki R. *et al.* (2004a) show that descending facilitatory control of mechanically evoked responses is enhanced in dorsal horn neurones after nerve injury. Local anesthesia of the PAG or RVM attenuates allodynia in rats receiving ligation of the two spinal nerves (Pertovaara, A. *et al.*, 1996; Kovelowski, C. J. *et al.*, 2000). The tactile allodynia and cold hypersensitivity after nerve injury depends upon activation of bulbospinal descending facilitatory pathways (Ossipov, M. H. *et al.*, 2000; Urban, M. O. *et al.*, 2003). Microinjection of the CCK_2 receptor antagonist L365,260 into the RVM reverses tactile allodynia and thermal hyperalgesia after L5/L6 spinal nerve ligation and intra-RVM CCK-8 produces allodynia and hyperalgesia (Kovelowski, C. J. *et al.*, 2000). These findings indicate that CCK may drive descending facilitation to maintain neuropathic pain in rats, which is consistent with the anti-opioid action of CCK in RVM. In nerve-injured rats, lesions of the DLF, local anesthetic block of the RVM, and lesions of RVM mu-opioid receptor expressing cells do not prevent the onset, but reverse the later maintenance of tactile and thermal hypersensitivity (Porreca, F. *et al.*, 2001; Burgess, S. E. *et al.*, 2002). These observations point to an ascending–descending loop that is activated in response to prolonged stimulation to facilitate spinal nociceptive processing. The ascending dorsal column pathways likely contribute to this pain facilitatory circuitry in neuropathic pain models (Miki, K. *et al.*, 2000). There is no evidence of activation of the dorsal column pathway in models of inflammatory primary and secondary hyperalgesia.

49.3.1.4 Descending facilitation versus inhibition

Clearly, descending modulation of persistent pain involves both inhibition and facilitation. Although there may exist selective circuitry that is responsible for multiple phases of modulation, it is rather surprising that the facilitatory sites in the rostral medulla generally overlap with the sites that also produce inhibition. The inhibition and facilitation may also share some synaptic mechanisms with subtle difference in sensitivity. Although many of the above cited studies have concluded that the hyperalgesia is dependent completely on facilitatory influences from the brainstem, it should be noted that the same effects could be produced by a reduction in descending inhibition leading to a dominance of descending facilitation. The serotonin concentration in the NRM is reduced in mice with diabetic neuropathy and sciatic nerve ligation, which is associated with a reduced opioid analgesia (Sounvoravong, S. *et al.*, 2004). Thus, neuropathic hyperalgesia may be dependent on a net descending facilitatory effect wherein a simultaneous descending inhibition is

dominated by the facilitation. After tissue or nerve injury, there is an increase in synaptic strength for both descending inhibition and facilitation. Both facilitatory and inhibitory circuitry may be activated by ascending input after injury (Herrero, J. F. and Cervero, F., 1996; Gozariu, M. *et al.*, 1998). There is an increase in ON- and OFF-cell activity after inflammation (Miki, K. *et al.*, 2002). Thus, the balance between synaptic excitation and inhibition under different conditions appears important. The Gi plays a role in descending facilitation of nociceptive transmission after transient noxious stimuli (Zhuo, M. and Gebhart, G. F., 1992). Lesions of the Gi produce an attenuation of hyperalgesia and spinal Fos expression after inflammation (Wei, F. *et al.*, 1999). However, combined Gi and NRM lesions reverse the opposite Gi or NRM lesion-induced effects. It is possible that severe persistent pain may be enhanced when the facilitatory network overrides the inhibition. In cases of neuropathic pain and secondary hyperalgesia, descending facilitation provides a driving force for the maintenance of persistent pain (also see Chapter Neuropathic Pain: Basic Mechanisms (Animal)).

49.3.2 Dynamics of Descending Pain Modulation

49.3.2.1 Time-dependent shift in descending control

There is a gradual build-up of descending inhibition in the early phase of tissue inflammation (Schaible, H. G. *et al.*, 1991; Ren, K. and Dubner, R., 1996; Hurley, R. W. and Hammond, D. L., 2000). The difference in inflammation-induced Fos expression between transected and sham controls is clear at 3 days but not apparent shortly (2 h) after inflammation (Ren, K. and Ruda, M. A., 1996). In early stages of inflammation, the dynamic response is similar in both the spinally intact and transected rats, suggesting that descending inhibitory inputs play less of a role in the early response to peripheral inflammation. In contrast, at 3 days after inflammation, descending inhibition is enhanced to counteract continuous noxious input. The results from electrophysiological studies also suggest a progressive enhancement of descending inhibition of dorsal horn neurons during the development of inflammation (Schaible, H. G. *et al.*, 1991). An effect of Freund's adjuvant on the contralateral spinal cord may also be subject to the inhibitory control of descending inputs as suggested by a clear contralateral increase in Fos-positive cells in transected rats after inflammation (Ren, K. and Ruda, M. A., 1996). Williams F. G. and Beitz A. J. (1993) demonstrate that after peripheral inflammation or nerve injury, the increase in neurotensin mRNA expression was confined to the ventromedial PAG and the dorsal raphe nucleus initially and spreads to the structures neighboring the PAG including the cuneiform nucleus and the PPTg as the injury persists. The enhanced descending inhibition in inflamed animals has been shown to subside in the late stage at 2 weeks (Guan, Y. *et al.*, 2003). In the ankle monoarthritis rat model, tonic descending inhibition of convergent neurons with input from the inflamed ankle was enhanced during the acute stage (1–2 d) and then decreased in the chronic stage (3–4 weeks) of monoarthritis (Danziger, N. *et al.*, 1999; 2001).

The dynamic changes in descending inhibition after inflammation can be examined over time by monitoring antinocifensive responses in lightly anesthetized rats during RVM stimulation (Figure 4). Terayama R. *et al.* (2000) directly compared the potency of descending inhibition during the development of inflammation in the lightly anesthetized rat preparation. Following a unilateral hind paw inflammation, the stimulus-response function (S-R) curve for the inflamed paw was initially shifted to the right of the noninflamed paw at 3 h and then gradually shifted to the left and reached the maximal potency at 24 h after inflammation (Figure 4(b)). These findings indicate that inflammation induces dramatic changes in the excitability of RVM pain-modulating circuitry. Early (up to 3 h) in the development of inflammation there is an increased descending facilitation also shown by Urban M. O. and Gebhart G. F. (1999), which reduces the net effect of the inhibition. Over time, the level of descending inhibition increases, or descending facilitation decreases, leading to a net enhancement of antinocifensive behavior. Direct stimulation of the DLF that bypasses brainstem synaptic mechanisms does not produce a change in excitability, indicating that the changes are due to supraspinal mechanisms at the level of the RVM or higher.

Thus the activity of descending pathways exhibits considerable plasticity. Brainstem descending pathways become progressively more involved in suppressing incoming nociceptive signals in primary hyperalgesic zones following inflammation. Injury-related, persistent primary afferent input is most likely responsible for triggering this ascending–descending feedback circuit.

(a)

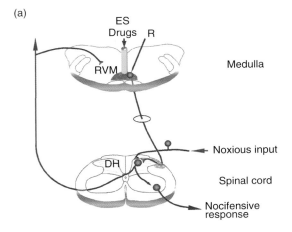

(b)

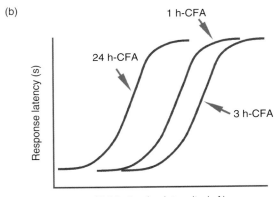

Figure 4 Time- and stimulus intensity-dependent changes in RVM electrical stimulation-produced suppression of nocifensive responses. (a) Schematic representation of the experimental setup. In lightly anesthetized animals, nocifensive responses such as paw withdrawal can be evoked by noxious stimulus. Electrical stimulation (ES) or drugs can be delivered through a guide cannula placed in the RVM. Neuronal activity can be recorded (R) in this preparation (see Figure 5). (b) The nocifensive response latency is plotted against the RVM ES intensity. The stimulus-response function (S-R) curve illustrates intensity-dependent inhibition of nocifensive responses, as indicated by an increase in response latency. Animals were inflamed with complete Freund's adjuvant (CFA), an inflammatory agent. The S-R curve at 1 h post-CFA is located in the middle. At 3 h post-CFA, the curve shifted to the right, suggesting a decrease in descending inhibition at this time. At 24 h post-CFA, the S-R curve shifted to the left, suggesting an enhanced descending inhibition. These results demonstrate dynamic changes in descending pain control during the development of inflammatory hyperalgesia. Adapted from Terayama, R., Guan, Y., Dubner, R., and Ren, K. 2000. Activity-induced plasticity in brain stem pain modulatory circuitry after inflammation. Neuroreport 11, 1915–1919.

49.3.2.2 Phenotypic changes in pain modulatory neurons

The time-dependent plasticity in descending pain modulatory circuitry involves changes in the response profiles of RVM neurons. Montagne-Clavel J. and Oliveras J. L. (1994) have shown changes in RVM neuronal properties after inflammation in the awake, freely moving rat, which suggests an increase in the population of neurons involved in pain modulatory activity. Using paw withdrawal as a behavioral correlate to assess the relationship between nocifensive behavior and RVM neural activity, ON-, OFF-, and NEUTRAL-like cells can be identified (Miki, K. *et al.*, 2002). Interestingly, some NEUTRAL-like cells change their response profile and can be reclassified as ON- or OFF-like cells through continuous recordings during the development of inflammation (Figure 5). These findings suggest that NEUTRAL cells have potential to become involved in pain modulation after injury. Another study by Schnell C. *et al.* (2002) shows that rat RVM neurons may display different properties depending on the site of peripheral stimulation. A NEUTRAL cell responds in an ON or OFF fashion to stimuli applied to the craniofacial region, although it is nonresponsive to noxious stimulation of the tail.

The switch in the response profile of RVM neurons correlates with the temporal changes in excitability in the RVM after inflammation. This phenotypic change of RVM neurons is verified in a population study that shows a significant increase in the percentage of ON- and OFF-like cells, and a decrease in the NEUTRAL-like cell population 24 h after inflammation as compared to control animals (Miki, K. *et al.*, 2002). The studies of RVM neurons support the view that enhanced descending modulation after inflammation involves both facilitation and inhibition since there are changes in the responses of both ON and OFF cells (Miki, K. *et al.*, 2002; de Novellis, V. *et al.*, 2005). After inflammation, there is a greater increase in ON-like cell responses before the onset of paw withdrawal as compared to that in naïve controls. ON-cell activity is associated with facilitation of nocifensive behavior (Fields, H. L. and Basbaum, A. I., 1999). In contrast, OFF-like cell responses are reduced after inflammation indicated

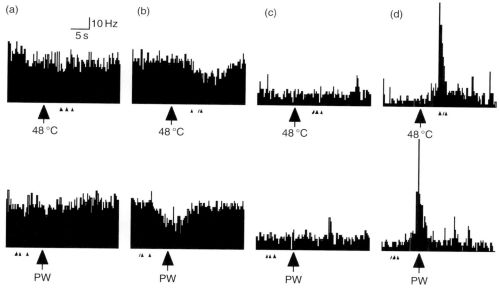

Figure 5 Examples of phenotypic changes of RVM neurons following inflammation. (a) Neutral-like cell, 9 h post-CFA. (b) Panels illustrate that at 14.5 h post-CFA, the same neuron shown in A exhibited a reduction of activity immediately before the paw withdrawal, characteristic of an OFF-like cell. (c) Neutral-like cell, 2 h post-CFA. (d) The same cell shown in (c) became an ON-like cell at 5.5 h post-CFA when the cell showed a burst of activity immediately before the paw withdrawal. All histograms were averaged from three repeated trials (200 ms bin width). The histograms are time-locked to the thermal stimuli (upper row) or the onset of paw withdrawal (lower row), respectively. Small arrowheads indicate the onset of paw withdrawal s (upper row) or onset of thermal stimuli (48 °C, lower row) for individual trials, respectively. Note that the changes in neuronal activity are best correlated with the paw withdrawal response. Adapted from Miki, K., Zhou, Q. Q., Guo, W., Guan, Y., Terayama, R., Dubner, R., and Ren, K. 2002. Changes in gene expression and neuronal phenotype in brain stem pain modulatory circuitry after inflammation. J. Neurophysiol. 87, 750–760.

by a less reduction in neuronal activity after the noxious stimulus, and a lack of a complete pause. The pause in OFF-cell activity is associated with disinhibition; the lack of the pause and the less reduction in neuronal activity suggest an increased inhibition of nocifensive behavior.

49.3.2.3 Involvement of excitatory amino acids and receptors

What are the cellular mechanisms that underlie these changes? Excitatory amino acids (EAAs) have been shown to mediate descending modulation in response to transient noxious stimulation and early inflammation (Urban, M. O. and Gebhart, G. F., 1999; Heinricher, M. M. *et al.*, 1999), and they are involved in the development of RVM excitability associated with inflammatory hyperalgesia (Terayama, R. *et al.*, 2000; Guan, Y. *et al.*, 2002; Miki, K. *et al.*, 2002). Microinjection of NMDA into the RVM produces effects that are dependent upon the postinflammatory time period. At 3 h post inflammation, low doses of NMDA produce facilitation of the response to noxious heat of the inflamed hind paw and the non-inflamed hind paw and tail, supporting findings that descending facilitatory effects are NMDA receptor dependent and occur early after inflammation (Urban, M. O. and Gebhart, G. F., 1999). Higher doses of NMDA at 3 h post inflammation only produce inhibition. At 24 h post inflammation, NMDA produces only inhibition. All of these effects are blocked by administration of NMDA receptor antagonists. AMPA, a selective AMPA receptor agonist, produces dose- and time-dependent inhibition that is significantly greater at 24 h compared to that at 3 h post inflammation. There is a leftward shift of the dose-response curves of NMDA- and AMPA-produced inhibition at 24 h post inflammation as compared to 3 h. The leftward shift of the dose-response curves of EAA receptor agonists parallels the time-dependent enhancement of net descending inhibition produced by RVM electrical stimulation, which is also attenuated by NMDA receptor antagonists (Terayama, R. *et al.*, 2000). These results suggest that the enhanced EAA neurotransmission contributes to the time-dependent functional changes in descending modulation.

49.3.2.4 Molecular mechanisms of activity-dependent plasticity in descending pathways

Correlate changes in gene transcription and protein translation have been found in the descending circuitry after injury. Williams F. G. and Beitz A. J. (1993) show that adjuvant-induced hindpaw inflammation increased neurotensin mRNA expression in the PAG and the lateral tegmental nuclei, from which neurotensinergic neurons project to the RVM. The increased sensitivity of EAA receptors in the descending circuitry during the development of inflammatory hyperalgesia is related to transcriptional and translational modulation of the receptors (Ren, K. and Dubner, R., 2002). Examination of the mRNA expression of the NR1, NR2A, and NR2B subunits of the NMDA receptor in the RVM reveals an upregulation that parallels the time course of the RVM excitability changes. This is accompanied by an increase in NMDA receptor protein. Protein phosphorylation is a major mechanism for regulation of receptor function. The native NMDA receptor is likely a tetramer that consists of two NR1 and two NR2 subunits. Phosphorylation of multiple sites in the cytoplasmic C-termini of the NR1 and NR2 subunits, including tyrosine, serine, and threonine residues, is known to modulate NMDA receptor activity and affect synaptic transmission. There was an increase in NR2A, but not NR2B, tyrosine phosphorylation in the RVM after inflammation (Guo, W. *et al.*, 2006). In contrast, inflammation induces an increase in NR2B, but not NR2A subunit tyrosine phosphorylation in the spinal cord. A time-dependent increase in NR1 serine phosphorylation has also been identified in the RVM after inflammation.

There is also an increase in the AMPA receptor GluR1 subunit levels in the RVM post inflammation (Guan, Y. *et al.*, 2003). AMPA receptor subunits (GluR1–4) exist in the two flip and flop isoforms that differentially affect the desensitization properties of the receptor. Reverse transcription polymerase chain reaction analysis indicates that inflammation induces a significant upregulation of mRNAs encoding the GluR1-flip (5–24 h post inflammation), GluR2-flip (24 h post inflammation) and GluR2-flop (24 h post inflammation) isoforms in the RVM, whereas the levels of GluR1-1 flop mRNAs do not exhibit significant changes (Guan, Y. *et al.*, 2003). GluR1 serine 831 phosphoprotein levels are also increased as early as 30 min after inflammation. The increase in GluR1 phosphorylation depends on primary afferent input and interestingly, is blocked by NMDA receptor antagonists, suggesting an NMDA receptor-mediated modification of AMPA receptor function in the RVM circuitry.

These findings indicate that EAA receptors in the RVM undergo selective transcriptional, translational, and posttranslational modulation following inflammation. The activity-induced plasticity in descending circuitry complements the hyperexcitability in ascending pain transmission pathways (Dubner, R. and Ruda, M. A., 1992). The increased neuronal barrage at the spinal level activates spinal projection neurons that then activate glutamatergic, opioidergic, and presumably GABAergic neurons at the brainstem level leading to a similar but not necessarily identical form of neuronal plasticity.

49.3.2.5 Changes in sensitivity to opioids

49.3.2.5.(i) Increased analgesic potency Rats with inflammatory hyperalgesia exhibit an increased sensitivity to opioid analgesics (Neil, A. *et al.*, 1986). Typically, there is a leftward shift of the dose-response curve for opioids from the inflamed hyperalgesic paw when compared to the noninflamed paw (Hylden, J. L. K. *et al.*, 1991). Kayser V. *et al.* (1991) suggest that this increased opioid sensitivity in inflamed animals is related to a peripheral mechanism as it is significantly attenuated after local injections of very low doses (0.5–1 µg) of naloxone. Ossipov M. H. *et al.* (1995) demonstrate that the increased morphine analgesic potency is mediated through delta opioid receptors. This delta-mediated effect may require mu-opioid receptor-induced trafficking of delta-receptors to neuronal membranes (Morinville, A. *et al.*, 2003). The increased opioid sensitivity after inflammation may also be explained by changes in central pain modulating pathways. Hurley R. W. and Hammond D. L. (2000; 2001) demonstrate enhancement and plasticity of the descending inhibitory effects of mu and delta 2 opioid receptor agonists microinjected into the RVM during the development and maintenance of inflammatory hyperalgesia.

49.3.2.5.(ii) Opioid tolerance and hyperalgesia Hyperalgesia may occur after long exposure to opioids as opioid tolerance develops (Mao, J., 2002). The underlying mechanisms may involve activity of the pain modulatory circuitry. Morphine-tolerant rats exhibit an increased glutamate synaptic transmission in RVM neurons (Bie, B. and Pan, Z. Z., 2005). Chronic morphine treatment induces delta receptor-mediated synaptic inhibition in PAG neurons that is not seen in the untreated mice (Hack, S. P. *et al.*, 2005). The activation of descending pain facilitatory

pathways from the RVM and an increase in CCK activity may contribute to hyperalgesia after prolonged morphine exposure (Ossipov, M. H. *et al.*, 2005).

49.3.2.6 Aging-related alteration in descending pain modulation

Iwata K. *et al.* (2002) have reported age-related plasticity in descending pathways in rats. It is found that, in aged rats, there is an increased dorsal horn nociceptive neuronal activity and nocifensive behavior associated with a significant loss of serotonergic and noradrenergic fibers in the spinal dorsal horn. These results suggest a reduced descending inhibition with advancing age. Human studies concur that the DNIC, a form of endogenous pain inhibition, is reduced in the older adults (Edwards, R. R. *et al.*, 2003).

49.3.3 Significance of Altered Descending Modulation in Persistent Pain

Descending pathways modulate the intensity of perceived pain under normal conditions and persistent pain as a result of injury. The activity-dependent plasticity within the pain modulatory circuitry is a normal function of the brain and presumably is activated to protect the organism from further environmental injury. The dynamic changes in descending modulation after inflammation and primary hyperalgesia and allodynia are protective. The early facilitation may function to enhance nocifensive escape behavior whereas the dominant late inhibition may provide a mechanism by which movement of the injured site is suppressed or reduced to aid in healing and recuperation. Gebhart G. F. (2004) proposes that the need to escape from a predator requires enhanced control of pain and thus more descending inhibition, which is supported by opioid and nonopioid mechanisms of stress-induced analgesia seen in animals (Hayes, R. L. *et al.*, 1978). However, the significance of the enhanced descending facilitation found at sites of secondary hyperalgesia and after nerve injury is unclear. Gebhart G. F. (2004) hypothesizes that descending facilitation is necessary to maintain hyperalgesia as the tissue heals and to protect the injured tissue from further insult. The enhanced descending pain modulation clearly includes shifts in the balance between inhibitory and facilitatory components. Present evidence suggests that there is a different balance in neural networks receiving input from zones of secondary hyperalgesia where there is no primary injury. The balance toward facilitatory influences appears to be maintained for longer periods. Activation of these sites would lead to an enhancement of movement behavior that could also be protective.

Injury to neural tissues will also upset the balance between facilitation and inhibition. The dynamic plasticity of descending pathways after peripheral nerve injury leading to neuropathic pain may render the system vulnerable and lead to pathological consequences. In this situation, the initiation of hyperalgesia is not dependent upon descending facilitation, whereas the maintenance of the hyperalgesia for long periods of time is dependent on such descending facilitation. Nerve injury may activate a descending nociceptive system that normally is protective but can become a source of persistent pain after pathology in the nervous system.

49.4 Therapeutic Implications: Activating Descending Pathways

The mechanisms underlying various clinical approaches for analgesia and pain control involve activation of descending pathways. One important area of research is to find efficient and selective ways to activate endogenous descending pathways for pain control.

49.4.1 Electrical Stimulation

Electrical stimulation has been applied to the all levels of the nervous system to produce analgesia or pain attenuation. Presumably, this approach activates endogenous pain control including the descending pathways to the spinal cord. However, the underlying mechanisms for the stimulation-produced analgesia, particularly the selectivity of the stimulation against certain types of chronic pain in humans, remain to be elucidated.

49.4.1.1 Brain stimulation
Stimulation-produced analgesia has been confirmed in humans (Kumar, K. *et al.*, 1997; Wallace, B. A. *et al.*, 2004). For pain control, the most commonly used sites for deep brain stimulation have been the periventricular gray and PAG regions, which is consistent with findings in animal studies that these areas are involved in opioid analgesia and descending pain inhibition. Practically, this approach is only

considered on patients suffering from intractable pain refractory to other treatments.

49.4.1.2 Spinal cord stimulation
Effective pain control can be achieved by direct spinal cord stimulation (SCS) through an implanted stimulator in the epidural space (Shealy, C. N. *et al.*, 1967; Janfaza, D. R. *et al.*, 1998). In theory, spinal stimulation can bypass brainstem circuitry to directly activate descending fibers or antidromically excite collaterals of ascending fibers that indirectly inhibit projection neurons, leading to inhibition of pain transmission at the spinal level. However, the mechanisms of SCS-produced pain relief are more complex (Meyerson, B. A. and Linderoth, B., 2000). One interesting aspect is that SCS is selectively effective against neuropathic pain and hyperalgesia, leaving normal nociceptive threshold unaffected (Meyerson, B. A. and Linderoth, B., 2000).

49.4.1.3 Vagal stimulation
Animal studies have provided evidence that stimulation of vagal afferent fibers produces antinociception, although low intensity may facilitate pain (Randich, A. and Gebhart, G. F., 1992). Taking advantage of the approved vagal stimulation approach for epilepsy, the effect of vagal stimulation on pain has been confirmed in humans (Kirchner, A. *et al.*, 2000; Ness, T. J. *et al.*, 2000). Vagal stimulation may offer another alternative method for activating the body's pain control system.

49.4.1.4 Transcutaneous electrical nerve stimulation
Peripheral stimulation can be applied through transcutaneous electrical nerve stimulation (TENS), an alternative modality for pain relief (Shealy, C. N., 2003). Although the results have been controversial, the clinical efficacy of this approach has been demonstrated by many studies. The effect of TENS is likely produced through modulation of the intrinsic spinal pain processing and release of endogenous opioids (Sluka, K. A. and Walsh, D., 2003).

49.4.2 Pharmacological Approaches

49.4.2.1 Opioids
Opioid agents have been used extensively for analgesia. Even applied peripherally, a central action involving activation of the endogenous pain control system is required for the analgesic effect of the opioids. The opioids are potent and efficacious

analgesics. However, their side effects have limited their use in chronic pain conditions.

49.4.2.2 Alpha 2 adrenoceptor agonists

Extensive documentation of the role of descending noradrenergic pathways in pain inhibition provides the bases for the use of α_2 adrenoceptor agonists for analgesic purpose. Epidural clonidine for cancer pain is the approved application (Kamibayashi, T. and Maze, M., 2000). The use of the α_2 agonist can reduce the dose of fentanyl because of their synergistic action. Intrathecal clonidine has been shown to improve the effect of SCS (Schechtmann, G. *et al.*, 2004).

49.4.3 Peripheral Stimulation

49.4.3.1 Diffuse noxious inhibitory controls

Heterotopic noxious stimulation produces analgesia in humans (Reinert, A. *et al.*, 2000), which can be understood by the DNIC hypothesis proposed by Le Bars and colleagues (Le Bars, D. *et al.*, 1979a; 1979b). DNIC provides a rationale for understanding counterirritation-produced hypoalgesia and help in the design of nonpharmacological approaches for pain relief (see Chapter Diffuse Noxious Inhibitory Controls (DNIC)). Additionally, non-noxious manipulations also provide pain relief (Simkin, P. P. and O'Hara, M., 2002), presumably through modulation of pain processing by interfering stimuli.

49.4.3.2 Acupuncture

Acupuncture and moxibustion have been used for pain relief for thousands of years. The analgesic effect of acupuncture is related to the recruitment of endogenous opioids (Han, J. S., 2004; Zhang, R. X. *et al.*, 2004) and pain modulatory circuitry (Ma, F. *et al.*, 2004). Since a typical needling during the acupuncture treatment does not induce pain in humans, the mechanisms of acupuncture analgesia are likely different from that of DNIC.

49.4.4 Psychological Factors

49.4.4.1 Stress

Analgesia can be produced by stress through both opioid and nonopioid mechanisms (Hayes, R. L. *et al.*, 1978; Watkins, L. R. and Mayer, D. J., 1982). Stress activates the PAG–RVM pathway, which apparently is a mechanism of stress-induced antinociception (Bellchambers, C. E. *et al.*, 1998). Fear conditioning

induces antinociception involving the amygdala, PAG, and RVM (Helmstetter, F. J. and Tershner, S. A., 1994). The relative painless state in the injured athlete or the wounded soldiers may be explained by stress-induced pain control related to a particular setting.

49.4.4.2 Placebo

Placebo analgesia may be mediated by endogenous opioids (Levine, J. D. *et al.*, 1978; Benedetti, F. and Amanzio, M., 1997). The imaging studies show that the neural substrate related to placebo effect on pain lies in the prefrontal and dorsal anterior cingulate cortex (Lieberman, M. D. *et al.*, 2004). It appears that expectation is a factor for triggering the placebo response (see Chapter The Placebo Effect) (Benedetti, F. *et al.*, 2003).

49.5 Concluding Remarks

Tremendous progress has been made in the past four decades on descending pain control mechanisms. In addition to well-established descending pain inhibition, the demonstration of descending pain facilitation provides a more complete picture of the body's endogenous pain modulatory system. The simultaneous descending excitatory and inhibitory inputs contribute to a balance between perception and suppression of noxious events. We now appreciate that the pain modulatory system undergoes dramatic changes after injury. There is a concurrent activation of brainstem descending inhibitory and facilitatory pathways and the interaction between these pathways will dictate, or affect, the development of neuronal hyperexcitability and behavioral hyperalgesia. The dynamic plasticity of descending pathways may sometimes render the system vulnerable and lead to pathological consequences. The imbalance between these modulatory pathways may be one mechanism underlying variability in acute and chronic pain conditions. For patients suffering from deep pains such as temporomandibular disorders, fibromyalgia, and low back pain, the diffuse nature and amplification of persistent pain, in part, may be the result of a net increase in endogenous facilitation.

References

Aarnisalo, A. A. and Panula, P. 1995. Neuropeptide FF-containing efferent projections from the medial hypothalamus of rat: a *Phaseolus vulgaris* leucoagglutinin study. Neuroscience 65, 175–192.

Abbott, F. V., Hong, Y., and Franklin, K. B. 1996. The effect of lesions of the dorsolateral funiculus on formalin pain and morphine analgesia: a dose–response analysis. Pain 65, 17–23.

Ackley, M. A., Hurley, R. W., Virnich, D. E., and Hammond, D. L. 2001. A cellular mechanism for the antinociceptive effect of a kappa opioid receptor agonist. Pain 91, 377–388.

Ahmadi, S., Muth-Selbach, U., Lauterbach, A., Lipfert, P., Neuhuber, W. L., and Zeilhofer, H. U. 2003. Facilitation of spinal NMDA receptor currents by spillover of synaptically released glycine. Science 300, 2094–2097.

Aicher, S. A., Saravay, R. H., Cravo, S., Jeske, I., Morrison, S. F., Reis, D. J., and Milner, T. A. 1996. Monosynaptic projections from the nucleus tractus solitarii to C1 adrenergic neurons in the rostral ventrolateral medulla: comparison with input from the caudal ventrolateral medulla. J. Comp. Neurol. 373, 62–75.

Aimone, L. D. and Gebhart, G. F. 1986. Stimulation-produced spinal inhibition from the midbrain in the rat is mediated by an excitatory amino acid neurotransmitter in the medial medulla. J. Neurosci. 6, 1803–1813.

Aimone, L. D., Bauer, C. A., and Gebhart, G. F. 1988. Brain-stem relays mediating stimulation-produced antinociception from the lateral hypothalamus in the rat. J. Neurosci. 8, 2652–2663.

Akil, H., Mayer, D. J., and Liebeskind, J. C. 1976. Antagonism of stimulation-produced analgesia by naloxone, a narcotic antagonist. Science 191, 961–962.

Antal, M., Petko, M., Polgar, E., Heizmann, C. W., and Storm-Mathisen, J. 1996. Direct evidence of an extensive GABAergic innervation of the spinal dorsal horn by fibres descending from the rostral ventromedial medulla. Neuroscience 73, 509–518.

Azami, J., Green, D. L., Roberts, M. H., and Monhemius, R. 2001. The behavioural importance of dynamically activated descending inhibition from the nucleus reticularis gigantocellularis pars alpha. Pain 92, 53–62.

Azkue, J. J., Mateos, J. M., Elezgarai, I., Benitez, R., Lazaro, E., Streit, P., and Grandes, P. 1998. Glutamate-like immunoreactivity in ascending spinofugal afferents to the rat periaqueductal grey. Brain Res. 790, 74–81.

Bandler, R. and Shipley, M. T. 1994. Columnar organization in the midbrain periaqueductal gray: modules for emotional expression? Trends Neurosci. 17, 379–389.

Barbaresi, P., Gazzanelli, G., and Malatesta, M. 1997. Glutamate-positive neurons and terminals in the cat periaqueductal gray matter (PAG): a light and electron microscopic immunocytochemical study. J. Comp. Neurol. 383, 381–396.

Basbaum, A. I. and Fields, H. L. 1979. The origin of descending pathways in the dorsolateral funiculus of the spinal cord of the cat and rat: further studies on the anatomy of pain modulation. J. Comp. Neurol. 187, 513–531.

Basbaum, A. I. and Fields, H. L. 1984. Endogenous pain control systems: brainstem spinal pathways and endorphin circuitry. Annu. Rev. Neurosci. 7, 309–338.

Basbaum, A. I., Marley, N. J., O'Keefe, J., and Clanton, C. H. 1977. Reversal of morphine and stimulus-produced analgesia by subtotal spinal cord lesions. Pain 3, 43–56.

Battaglia, G. and Rustioni, A. 1988. Coexistence of glutamate and substance P in dorsal root ganglion neurons of the rat and monkey. J. Comp. Neurol. 277, 302–312.

Bederson, J. B., Fields, H. L., and Barbaro, N. M. 1990. Hyperalgesia following naloxone-precipitated withdrawal from morphine is associated with increased on-cell activity in the rostral ventromedial medulla. Somatosens. Motor Res. 7, 185–203.

Behbehani, M. M. 1995. Functional characteristics of the midbrain periaqueductal gray. Prog. Neurobiol. 46, 575–605.

Behbehani, M. M. and Fields, H. L. 1979. Evidence that an excitatory connection between the periaqueductal gray and nucleus raphe magnus mediates stimulation produced analgesia. Brain Res. 170, 85–93.

Behbehani, M. M. and Pomeroy, S. L. 1978. Effect of morphine injected in periaqueductal gray on the activity of single units in nucleus raphe magnus of the rat. Brain Res. 149, 266–269.

Beitz, A. J. 1982a. The nuclei of origin of brain stem enkephalin and substance P projections to the rodent nucleus raphe magnus. Neuroscience 7, 2753–2768.

Beitz, A. J. 1982b. The organization of afferent projections to the midbrain periaqueductal gray of the rat. Neuroscience 7, 133–159.

Beitz, A. J. 1982c. The sites of origin brain stem neurotensin and serotonin projections to the rodent nucleus raphe magnus. J. Neurosci. 2, 829–842.

Beitz, A. J., Shepard, R. D., and Wells, W. E. 1983. The periaqueductal gray-raphe magnus projection contains somatostatin, neurotensin and serotonin but not cholecystokinin. Brain Res. 261, 132–137.

Bellavance, L. L. and Beitz, A. J. 1996. Altered c-fos expression in the parabrachial nucleus in a rodent model of CFA-induced peripheral inflammation. J. Comp. Neurol. 366, 431–447.

Bellchambers, C. E., Chieng, B., Keay, K. A., and Christie, M. J. 1998. Swim-stress but not opioid withdrawal increases expression of c-fos immunoreactivity in rat periaqueductal gray neurons which project to the rostral ventromedial medulla. Neuroscience 83, 517–524.

Benedetti, F. and Amanzio, M. 1997. The neurobiology of placebo analgesia: from endogenous opioids to cholecystokinin. Prog. Neurobiol. 52, 109–125.

Benedetti, F., Pollo, A., Lopiano, L., Lanotte, M., Vighetti, S., and Rainero, I. 2003. Conscious expectation and unconscious conditioning in analgesic, motor, and hormonal placebo/nocebo responses. J. Neurosci. 23, 4315–4323.

Bernard, J. F., Alden, M., and Besson, J-M. 1993. The organization of the efferent projections from the pontine parabrachial area to the amygdaloid complex: a Phaseolus vulgaris leucoagglutinin (PHA-L) study in the rat. J. Comp. Neurol. 329, 201–229.

Berrino, L., Oliva, P., Rossi, F., Palazzo, E., Nobili, B., and Maione, S. 2001. Interaction between metabotropic and NMDA glutamate receptors in the periaqueductal grey pain modulatory system. Naunyn Schmiedebergs Arch. Pharmacol. 364, 437–443.

Besson, J. M., Guilbaud, G., and Le Bars, D. 1975. Descending inhibitory influences exerted by the brain stem upon the activities of dorsal horn lamina V cells induced by intra-arterial injection of bradykinin into the limbs. J. Physiol. 248725–739.

Bester, H., Besson, J.-M., and Bernard, J. F. 1997. Organization of efferent projections from the parabrachial area to the hypothalamus: a Phaseolus vulgaris-leucoagglutinin study in the rat. J. Comp. Neurol 383, 245–281.

Bester, H., Chapman, V., Besson, J-M., and Bernard, J. F. 2000. Physiological properties of the lamina I spinoparabrachial neurons in the rat. J. Neurophysiol. 83, 2239–2259.

Bie, B. and Pan, Z. Z. 2003. Presynaptic mechanism for anti-analgesic and anti-hyperalgesic actions of kappa-opioid receptors. J. Neurosci. 23, 7262–7268.

Bie, B. and Pan, Z. Z. 2005. Increased glutamate synaptic transmission in the nucleus raphe magnus neurons from morphine-tolerant rats. Mol. Pain. 1, 7.

Bie, B., Fields, H. L., Williams, J. T., and Pan, Z. Z. 2003. Roles of alpha1- and alpha2-adrenoceptors in the nucleus raphe magnus in opioid analgesia and opioid abstinence-induced hyperalgesia. J. Neurosci. 23, 7950–7957.

Bitner, R. S., Nikkel, A. L., Curzon, P., Arneric, S. P., Bannon, A. W., and Decker, M. W. 1998. Role of the nucleus raphe magnus in antinociception produced by ABT-594: immediate early gene responses possibly linked to neuronal nicotinic acetylcholine receptors on serotonergic neurons. J. Neurosci. 18, 5426–5432.

Bohn, L. M., Xu, F., Gainetdinov, R. R., and Caron, M. G. 2000. Potentiated opioid analgesia in norepinephrine transporter knock-out mice. J. Neurosci. 20, 9040–9045.

Bouhassira, D., Chitour, D., Villanueva, L., and Le Bars, D. 1993. Morphine and diffuse noxious inhibitory controls in the rat: effects of lesions of the rostral ventromedial medulla. Eur. J. Pharmacol. 232, 207–215.

Bouhassira, D., Villanueva, L., Bing, Z., and Le Bars, D. 1992. Involvement of the subnucleus reticularis dorsalis in diffuse noxious inhibitory controls in the rat. Brain Res. 595, 353–357.

Bowker, R. M., Abbott, L. C., and Dilts, R. P. 1988. Peptidergic neurons in the nucleus raphe magnus and the nucleus gigantocellularis: their distributions, interrelationships, and projections to the spinal cord. Prog. Brain Res. 77, 95–127.

Bowker, R. M., Westlund, K. N., Sullivan, M. C., Wilber, J. F., and Coulter, J. D. 1983. Descending serotonergic, peptidergic and cholinergic pathways from the raphe nuclei: a multiple transmitter complex. Brain Res. 288, 33–48.

Buhler, A. V., Proudfit, H. K., and Gebhart, G. F. 2004. Separate populations of neurons in the rostral ventromedial medulla project to the spinal cord and to the dorsolateral pons in the rat. Brain Res. 1016, 12–19.

Buhler, A. V., Choi, J., Proudfit, H. K., and Gebhart, G. F. 2005. Neurotensin activation of the NTR1 on spinally-projecting serotonergic neurons in the rostral ventromedial medulla is antinociceptive. Pain 114, 285–294.

Burgess, S. E., Gardell, L. R., Ossipov, M. H., Malan, T. P., Jr., Vanderah, T. W., Lai, J., and Porreca, F. 2002. Time-dependent descending facilitation from the rostral ventromedial medulla maintains, but does not initiate, neuropathic pain. J. Neurosci. 22, 5129–5136.

Buritova, J., Besson, J. M., and Bernard, J. F. 1998. Involvement of the spinoparabrachial pathway in inflammatory nociceptive processes: a c-Fos protein study in the awake rat. J. Comp. Neurol. 397, 10–28.

Burnett, A. and Gebhart, G. F. 1991. Characterization of descending modulation of nociception from the A5 cell group. Brain Res. 546, 271–281.

Camarata, P. J. and Yaksh, T. L. 1985. Characterization of the spinal adrenergic receptors mediating the spinal effects produced by the microinjection of morphine into the periaqueductal gray. Brain Res. 336, 133–142.

Cameron, A. A., Khan, I. A., Westlund, K. N., and Willis, W. D. 1995b. The efferent projections of the periaqueductal gray in the rat: a Phaseolus vulgaris-leucoagglutinin study. II. Descending projections. J. Comp. Neurol. 351, 585–601.

Cameron, A. A., Khan, I. A., Westlund, K. N., Cliffer, K. D., and Willis, W. D. 1995a. The efferent projections of the periaqueductal gray in the rat: a Phaseolus vulgaris-leucoagglutinin study. I. Ascending projections. J. Comp. Neurol. 351, 568–584.

Calejesan, A. A., Kim, S. J., and Zhuo, M. 2000. Descending facilitatory modulation of a behavioral nociceptive response by stimulation in the adult rat anterior cingulate cortex. Eur. J. Pain 4, 83–96.

Cannon, K. E., Nalwalk, J. W., Stadel, R., Ge, P., Lawson, D., Silos-Santiago, I., and Hough, L. B. 2003. Activation of spinal histamine H3 receptors inhibits mechanical nociception. Eur. J. Pharmacol. 2003 470, 139–147.

Carlson, J. D., Iacono, R. P., and Maeda, G. 2004. Nociceptive excited and inhibited neurons within the pedunculopontine

tegmental nucleus and cuneiform nucleus. Brain Res. 1013, 182–187.

Carstens, E., Klumpp, D., and Zimmermann, M. 1980. Differential inhibitory effects of medial and lateral midbrain stimulation on spinal neuronal discharges to noxious skin heating in the cat. J Neurophysiol. 43, 332–342.

Cejas, P. J., Martinez, M., Karmally, S., McKillop, M., McKillop, J., Plunkett, J. A., Oudega, M., and Eaton, M. J. 2000. Lumbar transplant of neurons genetically modified to secrete brain-derived neurotrophic factor attenuates allodynia and hyperalgesia after sciatic nerve constriction. Pain 86, 195–210.

Cervero, F. and Laird, J. M. 1996. Mechanisms of touch-evoked pain (allodynia): a new model. Pain 68, 13–23.

Cervero, F. and Wolstencroft, J. H. 1984. A positive feedback loop between spinal cord nociceptive pathways and antinociceptive areas of the cat's brain stem. Pain 20, 125–138.

Cervero, F., Iggo, A., and Ogawa, H. 1976. Nociceptor-driven dorsal horn neurones in the lumbar spinal cord of the cat. Pain 2, 5–24.

Cervero, F., Molony, V., and Iggo, A. 1979. Supraspinal linkage of substantia gelatinosa neurones: effects of descending impulses. Brain Res. 175, 351–355.

Chen, H. and Weber, A. J. 2004. Brain-derived neurotrophic factor reduces TrkB protein and mRNA in the normal retina and following optic nerve crush in adult rats. Brain Res. 1011, 99–106.

Chen, Y., Mestek, A., Liu, J., Hurley, J. A., and Yu, L. 1993. Molecular cloning and functional expression of a mu-opioid receptor from rat brain. Mol. Pharmacol. 44, 8–12.

Chiang, C. Y., Hu, J. W., and Sessle, B. J. 1994. Parabrachial area and nucleus raphe magnus-induced modulation of nociceptive and nonnociceptive trigeminal subnucleus caudalis neurons activated by cutaneous or deep inputs. J. Neurophysiol. 71, 2430–2445.

Cho, H. J. and Basbaum, A. I. 1991. GABAergic circuitry in the rostral ventral medulla of the rat and its relationship to descending antinociceptive controls. J. Comp. Neurol. 303, 316–328.

Clark, F. M. and Proudfit, H. K. 1991. The projection of locus coeruleus neurons to the spinal cord in the rat determined by anterograde tracing combined with immunocytochemistry. Brain Res. 538, 231–245.

Clark, F. M. and Proudfit, H. K. 1993. The projections of noradrenergic neurons in the A5 catecholamine cell group to the spinal cord in the rat: anatomical evidence that A5 neurons modulate nociception. Brain Res. 616, 200–210.

Collins, J. G. and Ren, K. 1987. WDR response profiles of spinal dorsal horn neurons may be unmasked by barbiturate anesthesia. Pain 28, 369–378.

Commons, K. G. and Valentino, R. J. 2002. Cellular basis for the effects of substance P in the periaqueductal gray and dorsal raphe nucleus. J. Comp. Neurol. 447, 82–97.

Cordero-Erausquin, M. and Changeux, J. P. 2001. Tonic nicotinic modulation of serotoninergic transmission in the spinal cord. Proc. Natl. Acad. Sci. U. S. A. 98, 2803–2807.

Cordero-Erausquin, M., Pons, S., Faure, P., and Changeux, J. P. 2004. Nicotine differentially activates inhibitory and excitatory neurons in the dorsal spinal cord. Pain 109, 308–318.

Cravatt, B. F. and Lichtman, A. H. 2004. The endogenous cannabinoid system and its role in nociceptive behavior. J Neurobiol. 61, 149–160.

Cronin, J. N., Bradbury, E. J., and Lidierth, M. 2004. Laminar distribution of GABAA- and glycine-receptor mediated tonic inhibition in the dorsal horn of the rat lumbar spinal cord: effects of picrotoxin and strychnine on expression of Fos-like immunoreactivity. Pain 112, 156–163.

Cucchiaro, G., Chaijale, N., and Commons, K. G. 2005. The dorsal raphe nucleus as a site of action of the antinociceptive and behavioral effects of the alpha4 nicotinic receptor agonist epibatidine. J. Pharmacol. Exp. Ther. 313, 389–394.

Cui, J. G., O'Connor, W. T., Ungerstedt, U., Linderoth, B., and Meyerson, B. A. 1997. Spinal cord stimulation attenuates augmented dorsal horn release of excitatory amino acids in mononeuropathy via a GABAergic mechanism. Pain 73, 87–95.

Cui, M., Feng, Y., McAdoo, D. J., and Willis, W. D. 1999. Periaqueductal gray stimulation-induced inhibition of nociceptive dorsal horn neurons in rats is associated with the release of norepinephrine, serotonin, and amino acids. J. Pharmacol. Exp. Ther. 289, 868–876.

Dahlström, A. and Fuxe, K. 1964. Evidence for the existence of monoamine-containing neurons in the central nervous system. I. Demonstration of monoamines in the cell bodies of brain stem neurons. Acta Physiol. Scand. 62(Suppl 232), 1–55.

Danzebrink, R. M. and Gebhart, G. F. 1991. Intrathecal coadministration of clonidine with serotonin receptor agonists produces supra-additive visceral antinociception in the rat. Brain Res. 555, 35–42.

Danziger, N., Weil-Fugazza, J., Le Bars, D., and Bouhassira, D. 1999. Alteration of descending modulation of nociception during the course of monoarthritis in the rat. J. Neurosci. 19, 2394–2400.

Danziger, N., Weil-Fugazza, J., Le Bars, D., and Bouhassira, D. 2001. Stage-dependent changes in the modulation of spinal nociceptive neuronal activity during the course of inflammation. Eur J Neurosci. 13, 230–240.

de Novellis, V., Mariani, L., Palazzo, E., Vita, D., Marabese, I., Scafuro, M., Rossi, F., and Maione, S. 2005. Periaqueductal grey CB1 cannabinoid and metabotropic glutamate subtype 5 receptors modulate changes in rostral ventromedial medulla neuronal activities induced by subcutaneous formalin in the rat. Neuroscience 134, 269–281.

Depaulis, A., Morgan, M. M., and Liebeskind, J. C. 1987. GABAergic modulation of the analgesic effects of morphine microinjected in the ventral periaqueductal gray matter of the rat. Brain Res. 436, 223–228.

Dickenson, A. H., Oliveras, J. L., and Besson, J. M. 1979. Role of the nucleus raphe magnus in opiate analgesia as studied by the microinjection technique in the rat. Brain Res. 170, 95–111.

Dickhaus, H., Pauser, G., and Zimmermann, M. 1985. Tonic descending inhibition affects intensity coding of nociceptive responses of spinal dorsal horn neurones in the cat. Pain 23, 145–158.

Dostrovsky, J. O., Shah, Y., and Gray, B. G. 1983. Descending inhibitory influences from periaqueductal gray, nucleus raphe magnus, and adjacent reticular formation. II. Effects on medullary dorsal horn nociceptive and nonnociceptive neurons. J. Neurophysiol. 49, 948–960.

Dubner, R. and Bennett, G. J. 1983. Spinal and trigeminal mechanisms of nociception. Annu. Rev. Neurosci. 6, 381–418.

Dubner, R. and Ruda, M. A. 1992. Activity-dependent neuronal plasticity following tissue injury and inflammation. Trends Neurosci. 15, 96–103.

Dubuisson, D. and Wall, P. D. 1979. Medullary raphe influences on units in laminae 1 and 2 of cat spinal cord. J. Physiol. (Lond) 300, 33P.

Dubuisson, D. and Wall, P. D. 1980. Descending influences on receptive fields and activity of single units recorded in laminae 1,2 and 3 of cat spinal cord. Brain Res. 199, 283–298.

Duggan, A. W. 1985. Pharmacology of descending control systems. Phil. Trans. R. Soc. Lond. B 308, 375–391.

Eaton, M. J., Blits, B., Ruitenberg, M. J., Verhaagen, J., and Oudega, M. 2002. Amelioration of chronic neuropathic pain after partial nerve injury by adeno-associated viral (AAV) vector-mediated over-expression of BDNF in the rat spinal cord. Gene Ther. 9, 1387–1395.

Edwards, R. R., Fillingim, R. B., and Ness, T. J. 2003. Age-related differences in endogenous pain modulation: a comparison of diffuse noxious inhibitory controls in healthy older and younger adults. Pain 101, 155–165.

Ennis, M., Behbehani, M., Shipley, M. T., Van Bockstaele, E. J., and Aston-Jones, G. 1991. Projections from the periaqueductal gray to the rostromedial pericoerulear region and nucleus locus coeruleus: anatomic and physiologic studies. J. Comp. Neurol. 306, 480–494.

Evans, C. J., Keith, D. E., Jr., Morrison, H., Magendzo, K., and Edwards, R. H. 1992. Cloning of a delta opioid receptor by functional expression. Science 258, 1952–1955.

Fairbanks, C. A., Stone, L. S., Kitto, K. F., Nguyen, H. O., Posthumus, I. J., and Wilcox, G. L. 2002. alpha(2C)-Adrenergic receptors mediate spinal analgesia and adrenergic-opioid synergy. J. Pharmacol. Exp. Ther. 300, 282–290.

Fardin, V., Oliveras, J. L., and Besson, J. M. 1984a. A reinvestigation of the analgesic effects induced by stimulation of the periaqueductal gray matter in the rat. II. Differential characteristics of the analgesia induced by ventral and dorsal PAG stimulation. Brain Res. 306, 125–139.

Fardin, V., Oliveras, J. L., and Besson, J. M. 1984b. Projections from the periaqueductal gray matter to the B3 cellular area (nucleus raphe magnus and nucleus reticularis paragigantocellularis) as revealed by the retrograde transport of horseradish peroxidase in the rat. J. Comp. Neurol. 223, 483–500.

Fields, H. L. 1988. Sources of variability in the sensation of pain. Pain 33, 195–200.

Fields, H. L. and Anderson, S. D. 1978. Evidence that raphe-spinal neurons mediate opiate and midbrain stimulation-produced analgesia. Pain 5, 333–349.

Fields, H. L. and Basbaum, A. I. 1978. Brainstem control of spinal pain-transmission neurons. Annu. Rev. Physiol. 40, 217–248.

Fields, H. L. and Basbaum, A. I. 1999. Central Nervous System Mechanisms of Pain Modulation. In: Textbook of Pain (eds. P. D. Wall and R. Melzack R) pp. 309–329. London: Churchill Livingstone.

Fields, H. L., Heinricher, M. M., and Mason, P. 1991. Neurotransmitters in nociceptive modulatory circuits. Annu. Rev. Neurosci. 14, 219–245.

Fields, H. L., Bry, J., Hentall, I., and Zorman, G. 1983a. The activity of neurons in the rostral medulla of the rat during withdrawal from noxious heat. J. Neurosci. 3, 2545–2552.

Fields, H. L., Malick, A., and Burstein, R. 1995. Dorsal horn projection targets of ON and OFF cells in the rostral ventromedial medulla. J. Neurophysiol. 74, 1742–1759.

Fields, H. L., Vanegas, H., Hentall, I. D., and Zorman, G. 1983b. Evidence that disinhibition of brain stem neurones contributes to morphine analgesia. Nature 306, 684–686.

Finley, J. C., Lindstrom, P., and Petrusz, P. 1981. Immunocytochemical localization of beta-endorphin-containing neurons in the rat brain. Neuroendocrinology 33, 28–42.

Flores, J. A., El Banoua, F., Galan-Rodriguez, B., and Fernandez-Espejo, E. 2004. Opiate anti-nociception is attenuated following lesion of large dopamine neurons of the periaqueductal grey: critical role for D1 (not D2) dopamine receptors. Pain 110, 205–214.

Fowler, C. J. and Fraser, G. L. 1994. Mu-, delta-, kappa-opioid receptors and their subtypes. A critical review with emphasis on radioligand binding experiments. Neurochem. Int. 24, 401–426.

Frank, L., Wiegand, S. J., Siuciak, J. A., Lindsay, R. M., and Rudge, J. S. 1997. Effects of BDNF infusion on the regulation of TrkB protein and message in adult rat brain. Exp. Neurol. 145, 62–70.

Friedrich, A. E. and Gebhart, G. F. 2003. Modulation of visceral hyperalgesia by morphine and cholecystokinin from the rat rostroventral medial medulla. Pain 104, 93–101.

Fritschy, J. M. and Grzanna, R. 1990. Demonstration of two separate descending noradrenergic pathways to the rat spinal cord: evidence for an intragriseal trajectory of locus coeruleus axons in the superficial layers of the dorsal horn. J. Comp. Neurol. 291, 553–582.

Fu, J. J., Tang, J. S., Yuan, B., and Jia, H. 2002. Response of neurons in the thalamic nucleus submedius (Sm) to noxious stimulation and electrophysiological identification of on- and off-cells in rats. Pain 99, 243–251.

Galeotti, N., Malmberg-Aiello, P., Bartolini, A., Schunack, W., and Ghelardini, C. 2004. H1-receptor stimulation induces hyperalgesia through activation of the phospholipase C-PKC pathway. Neuropharmacology 47, 295–303.

Gao, K. and Mason, P. 2000. Serotonergic Raphe magnus cells that respond to noxious tail heat are not ON or OFF cells. J. Neurophysiol. 84, 1719–1725.

Gao, K., Kim, Y. H., and Mason, P. 1997. SEROTONERGIC pontomedullary neurons are not activated by antinociceptive stimulation in the periaqueductal gray. J. Neurosci. 17, 3285–3292.

Gao, Y. J., Ren, W. H., Zhang, Y. Q., and Zhao, Z. Q. 2004. Contributions of the anterior cingulate cortex and amygdala to pain- and fear-conditioned place avoidance in rats. Pain 110, 343–353.

Gauriau, C. and Bernard, J. F. 2002. Pain pathways and parabrachial circuits in the rat. Exp. Physiol. 87, 251–258.

Gaveriaux-Ruff, C. and Kieffer, B. L. 2002. Opioid receptor genes inactivated in mice: the highlights. Neuropeptides 36, 62–71.

Gebhart, G. F. 1982. Opiate and opioid peptide effects on brain stem neurons: relevance to nociception and antinociceptive mechanisms. Pain 12, 93–140.

Gebhart, G. F. 1988. Descending Inhibition of Nociceptive Transmission. In: Basic Mechanisms of Headache (eds. J. Olesen and L. Edvinsson), pp. 202–212. Elsevier.

Gebhart, G. F. 2004. Descending modulation of pain. Neurosci. Biobehav. Rev. 27, 729–737.

Gebhart, G. F., Sandkuhler, J., Thalhammer, J. G., and Zimmermann, M. 1983. Inhibition of spinal nociceptive information by stimulation in midbrain of the cat is blocked by lidocaine microinjected in nucleus raphe magnus and medullary reticular formation. J. Neurophysiol. 50, 1446–1459.

Giesler, G. J., Jr., Gerhart, K. D., Yezierski, R. P., Wilcox, T. K., and Willis, W. D. 1981. Postsynaptic inhibition of primate spinothalamic neurons by stimulation in nucleus raphe magnus. Brain Res. 204, 184–188.

Girardot, M. N., Brennan, T. J., Martindale, M. E., and Foreman, R. D. 1987. Effects of stimulating the subcoeruleus-parabrachial region on the non-noxious and noxious responses of T1–T5 spinothalamic tract neurons in the primate. Brain Res. 409, 19–30.

Gogas, K. R., Presley, R. W., Levine, J. D., and Basbaum, A. I. 1991. The antinociceptive action of supraspinal opioids results from an increase in descending inhibitory control: correlation of nociceptive behavior and c-fos expression. Neuroscience 42, 617–628.

Gogas, K. R., Cho, H. J., Botchkina, G. I., Levine, J. D., and Basbaum, A. I. 1996. Inhibition of noxious stimulus-evoked pain behaviors and neuronal fos-like immunoreactivity in the spinal cord of the rat by supraspinal morphine. Pain 65, 9–15.

Goldstein, A., Tachibana, S., Lowney, L. I., Hunkapiller, M., and Hood, L. 1979. Dynorphin-(1–13), an extraordinarily potent opioid peptide. Proc. Natl. Acad. Sci. U. S. A. 76, 6666–6670.

Gozariu, M., Bouhassira, D., Willer, J. C., and Le Bars, D. 1998. The influence of temporal summation on a C-fibre reflex in the rat: effects of lesions in the rostral ventromedial medulla (RVM). Brain Res. 792, 168–172.

Graff-Guerrero, A., Gonzalez-Olvera, J., Fresan, A., Gomez-Martin, D., Carlos Mendez-Nunez, J., and Pellicer, F. 2005. Repetitive transcranial magnetic stimulation of dorsolateral prefrontal cortex increases tolerance to human experimental pain. Brain Res. Cogn. Brain Res. May 31; (Epub ahead of print).

Grudt, T. J., van den Pol, A. N., and Perl, E. R. 2002. Hypocretin-2 (orexin-B) modulation of superficial dorsal horn activity in rat. J. Physiol. 538(Pt 2), 517–525.

Guan, J. L., Wang, Q. P., and Shioda, S. 2003. Immunoelectron microscopic examination of orexin-like immunoreactive fibers in the dorsal horn of the rat spinal cord. Brain Res. 987, 86–92.

Guan, Y., Terayama, R., Dubner, R., and Ren, K. 2002. Plasticity in excitatory amino acid receptor-mediated descending pain modulation after inflammation. J. Pharmacol. Exp. Ther. 300, 513–520.

Guan, Y., Guo, W., Zou, S-P., Dubner, R., and Ren, K. 2003. Inflammation–induced upregulation of AMPA receptor subunit expression in brain stem pain modulatory circuitry. Pain 104, 401–413.

Guo, W., Robbins, M. T., Wei, F., Zou, S., Dubner, R., and Ren, K. 2006. Supraspinal brain-derived neurotrophic factor signaling: a novel mechanism for descending pain facilitation. J. Neurosci. 26, 126–137.

Haber, L. H., Martin, R. F., Chung, J. M., and Willis, W. D. 1980. Inhibition and excitation of primate spinothalamic tract neurons by stimulation in region of nucleus reticularis gigantocellularis. J. Neurophysiol. 43, 1578–1593.

Hack, S. P., Bagley, E. E., Chieng, B. C., and Christie, M. J. 2005. Induction of delta-opioid receptor function in the midbrain after chronic morphine treatment. J. Neurosci. 25, 3192–3198.

Hagbarth, K. E. and Kerr, D. I. 1954. Central influences on spinal afferent conduction. J. Neurophysiol. 17, 295–307.

Hammond, D. L., Levy, R. A., and Proudfit, H. K. 1980. Hypoalgesia following microinjection of noradrenergic antagonists in the nucleus raphe magnus. Pain 9, 85–101.

Hammond, D. L., Nelson, V., and Thomas, D. A. 1998. Intrathecal methysergide antagonizes the antinociception, but not the hyperalgesia produced by microinjection of baclofen in the ventromedial medulla of the rat. Neurosci. Lett. 244, 93–96.

Han, J. S. 1995. Cholecystokinin octapeptide (CCK-8): a negative feedback control mechanism for opioid analgesia. Prog. Brain Res. 105, 263–271.

Han, J. S. 2004. Acupuncture and endorphins. Neurosci. Lett. 361, 258–261.

Haws, C. M., Williamson, A. M., and Fields, H. L. 1989. Putative nociceptive modulatory neurons in the dorsolateral pontomesencephalic reticular formation. Brain Res. 483, 272–282.

Hayes, R. L., Bennett, G. J., Newlon, P. G., and Mayer, D. J. 1978. Behavioral and physiological studies of non-narcotic analgesia in the rat elicited by certain environmental stimuli. Brain Res. 155, 69–90.

Hedo, G. and Lopez-Garcia, J. A. 2001. Alpha-1A adrenoceptors modulate potentiation of spinal nociceptive pathways in the rat spinal cord in vitro. Neuropharmacology 41, 862–869.

Heinricher, M. M. 2005. Nociceptin/orphanin FQ: pain, stress and neural circuits. Life Sci. 77, 3127–3132.

Heinricher, M. M., Cheng, Z. F., and Fields, H. L. 1987. Evidence for two classes of nociceptive modulating neurons in the periaqueductal gray. J. Neurosci. 7, 271–278.

Heinricher, M. M., Martenson, M. E., and Neubert, M. J. 2004a. Prostaglandin E2 in the midbrain periaqueductal gray produces hyperalgesia and activates pain-modulating circuitry in the rostral ventromedial medulla. Pain 110, 419–426.

Heinricher, M. M., McGaraughty, S., and Farr, D. A. 1999. The role of excitatory amino acid transmission within the rostral ventromedial medulla in the antinociceptive actions of systemically administered morphine. Pain 81, 57–65.

Heinricher, M. M., McGaraughty, S., and Tortorici, V. 2001. Circuitry underlying antiopioid actions of cholecystokinin within the rostral ventromedial medulla. J. Neurophysiol. 85, 280–286.

Heinricher, M. M. and Neubert, M. J. 2004. Neural basis for the hyperalgesic action of cholecystokinin in the rostral ventromedial medulla. J. Neurophysiol. 92, 1982–1989.

Heinricher, M. M., Neubert, M. J., Martenson, M. E., and Goncalves, L. 2004b. Prostaglandin E2 in the medial preoptic area produces hyperalgesia and activates pain-modulating circuitry in the rostral ventromedial medulla. Neuroscience 128, 389–398.

Helmstetter, F. J. and Tershner, S. A. 1994. Lesions of the periaqueductal gray and rostral ventromedial medulla disrupt antinociceptive but not cardiovascular aversive conditional responses. J. Neurosci. 14, 7099–7108.

Henderson, G. and McKnight, A. T. 1997. The orphan opioid receptor and its endogenous ligand–nociceptin/orphanin FQ. Trends Pharmacol. Sci. 18, 293–300.

Hernandez, N. and Vanegas, H. 2001. Encoding of noxious stimulus intensity by putative pain modulating neurons in the rostral ventromedial medulla and by simultaneously recorded nociceptive neurons in the spinal dorsal horn of rats. Pain 91, 307–315.

Hernandez, N., Lopez, Y., and Vanegas, H. 1989. Medullary on- and off-cell responses precede both segmental and thalamic responses to tail heating. Pain 39, 221–230.

Hernandez-Peon, R. and Hagbarth, K. E. 1955. Interaction between afferent and cortically induced reticular responses. J. Neurophysiol. 18, 44–55.

Herrero, J. F. and Cervero, F. 1996. Supraspinal influences on the facilitation of rat nociceptive reflexes induced by carrageenan monoarthritis. Neurosci. Lett. 209, 21–24.

Hohmann, A. G., Suplita, R. L., Bolton, N. M., Neely, M. H., Fegley, D., Mangieri, R., Krey, J. F., Walker, J. M., Holmes, P. V., Crystal, J. D., Duranti, A., Tontini, A., Mor, M., Tarzia, G., and Piomelli, D. 2005. An endocannabinoid mechanism for stress-induced analgesia. Nature 435, 1108–1112.

Holden, J. E. and Naleway, E. 2001. Microinjection of carbachol in the lateral hypothalamus produces opposing actions on nociception mediated by alpha(1)- and alpha(2)-adrenoceptors. Brain Res. 911, 27–36.

Holden, J. E., Schwartz, E. J., and Proudfit, H. K. 1999. Microinjection of morphine in the A7 catecholamine cell group produces opposing effects on nociception that are mediated by alpha1- and alpha2-adrenoceptors. Neuroscience 91, 979–990.

Holden, J. E., Van Poppel, A. Y., and Thomas, S. 2002. Antinociception from lateral hypothalamic stimulation may be mediated by NK(1) receptors in the A7 catecholamine cell group in rat. Brain Res. 953, 195–204.

Hough, L. B., Nalwalk, J. W., Li, B. Y., Leurs, R., Menge, W. M., Timmerman, H., Carlile, M. E., Cioffi, C., and Wentland, M. 1997. Novel qualitative structure-activity relationships for the antinociceptive actions of H2 antagonists, H3

antagonists and derivatives. J. Pharmacol. Exp. Ther. 283, 1534–1543.

Hoyer, D., Clarke, D. E., Fozard, J. R., Hartig, P. R., Martin, G. R., Mylecharane, E. J., Saxena, P. R., and Humphrey, P. P. 1994. International Union of Pharmacology classification of receptors for 5-hydroxytryptamine (Serotonin). Pharmacol. Rev. 46, 157–203.

Hughes, J., Smith, T. W., Kosterlitz, H. W., Fothergill, L. A., Morgan, B. A., and Morris, H. R. 1975. Identification of two related pentapeptides from the brain with potent opiate agonist activity. Nature 258, 577–580.

Hung, K. C., Wu, H. E., Mizoguchi, H., Leitermann, R., and Tseng, L. F. 2003. Intrathecal treatment with 6-hydroxydopamine or 5,7-dihydroxytryptamine blocks the antinociception induced by endomorphin-1 and endomorphin-2 given intracerebroventricularly in the mouse. J. Pharmacol. Sci. 93, 299–306.

Hunter, J. C., Fontana, D. J., Hedley, L. R., Jasper, J. R., Lewis, R., Link, R. E., Secchi, R., Sutton, J., and Eglen, R. M. 1997. Assessment of the role of alpha2-adrenoceptor subtypes in the antinociceptive, sedative and hypothermic action of dexmedetomidine in transgenic mice. Br. J. Pharmacol. 122, 1339–1344.

Hurley, R. W. and Hammond, D. L. 2000. The analgesic effects of supraspinal mu and delta opioid receptor agonists are potentiated during persistent inflammation. J. Neurosci. 20, 1249–1259.

Hurley, R. W. and Hammond, D. L. 2001. Contribution of endogenous enkephalins to the enhanced analgesic effects of supraspinal mu opioid receptor agonists after inflammatory injury. J. Neurosci. 21, 2536–2545.

Hylden, J. L. K., Anton, F., and Nahin, R. L. 1989. Spinal lamina I projection neurons in the rat: collateral innervation of parabrachial area and thalamus. Neuroscience 28, 27–37.

Hylden, J. L. K., Hayashi, H., Dubner, R., and Bennett, G. J. 1986. Physiology and morphology of the lamina I spinomesencephalic projection. J. Comp. Neurol. 247, 505–515.

Hylden, J. L. K., Thomas, D. A., Iadarola, M. J., and Dubner, R. 1991. Spinal opioid analgesic effects are enhanced in a model of unilateral inflammation/hyperalgesia: possible involvement of noradrenergic mechanisms. Eur. J. Pharmacol. 194, 135–143.

Ikeda, K., Ide, S., Han, W., Hayashida, M., Uhl, G. R., and Sora, I. 2005. How individual sensitivity to opiates can be predicted by gene analyses. Trends Pharmacol. Sci. 26, 311–317.

Iwamoto, E. T. 1991. Characterization of the antinociception induced by nicotine in the pedunculopontine tegmental nucleus and the nucleus raphe magnus. J. Pharmacol. Exp. Ther. 257, 120–133.

Iwata, K., Fukuoka, T., Kondo, E., Tsuboi, Y., Tashiro, A., Noguchi, K., Masuda, Y., Morimoto, T., and Kanda, K. 2002. Plastic changes in nociceptive transmission of the rat spinal cord with advancing age. J. Neurophysiol. 87, 1086–1093.

Janfaza, D. R., Michna, E., Pisini, J. V., and Ross, E. L. 1998. Bedside implantation of a trial spinal cord stimulator for intractable anginal pain. Anesth. Analg. 87, 1242–1244.

Janss, A. J. and Gebhart, G. F. 1987. Spinal monoaminergic receptors mediate the antinociception produced by glutamate in the medullary lateral reticular nucleus. J. Neurosci. 7, 2862–2873.

Janss, A. J. and Gebhart, G. F. 1988. Quantitative characterization and spinal pathway mediating inhibition of spinal nociceptive transmission from the lateral reticular nucleus in the rat. J. Neurophysiol. 59, 226–47.

Jasmin, L., Rabkin, S. D., Granato, A., Boudah, A., and Ohara, P. T. 2003b. Analgesia and hyperalgesia from GABA-mediated modulation of the cerebral cortex. Nature 424, 316–320.

Jensen, T. S. and Gabhart, G. F. 1988. General anatomy of antinociceptive systems. In: Basic mechanisms of headache (eds. J. Olesen and L. Edvinsson L), pp. 189–200. Elsevier.

Jensen, T. S. and Yaksh, T. L. 1992. Brainstem excitatory amino acid receptors in nociception: microinjection mapping and pharmacological characterization of glutamate-sensitive sites in the brainstem associated with algogenic behavior. Neuroscience 46, 535–547.

Jones, S. L. 1991. Descending noradrenergic influences on pain. Prog. Brain Res. 88, 381–394.

Jones, S. L. and Light, A. R. 1990. Electrical stimulation in the medullary nucleus raphe magnus inhibits noxious heat-evoked fos protein-like immunoreactivity in the rat lumbar spinal cord. Brain Res. 530, 335–338.

Jones, S. L. and Gebhart, G. F. 1986. Quantitative characterization of ceruleospinal inhibition of nociceptive transmission in the rat. J. Neurophysiol. 56, 1397–1410.

Jones, S. L. and Light, A. R. 1992. Serotoninergic medullary raphespinal projection to the lumbar spinal cord in the rat: a retrograde immunohistochemical study. J. Comp. Neurol. 322, 599–610.

Jones, B. E., Pare, M., and Beaudet, A. 1986. Retrograde labeling of neurons in the brain stem following injections of [3H]choline into the rat spinal cord. Neuroscience 18, 901–916.

Kamibayashi, T. and Maze, M. 2000. Clinical uses of alpha2 - adrenergic agonists. Anesthesiology 93, 1345–1349.

Kauppila, T., Kontinen, V. K., and Pertovaara, A. 1998. Influence of spinalilzation on spinal withdrawal reflex responses varies depending on the submodality of the test stimulus and the experimental pathophysiological condition in the rat. Brain Res. 797, 234–242.

Kayser, V., Chen, Y. L., and Guilbaud, G. 1991. Behavioural evidence for a peripheral component in the enhanced antinociceptive effect of a low dose of systemic morphine in carragenin-induced hyperalgesic rats. Brain Res. 560, 237–244.

Kieffer, B. L., Befort, K., Gaveriaux-Ruff, C., and Hirth, C. G. 1992. The delta-opioid receptor: isolation of a cDNA by expression cloning and pharmacological characterization. Proc. Natl. Acad. Sci. U. S. A. 89, 12048–12052.

Kim, S. J., Calejesan, A. A., and Zhuo, M. 2002. Activation of brainstem metabotropic glutamate receptors inhibits spinal nociception in adult rats. Pharmacol. Biochem. Behav. 73, 429–437.

King, M. A., Bradshaw, S., Chang, A. H., Pintar, J. E., and Pasternak, G. W. 2001. Potentiation of opioid analgesia in dopamine2 receptor knock-out mice: evidence for a tonically active anti-opioid system. J. Neurosci. 21, 7788–7792.

Kingery, W. S., Guo, T. Z., Davies, M. F., Limbird, L., and Maze, M. 2000. The alpha(2A) adrenoceptor and the sympathetic postganglionic neuron contribute to the development of neuropathic heat hyperalgesia in mice. Pain 85, 345–358.

Kingery, W. S., Agashe, G. S., Guo, T. Z., Sawamura, S., Davies, M. F., Clark, J. D., Kobilka, B. K., and Maze, M. 2002. Isoflurane and nociception: spinal alpha2A adrenoceptors mediate antinociception while supraspinal alpha1 adrenoceptors mediate pronociception. Anesthesiology 96, 367–374.

Kirchner, A., Birklein, F., Stefan, H., and Handwerker, H. O. 2000. Left vagus nerve stimulation suppresses experimentally induced pain. Neurology 55, 1167–1171.

Kivipelto, L. and Panula, P. 1991. Origin and distribution of neuropeptide-FF-like immunoreactivity in the spinal cord of rats. J. Comp. Neurol. 307, 107–119.

Kovelowski, C. J., Ossipov, M. H., Sun, H., Lai, J., Malan, T. P., and Porreca, F. 2000. Supraspinal cholecystokinin may drive tonic descending facilitation mechanisms to maintain neuropathic pain in the rat. Pain 87, 265–273.

Krout, K. E., Jansen, A. S., and Loewy, A. D. 1998. Periaqueductal gray matter projection to the parabrachial nucleus in rat. J. Comp. Neurol. 401, 437–454.

Kumar, K., Toth, C., and Nath, R. K. 1997. Deep brain stimulation for intractable pain: a 15-year experience. Neurosurgery. 40, 736–746, Discussion 746–747.

Lahdesmaki, J., Scheinin, M., Pertovaara, A., and Mansikka, H. 2003. The alpha2A-adrenoceptor subtype is not involved in inflammatory hyperalgesia or morphine-induced antinociception. Eur. J. Pharmacol. 468, 183–189.

Lakhlani, P. P., MacMillan, L. B., Guo, T. Z., McCool, B. A., Lovinger, D. M., Maze, M., and Limbird, L. E. 1997. Substitution of a mutant alpha2a-adrenergic receptor via "hit and run" gene targeting reveals the role of this subtype in sedative, analgesic, and anesthetic-sparing responses in vivo. Proc. Natl. Acad. Sci. U. S. A. 94, 9950–9955.

LaMotte, C. C. 1988. Lamina X of primate spinal cord: distribution of five neuropeptides and serotonin. Neuroscience 25, 639–658.

Lavezzi, A. M., Ottaviani, G., Rossi, L., and Matturri, L. 2004. Cytoarchitectural organization of the parabrachial/Kolliker-Fuse complex in man. Brain Dev. 26, 316–320.

Le Bars, D., Dickenson, A. H., and Besson, J. M. 1980. Microinjection of morphine within nucleus raphe magnus and dorsal horn neurone activities related to nociception in the rat. Brain Res. 189, 467–481.

Le Bars, D., Dickenson, A. H., and Besson, J. M. 1979a. Diffuse noxious inhibitory controls (DNIC). I. Effects on dorsal horn convergent neurones in the rat. Pain 6, 283–304.

Le Bars, D., Dickenson, A. H., and Besson, J. M. 1979b. Diffuse noxious inhibitory controls (DNIC). II. Lack of effect on non-convergent neurones, supraspinal involvement and theoretical implications. Pain 6, 305–327.

Lee, H. S. and Mihailoff, G. A. 1999. Fluorescent double-label study of lateral reticular nucleus projections to the spinal cord and periaqueductal gray in the rat. Anat. Rec. 256, 91–98.

Leiphart, J. W., Dills, C. V., and Levy, R. M. 2003. Decreased spinal alpha2a- and alpha2c-adrenergic receptor subtype mRNA in a rat model of neuropathic pain. Neurosci. Lett. 349, 5–8.

Leung, C. G. and Mason, P. 1999. Physiological properties of raphe magnus neurons during sleep and waking. J. Neurophysiol. 81, 584–595.

Levine, J. D., Gordon, N. C., and Fields, H. L. 1978. The mechanism of placebo analgesia. Lancet. 2, 654–657.

Li, C. H. and Chung, D. 1976. Isolation and structure of an untriakontapeptide with opiate activity from camel pituitary glands. Proc. Natl. Acad. Sci. U. S. A. 73, 1145–1148.

Li, Y. Q., Takada, M., Matsuzaki, S., Shinonaga, Y., and Mizuno, N. 1993. Identification of periaqueductal gray and dorsal raphe nucleus neurons projecting to both the trigeminal sensory complex and forebrain structures: a fluorescent retrograde double-labeling study in the rat. Brain Res. 623, 267–277.

Li, P. and Zhuo, M. 1998. Silent glutamatergic synapses and nociception in mammalian spinal cord. Nature 393, 695–698.

Lieberman, M. D., Jarcho, J. M., Berman, S., Naliboff, B. D., Suyenobu, B. Y., Mandelkern, M., and Mayer, E. A. 2004. The neural correlates of placebo effects: a disruption account. Neuroimage 22, 447–455.

Light, A. R., Casale, E. J., and Menetrey, D. M. 1986. The effects of focal stimulation in nucleus raphe magnus and periaqueductal gray on intracellularly recorded neurons in spinal laminae I and II. J. Neurophysiol 56, 555–571.

Lim, G., Sung, B., Ji, R. R., and Mao, J. 2003. Upregulation of spinal cannabinoid-1-receptors following nerve injury enhances the effects of Win 55,212-2 on neuropathic pain behaviors in rats. Pain 105, 275–283.

Lima, D. and Almeida, A. 2002. The medullary dorsal reticular nucleus as a pronociceptive centre of the pain control system. Prog. Neurobiol. 66, 81–108.

Lin, Q., Wu, J., Peng, Y. B., Cui, M., and Willis, W. D. 1999. Inhibition of primate spinothalamic tract neurons by spinal glycine and GABA is modulated by guanosine 3′,5′-cyclic monophosphate. J. Neurophysiol. 81, 1095–1103.

Liu, R. H., Fung, S. J., Reddy, V. K., and Barnes, C. D. 1995. Localization of glutamatergic neurons in the dorsolateral pontine tegmentum projecting to the spinal cord of the cat with a proposed role of glutamate on lumbar motoneuron activity. Neuroscience 64, 193–208.

Lu, Y., Sweitzer, S. M., Laurito, C. E., and Yeomans, D. C. 2004. Differential opioid inhibition of C- and A delta- fiber mediated thermonociception after stimulation of the nucleus raphe magnus. Anesth. Analg. 98, 414–419.

Ma, F., Xie, H., Dong, Z. Q., Wang, Y. Q., and Wu, G. C. 2004. Effects of electroacupuncture on orphanin FQ immunoreactivity and preproorphanin FQ mRNA in nucleus of raphe magnus in the neuropathic pain rats. Brain Res. Bull. 63, 509–513.

Ma, W. and Eisenach, J. C. 2003. Chronic constriction injury of sciatic nerve induces the up-regulation of descending inhibitory noradrenergic innervation to the lumbar dorsal horn of mice. Brain Res. 970, 110–118.

Maione, S., Marabese, I., Leyva, J., Palazzo, E., de Novellis, V., and Rossi, F. 1998. Characterisation of mGluRs which modulate nociception in the PAG of the mouse. Neuropharmacology 37, 1475–1483.

Maione, S., Marabese, I., Oliva, P., de Novellis, V., Stella, L., Rossi, F., Filippelli, A., and Rossi, F. 1999. Periaqueductal gray matter glutamate and GABA decrease following subcutaneous formalin injection in rat. Neuroreport 10, 1403–1407.

Malick, A., Strassman, A. M., and Burstein, R. 2000. Trigeminohypothalamic and reticulohypothalamic tract neurons in the upper cervical spinal cord and caudal medulla of the rat. J. Neurophysiol. 84, 2078–2112.

Malmberg, A. B., Hedley, L. R., Jasper, J. R., Hunter, J. C., and Basbaum, A. I. 2001. Contribution of alpha(2) receptor subtypes to nerve injury-induced pain and its regulation by dexmedetomidine. Br. J. Pharmacol. 132, 1827–1836.

Malmberg-Aiello, P., Lamberti, C., Ghelardini, C., Giotti, A., and Bartolini, A. 1994. Role of histamine in rodent antinociception. Br. J. Pharmacol. 111, 1269–1279.

Mansikka, H. and Pertovaara, A. 1997. Supraspinal influence on hindlimb withdrawal thresholds and mustard oil-induced secondary allodynia in rats. Brain Res. Bull. 42, 359–365.

Mansikka, H., Lahdesmaki, J., Scheinin, M., and Pertovaara, A. 2004. Alpha(2A) adrenoceptors contribute to feedback inhibition of capsaicin-induced hyperalgesia. Anesthesiology 101, 185–190.

Mansour, A., Fox, C. A., Akil, H., and Watson, S. J. 1995. Opioid-receptor mRNA expression in the rat CNS: anatomical and functional implications. Trends Neurosci. 18, 22–29.

Mantyh, P. W. 1983. Connections of midbrain periaqueductal gray in the monkey. II. Descending efferent projections. J. Neurophysiol. 49, 582–594.

Mantyh, P. W. and Peschanski, M. 1982. Spinal projections from the periaqueductal grey and dorsal raphe in the rat, cat and monkey. Neuroscience 7, 2769–2776.

Mao, J. 2002. Opioid-induced abnormal pain sensitivity: implications in clinical opioid therapy. Pain 100, 213–217.

Mao, J., Mayer, D. J., and Price, D. D. 1993. Patterns of increased brain activity indicative of pain in a rat model of peripheral mononeuropathy. J. Neurosci. 13, 2689–2702.

Marinelli, S., Vaughan, C. W., Schnell, S. A., Wessendorf, M. W., and Christie, M. J. 2002. Rostral ventromedial medulla neurons that project to the spinal cord express multiple opioid receptor phenotypes. J. Neurosci. 22, 10847–10855.

Martin, R. F., Haber, L. H., and Willis, W. D. 1979. Primary afferent depolarization of identified cutaneous fibers following stimulation in medial brain stem. J. Neurophysiol. 42, 779–790.

Martin, W. R., Eades, C. G., Thompson, J. A., Huppler, R. E., and Gilbert, P. E. 1976. The effects of morphine- and nalorphine- like drugs in the nondependent and morphine-dependent chronic spinal dog. J. Pharmacol. Exp. Ther. 197, 517–532.

Mason, P. 2001. Contributions of the medullary raphe and ventromedial reticular region to pain modulation and other homeostatic functions. Annu. Rev. Neurosci. 24, 737–777.

Matsutani, K., Tsuruoka, M., Shinya, A., Furuya, R, Kawawa, T., and Inoue, T. 2003. Coeruleotrigeminal suppression of nociceptive sensorimotor function during inflammation in the craniofacial region of the rat. Brain Res. Bull. 61, 73–80.

Matthes, H. W., Maldonado, R., Simonin, F., Valverde, O., Slowe, S., Kitchen, I., Befort, K., Dierich, A., Le Meur, M., Dolle, P., Tzavara, E., Hanoune, J., Roques, B. P., and Kieffer, B. L. 1996. Loss of morphine-induced analgesia, reward effect and withdrawal symptoms in mice lacking the mu-opioid-receptor gene. Nature 383, 819–823.

Mayer, D. J. and Liebeskind, J. C. 1974. Pain reduction by focal electrical stimulation of the brain: an anatomical and behavioral analysis. Brain Res. 68, 73–93.

Mayer, D. J., Wolfle, T. L., Akil, H., Carder, B., and Liebeskind, J. C. 1971. Analgesia from electrical stimulation in the brainstem of the rat. Science 174, 1351–1354.

McMahon, S. B. and Wall, P. D. 1988. Descending excitation and inhibition of spinal cord lamina I projection neurons. J. Neurophysiol. 59, 1204–1219.

Melzack, R. and Wall, P. D. 1965. Pain mechanisms: a new theory. Science 150, 971–979.

Menetrey, D., Chaouch, A., Binder, D., and Besson, J. M. 1982. The origin of the spinomesencephalic tract: an anatomical study using the retrograde transport of horseradish peroxidase. J. Comp. Neurol. 206, 193–207.

Meng, F., Xie, G. X., Thompson, R. C., Mansour, A., Goldstein, A., Watson, S. J., and Akil, H. 1993. Cloning and pharmacological characterization of a rat kappa opioid receptor. Proc. Natl. Acad. Sci. U. S. A. 90, 9954–9958.

Meng, I. D., Manning, B. H., Martin, W. J., and Fields, H. L. 1998. An analgesia circuit activated by cannabinoids. Nature 395, 381–383.

Meng, X. W., Budra, B., Skinner, K., Ohara, P. T., and Fields, H. L. 1997. Noradrenergic input to nociceptive modulatory neurons in the rat rostral ventromedial medulla. J. Comp. Neurol. 377, 381–391.

Meyerson, B. A. and Linderoth, B. 2000. Mechanisms of spinal cord stimulation in neuropathic pain. Neurol. Res. 22, 285–292.

Meunier, J. C. 1997. Nociceptin/orphanin FQ and the opioid receptor-like ORL1 receptor. Eur. J. Pharmacol. 340, 1–15.

Meunier, J. C., Mollereau, C., Toll, L., Suaudeau, C., Moisand, C., Alvinerie, P., Butour, J. L., Guillemot, J. C., Ferrara, P., Monsarrat, B., Mazarguil, H., Vassart, G., Parmentier, M., and Costentin, J. 1995. Isolation and structure of the endogenous agonist of opioid receptor-like ORL1 receptor. Nature 377, 532–535.

Miki, K., Zhou, Q. Q., Guo, W., Guan, Y., Terayama, R., Dubner, R., and Ren, K. 2002. Changes in gene expression and neuronal phenotype in brain stem pain modulatory circuitry after inflammation. J. Neurophysiol. 87, 750–760.

Miki, K., Iwata, K., Tsuboi, Y., Morimoto, T., Kondo, E., Dai, Y., Ren, K., and Noguchi, K. 2000. Dorsal column-thalamic pathway is involved in thalamic hyperexcitability following peripheral nerve injury: a lesion study in rats with experimental mononeuropathy. Pain 85, 263–271.

Millan, M. J. 2002. Descending control of pain. Prog. Neurobiol. 66, 355–474.

Mobarakeh, J. I., Takahashi, K., Sakurada, S., Nishino, S., Watanabe, H., Kato, M., and Yanai, K. 2005. Enhanced antinociception by intracerebroventricularly and intrathecally-administered orexin A and B (hypocretin-1 and -2) in mice. Peptides 26, 767–777.

Mokha, S. S., McMillan, J. A., and Iggo, A. 1986. Pathways mediating descending control of spinal nociceptive transmission from the nuclei locus coeruleus (LC) and raphe magnus (NRM) in the cat. Exp. Brain Res. 61, 597–606.

Mollereau, C., Roumy, M., and Zajac, J. M. 2005. Opioid-modulating peptides: mechanisms of action. Curr. Top Med Chem. 5, 341–355.

Montagne-Clavel, J. and Oliveras, J. L. 1994. Are ventromedial medulla neuronal properties modified by chronic peripheral inflammation? A single-unit study in the awake, freely moving polyarthritic rat. Brain Res. 657, 92–104.

Morgan, M. M., Gold, M. S., Liebeskind, J. C., and Stein, C. 1991. Periaqueductal gray stimulation produces a spinally mediated, opioid antinociception for the inflamed hindpaw of the rat. Brain Res. 545, 17–23.

Morinville, A., Cahill, C. M., Esdaile, M. J., Aibak, H., Collier, B., Kieffer, B. L., and Beaudet, A. 2003. Regulation of delta-opioid receptor trafficking via mu-opioid receptor stimulation: evidence from mu-opioid receptor knock-out mice. J. Neurosci. 23, 4888–4898.

Murphy, A. Z., Ennis, M., Rizvi, T. A., Behbehani, M. M., and Shipley M, T. 1995. Fos expression induced by changes in arterial pressure is localized in distinct, longitudinally organized columns of neurons in the rat midbrain periaqueductal gray. J. Comp. Neurol. 360, 286–300.

Nalwalk, J. W., Svokos, K., Taraschenko, O., Leurs, R., Timmerman, H., and Hough, L. B. 2004. Activation of brain stem nuclei by improgan, a non-opioid analgesic. Brain Res. 1021, 248–255.

Neil, A., Kayser, V., Gacel, G., Besson, J-M., and Guilbaud, G. 1986. Opioid receptor types and antinociceptive activity in chronic inflammation: both kappa and mu opiate agonistic effects are enhanced in arthritic rats. Eur. J. Pharmacol. 130, 203–208.

Ness, T. J., Fillingim, R. B., Randich, A., Backensto, E. M., and Faught, E. 2000. Low intensity vagal nerve stimulation lowers human thermal pain thresholds. Pain 86, 81–85.

Neubert, J. K., Widmer, C. G., Malphurs, W., Rossi, H. L., Vierck, C. J., Jr., and Caudle, R. M. 2005. Use of a novel thermal operant behavioral assay for characterization of orofacial pain sensitivity. Pain 116, 386–395.

North, R. A. and Yoshimura, M. 1984. The actions of noradrenaline on neurones of the rat substantia gelatinosa in vitro. J. Physiol. 349, 43–55.

Nuseir, K. and Proudfit, H. K. 2000. Bidirectional modulation of nociception by GABA neurons in the dorsolateral pontine tegmentum that tonically inhibit spinally projecting noradrenergic A7 neurons. Neuroscience 96, 773–783.

Okuda-Ashitaka, E., Minami, T., Tachibana, S., Yoshihara, Y., Nishiuchi, Y., Kimura, T., and Ito, S. 1998. Nocistatin, a peptide that blocks nociceptin action in pain transmission. Nature 392, 286–289.

Olave, M. J. and Maxwell, D. J. 2003. Neurokinin-1 projection cells in the rat dorsal horn receive synaptic contacts from

axons that possess alpha2C-adrenergic receptors. J. Neurosci. 23, 6837–6846.

Oliveras, J. L. and Besson, J. M. 1988. Stimulation-produced analgesia in animals: behavioural investigations. Prog. Brain Res. 77, 141–157.

Oliveras, J. L., Redjemi, F., Guilbaud, G., and Besson, J. M. 1975. Analgesia induced by electrical stimulation of the inferior centralis nucleus of the raphe in the cat. Pain 1, 139–145.

Omote, K., Kawamata, T., Kawamata, M., and Namiki, A. 1998. Formalin-induced nociception activates a monoaminergic descending inhibitory system. Brain Res. 814, 194–198.

Ossipov, M. H., Kovelowski, C. J., and Porreca, F. 1995. The increase in morphine antinociceptive potency produced by carrageenan-induced hindpaw inflammation is blocked by naltrindole, a selective delta-opioid antagonist. Neurosci. Lett. 184, 173–176.

Ossipov, M. H., Lai, J., Malan, T. P., Jr., and Porreca, F. 2000. Spinal and supraspinal mechanisms of neuropathic pain. Ann. N. Y. Acad. Sci. U. S. A. 909, 12–24.

Ossipov, M. H., Lai, J., King, T., Vanderah, T. W., and Porreca, F. 2005. Underlying mechanisms of pronociceptive consequences of prolonged morphine exposure. Biopolymers. 80, 319–324.

Ozdogan, U. K., Lahdesmaki, J., Mansikka, H., and Scheinin, M. 2004. Loss of amitriptyline analgesia in alpha 2A-adrenoceptor deficient mice. Eur. J. Pharmacol. 485, 193–196.

Pan, L., Xu, J., Yu, R., Xu, M. M., Pan, Y. X., and Pasternak, G. W. 2005. Identification and characterization of six new alternatively spliced variants of the human mu opioid receptor gene, Oprm. Neuroscience 133, 209–220.

Pan, Z. Z., Hirakawa, N., and Fields, H. L. 2000. A cellular mechanism for the bidirectional pain-modulating actions of orphanin FQ/nociceptin. Neuron 26, 515–522.

Pan, Z. Z., Tershner, S. A., and Fields, H. L. 1997. Cellular mechanism for anti-analgesic action of agonists of the kappa-opioid receptor. Nature 389, 382–385.

Pan, Z. Z., Williams, J. T., and Osborne, P. B. 1990. Opioid actions on single nucleus raphe magnus neurons from rat and guinea-pig in vitro. J. Physiol. 427, 519–532.

Pasternak, G. W. 2001. Incomplete cross tolerance and multiple mu opioid peptide receptors. Trends Pharmacol. Sci. 22, 67–70.

Paxinos, G. and Watson, C. 2005. The Rat Brain in Stereotaxic Coordinates, 5th edn. Academic Press.

Peng, Y. B., Lin, Q., and Willis, W. D. 1996. Involvement of alpha-2 adrenoceptors in the periaqueductal gray-induced inhibition of dorsal horn cell activity in rats. J. Pharmacol. Exp. Ther. 278, 125–135.

Pert, C. B. and Snyder, S. H. 1973. Opiate receptor: demonstration in nervous tissue. Science 179, 1011–1014.

Pertovaara, A., Keski-Vakkuri, U., Kalmari, J., Wei, H., and Panula, P. 2001. Response properties of neurons in the rostroventromedial medulla of neuropathic rats: attempted modulation of responses by [1DMe]NPYF, a neuropeptide FF analogue. Neuroscience 105, 457–68.

Pertovaara, A., Wei, H., and Hamalainen, M. M. 1996. Lidocaine in the rostroventromedial medulla and the periaqueductal gray attenuates allodynia in neuropathic rats. Neurosci. Lett. 218, 127–130.

Petrovic, P., Carlsson, K., Petersson, K. M., Hansson, P., and Ingvar, M. 2004. Context-dependent deactivation of the amygdala during pain. J. Cogn. Neurosci. 16, 1289–1301.

Porreca, F., Ossipov, M. H., and Gebhart, G. F. 2002. Chronic pain and medullary descending facilitation. Trends Neurosci. 25, 319–325.

Porreca, F., Burgess, S. E., Gardell, L. R., Vanderah, T. W., Malan, T. P., Jr., Ossipov, M. H., Lappi, D. A., and Lai, J. 2001. Inhibition of neuropathic pain by selective ablation of brainstem medullary cells expressing the micro-opioid receptor. J. Neurosci. 21, 5281–5288.

Price, D. D. 2002. Central neural mechanisms that interrelate sensory and affective dimensions of pain. Mol. Interv. 2, 392–403.

Proudfit, H. K. 2002. The challenge of defining brainstem pain modulation circuits. J. Pain 3, 350–354.

Randich, A. and Gebhart, G. F. 1992. Vagal afferent modulation of nociception. Brain Res. Rev 17, 77–99.

Randich, A. and Maixner, W. 1984. Interactions between cardiovascular and pain regulatory systems. Neurosci. Biobehav. Rev. 8, 343–367.

Reddy, S. V., Maderdrut, J. L., and Yaksh, T. L. 1980. Spinal cord pharmacology of adrenergic agonist-mediated antinociception. J. Pharmacol. Exp. Ther. 213, 525–533.

Reichling, D. B. and Basbaum, A. I. 1990. Contribution of brainstem GABAergic circuitry to descending antinociceptive controls: II. Electron microscopic immunocytochemical evidence of GABAergic control over the projection from the periaqueductal gray to the nucleus raphe magnus in the rat. J. Comp. Neurol. 302, 378–393.

Reinert, A., Treede, R., and Bromm, B. 2000. The pain inhibiting pain effect: an electrophysiological study in humans. Brain Res. 862, 103–110.

Reinscheid, R. K., Nothacker, H. P., Bourson, A., Ardati, A., Henningsen, R. A., Bunzow, J. R., Grandy, D. K., Langen, H., Monsma, F. J., Jr., and Civelli, O. 1995. Orphanin FQ: a neuropeptide that activates an opioidlike G protein-coupled receptor. Science 270, 792–794.

Ren, K. 2002. The medulla oblongata: the vital center for descending modulation. J. Pain 3, 355–357.

Ren, K. and Ruda, M. A. 1996. Descending modulation of Fos expression after persistent peripheral inflammation. Neuroreport 7, 2186–2190.

Ren, K. and Dubner, R. 1996. Enhanced descending modulation of nociception in rats with persistent hindpaw inflammation. J. Neurophysiol. 76, 3025–3037.

Ren, K. and Dubner, R. 2002. Descending modulation in presistent pain: an update. Pain 100, 1–6.

Ren, K., Randich, A., and Gebhart, G. F. 1990. Modulation of spinal nociceptive transmission from nuclei tractus solitarii: a relay for effects of vagal afferent stimulation. J. Neurophysiol. 63, 971–986.

Ren, K., Randich, A., and Gebhart, G. F. 1991. Spinal serotonergic and kappa opioid receptors mediate facilitation of the tail flick reflex produced by vagal afferent stimulation. Pain 45, 321–329.

Ren, K., Zhuo, M., and Willis, W. D. 2000. Multiplicity and Plasticity of Descending Modulation of Nociception: Implications for Persistent Pain. In: Proceedings of the 9th World Congress on Pain (eds. M. Devor, M. C. Rowbotham, and Z. Wiesenfeld-Hallin), pp. 387–400. IASP Press.

Renno, W. M. and Beitz, A. J. 1999. Peripheral inflammation is associated with decreased veratridine-induced release of GABA in the rat ventrocaudal periaqueductal gray: microdialysis study. J. Neurol. Sci. 163, 105–110.

Reynolds, D. V. 1969. Surgery in the rat during electrical analgesia induced by focal brain stimulation. Science 164, 444–445.

Richter, R. C. and Behbehani, M. M. 1991. Evidence for glutamic acid as a possible neurotransmitter between the mesencephalic nucleus cuneiformis and the medullary nucleus raphe magnus in the lightly anesthetized rat. Brain Res. 544, 279–286.

Rivot, J. P., Chaouch, A., and Besson, J. M. 1980. Nucleus raphe magnus modulation of response of rat dorsal horn

neurons to unmyelinated fiber inputs: partial involvement of serotonergic pathways. J. Neurophysiol. 44, 1039–1057.

Rivot, J. P., Chiang, C. Y., and Besson, J. M. 1982. Increase of serotonin metabolism within the dorsal horn of the spinal cord during nucleus raphe magnus stimulation, as revealed by in vivo electrochemical detection. Brain Res. 238, 117–126.

Rizvi, T. A., Ennis, M., Behbehani, M. M., and Shipley, M. T. 1991. Connections between the central nucleus of the amygdala and the midbrain periaqueductal gray: topography and reciprocity. J. Comp. Neurol. 303, 121–131.

Rossi, G. C., Mathis, J. P., and Pasternak, G. W. 1998. Analgesic activity of orphanin FQ2, murine prepro-orphanin FQ141–157 in mice. Neuroreport 9, 1165–1168.

Ruda, M. A. 1982. Opiates and pain pathways: demonstration of enkephalin synapses on dorsal horn projection neurons. Science 215, 1523–1525.

Ruda, M. A. 1988. Spinal dorsal horn circuitry involved in the brain stem control of nociception. Prog. Brain Res. 77, 129–140.

Ruda, M. A., Coffield, J., and Steinbusch, H. W. 1982. Immunocytochemical analysis of serotonergic axons in laminae I and II of the lumbar spinal cord of the cat. J. Neurosci. 2, 1660–1671.

Sagen, J. and Proudfit, H. K. 1984. Effect of intrathecally administered noradrenergic antagonists on nociception in the rat. Brain Res. 310, 295–301.

Sakurai, T., Amemiya, A., Ishii, M., Matsuzaki, I., Chemelli, R. M., Tanaka, H., Williams, S. C., Richardson, J. A., Kozlowski, G. P., Wilson, S., Arch, J. R., Buckingham, R. E., Haynes, A. C., Carr, S. A., Annan, R. S., McNulty, D. E., Liu, W. S., Terrett, J. A., Elshourbagy, N. A., Bergsma, D. J., and Yanagisawa, M. 1998. Orexins and orexin receptors: a family of hypothalamic neuropeptides and G protein-coupled receptors that regulate feeding behavior. Cell 92, 573–585.

Samad, T. A., Moore, K. A., Sapirstein, A., Billet, S., Allchorne, A., Poole, S., Bonventre, J. V., and Woolf, C. J. 2001. Interleukin-1beta-mediated induction of Cox-2 in the CNS contributes to inflammatory pain hypersensitivity. Nature 410, 471–475.

Sandkuhler, J. 1996. The organization and function of endogenous antinociceptive systems. Prog. Neurobiol. 50, 49–81.

Sawamura, S., Kingery, W. S., Davies, M. F., Agashe, G. S., Clark, J. D., Kobilka, B. K., Hashimoto, T., and Maze, M. 2000. Antinociceptive action of nitrous oxide is mediated by stimulation of noradrenergic neurons in the brainstem and activation of [alpha]2B adrenoceptors. J. Neurosci. 20, 9242–9251.

Sawynok, J. 1989. The 1988 Merck Frosst Award. The role of ascending and descending noradrenergic and serotonergic pathways in opioid and non-opioid antinociception as revealed by lesion studies. Can. J. Physiol. Pharmacol. 67, 975–988.

Schaible, H. G., Neugebauer, V., Cervero, F., and Schmidt, R. F. 1991. Changes in tonic descending inhibition of spinal neurons with articular input during the development of acute arthritis in the cat. J. Neurophysiol. 66, 1021–1032.

Schechtmann, G., Wallin, J., Meyerson, B. A., and Linderoth, B. 2004. Intrathecal clonidine potentiates suppression of tactile hypersensitivity by spinal cord stimulation in a model of neuropathy. Anesth. Analg. 99, 135–139.

Schnell, C., Ulucan, C., and Ellrich, J. 2002. Atypical on-, off- and neutral cells in the rostral ventromedial medulla oblongata in rat. Exp. Brain Res. 145, 64–75.

Sessle, B. J., Hu, J. W., Dubner, R., and Lucier, G. E. 1981. Functional properties of neurons in cat trigeminal subnucleus caudalis (medullary dorsal horn). II. Modulation of responses to noxious and nonnoxious stimuli by periaqueductal gray, nucleus raphe magnus, cerebral cortex, and afferent influences, and effect of naloxone. J. Neurophysiol. 45, 193–207.

Shealy, C. N. 2003. Transcutaneous electrical nerve stimulation: the treatment of choice for pain and depression. J. Altern. Complement. Med. 9, 619–623.

Shealy, C. N., Mortimer, J. T., and Reswick, J. B. 1967. Electrical inhibition of pain by stimulation of the dorsal columns: preliminary clinical report. Anesth. Analg. 46, 489–491.

Simkin, P. P. and O'hara, M. 2002. Nonpharmacologic relief of pain during labor: systematic reviews of five methods. Am. J. Obstet. Gynecol. 186, S131–159.

Simon, E. J., Hiller, J. M., and Edelman, I. 1973. Stereospecific binding of the potent narcotic analgesic (3H) Etorphine to rat-brain homogenate. Proc. Natl. Acad. Sci. U. S. A. 70, 1947–1949.

Siuciak, J. A., Wong, V., Pearsall, D., Wiegand, S. J., and Lindsay, R. M. 1995. BDNF produces analgesia in the formalin test and modifies neuropeptide levels in rat brain and spinal cord areas associated with nociception. Eur. J. Neurosci. 7, 663–670.

Skinner, K., Basbaum, A. I., and Fields, H. L. 1997. Cholecystokinin and enkephalin in brain stem pain modulating circuits. Neuroreport 8, 2995–2998.

Skirboll, L., Hokfelt, T., Dockray, G., Rehfeld, J., Brownstein, M., and Cuello, A. C. 1983. Evidence for periaqueductal cholecystokinin-substance P neurons projecting to the spinal cord. J. Neurosci. 3, 1151–1157.

Sluka, K. A. and Walsh, D. 2003. Transcutaneous electrical nerve stimulation: basic science mechanisms and clinical effectiveness. J. Pain 4, 109–121.

Sluka, K. A. and Westlund, K. N. 1992. Spinal projections of the locus coeruleus and the nucleus subcoeruleus in the Harlan and the Sasco Sprague-Dawley rat. Brain Res. 579, 67–73.

Sonohata, M., Furue, H., Katafuchi, T., Yasaka, T., Doi, A., Kumamoto, E., and Yoshimura, M. 2004. Actions of noradrenaline on substantia gelatinosa neurones in the rat spinal cord revealed by in vivo patch recording. J. Physiol. 555(Pt 2), 515–526.

Sora, I., Takahashi, N., Funada, M., Ujike, H., Revay, R. S., Donovan, D. M., Miner, L. L., and Uhl, G. R. 1997. Opiate receptor knockout mice define mu receptor roles in endogenous nociceptive responses and morphine-induced analgesia. Proc. Natl. Acad. Sci. U. S. A. 94, 1544–1549.

Sorkin, L. S., McAdoo, D. J., and Willis, W. D. 1993. Raphe magnus stimulation-induced antinociception in the cat is associated with release of amino acids as well as serotonin in the lumbar dorsal horn. Brain Res. 618, 95–108.

Sotgui, M. L. 1993. Descending influence on dorsal horn neuronal hyperactivity in a rat model of neuropathic pain. Neuroreport. 4, 21–24.

Sounvoravong, S., Nakashima, M. N., Wada, M., and Nakashima, K. 2004. Decrease in serotonin concentration in raphe magnus nucleus and attenuation of morphine analgesia in two mice models of neuropathic pain. Eur. J. Pharmacol. 484, 217–223.

Spinella, M., Cooper, M. L., and Bodnar, R. J. 1996. Excitatory amino acid antagonists in the rostral ventromedial medulla inhibit mesencephalic morphine analgesia in rats. Pain 64, 545–552.

Stamford, J. A. 1995. Descending control of pain. Br. J. Anaesth. 75, 217–227.

Stevens, R. T., Hodge, C. J., Jr., and Apkarian, A. V. 1982. Kolliker-Fuse nucleus: the principal source of pontine catecholaminergic cells projecting to the lumbar spinal cord of cat. Brain Res. 239, 589–594.

Stiller, C. O., Gustafsson, H., Fried, K., and Brodin, E. 1997. Opioid-induced release of neurotensin in the periaqueductal gray matter of freely moving rats. Brain Res. 774, 149–158.

Stone, L. S., MacMillan, L. B., Kitto, K. F., Limbird, L. E., and Wilcox, G. L. 1997. The alpha2a adrenergic receptor subtype mediates spinal analgesia evoked by alpha2 agonists and is necessary for spinal adrenergic-opioid synergy. J. Neurosci. 17, 7157–7165.

Stone, L. S., Vulchanova, L., Riedl, M. S., Wang, J., Williams, F. G., Wilcox, G. L., and Elde, R. 1999. Effects of peripheral nerve injury on alpha-2A and alpha-2C adrenergic receptor immunoreactivity in the rat spinal cord. Neuroscience 93, 1399–1407.

Sugiyo, S., Takemura, M., Dubner, R., and Ren, K. 2005. Trigeminal transition zone-rostral ventromedial medulla connections and facilitation of orofacial hyperalgesia after masseter inflammation in rats. J. Comp. Neurol. (in press).

Suzuki, R., Rahman, W., Hunt, S. P., and Dickenson, A. H. 2004a. Descending facilitatory control of mechanically evoked responses is enhanced in deep dorsal horn neurones following peripheral nerve injury. Brain Res. 1019, 68–76.

Suzuki, R., Rygh, L. J., and Dickenson, A. H. 2004b. Bad news from the brain: descending 5-HT pathways that control spinal pain processing. Trends Pharmacol. Sci. 25, 613–617.

Suzuki, R., Green, G. M., Millan, M. J., and Dickenson, A. H. 2002a. Electrophysiologic characterization of the antinociceptive actions of S18616, a novel and potent alpha 2-adrenoceptor agonist, after acute and persistent pain states. J. Pain 3, 234–243.

Suzuki, R., Morcuende, S., Webber, M., Hunt, S. P., and Dickenson, A. H. 2002b. Superficial NK1-expressing neurons control spinal excitability through activation of descending pathways. Nat. Neurosci. 5, 1319–1326.

Swanson, L. W. and Hartman, B. K. 1975. The central adrenergic system. An immunofluorescence study of the location of cell bodies and their efferent connections in the rat utilizing dopamine-beta-hydroxylase as a marker. J. Comp. Neurol. 163, 467–505.

Tanaka, M., Matsumoto, Y., Murakami, T., Hisa, Y., and Ibata, Y. 1996. The origins of catecholaminergic innervation in the rostral ventromedial medulla oblongata of the rat. Neurosci. Lett. 207, 53–56.

Tang, J., Ko, S., Ding, H. K., Qiu, C. S., Calejesan, A. A., and Zhuo, M. 2005. Pavlovian fear memory induced by activation in the anterior cingulate cortex. Mol. Pain 1, 6.

Tavares, I. and Lima, D. 2002. The caudal ventrolateral medulla as an important inhibitory modulator of pain transmission at the spinal cord. J. Pain 3, 337–346.

Terayama, R., Guan, Y., Dubner, R., and Ren, K. 2000. Activity-induced plasticity in brain stem pain modulatory circuitry after inflammation. Neuroreport 11, 1915–1919.

Terenius, L. 1973. Stereospecific interaction between narcotic analgesics and a synaptic plasm a membrane fraction of rat cerebral cortex. Acta Pharmacol. Toxicol. (Copenh). 32, 317–320.

Tattersall, J. E., Cervero, F., and Lumb, B. M. 1986. Effects of reversible spinalization on the visceral input to viscerosomatic neurons in the lower thoracic spinal cord of the cat. J. Neurophysiol. 56, 785–796.

Thomas, D. A., McGowan, M. K., and Hammond, D. L. 1995. Microinjection of baclofen in the ventromedial medulla of rats: antinociception at low doses and hyperalgesia at high doses. J Pharmacol. Exp. Ther. 275, 274–284.

Thurston, C. L. and Randich, A. 1995. Responses of on and off cells in the rostral ventral medulla to stimulation of vagal afferents and changes in mean arterial blood pressure in intact and cardiopulmonary deafferented rats. Pain 62, 19–38.

Tjolsen, A., Berge, O. G., and Hole, K. 1991. Lesions of bulbo-spinal serotonergic or noradrenergic pathways reduce nociception as measured by the formalin test. Acta Physiol. Scand. 142, 229–236.

Todd, A. J. and Millar, J. 1984. Antagonism of 5-hydroxytryptamine-evoked excitation in the superficial dorsal horn of the cat spinal cord by methysergide. Neurosci Lett. 48, 167–170.

Tseng, L. F. and Tang, R. 1990. Different mechanisms mediate beta-endorphin- and morphine-induced inhibition of the tail-flick response in rats. J. Pharmacol. Exp. Ther. 252, 546–551.

Tsou, K. and Jang, C. S. 1964. Studies on the site of analgesic action of morphine by intracerebral microinjections. Sci. Sin. 13, 1099–1109.

Tsou, K., Brown, S., Sanudo-Pena, M. C., Mackie, K., and Walker, J. M. 1998. Immunohistochemical distribution of cannabinoid CB1 receptors in the rat central nervous system. Neuroscience 83, 393–411.

Tsuruoka, M. and Willis, W. D. 1996. Bilateral lesions in the area of the nucleus locus coeruleus affect the development of hyperalgesia during carrageenan-induced inflammation. Brain Res. 726, 233–236.

Tsuruoka, M., Hitoto, T., Hiruma, Y., and Matsui, Y. 1999. Neurochemical evidence for inflammation-induced activation of the coeruleospinal modulation system in the rat. Brain Res. 821, 236–240.

Tsuruoka, M., Arai, Y. C., Nomura, H., Matsutani, K., and Willis, W. D. 2003. Unilateral hindpaw inflammation induces bilateral activation of the locus coeruleus and the nucleus subcoeruleus in the rat. Brain Res. Bull. 61, 117–123.

Urban, M. O. and Gebhart, G. F. 1997. Characterization of biphasic modulation of spinal nociceptive transmission by neurotensin in the rat rostral ventromedial medulla. J. Neurophysiol. 78, 1550–1562.

Urban, M. O. and Gebhart, G. F. 1999. Supraspinal contributions to hyperalgesia. Proc Natl Acad Sci USA 96, 7687–7692.

Urban, M. O., Jiang, M. C., and Gebhart, G. F. 1996. Participation of central descending nociceptive facilitatory system in secondary hyperalgesia produced by mustard oil. Brain Res. 737, 83–91.

Urban, M. O., Hama, A. T., Bradbury, M., Anderson, J., Varney, M. A., and Bristow, L. 2003. Role of metabotropic glutamate receptor subtype 5 (mGluR5) in the maintenance of cold hypersensitivity following a peripheral mononeuropathy in the rat. Neuropharmacology 44, 983–993.

Valverde, O., Ledent, C., Beslot, F., Parmentier, M., and Roques, B. P. 2000. Reduction of stress-induced analgesia but not of exogenous opioid effects in mice lacking CB1 receptors. Eur. J. Neurosci. 12, 533–539.

Van Bockstaele, E. J., Aston-Jones, G., Pieribone, V. A., Ennis, M., and Shipley, M. T. 1991. Subregions of the periaqueductal gray topographically innervate the rostral ventral medulla in the rat. J. Comp. Neurol. 309, 305–327.

Van den Pol, A. N. 1999. Hypothalamic hypocretin (orexin): robust innervation of the spinal cord. J. Neurosci. 19, 3171–3182.

Van den Pol, A. N. and Collins, W. F. 1994. Paraventriculospinal tract as a model for axon injury: spinal cord. J. Comp. Neurol. 349, 244–258.

Vanegas, H. and Schaible, H-G. 2004. Descending control of persistent pain: inhibitory or facilitatory? Brain Res. Rev. 46, 295–309.

Vanegas, H., Barbaro, N. M., and Fields, H. L. 1984. Midbrain stimulation inhibits tail-flick only at currents sufficient to excite rostral medullary neurons. Brain Res. 321, 127–133.

van Praag, H. and Frenk, H. 1990. The role of glutamate in opiate descending inhibition of nociceptive spinal reflexes. Brain Res. 524, 101–105.

Vaughan, C. W., Connor, M., Jennings, E. A., Marinelli, S., Allen, R. G., and Christie, M. J. 2001. Actions of nociceptin/orphanin FQ and other prepronociceptin products on rat rostral ventromedial medulla neurons in vitro. J. Physiol. 534(Pt 3), 849–859.

Veraksits, A., Runkorg, K., Kurrikoff, K., Raud, S., Abramov, U., Matsui, T., Bourin, M., Koks, S., and Vasar, E. 2003. Altered pain sensitivity and morphine-induced anti-nociception in mice lacking CCK2 receptors. Psychopharmacology (Berl). 166, 168–175.

Vierck, C. J., Jr., Kline, R., 4th, and Wiley, R. G. 2004. Comparison of operant escape and innate reflex responses to nociceptive skin temperatures produced by heat and cold stimulation of rats. Behav. Neurosci. 118, 627–635.

Walker, J. M. and Huang, S. M. 2002. Endocannabinoids in pain modulation. Prostaglandins Leukot. Essent. Fatty Acids. 66, 235–242.

Wall, P. D. 1967. The laminar organization of dorsal horn and effects of descending impulses. J. Physiol. 188, 403–423.

Wallace, B. A., Ashkan, K., and Benabid, A. L. 2004. Deep brain stimulation for the treatment of chronic, intractable pain. Neurosurg. Clin. N. Am. 15, 343–357, vii.

Wang, H. and Wessendorf, M. W. 1999. Mu- and delta-opioid receptor mRNAs are expressed in spinally projecting serotonergic and nonserotonergic neurons of the rostral ventromedial medulla. J. Comp. Neurol. 404, 183–96.

Wang, H. and Wessendorf, M. W. 2002. Mu- and delta-opioid receptor mRNAs are expressed in periaqueductal gray neurons projecting to the rostral ventromedial medulla. Neuroscience 109, 619–634.

Wang, Q., Mao, L. M., Shi, Y. S, and Han, J. S. 1990. Lumbar intrathecal administration of naloxone antagonizes analgesia produced by electrical stimulation of the hypothalamic arcuate nucleus in pentobarbital-anesthetized rats. Neuropharmacology 29, 1123–1129.

Watanabe, S., Kuwaki, T., Yanagisawa, M., Fukuda, Y., and Shimoyama, M. 2005. Persistent pain and stress activate pain-inhibitory orexin pathways. Neuroreport 16, 5–8.

Watkins, L. R. and Mayer, D. J. 1982. Organization of endogenous opiate and nonopiate pain control systems. Science 216, 1185–1192.

Watkins, L. R., Milligan, E. D., and Maier, S. F. 2003. Glial proinflammatory cytokines mediate exaggerated pain states: implications for clinical pain. Adv. Exp. Med. Biol. 521, 1–21.

Wei, F., Dubner, R., and Ren, K. 1999. Nucleus reticularis gigantocellularis and nucleus raphe magnus in the brain stem exert opposite effects on behavioral hyperalgesia and spinal Fos protein expression after peripheral inflammation. Pain 80, 127–141.

Westlund, K. N., Bowker, R. M., Ziegler, M. G., and Coulter, J. D. 1983. Noradrenergic projections to the spinal cord of the rat. Brain Res. 263, 15–31.

Westlund, K. N., Carlton, S. M., Zhang, D., and Willis, W. D. 1990. Direct catecholaminergic innervation of primate spinothalamic tract neurons. J. Comp. Neurol. 299, 178–86.

Wiertelak, E. P., Roemer, B., Maier, S. F., and Watkins, L. R. 1997. Comparison of the effects of nucleus tractus solitarius and ventral medial medulla lesions on illness-induced and subcutaneous formalin-induced hyperalgesias. Brain Res. 748, 143–150.

Willer, J. C., Boureau, F., and Albe-Fessard, D. 1979. Supraspinal influences on nociceptive flexion reflex and pain sensation in man. Brain Res. 179, 61–68.

Williams, F. G. and Beitz, A. J. 1993. Chronic pain increases brainstem proneurotensin/neuromedin-N mRNA expression:

a hybridization-histochemical and immunohistochemical study using three different rat models for chronic nociception. Brain Res. 611, 87–102.

Willis, W. D. and Westlund, K. N. 1997. Neuroanatomy of the pain system and of the pathways that modulate pain. J. Clin. Neurophysiol. 14, 2–31.

Willis, W. D., Westlund, K. N., and Carlton, S. M. 1995. Pain, In: The Rat Nervous System, 2nd edn. (*ed.* G. Paxinos), pp. 725–750. Academic Press.

Xie, J. Y., Herman, D. S., Stiller, C. O., Gardell, L. R., Ossipov, M. H., Lai, J., Porreca, F., and Vanderah, T. W. 2005. Cholecystokinin in the rostral ventromedial medulla mediates opioid-induced hyperalgesia and antinociceptive tolerance. J. Neurosci. 25, 409–416.

Yaksh, T. L. 1978. Opiate receptors for behavioral analgesia resemble those related to the depression of spinal nociceptive neurons. Science 199, 1231–1233.

Yaksh, T. L. 1979. Direct evidence that spinal serotonin and noradrenaline terminals mediate the spinal antinociceptive effects of morphine in the periaqueductal gray. Brain Res. 160, 180–185.

Yaksh, T. L. and Tyce, G. M. 1979. Microinjection of morphine into the periaqueductal gray evokes the release of serotonin from spinal cord. Brain Res. 1979 171, 176–181.

Yamamoto, T., Saito, O., Shono, K., and Hirasawa, S. 2003. Activation of spinal orexin-1 receptor produces anti-allodynic effect in the rat carrageenan test. Eur. J. Pharmacol. 481, 175–180.

Yamamoto, T., Nozaki-Taguchi, N, Sakashita, Y., and Kimura, S. 1999. Nociceptin/orphanin FQ: role in nociceptive information processing. Prog. Neurobiol. 57, 527–535.

Yang, H. Y., Fratta, W., Majane, E. A., and Costa, E. 1985. Isolation, sequencing, synthesis, and pharmacological characterization of two brain neuropeptides that modulate the action of morphine. Proc. Natl. Acad. Sci. U. S. A. 82, 7757–7761.

Yoshida, A., Chen, K., Moritani, M., Yabuta, N. H., Nagase, Y., Takemura, M., and Shigenaga, Y. 1997. Organization of the descending projections from the parabrachial nucleus to the trigeminal sensory nuclear complex and spinal dorsal horn in the rat. J. Comp. Neurol. 383, 94–111.

Young, R. F., Tronnier, V., and Rinaldi, P. C. 1992. Chronic stimulation of the Kolliker-Fuse nucleus region for relief of intractable pain in humans. J. Neurosurg. 76, 979–985.

Zadina, J. E., Hackler, L., Ge, L. J., and Kastin, A. J. 1997. A potent and selective endogenous agonist for the mu-opiate receptor. Nature 386, 499–502.

Zagon, A. 1995. Internal connections in the rostral ventromedial medulla of the rat. J. Auton. Nerv. Syst. 53, 43–56.

Zagon, A., Meng, X., and Fields, H. L. 1997. Intrinsic membrane characteristics distinguish two subsets of nociceptive modulatory neurons in rat RVM. J. Neurophysiol. 78, 2848–2858.

Zaki, P. A., Bilsky, E. J., Vanderah, T. W., Lai, J., Evans, C. J., and Porreca, F. 1996. Opioid receptor types and subtypes: the delta receptor as a model. Annu. Rev. Pharmacol. Toxicol. 36, 379–401.

Zambreanu, L., Wise, R. G., Brooks, J. C., Iannetti, G. D., and Tracey, I. 2005. A role for the brainstem in central sensitisation in humans. Evidence from functional magnetic resonance imaging. Pain 114, 397–407.

Zeilhofer, H. U. 2005. The glycinergic control of spinal pain processing. Cell. Mol. Life Sci. 2005 Jun 17 (Epub ahead of print).

Zeitz, K. P., Guy, N., Malmberg, A. B., Dirajlal, S., Martin, W. J., Sun, L., Bonhaus, D. W., Stucky, C. L., Julius, D., and Basbaum, A. I. 2002. The 5-HT3 subtype of serotonin

receptor contributes to nociceptive processing via a novel subset of myelinated and unmyelinated nociceptors. J. Neurosci. 22, 1010–1019.

Zhang, D. X., Owens, C. M., and Willis, W. D. 1991. Two forms of inhibition of spinothalamic tract neurons produced by stimulation of the periaqueductal gray and the cerebral cortex. J. Neurophysiol. 65, 1567–1579.

Zhang, Y., Yang, Z., Gao, X., and Wu, G. 2001. The role of 5-hydroxytryptamine1A and 5-hydroxytryptamine1B receptors in modulating spinal nociceptive transmission in normal and carrageenan-injected rats. Pain 92, 201–211.

Zhang, J., Hoffert, C., Vu, H. K., Groblewski, T., Ahmad, S., and O'Donnell, D. 2003. Induction of CB2 receptor expression in the rat spinal cord of neuropathic but not inflammatory chronic pain models. Eur. J. Neurosci. 17, 2750–2754.

Zhang, R. X., Lao, L., Wang, L., Liu, B., Wang, X., Ren, K., and Berman, B. M. 2004. Involvement of opioid receptors in electroacupuncture-produced anti-hyperalgesia in rats with peripheral inflammation. Brain Res. 1020, 12–17.

Zhong, F. X., Ji, X. Q., and Tsou, K. 1985. Intrathecal DSP4 selectively depletes spinal noradrenaline and attenuates morphine analgesia. Eur. J. Pharmacol. 116, 327–330.

Zhuo, M. and Gebhart, G. F. 1990. Spinal cholinergic and monoaminergic receptors mediate descending inhibition from the nuclei reticularis gigantocellularis and gigantocellularis pars alpha in the rat. Brain Res. 535, 67–78.

Zhuo, M. and Gebhart, G. F. 1991. Spinal serotonin receptors mediate descending facilitation of a nociceptive reflex from the nuclei reticularis gigantocellularis and gigantocellularis pars alpha in the rat. Brain Res. 550, 35–48.

Zhuo, M. and Gebhart, G. F. 1992. Characterization of descending facilitation and inhibition of spinal nociceptive transmission from the nuclei reticularis gigantocellularis and pars alpha in the rat. J. Neurophysiol. 67, 1599–1614.

Zhuo, M. and Gebhart, G. F. 1997. Biphasic modulation of spinal nociceptive transmission from the medullary raphe nuclei in the rat. J. Neurophysiol. 78, 746–758.

Zorman, G., Belcher, G., Adams, J. E., and Fields, H. L. 1982. Lumbar intrathecal naloxone blocks analgesia produced by microstimulation of the ventromedial medulla in the rat. Brain Res. 236, 77–84.

Further Reading

Calejesan, A. A., Ch'ang, M. H., and Zhuo, M. 1998. Spinal serotonergic receptors mediate facilitation of a nociceptive reflex by subcutaneous formalin injection into the hindpaw in rats. Brain Res. 798, 46–54.

Dubner, R. and Ren, K. 1999. Endogenous mechanisms of sensory modulation. Pain Suppl. 6, S45–53.

Guan, Y., Guo, W., Robbins, M. T., Dubner, R., and Ren, K, 2004. Changes in AMPA receptor phosphorylation in the rostral ventromedial medulla after inflammatory hyperalgesia. Neurosci. Lett. 366, 201–205.

Janss, A. J., Cox, B. F., Brody, M. J., and Gebhart, G. F. 1987. Dissociation of antinociceptive from cardiovascular effects of stimulation in the lateral reticular nucleus in the rat. Brain Res. 405, 140–149.

Jasmin, L., Boudah, A., and Ohara, P. T. 2003a. Long-term effects of decreased noradrenergic central nervous system innervation on pain behavior and opioid antinociception. J. Comp. Neurol. 460, 38–55.

Mason, P. 1997. Central mechanisms of pain modulation. Curr. Op. Neurobiol. 9, 436–441.

Rahman, W., Suzuki, R., Rygh, L. J., and Dickenson, A. H. 2004. Descending serotonergic facilitation mediated through rat spinal 5HT3 receptors is unaltered following carrageenan inflammation. Neurosci. Lett. 361, 229–231.

Vaccarino, A. L. and Chorney, D. A. 1994. Descending modulation of central neural plasticity in the formalin pain test. Brain Res. 666, 104–108.

Villanueva, L., Desbois, C., Le Bars, D., and Bernard, J. F. 1998. Organization of diencephalic projections from the medullary subnucleus reticularis dorsalis and the adjacent cuneate nucleus: a retrograde and anterograde tracer study in the rat. J. Comp. Neurol. 390, 133–160.

Wei, F., Ren, K., and Dubner, R. 1998. Inflammation-induced Fos protein expression in the rat spinal cord is enhanced following dorsolateral or ventrolateral funiculus lesions. Brain Res. 782, 136–141.

Woolf, C. J. and Salter, M. W. 2000. Neuronal plasticity: increasing the gain in pain. Science 288, 1765–1768.

50 Diffuse Noxious Inhibitory Controls (DNIC)

D Le Bars, INSERM U-713, Paris, France

J C Willer, INSERM U-731, Paris, France

Glossary

allodynia Pain caused by stimuli that would not normally cause pain (i.e., non-noxious stimuli).

basic somesthetic activity Description of the ongoing activity in somatosensory pathways in the absence of any deliberate stimuli but including stimuli provided by the environment.

blink reflex Exteroceptive reflex of the orbicularis oculi muscles usually evoked by stimuli such as a mechanical tap on the glabella or by electrical currents delivered to the skin in the periorbital region.

Brown-Séquard syndrome Patient with a hemisection of the spinal cord of traumatic origin.

bulbospinal control A neural control mechanism originating in the brainstem and projecting to the spinal cord.

dermatome Area of skin innervated by nerves from a given segment of the spinal cord.

descending inhibition Inhibition of spinal cord function produced by pathways originating in the brain.

diffuse noxious inhibitory controls (DNICs) Neural controls triggered by noxious stimulation of widespread areas of the body which exert an inhibitory influence on wide-dynamic-range (WDR) neurons.

electromyography (EMG) All electrophysiological methods for exploring the physiology and pathophysiology of peripheral nerves and muscles.

excitatory receptive field Area of the body (surface or interior) which when stimulated will produce excitation of a given neuron.

hemianalgesia Loss of pain sensation from one half of the body.

heterotopic Part of the body remote from the area of interest (e.g., the excitatory receptive field of a neuron).

Hoffmann reflex (H reflex) Monosynaptic reflex from the calf muscles elicited by electrical stimulation of the sciatic nerve.

inhibitory receptive field Area of the body (surface or interior) which when stimulated will produce inhibitory effects on the activity of a given neuron.

jaw-opening reflex Exteroceptive reflex produced by stimulation of facial or intraoral afferents and involving activation of jaw opening muscles (e.g., the digastric) and/or inhibition of activity in jaw closing muscles (e.g., the masseter).

jerk reflex Myotatic monosynaptic reflex usually elicited by a percussion of a muscle's tendon using a reflex testing hammer.

Lasègue's maneuver Elevation of a lower limb in an extended position on a patient lying on a bed. This technique explores the Lasègue's sign which is the painful limitation of the angle of elevation in cases of radicular compression of the sciatic nerve.

microelectrophoretic application Application of a substance in its ionic form from a microelectrode by the application of electrical current.

monoarthritis Inflammation of a single joint.

mononeuropathy A disturbance of function or pathological changes in a single nerve.

nociceptive stimulus Stimulus that activates nociceptive afferents (which could be a noxious stimulus or an electrical stimulus).

noxious stimulus Stimulus that causes or threatens to cause tissue damage.

polyarthritis Inflammation of several joints.

propriospinal mechanism Neural mechanism mediated entirely within the spinal cord.

R_{III} reflex Electromyographic response elicited by electrical painful stimulation of the (purely cutaneous) sural or ulnar nerve by recording from the biceps femoris (lower limb) or biceps brachialis (upper limb), respectively. So-named because it involves activation of group III ($A\delta$) afferents.

subnucleus reticularis dorsalis (SRD) Brainstem nucleus located ventral to the cuneate nucleus, between trigeminal nucleus caudalis and the nucleus of the solitary tract.

supraspinal mechanism Neural mechanism that involves structures above the spinal cord.

tetraplegic Patient with a complete transection of traumatic origin at a high level of the spinal cord thus affecting all four limbs.

transcutaneous electrical nerve stimulation (TENS) Technique for inducing analgesia/ hypoalgesia by electrically stimulating peripheral nerves through the skin (i.e., with electrodes placed on the skin).

trigeminal nucleus caudalis Most caudal part of the trigeminal sensory nuclear complex (or of the trigeminal spinal nucleus); also sometimes called the medullary dorsal horn.

ventrolateral quadrant Part of the spinal cord where spinoreticular and spinothalamic fibers travel.

Wallenberg's syndrome Patient with a unilateral lesion of the retro-olivary part of the brainstem resulting from the ischemia of a posterior cerebellar artery.

wide-dynamic-range (WDR) neurons Neurons of the dorsal horn activated by both noxious and non-noxious stimuli.

50.1 Introduction

Painful stimuli can diminish or even mask pain elicited by stimulation of a remote (extrasegmental) part of the body (see references in Melzack, R., 1989; Le Bars, D. and Willer, J. C., 2002). This phenomenon has been known since ancient times as illustrated by the Hippocrates' aphorism: "If a patient be subject to two pains arising in different parts of the body simultaneously, the stronger blunts the other." Numerous popular therapeutic methods for the alleviation of pain – some used spontaneously by patients – take advantage of this common clinical observation. It is often referred to as counter-irritation or counter-stimulation. In Kabylia, for example, healers treat rheumatic pain by placing a red-hot scythe close to the abdomen of the patient and then vibrating it at a frequency of about 3 Hz, in order to create a series of acute burning-type painful sensations.

A working hypothesis was developed that some of the neurons involved in the transmission of nociceptive signals might be inhibited by nociceptive stimulation of peripheral territories outside their own excitatory receptive fields. Such a hypothesis was found to be correct at as early a stage in the processing of nociceptive signals as the spinal cord. This phenomenon was termed diffuse noxious inhibitory control (DNIC). Since these spinal neurons are involved in mediating both pain and nociceptive reflexes, DNIC can be studied at three related endpoints: spinal neuronal activities, reflexes, and sensations. The first two are accessible in animals while the last two can be studied in human beings (Figure 1).

50.2 A Spinally Mediated Process

Combined psychophysical measurements and recordings of nociceptive reflexes in man have shown that the spinal cord is involved in the phenomenon of pain inhibiting pain (Willer, J. C. *et al.*, 1984). Electrical stimulation of the sural nerve at the ankle simultaneously induces a nociceptive reflex in a knee flexor muscle (the R_{III} reflex) and a pinprick type of painful sensation in the territory of the nerve (Willer, J. C., 1977). Both the reflex and the sensation have strong relationships with the stimulus intensities – with the thresholds for both parameters being practically identical.

One line of study involved recording these parameters (i.e., the reflex and pain) before, during, and after the application of heterotopic conditioning stimuli such as the immersion of a hand in a thermoregulated water bath at various temperatures (Figure 2, upper panels). When the temperature of the bath was lower than 45 °C, the immersion of the

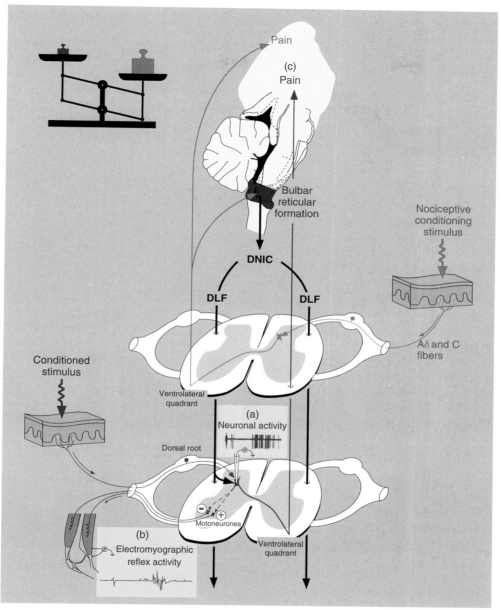

Figure 1 Nociceptive signals activate dorsal horn neurons, which in turn transfer the information to the brain to elicit pain and to spinal interneurons (broken blue line) to elicit somatic and vegetative reflexes. The effects on motoneurons through polysynaptic pathways trigger a movement that moves the stimulated area away from the stimulus. This movement results from both excitatory (e.g., contraction of flexor muscles) and inhibitory mechanisms – the latter affecting antagonist muscles. In animals, one can record from neurons (a) and from muscles (b) but the sensation (c) cannot be assessed. In man, on the other hand, one cannot record from neurons but it is possible to monitor both the sensation and the reflex activities. In an experimental paradigm designed for the study of pain inhibiting pain phenomena, one requires a conditioned stimulus (here in blue) with the corresponding endpoints (a, b, or c) and a conditioning stimulus (here in red). A painful focus activates dorsal horn neurons that send an excitatory signal through the ventrolateral quadrant toward higher centers, including the lower brainstem. This signal activates DNICs, which will inhibit WDR neurons through the dorsolateral funiculi (DLF). With two painful foci, the resulting sensation will depend on the balance of the nociceptive information elicited from the two sources. The power of such nociceptive information depends on the three dimensions of any nociceptive stimulus, namely strength, duration, and area. In an experimental situation designed to observe inhibitions of the response to the conditioned stimulus, the beam of the balance should tilt in favor of the conditioning stimulus. This is illustrated in Figure 2 where the response to a 20 ms duration series of electrical pulses is inhibited by the stimulation of the hand for 2 min.

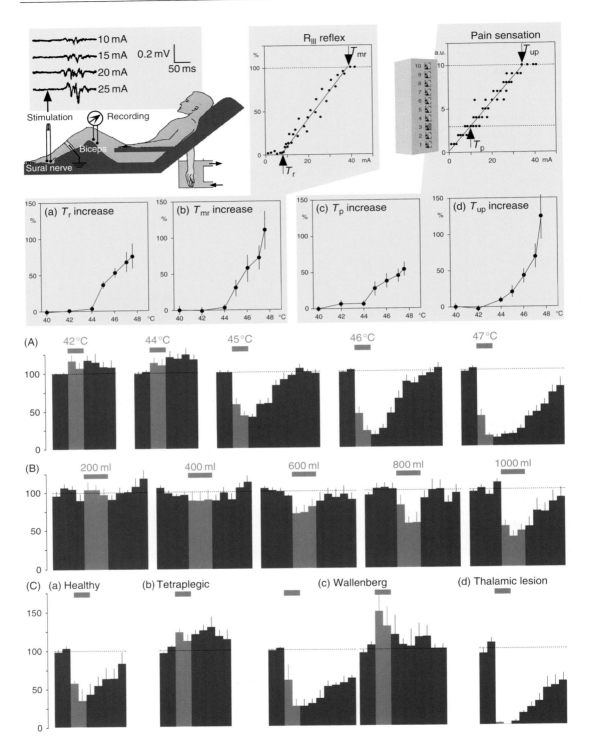

hand did not elicit any change in the stimulus–response relationships. By contrast, temperatures between 45 and 47.5 °C caused the stimulus–response curves to shift to the right as a direct function of the temperature. Similar effects were observed when other painful procedures were applied as conditioning stimuli (e.g., cold, pressure, exercise under ischemia, high-intensity transcutaneous electrical nerve stimulation) (Willer, J. C. *et al.*, 1984; 1989; Danziger, N. *et al.*, 1998; France, C. R. and Suchowiecki, S., 1999; Sandrini, G. *et al.*, 2000).

The temporal evolution of these phenomena can be studied by applying a heterotopic conditioning stimulus while the sural nerve is stimulated at a constant intensity (Figure 2, lower panels). This procedure revealed aftereffects of long duration. For example, a 2 min immersion of the hand in a 47 °C water bath completely abolished the R_{III} reflex and it took more than 9 min to achieve a full recovery

(Figure 2(A)). Very similar effects were produced by visceral stimulation (gastric or rectal distension) in healthy volunteers (Bouhassira, D. *et al.*, 1994; 1998). During gastric distension for example (Figure 2(B)), the inhibitory effects increased with the volume of distension and the resultant sensation. As with the inhibitions elicited by stimulation of somatic structures, these visceral-evoked inhibitions were correlated to the intensity of the conditioning stimuli. However, they could be triggered by stimuli which were unpleasant but not quite painful (Figure 2(C)). When they become clearly painful, the duration of the inhibitory effects outlasted the duration of the distension by several minutes.

Clinical pains can also activate this type of inhibition. Thus, in patients suffering from sciatic pain due to a herniated disk, a Lasègue's maneuver of the affected leg, which elicits a typical radicular pain, also blocks the R_{III} reflex, both in the affected and

Figure 2 (Upper panels) Experimental setup for recording the nociceptive R_{III} flexion reflex from the biceps femoris muscle in the lower limb using surface electrodes. The reflex is evoked by electrical stimulation of the ipsilateral sural nerve. The insert shows typical examples of electromyographic (EMG) responses elicited by increasing stimulus intensities (from top to bottom). The subjective sensation was estimated by the subject on a 10-level scale (box with switches shown in insert), with the pain threshold being defined as level 3. Examples of stimulus–response curves for the nociceptive reflex and subjective reports of sensations produced by a wide range of stimulation intensities applied randomly are shown on the left and right graphs, respectively. Note the close correlations between (1) the reflex threshold (T_r) and the pain threshold (T_p) around 10 mA; and (2) the stimulus intensities producing maximal responses, that is, the threshold of the maximal reflex recruitment (T_{mr}) versus the threshold of unbearable pain (T_{up}). Such recruitment curves were built before and during a 2 min period of immersion of the right hand in a thermoregulated water bath. No effects were seen at lower temperatures but a shift of the curves to the right was elicited when the temperature reached painful levels (45 °C and above). The shift applied to the reflex (a, b) and sensation (c, d) thresholds. (Lower panels) The temporal evolution of the reflex can be monitored by employing repeated, constant, stimulation of the sural nerve (1.2 times threshold every 6 s). The individual R_{III} reflex is expressed in terms of the percentages of the mean value recorded during a control period. Each vertical bar represents the mean of ten responses recorded during 1 min. Graphs show pooled data from several subjects. (A) Effects of immersion of the hand. The nonpainful temperature (42 and 44 °C, left) did not modify the reflex. By contrast, the painful temperatures (45, 46, and 47 °C) depressed the reflex during and after the period of conditioning. The extent of these depressions was temperature dependent. Note the long duration (around 10 min) of the inhibitory posteffects following the application of the highest temperature (47 °C, right). (B) Effects of visceral stimulation. Gastric distension by means of a balloon introduced into the proximal part of the stomach was applied for a 3 min period. The inflation of the balloon in volumes of 200 or 400 ml did not modify the R_{III} reflex, whereas volumes of 600, 800, and 1000 ml produced inhibitions correlated to the volume of distension. The 600 and 800 ml distensions were unpleasant while the 1000 ml volume was clearly painful. (C) Effects of nociceptive electrical stimuli (20–25 mA) applied to the upper limb of healthy volunteers or patients with lesions of the spinal cord or the brain. (a) In healthy volunteers; (b) in tetraplegic patients, the conditioning stimulus did not inhibit the R_{III} reflex; (c) in patients with a Wallenberg syndrome (unilateral lesion of the retro-olivary part of the brainstem) the conditioning stimuli did or did not inhibit the R_{III} reflex depending on whether it was applied to the normal hand (ipsilateral to the brain lesion) or the analgesic hand (contralateral to the brain lesion); and (d) in patients with thalamic lesions, the conditioning stimulation applied to the analgesic hand (contralateral to the brain lesion) did produce inhibition of the R_{III} reflex. (Upper panels) Adapted from Willer, J. C., Roby, A., and Le Bars, D. 1984. Psychophysical and electrophysiological approaches to the pain relieving effect of heterotopic nociceptive stimuli. Brain 107, 1095–1112. (A) Adapted from Willer, J. C., De Broucker, T., and Le Bars, D. 1989. Encoding of nociceptive thermal stimuli by diffuse noxious inhibitory controls in humans. J. Neurophysiol. 62, 1028–1038. (B) Adapted from Bouhassira, D., Chollet, R., Coffin, B., *et al.* 1994. Inhibition of a somatic nociceptive reflex by gastric distension in humans. Gastroenterology 107, 985–992. (C) Adapted from Roby-Brami, A., Bussel, B., Willer, J. C., and Le Bars, D. 1987. An electrophysiological investigation into the pain-relieving effects of heterotopic nociceptive stimuli: probable involvement of a supraspinal loop. Brain 110, 1497–1508; and De Broucker, T., Cesaro, P., Willer, J. C., Le Bars, D. 1990. Diffuse noxious inhibitory controls (DNIC) in man: involvement of the spino-reticular tract. Brain 113, 1223–1234.

in the nonaffected leg (Willer, J. C. *et al.*, 1987). When the Lasègue's maneuver is performed on the healthy limb, it is painless and has no effect on the reflex. In patients suffering from neuropathic pain, the R_{III} reflex is inhibited when pain is triggered by pressure ('static allodynia'), but not when it is triggered by light brushing ('dynamic allodynia') (Bouhassira, D. *et al.*, 2003). Interestingly, it is generally agreed that the former is due to the activation of nociceptive processes via $A\delta$- and C-fibers, while the latter is due to the activation of $A\beta$-fibers.

In summary, a painful conditioning stimulus can depress both a preexisting pain and its associated nociceptive reflex at the first spinal relays for the transmission of nociceptive information. Interestingly, nociceptive brainstem reflexes such as the blink or jaw-opening reflexes are also inhibited by remote painful stimulation (Pantaleo, T. *et al.*, 1988; Maillou, P. and Cadden, S. W., 1997; Ellrich, J. and Treede, R. D., 1998). By contrast, nonnociceptive reflexes such as jerk or Hoffmann reflexes remain unaffected during such procedures. One must stress here that there has to be an obvious imbalance between the conditioned and the conditioning stimuli for a clear depressive effect to be seen. For instance, in the examples shown in Figure 2, the conditioned stimuli were short (20 ms) trains of electrical shocks, whereas the conditioning stimuli were applied over time periods of several seconds to large areas of the body. Clearly, spatiotemporal summation is required to elicit visible inhibitory effects because of the reciprocal nature of the phenomenon triggered by two remote painful foci. A balance, the beam of which tips (or does not tip) in favor of one or other pain, could symbolize the net result of such a reciprocal process (Figure 1, upper left). Of course, clinical situations are always much more complicated than these experimental situations which are deliberately designed to simplify the problem under analysis. More particularly, it seems likely that the existence of multiple painful foci would produce very complicated interactions between excitatory and inhibitory processes.

50.3   A Spinally Mediated Process Involving Supraspinal Structures

Are the inhibitory mechanisms described above due to propriospinal mechanisms or do they involve supraspinal structures? To answer this question, the effects on the R_{III} reflex in the right leg, of nociceptive conditioning stimuli applied to the fourth and fifth fingers of the left hand were compared in normal subjects and tetraplegic patients with lesions of traumatic origin at the cervical-5–7 level (Roby-Brami, A. *et al.*, 1987). In the normal subjects (Figure 2(Ca)), the painful conditioning stimuli strongly depressed both the R_{III} reflex and the associated pain. By contrast, in the tetraplegic patients, nociceptive stimulation of the same territories, which, being in the cervical-8 and thoracic-1 dermatomes, were clinically unaffected by the spinal lesion, and did not produce any depression of the R_{III} reflex (Figure 2(Cb)). One can therefore conclude that the inhibitory effects triggered by heterotopic nociceptive stimuli are sustained by a loop that includes supraspinal structures.

In order to identify, or at least localize, these supraspinal structures, observations were made on patients with cerebral lesions causing contralateral hemianalgesia (De Broucker, T. *et al.*, 1990). These were patients with either a lesion of the retro-olivary part of the medulla (Wallenberg's syndrome, Figure 2(Cc)) or a unilateral thalamic lesion (Figure 2(Cd)). In the former group, no inhibitions were observed when the (not felt) nociceptive conditioning stimulus was applied to the affected side of the body, contralateral to the brain lesion. If this stimulus was applied to the normal side, ipsilateral to the brain lesion, it was felt as painful and triggered inhibitory effects very similar to those seen in normal subjects. By contrast, in the patients with a thalamic lesion, the R_{III} reflex was strongly depressed, as in normal subjects, by the nociceptive conditioning stimulus applied to the affected side (which was not felt as painful). Thalamic structures and consequently spinothalamic pathways, are therefore not involved in DNIC, whereas brainstem – probably reticular – structures seem to play a key role in these phenomena. These observations also demonstrate unambiguously that such phenomena can be elicited in the absence of a painful sensation, provided the nociceptive nature of the information reaches some brain structures. This suggests that the observed phenomenon does not result from a competition between the two attentional foci that the conditioned and the conditioning pain represent in the healthy volunteers. This is an important observation as it is known that the perception of pain is sensitive to attentional processes (Bushnell, M. C. *et al.*, 1985; Willer, J. C. *et al.*, 1979).

An exceptional case was also reported in a patient with a Brown-Séquard syndrome due to a hemisection of the spinal cord (left side, T6 level) produced by a knife-stab in the back (Bouhassira, D. *et al.*, 1993). The R$_{III}$ reflexes elicited by stimulation of cutaneous afferents in the ulnar and sural nerves were studied in the upper and lower limbs by recording from the biceps brachialis and biceps femoris muscles, respectively. For each limb, the R$_{III}$ reflex was elicited regularly before, during, and after periods of nociceptive electrical conditioning stimulation applied successively to the other three limbs. Inhibitions of around 90% followed by strong poststimulus effects were observed in all situations except that (1) no inhibition could be obtained when the conditioning stimuli were applied to the lower right limb (contralateral to the spinal lesion), and (2) the R$_{III}$ reflex in the lower left limb was completely insensitive to any of the conditioning stimuli. These results strongly suggest that in human beings (1) the ascending part of the loop subserving DNIC is completely crossed at the spinal level, and (2) the descending part is confined to the white matter ipsilateral to the limb being tested.

50.4   A Model in the Rat

Another nociceptive reflex – this time recorded in the anaesthetized rat – is the C-fiber reflex which can be elicited by electrical stimulation of the sural nerve and recorded from the biceps femoris muscle. This reflex can be strongly inhibited by both mechanical and thermal noxious heterotopic stimuli applied to the muzzle, a paw or the tail, and by colorectal distension. These inhibitory effects on the C-fiber reflex did not occur in spinal animals, or ipsilateral to a rostral unilateral lesion of the dorsolateral funiculus (DLF). Such observations are consistent with several reports showing the inhibition of reflexes or the increase of nociceptive thresholds elicited by heterotopic noxious conditioning. For example, the reflex discharge in the common peroneal nerve following electrical stimulation of the sural nerve in the rat was inhibited by pinching the muzzle or tail; the gastrocnemius medialis reflex evoked by sural nerve stimulation in the decerebrate rabbit was inhibited by electrical stimulation of the contralateral common peroneal or of the ipsi- or contralateral median nerves; the digastric reflex evoked by tooth pulp stimulation in the cat was inhibited by toe pinch, percutaneous electrical stimulation of a limb or electrical stimulation of the saphenous nerve. In behavioral experiments, intraperitoneal or cutaneous

injections of an irritant agent or noxious heat all produce an increase in nociceptive thresholds in distant somatic structures, notably the tail (e.g., tail flick test) and paws (e.g., paw withdrawal test). (See references in Le Bars, D. *et al.*, 2001.)

50.5   The Role of Wide-Dynamic-Range Neurons

In several species (rat, mouse, cat, and probably monkey), most WDR and some nociceptive-specific neurons are strongly inhibited by a noxious stimulus applied outside their excitatory receptive fields. Such effects do not appear to be somatotopically organized but apply to the whole body. Figure 3 shows examples of recordings from lumbar WDR neurons in the rat. The activities of these neurons – whether elicited by pinching their receptive field or by applying an excitatory amino acid to their membranes – were strongly inhibited by noxious mechanical (pinch of the muzzle, the contralateral hindpaw or the tail) or thermal stimuli (immersion of the tail in a 52 °C water bath) (Figure 3(A)). These examples illustrate the whole body inhibitory receptive fields of WDR neurons as recorded in intact rats. Interestingly, such inhibitory controls are exacerbated during clinical pains, for example, when an animal suffers from monoarthritis, polyarthritis or a peripheral mononeuropathy (Danziger, N. *et al.*, 2000; 2001).

These properties of dorsal horn neurons were observed at the level of various segments of the spinal cord and in both nucleus caudalis and nucleus oralis of the trigeminal system – thus suggesting a general organization. In keeping with the human studies shown in Figure 2, there is a clear relationship between the intensity of the conditioning stimulus and the strength of the resultant DNIC. With strong stimuli, the inhibitory effects are powerful and followed by long-lasting poststimulus effects, which can persist for several minutes. In some cases, the inhibitory effects can involve a complete abolition of activity for a long period of time following removal of the conditioning stimuli ('switch off') and the activity can be restored to preconditioning levels by further manipulations of the excitatory receptive field ('switch on') (Cadden, S. W., 1993). (See references in Le Bars, D., 2002.)

With regard to the viscera, some differences should be noted: visceral stimuli, for example, distension of the colon or urinary bladder, generally produce inhibitions with slower rates of onset and recovery but

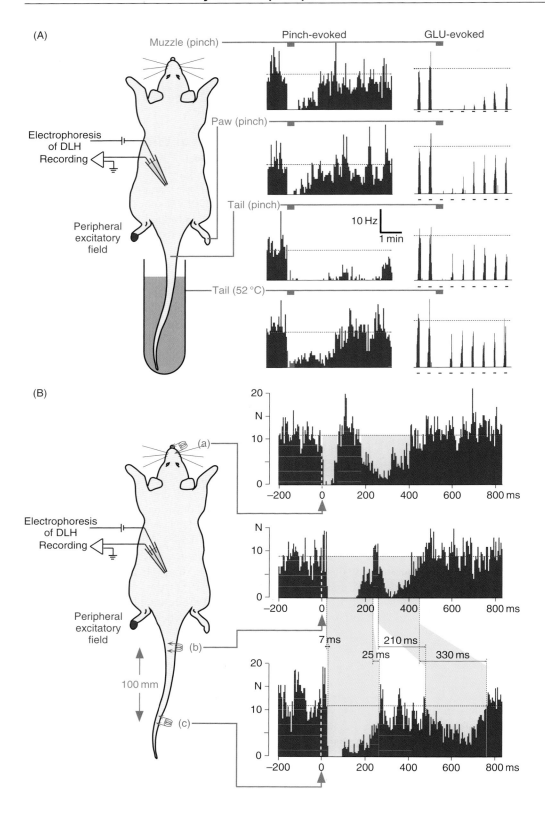

starting at intensities below a painful level (Cadden, S. W. and Morrison, J. F. B., 1991). It was proposed that these differences might have reflected different amounts and patterns of activity in the relevant primary afferent fibers rather than being due to different central neural mechanisms. Again, this is consistent with human studies (e.g., Figure 2(B)).

In any case, these data suggest that DNICs are triggered specifically by the activation of peripheral nociceptors whose signals are carried by Aδ- and C-fibers. In order to investigate further the types of peripheral fiber involved in DNIC, use was made of the facts that (1) trigeminal and spinal dorsal horn neurons respond with relatively steady discharges to the microelectrophoretic application of excitatory amino acids; and (2) DNICs act by a final postsynaptic inhibitory mechanism involving hyperpolarization of the neuronal membrane. It was found that when trigeminal WDR neurons were directly excited by the microelectrophoretic application of glutamate, the percutaneous application of single square-wave, electrical stimuli to the tail always induced a biphasic depression of the resultant activity. Both the early and late components of this inhibition occurred with shorter latencies when the base rather than the tip of the tail was stimulated. Such differences in latency were used to estimate the mean conduction velocities of the peripheral fibers triggering the inhibitions: the means were found to be

7.3 and 0.7 m s^{-1}, which fall into the Aδ- and C-fiber ranges, respectively. Such biphasic inhibitions could be elicited from any part of the body and recorded from any WDR neurons. Figure 3(B) shows a recording from a lumbar WDR neuron with an excitatory receptive field located on the extremity of the ipsilateral hind paw: two components of inhibition were induced by the activation of Aδ and C-fibers, respectively, when a single 2 ms duration shock of 10 mA was applied to the muzzle, the base or the tip of the tail.

DNICs are not observed in anesthetized or decerebrate animals in which the spinal cord has been sectioned. It is therefore obvious that the mechanisms underlying DNICs are not confined to the spinal cord and that supraspinal structures must be involved. Such a system is therefore completely different from segmental inhibitory systems, which work both in intact and in spinal animals, and can be triggered by the activation of low-threshold afferents. DNICs are also very different from the propriospinal inhibitory processes that can be triggered by noxious inputs. It should also be noted that DNIC are blocked by increasing the anesthetic regimen (Jinks, S. L. *et al.*, 2003).

The ascending and descending limbs of this loop travel through the ventrolateral and dorsolateral funiculi, respectively. Since thalamic lesions do not affect DNICs, it has been proposed that they result from a physiological activation of some of the brainstem structures that produce descending inhibition

---

Figure 3 (A) Example of inhibitions of a spinal WDR neuron elicited by noxious heterotopic stimuli. Recordings were made in the lumbar dorsal horn from a WDR neuron with a receptive field located on the ipsilateral hind paw. The neuron was activated either by a sustained pinch of its receptive field or by regular (once a minute) microelectrophoretic applications of the excitatory amino acid, glutamate (GLU, 20 nA, horizontal bars). Conditioning stimuli were pinch of the muzzle, the contralateral hind paw or the tail, and immersion of the tail in a 52 °C water bath (from top to bottom, respectively). Note that all these conditioning stimuli virtually blocked both the pinch-evoked and the DLH-induced firing. In the latter case, the inhibitions remained for several minutes in most cases, suggesting that WDR neurons were hyperpolarized for a long time following the conditioning stimulation. (B) Example of heterotopic activation of Aδ- and C-fibers triggering inhibitions in a spinal WDR neuron. Left: Schematic representation of the experimental design. Recordings were made in the lumbar dorsal horn from a WDR neuron with a receptive field located on the ipsilateral hind paw. The continuous microelectrophoretic application of the excitatory amino acid, DL-homocysteic acid (DLH) induced a steady discharge from the neuron under study. The repetitive application of individual percutaneous electrical stimuli of adequate intensities to the contralateral muzzle (a), the base (b), or the tip (c) of the tail induced biphasic depressions of the neuronal activity. Right: Peristimulus histograms (bin width 5 ms) prepared during the continuous microelectrophoretic application (15 nA) of DLH onto the membrane of the neuron. The broken white lines show the timing of percutaneous electrical stimulation (10 mA; 2 ms duration; 0.66 Hz; 200 ms delay; 100 sweeps). The broken black line represents the mean firing calculated during the prestimulation control period (−200 to 0 ms). Two waves of inhibition can be seen. They occurred earlier when the base of the tail (b) was stimulated instead of the tip (c). The time gaps are shown as yellow areas between the histograms, for both inhibitory components. The gap was 7 and 25 ms for the beginning and the end of the first component; it was 150 and 290 ms for the beginning and the end of the second component. Knowing that the distance between b and c was 100 mm, one can easily calculate the conduction velocities of fibers that elicited the first and second components: 4–14 m s^{-1} and 0.3–0.7 m s^{-1} respectively. These fibers therefore belong to the Aδ- and C- groups, respectively. (A) Adapted from Villanueva, L., Cadden, S. W., and Le Bars, D. 1984. Evidence that diffuse noxious inhibitory controls (DNIC) are mediated by a final post-synaptic inhibitory mechanism. Brain Res. 298, 67–74.

(see Chapter 5.49). Surprisingly, DNICs were not modified by lesions of the following structures: the periaqueductal gray (PAG), cuneiform nucleus, parabrachial area, locus coeruleus/subcoeruleus, rostral ventromedial medulla (RVM). By contrast, lesions of subnucleus reticularis dorsalis (SRD) in the caudal medulla strongly reduced DNICs. The SRD is located ventral to the cuneate nucleus, between trigeminal nucleus caudalis and the nucleus of the solitary tract and contains neurons with characteristics that suggest they have a key role in processing specifically nociceptive information (Villanueva, L. *et al.*, 1996). Indeed, they are preferentially or exclusively activated by nociceptive stimuli from a whole-body receptive field; they encode the intensity of cutaneous and visceral stimulation within noxious ranges; and they are activated exclusively by activity in $A\delta$- or $A\delta$- and C-peripheral fibers. In addition, they send descending projections through the dorsolateral funiculus that terminate in the dorsal horn at all levels of the spinal cord. The fact that the supraspinal loop sustaining DNICs is confined to the most caudal part of the medulla was confirmed in a series of experiments where the potency of DNIC was tested in animals with complete transections at different levels of the brainstem.

50.6 Summary and Conclusions

There is a body of evidence, based on both human and animal experiments, which strongly suggests that the phenomenon of 'pain inhibiting pain' is sustained by a well-defined neurological substratum based on a spino-reticulo-spinal loop. DNICs are in fact a very special and generalized case of lateral inhibition and it is very likely that they subserve a related function. Although lateral inhibition is sometimes thought of in terms of creating illusions, such phenomena play an important role in other senses. It is very probable that DNICs, as revealed by the empiric observation that pain inhibits pain, also play an important role in pain processing, probably by focusing the pain network onto a particular part of the body. The DNIC system could be understood as a filter allowing the extraction by the brain of a clear signal of pain from a basic somesthetic activity provided by dorsal horn WDR neurons (see Chapter 5.25). If these statements are correct, then defocusing should alleviate the unpleasantness of pain. Interestingly, the reference analgesic drug, morphine, blocks DNICs in both man and animals (Le Bars, D. *et al.*, 1992).

Many sources of descending inhibition from the brain that modulate the spinal transmission of nociceptive information have been described in animals. To date, the only descending inhibitory mechanisms that have been described in man are DNICs.

References

Bouhassira, D., Chollet, R., Coffin, B., Lemann, M., Le Bars, D., Willer, J. C., and Jian, R. 1994. Inhibition of a somatic nociceptive reflex by gastric distention in humans. Gastroenterology 107, 985–992.

Bouhassira, D., Danziger, N., Guirimand, F., and Attal, N. 2003. Comparison of the pain suppressive effects of clinical and experimental painful conditioning stimuli. Brain. 126, 1068–1078.

Bouhassira, D, Le Bars, D., Bolgert, F., Laplane, D., and Willer, J. C. 1993. Diffuse noxious inhibitory controls (DNIC) in man. A neurophysiological investigation of a patient with a form of Brown-Séquard syndrome. Ann. Neurol. 34, 536–543.

Bouhassira, D., Sabate, J. M., Coffin, B., Le Bars, D., Willer, J. C., and Jian, R. 1998. Effects of rectal distensions on nociceptive flexion reflexes in humans. Am. J. Physiol. 275, G410–G417.

Bushnell, M. C., Duncan, G. H., Dubner, R., Jones, R. L., and Maixner, W. 1985. Attentional influences on noxious and innocuous cutaneous heat detection in humans and monkeys. J. Neurosci. 5, 1103–1110.

Cadden, S. W. 1993. The ability of inhibitory controls to 'switch-off' activity in dorsal horn convergent neurones in the rat. Brain Res. 628, 65–71.

Cadden, S. W. and Morrison, J. F. B. 1991. Effects of visceral distension on the activities of neurones receiving cutaneous inputs in the rat lumbar dorsal horn; comparison with effects of remote noxious somatic stimuli. Brain Res. 558, 63–74.

Danziger, N., Gautron, M., Le Bars, D., and Bouhassira, D. 2001. Activation of diffuse noxious inhibitory controls (DNIC) in rats with an experimental peripheral mononeuropathy. Pain 91, 287–296.

Danziger, N., Rozenberg, S., Bourgeois, P., Charpentier, G., and Willer, J. C. 1998. Depressive effects of segmental and heterotopic application of transcutaneous electrical nerve stimulation and piezo-electric current on lower limb nociceptive flexion reflex in human subjects. Arch. Phys. Med. Rehabil. 79, 191–200.

Danziger, N., Le Bars, D., and Bouhassira, D. 2000. Diffuse Noxious Inhibitory Controls and Arthritis in the Rat. In: Pain and Neuroimmune Interactions (eds. N. E. Saadé, A. V. Apkarian, and S. J. Jabbur), pp. 69–78. Kluwer Academic Press.

De Broucker, T., Cesaro, P., Willer, J. C., and Le Bars, D. 1990. Diffuse noxious inhibitory controls (DNIC) in man: involvement of the spino-reticular tract. Brain 113, 1223–1234.

Ellrich, J. and Treede, R. D. 1998. Characterization of blink reflex interneurons by activation of diffuse noxious inhibitory controls in man. Brain Res. 803, 161–168.

France, C. R. and Suchowiecki, S. 1999. A comparison of diffuse noxious inhibitory controls in men and women. Pain 81, 77–84.

Jinks, S. L., Antognini, J. F., and Carstens, E. 2003. Isoflurane depresses diffuse noxious inhibitory controls in rats between

0.8 and 1.2 minimum alveolar anesthetic concentration. Anesth. Analg. 97, 111–116.

Le Bars, D. 2002. The whole body receptive field of dorsal horn multireceptive neurones. Brain Res. Rev. 40, 29–44.

Le Bars, D. and Willer, J. C. 2002. Pain Modulation Triggered by High-Intensity Stimulation: Implication for Acupuncture Analgesia. In: Acupuncture: Is there a Physiological Basis? (eds. P. Li, A. Sato, and J. Campbell), International Congress Serie 1238, pp. 11–29. Elsevier.

Le Bars, D., De Broucker, T., and Willer, J. C. 1992. Morphine blocks pain inhibitory controls in humans. Pain 48, 13–20.

Le Bars, D., Gozariu, M., and Cadden, S. W. 2001. Animal models of nociception. Pharmacol. Rev. 53, 597–652.

Maillou, P. and Cadden, S. W. 1997. Effects of remote deep somatic noxious stimuli on a jaw reflex in man. Arch. Oral. Biol. 42, 323–327.

Melzack, R. 1989. Folk Medicine and the Sensory Modulation of Pain. In: Textbook of Pain (eds. P. D. Wall and R. Melzack), pp. 897–905. Churchill Livingstone.

Pantaleo, T., Duranti, R., and Bellini, F. 1988. Effects of heterotopic ischemic pain on muscular pain threshold and blink reflex in humans. Neurosci. Lett. 85, 56–60.

Roby-Brami, A, Bussel, B., Willer, J. C., and Le Bars, D. 1987. An electrophysiological investigation into the pain-relieving effects of heterotopic nociceptive stimuli: probable involvement of a supraspinal loop. Brain 110, 1497–1508.

Sandrini, G., Milanov, I., Malaguti, S., Nigrelli, M. P., Moglia, A., and Nappi, G. 2000. Effects of hypnosis on diffuse noxious inhibitory controls. Physiol. Behav. 69, 295–300.

Villanueva, L., Bouhassira, D., and Le Bars, D. 1996. The medullary subnucleus reticularis dorsalis (SRD) as a key link in both the transmission and modulation of pain signals. Pain 67, 231–240.

Villanueva, L., Cadden, S. W., and Le Bars, D. 1984. Evidence that diffuse noxious inhibitory controls (DNIC) are mediated by a final post-synaptic inhibitory mechanism. Brain Res. 298, 67–74.

Willer, J. C. 1977. Comparative study of perceived pain and nociceptive flexion reflex in man. Pain 3, 69–80.

Willer, J. C., Boureau, F., and Albe-Fessard, D. 1979. Supraspinal influences on nociceptive flexion reflex and pain sensation in man. Brain Res. 179, 61–68.

Willer, J. C., De Broucker, T., and Le Bars, D. 1989. Encoding of nociceptive thermal stimuli by diffuse noxious inhibitory controls in humans. J. Neurophysiol. 62, 1028–1038.

Willer, J. C., Roby, A., and Le Bars, D. 1984. Psychophysical and electrophysiological approaches to the pain relieving effect of heterotopic nociceptive stimuli. Brain 107, 1095–1112.

Willer, J. C., Barranquero, A., Kahn, M. F., and Salliere, D. 1987. Pain in sciatica depresses lower limb cutaneous reflexes to sural nerve stimulation. J. Neurol. Neurosurg. Psych. 50, 1–5.

51 Fibromyalgia

R Staud, University of Florida, Gainesville, FL, USA

Glossary

chronic pain Pain for more than 3 months.

central sensitization Increased sensitivity to painful stimuli due to functional changes of the central nervous system (neuroplasticity).

peripheral sensitization Increased sensitivity to painful stimuli due to decreased thresholds of primary nociceptive afferents.

tender point (TP) Nine paired body areas at tendon insertion points of the neck, chest, buttocks, as well as all upper and lower extremities as defined by the 1990 American College of Rheumatology.

TPs represent areas of mechanical allodynia (pain threshold $\leq 4\,kg\,cm^{-2}$). Eleven or more TPs as well as chronic widespread pain are required to fulfill fibromyalgia criteria.

trigger point (TrP) Painful muscle areas that show increased muscle tension (taut bands), a twitch response to palpation, and referred pain.

wind-up Central pain mechanisms that depend on N-methyl-D-aspartate and substance P receptor mechanisms.

51.1 Introduction

Chronic nonmalignant pain is very frequent in the general population (Croft, P. *et al.*, 1993). Fibromyalgia (FM) represents the combination of chronic widespread pain with mechanical allodynia and disproportionably affects women (ratio women:men 9:1).

Like many other syndromes, FM has no single specific feature but is a symptom complex of self-reported or elicited findings. In 1990, the American College of Rheumatology (ACR) published diagnostic criteria for FM which include chronic widespread pain (>3 months) and mechanical tenderness in at least 11 out of 18 tender points (TP; Wolfe, F. *et al.*, 1990). Most

TP sites are located at tendon insertion areas and have shown few detectable tissue abnormalities. Besides musculoskeletal pain and mechanical tenderness most FM patients also complain of insomnia, fatigue, and distress. In addition FM coaggregates with major mood disorders (Raphael, K. G. *et al.*, 2004).

51.2   Pathogenesis of Fibromyalgia

FM can be characterized as a pain amplification syndrome of patients who often display widespread hyperalgesia and allodynia to mechanical, thermal, and chemical stimuli. However, the hypersensitivity of FM patients is not limited to pain, but also includes light, sound, and smell. The cause for this heightened sensitivity is unknown, but central nervous system (CNS) sensory processing abnormalities have been reported in several studies (Desmeules, J. A. *et al.*, 2003; Staud, R., 2004). Most of these studies found evidence of central sensitization of dorsal horn neurons of the spinal cord and the brain including those that are related to pain. The pathogenesis of central sensitization in FM is unclear but may be related to prior stressors, including infections and traumas. Because most FM patients relate their pain to deep tissues, particularly muscles, these structures may play an important role in the initiation and maintenance of central sensitization and pain. No consistent tissue abnormalities have been reported in FM patients that would explain persistent pain. However, focal muscle areas, like trigger points (TrPs), may play an important role for FM pain because little nociceptive input from peripheral tissues is required for the maintenance of central sensitization (Bengtsson, M. *et al.*, 1989).

51.3   Muscle Nociception

Nociceptive input from muscle may be particularly relevant for FM pain. Muscle pain travels in myelinated A-delta fibers (group III) and in unmyelinated type C fibers (group IV). Pain receptors are mostly located at free nerve endings which are concentrated around small arterioles and capillaries between the muscle fibers. The cell bodies for these nerves are found in the dorsal root ganglion and synapse in lamina I and IV–V of the dorsal horn of the spinal cord. Activation of peripheral nociceptors leads to a release of neurotransmitters in the spinal cord,

mostly substance P and calcitonin gene-related peptide (CGRP). There is also a retrograde migration of substance P from cell bodies which is then released in the region of the free nerve endings. This retrograde flow can sensitize tissues and increase their responsiveness to less intense stimuli. Muscle nociceptors are activated by mechanical stimuli (stretching or pressure), bradykinin, serotonin, and potassium ions, but are not activated by normal muscle movements or even increased muscle tension. Sensory input from muscle, as opposed to skin, is a much more potent effector of central sensitization and this may be particularly relevant for FM (Wall, P. D. and Woolf, C. J., 1984).

51.4   Triggering Events for Fibromyalgia

The onset of FM has frequently been associated with certain triggers or stressors include physical trauma, infections, emotional distress, endocrine disorders, and immune activation (Greenfield, S. *et al.*, 1992; Middleton, G. D. *et al.*, 1994; Waylonis, G. W. and Perkins, R. H., 1994). These stressors have been associated with the degree of pain, disability, life interference, and affective distress of FM patients (Turk, D. C. *et al.*, 1996). Strong evidence for trauma as a trigger of FM symptoms comes from studies of patients with acute injuries who developed chronic widespread pain at much higher rates than controls (Al Allaf, A. W. *et al.*, 2002; Buskila, D. and Neumann, L., 2002). Additional evidence for such an association include postinjury reported sleep abnormalities (Berglund, A. *et al.*, 2001), local injury sites as a source of chronic distant regional pain (Arendt-Nielsen, L. and Svensson, P., 2001), and recent evidence of injury-related CNS neuroplasticity in FM (Carli, G., 2000). Further prospective studies, however, are needed to confirm these associations and to identify the mechanisms that are relevant for posttraumatic FM pain (White, K. P. *et al.*, 2000).

51.5   Abnormal Response to Stressors

The biological response to stressors appears predictable in animals and humans. Particularly, events that are perceived as inescapable or unavoidable, or which appear unpredictable, evoke the strongest adverse biological responses (Gold, P. W. *et al.*, 1988a;

1988b; Chrousos, G. P. and Gold, P. W., 1992; Viau, V. *et al.*, 1993; Sapolsky, R. M. *et al.*, 2000). This may explain why victims of trauma appear to develop much higher rates of FM than persons with self-inflicted injuries (Greenfield, S. *et al.*, 1992). In addition, early life stressors can have a permanent and profound impact on subsequent biological responses to stress in animals and humans. Studies in rodents have demonstrated that exposure to multiple stressors including trauma or separation in the neonatal period can lead to permanent changes in the biological response to stress (Sapolsky, R. M., 1996; McNamara, R. K. *et al.*, 2002). This permanent effect of early stressors could explain the higher than expected incidence of traumatic childhood events in individuals who later develop chronic pain (Anderberg, U. M. *et al.*, 2000; Bailey, B. E. *et al.*, 2003).

51.6 Posttraumatic Stress Disorder and Fibromyalgia

Posttraumatic stress disorder (PTSD) often occurs after a significant traumatic event and is characterized by behavioral, emotional, functional, and physiological symptoms. Relevant traumatic events related to PTSD are usually perceived as threatening one's life or physical integrity and can lead to emotional responses including horror, helplessness, or intense fear. The psychological symptoms of PTSD include re-experience of the traumatic event, avoidance, and increased arousal. It has been shown that the experience of trauma is associated with increased somatic and physical complaints, including pain (Beckham, J. C. *et al.*, 1997; 1998). More than 50% of patients with FM suffer from PTSD in the USA and Israel (Sherman, J. J. *et al.*, 2000; Cohen, H. *et al.*, 2002). Compared to the prevalence of PTSD in the general population (6%), FM patients show greatly increased rates similar to Vietnam veterans and victims of natural disasters (Shore, J. H. *et al.*, 1986; Green, M. M. *et al.*, 1993). Not surprisingly, the incidence of FM is also increased in patients with PTSD (21%) and often associated with increased pain ratings, more distress, and higher functional impairment (Amir, M. *et al.*, 1997). As with several other disorders, however, it is unclear whether PTSD is the cause or consequence for FM.

51.7 Fibromyalgia and Affective Spectrum Disorders

Several studies have reported that FM is comorbid with major depressive disorder (MDD; Arnold, L. M. *et al.*, 2000; White, K. P. *et al.*, 2002). In addition, anxiety is very prevalent in patients with FM (Thieme, K. *et al.*, 2004). The outcome of a recent large family study of FM probands was consistent with the hypothesis that FM and MDD are characterized by shared, genetically mediated risk factors (Raphael, K. G. *et al.*, 2004). Although the findings of this study should not be interpreted that MDD and FM represent different forms of the same syndrome, they strongly suggest that FM and MDD share important CNS mechanisms.

51.8 Central Sensitization in Fibromyalgia

Whereas peripheral sensitization is related to changes of primary nociceptive afferent properties, central sensitization requires functional changes in the CNS (neuroplasticity). Behaviorally, centrally sensitized patients, like FM sufferers report abnormal or heightened pain sensitivity with spreading of hypersensitivity to uninjured sites and the generation of pain by low threshold mechanoreceptors that are normally silent in pain processing (Torebjork, H. E. *et al.*, 1992). Thus, tissue injury may not only cause pain but also an expansion of dorsal horn receptive fields and central sensitization.

51.9 Temporal Summation of Second Pain (Wind-Up)

In 1965, Mendell and Wall described for the first time, that repetitive C-fiber stimulation can result in a progressive increase of electrical discharges from second-order neurons in the spinal cord (Mendell, L. M. and Wall, P. D., 1965). This important mechanism of pain amplification in the dorsal horn neurons of the spinal cord is related to temporal summation of second pain or wind-up (WU) and has been used in FM patients for the evaluation of central sensitization (Staud, R. *et al.*, 2001). This technique reveals sensitivity to input from unmyelinated (C) afferents and the status of the N-methyl-D-aspartate (NMDA) receptor systems that are implicated in a variety of chronic pain conditions. Temporal summation depends upon activation of NMDA transmitter systems by C nociceptors, and chronic central pain states like FM can result from excessive temporal summation of pain (Staud, R. and Smitherman, M. L., 2002).

51.10 Abnormal Wind-Up of Fibromyalgia Patients

Several recent studies have demonstrated psychophysical evidence that input to central nociceptive pathways is abnormally processed in patients with FM (Staud, R. and Domingo, M., 2001a; 2001b; Staud, R. et al., 2001; Vierck, C. J. et al., 2001; Price, D. D. et al., 2002; Staud, R., 2002). When WU pain is evoked both in normal controls and FM patients, the perceived magnitude of the experimental stimuli (heat, cold, pressure, electricity) is greater for FM patients compared to controls, as is the amount of temporal summation within a series of stimuli. Following a series of stimuli, after-sensations are greater in magnitude, last longer, and are more frequently painful in FM subjects. These results indicate both augmentation and prolonged decay of nociceptive input in FM patients and provide convincing evidence for the presence of central sensitization.

51.11 Wind-Up Measures as Predictors of Clinical Pain Intensity in Fibromyalgia Patients

The important role of central pain mechanisms, like WU, for FM patients is also supported by their ability to predict clinical pain intensity. Thermal WU ratings correlate with clinical pain intensity (Pearson's $r = 0.529$), thus emphasizing the important role of these pain mechanisms for FM. In addition, statistical prediction models that includes TP count, pain-related negative affect, and WU ratings have been shown to account for nearly 50% of the variance in FM clinical pain intensity (Staud, R. et al., 2003; 2004).

51.12 Fibromyalgia and Other Chronic Pain Conditions

Many systemic illnesses can also present with diffuse pain similar to FM, including polymyalgia rheumatica (Gowin, K. M., 2000), rheumatoid arthritis,

Sjogren's syndrome, inflammatory myopathies (Sultan, S. M. et al., 2002), systemic lupus erythematosus (Tench, C. et al., 2002), multiple sclerosis, and joint hypermobility syndrome (Nef, W. and Gerber, N. J., 1998). Furthermore, several infectious diseases including hepatitis C, Lyme disease, coxsackie B infection, HIV, and parvovirus infection, frequently result in chronic pain states (Barkhuizen, A., 2002). Although the majority of FM patients report the insidious onset of pain and fatigue, approximately half of all patients describe the start of chronic pain after a traumatic or infectious event.

51.13 Myofascial Pain Syndrome

Myofascial pain or regional musculoskeletal pain is one of the most common pain syndromes encountered in clinical practice. Myofascial pain represents the most common cause of chronic pain, including neck and shoulder pain, tension headaches, and lower back pain (Granges, G. and Littlejohn, G., 1993; Fricton, J. R., 1994; Macfarlane, G. J. et al., 1996; Borg-Stein, J. and Simons, D. G., 2002). The term myofascial pain was introduced in the early 1950s by Janet Travell. She also defined the term myofascial TrP and demonstrated with David Simons that individual muscles have specific nondermatomal patterns of TrP pain referral (Simons, D. G. et al., 1999). In 1983, both authors first described the clinical picture and pathophysiology of a new syndrome which they named myofascial pain syndrome (MPS; Long, S. P. and Kephart, W., 1998; Simons, D. G. et al., 1999). MPS has been defined as a chronic pain syndrome accompanied by TrPs in one or more muscles or groups of muscles. Similar to FM, it is more frequently found in women compared to men. Besides the presence of TrPs and referred pain, MPS is frequently associated with limitation of movement, weakness, and autonomic dysfunction (Long, S. P. and Kephart, W., 1998) similar to FM.

51.14 Trigger Points

TrPs represent areas of local mechanical hyperalgesia that can be found in MPS and several chronic pain conditions, including FM, osteoarthritis, and rheumatoid arthritis. They are defined as specific areas of hyperirritability in muscle, but can also be detected in ligaments, tendons, periosteum, scar tissue, or skin (Han, S. C. and Harrison, P., 1997; Simons, D. G. et al.,

1999). TrPs are located in palpable taut bands and produce local and referred pain, which is specific for the particular muscle. When TrPs are mechanically stimulated, so-called taut bands within a muscle, rather than the entire muscle, will contract (Chu, J., 1998). They are often associated with a local muscle twitch response, which can easily be elicited by needling or palpation of the TrP (Chu, J., 1998; 1999). Latent TrPs are similar to active TrPs, but they are not associated with spontaneous pain and no referral of pain occurs. However, latent TrPs are painful when palpated.

51.15 Relationship between Myofascial Pain and Fibromyalgia

Approximately 70% of patients with FM have TrPs (Granges, G. and Littlejohn, G., 1993). A TP is considered to be different from a TrP because of the absence of referred pain, local twitch response, and a taut band in the muscle. The distinction between TPs and TrPs requires careful physical examination. TrPs are frequently located in areas of muscular TPs of patients with FM (Wolfe, F. *et al.*, 1992; Borg-Stein, J. and Stein, J., 1996) suggesting that some muscular TPs in patients with FM may actually be TrPs (Inanici, F. *et al.*, 1999). The presence of TrPs in most if not all FM patients represents evidence for local muscle abnormalities in this chronic musculoskeletal pain syndrome. Although it is unclear if TrPs are the cause or effect of muscle injury, they represent abnormally contracted muscle fibers. This muscle contraction can lead to accumulation of histamine, serotonin, tachykinins, and prostaglandins, which may result in the activation of local nociceptors. Prolonged muscle contractions may also result in local hypoxemia and energy depletion (Simons, D. G. *et al.*, 1999).

51.16 Inflammatory Connective Tissue Diseases

Many patients (up to 25%) with chronic arthritis have also wide-spread pain similar to FM. These patients may also present with symptoms of chronic fatigue, impaired memory and concentration, and mood abnormalities. Most of these patients, however, will have findings suggestive of inflammation, including joint pain/swelling, rashes, muscle weakness, as well as laboratory abnormalities, like elevated erythrocyte sedimentation rate (ESR), C-reactive protein (CRP), anemia, autoantibodies (rheumatoid factor, cyclic citrullinated peptide antibodies, anti-nuclear antibodies (ANA)), etc.

51.17 Treatment of Fibromyalgia

Treatment of patients with chronic widespread pain like FM needs to be individually tailored to each patient's needs. This includes the assessment of biopsychosocial abnormalities which are readily detectable in most FM patients. Importantly the identification of pain generators is essential for an effective treatment plan. Thus patients with arthritis, particularly osteoarthritis of the spine will benefit from muscle relaxants, physical therapy, and massage. In addition, these patients may respond well to therapy with cyclo-oxygenase (COX) inhibitors. Identification and treatment of mood abnormalities is crucial because affective spectrum disorders seem to share important mechanisms with FM.

Pharmacotherapy for FM has been most successful with antidepressant, muscle relaxant, or anticonvulsant drugs. These drugs affect the release of various neurochemicals (e.g., serotonin, norepinephrine, substance P) that have a broad range of activities in the brain and spinal cord, including modulation of pain sensation and tolerance. None of these drugs, however, is currently approved by the US Food and Drug Administration for the treatment of FM.

Most FM patients will respond to low-dose tricyclic medications, such as amitriptyline and cyclobenzaprine, as well as cardiovascular exercise, cognitive behavioral therapy (CBT), patient education, or a combination of these for the management of FM. Also tramadol, selective serotonin reuptake inhibitors (SSRI), selective norepinephrine re-uptake inhibitors (SNRI), and anticonvulsants have been found to be moderately effective. There is some evidence for the efficacy of strength training exercise, acupuncture, hypnotherapy, biofeedback, massage, and warm water baths. However, many commonly used FM therapies like guaifenesine have been found to be ineffective.

The efficacy of FM patients to manage their pain seems to correlate with their functional status. Brain imaging and psychological profiles have identified at least three FM subgroups, that is, patients who are highly dysfunctional, interpersonally distressed, or are effective copers (Giesecke, T. *et al.*, 2003). Such studies provide an explanation why some treatments seem to be differentially effective in individual patients. Thus optimal FM management will require a combination of pharmacological and nonpharmacological therapies with patients and healthcare professionals working as a team.

51.18 Prognosis of Fibromyalgia

FM can be mild or disabling, but often has substantial emotional and social consequences. About 50% of all patients have difficulty with or are unable to perform routine daily activities. Estimates of patients who have had to stop working or change jobs range from 30% to 40%. Patients with FM suffer job losses and social abandonment more often than people with other conditions that cause pain and fatigue. Although FM symptoms seem to remain stable, several long-term studies indicate that physical function and pain worsen over time (Forseth, K. O. *et al.*, 1999; Baumgartner, E., 2002). Significant life stressors often result in a poor outcome, including diminished capacity to work, poor self-efficacy, increased pain sensations, disturbed sleep, fatigue, and depression. A recent study reported increased mortality in patients with widespread pain compared to those without chronic pain (McBeth, J. *et al.*, 2003). Although this study did not specifically evaluate FM, these findings may be relevant for FM patients. The higher mortality was mostly associated with malignancies, although the cause for this association is currently unknown.

51.19 Conclusions

The pathogenesis of FM pain is currently unknown, but up to 50% of the variance can be explained by central sensitization and distress. However, additional mechanisms like tonic peripheral nociceptive input may play an important role for the initiation and maintenance of the increased pain sensitivity of FM patients. FM, like many other chronic pain syndromes, is treatable, and remission can occur in many patients who actively participate in effective disease management programs. New treatment strategies, however, may benefit from interventions that improve peripheral and central pain sensitization as well as the affective disorders of FM patients.

References

Al Allaf, A. W., Dunbar, K. L., Hallum, N. S., Nosratzadeh, B., Templeton, K. D., and Pullar, T. 2002. A case-control study examining the role of physical trauma in the onset of fibromyalgia syndrome. Rheumatology 41, 450–453.

Amir, M., Kaplan, Z., Neumann, L., Sharabani, R., Shani, N., and Buskila, D. 1997. Posttraumatic stress disorder, tenderness and fibromyalgia. J. Psychosom. Res. 42, 607–613.

Anderberg, U. M., Marteinsdottir, I., Theorell, T., and von Knorring, L. 2000. The impact of life events in female patients with fibromyalgia and in female healthy controls. Eur. Psychiatry 15, 295–301.

Arendt-Nielsen, L. and Svensson, P. 2001. Referred muscle pain: basic and clinical findings. Clin. J. Pain 17, 11–19.

Arnold, L. M., Keck, P. E., Jr., and Welge, J. A. 2000. Antidepressant treatment of fibromyalgia: a meta-analysis and review. Psychosomatics 41, 104–113.

Bailey, B. E., Freedenfeld, R. N., Kiser, R. S., and Gatchel, R. J. 2003. Lifetime physical and sexual abuse in chronic pain patients: psychosocial correlates and treatment outcomes. Disabil. Rehabil. 25, 331–342.

Barkhuizen, A. 2002. Rational and targeted pharmacologic treatment of fibromyalgia. Rheum. Dis. Clin. North Am. 28, 261–290.

Baumgartner, E., Finckh, A., Cedraschi, C., and Vischer, T. L. 2002. A six year prospective study of a cohort of patients with fibromyalgia. Ann. Rheum. Dis. 61, 644–645.

Beckham, J. C., Crawford, A. L., Feldman, M. E., Kirby, A. C., Hertzberg, M. A., Davidson, J. R., and Moore, S. D. 1997. Chronic posttraumatic stress disorder and chronic pain in Vietnam combat veterans. J. Psychosom. Res. 43, 379–389.

Beckham, J. C., Moore, S. D., Feldman, M. E., Hertzberg, M. A., Kirby, A. C., and Fairbank, J. A. 1998. Health status, somatization, and severity of posttraumatic stress disorder in Vietnam combat veterans with posttraumatic stress disorder. Am. J. Psychiatry 155, 1565–1569.

Bengtsson, M., Bengtsson, A., and Jorfeldt, L. 1989. Diagnostic epidural opioid blockade in primary fibromyalgia at rest and during exercise. Pain 39, 171–180.

Berglund, A., Alfredsson, L., Jensen, I., Cassidy, J. D., and Nygren, A. 2001. The association between exposure to a rear-end collision and future health complaints. J. Clin. Epidemiol. 54, 851–856.

Borg-Stein, J. and Simons, D. G. 2002. Myofascial pain. Arch. Phys. Med. Rehabil. 83, S40–S47.

Borg-Stein, J. and Stein, J. 1996. Trigger points and tender points: one and the same? Does injection treatment help? Rheum. Dis. Clin. North Am. 22, 305–322.

Buskila, D. and Neumann, L. 2002. The development of widespread pain after injuries. J. Musculoskel. Pain 10, 261–267.

Carli, G. 2000. Neuroplasticity and clinical pain. Prog. Brain Res. 129, 325–330.

Chrousos, G. P. and Gold, P. W. 1992. The concepts of stress and stress system disorders. Overview of physical and behavioral homeostasis. JAMA 267, 1244–1252.

Chu, J. 1998. Twitch response in myofascial trigger points. J. Musculoskel. Pain 6, 99–116.

Chu, J. 1999. Twitch-obtaining intramuscular stimulation: observations in the management of radiculopathic chronic low back pain. J. Musculoskel. Pain 7, 131–146.

Cohen, H., Neumann, L., Haiman, Y., Matar, M. A., Press, J., and Buskila, D. 2002. Prevalence of post-traumatic stress disorder in fibromyalgia patients: overlapping syndromes or post-traumatic fibromyalgia syndrome? Semin. Arthritis Rheum. 32, 38–50.

Croft, P., Rigby, A. S., Boswell, R., Schollum, J., and Silman, A. 1993. The prevalence of chronic widespread pain in the general population. J. Rheumatol. 20, 710–713.

Desmeules, J. A., Cedraschi, C., Rapiti, E., Baumgartner, E., Finckh, A., Cohen, P., Dayer, P., and Vischer, T. L. 2003. Neurophysiologic evidence for a central sensitization in patients with fibromyalgia. Arthritis Rheum. 48, 1420–1429.

Forseth, K. O., Forre, O., and Gran, J. T. 1999. A 5.5 year prospective study of self-reported musculoskeletal pain and of fibromyalgia in a female population: significance and natural history. Clin. Rheumatol. 18, 114–121.

Fricton, J. R. 1994. Myofascial pain. Clin. Rheumatol. 8, 857–880.

Giesecke, T., Williams, D. A., Harris, R. E., Cupps, T. R., Tian, X., Tian, T. X., Gracely, R. H., and Clauw, D. J. 2003. Subgrouping of fibromyalgia patients on the basis of pressure-pain thresholds and psychological factors. Arthritis Rheum. 48, 2916–2922.

Gold, P. W., Goodwin, F. K., and Chrousos, G. P. 1988a. Clinical and biochemical manifestations of depression. Relation to the neurobiology of stress (1). N. Engl. J. Med. 319, 348–353.

Gold, P. W., Goodwin, F. K., and Chrousos, G. P. 1988b. Clinical and biochemical manifestations of depression. Relation to the neurobiology of stress (2). N. Engl. J. Med. 319, 413–420.

Gowin, K. M. 2000. Diffuse pain syndromes in the elderly. Rheum. Dis. Clin. North Am. 26, 673–682.

Granges, G. and Littlejohn, G. 1993. Prevalence of myofascial pain syndrome in fibromyalgia syndrome and regional pain syndrome: a comparative study. J. Musculoskel. Pain 1, 19–35.

Green, M. M., McFarlane, A. C., Hunter, C. E., and Griggs, W. M. 1993. Undiagnosed post-traumatic stress disorder following motor vehicle accidents. Med. J. Aust. 159, 529–534.

Greenfield, S., Fitzcharles, M. A., and Esdaile, J. M. 1992. Reactive fibromyalgia syndrome. Arthritis Rheum. 35, 678–681.

Han, S. C. and Harrison, P. 1997. Myofascial pain syndrome and trigger-point management. Reg. Anesth. 22, 89–101.

Inanici, F., Yunus, M. B., and Aldag, J. C. 1999. Clinical features and psychologic factors in regional soft tissue pain: comparison with fibromyalgia syndrome. J. Musculoskelet. Pain 7, 293–301.

Long, S. P. and Kephart, W. 1998. Myofascial Pain Syndrome. In: The Management of Pain (eds. M. A. Ashburn and L. J. Rice), pp. 299–321. Churchill Livingstone.

Macfarlane, G. J., Thomas, E., Papageorgiou, A. C., Schollum, J., Croft, P. R., and Silman, A. J. 1996. The natural history of chronic pain in the community: a better prognosis than in the clinic? J. Rheumatol. 23, 1617–1620.

McBeth, J., Silman, A. J., and Macfarlane, G. J. 2003. Association of widespread body pain with an increased risk of cancer and reduced cancer survival: a prospective, population-based study. Arthritis Rheum. 48, 1686–1692.

McNamara, R. K., Huot, R. L., Lenox, R. H., and Plotsky, P. M. 2002. Postnatal maternal separation elevates the expression of the postsynaptic protein kinase c substrate rc3, but not presynaptic GAP-43, in the developing rat hippocampus. Dev. Neurosci. 24, 485–494.

Mendell, L. M. and Wall, P. D. 1965. Responses of single dorsal cord cells to peripheral cutaneous unmyelinated fibres. Nature 206, 97–99.

Middleton, G. D., McFarlin, J. E., and Lipsky, P. E. 1994. The prevalence and clinical impact of fibromyalgia in systemic lupus erythematosus. Arthritis Rheum. 37, 1181–1188.

Nef, W. and Gerber, N. J. 1998. Hypermobility syndrome. When too much activity causes pain. Schweiz. Med. Wochenschr. 128, 302–310.

Price, D. D., Staud, R., Robinson, M. E., Mauderli, A. P., Cannon, R. L., and Vierck, C. J. 2002. Enhanced temporal summation of second pain and its central modulation in fibromyalgia patients. Pain 99, 49–59.

Raphael, K. G., Janal, M. N., Nayak, S., Schwartz, J. E., and Gallagher, R. M. 2004. Familial aggregation of depression in fibromyalgia: a community-based test of alternate hypotheses. Pain 110, 449–460.

Sapolsky, R. M. 1996. Why stress is bad for your brain. Science 273, 749–750.

Sapolsky, R. M., Romero, L. M., and Munck, A. U. 2000. How do glucocorticoids influence stress responses? Integrating permissive, suppressive, stimulatory, and preparative actions. Endocr. Rev. 21, 55–89.

Sherman, J. J., Turk, D. C., and Okifuji, A. 2000. Prevalence and impact of posttraumatic stress disorder-like symptoms on patients with fibromyalgia syndrome. Clin. J. Pain 16, 127–134.

Shore, J. H., Tatum, E. L., and Vollmer, W. M. 1986. Psychiatric reactions to disaster: the Mount St. Helens experience. Am. J. Psychiatry 143, 590–595.

Simons, D. G., Travell, J. G., Simons, L. S., and Cummings, B. D. 1999. Myofascial Pain and Dysfunction: The Trigger Point Manual. Upper Half of the Body. 2nd edn. Lippincott Williams & Wilkins.

Staud, R. 2002. Evidence of involvement of central neural mechanisms in generating fibromyalgia pain. Curr. Rheumatol. Rep. 4, 299–305.

Staud, R. 2004. New evidence for central sensitization in patients with fibromyalgia. Curr. Rheumatol. Rep. 6, 259.

Staud, R. and Domingo, M. 2001a. Evidence for abnormal pain processing in fibromyalgia syndrome. Pain Med. 2, 208–215.

Staud, R. and Domingo, M. 2001b. New insights into the pathogenesis of fibromyalgia syndrome. Med. Aspects Hum. Sex. 1, 51–57.

Staud, R. and Smitherman, M. L. 2002. Peripheral and central sensitization in fibromyalgia: pathogenetic role. Curr. Pain Headache Rep. 6, 259–266.

Staud, R., Price, D. D., Robinson, M. E., and Vierck, C. J. 2004. Body pain area and pain-related negative affect predict clinical pain intensity in patients with fibromyalgia. J. Pain 5, 338–343.

Staud, R., Robinson, M. E., Vierck, C. J., Cannon, R. L., Mauderli, A. P., and Price, D. D. 2003. Ratings of experimental pain and pain-related negative affect predict clinical pain in patients with fibromyalgia syndrome. Pain 105, 215–222.

Staud, R., Vierck, C. J., Cannon, R. L., Mauderli, A. P., and Price, D. D. 2001. Abnormal sensitization and temporal summation of second pain (wind-up) in patients with fibromyalgia syndrome. Pain 91, 165–175.

Sultan, S. M., Ioannou, Y., Moss, K., and Isenberg, D. A. 2002. Outcome in patients with idiopathic inflammatory myositis: morbidity and mortality. Rheumatology (Oxf.) 41, 22–26.

Tench, C., Bentley, D., Vleck, V., McCurdie, I., White, P., and D'Cruz, D. 2002. Aerobic fitness, fatigue, and physical disability in systemic lupus erythematosus. J. Rheumatol. 29, 474–481.

Thieme, K., Turk, D. C., and Flor, H. 2004. Comorbid depression and anxiety in fibromyalgia syndrome: relationship to somatic and psychosocial variables. Psychosom. Med. 66, 837–844.

Torebjork, H. E., Lundberg, L. E., and LaMotte, R. H. 1992. Central changes in processing of mechanoreceptive input in capsaicin-induced secondary hyperalgesia in humans. J. Physiol. 448, 765–780.

Turk, D. C., Okifuji, A., Starz, T. W., and Sinclair, J. D. 1996. Effects of type of symptom onset on psychological distress and disability in fibromyalgia syndrome patients. Pain 68, 423–430.

Viau, V., Sharma, S., Plotsky, P. M., and Meaney, M. J. 1993. Increased plasma ACTH responses to stress in nonhandled compared with handled rats require basal levels of corticosterone and are associated with increased levels of ACTH secretagogues in the median eminence. J. Neurosci. 13, 1097–1105.

Vierck, C. J., Staud, R., Price, D. D., Cannon, R. L., Mauderli, A. P., and Martin, A. D. 2001. The effect of maximal exercise on temporal summation of second pain (wind-up) in patients with fibromyalgia syndrome. J. Pain 2, 334–344.

Wall, P. D. and Woolf, C. J. 1984. Muscle but not cutaneous C-afferent input produces prolonged increases in the excitability of the flexion reflex in the rat. J. Physiol. Lond. 356, 443–458.

Waylonis, G. W. and Perkins, R. H. 1994. Post-traumatic fibromyalgia. A long-term follow-up. Am. J. Phys. Med. Rehabil. 73, 403–412.

White, K. P., Harth, M., Speechley, M., and Ostbye, T. 2000. A general population study of fibromyalgia tender points in noninstitutionalized adults with chronic widespread pain. J. Rheumatol. 27, 2677–2682.

White, K. P., Nielson, W. R., Harth, M., Ostbye, T., and Speechley, M. 2002. Chronic widespread musculoskeletal pain with or without fibromyalgia: psychological distress in a representative community adult sample. J. Rheumatol. 29, 588–594.

Wolfe, F., Simons, D. G., Fricton, J. R., Bennett, R. M., Goldenberg, D. L., Gerwin, R., Hathaway, D., McCain, G. A., Russell, I. J., and Sanders, H. O. 1992. The fibromyalgia and myofascial pain syndromes: a preliminary study of tender points and trigger points in persons with fibromyalgia, myofascial pain syndrome and no disease. J. Rheumatol. 19, 944–951.

Wolfe, F., Smythe, H. A., Yunus, M. B., Bennett, R. M., Bombardier, C., Goldenberg, D. L., Tugwell, P., Campbell, S. M., Abeles, M., and Clark, P. 1990. The American College of Rheumatology 1990 criteria for the classification of fibromyalgia. Report of the Multicenter Criteria Committee. Arthritis Rheum. 33, 160–172.

52 Pain Perception – Nociception during Sleep

G J Lavigne, Université de Montréal, Montreal, QC, Canada

K Okura, Tokushima Graduate School, Tokushima, Japan

M T Smith, John Hopkins Medical School, Baltimore, MD, USA

Glossary

alpha EEG wave intrusion The intrusion of fast-frequency electroencephalographic (EEG) alpha (7.5–11 Hz) activity into slow wave sleep (SWS) or into deep sleep (stages 3 and 4). SWS is dominated by large and slow EEG waves of delta type (0.5–4.0 or 0.75–4.5 Hz); it also characterizes sleep stages 3 and 4. SWS is not a specific biological marker of pain during sleep.

sleep awakening (major arousal) A sudden increase in EEG and heart rate frequency with a rise in muscle tone that lasts more than 10 or 15 s. Subject is not usually aware of external influences. An excessive number of major arousals during sleep is frequently followed by a complaint of unrefreshing sleep and cognitive impairment the following day.

cyclic alternating pattern Repetition of micro-arousals every 20–60 s as a sentinel that allows for a reset of physiological functions (e.g., heart rate, respiration, muscle tone) or prepares the body for an appropriate response in relation to potential disrupting events.

micro-arousal A sudden increase, during sleep, in EEG activity and heart rate frequency under a cardiac sympathetic dominance with a possible rise in muscle tone. It should last more than 3 s but less than 10 or 15 s depending on scoring method. A sleeping subject is normally unaware of this physiological activity; it tends to be repeated 8–15 times per hour of sleep.

nociception The receiving and transmitting of noxious sensory signals from the periphery to the central nervous system.

pain An unpleasant and sensory, cognitive, and emotional experience normally associated or not with tissue damage, which usually requires a minimal level of consciousness for expression.

pain threshold The lowest experimental stimulation perceived as being painful by a conscious subject in 50% of the trials.

sleep A natural physiological and behavioral state usually characterized by a partial isolation from the environment except when an unpleasant or potentially harmful or life-threatening event is present.

sleep stage A division of the specific sleep period into sleep stages (St): light sleep (St 1 or 2), deep sleep (St 3 and 4) are both described as non-REM sleep, and the REM or paradoxical sleep. Most body movements occur in light non-REM sleep (periodic limb movement, sleep bruxism). Sleep apnea (cessation of breathing for more than 10 s) is dominant in deep sleep and in REM sleep. REM behavior disorder (RBD) is characterized by sudden movement in REM sleep that is normally associated with muscle hypotonia (behaviorally described as motor paralysis).

sleep fragmentation Interruption of any sleep stage by isolated or repetitive events such as sleep stage shifts (deeper to lighter), micro-arousals, frank awakenings, with or without low duration of deep St 3 and 4 sleep and possible alpha EEG wave intrusion. As a result, sleep continuity may be impaired and poor sleep quality (e.g., unrefreshing) is frequently reported upon awakening.

52.1 Pain in Relation to Sleep

52.1.1 Definition, Epidemiology, and Relevance

Pain is a sensation that usually requires a certain level of consciousness before it is interpreted as an unpleasant sensory experience. Most persistent pain states can trigger complaints of poor sleep quality. Poor sleep quality is reported in 50–90% of patients with the following chronic pain conditions: arthritis, cervical and orofacial pain, low back pain, cancer pain, fibromyalgia. (Atkinson, J. H. *et al.*, 1988; Dao, T. T. T. *et al.*, 1994; Morin, C. M. *et al.*, 1998; Smith, M. T. *et al.*, 2000; Dauvilliers, Y. and Touchon, J., 2001; Riley, J. L. III *et al.* 2001; Roizenblatt, S. *et al.*, 2001; McCracken, L. M. and Iverson, G. L., 2002). Sleep quality is also impaired in the presence of other medical conditions of which pain is a common symptom, including: headaches, irritable bowel syndrome, spinal cord injury, and metastatic breast cancer (Cohen, M. *et al.*, 2000; Menefee, L. A. *et al.*, 2000; Moldofsky, H., 2001; Widerstrom-Noga, E. G. *et al.*, 2001; Rains, J. C. and Penzien, D. B., 2002; Koopman, C. *et al.*, 2002; Okifuji, A. and Turk, D. C., 2002; Lavigne, G. J. *et al.*, 2005). According to a recent survey from the National Sleep Foundation (USA), up to 20% of the adult population reports that their sleep is disrupted by somatic discomfort and pain.

The interaction between pain and poor sleep is supported by cohort and case control studies. Patients with orofacial pain are 3.7 more times at risk of reporting poor sleep (Macfarlane, T. V. and Worthington, H. V., 2004). In a survey done in the USA with 1506 community women and men aged between 55 and 84, it was found that bodily pain was associated with a risk for complaints of insomnia (e.g., difficulty falling asleep, middle of the night/ early morning awakenings, or nonrestorative sleep). Odds ratios for insomnia among those reporting bodily pain were estimated to be between 1.88 and 2.68 (Foley, D. *et al.*, 2004). It has also been reported that the prevalence of insomnia is 2 times higher in Canadian chronic pain patients in comparison to the general population without pain (Sutton, D. A. *et al.*, 2001; Moldofsky, H., 2001). A cross-sectional study done in Hungary with over 12 643 adults showed that insomnia-related complaints were 3 times higher in chronic pain patients than in nonpain subjects (Novak, M. *et al.*, 2004). Finally, the concomitance of sleep disorders and pain may modify the subjective sensation related to sleep quality (e.g., refreshing) since 50% of insomnia patients with pain reported that pain interfered with daily activities (Novak, M. *et al.*, 2004). Moreover, in the presence of bodily pain there is an increase (OR of 1.73) in self-reports of daytime sleepiness (Foley, D. *et al.*, 2004) a topic further described in the next section.

In explorations of the interaction between high pain intensity and poor sleep, it is important to consider that chronic pain is frequently concomitant with anxiety, fatigue, mood disturbance, depression, and poor physical fitness (Menefee, L. A. *et al.*, 2000b; Riley, III. J. L. *et al.*, 2001; Smith, M. T. *et al.*, 2000; McCraken, L. M. and Iverson, G. L. 2002b; Nicassio, P. M. *et al.*, 2002; Sayar, K. *et al.*, 2002; Yatani, H. *et al.*, 2002). For example, fatigue is frequently reported in chronic pain patients and can explain 6.5% of the variability between pain intensity and complaints of poor sleep. Depression may further explain up to 40% of this variability. The presence of frank sleep apnea, a major risk for fatigue and daytime sleepiness, seems to be relatively low in chronic pain patients, as reported in a recent review (Dauvilliers, Y. and Touchon, J. 2001; Donald, F. *et al.*, 1996). However, the cumulative effect of poor sleep over a given night and possibly over several nights also needs to be considered. While it is recognized that daytime sleepiness may result from long interruptions of respiration (more than 10 s is defined as sleep apnea), less information is available on the cumulative consequence of several briefer hypoxic events, which may alter cognitive function and brain metabolism (Bonnet, M. H. 1985; Wesenten, N. J. *et al.*, 1999; Stepanski, E. J., 2002; Martin, S. E. *et al* 1996; Van Dogen, H. P. *et al.*, 2003; Gozal, D. and O'Brien, L. M., 2004). Recent work in fibromyalgia, for instance, suggests the possibility of a high rate of sleep-related upper airway resistance syndrome that when reversed by continuous positive airway pressure therapy, may improve daytime function (Gold, A. R. *et al.*, 2004). The impact of the above-mentioned variables on pain and sleep interaction and their reversal by continuous positive airway pressure (CPAP) needs to be further assessed in controlled and blind studies.

52.1.2 Cognitive Disturbances

Very few studies have investigated the effect of pain on cognitive functions such as memory and attention. Three studies using chronic musculoskeletal pain and fibromyalgia patients suggest a causative association (Kewman, D. G. *et al.*, 1991; Landro, N. I. *et al.*, 1997; Côté, K. A. and Moldofsky, H., 1997). The impact of cognitive deficits and their potential aggravation by disturbed sleep needs to be further studied. The potential impact of several classes of pain medication (e.g., opioids, anticonvulsants, muscle relaxants, antidepressants) on sleep and daytime alertness is also a particularly important line of research with real-life consequences ranging from sleep-related industrial accidents to impaired driving ability (George, C .F. P., 2003; Chapman, S., 2001). A recent experimental study done in normal volunteers showed that a one night infusion of morphine during sleep time can alter sleep quality by reducing deep-sleep duration and increasing sleep stage 2 duration (Shaw, I. R., *et al.*, 2005). Evidence-based data on long-term use of morphine on sleep homeostasis are missing.

52.1.3 Sleep Fragmentation and Deprivation

Sleep fragmentation is usually characterized by a rapid shift from a deeper sleep stage to a lighter one or by an increase in the number of brief arousals up to longer awakenings during the night (see Glossary for definitions). The number of body movements and respiratory disturbance events, which are often associated with arousals, seems to have an additive effect on daytime functioning over the following day or days. Sleep deprivation is a more general term that either characterizes the absence or very short duration of a given sleep stage and/or the near absence or reduction of sleep as tested by experimental manipulations. As recently demonstrated, measures of the impact of chronic sleep loss fall into a clear hierarchy: mood measures being more sensitive than cognitive tasks, which are in turn more sensitive than performance of motor tasks (Bonnet, M. H., 2005).

The increase in sleep fragmentation in normal subjects or in patients with sleep apnea seems to be clearly associated with a reduction in next-day performance, sleepiness, and impaired alertness (Bonnet, M. H., 1985; Wesenten, N. J. *et al.*, 1999; Stepanski, E. J., 2002; Martin, S. E. *et al.*, 1996; Van Dogen, H. P. *et al.*, 2003). A minimum of 5–6 h of sleep seems to be the threshold below which daytime performance is altered and, interestingly, several days are required before a return to baseline levels (Van Dogen, H. P. *et al.*, 2003). There is still no definitive evidence indicating that lack of deep sleep (St 3 and 4) or of slow wave EEG activity (SWA) increases pain sensitivity, although some animal and human experimental studies suggest such a link (see review from Kundermann, B. *et al.*, 2004a; Kundermann, B. and Lautenbacher, S., 2007). To our knowledge only one study has compared deep-sleep deprivation with other sleep stages or total sleep deprivation (Onen, A. H. *et al.*, 2001). Moreover, the absence of deep sleep is a feature of all chronic-pain conditions (Okura, K. *et al.*, 2004).

Before summarizing the studies on sleep deprivation and changes in pain perception, it is important to reiterate that study design and data analysis need to control for: (1) the differences between groups for age, since after 40 years of age, slow wave density, a dominant feature of deep sleep, tends to decline; (2) presence of respiratory disturbances during sleep, since patients with fibromyalgia who complain of daytime hypersomnolence tend to present with more micro-arousals, a higher number of short respiratory disturbances and longer oxygen desaturation periods than matched normal controls, an effect reversed by CPAP (Sergi, M. *et al.*, 1999; Gold, A. R. *et al.*, 2004); (3) the concomitant effect of fatigue, mood depression, anxiety, poor lifestyle habits, etc., as described above. In addition to the above issues, it is also important to note that the literature of sleep deprivation and pain is generally limited by experimental studies done with small sample sizes and a lack of adequate control groups.

Several studies have been conducted in normal subjects to assess the effect of deep-sleep deprivation on next-day pain perception. The disruption of sleep stage 4 or SWS by sound was associated with an increase in muscle tenderness and a reduction in pain threshold (mechanical pressure) the next morning, but this effect was less dominant when studies controlled for concomitant influences of fatigue (Moldofsky, H. and Scarisbrick, P., 1976; Lentz, M. J. *et al.*, 1999; Older, S. A. *et al.*, 1998). Negative findings have been reported by others (Drewes, A. M. *et al.*, 1997; Kundermann, B.

et al., 2004a; Older, S. A. *et al.*, 1998; Arima, T. *et al.*, 2001; Smith, M. T., *et al.*, 2007). Surprisingly, when normal subjects were deprived of SWA sleep, REM sleep or total sleep duration, no change in thermal pain threshold was noted during any of the sleep interruptions (Onen, A. H. *et al.*, 2001). In this study, however, total sleep deprivation slightly reduced mechanical pain tolerance scores (8%) and in the period after sleep recovery, a significant rebound effect was noted, such that mechanical pain tolerance was 15% higher. A recent study done in normal volunteers showed that two nights of total sleep deprivation reduced both heat and cold pain thresholds, but had no effect on warmth and cold detection threshold; results suggested that change in pain threshold was not specific to alterations in somatosensory function (Kundermann, B. *et al.*, 2004b). Kundermann and colleagues also suggested, in a comprehensive review, that the role of stress and changes in endogenous serotonin and opioid activity can also play an important role in arousal and changes in pain perception (Kundermann, B. *et al.*, 2004a).

It is important to clarify that experimental sleep deprivation, which has been induced in the above studies, is an extreme manipulation that may be different from the natural ongoing sleep disruption experienced by chronic pain patients. As described above, fragmentation or deprivation, not necessarily limited to St 3 and 4, seems to alter cognitive function and memory consolidation; all sleep stages seem to be equally important for the preservation and maintenance of daytime functions (Stickgold, R. *et al.*, 2000; Mednick, S. *et al.*, 2003; Gais, S. *et al.*, 2000).

52.1.4 Circadian Variation in Pain Perception

Surprisingly, most studies using normal nonpain subjects without mood alteration have failed to demonstrate a clear circadian variation in experimental pain threshold or pain intensity reports over a 24 h schedule or when comparing evening and morning data (Rogers, E. J. and Vilkin, B. 1978; Strian, F. *et al.*, 1989; Koltyn, K. F. *et al.*, 1999; Lavigne, G. J. *et al.*, 2000; Bentley, A. J. *et al.*, 2003; Lavigne, G. J. *et al.*, 2004). In patients with chronic pain, however, symptom presentation often has a typical circadian pattern. Arthritic pain is often worse in the morning, while pain levels related to fibromyalgia, myofascial orofacial pain, and some headache conditions are typically higher in the afternoon (Bellamy, N. *et al.*, 1991; Reilly, P. A. and Littlejohn, G. O., 1993; Dao, T. T. T. *et al.*, 1994). Furthermore, not all pain conditions

involve pain that directly interferes with sleep; in patients with torticolis, pain is rapidly reduced by adopting the supine posture (Lobbezoo, F. *et al.*, 1996; Bentley, A. J., 2007).

52.2 Perception of Pain during Sleep: Nociception or Sleep Arousal

52.2.1 Nociception – Pain Perception during Sleep

While awake, tactile and thermal sensory input of sufficient intensity can activate A delta, A beta, and C sensory fibres projecting to the thalamus and several cortical areas (Bromm, B. and Lorenz, J. 1998; Coghill, R.C. *et al.*, 1994; Le Pera, D. *et al.*, 2000; Peyron, R. *et al.*, 2000). The long duration of painful hypertonic saline infusions seems to activate A delta and C afferent fibers that in the wake state reach the thalamus, amygdale, and other cortical areas associated with pain processing (Le Pera, D. *et al.*, 2000; Zubieta, J. K. *et al.*, 2001). During sleep, however, a putative gating process of somatosensory inputs is thought to prevent awakening resulting from irrelevant input or non-life-threatening events. Several studies suggest that sleep-preserving sensory gating mechanisms are likely to be located at the upper brain stem area (see Figure 1) in the so-called reticular activating system (Pompeiano, O. and Swett, J. E. 1963; Hernandez-Peon, R. *et al.*, 1965; Soja, P. J. *et al.*, 2001; Soja, P. J., 2007; Nofzinger, E. and Derbyshire, S. W. G., 2007). However, the balance between wakefulness system(s) and sleep-promoting neuronal network(s) in relation to sensory discrimination and pain during sleep is poorly understood and requires further investigation.

Regarding sensory perception during sleep, it is crucial to discriminate between studies made using sound and those using tactile stimulations. The results of sound studies have made it apparent that acoustic stimulation triggers more sleep arousals and awakenings in non-REM light sleep stages (St 1 and 2) than in deep sleep (St 3 and 4) or in REM sleep (Rechstaffen, A. and Kales, A., 1968; Carley, D. W. *et al.*, 1997; Kato, T. *et al.*, 2004). Sound is a different sensory modality from pain: its main function with respect to sleep is to alert the individual to wake up (alarm clock) or to parental duties (a baby's cry). Tactile stimulation, however, is more likely to be interpreted as a touch to the body, which may or may not enforce the awakening process and is likely to be interpreted differently than sound by a sleeping

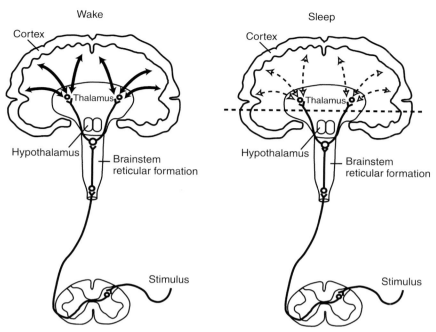

Figure 1 During wake state, there is a free flow in neuronal activity from spinal cord or/and brainstem to and from thalamus and cortical networks. During sleep, a gating mechanism seems to isolate the upper brain from most incoming inputs originating at somatosensory level. However, some circuits, for example, auditory processing, remain fully active. The exact anatomical site of the separation (see dashed line) between spinal and brainstem networks and thalamo-cortical pathways remain to be demonstrated.

brain (Velluti, R. A. *et al.*, 2000; Kato, T. *et al.*, 2004; Soja, P. J. *et al.*, 2001) Moreover, during sleep the term nociception is probably more accurate to describe the process of potentially harmful inputs, interpreted in conscious and awake subjects as pain (Bromm, B. and Lorenz, J., 1998; Lavigne, G. J. *et al.*, 2000).

Studies done in the 1960s, using tactile-cutaneous sensory stimulations during wake/sleep states, are a rich source of information (Pompeiano, O. and Swett, J. E., 1963; Hernandez-Peon, R. *et al.*, 1965) It was suggested that a filtering mechanism is active during synchronized or non-REM sleep at the first relay of neurons in the brainstem mesencephalic-reticular formation and in the trigeminal sensory nucleus. More recent work lends further support to the concept that sleep involves a selective isolation from the external milieu. Recordings of ascending spinoreticular and trigemino-thalamic tract neurons show that a gating mechanism seems to be present during sleep when brief sensory or pain stimulations are applied to the skin of the animal. Furthermore, the neurons behaved differently across sleep stages, that is, during non REM or REM sleep: air puff activated trigemino-thalamic tract neurons during REM sleep whereas neuronal activity following tooth pulp stimulation was

suppressed (Cairns, B. E. *et al.*, 1996; Soja, P. J. *et al.*, 2001). It was further suggested that the serotoninergic and nonserotoninergic brainstem raphe magnus cells are involved to an important extent in the sensory modulation of pain inputs (Foo, H. and Mason, P., 2003). Interestingly, recording of thalamic cells in one human has confirmed previous animal findings suggesting that thalamic neurons discharge differently during states of waking and sleeping. It was observed that the discharge of thalamic neurons was in a tonic mode while the patient was awake and in a bursting mode when the patient was in light sleep; nevertheless none of these cells had a clear sensory receptive field in the periphery (Tsoukatos, J. *et al.*, 1997). These findings are similar to those from animal studies showing that thalamic neurons are under a global inhibition during sleep while cortical neurons, by contrast, are highly active (Steriade, M. and Timofeev, I., 2003). The exact location of the mechanism(s) supporting the somatosensory dissociation during sleep remains under investigation (see Figure 1). The balance between mechanisms that maintain vigilance or promote sleep continuity, however, seems to be state-dependent and related to the modulation of neuronal networks at the level of the reticular activating system.

The initiation of an awakening and conscious reaction during sleep may be secondary to a decrease in the sensory gating in sleep resulting in the recovery of the free flow between ascending and descending brainstem, thalamus and cortical activities (Siegel, J., 2004; Skinner, R. D. *et al.*, 2004). In other words, it is as if the reptilian brain maintains basic survival function during sleep, and interaction with thalamo-cortical networks depends on the nature (sound, touch, and harmful stimulus), magnitude, and duration of the stimulation. Obviously, this hypothesis needs to be supported by further evidence.

A recent paper summarizing the evidence on brain activation during sleep provides relevant information (Peigneux, P. *et al.*, 2003). During light sleep, pontine tegmentum and thalamus are deactivated while mesencephalic regions remain active. In deep sleep, all three structures are relatively deactivated while activity is largely maintained at the cortical temporal and insular lobe levels. Paradoxically, all the above structures are activated during REM sleep, although a strong dissociation is present: there is no peripheral efferent activity (e.g., the motor cortex does not trigger a movement except in the presence of sleep pathology: the REM behaviour disorder). There is little evidence strongly supporting a clear-cut gating of sensory information at the lower brain level, to prevent awakening from irrelevant sensory inputs. However, the role of the amygdala and the hypothalamus is discussed in the interpretation of sensory gating outside the higher thalamocortical levels (Morrison, A. R. *et al.*, 2000; Lee, R. S. *et al.*, 2001). A review of behavioral animal and human studies is also useful for a better understanding of the modulation of sensory inputs during sleep.

The interaction between the pain system and vigilance versus sleep is further supported by a cat study using the classical tail flick sensory motor reflex (Kshatri, A. M. *et al.*, 1998). It was suggested that the gating of sensory inputs during sleep depends on the integrity of the cholinergic system at the medial pontine–reticular level. This work also clearly showed, in comparison to the wake state and in the absence of any medication, that the reflex latency was 3 times longer in non-REM sleep (quiet sleep in animal) and 5 times longer in REM sleep. This finding again supports the concept of higher responsiveness to brief sensory inputs during non-REM in comparison to REM sleep. However, another cat study revealed that under tonic sensory nociceptive influences (formalin injection in the foot) the animals tended to remain awake with delayed sleep onset for

the first 2 days, and a reduction in duration of deep sleep (St 3 and 4 in humans) and REM sleep stages (Carli, G. *et al.*, 1987) Interestingly, after 2 days, the animals recovered, exhibiting total sleep time and deep sleep of longer duration; a typical pattern seen after sleep deprivation (Onen, A. H. *et al.*, 2001; Kundermann, B. *et al.*, 2004a). Another sleep and pain study in rats, using the experimental adjuvant arthritis model in both hind paws, also showed fragmentation of the typical sleep and wake pattern (Landis, C. A. *et al.*, 1988). The number of brief sleep periods was increased over the circadian rhythm; total sleep time and deep sleep duration were also shorter in rats with arthritis. The above-mentioned animal studies suggest that a sleeping brain is not completely isolated from the external milieu and that pain-nociception can activate networks related to vigilance and body protection. This activation appears to be related to the type (phasic versus tonic, intensity, etc.) and duration of the painful stimuli (Velluti R. A. *et al.*, 2000; Soja, P. J. *et al.*, 2001; Kato, T. *et al.*, 2004; Lavigne, G. J. *et al.*, 2004).

In humans, the perception of pain during sleep also depends on the duration and type of stimulus used; brief sensory nerve-related or mild duration thermal or long duration chemical or mechanical stimulations initiate different types of responses that need to be interpreted accordingly. In a study using the spinal motor responses, following brief stimulation of the leg sural nerve to evoke a flexion reflex, it was noted that the reflex latency was prolonged during sleep in comparison to wake state (Sandrini, G. *et al.*, 2001). In parallel, the threshold was 60% higher in non-REM and 200% higher in REM sleep than during wake state. This study supports the above-mentioned animal findings, suggesting that during sleep lower brain neurons modulate pain inputs in such a way that a hypoalgesic response seems to be present. In contrast, when sensory processing from periphery to cortex was estimated with extremely brief (<60 ms) sensory-laser or electrical evoked responses to pain, the somatosensory cortically evoked potential completely disappeared during light sleep (Bedyoun, A. *et al.*, 1993; Wang, X. *et al.*, 2003). This absence of somatosensory evoked potentials during sleep stands in contrast to similarly present auditory stimuli in which evoked potentials are largely preserved during sleep. Sound appears to play a more sensitive role in preserving bodily integrity during sleep than tactile-somatosensory input (Kato, T. *et al.*, 2004). Alternatively, in collaboration with the Italian Milano sleep group, we found that longer (6–12 s) moderate heat pain stimulation (46 °C) applied over the upper arm can evoke a clear sleep behavioral response: nearly twice the number of micro-arousals (see Glossary for definition) in light non-REM sleep compared to deep or REM sleep (Lavigne,G. J. et al., 2000). These results were confirmed by another group that found that higher thermal pain temperatures were required to elicit arousal during deep sleep and REM in comparison to light sleep. These temperatures were closer to waking pain tolerance levels instead of pain threshold level (Bentley, A. J. et al., 2003). More recently, with the objective of challenging sensory pain processing across sleep stages, we used a hypertonic saline infusion in the arm muscle to mimic longer clinical pain episodes (>100 s). A similar procedure has been used by others during sleep (Drewes, A. M. et al., 1997). Interestingly, we found that micro-arousal responses were also more dominant in light sleep in comparison to deep and REM sleep, and a clear and dominant sleep awakening response was present with an equipotent response rate in all sleep stages (Lavigne, G. J. et al., 2004).

This evidence suggests that when a painful stimulus or a clinical pain episode lasts long enough, the protective mechanism that maintains sleep continuity (the gating barrier to irrelevant input) is released and a clear behavioral response may occur with a potential return to consciousness. Recent clinical evidence for impaired somatosensory gating during sleep in chronic pain conditions has been suggested by findings of decreased sleep spindle activity in chronic low back pain and fibromyalgia patients (Harman, K. et al., 2002; Landis, C. A. et al., 2004). Thus it can be hypothesized that brief sensory stimuli, such as electrical or thermal pain, are too brief to be processed by a sleeping brain as potentially harmful, whereas longer-lasting or tonic painful stimuli akin to clinical pain states are more likely to elicit a full-blown arousal response.

52.2.2 Sleep Arousal: Brain and Autonomic Activation

The above arousal and awakening responses observed with thermal and deep muscle infusion are not only restricted to a cortical activation as seen on electroencephalogram. They are also associated with a clear increase in heart rate and a change in respiration that is frequently interrupted with the onset of the pain stimulation (Lavigne, G. J. et al., 2000;

Lavigne, G. J. et al., 2001; Bentley, A. J. et al., 2003; Lavigne, G. J. et al., 2004). It is interesting to note that the so-called fast alpha EEG intrusion, which for a long time was believed to be a marker of pain in relation to sleep, is not considered to provide the sole explanation for poor sleep in pain patients (Moldofsky, H., 2001; Mahowald, M. L. and Mahowald M. W., 2000; Roizenblat, S. et al., 2001; Rains, J. C. and Penzien, D. B., 2003). More recent thinking conceptualizes pain as a disruptor of sleep homeostasis that may increase sleep arousal instability, described as an augmentation in the normal cyclical alternating pattern in which brain and cardiac activation tends to recur in clusters (Staedt, J. et al., 1993; Moldofsky, H., 2001; Roizenblatt, S. et al., 2001; Parrino, L., Zucconi, M. and Terzano, M. G., 2007).

Cardiac activation during sleep is used as an indirect estimate of sympathetic (e.g., accelerator) and parasympathetic (decelerator or brake) balance during wake and sleep states. During wake state, a sympathetic dominance predominates, maintaining a high level of vigilance. Sympathetic-cardiac dominance is also present during REM sleep, a state with high cortical and cardiac activity but with a behavioral muscle paralysis. Conversely, during non-REM sleep, parasympathetic-cardiac activity is dominant and this may provide a recuperative function since there is an inverse coupling: a low sympathetic dominance is inversely coupled with a high degree of SWA in each consecutive non-REM to REM cycle (Brandenberger, G. et al., 2001). Interestingly, in insomnia and fibromyalgia patients, a sympathetic dominance persists across NREM sleep cycles at levels similar to wake and REM sleep states (Bonnet, M. H. and Arand D. L., 1998; Martínez-Lavín, M. et al., 1998). It has been hypothesized that enhanced sympathetic activity during sleep may over-facilitate arousal mechanisms and cyclical patterns of brain activation. However, this hypothesis was not supported in a study of medication-free women suffering from fibromyalgia in comparison to women with efficient sleep and no pain (McMillan, D. et al., 2004). By contrast, when patients with chronic fibromyalgia (widespread pain) and pain (low back, etc.) presenting with poor sleep efficiency were matched for age and gender to normal sleepers, we found (see Figure 2) that the normal parasympathetic and sympathetic balance was lost in both pain groups during non-REM sleep with an inverse coupling of SWA, that is, a lower density of SWA was observed during consecutive sleep stages compared to control

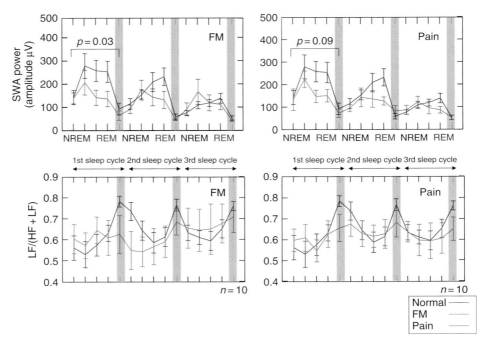

Figure 2 Dynamics of SWA and HRV over three consecutive sleep cycles. In normal subjects, SWA shows an inverse coupling with sympathetic activity (low-frequency divided by high-frequency plus low-frequency. One of the heart rate variability: HRV). During non-REM sleep, SWA was high when the sympathetic cardiac activity was low, and, conversely, SWA was low when HRV was high in REM sleep. In comparison to normal subjects, fibromyalgia (FM) chronic widespread pain patients showed a significant reduction in SWA for the first sleep cycle only (t-test: $p = 0.03$). Chronic pain patients (PAIN) also presented a similar trend in SWA, but not statistically significant (t-test: $p = 0.09$). PAIN patients did not show a rise in sympathetic activity during REM sleep ($p = $ NS: repeated ANOVA); FM patients showed a trend toward no rise in REM sleep.

subjects (Okura, K. *et al.*, 2005). Furthermore, like previous studies of men with fibromyalgia in a wake state, our preliminary analysis suggests that men maintain a high sympathetic activity across all sleep stages (Cohen, H. *et al.*, 2001). Women, however, do not show the usual rise in sympathetic activity during REM sleep. Our data differ from those of (Martinez-Lavin, M., *et al.*, 1998) since we separated data into consecutive non-REM and REM cycles. Caution is needed when interpreting the above-mentioned observations since parasympathetic and sympathetic balance tends to be altered with age and acute stress in relation to sleep (Brandenberger, G. *et al.*, 2003; Hall, M. *et al.*, 2004).

52.3 Circular Relation between Pain and Poor Sleep and Putative Consequence on Health Cost

The initiation of pain (e.g., acute post-op or trauma) usually precedes or coincides with the onset of poor sleep in close to 53–89% of patients (Morin, C. M. *et al.*,

1998; Smith, M. T. *et al.*, 2000; Riley, J. L., III *et al.*, 2001). In other words, in retrospective accounts, sleep disturbance is generally not considered a major problem for many patients until an injury is experienced; then pain and sleep become salient interacting issues. Intensive, time-series diary studies using multilevel modeling statistical techniques consistently show that over time, a day with intense pain is frequently followed by a poor sleep quality and subsequent ratings of poor sleep are in turn linked to next-day increases in clinical pain. This circular relationship, also described as a vicious circle, has been observed in patients with fibromyalgia, severe skin burns, and rheumatoid arthritis (Affleck, G. *et al.*, 1996; Raymond, I. *et al.*, 2001; Stone, A. A. *et al.*, 1997; Nicassio, P. M. *et al.*, 2002; Lavigne, G. J. *et al.*, 2005). As has been mentioned above, the influences of fatigue, mood alteration, other sleep disorders (such as sleep apnea or periodic limb movement), and medication use have not been well controlled for in these studies.

No direct data, to our knowledge, reports on the cost of pain and its interaction with poor sleep. However, an extensive study done in a Canadian population (125 574 respondents) estimated that for every four patients seen by a family physician, three will report chronic pain (Meana, M. *et al.*, 2004). When the potential cognitive disruption of pain medications is considered together with the rates at which analgesic, antidepressant, codeine, and hypnotics medications are used (3–4 times more often by women with chronic pain), it becomes apparent that the reduced ability of chronic pain patients with poor sleep to perform daily tasks is a societal burden (Meana, M. *et al.*, 2004b). Moreover, given the aging population and the high prevalence of chronic pain (reported by over 50% of elderly of both genders), it is clear that health managers have a major challenge in planning service use (Harstall, C. and Ospina, M., 2003; Meana, M. *et al.*, 2004).

The economic cost of chronic pain can probably be estimated in the order of billion dollars from headaches to pelvic pain; obviously specific estimation needs to be done. The large range of costs, for various pain symptoms and syndromes, reflects a lack of reliable information, although it is clear that chronic pain conditions are extremely costly (Latham, J. and Davis B. D., 1994). To date, an approximate estimate of the additional burden that poor sleep places on chronic pain costs can be only derived from cost studies of insomnia, which have yielded estimates ranging from 77 to 92 billion dollars (Stoller, M. K., 1994). If we agree that the suggested prevalence of chronic pain in adult population stands at 11% (Harstall, C. and Ospina, M., 2003) and that insomnia (defined as difficulties in sleep onset or in maintaining sleep and/or perception of nonrestorative sleep) is reported conservatively at a similar prevalence (Martin, S. A. *et al.*, 2004) we can extrapolate, based on the assumption that one chronic pain patient in two will report similar insomnia complaints, that the overall direct cost is probably over 1 billion dollars. Indirect costs arising from as sick leave, hospitalization, and vehicular accident also need to be estimated, as has been done for other sleep disorders (Barbé, F. *et al.*, 1998; George, C. F. P., 2003).

There is an obvious need for prospective and systematic studies of pain and economics that will estimate cost of health care, corrected for concomitant conditions such as aging, diabetes, obesity, cardiovascular problems, sleep disorders, depression, and anxiety (Lavigne, G. and Manzini, C., 2007). Moreover, indirect cost of absenteeism, low working performance, accident, reduced familial activities and risk of divorce also need to be factored into estimates of the cost of pain in relation to poor sleep.

References

Affleck, G., Urrows, S., and Tennen, H. 1996. Sequential daily relations of sleep, pain intensity, and attention to pain among women with fibromyalgia. Pain 68, 363–368.

Arima, T., Arendt-Nielsen, L., and Svensson, P. 2001. Effect of jaw muscle pain and soreness evoked by capsaicin before sleep on orofacial motor activity during sleep. J. Orofac. Pain 15(3), 245–256.

Atkinson, J. H., Ancoli-Israel, S., and Slater, M. A. 1988. Subjective sleep disturbance in chronic back pain. Clin. J. Pain 4, 225–232.

Barbé, F., Pericas, J., Munoz, A., Findley, L., Anto, J. M., and Agusti, A. G. N. 1998. Automobile accidents in patients with sleep apnea syndrome. Am. J. Respir. Crit. Care Med. 158, 18–22.

Bedyoun, A., Morrow, T. J., and Shen, J. 1993. Variability of laser-evoked potentials: attention, arousal and laterizes differences. Electroenceph Clin Neurophysiol 88, 173–181.

Bentley, A. J. 2007. Pain Perception during Sleep and Circadian Influences: The Experimental Evidence. In: *Sleep and Pain* (eds. G. Lavigne, B. J. Sessle, M. Choinière, P. J. Soja), pp. 123–136. IASP Press..

Bellamy, N., Sothern, R. B., and Campbell, J. 1991. Circadian rhythm in pain, stiffness, and manual dexterity in rheumatoid arthritis: relation between discomfort and disability. Annals. Rheum Dis 50, 243–248.

Bentley, A. J., Newton, S., and Zio, C. D. 2003. Sensitivity of sleep stages to painful thermal stimuli. J. Sleep Res. 12, 143–147.

Bonnet, M. H. 2005. Acute Sleep Deprivation. In: Principles and Practice Sleep of Medicine (eds. M. H. Kryger, T. Roth, W. C. Dement), 5, pp. 51–66. Elsevier Saunders.

Bonnet, M. H. 1985. Effect of sleep disruption on sleep, performance, and mood. Sleep 8(1), 11–19.

Bonnet, M. H. and Arand, D. L 1998. Heart rate variability in insomniacs and matched normal sleepers. Psychosom. Med. 60, 610–615.

Brandenberger, G., Ehrhart, J., Piquard, F., and Simon, C. 2001. Inverse coupling between ultradian oscillations in delta wave activity and heart rate variability during sleep. Clin. Neurophysiol. 112, 992–996.

Brandenberger, G., Viola, A. U., Ehrhart, J., Charloux, A., Geny, B., Piquard, F., and Simon, C. 2003. Age related changes in cardiac autonomic control during sleep. J. Sleep Res. 12, 173–180.

Bromm, B. and Lorenz, J. 1998. Neurophysiological evaluation of pain. Electroenceph. Clin. Neurophysiol. 107, 227–253.

Cairns, B. E., McErlane, S. A., and Fragoso, M. C. 1996. Spontaneous discharge and peripherally evoked orofacial responses of trigemino-thalamic tract neurons during wakefulness and sleep. J. Neurosci. 16, 8149–8159.

Carley, D. W., Applebaum, R., Basner, R. C., Onal, E., and Lopata, M. 1997. Respiratory and arousal responses to acoustic stimulation. Chest 112(6), 1567–1571.

Carli, G., Montesano, A., Rapezzi, S., and Paluffi, G. 1987. Differential effects of persistent nociceptive stimulation on sleep stages. Behav. Brain Res. 26, 89–98.

Chapman, S. 2001. The effects of opioids on driving ability in patients with chronic pain. APS Bull. 11, 5–9.

Coghill, R. C., Talbot, J. D., Evans, A. C., Meyer, E., Gjedde, A., Bushnell, M. C., and Duncan, G. H. 1994. Distributed processing of pain and vibration by the human brain. J. Neurosci. 14, 4095–4108.

Cohen, M., Menefee, L. A., and Doghramji, K. 2000. Sleep in chronic pain: problems and treatments. Int. Rev. Psychiat 12, 115–126.

Cohen, H., Neumann, L., Alhosshle, A., Kotler, A., Abu-Shakra, M., and Buskila, D. 2001. Abnormal sympathovagal balance in men with fibromyalgia. J. Rheumatol. 28, 581–589.

Côté, K. A and Moldofsky, H. 1997. Sleep, daytime symptoms, and cognitive performance in patients with fibromyalgia. J. Rheumatol. 24, 2014–2023.

Dao, T. T. T., Lund, J. P., and Lavigne, G. J. 1994. Comparison of pain and quality of life in bruxers and patients with myofascial pain of the masticatory muscles. Orofac. Pain. 8, 350–356.

Dauvilliers, Y. and Touchon, J. 2001. Le sommeil du fibromyalgique: revue des données cliniques et polygraphiques. Neurophysiol. Clin. 2001 31, 18–33.

Donald, F., Esdaile, J. M., Kimoff, J. R., and Fitzcharles, M. A. 1996. Musculoskeletal complaints and fibromyalgia in patients attending a respiratory sleep disorders clinic. J. Rheumatol. 23(9), 1612–1616.

Drewes, A. M., Nielsen, K. M., and Arendt-Nielsen, L. 1997. The effect of cutaneous and deep pain on the electroencephalogram during sleep – An experimental study. Sleep 20, 623–640.

Foley, D., Ancoli-Israel, S., Britz, P., and Walsh, J. 2004. Sleep disturbances and chronic disease in older adults: results of the 2003 National Sleep Foundation Sleep in America Survey. J. Psychosom. Res. 56(5), 497–502.

Foo, H. and Mason, P. 2003. Brainstem modulation of pain during sleep and waking. Sleep Med. Rev. 7, 145–154.

Gais, S., Plihal, W., and Wagner, U. 2000. Early sleep triggers memory for early visual discrimination skills. Nature Neurosci 3, 1335–1339.

George, C. F. P. 2003. Driving simulators in clinical practice. Sleep Med. Rev. 7, 311–320.

Gold, A. R., Dipalo, F., Gold, M. S., and Broderick, J. 2004. Inspiratory airflow dynamics during sleep in women with fibromyalgia. Sleep 27(3), 459–466.

Gozal, D. and O'Brien, L. M. 2004. Snoring and obstructive sleep apnoea in children: why should we treat? Paediatr. Respir. Rev. 5 (Suppl A), S371–S376.

Hall, M., Vasko, R., Buysse, D., Ombao, H., Chen, Q., Cashmere, J. D., Kupper, D., and Thayer, J. F. 2004. Acute stress affects heart rate variability during sleep. Psychosomatic Med 66, 56–62.

Harstall, C. and Ospina, M. 2003. How prevalent is chronic pain. Pain Clin. Update 11(2), 1–4.

Harman, K., Pivik, R. T., D'Eon, J. L., Wilson, K. G., Swenson, J. R., and Matsunaga, L. 2002. Sleep in depressed and nondepressed participants with chronic low back Pain: electroencephalographic and behavior findings. Sleep 25(7), 775–783.

Hernandez-Peon, R., O'Flaherty, J. J., and Mazzuchelli-O'Flaherty, A. L. 1965. Modifications of tactile evoked potentials at the spinal trigeminal sensory nucleus during wakefulness and sleep. Experimental Neurol. 13, 40–57.

Kato, T., Guitard, F., Rompré, P. H., Lund, J. P., Montplaisir, J. Y., and Lavigne, G. J. 2004. Experimentally induced arousals during sleep: a cross-modality matching paradigm. 2004. J. Sleep Res. 13, 1–10.

Kewman, D. G., Vaishampayan, N., and Zald, D. 1991. Cognitive impairment in musculoskeletal pain patients. Int'l. J. Psychiat. Med. 21, 253–262.

Koltyn, K. F., Focht, B. C., and Ancker, J. M. 1999. Experimentally induced pain perception in men and women in the morning and evening. Int. J. Neurosci. 98, 1–11.

Koopman, C., Nouriani, B., and Erickson, V. E. 2002. Sleep disturbances in women with metastatic breast cancer. Breast Journal 6, 362–370.

Kshatri, A. M., Baghdoyan, H. A., and Lydic, R. 1998. Cholinomimetics, but not morphine, increase antinociceptive behavior from pontine reticular regions regulating rapid eye movement sleep. Sleep 21(7), 677–685.

Kundermann, B. and Lautenbacher, S. 2007. Effects of Impaired Sleep Quality and Sleep Deprivation on Diurnal Pain Perception. In: Sleep and Pain (eds. C. Lavigne, B. J. Sessle, M. Choinière, P. J. Soja), Seattle IASP Press.

Kundermann, B., Krieg, J. C., Schreiber, W., and Lautenbacher, S. 2004a. The effect of sleep deprivation on pain. Pain Res. Manage. 39(1), 25–32.

Kundermann, B., Spernal, J., Huber, M. T., Krieg, J. C., and Lautenbacher, S. 2004b. Sleep deprivation affects thermal pain thresholds but not somato-sensory thresholds in healthy volunteers. Psychosomatic. Med. 66, 932–937.

Landis, C. A., Robinson, C. R., and Levine, J. D. 1988. Sleep fragmentation in the arthritic rat. Pain 34, 93–99.

Landis, C. A., Lentz, M. J., Rothermel, J., Buchwald, D., and Shaver, J. L. 2004. Decreased sleep spindles and spindle activity in midlife women with fibromyalgia and pain. Sleep 27(4), 741–750.

Landro, N. I., Stiles, T. C., and Sletvold, H. 1997. Memory functioning in patients with primary fibromyalgia and major depression and healthy controls. J. Psychosom Res. 42, 297–306.

Latham, J. and Davis, B. D. 1994. The socioeconomic impact of chronic pain. Disabil. Rehabil. 16(1), 39–44.

Lavigne, G. J. and Manzini, C. 2007. Pain and Poor Sleep: Impact on Public Health. In: The Public Health Impact of Sleep Disorders (eds. S. R. Pandi-Perumal and D. Leger), pp. 193–208.

Lavigne, G. J., Brousseau, M., Kato, T., Mayer, P., Manzini, C., Guitard, F., and Montplaisir, J. Y. 2004. Experimental pain perception remains equally active over all sleep stages. Pain 110, 646–655.

Lavigne, G. J., McMillan, D., and Zucconi, M. 2005. Pain and Sleep. In: principles and practice sleep of medicine (eds. M. H. Kryger, T. Roth, W. C. Dement), Vol. 106, pp. 1246–1255. Elsevier Saunders.

Lavigne, G. J., Zucconi, M., and Castronovo, C. 2000. Sleep arousal response to experimental thermal stimulation during sleep in human subjects free of pain and sleep problems. Pain 84, 283–290.

Lavigne, G. J., Zucconi, M., Castronovo, C., Manzini, C., Veglia, F., Smirne, S., and Ferini-Strambi, L. 2001. Heart rate changes during sleep in response to experimental thermal (nociceptive) stimulations in healthy subjects. Clin. Neurophysiol. 112, 532–535.

Lee, R. S., Steffenson, S. C., and Henriksen, S. J. 2001. Discharge profiles of ventral tegmental area GABA neurons during movement, anesthesia, and the sleep-wake cycle. J. Neurosci. 21(5), 1757–1766.

Lentz, M. J., Landis, C. A., and Rothermel, J. 1999. Effects of selective slow wave sleep disruption on musculoskeletal pain and fatigue in middle aged women. J. Rheumatol. 26, 1586–1592.

Le Pera, D., Svensson, P., Valeriani, M., Watanabe, I., Arendt-Nielsen, L., and Chen, A. C. 2000. Long-lasting effect evoked by tonic muscle pain on parietal EEG activity in humans. Clin. Neurophysiol. 111(12), 2130–2137.

Lobbezoo, F., Thu Thon, M., and Montplaisir, J. 1996. Relationship between sleep, neck muscle activity, and pain in cervical dystonia. Can. J. Neurol. Sci. 23, 285–290.

Macfarlane, T. V. and Worthington, H. V. 2004. Association between orofacial pain and other symptoms: a population based study. Oral. Biosci and Med. 1, 45–54.

Mahowald, M. L. and Mahowald, M. W. 2000. Nighttime sleep and daytime functioning (sleepiness and fatigue) in less well-defined chronic rheumatic diseases with particular reference to the 'alpha-delta NREM sleep anomaly'. Sleep Med. 1, 195–207.

Martin, S. A., Aikens, J. E., and Chervin, R. D. 2004. Toward cost-effectiveness analysis in the diagnosis and treatment of insomnia. S. Med. Rev. 8(1), 63–72.

Martin, S. E., Engleman, M. H., Deary, I. J., and Douglas, N. J. 1996. The effect of sleep fragmentation on daytime function. Am. J. Respir. Crit. Care Med. 153, 1328–1332.

Martínez-Lavín, M., Hermosillo, A. G., and Rosas, M. 1998. Circadian studies of autonomic nervous balance in patients with fibromyalgia: a heart rate variability analysis. Arthri. Rheum. 41, 1966–1971.

McCracken, L. M. and Iverson, G. L. 2002. Disrupted sleep patterns and daily functioning in patients with chronic pain. Pain Res. Manage. 7, 75–79.

McMillan, D., Landis, C., Lentz, M., Shaver, J., Heitkemper, M., Rothermel, J., Blunden, P., and Buchwald, D. 2004. Hearth rate variability during sleep in women with fibromyalgia. Sleep. 27, A324 (abstract Suppl.).

Meana, M., Cho, R., and DesMeules, M. 2004. Chronic pain: the extra burden on Canadian women. BMC Women's Health 4 (Suppl 1), S17.

Mednick, S., Nakayama, K., and Stickgold, R. 2003. Sleep dependent learning: a nap is as good as a night. Nature Neurosci. 6, 697–698.

Menefee, L. A., Cohen, M., and Anderson, W. R. 2000. Sleep disturbance and nonmalignant chronic pain: a comprehensive review of the literature. Pain Med 1, 156–172.

Moldofsky, H. 2001. Sleep and pain: clinical review. Sleep Med. Rev. 5, 387–398.

Moldofsky, H. and Scarisbrick, P. 1976. Induction of neurasthenic musculoskeletal pain syndrome by selective sleep stage deprivation. Psychosom. Med. 38, 35–44.

Morin, C. M., Gibson, D., and Wade, J. 1998. Self-reported sleep and mood disturbance in chronic pain patients. Clin. J. Pain. 14, 311–314.

Morrison, A. R., Sanford, L. D., and Ross, R. J. 2000. The amygdale: A critical modulator of sensory influence on sleep. Biol. Signals Recept. 9, 283–296.

Nicassio, P. M., Moxham, E. G., and Schuman, C. E. 2002. The contribution of pain, reported sleep quality, and depressive symptoms to fatigue in fibromyalgia. Pain. 100, 271–279.

Nofzinger, E. A. and Derbyshire, S. W. G. 2007. Pain Imaging in Relation to Sleep. In: Sleep and Pain (eds. G. Lavigne, B. J. Sessle, M. Choinière, P. J. Soja), pp. 153–173. IASP Press.

Novak, M., Mucsi, I., Shapiro, C. M., Rethelyi, J., and Kopp, M. S. 2004. Increased Utilization of health services by insomniacs – an epidemiological perspective. J. Psychosom. Res. 56, 527–536.

Okifuji, A. and Turk, D. C. 2002. Stress and psychophysiological dysregulation in patients with fibromyalgia syndrome. Appl. Psychophysiol. Biofeed 27(2), 129–141.

Okura, K., Lavigne, G. J., Montplaisir, J. Y., Rompré, P. H., Huynh, N., and Lafranchi, P. 2005. Slow wave activity and heart rate variation are less dominant in fibromyalgia and chronic pain patients than in controls. Sleep 28, 0894.

Okura, K., Rompré, S., Manzini, C., Rompré, P. H., Huynh, N., Montplaisir, J. Y., and Lavigne, G. J. 2004. Sleep duration and quality in chronic pain patients in comparison to the sleep of fibromyalgia, insomnia, RLS and control subjects. Sleep. 27: (abstract Suppl.) A329 No. 736.

Older, S. A., Battafarano, D. F., and Danning, C. L. 1998. The effects of delta wave sleep interruption on pain thresholds and fibromyalgia like symptoms in healthy subjects; correlations with insulin like growth factor 1. J. Rheum. 25, 1180–1186.

Onen, A. H., Alloui, A., and Gross, A. 2001. The effects of total sleep deprivation, selective sleep interruption and sleep

recovery on pain tolerance thresholds in healthy subjects. J. Sleep Res. 10, 35–42.

Parrino, L., Zucconi, M., and Terzano, M. G. Sleep Fragmentation and Arousal in the Pain Patient. In: Sleep and Pain (eds G. Lavigne, B. J. Sessle, M. Choinière, P. J. Soja,) pp. 213–231. IASP Press.

Peigneux, P., Laureys, S., Cleeremans, A., and Maquet, P. 2003. Cerebral Correlates of Memory Consolidation during Human Sleep : Contribution of Functional Neuroimaging. In: Sleep and Brain Plasticity. (eds. P. Maquet, C. Smith, R. Stickgold), Chap. 11,209–224.

Peyron, R., Laurent, B., and Garcia-Larrea, L. 2000. Functionnal imaging of brain responses to pain. A review and meta-analysis. Neurophysiol. Clin. 30, 263–288.

Pompeiano, O. and Swett, J. E. 1963. Actions of grated cutaneous and muscular afferent volleys on brain stem units in the decerebrate, cerebellectomized cat. Arch. Ital. Biol. 101, 552–583.

Rains, J. C. and Penzien, D. B. 2002. Chronic headache and sleep disturbance. Curr. Pain Headache Rep. 6, 498–504.

Rains, J. C. and Penzien, D. B. 2003. Sleep and chronic pain challenges to the αEEG sleep pattern as a pain specific sleep anormaly. J. Psychosom. Res. 54, 77–83.

Raymond, I., Nielsen, T. A., and Lavigne, G. J. 2001. Quality of sleep and its daily relationship to pain intensity in hospitalized adult burn patients. Pain 92, 381–388.

Rechtschaffen, A. and Kales, A. 1968. A manual of standardized terminology, techniques and scoring techniques for sleep stages of human subjects. Brain Res. Inst., University of California.

Reilly, P. A. and Littlejohn, G. O. 1993. Diurnal variation in the symptoms and signs of the fibromyalgia syndrome (FS). J. Musculoskel. Pain 1, 237–243.

Riley, III, J. L., Benson, M. B., and Gremillion, H. A. 2001. Sleep disturbances in orofacial pain patients: pain-related or emotional distress? J. Craniomandib. Pract. 19, 106–113.

Rogers, E. J. and Vilkin, B. 1978. Diurnal variation in sensory and pain thresholds correlated with mood states. J. Clin. Psychiat. 39(5), 431–438.

Roizenblatt, S., Moldofsky, H., and Benedito-Silva, A. A. 2001. Alpha sleep characteristics in fibromyalgia. Arthritis. Rheum. 44, 222–230.

Sandrini, G., Milanov, I., and Rossi, B. 2001. Effects of sleep on spinal nociceptive reflexes in humans. Sleep 24, 13–17.

Sayar, K., Arikan, M., and Yontem, T. 2002. Sleep quality in chronic pain patients. Can. J. Psychiatry. 47, 844–848.

Sergi, M., Rizzi, M., Braghiroli, A., Puttini, P. S., Greco, M., Cazzola, M., and Andreoli, A. 1999. Periodic breathing during sleep in patients affected by fibromyalgia syndrome. Eur. Respir. J. 14, 203–208.

Shaw, I. R., Lavigne, G., Mayer, P., and Choinière, M. 2005. Acute intravenous administration of morphine perturbs sleep architecture in healthy pain-free young adults: a preliminary study. Sleep. 28(6), 677–682.

Siegel, J. 2004. Brain mechanisms that control sleep and waking. Naturwissens Chaften 91, 355–365.

Skinner, R. D., Homma, Y., and Garcia-Rill, E. 2004. Arousal mechanisms related to posture and locomotion. 2. Ascending modulation. Prog. Brain Res. 143, 291–298.

Smith, M. T. and Haythornthwaite, J. A. 2007. Cognitive-Behavioral Treatment for Insomnia and Pain. In: Sleep and Pain (eds. G. Lavigne, B. J. Sessle, M. Choinière, P. J. Soja), pp. 439–457. IASP Press.

Smith, M. T., Perlis, M. L., and Smith, M. S. 2000. Sleep quality and presleep arousal in chronic pain. J. Behav. Med. 23, 1–13.

Soja, P. J. 2007. Modulation of Prethalamic Sensory Inflow during Sleep versus Wakefulness. In: Sleep and Pain (eds. G. Lavigne, B. J. Sessle, M. Choinière, P. J. Soja), pp. 45–76. IASP Press.

Soja, P. J., Pang, W., Taepavarapruk, N., and McErlane, S. A. 2001. Spontaneous spike activity of spinoreticular tract neurons during sleep and wakefulness. Sleep 24, 18–25.

Staedt, J., Windt, H., and Hajak, G. 1993. Cluster arousal analysis in chronic pain- disturbed sleep. J. Sleep Res. 2, 134–137.

Stepanski, E. J. 2002. The effect of sleep fragmentation on daytime function. Sleep 25(3), 268–276.

Steriade, M. and Timofeev, I. 2003. Neuronal Plasticity during Sleep Oscillations in Corticothalamic Systems. In: Sleep and Brain Plasticity (eds. P. Maquet, C. Smith, and R. Stickgold), Chap. 14,270–291.

Stickgold, R., James, L. T., and Hobson, J. A. 2000. Visual discrimination learning requires sleep after training. Nature Neurosci. 3(12), 1237–1238.

Stoller, M. K. 1994. Economic effects of insomnia. Clini. Ther. 16(5), 873–897.

Stone, A. A., Brokerick, J. E., Porter, L. S., and Kaell, A. T. 1997. The experience of rheumatoid arthritis pain and fatigue: examining momentary reports and correlates over one week. Arthritis. Care Res. 10(3), 185–193.

Strian, F., Lautenbacher, S., and Galfe, G. 1989. Diurnal variations in pain perception and thermal sensitivity. Pain 36, 125–131.

Sutton, D. A., Moldofsky, H., and Badley, E. M. 2001. Insomnia and health problems in Canadians. Sleep 24, 665–670.

Tsoukatos, J., Kiss, Z. H. T., Davis, K. D., Tasker, R. R., and Dostrovsky, J. O. 1997. Patterns of neuronal firing in the human lateral thalamus during sleep and wakefulness. Exp. Brain Res. 113, 273–282.

Van Dongen, H. P., Maislin, P., Mullington, J. M., and Dinges, D. F. 2003. The cumulative cost of additional wakefulness: dose–response effects on neurobehavioral functions and sleep physiology from chronic sleep restriction and total sleep deprivation. Sleep 26(2), 117–126.

Velluti, R. A., Pena, J. L., and Pedemonte, M. 2000. Reciprocal actions between sensory signals and sleep. Biol. Signals Recept. 9, 297–308.

Wang, X., Inui, K., Qiu, Y., Hoshiyama, M., Tran, T. D., and Kakigi, R. 2003. Effects of sleep on pain related somato sensory evoked potentials in humans. Neurosci. Res. 45, 53–57.

Wesenten, N. J., Balkin, T. J., and Belenky, G. 1999. Does sleep fragmentation impact recuperation? A review and reanalysis. J. Sleep Res. 8, 237–245.

Widerstrom-Noga, E. G., Felipe-Cuervo, E., and Yezierski, R. P. 2001. Chronic pain after spinal injury: interference with sleep and daily activities. Arch. Phys. Med. Rehabil. 82, 1571–1577.

Yatani, H., Studts, J., and Cordova, M. 2002. Comparison of sleep quality and clinical and psychologic characteristics in patients with temporomandibular disorders. J. Orofac. Pain 16, 221–228.

Zubieta, J. K., Smith, Y. R., Bueller, J. A., Xu, Y., Kilbourn, M. R., Jewett, D. M., Meyer, C. R., Koeppe, R. A., and Stohler, C. S. 2001. Regional Mu opioid receptor regulation of sensory and affective dimensions of pain. Science 293, 311–315.

Relevant Website

http://www.sleepfoundation.org – National Sleep Foundation.

53 Pharmacological Modulation of Pain

A Dray, AstraZeneca Research and Development, Montreal, PQ, Canada

53.1 Introduction

Chronic pain is a complex integration of sensory, affective, and cognitive processes encompassing a variety of clinical conditions that have been generalized as nociceptive (rheumatoid and osteoarthritis (OA)), neuropathic (postdiabetic neuropathy, posttraumatic neuralgia), or etiologically mixed (low back pain, cancer pain; for references see Merskey, H. *et al.*, 2005; Campbell, J. N. *et al.*, 2006). Pain conditions are etiologically diverse and there is limited understanding of clinical pain mechanisms and of environmental and social factors that aggravate pain such as stress and anxiety. In addition, it has been difficult to provide a unifying linkage of pain symptoms with the cellular mechanistic processes underlying these symptoms that would rationalize chronic pain therapy. The complexity of chronic pain is further reflected in the diversity of molecular targets considered to drive pain

processes. Furthermore there is the evolving view that the pharmacology of pain may alter with time-dependent expression of targets, in the initiation versus the maintenance phases of pain, and that target expression is state-dependent (related to relative activity of the pain process or pathway). Thus poor understanding of pain pharmacology has seriously hampered the introduction of new drugs as there have been few drugs introduced with a new mechanism of action within living memory.

A recent step to building rational pain therapy has embraced a mechanism-based approach guided by advances in pain pharmacology. Thus major symptoms such as hyperalgesia, allodynia, and spontaneous pain have been linked with some of the identifiable, and often overlapping, neural cellular processes. The key processes are considered to be sensitization (reduced threshold for stimulation of pain pathways), hyperexcitability (amplification or prolongation of nerve discharges in pain pathways), and spontaneous nerve activity (ectopic and/or spontaneous discharges in pain pathways). These processes or mechanisms are driven by molecular changes that provide the targets for pharmacological characterization and analgesia intervention (summarized in Figure 1). It is important to remember that for the most part these neural mechanisms and their molecular drivers have been explored in animal models and rarely consolidated to the same degree in human studies.

There have been a number of excellent reviews of pain mechanisms and molecules (Hunt, S. P. and Mantyh, P., 2001; Scholz, J. and Woolf, C. J., 2002; Watkins, L. R. and Maier, S., 2002) in the peripheral nervous system and central nervous system (CNS). Causative factors for pain are inflammatory mediators, factors released and made in response to nerve injury, phenotypic changes in neural pathways as well as supporting glial cells, structural modifications including nerve sprouting (sympathetic and sensory fibers), nerve rewiring (sensory and CNS fibers), and neurodegeneration of specific neurons in the CNS resulting in a hyperexcitable nervous system. Specific molecular interactions that drive the aforementioned excitability changes may differ according to the specific injury and consequent chemical environments that are created. The molecular interactions involve all major families of regulatory proteins (G-protein-coupled receptors (GPCRs), ion channels, regulatory enzymes, neurotrophins, kinases), offering an abundance of analgesic targets and therapeutic opportunities. The pharmacology of these targets is evolving and is dependent on the availability of reliable chemical and biopharmaceutical tools.

In this chapter I have selected from the emerging pharmacology of pain molecular approaches (summarized in Table 1) that will provide direction for future pain treatments.

53.2 G-Protein-Coupled Receptors

GPCRs have been a relatively tractable family of drug targets in a number of therapeutic arenas. A variety of GPCRs are involved in regulating neural excitability in pain pathways, often through the action of a variety of inflammatory mediators. These mediators act peripherally or centrally, via an equally large variety of specific membrane receptors (Scholz, J. and Woolf, C. J., 2002). In general, GPCRs signal via the production of second messengers (cyclic adenosine monophosphate (cAMP), cyclic guanosine monophosphate (cGMP), diacylglycerol (DAG), phospholipase C (PLC)) coupled with intracellular protein kinases and phosphatases (protein kinase A, several isotypes of protein kinase C (PKC)) which in their turn, phosphorylate and dephosphorylate specific cellular proteins including ligand- (e.g., TRPV1) and voltage-gated membrane ion channels (e.g., NaV1.8) which are important in regulating nerve excitability.

53.3 Opioids and Their Receptors

Opioids, and morphine in particular, are among the most effective pain medications. They act at peripheral, spinal, and supraspinal sites through a variety of opioid receptors (μ-, δ-, and κ-opioid receptors). These receptors are considered to be targets for an endogenous opioid system that has been extensively reviewed (Yaksh, T. L., 1997). μ-Opioid receptor activation commonly causes a variety of well-documented, target-related, CNS side effects including sedation, dysphoria, respiratory depression, and constipation. Thus there has been a vigorous attempt to exploit the peripheral antinociceptive actions of opioids as a means of avoiding CNS side effects. In support of this approach, sensory neurons express and transport opioid receptors to both the central and peripheral terminals. At central terminals, opioids reduce transmitter release from primary afferent nociceptors, thus blocking synaptic transmission while in the periphery opioid receptor activation directly hyperpolarize sensory neurons and attenuate nerve sensitization or hyperexcitability induced by inflammation or injury (Hurley, R. W. and Hammond, D. L., 2000; Sawynok, J., 2003).

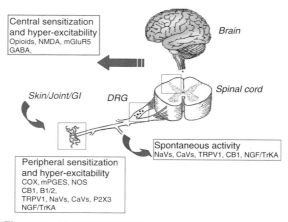

Figure 1 Mechanistic processes in chronic pain and key molecular drivers.

Table 1 Pharmacological modulators of pain and emerging mechanisms of action and clinical indications.

Molecular Target	Pharmacological tools		Analgesia mechanism and targeted tissue	Pain indications
	Agonist	Antagonist		
μ-Opioid receptors	Morphine, fentanyl	Naloxone, naltrexone	Agonist: reduced peripheral and CNS excitability	Acute and chronic pain states
δ-Opioid receptor	Enkephalin, DPDPE, SNC80	Naltrindole	Agonist: Modulation of CNS pathways	Nociceptive and inflammatory
B1 receptor	Des Arg9 bradykinin	des Arg10, HOE-140; SSR240612, NVP-SAA 164	Antagonist: inflammatory tissue: pain pathway: PNS, CNS	Nociceptive and inflammatory
B2 receptor	Bradykinin	Icatibant, bradyzide	Antagonist: inflammatory tissue: pain pathway: PNS, CNS	Nociceptive and inflammatory
PAR2	Proteases	Unknown	Antagonist: PNS	Nociceptive and inflammatory
CB1	THC, anandamide, 2-arachidonylglycerol, palmitoylethanolamide, WIN55, 212-2, ajulemic acid	SR141716A (rimonabant), SR147778,	Agonist: Pain pathways PNS and CNS	Nociceptive and neuropathic
CB2	THC, anandamide, 2-arachidonylglycerol, palmitoylethanolamide, WIN55, 212-2, ajulemic acid, GW405833, HU-308, AM 1241	SR144428	Agonist: Pain pathways PNS and CNS	Nociceptive and neuropathic
SNSR	BAM8-22, γ2-MSH	Unknown	Antagonist: PNS	Nociceptive
COX-1 constitutive		Ibuprofen, naproxene	Inhibition of PG synthesis	Nociceptive and inflammatory
COX-2 induced		Ibuprofen, naproxene, celecoxib, rofecoxib	Inhibition of PG synthesis in PNS and CNS	Nociceptive and inflammatory
mPGES		Unknown	Inhibition of PG synthesis in PNS and CNS	Nociceptive and inflammatory
IL-1β	IL-1β	Unknown	Antibody versus ligand	Nociceptive and neuropathic
TNF-α	TNF-α	Anti-TNF, etanercept, adalimumab	Antibody versus ligand	Nociceptive and neuropathic
α-2	Clonidine, dexmedetomidine		Inhibition of neuroexcitability in PNS and CNS	Nociceptive and neuropathic
iGluR	Glutamate, kainate, AMPA, NMDA	LY293558, GV196771, ifenprodil, CP-101, 606	Antagonist: CNS pathways	Nociceptive and neuropathic
mGluR (5)	Glutamate	MPEP, SIB1757	Antagonist: CNS pathways	Nociceptive and neuropathic
GABA$_A$	Muscimol	Bicuculline	Agonist: reduced CNS excitability	Nociceptive and neuropathic
GABA$_B$	Baclofen	CGP35348	Agonist: reduced CNS excitability	Nociceptive and neuropathic
TRPV1	Capsaicin	Capsazepine, DD161515, SB705498	Antagonist: PNS excitability	Nociceptive and neuropathic
TRPA1	Cinnamaldehyde	Unknown	Antagonist: PNS	Neuropathic

(Continued)

Table 1 (Continued)

Molecular Target	Pharmacological tools		Analgesia mechanism and targeted tissue	Pain indications
	Agonist	Antagonist		
TRPM1	Menthol, icilin	Unknown	Antagonist: PNS	Neuropathic
P2X3	ATP, α, β-methy ATP	A-3174919	Antagonist: PNS	Nociceptive
P2X4	ATP	TNP-ATP	Antagonist: CNS	Neuropathic
P2X7	ATP	Unknown	Antagonist: CNS and PNS excitability	Nociceptive and neuropathic
ASIC1-3	Low pH	A-317567	Antagonist: PNS	Nociceptive
NaV1.3, NaV1.8, NaV1.7, NaV1.9		Lidocaine, mexiletine, lamotrigine	Blocker: PNS excitability	Nociceptive and neuropathic
CaV2.2		SNX-111, NMED-160	Blockers: PNS and CNS excitability	Nociceptive and neuropathic
CaV3.2		Ethosuximide	Blocker: PNS	Neuropathic
NGF	NGF	ALE0540, PD90780, RN624 (mAb)	Antiligand: PNS excitability	Nociceptive and neuropathic
P38-kinase	ATP	SB203580, CNI-14930	Blocker: PNS and CNS excitability	Nociceptive and neuropathic
iNOS		L-NAME, GW27415	Blocker: PNS and CNS excitability	Nociceptive and neuropathic

AMPA, α-amino-3-hydroxy-5-methyl-4-isoxazolepropionic acid; ATP, adenosine triphosphate; B, bradykinin; CaV, voltage-gated calcium channels; CB, cannabinoid; CNS, central nervous system; COX, cyclo-oxygenase; DPDPE, [D-Pen2,D-Pen5]-enkephalin; GABA, gamma-aminobutyric acid; iNOS, inducible nitric oxide synthase; L-NAME, L-N-nitro-L-arginine methyl ester; mGluR, metabotropic glutamate receptor; MPEP, 2-methyl-6-(phenylethynyl)-pyridine; mPGES, microsomal-associated prostaglandin E synthase; NaV, voltage-gated sodium channel; NGF, nerve growth factor; NMDA, N-methyl-D-aspartic acid; PAR, protease-activated receptor; PG, prostaglandin; PNS, peripheral nervous system; SNSR, sensory neuron specific receptor; THC, tetrahydrocannabinol; mAb, monoclonal antibody; TNP-ATP, 2′,3′-O-(2,4,6-trinitrophenyl) adenosine 5′-triphosphate.

In addition peripherally mediated opioid analgesia may be enhanced because μ-opioid receptor expression is increased by inflammation and nerve injury (Stein, C. *et al.*, 2003). This is also accompanied by a peripheral increase in the expression of opioids peptides. Novel opioids such as [8-(3,3-diphenyl-propyl)-4-oxo-1-phenyl-1,3,8-triazaspiro[4.5]dec-3-yl]-acetic acid (DiPOA) and the antidiarrheal drug loperamide, whose distribution is peripherally restricted, have shown efficacy in a number of post-operative, inflammatory, and neuropathic pain models (Whiteside, G. T. *et al.*, 2005a; Shidoda, K. *et al.*, 2006).

The pharmacology and analgesic efficacy of δ-opioid receptors (DOR) have also been explored since there is a potential for analgesic efficacy without the confounding side effects of other opioid (μ and κ) receptor therapies. A variety of studies have supported the concept that δ-selective ligands produce analgesia without notable sedation, respiratory depression, or inhibition of gastrointestinal motility. These studies have used genetic methods ranging from the knockdown of receptors to deletion of the DOR gene as well as pharmacological tools such as selective peptides and nonpeptide agonists ([D-Pen2,D-Pen5]-enkephalin (DPDPE), SNC80, AM-390) and antagonists (e.g., naltrindole; see Dray, A., 1999 for references).

δ Ligands may provide less analgesic efficacy than morphine but their effectiveness appears to depend on the pain stimulus, the type of injury, and the influence of their neurochemical environment. Thus, systemically administered δ ligands have low analgesic efficacy in acute pain models but show robust analgesia efficacy in a variety of chronic pain conditions accompanies by inflammation (Cahill, C. M. *et al.*, 2001). Enhanced activity has been attributed to stimulus-induced trafficking of DOR from the cytoplasm to nerve membranes in CNS neurons (Cahill, C. M. *et al.*, 2001). Activity and environment-dependent receptor trafficking also appears as an important regulatory mechanism in sensory neurons (Bao, L. *et al.*, 2003) but further studies are required to understand the significance of this in clinical pain and analgesia.

53.4 Kinins and Their Receptors

Bradykinin is an important peptidic mediator of inflammatory pain causing nociceptor activation and sensitization via constitutively expressed bradykinin B2 receptors (BK2) on peripheral nociceptors (Dray, A., 1997). It also stimulates the production and release of prostaglandin (PGs) cytokines and nitric oxide (NO). The abundant metabolite of bradykinin, des Arg9 bradykinin (kallidin), activates BK1 receptors which also occur constitutively, but in low abundance, in the periphery (Dray, A. and Perkins, M. N., 1993; Wotherspoon, G. and Winter, J., 2000) and in the primate CNS (Shughrue, P. J. *et al.*, 2003). Both receptors are coupled to similar transduction mechanism including G-αi and G-αq causing stimulation of DAG, PLC, and the release of intracellular calcium.

BK2 receptors are considered to undergo desensitization following prolonged kinin exposure, whereas BK1 receptors do not desensitize rapidly and thus may promote more prolonged pain and inflammation. In keeping with this, BK1 receptors are dramatically unregulated in many tissues, including sensory neurons (Levy, D. and Zochodne, D. W., 2000) and spinal cord (Fox, A. *et al.*, 2003; Ferreira, J. *et al.*, 2005), following tissue or ultraviolet injury (Eisenbarth, H. *et al.*, 2004) or traumatic nerve injury and by mediators such as interleukin (IL)-1β or the neurotrophin glial cell line-derived neurotrophic factor (GDNF; Fox, A. *et al.*, 2003; Valleni, V. *et al.*, 2004). These kinins cause a cascade of secondary changes, including prostanoid production, phosphorylation of signaling proteins such as PKC, and the sensitization of sensory transducers such as the TRPV1 receptor (Marceau, F. *et al.*, 1998). These events are linked with heat and mechanical hyperalgesia (Liang, Y. F. *et al.*, 2001; Fox, A. *et al.*, 2003).

There is an abundance of evidence supporting the analgesic potential of BK2 receptor antagonists (Stewart, J. M., 2004). For example, BK2 antagonists (icatibant [HOE-140], bradyzide) and BK1 antagonist (des Arg10 HOE-140; SSR240612) produce robust antihyperalgesic effects in a variety of animal models of inflammatory hyperalgesia (Burgess, G. M. *et al.*, 2000; Fox, A. *et al.*, 2003, Gougat, J. *et al.*, 2004) as well as in models of nerve injury-induced hyperalgesia (Gabra, B. H. and Sirois, P., 2003). In keeping with this deletion of the BK receptor genes reduces immune cell chemotaxis, spinal sensitization, as well as heat and mechanical allodynia following traumatic and diabetic nerve lesions (Pesquero, J. B. *et al.*, 2000; Ferreira, J. *et al.*, 2005; Gabra, B. H. *et al.*, 2005). To strengthen the translation of BK1 receptor therapy to humans, transgenic mice expressing human BK1 receptors, show robust inflammatory pain that was attenuated by the oral administration of the human-selective BK1 antagonist NVP-SAA 164 (Fox, A. *et al.*, 2005).

Kinin antagonists have not been extensively evaluated in clinical pain although they inhibit other kinin-mediated changes in humans, for example, cancer, allergy, and vasodilatation. However, recent clinical study using a single dose intra-articular joint administration of Icatibant (Flechtenmacher, J. *et al.*, 2004) produced long-lasting (up to 1 week) relief of movement-related pain in patients suffering from OA. This study highlights a strong proof of principle for BK2 involvement in OA and that localized monotherapy with a selective antagonist can produce significant pain relief.

53.5 Protease-Activated Receptors

Several proteases, from circulating inflammatory cells and vascular epithelium, cleave a variety of protease-activated receptors (PARs) that are associated with sensory neurons (Vergnolle, N. *et al.*, 2003). PAR2 in particular has been highlighted as a regulator of sensory nerve excitability and plays an important role in inflammatory hyperalgesia. Thus PAR2 agonists contribute to neurogenic inflammation through the stimulation of peripheral substance P and calcitonin gene-related peptide (CGRP) release and to spinal hyperexcitability via release of these neuropeptides in the spinal cord. PAR2 receptor activation directly increases dorsal root ganglion (DRG) excitability and sensitizes TRPV1 receptors via a PKC mechanism (Amadesi, S. *et al.*, 2004); thus contributing to inflammation-induced hyperalgesia. In keeping with this, deletion of the *TRPVR1* gene or pretreatment with the TRPV1 antagonist capsazepine, abolished PAR2-induced thermal hyperalgesia, suggesting that TRPV1 is required for PAR2-induced hyperalgesia. PAR1 is also coexpressed on peptide-containing DRGs, and although PAR1 agonists induce neurogenic inflammation they do not sensitize TRPV1 receptors, nor do they induce hyperalgesia. Overall these data suggest that PAR2 antagonists would have antihyperalgesic activity (Mantyh, P. W. and Yaksh, T. L., 2001)

53.6 Cannabinoids and Their Receptors

Two major cannabinoid receptor subtypes, CB1 and CB2 are associated with pain modulation and have received much attention in recent years (reviewed by Fox, A. and Bevan, S., 2005). These receptors are widely distributed in the nervous system but CB1 receptors are located in greatest abundance in the CNS and on peripheral sensory neurons. CB2 receptors have been found mainly in peripheral tissues including immune tissues and keratinocytes (Hohmann, A. G. and Herkenham, M., 1999; Rice, A. S. C. *et al.*, 2002) but more recently on some sensory and CNS cells. Several fatty acids, for example, anandamide, 2-arachidonylglycerol, and palmitoylethanolamide, have been identified as endogenous ligands for these receptors and are termed endocannabinoids. They are released from postsynaptic nerve terminals to modulate postsynaptic excitability and transmitter release at presynaptic elements via Gi/Go-coupled activity. Specific antagonists such as SR141716A and SR147778 for CB1 and SR144428 for CB2 have been used to characterize receptor functions.

For the most part CB1 receptors are located on neurons where activation causes attenuation of presynaptic N-calcium channel activity and inhibition of transmitter release (Kreitzer, A. C. and Regeher, W. G., 2002). In contrast, activation of neuroglia (astrocytes, microglia) is accompanied by *de novo* expression of both CB1 and CB2 and increased release of endocannabinoids (Stella, N. 2004). Specifically CB2 receptors are expressed in spinal microglia following peripheral nerve lesions but not by peripheral inflammation (Zhang, J. M. *et al.*, 2003). The significance of this is presently unclear although prolonged CB receptor activation has been associated with increased expression of neuroprotective modulators such as brain-derived neurotrophic factor (BDNF; Marsicano, G. *et al.*, 2003).

Both CB1 and CB2 receptor types have been shown to have a key role in the modulation of nociceptors. Although CB1 effects have been best characterized, the mechanism and sites of action are incompletely understood. It is likely that an interplay of several CB1-driven systems occurs during acute or chronic pain. Brainstem structures appear to be important. For example, CB1-induced local activation of the periaqueductal gray (PAG; by stress-induced release of endocannabinoids) has been suggested to be involved in aversive responses and acute stress-induced analgesia (Finn, D. P. *et al.*, 2004; Hohmann, A. G. *et al.*, 2005). This was blocked by SR141716A (rimonabant, the CB1 antagonist) but not by naltrexone (μ-opioid antagonist). Paradoxically, recent data suggest that deletion of the CB1 receptor in mice is associated with a greater vulnerability to stress (Fride, E. *et al.*, 2005). Finally CB1-induced

analgesia has been proposed to be driven by activation of descending inhibitory mechanisms. In keeping with this, block of the descending noradrenergic systems (Gutierrez, T. *et al.*, 2003) has been proposed to account, in part, for CB1-induced antinociception.

Several studies have now shown that CB2 agonists modulation acute pain, for example, AM1241 and GW405833 (incision model) (LaBuda, C. J. *et al.*, 2005) while GW405833 has been reported to exhibit efficacy in neuropathic and inflammatory pain models (Valenzano, K. J. *et al.*, 2005). Interestingly another CB2 agonist, JWH-133 shows antihyperalgesia as well as anti-inflammatory activity (Elmes, S. J. R. *et al.*, 2005). The effects of CB2 agonists appear not to be due to the release of endogenous opioids as actions were unaffected by naltrexone (Whiteside, G. T. *et al.*, 2005b).

Several clinical studies have supported the use of cannabinoids, such as tetrahydrocannabinol (THC), to reduce pain. However this action is commonly accompanied by adverse effect such as euphoria, dizziness, and sedation (Campbell, F. A. *et al.*, 2001; Wade, D. T. *et al.*, 2003; Svenden, K. B. *et al.*, 2004). The reduction of CNS side effects has been approached through targeting peripheral CB receptors, since both CB1 and CB2 receptor activation produce antinociception. In support of this approach, the systemic administration of WIN55,212-2, a nonselective CB agonist, attenuated inflammatory hyperalgesia and this action was reversed by peripheral (intraplantar) but not spinal administration of a CB1 antagonist (Fox, A. and Bevan, S., 2005). Moreover a peripherally restricted CB1 agonist (e.g., NVP-001, personal communication) was shown to reduced inflammatory hyperalgesia by directly reducing nociceptor excitability (Richardson, J. D. *et al.*, 1998). Finally, CT-3 (ajulemic acid) a nonspecific CB1 and CB2 agonist, with limited CNS availability produced analgesia in a number of inflammatory and neuropathic pain models (Burstein, S. H. *et al.*, 2004; Dyson, A. *et al.*, 2005) as well as in human neuropathic pain, at doses which cause minimal CNS side effects (Karst, M. *et al.*, 2003). The mechanism of action of CT-3 appears to involve mainly CB1 receptors and a number of other indirect actions including activation of peroxisome proliferator-activated receptor (PPAR)-γ receptors and inhibition of inflammatory cell activity and mediator release (Liu, J. *et al.*, 2003).

Selective CB2 ligands (HU-308, AM-1241, and GW405833) as well as nonselective compounds (HU-210 and CP55940) have shown antinociceptive effects in a variety of inflammatory and neuropathic pain models (Malan, T. P. *et al.*, 2003, Valenzano, K. J. *et al.*, 2005). These effects are still present in CB1-knockout (KO) mice, absent in CB2-KO animals, can be revered by CB2 antagonist such as SR144528, and occur without major CNS side effects (sedation, catalepsy, or motor impairments). It is unclear how these effects are produced since CB2 receptors have not been found on sensory neurons (Sokal, D. M. *et al.*, 2003). Rather an indirect effect via immune cell modulation and the naloxone-reversible release of β-endorphin from keratinocytes has been proposed (Ibrahim, M. M. *et al.*, 2005). In contrast, a number of reports have suggested that CB2 agonists, for example, JWH-133, may reduce C-fiber excitability directly (Ross, R. A. *et al.*, 2001; Sagar, D. R. *et al.*, 2005). There is also a possibility that CB2 agonists cause antinociception via the CNS as CB2 receptors are expressed in spinal microglia after nerve injury (Zhang, J. M. *et al.*, 2003), and CB2 activation may modulate microglial activity (Water, L. *et al.*, 2003). Indeed the direct spinal administration of JWH-133 attenuated mechanical allodynia (Sagar, D. R. *et al.*, 2005). Clearly greater understanding is still required with respect to the mechanisms of CB2-mediated analgesia.

Other approaches to address the role of endogenous cannabinoid systems in pain and analgesia have targeted fatty acid amide hydrolysis (FAAH), the major degradation pathway for endogenous cannabinoids (Cravatt, B. F. and Lichtman, A. H., 2004). Thus in mice lacking this enzyme (Cravatt, B. J. *et al.*, 2001), or treatment of naïve mice with a novel FAAH inhibitor such as OL135 (Lichtman, A. H. *et al.*, 2004), increases the analgesic efficacy of anandamide and significantly elevated brain anandamide levels were measured. In addition, OL135 increased pain threshold in acute pain models, decreased the mechanical allodynia of neuropathic pain, and decreased thermal hyperalgesia in models of persistent pain (Chang, L. *et al.*, 2006). Surprisingly a CB2 rather than CB1 antagonist attenuated the effect of OL135 in the persistent pain models and its effects appeared to be mediated by a naloxone-reversible mechanism. In this latter respect several reports have indicated analgesic synergy, between μ-opioid and CB receptors. Thus combinations of these agonist have been shown to provide pain reduction with minimal side effects in acute pain models (Cichewicz, D. L. and McCarthy, E. A., 2003). However, it is still unclear whether such synergy can be exploited in chronic pain treatment.

53.7 Mrg-Related GPCR and Their Ligands

The mas-related gene (Mrg) family is a large family of GPCRs which vary in numbers depending on species, are rather exclusively expressed in subsets of small sensory neurons (Dong, X. *et al.*, 2001), and may be co-localized with markers for nociceptive neurons such as TRPV1 (Lembo, P. M. C. *et al.*, 2002). Selective manipulations of individual Mrgs have provided exquisite markers for subsets of small sensory neurons showing selective epidermal and spinal innervation, suggesting a close anatomical and functional relationship (Zylka, M. J. *et al.*, 2005).

Focus on the human mrgX subfamily (six genes in humans), also called sensory neuron-specific receptors (SNSR), has revealed a complex pharmacology. The endogenous ligand for SNSR is unconfirmed, but several mammalian peptide ligands have been identified with high affinity for this receptor. These peptides include bovine adrenal medullary peptide 22 (BAM22), and its fragment BAM8-22 derived from the endogenous opioid precursor pro-enkephalin A, as well as two unrelated peptides γ2-melanocytostimulating hormone (MSH) and CT-γ2-MSH. Apart from BAM22, these substances do not activate opioid receptors, rather they increase peripheral and spinal excitability to potentiate thermal and mechanical nociception (Grazzini, E. *et al.*, 2004). This suggests that the same gene products can facilitate as well as inhibit sensory neurons and that an SNSR antagonist should be sought to inhibit hyperalgesia. Although these data strongly support a role for SNSRs in the modulation of pain, the endogenous ligand for this receptor has not been confirmed and there is little direct evidence linking SNSRs in the etiology of acute or chronic pain.

53.8 Prostanoids and Receptors

A variety of prostanoid cyclo-oxygenase (COX) enzyme products (PGE_2, PGD_2, $PGF_{2\alpha}$, thromboxane, PGI_2) occur during inflammation but PGE_2 is considered to be the major contributor to inflammatory pain. Thus blocking the major synthetic enzymes, COX-1 (constitutive) and COX-2 (inducible), or inhibition of prostanoid receptors continue to be important approaches for reducing inflammatory pain (Flower, R. J., 2003). Experience with selective inhibitors of COX-2 (celecoxib, rofecoxib) shows improved gastrointestinal tract safety but little improvement in analgesic efficacy. A relatively infrequent though increased cardiovascular risk (coronary constriction) has been observed with some COX inhibitors. The mechanism is not understood but has led to Vioxx (rofecoxib) being withdrawn from therapy and considerable clinical debate about the safety of future approaches to COX inhibition.

An alternative approach for the clinical improvement of COX inhibitors has been to combine COX inhibition with NO donation (CINOD). Such molecules (NO-naproxene, NO-ibuprofen) have been claimed to show improved efficacy and safety over the parent nonsteroidal anti-inflammatory drugs (NSAIDs) due to actions (improved side effects and efficacy) of cleaved NO (Fiorucci, S. *et al.*, 2001).

Recently a splice variants of COX-1 has been identified, COX-3 (Chandrasekharan, N. V. *et al.*, 2002). However it has low enzymic capability and its distribution and low abundance in the CNS and in periphery does not make this a compelling target for analgesia. However several NSAIDs (acetaminophen, diclofenac, phenacetin) show low efficacy but some degree of selectivity for COX-3. Overall however the analgesic efficacy of acetaminophen is poorly correlated with COX inhibition. Interestingly endogenous acetaminophen metabolism produces a conjugated arachidonic acid derivative (AM404) that shows cannabinoidlike analgesia properties (Hogestatt, E. D. *et al.*, 2005; Ottani, A. *et al.*, 2006). Furthermore gene deletion studies suggested that COX-3 inhibition may be linked with the antipyretic and analgesia properties of NSAIDs. Clearly further studies are needed to link this enzyme to inflammatory pain pathophysiology.

Another route of inhibiting PGE_2 synthesis is via the blockade of PGE synthase (PGES), a major route of conversion of PGH_2 to PGE_2. Two isoforms of the enzyme have been identified, membrane or microsomal-associated PGES (mPGES-1) and a cytosolic enzyme (cPGES/p23) which are linked with COX-2- and COX-1-dependent PGE_2 production, respectively (Jakobsson, P. J. *et al.*, 1999; Claveau, D. *et al.*, 2003). Both isoforms are upregulated by inflammatory mediators and gene deletion studies in mice indicate an important role for mPGES in acute and chronic inflammation and inflammatory pain (Trebino, C. E. *et al.*, 2003).

Apart from prominent roles in regulating peripheral excitability during inflammatory pain, COX-1 (glia) and COX-2 (ventral horn cells) are constitutively present in spinal cord and are increased by inflammation, peripheral nerve injury, or by administration of cytokines, leading to increased production of spinal PGE_2. In keeping with this, several NSAIDs have been shown to reduce inflammatory hyperalgesia via inhibition of spinal COX activity (Yaksh, T. L. *et al.*, 2001). Several mechanisms have been proposed for the PGE_2-induced increase in spinal excitability. Prominent are the contribution of EP1 receptors and spinal release of glutamate. Recently however the spinal effects of PGE_2 have been linked with a glycine receptor. Thus deletion of the GlyR3α subunit gene, reduced pain sensitivity caused by PGE_2 administration or inflammation (Harvey, R. J. *et al.*, 2004).

Finally, PGE_2 exerts its effects via a variety of EP receptors (EP1, 2, 3 4) present both in peripheral sensory neurons and in the spinal cord. Activation of these receptors produces a complexity of effects, ranging from Ca^{2+} influx to cAMP activation or inhibition. In the periphery, sensitization by PGE_2 as been shown to be cAMP mediated causing the enhancement of tetrodotoxin-resistant (TTX-r) sodium currents in nociceptors via channel phosphorylation (England, S. *et al.*, 1996; Gold, M. S. *et al.*, 1998). However in the spinal cord, excitability was enhanced by EP1 receptors but reduced by an EP3α agonist (ONO-AE-248) suggesting further complexity in the prostanoid regulation of pain mechanisms (Bar, K. J. *et al.*, 2004).

53.9 Cytokines, Chemokines, and Their Receptors

Cytokines are produced by a variety of immune cells as well as brain neuroglial cells in response to injury and inflammation. They are powerful mediators of hyperalgesia (see Abbadie, C., 2005).

Probably the most characterized are IL-1β and tumor necrosis factor (TNF)-α that act via specific receptors on sensory neurons. Their effects can be attenuated by receptor antagonists (IL-1-r) that sequester the ligand as well as by neutralizing antibodies. Indeed the TNF antibody, etanercept, has been developed for the treatment of chronic inflammation and the presence of TNF-α has been correlated with a number of painful inflammatory clinical conditions (Lindenlaub, T. and Sommer, C., 2003). Cytokines induce hyperalgesia by a number of direct and indirect actions. Thus IL-1β activates nociceptors directly and produces heat sensitization via intracellular kinase activation, but it may also cause indirect nociceptor sensitization via the production of kinins and prostanoids (Sommer, C. and Kress, M., 2004). TNF-α also activates sensory neurons directly via TNFR1 and TNFR2 receptors, (Pollock, J. *et al.*, 2002; Ohtori, S. *et al.*, 2004) and initiates a cascade of inflammatory reactions through the production of IL-1, IL-6, and IL-8. It is significant that direct TNF-α application in the periphery induces neuropathic pain behavior that is blocked by ibuprofen and celecoxib (Schafers, M. *et al.*, 2004), while nerve ligation causes increased TNF-α in damaged as well as adjacent undamaged axons (Schafers, M. *et al.*, 2003a). Interestingly anti-TNF-α treatment with TNF antibody, adalimumab, produced a prolonged reduction of pain symptoms in OA (Grunke, M. and Schulze-Koops, H., 2006).

There appears to be a complex interplay between cytokines, and it should be noted that like other mediators of pain not all should be considered as detrimental. Thus IL-6 has been associated with potential beneficial effects after nerve injury including protection against cell death and the promotion of growth. However there is also compelling evidence of a strong association with this cytokine in several human pain conditions including herniated lumbar disc pain and pain caused by failed back surgery (De Jongh, R. F. *et al.*, 2003).

A variety of studies have demonstrated an important role for spinal inflammatory and neuroimmune processes triggered by peripheral inflammation and nerve injuries. These processes involve the regulation of a variety of receptors, channels, and enzymes with patterns that are likely to differentiate one pain state from another (Honore, P. *et al.*, 2000). In addition, activation of spinal neuroglial cells (microglia, astrocytes, satellite cells) stimulates a cascade of secondary excitability changes (Watkins, L. R. and Maier, S., 2002). Neuroglia make close-junctional connections with other cells, providing a means of spreading excitability changes beyond the boundaries of spinal segmental input. Neuroglia also secrete a number of mediators such NO, neurotrophins, IL-1β, TNF-α, free radicals, and glutamate. In addition, neuroglial mediators may contribute to spinal excitability by causing dysfunction or degeneration of inhibitory spinal neurons (Moore, A. K. *et al.*, 2002). In keeping with an important role for neuroglial mediators, treatments with anti-inflammatory agents or modulators of neuroglial activity such as propentofylline and minocycline inhibit glial activation and the release of glial products with reduced

behavioral signs of hyperexcitability (Watkins, L. R. and Maier, S., 2002; Raghavendra, V. *et al.*, 2003).

Chemokines are important peripheral and central mediators of inflammation. The major chemokines and their respective receptors are macrophage-derived chemokine (MDC)/CCR4, regulated upon activation, normal T-cell expressed, and secreted (RANTES)/CCR5, fractalkine/CX3CR1, and SDF-1α/CXCR4. Receptors are located on leukocytes, on central neurons, sensory neurons, and neuroglial cells (Watkins, L. R. and Maier, S., 2002). Apart from their major chemoattractant effects, chemokines contribute directly to inflammatory hyperalgesia through G-protein-coupled activation and sensitization of sensory neurons (Oh, S. B. *et al.*, 2001). Block of CX3CR1 by a fractalkine receptor-neutralizing antibody induces antiallodynic effects in models of peripheral nerve inflammation (Milligan, E. D. *et al.*, 2005b). However some chemokines appear beneficial. Thus gene therapy (AV-333) has been used to increase IL-10 production through the delivery of viral and nonviral vectors by acute spinal intrathecal delivery. This experimental treatment has been shown to reverse mechanical allodynia in the chronic constriction injury (CCI) model of neuropathic pain (Milligan, E. D. *et al.*, 2005a).

53.10 Adrenoceptors

A number of chronic pain disorders termed sympathetically maintained pain have highlighted the importance of the release of sympathetic transmitters (epinephrine or norepinephrine) from sympathetic varicosities and the involvement of adrenergic receptors in pain etiology. For example following peripheral nerve injury sprouting of sympathetic nerve endings occurs at several sites. Thus sympathetic/sensory coupling at the level of the DRG (Zhang, J. M. *et al.*, 2004), at the site of injury (neuroma), and in the periphery (Shinder, V. *et al.*, 1999) have been demonstrated. In keeping with this some neuropathic pain symptoms have been attenuated by sympathetic blocks or surgical sympathectomy.

Other studies have shown the expression of α1- and α2-adrenoceptors on sensory neurons or postganglionic sympathetic terminals (Sato, J. and Perl, E. R., 1991; Tracey, D. J. *et al.*, 1995; Lee, D. H. *et al.*, 1999) after nerve injuries. Under these conditions sensory neurons can be directly activated by the endogenous release of sympathetic transmitters (via

$\alpha1$ receptors) or in the clinic by intradermal injection of norepinephrine (Ali, Z. *et al.*, 2000).

On the other hand, transdermal applications of the $\alpha2$-agonist, clonidine via transdermal patches or creams, has proven efficacious in a variety of neuropathic pain conditions (Byas-Smith, M. G. *et al.*,1995). Since clonidine has been shown to decrease the excitability of small sensory neurons from nerve injured animals (Ma, W. *et al.*, 2005) the antinociceptive efficacy of clonidine is considered to be due to $\alpha2$ receptor-coupled inhibition. Clonidine and other $\alpha2$-agonists such as dexmedetomidine have also been used systemically to inhibit of sensory transmission in the spinal cord by block of pre- and postsynaptic membrane excitability. Unfortunately sedation and hypotension are major target-related side effects of these compounds. Great efforts have been made to identify ligands with improved $\alpha2$-receptor subtype selectivity, to avoid side effects, but thus far this has not been particularly successful.

53.11 Glutamate Regulation and Glutamate Receptors

Glutamate is the major excitatory neurotransmitter in the CNS with important regulatory roles for pain transmission. However there is also considerable evidence that it plays a role in pain processing in the periphery (Carlton, S. M. *et al.*, 1995; Bahave, G., *et al.*, 2001). Glutamate acts through receptor-coupled ligand-gated ion channels (α-amino-3-hydroxy-5-methyl-4-isoxazolepropionic acid (AMPA)/kinate receptors: iGluRs) and GPCR-coupled receptors (metabotropic glutamate receptors (mGluRs)). Injections of glutamate or metabolically stable receptor-selective agonists such as *N*-methyl-D-aspartic acid (NMDA), AMPA, and kainate, cause a reduction in thermal and mechanical thresholds for pain, while application of iGluR and mGluR antagonists attenuate pain in acute models (Zhou *et al.*,1996; Bhave, G., *et al.*, 2001).

Building upon this has been significant progress in development of a glutamate antagonist optimized for clinic testing. For example, the AMPA-kinate receptor antagonist, LY293558 has been shown to have efficacy in models of neuropathic pain (Blackburn-Munro, G. *et al.*, 2004) and to reduce capsaicin-induced pain in human volunteers as well as clinical dental pain with minimal side effects (Sang, C. N. *et al.*, 1998).

There has been a great deal of information linking NMDA receptors in central sensitization and CNS excitability during chronic pain and repetitive stimulation of pain pathways. Indeed NMDA antagonists show robust attenuation of pain behaviors but provoke a number of side effects (sedation, confusion, and motor incoordination) and thus have an insufficient therapeutic window. There has been a refocus on more specific NMDA-receptor subtype blockers (NR1 and NR2) directed toward the glycine modulatory site to avoid side effects. This site actively modulates the NMDA channel only during the sustained stimulation of the receptor, which is considered to occur during chronic pain. Selective NR1-Gly antagonists have been claimed to reduce pain with reduced side effects (Danysz, W. and Parsons, C. G., 1998; Quartaroli, M. *et al.*, 2001). However clinical experience has not yet confirmed this. GV196771 did not show efficacy against clinical pain, possible due to inadequate penetration into the CNS (Wallace, M. S. *et al.*, 2002).

Alternative initiatives have targeted other NMDA receptor subtypes such as the NR2B receptor, which has a specific distribution in sensory pathways. Blockade of this receptor has also been claimed to produce antinociception (ifenprodil, CP-101,606) with reduced side effects (Taniguchi, K. *et al.*, 1997). This concept has yet to be evaluated in the clinic.

Metabotropic glutamate receptors, particularly mGluR1 and mGluR5, have been reported to play a key role in sustaining heightened central excitability in chronic pain with minimal involvement in acute nociception. Thus spinal administration of selective agonists such as dihydroxy phenyl glycine produced allodynia, while mGluR5 was shown to be significantly overexpressed in some, but not all, chronic pain models (Hudson, L. J. *et al.*, 2002). Furthermore antisense-oligonucleotide and antibody treatments, directed at these receptors, reduce chronic inflammatory and neuropathic pain behaviors (Fundytus, M. E. *et al.*, 2002). In keeping with these observations, the mGluR5-selective antagonists 2-methyl-6-(phenylethynyl)-pyridine (MPEP) did not reduce acute physiological pain responses but reduced heat hyperalgesia and allodynia in models of inflammatory and neuropathic pain (Walker, K. *et al.*, 2001; Hudson, L. J. *et al.*, 2002; Urban, M. O. *et al.*, 2003).

Peripheral mGluR5 receptors have also been claimed to modulate pain (Jang, J. H. *et al.*, 2004). Thus local administration of mGluR5 (MPEP, SIB1757) have been effective in reducing pain behavior (Dogrul, A. *et al.*, 2000; Zhu, C. Z. *et al.*, 2004) suggesting a potential for using peripheral mGlu5 antagonists in pain therapy.

Metabotropic group II receptors (mGluR2 and mGluR3) also modulate pain transmission. The mGluR2 is located in sensory neurons and presynaptic nerve terminals whereas mGluR3 is found all over the brain. MGluR3 can be selectively increased in the spinal dorsal horn neurons after peripheral ultraviolet injury (Boxall, S. J. *et al.*, 1998). mGluR2/3 receptor activation appears necessary to reduce nerve terminal excitability and modulate pain transmission since treatment with L-acetyl-carnitine reduced inflammatory hyperalgesia and mechanical allodynia and increased the expression of mGlur2/3. The effect of L-acetyl-carnitine were attenuated by LY379268, a mGlu2/3 antagonist (Chiechio, S. *et al.*, 2002).

53.12 Gamma-Aminobutyric Acid Receptors

Gamma-aminobutyric acid (GABA) receptors are emerging as important regulators of pain, particularly in the spinal cord as they are abundantly expressed both on spinal afferent nerve terminals and spinal neurons. Activation of both subtypes of GABA receptor by muscimol (GABA$_A$ agonist) or baclofen (GABA$_B$ agonist) reduced pain behavior. In contrast, GABA$_A$ (bicuculline) and GABA$_B$ (CGP35348) antagonists cause pain when injected into the spinal intrathecal space (Malan, T. P. *et al.*, 2002). Loss of GABA (Ibuki, T. *et al.*, 1997) and impaired GABA-mediated inhibition through loss of spinal interneurons has been demonstrated in models of neuropathic pain (Moor, A. K., *et al.*, 2002).

Improvement of GABA synthesis through promotion of its synthetic enzyme, glutamic acid decarboxylase (GAD), has been achieved by gene transfer of a GAD promoter (in herpes simplex virus vector) via peripheral injection. This was taken up into sensory neurons and transported to DRGs to reduce spinal excitability, as well as mechanical and thermal allodynia caused by nerve injury (Hao, S. *et al.*, 2005).

Other important changes in ionic regulation have been highlighted to occur in chronic pain which indirectly affects GABA- and glycine-mediated inhibition and excitability in the spinal cord. In particular changes in transmembrane ion transporters have been associated with modulation of primary afferent excitability and related to neuropathic pain disorders. Thus the chloride transporter NKCC1, localized in primary sensory neurons, is responsible for maintaining the high chloride

ion gradient in afferent terminals. This allows the outflow of chloride ions associated with the primary afferent depolarization (PAD) and presynaptic inhibition, following GABA$_A$ receptor activation of afferent terminals. Deletion of the NKCC1 gene induced touch-evoked pain caused by light stroking. This was shown to be due to increased nerve terminal excitability caused by a reduced capability to generate presynaptic inhibition (Laird, J. M. A. *et al.*, 2004).

On the other hand increasing the expression of this cotransporter or increasing its efficacy through AMPA mediated Ca/CaM-kinase II phosphorylation has been postulated (Cervero, F. *et al.*, 2003) to increase PAD sufficient to reach the firing threshold of the nerve terminal. This mechanism may also contribute to spinal sensitization following nerve injury and may indeed generate antidromic action potentials triggering a dorsal root reflex or the ectopic discharges in sensory nerves which cause, or contribute to spontaneous pain.

Spinal sensitization that occurs following peripheral nerve injury may also arise by other mechanisms of spinal disinhibition. This has been hypothesized to involve the reduced expression of the potassium-chloride exporter KCC2, seen only in superficial spinal lamina 1 neurons after injury. In addition, knockdown of KCC2 with antisense or block of this transporter with ((dihydronindenyl)oxy) alkanoic acid (DIOA) causes a similar shift in the transmembrane ionic gradient in spinal lamina 1 neurons and a consequent behavioral hyperalgesia resembling that seen after nerve injury (Coull, J. A. M. *et al.*, 2003). In addition, normal spinal inhibitory currents, mediated by GABA and glycine interneurons, are reversed in the absence of KCC2, so that the effects of GABA become predominant excitatory, due to the outward, rather than inward, flow of chloride ions. The emerging view suggests that a variety of changes in ionic regulation occur during chronic pain. This adds complexity to the puzzle of pain but opportunity for intervention. It will be important to learn how these disinhibitory mechanisms contribute in different chronic pain conditions and whether they are critical for pain initiation as well as its maintenance.

53.13 Ion Channels

53.13.1 Ligand-Gated Channels

Transient receptor potential (TRP) channels form a large family of sensory transducers involved in cellular calcium regulation. Many are localized to mammalian sensory nerves and are involved in the transduction of temperature and chemical signals that result in pain. The TRPV family has received considerable attention and is considered a good targeted for developing analgesics (Krause, J. E. *et al.*, 2005). TRPV1, originally called the vanilloid receptor because of its selective activation by capsaicin and pungent vanilloids, has the properties of a noxious heat transducer. The effects of capsaicin on sensory neurons are complex, involving activation through ligand-operated cation channels, and the extracellular influx as well as the intracellular increase in calcium ions. Calcium-induced stimulation of phosphatase such as cacineurin then reduces excitability via inactivation voltage-gated calcium (CaV) channels as well as rapid CaV2.2 internalization (Wu, Z. Z. *et al.*, 2005). This mechanism explains the transient and reversible analgesia effects of capsaicin either on peripheral or spinal afferent nerve terminals. TRP channel regulation of heat transduction is complex as animals with deleted *TRPV1* and *TRPV2* genes appear to have normal heat sensitivity (Woodbury, J. *et al.*, 2004) whereas gene deletion of *TRPV1* prevented inflammatory heat hyperalgesia (Caterina, M. J. *et al.*, 2000).

The TRPV1 receptor has assumed an increasingly prominent role in regulating sensory neuronal excitability in inflammatory pain. A variety of inflammatory agents including protons, bradykinin, adenosine triphosphate (ATP), PGE2, 12-lipoxygenase products, PAR2, anandamide, and nerve growth factor (NGF) indirectly sensitize TRPV1, or regulate its expression, to cause thermal hyperalgesia. A common mechanism appears to be GPCR-coupled stimulation of PLC, the formation of intermediates inositol triphosphate (IP3) and DAG and the activation and mobilization of PKC (Vellani, V. *et al.*, 2001) which can phosphorylate the TRPV1 receptor.

Earlier analgesia strategies, targeting TRPV1, were focused on capsaicin-like agonists that induced functional inactivation of sensory fibers by causing reversible subepidermal degeneration. This has been successfully translated into the clinic with the introduction of a number of topical capsaicin therapies for inflammatory pain. Currently there is a focus on TRPV1 channel blockers or selective antagonists against the TRPV1 receptor (Garcia-Martiez, C. *et al.*, 2002). Supporting these approaches, competitive (AMG-9810, personal communication) and noncompetitive vanilloid receptor-1 (VR1) antagonists (DD161515; Sachez-Baez, F. *et al.*, 2002) block chemical and thermal pain sensitivity, heralding the

emergence of a novel therapy. Indeed recent clinical studies in human volunteers have shown that oral SB705498 attenuated capsaicin and ultraviolet C (UVC)-induced pain and hyperalgesia (Chizh, B. *et al.*, 2006).

Other TRP channels (TRPV3, TRPV4, TRPA1) have been suggested to be involved in transduction. Thus TRPA1 (ANKTM1) is colocalized with TRPV1, is activated by capsaicin and mustard oil but can also be sensitized by inflammatory mediators including bradykinin to produce cold-induced burning pain (Bandell, M. *et al.*, 2004). In addition TRPV1 can oligomerize with other TRP family members including TRPV3. TRPV3 is also a heat transducer that it sensitive in the physiological temperature range and responds to noxious heat, but is insensitive to capsaicin. This is found in keratinocytes and appears to be upregulated in inflammatory pain conditions. It is unclear how activity of TRPV3 leads to nociceptor activation. So far there are few chemical tools to help characterize the functions of these TRP receptors and support their value as analgesia targets.

53.14 Purinergic Receptors

53.14.1 P2X Receptors

The unique localization of the P2X3 receptor to small sensory fibers has highlighted its importance in pain. Large amounts of the endogenous ligand, ATP are released after tissue injury and during inflammatory injuries while both ATP and a stable analogue α,β-Me ATP, induce pain and are pronociceptive when administered intradermally in human volunteers.

In chronic inflammatory pain, P2X3-mediated excitability is enhanced while reduction of P2X3 receptors, by antisense oligonucleotide administrations, reduced inflammatory hyperalgesia as well as that evoked by α,β-Me ATP (Honore, P. *et al.*, 2002). In keeping with this a number of antagonists including 2′,3′-O-(2,4,6-trinitrophenyl) adenosine 5′-triphosphate (TNP-ATP), pyridoxal-phosphate-6-azophenyl-2′,4′-disulfonate (PPADS), and suramin, reduce pain behavior. More selective, and druglike, antagonists such as A-3174919 reduced pain in a number of acute and chronic pain models supporting the possibility for future analgesia therapy (Jarvis, M. F. *et al.*, 2002).

It should be noted that a number of other purinergic receptor subtypes, for example, P2X4 and P2X7, have also been suggested to modulate pain through altered central excitability and the release of neuroglial-cell products. Thus activated microglial, astrocytes, and satellite cells release a variety of inflammatory mediators including IL-1β, TNF-α, prostanoids, and NO upon ATP stimulation. Indeed increased expression of P2X4 has been shown to occur in spinal microglial after peripheral nerve lesions and this was related to painful mechanical allodynia. This behavior was blocked by spinal administrations of the selective P2X4 antagonist TNP-ATP (Tsuda, M. *et al.*, 2003). Remarkably spinal administration of activated microglia reproduced TNP-ATP sensitive mechanical allodynia in naïve animals.

Increased P2X7 expression has been found in peripheral macrophages following inflammation but this receptor is also expression in spinal neurons and microglia following peripheral nerve injury (Deuchars, S. A. *et al.*, 2001). In keeping with an important role in chronic pain both microglia and P2X7 receptors are upregulated in human chronic pain patients while deletion of the P2X7 receptor gene produced a complete absence of mechanical and thermal pain in mice (Chessell, I. P. *et al.*, 2005).

It is worth noting that other nucleotide-gated ion channels have also been shown to be important for regulating peripheral excitability. Thus the sodium/potassium repolarizing pacemaker current, Ih, which is activated during membrane hyperpolarization is important for generation of rhythmic and spontaneous action potentials in sensory neurons following nerve injury. Ih currents are controlled by cyclic nucleotides (cAMP and cGMP) via a family (HCN1–4 channels) of ligand-gated ion channels that are constitutively expressed in sensory nerves and differentially distributed after crush or inflammatory nerve injuries (Chaplan, S. R. *et al.*, 2003; Yao, H. *et al.*, 2003). Nerve injury has been shown to enhance the Ih and this can be blocked with ZD7288 which also prevents repetitive firing in damaged sensory neurons and reverses touch hypersensitivity in neuropathic pain models (Chaplan, S. R. *et al.*, 2003). This approach clearly has great potential for addressing peripheral excitability in neuropathic pain disorders.

53.14.2 Acid-Sensing Channels

Proton production is increased in inflammation and is likely to be involved in inflammatory hyperalgesia and in the sensation of muscle aching and discomfort due to the hypoxia/anoxia of muscle exercise. Indeed direct activation of nociceptors accounts for the sharp stinging pain produced by intradermal injections of acidic solutions and low extracellular pH enhancing the effects of other inflammatory mediators (Krishtal, O., 2003; Mamet, J. *et al.*, 2003).

Exogenously administered acidic solutions produce a rapid but transient increase in membrane cation permeability as well as a more prolonged permeability increase in sensory neurons. This can give rise to sustained nerve activation as well as an enhanced mechanosensitivity. The mechanism of proton-induced activation of sensory neurons underlying pain has not been fully elucidated, but appears to be driven by a number of acid-sensing ion channel (ASICs) particularly ASIC1 and ASIC3 (also called DRASIC). ASIC3 has been shown to have a strong antinociceptive phenotype following deletion of this gene in KO mice (Sluka, K. A. *et al.*, 2003).

A novel blocker (A-317567) of peripheral ASIC1, ASIC2, and ASIC3 channels has been described (Dube, G. R. *et al.*, 2005). This produces antihyperalgesia in models of inflammatory and postoperative pain but there have been no reports of therapeutic advances with more selective inhibitors.

53.15 Sodium Channels

Voltage-gated sodium channels are characterized by their primary structure and sensitivity to tetrodotoxin (TTX). A variety of TTX sensitive (NaV1.1, Nav1.2, NaV1.6, and Nav1.7) and TTX insensitive (Nav1.8, NaV1.9) channels are involved in regulating sensory neural excitability (Matzner, O. and Devor, M., 1994; Eglen, R. M. *et al.*, 1999). Changes in the expression, trafficking, and redistribution of NaVs, following inflammation or nerve injury is considered to account for the abnormal firing (ectopic generators) of afferent nerves (Devor, M., 2005). However channel redistribution appears complex as most channels appear to be downregulated in DRGs after nerve injury whereas, for example, NaV1.8 is redistributed along small axons. It has also been important to recognize that channel expression and nerve excitability are changed dramatically in uninjured axons that are closely apposed to injury (Gold, M. S. *et al.*, 2003).

The TTX-R sodium channel, NaV1.8, is uniquely expressed in all sensory neurons (IB4 positive and NGF sensitive) and appears to be an important contributor in the generation of abnormal excitability in sensory axons. Thus knockdown of NaV1.8 produces a marked reduction in abnormal pain responsiveness in pain models (Lai, J. *et al.*, 2003).

Inflammation also causes the overexpression of NaV1.7 in several types of sensory neurons in models of inflammatory pain (Gould, H. J. *et al.*, 2004) and in inflamed human tooth pulp. Interestingly NaV1.7 overexpression can be prevented by pretreatment with COX-1 and COX-2 inhibitors (ibuprofen, NS-398).

The clinical utility of nonselective sodium channel blockade in pain treatment has been well established with the use of local anesthetics such as lidocaine. Clinical experience has confirmed that local block of abnormally active afferents reduces secondary pain in established neuropathic pain conditions (Gracely, R. *et al.*, 1992). Interestingly this can be achieved at lower concentrations than are required for block of conduction in quiescent fibers highlighting a potential therapeutic advantage for selective fiber block (Devor, M. *et al.*, 1992). The basis of this is due to the state-dependent increase in the affinity of channel blockers for ion channel proteins (Devor, M., 2005). An additional utility of lidocaine is that intravenous administration has been reported to produce long-lasting pain relief both in animal models (Araujo, M. C. *et al.*, 2003) and in intractable neuropathic pain (Kastrup, J. *et al.*, 1987). The major disadvantages of nonselective sodium channel blockers are cardiotoxicity and CNS sedation and confusion produced by NaV1.5 and NaV1.2 channel block, respectively. Thus great activity is currently focused on discovering novel but selective sodium channel blockers.

In this regard novel activity-dependent sodium channel blockers (e.g., NW-1029) have been shown to reduce excitability of peripheral nerves and cause antihyperalgesia in models of inflammatory and neuropathic pain (Veneroni, O. *et al.*, 2003). Overall however, channel blockers still appear insufficiently selective to avoid cardiac and CNS side effects.

An alternative approach to selectively regulate ion channels is to block the trafficking of channels to the nerve membrane. For example the functioning of NaV1.8 may be reduced by preventing its interaction with p-11, an annexin II-related protein that tethers the channel to the nerve membrane (Okuse, K. *et al.*, 2002). In addition, channel-associated cell surface glycoproteins such as contactin may be involved in concentrating specific channel subtypes, e.g., NaV1.8 and NaV1.9 (IB4 positive) but not NaV1.6 and NaV1.7 (IB4 negative) in DRG nerve membranes, with an associated increased in current density (Rush, A. M. *et al.*, 2005). Although these approaches are attractive it

is unclear whether they will impact on nerve excitability associated with specific pain etiology.

NaV1.3 is another ion channel that is dramatically regulated after injury. This channel has been found abundantly in fetal but not in adult tissue. However NaV1.3 is dramatically upregulated in adult sensory neurons and spinal cord following peripheral and CNS injury (Hains, B. C. *et al.*, 2004). The expression of NaV1.3 has been suggested to make an important contribution to the sustained and high-frequency firing found in injured afferents related to neuropathic pain (Cummins, T. R. and Waxman, S. G., 1997; Kim, C. H. *et al.*, 2002). Further it has been proposed that NaV1.3 expression may not be directly related to neural injury as it was not correlated with activating transcription factor (ATF)-3 expression, a neuronal cell death marker (Lindia, J. A. *et al.*, 2005), but caused by a deficiency of growth factor. In this respect, GDNF administration reduced NaV1.3 expression and attenuated neuropathic pain behavior (Boucher, T. J. *et al.*, 2000). In contrast, antisense treatments that significantly reduced sensory neuronal NaV1.3 expression in the spared nerve injury (SNI) model of neuropathic pain did not attenuate nerve injury-induced allodynia (mechanical or cold; Lindia, J. A. *et al.*, 2005) whereas this was reduced by the nonselective channel blockers mexiletine and lamotrigine.

53.16 Calcium Channels

A variety of calcium channels (CaV) have been identified and characterized. Several have been shown to be prominently involved in pain regulation (Yaksh, T. L., 2006). The N-type calcium channel CaV2.2 is an important regulator of nerve terminal excitability and neurotransmitter release. It has been well established that there is complex chemical regulation of N-channels, particularly through GPCR signaling. For example, N-channel activity is attenuated by analgesic drugs such as opioids, with a resultant modulation of sensory transmitter release, e.g., substance P, CGRP, and glutamate, at both spinal and at peripheral sensory nerve terminals. N-channel trafficking may also be affected by GPCRs. For example, activation of the orphanin FQ receptor (ORL) by nociceptin causes channel internalization and downregulation of calcium entry (Altier, C. *et al.*, 2005).

Of additional importance is the fact that deletion of the N-channel gene reduces inflammatory and neuropathic pain (Kim, C. *et al.*, 2001; Saegusa, H. *et al.*, 2001). Moreover selective blockers such as ziconotide, (a naturally occurring conopeptide) and verapamil have been used to characterize channel activity while ziconitide (SNX-111, Prialt) has been used experimentally and clinically by intrathecal administration, to show utility for pain relief (Xiao, W. H. and Bennett, G. J., 1995; Snutch, T. P., 2005). Building on this concept, small molecule CaV2.2 channel blockers with oral availability are now reported to have been developed, e.g., NMED-160, showing efficacy in chronic pain models (Snutch, T. P., 2005).

While it is believed that blockers of CaV2.2 exert their major analgesic effects via the spinal cord, peripheral actions are also possible. Thus N-channels expression also occurs in peripheral terminals and has been shown to increase following axotomy (Baccei, M. L. and Kocsis, J. D., 2000). This suggests that peripheral N-channels can be targeted for neuropathic pain treatments; thus avoiding potential CNS side effects such as sedation.

Low-voltage-activated T channels (CaV3.1, CaV3.2, and CaV3.3) also appear important for pain transmission and as targets for pain therapy. Thus they are expressed in superficial laminas of the spinal cord and in DRG neurons (for references see Altier, C. and Zamponi, G. W., 2004; Yaksh, T. L., 2006). T-channels appear unaffected in DRGs after axotomy (Baccei, M. L. and Kocsis, J. D., 2000) but may play a more prominent role in regulating spinal excitability and spinal sensitization following repetitive C-fiber stimulation (Ikeda, H. *et al.*, 2003). Moreover nerve injury-induced hyper-responsiveness was blocked by the T-channel blocker ethosuximide (Matthews, E. A. and Dickenson, A. H., 2001) which also attenuated mechanical allodynia in animal models of vincristine- and paclitaxel-induced neuropathic pain (Flatters, S. J. and Bennett, G. J., 2004).

Finally a strongly validated approach for neuropathic pain has targeted the $\alpha2,\delta1$ calcium channel subunit, the substrate for the antiallodynic drugs, gabapentin and pregabalin. This subunit is important for channel assembly, is expressed in small DRGs and in spinal neurons, and its overexpression has been associated with allodynia in a number of specific pain models (Luo, Z. D. *et al.*, 2002).

53.17 Neurotrophins and Their Receptors

The neurotrophins represent an important family of regulatory proteins essential for sensory nerve development, survival, and determination of chemical phenotype including excitability (Sah, D. W. H. *et al.*, 2003; Zweifel, L. S. *et al.*, 2005). Several neurotrophins have been identified including NGF, brain-derived growth factor (BDNF) and neurotrophins (NT) 3, neurotrophin 4/5. Each neurotrophin binds with high affinity to receptor tyrosine kinases (Trk): NGF to TrkA, BDNF and NT4/5 to TrkB, and NT3 to TrkC. NT3 also binds with TrkA and TrkB. Mature neurotrophins also bind to a structurally distinct receptor p75 which affects neuronal development through downstream signaling. Neurotrophins arise from proneurotrophin precursors following extracellular cleavage by metalloproteinases and plasmin. It should be observed that proneurotrophins also signal through the p75 receptor and may produce opposite effects from neurotrophins, e.g., apoptosis rather than cell survival (Lu, B. *et al.*, 2005).

NGF has been most studies with respect to inflammatory hyperalgesia as its production is unregulated by inflammation in macrophages, fibroblasts, and Schwann cells. NGF has emerged as a key regulator of sensory neuron excitability and as an important mediator of injury-induced nociceptive and neuropathic pain (Ro, L. S. *et al.*, 1999; Theodosiou, M. *et al.*, 1999; Hefti, F. F. *et al.*, 2005). NGF acts via TrkA and p75 to activate a number of kinase pathways (e.g., p38 kinase; Ji, R. R. *et al.*, 2002) leading to altered gene transcription and the increased synthesis of sensory neuropeptides (substance P, CGRP), ion channels (TRPV1, NaV1.8, ASIC3; Fjell, J. *et al.*, 1999; Ji, R. R. *et al.*, 2002; Mamet, J. *et al.*, 2003), membrane receptors such as bradykinin and P2X3 (Petersen, M. *et al.*, 1998; Ramer, M. S. *et al.*, 2001), and structural molecules including neurofilament and channel anchoring proteins such as the annexin light chain p11 (Okuse, K. *et al.*, 2002).

Increased expression and release of NGF have been demonstration in several painful conditions in animal models (e.g., ultraviolet injury, surgical injury; Oddiah, D. *et al.*, 1998; Miller, L. J. *et al.*, 2002) and in human conditions including arthritis, cystitis, prostitis, and headache (Aloe, L. *et al.*, 1992; Halliday, D. A. *et al.*, 1998; Sarchielli, P. *et al.*, 2001). Administration of exogenous NGF induces thermal and mechanical hyperalgesia in animals and humans (Andreev, N. *et al.*, 1995; Apfel, S. C., 2002), which is considered to be due in part to mast cell degranulation and by directly increasing sensory neuronal excitability (Sah, D. W. H. *et al.*, 2003).

Few small molecule NGF antagonists are available but ALE0540, which inhibits the binding of

NGF to TrkA and p75, and PD90780 which inhibits NGF binding to p75, have been proposed to have efficacy in chronic pain models (Owolabi, J. B. *et al.*, 1999; Colquhoun, A. *et al.*, 2004). In contrast, the use of TrkA-immunoglobulin G (IgG) and NGF monoclonal antisera have confirmed the importance of NGF in pain. Thus anti-NGF treatment have provided positive behavioral and biochemical readouts in several models of chronic nociceptive (including visceral) and neuropathic pain (Obata, K. *et al.*, 2004; Hefti, F. F. *et al.*, 2005; Sevcik, M. A. *et al.*, 2005; Shelton, D. L. *et al.*, 2005). These studies have also indicted a lack of effect on acute nociceptive processing or on sympathetic nerves whose phenotype is also regulated by NGF.

The importance of NGF has also received clinical confirmations since RN624, a humanized anti-NGF monoclonal antibody (mAb), has been reported to be efficacious in reducing pain and improved mobility in OA (Lane, N. *et al.*, 2005). Anti-NGF mAb therapy appears as an attractive therapeutic approach with the potential for long-lasting pain treatment, similar in efficacy to morphine, without necessarily compromising physiological nociception.

NGF also induces the synthesis of another neurotrophin, BDNF, from peptide-containing sensory neurons. BDNF also accumulates in sensory neurons following painful nerve injury (see Sah, D. W. H. *et al.*, 2003). Release of BDNF in the spinal dorsal horn increasing spinal excitability and pain sensitization via TrkB receptors. This initiates a variety of effects, including direct neural excitation, activation of a signaling cascade via the phosphorylation of NMDA receptors, and an altered regulation of the neural chloride transporter that contributes to pain hypersensitivity (Coull, J. A. M. *et al.*, 2005).

In keeping with these observations, spinal BDNF administration induces thermal and mechanical allodynia whereas anti-BNDF neutralization or TrkB IgG administration reduces inflammation or nerve injury-mediated hypersensitivity pain in a number of animal models (Kerr, B. J. *et al.*, 1999; Theodosiou, M. *et al.*, 1999; Deng, Y. S. *et al.*, 2000).

Finally glial cell-line-derived neurotrophic factor, GDNF represents an extensive family of ligands and membrane receptor complexes which have an important role in regulating peripheral and central neural phenotypes. GDNF-related ligands include neurturin and artemin, which act via the complex RET Trk receptor and co-receptors GFRα1, GFRα2, GFRα3, and GFRα4.

Although there appears not to be a specific role in inflammation, GDNF has been shown to have neuroprotective and restorative properties in a number of neurodegenerative and neuropathic pain states (Sah, D. W. H. *et al.*, 2003). Specifically GDNF treatment has been shown to restore peripheral sensory neuron function, including peptide and ion channel expression patterns, following painful peripheral nerve injury accompanied by an attenuation of pain behaviors. Unfortunately clinical observations using GDNF have shown unacceptable side effects such as weight loss and allodynia, which has discouraged therapeutic developments (Nutt, J. G. *et al.*, 2003).

53.18 Kinases

As mentioned earlier inflammatory mediators also activate a number of protein kinases in sensory neurons and in the spinal cord. These include PKA, PKC, and mitogen-activated protein kinases (MAPK) considered to be important downstream regulators of excitability through altering gene transcription and posttranslational modification of target proteins (Woolf, C. J. and Salter, M. W., 2000). There are several types of MAPKs including extracellular signal-regulated kinases (ERK), cJUN, N-terminal kinase (JNK), and p38 kinase that are considered as targets for inflammatory pain. For example, several inhibitors of p38 kinase (e.g., SB203580, CNI-14930) posses antiinflammatory as well as antihyperalgesic properties in a variety of animal models (Schafers *et al.*, 2003b).

53.19 Botulinum Toxin

Another approach to pain modulation has been the use of botulinum toxins (BoTNs). This family of neurotoxins has been traditionally used as an experimental tool to study muscle nerve interactions. Recently BoTN-A has been approved for clinical use to induce muscle relaxation. The mechanism of action of BoTN is related to inhibition of transmitter release from motor fibers through proteolytic cleavage of a number of synaptosomal regulatory proteins (SNARE, syntaxin, SNAP-25, synaptobrevin). More recent studies also indicated potential for inhibition of neuropeptide transmitter release from small afferent neurons (Welch, M. J. *et al.*, 2000; Mense, S., 2004). In keeping with this BoTN has been shown to provide long-lasting pain relief following administration into human osteoarthritic joints and improve bladder dysfunction in overactive bladder patients. This was correlated with loss of both P2X3 and VR1 receptors in the bladder (Apostolidis, A. *et al.*, 2005).

53.20 Nitric Oxide

NO induces a delayed burning pain upon intradermal injection (Holthusen, H. and Arndt, J. O., 1994) and NO donors have been postulated to activate cerebral sensory fibers directly, causing release of the sensory vasodilator CGRP (Wei, P. *et al.*, 1992). Indeed NO has been suggested to contribute to migraine and other types of head pain (Olesen, J. *et al.*, 1994) and there is an abundance of evidence linking NO in the etiology of a number of chronic pain conditions (Millan, M. J., 1999).

Peripheral nerve injury and inflammation increases the expression of NO in sensory neurons and causes the upregulation of the nitric oxide synthase (NOS) isozymes neuronal NOS (nNOS) and inducible NOS (iNOS) in the spinal cord (Gordh, T. *et al.*, 1998) as a result of axonal neurodegeneration, neuroglial cell activation and inflammation (Levy, D. *et al.*, 2001). In keeping with these observations, treatment with nonspecific NOS inhibitors such as L-N-nitro-L-arginine methyl ester (L-NAME) has been shown to prevent or reduce pain hypersensitivity (Meller, S. T. *et al.*, 1992; Hao, J. X. and Xu, X. J., 1996; Handy, R. L. and Moore, P. K., 1998). Indeed particular interest has been devoted to producing selective blockers of iNOS, to avoid side effects associated with nonspecific block of nNOS and endothelial NOS (eNOS). Recently GW27415, a selective iNOS inhibitor has been shown to partially reduce Freund's complete adjuvant (FCA)-induced inflammatory pain as well the hypersensitivity associated with the CCI neuropathy model. This action was most likely to have been caused via a peripheral mechanism as no iNOS expression was detectable in sensory ganglia or spinal cord (Alba, J. D. *et al.*, 2006).

53.21 Summary and Conclusions

Pharmacological understanding of pain is the foundation on which new and effective therapies for pain can emerge. Great advances have been made with greater availability of improved chemical and biological tools. This has fueled greater understanding of the mechanistic and molecular substrates of pain,

derived for the most part from our studies of animal models (see Figure 1). The complexity and heterogeneity of chronic pain states will clearly remain an enormously challenging area for future research. There is still a great need for human data showing the regulated expression of pain molecules and clinical data that provides more rapid evaluation of emerging pharmacological concepts. This will allow the most valuable targets to be selected for analgesia development in the clinic.

Clearly the improvements in our pharmacological understand of pain has also provided an abundance of opportunities for developing analgesics. Among the most comprehensively studied molecules in chronic pain are inflammatory mediators and their key receptors. These include PGE2/EP1, bradykinin/BK1 and BK2, ATP/P2X3, cytokines/ IL-1β, chemokines/ CCr2 and neurotrophins/NGF/TrkA that sensitize and increase excitability in peripheral and central pain pathways. Many of these mediators act through GPCRs and there appears to be some convergence of mechanisms with changes in expression and regulation of ion channel. Key channels for pain are either ligand gated, for example, TRPV1, P2X3, P2X4, P2X7, or voltage gated, for example, NaV1.8, NaV1.7, NaV1.3, CaV2.2.

A variety of additional cellular process including protein phosphorylation of membrane receptors and channels via key kinases, for example, MAPK, as well as complex phenotype change regulated by neurotrophins require further understanding. In addition, the emerging importance of neuroglia and their pharmacology needs exploration and consolidation.

Finally despite the great advances in pain pharmacology, many pain molecules remain poorly validated as good targets for analgesia. Stronger clinical translation is highly likely with the progression of pharmacological opportunities into humans.

References

Abbadie, C. 2005. Chemokines, chemokine receptors and pain. Trends Immunol. 26, 529–534.

Alba, J. D., Clayton, N. M., Collins, S. D., Colthup, P., Chessell, I., and Knowles, R. G. 2006. GW274150, a novel and highly selective inhibitor of the inducible isoform of nitric oxide synthase (iNOS), shows analgesic effects in rat models of inflammatory and neuropathic pain. Pain 120, 170–181.

Ali, Z., Raja, S. N., Wesselmann, U., Fuchs, P. N., Meyer, R. A., and Campbell, J. N. 2000. Intradermal injection of norepinephrine evokes pain in patients with sympathetically maintained pain. Pain 88, 161–168.

Aloe, L., Tuveri, M. A., Carcassi, U., and Levi-Montalcini, R. 1992. Nerve growth factor in the synovial fluid of patients with chronic arthritis. Arthritis Rheum. 35, 3351–3355.

Altier, C. and Zamponi, G. W. 2004. Targeting Ca channels to treat pain: T type versus N-type. Trends Pharmacol. Sci. 26, 465–470.

Altier, C., Khosravani, H., Evans, R. M. Hameed, S., Peloquin, J. B., Vartian, B. A., Chen, L., Beedle, A. M., Ferguson, S. S., Mezghrani, A., Dubel, S. J., Bourinet, E., McRory, J. E., and Zamponi, G. W. 2005. ORL-1 receptor mediated internalization of N-type calcium channels. Nature Neurosci. 9, 31–40.

Amadesi, S., Nie, J., Vergnolle, N., Cottrell, G. S., Grady, E. F., Trevisani, M., Manni, C., Geppetti, P., McRoberts, J. A., Ennes, H., Davis, J. B., Mayer, E. A., and Bunnett, N. W. 2004. Protease-activated receptor 2 sensitizes the capsaicin receptor transient receptor potential vanilloid receptor 1 to induce hyperalgesia. J. Neurosci. 24, 4300–4312.

Andreev, N., Dimitrieva, N., Kolzenburg, M., and McMahon, S. B. 1995. peripheral administration of nerve growth factor in the adult rat produces a thermal hyperalgesia that requires the presence of sympathetic post-ganglionic neurones. Pain 63, 109–115.

Apfel, S. C. 2002. Nerve growth factor for the treatment of diabetic neuropathy: what went wrong, what went right, and what does the future hold? Int. Rev. Neurobiol. 50, 393–413.

Apostolidis, A., Popat, R., Yiangou, Y., Cockyne, D., Ford, A. B. D. W., Davis, J. B., Dasgupta, P., Fowler, C. J., and Anand, P. 2005. Decreased sensory receptors P2X (3) and TRPV1 in suburothelial nerve fibers following intradetrusor injection of botulinum toxin for human detrusor overactivity. J. Urol. 174, 977–982.

Araujo, M. C., Sinnott, C. J., and Strichartz, G. R. 2003. Multiple phases of relief from experimental mechanical allodynia by systemic lidocaine: responses to early and late infusions. Pain 103, 21–29.

Baccei, M. L. and Kocsis, J. D. 2000. Voltage gated calcium currents in axotomized rat cutaneous afferent neurons. spinal dorsal horn of rats. Eur. J. Neurosci. 419, 1336–1342.

Bandell, M., Story, G. M., Hwang, S. W., Viswanath, V., Eid, S. R., Petrus, M. J., Earley, T. J., and Pataapouian, A. 2004. Noxious cold ion channel TRPA1 is activated by pungent compounds and bradykinin. Neuron 41, 849–853.

Bao, L., Jin, S. X., Zhang, C., Wang, L. H., Xu, Z. Z., Zhang, F. X., Wang, L. C., Ning, F. S., Cai, H. J., Guan, J. S., Xiao, H. S., Xu, Z. Q., He, C., Hokfelt, T., Zhou, Z., and Zhang, X. 2003. Activation of delta opioid receptors induces receptor induces receptor insertion and neuropeptide secretion. Neuron 37, 121–133.

Bar, K. J., Natura, G., Telleria-Diaz, A., Tascher, P., Vogel, R., Vasques, E., Schaible, H. G., and Ebersberger, A. 2004. Changes in the effect of spinal prostaglandin E2 during inflammation: prostaglandin E9EP1-EP4) receptors in spinal nociceptive processing of input from the normal and inflamed knee joint. J. Neurosci. 24, 642–651.

Bhave, G. and Gereau, R. W. 2004. Posttranslational mechanisms of peripheral sensitization. J. Neurobiol. 6188–6106.

Bhave, G., Karim, F., Carlton, S. M., and Gereau, R. W. 2001. Peripheral group I metabotropic glutamate receptors modulate nociception in mice. Nat. Neurosci. 4, 417–423.

Blackburn-Munro, G., Bomholt, S. F., and Erichsen, H. K. 2004. Behavioural effects of the novel AMPA/GluR5 selective receptor antagonist NS1209 after systemic administration in animal models of experimental pain. Neuropharmacology 47, 351–362.

Boucher, T. J., Okuse, K., Bennett, D. L., Munson, J. B., Wood, J. N., and McMahon, S. B. 2000. Potent analgesic effects of GDNF in neuropathic pain states. Science 290, 124–127.

Boxall, S. J., Berthele, A., Laurie, D. J., Sommer, B., Zieglgansberger, W., Urban, L., and Tolle, T. R. 1998. Enhanced expression of metabotropic glutamate receptor 3 messenger RNA in the rat spinal cord during ultraviolet irradiation induced peripheral inflammation. Neuroscience 82, 591–602.

Burgess, G. M., Perkins, M. N., Rang, H. P., Campbell, E. A., Browen, M. C., McIntyre, P., Urban, L., Dziadulewicz, E. K., Ritchie, T. J., Hallet, A., Snell, C. R., Wrigglesworth, R., Lee, W., Davis, C., Phagoo, S. B., Davis, A. J., Phillips, E., Drake, G. S., Hughes, G. A., Dunstan, A., and Bloomfield, G. C. 2000. Bradyzide, a potent nonpeptide B2 bradykinin receptors antagonist with long lasting oral activity in animal models of inflammatory hyperalgesia. Br. J. Pharmacol. 129, 77–86.

Burstein, S. H., Karst, M., Schneider, U., and Zurier, R. B. 2004. Ajulemic acid: a novel cannabinoid produces analgesia without a "high". Life Sci. 75, 1513–1522.

Byas-Smith, M. G., Max, M. B., Muir, J., and Kingman, A. 1995. Transdermal clonidine compared to placebo in painful diabetic neuropathy using a two-stage 'enriched enrollment' design. Pain 60, 267–274.

Cahill, C. M., Morinville, A., Lee, M. C., Vincent, J. P., Collier, B., and Beaudet, A. 2001. Prolonged morphine treatment targets delta opioid receptors to neuronal plasma membranes and enhances delta-mediated antinociception. J. Neurosci. 21, 7598–7607.

Campbell, F. A., Tramer, M. R., Carroll, D., Reynolds, D. J., Moore, R. A., and McQuay, H. J. 2001. Are cannabinoids an effective and safe treatment option in the management of pain? A quantitative systematic review. BMJ 323, 13–16.

Campbell, J. N., Basbaum, A. I., Dray, A., Dubner, R., Dworkin, R. H., and Sang, C. N. (eds.) 2006. Emerging Strategies for the Treatment of Neuropathic Pain. IASP Press.

Carlton, S. M., Hargett, G. L., and Coggeshall, R. E. 1995. Localization and activation of glutamate receptors in unmyelinated axons of rat glabrous skin. Neurosci. Lett. 197, 25–28.

Caterina, M. J., Leffer, A., Malmberg, A. B., Martin, W. J., Trafton, J., Petersen-Zeitz, K. R., Koltzenburg, M., Basbaum, A. I., and Julius, D. 2000. Impaired nociception and pain sensation I mice lacking the capsaicin receptor. Science 288, 306–313.

Cervero, F., Laird, J. M. A., and Garcia-Nicas, E. 2003. Secondary hyperalgesia and presynaptic inhibition: an update. Eur. J. Pharmacol. 7, 345–351.

Chandrasekharan, N. V., Dai, H., Roos, K. L. T., Evanson, N. K., Tomsik, J., Elton, T. S., and Simmons, D. L. 2002. COX-3, a cyclooxygenase-1 variant inhibited by acetaminophen and other analgesic/antipyretic drugs: cloning, structure, and expression. Proc. Natl. Acad. Sci. U. S. A. 99, 13926–13931.

Chang, L., Luo, L., Palmer, J. A., Sutton, S., Wilson, S. J., Barbier, A. J., Breitenbucher, J. G., Chaplan, S. R., and Webb, M. 2006. Inhibition of fatty acid amide hydrolase produces analgesia by multiple mechanisms. Br. J. Pharmacol. 148, 102–113.

Chaplan, S. R., Guo, H. Q., Lee, D. H., Luo, L., Liu, C., Kuei, C., Velumian, A. A. L., Butler, M. P., Brown, S. M., and Dubin, A. E. 2003. Neuronal hyperpolarization-activated pacemaker channels drive neuropathic pain. J. Neurosci. 23, 1169–1178.

Chessell, I. P., Hatcher, J. P., Bountra, C., Michel, A. D., Hughes, J. P., Green, P., Egerton, J., Murfin, M., Richardson, J., Peck, W. L., Grahames, C. B., Casula, M. A., Yiangou, Y., Birch, R., Anand, P., and Buell, G. N. 2005. Disruption of the P2x7 purinoceptor gene abolishes chronic inflammatory and neuropathic pain. Pain 114, 386–396.

Chiechio, S., Caricasole, A., Barletta, E., Storto, M., Catania, M. V., Copani, A., Vertechy, M., Nicolai, R., Calvani, M., Melchiorri, D., and Nicoletti, F. 2002. L-Acetylcarnitine induces analgesia by selectively up-regulating mGlu2 metabotropic glutamate receptors. Mol. Pharmacol. 61, 989–996.

Chizh, B., Napolitano, A., O'Donnell, M., Wang, J., Brooke, A., Lai, R., Aylott, M., Bullman, J., Gray, E., Williams, P., and Appleby, J. 2006. The TRPV1 antagonist SB705498 attenuates TRPV1 receptor-mediated activity and inhibits inflammatory hyperalgesia in humans: results from a Phase 1 study. Am. Pain Soc. 765.

Cichewicz, D. L. and McCarthy, E. A. 2003. Antinociceptive synergy between delta9 tetrahydrocannabinol and opioids after oral administration J. Pharmacol. Exp. Ther. 304, 1010–1015.

Claveau, D., Sirinyan, M., Guay, J., Gordon, R., Chan, C. C., Bureau, Y., Riendeau, D., and Mancini, J. A. 2003. Microsomal prostaglandin synthase-1 is a major terminal synthase that is selectively up-regulated during cyclooxygenase-2-dependent prostaglandin E2 production in the rat adjuvant-arthritis model. J. Immunol. 170, 4738–4744.

Colquhoun, A., Lawrance, G. M., Shamovsky, I. L., Riopelle, R. J., and Ross, G. M. 2004. Differential activity of the nerve growth factor (NGF) antagonist PD90780 [7-benzoylamino)-4,9-dihydro-4methyl-8-oxo-pyrazolo-[5,1-b]-quinazoline-2-carboxylic acid] suggest altered NGF-p75NTR interactions in the presence of TrkA. J. Pharmacol. Exp. Ther. 310, 505–511.

Coull, J. A. M., Boudreau, R., Bachand, K., Prescott, S. A., Nault, F., Sik, A., de Koninck, P., and de Koninck, Y. 2003. Trans-synaptic shift in ionic gradient in spinal lamina 1 neurons as a mechanism of neuropathic pain. Nature 424, 938–942.

Cravatt, B. F. and Lichtman, A. H. 2004. The endogenous cannabinoid system and its role in nociceptive behavior. J. Neurobiol. 61, 149–160.

Cravatt, B. J., Demarest, K., Patricelli, M. P., Bracey, M. H., Giang, D. K., Martin, B. R., and Lichtman, A. H. 2001. Supersensitivity to anandamide and enhanced endogenous cannabinoid signaling in mice lacking fatty acid amide hydrolyse. Proc. Natl. Acad. Sci. U. S. A. 98, 9371–9376.

Cummins, T. R. and Waxman, S. G. 1997. Downregulation of tetrodotoxin-resistant sodium currents and upregulation of a rapidly repriming tetrodotozin-sensitive sodium current in small spinal sensory neurons after nerve injury. J. Neurosci. 17, 3503–3514.

Danysz, W. and Parsons, C. G. 1998. GlycineB recognition site of NMDA receptors and its antagonists. Amino Acids 14, 205–206.

De Jongh, R. F., Vissers, K., Meert, T. F., Booij, L. H. D. J., de Deyne, C. S., and Heylen, R. J. 2003. The role of interleukin-6 in nociception and pain. Anaesth. Analg. 96, 1096–1103.

Deng, Y. S., Zhong, J. H., and Zhou, X. F. 2000. Effects of endogenous neurotrophin on sympathetic sprouting in the dorsal root ganglia and allodynia following spinal nerve injury. Exp. Neurol. 164, 344–350.

Devor, M. 2005. Sodium channels and mechanisms of neuropathic pain. J. Pain 7, S3–S12.

Devor, M., Wall, P. D., and Catalan, N. 1992. Systemic lidocaine silences ectopic neuroma and DRG discharges without blocking nerve conduction. Pain 48, 261–268.

Deuchars, S. A., Atkinson, L., Brooke, R. E., Musa, H., Milligan, C. J., Batten, T. F. C., Buckley, N. J., Parson, S. H., and Deuchars, J. 2001. Neuronal P2X7 receptors are targeted to presynaptic terminals in the central and peripheral nervous systems. J. Neurosci. 21, 7143–7152.

Dogrul, A., Ossipov, M. H., Lai, J., Malan, T. P., and Porecca, F. 2000. Peripheral and spinal antihyperalgesic activity of SIB-1757, a metabotropic glutamate receptor (mGluR5) antagonist, in experimental neuropathic pain in rats. Neurosci. Lett. 292, 115–118.

Dong, X., Han, S., Zylka, M. J., Simon, M. I., and Anderson, D. J. 2001. A diverse family of GPCRs expressed in specific subsets of nociceptive sensory neurons. Cell 106, 619–632.

Dray, A. 1997. Kinins and their receptors in hyperalgesia. Can. J. Pharmacol. 75, 704–712.

Dray, A. 1999. Alternatives to mu-Opioid Analgesics: delta-Opioid and Galanin-Receptor Selective Compounds. In: Progress in Pain research and Management (eds. E. Kalso, H. J. McQuay, and Z. Wiesenfeld-Hallin), pp. 269–280. IASP Press.

Dray, A. and Perkins, M. N. 1993. Bradykinin and inflammatory pain. Trends Neurosci. 16, 99–104.

Dube, G. R., Lehto, S. G., Breese, N. M., Baker, S. J., Wang, X., Matulenko, M. A., Honore, P., Stewart, A. O., Moreland, R. B., and Brioni, J. D. 2005. Electrophysiological and in vivo characterization of A-317567, a novel blocker of acid sensing ion channels. Pain 117, 88–96.

Dyson, A., Peacock, M., Chen, A., Courade, J. P., Yaqoob, M., Groark, A., Brain, C., Loong, Y., and Fox, A. 2005. Antihyperalgesic properties of the cannabinoid CT-3 in chronic neuropathic and inflammatory pain states in the rat. Pain 116, 129–137.

Eglen, R. M., Hunter, J. C., and Dray, A. 1999. Ions in the fire: recent ion-channel research and approaches to pain therapy. Trends Pharmacol. Sci. 8, 337–342.

Eisenbarth, H., Rukwied, R., Petersen, M., and Schmelz, M. 2004. Sensitization to bradykinin B1 and B2 receptor activation in UV-B irradiated human skin. Pain 110, 197–204.

Elmes, S. J. R., Winyard, L. A., Medhurst, L. A., Clayton, S. J., Wilson, N. M., Kendall, A. W., and Chapman, V. 2005. Activation of CB1 and CB2 receptors attenuates the induction and maintenance of inflammatory pain in the rat. Pain 118, 327–335.

England, S., Bevan, S., and Dougherty, R. J. 1996. PGE2 modulate the tetrodotoxin-resistant sodium current in neonatal dorsal root ganglion neurons via the cyclic AMP-protein kinase A cascade. J. Physiol. 495, 429–440.

Ferreira, J., Beirith, A., Mori, M. A. S., Araujo, R. C., Bader, M., Pesqero, J. B., and Calixto, J. B. 2005. Reduced nerve injury induced neuropathic pain in kinin B1 receptor knock-out mice. J. Neurosci. 25, 2405–2412.

Fiorucci, S., Antonelli, E., Burgand, J. L., and Morelli, A. 2001. Nitric-oxide releasing NSAIDs: a review of their current status. Drug Saf. 24, 801–811.

Finn, D. P., Jhaveri, M. D., Beckett, S. R., Kendall, D. A., Marsden, C. A., and Chapman, V. 2004. Cannabinoids modulate ultrasound-induced aversive responses in rats. Psychopharmacology 172, 41–51.

Flatters, S. J. and Bennett, G. J. 2004. Ethosuximaide reverses paclitaxel and vincristine-induced painful peripheral neuropathy. Pain 109, 150–161.

Flechtenmacher, J., Talke, M., Veith, D., Heil, K., Gebauer, A., and Schoenharting, M. 2004. Icatibant induces pain relief in patients with osteoarthritis of the knee. 9th World Cong. Osteoarthritis Res. Soc. Intl. 12, Abst. P332.

Fjell, J., Cummings, T. R., Fried, K., Black, J. A., and Waxman, S. G. 1999. In vivo NGF deprivation reduces SNS expression and TTX-R sodium currents in IB4-negative DRG neurons. J Neurophysiol. 81, 803–810.

Flower, R. J. 2003. The development of COX2 inhibitors. Nat. Rev. 2, 179–191.

Fox, A. and Bevan, S. 2005. Therapeutic potential of cannabinoid receptor agonists as analgesic agents. Exp. Opin. Invest. Drugs 14, 695–703.

Fox, A., Kaur, S., Li, B., Panasar, M., Saha, U., Davis, C., Dragoni, I., Colley, S., Ritchie, T., Bevan, S., Burgess, G., and McIntyre, P. 2005. Antihyperalgesic activity of a novel nonpeptide bradykinin B1 receptor antagonist in transgenic mice expressing the human B1 receptor. Br. J. Pharmacol. 144, 889–899.

Fox, A., Wotherspoon, G., McNair, K., Hudson, L., Patel, S., Gentry, C., and Winter, L. 2003. Regulation and function of spinal and peripheral neuronal B(1) bradykinin receptors in inflammatory mechanical hyperalgesia. Pain 104, 683–691.

Fride, E., Suris, R., Weidenfeld, J., and Mechoulam, R. 2005. Differential response to acute and repeated stress in cannabinoid cB1 receptor knockout newborn and adult mice. Behav. Pharmacol. 16, 431–440.

Fundytus, M. E., Osborne, M. G., Henry, J. L., Coderre, T. J., and Dray, A. 2002. Antisense oligonucleotide knockdown of mGluR(1) alleviates hyperalgesia and allodynia associated with chronic inflammation. Pharmacol. Biochem. Behav. 73, 401–410.

Gabra, B. H. and Sirois, P. 2003. Beneficial effects of chronic treatment with the selective bradykinin B1 receptor antagonists, R-715 and R-953, in attenuating streptozotocin-diabetic thermal hyperalgesia. Peptides 24, 1131–1139.

Gabra, B. H., Merino, V. F., Bader, M., Pasquero, J. B., and Sirois, P. 2005. Absence of diabetic hyperalgesa in bradykinin B1 receptor-knockout mice. Reg. Pept. 127, 245–248.

Garcia-Martiez, C., Humet, M., Planells-Casas, R., Gomis, A., Capri, M., Viaa, F., de la Pena, E., Sachez-Baez, F., Carbonell, T., de Felipe, C., Perez-Paya, E., Belmote, C., Messeguer, A., and Ferrer-Motiel, A. 2002. Attenuation of thermal nociception and hyperalgesia by VR1 blockers. Proc. Natl. Acad. Sci. U. S. A. 99, 2374–2379.

Gold, M. S., Levine, J. D., and Correa, M. 1998. Modulation of TTX-R INa by PKC and PKA and their role in PGE2-induced sensitization of rat sensory neurons *in vitro*. J. Neurosci. 18, 10345–10355.

Gold, M. S., Weinreich, D., Kim, C. S., Wang, R., Treanor, J., Porecca, F., and Lai, J. 2003. Redistribution of Na(V) 1.8 in uninjured axons enables neuropathic pain. J. Neurosci. 23, 158–166.

Gordh, T., Sharma, H. S., Alm, P., and Westman, J. 1998. Spinal nerve lesion induces upregulation of neuronal nitric oxide synthase in the spinal cord. An immunohistochemical investigation in the rat. Amino Acids 14, 105–112.

Gougat, J., Ferrari, B., Sarran, L., Planchenault, C., Poncelet, M., Maruani, J., Alonso, R., Cudennec, A., Croci, T., Guagnini, F., Urban-Szabo, K., Martinolle, J. P., Soubrié, P., Finance, O., and Le Fur, G. 2004. SSR240612 [(2R)-2-[((3R)-3-(1,3-benzodioxol-5-yl)-3-[[(6-methoxy-2-naphthyl)sulfonyl]amino]propanoyl)amino]-3-(4-[[2R,6S)-2,6-dimethylpiperidinyl]methyl]phenyl)-N-isopropyl-N-methylpropanamide hydrochloride], a new nonpeptide antagonist of the bradykinin B₁ receptor: biochemical and pharmacological characterization. J. Pharmacol. Exp. Ther. 309, 661–669.

Gould, H. J., England, J. D., Soignier, R. D., Nolan, P., Minor, L. D., Liu, Z. P., Levinson, S. R., and Paul, D. 2004. Ibuprofen block changes in NaV1.7 and 1.8 sodium channels associated with complete Freund's adjuvant-induced inflammation in rat. J. Pain 5, 270–280.

Gracely, R., Lynch, S., and Bennett, G. 1992. Painful neuropathy: altered central processing, maintained dynamically by peripheral input. Pain 51, 175–194.

Grazzini, E., Puma, C., Roy, M. O., Yu, X. H., O'Donnell, D., Schmidt, R., Dautry, S., Ducharme, J., Perkins, M., Panetta, R., Laird, J. M. A., Ahmad, S., and Lembo, P. M. 2004. Sensory neuron-specific receptor activation elicits central and peripheral nociceptive effects in rats. Proc. Nat. Acad. Sci. U. S. A. 101, 7171–7180.

Grunke, M. and Schulze-Koops, H. 2006. Successful treatment of inflammatory knee osteoarthritis with tumour necrosis factor blockade. Annu. Rheum. Dis. 65, 555–556.

Gutierrez, T., Nackley, A. G., Neely, M. H., Freeman, K. G., Edwards, G. L., and Hohmann, A. G. 2003. Effects of neurotoxic destruction of descending noradrenergic pathways on cannabinoid antinociception in models of acute and tonic nociception. Brain Res. 987, 176–185.

Hains, B. C., Saab, C. Y., Klein, J. P., Craner, M. J., and Waxman, S. G. 2004. Altered sodium channel expression in second-order spinal sensory neurons contributes to pain after peripheral nerve injury. J. Neurosci. 24, 4832–4839.

Halliday, D. A., Zettler, C., Rush, R. A., Scicchitano, R., and McNeil, J. D. 1998. Elevated growth factor levels in the synovial fluid of patients with inflammatory joint disease. Neurochem. Res. 23, 919–922.

Handy, R. L. and Moore, P. K. 1998. A comparison of the effects of L-NAME, 7NI and l-NIL on carrageenan induced hindpaw edema and NOS activity. Br. J. Pharmacol. 12, 1119–1126.

Hao, J. X. and Xu, X. J. 1996. Treatment of a chronic allodynia-like response in spinally injured rats: effects of systemically administered nitric oxide synthase inhibitors. Pain 66, 313–319.

Hao, S., Mata, M., Wolfe, D., Glorioso, J. C., and Fink, D. J. 2005. Gene transfer of glutamic acid decarboxylase reduces neuropathic pain. Ann. Neurol. 57, 914–918.

Harvey, R. J., Depner, U. B., Wassle, H., Ahmadi, S., Heidl, C., Reinold, H., Smart, T. G., Harvey, K., Schultz, B., Abo-Salem, O. M., Zimmer, A., Poisbeau, P., Welzl, H., Wolfer, D. P., Betz, H., Zeilhofer, H. U., and Muller, U. 2004. GlyRα3: an essential target for spinal PGE2-mediated inflammatory pain sensitization. Science 304, 884–887.

Hefti, F. F., Rosenthal, A., Walicka, P. A., Wyatt, S., Vergara, G., Shelton, D. L., and Davis, A. M. 2005. Novel class of pain drug based on antagonism of NGF. Trends Pharmacol. Sci. 27, 85–91.

Hogestatt, E. D., Jonsson, B. A. G., Ermund, A., Andersson, D. A., Bjork, H., Alexander, J. P., Cravatt, B. J., Basbaum, A. I., and Zygmunt, P. M. 2005. Conversion of acetaminophen to bioactive N-acylphenolamine AM4040 via fatty acid amide hydrolase-dependent arachidonic acid conjugation in the nervous system. J. Biol. Chem. 280, 31405–31412.

Hohmann, A. G. and Herkenham, M. 1999. Cannabinoid receptors undergo axonal flow in sensory nerves. Neuroscience 92, 1171–1175.

Hohmann, A. G., Suplita, R. L., Bolton, N. M., Neely, M. H., Fegley, D., Mangieri, R., Krey, J. F., Walker, J. M., Holmes, P. V., Crystal, J. D., Duranti, A., Tonini, A., Mor, M., Tarzia, G., and Piomelli, D. 2005. An endocannabinoid mechanism for stress-induced analgesia. Nature 435, 1108–1112.

Holthusen, H. and Arndt, J. O. 1994. Nitric oxide evokes pain in humans on intracutaneous injection. Neurosci. Letts. 165, 71–74.

Honore, P., Kage, K., Mikusa, J., Watt, A. T., Johnston, J. F., Wyatt, J. R., Faltynek, C. R., Jarvis, M. F., and Lynch, K. 2002. Analgesic profile of intrathecal P2X(3) antisense oligonucleotide treatment in chronic inflammatory and neuropathic pain states in rats. Pain 99, 11–19.

Honore, P., Rogers, S. D., Schwei, M. J., Salak-Johnson, J. L., Luger, N. M., Sabino, M. C., Clohisy, D. R., and Mantyh, P. W. 2000. Murine models of inflammatory, neuropathic and cancer pan each generate a unique set of neurochemical changes in the spinal cord and sensory neurons. J. Neurosci. 98, 585–598.

Hudson, L. J., Bevan, S., McNair, K., Gentry, C., Fox, A., Kuhn, R., and Winter, J. 2002. Metabotropic glutamate receptor 5 upregulation in A-fibers after spinal nerve injury: 2-Methyl-6-(phenylethynyl)-pyridine (MPEP) reverses the induced thermal hyperalgesia. J. Neurosci. 22, 2660–2668.

Hunt, S. P. and Mantyh, P. 2001. The molecular dynamics of pain control. Nat. Rev. 2, 83–91.

Hurley, R. W. and Hammond, D. L. 2000. The analgesic effects of supraspinal mu and delta opioid receptor agonists are potentiated during persistent inflammation J. Neurosci. 20, 1249–1259.

Ibrahim, M. M., Porreca, F., Lai, J., Albrecht, P. J., Rise, F. L., Khodorova, A., Davar, G., Makriyannis, A., Vanderah, T. W., Mata, H. P., and Malan, T. P. 2005. CB2 cannabinoid receptor activation produces antinociception by stimulating peripheral release of endogenous opioids. Proc. Natl. Acad. Sci. U. S. A. 102, 3093–3098.

Ibuki, T., Hama, A. T., Wang, X. T., Pappas, G. D., and Sagen, J. 1997. Loss of GABA-immunoreactivity in the spinal dorsal horn of rats with peripheral nerve injury and promotion of recovery by adrenal medullary grafts. Neuroscience 76, 845–858.

Ikeda, H., Heinke, B., Ruscheweyh, R., and Sandkuhler, J. 2003. Synaptic plasticity in spinal lamina 1 projection neurons that mediate hyperalgesia. Science 299, 1237–1240.

Jakobsson, P. J., Thoen, S., Morgenstern, R., and Samuelsson, B. 1999. Identification of human prostaglandin E synthase: a microsomal, glutathione-dependent, inducible enzyme, constituting a potential novel drug target. Proc. Natl. Acad. Sci. U. S. A. 96, 7220–7225.

Jang, J. H., Kim, D. W., Sang Nam, T., Se Paik, K., and Leem, J. W. 2004. Peripheral glutamate receptors contribute to mechanical hyperalgesia in a neuropathic pain model of the rat. Neuroscience 128, 169–176.

Jarvis, M. F., Burgard, E. C., McGaraughty, S., Honore, P., Lynch, K., Brennan, T. J., Subieta, A., van Biesen, T., Cartmell, J., Bianchi, B., Niforatos, W., Kage, K., Yu, H. X., Mikusa, J., Wismer, C. T., Zhu, C. Z., Chu, K., Lee, C. H., Stewart, A. O., Polakowski, J., Cox, B. F., Kowaluk, E., Williams, M., Sullivan, J., and Faltynek, C. 2002. A-3174919, a novel potent and selective non nucleotide antagonist of P2X(3) and P2X(2/3) receptors, reduces chronic inflammatory and neuropathic pain in the rat. Proc. Natl. Acad. Sci. U. S. A. 99, 17179–17184.

Ji, R. R., Samad, T. A., Jin, S. X., Schmoll, R., and Woolf, C. J. 2002. P38 MAPK activation by NGF in primary sensory neurons after inflammation increases TRPV1 levels and maintains heat hyperalgesia. Neuron 36, 57–68.

Karst, M., Salim, K., Burstein, S., Conrad, I., Hoy, L., and Schneider, U. 2003. Analgesic effect of the synthetic cannabinoid CT-3 on chronic neuropathic pain. JAMA 290, 1757–1762.

Kastrup, J., Petersen, P., Dejgard, A., and Angelo, F. R. 1987. Intravenous lidocaine infusion – a new treatment of chronic painful diabetic neuropathy. Pain 28, 69–75.

Kerr, B. J., Bradbury, E. J., Bennett, D. L., Trivedi, P. M., Dassan, P., French, J., Shelton, D. B., McMahon, S. B., and Thompson, S. W. 1999. Brain derived neurotrophic factor modulates nociceptive sensory neurone input and NMDA-evoked responses in the rat spinal cord. J. Neurosci. 19, 5136–5149.

Kim, C., Jun, K., Lee, T., Kim, S. S., et al., 2001. Altered nociceptive responses in mice deficient in the α1B subunit of the voltage dependent calcium channel. Mol. Cell Neurosci. 18, 235–245.

Kim, C. H., Oh, Y., Chung, J. M., and Chung, K. 2002. Changes in three subtypes of tetrodotoxin sensitive sodium channel expression in the axotomized dorsal root ganglion in the rat. Neurosci. Letts. 323, 125–128.

Krause, J. E., Chenard, B. L., and Cortright, D. N. 2005. Transient receptor potential ion channels for the discovery of pain therapeutics Curr. Opin. Inv. Drugs 6, 48–57.

Kreitzer, A. C. and Regeher, W. G. 2002. Retrograde signaling by endocannabinoids. Curr. Opin. Neurobiol. 12, 324–330.

Krishtal, O. 2003. The ASICs: signaling molecules? Modulators? TINS 26, 477–483.

LaBuda, C. J., Koblish, M., and Little, P. J. 2005. Cannabinoid CB2 receptor agonist activity in the hindpaw incision model of postoperative pain. Eur. J. Pharmacol. 527, 172–175.

Lai, J., Hunter, J. C., and Porreca, F. 2003. The role of voltage-gated sodium channels in neuropathic pain. Curr. Opin. Neurobiol. 13, 291–297.

Laird, J. M. A., García-Nicas, E., Delpire, E. J., and Cervero, F. 2004. Presynaptic inhibition and spinal pain processing in mice: a possible role of the NKCC1 cation-chloride co-transporter in hyperalgesia. Neurosci. Letts. 361, 200–203.

Lane, N., Webster, L., Shiao-Ping, L., Gray, M., Hefti, F., and Walicke, P. 2005. RN624 (Anti-NGF) improves pain and function in subjects with moderate knee osteoarthritis: a phase I study. Arthritis Rheumat. 52, S461.

Lee, D. H., Liu, X., Kim, H. T., Chung, K., and Chung, J. M. 1999. Receptor subtype mediating the adrenergic sensitivity of pain behavior and ectopic discharges in neuropathic Lewis rats. J. Neurophysiol. 81, 2226–2233.

Lembo, P. M. C., Grazzini, E., Groblewski, T., O'Donnell, D., Roy, M. O., Zhang, J., Hoffert, C., Cao, J., Schmidt, R., Pelletier, M., Labarre, M., Gosselin, M., Fortin, Y., Banville, D., Shen, S. H., Strom, P., Payza, K., Dray, A., Walker, P., and Ahmad, S. 2002. Proenkephalin A gene products activate a new family of sensory neuron-specific GPCRs. Nat. Neurosci. 5, 201–209.

Levy, D. and Zochodne, D. W. 2000. Increased mRNA expression of the B1 and B2 bradykinin receptors and antinociceptive effects of their antagonists in an animal model of neuropathic pain. Pain 86, 265–271.

Levy, D., Kubes, P., and Zachodne, D. W. 2001. Delayed peripheral nerve degeneration, regeneration and pain in mice lacking inducible nitric oxide synthase. J. Neuropathol. Exp. Neurol. 60, 411–421.

Liang, Y. F., Haake, B., and Reeh, P. W. 2001. Sustained sensitization and recruitment of rat cutaneous nociceptors by bradykinin and a novel theory of its excitatory action. J. Physiol. 532, 229–239.

Lichtman, A. H., Leung, D., Shelton, C. C., Saghatelian, A., Hardouin, C., Boger, D. L., Cravatt, B. J. 2004. Reversible inhibitors of fatty acid amide hydrolase that promote analgesia: evidence for an unprecedented combination of potency and selectivity. J. Pharmacol. Exp. Ther. 311, 441–448.

Lindenlaub, T. and Sommer, C. 2003. Cytokines in sural nerve biopsies from inflammatory and non-inflammatory neuropathies. Acta Neuropathol. 105, 593–602.

Lindia, J. A., Kohler, M. G., Martin, W. J., and Abbadie, C. 2005. Relationship between sodium channel Nav1.3 expression and neuropathic pain behavior in rats. Pain 117, 145–153.

Liu, J., Li, H., Bursten, S. H., Zurier, R. B., and Chen, J. D. 2003. Activation and binding of peroxisomal proliferator-activated receptor gamma by synthetic cannabinoid ajulemic acid. Mol. Pharmacol. 63, 983–992.

Lu, B., Pang, P. T., and Woo, N. H. 2005. The Yin and Yang of neurotrophin action. Nat. Rev. Neurosci. 6, 603–614.

Luo, Z. D., Calcutt, N. A., Higuera, E. S., Valder, C. R., Song, Y. H., Svensson, C. I., and Myers, R. R. 2002. Injury type-specific calcium channel alpha (2) delta-1 subunit up-regulation in rat neuropathic pain models correlates with antiallodynic effects of gabapentin. J. Pharmacol. Exp. Ther. 303, 1199–1205.

Ma, W., Zhang, Y., Bantel, C., and Eisenach, J. C. 2005. Medium and large injured dorsal root ganglion cells increase TRPV-1, accompanied by increased alpha2C-adrenoceptor co-expression and functional inhibition by clonidine. Pain 113, 386–394.

Malan, T. P., Ibrahim, M. M., Lai, J., Vanderah, T. W., Makriyannis, A., and Porreca, F. 2003. CB2 cannabinoid receptor agonists: pain relief without psychoactive effects?. Curr. Opin. Pharmacol. 3, 62–67.

Malan, T. P., Mata, H. P., and Porreca, F. 2002. Spinal GABA(A) and GABA(B) receptor pharmacology in a rat model of neuropathic pain. Anesthesiology 96, 1161–1167.

Mamet, J., Lazdunski, M., and Voilley, N. 2003. How nerve growth factor drives physiological and inflammatory expression of acid-sensing ion channels 3 in sensory neurons. J. Biol. Chem. 278, 48907–48913.

Mantyh, P. W. and Yaksh, T. L. 2001. Sensory neurons are PARtial to pain. Nat. Med. 7, 772–773.

Marceau, F., Hess, J. F., and Bachvarov, D. R. 1998. The B1 receptors for kinins. Pharmacol. Rev. 50, 357–386.

Marsicano, G., Goodenough, S., Monory, K., Hermann, H., Eder, M., Cannich, A., Azad, S. C., Cascio, M., Ortega Gutierrez, S., Van der Stelt, M., Lopez-Rodriguez, M. L., Casanova, E., Schutz, G., Zieglgansberger, W., Di Marzo, V., Behl, C., and Lutz, B. 2003. CB1 cannabinoid receptors and on demand defense against excitotoxicity. Science 302, 84–88.

Matthews, E. A. and Dickenson, A. H. 2001. Effects of spinally delivered N- and P-type voltage dependent calcium channel antagonists on dorsal horn neuronal responses in a rat model of neuropathy. Eur. J. Pharmacol. 415, 141–149.

Matzner, O. and Devor, M. 1994. Hyperexcitability at sites of nerve injury depends on voltage sensitive – sodium channels. J. Neurophysiol. 72, 349–359.

Meller, S. T., Pechman, P. S., Gebhart, G. F., and Maves, T. J. 1992. Nitric oxide mediates the thermal hyperalgesia produced in a model of neuropathic pain in the rat. Neuroscience 50, 7–10.

Mense, S. 2004. Neurobiological basis for the use of botulinum toxin in pain therapy. J. Neurol. 251, S11–S17.

Merskey, H., Loeser, J.D., and Dubner, R. (eds.) 2005. The Paths of Pain: 1975–2005. IASP Press.

Millan, M. J. 1999. The induction of pain: an integrative review. Prog. Neurobiol. 57, 1–164.

Miller, L. J., Fischer, K. A., Goralnick, S. J., Litt, M., Burleson, J. A., Albertsen, P., and Kreutzer, D. L. 2002. Nerve growth factor and chronic prostatis/chronic pelvic pain syndrome. Urology 59, 603–608.

Milligan, E., Zapata, V., Schoeniger, D., Chacur, M., Green, P., Poole, S., Martin, D., Maier, S. F., and Watkins, L. R. 2005b. An initial investigation of spinal mechanisms underlying pain enhancement induced by fractalkine, a neuronally released chemokine. Eur. J. Neurosci. 22, 2775–2782.

Milligan, E. D., Langer, S. J., Sloane, E. M., He, L., Wieseler-Frank, J., O'Connor, K., Martin, D., Forsayeth, J. R., Maier, S. F., Johnson, K., Chavez, R. A., Leinwand, L. A., and Watkins, L. R. 2005a. Controlling pathological pain by adenovirally driven spinal production of the anti-inflammatory cytokine, interleukin-10. Eur. J. Neurosci. 21, 2136–2148.

Moore, A. K., Kohno, T., Karchewski, L. A., Scholz, J., Baba, H., and Woolf, C. J. 2002. Partial peripheral nerve injury promotes a selective loss of GABAergic inhibition in the superficial dorsal horn of the spinal cord. J. Neurosci. 22, 6724–6731.

Nutt, J. G., Burchiel, K. J., Comella, C. L., Jankovic, J., Lang, A. E., Laws, E. R., Jr., Lozano, A. M., Penn, R. D., Simpson, R. K., Jr., Stacy, M., and Wooten, G. F. ICV GDNF Study Group. 2003. Implanted intracerebroventricular. Glial cell line-derived neurotrophic factor. Randomized, double-blind trial of glial cell line-derived neurotrophic factor (GDNF) in PD. Neurology 60, 69–73.

Obata, K., Yamanaka, H., Kobayashi, K., Dai, Y., Mizushima, T., Katsura, H., Fukuoka, T., Tokunaga, A., and Noguchi, K. 2004. Role of mitogen-activated protein kinase activation in injured and intact primary afferent neurons for mechanical and heat hypersensitivity after spinal nerve ligation. J. Neurosci. 24, 10211–10222.

Oddiah, D., Anand, P., McMahon, S. B., and Rattray, M. 1998. Rapid increase of NGF, BDNF and NT3 mRNAs in inflamed bladder. Neuroreport 9, 603–608.

Oh, S. B., Tran, P. B., Gillard, S. E., Hurley, R. W., Hammond, D. L., and Miller, R. J. 2001. Chemokines and glycoprotein 120 produce pain hypersensitivity by directly exciting primary nociceptive neurons. J. Neurosci. 21, 5027–5035.

Ohtori, S., Takahashi, K., Moriya, H., and Myers, R. R. 2004. TNF-alpha and TNF-alpha receptor type 1 upregulation in glia and neurons after peripheral nerve injury: studies in murine DRG and spinal cord. Spine 29, 1082–1088.

Okuse, K., Malik-Hall, M., Baker, M. D., Poon, W. Y., Kong, H., Chao, M. V., and Wood, J. N. 2002. Annexin II light chain regulates sensory neurone-specific sodium channel expression. Nature 417, 653–656.

Olesen, J., Thomsen, L. L., and Iversen, H. 1994. Nitric oxide is a key molecule in migraine and other vascular headaches. Trend Pharmacol. Sci. 15, 149–153.

Ottani, A., Leone, S., Sandrini, M., Ferrari, A., and Bertolinin, A. 2006. The analgesic activity of paracetamol is prevented by the blockade of cannabinoid CB1 receptors. Eur. J. Pharmacol. 531, 280–281.

Owolabi, J. B., Rizkalla, G., Tehim, A., Ross, G. M., Riopelle, R. J., Kamboj, R., Ossipov, M., Bian, D., Wegert, S., Porreca, F., and Lee, DK. 1999. Characterization of antiallodynic actions ALE-0540, a novel nerve growth factor receptor antagonist, in the rat. J. Pharmacol. Exp. Ther. 289, 1271–1276.

Pesquero, J. B., Araujo, R. C., Heppenstall, P. A., Stucjy, C. L., Silva, J. A., Walther, T., Oliveira, S. M., Pasquero, J. L., Paiva, A. C. M., Calixto, J. B., Lewin, G. R., and Bader, M. 2000. Hypoalgesia and altered inflammatory responses in mice lacking kinin B1 receptors. Proc. Natl. Acad. Sci. U. S. A. 97, 8140–8145.

Petersen, M., Segond von Bachet, G., Heppelmann, B., and Kolzenburg, M. 1998. Nerve growth factor regulates the expression of bradykinin binding sites on adult sensory neurons via the neurotrophin p75. Neuroscience 83, 161–169.

Pollock, J., McFarlane, S. M., Connell, M. C., Zehavi, U., Vandenabeele, P., MacEwan, D. J., and Scott, R. H. 2002. TNF-alpha receptors simultaneously activate Ca^{2+} mobilisation and stress kinases in cultured sensory neurons. Neuropharmacology 42, 93–106.

Quartaroli, M., Fasdelli, N., Bettelini, L., Maraia, G., and Corsi, M. 2001. GV196771A, an NMDA receptor/glycine site antagonist, attenuates mechanical allodynia in neuropathic rats and reduces tolerance induced by morphine in mice. Eur. J. Pharmacol. 430, 219–227.

Raghavendra, V., Tanga, F., Rutkowski, M. D., and DeLeo, J. A. 2003. Anti-hyperalgesic and morphine-sparing actions pf propentofylline following peripheral nerve injury in rats: mechanistic implications of spinal glia and proinflammatory cytokines. Pain 104, 655–664.

Ramer, M. S., Bradbury, E. J., and McMahon, S. B. 2001. Nerve growth factor induces P2X(3) expression in sensory neurons. J. Neurochem. 77, 864–875.

Rice, A. S. C., Farquhar-Smith, W. P., and Nagy, I. 2002. Endocannabinoids and pain: spinal and peripheral analgesia in inflammation and neuropathy. Prostaglandins Leukot. Essent. Fatty Acids 66, 243–256.

Richardson, J. D., Kilo, S., and Hargreaves, K. M. 1998. Cannabinoids reduce hyperalgesia and inflammation via interactions with peripheral CB1 receptors. Pain 75, 111–119.

Ro, L. S., Chen, S. T., Tang, L. M., and Jacobs, J. M. 1999. Effect of NGF and anti-NGF on neuropathic pain in rats following chronic constriction injury of the sciatic nerve. 79, 265–274.

Ross, R. A., Coutts, A. A., McFarlane, S. M., Anavi-Goffer, S., Irving, A. J., Pertwee, R. G., MacEwan, D. J., and Scott, R. H. 2001. Actions of cannabinoid receptor ligands on rat cultured sensory neurones: implications of antinociception. Neuropharmacology 40, 221–232.

Rush, A. M., Craner, M. J., Kageyama, T., Dib-Haj, S. D., and Waxman, S. G. 2005. Contactin regulates the current density and axonal expression of tetrodotoxin-resistant but not tetrodotoxin- sensitive sodium channels in DRG neurons. Eur. J. Neurosci. 22, 39–49.

Sachez-Baez, F., Carbonell, T., de Felipe, C., Perez-Paya, E., Belmote, C., Messeguer, A. and Ferrer-Motiel, A. 2002. Attenuation of thermal nociception and hyperalgesia by VR1 blockers. Proc. Natl. Acad. Sci. U. S. A. 99, 2374–2379.

Saegusa, H., Kurihara, T., Zong, S., Kazuno, A., Matsuda, Y., Nonaka, T., Han, W., Toriyama, H., and Tanabe, T. 2001. Suppression of inflammation and neuropathic pain symptoms in mice lacking the N-type Ca^{2+} channel. EMBO J. 20, 2349–2356.

Sagar, D. R., Kelly, S., Millns, P. J., O'Shaugnessey, C. T., Kendall, D. A., and Chapman, V. 2005. Inhibitory effects of CB1 and CB2 receptor agonist on responses of DRG neurons and dorsal horn neurons in neuropathic rats. Eur. J. Neurosci. 22, 371–379.

Sah, D. W. H., Ossipov, M. H., and Porreca, F. 2003. Neurotrophic factors as novel therapeutics for neuropathic pain. Nat. Rev. Drug Disc. 2, 460–472.

Sang, C. N., Hostetter, M. P., Gracely, R. H., Chappell, A. S., Schoepp, D. D., Lee, G., Whitcup, S., Caruso, R., and Max, M. B. 1998. AMPA/kainate antagonist LY293558 reduces capsaicin-evoked hyperalgesia but not pain in normal skin in humans. Anesthesiology 89, 1060–1067.

Sarchielli, P., Alberti, A., Floridi, A., and Gallai, V. 2001. Levels of nerve growth factor in cerebrospinal fluid of chronic headache patients. Neurology 57, 132–134.

Sato, J. and Perl, E. R. 1991. Adrenergic excitation of cutaneous pain receptors induced by peripheral nerve injury. Science 251, 1608–1610.

Sawynok, J. 2003. Topical and peripherally acting analgesics. Pharmacol. Rev. 55, 1–20.

Schafers, M., Marziniak, M., Sorkin, L. S., Yaksh, T. L., and Sommer, C. 2004. Cyclooxygenase inhibition in nerve-injury and TNF-induced hyperalgesia in the rat. Exp. Neurol. 185, 160–168.

Schafers, M., Sorkin, L. S., Geis, C., and Shubayev, V. I. 2003a. Spinal ligation induces transient upregulation of tumor necrosis factor receptors 1 and 2 in injured and adjacent uninjured dorsal root ganglia of rat. Neurosci. Letts. 347, 179–182.

Schafers, M., Svensson, C. I., Sommer, C., and Sorkin, L. S. 2003b. Tumor necrosis factor-alpha induces mechanical allodynia after spinal nerve ligation by activation of p38 MAPK in primary sensory neurons. J. Neurosci. 23, 2517–2521.

Scholz, J. and Woolf, C. J. 2002. Can we conquer pain? Nature Neurosci. 5, 1062–1065.

Sevcik, M. A., Ghilardi, J. R., Peters, C. M., Lindsay, T. H., Halvorson, K. G., Jonas, BM, Kubota, K., Kuskowski, M. A., Boustany, L., Shelton, D. L., and Mantyh, P. W. 2005. Anti-NGF therapy profoundly reduces bone cancer pain and the accompanying increase in markers for peripheral and central sensitisation. Pain 115, 128–141.

Shelton, D. L., Zeller, J., Ho, W. H., Pons, J., and Rosenthal, A. 2005. Nerve growth factor mediates hyperalgesia and cachexia in auto-immune arthritis. Pain 116, 8–16.

Shidoda, K., Hruby, V. J., and Porecca, F. 2006. Antihyperalgesic effects of loperamide in a model of rat neuropathic pain are mediated by peripheral δ-opioid receptors. Neurosci Letts. 411, 143–146.

Shinder, V., Govrin-Lippmann, R., Cohen, S., Belenky, M., Ilin, P., Fried, K., Wilkinson, H. A., and Devor, M. 1999. Structural basis of sympathetic-sensory coupling in rat and human dorsal root ganglia following peripheral nerve injury. J. Neurocytol. 28, 743–761.

Shughrue, P. J., Ky, B., and Austin, C. P. 2003. Localization of B1 bradykinin receptors mRNA in the primate brain and spinal cord: an in situ hybridization study. J. Comp. Neurol. 46, 372–384.

Sluka, K. A., Price, M. P., Breese, N. M., Stucky, C. L., Wemmie, J. A., and Welsh, M. J. 2003. Chronic hyperalgesia induced by repeated acid injections in muscle is abolished by loss of ASIC3, but not ASIC1. Pain 106, 229–239.

Snutch, T. P. 2005. Targeting chronic and neuropathic pain: the N-type calcium channel comes of age. NeuroRx. 2, 662–670.

Sokal, D. M., Elmes, S. J. R., Kendall, D. A., and Chapman, V. 2003. Intraplantar injection of anandamide inhibits mechanically-evoked responses of spinal neurons via activation of Cb2 receptors in anaesthetized rats. Neuropharmacology 45, 404–411.

Sommer, C. and Kress, M. 2004. Recent findings on how proinflammatory cytokines cause pain: peripheral mechanisms in inflammatory and neuropathic hyperalgesia. Neurosci. Letts. 361, 184–187.

Stein, C., Schafer, M., and Machelska, H. 2003. Attacking pain at its source: new perspectives on opioids. Nature Med. 9, 1003–1008.

Stella, N. 2004. Cannabinoid signaling in glial cells. Glia 48, 267–277.

Stewart, J. M. 2004. Bradykinin antagonists: discovery and development. Peptides 25, 527–732.

Svenden, K. B., Jensen, T. S., and Bach, F. W. 2004. Does the cannabinoid dronabilon reduce central pain in multiple sclerosis? Randomized double blind placebo controlled crossover trial. BMJ 329, 253.

Taniguchi, K., Shinjo, K., Mizutani, M., Shimada, K., Ishikawa, T., Menniti, F., and Nagahisa, A. 1997. Antinociceptive actions of CP-101,606, an NMDA receptor NR2B subunit antagonist. Br. J. Pharmacol. 12, 809–812.

Theodosiou, M., Rush, R. A., Zhou, X. F., Hu, D., Walker, J. S., and Tracey, D. J. 1999. Hyperalgesia due to nerve damage: role of nerve growth factor. Pain 81, 245–255.

Tracey, D. J., Cunningham, J. E., and Romm, M. A. 1995. Peripheral hyperalgesia in experimental neuropathy: mediation by alpha 2-adrenoreceptors on post-ganglionic sympathetic terminals. Pain 60, 317–327.

Trebino, C. E., Stock, J. L., Gibbons, C. P., Naiman, B. M., Wachtmann, T. S., Umland, J. P., Pandher, K., Lapointe, J. M., Saha, S., Roach, M. L., Carter, D., Thomas, N. A., Durtschi, B. A., McNeish, J. D., Hambor, J. E. Jakobsson, P. J., Carty, T. J., Perez, J. R. and Audoly, L. P. 2003. Impaired inflammatory and pain responses in mice lacking an inducible prostaglandin E synthase. Proc. Natl. Acad. Sci. U. S. A. 100, 9044–9049.

Tsuda, M., Shigemoto-Mogami, Y., Koizumi, S., Mizokoshi, A., Kohsaka, S., Salter, M. W., and Inoue, K. 2003. P2X4 receptors induced in spinal microglia gate tactile allodynia after nerve injury. Nature 424, 778–783.

Urban, M. O., Hama, A. T., Bradbury, M., Anderson, J., Varney, M. A., and Bristow, L. 2003. Role of metabotropic glutamate receptor subtype 5 (mGluR5) in the maintenance of cold hypersensitivity following a peripheral mononeuropathy in the rat. Neuropharmacology 44, 983–993.

Valenzano, K. J., Tafesse, L., Lee, G., Harrison, J. E., Bulet, J. M., Gottshall, S. L., Mark, L., Pearson, M. S., Miller, W., Shan, S., Rabadi, L., Rotshteyn, Y., Chaffer, S. M.,

Turchin, P. I., Elsmore, D. A., Toth, M., Koetzner, L., and Whiteside, G. T. 2005. Pharmacological and pharmacokinetic characterization of the cannabinoid receptor 2 agonist GW405833, utilizing rodent models of acute and chronic pain, anxiety, ataxia and catalepsy. Neuropharmacology 48, 658–672.

Vellani, V., Mappleback, S., Moriondo, A., Davis, J. B., and McNaughton, P. A. 2001. Protein kinase C activation gating of the vanilloid receptor VR1 by capsaicin, protons, heat and anandamide. J. Physiol. 534, 813–825.

Valleni, V., Zachrisson, O., and McNaughton, P. A. 2004. Functional bradykinin B1 receptors are expressed in nociceptive neurons and are upregulated by the neurotrophin GDNF. J. Physiol. 560, 391–401.

Veneroni, O., Maj, R., Calbresi, M., Faravelli, L., Farielo, R. G., and Salvati, P. 2003. Anti-allodynic effect of NW-1029, a novel Na channel blocker, in experimental animal models of inflammatory and neuropathic pain. Pain 102, 17–25.

Vergnolle, N., Ferazzini, M., D'Andrea, M. R., Buddenkotte, J., and Steinhoff, M. 2003. Proteinase-activated receptors: novel signals for peripheral nerves. Trends Neurosci. 26, 496–500.

Wade, D. T., Robson, P., House, H., Makela, P., Aram, J. A. 2003. Preliminary controlled study to determine whether whole-plant cannabis extracts can improve intractable neurogenic symptoms. Clin. Rehabil. 17, 21–29.

Walker, K., Bowes, M., Panesar, M., Davis, A., Gentry, C., Kesingland, A., Gasparini, F., Spooren, W., Stoehr, N., Pagano, A., Flor, P. J., Vranesic, I., Lingenhoehl, K., Johnson, E. C., Varney, M., Urban, L., and Kuhn, R. 2001. Metabotropic glutamate receptor subtype 5 (mGlu5) and nociceptive function. I. Selective blockade of mGlu5 receptors in models of acute, persistent and chronic pain. Neuropharmacology 40, 1–9.

Wallace, M. S., Rowbotham, M. C., Katz, N. P., Dworkin, R. H., Dotson, R. M., Galer, B. S., Rauck, R. L., Backonja, M. M., Quessy, S. N., and Meisner, P. D. 2002. A randomized, double blind, placebo-controlled trial of a glycine antagonist in neuropathic pain. Neurology 59, 1694–1700.

Water, L., Franklin, A., Witting, A., Wade, C., Xie, Y., Kunos, G., Mackie, K., and Stella, N. 2003. Nonpsychotropic cannabinoid receptors regulate microglial cell migration. J. Neurosci. 23, 1398–1405.

Watkins, L. R. and Maier, S. 2002. Beyond neurons: evidence that immune and glial cells contribute to pathological pain states. Physiol. Rev. 82, 981–1011.

Wei, P., Moskowitz, M. A., Boccalini, P., and Kontos, H. A. 1992. Calcitonin gene-related peptide mediates nitroglycerin and sodium nitroprusside-induced vasodilatation in feline cerebral arterioles. Circulation Res. 70, 1313–1319.

Welch, M. J., Purkiss, J. R., and Foster, K. A. 2000. Sensitivity of embryonic rat dorsal root ganglia neurons to Clostridium botulinum neurotoxins. Toxicon 38, 245–258.

Whiteside, G. T., Boulet, J. M., and Walker, K. 2005a. The role of central and peripheral mu opioid receptors in inflammatory pain and edema: a study using morphine and DiPOA ([8-(3,3-diphenyl-propyl)-4-oxo-1-phenyl-1,3,8-triaza-spiro[4.5]dec-3-yl]-acetic acid). J. Pharmacol. Exp. Ther. 314, 1234–1240.

Whiteside, G. T., Gottshall, S., Boulet, J. M., Chaffer, S. M., Harrison, J. E., Pearson, J. E., Turchin, P. I., Mark, L., Garrison, A. E., and Valenzano, K. J. 2005b. A role for cannabinoid receptors, but not endogenous opioids, in the antinociceptive activity of the CB(2) selective agonist GW405833. Eur. J. Pharmacol. 528, 65–72.

Woodbury, J., Zwick, M., Wang, S., Lawson, J. J., Caterina, M. J., Koltzenburg, M., Albers, K. M., Koerber, H. R., and Davis, B. M. 2004. Nociceptors Lacking TRPV1 and TRPV2 Have Normal Heat Responses. J. Neurosci. 24, 6410–6415.

Woolf, C. J. and Salter, M. W. 2000. Neuronal plasticity: increasing the gain in pain. Science 288, 1765–1769.

Wotherspoon, G. and Winter, J. 2000. Bradykinin B1 receptor is constitutively expressed in the rat sensory nervous system. Neurosci. Letts. 294, 175–178.

Wu, Z. Z., Chen, S. R., and Pan, H. L. 2005. Transient receptor potential vanilloid type 1 activation down-regulates voltage-gated calcium channels through calcium dependent calcineurin in sensory neurons. J. Biol. Chem. 280, 18142–18151.

Yaksh, T. L. 1997. Pharmacology and mechanisms of opioid analgesic activity. Acta Anaesth. Scand. 41, 94–111.

Yaksh, T. L. 2006. Calcium channels as therapeutic targets in neuropathic pain. J. Pain 7, S13–S30.

Yaksh, T. L., Dirig, D. M., Conway, C. M., Svensson, C., Luo, Z. D., and Isakson, P. C. 2001. The acute hyperalgesic action of non-steroidal, anti-inflammatory drugs and release of spinal prostaglandin E2 is mediated by the inhibition of constitutive spinal cyclooxygenase-2 (COX-2) but not COX-1. J. Neurosci. 21, 5847–5853.

Yao, H., Donnelly, D. F., Ma, C., and LaMotte, R. H. 2003. Upregulation of the hyperpolarization-activation cation current after chronic compression of the dorsal root ganglion. J. Neurosci. 23, 2069–2074.

Xiao, W. H. and Bennett, G. J. 1995. Synthetic omega-conopeptides applied to the site of nerve injury suppress neuropathic pains in rats. J. Pharmacol. Exp. Ther. 274, 666–672.

Zhang, J. M., Li, H., and Munir, M. A. 2004. Decreasing sympathetic sprouting in pathological sympathetic ganglia: a new mechanism for treating neuropathic pain using lidocaine. Pain 109, 143–149.

Zhang, J. M., Hoffert, C., Vu, K., Groblewski, T., Ahmad, S., and O'Donnell, D. 2003. Induction of CB2 receptor expression in the rat spinal cord of neuropathic but not inflammatory chronic pain models. Eur. J. Neurosci. 17, 2750–2754.

Zhu, C. Z., Wilson, S. G., Mikusa, J. P., Wismer, C. T., Gauvin, D. M., Lynch, J. J., 3rd., Wade, C. L., Decker, M. W., and Honore, P. 2004. Assessing the role of metabotropic glutamate receptor 5 in multiple nociceptive modalities. Eur. J. Pharmacol. 506, 107–118.

Zweifel, L. S., Kuruvilla, R., and Ginty, D. D. 2005. Functions and mechanisms of retrograde neurotrophin signaling. Nat. Rev. Neurosci. 6, 615–625.

Zylka, M. J., Rise, F. L., and Anderson, D. J. 2005. Topographically distinct epidermal nociceptive circuits revealed by axonal tracers targeted to Mrgprd. Neuron 45, 17–25.

Further Reading

Abdulla, F. A. and Smith, P. A. 2001. Axotomy- and autotomy-induced changes in the excitability of rat dorsal root ganglion neurons. Am. J. Physiol. 85, 630–643.

Cahill, C. M., Morinville, A., Hoffert, C., O'Donnell, D., and Beaudet, A. 2003. Up-regulation and trafficking of delta opioid receptor in a model of chronic inflammation: implications for pain control. Pain 101, 199–208.

Davis, K. D., Treede, R. D., Raja, S. N., Meyer, R. A., and Campbell, J. N. 1991. Topical application of clonidine relieves hyperalgesia in patients with sympathetically maintained pain. Pain 47, 309–317.

Dray, A. 1994. Tasting the inflammatory soup: the role of peripheral neurones. Pain Rev. 1, 153–171.

Ferreira, J., Campos, M. M., Pesquero, J. B., Araujo, R. C., Bader, M., and Calixto, J. B. 2001. Evidence for the participation of kinins in Freund's adjuvant induced inflammatory and nociceptive responses in kinin B1 and B2 knockout mice. Neuropharmacology 41, 1006–1012.

Hao, S., Mta, M., and Goins, W. 2003. Transgene-mediated enkephaline release enhances the effect of morphine and evades tolerance to produce a sustained antiallodynic effect. Pain 102, 135–142.

Kim, S. F., Huri, D. A., and Snyder, S. H. 2005. Inducible nitric oxide synthase binds, N-nitrosylates and activates cyclooxygenase-2. Science 310, 1966–1970.

Kress, M., Izydorcyk, I., and Kuhn, A. 2001. N- and L-but not P/Q-type calcium channels contribute to neuropeptide release from rat skin in vitro. Neuroreport 12, 867–870.

Ossovskaya, V. S. and Bunnett, N. W. 2003. Protease-activated receptors: contribution to physiology and disease. J. Physiol. 552, 589–601.

54 Forebrain Opiates

J-K Zubieta, University of Michigan, Ann Arbor, MI, USA

54.1 Introduction

The initial studies on the neurobiology of pain and the discovery and understanding of the endogenous opioid systems have been historically intrinsically linked. In the 1960s, early research by Liebeskind and coworkers (Mayer, D. J. *et al.*, 1971) demonstrated that the electrical stimulation of the ventral periaqueductal and periventricular gray produced profound analgesia, an effect that was independently observed by Reynolds D. V. (1969). During the same time, Tsou K. and Jang C. S. (1964) observed similar analgesic effects when morphine was microinjected into these locations. As a consequence, and before opioid receptors or their ligands had been isolated, Mayer D. J. *et al.* (1971) proposed that electrical stimulation induced analgesia by activating neural substrates that are involved in the blockage of pain and that were related to the analgesic action of morphine. These observations were followed by the subsequent use of electrical stimulation of the posterior aspect of the paraventricular nucleus of the thalamus for the treatment of pain (Richardson, D. E. and Akil, H., 1977a; 1977b). In 1973, Snyder and Pert published the first manuscript demonstrating the existence of opioid receptors (Pert, C. B. and Snyder, S. H., 1973), a development that was paralleled by the work of Eric Simon (Simon, E. J. *et al.*, 1973) and Lars Terenius (Terenius, L., 1973). These initial findings were followed by the identification of met- and leu-enkephalin (Hughes, J. *et al.*, 1975), β-endorphin (Li, C. H. *et al.*, 1976; Loh, H. H. *et al.*, 1976), and the dynorphin peptides (Goldstein, A. *et al.*, 1979). The 1990s saw the cloning of the opioid receptors delta (δ) (Evans, C. J. *et al.*, 1992; Kieffer, B. L. *et al.*, 1992), mu (μ) (Chen, Y. *et al.*, 1993; Thompson, R. C. *et al.*, 1993), and kappa (κ) (Li, S. *et al.*, 1993; Meng, F. *et al.*, 1993). These are seven transmembrane domain G-protein-coupled receptors inhibiting adenyl cyclase and modulating calcium and potassium conductance.

54.2 Endogenous Opioid Mechanisms in the Regulation of Pain

54.2.1 μ-Opioid-Receptor-Mediated Pain Processing

This chapter will cover the regulation of pain by supraspinal opioid systems. Investigation of these mechanisms has been most extensive for those pertaining to the μ-opioid receptor, as this is the site of action of opiate analgesics and the receptor most consistently associated with pain suppression. Knockout animals devoid of these receptors display shorter latencies in the tail flick and hot plate tests, thought to involve spinal and supraspinal mechanisms, respectively, as well as a lack of morphine effects (Sora, I. *et al.*, 1997). Its endogenous ligands include β-endorphin, derived from the peptide precursor proopiomelanocortin (POMC), the recently described endomorphins 1 and 2, and in some circuits, the enkephalins (cleaved from the larger preproenkephalin precursor). POMC-containing neurons are located in the arcuate nucleus of the hypothalamus and periarcuate regions of the medial-basal hypothalamus, with a smaller group localized in the nucleus of the tractus solitarius of the medulla. Projections are extensive to the periventricular thalamus, septum, and amygdala, dorsally descending to the periaqueductal gray (PAG), midline raphe, and reticular formation, and laterally to the anterior and lateral hypothalamus. Lighter innervation is observed in the medial portion of the nucleus accumbens and olfactory cortex (Khachaturian, H. *et al.*, 1985). Animals devoid of a functioning POMC gene show a lack of naloxone-reversible analgesia induced by mild swim stress, but greater nonopioid (naloxone nonreversible) analgesic effects and normal effects of morphine. These data demonstrated the involvement of β-endorphin in stress-induced analgesic responses, while also highlighting the presence of alternate mechanisms not mediated by β-endorphin and μ-opioid receptors (Rubinstein, M. *et al.*, 1996).

In contrast to the narrow localization of POMC-containing cells, enkephalinergic cells and projections are widely distributed, forming both local circuits and long projections, with similar distributions for met- and leu-enkephalin, albeit with a higher level of met-enkephalin in all regions. The highest concentrations of terminals with these peptides are encountered in the globus pallidus and central and lateral amygdala nuclei. Globus pallidus and ventral pallidal projections form part of the enkephalinergic striatopallidal pathway, with cell bodies arising primarily in the nucleus accumbens and to a lesser extent in the ventromedial hypothalamus. Cell bodies containing preproenkephalin are also encountered in the olfactory bulb, septal nuclei, bed nucleus of the stria terminalis, throughout the basal ganglia and in the cortex. Terminal densities are moderate to dense in the cingulate, piriform, and entorhinal cortex, anterior and periventricular thalamus, and preoptic hypothalamus (the latter from both local projections and terminals arising from the amygdala and periamygdalar area). Both cell bodies and rich local projections are encountered in the PAG, raphe, locus coeruleus, nucleus ambiguus, and nucleus of the solitary tract (Petrusz, P. *et al.*, 1985). Enkephalin-deficient mice display large increases in the concentration of μ-opioid receptors in the globus pallidus, central nucleus of the amygdala, bed nucleus of the stria terminalis and substantia innominata, preoptic, medial and lateral hypothalamus, medial thalamus, PAG, raphe, nucleus ambiguus, and nucleus of the tractus solitarius, suggesting substantial enkephalinergic activity on μ-opioid receptors in these regions, with compensatory upregulation in the knockouts (Brady, L. S. *et al.*, 1999). Enkephalin-deficient mice display reductions in supraspinal analgesic responses as measured with the hot-plate test and rapid nocifensive responses in the formalin assay, but unaltered spinal analgesic responses (tail-flick test) and stress-induced analgesia (Konig, M. *et al.*, 1996).

Last, the most recently identified family of opioid peptides, the endomorphins 1 and 2, and in a manner similar to that for β-endorphin, is synthesized in the brain only in hypothalamic cells (periventricular, arcuate, and ventromedial and dorsomedial nuclei), and project to pain-processing regions of the brainstem and pons (trigeminal tract, parabrachial nucleus, nucleus of the tractus solitarious, PAG, and locus coeruleus) as well as to midline thalamic nuclei and the amygdala. Besides the hypothalamic cell bodies, the only other known areas containing endomorphin-positive cell bodies include the nucleus of the tractus solitarius and the dorsal root ganglia. Consistent with its actions at μ-opioid receptors, intracerebroventricular administration of endomorphins induces profound analgesia that may be dissociated from the rewarding properties associated with μ-opioid receptor agonists (Zadina, J. E., 2002).

Initial studies on the role of supraspinal μ-opioid-receptor-mediated antinociception focused on the role of the PAG and its connections with the rostroventral medulla (RVM). These are regions that, together with the dorsolateral pontine tegmentum, control nociceptive transmission via projections through the spinal cord dorsolateral funiculus to the dorsal horn laminae (Fields, H. L. and Basbaum, A. I., 1978). Much of the focus during the following decades was then dedicated to the understanding of the mechanisms underlying this modulation. Both pain facilitatory and inhibitory cells have been described at the level of the RVM (Moreau, J. L. and Fields, H. L., 1986; Heinricher, M. M. *et al.*, 1987; Urban, M. O. and Smith, D. J., 1994; Zhuo, M. and Gebhart, G. E., 1997). μ-Opioid receptors, through the activation of enkephalinergic neurotransmission and GABA interneurons in the PAG and RVM (al-Rodhan, N. *et al.*, 1990), increase the activity of inhibitory (off) cells reducing pain transmission. The administration of μ-opioid receptor agonists systemically, or directly into the PAG or RVM, increases the activity of these off cells, with blockade of the activation of these neurons preventing morphine's antinociceptive effects (Heinricher, M. M. *et al.*, 1999). Not surprisingly, in view of these results, the possible role of μ-opioid neurotransmission beyond the PAG, RVM, and descending pathways into the spinal cord (e.g., supraspinal mechanisms) was thought to be less critical and mostly regarded as secondary to the analgesia mediated primarily by brainstem and spinal cord regions.

Nevertheless, examination of the distribution of μ-opioid receptors, mRNA in the rodent (Mansour,

A. *et al.*, 1995) and *in vivo* μ-opioid receptor availability in the human brain (Frost, J. *et al.*, 1985) demonstrate a broad distribution of these sites in telencephalic areas. Administration of μ-opioid receptor agonists increases the blood flow of brain regions rich in μ-opioid receptors (cingulate, prefrontal, temporal and insular cortices, thalamus, hypothalamus, basal ganglia, amygdala, and brainstem) (Adler, L. J. *et al.*, 1997; Schlaepfer, T. *et al.*, 1998; Casey, K. *et al.*, 2000; Wagner, K. J. *et al.*, 2001), and reduces pain-induced activation in the thalamus (Casey, K. *et al.*, 2000). Some of these areas, such as thalamic and hypothalamic nuclei, the central nucleus of the amygdala, the agranular insular cortex and lateral orbitofrontal cortex, have been associated with the suppression of pain behavior in animal models (Mayer, D. J. and Liebeskind, J. C., 1974; Yaksh, T. L. and Rudy, T. A., 1978; Bodnar, R. J. *et al.*, 1980; Coffield, J. A. *et al.*, 1992; Burkey, A. R. *et al.*, 1996; Manning, B., 1998; Manning, B. and Franklin, K., 1998; Harte, S. *et al.*, 2000; Manning, B. H. *et al.*, 2001). However, the inhibition of pain signaling elicited by the activation of μ-opioid receptors in these supraspinal regions is thought to take place through their connections with the PAG, since inactivation of the PAG or the administration of opioid antagonists in this area abolishes the analgesic response. In turn, PAG-induced analgesia is blocked by μ-opioid receptor antagonists microinjected in the RVM, then reducing nociceptive transmission from the spinal cord (Kiefel, J. M. *et al.*, 1993; Roychowdhury, S. M. and Fields, H. L., 1996).

In view of the critical role of the PAG and RVM on μ-opioid-receptor-mediated antinociception, perhaps the question to be answered regarding the involvement of opioid systems in supraspinal nuclei is their respective role in the experience of pain as a complex phenomenon. Opioid neurotransmission is activated in humans during clinical pain, or experimental pain of some duration, as evidenced by enhancements in pain ratings after the administration of naloxone, a nonselective opioid receptor antagonist (Levine, J. *et al.*, 1978). Conversely, experimental pain of relatively short duration (e.g., <10 min) does not elicit an apparent activation of opioid neurotransmission, evidenced by a lack of increases in pain ratings after naloxone administration (Grevert, P., and Goldstein, A., 1977). These findings are similar to those observed in response to nonpainful stressors, whereby prolonged or unpredictable stimuli activate opioid neurotransmission, leading to the term stress-induced analgesia (Madden, J. T. *et al.*,

1977; Watkins, L. and Mayer, D., 1982). It is therefore possible that the progressive engagement of supraspinal μ-opioid mechanisms may be of greater importance for the control of noxious stimuli as it progresses from a relatively simple sensory experience to one that requires the assessment of its significance and an appropriate response by the organism.

Two studies have examined the activity of μ-receptor-mediated opioid neurotransmission in humans during the experience of pain. Both studies employed the μ-opioid-receptor-selective radiotracer [^{11}C]carfentanil and positron emission tomography. Bencherif B. *et al.* (2002) utilized the capsaicin model of experimental pain, while in Zubieta J. K. *et al.* (2001), jaw muscle pain was induced by the intramuscular infusion of hypertonic saline. The former, in a smaller sample, showed evidence of activation of μ-opioid-receptor-mediated neurotransmission in the thalamus. In the latter, engagement of this neurotransmitter system was observed in a number of brain areas, including the PAG, dorsal and rostral anterior cingulate, dorsolateral prefrontal cortex, anterior and posterior insular cortex, medial and lateral thalamus, hypothalamus, ventral basal ganglia, and amygdala. Invariably, the relationship between μ-opioid activation and subjective ratings of pain was in the negative direction (the larger the activation the lower the pain ratings), with sensory or pain affect elements being modulated by different regions. Some of these regions had been previously implicated in opioid-mediated suppression of nociceptive responses in animal models (e.g., PAG, thalamus, hypothalamus, amygdala, and to a lesser extent areas of the insular cortex), as noted above. Conversely, while pain-induced increases in synaptic activity of the prefrontal cortex (Baron, R. *et al.*, 1999; Lorenz, J. *et al.*, 2002) and dorsal anterior cingulate cortex (Rainville, P. *et al.*, 1997) had been described and associated with hyperalgesia and pain affect, respectively, the involvement of the endogenous opioid system on these processes had not been previously explored. Similarly, the ventral basal ganglia (nucleus accumbens, ventral pallidum) have been typically associated with either responses to natural rewards or drugs of abuse, stress, or, as is more recently postulated, with salient stimuli irrespective of its valence (Horvitz, J., 2000). These data then appear to indicate that, besides its traditional role in the modulation of descending inhibitory pathways into the PAG and RVM, supraspinal μ-opioid receptors are additionally involved in the local processing of complex assessments of the pain experience. In this regard, μ-opioid receptors appear implicated in regulating emotional responses, as examined in knockout rodents devoid of these receptors. These animals display an enhancement of anxiety-like responses in the elevated arm maze (Filliol, D. *et al.*, 2000) and deficits in attachment behavior (Moles, A. *et al.*, 2004). Proenkephalin knockouts also show a lack of conditioned place aversion to naloxone (Skoubis, P. D. *et al.*, 2005), increased aggressive behavior to an intruder and anxiety-like behavior in the open arm maze, as well as alterations in supraspinal but not spinal pain processing (Konig, M. *et al.*, 1996), additionally involving this neurotransmitter system in emotional behavioral responses.

Work in humans has also shown that in the case of some of the brain regions in which μ-opioid systems become active during pain (e.g., rostral anterior cingulate, dorsolateral prefrontal cortex, nucleus accumbens, and insular cortex), cognitive expectations associated with the administration of a placebo with expected analgesic properties are also capable of activating this neurotransmitter system to mediate the placebo analgesic effect (Benedetti, F. *et al.*, 2005; Zubieta, J. K. *et al.*, 2005).

A relatively less explored area in the understanding of μ-opioid receptor modulation of pain is the additional influences of age and gonadal steroids (e.g., sex effects). In rodents, age-associated declines in μ-opioid receptor concentrations have been reported (Piva, F. *et al.*, 1987). However, humans demonstrate opposite effects, with increases in most brain regions shown in postmortem material (Gross-Isseroff, R. *et al.*, 1990; Gabilondo, A. *et al.*, 1995) and in neuroimaging studies (Zubieta, J. K. *et al.*, 1999), albeit the psychophysical correlates of these changes in receptor concentrations have not been determined. In human subjects, women further demonstrate declines in μ-opioid receptor levels after menopause (Zubieta, J. K. *et al.*, 1999). In this regard, trophic effects of estradiol have been shown at the level of μ-opioid receptor levels and their mRNA, POMC content, and β-endorphin release in animal models (Hammer, R. P. and Bridges, R. S., 1987; Dondi, D. *et al.*, 1992; Quinones-Jenab, V. *et al.*, 1997; Eckersell, C. *et al.*, 1998). The clinical implications of these effects have not been well developed, albeit sex differences have been observed in the effects of μ-opioid receptor agonists, including their side effects and cardiovascular parameters, which appear to be more prominently affected in women (Zacny, J.

P., 2001; Fillingim, R. B. and Gear, R. W., 2004). A less prominent activation of μ-opioid-receptor-mediated neurotransmission, as measured with neuroimaging techniques, has been shown in women studied under conditions of low estradiol, low progesterone during an experimental pain challenge (Zubieta, J. K. *et al.*, 2002). This effect, observed in the thalamus, amygdala, and nucleus accumbens, was reversed after selectively increasing estradiol plasma levels, which also increased μ-opioid receptor concentrations in these regions (Smith, Y. R. *et al.*, 2006). These data then suggest an active regulation of μ-opioid-receptor-mediated mechanisms by gonadal steroids.

54.2.2 κ-Opioid-Receptor-Mediated Processing

The prodynorphin-derived peptides, which present the highest affinity and colocalization with κ-opioid receptors, consist of extended forms of leu-enkephalin coded by a different gene and mRNA. These peptides include dynorphin A, B, and neoendorphins α and β (Goldstein, A. *et al.*, 1979). Their involvement in pain regulatory mechanisms has been controversial, as both anti- and pronociceptive responses have been described in response to κ-agonists depending on the brain regions studied. Similarly to β-endorphin and endomorphin systems, there is a dense accumulation of dynorphin-containing cells in the multiple regions of the hypothalamus, forming both local circuits and longer projections, particularly to the amygdala and descending to brainstem, pontine, and medullary/spinal regions (e.g., PAG, but also substantia nigra). However, and closer to the distribution of the enkephalin peptides, cell bodies are also encountered in limbic areas of the cortex, such as the anterior cingulate, orbitofrontal, insular, periamygdaloid, and entorhinal cortex, in the basal ganglia (caudate, putamen, globus pallidus), nucleus accumbens, amygdala, particularly in its central nucleus, PAG, raphe nuclei, locus coeruleus, nucleus ambiguus, and nucleus of the solitary tract. Long pathways have been described from the striatum to the substantia nigra and ventral pallidum, as well as from the amygdala to brainstem and pons regions. Minimal innervation of the thalamus has been observed, which is confined to midline and intralaminar nuclei (Fallon, J. and Ciofi, P., 1990). Mice devoid of the dynorphin gene show an upregulation of κ-receptors in the nucleus accumbens shell and core, hypothalamus, amygdala, and substantia nigra, consistent with their high level of dynorphin innervation, and suggesting an active role of this peptide in the modulation of the κ-receptors in these brain regions (Clarke, S. *et al.*, 2003).

Intrathecal administration of κ-opioid agonists or their microinjection in the RVM has been shown to result in antinociceptive responses in various rat models (Harada, Y. *et al.*, 1995; Ackley, M. A. *et al.*, 2001), albeit modest, and typically requiring high doses. These effects have also been demonstrated to be sex dimorphic, with female rats demonstrating a more profound analgesic effect than the males in the tail-flick test (Tershner, S. A. *et al.*, 2000). In κ-receptor knockout animals, female but not male rats demonstrate decreased withdrawal latencies in the tail-immersion test, compared to wild-type animals (Martin, M. *et al.*, 2003). Consistent with these findings, one study has shown a preferential antinociceptive effect of opiates with κ-opioid activity in women, with respect to men, after surgical extraction of the third molar (Gear, R. *et al.*, 1996).

κ-Opioid agonists, however, have also been shown to antagonize morphine analgesia, by the hyperpolarization of pain-inhibitory cells activated by μ-opioid receptors in the RVM (Pan, Z. Z. *et al.*, 1997; Meng, I. D. *et al.*, 2005). Therefore pronociceptive, antianalgesic effects of κ-opioids may exist simultaneously with antinociceptive activity, depending on the brain regions involved or the existing level of μ-opioid receptor activity. Opposing effects of μ- and κ-receptors have also been shown at the level of the medial thalamus, whereby microinjection of μ-agonists increased, and κ-agonists reduced, the thresholds for pain-induced vocalization and motor responses in male rodents (Carr, K. D. and Bak, T. H., 1988). Dynorphin also appears to be involved in the maintenance of hyperalgesia and persistent pain, with only spinal mechanisms having been investigated at this time (Wang, Z. *et al.*, 2001).

Dynorphin and κ-opioid agonists further antagonize dopamine release in the basal ganglia (Zhang, Y. *et al.*, 2004), whether induced by μ-opioid receptor agonists, psychostimulants, or stressors, by the inhibition of dopamine (DA) cells in the ventral tegmental area and substantia nigra or possibly indirectly through the regulation of DA reuptake at the level of the nucleus accumbens (Thompson, A. C. *et al.*, 2000; Margolis, E. B. *et al.*, 2003; McLaughlin, J. P. *et al.*, 2003; Narita, M. *et al.*, 2005). Dopaminergic mechanisms, typically associated with responses to drugs of abuse and more generally to salient stimuli regardless of its valence, have been found dysregulated in chronic pain

patients, albeit the clinical consequences of these alterations are presently unknown (Hagelberg, N. *et al.*, 2004).

54.2.3 δ-Opioid-Receptor-Mediated Processing

The enkephalins are thought to be the endogenous ligand for the δ-opioid receptor, in addition to acting on μ-opioid receptors, as noted above. Their distribution was described in this chapter under Section 54.2.1. Knockout mice devoid of the enkephalin gene display increases in the regional concentration of μ- and δ-opioid receptors (Brady, L. S. *et al.*, 1999), suggesting an effect of this peptide in both receptor systems. In the case of the δ-receptor, these increases were observed in the caudate–putamen, nucleus accumbens core and shell, ventral pallidum, globus pallidus, the central nucleus of the amygdala, preoptic area, and lateral hypothalamus. Microinjection of δ-receptor agonists in the PAG and medullary reticular formation has been shown to induce antinociception (Jensen, T. S. and Yaksh, T. L., 1986). Their intracerebroventricular administration has also been shown to have antinociceptive properties in tests of spinal and supraspinal nociceptive mechanisms in wild-type and μ-opioid-receptor-deficient mice (Porreca, F. *et al.*, 1984; Vaught, J. L. *et al.*, 1988; Ossipov, M. H. *et al.*, 1995). Rodent, primate, and human δ-opioid receptors are uniformly distributed throughout the cortex, diffusely in the striatum, and moderately to densely in the nucleus accumbens, ventral pallidum, hippocampus, amygdala, medial thalamus, hypothalamus, PAG, and raphe nuclei. Low concentrations are observed in lateral thalamic nuclei and the cerebellum (Lewis, M. E. *et al.*, 1981; 1983; Moskowitz, A. S. and Goodman, R. R., 1985; Smith, J. S. *et al.*, 1999; Cahill, C. M. *et al.*, 2001b). This distribution, which emphasizes a limbic and paralimbic organization as well as its relationship with pain circuitry, has suggested the involvement of this receptor and neurotransmitter system in emotional and stress regulation in addition to antinociceptive mechanisms. Regarding the latter, δ-opioid-receptor-agonist-induced analgesia and respiratory depression has been shown either reduced (Matthes, H. W. *et al.*, 1998) or in some cases abolished (Scherrer, G. *et al.*, 2004), depending on the assays employed, in animals devoid of μ-opioid receptors. These results then suggested the presence of functional interactions between μ- and δ-opioid-receptor types. Conversely, δ-opioid stress-induced analgesia after forced swim stress was found maintained in μ-opioid receptor knockout mice. These data supported a lack of interactions between μ- and δ-opioid receptors when endogenous opioid transmission was activated by a stressor, as opposed to exogenously administered agonists. However, and in the same studies, two phases of the stress-induced analgesia were observed, one early response mediated by δ-opioid receptors, and a later response that was abolished in the μ-opioid receptor knockouts and unaffected by δ-opioid receptor antagonists (LaBuda, C. J. et al., 2000).

Utilizing single, double, and triple knockouts for μ-, κ-, and δ-opioid receptors Martin M. et al. (2003) have suggested that δ-opioid receptors, together with μ-opioid receptors, were involved in mechanical nociception and some, but not other phases of inflammatory pain. Besides their independent functions, interactions between μ- and δ-receptors have also been demonstrated at various levels. Treatment with μ-opioid agonists has been shown to increase the concentration of membrane-bound δ-receptors in the spinal cord by increasing their trafficking from intracellular sites, an effect that was abolished in μ-opioid receptor knockout animals (Cahill, C. M. et al., 2001a; Morinville, A. et al., 2003). Furthermore, interacting complexes of μ- and δ-receptors (heterodimers) have been demonstrated in spinal cord and neuroblastoma cell lines, as well as synergistic interactions in analgesic effects between the two receptor types (Gomes, I. et al., 2004).

Regarding other possible effects of δ-opioid neurotransmission, stressors have been shown to decrease the concentration of δ-opioid receptors in limbic areas of the rat (Fadda, P. et al., 1991) and increase that of enkephalin mRNA in the ventral medulla (Mansi, J. A. et al., 2000) and hypothalamus (Helmreich, D. L. et al., 1999; Dumont, E. C. et al., 2000), confirming their involvement in the modulation of the stress response, albeit not in its neuroendocrine components, at least in primates (Williams, K. L. et al., 2003). Antidepressant-like effects have also been demonstrated in animal models of depression, suggesting the participation of this system in complex behavioral functions (Jutkiewicz, E. M. et al., 2003; Torregrossa, M. M. et al., 2004).

54.2.4 Opioid Receptor-like (ORL-1, NOP Receptor)-Mediated Processing

The opioid receptor-like (ORL1) receptor (now denominated NOP) is the most recently identified opioid receptor, with high homology with other opioid receptors, but does not bind to traditional opioid peptides. It is present in high concentrations in the septum, diagonal band of Broca, hypothalamus (preoptic, arcuate, paraventricular, ventromedial), hippocampus, medial amygdala, substantia nigra, locus coeruleus, raphe nuclei, PAG, as well as in the dorsal and ventral horns of the spinal cord and in the trigeminal nucleus. It is also moderately concentrated in the thalamus, cortex (olfactory, cingulate, claustrum, and frontoparietal areas), the shell of the nucleus accumbens, and islets of Calleja. Low levels are observed in the caudate–putamen, core of the nucleus accumbens, and cerebellum (Mollereau, C. et al., 1994). The 17-amino acid peptide orphanin FQ/nociceptin (OFQ/N), the endogenous ligand for this receptor, was subsequently isolated. It presented high homology with opioid peptide transmitters, particularly dynorphin (Meunier, J. C. et al., 1995; Reinscheid, R. K. et al., 1995). This peptide is synthesized as part of a larger precursor, preproOFQ/N, with a distribution largely overlapping with that of the NOP receptor in the rat brain (Mollereau, C. et al., 1996; Darland, T. et al., 1998). Like opioid agonists, OFQ/N inhibits adenylate cyclase and calcium channels, and induces an inhibitory effect on postsynaptic neurons by opening potassium channels and hyperpolarization (Hawes, B. E. et al., 2000; Moran, T. D. et al., 2000). This inhibitory effect has been invoked in explaining the often-times contradictory effects of generalized routes of OFQ/N peptide administration (e.g., intracerebroventricular) (Darland, T. et al., 1998). For example, initial reports described increases in pain behavior (Meunier, J. C. et al., 1995; Reinscheid, R. K. et al., 1995), no effect on pain responses (Mogil, J. S. et al., 1996a; 1996b; Tian, J. H. et al., 1997), or even hyperalgesia (Rossi, G. C. et al., 1996; 1997), and a reversal of the analgesic actions of morphine and endogenous opioid peptides (Mogil, J. S. et al., 1996a; 1996b). It is thought that these effects depend on the populations of cells being inhibited, as OFQ/N can hyperpolarize neurons that both facilitate and inhibit pain signal transmission at the level of the RVM (Heinricher, M. M. et al., 1997), as well as most of the neuronal populations rich in the NOP receptor (e.g., hypothalamus, medulla, hippocampus, amygdala, ventral tegmental area, locus coeruleus, and PAG) (see Heinricher, M. M., 2003 for a recent review). Using data acquired by the microinjection of OFQ/N in the RVM, Heinricher M. M. et al. (1997) proposed that the behavioral effects of this peptide depend on the state of activity of the output neurons affected by the peptide. Under conditions of stimulation (opioid withdrawal, stress, and opioid agonist effects) the effects of OFQ/ N would oppose the state of activation in the relevant circuits, inducing an opposing effect, with additional influencing variables, such as sex or genotype, also suggested to influence its effects (Mogil, J. S. and Pasternak, G. W., 2001; Flores, C. A. *et al.*, 2003). On the other hand, and during basal states, OFQ/N effects would be minimal (Heinricher, M. M. *et al.*, 1997). Consistent with these findings, OFQ/N enhanced stress-induced ACTH production at the level of the pituitary and plasma corticosterone, an effect that in this case was also observed to a lesser extent in nonstressed animals (Devine, D. P. *et al.*, 2001). Knockout rodents devoid of either OFQ/N or NOP receptors do not appear to have significant changes in nociceptive responses or changes in opiate analgesia under nonstimulated conditions (Nishi, M. *et al.*, 1997; Koster, A. *et al.*, 1999; Kest, B. *et al.*, 2001; Mamiya, T. *et al.*, 2001). In a neuropathic pain model, NOP mRNA expression increased in pain regulatory circuitry (raphe magnus, PAG, and RVM), albeit the behavioral consequences of these increases have not been explored (Ma, F. *et al.*, 2005).

References

Ackley, M. A., Hurley, R. W., Virnich, D. E., and Hammond, D. L. 2001. A cellular mechanism for the antinociceptive effect of a kappa opioid receptor agonist. Pain 91, 377–388.

Adler, L. J., Gyulai, F. E., Diehl, D. J., Mintun, M. A., Winter, P. M., and Firestone, L. L. 1997. Regional brain activity changes associated with fentanyl analgesia elucidated by positron emission tomography. Anesth. Analg. 8, 4120–126.

Baron, R., Baron, Y., Disbrow, E., and Roberts, T. P. 1999. Brain processing of capsaicin-induced secondary hyperalgesia: a functional MRI study. Neurology 53, 548–557.

Bencherif, B., Fuchs, P. N., Sheth, R., Dannals, R. F., Campbell, J. N., and Frost, J. J. 2002. Pain activation of human supraspinal opioid pathways as demonstrated by [^{11}C]-carfentanil and positron emission tomography (PET). Pain 99, 589–598.

Benedetti, F., Mayberg, H. S., Wager, T. D., Stohler, C. S., and Zubieta, J. K. 2005. Neurobiological mechanisms of the placebo effect. J. Neurosci. 25, 10390–10402.

Bodnar, R. J., Kelly, D. D., Brutus, M., and Glusman, M. 1980. Stress-induced analgesia: neural and hormonal determinants. Neurosci. Biobehav. Rev. 4, 87–100.

Brady, L. S., Herkenham, M., Rothman, R. B., Partilla, J. S., Konig, M., Zimmer, A. M., and Zimmer, A. 1999. Region-specific up-regulation of opioid receptor binding in enkephalin knockout mice. Brain Res. Mol. Brain Res. 68, 193–197.

Burkey, A. R., Carstens, E., Wenniger, J. J., Tang, J., and Jasmin, L. 1996. An opioidergic cortical antinociception triggering site in the agranular insular cortex of the rat that contributes to morphine antinociception. J. Neurosci. 16, 6612–6623.

Cahill, C. M., Morinville, A., Lee, M. C., Vincent, J. P., Collier, B., and Beaudet, A. 2001a. Prolonged morphine treatment targets delta opioid receptors to neuronal plasma membranes and enhances delta-mediated antinociception. J. Neurosci. 21, 7598–7607.

Cahill, C. M., McClellan, K. A., Morinville, A., Hoffert, C., Hubatsch, D., O'Donnell, D., and Beaudet, A. 2001b. Immunohistochemical distribution of delta opioid receptors in the rat central nervous system: evidence for somatodendritic labeling and antigen-specific cellular compartmentalization. J. Comp. Neurol. 440, 65–84.

Carr, K. D. and Bak, T. H. 1988. Medial thalamic injection of opioid agonists: mu-agonist increases while kappa-agonist decreases stimulus thresholds for pain and reward. Brain Res. 441, 173–184.

Casey, K., Svensson, P., Morrow, T., Raz, J., Jone, C., and Minoshima, S. 2000. Selective opiate modulation of nociceptive processing in the human brain. J. Neurophysiol. 84, 525–533.

Chen, Y., Mestek, A., Liu, J., Hurley, J. A., and Yu, L. 1993. Molecular cloning and functional expression of a mu-opioid receptor from rat brain. Mol. Pharmacol. 44, 8–12.

Clarke, S., Zimmer, A., Zimmer, A. M., Hill, R. G., and Kitchen, I. 2003. Region selective up-regulation of micro-, delta- and kappa-opioid receptors but not opioid receptor-like 1 receptors in the brains of enkephalin and dynorphin knockout mice. Neuroscience 122, 479–489.

Coffield, J. A., Bowen, K. K., and Miletic, V. 1992. Retrograde tracing of projections between the nucleus submedius, the ventrolateral orbital cortex, and the midbrain in the rat. J. Comp. Neurol. 321, 488–499.

Darland, T., Heinricher, M. M., and Grandy, D. K. 1998. Orphanin FQ/nociceptin: a role in pain and analgesia, but so much more. Trends Neurosci. 21, 215–221.

Devine, D. P., Watson, S. J., and Akil, H. 2001. Nociceptin/ orphanin FQ regulates neuroendocrine function of the limbic–hypothalamic–pituitary–adrenal axis. Neuroscience 102, 541–553.

Dondi, D., Limonta, P., Maggi, R., and Piva, F. 1992. Effects of ovarian hormones on brain opioid binding sites in castrated female rats. Am. J. Physiol. 263, E507–E511.

Dumont, E. C., Kinkead, R., Trottier, J. F., Gosselin, I., and Drolet, G. 2000. Effect of chronic psychogenic stress exposure on enkephalin neuronal activity and expression in the rat hypothalamic paraventricular nucleus. J. Neurochem. 75, 2200–2211.

Eckersell, C., Popper, P., and Micevych, P. 1998. Estrogen-induced alteration of μ-opioid receptor immunoreactivity in the medial preoptic nucleus and medial amygdala. J. Neurosci. 18, 3967–3976.

Evans, C. J., Keith, D. E., Jr., Morrison, H., Magendzo, K., and Edwards, R. H. 1992. Cloning of a delta opioid receptor by functional expression. Science 258, 1952–1955.

Fadda, P., Tortorella, A., and Fratta, W. 1991. Sleep deprivation decreases mu and delta opioid receptor binding in the rat limbic system. Neurosci. Lett. 129, 315–317.

Fallon, J. and Ciofi, P. 1990. Dynorphin Containing Neurons. In: Handbook of Chemical Neuroanatomy (eds. A. Björklund, T. Hökfelt, and M. Kuhar), pp. 1–130. Elsevier.

Fields, H. L. and Basbaum, A. I. 1978. Brainstem control of spinal pain-transmission neurons. Annu. Rev. Physiol. 40, 217–248.

Fillingim, R. B. and Gear, R. W. 2004. Sex differences in opioid analgesia: clinical and experimental findings. Eur. J. Pain 8, 413–425.

Filliol, D., Ghozland, S., Chluba, J., Martin, M., Matthes, H., Simonin, F., Befort, K., Gaveriaux-Ruff, C., Dierich, A., Le, M. M., Valverde, O., Maldonado, R., and Kieffer, B. 2000. Mice deficient for delta- and mu-opioid receptors exhibit opposing alterations of emotional responses. Nat. Genet. 25, 195–200.

Flores, C. A., Shughrue, P., Petersen, S. L., and Mokha, S. S. 2003. Sex-related differences in the distribution of opioid receptor-like 1 receptor mRNA and colocalization with estrogen receptor mRNA in neurons of the spinal trigeminal nucleus caudalis in the rat. Neuroscience 118, 769–778.

Frost, J., Wagner, H. J., Dannals, R., Ravert, H., Links, J., Wilson, A., Burns, H., Wong, D., McPherson, R., and Rosenbaum, A., et al. 1985. Imaging opiate receptors in the human brain by positron tomography. J. Comput. Assist. Tomogr. 9, 231–236.

Gabilondo, A., Meana, J., and Garcia-Sevilla, J. 1995. Increased density of mu-opioid receptors in the postmortem brain of suicide victims. Brain Res. 682, 245–250.

Gear, R., Miaskowski, C., Gordon, N., Paul, S., Heller, P., and Levine, J. 1996. Kappa-opioids produce significantly greater analgesia in women than in men. Nat. Med. 2, 1248–1250.

Goldstein, A., Tachibana, S., Lowney, L. I., Hunkapiller, M., and Hood, L. 1979. Dynorphin-(1–13), an extraordinarily potent opioid peptide. Proc. Natl. Acad. Sci. U. S. A. 76, 6666–6670.

Gomes, I., Gupta, A., Filipovska, J., Szeto, H. H., Pintar, J. E., and Devi, L. A. 2004. A role for heterodimerization of mu and delta opiate receptors in enhancing morphine analgesia. Proc. Natl. Acad. Sci. U. S. A. 101, 5135–5139.

Grevert, P. and Goldstein, A. 1977. Effects of naloxone on experimentally induced ischemic pain and on mood in human subjects. Proc. Natl. Acad. Sci. U. S. A. 74, 1291–1294.

Gross-Isseroff, R., Dillon, K., Israeli, M., and Biegon, A. 1990. Regionally selective increases in mu opioid receptor density in the brains of suicide victims. Brain Res. 530, 312–316.

Hagelberg, N., Jaaskelainen, S. K., Martikainen, I. K., Mansikka, H, Forssell, H., Scheinin, H., Hietala, J., and Pertovaara, A. 2004. Striatal dopamine D2 receptors in modulation of pain in humans: a review. Eur. J. Pharmacol. 500, 187–192.

Hammer, R. P., Jr. and Bridges, R. S. 1987. Preoptic area opioids and opiate receptors increase during pregnancy and decrease during lactation. Brain Res. 420, 48–56.

Harada, Y., Nishioka, K., Kitahata, L. M., Nakatani, K., and Collins, J. G. 1995. Contrasting actions of intrathecal U50,488H, morphine, or [D-Pen2, D-Pen5] enkephalin or intravenous U50,488H on the visceromotor response to colorectal distension in the rat. Anesthesiology 83, 336–343.

Harte, S., Lagman, A., and Borszcz, G. 2000. Antinociceptive effects of morphine injected into the nucleus parafascicularis thalami of the rat. Brain Res. 874, 78–86.

Hawes, B. E., Graziano, M. P., and Lambert, D. G. 2000. Cellular actions of nociceptin: transduction mechanisms. Peptides 21, 961–967.

Heinricher, M. M. 2003. Orphanin FQ/nociceptin: from neural circuitry to behavior. Life Sci. 73, 813–822.

Heinricher, M. M., Cheng, Z. F., and Fields, H. L. 1987. Evidence for two classes of nociceptive modulating neurons in the periaqueductal gray. J. Neurosci. 7, 271–278.

Heinricher, M. M., McGaraughty, S., and Grandy, D. K. 1997. Circuitry underlying antiopioid actions of orphanin FQ in the rostral ventromedial medulla. J. Neurophysiol. 78, 3351–3358.

Heinricher, M. M., McGaraughty, S., and Farr, D. A. 1999. The role of excitatory amino acid transmission within the rostral ventromedial medulla in the antinociceptive actions of systemically administered morphine. Pain 81, 57–65.

Helmreich, D. L., Watkins, L. R., Deak, T., Maier, S. F., Akil, H., and Watson, S. J. 1999. The effect of stressor controllability on stress-induced neuropeptide mRNA expression within

the paraventricular nucleus of the hypothalamus. J. Neuroendocrinol. 11, 121–128.

Horvitz, J. 2000. Mesolimbic and nigrostriatal dopamine responses to salient non-rewarding stimuli. Neuroscience 96, 651–656.

Hughes, J., Smith, T. W., Kosterlitz, H. W., Fothergill, L. A., Morgan, B. A., and Morris, H. R. 1975. Identification of two related pentapeptides from the brain with potent opiate agonist activity. Nature 258, 577–580.

Jensen, T. S. and Yaksh, T. L. 1986. Comparison of the antinociceptive action of mu and delta opioid receptor ligands in the periaqueductal gray matter, medial and paramedial ventral medulla in the rat as studied by the microinjection technique. Brain Res. 372, 301–312.

Jutkiewicz, E. M., Rice, K. C., Woods, J. H., and Winsauer, P. J. 2003. Effects of the delta-opioid receptor agonist SNC80 on learning relative to its antidepressant-like effects in rats. Behav. Pharmacol. 14, 509–516.

Kest, B., Hopkins, E., Palmese, C. A., Chen, Z. P., Mogil, J. S., and Pintar, J. E. 2001. Morphine tolerance and dependence in nociceptin/orphanin FQ transgenic knock-out mice. Neuroscience 104, 217–222.

Khachaturian, H., Lewis, M. E., Tsou, K., and Watson, S. J. 1985. Beta-Endorphin, Alpha-MSH, ACTH, and Related Peptides. In: Handbook of Chemical Neuroanatomy (eds. A. Bjorklundand and T. Hokfelt), pp. 216–272. Elsevier Science.

Kiefel, J. M., Rossi, G. C., and Bodnar, R. J. 1993. Medullary mu and delta opioid receptors modulate mesencephalic morphine analgesia in rats. Brain Res. 624, 151–161.

Kieffer, B. L., Befort, K., Gaveriaux-Ruff, C., and Hirth, C. G. 1992. The delta-opioid receptor: isolation of a cDNA by expression cloning and pharmacological characterization. Proc. Natl. Acad. Sci. U. S. A. 89, 12048–12052.

Konig, M., Zimmer, A. M., Steiner, H., Holmes, P. V., Crawley, J. N., Brownstein, M. J, and Zimmer, A. 1996. Pain responses, anxiety and aggression in mice deficient in pre-proenkephalin. Nature 383, 535–538.

Koster, A., Montkowski, A., Schulz, S., Stube, E. M., Knaudt, K., Jenck, F., Moreau, J. L., Nothacker, H. P., Civelli, O., and Reinscheid, R. K. 1999. Targeted disruption of the orphanin FQ/nociceptin gene increases stress susceptibility and impairs stress adaptation in mice. Proc. Natl. Acad. Sci. U. S. A. 96, 10444–10449.

LaBuda, C. J., Sora, I., Uhl, G. R., and Fuchs, P. N. 2000. Stress-induced analgesia in mu-opioid receptor knockout mice reveals normal function of the delta-opioid receptor system. Brain Res. 869, 1–5.

Levine, J., Gordon, N., Jones, R., and Fields, H. 1978. The narcotic antagonist naloxone enhances clinical pain. Nature 272, 826–827.

Lewis, M. E., Pert, A., Pert, C. B., and Herkenham, M. 1983. Opiate receptor localization in rat cerebral cortex. J. Comp. Neurol. 216, 339–358.

Lewis, M. E., Mishkin, M., Bragin, E., Brown, R. M., Pert, C. B., and Pert, A. 1981. Opiate receptor gradients in monkey cerebral cortex: correspondence with sensory processing hierarchies. Science 211, 1166–1169.

Li, C. H., Lemaire, S., Yamashiro, D., and Doneen, B. A. 1976. The synthesis and opiate activity of beta-endorphin. Biochem. Biophys. Res. Commun. 71, 19–25.

Li, S., Zhu, J., Chen, C., Chen, Y. W., Deriel, J. K., Ashby, B., and Liu-Chen, L. Y. 1993. Molecular cloning and expression of a rat kappa opioid receptor. Biochem. J. 295(Pt 3), 629–633.

Loh, H. H., Tseng, L. F., Wei, E., and Li, C. H. 1976. Beta-endorphin is a potent analgesic agent. Proc. Natl. Acad. Sci. U. S. A. 73, 2895–2898.

Lorenz, J., Cross, D., Minoshima, S., Morrow, T., Paulson, P., and Casey, K. 2002. A unique representation of heat allodynia in the human brain. Neuron 35, 383–393.

Ma, F., Xie, H., Dong, Z. Q., Wang, Y. Q., and Wu, G. C. 2005. Expression of ORL1 mRNA in some brain nuclei in neuropathic pain rats. Brain Res. 1043, 214–217.

Madden, J. T., Akil, H., Patrick, R., and Barchas, J. 1977. Stress-induced parallel changes in central opioid levels and pain responsiveness in the rat. Nature 265, 358–360.

Mamiya, T., Noda, Y., Ren, X., Nagai, T., Takeshima, H., Ukai, M., and Nabeshima, T. 2001. Morphine tolerance and dependence in the nociceptin receptor knockout mice. J. Neural. Transm. 108, 1349–1361.

Manning, B. 1998. A lateralized deficit in morphine antinociception after unilateral inactivation of the central amygdala. J. Neurosci. 18, 9453–9470.

Manning, B. and Franklin, K. 1998. Morphine analgesia in the formalin test: reversal by microinjection of quaternary naloxone into the posterior hypothalamic area or periaqueductal gray. Behav. Brain Res. 92, 97–102.

Manning, B. H., Merin, N. M., Meng, I. D., and Amaral, D. G. 2001. Reduction in opioid- and cannabinoid-induced antinociception in rhesus monkeys after bilateral lesions of the amygdaloid complex. J. Neurosci. 21, 8238–8246.

Mansi, J. A., Laforest, S., and Drolet, G. 2000. Effect of stress exposure on the activation pattern of enkephalin-containing perikarya in the rat ventral medulla. J. Neurochem. 74, 2568–2575.

Mansour, A., Fox, C. A., Akil, H., and Watson, S. J. 1995. Opioid-receptor mRNA expression in the rats CNS: anatomical and functional implications. Trends Neurosci. 18, 22–29.

Margolis, E. B., Hjelmstad, G. O., Bonci, A., and Fields, H. L. 2003. Kappa-opioid agonists directly inhibit midbrain dopaminergic neurons. J. Neurosci. 23, 9981–9986.

Martin, M., Matifas, A., Maldonado, R., and Kieffer, B. L. 2003. Acute antinociceptive responses in single and combinatorial opioid receptor knockout mice: distinct mu, delta and kappa tones. Eur. J. Neurosci. 17, 701–708.

Matthes, H. W., Smadja, C., Valverde, O., Vonesch, J. L., Foutz, A. S., Boudinot, E., Denavit-Saubie, M., Severini, C., Negri, L., Roques, B. P., Maldonado, R., and Kieffer, B. L. 1998. Activity of the delta-opioid receptor is partially reduced, whereas activity of the kappa-receptor is maintained in mice lacking the mu-receptor. J. Neurosci. 18, 7285–7295.

Mayer, D. J. and Liebeskind, J. C. 1974. Pain reduction by focal electrical stimulation of the brain: an anatomical and behavioral analysis. Brain Res. 68, 73–93.

Mayer, D. J., Wolfle, T. L., Akil, H., Carder, B., and Liebeskind, J. C. 1971. Analgesia from electrical stimulation in the brainstem of the rat. Science 174, 1351–1354.

McLaughlin, J. P., Marton-Popovici, M., and Chavkin, C. 2003. Kappa opioid receptor antagonism and prodynorphin gene disruption block stress-induced behavioral responses. J. Neurosci. 23, 5674–5683.

Meng, I. D., Johansen, J. P., Harasawa, I., and Fields, H. L. 2005. Kappa opioids inhibit physiologically identified medullary pain modulating neurons and reduce morphine antinociception. J. Neurophysiol. 93, 1138–1144.

Meng, F., Xie, G. X., Thompson, R. C., Mansour, A., Goldstein, A., Watson, S. J., and Akil, H. 1993. Cloning and pharmacological characterization of a rat kappa opioid receptor. Proc. Natl. Acad. Sci. U. S. A. 90, 9954–9958.

Meunier, J. C., Mollereau, C., Toll, L., Suaudeau, C., Moisand, C., Alvinerie, P., Butour, J. L., Guillemot, J. C., Ferrara, P., and Monsarrat, B., et al. 1995. Isolation and structure of the endogenous agonist of opioid receptor-like ORL1 receptor. Nature 377, 532–535.

Mogil, J. S. and Pasternak, G. W. 2001. The molecular and behavioral pharmacology of the orphanin FQ/nociceptin peptide and receptor family. Pharmacol. Rev. 53, 381–415.

Mogil, J. S., Grisel, J. E., Reinscheid, R. K., Civelli, O., Belknap, J. K., and Grandy, D. K. 1996a. Orphanin FQ is a functional anti-opioid peptide. Neuroscience 75, 333–337.

Mogil, J. S., Grisel, J. E., Zhangs, G., Belknap, J. K., and Grandy, D. K. 1996b. Functional antagonism of mu-, delta- and kappa-opioid antinociception by orphanin FQ. Neurosci. Lett. 214, 131–134.

Moles, A., Kieffer, B. L., and D'Amato, F. R. 2004. Deficit in attachment behavior in mice lacking the mu-opioid receptor gene. Science 304, 1983–1986.

Mollereau, C., Parmentier, M., Mailleux, P., Butour, J. L., Moisand, C., Chalon, P., Caput, D., Vassart, G., and Meunier, J. C. 1994. ORL1, a novel member of the opioid receptor family. Cloning, functional expression and localization. FEBS Lett. 341, 33–38.

Mollereau, C., Simons, M. J., Soularue, P., Liners, F., Vassart, G., Meunier, J. C., and Parmentier, M. 1996. Structure, tissue distribution, and chromosomal localization of the prepronociceptin gene. Proc. Natl. Acad. Sci. U. S. A. 93, 8666–8670.

Moran, T. D., Abdulla, F. A., and Smith, P. A. 2000. Cellular neurophysiological actions of nociceptin/orphanin FQ. Peptides 21, 969–976.

Moreau, J. L. and Fields, H. L. 1986. Evidence for GABA involvement in midbrain control of medullary neurons that modulate nociceptive transmission. Brain Res. 397, 37–46.

Morinville, A., Cahill, C. M., Esdaile, M. J., Aibak, H., Collier, B., Kieffer, B. L., and Beaudet, A. 2003. Regulation of delta-opioid receptor trafficking via mu-opioid receptor stimulation: evidence from mu-opioid receptor knock-out mice. J. Neurosci. 23, 4888–4898.

Moskowitz, A. S. and Goodman, R. R. 1985. Autoradiographic analysis of mu1, mu2, and delta opioid binding in the central nervous system of C57BL/6BY and CXBK (opioid receptor-deficient) mice. Brain Res. 360, 108–116.

Narita, M., Kishimoto, Y., Ise, Y., Yajima, Y., Misawa, K., and Suzuki, T. 2005. Direct evidence for the involvement of the mesolimbic kappa-opioid system in the morphine-induced rewarding effect under an inflammatory pain-like state. Neuropsychopharmacology 30, 111–118.

Nishi, M., Houtani, T., Noda, Y., Mamiya, T., Sato, K., Doi, T., Kuno, J., Takeshima, H., Nukada, T., Nabeshima, T., Yamashita, T., Noda, T., and Sugimoto, T. 1997. Unrestrained nociceptive response and disregulation of hearing ability in mice lacking the nociceptin/orphaninFQ receptor. Embo. J. 16, 1858–1864.

Ossipov, M. H., Kovelowski, C. J., Nichols, M. L., Hruby, V. J., and Porreca, F. 1995. Characterization of supraspinal antinociceptive actions of opioid delta agonists in the rat. Pain 62, 287–293.

Pan, Z. Z., Tershner, S. A., and Fields, H. L. 1997. Cellular mechanism for anti-analgesic action of agonists of the kappa-opioid receptor. Nature 389, 382–385.

Pert, C. B. and Snyder, S. H. 1973. Opiate receptor: demonstration in nervous tissue. Science 179, 1011–1014.

Petrusz, P., Merchenthaler, I., and Maderdrut, J. 1985. Distribution of Enkephalin-Containing Neurons in the Central Nervous System. In: Handbook of Chemical Neuroanatomy (eds. A. Björklundand and T. Hökfelt), pp. 273–334. Elsevier.

Piva, F., Maggi, R., Limonta, P., Dondi, D., and Martini, L. 1987. Decrease of mu opioid receptors in the brain and in the hypothalamus of the aged male rat. Life Sci. 40, 391–398.

Porreca, F., Mosberg, H. I., Hurst, R., Hruby, V. J., and Burks, T. F. 1984. Roles of mu, delta and kappa opioid receptors in spinal and supraspinal mediation of gastrointestinal transit effects and hot-plate analgesia in the mouse. J. Pharmacol. Exp. Ther. 230, 341–348.

Quinones-Jenab, V., Jenab, S., Ogawa, S., Inturrisi, C., and Pfaff, D. W. 1997. Estrogen regulation of mu-opioid receptor mRNA in the forebrain of female rats. Brain Res. Mol. Brain Res. 47, 134–138.

Rainville, P., Duncan, G., Price, D., Carrier, B., and Bushnell, M. 1997. Pain affect encoded in human anterior cingulate but not somatosensory cortex. Science 277, 968–971.

Reinscheid, R. K., Nothacker, H. P., Bourson, A., Ardati, A., Henningsen, R. A., Bunzow, J. R., Grandy, D. K., Langen, H., Monsma, F. J., Jr, and Civelli, O. 1995. Orphanin FQ: a neuropeptide that activates an opioidlike G protein-coupled receptor. Science 270, 792–794.

Reynolds, D. V. 1969. Surgery in the rat during electrical analgesia induced by focal brain stimulation. Science 164, 444–445.

Richardson, D. E. and Akil, H. 1977a. Pain reduction by electrical brain stimulation in man. Part 1: Acute administration in periaqueductal and periventricular sites. J. Neurosurg. 47, 178–183.

Richardson, D. E. and Akil, H. 1977b. Pain reduction by electrical brain stimulation in man. Part 2: Chronic self-administration in the periventricular gray matter. J Neurosurg. 47, 184–194.

al-Rodhan, N., Chipkin, R., and Yaksh, T. L. 1990. The antinociceptive effects of SCH-32615, a neutral endopeptidase (enkephalinase) inhibitor, microinjected into the periaqueductal, ventral medulla and amygdala. Brain Res. 520, 123–130.

Rossi, G. C., Leventhal, L., Bolan, E., and Pasternak, G. W. 1997. Pharmacological characterization of orphanin FQ/nociceptin and its fragments. J. Pharmacol. Exp. Ther. 282, 858–865.

Rossi, G. C., Leventhal, L., and Pasternak, G. W. 1996. Naloxone sensitive orphanin FQ-induced analgesia in mice. Eur. J. Pharmacol. 311, R7–R8.

Roychowdhury, S. M. and Fields, H. L. 1996. Endogenous opioids acting at a medullary mu-opioid receptor contribute to the behavioral antinociception produced by GABA antagonism in the midbrain periaqueductal gray. Neuroscience 74, 863–872.

Rubinstein, M., Mogil, J. S., Japon, M., Chan, E. C., Allen, R. G., and Low, M. J. 1996. Absence of opioid stress-induced analgesia in mice lacking β-endorphin by site directed mutagenesis. Proc. Natl. Acad. Sci. U. S. A. 93, 3995–4000.

Scherrer, G., Befort, K., Contet, C., Becker, J., Matifas, A., and Kieffer, B. L. 2004. The delta agonists DPDPE and deltorphin II recruit predominantly mu receptors to produce thermal analgesia: a parallel study of mu, delta and combinatorial opioid receptor knockout mice. Eur. J. Neurosci. 19, 2239–2248.

Schlaepfer, T., Strain, E., Greenberg, B., Preston, K., Lancaster, E., Bigelow, G., Barta, P., and Pearlson, G 1998. Site of opioid action in the human brain: mu and kappa agonists' subjective and cerebral blood flow effects. Am. J. Psychiatry. 155, 470–473.

Simon, E. J., Hiller, J. M., and Edelman, I. 1973. Stereospecific binding of the potent narcotic analgesic (3H) Etorphine to rat-brain homogenate. Proc. Natl. Acad. Sci. U. S. A. 70, 1947–1949.

Skoubis, P. D., Lam, H. A., Shoblock, J., Narayanan, S., and Maidment, N. T. 2005. Endogenous enkephalins, not endorphins, modulate basal hedonic state in mice. Eur. J. Neurosci. 21, 1379–1384.

Smith, Y. R., Stohler, C. S., Nichols, T. E., Bueller, J. A., Koeppe, R. A., and Zubieta, J. K. 2006. Pronociceptive and antinociceptive effects of estradiol through endogenous opioid neurotransmission in women. J. Neurosci. 26, 5777–5785.

Smith, J. S., Zubieta, J. K., Price, J. C., Flesher, J. E., Madar, I., Lever, J. R., Kinter, C. M., Dannals, R. F., and Frost, J. J. 1999. Quantification of delta-opioid receptors in human brain with N1'-([^{11}C]methyl) naltrindole and positron emission tomography. J. Cereb. Blood Flow Metab. 19, 956–966.

Sora, I., Takahashi, N., Funada, M., Ujike, H., Revay, R. S., Donovan, D. M., Miner, L. L., and Uhl, G. R. 1997. Opiate receptor knockout mice define μ receptor roles in endogenous nociceptive responses and morphine-induced analgesia. Proc. Natl. Acad. Sci. U. S. A. 94, 1544–1549.

Terenius, L. 1973. Characteristics of the "receptor" for narcotic analgesics in synaptic plasma membrane fraction from rat brain. Acta Pharmacol. Toxicol. (Copenh) 33, 377–384.

Tershner, S. A., Mitchell, J. M., and Fields, H. L. 2000. Brainstem pain modulating circuitry is sexually dimorphic with respect to mu and kappa opioid receptor function. Pain 85, 153–159.

Thompson, R. C., Mansour, A., Akil, H., and Watson, S. J. 1993. Cloning and pharmacological characterization of a rat mu opioid receptor. Neuron 11, 903–913.

Thompson, A. C., Zapata, A., Justice, J. B., Jr., Vaughan, R. A., Sharpe, L. G., and Shippenberg, T. S. 2000. Kappa-opioid receptor activation modifies dopamine uptake in the nucleus accumbens and opposes the effects of cocaine. J. Neurosci. 20, 9333–9340.

Tian, J. H., Xu, W., Fang, Y., Mogil, J. S., Grisel, J. E., Grandy, D. K., and Han, J. S. 1997. Bidirectional modulatory effect of orphanin FQ on morphine-induced analgesia: antagonism in brain and potentiation in spinal cord of the rat. Br. J. Pharmacol. 120, 676–680.

Torregrossa, M. M., Isgor, C., Folk, J. E., Rice, K. C., Watson, S. J., and Woods, J. H. 2004. The delta-opioid receptor agonist (+)BW373U86 regulates BDNF mRNA expression in rats. Neuropsychopharmacology 29, 649–659.

Tsou, K. and Jang, C. S. 1964. Studies on the site of analgesic action of morphine by intracerebral micro-injection. Sci. Sin. 13, 1099–1109.

Urban, M. O. and Smith, D. J. 1994. Nuclei within the rostral ventromedial medulla mediating morphine antinociception from the periaqueductal gray. Brain Res. 652, 9–16.

Vaught, J. L., Mathiasen, J. R., and Raffa, R. B. 1988. Examination of the involvement of supraspinal and spinal mu and delta opioid receptors in analgesia using the mu receptor deficient CXBK mouse. J. Pharmacol. Exp. Ther. 245, 13–16.

Wang, Z., Gardell, L. R., Ossipov, M. H., Vanderah, T. W., Brennan, M. B., Hochgeschwender, U., Hruby, V. J., Malan, T. P., Jr., Lai, J., and Porreca, F. 2001. Pronociceptive actions of dynorphin maintain chronic neuropathic pain. J. Neurosci. 21, 1779–1786.

Wagner, K. J., Willoch, F., Kochs, E. F., Siessmeier, T., Tolle, T. R., Schwaiger, M., and Bartenstein, P. 2001. Dose-dependent regional cerebral blood flow changes during remifentanil infusion in humans: a positron emission tomography study. Anesthesiology 94, 732–739.

Watkins, L. and Mayer, D. 1982. Organization of endogenous opiate and nonopiate pain control systems. Science 216, 1185–1192.

Williams, K. L., Ko, M. C., Rice, K. C., and Woods, J. H. 2003. Effect of opioid receptor antagonists on hypothalamic-pituitary–adrenal activity in rhesus monkeys. Psychoneuroendocrinology 28, 513–528.

Yaksh, T. L. and Rudy, T. A. 1978. Narcotic analgesics: CNS sites and mechanisms of action as revealed by intracerebral injection techniques. Pain 4, 299–359.

Zacny, J. P. 2001. Morphine responses in humans: a retrospective analysis of sex differences. Drug Alcohol Depend. 63, 23–28.

Zadina, J. E. 2002. Isolation and distribution of endomorphins in the central nervous system. Jpn. J. Pharmacol. 89, 203–208.

Zhang, Y., Butelman, E. R., Schlussman, S. D., Ho, A., and Kreek, M. J. 2004. Effect of the endogenous kappa opioid agonist dynorphin A(1–17) on cocaine-evoked increases in striatal dopamine levels and cocaine-induced place preference in C57BL/6J mice. Psychopharmacology (Berl) 172, 422–429.

Zhuo, M. and Gebhart, G. F. 1997. Biphasic modulation of spinal nociceptive transmission from the medullary raphe nuclei in the rat. J. Neurophysiol. 78, 746–758.

Zubieta, J. K., Dannals, R., and Frost, J. 1999. Gender and age influences on human brain mu opioid receptor binding measured by PET. Am. J. Psychiatry. 156, 842–848.

Zubieta, J. K., Bueller, J. A., Jackson, L. R., Scott, D. J., Xu, Y., Koeppe, R. A., Nichols, T. E., and Stohler, C. S. 2005.

Placebo effects mediated by endogenous opioid activity on mu-opioid receptors. J. Neurosci. 25, 7754–7762.

Zubieta, J. K., Smith, Y. R., Bueller, J. A., Xu, Y., Kilbourn, M. R., Jewett, D. M., Meyer, C. R., Koeppe, R. A., and Stohler, C. S. 2002. mu-Opioid receptor-mediated antinociceptive responses differ in men and women. J. Neurosci. 22, 5100–5107.

Zubieta, J. K., Smith, Y., Bueller, J., Xu, Y., Kilbourn, M., Meyer, C., Koeppe, R., and Stohler, C. 2001. Regional mu opioid receptor regulation of sensory and affective dimensions of pain. Science 293, 311–315.

55 Neuropathic Pain: Basic Mechanisms (Animal)

M H Ossipov and F Porreca, University of Arizona, Tucson, AZ, USA

Published by Elsevier Inc.

55.1 Introduction

Neuropathic pain is defined by the International Association for the Study of Pain (IASP) as "pain initiated or caused by a primary lesion or dysfunction in the nervous system". Although this definition functions well as a guide for the clinical assessment of neuropathic pain, a definitive categorization of neuropathic pain is difficult because there are many descriptions of the pain syndrome. Neuropathic pain may arise from an actual injury to a peripheral nerve or it may arise in the absence of any obvious nerve damage (e.g., trigeminal neuralgia and complex regional pain syndrome (CRPS), type I). Disease processes, such as diabetic neuropathy, postherpetic neuralgia resulting from varicella zoster infection (shingles), and complications from AIDS also may cause neuropathic pain, as may alterations in peripheral nerve function induced by chemotherapeutic regimens. Neuropathic pain may include spontaneous, or stimulus-independent, pain that has been described as shooting, burning, lancinating, prickling, and electrical. Evoked, or stimulus-dependent, neuropathic pain includes allodynia, defined by IASP as "pain due to a stimulus which does not normally provoke pain." These stimuli may be nonnoxious heat, light touch, or even a puff of cool air. Moreover, mechanical allodynia may be static, as evoked by a light touch, or dynamic, as evoked by a light brush of the skin. Hyperalgesia is identified when a stimulus that normally produces a nociceptive responses produces an exaggerated responses. The different forms through which neuropathic pain manifests suggest that differing mechanisms are likely to mediate the different features of the condition. The numerous features of neuropathic pain can complicate an accurate clinical diagnosis. For this reason, efforts have emphasized the development of highly specific questionnaires emphasizing a unique set of symptoms that can aid clinical diagnosis with relatively little effort. Features that appear to be most often related to neuropathic pain include tingling or numbness, pain evoked by heat or cold, and especially, a sensation of heat or a burning-like quality is associated with the neuropathic pain state (Krause, S. J. and Backonja, M. M., 2003; Backonja, M. M. and Stacey, B., 2004; Bennett, M. I. et al., 2007). More recently, a questionnaire identified at DN4 was designed to assess the presence of pain combined with parasthesias or dysesthesias (Bouhassira, D. et al., 2005).

55.2 Injury-Induced Sensitization of Peripheral Nerves and Neuropathic Pain

There are several factors common to the neuropathic pain state, regardless of the etiology that may initiate the condition. It has been repeatedly demonstrated that peripheral nerve injury is associated with a heightened sensitivity of nerves to sensory stimuli, a phenomenon that has been termed peripheral sensitization. Importantly, peripheral sensitization is not restricted to the injured nerve fibers, but also includes adjacent, uninjured (but abnormal) nerves. Injured nerves may form a tangled mass of connective tissue termed a neuroma with distally sprouting regenerating axons. Physical manipulation of the neuroma elicits sensations ranging from minor dysesthesias to intense pain. Electrophysiologic studies showed that both myelinated and unmyelinated fibers in neuromas show a basal spontaneous activity as well as enhanced discharges provoked by touch or movement, indicating that the neuroma may serve as an ectopic generator of action potentials. In addition, cell bodies of the dorsal root ganglia (DRG) may also generate ectopic discharges, thus producing spontaneous hyperesthesias and pain as well as enhanced sensitivity to stimuli. Traumatic injury of the rat sciatic nerve produced by loose ligatures, the chronic constriction injury or CCI model, resulted in spontaneous discharges generated from the DRG. In this study, discharges were recorded from 15% to 35% of myelinated and 3% of unmyelinated fibers (Kajander, K. C. et al., 1992). Electrophysiological recordings performed on sciatic nerve bundles after CCI also demonstrated spontaneous and evoked ectopic discharges from the region of injury (Tal, M. and Bennett, G. J., 1994). Surprisingly, these spontaneous discharges of DRG neurons occurred from cells of both injured as well as intact nerves (Michaelis, M. et al., 1996). Additionally, electrophysiologic studies showed spinal nerve transection enhanced the

spontaneous activity of nerve fibers in adjacent, uninjured nerve roots and in peripheral nerves (Ali, Z. et al., 1999; Wu, G., et al., 2001). Such studies raised the question of relative contributions of enhanced excitability from injured and uninjured fibers in eliciting neuropathic pain. Yoon Y. W. et al. (1996) used a model of neuropathic pain in which the L5 spinal nerve was injured (the spinal nerve ligation or SNL model, Kim, S. H. and Chung, J. M., 1992) and showed that the behavioral signs of pain were abolished by dorsal rhizotomy or bupivacaine applied to the injured spinal nerves emphasizing the importance of afferent drive. Moreover, dorsal rhizotomy or bupivacaine at L4 and L5 also abolished behavioral signs of mechanical and cold allodynia, but not signs of ongoing pain (Yoon, Y. W. et al., 1996). The contribution of uninjured nerves, or uninjured adjacent fibers, to neuropathic pain has been debated. In an insightful study, Belzberg and coworkers (Li, Y. et al., 2000) used the SNL model and found that the behavioral signs of experimental neuropathic pain produced by ligation of the L5 spinal nerve were not abolished by rhizotomy at L5, but rather were abolished by rhizotomy at L4. These findings suggested a prominent role for uninjured but abnormal adjacent nerve fibers following nerve injury and were consistent with the suggested importance of uninjured, but abnormal, L4 afferents; these fibers show significant redistribution of sodium channels such as $Na_V1.8$ following L5 injury (Lai, J. et al., 2002) and this may reflect a mechanism which drives neuropathic pain (see below). It has also been suggested that sensitization of uninjured peripheral nerves may result, in part, from ephaptic excitation caused by the close proximity of injured and noninjured nerves within the nerve trunks or in the DRG (Devor, M. and Wall, P. D., 1990; Attal, N. and Bouhassira, D., 1999). Sensitization of peripheral nerves may mediate SNL-induced thermal hyperalgesia, but may not mediate tactile allodynia. Unlike tactile allodynia, thermal hyperalgesia was abolished by the intravenous infusion of the quaternary lidocaine analogs QX-314 and QX-222, which do not cross the blood–brain barrier (Chen, Q. et al., 2004). In contrast, both tactile and thermal parameters of enhanced pain were abolished by systemic lidocaine, or microinjection of lidocaine or QX-222 into the rostralventromedial medulla (RVM) (Chen, Q. et al., 2004). These observations emphasize the contribution of a central mechanism (addressed below) which may act in addition to peripheral mechanisms to elicit neuropathic pain states.

55.2.1 Role of Peripheral Nerve Degeneration in Neuropathic Pain

Degeneration of peripheral nerves provokes the release of proinflammatory cytokines such as interleukins and tumor necrosis factor (TNF), and neurotrophic factors in the region of the target tissue. The initial reaction is followed by macrophage infiltration that further mediate proinflammatory signals. This accumulation of proinflammatory substances may sensitize adjacent nerve terminals of uninjured fibers. Myelinated peripheral nerve fibers undergoing Wallerian degeneration elicit the release of several pronociceptive factors in the target tissue and cytokines along the peripheral nerve, thereby sensitizing the adjacent nerve endings of uninjured nerve fibers, and exciting and potentially producing an ectopic focus in the uninjured adjacent C fibers (Li, Y. et al., 2000). Wallerian degeneration of the injured, spontaneously active myelinated fibers may excite unmyelinated fibers through the comingling of injured and uninjured nerves that occurs within the Remak bundles of the sciatic nerve trunk (Wu, G. et al., 2001; 2002). Furthermore, Wallerian degeneration associated with injured nerves causes alterations in the Schwann cells that provoke them to release proinflammatory mediators, including glutamate as well as nerve growth factor (NGF) (Shamash, S. et al., 2002; MacInnis, B. L. and Campenot, R. B., 2005). Within the nerve trunk, these substances may diffuse to adjacent uninjured nerve fibers and enhance sensitivity such that an ectopic locus may develop. Release of proinflammatory mediators including NGF in the peripheral terminal fields that include uninjured neurons may sensitize these fibers to ensuing stimulation, thus producing peripheral sensitization. It has been shown that nerve injury causes significant elevations in NGF and its mRNA in Schwann cells, macrophages, and fibroblasts at the site of injury (Shamash, S. et al., 2002; MacInnis, B. L. and Campenot, R. B., 2005). Treatment of cultured fibroblasts or keratinocytes with inflammatory mediators, including histamine and interleukin-1 also provoke upregulation of NGF (Kanda, N. and Watanabe, S., 2003). Moreover, NGF promotes degranulation of mast cells, which release inflammatory mediators and thereby promotes peripheral sensitization of nociceptors as well as further release of NGF (Lewin, G. R. et al., 1994; Stempelj, M. and Ferjan, I., 2005). Denervation of the skin due to degeneration of peripheral nerves produce regions that are insensate surrounded by regions of increased sensitivity to noxious stimuli perhaps as a consequence of upregulation of NGF by keratinocytes (Taherzadeh, O. et al., 2003). Increased NGF production promotes sprouting of peripheral nerve terminals into the region into the boundary between the denervated region and that with normal innervation (Rajan, B. et al., 2003). Collateral sprouting of peripheral axons together with the pronociceptive actions of peripheral NGF can promote nociceptive hypersensitivity in the region immediately surrounding the zone of denervation (Griffin, J. W., 2006).

Other growth factors, such as those in the glial-cell-line-derived neutrotropic factor (GDNF) family including GDNF and artemin, may play a significant role in expression of neuropathic pain. NGF and GDNF family members may act at peripheral nociceptive primary afferents, leading to enhancement of sensitivity to noxious inputs (Zwick, M. et al., 2002; Elitt, C. M. et al., 2006; Griffin, J. W., 2006; Malin, S. A. et al., 2006). Injury-induced upregulation of growth factors such as NGF results in a situation where intact, uninjured nociceptive C fibers expressing the trkA receptor are exposed to a surplus of NGF and may become overtrophed in response to nerve injury (Griffin, J. W., 2006). In these fibers, NGF would promote nociceptive transmission through phosphorylation of TRPV1 ion channels, as well as posttranslational mechanisms that include upregulation of ion channels, redistribution of $Na_V1.8$ and increased synthesis of substance P and CGRP (Boucher, T. J. et al., 2000; Boucher, T. J. and McMahon, S. B., 2001; Griffin, J. W., 2006). In contrast, the axons of the injured peripheral fibers, which include C fiber nociceptors as well as some myelinated sensory fibers, are not able to traffic target-derived growth factors to the cell body. These peripheral nerves may be considered to be undertrophed (Griffin, J. W., 2006) and can demonstrate increased expression of tetrodotoxin (TTX)-sensitive (TTX-S) sodium channels and enhanced spontaneous activity, adaptations which may underlie such painful conditions as postamputation pain and neuropathic pain associated with mastectomy and thoracotomy scars (Griffin, J. W., 2006). This hypothesis was tested by providing support with exogenous trophic factors including GDNF and artemin. Administration of exogenous GDNF normalized expression of sodium channels and the reversed the behavioral signs of neuropathic pain (Boucher, T. J. et al., 2000; Boucher, T. J. and McMahon, S. B., 2001; Wang, R. et al., 2003). In addition, exogenous GDNF delivered by the intrathecal route also abolished electrophysiologic signs of enhanced firing, ectopic discharges, and normalized sodium channel currents in injured peripheral nerves (Boucher, T. J. *et al.*, 2000; Boucher, T. J. and McMahon, S. B., 2001) as well as multiple biochemical markers in the DRG (Wang, R. *et al.*, 2003). Systemic treatment with artemin to rats with SNL also normalized the expression of biochemical markers of neuropathy, including the expression and redistribution of $Na_V1.8$ and abolished signs of neuropathic pain (Gardell *et al.*, 2003b). The pronociceptive role of neurotrophic factors is discussed below.

55.2.2   Role of Nerve Growth Factor in Immediate Pronociceptive Function

NGF binds preferentially to the high-affinity tyrosine kinase receptor subtype trkA and to the low-affinity, nonselective neurotrophin receptor p75 (Sah, D. W. *et al.*, 2005a; 2005b). The neuronal distribution of the trkA receptor indicates a substantial role of NGF in nociceptive processes, since trkA is found almost exclusively on small-diameter unmyelinated peptidergic DRG neurons (Sah, D. W. *et al.*, 2003). Double-labeling studies indicate that 92% of trkA-expressing DRG neurons also express CGRP and substance P, whereas it is present on about 20% of myelinated DRG neurons (Bennett, D. L. *et al.*, 2000; Averill, S. *et al.*, 2004). Investigations performed with a skin–nerve preparation showed that NGF increased mechanical hyperalgesia mediated by thin myelinated fibers and thermal hyperalgesia mediated by unmyelinated nociceptors (Stucky, C. L. *et al.*, 2002). Electrophysiologic studies revealed that endogenous NGF regulates the sensitivity of unmyelinated nociceptors to chemical and thermal stimuli (Bennett, D. L. *et al.*, 1998). Sensitization to heat stimulus or to capsaicin application is induced rapidly by NGF (Shu, X. and Mendell, L. M., 2001). Within 10 min of exposure to NGF, there is a rapid potentiation of capsaicin-evoked ion currents in cultured DRG neurons gated by TRPV1 (Shu, X. and Mendell, L. M., 2001). Activation of the trkA receptor by NGF leads to, among other events, phosphorylation of TRPV1, in part through the signaling cascades associated with PKA, PKC, MEK, and the MAP kinases (Shu, X. and Mendell, L. M., 1999; 2001). Moreover, NGF increases nociceptor activity by a rapid facilitation of sodium currents and suppression of an outward potassium current. Behavioral hyperalgesia is evoked by NGF administered systemically or into the endoneurium of a peripheral nerve (Ruiz, G. *et al.*, 2004; Ruiz, G. and Banos, J. E., 2005). Clinical

studies also showed that exogenous NGF produced long-lasting mechanical allodynia and hyperalgesia and mild-to-moderate myalgia (Apfel, S. C., 2000; Svensson, P. *et al.*, 2003). Conversely, manipulations that block NGF activity alleviate signs of hyperalgesia. Blockade of NGF activity with the trkA–IgG fusion protein, that antagonist ALE-540 or with antiserum to NGF abolished mechanical and thermal hyperalgesia in animal models of inflammation, incision, or nerve injury (Owolabi J. B. *et al.*, 1999; Banik, R. K. *et al.*, 2005). More recently, intrathecal injection of antibody to NGF blocked behavioral signs of cold allodynia after nerve ligation (Obata, K. *et al.*, 2005). In addition, anti-NGF given by intrathecal injection or local application onto the L4 spinal nerve abolished signs of thermal and mechanical hyperalgesia in rats with peripheral nerve injury (Obata, K. *et al.*, 2004; 2006). The prominent role of NGF in the expression of chronic pain states has led a number of pharmaceutical companies, including Pfizer/Rinat, Amgen, and PainCeptor Pharma to initiate drug discovery and development efforts targeting NGF or trkA (Shelton, D. L. *et al.*, 2005; Hefti, F. F. *et al.*, 2006). Drug development approaches include sequestering or binding NGF so that it does not interact with the trkA receptor, trkA receptor antagonists, or disruption of the signaling pathways linked to trkA receptor activation (Hefti, F. F. *et al.*, 2006).

55.2.3   Nerve Growth Factor May Mediate Posttranslational Pronociceptive Functions

In addition to its immediate actions in the periphery, NGF can also initiate long-lasting posttranscriptional changes in peripheral nerves to promote and sustain an enhanced pain state. When NGF binds to the trkA receptor at the peripheral terminal, it is rapidly internalized and the phosphorylated trkA–NGF complex is accumulated in signaling endosomes that are transported to the DRG, where it triggers the activation of several intracellular signaling cascades, including expression of the MAP kinases, ERK, and p38 (Ji, R. R. *et al.*, 2002; McMahon, S. B. and Cafferty, W. B., 2004). Inflammation or exogenous NGF administered spinally causes phosphorylation of p38 in nociceptors in the DRG, but not the dorsal horn, along with a substantial upregulation of TRPV1 in the DRG (Ji, R. R. *et al.*, 2002; McMahon, S. B. and Cafferty, W. B., 2004; Malik-Hall, M. *et al.*, 2005). Moreover, there is evidence that the TRPV1 receptor is preferentially trafficked to the peripheral terminals of the nociceptors, where it can contribute to peripheral sensitization and hyperalgesia (Ji, R. R. et al., 2002). Furthermore, hyperalgesia may also be maintained by the increased expression of peptidic neurotransmitters specific to primary afferent nociceptors. The exposure of DRG cultures to NGF produced 15- to 60-fold increases in mRNA for substance P and CGRP in trkA-expressing cells, and increased capsaicin-evoked release of CGRP or substance P from cultured trigeminal neurons and from central terminals of primary afferent nociceptors in spinal cord sections (McMahon, S. B. and Cafferty, W. B., 2004; Puntambekar, P. et al., 2005). Animals with diabetic or chemotherapy-induced neuropathy treated with NGF showed abnormally elevated levels of substance P and CGRP protein and mRNA (Tomlinson, D. R. et al., 1997). These functional consequences of NGF expression, including stimulation of BDNF production, may enable transition from peripheral to central sensitization after injury, resulting in chronic pain (Sah, D. W. et al., 2003; McMahon, S. B., et al., 2005; Sah, D. W. et al., 2005a; 2005b). Moreover, retrogradely transported NGF promotes the upregulation of ion channels that promote activation of peripheral nerves and enhances nociceptive transmission. For example, hyperalgesia is also maintained by NGF-mediated upregulation of ASIC3 and upregulation and redistribution of Na$_V$1.8 on primary afferent nociceptors (McMahon, S. B. and Jones, N. G., 2004). Importantly, it was shown that peripheral nerve injury is associated with enhanced retrograde transport of NGF to the cell bodies of spared, or uninjured adjacent, peripheral nerves, thus promoting pronociceptive transcriptional changes in these nerves as well. For example, ligation of the L5 spinal nerve increased the expression of p75NTR in the DRG that was abolished by spinal administration of anti-NGF (Obata, K. et al., 2006). Moreover, the application of anti-NGF onto the L4 spinal nerve after ligation of the L5 spinal nerve blocked the injury-induced upregulation of p75NTR (Obata, K. et al., 2006). The spinal injection of the NGF/TrkA antagonist K252a blocked injury-induced thermal and cold hyperalgesia and the upregulation of phosphorylated trkA, p38, and of TRPV1 and TRPA1 in L4 DRG (Obata, K. et al., 2006).

55.2.4 Sodium Channels and Enhanced Peripheral Nerve Activity

Because of their critical role in the generation of action potentials and neuronal excitation, the increased abnormal spontaneous and ectopic discharges of injured peripheral nerves are related to increased activity of voltage-gated sodium channels (VGSCs) (Lai, J. et al., 2002; 2003). The VGSCs have been linked to excitability of primary afferent nociceptors and may mediate sensitized pain states (Lai, J. et al., 2003). Peripheral nerve injury is also associated with changes in the expression and distribution of isoforms of the VGSCs (Novakovic, S. D. et al., 2001). Three sodium channels merit special consideration.

The TTX-S sodium channel Na$_V$1.3 is fast-activating channel normally present in brain tissue but scarce in the DRG. However, peripheral nerve injury results in a substantial expression this channel in peripheral nerves, and it is believed that this novel expression may be partly responsible for the enhanced excitability of the primary afferents (Waxman, S. G. et al., 1999; Lai, J. et al., 2003). However, although its biophysical properties favor generation of spontaneous discharges, a causal relationship between expression of Na$_V$1.3 in the DRG and neuronal discharges has not been established. It had been hypothesized that upregulation of Na$_V$1.3 in peripheral nerves might mediate neuropathic pain as intrathecal treatment with GDNF abolished behavioral signs of enhanced pain, injury-induced enhanced peripheral neuronal excitability, and the upregulation of Na$_V$1.3 (Boucher, T. J. et al., 2000). Other studies, however, showed that intrathecal GDNF normalized many changes observed in the DRG following peripheral nerve injury (Wang, R. et al., 2003) making specific conclusions about Na$_V$1.3 difficult. Observations that knockdown of Na$_V$1.3 by antisense treatment abolished both the enhanced neuronal activity and behavioral signs of neuropathic pain in nerve-injured rats seemed to support this hypothesis (Hains, B. C. et al., 2004). However, more recent evidence has clearly shown that this channel may not be relevant to enhanced pain after nerve injury. Studies employing two different conditional mutant mice lines, one with knockout of Na$_V$1.3 in nociceptors and another with global knockout of Na$_V$1.3 throughout the nervous system, showed that behavioral signs of neuropathic pain developed normally in both mutant lines after peripheral nerve injury (Nassar, M. A. et al., 2006). Knockout of Na$_V$1.3 also did not alter responses to acute nociceptive stimuli or to inflammation (Nassar, M. A. et al., 2006). Accordingly, it is believed that the Na$_V$1.3 channel is neither sufficient nor necessary for the expression of neuropathic pain (Nassar, M. A. et al., 2006).

The TTX-insensitive sodium channel Na$_V$1.8 exhibits properties consistent with a role in neuropathic pain (Sangameswaran, L. et al., 1997; Lai, J. et al., 2002). It is expressed primarily in small-diameter, unmyelinated peripheral nerve fibers (C fibers) and in other cells in the DRG, and is not found in other peripheral or central neurons or nonneuronal tissue (Sangameswaran, L. *et al.*, 1997; Akopian, A. N. *et al.*, 1999). Studies employing mutant mice lacking the $Na_V1.8$ channel or antisense knockdown techniques have shown that this channel is not essential for resting membrane potentials, neuronal excitation thresholds, or basal nociceptive thresholds (Akopian, A. N. *et al.*, 1999; Renganathan, M., *et al.*, 2001; Lai, J. *et al.*, 2003). In contrast, knockdown of $Na_V1.8$ in rats with SNL reversed behavioral signs of neuropathic pain (Porreca, F. *et al.*, 2000; Lai, J. *et al.*, 2003), which is consistent with the ability of $Na_V1.8$ to enable DRG neurons to fire repetitively after stimulation (Akopian, A. N. *et al.*, 1999; Renganathan, M., *et al.*, 2001). There is evidence that suggests that behavioral neuropathic pain may be partly mediated through an abnormal redistribution of $Na_V1.8$ after nerve injury (Lai, J. *et al.*, 2003). Recent investigations found that the μO-conotoxin MrVIB produced a selective block of the $Na_V1.8$ current (Ekberg, J. *et al.*, 2006). Intrathecal injections of MrVIB produced dose-dependent attenuation of tactile allodynia and thermal hyperalgesia in rats with peripheral nerve injury, leading to the conclusion that blockade of $Na_V1.8$ may provide one therapeutic avenue for the alleviation of neuropathic pain (Ekberg, J. *et al.*, 2006).

Interestingly, immunological and electrophysiological studies showed that SNL resulted in a downregulation of $Na_V1.8$ protein or conductance in the DRG of injured neurons but upregulation in adjacent, uninjured unmyelinated, and myelinated DRG neurons (Porreca, F. *et al.*, 1998; Gold, M. S. *et al.*, 2003). This redistribution of $Na_V1.8$ after SNL was blocked by antisense knockdown techniques (Gold, M. S. *et al.*, 2003; Lai, J. *et al.*, 2003). The injury-induced changes in immunoreactivity of $Na_V1.8$ was correlated with an increase in tetrodotoxin-resistant (TTX-R) compound action potential at C fiber conduction velocity along with a minor contribution at A fiber conduction velocity, suggesting a functional reorganization of $Na_V1.8$ along unmyelinated fibers and in some myelinated fibers (Gold, M. S. *et al.*, 2003). Antisense-mediated knockdown of $Na_V1.8$ immunoreactivity and TTX-R current in these uninjured axons correlates with the reversal of both mechanical and thermal hypersensitivity, suggesting that this reorganization of $Na_V1.8$ activity along the uninjured axons may be necessary for expression of neuropathic pain in the injured rat (Gold, M. S. *et al.*, 2003; Lai, J. *et al.*, 2003). A redistribution of $Na_V1.8$ along the injured sciatic

nerve has been also observed in the chronic constriction injury model of neuropathic pain, and $Na_V1.8$ immunoreactivity is evident in peripheral nerve tissues from patients with chronic neuropathic pain (Novakovic, S. D. *et al.*, 1998; Coward, K. *et al.*, 2000). Immunohistochemical studies of human tooth pulp obtained from painful and nonpainful teeth demonstrated that fine unmyelinated nerve fibers expressed $Na_V1.8$ (Renton, T. *et al.*, 2005). In addition, there was an increased expression of $Na_V1.8$ in nerve fibers obtained from painful teeth (Renton, T. *et al.*, 2005).

$Na_V1.7$, a TTX-S VGSC is found predominantly in small-diameter sensory DRG neurons and in sympathetic nerves and plays a prominent role in enhanced pain states (Wood, J. N. *et al.*, 2004b). Recent investigations confirm that this channel mediates the intense burning pain associated with familial erythermalgia. Primary erythermalgia is a rare, autosomal-dominant condition that manifests itself by intermittent bouts of swelling of the feet or hands, intense burning pain, and erythemia (Yang, Y. *et al.*, 2004). The syndrome may be provoked by warmth or exercise and is refractory to therapeutic interventions (Yang, Y. *et al.*, 2004). It was found that individuals afflicted with erythromelalgia expressed a T to A transversion mutation in the gene *SCN9A*, which codes the alpha-subunit of the $Na_V1.7$ sodium channel (Yang, Y. *et al.*, 2004). These mutations of the $Na_V1.7$ sodium channel were found to promote abnormal activity of small-diameter nociceptors, which would be consistent with a small-fiber neuropathy (Drenth, J. P. *et al.*, 2005). Electrophysiologic studies demonstrated that the mutated channel produced a hyperpolarizing shift in activation, slowed deactivation of the channel and markedly increased the ramp current, with the net effect being a marked hyperexcitation of the cells (Dib-Hajj, S. D. *et al.*, 2005). The $Na_V1.7$ sodium channel is upregulated after carrageenan-induced inflammation, and genetically modified mice with knockdown of the $Na_V1.7$ channel of sensory neurons showed an attenuated inflammation-induced hyperalgesia (Wood, J. N. *et al.*, 2004a). More recently, voltage clamp studies performed on transfected mouse DRG cells showed that the human mutations lowered the thresholds for generation of action potentials and repetitive firing, thus provoking high-frequency firing of nociceptive small-diameter sensory neurons in the DRG (Dib-Hajj, S. D. *et al.*, 2005). It is now believed that enhanced sensitivity of this sodium channel may exacerbate activation of the TRPV1 channels, thus driving nociceptor hyperexcitability.

More recent studies were undertaken with patients with the autosomal dominant spontaneous pain condition, paroxysmal extreme pain disorder (PEPD), which manifests as paroxysmal episodes of burning pain in ocular, mandibular, and rectal regions. Mutational analysis of SCN9A obtained from individuals with PEPD and unafflicted family members revealed that the mutation was associated with enhanced $Na_V1.7$ current activity (Fertleman, C. R. et al., 2006). Although primary erythermalgia and PEPD are due to mutations of *SCN9A*, these familial disorders are considered allelic variants that are clinically and mechanistically distinct, related to the specific effects of the mutations on the sodium channel activity (Fertleman, C. R. et al., 2006). On the other side of the spectrum, individuals with one of three distinct nonsense mutations of *SCN9A* show a loss of function of the $Na_V1.7$ channel activity show a marked insensitivity to pain (Cox, J. J. et al., 2006). This sodium channel appears to be critical to nociceptive processing, both acutely and in chronic pain states (Cox, J. J. et al., 2006; Fertleman, C. R. et al., 2006). In spite of the evidence pointing to a prominent role of voltage-gated sodium channels in nociception and enhanced pain due to nerve injury, therapeutic interventions based on blockade of the sodium channels have, to this point, been disappointing. A large multicenter clinical investigation that was recently concluded showed that lamotrigine, which is a nonselective sodium channel blocker, failed to produce changes that were different from placebo in patients with diabetic neuropathy (Vinik, A. I. et al., 2007). A multicenter clinical investigation with oxcarbazepine for diabetic neuropathy suggested that this nonselective sodium channel blocker produced a moderate effect that may be clinically meaningful (Dogra, S. et al., 2005). Though currently unclear, one conclusion is that sodium channel blockers may not be suitable for routine use against neuropathic pain, but may be appropriate for responsive subgroups of neuropathic pain patients (Sindrup, S. H. and Jensen, T. S., 2007). One such example may be patients with postherpetic neuralgia mediated by irritable nociceptors (Sindrup, S. H. and Jensen, T. S., 2007).

55.2.5 Role of Calcium Channels

Calcium channels also are critical elements that modulate neuronal excitation (Snutch, T. P. et al., 2001). The N-type calcium channel, which is blocked by omega-conotoxin, has been identified in DRG neurons (Nowycky, M. C. et al., 1985; McCleskey, E. W. et al.,

1987). This channel is predominant on nerve terminals of primary afferent terminals in the superficial lamina of the spinal dorsal horn and participates in nociceptive transmission since it is critical to depolarization-coupled release of neurotransmitters, including substance P, CGRP, and glutamate, from these terminals (Santicioli, P. et al., 1992). It was suggested that nerve injury results in either increased frequency of opening of the N-type calcium channel, or an increase in the population (see Snutch, T. P. et al., 2001; Snutch T. P., 2005, for reviews). Blocking the activity of the N-type calcium channel with the omega-conotoxin–GVIA abolished enhanced electrically evoked responses of dorsal horn neurons as well as windup and post-discharge phenomena in nerve-injured rats, which is consistent with inhibition of signs of enhanced pain (Matthews, E. A. and Dickenson, A. H., 2001b). The spinal administration of a novel N-type calcium channel blocker ziconitide produced antinociception in uninjured rats and abolished tactile and thermal hyperesthesias in those with peripheral nerve injuries (Scott, D. A. et al., 2002; Wang, Y. X. et al., 2000a; 2000b). Recent clinical trials with ziconotide (PRIALT) provide strong evidence that blockade of the N-type calcium channel may provide a significant clinical benefit in the management of severe, refractory pain, such as deafferentation pain or pain due to cancer or neuropathies associated with HIV/AIDS and postherpetic neuralgia (Wang, Y. X., et al., 2000a; 2000b; Miljanich, G. P., 2004; Prommer, E. E., 2005).

The T-type calcium channel is present in DRG neurons and on second-order neurons of the dorsal horn of the spinal cord (Talley, E. M. et al., 1999; see Hildebrand, M. E. and Snutch, T. P., 2006, for a review). Activation of the NK1 receptor and of the T-type voltage-gated calcium channels (VGCCs) acts synergistically with activation of the NMDA/calcium channel complex to facilitate firing of the NK1-expressing cells of lamina I, thus promoting central sensitization (Ikeda, H. et al., 2003). This observation is consistent with observations that blockade of the T-type VGCC produced dose-dependent reversal of behavioral signs of neuropathic pain in rats with peripheral nerve injury (Dogrul, A. et al., 2003). Moreover, pharmacologic blockade of T-type VGCCs also diminished electrophysiologic signs of postsynaptic sensitization indicated by windup and after-discharges in nerve-injured rats (Matthews, E. A. and Dickenson, A. H., 2001a). The role of this VGCC in pain modulation makes the T-type calcium channel an attractive target for drug development. Although several compounds are under preclinical investigation as T-type channel blockers, no clinical data exist with which to gauge their clinical utility against neuropathic pain (Hildebrand, M. E. and Snutch, T. P., 2006).

55.3 Enhanced Afferent Discharges Lead to Central Sensitization in the Spinal Cord

The pronociceptive changes that occur in peripheral afferent fibers after nerve injury promote enhanced activity of these fibers, and this afferent drive is thought to result in hyperalgesia and allodynia (Amir, R. et al., 2002). However, although although onset of neuropathic pain is consistent with the initiation of the increased afferent discharges, the enhanced activity of peripheral fibers diminishes fairly rapidly, whereas the behavioral signs of enhanced pain persist long after the initial insult, indicating that mechanisms which sustain neuropathic pain may differ from those which initiate such pain (Liu, C. N. et al., 2001; Ossipov, M. H. et al., 2005; Sun, Q. et al., 2005). There is general agreement that these behavioral manifestations of enhanced pain are mediated in large part by an enhanced responsiveness of the spinal cord to sensory inputs, or central sensitization, which may be both initiated, and in part, sustained by the primary afferent barrage. Sensitization of second-order neurons of the spinal dorsal horn has been illustrated experimentally by the phenomena of windup. Windup is characterized by progressively increasing responses of spinal dorsal horn neurons in response to repetitive electrical stimulation of C fibers (Mendell, L. M., 1966; Woolf, C. J., 1996). Windup is relatively short term in duration and occurs only in response to stimulation of C fibers, and, in the uninjured state, not of A-beta fibers, and is nociceptive specific (Woolf, C. J., 1996; Ji, R. R. et al., 2003; Ji, R. R. and Strichartz, G., 2004). Conditioning stimuli in the form of electrically evoked trains of C fiber activity has been found to produce long-term responsiveness of dorsal horn units to additional stimuli, a profile that appears similar to long-term potentiation (LTP) (Woolf, C. J., 1996; Ji, R. R. et al., 2003; Ji, R. R. and Strichartz, G., 2004). Although these dorsal horn neurons are unreactive to A-beta fiber inputs in the resting state, they become sensitized through repetitive C fiber stimulation which can render them excitable by A-beta fiber inputs, and this may represent a mechanism which contributes to touch-evoked allodynia (Woolf, C. J., 1996; Ji, R. R. et al., 2003; Ji, R. R. and Strichartz, G., 2004).

The state of central sensitization is generally characterized by electrophysiologic parameters that include a reduction in sensory thresholds, increased responsiveness of second-order neurons to afferent inputs along with after-discharges, and expansion of the receptive field along with recruitment of afferent fibers that are not normally nociceptive (Cook, A. J. et al., 1987). Behaviorally, a reduction in sensory threshold is demonstrated by the significantly reduced withdrawal latencies to noxious thermal stimuli or reduced thresholds to noxious mechanical stimuli (Woolf, C. J. and Thompson, S. W., 1991; Hedo, G. et al., 1999). In addition, light tactile stimuli that do not elicit behavioral responses in normal animals produce nocifensive behaviors in the sensitized state (Woolf, C. J. and Thompson, S. W., 1991; Hedo, G. et al., 1999; Ossipov, M. H. and Porreca, F., 2005). For example, sensitization induced by the intradermal injection of capsaicin reduced pain thresholds to noxious heat and elicited allodynia to light brush in human volunteers (Simone, D. A. et al., 1989). Similarly, intradermal injection of capsaicin reduced the stimulus threshold required to elicit responses of the spinothalamic tract (STT) neurons to light brush and also increased the response intensity of these second-order neurons (Simone, D. A. et al., 1991). Peripheral nerve injury was shown to increase the spontaneous activity of dorsal horn neurons and to enhance the responses of these neurons to noxious heat or light touch applied to the injured hindpaw (Suzuki, R. and Dickenson, A., 2005a; 2005b). Moreover, nerve injury also provoked pronounced after-discharges and significantly enlarged the receptive field of dorsal horn neurons in response to noxious heat or pinch (Suzuki, R. et al., 2004a; Suzuki, R. and Dickenson, A., 2005a; 2005b). Peripheral nerve injury also caused a shift to the left of the stimulus–response function of wide dynamic range (WDR) neurons in response to tactile, but not thermal, stimuli (Kauppila, T. et al., 1998; Pertovaara, A. et al., 2001).

In addition to electrophysiologic means, immunohistochemical detection of the proto-oncogene product FOS in the spinal cord has been widely employed as an indicator for neurons that are activated by noxious inputs, and its enhanced expression in the spinal dorsal horn is suggestive of sensitization (Hunt, S. P. et al., 1987; Shortland, P. and Molander, C., 1998). Peripheral nerve injury produced significant increases in noxious and touch-evoked FOS expression in the superficial and intermediate laminae of the spinal dorsal horn (Catheline, G. et al., 1999; Vera-Portocarrero, L. P. et al., 2006; Zhang, E. T. et al., 2007). Electrical stimulation at A-beta fiber intensity or application of light touch evoked FOS expression, but only after nerve injury or inflammation (Ma, Q. P. and Woolf, C. J., 1996; Shortland, P. and Molander, C., 1998; Nomura, H. *et al.*, 2002). Recently, gentle stroking of the hindpaw, which did not evoke FOS expression in normal rats, but caused significant increases in FOS expression after nerve injury (Catheline, G. *et al.*, 1999; Zhang, E. T. *et al.*, 2007). The enhanced expression of FOS was found in the superficial and deep laminae of the spinal dorsal horn (Zhang, E. T. *et al.*, 2007). Together, these studies indicate that peripheral nerve injury can promote a state of sensitization of spinal dorsal horn neurons.

55.4 Excitatory Transmitters Promote Central Sensitization

Sensitization of the spinal cord elicited by enhanced peripheral afferent inputs results in part from the recruitment of subthreshold stimuli to evoke a neuronal response that is related to increased release of excitatory neurotransmitters form the primary afferent terminals (Li, Q. J. and Lu, G. W., 1989; Li, J. *et al.*, 1999). Microdialysis studies demonstrated the release of glutamate and aspartate from primary afferents in response to the capsaicin injected intradermally in the plantar aspect of the hindpaw of the rat or electrical stimulation of the sciatic nerve at C fiber intensity (Paleckova, V. *et al.*, 1992; Sluka, K. A. and Willis, W. D., 1998). Similarly, behavioral hypersensitivity and central sensitization due to peripheral nerve injury or inflammation are associated with increased evoked release of glutamate and aspartate from primary afferent terminals (Paleckova, V. *et al.*, 1992; Kawamata, M. and Omote, K., 1996; Sluka, K. A. and Willis, W. D., 1998; Kawamata, M. and Omote, K., 1999). Spontaneous and stimulus-evoked release of substance P and CGRP from primary afferent terminals is increased after peripheral nerve injury (Paleckova, V. *et al.*, 1992; Sluka, K. A. and Willis, W. D., 1998; Wallin, J. and Schott, E., 2002; Gardell, L. R. *et al.*, 2003a; 2003b). Increased release of excitatory neurotransmitters, including glutamate, substance P, and CGRP is believed to contribute to central sensitization and to hyperalgesia. For example, CGRP release after intradermal capsaicin increased WDR activity in the dorsal horn in response to brush or pinch (Sun, R. Q. *et al.*, 2003; Sun, R. Q. *et al.*, 2004).

The existence of excitatory NMDA autoreceptors on central terminals of primary afferents has been described (Liu, H. *et al.*, 1997). Nociception and release of substance P and glutamate from primary

afferent terminals evoked by spinal NMDA was abolished by capsaicin, indicating that presynaptic NMDA receptors provoke further release of excitatory transmitters (Liu, H. *et al.*, 1997). Glutamate may also act at presynaptic metabotropic receptors to promote sensitization (Ohishi, H. *et al.*, 1995). Prostacyclin and PGE2 enhanced capsaicin-evoked release of substance P and CGRP through presynaptic excitation (Southall, M. D. *et al.*, 1998; Southall, M. D. and Vasko, M. R., 2001; Southall, M. D. *et al.*, 2002). These observations are consistent with the fact that central sensitization elicited by C fiber stimulation or inflammation was blocked by NMDA antagonists (Woolf, C. J. and Salter, M. W., 2000).

A measure indicative of the level of primary afferent drive into the spinal dorsal horn is the determination of internalization of the NK1 receptor on second-order neurons that are activated by substance P released from primary afferent nociceptors (Allen, B. J. *et al.*, 1999; Honore, P. *et al.*, 1999). Under nonsensitized conditions, noxious stimulation, such as electrical stimulation at C fiber intensity, subcutaneous capsaicin or noxious heat or pinch, will evoke rapid internalization of the NK1 receptor almost exclusively in the outer lamina of the spinal cord (Allen, B. J. *et al.*, 1999; Honore, P. *et al.*, 1999). Moreover, nonnoxious stimulation does not evoke internalization of NK1. However, when sensitization occurs due to inflammation or peripheral nerve injury, there is a marked enhancement of noxious-induced NK1 internalization, and NK1 internalization is found in the deeper laminae of the spinal cord as well (Allen, B. J. *et al.*, 1999; Honore, P. *et al.*, 1999). In addition, nonnoxious stimuli such as light brush also evoke NK1 internalization in the intermediate and deep laminae of the spinal dorsal horn (Allen, B. J. *et al.*, 1999; Honore, P. *et al.*, 1999; Khasabov, S. G. *et al.*, 2002). Critically, electrical stimulation at A-beta fiber intensity does not elicit NK1 internalization after nerve injury, indicating that the effect produced by light brush is not a direct result of large-diameter fiber stimulation, but is likely to be secondary to other mechanisms driving sensitization (Allen, B. J. *et al.*, 1999; Khasabov, S. G. *et al.*, 2002). NK1 internalization observed in the deeper laminae may be due in part to diffusion of substance P to NK1 receptors on sensitized deep lamina neurons, or partly due to the small number of direct synaptic contacts between C fibers and deep lamina neurons (Allen, B. J. *et al.*, 1999; Khasabov, S. G. *et al.*, 2002).

Recent studies demonstrated that the postsynaptic second-order neurons of the spinal dorsal horn express message for the bradykinin B2 receptor (Wang, H. B. *et al.*, 2005). Patch-clamp studies showed that activation of the B2 receptor produces a dose-dependent potentiation of AMPA and NMDA currents in dorsal horn neurons. Excitation of the B2 receptor mediates increased frequency and amplitude of AMPA-mediated excitatory postsynaptic potentials, whereas B2 antagonists abolished electrophysiologic signs of central sensitization in dorsal horn neurons (Wang, H. B. *et al.*, 2005). These studies suggest that spinal bradykinin B2 receptors represent an important postsynaptic mechanism in promoting central sensitization and enhanced pain through the potentiation of glutaminergic synaptic transmission (Wang, H. B. *et al.*, 2005). While the importance of bradykinin receptors in enhancing pain, including chronic neuropathic pain (see below) seem clear, the mechanism by which the receptors are activated is not known.

55.5 Descending Facilitation Promotes Enhanced Pain

The existence of a pronociceptive pain facilitatory system projecting from the RVM has been well established (see Porreca, F. *et al.*, 2002; Heinricher, M. M. *et al.*, 2003; Ossipov, M. H. and Porreca, F., 2005 for reviews). It has been demonstrated that the microinjection of lidocaine into the RVM abolished behavioral signs of enhanced pain after nerve injury or inflammation, and abolished the sensitization of spinal WDR neurons to mustard oil induced inflammation (Mansikka, H. and Pertovaara, A., 1997; Burgess, S. E. *et al.*, 2002; Pertovaara, A. and Almeida, A., 2006). Lesions of the descending projections from the RVM in the dorsolateral funiculus also abolished behavioral hypersensitivity after inflammation or nerve injury (Wei, F. *et al.*, 1998; 1999; Ossipov, M. H. *et al.*, 2000; Burgess, S. E. *et al.*, 2002). Finally, it was found that selective ablation of RVM neurons that express the mu-opioid receptor, and which may label cells which contribute to descending facilitation, also abolished behavioral signs of sensitization and the enhanced capsaicin-evoked release of CGRP (Burgess, S. E. *et al.*, 2002; Gardell, L. R. *et al.*, 2003a; 2003b; Porreca, F. *et al.*, 2002). Importantly, disruption of descending facilitation abolished behavioral hypersensitization during later, but not initial, phases after nerve injury, indicating that it is essential to the maintenance of a sensitized state (Burgess, S. E. *et al.*, 2002; Porreca, F. *et al.*, 2002;

Gardell, L. R. *et al.*, 2003a; 2003b). Most recently, it was shown that activation of descending facilitation from the RVM was subsequent to stimulation of NK1-expressing lamina I postsynaptic cells and increased the neuronal responses of WDR neurons of the deeper laminae of the dorsal horn after nerve injury or inflammation (Suzuki, R. *et al.*, 2002; 2004a; Rahman, W. *et al.*, 2004; 2006). Spinal ondansetron blocked the sensitization of these WDR neurons, which is consistent with a descending serotonergic facilitatory pathway from the RVM (Suzuki, R. *et al.*, 2002; Rahman, W. *et al.*, 2004; Suzuki, R. *et al.*, 2004a; Rahman, W. *et al.*, 2006).

55.5.1 Role of the Rostralventromedial Medulla in Descending Facilitation of Pain

The RVM and surrounding tissue has long been recognized as a region critical to nociceptive processing and receives inputs from ascending nociceptive pathways as well as from descending pain modulatory sites. The RVM is considered to be the final common relay for descending inhibition of nociceptive inputs. More recently, this region has also been demonstrated to mediate a significant descending pain facilitatory system as well (Fields, H. L. and Basbaum, A. I., 1999; Porreca, F. *et al.*, 2002; Heinricher, M. M. *et al.*, 2003; Ossipov, M. H. and Porreca, F., 2005). Stimulation of this region, either through electrical stimulating electrodes or through the microinjection of glutamate or neurotensin, resulted in biphasic effects on behavioral and electrophysiologic responses to noxious stimuli (Zhuo, M. and Gebhart, G. F., 1997; Fields, H. L. and Basbaum, A. I., 1999). Whereas high levels of stimulation inhibited nociceptive responses, lower levels of stimulation facilitated the same responses (Zhuo, M. and Gebhart, G. F., 1997; Urban, M. O. *et al.*, 1999). Similarly, microinjection of the neurotensin antagonist SR48692 into the RVM also produced a biphasic effect on nociceptive responses. Taken together, these observations indicate that the RVM is capable of mediating both pain inhibitory functions, which are likely to predominate under basal conditions, and facilitation of pain, which would predominate under conditions of enhanced pain (Ossipov, M. H. *et al.*, 2001). Pharmacologic manipulations that attenuate RVM activity also block enhanced nociception. For example, thermal hyperalgesia induced by naloxone-precipitated withdrawal or by previous prolonged exposure to a nociceptive stimulus can be blocked by the microinjection of lidocaine into the RVM (Kaplan, H. and Fields, H. L., 1991; Morgan, M. M. and Heinricher, M. M., 1992). It is likely that activation of these pronociceptive circuits in the RVM in response to persistent or prolonged noxious inputs leads to a facilitation of more pain (Porreca, F. et al., 2002). Tail withdrawal responses to noxious thermal or mechanical stimuli can be facilitated by the subcutaneous injection of formalin into a hindpaw, and this effect can be reversed by local infusion of the hydrophilic lidocaine derivative QX-314 injected at the site of formalin (Calejesan, A. A. et al., 1998). It is currently understood that descending pain facilitatory systems arising from the RVM are likely mediated through spinopetal serotonergic projections that enhance nociceptive inputs at the level of the spinal cord. The relationship among noxious inputs, the RVM, and descending serotonergic projections to chronic pain states is described in greater detail below.

55.5.2 The ON Cells of the Rostralventromedial Medulla Promote Pain Through Descending Facilitation

Neuronal populations within the RVM have been described as ON cells, OFF cells and neutral cells based upon electrophysiologic responses to noxious heat applied to the tail of lightly anesthetized rats (Fields, H. L., 2000; Ossipov, M. H. et al., 2001; Neubert, M. J. et al., 2004). By definition, the OFF cells are tonically active and pause in firing immediately before a withdrawal response from a noxious thermal stimulus, whereas the ON cells accelerate firing immediately before the nociceptive reflex occurs (Fields, H. L. and Heinricher, M. M., 1985; Neubert, M. J. et al., 2004). As the name implies, the activity of the neutral cells does not correlate with nociceptive inputs. Characterization of the neurons of the RVM were initially determined in lightly anesthetized rat preparations, and it was found that tail flick latency was longer during periods of increased OFF cell activity and shorter when the ON cells were active. The considerable body of evidence collected subsequent to these studies confirmed the role of these neurons in the modulation of nociception and extended the findings to other models of acute and enhanced pain. It is now generally accepted that the activity of OFF cells correlates with inhibition of nociceptive inputs and of behavioral nocifensive responses to nociception. In the presence of opioids, OFF cells continue firing after a noxious stimulus is applied and inhibit electrophysiologic

and behavioral parameters of nociception through descending inhibition from the RVM (Fields, H. L. and Heinricher, M. M., 1985; Neubert, M. J. et al., 2004). In contrast, the electrophysiologically determined response characteristics of the ON cells are consistent with a pronociceptive function, and it is believed that these neurons serve as the source of descending facilitation from the RVM (Fields, H. L. and Heinricher, M. M., 1985; Heinricher, M. M. et al., 2003; Neubert, M. J. et al., 2004). Accordingly, manipulations that promote nociceptive responsiveness also increase ON cell activity, which is consistent with pain facilitation.

The enhanced behavioral responses to noxious stimuli elicited during naloxone-precipitated withdrawal from morphine correlates with increased spontaneous activity of the RVM ON cells, and both behavioral hyperalgesia and enhanced on cell activity are abolished by lidocaine microinjected into the RVM (Bederson, J. B. et al., 1990; Kaplan, H. and Fields, H. L., 1991). Prolonged application of a noxious thermal stimulus increased ON cell activity and facilitated behavioral nociceptive responses, and both the electrophysiologic and behavioral responses were abolished by microinjection of lidocaine into the RVM (Kaplan, H. and Fields, H. L., 1991; Morgan, M. M. and Fields, H. L., 1994). Electrophysiologic monitoring of RVM neurons demonstrated that the microinjection of either cholecystokinin (CCK) or neurotensin at doses that selectively activated on cells produced facilitated behavioral responses to nociceptive stimuli in lightly anesthetized rats (Heinricher, M. M. and Neubert, M. J., 2004; Neubert, M. J. et al., 2004). Application of mustard oil to the skin of the rat produced increased firing of ON cells and thermal hyperalgesia that were both blocked by NMDA antagonist application into the RVM (Kincaid, W. et al., 2006; Xu, M. et al., 2007). It was also shown that inflammation induces an upregulation of NMDA receptors on the ON cells, along with increased phosphorylation of AMPA receptors, thus enhancing ON cell activity and facilitating nociceptive responses (Guan, Y. et al., 2003; 2004). Finally, recent studies showed that inflammation results in upregulation of BDNF, which facilitates ON cell firing through an NMDA-dependent mechanism (Guo, W. et al., 2006). Most recently, it was discovered that capsaicin injected into a hindpaw or iontophoretic application of $Sar^9,Met(O_2)^{11}$-substance P into the RVM enhanced ON cell activity and potentiated the responses of ON cells to iontophoretic NMDA (Budai, D. et al., 2007). Conversely, iontophoretic administration of an NK1 antagonist into the RVM abolished the enhanced responses of ON cells (Budai, D. *et al.*, 2007). Taken together, these studies indicate that ON cells promote nociception, and that enhanced ON cell activity underlies hyperalgesia associated with chronic pain states. In addition, it is suggested that substance P in the RVM may modulate enhanced ON cell activity, and consequently, enhanced pain states.

In light of these observations, it has been proposed that neuropathic pain due to peripheral nerve injury may be initially mediated through enhanced afferent activity, but that the enhanced discharges lead to a sensitized spinal cord and neuroplastic changes culminating in an enhancement of descending facilitation mediated through ON cells of the RVM (Heinricher, M. M. *et al.*, 2003; Ossipov, M. H. *et al.*, 2005; Ossipov, M. H. and Porreca, F., 2006). Considerable electrophysiologic evidence exists to indicate that the population of RVM neurons that express the mu-opioid receptor, or a subset of these cells, are likely to drive descending facilitation (Heinricher, M. M. *et al.*, 1994; Neubert, M. J. *et al.*, 2004). Based on this evidence, presumptive pain facilitatory neurons of the RVM were selectively ablated by using the mu-opioid receptor as a portal of entry into these cells for dermorphin conjugated with the cytotoxin saporin (Porreca, F. *et al.*, 2001; Burgess, S. E. *et al.*, 2002; Vera-Portocarrero, L. P. *et al.*, 2006). With this technique, when the agonist binds to the receptor, internalization occurs, bringing the conjugated agonist and saporin complex into the cell body, where saporin is released and leads to inhibition of ribosome activity, thus causing cell death within 2–3 weeks (Mantyh, P. W. *et al.*, 1997). Selective ablation of pain facilitatory neurons of the RVM by microinjection of the dermorphin–saporin conjugate either 7 days prior to SNL or once tactile and thermal hyperesthesias were well established, respectively, prevented and reversed the behavioral signs of neuropathic pain (Porreca, F. *et al.*, 2001; Burgess, S. E. *et al.*, 2002). Importantly, these behavioral signs of neuropathic pain were abolished 1 week after SNL, and not earlier. Similarly, physical disruption of descending facilitatory tracts from the RVM or microinjection of lidocaine at different time points after SNL produced similar results (Porreca, F. *et al.*, 2001; Burgess, S. E. *et al.*, 2002). Enhanced expression of FOS in the spinal cord is indicative of sensitization. Recently, it was shown that SNL enhanced both the intensity and laminar distribution of FOS evoked by noxious stimuli to include the intermediate laminae of the dorsal horn and promoted novel expression of FOS in response to light brush (Vera-Portocarrero, L. P. *et al.*, 2006; Zhang, E. T. *et al.*, 2007). Microinjection of dermorphin–saporin conjugate into the RVM abolished enhanced FOS expression due to SNL (Vera-Portocarrero, L. P. *et al.*, 2006). These observations indicate that the maintenance phase of the neuropathic pain state is dependent on the certain neuroplastic changes that result in activation of a tonic descending pain facilitation from the RVM, whereas the earlier initiation phase is independent of descending systems.

55.5.3 Cholecystokinin in the Rostralventromedial Medulla May Mediate Pain Facilitation

It was recently shown that CCK opposes the antinociceptive activity of morphine by attenuating the morphine-induced activation of OFF cells, which constitute part of the descending inhibition of nociceptive inputs (Heinricher, M. M. *et al.*, 2001; Heinricher, M. M. and Neubert, M. J., 2004). The microinjection of CCK into the RVM produces enhancement of responses to light tactile or noxious thermal stimuli similar to that observed after SNL (Kovelowski, C. J. *et al.*, 2000; Xie, J. Y. *et al.*, 2005). Moreover, the microinjection of CCK2 antagonists into the RVM attenuated behavioral signs of neuropathic pain in rats with SNL (Kovelowski, C. J. *et al.*, 2000; Xie, J. Y. *et al.*, 2005). The behavioral hyperesthesias and hyperalgesia elicited by RVM CCK were abolished by dorsolateral funiculus (DLF) lesions or microinjection of L365,260 (Kovelowski, C. J. *et al.*, 2000; Xie, J. Y. *et al.*, 2005). In addition to its activities in inhibiting morphine-induced activation of OFF cells, it was also recently demonstrated that CCK provokes a direct and selective activation of ON cells of the RVM and that this activation is directly related to the production of hyperalgesia (Heinricher, M. M. and Neubert, M. J., 2004).

55.6 Second-Order Projection Neurons Expressing NK1 Receptors May Result in Activation of Descending Pain Facilitation

Converging evidence strongly indicates that central sensitization and hypersensitivity to sensory stimuli is dependent predominantly on the long-term sensitization of the lamina I cells that express the NK1 receptor (Ikeda, H. *et al.*, 2003). Electrophysiologic studies have revealed that approximately 75% of the neurons of lamina I are nociceptive specific, and the remainder are either WDR neurons or polymodal nociceptors that also respond to noxious cold (Suzuki, R. and Dickenson, A. H., 2005a; 2005b). Approximately one-half of the lamina I neurons express the NK1 receptor and project to supraspinal sites that process nociception, including the parabrachial region and the thalamus (Suzuki, R. and Dickenson, A. H., 2005a; 2005b). Recent studies showed that selective ablation of lamina I neurons that express the NK1 receptor abolish behavioral signs of neuropathic pain in rats with SNL (Nichols, M. L. *et al.*, 1999; Honore, P. *et al.*, 1999; Suzuki, R. *et al.*, 2002). In these studies, ablation was performed by spinal injection of substance P conjugated to saporin. Internalization of the NK1 receptor stimulated by substance P introduced the conjugate into the cell body, where saporin interfered with protein synthesis, causing eventual death of the cell (Honore, P. *et al.*, 1999; Nichols, M. L. *et al.*, 1999). Electrophysiologic studies indicated that signs of spinal sensitization induced by intraplantar capsaicin or peripheral inflammation were abolished by selective ablation of the NK1-expressing neurons of lamina I (Khasabov, S. G. *et al.*, 2002; 2005). Additional electrophysiologic studies demonstrated that SNL-induced enhancement of neuronal responses to noxious thermal stimuli or light brush, neuronal after-discharge, and enlargement of the receptive field were abolished in rats with SNL and in which NK1-expressing spinal neurons were selectively ablated with spinal injections of substance P conjugated to saporin (Suzuki, R. *et al.*, 2004a; 2004b; Suzuki, R. and Dickenson, A. H., 2005a; 2005b). Moreover, evoked expression of FOS in the RVM and in neurons in the deep laminae of the spinal cord was also abolished by selective ablation of NK1-expressing neurons of lamina I (Suzuki, R. *et al.*, 2002; Suzuki, R. and Dickenson, A. H., 2005a; 2005b). Based on these observations, especially the critical finding that the projection neurons of the superficial lamina of the spinal cord appear to promote sensitization of dorsal horn neurons in the deep lamina through an indirect mechanism, led to the hypothesis that injury-induced enhanced noxious inputs are transmitted to supraspinal sites, leading to the activation of a pronociceptive descending pain facilitatory system that maintains the state of spinal sensitization and maintains the neuropathic pain state.

55.7 Descending Facilitation Maintains a Sensitized Spinal Cord: Upregulation of Spinal Dynorphin and Enhanced Release of Primary Afferent Transmitters

The enhancement of descending facilitation from the RVM is believed to lead to neuroplastic changes in the spinal cord that favor a sensitized state and enhanced nociceptive inputs. A number of studies have demonstrated that enhanced descending facilitation elicited by peripheral nerve injury is associated with an upregulation of spinal dynorphin content and enhanced release of excitatory neurotransmitters from primary afferent neurons. Based on these observations, we formulated the hypothesis that neuropathic pain is initially driven by enhanced primary afferent the ectopic discharge but is then maintained in large part through descending facilitation and upregulation of spinal dynorphin (Lai, J. *et al.*, 2001; Porreca, F. *et al.*, 2002; Ossipov, M. H. and Porreca, F., 2006). Dynorphin upregulation occurs in the same postsynaptic lamina I cells that also express FOS in response to sensitization (Ji, R. R., 2004; Kawasaki, Y. *et al.*, 2004). Elevated levels of spinal dynorphin enhance capsaicin-evoked release of CGRP and substance P from primary afferent terminals and of PGE2, presumably from postsynaptic cells, which further enhances the release of excitatory amino acids from afferent terminals (Gardell, L. R. *et al.*, 2003a; 2003b; Koetzner, L. *et al.*, 2004). Sequestration of dynorphin with antiserum or surgical and neurochemical manipulations that blocked the upregulation of spinal dynorphin also abolished enhanced transmitter release and behavioral neuropathic pain (Burgess, S. E. *et al.*, 2002; Porreca, F. *et al.*, 2002; Gardell, L. R. *et al.*, 2003a; 2003b).

It is well established that elevated levels of spinal dynorphin and its fragments have a prominent pronociceptive role in chronic pain states (Lai, J. *et al.*, 2001). The spinal injection of antiserum to dynorphin did not abolish behavioral signs of neuropathic pain in mice with SNL when administered 2 days after the injury, but was effective 10 days after SNL (Wang, Z. *et al.*, 2001; Gardell, L. R. *et al.*, 2004). Moreover, knockout mice that do not express dynorphin still develop behavioral signs of neuropathic pain after SNL, but these signs resolve within 1 week (Wang, Z. *et al.*, 2001; Gardell, L. R. *et al.*, 2004). The dynorphin-dependent maintenance phase is dependent on descending facilitation form the RVM. Surgical and pharmacologic manipulations that disrupt descending facilitation abolish both dynorphin upregulation and the maintenance, but not the initiation, of behavioral signs of neuropathic pain. Rats with SNL and either lesions of the DLF or selective ablation of facilitatory RVM neurons after microinjection of the dermorphin–saporin conjugate did not show an upregulation of spinal dynorphin levels and the behavioral signs of neuropathic pain resolved over the time course during which dynorphin upregulation would be expected after SNL (Burgess, S. E. *et al.*, 2002; Gardell, L. R. *et al.*, 2003a). Thus, in the absence of elevated spinal dynorphin, neuropathic pain is not maintained.

Elevated levels of spinal dynorphin promote the release of transmitters from primary afferent terminals, thereby enhancing nociceptive input. The time course and profile of the effect of spinal dynorphin on capsaicin-evoked release of CGRP from primary afferent terminals correlates closely with the behavioral studies. Addition of antiserum to dynorphin abolishes the enhanced capsaicin-evoked release of CGRP 10 days after SNL, but not 2 days after SNL (Burgess, S. E. *et al.*, 2002; Gardell, L. R. *et al.*, 2003a). Disruption of descending facilitation by microinjection of dermorphin–saporin conjugate into the RVM or by lesioning the DLF also abolishes SNL-induced enhancement of capsaicin-evoked release of CGRP (Burgess, S. E. *et al.*, 2002; Gardell, L. R. *et al.*, 2003a). The significance of this line of investigation requires additional examination, however.

55.7.1 Descending Facilitation Maintains a Sensitized Spinal Cord: Upregulation of Spinal Dynorphin and Enhanced Release of Primary Afferent Transmitters

The mechanism through which dynorphin may exert a pronociceptive function in the spinal cord is not fully understood. It was once suggested that dynorphin may act at the glycine coagonist site to potentiate the effect of glutamate at the NMDA receptor (Zhang, L. *et al.*, 1997). However, other studies provided evidence that dynorphin could inhibit opening of ion channels mediated by the NMDA receptor (Chen, L. and Huang, L. Y., 1998). Bakalkin and coworkers suggested that dynorphin could promote the formation of pores through the cell membrane, allowing the leakage of sodium and calcium ions, thus sensitizing neurons in this fashion, but this mechanism was not explicitly demonstrated (Hugonin, L. *et al.*, 2006). The mechanism by which dynorphin may promote nociception has

recently been suggested to be mediated through bradykinin receptors (Lai, J. *et al.*, 2006). Bradykinin B1 and B2 receptors are present on nerve terminals of primary afferent fibers (Botticelli, L. J. *et al.*, 1981; Cloutier, F. *et al.*, 2002), and are in close proximity to interneurons that express dynorphin (Botticelli, L. J. *et al.*, 1981). Moreover, the B1 and B2 receptors are upregulated after SNL (Lai, J. *et al.*, 2006). A pronociceptive role of these receptors is indicated since intrathecal administration of bradykinin produced behavioral signs of hyperalgesia (Wang, H. B. *et al.*, 2005). Behavioral signs of enhanced pain after nerve injury were attenuated in B1 or B2 knockout mice (Ferreira, J. *et al.*, 2002; Wang, H. B. *et al.*, 2005). Curiously, although the bradykinin receptors are present in the spinal cord, there is little evidence of bradykinin synthesis within the spinal cord, suggesting the either bradykinin must diffuse from the circulatory system to act at the receptors, or there may be another endogenous ligand (Lai, J. *et al.*, 2006). It was shown that dynorphin induced a transient increase in calcium conductance in embryonic DRG cells that was blocked by B2 antagonists (Lai, J. *et al.*, 2006). Dynorphin-induced calcium currents were also blocked by B1 or B2 antagonists in transfected cells expressing these receptors (Lai, J. *et al.*, 2006). In the same study, it was shown that dynorphin demonstrated specific binding to the B1 and B2 receptors (Lai, J. *et al.*, 2006). Finally, antagonists of the B1 or B2 receptors given intrathecally produced dose-dependent attenuation of behavioral parameters indicative of neuropathic pain in rats with SNL, but only at the time points when spinal dynorphin was elevated (Lai, J. *et al.*, 2006). The antagonists were without effect during the dynorphin-independent early phase of nerve injury. Taken together, these studies indicate that dynorphin may mediate its pronociceptive enhancement of neuropathic pain through actions at the bradykinin receptors. The activity of dynorphin at bradykinin receptors could underlie the sensitization of dorsal horn neurons as described above (Wang, H. B. *et al.*, 2005).

55.7.2 Descending Facilitation Maintains a Sensitized Spinal Cord: Serotonergic Contributions

More recent studies suggest that descending serotonergic projections from the RVM may promote spinal sensitization through an action at the 5-HT$_3$ receptor. The 5-HT$_3$ receptor facilitates release or transmitter from nerve terminals through a ligand-gated Ca^{2+} channel (Miquel, M. C. *et al.*,

2002; Zeitz, K. P. *et al.*, 2002). This receptor is predominant in the superficial laminae of the spinal dorsal horn (Doucet, E. *et al.*, 2000; Miquel, M. C. *et al.*, 2002). Of this population, the major portion, approximately 85%, are found on primary afferent nerve terminals and the remainder at postsynaptic sites (Doucet, E. *et al.*, 2000; Miquel, M. C. *et al.*, 2002). Electrophysiologic studies indicated that primary afferent nerves that express the 5-HT$_3$ receptor on their terminals include C fiber nociceptors and myelinated A-delta nociceptors sensitive to mechanical stimuli (Zeitz, K. P. *et al.*, 2002). This distribution is consistent with the modulation of mechanical-evoked nociceptive signals, and the 5-HT$_3$ receptor has been shown to facilitate nociceptive transmission at the level of the spinal cord (Zeitz, K. P. *et al.*, 2002). Double-labeling studies in the nucleus raphe magnus (NRM) detected a population of serotonergic neurons that express FOS in response to noxious stimulation, suggesting the existence of a pronociceptive serotonergic projection from the RVM (Suzuki, R. *et al.*, 2002). Moreover, it was found that serotonergic terminals from the RVM are juxtaposed with cell bodies in the intermediate laminae of the dorsal horn, some of which also express the NK1 receptor (Todd, A. J. *et al.*, 2000; Todd, A. J., 2002). Antagonists of the 5-HT$_3$ receptor or depletion of serotonin from the spinal cord abolished the enhanced responses of dorsal horn neurons to mechanical stimulation in rats with SNL (Rahman, W. *et al.*, 2006). The 5-HT$_3$ antagonist ondansetron abolished behavioral signs of neuropathic pain and electrophysiologic parameters indicative of sensitization of superficial and deep dorsal horn neurons in rats with SNL (Suzuki, R. *et al.*, 2004a; 2004b). The effects of ondansetron were blocked by disrupting enhanced RVM activity through the ablation of NK1-expressing dorsal horn neurons with substance P–saporin conjugate, indicating that the 5-HT$_3$ receptors were activated by a descending serotonergic pain facilitatory system from the RVM (Suzuki, R.*et al.*, 2004a; 2004b).

Collectively, these observations paint a picture suggesting that neuropathic pain states are maintained, at least in part, by increased sensitivity of primary afferent neurons to noxious stimuli. In addition, it is hypothesized that persistent nociceptive inputs elicits supraspinal neuroplastic changes that results in the activation of a tonic descending facilitation of nociceptive inputs, which is mediated by an upregulation of spinal dynorphin content. Thus, increased spinal dynorphin serves to promote excitatory transmitter release, which in turn promotes the maintenance of the tonic descending facilitatory system, thus perpetuating a chronic pain state. The existence of a pronociceptive spinal–supraspinal loop that maintains a sensitized spinal cord may also provide a mechanism through which tactile hyperesthesia might be mediated.

55.8 Summary

Tremendous progress in our understanding of mechanisms that initiate and maintain neuropathic pain has been made over the past two decades. Importantly, such increased understanding has led to programs aimed at the development of novel therapeutics for neuropathic pain. It has been well appreciated that peripheral nerve injury produces increased activity of the injured and adjacent uninjured peripheral nerve fibers. This enhanced activity, or peripheral sensitization, is maintained in part by increased or relocation of ion channels that promote axonal firing. Programs are in place in the pharmaceutical industry to develop subtype selective sodium channel blockers that may diminish enhance activities of injured or adjacent nerves. Increase in pronociceptive mediators, including cytokines, chemokines, and growth factors have been noted in multiple experimental models of neuropathic pain. The enhanced primary afferent activity leads to a state of sensitization of the spinal cord neurons, such that they are more responsive to sensory inputs. Approaches aimed at decreasing sensitization of nociceptors through anti-NGF strategies, or through the administration of exogenous growth factors of the GDNF family, are either in development or under consideration by the pharmaceutical industry. Enhanced activity of sensitized nociceptors can increase ascending sensory inputs to supraspinal centers and neuroplastic changes that help maintain a sensitized state. One prominent change is increased availability of CCK in the RVM and enhanced descending facilitation of pain from this region. The enhanced pain facilitation leads to further changes in the spinal cord, notably upregulation of spinal dynorphin to pathological levels. Dynorphin, along with pronociceptive transmitters such as PGEs and glutamate, enhance the release of excitatory transmitters from primary afferent terminals. The result is a net increase in nociceptive inputs that help perpetuate this pronociceptive spinal/supraspinal/spinal pain facilitatory loop. Such knowledge underlies efforts to development of N-type calcium channel blockers. Understanding the changes that occur at each of these nodes in this cycle could lead to means through which this self-perpetuating chain of events may be broken, thus abolishing the mechanisms that perpetuate chronic pain. The accelerating understanding of pain regulatory systems advanced by the discovery of underlying mechanisms driving enhanced abnormal pain will be continuously validated by clinical trials of drugs with novel mechanisms of action. Critically, new mechanisms of action will be tested clinically with the expectation that improved therapy can be provided to patients.

References

Akopian, A. N., Souslova, V., England, S., Okuse, K., Ogata, N., Ure, J., Smith, A., Kerr, B. J., McMahon, S. B., Boyce, S., Hill, R., Stanfa, L. C., Dickenson, A. H., and Wood, J. N. 1999. The tetrodotoxin-resistant sodium channel SNS has a specialized function in pain pathways. Nat. Neurosci. 2, 541–548.

Ali, Z., Ringkamp, M., Hartke, T. V., Chien, H. F., Flavahan, N. A., Campbell, J. N., and Meyer, R. A. 1999. Uninjured C-fiber nociceptors develop spontaneous activity and alpha-adrenergic sensitivity following L6 spinal nerve ligation in monkey. J. Neurophysiol. 81, 455–466.

Allen, B. J., Li, J., Menning, P. M., Rogers, S. D., Ghilardi, J., Mantyh, P. W., and Simone, D. A. 1999. Primary afferent fibers that contribute to increased substance P receptor internalization in the spinal cord after injury. J. Neurophysiol. 81, 1379–1390.

Amir, R., Michaelis, M., and Devor, M. 2002. Burst discharge in primary sensory neurons: triggered by subthreshold oscillations, maintained by depolarizing afterpotentials. J. Neurosci. 22, 1187–1198.

Apfel, S. C. 2000. Neurotrophic factors and pain. Clin. J. Pain 16, 7–11.

Attal, N. and Bouhassira, D. 1999. Mechanisms of pain in peripheral neuropathy. Acta Neurol. Scand. Suppl. 173, 12–24.

Averill, S., Michael, G. J., Shortland, P. J., Leavesley, R. C., King, V. R., Bradbury, E. J., McMahon, S. B., and Priestley, J. V. 2004. NGF and GDNF ameliorate the increase in ATF3 expression which occurs in dorsal root ganglion cells in response to peripheral nerve injury. Eur. J. Neurosci. 19, 1437–1445.

Backonja, M. M. and Stacey, B. 2004. Neuropathic pain symptoms relative to overall pain rating. J. Pain 5, 491–497.

Banik, R. K., Subieta, A. R., Wu, C., and Brennan, T. J. 2005. Increased nerve growth factor after rat plantar incision contributes to guarding behavior and heat hyperalgesia. Pain 117, 68–76.

Bederson, J. B., Fields, H. L., and Barbaro, N. M. 1990. Hyperalgesia during naloxone-precipitated withdrawal from morphine is associated with increased on-cell activity in the rostral ventromedial medulla. Somatosens. Mot. Res. 7, 185–203.

Bennett, M. I., Attal, N., Backonja, M. M., Baron, R., Bouhassira, D., Freynhagen, R., Scholz, J., Tolle, T. R., Wittchen, H. U., and Jensen, T. S. 2007. Using screening tools to identify neuropathic pain. Pain 127, 199–203.

Bennett, D. L., Boucher, T. J., Armanini, M. P., Poulsen, K. T., Michael, G. J., Priestley, J. V., Phillips, H. S., McMahon, S. B., and Shelton, D. L. 2000. The glial cell line-derived neurotrophic factor family receptor components are differentially regulated within sensory neurons after nerve injury. J. Neurosci. 20, 427–437.

Bennett, D. L., Koltzenburg, M., Priestley, J. V., Shelton, D. L., and McMahon, S. B. 1998. Endogenous nerve growth factor regulates the sensitivity of nociceptors in the adult rat. Eur. J. Neurosci. 10, 1282–1291.

Botticelli, L. J., Cox, B. M., and Goldstein, A. 1981. Immunoreactive dynorphin in mammalian spinal cord and dorsal root ganglia. Proc. Natl. Acad. Sci. U. S. A. 78, 7783–7786.

Boucher, T. J. and McMahon, S. B. 2001. Neurotrophic factors and neuropathic pain. Curr. Opin. Pharmacol. 1, 66–72.

Boucher, T. J., Okuse, K., Bennett, D. L., Munson, J. B., Wood, J. N., and McMahon, S. B. 2000. Potent analgesic effects of GDNF in neuropathic pain states. Science 290, 124–127.

Bouhassira, D., Attal, N., Alchaar, H., Boureau, F., Brochet, B., Bruxelle, J., Cunin, G., Fermanian, J., Ginies, P., Grun-Overdyking, A., Jafari-Schluep, H., Lanteri-Minet, M., Laurent, B., Mick, G., Serrie, A., Valade, D., and Vicaut, E. 2005. Comparison of pain syndromes associated with nervous or somatic lesions and development of a new neuropathic pain diagnostic questionnaire (DN4). Pain 114, 29–36.

Budai, D., Khasabov, S. G., Mantyh, P. W., and Simone, D. A. 2007. NK-1 receptors modulate the excitability of ON cells in the rostral ventromedial medulla. J. Neurophysiol. 97, 1388–1395.

Burgess, S. E., Gardell, L. R., Ossipov, M. H., Malan, T. P., Jr., Vanderah, T. W., Lai, J., and Porreca, F. 2002. Time-dependent descending facilitation from the rostral ventromedial medulla maintains, but does not initiate, neuropathic pain. J. Neurosci. 22, 5129–5136.

Calejesan, A. A., Ch'ang, M. H., and Zhuo, M. 1998. Spinal serotonergic receptors mediate facilitation of a nociceptive reflex by subcutaneous formalin injection into the hindpaw in rats. Brain Res. 798, 46–54.

Catheline, G., Le Guen, S., Honore, P., and Besson, J. M. 1999. Are there long-term changes in the basal or evoked Fos expression in the dorsal horn of the spinal cord of the mononeuropathic rat? Pain 80, 347–357.

Chen, Q. and Huang, L. Y. 1998. Dynorphin block of N-methyl-D-aspartate channels increases with the peptide length. J. Pharmacol. Exp. Them. 284, 824–831.

Chen, Q., King, T., Vanderah, T. W., Ossipov, M. H., Malan, T. P., Jr., Lai, J., and Porreca, F. 2004. Differential blockade of nerve injury-induced thermal and tractile hypersensitivity by systemically adminstered brain-penetrating and peripherally restricted local anesthetics. J. Pain 5, 281–289.

Cloutier, F., de Sousa Buck, H., Ongali, B., and Couture, R. 2002. Pharmacologic and autoradiographic evidence for an up-regulation of kinin B2 receptors in the spinal cord of spontaneously hypertensive rats. Br. J. Pharmacol. 135, 1641–1654.

Cook, A. J., Woolf, C. J., Wall, P. D., and McMahon, S. B. 1987. Dynamic receptive field plasticity in rat spinal cord dorsal horn following C-primary afferent input. Nature 325, 151–153.

Coward, K., Plumpton, C., Facer, P., Birch, R., Carlstedt, T., Tate, S., Bountra, C., and Anand, P. 2000. Immunolocalization of SNS/PN3 and NaN/SNS2 sodium channels in human pain states. Pain 85, 41–50.

Cox, J. J., Reimann, F., Nicholas, A. K., Thornton, G., Roberts, E., Springell, K., Karbani, G., Jafri, H., Mannan, J., Raashid, Y., Al-Gazali, L., Hamamy, H., Valente, E. M., Gorman, S., Williams, R., McHale, D. P., Wood, J. N., Gribble, F. M., and Woods, C. G. 2006. An SCN9A channelopathy causes congenital inability to experience pain. Nature 444, 894–898.

Devor, M. and Wall, P. D. 1990. Cross-excitation in dorsal root ganglia of nerve-injured and intact rats. J. Neurophysiol. 64, 1733–1746.

Dib-Hajj, S. D., Rush, A. M., Cummins, T. R., Hisama, F. M., Novella, S., Tyrrell, L., Marshall, L., and Waxman, S. G. 2005. Gain-of-function mutation in Nav1.7 in familial erythromelalgia induces bursting of sensory neurons. Brain 128, 1847–1854.

Dogra, S., Beydoun, S., Mazzola, J., Hopwood, M., and Wan, Y. 2005. Oxcarbazepine in painful diabetic neuropathy: a randomized, placebo-controlled study. Eur. J. Pain 9, 543–554.

Dogrul, A., Gardell, L. R., Ossipov, M. H., Tulunay, F. C., Lai, J., and Porreca, F. 2003. Reversal of experimental neuropathic pain by T-type calcium channel blockers. Pain 105, 159–168.

Doucet, E., Miquel, M. C., Nosjean, A., Verge, D., Hamon, M., and Emerit, M. B. 2000. Immunolabeling of the rat central nervous system with antibodies partially selective of the short form of the 5-HT$_3$ receptor. Neuroscience 95, 881–892.

Drenth, J. P., Te Morsche, R. H., Guillet, G., Taieb, A., Kirby, R. L., and Jansen, J. B. 2005. SCN9A mutations define primary erythromelalgia as a neuropathic disorder of voltage gated sodium channels. J. Invest. Dermatol. 124, 1333–1338.

Ekberg, J., Jayamanne, A., Vaughan, C. W., Aslan, S., Thomas, L., Mould, J., Drinkwater, R., Baker, M. D., Abrahamsen, B., Wood, J. N., Adams, D. J., Christie, M. J., and Lewis, R. J. 2006. muO-conotoxin MrVIB selectively blocks Nav1.8 sensory neuron specific sodium channels and chronic pain behavior without motor deficits. Proc. Natl. Acad. Sci. U. S. A. 103, 17030–17035.

Elitt, C. M., McIlwrath, S. L., Lawson, J. J., Malin, S. A., Molliver, D. C., Cornuet, P. K., Koerber, H. R., Davis, B. M., and Albers, K. M. 2006. Artemin overexpression in skin enhances expression of TRPV1 and TRPA1 in cutaneous sensory neurons and leads to behavioral sensitivity to heat and cold. J. Neurosci. 26, 8578–8587.

Ferreira, J., Campos, M. M., Araujo, R., Bader, M., Pesquero, J. B., and Calixto, J. B. 2002. The use of kinin B-1 and B-2 receptor knockout mice and selective antagonists to characterize the nociceptive responses caused by kinins at the spinal level. Neuropharmacology 43, 1188–1197.

Fertleman, C. R., Baker, M. D., Parker, K. A., Moffatt, S., Elmslie, F. V., Abrahamsen, B., Ostman, J., Klugbauer, N., Wood, J. N., Gardiner, R. M., and Rees, M. 2006. SCN9A mutations in paroxysmal extreme pain disorder: allelic variants underlie distinct channel defects and phenotypes. Neuron 52, 767–774.

Fields, H. L. 2000. Pain modulation: expectation, opioid analgesia and virtual pain. Prog. Brain Res. 122, 245–253.

Fields, H. L. and Basbaum, A. I. 1999. Central Nervous System Mechanisms of Pain Modulation. In: Textbook of Pain (eds. P. D. Wall and R. Melzack), pp. 309–329. Churchill Livingstone.

Fields, H. L. and Heinricher, M. M. 1985. Anatomy and physiology of a nociceptive modulatory system. Philos. Trans. R. Soc. Lond. B Biol. Sci. 308, 361–374.

Gardell, L. R., Ibrahim, M., Wang, R., Wang, Z., Ossipov, M. H., Malan, T. P., Porreca, F., and Lai, J. 2004. Mouse strains that lack spinal dynorphin upregulation after peripheral nerve injury do not develop neuropathic pain. Neuroscience 123, 43–52.

Gardell, L. R., Vanderah, T. W., Gardell, S. E., Wang, R., Ossipov, M. H., Lai, J., and Porreca, F. 2003a. Enhanced evoked excitatory transmitter release in experimental neuropathy requires descending facilitation. J. Neurosci. 23, 8370–8379.

Gardell, L. R., Wang, R., Ehrenfels, C., Ossipov, M. H., Rossomando, A. J., Miller, S., Buckley, C., Cai, A. K., Tse, A., Foley, S. F., Gong, B., Walus, L., Carmillo, P., Worley, D.,

Huang, C., Engber, T., Pepinsky, B., Cate, R. L., Vanderah, T. W., Lai, J., Sah, D. W., and Porreca, F. 2003b. Multiple actions of systemic artemin in experimental neuropathy. Nat. Med. 9, 1383–1389.

Gold, M. S., Weinreich, D., Kim, C. S., Wang, R., Treanor, J., Porreca, F., and Lai, J. 2003. Redistribution of Na(V)1.8 in uninjured axons enables neuropathic pain. J. Neurosci. 23, 158–166.

Griffin, J. W. 2006. The Roles of Growth Factors in Painful Length-Dependent Axonal Neuropathies. In: Emerging Strategies for the Treatment of Neuropathic Pain (eds. J. N. Campbell et al.), pp. 271–290. IASP Press.

Guan, Y., Guo, W., Robbins, M. T., Dubner, R., and Ren, K. 2004. Changes in AMPA receptor phosphorylation in the rostral ventromedial medulla after inflammatory hyperalgesia in rats. Neurosci. Lett. 366, 201–205.

Guan, Y., Guo, W., Zou, S. P., Dubner, R., and Ren, K. 2003. Inflammation-induced upregulation of AMPA receptor subunit expression in brain stem pain modulatory circuitry. Pain 104, 401–413.

Guo, W., Robbins, M. T., Wei, F., Zou, S., Dubner, R., and Ren, K. 2006. Supraspinal brain-derived neurotrophic factor signaling: a novel mechanism for descending pain facilitation. J. Neurosci. 26, 126–137.

Hains, B. C., Saab, C. Y., Klein, J. P., Craner, M. J., and Waxman, S. G. 2004. Altered sodium channel expression in second-order spinal sensory neurons contributes to pain after peripheral nerve injury. J. Neurosci. 24, 4832–4839.

Hedo, G., Laird, J. M., and Lopez-Garcia, J. A. 1999. Time-course of spinal sensitization following carrageenan-induced inflammation in the young rat: a comparative electrophysiological and behavioural study in vitro and in vivo. Neuroscience 92, 309–318.

Hefti, F. F., Rosenthal, A., Walicke, P. A., Wyatt, S., Vergara, G., Shelton, D. L., and Davies, A. M. 2006. Novel class of pain drugs based on antagonism of NGF. Trends Pharmacol. Sci. 27, 85–91.

Heinricher, M. M. and Neubert, M. J. 2004. Neural basis for the hyperalgesic action of cholecystokinin in the rostral ventromedial medulla. J. Neurophysiol. 92, 1982–1989.

Heinricher, M. M., McGaraughty, S., and Tortorici, V. 2001. Circuitry underlying antiopioid actions of cholecystokinin within the rostral ventromedial medulla. J. Neurophysiol. 85, 280–286.

Heinricher, M. M., Morgan, M. M., Tortorici, V., and Fields, H. L. 1994. Disinhibition of off-cells and antinociception produced by an opioid action within the rostral ventromedial medulla. Neuroscience 63, 279–288.

Heinricher, M. M., Pertovaara, A., and Ossipov, M. H. 2003. Descending Modulation After Injury. In: Proceedings of the 10th World Congress on Pain (eds. D. O. Dostrovsky, D. B. Carr, and M. Koltzenburg), pp. 251–260. IASP Press.

Hildebrand, M. E. and Snutch, T. P. 2006. Contributions of T-type calcium channels to the pathophysiology of pain signaling. Drug Discov. Today: Dis. Mech. 3, 335–341.

Honore, P., Menning, P. M., Rogers, S. D., Nichols, M. L., Basbaum, A. I., Besson, J. M., and Mantyh, P. W. 1999. Spinal substance P receptor expression and internalization in acute, short-term, and long-term inflammatory pain states. J. Neurosci. 19, 7670–7678.

Hugonin, L., Vukojevic, V., Bakalkin, G., and Graslund, A. 2006. Membrane leakage induced by dynorphins. FEBS Lett. 580, 3201–3205.

Hunt, S. P., Pini, A., and Evan, G. 1987. Induction of c-fos-like protein in spinal cord neurons following sensory stimulation. Nature 328, 632–634.

Ikeda, H., Heinke, B., Ruscheweyh, R., and Sandkuhler, J. 2003. Synaptic plasticity in spinal lamina I projection neurons that mediate hyperalgesia. Science 299, 1237–1240.

Ji, R. R. 2004. Peripheral and central mechanisms of inflammatory pain, with emphasis on MAP kinases. Curr. Drug Targets Inflamm. Allergy 3, 299–303.

Ji, R. R. and Strichartz, G. 2004. Cell signaling and the genesis of neuropathic pain. Sci. STKE, 252 reE14.

Ji, R. R., Befort, K., Brenner, G. J., and Woolf, C. J. 2002. ERK MAP kinase activation in superficial spinal cord neurons induces prodynorphin and NK-1 upregulation and contributes to persistent inflammatory pain hypersensitivity. J. Neurosci. 22, 478–485.

Ji, R. R., Kohno, T., Moore, K. A., and Woolf, C. J. 2003. Central sensitization and LTP: do pain and memory share similar mechanisms? Trends Neurosci. 26, 696–705.

Ji, R. R., Samad, T. A., Jin, S. X., Schmoll, R., and Woolf, C. J. 2002. p38 MAPK activation by NGF in primary sensory neurons after inflammation increases TRPV1 levels and maintains heat hyperalgesia. Neuron 36, 57–68.

Kajander, K. C., Wakisaka, S., and Bennett, G. J. 1992. Spontaneous discharge originates in the dorsal root ganglion at the onset of a painful peripheral neuropathy in the rat. Neurosci. Lett. 138, 225–228.

Kanda, N. and Watanabe, S. 2003. Histamine enhances the production of nerve growth factor in human keratinocytes. J. Invest. Dermatol. 121, 570–577.

Kaplan, H. and Fields, H. L. 1991. Hyperalgesia during acute opioid abstinence: evidence for a nociceptive facilitating function of the rostral ventromedial medulla. J. Neurosci. 11, 1433–1439.

Kauppila, T., Kontinen, V. K., and Pertovaara, A. 1998. Influence of spinalization on spinal withdrawal reflex responses varies depending on the submodality of the test stimulus and the experimental pathophysiological condition in the rat. Brain Res. 797, 234–242.

Kawamata, M. and Omote, K. 1996. Involvement of increased excitatory amino acids and intracellular Ca^{2+} concentration in the spinal dorsal horn in an animal model of neuropathic pain. Pain 68, 85–96.

Kawamata, T. and Omote, K. 1999. Activation of spinal N-methyl-D-aspartate receptors stimulates a nitric oxide/cyclic guanosine 3,5-monophosphate/glutamate release cascade in nociceptive signaling. Anesthesiology 91, 1415–1424.

Kawasaki, Y., Kohno, T., Zhuang, Z. Y., Brenner, G. J., Wang, H., Van Der Meer, C., Befort, K., Woolf, C. J., and Ji, R. R. 2004. Ionotropic and metabotropic receptors, protein kinase A, protein kinase C, and Src contribute to C-fiber-induced ERK activation and cAMP response element-binding protein phosphorylation in dorsal horn neurons, leading to central sensitization. J. Neurosci. 24, 8310–8321.

Khasabov, S. G., Rogers, S. D., Ghilardi, J. R., Peters, C. M., Mantyh, P. W., and Simone, D. A. 2002. Spinal neurons that possess the substance P receptor are required for the development of central sensitization. J. Neurosci. 22, 9086–9098.

Khasabov, S. G., Ghilardi, J. R., Mantyh, P. W., and Simone, D. A. 2005. Spinal neurons that express NK-1 receptors modulate descending controls that project through the dorsolateral funiculus. J. Neurophysiol. 93, 998–1006.

Kincaid, W., Neubert, M. J., Xu, M., Kim, C. J., and Heinricher, M. M. 2006. Role for medullary pain facilitating neurons in secondary thermal hyperalgesia. J. Neurophysiol. 95, 33–41.

Kim, S. H. and Chung, J. M. 1992. An experimental model for peripheral neuropathy produced by segmental spinal nerve ligation in the rat. Pain 50, 355–363.

Koetzner, L., Hua, X. Y., Lai, J., Porreca, F., and Yaksh, T. 2004. Nonopioid actions of intrathecal dynorphin evoke spinal excitatory amino acid and prostaglandin E2 release mediated by cyclooxygenase-1 and -2. J. Neurosci. 24, 1451–1458.

Kovelowski, C. J., Ossipov, M. H., Sun, H., Lai, J., Malan, T. P., and Porreca, F. 2000. Supraspinal cholecystokinin may drive tonic descending facilitation mechanisms to maintain neuropathic pain in the rat. Pain 87, 265–273.

Krause, S. J. and Backonja, M. M. 2003. Development of a neuropathic pain questionnaire. Clin. J. Pain 19, 306–314.

Lai, J., Gold, M. S., Kim, C. S., Bian, D., Ossipov, M. H., Hunter, J. C., and Porreca, F. 2002. Inhibition of neuropathic pain by decreased expression of the tetrodotoxin-resistant sodium channel, NaV1.8. Pain 95, 143–152.

Lai, J., Hunter, J. C., and Porreca, F. 2003. The role of voltage-gated sodium channels in neuropathic pain. Curr. Opin. Neurobiol. 13, 291–297.

Lai, J., Luo, M. C., Chen, Q., Ma, S., Gardell, L. R., Ossipov, M. H., and Porreca, F. 2006. Dynorphin A activates bradykinin receptors to maintain neuropathic pain. Nat. Neurosci. 9, 1534–1540.

Lai, J., Ossipov, M. H., Vanderah, T. W., Malan, T. P., Jr., and Porreca, F. 2001. Neuropathic pain: the paradox of dynorphin. Mol. Interv. 1, 160–167.

Lewin, G. R., Rueff, A., and Mendell, L. M. 1994. Peripheral and central mechanisms of NGF-induced hyperalgesia. Eur. J. Neurosci. 6, 1903–1912.

Li, Q. J. and Lu, G. W. 1989. The property of peripheral input of projection neurons in the spinal cord dorsal horn. Sci. China B 32, 1224–1232.

Li, Y., Dorsi, M. J., Meyer, R. A., and Belzberg, A. J. 2000. Mechanical hyperalgesia after an L5 spinal nerve lesion in the rat is not dependent on input from injured nerve fibers. Pain 85, 493–502.

Li, J., Simone, D. A., and Larson, A. A. 1999. Windup leads to characteristics of central sensitization. Pain 79, 75–82.

Liu, H., Mantyh, P. W., and Basbaum, A. I. 1997. NMDA-receptor regulation of substance P release from primary afferent nociceptors. Nature 386, 721–724.

Liu, C. N., Raber, P., Ziv-Sefer, S., and Devor, M. 2001. Hyperexcitability in sensory neurons of rats selected for high versus low neuropathic pain phenotype. Neuroscience 105, 265–275.

Ma, Q. P. and Woolf, C. J. 1996. Basal and touch-evoked fos-like immunoreactivity during experimental inflammation in the rat. Pain 67, 307–316.

MacInnis, B. L. and Campenot, R. B. 2005. Regulation of Wallerian degeneration and nerve growth factor withdrawal-induced pruning of axons of sympathetic neurons by the proteasome and the MEK/Erk pathway. Mol.Cell Neurosci. 28, 430–439.

Malik-Hall, M., Dina, O. A., and Levine, J. D. 2005. Primary afferent nociceptor mechanisms mediating NGF-induced mechanical hyperalgesia. Eur. J. Neurosci. 21, 3387–3394.

Malin, S. A., Molliver, D. C., Koerber, H. R., Cornuet, P., Frye, R., Albers, K. M., and Davis, B. M. 2006. Glial cell line-derived neurotrophic factor family members sensitize nociceptors in vitro and produce thermal hyperalgesia in vivo. J. Neurosci. 26, 8588–8599.

Mansikka, H. and Pertovaara, A. 1997. Supraspinal influence on hindlimb withdrawal thresholds and mustard oil-induced secondary allodynia in rats. Brain Res. Bull. 42, 359–365.

Mantyh, P. W., Rogers, S. D., Honore, P., Allen, B. J., Ghilardi, J. R., Li, J., Daughters, R. S., Lappi, D. A., Wiley, R. G., and Simone, D. A. 1997. Inhibition of hyperalgesia by ablation of lamina I spinal neurons expressing the substance P receptor. Science 278, 275–279.

Matthews, E. A. and Dickenson, A. H. 2001a. Effects of ethosuximide, a T-type Ca(2+) channel blocker, on dorsal horn neuronal responses in rats. Eur. J. Pharmacol. 415, 141–149.

Matthews, E. A. and Dickenson, A. H. 2001b. Effects of spinally delivered N- and P-type voltage-dependent calcium channel antagonists on dorsal horn neuronal responses in a rat model of neuropathy. Pain 92, 235–246.

McCleskey, E. W., Fox, A. P., Feldman, D. H., Cruz, L. J., Olivera, B. M., Tsien, R. W., and Yoshikami, D. 1987. Omega-conotoxin: direct and persistent blockade of specific types of calcium channels in neurons but not muscle. Proc. Natl. Acad. Sci. U. S. A. 84, 4327–4331.

McMahon, S. B. and Cafferty, W. B. 2004. Neurotrophic influences on neuropathic pain. Novartis Found. Symp. 261, 68–92.

McMahon, S. B. and Jones, N. G. 2004. Plasticity of pain signaling: role of neurotrophic factors exemplified by acid-induced pain. J. Neurobiol. 61, 72–87.

McMahon, S. B., Cafferty, W. B., and Marchand, F. 2005. Immune and glial cell factors as pain mediators and modulators. Exp. Neurol. 192, 444–462.

Mendell, L. M. 1966. Physiological properties of unmyelinated fiber projection to the spinal cord. Exp. Neurol. 16, 316–332.

Michaelis, M., Devor, M., and Janig, W. 1996. Sympathetic modulation of activity in rat dorsal root ganglion neurons changes over time following peripheral nerve injury. J. Neurophysiol. 76, 753–763.

Miljanich, G. P. 2004. Ziconotide: neuronal calcium channel blocker for treating severe chronic pain. Curr. Med. Chem. 11, 3029–3040.

Miquel, M. C., Emerit, M. B., Nosjean, A., Simon, A., Rumajogee, P., Brisorgueil, M. J., Doucet, E., Hamon, M., and Verge, D. 2002. Differential subcellular localization of the 5-HT$_3$-As receptor subunit in the rat central nervous system. Eur. J. Neurosci. 15, 449–457.

Morgan, M. M. and Fields, H. L. 1994. Pronounced changes in the activity of nociceptive modulatory neurons in the rostral ventromedial medulla in response to prolonged thermal noxious stimuli. J. Neurophysiol. 72, 1161–1170.

Morgan, M. M. and Heinricher, M. M. 1992. Activity of neurons in the rostral medulla of the halothane-anesthetized rat during withdrawal from noxious heat. Brain Res. 582, 154–158.

Nassar, M. A., Baker, M. D., Levato, A., Ingram, R., Mallucci, G., McMahon, S. B., and Wood, J. N. 2006. Nerve injury induces robust allodynia and ectopic discharges in Nav1.3 null mutant mice. Mol. Pain 2, 33.

Neubert, M. J., Kincaid, W., and Heinricher, M. M. 2004. Nociceptive facilitating neurons in the rostral ventromedial medulla. Pain 110, 158–165.

Nichols, M. L., Allen, B. J., Rogers, S. D., Ghilardi, J. R., Honore, P., Luger, N. M., Finke, M. P., Li, J., Lappi, D. A., Simone, D. A., and Mantyh, P. W. 1999. Transmission of chronic nociception by spinal neurons expressing the substance P receptor. Science 286, 1558–1561.

Nomura, H., Ogawa, A., Tashiro, A., Morimoto, T., Hu, J. W., and Iwata, K. 2002. Induction of Fos protein-like immunoreactivity in the trigeminal spinal nucleus caudalis and upper cervical cord following noxious and non-noxious mechanical stimulation of the whisker pad of the rat with an inferior alveolar nerve transection. Pain 95, 225–238.

Novakovic, S. D., Eglen, R. M., and Hunter, J. C. 2001. Regulation of Na$^+$ channel distribution in the nervous system. Trends Neurosci. 24, 473–478.

Novakovic, S. D., Tzoumaka, E., McGivern, J. G., Haraguchi, M., Sangameswaran, L., Gogas, K. R., Eglen, R. M., and Hunter, J. C. 1998. Distribution of the tetrodotoxin-resistant sodium channel PN3 in rat sensory neurons in normal and neuropathic conditions. J. Neurosci. 18, 2174–2187.

Nowycky, M. C., Fox, A. P., and Tsien, R. W. 1985. Three types of neuronal calcium channel with different calcium agonist sensitivity. Nature 316, 440–443.

Obata, K., Katsura, H., Mizushima, T., Yamanaka, H., Kobayashi, K., Dai, Y., Fukuoka, T., Tokunaga, A., Tominaga, M., and Noguchi, K. 2005. TRPA1 induced in

sensory neurons contributes to cold hyperalgesia after inflammation and nerve injury. J. Clin. Invest. 115, 2393–2401.

Obata, K., Katsura, H., Sakurai, J., Kobayashi, K., Yamanaka, H., Dai, Y., Fukuoka, T., and Noguchi, K. 2006. Suppression of the p75 neurotrophin receptor in uninjured sensory neurons reduces neuropathic pain after nerve injury. J. Neurosci. 26, 11974–11986.

Obata, K., Yamanaka, H., Kobayashi, K., Dai, Y., Mizushima, T., Katsura, H., Fukuoka, T., Tokunaga, A., and Noguchi, K. 2004. Role of mitogen-activated protein kinase activation in injured and intact primary afferent neurons for mechanical and heat hypersensitivity after spinal nerve ligation. J. Neurosci. 24, 10211–10222.

Ohishi, H., Nomura, S., Ding, Y. Q., Shigemoto, R., Wada, E., Kinoshita, A., Li, J. L., Neki, A., Nakanishi, S., and Mizuno, N. 1995. Presynaptic localization of a metabotropic glutamate receptor, mGluR7, in the primary afferent neurons: an immunohistochemical study in the rat. Neurosci. Lett. 202, 85–88.

Ossipov, M. H. and Porreca, F. 2005. Endogenous Pain Modulation: Descending Excitatory Systems. In: Handbook of Clinical Neurology (eds. F. Cervero and T. S. Jensen), pp. 211–238. Elsevier.

Ossipov, M. H. and Porreca, F. 2006. Role of Descending Facilitation in Neuropathic Pain States. In: Emerging Strategies for the Treatment of Neuropathic Pain (eds. J. N. Campbell et al.), pp. 211–238. IASP Press.

Ossipov, M. H., Hong Sun, T., Malan, P., Jr., Lai, J., and Porreca, F. 2000. Mediation of spinal nerve injury induced tactile allodynia by descending facilitatory pathways in the dorsolateral funiculus in rats. Neurosci. Lett. 290, 129–132.

Ossipov, M. H., Lai, J., Malan, T. P., Jr., Vanderah, T. W., and Porreca, F. 2001. Tonic Descending Facilitation as a Mechanism of Neuropathic Pain. In: Neuropatic Pain: Pathophysiology and Treatment (eds. P. T. Hansson et al.), pp. 107–124. IASP Press.

Ossipov, M. H., Lai, J., and Porreca, F. 2005. Mechanisms of Experimental Neuropathic Pain: Integration from Animal Models. In: Wall and Melzack's Textbook of Pain (eds. S. B. McMahon and M. Koltzenburg), pp. 929–946. Elsevier.

Owolabi, J. B., Rizkalla, G., Tehim, A., Ross, G. M., Riopelle, R. J., Kamboj, R., Ossipov, M., Bian, D., Wegert, S., Porreca, F., and Lee, D. K. 1999. Characterization of antiallodynic actions of ALE-0540, a novel nerve growth factor receptor antagonist, in the rat. J. Pharmacol. Exp. Ther. 289, 1271–1276.

Paleckova, V., Palecek, J., McAdoo, D. J., and Willis, W. D. 1992. The non-NMDA antagonist CNQX prevents release of amino acids into the rat spinal cord dorsal horn evoked by sciatic nerve stimulation. Neurosci. Lett. 148, 19–22.

Pertovaara, A. and Almeida, A. 2006. Endogenous Pain Modulation: Descending Inhibitory Systems. In: Handbook of Clinical Neurology (eds. F. Cervero and T. S. Jensen), pp. 179–192. Elsevier.

Pertovaara, A., Wei, H., Kalmari, J., and Ruotsalainen, M. 2001. Pain behavior and response properties of spinal dorsal horn neurons following experimental diabetic neuropathy in the rat: modulation by nitecapone, a COMT inhibitor with antioxidant properties. Exp. Neurol. 167, 425–434.

Porreca, F., Burgess, S. E., Gardell, L. R., Vanderah, T. W., Malan, T. P., Jr., Ossipov, M. H., Lappi, D. A., and Lai, J. 2001. Inhibition of neuropathic pain by selective ablation of brainstem medullary cells expressing the mu-opioid receptor. J. Neurosci. 21, 5281–5288.

Porreca, F., Ossipov, M. H., and Gebhart, G. F. 2002. Chronic pain and medullary descending facilitation. Trends Neurosci. 25, 319–325.

Porreca, F., Ossipov, M. H., Lai, J., Wegert, S., Bian, D., Rogers, S., Mantyh, P., Novakovich, S., and Hunter, J. C. 2000. Blockade of Diabetic, Chemotherapeutic, and NGF-Induced Pain by Antisense Knockdown of PN3/SNS, a TTX-Resistant Sodium Channel. In: Proceedings of the 9th World Congress on Pain (eds. M. Devor, M. C. Rowbotham, and Z. Wiesenfeld-Hallin), pp. 273–285. IASP Press.

Porreca, F., Tang, Q. B., Bian, D., Riedl, M., Elde, R., and Lai, J. 1998. Spinal opioid mu receptor expression in lumbar spinal cord of rats following nerve injury. Brain Res. 795, 197–203.

Prommer, E. E. 2005. Ziconotide: can we use it in palliative care? Am. J. Hosp. Palliat. Med. 22, 369–374.

Puntambekar, P., Mukherjea, D., Jajoo, S., and Ramkumar, V. 2005. Essential role of Rac1/NADPH oxidase in nerve growth factor induction of TRPV1 expression. J. Neurochem. 95, 1689–1703.

Rahman, W., Suzuki, R., Rygh, L. J., and Dickenson, A. H. 2004. Descending serotonergic facilitation mediated through rat spinal 5HT$_3$ receptors is unaltered following carrageenan inflammation. Neurosci. Lett. 361, 229–231.

Rahman, W., Suzuki, R., Webber, M., Hunt, S. P., and Dickenson, A. H. 2006. Depletion of endogenous spinal 5-HT attenuates the behavioural hypersensitivity to mechanical and cooling stimuli induced by spinal nerve ligation. Pain 123, 264–274.

Rajan, B., Polydefkis, M., Hauer, P., Griffin, J. W., and McArthur, J. C. 2003. Epidermal reinnervation after intracutaneous axotomy in man. J. Comp. Neurol. 457, 24–36.

Renganathan, M., Cummins, T. R., and Waxman, S. G. 2001. Contribution of Na(v)1.8 sodium channels to action potential electrogenesis in DRG neurons. J. Neurophysiol. 86, 629–640.

Renton, T., Yiangou, Y., Plumpton, C., Tate, S., Bountra, C., and Anand, P. 2005. Sodium channel Nav1.8 immunoreactivity in painful human dental pulp. BMC Oral Health 5, 5.

Ruiz, G. and Banos, J. E. 2005. The effect of endoneurial nerve growth factor on calcitonin gene-related peptide expression in primary sensory neurons. Brain Res. 1042, 44–52.

Ruiz, G., Ceballos, D., and Banos, J. E. 2004. Behavioral and histological effects of endoneurial administration of nerve growth factor: possible implications in neuropathic pain. Brain Res. 1011, 1–6.

Sah, D. W., Ossipov, M. H., and Porreca, F. 2003. Neurotrophic factors as novel therapeutics for neuropathic pain. Nat. Rev. Drug Discov. 2, 460–472.

Sah, D. W., Ossipov, M. H., Rossomando, A., Silvian, L., and Porreca, F. 2005a. New approaches for the treatment of pain: the GDNF family of neurotrophic growth factors. Curr. Top. Med. Chem. 5, 577–583.

Sah, D. W., Porreca, F., and Ossipov, M. H. 2006. Modulation of Neurotrophic Growth Factors as a Therapeutic Strategy for Neuropathic Pain. Drug Dev. Res. 67, 389–403.

Sangameswaran, L., Fish, L. M., Koch, B. D., Rabert, D. K., Delgado, S. G., Ilnicka, M., Jakeman, L. B., Novakovic, S., Wong, K., Sze, P., Tzoumaka, E., Stewart, G. R., Herman, R. C., Chan, H., Eglen, R. M., and Hunter, J. C. 1997. A novel tetrodotoxin-sensitive, voltage-gated sodium channel expressed in rat and human dorsal root ganglia. J. Biol. Chem. 272, 14805–14809.

Santicioli, P., Del, B. E., Tramontana, M., Geppetti, P., and Maggi, C. A. 1992. Release of calcitonin gene-related peptide like-immunoreactivity induced by electrical field stimulation from rat spinal afferents is mediated by conotoxin-sensitive calcium channels. Neurosci. Lett. 136, 161–164.

Scott, D. A., Wright, C. E., and Angus, J. A. 2002. Actions of intrathecal omega-conotoxins CVID, GVIA, MVIIA, and morphine in acute and neuropathic pain in the rat. Eur. J. Pharmacol. 451, 279–286.

Shamash, S., Reichert, F., and Rotshenker, S. 2002. The cytokine network of Wallerian degeneration: tumor necrosis factor-alpha, interleukin-1alpha, and interleukin-1beta. J. Neurosci. 22, 3052–3060.

Shelton, D. L., Zeller, J., Ho, W. H., Pons, J., and Rosenthal, A. 2005. Nerve growth factor mediates hyperalgesia and cachexia in auto-immune arthritis. Pain 116, 8–16.

Shortland, P. and Molander, C. 1998. The time-course of abeta-evoked c-fos expression in neurons of the dorsal horn and gracile nucleus after peripheral nerve injury. Brain Res. 810, 288–293.

Shu, X. Q. and Mendell, L. M. 1999. Neurotrophins and hyperalgesia. Proc. Natl. Acad. Sci. U. S. A. 96, 7693–7696.

Shu, X. and Mendell, L. M. 2001. Acute sensitization by NGF of the response of small-diameter sensory neurons to capsaicin. J. Neurophysiol. 86, 2931–2938.

Simone, D. A., Baumann, T. K., and LaMotte, R. H. 1989. Dose-dependent pain and mechanical hyperalgesia in humans after intradermal injection of capsaicin. Pain 38, 99–107.

Simone, D. A., Sorkin, L. S., Oh, U., Chung, J. M., Owens, C., LaMotte, R. H., and Willis, W. D. 1991. Neurogenic hyperalgesia: central neural correlates in responses of spinothalamic tract neurons. J. Neurophysiol. 66, 228–246.

Sindrup, S. H. and Jensen, T. S. 2007. Are sodium channel blockers useless in peripheral neuropathic pain? Pain 128, 6–7.

Sluka, K. A. and Willis, W. D. 1998. Increased spinal release of excitatory amino acids following intradermal injection of capsaicin is reduced by a protein kinase G inhibitor. Brain Res. 798, 281–286.

Snutch, T. P. 2005. Targeting chronic and neuropathic pain: the N-type calcium channel comes of age. NeuroRx 2, 662–670.

Snutch, T. P., Sutton, K. G., and Zamponi, G. W. 2001. Voltage-dependent calcium channels – beyond dihydropyridine antagonists. Curr. Opin. Pharmacol. 1, 11–16.

Southall, M. D. and Vasko, M. R. 2001. Prostaglandin receptor subtypes, EP3C and EP4, mediate the prostaglandin E2-induced cAMP production and sensitization of sensory neurons. J. Biol. Chem. 276, 16083–16091.

Southall, M. D., Bolyard, L. A., and Vasko, M. R. 2002. Twenty-four hour exposure to prostaglandin downregulates prostanoid receptor binding but does not alter PGE(2)-mediated sensitization of rat sensory neurons. Pain 96, 285–296.

Southall, M. D., Michael, R. L., and Vasko, M. R. 1998. Intrathecal NSAIDS attenuate inflammation-induced neuropeptide release from rat spinal cord slices. Pain 78, 39–48.

Stempelj, M. and Ferjan, I. 2005. Signaling pathway in nerve growth factor induced histamine release from rat mast cells. Inflamm. Res. 54, 344–349.

Stucky, C. L., Shin, J. B., and Lewin, G. R. 2002. Neurotrophin-4: a survival factor for adult sensory neurons. Curr. Biol. 12, 1401–1404.

Sun, R.-Q., Lawand, N. B., Lin, Q., and Willis, W. D. 2004. Role of calcitonin gene-related peptide in the sensitization of dorsal horn neurons to mechanical stimulation after intradermal injection of capsaicin. J. Neurophysiol. 92, 320–326.

Sun, R. Q., Lawand, N. B., and Willis, W. D. 2003. The role of calcitonin gene-related peptide (CGRP) in the generation and maintenance of mechanical allodynia and hyperalgesia in rats after intradermal injection of capsaicin. Pain 104, 201–208.

Sun, Q., Tu, H., Xing, G. G., Han, J. S., and Wan, Y. 2005. Ectopic discharges from injured nerve fibers are highly correlated with tactile allodynia only in early, but not late, stage in rats with spinal nerve ligation. Exp. Neurol. 191, 128–136.

Suzuki, R. and Dickenson, A. 2005a. Spinal and supraspinal contributions to central sensitization in peripheral neuropathy. Neurosignals 14, 175–181.

Suzuki, R. and Dickenson, A. H. 2005b. Differential pharmacological modulation of the spontaneous stimulus-independent activity in the rat spinal cord following peripheral nerve injury. Exp. Neurol. 198, 72–80.

Suzuki, R., Morcuende, S., Webber, M., Hunt, S. P., and Dickenson, A. H. 2002. Superficial NK1-expressing neurons control spinal excitability through activation of descending pathways. Nat. Neurosci. 5, 1319–1326.

Suzuki, R., Rahman, W., Hunt, S. P., and Dickenson, A. H. 2004a. Descending facilitatory control of mechanically evoked responses is enhanced in deep dorsal horn neurones following peripheral nerve injury. Brain Res. 1019, 68–76.

Suzuki, R., Rygh, L. J., and Dickenson, A. H. 2004b. Bad news from the brain: descending 5-HT pathways that control spinal pain processing. Trends Pharmacol. Sci. 25, 613–617.

Svensson, P., Cairns, B. E., Wang, K., and Arendt-Nielsen, L. 2003. Injection of nerve growth factor into human masseter muscle evokes long-lasting mechanical allodynia and hyperalgesia. Pain 104, 241–247.

Tal, M. and Bennett, G. J. 1994. Extra-territorial pain in rats with a peripheral mononeuropathy: mechano-hyperalgesia and mechano-allodynia in the territory of an uninjured nerve. Pain 57, 375–382.

Talley, E. M., Cribbs, L. L., Lee, J. H., Daud, A., Perez-Reyes, E., and Bayliss, D. A. 1999. Differential distribution of three members of a gene family encoding low voltage-activated (T-type) calcium channels. J. Neurosci. 19, 1895–1911.

Taherzadeh, O., Otto, W. R., Anand, U., Nanchahal, J., and Anand, P. 2003. Influence of human skin injury on regeneration of sensory neurons. Cell Tissue Res. 312, 275–280.

Todd, A. J. 2002. Anatomy of primary afferents and projection neurones in the rat spinal dorsal horn with particular emphasis on substance P and the neurokinin 1 receptor. Exp. Physiol. 87, 245–249.

Todd, A. J., McGill, M. M., and Shehab, S. A. 2000. Neurokinin 1 receptor expression by neurons in laminae I, III and IV of the rat spinal dorsal horn that project to the brainstem. Eur. J. Neurosci. 12, 689–700.

Tomlinson, D. R., Fernyhough, P., and Diemel, L. T. 1997. Role of neurotrophins in diabetic neuropathy and treatment with nerve growth factors. Diabetes 46(Suppl. 2), 43–49.

Urban, M. O., Coutinho, S. V., and Gebhart, G. F. 1999. Biphasic modulation of visceral nociception by neurotensin in rat rostral ventromedial medulla. J. Pharmacol. Exp. Ther. 290, 207–213.

Vera-Portocarrero, L. P., Zhang, E. T., Ossipov, M. H., Xie, J. Y., King, T., Lai, J., and Porreca, F. 2006. Descending facilitation from the rostral ventromedial medulla maintains nerve injury-induced central sensitization. Neuroscience 140, 1311–1320.

Vinik, A. I., Tuchman, M., Safirstein, B., Corder, C., Kirby, L., Wilks, K., Quessy, S., Blum, D., Grainger, J., White, J., and Silver, M. 2007. Lamotrigine for treatment of pain associated with diabetic neuropathy: results of two randomized, double-blind, placebo-controlled studies. Pain 128, 169–179.

Wallin, J. and Schott, E. 2002. Substance P release in the spinal dorsal horn following peripheral nerve injury. Neuropeptides 36, 252–256.

Wang, Y. X., Gao, D., Pettus, M., Phillips, C., and Bowersox, S. S. 2000a. Interactions of intrathecally administered ziconotide, a selective blocker of neuronal N-type voltage-sensitive calcium channels, with morphine on nociception in rats. Pain 84, 271–281.

Wang, Z., Gardell, L. R., Ossipov, M. H., Vanderah, T. W., Brennan, M. B., Hochgeschwender, U., Hruby, V. J., Malan, T. P., Jr., Lai, J., and Porreca, F. 2001. Pronociceptive actions of dynorphin maintain chronic neuropathic pain. J. Neurosci. 21, 1779–1786.

Wang, R., Guo, W., Ossipov, M. H., Vanderah, T. W., Porreca, F., and Lai, J. 2003. Glial cell line-derived neurotrophic factor normalizes neurochemical changes in injured dorsal root ganglion neurons and prevents the expression of experimental neuropathic pain. Neuroscience 121, 815–824.

Wang, H. B., Kohno, T., Amaya, F., Brenner, G. J., Ito, N., Allchorne, A., Ji, R. R., and Woolf, C. J. 2005. Bradykinin produces pain hypersensitivity by potentiating spinal cord glutamatergic synaptic transmission. J. Neuroscience. 25, 7986–7992.

Wang, Y. X., Pettus, M., Gao, D., Phillips, C., and Scott Bowersox, S. 2000b. Effects of intrathecal administration of ziconotide, a selective neuronal N-type calcium channel blocker, on mechanical allodynia and heat hyperalgesia in a rat model of postoperative pain. Pain 84, 151–158.

Waxman, S. G., Cummins, T. R., Dib-Hajj, S., Fjell, J., and Black, J. A. 1999. Sodium channels, excitability of primary sensory neurons, and the molecular basis of pain. Muscle Nerve 22, 1177–1187.

Wei, F., Dubner, R., and Ren, K. 1999. Dorsolateral funiculus-lesions unmask inhibitory or disfacilitatory mechanisms which modulate the effects of innocuous mechanical stimulation on spinal Fos expression after inflammation. Brain Res. 820, 112–116.

Wei, F., Ren, K., and Dubner, R. 1998. Inflammation-induced Fos protein expression in the rat spinal cord is enhanced following dorsolateral or ventrolateral funiculus lesions. Brain Res. 782, 136–141.

Wood, J. N., Abrahamsen, B., Baker, M. D., Boorman, J. D., Donier, E., Drew, L. J., Nassar, M. A., Okuse, K., Seereeram, A., Stirling, C. L., and Zhao, J. 2004a. Ion channel activities implicated in pathological pain. Novartis Found Symp. 261, 32–40.

Wood, J. N., Boorman, J. P., Okuse, K., and Baker, M. D. 2004b. Voltage-gated sodium channels and pain pathways. J Neurobiol. 61, 55–71.

Woolf, C. J. 1996. Windup and central sensitization are not equivalent. Pain 66, 105–108.

Woolf, C. J. and Salter, M. W. 2000. Neuronal plasticity: increasing the gain in pain. Science 288, 1765–1769.

Woolf, C. J. and Thompson, S. W. 1991. The induction and maintenance of central sensitization is dependent on N-methyl-D-aspartic acid receptor activation; implications for the treatment of post-injury pain hypersensitivity states. Pain 44, 293–299.

Wu, G., Ringkamp, M., Hartke, T. V., Murinson, B. B., Campbell, J. N., Griffin, J. W., and Meyer, R. A. 2001. Early onset of spontaneous activity in uninjured C-fiber nociceptors after injury to neighboring nerve fibers. J. Neurosci. 21, RC140.

Wu, G., Ringkamp, M., Murinson, B. B., Pogatzki, E. M., Hartke, T. V., Weerahandi, H. M., Campbell, J. N., Griffin, J. W., and Meyer, R. A. 2002. Degeneration of myelinated efferent fibers induces spontaneous activity in uninjured C-fiber afferents. J. Neurosci. 22, 7746–7753.

Xie, J. Y., Herman, D. S., Stiller, C. O., Gardell, L. R., Ossipov, M. H., Lai, J., Porreca, F., and Vanderah, T. W. 2005. Cholecystokinin in the rostral ventromedial medulla mediates opioid-induced hyperalgesia and antinociceptive tolerance. J. Neurosci. 25, 409–416.

Xu, M., Kim, C. J., Neubert, M. J., and Heinricher, M. M. 2007. NMDA receptor-mediated activation of medullary pronociceptive neurons is required for secondary thermal hyperalgesia. Pain 127, 253–262.

Yang, Y., Wang, Y., Li, S., Xu, Z., Li, H., Ma, L., Fan, J., Bu, D., Liu, B., Fan, Z., Wu, G., Jin, J., Ding, B., Zhu, X., and Shen, Y. 2004. Mutations in SCN9A, encoding a sodium channel alpha subunit, in patients with primary erythromelalgia. J. Med. Genet. 41, 171–174.

Yoon, Y. W., Na, H. S., and Chung, J. M. 1996. Contributions of injured and intact afferents to neuropathic pain in an experimental rat model. Pain 64, 27–36.

Zeitz, K. P., Guy, N., Malmberg, A. B., Dirajlal, S., Martin, W. J., Sun, L., Bonhaus, D. W., Stucky, C. L., Julius, D., and Basbaum, A. I. 2002. The 5-HT$_3$ subtype of serotonin receptor contributes to nociceptive processing via a novel subset of myelinated and unmyelinated nociceptors. J. Neurosci. 22, 1010–1019.

Zhang, E. T., Ossipov, M. H., Zhang, D. Q., Lai, J., and Porreca, F. 2007. Nerve injury-induced tactile allodynia is present in the absence of FOS labeling in retrogradely labeled post-synaptic dorsal column neurons. Pain, 128, 143–154.

Zhang, L., Peoples, R. W., Oz, M., Harvey-White, J., Weight, F. F., and Brauneis, U. 1997. Potentiation of NMDA receptor-mediated responses by dynorphin at low extracellular glycine concentrations. J. Neurophysiol. 78, 582–590.

Zhuo, M. and Gebhart, G. F. 1997. Biphasic modulation of spinal nociceptive transmission from the medullary raphe nuclei in the rat. J. Neurophysiol. 78, 746–758.

Zwick, M., Davis, B. M., Woodbury, C. J., Burkett, J. N., Koerber, H. R., Simpson, J. F., and Albers, K. M. 2002. Glial cell line-derived neurotrophic factor is a survival factor for isolectin B4-positive, but not vanilloid receptor 1-positive, neurons in the mouse. J Neurosci. 22, 4057–4065.

Further Reading

Ali, Z., Meyer, R. A., and Campbell, J. N. 1996. Secondary hyperalgesia to mechanical but not heat stimuli following a capsaicin injection in hairy skin. Pain 68, 401–411.

Backonja, M. M. and Krause, S. J. 2003. Neuropathic pain questionnaire – short form. Clin. J. Pain 19, 315–316.

Chapman, V., Suzuki, R., and Dickenson, A. H. 1998. Electrophysiological characterization of spinal neuronal response properties in anaesthetized rats after ligation of spinal nerves L5–L6. J. Physiol. 507(Pt 3), 881–894.

Cunha, F. Q., Teixeira, M. M., and Ferreira, S. H. 1999. Pharmacological modulation of secondary mediator systems – cyclic AMP and cyclic GMP – on inflammatory hyperalgesia. Br. J. Pharmacol. 127, 671–678.

Eschenfelder, S., Habler, H. J., and Janig, W. 2000. Dorsal root section elicits signs of neuropathic pain rather than reversing them in rats with L5 spinal nerve injury. Pain 87, 213–219.

Han, H. C., Lee, D. H., and Chung, J. M. 2000. Characteristics of ectopic discharges in a rat neuropathic pain model. Pain 84, 253–261.

Hunt, S. P. and Mantyh, P. W. 2001. The molecular dynamics of pain control. Nat. Rev. Neurosci. 2, 83–91.

Ikeda, H., Stark, J., Fischer, H., Wagner, M., Drdla, R., Jager, T., and Sandkuhler, J. 2006. Synaptic amplifier of inflammatory pain in the spinal dorsal horn. Science 312, 1659–1662.

Ji, R. R. 2004. Mitogen-activated protein kinases as potential targets for pain killers. Curr. Opin. Invest. Drug. 5, 71–75.

Li, W. P., Xian, C., Rush, R. A., and Zhou, X. F. 1999b. Upregulation of brain-derived neurotrophic factor and neuropeptide Y in the dorsal ascending sensory pathway following sciatic nerve injury in rat. Neurosci. Lett. 260, 49–52.

Ma, W., Peschanski, M., and Ralston, H. J.3. 1987. The differential synaptic organization of the spinal and lemniscal projections to the ventrobasal complex of the rat thalamus. Evidence for convergence of the two systems upon single thalamic neurons. Neuroscience 22, 925–934.

Malmberg, A. B. and Yaksh, T. L. 1993. Spinal nitric oxide synthesis inhibition blocks NMDA-induced thermal hyperalgesia and produces antinociception in the formalin test in rats. Pain 54, 291–300.

Malmberg, A. B. and Yaksh, T. L. 1995. Cyclooxygenase inhibition and the spinal release of prostaglandin E2 and amino acids evoked by paw formalin injection: a microdialysis study in unanesthetized rats. J. Neurosci. 15, 2768–2776.

Malmberg, A. B., Chen, C., Tonegawa, S., and Basbaum, A. I. 1997. Preserved acute pain and reduced neuropathic pain in mice lacking PKCgamma (see comments). Science 278, 279–283.

Malmberg, A. B., Hamberger, A., and Hedner, T. 1995. Effects of prostaglandin E2 and capsaicin on behavior and cerebrospinal fluid amino acid concentrations of unanesthetized rats: a microdialysis study. J. Neurochem. 65, 2185–2193.

Mayer, D. J., Mao, J., Holt, J., and Price, D. D. 1999. Cellular mechanisms of neuropathic pain, morphine tolerance, and their interactions. Proc. Natl. Acad. Sci. U. S. A. 96, 7731–7736.

Miki, K., Fukuoka, T., Tokunaga, A., and Noguchi, K. 1998. Calcitonin gene-related peptide increase in the rat spinal dorsal horn and dorsal column nucleus following peripheral nerve injury: up-regulation in a subpopulation of primary afferent sensory neurons. Neuroscience 82, 1243–1252.

Miki, K., Iwata, K., Tsuboi, Y., Morimoto, T., Kondo, E., Dai, Y., Ren, K., and Noguchi, K. 2000. Dorsal column–thalamic pathway is involved in thalamic hyperexcitability following peripheral nerve injury: a lesion study in rats with experimental mononeuropathy. Pain 85, 263–271.

Minami, T., Nishihara, I., Ito, S., Sakamoto, K., Hyodo, M., and Hayaishi, O. 1995. Nitric oxide mediates allodynia induced by intrathecal administration of prostaglandin E2 or prostaglandin F2 alpha in conscious mice. Pain 61, 285–290.

Minami, T., Nishihara, I., Uda, R., Ito, S., Hyodo, M. and Hayaishi, O., 1994. Involvement of glutamate receptors in allodynia induced by prostaglandins E2 and F2 alpha injected into conscious mice. Pain 57, 225–231.

Noguchi, K., Kawai, Y., Fukuoka, T., Senba, E., and Miki, K. 1995. Substance P induced by peripheral nerve injury in primary afferent sensory neurons and its effect on dorsal column nucleus neurons. J. Neurosci. 15, 7633–7643.

Ossipov, M. H. and Porreca, F. 2005. Descending Modulation of Pain. In: The Paths of Pain (eds. H. Merskey, J. D. Loeser, and R. Dubner). pp. 117–130. IASP Press.

Ossipov, M. H. and Porreca, F. 2007. Neuropathic Pain: Basic Mechanisms (Animal). In: Handbook of the Senses (eds. A. I. Basbaum, C. Bushnell, and D. Julius). Elsevier, in press.

Ossipov, M. H., Zhang, E. T., Carvajal, C., Gardell, L., Quirion, R., Dumont, Y., Lai, J., and Porreca, F. 2002. Selective mediation of nerve injury-induced tactile hypersensitivity by neuropeptide Y. J. Neurosci. 22, 9858–9867.

Seybold, V. S., McCarson, K. E., Mermelstein, P. G., Groth, R. D., and Abrahams, L. G. 2003. Calcitonin gene-related peptide regulates expression of neurokinin1 receptors by rat spinal neurons. J. Neurosci. 23, 1816–1824.

Sun, H., Ren, K., Zhong, C. M., Ossipov, M. H., Malan, T. P., Lai, J., and Porreca, F. 2001. Nerve injury-induced tactile allodynia is mediated via ascending spinal dorsal column projections. Pain 90, 105–111.

Sun, R. Q., Tu, Y. J., Lawand, N. B., Yan, J. Y., Lin, Q., and Willis, W. D. 2004. Calcitonin gene-related peptide receptor activation produces pka- and pkc-dependent mechanical hyperalgesia and central sensitization. J. Neurophysiol. 92, 2859–2866.

Suzuki, R. and Dickenson, A. H. 2000. Neuropathic pain: nerves bursting with excitement. Neuroreport 11, 17–21.

Suzuki, R., Kontinen, V. K., Matthews, E., Williams, E., and Dickenson, A. H. 2000. Enlargement of the receptive field size to low intensity mechanical stimulation in the rat spinal nerve ligation model of neuropathy. Exp. Neurol. 163, 408–413.

Vetter, G., Geisslinger, G., and Tegeder, I. 2001. Release of glutamate, nitric oxide and prostaglandin E2 and metabolic activity in the spinal cord of rats following peripheral nociceptive stimulation. Pain 92, 213–218.

Willis, W. D. and Coggeshall, R. E. 2004. Sensory Mechanisms of the Spinal Cord, Vol. 1, Primary Afferent Neurons and the Spinal Dorsal Horn, 3rd edn. Kluwer Academic/Plenum Publishers.

Willis, W. D., Jr. 2001. Role of neurotransmitters in sensitization of pain responses. Ann. N. Y. Acad. Sci. 933, 142–156.

Woolf, C. J. 1996. Phenotypic modification of primary sensory neurons: the role of nerve growth factor in the production of persistent pain. Philos. Trans. R. Soc. Lond. B Biol. Sci. 351, 441–448.

Relevant Website

www.iasp-pain.org – International Association for the Study of Pain.

56 Animal Models and Neuropathic Pain

I Decosterd and T Berta, University of Lausanne, Lausanne, Switzerland

Glossary

allodynia Pain due to a stimulus which does not normally provoke pain.

autotomy Self amputation by animal that severs its own appendages (usually distal phalanges).

hyperalgesia An increased response to a stimulus which is normally painful.

neuropathic pain Pain initiated or caused by a primary lesion or dysfunction in the nervous system.

neuropathy A disturbance of function or pathological change in a nerve.

spontaneous pain Pain without an evident stimulus.

stimulus-evoked pain pain evoked by a mechanical/thermal/chemical stimulus.

Wallerian degeneration The degeneration of a nerve fiber induces by injury or disease, characterized by segmentation of the myelin and resulting in atrophy and destruction of the axon.

56.1 Introduction

Animal models in pain research offer great promise for both the identification of pain mechanisms, and the investigation of possible therapeutic applications. Our understanding of neuropathic pain (NP) has progressed during last few decades and although animal models have provided important information, in humans a clear pattern of the mechanisms responsible for its generation and maintenance is still unclear. Therefore, it is essential to put forward advanced animal models based, on the investigation of specific mechanisms, as well as refining the way in which the data is interpreted. Moreover, it is clear that the quest of a unique and universal animal model of all symptoms in patients ignores the complexity of the mechanisms involved. Therefore, it is important to select the best model for each specific research interest or a combination of appropriate but distinct models. In this chapter we will argue for the most frequently used animal models of NP, compare their advantages and disadvantages, and summarize the behavioral characterization.

56.2 Animal Models of Neuropathic Pain

At present, animal models of NP cover various etiologies and are related to symptoms leading to an extensive picture of clinical NP manifestations. The majority of animal models of NP involve traumatic injuries to peripheral nerves, nerve roots, or spinal cord by transection, ligation, or compression (Figure 1). Other experimental models are related to direct or indirect nerve inflammation, ischemia, drug toxicity, or systemic metabolic disorders. Each animal model has been characterized by precise behavioral-based evaluations using different modalities of sensory stimulus. Many temporal and spatial profiles of molecular, physiological, and structural changes occurring during the development of NP have been investigated and most of the rodent models are compatible with actual electrophysiological measurements, functional imaging techniques, as well as genomic and proteomic screenings.

56.2.1 Models of Central Neuropathic Pain: Spinal Cord Injuries

Partial or complete spinal cord injuries result in loss of function in varying degrees below the level of injury, and are often associated with central NP pain syndromes (Christensen, M. D. and Hulsebosch, C. E., 1997a; Siddall, P. J. and Loeser, J. D., 2001). This type of injury involves functional and structural changes, partial loss of normal descending pathway, and hyperexcitability of dorsal horn neurons (Woolf, C. J., 1983; Woolf, C. J. and Fitzgerald, M., 1983; Hulsebosch, C. E., 2002). Spinal cord transection, contusion, compression, and chemical toxicity induce all at-level or below-level signs of pain hypersensitivity (Vierck, C. J. et al., 1999; Sjolund, B. H., 2002; Rosenzweig, E. S. and McDonald, J. W., 2004). In the hemisection model (Christensen, M. D. et al., 1996), a lateral spinal hemisection is performed at T13 level. After a period of motor dysfunction, ipsilateral to the injury, rats display persistent stimulus-evoked hypersensitivity of the four limbs associated with vocalization. This model was successfully used to observe molecular and structural reorganization of the spinal cord in NP (Christensen, M. D. and Hulsebosch, C. E., 1997b; Bennett, A. D. et al., 2000; Hains, B. C. et al., 2003b). The anterolateral cordotomy model (Vierck, C. J. and Light, A. R., 1999) consists in an unilateral injury of the anterolateral column at the T9–T11 spinal segments. The injury is associated with overgrooming and signs of allodynia/hyperalgesia. The model has been adapted

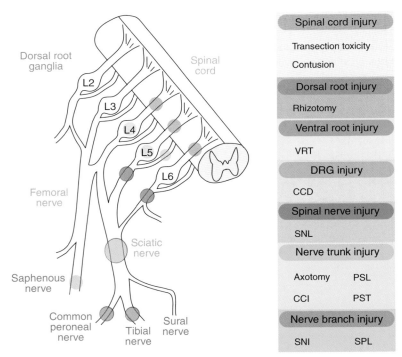

Figure 1 Anatomy of the rat lumbar plexus and the location of the different injuries used to create the different neuropathic pain models. VRT, ventral root transection; CCD, chronic compression of dorsal root ganglia; SNL, spinal nerve ligation; CCI, chronic constriction injury; PSL, partial sciatic ligation; PST, partial sciatic nerve transection; SNI, spared nerve injury (tibial and c. peroneal nerves are injured, sural nerve is intact); SPL, saphenous nerve partial ligation (30–50% of the saphenous is transected).

to primates and much information has been acquired on thalamus circuitry in NP (Weng, H.-R. *et al.*, 2000). The model of spinal cord contusion (Gruner, J. A., 1992; Siddall, P. *et al.*, 1995) is produced by an impact injury on the spinal cord at the level of T12/L1 using a weight dropped with a standard impact device (MASCIS, Multi-Center Animal Spinal Cord Injury Study). The incidence of pain hypersensitivity is correlated to the extent of the injury (Hulsebosch, C. E. *et al.*, 2000; Mills, C. D. *et al.*, 2001; Gwak, Y. S. *et al.*, 2004) and evidence of hyperexcitability of dorsal horn neurons and regulation of voltage-gated sodium channels have been shown (Hains, B. C. *et al.*, 2003a). Squeezing the spinal cord is the basis of the compression injury model (Fehlings, M. G. and Nashmi, R., 1995). A 35- or 50-g clip compression is placed on the dura mater at the juncture of T12 and T13 spinal segments and leads to microvasculature disruption, ischemia, and progressive axonal injury associated with the development of tactile allodynia (Bruce, J. C. *et al.*, 2002). Other central models of NP use toxic substances. The photochemically induced lesion of the spinal cord (Hao, J. X. *et al.*, 1991) is produced by the intravenous injection of erythrosine B (a photosensitizing dye) immediately followed by irradiation with an argon laser beam at T13 level. This procedure results in an ischemic lesion of thoracic spinal cord by occlusion of spinal cord vessels and produces a particularly severe mechanical and cold allodynia-like behavior in skin territories innervated by the rostral border of the injury (Hao, J. X. *et al.*, 1991; 1992). Excitotoxic spinal cord injury (ESCI) is directly mediated by the intraspinal injection at levels ranging from T10 to L4 of quisqualic acid (QUIS), an α-amino-3-hydroxy-5-methylisoxazole-4-propionic acid (AMPA) metabotropic receptor agonist (Yezierski, R. P. *et al.*, 1993; 1998). This model offers the considerable advantage of avoiding surgical laminectomy, and although less representative of a clinical pathology, it induces sustained NP-like symptoms related to a precise mechanism. The QUIS model leads to a gradual and specific loss of neurons associated to NP-like behaviors and the extension of the injury and the intensity of changes are dependent on the depth and volume of the QUIS injections.

56.2.2 Models of Peripheral Neuropathic Pain: Injury to Roots, Dorsal Root Ganglia, and Spinal Nerves

Surgical procedures on selective root, dorsal root ganglia, and spinal nerves allows direct access to sensory and motor fibers and a clear segmental location of injured and noninjured primary afferents. Spontaneous and stimulus-evoked pain are sustained by different mechanisms such as deafferentation, ectopic activities in dorsal root ganglion (DRG), and spinal cord neurons, Wallerian degeneration of the peripheral axons and central reorganization (Eschenfelder, S. *et al.*, 2000; Li, Y. *et al.*, 2000). For dorsal and ventral root rhizotomies, typically at L5 level, the root is ligated and transected 3–4 mm proximal to the DRG (Colburn, R. W. *et al.*, 1999; Li, L. *et al.*, 2002). The ventral root transection model provides an additional feature to investigate NP, in the sense that lesion of a motor nerve induces alteration of pain sensitivity (Li, L. *et al.*, 2002; Sheth, R. N. *et al.*, 2002). Robust mechanical and cold allodynia-like behaviors have been documented (Sheen, K. and Chung, J. M., 1993), whereas loose ligation of L4–L6 dorsal roots produces only mechanical allodynia (Tabo, E. *et al.*, 1999). The chronic compression of dorsal root ganglia (CCD), proposed as a model of lower back pain/disc herniation, consists of the placement of a fine stainless-steel rod into the L5/L4 intervertebral foramen (Hu, S. J. and Xing, J. L., 1998). Heat hyperalgesia and a tactile allodynia are the resulting behaviors encountered in this model (Zhang, J. M. *et al.*, 1999; Ma, C. and LaMotte, R. H., 2005). The widely used spinal nerve ligation (SNL), where L5/L6 spinal nerve are tightly ligated (Kim, S. H. and Chung, J. M., 1992), produces robust and consistent sympathetic-dependent NP-related behaviors including indirect signs of spontaneous pain, heat hyperalgesia, mechanical allodynia/hyperalgesia, and cold allodynia (Choi, Y. *et al.*, 1994). This type of damage has also been applied successfully to mice and primates (Choi, Y. *et al.*, 1994; Ali, Z. *et al.*, 1999; Mogil, J. S. *et al.*, 1999). The additional and considerable advantage of SNL is the possibility to distinguish injured (L5/L6) and noninjured (L4) DRG and spinal segments.

56.2.3 Models of Peripheral Neuropathic Pain: Injury to Peripheral Nerves

The first models of NP used complete transection of major peripheral nerve. Wall and co-workers produce an entire denervation of the distal hindlimb, damaging the sciatic and saphenous nerves, to model anesthesia dolorosa, a type of NP referred to denervated zone without any sensory input from that area. Nerves were transected at midthigh level and the proximal stump was either tightly ligated or encapsulated into a polyethylene tube (Wall, P. D. *et al.*, 1979a; 1979b). The total sciatic transection prevents stimulus-evoked assessment of the hindlimb but self-mutilation (autotomy) was the major feature of the experimental anesthesia dolorosa mode. Degrees of autotomy have been defined according to the extension of self-mutilation in the hindpaw and are used to evaluate the progression of the neuropathy. The subsequent chronic constriction injury (CCI) model, four catgut ligations loosely placed around the sciatic nerve (Bennett, G. J. and Xie, Y. K., 1988) or the partial sciatic ligation (PSL) model, ligation of one-third to one-half of the sciatic nerve in rats (Seltzer, Z. *et al.*, 1990) or mice (Malmberg, A. B. and Basbaum, A. I., 1998), allow sensory testing in the hindpaw (Figure 2). Moderate autotomy, guarding, and excessive grooming of the injured limb were noticed as well as thermal/mechanical hyperalgesia- and allodynia-like behavior were recorded (Bennett, G. J., 1993). Alternative to the CCI models include application of fixed-diameter polyethylene cuffs around the sciatic nerve, favoring a standardization of the loose, but constrictive ligatures in CCI (Mosconi, T. and Kruger, L., 1996). Another valuable alternative is the partial sciatic nerve transection (PST) model where half of the diameter of the sciatic nerve is only transected, avoiding inflammation caused by suture material with PSL. Chronic constriction of the trigeminal nerve (infraorbital nerve CCI, ION–CCI) was designed by Vos B. P. *et al.* (1994) as a model of trigeminal pain. ION–CCI model elicits excessive grooming of the face, mechanical allodynia, and thermal hyperalgesia (Imamura, Y. *et al.*, 1997). Sensory assessment of the animal face is difficult; however, the recent introduction of an automated thermal behavioral assay should provide valuable characterizations of orofacial pain models (Neubert, J. K. *et al.*, 2005).

56.2.4 Models of Peripheral Neuropathic Pain: Distal Nerve Branch Injuries

Injuries to distal nerves branches offer the advantage that they can be anatomically defined, helping the standardization of surgical procedures. The spared nerve injury (SNI) in rats and mice is based on the ligation and transection of the tibial and common peroneal nerves leaving the sural nerve intact (Decosterd, I. and Woolf, C. J., 2000; Bourquin, A. F. *et al.*, 2006). This model results in early, robust, and intense mechanical allodynia/hyperalgesia and cold allodynia in the territory of the sural nerve (Figure 3). It offers the advantage of sparing an almost sensory nerve in combination with a clear separation of injured and noninjured axons distal to the injury. SNI has also been adapted to neonatal animals (Howard, R. F. *et al.*, 2005) and variants of the model have been described in rats and mice (Lee, B. H. *et al.*, 2000; Shields, S. D. *et al.*, 2003). Recently, the saphenous nerve partial ligation (SPL) model (30–50% of the saphenous nerve, a branch of the femoral nerve that innervates the medial part of the paw, is tightly ligated) leads to the development of NP-related symptoms (Walczak, J. S. *et al.*, 2005). The relevance in this model is that the injury of an exclusively sensory nerve is easily transferable for investigations with skin–nerve preparations (Reeh, P. W., 1986). Injury to peripheral branches has also been

Figure 2 In many experimental models of neuropathic pain (NP), stimulus-evoked pain sensitivity is assessed in the plantar side of the hindpaw with various stimuli, usually in the territory of spared or regenerated afferents.

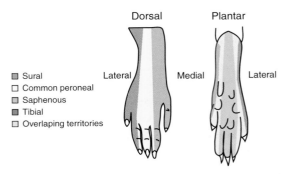

Figure 3 Areas of the dorsal and plantar paw innervated by the sciatic and femoral nerves in the rat. Some overlap between territories may occur (pink).

applied to the trigeminal nerve to study orofacial pain and similar to the SNI model, transection of a branch of the trigeminal nerve (infra-alveolar nerve) leads to hypersensitivity in the whisker pad area, which is innervated by a spared nerve (ION) (Tsuboi, Y. *et al.*, 2004).

56.2.5 Nontraumatic Models of Neuropathic Pain

Vincristine, taxol, and cisplatin may lead to peripheral neuropathy and NP in cancer patients. Models based on these substances have been created with different features of NP-related symptoms, usually in a dose-dependent manner (de Koning, P. *et al.*, 1987; Cavaletti, G. *et al.*, 1995; Aley, K. O. *et al.*, 1996; Authier, N. *et al.*, 2000; Mimura, Y. *et al.*, 2000; Nozaki-Taguchi, N. *et al.*, 2001). Diabetes is a major cause of painful neuropathy in humans. The diabetic model was induced by intraperitoneal injections of streptozotocin (STZ), which leads to pancreatic islet destruction and chronic hyperglycemia (Courteix, C. *et al.*, 1993; 1994). NP symptoms such as hyperalgesia- and allodynia-like behavior appear 2 weeks after. STZ produces functional, biochemical, and structural abnormalities in the sciatic nerve similar to those seen in human diabetic neuropathy (Sima, A. A. and Sugimoto, K., 1999). Prolonged ischemia of the hindpaw was proposed as a surrogate to complex regional pain syndrome (CRPS) type I (Coderre, T. J. *et al.*, 2004). After 8 h and lasting for 4 weeks, mechanical hyperalgesia and bilateral mechanical/thermal allodynia was observed along with excessive shaking and licking of hindpaws (Coderre, T. J. *et al.*, 2004). Axonal damage is combined with inflammation and nerve inflammatory models, without direct nerve injury were developed by wrapping the sciatic nerve with cellulose-containing inflammatory reagents such as carrageenan or complete Freund's adjuvant (CFA) (Eliav, E. *et al.*, 1999). In these cases, the course of pain hypersensitivity is very short (2–5 days only).

56.3 Behavioral Assessment of Neuropathic Pain in Animal Models: Spontaneous Pain, Stimulus-Evoked Pain, and Stimulus-Induced Pain

Tests based on pain sensitivity consider the type of injury, the remaining capacity of withdrawal reflexes (e.g., when neuromuscular defect is caused by the nerve injury), plus the localization, the intensity, and the time course of symptoms. NP in humans is characterized by modification of basal pain sensitivity with ongoing/spontaneous pain, and stimulus-evoked hyper- or hyposensitivity (Woolf, C. J. and Decosterd, I., 1999). Spontaneous pain appears independently of any external stimulus and cannot be assessed directly in animals. Only indirect physiological or behavioral surrogates of spontaneous pain are measurable: spontaneous discharges in sensory neurons, autotomy, posture (e.g., guarding behavior) movements (e.g., brisk paw flinches), or changes in general behavior (e.g., grooming, sleep patterns, vocalization, exploratory behavior) (Mogil, J. S. and Crager, S. E., 2004). All raised substantial issues that have meant laboratories favoring the recording of stimulus-evoked pain.

Stimulus-evoked pain is elicited by distinct stimuli like touch, pinprick, or temperature (heat or cold). Change in the threshold of response and/or quality of response are generally used as outcomes of allodynia/hyperalgesia-like behavior.

Mechanical allodynia-like behavior is usually measured by using a series of calibrated monofilaments of logarithmic incremental force (von Frey hairs) (Bennett, G. J., 2001). Monofilaments are applied in various territories of the animal's body, and positive responses are recorded when the application induces a withdrawal, movement, or vocalization. The evaluation of nociceptive threshold in sensible regions is the result of different procedures, such as the up–down method of Dixon (Chaplan, S. R. *et al.*, 1994). Cold allodynia-like behavior is assessed by placing an acetone drop on the area of interest, the acetone drop quickly evaporates eliciting a cold sensation (Choi, Y. *et al.*, 1994). Frequency of response to multiple applications as well as the type and duration of response have been used to score cold allodynia. Another method for measuring response to cold is the use of a cooled metal floor, usually at noxious low temperatures, thus evaluating cold hyperalgesia (Bennett, G. J. and Xie, Y. K., 1988; Choi, Y. *et al.*, 1994). Thermal hyperalgesia-like behavior is typically evaluated by measuring the latency and duration of reaction toward a heat noxious stimulus induced by a radial infrared beam (Hargreaves, K. *et al.*, 1988). Mechanical hyperalgesia-like behavior is evoked by a pinprick test, using a safety pin stimulus that provokes, in a normal animal, a small brisk withdrawal reflex, while abnormal responses are evoked in experimental NP rats. A large duration of withdrawal and paw guarding/shaking/licking follows the stimulus (Tal, M. and Bennett, G. J., 1994).

Stimulus-induced pain refers to the capacity of external stimuli not only to directly evoke pain, but also to alter sensory processing such as to generate a state of pain hypersensitivity that outlast the primary stimuli. In animals, stimulus-evoked pain after capsaicin, inflammation, or nerve crush (Bennett, G. J. and Xie, Y. K., 1988; Choi, Y. *et al.*, 1994; Ma, Q. P. and Woolf, C. J., 1996; Kim, H. T. *et al.*, 2001; Decosterd, I. *et al.*, 2002) has been investigated after repeated low-threshold mechanical stimuli and further measurement of mechanical sensitivity is carried out using von Frey hairs. In this case, stimulus-induced pain has therefore been termed progressive tactile hypersensitivity.

Internal and external factors influence behavioral outcomes. These include: experimenter, genetic background, and gender of the animals, food, sawdust, noise, transportation, and so on (Chesler, E. J. *et al.*, 2002). Harmonized environments, as well as studies conducted by an investigator unaware of the treatment applied, facilitate this standardization of behavioral assessment.

56.4 Conclusions

Many factors need to be considered in any particular model such as the location, form, extension and evolution of the injury. At present, animal models of NP display unique and indispensable features and they are:

1. representative of a clinical entity or a fundamental mechanism;
2. ease in performance and reproducibility from investigator to investigator and laboratory to laboratory;
3. related to a development of symptoms consistent with clinical NP;
4. allowing well-characterized measure of pain sensitivity by behavioral assessments;
5. presenting robust alterations of nociceptive thresholds; and
6. compatible with reproducible pharmacological response.

Future challenges remain open for the development of new animal models as well as methods of measurement that include for instance measuring devices independent of the experimenter (Neubert, J. K. *et al.*, 2005). Pain symptoms are not equivalent to

pain mechanisms, but they may reflect them. Similarly, all outcome measures in experimental models are not equivalent. A sophisticated approach using multiple models (Decosterd, I. *et al.*, 2004) is then required in order to correlate information obtained in experimental models with clinical syndromes of NP. Although all models of NP have weak points in modeling only partially human NP and the impossibility to have direct behavioral measurement of spontaneous pain, we believe that current and future models of NP will strengthen our knowledge of the underlying mechanisms and favor the development of new treatment strategies for this currently intractable form of pain.

References

Aley, K. O., Reichling, D. B., and Levine, J. D. 1996. Vincristine hyperalgesia in the rat: a model of painful vincristine neuropathy in humans. Neuroscience 73, 259–265.

Ali, Z., Ringkamp, M., Hartke, T. V., Chien, H. F., Flavahan, N. A., Campbell, J. N., and Meyer, R. A. 1999. Uninjured C-fiber nociceptors develop spontaneous activity and alpha-adrenergic sensitivity following L6 spinal nerve ligation in monkey. J. Neurophysiol. 81, 455–466.

Authier, N., Fialip, J., Eschalier, A., and Coudore, F. 2000. Assessment of allodynia and hyperalgesia after cisplatin administration to rats. Neurosci. Lett. 291, 73–76.

Bennett, G. J. 1993. An animal model of neuropathic pain: a review. Muscle Nerve 16, 1040–1048.

Bennett, G. J. 2001. Animal Models of Pain. In: Methods in Pain Research (ed. L. Kruger), pp. 67–91. CRC Press LLC.

Bennett, G. J. and Xie, Y. K. 1988. A peripheral mononeuropathy in rat that produces disorders of pain sensation like those seen in man. Pain 33, 87–107.

Bennett, A. D., Everhart, A. W., and Hulsebosch, C. E. 2000. Intrathecal administration of an NMDA or a non-NMDA receptor antagonist reduces mechanical but not thermal allodynia in a rodent model of chronic central pain after spinal cord injury. Brain Res. 859, 72–82.

Bourquin, A. F., Suveger, M., Pertin, M., Gilliard, N., Sardy, S., Davison, A. C., Spahn, D. R., and Decosterd, I. 2006. Assessment and analysis of mechanical allodynia- like behavior induced by spared nerve injury (SNI) in the mouse. Pain 122, 14.e1–14. e14.

Bruce, J. C., Oatway, M. A., and Weaver, L. C. 2002. Chronic pain after clip-compression injury of the rat spinal cord. Exp. Neurol. 178, 33–48.

Cavaletti, G., Tredici, G., Braga, M., and Tazzari, S. 1995. Experimental peripheral neuropathy induced in adult rats by repeated intraperitoneal administration of taxol. Exp. Neurol. 133, 64–72.

Chaplan, S. R., Bach, F. W., Pogrel, J. W., Chung, J. M., and Yaksh, T. L. 1994. Quantitative assessment of tactile allodynia in the rat paw. J. Neurosci. Methods 53, 55–63.

Chesler, E. J., Wilson, S. G., Lariviere, W. R., Rodriguez-Zas, S. L., and Mogil, J. S. 2002. Influences of laboratory environment on behavior. Nat. Neurosci. 5, 1101–1102.

Choi, Y., Yoon, Y. W., Na, H. S., Kim, S. H., and Chung, J. M. 1994. Behavioral signs of ongoing pain and cold allodynia in a rat model of neuropathic pain. Pain 59, 369–376.

Christensen, M. D. and Hulsebosch, C. E. 1997a. Chronic central pain after spinal cord injury. J. Neurotrauma 14, 517–537.

Christensen, M. D. and Hulsebosch, C. E. 1997b. Spinal cord injury and anti-NGF treatment results in changes in CGRP density and distribution in the dorsal horn in the rat. Exp. Neurol. 147, 463–475.

Christensen, M. D., Everhart, A. W., Pickelman, J. T., and Hulsebosch, C. E. 1996. Mechanical and thermal allodynia in chronic central pain following spinal cord injury. Pain 68, 97–107.

Coderre, T. J., Xanthos, D. N., Francis, L., and Bennett, G. J. 2004. Chronic post-ischemia pain (CPIP): a novel animal model of complex regional pain syndrome-Type I (CRPS-I; reflex sympathetic dystrophy) produced by prolonged hindpaw ischemia and reperfusion in the rat. Pain 112, 94–105.

Colburn, R. W., Rickman, A. J., and DeLeo, J. A. 1999. The effect of site and type of nerve injury on spinal glial activation and neuropathic pain behavior. Exp. Neurol. 157, 289–304.

Courteix, C., Bardin, M., Chantelauze, C., Lavarenne, J., and Eschalier, A. 1994. Study of the sensitivity of the diabetes-induced pain model in rats to a range of analgesics. Pain 57, 153–160.

Courteix, C., Eschalier, A., and Lavarenne, J. 1993. Streptozocin-induced diabetic rats: behavioural evidence for a model of chronic pain. Pain 53, 81–88.

Decosterd, I. and Woolf, C. J. 2000. Spared nerve injury: an animal model of persistent peripheral neuropathic pain. Pain 87, 149–158.

Decosterd, I., Allchorne, A., and Woolf, C. J. 2002. Progressive tactile hypersensitivity after a peripheral nerve crush: non-noxious mechanical stimulus-induced neuropathic pain. Pain 100, 155–162.

Decosterd, I., Allchorne, A., and Woolf, C. J. 2004. Differential analgesic sensitivity of two distinct neuropathic pain models. Anesth. Analg. 99, 457–463.

Eliav, E., Herzberg, U., Ruda, M. A., and Bennett, G. J. 1999. Neuropathic pain from an experimental neuritis of the rat sciatic nerve. Pain 83, 169–182.

Eschenfelder, S., Habler, H. J., and Janig, W. 2000. Dorsal root section elicits signs of neuropathic pain rather than reversing them in rats with L5 spinal nerve injury. Pain 87, 213–219.

Fehlings, M. G. and Nashmi, R. 1995. Assessment of axonal dysfunction in an in vitro model of acute compressive injury to adult rat spinal cord axons. Brain Res. 677, 291–299.

Gruner, J. A. 1992. A monitored contusion model of spinal cord injury in the rat. J. Neurotrauma 9, 123–126.

Gwak, Y. S., Hains, B. C., Johnson, K. M., and Hulsebosch, C. E. 2004. Effect of age at time of spinal cord injury on behavioral outcomes in rat. J. Neurotrauma 21, 983–993.

Hains, B. C., Klein, J. P., Saab, C. Y., Craner, M. J., Black, J. A., and Waxman, S. G. 2003a. Upregulation of sodium channel Nav1.3 and functional involvement in neuronal hyperexcitability associated with central neuropathic pain after spinal cord injury. J. Neurosci. 23, 8881–8892.

Hains, B. C., Willis, W. D., and Hulsebosch, C. E. 2003b. Serotonin receptors 5-HT1A and 5-HT3 reduce hyperexcitability of dorsal horn neurons after chronic spinal cord hemisection injury in rat. Exp. Brain Res. 149, 174–186.

Hao, J. X., Xu, X. J., Aldskogius, H., Seiger, A., and Wiesenfeld-Hallin, Z. 1991. Allodynia-like effects in rat after ischaemic spinal cord injury photochemically induced by laser irradiation. Pain 45, 175–185.

Hao, J. X., Xu, X. J., Aldskogius, H., Seiger, A., and Wiesenfeld-Hallin, Z. 1992. Photochemically induced transient spinal ischemia induces behavioral hypersensitivity to mechanical and cold stimuli, but not to noxious-heat stimuli, in the rat. Exp. Neurol. 118, 187–194.

Hargreaves, K., Dubner, R., Brown, F., Flores, C., and Joris, J. 1988. A new and sensitive method for measuring thermal nociception in cutaneous hyperalgesia. Pain 32, 77–88.

Howard, R. F., Walker, S. M., Mota, P. M., and Fitzgerald, M. 2005. The ontogeny of neuropathic pain: postnatal onset of mechanical allodynia in rat spared nerve injury (SNI) and chronic constriction injury (CCI) models. Pain 115, 382–389.

Hu, S. J. and Xing, J. L. 1998. An experimental model for chronic compression of dorsal root ganglion produced by intervertebral foramen stenosis in the rat. Pain 77, 15–23.

Hulsebosch, C. E. 2002. Recent advances in pathophysiology and treatment of spinal cord injury. Adv. Physiol. Educ. 26, 238–255.

Hulsebosch, C. E., Xu, G. Y., Perez-Polo, J. R., Westlund, K. N., Taylor, C. P., and McAdoo, D. J. 2000. Rodent model of chronic central pain after spinal cord contusion injury and effects of gabapentin. J. Neurotrauma 17, 1205–1217.

Imamura, Y., Kawamoto, H., and Nakanishi, O. 1997. Characterization of heat-hyperalgesia in an experimental trigeminal neuropathy in rats. Exp. Brain Res. 116, 97–103.

Kim, S. H. and Chung, J. M. 1992. An experimental model for peripheral neuropathy produced by segmental spinal nerve ligation in the rat. Pain 50, 355–363.

Kim, H. T., Park, S. K., Lee, S. E., Chung, J. M., and Lee, D. H. 2001. Non-noxious A fiber afferent input enhances capsaicin-induced mechanical hyperalgesia in the rat. Pain 94, 169–175.

de Koning, P., Neijt, J. P., Jennekens, F. G., and Gispen, W. H. 1987. Evaluation of cis-diamminedichloroplatinum (II) (cisplatin) neurotoxicity in rats. Toxicol. Appl. Pharmacol. 89, 81–87.

Lee, B. H., Won, R., Baik, E. J., Lee, S. H., and Moon, C. H. 2000. An animal model of neuropathic pain employing injury to the sciatic nerve branches. Neuroreport 11, 657–661.

Li, Y., Dorsi, M. J., Meyer, R. A., and Belzberg, A. J. 2000. Mechanical hyperalgesia after an L5 spinal nerve lesion in the rat is not dependent on input from injured nerve fibers. Pain 85, 493–502.

Li, L., Xian, C. J., Zhong, J. H., and Zhou, X. F. 2002. Effect of lumbar 5 ventral root transection on pain behaviors: a novel rat model for neuropathic pain without axotomy of primary sensory neurons. Exp. Neurol. 175, 23–34.

Ma, C. and LaMotte, R. H. 2005. Enhanced excitability of dissociated primary sensory neurons after chronic compression of the dorsal root ganglion in the rat. Pain 113, 106–112.

Ma, Q. P. and Woolf, C. J. 1996. Progressive tactile hypersensitivity: an inflammation-induced incremental increase in the excitability of the spinal cord. Pain 67, 97–106.

Malmberg, A. B. and Basbaum, A. I. 1998. Partial sciatic nerve injury in the mouse as a model of neuropathic pain: behavioral and neuroanatomical correlates. Pain 76, 215–222.

Mills, C. D., Hains, B. C., Johnson, K. M., and Hulsebosch, C. E. 2001. Strain and model differences in behavioral outcomes after spinal cord injury in rat. J. Neurotrauma 18, 743–756.

Mimura, Y., Kato, H., Eguchi, K., and Ogawa, T. 2000. Schedule dependency of paclitaxel-induced neuropathy in mice: a morphological study. Neurotoxicology 21, 513–520.

Mogil, J. S. and Crager, S. E. 2004. What should we be measuring in behavioral studies of chronic pain in animals? Pain 112, 12–15.

Mogil, J. S., Wilson, S. G., Bon, K., Lee, S. E., Chung, K., Raber, P., Pieper, J. O., Hain, H. S., Belknap, J. K., Hubert, L., Elmer, G. I., Chung, J. M., and Devor, M. 1999. Heritability of nociception I: responses of 11 inbred mouse strains on 12 measures of nociception. Pain 80, 67–82.

Mosconi, T. and Kruger, L. 1996. Fixed-diameter polyethylene cuffs applied to the rat sciatic nerve induce a painful neuropathy: ultrastructural morphometric analysis of axonal alterations. Pain 64, 37–57.

Neubert, J. K., Widmer, C. G., Malphurs, W., Rossi, H. L., Vierck, C. J., and Caudle, R. M. 2005. Use of a novel thermal operant behavioral assay for characterization of orofacial pain sensitivity. Pain 116, 386–395.

Nozaki-Taguchi, N., Chaplan, S. R., Higuera, E. S., Ajakwe, R. C., and Yaksh, T. L. 2001. Vincristine-induced allodynia in the rat. Pain 93, 69–76.

Reeh, P. W. 1986. Sensory receptors in mammalian skin in an *in vitro* preparation. Neurosci. Lett. 66, 141–146.

Rosenzweig, E. S. and McDonald, J. W. 2004. Rodent models for treatment of spinal cord injury: research trends and progress toward useful repair. Curr. Opin. Neurol. 17, 121–131.

Seltzer, Z., Dubner, R., and Shir, Y. 1990. A novel behavioral model of neuropathic pain disorders produced in rats by partial sciatic nerve injury. Pain 43, 205–218.

Sheen, K. and Chung, J. M. 1993. Signs of neuropathic pain depend on signals from injured nerve fibers in a rat model. Brain Res. 610, 62–68.

Sheth, R. N., Dorsi, M. J., Li, Y., Murinson, B. B., Belzberg, A. J., Griffin, J. W., and Meyer, R. A. 2002. Mechanical hyperalgesia after an L5 ventral rhizotomy or an L5 ganglionectomy in the rat. Pain 96, 63–72.

Shields, S. D., Eckert, W. A., and Basbaum, A. I. 2003. Spared nerve injury model of neuropathic pain in the mouse: a behavioral and anatomic analysis. J. Pain 4, 465–470.

Siddall, P. J. and Loeser, J. D. 2001. Pain following spinal cord injury. Spinal Cord 39, 63–73.

Siddall, P., Xu, C. L., and Cousins, M. 1995. Allodynia following traumatic spinal cord injury in the rat. Neuroreport 6, 1241–1244.

Sima, A. A. and Sugimoto, K. 1999. Experimental diabetic neuropathy: an update. Diabetologia 42, 773–788.

Sjolund, B. H. 2002. Pain and rehabilitation after spinal cord injury: the case of sensory spasticity? Brain Res. Rev. 40, 250–256.

Tabo, E., Jinks, S. L., Eisele, J. H., Jr., and Carstens, E. 1999. Behavioral manifestations of neuropathic pain and mechanical allodynia, and changes in spinal dorsal horn neurons, following L4–L6 dorsal root constriction in rats. Pain 80, 503–520.

Tal, M. and Bennett, G. J. 1994. Extra-territorial pain in rats with a peripheral mononeuropathy: mechano-hyperalgesia and mechano-allodynia in the territory of an uninjured nerve. Pain 57, 375–382.

Tsuboi, Y., Takeda, M., Tanimoto, T., Ikeda, M., Matsumoto, S., Kitagawa, J., Teramoto, K., Simizu, K., Yamazaki, Y., Shima, A., Ren, K., and Iwata, K. 2004. Alteration of the second branch of the trigeminal nerve activity following inferior alveolar nerve transection in rats. Pain 111, 323–334.

Vierck, C. J. and Light, A. R. 1999. Effects of combined hemotoxic and anterolateral spinal lesions on nociceptive sensitvity. Pain 83, 447–457.

Vierck, C. J., Siddall, P., and Yezierski, R. P. 2000. Pain following spinal cord injury: animal models and mechanistic studies. Pain 89, 1–5.

Vos, B. P., Strassman, A. M., and Maciewicz, R. J. 1994. Behavioral evidence of trigeminal neuropathic pain following chronic constriction injury to the rat's infraorbital nerve. J. Neurosci. 14, 2708–2723.

Walczak, J. S., Pichette, V., Leblond, F., Desbiens, K., and Beaulieu, P. 2005. Behavioral, pharmacological and molecular characterization of the saphenous nerve partial ligation: a new model of neuropathic pain. Neuroscience 132, 1093–1102.

Wall, P. D., Devor, M., Inbal, R., Scadding, J. W., Schonfeld, D., Seltzer, Z., and Tomkiewicz, M. M. 1979a. Autotomy following peripheral nerve lesions: experimental anaesthesia dolorosa. Pain 7, 103–111.

Wall, P. D., Scadding, J. W., and Tomkiewicz, M. M. 1979b. The production and prevention of experimental anesthesia dolorosa. Pain 6, 175–182.

Weng, H.-R, Lee, J. I., Lenz, F. A., Schwartz, A., Vierck, C., Rowland, L., and Dougherty, P. M. 2000. Functional plasticity in primate somatosensory thalamus following chronic lesion of the ventral lateral spinal cord. Neuroscience 101, 393–401.

Woolf, C. J. 1983. Evidence for a central component of post-injury pain hypersensitivity. Nature 306, 686–688.

Woolf, C. J. and Decosterd, I. 1999. Implications of recent advances in the understanding of pain pathophysiology for the assessment of pain in patients. Pain Suppl. 6, S141–S147.

Woolf, C. J. and Fitzgerald, M. 1983. The properties of neurones recorded in the superficial dorsal horn of the rat spinal cord. J. Comp. Neurol. 221, 313–328.

Yezierski, R. P., Liu, S., Ruenes, G. L., Kajander, K. J., and Brewer, K. L. 1998. Excitotoxic spinal cord injury: behavioral and morphological characteristics of a central pain model. Pain 75, 141–155.

Yezierski, R. P., Santana, M., Park, S. H., and Madsen, P. W. 1993. Neuronal degeneration and spinal cavitation following intraspinal injections of quisqualic acid in the rat. J. Neurotrauma 10, 445–456.

Zhang, J. M., Song, X. J., and LaMotte, R. H. 1999. Enhanced excitability of sensory neurons in rats with cutaneous hyperalgesia produced by chronic compression of the dorsal root ganglion. J. Neurophysiol. 82, 3359–3366.

57 Neuropathic Pain: Clinical

R Baron, Christian-Albrechts-Universität Kiel, Kiel, Germany

Glossary

CRPS Complex regional pain syndrome.
DRG Dorsal root ganglion, spinal ganglion.
MEG Magnetic encephalography.
NCF Nucleus cuneiformis.
PAG Periaqueductal gray.

PET Positron emission tomography.
TRP Ion channels of transient receptor potential family.
TNF Tumor necrosis factor.
QST Quantitative sensory testing.

57.1 Definition of Neuropathic Pain

The current NeuPSIG (Neuropathic pain special interest group of the International Association for the Study of Pain) taxonomy of chronic pain, defines neuropathic pain as a "Pain arising as a direct consequence of a lesion or disease affecting the somatosensory system." The term 'disease' refers to identifiable disease processes such as inflammatory, autoimmune conditions, or channelopathies, while lesions refer to macro- or microscopically identifiable damage. The restriction to the somatosensory system is necessary, because diseases and lesions of other parts of the nervous system may cause nociceptive pain. For example, lesions or diseases of the motor system may lead to spasticity or rigidity, and thus may indirectly cause muscle pain. The latter pain conditions are now explicitly excluded from the condition neuropathic pain. Where possible, neuropathic pain should be qualified as being of peripheral or central origin in terms of the location of the lesion or disease process. This distinction is important, as lesions or diseases of the central nervous system (CNS) and peripheral nervous system (PNS) are distinct in terms of clinical manifestations and underlying pathophysiology.

57.2 Epidemiology of Neuropathic Pain

Chronic neuropathic pain is common in clinical practice, greatly impairs the quality of life of patients and is a major economical health problem. Estimates of point prevalence for neuropathic pain in the general population are as high as 5% (Daousi, C. et al., 2004), a quarter of them suffering from severe intensity (McDermott, A. M. et al., 2006). Moreover, a recent prospective cross-sectional survey in more than 12 000 chronic pain patients with both

nociceptive and neuropathic pain types who were referred to pain specialists in Germany (Freynhagen, R. et al., 2005/2006) revealed that already 13% of these patients suffer from the two classical neuropathic disorders, postherpetic neuralgia (PHN) and painful diabetic neuropathy and 40% of all patients at least have a neuropathic component to their discomfort (especially patients with chronic back pain and radiculopathy). In neuropathic patients pain therapy was already performed for 2–4 years, one-third of the patients mentioned three or more doctor visits for pain therapy in the past month and 8–20% already applied for pension (Table 1).

Comorbidities such as poor sleep, depressed mood, and anxiety are common in neuropathic pain and have a significant impact on the global pain experience. The German survey revealed that about 60% of neuropathic patients were moderate or severely depressed compared with 25% in the nociceptive patient group. Optimal sleep quality was only scored by 27% of the neuropathic patients and in 49% of the nociceptive pain patients, respectively.

57.3 Classification

57.3.1 Disease/Anatomy-Based Classification

It is common clinical practice to classify neuropathic pain according to the underlying etiology of the disorder and the anatomical location of the specific lesion (Jensen, T. S. and Baron, R., 2003). The majority of patients fall into four broad classes (Table 2): painful peripheral neuropathies (focal, multifocal, or generalized, e.g., traumatic, ischemic, inflammatory, toxic, metabolic, and hereditary), central pain syndromes (e.g., stroke, multiple sclerosis (MS), and spinal cord injury (SCI)), complex painful neuropathic disorders (complex regional pain syndromes, CRPSs) and mixed pain syndromes (combination of

Table 1   Sociodemographic and clinical characteristics in patients with postherpetic neuralgia (PHN) and diabetic painful neuropathy (DPN) (%)

	PHN	DPN
Number of patients	498	1144
Male	199 (40.0)	596 (52.1)
Female	299 (60.0)	548 (47.9)
Age (years)		
Mean	60.6	61.7
SD ±	15.4	13.2
Duration of pain treatment (months)		
Mean	24	52
SD ±	41	63
No. of doctor visits for pain during the past 4 weeks		
One visit	188 (37.8)	348 (30.4)
Two visits	135 (27.1)	339 (29.6)
Three or more	132 (26.5)	348 (30.4)
Ongoing psychotherapy	8 (1.6)	34 (3.0)
Intended/submitted pension application	42 (8.2)	221 (19.3)
Pain intensity on VAS during the past 4 weeks		
Mild	154 (30.9)	492 (43.0)
Moderate	239 (48.0)	424 (37.1)
Severe	105 (21.1)	228 (19.9)
VAS worst pain (mean)	7.4	6.3
VAS average pain (mean)	5.5	4.9

VAS, visual analogue scale.

nociceptive and neuropathic pain, e.g., chronic low back pain with radiculopathy).

The anatomical distribution pattern of the affected nerves provides valuable differential diagnostic clues as to possible underlying causes. Therefore, painful peripheral neuropathies are grouped into symmetrical generalized polyneuropathies, disease affecting many nerves simultaneously and into asymmetrical neuropathies with a focal or multifocal distribution or processes affecting the brachial or lumbosacral plexuses.

One important subgroup of painful polyneuropathies is characterized by a predominant or in some cases even isolated affection of small afferent fibers, i.e., unmyelinated C fibers and small myelinated $A\delta$ fibers. In many cases, autonomic efferent small fiber systems are also affected. Several different etiologies may lead to small-fiber polyneuropathies (see below), up to 20% of cases, however, are of unknown cause. It is important to realize that conventional electrophysiological techniques like nerve conduction study (NCS), sensory-evoked potential (SEP), etc., only assess the function of myelinated peripheral axonal systems, the affection of small fibers are missed. Therefore, especially in small-fiber neuropathies alternative diagnostic procedures have to be used (see below).

57.3.1.1 Painful peripheral (focal, multifocal, generalized) neuropathies

Although most of the different neuropathic pain syndromes are common and well known to the pain specialists, there are certain entity-specific features that should be mentioned in detail (See also Table 3):

57.3.1.1.(i) Diabetic neuropathy

- High prevalence
- Several types of polyneuropathies, most prevalent symmetrical distally, sensory predominant neuropathy
- Numbness and paresthesias (sometimes burning quality) are common
- Sometimes spontaneous, deep, aching pain, and lightning pain
- Severe sensory neuropathy in diabetes may lead to painless perforating foot ulcers often associated with autonomic neuropathy
- Patients without pain tend to have areflexia with distal sensory loss particularly involving joint position sense
- Patients with severe burning pain and hyperesthesia tend to have sensory loss with a relative preservation of position sense and intact reflexes and often have accompanying autonomic neuropathy

Table 2   Disease-/anatomy-based classification of painful peripheral neuropathies

Painful peripheral neuropathies
  Focal, multifocal
    Phantom pain, stump pain, nerve transection pain (partial or complete)
    Neuroma (posttraumatic or postoperative)
    Posttraumatic neuralgia
    Entrapment syndromes
    Mastectomy
    Postthoracotomy
    Morton's neuralgia
    Painful scars
    Herpes zoster and postherpetic neuralgia
    Diabetic mononeuropathy, diabetic amyotrophy
    Ischemic neuropathy
    Borreliosis
    Connective tissue disease (vasculitis)
    Neuralgic amyotrophy
    Peripheral nerve tumors
    Radiation plexopathy
    Plexus neuritis (idiopathic or hereditary)
    Trigeminal or glossopharyngeal neuralgia
    Vascular compression syndromes

Generalized (polyneuropathies)
  Metabolic or nutritional
    Diabetic, often 'Burning feet syndrome'
    Alcoholic
    Amyloid
    Hypothyroidism
    Beri beri, pellagra
  Drugs
    Antiretrovirals, cisplatin, disulfiram, ethambutol, isoniazid, nitrofurantoin, thalidomid, thiouracil, vincristine,
    chloramphenicol, metronidazole, taxoids, gold
  Toxins
    Acrylamide, arsenic, clioquinol, dinitrophenol, ethylene oxide, pentachlorophenol, thallium
  Hereditary
    Amyloid neuropathy
    Fabry's disease
    Charcot–Marie–Tooth disease type 5, type 2B
    Hereditary sensory and autonomic neuropathy (HSAN) type 1, type 1B
  Malignant
    Carcinomatous (paraneoplastic), myeloma
  Infective or postinfective, immune
    Acute or inflammatory polyradiculoneuropathy (Guillain–Barré syndrome), borreliosis, HIV
  Other polyneuropathies
    Erythromelalgia
    Idiopathic small-fiber neuropathy
    Trench foot (cold injury)

Central pain syndromes
  Vascular lesions in the brain (especially brainstem and thalamus) and spinal cord
    Infarct
    Hemorrhage
    Vascular malformation
  Multiple sclerosis
  Traumatic spinal cord injury including iatrogenic cordotomy
  Traumatic brain injury
  Syringomyelia and syringobulbia
  Tumors
  Abscesses
  Inflammatory diseases other than multiple sclerosis; myelitis caused by viruses, syphilis
  Epilepsy
  Parkinson's disease

(Continued)

Table 2 (Continued)

Complex painful neuropathic disorders
  Complex regional pain syndromes type I and II (reflex sympathetic dystrophy, causalgia)

Mixed pain syndromes
  Chronic low back pain with radiculopathy
  Cancer pain with malignant plexus invasion
  Complex regional pain syndromes

Table 3 Special clinical features that are relevant for specific diagnoses

Special feature	Possible diagnosis
Cold allodynia	Traumatic nerve injury
	Trench foot syndrome
	Complex regional pain syndrome
	Oxaliplatin-induced polyneuropathy
	Central poststroke pain
Deep somatic allodynia	Complex regional pain syndrome
SMP	Complex regional pain syndrome, acute herpes
Isolated small-fiber affection	Diabetic polyneuropathy
	Amyloid polyneuropathy
	Fabry's disease
	Hereditary polyneuropathy
	Idiopathic small-fiber polyneuropathy
Painful polyneuropathy in several family members	Amyloid polyneuropathy,
	Fabry's disease
	Charcot–Marie–Tooth disease type 5, type 2B
	Hereditary sensory, autonomic polyneuropathy (HSAN) type 1, 1B

SMP, sympathetically maintained pain.

• Small sensory and autonomic fibers are involved early in diabetic neuropathy as well as in impaired glucose tolerance (IGT)

• Abnormal epidermal C fiber density is highly correlated with warm detection threshold (QST)

• Regeneration capacity of cutaneous C fibers (experimental challenge with topical capsaicin) is reduced

• Hyperglycemia in diabetic patients may itself be an important factor for acute exacerbations of pain

• Up 40% of patients with a so-called burning feet syndrome have a small-fiber neuropathy and many already demonstrate an IGT indicative of early diabetes

57.3.1.1.(ii) Alcoholic neuropathy

• Incidence of neuropathy in chronic alcoholism about 10%

• Paresthesias, burning pain, and tenderness of the feet

• Painful symptoms not related to the severity of somatosensory deficit

• Sensorimotor small- and large-fiber neuropathy
• Sometimes early small-fiber deficit

57.3.1.1.(iii) Amyloid neuropathy

• Predominant painful small fiber neuropathy in both inherited and sporadic forms

• Predominant distal sensory loss of pain and thermal sensations, often autonomic involvement

• Severe pain usually of a deep aching quality, sometimes shooting pains

57.3.1.1.(iv) Fabry's disease (angiokeratoma corporis diffusum)

• Lipid (glycosphingolipid) storage disorder, sex-linked recessive

• Heterozygous carrier females occasionally develop symptoms later in life

• Deficiency of α-galactosidase A lead to accumulation of glycosphingolipid (globotriaosylceramide) in tissues

- Painful small-fiber neuropathy early and presenting clinical feature
- Boys or young men present with tenderness of the feet and severe spontaneous burning pain in the legs
- Stroke in young adults due endothelial vasculopathy, about 5% of young male patients with cryptogenic stroke have Fabry's disease
- Extraneural manifestations: cutaneous telangiectasia, cardiac, renal, pulmonary failure
- Patients with preserved renal function involves exclusively small myelinated and unmyelinated fibers
- Normal results of nerve conduction studies
- Normal results of large-fiber quantitation by sural nerve biopsy

57.3.1.1.(v) Idiopathic small-fiber neuropathy

- Predominant or isolated loss of small fibers involving somatic and autonomic neurons
- Often elderly patients
- 40% of patients with a so-called burning feet syndrome have a small-fiber neuropathy suffering from burning sensation, dysesthesia and paresthesias in the feet, lancinating pains, and minimal or no distal weakness
- Abnormalities of epidermal skin innervation (skin biopsies)
- Normal results of nerve conduction studies
- Normal results of large-fiber quantitation by sural nerve biopsy

57.3.1.1.(vi) Human immunodeficiency virus- and highly active antiretroviral therapy–associated neuropathy

- Incidence about 15–30% in the human immunodeficiency virus (HIV) population
- Painful sensory polyneuropathy
- Distal painful paresthesias, spontaneous pain, evoked shock-like pain, and mechanical allodynia with symmetric stocking-like distribution associated with autonomic neuropathy
- Neuropathy is due to a combination of the virus itself and antiretroviral drugs (nucleoside analogue reverse transcriptase inhibitors, highly active antiretroviral therapy (HAART))
- Clinically, these two neuropathies are indistinguishable
- Nerve pathology in the painful sensory polyneuropathy includes axonal atrophy and endoneurial capillary thickening

57.3.1.1.(vii) Erythromelalgia (erythermalgia)

- Rare condition characterized by painful, red, hot extremities
- Acquired form sometimes associated with polycythemia rubra vera
- Rare hereditary form caused by a genetic defect of the sodium channel $Na_v1.7$
- Symptoms are aggravated by heat
- Immediate pain relief by cooling the affected limb
- Feet affected in more than 80%, hands in 25%
- Underlying small fiber neuropathy in some cases

57.3.1.1.(viii) Acute herpes zoster and postherpetic neuralgia

- Acute zoster: neurocutaneous disease with unilateral efflorescences within one or few dermatomes
- Reactivation of varicella zoster virus, which resides in latent form in sensory trigeminal and dorsal root ganglia following primary infection as varicella
- Preherpetic neuralgia (pain before cutaneous symptoms of acute zoster) of some days
- Every dermatome may be affected, thoracic 54%, especially Th5–10 (15%), trigeminal 20%, especially first branch (13%)
- PHN: chronic pain syndrome for more than 3 months after healing of acute zoster lesions
- Incidence of pain 9–15% after 1 month, 2–5% of all age groups after 1 year
- Main risk factor for PHN is age
- Spontaneous burning pain, stimulus-evoked pain, and shooting pains possible

57.3.1.1.(ix) Phantom limb pain

- Phantom limb: amputation of an extremity or other body part or transection of a peripheral nerve leads almost in all cases to a sensation of the missing or denervated limb
- Phantom limb pain: pain in the missing body part, up to 80% of patients, mainly after extremity amputation, cramp-like severe pain in distal phantom, sometime painful phantom movements
- Stump pain: the amputation stump itself shows mechanical allodynia especially close to the skin flap scar or near tender neuromas
- Telescoping: seemingly shrinking phantom
- Aggravation of both painful and nonpainful phantom sensations by external stimuli (noxious or

innocuous mechanical stimuli at the stump, sometimes at face or remote ipsilateral body parts)

• Aggravation of both painful and nonpainful phantom sensation by internal stimuli (micturition, yawning) or emotional stimuli

• Main reasons for amputations in times of peace: peripheral vascular disease (60%), half of the patients also suffer from additional diabetes mellitus, accidents (20%), rare causes like tumor, arterial thrombosis, and osteomyelitis (20%)

• Pathophysiological mechanisms: it seems obvious that the phantom limb image with its complex perceptual, emotional and cognitive qualities involves interaction of some cerebral structures. This, however, may not preclude the notion that neuronal mechanisms at a much lower level of the CNS or even within the severed peripheral nerves may furnish the brain with the information necessary for creating the phantom.

57.3.1.2 Central pain syndromes

Central pain is defined as chronic pain following a lesion or disease of the CNS. The cause of pain is a primary process within the CNS. All lesions that cause central pain affect the somatosensory pathways. They may be located at any level of the neuraxis. Thus lesions at the first synapse in the dorsal horn of the spinal cord or trigeminal nuclei, along the ascending pathways through the spinal cord (Figure 1) and brainstem, in the thalamus, in the

subcortical white matter, and in the cerebral cortex have all been reported (Leijon, G. et al., 1989).

The most common are cerebrovascular lesions, MS, and traumatic spinal cord injuries (SCIs). In these diseases the incidences are around 8%, 28%, and 40%, respectively (Siddall, P. J. et al., 2003). In stroke, there is a particularly high prevalence among patients with brainstem and thalamic lesions, whereas there is no difference in the prevalence after ischemic and hemorrhagic events. In patients with thalamic infarction only patients with lesions that includes the ventroposterior region, which receives a particularly dense spinothalamic projection develop central pain. In MS it has not been possible to determine the location of the lesions that cause central pain, but clinical observations indicate that many of them are located in the spinal cord.

In stroke the pain is most frequently a hemipain (75%). In MS it affects one or both sides more often in the legs (87%) than in the upper extremities (31%). Trigeminal neuralgia, caused by a lesion in the brain stem, occurs in 5% of all MS patients. In all conditions, there is a large variation in the quality of central pain.

Often central pain develops with a latency of weeks or month after the inciting event. The most frequent clinical characteristics are spontaneous burning pain, shooting pain and evoked pain within the affected body part. Allodynia to touch or light pressure as well as to heat or cold are common. Many internal and external events influence central pain,

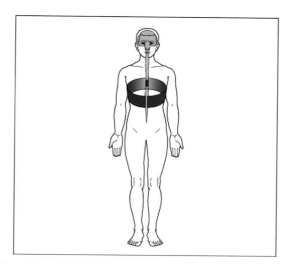
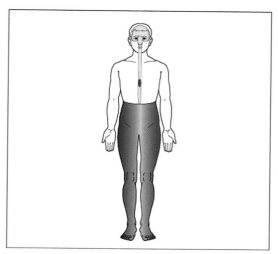

Figure 1 Different pains in spinal cord injury patients. (a) Typical pattern of below-level neuropathic pain following spinal cord injury (SCI) at T7. The dark shading represents the distribution of pain. (b) Typical pattern of at-level neuropathic pain following SCI at T7. The dark shading represents the distribution of pain. Adapted from Siddall, P. 2006. Pain Following Spinal Gord Injury. In: Walland Melzack's Textbook of Pain, 5th edn. (eds. McMahan, S. B. and Koltzenburg, M.), pp. 1043–1056. Elsevier.

such as cutaneous stimuli, body movements, visceral stimuli, emotions, and changes in mood.

In order to diagnose central pain it is necessary to ensure that the patient has a CNS disorder by a detailed history and a neurological examination, computed tomography (CT), magnetic resonance imaging (MRI), cerebrospinal fluid analysis, or electrophysiological techniques (in particular QST to assess spinothalamic function). It is necessary to exclude painful peripheral neuropathy as a cause. Polyneuropathy is not uncommon in, for instance stroke patients, a group with a high incidence of diabetes.

The pathophysiological mechanisms of central pain are unclear. An affection of the spinothalamo-cortical pathways seems to be crucial for the development of central pain, whereas isolated lesions of the lemniscal system are never associated with pain. Thus, the pain is always correlated with sensory disturbances, dominated by abnormalities in the sensibility to temperature and pain. Decreased sensibility to touch, vibration, and joint movements may be present, but is often less pronounced. It is hypothesized that central pain occurs only in patients who have lesions of the spinothalamocortical pathways. Lesioning of the spinothalamic tract leads to a loss of input to normal pain-modulating systems in particular in thalamic neurons. This loss may creates neuronal signs of hyperexcitability. Furthermore, infarction of the dorsolateral thalamus may lead to neuronal disinhibition in medial thalamic nuclei. Incomplete SCI may be associated with sensitization of spinal dorsal horn neurons.

57.3.1.2.(i) Spinal cord injury pain

- Prevalence of pain about 65% (Siddall, P. J. *et al.*, 2003)
- Four types of pain (Figure 1):

1. Nociceptive pain (musculoskeletal, spastic, visceral)
2. Neuropathic pain above level (secondary to nerve lesions: compressive mononeuropathies, CRPS)
3. Neuropathic pain at level (segmental, circumferential band-like distribution: nerve root compression, syringomyelia, spinal cord trauma, or ischemia)
4. Neuropathic pain below level (below the level of the spinal cord lesion perceived diffusely in anesthetic regions: spinal cord trauma, or ischemia)

- Musculoskeletal nociceptive pain is the most common (58%)
- At-level neuropathic pain 42%, sharp, shooting, electric, burning, and stabbing pain located in a region of sensory disturbance
- Below-level neuropathic pain 34%, develops months and even years following initial injury, spontaneous and/or evoked pain diffusely caudal to the level of SCI, burning, aching, stabbing, or electric shocks, often with hyperalgesia
- Fluctuation with mood, activity, infections
- Triggered by sudden noises, jarring movements
- Both complete and partial injuries may be associated with the diffuse, burning pain
- Incomplete injuries often show allodynia due to sparing of lemniscal tracts conveying touch sensations
- Mainly not painful phantom sensation present in 90%

57.3.1.3 *Complex painful neuropathic disorders*

In addition to the classical neuropathic syndromes like painful diabetic neuropathy, PHN, or phantom limb pain there are certain chronic painful conditions that share many clinical characteristics. These syndromes were formerly called reflex sympathetic dystrophy, M. Sudeck or causalgia and are now classified under the umbrella term CRPSs. CRPS are painful disorders that may develop as a disproportionate consequence of trauma typically affecting the limbs (Janig, W. and Baron, R., 2003; Baron, R. and Janig, W., 2004). CRPS type I usually develops after minor trauma with a small or no obvious nerve lesion at an extremity (e.g., bone fracture, sprains, bruises or skin lesions, surgeries). CRPS type II develops after trauma with a mostly large nerve lesion. The patients often report a burning spontaneous pain felt in the distal part of the affected extremity. Stimulus-evoked pains are a striking clinical feature; they include mechanical and thermal allodynia and/or hyperalgesia. These sensory abnormalities are most pronounced distally, and have no consistent spatial relationship to individual nerve territories or to the site of the inciting lesion. Typically, pain can be elicited by movement of and pressure on the joints (deep somatic allodynia), even if these are not directly affected by the inciting lesion.

Unlike other classical neuropathic pain syndromes CRPS show additional nonsensory features like signs of autonomic dysfunction (abnormal regulation of blood flow and sweating), edema of skin and subcutaneous tissues, active and passive movement disorders, and trophic changes of skin, appendages of skin and subcutaneous tissues. Especially for the acute phase, there is accumulating evidence that inflammatory processes might be involved. Thus, it is likely that nociceptive as well as neuropathic mechanisms are responsible for pain generation in CRPS, which would then fit into the classification if a mixed pain syndrome (see below).

57.3.1.4 Mixed pain syndromes

There is agreement that both nociceptive and neuropathic processes contribute to many chronic pain syndromes and that these different mechanisms may explain the qualitatively different symptoms and signs that patients experience. In particular, patients with chronic low back pain, cancer pain, and CRPS seem to fit into this theoretical construct (Baron, R. and Binder, A., 2004).

In chronic low back pain, the nociceptive component stems from activation of intact nociceptors that innervate ligaments, small joints, muscle, and tendons. The neuropathic pain component is caused by mechanical compression of the nerve root or by action of inflammatory mediators originating from the degenerative disc on the nerve root. Cytokines and chemokines have been implicated in the chemical pathomechanism of radicular pain. Increased levels of tumor necrosis factor alpha (TNF-α), interleukin (IL)-1, IL-6, and granulocyte-macrophage colony stimulating factor (GM-CSF) have been detected in herniated disc tissue compared with normal, nondegenerated disc tissue. TNF-α applied to nerve roots reproduces the pathologic changes observed with nerve root exposure to nucleus pulposus extracts. Finally, in an animal model of radiculopathy induced by intervertebral disc material, inhibitors of TNF-α reversed the radicular pathology. So far it is unclear whether these mediators are capable of activating intact nerve root fibers even without any mechanical compression or if the combination of mechanical nerve damage and cytokine attack is the important prerequisite.

57.3.2 Mechanism-Based Classification

As described above neuropathic pain represents heterogeneous conditions, which neither can be explained by one single etiology or disease nor by a specific anatomical lesion, i.e., a disease- and anatomy-based classification is often insufficient. First, despite obvious differences in etiology, many of these conditions share common clinical phenomena; for example, touch-evoked pain in PHN and painful diabetic neuropathy. Conversely, different signs and symptoms can be present in the same disease; for example, pain paroxysms and stimulus-evoked abnormalities in PHN. Second, classification on the basis of anatomical location also has its shortcomings, as neuroplastic changes following nervous system lesions often give rise to sensory and pain distributions that do not respect nerve, root, segmental or cortical territories. Third, and most importantly, decades of rather discouraging systematic research on chronic pain therapy have revealed that a disease based strategy is of no or little help to these patients and their pain.

These observations have raised the question of whether an entirely different strategy, in which pain is analyzed on the basis of underlying mechanisms (Jensen, T. S. and Baron, R., 2003), could provide an alternative approach for examining and classifying patients, with the ultimate aim of obtaining a better treatment outcome (Woolf, C. W. et al., 1998; Dworkin, R. H. et al., 2003). Our increasing understanding of the mechanisms that underlie chronic pain, together with the discovery of new molecular therapy targets, has strengthened the demand for alternative concepts.

One theoretical possibility to identify pain mechanisms in patients is to assess differences in the somatosensory phenotype as precisely as possible. These specific patterns of signs and symptoms could be compared with the knowledge derived from animal experiments where the association of signs and symptoms and underlying mechanisms has been elucidated. This concept has led to the development of a symptom-oriented diagnostic approach to neuropathic pain conditions that supplements the etiology-based classification scheme, which recognizes the fact that neuropathic pains are usually a composite of several pain symptoms. A symptom-oriented approach does not negate the fact that distinct neuropathies present differently clinically, and that some neuropathic disease states may predispose to certain constellations of pain symptoms (e.g., touch-evoked pain in PHN). The rationale of this approach recognizes several principles:

● Clinically distinct pain symptoms such as ongoing stimulus-independent pain may be caused by similar, if not identical, neural mechanisms, even if the underlying neuropathies differ. For example, nociceptive afferents that develop ongoing activity after axon injury could mediate ongoing pain regardless of the precipitating or sustaining neuropathic cause.

- More than one pain mechanism is usually present in an individual patient.
- Some symptoms, such as mechanical hypersensitivity, can be explained by several distinct neural mechanisms that may even coexist in an individual patient.

As said, a symptom-based approach to painful neuropathies can be useful for dissecting the underlying neural mechanisms, and this knowledge may eventually be harnessed for the development of novel analgesic drugs that differentially target these mechanisms. A very promising but still hypothetical approach is summarized in Table 5.

57.4 Signs and Symptoms in Neuropathic Pain

Pain associated with nerve injury has several clinical characteristics (Baron, R., 2006) (Tables 4 and 6). If a mixed peripheral nerve with a cutaneous branch or a central somatosensory pathway are involved, there is almost always an area of abnormal sensation and the patient's maximum pain is coextensive with or within an area of sensory deficit. This is a key diagnostic feature for neuropathic pain. The sensory deficit is usually to noxious and thermal stimuli, indicating damage to small-diameter afferent fibers.

Beside these negative somatosensory signs (deficit in function), which are bothering but not painful also positive signs are characteristic for neuropathic conditions. Paresthesias (ant crawling, tingling) are not painful. Painful positive signs are spontaneous (not stimulus-induced) ongoing pain and spontaneous shooting, electric schock-like sensations. Many patients with peripheral neuropathic pain also have evoked types of pain (stimulus-induced pain, hypersensitivity), which are characterized by several sensory abnormalities. They may be adjacent to or intermingled with skin areas of sensory deficit. Most often, patients report mechanical hypersensitivity followed by hypersensitivity to heat and cold. Two types of hypersensitivity can be distinguished. First, allodynia is defined as pain in response to a nonnociceptive stimulus. In case of mechanical allodynia even gentle mechanical stimuli such that even slight bending of hairs may evoke severe pain. Second, hyperalgesia is defined as an increased pain sensitivity to a nociceptive stimulus. Another evoked feature is summation, which is the progressive worsening of pain evoked by slow repetitive stimulation with mildly noxious stimuli, for example, pinprick. A small percentage of patients with peripheral nerve injury have a nearly pure hypersensitive syndrome in which no sensory deficit is demonstrable.

The quality of the reported sensation may also be a clue; neuropathic pain commonly has a burning and/or shooting quality with unusual tingling, crawling or electrical sensations (dysesthesiae).

Although all these characteristics are neither universally present in, nor absolutely diagnostic of

Table 4   Symptoms and signs of neuropathic pain

	Peripheral neuropathic pain	Postherpetic neuralgia	Complex regional pain syndrome
Ongoing pain	100%	100%	100%
Brush-evoked pain (dynamic mechanical allodynia)	36%	60%	60%
Pinprick hyperalgesia	38%	37%	Present
Static allodynia	35%	52%	Present
Deep somatic allodynia	None	None	90%
Heat allodynia	29%	29%	30%
Cold allodynia	46%	21%	30%
Edema	Low	Low	50%
Trophic changes	Present	Scars	90%
Sympathetically maintained pain	Low	Low	90% (early)

Source: Data from Pappagallo M. *et al.* (2000) and unpublished data from Peters and Nurmikko (for a group of patients with mixed peripheral neuropathic pain) and Baron (for complex regional pain syndrome). Adapted from Scadding, J. W. and Koltzenburg, M. 2006. Painful peripheral neuropathies. In: Wall and Melzack's Textbook of Pain. 5th edition. (eds. S. B. McMahon and M. Koltzenburg), pp. 973-1000. Elsevier.

Table 5 Proposed model for the relationship between neuropathic pain mechanism and clinical symptoms and signs, and possible targets for therapeutic interventions

Symptom	Neuronal processes, mechanisms	Targets	Optimal compounds	Available
	Peripheral nociceptor hyperexcitability			
Spontaneous pain (shooting)	Ectopic impulse generation, oscillations in DRG	Na channels	Selective Na channel blocker	Lidocaine, carbamazepine, oxcarbazepine, lamotrigine, TCA
Spontaneous pain (ongoing)	**Peripheral nociceptor sensitization**			
	Inflammation within nerves Cytokine-release	Cytokines	Cytokine antagonists Cyclooxygenase blocker	TNF-α antagonists NSAIDS?
	Reduced activation threshold to:			
Heat hyperalgesia	Heat	TRPV1 receptor	TRPV1 receptor antagonists	Capsaicin cream
Cold hyperalgesia	Cold	TRPM8 receptor	TRPM8 receptor antagonists	Menthol?
Static mechanical hyperalgesia	Mechanical stimuli	ASCI receptor?	ASCI receptor antagonists	?
	Noradrenaline	α receptor	α receptor antagonists	Phentolamine, sympathetic block, TCA
SMP	Histamine	H1 receptor	H1 receptor antagonists	TCA
	Central dorsal horn hyperexcitability			
	Central sensitization on spinal level Ongoing C input induces increased synaptic transmission	Presynaptic: μ receptors Ca channels(α2–δ)	μ receptor agonists Ca channel blocker, α2-δ ligands	Opioids Gabapentin, pregabalin
	Amplification of C fiber input			
Dynamic mechanical hyperalgesia	Gating of Aβ fiber input (mechanical dynamic hyperalgesia)	Postsynaptic: NMDA receptors NK1 receptors NA channels Intracellular cascads	NMDA receptor antagonists NK1 receptor antagonists Selective Na channel blocker Mitogen activated protein kinase mediators	For example, ketamine, dextromethorphan ? For example, carbamazepine ?
Punctate mechanical hyperalgesia	Gating of Aδ fiber input (mechanical punctate hyperalgesia)			
	Central spinal disinhibition, Aδ cold fiber function ↓	?	?	?
	Intraspinal inhibitory interneurons ↓ (Functional, degeneration)			
	GABA-ergic	GABA-B receptors	GABA-B agonists	Baclofen
	Opioidergic	μ receptors	μ receptor agonists	Opioids
	Changes of supraspinal descending modulation			
	Inibitory control (noradrenaline, 5-HT) ↓	α2 receptors Serotonine receptors	α2 receptor agonists NA/5-HT reuptake blocker	Clonidine TCA, venlafaxine, duloxetine
	Faciliatory control ↑	?	?	?

5-HT, 5-hydroxytryptamine (serotonin); ASIC, acid-sensing ion channel; GABA, γ-aminobutyric acid; MAPK, mitogen-activated protein kinase; NK1, neurokinin 1; NMDA, N-methyl-D-aspartate; NSAIDS, nonsteroidal anti-inflammatory drugs; SMP, sympathetically maintained pain; TCA, tricyclic antidepressants; TNF-α, tumor necrosis factor-α.
Adapted from Baron R. 2006. Mechanism of disease: neuropathic pain – a clinical perspective. Nat. Clin. Pract. Neurol. 2, 95–106.

neuropathic pain, when they are present the diagnosis of neuropathic pain is likely (Table 4).

57.5   Pathophysiological Mechanisms in Neuropathic Pain

Most of the present pathophysiological ideas are derived from experimental work with animal models for neuropathic pain (Figure 2). This work has delineated a series of partially independent pathophysiological mechanisms presumed to be responsible for different types of neuropathic pain and different somatosensory abnormalities (Baron, R., 2006).

57.5.1   Peripheral Sensitization of Primary Afferent Nociceptors

Pain sensations are normally elicited by activity in unmyelinated (C) and thinly myelinated ($A\delta$) primary afferent neurons. After peripheral nerve lesion these neurons acquire an abnormal sensitization, i.e., an increased responsiveness of primary nociceptors to stimulation of their receptive field. The characteristic features of sensitized nociceptors are: pathological spontaneous discharge, a lowered activation threshold for thermal and mechanical stimuli, and an enhanced discharge to suprathreshold stimulation (hyperalgesia).

A large number of dramatic molecular and cellular changes at the level of the primary afferent nociceptor that are triggered by the nerve lesion underlie these pathological changes.

Nerve injury is matched by increased expression of messenger RNA for voltage gated sodium channels in primary afferent neurons. Two voltage-gated sodium channel genes ($Na_v1.8$ and $Na_v1.9$) are expressed selectively in nociceptive primary afferent neurons and an embryonic channel ($Na_v1.3$) is also upregulated in damaged peripheral nerves and associated with increased electrical excitability and spontaneous activity in neuropathic pain states. The accumulation of sodium channels at sites of ectopic impulse generation (sodium channel clusters) may be responsible for lowering of action potential threshold (Omana-Zapata, I. et al., 1997). There is evidence that small primary afferent fibers may acquire a unique sodium channel expression profile after nerve lesion (e.g., a specific relation of up- and downregulation of channel proteins), which makes them an interesting target (Wood, J. N. et al., 2004).

After peripheral nerve damage the accumulation of sodium channel clusters does not only occur at the site of the nerve lesion but also far proximally within the intact dorsal root ganglion (DRG). In the DRG a reciprocation between a phasically activating voltage-dependent, tetrodotoxin (TTX)-sensitive sodium conductance and a passive, voltage-independent potassium leak generates characteristic membrane potential oscillations. Ectopic firing is triggered when the amplitude of oscillation sinusoids reaches threshold (Amir, R. et al., 2002). Pathological membrane properties within the DRG that occur after nerve lesion are of particular therapeutic interest as the DRG is spared of the blood brain barrier and might be easily accessible for systemic therapies (Jacobs, J. M. et al., 1976). In addition to sodium currents, impaired potassium conductances in myelinated fibers were demonstrated in experimental diabetes and may underlie hyperexcitability in these fibers.

Damage to peripheral primary nociceptors also induces upregulation of a variety of receptor

(a)

(b)

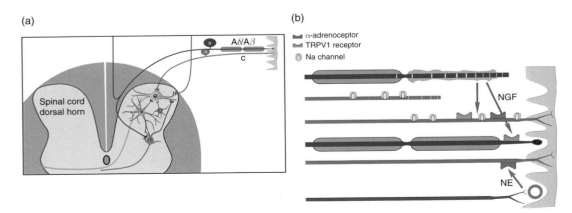

Figure 2   (continued)

(c)

Opioid receptor
Glutamate receptor
NA/5-HT receptor
GABA receptor
α-adrenoceptor
TRPV1 receptor
Kainate receptor

Na channel
Ca channel
(α2-δ subunit)

(d)

Opioid receptor
Glutamate receptor
NA/5-HT receptor
GABA receptor
α-adrenoceptor
TRPV1 receptor
Kainate receptor

Na channel
Ca channel
(α2-δ subunit)

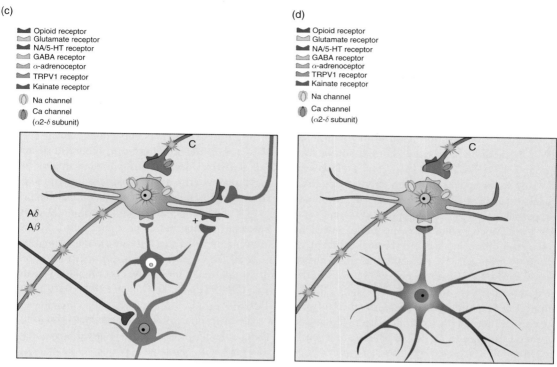

Figure 2 Mechanisms of peripheral sensitization and central sensitization in neuropathic pain. (a) Primary afferent pathways and their connections in the spinal cord dorsal horn. Nociceptive C fibers (red) terminate at spinothalamic projection neurons in upper laminae (orange neuron) whereas nonnociceptive myelinated A fibers project to deeper laminae. The second-order projection neuron is of wide dynamic range (WDR) type, that is, it receives direct synaptic input from nociceptive terminals and also multisynaptic input from myelinated A fibers (nonnoxions information, blue neuron system). γ-aminobutyric acid (GABA)-releasing interneurons (green neuron) normally exert inhibitory synaptic input on the WDR neuron. Furthermore, descending modulatory systems synapse at the WDR neuron (green descending terminal). Spinal cord glia cells (gray cell) also communicate with WDR neuron. (b) Peripheral changes at primary afferent neurons after partial nerve lesion leading to peripheral sensitization. Some axons are damaged and degenerate (upper two axons), whereas others (lower two axons) are still intact and connected with the peripheral end organ (skin). The lesion triggers the expression of sodium channels on damaged neurons. Furthermore, products such as nerve growth factor, which are associated with Wallerian degeneration, are released in the vicinity of spared fibers (arrows), triggering channel and receptor expression (sodium channels, TRPV1 receptors, adrenoceptors) on uninjured fibers. (c) Spontaneous activity in C nociceptors induces secondary changes in the central sensory processing, leading to spinal cord hyperexcitability (central sensitization of second-order WDR neurons indicated by star in orange neuron). This causes input from mechanoreceptive A fibers (light touch and punctate stimuli; blue neuron system) to be perceived as pain (dynamic and punctate mechanical allodynia; '+' indicates gating at synapse via AMPA/KA (α-amino-3-hydroxy-5-methyl-4-isoxazole propionic acid and kainate) receptors). Several presynaptic (opioid receptors, calcium channels) and postsynaptic molecular structures (glutamate receptors, NA (adrenceptors), 5-HT (serotonin) receptors, GABA receptors, sodium channels) are involved in central sensitization. Inhibitory interneurons and descending modulatory control systems (green neurons) are dysfunctional after nerve lesions, leading to disinhibition or facilitation of spinal cord dorsal horn neurons and to further central sensitization. (d) Peripheral nerve injury activates spinal cord nonneural glia cells (gray cell), which further enhances excitability in WDR neurons by releasing cytokines and glutamate. Adapted from Baron, R. 2006. Mechanisms of disease: neuropathic pain – a clinical perspective. Nat. Clin. Pract. Neurol. 2, 95–106.

proteins at the membrane some of them only marginally expressed under physiological conditions. Vanilloid receptors (TRPV1) are located predominantly on nociceptive afferent fibers and can be activated by the ingredient of hot chilli pepper (capsaicin). Physiologically, this receptor senses noxious heat (>43 °C) (Catarina, M. J. et al., 2000). Accordingly, TRPV1-deficient knockout (KO) mice have obvious deficits in chemical and thermal heat nociception but normal reactions to noxious mechanical and to noxious cold stimuli.

After partial nerve injury and in streptozotocin-induced diabetic rats the lesion triggers a TRPV1 downregulation on many damaged afferents but a novel expression of TRPV1 on uninjured C and A fibers (Figure 2(b)) (Hudson, L. J. *et al.*, 2001; Hong, S. and Wiley, J. W., 2005) which is likely involved in the development C nociceptor sensitization and the associated symptom of heat hyperalgesia. Recent studies also provides evidence for a an upregulation of TRPV1 in medium and large injured dorsal root ganglion cells (Ma, W. *et al.*, 2005). However, TRPV1 does not appear to be the only transduction mechanism for heat sensitization after nerve injury. After partial sciatic nerve ligation, wild-type and TRPV1-null mice exhibited comparable persistent enhancement of mechanical and thermal nociceptive responses (Catarina, M. J. *et al.*, 2000). In this context, a recent study examined strain differences in the normal sensitivity to noxious heat in mice. These differences reflect differential responsiveness of primary afferent thermal nociceptors to heat stimuli due to a genetic variance in CGRP expression and sensitivity (Mogil, J. S. *et al.*, 2005).

In taxol-induced small-fiber painful polyneuropathy TRPV4 which is normally activated by heat of more than $30\,°C$ seems to play a crucial role in producing taxol-induced mechanical hyperalgesia (Alessandri-Haber, N. *et al.*, 2004).

Investigations into temperature-sensitive excitatory ion channels also identified several cold-sensing ion channels in peripheral neurons. The cold- and menthol-sensitive TRP channel (TRPM8) is activated within the range of $8–28\,°C$ (Patapoutian, A. *et al.*, 2003) and sensitized by menthol. This receptor is expressed in around 10% of all afferent ganglion neurons of rats, primarily within small-diameter cells (McKemy, D. D. *et al.*, 2002). TRPA1 is activated at lower temperatures and its exogenous ligand is cinnamaldehyde, a constituent of cinnamon oil, mustard oil, and horseradish. Peripheral nerve lesions have been shown to upregulate the expression of the latter cold-sensing ion channels in DRG cells of rats that developed cold allodynia (Obata, K. *et al.*, 2005). Therefore, upregulation or gating of these channels after injury may lead to the peripheral sensitization of cold-sensitive C nociceptors, resulting in the sensory phenomenon of cold hyperalgesia.

Besides these temperature-sensitive receptors experimental nerve injury also triggers the expression of functional α_1 or α_2 adrenoceptors on cutaneous afferent fibers, these neurons develop adrenergic sensitivity (Figure 2(b)). I.v. epinephrine or physiological noradrenaline release after stimulation of sympathetic efferents that have regenerated into the neuroma can excite afferent nociceptors. After section and reanastomosis of peripheral nerves, electrical stimulation of the sympathetic trunk at physiological stimulus frequencies activates regenerated C nociceptors through an α_1 adrenoceptor mechanism (Habler, H. J. *et al.*, 1987). Furthermore, sympathetic activity can sensitize identified intact nociceptors following damage to the nerve in which they run via an α_2 adrenoceptor mechanism (Sato, J. and Perl, E. R., 1991). The concept of a pathological coupling between sympathetic postganglionic fibers and afferent neurons via noradrenaline forms the conceptual framework for the therapeutic application of sympathetic blocks in certain pain syndromes, e.g., CRPS (Price, D. D. *et al.*, 1998; Baron, R. *et al.*, 1999b).

There is increasing evidence that also uninjured fibers running in a partially lesioned nerve may take part in pain signaling (Wasner, G. *et al.*, 2005). Uninjured fibers comingle with degenerating fibers in the same nerves. Products associated with Wallerian degeneration released in the vicinity of spared fibers (e.g., nerve growth factor (NGF)) may be the trigger for channel and receptor expression and may alter the properties of uninjured afferents (Hudson, L. J. *et al.*, 2001; Wu, G. *et al.*, 2001). Expression of sodium channels, TRPV1 receptors, adrenoreceptors and an increase of TNF-α sensitivity has been shown to play a role in uninjured fibers adjacent to lesioned axons.

57.5.2 Central Sensitization

As a consequence of periphereal nociceptor hyperactivity also dramatic secondary changes in the spinal cord dorsal horn occur. Peripheral nerve injury leads to an increase in the general excitability of nociceptive and multireceptive spinal cord neurons (multiple synaptic input from C as well as A fibers, wide dynamic range neurons). This phenomenon is called central sensitization and is defined as an increased responsiveness of nociceptive neurons in the CNS to their normal afferent input. Central sensitization is manifested by at least three different modes:

1. increase of neuronal activity to noxious stimuli,
2. expansion of size of neuronal receptive fields, and
3. spread of spinal hyperexcitability to other segments.

Central sensitization is initiated and maintained by activity in pathologically sensitized C fibers, which sensitize second-order spinal cord dorsal horn neurons by releasing glutamate acting on N-methyl-D-aspartic acid (NMDA) receptors and the neuropeptide substance P. Furthermore, central neuronal voltage-gated N-calcium channels located at the presynaptic sites on terminals of primary afferent nociceptors are involved in central sensitization by facilitation of the release of glutamate and substance P. This channel is overexpressed after peripheral nerve lesion and in rats with streptozotocin-induced diabetes (Luo, Z. D. *et al.*, 2001). As a consequence of peripheral nerve lesion the dorsal horn neurons abnormally express $HNa_v1.3$ (Hains, B. C. *et al.*, 2004) also enhancing central sensitization. Several intracellular cascades contribute to central sensitization, in particular the mitogen-activated protein kinase (MAPK) system (Ji, R. R. and Woolf, C. J., 2001). If central sensitization is established, normally innocuous tactile stimuli become capable of activating spinal cord pain signalling neurons via Aδ and Aβ low-threshold mechanoreceptors (Tal, M. and Bennett, G. J., 1994). By this mechanism, light innocuous mechanical stimuli to the skin induces pain, i.e., punctuate and dynamic mechanical allodynia. Besides these dramatic changes in the spinal cord, sensitized neurons were also found in the thalamus and primary somatosensory cortex after partial peripheral nerve injury (Guilbaud, G. *et al.*, 1992).

There is increasing evidence that neuropathic pain is in part mediated by an interaction of non-neural spinal cord glia and nociceptive neurons (Figure 2(d)). In experimental pain states in animals, astrocytes and microglia are activated by neuronal signals including substance P, glutamate, and fractalkine (Wieseler-Frank, J. *et al.*, 2005). Activation of glia by these substances in turn leads to the release of mediators that then act on other glia and also on central nociceptive neurons. These include proinflammatory cytokines and most likely also other neuroexcitatory compounds like glutamate. By this interaction central sensitization is augmented. While traditional therapies for pathological pain have focused on neuronal targets, glia might be new therapeutic targets.

Some patients with neuropathic pain, in particular patients with CRPS characteristically report 'extra-territorial' and/or 'mirror-' image pain. The pain is experienced not only in the area of trauma but also in neighboring healthy tissues. In cases of mirror-image pain, the pain is perceived from the healthy, corresponding part on the opposite side of the body. New data suggest that communication of activated astrocytes via gap junctions may mediate such spread of pain.

57.5.3 Central Disinhibition and Fascilitation

Physiologically dorsal horn neurons receive a strong intraspinal inhibitory control by gamma-aminobutyric acid (GABA)-ergic interneurons. Partial peripheral nerve injury may promote a selective apoptotic loss of these GABA-ergic inhibitory neurons in the superficial dorsal horn of the spinal cord (Moore, K. A. *et al.*, 2002), a mechanism which would further increases central sensitization. Furthermore, there is a novel mechanism of disinhibition following peripheral nerve injury. This mechanism involves a trans-synaptic reduction in the expression of the potassium chloride exporter KCC2 in lamina I nociceptive neurons. This induces a consequent disruption of the anion homeostasis in these dorsal horn neurons. The resulting shift in the transmembrane anion gradient caused normally inhibitory anionic synaptic currents to be excitatory. Due to this mechanism the release of GABA from normally inhibitory interneurons now exerts an excitatory action on lamina I neurons via $GABA_A$ receptors (Coull, J. A. *et al.*, 2003). The changes in lamina I neurons is induced by brain-derived neurotrophic factor (BDNF) released from activated spinal cord glia (Coull, J. A. *et al.*, 2005).

Dorsal horn neurons receive a powerful descending modulating control from supraspinal brainstem centers (inhibitory as well as facilitatory) (Vanegas, H. and Schaible, H. G., 2004) (Figures 2(a) and 2(c)). It was hypothesized that a loss of function in descending inhibitory serotonergic and noradrenergic pathway contributes to central sensitization and pain chronicity. This idea nicely explained the efficacy of serotonin and noradrenaline reuptake blocking antidepressants in neuropathic pain. However, in animals, mechanical allodynia after peripheral nerve injury was dependent upon tonic activation of descending pathways that facilitate pain transmission indicating that structures in the mesencephalic reticular formation – possibly the nucleus cuneiformis (NCF) and the periaqueductal gray (PAG) – are involved in central sensitization in neuropathic pain (Ossipov, M. H. *et al.*, 2000). Interestingly, exactly the same brainstem structures were shown to be active in humans with allodynia using advanced functional MRI (fMRI) techniques

(Zambreanu, L. *et al.*, 2005). Because in most animal pain models descending facilitation and inhibition are triggered simultaneously, it will be important to elucidate why inhibition predominates in some neuronal pools and facilitation in others.

57.5.4   Inflammation and Neuropathic Pain

The connective tissue sheath of peripheral nerves is innervated by sensory fibers, the nervi nervorum, which enter the sheath with the nutrient blood vessels. Some of these fine diameter primary afferents are nociceptors (Zochodne, D. W., 1993). Nervi nervorum are a potential source of pain in diseases of peripheral nerve especially in those conditions with an inflammatory component.

A different mechanism involves the production of inflammatory mediators at the site of nerve injury (local inflammation) that might be a critical factor in the cascade of events leading to neuropathic pain. Activated macrophages infiltrating from endoneurial blood vessels have been demonstrated in experimentally injured nerves (Sommer, C. and Myers, R. R., 1996) and also the dorsal root ganglia after nerve transection (Lu, X. and Richardson, P. M., 1993). Proinflammatory cytokines and in particular TNF-α released by activated macrophages can induce ectopic activity in injured but also in adjacent uninjured primary afferent nociceptors at the lesion site (Sorkin, L. S. *et al.*, 1997) and thus is a potential cause of pain and hyperalgesia (Sommer, C. *et al.*, 1998; Wagner, R. *et al.*, 1998; Sommer, C., 2003; Marchand, F. *et al.*, 2005).

57.6   Pathophysiological Mechanisms in Patients in Relation to Their Somatosensory Profile

Although we have achieved enormous progress in the understanding of pathophysiological mechanisms generating neuropathic pain the most important question is whether it is possible to translate these concepts into the clinical situation. In the following section the findings of the correlation of mechanisms and somatosensory abnormalities in animal experiments will be matched by observations in human experimental and clinical pain conditions.

In order to get more insight into the puzzle of neuropathic pain and even solve the therapeutic dilemma this type of translational research should

be fostered in the future. The solution of the problem neuropathic pain can only be unraveled in the human experiment, that is, in patients that really feel all components of this complex sensation of neuropathic pain.

57.6.1   Peripheral Sensitization of Primary Afferent Neurons in Patients

Microneurographic single-fiber recordings support the idea of peripheral sensitzation of primary afferent neurons in patients with painful nerve lesions by demonstrating abnormal activity and reduced thresholds in cutaneous afferents. Abnormal ectopic activity of myelinated mechanosensitive fibers were found in traumatic nerve lesions, entrapment neuropathies, or radiculopathies (Figure 3). Because the ectopic nerve activity correlated in intensity and time course to the perceived paresthesias it is likely that pathological activity in A fibers is the underlying mechanism of positive nonpainful sensations.

Recordings from transected nerves in awake human amputees with phantom limb pain have demonstrated spontaneous ectopic activity as well as barrages of action potential firing in afferent A and C fibers projecting into the neuroma (Nystrom, B. and Hagbarth, K. E., 1981). Ectopic excitation occurred at multiple sites in damaged sensory neurons. Ongoing activity and mechanical sensitivity were recorded proximal to the nerve neuroma. Following local anesthetic blockade of the nerve distal to the recording site, impulses evoked by mechanical stimulation of the neuroma were abolished, but ongoing activity at the recording site continued, suggesting that this residual activity arose from the DRGs. As these patients suffered from spontaneous burning pain and electric shock-like sensations it is very likely that these symptoms are associated with ectopic firing in primary afferent C fibers.

In few patients with characteristic burning pain and heat hyperalgesia microneurographic recordings have provided evidence for sensitized C nociceptors. In patients with erythromelalgia nociceptors displayed ongoing activity, which is normally not observed in nociceptors, and there was a sensitization of mechanically insensitive afferents to nonpainful tactile stimuli. These abnormalities were only present in nociceptive fibers, but not of the sympathetic unmyelinated fibers, as an indicator of a neuropathic process (Figure 4) (Orstavik, K. *et al.*, 2003). In the

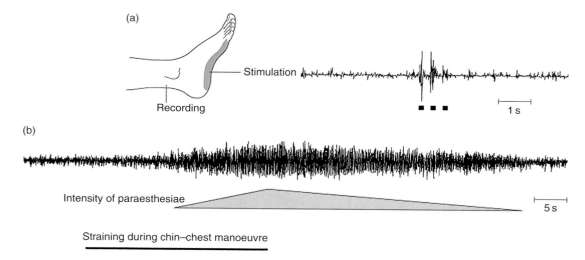

Figure 3 Microneurographic multiunit recording from the sural nerve in a patient with a compression of the S1 spinal root by a herniated disc. (a) Excitation of mechanosensitive units in the receptive field by tactile stimulation (bars). (b) Straining and chin–chest maneuver-provoked paresthesiae and an ectopic discharge of afferent fibers originating from the compressed root. Adapted from Nordin, M., Nyström, B., Wallin, U., and Hagbarth, K. E. 1984. Ectopic sensory discharges and paresthesiae in patients with disorders of peripheral nerve, dorsal roots and dorsal columns. Pain 20, 231–245. Elsevier Ltd.

autosomal dominantly inherited form of erythromelalgia a mutation in SCN9A was found, a gene that encodes the $Na_v1.7$ sodium channel, leading to an altered firing pattern in afferent neurons (Dib-Hajj, S. D. *et al.*, 2005). Hence, erythromelalgia is the first channelopathy associated with chronic pain.

Several clinical observations also support the concept of sensitized nociceptors in PHN patients. About 30% of patients with PHN do not show any loss of sensory function in the affected extremity indicating that in these particular group of patients loss of neurons is minimal or absent. Accordingly, thermal sensory thresholds in their region of greatest pain are either normal or even decreased (heat hyperalgesia) by up to 2–4 °C (Rowbotham, M. C. and Fields H. L., 1996; Pappagallo, M. *et al.*, 2000). The decrease of heat pain perception thresholds is a well-known phenomenon of peripheral nociceptor sensitization. Using skin punch biopsy, it was shown that thermal sensitivity is directly correlated with density of cutaneous innervation in the area of most severe pain (Rowbotham, M. C. *et al.*, 1996). Moreover, in PHN patients with heat hyperalgesia acute topical application of the vanilloid compound capsaicin (TRPV1 agonist), enhances pain, a sign that is indicative of an increased capsaicin sensitivity of nociceptors in the affected skin area and has been attributed to a sensitization of nociceptors as the driving element in some patients (Petersen, K. L. *et al.*, 2000). Furthermore, in a similar group of

PHN patients cutaneous iontophoresis of histamine evoked a burning pain sensation whereas only itch was elicited in normal skin. Again, this phenomenon indicates that nociceptive neurons in the affected skin are abnormally sensitive to histamine (Baron, R. *et al.*, 2001) probably due to expression of a novel receptor pattern.

These observations suggest that spontaneous burning pain and heat hyperalgesia at least in part are associated with sensitization of primary afferent C fibers to TRPV1 agonists and histamine or with an altered firing pattern in these fibers due to abnormal sodium channels.

Cutaneous hypersensitivity to cold, i.e., cold hyperalgesia, is particularly prominent in patients with posttraumatic neuralgias, some small-fiber polyneuropathies and chronic CRPS. Another condition of acute cold intolerance occurs after systemic injection of the cancer chemotherapeutic agent oxaliplatin, which is associated with paresthesiae and painful hypersensitivity aggravated by cold. Psychophysical studies of human volunteers using the topical menthol model suggest that sensitization of cold-sensitive nociceptors can produce cold hyperalgesia in normal volunteers (Wasner, G. *et al.*, 2004). Peripheral sensitization also appears to occur in acute oxaliplatin-induced peripheral neuropathy (Lehky, T. J. *et al.*, 2004). Therefore, it is likely that cold hyperalgesia in some patients is induced by sensitization of primary afferent cold-sensitive nociceptors.

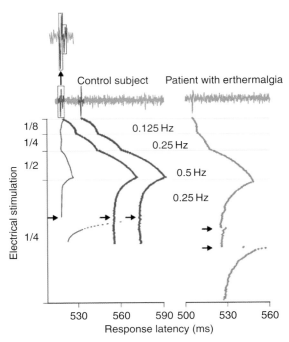

Figure 4 Microneurographic C fiber recordings from the peroneal nerve of a normal volunteer (left) and a patient suffering from erythromelalgia (right). The first trace shows the original nerve signal. The subsequent recordings are a falling leaf display in which each action potential is symbolized by a series of dots. Each row of dots represents the latency, at different electrical stimulation frequencies of the fiber, of its receptive field in the skin. Simultaneous recording of three afferent C fibers in a control subject shows two units with activity-dependent latency increases at low-frequency stimulation, and one unit that displays latency increases when excitation frequency increases to 1 impulse per 2 s. These biophysical features are characteristic for mechanically insensitive and mechanically sensitive nociceptive C fibers, respectively. Stimulation of the receptive field with a 750-mN von Frey hair (arrow) excited only the mechanosensitive C nociceptor, as evidenced by a strong latency shift to electrical stimulation. Mechanically insensitive afferents are not activated by this stimulus. The right panel shows a C fiber recording from a patient with erythromelalgia. The unit had the biophysical properties of a mechanically insensitive C fiber but responded reproducibly to mechanical stimulation (arrows). This is consistent with the hypothesis that mechanically insensitive afferents become sensitized in erythromelalgia. An alternative explanation would be that mechanically sensitive nociceptors start to display a different pattern of activity-dependent latency slowing in this disease. Adapted from Orstavik, K., Weidner, C., Schmidt, R., Schmelz, M., Hilliges, M., Jorum, E., Handwerker, H. and Torebjork, E. 2003. Pathological C-fibres in patients with a chronic painful condition. Brain 126, 567–578. Copyright 2003 Oxford University Press.

57.6.2 Sensitization to Catecholamines in Patients

Several clinical observations support the idea that nociceptors acquire a sensitivity to catecholamines that permits an abnormal excitation by either noradrenaline or by circulating catecholamines. This chemical sensitization makes the nociceptors susceptible for noradrenaline released from efferent sympathetic fibers in the periphery. The consecutive pathological sympathetic–afferent coupling forms the conceptual framework for sympathetically maintained pain states.

Noradrenergic sensitivity has been described in several human neuropathies and suggests that the ongoing pain can be caused or maintained by the sympathetic nervous system in selected patients: In amputees, perineuromal administration of physiological doses of norepinephrine induces intense pain as compared with saline injections (Raja, S. N. et al., 1998). Intraoperative stimulation of the sympathetic chain induces an increase of spontaneous pain in patients with causalgia (CRPS II) but not in patients with hyperhidrosis. In PHN, application of norepinephrine into a symptomatic skin area increased spontaneous pain and dynamic mechanical hyperalgesia (Choi, B. and Rowbotham, M. C., 1997). In CRPS II and posttraumatic neuralgias, intracutaneous norepinephrine rekindles pain and hyperalgesia that had been relieved by sympathetic blockade. Also intradermal norepenephrine, in physiologically relevant doses, was demonstrated to evoke greater pain in the affected regions of patients with sympathetically maintained pain (SMP) than in the contralateral unaffected limb, and in control subjects (Ali, Z. et al., 2000). Because noradrenaline-induced pain occurs during a differential blockade of myelinated fibers, unmyelinated fibers appear to signal sympathetically maintained pain (Torebjork, E. et al., 1995).

We performed a study in patients with CRPS I using physiological stimuli of the sympathetic nervous system (Baron, R. et al., 2002). Cutaneous sympathetic vasoconstrictor outflow to the painful extremity was experimentally activated to the highest possible physiological degree by whole body cooling. During the thermal challenge the affected extremity was clamped to 35 °C in order to avoid thermal effects at the nociceptor level. The intensity as well as area of spontaneous pain and mechanical allodynia (dynamic and punctate) increased significantly in patients that had been classified as having SMP by positive sympathetic blocks but not in SIP patients (Figure 5). The experimental setup used in the latter study selectively alters sympathetic cutaneous vasoconstrictor activity without influencing other sympathetic systems innervating the

extremities, i.e., piloarrector, sudomotor, and muscle vasoconstrictor neurons. Therefore, the interaction of sympathetic and afferent neurons measured here is likely to be located within the skin as

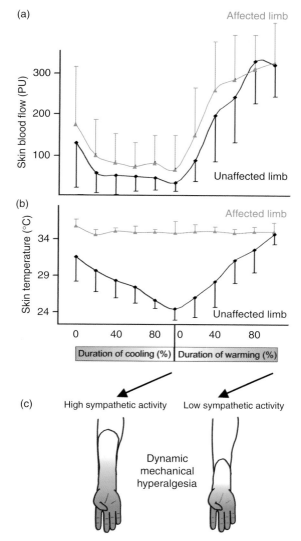

Figure 5 Experimental modulation of cutaneous sympathetic vasoconstrictor neurons by physiological thermoregulatory reflex stimuli in 13 complex regional pain syndrome (CRPS) patients. With the help of a thermal suit, whole-body cooling and warming was performed to alter sympathetic skin nerve activity. The subjects were lying in a suit supplied by tubes, in which running water of 12 °C and 50 °C, respectively (inflow temperature) was used to cool or warm the whole body. By these means sympathetic activity can be switched on and off. (a) High sympathetic vasoconstrictor activity during cooling induces considerable drop in skin blood flow on the affected and unaffected extremity (laser Doppler flowmetry). Measurements were taken at 5-min intervals (mean ± SD). (b) On the unaffected side, a secondary decrease of skin temperature was documented. On the affected side, the forearm temperature was clamped at 35 °C by a feedback-controlled heat lamp to exclude temperature effects on the sensory receptor level. Measurements were taken at 5-min intervals (mean ± SD). (c) Effect of cutaneous sympathetic vasoconstrictor activity on dynamic mechanical hyperalgesia in one CRPS patient with sympathetically maintained pain (SMP). Activation of sympathetic neurons (during cooling) leads to a considerable increase of the area of dynamic mechanical hyperalgesia. From Baron, R., Schattschneider, J., Binder, A., Siebrecht, D. and Wasner, G. 2002. Relation between sympathetic vasoconstrictor activity and pain and hyperalgesia in complex regional pain syndromes: a case–control study. Lancet 359, 1655–1660, with permission.

predicted by the pain-enhancing effect of intracutaneous norepinephrine injections (Ali, Z. *et al.*, 2000). Interestingly, the relief of spontaneous pain after sympathetic blockade was more pronounced than changes in spontaneous pain that could be induced experimentally by sympathetic activation. One explanation for this discrepancy might be that a complete sympathetic block affects all sympathetic outflow channels projecting to the affected extremity. It is very likely that in addition to a coupling in the skin, a sympathetic–afferent interaction may also occur in other tissues, in particular in the deep somatic domain such as bone, muscle, or joints. Supporting this view, especially these structures are extremely painful in some cases with CRPS (Baron, R. and Wasner, G., 2001). Furthermore, there may be patients who are characterized by a selective or predominant sympathetic–afferent interaction in deep somatic tissues sparing the skin (Wasner, G. *et al.*, 1999). In summary, it is likely that sympathetically maintained pain is a consequence of an adrenergic sensitization of C nociceptors.

57.6.3 Central Sensitization in Patients

One hallmark of central sensitization of spinal cord neurons in animals is that activity in A fiber mechanoreceptors is allowed to gain access to the nociceptive system and induces pain. These phenomena are called mechanical allodynia or hyperalgesia.

Mechanical hypersensitivity is a common phenomenon in neuropathic pain states in patients. There are several lines of evidence that also in patients central mechanisms contribute to these sensory phenomena. Two distinct types have been described in patients, dynamic mechanical and pinprick mechanical hypersensitivity.

There is consensus that dynamic mechanical allodynia is signaled out of the skin by afferent mechanoreceptors with large myelinated axons that normally encode nonpainful tactile stimuli:

1. Reaction time measurements show dynamic mechanical allodynia in patients to be signaled by afferents with conduction velocities appropriate for large myelinated axons (Lindblom, U. and Verrillo, R. T., 1979; Campbell, J. N. *et al.*, 1988),
2. transcutaneous or intraneural stimulation of nerves innervating the allodynic skin can evoke pain at stimulus intensities which only produce

tactile sensations in healthy skin (Gracely, R. H. *et al.*, 1992; Price, D. D. *et al.*, 1992), and
3. using differential nerve blocks dynamic allodynia is abolished at time points when tactile sensations is lost, but other modalities remain unaffected (Campbell, J. N. *et al.*, 1988; Ochoa, J. L. and Yarnitsky, D., 1993).

Therefore, patients with dynamic mechanical allodynia would be expected to have central sensitization as their underlying mechanism.

Hyperalgesia to pinprick stimuli, typically elicited by probing of the skin with a stiff von Frey hair is distinct from dynamic mechanical allodynia because of its different spatial and temporal profile and the fact that it is signaled by nonsensitized, heat-insensitive, $A\delta$ nociceptors.

Notably, in many neuropathic pain patients the mechanically sensitive skin area expands widely into the secondary zone, i.e., the area not affected by the primary nerve lesion, which is also indicative for CNS mechanisms involved. Furthermore, the area of secondary mechanical hypersensitivity is a dynamic phenomenon. In PHN patients with signs of peripheral nociceptor sensitization (heat hyperalgesia, see above) cutaneous capsaicin application into the primary skin area leads to an increase of the allodynic zone into previously nonallodynic and nonpainful skin that had normal sensory function and cutaneous innervation. These observations support the hypothesis that allodynia in a subgroup of PHN patients is a form of chronic secondary hyperalgesia dynamically maintained by input from intact and possibly sensitized ('irritable') primary afferent nociceptors to a sensitized CNS (Fields, H. L. *et al.*, 1998; Petersen, K. L. *et al.*, 2000). Because central sensitization involves the NMDA receptor, the fact that the NMDA receptor antagonist ketamine relieves some neuropathic pain disorders further supports the concept of central sensitization.

Besides these dramatic changes in the spinal cord, there is now evidence that higher centers of the neuraxis demonstrate an increased excitability as well as fundamental changes in the somatosensory representation. Magnetic encephalography (MEG), positron emission tomography (PET), and fMRI studies revealed cortical changes in patients with phantom limb pain, CRPS, and central pain syndromes (Flor, H. *et al.*, 1995; Willoch, F. *et al.*, 2000; Pleger, B. *et al.*, 2004; Willoch, F. *et al.*, 2004; Maihofner, C. *et al.*, 2005a) as well as experimental pain models (Baron, R. *et al.*, 1999a; Baron, R. *et al.*,

2000). Interestingly, these changes correlated with the intensity of the perceived pain and disappeared after successful treatment of the pain (Maihofner, C. *et al.*, 2004; Pleger, B. *et al.*, 2005).

57.6.4 Central Disinhibition Leading to Cold Hyperalgesia

Cutaneous hypersensitivity to cold, i.e., cold hyperalgesia is particularly prominent in patients with posttraumatic neuralgias, some polyneuropathies, and chronic CRPS. Some observations in neuropathic pain patients point to central disinhibition as the underlying mechanism in a subgroup of patients.

Cold stimuli are usually transmitted centrally by cold-sensitive Aδ fibers, whereas cold pain is conveyed via nociceptive cold-sensitive C fibers. In patients with polyneuropathies a disproportionate loss of Aδ axons and relative sparing of C fibers have been described (Ochoa, J. L. and Yarnitsky, D., 1994). These patients suffer from the so-called triple cold syndrome (CCC syndrome): cold hypoesthesia in combination with cold hyperalgesia and a cold skin. A selective damage of cold-sensitive Aδ fibers leads to a lack of inhibition (disinhibition) on C nociceptors transmission normally exerted by concomitant activation of myelinated cold Aδ fibers. This mechanism of a central interaction between cold specific afferents and nociceptors would nicely explain the combination of cold hyperalgesia and cold hypoesthesia in the above patients. Furthermore, it is suggested to play a crucial role in central pain syndromes after infarction of the dorsolateral thalamus with cold hyperalgesia and paradoxical heat sensation. In these patients the lesion in the lateral thalamus might disinhibit the nociceptive system projecting through the medial thalamus.

Recently, direct evidence was provided for the central disinhibhition theory of cold hyperalgesia in an experimental pain model in humans. Selective A fiber block induces the symptom combination of cold hyperalgesia and cold hypoesthesia and therefore mimics the situation in polyneuropathy patients with A fiber degeneration. With fMRI the disinhibition of the medial nociceptive system (medial thalamus, anterior cingulate cortex (ACC), and frontal cortices) could be demonstrated.

Thus, a loss of peripheral A fiber function disinhibits the nociceptive system centrally leading to cold hyperalgesia.

This is an entirely different mechanism as compared with the peripheral sensitization process of cold-sensitive C nociceptors described above leading to the same somatosensory phenomenon, cold hyperalgesia.

57.6.5 Deafferentation: Hyperactivity of Central Pain Transmission Neurons

The above data convincingly support a role for sensitization mechanisms in the peripheral as well as the CNS in the generation of neuropathic pain. However, in some patients there is a profound cutaneous deafferentation of the painful area. Up to 60% of PHN patients show considerable signs of neuronal degeneration and loss of sensory function within the affected tissues. Interestingly, some of these patients still suffer from severe dynamic mechanical allodynia although the function of nociceptors is diminished or absent in the same skin area (Wasner, G. *et al.*, 2005).

Punch skin biopsies and the anti-PGP 9.5 antibody, a panaxonal marker, were used in PHN patients and zoster patients without pain to quantify sensory nerve endings in the affected skin and compared the numbers with the homologous contralateral site (Rowbotham, M. C. et al., 1996; Oaklander, A.L., 1998; Oaklander, A.L., 2001). A severe loss could be demonstated on the affected side (20% as compared with the controls). Neurite loss was more prominent in the epidermis than in the dermis. Furthermore, the PHN group also had lost half of the neurites in the contralateral epidermis whereas distant areas where unaffected. However, these contralateral changes could not be observed with QST or axon reflex measurements (Baron, R. and Saguer, M., 1994).

Functional studies support the concept of degeneration of cutaneous C nociceptors. By using these C fiber axon reflex reactions it is possible to objectively assess cutaneous C fiber function in the human skin. In some patients the histamine evoked axon reflex vasodilatation and flare size were impaired or abolished in skin regions with intense dynamic allodynia (Baron, R. and Saguer, M., 1993). Similarly, during capsaicin stimulation, a subgroup did not experience worsening of pain (Petersen, K. L. *et al.*, 2000). Using quantitative sensory testing (QST) some chronic PHN patients have extremely high thermal thresholds in areas with marked dynamic allodynia (Nurmikko, T. and Bowsher, D., 1990; Nurmikko, T. *et al.*, 1994; Choi, B. and Rowbotham, M. C.,

1997). Thus, there is a subset of PHN patients with pain and loss of cutaneous C nociceptor function.

Assuming that the DRG cells and the central afferent connections are lost in such patients, their pain must be the result of intrinsic CNS changes. In animal studies, following complete primary afferent loss of a spinal segment, many dorsal horn cells begin to fire spontaneously at high frequencies (Lombard, M. C. and Larabi, Y., 1983; Fields, H. L. *et al.*, 1998). There is some evidence that a similar process may underlie the pain that follows extensive denervating injuries in human. Recordings of spinal neuron activity in a pain patient whose dorsal roots were injured by trauma to the cauda equina revealed high-frequency regular and paroxysmal bursting discharges (Loeser, J. D. *et al.*, 1967). That patient complained of spontaneous burning pain in a skin region that was anesthetic by the lesion (anesthesia dolorosa). Thus, extensive degeneration of primary afferents associated with severe somatosensory deficits points to an increased excitability of deafferented central neurons as underlying mechanism.

57.6.6 Inflammation in Patients

In patients with inflammatory demyelinating neuropathies such as acute Guillain–Barré syndrome or vasculitic neuropathies deep proximal aching pain in addition to paroxysmal types of pains is a characteristic phenomenon. Accordingly, COX2 was found to be upregulated in nerve biopsy specimens from patients with chronic inflammatory demyelinating neuropathy and an increased expression of proinflammatory cytokines have been demonstrated in peripheral nerves of most vasculitic neuropathies (Lindenlaub, T. and Sommer, C., 2003). About 15–50% of AIDS patients suffer from distal predominantly sensory neuropathy, which very often is painful. In sarcoidosis, painful small-fiber neuropathy may be present in a subgroup of patients. In the fluid of artificially produced skin blisters significantly higher levels of IL-6 and TNF-α were observed in CRPS-affected extremities as compared with the uninvolved extremity. In CRPS patients with hyperalgesia higher levels of the soluble TNF-α receptor type I were found (Maihofner, C., *et al.*, 2005b). Accordingly, a significant increases in IL-1β and IL-6, but not TNF-α was demonstrated in the CSF of individuals afflicted with CRPS as compared with controls (Alexander, G. M. *et al.*, 2005).

Acute zoster is accompanied by intense inflammation along the affected peripheral nerve that typically resolves in several weeks. However, a small subgroup of patients with PHN has inflammatory infiltrates throughout the affected peripheral nerve, DRG, and dorsal root (Watson, C. P. *et al.*, 1991). Moreover, Gilden D. H. *et al.* (1991) have described a subpopulation of PHN patients with evidence of continuing low-level viral expression whose pain responds to antiviral agents. Thus continuing VZV expression could produce sensitization and activity in primary afferents secondary to inflammation.

57.7 Diagnostic Tools for Neuropathic Pain

Modern research into the mechanisms of neuropathic pain clearly revealed that the nerve lesion leads to dramatic changes in the PNS and CNS that makes it distinct from other chronic pain types in which the nociceptive system is intact (chronic nociceptive pain, e.g., osteoarthritis). Furthermore, neuropathic pain states require different therapeutic approaches, e.g., anticonvulsants, that are not effective in nociceptive pain. To make the situation even more complex, many chronic pain states are characterized by a combination of both the pain types. Best examples for the so-called mixed pain syndromes are chronic radicular back pain, tumor pain, or CRPS.

For the clinician, it is, therefore, of utmost importance to have valid diagnostic tools that differentiate neuropathic from nociceptive pain or estimate the neuropathic pain component in mixed pain syndromes.

The easiest approach to this would be to use somatosensory symptoms assessed by questionnaires or history questions or simple signs testable at the bedside that are characteristic for neuropathic pain.

However, a recent study on this issue raised several caveats (Rasmussen, P. V. *et al.*, 2004): This study prospectively looked at symptoms and signs in 214 patients with suspected chronic neuropathic pain that were *a priori* classified by pain experts as having the so-called definite, possible or unlikely neuropathic pain. Pain symptoms including pain descriptors were recorded and sensory tests including repetitive pinprick stimulation, examination for cold-evoked pain by an acetone drop and brush-evoked pain were carried out in the maximal pain area and in a control area in order to determine if symptoms and signs cluster differentially in groups of patients with increasing evidence of neuropathic pain. Several symptoms (touch- or cold-provoked pain) and signs (brush-evoked allodynia) were more prominent in patients with definite or possible neuropathic pain; however, there was considerable overlap with the clinical presentation of patients with unlikely neuropathic pain. Even worse, the used pain descriptors could not at all distinguish between the three clinical categories.

57.7.1 Questionnaires

Other approaches to find easy screening tools for neuropathic pain gave more promising results: a clinician-administered 10-item questionnaire (DN4) consists of sensory descriptors (seven items) as well as signs related to bedside sensory examination (three items) (Bouhassira, D. et al., 2004). The interview questions address the quality of the pain (burning, painful cold, and electric shocks) and associated symptoms (tingling, pins and needles, numbness, and itching). The examination consists of the assessment of hypoesthesia to touch, prick and allodynia to brush. This questionnaire was validated in a prospective study of 160 patients presenting with pain of nociceptive and neuropathic origin and showed 86.0% of correctly identified patients (sensitivity 82.9%, specificity 89.9%).

A different approach to distinguish neuropathic from nonneuropathic pain uses a patient-based questionnaire with nine questions without the need of examinations by the physician (PainDetect) (Freynhagen, R. et al., 2006). The questionnaire consists of slightly different sensory descriptors (seven items, burning pain, tingling or prickling (electricity), sensitivity to touch (clothes, blanket), pain caused by light pressure (e.g., with finger), shooting pain or electric shock-like pain, occasional painful cold or heat (e.g., bath tub), and numbness), the question whether the pain is spatially radiating and a questions addressing the individual pain pattern. In the latter, the patients have to choose one of four graphically illustrated pain patters (permanent pain with a light variability, stable permanent pain with paroxysmal pain attacks, paroxysmal pain attacks with no permanent pain, and pain attacks with fluctuating permanent pain in between). This questionnaire was validated in 392 patients recruited at 10 highly specialized pain centers with either pain of pure predominant neuropathic origin ($n = 167$, e.g., PHN, painful polyneuropathies, and nerve trauma) or pure or predominant nociceptive origin ($n = 225$, e.g., ostheoathritis, mechanical low back pain, and inflammatory arthropathies). Patients were

diagnosed according to the results of two independent pain specialists who determined the predominant pain type (neuropathic versus nociceptive) by means of clinical experience as well as neurological examination, electrophysiologic, or imaging techniques. This questionnaire showed a correct classification rate of 82.5% with a sensitivity of 80.8% and a specificity 84.7%.

57.7.2 Bedside Assessment of Neuropathic Pain

Patients with neuropathic pain demonstrate a variety of distinct sensory symptoms and signs that can coexist in combinations (see above). Therefore, the sensory bedside examination should include the following qualities: touch, pinprick, pressure, cold, heat, vibration, temporal summation, and after sensations (Bouhassira, D. et al., 2004; Cruccu, G. et al., 2004, definition in Table 6). To assess either a loss (negative) or a gain of somatosensory function (positive sensory signs) the responses can be graded as normal, decreased or increased. The stimulus-evoked (positive) pain types are classified as hyperalgesic or allodynic, and according to the dynamic or static character of the stimulus (Rasmussen, P. V. et al., 2004).

Touch can be assessed by gently applying cotton wool to the skin, pinprick sensation by the response to sharp pinprick stimuli, deep pain by gentle pressure on muscle and joints, and cold and heat sensation by measuring the response to a thermal stimulus, for example, thermorollers kept at 20 or 45 °C. Cold sensation can also be assessed by the response to acetone spray. Vibration can be assessed by a tuning fork placed at strategic points (interphalangeal joints, etc.). Abnormal temporal summation is the clinical equivalent to increasing neuronal activity following repetitive C fiber stimulation >0.3 Hz. This windup-like pain can be produced by mechanical and thermal stimuli. After-sensations – the persistence of pain long after termination of a painful stimulus – is another characteristic feature of neuropathic pain, which is closely related to a coexistent dynamic or static hyperalgesia. When present, allodynia or hyperalgesia can be quantified by measuring intensity and area. At present, it is generally agreed that assessment should be carried out in the area of maximal pain using the contralateral area as control. However, contralateral segmental changes following a unilateral nerve or root lesion cannot be excluded, so an examination at mirror sites may not necessarily represent a true control site.

Table 6   Definition and assessment of negative and positive sensory symptoms or signs in neuropathic pain

Symptom/sign	Definition	Assessment Bedside exam	Expected pathological response
Negative signs and symptoms			
Hypoesthesia	Reduced sensation to nonpainful stimuli	Touch skin with painters brush, cotton swab, or gauze	Reduced perception, numbness
Pall hypoesthesia	Reduced sensation to vibration	Apply tuning fork on bone or joint	Reduced perception threshold
Hypoalgesia	Reduced sensation to painful stimuli	Prick skin with single pin stimulus	Reduced perception, numbness
Thermhypoesthesia	Reduced sensation to cold/warm stimuli	Contact skin with objects of 10 °C (metal roller, glass with water, coolants like acetone) Contact skin with objects of 45 °C (metal roller, glass with water)	Reduced perception
Spontaneous sensations/pain			
Paraesthesia	Nonpainful ongoing sensation (ant crawling)	Grade intensity (0–10) Area in cm^2	–
Paroxysmal pain	Shooting electrical attacks for seconds	Number per time Grade intensity (0–10) Threshold for evocation	–
Superficial pain	Painful ongoing sensation often of burning quality	Grade intensity (0–10) Area in cm^2	–
Evoked pain			
Mechanical dynamic allodynia	Normally nonpainful light moving stimuli on skin evoke pain	Stroking skin with painters brush, cotton swab, or gauze	Sharp burning superficial pain Present in the primary affected zone but spread beyond into unaffected skin areas (secondary zone)
Mechanical static allodynia	Normally nonpainful gentle static pressure stimuli at skin evoke pain	Manual gentle mechanical pressure at the skin	Dull pain Present in the area of affected (damaged or sensitized) primary afferent nerve endings (primary zone)
Mechanical punctuate, pinprick hyperalgesia	Normally not painful/slightly stinging stimuli evoke pain	Manual pricking the skin with a safety pin, sharp stick, or stiff von Frey hair	Sharp superficial pain Present in the primary affected zone but spread beyond into unaffected skin areas (secondary zone)
Temporal summation	Repetitive application of identical single noxious stimuli is perceived as increasing pain sensation (windup like pain)	Pricking skin with safety pin at interval <3 s for 30 s	Sharp superficial pain of increasing intensity
Cold allodynia hyperalgesia	Normally nonpainful/slightly painful cold stimuli evoke pain	Contact skin with objects of 20 °C (metal roller, glass with water, coolants like acetone) Control: contact skin with objects of skin temperature	Painful often burning temperature sensation Present in the area of affected (damaged or sensitized) primary afferent nerve endings (primary zone)
Heat allodynia hyperalgesia	Normally nonpainful/slightly painful heat stimuli evoke pain	Contact skin with objects of 40 °C (metal roller, glass with water)	Painful burning temperature sensation

(Continued)

Table 6 (Continued)

Symptom/sign	Definition	Assessment Bedside exam	Expected pathological response
		Control: contact skin with objects of skin temperature	Present in the area of affected (damaged or sensitized) primary afferent nerve endings (primary zone)
Mechanical deep somatic allodynia	Normally nonpainful pressure on deep somatic tissues evoke pain	Manual light pressure at joints or muscle	Deep pain at joints or muscles

It is important to realized the spatial distribution of abnormal sensations. In neuropathic conditions, the distinction between primary and secondary area corresponds to the tissue supplied by damaged nerves and the area outside this innervation territory. Mechanical hypersensitivity often expands into the secondary area.

In Table 6 several sensory signs that can be found in peripheral neuropathies are defined and the appropriate tests to assess these signs clinically are summarized.

57.7.3 Apparative Assessment of Somatosensory Phenotypes

It is important to realize that conventional electrophysiological techniques like NCS, SEP, etc., only assess the function of myelinated peripheral axonal systems, the affection of small fibers including nociceptors are missed. Similarly, SEP measures the conduction in the lemniscal system (dorsal columns) and not the spinothalamic tract, which conveys noxious information. Therefore, alternative diagnostic procedures have to be used to analyze nociceptive fibers and tracts in the peripheral or CNS, especially to diagnose small fiber neuropathies and central pain states that are associated with a isolated damage to central nociceptive systems.

A sophisticated neurophysiologic technique to test both systems in the PNS and the CNS is QST, which uses standardized mechanical and thermal stimuli (graded von Frey hairs, several pinprick stimuli, pressure algometers, quantitative thermotesting, etc.). Another advantage of QST is that it assesses a loss of function (minus signs) as well as a gain of function (positive signs). For example Allodynia or hyperalgesia, can be quantified by measuring intensity, threshold for elicitation, duration, and area.

A standardized protocol for QST was recently proposed by the nationwide German Network on Neuropathic Pain (Rolke R. *et al.*, 2006) including 13 parameters of sensory testing procedures for the analysis of the exact somatosensory phenotype of neuropathic pain patients (Figure 6). To evaluate plus or minus signs in patients, an age- and gender-matched database for absolute and relative QST reference data was established for healthy human subjects. Until now this nationwide multicenter trial comprises complete sensory profiles of 180 healthy human subjects and more than 1500 neuropathic pain patients of a variety of entities. Thermal detection and pain thresholds including a test for the presence of paradoxical heat sensations, mechanical detection thresholds to von Frey filaments and a 64-Hz tuning fork, mechanical pain thresholds to pinprick stimuli and blunt pressure, stimulus/response functions for pinprick and dynamic mechanical allodynia (pain to light touch), and pain summation (windup ratio) using repetitive pinprick stimulation are determined. Data of healthy human subjects were analyzed for the influence of body side and region, age, and gender. For most variables pathological values of positive and negative signs can be detected on the basis of reference data.

57.7.4 Skin Biopsies

The advent of skin biopsy as a diagnostic and investigative tool has changed the diagnostic picture in neuropathic pain; skin biopsies can be done at several sites to establish a spatial profile of nerve fiber involvement, and can be repeated over time. They have generally met good patient acceptance in neuropathies, and even in allodynic states like PHN. They can be used to assess epidermal nerve fibers, predominantly C fiber nociceptors, sympathetic fibers innervating the cutaneous sweat glands, and myelinated Aβ and Aδ fibers in the dermis.

Immunostaining of skin punch biopsies for the panaxonal marker protein gene product 9.5 (PGP9.5), a neuronal ubiquitin hydrolase is believed to identify axons of all categories of nerve fibers. In a

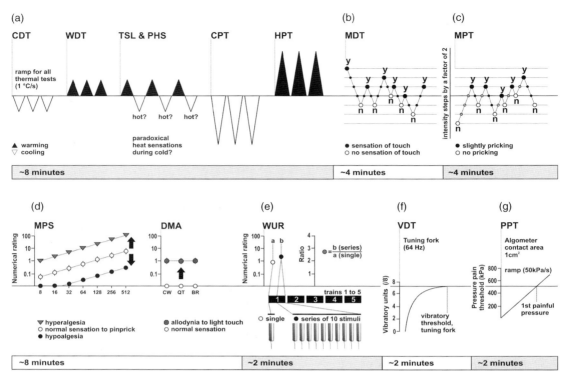

Figure 6 Quantitative sensory testing (QST) – a battery of sensory tests: figure of methods. The standardized QST protocol assesses 13 variables in seven test procedures (a–g). All procedures are presented including a time frame for testing over one area. (a) (thermal testing; CDT, cold detection threshold; WDT, warm detection threshold; TSL, thermal sensory limen for alternating warm and cold stimuli; PHS, number of paradoxical heat sensations during the TSL procedure; CPT, cold pain threshold; HPT, heat pain threshold) comprises detection and pain thresholds for cold, warm, or hot stimuli (C and Aδ fiber mediated). (b) (MDT, mechanical detection threshold) tests for Aβ fiber function using von Frey filaments. (c) (MPT, mechanical pain threshold) tests for Aδ-fiber-mediated hyper- or hypoalgesia to pinprick stimuli. (d) (stimulus/response functions: MPS for pinprick stimuli, and ALL for dynamic mechanical allodynia) assesses again Aδ-mediated sensitivity to sharp stimuli (pinprick), and also Aβ-fiber-mediated sensitivity to stroking light touch (CW, cotton wisp; QT, cotton wool tip; BR, brush). (e) (WUR, windup ratio) compares the verbal ratings within five trains of a single pinprick stimulus (a) with a series (b) of 10 repetitive pinprick stimuli to calculate WUR as the ratio: b/a. (f) (VDT, vibration detection threshold) tests again for Aβ fiber function using a Rydel–Seiffer 64-Hz tuning fork. (g) (PPT, pressure pain threshold) is the only test for deep pain sensitivity, most probably mediated by muscle C and Aδ fibers.

variety of neuropathies affecting small caliber nerve fibers, denervation has been documented to occur in a length-dependent manner. These include diabetes mellitus, HIV-associated sensory neuropathy (Polydefkis, M. *et al.*, 2002), Fabry's disease (Scott, L. J. *et al.*, 1999), restless legs syndrome, and idiopathic small-fiber sensory neuropathies (Holland, N. R. *et al.*, 1997).

Punch skin biopsies have more recently been used as a serial measure, to examine changes in epidermal innervation over time, or after therapeutic interventions in various neuropathic conditions. For example, epidermal reinnervation occured concomitant with symptomatic improvement in sensory neuropathy (Nodera, H. *et al.*, 2003), and impressive increases in epidermal, dermal, and sweat gland innervation

densities have been documented with improvement in neuropathic status spontaneously or after therapies (Hart, A. M. *et al.*, 2002).

Cutaneous innervation is usually reduced in PHN skin compared to mirror-image skin when assessed with PGP (Rowbotham, M. C. *et al.*, 1996; Oaklander, A.L. *et al.*, 1998; Petersen, K. L. *et al.*, 2000) (Figure 7). However, a significant proportion of PHN subjects have similar fiber densities in both PHN and mirror-image skin, suggesting that the symptoms of pain and sensory dysfunction is not due to a mere loss in overall innervation density (Rowbotham, M. C. *et al.*, 1996; Petersen, K. L. *et al.*, 2000). Oaklander demonstrated that the average cutaneous fiber density was lower in subjects with established PHN than in those who had recovered without PHN after herpes zoster.

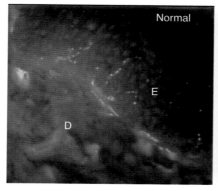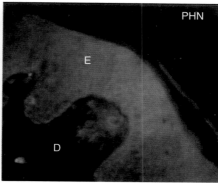

Figure 7 Representative skin biopsy samples from a postherpetic neuralgia (PHN) patient. PGP 9.5 immunofluorescence in normal and mirror-image painful PHN skin. The patient suffered from spontaneous pain and allodynia, but had a marked loss of thermal sensation (heat pain threshold elevated by $7.3 \,^\circ$C). D, dermis; E, epidermis. (Reprinted from Rowbotham, M. C., Yosipovitch, G., Connolly, M.K., Finlay, D., Forde, G. and Fields, H. L. 1996. Cutaneous innervation density in the allodynic form of postherpetic neuralgia. Neurobiol. Dis. 3, 205–214. Copyright 1996 Academic Press.)

Furthermore, mirror-image skin had lower fiber densities than distant control skin in subjects with established PHN (Oaklander, A.L. *et al.*, 1998).

57.7.5 Microneurography

Microneurography has been used to assess functional abnormalities in single small fibers. This technique makes it possible to subgroup C nociceptors in healthy controls into a variety of physiologically different classes. The technique requires huge technical skills from the researcher and is very time consuming.

Regarding pathological states it has been possible to identify spontaneously active C fibers in patients e.g., with erythromelalgia (Orstavik, K. *et al.*, 2003) (Figure 4). However, the results about abnormal firing patterns in diabetic neuropathy patients obtained so far have not been conclusive (Handwerker, personal communication). One major problem has been that normal aging changes the functional properties of C afferents dramatically. As the patient populations are generally older, normative data from a cohort of healthy older subjects have to be obtained. Furthermore, as recordings are performed from single afferent C fibers and not from whole nerves it is not possible to obtain reliable quantitative data about fiber numbers.

57.7.6 Indirect Test of Afferent Unmyelinated Fiber Function

The function of afferent unmyelinated fibers in the skin can be objectively assessed by using the efferent effector component of C fibers as indirect means (Baron, R. and Saguer, M., 1993; Schuller, T. B. *et al.*, 2000). Studies of the axon reflex vasodilatation (neurogenic flare) using visual inspection (Figure 8), thermography, or laser Doppler flowmetry in response to a chemical stimulus that activates cutaneous C fibers (histamine, capsaicin) are commonly used. While independent of patient cooperation, a major drawback of this technique is the dependence on a variety of other factors that affect the effector response.

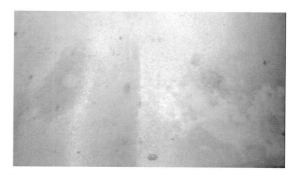

Figure 8 Histamine flares in a patient with severe postherpetic neuralgia in the right segment Th6. Histamine iontophoresis was performed within the allodynic transitional zone at the cranial border of the affected and adjacent segment and on the contralateral corresponding site. The formerly affected dermatome is characterized by heavy scar tissue, whereas the allodynic area shows no scars or secondary pigmentation. Note the difference in flare size (ipsilateral $260 \, mm^2$, contralateral $2060 \, mm^2$) indicating an impairment of nociceptive C fiber function in the allodynic region. Adapted from Baron, R. and Saguer, M. 1993. Copyright 1993 Oxford University Press.

57.7.7 Imaging Techniques

A new MRI technique, MR neurography makes it possible to identify small patches of inflammation in peripheral nerves. This might be of value in inflammatory neuropathies, acute zoster, or schwannoma and neuroma detection (Haanpaa, M. *et al.*, 1998; Bendszus, M. *et al.*, 2004) (Figure 9). More patients with visable lesions associated with acute herpes zoster still had pain after 3 months. No data are available to demonstrate changes that occur in chronic PHN. MRI can also identify changes in peripheral nerves and secondary neurogenic alterations in skeletal muscle (Koltzenburg, M. and Bendszus, M., 2004). Prolongation of the T2 relaxation time and gadolinium enhancement of denervated muscle develop in parallel with the development of pathological spontaneous activity on EMG that is a feature of axonal damage. Axonal nerve lesions cause a hyperintense signal on T2-weighted MRI of the affected nerve at and distal to the lesion site, which correlates with Wallerian degeneration and nerve edema. While the MRI changes do not distinguish between painful and painless nerve lesions, they supplement the differential diagnosis of peripheral nerve disease. Furthermore, diffusion tensor imaging technique is capable of detecting degeneration of fibers in the dorsal columns in severe peripheral neuropathies.

57.7.8 Do We Have Diagnostic Tool to Dissect Individual Mechanisms in Neuropathic Pain?

The modern theoretical concept of a mechanism-based therapy uses the assumption that a specific symptom predicts a specific underlying mechanism. It should be possible for a clinician to identify a particular

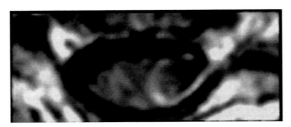

Figure 9 Magnetic resonance image of a patient with acute herpes zoster. Spinal cord in the cervical affected region. Note the contrast enhancement in the dorsal root, dorsal horn of the spinal cord, and lesser also in the ventral root. The patients was severely affected with pain and motor weakness at the left arm (T1 image with contrast medium). Photo courtesy of Prof. Dr. T. Tölle, Munich.

mechanism by assessing a specific sensory symptom. There are, however, some important caveats with such an approach. Recent clinical experimental studies indicate that it is not appropriate to link one single symptom with exactly one mechanism. It was shown that one specific symptom may be generated by several entire different underlying pathophysiological mechanisms. It became clear that only a specific symptom constellation, a symptom profile, i.e., a combination of negative and positive sensory phenomena, might be able to predict the mechanisms and not just one single symptom. In order to translate these ideas into the clinical framework of neuropathic pain the most important approach is to characterize the somatosensory phenotype of the patients as precisely as possible.

For example, standardized QST methods described above can nicely distinguish between phenotypic subtypes of PHN patients with distinct sensory symptom constellations that are likely correlated with different underlying mechanisms (sensitization type, deafferentation type, see above). Distinct pathophysiological changes in the excitability of peripheral and central neurons are likely involved in pain generation (for details, see Figure 10).

As attractive a subtype classification based on the nociceptor function and evoked pain types might be it should be emphasized that not all PHN patients fit exactly into one category or the other. It rather seems to be a continuum. Furthermore, in a large group of PHN patients many heterogeneous patterns of sensory dysfunction were detected (Pappagallo, M. *et al.*, 2000). Accordingly, detailed testing of sensory function, chemical stimulation, and cutaneous innervation in one PHN patient clearly showed areas of relative preservation in close vicinity to impaired thermal sensation, both within the affected dermatome (Petersen, K. L. *et al.*, 2000). Furthermore, the sensory patterns showed a variation over the time course of PHN. However, by classification of PHN patients due to sensory perception thresholds within the most painful skin area it is possible to detect the predominant individual sensory profile and the most likely underlying pain-generating mechanism.

Another caveat addressing the predictive value of QST measurements for therapy was recently revealed by two independent studies. It was hypothesized that topically applied lidocaine that is believed to act on ectopic discharges in nociceptive fibers would be in particular beneficial for patients with sensitized peripheral nociceptors as compared with patients with a loss of dermal nociceptors. In contrast to the hypothesis, however, skin biopsies, QST, histamine test as well nerve conduction studies could not identify

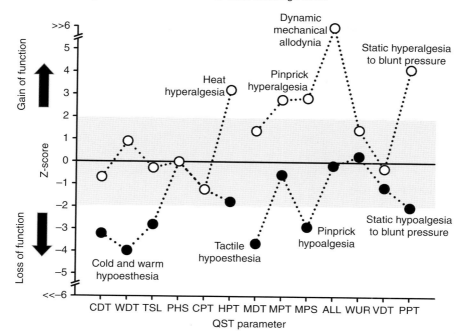

Figure 10 *Z*-score sensory profiles of two patients suffering from postherpetic neuralgia (PHN). Patient PHN I (open circles) presents the Quantitative sensory testing (QST) profile of a 70-year-old woman suffering from PHN for 8 years. Ongoing pain was 80 on a 0 to 100 numerical rating scale. The profile shows a predominant gain of sensory function in terms of heat pain hyperalgesia (HPT), pinprick mechanical hyperalgesia (MPS), dynamic mechanical allodynia (ALL), and static hyperalgesia to blunt pressure (PPT) outside the 95% confidence interval of the distribution of healthy subjects (=gray zone). This profile is consistent with a combination of peripheral and central sensitization. Patient PHN II (filled circles) shows the QST profile of a 71-year-old woman with pain for 8 months. Ongoing pain was 70 on a 0 to 100 numerical rating scale. The QST profile shows predominant loss of sensory function. Note the cold (CDT), warm detection thresholds (WDT), thermal sensory limen (TSL), heat pain thresholds (HPT), tactile detection thresholds (MDT), and mechanical pain thresholds to pinprick stimuli (MPT) outside the normal range as presented by the gray zone. This profile is consistent with a combined small- and large-fiber sensory deafferentation. *Z*-score: Numbers of standard deviations between patient data and group-specific mean value. Adapted from Rolke, R., Baron, R., Maire, C., Tölle, T. R., Treede, R. D., Binder, A., Birbaumer, N., Birklein, F., Bötefür, C. I., Braune, S., Flor, H., Huge, V., Klug, R., Landwehrmeyer, G. B., Magerl, W., Maihöfer, G., Rolko, C., Scherens, A., Sprenger, T., Valet, M., and Wasserka, B. 2006. Quantitative Sensory testing in the German Research Network on Neuropathic Pain (DFNS): standardized protocol and reference values. *Pain* 123, 231–243.

lidocaine responders in painful neuropathies (Herrmann, D. N. *et al.*, 2006) and PHN (Wasner, G. *et al.*, 2005). Alternatively, it might be possible that surviving A fibers in C-nociceptor-deprived skin may express sodium channels, develop ectopic firing and might therefore be the target for lidocaine.

57.8 Therapy

The number of trials for peripheral neuropathic pain has expanded greatly in the past few years. These have been summarized in several recent meta-analyses that are referred to in the following (Baron, R. and Wasner,

G., 2006;; McQuay H. *et al.*, 1995; McQuay, H. J. *et al.*, 1996; Sindrup, S. H. and Jensen, T. S., 1999; Dworkin, R. H. *et al.*, 2003; Finnerup, N. B. *et al.*, 2005; Siddall, P. J. and Middleton, J. W., 2005). For central neuropathic pain there are limited data. The treatments discussed below have all been demonstrated to provide statistically significant and clinically meaningful treatment benefits compared with placebo in multiple randomized controlled trials (Table 7).

Nonmedical treatments, i.e., interventional procedures, neurostimulation techniques, neurosurgical destructive techniques, psychological therapy as well as physiotherapy and occupational therapy are not the focus of this review.

Table 7 Pharmacological therapy of neuropathic pain syndromes

Compound	Evidence
Antidepressants	
Amitriptyline	PHN ⇑⇑⇑, PNP ⇑⇑⇑, PTN ⇑, STR ⇑
Venlafaxine	PNP ⇑⇑⇑
Duloxetine	PNP ⇑⇑⇑
Anticonvulsants (Na channel)	
Carbamazepine	PNP ⇑, TGN ⇑⇑⇑
Oxcarbazepine	PNP ⇑
Lamotrigine	HIV ⇑, PNP ⇑, STR ⇑
Anticonvulsants (Ca channel)	
Gabapentin	PHN ⇑⇑⇑, PNP ⇑⇑⇑, HIV ⇑, CRPS ⇑, PHAN ⇑, SCI ⇑, MIX ⇑, CANC ⇑
Pregabalin	PHN ⇑⇑⇑, PNP ⇑⇑⇑, SCI ⇑
Tramadol	PHN ⇑, PNP ⇑⇑⇑
Long-acting strong opioids	
Morphine	PHN ⇑, PHAN ⇑
Oxycodone	PHN ⇑, PNP ⇑⇑⇑
Cannabinoids	MS ⇑⇑⇑, PA ⇔, MIX ⇑
Topical therapy	
Capsaicin cream	PHN ⇑, PNP ⇑, PTN ⇑
Lidocaine patch	PHN ⇑⇑⇑, MIX ⇑

CANC, neuropathic cancer pain; CRPS, complex regional pain syndrome; HIV, HIV neuropathy; MIX, mixed neuropathic pain cohort; MS, central neuropathic associated with MS (multiple sclerosis); PA, central neuropathic pain after plexus avulsion; PHAN, phantom pain; PHN, postherpetic neuralgia; PNP, polyneuropathy (mainly diabetic); PTN, posttraumatic neuralgia; SCI, spinal cord injury; STR, poststroke pain; TGN, trigeminal neuralgia.
Levels of evidence: ⇑⇑⇑, several randomized clinical trials (RCTs) or meta-analyses; ⇑, at least 1 RCT; ⇔, unclear.

57.8.1 Antidepressants

The effectiveness of tricyclic antidepressants (TCAs) in neuropathic pain may account for their broad range of pharmacological actions. These compounds are inhibitors of the reuptake of monoaminergic transmitters. They are believed to potentiate the effects of biogenic amines in CNS pain-modulating pathways. In addition, they block voltage dependent sodium channels and α adrenergic receptors.

Venlafaxine and duloxetine that blocks both serotonin and norepinephrine reuptake (serotonin noradrenaline reuptake inhibitors (SNRI)) were efficacious in diabetic painful neuropathy (DPN). In a comparison of venlafaxine and imipramine in patients with painful polyneuropathy, both antidepressants demonstrated superior pain relief compared with placebo but did not differ from each other.

Selective serotonin reuptake inhibitors (SSRI) have fewer adverse effects and are generally better tolerated than TCAs. However, they did not show convincing efficacy in neuropathic pain states.

57.8.2 Anticonvulsants (Ca Channel Modulators)

There are extensive clinical trials of gabapentin for chronic neuropathic pain. These studies examined patients with PHN, DPN, mixed neuropathic pain syndromes, phantom limb pain, Guillain–Barré syndrome, and acute and chronic pain from SCI. Improvements in sleep, mood, and quality of life were also demonstrated. Pregabalin, the successor drug of gabapentin was shown to be efficacious in PHN, DPN, and SCI. There is growing evidence supporting the mechanism of action on the $\alpha2\delta$ subunit of neuronal calcium channels partly located at the presynaptic spinal terminals of primary afferent nociceptors. One advantage over gabapentin is its superior bioavailability, which makes it easier to use without the need of long titration periods. Gabapentin and pregabalin are generally well tolerated, safe, have no drug interactions, and no negative impact on cardiac function. This advantages makes them a suitable option especially for the elderly, a population very often suffering from several comorbidities that need multiple drug therapies.

57.8.3 Anticonvulsants (Na Channel Blockers)

For lamotrigine there is evidence of efficacy for HIV sensory neuropathy, DPN, and central poststroke pain. Carbamazepine is very effective in trigeminal neuralgia. However, the strength of evidence is much lower for the benefit of these drugs in other types of neuropathic pain. Oxcarbazepine which has fewer side effects and drug–drug interactions than carbamazepine was shown to be effective in painful diabetic neuropathy (Dogra, S. *et al.*, 2005).

57.8.4 Tramadol and Opioid Analgesics

Tramadol is a norepinephrine and serotonin reuptake inhibitor with a major metabolite that is a μ opioid agonist. Sustained efficacy for several weeks has been demonstrated for oral tramadol in PHN, DPN, and in patients with painful polyneuropathy of various causes.

Strong opioids are clearly effective in postoperative, inflammatory, and cancer pain. However, the use of narcotic analgesics for patients with chronic neuropathic pain was highly controversial, even among experts in the field of pain management. However, to date several positive trials of oral strong opioid analgesics in various neuropathic pain entities have been published. In an interesting three-period crossover study comparing treatment with opioid analgesics, TCAs, and placebo in patients with PHN, controlled-release morphine provided statistically significant benefits for pain and sleep but not for physical function and mood. In that trial, patients preferred treatment with opioid analgesics compared with TCAs and placebo despite a greater incidence of adverse effects and more dropouts during opioid treatment (Raja, S. N. *et al.*, 2002). Our experience is that many patients with pain due to central and peripheral nerve injury can be successfully and safely treated on a chronic basis with stable doses of narcotic analgesics. However, the use of opioids requires caution in patients with a history of chemical dependence or pulmonary disease. We recommend using long-acting opioid analgesics (e.g., sustained release preparation or transdermal applications) when alternative approaches to treatment have failed.

57.8.5 Topical Medications

57.8.5.1 *Topical capsaicin*

Capsaicin is an agonist of the vanilloid receptor (TRPV1) that is present on the sensitive terminals of primary nociceptive afferents. On initial application, it has an excitatory action and produces burning pain and hyperalgesia, but with repeated or prolonged application it inactivates the receptive terminals of nociceptors. Therefore, this approach is reasonable for those patients whose pain is maintained by anatomically intact sensitized nociceptors. Capsaicin has been reported to reduce the pain of PHN and DPN.

57.8.5.2 *Topical lidocaine*

A second topical medication for neuropathic pain is local anesthetics. Local anesthetics block voltage-dependent sodium channels. Although the site of action of membrane-stabilizing drugs for relief of pain has not been proven in patients, *in vitro* studies have shown that ectopic impulses generated by damaged primary afferent nociceptors are abolished by concentrations of local anesthetics much lower than that required for blocking normal axonal conduction. Studies report pain relief with topically applied special formulations of local anesthetic (lidocaine patches (5%)) in PHN and other neuropathies.

57.8.6 *N*-Methyl-D-Aspartic Acid Receptor Antagonists

These drugs block excitatory glutamate receptors in the CNS that are thought to be responsible for the increased central excitability (central sensitization) following noxious stimuli. Clinically available substances with NMDA receptor blocking properties include ketamine, dextromethorphane, memantine, and amantadine. Studies of small cohorts have generally confirmed the analgesic effects of ketamine in patients suffering from PHN. However, studies with oral NMDA antagonists formulations (e.g., dextromethorphane) showed positive results in DPN but the drug was without beneficial effect in PHN.

57.8.7 Cannabinoids

Cannabinoid receptors type 1 (CB1) have been demonstrated in upper laminae of the spinal dorsal horn intimately concerned with the processing of nociceptive information as well as on the cell bodies of primary afferent neurons. Relief of central pain was found with the oral tetrahydocannabinol dronabinol in MS patients and in plexus avulsion pain although in the latter the primary outcome measure defined in the hypothesis was failed. Cannabinoids were also effective in mixed peripheral neuropathic pain.

57.8.8 Treatment Guideline

In the previous section a variety of treatment regimens were reviewed that have shown to be effective. In summary, the relevant medical management of neuropathic pain consists of five main classes of oral medication (serotonin-/norepinephrine-modulating antidepressants, Na blocker anticonvulsants, Ca modulator anticonvulsants, tramadol, and opioids) and two categories of topical medications mainly for patients with cutaneous allodynia and hyperalgesia (capsaicin and local anesthetics).

A useful way to compare the efficacy of different treatments is the consultation of systematic reviews to determine the best available drugs (Sindrup, S. H. and Jensen, T. S., 1999; 2000; Dworkin, R. H. and Schmader, K. E., 2003). Up to now, more than 110 randomized controlled trials for the medical treatment of neuropathic pain have been published. In this respect, the measure: "Numbers needed to treat (NNT)" has been a useful measure. Numbers needed to treat is: the number of patients needed to treat with a certain drug to obtain one patient with a defined degree of pain relief. This method permit a comparison between different drugs and diseases and it allows generation of large numbers in order to provide reliable information about efficacy. Usually the NNT for more than 50% pain relief is used because it is easily understood and seems in many cases to be a relevant clinical effect. However, the methodological complexity of pooling data from small crossover and large parallel group trials and methodological shortcomings of some trials in a systematic review, remain as limitations.

In Figure 11 the NNT values of drugs for treatment of neuropathic pain is shown in relation to the number of patient that have been treated with each of the drug to get a sense of the validity of the different trials.

Because more than one mechanism is at work in most patients, a combination of two or more analgesic agents to cover multiple types of mechanisms will generally produce greater pain relief and fewer side effects. Therefore, in most patients, a stepwise proceeding with a successive monotherapy is not appropriate. Early combinations of two or three

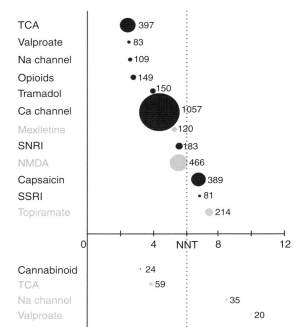

Figure 11 Numbers needed to treat in peripheral and central neuropathic pain. Combined numbers needed to treat (NNT) to obtain one patient with more than 50% pain relief in (a) peripheral neuropathic pain (painful polyneuropathy, postherpetic neuralgia, and peripheral nerve injury pain) and (b) central pain (central poststroke pain, pain following spinal cord injury and multiple sclerosis). SNRI, serotonin noradrenaline reuptake inhibitors; SSRI, selective serotonin reuptake inhibitor. Circle size and related numbers indicate number of patients who have received active treatment. Gray circle: at least half of conducted trials showed no significant effect. Right to Dotted line: clinical relevance should be questioned. (After Finnerup, N. B., Otto, M., McQuay, H. J., Jensen, T. S., Sindrup, S. H. 2005. Algorithm for neuropathic pain treatment: an evidence based proposal. Pain 118, 289–305.)

compounds out of different classes is the general practical approach. In fact, in a controlled four-period crossover trial, combined gabapentin and morphine achieved better analgesia at lower doses of each drug than either as a single agent, with constipation, sedation, and dry mouth as the most frequent adverse effects.

Drug-related adverse effects are common in the treatment of neuropathic pain, not only because of the specific medications used but also because many patients with this condition are older, take other medications, and have comorbid illnesses. Therefore, the drugs of first choice has to be judged on the basis of these data.

There are several exceptions from the above general treatment recommendations that should be emphasized: First, treatment guidelines for trigeminal neuralgia are distinct and include carbamazepine and baclofen. Second, the pharmacological therapy may be similar in CRPS, although controlled trials of first-line medications are lacking. However, it is very likely that antiinflammatory strategies in particular in the acute phase might be helpful. Third, although chronic neuropathic back pain (i.e., cervical and lumbar radiculopathic pain) is probably the most prevalent pain syndrome to which neuropathic mechanisms contribute, there are no accepted diagnostic criteria for identifying this neuropathic component. It is likely that a combination of neuropathic, skeletal, and myofascial mechanisms account for this type of pain in many patients.

Acknowledgments

This work was supported by the Deutsche Forschungsgemeinschaft (DFG Ba 1921/1-2), the German Ministry of Research and Education within the German Research Network on Neuropathic Pain (BMBF, 01EM01/04), and an unrestricted educational grant of Pfizer (Germany).

References

Alessandri-Haber, N., Dina, O. A., Yeh, J. J., Parada, C. A., Reichling, D. B., and Levine, J. D. 2004. Transient receptor potential vanilloid 4 is essential in chemotherapy-induced neuropathic pain in the rat. J. Neurosci. 24, 4444–4452.

Alexander, G. M., van Rijn, M. A., van Hilten, J. J., Perreault, M. J., and Schwartzman, R. J. 2005. Changes in cerebrospinal fluid levels of pro-inflammatory cytokines in CRPS. Pain 116, 213–219.

Ali, Z., Raja, S. N., Wesselmann, U., Fuchs, P., Meyer, R. A., and Campbell, J. N. 2000. Intradermal injection of norepinephrine evokes pain in patients with sympathetically maintained pain. Pain 88, 161–168.

Amir, R., Liu, C. N., Kocsis, J. D., and Devor, M. 2002. Oscillatory mechanism in primary sensory neurones. Brain 125, 421–435.

Baron, R. 2006. Mechanisms of disease: neuropathic pain – a clinical perspective. Nat. Clin. Pract. Neurol. 2, 95–106.

Baron, R. and Binder, A. 2004. How neuropathic is sciatica? The mixed pain concept. Orthopade 33, 568–575.

Baron, R. and Janig, W. 2004. Complex regional pain syndromes – how do we escape the diagnostic trap? Lancet 364, 1739–1741.

Baron, R. and Saguer, M. 1993. Postherpetic neuralgia. Are C-nociceptors involved in signalling and maintenance of tactile allodynia? Brain 116, 1477–1496.

Baron, R. and Saguer, M. 1994. Axon-reflex reactions in affected and homologous contralateral skin after unilateral peripheral injury of thoracic segmental nerves in humans. Neurosci. Lett. 165, 97–100.

Baron, R. and Wasner, G. 2001. Complex regional pain syndromes. Curr. Pain Headache Rep. 5, 114–123.

Baron, R. and Wasner, G. 2006. Prevention and treatment of postherpetic neuralgia. Lancet 367, 186–188.

Baron, R., Baron, Y., Disbrow, E., and Roberts, T. P. L. 1999a. Brain processing of capsaicin-induced secondary hyperalgesia: a functional MRI study. Neurology 53, 548–557.

Baron, R., Baron, Y., Disbrow, E., and Roberts, T. P. L. 2000. Activation of the somatosensory cortex during Aβ-fiber mediated hyperalgesia – a MSI study. Brain Res. 871, 75–82.

Baron, R., Levine, J. D., and Fields, H. L. 1999b. Causalgia and reflex sympathetic dystrophy: does the sympathetic nervous system contribute to the generation of pain? Muscle Nerve 22, 678–695.

Baron, R., Schattschneider, J., Binder, A., Siebrecht, D., and Wasner, G. 2002. Relation between sympathetic vasoconstrictor activity and pain and hyperalgesia in complex regional pain syndromes: a case–control study. Lancet 359, 1655–1660.

Baron, R., Schwarz, K., Kleinert, A., Schattschneider, J., and Wasner, G. 2001. Histamine-induced itch converts into pain in neuropathic hyperalgesia. Neuroreport 12, 3475–3478.

Bendszus, M., Wessig, C., Solymosi, L., Reiners, K., and Koltzenburg, M. 2004. MRI of peripheral nerve degeneration and regeneration: correlation with electrophysiology and histology. Exp. Neurol. 188, 171–177.

Bouhassira, D., Attal, N., Fermanian, J., Alchaar, H., Gautron, M., Masquelier, E., Rostaing, S., Lanteri-Minet, M., Collin, E., Grisart, J., and Boureau, F. 2004. Development and validation of the neuropathic pain symptom inventory. Pain 108, 248–257.

Campbell, J. N., Raja, S. N., Meyer, R. A., and Mackinnon, S. E. 1988. Myelinated afferents signal the hyperalgesia associated with nerve injury. Pain 32, 89–94.

Catarina, M. J., Leffler, A., Malmberg, A. B., Martin, W. J., Trafton, J., Petersen-Zeitz, K. R., Koltzenburg, M., Basbaum, A. I., and Julius, D. 2000. Impaired nociception and pain sensation in mice lacking the capsaicin receptor. Science 288, 306–313.

Choi, B. and Rowbotham, M. C. 1997. Effect of adrenergic receptor activation on post-herpetic neuralgia pain and sensory disturbances. Pain 69, 55–63.

Coull, J. A., Beggs, S., Boudreau, D., Boivin, D., Tsuda, M., Inoue, K., Gravel, C., Salter, M. W., and De Koninck, Y. 2005. BDNF from microglia causes the shift in neuronal anion gradient underlying neuropathic pain. Nature 438, 1017–1021.

Coull, J. A., Boudreau, D., Bachand, K., Prescott, S. A., Nault, F., Sik, A., De Koninck, P., and De Koninck, Y. 2003. Trans-synaptic shift in anion gradient in spinal lamina I neurons as a mechanism of neuropathic pain. Nature 424, 938–942.

Cruccu, G., Anand, P., Attal, N., Garcia-Larrea, L., Haanpaa, M., Jorum, E., Serra, J., and Jensen, T. S. 2004. EFNS guidelines on neuropathic pain assessment. Eur. J. Neurol. 11, 153–162.

Daousi, C., MacFarlane, I. A., Woodward, A., Nurmikko, T. J., Bundred, P. E., and Benbow, S. J. 2004. Chronic painful peripheral neuropathy in an urban community: a controlled comparison of people with and without diabetes. Diabet. Med. 21, 976–982.

Dib-Hajj, S. D., Rush, A. M., Cummins, T. R., Hisama, F. M., Novella, S., Tyrrell, L., Marshall, L., and Waxman, S. G. 2005. Gain-of-function mutation in Nav1.7 erythromelalgia induces bursting of sensory neurons. Brain.128, 1847–1854.

Dogra, S., Beydoun, S., Mazzola, J., Hopwood, M., and Wan, Y. 2005. Oxcarbazepine in painful diabetic neuropathy: a randomized, placebo-controlled study. Eur. J. Pain 9, 543–554.

Dworkin, R. H. and Schmader, K. E. 2003. Treatment and prevention of postherpetic neuralgia. Clin. Infect. Dis. 36, 877–882.

Dworkin, R. H., Backonja, M., Rowbotham, M. C., Allen, R. R., Argoff, C. R., Bennett, G. J., Bushnell, M. C., Farrar, J. T., Galer, B. S., Haythornthwaite, J. A., Hewitt, D. J., Loeser, J. D., Max, M. B., Saltarelli, M., Schmader, K. E., Stein, C., Thompson, D., Turk, D. C., Wallace, M. S., Watkins, L. R., and Weinstein, S. M. 2003. Advances in neuropathic pain: diagnosis, mechanisms, and treatment recommendations. Arch. Neurol. 60, 1524–1534.

Fields, H. L., Rowbotham, M., and Baron, R. 1998. Postherpetic neuralgia: irritable nociceptors and deafferentation. Neurobiol. Dis. 5, 209–227.

Finnerup, N. B., Otto, M., McQuay, H. J., Jensen, T. S., and Sindrup, S. H. 2005. Algorithm for neuropathic pain treatment: an evidence based proposal. Pain 118, 289–305.

Flor, H., Elbert, T., Knecht, S., Wienbruch, C., Pantev, C., Birbaumer, N., Larbig, W., and Taub, E. 1995. Phantom-limb pain as a perceptual correlate of cortical reorganization following arm amputation. Nature 375, 482–484.

Freynhagen, R., Tölle, T. R., and Baron, R. 2005. painDETECT – ein Palmtop-basiertes Verfahren für Versorgungsforschung, Qualitätsmanagement und Screening bei chronischen Schmerzen. Akt. Neurol 54, P641.

Gilden, D. H., Dueland, A. N., Cohrs, R., Martin, J. R., Kleinschmidt-DeMasters, B. K., and Mahalingam, R. 1991. Preherpetic neuralgia. Neurology 41, 1215–1218.

Gracely, R. H., Lynch, S. A., and Bennett, G. J. 1992. Painful neuropathy: altered central processing maintained dynamically by peripheral input (published erratum appears in Pain (1993, Feb.) 52(2), 251–253) (see comments). Pain 51, 175–194.

Guilbaud, G., Benoist, J. M., Levante, A., Gautron, M., and Willer, J. C. 1992. Primary somatosensory cortex in rats with pain-related behaviours due to a peripheral mononeuropathy after moderate ligation of one sciatic nerve: neuronal responsivity to somatic stimulation. Exp. Brain Res. 92, 227–245.

Haanpaa, M., Dastidar, P., Weinberg, A., Levin, M., Miettinen, A., Lapinlampi, A., Laippala, P., and Nurmikko, T. 1998. CSF and MRI findings in patients with acute herpes zoster. Neurology 51, 1405–1411.

Habler, H. J., Janig, W., and Koltzenburg, M. 1987. Activation of unmyelinated afferents in chronically lesioned nerves by adrenaline and excitation of sympathetic efferents in the cat. Neurosci. Lett. 82, 35–40.

Hains, B. C., Saab, C. Y., Klein, J. P., Craner, M. J., and Waxman, S. G. 2004. Altered sodium channel expression in second-order spinal sensory neurons contributes to pain after peripheral nerve injury. J. Neurosci. 24, 4832–4839.

Hart, A. M., Wiberg, M., Youle, M., and Terenghi, G. 2002. Systemic acetyl-L-carnitine eliminates sensory neuronal loss after peripheral axotomy: a new clinical approach in the management of peripheral nerve trauma. Exp. Brain Res. 145, 182–189.

Herrmann, D. N., Pannoni, V., Barbano, R. L., Pennella-Vaughan, J., and Dworkin, R. H. 2006. Skin biopsy and quantitative sensory testing do not predict response to lidocaine patch in painful neuropathies. Muscle Nerve 33, 42–48.

Holland, N. R., Stocks, A., Hauer, P., Cornblath, D. R., Griffin, J. W., and McArthur, J. C. 1997. Intraepidermal nerve fiber density in patients with painful sensory neuropathy. Neurology 48, 708–711.

Hong, S. and Wiley, J. W. 2005. Early painful diabetic neuropathy is associated with differential changes in the expression and function of vanilloid receptor 1. J. Biol. Chem. 280, 618–627.

Hudson, L. J., Bevan, S., Wotherspoon, G., Gentry, C., Fox, A., and Winter, J. 2001. VR1 protein expression increases in undamaged DRG neurons after partial nerve injury. Eur. J. Neurosci. 13, 2105–2114.

Jacobs, J. M., Macfarlane, R. M., and Cavanagh, J. B. 1976. Vascular leakage in the dorsal root ganglia of the rat, studied with horseradish peroxidase. J. Neurol. Sci. 29, 95–107.

Janig, W. and Baron, R. 2003. Complex regional pain syndrome: mystery explained? Lancet Neurol. 2, 687–697.

Jensen, T. S. and Baron, R. 2003. Translation of symptoms and signs into mechanisms in neuropathic pain. Pain 102, 1–8.

Ji, R. R. and Woolf, C. J. 2001. Neuronal plasticity and signal transduction in nociceptive neurons: implications for the initiation and maintenance of pathological pain. Neurobiol. Dis. 8, 1–10.

Koltzenburg, M. and Bendszus, M. 2004. Imaging of peripheral nerve lesions. Curr. Opin. Neurol. 17, 621–626.

Lehky, T. J., Leonard, G. D., Wilson, R. H., Grem, J. L., and Floeter, M. K. 2004. Oxaliplatin-induced neurotoxicity: acute hyperexcitability and chronic neuropathy. Muscle Nerve 29, 387–392.

Leijon, G., Boivie, J., and Johansson, I. 1989. Central post-stroke pain – neurological symptoms and pain characteristics. Pain 36, 13–25.

Lindblom, U. and Verrillo, R. T. 1979. Sensory functions in chronic neuralgia. J. Neurol. Neurosurg. Psychiatr. 42, 422–435.

Lindenlaub, T. and Sommer, C. 2003. Cytokines in sural nerve biopsies from inflammatory and non-inflammatory neuropathies. Acta Neuropathol. (Berl.) 105, 593–602.

Loeser, J. D., Ward, A. A., and White, L. E. 1967. Chronic deafferentation of human spinal cord neurons. J. Neurosurg. 29, 48–50.

Lombard, M. C. and Larabi, Y. 1983. Electrophysiological Study of Cervical Dorsal Horn Cells in Partially Deafferented Rats. In: Advances in Pain Research and Therapy (ed. J. J. Bonica), pp. 147–154. Raven Press.

Lu, X. and Richardson, P. M. 1993. Responses of macrophages in rat dorsal root ganglia following peripheral nerve injury. J. Neurocytol. 22, 334–341.

Luo, Z. D., Chaplan, S. R., Higuera, E. S., Sorkin, L. S., Stauderman, K. A., Williams, M. E., and Yaksh, T. L. 2001. Upregulation of dorsal root ganglion (alpha)2(delta) calcium channel subunit and its correlation with allodynia in spinal nerve-injured rats. J. Neurosci. 21, 1868–1875.

Ma, W., Zhang, Y., Bantel, C., and Eisenach, J. C. 2005. Medium and large injured dorsal root ganglion cells increase TRPV-1, accompanied by increased alpha2C-adrenoceptor co-expression and functional inhibition by clonidine. Pain 113, 386–394.

Maihofner, C., Forster, C., Birklein, F., Neundorfer, B., and Handwerker, H. O. 2005a. Brain processing during mechanical hyperalgesia in complex regional pain syndrome: a functional MRI study. Pain 114, 93–103.

Maihofner, C., Handwerker, H. O., Neundorfer, B., and Birklein, F. 2004. Cortical reorganization during recovery from complex regional pain syndrome. Neurology 63, 693–701.

Maihofner, C., Handwerker, H. O., Neundorfer, B., and Birklein, F. 2005b. Mechanical hyperalgesia in complex regional pain syndrome: a role for TNF-alpha? Neurology 65, 311–313.

Marchand, F., Perretti, M., and McMahon, S. B. 2005. Role of the immune system in chronic pain. Nat. Rev. Neurosci. 6, 521–532.

McDermott, A. M., Toelle, T. R., Rowbotham, D. J., Schaefer, C. P., and Dukes, E. M. 2006. The burden of neuropathic pain: results from a cross-sectional survey. Eur. J. Pain 10, 127–135.

McKemy, D. D., Neuhausser, W. M., and Julius, D. 2002. Identification of a cold receptor reveals a general role for TRP channels in thermosensation. Nature 416, 52–58.

McQuay, H., Carroll, D., Jadad, A. R., Wiffen, P., and Moore, A. 1995. Anticonvulsant drugs for management of pain: a systematic review [see comments]. BMJ 311, 1047–1052.

McQuay, H. J., Tramer, M., Nye, B. A., Carroll, D., Wiffen, P. J., and Moore, R. A. 1996. A systematic review of antidepressants in neuropathic pain. Pain 68, 217–227.

Mogil, J. S., Miermeister, F., Seifert, F., Strasburg, K., Zimmermann, K., Reinold, H., Austin, J. S., Bernardini, N., Chesler, E. J., Hofmann, H. A., Hordo, C., Messlinger, K., Nemmani, K. V., Rankin, A. L., Ritchie, J., Siegling, A., Smith, S. B., Sotocinal, S., Vater, A., Lehto, S. G., Klussmann, S., Quirion, R., Michaelis, M., Devor, M., and Reeh, P. W. 2005. Variable sensitivity to noxious heat is mediated by differential expression of the CGRP gene. Proc. Natl. Acad. Sci. U. S. A. 102, 12938–12943.

Moore, K. A., Kohno, T., Karchewski, L. A., Scholz, J., Baba, H., and Woolf, C. J. 2002. Partial peripheral nerve injury promotes a selective loss of GABAergic inhibition in the superficial dorsal horn of the spinal cord. J. Neurosci. 22, 6724–6731.

Nodera, H., Barbano, R. L., Henderson, D., and Herrmann, D. N. 2003. Epidermal reinnervation concomitant with symptomatic improvement in a sensory neuropathy. Muscle Nerve 27, 507–509.

Nordin, M, Nyström, B., Wallin, U., and Hagbarth, K. E. 1984. Ectopic sensory discharges and paresthesiae in patients with disorders of peripheral nerve, dorsal roots and dorsal columns. Pain 20, 231–245.

Nurmikko, T. and Bowsher, D. 1990. Somatosensory findings in postherpetic neuralgia. J. Neurol. Neurosurg. Psychiatr. 53, 135–141.

Nurmikko, T., Wells, C., and Bowsher, D. 1994. Sensory Dysfunction in Postherpetic Neuralgia. In: Touch, Temperature, and Pain in Health and Disease: Mechanisms and Assessments, Vol. 3 (eds. J. Boivie, P. Hansson, and U. Lindblom), pp. 133–141. IASP Press.

Nystrom, B. and Hagbarth, K. E. 1981. Microelectrode recordings from transected nerves in amputees with phantom limb pain. Neurosci. Lett. 27, 211–216.

Oaklander, A. L. 1998. Unilateral postherpetic neuralgia is associated with bilateral sensory neuron damage. Neurology 44, 789–795.

Oaklander, A. L. 2001. The density of remaining nerve endings in human skin with and without postherpetic neuralgia after shingles. Pain 92, 139–145.

Oaklander, A. L., Romans, K., Horasek, S., Stocks, A., Hauer, P., and Meyer, R. A. 1998. Unilateral postherpetic neuralgia is associated with bilateral sensory neuron damage. Ann. Neurol. 44, 789–795.

Obata, K., Katsura, H., Mizushima, T., Yamanaka, H., Kobayashi, K., Dai, Y., Fukuoka, T., Tokunaga, A., Tominaga, M., and Noguchi, K. 2005. TRPA1 induced in sensory neurons contributes to cold hyperalgesia after inflammation and nerve injury. J. Clin. Invest. 115, 2393–2401.

Ochoa, J. L. and Yarnitsky, D. 1993. Mechanical hyperalgesias in neuropathic pain patients: dynamic and static subtypes. Ann. Neurol. 33, 465–472.

Ochoa, J. L. and Yarnitsky, D. 1994. The triple cold syndrome. Cold hyperalgesia, cold hypoaesthesia and cold skin in peripheral nerve disease. Brain 117, 185–197.

Omana-Zapata, I., Khabbaz, M. A., Hunter, J. C., Clarke, D. E., and Bley, K. R. 1997. Tetrodotoxin inhibits neuropathic ectopic activity in neuromas, dorsal root ganglia and dorsal horn neurons. Pain 72, 41–49.

Orstavik, K., Weidner, C., Schmidt, R., Schmelz, M., Hilliges, M., Jorum, E., Handwerker, H., and Torebjork, E. 2003. Pathological C-fibres in patients with a chronic painful condition. Brain 126, 567–578.

Ossipov, M. H., Lai, J., Malan, T. P., Jr., and Porreca, F. 2000. Spinal and supraspinal mechanisms of neuropathic pain. Ann. N. Y. Acad. Sci. 909, 12–24.

Pappagallo, M., Oaklander, A. L., Quatrano-Piacentini, A. L., Clark, M. R., and Raja, S. N. 2000. Heterogenous patterns of sensory dysfunction in postherpetic neuralgia suggest multiple pathophysiologic mechanisms. Anesthesiology 92, 691–698.

Patapoutian, A., Peier, A. M., Story, G. M., and Viswanath, V. 2003. ThermoTRP channels and beyond: mechanisms of temperature sensation. Nat. Rev. Neurosci. 4, 529–539.

Petersen, K. L., Fields, H. L., Brennum, J., Sandroni, P., and Rowbotham, M. C. 2000. Capsaicin evoked pain and allodynia in post-herpetic neuralgia. Pain 88, 125–133.

Pleger, B., Tegenthoff, M., Ragert, P., Forster, A. F., Dinse, H. R., Schwenkreis, P., Nicolas, V., and Maier, C. 2005. Sensorimotor retuning [corrected] in complex regional pain syndrome parallels pain reduction. Ann. Neurol. 57, 425–429.

Pleger, B., Tegenthoff, M., Schwenkreis, P., Janssen, F., Ragert, P., Dinse, H. R., Volker, B., Zenz, M., and Maier, C. 2004. Mean sustained pain levels are linked to hemispherical side-to-side differences of primary somatosensory cortex in the complex regional pain syndrome I. Exp. Brain Res. 155, 115–119.

Polydefkis, M., Yiannoutsos, C. T., Cohen, B. A., Hollander, H., Schifitto, G., Clifford, D. B., Simpson, D. M., Katzenstein, D., Shriver, S., Hauer, P., Brown, A., Haidich, A. B., Moo, L., and McArthur, J. C. 2002. Reduced intraepidermal nerve fiber density in HIV-associated sensory neuropathy. Neurology 58, 115–119.

Price, D. D., Long, S., and Huitt, C. 1992. Sensory testing of pathophysiological mechanisms of pain in patients with reflex sympathetic dystrophy [see comments]. Pain 49, 163–173.

Price, D. D., Long, S., Wilsey, B., and Rafii, A. 1998. Analysis of peak magnitude and duration of analgesia produced by local anesthetics injected into sympathetic ganglia of complex regional pain syndrome patients. Clin. J. Pain 14, 216–226.

Raja, S. N., Abatzis, V., and Frank, S. M. 1998. Role of a-adrenoceptors in neuroma pain in amputees. Anesthesiology 89, A1083.

Raja, S. N., Haythornthwaite, J. A., Pappagallo, M., Clark, M. R., Travison, T. G., Sabeen, S., Royall, R. M., and Max, M. B. 2002. Opioids versus antidepressants in postherpetic neuralgia: a randomized, placebo-controlled trial. Neurology 59, 1015–1021.

Rasmussen, P. V., Sindrup, S. H., Jensen, T. S., and Bach, F. W. 2004. Symptoms and signs in patients with suspected neuropathic pain. Pain 110, 461–469.

Rolke, R., Baron, R., Maire, C., Tölle, T. R., Treede, R. D., Binder, A., Birbaumer, N., Birklein, F., Bötefür, C. I., Braune, S., Flor, H., Huge, V., Klug, R., Landwchrmeyer, G. B., Magerl, W., Maihöfner, G., Rolko, C., Scherens, A., Sprenger, T., Valet, M., and Wasserka, B. 2006. Quantitative sensory testing in the German Research Network on Neuropathic Pain (DFNS): standardized protocol and reference values. Pain 123, 231–243.

Rowbotham, M. C. and Fields, H. L. 1996. The relationship of pain, allodynia and thermal sensation in post-herpetic neuralgia. Brain 119, 347–354.

Rowbotham, M. C., Yosipovitch, G., Connolly, M. K., Finlay, D., Forde, G., and Fields, H. L. 1996. Cutaneous innervation density in the allodynic form of postherpetic neuralgia. Neurobiol. Dis. 3, 205–214.

Sato, J. and Perl, E. R. 1991. Adrenergic excitation of cutaneous pain receptors induced by peripheral nerve injury. Science 251, 1608–1610.

Scadding, J. W. and Koltzenburg, M. 2006. Painful peripheral neuropathies. In: Wall and Melzack's Textbook of Pain. 5th edition. (eds. S. B. McMahon and M. Koltzenburg), pp. 973–1000. Elsevier.

Schuller, T. B., Hermann, K., and Baron, R. 2000. Quantitative assessment and correlation of sympathetic, parasympathetic, and afferent small fiber function in peripheral neuropathy. J. Neurol. 247, 267–272.

Scott, L. J., Griffin, J. W., Luciano, C., Barton, N. W., Banerjee, T., Crawford, T., McArthur, J. C., Tournay, A., and Schiffmann, R. 1999. Quantitative analysis of epidermal innervation in Fabry disease. Neurology 52, 1249–1254.

Siddall, P. J. and Middleton, J. W. 2005. A proposed algorithm for the management of pain following spinal cord injury. Spinal Cord. 44, 67–77.

Siddall, P. J., McClelland, J. M., Rutkowski, S. B., and Cousins, M. J. 2003. A longitudinal study of the prevalence and characteristics of pain in the first 5 years following spinal cord injury. Pain 103, 249–257.

Sindrup, S. H. and Jensen, T. S. 1999. Efficacy of pharmacological treatments of neuropathic pain: an update and effect related to mechanism of drug action. Pain 83, 389–400.

Sindrup, S. H. and Jensen, T. S. 2000. Pharmacologic treatment of pain in polyneuropathy. Neurology 55, 915–920.

Sommer, C. 2003. Painful neuropathies. Curr. Opin. Neurol. 16, 623–628.

Sommer, C. and Myers, R. R. 1996. Vascular pathology in CCI neuropathy: a quantitative temporal study. Exp. Neurol. 141, 113–119.

Sommer, C., Marziniak, M., and Myers, R. R. 1998. The effect of thalidomide treatment on vascular pathology and hyperalgesia caused by chronic constriction injury. Pain 74, 83–91.

Sorkin, L. S., Xiao, W.-H., Wagner, R., and Myers, R. R. 1997. Tumor necrosis factor-alpha induces ectopic activity in nociceptive primary afferent fibres. Neuroscience 81, 255–262.

Tal, M. and Bennett, G. J. 1994. Extra-territorial pain in rats with a peripheral mononeuropathy: mechano-hyperalgesia and mechano-allodynia in the territory of an uninjured nerve. Pain 57, 375–382.

Torebjork, E., Wahren, L., Wallin, G., Hallin, R., and Koltzenburg, M. 1995. Noradrenaline-evoked pain in neuralgia (see comments). Pain 63, 11–20.

Vanegas, H. and Schaible, H. G. 2004. Descending control of persistent pain: inhibitory or facilitatory? Brain Res. Brain Res. Rev. 46, 295–309.

Wagner, R., Janjigian, M., and Myers, R. R. 1998. Anti-inflammatory interleukin-10 therapy in CCI neuropathy decreases thermal hyperalgesia, macrophage recruitment, and endoneurial TNF-alpha expression. Pain 74, 35–42.

Wasner, G., Heckmann, K., Maier, C., and Baron, R. 1999. Vascular abnormalities in acute reflex sympathetic dystrophy (CRPS I) – complete inhibition of sympathetic nerve activity with recovery. Arch. Neurol. 56, 613–620.

Wasner, G., Kleinert, A., Binder, A., Schattschneider, J., and Baron, R. 2005. Postherpetic neuralgia: topical lidocaine is effective in nociceptor-deprived skin. J. Neurol. 252, 677–686.

Wasner, G., Schattschneider, J., Binder, A., and Baron, R. 2004. Topical menthol – a human model for cold pain by activation and sensitization of C nociceptors. Brain. 127, 1159–1171.

Watson, C. P., Deck, J. H., Morshead, C., Van der Kooy, D., and Evans, R. J. 1991. Post-herpetic neuralgia: further post-mortem studies of cases with and without pain. Pain 44, 105–117.

Wieseler-Frank, J., Maier, S. F., and Watkins, L. R. 2005. Immune-to-brain communication dynamically modulates pain: physiological and pathological consequences. Brain Behav. Immun. 19, 104–111.

Willoch, F., Rosen, G., Tolle, T. R., Oye, I., Wester, H. J., Berner, N., Schwaiger, M., and Bartenstein, P. 2000. Phantom limb pain in the human brain: unraveling neural circuitries of phantom limb sensations using positron emission tomography. Ann. Neurol. 48, 842–849.

Willoch, F., Schindler, F., Wester, H. J., Empl, M., Straube, A., Schwaiger, M., Conrad, B., and Tolle, T. R. 2004. Central poststroke pain and reduced opioid receptor binding within pain processing circuitries: a [11C]diprenorphine PET study. Pain 108, 213–220.

Wood, J. N., Boorman, J. P., Okuse, K., and Baker, M. D. 2004. Voltage-gated sodium channels and pain pathways. J. Neurobiol. 61, 55–71.

Woolf, C. J., Bennett, G. J., Doherty, M., Dubner, R., Kidd, B., Koltzenburg, M., Lipton, R., Loeser, J. D., Payne, R., and Torebjork, E. 1998. Towards a mechanism-based classification of pain? Pain 77, 227–229.

Wu, G., Ringkamp, M., Hartke, T. V., Murinson, B. B., Campbell, J. N., Griffin, J. W., and Meyer, R. A. 2001. Early onset of spontaneous activity in uninjured C-fiber nociceptors after injury to neighboring nerve fibers. J. Neurosci. 21, RC140.

Zambreanu, L., Wise, R. G., Brooks, J. C., Iannetti, G. D., and Tracey, I. 2005. A role for the brainstem in central sensitisation in humans. Evidence from functional magnetic resonance imaging. Pain 114, 397–407.

Zochodne, D. W. 1993. Epineurial peptides: a role in neuropathic pain? Can. J. Neurol. Sci. 20, 69–72.

58 Neurogenic Inflammation in Complex Regional Pain Syndrome (CRPS)

F Birklein, University of Mainz, Mainz, Germany

M Schmelz, University of Heidelberg, Mannheim, Germany

Glossary

ACE Angiotensin-converting enzyme.

CGRP Calcitonin gene-related peptide is the most potent vasodilator neuropeptide released from primary afferent nociceptors.

C-nociceptors Free nerve endings with high activation thresholds.

dermal microdialysis Single hollow semipermeable fibers are inserted into the skin.

laser Doppler Monochromatic light is directed to the skin. The frequency shift of the backscattered is proportional to skin perfusion.

NEP Neutral endopeptidase.

NGF Nerve growth factor.

NK1-R Substance P receptors.

polymorphism Common genetic variability.

QSART Quantitative measurement of peripheral axon reflex sweating.

SP Substance P binds to NK1 receptors at venoles and mediates plasma protein extravasation.

TNF-α Tumor necrosis factor-alpha.

TST Sweating induced by body heating – thermoregulatory sweating.

58.1 Principles and Mechanisms of Neurogenic Inflammation – Basic Research

58.1.1 Concept of Neurogenic Inflammation

It has been more than 100 years since Bayliss W. M. (1901) described antidromic vasodilation following electrical stimulation of centrally cut dorsal roots. Meanwhile it has become clear that retrogradely conducted action potentials invade the arborizations of the primary afferent neurons (axon reflex) and release neuropeptides from their terminals (Szolcsanyi, J., 1996). The acute effects of neuropeptide release in rodent skin are vasodilation and protein extravasation, the combination of which has been termed neurogenic inflammation. Pivotal neuro-peptides in the induction of neurogenic inflammation comprise calcitonin gene-related peptide (CGRP) for vasodilation and substance P (SP) for the induction of protein extravasation (Holzer, P., 1998).

58.1.2 Neuronal Basis of Neurogenic Inflammation

Chemical, thermal, and electrical stimulation has been widely used to elicit neurogenic inflammation. As a result of previous studies, mechanoinsensitive, but heat and chemosensitive C-nociceptors have been found responsible for the neurogenic vasodilation in pig (Lynn, B. *et al.*, 1996) and human skin (Schmelz, M. *et al.*, 2000), whereas in rodent, skin polymodal noci-ceptors play a crucial role (Gee, M. D. *et al.*, 1997).

A few approaches so far have tried to directly measure neuropeptide release in the skin, for exam-ple, *ex vivo* analysis by superfusion of excised rat skin (Kilo, S. *et al.*, 1997; Kress, M. *et al.*, 1999). In most studies, neurogenic vasodilation and protein extra-vasation have been used to functionally assess nociceptor activation. Vasodilation has been mea-sured by laser Doppler techniques and infrared thermography. Assessment of protein extravasation requires more invasive techniques, for example, extravasation of albumin bound dyes like Evans blue, which can be photometrically quantified *ex vivo*. Recent advances in dermal microdialysis have opened new possibilities by allowing *in vivo* measure-ment of local mediator concentrations, including protein extravasation (Sauerstein, K. *et al.*, 2000).

58.1.3 Humoral Interactions between Neurons and Inflammatory Cells

In addition to the classical features of neurogenic inflammation, that is, vasodilation and protein extra-vasation, neuropeptides also directly interact with inflammatory cells. Primary afferent nociceptors release a variety of neuropeptides like neurokinin A, neurokinin B, somatostatin, corticotropin-releasing hormone (CRH), pituitary adenylate cyclase-activat-ing polypeptide (PACAP), vasoactive intestinal peptide (VIP), galanin, and endomorphins (Steinhoff, M. *et al.*, 2003). A summary of important activating (Figure 1(a)) and inhibitory interactions (Figure 1(b)) between nociceptive nerve endings, tissue cells, and inflammatory cells is shown below. As an example, nerve growth factor (NGF) derived from keratino-cytes and mast cells sensitizes nociceptors and also

increases expression of mast cell tryptase and hista-mine (Groneberg, D. A. *et al.*, 2005) in mast cells. Histamine and tryptase activate nociceptive terminals via H1 and proteinase-activated receptors (PAR2) and release SP from the terminals (Steinhoff, M. *et al.*, 2000). SP via activation of NK1 receptors on the mast cells (van der Kleij, H. P. *et al.*, 2003) increases expression of histamine and tumor necrosis factor-alpha (TNF-α) (Ansel, J. C. *et al.*, 1993; Cocchiara, R. *et al.*, 1999), which in return sensitizes nociceptive terminals via activation of the TNF receptor.

58.2 Animal Models of Neuropathic Pain, Which Focus on Neurogenic Inflammation

Different animal models have been used to study the nature of neuropathic pain. Unfortunately, most studies focused on mechanisms and symptoms of pain but did not look at symptoms that characterize neurogenic inflammation after nerve lesion. Nevertheless, these pain models indicate that cyto-kines (TNF-α and NGF) were upregulated in the injured nerve (Herzberg, U. *et al.*, 1997a; 1997b; Schafers, M. *et al.*, 2002) and that neutralizing TNF-α reduces pain behavior (Sommer, C. *et al.*, 2001). It is the merit of Wade Kingery and his group that these shortcomings were meanwhile eliminated.

58.2.1 Bone Fracture Model for Complex Regional Pain Syndrome I

Most cases of complex regional pain syndrome (CRPS) I develop after bone fractures, e.g. to the wrist or the calf. Therefore, the most interesting animal model may be a controlled fracture of a large extremity bone. Very recently, skin tempera-ture increase, edema, plasma protein extravasation (PPE), osteoporosis at the metaphysal parts of bones, and hyperalgesia-like behavior over weeks were observed in a rat model of tibial fracture (Guo, T. Z. *et al.*, 2004). As indicated above, these are signs of neurogenic inflammation, which can also be found in acute CRPS after bone fracture. These signs were even pronounced if the limbs of the rats were immobilized – as it is a common treatment in most human fractures. Supporting the neurogenic inflammation hypothesis of these symptoms, skin temperature difference, edema, and pain behavior can be reversed by NK1 antagonism (Kingery, W. S. *et al.*, 2003).

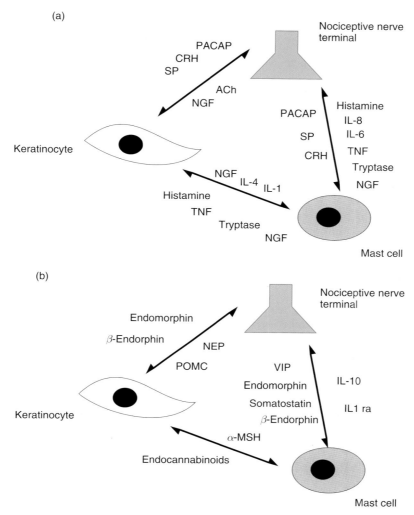

(a)

(b)

Figure 1 (a) Excitatory and (b) inhibitory neuroimmune interactions at peripheral nociceptors.

58.2.2 Nerve Transection Model for Complex Regional Pain Syndrome II

The sciatic nerve transection model was introduced as a surrogate for CRPS II. Soon after transection of the sciatic nerve, similar to the fracture model, rats develop increased skin temperature, paw edema, regional osteoporosis, and hyperalgesia-like behavior. Thus, symptoms after nerve transection also resemble acute CRPS. In this model, different treatments were tested. Steroids were able to reverse inflammatory symptoms (Kingery, W. S. *et al.*, 2001a; 2001b), and as described in the bone fracture model, NK1 antagonism was also effective (Kingery, W. S. *et al.*, 2003).

Taken together, both the bone fracture and the sciatic nerve transection model corroborates that many of the peripheral symptoms in acute posttraumatic CRPS may be due to C-fiber-mediated neurogenic inflammation.

58.3 Clinical Symptoms Suggesting Neurogenic Inflammation in Complex Regional Pain Syndrome

In general, neurogenic inflammation in CRPS can only assumed only if there has been an adequate limb trauma and if the initial clinical presentation corresponds to the primarily warm CRPS subtype (Veldman, P. H. J. M. *et al.*, 1993). In CRPS cases after minor trauma, which often start with a cold extremity (primarily cold), neurogenic inflammation may not be present. Corresponding to the animal models, CRPS I and CRPS II could not be distinguished without detailed neurological investigation. CRPS II can be diagnosed when there is obvious evidence of peripheral nerve lesion – clinically (sensory loss, muscle weakness, and reflex loss) or neurophysiologically (nerve conduction and electromyography) (Harden, N. and Bruehl, S., 2004). However, these peripheral nerve lesions must involve either major nerves or skin nerves to be clinically recognized. Nerve lacerations (from broken bones) or stretches (from joint distraction) are usually unrecognizable. Injuries to small distal nerve branches, especially to purely sensory branches that innervate skin or bone, may not cause weakness or loss of reflexes, and thus go undetected in most clinical settings (Oaklander, A. L. and Birklein, F., 2004). For instance, a seemingly minor knee sprain can trigger CRPS if it is evaluated by knee arthroscopy, a well-documented cause of painful damage to the infrapatellar branch of the saphenous nerve (Mochida, H. and Kikuchi, S., 1995). Seemingly trivial medical procedures such as routine venipuncture can cause CRPS-inducing nerve damage to forearm nerves (Horowitz, S. H., 2001). Thus, we will not continue the traditional but often virtual differentiation between CRPS I and II in this chapter.

58.3.1 Pain and Hyperalgesia

In CRPS, pain and hyperalgesia are the most distressing symptoms. Since the initial trauma for CRPS usually does not include a complete nerve lesion, there are no serious reasons to doubt that pain and hyperalgesia result, at least partially, from activation of peripheral nociceptors. If these nociceptors are sensitized, they might be spontaneously active causing pain at rest (Orstavik, K. et al., 2003) or they might respond more vigorously to physiological stimuli causing pain, for example, while moving a joint (Schaible, H. G. et al., 2002). Nociceptor sensitization might furthermore drive spinal sensitization, which reinforces mechanical hyperalgesia. Since neurogenic inflammation, for example, the release of neuropeptides from nociceptors, is inevitably linked to nociceptor activation (even subthreshold for pain might be enough; Magerl, W. et al., 1990), we regard pain and hyperalgesia as the one clinical symptom pointing to neurogenic inflammation in CRPS. In a study of 145 patients with untreated CRPS (Birklein, F.

et al., 2000), 75% reported pain even when the limb was completely at rest, but nearly all patients reported worsening of pain under certain circumstances – clinically indicating hyperalgesia. The most powerful stimulus that increased pain was joint movement. Accordingly, quantitative sensory testing revealed that on the affected skin there is a moderate decrease (about 4°C) of heat pain thresholds (Tahmoush, A. J. et al., 2000). Heat pain sensitization of skin nociceptors is regarded as a clinical sign for peripheral sensitization (Baron, R., 2000). Even more obvious was the decrease of mechanical pain thresholds on the affected side. Our group investigated 40 CRPS patients and regularly found the presence of hyperalgesia to brief mechanical impact stimuli (Sieweke, N. et al., 1999).

58.3.2 The Increase of Skin Perfusion and Temperature

In posttraumatic CRPS, skin temperature is increased usually during the first 6 months of the disease. The maximum skin temperature difference might be up to 10°C (Birklein, F. et al., 1997a; 1997b) Accordingly, skin color in this stage is reddish and the skin appears hyperemic. Only with disease chronification, skin temperature decreases on the affected side, the skin becomes bluish and appears thin and shiny (Birklein, F. et al., 1998).

58.3.3 Hyperhidrosis

In 50% of the patients with CRPS, increased sweating can be clinically observed (Birklein, F. et al., 1997a; 1997b); Two tests were used to localize sudomotor dysfunction along the neuraxis: thermoregulatory sweating test (TST) activates central thermoreceptors in the ventral hypothalamus (Jänig, W., 1990) and quantitative sudomotor axon reflex test (QSART) tests purely the peripheral sudomotor axon reflex (Low, P. A. et al., 1983). In acute CRPS both TST and QSART are increased pointing to sudomotor dysfunction in the peripheral nervous system. From basic experiments we know that the capacity of sweat glands increases with increasing skin temperature (Sato, K. et al., 1993) and that the neuropeptide CGRP, which might be responsible for increased skin temperature and perfusion, itself enhances sweat gland activity (Kumazawa, K. et al., 1994).

58.3.4 Limb Edema

Edema is the most striking clinical sign in CRPS. During acute stages 81% of patients have visible edema on the affected limb (Birklein, F. *et al.*, 2000). The edema is pronounced at backs of hands or feet. After exercise or during the course of a day, when repeated nociceptor activation has occurred, edema increases. Only if CRPS persists longer and becomes chronic, edema dwindles (Birklein, F. *et al.*, 1999).

58.3.5 Trophic Changes

Trophic changes can be found in more than 50% of patients. Increased hair and nail growth appears several days after the onset of symptoms (Birklein, F. *et al.*, 2000). The well-known periarticular patchy osteoporosis, which can bee seen on X-ray pictures, belongs to the same group of symptoms (Maresca, V., 1985). This osteoporosis is a so-called high-turnover osteoporosis with increased osteoclastic activity (Leitha, T. *et al.*, 1996). Interestingly, TNF-α and SP (Goto, T. *et al.*, 1998) activate osteoclasts, and CGRP promotes hair growth (Hagner, S. *et al.*, 2002).

58.4 Clinical Investigations Proving Neurogenic Inflammation in Complex Regional Pain Syndrome

58.4.1 Experimental Studies

Many clinical symptoms of acute CRPS resemble inflammation: pain, edema, increased skin temperature, and blood flow. Inflammation in the classical sense with positive blood markers has not been proven, but any inflammation has a neurogenic component (Brain, S. D. and Williams, T. J., 1988). Peripheral trauma, in particular if it is accompanied by partial peripheral nerve lesion, causes a rapid release of cytokines like NGF (Herzberg, U. *et al.*, 1997a; 1997b) and TNF-α (Grellner, W., 2002). Both cytokines are able to activate and sensitize primary afferents, locally at the injury site or proximally in the respective nerve trunk. Recently, increased TNF-α has been shown in CRPS skin by analysis of suction blister fluids (Huygen, F. J. *et al.*, 2002). Our own group was able to demonstrate increased TNF-α in plasma samples of two different independent CRPS patients groups (Maihofner, C. *et al.*, 2005). Increase of TNF-α was correlated to mechanical hyperalgesia, a clinical surrogate for nociceptive sensitization.

Cytokines also increase the neuropeptide content of primary afferent neurons (Opree, A. and Kress, M., 2000). Activation of sensitized primary afferents then causes an increased release of neuropeptides into the affected body region (SP and CGRP). Chronic release of neuropeptides might be responsible for the above-mentioned CRPS symptoms. In addition, central neuropeptide release facilitates nociceptive sensitization. In analogy to migraine studies, we therefore measured CGRP in serum samples from patients with acute CRPS. CGRP was significantly increased, in particular when clinical inflammatory signs were present and when there was evidence for trauma-related nerve lesion (Birklein, F. *et al.*, 2001). In order to verify that increased CGRP indeed comes from primary nociceptive afferents, neurogenic inflammation was elicited directly in the skin by transcutaneous electrical stimulation via intradermal microdialysis capillaries. We first investigated the flare by using laser Doppler scanning on the affected and the unaffected side in our CRPS patients. Neurogenic flare was significantly more intense in patients, surprisingly on both sides – the affected and the clinically unaffected one (Leis, S. *et al.*, 2004). Another characteristic of neurogenic inflammation in rodents is SP-mediated PPE. In healthy humans, C fibers usually contain too less SP to induce PPE. In CRPS, however, significant PPE could be shown in almost all patients investigated. In contrast to the flare response, this increased PPE was limited to the affected side (Weber, M. *et al.*, 2001). These results suggested two possible pathomechanisms leading to facilitated neurogenic inflammation in CRPS – either increased release, hampered inactivation of neuropeptides, or both. In order to further unravel these mechanisms, we perfused SP in ascending concentrations through dermal microdialysis fibers in CRPS patients and controls. We found SP to be significantly more effective at inducing PPE in CRPS patients. Similar to the increased flare response, this increased responsiveness to SP was present on both the affected and the unaffected limbs (Leis, S. *et al.*, 2003) (Figure 2).

To summarize, in this series of experiments we found a trauma-related upregulation of neuropeptide release from primary afferent on the affected side and in addition an impaired inactivation of neuropeptides on both sides, possibly as a predisposing factor. Neuropeptide action in human tissue is terminated by degradation of peptidases. The most familiar peptidase, which cleaves SP and CGRP, is the angiotensin-converting enzyme (ACE). A previous

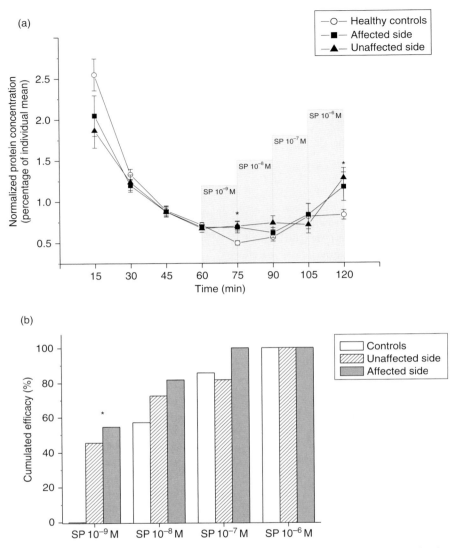

Figure 2 Substance P (SP) induced significantly increased protein extravasation in complex regional pain syndrome (CRPS) patients on the ipsilateral affected side (circles) and the contralateral unaffected side (triangles) as compared to healthy controls (squares). Significant differences at SP 10^{-9} M (receptor-dependent) and SP 10^{-6} M (receptor-independent) are marked. Results with intermediate SP concentrations might be false-negative due to SP receptor internalization.

report of an insertion/deletion (I/D) polymorphism in intron 16 within the ACE gene implicated an increased risk to develop CRPS I associated with the D allele (Kimura, T. *et al.*, 2000). However, in a linkage study in CRPS families, we failed to show any cosegregation of this polymorphism and the CRPS phenotype (Hühne, K. *et al.*, 2004). Therefore, we will focus on other enzymes that are involved in neuropeptide degradation. One promising example is the neutral endopeptidase (NEP). Blocking the NEP facilitates neurogenic

inflammation in a way, which resembles the findings on the unaffected side in CRPS patients (Krämer, H. H. *et al.*, 2005).

58.4.2 Treatment Studies

Another indication for neurogenic inflammation in warm posttraumatic CRPS cases is the effectiveness of anti-inflammatory treatment with steroids, which is based on small, but controlled studies (Christensen, K. *et al.*, 1982). Cortisone has multiple effects in

CRPS – it inhibits the production of inflammatory mediators, reduces the transcription rate in dorsal root ganglia cells, and thereby reduces neuropeptide content of sensory neurons (Kingery, W. S. *et al.*, 2001a; 2001b). Furthermore, it even facilitates the degradation of neuropeptides (Nadel, J. A., 1992) and inhibits ectopic impulse generation after nerve trauma by stabilizing nerve membranes. Thus, steroids may prevent the development of facilitated neurogenic inflammation and of neuropathic pain in warm CRPS.

References

Ansel, J. C., Brown, J. R., Payan, D. G., and Brown, M. A. 1993. Substance P selectively activates TNF-alpha gene expression in murine mast cells. J. Immunol. 150, 4478–4485.

Baron, R. 2000. Peripheral neuropathic pain: from mechanisms to symptoms. Clin. J. Pain 16, 12–20.

Bayliss, W. M. 1901. On the origin from the spinal cord of the vasodilator fibres of the hindlimb, and on the nature of these fibers. J. Physiol. (Lond.) 32, 1025–1043.

Birklein, F., Riedl, B., Claus, D., and Neundörfer, B. 1998. Pattern of autonomic dysfunction in time course of complex regional pain syndrome. Clin. Auton. Res. 8, 79–85.

Birklein, F., Riedl, B., Griessinger, N., and Neundörfer, B. 1999. Komplexes regionales schmerzsyndrom. Klinik und autonome störungen während akuter und chronischer krankheitsstadien. Nervenarzt 70, 335–341.

Birklein, F., Riedl, B., Handwerker, H. O., and Neundörfer, B. 1997a. Störung der thermoregulatorischen Hautdurchblutung bei sympathischer Reflexdystrophie. Akt. Neurol. 24, 61–66.

Birklein, F., Riedl, B., Sieweke, N., Weber, M., and Neundorfer, B. 2000. Neurological findings in complex regional pain syndromes – analysis of 145 cases. Acta Neurol. Scand. 101, 262–269.

Birklein, F., Schmelz, M., Schifter, S., and Weber, M. 2001. The important role of neuropeptides in complex regional pain syndrome. Neurology 57, 2179–2184.

Birklein, F., Sittl, R., Spitzer, A., Claus, D., Neundörfer, B., and Handwerker, H. O. 1997b. Sudomotor function in sympathetic reflex dystrophy. Pain 69, 49–54.

Brain, S. D. and Williams, T. J. 1988. Substance P regulates the vasodilator activity of calcitonin gene-related peptide. Nature 335, 73–75.

Christensen, K., Jensen, E. M., and Noer, I. 1982. The reflex sympathetic dystrophy syndrome response to treatment with systemic corticosteroids. Acta Chir. Scand. 148, 653–655.

Cocchiara, R., Lampiasi, N., Albeggiani, G., Bongiovanni, A., Azzolina, A., and Geraci, D. 1999. Mast cell production of TNF-alpha induced by substance P evidence for a modulatory role of substance P-antagonists. J. Neuroimmunol. 101, 128–136.

Gee, M. D., Lynn, B., and Cotsell, B. 1997. The relationship between cutaneous C fibre type and antidromic vasodilatation in the rabbit and the rat. J. Physiol. (Lond.) 503, 31–44.

Goto, T., Yamaza, T., Kido, M. A., and Tanaka, T. 1998. Light- and electron-microscopic study of the distribution of axons containing substance P and the localization of neurokinin-1 receptor in bone. Cell Tissue Res. 293, 87–93.

Grellner, W. 2002. Time-dependent immunohistochemical detection of proinflammatory cytokines (IL-1beta, IL-6, TNF-alpha) in human skin wounds. Forensic Sci. Int. 130, 90–96.

Groneberg, D. A., Serowka, F., Peckenschneider, N., Artuc, M., Grutzkau, A., Fischer, A., Henz, B. M., and Welker, P. 2005. Gene expression and regulation of nerve growth factor in atopic dermatitis mast cells and the human mast cell line-1. J. Neuroimmunol. 161, 87–92.

Guo, T. Z., Offley, S. C., Boyd, E. A., Jacobs, C. R., and Kingery, W. S. 2004. Substance P signaling contributes to the vascular and nociceptive abnormalities observed in a tibial fracture rat model of complex regional pain syndrome type I. Pain 108, 95–107.

Hagner, S., Haberberger, R. V., Overkamp, D., Hoffmann, R., Voigt, K. H., and McGregor, G. P. 2002. Expression and distribution of calcitonin receptor-like receptor in human hairy skin. Peptides 23, 109–116.

Harden, N. and Bruehl, S. 2004. Diagnostic Criteria: The Statistical Derivation of Four Criterion Factors. In: CRPS: Current Diagnosis and Therapy (eds. P. R. Wilson, M. Stanton-Hicks, and R. N. Harden), pp. 45–58. IASP Press.

Herzberg, U., Eliav, E., Dorsey, J. M., Gracely, R. H., and Kopin, I. J. 1997a. NGF involvement in pain induced by chronic constriction injury of the rat sciatic nerve. Neuroreport 8, 1613–1618.

Herzberg, U., Eliav, E., Dorsey, J. M., Gracely, R. H., and Kopin, I. J. 1997b. NGF involvement in pain induced by chronic constriction injury of the rat sciatic nerve. Neuroreport 8, 1613–1618.

Holzer, P. 1998. Neurogenic vasodilatation and plasma leakage in the skin. Gen. Pharmacol. 30, 5–11.

Horowitz, S. H. 2001. Venipuncture-induced neuropathic pain: the clinical syndrome, with comparisons to experimental nerve injury models. Pain 94, 225–229.

Hühne, K., Leis, S., Schmelz, M., Rautenstrauss, B., and Birklein, F. 2004. A polymorphic locus in the intron 16 of the human angiotensin-converting enzyme (ACE) gene is not correlated with complex regional pain syndrome I (CRPS I). Eur. J. Pain 8, 221–225.

Huygen, F. J., De Bruijn, A. G., De Bruin, M. T., Groeneweg, J. G., Klein, J., and Zijistra, F. J. 2002. Evidence for local inflammation in complex regional pain syndrome type 1. Mediators Inflamm. 11, 47–51.

Jänig, W. 1990. Functions of the Sympathetic Innervation of the Skin. In: Central Regulation of Autonomic Functions (eds. A. D. Loewy and K. M. Spyer), pp. 334–348. Oxford University Press.

Kilo, S., Harding-Rose, C., Hargreaves, K. M., and Flores, C. M. 1997. Peripheral CGRP release as a marker for neurogenic inflammation: a model system for the study of neuropeptide secretion in rat paw skin. Pain 73, 201–207.

Kimura, T., Komatsu, T., Hosada, R., Nishiwaki, K., and Shimada, Y. 2000. Angiotensin-Converting Enzyme Gene Polymorphism in Patients with Neuropathic Pain. In: Proceedings of the 9th World Congress in Pain (eds. M. Devor, M. Rowbotham, and Z. Wiesenfeld-Hallin), pp. 471–476. IASP Press.

Kingery, W. S., Davies, M. F., and Clark, J. D. 2003. A substance P receptor (NK1) antagonist can reverse vascular and nociceptive abnormalities in a rat model of complex regional pain syndrome type II. Pain 104, 75–84.

Kingery, W. S., Guo, T., Agashe, G. S., Davies, M. F., Clark, J. D., and Maze, M. 2001a. Glucocorticoid inhibition of neuropathic limb edema and cutaneous neurogenic extravasation. Brain Res. 913, 140–148.

Kingery, W. S., Guo, T., Agashe, G. S., Davies, M. F., Clark, J. D., and Maze, M. 2001b. Glucocorticoid inhibition of neuropathic limb edema and cutaneous neurogenic extravasation. Brain Res. 913, 140–148.

van der Kleij, H. P., Ma, D., Redegeld, F. A., Kraneveld, A. D., Nijkamp, F. P., and Bienenstock, J. 2003. Functional expression of neurokinin 1 receptors on mast cells induced by IL-4 and stem cell factor. J. Immunol. 171, 2074–2079.

Krämer, H. H., Schmidt, K., Leis, S., Schmelz, M., Sommer, C., and Birklein, F. 2005. Inhibition of neutral endopeptidase (NEP) facilitates neurogenic inflammation. Exp. Neurol. 195(1), 179–184.

Kress, M., Guthmann, C., Averbeck, B., and Reeh, P. W. 1999. Calcitonin gene-related peptide and prostaglandin E2 but not substance P release induced by antidromic nerve stimulation from rat skin, *in vitro*. Neuroscience 89, 303–310.

Kumazawa, K., Sobue, G., Mitsuma, T., and Ogawa, T. 1994. Modulatory effects of calcitonin gene-related peptide and substance P on human cholinergic sweat secretion. Clin. Auton. Res. 4, 319–322.

Leis, S., Weber, M., Isselmann, A., Schmelz, M., and Birklein, F. 2003. Substance-P-induced protein extravasation is bilaterally increased in complex regional pain syndrome. Exp. Neurol. 183, 197–204.

Leis, S., Weber, M., Schmelz, M., and Birklein, F. 2004. Facilitated neurogenic inflammation in unaffected limbs of patients with complex regional pain syndrome. Neurosci. Lett. 359, 163–166.

Leitha, T., Staudenherz, A., Korpan, M., and Fialka, V. 1996. Pattern recognition in five-phase bone scintigraphy: diagnostic patterns of reflex sympathetic dystrophy in adults. Eur. J. Nucl. Med. 23, 256–262.

Low, P. A., Caskey, P. E., Tuck, R. R., Fealey, R. D., and Dyck, P. J. 1983. Quantitative sudomotor axon reflex test in normal and neuropathic subjects. Ann. Neurol. 14, 573–580.

Lynn, B., Schutterle, S., and Pierau, F. K. 1996. The vasodilator component of neurogenic inflammation is caused by a special subclass of heat-sensitive nociceptors in the skin of the pig. J. Physiol. (Lond.) 494, 587–593.

Magerl, W., Geldner, G., and Handwerker, H. O. 1990. Pain and vascular reflexes elicited by prolonged noxious mechano-stimulation. Pain 43, 219–225.

Maihofner, C., Handwerker, H. O., Neundorfer, B., and Birklein, F. 2005. Mechanical hyperalgesia in complex-regional pain syndrome: a role for tnf-alpha. Neurology 65(2), 311–313.

Maresca, V. 1985. Human calcitonin in the management of osteoporosis: a multicentre study. J. Int. Med. Res. 13, 311–316.

Mochida, H. and Kikuchi, S. 1995. Injury to infrapatellar branch of saphenous nerve in arthroscopic knee surgery. Clin. Orthop. Relat. Res. 88–94.

Nadel, J. A. 1992. Neurogenic inflammation in airways and its modulation by peptidases. Ann. N. Y. Acad. Sci. 664, 408–414.

Oaklander, A. L. and Birklein, F. 2004. Peripheral Somatosensory Abnormalities – Pathology and Pathophysiology. In: CRPS: Current Diagnosis and Therapy (eds. P. R. Wilson, M. Stanton-Hicks, and R. N. Harden), pp. 59–79. IASP Press.

Opree, A. and Kress, M. 2000. Involvement of the proinflammatory cytokines tumor necrosis factor-alpha, IL-1 beta, and IL-6 but not IL-8 in the development of heat hyperalgesia: effects on heat-evoked calcitonin gene-related peptide release from rat skin. J. Neurosci. 20, 6289–6293.

Orstavik, K., Weidner, C., Schmidt, R., Schmelz, M., Hilliges, M., Jorum, E., Handwerker, H., and Torebjork, E. 2003. Pathological C-fibres in patients with a chronic painful condition. Brain 126, 567–578.

Sato, K., Ohtsuyama, M., and Sato, F. 1993. Normal and Abnormal Eccrine Sweat Gland Function. In: Clinical Autonomic Disorders (ed. P. A. Low), pp. 93–104, Little, Brown.

Sauerstein, K., Klede, M., Hilliges, M., and Schmelz, M. 2000. Electrically evoked neuropeptide release and neurogenic inflammation differ between rat and human skin. J. Physiol. 529, 803–810.

Schafers, M., Geis, C., Brors, D., Yaksh, T. L., and Sommer, C. 2002. Anterograde transport of tumor necrosis factor-alpha in the intact and injured rat sciatic nerve. J. Neurosci. 22, 536–545.

Schaible, H. G., Ebersberger, A., and von Banchet, G. S. 2002. Mechanisms of pain in arthritis. Ann. N. Y. Acad. Sci. 966, 343–354.

Schmelz, M., Michael, K., Weidner, C., Schmidt, R., Torebjörk, H. E., and Handwerker, H. O. 2000. Which nerve fibers mediate the axon reflex flare in human skin? Neuroreport 11, 645–648.

Sieweke, N., Birklein, F., Riedl, B., Neundorfer, B., and Handwerker, H. O. 1999. Patterns of hyperalgesia in complex regional pain syndrome. Pain 80, 171–177.

Sommer, C., Lindenlaub, T., Teuteberg, P., Schafers, M., Hartung, T., and Toyka, K. V. 2001. Anti-TNF-neutralizing antibodies reduce pain-related behavior in two different mouse models of painful mononeuropathy. Brain Res. 913, 86–89.

Steinhoff, M., Stander, S., Seeliger, S., Ansel, J. C., Schmelz, M., and Luger, T. 2003. Modern aspects of cutaneous neurogenic inflammation. Arch. Dermatol. 139, 1479–1488.

Steinhoff, M., Vergnolle, N., Young, S. H., Tognetto, M., Amadesi, S., Ennes, H. S., Trevisani, M., Hollenberg, M. D., Wallace, J. L., Caughey, G. H., Mitchell, S. E., Williams, L. M., Geppetti, P., Mayer, E. A., and Bunnett, N. W. 2000. Agonists of proteinase-activated receptor 2 induce inflammation by a neurogenic mechanism. Nat. Med. 6, 151–158.

Szolcsanyi, J. 1996. Capsaicin-sensitive sensory nerve terminals with local and systemic efferent functions: facts and scopes of an unorthodox neuroregulatory mechanism. Prog. Brain Res. 113, 343–359.

Tahmoush, A. J., Schwartzman, R. J., Hopp, J. L., and Grothusen, J. R. 2000. Quantitative sensory studies in complex regional pain syndrome type 1/RSD. Clin. J. Pain 16, 340–344.

Veldman, P. H. J. M., Reynen, H. M., Arntz, I. E., and Goris, R. J. A. 1993. Signs and symptoms of reflex sympathetic dystrophy: prospective study of 829 patients. Lancet 342, 1012–1016.

Weber, M., Birklein, F., Neundorfer, B., and Schmelz, M. 2001. Facilitated neurogenic inflammation in complex regional pain syndrome. Pain 91, 251–257.

59 Complex Regional Pain Syndromes

R Baron, Christian-Albrechts-Universität Kiel, Kiel, Germany

Glossary

CRPS Complex regional pain syndrome.
DRG Dorsal root ganglion, spinal ganglion.
MEG Magnetic encephalography.

PET Positron emission tomography.
QST Quantitative sensory testing.
TNF Tumor necrosis factor.

59.1 Introduction

The term complex regional pain syndrome (CRPS) describes a variety of painful conditions following injury, which appears regionally with a distal predominance of abnormal findings. The symptoms exceed in both magnitude and duration the expected clinical course of the inciting event and often result in significant impairment of motor function. The disorder shows a variable progression over time. These chronic pain syndromes comprise different additional clinical features including spontaneous pain, allodynia, hyperalgesia, edema, autonomic abnormalities, and trophic signs. In CRPS type I (reflex sympathetic dystrophy) minor injuries or fractures of a limb precede the onset of symptoms. CRPS type II (causalgia) develops after injury to a major peripheral nerve (Baron, R. *et al.*, 2002a; Janig, W. and Baron, R., 2003; Baron, R., 2004).

59.2 Complex Regional Pain Syndrome Type I (Reflex Sympathetic Dystrophy)

The most common precipitating event is a trauma affecting the distal part of an extremity (65%), especially fractures, postsurgical conditions, contusions, and strain or sprain. Less common occasions are central nervous system (CNS) lesions like spinal cord injuries and cerebrovascular accidents as well as cardiac ischemia (Wasner, G. *et al.*, 2003).

CRPS I patients develop asymmetrical distal extremity pain and edema without presenting an overt nerve lesion. These patients often report a burning spontaneous pain felt in the distal part of the affected extremity. Characteristically, the pain is disproportionate in intensity to the inciting event. The pain usually increases when the extremity is in a dependent position. Stimulus-evoked pains are a striking clinical feature. These sensory abnormalities often appear early, are most pronounced distally, and have no consistent spatial relationship to individual nerve territories or to the site of the inciting lesion. Typically, pain can be elicited by movement of and pressure on the joints (deep somatic allodynia), even if these are not directly affected by the inciting lesion.

Autonomic abnormalities include swelling and changes of sweating and skin blood flow. In the acute stages of CRPS I, the affected limb is often warmer than the contralateral limb. Sweating abnormalities, either hypohidrosis or, more frequently, hyperhidrosis are present in nearly all CRPS I patients. The acute distal swelling of the affected limb depends very critically on aggravating stimuli. Since it diminishes after sympathetic blocks, it is likely that it is maintained by sympathetic activity.

Trophic changes such as abnormal nail growth, increased or decreased hair growth, fibrosis, thin glossy skin, and osteoporosis may be present, particularly in chronic stages. Restrictions of passive movement are often present in long-standing cases and may be related to both functional motor disturbances and trophic changes of joints and tendons.

Weakness of all muscles of the affected distal extremity is often present. Small accurate movements are characteristically impaired. Nerve conduction and electromyography studies are normal, except in patients in very chronic and advanced stages. About half of the patients have a postural or action tremor representing an increased physiological tremor. In about 10% of the cases, dystonia of the affected hand or foot develops.

59.3   Complex Regional Pain Syndrome Type II (Causalgia)

The symptoms of CRPS II are similar to those of CRPS I. The only exception is that a lesion of peripheral nerve structures and subsequent focal deficits are mandatory for the diagnosis. The symptoms and signs spread beyond the innervation territory of the

injured peripheral nerve and often occur remote from the site of injury but a restriction to the territory is not in conflict with the current definition.

59.4   Important Issues that Are Unique in Complex Regional Pain Syndrome

59.4.1   Motor Abnormalities

About 50% of the patients with CRPS show motor abnormalities (Deuschl, G. et al., 1991; Bhatia, K. P. et al., 1993). It is unlikely that these motor changes are related to a peripheral process (e.g., influence of the sympathetic nervous system on neuromuscular transmission and/or contractility of skeletal muscle). These somatomotor changes are more likely generated by central changes of activity in the motoneurons. With kinematic analysis of target reaching as well as grip force analysis to quantitatively assess motor deficits in CRPS patients (Schattschneider, J. et al., 2001) abnormalities in the cerebral motor processing were revealed. A pathological sensorimotor integration located in the parietal cortex may induce an abnormal central programming and processing of motor tasks. Interestingly, the motor performance is also slightly impaired on the contralateral unaffected side (Ribbers, G. M. et al., 2002). According to this view, a neglect-like syndrome was clinically described to be involved in the disuse of the extremity (Galer, B. S. et al., 1995). Recent controlled studies also support an incongruence between central motor output and sensory input as underlying mechanism in CRPS. Using the method of mirror visual feedback, the visual input from a moving unaffected limb to the brain was able to reestablish the pain-free relationship between sensory feedback and motor execution. After 6 weeks of therapy, pain and function were improved as compared with the control group (McCabe, C. S. et al., 2003; Moseley, G. L., 2004). Furthermore, a sustained inhibition of the motor cortex was found in CRPS patients on the contralateral as well as ipsilateral hemisphere (Juottonen, K. et al., 2002; Schwenkreis, P. et al., 2003).

59.4.2   Inflammatory Processes

Some of the clinical features of CRPS particularly in its early phase, that is, calor, dolor, rubor, functio lesae, and swelling suggest that the pathological extremity is exhibiting an excessive inflammatory process (Leitha, T. *et al.*, 1996; van der Laan, L. and Goris, R. J., 1997; Calder, J. S. *et al.*, 1998). The idea of an inflammatory processes, in particular in the deep somatic tissues including bones, goes back to Sudeck who believed that this syndrome is an inflammatory bone atrophy ("entzündliche Knochenatrophie") (Sudeck, P., 1902; 1931). There are now several lines of evidence supporting this view.

59.4.2.1 *Immune cell-mediated inflammation and cytokine release*

Several studies in CRPS patients address the question to what extend an immune cell-mediated inflammation is involved. Skin biopsies in these patients showed a striking increase in the number of Langerhans cells that can release immune cell chemoattractants and proinflammatory cytokines (Calder, J. S. *et al.*, 1998). In accordance, in the fluid of artificially produced skin blisters significantly higher levels of IL-6 and tumor necrosis factor alpha (TNF-α) as well as tryptase (a measure of mast cell activity) were observed in the involved extremity as compared with the uninvolved extremity (Huygen, F. J. *et al.*, 2001; 2002). In CRPS patients with hyperalgesia, higher levels of the soluble TNF-α receptor type I were found (Maihofner, C. *et al.*, 2005b). Accordingly, a significant increases in IL-1β and IL-6, but not TNF-α was demonstrated in the cerebrospinal fluid (CSF) of individuals afflicted with CRPS as compared with controls (Alexander, G. M. *et al.*, 2005).

In this context, it is of particular interest that cutaneous denervation can cause rapid activation and proliferation of Langerhans cells that continues until reinnervation occurs. Thus, one might speculate that such cell proliferation may reflect also a functional denervation of cutaneous sympathetic outflow that occurs in CRPS (Wasner, G. *et al.*, 1999; 2001).

The patchy osteoporosis which is found in more advanced CRPS cases may also be consistent with a regional inflammatory process in deep somatic tissues. Both IL-1 and IL-6 cause proliferation and activation of osteoclasts and suppress the activity of osteoblasts.

Changes in hair growth can also be created by proinflammatory cytokines. TNF and IL-1 directly inhibit hair growth. Keratocyte-derived TNF and IL-6 cause retarded hair growth, signs of fibrosis, and in turn immune infiltration of the dermis, all present in CRPS patients.

59.4.2.2 *Neurogenic inflammation and nociceptor sensitization*

Because infection has so far not been shown to account for the inflammatory reactions in CRPS an exaggerated neurogenic inflammation has been proposed. Axon reflex vasodilatation was significantly increased on the affected side as measured with laser Doppler flowmetry after electrical C-fiber stimulation. Accordingly, systemic CGRP levels were found to be increased in acute CRPS but not in chronic stages (Birklein, F. *et al.*, 2001). Also, increased axon reflex sweating could be explained by release of CGRP from nociceptive terminals during nociceptor sensitization that act on peripheral sweat glands (Birklein, F. *et al.*, 2001).

In acute untreated CRPS I patients axon reflex activation was elicited by strong transcutaneous electrical stimulation of peptidergic unmyelinated afferents via intradermal microdialysis capillaries. Protein extravasation that was simultaneously assessed by the microdialysis system was only provoked on the affected extremity as compared with the normal side. The time course of electrically induced protein extravasation in the patients resembled the one observed following application of exogenous substance P (SP) (Weber, M. *et al.*, 2001).

Plasma extravasation could also be demonstrated with other experimental techniques: Bone scintigraphy demonstrated periarticular tracer uptake in acute CRPS (Leitha, T. *et al.*, 1996) and analysis of joint fluid and synovial biopsies in CRPS patients have shown an increase in protein concentration and synovial hypervascularity (Renier, J. C. *et al.*, 1983). Furthermore, synovial effusion is enhanced in affected joints as measured with magnetic resonance imaging (MRI) (Graif, M. *et al.*, 1998). Scintigraphic investigations with radiolabeled immunoglobulins show extensive plasma extravasation (Oyen, W. J. *et al.*, 1993).

59.4.2.3 *Autoimmune etiology*

There are several studies focusing at an autoimmune response in patients with CRPS. Autoantibodies against rat sympathetic neurons were detected in 40% of CRPS patients, whereas only 5% of other neuropathy patients showed these autoantibodies (Blaes, F. *et al.*, 2004). Immunoreactivity to *Campylobacter* was found in many early CRPS patients in particular associated with minimal trauma indicating a postinfectious autoimmune basis in some patients (Goebel, A. *et al.*, 2005).

59.4.3   Sensitization to Adrenergic Stimuli – Sympathetically Maintained Pain

After nerve lesion in animal experiments (Baron, R. *et al.*, 1999), afferent nociceptive and nonnociceptive neurons undergo dramatic functional and anatomical changes including upregulating of α-adrenoceptors. Noradrenaline released by the sympathetic nerve fibers may activate and/or sensitize the afferent neurons. This sympathetic–afferent coupling forms the theoretical basis for the clinical phenomenon of sympathetically maintained pain (SMP). The interaction may occur at the lesion site, along the lesioned nerve or even in the dorsal root ganglion (DRG).

Clinical studies in CRPS support the idea that nociceptors may develop catecholamine sensitivity (Baron, R. *et al.*, 1999): intraoperative stimulation of the sympathetic chain induces an increase of spontaneous pain in patients with causalgia (CRPS II) but not in patients with hyperhidrosis. In CRPS II, intracutaneous application of norepinephrine into a symptomatic area rekindles spontaneous pain and dynamic mechanical hyperalgesia that had been relieved by sympathetic blockade, supporting the idea that noradrenergic sensitivity of human nociceptors is present after partial nerve lesion. Also intradermal NE, in physiologically relevant doses, was demonstrated to evoke greater pain in the affected regions of patients with SMP, than in the contralateral unaffected limb, and in control subjects (Ali, Z. *et al.*, 2000).

We performed a study in patients with CRPS I using physiological stimuli of the sympathetic nervous system (Baron, R. *et al.*, 2002b). Cutaneous sympathetic vasoconstrictor outflow to the painful extremity was experimentally activated to the highest possible physiological degree by whole body cooling. The intensity as well as area of spontaneous pain and mechanical hyperalgesia (dynamic and punctate) increased significantly in patients that had been classified as having SMP by positive sympathetic blocks but not in sympathetically independent pain (SIP) patients. It is very likely that in addition to a coupling in the skin, a sympathetic–afferent interaction may also occur in other tissues, in particular in the deep somatic domain such as bone, muscle, or joints. Supporting this view, especially these structures are extremely painful in some cases with CRPS (Baron, R. and Wasner, G., 2001). Furthermore, there may be patients who are characterized by a selective or predominant sympathetic–afferent interaction in deep somatic tissues sparing the skin (Wasner, G. *et al.*, 1999).

59.5   Future Research, Diagnostic Classification, and Therapeutic Strategies

59.5.1   Integration of the Clinics and Basic Research

Recently a new animal model for CRPS I was introduced (Coderre, T. J. *et al.*, 2004). The clinical observation that signs of tissue inflammation and ischemia are present in many CRPS I patients led to the hypothesis that CRPS I might depend on an ischemia–reperfusion injury, which may at least in part induce the classical CRPS symptomatology. Consequently, an animal pain model was established that utilizes prolonged hindlimb ischemia for 3 h, followed by reperfusion, which produces a chronic postischemia pain syndrome. The reproducability of this model awaits further confirmation.

In the future an integrative research approach on animal and human models combined with clinical research on patients who suffer from CRPS is a necessity if we are to unravel the mechanisms operating in CRPS and if we are to find the organising pathophysiological principles that orchestrate the different changes. It is essential that research on animal and human models and clinical investigations of CRPS should be closely aligned.

59.5.2   The Long Path Approaching the Central Nervous System

In recent years, research concentrated on investigation of peripheral mechanisms underlying the symptoms of CRPS. We are just now at a turning point recognising that important parts of the CRPS pathophysiology are obviously located within the CNS.

There is much evidence that the changed somatosensory perceptions observed in CRPS patients are likely due to changes in the central representation in the thalamus and cortex (Maleki, J. *et al.*, 2000). Positron emission tomography (PET) studies demonstrated adaptive changes in the thalamus during the course of the disease (Fukumoto, M. *et al.*, 1999). Furthermore, recent magnetic encephalography (MEG) and functional MRI (fMRI) studies revealed a shortened distance between little finger and thumb representations in the primary somatosensory (S1) cortex on the painful side (Maihofner, C. *et al.*, 2003; Pleger, B. *et al.*, 2004; Maihofner, C. *et al.*, 2005a). The MEG SI responses were increased on the affected side, indicating processes of central sensitization (Juottonen, K. *et al.*, 2002). fMRI studies indicate that prefrontal cortical networks are involved in SMP in CRPS patients. Furthermore, in one CRPS patient a traumatic cerebral contusion of the left temporal lobe resolved the symptoms (Shibata, M. *et al.*, 1999; Apkarian, A. V. *et al.*, 2001).

We do not know the extent to which these central changes depend on continuous nociceptive input from the affected extremity; however, it has been shown that these generalized sensory changes may disappear after successful treatment of the pain (Maihofner, C. *et al.*, 2004; Pleger, B. *et al.*, 2005). In the future, we will fully understand that CRPS is a neurological disease affecting several functional systems that are closely integrated. This will include centers for the processing of cognitive and affective information.

59.5.3 A Possible Link between Inflammation and the Sympathetic Nervous System

In some patients sympatholytic procedures can ameliorate pain as well as inflammation and edema. Accordingly, animal studies have demonstrated that the sympathetic nervous system can influence the intensity of an inflammatory process (Levine, J. D. *et al.*, 1985; Jänig, W. *et al.*, 1996; Miao, F. J. *et al.*, 1996a; 1996b; Perl, E. R., 1996). The question arises whether the sympathetic nervous system might be involved in the inflammatory process in CRPS and whether it might interact with immune cells.

Under normal conditions, catecholamines act via $\beta2$-adrenoceptors on immune cells to inhibit the production and release of proinflammatory cytokines. These cells do not express α-adrenoceptors under basal conditions. However, the situation can dramatically change in chronic inflammation. Now, immune cells downregulate their expression of $\beta2$-adrenoceptors and upregulate their expression of $\alpha1$-adrenoceptors over time (Gazda, L. S. *et al.*, 2001). In contrast to $\beta2$-adrenoceptors, $\alpha1$-adrenoceptors stimulate the production and release of proinflammatory cytokines. If $\alpha1$-adrenoceptors were to become expressed by the resident and/or recruited immune or immunocompetent cells of the affected CRPS extremity, then sympathetic activation would be predicted to cause pain and other inflammatory signs via cytokine release.

Summarizing the hypothetical ideas described above Figure 1 illustrates the possible interactions between sympathetic fibers, afferent fibers, blood vessels, and nonneural cells related to the immune system (e.g., macrophages) leading theoretically to the inflammatory changes observed in CRPS patients.

59.5.4 Do the Genes Predispose for Complex Regional Pain Syndrome?

One of the unsolved features in human pain diseases is the fact that only a minority of patients develop chronic pain after identical inciting events. Similarly, in certain nerve lesion animal models, differences in pain susceptibility were found to be due genetic factors (Mogil, J. S. *et al.*, 1999). To address this question in CRPS patients, gene technology has been used to characterize the genetic pattern of patients at risk to develop CRPS (Mailis, A. and Wade, J., 1994; 2001). In a cohort of 52 CRPS patients, class I and II major histocompatibility antigens were typed. The frequency of HLA-DQ1 was found to be significantly increased compared with control frequencies (Kemler, M. A. *et al.*, 1999). In patients with CRPS who progressed towards multifocal or generalized tonic dystonia an association with HLA-DR13 was reported (van Hilten, J. J. *et al.*, 2000). Furthermore, a different locus, centromeric in HLA-class I, was found to be associated with spontaneous development of CRPS, suggesting an interaction between trauma severity and genetic factors that describe CRPS susceptibility (van de Beek, W. J. *et al.*, 2003).

59.5.5 The Future Need for Diagnostic Procedures

To design experiments in CRPS patients, it is important to define the population as precisely as possible. An exact diagnostic classification of the study cohort is essential to perform clinical treatment trials on a homogeneous group.

59.5.5.1 *Validation of clinical diagnostic criteria*

The definition of standardized diagnostic criteria for CRPS in 1995 was a major advance (Stanton-Hicks, M. *et al.*, 1995). However, these criteria were derived based upon the consensus opinion of a small group of expert clinicians. While this was an appropriate first step it is important to continuously improve the criteria, i.e., to validate and, if necessary, modify these initial consensus-based criteria based upon results of systematic validation research. The CRPS diagnostic criteria were adequately sensitive (i.e., rarely miss a case of actual CRPS). However, both internal and external

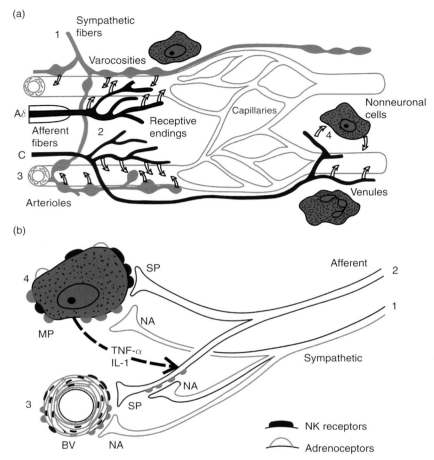

Figure 1 (a) The micromilieu of nociceptors. The microenvironment of primary afferents is thought to affect the properties of the receptive endings of myelinated (A) and unmyelinated (C) afferent fibers. This has been particularly documented for inflammatory processes, but one may speculate that pathological changes in the direct surroundings of primary afferents may contribute to other pain states as well. The vascular bed consists of arterioles (directly innervated by sympathetic and afferent fibers), capillaries (not innervated and not influenced by nerve fibers), and venules (not directly innervated but influenced by nerve fibers). The micromilieu depends on several interacting components: Neural activity in postganglionic noradrenergic fibers (1) supplying blood vessels (3, BV) causes release of noradrenaline (NA) and possibly other substances and vasoconstriction. Excitation of primary afferents (Aδ and C fibers) (2) causes vasodilation in precapillary arterioles and plasma extravasation in postcapillary venules (C fibers only) by the release of substance P (SP) and other vasoactive compounds (e.g., calcitonin gene-related peptide, CGRP). Some of these effects may be mediated by nonneuronal cells such as mast cells and macrophages (4). Other factors that affect the control of the microcirculation are the myogenic properties of arterioles (3) and more global environmental influences such as a change of the temperature and the metabolic state of the tissue. (b) Hypothetical relation between sympathetic noradrenergic nerve fibers (1), peptidergic afferent nerve fibers (2), blood vessels (3), and macrophages (4). The activated and sensitized afferent nerve fibers activate macrophages (MP) possibly via SP release. The immune cells start to release cytokines, such as tumor necrosis factor alpha (TNF-α) and interleukin-1 (IL-1), which further activate afferent fibers. SP (and CGRP) released from the afferent nerve fibers reacts with neurokinin 1 (NK1) receptors in the blood vessels (arteriolar vasodilation, venular plasma extravasation, neurogenic inflammation). The sympathetic nerve fibers interact with this system on three levels: (1) via adrenoceptors (mainly α) on the blood vessels (vasoconstriction); (2) via adrenoceptors (mainly β) on macrophages (further release of cytokines), and (3) via adrenoceptors (mainly α) on afferents (further sensitization of these fibers.) Adapted from Janig, W. and Baron, R. 2003. Complex regional pain syndrome: mystery explained? Lancet Neurol. 2, 687–697.

validation research suggests that CRPS was overdiagnosed (Bruehl, S. *et al.*, 1999; Harden, R. N. *et al.*, 1999). For example, an external validation of the International Association for the Study of Pain (IASP) criteria in 117

patients with CRPS and 43 patients with neuropathic pain without CRPS etiology demonstrated a sensitivity of 0.98 and a specificity of 0.36. The inclusion of a category motor and trophic signs and symptoms, for

Table 1 Revised diagnostic criteria for CRPS

A. Continuing pain, which is disproportionate to any inciting event
B. Categories of clinical signs/symptoms
1. Positive sensory abnormalities
2. Vascular abnormalities
   Vasodilation
   Vasoconstriction
   Skin temperature asymmetries
   Skin color changes
3. Edema, sweating abnormalities
   Swelling
   Hyperhidrosis
   Hypohidrosis
4. Motor, trophic changes
   Motor weakness
   Tremor
   Dystonia
   Coordination deficits
   Nail, hair changes
   Skin atrophy
   Joint stiffness
   Soft tissue
Interpretation
Clinical use
   ≥1 Symptoms in ≥3 of the categories AND ≥1 signs at time of evaluation in ≥2 of the categories
   Sensitivity = 0.85, specificity = 0.60
Research use
   ≥1 Symptoms in each of the categories each AND ≥1 signs at the time of evaluation in ≥2 of the categories each
   Sensitivity = 0.70, specificity = 0.96

CRPS, complex regional pain syndrome.

Table 2 Diagnostic tests in CRPS

	Sensitivity	Specificity
Plain radiograph (only chronic stages)	73	57
Bone scan (only acute stages)	97	86
Quantitative sensory testing (QST)	High	Low
Temperature differences (during sympathetic stimulation)	76	93
MRI (skin, joint, etc.)	91	17

CRPS, complex regional pain syndrome; MRI, magnetic resonance imaging.

somatosensory and motor abnormalities can be visualized using MEG and fMRI. Video-assisted kinematic analysis and tremor analysis is used to quantify and qualify motor abnormalities like weakness, coordination deficits and tremor. The acute inflammatory component of the disorder, in particular the role of specific inflammatory mediators, are explored using microdialysis techniques and analysing the content of suction blister fluid. These and other investigations will pave the way for a quantitative and objective analysis of these abnormalities. Furthermore, genetic patterns of patients at risk to develop CRPS have already been revealed and more genes determining the susceptible patient are to come.

In summary, we have to stimulate researchers from various fields to unravel the diagnostic puzzle of the CRPS mystery using an interdisciplinary research approach.

example, improves specificity considerably without losing sensitivity (Bruehl, S. et al., 2002).

Based on this validation research a novel diagnostic algorithm was recently proposed (Table 1, Baron, R. and Janig, W., 2004; Burton, A. W. et al., 2005). In addition to the improved clinical categories, it became clear to distinguish between criteria for clinical use and a classification for research purposes. For the clinician and in particular for the patients, it is important to have a high sensitivity value combined with a fair specificity (e.g., 0.85 vs. 0.60, Table 1). For research purposes, however, it is much more important to have a high specificity in order to perform studies in a precisely diagnosed population (e.g., 0.7 vs. 0.96, Table 1).

59.5.5.2 Tests which aid the diagnosis of complex regional pain syndrome

Several diagnostic procedures have been evaluated in recent years (Table 2). New techniques to assess specific signs objectively are in the experimental testing phase. Cortical reorganization underlying

Acknowledgments

This work was supported by the Deutsche Forschungsgemeinschaft (DFG Ba 1921/1-2), the German Ministry of Research and Education within the German Research Network on Neuropathic Pain (BMBF, 01EM01/04) and an unrestricted educational grant of Pfizer (Germany).

References

Alexander, G. M., van Rijn, M. A., van Hilten, J. J., Perreault, M. J., and Schwartzman, R. J. 2005. Changes in cerebrospinal fluid levels of pro-inflammatory cytokines in CRPS. Pain 116, 213–219.
Ali, Z., Raja, S. N., Wesselmann, U., Fuchs, P., Meyer, R. A., and Campbell, J. N. 2000. Intradermal injection of norepinephrine

evokes pain in patients with sympathetically maintained pain. Pain 88, 161–168.

Apkarian, A. V., Thomas, P. S., Krauss, B. R., and Szeverenyi, N. M. 2001. Prefrontal cortical hyperactivity in patients with sympathetically mediated chronic pain. Neurosci. Lett. 311, 193–197.

Baron, R. 2004. Mechanistic and clinical aspects of complex regional pain syndrome (CRPS). Novartis Found. Symp. 261, 220–233; discussion 233–238, 256–261.

Baron, R. and Janig, W. 2004. Complex regional pain syndromes – how do we escape the diagnostic trap? Lancet 364, 1739–1741.

Baron, R. and Wasner, G. 2001. Complex regional pain syndromes. Curr. Pain Headache Rep. 5, 114–123.

Baron, R., Fields, H. L., Janig, W., Kitt, C., and Levine, J. D. 2002a. National Institutes of Health Workshop: reflex sympathetic dystrophy/complex regional pain syndromes – state-of-the-science. Anesth. Analg. 95, 1812–1816.

Baron, R., Levine, J. D., and Fields, H. L. 1999. Causalgia and reflex sympathetic dystrophy: Does the sympathetic nervous system contribute to the generation of pain? Muscle Nerve 22, 678–695.

Baron, R., Schattschneider, J., Binder, A., Siebrecht, D., and Wasner, G. 2002b. Relation between sympathetic vasoconstrictor activity and pain and hyperalgesia in complex regional pain syndromes: a case–control study. Lancet 359, 1655–1660.

Bhatia, K. P., Bhatt, M. H., and Marsden, C. D. 1993. The causalgia–dystonia syndrome. Brain 116, 843–851.

Birklein, F., Schmelz, M., Schifter, S., and Weber, M. 2001. The important role of neuropeptides in complex regional pain syndrome. Neurology 57, 2179–2184.

Blaes, F., Schmitz, K., Tschernatsch, M., Kaps, M., Krasenbrink, I., Hempelmann, G., and Brau, M. E. 2004. Autoimmune etiology of complex regional pain syndrome (M. Sudeck). Neurology 63, 1734–1736.

Bruehl, S., Harden, R. N., Galer, B. S., Saltz, S., Backonja, M., and Stanton-Hicks, M. 2002. Complex regional pain syndrome: are there distinct subtypes and sequential stages of the syndrome? Pain 95, 119–124.

Bruehl, S., Harden, R. N., Galer, B. S., Saltz, S., Bertram, M., Backonja, M., Gayles, R., Rudin, N., Bhugra, M. K., and Stanton-Hicks, M. 1999. External validation of IASP diagnostic criteria for Complex Regional Pain Syndrome and proposed research diagnostic criteria. Pain 81, 147–154.

Burton, A. W., Bruehl, S., and Harden, R. N. 2005. Current diagnosis and therapy of complex regional pain syndrome: refining diagnostic criteria and therapeutic options. Expert Rev. Neurother. 5, 643–651.

Calder, J. S., Holten, I., and McAllister, R. M. 1998. Evidence for immune system involvement in reflex sympathetic dystrophy. J. Hand Surg. [Br.] 23, 147–150.

Coderre, T. J., Xanthos, D. N., Francis, L., and Bennett, G. J. 2004. Chronic post-ischemia pain (CPIP): a novel animal model of complex regional pain syndrome – Type I (CRPS-I; reflex sympathetic dystrophy) produced by prolonged hindpaw ischemia and reperfusion in the rat. Pain 112, 94–105.

Deuschl, G., Blumberg, H., and Lücking, C. H. 1991. Tremor in reflex sympathetic dystrophy. Arch. Neurol. 48, 1247–1252.

Fukumoto, M., Ushida, T., Zinchuk, V. S., Yamamoto, H., and Yoshida, S. 1999. Contralateral thalamic perfusion in patients with reflex sympathetic dystrophy syndrome. Lancet 354, 1790–1791.

Galer, B. S., Butler, S., and Jensen, M. P. 1995. Case reports and hypothesis: a neglect-like syndrome may be responsible for the motor disturbance in reflex sympathetic dystrophy (Complex Regional Pain Syndrome – 1). J. Pain Symptom Manage. 10, 385–391.

Gazda, L. S., Milligan, E. D., Hansen, M. K., Twining, C. M., Poulos, N. M., Chacur, M., O'Connor, K. A., Armstrong, C., Maier, S. F., Watkins, L. R., and Myers, R. R. 2001. Sciatic inflammatory neuritis (SIN): behavioral allodynia is paralleled by peri-sciatic proinflammatory cytokine and superoxide production. J. Peripher. Nerv. Syst. 6, 111–129.

Goebel, A., Vogel, H., Caneris, O., Bajwa, Z., Clover, L., Roewer, N., Schedel, R., Karch, H., Sprotte, G., and Vincent, A. 2005. Immune responses to Campylobacter and serum autoantibodies in patients with complex regional pain syndrome. J. Neuroimmunol. 162, 184–189.

Graif, M., Schweitzer, M. E., Marks, B., Matteucci, T., and Mandel, S. 1998. Synovial effusion in reflex sympathetic dystrophy: an additional sign for diagnosis and staging. Skeletal Radiol. 27, 262–265.

Harden, R. N., Bruehl, S., Galer, B. S., Saltz, S., Bertram, M., Backonja, M., Gayles, R., Rudin, N., Bhugra, M. K., and Stanton-Hicks, M. 1999. Complex regional pain syndrome: are the IASP diagnostic criteria valid and sufficiently comprehensive? Pain 83, 211–219.

Huygen, F. J., De Bruijn, A. G., De Bruin, M. T., Groeneweg, J. G., Klein, J., and Zijistra, F. J. 2002. Evidence for local inflammation in complex regional pain syndrome type 1. Mediators Inflamm. 11, 47–51.

Huygen, F. J., de Bruijn, A. G., Klein, J., and Zijlstra, F. J. 2001. Neuroimmune alterations in the complex regional pain syndrome. Eur. J. Pharmacol. 429, 101–113.

Janig, W. and Baron, R. 2003. Complex regional pain syndrome: mystery explained? Lancet Neurol. 2, 687–697.

Jänig, W., Levine, J. D., and Michaelis, M. 1996. Interactions of sympathetic and primary afferent neurons following nerve injury and tissue trauma. Prog. Brain Res. 113, 161–184.

Juottonen, K., Gockel, M., Silen, T., Hurri, H., Hari, R., and Forss, N. 2002. Altered central sensorimotor processing in patients with complex regional pain syndrome. Pain 98, 315–323.

Kemler, M. A., van de Vusse, A. C., van den Berg-Loonen, E. M., Barendse, G. A., van Kleef, M., and Weber, W. E. 1999. HLA-DQ1 associated with reflex sympathetic dystrophy. Neurology 53, 1350–1351.

Leitha, T., Korpan, M., Staudenherz, A., Wunderbaldinger, P., and Fialka, V. 1996. Five phase bone scintigraphy supports the pathophysiological concept of a subclinical inflammatory process in reflex sympathetic dystrophy. Q. J. Nucl. Med. 40, 188–193.

Levine, J. D., Dardick, S. J., Basbaum, A. I., and Scipio, E. 1985. Reflex neurogenic inflammation. I. Contribution of the peripheral nervous system to spatially remote inflammatory responses that follow injury. J. Neurosci. 5, 1380–1386.

Maihofner, C., Forster, C., Birklein, F., Neundorfer, B., and Handwerker, H. O. 2005a. Brain processing during mechanical hyperalgesia in complex regional pain syndrome: a functional MRI study. Pain 114, 93–103.

Maihofner, C., Handwerker, H. O., Neundorfer, B., and Birklein, F. 2003. Patterns of cortical reorganization in complex regional pain syndrome. Neurology 61, 1707–1715.

Maihofner, C., Handwerker, H. O., Neundorfer, B., and Birklein, F. 2004. Cortical reorganization during recovery from complex regional pain syndrome. Neurology 63, 693–701.

Maihofner, C., Handwerker, H. O., Neundorfer, B., and Birklein, F. 2005b. Mechanical hyperalgesia in complex regional pain syndrome: a role for TNF-alpha? Neurology 65, 311–313.

Mailis, A. and Wade, J. 1994. Profile of Caucasian women with possible genetic predisposition to reflex sympathetic dystrophy: a pilot study. Clin. J. Pain 10, 210–217.

Mailis, A. and Wade, J. 2001. Genetic Considerations in CRPS. In: Complex Regional Pain Syndrome (eds. R. N. Harden, R. Baron, and W. Jänig), Vol. 22, pp. 227–238. IASP Press.

Maleki, J., LeBel, A. A., Bennett, G. J., and Schwartzman, R. J. 2000. Patterns of spread in complex regional pain syndrome, type I (reflex sympathetic dystrophy). Pain 88, 259–266.

McCabe, C. S., Haigh, R. C., Ring, E. F., Halligan, P. W., Wall, P. D., and Blake, D. R. 2003. A controlled pilot study of the utility of mirror visual feedback in the treatment of complex regional pain syndrome (type 1). Rheumatology (Oxford) 42, 97–101.

Miao, F. J., Green, P. G., Coderre, T. J., Janig, W., and Levine, J. D. 1996a. Sympathetic-dependence in bradykinin-induced synovial plasma extravasation is dose-related. Neurosci. Lett. 205, 165–168.

Miao, F. J., Janig, W., Green, P. G., and Levine, J. D. 1996b. Inhibition of bradykinin-induced synovial plasma extravasation produced by intrathecal nicotine is mediated by the hypothalamopituitary adrenal axis. J. Neurophysiol. 76, 2813–2821.

Mogil, J. S., Wilson, S. G., Bon, K., Lee, S. E., Chung, K., Raber, P., Pieper, J. O., Hain, H. S., Belknap, J. K., Hubert, L., Elmer, G. I., Chung, J. M., and Devor, M. 1999. Heritability of nociception I: responses of 11 inbred mouse strains on 12 measures of nociception. Pain 80, 67–82.

Moseley, G. L. 2004. Graded motor imagery is effective for long-standing complex regional pain syndrome: a randomised controlled trial. Pain 108, 192–198.

Oyen, W. J., Arntz, I. E., Claessens, R. M., Van der Meer, J. W., Corstens, F. H., and Goris, R. J. 1993. Reflex sympathetic dystrophy of the hand: an excessive inflammatory response? Pain 55, 151–157.

Perl, E. R. 1996. Cutaneous polymodal receptors: characteristics and plasticity. Prog. Brain Res. 113, 21–37.

Pleger, B., Tegenthoff, M., Ragert, P., Forster, A. F., Dinse, H. R., Schwenkreis, P., Nicolas, V., and Maier, C. 2005. Sensorimotor retuning [corrected] in complex regional pain syndrome parallels pain reduction. Ann. Neurol. 57, 425–429.

Pleger, B., Tegenthoff, M., Schwenkreis, P., Janssen, F., Ragert, P., Dinse, H. R., Volker, B., Zenz, M., and Maier, C. 2004. Mean sustained pain levels are linked to hemispherical side-to-side differences of primary somatosensory cortex in the complex regional pain syndrome I. Exp. Brain Res. 155, 115–119.

Renier, J. C., Arlet, J., Bregeon, C., Basle, M., Seret, P., Acquaviva, P., Schiano, A., Serratrice, G., Amor, B., May, V., Delcambre, B., D'Eshoughes, J. R., Vincent, G., Ducastelle, F., and Pawlotsky, Y. 1983. The joint in algodystrophy. Joint fluid, synovium, cartilage. Rev. Rhum. Mal. Osteoartic. 50, 255–260.

Ribbers, G. M., Mulder, T., Geurts, A. C., and den Otter, R. A. 2002. Reflex sympathetic dystrophy of the left hand and motor impairments of the unaffected right hand: impaired central motor processing? Arch. Phys. Med. Rehabil. 83, 81–85.

Schattschneider, J., Wenzelburger, R., Deuschl, G., and Baron, R. 2001. Kinematic Analysis of the Upper Extremity in CRPS. In: Complex Regional Pain Syndrome, Vol. 22 (eds. R. N. Harden, R. Baron, and W. Jänig), pp. 119–128. IASP Press.

Schwenkreis, P., Janssen, F., Rommel, O., Pleger, B., Volker, B., Hosbach, I., Dertwinkel, R., Maier, C., and Tegenthoff, M. 2003. Bilateral motor cortex disinhibition in complex regional pain syndrome (CRPS) type I of the hand. Neurology 61, 515–519.

Shibata, M., Nakao, K., Galer, B. S., Shimizu, T., Taniguchi, H., and Uchida, T. 1999. A case of reflex sympathetic dystrophy (complex regional pain syndrome, type I) resolved by cerebral contusion. Pain 79, 313–315.

Stanton-Hicks, M., Janig, W., Hassenbusch, S., Haddox, J. D., Boas, R., and Wilson, P. 1995. Reflex sympathetic dystrophy: changing concepts and taxonomy. Pain 63, 127–133.

Sudeck, P. 1902. Über die akute (trophoneurotische) Knochenatrophie nach Entzündungen und Traumen der Extremitäten. Deut. Med. Wschr. 28, 336–342.

Sudeck, P. 1931. Die trophische Extremitätenstörung durch periphere (infektiöse und traumatische) Reize. Deutsch. Zeitschr. Chirurg. 234, 596–612.

van de Beek, W. J., Roep, B. O., van der Slik, A. R., Giphart, M. J., and van Hilten, B. J. 2003. Susceptibility loci for complex regional pain syndrome. Pain 103, 93–97.

van der Laan, L. and Goris, R. J. 1997. Reflex sympathetic dystrophy. An exaggerated regional inflammatory response? Hand Clin. 13, 373–385.

van Hilten, J. J., van de Beek, W. J., and Roep, B. O. 2000. Multifocal or generalized tonic dystonia of complex regional pain syndrome: a distinct clinical entity associated with HLA-DR13. Ann. Neurol. 48, 113–116.

Wasner, G., Heckmann, K., Maier, C., and Baron, R. 1999. Vascular abnormalities in acute reflex sympathetic dystrophy (CRPS I) – complete inhibition of sympathetic nerve activity with recovery. Arch. Neurol. 56, 613–620.

Wasner, G., Schattschneider, J., Binder, A., and Baron, R. 2003. Complex regional pain syndrome – diagnostic, mechanisms, CNS involvement and therapy. Spinal Cord 41, 61–75.

Wasner, G., Schattschneider, J., Heckmann, K., Maier, C., and Baron, R. 2001. Vascular abnormalities in reflex sympathetic dystrophy (CRPS I): mechanisms and diagnostic value. Brain 124, 587–599.

Weber, M., Birklein, F., Neundorfer, B., and Schmelz, M. 2001. Facilitated neurogenic inflammation in complex regional pain syndrome. Pain 91, 251–257.

60 Poststroke Pain

T S Jensen and N B Finnerup, Aarhus University Hospital, Aarhus, Denmark

Glossary

allodynia Pain due to a stimulus that does not normally provoke pain.

central pain Pain initiated or caused by a primary lesion or dysfunction in the central nervous system.

hyperalgesia An increased response to a stimulus that is normally painful.

60.1 Introduction

Central poststroke pain (PSP) is defined as pain directly caused by a cerebrovascular lesion, regardless of the anatomical location within the central nervous system (CNS).

Neuropathic pain has mainly been discussed in the context of peripheral neuropathic pain, that is, conditions where the injury or the dysfunction is located within the peripheral nervous system (Jensen, T. S. *et al.*, 2001b). Typical examples of diseases affecting the peripheral nervous system giving rise to pain are painful polyneuropathies, such as diabetic neuropathy and toxic neuropathies, postherpetic neuralgia, trigemenial neuralgia, and nerve injuries. Although probably less common and more complex in nature, central neuropathic pain conditions share some of the same phenomena seen in peripheral neuropathic pain, that is, sensory loss in part of the territory with pain, the paradoxical presence of lost sensibility, and hypersensitivity to one or several submodalities of pain (Tasker, R. R. *et al.*, 1991; Schott, G. D., 1995; Bowsher, D. *et al.*, 1998; Boivie, J., 1999). This chapter will briefly review central neuropathic pain after stroke: its underlying mechanisms, clinical picture, and management.

60.2 Characteristics of Central Pain After Stroke

Despite the heterogeneity in etiology and anatomical location, neuropathic pains share certain characteristics (Boivie, J., 1999; Woolf, C. J. and Mannion, R. J., 1999; Jensen, T. S. *et al.*, 2001b; Dworkin, R. H. *et al.*, 2003):

● Pain located in a neuroanatomically defined area with partial or complete sensory loss.

- Ongoing pain (stimulus independent).
- Evoked types of pain (stimulus dependent).
- Hypersensitivity.

These symptoms and signs may occur in various combinations but do not necessarily have to be present altogether at the same time and in all patients affected by the same disease. There are no universally accepted criteria for PSP, but the following seem to be minimal criteria:

- A confirmed cerebrovascular accident.
- Development of pain after stroke.
- Sensory abnormality (hypo- or hypersensitivity to temperature or pinprick) in a body area corresponding to the CNS region damaged by the stroke.
- Pain located within the body area of sensory abnormality.

60.2.1 Symptoms in Central Pain

As for peripheral neuropathic pain, central pains are characterized by the presence of spontaneous types of pain and various types of evoked pains (Boivie, J., 1999; Jensen, T. S. *et al.*, 2001b).

Ongoing pains. These pains are spontaneous and may be continuous or paroxysmal. The character differs, but they can be shooting, shock-like, aching, cramping, crushing, smarting, burning, etc. No specific pain character has been associated with PSP (Boivie, J., 1999).

Evoked pains. The stimulus-evoked pains are classified according to the type of stimulus that provokes them. Several types of evoked pains can be present; the most prominent types in PSP are mechanical and thermal.

In some patients ongoing pain is dominating, in others evoked pain. Some patients have several of these phenomena, others only one type of hyperalgesia. For example, in some patients with neuropathy, cold allodynia may be the only abnormality present. Therefore, a number of different stimuli have to be used to document or exclude abnormality. The evoked pains are usually brief, lasting only for the duration of stimulation, but sometimes they can persist even after cessation of stimulation causing aftersensations (see below), which can last for minutes, hours, or perhaps days. In such cases, the distinction between evoked and spontaneous pain can be difficult. Both evoked and spontaneous pains may be confined to deep structures.

60.2.2 Sensory Abnormalities

Sensory disturbances are key phenomena in patients with PSP and a variety of disturbances may be present.

Sensory deficit and pain. An essential part of neuropathic pain is a loss (partial or complete) of afferent sensory function. In some patients the sensory deficit may be gross, in others subtle and difficult to detect with bedside methods, but quantitative testing can usually disclose minor changes. The sensory loss may involve all sensory abnormalities, but a loss of spinothalamic functions (cold, warmth, and pinprick) appears to be crucial, and there is now some evidence to indicate that spinothalamic loss is a necessary, albeit not sufficient, requirement for the development and maintenance of pain (Andersen, G. *et al.*, 1995; Vestergaard, K. *et al.*, 1995). In the study by Andersen G. *et al.* (1995), decreased sensibility to all types of stimuli was seen in the same proportion of patients, regardless of the presence of pain. It can occasionally be difficult to document sensory loss because of the additional presence of hypersensitivity phenomena in the same area.

Sensory hypersensitivity and pain. Another characteristic of PSP is the paradoxical presence of hyperphenomena in the painful area. These hyperphenomena may be present as hypersensitivity, allodynia, or dysesthesia to one or several sensory modalities (touch, pinprick, cold, or warmth). Allodynia or hypersensitivity has mainly been noted as a cutaneous phenomenon, but it was originally suggested that allodynia could also be present in deep structures as a hyperalgesia to pressure (Vestergaard, K. *et al.*, 1995). Consistent with this observation, Bowsher D. *et al.* (1998) noted movement-induced allodynia in 9 of 73 patients, and in another study, 22% of 92 patients with central pain had movement-induced pain (Bowsher, D., 1996). A dissociation of cutaneous and deep sensibility has been noted in three case reports (Mailis, A. and Bennett, G. J., 2002). In the study by Andersen G. *et al.* (1995), cold allodynia/dysesthesia was present in 88% of PSP patients but only in 3% of stroke patients with nonpainful sensory abnormalities. Similarly, 75% had touch-evoked allodynia or dysesthesia while none in the group without pain had these abnormalities. In the study by Bowsher D. *et al.* (1998), 59% of the patients had allodynia mostly seen as a mechanical allodynia. Abnormal temperature and pain sensibility are the most consistent abnormalities in PSP. In two studies, it was found that 81%

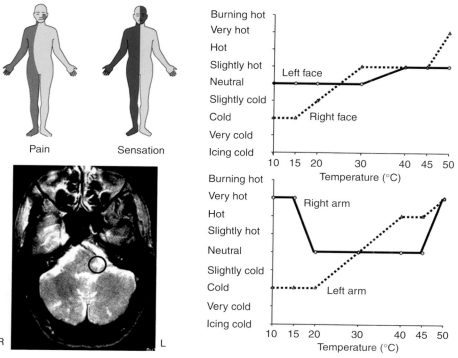

Figure 1 A case presenting a 75-year-old patient with lateral medullary infarction and central pain. At the age of 68, he experienced short-lasting vertigo. At the age of 72, he had a left brain infarction with left facial and right hemibody ongoing pain and burning pain evoked by touch. The magnetic resonance imaging (MRI) scan shows a left lateral medullary infarction. Stimulus–response functions of cold and warm stimuli in the patient using a thermotester showing decreased thermal sensation on the left side of the face and paradoxical sensitivity to cold on the left side of the body.

had reduced sensibility to temperature (Boivie, J., 1989; Andersen, G. *et al.*, 1995). In some studies, on the other hand, sensation to touch is only reduced in a smaller proportion of patients (Boivie, J., 1989).

Based on the observation that thermal sensation is also disturbed in a large proportion of stroke patients with no pain, it has been concluded that loss of spinothalamic function mediating thermal stimuli is a necessary, but probably not a sufficient condition for the development of PSP (Jensen, T. S. and Lenz, F. A., 1995). In addition to the sensory deficit, some patients may have a paradoxical sensitivity to cold and heat so that cold is perceived as hot and vice versa. An example of a patient with a lateral medullary infarction and such a sensory pattern is shown in Figure 1.

60.3 Epidemiology of Poststroke Pain

There is limited information on the frequency of PSP. A prospective study found that 8.4% of 191

unselected consecutive stroke patients developed PSP within the first year after stroke (Andersen, G. *et al.*, 1995). In an unpublished study from the UK, the frequency of PSP was found to be 11% in a group of stroke patients aged 80 years and above (Bowsher, D., personal communication).

In 63 patients with lateral medullary infarction, 16 (25%) developed central pain within 6 months after stroke. There was a significant relationship between the degree of sensory loss and the development of pain, suggesting to the authors that pain is a result of a lesion of the neospinothalamic system, which causes a denervation sensitivity of the paleospinothalamic system. In patients with sensory disturbances, the frequency of pain is much higher: thus 44% in one series (MacGowan, D. J. *et al.*, 1997) and 18% in another series (Andersen, G. *et al.*, 1995).

Stroke lesion. Both infarcts and hemorrhages may cause PSP (Andersen, G. *et al.*, 1995; Bowsher, D. *et al.*, 1998; Kim, J. S., 2003). Based on the fact that ischemic lesions are 7–8 times more frequent than are hemorrhages, infarction is the most common cause for PSP. Lesions are known to occur at any level of the spinothalamic pathways and its cortical projection. Previous studies using computed tomography (CT) imaging show that the lesion involves the thalamus in one-fourth to one-third of cases (Bogousslavsky, J. *et al.*, 1988; Leijon, G. *et al.*, 1989; Andersen, G. *et al.*, 1995). In one study of PSP, more than 60% of patients had a lesion of the ventroposterior thalamic nucleus demonstrated by magnetic resonance imaging (MRI) (Bowsher, D. *et al.*, 1998). In a series of 63 patients with lateral medullary infarctions, it was found that 16 patients (25%) developed a PSP within 6 months after their stroke (MacGowan, D. J. *et al.*, 1997). In another series of 55 patients with medullary infarctions, 24 patients had sensory complaints from the face (Kim, J. S. and Choi-Kwon, S., 1999).

60.4 Mechanisms of Pain

The pathophysiological mechanisms underlying peripheral neuropathic pain have been reviewed extensively in recent years (Besson, J. M., 1999; Woolf, C. J. and Salter, M. W., 2000; Koltzenburg, M. and Scadding, J., 2001; Jensen, T. S. *et al.*, 2001a; Bolay, H. and Moskowitz, M. A., 2002).

Evidence from a large number of studies indicates that neuronal hyperexcitability, either at the level of the peripheral receptor or centrally, plays a key role in generating pain after nerve injury. The mechanisms responsible for central pain after stroke are much less clear. The combination of symptoms and signs is compatible with a general sensitization of second- or third-order neurons in the CNS that have lost part of their normal patterned input and are substituted by a changed afferent influx. In fact, sensitization is today considered to be an essential phenomenon behind persistent neuropathic pains (Woolf, C. J. and Salter, M. W., 2000; Woolf, C. J. and Max, M. B., 2001; Dworkin, R. H. *et al.*, 2003; Jensen, T. S. and Baron, R., 2003).

Patients with central pain display ongoing spontaneous pain and abnormally evoked pain with hypersensitivity to noxious and nonnoxious stimuli in the painful area. These findings suggest that hypersensitive pools of cells are present in the nervous system and that they probably are involved in generating pain (Tasker, R. R. *et al.*, 1991; Jensen, T. S. and Lenz, F. A., 1995; Schott, G. D., 1995). Several hypotheses have been forwarded. The most important ones are summarized briefly below:

Spinothalamic tract lesion. Sensory examination and more elaborative quantitative sensory analysis show that patients with PSP have abnormal temperature and pain sensitivity, but normal epicritic sensibility (touch, vibration, and pressure). In addition, it is well known that dorsolateral infarctions of the lower medulla and cordotomies, two conditions characterized by loss of spinothalamic but preservation of lemniscal pathways, cause central pain (Boivie, J. *et al.*, 1989; Vestergaard, K. *et al.*, 1995; Nathan, P. W. *et al.*, 2001).

Subsequent studies have indicated that while a spinothalamic tract lesion and its cortical projection is necessary for the development of pain, it is not sufficient for the development of central pain (Jensen, T. S. and Lenz, F. A., 1995) since similar somatosensory disturbances may occur in stroke patients without pain (Andersen, G. *et al.*, 1995; Vestergaard, K. *et al.*, 1995).

Lemniscal disinhibition. In their original description Head H. and Holmes G. (1911) suggested that injury to the lateral part of thalamus sets free a medial thalamic center and the result is allodynia to all types of stimuli. A consequence of this theory is that sensory modalities such as touch, vibration, and pressure conveyed by the lemniscal system exert a control over pain and temperature, so that sensory deficit in discrimination causes spontaneous pain and allodynia.

Loss of normal cold inhibition. On the basis of experimental studies in animals and in humans, Craig A. D. (2000) has suggested that central pain is due to loss of a normal inhibitory effect exerted by cool-signaling pathways from lamina I projecting to the thalamus and insula. According to this hypothesis, a lesion of the lateral cool projection system disinhibits the medial system of heat–pinch–cold neurons passing from lamina I to the medial part of the thalamus. This disinhibition then results in the release of cold allodynia and burning and ongoing pain (Craig, A. D. and Bushnell, M. C., 1994; Craig, A. D., 2000). Disruption of thermosensory integrations leads to a disinhibition of noxious responding thalamocortical neurons and a sensation of burning pain (Craig, A. D., 2000). The hypothesis and the processing of cold proposed by Craig has been criticized by Willis W. D. Jr. *et al.* (2002), and more recently, Greenspan J. D. *et al.* (2004) suggested, based on studies in 13 patients with PSP, that cold allodynia is not caused by loss of normal cold pathway but is rather due to input via an intact cold pathway driving nociceptive neurons in the thalamus. So it is still unclear whether and to what extent an intact cold pathway plays a role for the cold allodynia seen in stroke patients.

Thalamus and PSP. In all instances of PSP, the thalamus is thought to play a role for the pain. The most critical parts of the thalamus are the ventroposterior part, the medial intralaminar nuclei, and the reticular nucleus, but the exact mechanisms of central pain are still unclear. Based on clinical and experimental observations, it has been suggested that the development of PSP may be related to an imbalance in the activity of the spinothalamic and reticulothalamic systems so that if the former is damaged, as for example in medullary infarction, this may result in pain from non-noxious stimuli via the reticulothalamic pathway (Tasker, R. R. *et al.*, 1991; MacGowan, D. J. *et al.*, 1997).

60.5 Treatment

There is limited amount of data on pharmacological treatment of PSP and no evidence to support treatment of the individual patient based on symptoms and signs. Pharmacological agents used to treat PSP share the ability to interfere with some of the processes involved in neuronal hyperexcitability either by decreasing excitatory or by increasing inhibitory transmission, thereby exerting a neuronal depressant effect.

60.5.1 Antidepressants

Tricyclic antidepressants inhibit the presynaptic reuptake of noradrenaline and serotonin and interfere with N-methyl-D-aspartate (NMDA) receptors and sodium channels (Sindrup, S. H. *et al.*, 2005). Amitriptyline 75 mg day^{-1} effectively relieved PSP (Leijon, G. and Boivie, J., 1989). The pain-relieving effect correlated with total plasma concentration, with a plasma concentration of amitriptyline and its metabolite nortriptyline exceeding the 300 nmol l^{-1} necessary for pain relief. Amitriptyline administered from the first day after a stroke and thereafter for 1 year in 39 stroke patients did not prevent PSP and allodynia (Lampl, C. *et al.*, 2002). Seven patients developed PSP (three receiving amitriptyline and four receiving placebo), and a large sample size is needed to exclude a type II error. Tricyclic antidepressants cannot be used in patients with heart failure, cardiac conduction disturbances, and epilepsy, and side effects including dry mouth, sweating, and sedation may limit their use. Recently, serotonin noradrenaline reuptake

inhibitors with a better side-effect profile have been shown to relieve peripheral neuropathic pain, but these antidepressants are not yet studied in central pain conditions. Selective serotonin reuptake inhibitors are less effective in relieving neuropathic pain (Sindrup, S. H. *et al.*, 2005), and citalopram was found not to relieve PSP (Vestergaard, K. *et al.*, 1996).

60.5.2 Sodium Channel Blockers

In peripheral neuropathic pain, increased and abnormal sodium channels are known to be a source of neuronal hyperexcitability (Waxman, S. G. *et al.*, 2000). A role for such sodium channels and neuronal hyperexcitability in central pain in humans is supported by randomized controlled trials using sodium channel blockers. In 16 patients with central pain (six with PSP), lidocaine 5 mg kg^{-1} given intravenously over 30 min alleviated spontaneous ongoing pain, brush-induced allodynia, and mechanical hyperalgesia (Attal, N. *et al.*, 2000). This effect in central pain may be related to suppression of ectopic discharges from hyperexcitable central neurons. This is supported by a recent experimental study demonstrating increased expression of abnormal sodium channels functionally linked to neuronal hyperexcitability and central pain behaviors on second-order dorsal horn neurons close to a spinal cord lesion (Hains, B. C. *et al.*, 2003).

Along this line, the sodium channel blocker lamotrigine, at a relatively low dosage of 200 mg day^{-1}, reduced ongoing pain with a mean reduction of 30% and cold allodynia in PSP patients (Vestergaard, K. *et al.*, 2001). Lamotrigine was generally well tolerated. Side effects associated with the use of lamotrigine include dizziness, nausea, headache, and fatigue. Rash and potentially life-threatening hypersensitivity reactions are rare but potential risks, and slow dose escalation is recommended to minimize these risks. Carbamazepine 800 mg day^{-1} showed a statistically nonsignificant trend toward a pain-relieving effect in PSP (Leijon, G. and Boivie, J., 1989).

60.5.3 Opioids

Opioids seem to effectively relieve peripheral neuropathic pain with an effect size in line with tricyclic antidepressants and gabapentin/pregabalin (Finnerup, N. B. *et al.*, 2005) but seem to be less effective in PSP, although a subgroup of patients may respond well to opioids (Attal, N. *et al.*, 2002; Rowbotham, M. C. *et al.*, 2003).

60.5.4 Gabapentin

Gabapentin, which binds to the $\alpha_2\delta$ subunits of the voltage-dependent calcium channels, has not been studied in PSP but has been found to relieve spinal cord injury pain (Levendoglu, F. *et al.*, 2004), and given its tolerability, it is a treatment option in PSP.

60.5.5 GABA Agonists

Increasing γ-aminobutyric acid (GABA)-mediated inhibition using intravenous propofol is reported to relieve central pain including PSP (Canavero, S. and Bonicalzi, V., 2004), but propofol is very sedating and oral GABA agonists have not proven effective in neuropathic pain.

60.5.6 Treatment Option

Based on a few randomized controlled trials, an evidence-based treatment algorithm for PSP is not possible, but lamotrigine, gabapentin, tricyclic antidepressants, or serotonin noradrenaline reuptake inhibitors may be suggested as first choice.

Acknowledgments

This paper is in part based on support from Karen Elise Jensen Foundation, Danish Medical Research Council (grant no. 22-04-0561), and the Elsass Foundation.

References

Andersen, G., Vestergaard, K., Ingeman, N. M., and Jensen, T. S. 1995. Incidence of central post-stroke pain. Pain 61, 187–193.

Attal, N., Gaude, V., Brasseur, L., Dupuy, M., Guirimand, F., Parker, F., and Bouhassira, D. 2000. Intravenous lidocaine in central pain: a double-blind, placebo-controlled, psychophysical study. Neurology 54, 564–574.

Attal, N., Guirimand, F., Brasseur, L., Gaude, V., Chauvin, M., and Bouhassira, D. 2002. Effects of IV morphine in central pain: a randomized placebo-controlled study. Neurology 58, 554–563.

Besson, J. M. 1999. The neurobiology of pain. Lancet 353, 1610–1615.

Bogousslavsky, J., Regli, F., and Uske, A. 1988. Thalamic infarcts: clinical syndromes, etiology, and prognosis. Neurology 38, 837–848.

Boivie, J. 1989. On central pain and central pain mechanisms. Pain 38, 121–122.

Boivie, J. 1999. Central Pain. In: Textbook of Pain (eds. P. D. Wall and R. Melzack), pp. 879–914. Churchill Livingstone.

Boivie, J., Leijon, G., and Johansson, I. 1989. Central post-stroke pain – a study of the mechanisms through analyses of the sensory abnormalities. Pain 37, 173–185.

Bolay, H. and Moskowitz, M. A. 2002. Mechanisms of pain modulation in chronic syndromes. Neurology 59, S2–S7.

Bowsher, D. 1996. Central pain: clinical and physiological characteristics. J. Neurol. Neurosurg. Psychiatry 61, 62–69.

Bowsher, D., Leijon, G., and Thuomas, K. A. 1998. Central poststroke pain: correlation of MRI with clinical pain characteristics and sensory abnormalities. Neurology 51, 1352–1358.

Canavero, S. and Bonicalzi, V. 2004. Intravenous subhypnotic propofol in central pain: a double-blind, placebo-controlled, crossover study. Clin. Neuropharmacol. 27, 182–186.

Craig, A. D. 2000. The Functional Anatomy of Lamina I and Its Role in Post-Stroke Central Pain. In: Progress in Brain Research (eds. J. Sandkuhler, B. Bromm, and G. F. Gebhart), pp. 137–151. Elsevier Science.

Craig, A. D. and Bushnell, M. C. 1994. The thermal grill illusion: unmasking the burn of cold pain. Science 265, 252–255.

Dworkin, R. H., Backonja, M., Rowbotham, M. C., Allen, R. R., Argoff, C. R., Bennett, G. J., Bushnell, M. C., Farrar, J. T., Galer, B. S., Haythornthwaite, J. A., Hewitt, D. J., Loeser, J. D., Max, M. B., Saltarelli, M., Schmader, K. E., Stein, C., Thompson, D., Turk, D. C., Wallace, M. S., Watkins, L. R., and Weinstein, S. M. 2003. Advances in neuropathic pain: diagnosis, mechanisms, and treatment recommendations. Arch. Neurol. 60, 1524–1534.

Finnerup, N. B., Otto, M., McQuay, H. J., Jensen, T. S., and Sindrup, S. H. 2005. Algorithm for neuropathic pain treatment: an evidence based proposal. Pain 118, 289–305.

Greenspan, J. D., Ohara, S., Sarlani, E., and Lenz, F. A. 2004. Allodynia in patients with post-stroke central pain (CPSP) studied by statistical quantitative sensory testing within individuals. Pain 109, 357–366.

Hains, B. C., Klein, J. P., Saab, C. Y., Craner, M. J., Black, J. A., and Waxman, S. G. 2003. Upregulation of sodium channel Nav1.3 and functional involvement in neuronal hyperexcitability associated with central neuropathic pain after spinal cord injury. J. Neurosci. 23, 8881–8892.

Head, H. and Holmes, G. 1911. Sensory disturbances from cerebral lesions. Brain 34, 102–254.

Jensen, T. S. and Baron, R. 2003. Translation of symptoms and signs into mechanisms in neuropathic pain. Pain 102, 1–8.

Jensen, T. S. and Lenz, F. A. 1995. Central post-stroke pain: a challenge for the scientist and the clinician [editorial]. Pain 61, 161–164.

Jensen, T. S., Gottrup, H., Kasch, H., Nikolajsen, L., Terkelsen, A. J., and Witting, N. 2001a. Has basic research contributed to chronic pain treatment? Acta Anaesthesiol. Scand. 45, 1128–1135.

Jensen, T. S., Gottrup, H., Sindrup, S. H., and Bach, F. W. 2001b. The clinical picture of neuropathic pain. Eur. J. Pharmacol. 429, 1–11.

Kim, J. S. 2003. Central post-stroke pain or paresthesia in lenticulocapsular hemorrhages. Neurology 61, 679–682.

Kim, J. S. and Choi-Kwon, S. 1999. Sensory sequelae of medullary infarction, differences between lateral and medial medullary syndrome. Stroke 30, 2697–2703.

Koltzenburg, M. and Scadding, J. 2001. Neuropathic pain. Curr. Opin. Neurol. 14, 641–647.

Lampl, C., Yazdi, K., and Roper, C. 2002. Amitriptyline in the prophylaxis of central poststroke pain. Preliminary results of 39 patients in a placebo-controlled, long-term study. Stroke 33, 3030–3032.

Leijon, G. and Boivie, J. 1989. Central post-stroke pain – a controlled trial of amitriptyline and carbamazepine. Pain 36, 27–36.

Leijon, G., Boivie, J., and Johansson, I. 1989. Central post-stroke pain – neurological symptoms and pain characteristics. Pain 36, 13–25.

Levendoglu, F., Ogun, C. O., Ozerbil, O., Ogun, T. C., and Ugurlu, H. 2004. Gabapentin is a first line drug for the treatment of neuropathic pain in spinal cord injury. Spine 29, 743–751.

MacGowan, D. J., Janal, M. N., Clark, W. C., Wharton, R. N., Lazar, R. M., Sacco, R. L., and Mohr, J. P. 1997. Central poststroke pain and Wallenberg's lateral medullary infarction: frequency, character, and determinants in 63 patients. Neurology 49, 120–125.

Mailis, A. and Bennett, G. J. 2002. Dissociation between cutaneous and deep sensibility in central post-stroke pain (CPSP). Pain 98, 331–334.

Nathan, P. W., Smith, M., and Deacon, P. 2001. The crossing of the spinothalamic tract. Brain 124, 793–803.

Rowbotham, M. C., Twilling, L., Davies, P. S., Reisner, L., Taylor, K., and Mohr, D. 2003. Oral opioid therapy for chronic peripheral and central neuropathic pain. N. Engl. J. Med. 348, 1223–1232.

Schott, G. D. 1995. From thalamic syndrome to central poststroke pain. J. Neurol. Neurosurg. Psychiatry 61, 560–564.

Sindrup, S. H., Otto, M., Finnerup, N. B., and Jensen, T. S. 2005. Antidepressants in the treatment of neuropathic pain. Basic Clin. Pharmacol. Toxicol. 96, 399–409.

Tasker, R. R., de Corvallho, G., and Dostrovsky, J. O. 1991. The History of Central Pain Syndromes, with Observations Concerning Pathophysiology and Treatment. In: Pain and Central Nervous System Disease: The Central Pain Syndromes (ed. K. L. Casey), pp. 31–58. Raven Press.

Vestergaard, K., Andersen, G., Gottrup, H., Kristensen, B. T., and Jensen, T. S. 2001. Lamotrigine for central poststroke pain: a randomized controlled trial. Neurology 56, 184–190.

Vestergaard, K., Andersen, G., and Jensen, T. S. 1996. Treatment of central post-stroke pain with a selective serotonin reuptake inhibitor. Clin. J. Pain 3(Suppl. 5), 169.

Vestergaard, K., Nielsen, J., Andersen, G., Ingeman-Nielsen, M., Arendt-Nielsen, L., and Jensen, T. S. 1995. Sensory abnormalities in consecutive, unselected patients with central post-stroke pain. Pain 61, 177–186.

Waxman, S. G., Cummins, T. R., Dib-Hajj, S. D., and Black, J. A. 2000. Voltage-gated sodium channels and the molecular pathogenesis of pain: a review. J. Rehabil. Res. Dev. 37, 517–528.

Willis, W. D., Jr., Zhang, X., Honda, C. N., and Giesler, G. J., Jr. 2002. A critical review of the role of the proposed VMpo nucleus in pain. J. Pain 3, 79–94.

Woolf, C. J. and Mannion, R. J. 1999. Neuropathic pain: aetiology, symptoms, mechanisms, and management. Lancet 353, 1959–1964.

Woolf, C. J. and Max, M. B. 2001. Mechanism-based pain diagnosis: issues for analgesic drug development. Anesthesiology 95, 241–249.

Woolf, C. J. and Salter, M. W. 2000. Neuronal plasticity: increasing the gain in pain. Science 288, 1765–1769.

Further Reading

Loeser, J. D., Max, M. B., Saltarelli, M., Schmader, K. E., Stein, C., Thompson, D., Turk, D. C., Wallace, M. S., Watkins, L. R., and Weinstein, S. M. 2003. Advances in neuropathic pain: diagnosis, mechanisms, and treatment recommendations. Arch. Neurol. 60, 1524–1534.

61 Psychophysics of Pain

R H Gracely, University of Michigan Health System, VAMC, Ann Arbor, MI, USA

E Eliav, UMDNJ-New Jersey Dental School, Newark, NJ, USA

Published by Elsevier Inc.

61.1 Measuring the Single Dimension of Pain in the Laboratory and in the Clinic

To the layman, the pain threshold is often synonymous with the concept of pain sensitivity. It is used to describe individual differences, such as, "she has a high pain threshold." In pain research, the threshold describes a distinction between sensory qualities of nonpainful and painful. As the magnitude of a noxious stimulus is increased, the pain threshold marks the transition from the absence of pain sensation to the presence of pain sensation, and is quantified as the amount of stimulus intensity needed to evoke a painful sensation.

In the clinic, gross changes in pain threshold can be assessed in a qualitative way by administering a standard stimulus (e.g., safety pin, manual palpation) and by asking if the evoked sensation is painful (e.g., "does this hurt?"). More precise, quantitative methods increase the intensity of the stimulus until the evoked sensation is described as painful. The amount of stimulus intensity at this point, or the average from a number of trials, is the pain threshold. Clinical examples of this method are the determination of the pain pressure threshold (PPT) to blunt pressure by mechanical or electrical dolorimeters for syndromes such as temporomandibular disorders or fibromyalgia (Petzke, F. *et al.*, 2003a; 2003b).

61.1.1 Measuring Pain in the Laboratory

In the laboratory, the pain threshold is determined by more extensive psychophysical procedures that are designed to control for a number of biases that may influence threshold assessments. These share a number of common features. They present a series of stimuli of varying intensity in a random sequence, or increasing from a randomly varied starting point. Subjects judge and rate the sensations evoked by each stimulus, and the different paradigms are designed to collect fresh judgments of the evoked sensations while minimizing extraneous information about stimulus intensity. They assume that the threshold is not succinct, but rather spans a range of stimulation in which painfulness is uncertain.

These laboratory methods include the classical Method of Limits, which uses alternating ascending and descending series of continuous or discrete stimuli, and the Method of Constant Stimuli that presents discrete stimuli of varying intensities that span the range from certain judgments of nonpainful to certain judgments of painful. Figure 1(a) shows an example of the method of limits. This method is usually modified for pain evaluation by using only ascending series to avoid the possibility of unacceptably painful stimulation at the beginning of a descending series. Biases are controlled by varying the starting point or step size, and traditionally in studies of nonpainful stimulation, the use of ascending and descending series to control for directional errors such as anticipation, of indicating the presence of pain before the sensation becomes painful, and errors of habituation, of continuing to use the nonpainful responses after the stimulus has become painful (Engen, T., 1971a). Figure 1(b) shows a histogram of a sequence of random stimuli that span the gray area between certain responding and an example of a probability function generated by this procedure. In addition to these classical methods, Figure 1(c) shows the logic of signal detection theory or sensory decision theory (SDT) evaluation,

which can be conceptualized in sensory detection as a 2×2 table of the four possibilities of whether a particular stimulus value was delivered on a trial (versus delivery of nothing, termed a blank stimulus) and whether a sensation was detected or not. The common SDT terms for the four possibilities are a hit (correctly identifying a delivered stimulus), a correct rejection (correctly indicating that no stimulus was given) and the two errors of a miss (not identifying a stimulus when it was given), and a false alarm, indicating that a stimulus was given when it was not. The reader can appreciate that this 2×2 table is common in statistics, where the correct responses are referred to as accepting or rejecting the null hypothesis, and the errors are referred to as the α or type I error (labeling an effect as present when it is not, corresponding to a false alarm), and β or type II error (labeling something as absent when it is present, corresponding to a miss). Probably the most self-explanatory labels for the four outcomes are true positive, true negative, false-positive, and false-negative. While the classical methods consider the concept of a threshold to be fuzzy, and therefore best described by a mean or probability, SDT extends the concept to include two variables. One is a measure of sensitivity of the system, directly analogous to a statistic such as t or Z, which can be thought of as a distance between determinations of no detection and detection (or no pain and pain) divided by the variability of these judgments. The second refers to the labeling behavior, and is directly related to the α level (such as $p < 0.05$) arbitrarily set by an experimenter in statistics. It describes the willingness to use specific responses, and it can be liberal or conservative. As in statistics, this behavior divides the gray area of indecision into two types of error. For a given sensitivity, changing this behavior decreases one error only at the expense of increasing the other. For example, adapting a liberal criterion and calling uncertain sensations painful increases the false alarm error (or α, type I, or false-positive) while decreasing the miss error (or β, type II, or false-negative).

One of the first applications of SDT was to assess the performance of radar operators (Swets, J. A., 1964). In this context, the sensitivity parameter d' describes the overall sensitivity of the system, which includes the visual acuity of the human operator and the physical capabilities of the radar receiver. The response criterion parameter is set by the operator and in this case determined by the demand characteristics. During wartime, operators must adapt a liberal criterion to label any suspicious blip as an enemy plane. This vigilant strategy minimizes the misses at the expense of increased false alarms.

Other situations such as capital punishment must adapt the opposite conservative criterion, to be as sure as possible of a hit, minimizing false alarms at the expense of increased misses.

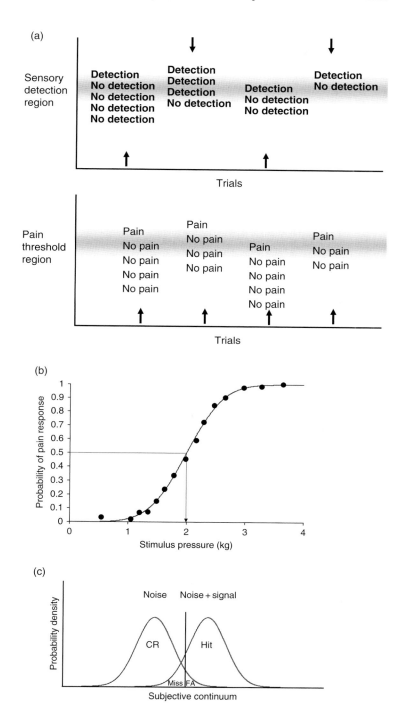

Figure 1 (Continued)

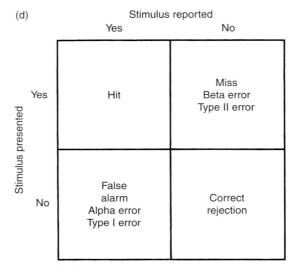

Figure 1 (a) The method of limits. Stimulus intensity is increased in successive discrete presentations on trial 1 until a positive response (yes) is made. Intensity is decreased on trial 2 until a negative response (no) is made. Several trials are run with varied initial starting intensities. The threshold is defined as the mean of the response transitions for each trial. A common modification of the method of limits to measure pain threshold delivers only ascending series to avoid excessively painful stimulation. (b and c) The method of constant stimuli. (b) Shows a range of stimulus intensities about the threshold that are presented in random sequence. (c) Shows the probability of a positive response over stimulus intensity. The result typically is an ogive shape (cumulative normal distribution) and the threshold is defined as the stimulus intensity corresponding to a specific response probability (in this case 0.5). This method emphasizes that the transition between no sensation, nonpainful sensation, and pain sensation shown in (a) is not distinct and varies over trials. (c and d) The logic of signal detection theory (SDT). (c) Shows the sensory distributions of a blank stimulus and an actual stimulus. Because of noise in the system, these distributions overlap. (d) Shows the four possibilities of signal presence and response choice. (d) This table is analyzed in terms of discrimination sensitivity and the relative frequency of the errors of false alarms or misses.

SDT methods are easily implemented by including the presentation of blank stimuli to determine the rate of false-positive errors. In addition to SDT, the robust method of two alternative forced choice (2ALFC) controls the labeling behavior rather than assessing it, while providing a measure of sensitivity similar to that provided by SDT. The 2ALFC is implemented by presenting two stimuli at the same time in different places (e.g., left and right hands) and asking which was the true stimulus or which had the greater sensation. The same method can also be used by applying two stimuli to the same (or different) place at different times and asking whether the first or second stimulus period contained the stimulus or had a greater sensation.

There are a number of issues that have been raised about the pain threshold, which assesses pain sensitivity only at the very bottom of the pain range. One issue is whether SDT methods are superior to the classical procedures. This is also an issue for suprathreshold scaling and this issue is presented below in the suprathreshold context. SDT methods usually require more trials than conventional procedures, which may be problematic if these trials are painful.

Another issue is whether the pain threshold predicts suprathreshold sensitivity. Figure 2 shows that the answer is "not always." In the clinical condition of

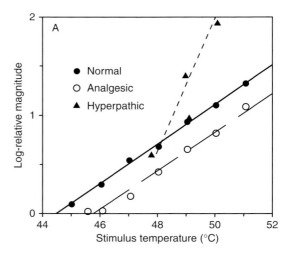

Figure 2 Pain thresholds may not indicate suprathreshold sensitivity. The triangles show the type of function that could be obtained in the clinical condition of hyperpathia in which the threshold is raised and suprathreshold sensitivity is increased.

hyperpathia, an elevated pain threshold (decreased pain sensitivity) is combined with significant suprathreshold sensitivity. Increasing the intensity of the stimulus does not evoke a sensation at usual levels and an explosive experience of pain at higher levels. In cases in which the pain threshold is predictive of suprathreshold sensitivity, an issue is whether it is more or less sensitive, and more or less responsive to pain interventions.

Another issue is the presence of prepain sensation. In studies of nonpainful modalities, the sensation is either not present or present, as shown in Figure 1(c). In pain studies, there are usually three alternatives: nothing, nonpainful sensation, and painful sensation. This leads to some operational difficulties with methods such as SDT, and changes detection tasks (with the control provided by blank stimulation) to rating tasks described below.

61.1.2 Measuring Pain in the Clinic

The requirements of clinical and laboratory assessments are usually quite different. Because of the limitations of time, the need to test multiple areas with multiple modalities, and the gross nature of many clinical abnormalities, clinical methods are necessarily brief and thus devoid of the protection against bias contained in the classical and more modern methods. This may not be an important issue in specific situations, but can significantly influence the outcome, especially when the comparisons are between individuals.

These issues of clinical versus laboratory methods were addressed in a recent study by Petzke F. et al. (2003a) that compared the influence of psychological distress on clinical and laboratory measures of the

PPT in fibromyalgia. This study compared the methods described above. The qualitative clinical method is the manual tender point count, which applies a standard stimulus (4 kg of thumb pressure) to nine bilateral pairs of defined tender points defined by the American College Rheumatology (1990). The tenderness part of the diagnosis of fibromyalgia is satisfied if the standard stimulus evokes pain in at least 11 of the 18 tender point sites. The reader can appreciate that patients with fibromyalgia know about these points and criteria, and know what to say if they want to satisfy this aspect of the diagnosis.

The quantitative clinical measure used a standard commercial dolorimeter at the same 18 tender point sites, slowly increasing the pressure of a 1 cm diameter probe until the sensation is reported to be painful. The laboratory procedure used the multiple random staircase (MRS) method (see below) that uses a computerized interactive logic to determine the PTT and also pain sensitivity at more intense levels between pain threshold and tolerance (Petzke, F. et al., 2003b).

Since one theory of tenderness in fibromyalgia is that it is the extreme tail of a normal distribution of tenderness, Petzke F. et al. (2003a) composed a group of 76 subjects that included both healthy controls and patients with fibromyalgia with a normally distributed range of PPTs determined by a dolorimeter. Table 1 shows the association of the different measures of pressure pain sensitivity with measures of psychological distress. The manual tender point count is significantly associated (23%) with distress, the more quantitative clinical dolorimeter is significantly but less associated (15% at tender points, 12% at thumb) with distress, while the random psychophysical method applied to the thumb is insignificantly associated with distress (4%).

Table 1 Results of regression analysis[a]

	R^2	B	SE	Significance
MTPC	0.23	0.24	0.06	0.001
DM-PT	0.15	−0.02	0.009	0.008
DM-THU	0.12	−0.08	0.033	0.019
RAN	0.04	0.59	0.42	0.169

[a]R-square, regression coefficient B, standard error of B, significance , for four dependent measures of tenderness measures with distress (GSI) as independent variable. MTPC is the manual tender point count, defined as the number of 18 predefined sites at which 4 kg of manual thumb pressure evokes pain. DM-PT is the mean of dolorimeter measures of the pain pressure threshold obtained at the 18 tender point sites. DM-THU is the dolorimeter-derived pain pressure threshold obtained at the left thumbnail. RAN is the pressure pain sensitivity at the left thumbnail determined by the multiple random staircase method. Adapted from Petzke, Petzke, F., Gracely, R. H., Park, K. M., Ambrose, K., and Clauw, D. J. 2003b. What do tender points measure? Influence of distress on 4 measures of tenderness. J. Rheumatol. 30, 567–574.

61.1.2.1 *Quantitative sensory testing methods*

The influence of distress may be less of an issue in other clinical situations. In cases in which multiple determinations are made in a single individual and compared between sites, trait characteristics of individual response styles may cancel out. In addition, in many cases the abnormalities that are the target of the evaluations are so dramatic that they can be accurately assessed by a variety of methods. Indeed, simple threshold measures form the core of a set of procedures referred to as quantitative sensory testing (QST). After a considerable history in which clinicians applied pinpricks to patients in the clinic and psychologists applied electrical or thermal stimuli to healthy students in the laboratory, QST bridged the gap, applying an increasing number of stimulus methods and modalities to patients with a variety of pain complaints. The section below briefly summarizes these methods, and provides a few examples of studies in which pain psychophysical threshold procedures have elucidated mechanisms and aided diagnosis. It focuses on the pain in pain psychophysics.

The goal of sensory assessment is to obtain information from the neural system, which may help in diagnosis and treatment of disorders in the nervous system itself and in surrounding tissue. Traditionally, this examination included the response to a brush, a cotton swab, a warm or a cold object, and a pinprick (Hansson, P., 2002). QST extends these methods.

For clinical applications, the methods can be organized into a protocol that is applied routinely to all patients. Determination of an abnormality can be made on the basis of comparison to a set of standard control values, although comparison to a contralateral unaffected site in the same patient may be the best approach when available (Kemler, M. A. *et al.*, 2000).

Common stimulus modalities used to assess the function of specific nerve fibers are:

1. Touch, electrical, and vibratory stimuli for the evaluation of the function of the thickly myelinated $A\beta$ fibers.
2. Cold detection threshold for the assessment of thinly myelinated $A\delta$ fibers.
3. Heat detection threshold for the assessment of thin unmyelinated C-fibers.

Touch and vibration sensitivity is usually determined by the detection and pain thresholds to both calibrated monofilaments and a vibrating probe. Four different thermal parameters (thresholds to warm and cold stimuli, heat pain, and cold pain) are typically assessed by Peltier element-based thermal stimulators. The cold pain threshold is a specific test for cold allodynia, a common finding in conditions involving nerve injury. Altered heat pain thresholds can reflect either the function of myelinated $A\delta$ fibers or unmyelinated C-fibers, depending on the rate, location, and sequence of stimulation (Collins, W. R. *et al.*, 1960; Price, D. D. *et al.*, 1989; Andersen, O. K. *et al.*, 1995; Koltzenburg, M., 1997).

Tests of the area of abnormality assess the threshold for sensory detection, and usually a higher threshold for pain. In some cases both may be altered, or the sensory detection threshold may be absent, and stimulation evokes only pain. The profile of altered sensory and pain thresholds across mechanical and thermal modalities provides information about the type of nerve abnormality, and can be supplemented by additional tests that examine pain processing at levels above the pain threshold (Hansson, P., 2002). Since it is important to ascertain whether the distribution of sensory abnormalities match the damaged nerve's dermatome, the borders of sensory dysfunction area should be carefully determined using the different modalities. The finding of extraterritorial pain that extends beyond the border of a dermatome is discussed below.

The response to electrical stimulation is usually not included in standard QST assessments, and several studies have found conflicting and inconsistent results using electrical tests (Hagberg, C. *et al.*, 1990; Vecchiet, L. *et al.*, 1991; Bendtsen, L. *et al.*, 1997). However, recent studies demonstrate a number of unique features of electrical stimulation. Unlike the other methods, which naturally stimulate nerve receptors, electrical stimuli may bypass the receptor to stimulate the axon of the primary afferent. Due to this property, electrical stimulation does not show the same temporal profile of sensitization, suppression, or fatigue as found in natural stimulation of receptors. In addition, at the threshold for detection, electrical stimuli exclusively activate the thickly myelinated $A\beta$ fibers. Thus, a comparison of the detection threshold to both mechanical and electrical stimuli can provide a differential method that isolates receptor and postreceptor processes of $A\beta$ fibers. Changes in both or in only electrical detection indicate a postreceptor process, while changes in only the results of mechanical stimulation indicate a receptor process. Thus, altered electrical sensitivity alone indicates a postreceptor process in large myelinated fibers. A single electrical test can identify a number of central processes, including spinal sensitization resulting in $A\beta$-mediated allodynia, and inflammatory processes that include the nerve in passage. One problem with electrical sensory assessment is inter-patient variability. This inconsistency can be reduced by expressing the electrical detection thresholds as the ratio between the pathological and the contral-ateral side, that is, the healthy side used as a control for the pathological side (Kemler, M. A. *et al.*, 2000).

The results of QST methods help distinguish between nociceptive and neuropathic pain, which dif-fer significantly in etiology, mechanism, and effective treatment. Nociceptive pain is the normal condition in which actual or impending tissue damage is sensed by the receptor organs, nociceptors, resulting in a range of normal pain conditions that may be treated with conventional (e.g., NSAID, opioids) analgesics. Neuropathic pain describes a special condition in which the sensing neural apparatus is damaged, lead-ing to a variety of pain sensations with unnatural characteristics of quality and distribution. Neuropathic pain can be particularly difficult to treat, and successful agents include drugs such as anticonvulsants that modify neural function.

61.1.2.2 Quantitative sensory testing signatures of pathological conditions

Mechanical nerve damage or total nerve transection is characterized by myelinated and unmyelinated nerve fiber hyposensitivity reflected in elevated detection thresholds to heat, electrical, and mechan-ical stimulation (Dao, T. T. and Mellor, A., 1998). Partial damage may be followed by either hypo- or hypersensitivity accompanied by ongoing neuro-pathic pain (Benoliel, R. *et al.*, 2001; 2002).

In contrast to the neuropathic process of mechan-ical nerve damage, specific nociceptive processes may provide a different, identifiable sensory signa-ture. For example, early perineural inflammation produces short-lasting large myelinated nerve fiber hypersensitivity that results in reduced detection to electrical and mechanical stimuli. This increased detection sensitivity has been demonstrated in clin-ical studies and in animal spinal nerve models (Eliav, E. and Gracely, R. H., 1998; Eliav, E. *et al.*, 1999; Chacur, M. *et al.*, 2001; Eliav, E. *et al.*, 2001; 2003) and reproduced in a model of inflammatory trigem-inal nerve neuropathy (Benoliel, R. *et al.*, 2002).

Although the mechanisms are not clear, muscle-induced pain may result in the opposite effect, that is, reduced large nerve fiber sensitivity reflected by ele-vated electrical or mechanical detection thresholds (Stohler, C. S. *et al.*, 2001; Eliav, E. *et al.*, 2003). In the orofacial region, there is both experimental and clin-ical evidence for this effect. Following injection of

hypertonic saline to the masseter muscle, mechanical stimuli applied into the overlying skin induced signif-icant increases in the verbal rating score (Svensson, P. *et al.*, 1998), and lowered pain thresholds to this type of stimulation have been found in patients with orofacial myofascial pain (Svensson, P. *et al.*, 1998; 2001).

Both neuropathic and intense nociceptive pain can initiate and maintain a condition of spinal central sensitization (CS). CS is characterized by sponta-neous pain and evoked pain abnormalities that can cross dermatomes. One striking abnormality is dynamic mechanical allodynia, pain to light brushing of the skin. Theoretically, this can be mediated by either sensitized pain fibers or a functional change in large-diameter Aβ fibers that usually only mediate the sense of touch. Mechanical stimulation cannot distinguish between these alternatives, but recent evidence shows that a ratio between the electrical pain and the detection threshold of <2.0 may indicate altered central nervous system processing of Aβ fiber input (Sang, C. N. *et al.*, 2003). This result is strong evidence of altered central pain modulation and CS.

Combining the above procedures into one meth-odological approach provides a clinical decision-making tool that facilitates diagnosis and choice of treatment. The following section illustrates how these tests have helped identify the underlying mechanisms of three painful syndromes.

61.1.2.3 Clinical studies

61.1.2.3.(i) Burning mouth syndrome Burning mouth syndrome (BMS) is a putative centrally mediated pain that is poorly understood and treated. BMS is diagnosed only when the burning sensation is not associated with a local or systemic pathology. In contrast, burning mouth symptoms (BMST) is diag-nosed in the presence of a known etiology for the altered sensation. BMS is an intraoral disorder most prevalent in postmenopausal women (Grushka, M., 1987; Grushka, M. *et al.*, 2002), characterized by a burning mucosal pain without major visible physical signs. Altered taste sensations have long been asso-ciated with BMS and nearly 70% of the patients complain of accompanying dysguesia (altered taste sensation) (Ship, J. A. *et al.*, 1995). Conventional QST in BMS patients identified defects in pain tolerance, altered chemosensory function, increased pain threshold to laser stimulation, and hypoesthesia of large and small nerves (Grushka, M. and Sessle, B., 1988; Grushka, M. and Sessle, B. J., 1991; Mott, A. E. *et al.*, 1993; Formaker, B. K. and Frank, M. E., 2000). Accumulating evidence suggests that BMS involves central and peripheral nervous system pathologies induced by the damage to the taste system at the level of the chorda tympani nerve. This damage results in reduced trigeminal inhibition that in turn leads to an intensified response to oral irritants and eventually to oral phantom pain (i.e., BMS) (Grushka, M. *et al.*, 2003).

Recent studies of BMS show how electrical tests can enhance a QST examination. Electrical stimulation of the tongue can provoke two different sensations; one is described as itch (tingling) and the other as an electrical taste (Bujas, Z. *et al.*, 1979; Lindemann, B., 1996). The electrical taste threshold in the tongue is easily recognized as a sensation usually described as a battery-like or sour taste. The taste sensation is assumed to be conducted via the chorda tympani nerve and the itch sensation via the lingual nerve. Thus electrical detection thresholds of the anterior two thirds of the tongue likely evaluate the role of chorda tympani and lingual nerve function in BMS.

Eliav E. *et al.* (2007) found no difference between electrical taste and electrical detection thresholds among 23 BMS patients, 14 BMS patients, or 10 asymptomatic controls at extraoral nerve mental and infraorbital control sites. In the control and BMST patients, the ratio between the electrical taste and the itch sensation on the tongue was <0.7, while this ratio was significantly higher (1.4) in the BMS patients. This result is consistent with the hypothesis that BMS is an oral phantom-type pain induced by damage to the taste system (Grushka, M. *et al.*, 2003), and suggests that the electrical taste and itch ratio may be a useful diagnostic test for BMS.

61.1.2.3.(ii) Painful temporomandibular disorder Painful temporomandibular disorder (TMD) can be due to arthralgia in the temporomandibular joint or from processes in nearby muscle. Eliav E. *et al.* (2003) classified a cohort of patients as suffering from joint arthralgia ($n = 44$) or myalgia ($n = 28$). Electrical thresholds for Aβ fibers and heat thresholds for C-fibers were assessed in these patients and in 33 healthy control subjects. These tests were performed bilaterally in three trigeminal nerve sites: the auriculotemporal nerve territory, the buccal nerve territory, and the mental nerve territory.

Electrical detection threshold ratios were calculated by dividing the affected side by the control side; thus reduced ratios indicate hypersensitivity of the affected side. In control patients, ratios obtained at all sites did not vary significantly from the expected value of one. In arthralgia patients, the mean ratio

obtained for the auriculotemporal nerve territory (0.63) was significantly lower than the ratio for the mental nerve (1.02) and buccal nerve (0.96) territories in patients and lower than the auriculotemporal nerve ratios in myalgia (1.27) and control subjects (1.00). In the myalgia group, the electrical detection threshold ratios in the auriculotemporal nerve territory were significantly elevated compared to the auriculotemporal nerve ratios in control subjects. There were no significant differences between and within the groups for electrical detection threshold ratios in the buccal and the mental nerve territories, and for the heat detection thresholds in all tested sites. Following arthrocentesis (fluid draining) delivered to 10 arthralgia patients, mean electrical detection threshold ratios in the auriculotemporal nerve territory were significantly elevated from 0.64 to 0.99, indicating resolution of the hypersensitivity. Thus, electrical stimulation distinguished between joint and muscle-related facial pain, and monitored the time course of the joint pain.

61.1.2.3.(iii) Sinusitis Sinusitis can be classified as acute (<1 month) or chronic (>3 months). The transition from acute to chronic likely involves a shift from acute inflammation to more persistent mechanisms that may involve nerve damage, changes that should be detectable by QST. In a pilot study by Eliav and coworkers (personal communication), neurosensory function was evaluated in seven patients and eight control subjects by measuring the electrical detection threshold for large myelinated nerve fibers and heat detection thresholds for the assessment of thinly myelinated nerve fibers. The sensory tests were conducted in the region of the infraorbital, supraorbital, and mental nerve distributions. Nine acute and five chronic sinusitis were diagnosed. Acute sinusitis was associated with bilateral large myelinated nerve hypersensitivity (to electrical stimulation) compared to healthy controls with no significant change in the unmyelinated nerve fiber detection threshold (to heat). Chronic sinusitis resulted in a large myelinated fiber hyposensitivity and a thin myelinated fiber bilateral hyposensitivity compared to healthy controls. These preliminary results suggest that QST may distinguish between an early inflammatory neuritis and a chronic condition likely accompanied by nerve damage.

61.1.2.3.(iv) Malignancy The ability of electrical stimulation to detect inflammation along the course of the nerve has also been applied to the detection of malignancy. Eliav E. *et al.* (2002) assessed the sensitivity of Aβ primary afferents to weak electrical currents applied bilaterally to regions innervated by three peripheral branches of the trigeminal nerve. For dermatomes containing lesions, ratios of electrical detection thresholds (affected/unaffected side) were <0.8 in 12 patients. Biopsy showed that the lesions in 10 of these 12 patients were malignant. No malignancy was found in the remaining 11 patients. These results suggest that Aβ primary afferent hypersensitivity is observed to occur in nerves exposed to early soft tissue oral malignancies, and that a simple electrical test may provide a nonivasive and inexpensive method of early detection.

61.1.3 Suprathreshold Pain Sensation

Although the pain threshold is very useful in QST examinations, in some cases it may be necessary to assess pain sensitivity at higher, suprathreshold levels. As shown in Figure 2, in conditions such as hyperpathia the pain threshold may not predict suprathreshold sensitivity. In this case the psychophysical function rotates; decreased sensitivity at threshold levels is accompanied by increased sensitivity at suprathreshold levels. In addition, suprathreshold pain stimulation is important for goals outside of QST testing in which the painful stimulus is designed to simulate some aspects of spontaneous pain. In the clinical situation, threshold-level pain sensations are less relevant and the focus is on treatment of intense acute pain following injury or medical procedures, and the management of relentless chronic pain.

The psychophysical scaling techniques used to evaluate suprathreshold sensations are reviewed in a number of sources (Torgerson, W. S., 1958; Bock, R. D. and Jones, L. V., 1968; Engen, T., 1971b; Anderson, N. H. *et al.*, 1974; Marks, L. E., 1974; Stevens, S. S., 1975; Gescheider, G. A., 1985; Falmagne, J. C., 1986; Bolanowski, S. J. and Gescheider, G. A., 1991; Atkinson, R. C. *et al.*, 1998). In terms of the task performed by the subjects, these methods can be divided into three classes: discrete numerical or verbal category scales, bounded or confined continuous-measure equivalents of bounded scales such as the 10 cm visual analog scale (VAS), and unbounded scales such as the well-known method of magnitude estimation (using any number).

These methods also differ in analysis, with some common rating tasks analyzed by distinctly different procedures. For example, discrete numerical or verbal category scales can be analyzed by considering the distances between categories to be psychologically equal (formally titled as the method of equal appearing intervals (Engen, E., 1971b). Many pain studies have used discrete category scales such as the 0–10 numerical scale (Brennum, J. *et al.*, 1992), the four-point category scales of pain (none, mild, moderate, severe) or pain relief (none, some, lots, complete) (Gracely, R. H. and Naliboff, B. D., 1996). The most widely used McGill Pain Questionnaire (MPQ) (Melzack, R., 1975) is usually analyzed by assigning integer ranks to each category subscale.

The actual category values can be determined by the response behavior of the subjects as they use the scale. Thurstone L. I. (1959) developed a number of methods that use this approach, in which the spacing of two categories is inversely related to the extent that they are confused, such as the relative proportion that each response is used to describe the same stimulus intensity. The method of successive categories (Thurstone, L. I., 1959; Engen, E., 1971b) is applied to category scaling data and a related analysis is used for the simpler task of paired comparisons, in which subjects are presented a series of stimulus pairs and they indicate which of the pair is more intense. Thurstone analyses are rarely applied to pain assessment (LaMotte, R. H. and Campbell, J. N., 1978), but could be applied to many more studies currently using the method of equal appearing intervals to analyze categorical data. A comparison of the two methods can actually be performed post hoc, either validating the assumption of equal intervals or providing an alternative that more closely resembles the values of the category responses.

Methods that use response distributions can also be applied to clinical pain ratings, when the underlying stimulus intensities are not known. An example is Rasch analysis, which has been applied to the assessment of low back pain. The results can be used for item weighting or for item selection during further scale development (McArthur, D. J. *et al.*, 1991).

A third class of methods also determine the appropriate value for each category response. However, in contrast to the Thurstone methods of determining these values simultaneous with scaling, these methods quantify category values in a separate procedure. The scales can be used for both experimental and clinical pain assessment. A notable example of this approach is the MPQ. The 2–6 items within each of the 20 subscales were quantified during the development of the scale (Melzack, R. and Torgerson, W. S., 1971). One of the most sophisticated examples of determining category scale values was performed by Tursky B. (1976), who requantified some of the MPQ items using a ratio-scaling paradigm advocated by Stevens S. S. (1975). The details of this method are described in Section 61.1.3.3.

Once category values are quantified, the issue is how these categories should be used to measure pain. Categorical scales can be presented in ordered sequences as a simple checklist such as the MPQ, or be placed at the appropriate locations on a spatial continuum to form a graphic rating scale (Heft, M. W. and Parker, S. R., 1984; Gracely, R. H., 1991). This continuum can be some type of VAS discussed below (Lembo, A. et al., 2000). Figure 3 shows an example of an alternative format in which descriptors are placed at the appropriate locations along a higher-resolution numerical category scale. This type of scale is useful in situations in which the subjects cannot mark a line, such as in neuroimaging experiments or in telephone assessments.

A novel method presents the categories in random order and subjects rate experimental or clinical pain magnitude by choosing a category from the randomized list (Gracely, R. H. et al., 1978a). This method is unique since it forces category choices based on the semantic content of the item rather than on its spatial location in a list. This method has two prominent advantages. First, basing choices on the meaning of a word or phrase may help subjects distinguish between different dimensions of pain experience (see below). Second, choices from a randomized list may be free of the rating biases associated with choosing responses from a bounded scale. These biases relate to the subject's tendency to spread repeated responses evenly over a category scale. This tendency results in scale responses that are influenced by the distribution of delivered stimulus intensities, including range, spacing, and frequency of presentation (Parducci, A., 1974). Varying these parameters distorts the resultant scale. Methodological improvements may reduce the effects of these distortions, such as iterative scaling techniques that systematically eliminate distortions (Pollack, I., 1965), and the use of open-ended scales described below (Stevens, S. S., 1975). The randomized descriptor method is likely the best improvement, since it eliminates the scaling space that forms the basis for the response spreading. The disadvantages of the random method is that resolution is limited to the number of categories (in contrast to a graphic rating scale) and that the task may be more difficult than scaling with an ordered set of categories.

Figure 3 Box scale of pain intensity. This scale combines a spatial range of verbal descriptors previously quantified and position on a logarithmic scale adjacent to a 0–20 category numerical category scale. Subjects use the descriptor space to determine an appropriate amount of intensity and to then choose the nearest numbered box to express this magnitude of pain intensity. This method increases the resolution beyond the number of descriptors, allows for semantic gaps such as mild to moderate, adds additional scaling space at the top, is used also for a scale of unpleasantness, does not require a motor response, and can be used for applications such as telephone surveys.

Although an undistorted scale is desirable for pain measurement, applied studies of pain may be more concerned with how these response behaviors adversely affect the evaluation of changes in pain sensitivity such as the effect of analgesic interventions. In this situation, the tendency to produce a uniform response distribution tends to diminish the response to a real pain treatment effect, reducing assay sensitivity. This is one of several normalizing effects; others include regression to the mean and the physiological effect termed the law of initial values, together forming a group of effects informally called floor-ceiling effects. For example, study using a 0–6 category scale showed a reduced response range immediately after a pain-reducing intervention, which quickly spread to the previous response distribution, diminishing the effect. In contrast, a randomized verbal descriptor scale, with no spatial response range, showed a persistent reduction (Gracely, R. H. et al., 1984).

61.1.3.1 Advantages of multiple category items

In contrast to scaling methods that average a large number of ratings, many clinical pain scales are based on a single score or only a few scores. One advantage of verbal categories is that they describe different magnitudes of similar or different pain dimensions, and a scale can consist of a number of items. Both the long (Melzack, R., 1975) and the short (Melzack, R., 1987) forms of the MPQ take advantage of this property. The use of composite scores from a number of items decreases the influence of random scaling errors. For example, the reliability and homogeneity of the overall score of the descriptor differential scale (DDS) is greater than the mean score obtained from each individual item (Gracely, R. H. and Kwilosz, D. M., 1988). This improvement obtained from the use of multiple items at a single testing time is similar to the improvement noted with repeated use of single items; averages of multiple responses to a single-item pain diary are superior to single responses or to the mean of only a few responses (Jensen, M. P. and McFarland, C. A., 1993). Multiple measures delivered at the same time must use different items, a requirement naturally filled by the use of verbal categories. The DDS uses this approach by presenting 12 quantified descriptors and instructing subjects to rate if their clinical pain is equal to that implied descriptors, or how much less or greater on a $+10$ to -10 scale. The DDS has been administered to patients in acute or chronic pain in several studies, revealing adequate reliability, internal consistency, validity, and sensitivity (Hargreaves, K. M. et al., 1986; Kwilosz, D. M. et al., 1986; Gracely, R. H. and Kwilosz, D. M., 1988; Doctor, J. N. et al., 1995).

In these applications, the different categories must describe different amounts or aspects of pain, but do not have to be quantified. Quantifying these verbal categories provides additional advantages. Comparisons between the pattern of item responses and the previously determined descriptor values provide a measure of scaling consistency. The DDS

shows high consistency that improves from the first to the second administration (Gracely, R. H. and Kwilosz, D. M., 1988). These consistency measures can be used at screening, or over the course of a study to identify subjects who obviously are not attending to the scaling task. Eliminating these subjects may improve psychometric properties and increase the sensitivity of an analgesic assay.

The use of multiple, quantified items also allows the construction of alternative forms, each with different descriptor items, that are theoretically equivalent. Kwilosz D. M. et al. (1984) validated such forms and used them to evaluate the accuracy of memory of acute postsurgical pain over a 1-week period. The use of alternative forms is critical in such studies. Without alternative forms, memory for pain cannot be distinguished from an alternative explanation, that it represents only the memory of a previous pain rating.

Finally, the methods used to quantify verbal categories can provide information about the stability of the meanings attached to the words, and the consistency for these meanings over the general population. The exact procedures used for the quantification are described in Section 61.1.3.3. The overall approach uses methods that reduce bias. Each subject uses two independent methods to rate the words and the comparison of these methods provides a check of internal consistency. The similarity of these scales validates both the methods and the general concept that verbal descriptors, with no physical metric of their own, can be rated as reliably as physical stimuli. After scales produced by different response methods are shown to be similar, they are combined into an overall scale of magnitude. Figure 4 shows an example of these separate and combined scales for descriptors of pain intensity (Gracely, R. H. et al., 1978a).

In addition to these measures of internal consistency, comparisons among individuals evaluate common usage. The Gracely R. H. et al. (1978a) quantification of pain sensory intensity descriptors found evidence for common usage, expressed as objectivity. Mean individual reliabilities were 0.96 over 1 week, and the mean of correlations between each individual's scale and mean scale from the remaining subjects in a group were also 0.96. The equivalence of these correlations indicates very high objectivity; an individual's quantification of the verbal descriptors was predicted equally well by that individual or by a group norm. This objectivity indicated general agreement about the ordering and spacing of English words describing magnitudes of sensory

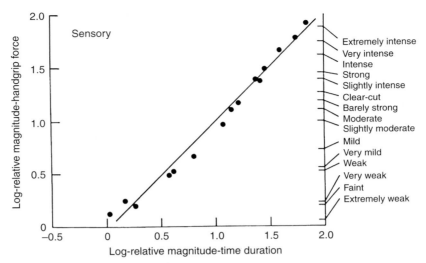

Figure 4 Relative magnitude of sensory intensity verbal descriptors derived from cross-modality matching to handgrip force (ordinate) and to time duration (abscissa). Each point is the geometric mean of 80 responses of 40 subjects to a single descriptor. The mean of the two magnitudes for each descriptor is shown on the right vertical axis. From Gracely, R. H., McGrath, P. A., and Dubner, R 1978a. Ratio scales of sensory and affective verbal pain descriptors. Pain 5, 5–18.

intensity, validating the use of a group normative scale. The results of these pain-related investigations were in general agreement with previous psychological studies of semantic meaning. For example, the adverbs very and slightly, when used to modify adjectives of pain magnitude (Gracely, R. H. and Dubner, R., 1987) or pain unpleasantness (Walther, D. J. and Gracely, R. H., 1986), appear to serve as multiplicative modifiers independent of the adjective modified (e.g., pain, distressing, unpleasantness). This property was described by Cliff N. (1959) and the constants found in that study using a Thurstone analysis of paired comparisons (1.30 for very, 0.55 for slightly) are in general agreement with those found with ratio-scaling procedures (1.44 for very, 0.42 for slightly; Gracely, R. H. and Dubner, R., 1987).

61.1.3.2 Bounded continuous scales: the visual analog scale

The VAS is likely the most widely used pain measurement tool. As shown in Figure 5, the VAS is usually a 10-cm line labeled at the ends with descriptors such as no pain and worst pain imaginable. Subjects indicate pain magnitude by marking the line, and a ruler is used to quantify the measurement on a 0- to 100-mm scale. Variations include vertical or horizontal alignments, placing descriptors along the scale, scales of different lengths, use of mechanical devices, and presentation on a computer or small

No pain   Worst pain
                 imaginable

Figure 5 A typical visual analog scale (VAS). A horizontal 10-cm line is labeled with extremes of no pain and pain as bad as it could be. Subjects indicate pain magnitude by a mark on the line. Other formats use vertical orientations, plastic slide rules, presentations on a computer screen or handheld PDA, with different or more numerous verbal labels.

personal digital assistant (PDS) device (Jensen, M. P. et al., 1986; Price, D. D., 1988; Jamison, R. N. et al., 2002).

The popularity of the VAS is due to its ease of use and evidence to support the reliability and validity of this method. It has become a standard in clinical pharmacology, a field long dominated by the consistent use of four-point categorical scales of pain intensity and pain relief. The continuous nature of the VAS provides a 0–100 response resolution, limited by the observer's discriminative capacity rather than by the scale. It also allows for increased independence of multiple responses in designs requiring repeated measures, since subjects cannot remember previous responses as well as with category scales.

Other bounded continuous scales have been applied to pain assessment, usually requiring some form of apparatus. These include squeezing a handgrip dynamometer (Gracely, R. H. et al., 1978b) using a finger span device (Cooper, B. Y. *et al.*, 1986), a mechanical slide rule (Price, D. D., 1988), a slider along the back of a meter stick (McGrath, P. A. *et al.*, 1983), and a VAS presented on a PDA (Jamison, R. N. *et al.*, 2002).

Since the VAS and similar scales are bounded, representing a confined response space, they should be vulnerable to the same spatial rating biases demonstrated with category scales. They should be sensitive to stimulus range, spacing, and frequency of presentation. Subjects may spread out their responses after an analgesic intervention, reducing sensitivity. A few studies have addressed these biases and limitations of the VAS (Fernandez, E. *et al.*, 1991; Yarnitsky, D. *et al.*, 1996), but generally these effects have not been investigated in any detail.

61.1.3.3 Unbounded scales: ratio-scaling methods

Ratio scales are defined as having both a true zero point and a defined interval, corresponding to physical scales such as length and mass. In contrast, interval scales have a defined unit but no true zero point (Fahrenheit temperature scale). Other scales are classified as ordinal, ranks with no information about distances between values, as finishing first, second, and third in a race. The use of numbers to connote identity without order, such a sport jersey numbers, is referred to as nominal. These distinctions were emphasized by the late Stevens S. S. (1975), who promoted the use of ratio scales and developed a set of methods to produce scales with ratio properties (Engen, E., 1971b; Carterette, E. C. and Friedman, M. P., 1974; Marks, L. E., 1974; Stevens, S. S., 1975; Gescheider, G. A., 1985; Falmagne, J. C., 1986; Bolanowski, S. J. and Gescheider, G. A., 1991).

The ratio-scaling methods advocated by Stevens used instructions to rate stimulus-evoked sensations in a proportional or ratio manner. In the most commonly used ratio method of magnitude estimation, subjects used any number to rate the sensations produced by a series of stimuli. The instructions emphasized ratios or proportions, such as if a subsequent stimulus-evoked sensation felt twice as intense as a previous sensation, it should be rated by a number twice that used to describe the previous sensation. Ratio scaling avoided the biases of a bounded scale described above by offering an unlimited response continuum. This is literally true for measures such as magnitude estimation (number choice) and time duration, and virtually true for brightness (13 log unit range). Other ratio responses, which can be any

adjustable continuum, may be relatively unbounded in practice but ultimately bounded due to physical (handgrip force) or safety (loudness) limitations (Gracely, R. H. and Naliboff, B. D., 1996).

The use of responses other than magnitude estimation is referred to as cross-modality matching, which describes the general case in which any adjustable response is used to match responses to any controllable stimulus modality. Magnitude estimation itself can be thought of cross-modality matching to number use.

The characteristic feature of the ratio-scaling methods advocated by Stevens S. S. (1975) is that the relationship between a particular response and a stimulus modality is described by a power function of the form $R = S^n$ in which R is the response, S the stimulus, and n the power function exponent (ignoring additive constants). This function is expressed conveniently in a logarithmic plot ($\log R = n + \log S$) since this format is linear and the exponent is equal to the slope of the function. The value of this slope, or exponent, is related to both the stimulus and the response modalities. Each modality is associated with its own specific exponent, and the theoretical exponent is the ratio of the response-specific exponent divided by the stimulus-specific exponent. Since the exponent for the very common response modality of magnitude estimation is 1, the theoretical exponent for any stimulus modality assessed by magnitude estimation would be equal to the reciprocal of the exponent of the psychophysical function, since the function exponent is equal to the response exponent (1) divided by the stimulus exponent.

The theoretical value computed from the modality-specific exponents will likely not be exactly the same since the empirical exponent is influenced by a number of other factors, including the influence of regression bias, which is a tendency to contract the response continuum, resulting in an empirical exponent always less than the theoretical exponent computed from the ratio of the response and stimulus-specific exponents. Other factors may also influence the empirical exponent. Although the open-ended nature of ratio-scaling methods such as magnitude estimation have been shown to reduce biases due to stimulus range, frequency, and spacing, there is considerable evidence that power function exponents can be influenced by factors such as stimulus range (Poulton, E. C., 1979), although the variation can be viewed as the factor that actually determines the modality-specific exponent (Teghtsoonian, R., 1971).

Ratio-scaling methods have been applied to experimental pain assessment, with resulting power function exponents of 1.0 for radiant heat, 2.1–2.2 for contact heat, and 1.8 for electrical skin stimulation (Hardy, J. D. *et al.*, 1952; Tursky, B., 1974; Gracely, R. H., 1977; Price, D. D., 1988). However, theses applications provide examples of how experimental parameters may influence the size of the exponent. For example, the stimulus range is usually restricted in pain studies to avoid excessively high levels of pain. This restriction will tend to increase the value of the exponent. Similarly, constriction of the response scale, such as using a bounded VAS scale, will usually lower the exponent by about 50%. These lowered exponents are termed virtual exponents, since they reflect both the exponent of the underlying unrestrained continuum and the compressive effect of limiting the response range (Stevens, S. S., 1975). Price and coworkers (Price, D. D. *et al.*, 1983; Price, D. D., 1988) obtained exponents for VAS measures of thermal pain intensity (2.1–2.2) that were similar to those found with line production. However, the compressive effect should have resulted in a VAS exponent that was reduced to approximately one-half of that observed for line production. Price D. D. *et al.* (1983) posit that this lack of reduced exponent was due to specific instruction to subjects, to the use of the longer (15 versus 10 cm) scales, and experience with the range of stimuli. According to the investigators, this training essentially converts the bounded VAS to a line production response. Whatever the mechanism, the results of Price D. D. *et al.* (1983) serve as an excellent example of how the power function exponent may be manipulated by subject instructions. This is a critical issue when exponents are compared across studies that differ in instructions and scale formats.

61.1.4 The Different Goals of Traditional and Pain Psychophysics

Traditional Steven's psychophysics and other scaling methods have been concerned primarily with the shape of the function relating sensory magnitude to stimulus intensity. Steven's ratio-scaling methods focus on the magnitude of the slope (in log coordinates), which describes the rate-perceived intensity increases with a given increase in stimulus intensity.

Most studies using these methods focus on the variation of slope with different stimulus modalities or response procedures. In addition, the function derived from ratio-scaling methods is compared to the functions obtained with different methods such as category scales in an effort to find the true function relating stimulus intensity to the magnitude of evoked sensation.

In contrast to the research focus of traditional psychophysical studies, pain psychophysics is necessarily concerned with the absolute level of sensation, addressing questions such as, "does this group experience greater pain than this group?" or "does this treatment reduce pain magnitude in comparison to a sham treatment?"

Since psychophysical methods generally focus on the shape and slope of the psychophysical function, they usually provide little information about the height of the function. In certain methods this height has little meaning. In the common use of a free modulus in ratio scaling, subjects choose their own personal unit of measurement, which arbitrarily determines the height of the psychophysical function. In another option in ratio scaling, the initial stimulus is described by a fixed response value (modulus) such as 10. The use of a modulus equalizes the units among subjects, eliminating differences in absolute magnitude between these subjects.

The emphasis on the shape of the function pervades much of the psychophysical literature, including comparisons of the different functions generated by bounded category scales and unbounded ratio-scaling procedures. Because of this emphasis on height rather than slope, much of this literature may have little applied utility to pain evaluation.

61.1.4.1 Adapting psychophysical methods to pain assessment

To provide absolute measurement useful for pain assessment, scales must be anchored to a subjective standard. Bounded scales include a degree of anchoring. In the bounded VAS scale, the possible range of pain from threshold to worst imaginable is represented by a short line. Thus a mark of 3 cm on a 10-cm scale represents a moderate amount of pain, a mark of 5 cm represents greater pain, and a mark of 6.7 cm represents more intense pain. Numerical category scales represent the range of possible pain magnitude in a similar response continuum. Verbal categories also present a limited response range; however, the use of verbal categories further defines subjective standards along the scale. In a sense, verbal categories provide the anchoring of a response space in a manner similar to VAS and numerical categories, and in addition add a second anchoring system defined by the meanings of the words along the category. Subjects could conceivably use only one of these systems; they could use the meaning of the words irrespective of the number or spacing of the categories or alternatively treat the scale as a numerical scale without attending to the meaning of the words. Studies of the relative contribution of these two sources of subjective standard suggest that subjects use both sources of information in making judgments. Shifting the position of quantified descriptors from an arithmetic to a logarithmic spacing changes the scale, but not enough to suggest that subjects are only attending to word values (Gracely, unpublished observations). One feature of randomized scales discussed above is that this ambiguity is reduced, since information provided by spatial location in a list is purposefully omitted.

The anchoring provided by the limited range of bounded scales is absent in conventional ratio scaling, but can be included in several ways. In one approach, a subjective standard is used in conjunction with a modulus. Hilgard, in a study of hypnotic modification of cold pressor pain, instructed subjects to use the modulus 10 to describe a critical level of pain, one that "they would very much wish to terminate" (Hildard, E. R., 1975). In a second approach, subjects rate both pain stimuli and intermingled verbal stimuli that anchor responses, such as moderate or intense, like childbirth or headache, or pain severe enough to require medication. This second paradigm simultaneously scales the stimuli, the subjective standards, and anchors the pain responses (Gracely, R. H. and Wolskee, P. J., 1983).

A third approach uses verbal categories quantified by ratio methods. This method combines the anchoring provided by category scales with the advantages of ratio scaling, for example, permitting inferences such as, "this drug reduced pain by 36%." In this method the subjective anchors are quantified in a session separate from scaling pain sensations. The descriptors are used directly to measure pain sensations, effectively anchoring each judgment to a subjective standard. The second and third approaches are functionally similar, each with specific advantages. Pain scaling with the third, descriptor method, is efficient, with each response providing a subjective anchor. However, choosing specific descriptors forces each response into a discrete

category. This method is also vulnerable to poor response behaviors, such as perseverance or use of a limited number of descriptors. Although longer, the use of the second approach that collects ratio-scaled responses to both pain and verbal stimuli may provide greater resolution and eliminate problems with the second approach.

61.1.5 Using Experimental Stimulation as an Anchor for Experimental and Clinical Pain Assessment

This approach uses evoked sensations as a standard anchor. Rather than match pain to a category or response space, pain is matched to pain evoked by methods such as contact heat, or matched to the intensity of a nonpainful stimulus such as the brightness of a light (Duncan, G. H. et al., 1988). Another approach for clinical pain assessment anchors judgments by using the same pain rating scale to rate both clinical and experimental pain sensations (Gracely, R. H., 1979; Heft, M. W. et al., 1980). The validity of these methods is dependent on little variance in experimental pain sensitivity across the population. Heat pain approaches this ideal. Other stimuli such as electrical tooth pulp stimuli are extremely variable. However, even in the case of heat there is considerable variation due to genetic (Kim, H. et al., 2004; Mogil, J. S. et al., 2005) and other factors. At this point it appears that the validity of an absolute pain scale must rely on the face validity of the meaning of words within a language and other scale properties such as number use and anchoring in a response space such as the VAS.

61.1.6 Other Suprathreshold Scaling Techniques: Discrimination and Single-Stimulus, Single-Response Designs in General

61.1.6.1 Discrimination
SDT evaluation of painful stimulation was presented in Section 61.1. Although usually used to assess detection, the applications to pain measurement have modified the procedure to assess category ratings of suprathreshold painful stimulation. In the usual detection paradigm, binary responses (detected, not detected) are made to random sequences of two stimulus conditions (present, blank). These stimulus and response pairs form a 2×2 table, with each of the four cells representing the outcomes (hit, miss, false alarm, and correct

rejection) used in the analysis. The analysis of detection examines the proportion of positive and negative responses to the presentations of a stimulus and blank stimulus, that is, between a stimulus pair and a response pair. The suprathreshold modification uses this set of stimuli and responses, and in addition, many more 2×2 tables formed from possible pairs of two stimuli and two category responses. In these cases, the lesser stimulus serves as a blank in the analysis, and the lesser response serves as a not detected. Among all of the stimulus and response possibilities, this analysis can be performed on any 2×2 table with sufficient responses in all four cells, that is, an adequate number of hits, misses, false alarms, and misses. Widely separated stimuli and response categories will usually not meet these criteria because of insufficient false alarms or misses.

The differences between direct scaling and SDT scaling are primarily in the analysis. A subject may not perceive any difference; both methods require category responses to randomly presented stimulus intensities. Each method presents multiple blocks of stimuli that vary in the number of stimuli, stimulus repetitions, and response categories. The two methods are actually similar in many respects. Direct scaling and SDT can be described as single-stimulus, single-response (SSSR) methods (Gracely, R. H. and Naliboff, B. D., 1996). All SSSR designs provide a mean response to each stimulus, and a measure of stimulus discrimination. It is clear that a mean response can be determined from any of the methods. Similarly, the evaluation of discrimination is not limited to the traditional SDT method of using a small number of stimulus intensities (e.g., four) and a large number of response categories (e.g., 12–14). A nonparametric pain-rating paradigm developed by Richard Coppola (Buchsbaum, M. S. *et al.*, 1981) uses the opposite approach; 4 responses categories are available to describe 31 stimulus intensities.

Discrimination can also be computed from the standard direct scaling paradigm in which any number of response formats are used to describe five to seven stimulus intensities presented two to four times each. The discrimination between any two stimulus pairs can be computed by parametric or nonparametric tests such as the t-test or the Mann–Whitney U test. This measure is similar to d'; both represent the ratio of difference in means of distributions and the variability of these distributions. All of the data can be used to compute overall discrimination by a parametric or nonparametric one-way analysis of variance with the stimuli as the factors and the

repetitions as pseudosubjects. The resultant statistic is related to discrimination; high slopes and/or small standard errors around each stimulus will produce a large value of the statistic (e.g., F), while low slopes and/or large standard errors will result in small values of the statistic. This method has been applied to a study of the effect of naloxone (Gracely, R. H. *et al.*, 1987) on pain evoked by electrical tooth pulp stimulation using an analysis of variance and the F statistic, and to taste discrimination using the inter-class correlation coefficient (ICC) as the measure of discrimination (Weiffenbach, J. M. *et al.*, 1986).

61.1.6.2 Sensory decision theory applications to pain assessment

A person suffering from acute or chronic pain lives with a burden that is difficult to understand unless it has been experienced. The burden is augmented by the fact that pain has no physical correlate; it is only defined and communicated by self-report. Mistrust of such reports adds to the burden of having pain; patients must fight for credibility as well as relief. Even in the laboratory, in which the level of a painful stimulus can be controlled, measures of pain are based on subjective reports. Although conventional psychophysical methods control for many biases, there is general agreement that any measure of pain can be clouded by extraneous factors that augment or reduce reported pain.

SDT was introduced to pain assessment as means of solving this problem, of separating out extraneous factors from a measure of actual pain. The method was used in studies examining the influence of placebo, nitrous oxide, the minor tranquilizer diazepam, the narcotic analgesic morphine, acupuncture, and trans-cutaneous electrical nerve stimulation (TENS) (Feather, B. W. *et al.*, 1972; Chapman, C. R. and Feather, B. W., 1973; Chapman, C. R. *et al.*, 1973; Clark, W. C. and Yang, Y. C., 1974; Chapman, C. R. *et al.*, 1975; Chapman, C. R. *et al.*, 1976; Yang, J. C. *et al.*, 1979). Subjects in these studies used category scales to rate the pain evoked by either electrical stimulation of the tooth pulp or heat applied to the skin. These investigations assumed that discriminative sensitivity, assessed by parameters such as d', represented true pain sensitivity, and a change in this parameter after an analgesic intervention would represent actual analgesia. Similarly, the response criterion parameter was assumed to represent response bias, and a change in only this parameter would represent a nonphysiological effect in which subjects changed the labels to describe the evoked pain sensation without any real physiological change in this sensation. The results of these initial studies were consistent with this interpretation. Administration of placebo resulted in only a shift in the response criterion without a change in d', while the active interventions reduced d' with variable effects on the response criterion.

These results focused attention on the issues of pain measurement and generated additional studies and a lively dialog between proponents of the SDT method and critics who questioned the interpretation of the SDT variables in terms of pain sensitivity and labeling bias (Chapman, C. R., 1977; Rollman, G. B., 1977; Clark, W. C. and Yang, Y. C., 1983; Coppola, R. and Gracely, R. H., 1983). This focus identified at least three theoretical issues that potentially limit the interpretation of SDT parameters for pain evaluation (Gracely, R. H. and Naliboff, B. D., 1996). First, discriminability among sensations is not equivalent to sensory magnitude (Rollman, G. B., 1977; Coppola, R. and Gracely, R. H., 1983). It is also influenced by variability in the afferent system from receptor to brain, and cognitive variability in choosing labels to describe evoked sensations. Second, discriminability between sensations could be based on sensory qualities unrelated to pain magnitude (Gracely, R. H., 1994) such as radiation or temporal characteristics. Third, a change in the response criterion may represent analgesia. Pain is defined as a feeling of hurt with both sensory and affective components. An analgesic intervention that reduces the affective, unpleasant aspect of a pain sensation could conceivably result in a change in the response criterion without a change in the discriminability of the stimulus-evoked sensations.

The extent to which these theoretical limitations apply in practice is not known. At present, these issues can be considered to be working hypotheses to be addressed by further studies. Additional issues include the role of prepain sensation in the determination of sensitivity, and the negative influence of pain unpleasantness on discrimination of painful or nonpainful sensations. Until disproved, it may be best to assume that the influence of extraneous biases cannot be eliminated by analytical method but rather controlled as in other psychophysical experiments, through careful experimental designs that consider the many factors that can influence response behavior.

61.1.6.3 Pragmatic problems with a fixed set of stimulus intensities

Both direct scaling and SDT methods deliver a predetermined set of fixed stimulus values in a randomized sequence. These methods are easily applied to the evaluation of nonpainful stimuli, but become problematic when evaluating pain. Care must be taken to avoid delivering unacceptable stimulation to any person. This unacceptable level varies between healthy control subjects. One method simply limits the maximum stimulus to a value that is tolerated by the most sensitive subject. A problem with this approach is that the least sensitive individuals will not experience the same level of subjective pain as experienced by the more sensitive individuals. Another problem is that in studies of analgesia, an active drug may be identified by the effects on stimulus range; a wide range before the intervention may be contracted afterward, providing a clue that an active treatment had been administered. In addition, insensitive subjects may show no effect, since the stimuli evoked minimal pain sensations before the intervention.

These effects are observed in a group of homogenous subjects and pronounced in comparisons between normal and hypersensitive groups. The obvious solution to this problem is to determine ranges of tolerable stimulation for each individual and compute separate stimulus sets for each individual. This can be done once or possibly a second time after an analgesic intervention. This method has implications for the analysis. Rather than use parametric t-tests or analyses of variance, the analysis must use multivariate or regression methods to evaluate both the response and the subject-specific stimulus intensity associated with the response.

Direct scaling and SDT methods share another common feature. Both procedures were developed prior to laboratory computers and performed using pencil and paper methodology. Adaptations to computer administration simply automated the paper and pencil process.

61.1.6.4 Utility of interactive adaptive methods

A relatively recent set of procedures solves the problems with fixed stimulus sets by using interactive, computerized paradigms. These adaptive methods extend a concept developed by Cornsweet T. N. (1962), in which the stimulus series is based on the behavior of the subject. In its simplest form, this method is a tracking task in which a response of painful would lower the intensity of the next presented stimulus while a response of not painful would increase the intensity of the next presented stimulus. However, since the next stimulus is always response-contingent, subjects, or even pigeons in the case of Cornsweet T. N. (1962), can become aware of the staircase rule and adapt simple strategies to reduce the painful stimulus (humans) or maximize positive reward (pigeons). Cornsweet's innovation minimized the contingency between the response and the change in the stimulus intensity by using two different and independent tracking paradigms and randomly alternating between them. This method of double random staircases was used in conjunction with a light detection task to demonstrate dark adaptation curves in the pigeon (Cornsweet, T. N., 1962). This method formed the basis for the method of MRS scaling of pain sensation. The MRS expanded on the double random staircase method in three ways. First, it increased the number of independent staircases, with the initial application using six staircases. Second, rather than assess only detection, the method was expanded to evaluate both detection-level and suprathreshold pain sensations. Third, the paradigm took advantage of computerization to become more interactive, with a dynamically changing amount of stimulus step size to match the method's performance to the discriminative ability of the subject.

Each staircase in the MRS is associated with a response interval between two categories or between two alternatives of a continuous rating scale (e.g., 20.5 mm on a 0- to 100-mm VAS scale). On each trial, a staircase is randomly chosen and stimulus intensity is delivered to the subject. The response determines the new stimulus intensity to be delivered the next time that particular staircase is again chosen. Different decision rules are possible; in the simplest, any response above the associated stimulus level reduces the intensity of the next delivered stimulus, while any response below the associated level increases the intensity of the next delivered stimulus. The amount the stimulus changes, the step size, is a dynamic variable particular to each staircase. It is usually large in the beginning and reduces if the direction of the response change (increase or decrease) changes too frequently (e.g., a very large step size that would administer either nonpainful or very painful stimuli). It can increase if the direction of the response change is the same over a large number of trials (e.g., an increment that is so small that the staircase never becomes painful).

The MRS method solves the problem of choosing appropriate stimulus ranges for each subject since it adjusts the range automatically. The initial stimulus values can be either set low for all staircases or set near the appropriate value by using the results of a simple ascending series. The method has a number of additional potential advantages (Gracely, R. H., 1988; Gracely, R. H., *et al.*, 1988). The results of the independent staircases can be compared to assess reliable scaling performance. Poor subject ability or compliance results in random walks for each staircase, while consistent responding produces characteristic response patterns. The method expresses the results in units of stimulus intensity, permitting simple comparisons across subjects or time. Figure 6 shows that the method can tract sensitivity over small intervals of time, demonstrating analgesia to fast-acting agents such as nitrous oxide or narcotic analgesics. Most importantly, the method results in a similar subjective range of pain in each subject regardless of individual pain sensitivities. This reduces confounds in which more sensitive subjects experience greater subjective levels of pain. In studies of pain control interventions, subjects experience the same level of pain before and after the intervention (i.e., after an efficacious analgesic the computer increases the stimulus intensities to evoke the previously experienced pain level). This minimizes biases associated with analgesic-induced reduction of the range

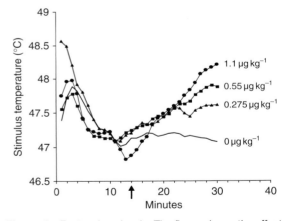

Figure 6 Fentanyl analgesia. The figure shows the effect of dose of the opioid fentanyl for the staircases titrated between moderate and intense. The double-random staircases titrated at this interval were combined into one mean staircase to show the results of four groups (*n* = 16 each). Stimulus intensity is shown on the ordinate, and time before, during, and after a 2-min double-blind intravenous infusion administered at 10 min (arrow) is shown on the abscissa. Subjects received either placebo, shown by solid line, or 0.275, 0.55, or 1.1 μg kg^{-1} fentanyl shown respectively by the triangles, squares, and circles. Fentanyl significantly increased staircase temperatures, indicating analgesia. Both rate of onset and final temperature were dose-related.

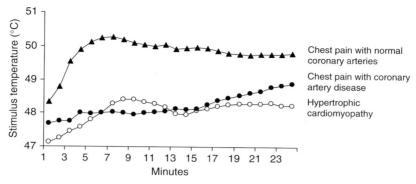

Figure 7 Multiple random staircase evaluation of thermal cutaneous pain sensitivity in cardiac patients. Threshold and suprathreshold values determined over a 34-min period are plotted for patients with chest pain and normal coronary arteries (triangles), hypertrophic cardiomyopathy (open circles), or coronary artery disease (closed circles). Stimulus temperatures required to produce these three subjective levels of pain sensation are plotted against time. These temperature differences indicate decreased cutaneous pain sensitivity in the chest pain, normal coronary artery group, suggesting that their chest pain does not result from a generalized pain sensitivity. Adapted from Cannon, R. C., Quyyumi, A. A., Schenke, W. H., Fananapazir, L., Tucker, E. E., Gaughan, A. M., Gracely, R. H., Cattan, E. L., and Epstein, S. 1990. Abnormal Cardiac Sensitivity in Patients with Chest Pain and Normal Coronary Arteries. J. Am. Coll. Cardiol. 16, 1359–1366.

of pain sensations that plagues methods that used fixed sets of stimulus intensities. Figure 7 suggests that the reductions of biases is also useful for the determination of stable values that can be compared among different groups. This figure shows the comparison of cutaneous pain sensitivity of patients suffering from angina pain, suggesting that patients with coronary artery disease are generally more sensitive than patients with pain but no evidence of coronary pathology (Cannon, R. C. *et al.*, 1990).

The last advantage is an ethical one; studies of analgesia at high stimulus intensities are achieved with less pain using the MRS method. For a simplistic example, that is, to study the analgesic effect at 10 kg of pressure, a conventional method delivers 10 kg, and an analgesic reduces the response to what would be evoked by 8 kg before the analgesic was administered. With the MRS, 8 kg is administered, which evokes a specific response. After the analgesic, the computer raises the pressure to 10 kg to evoke the same response. With the conventional method, the evaluation of the analgesic effect at 10 kg requires the subjects to experience the pain evoked by 10 kg. With the MRS method, the subjects only experience the pain produced by 8 kg of pressure.

The main principle of adaptive methods is that the intensity of stimulation delivered to the subjects is adjusted based on the subject's ratings of the sensations evoked by the stimulation. In addition to the staircase algorithm described above, which bases response change primarily on the immediate response history, an alternative method that computes probabilities from past responses provides a current best estimate of the stimulus–response function (Duncan, G. H. *et al.*, 1992). This method provides an excellent estimate of unvarying sensitivity, since all responses are used in the computation. However, these methods may not be optimal for the assessment of pain sensitivity that naturally varies, or is reduced by an analgesic intervention. In their present form, each of these procedures may be ideal for different purposes: the staircase algorithm for tracking changing pain sensitivity and the probability algorithm for evaluating unchanging sensitivity. However, the algorithm of each can be altered to approach the characteristic of the other. The sensitivity of probability methods to varying pain sensitivity can be increased by decreasing the time interval used for the computation of probabilities, while the use of a dynamic increment in staircase paradigms provides an influence of response history.

SSSR designs share common features and provide (1) information about the mean response applied to a stimulus and (2) the ability to discriminate stimulus-evoked sensations. The limitations of these parameters have been discussed above in Sensory Decision Theory Applications to Pain Assessment. The first parameter is a measure of perceived pain magnitude, influenced by a subject's response criterion. The second parameter is a measure of the intensity of the pain sensation, noise in the neural afferent system, and, most importantly, general cognitive ability in consistently choosing the same labels to describe the same sensory magnitude. Neither is a pure measure of pain magnitude.

61.1.6.5 Stimulus integration: double-stimulus, single response methods

In contrast to SSSR methods in which subjects make a single response to a single stimulus, another family of methods presents two or more stimuli and collects a single response. Usually two stimuli are delivered, which could be termed a double-stimulus, single-response (DSSR) design. In the usual case, a pair of stimuli is administered on each trial and the subject chooses a single response to describe some integrated impression of the two stimuli. The analysis produces an independent scale for each stimulus set. The statistical interaction between these two scales can be used to assess both individual and group scaling performance. This evaluation of performance is a unique feature of these methods. Rather than passively collect responses to stimuli and accept the resultant scale, these procedures treat the processes as an experiment that either fails or succeeds. If successful, the separate scales are accepted. A negative outcome indicates a failure in one or more of the three operations involved in the task: scaling, integrating, or choosing a response. Two different methods have been used, differing primarily in the analysis. Functional measurement treats the responses as interval-level data and performs a parametric analysis. Conjoint measurement assumes only ordinal properties in the responses, and performs a nonparametric analysis.

This description and the term DSSR are oversimplified. In most applications to pain measurement, a subject provides a single response to two stimuli using a specified integrating operation such as averaging. The subjects could also be asked to respond to ratios, subtract or summate the sensations evoked by the two different stimuli. In other applications, two attributes of a single stimulus, rather than two separate stimuli, are assessed. In this case the integration is implicit. For example, subjects may rate the heaviness of weights varying independently in size and mass, and the analysis assesses the separate influence of mass and size without specific instructions. Applications to painful stimuli could vary both intensity and another attribute such as stimulation site, area, or stimulus duration. The design is not limited to two stimulus sets or two attributes. Subjects may integrate any number of delivered stimuli or stimulus attributes.

The literature on these DSSR methods includes a few studies that integrated pain sensations. One of the first presented all possible pairs of two identical sets of four electrocutaneous stimuli varying from 1 to 4 mA (Jones, B., 1980). Responses to the average painfulness of these stimuli were analyzed by functional measurement. A two-way analysis of variance (ANOVA) validated the method by showing two main effects and no interaction. Algom D. et al., (1986) varied this design by asking subjects to integrate sensations evoked by painful and, aversive but not painful, stimuli. Subjects rated overall aversiveness of all possible pairs of six electrical stimuli and six loud auditory stimuli. Gracely R. H. and Wolskee P. J. (1983) extended the generality of this method for pain assessment (1983) by asking subjects to integrate the magnitude of pain sensations with the magnitude implied by a verbal descriptor. Subjects matched handgrip force to the average intensity of pain sensations produced by all possible pairs of five electrical tooth pulp stimuli and five verbal descriptors (weak, mild, moderate, strong, and intense). A functional measurement ANOVA analysis showed significant main effects for the tooth pulp and the verbal stimuli and a nonsignificant interaction, validating the method. This integration of somatic and verbal stimuli was replicated by Heft M. W. and Parker S. R. (1984) using electrical skin stimulation and a conjoint measurement analysis.

To summarize evaluation of a single pain dimension, SSSR methods that deliver single stimulus and collect a single response on each of multiple trials contain similar information regardless of how the response is analyzed. The responses contain information about the mean value of the stimulus and about the variability in response to the same stimulus and between different stimulus intensities. These are the most common methods for the evaluation of suprathreshold pain sensation. Measures of discrimination and response magnitude do not provide pure measures of analgesia, but are influenced by other factors such as cognitive ability and variable response criterion. These factors cannot be isolated by a SSSR method, but rather must be controlled by careful experimental design. Preliminary results with other procedures such as DSSR methods, however, may be capable of providing more information than SSSR methods, although the capabilities and limitations of these methods have not been examined in detail. Measures such as performance in DSSR designs or sensitivity measures in SDT designs may be considered as measures of performance, which may be independent of pain or provide a behavioral measure of pain. Measures in which subjects judge intensity or some other aspect of a sensation must rely on the reduction of biases through adequate experimental control, and the anchoring of judgments to a standard. Choice of anchoring method may be based in part on which method has greater face validity: a mark on a VAS scale, a numerical category scale on a scale of 1 to 10 or 20, or the use of language and the commonalities inherent in language.

61.2 Pain as Dual Variables of Intensity and Unpleasantness

Pain is usually treated as a single dimension of intensity, such as heaviness. However, the history of pain literature emphasizes a second dimension. Aristotle and Plato described pain as a "passion of the soul," and contemporary science has used terms such as the reaction component, unpleasantness, discomfort, distress, affective, or evaluative.

61.2.1 Psychophysical Studies of Feeling States: Hedonics

Pain shares this feeling state with other sensory systems. Modalities involved with maintaining homeostatic equilibrium such as temperature and the chemical senses of taste and olfaction possess two distinct dimensions. The first is a dimension of sensory intensity that is shared with all other senses and that has been the focus of the discussion of pain measurement above. The second is a feeling state that for most homeostatic modalities can be either pleasant or unpleasant. Together, these bivalent feeling states have been described as hedonic. Generally, departures from homeostatic neutrality result in an unpleasant hedonic for sensations consistent with that departure; warm water is unpleasant if one is overheated, that is, hyperthermic. Sensations associated with the restoration of homeostatic are usually pleasant. Cool objects, air, or water feel pleasant when overheated. Thus, sensations of sensory intensity can be associated with either pleasant or unpleasant feelings; these sensations are characterized by bivalent hedonics. Hedonic ratings of sugar solutions are initially pleasant with increasing concentrations, then decline and become unpleasant at very high concentrations. Extreme temperatures are unpleasant, while more moderate departures from thermal neutrality can be pleasant. In addition to this bivalent nature, the examples above indicate that the

sign and magnitude of the hedonic depends on the internal state. Hunger shifts the hedonic function for sugar solutions so that the formerly moderately pleasant concentrations become more pleasant, and formerly unpleasant concentrations become pleasant (Mower, G. et al., 1977). Satiety has the opposite effect. After overeating at a buffet table, even favorite deserts lose their appeal. Similarly, thermal pleasantness or unpleasantness depends on core temperature. Water of a specific temperature can feel either pleasant or unpleasant, depending if one is hypothermic or hyperthermic (Mower, G., 1976).

In contrast to these bivalent hedonics, a number of sensations appear to be associated with only unpleasant feelings. Vertigo, nausea, and tinnitus do not usually have a positive counterpart, and of course pain is a prime example of an unpleasant sensation. This group of sensations can be characterized as having a univalent hedonic; they are unpleasant throughout the range of sensory intensity.

Although pain does not parallel the homeostatic regulatory functions of hunger, thirst, or temperature regulation, it does serve to protect the integrity of the organism by both avoiding or minimizing injurious stimulation, and once injured, to promote behaviors that maximize recuperation.

While the distinction between the sensory intensity and the hedonic dimension is readily apparent for modalities that can feel either good or bad, this distinction is less obvious for modalities such as pain that have only a univalent hedonic. Pain unpleasantness increases with intensity, and the intertwined nature may make it difficult to distinguish between these dimensions. VAS or cross-modality scales have used different sets of instructions with mixed results. Extensive instructions have been sufficient to facilitate the distinction between intensity and unpleasantness (Price, D. D., 1988), while instructions to cross-modality match or use magnitude estimation to rate unpleasantness appeared to result in scales only of the salient dimension of sensory intensity (Gracely, R. H. and Naliboff, B. D., 1996).

One tactic to avoid the influence of variable instructions is to build in the instruction by using verbal categories. The most popular example of using language to facilitate the discrimination of unpleasantness and other qualitative dimensions was initiated by Melzack R. and Torgerson W. S. (1971), who quantified a list of words compiled by Dallenbach K. M. (1939). In this study, the verbal stimuli were first sorted into discrete categories of qualitative meaning. For each category, the values of the words within the category were computed using a Thurstone L. I. (1959) scaling analysis of stimulus rankings. The results of this study provided the core of the subsequent MPQ, which is undoubtedly the most widely used multidimensional pain assessment instrument (Melzack, R., 1975). This questionnaire and a short version are described more fully in The Need for Scaling Multiple Pain Dimensions.

Tursky B. (1976) applied Steven's ratio-scaling methods to a subset of the MPQ words. The selection of the specific words reduced the large number of categories of the Melzack and Torgerson study and the subsequent MPQ to a few dimensions common to all of pain experience. Gracely R. H. *et al.* (1978a; 1979) refined Tursky's methods by reducing the number of words further to describe two dimensions of sensory intensity and unpleasantness, and by omitting ambiguous descriptors (e.g., severe) to satisfy the criterion that the sensory intensity class could be used to described the intensity of any sensation. Initial studies demonstrated reliability and scaling consistency (Gracely, R. H., 1978a; 1978b) and subsequent studies evaluated the modulation of experimentally evoked pain sensations evoked by analgesic interventions.

61.2.2 Validity of Hedonic Scales

Development of separate scales to assess sensory intensity and hedonic tone does not ensure that such scales will actually assess these dimensions. It is conceivable that, regardless of the scale used, subjects may literally or figuratively take off their glasses and use the spatial properties of the scale to make responses. These responses may be made to the most salient dimension, perhaps the dimension of sensory intensity for a study assessing responses to experimental stimulation in healthy subjects and the dimension of unpleasantness for patients rating the magnitude of persistent chronic pain. This possibility must always be considered when a study finds that scales of sensory intensity and unpleasantness yield essentially the same result. Does such a finding indicate that the scales are indeed the same, or that they fail to discriminate between these dimensions?

Studies of hedonic scales establish validity in two ways. First, validity is assumed if the scales of sensory intensity and unpleasantness result in different psychophysical functions. The finding of similar functions does not invalidate the scales since the functions could be similar; thus similar scales are ambiguous for purposes of demonstrating validity. A related issue is that separate scales of intensity and hedonics should be physically similar, such as category scales or labeled VAS scales. Using different formats introduces method variance such that a finding of different psychophysical functions could represent the different scale formats and not the different subjective dimensions.

The second line of evidence for validity is that sensory and hedonic ratings are altered appropriately by interventions targeted at one or the other dimension. In the examples of temperature and taste described above, the validity of the scales is supported if manipulations that alter homeostasis alter hedonic ratings without changing ratings of sensory intensity.

Both types of evidence have been found for non-painful modalities such as temperature, taste, and olfaction. The hedonic functions are different than ratings of sensory intensity, and are shifted by changes in internal state by manipulation of core temperature or hunger (Mower, G., 1976; Mower, G. *et al.*, 1977).

61.2.3 Validity of Scales of Pain Unpleasantness

Similar types of evidence have also been shown for ratings of painful stimulation. Evidence for different functions is shown in Figure 8, which summarizes the results of studies that assessed pain sensations evoked by electrical stimulation by choosing quantified descriptors of sensory intensity or unpleasantness. The result was significantly different functions. In contrast, instructions to use cross-modality matching to rate either dimensions resulted in the same function that was coincident with the verbal descriptor function of sensory intensity. This result suggests that the instructions alone were not sufficient to discriminate between these dimensions, and subjects rated the salient dimension of sensory intensity regardless of instructions to rate either sensory intensity or unpleasantness.

The second line of evidence for the validity of scales of pain unpleasantness, differential and appropriate response to an intervention, has been demonstrated in a study in which different groups of healthy controls used randomized lists of verbal descriptors to rate either the intensity or the unpleasantness of sensations evoked by electrocutaneous stimuli before and after the intravenous

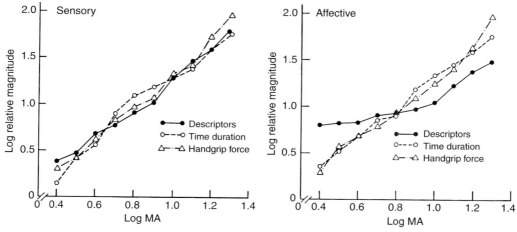

Figure 8 Verbal descriptor and cross-modality matching functions for sensory intensity and affective/unpleasantness responses. Log relative magnitude is plotted against stimulus intensity in log milliamperes. The *left graph* shows psychophysical functions derived from choosing sensory descriptors from a random list and from cross-modality matching time duration and handgrip force to the intensity of the pain sensations evoked by the stimuli. The *right graph* shows psychophysical functions derived from choosing unpleasantness descriptors from random lists and from cross-modality matching of time duration and handgrip force to the unpleasantness of the stimulus-evoked sensations. Each point is the mean of 4 observations from each of 16 subjects. All of the functions are similar except for the unpleasantness descriptor responses.

administration of the minor tranquilizer diazepam. Unpleasantness responses were reduced following diazepam while sensory intensity responses were not altered (Gracely, R. H. *et al.*, 1978b). Subsequent studies have demonstrated different functions for sensory intensity and unpleasantness (Greenspan, J. D. *et al.*, 2003) and differential effects following administration of the opioid fentanyl (Gracely, R. H. *et al.*, 1979), hypnosis (Malone, M. D., 1989; Rainville, P. *et al.*, 1997; Hofbauer, R. K. *et al.*, 2001), conditioning (Wunsh, A. *et al* 2003), and different experimental pain modalities (Rainville, P. *et al.*, 1992; Strigo, I. A. *et al.*, 2002; Dunckley, P. *et al.*, 2005). Tasks such as attending to either the localization or the unpleasantness of evoked pain have shown differential activation of lateral and medial structures, respectively (Bentley, D. E. *et al.*, 2004; Kulkarni, B. *et al.*, 2005).

61.2.4 The Concept of Affective Gain

Recent studies of supraspinal structures involved in sensory discrimination and unpleasantness may ultimately provide support for models of pain unpleasantness that posit an affective system that depends both on sensory input and what can be simply considered as the gain of the affective processing system (Gracely, R. H., 1992a; Price, D. D. and Harkins, S. W. 1992). Affective gain can be expressed

as the amount of unpleasantness associated with a specific sensory pain intensity. Interventions that reduce the amount of unpleasantness without altering sensory intensity can be considered to reduce affective gain, while interventions that reduce both sensory intensity and unpleasantness equally can be considered to reduce only sensory input. In this case affective gain has not been changed, only the input to the affective gain system.

One result of this affective gain system is that in studies in which clinical pain varies or experimental pain is varied, there is an expected correlation between sensory intensity and unpleasantness. This correlation has led to the suggestion that the concepts of sensory intensity and unpleasantness are not independent and should be simply combined into a measure of overall pain magnitude (Gracely, R. H., 1992b). However, the affective gain, which has been expressed as the ratio of unpleasantness to sensory intensity ratings, is relatively independent of sensory intensity and a meaningful variable that has been specifically altered by pharmacological and cognitive interventions (Gracely, R. H. *et al.*, 1978b; Gracely, R. H., 1979; Malone, M. D. *et al.*, 1989; Rainville, P. *et al.*, 1997; Wunsh, A. *et al.*, 2003; Bentley, D. E. *et al.*, 2004; Kulkarni, B. *et al.*, 2005) and it varies widely over the type of experimental or clinical pain (Price, D. D. *et al.*, 1987; Rainville, P. *et al.*, 1992).

It is interesting to note that interventions can reduce only the affective gain in both pain modalities (e.g., hypnosis) and nonpainful modalities (core temperature, satiety) such as taste or temperature. This selective attenuation of affect may be only one of the few options available for intractable pain syndromes such as poststroke pain that currently are not adequately managed by available treatments. The aversive aspect of pain is not a unitary construct and includes multiple levels of distress. The term unpleasantness has been used to refer to the immediate disagreeable nature of pain that prompts withdrawal and other avoidance and escape behaviors. This conceptualization is similar to the homeostatic feelings of hunger, thirst, and hypo- or hyperthermia; all are negative feeling states that prompt behaviors that reduce the negative feeling. They are present in both animals and man. In addition to this immediate feeling state of unpleasantness, pain is also associated with an emotional or affective state that includes cognitive appraisals (thought) and described in terms of distress, hopelessness, and despair (Gracely, R. H., 1992a; Price, D. D. and Harkins S. W., 1992). Implicit in this view is an emotional state resulting from thoughts about both the present and the future consequences. While the presence of this level in animals is at least debatable, it is clear that the human ability to think about the future contributes to this secondary level of pain affect. In addition to these levels is a tertiary level that is not specific to pain (Gracely, R. H., 1992a). It is shared by all medical conditions and results from the presence of a condition, interaction with the health care system, and general concern about the present and future.

In addition to the hedonic and affective dimensions, pain varies in locus and spatial extent, and in sensory quality. Pain can be experienced in mechanical terms of pulling, squeezing, pressing, and tearing. It can be sharp or dull, deep or superficial, and localized or diffuse. Pain can burn, throb, or ache. Pain can be localized to one constant site, or spread or move. The following section describes methods that extend the measurement of pain sensation and unpleasantness to consider all secondary levels of pain affect and all of these sensory dimensions.

61.3 Multidimensional Pain Assessment

Sensations can be described by four attributes: location, duration, intensity, and quality. Pain can be localized and its presence or absence simply noted. Location and duration are recognized as important aspects of the pain experience, and while they are not always easily quantifiable, they are at least conceptually straightforward. In most situations, pain is treated as a single dimension of sensory intensity, although there is increasing recognition of the need to also assess pain unpleasantness. A number of approaches have extended the evaluation of pain to include all of the qualitative dimensions, affective dimensions, and in some cases the reaction to pain, including emotions, cognitions, behaviors, and neural and autonomic responses.

61.3.1 The Need for Scaling Multiple Pain Dimensions

The utility of extending pain measurement beyond a single dimension depends on the incremental value of the increased information in terms of diagnosis, choice of treatment, evaluation of treatment efficacy, or investigation of underlying mechanisms. In the case in which multiple pain qualities are assessed, multidimensional pain scales that determined the mixture of pain qualities present, similar to a color mixture for house paint, must demonstrate incremental utility over the assignment of a single dimension, or in the case of paint, a primary color such as blue, red, or yellow (Gracely, R. H. and Naliboff, B. D., 1996).

Clark W.C. *et al.* (1989) distinguish between restrictive and nonrestrictive measurement. Restricted measurement systems determine the number of dimensions *a priori*, while nonrestrictive systems discover the number and type of dimensions for a particular type of subject and pain condition. Naliboff points out that this distinction becomes blurred as results of these methods accumulate (Gracely, R. H. and Naliboff, B. D., 1996), "In actuality this differentiation in approach is not so clear since the objective of even 'unrestrictive' methods is to discover the real pain dimensions and therefore lead to restrictive measurement. Conversely, the system of sensory and affective dimensions was at least in part 'discovered' through empirical categorizing of reports."

Multidimensional scaling can be applied to any scale of pain intensity, and, similar to quantified category scales described above, can be used to evaluate verbal response items independent of pain experience. The most popular and widely used multidimensional pain instrument is the MPQ. The MPQ presents 20 categories of verbal descriptors with 2 to 6 descriptors per category for a total of 78

Sensory

1.	2.	3.	4.	5.
Flickering	Jumping	Pricking	Sharp	Pinching
Quivering	Flashing	Boring	Cutting	Pressing
Pulsing	Shooting	Drilling	Lacerating	Gnawing
Throbbing		Stabbing		Cramping
Beating		Lancinating		Crushing
Pounding				

6.	7.	8.	9.	10.
Tugging	Hot	Tingling	Dull	Tender
Pulling	Burning	Itchy	Sore	Taut
Wrenching	Scalding	Smarting	Hurting	Rasping
	Searing	Stinging	Aching	Splitting
			Heavy	

Affective

11.	12.	13.	14.	15.
Tiring	Sickening	Fearful	Punishing	Wretched
Exhausting	Suffocating	Frightful	Grueling	Blinding
		Terrifying	Cruel	
			Vicious	
			Killing	

Evaluative **Miscellaneous**

16.	17.	18.	19.	20.
Annoying	Spreading	Tight	Cool	Nagging
Troublesome	Radiating	Numb	Cold	Nauseating
Miserable	Penetrating	Drawing	Freezing	Agonizing
Intense	Piercing	Squeezing		Dreadful
Unbearable		Tearing		Torturing

Figure 9 Pain descriptors from the McGill Pain Questionnaire (MPQ). The 20 word sets are grouped into 4 major classes; 1–10 are sensory descriptors, 11–15 are affective descriptors, 16 is evaluative, and 17–20 are miscellaneous. Descriptors are ordered within each set based on increasing intensity of the particular pain quality. (McGill Pain Questionnaire, copyright, Ronald Melzack, 1975.)

pain-related descriptors (Figure 9). The MPQ is based on the pioneering study of Melzack R. and Togerson W. S. (1971), which used physician and nonphysician groups that classified a list of 102 pain descriptors in terms of similarity of pain quality. The words were grouped into three general classes of sensory words describing pain in terms of temporal, spacial, pressure, and thermal features, affective words describing tension, fear, and autonomic qualities, and evaluative words describing the general intensity of the pain experience. The result included 16 subclasses. In the second step, the words of each subclass where scaled for intensity. Ultimately, four additional miscellaneous categories were added for descriptors that were not adequately described by the three general categories. In theory, a patient's pain can be described by a profile of the 20 subclasses although in practice the results of the MPQ are usually expressed by the three general dimensions (sensory, affective, and evaluative) by summing the scores of the respective subclasses.

One psychometric property of the MPQ scoring system is that a general score such as the sensory class, or the overall score, can be obtained by either relatively high scores on a few subclasses or lower scores on a larger number of subclasses. This system equates a low-intensity pain with many qualities with a higher-intensity pain described by one or only a few qualities. Scaling verbosity, the tendency to choose many categories, inflates the scores. The subsequent short form of the MPQ avoids these issues by requiring a rating response to each verbal item from a subset of the MPQ items (Melzack, R., 1987).

Responses from the two forms of the MPQ and other multidimensional pain scales can be analyzed by several methods, including factor analysis, ideal type analysis, and individual differences scaling (INDSCAL) (Gracely, R. H. and Naliboff, B. D., 1996). The factor analytic methods refine items into separate dimensions with maximal attainable internal consistency. These methods have been used to confirm 3D MPQ, and also in studies that aim to identify the true number of relevant pain dimensions (Melzack, R. and Katz, J., 1992; Holroyd, K. A. *et al.*, 1992).

However, there is an important distinction between studies that use factor analysis to analyze items in a questionnaire (the items are presented as stimuli and rated) and studies that use factor analysis to evaluate patients' actual use of a questionnaire. The analysis of items is a standard psychometric technique that indicates the number and homogeneity of the subscales in an instrument. Like the MPQ, the result is similar items grouped together by semantic similarity. In contrast, factor analysis of patient responses evaluates associative similarity (Torgerson, W. S. *et al.* 1988; Gracely, R. H., 1992b; Gracely, R. H. and Naliboff, B. D., 1996). These identify symptoms that coalesce in specific pain syndromes such as deep and throbbing or superficial and burning. They identify the constellation of symptoms found with the particular pain disorders present in the population using the scale. As Torgerson (private communication) has pointed out using a food analogy, factor analysis of items groups butter and margarine together and bread and cake together, but a factor analysis of responses to such a food scale would provide associate similarities, such as grouping bread and butter.

When factor analysis is used appropriately, the item analyses can also eliminate important information in the combinations of correlated items into a scale. For example, measures of both height and weight could be highly correlated in a sample of subjects and it might seem reasonable to combine these into an overall scale of body size. However, a consideration of the difference between the two yields potentially medically important information about conditions of overweight or underweight that would be missed by combining the similar items of height and weight into a single scale (Gracely, R. H., 1992b; Gracely, R. H. and Naliboff, B. D., 1996). Attempts to combine the correlated pain dimensions of intensity and unpleasantness into a single scale are similarly problematic because important information, the amount of unpleasantness associated with a particular pain intensity (affective gain), is lost in the process (Gracely, R. H., 1992b).

In addition to these issues, Naliboff notes that almost all factor analytic studies are based on the data collected at a single point in time and thus do not evaluate temporal properties of pain dimensions (Gracely, R. H. and Naliboff, B. D., 1996). Circadian and hormonal rhythms, pain interventions, behavioral patterns, and fluctuations in specific disease symptoms may alter the compositions of the multidimensional experience of pain that as yet is not captured by these instruments.

The subclass structure of the MPQ assumes that each descriptor represents a magnitude of a specific quality, for example that flickering, throbbing, and pounding represent different degrees of temporal variation. Torgerson and coworkers developed an alternative approach in which a descriptor could represent different amounts of various qualities (Torgerson, W. S. *et al.*, 1988). A specific descriptor could represent a pure quality, but this approach assumes that many of the pure qualities are not represented by a single word. These specific qualities were referred to as ideal types.

Torgerson's method required subjects to scale each descriptor by both its similarity to a set of ideal types and a quantitative dimension of magnitude, a study of 17 stimuli (16 from the original Melzack–Torgerson set), and identified four ideal types: bright, slow/rhythmic, thermal, and vibratory-arrhythmic. Figure 10 shows the results of studies using either sensory or affective descriptors, each in relation to three ideal types (Torgerson, W. S., 1988). Torgerson W. S. *et al.* (1988) used overlapping subsets of the original Melzack and Torgerson descriptors and ultimately derived 19 ideal types. Pain qualities as well as an overall quantitative pain intensity scale were obtained. The results generally confirmed the sensory intensity dimension of the MPQ organization with some refinement such as dividing the punctuate pressure MPQ class into separate dimensions of pricking, stabbing and drilling, boring, acknowledging the rotational character of the latter group. Words that closely matched the sensory ideal types include sharp, tugging, and pounding, and close affective types words included frightful, sickening, and cruel. Figure 11 shows the two pain descriptors closest to each of the 19 ideal types. Use of these descriptors to assess low back pain found an association between sensory qualities and diagnosis, and between the affective qualities and the psychological distress (BenDebba, M. *et al.*, 1993).

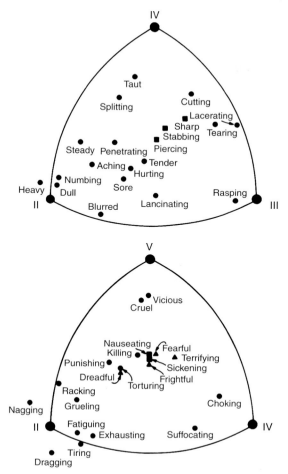

Figure 10 Surface plots illustrating the ideal type structures obtained from a set of sensory (*top*) and affective (*bottom*) words. The numbers at the vertices of the hyperspherical triangle correspond to three of the ideal types identified in each analysis. A total of four ideal types were identified for the sensory words and five for the affective words. From Torgerson, W. S., BenDebba, M., and Mason, K. J. 1988. Varieties of Pain. In: Proceedings of the Vth World Congress on Pain (*eds*. R. Dubner, G. R. Gebhart. and M. R. Bond), p. 368. Elsevier.

A third major approach to scaling multidimensional pain descriptors uses software developed to scale any type of multidimensional modality. Clark and coworkers use INDSCAL procedures combined with preference mapping (PREMAP) to define a multidimensional pain coordinate system and to place a particular judgment within this system. These software programs use judgments of proximity (sameness) to identify the number and type of dimensions inherent in the responses. As with other techniques, the meaning of the dimensions must be inferred from the configuration of the results, especially at the poles of the dimension. The PREMAP model aids in this process by having judges rate each stimuli on bipolar property scales, which are used to help interpret the meaning of the dimensions from the INDSCL analysis. This combination of INDSCAL/PREMAP method also provides a measure of the importance of each dimension to individual subjects.

Clark and coworkers have applied the INDSCAL/PREMAP method to judgments of pain descriptors, clinical pain ratings, and experimental pain stimuli. In the study of pain descriptors, 9 descriptors (mild pain, annoying, cramping, sickening, miserable, burning, unbearable pain, intense pain, and shooting) were presented to subjects in all 36 possible pairs and subjects judged the similarity of each pair of descriptors. The INDSCAL/ PREMAP analysis resulted in three dimensions. Clark W. C. *et al.* (1989) labeled these three dimensions as evaluative, aversive, and somatosensory, although Syrjala K. (1989) raises issues about this interpretation. The evaluative dimension includes both pain magnitude and affective qualities. The aversive dimension also contains descriptors of pain magnitude combined with affective descriptor of sickening and shooting. The somatosensory dimension described two qualities related to depth and temporal qualities of pain.

Clark W. C. *et al.* (1989) also collected ratings from a group of patients with cancer and a control group. Interestingly, the cancer group found the affective descriptors to be less important than did the control group. Analysis of the importance of the affective variation within the patient group revealed variability in the importance of the affective descriptors, suggesting the possibility that those who found the affective dimensions to be important might be responding to interventions targeting affect such as antianxiety agents, while those assigning more importance to the somatosensory dimension might be more responsive to analgesics.

The INDSCAL/PREMAP method has also been used to assess the perceptual dimensions of pain evoked by experimental stimulation (Figure 12). Janal M. N. *et al.* (1991; 1993)delivered electrocutaneous and thermal stimuli in pairs and subjects rated the similarity of the pairs. The software identified a robust dimension of intensity and additional stimulus-specific dimensions: a frequency dimension for the electrical stimuli and a warm–hot dimension for the thermal stimuli. The intensity dimension for the

I. Sensory domain

1. Sharp Cutting	2. Boring Drilling	3. Stabbing Pricking	4. Dull Numbing	5. Rasping Tearing
6. Taut Splitting	7. Tugging Pulling	8. Squeezing Crushing	9. Hot Burning	10. Flickering Quivering
11. Pounding Throbbing	12. Itching Tickling	13. Radiating Spreading	14. Flashing Shocking	

II. Affective domain

15. Frightful Fearful	16. Dragging Fatiguing	17. Sickening Nauseating	18. Choking Suffocating	19. Cruel Vicious

Figure 11 Two pain descriptors closest to each of the 19 ideal types identified by Torgerson *et al*. From Torgerson, W. S., BenDebba, M., and Mason, K. J. 1988. Varieties of Pain. In: Proceedings of the Vth World Congress on Pain (*eds*. R. Dubner, G. R. Gebhart. and M. R. Bond), p. 368. Elsevier.

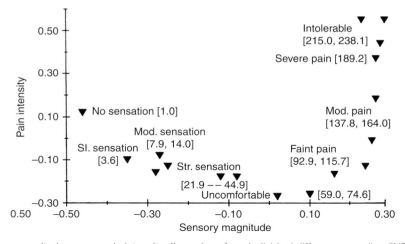

Figure 12 Sensory magnitude versus pain intensity dimensions from individual differences scaling (INDSCAL) analysis of electrocutaneous stimuli. Category labels for each stimulus are determined from the mean response from a method of limits procedure. Numbers in brackets refer to the mean power (mW) of the stimulus. Note that the stimuli labeled no sensation to faint pain order well on the abscissa (sensory magnitude dimension). Higher intensities are clustered together on the sensory magnitude dimension but separate out on the pain intensity dimension (the ordinate) from faint pain to intolerable. From Janal, M. N., Clark, W. C., and Carroll, J. D. 1991. Multidimensional scaling of painful and innocuous electrocutaneous stimuli: reliability and individual differences. Percept. Psychophys. 50, 108–116.

electrical stimuli distinguished between nonpainful and painful electrical stimulation. Together, these results suggest that this scaling approach scales the common dimension of intensity present for all stimuli and in addition produces separate scales specific to the unique characteristics of each stimulus modality.

61.3.2 Issues Concerning Multidimensional Scaling

The complex methods of multidimensional scaling have been shown to provide a sensitive assessment of common and stimulus-specific pain dimensions. When is this extra effort warranted? Naliboff

(Gracely, R. H. and Naliboff, B. D., 1996) identified four criteria: increase in reliability or accuracy, diagnostic sensitivity, improved evaluation of psychological and other variables that lead to improved treatment, and increasing basic knowledge of pain mechanisms.

There is at present little empirical evidence that the extra dimensions provided by these measures increase either reliability or accuracy. In most cases, the extra dimensions describe how different stimulus modalities are different, and add little to the dimensions common to all stimuli that are already adequately described by simple measures of pain intensity or possibly dual scales of pain intensity and unpleasantness. However, in certain instances, the distinction between different qualities may be important to describe pain changes in this quality (efficacy of nerve block, spontaneous change in pain qualities).

The use of these methods to determine pain quality as an aid to diagnosis has immediate face appeal. Current diagnostic decision making is based in large part on qualitative dimensions. Although these methods are promising, there are little data linking the outcome of these scales with diagnostic pain groupings. Studies have discriminated between heterogeneous groups such as back pain or headache. However, there is no evidence that these approaches provide any increase in utility over simple diagnostic tests and simple scales of one or two pain dimensions. Such evidence would have to be substantial, satisfying standard psychometric criteria for reliability, cross-validation in different setting, sensitivity, specificity, and predictive utility in relation to prevalence (Anastasi, A., 1968). If utility is demonstrated by these methods, it is likely to be in specific situations. One of the best examples of a more difficult discrimination aided by these methods is the study by Grushka M. and Sessle B. J. (1984)in which the MPQ distinguished a tooth with a reversibly inflamed pulp and a tooth in which the pulp inflammation was irreversible based on verbal pain reports.

Multidimensional analysis of pain ratings have also been proposed as a way to assess broader issues of psychological functioning in pain patients. Clark and coworkers have even suggested that questions such as "Is this patient really in pain or merely depressed?" can be addressed by an analysis of "What are the patient's coordinates in the multidimensional global pain space and which dimensions are most relevant for him?" The extension of this view is that treatment would follow from an analysis of an individual multidimensional diagnosis. A patient rating clinical pain high on a somatosensory burning dimension would receive a different medication than one with a similar pathophysiology but a greater score on a somatosensory tingling dimension or an affective dimension. To date there is no empirical support for the validity of using pain reports in this manner or for the assumption that assessment of subjective pain quality offers a more accurate and useful assessment of psychological functioning than more traditional mood or personality instruments.

Improved quantification of a patient's pain report might also increase health providers' ability to communicate, express empathy, and enlist cooperation from a suffering patient. Patients often complain that their pain report is misunderstood or dismissed by providers and perhaps a better common language could help this process. This may be especially relevant for changes in pain over the course of treatment and adjustment of pharmacological interventions. A potential difficulty with the current approach to multidimensional scaling is that the dimensions are often difficult for even the researcher to interpret. Application to the clinical setting requires translation into language with face validity for both clinicians and patients. This application of multidimensional assessment reinforces the need to examine pain as a dynamic process and to validate the ability of any multidimensional framework to monitor pain as it changes with time and treatment. In addition, the multidimensional analytical techniques applied to pain have to date emphasized the discovery and validation of painfulness dimensions, and not analytical tools for assessing pain in a clinical situation over time.

Perhaps the most promising area for application of multidimensional pain assessment is in basic research on neurophysiology and neuroanatomy of pain. There has been and will continue to be rapid growth in techniques for the study of ongoing central nervous system (CNS) processes such as functional brain imaging and CNS stimulation and recording techniques. Better quantification of a subject's subjective experience along with multiple dimensions may allow for specificity in the psychological domain corresponding to specific neurophysiological processes. For example, Lenz F. A. et al. (1993; 1994) have found that electrical stimulation of the thalamus in patients undergoing stereotactic neurosurgery evokes sensations that can be referred to distinct body locations. When these locations coincide with the locus of a chronic pain syndrome, the quality of the evoked pain is similar to the quality of the pain experienced chronically by the patient. Evoked sensations located at other locations either were not painful or did not resemble the patient's clinical pain. These studies rely heavily on valid multidimensional pain assessment. In this case, the investigators used a tree-branching questionnaire that efficiently evaluated pain quality in an intraoperative environment.

61.4 Conclusion

Pain is in part a sensory experience that can be assessed by the methods used to evaluate other sensory modalities. While the concept of pain threshold is well known and often used to imply pain sensitivity, adequate evaluation of pain sensitivity must include the range of pain from threshold to tolerable suprathreshold intensities. While many traditional psychophysical methods focus on the relation between different stimulus intensities, pain measures strive to provide absolute pain assessment that facilitates evaluation of slow-acting analgesics or comparing pain levels between groups. Pain also varies greatly in quality, and in temporal and spatial characteristics. New measures employing multidimensional scaling and other novel methods continue to address these dimensions. Finally, similar to some modalities, pain motivates behaviors of avoidance, escape, guarding, and recuperation. It is immediately unpleasant and commands attention. If pain persists, it is associated with both plastic changes in the nervous systems and secondary affective components associated with cognitions of distress and despair. Assessment of sensory function provides an important part of the evaluation, and the application of psychophysical methods continues to increase the understanding of pain mechanisms and provide diagnostic information in individual patients. This realm of pain assessment provides necessary information, but no single approach is completely sufficient. In the clinic, measures of pain sensation and unpleasantness must be supplemented by measures from other branches of psychology and medicine, including cognitive style, personality, mood, beliefs, and function.

References

Algom, D., Raphaeli, N., and Cohen-Raz, L. 1986. Integration of noxious stimulation across separate somatosensory communications systems: a functional theory of pain. J. Exp. Psychol. Hum. Percept. Perform. 12, 92–102.

Anastasi, A. 1968. Psychological Testing, 3rd ed. Macmillan.

Anderson, N. H. 1974. Algebraic Models in Perception. In Handbook of Perception, Vol. 2 (eds. E. C. Carterette and M. P. Friedman), pp. 251–298. Academic Press.

Andersen, O. K., Gracely, R. H., and Arendt-Nielsen, L. 1995. Facilitation of the human nociceptive reflex by stimulation of A beta-fibres in a secondary hyperalgesic area sustained by nociceptive input from the primary hyperalgesic area. Acta Physiol. Scand. 155, 87–97.

Atkinson, R. C., Herrnstein, R. J., Lindzey, G., and Luce, R. D. 1998. Stevens Handbook of Experimental Psychology. 2nd ed., Vol. 1, Perception and Motivation. Vol. 1. John Wiley and Sons.

BenDebba, M., Torgerson, W. S., and Long, D. M. 1993. Measurement of affective and sensory qualities of back pain. *Paper presented at the 7th World Congress on Pain, Paris, France.*

Bendtsen, L., Jensen, R. A., and Olesen, J. 1997. Decreased pain threshold and tolerance in patients with chronic tension headache. Ugeskr. Laeg. 159, 4521–4525.

Benoliel, R., Eliav, E., and Iadarola, M. J. 2001. Neuropeptide Y in trigeminal ganglion following chronic constriction injury of the rat infraorbital nerve: is there correlation to somatosensory parameters? Pain 91, 111–121.

Benoliel, R., Wilensky, A., Tal, M., and Eliav, E. 2002. Application of a pro-inflammatory agent to the orbital portion of the rat infraorbital nerve induces changes indicative of ongoing trigeminal pain. Pain 99, 567–578.

Bentley, D. E., Watson, A., Treede, R. D., Barrett, G., Youell, P. D., Kulkarni, B., and Jones, A. K. 2004. Differential effects on the laser evoked potential of selectively attending to pain localisation versus pain unpleasantness. Clin. Neurophysiol. 115, 1846–1856.

Bock, R. D. and Jones, L. V. 1968. The Measurement and Prediction of Judgment and Choice. Holden-Day.

Bolanowski, S. J. and Gescheider, G. A. 1991. Ratio Scaling of Psychological Magnitude. Lawrence Erlbaum Associates, Inc.

Brennum, J., Arendt-Nielsen, L., Secher, N. H., Jensen, T. S., and Bjerring, P. 1992. Quantitative sensory examination during epidural anesthesia in man: effects of lidocaine. Pain 51, 27–34.

Buchsbaum, M. S., Davis, G. C., Copopola, R., and Naber, D. 1981. Opiate pharmacology and individual differences. I. Psychophysical pain measurements. Pain 10, 357–366.

Bujas, Z., Frank, M., and Pfaffmann, C. 1979. Neural effects of electrical taste stimuli. Sens. Processes. 3, 353–365.

Cannon, R. C., Quyyumi, A. A., Schenke, W. H., Fananapazir, L., Tucker, E. E., Gaughan, A. M., Gracely, R. H., Cattan, E. L., and Epstein, S. 1990. Abnormal Cardiac Sensitivity in Patients with Chest Pain and Normal Coronary Arteries. J. Am. Coll. Cardiol. 16, 1359–1366.

Carterette, E. C. and Friedman, M. P. 1974. Handbook of Perception, Vol. 2. Academic Press.

Chacur, M., Milligan, E. D., Gazda, L. S., Armstrong, C., Wang, H., Tracey, K. J., Maier, S. F., and Watkins, L. R. 2001. A new model of sciatic inflammatory neuritis (SIN): induction of unilateral and bilateral mechanical allodynia following acute unilateral peri-sciatic immune activation in rats. Pain 94, 231–244.

Chapman, C. R. 1977. Sensory decision theory methods in pain research: a reply to Rollman. Pain 3, 295–305.

Chapman, C. R. and Feather, B. W. 1973. Effects of diazepam on human pain tolerance and sensitivity. Psychosom. Med. 35, 330–340.

Chapman, C. R., Gehrig, J. D., and Wilson, M. E. 1975. Acupuncture compared with 33 percent nitrous oxide for dental analgesia: a sensory decision theory evaluation. Anesthesiology 42, 532–537.

Chapman, C. R., Murphy, T. M., and Butler, S. H. 1973. Analgesic strength of 33 percent nitrous oxide: a signal detection evaluation. Science 179, 1246–1248.

Chapman, C. R., Wilson, M. E., and Gehrig, J. D. 1976. Comparative effects of acupuncture and transcutaneous stimulation on the perception of painful dental stimuli. Pain 2, 265–283.

Clark, W. C. and Yang, J. C. 1974. Acupunctural analgesia? Evaluation by signal detection theory. Science 184, 1096–1098.

Clark, W. C. and Yang, J. C. 1983. Applications of Sensory Decision Theory to Problems in Laboratory and Clinical Pain. In: Pain Measurement and Assessment (ed. R. Melzack), pp. 15–25. Raven Press.

Clark, W. C., Janal, M. N., and Carroll, J. D. 1989. Multidimensional Pain Requires Multidimensional Scaling. In: Issues in Pain Measurement (eds. C. R. Chapman and J. D. Loeser), p. 285. Raven Press.

Cliff, N. 1959. Adverbs as multipliers. Psychol. Rev. 66, 27–44.

Collins, W. R. Jr., Nulsen, F. E., and Randt, C. T. 1960. Relation of peripheral nerve fiber size and sensation in man. Arch. Neurol. 3, 381–385.

Cooper, B. Y., Vierck, C. J, Jr, and Yeomans, D. C. 1986. Selective reduction of second pain sensation by systemic morphine in humans. Pain 24, 93–116.

Coppola, R. and Gracely, R. H. 1983. Where is the noise in SDT pain assessment? Pain 17, 257–266.

Cornsweet, T. N. 1962. The staircase-method in psychophysics. Am. J. Psychol. 75, 485–491.

Dallenbach, K. M. 1939. Pain: history and present status. Am. J. Psychol. 52, 331–347.

Dao, T. T. and Mellor, A. 1998. Sensory disturbances associated with implant surgery. Int. J. Prosthodont. 11, 462–469.

Doctor, J. N., Slater, M. A., and Atkinson, J. H. 1995. The descriptor differential scale of pain intensity: an evaluation of item & scale properties. Pain 61, 251–260.

Duncan, G. H., Feine, J. S., Bushnell, M. C., and Boyer, M. 1988. Use of Magnitude Matching for Measuring Group Differences in Pain Perception. In: Proceedings of the Vth World Congress on Pain (eds. R. Dubner, G. R. Gebhart, and M. R. Bond), pp. 383–390. Elsevier.

Duncan, G. H., Miron, D., and Parker, S. R. 1992. Yet another adaptive scheme for tracking threshold. *Meeting of the International Society for Psychophysics*, July, Stockholm, July.

Dunckley, P., Wise, R. G., Aziz, Q., Painter, D., Brooks, J., Tracey, I., and Chang, L. 2005. Cortical processing of visceral and somatic stimulation: differentiating pain intensity from unpleasantness. Neuroscience 133, 533–542.

Eliav, E. and Gracely, R. H. 1998. Sensory changes in the territory of the lingual and inferior alveolar nerves following lower third molar extraction. Pain 77, 191–199.

Eliav, E., Benoliel, R., and Tal, M. 2001. Inflammation with no axonal damage of the rat saphenous nerve trunk induces ectopic discharge and mechanosensitivity in myelinated axons. Neurosci. Lett. 311, 49–52.

Eliav, E., Herzberg, U., Ruda, M. A., and Bennett, G. J. 1999. Neuropathic pain from an experimental neuritis of the rat sciatic nerve. Pain 83, 169–182.

Eliav, E., Kamran, B., Gracely, R. H., and Benoliel, R. 2007. Evidence for chorda Tympani Dysfunction in Buring Mouth Syndrome patients. J. Orofacial. Pain 138, 628–633.

Eliav, E., Teich, S., Benoliel, R., Nahlieli, O., Lewkowicz, A. A., Baruchin, A., and Gracely, R. H. 2002. Large myelinated nerve fiber hypersensitivity in oral malignancy. Oral Surg. Oral Med. Oral Pathol. Oral Radiol. Endod. 94, 45–50.

Eliav, E., Teich, S., Nitzan, D., El Raziq, D. A., Nahlieli, O., Tal, M., Gracely, R. H., and Benoliel, R. 2003. Facial arthralgia and myalgia: can they be differentiated by trigeminal sensory assessment? Pain 104, 481–490.

Engen, T. 1971a. Psychophysics I. Discrimination and Detection. In: Experimental Psychology 3rd ed.,. (eds. J. W. Kling and L. A. Riggs), 3rd ed., p. 11. Holt.

Engen, T. 1971b. Psychophysics II. Scaling Methods. In: Experimental Psychology 3rd ed.,. (eds. J. W. Kling and L. A. Riggs), 3rd ed., p. 47. Holt.

Falmagne, J. C. 1986. Psychophysical measurement and theory. In: Handbook of Perception and Human Performance, Vol. 1, Sensory Processes and Perception (eds. K. R. Boff, L. Kaufman, and J. P. Thomas), John Wiley & Sons.

Feather, B. W., Chapman, C. R., and Fisher, S. B. 1972. The effect of placebo on the perception of painful radiant heat stimuli. Psychosoma. Med. 34, 290–294.

Fernandez, E., Nygren, T. E., and Thorn, B. E. 1991. An "open-transformed scale" for correcting ceiling effects and enhancing retest reliability: the example of pain. Percept. Psychophys. 49, 572–578.

Formaker, B. K. and Frank, M. E. 2000. Taste function in patients with oral burning. Chem. Senses 25, 575–581.

Gescheider, G. A. 1985. Psychophysics: Method, Theory, and Application. Lawrence Erlbaum Associates, Inc.

Gracely, R. H. 1977. Pain psychophysics. Doctoral dissertation, Department of Psychology, Brown University.

Gracely, R. H. 1979. Psychophysical Assessment of Human Pain. In: Advances in pain research and therapy, Vol. 3 (eds. J. J. Bonica, J. C. Liebeskind, and D. G. Able-Fessard), pp. 805–824. Raven Press.

Gracely, R. H. 1988. Multiple Random Staircase Assessment of Thermal Pain Perception. In: Proceedings of the Fifth World Congress on Pain (eds. R. Dubner, M. Bond, and G. Gebhart), p. 391. Elsevier.

Gracely, R. H. 1991. Experimental Pain Models. In: Advances in Pain Research and Therapy. Vol. 18, The Design of Analgesic Clinical Trials, (eds. M. Max, R. Portenoy, and E. Laska), p. 33, Raven Press.

Gracely, R. H. 1992a. Affective dimensions of pain: how many and how measured? APS J. 71, 243–247.

Gracely, R. H. 1992b. Evaluation of multidimensional pain scales. Pain 48, 297–300.

Gracely, R. H. 1994. Methods of Testing Pain Mechanisms in Normal Man. In: Textbook of Pain, 3rd edition. (eds. P. D. Wall and R. Melzack), p. 315. Churchill Livingstone.

Gracely, R. H. and Dubner, R. 1987. Reliability and validity of verbal descriptor scales of painfulness. Pain 29, 175–185.

Gracely, R. H. and Kwilosz, D. M. 1988. The descriptor differential scale: applying psychophysical principles to clinical pain assessment. Pain 35, 279–288.

Gracely, R. H. and Naliboff, B. D. 1996. Measurement of Pain Sensation. In: Handbook of Perception and Cognition: Somatosensory Systems (ed. L. Kruger), p. 243. Raven Press.

Gracely, R. H. and Wolskee, P. J. 1983. Semantic functional measurement of pain: integrating perception and language. Pain 15, 389–398.

Gracely, R. H., Deeter, W. R., Wolskee, P. J., and Dubner, R. 1987. Does naloxone alter experimental pain perception? Soc. Neurosci. Abstr. 6, 246.

Gracely, R. H., Dubner, R., and McGrath, P. A. 1979. Narcotic analgesia: fentanyl reduces the intensity but not the unpleasantness of painful tooth pulp stimulation. Science 203, 1261–1263.

Gracely, R. H., Lota, L., Walther, D. J., and Dubner, R. 1988. A multiple random staircase method of psychophysical pain assessment. Pain 32, 55–63.

Gracely, R. H., McGrath, P. A., and Dubner, R 1978a. Ratio scales of sensory and affective verbal pain descriptors. Pain 5, 5–18.

Gracely, R. H., McGrath, P., and Dubner, R. 1978b. Validity and sensitivity of ratio scales of sensory and affective verbal pain descriptors: manipulation of affect by diazepam. Pain 5, 19–29.

Gracely, R. H., Taylor, F., Schilling, R. M., and Wolskee, P. J. 1984. The effect of a simulated analgesic on verbal descriptor and category responses to thermal pain. Pain Suppl. 2, 173.

Greenspan, J. D., Roy, E. A., Caldwell, P. A., and Farooq, N. S. 2003. Thermosensory intensity and affect throughout the perceptible range. Somatosens. Mot. Res. 20, 19–26.

Grushka, M. 1987. Clinical features of burning mouth syndrome. Oral Surg. Oral Med. Oral Pathol. 63, 30–36.

Grushka, M. and Sessle, B. J. 1984. Applicability of the McGill Pain Questionnaire to the differentiation of 'toothache' pain. Pain 19, 49–57.

Grushka, M. and Sessle, B. 1988. Taste dysfunction in burning mouth syndrome. Gerodontics 4, 256–258.

Grushka, M. and Sessle, B. J. 1991. Burning mouth syndrome. Dent. Clin. North Am. 35, 171–184.

Grushka, M., Epstein, J. B., and Gorsky, M. 2002. Burning mouth syndrome. Am. Fam. Physician 65, 615–620.

Grushka, M., Epstein, J. B., and Gorsky, M. 2003. Burning mouth syndrome and other oral sensory disorders: a unifying hypothesis. Pain Res. Manag. 8, 133–135.

Hagberg, C., Hellsing, G., and Hagberg, M. 1990. Perception of cutaneous electrical stimulation in patients with cranioman-dibular disorders. J. Craniomandib. Disord. 4, 120–125.

Hansson, P. 2002. Neuropathic pain: clinical characteristics and diagnostic workup. Eur. J. Pain 6(Suppl), A47–A50.

Hardy, J. D., Wolff, H. G., and Goodell, H. S. 1952. Pain Sensations and Reactions. Hafner.

Hargreaves, K. M., Dionne, R. A., Meuller, G. P., Goldstein, D. S., and Dubner, R. 1986. Naloxone, fentanyl and diazepam modify plasma beta-endorphin levels during surgery. Clin. Pharmacol. Ther. 40, 165–171.

Heft, M. W. and Parker, S. R. 1984. An experimental basis for revising the graphic rating scale. Pain 19, 153–161.

Heft, M. W., Gracely, R. H., Dubner, R., and McGrath, P. A. 1980. A validation model for verbal descriptor scaling of human clinical pain. Pain 9, 363–373.

Hildgard, E. R. 1975. The alleviation of pain by hypnosis. Pain 1, 213–231.

Hofbauer, R. K., Rainville, P., Duncan, G. H., and Bushnell, M. C. 2001. Cortical representation of the sensory dimension of pain. J. Neurophysiol. 86, 402–411.

Holroyd, K. A., Holm, F. J., Keefe, F. J., Turner, J. A., Bradley, L. A., Murphy, W. D., Johnson, P., Anderson, K., Hinkle, A. L., and O'Malley, W. B. 1992. A multi-center evaluation of the McGill Pain Questionnaire: results from more than 1700 chronic pain patients. Pain 48, 297–300.

Jamison, R. N., Gracely, R. H., Raymond, S. A., Levine, J. G., Marino, B., Hermann, T. J., Daly, M., Fram, D., and Katz, N. P. 2002. Comparative study of electronic vs. paper VAS ratings: a randomized, crossover trial using healthy volunteers. Pain 99, 341–347.

Janal, M. N., Clark, W. C., and Carroll, J. D. 1991. Multidimensional scaling of painful and innocuous electrocutaneous stimuli: reliability and individual differences. Percept. Psychophys. 50, 108–116.

Janal, M. N., Clark, W. C., and Carroll, J. D. 1993. Multidimensional scaling of painful electrocutaneous stimulation: INDSCAL dimensions, signal detection theory indices, and the McGill Pain Questionnaire. Somatosens. Mot. Res. 10, 31–39.

Jensen, M. P. and McFarland, C. A. 1993. Increasing the reliability and validity of pain intensity measurement in chronic pain patients. Pain 55, 195–203.

Jensen, M. P., Koroly, P., and Braver, S. 1986. The measurement of clinical pain intensity: a comparison of six methods. Pain 27, 117–126.

Jones, B. 1980. Algebraic models for integration of painful and nonpainful electric shocks. Percept. Psychophys. 28, 572–576.

Kemler, M. A., Schouten, H. J., and Gracely, R. H. 2000. Diagnosing sensory abnormalities with either normal values or values from contralateral skin: comparison of two approaches in complex regional pain syndrome I. Anesthesiology 93, 718–727.

Kim, H., Neubert, J. K., San Miguel, A., Xu, K., Krishnaraju, R. K., Iadarola, M. J., Goldman, D., and Dionne, R. A. 2004. Genetic influence on variability in human acute experimental pain sensitivity associated with gender, ethnicity and psychological temperament. Pain 109, 488–496.

Koltzenburg, M., Stucky, C. L., and Lewin, G. R. 1997. Receptive properties of mouse sensory neurons innervating hairy skin. J. Neurophysiol. 78, 1841–1850.

Kulkarni, B., Bentley, D. E., Elliott, R., Youell, P., Watson, A., Derbyshire, S. W. G., Frackowiak, R. S. J., Friston, K. J., and Jones, A. K. P. 2005. Attention to pain localization and unpleasantness discriminates the functions of the medial and lateral pain systems. Eur. J. Neurosci. 21, 3133–3142.

Kwilosz, D. M., Gracely, R. H., and Torgerson, W. S. 1984. Memory for post-surgical dental pain. Pain Suppl. 2, 426.

Kwilosz, D. M., Schmidt, E. A., and Gracely, R. H. 1986. Assessment of clinical pain: cross-validation of parallel forms of the descriptor differential scale. Am. Pain Soc. Abstr. 6, 81.

LaMotte, R. H. and Campbell, J. N. 1978. Comparison of responses of warm and nociceptive C-fiber afferents in monkey with human judgments of thermal pain. J. Neurophysiol. 41, 509–528.

Lembo, A., Naliboff, B., Matin, K., Munakata, J., Parker, R., Gracely, R. H., and Mayer, E. A. 2000. Irritable bowel syndrome patients show altered sensitivity to exogenous opioids. Pain 87, 137–147.

Lenz, F. A., Gracely, R. H., Hope, E. J., Baker, F. H., Rowland, L. H., Dougherty, P. M., and Richardson, R. T. 1994. The sensation of angina pectoris can be evoked by stimulation of the human thalamus. Pain 59, 119–125.

Lenz, F. A., Seike, M., Richardson, R. T., Lin, Y. C., Baker, F. H., Khoja, I., Yeager, C. J., and Gracely, R. H. 1993. Thermal and pain sensations evoked by microstimulation in the area of human ventrocaudal nucleus (Vc). J. Neurophysiol. 70, 200–212.

Lindemann, B. 1996. Taste reception. Physiol. Rev. 76, 718–766.

Malone, M. D., Kurtz, R. M., and Strube, M. J. 1989. The effects of hypnotic suggestion on pain report. Am. J. Clin. Hypn. 31, 221–230.

Marks, L. E. 1974. Sensory Processes: The New Psychophysics. Academic Press.

McArthur, D. L., Cohen, M. J., and Schandler, S. L. 1991. Rash analysis of functional assessment scales: and example using pain behaviors. Arch. Phys. Med. Rehabil. 72, 296–304.

McGrath, P. A., Gracely, R. H., Dubner, R., and Heft, M. W. 1983. Non-pain and pain sensations evoked by tooth pulp stimulation. Pain 15, 377–388.

Melzack, R. 1975. The McGill Pain Questionnaire: major properties and scoring methods. Pain 1, 277–299.

Melzack, R. 1987. The short-form McGill Pain Questionnaire. Pain 30, 191–197.

Melzack, R. and Katz, J. 1992. The McGill Pain Questionnaire: Appraisal and Current Status. In: Handbook of Pain Assessment (eds. D. C. Turk and R. Melzack), p. 152. Guilford Press.

Melzack, R. and Torgerson, W. S. 1971. On the language of pain. Anesthesiology 34, 50–59.

Mogil, J. S., Miermeister, F., Seifert, F., Strasburg, K., Zimmermann, K., Reinold, H., Austin, J. S., Bernardini, N., Chesler, E. J., Hofmann, H. A., Hordo, C., Messlinger, K., Nemmani, K. V., Rankin, A. L., Ritchie, J., Siegling, A.,

Smith, S. B., Sotocinal, S., Vater, A., Lehto, S. G., Klussmann, S., Quirion, R., Michaelis, M., Devor, M., and Reeh, P. W. 2005. Variable sensitivity to noxious heat is mediated by differential expression of the CGRP gene. Proc. Natl. Acad. Sci. U. S. A. 102, 12938–12943.

Mower, G. 1976. Perceived intensity of peripheral thermal stimuli is independent of internal body temperature. J. Comp. Physiol. Psychol. 90, 1152–1155.

Mower, G., Mair, R., and Engen, T. 1977. Influence of internal factors on the perception of olfactory and gustatory stimuli. In: The Chemical Senses and Nutrition. (eds. M. R. Hare and O. Maller), Academic Press.

Mott, A. E., Grushka, M., and Sessle, B. J. 1993. Diagnosis and management of taste disorders and burning mouth syndrome. Dent. Clin. North Am. 37, 33–71.

Parducci, A. 1974. Contextual Effects: A Range-Frequency Analysis. In: Handbook of Perception, Vol. 2 (eds. E. C. Carterette and M. P. Friedman), p. 127. Academic Press.

Petzke, F., Clauw, D. J., Ambrose, K., Khine, A., and Gracely, R. H. 2003a. Increased pain sensitivity in fibromyalgia: effects of stimulus type and mode of presentation. Pain 105, 403–413.

Petzke, F., Gracely, R. H., Park, K. M., Ambrose, K., and Clauw, D. J. 2003b. What do tender points measure? Influence of distress on 4 measures of tenderness. J. Rheumatol. 30, 567–574.

Pollack, I. 1965. Iterative techniques for unbiased rating scales. Q. J. Exp. Psychol. 17, 139–148.

Price, D. D. 1988. Psychological and Neural Mechanisms of Pain., Raven Press.

Price, D. D. and Harkins, S. W. 1992. The affective-motivational dimension of pain: a two stage model. APS J. 1, 229–239.

Price, D. D., Harkins, S. W., and Baker, C. 1987. Sensory-affective relationships among different types of clinical and experimental pain. Pain 28, 297–307.

Price, D. D., McGrath, P. A., Rafii, A., and Buckingham, B. 1983. The validation of visual analogue scales as ratio scale measures in for chronic and experimental pain. Pain 17, 45–56.

Price, D. D., McHaffie, J. G., and Larson, M. A. 1989. Spatial summation of heat-induced pain: influence of stimulus area and spatial separation of stimuli on perceived pain sensation intensity and unpleasantness. J. Neurophysiol. 62, 1270–1279.

Poulton, E. C. 1979. Models for biases in judging sensory magnitude. Psychol. Bull. 86, 777–803.

Rainville, P., Duncan, G. H., Price, D. D., Carrier, B., and Bushnell, M. C. 1997. Pain affect encoded in human anterior cingulated but not somatosensory cortex. Science 277, 968–971.

Rainville, P., Feine, J. S., Bushnell, M. C., and Duncan, G. H. 1992. A psychophysical comparison of sensory and affective responses to four modalities of experimental pain. Somatosens. Mot. Res. 9, 265–277.

Rollman, G. B. 1977. Signal detection theory measurement of pain: a review and critique. Pain 3, 187–211.

Sang, C. N., Max, M. B., and Gracely, R. H. 2003. Stability and reliability of detection thresholds for human A-Beta and A-delta sensory afferents determined by cutaneous electrical stimulation. J. Pain Symptom Manage. 25, 64–73.

Ship, J. A., Grushka, M., Lipton, J. A., Mott, A. E., Sessle, B. J., and Dionne, R. A. 1995. Burning mouth syndrome: an update. J. Am. Dent. Assoc. 126, 842–853.

Stevens, S. S. 1975. Psychophysics: Introduction to its Perceptual, Neural and Social Prospects. Wiley.

Stohler, C. S., Kowlski, C. J., and Lund, J. P. 2001. Muscle pain inhibits cutaneous touch perception. Pain 92, 327–333.

Strigo, I. A., Bushnell, M. C., Boivin, M., and Duncan, G. H. 2002. Psychophysical analysis of visceral and cutaneous pain in human subjects. Pain 97, 235–246.

Svensson, P., Graven-Nielsen, T., and Arendt-Nielsen, L. 1998. Mechanical hyperesthesia of human facial skin induced by tonic painful stimulation of jaw muscles. Pain 74, 93–100.

Svensson, P., List, T., and Hector, G. 2001. Analysis of stimulus-evoked pain in patients with myofascial temporomandibular pain disorders. Pain 92, 399–409.

Swets, J. A. 1964. Signal detection and recognition by human observers, Wiley.

Syrjala, K. 1989. Multidimensional Versus Unidimensional Scaling: Little To Debate, Much To Determine. In: Issues in Pain Measurement (eds. C. R. Chapman and J. D. Loeser), p. 327. Raven Press.

Teghtsoonian, R. 1971. On the exponents in Stevens' law and the constant in Ekman's law. Psychol. Rev. 78, 71–80.

Thurstone, L. I. 1959. The measurement of values. University of Chicago Press.

Torgerson, W. S. 1958. Theory and Methods of Scaling. John Wiley and Sons.

Torgerson, W. S., BenDebba, M., and Mason, K. J. 1988. Varieties of Pain. In: Proceedings of the Vth World Congress on Pain (eds. R. Dubner, G. R. Gebhart. and M. R. Bond), p. 368. Elsevier.

Tursky, B. 1974. Physical physiological and psychological factors that affect pain reaction to electric shock. Psychophysiology 11, 95–112.

Tursky, B. 1976. The Development of a Pain Perception Profile: A Psychophysical Approach. In: Pain: New Perspectives in Therapy and Research (eds. M. Weisenberg and B. Tursky), p. 171. Plenum Press.

Vecchiet, L., Giamberardino, M. A., and Saggini, R. 1991. Myofascial pain syndromes: clinical and pathophysiological aspects. Clin. J. Pain 7 (Suppl), 1, S16–S22.

Walther, D. J. and Gracely, R. H. 1986. Ratio scales of pleasantness and unpleasantness affective descriptors. Am. Pain Soc. Abstr. 6, 34.

Weiffenbach, J. M., Cowart, B. J., and Baum, B. J. 1986. Taste intensity perception in aging. J. Gerontol. 41, 460–468.

Wunsch, A., Philippopt, P, and Plaghki, L. 2003. Affective associative leaning modifies the sensory perception of nociceptive stimuli without participant' awareness. Pain 102, 27–38.

Yang, J. C., Clark, W. C., Ngai, S. H., Berkowitz, B. A., and Spector, S. 1979. Analgesic action and pharmaco-kinetics of morphine and diazepam in man: an evaluation by sensory decision theory. Anesthesiology 51, 495–502.

Yarnitsky, D., Sprecher, E., Zaslansky, R., and Hemli, J. A. 1996. Multiple session experimental pain measurement. Pain. 67, 327–333.

Further Reading

ACR: Report of the Multicenter Criteria Committee. 1990. 1990 classification criteria of fibromyalgia from the American College of Rheumatology. Union Med. Can. 119, 272.

Gracely, R. H. 1985. Pain psychophysics. In: Advances in Behavioral Medicine Vol. 1 (ed. S. Manuk), JAI Press.

Price, D. D., Bush, F. M., Long, S., and Harkins, S. W. 1994. A comparison of pain measurement characteristics of mechanical and simple numerical rating scales. Pain 56, 217–226.

Rainville, P., Carrier, B., Hofbauer, R. K., Bushnell, M. C., and Duncan, G. H. 1999. Dissociation of sensory and affective dimensions of pain using hypnotic modulation. Pain 82, 159–171.

Warren, R. M. and Warren, R. P. 1963. A critique of SS Stevens' "new psychophysics" Percept. Mot. Skills 16, 797–810.

62 Consciousness and Pain

M Devor, Hebrew University of Jerusalem, Jerusalem, Israel

Glossary

the hard problem Mechanism or algorithm that accounts for the emergence of conscious experience and willed action from neural activity. The expression was coined by Chalmers D. (1995).

locked in Neurological condition, usually due to severe brain injury, in which an individual is awake and alert but completely paralyzed and unable to express himself with words or movement.

MPTA (mesopontine tegmental anesthesia area) Restricted location in the reticular formation bordering the brainstem mesencephalon and pons at which microinjection of minute quantities of pentobarbital, and other $GABA_A$-receptor active drugs, induces a brief period of general anesthesia.

nocifensive response Adaptive behavioral responses intended to avoid, minimize, or escape from noxious, tissue-threatening stimuli. Examples are the flexor reflex as well as more complex escape, distress vocalization, and avoidance behaviors. The term nocifensive does not necessarily imply a reasoned, intentional response.

stress-induced analgesia The observation that stressful events such as electrical shock or forced cold-water swim may be followed by a period of reduced pain sensitivity.

62.1 Introduction

Pain is a sensory and emotional experience, a mental state (Merskey, H., 1986). As such, it cannot be properly addressed without reference to the conscious brain. Analysis of skin receptors, spinal pathways, and other substrates of pain without considering the resulting conscious experience is analysis of nociception, not pain *per se* (see below). The hard problem of how neural activity gives rise to conscious experience and willed action is probably intractable as a subject of philosophical analysis (McGinn, C., 1999). However, despite alleged semantic proofs to the contrary, it should ultimately be amenable to solution using neurobiological tools. Just as life is now understood without reference to supernatural vital spirits, the proposition that consciousness is a bodiless ghost temporarily resident in the gray matter is implausible. It is too obviously subject to conventional neurobiological variables such as circadian rhythmicity, pharmacological modifiers of membrane excitability and synaptic action, blood glucose levels, and so forth. Consciousness is a product of neural activity; there is nothing spooky about it (Searle, J. R., 1992). And while new principles might have to emerge before we fully understand it, they will be natural ones that play out within the cranial vault.

At present we do not know whether the hard problem is hard because of intrinsic complexity (like decoding the genome) or hard only because nobody has yet had the flash of insight necessary to see the essence of the problem and its simple solution. Think of a puzzle many of us faced as kids where six toothpicks were laid on the table and we were challenged to use them to construct four equilateral triangles. You can spend a lifetime working on this one, or solve it in a second by raising the toothpicks off the plane of the table to form a pyramid. Knowledge that the problem has a solution can help keep you from giving up, but it does not provide hints as to how to proceed, nor does it let you know in which of these two senses the problem is hard.

62.2   Definitions

Like alert awareness in general, the experience of pain is a causal consequence of neural activity in the brain – the execution of a neural algorithm. Unfortunately, we do not have the foggiest idea of how this algorithm operates, nor can we even conceive of an algorithm whose output would constitute pain experience. This is the reason that current attempts to define consciousness are basically futile. The numerous definitions that have been proposed are not definitions at all, but either metaphors (spotlight, stage, narrator, multiple drafts, etc.) or semantic restatements (e.g., alertness, subjective experience, self). Imagine some large-headed extraterrestrial arriving on earth and delivering the pain algorithm in the form of compiled code ready to run on a Pentium IV. He assures us that the computer is now conscious and experiences pain given an appropriate input signal. But would we have any way of knowing if this was in fact so unless he were willing to explain the principle in a way we could grasp? The red flashing icon that hurts would clearly prove nothing ... you don't need a large head to make a computer do that. The computer analogy itself may be flawed ... perhaps an entirely different sort of physical architecture is required to implement the algorithm.

The erstwhile conscious Pentium is not unlike humans who become locked in due to central nervous system (CNS) damage and are unable to feel, move, or communicate. However, reports of locked-in patients who found a way to make themselves understood (e.g., by an eye-blink code) leave no doubt that one can be fully awake and aware despite having no way of letting others know (Bauby, J.-D., 1997). Pain is subjective, inaccessible to others, not in its essence but because we do not yet know how to assess it objectively. This may change as noninvasive brain imaging technologies improve. It is important to realize that our inability to conceive of an algorithm that generates awareness in no way undermines the fact that such an algorithm is running in my brain right now, and in yours. To insist that research on the biology of consciousness must begin with a definition is misguided. It stymies research on the subject unnecessarily, before it has begun. The definition is the algorithm; it is the endpoint of the research, not the starting point. But where to begin?

62.3   Pain and the Neurobiology of Consciousness

Although they are private, first-person experiences, we can usually detect whether a person or an animal is conscious or unconscious, or in pain, by simple observations of behavior. This is enough to provide operational definitions for research. True, there may be gray areas (and deceit), but the problem is minor as long as one avoids subtle distinctions and sticks to major effects (see below). Thus, even without knowing the computation that generates conscious experience, practical experiments should be able to uncover some of its general characteristics. For example, it might be possible to learn whether the computation is carried out in a distributed manner or focally, and if focally, in what part of the brain. Beyond that, simply engaging in relevant biological research holds out the prospect that the imagination might be primed, increasing the likelihood of an intuitive leap taking place. Six toothpicks a pyramid makes.

Pain is a favorable point of entry into the neurobiology of consciousness for several reasons. First among them is its immediacy. In the visual system, extensive computation is required to extract meaning from contrasts falling on the retina. Is it a cloud passing by or a rhinoceros charging? Only after this image analysis is completed does the visual scene become a useful object of attention. We do not have conscious access to the intermediate stages of visual image analysis just as we are not aware of the detailed command sequences sent spinally by our motor cortex, or the workings of our autonomic nervous system. The system delivers to a conscious brain the end product of the analysis, a meaningful visual scene, smooth gait, or a feeling of wellness. Output of the analysis is input to the consciousness algorithm. Injury to image analysis or motor sequencing functions of the brain may be crippling, but they cause blindness and spasticity, not obtunded consciousness or coma. Essential meaning in the pain system is an answer to the question: "Is it good for me or bad for me?" But unlike in vision, this information is available without extensive signal processing. Entry into the circuitry of consciousness via the pain system probably takes place within a synapse or two of the dorsal horn. A painful stimulus captures attention like no other.

A second advantageous feature of the pain system is the variable relation between provoking stimuli and the resulting percept. Photons of 450 nm invariably evoke blue, but 48 °C heat pulses do not necessarily evoke pain. The stimulus-color response function adjusts somewhat to environmental contingencies (e.g., color constancy despite variable illumination), but not nearly so robustly as the stimulus-pain response function. For example, at times of danger, pain normally evoked by an intense noxious stimulus, for example, an injury, may be completely suppressed (Melzack, R. et al., 1982). It is generally presumed that this stress-induced analgesia evolved to increase the likelihood of successful fight or flight. Intense pain captures attention, occupying cerebral resources and degrading performance in other spheres. In an existential emergency, performance is critical; the wound can wait. In effect, the pain system disconnects the signal analysis (nociception) from the percept (pain). How is this done?

Most data suggest that the disconnect between injury and pain is executed primarily in the spinal cord. This is unlike color constancy, which is a cortical function. Disconnect reflects, at least in part, the operation of a specific neural circuit with a nodal point in the midbrain periaqueductal gray (PAG) and a variety of defined efferent effector pathways that employ known neurotransmitters, including bioamines and endogenous opiates (Fields, H. L.

and Basbaum, A. I., 1999). The intrinsic descending pain control system(s), or something like it, is probably also engaged in other situations of noxious stimulus-pain disconnect such as the placebo effect, distracted attention, and hypnosis (Rainville, P. et al., 1997; Benedetti, F. et al., 1999). Appreciation of behavioral context (danger, anticipation, belief, and distraction) is a complex, multisensory, cortical function to be sure, but the actual pain disconnect appears to be carried out in the spinal cord or brainstem. Situations in which defined stimulus configurations yield alternative percepts, for example, ambiguous figures like the Necker cube, are fascinating grist for consciousness research. But in the pain system some of the actual neural pathways that implement flip-flop between alternative percepts (viz pain or no pain to a fixed noxious stimulus) have been worked out. For example, under some circumstances pharmacological blockade of enkephalinergic synapses can block the effect of placebos (Benedetti, F. et al., 1999; Skoubis, P. D. et al., 2005). We are talking here about the neural circuitry of belief and anticipation.

A third salient feature of the pain system is that a qualitative, emotional vector is an essential feature of the sensory experience. Authorities on pain psychophysics insist that people can assign a separate value to pain intensity and pain unpleasantness (Rainville, P. et al., 1997). Moreover, drugs and lesions are said to be capable of dissociating between the two: "It hurts just as much as before, but it no longer bothers me" (i.e., no longer causes suffering). This is reflected in extensive involvement of the limbic system in functional brain imaging activations as well as in neuroanatomical connectivity (Peyron, R. et al., 2000). A surprisingly large fraction of the human brain, including extensive regions of the frontal neocortex, appears to be devoted to limbic functions. Affect is undoubtedly an essential, if enigmatic, part of the circuitry of consciousness.

Finally, assuming a willingness to grant that animals, mammals at least, have mental states and neural mechanisms homologous to those that subserve conscious experience in man, the analysis of pain in animals offers a useful experimental shortcut. Present a red triangle to a monkey and it undoubtedly has a corresponding visual percept. To be certain, however (the beast may be daydreaming), and to know that both color and shape information were perceived, it is usually necessary to train the monkey beforehand in order to get an informative behavioral response. In the pain system behavioral responses that mark pain perception, nocifensive responses, are built in, at least partly circumventing the need for extensive operant conditioning. Such responses include reflexes as well as complex escape behavior, distress vocalization, avoidance, and other such nonreflexive nocifensive behaviors. Granted, inferences about conscious perception in animals will always include an element of uncertainty. But progress can be made by taking as a given that mental states are present, and then later verifying in humans conclusions derived about biological mechanisms.

62.4   Pain versus Nociception

Nocifensive responses and avoidance learning could indicate the presence of perceived pain, or they could reflect automatic machine-like reactions of a nonconscious zombie (Woolf, C. J., 1984; Giacino, J. T. *et al.*, 2002; Kobylarz, E. J. and Schiff, N. D., 2004). Since Sherrington C. S. (1904) pain researchers have been trained to use the term nociception rather than pain when there is uncertainty that a conscious percept accompanies the nocifensive response. Nociception may occur when a person is unconscious, but by definition, pain cannot (Merskey, H., 1986). It is not obvious why conscious registration of noxious stimuli (i.e., pain) needs to occur at all. However, under normal circumstances nocifensive responses and pain experience go hand in hand. More than that, the pain system goes out of its way to ensure that noxious stimuli overcome distraction, and even arouse the individual from sleep and sedation. The experience of pain must have survival benefits that unconsciousness nociception cannot provide, perhaps in the realm of optimizing performance, learning, or empathy.

Investigators of other senses have not adopted a mnemonic equivalent to the pain–nociception dichotomy, but this distinction is not unique to pain. The monkey who responded to the red triangle could be machine-like, without awareness, like the door-opener mechanism in a supermarket. Monotonous visual or auditory impressions often do not register in awareness during routine tasks like driving. A particularly striking example is sensory processing during sleep. We toss and turn during sleep, unconscious, but continuously monitor the edge of the bed, and almost never fall out. Complex behavior does not necessarily require attention. And yet, despite periods on autopilot, most of the time we are aware and act with intent.

62.5   Neural Correlates of Pain Perception

62.5.1   Central Nervous System Activation

A noxious stimulus applied to the paw of an animal activates a large number of neurons in the spinal cord, many of which send ascending axons to the brain. A small minority, cells located in the most superficial layers of the dorsal horn, are nociceptive-selective in their responses. Most, however, have a wide dynamic range (WDR), and respond to weak as well as strong cutaneous stimuli. Still others show WDR responses combined with input from proprioceptive and visceral afferents (multireceptive neurons). The nociceptive-selective neurons have attracted the most attention, with some investigators insisting that their projections alone define the brain's pain-processing network (Craig, A. D., 2003). However, there is no rational basis for denying a role for the WDR and multireceptive neurons. These could well deliver a specific pain signal after extraction from the multiplexed mix (Price, D. D. *et al.*, 2003).

The spinal cord and corresponding brainstem trigeminal areas are sufficient alone to sustain simple nocifensive responses, and they play an important role in the early stages of pain processing. But it is a safe bet that the algorithm of conscious pain perception does not run in the spinal cord on the grounds that paraplegics are fully conscious, and may even report pain in their (phantom) legs. Pain happens supraspinally. Microelectrode recordings in animals and noninvasive functional imaging in humans show excitations in many brain areas following pain-provoking stimulation of the skin and internal organs (Peyron, R. *et al.*, 2000). These include structures that are well-known parts of the somatosensory system, such as the ventrobasal thalamus and the primary (S1) and secondary (S2) somatosensory cortices, as well as areas not classically thought of as somatosensory processors, such as the cerebellar cortex and the corpus striatum. Curiously, the classical somatosensory cortex (S1, S2) tends to show less robust and reliable activations than limbic cortical areas such as the anterior cingulate cortex (ACC), the insula, and parts of the frontal lobe. Noxious stimulation of different organs, skin versus viscera for example, reveals different if overlapping patterns of cortical activation appropriate to the different feels evoked (Rainville, P. *et al.*, 1997; Strigo, I. A. *et al.*, 2003). It is likely that intensity, affect, localization, and other aspects of sensory quality are processed as a mosaic, in distinct brain areas. But this does not necessarily mean that perception is a mosaic. The various outputs of cortical processing may be bound in a central location.

62.5.2 Central Nervous System Lesions

Lesions in the primary visual cortex cause perceptual blindness although some visual performance persists based on subcortical and extrastriate pathways (blindsight, Poppel, E. *et al.*, 1973). In contrast, no cerebral area has been discovered in which lesions reliably produce analgesia or provide relief from ongoing pathophysiological pain. Even in the occasional case report of stroke or surgery relieving a preexisting chronic pain, pain typically returns. Indeed, cortical lesions frequently trigger chronic pain (Boivie, J. *et al.*, 1989; Gybels, J. M. and Sweet, W. H., 1990).

Large right parietal lesions can evoke a state of mind (anosognosia) where the individual denies that a body part belongs to him or her despite the obvious visual evidence to the contrary ("Whose leg is it that you put into my bed!"). The leg is no longer an integral part of the body schema, the inverse, as it were, of an amputation phantom. Nonetheless, pain is registered when the leg is pinched; only the source of the pain is denied. Prolonged anoxia, for example, in drowning victims, can cause still more extensive cortical lesions destroying essentially all of the cerebral cortex. If the patient survives, he or she may be left in a persistent vegetative state (PVS). In PVS the patient is eyes-open awake (with retained sleep–wake cycles), but largely unresponsive, and with massive loss of cognitive function and self-initiated movement. However, he or she typically continues to show appropriate, adaptive nocifensive responses, for example, to pinch stimuli (Plum, F. and Posner, J. B., 1980; Kinney, H. C. and Samuels, M. A., 1994; Kobylarz, E. J. and Schiff, N. D., 2004). With no possibility of communicating, reasonable people can differ over whether the patient has any residual conscious pain experience. Some authorities are sufficiently confident that pain is absent in PVS that once the decision is made to permit ending life, they see it as ethically acceptable to withdraw feeding and end life by slow starvation, rather than by painless injection.

The unexpected preservation of pain response following cortical lesions in S1, S2, and beyond can be interpreted in a number of ways: redundancy of cortical areas that support pain perception, a

distributed architecture, or functional organization that is entirely different from the other senses. A final possibility, bordering on heresy, is that fundamental pain perception might not require the cerebral cortex (Dover, M., 2007).

62.5.3 Central Nervous System Stimulation

Expectations based on vision are violated even more dramatically in trials of direct cortical excitation. If pain function is multiply represented (redundant) focal lesions might not produce analgesia, but focal stimulation should nonetheless evoke pain. In fact, it has remained a mystery since the classic experiments of Penfield W. and Rasmussen T. (1955) why pain is rarely if ever evoked by stimulation of the cortical convexity (also see Libet, B., 1973), while visual, auditory, and (nonpainful) somatosensory percepts readily are. It may be argued that the relevant structures are buried in the midsagittal (cingulate gyrus) or Sylvian sulci (insula), or that unlike the other senses, multiple cortical areas must be activated simultaneously. However, it is very rare for epileptic seizures to include auras that are painful even though the underlying cortical discharge is frequently widespread and includes these buried cortices. Recent observations of pain evoked in a small number of epilepsy patients by depth electrodes on the insular cortex is a potential exception (Mazzola, L. *et al.*, 2006), but this needs detailed presentation and replication.

If not for the deeply held conviction that conscious perception is a high cortical function, the foregoing observations might have led to the tentative conclusion that the neural computations that generate pain experience do not, in fact, occur in or require the cortex. In contrast to the cortex, pain is readily evoked by focal (microelectrode) stimulation in areas of the thalamus and brainstem (Dostrovsky, J. O., 2000). Of course, this may simply reflect activation of ascending pathways, as for nerve or spinal cord stimulation. On the other hand, it might mean that the algorithm of pain experience runs subcortically. If this were the case, cortical (like cerebellar) activations evoked by noxious stimuli could reflect the contribution of noxious signals to functions such as memory formation, arousal, or the generation of motor sequences, rather than pain perception. Alternatively, the cortex could contribute qualitative detail to the pain percept such as precise localization and dynamics, which are not required for the raw feel of pain. Indeed, given the high prevalence of chronic poststroke pain (Boivie, J. *et al.*, 1989), it is even possible that overall, cortical activity modulates pain by means of inhibitory control of subcortical circuitry.

62.6 Contents of Consciousness versus Its Presence

62.6.1 Contents

Much research on the neural correlates of perceptual states focuses on the contents of ongoing narrative. For example, an investigator may ask whether cerebral activation (or synchronization) in response to a fixed stimulus configuration changes when the resulting percept changes (e.g., at state transitions of ambiguous figures). In the realm of pain, many studies have shown that certain cortical activations in response to fixed noxious stimuli change when the resulting percept changes, say as a result of hypnotic suggestion or placebo/expectation (Rainville, P. *et al.*, 1999; Wager, T. D. *et al.*, 2004; Raij, T. T. *et al.*, 2005). An optimistic inference is that the altered neural activity correlates with the pain percept because it actually constitutes the pain percept. An alternative possibility, however, is that the change in the percept and in the underlying neural activity both occur subcortically, with the change in the cortex only secondary reflecting the fundamental subcortical change.

62.6.2 Pain and Loss of Consciousness

Compared to contents (states) of consciousness, much less research is done on the transition from consciousness to unconsciousness. By unconscious, I refer to the absence of consciousness, as in coma, not subconscious material that may be inaccessible. This more basic transition may sound mundane, but it is probably a more practical entry point into the underlying neural circuitry. Without question the most potent modality of pain suppression is general anesthesia. Shortly after adequate numbers of anesthetic molecules are inhaled or injected systemically, the most extreme of noxious stimuli can be delivered with no pain sensation at all. It might be argued that this is not analgesia at all as sensory experience is eliminated globally. Of course one might equally say that with loss of consciousness, analgesia is present, and that it is accompanied by blindness, deafness, and anosmia. Aspects of unconscious nociception may

persist during anesthesia, such as cardiovascular and some spinal reflex response. Moreover, neurons at many levels of the neuraxis, from spinal cord to cerebral cortex, continue to respond to sensory input. This includes cortical responses to noxious and nonnoxious somatosensory stimuli. Indeed, nearly all of classical cortical neurophysiology was carried out in anesthetized animals. One could even make the case that consciousness is the thing that is lost by the action of general anesthetics. The mechanism of action of anesthetics is therefore of substantial interest with respect to pain – a unique experimental lead.

How do anesthetics obtund consciousness? The most widely held theory is that anesthetic agents distributed by the circulatory system directly access neurons throughout the CNS where they bind to membrane receptors and suppress either electrical excitability or synaptic action. For example, the anesthetic ketamine suppresses glutamatergic excitation while barbiturates such as pentobarbital enhance GABAergic inhibition. By this model, CNS suppression is global, quenched as if by a wet blanket. Anesthetics, however, can also act in a very different way. Microinjection of minute quantities of pentobarbital into a highly localized region of the mesopontine tegmentum, a region termed the mesopontine tegmental anesthesia area (MPTA), can induce all of the indicators of lost consciousness, assessed behaviorally and by electroencephalography. The same dose injected anywhere else in the brain has minimal or no effects (Devor, M. and Zalkind, V., 2001).

The discovery of the MPTA as a nodal site capable of switching the brain rapidly between alertness and unconsciousness is of considerable interest. Among other things, it indicates that neural circuitry exists that is capable of accessing and switching off pain perception, motor control, and consciousness. The difference between the wet blanket hypothesis and the MPTA circuit hypothesis is not in the molecular agents and receptors involved but in the system architecture. Tracing the connectivity of MPTA neurons provides a practical experimental avenue into the neural circuitry of pain as a conscious experience (Sukhotinsky, I. *et al.*, 2005; 2007). Are motor suppression, analgesia, and cognitive collapse inseparable consequences of loss of consciousness, or are these independent functions with independent circuitries that just happen to merge in the MPTA? Does loss of consciousness from fainting (syncope), concussion, and coma also involve MPTA circuitry?

Such questions have ceased to be topics for armchair speculation and can now be investigated using conventional neurobiological tools.

62.7 Conclusion

Pain is neither a noxious stimulus nor an adaptive nocifensive response. It is a mental state, a sensory and emotional percept experienced by a conscious brain by virtue of the operation of a neural algorithm, which is, in principle, knowable. Adaptive nocifensive responses and avoidance learning may occur in the absence of consciousness, but pain experience cannot. The pain system offers many advantages for the experimental discovery and analysis of the neural circuitry that subserves conscious perception.

References

Bauby, J.-D. 1997. The Diving Bell and the Butterfly: A Memoir of Life in Death, p. 131. Knopf.

Benedetti, F., Arduino, C., and Amanzio, M. 1999. Somatotopic activation of opioid systems by target-directed expectations of analgesia. J. Neurosci. 19, 3639–3648.

Boivie, J., Leijon, G., and Johansson, I. 1989. Central post-stroke pain – a study of the mechanisms through analysis of the sensory abnormalities. Pain 37, 173–185.

Chalmers, D. 1995. Facing up to the problem of consciousness. J. Consciousness Studies 2, 200–219.

Craig, A. D. 2003. Pain mechanisms: labeled lines versus convergence in central processing. Annu. Rev. Neurosci. 26, 1–30.

Devor, M. 2007. Pain, cortex, and consciousness. Behav. Brain. Sci. 30, 89–90.

Devor, M. and Zalkind, V. 2001. Reversible analgesia, atonia, and loss of consciousness on bilateral intracerebral microinjection of pentobarbital. Pain 94, 101–112.

Dostrovsky, J. O. 2000. Role of thalamus in pain. Prog. Brain Res. 129, 245–257.

Fields, H. L. and Basbaum, A. I. 1999. Central Nervous System Mechanisms of Pain Modulation. In: Textbook of Pain (eds. P. D. Wall and R. Melzack), pp. 309–329. Churchill Livingstone.

Giacino, J. T., Ashwal, S., Childs, N., Cranford, R., Jennett, B., Katz, D. I., Kelly, J. P., Rosenberg, J. H., Whyte, J., Zafonte, R. D., and Zasler, N. D. 2002. The minimally conscious state: definition and diagnostic criteria. Neurology 58, 349–353.

Gybels, J. M. and Sweet, W. H. 1990. Neurosurgical Treatment of Persistent Pain, p. 441. Karger.

Kinney, H. C. and Samuels, M. A. 1994. Neuropathology of the persistent vegetative state, a review. J. Neuropath. Exp. Neurol. 53, 548–558.

Kobylarz, E. J. and Schiff, N. D. 2004. Functional imaging of severely brain-injured patients: progress, challenges, and limitations. Arch. Neurol. 61, 1357–1360.

Libet, B. 1973. Electrical Stimulation of Cortex in Human Subjects, and Conscious Sensory Aspects. In: Handbook of Sensory Physiology, (ed. A. Iggo), Vol. II, pp. 743–790. Springer.

Mazzola, L., Isnard, J., and Mauguiere, F. 2006. Somatosensory and pain responses to stimulation of the second somatosensory area (SII) in humans. A comparison with SI and insular responses. Cereb. Cort. 16, 960–968.

McGinn, C. 1999. Can We Solve the Mind-Body Problem? In: The Nature of Consciousness: Philosophical Debates (eds. N. Block, O. Flanagan, and G. Guzeldere), pp. 529–542. MIT Press.

Melzack, R., Wall, P. D., and Ty, T. C. 1982. Acute pain in an emergency clinic: latency of onset and descriptor patterns related to different injuries. Pain 14, 33–43.

Merskey, H. 1986. Pain terms: a current list with definitions and notes on usage. Pain Suppl. 3, S215–S221.

Penfield, W. and Rasmussen, T. 1955. The Cerebral Cortex of Man. Macmillan.

Peyron, R., Laurent, B., and Garcia-Larrea, L. 2000. Functional imaging of brain responses to pain: a review and meta-analysis. Neurophysiol. Clin. 2000, 263–288.

Plum, F. and Posner, J. B. 1980. The Diagnosis of Stupor and Coma. Davis.

Poppel, E., Held, R., and Frost, D. 1973. Residual visual function after brain wounds involving the central visual pathways in man. Nature 243, 295–296.

Price, D. D., Greenspan, J. D., and Dubner, R. 2003. Neurons involved in the exteroceptive function of pain. Pain 106, 215–219.

Raij, T. T., Numminen, J., Narvanen, S., Hiltunen, J., and Hari, R. 2005. Brain correlates of subjective reality of physically and psychologically induced pain. Proc. Natl. Acad. Sci. U. S. A. 102, 2147–2151.

Rainville, P., Carrier, B., Hofbauer, R. K., Bushnell, M. C., and Duncan, G. H. 1999. Dissociation of sensory and affective dimensions of pain using hypnotic modulation. Pain 82, 159–171.

Rainville, P., Duncan, G. H., Price, D. D., Carrier, B., and Bushnell, M. C. 1997. Pain affect encoded in human anterior cingulate but not somatosensory cortex. Science 277, 968–971.

Searle, J. R. 1992. The Rediscovery of the Mind, p. 320. MIT Press.

Sherrington, C. S. 1904. Qualitative difference of spinal reflex corresponding with qualitative difference of cutaneous stimulus. J. Physiol. (Lond.) 30, 39–46.

Skoubis, P. D., Lam, H. A., Shoblock, J., Narayanan, S., and Maidment, N. T. 2005. Endogenous enkephalins, not endorphins, modulate basal hedonic state in mice. Eur. J. Neurosci. 21, 1379–1384.

Strigo, I. A., Duncan, G. H., Boivin, M., and Bushnell, M. C. 2003. Differentiation of visceral and cutaneous pain in the human brain. J. Neurophysiol. 89, 3294–3303.

Sukhotinsky, I., Hopkins, D. A., Lu, J., Saper, C. and Devor, M. 2005. Movement suppression during anesthesia: neural projections from the mesopontine tegmentum to areas involved in motor control. J. Comp. Neurol. 489, 425–448.

Sukhotinsky, I., Zalkind, V., Lu, J., Hopkins, D. A., Saper, C. B., and Devor, M. 2007. Neural pathways associated with loss of consciousness caused by intracerebral microinjection of $GABA_A$-active anesthetics. Eur. J. Neurosci. 25, 1417–1436.

Wager, T. D., Rilling, J. K., Smith, E. E., Sokolik, A., Casey, K. L., Davidson, R. J., Kosslyn, S. M., Rose, R. M., and Cohen, J. D. 2004. Placebo-induced changes in fMRI in the anticipation and experience of pain. Science 303, 1162–1167.

Woolf, C. J. 1984. Long term alterations in the excitability of the flexion reflex produced by peripheral tissue injury in the chronic decerebrate rat. Pain 18, 325–343.

63 Assessing Pain in Animals

S W G Derbyshire, University of Birmingham, Birmingham, UK

Glossary

anthropomorphism The tendency to attribute human motivations, characteristics, or other attributes to nonhuman things including inanimate objects, the weather and, especially, animals.

cortex The outer layer of an organ. Thus, cerebral cortex: the layers of nerve cells at the outer part of the brain.

nociception The biological response to a noxious stimulus or other stimulus that threatens tissue damage if prolonged.

pain The subjective experience that often, but not always, accompanies nociception in socially conscious beings.

representational memory Memories of a particular event or thing with explicit representational content that can be consciously accessed.

63.1 Introduction

There are few living creatures that cannot respond to a noxious stimulus such as a pinch or a burning flame. Strike a match next to the humble *Drosophila* (fruit fly) larvae, for example, and it will bend and roll away (Goodman, M. B., 2003). Responses such as these are dependent upon specialized sensory neurons that preferentially respond to stimuli with the potential to cause tissue damage. These sensory neurons can be found in most animals, including fishes (Sneddon, L. U. *et al.*, 2003a; 2003b), as well as human beings and larvae. Clearly, then, all of these creatures have the apparatus to detect and respond to potentially dangerous stimulation, but can they all be said to feel pain?

If we define pain as the response to noxious stimulation then the answer must be yes, the larvae, fishes, and human being all feel pain. This definition, however, would also mean that a thermostat feels pain because it responds to excessive heat by changing its internal state. Similarly, even rocks must feel pain as they respond to excessive force by shattering. Such a definition seems much too permissive, but it is also

flawed because it leads to a tautological understanding of pain. Pain is defined in terms of a stimulus that is deemed to be painful because it elicits the pain response. To put that more simply, pain is defined as pain (Derbyshire, S. W. G., 1999a). To escape from this tautology, a deeper examination of both the biology and the psychology of pain is required. Understanding the minimal necessary biology for pain will enable us to eliminate those animals that lack that biology from being able to experience pain. Similarly, understanding the psychological structure of pain will enable us to eliminate those animals that lack that structure from being able to experience pain.

63.2 The Biology of Pain

Modern imaging techniques have allowed unprecedented access to the workings of the brain and over the past decade a series of researchers have examined brain responses during pain experience (e.g., Craig,

A. D. *et al.*, 1996; Rosen, S. D. *et al.*, 1996; Coghill, R. C. *et al.*, 2003; Derbyshire, S. W. G. *et al.*, 2004). These studies have not identified a single center responsible for pain processing but have instead revealed a network or matrix of brain areas (see reviews by Treede, R. D. *et al.*, 1999; Derbyshire, S. W. G., 2000; Peyron, R. *et al.*, 2000; Vogt, B. A. *et al.*, 2003). This matrix includes subcortical structures, such as the thalamus, as well as cortical regions widely believed to be responsible for the higher processes of thought and feeling (Baars, B. J. *et al.*, 2003). The major areas of the cortex now demonstrated to be involved in pain processing include the anterior cingulate cortex, insular cortex, primary and secondary somatosensory cortices, and the prefrontal cortex (Treede, R. D. *et al.*, 1999; Derbyshire, S. W. G. 2000; Peyron, R. *et al.*, 2000; Vogt, B. A. *et al.*, 2003).

Evidence for the direct generation of pain by the cortex comes from an imaging study involving the so-called thermal grill illusion (Craig, A. D. *et al.*, 1996). The thermal grill illusion is the experience of a cold burn produced by interleaving bars that are cool (20 °C) and warm (40 °C) onto the palm of the hand. Instead of experiencing interleaved warmth and coolness, which is the actual stimulus being applied to the hand, the subject will report a feeling of pain that is similar to a cold burn. Explanation for this paradox rests upon the fact that the mammalian nervous system is generally designed to process warmth and coolness separately. Detection of coolness is suggested to proceed via the inhibition or deactivation of an intense cold pathway combined with the activation of a cool channel to produce the experience of coolness. The warm bars, however, act to inhibit the cool channel resulting in a net flow of information through the intense cold channel and the paradoxical experience of a burning cold (Casey, K. L., 1996). The nervous system is tricked, so to speak, into the experience of a burning cold pain sensation, and this trick results in the activation of anterior cingulate cortex and insular cortex. Thus, this activation of the cortex is directly associated with an experience of pain in the absence of an actual noxious stimulus.

In our laboratory, we have completed a study that also provides evidence for the direct involvement of the cortex in the generation of pain experience (Derbyshire, S. W. G. *et al.*, 2004). Eight subjects were selected for their ability to be hypnotized and to experience pain in the absence of any stimulation via hypnotic suggestion. Images of brain activation were recorded when actual noxious stimulation was provided to the subjects and when the pain was suggested in the absence of any stimulation. We demonstrated significant changes during this hypnotically induced pain experience within the anterior cingulate, insular, prefrontal, and parietal cortices. These findings compare well with the activation patterns during pain from an actual noxious source and provide direct experimental evidence in humans linking specific cortical activity with the immediate generation of a pain experience.

Thus, imaging studies demonstrate that the cortex is necessary for pain experience and, as rocks and thermostats do not have a cortex, they are excluded from the world of pain. The same is also true of fruit fly larvae. Other animals, however, are much more likely to share with humans the necessary biology for pain experience.

63.3 Animal Biology

In terms of gross anatomy and surface features, human cortex shares many features with the higher primates including the presence of neural structures widely associated with the pain neuromatrix. These similarities have facilitated the investigation of animal nociceptive systems with application to the human species (Vogt, B. A. *et al.*, 1987; Apkarian, A. V. and Hodge, C. J., 1989). At the same time, differences across species include how much the brain, and its constituent parts, weigh; the appearance of the individual cells in various brain regions; the destinations of axons from these cells; and the extent to which brain cells respond electrically when the sense organs are experimentally activated or when the animal is behaving. A review of each of these areas for all animals would be obviously vast and beyond the scope of this chapter (see Dehaene, S. *et al.*, 2005 for extensive comparison of humans and primates). For the purposes of expediency, I will focus only upon the higher primates and will consider only the comparative size and cytoarchitecture of the frontal cortex, which is often activated during pain and is more broadly involved in higher-cognitive abilities including the organization of mental contents that control creative thinking and language, and the artistic expression and planning of future actions (Damasio, A. R., 1985).

Human frontal cortex ranges in size from 239 to 330 cm^3, the range in the great apes is from 50 cm^3 (in a chimpanzee) to 112 cm^3 (in an orangutan) (Semendeferi, K. *et al.*, 2002). Within area 10 of the dorsolateral prefrontal cortex, the right human hemisphere has an estimated 254 million neurons, while the great apes have less than one-third of that amount (Semendeferi, K. *et al.*, 2001). Although human area 10 has more neurons, it is less densely packed than in the apes such that the supragranular layers (layers II and III) stand out more in the human brain compared to apes. Consequently, humans have more space available for extrinsic and intrinsic connections, which may indicate increased communication within and between area 10 and other higher-order association cortices. The relative volume of white matter underlying prefrontal association cortices is also larger in humans than in apes, compatible with the idea that neural connectivity is increased in the human brain (Zhang, K. and Sejnowski, T. J., 2000).

We cannot replace psychological investigations by the examination of brain tissue (Walker, S. F., 1983; Derbyshire, S. W. G., 2000), but if the assumption that mental states reflect brain activation is to be taken seriously, then it is clear that the physical structure of animal brains must be taken into account when considering mental activity in animals. The similarities in structure between the brains of apes and humans suggest some commonality of function, but the differences in size, connectivity, and cytoarchitecture equally suggest important diversion of function. Until the relationship between the structure of the brain and function is fully realized, it will remain unclear what the differences in neural biology between humans and animals mean in terms of perception and experience, but further examination of the similarities in experience can proceed via comparative investigation of the psychology of animals and humans.

63.4 The Psychology of Pain, the Psychology of Humans and the Psychology of Animals

To avoid the tautological use of the term pain, we need some sort of definition of what pain is. In the absence of a definition, pain is ascribed when there are pain behaviors or pain stimuli or, more simply, pain is present when there is pain. The circularity arises because the description of behaviors that follow noxious insult and the psychological description of pain experience are at two different levels: the link between them is at the heart of the problem of understanding animal suffering or other types of animal experience (Nagel, T., 1974).

To escape the inherent tautology of describing pain as the result of painful stimuli, pain is generally

defined as a sort of amalgam of cognition, sensation, and affective processes, commonly described under the rubric of the biopsychosocial model of pain (Waddell, G., 1987). Pain is no longer regarded as merely a physical sensation of noxious stimulus and disease, but is seen as a conscious experience that includes mental, emotional, and sensory mechanisms. This understanding is reflected in the current International Association for the Study of Pain (IASP) definition of pain as "an unpleasant sensory and emotional experience associated with actual or potential tissue damage, or described in terms of such damage" (Merskey, H., 1991).

By this definition, pain is a high-level process that makes no sense in the absence of sentience; pain accompanies injury in minds that are capable of subjectivity or consciousness. Without consciousness, there can be the response to noxious stimulation, technically referred to as nociception, but there cannot be pain. Thus, to understand how pain experience in particular becomes possible, it is necessary to understand the origin and course of development of conscious experience in general.

I suggest that the origin of human subjectivity lies within development. Before an infant can think about objects or events, or experience sensations and emotion, elements of thought must have their own independent existence in the infant's mind. This is something that is achieved via continued brain development in conjunction with discoveries made in action and in patterns of mutual adjustment and interactions with the infant's caregiver. For example, an important step toward subjectivity is the development of representational memory, allowing responses to learned information as well as material directly available. Representational memory begins to emerge as the frontal cortex develops between 2 and 4 months of age. From this point on, there is the possibility of tagging in memory, or labeling as a something, all the objects, emotions, and sensations that appear or are felt. When a primary caregiver points to a spot on the body and asks, "does that hurt?" she is providing content and enabling an internal discrimination. This type of interaction provides for content and symbols allowing the infant to locate and anchor emotions and sensations and to arrive at a particular state of being within her own mind. Although pain experience is clearly individual, it is created by a process that extends beyond the individual. If pain were an entirely private affair, no words would be able to express it because no external frame of reference would be comparable and therefore adequate to express the sensation. Pain is not like this because clearly human beings do express their painful experiences and these expressions have meaning that allow for diagnosis and treatment. In so far as human beings live in a community of thinking, feeling, talking beings, the privacy of experience is broken down and externalized for further analysis. As we are able to externalize our inner world, so we are able to reflect upon that world and become self-aware or self-conscious. Thus, the contents of our inner world come to mean something to us only in so far as they mean something to others, and the realization of that meaning occurs through a developmental process that is social as well as natural (Derbyshire, S. W. G., 1999b; Malik, K., 2000).

The evidence for this type of self-awareness within the animal world is limited and controversial (Heyes, C. M., 1993; Vauclair, J., 1997). There is some evidence of potentially reflective consciousness from behaviors such as giving of alarm calls and the use of deception, but any awareness or insight that these behaviors may indicate is heavily circumscribed and does not generalize. There is no evidence that the awareness, however defined or understood, provides any transformative impact within the individual animal or group. A chimpanzee, for example, behaves today in the same way as a chimpanzee did 100 000 years ago. When chimps forage for food, they do not ask themselves why or consider better alternatives any more than does a beaver consider better ways of building dams. When swallows fly south in the winter, they do not ask why it is hotter in Africa or what would happen if they flew further south or whether they could stay by creating warmth in the north. Humans do ask these kinds of questions and do engage in behavior that transforms their circumstances. We are not trapped inside a purely personal, solipsistic view of the world. We have a conscious insight that makes pain and all other experiences possible.

Pain is much more than working nociceptors connected to a brain. When we feel pain, it is an experience we have spent a good portion of our psychological development learning to recognize. This feeling is not given directly by our brain, but is made through repeated experience, categorization, memory, and reconnection. This development is obviously dependent upon the presence of a sufficiently developed neural system, the human brain, but the content of that experience is dependent upon elements that lie outside the human brain. Animals do not have a human brain and do not share in the psychological development of humans. Animal pain is therefore dependent upon a false equivalence of both biology and psychology and can be rejected.

63.5   Conclusion

The fact that some animal behaviors are specific to noxious stimulation and are particular defensive reactions driven by homologous neuronal centers is a long way from demonstrating pain experience. In their report on fishes, Sneddon L. U. *et al.* (2003a) concede that what a fish feels "is possibly nothing like the experience of humans" but is nevertheless "no less important in terms of biology or ethics" than human pain. This is similar to the argument of Beckoff M. (2002) who has suggested, "even though the experience of pain might not be the same across species, individuals of different species can still suffer their own type or version of pain."

It is not obvious why, if the experience is nothing like our experience, it still retains the same moral dimension and should still be called pain. Equally, there is no description of what the pain experience is like for the animals, beyond it being nothing like the human experience but still being important. What is perhaps being argued is that while animals are incapable of a broad awareness they are capable of experiencing something like pain or a narrow awareness. Whereas human awareness is a general phenomenon, carried along with every experience we have, from hitting ourselves with a hammer to enjoying a sunny day, that of the animals might be piecemeal with the individual pieces, including pain, available whole and fully formed.

What remains unexplained is how the animal can be aware of that particular piece of experience. As discussed, humans use symbolic labeling to differentiate pain, red, and so forth, and it is through this public process of externalizing our thoughts and feelings that those thoughts and feelings come to have meaning to us (Malik, K., 2000). We become conscious (Hobson, P., 2002).

As a subjective experience, pain is one of the many exceptional properties of human existence. Belief in the exceptional properties of human beings, however, is not currently popular. John Gray, professor of European thought at the London School of Economics, for example, has coined the phrase Homo rapiens to describe the human race, which he presents as no better than bacteria (Gray, J., 2002). Although certainly an extreme view, Gray is not alone in comparing humanity to bacteria (Fernandez-

Armesto, F., 2004), or in suggesting the planet would do well to rid itself of our existence (Preston, P., 1993).

Even though many will likely reject the extreme antihumanism of Gray and others, far fewer will reject the notion that animals are ultimately quite similar to us. Almost every day, we are presented with new revelations about how animals are more like us than we ever imagined. A selection of news headlines includes: "How animals kiss and make up"; "Male birds punish unfaithful females"; "Dogs experience stress at Christmas"; "Capuchin monkeys demand equal rights"; "Scientists prove fish intelligence"; "Birds going through divorce proceedings"; "Bees can think say scientists"; "Chimpanzees are cultured creatures"; "Psychic' parrot expected to ruffle scientific feathers"; and, "They're in love. They're gay. They're penguins."

While this kind of anthropomorphism can seem harmless and is often amusing, the ultimate effect is similar to that of the frank antihumanists: to denigrate human abilities (Schwab, A., 2003; Guldberg, H., 2004). The sheer staggering scale and richness of human culture are unlike anything in any other species. The development of medicine, industry, transportation, communication, clean water, a stable food supply, warm habitats, and so on, are the discernible signs of culture and progress that are evidently absent from the nonhuman world.

This does not mean we should ignore issues of animal welfare, but it does mean that any discussion of animal welfare is a human debate and is fundamentally about what it means to be human (Engelhardt, T. H., 2001; Derbyshire, S. W. G., 2002). The self-reflective ability of humans, which encapsulates feelings and experience and which made tangible material and cultural progress possible, has moral and political consequences. Humans are free agents, within the limits of any given political system, and organize their behavior according to rights and customs that they create, which includes weighing and judging the life of humans or animals (Engelhardt, T. H., 2001).

Human beings take a central moral position, not via some act of special pleading, but as a function of their exceptional and special abilities (Engelhardt, T. H., 2001). We have an unavoidable moral centrality that places us into a position of authority over animals and at liberty to do with our own animals whatever we wish. Arguably, where the aim is one of frank malevolence, the agents involved have rejected morality itself and are acting in an antisocial manner worthy of sanction. It is probably appropriate to distinguish between practices aimed at cruelty and those motivated by a scientific, educational, or cultural practice. Such distinction will lead to a condemnation of some animal uses but will not count against those that are part of sport, consumption, or research (Engelhardt, T. H., 2001).

For researchers involved in understanding pain, there is a simple message. The use of animals has application to human welfare because of the similarities between the nociceptive system of animals and humans. There is the real possibility of discovering new ways of preventing noxious transmission, and the processing of noxious information, in humans through the use of animal models. At the same time, the use of animals is morally acceptable, because animals precisely lack the insight and conscious appreciation that makes pain, moral agency, and personhood possible.

References

Apkarian, A. V. and Hodge, C. J. 1989. Primate spinothalamic pathways. III. Thalamic terminations of the dorsolateral and ventral spinothalamic pathways. J. Comp. Neurol. 288, 493–511.

Baars, B. J., Ramsoy, T. Z., and Laureys, S. 2003. Brain, conscious experience and the observing self. Trends Neurosci. 26, 671–675.

Beckoff, M. 2002. Animal reflections. Nature 419, 255.

Casey, K. L. 1996. Resolving a paradox of pain. Nature 384, 217–218.

Coghill, R. C., McHaffie, J. G., and Yen, Y. F. 2003. Neural correlates of interindividual differences in the subjective experience of pain. Proc. Natl. Sci. U. S. A. 100, 8538–8542.

Craig, A. D., Reiman, E. M., Evans, A., and Bushnell, M. C. 1996. Functional imaging of an illusion of pain. Nature 384, 258–260.

Dehaene, S., Duhamel, J.-R, Hauser, M. D., and Rizzolatti, G. 2005. From Monkey Brain to Human Brain. Bradford Books.

Damasio, A. R. 1985. The Frontal Lobes. In: Clinical Neuropsychology (eds. K. Heilman and E. Valenstein), pp. 339–375. Oxford University Press.

Derbyshire, S. W. G. 1999a. The IASP definition captures the essence of pain experience. Pain Forum 8, 106–109.

Derbyshire, S. W. G. 1999b. Locating the beginnings of pain. Bioethics 13, 1–31.

Derbyshire, S. W. G. 2000. Exploring the pain "neuromatrix." Curr Rev. Pain 6, 467–477.

Derbyshire, S. W. G. 2002. Why Animal Rights Are Wrong. In: Animal Experiments: Good or Bad? (ed. E. Lee), pp. 37–55. Hodder & Stoughton.

Derbyshire, S. W. G., Whalley, M. G., Stenger, V. A., and Oakley, D. A. 2004. Cerebral activation during hypnotically induced and imagined pain. NeuroImage 23, 392–401.

Engelhardt, T. H. 2001. Animals: Their Right to be Used. In: Why Animal Experimentation Matters: The Use of Animals in Medical Research (eds. E. F. Paul and J. Paul), pp. 175–195. Transaction Publishers.

Fernandez-Armesto, F. 2004. So You Think You're Human? Oxford University Press.

Goodman, M. B. 2003. Sensation is painless. Trends Neurosci. 26, 643–645.

Gray, J. 2002. Straw Dogs: Thoughts on Humans and Other Animals. Granta Press.

Guldberg, H. 2004. Why humans are superior to apes. Spiked-online (http://www.spiked-online.co.uk/Articles/0000000CA40E.htm).

Heyes, C. M. 1993. Anecdotes, trapping and triangulating: do animals attribute mental states? Anim. Behav. 46, 177–188.

Hobson, P. 2002. The Cradle of Thought: Exploring the Origins of Thinking. Macmillan Press.

Malik, K. 2000. Man, Beast and Zombie. Weidenfeld and Nicholson.

Merskey, H. 1991. The definition of pain. Eur. Psychiatry 6, 153–159.

Nagel, T. 1974. What is it like to be a bat? Philos. Rev. 83, 435–450.

Peyron, R., Laurent, B., and Garcia-Larrea, L. 2000. Functional imaging of brain responses to pain. A review and meta-analysis. Neurophysiol. Clin. 30, 263–288.

Preston, P. 1993. The Hot Zone. Anchor Publications.

Rosen, S. D., Paulesu, E., Nihoyannopoulos, P., Tousoulis, D., Frackowiak, R. S. J., Frith, C. D., Jones, T., and Camici, P. G. 1996. Silent ischemia as a central problem: regional brain activation compared in silent and painful myocardial ischemia. Ann. Intern. Med. 124, 939–949.

Schwab, A. 2003. Hook, Line and Thinker: Angling and Ethics. Merlin Unwin Books.

Semendeferi, K., Armstrong, E., Schleicher, A., Zilles, K., and Van Hoesen, G. W. 2001. Prefrontal cortex in humans and apes: a comparative study of area 10. Am. J. Phys. Anthropol. 114, 224–241.

Semendeferi, K., Lu, A., Schenker, N., and Damasio, H. 2002. Humans and great apes share a large frontal cortex. Nat. Neurosci. 5, 272–276.

Sneddon, L. U., Braithwaite, V. A., and Gentle, M. J. 2003a. Do fishes have nociceptors? Evidence for the evolution of a vertebrate sensory system. Proc. R. Soc. Lond. B 270, 1115–1121.

Sneddon, L. U., Braithwaite, V. A., and Gentle, M. J. 2003b. Novel object test: examining nociception and fear in the rainbow trout. J. Pain 4, 431–440.

Treede, R. D., Kenshalo, D. R., Gracely, R. H., and Jones, A. K. P., 1999. The cortical representation of pain. Pain 79, 105–111.

Vauclair, J. 1997. Mental states in animals: cognitive ethology. Trends Cogn. Sci. 1, 35–39.

Vogt, B. A., Berger, G. R., and Derbyshire, S. W. G. 2003. Structural and functional dichotomy of human midcingulate cortex. Eur. J. Neurosci. 18. 3134–3144.

Vogt, B. A., Pandya, D. N., and Rosene, D. L. 1987. Cingulate cortex of the rhesus monkey. 1. Cyto-architecture and thalamic afferents. J. Comp. Neurol. 262, 256–270.

Waddell, G. 1987. A new clinical model for the treatment of low-back pain. Spine 12, 632–644.

Walker, S. F. 1983. Animal Thought. Routledge & Kegan Paul, International Library of Psychology Series.

Zhang, K. and Sejnowski, T. J. 2000. A universal scaling law between gray matter and white matter of cerebral cortex. Proc. Natl. Acad. Sci. U. S. A. 97, 5621–5626.

64 Psychological Modulation of Pain

D D Price, A Hirsh, and M E Robinson, University of Florida, Gainesville, FL, USA

64.1   Overview of Pain

64.1.1   Overview of Stages/Dimensions of Pain

64.1.1.1   *Psychological stages of pain processing*

As a preliminary foundation for these discussions, this chapter begins with very general and somewhat simplified psychological and neurophysiological overviews of pain processing mechanisms. As a starting point for discussion, one can think of pain modulatory mechanisms as those that can intervene in the pathways outlined in Figures 1 and 2. Pain contains both sensory and emotional dimensions and is often accompanied by desires to terminate, reduce, or escape its presence (Buytendyck, F. J. J., 1961; Melzack, R. and Casey, K. L., 1968). By definition, pain is unpleasant and contains emotional feelings. This is largely due to the fact that the sensory qualities of physical pain dispose us to feel unpleasantness in most contexts. Thus, sensory qualities associated with pain are usually unpleasant for the same reasons that nausea and air hunger are perceived as unpleasant, as indicated by the path from nociceptive sensations to unpleasantness in Figure 1. In addition, there are parallel contributions to pain unpleasantness because the meaning of these sensory qualities is shaped by context and by a

person's ongoing anticipations and attitudes. These contextual and cognitive factors are partly the result of the fact that pain often occurs within a situation that is threatening, such as during physical trauma or disease. Part of the affective dimension of pain is the moment-by-moment unpleasantness of pain, comprised of emotional feelings that pertain to the present or short-term future, such as annoyance, fear, or distress. Pain-unpleasantness is often though not always closely linked to both the intensity and unique qualities of the painful sensation. Another component of pain affect, secondary pain affect, includes emotional feelings directed toward long term implications of having pain (e.g., suffering), as shown in Figure 1 by the path extending from pain unpleasantness to secondary pain affect.

64.1.1.2   *Factors contributing to pain unpleasantness*

Multiple factors contribute to pain unpleasantness. Several sensory attributes of pain dispose unpleasant emotional feelings. The foremost among these is that sensations of pain are often, though not necessarily, more intense than other types of somatic sensations. Similar to other sensory modalities, it is commonly accepted that perceived intensity is a salient factor that contributes to the unpleasantness of pain. It may be the reason that very bright lights and loud sounds are sometimes metaphorically referred to as "painful". Qualities of sensations evoked by tissue damaging stimuli or stimuli that would produce tissue damage if maintained dispose us to perceive pain as invasive and intrusive for both the body and consciousness (Buytendyck, F. J. J., 1961; Price, D. D., 1999; 2000). Both neural and psychological processes related to sensory qualities of pain can be conceived as important causal links in the production of pain-related emotional disturbance. The persistence of these sensory qualities over time enhances unpleasantness. Thus, there exists a serial interaction between pain sensation intensity and pain unpleasantness.

Nociceptive, exteroceptive (e.g., sight, sound), and interoceptive sensory processes (e.g., startle, increased visceral responses) may provide parallel contributions to pain affect (Price, D. D., 1999). Pain itself can be conceptualized as both an exteroceptive and interoceptive phenomenon depending on the type of tissue that is stimulated and/or the types of sensory qualities that are present during pain. Consistent with Damasio's (Damasio, A., 1994) neurological view of emotion mechanisms, pain unpleasantness reflects the

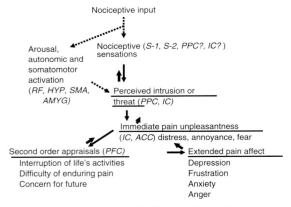

Figure 1 Schematic used to illustrate interactions between pain sensation, pain unpleasantness, and secondary pain affect (solid arrows). Neural structures likely to have a role in these dimensions are shown by abbreviations in adjacent parentheses and their full names are given in the legend of Figure 2. Dashed arrows indicate nociceptive or endogenous physiological factors that influence pain sensation and unpleasantness. From Price, D. D. and Bushnell, M. C. (*eds*.) 2004. Psychological Methods of Pain Control: Basic Science and Clinical Perspectives, pp. 322. IASP Press.

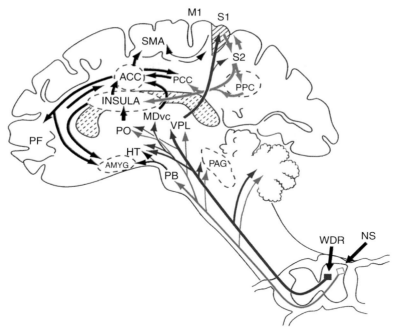

Figure 2 Schematic of ascending pathways, subcortical structures, and cerebral cortical structures involved in processing pain. PAG, periaqueductal gray; PB, parabrachial nucleus of the dorsolateral pons; VMpo, ventromedial part of the posterior nuclear complex; MDvc, ventrocaudal part of the medial dorsal nucleus; VPL, ventroposterior lateral nucleus; ACC, anterior cingulate cortex; PCC, posterior cingulate cortex; HY, hypothalamus; S-1 and S-2, first and second somatosensory cortical areas; PPC, posterior parietal complex; SMA, supplementary motor area; AMYG, amygdala; PF, prefrontal cortex. From Price, D. D. and Bushnell, M. C. (*eds.*) 2004. Pschological Methods of Pain Control: Basic Science and Clinical Perspectives, 322 pp. IASP Press.

contribution of several sources, including pain sensory qualities, arousal, visceral, and somatomotor responses. Thus, these psychophysiological sources of input, in combination with appraisal of the context in which they occur, give rise to a feeling of what is happening to the body and self, often but not necessarily accompanied by specific thoughts. This felt meaning derives largely from the experience of the body and constitutes the immediate unpleasantness of pain. For Damasio, "the essence of feeling an emotion is the experience of such changes in juxtaposition to the mental images that initiated the cycle" (Damasio, A., 1994, p. 145). In the case of pain, the mental image is that of a physically intrusive and qualitatively unique somatic perception.

Psychophysical studies demonstrate that pain sensation and pain unpleasantness represent two distinct dimensions of pain that demonstrate reliably different relationships to nociceptive stimulus intensity and are separately influenced by various psychological factors. Pain unpleasantness can be selectively influenced by emotions, changes in meaning of the pain, and factors related to the predictability,

controllability, and duration of pain (Buytendyck, F. J. J., 1961; Melzack, R. and Casey, K. L., 1968). Pain unpleasantness also often leads to and sustains pain-related emotions that are based on cognitive appraisals that are more reflective than is pain unpleasantness. We term these latter pain-related emotional feelings as secondary pain affect. Unlike pain unpleasantness, secondary pain affect is based on more elaborate reflection related to that which one remembers or imagines. This involves meanings such as perceived interference with one's life, difficulties of enduring pain over time, and the implications for the future (Buytendyck F. J. J., 1961; Price, D. D., 1999). Pain is often experienced not only as a threat to the present state of one's body, comfort, or activity, but also to one's future well being and life in general. The perceived implications that present pain-related distress or annoyance holds for future well being and functioning support the link between pain unpleasantness and secondary pain affect, as illustrated in Figure 1. Studies of pain patients show the distinction between immediate pain unpleasantness and secondary pain affect and their sequential interactions.

64.1.2 Overview of Neurophysiological Processing of Pain

The stages of pain processing just described are also related to our present understanding of the neurophysiology of pain. Underlying neural mechanisms of sensory as well as primary and secondary affective dimensions of pain include multiple ascending spinal pathways to the brain and both parallel and serial processing in the brain. The neural representation associated with the experience of pain is likely to be related to a distributed network of brain structures that participate in the different dimensions of pain, such as identification of unique sensory qualities and their intensity, response selection, arousal, emotional feelings, and finally emotional expression and motivation. This network of brain circuitry is itself controlled by inhibitory and facilitatory interactions within the brain, as well as descending inhibitory and facilitatory control within brain-to-spinal cord pathways (Basbaum, A. I. and Fields, H. L., 1984; Price, D. D., 2000; Rainville, P. *et al.*, 1997). Thus, pain can be enhanced or reduced by interactions that take place at any one of several points in the schematics in Figures 1 and 2. A major pathway for pain modulation, known since the 1970s, is that of a brainstem-to-spinal cord pathway that utilizes endogenous opioids (Figure 3).

64.1.2.1 Ascending nervous system pathways for pain

The principle ascending pathways for pain (Figure 2) originate in specialized receptors in body tissues, called nociceptors. These receptors are specialized to respond to stimuli to tissues that either cause

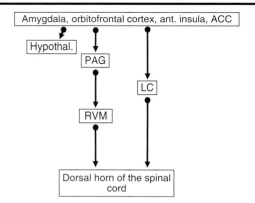

Figure 3 Schematic used to illustrate pathways and mechanisms that modulate pain. From Price, D. D. and Bushnell, M. C. (*eds.*) 2004. Pscyhological Methods of Pain Control: Basic Science and Clinical Perspectives, pp. 322. IASP Press.

damage or would cause damage if they were maintained for a sufficient length of time. Nociceptors are connected to primary afferent neuron axons that synapse onto dorsal horn neurons (Figure 2). The dorsal horn of the spinal cord receives somatosensory information from most of the body, and the dorsal horn of the medulla oblongata receives input from the orofacial area via the trigeminal nerve. There are different classes of primary nociceptive afferent neurons that innervate different types of tissues, such as skin, muscle, joints, and viscera. Dorsal horn neurons that receive primary nociceptive neuron input are at the origin of ascending pathways to the brain that have different functional roles in pain processing. Two types of dorsal horn sensory projection neurons include nociceptive-specific neurons (NS) that respond predominantly to nociceptive stimuli and wide-dynamic-range (WDR) neurons that respond differentially to gentle and nociceptive stimulation. Both WDR and NS neurons project in the spinothalamic tract, the main somatosensory pathway for pain. The spinothalamic tract projects to the ventro-posterior-lateral thalamic nucleus (VPL), some medial thalamic nuclei, and nuclei just posterior and inferior to VPL. VPL (and VPM in the case of the trigeminal system) projects to the somatosensory cortices (S1 and S2), whereas medial and posterior thalamic nuclei mainly project to limbic cortical areas (Figure 2). The pathway to VPL/VPM is critical for sensory qualities of pain and appreciation of its perceived sensory intensity. In addition to this major pathway, there are pathways that proceed from the dorsal horn directly to reticular formation nuclei of the medulla and midbrain and to the hypothalamus. There also is a pathway that ascends to the amygdala by way of a synapse in the pons (Bernard, J. F. *et al.*, 1996). These various target structures are involved in pain because they participate in such functions as arousal and diffuse cortical activation (reticular formation nuclei), body regulation (hypothalamus), and affective states such as fear (amygdala). The central nucleus of the amygdala has been strongly implicated in fear, emotional memory, and behavior, and autonomic and somatomotor responses to threatening stimuli (Bernard, J. F. *et al.*, 1996). Various hypothalamic nuclei have also been implicated in these functions. Based largely on their central targets, all ascending pathways described above are likely to participate in the affective dimension of pain. Thus, some pathways directly activate cortical and subcortical limbic structures and the pathway to VPL/VPM and somatosensory cortex activates these same limbic structures through a ventrally directed route, as shown in Figure 2. This latter pathway reflects the serial interaction between pain sensory qualities and intensity and pain unpleasantness. Thus, direct and indirect activation of emotion-related limbic structures is consistent with the pathways shown in the schematic of Figure 1. Emotions related to pain are a consequence of sensations related to pain (i.e., serial interrelationships) as well as other psychophysiological and contextual factors (i.e., parallel interrelationships). Some aspects of pain-related emotions may occur automatically as a result of direct input to subcortical limbic areas such as the amygdala, a structure implicated in fear.

64.1.2.2 Brain pathways and regions interrelating sensory and affective dimensions of pain

Once impulses in ascending pathways from the dorsal horn reach different targets within the brain, both serial and parallel circuitry is activated. The serial interconnections are established by the pathway to VPL and to somatosensory cortices that is anatomically interconnected with a ventrally directed corticolimbic pathway that integrates somatosensory input with other sensory modalities such as vision and audition and with learning and memory (Friedman, D. P. et al., 1986). This pathway proceeds from primary and secondary somatosensory cortices to posterior parietal cortical areas and to insular cortex (IC) and from the latter to subcortical limbic structures (e.g., amygdala, perirhinal cortex, and hippocampus). This corticolimbic pathway may be critical for pain affect (Friedman, D. P. et al., 1986). This serial corticolimbic somatosensory pathway ultimately converges on the same limbic and subcortical structures that may be more directly accessed by other ascending spinal pathways. The latter reflect parallel processing. This dual convergence may be related to a mechanism whereby multiple neural sources, including somatosensory cortices S1 and S2, contribute to pain affect and reflects both parallel and serial processing. This corticolimbic serial pathway directly implies that somatosensory cortices contribute to pain affect. This implication is consistent with psychophysical studies in which pain unpleasantness was shown to be the result of pain sensation intensity and with studies showing deficits in both pain sensation and pain unpleasantness as a result of somatosensory cortical lesions (see references in Price, D. D., 1999; 2000).

64.1.2.3 The pivotal role of the anterior cingulate cortex (ACC) in pain affect

As shown in Figure 2, ACC receives inputs from multiple sources, including thalamus, insular cortex, posterior cingulate cortex (PCC), and prefrontal cortex. Convergence of these inputs at the level of ACC would be consistent with a mechanism in which somatic perceptual and cognitive features of pain would be integrated with attentional and rudimentary emotion mechanisms. Based on neurological evidence, the ACC may have a complex pivotal role in interrelating attentional and evaluative functions with that of establishing emotional valence and response priorities (Devinsky, O. et al., 1995). Response priorities would be closely related to premotor functions that are integrally related to motivation and emotions and may be associated with immediate efforts to cope with, escape, or avoid the pain and pain-evoking situation. For example, the projection of ACC to the supplementary premotor area may reflect the motor side of an overall affective-motivational dimension of pain. In this view, cortical areas controlling sensory, attentional, pre-motor, and affective functions of pain are part of a distributed network that has serial and parallel connections. This network has implications for psychological modulation of pain because each of these connections represents a stage of pain processing that can, in principle, be modified by psychological factors.

Response priorities change over an extended period of time. Pain unpleasantness endured over time engages prefrontal cortical areas involved in reflection and rumination over the future implications of a persistent pain condition. The ACC may serve this function by coordinating somatosensory features of pain with prefrontal cerebral mechanisms involved in attaching significance and long term implications to pain, a function associated with secondary pain affect (Apkarian, A. V. et al. 2001; Devinsky, O. et al., 1995; Price, D. D., 1999; Verne, G. N. et al., 2003). Thus, ACC may be a region that coordinates inputs from parietal areas involved in perception of bodily threat with frontal cortical areas involved in plans and response priorities for pain-related behavior. Both functions would help explain observations on patients with prefrontal lobotomy and patients with pain asymbolia as a result of insular cortical lesions (Price, D. D., 1999). The former have deficits in spontaneous concern or rumination about their pain but can experience the immediate threat of pain once it is brought to their attention. In contrast, asymbolia patients appear incapable of perceiving the threat of nociceptive stimuli under any circumstances. The role of medial prefrontal cortical areas in secondary pain affect is also supported by recent brain imaging studies. These studies show increased prefrontal neural activity in clinical pain patients likely to have high levels of secondary pain affect (Apkarian, A. V. *et al.* 2001; Verne, G. N. *et al.*, 2003).

64.2 Modulation of Pain Sensation and Pain Affect by Attention and Distraction

There are neural circuits that provide top-down modulation of the pain experience. Central neural mechanisms associated with such phenomena as placebo/nocebo, hypnotic suggestion, attention, distraction, and even ongoing emotions are now thought to modulate pain by decreasing or increasing neural activity within many central neural structures, including the dorsal horn (Figure 3). This modulation includes endogenous pain-inhibitory and pain-facilitatory pathways that descend to spinal dorsal horn, the origin of ascending spinal pathways for pain (Basbaum, A. I. and Fields, H. L., 1978). Recent brain imaging studies have begun to reveal the cortical areas involved in pain modulation (Hofbauer, R. K. *et al.*, 2001; Koeppe, R. A. and Stohler, C. S., 2001; Petrovic, P. *et al.*, 2001, Rainville, P. *et al.*, 1997).

The modulation of pain by endogenous cortical and subcortical networks is consistent with other types of brain imaging experiments that show that anticipation of pain activates cortical nociceptive regions to some extent, though to a lesser degree than pain itself. Thus, the cortical networks shown in Figure 2 may be directly influenced by cognitive factors. These are likely to include not only anticipation of the presence of pain, but also anticipation of its reduction, as in the case of placebo or hypnotic analgesia. The modulatory mechanisms are likely to be diverse, extending from those which are confined only to neural interactions within the brain to those involving activation of brain-to-spinal cord circuits long known to potently reduce or enhance pain, as represented in Figure 3 (Basbaum, A. I. and Fields, H. L., 1978).

64.2.1 Psychophysical Studies of Effects of Attention and Distraction on Pain

Attentional manipulation can be achieved in numerous ways; however, distraction is the most general attentional manipulation for pain modification. An interesting stimulus can be presented to an individual, or the individual can engage in an attention-demanding task. Unfortunately, in addition to attention, other variables are often affected by these techniques. For example, mood and/or arousal may change in response to an interesting stimulus (e.g., music), and alteration of these states may contribute to changes in pain experience. Furthermore, simple distraction tasks often lack a measure of attentional state. This is problematic, as some individuals may attend to music during presentation of noxious stimuli, whereas others may focus only on the pain. In the case of attention-demanding tasks, some individuals who are particularly adept at performing the task (e.g., mental arithmetic) may devote much less attention to it than those who are less skilled.

A psychophysical paradigm with two balanced conditions provides a more controlled alteration of attentional state. An example is that of the cross-modality attention paradigm in which stimuli from two sensory modalities are presented simultaneously, and the participant must perform tasks that require attention to one modality. Attentional status is measured via task performance. These paradigms permit the control of level of arousal and vigilance, because the same stimuli are presented on all trials, task difficulty is equivalent between the two modalities, and only direction of attention is altered.

One variant of the balanced-condition psychophysical approach is the intra-modal spatial attention paradigm. In this paradigm, concurrent identical stimuli are presented to different receptive fields – or different body regions for somatosensory stimuli – and tasks are performed that require attention to one or the other location. Although used extensively with visual and auditory systems (Corbetta, M. and Shulman, G. L., 2002; Hikosaka, O. *et al.*, 1996), when applied to the pain system, this paradigm may engage diffuse noxious inhibitory controls (DNIC), a pain modulatory system evoked by simultaneous presentation of two noxious stimuli, that could alter pain perception (Le Bars, D. *et al.*, 1979a; 1979b; Price, D. D. and McHaffie, J. G., 1988). Consequently, this paradigm makes it difficult to distinguish changes in pain perception due to shifts of spatial attention from those due to other pain modulatory networks.

Finally, a balanced-condition psychophysical paradigm can require shifts of attention to different dimensions of the same stimulus within a single stimulus modality. For example, one can attend to the shape, color, texture, or motion of a visual stimulus.

Similarly, the quality of a painful stimulus – the burning, stinging, or aching sensations – can be the focus of attention. The unpleasantness of the sensation can also be attended to. Such manipulations can alter both the pain experience and pain-evoked neural activity.

64.2.1.1 Studies of basic distraction effects on pain

Many studies have compared experimental pain ratings in the presence and absence of a distracter (Bentsen, B. et al., 1999; Hodes, R. L. et al., 1990; Janssen, S. A. and Arntz, A., 1996; Johnson, M. H. and Petrie, S. M., 1997). Although not perfectly consistent, the vast majority of results indicate lower pain ratings during distraction. However, as noted above, caution is in order when attributing this pain modulation to directed attention, since uncontrolled other factors such as vigilance, mood, and anxiety, could contribute to the overall modulatory effect.

64.2.1.2 Studies of cross-modal attention effect on pain

Other studies have employed cross-modality psychophysical paradigms to specifically examine the modulatory effects of directed attention. For example, Miron (Miron, D. et al., 1989) had participants detect changes in intensity of heat pain and visual stimuli. As expected, the speed and accuracy of detection of changes in stimuli were decreased when attention was directed to another stimulus modality. Their results further confirmed that pain stimuli were more effective in commanding attention than other stimuli. Subsequent results from the same investigators suggest that cross-modality attention may exert greater effects on the sensory dimension of pain than the unpleasantness dimension (Bushnell, M. C. et al., 1999; Villemure, C. et al., 2003). A limitation of all these studies is their focus on the sensory features of the stimulus. Requiring a study participant to attend to the emotional aspects of a stimulus might alter the affective more than the sensory component of pain.

64.2.1.3 Studies of spatial attention effects on pain

In a spatial attention paradigm, Bushnell M. C. et al. (1985) found that detection of changes to a heat stimulus applied to each side of the face was more rapid when these changes occurred on the side to which attention was directed. These results also demonstrated the modulatory effects of spatial attention on perceived intensity of noxious stimuli, as stimuli changes occurring outside a subject's attentional focus were less accurately detected. Furthermore, this study suggests that mechanisms of directed spatial attention involve areas of the nervous system that display a fine somatotopic organization of nociceptive information. The perceptual effects of spatial attention to pain are consistent with those of other sensory modalities and are critical for an organism's orientation to important features of the environment (Posner, M. E., 1980).

64.2.1.4 Studies of intra-modal attention effects on pain

In addition to focussing attention on different stimulus modalities and spatial locations, attention can also be directed to different features or dimensions of a stimulus. Several clinical and experimental studies have demonstrated decreased pain intensity and unpleasantness by directing subject attention to the sensory characteristics of pain (e.g., the burning, pricking, or aching sensation). A study of burn patients undergoing dressing changes found that those who focus on the sensory aspects of their pain report less pain than those distracted by music (Haythornthwaite, J. A. et al., 2001). Another study of health-anxious chronic pain patients observed less anxiety and discomfort when attentional focus was directed only to the physical sensations of their pain (Hadjistavropoulos, H. D. et al., 2000).

64.2.1.5 Other accounts of pain modulation by attentional and cognitive processes

An alternative account of attentional processing has been provided by Wegner D. M. (1994). He proposed the provocative theory of ironic processing, which has been applied to a number of cognitive processes (e.g., mood control, thought suppression) and has received some attention in the pain literature. This theory posits two cognitive mechanisms (the consciously controlled operating system and the automatic monitoring system) for achieving a desired mental state. The operating system searches for mental contents consistent with this state, whereas the monitoring system searches for mental contents inconsistent with this state. When the monitoring system locates such material, this material is brought into consciousness and thus activates the operating system. According to ironic processing theory, the operating system is compromised when mental capacity is reduced (e.g., because of pain, fatigue, stress), thereby permitting the undesired thoughts to enter consciousness. A rebound effect then occurs where the undesired thoughts are more frequent than if they had not been initially suppressed (see Wenzlaff, R. M. and Wegner, D. M., 2000 for a review). As applied to pain, this theory holds that directed attention to pain would result in less pain compared to intentional distraction. Furthermore, pain suppression is suggested to have less antinociceptive effects than pain distraction. There is some evidence in the literature supporting both these hypotheses (e.g., Cioffi, D. and Holloway, J., 1993; Haythronthwaite, J. A. *et al.*, 2001).

What is to be made of the seemingly competing hypotheses of attentional modulation of pain and the divergent evidence that provides support for both these hypotheses? It may be that the antinociceptive effects of directed attention toward or away from pain is dependent upon the intensity of the pain and the cognitive and emotional state of the individual. For example, in situations of moderate pain intensity, where a patient is likely to be able to achieve/experience pain reduction, focused attention towards the pain may prove beneficial, à la ironic processing theory. How may this occur? Suppose a particular patient holds a moderate expectation for reduction in pain (due to some pharmacological agent, behavioral strategy, etc.) and a high desire for pain relief. Upon experiencing a small amount of pain relief – which the patient is able to perceive due to his/her focused attention on pain – some period of time following the treatment, the patient begins to notice that the treatment is working and, thus, expects further pain reduction. As a result of the experienced pain relief, the patient is also likely to be less fearful or anxious, and consequently reduces his/her desire for pain relief. As discussed more thoroughly below in the section on placebo, these changes in expectation (increase) and desire (decrease) can have powerful effects on the emotional state of the individual and, consequently, on the pain experience itself. In this manner, Wegner's contention that directed attention toward pain results in decreased pain is supported, although due to different mechanisms than outlined in his theory of ironic processing. Conversely, it may be the case that during episodes of extreme pain, directed attention away from the pain is the favored approach. During extreme pain, it is often the case that the patient has an intense desire for pain relief and a concomitant low expectation of achieving this relief. As this cognitive state is likely to produce intense negative emotions, which will be further reinforced by directed attention toward the pain and the

experience of little pain reduction, it is likely that distraction techniques will prove most beneficial. Even in cases of moderate pain, if the patient has low expectations and high desire for pain relief (thus resulting in negative emotions), that individual may respond more favorably to a distraction technique than a pain-focused one. These are interesting and testable predictions, and they highlight the interaction between pain, cognition, and emotions. Other variables may also play an important role in these processes, as described below in the discussion of sex differences and pain-related attention.

In a laboratory cold-pressor paradigm, Keogh E. *et al.* (2000) found that for males, but not females, directed attention toward the pain sensation resulted in reports of less pain compared to a pain avoidance attentional condition. As a possible explanation for the seemingly paradoxical finding that men experience less pain by attending to it, the instructions required participants to concentrate on the sensations experienced during the task, and this sensory focussing may have reduced the painfulness of the situation. Thus, it appears that the use of attention-based strategies for pain reduction may be more effective for men than women.

64.2.2 Separating the Influence of Attention and Emotions

The attitude and emotional state of an individual can influence his/her perception of experimental and clinical pain (De Wied, M. and Verbaten, M. N., 2001; Marchand, S. and Arsenault, P., 2002; Meagher, *et al.*, 2001a; Schanberg, L.E. *et al.*, 2000; Weisenberg, M. *et al.*, 1998; Zelman, D. C. *et al.*, 1991). Unfortunately, pleasant stimuli used in laboratory studies to improve mood and reduce pain perception typically alter both attentional state and mood, thus confounding their effects. However, a recent study by Villemure C. *et al.* (2003) demonstrated that the effects of emotion and attention can be separated, and that these variables differentially alter pain intensity and pain-related unpleasantness. Using a cross-modality attention task, they showed that attending to the odor rather than heat pain reduced pain sensation intensity but did not alter pain-related unpleasantness, mood, or anxiety. In contrast, mood, anxiety level, and pain unpleasantness, but not pain intensity, were altered by odor valence. As compared to pleasant odor, unpleasant odor increased pain unpleasantness, negative mood, and anxiety. Their results further suggest that the selective modulation of pain affect is a function of emotional changes, which were elicited by differences in odor valence.

64.2.3 Neurophysiological Studies of Effects of Attention on Pain Processing, Including Brain Imaging and Electrophysiological Studies

64.2.3.1 Effect of attention on pain-related activity in the dorsal horn

The attentional modulatory effects described above relate to studies of neural modulation of nociceptive transmission. Thus, dorsal horn sensory projection neurons, including WDR and NS neurons, undergo inhibition and facilitation from pathways that descend from the brain to the spinal cord (Figure 3). The stimulus-evoked impulse responses and the receptive field areas of these neurons can increase or decrease as a result of activity in various descending pathways under different behavioral conditions.

Descending modulatory influences and the behavioral conditions with which they are associated have been shown in experiments in awake trained monkeys (Hayes, R. L. *et al.*, 1981; Hoffman, D. S. *et al.*, 1981). These studies showed that responses of trigeminothalamic WDR and NS neurons of the medullary dorsal horn could be modified by different attentional sets and conditions of stimulus relevance. Nociceptive stimulus temperatures (45–49°) applied to the monkeys' face evoked impulse discharges in these neurons that were an increasing function of stimulus temperature. However, these neuronal impulse responses increased during a task in which the monkey discriminated a small temperature shift in the nociceptive range (a relevant attentional set) and decreased when they performed a similarly demanding visual discrimination task (an irrelevant attentional set). These opposing effects predict that the perceived intensity of pain should increase during thermal as compared to visual discrimination. A psychophysical study by Miron D. *et al.* (1989) confirmed this prediction.

The impulse frequencies elicited in WDR and NS trigeminothalamic neurons by the same range of nociceptive temperatures (45–49°) also can be modified by attentional factors related to anticipation. Impulse frequencies of these neurons were higher when a warning light preceded nociceptive stimuli as opposed to when these stimuli occurred unexpectedly. Thus, the magnitudes of the stimulus–response relationships are under dynamic modulatory control and may play as

direct a role in determining pain perception as do the peripheral stimulus events themselves. The dorsal horn is part of a network involved in shaping pain perception, a network that includes the first synapse in ascending pathways for pain. This type of modulation had been anticipated by existing theories of pain (e.g., the gate control theory), but these studies were the first to demonstrate this principle in a behavioral context.

64.2.3.2 Effect of attention and distraction on pain-evoked activity in the brain

This type of study was extended by a similarly designed study of effects of attentional state on isolated nociceptive neurons in the thalamus (Bushnell, M. C. *et al.*, 1984; Bushnell, M. C. and Duncan, G. H., 1989). Similar to studies of trigeminothalamic neurons just described, tasks that directed a monkey's attention to a painful stimulus enhanced responses of medial thalamic neurons to nociceptive temperature stimuli applied to the monkeys' face, shown in Figure 4.

Imaging studies of the effects of attention and distraction in humans likewise show modulation of pain-evoked activity in the thalamus and several cortical regions, such as the primary somatosensory cortex (S1), anterior cingulate cortex (ACC), and insular cortex (IC) (Bantick, S. J. *et al.*, 2002; Brooks, J. C. *et al.*, 2002; Bushnell, M. C. *et al.*, 1999; Hoffman, H. G. *et al.*, 2004; Seminowicz, D. A. *et al.*, 2004; Valet, M. *et al.*, 2004). In one study, several pain-related brain regions were more activated during a pain-attention condition compared to a distraction condition (Peyron, R. *et al.*, 1999). In a similar study, Bushnell M. C. *et al.* (1999) showed the greatest attentional modulation of pain-evoked activity in S1 cortex, perhaps because subject attention was directed toward a sensory feature of the stimulus. Reciprocally, the primary auditory cortex was more activated when attention was directed to the auditory stimulus and not the painful stimulus, demonstrating the generality of attentional modulation across sensory modalities. Other regions, including the periaqueductal gray matter (PAG), ACC, and orbitofrontal cortex, may also be involved in the modulatory circuitry related to attention, as they have been shown to be activated during pain-distraction tasks (Petrovic, P. *et al.*, 2000; Tracey, I. *et al.*, 2002).

Results of EEG and MEG studies suggest that attentional modulation of pain likely involves both early sensory processing in S2-IC (Legrain, V. *et al.*,

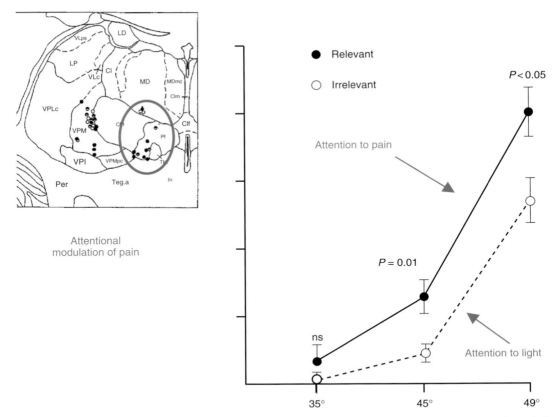

Figure 4 Stimulus–response functions showing single neuron responses in the medial thalamus (location shown in inset) during thermal stimuli of 35, 45, and 47 °C presented by a 1 cm² thermode on the maxillary facial area of a monkey. The solid line shows the responses when the monkey was required to attend to the thermal stimuli, and the dashed line shows responses to the same temperatures when the monkey was required to attend to a visual stimulus. Adapted from Bushnell, M. C and Duncan, G. H. 1989. Sensory and affective aspects of pain perception: Is medial thalamus restricted to emotional issues? Exp. Brain Res. 78, 415–418.

2002; Nakamura, Y. *et al.*, 2002) and later processing in ACC (Dowman, R., 2004; Garcia-Larrea, L. *et al.*, 1997; Siedenberg, R. and Treede, R. D., 1996). Changes in pain perception associated with attentional state appear to be due in part to changes in cortical processing and in part to a decrease in ascending afferent input from the spinal cord through activation of descending noxious inhibitory controls (DNIC).

64.3 Nonplacebo Effects of Expectation and Meaning

64.3.1 Effects on Pain Intensity

Studies attempting to modify either experimental or clinical pain intensity by manipulations designed to

change expectancy show either weak or negligible effects, as indicated in a meta-analysis by Fernandez and Turk (Fernandez, E and Turk, D. C., 1989). Interestingly, 10 of 12 studies in their analysis showed that expectancy manipulations were least effective in reducing pain compared to other psychological strategies. A limitation of nearly all these studies, however, is that the independent variable is a manipulation or instruction designed to persuade participants to develop a positive expectation rather than a measure of expectation itself. Therefore, whereas the weight of all of these studies indicates weak or negligible effects of expectancy *manipulations* on pain intensity, it is not at all clear that direct effects of actual expectations have such weak effects. In fact, data published after Fernandez and Turk's (Fernandez, E and Turk, D. C., 1989) meta-analysis indicate that experiential manipulations of expectancy have stronger effects than verbal manipulations of expectancy (Montgomery, G. H. and Kirsch, I., 1997; Voudouris, N. J. *et al.*, 1990).

64.3.2 Effects on the Affective Dimension of Experimental Pain

Studies of experimental pain wherein expectations can be directly manipulated provide the strongest evidence that desires and expectations are integral components of pain experience. Price D. D. *et al.* (1980) conducted a psychophysical analysis of experiential factors that influence the affective but not sensory discriminative dimension of heat pain. Trained participants made cross-modality matching judgments of both pain sensation intensity and unpleasantness. Non-noxious and noxious skin temperature stimuli were randomly interspersed during each experimental session. Changes in expectation of receiving painful stimulation were induced by preceding one-half of all the noxious stimuli by a warning signal. Results indicated that noxious temperatures were experienced as less unpleasant when preceded by a warning signal. In contrast, pain sensation magnitudes were unaffected by the warning signal. Moreover, the selective lowering of pain unpleasantness by the warning signal was greatest at the lowest end of the noxious temperature range (i.e. 45 °C) and was minimal at the highest end of the stimulus range (i.e., 51 °C). These data suggest that subjects prefer knowing that the next stimulus will be painful as compared to being surprised. Not knowing if the next stimulus will be painful is likely to produce anxiety; however, advanced warning has little effect if the intensity of stimulation is sufficiently high.

This interpretation of the anti-anxiety effects of a warning signal is supported by other experiments that reduce anxiety associated with experimental pain. Gracely *et al.* (1976) demonstrated that 5 mg diazepam significantly reduced affective descriptor responses to painful shock without altering sensory descriptor responses. Consistent with the Price D. D. *et al.* (1980) study, the reductions were greatest for low-intensity noxious stimuli. Another study by Gracely R. H. (1979) found a similar pattern for saline placebo administration. The results of all of these studies can be parsimoniously explained as a selective lowering of pain unpleasantness by reduction in anxiety. Part of pain unpleasantness is the anxiety associated with anticipating and receiving a noxious stimulus. Anxiety represents a state of wanting to avoid negative consequences combined with an experienced uncertainty

of avoiding them (Barrell, J. J. *et al.*, 1985; Price, D. D. *et al.*, 1985). Regardless of how anxiety reduction is achieved, the result is that of selectively reducing pain unpleasantness, with the largest effects occurring for mildly painful intensities.

Expectations about the qualitative nature of pain sensations have also been shown to selectively influence pain affect. Johnson J. E. (1973) found that subjects who received a description of the painful sensations produced by forearm ischemia had lower levels of distress compared to subjects who only received a description of the procedure. Similar to Price D. D. *et al.* (1980), pain sensation intensities were unaffected by this difference in description. Thus, it appears that different kinds of expectations, either about the time of occurrence of pain or the types of sensations that will occur, can alter experienced unpleasantness without changing the intensity of experimentally induced painful sensation.

In summary, based on studies of experimental pain, it is apparent that expectations about different aspects of pain and about the contextual conditions in which pain is present influence pain differently. Expectations about when the pain will occur and what it means once it does occur may have relatively selective effects on the unpleasantness or immediate emotional disturbance associated with pain. Expectations about pain sensation intensity itself appear to have a modest influence on perceived pain intensity. Finally, manipulations designed to produce a positive expectation appear to have the weakest effect on pain intensity, although it is difficult to reach a definitive conclusion about this possibility because of the questionable pain measures used among most available studies.

64.3.3 The Influence of Expectancy on the Immediate Unpleasantness of Clinical Pain

The factors of desire for relief and expectation would intuitively have a greater influence on pain-related affect in the case of clinical pain, because the implications of having clinical pain are likely to be perceived as more open-ended and threatening than experimental pain. Unfortunately, few attempts have been made to directly assess the role of expectancy in clinical pain. Nevertheless, such a role is strongly supported by at least indirect evidence largely consistent with the idea that anxiety is a significant component of clinical pain affect. One study compared affective ratings of clinical pain in patients whose pain is associated with a serious threat to health (cancer pain) to those of patients whose pain is of a less threatening nature (labor pain) (Price, D. D. *et al.*, 1987). Comparisons were also made between women in labor whose focus was primarily on the birth of their child versus those who focused mainly on pain or on avoiding pain. In cancer pain patients, unpleasantness ratings were higher than sensory ratings, whereas the reverse was true for labor pain patients. Furthermore, patients who focused primarily on having the baby rated the unpleasantness of their pain as approximately one-half that of patients who focused primarily on pain or avoiding pain (Figure 5(a)); this difference occurred for each stage of labor. In contrast, no significant differences in pain sensation intensity ratings occurred between patients with these two orientations at any stage of labor (Figure 5(b)).

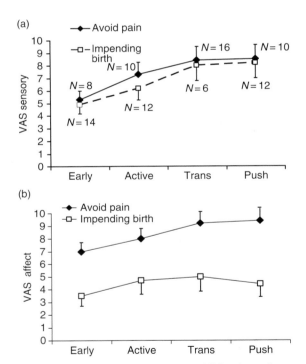

Figure 5 Mean sensory and affective VAS ratings of two groups of labor patients at various stages of labor. (a) Mean sensory ratings of the women who focused mainly on pain or avoiding pain (black diamonds) were very similar to women who focused mainly on the impending birth of the child. (b) Mean unpleasantness ratings of women focused on the impending birth were approximately half that of women who focused on pain or avoiding pain. Adapted from Price, D. D., Harkins, S. W., and Baker, C. 1987. Sensory-affective relationships among different types of clinical and experimental pain. Pain 28(3), 291–299.

The combination of all these results indirectly indicates that a person's goals, desires, and expectations about outcomes strongly influence emotional feelings associated with clinical pain. The influence of these factors is most apparent when divergent psychological orientations exist within a clinical pain condition. Thus, the unpleasantness generated by the immediate implications of cancer pain, including the reminder that pain sensation is a signal for the presence of disease, appears to add to that which is directly related to the pain sensations. One of the implications of having labor pain, on the other hand, is that birth of a baby is imminent. The positive emotional consequence of this implication appears to offset to some degree the unpleasantness of labor pain. This interpretation is further supported by the much greater degree of labor pain unpleasantness among women who focus on avoiding pain compared to those who focus on the birth of the baby. Part of what constitutes pain unpleasantness is the immediate implication of the pain condition. The implication, in turn, is related to a goal, desire, and an expectation associated with that goal.

Results from the studies of experimental pain described above indicate the existence of pharmacological and psychological manipulations that can selectively, and sometimes powerfully, reduce the immediate unpleasantness of pain. Some of these manipulations can be quite simple, such as stimulus duration or the presence or absence of a warning signal. Others are more complex and relate to restructuring the meanings associated with pain (e.g., some hypnotic suggestions). Manipulations that enhance pain unpleasantness also exist. The identification of experimental manipulations that selectively modify affective components of pain is critical for developing experimental strategies for identifying neurons, central pathways, and brain regions that are uniquely involved in the immediate affective dimension of pain.

64.4 Factors That Contribute to the Magnitude of Placebo Analgesia

For decades, placebo responses were mainly conceptualized as resulting from inert physical agents. Within the last few years, explanations of placebo responses have shifted from emphasis on environmental factors to human meanings and perceptions. We explore the roles of both external and experienced factors in producing placebo analgesia as well as the relationships between them (see Chapter The Placebo Effect for placebo vignette).

64.4.1 The Magnitudes of Placebo Analgesia and the General Contexts That Influence Them

Although this study raised considerable controversy, the effect of placebo treatment was compared to a natural history condition in a recent meta-analysis of 29 studies (Hrobjartsson, A. and Gøtzsche, P. C., 2001). The effect size of placebo analgesia, measured as Cohen's d (pooled standardized mean difference), was 0.27, reflecting a small effect that is possibly clinically significant. Hrobjartsson and Gøtzsche suggested that the small placebo analgesia effect could be the result of confounding variables such as response bias, since pain reflects a subjective state. Interestingly, the large majority of the studies (24 of 29) included in the meta-analysis used a clinical trial design in which placebo administration served as the control condition. Verbal suggestions for analgesia are typically avoided in such studies and the only reference to placebo is made in the consent form in which patients are informed of the possibility of receiving an inert and/or ineffective agent. In most nonresearch clinical settings or in experimental studies of placebo analgesia mechanisms, however, positive suggestions or conditioning with effective agents are likely to enhance placebo analgesia, as will be discussed below. Therefore, patients are more likely to expect to receive a pain relieving medication in studies containing positive verbal suggestions in comparison to most clinical studies.

To determine whether placebo effect size is systematically influenced by the two contexts just described, two contrasting meta-analyses were conducted (Vase, L. et al., 2002). In the first, twenty-three studies in which placebo treatments were used as a control condition were included, the majority of which were also included in the meta-analysis by Hrobjartsson and Gøtzsche. Most of these were clinical studies that tested an active drug or treatment. In the second, fourteen studies that were about placebo analgesic mechanisms were included. The mean effect size was 0.15 in the first meta-analysis of twenty-three studies, even smaller than Hrobjartsson and Gøtzsche's finding of 0.27. However, the mean effect size was 0.95 in the fourteen studies of placebo mechanisms, where strong suggestions for pain relief were typically given. Overall, these meta-analyses indicate that although the magnitude of placebo analgesia effects is highly variable, one of the potent factors influencing the placebo effect may be whether overt verbal suggestions are given for pain relief. The effect size of placebo analgesia in the studies where placebo analgesia was induced via either conditioning alone or suggestion alone was 0.83 and 0.85, respectively. However, the magnitude of placebo analgesia was 1.45 (almost twice as high as either factor alone) in the studies where the placebo analgesia effect was induced via a combination of conditioning and suggestion. Thus, it appears that the effect of external manipulations such as conditioning and suggestion are additive, implying that the placebo analgesia effect may be related to how the placebo agent is perceived, regardless of the exact external factors that lead to that perception.

64.4.2 Analyses of Factors That Contribute to the Magnitude of Placebo Analgesia

Given the variability in placebo analgesic effects and the systematic influence of the presence or absence of factors such as previous experience with effective treatments and verbal suggestions, an emerging goal of placebo analgesia research is to identify factors that contribute to the perceived efficacy of a therapeutic intervention. The contribution of placebo analgesia to the effectiveness of analgesic drugs was tested in a clinical study using hidden and open injections of traditional painkillers such as buprenorphine (Colloca, L. et al., 2004). The patients needed less medication for analgesia when a doctor administrated open injections of painkilling agents with verbal suggestions for pain relief, than when a hidden machine administrated the painkilling agents. Presumably, the difference in analgesic response between open and hidden injections directly reflects the placebo analgesia effect. The difference between open and hidden injections could be eliminated by hidden injections of naloxone in a subsequent experimental study, demonstrating that part of the response variability to analgesic drugs can be attributed to opioid-mediated placebo analgesic effects. It may be particularly useful to explore exactly how subjects experience open injections versus hidden injections in order to clarify the mediating and moderating factors in placebo analgesia. As discussed below, such an approach is beginning to be applied to placebo analgesic research in general.

64.4.2.1   External factors that contribute to placebo analgesia

64.4.2.1.(i) Conditioning Previous experiences with pain, the analgesic remedy, and the setting are likely to influence perception of pain treatments. Some investigators propose the idea that placebo effects are based on Pavlovian conditioning. When a patient receives an agent (unconditioned stimulus) that leads to analgesia (unconditioned response), contextual cues such as the medical setting, the white coat, or the pill/syringe (conditioned stimuli) may be associated with pain relief (Wickramasekera, I. A., 1985). These contextual factors, which represent the conditioned stimuli, elicit pain relief in the absence of active agents. Cross-over studies have supported this hypothesis; the magnitude of placebo analgesia follows the graded doses of the active drug when placebo is given as the second drug (Laska, E. and Sunshine, A., 1973). Voudouris N. J. *et al.* (1985; 1989; 1990) conducted the first studies to investigate the contribution of conditioning with a design that included a repeated baseline control condition. Subjects were tested in three sessions: (1) pretest, where a noxious electrical stimulus was applied to determine the subjects' threshold, (2) manipulation, where an inert cream was applied to the skin and the stimulus was surreptitiously reduced to provide an experience of the cream's analgesic effect, and (3) posttest, where the analgesic cream was applied and the original intensity stimulus was delivered to the same area of the skin. Compared to a group given the cream with no conditioning, the conditioning group showed significant pain reduction by the placebo cream. This suggests that a previous pairing of reduced stimulus intensity with an inert cream can result in large analgesic effects.

With subjects undergoing ischemic arm pain, the contribution of conditioning also has been tested (Amanzio, M. and Benedetti, F., 1999). On the first day, subjects were tested during a no-treatment session. On the second and third day, a group of 14 subjects was conditioned with the opioid agent morphine and another group (14 subjects) was conditioned with the nonopioid agent ketorolac. On the fourth day, both groups received open injections of saline and were told that the agent was an antibiotic to test effects of conditioning alone without any effect of overt placebo suggestion. Conditioning in both the morphine and ketorolac groups produced moderate and statistically reliable placebo effects. When the open injection secretly contained naloxone in a subsequent morphine-conditioned group, the placebo effect could be completely prevented. In contrast, the placebo effect was not antagonized by naloxone if ketorolac was used as the conditioning stimulus. Thus, the placebo analgesia effect induced by opioid conditioning (morphine) was mediated by endogenous opioids, whereas the placebo analgesia effect induced by nonopioid conditioning (ketorolac) was not mediated by endogenous opioids.

64.4.2.1.(ii) Suggestion Expectation and belief in future pain relief can be induced by nonverbal and verbal suggestions, and expectations of caregivers can be transferred to patients. For example, the influence of nonverbal suggestions was indirectly tested in a study where the clinician knew that one group of patients would receive placebo whereas the other group would receive placebo as well as the active analgesic agent fentanyl (Gracely, R. H. *et al.*, 1985). The placebo analgesia effect was significantly higher in the second group, indicating that even without intentional or verbal communication, simply the clinician's belief that a patient may receive a powerful painkiller influences the magnitude of placebo analgesia. Several studies have investigated the influence of verbal suggestions for pain relief.

The same paradigm of ischemic tourniquet pain tolerance was used to test whether suggestion in itself was enough to produce placebo analgesia (Amanzio, M. and Benedetti, F., 1999). After the first day's no-treatment session, on the second day, patients were given an open injection of saline and told that the agent was a powerful painkiller. A small but statistically reliable placebo effect was produced by this placebo treatment. When the experiment was repeated with an open injection of naloxone instead of an open injection of saline on the second day, the placebo analgesia effect was entirely eliminated, indicating that suggestion induces a placebo analgesia effect that is mediated by endogenous opioids. Recent studies have shown that both direct and indirect suggestions for pain relief lead to placebo analgesia of large magnitudes (Pollo, A. *et al.*, 2001; Price, D. D. *et al.*, 1999; Vase, L. *et al.*, 2002). Additionally, several studies have shown that verbal suggestions can lead to reduction of pain in highly specific areas of the body and that these specific placebo effects may be mediated by endogenous opioids (Benedetti, F. *et al.*, 1999; De Pascalis, V. *et al.*, 2002; Montgomery, G. H. and Kirsch, I., 1996; Price, D. D. *et al.*, 1999).

The majority of the studies presented so far have shown that conditioning in itself or suggestion in itself may lead to placebo analgesia. However, in most studies and especially in most clinical settings, previous exposure to analgesic agents and verbal suggestions for pain relief combine to produce placebo analgesia. The combination of morphine conditioning and verbal suggestions for inducing placebo analgesia was also examined in the study by Amanzio M. and Benedetti F. (1999). The magnitude of this placebo analgesia effect was approximately twice the size of the placebo effect induced via either conditioning or suggestion alone. Interestingly, this large placebo effect could be completely prevented by naloxone. Thus, even though conditioning and suggestion may make separate contributions to placebo analgesia effects, they are likely to both activate a common opioid analgesia network. A recent study suggests that conditioning and suggestion both influence conscious phenomena (e.g., pain) through expectations, whereas conditioning without conscious expectations can influence unconscious physiological processes such as cardiovascular responses and hormonal secretions (Benedetti, F. *et al.*, 2003).

64.4.2.2 Experienced factors that contribute to placebo analgesia

Patients are likely to perceive external factors such as conditioning and suggestion in different ways. Therefore, it is important to examine how these environmental factors and factors within the human experience relate to each other during placebo analgesia.

64.4.2.2.(i) Expectancy

In pain studies, expectancy (the experienced likelihood of an outcome or an expected effect) can be measured by asking people about the level of pain they expect to experience. One of the first studies to directly measure and manipulate expected pain levels used a design in which subjects were conditioned with reduced stimulus intensity (cutaneous electrophoresis) in the presence of an inert cream (Montgomery, G. and Kirsch, I., 1997), similar to studies by Voudouris N. J. and co-workers (Voudouris, N. J. *et al.*, 1985; 1989; 1990). Subjects rated expected pain levels immediately after the conditioning trials. Pain ratings were noticeably reduced by the conditioning procedure in a group of subjects that did not know about the stimulus manipulation (i.e., the placebo analgesic effect). However, regression analyses showed that expected pain levels mainly accounted for this effect. Thus, expectancy accounted for 49% of the variance in post manipulation pain ratings. Furthermore, when

another group of subjects was informed about the experimental design and learned that the cream was inert, their expected pain levels did not differ from expected pain levels during the baseline condition and the placebo analgesia effect disappeared. Thus, although conditioning can lead to placebo analgesia, it appears to be mediated by conscious expectations.

Using a similar paradigm, the extent to which expectations of pain relief can be graded and related to specific areas of the body was further tested (Price, D. D. *et al.*, 1999). In this study three 'strengths' of placebo cream were applied to the subjects forearm and external manipulations were performed to perpetuate the belief that cream A was a strong analgesic, cream B a weak analgesic, and cream C a control agent on three adjacent areas of the arm. Conditioning was accomplished by surreptitiously lowering the heat stimuli in areas A (large reduction) and B (small reduction) but not C (no reduction), after the creams were applied. Following these manipulation trials, subjects were asked to rate their expected pain levels for the next series of trials. The conditioning trials led to graded levels of expectancy for the three creams C, B, and A (of successively less intense pain). When identical stimulus intensities were applied during the postmanipulation trials, subjects rated pains in areas C, B, and A as progressively less intense, demonstrating a graded placebo effect. In this study, expected pain levels accounted for between 25% and 36% of the variance in postmanipulation pain ratings. These results provide further evidence for the somatotopic specificity of the placebo analgesic effect, given that these placebo effects were induced on three immediately adjacent areas of the arm.

64.4.2.2.(ii) Desire for pain reduction

While expectancy appears to be an important psychological mediator of placebo analgesia, it is unlikely to work alone. Since motivation is known to influence perception, desire (the experiential dimension of wanting something to happen or wanting to avoid something happening) is also likely to be involved in placebo analgesia. In the study just described, the subjects' level of desire for pain reduction also was manipulated (Price, D. D. *et al.*, 1999). In one group, presenting them the prospect of receiving a large number of painful stimuli successfully increased their desire for pain relief. In a second group, informing them that only a few stimuli would be presented decreased desire. Interestingly, ratings of desire for pain relief were not significantly associated with the magnitudes of placebo analgesia. One possible reason for this lack of association may be because pain was induced via brief heat stimuli in this experimental setting. Desire for pain relief may be more of a factor in clinical pain where the pain is threatening or has an uncertain duration, therefore likely inducing fear or anxiety. Consequently, it is important to investigate the contribution of expectations and desire for pain relief in more clinically relevant settings.

64.4.2.2.(iii) *Memory of pain*

This same study also showed that distorted memory of recent pain also contributes to the placebo effect (Price, D. D. *et al.*, 1999). The placebo analgesia effect was assessed in two ways, on concurrent ratings of pain obtained immediately after the stimuli were applied and on retrospective ratings of pain obtained approximately two minutes after the stimuli were applied. Based on retrospective ratings, the magnitude of placebo analgesic effects was three to four times greater than the effect based on concurrent ratings. The main reason for this difference was that subjects remembered their baseline pain intensity as being much larger than it actually was. Similar to placebo analgesia effects based on concurrent ratings, the placebo effects based on remembered ratings were strongly correlated with expected pain intensities ($R = 0.5$–0.6). In fact, there was a stronger correlation between expected pain and remembered pain than between expected pain and actual pain. Consequently, it is important to measure placebo analgesia effects with concurrent ratings since it may be distorted by remembered pain. Moreover, since remembered pain and expected pain are closely related, these factors appear to interact. These effects of distorted memory on placebo analgesia have been replicated (De Pascalis, V. *et al.*, 2002).

64.4.2.3 *Predicting placebo analgesia*

Two similarly designed studies further explain how factors within and external to human experience may relate in placebo analgesia (Vase, L. *et al.*, 2003; Verne, G. N. *et al.*, 2003). In both studies, patients diagnosed with Irritable Bowl Syndrome (IBS) were exposed to rectal distention via a balloon barostat (a type of visceral stimulation that simulates their clinical pain) and tested under natural history, rectal placebo and rectal lidocaine conditions. Both pain intensity and unpleasantness were rated immediately after each stimulus. In the first study, a standard clinical trial was conducted where the patients were given an informed consent form stating that they "may receive an active pain reducing medication or an inert placebo agent" (Verne, G. N. *et al.*, 2003). The rectal lidocaine had a significant pain relieving effect compared to rectal placebo ($p < 0.001$), and the rectal placebo had a significant pain relieving effect compared to the natural history condition (Figure 6, left panel). Based on meta-analyses discussed earlier, the second study added a suggestion to the placebo condition and lidocaine condition (Vase, L. *et al.*, 2003). Patients of the second study were told at the beginning of each treatment condition that "the agent you have just been given is known to significantly reduce pain in some patients" (rectal placebo, rectal lidocaine). A much larger placebo analgesic effect was found in the second study. In fact, the magnitude of placebo analgesia was so high that there was no longer a significant difference between the magnitude of rectal lidocaine and rectal placebo (Figure 6, right panel). Therefore, these two studies indicate that it is possible to increase the magnitude of placebo analgesia to a level that matches an active agent by adding an overt suggestion for pain relief.

Patients were asked to rate their expected pain level and desire for pain relief immediately after the agent was administered (before the onset of its effects) in both studies. In the second study, the combination of expected pain levels and desire for pain relief accounted for 77% of the variance in post placebo pain ratings (see Table 1); however, expected pain levels have accounted for between 25% to 49% of the variance in post placebo pain ratings in previous studies. Therefore, it is possible that the combination of expected pain level and desire for pain relief accounts for a greater amount of the variance in post placebo pain ratings than either factor alone. Nevertheless, although the interaction between expected pain level and desire for pain relief seems to have contributed, the majority of the contribution came from the expected pain level (Table 1). Interestingly, the combination of expected pain level and desire for pain relief accounted for 81% of the variance in pain ratings during the rectal lidocaine condition. These findings further support the conclusion that placebo factors make strong contributions to the efficacy of active treatments.

64.4.3 Placebo Analgesia As an Emotional Response

More recently, the role of emotion in placebo analgesia has been examined in a number of studies.

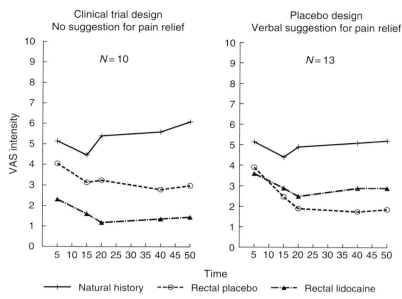

Figure 6 Comparisons of natural history, rectal placebo and rectal lidocaine scores on visceral pain intensity ratings (VAS) during a 50 min session within a clinical trial design, where no suggestions for pain relief are given (left) and within a placebo design with verbal suggestions for pain relief (right). From Vase, L., Robinson, M. E, Verne, G. N, and Price, D. D. 2003. The contribution of suggestion, expectancy and desire to placebo effect in irritable bowl syndrome patients. Pain 105, 17–25.

Table 1 The contribution of expectancy and desire to rectal placebo analgesia

Model	R^2 change	F	P
Expectancy[a] + desire	0.64	9.1	0.006
Expectancy X desire	0.12	5.0	0.05
Total model	0.77	10.2	0.003

[a]Denotes significant beta weight.
From Vase, L., Robinson, M. E, Verne, G. N, and Price, D. D. 2003. The contribution of suggestion, expectancy and desire to placebo effect in irritable bowl syndrome patients. Pain 105, 17–25.

Interactions between desire and expectation account for magnitudes of positive and negative feelings during hypothetical emotional situations and emotional feelings during actual events (Price, D. D. and Barrell, J. J., 2000). These same interactions also can account for direction and strength of choices during instances of decision making (Price, D. D. *et al.*, 2001). Placebo analgesia may be added to the list of phenomena that are associated with the desire-expectation model in showing that interactions between desire for relief and expected pain levels account for large and significant amounts of the variance in the placebo response.

64.4.3.1 Do placebo effects result from changes in emotional states?

Placebo effects also may change over time as a consequence of changes in emotional states. Similar to the studies on IBS patients described above, 26 IBS patients were exposed to rectal distension under natural history, rectal placebo, and rectal lidocaine conditions and ratings were obtained of expected pain, desire for pain relief, anxiety, and actual pain during each experimental condition (Vase, L. *et al.*, 2005). In this study, however, ratings of expected pain levels, desire for pain relief, and anxiety were obtained at two time points, at the onset of placebo administration (early) and midway through the 40 min observation period (late). Ratings of these three factors decreased from early to late time points as did pain ratings within the placebo condition. Decreases in these three factors across natural history versus placebo conditions accounted for 11% (not significant) of the variance in placebo responses (pain in natural history minus pain in placebo condition) at the onset of placebo administration and for 58% (significant at $P < 0.001$) of the variance in placebo responses in the late period of observation. These results suggest that the placebo effect can increase over time as a function of decreases in expected pain levels, desire for relief, and negative emotions such as anxiety. These findings support the view that placebo effects may be at least partly mediated by changes in emotional states. Equally interesting is that similar results occurred for the lidocaine condition, providing further evidence that placebo effects and the factors that evoke them are embedded in active treatments.

Placebo treatments can initiate a more positive or less negative emotional state for a patient. It is possible that the placebo analgesia effect can at least partly result from decreases in negative emotions or increases in positive emotions. This possibility is supported in part by the studies just described and also by a study, which shows that pain, unpleasantness, and pain intensity may be reduced by evoking positive emotions or increased by evoking negative emotions (Chretien, P. and Rainville, P., 2004). Since this study was not about placebo effects, it provides an independent line of evidence that emotional feelings modulate pain. If placebo manipulations alter pain, they may do so by altering emotional states. Alternatively, it is possible that desire and expectation are merely factors that are common to both placebo mechanisms and emotions. In a recent study, it was found that low desire for pain relief decreased pain intensity and unpleasantness, whereas a high desire for pain relief increased pain unpleasantness but not pain intensity, suggesting that a decrease in desire may actually lower pain intensity (Chretien, P. and Rainville, P., 2004). These results are consistent with placebo studies described above (Vase, L. *et al.*, 2003; Verne, G. N. *et al.*, 2003).

64.4.3.2   *Future directions*

Whereas past explanations of placebo analgesia focused mainly on environmental factors and classical conditioning models, modern explanations focus more on human meanings, conscious expectations, and emotions as mediators or moderators of this phenomenon. This shift in emphasis is exciting because it suggests that the placebo analgesic response can be understood as a mind–body interaction and as a phenomenon of human consciousness, which has parallels within pain-regulating areas of the brain. Thus, future studies that combine brain imaging with analyses of dimensions of human experience should provide important neuroscientific and psychological insights into the mechanisms of placebo effects, with both significant scientific benefits and consequent improvements in the care of patients with pain.

64.5   Modulation of Pain Intensity and Pain Unpleasantness by Emotions

Emotions are closely related to pain in several ways. The notion of pain-related suffering corresponds to the experience of strong negative emotion associated with pain. Emotions also share a number of features with pain, including output viscero-motor responses and several experiential dimensions. In addition to co-occurring with pain, increasing evidence suggests that emotions play an important role in pain modulation. This may reflect the overlap between pain and emotion-related neurophysiological processes. This section begins with a brief discussion of emotions and follows with a consideration of the effects of emotion on experimental and clinical pain.

64.5.1   Negative Emotions As a Constituent of Pain

64.5.1.1   *Pain definitions*

The International Association for the Study of Pain (IASP) defines pain as "an unpleasant sensory and emotional experience associated with actual or potential tissue damage, or described in terms of such damage" (Merskey, H. and Spear, F. G., 1967). A slightly different definition is offered by Price D. D. (1999) in which pain is "a somatic perception containing (1) a bodily sensation with qualities like those reported during tissue damaging stimulation, (2) an experienced threat associated with this sensation, and (3) a feeling of unpleasantness or other negative emotions based on this experienced threat" (p. 1–2). In both definitions, negative emotions are a component of the pain experience. Price's (1999) stages of pain processing described earlier also highlight the relationship between pain and emotions.

64.5.1.2   *Is pain an emotion?*

The components of emotion theories also relate to pain processes. Perhaps the primary difference between pain and emotion is that pain requires the presence of "a bodily sensation with qualities like those reported during tissue damaging stimulation" (Price, D. D., 1999; p. 1). Pain sensations then can be regarded as a specific inducer of a primitive emotional response, which is consistent with theoretical accounts of emotion systems as closely related to evolutionary processes (e.g., Izard, C. E., 1993; Plutchik, R., 1980). The prototypical "pain face" has been documented and can be differentiated from facial expressions of the basic emotions of fear, anger, and sadness (reviewed in Williams, A. C., 2002). Patterns of autonomic responses to nociceptive stimuli well documented in animals (Sato, A. *et al.*, 1997) may contribute to pain-related emotional responses as well as subjective feelings of unpleasantness (e.g. Fillingim, R. B. *et al.*, 1998; Rainville, P. *et al.*, 1999). The stages of pain processing described earlier include primary and secondary affective stages (Price, D. D., 1999), where the first corresponds to the basic experience of a threat, or the fear of tissue damage, and may include self-perception of autonomic and motor responses triggered automatically. The second stage reflects secondary evaluative processes consistent with modern views on the role of cognitive evaluative processes in emotions. This is also consistent with the experiential model in which emotional experiences are related to goals, desires, and expectations. A study by Price D. D. *et al.* (1980) indicated that this experiential model of emotions is predictive of variations in pain unpleasantness, but not pain intensity, which is consistent with the view that pain-related affective processes are emotional responses to pain sensations.

64.5.2 Effects of Emotion on Pain: Experimental Studies

In order to determine the effects of emotion on pain, one must first distinguish between emotions related to pain from emotions that co-occur with pain but are otherwise unrelated to pain. Emotions that are related to pain constitute secondary affective responses to pain. Emotions that are not related to pain may be considered distracters and/or they may produce brain–body responses that increase or decrease pain.

64.5.2.1 The experience of pain-related emotions

Hypnosis is one experimental paradigm that has been used to test the effect of pain-related emotions on pain. In a series of studies conducted by Rainville (reviewed in Rainville, P., 2004), positive and negative emotions were hypnotically induced in subjects undergoing an experimental pain task. Compared to baseline and hypnotic relaxation conditions, ratings of pain intensity and unpleasantness decreased in positive emotions conditions and increased in negative emotions conditions (Figure 7) (Huynh Bao, Q. V. and Rainville, P., 2002). Although affecting both pain intensity and unpleasantness, pain-related negative

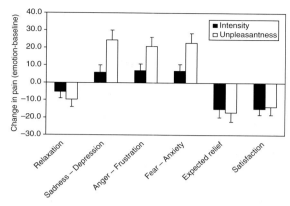

Figure 7 Changes in pain intensity and pain unpleasantness produced by 1 min immersions of the hand in hot water, from a baseline control condition and following the hypnotic induction of emotions. Significant increases in pain ($p < 0.05$) are found in response to negative emotions relative to the baseline ($y = 0$) and relaxation condition, with larger effects observed in pain unpleasantness than pain intensity. Decreases in pain were less pronounced and ns. in response to positive emotion induction. Adapted from Rainville, P. 2004. Pain and Emotions. In: Psychological Methods of Pain Control: Basic Science and Clinical Perspectives (*eds.* D. D. Price and M. C. Bushnell), pp. 117–141. IASP Press.

emotions primarily increased pain unpleasantness (Figure 7).

In a subsequent study examining the effects of anger and sadness, hypnotic suggestions for these emotions were associated with significant increases in desire to avoid and/or escape pain during the experimental task and reductions in expectations of avoiding and/or escaping pain (Figure 8(a)). Ratings of negative valence and arousal increased significantly in both negative emotion conditions, and subject ratings of their feeling of being in control (dominance) decreased significantly only in the sadness condition. In both negative emotion conditions, pain intensity and unpleasantness increased significantly, with pain unpleasantness showing a larger change (Figure 8(b)). Importantly, changes in pain unpleasantness, and pain intensity to a lesser degree, were predicted by changes in emotions (Table 2). The associations with pain unpleasantness remained significant after controlling for changes in pain intensity. However, after accounting for changes in pain unpleasantness, the associations with pain intensity were reduced to almost zero, save expectation that contributed some unique variance to changes in pain intensity. In a recent study, however, Chretien and Rainville (Chretien, P. and Rainville, P., 2004) found

(a)

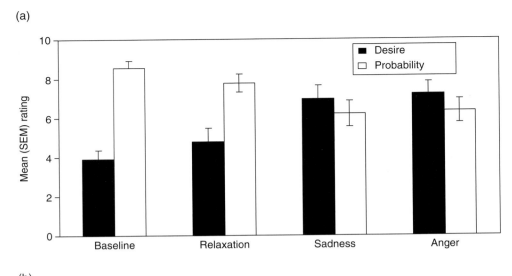

(b)

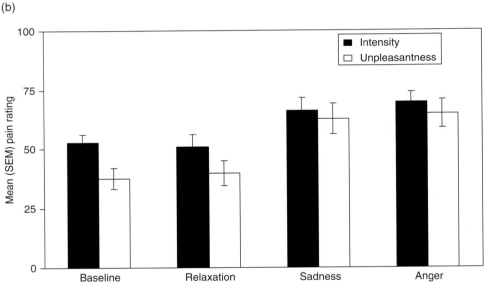

Figure 8 Effects of hypnotic relaxation and hypnotic induction of anger and sadness on subjective desires and expectations to avoid/escape pain (a) and on the pain intensity and unpleasantness produced by 1-min immersion of the hand in hot water (b). Desire increased and expectation decreased in the negative emotion conditions ($p < 0.05$). As a result, pain unpleasantness, and to a lesser extent, pain intensity increased in both conditions ($p < 0.05$). Figure adapted from Rainville, P. 2004. Pain and Emotions. In: Psychological Methods of Pain Control: Basic Science and Clinical Perspectives (*eds.* D. D. Price and M. C. Bushnell), pp. 117–141. IASP Press.

that the specific, hypnotic manipulation of desire to avoid/escape pain does produce a significant change in pain unpleasantness, independent of changes in pain intensity and expectation.

The previous study also included autonomic measures and found changes in heart rate to the experimental heat pain stimuli. Significant increases in heart rate were observed during negative emotions compared to baseline and control conditions, and this

increase was positively associated with changes in pain and emotional experience (Table 3). After controlling for changes in pain intensity and felt arousal, changes in pain unpleasantness remained positively correlated with changes in cardiac responses. Conversely, cardiac responses were not significantly correlated with pain intensity after controlling for pain unpleasantness, and were only slightly correlated with pain intensity after controlling for

Table 2 Correlation coefficients between changes in pain intensity and unpleasantness, and changes in several emotional dimensions

Pearson-R	Pain intensity	Pain unpleasantness	Partial intensity	Partial unpleasantness
Desire	0.74***	0.83***	0.16	0.56***
Expectation	−0.36**	−0.65***	0.43*	−0.68***
Valence	0.65***	0.76***	0.04	0.52***
Felt arousal	0.52***	0.67***	−0.09	0.50***
Dominance	−0.47***	−0.53***	−0.05	−0.29*

*, $p < 0.05$; **, $p < 0.01$; ***, $p < 0.001$.
From Rainville, P. 2004. Pain and Emotions. In: Psychological Methods of Pain control: Basic Science and Clinical Perspectives (eds. D. D. Price and M. C. Bushnell), pp. 117–141. IASP Press.

Table 3 Correlation between changes in pain-evoked cardiac responses and changes in pain- and emotion-related dimensions

Pearson-R	Modulation of RR changes	Residual (-INT)	Residual (-UNP)	Residual (-arousal)
Pain intensity	−0.48***		+0.02	−0.27*
Pain unpleasantness	−0.59***	−0.38**		−0.35**
Desire to avoid-escape	−0.47***	−0.19	+0.03	−0.26
Expectation to avoid-escape	+0.44***	+0.33	+0.10	+0.16
Valence	−0.37**	−0.09	+0.14	−0.13
Felt arousal	−0.55***	−0.40**	−0.27*	
Dominance	+0.30*	+0.10	−0.01	+0.10

*, $p < 0.05$; **, $p < 0.01$; ***, $p < 0.001$.
From Rainville, P. 2004. Pain and Emotions. In: Psychological Methods of Pain control: Basic Science and Clinical Perspectives (eds. D. D. Price and M. C. Bushnell), pp. 117–141. IASP Press.

arousal. As would be expected, changes in felt arousal were positively associated with cardiac responses; however, this association was attenuated when changes in pain unpleasantness were controlled. This suggests that feelings of arousal and pain unpleasantness are closely related.

Figure 9 summarizes the findings of these studies examining the effect of pain-related emotions on pain. In this model, the nociceptive stimulus contributes to pain intensity, which then contributes to pain unpleasantness (Price, D. D., 1999; 2000; Rainville, P. *et al.*, 1999). The nociceptive stimulus also elicits autonomic responses and somatic feedback, and these factors influence felt arousal and pain unpleasantness. Pain-related desire and expectation contribute to modulate pain. Desire to avoid or escape pain primarily affects pain unpleasantness, independent of pain intensity (Table 2). Alternatively, expectation appears to have a modulatory affect on both pain intensity and unpleasantness, which is consistent with the placebo literature discussed previously. Furthermore, pain-related

autonomic activity may be enhanced or diminished by descending neurophysiological mechanisms that are responsive to modulation of pain unpleasantness. When considered in their entirety, these results indicate that pain-related negative emotions have the primary effect of increasing pain unpleasantness and to a lesser degree pain intensity.

64.5.2.2 Pain anticipation and pain-related fear/anxiety

Other experimental studies have taken different approaches to examine the effects of emotions on pain perception. One phenomenon that has received particular attention is pain anticipation and the associated negative emotions. In addition to producing increased vigilance and attention towards future painful stimulation, anticipation of pain may also produce fearful responses and anxiety associated with the expectation of potential harm.

The effects of pain-related anxiety, fear, and fear-avoidance have been examined in clinical contexts as well as in animal studies. Meagher and Grau have

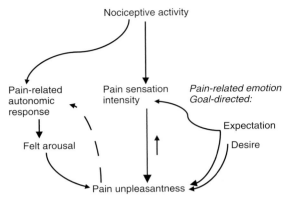

Figure 9 Pain-related emotions are suggested as affecting pain experience in several ways. Goal-directed desires appear to mainly influence pain unpleasantness, whereas goal-directed expectations affect both sensation intensity and pain unpleasantness. Physiological activity associated with autonomic activity that contributes to felt arousal. Adapted from Rainville, P. 2004. Pain and Emotions. In: Psychological Methods of Pain Control: Basic Science and Clinical Perspectives (*eds.* D. D. Price and M. C. Bushnell), pp. 117–141. IASP Press.

attempted to bridge the gap between these paradigms in a series of studies. Mild electric shocks in rats have been shown to produce hyperalgesia, whereas higher intensity stimuli typically results in stress-induced analgesia (e.g. Meagher, M. W. *et al.*, 2001b). In two human studies by Rhudy and Meagher (2000; 2001), pain threshold was increased in an anxiety condition and decreased in a fear condition, which they argued was consistent with the animal literature demonstrating anxiety-related hyperalgesia and fear-related analgesia. A subsequent study, however, found pain reductions in all conditions (Rhudy, J. L. and Meagher, M. W., 2003). Further complicating matters, the specific emotion experienced and its physiological features played important roles in pain modulation and these effects were influenced by gender. The mixed results of these and another study (Rhudy, J. L. and Meagher, M. W., 2003) highlight the importance of individual differences in the effect of the experimental manipulation and the effect of the emotion experienced on pain. A further complication arises when one considers attentional effects. The emotion-induction stimuli used in these studies, being distinct from the pain stimulus used to assess pain perception, might have competed with the less intense target pain stimulus for the subject's attentional resources. This possibility cannot be excluded based on the design of these studies.

Anticipation of pain is also suggested to be an important factor in pain modulation. Consistent with the previous discussion of the differential effects of pain-related and pain-unrelated emotions, Dougher M. J. *et al.* (1987) observed increased pain sensitivity in response to pain-specific, compared to non-specific, anxiety. Pain expectancies may also serve as a mediator of the effects of pain catastrophizing and depression on experimental pain in normals, as suggested recently by Sullivan M. J. *et al.* (2001a; 2001b). Interestingly, in this study expected mood was a stronger predictor of pain rating than expected pain, suggesting that the anticipation of an emotional response may contribute significantly to pain modulation.

64.5.2.3 *Emotions unrelated to pain*

Several experimental methods have been employed to induce emotions unrelated to pain in order to further examine the effect of emotional modulation on pain perception. Meagher M. W. *et al.* (2001b) found that presenting subjects with pictures that display fear and disgust before they underwent a cold-pressor test reduced threshold of pain intensity and unpleasantness; the fear stimuli also reduced pain tolerance. Conversely, erotic pictures increased pain intensity and unpleasantness threshold in men, but not women. A positive, graded increase in pain tolerance was observed with increases in the pleasantness of pictures presented during a cold-pressor test, whereas negative pictures that displayed pain-related images decreased pain tolerance (De Wied, M. and Verbaten, M. N., 2001). This is consistent with the distinction between pain-related and pain-unrelated emotions, pain-related emotions being generally associated with increased pain.

Reading positive and negative affectively valenced statements (Velten, E., 1968) has also been observed to increase and decrease experimental pain tolerance, respectively (Zelman, D. C. *et al.*, 1991). Willoughby S. G. *et al.* (2002), using this method, found decreases in pain tolerance related to pain catastrophizing in a depressed mood induction condition. Induction of positive affect via films has also been shown to reduce ratings of experimental pain (Weisenberg, M. *et al.*, 1998; Zillmann, D. *et al.*, 1996). Likewise, emotionally valenced music can exert pain modulatory effects (Roy, M. *et al.*, 2003; Whipple, B. and Glynn, N. J. 1992). Results of the studies by Roy and co-workers suggest that that the effects of emotions/moods are specific to pain processing and not merely due to non-specific distracting effects.

Sweet taste (Lewkowski, M. D. *et al.*, 2003) and pleasant odors (Marchand, S. and Arsenault, P., 2002) have also been shown to reduce experimental pain; however, the effect was only present in those with low arterial blood pressure in the study on taste and in women in the study on odors. Although providing further confirmation for the effects of positive and negative mood on pain perception, these findings highlight the importance of individual differences in these effects.

Studies using a cognitive-stress paradigm also demonstrate the effects of emotional modulation on pain perception and further highlight the importance of sex of the subject in this context. The cognitive stress of the stroop color-word task increased pain associated with capsaicin injection in women compared to men in one study (Logan, H. *et al.*, 2001). In this study, men showed a stronger correlation between pain increases and *physiological* indices of arousal, whereas women showed a stronger correlation between pain increases and *subjective* measures of arousal. This suggests that the effects of emotion-related physiological and subjective arousal on pain modulation may vary by sex. This study did not assess the specific emotion elicited during the cognitive task; however, a similar study found induced anger to be associated with increases in pain tolerance (Janssen, S. A. *et al.*, 2001).

64.5.2.4 *Emotion or attention?*

Emotion-induction stimuli naturally attract attention and consequently may compete with pain for attentional resources. This potential confound has long been recognized in the literature (e.g., Gardner, W. L. and Licklider, J. C., 1959; Gardner, W. L. *et al.*, 1960). As discussed earlier in section II.B. on attention, Villemure C. *et al.* (2003) demonstrated the pain modulatory effects of mood independent of attention. Although perhaps explaining the pain reduction induced by positive emotions in some instances, distraction does not account for increased pain ratings associated with negative emotions (e.g., Meagher, M. W. *et al.*, 2001b). Furthermore, attention usually exerts its effects on pain intensity and unpleasantness (e.g., Miron, D. *et al.*, 1989), whereas emotions – both pain-related and pain-unrelated – have a larger impact on pain unpleasantness (e.g., Huynh Bao, Q. V. and Rainville, P., 2002; 2003; Roy, M. *et al.*, 2003; Villemure, C. *et al.*, 2003). Specifically, negative emotions may increase pain unpleasantness by heightening an individual's awareness of his/her

viscero-somatic responses independent of pain intensity. This would be consistent with studies showing negative emotions to be associated with increased self-focus and attention to somatic signals (Mor, N. and Winquist, J., 2002; also see Salovey, P., 1992).

64.5.3 Effects of Emotions on Pain: Clinical Studies

Clinical studies of the effects of emotions on pain have largely focused on specific emotions or emotion disorders. These studies have not extensively investigated the effects of arousal, valence, or specific experiential dimensions of emotions. Depression and anxiety have received the most attention in this literature (see the reviews by Robinson, M. E. and Riley, J. L., 1998; Keefe, F. J. *et al.*, 2001; Huyser, B. A. and Parker, J. C., 1999; Fernandez, E. and Milburn, T. W., 1994; Fernandez, E. and Turk, D. C., 1995), and a large-scale study by the World Health Organization found that patients with chronic pain have higher rates of depressive and anxious disorders (Gureje, O. *et al.*, 1998). Comorbid negative mood in patients with pain has been observed to be a significant negative predictor of treatment outcome in several domains, including spine surgery (Junge, A. *et al.*, 1995; Riley, J. L. *et al.*, 1995), multidisciplinary treatment programs (Swimmer, G. *et al.*, 1992) and conservative therapies (Gerke, D. C. *et al.*, 1989). Negative emotions may also serve as a mediator in the pain, impairment, and disability nexus, as suggested by Banks S. M. and Kerns R. D. (1996) and demonstrated by Holzberg and colleagues (1996). A number of investigations have demonstrated that the chronicity of injury and related pain is predicted by measures of negative mood. In particular, a strong association has been observed in chronic pain patients between pain unpleasantness and feelings of depression, anxiety, frustration, fear, and anger (Wade, J. B. *et al.*, 1990; 1996). Patient care cost and health care utilization have also been associated with negative mood in chronic pain patients (Engel, C. C. *et al.*, 1996). As a whole, this literature suggests that negative emotions are extremely prevalent in patients with pain and are predictive of a number of key variables, including treatment efficacy and health care costs.

One theory on the nature of the relationship between pain and negative emotions is that the experience of negative emotion increases or maintains the report of chronic pain through some sensory process such as enhanced sensitivity to pain. By this account, the physiological perception of pain is modulated by one's emotions and cognitions, with depression and pain modulated by a similar process in the periaqueductal gray area of the dorsal horn. There is some evidence in the clinical literature for this contention, such as the study by Salovey P. and Birnbaum D. (1989) that found induction of negative mood to be associated with increased reports of aches and pains; however, this literature is thin and inconclusive.

Cognitive evaluations of the meaning of pain are an important component of the pain-emotion nexus. This is demonstrated in a study in which cancer patients who believed their pain to be due to the cancer reported more pain than patients who did not possess these beliefs. Pain-related emotions may mediate this effect. Other cognitive processes, such as attributional style (Love, A. W., 1988) and negative self-image (Holzberg, A. D. *et al.*, 1993), have been implicated in the relationship between pain and emotions. The constructs of pain catastrophizing and fear of pain have received considerable recent empirical attention and have been demonstrated to contribute to pain severity and distress in various patients (Crombez, G. *et al.*, 1999; Eccleston, C. *et al.*, 2001; Sullivan, M. J. *et al.*, 2001; Vlaeyen, J. W. and Linton, S. J., 2000); these effects may be mediated by increased attention to pain and/or changes in emotions. In fact, pain catastrophizing, fear, and other pain-related negative emotions, are consistently strong predictors of pain-related disability.

The effects of anger on pain also appear to depend on several factors, particularly the object of anger. Okifuji A. *et al.* (1999) found a positive relationship between pain severity and both anger directed at oneself and overall anger experienced. Depression and anger shared a similar relationship. In a recent review of this literature, Greenwood K. A. *et al.* (2003) found that gender, hostility, and anger management style, contribute to the effect of anger on pain. Interactions between these factors are complex (e.g., Burns, J. W. *et al.*, 1996) and beyond the scope of this discussion; however, anger remains of topic of increased interest in the pain literature.

In summary, negative emotions generally increase clinical pain, particularly pain unpleasantness, and are consistent predictors of pain behaviors and disability. Although the mechanisms by which emotions modulate clinical pain have not yet been entirely elucidated, the object of emotions appears to play a critical role. Emotions that are pain-related or directed toward the self appear to have the most deleterious effect.

64.6 Summary and Conclusions

This chapter has illustrated the potent modulating effects of psychological factors on pain. The evidence of these influences includes experimental studies that manipulate psychological factors such as mood, expectation, and desire for pain relief. Further evidence is provided by clinical studies that show strong predictive relationships between mood and pain, mood and disability, and placebo/expectancy manipulations and pain. Clearly, there is much overlap between different psychological mechanisms of pain modulation. For example, desires for relief and expectations are integral factors in placebo analgesia, yet they also play important roles in emotional influences on pain (Figure 9). Placebo analgesic mechanisms may well reflect modulation of pain by changes in emotions. On the other hand, attention and emotions have at least partly separate influences on pain. The neurophysiological basis of these psychological factors are partially understood, but much more work needs to be done to fully understand the basic mechanisms. Furthermore, the relationships and interactions of psychological factors to more traditional physiological and drug effects are also poorly understood and ripe for further study.

References

Amanzio, M. and Benedetti, F. 1999. Neuropharmacological dissection of placebo analgesia: expectation activated opioid systems versus conditioning-activated specific subsystems. J. Neurosci. 19. 484–494.

Apkarian, A. V., Thomas, P. S., Krauss, B. R., and Szeverenyi, N. M. 2001. Prefrontal cortical hyperactivity in patients with sympathetically mediated pain. Neurosci. Lett. 311(3), 193–197.

Banks, S. M. and Kerns, R. D. 1996. Explaining high rates of depression in chronic pain: a diathesis-stress framework. Psychol. Bull. 119, 96–110.

Bantick, S. J., Wise, R. G., Ploghaus, A., Clare, S., Smith, S. M., and Tracey, I. 2002. Imaging how attention modulates pain in humans using functional MRI. Brain 125, 310–319.

Barrell, J. J., Medieros, D., Barrell, J. E., and Price, D. D. 1985. Anxiety: an obstacle to performance. J. Hum. Psychol. 25, 106–122.

Basbaum, A. I. and Fields, H. L. 1978; Endogenous pain control mechanisms: review and hypothesis. Ann. Neurol. 4, 451–462.

Basbaum, A. I. and Fields, H. L. 1984. Endogenous pain control systems: brainstem spinal pathways and endorphin circuitry. Annu. Rev. Neurosci. 7, 309–338.

Beecher, H. K. 1959. Measurement of Subjective Responses: Quantitative effects of Drugs. Oxford University Press.

Benedetti, F., Arduino, C., and Amanzio, M. 1999. Somatotopic activation of opioid systems by target-directed expectations of analgesia. J. Neurosci. 19, 3639–3648.

Benedetti, F., Pollo, A., Lopiano, L., Lanotte, M., Vighetti, S., and Rainero, I. 2003. Conscious expectation and unconscious conditioning in analgesic, motor, and hormonal placebo/nocebo responses. J. Neurosci. 23, 4315–23.

Bentsen, B., Svensson, P., and Wenzel, A. 1999. The effect of a new type of video glasses on the perceived intensity of pain and unpleasantness evoked by a cold pressor test. Anesth. Prog. 46, 113–117.

Bernard, J. F., Bester, H., and Besson, J. M. 1996. Involvement of the spino-parabrachial amygdaloid and hypothalamic pathways in the autonomic and affective emotional aspects of pain. Prog. Brain Res. 107, 243–255.

Brooks, J. C., Nurmikko, T. J., Bimson, W. E., Singh, K. D., and Roberts, N. 2002. fMRI of thermal pain: effects of stimulus laterality and attention. Neuroimage 15, 293–301.

Burns, J. W., Johnson, B. J., Mahoney, N., Devine, J., and Pawl, R. 1996. Anger management style, hostility and spouse responses: gender differences in predictors of adjustment among chronic pain patients. Pain 64, 445–453.

Bushnell, M. C. and Duncan, G. H. 1989. Sensory and affective aspects of pain perception: Is medial thalamus restricted to emotional issues? Exp. Brain Res. 78, 415–418.

Bushnell, M. C., Duncan, G. H., Dubner, R., and He, L. F. 1984. Activity of trigeminothalamic neurons in medullary dorsal horn of awake monkeys trained in a thermal discrimination task. J. Neurophysiol. 52, 170–187.

Bushnell, M. C., Duncan, G. H., Dubner, R., Jones, R. L., and Maixner, W. 1985. Attentional influences on noxious and innocuous cutaneous heat detection in humans and monkeys. J. Neurosci. 5, 1103–1110.

Bushnell, M. C., Duncan, G. H., Hofbauer, R. K., Ha, B., Chen, J., and Carrier, B. 1999. Pain perception: is there a role for primary somatosensory cortex? Proc. Natl. Acad. Sci. USA 96, 7705–7709.

Buytendyck, F. J. J. 1961. Pain, Hutchinson.

Chretien, P. and Rainville, P. 2004. Desire for relief affects pain perception and autonomic nociceptive responses. J. Pain 3, 255.

Cioffi, D. and Holloway, 1993. J. Delayed costs of suppressed pain. J. Pers. Soc. Psychol. 64, 274–282.

Colloca, L., Lopiano, L., Lanotte, M., and Benedetti, F. 2004. Overt versus covert treatment for pain, anxiety, and Parkinson's disease. Lancet Neurol. 3, 679–684.

Corbetta, M. and Shulman, G. L. 2002. Control of goal-directed and stimulus-driven attention in the brain. Nat. Rev. Neurosci. 3, 201–215.

Crombez, G., Vlaeyen, J. W., Heuts, P. H., and Lysens, R. 1999. Pain-related fear is more disabling than pain itself: evidence on the role of pain-related fear in chronic back pain disability. Pain 80, 329–339.

Damasio, A. 1994. Descartes Error, Avon Books.

De Pascalis, V., Chiaradia, C., and Carotenuto, E. 2002. The contribution of suggestibility and expectation to placebo analgesia phenomenon in an experimental setting. Pain 96, 393–402.

Devinsky, O., Morrell, M. J., and Vogt, B. A. 1995. Contributions of anterior cingulate cortex to behavior. Brain 118(Pt 1), 279–306.

De Wied, M. and Verbaten, M. N. 2001. Affective pictures processing, attention, and pain tolerance. Pain 90, 163–172.

Dougher, M. J., Goldstein, D., and Leight, K. A. 1987. Induced anxiety and pain. J. Anxiety Disord. 1, 259–264.

Dowman, R. 2004. Distraction produces an increase in pain-evoked anterior cingulate activity. Psychophysiology 41, 613–624.

Eccleston, C., Crombez, G., Aldrich, S., and Stannard, C. 2001. Worry and chronic pain patients: a description and analysis of individual differences. Eur. J. Pain 5, 309–318.

Engel, C. C., von Korff, M., and Katon, W. J. 1996. Back pain in primary care: predictors of high health-care costs. Pain 65, 197–204.

Fernandez, E. and Milburn, T. W. 1994. Sensory and affective predictors of overall pain and emotions associated with affective pain. Clin. J. Pain 10, 3–9.

Fernandez, E. and Turk, D. C. 1989. The utility of cognitive coping strategies for altering pain perception: a meta-analysis. Pain 38, 123–135.

Fernandez, E. and Turk, D. C. 1995. The scope and significance of anger in the experience of chronic pain. Pain 61, 165–175.

Fillingim, R. B., Maixner, W., Bunting, S., and Silva, S. 1998. Resting blood pressure and thermal pain responses among females: effects on pain unpleasantness but not pain intensity. Int. J. Psychophysiol. 30, 313–318.

Friedman, D. P., Murray, E. A., and O'Neill, J. R. 1986. Cortical connections of the somatosensory fields of the lateral sulcus of macaques: evidence for a corticolimbic pathway for touch. J. Comp. Neurol. 252(3), 323–347.

Garcia-Larrea, L., Peyron, R., Laurent, B., and Mauguiere, F. 1997. Association and dissociation between laser-evoked potentials and pain perception. Neuroreport 8, 3785–3789.

Gardner, W. L. and Licklider, J. C. 1959. Auditory analgesia in dental operations. J. Am. Dent. Assoc. 59, 1144–1149.

Gardner, W. L., Licklider, J. C., and Weisz, A. Z. 1960. Suppression of pain by sound. Science 132, 32–33.

Gerke, D. C., Richards, L. C., and Goss, A. N. 1989. Discriminant function analysis of clinical and psychological variables in temporomandibular joint pain dysfunction. Aust. Dent. J. 34, 44–48.

Gracely, R. H. 1979. Psychophysical Assessment of Human Pain. In: Advances in Pain Research and Therapy (eds. J. J. Bonica, J. C. Liebeskind, and D. G. Albe-Fessard), Vol. 3, pp. 781–790. Raven Press.

Gracely, R. H., Dubner, R., Deeter, W. D., and Wolskee, P. J. 1985. Clinicians' expectations influence placebo analgesia. Lancet 5, 43.

Gracely, R. H., McGrath, P., and Dubner, R. 1976. Validity and sensitivity of ratio scales of sensory and affective verbal pain descriptors: manipulation of affect by diazepam. Pain 2, 19–29.

Greenwood, K. A., Thurston, R., Rumble, M., Waters, S. J., and Keefe, F. J. 2003. Anger and persistent pain: current status and future directions. Pain 103, 1–5.

Gureje, O., Von Korff, M., Simon, G., and Gater, R. 1998. Persistent pain and well-being: a world health organization study in primary care. JAMA 280, 147–151.

Hadjistavropoulos, H. D., Hadjistavropoulos, T., and Quine, A. 2000. Health anxiety moderates the effects of distraction versus attention to pain. Behav. Res. Ther. 38, 425–438.

Hayes, R. L., Dubner, R., and Hoffman, D. S. 1981. Neuronal activity in medullary dorsal horn of awake monkeys trained in a thermal discrimination task. II. Behavioral modulation of responses to thermal and mechanical stimuli. J. Neurophysiol. 46, 428–443.

Haythronthwaite, J. A., Lawrence, J. W., and Fauerbach, J. A. 2001. Brief cognitive interventions for burn pain. Ann. Behav. Med. 23, 42–49.

Hikosaka, O., Miyauchi, S., and Shimojo, S. 1996. Orienting a spatial attention – its reflexive, compensatory, and voluntary mechanisms. Brain Res. Cogn. Brain Res. 5, 1–9.

Hodes, R. L., Howland, E. W., Lightfoot, N., and Cleeland, C. S. 1990. The effects of distraction on responses to cold pressor pain. Pain 41, 109–114.

Hofbauer, R. K., Rainville, P., Duncan, G. H., and Bushnell, M. C. 2001. Cognitive modulation of pain sensation alters activity in human cerebral cortex. Neurophysiol. 86(1), 402–411.

Hoffman, D. S., Dubner, R., Hayes, R. L., and Medlin, T. P. 1981. Neuronal activity in medullary dorsal horn of awake monkeys trained in a thermal discrimination task. I. Responses to innocuous and noxious thermal stimuli. J. Neurophysiol. 46, 409–427.

Hoffman, H. G., Richards, T. L., Coda, B., Bills, A. R., Blough, D., Richards, A. L., and Sharar, S. R. 2004. Modulation of thermal pain-related brain activity with virtual reality: evidence from fMRI. Neuroreport 15, 1245–1248.

Holzberg, A. D., Robinson, M. E., and Geisser, M. E. 1993. The relationship of cognitive distortion to depression in chronic pain: the role of ambiguity and desirability in self-ratings. Clin. J. Pain 9, 202–206.

Holzberg, A. D., Robinson, M. E., Geisser, M. E., and Gremillion, H. A. 1996. The effects of depression and chronic pain on psychosocial and physical functioning. Clin. J. Pain 12, 118–125.

Huynh Bao, Q. V and Rainville, P. 2002. Emotional modulation of experimental pain under hypnosis. International Association for Dental Research Abstracts.

Huynh Bao, Q. V. and Rainville, P. 2003. Modulation of experimental pain by emotion induced using hypnosis. Pain Res. Manag. (Suppl 8), 35B.

Huyser, B. A. and Parker, J. C. 1999. Negative affect and pain in arthritis. Rheum. Dis. Clin. North. Am. 25, 105–121.

Izard, C. E. 1993. Four systems for emotion activation: cognitive and noncognitive processes. Psychol. Rev. 100, 68–90.

Janssen, S. A. and Arntz, A. 1996. Anxiety and pain: attentional and endorphinergic influences. Pain 66, 145–150.

Janssen, S. A., Spinhoven, P., and Brosschot, J. F. 2001. Experimentally induced anger, cardiovascular reactivity, and pain sensitivity. J. Psychosom. Res. 51, 479–485.

Johnson, J. E. 1973. Effects of accurate expectations about sensations on the sensory and distress components of pain. J. Pers. Soc. Psychol. 27, 261–275.

Johnson, M. H. and Petrie, S. M. 1997. The effects of distraction on exercise and cold pressor tolerance for chronic low back pain sufferers. Pain 69, 43–48.

Junge, A., Dvorak, J., and Ahrens, S. 1995. Predictors of bad and good outcomes of lumbar disc surgery. A prospective clinical study with recommendations for screening to avoid bad outcomes. Spine 20, 460–468.

Keefe, F. J., Lumley, M., Anderson, T., Lynch, T., and Carson, K. L. 2001. Pain and emotion: new research directions. J. Clin. Psychol. 57, 587–607.

Keogh, E., Hatton, K., and Ellery, D. 2000. Avoidance versus focused attention and the perception of pain: differential effects for men and women. Pain 85, 225–230.

Koeppe, R. A. and Stohler, C. S. 2001. Regional mu opioid receptor regulation of sensory and affective dimensions of pain. Science 293(5528), 311–315.

Laska, E., and Sunshine, A. 1973. Anticipation of analgesia. A placebo effect. Headache 13(1), 1–11.

Le Bars, D., Dickenson, A. H., and Besson, J. M. 1979a. Diffuse noxious inhibitory controls (DNIC). I. Effects on dorsal horn convergent neurones in the rat. Pain 6, 283–304.

Le Bars, D., Dickenson, A. H., and Besson, J. M. 1979b. Diffuse noxious inhibitory controls (DNIC). II. Lack of effect on non-convergent neurones, supraspinal involvement and theoretical implications. Pain 6, 305–327.

Legrain, V., Guerit, J. M., Bruyer, R., and Plaghki, L. 2002. Attentional modulation of the nociceptive processing into the human brain: selective spatial attention, probability of stimulus occurrence, and target detection effects on laser evoked potentials. Pain 99, 21–39.

Lewkowski, M. D., Ditto, B., Roussos, M., and Young, S. N. 2003. Sweet taste and blood pressure-related analgesia. Pain 106, 181–186.

Logan, H., Lutgendorf, S., Rainville, P., Sheffield, D., Iverson, K., and Lubaroff, D. 2001. Effects of Stress and Relaxation on capsaicin-induced pain. J. Pain. 2, 160–170.

Love, A. W. 1988. Attributional style of depressed chronic low back patients. J. Clin. Psychol. 44, 317–321.

Marchand, S. and Arsenault, P. 2002. Odors modulate pain perception: a gender-specific effect. Physiol Behav 76, 251–256.

Meagher, M. W., Arnau, R. C., and Rhudy, J. L. 2001a. Pain and emotion: effects of affective picture modulation. Psychosom. Med. 63, 79–90.

Meagher, M. W., Ferguson, A. R., Crown, E. D., McLemore, S., King, T. E., Sieve, A. N., and Grau, J. W. 2001b. Shock-induced hyperalgesia. IV. Generality. J. Exp. Psychol. Anim. Behav. Process. 27, 219–238.

Melzack, R. and Casey, K. L. 1968. Sensory, Motivational, and Central Control Determinants of Pain. In: The Skin Senses (ed. D. Kenshalo), pp. 423–443. CC Thomas.

Merskey, H. and Spear, F. G. 1967. The concept of pain. J. Psychosom. Res. 11, 59–67.

Miron, D., Duncan, G. H., and Bushnell, M. C. 1989. Effects of attention on the intensity and unpleasantness of thermal pain. Pain 39, 345–352.

Montgomery, G. H. and Kirsch, I. 1996. Mechanisms of placebo pain reduction: an empirical investigation. Psychol. Sci. 7, 174–176.

Montgomery, G. and Kirsch, I. 1997. Classical conditioning and the placebo effect. Pain 72, 107–113.

Mor, N. and Winquist, J. 2002. Self-focused attention and negative affect: a meta-analysis. Psych. Bull. 128, 638–662.

Nakamura, Y., Paur, R., Zimmermann, R., and Bromm, B. 2002. Attentional modulation of human pain processing in the secondary somatosensory cortex: a magnetoencephalographic study. Neurosci. Lett. 328, 29–32.

Okifuji, A., Turk, D. C., and Curran, S. L. 1999. Anger in chronic pain: investigations of anger targets and intensity. J. Psychosom. Res. 47, 1–12.

Petrovic, P., Kalso, E., Petersson, K. M., and Ingvar, M. 2001. Placebo and opioid analgesia–imaging a shared neuronal network. Science 295(5560), 1737–1740.

Petrovic, P., Petersson, K. M., Ghatan, P. H., Stone-Elander, S., and Ingvar, M. 2000. Pain-related cerebral activation is altered by a distracting cognitive task. Pain 85, 19–30.

Peyron, R., Garcia-Larrea, L., Gregoire, M. C., Costes, N., Convers, P., Lavenne, F., Mauguire, F., Michel, D., and Laurent, B. 1999. Haemodynamic brain responses to acute pain in humans: sensory and attentional networks. Brain 122, 1765–1780.

Plutchik, R. 1980. A General Psychoevolutionary Theory of Emotion. In: Emotion: Theory, Research, and Experience, Vol. 1. Theories of Emotion (eds. R. Plutchik and H. Kellerman), pp. 3–33. Academic Press.

Pollo, A., Amanzio, M., Arslanian, A., Casadio, C., Maggi, G., and Benedetti, F. 2001. Response expectancies in placebo analgesia and their clinical relevance. Pain 93, 77–84.

Posner, M. E. 1980. Orienting of attention. Quart. J. Exp. Psychol. 32, 3–25.

Price, D. D. 1999. Psychological Mechanisms of Pain and Analgesia. IASP Press.

Price, D. D. 2000. Psychological and neural mechanisms of the affective dimension of pain. Science 288, 1769–1772.

Price, D. D. and Barrell, J. J. 2000. Mechanisms of analgesia produced by hypnosis and placebo suggestions. Prog. Brain Res. 122, 255–71.

Price, D. D. and Bushnell, M. C. (eds.) 2004. Psychological Methods of Pain Control: Basic Science and Clinical Perspectives pp. 322. IASP Press.

Price, D. D. and McHaffie, J. G. 1988. Effects of heterotopic conditioning stimuli on first and second pain: a psychophysical evaluation in humans. Pain 34, 245–252.

Price, D. D., Barrell, J. E., and Barrell, J. J. 1985. A quantitative-experiential analysis of human emotions. Motiv. Emot. 9, 19–38.

Price, D. D., Barrell, J. J., and Gracely, R. H. 1980. A psychophysical analysis of experimental factors that selectively influence the affective dimension of pain. Pain 8, 137–149.

Price, D. D., Harkins, S. W., and Baker, C. 1987. Sensory-affective relationships among different types of clinical and experimental pain. Pain 28(3), 291–299.

Price, D. D., Milling, L. S., Kirsch, I., Duff, A., Montgomery, G. H., and Nicholls, S. S. 1999. An analysis of factors that contribute to the magnitude of placebo analgesia in an experimental paradigm. Pain 83, 147–156.

Price, D. D., Riley, J., and Barrell, J. J. 2001. Are lived choices based on emotional processes? Cogn. Emot. 15(3), 365–379.

Rainville, P. 2004. Pain and Emotions. In: Psychological Methods of Pain Control: Basic Science and Clinical Perspectives (eds. D. D. Price and M. C. Bushnell), pp. 117–141. IASP Press.

Rainville, P., Carrier, B., Hofbauer, R. K., Bushnell, M. C., and Duncan, G. H. 1999. Dissociation of pain sensory and affective dimensions using hypnotic modulation. Pain 82, 159–171.

Rainville, P., Duncan, G. H., Price, D. D., Carrier, B., and Bushnell, M. C. 1997. Pain affect encoded in human anterior cingulate but not somatosensory cortex. Science 277, 968–971.

Rhudy, J. L. and Meagher, M. W. 2000. Fear and anxiety: divergent effects on human pain thresholds. Pain 84, 65–75.

Rhudy, J. L. and Meagher, M. W. 2001. Noise stress and human pain thresholds: divergent effects in men and women. J. Pain 2, 57–64.

Rhudy, J. L. and Meagher, M. W. 2003. Negative affect: effects on an evaluative measure of human pain. Pain 104, 617–626.

Riley, J. L., Robinson, M. E., Geisser, M. E., Wittmer, V. T., and Graham-Smith, A. 1995. Relationship between MMPI-2 cluster profiles and surgical outcome in low-back pain patients. J. Spinal Disord. 8, 213–219.

Robinson, M. E. and Riley, J. L. 1998. Negative Emotion in Pain. In: Psychosocial Factors in Pain (eds. R. Gatchel and D. Turk), pp. 221–243. Guilford Press.

Roy, M., Peretz, I., and Rainville, P. 2003. Effects of emotion induced by music on experimental pain. Pain Res. Manag. (suppl 8), 33B.

Salovey, P. 1992. Mood-induced self-focused attention. J. Pers. Soc. Psychol. 62, 699–707.

Salovey, P. and Birnbaum, D. 1989. Influence of mood on health-relevant cognitions. J. Pers. Soc. Psychol. 57, 539–551.

Sato, A., Sato, Y., and Schmidt, R. F. 1997. The Impact of Somatosensoty Input on Autonomic Functions.In Reviews of Physiology Biochemistry and Pharmacology (eds. M. P. Blaustein, H. Grunicke, D. P. Konstanz, G. Schultz, and M. Schweiger) Vol. 130, pp. 119–134. Springer-Verlag.

Schanberg, L. E., Sandstrom, M. J., Starr, K., Gil, K. M., Lefebvre, J. C., Keefe, F. J., Affleck, G., and Tennen, H. 2000. The relationship of daily mood and stressful events to symptoms in juvenile rheumatic disease. Arthritis Care Res. 13, 33–41.

Seminowicz, D. A., Mikulis, D. J., and Davis, K. D. 2004. Cognitive modulation of pain-related brain responses depends on behavioral strategy. Pain 112, 48–58.

Siedenberg, R. and Treede, R. D. 1996. Laser-evoked potentials: exogenous and endogenous components. Electroencephalogr. Clin. Neurophysiol. 100, 240–249.

Sullivan, M. J., Rodgers, W. M., and Kirsch, I. 2001a. Catastrophizing, depression and expectancies for pain and emotional distress. Pain 91, 147–154.

Sullivan, M. J., Thorn, B., Haythornthwaite, J. A., Keefe, F., Martin, M., Bradley, L. A., and Lefebvre, J. C. 2001b. Theoretical perspectives on the relation between catastrophizing and pain. Clin. J. Pain 17, 52–64.

Swimmer, G., Robinson, M., and Geisser, M. 1992. The relationship of MMPI cluster type, pain coping strategy, and treatment outcome. Clin. J. Pain 8, 131–137.

Tracey, I., Ploghaus, A., Gati, J. S., Clare, S., Smith, S., Menon, R. S., and Matthews, P. M. 2002. Imaging attentional modulation of pain in the periaqueductal gray in humans. J. Neurosci. 22, 2748–2752.

Valet, M., Sprenger, T., Boecker, H., Wiloch, F., Rummeny, E., Conrad, B., Erhard, P., and Tolle, T. R. 2004. Distraction modulates connectivity of the cingulo-frontal cortex and the midbrain during pain – an fMRI analysis. Pain 109, 399–408.

Vase, L., Riley, J. L., and Price, D. D. 2002. A comparison of placebo effects in clinical analgesic trials versus studies of placebo analgesia. Pain 99, 443–452.

Vase, L., Robinson, M. E., Verne, G. N., and Price, D. D. 2003. The contribution of suggestion, expectancy and desire to placebo effect in irritable bowl syndrome patients. Pain 105, 17–25.

Vase, L., Robinson, M. E., Verne, G. N., and Price, D. D. 2005. Increased placebo analgesia over time in irritable bowel syndrome (IBS) patients is associated with desire and expectation but not endogenous opioid mechanisms. Pain 115(3), 338–347.

Velten, E. 1968. A laboratory task for induction of mood states. Behav. Res. Ther. 6, 473–482.

Verne, G. N., Robinson, M. E., Vase, L., and Price, D. D. 2003. Reversal of visceral and cutaneous hyperalgesia by local rectal anesthesia in irritable bowl syndrome (IBS) patients. Pain 105, 223–230.

Villemure, C., Slotnick, B. M., and Bushnell, M. C. 2003. Effects of odors on pain perception: deciphering the roles of emotion and attention. Pain 106, 101–108.

Vlaeyen, J. W. and Linton, S. J. 2000. Fear-avoidance and its consequences in chronic musculoskeletal pain: a state of the art. Pain 85, 317–332.

Voudouris, N. J., Peck, C. L., and Coleman, G. 1985. Conditioned placebo responses. J. Pers. Soc. Psychol. 48, 47–53.

Voudouris, N. J., Peck, C. L., and Colemann, G. 1998. Conditioned response models of placebo phenomena: further support. Pain 38, 109–116.

Voudouris, N. J., Peck, C. L., and Coleman, G. 1989. Conditioned response models of placebo phenomena: further support. Pain 38(1), 109–116.

Voudouris, N. J., Peck, C. L., and Coleman, G. 1990. The role of conditioning and verbal expectancy in the placebo response. Pain 43, 121–128.

Wade, J. B., Dougherty, L. M., Archer, R., and Price, D. D. 1996. Assessing the stages of pain processing: a multivariate analytical approach. Pain 68, 157–167.

Wade, J. B., Price, D. D., Hamer, R. M., Schwartz, S. M., and Hart, R. P. 1990. An emotional component analysis of chronic pain. Pain 40, 303–310.

Wegner, D. M. 1994. Ironic processes of mental control. Psychol. Rev. 101, 34–52.

Weisenberg, M., Raz, T., and Hener, T. 1998. The influence of film-induced mood on pain perception. Pain 76, 365–375.

Wenzlaff, R. M. and Wegner, D. M. 2000. Thought Suppression. In: Annual Review of Psychology (*ed*. S. T. Fiske), Vol. 51, pp. 51–91. Annual Reviews.

Whipple, B. and Glynn, N. J. 1992. Quantification of the effects of listening to music as a noninvasive method of pain control. Sch. Inq. Nurs. Pract. 6, 43–58.

Wickramasekera, I. A. 1985. Conditioned Response Model of the Placebo Effect: Predictions of the Model. In: Placebo: Theory, Research and Mechanisms (*eds*. L. White, B. Tursky, and G. E. Schwartz), pp. 255–287. Guilford Press.

Williams, A. C. 2002. Facial expression of pain: an evolutionary account. Behav. Brain Sci. 25, 439–55.

Willoughby, S. G., Hailey, B. J., Mulkana, S., and Rowe, J. 2002. The effect of laboratory-induced depressed mood state on responses to pain. Behav. Med. 28, 23–31.

Zelman, D. C., Howland, E. W., Nichols, S. N., and Cleeland, C. S. 1991. The effects of induced mood on laboratory pain. Pain 46, 105–111.

Zillmann, D., de Wied, M., King-Jablonski, C., and Jenzowsky, S. 1996. Drama-induced affect and pain sensitivity. Psychosom. Med. 58, 333–341.

Further Reading

Hrojartsson, A. and Gøtzsche, P. C. 2001. Is the placebo effect powerless? An analysis of clinical trials comparing placebo with no- treatment. N. Engl. J. Med. 344, 1594–1602.

Verne, G. N., Himes, N. C., Robinson, M. E., Gopinath, K. S., Briggs, R. W., Crosson, B., and Price, D. D. 2003. Central representation of visceral and cutaneous hypersensitivity in the irritable bowel syndrome. Pain 103, 99–110.

65 The Placebo Effect

F Benedetti, University of Turin Medical School, Turin, Italy

Glossary

immunosuppressive Adjective that indicates an inhibitory action on the immune system.

μ-opioid receptor Subtype of opioid receptors that bind μ-agonist drugs, such as morphine.

naloxone Antagonist drug of opioids.

natural history Spontaneous time course of a symptom, e.g., pain.

nocebo It has the opposite meaning of placebo, which indicates that a symptom may also increase after suggestions of increase.

Parkinson's disease Motor disorder characterized by three typical symptoms: rest tremor, muscle rigidity, and bradykinesia (reduction of movement velocity).

Pavlovian conditioning Conditioning according to Pavlov, in which the temporal contiguity between a neutral (conditioned) and an unconditioned stimulus leads to a conditioned response.

proglumide Antagonist drug of cholecystokinin.

subthalamic nucleus Nucleus located under the thalamus, which represents a major target of the surgical treatment of Parkinson's disease.

sumatriptan Agonist drug of serotonin 5-HT1 receptors that is used in the treatment of migraine.

65.1 Top-Down Modulation of Pain

The input coming from a damaged tissue and traveling along the pain pathways up to the brain is not always experienced in the same way. A complex modulation occurs at the supraspinal level and may either increase or decrease the global experience of pain. Many psychological factors, for example, attention, emotions, mood, stress, expectation, anticipation, distraction, anxiety, depression, and fear, all modulate the global experience of pain, although the underlying mechanisms are poorly understood. In recent times, the placebo effect, particularly placebo analgesia, has emerged as an interesting model to understand the psychological and physiological mechanisms through which this intricate top-down modulation occurs. Of course, the understanding of the placebo effect can only explain some of these factors, such as expectation and anticipation.

Nonetheless, its investigation has yielded new insights into the biological mechanisms that link a complex mental activity to different body functions. This chapter is a brief overview of what we know today about the neural mechanisms underlying the placebo effect. It will appear clear that this complex phenomenon can give us important information on the intricate mechanisms that link mind, brain, and body.

65.2 Methodological Aspects

The investigation of the placebo effect is full of pitfalls and drawbacks since, in order to identify a real psychobiological placebo response, several other phenomena have to be ruled out. In fact, the placebo itself is not always the cause of the effect that is observed (Colloca, L. and Benedetti, F., 2005). For example, most painful conditions show a spontaneous temporal variation that is known as natural history. If subjects take a placebo just before their discomfort starts decreasing, they may believe that the placebo is effective, although that decrease would have occurred anyway. Clearly, this is not a placebo effect but a spontaneous remission that leads to a misinterpretation of the cause–effect relationship. Another example is regression to the mean, a statistical phenomenon assuming that individuals tend to receive their initial clinical assessment when their pain is near its greatest intensity, and that their pain level is likely to be lower when they return for a second pain assessment. In this case also, the improvement cannot be attributed to any intervention they might have undergone. A further source of confusion is represented by the fact that a particular type of error made by the patient, a false positive error, may explain the placebo effect in some circumstances. This phenomenon is based on signal detection theory, and is due to the occurrence of errors in the detection of ambiguous signals. It also happens that a co-intervention actually is responsible for the reduction of a symptom, such as the analgesic effect following the mechanical insertion of a needle to inject an inert solution. All these examples show that, although an improvement may occur after the administration of a placebo, the placebo is not necessarily the cause of the effect that is observed. Since all these phenomena are sometimes difficult to identify, the mechanisms of the placebo response must be investigated under strictly controlled experimental conditions (Vase, L. *et al.*, 2002; Colloca, L. and Benedetti, F., 2005). For example, in order to rule out spontaneous remission, a group taking the placebo is compared to a group receiving no treatment, the latter giving information on the natural course of the symptom. The difference between the placebo group and the no-treatment group represents the real psychobiological placebo response.

65.3 Psychological Explanations

The placebo effect involves both cognitive factors and conditioning mechanisms. The deceptive administration of a placebo treatment can lead the subjects to believe that the treatment is effective, so that anticipation and expectation of analgesia lead to a placebo analgesic response. Some studies show that different verbal instructions lead to different expectations and thus to different responses, and this plays a fundamental role in the placebo effect (Amanzio, M., and Benedetti, F., 1999; Benedetti, F. *et al.*, 1999b; Price, D. D. *et al.*, 1999). The context around a therapy may act not only through expectation and conscious anticipatory processes. In fact, there are some lines of evidence indicating that the placebo response is sometimes a conditioned response due to repeated associations between a conditioned stimulus (e.g., shape and color of aspirin pills) and an unconditioned stimulus (the active substance of aspirin) (Amanzio, M., and Benedetti, F., F., 1999; Siegel, S., 2002; Benedetti, F. *et al.*, 2003). In this case, it is the context itself that is the conditioned stimulus. However, even by considering a typical conditioning procedure, it has been shown that a conditioned placebo analgesic response can result from conditioning but is actually mediated by expectation. In other words, conditioning would lead to the expectation that a given event will follow another event, and this occurs on the basis of the information that the conditioned stimulus provides about the unconditioned stimulus (Benedetti, F. *et al.*, 2003).

There is experimental evidence that some physiological functions are affected by placebos through anticipatory conscious processes whereas some other functions undergo an unconscious mechanism of conditioning (Benedetti, F. *et al.*, 2003). For example, verbally induced expectations of either analgesia or hyperalgesia antagonize completely the effects of a conditioning procedure in experimentally induced pain. By contrast, verbally induced expectations of either increase or decrease of growth hormone and cortisol do not have any effect on the secretion of these hormones. However, if a preconditioning is performed with sumatriptan, a $5-HT_{1B/1D}$ agonist that stimulates growth hormone and inhibits cortisol secretion, a significant increase of growth hormone and decrease of cortisol plasma concentrations can be found after placebo administration, even though opposite verbal suggestions are given. These findings suggest that placebo responses are mediated by conditioning when unconscious physiological functions, like hormonal secretion, are involved, whereas they are mediated by expectation when conscious physiological processes, like pain, come into play. Thus the placebo effect seems to be a phenomenon which can be learned either consciously or unconsciously, depending on the system that is involved (e.g., pain or hormone secretion).

65.4 Physiological Mechanisms

The complex cascade of events that may occur after placebo administration is shown in Figure 1. Several studies show that placebo-induced analgesia is antagonized by the opioid antagonist, naloxone, thus suggesting mediation by endogenous opioids (Levine, J. D. *et al.*, 1978; Amanzio, M., and Benedetti, F., 1999; Benedetti, F. *et al.*, 1999b). The cholecystokinin (CCK) antagonist, proglumide, has been found to enhance the placebo analgesic effect (Benedetti, F. *et al.*, 1995), which indicates that CCK has an inhibitory role in placebo-induced analgesia. The placebo analgesic responses are thus the result of the balance between endogenous opioids and endogenous CCK.

Placebo analgesia is not always mediated by endogenous opioids. In fact, if the placebo response is induced by means of strong expectation cues, it can be blocked by naloxone. Conversely, if the placebo response is induced by means of prior conditioning with a nonopioid drug, like ketorolac, it is naloxone-insensitive (Amanzio, M., and Benedetti, F., 1999). Today we know that specific placebo analgesic responses can be obtained in different parts of the body, and that these responses are naloxone-reversible (Benedetti, F. *et al.*, 1999b), which suggests that the placebo-activated endogenous opioid systems have a somatotopic organization.

The investigation of placebo analgesia by means of positron emission tomography found that similar

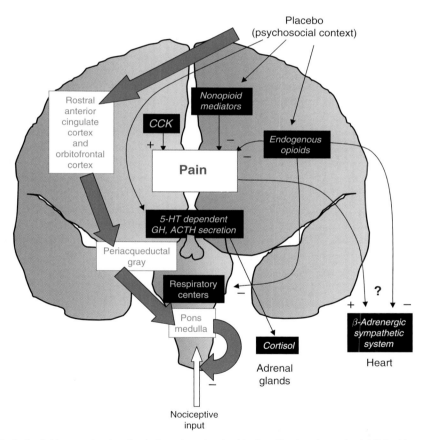

Figure 1 Events that might occur in placebo-induced analgesia. Nociceptive input may be inhibited by a descending network that involves the rostral anterior cingulate cortex, the orbitofrontal cortex, the periaqueductal gray, and the pons/medulla. Endogenous opioids might inhibit pain through this descending network and/or other mechanisms. The endogenous opioids also inhibit the respiratory centers. The β-adrenergic sympathetic system is also inhibited during placebo analgesia, although the mechanism is not known (either reduction of the pain itself or direct action of endogenous opioids). Nonopioid mechanisms are also involved. Cholecystokinin (CCK) counteracts the effects of the endogenous opioids thus reducing the placebo analgesic response. Placebo can also act on serotonin-dependent hormone secretion, mimicking the effect of the analgesic drug sumatriptan. 5-HT, serotonin; ACTH, adrenocorticotropic hormone; GH, growth hormone.

regions of the brain are affected by both a placebo and a narcotic drug, which indicates a related mechanism in placebo-induced and opioid-induced analgesia (Petrovic, P. *et al.*, 2002). In fact, the administration of a placebo induced the activation of the rostral anterior cingulate cortex (rACC) and the orbitofrontal cortex (OrbC). Moreover, there was a significant covariation in activity between the rACC and the lower pons/medulla corresponding to the rostral ventromedial medulla (RVM), and a subsignificant covariation between the rACC and the periaqueductal gray (PAG), thus suggesting that a descending rACC/PAG/RVM pain-modulating circuit is involved in placebo-induced analgesia. It is worth remembering that ACC and PAG are rich with opioid receptors, thus confirming the pharmacological studies with naloxone described above. By using functional magnetic resonance imaging to analyze the brain regions that are involved in placebo analgesia, another study showed that the activity of different regions involved in pain transmission, such as the thalamus, the anterior insula (aINS), and the caudal rACC, was decreased by a placebo treatment, which indicates a reduction of nociceptive transmission along the pain pathways (Wager, T. D. *et al.*, 2004). Furthermore, during the anticipatory phase of the placebo analgesic response, an activation of the dorsolateral prefrontal cortex (DLPFC), OrbC, superior parietal cortex (SPC), PAG, and other frontal regions occurs, suggesting the activation of a cognitive-evaluative network just before the placebo response. A recent attempt to identify the regions where endogenous opioids are released has been performed by using *in vivo* receptor binding techniques (Zubieta, J. K., 2005). A placebo-induced activation of μ-opioid receptors has also been found in different brain regions, such as the pregenual rostral anterior cingulate, DLPFC, INS, and the nucleus accumbens, which confirms once again the pharmacological blockade of placebo analgesia by opioid antagonists.

Placebo-activated endogenous opioids have also been shown to induce respiratory depression (Benedetti, F. *et al.*, 1999a), indicating that they act not only on pain mechanisms, but also on the respiratory centers. Also β-adrenergic sympathetic activity is reduced in placebo analgesia, and this might be due to either pain reduction itself or a direct action of placebo-activated endogenous opioids (Pollo, A. *et al.*, 2003). Some nonopioid mechanisms, such as the serotonin 5-HT$_{1B/1D}$ receptors, have also been investigated. For example, placebo-induced increase

of growth hormone secretion and decrease of cortisol secretion have been described after pharmacological preconditioning with the serotonin agonist sumatriptan (Benedetti, F. *et al.*, 2003).

The placebo response is not limited to pain and analgesia, but it also occurs in many other conditions. The integration of the understanding of the placebo mechanisms in the field of pain and in other diseases is crucial and essential to identify similarities and differences that might help us appreciate the complexity of the placebo effect better. For example, as described above, placebo-induced hormonal responses can be obtained after repeated administrations of a hormone-stimulating drug (Benedetti, F. *et al.*, 2003), so can placebo-induced immunosuppressive responses after repeated administrations of an immunosuppressive drug (Goebel, M. U. *et al.*, 2002), which suggests a mechanism of Pavlovian conditioning in the endocrine and immune systems. Parkinson's disease has also been used as an interesting model to understand the neurobiological mechanisms of the placebo response, which might help understand placebo analgesia better. It has been shown that placebo administration in patients with Parkinson's disease activates endogenous dopamine in the striatum (de la Fuente-Fernandez, R. *et al.*, 2001) and modifies the pattern of activity of the neurons in the subthalamic nucleus (Benedetti, F. *et al.*, 2004). The placebo-induced release of dopamine might represent a mechanism of reward, whereby dopamine release by expectation of reward (in this case the expectation of clinical benefit) could represent a common biochemical substrate in many conditions, including pain.

65.5 The Nocebo Effect

The nocebo effect, or response, is a placebo effect in the opposite direction. In fact, expectation of pain increase may induce a hyperalgesic effect. In this case, anticipatory anxiety may play a fundamental role. In one study, negative expectations were induced by injecting an inert substance (saline solution) along with the instructions that pain was going to increase in a few minutes (Benedetti, F. *et al.*, 1997). As a consequence of this procedure, a pain increase was observed, and this increase was blocked by the CCK antagonist, proglumide. This indicates that expectation-induced hyperalgesia is mediated, at least in part, by CCK. These effects of proglumide are not antagonized by naloxone, thus endogenous opioids are not involved. Since CCK plays a role in anxiety and negative expectations themselves are anxiogenic, proglumide is likely to act on a CCK-dependent increase of anxiety during the verbally induced negative expectations. Although, mainly due to ethical constraints, the nocebo effect has not been investigated in detail as has been done for the placebo effect, it shows the powerful effect of the top-down modulation of pain. In other words, cognitive and emotional factors can modulate pain perception in opposite directions.

65.6 Clinical Implications

One of the best evidences that suggest the important role of expectations and the top-down modulation of pain and analgesia is the decreased effectiveness of hidden therapies. In fact, it is possible to eliminate the placebo (psychosocial) component and to analyze the pharmacodynamic effects of an analgesic treatment, free of any psychological contamination. To eliminate the patient's expectations, the patient is made completely unaware that a medical therapy is being carried out. To do this, drugs are administered through hidden infusions by computer-controlled machines. The crucial point here is that the patients do not know that any analgesic is being injected, so that they do not expect anything. In postoperative pain, it was found that a hidden injection of different painkillers, in which the patients do not expect any outcome, is significantly less effective than an open one, in which the patients know that a pain reduction will occur (Colloca, L. *et al.*, 2004).

The difference between the open and hidden administration represents the placebo component, and underscores the importance of the placebo response in clinical practice. In fact, it basically shows that the specific effect of a treatment and the placebo response are additive. Since no placebo is given, the difference between an open and hidden injection cannot be called a placebo effect. Nevertheless it strongly indicates the important role of the psychosocial component of a therapy and the importance of the patient's perception that a therapy is being received. This new approach to the identification of the placebo effect might also have an important impact on the design of clinical trials (Colloca, L. and Benedetti, F., 2005; Finniss, D. G. and Benedetti, F., 2005).

References

Amanzio, M. and Benedetti, F. 1999. Neuropharmacological dissection of placebo analgesia: expectation-activated opioid systems versus conditioning-activated specific sub-systems. J. Neurosci. 19, 484–494.

Benedetti, F., Amanzio, M., Baldi, S., Casadio, C., and Maggi, G. 1999a. Inducing placebo respiratory depressant responses in humans via opioid receptors. Eur. J. Neurosci. 11, 625–631.

Benedetti, F., Amanzio, M., Casadio, C., Oliaro, A., and Maggi, G. 1997. Blockade of nocebo hyperalgesia by the cholecystokinin antagonist proglumide. Pain 71, 135–140.

Benedetti, F., Amanzio, M., and Maggi, G. 1995. Potentiation of placebo analgesia by proglumide. Lancet 346, 1231.

Benedetti, F., Arduino, C., and Amanzio, M. 1999b. Somatotopic activation of opioid systems by target-directed expectations of analgesia. J. Neurosci. 19, 3639–3648.

Benedetti, F., Colloca, L., Torre, E., Lanotte, M., Melcarne, A., Pesare, M., Bergamasco, B., and Lopiano, L. 2004. Placebo-responsive Parkinson patients show decreased activity in single neurons of subthalamic nucleus. Nat. Neurosci. 7, 587–588.

Benedetti, F., Pollo, A., Lopiano, L., Lanotte, M., Vighetti, S., and Rainero, I. 2003. Conscious expectation and unconscious conditioning in analgesic; motor and hormonal placebo/nocebo responses. J. Neurosci. 23, 4315–4323.

Colloca, L. and Benedetti, F. 2005. Placebos and painkillers: is mind as real as matter? Nat. Rev. Neurosci. 6, 545–552.

Colloca, L., Lopiano, L., Lanotte, M., and Benedetti, F. 2004. Overt versus covert treatment for pain, anxiety and Parkinson's disease. Lancet Neurol. 3, 679–684.

de la Fuente-Fernandez, R., Ruth, T. J., Sossi, V., Schulzer, M., Calne, D. B., and Stoessl, A. J. 2001. Expectation and dopamine release: mechanism of the placebo effect in Parkinson's disease. Science 293, 1164–1166.

Finniss, D. G. and Benedetti, F. 2005. Mechanisms of the placebo response and their impact on clinical trails and clinical practice. Pain 114, 3–6.

Goebel, M. U., Trebst, A. E., Steiner, J., Xie, Y. F., Exton, M. S., Frede, S., Canbay, A. E., Michel, M. C., Heemann, U., and Schedlowski, M. 2002. Behavioral conditioning of immunosuppression is possible in humans. FASEB J. 16, 1869–1873.

Levine, J. D., Gordon, N. C., and Fields, H. L. 1978. The mechanisms of placebo analgesia. Lancet 2, 654–657.

Petrovic, P., Kalso, E., Petersson, K. M., and Ingvar, M. 2002. Placebo and opioid analgesia – imaging a shared neuronal network. Science 295, 1737–1740.

Pollo, A., Vighetti, S., Rainero, I., and Benedetti, F. 2003. Placebo analgesia and the heart. Pain 102, 125–133.

Price, D. D., Milling, L. S., Kirsch, I., Duff, A., Montgomery, G. H., and Nicholls, S. S. 1999. An analysis of factors that contribute to the magnitude of placebo analgesia in an experimental paradigm. Pain 83, 147–156.

Siegel, S. 2002. Explanatory Mechanisms for Placebo Effects: Pavlovian Conditioning. In: The Science of the Placebo: Toward an Interdisciplinary Research Agenda (*eds*. H. A. Guess, A. Kleinman, J. W. Kusek, and L. W. Engel), pp. 133–157. BMJ Books.

Vase, L., Riley, J. L., and Price, D. D. 2002. A comparison of placebo effects in clinical analgesic trials versus studies of placebo analgesia. Pain 99, 443–452.

Wager, T. D., Rilling, J. K., Smith, E. E., Sokolik, A., Casey, K. L., Davidson, R. J., Kosslyn, S. M., Rose, R. M., and Cohen, J. D. 2004. Placebo-induced changes in fMRI in the anticipation and experience of pain. Science 303, 1162–1166.

Zubieta, J. K., Bueller, J. A., Jackson, L. R., Scott, D. J., Xu, Y., Koeppe, R. A., Nichols, T. E., and Stohler, C. S. 2005. Placebo effects mediated by endogenous. J. Neurosci. 25, 7754–7762.

66 Hypnotic Analgesia

P Rainville, Université de Montréal, Montreal, QC, Canada

I Marc, Université Laval, Quebec City, QC, Canada

Glossary

affective dimension of pain The feelings of unpleasantness and/or other negative emotions associated with the sensory experience of pain. This dimension of the pain experience encompasses both the basic emotional and the motivational responses produced by the actual or potential threat associated with pain.

Automaticity Behavioral responses and changes in perception and subjective experience produced without a feeling voluntary control (self-agency), that is, changes induced by hypnotic suggestions may be felt passively, as if they were induced by an external cause/agent.

hypnotic state The coordinated changes in activity within networks of brain regions involved in the regulation of consciousness (vigilance, attention, and self-monitoring), associated with self-experienced increases in mental relaxation, mental absorption, and feelings of automaticity, and typically (although not necessarily) induced by suggestions.

hypnotic suggestion Suggestion given during a context of hypnosis for an experience to occur or a behavior to be carried out.

nociceptive processes Biological responses induced by actual or potential tissue damage and normally (although not always) associated with the experience of pain in humans.

nociceptive withdrawal reflex (RIII reflex) The RIII reflex is a spinally mediated multisynaptic cutaneous reflex involving the flexion of the stimulated limb and the extension of the opposite limb. It can be induced experimentally by a transcutaneous electrical stimulation of the sural nerve at the level of the ankle and measured by the amplitude of the electromyographic response of the ipsilateral bicep femoris (flexor of the leg).

positron emission tomography (PET) A nuclear medical imaging technique where a radioactive tracer is injected into the blood circulation of a living organism. Applied to brain imaging, this technique allows for the measurement of the distribution of blood flow or the availability of various receptors in the brain. In functional brain imaging studies, this technique is used to test the changes in local brain activity in various experimental conditions.

self-monitoring Processes by which intentions, actions, and somatic activity are registered by the brain and mind. This process is central to the representation of self and is thought to be altered during hypnosis.

somatosensory-evoked potential Changes in regional electrical activity of the cerebral cortex in response to acute somatosensory stimuli and typically measured using surface electroencephalography.

66.1 Introduction

Scientific conceptions of hypnosis can be traced back to the end of the eighteenth century, when the pseudoscientific theory of Franz Anton Mesmer on *Animal Magnetism* (Bouleur, Jt., 1998) was formally examined by a Royal Commission ordered by King Louis XVI and chaired by Benjamin Franklin (Salas, C. and Salas, D., 1996; Laurence, J. R., 2002). Although the commissioners noted that the interventions (i.e., magnetization) used by Mesmer did produce dramatic responses in some patients (including analgesia), they concluded that the subjects' imagination was the main mediating mechanisms activated under the power of suggestion. This explanation based on the subject's imaginative resources is consistent with most contemporary theories of hypnosis.

Modern theories of hypnosis are based on complementary psychosocial, cognitive, and neurobiological interpretations (Fromm, E. and Nash, M. R., 1992). Experimental research has provided objective evidence that hypnotic interventions have robust, and in many cases specific, physiological effects. Furthermore, recent brain imaging methods have provided insight into the mechanisms underlying hypnotic states and into the specific responses to various types of hypnotic suggestions, including those aimed at reducing pain. Those advances are paralleled by well-controlled clinical studies, providing conclusive evidence on the efficacy of hypnosis in the control of pain.

66.2 Hypnosis Affects Consciousness-Related Brain Mechanisms

At the core of hypnosis is a social interaction where a hypnotist gives suggestions to another person, thus leading some authors to explain hypnosis-related effects primarily as suggestion-induced changes in expectancy (Kirsch, I. and Lynn, S. R., 1995). Hypnosis may also be defined as basic changes in subjective experience typically induced by standard hypnosis induction procedures and sometimes described as a hypnotic state or an altered state of consciousness. As those changes may be self-induced (e.g., in autohypnosis or some forms of meditation), an explanation strictly based on social interactions and suggestions may be limited. The analysis of experiential changes associated with the induction of hypnotic

states has revealed increases in mental ease and absorption, followed by reductions in self-monitoring and orientation, which in turn lead to feelings of automaticity (Price, D. D., 1996; Rainville, P. and Price, D. D., 2004). Increases in absorption (e.g., concentration and fascination) are consistent with the central role of attention processes in hypnotic responsiveness (Tellegen, A. and Atkinson, G., 1974; Nadon, R. *et al.*, 1987; Radtke, H. L. and Stam, H. J., 1991; Balthazard, C. G., and Woody, E. Z., 1992; Crawford, H. J., 1994). However, the subjective increase in automaticity has been identified as the classical suggestion effect characterizing hypnosis (Weitzenhoffer, A. M., 1980). This sense of automaticity may also relate to dissociative processes posited to explain the dual streams of consciousness reported by some subjects during hypnosis (Hilgard, E. R., 1992).

In agreement with the involvement of basic consciousness-related processes, increases in mental ease and absorption are associated with the changes in activity within brain networks underlying the regulation of vigilance and attention, including brainstem and medial thalamic nuclei, the anterior cingulate cortex, and the prefrontal and posterior parietal cortices (Maquet, P. *et al.*, 1999; Rainville, P. *et al.*, 1999b; 2002). Those findings are consistent with increases in rhythmic electroencephalographic activity within the theta frequency band, associated with attention and mental imagery (Sabourin, M. E. *et al.*, 1991; Crawford, H. J., 1994; Ray, W. J., 1997). Neurophysiological effects have been suggested to reduce the reciprocal inhibition between competing mental representations and processes, thereby facilitating the consideration of alternative experiences and allowing for a reinterpretation of sensory inputs during hypnosis (Rainville, P. and Price, D. D., 2004). In addition, feelings of automaticity relative to movement produced in response to hypnotic suggestions have been suggested to reflect a disruption of the coordinated activity of brain systems underlying the production and monitoring of self-generated actions within the frontoparietal cortices, thus leading to an altered sense of self-agency (Blakemore, S. J. and Frith, C., 2003; Blakemore, S. J., *et al.*, 2003). Generalizing those findings to the generation and regulation of mental images and thoughts, this suggests that hypnosis-related changes in perceptual and cognitive processes are felt as automatic because of the alteration in self-monitoring processes or the disruption in the communication between brain systems involved in the production of the perceptual–cognitive–behavioral changes and those involved in self-monitoring of those changes (also see Woody, E. and Szechtman, H., 2000; Rainville, P. and Price, D. D., 2004; Raz, A. *et al.*, 2005; Egner, T. *et al.*, 2005). These basic changes provide the background state upon which suggestions are given to further modulate specific perceptual processes such as those involved in pain.

66.3 Hypnotic Analgesia in the Brain

There are a number of psychophysiological studies providing convincing evidence that hypnotic suggestions to modulate pain perception affect nociceptive processes in the central nervous system (CNS) (Arendt-Nielsen, L. *et al.*, 1990; Zachariae, R. and Bjerring, P., 1994; Crawford, H. J. *et al.*, 1998; De Pascalis, V. *et al.*, 1999; 2001). New evidence based on modern functional brain imaging methods has further demonstrated that the hypnotic modulation of pain affects responses within pain-related brain network. In normal conditions, brain responses to

acute experimental pain tests involve the activation of primary and secondary somatosensory cortical areas, the insular cortex, and the anterior cingulate cortex (Apkarian, A. V. *et al.*, 2005). In a positron emission tomography (PET) experiment (Hofbauer, R. K. *et al.*, 2001), hypnotic suggestions were given to modulate the sensory intensity of pain before subjects immersed their hand in painfully hot water. In those conditions, the subjects' ratings of pain intensity confirmed the expected effects on pain experiences (Figure 1). Brain activity measured by PET further demonstrated significant changes in the amplitude of the responses within the primary somatosensory cortex, consistent with a role of this area in the perception of pain sensory intensity. In an earlier PET study using the same experimental model (immersion of the hand in hot water) (Rainville, P. *et al.*, 1997), hypnotic suggestions were given to modulate specifically the affective dimension of pain experience. In those conditions, the subjects reported changes in pain unpleasantness, but not pain intensity, and brain responses were modulated within the anterior cingulate cortex, but not the somatosensory cortex (Rainville, P. *et al.*, 1997). Furthermore, activity within the anterior cingulate cortex was found to correlate significantly with subjective reports of pain

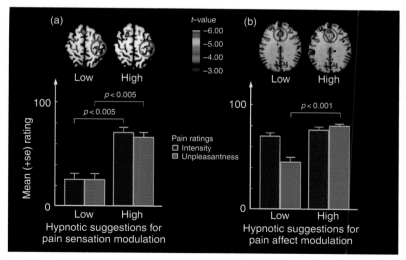

Figure 1 (a) Modulation of pain intensity and S1 activity. (b) Modulation of pain unpleasantness and anterior cingulate cortex activity. Horizontal slices of the brain are shown in conditions of decreased and increased pain intensity (a) and unpleasantness (b). Bar graph represent ratings of pain intensity and pain unpleasantness obtained in those conditions and demonstrate the significant modulation of both pain intensity and unpleasantness in the sensation-modulation experiment (a) and the significant modulation of pain unpleasantness only in the affect-modulation experiment (b). Reproduced from Rainville, P. 2002. Brain mechanisms of pain affect and pain modulation. Curr. Opin. Neurobiol. 12, 195–204.

unpleasantness, consistent with a contribution of this area to the affective dimension of pain. These studies

1. confirmed the efficacy of hypnosis to modulate pain-related brain responses,
2. demonstrated a functional segregation of brain activity associated more specifically with the sensory or affective dimensions of pain, and
3. provide further evidence supporting the specificity of hypnosis-related effects as a function of the precise content of the suggestions.

Importantly, such anatomofunctional specificity of the effects of hypnotic suggestions on perceptual processes has been observed in both the auditory modality (Szechtman, H. *et al.*, 1998) and in the visual modality (Kosslyn, S. M. *et al.*, 2000). This implies that the effect of hypnotic suggestions does not simply reflect a nonspecific response but involve the precise top-down modulation of targeted sensory–perceptual or affective–emotional processes. In the specific case of pain, this process may involve descending modulatory mechanisms affecting lower-level nociceptive activity in the CNS.

In addition to changes specifically found in pain-related areas, suggestions also produce widespread increases in lateral and medial prefrontal cortices including the dorsal aspect of the medial prefrontal cortex, as well as in parietal cortices (Rainville, P. *et al.*, 1999b; Faymonville, M. E. *et al.*, 2000). Interestingly, the activation found in the frontal and parietal cortices in response to hypnotic suggestions for pain modulation is consistent with the observations of the activation of the medial prefrontal cortex during hypnotic auditory hallucination (Szechtman, H. *et al.*, 1998), and of the lateral parietal cortex during hypnotically induced delusion of movement (Blakemore, S. J. *et al.*, 2003), as discussed above. In the case of pain modulation, those changes may also reflect the activation of modulatory circuits that affect pain-related areas directly through corticocortical connections or indirectly through descending projections affecting thalamic or brainstem nuclei. The medial prefrontal cortex, including the more dorsal and rostral parts of the anterior cingulate cortex may interact with other cortical and subcortical structures of the cerebral network involved in pain perception, including the insular cortices, the pain-related cingulate area, right prefrontal cortices, striatum, thalamus, and brainstem. Coactivation of the brainstem with the medial prefrontal cortex may

further reflect the activation of descending mechanisms involved in the regulation of ascending spinothalamic nociceptive pathways (Faymonville, M. E. *et al.*, 2003). In addition to their contribution to the regulation of ascending pathways, descending modulatory processes are likely to influence the reflex nociceptive responses generally observed in the periphery.

66.4 Somatic Consequences of Hypnotic Analgesia

Descending mechanisms have played a central role in the conceptualization of pain modulation over the past 40 years (Melzack, R. and Wall, P. D., 1965; Melzack, R. and Casey, K. L., 1968). In relation to descending inhibitory mechanism, there are abundant research reports examining the effects of hypnotic analgesia on peripheral physiological responses to acute pain stimuli. One of the most interesting effects reported in that literature is the effect of hypnotic analgesia on the nociceptive withdrawal reflex (the RIII reflex). Various experimental conditions involving pain modulation by conditioning stimuli, contextual factors, and psychological variables have demonstrated a robust modulation of the RIII reflex (Willer, J. C. *et al.*, 1979; 1981; Sandrini, G. *et al.*, 2005). The modulation of pain by hypnotic suggestion has been shown to affect the amplitude of the RIII reflex, consistent with the involvement of descending modulatory processes during hypnotic analgesia (Kiernan, B. D. *et al.*, 1995; Sandrini, G. *et al.*, 2005). However, a subsequent study also showed that some subjects demonstrated an increase in the RIII response during effective hypnotic analgesia as reflected in increased pain threshold and reduced cortical response (somatosensory-evoked potential), suggesting that the activation of descending modulatory processes during hypnotic analgesia may present considerable interindividual variability (Danziger, N. *et al.*, 1998).

In addition to studies examining the RIII response, several reports have demonstrated that hypnotic analgesia affects autonomic response. Those include studies showing reduced cardiovascular response and pain ratings evoked by experimental ischemia (Lenox, J. R., 1970) and the cold pressor test (Hilgard, E. R. *et al.*, 1974; Hilgard, E. R. and Hilgard, J. R., 1994), and reduced cutaneous and cardiac sympathetic responses evoked by painful electric shocks (De Pascalis, V. *et al.*, 2001). In addition, a study examining the effect of hypnotic suggestions focusing on the affective dimension of pain has found changes in heart rate responses correlated with changes in ratings of pain unpleasantness (Rainville, P. *et al.*, 1999a). Importantly, those effects were independent from changes in ratings of pain intensity, further supporting the distinction between pain sensation and pain affect and the stronger functional relation between autonomic responses and affective responses.

Taken together, studies on the modulation of motor and autonomic responses to noxious stimuli imply that the central neurophysiological mechanisms engaged during hypnotic analgesia modulate low-level brainstem and spinal systems responsible for the regulation of acute somatic responses to nociceptive stimuli. Combined with the studies demonstrating changes in pain-related brain activity, this strongly supports the specificity of hypnosis-related effects for the control of pain, providing a convincing demonstration that hypnotic analgesia can modulate both central and peripheral pain-related responses. This experimental demonstration of the neurophysiological mechanisms underlying hypnotic analgesia argues strongly in favor of the application of hypnosis for the control pain.

66.5 Clinical Applications of Hypnotic Analgesia

The increasing number of sophisticated neurophysiologic studies on hypnosis and pain has largely contributed to the renewed interest in hypnosis within healthcare settings. Consistent with the validation of hypnotic analgesia in experimental studies, nonpharmacological approaches including hypnosis have been strongly recommended in the management of pain. Among psychological strategies, there is accumulating evidence suggesting that hypnotic interventions are helpful for the management of acute procedural pain and anxiety. Chronic pain may also benefit from hypnosis interventions that must be integrated to the psychological, social, and environmental context.

Since the meta-analysis of Montgomery G. H. *et al.* (2000), which has focused on hypnotic analgesia in both clinical and experimental settings, some new randomized clinical trials have further documented the effectiveness of hypnosis in reducing pain and anxiety in various conditions. A more recent review of the literature has focused on randomized clinical

trials to assess the efficacy of hypnotic interventions (Patterson, D. R. and Jensen, M. P., 2003). Authors have concluded that hypnosis has a reliable and significant impact on acute procedural as well as chronic pain conditions in adults. Evidence is also presented supporting the successful application of hypnosis in children to decrease pain and distress during needle-related procedures. Nevertheless, authors recommend that more high quality studies have to be conducted to specify the scope and limits of hypnotic interventions. Furthermore, they underline that reviews on hypnosis effectiveness are limited by the heterogeneity of the type of hypnotic interventions, emphasizing the importance of a more detailed description of the specific suggestions used (Patterson, D. R., 2004; Jensen, M. P. and Patterson, D. R., 2005). This also appears to be critical in view of the specificity of the effects of different types of hypnotic suggestions as described above for suggestions focusing on sensory or affective dimensions of the pain experience.

To illustrate the potential benefit of hypnotic analgesia, some clinical studies provide particularly convincing evidence. (Note: the readers should refer to Medline or Cochrane Review sites for an extensive search of clinical trials on hypnosis; here we describe a few original reports to illustrate the potential efficacy of hypnosis in a clinical setting.) The potential benefits of hypnosis in patients receiving conscious sedation for plastic surgery have been demonstrated in a randomized clinical trial ($n = 60$) (Faymonville, M. E. *et al.*, 1997). In this study, hypnosis did not only provide better perioperative pain and anxiety relief, but also decreased medication used and improved patient's satisfaction and surgical conditions as compared with conventional stress-reducing strategies support. In another randomized clinical trial ($n = 241$), the effects of structured attention and self-hypnotic relaxation on pain and anxiety during percutaneous vascular and renal procedures in radiology were compared to standard care (Lang, E. V. *et al.*, 2000). The hypnotic intervention was performed during the surgery with suggestions for relaxation, eye closing, deep breathing, sensation of floating, and finally suggestion to self-generate imagery and focus on a safe and pleasant experience. As pain increased with the procedure time in the standard care and attention groups, it remained stable or decreased slightly in the hypnosis group even if the patients in the hypnosis group had requested less intravenous analgesia during the procedure. Mean anxiety scores also decreased during the procedure in the hypnosis group, while it remained stable in the structured attention group and increased in the standard care group. Finally, hypnosis improved hemodynamic stability significantly during the procedure consistent with the notion that the improvement in subjective reports of pain and negative emotions are accompanied by beneficial impact on peripheral responses to the noxious stimuli (here the surgical procedure). These results are highly consistent with the experimental studies reported above showing that the improvements in pain reports reflect a reduction in pain-related brain responses and are paralleled by an attenuation of peripheral responses as a consequence of active inhibitory processes acting on low-level nociceptive processes in the brain and spinal cord.

The benefits of hypnotic intervention on procedural pain and distress have also been demonstrated in children (Woodgate, R. L. and Degner, L. F., 2003; Wild, M. R. and Espie, C. A., 2004; Richardson, J. et al., 2006). In a recently published randomized clinical study in children with cancer ($n = 45$), Liossi and co-workers have illustrated the benefit of teaching self-hypnosis to children undergoing repeated lumbar punctures by comparing the effects of a local anesthetic plus self-hypnosis, to the effects of a local anesthetic or a local anesthetic plus attention. In addition to showing hypnosis-related pain relief, this study also demonstrated a decrease in the child's anticipatory anxiety before the procedure (Liossi, C. et al., 2006). Repeated invasive procedures such as cystourethrography are also painful and frightening for young children. Forty-four children (age range: 4–15 years) who were scheduled for at least the second VCUG were randomized to receive hypnosis ($n = 21$) or routine care with support ($n = 23$) while undergoing the procedure (Butler, L. D. et al., 2005). Hypnosis was found to be beneficial for children, because it may decrease distress, improve the collaboration of patients, and facilitate the procedure. However, as children are often receiving light sedation and analgesia in addition to local anesthetic, more studies have to be conducted to determine the specific contribution of hypnosis in pain and anxiety management strategies and, more critically, the potential interaction between multiple interventions.

Another clinical application for hypnosis is labor pain. A recent Cochrane review on complementary and alternative therapies for pain management in labor concluded that women who were taught self-hypnosis (five clinical trials for a total of 749 women)

had decreased requirements for pharmacological analgesia including epidural analgesia and were more satisfied with their pain management in labor compared with controls (Smith, C. A. et al., 2006). In a preliminary study, Marc I. et al. (2007) also suggested that hypnosis significantly reduces the need for analgesic/sedative medication without increasing pain or distress in patients undergoing a first-trimester surgical abortion. These effects further validate the use of hypnosis as an effective means to complement or replace the standard pharmacological approaches without reducing the quality of care, and sometimes by improving it.

There are several other conditions in which hypnosis may provide an advantageous approach to treat pain. In recent years, considerable work has been done by several teams on the use of hypnotic intervention in the treatment of mind–body dysfunction especially when stress and anxiety are involved. For example, irritable bowel syndrome is a functional gastrointestinal disorder characterized by chronic abdominal pain and altered bowel habits in the absence of any organic cause. Hypnosis as part of a structured intervention has been found to improve the status of all major symptoms in this condition (Gonsalkorale, W. M. and Whorwell, P. J., 2005). However, as is the case of many uncontrolled clinical studies, the use of hypnosis is associated with significant improvement but the specific contribution of hypnosis-related phenomena remains unclear. Of course, from a practical clinical perspective, the available evidence may be sufficient for the integration of hypnosis to clinical interventions, especially given the fact that adverse effects are unlikely to occur when hypnosis is proposed as an adjunct to standard care. Nevertheless, additional randomized controlled trials must be performed to extend the available evidence on the efficacy of hypnotic analgesia for the relief of acute pain and chronic pain condition (Stewart, J. H., 2005).

66.6 Conclusion

The effects of hypnosis and analgesic suggestions on brain activity are highly consistent with a modification of sensory and affective processes underlying the experience of pain. The available evidence clearly demonstrates the specificity of the effects on pain-related physiological processes both in the CNS and in the periphery. Hypnosis may also be used to improve relief during painful procedures and may be integrated to the arsenal of strategies to relieve chronic pain conditions, especially if stress and anxiety are involved. Although many questions on the neurobiological mechanisms underlying hypnosis-related effects remain to be answered, the available evidence strongly supports the instrumental use of hypnosis to study conscious perceptual processes and the clinical use of hypnosis to reduce pain and suffering.

References

Apkarian, A. V., Bushnell, M. C., Treede, R. D., and Zubieta, J. K. 2005. Human brain mechanisms of pain perception and regulation in health and disease. Eur. J. Pain 9, 463–484.

Arendt-Nielsen, L., Zachariae, R., and Bjerring, P. 1990. Quantitative evaluation of hypnotically suggested hyperaesthesia and analgesia by painful laser stimulation. Pain 42, 243–251.

Balthazard, C. G. and Woody, E. Z. 1992. The spectral analysis of hypnotic performance with respect to "absorption". Int. J. Clin. Exp. Hypn. 40, 21–43.

Blakemore, S. J. and Frith, C. 2003. Self-awareness and action. Curr. Opin. Neurobiol. 13, 219–224.

Blakemore, S. J., Oakley, D. A., and Frith, C. D. 2003. Delusions of alien control in the normal brain. Neuropsychologia 41, 1058–1067.

Bouleur, Jt. 1998. Mesmerism: The Discovery of Animal Magnetism (1779). A new translation. Holmes Publishing Group.

Butler, L. D., Symons, B. K., Henderson, S. L., Shortliffe, L. D., and Spiegel, D. 2005. Hypnosis reduces distress and duration of an invasive medical procedure for children. Pediatrics 115, e77–e85.

Crawford, H. J. 1994. Brain dynamics and hypnosis: attentional and disattentional processes. Int. J. Clin. Exp. Hypn. 42, 204–232.

Crawford, H. J., Knebel, T., Kaplan, L., Vendemia, J. M. C., Xie, M., Jamison, S., and Pribram, K. H. 1998. Hypnotic analgesia. 1. Somatosensory event-related potential changes to noxious stimuli. 2. Transfer learning to reduce chronic low back pain. Int. J. Clin. Exp. Hypn. 46, 92–132.

Danziger, N., Fournier, E., Bouhassira, D., Michaud, D., De Broucker, T., Santarcangelo, E., Carli, G., Chertock, L., and Willer, J. C. 1998. Different strategies of modulation can be operative during hypnotic analgesia: a neurophysiological study. Pain 75, 85–92.

De Pascalis, V., Magurano, M. R., and Bellusci, A. 1999. Pain perception, somatosensory event-related potentials and skin conductance responses to painful stimuli in high, mid, and low hypnotizable subjects: effects of differential pain reduction strategies. Pain 83, 499–508.

De Pascalis, V., Magurano, M. R., Bellusci, A., and Chen, A. C. 2001. Somatosensory event-related potential and autonomic activity to varying pain reduction cognitive strategies in hypnosis. Clin. Neurophysiol. 112, 1475–1485.

Egner, T., Jamieson, G., and Gruzelier, J. 2005. Hypnosis decouples cognitive control from conflict monitoring processes of the frontal lobe. Neuroimage 27, 969–978.

Faymonville, M. E., Laureys, S., Degueldre, C., Delfiore, G., Luxen, A., Franck, G., Lamy, M., and Maquet, P. 2000. Neural mechanisms of antinociceptive effects of hypnosis. Anesthesiology 92, 1257–1267.

Faymonville, M. E., Mambourg, P. H., Joris, J., Vrijens, B., Fissette, J., Albert, A., and Lamy, M. 1997. Psychological approaches during conscious sedation. Hypnosis versus stress reducing strategies: a prospective randomized study. Pain 73, 361–367.

Faymonville, M. E., Roediger, L., Del Fiore, G., Delgueldre, C., Phillips, C., Lamy, M., Luxen, A., Maquet, P., and Laureys, S. 2003. Increased cerebral functional connectivity underlying the antinociceptive effects of hypnosis. Brain Res. Cogn. Brain Res. 17, 255–262.

Fromm, E. and Nash, M. R. 1992. Contemporary Hypnosis Research Guilford Press.

Gonsalkorale, W. M. and Whorwell, P. J. 2005. Hypnotherapy in the treatment of irritable bowel syndrome. Eur. J. Gastroenterol. Hepatol. 17, 15–20.

Hilgard, E. R. 1992. Dissociation and Theories of Hypnosis. In: Contemporary Hypnosis Research (eds. E. Fromm and M. R. Nash), pp. 69–101. Guilford Press.

Hilgard, E. R. and Hilgard, J. R. 1994. Hypnosis in the Relief of Pain, revised edn. Brunner/Mazel.

Hilgard, E. R., Morgan, A. H., Lange, A. F., Lenox, J. R., MacDonald, H., Marshall, G. D., and Sachs, L. B. 1974. Heart rate changes in pain and hypnosis. Psychophysiology 11, 69–702.

Hofbauer, R. K., Rainville, P., Duncan, G. H., and Bushnell, M. C. 2001. Cortical representation of the sensory dimension of pain. J. Neurophysiol. 86, 402–411.

Jensen, M. P. and Patterson, D. R. 2005. Control conditions in hypnotic-analgesia clinical trials: challenges and recommendations. Int. J. Clin. Exp. Hypn. 53, 170–197.

Kiernan, B. D., Dane, J. R., Philips, L. H., and Price, D. D. 1995. Hypnotic analgesia reduces R-III nociceptive reflex: further evidence concerning the multifactorial nature of hypnotic analgesia. Pain 60, 39–47.

Kirsch, I. and Lynn, S. J. 1995. The altered state of hypnosis: changes in the theoretical landscape. Am. Psychol. 50, 846–858.

Kosslyn, S. M., Thompson, W. L., Costantini-Ferrando, M. F., Alpert, N. M., and Spiegel, D. 2000. Hypnotic visual illusion alters color processing in the brain. Am. J. Psychiatry 157, 1279–1284.

Lang, E. V., Benotsch, E. G., Fick, L. J., Lutgendorf, S., Berbaum, M. L., Berbaum, K. S., Logan, H., and Spiegel, D. 2000. Adjunctive non-pharmacological analgesia for invasive medical procedures: a randomized trial. Lancet 355, 1486–1490.

Laurence, J. R. 2002. 1784. Int. J. Clin. Exp. Hypn. 50, 309–319.

Lenox, J. R. 1970. Effect of hypnotic analgesia on verbal report and cardiovascular responses to ischemic pain. J. Abnormal Psychol. 75, 199–206.

Liossi, C., White, P., and Hatira, P. 2006. Randomized clinical trial of local anesthetic versus a combination of local anesthetic with self-hypnosis in the management of pediatric procedure-related pain. Health Psychol. 25, 307–315.

Maquet, P., Faymonville, M. E., Degueldre, C., Delfiore, G., Franck, G., Luxen, A., and Lamy, M. 1999. Functional neuroanatomy of hypnotic state. Biol. Psychiatry 45, 327–333.

Marc, I., Rainville, P., Verreault, R., Vaillancourt, L., Masse, B., and Dodin, S. 2007. The use of hypnosis to improve pain management during voluntary interruption of pregnancy: an open randomized preliminary study. Contraception 75, 52–58.

Melzack, R. and Casey, K. L. 1968. Sensory, Motivational, and Central Control Determinants of Pain: A New Conceptual Model. In: The Skin Senses (ed. D. Kenshalo), pp. 423–443. Thomas.

Melzack, R. and Wall, P. D. 1965. Pain mechanisms: a new theory. Science 150, 971–978.

Montgomery, G. H., DuHamel, K. N., and Redd, W. H. 2000. A meta-analysis of hypnotically induced analgesia: how effective is hypnosis? Int. J. Clin. Exp. Hypn. 48, 138–153.

Nadon, R., Laurence, J. R., and Perry, C. 1987. Multiple predictors of hypnotic susceptibility. J. Pers. Soc. Psychol. 53, 948–960.

Patterson, D. R. 2004. Treating pain with hypnosis. Curr. Dir. Psychol. Sci. 13, 252–255.

Patterson, D. R. and Jensen, M. P. 2003. Hypnosis and clinical pain. Psychol. Bull. 129, 495–521.

Price, D. D. 1996. Hypnotic Analgesia: Psychological and Neural Mechanisms. In: Hypnosis and Suggestions in the Treatment of pain (ed. J. Barber), pp. 67–84. Norton.

Radtke, H. L. and Stam, H. J. 1991. The relationship between absorption, openness to experience, anhedonia, and susceptibility. Int. J. Clin. Exp. Hypn. 39, 39–56.

Rainville, P. 2002. Brain mechanisms of pain affect and pain modulation. Curr. Opin. Neurobiol. 12, 195–204.

Rainville, P. and Price, D. D. 2004. The Neurophenomenology of Hypnosis and Hypnotic Analgesia. In: Psychological Methods of Pain Control: Basic Science and Clinical Perspectives (eds. D. D. Price and M. C. Bushnell), Vol. 29, pp. 235–267. IASP Press.

Rainville, P., Carrier, B., Hofbauer, R. K., Bushnell, M. C., and Duncan, G. H. 1999a. Dissociation of pain sensory and affective dimensions using hypnotic modulation. Pain 82, 159–171.

Rainville, P., Duncan, G. H., Price, D. D., Carrier, B., and Bushnell, M. C. 1997. Pain affect encoded in human anterior cingulate but not somatosensory cortex. Science 277, 968–971.

Rainville, P., Hofbauer, R. K., Bushnell, M. C., Duncan, G. H., and Price, D. D. 2002. Hypnosis modulates activity in brain structures involved in the regulation of consciousness. J. Cogn. Neurosci. 14, 887–901.

Rainville, P., Hofbauer, R. K., Paus, T., Duncan, G. H., Bushnell, M. C., and Price, D. D. 1999b. Cerebral mechanisms of hypnotic induction and suggestion. J. Cogn. Neurosci. 11, 110–125.

Ray, W. J. 1997. EEG concomitants of hypnotic susceptibility. Int. J. Clin. Exp. Hypn. 45, 301–313.

Raz, A., Fan, J., and Posner, M. I. 2005. Hypnotic suggestion reduces conflict in the human brain. Proc. Natl. Acad. Sci. U. S. A. 102, 9978–9983.

Richardson, J., Smith, J. E., McCall, G., and Pilkington, K. 2006. Hypnosis for procedure-related pain and distress in pediatric cancer patients: a systematic review of effectiveness and methodology related to hypnosis interventions. J. Pain Symptom Manage. 31, 70–84.

Sabourin, M. E., Cutcomb, S. D., Crawford, H. J., and Pribram, K. 1991. EEG correlates of hypnotic susceptibility and hypnotic trance: spectral analysis and coherence. Int. J. Psychophysiol. 10, 125–142.

Salas, C. and Salas, D. 1996. The first scientific investigation of the paranormal ever conducted: testing the claims of Mesmerism, Commissioned by King Louis XVI, Designed conducted and written by Benjamin Franklin, Antoine Lavoisier & Others. Translated from "Rapport des Comissaires chargés par le Roi de l'Examen du Magnétirme animal. Imprimé par Ordre du Roi No 4 à Paris de l'Imprimerie Royale (1784)". Introduction by Shermer, M. Skeptic 4, 66–83.

Sandrini, G., Serrao, M., Rossi, P., Romaniello, A., Cruccu, G., and Willer, J. C. 2005. The lower limb flexion reflex in humans. Prog. Neurobiol. 77, 353–395.

Smith, C. A., Collins, C. T, Cyna, A. M., and Crowther, C. A. 2006. Complementary and alternative therapies for pain management in labour. Cochrane Database Syst. Rev. CD003521.

Stewart, J. H. 2005. Hypnosis in contemporary medicine. Mayo Clin. Proc. 80, 511–524.

Szechtman, H., Woody, E., Bowers, K. S, and Nahmias, C. 1998. Where the imaginal appears real: a positron emission tomography study of auditory hallucinations. Proc. Natl. Sci. U. S. A. 95, 1956–1960.

Tellegen, A. and Atkinson, G. 1974. Openness to absorbing and self-altering experiences ("absorption"), a trait related to hypnotic susceptibility. J. Abnorm. Psychol. 83, 268–277.

Weitzenhoffer, A. M. 1980. Hypnotic susceptibility revisited. Am. J. Clin. Hypn. 22, 130–146.

Wild, M. R. and Espie, C. A. 2004. The efficacy of hypnosis in the reduction of procedural pain and distress in pediatric oncology: a systematic review. J. Dev. Behav. Pediatr. 25, 207–213.

Willer, J. C., Boureau, F., and Albe-Fessard, D. 1979. Supraspinal influences on nociceptive flexion reflex and pain sensation in man. Brain Res. 179, 61–68.

Willer, J. C., Dehen, H., and Cambier, J. 1981. Stress-induced analgesia in humans: endogenous opioids and naloxone-reversible depression of pain reflexes. Science 212, 689–691.

Woodgate, R. L. and Degner, L. F. 2003. Expectations and beliefs about children's cancer symptoms: perspectives of children with cancer and their families. Oncol. Nurs. Forum 30, 479–491.

Woody, E. and Szechtman, H. 2000. Hypnotic hallucinations: towards a biology of epistemology. Contemp. Hypn. 17, 4–14.

Zachariae, R. and Bjerring, P. 1994. Laser-induced pain-related brain potentials and sensory pain ratings in high and low hypnotizable subjects during hypnotic suggestions of relaxation, dissociated imagery, focused analgesia, and placebo. Int. J. Clin. Exp. Hypn. 42, 56–80.

Index

Cross-reference terms in *italics* are general cross-references, or refer to subentry terms within the main entry (the main entry is not repeated to save space). Readers are also advised to refer to the end of each article for additional cross-references - not all of these cross-references have been included in the index cross-references.

The index is arranged in set-out style with a maximum of three levels of heading. Major discussion of a subject is indicated by bold page numbers. Page numbers suffixed by *t* and *f* refer to Tables and Figures respectively. vs. indicates a comparison.

This index is in letter-by-letter order, whereby hyphens and spaces within index headings are ignored in the alphabetization. Prefixes and terms in parentheses are excluded from the initial alphabetization.